Handbook of Parallel Computing

Models, Algorithms and Applications

CHAPMAN & HALL/CRC
COMPUTER and INFORMATION SCIENCE SERIES

Series Editor: Sartaj Sahni

PUBLISHED TITLES

ADVERSARIAL REASONING: COMPUTATIONAL APPROACHES TO READING THE OPPONENT'S MIND
Alexander Kott and William M. McEneaney

DISTRIBUTED SENSOR NETWORKS
S. Sitharama Iyengar and Richard R. Brooks

DISTRIBUTED SYSTEMS: AN ALGORITHMIC APPROACH
Sukumar Ghosh

FUNDEMENTALS OF NATURAL COMPUTING: BASIC CONCEPTS, ALGORITHMS, AND APPLICATIONS
Leandro Nunes de Castro

HANDBOOK OF ALGORITHMS FOR WIRELESS NETWORKING AND MOBILE COMPUTING
Azzedine Boukerche

HANDBOOK OF APPROXIMATION ALGORITHMS AND METAHEURISTICS
Teofilo F. Gonzalez

HANDBOOK OF BIOINSPIRED ALGORITHMS AND APPLICATIONS
Stephan Olariu and Albert Y. Zomaya

HANDBOOK OF COMPUTATIONAL MOLECULAR BIOLOGY
Srinivas Aluru

HANDBOOK OF DATA STRUCTURES AND APPLICATIONS
Dinesh P. Mehta and Sartaj Sahni

HANDBOOK OF DYNAMIC SYSTEM MODELING
Paul A. Fishwick

HANDBOOK OF PARALLEL COMPUTING: MODELS, ALGORITHMS AND APPLICATIONS
Sanguthevar Rajasekaran and John Reif

HANDBOOK OF REAL-TIME AND EMBEDDED SYSTEMS
Insup Lee, Joseph Y-T. Leung, and Sang H. Son

HANDBOOK OF SCHEDULING: ALGORITHMS, MODELS, AND PERFORMANCE ANALYSIS
Joseph Y.-T. Leung

HIGH PERFORMANCE COMPUTING IN REMOTE SENSING
Antonio J. Plaza and Chein-I Chang

THE PRACTICAL HANDBOOK OF INTERNET COMPUTING
Munindar P. Singh

SCALABLE AND SECURE INTERNET SERVICES AND ARCHITECTURE
Cheng-Zhong Xu

SPECULATIVE EXECUTION IN HIGH PERFORMANCE COMPUTER ARCHITECTURES
David Kaeli and Pen-Chung Yew

Handbook of Parallel Computing

Models, Algorithms and Applications

Edited by
Sanguthevar Rajasekaran
University of Connecticut
Storrs, U.S.A.

John Reif
Duke University
Durham, North Carolina, U.S.A.

Boca Raton   London   New York

Chapman & Hall/CRC is an imprint of the
Taylor & Francis Group, an **informa** business

Chapman & Hall/CRC
Taylor & Francis Group
6000 Broken Sound Parkway NW, Suite 300
Boca Raton, FL 33487-2742

© 2008 by Taylor & Francis Group, LLC
Chapman & Hall/CRC is an imprint of Taylor & Francis Group, an Informa business

No claim to original U.S. Government works
Printed in the United States of America on acid-free paper
10 9 8 7 6 5 4 3 2 1

International Standard Book Number-13: 978-1-58488-623-5 (Hardcover)

This book contains information obtained from authentic and highly regarded sources. Reprinted material is quoted with permission, and sources are indicated. A wide variety of references are listed. Reasonable efforts have been made to publish reliable data and information, but the author and the publisher cannot assume responsibility for the validity of all materials or for the consequences of their use.

Except as permitted under U.S. Copyright Law, no part of this book may be reprinted, reproduced, transmitted, or utilized in any form by any electronic, mechanical, or other means, now known or hereafter invented, including photocopying, microfilming, and recording, or in any information storage or retrieval system, without written permission from the publishers.

For permission to photocopy or use material electronically from this work, please access www.copyright.com (http://www.copyright.com/) or contact the Copyright Clearance Center, Inc. (CCC) 222 Rosewood Drive, Danvers, MA 01923, 978-750-8400. CCC is a not-for-profit organization that provides licenses and registration for a variety of users. For organizations that have been granted a photocopy license by the CCC, a separate system of payment has been arranged.

Trademark Notice: Product or corporate names may be trademarks or registered trademarks, and are used only for identification and explanation without intent to infringe.

Library of Congress Cataloging-in-Publication Data

Rajasekaran, Sanguthevar.
   Handbook of parallel computing : models, algorithms and applications / Sanguthevar Rajasekaran and John Reif.
      p. cm. -- (Chapman & Hall/CRC computer & information science)
   Includes bibliographical references and index.
   ISBN-13: 978-1-58488-623-5
   ISBN-10: 1-58488-623-4
   1. Parallel processing (Electronic computers)--Handbooks, manuals, etc. 2. Computer algorithms--Handbooks, manuals, etc. I. Reif, J. H. (John H.) II. Title. III. Series.

QA76.58.R34 2007
005.1--dc22
                                      2007025276

Visit the Taylor & Francis Web site at
http://www.taylorandfrancis.com

and the CRC Press Web site at
http://www.crcpress.com

Dedication

To
Pandi, Pandiammal, and Periyasamy
                            Sanguthevar Rajasekaran

To
Jane, Katie, and Emily
                    John H. Reif

தொட்டனைத் தூறும் மணற்கேணி மாந்தர்க்குக்
கற்றனைத் தூறும் அறிவு

திருவள்ளுவர்
(திருக்குறள்; அதிகாரம் நாற்பது: கல்வி)

The deeper you dig in sand, the larger will be the water flow;
The more you learn, the more will be the knowledge flow

Thiruvalluvar (circa 100 B.C.)
(*Thirukkural*; Chapter 40: Education)

Contents

Preface xiii

Acknowledgments xxi

Editors xxiii

Contributors xxv

Models

1   Evolving Computational Systems
    Selim G. Akl 1-1

2   Decomposable BSP: A Bandwidth-Latency Model for Parallel and Hierarchical Computation
    Gianfranco Bilardi, Andrea Pietracaprina, and Geppino Pucci 2-1

3   Membrane Systems: A "Natural" Way of Computing with Cells
    Oscar H. Ibarra and Andrei Păun 3-1

4   Optical Transpose Systems: Models and Algorithms
    Chih-Fang Wang and Sartaj Sahni 4-1

5   Models for Advancing PRAM and Other Algorithms into Parallel Programs for a PRAM-On-Chip Platform
    Uzi Vishkin, George C. Caragea, and Bryant C. Lee 5-1

6 Deterministic and Randomized Sorting Algorithms for Parallel Disk Models
   Sanguthevar Rajasekaran **6**-1

7 A Programming Model and Architectural Extensions for Fine-Grain Parallelism
   Alex Gontmakher, Assaf Schuster, Gregory Shklover, and Avi Mendelson **7**-1

8 Computing with Mobile Agents in Distributed Networks
   Evangelos Kranakis, Danny Krizanc, and Sergio Rajsbaum **8**-1

9 Transitional Issues: Fine-Grain to Coarse-Grain Multicomputers
   Stephan Olariu **9**-1

10 Distributed Computing in the Presence of Mobile Faults
   Nicola Santoro and Peter Widmayer **10**-1

11 A Hierarchical Performance Model for Reconfigurable Computers
   Ronald Scrofano and Viktor K. Prasanna **11**-1

12 Hierarchical Performance Modeling and Analysis of Distributed Software Systems
   Reda A. Ammar **12**-1

13 Randomized Packet Routing, Selection, and Sorting on the POPS Network
   Jaime Davila and Sanguthevar Rajasekaran **13**-1

14 Dynamic Reconfiguration on the R-Mesh
   Ramachandran Vaidyanathan and Jerry L. Trahan **14**-1

15 Fundamental Algorithms on the Reconfigurable Mesh
   Koji Nakano **15**-1

16 Reconfigurable Computing with Optical Buses
   Anu G. Bourgeois **16**-1

Algorithms

17 Distributed Peer-to-Peer Data Structures
   Michael T. Goodrich and Michael J. Nelson **17**-1

18 Parallel Algorithms via the Probabilistic Method
   Lasse Kliemann and Anand Srivastav **18**-1

19 Broadcasting on Networks of Workstations
   Samir Khuller, Yoo-Ah Kim, and Yung-Chun (Justin) Wan **19**-1

20 Atomic Selfish Routing in Networks: A Survey
   Spyros Kontogiannis and Paul G. Spirakis **20**-1

Contents

21 Scheduling in Grid Environments
   Young Choon Lee and Albert Y. Zomaya 21-1

22 QoS Scheduling in Network and Storage Systems
   Peter J. Varman and Ajay Gulati 22-1

23 Optimal Parallel Scheduling Algorithms in WDM Packet Interconnects
   Zhenghao Zhang and Yuanyuan Yang 23-1

24 Real-Time Scheduling Algorithms for Multiprocessor Systems
   Michael A. Palis 24-1

25 Parallel Algorithms for Maximal Independent Set and Maximal Matching
   Yijie Han 25-1

26 Efficient Parallel Graph Algorithms for Multicore and Multiprocessors
   David A. Bader and Guojing Cong 26-1

27 Parallel Algorithms for Volumetric Surface Construction
   Joseph JaJa, Amitabh Varshney, and Qingmin Shi 27-1

28 Mesh-Based Parallel Algorithms for Ultra Fast Computer Vision
   Stephan Olariu 28-1

29 Prospectus for a Dense Linear Algebra Software Library
   James Demmel, Beresford Parlett, William Kahan, Ming Gu, David Bindel, Yozo Hida, E. Jason Riedy, Christof Voemel, Jakub Kurzak, Alfredo Buttari, Julie Langou, Stanimire Tomov, Jack Dongarra, Xiaoye Li, Osni Marques, Julien Langou, and Piotr Luszczek 29-1

30 Parallel Algorithms on Strings
   Wojciech Rytter 30-1

31 Design of Multithreaded Algorithms for Combinatorial Problems
   David A. Bader, Kamesh Madduri, Guojing Cong, and John Feo ... 31-1

32 Parallel Data Mining Algorithms for Association Rules and Clustering
   Jianwei Li, Wei-keng Liao, Alok Choudhary, and Ying Liu 32-1

33 An Overview of Mobile Computing Algorithmics
   Stephan Olariu and Albert Y. Zomaya 33-1

Applications

34 Using FG to Reduce the Effect of Latency in Parallel Programs Running on Clusters
   Thomas H. Cormen and Elena Riccio Davidson 34-1

35 High-Performance Techniques for Parallel I/O
Avery Ching, Kenin Coloma, Jianwei Li, Wei-keng Liao, and Alok Choudhary . . . **35**-1

36 Message Dissemination Using Modern Communication Primitives
Teofilo F. Gonzalez **36**-1

37 Online Computation in Large Networks
Susanne Albers **37**-1

38 Online Call Admission Control in Wireless Cellular Networks
Ioannis Caragiannis, Christos Kaklamanis, and Evi Papaioannou **38**-1

39 Minimum Energy Communication in Ad Hoc Wireless Networks
Ioannis Caragiannis, Christos Kaklamanis, and Panagiotis Kanellopoulos **39**-1

40 Power Aware Mapping of Real-Time Tasks to Multiprocessors
Dakai Zhu, Bruce R. Childers, Daniel Mossé, and Rami Melhem **40**-1

41 Perspectives on Robust Resource Allocation for Heterogeneous Parallel and Distributed Systems
Shoukat Ali, Howard Jay Siegel, and Anthony A. Maciejewski **41**-1

42 A Transparent Distributed Runtime for Java
Michael Factor, Assaf Schuster, and Konstantin Shagin **42**-1

43 Scalability of Parallel Programs
Ananth Grama and Vipin Kumar **43**-1

44 Spatial Domain Decomposition Methods in Parallel Scientific Computing
Sudip Seal and Srinivas Aluru **44**-1

45 Game Theoretical Solutions for Data Replication in Distributed Computing Systems
Samee Ullah Khan and Ishfaq Ahmad **45**-1

46 Effectively Managing Data on a Grid
Catherine L. Ruby and Russ Miller **46**-1

47 Fast and Scalable Parallel Matrix Multiplication and Its Applications on Distributed Memory Systems
Keqin Li **47**-1

Index **I**-1

Preface

We live in an era in which parallel computing has become mainstream and very affordable. This is mainly because hardware costs have come down rapidly. With the advent of the Internet, we also experience the phenomenon of data explosion in every application of interest. Processing voluminous datasets is highly computation intensive. Parallel computing has been fruitfully employed in numerous application domains to process large datasets and handle other time-consuming operations of interest. As a result, unprecedented advances have been made in such areas as biology, scientific computing, modeling and simulations, and so forth. In this handbook we present recent developments in the areas of parallel models, algorithms, and applications.

An Introduction to Parallel Computing

There are many ways of achieving parallelism. Some examples include the use of supercomputers, clusters, network of workstations, and grid computing. In sequential computing, the *Random Access Machine (RAM)*, which executes arithmetic and Boolean operations as well as read and write memory operations, has been universally accepted as an appropriate model of computing. On the other hand, numerous parallel models of computing have been proposed in the literature. These models differ in the way the processors communicate among themselves.

If we use P processors to solve a problem, then there is a potential of reducing the (sequential) runtime by a factor of up to P. If S is the best known sequential runtime and if T is the parallel runtime using P processors, then $PT \geq S$. Otherwise, we can simulate the parallel algorithm using a single processor and get a runtime better than S (which will be a contradiction). We refer to PT as the *work done* by the parallel algorithm. Any parallel algorithm for which $PT = O(S)$ will be referred to as an *optima work* algorithm.

In most parallel machines we can still think of each processor as a RAM. Variations among different architectures arise from the ways in which they implement interprocessor communications. Parallel models can be categorized broadly into *parallel comparison trees, shared memory models,* and *fixed connection networks*.

A *parallel comparison tree* is analogous to the sequential comparison (or decision) tree. It is typically employed for the study of comparison problems such as sorting, selection, merging, and so forth. An algorithm under this model is represented as a tree. The computation starts at the root. P pairs of input keys are compared in parallel at the root (P being the number of processors). On the basis of the outcomes of these comparisons, the computation branches to an appropriate child of the root. Each node of the tree corresponds to comparison of P pairs of input keys. The computation terminates at a leaf node that has enough information to output the correct answer. Thus, there is a tree corresponding to every input size. For a given instance of the problem, a branch in the tree is traversed. The worst case runtime is proportional to the depth of the tree. This model takes into account only the comparison operations performed.

The *Parallel Random Access Machine (PRAM)* is a shared memory model for parallel computation consisting of a collection of RAMs working in synchrony where communication takes place with the help of a common block of shared memory. If, for example, processor i wants to communicate with processor j, it can do so by writing a message in memory cell j, which can then be read by processor j.

More than one processor may want to access the same cell at the same time for either reading from or writing into. Depending on how these conflicts are resolved, a PRAM can further be classified into three. In an *Exclusive Read and Exclusive Write (EREW)* PRAM, neither concurrent reads nor concurrent writes are allowed. In a *Concurrent Read and Exclusive Write (CREW)* PRAM, concurrent reads are permitted but not concurrent writes. Finally, a *Concurrent Read and Concurrent Write (CRCW)* PRAM allows both concurrent reads and concurrent writes. A mechanism for handling write conflicts is needed for a CRCW PRAM, since the processors trying to write at the same time in the same cell can possibly have different data to write and we should determine which data gets written. This is not a problem in the case of concurrent reads since the data read by different processors will be the same. In a *Common-CRCW PRAM*, concurrent writes are permissible only if the processors trying to access the same cell at the same time have the same data to write. In an *Arbitrary-CRCW PRAM*, if more than one processor tries to write in the same cell at the same time, an arbitrary one of them succeeds. In a *Priority-CRCW PRAM*, processors have assigned priorities. Write conflicts are resolved using these priorities.

Also, we consider *fixed connection network* machine models. A directed graph is used to represent the fixed connection network. The nodes of this graph correspond to processing elements, and the edges correspond to communication links. If two processors are connected by an edge, they can communicate in a unit step. Two processors not connected by an edge can communicate by sending a message along a path that connects the two processors. Each processor in a fixed connection machine is a RAM. Examples of fixed connection machines include the mesh, the hypercube, the star graph, and so forth.

A mesh is an $n \times n$ square grid whose nodes are processors and whose edges are communication links. The *diameter* of a mesh is $2n-2$. (*Diameter* of a graph is defined to be the maximum of the shortest distance between any two nodes in the graph.) Diameter of a fixed connection machine is often a lower bound on the solution time of any nontrivial problem on the machine. The *degree* of a fixed connection network should be as small as possible for it to be physically realizable. (The *degree* of a fixed connection machine is defined to be the maximum number of neighbors for any node.) The degree of a mesh is four. Many variants of the mesh have also been proposed in the literature. Two examples are meshes with fixed buses and meshes with reconfigurable buses. In a mesh with fixed buses, in addition to the edge connections of a standard mesh, each row and each column has an associated bus. In any time unit, one of the processors connected to a bus can send a message along the bus (which can be read by the others in the same time unit). In a mesh with reconfigurable buses also, each row and each column has an associated bus, but the buses can be dynamically reconfigured.

A hypercube of dimension n has 2^n nodes. Any node in a hypercube can be denoted as an n-bit binary number. Let x and y be the binary representations of any two nodes in a hypercube. Then, these two nodes will be connected by an edge if and only if the Hamming distance between x and y is one, that is, x and y differ in exactly one bit position. Thus, the degree of a hypercube with 2^n nodes is n. The diameter of a 2^n-noded hypercube can also be seen to be n. Butterfly, CCC, de Bruijn, and so forth are networks that are very closely related to a hypercube.

The chapters in this book span all of the above models of computing and more. They have been loosely categorized into three sections, namely, Parallel Models, Parallel Algorithms, and Parallel Applications. However, many chapters will cover more than one of these three aspects.

Parallel Models

Chapter 1 explores the concept of *evolving computational systems*. As an example, it considers the computation of n functions $F_0, F_1, \ldots, F_{n-1}$ on the n variables $x_0, x_1, \ldots, x_{n-1}$, where the variables themselves change with time. Both sequential and parallel solutions are considered for this problem.

An important consideration in adapting a model of computation is that it should provide a framework for the design and analysis of algorithms that can be executed efficiently on physical machines. In the area of parallel computing, the *Bulk Synchronous Model (BSP)* has been widely studied. **Chapter 2** deals with a variant of this model, called *Decomposable Bulk Synchronous Model (D-BSP)*.

In **Chapter 3**, basic elements of *membrane computing*, a recent branch of natural computing, are presented. Membrane computing is a model abstracted from the way living cells function. The specific version of membrane computing considered in this chapter is called a *P system*. A *P* system can be thought of as a distributed computing device.

In any standard fixed connection machine (like a mesh), it is assumed that all the interconnects are electric. It is known that when communication distances exceed a few millimeters, optical interconnects are preferable since they provide bandwidth and power advantages. Thus, it is wise to use both optical (for long distances) and electrical (for short distances) interconnects. **Chapter 4** deals with one such model called the *Optical Transpose Interconnection System (OTIS)*.

Chapter 5 proposes models and ways for converting PRAM and other algorithms onto a so-called PRAM-on-chip architecture. PRAM-on-chip is a commodity high-end multicore computer architecture. This chapter provides the missing link for upgrading the standard theoretical PRAM algorithms class to a parallel algorithms and programming course. It also guides future compiler work by showing how to generate performance-tuned programs from simple PRAM-like programs. Such program examples can used by compiler experts in order to "teach the compiler" about performance-tuned programs.

When the amount of data to be processed is very large, it may not be possible to fit all the data in the main memory of the computer used. This necessitates the use of secondary storage devices (such as disks) and the employment of out-of-core algorithms. In the context of parallel computing, the *Parallel Disk Model (PDM)* has been studied as a model for out-of-core computing. In **Chapter 6**, deterministic and randomized algorithms for sorting on the PDM model are surveyed.

Chapter 7 introduces a new threading model called *Inthreads*. This model is sufficiently lightweight to use parallelization at very fine granularity. The Inthreads model is based on microthreads that operate within the context of a conventional thread.

Mobile agents are software entities with the capacity for motion that can act on behalf of their user with a certain degree of autonomy in order to accomplish a variety of computing tasks. They find applications in numerous computer environments such as operating system daemons, data mining, web crawlers, and so forth. In **Chapter 8**, an introduction to mobile agents is provided.

The main focus of **Chapter 9** is to investigate transitional issues arising in the migration from fine- to coarse-grain multicomputers, as viewed from the perspective of the design of scalable parallel algorithms. A number of fundamental computational problems whose solutions are fairly well understood in the fine-grain model are treated as case studies.

A *distributed computing environment* is a collection of networked computational *entities* communicating with each other by means of *messages*, in order to achieve a common goal, for example, to perform a given task, to compute the solution to a problem, or to satisfy a request. Examples include data communication networks, distributed databases, transaction processing systems, and so forth. In **Chapter 10**, the problem of fault-tolerant computing (when communication faults are present) in a distributed computing environment is studied.

Reconfigurable hardware has been shown to be very effective in providing application acceleration. For any scientific computing application, certain components can be accelerated using reconfigurable hardware and the other components may benefit from general purpose processors. In **Chapter 11**, a hierarchical performance model is developed to aid in the design of algorithms for a hybrid architecture that has both reconfigurable and general purpose processors.

In **Chapter 12**, a general purpose-methodology called *Hierarchical Performance Modeling (HPM)* to evaluate the performance of distributed software systems is provided. It consists of a series of separate modeling levels to match abstraction layers present in distributed systems. Each level represents a different view of the system at a particular amount of details.

A partitioned optical passive star (POPS) network is closely related to the OTIS network of Chapter 5. In a POPS(d, g) there are $n = dg$ processors. These processors are divided into g groups with d nodes in every group. There is an optical coupler between every pair of groups, so that at any given time step any coupler can receive a message from a processor of the source group and broadcast it to the processors of the destination group. In **Chapter 13**, algorithms are presented on the POPS network for several fundamental problems such as packet routing and selection.

Chapter 14 studies the power of the reconfigurable mesh (R-Mesh) model. Many variants of the R-Mesh are described. Algorithms for fundamental problems such as permutation routing, neighbor localization, prefix sums, sorting, graph problems, and so forth are presented. The relationship between R-Mesh and other models such as the PRAM, Boolean circuits, and Turing machines is brought out.

In **Chapter 15** the subject matter is also the reconfigurable mesh. Algorithms for fundamental problems such as finding the leftmost 1, compression, prime numbers selection, integer summing, and so forth are presented.

Chapter 16 considers reconfigurable models that employ optical buses. The one-dimensional version of this model is called a *Linear Array with a Reconfigurable Pipelined Bus System (LARPBS)*. This model is closely related to the models described in Chapters 4, 11, 13, 14, and 15. Fundamental problems such as compression, prefix computation, sorting, selection, PRAM simulations, and so forth are considered. A comparison of different optical models is also presented in this chapter.

Parallel Algorithms

Peer-to-peer networks consist of a collection of machines connected by some networking infrastructure. Fundamental to such networks is the need for processing queries to locate resources. The focus of **Chapter 17** is on the development of distributed data structures to process these queries effectively.

Chapter 18 gives an introduction to the design of parallel algorithms using the probabilistic method. Algorithms of this kind usually possess a randomized sequential counterpart. Parallelization of such algorithms is inherently linked with derandomization, either with the Erdős–Spencer method of conditional probabilities or exhaustive search in a polynomial-sized sample space.

Networks of Workstations (NOWs) are a popular alternative to massively parallel machines and are widely used. By simply using off-the-shelf PCs, a very powerful workstation cluster can be created, and this can provide a high amount of parallelism at relatively low cost. **Chapter 19** addresses issues and challenges in implementing broadcast and multicast on such platforms.

Chapter 20 presents a survey of some recent advances in the *atomic congestion games* literature. The main focus is on a special case of congestion games, called *network congestion games*, which are of particular interest to the networking community. The algorithmic questions of interest include the *existence* of pure Nash equilibria.

Grid scheduling requires a series of challenging tasks. These include searching for resources in collections of geographically distributed heterogeneous computing systems and making scheduling decisions, taking into consideration quality of service. In **Chapter 21**, key issues related to grid scheduling are described and solutions for them are surveyed.

Chapter 22 surveys a set of scheduling algorithms that work for different models of sharing, resource, and request requirements. An idealized fluid flow model of fair service is described, followed by practical schedulers using discrete service models such as WFQ, SC, and WF^2Q. Algorithms for parallel storage servers are also given.

In **Chapter 23**, the problem of packet scheduling in wavelength-division-multiplexing (WDM) optical interconnects with limited range wavelength conversion and shared buffer is treated. Parallel algorithms

Preface

for finding an optimal packet schedule are described. Finding an optimal schedule is reduced to a matching problem in a bipartite graph.

Chapter 24 presents a summary of the past and current research on *multiprocessor scheduling algorithms* for *hard* real-time systems. It highlights some of the most important theoretical results on this topic over a 30-year period, beginning with the seminal paper of Liu and Layland in 1973, which provided impetus to much of the subsequent research in real-time scheduling theory.

Maximal independent set and maximal matching are fundamental problems studied in computer science. In the sequential case, a greedy linear time algorithm can be used to solve these problems. However, in parallel computation these problems are not trivial. **Chapter 25** provides details of the *derandomization technique* and its applications to the maximal independent set and maximal matching problems.

Chapter 26 focuses on algorithm design and engineering techniques that better fit current (or near future) architectures and help achieve good practical performance for solving arbitrary, sparse instances of fundamental graph problems. The target architecture is shared-memory machines, including symmetric multiprocessors (SMPs) and the multithreaded architecture.

Large-scale scientific data sets are appearing at an increasing rate, whose sizes can range from hundreds of gigabytes to tens of terabytes. Isosurface extraction and rendering is an important visualization technique that enables the visual exploration of such data sets using surfaces. In **Chapter 27**, basic sequential and parallel techniques used to extract and render isosurfaces with a particular focus on out-of-core techniques are presented.

Chapter 28 surveys computational paradigms offered by the reconfigurability of bus systems to obtain fast and ultra fast algorithms for a number of fundamental low and mid-level vision tasks. Applications considered range from the detection of man-made objects in a surrounding environment to automatic target recognition.

There are several current trends and associated challenges in Dense Linear Algebra (DLA) that influence the development of DLA software libraries. **Chapter 29** identifies these trends, addresses the new challenges, and consequently outlines a prospectus for new releases of the LAPACK and ScaLAPACK libraries.

Chapter 30 presents several *polylogarithmic*-time parallel string algorithms for the *Parallel Random Access Machine* (PRAM) models. Problems considered include *string matching*, construction of the *dictionary of basic subwords*, *suffix arrays*, and *suffix trees*. Suffix tree is a very useful data structure in string algorithmics.

Graph theoretic and combinatorial problems arise in several traditional and emerging scientific disciplines such as VLSI design, optimization, databases, and computational biology. Some examples include phylogeny reconstruction, protein–protein interaction network, placement and layout in VLSI chips, and so forth. **Chapter 31** presents fast parallel algorithms for several fundamental graph theoretic problems, optimized for multithreaded architectures such as the Cray MTA-2.

Volumes of data are exploding in both scientific and commercial domains. Data mining techniques that extract information from huge amount of data have become popular in many applications. In **Chapter 32**, parallel algorithms for association rule mining and clustering are presented.

Algorithmics research in mobile computing is still in its infancy, and dates back to only a few years. The elegance and terseness that exist today in algorithmics, especially parallel computing algorithmics, can be brought to bear on research on "mobile computing algorithmics." **Chapter 33** presents an overview in this direction.

Parallel Applications

One of the most significant obstacles to high-performance computing is latency. Working with massive data on a cluster induces latency in two forms: accessing data on disk and interprocessor communication. **Chapter 34** explains how pipelines can mitigate latency and describes FG, a programming environment that improves pipeline-structured programs.

An important aspect of any large-scale scientific application is data storage and retrieval. I/O technology lags behind other computing components by several orders of magnitude with a performance gap that is still growing. **Chapter 35** presents many powerful I/O techniques applied to each stratum of the parallel I/O software stack.

Chapter 36 surveys algorithms, complexity issues, and applications for message dissemination problems defined over networks based on modern communication primitives. More specifically, the problem of disseminating messages in parallel and distributed systems under the multicasting communication mode is discussed.

In **Chapter 37**, fundamental network problems that have attracted considerable attention in the algorithms community over the past 5–10 years are studied. Problems of interest are related to communication and data transfer, which are premier issues in high-performance networks today.

An important optimization problem that has to be solved by the base stations in wireless cellular networks that utilize frequency division multiplexing (FDM) technology is the *call admission control problem*. Given a spectrum of available frequencies and users that wish to communicate with their base station, the problem is to maximize the benefit, that is, the number of users that communicate without signal interference. **Chapter 38** studies the online version of the call admission control problem under a competitive analysis perspective.

In ad hoc wireless networks, the establishment of typical communication patterns like broadcasting, multicasting, and group communication is strongly related to energy consumption. Since energy is a scarce resource, corresponding *minimum energy communication problems* arise. **Chapter 39** considers a series of such problems on a suitable combinatorial model for ad hoc wireless networks and surveys recent approximation algorithms and inapproximability results.

Chapter 40 investigates the problem of *power aware scheduling*. Techniques that explore different degrees of parallelism in a schedule and generate an energy-efficient *canonical schedule* are surveyed. A canonical schedule corresponds to the case where all the tasks take their worst case execution times.

Parallel and distributed systems may operate in an environment that undergoes unpredictable changes causing certain system performance features to degrade. Such systems need robustness to guarantee limited degradation despite some fluctuations in the behavior of their component parts or environment. **Chapter 41** investigates the robustness of an allocation of resources to tasks in parallel and distributed systems.

Networks of workstations are widely considered a cost-efficient alternative to supercomputers. Despite this consensus and high demand, there is no general parallel processing framework that is accepted and used by all. **Chapter 42** discusses such a framework and addresses difficulties that arise in the use of NOWs.

Chapter 43 focuses on analytical approaches to scalability analysis of multicore processors. In relatively early stages of hardware development, such studies can effectively guide hardware design. For many applications, parallel algorithms scaling to very large configurations may not be the same as those that yield high efficiencies on smaller configurations. For these desired predictive properties, analytical modeling is critical.

Chapter 44 studies spatial locality-based parallel domain decomposition methods. An overview of some of the most widely used techniques such as orthogonal recursive bisection (ORB), space filling curves (SFCs), and parallel octrees is given. Parallel algorithms for computing the specified domain decompositions are surveyed as well.

Data replication is an essential technique employed to reduce the user access time in distributed computing systems. Algorithms available for the data replication problem (DRP) range from the traditional mathematical optimization techniques, such as linear programming, dynamic programming, and so forth, to the biologically inspired metaheuristics. **Chapter 45** aims to introduce game theory as a new oracle to tackle the data replication problem.

In **Chapter 46**, three initiatives that are designed to present a solution to the data service requirements of the *Advanced Computational Data Center Grid (ACDC Grid)* are presented. Discussions in this chapter also include two data grid file utilization simulation tools, the Scenario Builder and the Intelligent Migrator, which examine the complexities of delivering a distributed storage network.

Chapter 47 presents fast and scalable parallel matrix multiplication algorithms and their applications to distributed memory systems. This chapter also describes a processor-efficient parallelization of the best-known sequential matrix multiplication algorithm on a distributed memory system.

Intended Use of the Book

This book is meant for use by researchers, developers, educators, and students in the area of parallel computing. Since parallel computing has found applications in a wide variety of domains, the ideas and techniques illustrated in this book should prove useful to a very wide audience.

This book can also be used as a text in a graduate course dealing with parallel computing. Graduate students who plan to conduct research in this area will find this book especially invaluable. The book can also be used as a supplement to any course on algorithms, complexity, or applied computing.

Other Reading

A partial list of books that deal with randomization is given below. This is by no means an exhaustive list of all the books in the area.

1. T.H. Cormen, C.E. Leiserson, R.L. Rivest, and C. Stein, *Introduction to Algorithms*, Second edition, MIT Press, Cambridge, MA, 2001.
2. E. Horowitz, S. Sahni, and S. Rajasekaran, *Computer Algorithms*, W. H. Freeman, New York, 1998.
3. J. Já Já, *An Introduction to Parallel Algorithms*, Addison-Wesley Publishers, Reading, 1992.
4. T. Leighton, *Introduction to Parallel Algorithms and Architectures: Arrays–Trees–Hypercube*, Morgan-Kaufmann Publishers, San Mateo, CA, 1992.
5. P.M. Pardalos and S. Rajasekaran, editors, *Advances in Randomized Parallel Computing*, Kluwer Academic Publishers, Boston, 1999.
6. J.H. Reif, editor, *Synthesis of Parallel Algorithms*, Morgan-Kaufmann Publishers, San Mateo, CA, 1992.
7. M.T. Goodrich and R. Tamassia, *Algorithm Design: Foundations, Analysis, and Internet Examples*, John Wiley & Sons, Inc., New York, 2002.

Acknowledgments

We are very thankful to the authors who have written these chapters under a very tight schedule. We thank the staff of Chapman & Hall/CRC, in particular Bob Stern and Jill Jurgensen. We also gratefully acknowledge the partial support from the National Science Foundation through grants CCR-9912395 and ITR-0326155.

Sanguthevar Rajasekaran
John H. Reif

Editors

Sanguthevar Rajasekaran received his ME degree in automation from the Indian Institute of Science, Bangalore, in 1983, and his PhD degree in computer science from Harvard University in 1988. Currently he is the UTC chair professor of computer science and engineering at the University of Connecticut and the director of Booth Engineering Center for Advanced Technologies (BECAT). Before joining UConn, he has served as a faculty member in the CISE Department of the University of Florida and in the CIS Department of the University of Pennsylvania. During 2000–2002 he was the chief scientist for Arcot Systems. His research interests include parallel algorithms, bioinformatics, data mining, randomized computing, computer simulations, and combinatorial optimization. He has published more than 150 articles in journals and conferences. He has coauthored two texts on algorithms and coedited four books on algorithms and related topics. He is an elected member of the Connecticut Academy of Science and Engineering.

John Reif is Hollis Edens distinguished professor at Duke University. He developed numerous parallel and randomized algorithms for various fundamental problems including the solution of large sparse systems, sorting, graph problems, data compression, and so forth. He also made contributions to practical areas of computer science including parallel architectures, data compression, robotics, and optical computing. He is the author of over 200 papers and has edited four books on synthesis of parallel and randomized algorithms. He is a fellow of AAAS, IEEE, and ACM.

Contributors

Ishfaq Ahmad
Department of Computer Science and Engineering
University of Texas
Arlington, Texas

Selim G. Akl
School of Computing
Queen's University
Kingston, Ontario, Canada

Susanne Albers
Department of Computer Science
University of Freiburg
Freiburg, Germany

Shoukat Ali
Platform Validation Engineering
Intel Corporation
Sacramento, California

Srinivas Aluru
Department of Electrical and Computer Engineering
Iowa State University
Ames, Iowa

Reda A. Ammar
Computer Science and Engineering Department
University of Connecticut
Storrs, Connecticut

David A. Bader
Georgia Institute of Technology
Atlanta, Georgia

Gianfranco Bilardi
Department of Information Engineering
University of Padova
Padova, Italy

David Bindel
University of California
Berkeley, California

Anu G. Bourgeois
Department of Computer Science
Georgia State University
Atlanta, Georgia

Alfredo Buttari
University of Tennessee
Knoxville, Tennessee

George C. Caragea
Department of Computer Science
University of Maryland
College Park, Maryland

Ioannis Caragiannis
University of Patras and Research Academic Computer Technology Institute
Patras, Greece

Bruce R. Childers
Department of Computer Science
University of Pittsburgh
Pittsburgh, Pennsylvania

Avery Ching
Northwestern University
Evanston, Illinois

Alok Choudhary
Northwestern University
Evanston, Illinois

Kenin Coloma
Northwestern University
Evanston, Illinois

Guojing Cong
IBM T.J. Watson Research Center
Yorktown Heights, New York

Thomas H. Cormen
Dartmouth College
Hanover, New Hampshire

Elena Riccio Davidson
Dartmouth College
Hanover, New Hampshire

Jaime Davila
Department of Computer Science and Engineering
University of Connecticut
Storrs, Connecticut

xxv

James Demmel
University of California
Berkeley, California

Jack Dongarra
University of Tennessee
Knoxville, Tennessee
and
Oak Ridge National Laboratory
Oak Ridge, Tennessee

Michael Factor
IBM Research Laboratory
Haifa, Israel

John Feo
Microsoft Corporation
Redmond, Washington

Alex Gontmakher
Technion–Israel Institute of Technology
Haifa, Israel

Teofilo F. Gonzalez
University of California
Santa Barbara, California

Michael T. Goodrich
University of California
Irvine, California

Ananth Grama
Purdue University
West Lafayette, Indiana

Ming Gu
University of California
Berkeley, California

Ajay Gulati
Department of Electrical and Computer Engineering
Rice University
Houston, Texas

Yijie Han
University of Missouri–Kansas City
Kansas City, Missouri

Yozo Hida
University of California
Berkeley, California

Oscar H. Ibarra
Department of Computer Science
University of California
Santa Barbara, California

Joseph JaJa
Institute for Advanced Computer Studies and Department of Electrical and Computer Engineering
University of Maryland
College Park, Maryland

William Kahan
University of California
Berkeley, California

Christos Kaklamanis
University of Patras and Research Academic Computer Technology Institute
Patras, Greece

Panagiotis Kanellopoulos
University of Patras and Research Academic Computer Technology Institute
Patras, Greece

Samee Ullah Khan
Department of Electrical and Computer Engineering
Colorado State University
Fort Collins, Colorado

Samir Khuller
University of Maryland
College Park, Maryland

Yoo-Ah Kim
Computer Science and Engineering Department
University of Connecticut
Storrs, Connecticut

Lasse Kliemann
Department of Computer Science
Christian Albrechts University
Kiel, Germany

Spyros Kontogiannis
Computer Science Department
University of Ioannina
Ioannina, Greece

Evangelos Kranakis
School of Computer Science
Carleton University
Ottawa, Canada

Danny Krizanc
Department of Mathematics and Computer Science
Wesleyan University
Middletown, Connecticut

Vipin Kumar
Department of Computer Science
University of Minnesota
Minneapolis, Minnesota

Jakub Kurzak
University of Tennessee
Knoxville, Tennessee

Julie Langou
University of Tennessee
Knoxville, Tennessee

Julien Langou
University of Colorado
Denver, Colorado

Bryant C. Lee
Computer Science Department
Carnegie Mellon University
Pittsburgh, Pennsylvania

Young Choon Lee
School of Information Technologies
The University of Sydney
Sydney, Australia

Jianwei Li
Northwestern University
Evanston, Illinois

Contributors

Keqin Li
Department of Computer Science
State University of New York
New Paltz, New York

Xiaoye Li
Lawrence Berkeley National Laboratory
Berkeley, California

Wei-keng Liao
Northwestern University
Evanston, Illinois

Ying Liu
DTKE Center
Graduate University of the Chinese Academy of Sciences
Beijing, China

Piotr Luszczek
MathWorks
Natick, Massachusetts

Anthony A. Maciejewski
Department of Electrical and Computer Engineering
Colorado State University
Fort Collins, Colorado

Kamesh Madduri
Computational Science and Engineering Division
College of Computing
Georgia Institute of Technology
Atlanta, Georgia

Osni Marques
Lawrence Berkeley National Laboratory
Berkeley, California

Rami Melhem
Department of Computer Science
University of Pittsburgh
Pittsburgh, Pennsylvania

Avi Mendelson
Intel Israel
MTM Scientific Industries Center
Haifa, Israel

Russ Miller
Center for Computational Research
Department of Computer Science and Engineering
State University of New York
Buffalo, New York

Daniel Mossé
Department of Computer Science
University of Pittsburgh
Pittsburgh, Pennsylvania

Koji Nakano
Department of Information Engineering
Hiroshima University
Higashi Hiroshima, Japan

Michael J. Nelson
University of California
Irvine, California

Stephan Olariu
Department of Computer Science
Old Dominion University
Norfolk, Virginia

Michael A. Palis
Rutgers University
Camden, New Jersey

Evi Papaioannou
University of Patras and Research Academic Computer Technology Institute
Patras, Greece

Beresford Parlett
University of California
Berkeley, California

Andrei Păun
Department of Computer Science/IfM
Louisiana Tech University
Ruston, Louisiana

Andrea Pietracaprina
Department of Information Engineering
University of Padova
Padova, Italy

Viktor K. Prasanna
Department of Electrical Engineering
University of Southern California
Los Angeles, California

Geppino Pucci
Department of Information Engineering
University of Padova
Padova, Italy

Sanguthevar Rajasekaran
Department of Computer Science and Engineering
University of Connecticut
Storrs, Connecticut

Sergio Rajsbaum
Instituto de Matemáticas
Universidad Nacional Autónoma de México
Ciudad Universitaria, Mexico

E. Jason Riedy
University of California
Berkeley, California

Catherine L. Ruby
Center for Computational Research
Department of Computer Science and Engineering
State University of New York
Buffalo, New York

Wojciech Rytter
Warsaw University
Warsaw, Poland

Sartaj Sahni
University of Florida
Gainesville, Florida

Nicola Santoro
School of Computer Science
Carleton University
Ottawa, Canada

Assaf Schuster
Technion–Israel Institute of Technology
Haifa, Israel

Ronald Scrofano
Computer Science Department
University of Southern California
Los Angeles, California

Sudip Seal
Department of Electrical and Computer Engineering
Iowa State University
Ames, Iowa

Konstantin Shagin
Technion–Israel Institute of Technology
Haifa, Israel

Qingmin Shi
Institute for Advanced Computer Studies
University of Maryland
College Park, Maryland

Gregory Shklover
Technion–Israel Institute of Technology
Haifa, Israel

Howard Jay Siegel
Department of Electrical and Computer Engineering
Department of Computer Science
Colorado State University
Fort Collins, Colorado

Paul G. Spirakis
Research Academic Computer Technology Institute
University of Patras
Patras, Greece

Anand Srivastav
Department of Computer Science
Christian Albrechts University
Kiel, Germany

Stanimire Tomov
Computer Science Department
University of Tennessee
Knoxville, Tennessee

Jerry L. Trahan
Department of Electrical and Computer Engineering
Louisiana State University
Baton Rouge, Louisiana

Ramachandran Vaidyanathan
Department of Electrical and Computer Engineering
Louisiana State University
Baton Rouge, Louisiana

Peter J. Varman
Department of Electrical and Computer Engineering
Rice University
Houston, Texas

Amitabh Varshney
Department of Computer Science and Institute for Advanced Computer Studies
University of Maryland
College Park, Maryland

Uzi Vishkin
University of Maryland Institute for Advanced Computer Studies
Department of Electrical and Computer Engineering
University of Maryland
College Park, Maryland

Christof Voemel
University of California
Berkeley, California

Yung-Chun (Justin) Wan
Google Inc.
Mountain View, California

Chih-Fang Wang
Department of Computer Science
Southern Illinois University at Carbondale
Carbondale, Illinois

Peter Widmayer
Institute for Theoretical Informatics
ETH Zurich
Zurich, Switzerland

Yuanyuan Yang
Departmental of Electrical and Computer Engineering
State University of New York at Stony Brook
Stony Brook, New York

Zhenghao Zhang
Departmental of Electrical and Computer Engineering
State University of New York at Stony Brook
Stony Brook, New York

Dakai Zhu
University of Texas at San Antonio
San Antonio, Texas

Albert Y. Zomaya
Department of Computer Science
School of Information Technologies
The University of Sydney
Sydney, Australia

Models

1
Evolving Computational Systems

1.1	Introduction..	1-1
1.2	Computational Models	1-2
	Time and Speed • What Does it Mean to Compute? • Sequential Model • Parallel Model • A Fundamental Assumption	
1.3	Time-Varying Variables	1-4
	Quantum Decoherence • Sequential Solution • Parallel Solution	
1.4	Time-Varying Computational Complexity	1-6
	Examples of Increasing Functions $C(t)$ • Computing with Deadlines • Accelerating Machines	
1.5	Rank-Varying Computational Complexity	1-9
	An Algorithmic Example: Binary Search • The Inverse Quantum Fourier Transform	
1.6	Interacting Variables..................................	1-12
	Disturbing the Equilibrium • Solutions • Distinguishability in Quantum Computing	
1.7	Computations Obeying Mathematical Constraints ...	1-17
	Mathematical Transformations • Sequential Solution • Parallel Solution	
1.8	The Universal Computer is a Myth	1-19
1.9	Conclusion	1-20
	Acknowledgment...	1-20
	References	1-21

Selim G. Akl
Queen's University

1.1 Introduction

The universe in which we live is in a constant state of evolution. People age, trees grow, the weather varies. From one moment to the next, our world undergoes a myriad of transformations. Many of these changes are obvious to the naked eye, others more subtle. Deceptively, some appear to occur independently of any direct external influences. Others are immediately perceived as the result of actions by other entities.

In the realm of computing, it is generally assumed that the world is static. The vast majority of computations take place in applications where change is thought of, rightly or wrongly, as inexistent or irrelevant. Input data are read, algorithms are applied to them, and results are produced. The possibility that the data, the algorithms, or even the results sought may vary *during* the process of computation is rarely, if ever, contemplated.

In this chapter we explore the concept of *evolving computational systems*. These are systems in which everything in the computational process is subject to change. This includes inputs, algorithms, outputs, and even the computing agents themselves. A simple example of a computational paradigm that meets this definition of an evolving system to a limited extent is that of a computer interacting in real time with a user while processing information. Our focus here is primarily on certain changes that may affect the data required to solve a problem. We also examine changes that affect the complexity of the algorithm used in the solution. Finally, we look at one example of a computer capable of evolving with the computation.

A number of evolving computational paradigms are described for which a parallel computing approach is most appropriate. In Sections 1.3, 1.4, and 1.5, time plays an important role either directly or indirectly in the evolution of the computation. Thus, it is the passage of time that may cause the change in the data. In another context, it may be the order in which a stage of an algorithm is performed that determines the number of operations required by that stage. The effect of time on evolving computations is also used to contrast the capabilities of a parallel computer with those of an unconventional model of computation known as the *accelerating machine*. In Sections 1.6 and 1.7, it is not time but rather external agents acting on the data that are responsible for a variable computation. Thus, the data may be affected by a measurement that perturbs an existing equilibrium, or by a modification in a mathematical structure that violates a required condition. Finally, in Section 1.8 evolving computations allow us to demonstrate the impossibility of achieving universality in computing. Our conclusions are offered in Section 1.9.

1.2 Computational Models

It is appropriate at the outset that we define our models of computation. Two such models are introduced in this section, one sequential and one parallel. (A third model, the accelerating machine, is defined in Section 1.4.3.) We begin by stating clearly our understanding regarding the meaning of time, and our assumptions in connection with the speed of processors.

1.2.1 Time and Speed

In the classical study of algorithms, whether sequential or parallel, the notion of a *time unit* is fundamental to the analysis of an algorithm's running time. A time unit is the smallest discrete measure of time. In other words, time is divided into consecutive time units that are indivisible. All events occur at the beginning of a time unit. Such events include, for example, a variable changing its value, a processor undertaking the execution of a step in its computation, and so on.

It is worth emphasizing that the length of a time unit is not an absolute quantity. Instead, the duration of a time unit is specified in terms of a number of factors. These include the parameters of the computation at hand, such as the rate at which the data are received, or the rate at which the results are to be returned. Alternatively, a time unit may be defined in terms of the speed of the processors available (namely, the single processor on a sequential computer and each processor on a parallel computer). In the latter case, a faster processor implies a smaller time unit.

In what follows, the standard definition of time unit is adopted, namely: A time unit is the length of time required by a processor to perform a *step* of its computation. Specifically, during a time unit, a processor executes a step consisting of

1. A *read* operation in which it receives a constant number of fixed-size data as input
2. A *calculate* operation in which it performs a fixed number of constant-time *arithmetic* and *logical* calculations (such as adding two numbers, comparing two numbers, and so on)
3. A *write* operation in which it returns a constant number of fixed-size data as output

Evolving Computational Systems

All other occurrences external to the processor (such as the data arrival rate, for example) will be set and measured in these terms. Henceforth, the term *elementary operation* is used to refer to a read, a calculate, or a write operation.

1.2.2 What Does it Mean to Compute?

An important characteristic of the treatment in this chapter is the broad perspective taken to define what it means *to compute*. Specifically, *computation* is a process whereby information is manipulated by, for example, acquiring it (input), transforming it (calculation), and transferring it (output). Any form of information processing (whether occurring spontaneously in nature, or performed on a computer built by humans) is a computation. Instances of computational processes include

1. Measuring a physical quantity
2. Performing an arithmetic or logical operation on a pair of numbers
3. Setting the value of a physical quantity

to cite but a few. These computational processes themselves may be carried out by a variety of means, including, of course, conventional (electronic) computers, and also through physical phenomena [35], chemical reactions [1], and transformations in living biological tissue [42]. By extension, *parallel computation* is defined as the execution of several such processes of the same type simultaneously.

1.2.3 Sequential Model

This is the conventional model of computation used in the design and analysis of sequential (also known as *serial*) algorithms. It consists of a single processor made up of circuitry for executing arithmetic and logical operations and a number of registers that serve as internal memory for storing programs and data. For our purposes, the processor is also equipped with an input unit and an output unit that allow it to receive data from, and send data to, the outside world, respectively.

During each time unit of a computation the processor can perform

1. A read operation, that is, receive a constant number of fixed-size data as input
2. A calculate operation, that is, execute a fixed number of constant-time calculations on its input
3. A write operation, that is, return a constant number of fixed-size results as output

It is important to note here that the read and write operations can be, respectively, from and to the model's internal memory. In addition, both the reading and writing may be, on occasion, from and to an external medium in the environment in which the computation takes place. Several incarnations of this model exist, in theory and in practice [40]. In what follows, we refer to this model of computation as a *sequential computer*.

1.2.4 Parallel Model

Our chosen model of parallel computation consists of n processors, numbered 1 to n, where $n \geq 2$. Each processor is of the type described in Section 1.2.3. The processors are connected in some fashion and are able to communicate with one another in order to exchange data and results. These exchanges may take place through an *interconnection network*, in which case selected pairs of processors are directly connected by two-way communication links. Pairs of processors not directly connected communicate indirectly by creating a route for their messages that goes through other processors. Alternatively, all exchanges may take place via a global *shared memory* that is used as a bulletin board. A processor wishing to send a datum to another processor does so by writing it to the shared memory, from where it is read by the other processor. Several varieties of interconnection networks and modes of shared memory access exist, and any of them would be adequate as far as we are concerned here. The exact nature of the communication medium among the processors is of no consequence to the results described in this chapter. A study of various ways of connecting processors is provided in [2].

During each time unit of a computation a processor can perform

1. A read operation, that is, receive as input a constant number of fixed-size data
2. A calculate operation, that is, execute a fixed number of constant-time calculations on its input
3. A write operation, that is, return as output a constant number of fixed-size results

As with the sequential processor, the input can be received from, and the output returned to, either the internal memory of the processor or the outside world. In addition, a processor in a parallel computer may receive its input from and return its output to the shared memory (if one is available), or another processor (through a direct link or via a shared memory, whichever is available). Henceforth, this model of computation will be referred to as a *parallel computer* [3].

1.2.5 A Fundamental Assumption

The analyses in this chapter assume that all models of computation use the fastest processors possible (within the bounds established by theoretical physics). Specifically, no sequential computer exists that is faster than the one of Section 1.2.3, and similarly no parallel computer exists whose processors are faster than those of Section 1.2.4. Furthermore, no processor on the parallel computer of Section 1.2.4 is faster than the processor of the sequential computer of Section 1.2.3. This is the *fundamental assumption in parallel computation*. It is also customary to suppose that the sequential and parallel computers use identical processors. We adopt this convention throughout this chapter, with a single exception: In Section 1.4.3, we assume that the processor of the sequential computer is in fact capable of increasing its speed at every step (at a pre-established rate, so that the number of operations executable at every consecutive step is known a priori and fixed once and for all).

1.3 Time-Varying Variables

For a positive integer n larger than 1, we are given n functions, each of one variable, namely, $F_0, F_1, \ldots, F_{n-1}$, operating on the n variables $x_0, x_1, \ldots, x_{n-1}$, respectively. Specifically, it is required to compute $F_i(x_i)$, for $i = 0, 1, \ldots, n-1$. For example, $F_i(x_i)$ may be equal to x_i^2.

What is unconventional about this computation is the fact that the x_i are themselves functions that vary with time. It is therefore appropriate to write the n variables as

$$x_0(t), x_1(t), \ldots, x_{n-1}(t),$$

that is, as functions of the time variable t. It is important to note here that, while it is known that the x_i change with time, the actual functions that effect these changes are not known (e.g., x_i may be a true random variable).

All the physical variables exist in their natural environment within which the computation is to take place. They are all available to be operated on at the beginning of the computation. Thus, for each variable $x_i(t)$, it is possible to compute $F_i(x_i(t))$, provided that a computer is available to perform the calculation (and subsequently return the result).

Recall that time is divided into intervals, each of duration one time unit. It takes one time unit to evaluate $F_i(x_i(t))$. The problem calls for computing $F_i(x_i(t))$, $0 \leq i \leq n-1$, at time $t = t_0$. In other words, once all the variables have assumed their respective values at time $t = t_0$, the functions F_i are to be evaluated for all values of i. Specifically,

$$F_0(x_0(t_0)), F_1(x_1(t_0)), \ldots, F_{n-1}(x_{n-1}(t_0))$$

are to be computed. The fact that $x_i(t)$ changes with the passage of time should be emphasized here. Thus, if $x_i(t)$ is not operated on at time $t = t_0$, then after one time unit $x_i(t_0)$ becomes $x_i(t_0 + 1)$, and after two time units it is $x_i(t_0 + 2)$, and so on. Indeed, time exists as a fundamental fact of life. It is real,

Evolving Computational Systems

relentless, and unforgiving. Time cannot be stopped, much less reversed. (For good discussions of these issues, see [28, 45].) Furthermore, for $k > 0$, not only is each value $x_i(t_0 + k)$ different from $x_i(t_0)$ but also the latter cannot be obtained from the former. We illustrate this behavior through an example from physics.

1.3.1 Quantum Decoherence

A binary variable is a mathematical quantity that takes exactly one of a total of two possible values at any given time. In the base 2 number system, these values are 0 and 1, and are known as *binary digits* or *bits*. Today's conventional computers use electronic devices for storing and manipulating bits. These devices are in either one or the other of two physical states at any given time (e.g., two voltage levels), one representing 0, the other 1. We refer to such a device, as well as the digit it stores, as a *classical bit*.

In *quantum computing*, a bit (aptly called a quantum bit, or *qubit*) is both 0 and 1 at the same time. The qubit is said to be in a *superposition* of the two values. One way to implement a qubit is by encoding the 0 and 1 values using the spin of an electron (e.g., clockwise, or "up" for 1, and counterclockwise, or "down" for 0). Formally, a qubit is a unit vector in a two-dimensional state space, for which a particular orthonormal basis, denoted by $\{|0\rangle, |1\rangle\}$ has been fixed. The two basis vectors $|0\rangle$ and $|1\rangle$ correspond to the possible values a classical bit can take. However, unlike classical bits, a qubit can also take many other values. In general, an arbitrary qubit can be written as a linear combination of the computational basis states, namely, $\alpha|0\rangle + \beta|1\rangle$, where α and β are complex numbers such that $|\alpha|^2 + |\beta|^2 = 1$.

Measuring the value of the qubit (i.e., reading it) returns a 0 with probability $|\alpha|^2$ and a 1 with a probability $|\beta|^2$. Furthermore, the measurement causes the qubit to undergo *decoherence* (literally, to lose its coherence). When decoherence occurs, the superposition is said to collapse: any subsequent measurement returns the same value as the one obtained by the first measurement. The information previously held in the superposition is lost forever. Henceforth, the qubit no longer possesses its quantum properties and behaves as a classical bit [33].

There is a second way, beside measurement, for decoherence to take place. A qubit loses its coherence simply through prolonged exposure to its natural environment. The interaction between the qubit and its physical surroundings may be thought of as an external action by the latter causing the former to behave as a classical bit, that is, to lose all information it previously stored in a superposition. (One can also view decoherence as the act of the qubit making a mark on its environment by adopting a classical value.) Depending on the particular implementation of the qubit, the time needed for this form of decoherence to take place varies. At the time of this writing, it is well below 1 s (more precisely, in the vicinity of a nanosecond). The information lost through decoherence cannot be retrieved. For the purposes of this example, the time required for decoherence to occur is taken as one time unit.

Now suppose that a quantum system consists of n independent qubits, each in a state of superposition. Their respective values at some time t_0, namely, $x_0(t_0), x_1(t_0), \ldots, x_{n-1}(t_0)$, are to be used as inputs to the n functions $F_0, F_1, \ldots, F_{n-1}$, in order to perform the computation described at the beginning of Section 1.3, that is, to evaluate $F_i(x_i(t_0))$, for $0 \leq i \leq n - 1$.

1.3.2 Sequential Solution

A sequential computer fails to compute all the F_i as desired. Indeed, suppose that $x_0(t_0)$ is initially operated upon. It follows that $F_0(x_0(t_0))$ can be computed correctly. However, when the next variable, x_1, for example, is to be used (as input to F_1), the time variable would have changed from $t = t_0$ to $t = t_0 + 1$, and we obtain $x_1(t_0 + 1)$, instead of the $x_1(t_0)$ that we need. Continuing in this fashion, $x_2(t_0 + 2)$, $x_3(t_0 + 3), \ldots, x_{n-1}(t_0 + n - 1)$, represent the sequence of inputs. In the example of Section 1.3.1, by the time $F_0(x_0(t_0))$ is computed, one time unit would have passed. At this point, the $n - 1$ remaining qubits would have undergone decoherence. The same problem occurs if the sequential computer attempts to first read all the x_i, one by one, and store them before calculating the F_i.

Since the function according to which each x_i changes with time is not known, it is impossible to recover $x_i(t_0)$ from $x_i(t_0 + i)$, for $i = 1, 2, \ldots, n - 1$. Consequently, this approach cannot produce $F_1(x_1(t_0))$, $F_2(x_2(t_0)), \ldots, F_{n-1}(x_{n-1}(t_0))$, as required.

1.3.3 Parallel Solution

For a given n, *any* computer capable of performing n calculate operations per step can easily evaluate the $F_i(x_i(t_0))$, all simultaneously, leading to a successful computation.

Thus, a parallel computer consisting of n independent processors may perform all the computations at once: For $0 \le i \le n - 1$, and all processors working at the same time, processor i computes $F_i(x_i(t_0))$. In the example of Section 1.3.1, the n functions are computed in parallel at time $t = t_0$, before decoherence occurs.

1.4 Time-Varying Computational Complexity

In traditional computational complexity theory, the *size* of a problem $\mathcal{P}$ plays an important role. If $\mathcal{P}$ has size n, for example, then the number of operations required in the worst case to solve $\mathcal{P}$ (by any algorithm) is expressed as a function of n. Similarly, the number of operations executed (in the best, average, and worst cases) by a specific algorithm that solves $\mathcal{P}$ is also expressed as a function of n. Thus, for example, the problem of sorting a sequence of n numbers requires $\Omega(n \log n)$ comparisons, and the sorting algorithm Quicksort performs $O(n^2)$ comparisons in the worst case.

In this section, we depart from this model. Here, the size of the problem plays a secondary role. In fact, in most (though not necessarily all) cases, the problem size may be taken as constant. The computational complexity now depends on *time*. Not only science and technology but also everyday life provide many instances demonstrating time-varying complexity. Thus, for example,

1. An illness may get better or worse with time, making it more or less amenable to treatment.
2. Biological and software viruses spread with time making them more difficult to cope with.
3. Spam accumulates with time making it more challenging to identify the legitimate email "needles" in the "haystack" of junk messages.
4. Tracking moving objects becomes harder as they travel away from the observer (e.g., a spaceship racing toward Mars).
5. Security measures grow with time in order to combat crime (e.g., when protecting the privacy, integrity, and authenticity of data, ever stronger cryptographic algorithms are used, i.e., ones that are more computationally demanding to break, thanks to their longer encryption and decryption keys).
6. Algorithms in many applications have complexities that vary with time from one time unit during the computation to the next. Of particular importance here are
   a. *Molecular dynamics* (the study of the dynamic interactions among the atoms of a system, including the calculation of parameters such as forces, energies, and movements) [18,39].
   b. *Computational fluid dynamics* (the study of the structural and dynamic properties of moving objects, including the calculation of the velocity and pressure at various points) [11].

Suppose that we are given an algorithm for solving a certain computational problem. The algorithm consists of a number of stages, where each stage may represent, for example, the evaluation of a particular arithmetic expression such as

$$c \leftarrow a + b.$$

Further, let us assume that a computational stage executed at time t requires a number $C(t)$ of constant-time operations. As the aforementioned situations show, the behavior of C varies from case to case.

Evolving Computational Systems

Typically, C may be an increasing, decreasing, unimodal, periodic, random, or chaotic function of t. In what follows, we study the effect on computational complexity of a number of functions $C(t)$ that grow with time.

It is worth noting that we use the term *stage* to refer to a component of an algorithm, hence a variable entity, in order to avoid confusion with a *step*, an intrinsic property of the computer, as defined in Sections 1.2.1 and 1.4.3. In conventional computing, where computational complexity is invariant (i.e., oblivious to external circumstances), a *stage* (as required by an algorithm) is exactly the same thing as a *step* (as executed by a computer). In *unconventional computing* (the subject of this chapter), computational complexity is affected by its environment and is therefore variable. Under such conditions, one or more steps may be needed in order to execute a stage.

1.4.1 Examples of Increasing Functions $C(t)$

Consider the following three cases in which the number of operations required to execute a computational stage increases with time. For notational convenience, we use $S(i)$ to express the number of operations performed in executing stage i, at the time when that stage is in fact executed. Denoting the latter by t_i, it is clear that $S(i) = C(t_i)$.

1. For $t \geq 0$, $C(t) = t + 1$. Table 1.1 illustrates t_i, $C(t_i)$, and $S(i)$, for $1 \leq i \leq 6$.
   It is clear in this case that $S(i) = 2^{i-1}$, for $i \geq 1$. It follows that the total number of operations performed when executing all stages, from stage 1 up to and including stage i, is

$$\sum_{j=1}^{i} 2^{j-1} = 2^i - 1.$$

It is interesting to note that while $C(t)$ is a linear function of the time variable t, the quantity $S(i)$ grows exponentially with $i - 1$, where i is the number of stages executed so far. The effect of this behavior on the total number of operations performed is appreciated by considering the following example. When executing a computation requiring $\log n$ stages for a problem of size n,

$$2^{\log n} - 1 = n - 1$$

operations are performed.

2. For $t \geq 0$, $C(t) = 2^t$. Table 1.2 illustrates t_i, $C(t_i)$, and $S(i)$, for $1 \leq i \leq 5$.
   In this case, $S(1) = 1$, and for $i > 1$, we have

$$S(i) = 2^{\sum_{j=1}^{i-1} S(j)}.$$

TABLE 1.1 Number of Operations Required to Complete Stage i When $C(t) = t + 1$

Stage i	t_i	$C(t_i)$	$S(i)$
1	0	$C(0)$	1
2	$0 + 1$	$C(1)$	2
3	$1 + 2$	$C(3)$	4
4	$3 + 4$	$C(7)$	8
5	$7 + 8$	$C(15)$	16
6	$15 + 16$	$C(31)$	32
7	$31 + 32$	$C(63)$	64

TABLE 1.2 Number of Operations Required to Complete Stage i When $C(t) = 2^t$

Stage i	t_i	$C(t_i)$	$S(i)$
1	0	$C(0)$	2^0
2	$0 + 1$	$C(1)$	2^1
3	$1 + 2$	$C(3)$	2^3
4	$3 + 8$	$C(11)$	2^{11}
5	$11 + 2048$	$C(2059)$	2^{2059}

TABLE 1.3 Number of Operations Required to Complete Stage i When $C(t) = 2^{2^t}$

Stage i	t_i	$C(t_i)$	$S(i)$
1	0	$C(0)$	2^{2^0}
2	$0 + 2$	$C(2)$	2^{2^2}
3	$2 + 16$	$C(18)$	$2^{2^{18}}$

Since $S(i) > \sum_{j=1}^{i-1} S(j)$, the total number of operations required by i stages is less than $2S(i)$, that is, $O(S(i))$.

Here we observe again that while $C(t) = 2C(t-1)$, the number of operations required by $S(i)$, for $i > 2$, increases significantly faster than double those required by all previous stages combined.

3. For $t \geq 0$, $C(t) = 2^{2^t}$. Table 1.3 illustrates t_i, $C(t_i)$, and $S(i)$, for $1 \leq i \leq 3$.

Here, $S(1) = 2$, and for $i > 1$, we have

$$S(i) = 2^{2^{\sum_{j=1}^{i-1} S(j)}}.$$

Again, since $S(i) > \sum_{j=1}^{i-1} S(j)$, the total number of operations required by i stages is less than $2S(i)$, that is, $O(S(i))$.

In this example, the difference between the behavior of $C(t)$ and that of $S(i)$ is even more dramatic. Obviously, $C(t) = C(t-1)^2$, where $t \geq 1$ and $C(0) = 2$, and as such $C(t)$ is a fast growing function ($C(4) = 65,536$, while $C(7)$ is represented with 39 decimal digits). Yet, $S(i)$ grows at a far more dizzying pace: Already $S(3)$ is equal to 2 raised to the power $4 \times 65,536$.

The significance of these examples and their particular relevance in parallel computation are illustrated by the paradigm in the following section.

1.4.2 Computing with Deadlines

Suppose that a certain computation requires that n functions, each of one variable, be computed. Specifically, let $f_0(x_0), f_1(x_1), \ldots, f_{n-1}(x_{n-1})$ be the functions to be computed. This computation has the following characteristics:

1. The n functions are entirely independent. There is no precedence whatsoever among them; they can be computed in any order.
2. Computing $f_i(x_i)$ at time t requires $C(t) = 2^t$ operations, for $0 \leq i \leq n-1$ and $t \geq 0$.
3. There is a deadline for reporting the results of the computations: All n values $f_0(x_0), f_1(x_1), \ldots, f_{n-1}(x_{n-1})$ must be returned by the end of the third time unit, that is, when $t = 3$.

It should be easy to verify that no sequential computer, capable of exactly one constant-time operation per step (i.e., per time unit), can perform this computation for $n \geq 3$. Indeed, $f_0(x_0)$ takes $C(0) = 2^0 = 1$

Evolving Computational Systems

time unit, $f_1(x_1)$ takes another $C(1) = 2^1 = 2$ time units, by which time three time units would have elapsed. At this point none of $f_2(x_2), f_3(x_3), \ldots, f_{n-1}(x_{n-1})$ would have been computed.

By contrast, an n-processor parallel computer solves the problem handily. With all processors operating simultaneously, processor i computes $f_i(x_i)$ at time $t = 0$, for $0 \leq i \leq n - 1$. This consumes one time unit, and the deadline is met.

The example in this section is based on one of the three functions for $C(t)$ presented in Section 1.4.1. Similar analyses can be performed in the same manner for $C(t) = t + 1$ and $C(t) = 2^{2^t}$, as well as other functions describing time-varying computational complexity.

1.4.3 Accelerating Machines

In order to put the result in Section 1.4.2 in perspective, we consider a variant on the sequential model of computation described in Section 1.2.3. An *accelerating machine* is a sequential computer capable of increasing the number of operations it can do at each successive step of a computation. This is primarily a theoretical model with no existing implementation (to date!). It is widely studied in the literature on unconventional computing [10,12,14,43,44,46]. The importance of the accelerating machine lies primarily in its role in questioning some long-held beliefs regarding uncomputability [13] and universality [7].

It is important to note that the rate of acceleration is specified at the time the machine is put in service and remains the same for the lifetime of the machine. Thus, the number of operations that the machine can execute during the ith step is known in advance and fixed permanently, for $i = 1, 2, \ldots$.

Suppose that an accelerating machine that can *double* the number of operations that it can perform at each step is available. Such a machine would be able to perform one operation in the first step, two operations in the second, four operations in the third, and so on. How would such an extraordinary machine fare with the computational problem of Section 1.4.2?

As it turns out, an accelerating machine capable of doubling its speed at each step, is unable to solve the problem for $n \geq 4$. It would compute $f_0(x_0)$, at time $t = 0$ in one time unit. Then it would compute $f_1(x_1)$, which now requires two operations at $t = 1$, also in one time unit. Finally, $f_2(x_2)$, requiring four operations at $t = 2$, is computed in one time unit, by which time $t = 3$. The deadline has been reached and none of $f_3(x_3), f_4(x_4), \ldots, f_{n-1}(x_{n-1})$ has been computed.

In closing this discussion of accelerating machines we note that once an accelerating machine has been defined, a problem can always be devised to expose its limitations. Thus, let the acceleration function be $\Phi(t)$. In other words, $\Phi(t)$ describes the number of operations that the accelerating machine can perform at time t. For example, $\Phi(t) = 2\Phi(t - 1)$, with $t \geq 1$ and $\Phi(0) = 1$, as in the case of the accelerating machine in this section. By simply taking $C(t) > \Phi(t)$, the accelerating machine is rendered powerless, *even in the absence of deadlines*.

1.5 Rank-Varying Computational Complexity

Unlike the computations in Section 1.4, the computations with which we are concerned here have a complexity that does not vary with time. Instead, suppose that a computation consists of n stages. There may be a certain precedence among these stages, that is, the order in which the stages are performed matters since some stages may depend on the results produced by other stages. Alternatively, the n stages may be totally independent, in which case the order of execution is of no consequence to the correctness of the computation.

Let the *rank* of a stage be the order of execution of that stage. Thus, stage i is the ith stage to be executed. In this section, we focus on computations with the property that the number of operations required to execute a stage whose rank is i is a function of i only. For example, as in Section 1.4, this function may be increasing, decreasing, unimodal, random, or chaotic. Instances of algorithms whose computational complexity varies from one stage to another are described in Reference 15. As we did before, we concentrate here on the case where the computational complexity C is an increasing function of i.

When does rank-varying computational complexity arise? Clearly, if the computational requirements grow with the rank, this type of complexity manifests itself in those circumstances where it is a disadvantage, whether avoidable or unavoidable, to being ith, for $i \geq 2$. For example

1. A penalty may be charged for missing a deadline, such as when a stage s must be completed by a certain time d_s.
2. The precision and/or ease of measurement of variables involved in the computation in a stage s may decrease with each stage executed before s.
3. Biological tissues may have been altered (by previous stages) when stage s is reached.
4. The effect of $s - 1$ quantum operations may have to be reversed to perform stage s.

1.5.1 An Algorithmic Example: Binary Search

Binary search is a well-known (sequential) algorithm in computer science. It searches for an element x in a sorted list L of n elements. In the worst case, binary search executes $O(\log n)$ stages. In what follows, we denote by $B(n)$ the total number of elementary operations performed by binary search (on a sequential computer), and hence its running time, in the worst case.

Conventionally, it is assumed that $C(i) = O(1)$, that is, each stage i requires the same constant number of operations when executed. Thus, $B(n) = O(\log n)$. Let us now consider what happens to the computational complexity of binary search when we assume, unconventionally, that the computational complexity of every stage i increases with i. Table 1.4 shows how $B(n)$ grows for three different values of $C(i)$.

In a parallel environment, where n processors are available, the fact that the sequence L is sorted is of no consequence to the search problem. Here, each processor reads x, compares one of the elements of L to x, and returns the result of the comparison. This requires one time unit. Thus, regardless of $C(i)$, the running time of the parallel approach is always the same.

1.5.2 The Inverse Quantum Fourier Transform

Consider a quantum register consisting of n qubits. There are 2^n computational basis vectors associated with such a register, namely,

$$\begin{aligned} |0\rangle &= |000\cdots 00\rangle, \\ |1\rangle &= |000\cdots 01\rangle, \\ &\vdots \\ |2^n - 1\rangle &= |111\cdots 11\rangle. \end{aligned}$$

TABLE 1.4 Number of Operations Required by Binary Search for Different Functions $C(i)$

$C(i)$	$B(n)$
i	$O(\log^2 n)$
2^i	$O(n)$
2^{2^i}	$O(2^n)$

Evolving Computational Systems

Let $|j\rangle = |j_1 j_2 j_3 \cdots j_{n-1} j_n\rangle$ be one of these vectors. For $j = 0, 1, \ldots, 2^n - 1$, the quantum Fourier transform of $|j\rangle$ is given by

$$\frac{\left(|0\rangle + e^{2\pi i 0.j_n}|1\rangle\right) \otimes \left(|0\rangle + e^{2\pi i 0.j_{n-1}j_n}|1\rangle\right) \otimes \cdots \otimes \left(|0\rangle + e^{2\pi i 0.j_1 j_2 \cdots j_n}|1\rangle\right)}{2^{n/2}},$$

where

1. Each transformed qubit is a balanced superposition of $|0\rangle$ and $|1\rangle$.
2. For the remainder of this section $i = \sqrt{-1}$.
3. The quantities $0.j_n$, $0.j_{n-1}j_n$, $\ldots$, $0.j_1 j_2 \cdots j_n$, are binary fractions, whose effect on the $|1\rangle$ component is called a *rotation*.
4. The operator $\otimes$ represents a tensor product; for example.

$$(a_1|0\rangle + b_1|1\rangle) \otimes (a_2|0\rangle + b_2|1\rangle) = a_1 a_2 |00\rangle + a_1 b_2 |01\rangle + b_1 a_2 |10\rangle + b_1 b_2 |11\rangle.$$

We now examine the inverse operation, namely, obtaining the original vector $|j\rangle$ from its given quantum Fourier transform.

1.5.2.1 Sequential Solution

Since the computation of each of $j_1, j_2, \ldots j_{n-1}$ depends on j_n, we must begin by computing the latter from $|0\rangle + e^{2\pi i 0.j_n}|1\rangle$. This takes one operation. Now j_n is used to compute j_{n-1} from $|0\rangle + e^{2\pi i 0.j_{n-1}j_n}|1\rangle$ in two operations. In general, once j_n is available, j_k requires knowledge of $j_{k+1}, j_{k+2}, \ldots, j_n$, must be computed in $(n-k+1)$st place, and costs $n-k+1$ operations to retrieve from $|0\rangle + e^{2\pi i 0.j_k j_{k+1} \cdots j_n}|1\rangle$, for $k = n-1, n-2, \ldots, 1$. Formally, the sequential algorithm is as follows:

for $k = n$ **downto** 1 **do**
$\quad |j_k\rangle \leftarrow \frac{1}{\sqrt{2}} \begin{pmatrix} |0\rangle \\ e^{2\pi i 0.j_k j_{k+1} \cdots j_n}|1\rangle \end{pmatrix}$
$\quad$ **for** $m = k+1$ **to** n **do**
$\quad\quad$ **if** $j_{n+k+1-m} = 1$ **then**
$\quad\quad\quad |j_k\rangle \leftarrow |j_k\rangle \begin{pmatrix} 1 & 0 \\ 0 & e^{-2\pi i/2^{n-m+2}} \end{pmatrix}$
$\quad\quad$ **end if**
$\quad$ **end for**
$\quad |j_k\rangle \leftarrow |j_k\rangle \frac{1}{\sqrt{2}} \begin{pmatrix} 1 & 1 \\ 1 & -1 \end{pmatrix}$
end for. ∎

Note that the inner **for** loop is not executed when $m > n$. It is clear from the above analysis that a sequential computer obtains $j_1, j_2, \ldots, j_n$ in $n(n+1)/2$ time units.

1.5.2.2 Parallel Solution

By contrast, a parallel computer can do much better in two respects. First, for $k = n, n-1, \ldots, 2$, once j_k is known, all operations involving j_k in the computation of $j_1, j_2, \ldots, j_{k-1}$ can be performed simultaneously, each being a rotation. The parallel algorithm is given below:

for $k = 1$ **to** n **do in parallel**
$\quad |j_k\rangle \leftarrow \frac{1}{\sqrt{2}} \begin{pmatrix} |0\rangle \\ e^{2\pi i 0.j_k j_{k+1} \cdots j_n}|1\rangle \end{pmatrix}$
end for

$$|j_n\rangle \leftarrow |j_n\rangle \frac{1}{\sqrt{2}} \begin{pmatrix} 1 & 1 \\ 1 & -1 \end{pmatrix}$$

    for $k = n - 1$ **downto** 1 **do**
        if $j_{k+1} = 1$ **then**
            for $m = 1$ **to** k **do in parallel**

$$|j_m\rangle \leftarrow |j_m\rangle \begin{pmatrix} 1 & 0 \\ 0 & e^{-2\pi i/2^{n-m+1}} \end{pmatrix}$$

            end for
        end if

$$|j_k\rangle \leftarrow |j_k\rangle \frac{1}{\sqrt{2}} \begin{pmatrix} 1 & 1 \\ 1 & -1 \end{pmatrix}$$

    end for. ∎

The total number of time units required to obtain $j_1, j_2, \ldots, j_n$ is now $2n - 1$.

Second, and more important, if decoherence takes place within δ time units, where $2n - 1 < \delta < n(n+1)/2$, the parallel computer succeeds in performing the computation, while the sequential computer fails [34].

1.6 Interacting Variables

So far, in every one of the paradigms that we have examined, the unconventional nature of the computation was due either to the passage of time or to the order in which an algorithmic stage is performed. In this and the next section, we consider evolving computations that occur in computational environments where time and rank play no role whatsoever either in the outcome or the complexity of the computation. Rather, it is the interactions among mutually dependent variables, caused by an interfering agent (performing the computation) that is the origin of the evolution of the system under consideration.

The computational paradigm to be presented in this section does have one feature in common with those discussed in the previous sections, namely, the central place occupied by the physical environment in which the computation is carried out. Thus, in Section 1.3, for example, the passage of time (a physical phenomenon, to the best of our knowledge) was the reason for the variables acquiring new values at each successive time unit. However, the attitude of the physical environment in the present paradigm is a passive one: Nature will not interfere with the computation until it is disturbed.

Let S be a physical system, such as one studied by biologists (e.g., a living organism), or one maintained by engineers (e.g., a power generator). The system has n variables each of which is to be measured or set to a given value at regular intervals. One property of S is that measuring or setting one of its variables modifies the values of any number of the system variables unpredictably. We show in this section how, under these conditions, a parallel solution method succeeds in carrying out the required operations on the variables of S, while a sequential method fails. Furthermore, it is principles governing such fields as physics, chemistry, and biology that are responsible for causing the inevitable failure of any sequential method of solving the problem at hand, while at the same time allowing a parallel solution to succeed. A typical example of such principles is the uncertainty involved in measuring several related variables of a physical system. Another principle expresses the way in which the components of a system in equilibrium react when subjected to outside stress.

1.6.1 Disturbing the Equilibrium

A physical system S possesses the following characteristics:

1. For $n > 1$, the system possesses a set of n variables (or properties), namely, $x_0, x_1, \ldots, x_{n-1}$. Each of these variables is a physical quantity (such as, e.g., temperature, volume, pressure, humidity, density, electric charge, etc.). These quantities can be measured or set independently, each at a

given discrete location (or point) within $\mathcal{S}$. Henceforth, x_i, $0 \leq i \leq n-1$ is used to denote a variable as well as the discrete location at which this variable is measured or set.
2. The system is in a state of *equilibrium*, meaning that the values $x_0, x_1, \ldots, x_{n-1}$ satisfy a certain global condition $\mathcal{G}(x_0, x_1, \ldots, x_{n-1})$.
3. At regular intervals, the state of the physical system is to be recorded and possibly modified. In other words, the values $x_0, x_1, \ldots, x_{n-1}$ are to be measured at a given moment in time where $\mathcal{G}(x_0, x_1, \ldots, x_{n-1})$ is satisfied. New values are then computed for $x_0, x_1, \ldots, x_{n-1}$, and the variables are set to these values. Each interval has a duration of $\mathcal{T}$ time units; that is, the state of the system is measured and possibly updated every $\mathcal{T}$ time units, where $\mathcal{T} > 1$.
4. If the values $x_0, x_1, \ldots, x_{n-1}$ are measured or set *one by one*, each separately and independently of the others, this disturbs the equilibrium of the system. Specifically, suppose, without loss of generality, that all the values are first measured, and later all are set, in the order of their indices, such that x_0 is first and x_{n-1} last in each of the two passes. Thus
   a. When x_i is measured, an arbitrary number of values x_j, $0 \leq j \leq n-1$ will change unpredictably shortly thereafter (within one time unit), such that $\mathcal{G}(x_0, x_1, \ldots, x_{n-1})$ is no longer satisfied. Most importantly, when $i < n-1$, the values of $x_{i+1}, x_{i+2}, \ldots, x_{n-1}$, none of which has yet been registered, may be altered irreparably.
   b. Similarly, when x_i is set to a new value, an arbitrary number of values x_j, $0 \leq j \leq n-1$ will change unpredictably shortly thereafter (within one time unit), such that $\mathcal{G}(x_0, x_1, \ldots, x_{n-1})$ is no longer satisfied. Most importantly, when $i > 0$, the values of $x_0, x_1, \ldots, x_{i-1}$, all of which have already been set, may be altered irreparably.

This last property of $\mathcal{S}$, namely, the way in which the system reacts to a sequential measurement or setting of its variables, is reminiscent of a number of well-known phenomena that manifest themselves in many subfields of the physical and natural sciences and engineering [8]. Examples of these phenomena are grouped into two classes and presented in what follows.

1.6.1.1 Uncertainty in Measurement

The phenomena of interest here occur in systems where measuring one variable of a given system affects, interferes with, or even precludes the subsequent measurement of another variable of the system. It is important to emphasize that the kind of uncertainty of concern in this context is in no way due to any errors that may be introduced by an imprecise or not sufficiently accurate measuring apparatus.

1. In quantum mechanics, *Heisenberg's uncertainty principle* puts a limit on our ability to measure pairs of "complementary" variables. Thus, the *position* and *momentum* of a subatomic particle, or the *energy* of a particle in a certain state and the *time* during which that state existed, cannot be defined at the same time to arbitrary accuracy [9]. In fact, one may interpret this principle as saying that once *one* of the two variables is measured (however accurately, but independently of the other), the act of measuring itself introduces a disturbance that affects the value of the *other* variable. For example, suppose that at a given moment in time t_0 the position p_0 of an electron is measured. Assume further that it is also desired to determine the electron's momentum m_0 at time t_0. When the momentum is measured, however, the value obtained is not m_0, as it would have been changed by the previous act of measuring p_0.
2. In digital signal processing, the *uncertainty principle* is exhibited when conducting a Fourier analysis. Complete resolution of a signal is possible either in the time domain t or the frequency domain w, but not both simultaneously. This is due to the fact that the Fourier transform is computed using e^{iwt}. Since the product wt must remain constant, narrowing a function in one domain causes it to be wider in the other [19, 41]. For example, a pure sinusoidal wave has no time resolution, as it possesses nonzero components over the infinitely long time axis. Its Fourier transform, on the other hand, has excellent frequency resolution: it is an impulse function with a single positive frequency component. By contrast, an impulse (or *delta*) function has only one

value in the time domain, and hence excellent resolution. Its Fourier transform is the constant function with nonzero values for all frequencies and hence no resolution.

Other examples in this class include image processing, sampling theory, spectrum estimation, image coding, and filter design [49]. Each of the phenomena discussed typically involves *two* variables in equilibrium. Measuring one of the variables has an impact on the value of the other variable. The system S, however, involves *several* variables (two or more). In that sense, its properties, as listed at the beginning of this section, are extensions of these phenomena.

1.6.1.2 Reaction to Stress

Phenomena in this class arise in systems where modifying the value of a parameter causes a change in the value of another parameter. In response to stress from the outside, the system automatically reacts so as to relieve the stress. Newton's third law of motion ("For every action there is an equal and opposite reaction") is a good way to characterize these phenomena.

1. In chemistry, *Le Châtelier's principle* states that if a system at equilibrium is subjected to a stress, the system will shift to a new equilibrium in an attempt to reduce the stress. The term *stress* depends on the system under consideration. Typically, stress means a change in pressure, temperature, or concentration [36]. For example, consider a container holding gases in equilibrium. Decreasing (increasing) the volume of the container leads to the pressure inside the container increasing (decreasing); in response to this external stress the system favors the process that produces the sleast (most) molecules of gas. Similarly, when the temperature is increased (decreased), the system responds by favoring the process that uses up (produces) heat energy. Finally, if the concentration of a component on the left (right) side of the equilibrium is decreased (increased), the system's automatic response is to favor the reaction that increases (decreases) the concentration of components on the left (right) side.
2. In biology, the *homeostatic principle* is concerned with the behavior displayed by an organism to which stress has been applied [37,48]. An automatic mechanism known as *homeostasis* counteracts external influences in order to maintain the equilibrium necessary for survival, at all levels of organization in living systems. Thus, at the molecular level, homeostasis regulates the amount of enzymes required in metabolism. At the cellular level, it controls the rate of division in cell populations. Finally, at the organismic level, it helps maintain steady levels of temperature, water, nutrients, energy, and oxygen. Examples of homeostatic mechanisms are the sensations of hunger and thirst. In humans, sweating and flushing are automatic responses to heating, while shivering and reducing blood circulation to the skin are automatic responses to chilling. Homeostasis is also seen as playing a role in maintaining population levels (animals and their prey), as well as steady state conditions in the Earth's environment.

Systems with similar behavior are also found in cybernetics, economics, and the social sciences [25]. Once again, each of the phenomena discussed typically involves *two* variables in equilibrium. Setting one of the variables has an impact on the value of the other variable. The system S, however, involves *several* variables (two or more). In that sense, its properties, as listed at the beginning of this section, are extensions of these phenomena.

1.6.2 Solutions

Two approaches are now described for addressing the problem defined at the beginning of Section 1.6.1, namely, to measure the state of S while in equilibrium, thus disturbing the latter, then setting it to a new desired state.

1.6.2.1 Simplifying Assumptions

In order to perform a concrete analysis of the different solutions to the computational problem just outlined, we continue to assume in what follows that the time required to perform all three operations below (in the given order) is one time unit:

1. Measuring one variable x_i, $0 \leq i \leq n-1$
2. Computing a new value for a variable x_i, $0 \leq i \leq n-1$
3. Setting one variable x_i, $0 \leq i \leq n-1$

Furthermore, once the new values of the parameters $x_0, x_1, \ldots, x_{n-1}$ have been applied to $\mathcal{S}$, the system requires one additional time unit to reach a new state of equilibrium. It follows that the smallest $\mathcal{T}$ can be is two time units; we therefore assume that $\mathcal{T} = 2$.

1.6.2.2 A Mathematical Model

We now present a mathematical model of the computation in Section 1.6.1. Recall that the physical system has the property that all variables are related to, and depend on, one another. Furthermore, measuring (or setting) one variable disturbs any number of the remaining variables unpredictably (meaning that we cannot tell which variables have changed value, and by how much). Typically, the system evolves until it reaches a state of equilibrium and, in the absence of external perturbations, it can remain in a stable state indefinitely.

Formally, the interdependence among the n variables can be modeled using n functions, $g_0, g_1, \ldots, g_{n-1}$, as follows:

$$x_0(t+1) = g_0(x_0(t), x_1(t), \ldots, x_{n-1}(t))$$
$$x_1(t+1) = g_1(x_0(t), x_1(t), \ldots, x_{n-1}(t))$$
$$\vdots$$
$$x_{n-1}(t+1) = g_{n-1}(x_0(t), x_1(t), \ldots, x_{n-1}(t)).$$

These equations describe the evolution of the system from state $(x_0(t), x_1(t), \ldots, x_{n-1}(t))$ at time t to state $(x_0(t+1), x_1(t+1), \ldots, x_{n-1}(t+1))$, one time unit later. While each variable is written as a function of time, there is a crucial difference between the present situation and that in Section 1.3: When the system is in a state of equilibrium, its variables do not change over time. It is also important to emphasize that, in most cases, the dynamics of the system are very complex, so the mathematical descriptions of functions $g_0, g_1, \ldots, g_{n-1}$ are either not known to us or we only have rough approximations for them.

Assuming the system is in an equilibrium state, our task is to measure its variables (in order to compute new values for these variables and set the system to these new values). In other words, we need the values of $x_0(t_0), x_1(t_0), \ldots, x_{n-1}(t_0)$ at moment $t = t_0$, when the system is in a stable state.

We can obtain the value of $x_0(t_0)$, for instance, by measuring that variable at time t_0 (noting that the choice of x_0 here is arbitrary; the argument remains the same regardless of which of the n variables we choose to measure first). Although we can acquire the value of $x_0(t_0)$ easily in this way, the consequences for the entire system can be dramatic. Unfortunately, any measurement is an external perturbation for the system, and in the process, the variable subjected to measurement will be affected unpredictably.

Thus, the measurement operation will change the state of the system from $(x_0(t_0), x_1(t_0), \ldots, x_{n-1}(t_0))$ to $(x'_0(t_0), x_1(t_0), \ldots, x_{n-1}(t_0))$, where $x'_0(t_0)$ denotes the value of variable x_0 after measurement. Since the measurement process has a nondeterministic effect on the variable being measured, we cannot estimate $x'_0(t_0)$ in any way. Note also that the transition from $(x_0(t_0), x_1(t_0), \ldots, x_{n-1}(t_0))$, that is, the state before measurement, to $(x'_0(t_0), x_1(t_0), \ldots, x_{n-1}(t_0))$, that is, the state after measurement, does not correspond to the normal evolution of the system according to its dynamics described by functions g_i, $0 \leq i \leq n-1$.

However, because the equilibrium state was perturbed by the measurement operation, the system will react with a series of state transformations, governed by equations defining the g_i. Thus, at each time unit after t_0, the parameters of the system will evolve either toward a new equilibrium state or perhaps fall into a chaotic behavior. In any case, at time $t_0 + 1$, all n variables have acquired new values, according to the functions g_i:

$$x_0(t_0 + 1) = g_0(x_0'(t_0), x_1(t_0), \ldots, x_{n-1}(t_0))$$
$$x_1(t_0 + 1) = g_1(x_0'(t_0), x_1(t_0), \ldots, x_{n-1}(t_0))$$
$$\vdots$$
$$x_{n-1}(t_0 + 1) = g_{n-1}(x_0'(t_0), x_1(t_0), \ldots, x_{n-1}(t_0)).$$

Consequently, unless we are able to measure all n variables, in parallel, at time t_0, some of the values composing the equilibrium state

$$(x_0(t_0), x_1(t_0), \ldots, x_{n-1}(t_0))$$

will be lost without any possibility of recovery.

1.6.2.3 Sequential Approach

The sequential computer measures *one* of the values (x_0, for example) and by so doing it disturbs the equilibrium, thus losing all hope of recording the state of the system within the given time interval. Any value read afterward will not satisfy $\mathcal{G}(x_0, x_1, \ldots, x_{n-1})$.

Similarly, the sequential approach cannot update the variables of $\mathcal{S}$ properly: once x_0 has received its new value, setting x_1 disturbs x_0 unpredictably.

1.6.2.4 Parallel Approach

A parallel computer with n processors, by contrast, will measure *all* the variables $x_0, x_1, \ldots, x_{n-1}$ simultaneously (one value per processor), and therefore obtain an accurate reading of the state of the system within the given time frame. Consequently,

1. A snapshot of the state of the system that satisfies $\mathcal{G}(x_0, x_1, \ldots, x_{n-1})$ has been obtained.
2. The new variables $x_0, x_1, \ldots, x_{n-1}$ can be computed in parallel (one value per processor).
3. These new values can also be applied to the system simultaneously (one value per processor).

Following the resetting of the variables $x_0, x_1, \ldots, x_{n-1}$, a new equilibrium is reached. The entire process concludes within $\mathcal{T}$ time units successfully.

1.6.3 Distinguishability in Quantum Computing

We conclude our study of interacting variables with an example from quantum computation. In Section 1.3.1 we saw that a single qubit can be in a superposition of two states, namely $|0\rangle$ and $|1\rangle$. In the same way, it is possible to place an entire quantum register, made up of n qubits, in a superposition of two states. The important point here is that, unlike the case in Section 1.3.1, it is not the individual qubits that are in a superposition, but rather the entire register (viewed as a whole).

Thus, for example, the register of n qubits may be put into any one of the following 2^n states:

$$\frac{1}{\sqrt{2}}(|000 \cdots 0\rangle \pm |111 \cdots 1\rangle)$$
$$\frac{1}{\sqrt{2}}(|000 \cdots 1\rangle \pm |111 \cdots 0\rangle)$$
$$\vdots$$
$$\frac{1}{\sqrt{2}}(|011 \cdots 1\rangle \pm |100 \cdots 0\rangle).$$

These vectors form an orthonormal basis for the state space corresponding to the n-qubit system. In such superpositions, the n qubits forming the system are said to be *entangled*: Measuring any one of them causes the superposition to collapse into one of the two basis vectors contributing to the superposition. Any subsequent measurement of the remaining $n-1$ qubits will agree with that basis vector to which the superposition collapsed. This implies that it is impossible through single measurement to distinguish among the 2^n possible states. Thus, for example, if after one qubit is read the superposition collapses to $|000\cdots 0\rangle$, we will have no way of telling which of the two superpositions, $\frac{1}{\sqrt{2}}(|000\cdots 0\rangle + |111\cdots 1\rangle)$ or $\frac{1}{\sqrt{2}}(|000\cdots 0\rangle - |111\cdots 1\rangle)$, existed in the register prior to the measurement.

The only chance to differentiate among these 2^n states using quantum measurement(s) is to observe the n qubits simultaneously, that is, perform a single joint measurement of the entire system. In the given context, *joint* is really just a synonym for *parallel*. Indeed, the device in charge of performing the joint measurement must possess the ability to "read" the information stored in each qubit, in parallel, in a perfectly synchronized manner. In this sense, at an abstract level, the measuring apparatus can be viewed as having n probes. With all probes operating in parallel, each probe can "peek" inside the state of one qubit, in a perfectly synchronous operation. The information gathered by the n probes is seen by the measuring device as a single, indivisible chunk of data, which is then interpreted to give one of the 2^n entangled states as the measurement outcome.

It is perhaps worth emphasizing that if such a measurement cannot be applied then the desired distinguishability can no longer be achieved, regardless of how many other measuring operations we are allowed to perform. In other words, even an infinite sequence of measurements touching at most $n-1$ qubits at the same time cannot equal a single joint measurement involving all n qubits. Furthermore, with respect to the particular distinguishability problem that we have to solve, a single joint measurement capable of observing $n-1$ qubits simultaneously offers no advantage whatsoever over a sequence of $n-1$ consecutive *single* qubit measurements [31, 32].

1.7 Computations Obeying Mathematical Constraints

In this section, we examine computational problems in which a certain mathematical condition must be satisfied throughout the computation. Such problems are quite common in many subareas of computer science, such as numerical analysis and optimization. Thus, the condition may be a local one; for example, a variable may not be allowed to take a value larger than a given bound. Alternatively, the condition may be global, as when the average of a set of variables must remain within a certain interval. Specifically, for $n > 1$, suppose that some function of the n variables, $x_0, x_1, \ldots, x_i, \ldots, x_{n-1}$, is to be computed. The requirement here is that the variables satisfy a stated condition at each step of the computation. In particular, if the effect of the computation is to change x_i to x_i' at some point, then the condition must remain true, whether it applies to x_i alone or to the entire set of variables, whatever the case may be. If the condition is not satisfied at a given moment of the computation, the latter is considered to have failed.

Our concern in what follows is with computations that fit the broad definition just presented, yet can only be performed successfully in parallel (and not sequentially). All n variables, $x_0, x_1, \ldots, x_i, \ldots, x_{n-1}$, are already stored in memory. However, modifying any *one* of the variables from x_i to x_i', to the exclusion of the others, causes the required condition (whether local or global) to be violated, and hence the computation to fail.

1.7.1 Mathematical Transformations

There exists a family of computational problems where, given a mathematical object satisfying a certain property, we are asked to transform this object into another, which also satisfies the same property. Furthermore, the property is to be maintained throughout the transformation, and be satisfied by every intermediate object, if any. Three examples of such transformations are now described.

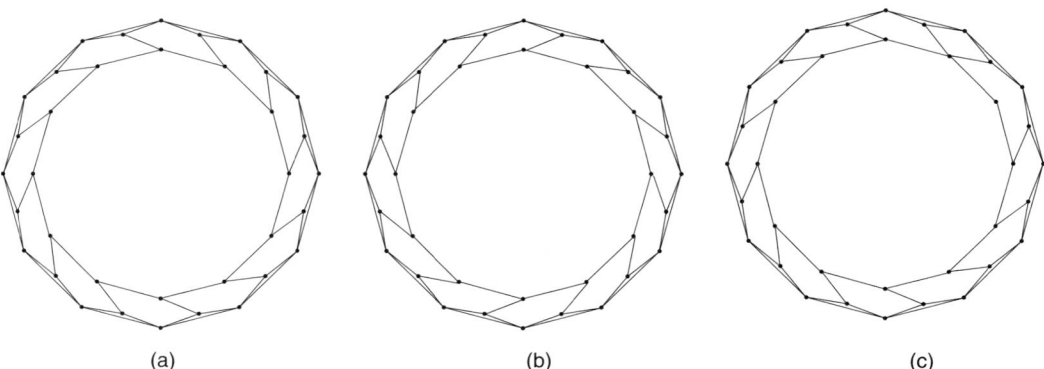

FIGURE 1.1 Subdivision: (a) origin, (b) destination, (c) with a concavity.

1.7.1.1 Geometric Flips

The object shown in Figure 1.1a is called a *convex subdivision*, as each of its faces is a convex polygon. This convex subdivision is to be transformed into that in Figure 1.1b.

The transformation can be effected by removing edges and replacing them with other edges. The condition for a successful transformation is that each intermediate figure (resulting from a replacement) be a convex subdivision as well. There are n edges in Figure 1.1a that can be removed and replaced with another n edges to produce Figure 1.1b, where $n = 12$ for illustration. These are the "spokes" that connect the outside "wheel" to the inside one. However, as Figure 1.1c illustrates, removing any *one* of these edges and replacing it with another creates a concavity, thus violating the condition [6, 29].

1.7.1.2 Map Coloring

A simple map is given consisting of n contiguous regions, where $n > 1$. Each region is a vertical strip going from the top edge to the bottom edge of the map. The regions are colored using two colors, red (R) and blue (B), in alternating fashion, thus

$$\ldots RBRBRBRBRBRBRBRBRB \ldots$$

It is required to recolor this map, such that each region previously colored R is now colored B, and conversely, each region previously colored B is now colored R, thus

$$\ldots BRBRBRBRBRBRBRBRBR \ldots$$

The condition to be satisfied throughout the recoloring is that no two adjacent regions are colored using the same color, and no third color (beside R and B) is ever used. It is clear that changing any *one* color at a time violates this requirement [24].

1.7.1.3 Rewriting Systems

From an initial string ab, in some formal language consisting of the two symbols a and b, it is required to generate the string $(ab)^n$, for $n > 1$. The rewrite rules to be used are

$$\begin{aligned} a &\to ab \\ b &\to ab. \end{aligned}$$

Thus, for $n = 3$, the target string is *ababab*. Throughout the computation, no intermediate string should have two adjacent identical characters. Such rewrite systems (also known as $\mathcal{L}$-systems) are used to draw fractals and model plant growth [38]. Here we note that applying any *one* of the two rules at a time causes the computation to fail (e.g., if *ab* is changed to *abb*, by the first rewrite rule, or to *aab* by the second) [24].

1.7.2 Sequential Solution

With all the x_i in its memory, suppose without loss of generality that the sequential computer obtains x'_0. This causes the computation to fail, as the set of variables $x'_0, x_1, x_2, \ldots, x_{n-1}$ does not satisfy the global condition. Thus, in Section 1.7.1.1, only one edge of the subdivision in Figure 1.1(a) can be replaced at a time. Once any one of the n candidate edges is replaced, the global condition of convexity no longer holds. The same is true in Sections 1.7.1.2 and 1.7.1.3, where the sequential computer can change only one color or one symbol at once, respectively, thereby causing the adjacency conditions to be violated.

1.7.3 Parallel Solution

For a given n, a parallel computer with n processors can easily perform a transformation on all the x_i collectively, with processor i computing x'_i. The required property in each case is maintained leading to a successful computation. Thus, in Section 1.7.1.1, n edges are removed from Figure 1.1a and n new edges replace them to obtain Figure 1.1b, all in one step. Similarly in Section 1.7.1.2, all colors can be changed at the same time. Finally, in Section 1.7.1.3, the string $(ab)^n$ is obtained in log n steps, with the two rewrite rules being applied simultaneously to all symbols in the current intermediate string, in the following manner: *ab, abab, abababab*, and so on. It is interesting to observe that a successful generation of $(ab)^n$ also provides an example of a rank-varying computational complexity (as described in Section 1.5). Indeed, each legal string (i.e., each string generated by the rules and obeying the adjacency property) is twice as long as its predecessor (and hence requires twice as many operations to be generated).

1.8 The Universal Computer is a Myth

The principle of simulation is the cornerstone of computer science. It is at the heart of most theoretical results and practical implements of the field such as programming languages, operating systems, and so on. The principle states that any computation that can be performed on any one general-purpose computer can be equally carried out through simulation on any other general-purpose computer [17, 20, 30]. At times, the imitated computation, running on the second computer, may be faster or slower depending on the computers involved. In order to avoid having to refer to different computers when conducting theoretical analyses, it is a generally accepted approach to define a model of computation that can simulate all computations by other computers. This model would be known as a universal computer $\mathcal{U}$. Thus, universal computation, which clearly rests on the principle of simulation, is also one of the foundational concepts in the field [16, 21, 22].

Our purpose here is to prove the following general statement: There does not exist a *finite* computational device that can be called a universal computer. Our reasoning proceeds as follows. Suppose there exists a universal computer capable of n elementary operations per step, where n is a finite and fixed integer. This computer will fail to perform a computation *requiring* n' operations per step, for any $n' > n$, and consequently lose its claim of universality. Naturally, for each $n' > n$, another computer capable of n' operations per step will succeed in performing the aforementioned computation. However, this new computer will in turn be defeated by a problem requiring $n'' > n'$ operations per step.

This reasoning is supported by each of the computational problems presented in Sections 1.3 through 1.7. As we have seen, these problems *can* easily be solved by a computer capable of executing n operations at every step. Specifically, an n-processor parallel computer led to a successful computation in each case. However, *none* of these problems is solvable by any computer capable of at most $n - 1$ operations per step, for any integer $n > 1$. Furthermore, the problem size n itself is a variable that changes with each problem instance. As a result, *no* parallel computer, regardless of how many processors it has available, can cope with a growing problem size, as long as the number of processors is finite and fixed. This holds even if the finite computer is endowed with an unlimited memory and is allowed to compute for an indefinite period of time.

The preceding reasoning applies to any computer that obeys the *finiteness condition*, that is, a computer capable of only a finite and fixed number of operations per step. It should be noted that computers obeying the finiteness condition include all "reasonable" models of computation, both theoretical and practical, such as the Turing machine [26], the random access machine [40], and other idealized models, as well as all of today's general-purpose computers, including existing conventional computers (both sequential and parallel), as well as contemplated unconventional ones such as biological and quantum computers [5]. It is clear from Section 1.4.3 that even accelerating machines are not universal.

Therefore, the universal computer $\mathcal{U}$ is clearly a myth. As a consequence, the principle of simulation itself (though it applies to most *conventional* computations) is, in general, a fallacy. In fact, the latter principle is responsible for many other myths in the field. Of particular relevance to parallel computing are the myths of the *Speedup Theorem* (speedup is at most equal to the number of processors used in parallel), the *Slowdown Theorem*, also known as *Brent's Theorem* (when q instead of p processors are used, $q < p$, the slowdown is at most p/q), and Amdahl's Law (maximum speedup is inversely proportional to the portion of the calculation that is sequential). Each of these myths can be dispelled using the same computations presented in this chapter. Other computations for dispelling these and other myths are presented in Reference 4.

1.9 Conclusion

An evolving computation is one whose characteristics vary during its execution. In this chapter, we used evolving computations to identify a number of computational paradigms involving problems whose solution necessitates the use of a parallel computer. These include computations with variables whose values change with the passage of time, computations whose computational complexity varies as a function of time, computations in which the complexity of a stage of the computation depends on the order of execution of that stage, computations with variables that interact with one another and hence change each other's values through physical processes occurring in nature, and computations subject to global mathematical constraints that must be respected throughout the problem-solving process. In each case, n computational steps must be performed simultaneously in order for the computation to succeed. A parallel computer with n processors can readily solve each of these problems. No sequential computer is capable of doing so. Interestingly, this demonstrates that one of the fundamental principles in computing, namely, that any computation by one computer can be simulated on another, is invalid. None of the parallel solutions described in this chapter can be simulated on a sequential computer, regardless of how much time and memory are allowed.

Another consequence of our analysis is that the concept of universality in computing is unachievable. For every putative universal computer $\mathcal{U}_1$ capable of $V(t)$ operations at time unit t, it is always possible to define a computation $\mathcal{P}_1$ requiring $W(t)$ operations at time unit t to be completed successfully, where $W(t) > V(t)$, for all t. While $\mathcal{U}_1$ fails, another computer $\mathcal{U}_2$ capable of $W(t)$ operations at time unit t succeeds in performing $\mathcal{P}_1$ (only to be defeated, in turn, by a computation $\mathcal{P}_2$ requiring more than $W(t)$ operations at time unit t). Thus, no finite computer can be universal. That is to say, no machine, defined once and for all, can do all computations possible on other machines. This is true regardless of how $V(t)$ is defined, so long as it is fixed: It may be a constant (as with all of today's computers), or grow with t (as with accelerating machines). The only possible universal computer would be one that is capable of an *infinite* number of operations per step. As pointed out in Reference 5 the Universe satisfies this condition. This observation agrees with recent thinking to the effect that the Universe is a computer [23, 27, 47, 50]. As stated in Reference 17: "[T]hink of all our knowledge-generating processes, our whole culture and civilization, and all the thought processes in the minds of every individual, and indeed the entire evolving biosphere as well, as being a gigantic *computation*. The whole thing is executing a self-motivated, self-generating computer program."

Acknowledgment

This research was supported by the Natural Sciences and Engineering Research Council of Canada.

References

[1] A. Adamatzky, B. DeLacy Costello, and T. Asai, *Reaction-Diffusion Computers*, Elsevier Science, New York, 2005.
[2] S.G. Akl, *Parallel Computation: Models and Methods*, Prentice Hall, Upper Saddle River, NJ, 1997.
[3] S.G. Akl, The design of efficient parallel algorithms, in: *Handbook on Parallel and Distributed Processing*, Blazewicz, J., Ecker, K., Plateau, B., and Trystram, D., Eds., Springer-Verlag, Berlin, 2000, pp. 13–91.
[4] S.G. Akl, Superlinear performance in real-time parallel computation, *The Journal of Supercomputing*, Vol. 29, No. 1, 2004, pp. 89–111.
[5] S.G. Akl, The myth of universal computation, in: *Parallel Numerics*, Trobec, R., Zinterhof, P., Vajteršic, M., and Uhl, A., Eds., Part 2, Systems and Simulation, University of Salzburg, Austria and Jožef Stefan Institute, Ljubljana, Slovenia, 2005, pp. 211–236.
[6] S.G. Akl, Inherently parallel geometric computations, *Parallel Processing Letters*, Vol. 16, No. 1, March 2006, pp. 19–37.
[7] S.G. Akl, Even accelerating machines are not universal, Technical Report No. 2006-508, School of Computing, Queen's University, Kingston, Ontario, Canada, 2006.
[8] S.G. Akl, B. Cordy, and W. Yao, An analysis of the effect of parallelism in the control of dynamical systems, *International Journal of Parallel, Emergent and Distributed Systems*, Vol. 20, No. 2, June 2005, pp. 147–168.
[9] B.H. Bransden and C.J. Joachain, *Quantum Mechanics*, Pearson Education, Harlow (Essex), England, 2000.
[10] C.S. Calude and G. Păun, Bio-steps beyond Turing, *BioSystems*, Vol. 77, 2004, pp. 175–194.
[11] T.J. Chung, *Computational Fluid Dynamics*, Cambridge University Press, Cambridge, United Kingdom, 2002.
[12] B.J. Copeland, Super Turing-machines, *Complexity*, Vol. 4, 1998, pp. 30–32.
[13] B.J. Copeland, Even Turing machines can compute uncomputable functions, in: *Unconventional Models of Computation*, Calude, C.S., Casti, J., and Dinneen, M.J., Eds., Springer-Verlag, Singapore, 1998, pp. 150–164.
[14] B.J. Copeland, Accelerating Turing machines, *Mind and Machines*, Vol. 12, No. 2, 2002, pp. 281–301.
[15] T.H. Cormen, C.E. Leiserson, R.L. Rivest, and C. Stein, *Introduction to Algorithms*, MIT Press, Cambridge, MA, 2001.
[16] M. Davis, *The Universal Computer*, W.W. Norton, New York, 2000.
[17] D. Deutsch, *The Fabric of Reality*, Penguin Books, London, England, 1997.
[18] P.L. Freddolino, A.S. Arkhipov, S.B. Larson, A. McPherson, and K. Schulten, Molecular dynamics simulation of the complete satellite tobacco mosaic virus, *Structure*, Vol. 14, No. 3, March 2006, pp. 437–449.
[19] D. Gabor, Theory of communication, *Proceedings of the Institute of Electrical Engineers*, Vol. 93, No. 26, 1946, pp. 420–441.
[20] D. Harel, *Algorithmics: The Spirit of Computing*, Addison-Wesley, Reading, MA, 1992.
[21] D. Hillis, *The Pattern on the Stone*, Basic Books, New York, 1998.
[22] J.E. Hopcroft and J.D. Ullman, *Formal Languages and their Relations to Automata*, Addison-Wesley, Reading, MA, 1969.
[23] K. Kelly, God is the machine, *Wired*, Vol. 10, No. 12, December 2002, pp. 170–173.
[24] A. Koves, personal communication, 2005.
[25] R. Lewin, *Complexity*, The University of Chicago Press, Chicago, 1999.

[26] H.R. Lewis and C.H. Papadimitriou, *Elements of the Theory of Computation*, Prentice Hall, Englewood Cliffs, NJ, 1981.
[27] S. Lloyd, *Programming the Universe*, Knopf, New York, 2006.
[28] M. Lockwood, *The Labyrinth of Time: Introducing the Universe*, Oxford University Press, Oxford, England, 2005.
[29] H. Meijer and D. Rappaport, Simultaneous Edge Flips for Convex Subdivisions, *16th Canadian Conference on Computational Geometry*, Montreal, August 2004, pp. 57–69.
[30] M.L. Minsky, *Computation: Finite and Infinite Machines*, Prentice-Hall, Englewood Cliffs, NJ, 1967.
[31] M. Nagy and S.G. Akl, On the importance of parallelism for quantum computation and the concept of a universal computer, *Proceedings of the Fourth International Conference on Unconventional Computation*, Sevilla, Spain, October 2005, pp. 176–190.
[32] M. Nagy and S.G. Akl, Quantum measurements and universal computation, *International Journal of Unconventional Computing*, Vol. 2, No. 1, 2006, pp. 73–88.
[33] M. Nagy and S.G. Akl, Quantum computation and quantum information, *International Journal of Parallel, Emergent and Distributed Systems*, Vol. 21, No. 1, February 2006, pp. 1–59.
[34] M. Nagy and S.G. Akl, Coping with decoherence: Parallelizing the quantum Fourier transform, Technical Report No. 2006-507, School of Computing, Queen's University, Kingston, Ontario, Canada, 2006.
[35] M.A. Nielsen and I.L. Chuang, *Quantum Computation and Quantum Information*, Cambridge University Press, Cambridge, United Kingdom, 2000.
[36] L.W. Potts, *Quantitative Analysis*, Harper & Row, New York, 1987.
[37] D.B. Pribor, *Functional Homeostasis: Basis for Understanding Stress*, Kendall Hunt, Dubuque, IA, 1986.
[38] P. Prusinkiewicz and A. Lindenmayer, *The Algorithmic Beauty of Plants*, Springer Verlag, New York, 1990.
[39] D.C. Rapaport, *The Art of Molecular Dynamics Simulation*, Cambridge University Press, Cambridge, United Kingdom, 2004.
[40] J.E. Savage, *Models of Computation*, Addison-Wesley, Reading, MA, 1998.
[41] C.E. Shannon, Communication in the presence of noise, *Proceedings of the IRE*, Vol. 37, 1949, pp. 10–21.
[42] T. Sienko, A. Adamatzky, N.G. Rambidi, M. Conrad, Eds., *Molecular Computing*, MIT Press, Cambridge, MA, 2003.
[43] I. Stewart, Deciding the undecidable, *Nature*, Vol. 352, August 1991, pp. 664–665.
[44] I. Stewart, The dynamics of impossible devices, *Nonlinear Science Today*, Vol. 1, 1991, pp. 8–9.
[45] G. Stix, Real time, *Scientific American Special Edition: A Matter of Time*, Vol. 16, No. 1, February 2006, pp. 2–5.
[46] K. Svozil, The Church-Turing thesis as a guiding principle for physics, in: *Unconventional Models of Computation*, Calude, C.S., Casti, J., and Dinneen, M.J., Eds., Springer-Verlag, Singapore, 1998, pp. 150–164.
[47] F.J. Tipler, *The Physics of Immortality*, Macmillan, London, 1995.
[48] P.A. Trojan, *Ecosystem Homeostasis*, Dr. W. Junk, Dordrecht, The Netherlands, 1984.
[49] R. Wilson and G.H. Granlund, The uncertainty principle in image processing, *IEEE Transactions on Pattern Analysis and Machine Intelligence*, Vol. PAMI-6, No. 6, 1984, pp. 758–767.
[50] S. Wolfram, *A New Kind of Science*, Wolfram Media, Champaign, IL, 2002.

2
Decomposable BSP: A Bandwidth-Latency Model for Parallel and Hierarchical Computation

Gianfranco Bilardi
Andrea Pietracaprina
Geppino Pucci
University of Padova

2.1	Reflections on Models of Parallel Computation	2-1
2.2	Model Definition and Basic Algorithms	2-6
2.3	Effectiveness of D-BSP	2-9
	Methodology for the Quantitative Assessment of Effectiveness • Effectiveness of D-BSP with Respect to Processor Networks • Effectiveness of D-BSP versus BSP on Multidimensional Arrays	
2.4	D-BSP and the Memory Hierarchy	2-13
	Models for Sequential Hierarchies • Translating Submachine Locality into Temporal Locality of Reference • Extension to Space Locality	
2.5	Conclusions	2-18
	Acknowledgments......................................	2-19
	References	2-19

2.1 Reflections on Models of Parallel Computation

One important objective of models of computation [1] is to provide a framework for the design and the analysis of algorithms that can be executed efficiently on physical machines. In view of this objective, this chapter reviews the *Decomposable Bulk Synchronous Parallel* (D-BSP) model, introduced in Reference 2 as an extension to BSP [3], and further investigated in References 4–7. In the present section, we discuss a number of issues to be confronted when defining models for parallel computation and briefly outline some historical developments that have led to the formulation of D-BSP. The goal is simply to put D-BSP in perspective, with no attempt to provide a complete survey of the vast variety of models proposed in the literature.

Broadly speaking, three properties of a parallel algorithm contribute to its efficiency: a small number of operations, a high degree of parallelism, and a small amount of communication. The *number of operations* is a metric simple to define and is clearly machine independent. Most sequential algorithmics (see, e.g., Reference 8) developed on the random access machine (RAM) model of computation [9] have been concerned mainly with the minimization of execution time which, in this model, is basically proportional to the number of operations.

The degree of parallelism of an algorithm is a somewhat more sophisticated metric. It can be defined, in a machine-independent manner, as the ratio between the number of operations and the length of the critical path, that is, the length of the longest sequence of operations, each of which takes the result of the previous one as an operand. Indeed, by adapting arguments developed in Reference 10, one can show that any algorithm can be executed in a nearly optimal number of parallel steps (assuming that operations can be scheduled off-line and ignoring delays associated with data transfers) with a number of operations per step equal to the degree of parallelism. As defined above, the degree of parallelism measures *implicit* parallelism. In fact, the metric is properly defined even for an algorithm expressed as a program for a sequential RAM. At the state of the art, automatic exposure of implicit parallelism by both compilers and machines is achievable only to a limited extent. Hence, it is desirable to work with models of computation that afford an *explicit* formulation of algorithmic parallelism. In this spirit, particularly during the eighties and the nineties, much attention has been devoted to the parallel random access machine (PRAM). Essentially, the PRAM consists of synchronous RAMs accessing a common memory; a number of formal variants have been proposed, for example, in References 11 and 12. A substantial body of PRAM algorithms (see, e.g., References 13 and 14) has been developed, often targeting the minimization of PRAM time, which closely corresponds to the length of the critical path of the underlying computation. More generally, one could view the PRAM complexity of an algorithm as a processor-time tradeoff, that is as a function which, for each number of available processors, gives the minimum time to execute the algorithm on a PRAM with those many processors. The number of processors used by a PRAM algorithm can be taken as a measure of the amount of parallelism that has been made explicit.

Neither the RAM nor the PRAM models reflect the fact that, in a physical system, moving a data item from a memory location to a processor takes time. This time is generally a function of both the processor and the memory location and grows, on average, with machine size. At a fundamental level, this fact is a corollary of the principle that information cannot travel faster than light. In engineering systems, further limitations can arise from technological constraints as well as from design choices. In any case, a *latency* is associated with each data access. The available *bandwidth* across suitable cuts of the system further constrains the timing of sets of concurrent data accesses. At a fundamental level, bandwidth limitations arise from the three-dimensional nature of physical space, in conjunction with the maximum number of bits that can be transferred across a unit area in unit time. Engineering systems are further constrained, for example, by the inability to fully integrate devices or remove heat in three dimensions.

In principle, for a given machine structure, the time required to perform any set of data movements is well defined and can be taken into account when studying execution time. However, two levels of complex issues must be faced. First, defining algorithms for a specific machine becomes more laborious, since the mapping of data to machine locations must be specified as part of the program. Second, the relative time taken by different data movements may differ substantially in different machines. Then, it is not clear a priori whether a single model of computation can adequately capture the communication requirements of a given algorithm for all machines, nor is it clear whether the communication requirements of an algorithm can be meaningfully expressed in a machine-independent fashion. From this perspective, it is not surprising that a rich variety of models of parallel computation have been proposed and explored. Indeed, a tradeoff arises between the goal of enabling the design and analysis of algorithms portable across several different platforms, on the one side, and the goal of achieving high accuracy in performance estimates, on the other side.

During the two decades from the late sixties to the late eighties, although a substantial body of algorithms have been developed for PRAM-like models [14], essentially ignoring data movement costs, an equally substantial effort has been devoted to the design of parallel algorithms on models that explicitly account for these costs, namely the processor network (PN) models [15]. In a PN model, a machine is viewed as a set of nodes, each equipped with processing and storage capabilities and connected through a set of point-to-point communication links to a subset of other nodes. Typically, the PN is assumed to be synchronized by a global clock. During a clock cycle, a node can execute one functional or memory operation on local data and send/receive one word on its communication links. A specific PN is characterized by the interconnection pattern among its nodes, often referred to as the network topology. One can argue that PNs are more realistic models than PRAMs, with respect to physical machines. Unfortunately, from the perspective of a unified theory of parallel algorithms, quite a number of network topologies have been proposed in the literature and many have also been adopted in experimental or commercial systems. Among the candidates that have received more attention, it is worth mentioning the *hypercube* [16] and its constant-degree derivatives such as the *shuffle-exchange* [17] and the *cube-connected-cycles* [18], *arrays* and *tori* of various dimensions [15], and *fat-trees* [19]. Although the popularity enjoyed by the hypercube family in the eighties and in the early nineties has later declined in favor of multidimensional (particularly, three-dimensional) array and fat-tree interconnections, no topology has yet become the undisputed choice. In this scenario, it is not only natural but also appropriate that a variety of topologies be addressed by the theory of parallel computation. At the same time, dealing with this variety translates into greater efforts from the algorithm developers and into a more restricted practical applicability of individual results.

In the outlined context, beginning from the late eighties, a number of models have been formulated with the broad goal of being more realistic than PRAMs and yet providing reasonable approximations for a variety of machines, including processor networks. The pursuit of wide applicability has usually led to models that (like the PRAM and unlike the PNs) are symmetric with respect to arbitrary permutations of the processors (or processor/memory nodes). In the direction of a more realistic description, these models can be viewed as evolutions of the PRAM aiming at better capturing at least some of the following aspects of physical machines: the *granularity of memory*, the *nonuniformity of memory-access time*, the *time of communication*, and the *time of synchronization*. Next, we briefly review each of these aspects.

A large memory is typically realized as a collection of memory modules; while a memory module contains many words (from millions to billions in current machines), only one or a few of these words can be accessed in a given cycle. If, in a parallel step, processors try to access words in the same memory module, the accesses are serialized. The PRAM memory can be viewed as made of modules with just one word. The problem of automatically simulating a memory with one-word modules on a memory with m-word modules has been extensively studied, in the context of distributed realizations of a logically shared storage (e.g., see References 20, 21, and references therein). While the rich body of results on this topic cannot be summarized here, they clearly indicate that such simulations incur non-negligible costs (typically, logarithmic in m), in terms of both storage redundancy and execution slowdown. For most algorithms, these costs can be avoided if the mapping of data to modules is explicitly dealt with. This has motivated the inclusion of memory modules into several models.

Another characteristic of large memories is that the time to access a memory location is a function of such location; the function also varies with the accessing processor. In a number of models of parallel computation, the nonuniform nature of memory is partially reflected by assuming that each processor has a local memory module: local accesses are charged one-cycle latencies; nonlocal accesses (in some models viewed as accesses to other processors' local modules, in others viewed as accesses to a globally shared set of modules) are charged multiple-cycle latencies.

As already emphasized, communication, both among processors and between processors and memory modules, takes time. For simplicity, most models assume that program execution can be clearly decomposed as a sequence of computation steps, where processors can perform operations on local data, and communication steps, where messages can be exchanged across processors and memory modules. If the structure of the communication network is fully exposed, as it is the case in PN models, the routing path of each message and the timing of motion along such path can in principle be specified by the algorithm. However, in several models, it is assumed that the program only specifies source, destination, and body of a message, and that a "hardware router" will actually perform the delivery, according to some routing algorithm. This is indeed the case in nearly all commercial machines.

In order to attribute an execution time to a communication step, it is useful to characterize the nature of the message set with suitable parameters. When the intent of the model is to ignore the details of the network structure, a natural metric is the degree h of the message set, that is, the maximum number of messages either originating at or destined to the same node. In the literature, a message set of degree h is also referred to as an *h-relation*. Routing time for a message set is often assumed to be proportional to its degree, the constant of proportionality being typically left as an independent parameter of the model, which can be adjusted to capture different communication capabilities of different machines. Some theoretical justification for the proportionality assumption comes from the fact that, as a corollary of Hall's theorem [22], any message set of degree h can be decomposed into h message sets of degree 1, although such a decomposition is not trivial to determine online and, in general, it is not explicitly computed by practical routers.

During the execution of parallel programs, there are crucial times where all processes in a given set must have reached a given point, before their execution can correctly continue. In other words, these processes must synchronize with each other. There are several software protocols to achieve synchronization, depending on which primitives are provided by the hardware. In all cases, synchronization incurs a nontrivial time penalty, which is explicitly accounted for in some models.

The *Bulk Synchronous Parallel* (BSP) model, introduced in Reference 3, provides an elegant way to deal with memory granularity and nonuniformity as well as with communication and synchronization costs. The model includes three parameters, respectively denoted by n, g, and ℓ, whose meaning is given next. $BSP(n, g, \ell)$ assumes n nodes, each containing a processor and a memory module, interconnected by a communication medium. A BSP computation is a sequence of phases, called *supersteps*: in one superstep, each processor can execute operations on data residing in the local memory, send messages and, at the end, execute a global synchronization instruction. A messages sent during a superstep becomes visible to the receiver only at the beginning of the next superstep. If the maximum time spent by a processor performing local computation is τ cycles and the set of messages sent by the processors forms an h-relation, then the superstep execution time is defined to be $\tau + gh + \ell$.

A number of other models, broadly similar to BSP, have been proposed by various authors. As BSP, the *LogP* model [23] also assumes a number of processor/memory nodes that can exchange messages through a suitable communication medium. However, there is no global synchronization primitive, so that synchronization, possibly among subsets of processors, must be explicitly programmed. A delay of g time units must occur between subsequent sends by the same node; if the number of messages in flight toward any given node never exceeds L/g, then the execution is defined to be nonstalling and any message is delivered within L time units. As shown in Reference 24 by suitable cross simulations, LogP in nonstalling mode is essentially equivalent to BSP, from the perspective of asymptotic running time of algorithms. The possibility of stalling detracts both simplicity and elegance from the model; it also has a number of subtle, probably unintended consequences, studied in Reference 25.

The *LPRAM* [26] is based on a set of nodes, each with a processor and a local memory, private to that processor; the nodes can communicate through a globally shared memory. Two types of steps are defined and separately accounted for: computation steps, where each processor performs one operation on local data, and communication steps, where each processor can write, and then read a word from global memory (technically, concurrent read is allowed, but not concurrent write). The key differences with BSP are that, in the LPRAM, (i) the granularity of the global memory is not modeled; (ii) synchronization is automatic after each step; (iii) at most one message per processor is outstanding at any given time. Although these differences are sufficiently significant to make $O(1)$ simulations of one model on the other unlikely, the mapping and the analysis of a specific algorithm in the two models typically do carry a strong resemblance.

Similar to the LPRAM, the *Queuing Shared Memory* (QSM) [27] also assumes nodes, each with a processor and a local private memory; the computation is instead organized essentially in supersteps, like in BSP. A superstep is charged with a time cost $\max(\tau, gh, g'\kappa)$, where τ is the maximum time spent by a processor on local computation, h is the maximum number of memory requests issued by the same processor, κ is the maximum number of processors accessing the same memory cell, and $g' = 1$ (a variant where $g' = g$ is also considered and dubbed the symmetric QSM). We may observe that contention at individual cell does not constitute a fundamental bottleneck, since suitable logic in the routing network can in principle combine, in a tree-like fashion, requests directed to or returning from the same memory cell [28]. However, the memory contention term in the time of a QSM superstep can be relevant in the description of machines without such combining capabilities.

The models reviewed above take important steps in providing a framework for algorithm design that leads to efficient execution on physical machines. All of them encourage exploiting some locality in order to increase the computation over communication ratio of the nodes. BSP and LogP address the dimension of memory granularity, considered only in part in the QSM and essentially ignored in the LPRAM. BSP and LogP also stress, albeit in different ways, the cost of synchronization, which is instead ignored in the other two models, although in QSM there is still an indirect incentive to reduce the number of supersteps owing to the subadditive nature of the parameters τ, h, and κ. When the complications related to stalling in LogP are also considered, BSP does appear as the model of choice among these four, although, to some extent, this is bound to be a subjective judgment.

Independently of the judgment on the relative merits of the models we have discussed thus far, it is natural to ask whether they go far enough in the direction of realistic machines. Particularly drastic appears the assumption that all the nonlocal memory is equally distant from a given node, especially when considering today's top supercomputers with tens or hundred thousand nodes (e.g., see Reference 29). Indeed, most popular interconnection networks, for example arrays and fat-trees, naturally lend themselves to be recursively partitioned into smaller subnetworks, where both communication and synchronization have smaller cost. Subcomputations that require interactions only among nodes of the same subnetwork, at a certain level of the partition, can then be executed considerably faster than those that require global interactions.

These types of considerations have motivated a few proposals in terms of models of computations that help exposing what has been dubbed as submachine locality. One such model, formulated in Reference 2, is the D-BSP model, which is the main focus of the present chapter. In its more common formulation, the D-BSP model is a variant of BSP where the nodes are conceptually placed at the leaves of a rooted binary tree (for convenience assumed to be ordered and complete). A D-BSP computation is a sequence of *labeled* supersteps. The label of a superstep identifies a certain level in the tree and imposes that message exchange and synchronization be performed independently within the groups of nodes associated with the different subtree rooted at that level. Naturally, the parameters g and ℓ, used in BSP to estimate the execution time, vary, in D-BSP, according to the size of groups where communication takes place, hence according to the superstep labels. A more formal definition of the model is given in Section 2.2, where a number of D-BSP algorithmic results are also discussed for basic operations such as broadcast, prefix, sorting, routing, and shared memory simulation.

Intuitively, one expects algorithm design to entail greater complexity on D-BSP than on BSP, since supersteps at logarithmically many levels must be managed and separately accounted for in the analysis. In exchange for the greater complexity, one hopes that algorithms developed on D-BSP will ultimately run more efficiently on real platforms than those developed on BSP. We consider this issue systematically in Section 2.3, where we begin by proposing a quantitative formulation of the notion, which we call *effectiveness*, that a model M provides a good framework to design algorithms that are translated for and executed on a machine M'. Often, effectiveness is mistakenly equated with *accuracy*, the property by which the running time T of a program for M is a good estimate for the running time T' of a suitable translation of that program for M'. Although accuracy is sufficient to guarantee effectiveness, it is not necessary. Indeed, for effectiveness, all that is required is that the relative performance of two programs on M be a good indicator of the relative performance of their translations on M'. On the basis of these considerations, we introduce a quantity $\eta(M, M') \geq 1$, whose meaning is that if (i) two programs for M have running times T_1 and T_2, respectively, and (ii) their suitable translations for M' have running times T_1' and T_2', respectively, with $T_2' \geq T_1'$, then $T_2'/T_1' \leq \eta(M, M') T_2/T_1$. Furthermore, at least for one pair of programs, $T_2'/T_1' = \eta(M, M') T_2/T_1$. Intuitively, the closer $\eta(M, M')$ is to 1, the better the relative performance of two programs on M is an indicator of their relative performance on M'. The established framework lets us make a quantitative comparison between BSP and D-BSP, for example, when the target machine is a d-dimensional array M_d. Specifically, we show D-BSP is more effective than BSP by a factor $\Omega(n^{1/d(d+1)})$, technically by proving that $\eta(BSP, M_d) = \Omega(n^{1/d(d+1)}\eta(D\text{-}BSP, M_d))$.

The models we have discussed thus far mostly focus on the communication requirements arising from the distribution of the computation across different processors. As well known, communication also plays a key role within the memory system of a uniprocessor. Several models of computation have been developed to deal with various aspects of the memory hierarchy. For example, the *Hierarchical Memory Model* (HMM) of Reference 30 captures the dependence of access time upon the address, assuming nonoverlapping addresses; the *Block Transfer* (BT) model [31] extends HMM with the capability of pipelining accesses to consecutive addresses. The *Pipelined Hierarchical RAM* (PH-RAM) of Reference 32 goes one step further, assuming the pipelinability of accesses to arbitrary locations, whose feasibility is shown in Reference 33. Although D-BSP does not incorporate any hierarchical assumption on the structure of the processors local memory, in Section 2.4, we show how, under various scenarios, efficient D-BSP algorithms can be automatically translated into efficient algorithms for the HMM and the BT models. These results provide important evidence of a connection between the communication requirements of distributed computation and those of hierarchical computation, a theme that definitely deserves further exploration.

Finally, in Section 2.5, we present a brief assessment of D-BSP and suggest some directions for future investigations.

2.2 Model Definition and Basic Algorithms

In this section, we give a formal definition of the D-BSP model used throughout this chapter and review a number of basic algorithmic results developed within the model.

Definition 2.2.1 *Let n be a power of two. A D-BSP$(n, g(x), \ell(x))$ is a collection of n processor/memory pairs $\{P_j : 0 \leq j < n\}$ (referred to as processors, for simplicity), communicating through a router. The n processors are grouped into clusters $C_j^{(i)}$, for $0 \leq i \leq \log n$ and $0 \leq j < 2^i$, according to a binary decomposition tree of the D-BSP machine. Specifically, $C_j^{\log n} = \{P_j\}$, for $0 \leq j < n$, and $C_j^{(i)} = C_{2j}^{(i+1)} \cup C_{2j+1}^{(i+1)}$, for $0 \leq i < \log n$ and $0 \leq j < 2^i$. For $0 \leq i \leq \log n$, the disjoint i-clusters $C_0^{(i)}, C_1^{(i)}, \cdots, C_{2^i-1}^{(i)}$, of $n/2^i$ processors each, form a partition of the n processors. The base of the logarithms is assumed to be 2. The processors of an i-cluster are capable of communicating and synchronizing among themselves, independently of the other processors, with bandwidth and latency characteristics given by the functions $g(n/2^i)$ and $\ell(n/2^i)$ of the cluster size.*

A D-BSP computation consists of a sequence of labeled supersteps. In an i-superstep, $0 \leq i \leq \log n$, each processor computes on locally held data and sends (constant-size) messages exclusively to processors within its i-cluster. The superstep is terminated by a barrier, which synchronizes processors within each i-cluster, independently. A message sent in a superstep is available at the destination only at the beginning of the next superstep. If, during an i-superstep, $\tau \geq 0$ is the maximum time spent by a processor performing local computation and $h \geq 0$ is the maximum number of messages sent by or destined to the same processor (i.e., the messages form an h-relation), then the time of the i-superstep is defined as $\tau + hg(n/2^i) + \ell(n/2^i)$.

The above definition D-BSP is the one adopted in References 4 and later used in Reference 5–7. We remark that, in the original definition of D-BSP [2], the collection of clusters that can act as independent submachines is itself a parameter of the model. Moreover, arbitrarily deep nestings of supersteps are allowed. As we are not aware of results that are based on this level of generality, we have chosen to consider only the simple, particular case where the clusters correspond to the subtrees of a tree with the processors at the leaves, superstep nesting is not allowed, and, finally, only clusters of the same size are active at any given superstep. This is in turn a special case of what is referred to as *recursive D-BSP* in Reference 2. We observe that most commonly considered processor networks do admit natural tree decompositions. Moreover, binary decomposition trees derived from layouts provide a good description of bandwidth constraints for any network topology occupying a physical region of a given area or volume [19].

We also observe that any BSP algorithm immediately translates into a D-BSP algorithm by regarding every BSP superstep as a 0-superstep, whereas any D-BSP algorithm can be made into a BSP algorithm by simply ignoring superstep labels. The key difference between the models is in their cost functions. In fact, D-BSP introduces the notion of proximity in BSP through clustering, and groups h-relations into a logarithmic number of classes associated with different costs.

In what follows, we illustrate the use of the model by presenting efficient D-BSP algorithms for a number of key primitives. For concreteness, we will focus on D-BSP(n, x^α, x^β), where α and β are constants, with $0 < \alpha, \beta < 1$. This is a significant special case, whose bandwidth and latency functions capture a wide family of machines, including multidimensional arrays.

Broadcast and Prefix *Broadcast* is a communication operation that delivers a constant-sized item, initially stored in a single processor, to every processor in the parallel machine. Given n constant-sized operands, $a_0, a_1, \ldots, a_{n-1}$, with a_j initially in processor P_j, and given a binary, associative operator $\oplus$, *n-prefix* is the primitive that, for $0 \leq j < n$, computes the prefix $a_0 \oplus a_1 \oplus \cdots \oplus a_j$ and stores it in P_j. We have the following:

Proposition 2.2.2([2]) *Any instance of broadcast or n-prefix can be accomplished in optimal time $O\left(n^\alpha + n^\beta\right)$ on D-BSP(n, x^α, x^β).*

Together with an $\Omega\left((n^\alpha + n^\beta) \log n\right)$ time lower bound established in Reference 34 for n-prefix on BSP(n, n^α, n^β) when $\alpha \geq \beta$, the preceding proposition provides an example where the flat BSP, unable to exploit the recursive decomposition into submachines, is slower than D-BSP.

Sorting In *k-sorting*, k keys are initially assigned to each one of the n D-BSP processors and are to be redistributed so that the k smallest keys will be held by processor P_0, the next k smallest ones by processor P_1, and so on. We have the following:

Proposition 2.2.3([35]) *Any instance of k-sorting, with k upper bounded by a polynomial in n, can be executed in optimal time $T_{\text{SORT}}(k, n) = O\left(kn^\alpha + n^\beta\right)$ on D-BSP(n, x^α, x^β).*

Proof. First, sort the k keys inside each processor sequentially, in time $O(k \log k)$. Then, simulate bitonic sorting on a hypercube [15] using the *merge-split* rather than the *compare-swap* operator [36]. Now, in bitonic sorting, for $q = 0, 1, \ldots, \log n - 1$, there are $\log n - q$ merge-split phases between processors whose binary indices differ only in the coefficient of 2^q and hence belong to the same cluster of size 2^{q+1}. Thus, each such phase takes time $O(k(2^q)^\alpha + (2^q)^\beta)$. In summary, the overall running time for the k-sorting algorithm is

$$O\left(k \log k + \sum_{q=0}^{\log n - 1} (\log n - q)\left(k\left(2^q\right)^\alpha + \left(2^q\right)^\beta\right)\right) = O\left(k \log k + kn^\alpha + n^\beta\right),$$

which simplifies to $O\left(kn^\alpha + n^\beta\right)$ if k is upper bounded by a polynomial in n. Straightforward bandwidth and latency considerations suffice to prove optimality. ∎

A similar time bound appears hard to achieve in BSP. In fact, results in Reference 34 imply that any BSP sorting strategy where each superstep involves a k-relation requires time $\Omega\left((\log n / \log k)(kn^\alpha + n^\beta)\right)$, which is asymptotically larger than the D-BSP time for small k.

Routing We call (k_1, k_2)-*routing* a routing problem where each processor is the source of at most k_1 packets and the destination of at most k_2 packets. Greedily routing all the packets in a single superstep results in a $\max\{k_1, k_2\}$-relation, taking time $\Theta\left(\max\{k_1, k_2\} \cdot n^\alpha + n^\beta\right)$ both on D-BSP(n, x^α, x^β) and on

BSP(n, n^α, n^β). However, while on BSP this time is trivially optimal, a careful exploitation of submachine locality yields a faster algorithm on D-BSP.

Proposition 2.2.4 ([35]) *Any instance of (k_1, k_2)-routing can be executed on D-BSP(n, x^α, x^β) in optimal time*

$$T_{\text{rout}}(k_1, k_2, n) = O\left(k_{\min}^\alpha k_{\max}^{1-\alpha} n^\alpha + n^\beta\right),$$

where $k_{\min} = \min\{k_1, k_2\}$ and $k_{\max} = \max\{k_1, k_2\}$.

Proof. We accomplish (k_1, k_2)-routing on D-BSP in two phases as follows:

1. For $i = \log n - 1$ down to 0, in parallel within each $C_j^{(i)}$, $0 \le j < 2^i$: evenly redistribute messages with origins in $C_j^{(i)}$ among the processors of the cluster.
2. For $i = 0$ to $\log n - 1$, in parallel within each $C_j^{(i)}$, $0 \le j < 2^i$: send the messages destined to $C_{2j}^{(i+1)}$ to such cluster, so that they are evenly distributed among the processors of the cluster. Do the same for messages destined to $C_{2j+1}^{(i+1)}$.

Note that the above algorithm does not require that the values of k_1 and k_2 be known a priori. It is easy to see that at the end of iteration i of the first phase each processor holds at most $a_i = \min\{k_1, k_2 2^i\}$ messages, while at the end of iteration i of the second phase, each message is in its destination $(i+1)$-cluster, and each processor holds at most $d_i = \min\{k_2, k_1 2^{i+1}\}$ messages. Note also that iteration i of the first (resp., second) phase can be implemented through a constant number of prefix operations and one routing of an a_i-relation (resp., d_i-relation) within i-clusters. Putting it all together, the running time of the above algorithm on D-BSP(n, x^α, x^β) is

$$O\left(\sum_{i=0}^{\log n - 1} \left(\max\{a_i, d_i\}\left(\frac{n}{2^i}\right)^\alpha + \left(\frac{n}{2^i}\right)^\beta\right)\right). \tag{2.1}$$

The theorem follows by plugging in Equation 2.1 the bounds for a_i and d_i derived above.

The optimality of the proposed (k_1, k_2)-routing algorithm is again based on a bandwidth argument. If $k_1 \le k_2$, consider the case in which each of the n processors has exactly k_1 packets to send; the destinations of the packets can be easily arranged so that at least one 0-superstep is required for their delivery. Moreover, suppose that all the packets are sent to a cluster of minimal size, that is a cluster containing $2^{\lceil \log(k_1 n/k_2) \rceil}$ processors. Then, the time to make the packets enter the cluster is

$$\Omega\left(k_2\left(\frac{k_1}{k_2}n\right)^\alpha + \left(\frac{k_1}{k_2}n\right)^\beta\right) = \Omega\left(k_1^\alpha k_2^{1-\alpha} n^\alpha\right). \tag{2.2}$$

The lower bound is obtained by summing up the quantity in Equation 2.2 and the time required by the 0-superstep. The lower bound for the case $k_1 > k_2$ is obtained in a symmetric way; this time each processor receives exactly k_2 packets, which come from a cluster of minimal size $2^{\lceil \log(k_2 n/k_1) \rceil}$. ∎

As a corollary of Proposition 2.2.4, we can show that, unlike the standard BSP model, D-BSP is also able to handle unbalanced communication patterns efficiently, which was the main objective that motivated the introduction of the E-BSP model [37]. Let an (h, m)-relation be a communication pattern where each processor sends/receives at most h messages, and a total of $m \le hn$ messages are exchanged. Although greedily routing the messages of the (h, m)-relation in a single superstep may require time $\Theta\left(hn^\alpha + n^\beta\right)$ in the worst case on both D-BSP and BSP, the exploitation of submachine locality in D-BSP allows us to route any (h, m)-relation in optimal time $O\left(\lceil m/n \rceil^\alpha h^{1-\alpha} n^\alpha + n^\beta\right)$, where optimality follows by adapting the argument employed in the proof of Proposition 2.2.4.

PRAM Simulation A very desirable primitive for a distributed-memory model such as D-BSP is the ability to support a shared memory abstraction efficiently, which enables the simulation of PRAM algorithms [14] at the cost of a moderate time penalty. Implementing shared memory calls for the development of a *scheme* to represent m shared cells (*variables*) among the n processor/memory pairs of a distributed-memory machine in such a way that any n-tuple of variables can be read/written efficiently by the processors.

Numerous randomized and deterministic schemes have been developed in the literature for a number of specific processor networks. Randomized schemes (see e.g., References 38 and 39) usually distribute the variables randomly among the memory modules local to the processors. As a consequence of such a scattering, a simple routing strategy is sufficient to access any n-tuple of variables efficiently, with high probability. Following this line, we can give a simple, randomized scheme for shared memory access on D-BSP. Assume, for simplicity, that the variables be spread among the local memory modules by means of a totally random function. In fact, a polynomial hash function drawn from a $\log n$-universal class [40] suffices to achieve the same results [41], and it takes only poly$(\log n)$ rather than $O(n \log n)$ random bits to be generated and stored at the nodes. We have the following:

Theorem 2.2.5 *Any n-tuple of shared memory cells can be accessed in optimal time $O\left(n^\alpha + n^\beta\right)$, with high probability, on a D-BSP(n, x^α, x^β).*

Proof. Consider first the case of write accesses. The algorithm consists of $\lfloor \log(n/\log n) \rfloor + 1$ steps. More specifically, in Step i, for $1 \leq i \leq \lfloor \log(n/\log n) \rfloor$, we send the messages containing the access requests to their destination i-clusters, so that each node in the cluster receives roughly the same number of messages. A standard occupancy argument suffices to show that, with high probability, there will be no more than $\lambda n/2^i$ messages destined to the same i-cluster, for a given small constant $\lambda > 1$, hence each step requires a simple prefix and the routing of an $O(1)$-relation in i-clusters. In the last step, we simply send the messages to their final destinations, where the memory access is performed. Again, the same probabilistic argument implies that the degree of the relation in this case is $O\left(\log n/\log \log n\right)$, with high probability. The claimed time bound follows by using the result in Proposition 2.2.2.

For read accesses, the return journey of the messages containing the accessed values can be performed by reversing the algorithm for writes, thus, remaining within the same time bound. ∎

Under a uniform random distribution of the variables among the memory modules, $\Theta\left(\log n/\log \log n\right)$ out of *any* set of n variables will be stored in the same memory module, with high probability. Thus, any randomized access strategy without replication requires at least $\Omega\left(n^\alpha \log n/\log \log n + n^\beta\right)$ time on BSP(n, n^α, n^β).

Finally, we point out that a deterministic strategy for PRAM simulation on D-BSP(n, x^α, x^β) was presented in Reference 35 attaining an access time slightly higher, but still optimal when $\alpha < \beta$, that is, when latency overheads dominate those due to bandwidth limitations.

2.3 Effectiveness of D-BSP

In this section, we provide quantitative evidence that D-BSP is more effective than the flat BSP as a bridging computational model for parallel platforms. First, we present a methodology introduced in Reference 4 to quantitatively measure the effectiveness of a model M with respect to a platform M'. Then, we apply this methodology to compare the effectiveness of D-BSP and BSP with respect to processor networks, with particular attention to multidimensional arrays. We also discuss D-BSP's effectiveness for specific key primitives.

2.3.1 Methodology for the Quantitative Assessment of Effectiveness

Let us consider a model M where designers develop and analyze algorithms, which we call M-*programs* in this context, and a machine M' onto which M-programs are translated and executed. We call M'-*programs* the programs that are ultimately executed on M'. During the design process, choices between different programs (e.g., programs implementing alternative algorithms for the same problem) will be clearly guided by model M, by comparing their running times as predicted by the model's cost function. Intuitively, we consider M to be effective with respect to M' if the choices based on M turn out to be good choices for M' as well. In other words, effectiveness means that the relative performance of any two M-programs reflects the relative performance of their translations on M'. In order for this approach to be meaningful, we must assume that the translation process is a reasonable one, in that it will not introduce substantial algorithmic insights, whereas at the same time fully exploiting any structure exposed on M, in order to achieve performance on M'.

Without attempting a specific formalization of this notion, we abstractly assume the existence of an equivalence relation ρ, defined on the set containing both M- and M'-programs, such that automatic optimization is considered to be reasonable within ρ-equivalence classes. Therefore, we can restrict our attention to ρ-*optimal* programs, where a ρ-optimal M-program (resp., ρ-optimal M'-program) is fastest among all M-programs (resp., M'-programs) ρ-equivalent to it. Examples of ρ-equivalence relations are, in order of increasing reasonableness, (i) realizing the same input–output map, (ii) implementing the same algorithm at the functional level, and (iii) implementing the same algorithm with the same schedule of operations. We are now ready to propose a formal definition of effectiveness, implicitly assuming the choice of some ρ.

Definition 2.3.1 *Let Π and Π', respectively, denote the sets of ρ-optimal M-programs and M'-programs and consider a translation function $\sigma : \Pi \to \Pi'$ such that $\sigma(\pi)$ is ρ-equivalent to π. For $\pi \in \Pi$ and $\pi' \in \Pi'$, let $T(\pi)$ and $T'(\pi')$ denote their running times on M and M', respectively. We call* inverse effectiveness *the metric*

$$\eta(M, M') = \max_{\pi_1, \pi_2 \in \Pi} \frac{T(\pi_1)}{T(\pi_2)} \cdot \frac{T'(\sigma(\pi_2))}{T'(\sigma(\pi_1))}. \tag{2.3}$$

Note that $\eta(M, M') \geq 1$, as the argument of the max function takes reciprocal values for the two pairs (π_1, π_2) and (π_2, π_1). If $\eta(M, M')$ is close to 1, then the relative performance of programs on M closely tracks relative performance on M'. If η is large, then the relative performance on M' may differ considerably from that on M, although not necessarily, η being a worst-case measure.

Next, we show that an upper estimate of $\eta(M, M')$ can be obtained on the basis of the ability of M and M' to simulate each other. Consider an algorithm that takes any M-program π and simulates it on M' as an M'-program π' ρ-equivalent to π. (Note that in neither π nor π' needs be ρ-optimal.) We define the *slowdown* $S(M, M')$ of the simulation as the ratio $T'(\pi')/T(\pi)$ maximized over all possible M-programs π. We can view $S(M, M')$ as an upper bound to the cost for supporting model M on M'. $S(M', M)$ can be symmetrically defined. Then, the following key inequality holds:

$$\eta(M, M') \leq S(M, M')S(M', M). \tag{2.4}$$

Indeed, since the simulation algorithms considered in the definition of $S(M, M')$ and $S(M', M)$ preserve ρ-equivalence, it is easy to see that for any $\pi_1, \pi_2 \in \Pi$, $T'(\sigma(\pi_2)) \leq S(M, M')T(\pi_2)$ and $T(\pi_1) \leq S(M', M)T'(\sigma(\pi_1))$. Thus, we have that $(T(\pi_1)/T(\pi_2)) \cdot (T'(\sigma(\pi_2))/T'(\sigma(\pi_1))) \leq S(M, M')S(M', M)$, which by Relation 2.3, yields Bound 2.4.

2.3.2 Effectiveness of D-BSP with Respect to Processor Networks

By applying Bound 2.4 to suitable simulations, in this subsection we derive an upper bound on the inverse effectiveness of D-BSP with respect to a wide class of processor networks. Let G be a connected processor network of n nodes. A computation of G is a sequence of *steps*, where in one step each node may execute a constant number of local operations and send/receive one message to/from each neighboring node (multiport regimen). We consider those networks G with a decomposition tree $\{G_0^{(i)}, G_1^{(i)}, \ldots, G_{2^i-1}^{(i)} :$ $\forall i, 0 \leq i \leq \log n\}$, where each $G_j^{(i)}$ (*i-subnet*) is a connected subnetwork with $n/2^i$ nodes and there are most b_i links between nodes of $G_j^{(i)}$ and nodes outside $G_j^{(i)}$; moreover, $G_j^{(i)} = G_{2j}^{(i+1)} \cup G_{2j+1}^{(i+1)}$. Observe that most prominent interconnections admit such a decomposition tree, among the others, multidimensional arrays, butterflies, and hypercubes [15].

Let us first consider the simulation of D-BSP onto any such network G. By combining the routing results of [42, 43] one can easily show that for every $0 \leq i \leq \log n$ there exist suitable values g_i and ℓ_i related, resp., to the bandwidth and diameter characteristics of the *i*-subnets, such that an h-relation followed by a barrier synchronization within an *i*-subnet can be implemented in $O(hg_i + \ell_i)$ time. Let M be any D-BSP$(n, g(x), \ell(x))$ with $g(n/2^i) = g_i$ and $\ell(n/2^i) = \ell_i$, for $0 \leq i \leq \log n$. Clearly, we have that $S(M, G) = O(1)$.

In order to simulate G on a n-processor D-BSP, we establish a one-to-one mapping between nodes of G and D-BSP processors so that the nodes of $G_j^{(i)}$ are assigned to the processors of *i*-cluster $C_j^{(i)}$, for every i and j. The simulation proceeds step-by-step as follows. Let $M_{i,j}^{\text{out}}$ (resp., $M_{i,j}^{\text{in}}$) denote the messages that are sent (resp., received) in a given step by nodes of $G_j^{(i)}$ to (resp., from) nodes outside the subnet. Since the number of boundary links of an *i*-subnet is at most b_i, we have that $|M_{i,j}^{\text{out}}|, |M_{i,j}^{\text{in}}| \leq b_i$. Let also $\bar{M}_{i,j}^{\text{out}} \subseteq M_{i,j}^{\text{out}}$ denote those messages that from $G_j^{(i)}$ go to nodes in its sibling $G_{j'}^{(i)}$, with $j' = j \pm 1$ depending on whether j is even or odd. The idea behind the simulation is to guarantee, for each cluster, that the outgoing messages be balanced among the processors of the cluster before they are sent out, and, similarly, that the incoming messages destined to any pair of sibling clusters be balanced among the processors of the cluster's father before they are acquired. More precisely, after a first superstep where each D-BSP processor simulates the local computation of the node assigned to it, the following two cycles are executed:

1. For $i = \log n - 1$ down to 0 do in parallel within each $C_j^{(i)}$, for $0 \leq j < 2^i$
   a. Send the messages in $\bar{M}_{i+1,2j}^{\text{out}}$ (resp., $\bar{M}_{i+1,2j+1}^{\text{out}}$) from $C_{2j}^{(i+1)}$ (resp., $C_{2j+1}^{(i+1)}$) to $C_{2j+1}^{(i+1)}$ (resp., $C_{2j}^{(i+1)}$), so that each processor receives (roughly) the same number of messages.
   b. Balance the messages in $M_{i,j}^{\text{out}}$ among the processors of $C_j^{(i)}$. (This step is vacuous for $i = 0$.)
2. For $i = 1$ to $\log n - 1$ do in parallel within each $C_j^{(i)}$, for $0 \leq j < 2^i$
   a. Send the messages in $M_{i,j}^{\text{in}} \cap M_{i+1,2j}^{\text{in}}$ (resp., $M_{i,j}^{\text{in}} \cap M_{i+1,2j+1}^{\text{in}}$) to the processors of $C_{2j}^{(i+1)}$ (resp., $C_{2j+1}^{(i+1)}$), so that each processor receives (roughly) the same number of messages.

It is easy to see that the above cycles guarantee that each message eventually reaches its destination, hence the overall simulation is correct. As for the running time, consider that $h_i = \lceil b_i/(n/2^i) \rceil$, for $0 \leq i \leq \log n$, is an upper bound on the average number of incoming/outgoing messages for an *i*-cluster. The balancing operations performed by the algorithm guarantee that iteration i of either cycle entails a $\max\{h_i, h_{i+1}\}$-relation within *i*-clusters.

As a further refinement, we can optimize the simulation by running it entirely within a cluster of $n' \leq n$ processors of the n-processor D-BSP, where n' is the value that minimizes the overall slowdown. When $n' < n$, in the initial superstep before the two cycles, each D-BSP processor will simulate all the local computations and communications internal to the subnet assigned to it. The following theorem summarizes the above discussion.

Theorem 2.3.2 *For any n-node network G with a decomposition tree of parameters* $(b_0, b_1, \ldots, b_{\log n})$, *one step of G can be simulated on a* $M = \text{D-BSP}(n, g(x), \ell(x))$ *in time*

$$S(G, M) = O\left(\min_{n' \leq n}\left\{\frac{n}{n'} + \sum_{i=\log(n/n')}^{\log n - 1} \left(g\left(n/2^i\right)\max\{h_i, h_{i+1}\} + \ell\left(n/2^i\right)\right)\right\}\right), \tag{2.5}$$

where $h_i = \lceil b_{i-\log(n/n')}/(n/2^i) \rceil$, *for* $\log(n/n') \leq i < \log n$.

We remark that the simulations described in this subsection transform D-BSP programs into ρ-equivalent G-programs, and vice versa, for most realistic definitions of ρ (e.g., same input-output map, or same high-level algorithm). Let M be any D-BSP$(n, g(x), \ell(x))$ machine whose parameters are adapted to the bandwidth and latency characteristics of G as discussed above. Since $S(M, G) = O(1)$, from Equation 2.4, we have that the inverse effectiveness satisfies $\eta(M, G) = O(S(G, M))$, where $S(G, M)$ is bounded as in Equation 2.5.

2.3.3 Effectiveness of D-BSP versus BSP on Multidimensional Arrays

In this subsection, we show how the richer structure exposed in D-BSP makes it more effective than the "flat" BSP, for multidimensional arrays. We begin with the effectiveness of D-BSP.

Proposition 2.3.3 *Let* G_d *be an n-node d-dimensional array. Then,* $\eta(\text{D-BSP}(n, x^{1/d}, x^{1/d}), G_d) = O\left(n^{1/(d+1)}\right)$.

Proof. Standard routing results [15] imply that such a D-BSP$(n, x^{1/d}, x^{1/d})$ can be simulated on G_d with constant slowdown. Since G_d has a decomposition tree with connected subnets $G_j^{(i)}$ that have $b_i = O\left((n/2^i)^{(d-1)/d}\right)$, the D-BSP simulation of Theorem 2.3.2 yields a slowdown of $O\left(n^{1/(d+1)}\right)$ per step, when entirely run on a cluster of $n' = n^{d/(d+1)}$ processors. In conclusion, letting $M = \text{D-BSP}(n, x^{1/d}, x^{1/d})$, we have that $S(M, G_d) = O\left(n^{1/(d+1)}\right)$, which, combined with Equation 2.4, yields the stated bound $\eta(M, G_d) = O\left(n^{1/(d+1)}\right)$. ∎

Next, we turn our attention to BSP.

Proposition 2.3.4 *Let* G_d *be an n-node d-dimensional array and let* $g = O(n)$, *and* $\ell = O(n)$. *Then,* $\eta(\text{BSP}(n, g, \ell), G_d) = \Omega\left(n^{1/d}\right)$.

The proposition is an easy corollary of the following two lemmas which, assuming programs with the same input-output map to be ρ-equivalent, establish the existence of two ρ-optimal BSP programs π_1 and π_2 such that

$$\frac{T_{\text{BSP}}(\pi_1)}{T_{\text{BSP}}(\pi_2)} \frac{T_{G_d}(\sigma(\pi_2))}{T_{G_d}(\sigma(\pi_1))} = \Omega\left(n^{1/d}\right),$$

where $T_{\text{BSP}}(\cdot)$ and $T_{G_d}(\cdot)$ are the running time functions for BSP and G_d, respectively.

Lemma 2.3.5 *There exists a BSP program* π_1 *such that* $T_{\text{BSP}}(\pi_1)/T_{G_d}(\sigma(\pi_1)) = \Omega(g)$.

Proof. Let $\Delta_T(A_n)$ be a dag modeling an arbitrary T-step computation of an n-node linear array A_n. The nodes of $\Delta_T(A_n)$, which represent operations executed by the linear array nodes, are all pairs (v, t), with v being a node of A_n, and $0 \leq t \leq T$; while the arcs, which represent data dependencies, connect nodes (v_1, t) and $(v_2, t+1)$, where $t < T$ and v_1 and v_2 are either the same node or are adjacent in A_n. During a computation of $\Delta_T(A_n)$, an operation associated with a dag node can be executed by a processor if and only if the processor knows the results of the operations associated with the node's predecessors. Note that, for $T \leq n$, $\Delta_T(A_n)$ contains the $\lceil T/2 \rceil \times \lceil T/2 \rceil$ *diamond dag* $D_{T/2}$ as a subgraph. The result in Reference 26, Theorem 5.1 implies that any BSP program computing $D_{T/2}$ either requires an $\Omega(T)$-relation or an $\Omega(T^2)$-time sequential computation performed by some processor. Hence, such a program requires $\Omega(T \min\{g, T\})$ time.

Since any connected network embeds an n-node linear array with constant load and dilation, then $\Delta_T(A_n)$ can be computed by G_d in time $O(T)$. The lemma follows by choosing $n \geq T = \Theta(g)$ and any ρ-optimal BSP-program π_1 that computes $\Delta_T(A_n)$. ∎

Lemma 2.3.6 *There exists a BSP program π_2 such that $T_{G_d}(\sigma(\pi_2))/T_{\text{BSP}}(\pi_2) = \Omega(n^{1/d}/g)$.*

Proof. Consider the following problem. Let Q be a set of n^2 4-tuples of the kind (i, j, k, a), where i, j, and k are indices in $[0, n-1]$, while a is an arbitrary integer. Q satisfies the following two properties: (i) if $(i, j, k, a) \in Q$ then also $(j, i, k, b) \in Q$, for some value b; and (ii) for each k there are exactly n 4-tuples with third component equal to k. The problem requires to compute, for each pair of 4-tuples $(i, j, k, a), (j, i, k, b)$, the product $a \cdot b$. It is not difficult to see that there exists a BSP program π_2 that solves the problem in time $O(n \cdot g)$. Consider now an arbitrary program that solves the problem on the d-dimensional array G_d. If initially one processor holds $n^2/4$ 4-tuples then either this processor performs $\Omega(n^2)$ local operations or sends/receives $\Omega(n^2)$ 4-tuples to/from other processors. If instead every processor initially holds at most $n^2/4$ 4-tuples then one can always find a subarray of G_d connected to the rest of the array by $O(n^{1-1/d})$ links, whose processors initially hold a total of at least $n^2/4$ and at most $n^2/2$ 4-tuples. An adversary can choose the components of such 4-tuples so that $\Omega(n^2)$ values must cross the boundary of the subarray, thus requiring $\Omega(n^{1+1/d})$ time. Thus, *any* G_d-program for this problem takes $\Omega(n^{1+1/d})$ time, and the lemma follows. ∎

Propositions 2.3.3 and 2.3.4 show that D-BSP is more effective than BSP, by a factor $\Omega(n^{1/d(d-1)})$, in modeling multidimensional arrays. A larger factor might apply, since the general simulation of Theorem 2.3.2 is not necessarily optimal for multidimensional arrays. In fact, in the special case where the nodes of G_d only have constant memory, an improved simulation yields a slowdown of $O\left(2^{O(\sqrt{\log n})}\right)$ on a D-BSP$(n, x^{1/d}, x^{1/d})$ [4]. It remains an interesting open question whether improved simulations can also be achieved for arrays with nonconstant local memories.

As we have noted, due to the worst case nature of the η metric, for specific programs the effectiveness can be considerable better than what is guaranteed by η. As an example, the optimal D-BSP algorithms for the key primitives discussed in Section 2.2 can be shown to be optimal for multidimensional arrays, so that the effectiveness of D-BSP is maximum in this case.

As a final observation, we underline that the greater effectiveness of D-BSP over BSP comes at the cost of a higher number of parameters, typically requiring more ingenuity in the algorithm design process. Whether the increase in effectiveness is worth the greater design effort remains a subjective judgment.

2.4 D-BSP and the Memory Hierarchy

Typical modern multiprocessors are characterized by a multilevel hierarchical structure of both the memory system and the communication network. As well known, performance of computations is considerably enhanced when the relevant data are likely to be found close to the unit (or the units) that must process it, and expensive data movements across several levels of the hierarchy are minimized or, if required, they are orchestrated so to maximize the available bandwidth. To achieve this objective the computation must exhibit a property generally referred to as *data locality*. Several forms of locality exist.

On a uniprocessor, data locality, also known as *locality of reference*, takes two distinct forms, namely, the frequent reuse of data within short time intervals (*temporal locality*), and the access to consecutive data in subsequent operations (*spatial locality*). On multiprocessors, another form of locality is *submachine* or *communication locality*, which requires that data be distributed so that communications are confined within small submachines featuring high per-processor bandwidth and small latency.

A number of preliminary investigations in the literature have pointed out an interesting relation between parallelism and locality of reference, showing that efficient sequential algorithms for two-level hierarchies can be obtained by simulating parallel ones [44–47]. A more general study on the relation and interplay of the various forms of locality has been undertaken in some recent works [7,48] on the basis of the D-BSP model, which provides further evidence of the suitability of such a model for capturing several crucial aspects of high-performance computations. The most relevant results of these latter works are summarized in this section, where for the sake of brevity several technicalities and proofs are omitted (the interested reader is referred to Reference 7 for full details).

In Subsection 2.4.1 we introduce two models for sequential memory hierarchies, namely, HMM and BT, which explicitly expose the two different kinds of locality of reference. In Subsection 2.4.2 we then show how the submachine locality exposed in D-BSP computations can be efficiently and automatically translated into temporal locality of reference on HMM, whereas in Subsection 2.4.3 we extend the results to the BT model to encompass both temporal and spatial locality of reference.

2.4.1 Models for Sequential Hierarchies

HMM The hierarchical memory model (HMM) was introduced in Reference 30 as a RAM where access to memory location x requires time $f(x)$, for a given nondecreasing function $f(x)$. We refer to such a model as $f(x)$-HMM. As most works in the literature, we will focus our attention on nondecreasing functions $f(x)$ for which there exists a constant $c \geq 1$ such that $f(2x) \leq cf(x)$, for any x. As in Reference 49, we will refer to these functions as $(2, c)$-*uniform* (in the literature these functions have also been called *well behaved* [50] or, somewhat improperly, *polynomially bounded* [30]). Particularly interesting and widely studied special cases are the polynomial function $f(x) = x^\alpha$ and the logarithmic function $f(x) = \log x$, where the base of the logarithm is assumed to be any constant greater than 1.

BT The *Hierarchical Memory Model with Block Transfer* was introduced in Reference 31 by augmenting the $f(x)$-HMM model with a block transfer facility. We refer to this model as the $f(x)$-BT. Specifically, as in the $f(x)$-HMM, an access to memory location x requires time $f(x)$, but the model makes also possible to copy a block of b memory cells $[x-b+1, x]$ into a *disjoint* block $[y-b+1, y]$ in time $\max\{f(x), f(y)\} + b$, for arbitrary $b > 1$. As before, we will restrict our attention to the case of $(2, c)$-uniform access functions.

It must be remarked that the block transfer mechanism featured by the model is rather powerful since it allows for the pipelined movement of arbitrarily large blocks. This is particularly noticeable if we look at the fundamental *touching* problem, which requires to bring each of a set of n memory cells to the top of memory. It is easy to see that on the $f(x)$-HMM, where no block transfer is allowed, the touching problem requires time $\Theta(nf(n))$, for any $(2, c)$-uniform function $f(x)$. Instead, on the $f(x)$-BT, a better complexity is attainable. For a function $f(x) < x$, let $f^{(k)}(x)$ be the iterated function obtained by applying f k times, and let $f^*(x) = \min\{k \geq 1 : f^{(k)}(x) \leq 1\}$. The following fact is easily established from Reference 31.

Fact 2.4.1 *The touching problem on the $f(x)$-BT requires time $T_{\text{TCH}}(n) = \Theta(nf^*(n))$. In particular, we have that $T_{\text{TCH}}(n) = \Theta(n \log^* n)$ if $f(x) = \log x$, and $T_{\text{TCH}}(n) = \Theta(n \log \log n)$ if $f(x) = x^\alpha$, for a positive constant $\alpha < 1$.*

The above fact gives a nontrivial lower bound on the execution time of many problems where all the inputs, or at least a constant fraction of them, must be examined. As argued in Reference 7, the memory transfer capabilities postulated by the BT model are already (reasonably) well approximated by current hierarchical designs, and within reach of foreseeable technology [33].

2.4.2 Translating Submachine Locality into Temporal Locality of Reference

We now describe how to simulate D-BSP programs on the HMM model with a slowdown that is merely proportional to the loss of parallelism. The simulation is able to hide the memory hierarchy costs induced by the HMM access function by efficiently transforming submachine locality into temporal locality of reference.

We refer to D-BSP and HMM as the *guest* and *host* machine, respectively, and restrict our attention to the simulation of D-BSP programs that end with a global synchronization (i.e., a 0-superstep), a reasonable constraint. Consider the simulation of a D-BSP program $\mathcal{P}$ and let μ denote the size of each D-BSP processor's local memory, which we refer to as the processor's *context*. We assume that a processor's context also comprises the necessary buffer space for storing incoming and outgoing messages. The memory of the host machine is divided into *blocks* of μ cells each, with block 0 at the top of memory. At the beginning of the simulation, block j, $j = 0, 1, \ldots, n-1$, contains the *context* (i.e., the local memory) of processor P_j, but this association changes as the simulation proceeds.

Let the supersteps of $\mathcal{P}$ be numbered consecutively, and let i_s be the label of the s-th superstep, with $s \geq 0$ (i.e., the s-th superstep is executed independently within i_s-clusters). At some arbitrary point during the simulation, an i_s-cluster C is said to be s-ready if, for all processors in C, supersteps $0, 1, \ldots, s-1$ have been simulated, while Superstep s has not been simulated yet. The simulation, whose pseudocode is given in Figure 2.1, is organized into a number of *rounds*, corresponding to the iterations of the while loop in the code. A round simulates the operations prescribed by a certain Superstep s for a certain s-ready cluster C, and performs a number of context swaps to prepare for the execution of the next round.

The correctness of the simulation follows by showing that the two invariants given below are maintained at the beginning of each round. Let s and C be defined as in Step 1 of the round.

Invariant 2.4.2 *C is s-ready.*

Invariant 2.4.3 *The contexts of all processors in C are stored in the topmost $|C|$ blocks, sorted in increasing order by processor number. Moreover, for any other cluster C', the contexts of all processors in C' are stored in consecutive memory blocks (although not necessarily sorted).*

It is important to observe that in order to transform the submachine locality of the D-BSP program into temporal locality of reference on the HMM, the simulation proceeds unevenly on the different D-BSP clusters. This is achieved by suitably selecting the next cluster to be simulated after each round, which, if needed, must be brought on top of memory. In fact, the same cluster could be simulated for several consecutive supersteps so to avoid repeated, expensive relocations of its processors' contexts

```
        while true do
1           P ← processor whose context is on top of memory
            s ← superstep number to be simulated next for P
            C ← i_s-cluster containing P
2           Simulate Superstep s for C
3           if P has finished its program then exit
4           if i_{s+1} < i_s then
                b ← 2^{i_s - i_{s+1}}
                Let Ĉ be the i_{s+1}-cluster containing C,
                    and let Ĉ_0 … Ĉ_{b-1} be its component i_s-clusters,
                    with C = Ĉ_j for some index j
                if j > 0 then swap the contexts of C with those of Ĉ_0
                if j < b - 1 then
                    swap the contexts of Ĉ_0 with those of Ĉ_{j+1}
```

FIGURE 2.1 The simulation algorithm.

in memory. More precisely, consider a generic round where Superstep s is simulated for an i_s-cluster C. If $i_{s+1} \geq i_s$ then no cluster swaps are performed at the end of the round, and the next round will simulate Superstep $s + 1$ for the topmost i_{s+1}-cluster contained in C and currently residing on top of memory. Such a cluster is clearly $(s + 1)$-ready. Instead, if $i_{s+1} < i_s$, Superstep $s + 1$ involves a coarser level of clustering, hence the simulation of this superstep can take place only after Superstep s has been simulated for *all* i_s-clusters that form the i_{s+1}-cluster $\hat{C}$ containing C. Step 4 is designed to enforce this schedule. In particular, let $\hat{C}$ contain $b = 2^{i_s - i_{s+1}}$ i_s-clusters, including C, which we denote by $\hat{C}_0, \hat{C}_1, \ldots, \hat{C}_{b-1}$, and suppose that $C = \hat{C}_0$ is the first such i_s-cluster for which Superstep s is simulated. By Invariant 2.4.3, at the beginning of the round under consideration the contexts of all processors in $\hat{C}$ are at the top of memory. This round starts a *cycle* of b phases, each phase comprising one or more simulation rounds. In the k-th phase, $0 \leq k < b$, the contexts of the processors in $\hat{C}_k$ are brought to the top of memory, then all supersteps up to Superstep s are simulated for these processors, and finally the contexts of $\hat{C}_k$ are moved back to the positions occupied at the beginning of the cycle.

If the two invariants hold, then the simulation of cluster C in Step 2 can be performed as follows. First the context of each processor in C is brought in turn to the top of memory and its local computation is simulated. Then, message exchange is simulated by scanning the processors' outgoing message buffers sequentially and delivering each message to the incoming message buffer of the destination processor. The location of these buffers is easily determined since, by Invariant 2.4.3, the contexts of the processors are sorted by processor number.

The running time of the simulation algorithm is summarized in the following theorem.

Theorem 2.4.4 ([7, Thm.5]) *Consider a D-BSP$(n, g(x), \ell(x))$ program $\mathcal{P}$, where each processor performs local computation for $O(\tau)$ time, and there are λ_i i-supersteps for $0 \leq i \leq \log n$. If $f(x)$ is $(2, c)$-uniform, then $\mathcal{P}$ can be simulated on a $f(x)$-HMM in time $O\left(n\left(\tau + \mu \sum_{i=0}^{\log n} \lambda_i f(\mu n / 2^i)\right)\right)$.*

Since our main objective is to assess to what extent submachine locality can be transformed into locality of reference, we now specialize the above result to the simulation of *fine-grained* D-BSP programs where the local memory of each processor has constant size (i.e., $\mu = O(1)$). In this manner, submachine locality is the only locality that can be exhibited by the parallel program. By further constraining the D-BSP bandwidth and latency functions to reflect the HMM access function, we get the following corollary that states the linear slowdown result claimed at the beginning of the subsection.

Corollary 2.4.5 *If $f(x)$ is $(2, c)$-uniform then any T-time fine-grained program for a D-BSP$(n, g(x), \ell(x))$ with $g(x) = \ell(x) = f(x)$ can be simulated in optimal time $\Theta(T \cdot n)$ on $f(x)$-HMM.*

We note that the simulation is *online* in the sense that the entire sequence of supersteps need not be known by the processors in advance. Moreover, the simulation code is totally oblivious to the D-BSP bandwidth and latency functions $g(x)$ and $\ell(x)$.

Case studies On a number of prominent problems the simulation described above can be employed to transform efficient fine-grained D-BSP algorithms into optimal HMM strategies. Specifically, we consider the following problems:

- *n-MM*: the problem of multiplying two $\sqrt{n} \times \sqrt{n}$ matrices on an n-processor D-BSP using only semiring operations
- *n-DFT*: the problem of computing the discrete fourier transform of an n-vector
- *1-sorting*: the special case of k-sorting, with $k = 1$, as defined in Section 2.2

For concreteness, we will consider the HMM access functions $f(x) = x^\alpha$, with $0 < \alpha < 1$, and $f(x) = \log x$. Under these functions, upper and lower bounds for our reference problems have been developed directly for the HMM in Reference 30. The theorem below follows from the results in Reference 7.

D-BSP: A Bandwidth-Latency Model for Parallel and Hierarchical Computation

Theorem 2.4.6 *There exist D-BSP algorithms for the n-MM and n-DFT problems whose simulations on the $f(x)$-HMM result into optimal algorithms for $f(x) = x^\alpha$, with $0 < \alpha < 1$, and $f(x) = \log x$. Also, there exists a D-BSP algorithm for 1-sorting whose simulation on the $f(x)$-HMM results into an optimal algorithm for $f(x) = x^\alpha$, with $0 < \alpha < 1$, and into an algorithm with a running time that is a factor $O\left(\log n / \log \log n\right)$ away from optimal for $f(x) = \log x$.*

It has to be remarked that the nonoptimality of the $\log(x)$-HMM 1-sorting is not due to an inefficiency in the simulation, but, rather, to the lack of an optimal 1-sorting algorithm for the D-BSP$(n, \log(x), \log(x))$, the best strategy known so far requiring $\Omega(\log^2 n)$ time. In fact, the results in Reference 30 and our simulation imply an $\Omega\left(\log n \log \log n\right)$ lower bound for 1-sorting on D-BSP$(n, \log(x), \log(x))$, which is tighter than the previously known, trivial bound of $\Omega\left(\log n\right)$.

Theorem 2.4.6 provides evidence that D-BSP can be profitably employed to obtain efficient, portable algorithms for hierarchical architectures.

2.4.3 Extension to Space Locality

In this subsection, we modify the simulation described in the previous subsection to run efficiently on the BT model that rewards both temporal and spatial locality of reference.

We observe that the simulation algorithm of Figure 2.1 yields a valid BT program, but it is not designed to exploit block transfer. For example, in Step 2 the algorithm brings one context at a time to the top of memory and simulates communications touching the contexts in a random fashion, which is highly inefficient in the BT framework. Since the BT model supports block copy operations only for nonoverlapping memory regions, additional buffer space is required to perform swaps of large chunks of data; moreover, in order to minimize access costs, such buffer space must be allocated close to the blocks to be swapped. As a consequence, the required buffers must be interspersed with the contexts. Buffer space can be dynamically created or destroyed by *unpacking* or *packing* the contexts in a cluster. More specifically, unpacking an i-cluster involves suitably interspersing $\mu n / 2^i$ empty cells among the contexts of the cluster's processors so that block copy operations can take place.

The structure of the simulation algorithm is identical to the one in Figure 2.1, except that the simulation of the s-th superstep for an i_s-cluster C is preceded (respectively, followed) by a packing (resp., unpacking) operation on C's contexts. The actual simulation of the superstep is organized into two phases: first, local computations are executed in a recursive fashion, and then the communications required by the superstep are simulated.

In order to exploit both temporal and spatial locality in the simulation of local computations, processor contexts are iteratively brought to the top of memory in chunks of suitable size, and the prescribed local computation is then performed for each chunk recursively. To deliver all messages to their destinations, we make use of sorting. Specifically, the contexts of C are divided into $\Theta\left(\mu|C|\right)$ constant-sized elements, which are then sorted in such a way that after the sorting, contexts are still ordered by processor number and all messages destined to processor P_j of C are stored at the end of P_j's context. This is easily achieved by sorting elements according to suitably chosen tags attached to the elements, which can be produced during the simulation of local computation without asymptotically increasing the running time.

The running time of the simulation algorithm is summarized in the following theorem.

Theorem 2.4.7 ([7, Thm.12]) *Consider a D-BSP$(n, g(x), \ell(x))$ program $\mathcal{P}$, where each processor performs local computation for $O(\tau)$ time, and there are λ_i i-supersteps for $0 \leq i \leq \log n$. If $f(x)$ is $(2, c)$-uniform, and $f(x) \in O(x)^\alpha$ with constant $0 < \alpha < 1$ then $\mathcal{P}$ can be simulated on a $f(x)$-BT in time $O\left(n\left(\tau + \mu \sum_{i=0}^{\log n} \lambda_i \log(\mu n/2^i)\right)\right)$.*

We remark that, besides the unavoidable term $n\tau$, the complexity of the sorting operations employed to simulate communications is the dominant factor in the running time. Moreover, it is important to observe that, unlike the HMM case, the complexity in the above theorem *does not* depend on the access function $f(x)$, neither does it depend on the D-BSP bandwidth and latency functions $g(x)$ and $\ell(x)$.

This is in accordance with the findings of Reference 31, which show that an efficient exploitation of the powerful block transfer capability of the BT model is able to hide access costs almost completely.

Case Studies As for the HMM model, we substantiate the effectiveness of our simulation by showing how it can be employed to obtain efficient BT algorithms starting from D-BSP ones. For the sake of comparison, we observe that Fact 2.4.1 implies that for relevant access functions $f(x)$, any straightforward approach simulating one entire superstep after the other would require time $\omega(n)$ per superstep just for touching the n processor contexts, whereas our algorithm can overcome such a barrier by carefully exploiting submachine locality.

First consider the n-MM problem. It is shown in Reference 7 that the same D-BSP algorithm for this problem underlying the result of Theorem 2.4.6 also yields an optimal $O\left(n^{3/2}\right)$-time algorithm for $f(x)$-BT, for both $f(x) = \log x$ and $f(x) = x^\alpha$. In general, different D-BSP bandwidth and latency functions $g(x)$ and $\ell(x)$ may promote different algorithmic strategies for the solution of a given problem. Therefore, without a strict correspondence between these functions and the BT access function $f(x)$ in the simulation, the question arises of which choices for $g(x)$ and $\ell(x)$ suggest the best "coding practices" for BT. Unlike the HMM scenario (see Corollary 2.4.5), the choice $g(x) = \ell(x) = f(x)$ is not always the best. Consider, for instance, the n-DFT problem. Two D-BSP algorithms for this problem are applicable. The first algorithm is a standard execution of the n-input FFT dag and requires one i-superstep, for $0 \leq i < \log n$. The second algorithm is based on a recursive decomposition of the same dag into two layers of $\sqrt{n}$ independent $\sqrt{n}$-input subdags, and can be shown to require 2^i supersteps with label $(1 - 1/2^i) \log n$, for $0 \leq i < \log \log n$. On a D-BSP(n, x^α, x^α) both algorithms yield a running time of $O(n^\alpha)$, which is clearly optimal. However, the simulations of these two algorithms on the x^α-BT take time $O\left(n \log^2 n\right)$ and $O\left(n \log n \log \log n\right)$, respectively. This implies that the choice $g(x) = \ell(x) = f(x)$ is not *effective*, since the D-BSP(n, x^α, x^α) does not reward the use of the second algorithm over the first one. On the other hand, D-BSP$(n, \log(x), \log(x))$ correctly distinguishes among the two algorithms, since their respective parallel running times are $O\left(\log^2 n\right)$ and $O\left(\log n \log \log n\right)$.

The above example is a special case of the following more general consideration. It is argued in Reference 7 that given two D-BSP algorithms A_1, A_2 solving the same problem, if the simulation of A_1 on $f(x)$-BT runs faster than the simulation of A_2, then A_1 exhibits a better asymptotic performance than A_2 also on the D-BSP$(n, g(x), \ell(x))$, with $g(x) = \ell(x) = \log x$, which may not be the case for other functions $g(x)$ and $\ell(x)$. This proves that D-BSP$(n, \log x, \log x)$ is the most effective instance of the D-BSP model for obtaining sequential algorithms for the class of $f(x)$-BT machines through our simulation.

2.5 Conclusions

In this chapter, we have considered the D-BSP model of computation as an *effective* framework for the design of algorithms that can run efficiently on realistic parallel platforms. Having proposed a quantitative notion of effectiveness, we have shown that D-BSP is more effective than the basic BSP when the target platforms have decomposition trees with geometric bandwidth progressions. This class includes multidimensional arrays and tori, as well as fat-trees with area-universal or volume-universal properties. These topologies are of particular interest, not only because they do account for the majority of current machines, but also because physical constraints imply convergence toward these topologies in the limiting technology [51]. The greater effectiveness of D-BSP is achieved by exploiting the hierarchical structure of the computation and by matching it with the hierarchical structure of the machine. In general, describing these hierarchical structures requires logarithmically many parameters in the problem and in the machine size, respectively. However, this complexity is considerably reduced if we restrict our attention to the case of geometric bandwidth and latency progressions, essentially captured by D-BSP(n, x^α, x^β), with a small, constant number of parameters. Although D-BSP is more effective than BSP with respect to multidimensional arrays, the residual loss of effectiveness is not negligible, not just in quantitative terms, but also in view of the relevance of some of the algorithms for which D-BSP is less effective. In fact, these include the d-dimensional near-neighbor algorithms frequently arising in technical and scientific computing.

The preceding observations suggest further investigations to evolve the D-BSP model both in the direction of greater effectiveness, perhaps by enriching the set of partitions into clusters that can be the base of a superstep, and in the direction of greater simplicity, perhaps by suitably restricting the space of bandwidth and latency functions. In all cases, particularly encouraging are the results reported in the previous section, showing how D-BSP is well poised to incorporate refinements for an effective modeling of the memory hierarchy and providing valuable insights toward a unified framework for capturing communication requirements of computations.

Acknowledgments

The authors wish to thank Carlo Fantozzi, who contributed to many of the results surveyed in this work. Support for the authors was provided in part by MIUR of Italy under Project *ALGO-NEXT: ALGOrithms for the NEXT generation Internet and Web* and by the European Union under the FP6-IST/IP Project 15964 *AEOLUS: Algorithmic Principles for Building Efficient Overlay Computers*.

References

[1] J.E. Savage. *Models of Computation—Exploring the Power of Computing*. Addison Wesley, Reading, MA, 1998.
[2] P. De la Torre and C.P. Kruskal. Submachine locality in the bulk synchronous setting. In *Proc. of EUROPAR 96*, LNCS 1124, pp. 352–358, August 1996.
[3] L.G. Valiant. A bridging model for parallel computation. *Communications of the ACM*, 33(8):103–111, August 1990.
[4] G. Bilardi, A. Pietracaprina, and G. Pucci. A quantitative measure of portability with application to bandwidth-latency models for parallel computing. In *Proc. of EUROPAR 99*, LNCS 1685, pp. 543–551, September 1999.
[5] G. Bilardi, C. Fantozzi, A. Pietracaprina, and G. Pucci. On the effectiveness of D-BSP as a bridging model of parallel computation. In *Proc. of the Int. Conference on Computational Science*, LNCS 2074, pp. 579–588, 2001.
[6] C. Fantozzi, A. Pietracaprina, and G. Pucci. A general PRAM simulation scheme for clustered machines. *Intl. Journal of Foundations of Computer Science*, 14(6):1147–1164, 2003.
[7] C. Fantozzi, A. Pietracaprina, and G. Pucci. Translating submachine locality into locality of reference. *Journal of Parallel and Distributed Computing*, 66:633–646, 2006. Special issue on 18th IPDPS.
[8] T.H. Cormen, C.E. Leiserson, R.L. Rivest, and C. Stein. *Introduction to Algorithms*. McGraw-Hill/MIT Press, Cambridge, MA, USA, 2nd edition, 2001.
[9] S.A. Cook and R.A. Reckhow. Time bounded random access machines. *Journal of Computer and System Sciences*, 7:354–375, 1973.
[10] R.P. Brent. The parallel evaluation of general arithmetic expressions. *Journal of the ACM*, 21(2):201–208, 1974.
[11] S. Fortune and J. Wyllie. Parallelism in random access machines. In *Proc. of the 10th ACM Symp. on Theory of Computing*, pp. 114–118, 1978.
[12] L.M. Goldschlager. A unified approach to models of synchronous parallel machines. *Journal of the ACM*, 29:1073–1086, 1982.
[13] R.M. Karp and V. Ramachandran. Parallel algorithms for shared-memory machines. In *Handbook of Theoretical Computer Science, Volume A: Algorithms and Complexity*, p. 869–942. Elsevier and MIT Press, Amsterdam, The Netherlands, 1990.
[14] J. JáJá. *An Introduction to Parallel Algorithms*. Addison Wesley, Reading, MA, 1992.
[15] F.T. Leighton. *Introduction to Parallel Algorithms and Architectures: Arrays, Trees, Hypercubes*. Morgan Kaufmann, San Mateo, CA, 1992.

[16] M.C. Pease. The indirect binary *n*-cube microprocessor array. *IEEE Trans. on Computers*, C-26(5):458–473, May 1977.

[17] H.S. Stone. Parallel processing with the perfect shuffle. *IEEE Trans. on Computers*, C-20(2):153–161, February 1971.

[18] F.P. Preparata and J. Vuillemin. The cube-connected-cycles: A versatile network for parallel computation. *Communications of the ACM*, 24(5):300–309, May 1981.

[19] C.E. Leiserson. Fat-trees: universal networks for hardware-efficient supercomputing. *IEEE Trans. on Computers*, C-34(10):892–901, October 1985.

[20] E. Upfal and A. Widgerson. How to share memory in a distributed system. *Journal of the ACM*, 34(1): 116–127, 1987.

[21] A. Pietracaprina, G. Pucci, and J. Sibeyn. Constructive, deterministic implementation of shared memory on meshes. *SIAM Journal on Computing*, 30(2):625–648, 2000.

[22] M. Hall Jr. *Combinatorial Theory*. John Wiley & Sons, New York, 2nd edition, 1986.

[23] D.E. Culler, R. Karp, D. Patterson, A. Sahay, E. Santos, K.E. Schauser, R. Subramonian, and T.V. Eicken. LogP: A practical model of parallel computation. *Communications of the ACM*, 39(11): 78–85, November 1996.

[24] G. Bilardi, K.T. Herley, A. Pietracaprina, G. Pucci, and P. Spirakis. BSP vs LogP. *Algorithmica*, 24:405–422, 1999. Special Issue on Coarse Grained Parallel Algorithms.

[25] G. Bilardi, K. Herley, A. Pietracaprina, and G. Pucci. On stalling in LogP. *Journal of Parallel and Distributed Computing*, 65:307–312, 2005.

[26] A. Aggarwal, A.K. Chandra, and M. Snir. Communication complexity of PRAMs. *Theoretical Computer Science*, 71:3–28, 1990.

[27] P.B. Gibbons, Y. Matias, and V. Ramachandran. Can a shared-memory model serve as a bridging-model for parallel computation? *Theory of Computing Systems*, 32(3):327–359, 1999.

[28] M. Snir and J.A. Solworth. Ultracomputer note 29. The Ultraswitch—A VLSI network node for parallel processing. Technical report, Courant Institute, New York University, 1984.

[29] G.L.-T. Chiu, M. Gupta, and A.K. Royyuru, editors. *Blue Gene*. Special issue of *IBM Journal of Research and Development* 49(2/3), 2005.

[30] A. Aggarwal, B. Alpern, A.K. Chandra, and M. Snir. A model for hierarchical memory. In *Proc. of the 19th ACM Symp. on Theory of Computing*, pp. 305–314, 1987.

[31] A. Aggarwal, A.K. Chandra, and M. Snir. Hierarchical memory with block transfer. In *Proc. of the 28th IEEE Symp. on Foundations of Computer Science*, pp. 204–216, 1987.

[32] G. Bilardi, K. Ekanadham, and P. Pattnaik. Computational power of pipelined memory hierarchies. In *Proc. of the 13th ACM Symp. on Parallel Algorithms and Architectures*, pp. 144–152, 2001.

[33] G. Bilardi, K. Ekanadham, and P. Pattnaik. Optimal organizations for pipelined hierarchical memories. In *Proc. of the 14th ACM Symp. on Parallel Algorithms and Architectures*, pp. 109–116, 2002.

[34] M.T. Goodrich. Communication-efficient parallel sorting. *SIAM Journal on Computing*, 29(2):416–432, 1999.

[35] C. Fantozzi, A. Pietracaprina, and G. Pucci. Implementing shared memory on clustered machines. In *Proc. of 2nd International Parallel and Distributed Processing Symposium*, 2001.

[36] G. Baudet and D. Stevenson. Optimal sorting algorithms for parallel computers. *IEEE Trans. on Computers*, C-27(1):84–87, January 1978.

[37] B.H.H. Juurlink and H.A.G. Wijshoff. A quantitative comparison of parallel computation models. In *Proc. of the 8th ACM Symp. on Parallel Algorithms and Architectures*, pp. 13–24, June 1996.

[38] A. Czumaj, F. Meyer auf der Heide, and V. Stemann. Shared memory simulations with triple-logarithmic delay. In *Proc. of the 3rd European Symposium on Algorithms*, pp. 46–59, 1995.

[39] A.G. Ranade. How to emulate shared memory. *Journal of Computer and System Sciences*, 42:307–326, 1991.

[40] J.L. Carter and M.N. Wegman. Universal classes of hash functions. *Journal of Computer and System Sciences*, 18:143–154, 1979.

[41] K. Mehlhorn and U. Vishkin. Randomized and deterministic simulations of PRAMs by parallel machines with restricted granularity of parallel memories. *Acta Informatica*, 21:339–374, 1984.

[42] F.T. Leighton and S. Rao. Multicommodity max-flow min-cut theorems and their use in designing approximation algorithms. *Journal of the ACM*, 46(6):787–832, 1999.

[43] F.T. Leighton, B.M. Maggs, and A.W. Richa. Fast algorithms for finding O(congestion + dilation) packet routing schedules. *Combinatorica*, 19(3):375–401, 1999.

[44] U. Vishkin. Can parallel algorithms enhance serial implementation? *Communications of the ACM*, 39(9):88–91, 1996.

[45] J.F. Sibeyn and M. Kaufmann. BSP-like external-memory computation. In *Proc. of 3rd CIAC*, LNCS 1203, pp. 229–240, 1999.

[46] F. Dehne, W. Dittrich, and D. Hutchinson. Efficient external memory algorithms by simulating coarse-grained parallel algorithms. *Algorithmica*, 36(2):97–122, 2003.

[47] F. Dehne, D. Hutchinson, D. Maheshwari, and W. Dittrich. Bulk synchronous parallel algorithms for the external memory model. *Theory of Computing Systems*, 35(6):567–597, 2002.

[48] C. Fantozzi, A. Pietracaprina, and G. Pucci. Seamless integration of parallelism and memory hierarchy. In *Proc. of 29th Int. Colloquium on Automata, Languages and Programming*, LNCS 2380, pages 856–867, July 2002.

[49] G. Bilardi and E. Peserico. A characterization of temporal locality and its portability across memory hierarchies. In *Proc. of 28th Int. Colloquium on Automata, Languages and Programming*, LNCS 2076, pp. 128–139, 2001.

[50] J.S. Vitter and E.A.M. Shriver. Algorithms for parallel memory II: Hierarchical multilevel memories. *Algorithmica*, 12(2/3):148–169, 1994.

[51] G. Bilardi and F.P. Preparata. Horizons of parallel computing. *Journal of Parallel and Distributed Computing*, 27:172–182, 1995.

3
Membrane Systems: A "Natural" Way of Computing with Cells

3.1	Introduction...	3-2
3.2	From Cells to Computers	3-3
3.3	The Basic Model of a P System.....................	3-5
3.4	Extensions and Variants	3-7
3.5	P Automata and P Transducers	3-8
3.6	P Systems with String Objects	3-9
3.7	Universality (for the Maximal Parallelism)	3-10
3.8	Solving Computationally Hard Problems in Polynomial Time...................................	3-12
3.9	Classes of P Systems with Nonmaximal Parallelism... Bounded Parallelism • Minimal Parallelism • P Systems with Proteins on Membranes	3-15
3.10	On Determinism versus Nondeterminism in P Systems	3-17
3.11	Other Classes of P Systems	3-18
3.12	Further Topics...	3-19
3.13	Closing Remarks	3-20
	References	3-20

Oscar H. Ibarra
University of California

Andrei Păun
Louisiana Tech University

This chapter briefly presents a series of basic elements of membrane computing, a recent branch of natural computing. The main goal of membrane computing is to abstract computing ideas and models from the structure and the function of living cells, as well as from the way the cells are organized in tissues or higher order structures. The corresponding models, called P systems, are distributed computing devices with various degrees of parallelism depending on the variant considered, ranging from maximal parallelism, to asynchronous systems, through bounded parallelism, minimal parallelism, sequential systems, sometimes with several levels of parallelism. Short descriptions of several of these possibilities are presented, with an overview of basic results (especially for cell-like P systems).

3.1 Introduction

Membrane computing is a part of the general research effort of describing and investigating computing models, ideas, architectures, and paradigms from the processes taking place in nature, especially in biology. This was a concern of computer science since "old times." Turing himself has tried to model the way that human beings compute, while finite automata were directly originated in investigations related to neurons and their cooperation—see McCulloch, Pitts, Kleene. More recently, neural computing, genetic algorithms, and evolutionary computing in general are well-known bioinspired areas of computer science with many applications. These areas try to find in biology suggestions for the better use of the existing electronic computers. DNA computing has a greater ambition, which is to find new hardware, able to support massive parallelism, by using DNA (and other) molecules as a medium for computing. All these directions of research—and several others less elaborated—now form a general area known under the name of *natural computing*, which is rapidly developing and rather promising both at the theoretical and practical level.

Membrane computing is a part of natural computing initiated in Reference 52 and vividly investigated in the past few years. Already in 2003, membrane computing was selected by the Institute for Scientific Information (ISI) as a fast "Emerging Research Front" in Computer Science (see `http://esi-topics.com/erf/october2003.html`).

In short, membrane computing abstracts computing models, starting from the structure and functioning of living cells and of their organization in tissues, organs, and so forth. These models are known as P systems, which are distributed and parallel computing models, processing multisets of symbol-objects in a localized manner (evolution rules and evolving objects are encapsulated into compartments delimited by membranes), with an essential role played by the communication among compartments (with the environment as well). We will mainly present the cell-like P systems, while only briefly describing the tissue-like and neural-like P systems (spiking neural P systems). For a comprehensive presentation of the domain (at the level of year 2002), we refer the reader to the monograph [54]; complete bibliographical information as well as many downloadable papers and proceedings volumes can be found in Reference 70. At this web site one can also find information about implementations of P systems on silicon computers and applications of membrane computing, especially in biology/medicine, linguistics, economics, and computer science. Such applications can be also found in Reference 12.

Membrane computing can be seen as an extension of DNA computing (or molecular computing, in general) because of the distributed nature of P systems; as we will see below, the basic ingredients of DNA computing (operations, data structures) were also used in this area. Thus membrane computing inherits most of the features of DNA computing, adding several others of interest for computing. It is important to mention that at this moment there is no lab implementation of a P system, while DNA computing, initiated theoretically by research related to the splicing operation (see Reference 28), was proved to be practically possible in 1994 (see Reference 1), although not yet proved to be practically useful.

The essential ingredients of a P system are the *membranes*, which delimit compartments where *multisets* of objects evolve according to specific *evolution rules*. The membranes correspond to the biological membranes, the objects correspond to the chemicals present in the compartments of a cell, and the evolution rules are the counterpart of reactions among these chemicals. There are also rules used for evolving the membranes (e.g., dividing them), and for moving objects across membranes (e.g., symport and antiport rules). Recently, models have been considered where chemicals (proteins) are also placed on the membranes, not only in the compartments defined by them—we will describe below these types of systems.

Inspired by the biochemistry taking place in the cells, the evolution rules are used in general in parallel, with various degrees of parallelism; we will discuss several possibilities in the following sections, pointing to their influence on the power and efficiency of the respective classes of P systems.

We end this introductory discussion by stressing the fact that while membrane computing started as an attempt to find computationally useful ideas in cellular biology (and not to provide models of cells useful to the biologist) and was impressively developed at the theoretical level (mainly investigating the computing power and the computing efficiency of P systems, that is, comparing them with Turing machines and their restrictions and trying to solve computationally hard problems in a feasible time, by a time-space trade-off made possible by the massive parallelism available), recently the domain turned back to biology as a framework for devising models of processes taking place in cells or populations of cells. Also, although no bioimplementation was attempted yet, there are many implementations on the usual electronic computers, sometimes on distributed/parallel architectures.

In this framework, several features important for computer science were touched upon—distribution, parallelism, communication, synchronization, decentralization, nondeterminism—with a specific definition and materialization, as suggested by the biological reality. Whether the suggestions coming from biology via membrane computing would be practically useful for computer science is not yet fully proven, although there are a series of papers that contain rather encouraging results from this point of view References 5, 24, 47, 48, 67, and so forth.

3.2 From Cells to Computers

The cell is the smallest "thing" unanimously considered as *alive*, but still the cell, whether prokaryotic or eukaryotic, is a very intricate small "factory," with a complex structure, function, and interaction with the environment (e.g., Reference 2).

Essential for a cell is the notion of a membrane. The cell itself is defined—separated from its environment—by its external membrane. Each membrane in the cell has the function of keeping the water molecules and all the other (large) molecules that are floating in the aqueous solution from passing to the other region. In this way, several membranes inside the cell enclose "protected reactors," which are compartments where specific biochemical processes take place, in many cases under the control of enzymes placed on the membranes. The membranes separating the compartments also selectively allow passage of substances, either depending on the size of substances or through protein channels in a much more intricate manner, even moving molecules from a low concentration to a higher concentration, and sometimes coupling molecules through so-called symport and antiport processes.

Some cells live alone (unicellular organisms), but in general cells are organized in tissues, organs, and organisms, with the special case of neurons that are organized in the best "computer" ever known, the brain. All these cells assume a specific organization, starting with the direct communication/cooperation among neighboring cells, and ending with the interaction with the environment, at various levels. Together with the internal structure and organization of the cell, all these levels of organization suggest a lot of ideas, exciting from a mathematical point of view, and potentially useful from a computability point of view. Part of them were already explored in membrane computing, though much more still wait for research efforts.

Not too many biological details are necessary in order to start a study of cell-like or tissue-like computing devices. In particular, we do not recall here any biological information, and we enter directly into discussing the basic ideas of membrane computing.

As we have already mentioned, the fundamental ingredients of a P system are (i) membranes, (ii) multisets of objects, and (iii) evolution rules. They correspond to biological membranes (3D vesicles, defining an inside and an outside), solutions of chemicals placed in the compartments of a cell, and reactions that make these chemicals evolve. In what follows, the notion of a membrane should be understood at an abstract level as a virtual border, with its dual role of both delimiting compartments and selecting chemicals for passage from one compartment to another. Let us now proceed to the *objects* that can be found in these regions delimited by membranes; we use the term objects as a generic name for all chemicals, from ions to large macromolecules. In the case when the structure of these molecules does not play any role in our investigations, such objects are represented by symbols from a specified alphabet. When the structure matters, we can also consider objects in the form of strings (or even more complex structures, such as trees, arrays, etc.). Because the concentration of molecules in the compartments of a cell is important for the cell's evolution, but all these chemicals (objects in our terminology) are swimming in an aqueous solution, we use the multiset as the main data structure. A multiset has a multiplicity (in the form of a natural number) associated with each element.

It is usual in membrane computing to represent the objects by symbols and the multisets by strings. For instance, a multiset in which the objects a, b, c are present in, respectively, 5, 2, and 6 copies each can be represented by the string $a^5 b^2 c^6$ or by any permutations of this string (mathematically, we can say that we work with strings modulo permutation, or that from a string we keep only the information provided by its Parikh mapping; for any formal language notions or results we invoke here, we refer the reader to Reference 63).

Now, having in mind the biochemical background, the multisets from the compartments of our computing device are processed by "reactions" encoded as rewriting-like rules. This means that rules are of the form $u \to v$, where u and v are multisets of objects (represented by strings). As an example, consider the rule $aab \to abcc$. It indicates the fact that two copies of object a together with a copy of object b react, and, as a result of this reaction, we get back a copy of a as well as the copy of b (hence b behaves here as a catalyst), and we produce two new copies of c. If this rule is applied to the multiset $a^5 b^2 c^6$, then, because aab are "consumed" and then $abcc$ are "produced," we pass to the multiset $a^4 b^2 c^8$. Similarly, by using the rule $bb \to aac$, we will get the multiset $a^7 c^7$, which contains no occurrence of object b.

We now arrive at two notions central to our discussion, that of nondeterminism and that of parallelism, both of which are also fundamental for computer science in general. Addressing problems in a nondeterministic way (guessing a solution and then checking it) is a fast way to solve a problem, but it is not an acceptable one from a practical point of view, because the guess can be wrong. Parallelism can be used for simulating the nondeterminism, especially in the case of biochemical-like frameworks, where reactions take place in a massively parallel way.

This is the case also in membrane computing, where both nondeterminism and parallelism are present. The rules to be used and the object to which they are applied are chosen randomly, while in a given step one applies in parallel as many rules as possible. One could say that the parallelism is maximal—and this was the way the evolution rules were used in Reference 52 and in most papers in membrane computing dealing with theoretical issues.

This is an important point that deserves some clarification. First, the P systems (again, as initially introduced and as investigated in most papers in the area) are synchronized computing devices. A global clock is assumed, marking the time in the same way for all compartments of the system. The use of a rule takes one time unit, and in each time unit all objects that can evolve should do it (more precisely, in each time unit, the multiset from each compartment must evolve by means of a maximal multiset of applicable rules). We illustrate this idea with the previous multiset and pair of rules. Using these rules in the maximally parallel manner means to either use the first rule twice (thus involving four copies of a and both copies of b) or the second rule once (it consumes both copies of b, hence the first rule cannot be used at the same time). In the first case, one copy of a remains unused (and the same with all copies of c), and the resulting multiset is $a^3 b^2 c^{10}$; in the second case, all copies of a and c remain unused, and the resulting multiset is $a^7 c^7$. It deserves to be noted that in the latter case the maximally parallel application of rules corresponds to the *sequential* (one at a time) application of the second rule.

Using the rules as suggested above, one obtains transitions between configurations of a P system. A sequence of transitions forms a computation, and with a halting computation (one reaching a configuration where no rule can be applied) we associate a result. Because in the system we have numbers (multiplicities of objects), the result will be a number, for example, the total number of objects in a specified compartment, or the vectors of numbers, in the case where we distinguish among the objects from that compartment. In this way, we get a number (or vector) generating device.

There are many variants of the architecture and functioning of the computing device sketched above, and we will mention some of them after introducing in some detail the basic type of P systems, the cell-like one working with symbol objects processed by multiset rewriting rules.

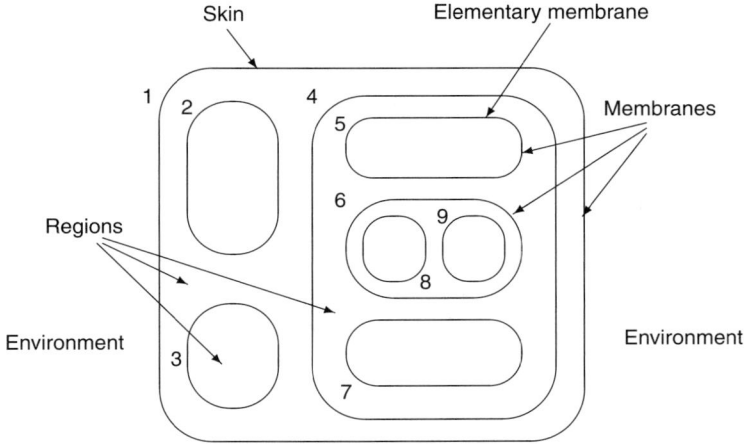

FIGURE 3.1 A membrane structure.

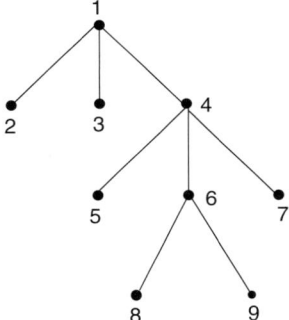

FIGURE 3.2 The tree describing the membrane structure from Figure 3.1.

3.3 The Basic Model of a P System

As mentioned above, the fundamental ingredient of a P system is that of a *membrane structure*.

The meaning of this notion is illustrated in Figure 3.1, where also the related terminology is presented.

Thus, as suggested by Figure 3.1, a membrane structure is a hierarchically arranged set of (labeled) membranes, contained in a distinguished external membrane (corresponding to the plasma membrane and usually called the *skin* membrane). Several membranes can be placed inside the skin membrane; a membrane without any other membrane inside it is said to be *elementary*. Each membrane determines a compartment, also called a *region*, which is the space delimited from above by the membrane itself and from below by the membranes placed directly inside, if any exist. Clearly, the correspondence membrane region is one-to-one; that is why we sometimes use these terms interchangeably, using the labels of membranes for identifying the regions.

A hierarchical structure of membranes can be represented by a rooted unordered tree. Figure 3.2 gives the tree that describes the membrane structure from Figure 3.1. The root of the tree is associated with the skin membrane and the leaves with the elementary membranes. Directly suggested by the tree representation is the symbolic representation of a membrane structure, which is given by strings of labeled matching parentheses. For instance, a string corresponding to the structure from Figure 3.1 is the following one:

$$[_1 [_2]_2 [_3]_3 [_4 [_5]_5 [_6 [_8]_8 [_9]_9]_6 [_7]_7]_4]_1.$$

Of course, membranes from the same level can float around (the tree is not ordered), hence two subexpressions placed at the same level represent the same membrane structure.

The objects present in the regions of a membrane structure evolve by means of *evolution rules*, which are also localized, meaning they are associated with the regions of the membrane structure. There are three main types of rules: (i) multiset-rewriting rules (one could simply call them evolution rules), (ii) communication rules, and (iii) rules for handling membranes.

As already mentioned in the previous section, the first type of rules are of the form $u \to v$, where u and v are multisets of objects. In order to make the compartments cooperate, we have to move objects across membranes, and to this aim we add *target indications* to the objects produced by a rule as above (to the objects from multiset v). These indications are *here, in, out*. An object associated with the indication *here* remains in the same region, one associated with the indication *in* goes immediately into a directly lower membrane, nondeterministically chosen, and *out* indicates that the object has to exit the membrane, thus becoming an element of the region surrounding it. An example of an evolution rule is $aab \to (a, here)(b, out)(c, here)(c, in)$ (this is the first of the rules considered in Section 3.2, where target indications are associated with the objects produced by the rule application). After using this rule in a given region of a membrane structure, two copies of a and one b are consumed (removed from the multiset of that region), and one copy of a, one of b, and two of c are produced; the resulting copy of a remains in the same region, and the same happens with one copy of c (indications *here*), while the new copy of b exits the membrane, going to the surrounding region (indication *out*), and one of the new copies of c enters one of the child membranes, nondeterministically chosen. If no such child membrane exists, that is, the membrane with which the rule is associated is elementary, then the indication *in* cannot be followed, and the rule cannot be applied. In turn, if the rule is applied in the skin region, then b will exit into the environment of the system (and it is "lost" there, as it can never come back). In general, the indication *here* is not specified (an object without an explicit target indication is supposed to remain in the same region where the rule is applied).

A rule as the one above, with at least two objects in its left-hand side, is said to be *cooperative*; a particular case is that of *catalytic* rules, of the form $ca \to cv$, where c is an object (called catalyst) that assists the object a to evolve into the multiset v. Rules of the form $a \to v$, where a is an object, are called *noncooperative*.

Rules can also have the form $u \to v\delta$, where δ denotes the action of *membrane dissolving*: if the rule is applied, then the corresponding membrane disappears and its contents, object and membranes alike, are left free in the surrounding membrane. The rules of the dissolved membrane disappear at the same time with the membrane. The skin membrane is never dissolved.

When choosing which rules to use, the key phrase is: *in the maximally parallel manner, nondeterministically choosing the rules and the objects.*

More specifically, this means that we assign objects to rules, nondeterministically choosing the objects and the rules, until no further assignment is possible. More mathematically stated, we look to the *set* of rules, and try to find a *multiset* of rules, by assigning multiplicities to rules, with two properties: (i) the multiset of rules is *applicable* to the multiset of objects available in the respective region, that is, there are enough objects in order to apply the rules as many times as indicated by their multiplicities, and (ii) the multiset is *maximal*, meaning no further rule can be added to it (because of the lack of available objects).

Thus, an evolution step in a given region consists of finding a maximal applicable multiset of rules, removing from the region all objects specified in the left hand of the chosen rules (with the multiplicities as indicated by the rules and by the number of times each rule is used), producing the objects from the right-hand sides of rules, and then distributing these objects as indicated by the targets associated with them. If at least one of the rules introduces the dissolving action δ, then the membrane is dissolved, and its contents become part of the immediately upper membrane—provided that this membrane was not dissolved at the same time, a case where we stop in the first upper membrane that was not dissolved (at least the skin remains intact).

It is worth pointing out in this context the role of catalysts as inhibitors/controllers of parallelism: a rule of the type $ca \to cv$ can be used in parallel, but at any given step the number of applications of such a rule is upper bounded by the number of copies of the catalysts c that are present. In most proofs, this number is rather limited, even one in most cases, which means that the respective rule is used only once (sequentially). This is essential when trying to simulate a sequential computing device, for example, a Turing machine, by means of a P system, which is intrinsically a parallel device.

We are now able to introduce formally the cell-like P systems with symbol objects; they are customarily called *transition P systems*. Such a device is a construct of the form

$$\Pi = (O, C, \mu, w_1, w_2, \ldots, w_m, R_1, R_2, \ldots, R_m, i_o),$$

where

1. O is the (finite and nonempty) alphabet of *objects*.
2. $C \subset O$ is the set of *catalysts*.
3. μ is a membrane structure, consisting of m membranes, labeled with $1, 2, \ldots, m$; one says that the membrane structure, and hence the system, is *of degree m*.
4. $w_1, w_2, \ldots, w_m$ are strings over O representing the *multisets of objects* present in the regions $1, 2, \ldots, m$ of the membrane structure.
5. $R_1, R_2, \ldots, R_m$ are finite *sets of evolution rules* associated with the regions $1, 2, \ldots, m$ of the membrane structure.
6. i_o is either one of the labels $1, 2, \ldots, m$, and the respective region is the *output region* of the system, or it is 0, and the result of a computation is collected in the environment of the system.

The rules are of the form $u \to v$ or $u \to v\delta$, with $u \in O^+$ and $v \in (O \times \text{Tar})^*$, $\text{Tar} = \{here, in, out\}$.

A possible restriction for the region i_o in the case when it is an internal one is to consider only regions enclosed by elementary membranes for output (i.e., i_o should be the label of an elementary membrane of μ).

The *initial configuration* is given by specifying the membrane structure and the multisets of objects available in its compartments at the beginning of a computation, hence $(\mu, w_1, \ldots, w_m)$. During the evolution of the system, by means of applying the rules, both the multisets of objects and the membrane structure can change.

Note also that in a transition P system we have two levels of parallelism: all regions evolve simultaneously, and, inside each region, the objects evolve in a maximally parallel manner.

Because of the nondeterminism of the application of rules, starting from an initial configuration, we can get several successful computations, hence several results. Thus, a P system *computes* (or *generates*) a set of numbers (we consider here only this case). For a given system Π we denote by $N(\Pi)$ the set of numbers computed by Π in the above way.

We do not give here examples of P systems, because such examples can be found in many places—in particular, in Reference 55, where the reader can find a friendly introduction to membrane computing.

3.4 Extensions and Variants

The previous type of P system is only one of many types already considered in literature. Actually, it was said several times that membrane computing does not deal with a model, it is a framework for devising models of a specific type. The motivations for considering extensions and variants come from different directions: from biology (with a lot of suggestions, some of them still waiting to be considered), mathematics, and computer science.

For instance, the rules as considered above can be changed in various ways: using only the targets *here*, *out*, *in* (hence not *in_j*, with the indication of the label of the target membrane); allowing catalysts to move across membranes; using "pseudo-catalysts," that are allowed to flip from a state c to a state c' and back; and using an operation opposite to membrane dissolving, such as membrane thickening (making it nonpermeable to object communication), or even creating new membranes. Similarly, one can control the use of rules, for example, by means of promoters (objects that must be present in order to apply a rule, but that are not modified by a rule and that can promote at the same time several rules—note that the promoters do not inhibit the parallelism), or inhibitors, by considering a priority relation among rules, and so forth. Note that these controls decrease the degree of nondeterminism of the system.

An important class of P systems is that based on symport/antiport rules. These rules correspond to the biological processes by which molecules are transported across membranes in a coupled way, with two molecules passing together across a membrane (through a specific protein channel) either in the same direction (and this is called *symport*) or in opposite directions (this process is called *antiport*).

These processes were formalized in membrane computing in the form of symport rules of the form (x, in), (x, out), and antiport rules of the form $(z, out; w, in)$, where x, z, w are multisets of arbitrary size. In the first case, all objects of multiset x pass together across the membrane with which the rule is associated; in the second case the objects of z are sent out and those of w are brought into the membrane (the length of x, denoted by $|x|$, is called the *weight* of a symport rule as above, and $\max(|z|, |w|)$ is the *weight* of the antiport rule).

Such rules can be used in a *symport/antiport P system*, which is a construct of the form

$$\Pi = (O, \mu, w_1, \ldots, w_m, E, R_1, \ldots, R_m, i_o),$$

where

1. O is the alphabet of objects.
2. μ is the membrane structure (of degree $m \geq 1$, with the membranes labeled in a one-to-one manner with $1, 2, \ldots, m$).
3. $w_1, \ldots, w_m$ are strings over O representing the multisets of objects present in the m compartments of μ in the initial configuration of the system.
4. $E \subseteq O$ is the set of objects supposed to appear in the environment in arbitrarily many copies.
5. $R_1, \ldots, R_m$ are the (finite) sets of rules associated with the m membranes of μ.
6. $i_o \in \{1, 2, 3, \ldots, m\}$ is the label of a membrane of μ, which indicates the *output* region of the system.

The rules from R can be of two types, symport rules and antiport rules, of the forms as specified above.

Again, the rules are used in the nondeterministic maximally parallel manner. In the usual way, we define transitions, computations, and halting computations. The number (or the vector of multiplicities) of objects present in region i_o in the halting configuration is said to be computed by the system along that computation; the set of all numbers computed in this way by Π is denoted by $N(\Pi)$.

There also are other variants considered in literature, both in what concerns the form of rules (also combinations of types of rules) and in what concerns the way of controlling the use of the rules, but we do not continue here in this direction. An important class of P systems, which together with the basic transition systems and the symport/antiport systems is one of the three central types of cell-like P systems considered in membrane computing, is that of P systems with active membranes, which will be presented separately, in Section 3.8.

3.5 P Automata and P Transducers

Besides the generative mode introduced above, there exist two other important modes of P systems: the *accepting* mode and the *transducer* mode. In both cases, an input is provided to the system in two possible ways: either a multiset w_0 is added to the initial configuration of a system and we say that it is accepted if the computation halts or (e.g., in the case of symport/antiport P systems) the objects taken by the system from the environment form the input of a computation. In the first case, we accept a number (the size of w_0), in the latter case we accept a string: the sequence of objects entering the system form a string that is accepted if the computation halts. In this latter case we can have several variants: the input objects are introduced one by one in the first steps of the computation or when necessary, with or without an end-marker, or we can associate labels with input multisets and the accepted string is that of labels, and so forth.

It is also possible to consider an output of a computation that starts by having an input, and in this way we obtain a transducer.

It is important to note that in both the accepting and the transducing case we can impose the system to work deterministically: in each step, at most one multiset of rules can be applied, hence either the computation halts or the next configuration is unique. Between deterministic systems and nondeterministic systems there is an important intermediate case, that of confluent systems; the computations are nondeterministic, but in the end, all computations starting from the same initial configuration (the same input) provide the same result (output). This is very useful when solving problems by means of P systems; we are interested in the answer to the input problem, not in the behavior of the system, hence we can relax the construction of the system in such a way to behave in a confluent way, not necessarily in a deterministic way.

Given a P system Π (it can be a transition or a symport/antiport one), let us denote by $N_a(\Pi)$ the set of all numbers accepted by Π, in the following sense: we introduce a^n, for a specified object a, into a specified region of Π, and we say that n is accepted if and only if there is a computation of Π starting from this augmented initial configuration that halts. In the case of systems taking objects from the environment, such as the symport/antiport ones, we can consider that the system accepts/recognizes the sequence of objects taken from the environment during a halting computation (if several objects are brought into the system at the same time, then all their permutations are accepted as substrings of the accepted string). This language is denoted by $L_a(\Pi)$.

We use the subscript a in order to distinguish this language from other types of languages that can be associated with a P system, when used in the generative mode: for any type of system that can send objects into the environment, we can consider the sequence of objects that leaves the system, and their string can be considered as generated by the halting computations of the system.

It is worth noting that in the case of string generating or accepting, there is an essential difference between the data structure used inside the system (multisets, hence numbers), and the data structure of the result of the computation (strings, hence carrying positional information).

3.6 P Systems with String Objects

Strings can be introduced directly into the regions of the system by considering the objects as structured—as it is the case with most chemicals from a cell, starting with the DNA molecules. In this way, we obtain P systems with string objects, which are constructs of the form

$$\Pi = (V, T, \mu, M_1, \ldots, M_m, R_1, \ldots, R_m),$$

where V is the alphabet of the system, $T \subseteq V$ is the terminal alphabet, μ is the membrane structure (of degree $m \geq 1$), $M_1, \ldots, M_m$ are finite sets of strings present in the m regions of the membrane structure, and $R_1, \ldots, R_m$ are finite sets of string-processing rules associated with the m regions of μ.

In turn, the rules can be of several types: rewriting, splicing, insertion–deletion, and so forth. In the case of *rewriting P systems*, the string objects are processed by rules of the form $a \to u(tar)$, where $a \to u$ is a context-free rule over the alphabet V and *tar* is one of the target indications *here, in, out*. When such a rule is applied to a string $x_1 a x_2$ in a region i, we obtain the string $x_1 u x_2$, that is placed in region i, in any inner region, or in the surrounding region, depending on whether *tar* is *here, in,* or *out*, respectively. The strings that leave the system do not come back; if they are composed only of symbols from T, then they are considered as generated by the system, and included in the language $L(\Pi)$ (note that we do not need to consider only the halting computations as successful, because the strings that exit the system cannot be further processed).

Note that the rules are applied sequentially (each string is processed by only one rule at a time), but all strings from a compartment and all compartments evolve simultaneously. Of course, we can also consider the case of applying several rules to the same string (e.g., like in Lindenmayer systems), and this possibility was explored in Reference 7.

We do not present here other types of rules, but we introduce an important extension of rewriting rules, namely, *rewriting with replication* (see Reference 41), which provides a way to enhance the parallelism of rewriting P systems. Specifically, one uses rules of the form $a \to (u_1, tar_1)||(u_2, tar_2)|| \ldots ||(u_n, tar_n)$, with the meaning that by rewriting a string $x_1 a x_1$ we get n strings, $x_1 u_1 x_2, x_1 u_2 x_2, \ldots, x_1 u_n x_2$, which have to be moved in the regions indicated by targets $tar_1, tar_2, \ldots, tar_n$, respectively. In this case we work again with halting computations, and the motivation is that if we do not impose the halting condition, then the strings $x_1 u_i x_2$ evolve completely independently, hence we can replace the rule $a \to (u_1, tar_1)||(u_2, tar_2)|| \ldots ||(u_n, tar_n)$ with n rules $a \to (u_i, tar_i), 1 \le i \le n$, without changing the language; that is, replication makes a difference only in the halting case.

Replicated rewriting is one of the ways to generate an exponential workspace in linear time. It is used for solving computationally hard problems in polynomial time.

3.7 Universality (for the Maximal Parallelism)

After having defined a computing model, it is a standard issue to compare it with existing computing models, in particular, with the standard one—the Turing machine and its restrictions. As the title of this section suggests, in the case of P systems the most frequent result in this respect is the equivalence with the Turing machine. Formally, this means "Turing completeness," but because the proofs are constructive, starting from a universal Turing machine one obtains a universal P system of the type dealt with, hence we can also speak about "universality" in the rigorous sense.

We present here only a few basic universality results, without proofs, and only for the classes of P systems introduced above. Similar results are reported for other classes of systems, in particular, for tissue-like P systems that are not recalled here.

The proofs are based on simulations of computing devices known to be universal by means of P systems of various types. In general, there are two difficulties in this framework: (i) to control/inhibit the parallelism of P systems so that the sequential behavior of classic devices (grammars and automata of various types) can be simulated by the inherently parallel machineries of P systems, and (ii) to avoid/encode the positional information provided by tapes of automata and strings of grammars, so that we can simulate a classic device with multiset processing devices such as P systems.

Because of these two difficulties/criteria, most of the universality proofs in membrane computing are based on simulations of register machines (already working with numbers), or context-free grammars with regulated rewriting—in general, in special normal forms. Thus, the first universality proofs in this area were based on using matrix grammars with appearance checking in the binary normal form (this normal form was improved several times with the motivation coming from membrane computing), but later also programmed grammars, random context grammars, and other regulated grammars were used—with an abundance of recent proofs based on register machines.

In what concerns the notations, the family of sets $N(\Pi)$ of numbers generated by P systems of a specified type, working with symbol objects, having at most m membranes, and using features/ingredients from a given *list* is denoted by $NOP_m(list\text{-}of\text{-}features)$. When the number of membranes is not bounded, the subscript m is replaced by $*$, and this is a general convention used also for other parameters.

In what concerns the list of features, there are a lot of possibilities: using cooperative rules is indicated by *coo*, catalytic rules are indicated by *cat*, while noting that the number of catalysts matters, hence we use cat_r in order to indicate that we use systems with at most r catalysts; bistable catalysts are indicated by $2cat$

($2cat_r$, if at most r catalysts are used); when using a priority relation, we write pri, membrane creation is represented by $mcre$, and endocytosis and exocytosis operations are indicated by $endo$, exo, respectively. In the case of P systems with active membranes, which will be introduced in the next section, one directly lists the types of rules used, from (a) to (e), as defined and denoted in Section 3.8. For systems with string objects, one writes rew, $repl_d$, spl to indicate that one uses rewriting rules, replicated rewriting rules (with at most d copies of each string produced by replication), and splicing rules, respectively.

In the case of systems using symport/antiport rules, we have to specify the maximal weight of the used rules, and this is done by writing sym_p, $anti_q$, meaning that symport rules of weight at most p and antiport rules of weight at most q are allowed.

Also, for a family FL of languages, we denote by NFL the family of length sets of languages in FL. Thus, NRE denotes the family of Turing computable sets of numbers (which is the same with the family of length sets of recursively enumerable languages, those generated by Chomsky type-0 grammars, or by many types of regulated rewriting grammars, and recognized by Turing machines). The family NRE is also the family of sets of numbers generated/recognized by register machines.

Here are some universality results (we mention some papers for proofs):

1. $NRE = NOP_1(cat_2)$, [18].
2. $NRE = NOP_3(sym_1, anti_1) = NOP_3(sym_2, anti_0)$, [4].
3. $NRE = NOP_3((a), (b), (c))$, [43].
4. $NRE = NSP_3(repl_2)$, [42].

Similar results are obtained in the accepting case. We do not present here other details than the important one concerning determinism versus nondeterminism. Both transition P systems and symport/antiport P systems working in the accepting mode are universal, with the following surprising difference between the two types of systems: nondeterministic systems are universal in both cases, but the deterministic version is universal only for symport/antiport systems, while deterministic catalytic P systems are not universal, see Reference 31.

In all these universality results, the number of membranes sufficient for obtaining universality is pretty small, and this has an interesting consequence about the hierarchy induced by the number of membranes: these hierarchies collapse at rather low levels. This is true for all universal classes of P systems where the numbers to be computed are encoded in objects and not in the membranes.

Still, "the number of membranes matters," as we read already in the title of Reference 29: there are (subuniversal) classes of P systems for which the number of membranes induces an infinite hierarchy of families of sets of numbers (see also Reference 30).

The intuition behind these universality results (specifically, behind their proofs) is the fact that the maximal parallelism together with the halting condition in defining successful computations is a powerful tool for "programming" the work of a P system. The maximal parallelism ensures the possibility to have global information about the multisets placed in each region, so that in the case of "wrong" steps (when nondeterministically choosing a transition that does not correspond to a correct step in the simulated grammar or automaton) a trap object that evolves forever can be introduced, thus preventing the halting of the computation.

Actually, the power of maximal parallelism, combined with the halting condition, was explicitly considered in Reference 23, and found to be directly related (equivalent, in a certain sense) with the appearance checking from matrix grammars and other grammars with regulated rewriting, and with the check for zero in register machines. It is known that both the appearance checking and the check for zero are essential for the universality of the respective devices; grammars without appearance checking and register machines that cannot check whether a register is zero are not universal.

As said above, universality is obtained only if the parallelism is restricted at least for part of the rules of the considered P system, and this is achieved by means of catalysts, by using objects with a small number of copies in symport/antiport systems, and by definition in P systems with active membranes (each membrane can enter only one rule) and in P systems with string objects. When the parallelism is not restricted, then in general we obtain subfamilies of NRE, often only $NREG = NCF$ (the length sets of regular languages, which is the same as the length sets of context-free languages), *NMAT* (the length sets of languages generated by matrix grammars without appearance checking), or *NET0L* (the length sets of languages generated by extended tabled Lindenmayer systems without interaction).

3.8 Solving Computationally Hard Problems in Polynomial Time

We still remain in the case of maximal parallelism, and we consider an important direction of research in membrane computing, dealing not with computing sets of numbers, but with solving computationally hard problems (typically, **NP**-complete, but recently also **PSPACE**-complete problems were considered) in a feasible amount of time (polynomial, but often the time is even linear). A lot of hard problems were considered in this respect and "solved" in a polynomial time. Initially, decidability problems were addressed (starting with SAT), but then also numerical problems were considered—see References 60 and 58. This later book also introduces a rigorous framework for dealing with complexity matters in the membrane computing area, based on recognizing/accepting P systems with input. The idea is to construct by a Turing machine, in a uniform manner and polynomial time, a family of P systems associated with a given problem, to introduce any given instance of the problem in the initial configuration of a system associated with the size of the instance, and let the obtained system work, providing an answer to the problem in a specified number of steps. The system can be deterministic or confluent. In order to have the answer provided in polynomial time, an exponential workspace is constructed, in general, in linear time. Owing to the parallelism of P systems, this space is used in such a way to simultaneously check exponentially many computations (and this is the meaning of the idea that the massive parallelism can simulate nondeterminism, by exploring an exponential computation tree in the same amount of time as exploring any single path in this tree).

Thus, essentially, the solution to the problem is based on a time-space trade-off, which raises two issues: (i) how to obtain this space in a "natural" way, and (ii) how to implement such a procedure in order to perform a real computation. At this moment, there is no answer to the second question, neither in lab nor in silico.

In turn, the first question has three basic answers in membrane computing literature, with a series of variants of them: the necessary exponential space is obtained by (i) membrane division, (ii) membrane creation, and (iii) string replication. The last one was already introduced. Membrane creation is based on rules of the form $a \to [_h v]_h$, with the meaning that object a creates a new membrane, with label h and the contents specified by the multiset v. Membrane division was the first way to address computationally hard problems in terms of P systems [53], and the most investigated, which is why it will be presented below with some details. Before that, it is worth discussing an interesting point.

Already in transition P systems (and the same in symport/antiport systems) we have the ability to exponentially grow the workspace, in the sense that we can exponentially grow the number of objects present in the system. For instance, by using n steps of the rule $a \to aa$ we get 2^n copies of a (similarly for the antiport rule $(a, out; aa, in)$). Still, this space is not sufficient for essentially speeding-up computations: the so-called *Milano theorem* from References 68 and 69 says that a deterministic transition P system can be simulated with a polynomial slow-down by a Turing machine (a similar theorem for symport/antiport systems was not explicitly proven, but it is expected that it has a proof similar to that of the Milano theorem). There is an intriguing fact here: we have exponentially many objects, but we cannot solve **NP**-complete problems in polynomial time (unless if **NP** = **P**); structuring this space, by placing the objects in membranes, makes the solution possible. This has something to do with the localization of rules; specific membranes/compartments have specific rules for evolving the local objects, which makes the parallelism effective.

Let us have an illustration of this approach by considering the *P systems with active membranes*, which are constructs of the form

$$\Pi = (O, H, \mu, w_1, \ldots, w_m, R),$$

where

1. $m \geq 1$ (the initial degree of the system).
2. O is the alphabet of *objects*.
3. H is a finite set of *labels* for membranes.
4. μ is a *membrane structure*, consisting of m membranes having initially neutral polarizations, labeled (not necessarily in a one-to-one manner) with elements of H.
5. $w_1, \ldots, w_m$ are strings over O, describing the *multisets of objects* placed in the m regions of μ.
6. R is a finite set of *developmental rules*, of the following forms:
   a. $[_h a \rightarrow v]_h^e$,
      for $h \in H, e \in \{+, -, 0\}, a \in O, v \in O^*$
      (object evolution rules, associated with membranes and depending on the label and the charge of the membranes, but not directly involving the membranes, in the sense that the membranes are neither taking part in the application of these rules nor are they modified by them).
   b. $a[_h]_h^{e_1} \rightarrow [_h b]_h^{e_2}$,
      for $h \in H, e_1, e_2 \in \{+, -, 0\}, a, b \in O$
      (*in* communication rules; an object is introduced in the membrane, possibly modified during this process; also the polarization of the membrane can be modified, but not its label)
   c. $[_h a]_h^{e_1} \rightarrow [_h]_h^{e_2} b$,
      for $h \in H, e_1, e_2 \in \{+, -, 0\}, a, b \in O$
      (*out* communication rules; an object is sent out of the membrane, possibly modified during this process; also the polarization of the membrane can be modified, but not its label).
   d. $[_h a]_h^e \rightarrow b$,
      for $h \in H, e \in \{+, -, 0\}, a, b \in O$
      (dissolving rules; in reaction with an object, a membrane can be dissolved, while the object specified in the rule can be modified).
   e. $[_h a]_h^{e_1} \rightarrow [_h b]_h^{e_2}[_h c]_h^{e_3}$,
      for $h \in H, e_1, e_2, e_3 \in \{+, -, 0\}, a, b, c \in O$
      (division rules for elementary membranes; in reaction with an object, the membrane is divided into two membranes with the same label, possibly of different polarizations; the object specified in the rule is replaced in the two new membranes by possibly new objects; the remaining objects are duplicated and may evolve in the same step by rules of type (a)).

The objects evolve in the maximally parallel manner, used by rules of type (a) or by rules of the other types, and the same is true at the level of membranes, which evolve by rules of types (b)–(e). Inside each membrane, the rules of type (a) are applied in the parallel way, with each copy of an object being used by only one rule of any type from (a) to (e). Each membrane can be involved in only one rule of types (b), (c), (d), (e) (the rules of type (a) are not considered to involve the membrane where they are applied). Thus, in total, the rules are used in the usual nondeterministic maximally parallel manner, in a bottom-up way (first we use the rules of type (a), and then the rules of other types; in this way, in the case of dividing membranes, in the newly obtained membranes we duplicate the result of using first the rules of type (a)). Also as usual, only halting computations give a result, in the form of the number (or the vector) of objects expelled into the environment during the computation.

The set H of labels has been specified because it is also possible to allow the change of membrane labels. For instance, a division rule can be of the more general form

(e') $[_{h_1} a]_{h_1}^{e_1} \rightarrow [_{h_2} b]_{h_2}^{e_2}[_{h_3} c]_{h_3}^{e_3}$,
for $h_1, h_2, h_3 \in H, e_1, e_2, e_3 \in \{+, -, 0\}, a, b, c \in O$.

The change of labels can also be considered for rules of types (b) and (c). Also, we can consider the possibility of dividing membranes in more than two copies, or even of dividing nonelementary membranes (in such a case, all inner membranes are duplicated in the new copies of the membrane).

It is important to note that in the case of P systems with active membranes, the membrane structure evolves during the computation, not only by decreasing the number of membranes due to dissolution operations (rules of type (d)) but also by increasing the number of membranes through division. This increase can be exponential in a linear number of steps: using successively a division rule, due to the maximal parallelism, in n steps we get 2^n copies of the same membrane.

We illustrate here the way of using membrane division in such a framework by an example dealing with the generation of all 2^n truth-assignments possible for n propositional variables.

Assume that we have the variables $x_1, x_2, \ldots, x_n$; we construct the following system, of degree 2:

$$\Pi = (O, H, \mu, w_1, w_2, R),$$

$$O = \{a_i, c_i, t_i, f_i \mid 1 \leq i \leq n\} \cup \{\text{check}\},$$

$$H = \{1, 2\},$$

$$\mu = [\,_1[\,_2\ ]_2\,]_1,$$

$$w_1 = \lambda,$$

$$w_2 = a_1 a_2 \ldots a_n c_1,$$

$$R = \{[\,_2 a_i]_2^0 \rightarrow [\,_2 t_i]_2^0 [\,_2 f_i]_2^0 \mid 1 \leq i \leq n\}$$

$$\cup \{[\,_2 c_i \rightarrow c_{i+1}]_2^0 \mid 1 \leq i \leq n-1\}$$

$$\cup \{[\,_2 c_n \rightarrow \text{check}]_2^0, [\,_2 \text{check}]_2^0 \rightarrow \text{check}[\,_2\ ]_2^+\}.$$

We start with the objects $a_1, \ldots, a_n$ in the inner membrane and we divide this membrane, repeatedly, by means of rules $[\,_2 a_i]_2^0 \rightarrow [\,_2 t_i]_2^0 [\,_2 f_i]_2^0$. Note that the object a_i used in each step is nondeterministically chosen, but each division replaces that object by t_i (for *true*) in one membrane and with f_i (for *false*) in the other membrane; hence, after n steps the obtained configuration is the same irrespective of the order of expanding the objects. Specifically, we get 2^n membranes with label 2, each one containing a truth-assignment for the n variables. Actually, simultaneously with the division, we have to use the rules of type (a) that evolve the "counter" c, hence at each step we increase by one the subscript of c. Therefore, when all variables are expanded, we get the object check in all membranes (the rule of type (a) is used first, and after that the result is duplicated in the newly obtained membranes). In step $n + 1$, this object exits each copy of membrane 2, changing its polarization to positive—this object is meant to signal the fact that the generation of all truth-assignments is completed, and we can start checking the truth values of the (clauses of a) propositional formula.

The previous example was chosen also for showing that the polarizations of membranes are not used during generating the truth-assignments, but they might be useful after that—and, up to now, this is the case in all polynomial time solutions to **NP**-complete problems obtained in this framework, in particular, for solving SAT (satisfiability of propositional formulas in the conjunctive normal form). This is an important *open problem* in this area: whether or not the polarizations can be avoided. This can be done if other ingredients are considered, such as label changing or division of nonelementary membranes, but without adding such features the best result obtained so far is that from Reference 3 where it is proven that the number of polarizations can be reduced to two.

As said above, using membrane division (or membrane creation, or string replication—but we do not give further details for these two last possibilities) polynomial or pseudo-polynomial solutions to **NP**-complete problems were obtained. The first problem addressed in this context was SAT [53] (the solution was improved in several respects in other subsequent papers), but similar solutions are reported in literature for the Hamiltonian Path problem, the Node Covering problem, the problem of inverting one-way functions, the Subset sum, the Knapsack problems (note that the last two are numerical problems, where the answer is not of the yes/no type, as in decidability problems), and for several other problems. Details can be found in References 54 and 58, as well as in the web page of the domain [70].

This direction of research is very active at the present moment. More and more problems are considered, the membrane computing complexity classes are refined, and characterizations of the **P≠NP** conjecture as well as of classic complexity classes were obtained in this framework. Two important recent results concern the surprising role of the operation of membrane dissolution (with characterizations of class **P** when this operation is not allowed [27]), and the characterization of class **PSPACE** in terms of P systems with the possibility of dividing both elementary and nonelementary membranes (see References 61 and 62).

3.9 Classes of P Systems with Nonmaximal Parallelism

As we have seen above, the maximal parallelism is rather useful from a theoretical point of view, both in what concerns the computing power and the efficiency of P systems, but it is mainly mathematically motivated (powerful, elegant) and not so much biologically grounded. The discussion is not trivial: is there a unique clock in a cell, marking uniformly the time of each compartment? Otherwise stated, is the cell a synchronized system or an asynchronous one? Some people answer affirmatively, other people answer negatively this question. What seems to be clear is that there is a high degree of parallelism in a cell. When several reactions can take place in the same compartment, many of them actually take place, in parallel. If we wait long enough, then *all* possible reactions happen. Well, "waiting long enough" means changing the time unit, hence we return again to the question whether the time is uniform in the whole cell.

In what follows we avoid this discussion (and other related ones), and we confine ourselves to a more technical discussion of the versions of parallelism considered so far in membrane computing. We do not present details concerning the sequential systems—this looks similarly "nonrealistic" from a biological point of view as the maximal parallelism (there are however several papers investigating sequential P systems of various types; details and references can be found, e.g., in References 17, 16, and 34).

3.9.1 Bounded Parallelism

Now we consider the following definition (different from the standard definition in the literature of "maximal parallelism" in the application of evolution rules in a P system Π: Let $R = \{r_1, \ldots, r_k\}$ be the set of (distinct) rules in the system. Π operates in maximally parallel mode if at each step of the computation, a maximal subset of R is applied, and at most one instance of any rule is used at every step (thus at most k rules are applicable at any step). We refer to this system as a maximally parallel system. We can define three semantics of parallelism. For a positive integer $n \leq k$, define

n-**Max-Parallel**: At each step, nondeterministically select a maximal subset of at most n rules in R to apply (this implies that no larger subset is applicable).

$\leq n$-**Parallel**: At each step, nondeterministically select any subset of at most n rules in R to apply.

n-**Parallel**: At each step, nondeterministically select any subset of exactly n rules in R to apply.

In all three cases, if any rule in the subset selected is not applicable, then the whole subset is not applicable. When $n = 1$, the three semantics reduce to the **Sequential** mode.

The three modes of parallelisms were studied in Reference 36 for two popular model P systems: multimembrane catalytic systems and communicating P systems. It was shown that for these systems, n-**Max-Parallel** mode is strictly more powerful than any of the following three modes: **Sequential**, $\leq n$-**Parallel**, or n-**Parallel**. For example, it follows from the result in Reference 20 that a maximally parallel communicating P system is universal for $n = 2$. However, under the three limited modes of parallelism, the system is equivalent to a vector addition system, which is known to only define a recursive set. Some results in Reference 36 are rather surprising. For example, it was shown that a **Sequential** 1-membrane communicating P system can only generate a semilinear set, whereas with k membranes, it is equivalent to a vector addition system for any $k \geq 2$ (thus the hierarchy collapses at 2 membranes—a rare collapsing result for nonuniversal P systems).

A simple cooperative system (SCO) is a P system where the only rules allowed are of the form $a \to v$ or of the form $aa \to v$, where a is a symbol and v is a (possibly null) string of symbols not containing a. It was also shown in Reference 36 that a 9-**Max-Parallel** 1-membrane SCO is universal.

3.9.2 Minimal Parallelism

The notion of minimal parallelism looks rather "biorealistic" and relaxed: if a compartment of a cell can evolve, then it has to evolve at least in the minimal way, by using at least one reaction. In terms of P systems, the condition is the following: we have sets of rules associated with compartments (as in the case of transition P systems) or with membranes (as in symport/antiport systems or in P systems with active membranes); then, if in a given configuration from such a set at least one rule *can* be applied, then at least one *must* be applied. How many rules are actually applied is not prescribed; exactly one rule or more than one rules, even all rules that can be used, all these possibilities are allowed, we only know that for sure at least one is effectively applied.

It is clear that this way of using the rules is much weaker than the maximal parallelism, because we have a rather limited knowledge about the evolution of the system (the degree of nondeterminism is increased, we cannot get an information about all objects from a compartment, etc.). Still, we have one level of parallelism, because the compartments/membranes evolve simultaneously, synchronized, and with a minimal action taking place in each of them.

It is somewhat surprising that this minimal control on the evolution of "subsystems" of a P system still ensures the computational universality. This was proved in Reference 13 for symport/antiport P systems working in the generating or the accepting mode (in the second case also for deterministic systems), and for P systems with active membranes. Thus, results as those recalled in Section 3.7 were obtained (not yet also for catalytic systems), but in all cases with some of the parameters (especially, the number of membranes) being larger, thus compensating for the loss of control when passing from maximal parallelism to minimal parallelism.

The "explanation" of the previously mentioned results lies in the way the proof is conducted: because we have to simulate a sequential computing device (a register machine), the parallelism is low anyway, hence the construction can be arranged in such a way that either exactly one rule is applicable for each membrane or only a small number of rules can be applied, hence for the constructed systems minimal parallelism is very close (if not identical) to maximal parallelism.

It was shown in Reference 13 that efficiency results can also be obtained in the case of minimal parallelism like in the case of maximal parallelism: the satisfiability of any propositional formula in the conjunctive normal form, using n variables, can be decided in linear time with respect to n by a P system with active membranes, using rules of types (a), (b), (c), and (e), while working in the minimally parallel mode.

Results as above (both universality and efficiency) were recently reported in Reference 38 for other types of P systems, on the basis of operations similar to membrane division, such as membrane separation (rules $[_h Q]_h \to [_h K]_h [_h Q - K]_h$, with the meaning that membrane h is split into two membranes, one containing the objects from the set K and the other one containing the complement of K with respect to Q).

3.9.3 P Systems with Proteins on Membranes

We have mentioned in the Introduction that part of the biochemistry taking place in a cell is controlled by proteins (enzymes) bound on membranes. In particular, the symport/antiport processes depend on the protein channels embedded in the membranes. Thus, it is natural to also consider objects (proteins) placed on membranes, not only in the compartments of a P system. This is the approach of so-called brane calculi introduced in Reference 8, which deals only with operations with membranes controlled

Membrane Systems

by their proteins. Perhaps more realistic is to combine the two approaches, thus considering objects both in the compartments of a membrane structure and on the membranes. There are already several papers dealing with such a bridge between membrane computing and brane calculi; see References 22 and 25.

An idea relevant for our discussion of the parallelism in P system is that considered in Reference 50: let us consider objects in compartments and proteins on membranes; the objects can pass across membranes (like in symport/antiport systems) and can also evolve (like in transition P systems) under the control of proteins placed on membranes. In turn, these proteins can change, either without restrictions or only in a flip-flop manner (like bistable catalysts). The most general form of rules in this framework is the following one:

$$a[_i p | b \to c[_i p' | d, \tag{3.1}$$

where a, b, c, d are objects and p, p' are proteins; the notation $[_i p|$ means "membrane i has protein p placed on it." When applying this rule, the object a placed outside membrane i and object b placed inside membrane i are simultaneously changed in c, d, respectively, assisted by protein p, placed on the membrane, which evolves to protein p'.

A P system using rules of this form starts with multisets of objects placed in the compartments, in the environment, and on membranes (called proteins). Because the proteins cannot leave the membranes and cannot multiply, their number is constant during the computation—hence finite. This means that the parallelism of the system is also bounded; in each step at most as many rules can be used as the number of proteins appearing on the membranes. Thus, the protein behaves like catalysts for symport/antiport-like rules, where the objects may also evolve.

The generality, hence the power of the rules in the form given above is clear, so a series of restrictions were considered in Reference 50: the protein should not evolve, or it should only oscillate between two states, p, p'; the objects cannot change, but only move across the membrane; only one of a and b is present, and it either evolves and remains in the same region, or moves to another region without evolving, or both evolves and moves to another region. A large number of types of rules are obtained, and a large number of combinations of such rules can then be considered and investigated in what concerns the computing power. In most cases, universality results were again obtained. Details can be found in Reference 50 and in Reference 40.

3.10 On Determinism versus Nondeterminism in P Systems

Two popular models of P systems are the catalytic system [54] and the symport/antiport system [49]. An interesting subclass of the latter was studied in Reference 21—each system is *deterministic* in the sense that the computation path of the system is unique, that is, at each step of the computation, the maximal multiset of rules that is applicable is unique. It was shown in Reference 21 that any recursively enumerable unary language $L \subseteq o^*$ can be accepted by a deterministic 1-membrane symport/antiport system. Thus, for symport/antiport systems, the deterministic and nondeterministic versions are equivalent and they are universal. It also follows from the construction in Reference 65 that for another model of P systems, called communicating P systems, the deterministic and nondeterministic versions are equivalent as both can accept any unary recursively enumerable language. However, the deterministic-versus-nondeterministic question was left open in Reference 21 for the class of catalytic systems (these systems have rules of the form $Ca \to Cv$ or $a \to v$), where the proofs of universality involve a high degree of parallelism [18, 65].

For a catalytic system serving as a *language acceptor*, the system starts with an initial configuration wz, where w is a fixed string of catalysts and noncatalysts not containing any symbol in z, and $z = a_1^{n_1} \ldots a_k^{n_k}$ for some nonnegative integers $n_1, \ldots, n_k$, with $\{a_1, \ldots, a_k\}$ a distinguished subset of noncatalyst symbols (the input alphabet). At each step, a maximal multiset of rules are nondeterministically selected and applied in parallel to the current configuration to derive the next configuration (note that the next configuration is not unique, in general). The string z is accepted if the system eventually halts. Unlike nondeterministic 1-membrane catalytic system acceptors (with 2 catalysts) that are universal, it was shown in Reference 35 using a graph-theoretic approach that the Parikh map of the language $\subseteq a_1^* \cdots a_k^*$ accepted by any deterministic catalytic system is a simple semilinear set that can also be effectively constructed. This result gives the first example of a P system for which the nondeterministic version is universal, but the deterministic version is not. For deterministic 1-membrane catalytic systems using only rules of type $Ca \to Cv$, it was shown that the set of reachable configurations from a given initial configuration is effectively semilinear. In contrast, the reachability set is no longer semilinear in general if rules of type $a \to v$ are also used. This result generalizes to multimembrane catalytic systems.

Deterministic catalytic systems that allow rules to be prioritized were also considered in Reference 35. Three such systems, namely, *totally prioritized*, *strongly prioritized*, and *weakly prioritized* catalytic systems, are investigated. For totally prioritized systems, rules are divided into different priority groups, and if a rule in a higher priority group is applicable, then no rules from a lower priority group can be used. For both strongly prioritized and weakly prioritized systems, the underlying priority relation is a *strict partial order* (i.e., irreflexive, asymmetric, and transitive). Under the semantics of strong priority, if a rule with higher priority is used, then no rule of a lower priority can be used even if the two rules do not compete for objects. For weakly prioritized systems, a rule is applicable if it cannot be replaced by a higher priority one. For these three prioritized systems, the results were contrasting: deterministic strongly and weakly prioritized catalytic systems are universal, whereas totally prioritized systems only accept semilinear sets.

Finally, we mention that in Reference 31, two restricted classes of communicating P system (CPS) acceptors [65] were considered, and the following shown:

1. For the first class, the deterministic and nondeterministic versions are equivalent if and only if deterministic and nondeterministic linear bounded automata are equivalent. The latter problem is a long-standing open question in complexity theory.
2. For the second class, the deterministic version is strictly weaker than the nondeterministic version.

Both classes are nonuniversal, but can accept fairly complex languages. Similar results were later shown for restricted classes of symport/antiport systems [32].

3.11 Other Classes of P Systems

Besides cell-like P systems, there are also investigated P systems whose compartments are not arranged in a hierarchical structure (described by a tree), but in a general graph-like structure. This corresponds to the arrangement of cells in tissues, and this was the name under which such systems were first considered and investigated: tissue-like P systems. The idea is simple: elementary membranes (called cells) are placed in the nodes of a graph with multisets of objects inside them; on the edges, there are antiport-like rules, controlling the passage of objects from a cell to another one or the exchange of objects between cells and the environment. More specifically, the rules are of the form $(i, x/y, j)$, with the meaning that the multiset of objects x is moved from cell i to cell j, in exchange for objects from multiset y, which pass from cell j to cell i. One of i, j can indicate the environment, and then we have an antiport-like operation between a cell and the environment. A chapter of Reference 54 is devoted to these systems. Universality is obtained for various combinations of ingredients (number of cells, weight of rules, etc.).

In this type of system, the edges of the graph are actually specified by the rules, hence the communication graph is dynamically defined. This aspect is stressed and made more explicit in population P systems, a variant of tissue-like P systems introduced in Reference 6, where there are links between cells that are created or destroyed during the evolution of the system and communication is allowed only along existing links. This kind of system is motivated by and applications in the study of populations of bacteria.

A class of P systems related to tissue-like systems, but with ingredients inspired from the way the neurons are organized in the brain, is that of neural-like P systems. In this case, the cells (neurons) both contain objects and also have a *state*, from a finite set; the rules are of the form $su \to s'v$, where s, s' are states and u, v are multisets of objects, with v containing objects with target indications *here* (they remain in the same neuron), *go* (they pass to the neurons to which there are edges in the graph—synapses—either in a replicative mode or not), and *out* (this indication is allowed only in a distinguished neuron called output neuron, and the respective objects exit into the environment, thus providing a result of the computation). The control by means of states is rather powerful (hence universality is again obtained, even for systems with a small number of neurons), while the replications of objects that exit a neuron can be used for producing an exponential workspace that can then lead to polynomial solutions to **NP**-complete problems.

Recently, a class of neural-like P systems was introduced still closer to neurobiology, namely the spiking neural P systems. They start from the way the neurons communicate: by means of spikes, identical to electrical impulses, with the information encoded mainly in the distance between consecutive impulses. In terms of P systems, this leads to systems where only one object is considered, the spike (we represent it with symbol a). Neurons (elementary membranes) are placed in the nodes of a graph whose edges represent the synapses. The rules for handling the spikes are of two types: firing rules, of the form $E/a^c \to a; d$, and forgetting rules, of the form $a^s \to \lambda$, where E is a regular expression using the unique symbol a. A rule $E/a^c \to a; d$ is used only if the neuron contains n spikes such that a^n is a string in the language represented by the expression E. Using this rule means to consume c spikes (hence only $n - c$ remains in the neuron) and to produce a spike after a delay of d time units. During these d time units, the neuron is closed, it cannot use another rule, and cannot receive spikes from other neurons. The spike produced by the rule is replicated in as many copies as there are synapses, leaving the neuron where the rule is applied, and one spike is passed to each neuron that is placed at the end of such a synapse and is open. In turn, using a forgetting rule $a^s \to \lambda$ means to erase all s spikes of the neuron.

It is important to note that in each neuron only one rule can be used and in such a way that all spikes present in the neuron are taken into consideration (either "covered" by the regular expression of a firing rule or by the left-hand member of a firing rule). Thus, in each neuron we work sequentially, but the system as a whole is synchronized, and all neurons evolve in parallel.

Starting from an initial distribution of spikes in the neurons, the system evolves. One neuron is designated as the output one, and its spikes also exit into the environment. The sequence of time units when spikes are sent into the environment forms a spike train, and with such a spike train we can associate various results from the computation: a set of numbers (the distance between the first two spikes, the distance between any two consecutive spikes) or a string (e.g., the binary description of the spike train, with 0 associated with a step when no spike is sent out of the system and 1 with a step when a spike is sent out). A series of possibilities is considered in References 33, 37, 56, and in other papers, and the obtained results are of two general types: Turing completeness is achieved if no bound is imposed on the number of spikes present in the neurons, and characterizations of semilinear sets of numbers is obtained if a bound is imposed on the number of spikes present in neurons during a computation. Interesting results are obtained when considering the spiking neural P systems as string generators, acceptors, or transducers, with a series of topics still waiting for research efforts.

3.12 Further Topics

There are many other issues that deserve to be considered in relation with parallelism in membrane computing, but due to space restrictions we will only mention some of them in what follows.

First, a close comparison with classic parallel architectures in computer science is possible and worth carrying out, for instance, starting with the investigations in Reference 11, where a series of known processor organizations (two-dimensional mesh, pyramid network, shuffle-exchange network, butterfly network, cube-connected network, etc.) were formalized in terms of P systems.

Then, an important topic is that of applications. Although initiated as a branch of natural computing, hence with computer science motivations, membrane computing turned out to be a useful framework for devising models of interest for biology, somewhat related to the important goal of systems biology to find models of the whole cell (e.g., References 39 and [66]). Indeed, a series of applications were reported in the last years, especially in modeling and simulating processes in the cell, but also related to populations of cells (e.g., of bacteria). The attractiveness of membrane computing from this point of view comes from several directions: P systems are directly inspired from biology, are easy to understand by biologists, easy to extend (scale up), modify, program, are nonlinear in behavior, and suitable for situations where we deal with small populations of molecules and/or with slow reactions (in such cases the traditional models based on differential equations are not adequate). Many applications of P systems in biology/medicine (and economics, which follow the same pattern as biology) can be found in References 70 or 12, hence we do not enter here into details. What is relevant for our discussion is that in these models one does not work with any of the types of parallelism considered above, but with rules having either probabilities or reaction rates like in biochemistry, and thus the parallelism is controlled by these probabilities and coefficients.

These applications are supported by an increasing number of simulations/implementations of various classes of P systems—see Reference 26 for an overview of membrane computing software available at the beginning of 2005 and Reference 70 for recent programs. There is an important detail: most of the membrane computing programs are not "implementations" of P systems, because of the inherent nondeterminism and the massive parallelism of the basic model, and these features cannot be implemented on the usual electronic computer. However, there are proposals for implementing P systems on a dedicated, reconfigurable hardware, as done in Reference 59, or on a local network, as reported in References 14 and 64.

3.13 Closing Remarks

The present chapter has presented only the main ideas of membrane computing, focusing on basic notions and results, with emphasis on the role of parallelism in this area, the levels/types of parallelism, and their influence on the power and the efficiency of the respective classes of P systems. A lot of further topics and results were considered in this area of literature, and we refer the reader to the web site from Reference 70 for details and recent developments. Also, the next bibliographical list contains some items that were not cited in the text, but are relevant for the topic of our discussion.

References

[1] L.M. Adleman, Molecular computation of solutions to combinatorial problems. *Science*, 226 (November 1994), 1021–1024.

[2] B. Alberts, A. Johnson, J. Lewis, M. Raff, K. Roberts, P. Walter, *Molecular Biology of the Cell*, 4th ed. Garland Science, New York, 2002.

[3] A. Alhazov, R. Freund, On the efficiency of P systems with active membranes and two polarizations, G. Mauri, Gh. Păun, M.J. Pérez-Jiménez, G. Rozenberg, A. Salomaa, eds., *Membrane Computing. International Workshop WMC5, Milan, Italy, 2004. Revised Papers*, Lecture Notes in Computer Science, 3365, Springer, Berlin, 2005, 147–161.

[4] A. Alhazov, R. Freund, Y. Rogozhin, Computational power of symport/antiport: history, advances and open problems, R. Freund, Gh. Păun, G. Rozenberg, A. Salomaa, eds., *Membrane Computing, International Workshop, WMC6, Vienna, Austria, 2005, Selected and Invited Papers*, LNCS 3850, Springer-Verlag, Berlin, 2006, 1–30.

[5] A. Alhazov, D. Sburlan, Static sorting algorithms for P systems, C. Martín-Vide, G. Mauri, Gh. Păun, G. Rozenberg, A. Salomaa, eds., *Membrane Computing. International Workshop, WMC2003, Tarragona, Spain, Revised Papers*. Lecture Notes in Computer Science, 2933, Springer, Berlin, 2004, 17–40.

[6] F. Bernardini, M. Gheorghe, Population P systems, *Journal of Universal Computer Science*, 10, 5 (2004), 509–539.

[7] D. Besozzi, *Computational and Modeling Power of P Systems*. PhD Thesis, Univ. degli Studi di Milano, 2004.

[8] L. Cardelli, Brane calculi. Interactions of biological membranes, V. Danos, V. Schachter, eds., *Computational Methods in Systems Biology. International Conference CMSB 2004. Paris, France, May 2004. Revised Selected Papers*, Lecture Notes in Computer Science, 3082, Springer-Verlag, Berlin, 2005, 257–280.

[9] M. Cavaliere, C. Martín-Vide, Gh. Păun, eds., *Proceedings of the Brainstorming Week on Membrane Computing, Tarragona, February 2003*. Technical Report 26/03, Rovirai Virgili University, Tarragona, 2003.

[10] M. Cavaliere, D. Sburlan, Time and synchronization in membrane systems. *Fundamenta Informaticae*, 64 (2005), 65–77.

[11] R. Ceterchi, M.J. Pérez-Jiménez, Simulating a class of parallel architectures: A broader perspective, M.A. Gutiérrez-Naranjo, Gh. Păun, M.J. Pérez-Jiménez, eds., *Cellular Computing. Complexity Aspects*. Fenix Editora, Sevilla, 2005, 131–148.

[12] G. Ciobanu, Gh. Păun, M.J. Pérez-Jiménez, eds., *Applications of Membrane Computing*. Springer-Verlag, Berlin, 2006.

[13] G. Ciobanu, L. Pan, Gh. Păun, M.J. Pérez-Jiménez, P systems with minimal parallelism. Submitted, 2005.

[14] G. Ciobanu, G. Wenyuan, A P system running on a cluster of computers, C. Martín-Vide, G. Mauri, Gh. Păun, G. Rozenberg, A. Salomaa, eds., *Membrane Computing. International Workshop, WMC2003, Tarragona, Spain, Revised Papers*. Lecture Notes in Computer Science, 2933, Springer, Berlin, 2004, 123–139.

[15] E. Csuhaj-Varju, G. Vaszil, P automata or purely communicating accepting P systems, Gh. Păun, G. Rozenberg, A. Salomaa, C. Zandron, eds., *Membrane Computing. International Workshop, WMC-CdeA 2002, Curtea de Argeş, Romania, Revised Papers*. Lecture Notes in Computer Science, 2597, Springer, Berlin, 2003, 219–233.

[16] Z. Dang, O.H. Ibarra, On P systems operating in sequential and limited parallel modes, *Pre-proc. of the Workshop on Descriptional Complexity of Formal Systems, DCFS 2004*, London, Ontario, 164–177.

[17] R. Freund, Asynchronous P systems and P systems working in the sequential mode, G. Mauri, Gh. Păun, M.J. Pérez-Jiménez, G. Rozenberg, A. Salomaa, eds., *Membrane Computing. International Workshop WMC5, Milan, Italy, 2004. Revised Papers*, Lecture Notes in Computer Science, 3365, Springer, Berlin, 2005, 36–62.

[18] R. Freund, L. Kari, M. Oswald, P. Sosik, Computationally universal P systems without priorities: two catalysts suffice. *Theoretical Computer Science*, 2004.

[19] R. Freund, M. Oswald, A short note on analysing P systems. *Bulletin of the EATCS*, 78 (2003), 231–236.

[20] R. Freund and A. Paun, Membrane systems with symport/antiport rules: universality results, *Proc. WMC-CdeA2002*, volume 2597 of Lecture Notes in Computer Science, pp. 270–287. Springer, 2003.

[21] R. Freund and Gh. Paun, On deterministic P systems, See P systems web page at http://psystems.disco.unimib.it, 2003.

[22] R. Freund, Gh. Păun, G. Rozenberg, A. Salomaa, eds., *Membrane Computing, International Workshop, WMC6, Vienna, Austria, 2005, Selected and Invited Papers*, LNCS 3850, Springer-Verlag, Berlin, 2006.

[23] P. Frisco, P systems, Petri nets and program machines, R. Freund, Gh. Păun, G. Rozenberg, A. Salomaa, eds., *Membrane Computing, International Workshop, WMC6, Vienna, Austria, 2005, Selected and Invited Papers*, LNCS 3850, Springer-Verlag, Berlin, 2006, 209–223.

[24] A. Georgiou, M. Gheorghe, F. Bernardini, Membrane-based devices used in computer graphics, G. Ciobanu, Gh. Păun, M.J. Pérez-Jiménez, eds., *Applications of Membrane Computing*. Springer-Verlag, Berlin, 2006, 253–282.

[25] M.A. Gutiérrez-Naranjo, Gh. Păun, M.J. Pérez-Jiménez, eds., *Cellular Computing. Complexity Aspects*. Fenix Editora, Sevilla, 2005.

[26] M.A. Gutiérrez-Naranjo, M.J. Pérez-Jiménez, A. Riscos-Núñez, Available membrane computing software, G. Ciobanu, Gh. Păun, M.J. Pérez-Jiménez, eds., *Applications of Membrane Computing*. Springer-Verlag, Berlin, 2006, 411–436.

[27] M.A. Guttierez-Naranjo, M.J. Perez-Jimenez, A. Riscos-Nunez, F.J. Romero-Campero, On the power of dissolution in P systems with active membranes, R. Freund, Gh. Păun, G. Rozenberg, A. Salomaa, eds., *Membrane Computing, International Workshop, WMC6, Vienna, Austria, 2005, Selected and Invited Papers*, LNCS 3850, Springer-Verlag, Berlin, 2006, 224–240.

[28] T. Head, Formal language theory and DNA: an analysis of the generative capacity of specific recombinant behaviors. *Bulletin of Mathematical Biology*, 49 (1987), 737–759.

[29] O.H. Ibarra, The number of membranes matters, C. Martín-Vide, G. Mauri, Gh. Păun, G. Rozenberg, A. Salomaa, eds., *Membrane Computing. International Workshop, WMC2003, Tarragona, Spain, Revised Papers*. Lecture Notes in Computer Science, 2933, Springer, Berlin, 2004, 218–231.

[30] O.H. Ibarra, On membrane hierarchy in P systems. *Theoretical Computer Science*, 334, 1–3 (2005), 115–129.

[31] O.H. Ibarra, On determinism versus nondeterminism in P systems. *Theoretical Computer Science*, 344 (2005), 120–133.

[32] O.H. Ibarra, S. Woodworth, On bounded symport/antiport P systems. *Proc. DNA11*, Lecture Notes in Computer Science, 3892, Springer, 2006, 129–143.

[33] O.H. Ibarra, A. Păun, Gh. Păun, A. Rodriguez-Paton, P. Sosik, S. Woodworth, Normal forms for spiking neural P systems. Submitted, 2006.

[34] O.H. Ibarra, S. Woodworth, H.-C. Yen, Z. Dang, On the computational power of 1-deterministic and sequential P systems. *Fundamenta Informaticae*, 2006.

[35] O.H. Ibarra, H.C. Yen, On deterministic catalytic P systems. *Proc. CIAA05*, LNCS 3845, Springer, 2006, 164–176.

[36] O.H. Ibarra, H.-C. Yen, Z. Dang, On various notions of parallelism in P systems. *International Journal of Foundations of Computer Science*, 16(4) 2005, 683–705.

[37] M. Ionescu, Gh. Păun, T. Yokomori, Spiking neural P systems. *Fundamenta Informaticae*, 71 (2006).

[38] T.-O. Ishdorj, Minimal parallelism for polarizationless P systems. *LNCS* 4287, 2006, 17–32, http://dx.doi.org/10.1007/11925903_2

[39] H. Kitano, Computational systems biology. *Nature*, 420, 14 (2002), 206–210.

[40] S.N. Krishna, Combining brane calculus and membrane computing. Conf. Bio-Inspired Computing—Theory and Applications, BIC-TA, Wuhan, China, 2006; *Progress in Natural Science*, 17, 4 (2007).

[41] S.N. Krishna, R. Rama, P systems with replicated rewriting. *Journal of Automata, Languages and Combinatorics*, 6, 3 (2001), 345–350.

[42] S.N. Krishna, R. Rama, H. Ramesh, Further results on contextual and rewriting P systems. *Fundamenta Informaticae*, 64 (2005), 235–246.

[43] M. Madhu, K. Krithivasan, Improved results about the universality of P systems. *Bulletin of the EATCS*, 76 (2002), 162–168.

[44] V. Manca, L. Bianco, F. Fontana, Evolution and oscillation in P systems: applications to biological phenomena, G. Mauri, Gh. Păun, M.J. Pérez-Jiménez, G. Rozenberg, A. Salomaa, eds., *Membrane Computing. International Workshop WMC5, Milan, Italy, 2004. Revised Papers*, Lecture Notes in Computer Science, 3365, Springer, Berlin, 2005, 63–84.

[45] C. Martín-Vide, G. Mauri, Gh. Păun, G. Rozenberg, A. Salomaa, eds., *Membrane Computing. International Workshop, WMC2003, Tarragona, Spain, Revised Papers*. Lecture Notes in Computer Science, 2933, Springer, Berlin, 2004.

[46] G. Mauri, Gh. Păun, M.J. Pérez-Jiménez, G. Rozenberg, A. Salomaa, eds., *Membrane Computing. International Workshop WMC5, Milan, Italy, 2004. Revised Papers*, Lecture Notes in Computer Science, 3365, Springer, Berlin, 2005.

[47] O. Michel, F. Jaquemard, An analysis of a public-key protocol with membranes, G. Ciobanu, Gh. Păun, M.J. Pérez-Jiménez, eds., *Applications of Membrane Computing*. Springer-Verlag, Berlin, 2006, 283–302.

[48] T.Y. Nishida, An application of P system: a new algorithm for NP-complete optimization problems, N. Callaos, et al., eds., *Proceedings of the 8th World Multi-Conference on Systems, Cybernetics and Informatics*, vol. V, 2004, 109–112.

[49] A. Păun, Gh. Păun, The power of communication: P systems with symport/antiport. *New Generation Computing*, 20, 3 (2002), 295–306.

[50] A. Păun, B. Popa, P systems with proteins on membranes. *Fundamenta Informaticae*, 72, 4 (2006), 467–483.

[51] A. Păun, B. Popa, P systems with proteins on membranes and membrane division. *Preproceedings of the Tenth International Conference on Developments in Language Theory (DLT)*, Santa Barbara, 2006.

[52] Gh. Păun, Computing with membranes. *Journal of Computer and System Sciences*, 61, 1 (2000), 108–143 (and Turku Center for Computer Science-TUCS Report 208, November 1998, www.tucs.fi).

[53] Gh. Păun, P systems with active membranes: attacking NP-complete problems. *Journal of Automata, Languages and Combinatorics*, 6, 1 (2001), 75–90.

[54] Gh. Păun, *Computing with Membranes: Introduction*. Springer, Berlin, 2002.

[55] Gh. Păun, Introduction to membrane computing. Chapter 1, G. Ciobanu, Gh. Păun, M.J. Pérez–Jiménez, eds., *Applications of Membrane Computing*. Springer-Verlag, Berlin, 2006, 1–42.

[56] Gh. Păun, M.J. Pérez-Jiménez, G. Rozenberg, Spike trains in spiking neural P systems. *International Journal of Foundations of Computer Science*, 2006.

[57] Gh. Păun, G. Rozenberg, A. Salomaa, C. Zandron, eds., *Membrane Computing. International Workshop, WMC-CdeA 2002, Curtea de Argeș, Romania, Revised Papers*. Lecture Notes in Computer Science, 2597, Springer, Berlin, 2003.

[58] M. Pérez-Jiménez, A. Romero-Jiménez, F. Sancho-Caparrini, *Teoría de la Complejidad en Modelos de Computatión Celular con Membranas*. Editorial Kronos, Sevilla, 2002.

[59] B. Petreska, C. Teuscher, A hardware membrane system, C. Martín-Vide, G. Mauri, Gh. Păun, G. Rozenberg, A. Salomaa, eds., *Membrane Computing. International Workshop, WMC2003, Tarragona, Spain, Revised Papers*. Lecture Notes in Computer Science, 2933, Springer, Berlin, 2004, 269–285.

[60] A. Riscos-Núñez, *Programacion celular. Resolucion eficiente de problemas numericos NP-complete*. PhD Thesis, Univ. Sevilla, 2004.

[61] A. Rodriguez-Paton, P Sosik, Membrane computing and complexity theory: characterization of PSPACE. *Journal of Computer and System Sciences, February 2007*, 73, 1 (2007), 137–152.

[62] P. Sosik, The computational power of cell division in P systems: beating down parallel computers? *Natural Computing*, 2, 3 (2003), 287–298.

[63] G. Rozenberg, A. Salomaa, eds., *Handbook of Formal Languages*, 3 vols. Springer-Verlag, Berlin, 1997.

[64] A. Syropoulos, P.C. Allilomes, E.G. Mamatas, K.T. Sotiriades, A distributed simulation of P systems, C. Martín-Vide, G. Mauri, Gh. Păun, G. Rozenberg, A. Salomaa, eds., *Membrane Computing. International Workshop, WMC2003, Tarragona, Spain, Revised Papers*. Lecture Notes in Computer Science, 2933, Springer, Berlin, 2004, 355–366.

[65] P. Sosik, P systems versus register machines: two universality proofs. In *Preproceedings of Workshop on Membrane Computing (WMC-CdeA2002)*, Curtea de Arges, Romania, pp. 371–382, 2002.

[66] M. Tomita, Whole-cell simulation: a grand challenge of the 21st century. *Trends in Biotechnology*, 19 (2001), 205–210.

[67] K. Ueda, N. Kato, LNMtal—a language model with links and membranes, G. Mauri, Gh. Păun, M.J. Pérez-Jiménez, G. Rozenberg, A. Salomaa, eds., *Membrane Computing. International Workshop WMC5, Milan, Italy, 2004. Revised Papers*, Lecture Notes in Computer Science, 3365, Springer, Berlin, 2005, 110–125.

[68] C. Zandron, *A Model for Molecular Computing, Membrane Systems.* PhD Thesis, Univ. degli Studi di Milano, 2001.
[69] C. Zandron, C. Ferretti, G. Mauri, Solving NP-complete problems using P systems with active membranes, I. Antoniou, C.S. Calude, M.J. Dinneen, eds., *Unconventional Models of Computation*, Springer, London, 2000, 289–301.
[70] The Web Page of Membrane Computing: http://psystems.disco.unimib.it

4
Optical Transpose Systems: Models and Algorithms

4.1	Introduction..	4-1
4.2	Optical Transpose Interconnection System Model ...	4-2
4.3	Optical Transpose Interconnection System Parallel Computers..	4-3
	Permutation Routing on Optical Transpose Interconnection System Computers • OTIS-Mesh • OTIS-Hypercube	
4.4	Summary of Results for Optical Transpose Interconnection System Computers	4-12
	General Properties • Algorithms • Routing	
4.5	Summary	4-15
	References	4-15

Chih-Fang Wang
Southern Illinois University at Carbondale

Sartaj Sahni
University of Florida

4.1 Introduction

It is well known that when communication distances exceed a few millimeters, optical interconnects provide speed (bandwidth) and power advantage over electronic interconnects Feldman et al. [1988]; Krishnamoorthy et al. [1992]. Therefore, in the construction of very large multiprocessor computers it is prudent to interconnect processors that are physically close using electronic interconnect, and to connect processors that are far apart using optical interconnect. We shall assume that physically close processors are in the same physical package (i.e., chip, wafer, board) and processors not physically close are in different packages. As a result, electronic interconnects are used for intra package communications while optical interconnect is used for inter package communication.

Various combinations of interconnection networks for intra package (i.e., electronic) communications and inter package (i.e., optical) communications have been proposed. In optical transpose interconnection system (OTIS) computers Hendrick et al. [1995]; Marsden et al. [1993]; Zane et al. [2000], optical interconnects are realized via a free space optical interconnect system that interconnects pairs of processors in different packages.

In this chapter, we begin by describing the OTIS model. Next, we describe the classes of computers that result when the OTIS optical interconnect system is used for inter package communication and different topologies are used for intra package communication, in particular, OTIS-Mesh and OTIS-Hypercube

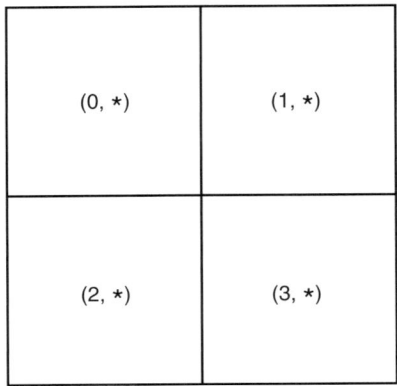 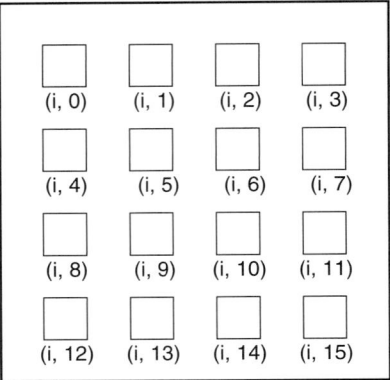

FIGURE 4.1 Two-dimensional arrangement of $L = 64$ inputs when $M = 4$ and $N = 16$ (a) $\sqrt{M} \times \sqrt{M} = 2 \times 2$ grouping of inputs (b) the $(i, *)$ group, $0 \leq i < M = 4$.

computers are the two classes researched frequently (i.e., a mesh and a hypercube are used for intra package communication respectively). Properties of OTIS-Mesh and OTIS-Hypercube computers are developed. We then summarize various algorithms developed on OTIS-Mesh and OTIS-Hypercube computers, as well as other classes of OTIS computers.

4.2 Optical Transpose Interconnection System Model

The OTIS was proposed by Marsden et al. [1993]. The OTIS connects $L = MN$ inputs to L outputs using free space optics and two arrays of lenslets. The first lenslet array is a $\sqrt{M} \times \sqrt{M}$ array and the second one is of dimension $\sqrt{N} \times \sqrt{N}$. Thus, a total of $M + N$ lenslets are used. The L inputs and outputs are arranged to form a $\sqrt{L} \times \sqrt{L}$ array. The L inputs are arranged into $\sqrt{M} \times \sqrt{M}$ groups with each group containing N inputs, arranged into a $\sqrt{N} \times \sqrt{N}$ array. Figure 4.1 shows the arrangement for the case of $L = 64$ inputs when $M = 4$ and $N = 16$. The $M \times N$ inputs are indexed (i, j) with $0 \leq i < M$ and $0 \leq j < N$. Inputs with the same i value are in the same $\sqrt{N} \times \sqrt{N}$ block. The notation $(2, *)$, for example, refers to all inputs of the form $(2, j), 0 \leq j < N$.

In addition to using the two-dimensional notation (i, j) to refer to an input, we also use a four-dimensional notation (i_r, i_c, j_r, j_c), where (i_r, i_c) gives the coordinates (row,column) of the $\sqrt{N} \times \sqrt{N}$ block that contains the input (see Figure 4.1a) and (j_r, j_c) gives coordinates of the element within a block (see Figure 4.1b). So all inputs $(i, *)$ with $i = 0$ have $(i_r, i_c) = (0, 0)$; those with $i = 2$ have $(i_r, i_c) = (1, 0)$, and so on. Likewise, all inputs with $j = 5$ have $(j_r, j_c) = (1, 1)$, and those with $j = 13$ have $(j_r, j_c) = (3, 1)$.

The L outputs are also arranged into a $\sqrt{L} \times \sqrt{L}$ array. This time, however, the $\sqrt{L} \times \sqrt{L}$ array is composed of $\sqrt{N} \times \sqrt{N}$ blocks, with each block containing M outputs that are arranged as a $\sqrt{M} \times \sqrt{M}$ array. The $L = MN$ outputs are indexed (i, j) with $0 \leq i < N$ and $0 \leq j < M$. All outputs of the form $(i, *)$ are in the same block i, with block i in position (i_r, i_c) with $i = i_r\sqrt{N} + i_c$, $0 \leq i_r, i_c < \sqrt{N}$, of the $\sqrt{N} \times \sqrt{N}$ block arrangement. Outputs of the form $(*, j)$ are in position (j_r, j_c) of their block, where $j = j_r\sqrt{M} + j_c$, $0 \leq j_r, j_c < \sqrt{M}$.

In the physical realization of OTIS, the $\sqrt{L} \times \sqrt{L}$ output arrangement is rotated $180°$. We have four two-dimensional planes: the first is the $\sqrt{L} \times \sqrt{L}$ input plane, the second is a $\sqrt{M} \times \sqrt{M}$ lenslet plane, the third is a $\sqrt{N} \times \sqrt{N}$ lenslet plane, and the fourth is the $\sqrt{L} \times \sqrt{L}$ plane of outputs rotated $180°$. When the OTIS is viewed from the side, only the first column of each of these planes is visible. Such a side view for the case $64 = L = M \times N = 4 \times 16$ is shown in Figure 4.2. Notice

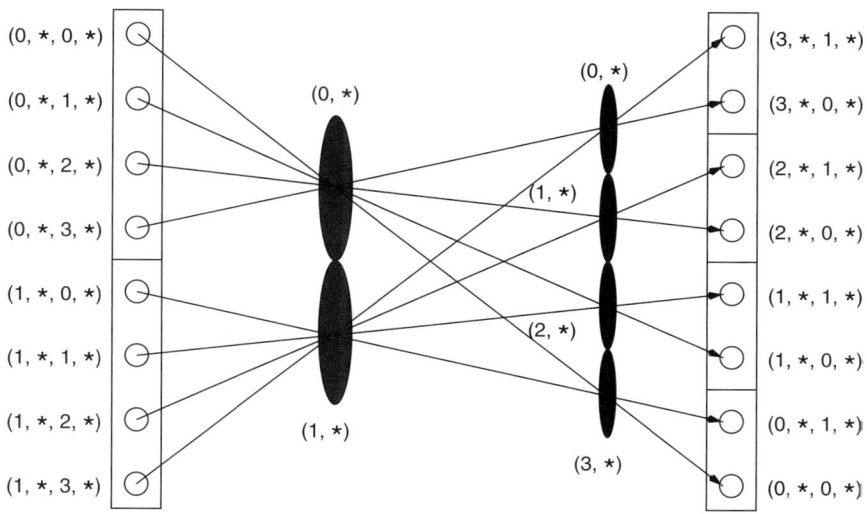

FIGURE 4.2 Side view of the OTIS with $M = 4$ and $N = 16$.

that the first column of the input plane consists of the inputs $(0,0)$, $(0,4)$, $(0,8)$, $(0,12)$, $(2,0)$, $(2,4)$, $(2,8)$, and $(2,12)$, which in four-dimensional notation are $(0,0,0,0)$, $(0,0,1,0)$, $(0,0,2,0)$, $(0,0,3,0)$, $(1,0,0,0)$, $(1,0,1,0)$, $(1,0,2,0)$, and $(1,0,3,0)$, respectively. The inputs in the same row as $(0,0,0,0)$ are $(0,*,0,*)$; in general, those in the same row as (i_r, i_c, j_r, j_c) are $(i_r, *, j_r, *)$. The (i_r, j_r) values from top to bottom are $(0,0)$, $(0,1)$, $(0,2)$, $(0,3)$, $(1,0)$, $(1,1)$, $(1,2)$, and $(1,3)$. The first column in the output plane (after the 180° rotation) has the outputs $(15,3)$, $(15,1)$, $(11,3)$, $(11,1)$, $(7,3)$, $(7,1)$, $(3,3)$, and $(3,1)$, which in four-dimensional notation are $(3,3,1,1)$, $(3,3,0,1)$, $(2,3,1,1)$, $(2,3,0,1)$, $(1,3,1,1)$, $(1,3,0,1)$, $(0,3,1,1)$, and $(0,3,0,1)$, respectively. The outputs in the same row as $(3,3,1,1)$ are $(3,*,1,*)$; generally speaking, those in the same row as (i_r, i_c, j_r, j_c) are $(i_r, *, j_r, *)$. The (i_r, j_r) values from top to bottom are $(3,1)$, $(3,0)$, $(2,1)$, $(2,0)$, $(1,1)$, $(1,0)$, $(0,1)$, and $(0,0)$.

Each lens of Figure 4.2 denotes a row of lenslets and each ○ a row of inputs or outputs. The interconnection pattern defined by the given arrangement of inputs, outputs, and lenslets connects input $(i,j) = (i_r, i_c, j_r, j_c)$ to output $(j,i) = (j_r, j_c, i_r, i_c)$. The connection is established via an optical ray that originates at input position (i_r, i_c, j_r, j_c), goes through lenslet (i_r, i_c) of the first lenslet array, then through lenslet (j_r, j_c) of the second lenslet array, and finally arrives at output position (j_r, j_c, i_r, i_c). The best connectivity provided by the OTIS is an optical connection between input (i,j) and output (j,i), $0 \leq i < M$, $0 \leq j < N$.

4.3 Optical Transpose Interconnection System Parallel Computers

Marsden et al. [1993] have proposed several parallel computer architectures in which OTIS is used to connect processors in different groups (packages), and an electronic interconnection network is used to connect processors in the same group. Since the bandwidth is maximized and power consumption minimized when the number of groups is the same as the number of processors in a group Krishnamoorthy et al. [1992], that is, an $L = N^2$ processor OTIS computer is partitioned into N groups of N processors

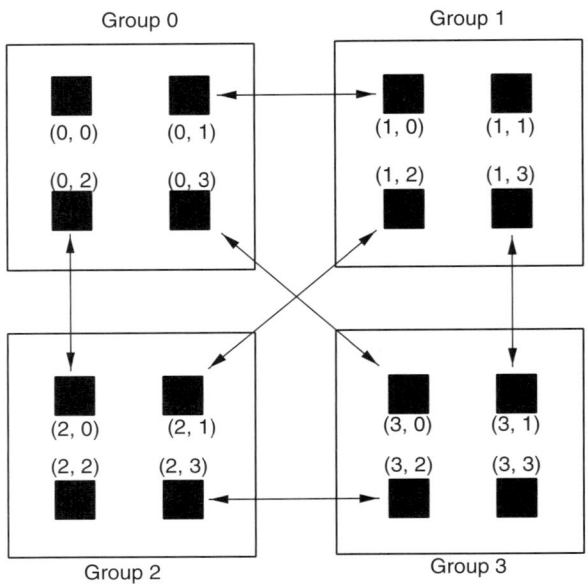

FIGURE 4.3 Example of OTIS connections with 16 processors.

each, Zane et al. Zane et al. [2000] limit the study of OTIS parallel computers to such an arrangement. Let (i, j) denote processor j of package i, $0 \leq i, j < N$. Processor (i, j), $i \neq j$, is connected to processor (j, i) using free space optics (i.e., OTIS). The only other connections available in an OTIS computer are the electronic intra group connections.

A generic 16-processor OTIS computer is shown in Figure 4.3. The solid boxes denote processors. Each processor is labeled (g, p) where g is the group index and p is the processor index. OTIS connections are shown by arrows. Intra group connections are not shown.

By using different topologies in each package (group), different classes of OTIS computers are defined. In an OTIS-Mesh, for example, processors in the same group are connected as a 2D mesh Marsden et al. [1993]; Zane et al. [2000]; Sahni and Wang [1997]; and in an OTIS-Hypercube, processors in the same group are connected using the hypercube topology Marsden et al. [1993]; Zane et al. [2000]; Sahni and Wang [1998]. OTIS-Mesh of trees Marsden et al. [1993], OTIS-Perfect shuffle, OTIS-Cube connected cycles, and so forth, may be defined in an analogous manner.

When analyzing algorithms for OTIS architectures, we count data moves along electronic interconnects (i.e., electronic moves) and those along optical interconnects (i.e., OTIS moves) separately. This allows us to later account for any differences in the speed and bandwidth of these two types of interconnects.

In the following sections, we first discuss the general property of an OTIS computer. In particular, the ability to route data from one processor to another. Then we discuss the properties of OTIS-Mesh and OTIS-Hypercube, two of the most frequently referenced classes of OTIS computers.

4.3.1 Permutation Routing on Optical Transpose Interconnection System Computers

Suppose we wish to rearrange the data in an N^2-processor OTIS computer according to the permutation $\Pi = \Pi[0] \cdots \Pi[N^2 - 1]$. That is, data from processor $i = gN + p$ is to be sent to processor $\Pi[i]$, $0 \leq i < N^2$. We assume that the interconnection network in each group is able to sort the data in its

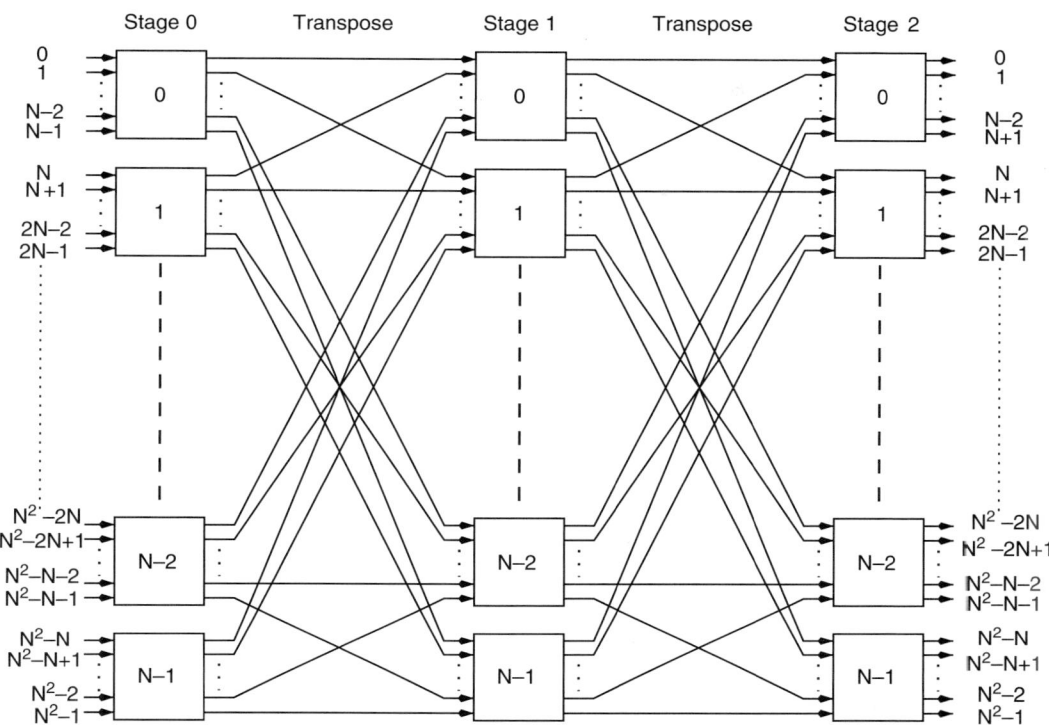

FIGURE 4.4 Multistage interconnection network (MIN) defined by OTIS.

N processors (equivalently, it is able to perform any permutation of the data in its N processors). This assumption is certainly valid for the mesh, hypercube, perfect shuffle, cube-connected cycles, and mesh of trees interconnections mentioned earlier.

Theorem 4.1 *Wang and Sahni [1998b] Every OTIS computer in which each group can sort can perform any permutation Π using at most 2 OTIS moves.*

Proof. When 2 OTIS moves are permitted, the data movement can be modeled by a 3-stage MIN (multistage interconnection network) as in Figure 4.4. Each switch represents a processor group that is capable of performing any N input to N output permutation. The OTIS moves are represented by the connections from one stage to the next.

The OTIS interstage connections are equivalent to the interstage connections in a standard MIN that uses $N \times N$ switches. From MIN theory Koppelman and Oruç [1994], we know that when $k \times k$ switches are used, $2\log_k N^2 - 1$ stages of switches are sufficient to make an N^2 input N^2 output network that can realize every input to output permutation. In our case (Figure 4.4), $k = N$. Therefore, $2\log_N N^2 - 1 = 3$ stages are sufficient. Hence 2 OTIS moves suffice to realize any permutation.

An alternative proof comes from an equivalence with the preemptive open shop scheduling problem (POSP) Gonzalez and Sahni [1976]. In the POSP, we are given n jobs that are to be scheduled on m machines. Each job i has m tasks. The task length of the jth task of job i is the integer $t_{ij} \geq 0$. In a preemptive schedule of length T, the time interval from 0 to T is divided into slices of length 1 unit each. A time slice is divided into m slots with each slot representing a unit time interval on one machine. Time slots on each machine are labeled with a job index. The labeling is done in such a way that (I) Each job (index) i is assigned to exactly t_{ij} slots on machine j, $0 \leq j < m$, and (II) No job is assigned to two or more machines in any time slice. T is the schedule length. The objective is to find the smallest T for which a schedule exists. Gonzalez and Sahni [1976] have shown that the length $T_{\min}$ of an optimal schedule is

$$T_{\min} = \max\{J_{\max}, M_{\max}\},$$

where

$$J_{\max} = \max_i \left\{ \sum_{j=0}^{m-1} t_{ij} \right\}$$

that is, $J_{\max}$ is the maximum job length; and

$$M_{\max} = \max_j \left\{ \sum_{i=0}^{n-1} t_{ij} \right\}$$

that is, $M_{\max}$ is the maximum processing to be done by any machine.

We can transform the OTIS permutation routing problem into a POSP. First, note that to realize a permutation Π with 2 OTIS moves, we must be able to write Π as a sequence of permutations $\Pi_0 T \Pi_1 T \Pi_2$, where Π_i is the permutation realized by the switches (i.e., processor groups) in stage i and T denotes the OTIS (transpose) interstage permutation. Let (g_q, p_q) denote processor p_q of group g_q where $q \in \{i_0, o_0, i_1, o_1, i_2, o_2\}$ (i_0 = input of stage 0, o_0 = output of stage 0, etc.). Then, the data path is

$$(g_{i_0}, p_{i_0}) \xrightarrow{\Pi_0} (g_{o_0}, p_{o_0}) \xrightarrow{T} (p_{o_0}, g_{o_0})$$
$$= (g_{i_1}, p_{i_1}) \xrightarrow{\Pi_1} (g_{o_1}, p_{o_1}) \xrightarrow{T} (p_{o_1}, g_{o_1})$$
$$= (g_{i_2}, p_{i_2}) \xrightarrow{\Pi_2} (g_{o_2}, p_{o_2})$$

We observe that to realize the permutation Π, the following must hold:

i. Switch i of stage 1 should receive exactly one data item from each switch of stage 0, $0 \le i < N$.
ii. Switch i of stage 1 should receive exactly one data item destined for each switch of stage 2, $0 \le i < N$.

Once we know which data items are to get to switch i, $0 \le i < N$, we can easily compute Π_0, Π_1, and Π_2. Therefore, it is sufficient to demonstrate the existence of an assignment of the N^2 stage 0 inputs to the switches in stage 1 satisfying conditions (i) and (ii). For this, we construct an N job N machine POSP instance. Job i represents switch i of stage 0 and machine j represents switch j of stage 2. The task time t_{ij} equals the number of inputs to switch i of stage 0 that are destined for switch j of stage 2 (i.e., t_{ij} is the number of group i data that are destined for group j). Since Π is a permutation, it follows that $\sum_{j=0}^{N-1} t_{ij}$ = total number of inputs to switch i of stage 0 = N and $\sum_{i=0}^{N-1} t_{ij}$ = total number of inputs destined for switch j of stage 2 = N. Therefore, $J_{\max} = M_{\max} = N$ and the optimal schedule length is N. Since $\sum_{i=0}^{N-1} \sum_{j=0}^{N-1} t_{ij} = N^2$ and the optimal schedule length is N, every slot of every machine is assigned a task in an optimal schedule. From the property of a schedule, it follows that in each time slice all N job labels occur exactly once. The N labels in slice i of the schedule define the inputs that are to be assigned to switch i of stage 1, $0 \le i < N$. From properties (a) and (b) of a schedule, it follows that this assignment satisfies the requirements (i) and (ii) for an assignment to the stage 1 switches. ∎

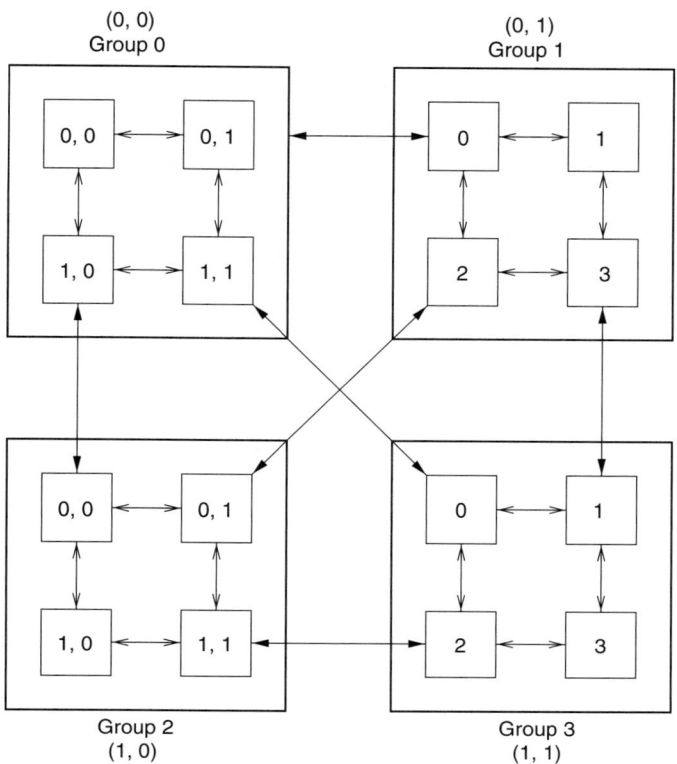

FIGURE 4.5 16-Processor OTIS-Mesh.

Even though every permutation Π can be realized with just 2 OTIS moves, simulating the 3-stage MIN of Figure 4.4 does not result in an efficient algorithm to perform permutation routing, since it takes many more OTIS moves to compute the decomposition $\Pi = \Pi_0 T \Pi_1 T \Pi_2$. Instead, it is more efficient to develop customized routing algorithms for specific permutations. Also, general permutations may be realized more efficiently using a sorting algorithm.

4.3.2 OTIS-Mesh

In an N^2 processor OTIS-Mesh, each group is a $\sqrt{N} \times \sqrt{N}$ mesh and there is a total of N groups. Figure 4.5 shows a 16-processor OTIS-Mesh. The processors of groups 0 and 2 are labeled using two-dimensional local mesh coordinates while the processors in groups 1 and 3 are labeled in row-major fashion. We use the notation (g, p) to refer to processor p of group g.

4.3.2.1 Diameter of the OTIS-Mesh

Let (g_1, p_1) and (g_2, p_2) be two OTIS-Mesh processors. The shortest path between these two processors is of one of the following forms:

a. The path involves only electronic moves. This is possible only when $g_1 = g_2$.
b. The path involves an even number of optical moves. In this case, the path is of the form $(g_1, p_1) \xrightarrow{E^*} (g_1, p'_1) \xrightarrow{O} (p'_1, g_1) \xrightarrow{E^*} (p'_1, g'_1) \xrightarrow{O} (g'_1, p'_1) \xrightarrow{E^*} (g'_1, p''_1) \xrightarrow{rmO} (p''_1, g'_1) \xrightarrow{E^*} (p''_1, g''_1) \xrightarrow{O} \cdots \xrightarrow{E^*} (g_2, p_2)$.

Here E^* denotes a sequence (possibly empty) of electronic moves and O denotes a single OTIS move. If the number of OTIS moves is more than two, we may compress paths of this form into the shorter path $(g_1, p_1) \xrightarrow{E^*} (g_1, p_2) \xrightarrow{O} (p_2, g_1) \xrightarrow{E^*} (p_2, g_2) \xrightarrow{O} (g_2, p_2)$. So we may assume that the path is of the above form with exactly two OTIS moves.

c. The path involves an odd number of OTIS moves. In this case, it must involve exactly one OTIS move (as otherwise it may be compressed into a shorter path with just one OTIS move as in (b)) and may be assumed to be of the form $(g_1, p_1) \xrightarrow{E^*} (g_1, g_2) \xrightarrow{O} (g_2, g_1) \xrightarrow{E^*} (g_2, p_2)$.

Let $D(i, j)$ be the shortest distance between processors i and j of a group using a path comprised solely of electronic moves. So, $D(i, j)$ is the Manhattan distance between the two processors of the local mesh group. Shortest paths of type (a) have length $D(p_1, p_2)$ while those of types (b) and (c) have length $D(p_1, p_2) + D(g_1, g_2) + 2$ and $D(p_1, g_2) + D(p_2, g_1) + 1$, respectively.

From the preceding discussion we have the following theorems:

Theorem 4.2 *Sahni and Wang [1997] The length of the shortest path between processors (g_1, p_1) and (g_2, p_2) is $D(p_1, p_2)$ when $g_1 = g_2$ and $\min\{D(p_1, p_2) + D(g_1, g_2) + 2, D(p_1, g_2) + D(p_2, g_1) + 1\}$ when $g_1 \neq g_2$.*

Proof. When $g_1 = g_2$, there are three possibilities for the shortest path. It may be of types (a), (b), or (c). If it is of type (a), its length is $D(p_1, p_2)$. If it is of type (b), its length is $D(p_1, p_2) + D(g_1, g_2) + 2 = D(p_1, p_2) + 2$. If it is of type (c), its length is $D(p_1, g_2) + D(p_2, g_1) + 1 = D(p_1, g_1) + D(p_2, g_1) + 1 D(p_1, g_1) + D(g_1, p_2) + 1 \geq D(p_1, p_2) + 1$. So, the shortest path has length $D(p_1, p_2)$. When $g_1 \neq g_2$, the shortest path is either of type (b) or (c). From our earlier development it follows that its length is $\min\{D(p_1, p_2) + D(g_1, g_2) + 2, D(p_1, g_2) + D(p_2, g_1) + 1\}$. ∎

Theorem 4.3 *Sahni and Wang [1997] The diameter of the OTIS-Mesh is $4\sqrt{N} - 3$.*

Proof. Since each group is a $\sqrt{N} \times \sqrt{N}$ mesh, $D(p_1, p_2)$, $D(p_2, g_1)$, $D(p_1, g_2)$, and $D(g_1, g_2)$ are all less than or equal to $2(\sqrt{N} - 1)$. From Theorem 4.2, it follows that no two processors are more than $4(\sqrt{N} - 1) + 1 = 4\sqrt{N} - 3$ apart. Hence, the diameter is $\leq 4\sqrt{N} - 3$. Now consider the processors (g_1, p_1), (g_2, p_2) such that p_1 is in position $(0, 0)$ of its group and p_2 is in position $(\sqrt{N} - 1, \sqrt{N} - 1)$ (i.e., p_1 the top left processor and p_2 the bottom right one of its group). Let g_1 be 0 and g_2 be $N - 1$. So, $D(p_1, p_2) = D(g_1, g_2) = D(p_1, g_2) = D(p_2, g_1) = \sqrt{N} - 1$. Hence, the distance between (g_1, p_1) and (g_2, p_2) is $4\sqrt{N} - 3$. As a result, the diameter of the OTIS-Mesh is exactly $4\sqrt{N} - 3$. ∎

4.3.2.2 Simulation of a 4D Mesh

Zane et al. [2000] have shown that the OTIS-Mesh can simulate each move of a $\sqrt{N} \times \sqrt{N} \times \sqrt{N} \times \sqrt{N}$ four-dimensional mesh by using either a single electronic move local to a group or using one local electronic move and two inter group OTIS moves. For the simulation, we must first embed the 4D mesh into the OTIS-Mesh. The embedding is rather straightforward with processor (i, j, k, l) of the 4D mesh being identified with processor (g, p) of the OTIS-Mesh. Here, $g = i\sqrt{N} + j$ and $p = k\sqrt{N} + l$.

The mesh moves $(i, j, k \pm 1, l)$ and $(i, j, k, l \pm 1)$ can be performed with one electronic move of the OTIS-Mesh while the moves $(i, j \pm 1, k, l)$ and $(i \pm 1, j, k, l)$ require one electronic and two optical moves. For example, the move $(i, j + 1, k, l)$ may be done by the sequence $(i, j, k, l) \xrightarrow{O} (k, l, i, j) \xrightarrow{E} (k, l, i, j + 1) \xrightarrow{O} (i, j + 1, k, l)$.

The above efficient embedding of a 4D mesh implies that 4D mesh algorithms can be run on the OTIS-Mesh with a constant factor (at most 3) slowdown Zane et al. [2000]. Unfortunately, the body of known 4D mesh algorithms is very small compared to that of 2D mesh algorithms. So, it is desirable to consider a 2D mesh embedding. Such an embedding will enable one to run 2D mesh algorithms on the

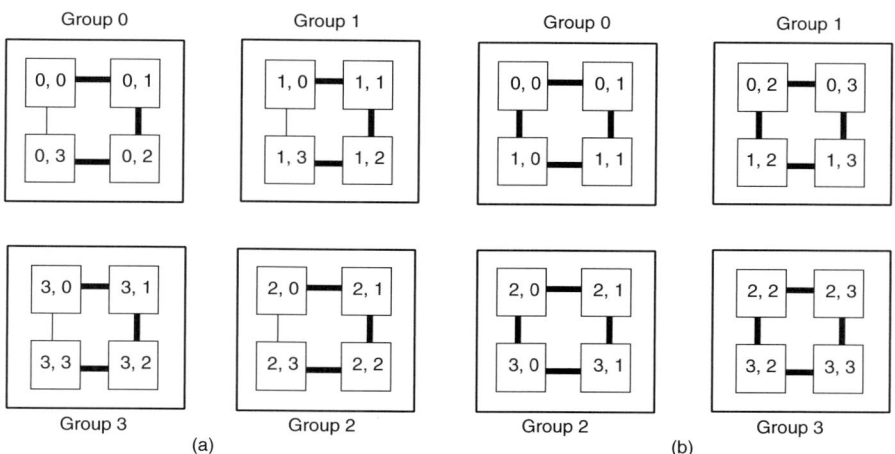

FIGURE 4.6 Mapping a 4 × 4 mesh onto a 16-processor OTIS-Mesh (a) group row mapping (GRM) (b) group submesh mapping (GSM).

OTIS-Mesh. Naturally, one would do this only for problems for which no 4D algorithm is known or for which the known 4D mesh algorithms are not faster than the 2D algorithms.

4.3.2.3 Simulation of a 2D Mesh

There are at least two intuitively appealing ways to embed an $N \times N$ mesh into the OTIS-Mesh. One is the *group row mapping* (GRM) in which each group of the OTIS-Mesh represents a row of the 2D mesh. The mapping of the mesh row onto a group of OTIS processors is done in a snake-like fashion as in Figure 4.6a. The pair of numbers in each processor of Figure 4.6a gives the (row, column) index of the mapped 2D mesh processor. The thick edges show the electronic connections used to obtain the 2D mesh row. Note that the assignment of rows to groups is also done in a snake-like manner. Let (i, j) denote a processor of a 2D mesh. The move to $(i, j+1)$ (or $(i, j-1)$) can be done with one electronic move as (i, j) and $(i, j+1)$ are neighbors in a processor group. If all elements of row i are to be moved over one column, then the OTIS-Mesh would need one electronic move in case of an MIMD mesh and three in case of an SIMD mesh as the row move would involve a shift by one left, right, and down within a group. A column shift can be done with two additional OTIS moves as in the case of a 4D mesh embedding. GRM is particularly nice for the matrix transpose operation. Data from processor (i, j) can be moved to processor (j, i) with one OTIS and zero electronic moves.

The second way to embed an $N \times N$ mesh is to use the *group submesh mapping* (GSM). In this, the $N \times N$ mesh is partitioned into N $\sqrt{N} \times \sqrt{N}$ submeshes. Each of these is mapped in the natural way onto a group of OTIS-Mesh processors. Figure 4.6b shows the GSM of a 4 × 4 mesh. Moving all elements of row or column i over by one is now considerably more expensive. For example, a row shift by -1 would be accomplished by the following data movements (a boundary processor is one on the right boundary of a group):

Step 1: Shift data in nonboundary processors right by one using an electronic move.
Step 2: Perform an OTIS move on boundary processor data. So, data from (g, p) moves to (p, g).
Step 3: Shift the data moved in Step 2 right by one using an electronic move. Now, data from (g, p) is in $(p, g+1)$.
Step 4: Perform an OTIS move on this data. Now data originally in (g, p) is in $(g+1, p)$.
Step 5: Shift this data left by $\sqrt{N} - 1$ using $\sqrt{N} - 1$ electronic moves. Now, the boundary data originally in (g, p) is in the processor to its right but in the next group.

The above five step process takes $\sqrt{N}$ electronic and two OTIS moves. Note, however, that if each group is a wraparound mesh in which the last processor of each row connects to the first and the bottom processor of each column connects to the top one, then row and column shift operations become much simpler as Step 1 may be eliminated and Step 5 replaced by a right wraparound shift of 1. The complexity is now two electronic and two OTIS moves.

GSM is also inferior on the transpose operation that now requires $8(\sqrt{N} - 1)$ electronic and 2 OTIS moves.

Theorem 4.4 *Sahni and Wang [1997] The transpose operation of an $N \times N$ mesh requires $8(\sqrt{N} - 1)$ electronic and 2 OTIS moves when the GSM is used.*

Proof. Let $g_x g_y$ and $p_x p_y$ denote processor (g, p) of the OTIS-Mesh. This processor is in position (p_x, p_y) of group (g_x, g_y) and corresponds to processor $(g_x p_x, g_y p_y)$ of the $N \times N$ embedded mesh. To accomplish the transpose, data is to be moved from the $N \times N$ mesh processor $(g_x p_x, g_y p_y)$ (i.e., the OTIS-Mesh processor $(g, p) = (g_x g_y, p_x p_y)$) to the mesh processor $(g_y p_y, g_x p_x)$ (i.e., the OTIS-Mesh processor $(g_y g_x, p_y p_x)$). The following movements do this: $(g_x p_x, g_y p_y) \xrightarrow{E^*} (g_x p_y, g_y p_x) \xrightarrow{O} (p_y g_x, p_x g_y) \xrightarrow{E^*} (p_y g_y, p_x g_x) \xrightarrow{O} (g_y p_y, g_x p_x)$. Once again E^* denotes a sequence of electronic moves local to a group and O denotes a single OTIS move. The E moves in this case perform a transpose in a $\sqrt{N} \times \sqrt{N}$ mesh. Each of these transposes can be done in $4(\sqrt{N} - 1)$ moves Nassimi and Sahni [1980]. So, the above transpose method uses $8(\sqrt{N} - 1)$ electronic and 2 OTIS moves.

To see that this is optimal, first note that every transpose algorithm requires at least two OTIS moves. For this, pick a group $g_x g_y$ such that $g_x \neq g_y$. Data from all N processors in this group are to move to the processors in group $g_y g_x$. This requires at least one OTIS move. However, if only one OTIS move is performed, data from $g_x g_y$ is scattered to the N groups. So, at least two OTIS moves are needed if the data ends up in the same group.

Next, we shall show that independent of the OTIS moves, at least $8(\sqrt{N} - 1)$ electronic moves must be performed. The electronic moves cumulatively perform one of the following two transforms (depending on whether the number of OTIS moves is even or odd, see Section 4.5 about the diameter):

    a. Local moves from (p_x, p_y) to (p_y, p_x); local moves from (g_x, g_y) to (g_y, g_x)
    b. Local moves from (p_x, p_y) to (g_y, g_x); local moves from (g_x, g_y) to (p_y, p_x)

For $(p_x, p_y) = (g_x, g_y) = (0, \sqrt{N} - 1)$, (a) and (b) require $2(\sqrt{N} - 1)$ left and $2(\sqrt{N} - 1)$ down moves. For $(p_x, p_y) = (g_x, g_y) = (\sqrt{N} - 1, 0)$, (a) and (b) require $2(\sqrt{N} - 1)$ right and $2(\sqrt{N} - 1)$ up moves. The total number of moves is thus $8(\sqrt{N} - 1)$. So, $8(\sqrt{N} - 1)$ is a lower bound on the number of electronic moves needed. ∎

4.3.3 OTIS-Hypercube

In an N^2-processor OTIS-Hypercube, each group is a hypercube of dimension $\log_2 N$. Figure 4.7 shows a 16-processor OTIS-Hypercube. The number inside a processor is the processor index within its group.

4.3.3.1 Diameter of the OTIS-Hypercube

Let $N = 2^d$ and let $D(i, j)$ be the length of the shortest path from processor i to processor j in a hypercube. Let (g_1, p_1) and (g_2, p_2) be two OTIS-Hypercube processors. Similar to the discussion of the diameter of OTIS-Mesh in Section 4.5, the shortest path between these two processors fits into one of the following categories:

    a. The path employs electronic moves only. This is possible only when $g_1 = g_2$.
    b. The path employs an even number of OTIS moves. If the number of OTIS moves is more than two, we may compress the path into a shorter path that uses two OTIS moves only: $(g_1, p_1) \xrightarrow{E^*} (g_1, p_2) \xrightarrow{O} (p_2, g_1) \xrightarrow{E^*} (p_2, g_2) \xrightarrow{O} (g_2, p_2)$.

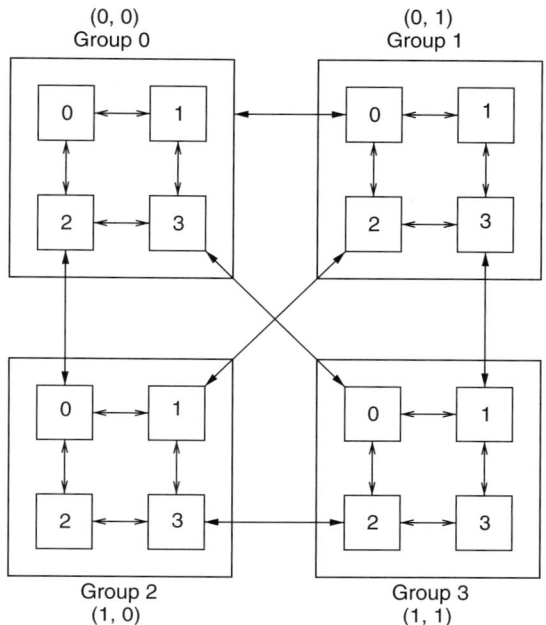

FIGURE 4.7 16-Processor OTIS-Hypercube.

c. The path employs an odd number of OTIS moves. Again, if the number of moves is more than one, we can compress the path into a shorter one that employs exactly one OTIS move. The compressed path looks like $(g_1, p_1) \xrightarrow{E^*} (g_1, g_2) \xrightarrow{O} (g_2, g_1) \xrightarrow{E^*} (g_2, p_2)$.

Shortest paths of type (a) have length exactly $D(p_1, p_2)$ (which equals the number of ones in the binary representation of $p_1 \oplus p_2$). Paths of type (b) and type (c) have length $D(p_1, p_2) + D(g_1, g_2) + 2$ and $D(p_1, g_2) + D(p_2, g_1) + 1$, respectively.

The following theorems follow directly from the above discussion.

Theorem 4.5 *Sahni and Wang [1998] The length of the shortest path between processors (g_1, p_1) and (g_2, p_2) is $d(p_1, p_2)$ when $g_1 = g_2$ and $\min\{D(p_1, p_2) + D(g_1, g_2) + 2, D(p_1, g_2) + D(p_2, g_1) + 1\}$ when $g_1 \neq g_2$.*

Proof. Similar to the proof in Theorem 4.2. ∎

Theorem 4.6 *Sahni and Wang [1998] The diameter of the OTIS-Hypercube is $2d + 1$.*

Proof. Since each group is a d-dimensional hypercube, $D(p_1, p_2)$, $D(g_1, g_2)$, $D(p_1, g_2)$, and $D(p_2, g_1)$ are all less than or equal to d. From Theorem 4.5, we conclude that no two processors are more than $2d + 1$ apart. Now consider the processors (g_1, p_1), (g_2, p_2) such that $p_1 = 0$ and $p_2 = N - 1$. Let $g_1 = 0$ and $g_2 = N - 1$. So $D(p_1, p_2) = D(g_1, g_2) = D(p_1, g_2) = D(p_2, g_1) = d$. Hence, the distance between (g_1, p_1) and (g_2, p_2) is $2d + 1$. As a result, the diameter of the OTIS-Mesh is exactly $2d + 1$. ∎

4.3.3.2 Simulation of an N^2-Processor Hypercube

Zane et al. [2000] have shown that each move of an N^2-processor hypercube can be simulated by either a single electronic move or by one electronic and two OTIS moves in an N^2-processor OTIS-Hypercube. For the simulation, processor q of the hypercube is mapped to processor (g, p) of the OTIS-Hypercube. Here $gp = q$ (i.e., g is obtained from the most significant $\log_2 N$ bits of q and p comes from the least significant $\log_2 N$ bits). Let $g_{d-1} \cdots g_0$ and $p_{d-1} \cdots p_0$, $d = \log_2 N$, be the binary representations of g and p, respectively. The binary representation of q is $q_{2d-1} \cdots q_0 = g_{d-1} \cdots g_0 p_{d-1} \cdots p_0$. A hypercube move moves data from processor q to processor $q^{(k)}$ where $q^{(k)}$ is obtained from q by complementing bit k in the binary representation of q. When k is in the range $c(0, d)$, the move is done in the OTIS-Hypercube by a local intra group hypercube move. When $k \geq d$, the move is done using the steps:

$$(g_{d-1} \cdots g_j \cdots g_0 p_{d-1} \cdots p_0)$$
$$\xrightarrow{O} (p_{d-1} \cdots p_0 g_{d-1} \cdots g_j \cdots g_0)$$
$$\xrightarrow{E} (p_{d-1} \cdots p_0 g_{d-1} \cdots \overline{g_j} \cdots g_0)$$
$$\xrightarrow{O} (g_{d-1} \cdots \overline{g_j} \cdots g_0 p_{d-1} \cdots p_0)$$

where $j = k - d$.

4.4 Summary of Results for Optical Transpose Interconnection System Computers

Algorithms for various operations on OTIS computers have been developed, also the properties of different classes of OTIS computers have been analyzed and extended. In this section, we categorize the papers that contributed to the development of various classes of OTIS computers.

4.4.1 General Properties

It is very important to know the properties of the underlining model of computation before the model is realized and algorithms are developed. Besides the discussions in Feldman et al. [1988]; Zane et al. [2000]; Wang and Sahni [1998b], Day and Al-Ayyoub [2002]; Coudert et al. [2000b]; Parhami [2004] discuss the general (topological) properties of OTIS networks. Parhami [2005]; Farahabady and Sarbazi-Azad [2005] demonstrate the relationship between the OTIS network and its Hamiltonicity, which is important for both parallel algorithm development and network routing. In specific classes of OTIS computers, Sahni and Wang [1997]; Mondal and Jana [2004] provide analysis on the properties of OTIS-Mesh; Sahni and Wang [1998] discuss some basic properties of OTIS-Hypercube; Coudert et al. [2002]; Ogata et al. [2005]; Wu and Deng [2006] present the relation between OTIS and de Bruijn network and its equivalent classes; Coudert et al. [2000a] present a multihop multi-OPS (optical passive star) network based on OTIS, Day [2004] discusses using k-ary n-cube network; and Al-Sadi [2005] presents an extended OTIS-Cube network.

4.4.2 Algorithms

Various algorithms have been developed for the OTIS-Mesh and OTIS-Hypercube. The complexities of these algorithms are governed by the number of data moves performed. Here we present a summary of algorithms that are developed for these two OTIS families by providing the number of electronic and OTIS moves they require.

For the OTIS-Mesh, Sahni and Wang [1997] develop efficient algorithms for frequently used permutations such as transpose and perfect shuffle, and for the general class of BPC permutations Nassimi and Sahni [1980]. The results are given in Table 4.1.

Table 4.2 presents algorithms for basic operations—broadcast, prefix sum, rank, concentrate, and so forth sorting—on the OTIS-Mesh with both SIMD and MIMD models Wang and Sahni [1998a]. Note that w in window broadcast indicates the window size, where $0 < w \leq \sqrt{N}/2$; and M in data accumulation, consecutive sum, and adjacent sum is the block size, $0 \leq M < \sqrt{N}/2$. Later Osterloh [2000] presents a

TABLE 4.1 Performance of Permutations on N^2-Processor OTIS-Mesh

Operation	Electronic	OTIS
Transpose	0	1
Perfect shuffle	$4\sqrt{N}+6$	2
Unshuffle	$4\sqrt{N}+6$	2
Bit reversal	$8(\sqrt{N}-1)$	1
Vector reversal	$8(\sqrt{N}-1)$	2
Bit shuffle	$\frac{28}{3}\sqrt{N}-4$	$\log_2 N + 2$
Shuffled row-major	$\frac{28}{3}\sqrt{N}-4$	$\log_2 N + 2$
BPC	$12(\sqrt{N}-1)$	$\log_2 N + 2$

Source: Sahni, S. and Wang, C.-F. (1997). *MPPOI'97*, pp. 130–135.

TABLE 4.2 Performance of Basic Operations on N^2-Processor OTIS-Mesh

Operation	SIMD Electronic	SIMD OTIS	MIMD Electronic	MIMD OTIS
Broadcast	$4(\sqrt{N}-1)$	1	$4(\sqrt{N}-1)$	
Window broadcast	$4\sqrt{N}-2w-2$	2	$4\sqrt{N}-2w-2$	2
Prefix sum	$7(\sqrt{N}-1)$	2	$7(\sqrt{N}-1)$	2
Data sum	$8(\sqrt{N}-1)$	1	$4\sqrt{N}$	1
Rank	$7(\sqrt{N}-1)$	2	$7(\sqrt{N}-1)$	2
Regular shift	s	2	s	2
Circular shift	$\sqrt{N}$	2	s	2
Data accumulation	$\sqrt{N}$	2	M	2
Consecutive sum	$4(M-1)$	2	$2(M-1)$	2
Adjacent sum	$\sqrt{N}$	2	M	2
Concentrate	$7(\sqrt{N}-1)$	2	$4(\sqrt{N}-1)$	2
Distribute	$7(\sqrt{N}-1)$	2	$4(\sqrt{N}-1)$	2
Generalize	$7(\sqrt{N}-1)$	2	$4(\sqrt{N}-1)$	2
Sorting	$22\sqrt{N}+o(\sqrt{N})$	$o(\sqrt{N})$	$11\sqrt{N}+o(\sqrt{N})$	$o(\sqrt{N})$

Source: Wang, C.-F. and Sahni, S. (1998a). *IEEE Transaction on Parallel and Distributed Systems*, 9(12): 1226–1234.

TABLE 4.3 Performance of Randomized Algorithms on N^2-Processor OTIS-Mesh

Operation	Complexity
Randomized routing	$4\sqrt{N}+\tilde{o}(\sqrt{N})$
Randomized sorting	$8\sqrt{N}+\tilde{o}(\sqrt{N})$
Randomized selection	$6\sqrt{N}+\tilde{o}(\sqrt{N})$

Source: Rajasekaran, S. and Sahni, S. (1998). *IEEE Transaction on Parallel and Distributed Systems*, 9(9): 833–840.

more efficient k-k sorting algorithm, on the basis of the principle of sorting in Kunde [1993], that uses $2k\sqrt{N} + O\left(kN^{\frac{1}{3}}\right)$ moves.

Rajasekaran and Shani, (1998) develop randomized algorithms for routing, selection, and sorting, and the results are in Table 4.3.

Table 4.4 gives the number of data moves for various matrix/vector multiplication operations Wang and Sahni [2001]. Since the mapping of the matrix and/or vector is a determining factor

TABLE 4.4 Performance of Matrix Multiplication on N^2-Processor OTIS-Mesh

Scheme	GRM		GSM	
Operation[a]	Electronic	OTIS	Electronic	OTIS
C × R	$2\sqrt{N}$	2	$4(\sqrt{N}-1)$	2
R × C	$2(\sqrt{N}-1)$	1	$5(\sqrt{N}-1)$	2
R × M	$4(\sqrt{N}-1)$	3	$8(\sqrt{N}-1)$	2
M × C	$4(\sqrt{N}-1)$	1	$8(\sqrt{N}-1)$	2
M × M $O(N)$	$4(N-1)$	$N/K+1$	$4(N-1)$	$2\sqrt{N}/K+1$
M × M $O(1)$	$8N+O(\sqrt{N})$	$N+1$	$4N+O(\sqrt{N})$	$\sqrt{N}$

[a] C, R, and M denote column vector, row vector, and matrix respectively.
Source: Wang, C.-F. and Sahni, S. (2001). *IEEE Transaction on Computers*, 50(7): 635–646.

TABLE 4.5 Performance of Permutations on N^2-Processor OTIS-Hypercube

Operation	Electronic	OTIS
Transpose	0	1
Perfect shuffle	$2d$	2
Unshuffle	$2d$	2
Bit reversal	$2d$	1
Vector reversal	$2d$	2
Bit shuffle	$3d$	$d/2+2$
Shuffled row-major	$3d$	$d/2+2$
BPC	$3d$	$d/2+2$

$d = \log_2 N$.
Source: Sahni, S. and Wang, C.-F. (1998). *Informatica*, 22: 263–269.

on the outcome, results for both GRM and GSM mappings (see Section 4.5) are listed. The big O besides the matrix × matrix operation denotes the memory capacity constraints for each processor and K denotes the maximum amount of data that can be transferred in a single OTIS move. (In an electronic move, 1 unit of data may be transferred. Since the optical bandwidth is larger than the electronic bandwidth, $K > 1$ units of data may be transferred in each OTIS move.)

Other than the results above, Wang and Sahni [2000] develop algorithms for image processing problems such as histogramming and its modification, image expanding and shrinking, and Hough transform, each having the complexity of $O(\sqrt{N})$, except for Hough transform, which uses $O(N)$ moves. Wang and Sahni [2002] present algorithms for various computational geometry problems, such as convex hull, smallest enclosing box, ECDF, two-set dominance, maximal points, all-nearest neighbor, and closest pair of points, each running in $O(\sqrt{N})$ time. Finally, Jana [2004] presents methods to perform polynomial interpolation that uses $6\sqrt{N} - 4$ electronic moves and 2 OTIS moves.

For the OTIS-Hypercube, Sahni and Wang [1998] develop algorithms for frequently used permutations as well as for general BPC permutations Nassimi and Sahni [1980]. The results are listed in Table 4.5.

4.4.3 Routing

Routing is the foundation for communications, whether it is between two processors in a computer, or two computers in a network. Efficient routing algorithms are necessary for many higher-level applications.

For OTIS-Hypercube, the study of routing algorithms is extensive. Najaf-Abadi and Sarbazi-Azad [2004a] evaluate deterministic routing, Najaf-Abadi and Sarbazi-Azad [2004b] discuss performance under different traffic patterns, and Najaf-Abadi and Sarbazi-Azad [2006] do an analytical performance evaluation. Najaf-Abadi and Sarbazi-Azad [2004c]; Al-Sadi and Awwad [2004], Awwad and Al-Sadi [2005] discuss various aspects of routing on the OTIS-Cube, a generalization of OTIS-Hypercube, and consider capacity and fault tolerance. Najaf-Abadi and Sarbazi-Azad [2005] present a comparison of deterministic routing between the OTIS-Mesh and OTIS-Hypercube models, and Al-Sadi et al. [2004] present efficient routing algorithms for the OTIS-Star.

4.5 Summary

In this chapter, we have described the OTIS family of computer architectures. This family accounts for the fact that computers with a very large number of processors cannot be built using very short connections (i.e., less than a few millimeters) alone. Therefore, at some level of the packaging hierarchy, the interprocessor distance will be such as to favor an optical interconnect over an electronic one. The OTIS system is a simple and easy to implement free space optical interconnect system. Using OTIS for the long connections and electronic mesh (say) for the short interconnects results in a computer with a hybrid optoelectronic interconnection network. The development of algorithms for computers using a hybrid interconnection network poses a challenge. Using the simulation methods developed by Zane et al. [2000], it is possible to obtain efficient OTIS algorithms from efficient algorithms for the 4D-mesh and hypercube architectures. As one might expect, we can do better than this simulation by developing algorithms specifically for the OTIS architecture.

References

Al-Sadi, J. (2005). An extended OTIS-Cube interconnection network. In *Proceedings of the IADIS International Conference on Applied Computing (AC'05)*, volume 2, pp. 167–172.

Al-Sadi, J. and Awwad, A. M. (2004). A new fault-tolerant routing algorithm for OTIS-Cube using unsafety vectors. In *Proceedings of the 2004 ACM Symposium on Applied Computing (SAC'04)*, pp. 1426–1430.

Al-Sadi, J. A., Awwad, A. M., and AlBdaiwi, B. F. (2004). Efficent routing algorithm on OTIS-star network. In *Proceedings of Advances in Computer Science and Technology (ACST'04)*, pp. 431–040.

Awwad, A. and Al-Sadi, J. (2005). On the routing of the OTIS-Cube network in presence of faults. *The International Arab Journal of Information Technology*, 2(1):17–23.

Coudert, D., Ferreira, A., and Muñoz, X. (2000a). A multihop-multi-OPS optical interconnection network. *Journal of Lightwave Technology*, 18(12):2076–2085.

Coudert, D., Ferreira, A., and Muñoz, X. (2000b). Topologies for optical interconnection networks based on the optical transpose interconnection system. *Applied Optics—Information Processing*, 39 17):2695–2974.

Coudert, D., Ferreira, A., and Pérennes, S. (2002). Isomorphisms of the de bruijn digraph and free-space optical networks. *Network*, 40(3):155–164.

Day, K. (2004). Optical transpose k-ary n-cube networks. *Journal of Systems Architecture: The EUROMICRO Journal*, 50(11):697–705.

Day, K. and Al-Ayyoub, A.-E. (2002). Topological properties of OTIS-networks. *IEEE Transactions on Parallel and Distributed Systems*, 13(4):359–366.

Farahabady, M. H. and Sarbazi-Azad, H. (2005). Algorithmic construction of Hamiltonian cycles in OTIS-networks. In *Proceedings of Networks and Communication Systems (NCS'05)*, pp. 464–110.

Feldman, M., Esener, S., Guest, C., and Lee, S. (1988). Comparison between electrical and free-space optical interconnects based on power and speed considerations. *Applied Optics*, 27(9):1742–1751.

Gonzalez, T. and Sahni, S. (1976). Open shop scheduling to minimize finish time. *Journal of the Association for Computing Machinery*, 23(4):665–679.

Hendrick, W., Kibar, O., Marchand, P., Fan, C., Blerkom, D. V., McCormick, F., Cokgor, I., Hansen, M., and Esener, S. (1995). Modeling and optimization of the optical transpose interconnection system. In *Optoelectronic Technology Center, Program Review, Cornell University*.

Jana, P. K. (2004). Polynomial interpolation on OTIS-Mesh optoelectronic computers. In *Proceedings of the 6th International Workshop on Distributed Computing*, pp. 373–378.

Koppelman, D. M. and Oruç, A. Y. (1994). The complexity of routing in Clos permutation networks. *IEEE Transactions on Information Theory*, 40(1):278–284.

Krishnamoorthy, A., Marchand, P., Kiamilev, F., and Esener, S. (1992). Grain-size considerations for optoelectronic multistage interconnection networks. *Applied Optics*, 31(26):5480–5507.

Kunde, M. (1993). Block gossiping on grids and tori: Deterministic sorting and routing match the bisection bound. In *Proceedings of European Symposium on Algorithms*, pp. 272–283.

Marsden, G. C., Marchand, P. J., Harvey, P., and Esener, S. C. (1993). Optical transpose interconnection system architectures. *Optics Letters*, 18(13):1083–1085.

Mondal, D. and Jana, P. K. (2004). Neighborhood property of OTIS-Mesh optoelectronic computer. In *Proceedings of the 7th International Symposium on Parallel Architectures, Algorithms, and Networks (I-SPAN'04)*, pp. 458–462.

Najaf-Abadi, H. H. and Sarbazi-Azad, H. (2004a). Comparative evaluation of adaptive and deterministic routing in the OTIS-Hypercube. In *Proceedings of the 9th Asia-Pacific Computer System Architecture Conference (ACSAC'04)*, pp. 349–362.

Najaf-Abadi, H. H. and Sarbazi-Azad, H. (2004b). The effect of adaptivity on the performance of the OTIS-Hypercube under different traffic patterns. In *Proceedings of the International Conference on Network and Parallel Computing (NPC'04)*, pp. 390–398.

Najaf-Abadi, H. H. and Sarbazi-Azad, H. (2004c). Routing capabilities and performance of cube-based OTIS systems. In *Proceedings of the International Symposium on Performance Evaluation of Computer and Telecommunication Systems (SPECTS'04)*, pp. 130–134.

Najaf-Abadi, H. H. and Sarbazi-Azad, H. (2005). An empirical comparison of OTIS-Mesh and OTIS-Hypercube multicomputer systems under deterministic routing. In *Proceedings of the 19th International Parallel and Distributed Processing Symposium (IPDPS'05)*. CD-ROM.

Najaf-Abadi, H. H. and Sarbazi-Azad, H. (2006). Analytic performance evaluation of OTIS-Hypercube. *IEICE Transactions on Information and Systems*, E89-D(2):441–451.

Nassimi, D. and Sahni, S. (1980). An optimal routing algorithm for mesh-connected parallel computers. *Journal of the Association for Computing Machinery*, 27(1):6–29.

Ogata, K., Yamada, T., and Ueno, S. (2005). A note on the implementation of De Bruijn networks by the optical transpose interconnection system. *IEICE Transactions on Fundamentals of Electronics, Communications and Computer Sciences*, E88-A(12):3661–3662.

Osterloh, A. (2000). Sorting on the OTIS-Mesh. In *Proceedings of the 14th International Parallel and Distributed Processing Symposium (IPDPS'00)*, pp. 269–274.

Parhami, B. (2004). Some properties of swapped interconnection networks. In *Proceedings of the International Conference on Communications in Computing (CIC'04)*, pp. 93–99.

Parhami, B. (2005). The hamiltonicity of swapped (OTIS) networks built of Hamiltonian component networks. *Imformation Processing Letters*, 95(4):441–445.

Rajasekaran, S. and Sahni, S. (1998). Randomized routing, selection, and sorting on the OTIS-Mesh optoelectronic computer. *IEEE Transactions on Parallel and Distributed Systems*, 9(9):833–840.

Sahni, S. and Wang, C.-F. (1997). BPC permutations on the OTIS-Mesh optoelectronic computer. In *Proceedings of the Fourth International Conference on Massively Parallel Processing Using Optical Interconnections (MPPOI'97)*, pp. 130–135.

Sahni, S. and Wang, C.-F. (1998). BPC permutations on the OTIS-Hypercube optoelectronic computer. *Informatica*, 22:263–269.

Wang, C.-F. and Sahni, S. (1998a). Basic operations on the OTIS-Mesh optoelectronic computer. *IEEE Transactions on Parallel and Distributed Systems*, 9(12):1226–1236.

Wang, C.-F. and Sahni, S. (1998b). OTIS optoelectronic computers. In Li, K., Pan, Y., and Zheng, S. Q., editors, *Parallel Computing Using Optical Interconnections*, Chapter 5, pp. 99–116. Kluwer Academic Publication, Norwell, MA.

Wang, C.-F. and Sahni, S. (2000). Image processing on the OTIS-Mesh optoelectronic computer. *IEEE Transactions on Parallel and Distributed Systems*, 11(2):97–109.

Wang, C.-F. and Sahni, S. (2001). Matrix multiplication on the OTIS-Mesh optoelectronic computer. *IEEE Transactions on Computers*, 50(7):635–646.

Wang, C.-F. and Sahni, S. (2002). Computational geometry on the OTIS-Mesh optoelectronic computer. In *Proceedings of the 2002 International Conference on Parallel Processing (ICPP'02)*, pp. 501–507.

Wu, Y. and Deng, A. (2006). OTIS layouts of De Bruijn digraphs. e-submission.

Zane, F., Marchand, P., Paturi, R., and Esener, S. (2000). Scalable network architectures using the optical transpose interconnection system (OTIS). *Journal of Parallel and Distributed Computing*, 60(5): 521–538.

5
Models for Advancing PRAM and Other Algorithms into Parallel Programs for a PRAM-On-Chip Platform

Part I: Extended Summary		5-2
5.1	Introduction...	5-2
5.2	Model Descriptions...................................	5-4
	PRAM Model • The Work-Depth Methodology • XMT Programming Model • XMT Execution Model • Clarifications of the Modeling	
5.3	An Example for Using the Methodology: Summation	5-13
5.4	Empirical Validation of the Performance Model	5-14
	Performance Comparison of Implementations • Scaling with the Architecture	
5.5	Conclusion of Extended Summary	5-19
Part II: Discussion, Examples, and Extensions		5-20
5.6	Compiler Optimizations	5-20
	Nested Parallel Sections • Clustering • Prefetching	
5.7	Prefix-Sums	5-23
	Synchronous Prefix-Sums • No-Busy-Wait Prefix-Sums • Clustering for Prefix-Sums • Comparing Prefix-Sums Algorithms	
5.8	Programming Parallel Breadth-First Search Algorithms	5-27
	Nested Spawn BFS • Flattened BFS • Single-Spawn and k-Spawn BFS	
5.9	Execution of Breadth-First Search Algorithms	5-30
	Flattened BFS • Single-Spawn BFS • k-Spawn BFS • Comparison	

5.10	Adaptive Bitonic Sorting	5-32
5.11	Shared Memory Sample Sort	5-34
5.12	Sparse Matrix—Dense Vector Multiplication	5-35
5.13	Speed-Ups over Serial Execution	5-36
5.14	Conclusion	5-37
	Acknowledgments	5-37
	Appendix: XMTC Code Examples	5-40
	K-Ary Summation	5-40
	Synchronous K-Ary Prefix-Sums	5-42
	No-Busy-Wait K-Ary Prefix-Sums	5-46
	Serial Summation	5-50
	Serial Prefix-Sums	5-50
	Flattened BFS Algorithm	5-51
	Single-Spawn BFS Algorithm	5-55
	k-Spawn BFS Algorithm	5-57
	References	5-57

Uzi Vishkin
George C. Caragea
University of Maryland

Bryant C. Lee
Carnegie Mellon University

It is better for Intel to get involved in this now so when we get to the point to having 10s and 100s of cores we will have the answers. There is a lot of architecture work to do to release the potential, and we will not bring these products to market until we have good solutions to the *programming problem*.

(Justin Rattner, CTO, Intel, *Electronic News*, March 13, 2006)

Part I: Extended Summary
5.1 Introduction

Parallel programming is currently a difficult task. However, it does not have to be that way. Current methods tend to be coarse-grained and use either a shared memory or a message passing model. These methods often require the programmer to think in a way that takes into account details of memory layout or architectural implementation, leading the 2003 NSF Blue-Ribbon Advisory Panel on Cyberinfrastructure to opine that: to many users, programming existing parallel computers is still "as intimidating and time consuming as programming in assembly language". Consequently, to date, the outreach of parallel computing has fallen short of historical expectations. With the ongoing transition to chip multiprocessing (multi-cores), the industry is considering the parallel programming problem a key bottleneck for progress in the commodity processor space. In recent decades, thousands of papers have been written on algorithmic models that accommodate simple representation of concurrency. This effort brought about a fierce debate between several schools of thought, with one of the approaches, the "PRAM approach," emerging as a clear winner in this "battle of ideas." Three of the main standard undergraduate computer science texts that came out in 1988-90 [5,18,43] chose to include large chapters on PRAM algorithms. The PRAM was the model of choice for parallel algorithms in all major algorithms/theory communities and was taught everywhere. This win did not register in the collective memory as the clear and decisive victory it was since (i) at about the same time (early 1990s), it became clear that it will not be possible to build a machine that can look to the performance programmer as a PRAM using 1990s technology, (ii) saying that the PRAM approach can never become useful became a mantra, and PRAM-related work came to a near halt, and (iii) apparently, most people made an extra leap into accepting this mantra.

The Parallel Random Access Model (PRAM) is an easy model for parallel algorithmic thinking and for programming. It abstracts away architecture details by assuming that many memory accesses to a shared memory can be satisfied within the same time as a single access. Having been developed mostly during the 1980s and early 1990s in anticipation of a parallel programmability challenge, PRAM algorithmics provides the second largest algorithmic knowledge base right next to the standard serial knowledge base.

With the continuing increase of silicon capacity, it has become possible to build a single-chip parallel processor. This has been the purpose of the Explicit Multi-Threading (XMT) project [47,60] that seeks to prototype a PRAM-On-Chip vision, as on-chip interconnection networks can provide enough bandwidth for connecting processors to memories [7,62].

The XMT framework, reviewed briefly in the current chapter, provides quite a broad computer system platform. It includes an Single Program, Multiple Data (SPMD) programming model that relaxes the lock-step aspect of the PRAM model, where each parallel step consists of concurrent operations all performed before proceeding to the next parallel step.

The XMT programming model uses thread-level parallelism (TLP), where threads are defined by the programming language and handled by its implementation. The threads tend to be short and are not operating system threads. The overall objective for multi-threading is reducing single-task completion time. While there have been some success stories in compiler effort to automatically extract parallelism from serial code [2, 4], it is mostly agreed that compilers alone are generally insufficient for extracting parallelism from "performance code" written in languages such as C. Henceforth, we assume that the performance programmer is responsible to extract and express the parallelism from the application.

Several multichip multiprocessor architectures targeted implementation of PRAM algorithms, or came close to that. (i) The NYU Ultracomputer project viewed the PRAM as providing theoretical yardstick for limits of parallelism as opposed to a practical programming model [50]. (ii) The Tera/Cray Multithreaded Architecture (MTA) advanced Burton Smith's 1978 HEP novel hardware design. Some authors have stated that an MTA with large number of processors looks almost like a PRAM; see References 6 and 15. (iii) The SB-PRAM may be the first whose declared objective was to provide emulation of the PRAM [39]. A 64-processor prototype has been built [21]. (iv) Although only a language rather than an architecture, NESL [10] sought to make PRAM algorithms easier to express using a functional program that is compiled and run on standard multichip parallel architectures. However, PRAM theory has generally not reached out beyond academia and it is still undecided whether a PRAM can provide an effective abstraction for a proper design of a multichip multiprocessor, due to limits on the bandwidth of such an architecture [19]. While we did not find an explicit reference to the PRAM model in MIT-Cilk related papers, the short tutorial [42] presents similar features as to how we could have approached a first draft of divide-and-conquer parallel programs. This applies especially to the incorporation of "work" and "depth" in such first draft programs. As pointed out in Section 5.6.2, we also do not claim in the current chapter original contributions beyond the MIT-Cilk on issues concerning clustering, and memory utilization related to implementation of nested spawns.

Guided by the fact that the number of transistors on a chip keeps growing and already exceeds one billion, up from less than 30,000 circa 1980, the main insight behind XMT is as follows. Billion transistor chips allow the introduction of a high-bandwidth low-overhead on-chip multiprocessor. It also allows an evolutionary path from serial computing. The drastic slow down in clock rate improvement for commodity processors since 2003 is forcing vendors to seek single task performance improvements through parallelism. While some have already announced growth plans to 100-core chips by the mid-2010s, they are yet to announce algorithms, programming, and machine organization approaches for harnessing these enormous hardware resources toward single task completion time. XMT addresses these issues.

Some key differences between XMT and the above multichip approaches are: (i) its larger bandwidth, benefiting from the on-chip environment; (ii) lower latencies to shared memory, since an on-chip approach allows on-chip shared caches; (iii) effective support for serial code; this may be needed for backward compatibility for serial programs, or for serial sections in PRAM-like programs; (iv) effective support for parallel execution where the amount of parallelism is low; certain algorithms (e.g., breadth first-search (BFS) on graphs presented later) have particularly simple parallel algorithms; some are only a minor variation of the serial algorithm; since they may not offer sufficient parallelism for some multichip architectures, such important algorithms had no merit for these architectures; and (v) XMT introduced a so-called *Independence of Order Semantics* (IOS), which means that each thread executes at its own pace and any ordering of interactions among threads is valid. If more than one thread may seek to write to the same shared variable this would be in line with the PRAM "arbitrary CRCW" convention (see Section 5.2.1). This IOS feature improves performance as it allows processing with whatever data is available at the processing elements and saves power as it reduces synchronization needs. An IOS feature could have been added to multichip approaches providing some, but apparently not all the benefits.

While most PRAM-related approaches also tended to emphasize competition with (massively parallel) parallel computing approaches, not falling behind modern serial architectures has been an additional objective for XMT.

XMT can also support standard application programming interfaces (APIs) such as those used for graphics (e.g., OpenGL) or circuit design (e.g., VHDL). For example, Reference 30: (i) demonstrated speedups exceeding a hundred fold over serial computing for gate-level VHDL simulations implemented on XMT, and (ii) explained how these results can be achieved by automatic compiler extraction of parallelism. Effective implementation of such APIs on XMT would allow an application programmer to take advantage of parallel hardware with few or no changes to an existing API.

Contributions. The main contributions of this chapter are as follows. (i) Presenting a programming methodology for converting PRAM algorithms to PRAM-On-Chip programs. (ii) Performance models used in developing a PRAM-On-Chip program are introduced, with a particular emphasis on a certain complexity metric, the length of the sequence of round trips to memory (LSRTM). While the PRAM algorithmic theory is pretty advanced, many more practical programming examples need to be developed. For standard serial computing, examples for bridging the gap between algorithm theory and practice of serial programming abound, simply because it has been practiced for so long. See also Reference 53. A similar knowledge base needs to be developed for parallel computing. (iii) The current paper provides a few initial programming examples for the work ahead. (iv) Alternatives to the strict PRAM model that by further suppressing details provide (even) easier-to-think frameworks for parallel algorithms and programming development are also discussed. And last, but not least, these contributions have a much broader reach than the context in which they are presented.

To improve readability of this long chapter the presentation comes in two parts. Part I is an extended summary presenting the main contributions of the chapter. Part II supports Part I with explanations and examples making the chapter self-contained.

Performance models used in developing a PRAM-On-Chip program are described in Section 5.2. An example using the models is given in Section 5.3. Some empirical validation of the models is presented in Section 5.4. Section 5.5 concludes Part I—the extended summary of the paper. Part II begins with Section 5.6, where compiler optimizations that could affect the actual execution of programs are discussed. Section 5.7 gives another example for applying the models to the prefix sums problem. Section 5.8 presents BFS in the PRAM-On-Chip Programming Model. Section 5.9 explains the application of compiler optimizations to BFS and compares performance of several BFS implementations. Section 5.10 discusses the Adaptive Bitonic Sorting algorithm and its implementation while Section 5.11 introduces a variant of Sample Sort that runs on a PRAM-On-Chip. Section 5.12 discusses sparse matrix—dense vector multiplication. The discussion on speedups over serial code from the section on empirical validation of the models was deferred to Section 5.13. A conclusion section is followed by a long appendix with quite a few XMTC code examples in support of the text.

5.2 Model Descriptions

Given a problem, a "recipe" for developing an efficient XMT program from concept to implementation is proposed. In particular, the stages through which such development needs to pass are presented.

Figure 5.1 depicts the proposed methodology. For context, the figure also depicts the widely used work-depth methodology for advancing from concept to a PRAM algorithm, namely, the sequence of models $1 \to 2 \to 3$ in the figure. For developing an XMT implementation, we propose following the sequence of models $1 \to 2 \to 4 \to 5$, as follows. Given a specific problem, an algorithm design stage will produce a high-level description of the parallel algorithm, in the form of a sequence of steps, each comprising a *set* of concurrent operations (box 1). In a first draft, the set of concurrent operations can be implicitly defined. See the BFS example in Section 5.2.2. This first draft is refined to a sequence of steps

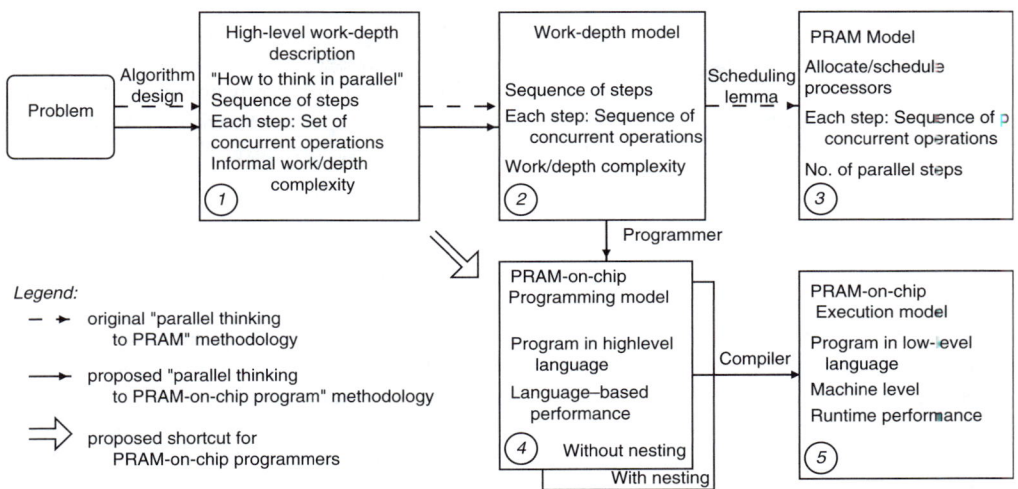

FIGURE 5.1 Proposed Methodology for Developing XMT Programs in view of the Work-Depth Paradigm for Developing PRAM algorithms.

each comprising now an *ordered sequence* of concurrent operations (box 2). Next, the programming effort amounts to translating this description into a single-program multiple-data (SPMD) program using a high-level XMT programming language (box 4). From this SPMD program, a compiler will transform and reorganize the code to achieve the best performance in the target XMT execution model (box 5). As an XMT programmer gains experience, he/she will be able to skip box 2 (the Work-Depth model) and directly advance from box 1 (high-level work-depth description) to box 4 (high-level XMT program). We also demonstrate some instances where it may be advantageous to skip box 2 because of some features of the programming model (such as some ability to handle nesting of parallelism). In Figure 5.1 this shortcut is depicted by the arrow 1 → 4. Much of the current chapter is devoted to presenting the methodology and demonstrating it. We start with elaborating on each model.

5.2.1 PRAM Model

Parallel Random Access Machine, or Model (PRAM) consists of p synchronous processors communicating through a global shared memory accessible in unit time from each of the processors. Different conventions exist regarding concurrent access to the memory. For brevity, we only mention arbitrary concurrent-read concurrent-write (CRCW) where simultaneous access to the same memory location for reads or writes are permitted, and concurrent writes to the same memory location result in an arbitrary one among the values sought to be written, but it is not known in advance which one.

The are quite a few sources for PRAM algorithms including References [23, 37–39, 59]. An algorithm in the PRAM model is described as a sequence of parallel time units, or rounds; each round consists of exactly p instructions to be performed concurrently, one per each processor. While the PRAM model is simpler than most parallel models, producing such a description imposes a significant burden on the algorithm designer. Luckily this burden can be somewhat mitigated using the work-depth methodology, presented next.

5.2.2 The Work-Depth Methodology

The work-depth methodology for designing PRAM algorithms, introduced in Reference 52, has been quite useful for describing parallel algorithms and reasoning about their performance. For example, it was used as the description framework in References 37 and 39. The methodology guides algorithm designers to optimize two quantities in a parallel algorithm: *depth* and *work*. Depth represents the number of steps the algorithm would take if unlimited parallel hardware was available, while work is the total number of operations performed, over all parallel steps.

The methodology comprises of two steps: (i) first, produce an informal description of the algorithm in a high-level work-depth model (HLWD), and (ii) refine this description into a fuller presentation in a model of computation called work-depth. These two models are described next.

5.2.2.1 High-Level Work-Depth Description

A HLWD description consists of a succession of parallel rounds, each round being a *set* of any number of instructions to be performed concurrently. Descriptions can come in several flavors and even implicit, where the number of instructions is not obvious, are acceptable.

Example: Input: An undirected graph $G(V, E)$ and a source node $s \in V$; the length of every edge in E is 1. Find the length of the shortest paths from s to every node in V. An informal work-depth description of a parallel *BFS* algorithm can look as follows. Suppose that the set of vertices V is partitioned into layers. Layer L_i includes all vertices of V whose shortest path from s have i edges. The algorithm works in iterations. In iteration i, layer L_i is found. Iteration 0: node s forms layer L_0. Iteration $i, i > 0$: Assume inductively that layer L_{i-1} was found. In parallel, consider all the edges (u, v) that have an endpoint u in layer L_{i-1}; if v is not in a layer L_j, $j < i$, it must be in layer L_i. As more than one edge may lead from a vertex in layer L_{i-1} to v, vertex v is marked as belonging to layer L_i based on one of these edges using the arbitrary concurrent write convention. This ends an informal, high-level work-depth verbal description.

A pseudocode description of an iteration of this algorithm could look as follows:

```
for all vertices v in L(i) pardo
   for all edges e=(v,w) pardo
      if w unvisited
         mark w as part of L(i+1)
```

The above HLWD descriptions challenge us to find an efficient PRAM implementation for an iteration. In particular, given a p-processor PRAM how to allocate processors to tasks to finish all operations of an iterations as quickly as possible? A more detailed description in the work-depth model would address these issues.

5.2.2.2 Work-Depth Model

In the work-depth model an algorithm is described in terms of successive time steps, where the concurrent operations in a time step form a sequence; each element in the sequence is indexed from 1 to the number of operations in the step. The work-depth model is formally equivalent to the PRAM. For example, a work-depth algorithm with $T(n)$ depth (or time) and $W(n)$ work runs on a p processor PRAM in at most $T(n) + \lfloor \frac{W(n)}{p} \rfloor$ time steps. The simple equivalence proof follows Brent's scheduling principle, which was introduced in Reference 13 for a model of parallel model of computation that was much more abstract than the PRAM (counting arithmetic operations, but suppressing anything else).

Example (continued): We only note here the challenges for coming up with a work-depth description for the BFS algorithm: to find a way for listing in a single sequence all the edges whose endpoint is a vertex at layer L_i. In other words, the work-depth model does not allow us to leave nesting of parallelism, such as in the pseudocode description of BFS above, unresolved. On the other hand XMT programming should allow nesting of parallel structures, since such nesting provides an easy way for parallel programming. It is also important to note that the XMT architecture includes some limited support for nesting of parallelism: a nested spawn can only spawn k extra threads, where k is a small integer (e.g., $k = 1, 2, 4$, or 8); nested spawn commands are henceforth called either k-spawn, or sspawn (for single spawn). We suggest the following to resolve this problem. The *ideal* long term solution is to (i) allow the programmer free unlimited use of nesting, (ii) have it implemented as efficiently as possible by a compiler, and (iii) make the programmer (especially the performance programmer) aware of the added cost of using nesting. However, since our compiler is not yet mature enough to handle this matter, our *tentative* short term solution is presented in Section 5.8, which shows how to build on the support for nesting provided by

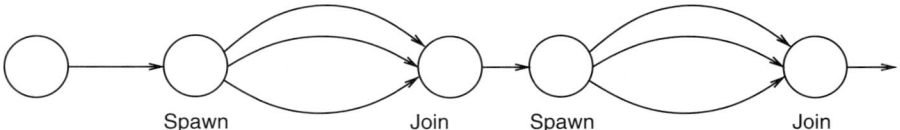

FIGURE 5.2 Switching between serial and parallel execution modes in the XMT programming model. Each parallel thread executes at its own speed, without ever needing to synchronize with another thread

the architecture. There is merit to this "manual solution" beyond its tentative role until the compiler matures. Such a solution still needs to be understood (even after the ideal compiler solution is in place) by performance programmers, so that the impact of nesting on performance is clear to them.

The reason for bringing this issue up this early in the discussion is that it actually demonstrates that our methodology can sometimes proceed directly to the PRAM-like programming methodology, rather than make a "stop" at the work-depth model.

5.2.3 XMT Programming Model

A framework for a high-level programming language, the XMT programming model seeks to mitigate two goals: (i) *Programmability*: given an algorithm in the HLWD or work-depth models, the programmer's effort in producing a program should be minimized; and (ii) *Implementability*: effective compiler translation of the program into the XMT execution model should be feasible.

The XMT programming model is fine-grained and SPMD type. Execution can frequently alternate between serial and parallel execution modes. A Spawn command prompts a switch from serial mode to parallel mode (see Figure 5.2). The Spawn command can specify any number of threads. Ideally, each such thread can proceed until termination (a Join command) without ever having to busy-wait or synchronize with other threads. To facilitate that, an *independence of order semantics (IOS)* was introduced. Inspired by the arbitrary concurrent-write convention of the PRAM model, commands such as "prefix-sum" permit threads to proceed even if they try to write into the same memory location.

Some primitives in the XMT programming model follows:

Spawn Instruction. Starts a parallel section. Accepts as parameter the number of parallel threads to start.

Thread-id. A reserved parameter inside a parallel section. Evaluates to the unique thread index. This allows SPMD style programming.

Prefix-sum Instruction. The prefix-sum instruction defines an atomic operation. First assume a global variable B, called base, and a local variable R, called increment, the result of a prefix-sum is (i) B gets the value $B + R$, and (ii) R gets the original value of B.

Example: Suppose that threads 2, 3, and 5, respectively, execute concurrently the commands $ps(B, R_2)$, $ps(B, R_3)$, and $ps(B, R_5)$, respectively, all relating to the same base B and the original values are $B = 0, R_2 = R_3 = R_5 = 1$. IOS allows any order of execution among the 3 prefix-sums commands; namely, any of the 6 possible permutations. The result of all 6 permutations is $B = 3$. If thread 5 precedes thread 2 that precedes thread 2, we will get $R_5 = 0, R_2 = 1$, and $R_3 = 2$, and if the thread order is 2, 3, and 5 then $R_2 = 0, R_3 = 1, R_5 = 2$. Two examples for the use of the prefix sums command are noted next. (i) In the array compaction code in Table 5.1a, a code example where the prefix-sum command is used is demonstrated. This code example is referenced later in the text. (ii) To implement the PRAM arbitrary concurrent write convention, the programmer is guided to do the following: Each location that might be written by several threads has an auxiliary "gatekeeper" location associated with it, initialized with a known value (say 0). When a thread wants to write to the shared location, it first executes a Prefix-sum instruction (e.g., with an increment of 1) on the gatekeeper location. Only one thread gets 0 as its result; this thread is allowed to write to the shared location, while the other threads advance to their next instruction without writing.

While, the basic definition of prefix-sum follows the fetch-and-add of the NYU-Ultracomputer [28], XMT uses a fast parallel hardware implementation if R is from a small range (e.g., one bit) and B can fit one of a small number of global registers [57]; otherwise, prefix-sums are obtained using a prefix-sum-to-memory instruction; in the latter case, prefix-sum implementation involves queuing in memory.

Nested parallelism. A parallel thread can be programmed to initiate more threads. However, as noted in Section 5.2.2, this comes with some (tentative) restrictions and cost caveats, due to compiler and hardware support issues. As illustrated with the breadth-first search example, nesting of parallelism could improve the programmer's ability to describe algorithms in a clear and concise way.

Note that Figure 5.1 depicts two alternative XMT programming models: without nesting and with nesting. The work-depth model maps directly into the programming model without nesting. Allowing nesting could make it easier to turn a description in the high-level work-depth model into a program.

We call the resulting programming language XMTC. XMTC is a superset of the language C which includes statements implementing the above primitives.

Examples of XMTC Code Table 5.1 provides 5 XMTC programs, excluding constructs such as variable and function declarations, which were derived from PRAM algorithms. The table demonstrates compactness of code, as may be appreciated by readers familiar with other parallel programming frameworks.

In Table 5.1a, the language features of XMTC are demonstrated using the *array compaction problem*: given an array of integers $T[0 \ldots n-1]$, copy all its nonzero elements into another array S; any order will do. The special variable $ denotes the thread-id. Command spawn(0,n-1) spawns n threads whose id's are the integers in the range $0 \ldots n-1$. The ps(increment,length) statement executes an atomic prefix-sum command using length as the base and increment as the increment value. The variable increment is local to a thread while length is a shared variable that holds the number of nonzero elements copied at the end of the spawn block. Variables declared inside a spawn block are local to a thread, and are typically much faster to access than shared memory. Note that on XMT, local thread variables are typically stored into local registers of the executing hardware thread control unit (TCU). The programmer is encouraged to use local variables to store frequently used values. Often, this type of optimization can also be performed by an optimizing compiler.

To evaluate the performance in this model, a *Language-Based Performance Model* is used, performance costs are assigned to each primitive instruction in the language with some accounting rules added for the performance cost of computer programs (e.g., depending on execution sequences). Such performance modeling was used by Aho and Ullman [1] and was generalized for parallelism by Blelloch [10]. The article [20] used language-based modeling for studying parallel list ranking relative to an earlier performance model for XMT.

5.2.4 XMT Execution Model

The execution model depends on XMT architecture choices. However, as can be seen from the modeling itself, this dependence is rather minimal and should not compromise the generality of the model for other future chip-multiprocessing architectures whose general features are similar. We are only reviewing the architecture below and refer to Reference 47 for a fuller description. Class presentation of the overall methodology proposed in the current chapter breaks here to review, based on Reference 47, the way in which (i) the XMT apparatus of the program counters and stored program extending the (well-known) von-Neumann serial apparatus, (ii) the switch from serial to parallel mode and back to serial mode is being implemented, (iii) virtual threads coming from an XMTC program are allocated dynamically at run time, for load balancing, to thread control units (TCUs), (iv) hardware implementation of prefix-sum enhances the computation, and (v) independence of order semantics (IOS), in particular, and the overall design principle of no-busy-wait finite-state-machines (NBW FSM), in general, allow making as much progress as possible with whatever data and instructions are available.

Advancing PRAM Algorithms into Parallel Programs 5-9

TABLE 5.1 Implementation of Some PRAM Algorithms in the XMT PRAM-on-Chip Framework to Demonstrate Compactness

---

(a) **Array compaction**

```
/* Input: N
numbers in T[0..N-1]                                              *
* Output: S contains the non-zero values from T                   */
length = 0; // shared by all threads
spawn(0,n-1) { //start one thread per array element
    int increment = 1; // local to each thread
    if(T[$] != 0) {
        ps(increment, length); //execute prefix-sum to allocate one entry in S
        S[increment] = T[$];
}}
```

(b) **k-ary Tree Summation**

```
/* Input: N
numbers in the leaves of a k-ary tree in a 1D array
representation*
* Output: The sum of the numbers in sum[0]                        */
level = 0;
while(level < log_k(N) ) {
// process levels of tree from leaves to root
level++; /* also compute current_level_start and end_index */
    spawn(current_level_start_index, current_level_end_index) {
        int count, local_sum=0;
        for(count = 0; count < k; count++)
            temp_sum += sum[k * $ + count + 1];
        sum[$] = local_sum;
} }
```

(c) **k-ary Tree Prefix-Sums**

```
/* Input: N
numbers in sum[0..N-1]                                            *
                                                                  *
* Output: the prefix-sums of the numbers in                       *
* prefix_sum[offset_to_1st_leaf..offset_to_1st_leaf+N-1]          *
* The prefix_sum array is a 1D complete tree representation (See Summation)  */
kary_tree_summation(sum); // run k-ary tree summation algorithm
prefix_sum[0] = 0; level = log_k(N); while(level > 0) { // all
levels from root to leaves
    spawn(current_level_start_index, current_level_end_index) {
        int count, local_ps = prefix_sum[$];
        for(count = 0; count < k; count++) {
            prefix_sum[k*$ + count + 1] = local_ps;
            local_ps += sum[k*$ + count + 1]; }
    }
    level--; /* also compute current_level_start and end_index */
}
```

(d) **Breadth-First Search**

```
/* Input: Graph
G=(E,V) using adjacency lists (See Programming BFS section)       *
* Output: distance[N] - distance from start vertex for each vertex *
* Uses: level[L][N] - sets of vertices at each BFS level.          */
//run prefix sums on degrees to determine position of start edge for each vertex
start_edge = kary_prefix_sums(degrees); level[0]=start_node; i=0;
while (level[i] not empty) {
    spawn(0,level_size[i] - 1) { // start one thread for each vertex in level[i]
        v = level[i][$]; // read one vertex
        spawn(0,degree[v]-1) {    // start one thread for each edge of each vertex
            int w = edges[start_edge[v]+$][2]; // read one edge (v,w)
            psm(gatekeeper[w],1);//check the gatekeeper of the end-vertex w
            if gakeeper[w] was 0 {
                psm(level_size[i+1],1);//allocate one entry in level[i+1]
                store w in level[i+1]; }
    } }    // join
    i++; }
```

(e) **Sparse Matrix—Dense Vector Multiplication**

```
/* Input: Vector
b[n], sparse matrix A[m][n] given in Compact Sparse Row form      *
* (See Sparse Matrix - Dense Vector Multiplication section)       *
* Output: Vector c[m] = A*b                                       */
spawn(0,m) {  // start one thread for each row in A
    int row_start=row[$], elements_on_row = row[$+1]-row_start;
    spawn(0,elements_on_row-1) {//start one thread for each non-zero element on row
        // compute A[i][j] * b[j] for all non-zero elements on current row
        tmpsum[$]=values[row_start+$]*b[columns[row_start+$]];
    }
    c[$] = kary_tree_summation(tmpsum[0..elts_on_row-1]); // sum up
}
```

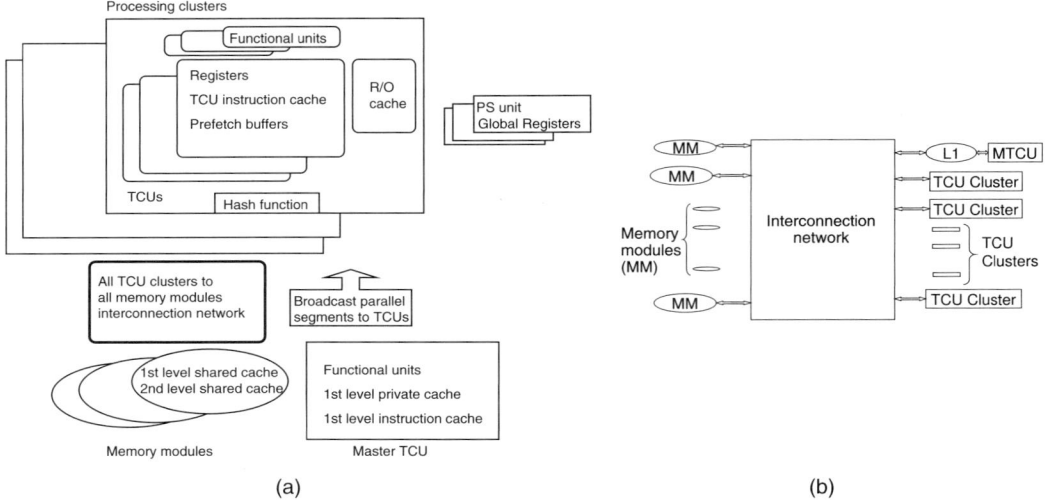

FIGURE 5.3 (a) An overview of the XMT architecture. (b) Block diagram of memories, TCUs and interconnection network.

Possible Architecture Choices   A bird eye's view of a slightly more recent version of XMT is presented in Figure 5.3. A number of (say 1024) TCUs are grouped into (say 64) clusters. Clusters are connected to the memory subsystem by a high-throughput, low-latency interconnection network, (see e.g., Reference 7: they also interface with specialized units such as prefix-sum unit and global registers. A hash function is applied to memory addresses to provide better load balancing at the shared memory modules. An important component of a cluster is the read-only cache; this is used to store values read from memory by all TCUs in the cluster. In addition, TCUs have prefetch buffers to hold values read by special prefetch instructions. The memory system consists of memory modules each having several levels of cache memories. In general each logical memory address can reside in only one memory module, alleviating cache coherence problems. This connects to including only read-only caches at the clusters. The master TCU runs serial code, or the serial mode for XMT and is the only TCU that has a local writeable cache. When it hits a spawn command it initiates a parallel mode by broadcasting the same SPMD parallel code segment to all the TCUs. As each TCU captures its copy, it executes it based on a thread-id assigned to it. A separate distributed hardware system, reported in Reference 47 but not shown in Figure 5.3, ensures that all the thread id's mandated by the current Spawn command are allocated to the TCUs. A sufficient part of this allocation is done dynamically to ensure that no TCU needs to execute more than one thread id, once another TCU is already idle.

Possible Role of Compiler   To take advantage of features of the architecture, an XMTC program needs to be translated by an optimizing compiler. In the execution model, a program could include (i) Prefetch instructions to bring data from the lower memory hierarch levels either into the shared caches or into the prefetch buffers located at the TCUs; (ii) Broadcast instructions, where some values needed by all, or nearly all TCUs, are broadcasted to all; (iii) Thread clustering: combining shorter virtual threads into a longer thread; and (iv) If the programming model allows nested parallelism, the compiler will use the mechanisms supported by the architecture to implement or emulate it. Compiler optimizations and issues such as nesting and thread clustering are discussed in Section 5.6.

Performance Modeling   The performance modeling of a program first extends the known PRAM notions of *work* and *depth*. Later, a formula for estimating execution time based on these extensions is provided.

The depth of an application in the XMT Execution model must include the following three quantities: (i) *Computation Depth*, given by the number of operations that have to be performed sequentially, either by a thread or while in serial mode. (ii) *Length of Sequence of Round-Trips to Memory (or LSRTM)*, which represents the number of cycles on the critical path spent by execution units waiting for data from memory. For example, a read request or prefix-sum instruction from a TCU usually causes a round-trip to memory (or RTM); memory writes in general proceed without acknowledgment, thus not being counted as round-trips, but ending a parallel section implies one RTM used to flush all the data still in the interconnection network to the memory. (iii) *Queuing delay (or QD)*, which is caused by concurrent requests to the same memory location; the response time is proportional to the size of the queue.

A prefix-sum (ps) statement is supported by a special hardware unit that combines calls from multiple threads into a single multioperand prefix-sum operation. Therefore, a ps statement to the same base over several concurrent threads causes one roundtrip through the interconnection network to the global register file (1 RTM) and 0 queuing delay, in each of the threads.

The syntax of a prefix-sum to memory (psm) statement is similar to ps except the base variable is a memory location instead of a global register. As updates to the memory location are queued, the psm statement costs 1 RTM and additionally has a queuing delay (QD) reflecting the plurality of concurrent threads executing psm with respect to the same memory location.

We can now define the XMT "execution depth" and "execution time." XMT execution depth represents the time spent on the "critical path" (i.e., the time assuming unlimited amount of hardware) and is the sum of the computation depth, LSRTM, and QD on the critical path. Assuming that a round-trip to memory takes $\mathcal{R}$ cycles:

$$\text{Execution depth} = \text{Computation depth} + \text{LSRTM} \times \mathcal{R} + \text{QD} \tag{5.1}$$

Sometimes, a step in the application contains more *Work* (the total number of instructions executed) to be executed in parallel than what the hardware can handle concurrently. For the additional time spent executing operations outside the critical path (i.e., beyond the execution depth), the work of each parallel section needs to be considered separately. Suppose that one such parallel section could employ in parallel up to p_i TCUs. Let $Work_i = p_i * ComputationDepth_i$ be the total computation work of parallel section i. If our architecture has p TCUs and $p_i < p$, we will be able to use only p_i of them, while if $p_i \geq p$, only p TCUs can be used to start the threads, and the remaining $p_i - p$ threads will be allocated to TCUs as they become available; each concurrent allocation of p threads to p TCUs is charged as one RTM to the execution time, as per Equation 5.2. The total time spent executing instructions outside the critical path over all parallel sections is given in Equation 5.3.

$$\text{Thread start overhead}_i = \left\lceil \frac{p_i - p}{p} \right\rceil \times \mathcal{R} \tag{5.2}$$

$$\text{Time additional work} = \sum_{\text{spawn block } i} \left(\frac{Work_i}{\min(p, p_i)} + Threadstartoverhead_i \right) \tag{5.3}$$

In the last equation we do not subtract the quantity that is already counted as part of the execution depth. The reason is that our objective in the current chapter is limited to extending the work-depth upper bound $T(n) + \lfloor \frac{W(n)}{p} \rfloor$, and such double count is possible in that original upper bound as well. Adding up, the execution time of the entire program is:

$$\text{Execution time} = \text{Execution depth} + \text{Time additional work} \tag{5.4}$$

5.2.5 Clarifications of the Modeling

The current chapter provides performance modeling that allows weighing alternative implementations of the same algorithm where asymptotic analysis alone is insufficient. Namely, a more refined measure than the asymptotic number of parallel memory accesses was needed.

Next, we point out a somewhat subtle point. Following the path from the HLWD model to the XMT models in Figure 5.1 may be important for optimizing performance, and not only for the purpose of developing a XMT program. Bandwidth is not accounted for in the XMT performance modeling, since the on-chip XMT architecture should be able to provide sufficient bandwidth for an efficient algorithm in the work-depth model. The only way in which our modeling accounts for bandwidth is indirect: by first screening an algorithm through the work-depth performance modeling, where we account for work. Now, consider what could have happened had XMT performance modeling not been coupled with work-time performance modeling. The program could include excessive speculative prefetching to supposedly improve performance (reduce LSRTM). The subtle point is that the extra prefetches add to the overall work count. Accounting for them in the Work-Depth model prevents this "loophole."

It is also important to recognize that the model abstracts away some significant details. The XMT hardware has a limited number of memory modules. If multiple requests attempt to access the same module, queuing will occur. While accounting for queuing to the same memory location, the model does not account for queuing accesses to different locations in the same module. Note that hashing memory addresses among modules lessens problems that would occur for accesses with high spatial locality and generally mitigates this type of "hot spots." If functional units within a cluster are shared between the TCUs, threads can be delayed while waiting for functional units to become available. The model also does not account for these delays.

Our modeling is a first-order approximation of run time. Such analytic results are not a substitute for experimental results, since the latter will not be subject to the approximations described earlier. In fact, some experimental results as well as a comparison between modeling and simulations are discussed in Section 5.4.

Similar to some serial performance modeling, the above modeling assumes that data is found in the (shared) caches. This allows proper comparison of serial computing where data is found in the cache, as the number of clocks to reach the cache for XMT is assumed to be significantly higher than in serial computing; for example, our prototype XMT architecture suggests values that range between 6 and 24 cycles for a round-trip to the first level of cache, depending on the characteristics of the interconnection network and its load level; we took the conservative approach to use the value $\mathcal{R} = 24$ cycles for one RTM for the rest of this chapter. We expect that the number of clocks to access main memory should be similar to serial computing and that both for serial computing and for XMT large caches will be built. However, this modeling is inappropriate, if XMT is to be compared to Cray MTA where no shared caches are used for the MTA the number of clocks to access main memory is important and for a true comparison this figure for cache misses on XMT would have to be included as well.

As pointed out earlier, some of the computation work is counted twice in our execution time, once as part of the critical path under execution depth and once in the additional work factor. Future work by us, or others, could refine our analysis into a more accurate model, but with much more involved formulae. In the current chapter, we made the choice to stop at this level of detail allowing a concise presentation while still providing relevant results.

Most other researchers that worked on performance modeling of parallel algorithms were concerned with other platforms. They focused on different factors than us. Helman and JáJá [32] measured the complexity of algorithms running on SMPs using the triplet of maximum number of noncontiguous accesses by any processor to main memory, number of barrier synchronizations, and local computation cost. These quantities are less important in a PRAM-like environment. Bader et al. [6] found that in some experiments on the Cray MTA, the costs of noncontiguous memory access and barrier synchronization were reduced almost to zero by multithreading and that performance was best modeled by computation alone. For the latest generation of the MTA architecture, a calculator for performance that

Advancing PRAM Algorithms into Parallel Programs

includes the parameters of count of trips to memory, number of instructions, and number of accesses to local memory [24] was developed. The RTMs that we count are round trips to the shared cache, which is quite different, as well as queuing at the shared cache. Another significant difference is that we consider the effect of optimizations such as prefetch and thread clustering. Nevertheless, the calculator should provide an interesting basis for comparison between performance of applications on MTA and XMT. The incorporation of queuing follows the well-known QRQW PRAM model of Gibbons, Matias and Ramachandran [27]. A succinct way to summarize the modeling contribution of the current chapter is that unlike previous practice, QRQW becomes secondary, though still quite important, to LSRTM.

5.3 An Example for Using the Methodology: Summation

Consider the problem of computing in parallel the sum of N values stored in array A. A high-level work-depth description of the algorithm is as follows: in parallel add groups of k values; apply the algorithm recursively on the $\lceil N/k \rceil$ partial sums until the total sum is computed. This is equivalent to climbing (from leaves towards root) a balanced k-ary tree. An iterative description of this algorithm that fits the work-depth model can be easily derived from this. The parameter k is a function of the architecture parameters and the problem size N, and is chosen to minimize the estimated running time.

An implementation of this algorithm in the XMT programming model is presented in Table 5.1b. Note that a unidimensional array is used to store the complete k-ary tree, where the root is stored at element 0, followed by the k elements of the second level from left to right, then the k^2 elements of the second level and so on.

For each iteration of the algorithm, the k partial sums from a node's children have to be read and summed. Prefetching can be used to pipeline the memory accesses for this operation, thus requiring only one round-trip to memory (RTM). An additional RTM is needed to flush all the data to memory at the end of each step. There are no concurrent accesses to the same memory location in this algorithm, thus the queuing delay (QD) is zero. By accounting for the constant factors in our own XMTC implementation, we determined the computation depth to be $(3k + 9) \log_k N + 2k + 33$, given that $\log_k N$ iteration are needed. To compute the additional time spent executing outside the critical path (in saturated regime), we determined the computation per tree node to be $C = 3k + 2$ and the total number of nodes processed under this regime to be $Nodes_{sat}$ as in Figure 5.4. An additional step is required to copy the data into the tree's leaves at the start of the execution.

This determines the execution work, additional work an the thread start overhead terms of the XMT execution time

$$Execution\ depth = (2 \log_k N + 1) \times \mathcal{R} + (3k + 9) \log_k N + 2k + 33 \qquad (5.5)$$

$$Additional\ work = \frac{2N + (3k + 2) Nodes_{sat}}{p} + (3k + 2) \log_k p +$$

$$+ \left\lceil \frac{Nodes_{sat}}{p} - \log_k \frac{N}{p} \right\rceil \times \mathcal{R} \qquad (5.6)$$

To avoid starting too many short threads, an optimization called thread clustering can be applied either by an optimizing compiler or a performance programmer: let c be a constant; start c threads that each run a serial summation algorithm on a contiguous subarray of N/c values from the input array. Each thread writes its computed partial sum into an array B. To compute the total sum, run the parallel summation algorithm described already on the array B.

We now consider how clustering changes the execution time. $SerSum(N)$ and $ParSum(N)$ denote the execution time for serial and the parallel summation algorithms over N values, respectively. The serial

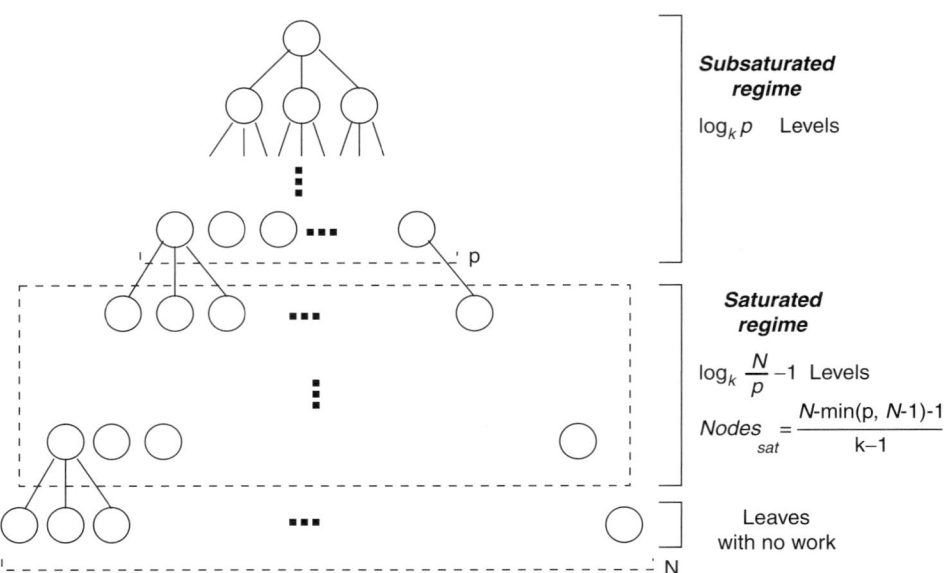

FIGURE 5.4 The $\log_k p$ levels of the tree closest to the root are processed in subsaturated regime, that is there is not enough parallelism in the application to run all the TCUs. The next $\log(N/p) - 1$ levels have more parallelism than the hardware can execute in parallel and some TCUs will run more than one thread (saturated regime). The computation starts at the first level above the leaves.

algorithm loops over N elements and, by using prefetching always have the next value available before it is needed, it has a running time of $SerSum(N) = 2N + 1 \times \mathcal{R}$.

The first part of the algorithm uses a total of $N - c$ additions evenly divided among p processors, while the second requires the parallel summation to be applied on an input of size c. This gives an execution time for the clustered algorithm of

$$\text{Execution time} = SerSum\left(\frac{N-c}{p}\right) + ParSum(c) \qquad (5.7)$$

The value of c, where $p \leq c \leq N$, that minimizes the execution time determines the best crossover point for clustering. Suppose $p = 1024$. To allow numerical comparison, we need to assign a value to $\mathcal{R}$, the number of cycles in one RTM. As noted in Section 5.2.5, for the prototype XMT architecture this value is upper bounded by 24 cycles under the assumption that the data is already in the on-chip cache and there is no queuing in the interconnection network or memory.

Since all the clustered threads are equal in length and in the XMT execution model they run at the same speed, we found, that when $N \gg p$, the optimal value for c is 1024.

The optimum value for k can be determined by minimizing execution time for a fixed N. For the interesting case when $N \geq p$ (where $p = 1024$), the parallel summation algorithm is only run on $c = 1024$ elements and in this case we found (see Figure 5.5.a) that $k = 8$ is optimal.

5.4 Empirical Validation of the Performance Model

Some empirical validation of the analytic performance model of the earlier sections are presented in this section. Given an XMTC program, estimated run-times using the analytic model are compared against simulated run-times using our XMT simulator. Derived from a synthesizable gate-level description of

Advancing PRAM Algorithms into Parallel Programs

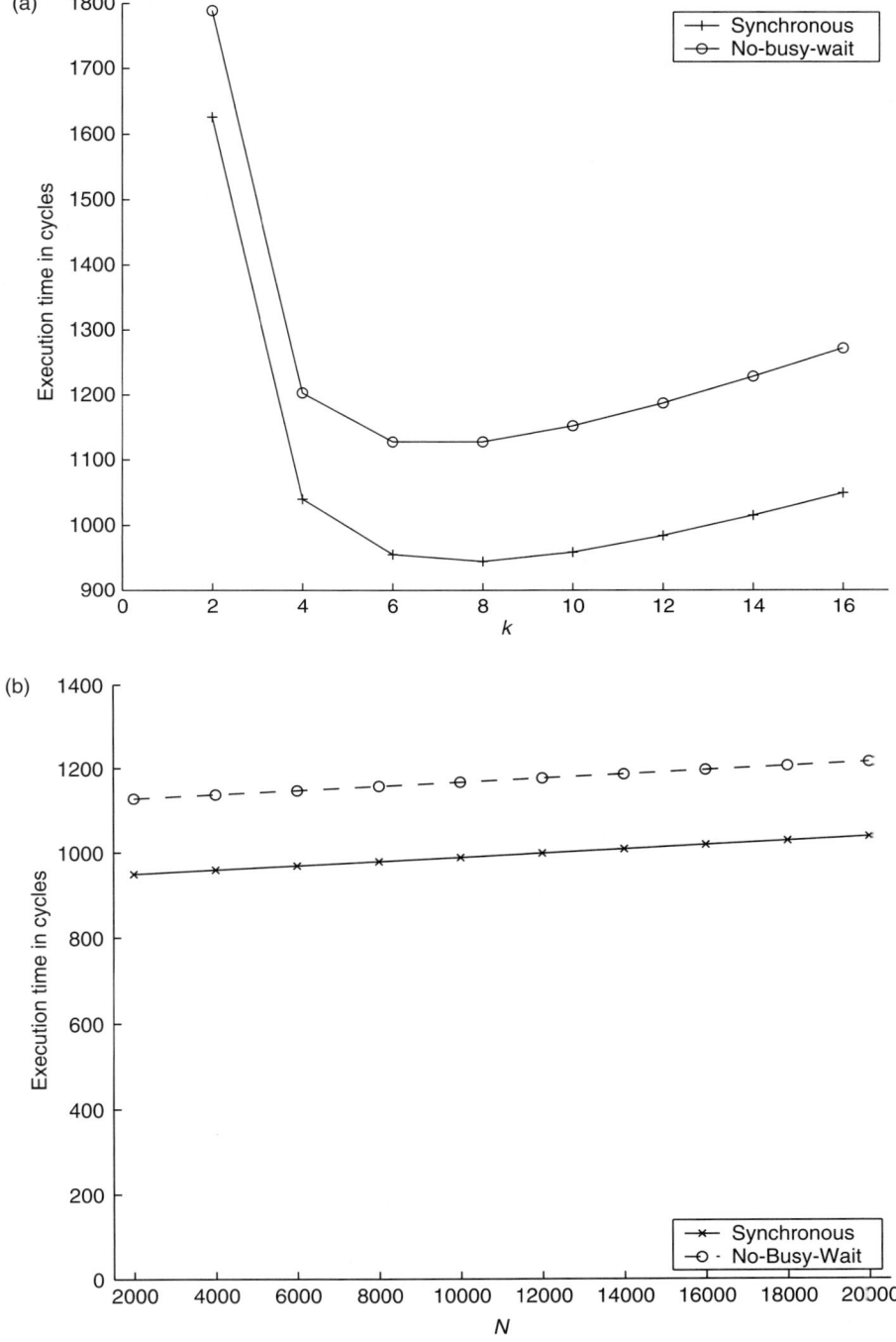

FIGURE 5.5 Estimated run times obtained using the analytic model: (a) Determining the optimum arity of the tree k for the two implementations of the prefix-sums algorithm for $N = 1024$. (b) Execution times for the two implementations of the k-ary prefix-sums algorithms. The optimum k is chosen for each case.

the XMT architecture, the XMT simulation engine aims at being cycle-accurate. A typical configuration includes 1024 thread control units (TCUs) grouped in 64 clusters and one Master TCU.

A gap between the simulations and the analytic model is to be expected. The analytic model as presented not only makes some simplifying assumptions, such as counting each XMTC statement as one cycle, and ignoring contention at the functional units, but also provides the same (worst-case) run-time estimates for different input data as long as the input has the same size. On the other hand, at the present time the XMTC compiler and the XMT cycle-accurate simulator lack a number of features that were assumed for the analytic performance model. More specifically, there is no compiler support for prefetching and thread clustering, and only limited broadcasting capabilities are included. (Also, a sleep-wait mechanism noted in Section 5.6.1 is not fully implemented in the simulator, making the single-spawning mechanism less efficient.) All these factors could cause the simulated run-times to be higher than the ones computed by the analytic approach. For that reason, we limit our focus to just studying the relative performance of two or more implementations that solve the same problem. This will enable evaluating programming implementation approaches against the relative performance gain—one of the goals of the full paper.

More information about the XMTC compiler and the XMT cycle-accurate simulator can be found in Reference 8. An updated version of the document which reflects the most recent version of the XMT compiler and simulator is available from http://www.umiacs.umd.edu/users/vishkin/XMT/XMTCManual.pdf and http://www.umiacs.umd.edu/users/vishkin/XMT/XMTCTutorial.pdf.

We proceed to describe our experimental results.

5.4.1 Performance Comparison of Implementations

The first experiment consisted of running the implementations of the Prefix-Sums [41,58] and the parallel breadth-first search algorithms discussed in Sections 5.7 and 5.8 using the XMT simulator.

Figure 5.6a presents the results reported for k-ary prefix-sums ($k = 2$) by the XMT cycle-accurate simulator for input sizes in the range of 2,000–20,000. The results show the synchronous program outperforming the no-busy-wait program and the execution times increasing linearly with N. This is in agreement with the analytic model results in Figure 5.5. For lack of space, we had to defer the text leading to that figure to Section 5.7.

However, there are some differences, as well. For example, the larger gap that the simulator shows between the synchronous program and no-busy-wait execution times. This can be explained by the relatively large overheads of the single-spawn instruction, which will be mitigated in future single-spawn implementations by mechanisms such as sleep-wait.

Figure 5.6b depicts the number of cycles obtained by running two XMTC implementations of the BFS algorithm, single-spawn and flattened, on the simulator. The results show the execution times for only one iteration of the BFS algorithm, that is, building BFS tree level $L(i + 1)$ given level $L(i)$. To generate the graphs, we pick a value M and choose the degrees of the vertices uniformly at random from the range $[M/2, 3M/2]$, which gives a total number of edges traversed of $|E(i)| = M * N(i)$ on average. Only the total number of edges is shown in Figure 5.7. Another arbitrary choice was to set $N(i + 1) = N(i)$, which gives gatekeeper queuing delay (GQD) of $\frac{|E(i)|}{N(i)} = M$ on average.

The number of vertices in levels $L(i)$ and $L(i + 1)$ was set to 500. By varying the average degree per vertex M, we generated graphs with the expected number of edges between $L(i)$ and $L(i + 1)$ in the range [12,500..125,000]. We observe that the single-spawn implementation outperforms the Flattened BFS for smaller problem sizes, but the smaller work factor makes the latter run faster when the number of edges traversed increases above a certain threshold.

By comparing these experimental results with the outcome of the analysis presented in Figure 5.7, we observe the same performance ranking between the different implementations providing a second example where the outcome of the analytic performance model is consistent with the cycle-accurate simulations. The text supporting Figure 5.7 is in Section 5.9.4.

Advancing PRAM Algorithms into Parallel Programs

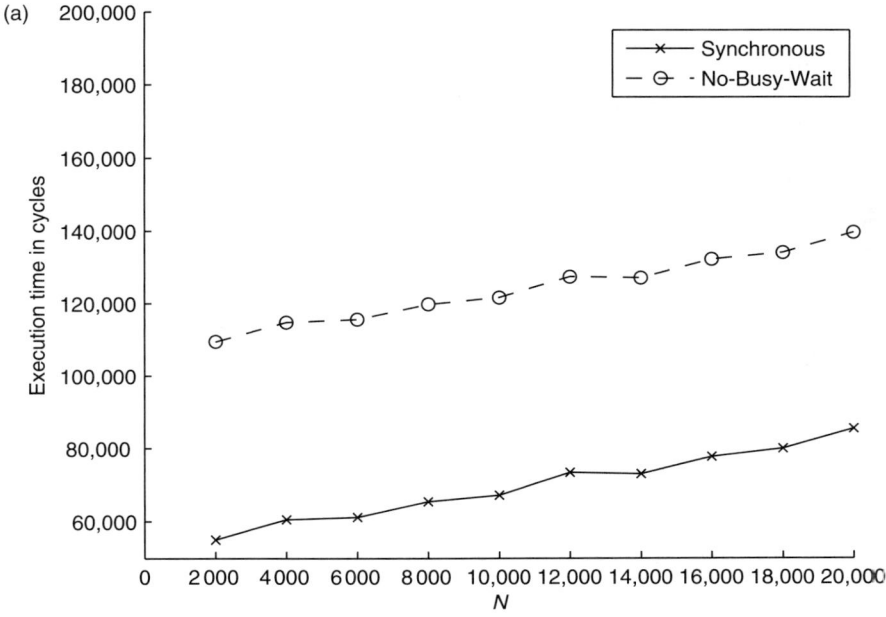

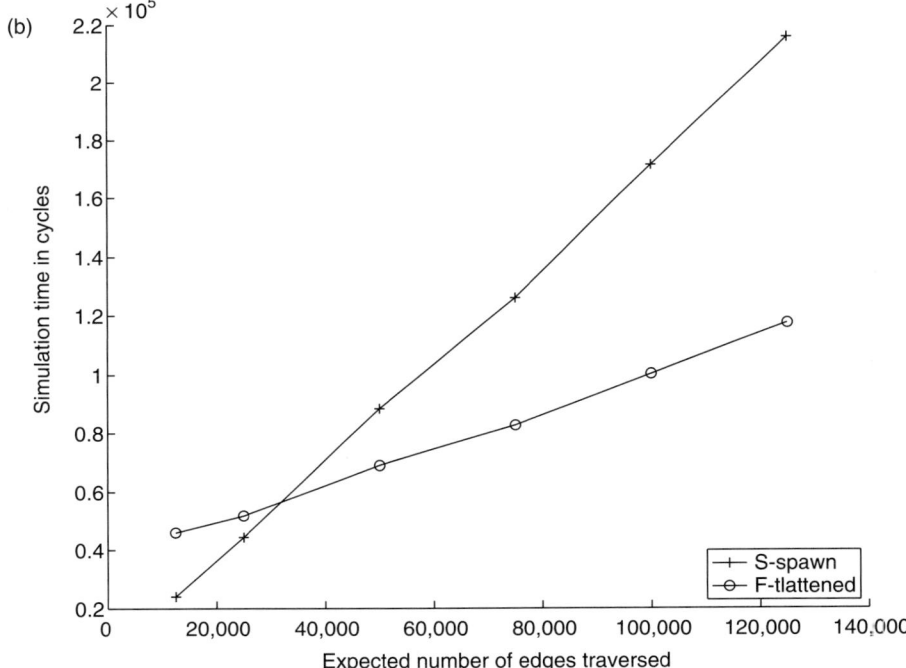

FIGURE 5.6 Cycle counts reported by the XMT cycle-accurate simulator. (a) Synchronous and no-busy-wait prefix-sums; (b) Flattened and single-spawn BFS.

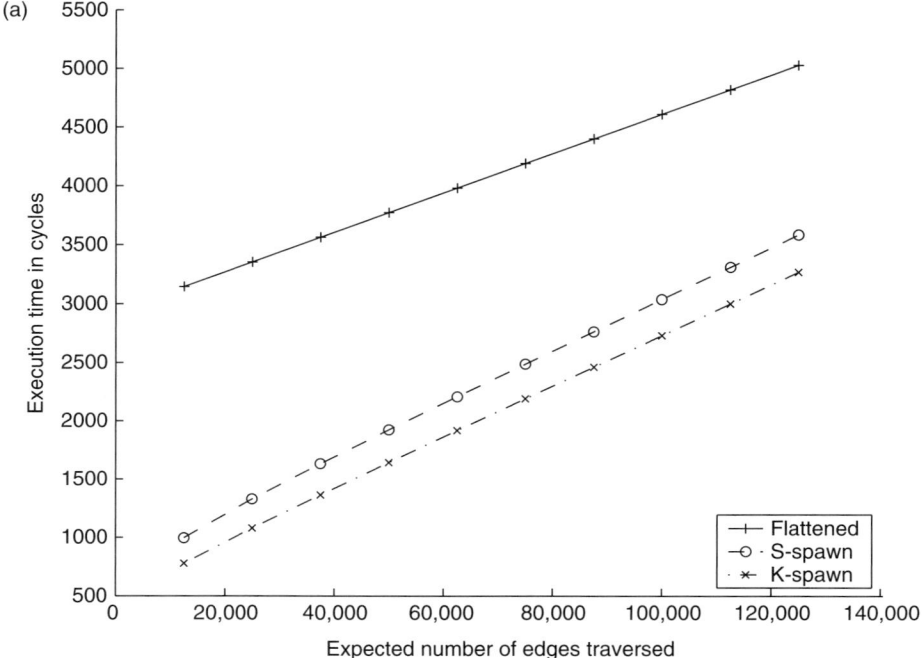

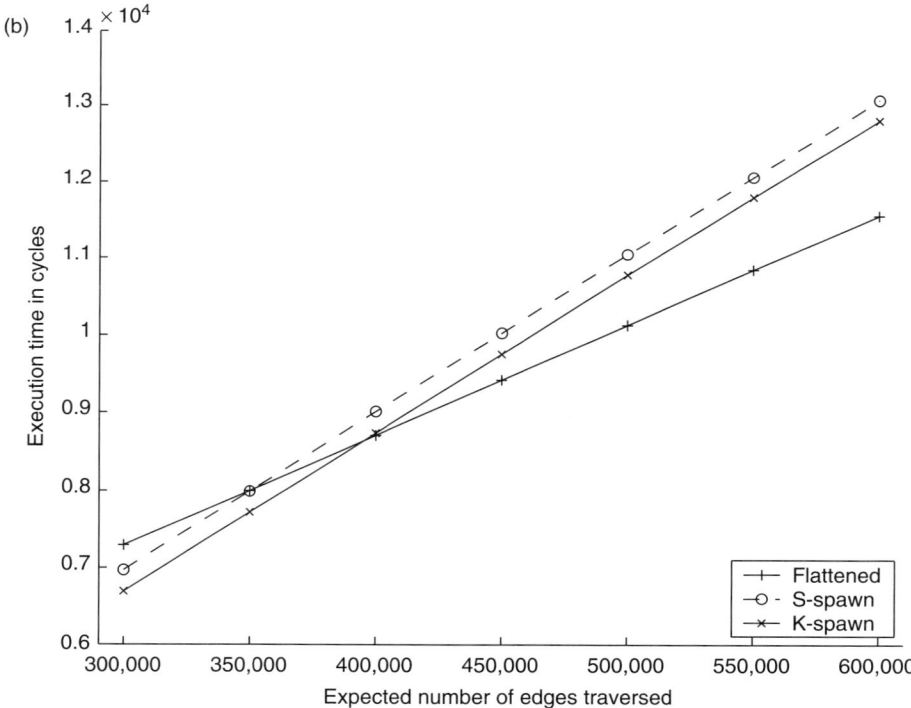

FIGURE 5.7 Analytic execution times for one iteration of BFS when the number of vertices at current level is (a) $N(i) = 500$ and (b) $N(i) = 2000$. The optimal value for k was calculated for each dataset.

TABLE 5.2 Scalability of Algorithms with the Number of Parallel TCUs

Algorithm	Method	16 TCUs	32 TCUs	128 TCUs	512 TCUs	1024 TCUs
Matvec	Simulation	11.23	22.61	90.32	347.64	646.97
	Analytic	11.42	22.79	84.34	306.15	548.92
	Difference(%)	1.67	0.84	6.63	11.94	15.16

5.4.2 Scaling with the Architecture

In this section, we report experiments with the scalability of implementations relative to the number of parallel processing units physically available. Both the XMT simulator and our analytic model are fully parametrized, allowing us to obtain running times for many XMT configurations.

For both analytic and experimental scaling results, we computed the ratio of running times when using one single TCU divided by the time when using 16, 32, 128, 512, or 1024 TCUs. Note that such ratio cancels out constant factors present in the two approaches.

We applied this method to the parallel sparse matrix—dense vector multiplication (or Matvec) algorithm [24, 51] presented in Section 5.12. Results are presented in Table 5.2. The fact that we were able to obtain analytic results that are under 1% different from the simulation cycle-counts for some configurations suggests that the dry analysis can be quite accurate. When the number of TCUs was increased, the larger discrepancy (up to 15.16% for a "full" 1024-TCU XMT configuration) can be explained by the fact that in the analytic model the queing delay is computed under worst-case assumptions, that is proportional to the number of TCUs, while in the simulator the threads are running at different speeds, causing the read requests to spread in time and imply less queuing.

It is worth noting that when scaling down the architecture, the interconnection network becomes simpler, causing the time for a round-trip to first level of cache to decrease; this factor was reflected accordingly in the analytic model.

5.5 Conclusion of Extended Summary

While innovation in education is often not a focus for technical research papers, it is worth noting that this chapter provides the missing link for upgrading a standard theoretical PRAM algorithms class to a parallel algorithms and programming class. Specifically, the programming assignments in a one-semester parallel algorithms class taught in 2006 at the University of Maryland included parallel MATVEC, general deterministic (Bitonic) sort, sample sort, BFS on graphs and parallel graph connectivity. A first class on serial algorithms coupled with serial programming typically does not require more demanding programming assignments. This is evidence that the PRAM theory coupled with XMT programming could be on par with serial algorithms and programming. An important purpose of the full chapter is to provide an initial draft for augmenting a PRAM algorithms textbook with an understanding of how to effectively program an XMT computer system to allow such teaching elsewhere. On the tool-chain side, the following progress has been made (i) a much more stable compiler has been developed [56]; (ii) from a synthesizable simulator written in the Verilog hardware description language, a cycle-accurate simulator in Java was derived; while more portable and not as slow, such simulators tend to still be rather slow; and (iii) a 75 MHz 64-TCU computer based on an FPGA programmable hardware platform has been built using a board comprising 3 Xilinx FPGA chips [61]. This XMT computer is tens of thousands times faster than the simulator. The XMT computer could potentially have a profound psychological effect by asserting that XMT is real and not just a theoretical dream. However, it is also important to realize that for some computer architecture aspects (e.g., relative speed of processors and main memory) a simulator may actually better reflect the anticipated performance of hard-wired (ASIC) hardware that will hopefully be built in the near future.

More generally (i) The paper suggests that given a proper chip multiprocessor design, parallel programming in the emerging multicore era does not need to be significantly more difficult than serial programming. If true, a powerful answer to the so-called parallel programming open problem is being provided. This open problem is currently the main stumbling block for the industry in getting the upcoming generation of multicore architectures to improve single task completion time using easy-to-program application programmer interfaces. Known constraints of this open problem, such as backwards compatibility on serial code, are also addressed by the overall approach. (ii) The chapter seeks to shed new light on some of the more robust aspects of the new world of chip-multiprocessing as we have envisioned it, and open new opportunities for teaching, research and practice. This work-in-progress is a significant chapter in an unfolding story.

Part II: Discussion, Examples, and Extensions
5.6 Compiler Optimizations

Given a program in the XMT programming model, an optimizing compiler can perform various transformations on it to better fit the target XMT execution model and reduce execution time. We describe several possible optimizations and demonstrate their effect using the summation algorithm described earlier.

5.6.1 Nested Parallel Sections

Quite a few PRAM algorithms can be expressed with greater clarity and conciseness when nested parallelism is allowed [10]. For this reason, it is desired to allow nesting parallel sections with arbitrary numbers of threads in the XMT programming model. Unfortunately, hardware implementation of nesting comes at a cost. The performance programmer needs to be aware of the implementation overheads. To explain a key implementation problem we need to review the *hardware mechanism that allocates threads to the p physical TCUs*. Consider an SMPD parallel code section that starts with a spawn(1,n) command, and each of the n threads ends with a join command without any nested spawns. As noted earlier, the master TCU broadcasts the parallel code section to all p TCUs. In addition it broadcasts the number n to all TCUs. TCU i, $1 \leq i \leq p$, will check whether $i > n$, and if not it will execute thread i; once TCU i hits a join, it will execute a special "system" ps instruction with an increment of 1 relative to a counter that includes the number of threads started so far; denote the result it gets back by j; if $j > n$ TCU i is done; if not TCU i executes thread j; this process is repeated each time a TCU hits a join until all TCUs are done, when a transition back into serial mode occurs.

Allowing the nesting of spawn blocks would require (i) Upgrading this thread allocation mechanism; the number n representing the total number of threads will be repeatedly updated and broadcast to the TCUs. (ii) Facilitating a way for the parent (spawning) thread to pass initialization data to each child (spawned) thread.

In our prototype XMT programming model, we allow nested spawns of a small fixed number of threads through the single-spawn and k-spawn instructions; sspawn starts one single additional thread while kspawn starts exactly k threads, where k is a small constant (such as 2 or 4). Each of these instructions causes a delay of one RTM before the parent can proceed, and an additional delay of 1-2 RTMs before the child thread can proceed (or actually get started). Note that the threads created with a sspawn or kspawn command will execute the same code as the original threads, starting with the first instruction in the current parallel section. Suppose that a parent thread wants to create another thread whose virtual thread number (as referenced from the SPMD code) is v. First, the parent uses a system ps instruction to a global thread-counter register to create a unique thread ID i for the child. The parent then enters the value v in $A(i)$, where A is a specially designated array in memory. As a result of executing an sspawn or kspawn command by the parent thread (i) n will be incremented, and at some point in the future (ii) the thread allocation mechanism will generate virtual thread i. The program for thread i starts with reading v through $A(i)$. It then can be programmed to use v as its "effective" thread ID.

An algorithm that could benefit from nested spawns is the BFS algorithm. Each iteration of the algorithm takes as input L_{i-1}, the set of vertices whose distance from starting vertex s is $i-1$ and outputs L_i. As noted in Section 5.2.2, a simple way to do this is to spawn one thread for each vertex in L_{i-1}, and have each thread spawn as many threads as the number of its edges.

In the BFS example, the parent thread needs to pass information, such as which edge to traverse, to child threads. To pass data to the child, the parent writes data in memory at locations indexed by the child's ID, using nonblocking writes (namely, the parent sends out a write request, and can proceed immediately to its next instruction without waiting for any confirmation). Since it is possible that the child tries to read this data before it is available, it should be possible to recognize that the data is not yet there and wait until it is committed to memory. One possible solution for that is described in the next paragraph. The kspawn instruction uses a prefix-sum instruction with increment k to get k thread IDs and proceeds, similarly; the delays on the parent and children threads are similar, though a few additional cycles being required for the parent to initialize the data for all k children.

When starting threads using single-spawn or k-spawn, a synchronization step between the parent and the child is necessary to ensure the proper initialization of the latter. Since we would rather not use a "busy-wait" synchronization technique that could overload the interconnection network and waste power, our envisioned XMT architecture would include a special primitive, called *sleep-waiting*: the memory system holds the read request from the child thread until the data is actually committed by the parent thread, and only then satisfies the request.

When advancing from the programming to the execution model, a compiler can automatically transform a nested spawn of n threads, and n can be any number, into a recursive application of single-spawns (or k-spawns). The recursive application divides much of the task of spawning n thread among the newly spawned threads. When a thread starts a new child, it assigns to it half (or $1/(k+1)$ for k-spawn) of the $n-1$ remaining threads that need to be spawned. This process proceeds in a recursive manner.

5.6.2 Clustering

The XMT programming model allows spawning an arbitrary number of virtual threads, but the architecture has only a limited number of TCUs to run these threads. In the progression from the programming model to the execution model, we often need to make a choice between two options: (i) spawn fewer threads each effectively executing several shorter threads, and (ii) run the shorter threads as is. Combining short threads into a longer thread is called clustering and offers several advantages: (a) *reduce RTMs and QDs*: we can pipeline memory accesses that had previously been in separate threads; this can reduce extra costs from serialization of RTMs and QDs that are not on the critical path; (b) *reduce thread initiation overhead*: spawning fewer threads means reducing thread initiation overheads, that is the time required to start a new thread on a recently freed TCU; (c) *reduce memory needed*: each spawned thread (even those that are waiting for a TCU) usually takes up space in the system memory to store initialization and local data.

Note that if the code provides fewer threads than the hardware can support, there are few advantages if any to using fewer longer threads. Also, running fewer, longer threads can adversely affect the automatic load balancing mechanism. Thus, as discussed later, the granularity of the clustering is an issue that needs to be addressed.

In some cases, clustering can be used to group the work of several threads and execute this work using a serial algorithm. For example, in the summation algorithm the elements of the input array are placed in the leaves of a k-ary tree, and the algorithm climbs the tree computing for each node the sum of its children. However, we can instead start with an embarrassingly parallel algorithm in which we spawn p threads that each serially sum N/p elements and then sum the p sums using the parallel summation algorithm. See Figure 5.8.

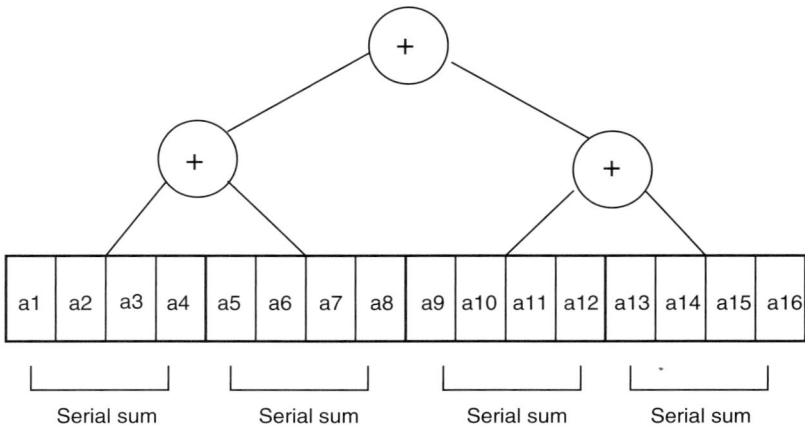

FIGURE 5.8 The sums algorithm with thread clustering.

With such switch to a serial algorithm, clustering is nothing more than a special case of the accelerating cascades technique [16]. For applying accelerating cascades, two algorithms that solve the same problem are used. One of the algorithms is slower than the other, but requires less work. If the slower algorithm progresses in iterations where each iteration reduces the size of the problem considered, the two algorithms can be assembled into a single algorithm for the original problem as follows: (i) start with the slower algorithm and (ii) switch to the faster one once the input size is below some threshold. This often leads to faster execution than by each of the algorithms separately.

Finding the optimal crossover point between the slow (e.g., serial) and faster algorithms is needed. Also, accelerating cascades can be generalized to situations where more than two algorithms exist for the problem at hand.

Though our current compiler does not yet do that, a compiler should some day be able to do the clustering automatically. When the number of threads is known statically (i.e., where there are no nested spawns), clustering is simpler. However, even with nested spawns, our limited experience is that methods of clustering tend not to be too difficult to implement. Both cases are described below.

Clustering without Nested Spawns Suppose we want to spawn N threads, where $N \gg p$. Instead of spawning each as a separate thread, we could trivially spawn only c threads, where c is a function of the number of TCUs, and have each one complete N/c threads in a serial manner. Sometimes an alternative serial algorithm can replace running the N/c threads. Applying this mechanism can create a situation where most TCUs have already finished, but a few long threads are still running. To avoid this, shorter threads can be ran as execution progresses toward completion of the parallel section [47].

Clustering for single-spawn and k-spawn In the hardware, the number of current virtual threads (either running or waiting) is broadcasted to TCUs as it is updated. Assuming some system threshold, each running thread can determine whether the number of (virtual) threads scheduled to run is within a certain range. When a single-spawn is encountered, if below the threshold, the single-spawn is executed; otherwise, the thread enters a temporary spawning suspension mode and continues execution of the original thread; the thread will complete its own work and can also serially do the work of the threads whose spawning it has suspended. However, the suspension decision can be revoked once the number of threads falls below a threshold. If that occurs, then a new thread is single-spawned. Often, half the remaining work is delegated to the new thread. Clustering with k-spawn is similar. If several threads complete at the same time it will take some time to reach p running threads again, causing a gap in parallel hardware usage. This can be avoided by having a larger threshold, which would keep a set of threads ready to be started as soon as hardware becomes available. Related implementation considerations have been reported by the MIT Cilk project. As no original insights are claimed here, we avoid for brevity a literature review.

5.6.3 Prefetching

Special data prefetch instructions can be used to issue read requests for data values before they are needed; this can prevent long waits due to memory and interconnection network latencies. Prefetched values are stored in read-only buffers at the cluster level. On the basis of our experiments with different applications, the interconnection network between TCUs and memory is expected to be powerful enough to serve all read requests but perhaps not all prefetch requests. In particular, this suggests avoiding speculative prefetches.

Advanced prefetch capabilities are supported by modern serial compilers and architectures, and the parallel domain is adopting them as well. Prefetching has been demonstrated to improve performance on SMPs [25, 55]. Pai and Adve [48] advocate both grouping read misses and using prefetch. Our approach builds on these results, using thread clustering to group large numbers of read requests, and possibly prefetching them as well. Grouping read requests allows overlapping memory latencies.

5.7 Prefix-Sums

Computing the prefix-sums for n values is a basic routine underlying many parallel algorithms. Given an array $A[0..n-1]$ as input, let $prefix\_sum[j] = \sum_{i=0}^{j-1} A[i]$ for j between 1 and n and $prefix\_sum[0] = 0$. Two prefix-sums implementation approaches are presented and compared. The first algorithm considered is closely tied to the synchronous (textbook) PRAM prefix-sums algorithm while the second one uses a no-busy-wait paradigm [58]. The main purpose of the current section is to demonstrate designs of efficient XMT implementation and the reasoning that such a design may require. It is perhaps a strength of the modeling in the current chapter that it provides a common platform for evaluating rather different algorithms. Interestingly enough, our analysis suggests that when it comes to addressing the most time consuming elements in the computation, they are actually quite similar.

Owing to Reference 41, the basic routine works in two stages each taking $O(\log n)$ time. The first stage is the summation algorithm presented earlier, namely, the computation advances up a balanced tree computing sums. The second stage advances from root to leaves. Each internal node has a value $C(i)$, where $C(i)$ is the prefix-sum of its rightmost descendant leaf. The $C(i)$ value of the root is the sum computed in the first stage, and the $C(i)$ for other nodes is computed recursively. Assuming that the tree is binary, any right child inherits the $C(i)$ value from its parent, and any left child takes $C(i)$ equal to the $C(i)$ of its left uncle plus this child's value of sum. The values of $C(i)$ for the leaves are the desired prefix-sums. See Figure 5.9a.

5.7.1 Synchronous Prefix-Sums

The implementation of this algorithm in the XMT programming model is presented in Table 5.1c using XMTC pseudocode. Similar to the summation algorithm, we use a k-ary tree instead of a binary one. The two overlapped k-ary trees are stored using two one-dimensional arrays sum and $prefix\_sum$ by using the array representation of a complete tree as discussed in Section 5.3.

The XMT algorithm works by first advancing up the tree using a summation algorithm. Then the algorithm advances down the tree to fill in the array $prefix\_sum$. The value of $prefix\_sum$ is defined as follows: (i) for a leaf, $prefix\_sum$ is the prefix-sum and (ii) for an internal node, $prefix\_sum$ is the prefix-sum for its leftmost descendant leaf (not the rightmost descendant leaf as in the PRAM algorithm—this is a small detail that turns out to make things easier for generalizing from binary to k-ary trees).

Analysis of Synchronous Prefix-Sums We analyze the algorithm in the XMT execution model. The algorithm has 2 round-trips to memory for each level going up the tree. One is to read sum from the children of a node, done in one RTM by prefetching all needed values in one round-trip. The other is to join the spawn at the current level. Symmetrically, there are 2 RTMs for each level going down the tree.

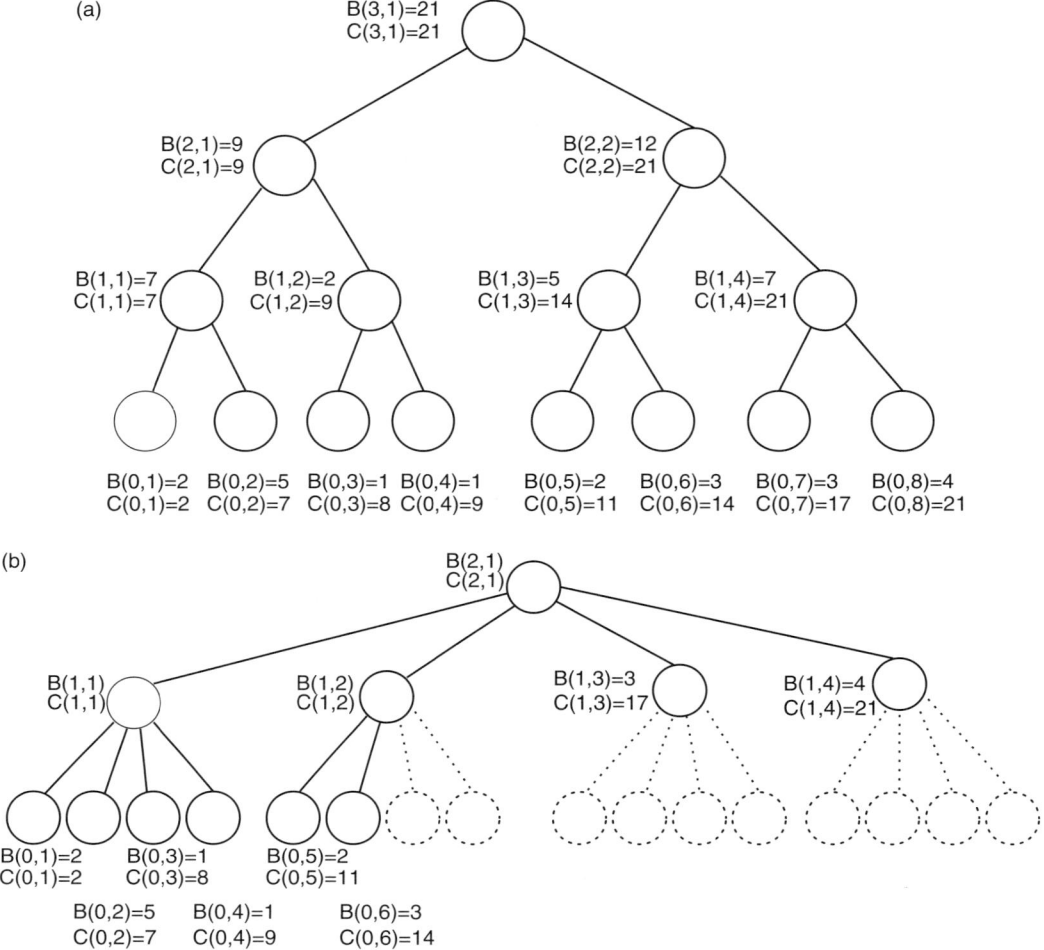

FIGURE 5.9 (a) PRAM prefix-sums algorithm on a binary tree and (b) PRAM prefix-sums algorithm on a k-ary tree ($k = 4$).

One to read *prefix_sum* of the parent and *sum* of all the children of a node. Another to join the spawn at the current level. This gives a total of $4 * \log_k N$ RTMs. There is no queuing.

In addition to RTMs, there is a computation cost. The depth is $O(\log_k N)$ due to ascending and descending a logarithmic depth tree. By analyzing our own XMTC implementation we observed this term to be $(7k + 18) \log_k N + 2k + 39$ portion of the depth formula. The *Additional work* is derived similarly to the summation algorithm. It contains a $3N/p$ term for copying data to the tree's leaves and a $\frac{C*(N-\min(p,N-1)-1)/(k-1)}{p} + C * \log_k p$ term to advance up and down the tree. This is derived by using the geometric series to count the number of internal nodes in the tree (because each internal node is touched by one thread and $C = (7k + 4)$ is the work per node) and considering that processing any level of the tree with fewer than p nodes has *Additional work* $= \frac{(C \times p_i)}{p_i} = C$. The overhead to start threads in oversaturated conditions is computed analogously.

Advancing PRAM Algorithms into Parallel Programs

For the moment, we do not consider how clustering will be applied. Assuming that a round-trip to memory takes $\mathcal{R}$ cycles, the performance of this implementation is as follows:

$$\text{Execution depth} = (4 \log_k N + 3) \times \mathcal{R} + (7k + 18) \log_k N + 2k + 39 \quad (5.8)$$

$$\text{Additional work} = \frac{3N + (7k + 4)(N - \min(p, N-1) - 1)/(k-1)}{p}$$
$$+ (7k + 4) \log_k p + \left\lceil \frac{(N - \min(p, N-1) - 1)/(k-1)}{p} - \log_k \frac{N}{p} \right\rceil \times 2\mathcal{R} \quad (5.9)$$

5.7.2 No-Busy-Wait Prefix-Sums

A less-synchronous XMT algorithm is presented. The synchronous algorithm presented earlier processes each level of the tree before moving to the next, but this algorithm has no such restriction. The algorithm is based on the no-busy-wait balanced tree paradigm [58]. As before, we use k-ary rather than binary trees.

The input and data structures are the same as earlier, with the addition of the *gatekeeper* array, providing a "gatekeeper" variable per tree node. The computation advances up the tree using a no-busy-wait summation algorithm. Then it advances down the tree using a no-busy-Wait algorithm to fill in the prefix-sums.

The pseudocode of the algorithm in the XMT Programming Model is as follows:

```
Spawn(first_leaf,last_leaf)
    Do while alive
    Perform psm on parent's gatekeeper
    If last to arrive at parent
        Move to parent and sum values from children
    Else
        Join
    If at root
        Join
prefix_sum[0] = 0         //set prefix_sum of root to 0
Spawn(1,1)                //spawn one thread at the root
    Let prefix_sum value of left child = prefix_sum of parent
    Proceed through children left to right where each child is
    assigned prefix_sum value equal to prefix_sum + sum of left
    sibling    Use a nested spawn command to start a thread to
    recursively handle each child thread except the leftmost.
    Advance to leftmost child and repeat.
```

Analysis of No-Busy-Wait Prefix-Sums When climbing the tree, the implementation executes 2 RTMs per level, just as in the earlier algorithm. One RTM is to read values of *sum* from the children, and the other is to use an atomic prefix-sum instruction on the gatekeeper. The LSRTM to descend the tree is also 2 RTMs per level. First, a thread reads the thread ID assigned to it by the parent thread, in one RTM. The second RTM is used to read *prefix_sum* from the parent and *sum* from the children in order to do the necessary calculations. This is an LSRTM of $4 \log_k N$. Also, there are additional $O(1)$ RTMs. Examining our own XMTC implementation, we have computed the constants involved.

Queuing is also a factor. In the current algorithm, up to k threads can perform a prefix-sum-to-memory operation concurrently on the same gatekeeper and create a $k - 1$ queuing delay (since the first access does not count towards queuing delay). The total QD on the critical path is $(k - 1) \log_k N$.

In addition to RTMs and QD, we count computation depth and work. The computation depth is $O(\log_k N)$. Counting the constants our implementation yields $(11+8k)*\log_k N + 2k + 55$. The $\Sigma \frac{\text{Work}}{\min(p,p_i)}$ part of the complexity is derived similarly as in the synchronous algorithm. It contains an $18N/p$ term, which is due to copying data to the tree's leaves and also for some additional work at the leaves. There is a $\frac{C*(N-\min(p,N-1)-1)/(k-1)}{p} + C * \log_k p$ term to traverse the tree both up and down. This value is derived by using the geometric series to count the number of internal nodes in the tree and multiplying by the work per internal node ($C = (11 + 8k)$) as well as considering that processing any level of the tree with fewer than p nodes has $\frac{\text{Work}}{\min(p,p_i)} = C$. Without considering clustering, the running time is given by

$$\text{Execution depth} = (4 \log_k N + 6) \times \mathcal{R} + (11 + 9k) * \log_k N + 2k + 55 \quad (5.10)$$

$$\text{Additional work} = \frac{6 + 18N + (11 + 8k)(N - \min(p, N - 1) - 1)/(k - 1)}{p}$$

$$+ (11 + 8k) \log_k p + \left\lceil \frac{(N - \min(p, N - 1) - 1)/(k - 1)}{p} - \log_k \frac{N}{p} \right\rceil \times 2\mathcal{R} \quad (5.11)$$

5.7.3 Clustering for Prefix-Sums

Clustering may be added to the synchronous k-ary prefix-sums algorithm to produce the following algorithm. The algorithm begins with an embarrassing parallel section, using the parallel prefix-sums algorithm to combine results, and ends with another embarrassing parallel section.

```
1.Let c be a constant.
  Spawn c threads that run the serial summation algorithm on a
  contiguous sub-array of N/c values from the input array. The
  threads write the resulting sum values into a temporary array B.
2.Invoke the parallel prefix-sums algorithm on array B.
3.Spawn c threads. Each thread retrieves a prefix-sum value from B.
  The thread then executes the serial prefix-sum algorithm on the
  appropriate sub-array of N/c values from the original array.
```

The no-busy-wait prefix-sums algorithm can be clustered in the same way.

We now present the formulas for execution time using clustering. Let c be the number of threads that are spawned in the embarrassing parallel portion of the algorithm. Let *SerSum* be the complexity of the serial summation algorithm, *SerPS* be the complexity of the serial PS algorithm, and *ParPS* be the complexity of the parallel PS algorithm (dependent on whether the synchronous or no-busy-wait is used). The serial sum and prefix-sum algorithms loop over N elements and from the serial code it is derived that $SerSum(N) = 2N + 1 \times \mathcal{R}$ and $SerPS(N) = 3N + 1 \times \mathcal{R}$. The following formula calculates the cost of performing the serial algorithms on a set of $N - c$ elements divided evenly among p processors and then adds the cost of the parallel step:

$$\text{Execution depth} = SerSum(\frac{N-c}{p}) + SerPS(\frac{N-c}{p}) + ParPS(c) \quad (5.12)$$

Optimal k and Optimal Parallel-Serial Crossover The value c, where $p \leq c \leq N$, that minimizes the formula determines the best crossover point for clustering. Let us say $p = 1024$ and $\mathcal{R} = 24$. Similar to the summation problem, we have concluded that in the XMT execution model for many values $N \geq p$, the best c is 1024. This is the case for both algorithms. A different value of c may be optimal for other applications, for example if the threads do not have equal work.

The optimal k value, where k denotes the arity of the tree, to use for either of the prefix-sums algorithms can be derived from the formulas. As shown in Figure 5.5.a, for $N \geq p$ (where $p = 1024$), the parallel sums algorithm is only run on $c = 1024$ elements and in this case $k = 8$ is optimal for the synchronous algorithm and $k = 7$ is optimal for the no-busy-wait algorithm. When $N < p$, clustering does not take effect, and the optimal value of k varies with N, for both algorithms.

5.7.4 Comparing Prefix-Sums Algorithms

Using the performance model presented earlier with these optimizations allows comparison of the programs in the XMT Execution model. The execution time for various N was calculated for both prefix-sums algorithms using the formula with clustering. This is plotted in Figure 5.5.b on page 5-15.

The synchronous algorithm performs better, due to the smaller computation constants. The LSRTM of both algorithms is the same, indicating that using gatekeepers and nesting is equivalent in RTMs to using synchronous methods. The no-busy-Wait algorithm has slightly longer computation depth and more computation work due to the extra overhead.

We note that in an actual XMT system, an implementation of the prefix-sums algorithm would be likely to be included as a library routine.

5.8 Programming Parallel Breadth-First Search Algorithms

As noted earlier, BFS provides an interesting example for XMT programming. We assume that the graph is provided using the incidence list representation, as pictured in Figure 5.10a.

Let $L(i)$ be the set of $N(i)$ nodes in level i and $E(i)$ the set of edges adjacent to these nodes. For brevity, we will only illustrate how to implement one iteration. Developing the full program based on this is straightforward.

As described in Section 5.2.2.1, the high-level work-depth presentation of the algorithm starts with all the nodes in parallel, and then using nested parallelism ramps up more parallelism to traverse all their adjacent edges in one step. Depending on the extent that the target programming model supports nested parallelism, the programmer needs to consider different implementations. We discuss these choices in the following paragraphs, laying out assumptions regarding the target XMT model.

We noted before that the work-depth model is not a direct match for our proposed programming model. With this in mind, we will not present a full work-depth description of the BFS algorithm; as will be shown, the "ideal" implementation will be closer to the high-level work-depth presentation.

5.8.1 Nested Spawn BFS

In a XMT programming model that supports nested parallel sections, the high-level XMT program can be easily derived from the HLWD description:

```
For every vertex v of current layer L(i) spawn a thread
    For every edge e=(v,w) adjacent on v spawn a thread
        Traverse edge e
```

A more detailed implementation of this algorithm using the XMTC programming language is included in Table 5.2.3d. To traverse an edge, threads use an atomic prefix-sum instruction on a special "gatekeeper" memory location associated with the destination node. All gatekeepers are initially set to 0. Receiving a 0 from the prefix-sum instruction means the thread was the first to reach the destination node. The newly discovered neighbors are added to layer $L(i+1)$ using another prefix-sum operation on the size of $L(i+1)$. In addition, the edge antiparallel to the one traversed is marked to avoid needlessly traversing it again (in

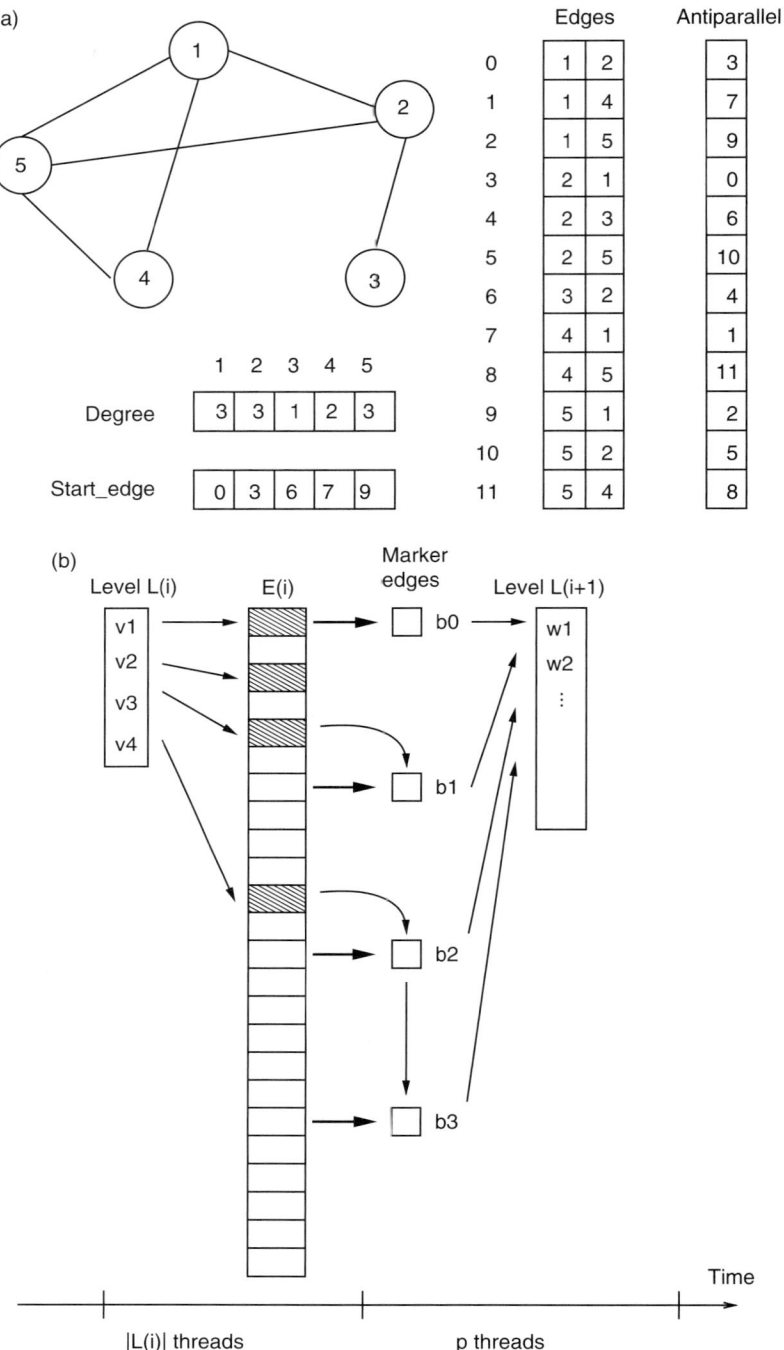

FIGURE 5.10 (a) The incidence list representation for a graph. (b) Execution of flattened BFS algorithm. First allocate $E[i]$ to hold all edges adjacent to $level[i]$. Next, identify marker edges b_i, which give the first edge per each subarray. Running one thread per subarray, all edges are traversed to build $level[i+1]$.

the opposite direction) in later BFS layers. Note that this matches the PRAM arbitrary concurrent write convention, as mentioned in Section 5.2.3.

The nested spawn algorithm bears a natural resemblance to the HLWD presentation of the BFS algorithm and in this sense, is the ideal algorithm to program. Allowing this type of implementations to be written and efficiently executed is the desired goal of a XMT framework.

Several other XMT BFS algorithms will be presented to demonstrate how BFS could be programmed depending on the quantitative and qualitative characteristics of a XMT implementation.

5.8.2 Flattened BFS

In this algorithm, the total amount of work to process one layer (i.e., the number of edges adjacent to its vertices) is computed, and it is evenly divided among a predetermined number of threads p, a value that depends on architecture parameters. For this, a prefix-sums subroutine is used to allocate an array of size $|E(i)|$. The edges will be laid out flat in this array, located contiguously by source vertex. p threads are then spawned, each being assigned one subarray of $|E(i)|/p$ edges and traversing these edges one by one. An illustration of the steps in this algorithm can be found in Figure 5.10b.

To identify the edges in each subarray, it is sufficient to find the first (called *marker*) edge in such an interval; we can then use the natural order of the vertices and edges to find the rest. We start by identifying first (if any) marker edge adjacent to v_j for all vertices $v_j \in L(i)$ in parallel, then use a variant of the pointer jumping technique [37, 59] to identify the rest of the marker edges (if any) adjacent to v_j using at most $\log_2 p$ steps.

5.8.3 Single-Spawn and k-Spawn BFS

Although the programming model can allow nested parallelism, the execution model might include limited or no support for nesting. To provide insight into the transformations applied by the compiler, and how to reason about the efficiency of the execution, we present two implementations of the nested spawn BFS algorithm that directly map into an execution model with limited support for nesting.

The **Single-spawn BFS Algorithm** uses sspawn() and a binary tree type technique to allow the nested spawning of any number T of threads in $\log_2 T$ steps. The algorithm spawns one thread for each vertex in the current level, and then uses each thread to start *degree*(*vertex*) − 1 additional threads by iteratively using the sspawn() instruction to delegate half a thread's work to a new thread. When one edge per thread is reached, the edges are traversed.

The pseudocode for a single layer is as follows:

```
For every vertex v of current layer L spawn a thread
    While a thread needs to handle s > 1 edges
        Use sspawn() to start another thread and
            delegate floor(s/2) edges to it
    Traverse one edge
```

The k-**spawn BFS Algorithm** follows the same principle as single-spawn BFS, but uses the kspawn() instruction to start the threads faster. By using a k-ary rather than binary tree to emulate the nesting, the number of steps to start T threads is reduced to $\log_k T$.

The k-spawn BFS pseudocode for processing one layer is

```
For every vertex v of current layer L spawn a thread
    While a thread needs to handle s > k edges
        Use kspawn() to spawn k threads and
            delegate to each floor(s/(k+1)) edges
    Traverse (at most) k edges
```

5.9 Execution of Breadth-First Search Algorithms

In this section, we examine the execution of the BFS algorithms presented, and analyze the impact compiler optimizations could have on their performance using the XMT execution model as a framework to estimate running times.

5.9.1 Flattened BFS

When each of p threads traverses the edges in its subarray serially, a simple optimization would prefetch the next edge data from memory and overlap the prefix-sum operations, thus reducing the number of round-trips to memory from $O(\frac{|E(i)|}{p})$ to a small constant. Such an improvement can be quite significant.

The flattened BFS algorithm uses the prefix sums algorithm as a procedure; we will use the running time computed for this routine in Section 5.7. Analyzing our implementation, we observed that identifying the marker edges uses 3 RTMs to initialize one marker edge per vertex, and then $4 \log_2 p$ RTMs to do $\log_2 p$ rounds of pointer jumping and find the rest of the adjacent marker edges. Finally, p threads cycle through their subarrays of $\frac{|E(i)|}{p}$ edges. By using the optimizations described earlier, the only round-trip to memory penalties paid in this step are that of traversing a single edge. A queuing delay occurs at the node gatekeeper level if several threads reach the same node, simultaneously. This delay depends on the structure of the graph, and is denoted GQD in the formula later.

In addition to LSRTM and QD, the computation depth also appears in the depth formula. The $10 \log_2 p$ term is the computation depth of the binary tree approach to identifying marker edges. The computation depth of the call to prefix-sums is included.

The dominating term of the additional work is $7|E(i)|/p + 28N(i)/p$, which comes from the step at the end of the algorithm in which all the edges are traversed in parallel by p threads, and the new found vertices are added to level $L(i+1)$. The *Additional Work* portion of the complexity also contains the work for the call to prefix-sums. The performance is

$$\text{Execution depth} = (4 \log_k N(i) + 4 \log_2 p + 14) \times \mathcal{R} + 38 + 10 \log_2 p + 7|E(i)|/p$$
$$+ 16N(i)/p + GQD + \text{Computation depth}(\text{Prefix-sums}) \qquad (5.13)$$

$$\text{Additional work} = \frac{7|E(i)| + 28N(i) + 15p + 17}{p} + \lceil \frac{N(i) - p}{p} \rceil \times \mathcal{R}$$
$$+ \text{Additional work}(\text{Prefix-sums}) \qquad (5.14)$$

As before, $N(i)$ is the number of nodes in layer $L(i)$, $|E(i)|$ is the number of edges adjacent to $L(i)$ and $\mathcal{R}$ is the number of cycles in one RTM.

The second term of relation 5.14 denotes the overhead of starting additional threads in over-saturated conditions. In the flattened algorithm, this can occur only in the initial phase, when the set of edges $E(i)$ is filled in. To reduce this overhead, we can apply clustering to the relevant parallel sections.

Note the following special case, when the number of edges adjacent to one layer is relatively small, there is no need to start p threads to traverse them. We choose a threshold θ, and if $\frac{|E(i)|}{p} < \theta$, then we use $p' = \frac{|E(i)|}{\theta}$ threads. Each will process θ edges. In this case, the running time is found by taking the formulas above and replacing p with $p' = \frac{|E(i)|}{\theta}$.

5.9.2 Single-Spawn BFS

In this algorithm, one thread per each vertex v_i is started, and each of these threads then repeatedly uses single-spawn to get $deg(v_i) - 1$ threads started.

Advancing PRAM Algorithms into Parallel Programs

To estimate the running time of this algorithm, we proceed to enumerate the operations that take place on the critical path during the execution:

- Start and initialize the original set of $N(i)$ threads, which in our implementation takes 3 RTMs to read the vertex data.
- Let d_{max} be the largest degree among the nodes in current layer. Use single-spawn and $\log_2 d_{max}$ iterations to start d_{max} threads using a balanced binary tree approach. Starting a new thread at each iteration takes 2 RTMs (as described in Section 5.6.1), summing up $2 \log_2 d_{max}$ RTM on the critical path.
- The final step of traversing edges implies using one prefix-sum instruction on the gatekeeper location and another one to add the vertex to the new layer.

The cost of queuing at gatekeepers is represented by GQD. In our implementation, the computation depth was $18 + 7 \log_2 d_{max}$.

Up to $|E(i)|$ threads are started using a binary tree, and when this number exceeds the number of TCUs p, we account for the additional work and the thread starting overhead. We estimate these delays by following the same reasoning as with the k-ary summation algorithm in Section 5.3 using a constant of $C = 19$ cycles work per node as implied by our implementation.

The performance is

$$\text{Execution depth} = (7 + 2 \log_2 d_{max})\mathcal{R} + (18 + 7 \log_2 d_{max}) + GQD \qquad (5.15)$$

$$\text{Additional work} = \frac{19(|E(i)| - \min(p, |E(i)| - 1) - 1) + 2}{p}$$

$$+ 19 \log_2 |E(i)| + \left\lceil \frac{|E(i)| - p}{p} \right\rceil \times \mathcal{R} \qquad (5.16)$$

To avoid starting too many threads, the clustering technique presented in Section 5.6.2 can be applied. This will reduce the additional work component since the cost of allocating new threads to TCUs will no longer be paid for every edge.

5.9.3 k-Spawn BFS

The threads are now started using k-ary trees and are therefore shorter. The LSRTM is $2 \log_k d_{max}$. The factor of 2 is due to the 2 RTMs per k-spawn, as per Section 5.6.1.

The computation depth in our XMTC implementation is $(5 + 4k) \log_k d_{max}$. This is an $O(k)$ cost per node, where $\log_k d_{max}$ nodes are on the critical path. The queuing cost at the gatekeepers is denoted by GQD. The *Additional Work* is computed as in single-spawn BFS with the constant $C = 4k + 3$ denoting the work per node in the k-ary tree used to spawn the $|E(i)|$ threads.

The performance is

$$\text{Execution depth} = (7 + 2 \log_k d_{max})\mathcal{R} + (5 + 4k) \log_k d_{max} + 15 + 4k + GQD \qquad (5.17)$$

$$\text{Additional work} = \frac{14|E(i)| + (4k + 3)(|E(i)| - \min(p, |E(i)| - 1) - 1)/(k - 1)}{p}$$

$$+ (4k + 3) \log_k |E(i)| + \left\lceil \frac{|E(i)| - p}{p} \right\rceil \times \mathcal{R} \qquad (5.18)$$

Similar to the case of the single-spawn BFS algorithm, thread clustering can be used in the k-spawn BFS algorithm by checking the number of virtual threads to determine whether the k-spawn instruction should continue to be used or if additional spawning is to be temporarily suspended.

5.9.4 Comparison

We calculated execution time for one iteration (i.e., processing one BFS level) of the BFS algorithms presented here and the results are depicted in Figure 5.7 on page 5-18. This was done for two values for the number of vertices $N(i)$ in current level $L(i)$, 500 and 2000. The analysis assumes that all edges with one end in $L(i)$ lead to vertices that have not been visited in an earlier iteration; since there is more work to be done for a "fresh" vertex, this constitutes a worst-case analysis. The same graphs are also used in Section 5.4 for empirically computing speedups over serial code as we do not see how they could significantly favor a parallel program over the serial one. To generate the graphs, we pick a value M and choose the degrees of the vertices uniformly at random from the range $[M/2, 3M/2]$, which gives a total number of edges traversed of $|E(i)| = M * N(i)$ on average. Only the total number of edges is shown in Figure 5.7. We arbitrarily set $N(i+1) = N(i)$, which gives a queuing delay at the gatekeepers (GQD) of $\frac{|E(i)|}{N(i)} = M$ on average. As stated in Section 5.2.5, we use the value $\mathcal{R} = 24$ for the number of cycles needed for one roundtrip to memory.

For small problems, the k-spawn algorithm came ahead and the single-spawn one was second best. For large problems, the flattened algorithm performs best, followed by k-spawn and single-spawn. When the hardware is subsaturated, the k-spawn and single-spawn algorithms do best because their depth component is short. These algorithms have an advantage on smaller problem sizes due to their lower constant factors. The k-spawn implementation performs better than single-spawn due to the reduced height of the "Spawn tree." The flattened algorithm has a larger constant factor for the number of RTMS, mostly due to the introduction of a setup phase which builds and partitions the array of edges. For supersaturated situations, the flattened algorithm does best due to a smaller work component than the other algorithms.

Note that using the formula ignores possible gaps in parallel hardware usage. In a highly unbalanced graph, some nodes have high degree while others have low degree. As many nodes with small degree finish, it may take time before the use of parallel hardware can be ramped up again. For example, in the single-spawn and k-spawn algorithms, the virtual threads from the small trees can happen to take up all the physical TCUs and prevent the deep tree from getting started. The small trees may all finish before the deep one starts. This means we are paying the work of doing the small trees plus the depth of the deep tree. A possible workaround would be to label threads according to the amount of work they need to accomplish and giving threads with more work a higher priority (e.g., by scheduling them to start as soon as possible). Note that this issue does not affect the flattened BFS algorithm, since the edges in a layer are evenly distributed among the running threads.

5.10 Adaptive Bitonic Sorting

Bilardi and Nicolau's [9] adaptive bitonic Sorting algorithm is discussed next. The key component of this sorting algorithm is a fast, work-optimal merging algorithm based on Batcher's bitonic network and adapted for shared memory parallel machines. An efficient, general purpose Merge-Sort type algorithm is derived using this merging procedure.

The main advantages of the bitonic merging and sorting algorithms are the small constants involved and the fact that they can be implemented on an EREW PRAM. This is an important factor for implementing an algorithm on our proposed XMT model, since it guarantees that no queuing delays will occur. We will only consider the case where the problem size is $N = 2^n$; the general case is treated in [9]. We note that XMT caters well to the heavy reliance on pointer manipulation in the algorithm, which tends to be a weakness for other parallel architectures.

For conciseness, we focus on presenting the merging algorithm here. The purpose of this algorithm is to take two sorted sequences and merge them into one single sorted sequence. To start, one of the input sequences is reversed and concatenated with the other one. The result is what is defined as a *bitonic sequence*, and it is stored in a *bitonic tree*—a fully balanced binary tree of height $\log_2 N$ whose in-order traversal yields the bitonic sequence, plus an extra node (called spare) to store the Nth element. The goal of the bitonic merging algorithm is to transform this tree into a binary tree whose in-order traversal, followed by the spare node, gives the elements in sorted order.

The key observation of the algorithm is that a single traversal of a bitonic tree from the root to the leaves is sufficient to have all the elements in the left subtree of the root smaller than the ones in the right subtree. At each of the $\log_2 N$ steps of one such traversal, one comparison is performed, and at most two pairs of values and pointers are exchanged. After one such traversal, the algorithm is applied recursively on the left and right children of the root, and after $\log_2 N$ rounds (that can be pipelined, as explained below) the leaves are reached and the tree is a binary search tree.

The full description of the algorithm can be found in Reference 9, and can be summarized by the following recursive function, called with the root and spare nodes of the bitonic tree and the direction of sorting (increasing or decreasing) as arguments:

```
procedure bimerge(root,spare,direction)
  1.   compare root and spare values to find direction of swapping
  2.   swap root and spare values if necessary
  3.   pl = root.left, pr = root.right
  4.   while pr not nil
  5.     compare pl, pr
  6.     swap values of pl, pr and two subtree pointers if necessary
  7.     advance pl,pr toward leaves
       end
  8.   in parallel run bimerge(root.left,root) and
           bimerge(root.right,spare)
end
```

In this algorithm, lines 4–7 traverse the tree from current height to the leaves in $O(\log N)$ time. The procedure is called recursively in line 8, starting at the next lowest level of the tree. This leads to an overall time of $O(\log^2 N)$.

We call *stage(k)* the set of tree traversals that start at level k (the root being at level 0). There are 2^k parallel calls in such a stage. Call *phase(0)* of a stage the execution of lines 1–3 in the above algorithm, and *phase(i)*, $i = 1..\log(N) - k - 1$ the iterations of lines 4–7.

To obtain a faster algorithm, we note that the traversals of the tree can be pipelined. In general, we can start stage $k + 1$ as soon as stage k has reached its *phase(2)*. On a synchronous PRAM model, all stages advance at the same speed and thus they will never overlap; on a less-asynchronous PRAM implementation, such as PRAM-on-chip, this type of lockstep execution can be achieved by switching from parallel to serial mode after each phase. With this modification, the bitonic merging has a running time of $O(\log N)$.

We have experimented with two implementations of the bitonic merging algorithm:

Pipelined This is an implementation of the $O(\log N)$ algorithm that pipelines the stages. We start with an active workset containing only one thread for *stage(0)* and run one phase at a time, joining threads after one phase is executed. Every other iteration, we initialize a new stage by adding a corresponding number of threads to the active workset. At the same time, the threads that have reached the leaves are removed from the workset. When the set of threads is empty, the algorithm terminates.

Nonpipelined The previous algorithm has a lock-step type execution, with one synchronization point after each phase. An implementation with fewer synchronization points is evaluated, where all phases of a stage are ran in one single parallel section with no synchronizations, followed by a `join` command and then the next stage is started. This matches the $O(\log^2 N)$ algorithm described above.

For the limited input sizes we were able to test at this time on the XMT cycle-accurate simulator, the performance of the pipelined version fell behind the simpler nonpipelined version. This is mainly caused by the overheads required by the implementation of pipelining. Namely, some form of added synchronization, such as using a larger number of spawn blocks, was added.

5.11 Shared Memory Sample Sort

The sample sort algorithm [36, 49] is a commonly used randomized sorting algorithm designed for multiprocessor architectures; it follows a "decomposition first" pattern, making it a good match for distributed memory machines. Being a randomized algorithm, its running time depends on the output of a random number generator. Sample sort has been proved to perform well on very large arrays, with high probability.

The idea behind sample sort is to find a set of $p - 1$ elements from the array, called *splitters*, which are then used to partition the n input elements into p buckets $bucket_0 \ldots bucket_{p-1}$ such that every element in $bucket_i$ is smaller than each element in $bucket_{i+1}$. The buckets are then sorted independently.

One key step in the standard sample sort algorithm is the distribution of the elements to the appropriate bucket. This is typically implemented using "one-to-one" and broadcasting communication primitives usually available on multiprocessor architectures. This procedure can create delays due to queuing at the destination processor [11, 22, 49].

In this section we discuss the shared memory sample sort algorithm, which is an implementation of sample sort for shared memory machines. The solution presented here departs slightly from the sample sorting algorithm and consists of an CREW PRAM algorithm that is better suited for an XMT implementation.

Let the input be the unsorted array A and let p the number of hardware thread control units (TCUs) available. An overview of the shared memory sample sort algorithms is as follows:

Step 1. In parallel, a set S of $s \times p$ random elements from the original array A is collected, where p is the number of TCUs available and s is called the oversampling ratio. Sort the array S, using an algorithm that performs well for the size of S (e.g. adaptive bitonic sorting). Select a set of $p - 1$ evenly spaced elements from it into S': $S' = \{S[s], S[2s], \ldots, S[(p-1) \times s]\}$ These elements are the splitters, which will be used to partition the elements of A into p sets $bucket_i$, $0 \leq i < p$ as follows: $bucket_0 = \{A[i] \mid A[i] < S'[0]\}$, $bucket_1 = \{A[i] \mid S'[0] < A[i] < S'[1]\}$, ..., $bucket_{p-1} = \{A[i] \mid S'[p-1] < A[i]\}$.

Step 2. Consider the the input array A divided into p subarrays, $B_0 = A[0, \ldots, N/p - 1]$, $B_1 = A[N/p, \ldots, 2N/p - 1]$ and so forth. The ith TCU iterates through the subarray B_i and for each element executes a binary search on the array of splitters S', for a total of N/p binary searches per TCU. The following quantities are computed (i) c_{ij}—the number of elements from B_i that belong in $bucket_j$. The c_{ij} make up the matrix C as in Figure 5.11, (ii) $bucket\_idx[k]$—the bucket in which element $A[k]$ belongs. Each element is tagged with such an index and (iii) $serial[k]$—the number of elements in B_i that belong in $bucket_{bucket\_idx[k]}$ but are located before $A[k]$ in B_i.

For example, if $B_0 = [105, 101, 99, 205, 75, 14]$ and we have $S' = [100, 150, \ldots]$ as splitters, we will have $c_{0,0} = 3$, $c_{0,1} = 2$ etc., $bucket[0] = 1$, $bucket[1] = 1$ and so on and $serial[0] = 0$, $serial[1] = 1$, $serial[5] = 2$.

Step 3. Compute the prefix-sums $ps[i, j]$ for each **column** of the matrix C. For example, $ps[0, j], ps[1, j], \ldots, ps[p-1, j]$ are the prefix-sums of $c[0, j], c[1, j], \ldots, c[p-1, j]$. In addition, compute the sum of column i, which is stored in sum_i.

Compute the prefix sums of the $sum_1, \ldots, sum_p$ into $global\_ps[0, \ldots, p-1]$ and the total sum of sum_i in $global\_ps[p]$.

Step 4. Each TCU i computes for each element $A[j]$ in segment B_i, $iN/p \leq j < (i+1)N/p - 1$:

$$pos[j] = global\_ps[bucket[j]] + ps[i, bucket[j]] + serial[j]$$

Copy $Result[pos[j]] = A[j]$.

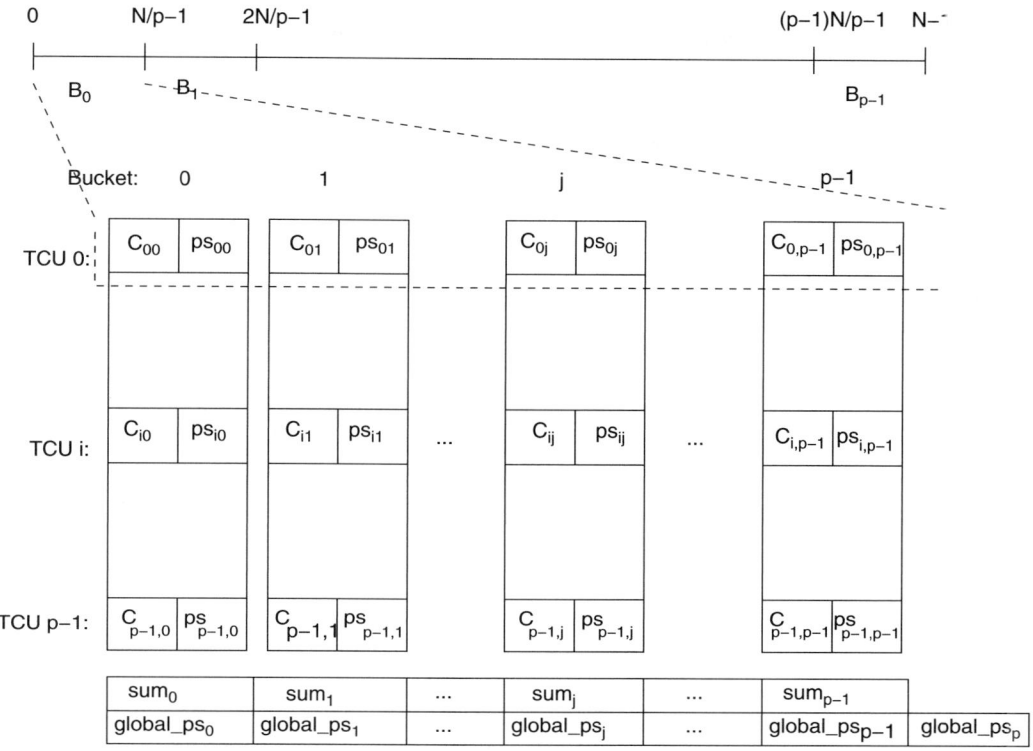

FIGURE 5.11 The helper matrix used in Shared Memory Sample Sort. c_{ij} stores the number of elements from B_i that fall in $bucket_j$. ps_{ij} represent the prefix-sums computed for each column, and $global\_ps_{0..p}$ are the prefix-sums and the sum of $sum_{0..p-1}$.

Step 5. TCU i executes a (serial) sorting algorithm on the elements of $bucket_i$, which are now stored in $Result[global\_ps[i], \ldots, global\_ps[i+1]-1]$.

At the end of Step 5, the elements of A are stored in sorted order in $Result$.

An implementation for the shared memory sample sort in the XMT programming model can be directly derived from the earlier description. Our preliminary experimental results show that this sorting algorithm performs well on average; due to the nature of the algorithm, it is only relevant for problem sizes $N \gg p$, and its best performance is for $N \geq p^2$. Current limitations on the cycle-accurate simulator have prevented us from running the sample-sort algorithm on datasets with such a N/p ratio. One possible work around that we are currently investigating could be to scale down the architecture parameters by reducing the number of TCUs p, and estimate performance by extrapolating the results.

5.12 Sparse Matrix—Dense Vector Multiplication

Sparse matrices, in which a large portion of the elements are zeros, are commonly used in scientific computations. Many software packages used in this domain include specialized functionality to store and perform computation on them. In this section we discuss the so-called *Matvec* problem of multiplying a sparse matrix by a dense vector. Matvec is the kernel of many matrix computations. Parallel implementations of this routine have been used to evaluate the performance of other parallel architectures [24, 51].

To save space, sparse matrices are usually stored in a compact form, for example using a compressed sparse row (CSR) data structure: for a matrix A of size $n \times m$ all the nz non-zero elements are stored in

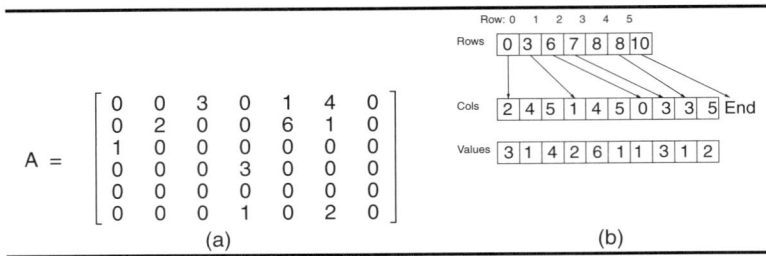

FIGURE 5.12 Compressed Sparse Row representation for a matrix. (a) A sparse matrix. (b) The "packed" CSR representation for matrix A.

an array *values*, and two new vectors *rows* and *cols* are used to store the start of each row in A and the column index of each nonzero element. An example is shown in Figure 5.12.

An "embarrassing parallel" solution to this problem exists the rows of the matrix can all be processed in parallel, each being multiplied with the dense vector. If the nonzero elements of the sparse matrix are relatively well distributed among the rows, then the work is well balanced and the algorithm performs well.

One factor that influences the performance of this algorithm on a XMT platform is the fact that the dense vector is read by all the parallel threads. On the XMT architecture, as TCUs request elements of the vector from memory, they are stored in the read-only caches at the cluster level, and subsequent requests from TCUs in the same cluster will not cause round-trips to memory and queuing.

Applying clustering to this algorithm requires a more in-depth analysis, since the threads are no longer symmetrical and can have significantly different lengths. In general, if such information about the lengths of the threads is known before the start of the parallel section, a compiler coupled with a runtime system could use this information to apply thread clustering, as described in Section 5.6.2. We note that the total length (number of nonzero elements) and the prefix sums of the number of elements each thread will process is actually provided as part of the input, making this task feasible.

In Spring 2005, an experiment to compare development time between two approaches to parallel programming of Matvec was conducted by software engineering researchers funded by the DARPA (High Productivity Computer Systems, HPCS). One approach was based on MPI and was taught by John Gilbert, a professor at the University of California, Santa Barbara. The second implemented two versions of the above algorithm using XMTC in a course taught by Uzi Vishkin at the University of Maryland. Both courses were graduate courses. For the UCSB course this was the forth programming assignment and for the UMD course it was the second. The main finding [34,35] was that XMTC programming required less than 50% of the time than for MPI programming.

5.13 Speed-Ups over Serial Execution

We present speed-up results over serial algorithms as observed empirically for some of the applications discussed in this chapter. The speed-up coefficients are computed by comparison with a "best serial" algorithm for each application. In our view such comparison is more useful than comparing with the parallel algorithm run on a single processing unit. In order not to give an unfair advantage to the parallel algorithm, we sought to use state-of-the-art compiler optimizations and architectural features when collecting serial running times. To this end, we used the SimpleScalar [14] toolset in the following manner. An "architecture serial speedup coefficient" was computed by running the same serial implementation using two different configurations of the SimpleScalar simulator (i) one resembling the Master TCU of the XMT platform, and (ii) one using the most advanced architecture simulated by SimpleScalar and aggressive compiler optimizations. We computed coefficients for each application and applied them to calibrate all the speedup results.

Advancing PRAM Algorithms into Parallel Programs 5-37

To obtain the same speedup results using the analytic model, we would have needed to present a performance model describing a "best serial" architecture and use it to estimate running times for the "best serial" algorithm. We considered this to be outside the scope of this work, and chose to present only experimental speedup results.

Figure 5.13 presents speedup results of parallel prefix-sums and BFS implementations. The XMT architecture simulated included 1024 parallel TCUs grouped into 64 clusters.

The current chapter represents work-in-progress both on the analytic model and on the XMT architecture as a whole. For the future, comparing the outcome of our dry analysis with the experimental results serves a double purpose: (i) It exposes the strengths and limitations of the analytic model, allowing further refinements to more closely match the architecture. (ii) Proposals for new architecture features can be easily evaluated using the analytic model and then confirmed in the simulator. This would be more cost-effective than testing such proposals by committing them to hardware architecture.

5.14 Conclusion

As pointed out earlier, the programming assignments in a one semester parallel algorithms class taught recently at the University of Maryland included parallel MATVEC, general deterministic (Bitonic) sort, sample sort, BFS on graphs and parallel graph connectivity. A first class on serial algorithms and serial programming typically does not require more demanding programming assignments. This fact provides a powerful demonstration that the PRAM theory coupled with XMT programming are on par with serial algorithms and programming. The purpose of the current chapter is to augment a typical textbook understanding of PRAM algorithms with an understanding of how to effectively program a XMT computer system to allow such teaching elsewhere.

It is also interesting to compare the XMT approach with other parallel computing approaches from the point of view of the first course to be taught. Other approaches tend to push the skill of parallel programming ahead of parallel algorithms. In other words, unlike serial computing and the XMT approach, where much of the intellectual effort of programming is taught in algorithms and data structure classes and programming itself is deferred to self-study and homework assignments, the art of fitting a serial algorithm to a parallel programming language such as MPI or OpenMP becomes the main topic. This may explain why parallel programming is currently considered difficult. However, if parallel computing is ever to challenge serial computing as a main stream paradigm, we feel that it should not fall behind serial computing in any aspects and in particular, in the way it is taught to computer science and engineering majors.

Finally, Figure 5.14 gives a bird's eye view on the productivity of both performance programming and application programming (using APIs). By productivity we mean the combination of run time and development time. For performance programming, we contrast the current methodology, where a serial version of an application is first considered and parallelism is then extracted from it using the rather involved methodology outlined for example by Culler and Singh [19], with the XMT approach where the parallel algorithm is the initial target and the way from it to a parallel program is more a matter of skill than an inventive step. For application programming, standard serial execution is automatically derived from APIs. A similar automatic process has already been demonstrated, though much more work remains to be done, for the XMT approach.

Acknowledgments

Contributions and help by the UMD XMT group at and, in particular, N. Ba, A. Balkan, F. Keceli, A. Kupershtok, P. Mazzucco, and X. Wen, as well as the UMD Parallel Algorithms class in 2005 and 2006 are gratefully acknowledged.

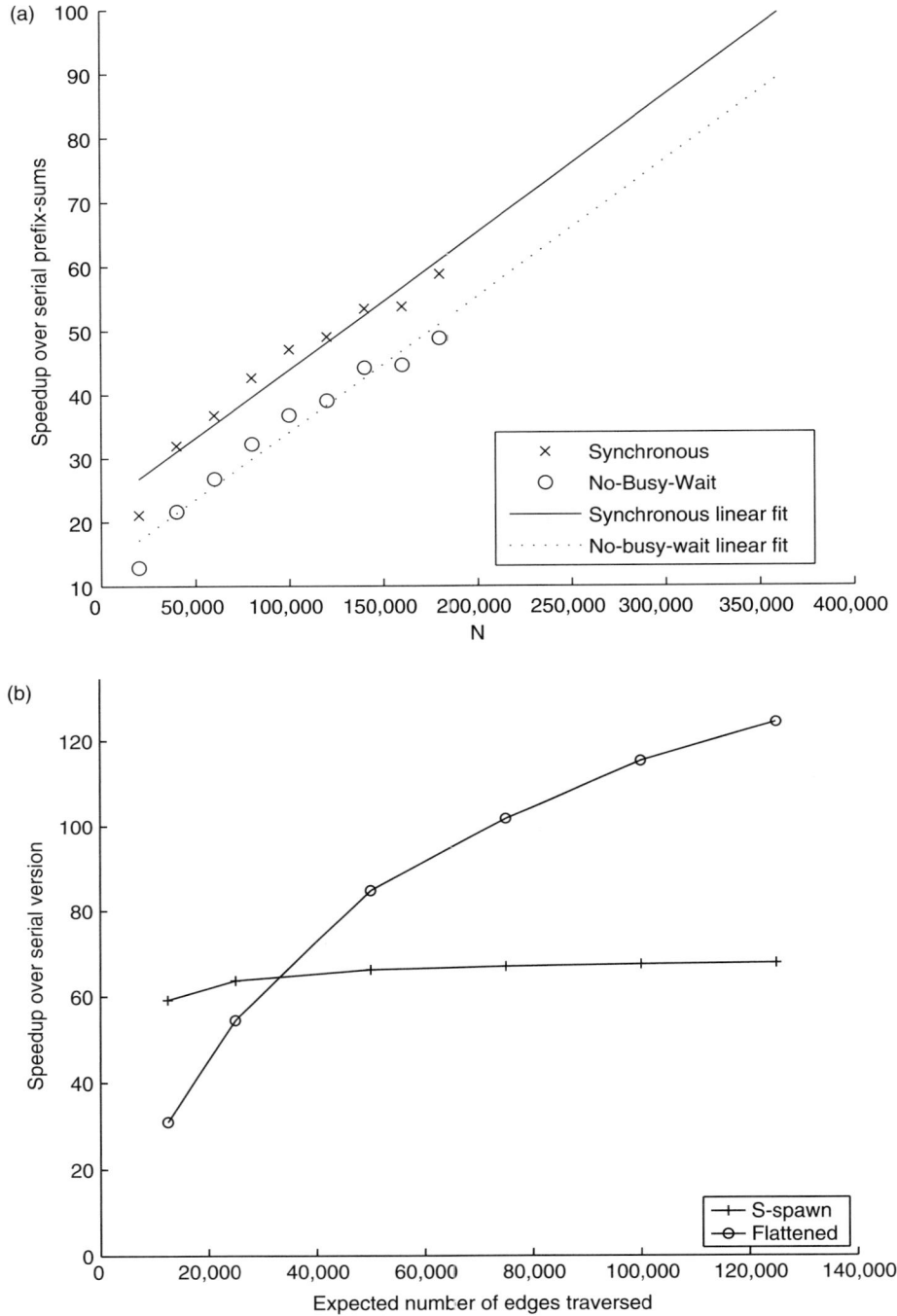

FIGURE 5.13 (a) Speedup over serial prefix-sums. Owing to current limitations of our simulator we could only run datasets of size up to 200,000. However, fitting a linear curve to the data indicates that the speedups would exceed 100 for problem sizes above 350,000 and (b) Speedups of BFS algorithms relative to a serial implementation for $N(i) = 500$ vertices in one BFS level.

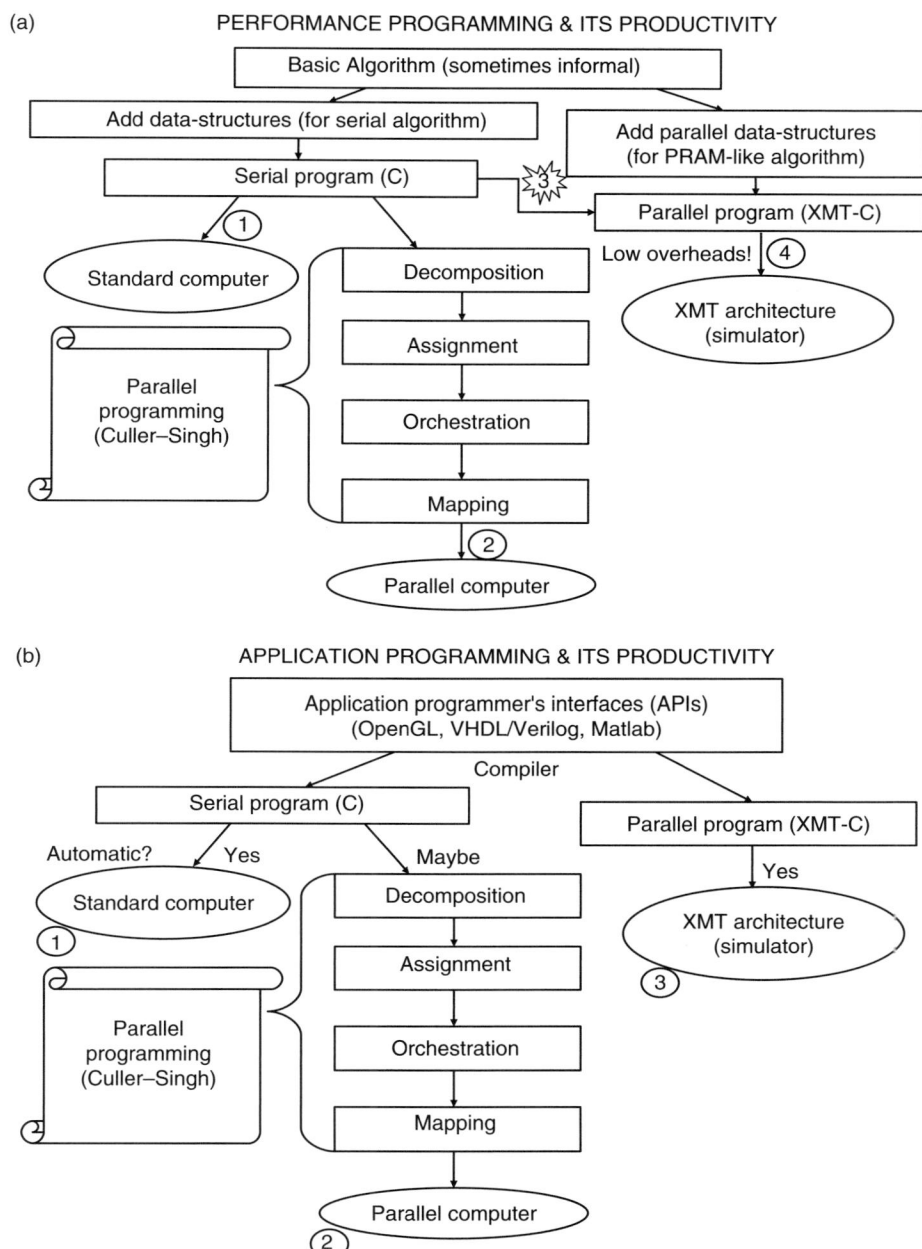

FIGURE 5.14 (a) Productivity of performance programming. Note that the path leading to (4) is much easier than the path leading to (2), and transition (3) is quite problematic. We conjecture that (4) is competitive with (1). (b) Productivity of High-Performance Application Programming. The chain (3) leading to XMT is similar in structure to serial computing (1), unlike the chain to standard parallel computing (2).

Appendix: XTMC Code Examples
K-Ary Summation

```
/*
 * void sum(...)
 *
 * The function computes sums by using a k-ary tree.
 * k is defined by the parameter k to the function.
 *
 * Input:
 * increment[] - an array of increment values
 * k - the value of k to use for the k-ary tree
 * size - the size of the increment[] array
 *
 * Output:
 * result[] - element 0 of the array is filled with the sum
 *
 */
void sum(int increment[], int result[], int k, int size) {
    register int low, high;
    int height = 20; //note: height should be as large as log_k(size)
    //int layersize[height]; //number of nodes in layer i
    int base[height*size]; //base of leaf is its value after PS
                    //base of internal node is the base of its leftmost leaf
    int sum[height*size]; //the value of sum for a node is the sum of the values of
                    //increment for all its leaves

    int iteration = 0; //determines the current height in the tree

    int temp;
    int done; //a loop control variable

    int l,   //level where leaves start
        sb,  //index where leaves would start if size is a power of 2
        d,   //size - k^l
        offset, //how much to offset due to size not being power of 2
        sr,  //sb + offset
        over,  //number of leaves at level l
        under, //number of leaves at level l + 1
        sbp1;  //index of one level higher from sb
    int fill; //nodes to fill in with 0 to make all nodes have k children
    int level, startindex, layersize;

    int i;

    /*
     * With non-blocking writes 0 RTM is required to initialize
     * the function parameters: k and size
     * 0 RTM is required to initialize local variables such as height
     */

    //Special case if size == 1
    if(size == 1) { //the check has 0 RTM because size is cached.
        result[0] = 0;
        return;
    }

    /*
     * 18 lines of code above, means computation cost = 18 up to this point.
     */

    //calculate location for leaves in the complete representation
    l = log(size) / log(k);

    sb = (pow(k,l) - 1) / (k - 1); //this is derived from geometric series
    sbp1 = (pow(k,l+1) - 1) / (k - 1);

    d = size - pow(k,l);
    offset = CEIL(((double) d) / (k - 1));
    sr = sb + offset;
    over = pow(k,l) - offset;
    under = size - over;
```

```
        /*
         * Computation cost = 8
         */

    //printf("l = %d, sb = %d, d = %d, offset = %d,
    //sr = %d, over = %d\n", l, sb, d, offset, sr, over);


        // Copy increment[...] to leaves of sum[...]

        low = 0;
        high = size - 1;
        spawn(low, high) {
                if($ < under) {
                        sum[$ + sbp1] = increment[$]; //1 RTM
                }
                else {
                        sum[($ - under) + sb + offset] = increment[$]; //1 RTM
                }
        } //1 RTM join

        /*
         * LSRTM = 2
         * QD = 0
         * Computation Depth = 5
         * Computation Work = 2N
         */

        // Make some 0 leaves at level l+1 so all nodes have exactly
        // k children

        fill = (k - (under % k)) % k;
        for(i = 0; i < fill; i++) {
                        sum[sbp1 + under + i] = 0;
        }

        /*
         * Computation Cost = 2k + 1
         */

        // Iteration 1: fill in all nodes at level l
        low = sb;
        high = sb + offset - 1;
        if(high >= low) {
            spawn(low, high) {
                int count;

                sum[$] = 0;
                for(count = 0; count < k; count++) {
                    sum[$] += sum[k * $ + count + 1];
                }
            }
        }

        /*
         * We will count the above "Iteration 1" as 1 iteration in
         * the climbing the tree loop below, for simplicity.
         * This gives an upper bound, since the "Iteration 1"
         * section above does slightly less.
         */

        // Climb the tree
        level = l;
        while(level > 0) {
            level --;
            startindex = (pow(k,level) - 1) / (k - 1);
            layersize = pow(k,level);

            low = startindex;
            high = startindex + layersize - 1;
            spawn(low, high) {
                int count;

                /*
                 * All the sum[X] elements are read at once
```

```
                     * for the below loop using prefetching.
                     *
                     * RTMs = 1
                     * (prefetch) Computation depth = k
                     */

                    sum[$] = 0;
                    for(count = 0; count < k; count++) {
                        sum[$] += sum[k * $ + count + 1];
                    }

                    /*
                     * Computation Depth = 2k + 1
                     */
                }
            } // 1 RTM join
            /*
             * For the above stage of climbing the tree:
             * LSRTM = 2 * logN
             * Computation Depth = (3k + 9) * logN + 1
             * Computation Work = (3k + 2) * (N - 1) / (k - 1)
             *
             * The (N - 1) / (k - 1) factor of the work is the
             * number of nodes in a k-ary tree of depth logN - 1
             * [there is no work for the leaves at depth logN]
             *
             * Computation Work / min(p, p_i) =
             * ((3k + 2) * (N - min(p, N-1) - 1) / (k - 1)) / p
             * + (3k + 2) * log_k(p)
             *
             * For each level where number of nodes < p, the denominator is p_i.
             * Otherwise the denominator is p.  This gives the above formula.
             */

            result[0] = sum[0];

            /*
             * For the whole algorithm:
             *
             * LSRTM = 2 * logN + 1
             * QD = 0
             * Computation Depth = (3k + 9) * logN + 2k + 33
             * Computation Work / min(p, p_i) =
             * ((3k + 2)(N - min(p, N-1) - 1) / (k - 1) + 2N) / p
             * + (3k + 2)log_k(p)
             */
}
```

Synchronous K-Ary Prefix-Sums

```
/*
 * void kps(...)
 *
 * The function computes prefix sums by using a k-ary tree.
 * k is defined by the parameter k to the function.
 *
 * Input:
 * increment[] - an array of increment values
 * k - the value of k to use for the k-ary tree
 * size - the size of the increment[] array
 *
 * Output:
 * result[] - this array is filled with the prefix sum on the values
 * of the array increment[]
 *
 */
void kps(int increment[], int result[], int k, int size) {
    register int low, high;
    int height = 20; //note: height should be as large as log_k(size)
    //int layersize[height]; // number of nodes in layer i
    int base[height*size]; // base of leaf is its value after PS
                           // base of internal node is the base of leftmost leaf
```

Advancing PRAM Algorithms into Parallel Programs

```
int sum[height*size]; // the value of sum for a node is the sum of the values
            // of increment for all its leaves

int iteration = 0; // determines the current height in the tree

int temp;
int done; //a loop control variable

int l,     //level where leaves start
    sb,    //index where leaves would start if size is a power of 2
    d,     //size - k^l
    offset, //how much to offset due to size not being power of 2
    sr,    //sb + offset
    over,  //number of leaves at level l
    under, //number of leaves at level l + 1
    sbp1;  //index of one level higher from sb
int fill; //nodes to fill in with 0 to make all nodes have k children
int level, startindex, layersize;
int i;

/*
 * With non-blocking writes 0 RTM is required to initialize
 * the function parameters: k and size
 * 0 RTM is required to initialize local variables such as height
 */

//Special case if size == 1
if(size == 1) { //the check has 0 RTM because size is cached.
        result[0] = 0;
        return;
}

/*
 * 18 lines of code above, means computation cost = 18 up to this point.
 */

//calculate location for leaves in the complete representation
l = log(size) / log(k);

sb = (pow(k,l) - 1) / (k - 1); //this is derived from geometric series
sbp1 = (pow(k,l+1) - 1) / (k - 1);

d = size - pow(k,l);
offset = CEIL(((double) d) / (k - 1));
sr = sb + offset;
over = pow(k,l) - offset;
under = size - over;

/*
 * Computation cost = 8
 */

//printf("l = %d, sb = %d, d = %d, offset = %d,
//sr = %d, over = %d\n", l, sb, d, offset, sr, over);

// Copy increment[...] to leaves of sum[...]
low = 0;
high = size - 1;
spawn(low, high) {
    if($ < under) {
        sum[$ + sbp1] = increment[$]; //1 RTM
    }
    else {
        sum[($ - under) + sb + offset] = increment[$]; //1 RTM
    }
} //1 RTM join

/*
 * LSRTM = 2
 * QD = 0
 * Computation Depth = 5
 * Computation Work = 2N
 */

// Make some 0 leaves at level l+1 so all nodes have exactly
```

```
// k children
fill = (k - (under % k)) % k;
for(i = 0; i < fill; i++) {
    sum[sbp1 + under + i] = 0;
}

/*
 * Computation Cost = 2k + 1
 */

// Iteration 1: fill in all nodes at level 1
low = sb;
high = sb + offset - 1;
if(high >= low) {
    spawn(low, high) {
        int count;

        sum[$] = 0;
        for(count = 0; count < k; count++) {
            sum[$] += sum[k * $ + count + 1];
        }
    }
}

/*
 * We will count the above "Iteration 1" as 1 iteration in
 * the climbing the tree loop below, for simplicity.
 * This gives an upper bound, since the "Iteration 1"
 * section above does slightly less.
 */

// Climb the tree
level = 1;
while(level > 0) {
    level --;
    startindex = (pow(k,level) - 1) / (k - 1);
    layersize = pow(k,level);

    low = startindex;
    high = startindex + layersize - 1;
    spawn(low, high) {
        int count;

        /*
         * All the sum[X] elements are read at once
         * for the below loop using prefetching.
         *
         * RTMs = 1
         * (prefetch) Computation Depth = k
         */

        sum[$] = 0;
        for(count = 0; count < k; count++) {
            sum[$] += sum[k * $ + count + 1];
        }

        /*
         * Computation Depth of loop = 2k + 1
         */
    }
} // 1 RTM join

/*
 * For the above stage of climbing the tree:
 * LSRTM = 2 * logN
 * Computation Depth = (3k + 9) * logN + 1
 * Computation Work = (3k + 2) * (N - 1) / (k - 1)
 *
 * The (N - 1) / (k - 1) factor of the work is the
 * number of nodes in a k-ary tree of depth logN - 1
 * [there is no work for the leaves at depth logN]
 *
 * Computation Work / min(p, p_i) =
 * ((3k + 2) * (N - min(p, N-1) - 1) / (k - 1)) / p
 * + (3k + 2) * log_k(p)
```

```
 *
 * For each level where number of nodes < p, the denominator is p_i.
 * Otherwise the denominator is p.  This gives the above formula.
 */
base[0] = 0;  //set root base = 0

// Descend the tree
startindex = 0;
while(level < l) {
    layersize = pow(k, level);

    low = startindex;
    high = startindex + layersize - 1;
    spawn(low, high) {
        int count, tempbase;
        tempbase = base[$];

        /*
         * All the sum[X] elements are read at once
         * for the below loop using prefetching.
         *
         * RTMs = 1
         * (prefetch) Computation Depth = k
         */

        for(count = 0; count < k; count++) {
                base[k*$ + count + 1] = tempbase;
                tempbase += sum[k*$ + count + 1];
        }

        /*
         *  Computation Depth = 3k;
         */

    } //1 RTM join

    startindex += layersize;
    level++;
}

// Iteration h: fill in all nodes at level l+1
low = sb;
high = sb + offset - 1;
if(high >= low) {
        spawn(low, high) {
                int count, tempbase;

                tempbase = base[$];

            for(count = 0; count < k; count++) {
                base[k*$ + count + 1] = tempbase;
                tempbase += sum[k*$ + count + 1];
            }
        }
}

/*
 * For simplicity count "Iteration h" as part of
 * the loop to descend the tree.  This gives
 * an upper bound.
 *
 * For the stage of descending the tree:
 * LSRTM = 2 * logN
 * Computation Depth = (4k + 9) * logN + 2
 * Computation Work = (4k + 2) * (N - 1) / (k - 1)
 *
 * The (N - 1) / (k - 1) factor of the work is the
 * number of nodes in a k-ary tree of depth logN - 1
 * [there is no work for the nodes at depth logN]
 *
 * Computation Work / min(p, p_i) =
 * ((4k + 2) * (N - min(p, N-1) - 1) / (k - 1)) / p
 * + (4k + 2) * log_k p
 *
 * For each level where number of nodes < p, the denominator is p_i.
```

```
             * Otherwise the denominator is p.  This gives the above formula.
             */

            //Copy to result matrix
            low = 0;
            high = size - 1;
            spawn(low, high) {
                    result[$] = base[sr + $]; //1 RTM
            }

            /*
             * For above code:
             * LSRTM = 1
             * Computation Depth = 4
             * Computation Work = N
             */
            /*
             * For the whole algorithm:
             *
             * LSRTM = 4 * logN + 3
             * QD = 0
             * Computation Depth = (7k + 18) * logN + 2k + 39
             * Computation Work = 3N + (7k + 4) * (N - 1) / (k - 1)
             *
             * Computation Work / min(p, p_i) =
             * (3N + (7k + 4) * (N - min(p, p_i) - 1) / (k - 1)) / p
             * + (7k +4) * log_k p
             */
}
```

No-Busy-Wait K-Ary Prefix-Sums

```
/*
 * void kps(...)
 *
 * The function computes prefix sums by using a k-ary tree.
 * k is defined by the parameter k to the function.
 *
 * Input:
 * increment[] - an array of increment values
 * k - the value of k to use for the k-ary tree
 * size - the size of the increment[] array
 *
 * Output:
 * result[] - this array is filled with the prefix sum on
 * the values of the array increment[]
 *
 */
void kps(int increment[], int result[], int k, int size) {
    register int low, high;
    int height = 20; //note: height should be as large as log_k(size)
    //int layersize[height]; // number of nodes in layer i
    int base[height*size]; // base of leaf is its value after PS,
                    // base of internal node is the base of leftmost leaf
    int sum[height*size]; // the value of sum for a node is the sum
                    // of the values of increment for all its leaves
    int isLeaf[height * size]; // if a leaf: 1; if not a leaf: 0
    int passIndex[height * size]; //array for passing index to child threads

    int iteration = 0; // determines the current height in the tree

    int temp;
    int done; //a loop control variable

    int l,    //level where leaves start
        sb,   //index where leaves would start if size is a power of 2
        d,    //size - k^l
        offset, //how much to offset due to size not being power of 2
        sr,   //sb + offset
        over, //number of leaves at level l
        under, //number of leaves at level l + 1
        sbp1; //index of one level higher from sb
```

```
    int fill; //nodes to fill in with 0 to make all nodes have k children

    int level, startindex, layersize;

    int i;

    /*
     * With non-blocking writes 0 RTM is required to initialize
     * the function parameters: k and size
     * 0 RTM is required to initialize local variables such as height
     */

    //Special case if size == 1
    if(size == 1) { //the check has 0 RTM because size is cached.
        result[0] = 0;
        return;
    }

    /*
     * 21 lines of code above, means computation cost = 21 up to this point.
     */

    //calculate location for leaves in the complete representation
    l = log(size) / log(k);

    sb = (pow(k,l) - 1) / (k - 1); //this is derived from geometric series
    sbp1 = (pow(k,l+1) - 1) / (k - 1);

    d = size - pow(k,l);
    offset = CEIL(((double) d) / (k - 1));
    sr = sb + offset;
    over = pow(k,l) - offset;
    under = size - over;

    /*
     * Computation cost = 8
     */

    //printf("l = %d, sb = %d, d = %d, offset = %d, sr = %d, over = %d\",
    //        l, sb, d, offset, sr, over);

    // Copy increment[...] to leaves of sum[...]

    low = 0;
    high = size - 1;
    spawn(low, high) {
        if($ < under) {
            sum[$ + sbp1] = increment[$]; // 1 RTM
            isLeaf[$ + sbp1] = 1;
        }
        else {
            sum[($ - under) + sb + offset] = increment[$]; //1 RTM
            isLeaf[($ - under) + sb + offset] = 1;
        }
    } // 1 RTM join

    /*
     * For code above:
     *
     * LSRTM = 2
     * Computation Depth = 6
     * Computation Work = 3N
     */

    // Make some 0 leaves at level l+1 so all nodes have exactly
    // k children

    fill = (k - (under % k)) % k;
    for(i = 0; i < fill; i++) {
        sum[sbp1 + under + i] = 0;
    }

    /*
     * Computation Cost = 2k + 1
     */
```

```
//Climb tree
low = sr;
high = sr + size + fill - 1;
spawn(low, high) {
    int gate, count, alive;
    int index= $;
    alive = 1;

    while(alive) {
        index = (index - 1) / k;

        gate = 1;
        psm(gate, &gatekeeper[index]); //1 RTM

        if(gate == k - 1) {
            /*
             * Using prefetching, the sum[X] elements
             * in the following loop are read all at once
             * LSRTM = 1
             * (prefetching) Computation Depth = k
             */

            sum[index] = 0;
            for(count = 0; count < k; count++) {
                sum[index] += sum[k*index + count + 1];
            }

            if(index == 0) {
                alive = 0;
            }

            /*
             * Computation Depth = 2k + 3;
             */
        }
        else {
            alive = 0;
        }
    }
} // 1 RTM join

/*
 * For code above:
 *
 * LSRTM = 2 * logN + 1
 * QD = k * logN
 * Computation Depth = (8 + 2k) * (logN + 1) + 6
 * Computation Work = (8 + 2k) * (N - 1) / (k - 1) + 8N
 *
 * The (N - 1) / (k - 1) factor of the work comes
 * from counting the total nodes in a tree with logN - 1
 * levels.  Each of the leaves at level logN only
 * executes the first 8 lines inside the spawn block
 * (that is, up to the check of the gatekeeper) before
 * most die and only 1 thread per parent continues.  This
 * gives the 8N term.
 *
 * Computation Work / min(p, p_i) =
 * ((8 + 2k)*(N - min(p, N-1) - 1)/(k-1) + 8N) / p
 * + (8 + 2k) * log_k p
 *
 * For each level where number of nodes < p, the denominator is p_i.
 * Otherwise the denominator is p.  This gives the above formula.
 */

base[0] = 0; //set root base = 0

low = 0;
high = 0;
spawn(low, high) {
    int count, tempbase;
    int index = $;
    int newID;

    if($ != 0) {
        index = passIndex[$];
```

```
            }
        while(isLeaf[index] == 0) {
            tempbase = base[index];

            /*
             * The k - 1 calls to sspawn can be executed with
             * a single kspawn instruction.
             * The elements sum[X] are read all at once using
             * prefetching.
             *
             * LSRTM = 2
             * (kspawn and prefetching) Computation Depth = k +1
             */

            for(count = 0; count < k; count++) {
                base[k*index + count + 1] = tempbase;
                tempbase += sum[k*index + count + 1];

                if(count != 0) {
                    sspawn(newID) {
                        passIndex[newID] = k*index + count + 1;
                    }
                }
            }

            index = k*index + 1;

            /*
             * Computation Depth = 6k + 1
             */
        }
} //1 RTM join

/*
 * For code above:
 *
 * LSRTM = 2 * logN + 1
 * Computation Depth = (3 + 6k) * logN + 9
 * Computation Work = (3 + 6k) * (N - 1) / (k - 1) + 6N + 6
 *
 * The (N - 1) / (k - 1) factor of the work comes
 * from counting the total nodes in a tree with logN - 1
 * levels.  Each of the leaves at level logN only
 * executes the first 6 lines inside the spawn block
 * (up to the check of isLeaf) before dying. This
 * gives the 6N term.
 *
 * Computation Work / min(p, p_i) =
 * ((3 + 6k)*(N - min(p, N-1) - 1) / (k-1) + 6N + 6)/p
 * + (3 + 6k) * log_k p
 */

//Copy to result matrix
low = 0;
high = size - 1;
spawn(low, high) {
    result[$] = base[sr + $]; //1 RTM
} //1 RTM join

/*
 * LSRTM = 2
 * Computation Depth = 4
 * Computation Work = N
 */

/*
 * For the whole algorithm:
 *
 * LSRTM = 4 * logN + 6
 * QD = k * logN
 * Computation Depth = (11 + 8k) * logN + 2k + 55
 * Computation Work = (11 + 8k) * (N - 1) / (k - 1) + 18N + 6
 *
 * Computation Work / min(p, p_i) =
 * ((11 + 8k)*(N - min(p, N-1) - 1) / (k-1) + 18N + 6) / p
```

```
     * + (11 + 8k)*log_k p
     */
}
```

Serial Summation

```
/*
 * void sum(...)
 * Function computes a sum
 *
 * Input:
 * increment[] - an array of increment values
 * k - the value of k to use for the k-ary tree
 * size - the size of the increment[] array
 *
 * Output:
 * sum
 *
 */
void sum(int increment[], int *sum, int k, int size) {
    int i;
    *sum = 0;

    for(i = 0; i < size; i++) {
        *sum += increment[i];
    }

    /*
     * LSRTM = 1
     * At first, 1 RTM is needed to read increment.  However, later reads
     * to increment are accomplished with prefetch.
     *
     * QD = 0
     * Computation = 2N
     */
}
```

Serial Prefix-Sums

```
/*
 * void kps(...)
 *
 * The function computes prefix sums serially.
 *
 * Input:
 * increment[] - an array of increment values
 * k - the value of k to use for the k-ary tree (not used)
 * size - the size of the increment[] array
 *
 * Output:
 * result[] - this array is filled with the prefix sum on the values of the array increment[]
 *
 */
void kps(int increment[], int result[], int k, int size) {
    int i;
    int PS = 0;

    for(i = 0; i < size; i++) {
        result[i] = PS;
        PS += increment[i];
    }

    /*
     * LSRTM = 1
     * At first, 1 RTM is needed to read increment.  However, later reads
     * to increment are accomplished with prefetch.
     *
     * QD = 0
```

```
         * Computation = 3N
         */
}
```

Flattened BFS Algorithm

```
/* Flattened BFS implementation
 */
psBaseReg newLevelGR, notDone; // global register for ps()

int * currentLevelSet, * newLevelSet, *tmpSet; // pointers to vertex sets
main() {

  int currentLevel;
  int currentLevelSize;
  register int low,high;
  int i;
  int nIntervals;

  /* variables for the edgeSet filling algorithm */
  int workPerThread;
  int maxDegree,nMax; // hold info about heaviest node

  /* initialize for first level */
  currentLevel = 0;
  currentLevelSize = 1;
  currentLevelSet = temp1;
  newLevelSet = temp2;

  currentLevelSet[0] = START_NODE;
  level[START_NODE]=0;
  gatekeeper[START_NODE]=1; // mark start node visited

  /* All of the above initializations can be done with non-blocking writes.
   * using 0 RTM
   * 7 lines of code above, cost = 9 up to this point
   */

  // 0 RTM, currentLevelSize in cache
  while (currentLevelSize > 0) {  // while we have nodes to explore

    /* clear the markers array so we know which values are uninitialized
     */
    low = 0;
    high = NTCU - 1; // 0 RTM, NTCU in cache
    spawn(low,high) {
      markers[$] = UNINITIALIZED; // 0 RTM, non-blocking write. UNITIALIZED is a constant
      // the final non-blocking write is overlapped with the RTM of the join
    } // 1 RTM for join

    /* Total for this spawn block + initialization steps before:
     * RTM Time = 1
     * Computation time = 1
     * Computation work = NTCU, number of TCUs.
     */

    /**********************************************************************
     *  Step 1:
     *  Compute prefix sums of the degrees of vertices in current level set
     **********************************************************************/

    /*
     * We use the k-ary tree Prefix_sums function.
     * Changes from "standard" prefix_sums:
     *    - also computes maximum element. this adds to computation time of
     *      upward traversal of k-ary tree
     */

    // first get all the degrees in an array
    low = 0;
    high = currentLevelSize-1;
```

```
spawn(low,high) {
  register int LR;
  /* prefetch crtLevelSet[$]
   * this can be overlaped with the ps below,
   * so it takes 0 RTM and 1 computation
   */
  LR = 1;
  ps(LR,GR); // 1 RTM
  degs[GR] = degrees[crtLevelSet[$]];
  // 1 RTM to read degrees[crtLevelSet[$]]. using non-blocking write
  // last write is overlapped with the RTM of the join
} // 1 RTM for join

/* the above spawn block:
 *  RTM Time = 3
 *  Computation Time = 3
 *  Computation Work = 3*Ni
 */

 kary_psums_and_max(degs,prefix_sums,k,currentLevelSize, maxDegree);

/*
 * this function has:
 *  RTM Time = 4 log_k (Ni)
 *  Computation Time = (17 + 9k) log_k(Ni) + 13
 *  Computation Work = (17 + 9k) Ni + 13
 */
outgoingEdgesSize = prefix_sums[currentLevelSize + 1]; // total sum. 0 RTM (cached)

/* compute work per thread and number of edge intervals
 * cost = 3 when problem is large enough, cost = 5 otherwise
 * no RTMs,everything is in cache and using non-blocking writes
 */
nIntervals = NTCU; // constant
workPerThread = outgoingEdgesSize / NTCU + 1;
if (workPerThread < THRESHOLD) {
  workPerThread = THRESHOLD;
  nIntervals = (outgoingEdgesSize / workPerThread) + 1;
}
/* Total Step 1:
 * RTM Time: 4 log_k Ni + 4
 * Computation Time: (17+9k) log_k Ni + 23
 * Computation Work: (19+9k) Ni + 21
 */

/*************************************************************************
 * Step 2:
 *  Apply parallel pointer jumping algorithm to find all marker edges
 *************************************************************************/

nMax = maxDegree / workPerThread; // 0 RTM, all in cache

/* Step 2.1 Pointer jumping - Fill in one entry per vertex */
low = 0;
// one thread for each node in current layer
high = currentLevelSize - 1;
spawn(low,high) {
  int crtVertex;
  int s,deg;
  int ncrossed;

  /*
   * prefetch currentLevelSet[$], prefix_sums[$]
   * 1 RTM, computation cost = 2
   */

  crtVertex = currentLevelSet[$]; // 0 RTM, value is in cache
  s = prefix_sums[$] / workPerThread + 1; // 0 RTM, values in cache
  // how many (if any) boundaries it crosses.
  ncrossed = (prefix_sums[$] + degrees[crtVertex]) / workPerThread - s;
    // above line has 1 RTM, degrees[] cannot be prefetched above, depends on crtVertex
  if (ncrossed>0) { // crosses at least one boundary
markers[s] = s * workPerThread - prefix_sums[$]; // this is the edge index (offset)
markerNodes[s] = $;  // this is the vertex
  }
  // last write is overlapped with the RTM of the join
```

Advancing PRAM Algorithms into Parallel Programs

```
    } // 1 RTM for join


/*
 * Total for the above spawn block
 *  RTM Time = 3
 *
 * Computation Time = 9
 * Computation Work <= 9 Ni
 */

/* Step  2.2 Actual pointer jumping */

jump = 1; notDone = 1;
while (notDone) { // is updated in parallel mode, 1 RTM to read it
  notDone = 0; // reset
  low=0; high = NTCU-1;
  spawn(low,high) {
register int LR;
// will be broadcasted: jump, workPerThread, UNINITIALIZED constant
/* Prefetch: markers[$], markers[$-jump]
 *  1 RTM, 2 Computation, 1 QD
 */
if (markers[$] == UNINITIALIZED) { // 0 RTM, cached
  if (markers[$-jump] != UNINITIALIZED) { // 0 RTM, cached
    // found one initialized marker
    markers[$] = markers[$-jump] + s * workPerThread;
    markerNodes[$] = markerNodes[$-jump];
  }
  else { // marker still not initialized. mark notDone
    LR = 1;
    ps(LR,notDone);   // 1 RTM
  }
}

  } // 1 RTM for join
  /* Total for the above spawn block + setup
   *  RTM Time = 3
   *  Computation time =  6
   *  Computation work = 6
   *
   */
  jump = jump * 2; // non-blocking write
}

/* above loop executes at most log NTCU times
 *  Total:
 *    RTM Time = 4 log NTCU
 *    Computation time = 10 log NTCU (includes serial code)
 *    Computation work = 6 NTCU
 */


/* Total step 2:
 *  RTM = 4 log NTCU + 3
 *  Computation depth = 10 log NTCU + 9
 *  Computation work. section 1: 9Ni, section 2=10 NTCU
 */

/************************************************************************
 * Step 3.
 * One thread per edge interval.
 * Do work for each edge, add it to new level if new
 ************************************************************************/

low = 0;
high = nIntervals;     // one thread for each interval
newLevelGR = 0; // empty set of nodes
spawn(low,high) {
  int crtEdge,freshNode,antiParEdge;
  int crtNode,i3;
  int gatekLR; // local register for gatekeeper psm
  int newLevelLR; // local register for new level size

  /*
   * Prefetch markerNodes[$], markers[$]
```

```
     *   1 RTM, computation cost 2
     */

    crtNodeIdx = markerNodes[$]; // cached, 0 RTM
    crtEdgeOffset = markers[$]; // cached, 0 RTM

    /* prefetch currentLevelSet[crtNodeIdx],
           vertices[currentLevelSet[crtNodeIdx]],
        degrees[currentLevelSet[crtNodeIdx]]
     * 2 RTM, cost = 2
     */

    // workPerThread is broadcasted, 0 RTM to read it
    for (i3=0;i3<workPerThread;i3++) {
 crtEdge = vertices[currentLevelSet[crtNodeIdx]] + crtEdgeOffset; // cached, 0 RTM
 // traverse edge and get new vertex
 freshNode = edges[crtEdge][1]; // 1 RTM
 if (freshNode!= -1) { // edge could be marked removed
   gatekLR = 1;
   psm(gatekLR,&gatekeeper[freshNode]); // 1 RTM, queuing for the indegree

   if (gatekLR == 0) { // destination vertex unvisited
     newLevelLR = 1;
     // increase size of new level set
     ps(newLevelLR,newLevelGR); // 1 RTM
     // store fresh node in new level. next two lines are 0 RTM, non-blocking writes
     newLevelSet[newLevelLR] = freshNode;
     level[freshNode] = currentLevel + 1;
     // now mark antiparallel edge as deleted
     antiParEdge = antiParallel[crtEdge]; // 0 RTM, prefetched
     edges[antiParEdge][1] =-1; edges[antiParEdge][0] =-1; // 0 RTM, non-blocking writes

   } // end if
 } // end if freshNode

 /* Previous if block costs:
  * 2 RTM, computation 8 for a "fresh" vertex
  * or
  * 1 RTM, computation 2 for a "visited" vertex
  */

 crtEdgeOffset++;
 if (crtEdgeOffset>=degrees[currentLevelSet[crtNodeIdx]]) { // exhausted all the edges?
   // 0 RTM, value is in cache
   crtNodeIdx++;
   crtEdgeOffset = 0;
   /* We have new current node. prefetch its data
      prefetch currentLevelSet[crtNodeIdx],
    *   vertices[currentLevelSet[crtNodeIdx]],
    *      degrees[currentLevelSet[crtNodeIdx]]
    * 2 RTM, cost = 2
    */
 }

 /* This if and instruction before it cost:
  * 2 RTM, 6 computation for each new marker edge in interval
  * or
  * 2 computation for all other edges
  */

 if (crtNodeIdx>= currentLevelSet)
   break;
 // this if is 0 RTM, 1 computation.

    } // end for

    /* Previous loop is executed C = Ei/p times.
     * We assume Ni nodes are "fresh", worst case analysis
     * Total over all iterations. AA is the number of marker edges in interval.
     *  WITHOUT PREFETCHING:
     *    RTM: 3*C + 2 AA
     *    Computation: 11*C + 4 AA
     */
    // last write is overlapped with the RTM of the join
 } // 1 RTM for join
```

```
    /*
     * Total for above spawn block + initialization: (C=Ei/p, AA = N/p = # marker edges)
     * WITHOUT PREFETCHING for multiple edges: RTM Time = 3*C + 3 + 2 AA
     * WITH PREFETCHING for multiple edges: RTM Time = 3 + 3 + 2
     * Computation Time = 8 + 7*C + 16 AA
     * Computation Work = 8p + 7E + 16N
     */

    // move to next layer
    currentLevel++;
    currentLevelSize = newLevelGR; // from the prefix-sums
    // "swap" currentLevelSet with newLevelSet
    tmpSet = newLevelSet;
    newLevelSet = currentLevelSet;
    currentLevelSet = tmpSet;

    /* all these above steps: 0 RTM, 5 computation */

  } // end while
  /*
   * Total for one BFS level (one iteration of above while loop):
   *    W/O PRE: RTM Time = 4 log_k Ni + 4 |Ei|/p + 11 + LSRTM of PSUMS
   *    W PRE  : RTM Time = 4 log_k Ni + 4     + 11 + LSRTM of PSUMS
   *    Computation Time =
   *    Comp Work =
   */
}
```

Single-Spawn BFS Algorithm

```
/* BFS implementation using single-spawn operation
 * for nesting
 */
psBaseReg newLevelGR; // global register for new level
set

int * currentLevelSet, * newLevelSet, *tmpSet; //
pointers to level sets

main() {

  int currentLevel;
  int currentLevelSize;
  int low, high;
  int i;

  currentLevel = 0;
  currentLevelSize = 1;
  currentLevelSet = temp1;
  newLevelSet = temp2;

  currentLevelSet[0] = START_NODE; // store the vertex# this thread will handle

  /*
   * 0 RTMs, 5 computation
   */

  while (currentLevelSize > 0) { // while we have nodes to explore
    newLevelGR = 0;
    low = 0;
    high = currentLevelSize - 1; // one thread for each node in current layer

    spawn(low,high) {
      int gatekLR, newLevelLR, newTID;
      int freshNode, antiParEdge;

      /*
       * All threads need to read their initialization data
       *   nForks[$] and currentEdge[$]
       */
      if ($ < currentLevelSize ) { // 0 RTM
      /*
```

```
     * "Original" threads read it explicitly from the graph
     */
    // only start degree-1 new threads, current thread will handle one edge
    nForks[$] = degrees[currentLevelSet[$]] - 1;  // 2 RTM
    // this thread will handle first outgoing edge
    currentEdge[$] = vertices[currentLevelSet[$]];  // 1 RTM
       }
       else {
    /*
     * Single-spawned threads, need to "wait" until init values
     * from the parent are written
     */

    while (locks[$]!=1) ;   // busy wait until it gets the signal

       } // end if
       /* The above if block takes
        *  3 RTM, 3 computation for "original" threads
        *  for child threads: 1 RTM for synchronization. 2 computation
        */

       while (nForks[$] > 0) { // 1 computation
    // this is executed for each child thread spawned
     sspawn(newTID) { // 1 RTM
       /*
        * writing initialization data for child threads.
        * children will wait till this data is commited
        */
       nForks[newTID] = (nForks[$]+1)/2 -1;
       nForks[$] = nForks[$] - nForks[newTID]-1;
       currentEdge[newTID] = currentEdge[$] + nForks[$]+1;
       locks[newTID] = 1; // GIVE THE GO SIGNAL!

         /*
          *  0 RTM
          *  4 computation
          */
     }
     /* For each child thread:
      *  1 RTM
      *  5 computation
      */
       } // done with forking

       /*
        * Prefetch edges[currentEdge[$]][1], antiParallel[currentEdge[$]]
        *  1 RTM, 2 computation
        */

       // let's handle one edge
       freshNode = edges[currentEdge[$]][1]; // 0 RTM, value was prefetched
       if (freshNode != -1 ) { // if edge hasn't been deleted
    gatekLR = 1;
    // test gatekeeper
    psm(gatekLR,&gatekeeper[freshNode]);   // 1 RTM. GQD queuing

    if (gatekLR == 0) { // destination vertex unvisited!
      newLevelLR = 1;
      // increase size of new level set
      ps(newLevelLR,newLevelGR);  // 1 RTM
      // store fresh node in new level
      newLevelSet[newLevelLR] = freshNode;
      level[freshNode] = currentLevel + 1;
      // now mark antiparallel edge as deleted
      antiParEdge = antiParallel[currentEdge[$]]; // 0 RTM, value was prefetched
      edges[antiParEdge][1] = -1;
      edges[antiParEdge][0] = -1;
    } // end if
       } // end if

       /*
        * Previous if block costs:
        *  2 RTM, 10 computation for "fresh" vertex/
        *  0 RTM, 2 computation for visited vertex
        */
```

```
    /*
     * Final write is blocking, but the RTM overlaps the join.
     */
    } // 1 RTM join

    /* Computation for a child thread that starts one single child: 19 */

    // move to next layer
    currentLevel++;
    currentLevelSize = newLevelGR; // from the prefix-sums
    // "swap" currentLevelSet with newLevelSet
    tmpSet = newLevelSet;
    newLevelSet = currentLevelSet;
    currentLevelSet = tmpSet;
    /* the above 5 lines of code: 0 RTM, 5 computation */

  } // end while
}
```

k-Spawn BFS Algorithm

The only difference between the single-spawn BFS algorithm and the k-spawn is the while loop that is starting children threads. We are including only that section of the code here, the rest is identical with the code in the BFS Single-Spawn implementation.

```
        while (nForks[$] > 0) { // 1 computation
    // this is executed for each child thread spawned
     kspawn(newTID) { // 1 RTM for kSpawn
        // newTID is the lowest of the k TIDs allocated by k-spawn
        // The other ones are newTID+1, newTID+2,..., newTID+(k-1)
        /*
         * writing initialization data for child threads.
         * children will wait till this data is commited
         */

        slice = nForks[$] / k;
        nForks[$] = nForks[$] - slice; // substract a slice for parent thread

        for (child=0;child<k;child++) {
           // initialize nForks[newTid + child] and currentEdge[newTid + child]
           nForks[newTID + child] = max(slice,nForks[$]); // for rounding
           currentEdge[newTID] = currentEdge[$] + child * slice;
           nForks[$] = nForks[$] - nForks[newTID + child];
           locks[newTID + child] = 1; // GIVE THE GO SIGNAL!
        }
          /*
           * loop is executed k times.
           * Each iterration:
           *    0 RTM
           *    4 computation
           */
      }
       /* For each k child threads:
        *  1 RTM
        *  2+4*k computation
        */
       } // done with forking
```

References

[1] A. V. Aho and J. D. Ullman. *Foundations of Computer Science*. W. H. Freeman & Co., New York, 1994.

[2] R. Allen, D. Callahan, and K. Kennedy. Automatic decomposition of scientific programs for parallel execution. In *POPL '87: Proceedings of the 14th ACM SIGACT-SIGPLAN Symposium on Principles of Programming Languages*, pages 63–76, New York, 1987. ACM Press.

[3] G. S. Almasi and A. Gottlieb. *Highly Parallel Computing*. Benjamin Cummings, Redwood City, CA, 1994.

[4] S. P. Amarasinghe, J. M. Anderson, M. S. Lam, and C. W. Tseng. The SUIF compiler for scalable parallel machines. In *Proceedings of the Seventh SIAM Conference on Parallel Processing for Scientific Computing*, 1995.

[5] S. Baase. *Computer Algorithms: Introduction to Design and Analysis*. Addison Wesley, Boston, MA, 1988.

[6] D. A. Bader, G. Cong, and J. Feo. On the architectural requirements for efficient execution of graph algorithms. In *ICPP '05: Proceedings of the 2005 International Conference on Parallel Processing (ICPP'05)*, pages 547–556, Washington, DC, 2005. IEEE Computer Society.

[7] A. Balkan, G. Qu, and U. Vishkin. Mesh-of-trees and alternative interconnection networks for single-chip parallel processing. In *ASAP 2006: 17th IEEE Internation Conference on Application-Specific Systems, Architectures and Processors*, pages 73–80, Steamboat Springs, Colorado, 2006. Best Paper Award.

[8] A. O. Balkan and U. Vishkin. Programmer's manual for XMTC language, XMTC compiler and XMT simulator. Technical Report UMIACS-TR 2005-45, University of Maryland Institute for Advanced Computer Studies (UMIACS), February 2006.

[9] G. Bilardi and A. Nicolau. Adaptive bitonic sorting: an optimal parallel algorithm for shared-memory machines. *SIAM J. Comput.*, 18(2):216–228, 1989.

[10] G. E. Blelloch. Programming parallel algorithms. *Commun. ACM*, 39(3):85–97, 1996.

[11] G. E. Blelloch, C. E. Leiserson, B. M. Maggs, C. G. Plaxton, S. J. Smith, and M. Zagha. A comparison of sorting algorithms for the connection machine CM-2. In *SPAA '91: Proceedings of the Third Annual ACM Symposium on Parallel Algorithms and Architectures*, pages 3–16, New York, 1991. ACM Press.

[12] R. D. Blumofe and C. E. Leiserson. Space-efficient scheduling of multithreaded computations. In *25th Annual ACM Symposium on the Theory of Computing (STOC '93)*, pages 362–371, 1993.

[13] R. P. Brent. The parallel evaluation of general arithmetic expressions. *J. ACM*, 21(2):201–206, 1974.

[14] D. Burger and T. M. Austin. The simplescalar tool set, version 2.0. *SIGARCH Comput. Archit. News*, 25(3):13–25, 1997.

[15] L. Carter, J. Feo, and A. Snavely. Performance and programming experience on the tera mta. In *Proceedings SIAM Conference on Parallel Processing*, 1999.

[16] R. Cole and U. Vishkin. Deterministic coin tossing and accelerating cascades: micro and macro techniques for designing parallel algorithms. In *STOC '86: Proceedings of the Eighteenth Annual ACM Symposium on Theory of Computing*, pages 206–219, New York, 1986. ACM Press.

[17] R. Cole and O. Zajicek. The APRAM: incorporating asynchrony into the PRAM model. In *SPAA '89: Proceedings of the First Annual ACM Symposium on Parallel Algorithms and Architectures*, pages 169–178, New York, 1989. ACM Press.

[18] T. H. Cormen, C. E. Leiserson, and R. L. Rivest. *Introduction to Algorithms*. 1st ed., MIT Press, Cambridge, MA, 1990.

[19] D. E. Culler and J. P. Singh. *Parallel Computer Architecture: A Hardware/Software Approach*. Morgan Kaufmann Publishers Inc., San Francisco, CA, 1999.

[20] S. Dascal and U. Vishkin. Experiments with list ranking for explicit multi-threaded (xmt) instruction parallelism. *J. Exp. Algorithmics*, 5:10, 2000. Special issue for the 3rd Workshop on Algorithms Engineering (WAE'99), London, UK, July 1999.

[21] R. Dementiev, M. Klein, and W. J. Paul. Performance of MP3D on the SB-PRAM prototype (research note). In *Euro-Par '02: Proceedings of the 8th International Euro-Par Conference on Parallel Processing*, pages 132–136, London, UK, 2002. Springer-Verlag.

[22] A. C. Dusseau, D. E. Culler, K. E. Schauser, and R. P. Martin. Fast parallel sorting under Logp: Experience with the CM-5. *IEEE Trans. Parallel Distrib. Syst.*, 7(8):791–805, 1996.

[23] D. Eppstein and Z. Galil. Parallel algorithmic techniques for combinatorial computation. *Ann. Rev. of Comput. Sci.*, 3:233–283, 1988.

[24] J. Feo, D. Harper, S. Kahan, and P. Konecny. Eldorado. In *CF '05: Proceedings of the 2nd Conference on Computing Frontiers*, pages 28–34, New York, 2005. ACM Press.

[25] M. J. Garzaran, J. L. Briz, P. Ibanez, and V. Vinals. Hardware prefetching in bus-based multiprocessors: pattern characterization and cost-effective hardware. In *Proceedings of the Euromicro Workshop on Parallel and Distributed Processing*, pages 345–354, 2001.

[26] P. B. Gibbons. A more practical PRAM model. In *SPAA '89: Proceedings of the First Annual ACM Symposium on Parallel Algorithms and Architectures*, pages 158–168, New York, 1989. ACM Press.

[27] P. B. Gibbons, Y. Matias, and V. Ramachandran. The queue-read queue-write asynchronous PRAM model. *Theor. Comput. Sci.*, 196(1-2):3–29, 1998.

[28] A. Gottlieb, R. Grishman, C. P. Kruskal, K. P. McAuliffe, L. Rudolph, and M. Snir. The NYU ultracomputer: designing a mimd, shared-memory parallel machine (extended abstract). In *ISCA '82: Proceedings of the 9th Annual Symposium on Computer Architecture*, pages 27–42, Los Alamitos, CA, 1982. IEEE Computer Society Press.

[29] A. Gress and G. Zachmann. GPU-BiSort: Optimal parallel sorting on stream architectures. In *IPDPS '06: Proceedings of the 20th IEEE International Parallel and Distributed Processing Symposium (IPDPS'06)*, 2006. To appear.

[30] P. Gu and U. Vishkin. Case study of gate-level logic simulation on an extermely fine-grained chip multiprocessor. *Journal of Embedded Computing, Special Issue on Embedded Single-Chip Multicore Architectures and Related Research—from System Design to Application Support*, 2006. To appear.

[31] M. J. Harris, W. V. Baxter, T. Scheuermann, and A. Lastra. Simulation of cloud dynamics on graphics hardware. In *HWWS '03: Proceedings of the ACM SIGGRAPH/EUROGRAPHICS Conference on Graphics Hardware*, pages 92–101, Aire-la-Ville, Switzerland, Switzerland, 2003. Eurographics Association.

[32] D. R. Helman and J. JáJá. Designing practical efficient algorithms for symmetric multiprocessors. In *ALENEX '99: Selected papers from the International Workshop on Algorithm Engineering and Experimentation*, pages 37–56, London, UK, 1999. Springer-Verlag.

[33] J. L. Hennessy and D. A. Patterson. *Computer Architecture: A Quantitative Approach*. Morgan Kaufmann Publishers Inc., San Francisco, CA, 2003.

[34] L. Hochstein and V. R. Basili. An empirical study to compare two parallel programming models. *18th ACM Symposium on Parallelism in Algorithms and Architectures (SPAA '06)*, July 2006. Position Paper and Brief Announcement.

[35] L. Hochstein, U. Vishkin, J. Gilbert, and V. Basili. An empirical study to compare the productivity of two parallel programming models. Preprint, 2005.

[36] J. S. Huang and Y. C. Chow. Parallel sorting and data partitioning by sampling. In *Proceedings of the IEEE Computer Society's Seventh International Computer Software and Applications Conference*, pages 627–631, November 1983.

[37] J. JáJá. *An Introduction to Parallel Algorithms*. Addison Wesley Longman Publishing Co., Inc., Redwood City, CA, 1992.

[38] R. M. Karp and V. Ramachandran. Parallel algorithms for shared-memory machines. *Handbook of Theoretical Computer science (vol. A): Algorithms and Complexity*, pages 869–941, 1990.

[39] J. Keller, C. W. Kessler, and J. L. Traff. *Practical PRAM Programming*. Wiley, New York, 2000.

[40] P. Kipfer, M. Segal, and R. Westermann. Uberflow: a GPU-based particle engine. In *HWWS '04: Proceedings of the ACM SIGGRAPH/EUROGRAPHICS Conference on Graphics Hardware*, pages 115–122, New York, 2004. ACM Press.

[41] R. E. Ladner and M. J. Fischer. Parallel prefix computation. *J. ACM*, 27(4):831–838, 1980.

[42] C. E. Leiserson and H. Prokop. Minicourse on multithreaded programming. http://supertech.csail.mit.edu/cilk/papers/index.html, 1998.

[43] U. Manber. *Introduction to Algorithms—A Creative Approach*. Addison Wesley, Boston, MA, 1989.

[44] K. Moreland and E. Angel. The FFT on a GPU. In *HWWS '03: Proceedings of the ACM SIGGRAPH/EUROGRAPHICS Conference on Graphics Hardware*, pages 112–119, Aire-la-Ville, Switzerland, 2003. Eurographics Association.

[45] D. Naishlos, J. Nuzman, C. W. Tseng, and U. Vishkin. Evaluating multi-threading in the prototype XMT environment. In *Proceedings 4th Workshop on Multithreaded Execution, Architecture, and Compilation*, 2000.

[46] D. Naishlos, J. Nuzman, C. W. Tseng, and U. Vishkin. Evaluating the XMT programming model. In *Proceedings of the 6th Workshop on High-Level Parallel Programming Models and Supportive Environments*, pages 95–108, 2001.

[47] D. Naishlos, J. Nuzman, C. W. Tseng, and U. Vishkin. Towards a first vertical prototyping of an extremely fine-grained parallel programming approach. In *Invited Special Issue for ACM-SPAA'01: TOCS 36,5*, pages 521–552, New York, 2003. Springer-Verlag.

[48] V. S. Pai and S. V. Adve. Comparing and combining read miss clustering and software prefetching. In *PACT '01: Proceedings of the 2001 International Conference on Parallel Architectures and Compilation Techniques*, page 292, Washington, DC, 2001. IEEE Computer Society.

[49] J. H. Reif and L. G. Valiant. A logarithmic time sort for linear size networks. *J. ACM*, 34(1):60–76, 1987.

[50] J. T. Schwartz. Ultracomputers. *ACM Trans. Program. Lang. Syst.*, 2(4):484–521, 1980.

[51] V. Shah and J. R. Gilbert. Sparse matrices in MATLAB: Design and implementation. In *HiPC*, pages 144–155, 2004.

[52] Y. Shiloach and U. Vishkin. An $o(n^2 \log n)$ parallel max-flow algorithm. *J. Algorithms*, 3(2):128–146, 1982.

[53] S. Skiena. *The Algorithm Design Manual*. Springer, November 1997.

[54] A. Snavely, L. Carter, J. Boisseau, A. Majumdar, K. S. Gatlin, N. Mitchell, J. Feo, and B. Koblenz. Multi-processor performance on the Tera MTA. In *Supercomputing '98: Proceedings of the 1998 ACM/IEEE Conference on Supercomputing (CDROM)*, pages 1–8, Washington, DC, 1998. IEEE Computer Society.

[55] X. Tian, R. Krishnaiyer, H. Saito, M. Girkar, and W. Li. Impact of compiler-based data-prefetching techniques on SPEC OMP application performance. In *IPDPS '05: Proceedings of the 19th IEEE International Parallel and Distributed Processing Symposium (IPDPS'05)—Papers*, page 53.1, Washington, DC, 2005. IEEE Computer Society.

[56] A. Tzannes, R. Barua, G. C. Caragea, and U. Vishkin. Issues in writing a parallel compiler starting from a serial compiler. Draft, 2006.

[57] U. Vishkin. From algorithm parallelism to instruction-level parallelism: An encode-decode chain using prefix-sum. In *SPAA '97: Proceedings of the 9th Annual ACM Symposium on Parallel Algorithms and Architectures*, pages 260–271, New York, 1997. ACM Press.

[58] U. Vishkin. A no-busy-wait balanced tree parallel algorithmic paradigm. In *SPAA '00: Proceedings of the Twelfth Annual ACM Symposium on Parallel Algorithms and Architectures*, pages 147–155, New York, 2000. ACM Press.

[59] U. Vishkin. Thinking in parallel: Some basic data-parallel algorithms and techniques. In use as class notes since 1993. http://www.umiacs.umd.edu/users/vishkin/PUBLICATIONS/classnotes.ps, February 2002.

[60] U. Vishkin, S. Dascal, E. Berkovich, and J. Nuzman. Explicit multi-threading (XMT) bridging models for instruction parallelism (extended abstract). In *SPAA '98: Proceedings of the Tenth Annual ACM Symposium on Parallel Algorithms and Architectures*, pages 140–151, New York, 1998. ACM Press.

[61] X. Wen and U. Vishkin. Pram-on-chip: First commitment to silicon. In *SPAA '07: Proceedings of the Nineteenth Annual ACM Symposium on Parallel Algorithms and Architectures*, pages 301–302, 2007, ACM Press.

[62] Balkan, M. Horak, G. Qu, and U. Vishkin. Layout-accurate design and implementation of a high-throughput interconnection network for single-chip parallel processing. In *Processings of Hot Interconnects 15*, pages 21–28, Stanford, CA, August 22–24, 2007. IEEE.

6
Deterministic and Randomized Sorting Algorithms for Parallel Disk Models

6.1	Introduction...	6-1
6.2	Parallel Disks Model	6-2
6.3	A Summary of Sorting Results on the Parallel Disks Model	6-2
6.4	The (ℓ, m)-Merge Sort (LMM)	6-4
	A Merging Algorithm • LMM on PDM	
6.5	A Practical Realization of Parallel Disks	6-8
6.6	Sorting $M\sqrt{M}$ Keys	6-9
	A Lower Bound • A Three-Pass Algorithm	
6.7	Algorithms with Good Expected Performance	6-10
	A Useful Lemma • An Expected Two-Pass Algorithm • An Expected Three-Pass Algorithm	
6.8	Optimal Integer Sorting	6-14
6.9	A Simple Optimal Randomized Sorting Algorithm ...	6-15
6.10	Conclusions	6-16
	Acknowledgments...	6-16
	References	6-16

Sanguthevar Rajasekaran
University of Connecticut

6.1 Introduction

Today's applications have to deal with voluminous data sets. The use of secondary storage devices such as disks is inevitable. Even a single disk may not be enough to handle I/O operations efficiently. Thus, researchers have proposed models with multiple disks.

One of the models (which refines prior models) that has been studied extensively is the parallel disks model (PDM) [40]. In this model, there are D disks and an associated sequential or parallel computer. In a single parallel I/O operation, a block of data from each of the D disks can be fetched into the main memory of the computer. A block consists of B records. We usually require that $M \geq 2DB$, where M is the internal memory size. At least DB amount of memory is needed to store the data fetched from the disks and the remaining part of the main memory can be used to overlap local computations with I/O operations. Algorithms for PDM have been devised for various fundamental problems. In the analysis of these algorithms, only I/O operations are counted since the local computations are usually very fast.

In this chapter, we survey sorting algorithms that have been proposed for models with multiple disks. We also investigate the issue of implementing such models in practice.

6.2 Parallel Disks Model

In this section, we provide details on PDM. A PDM consists of a sequential or parallel computer together with D disks. For any given problem, the input will be given in the disks and the output also is expected to be written in the disks. In one I/O operation, a block of B records can be brought into the core memory of the computer from each one of the D disks. It is typically assumed that the time for I/O operations is much more than that for local computations and, hence, the latter is ignored in any analysis.

If M is the internal memory size of the computer, then one usually requires that $M \geq 2DB$. A portion of this memory is used to store operational data whereas the other portion is used for storing prefetched data that enables overlap of local computations with I/O operations. From hereon, we use M to refer to only DB.

The sorting problem on the PDM can be defined as follows. There are a total of N records to begin with so that there are N/D records in each disk. The problem is to rearrange the records such that they are in either ascending order or descending order with N/D records ending up in each disk.

6.3 A Summary of Sorting Results on the Parallel Disks Model

Sorting has been studied on a variety of models owing to its great importance. It is easy to show that a lower bound on the number of I/O read steps for parallel disk sorting is* $\Omega\left(\frac{N}{DB}\left[\frac{\log(N/B)}{\log(M/B)}\right]\right)$. Here N is the number of records to be sorted and M is the internal memory size of the computer. B is the block size and D is the number of parallel disks used. Numerous asymptotically optimal algorithms that make $O\left(\frac{N}{DB}\left[\frac{\log(N/B)}{\log(M/B)}\right]\right)$ I/O read steps have been proposed (see, e.g., References 1, 3, and 25).

Many asymptotically optimal algorithms have been developed for sorting on the PDM. Nodine and Vitter's optimal algorithm [24] involves the solution of certain matching problems. Aggarwal and Plaxton's optimal algorithm [1] is based on the Sharesort algorithm of Cypher and Plaxton (that was originally offered for the hypercube model). Vitter and Hutchinson's algorithm is also asymptotically optimal [39]. An optimal randomized algorithm was given by Vitter and Shriver for disk sorting [40]. Though these algorithms are highly nontrivial and theoretically interesting, the underlying constants in their time bounds are high.

One of the algorithms that people use in practice is the simple disk-striped mergesort (DSM) [6], even though it is not asymptotically optimal. DSM is simple and the underlying constant is small. In any I/O operation, DSM accesses the same portions of the D disks. This has the effect of having a single disk that can transfer DB records in a single I/O operation. DSM is basically an M/DB-way mergesort. To start with, initial runs are formed in one pass through the data. After this, the disks have N/M runs each of length M. Next, M/DB runs are merged at a time. Blocks of any run are uniformly striped across the disks so that in future they can be accessed in parallel utilizing the full bandwidth.

Each phase of merging can be done with one pass through the data. There are $\log(N/M)/\log(M/DB)$ phases and hence the total number of passes made by DSM is $\log(N/M)/\log(M/DB)$, that is, the total number of I/O read operations called for by DSM is $\frac{N}{DB}\left(1 + \frac{\log(N/M)}{\log(M/DB)}\right)$. Note that the constant here is just 1.

If N is a polynomial in M and B is small (that are readily satisfied in practice), the lower bound simply yields $\Omega(1)$ passes. All the optimal algorithms mentioned above make only $O(1)$ passes. Thus, the

---

*Throughout this chapter we use log to denote logarithms to the base 2 and ln to denote natural logarithms.

challenge in the design of sorting algorithms lies in reducing this constant. If $M = 2DB$, the number of passes made by DSM is $1 + \log(N/M)$, which is $\omega(1)$.

Two kinds of sorting algorithms can be found in the literature: merge- and distribution-based. The first kind of algorithms employ R-way merge for some suitable choice of R. Distribution-based algorithms use ideas similar to bucket sort.

Some of the works specifically focus on the design of practical sorting algorithms. For example, Pai et al. [26] analyzed the average case performance of a simple merging algorithm, with the help of an approximate model of average case inputs. Barve et al. [6] have proposed a simple randomized algorithm (SRM) and analyzed its performance. In a nutshell this algorithm works as follows. It employs an R-way merge algorithm (for an appropriate value of R). For each run, the disk for the placement of the first block is chosen randomly. The analysis of SRM involves the solution of certain occupancy problems. The expected number Reads$_{SRM}$ of I/O read operations needed in their algorithm is such that

$$\text{Reads}_{SRM} \leq \frac{N}{DB} + \frac{N}{DB} \frac{\log(N/M)}{\log kD} \frac{\log D}{k \log \log D} \left(1 + \frac{\log \log \log D}{\log \log D} + \frac{1 + \log k}{\log \log D} + O(1)\right).$$

SRM merges $R = kD$ runs at a time, for some integer k. The expected performance of SRM is optimal when $R = \Omega(D \log D)$. However, in this case, the internal memory needed is $\Omega(BD \log D)$. They have also compared SRM with DSM through simulations and shown that SRM performs better than DSM.

The LMM sort of Rajasekaran [29] is optimal when N, B, and M are polynomially related and is a generalization of Batcher's odd–even merge sort [7], Thompson and Kung's s^2-way merge sort [38], and Leighton's columnsort [19]. The algorithm is as simple as DSM. LMM makes less number of passes through the data than DSM when D is large.

Problems such as FFT computations (see, e.g., Reference 12), selection (see, e.g., Reference 30), and so forth have also been studied on the PDM.

Recently, many algorithms have been devised for problem sizes of practical interest. For instance, Dementiev and Sanders [14] have developed a sorting algorithm based on multiway merge that overlaps I/O and computation optimally. Their implementation sorts gigabytes of data and competes with the best practical implementations. Chaudhry, Cormen, and Wisniewski [11] have developed a novel variant of columnsort that sorts $M\sqrt{M}$ keys in three passes over the data (assuming that $B = M^{1/3}$). Their implementation is competitive with NOW-Sort. (By a pass we mean N/DB read I/O operations and the same number of write operations.) In Reference 9, Chaudhry and Cormen introduce some sophisticated engineering tools to speed up the algorithm of Reference 11 in practice. They also report a three-pass algorithm that sorts $M\sqrt{M}$ keys in this paper (assuming that $B = \Theta(M^{1/3})$). In [10], Chaudhry, Cormen, and Hamon present an algorithm that sorts $M^{5/3}/4^{2/3}$ keys (when $B = \Theta(M^{2/5})$). They combine columnsort and Revsort of Schnorr and Shamir [37] in a clever way. This paper also promotes the need for oblivious algorithms and the usefulness of mesh-based techniques in the context of out-of-core sorting. In fact, the LMM sort of Rajasekaran [29] is oblivious.

In Reference 34, Rajasekaran and Sen present several algorithms for sorting $\leq M^2$ keys. For most applications of practical interest, it seems safe to assume that $N \leq M^2$. For instance, if $M = 10^8$ (integers), then M^2 is 10^{16} (integers) (i.e. around 100,000 terabytes). In Reference 34 algorithms that have good expected performance have been presented. In particular, they are interested in algorithms that take only a small number of passes on an **overwhelming fraction** of all possible inputs. As an example, consider an algorithm $\mathcal{A}$ that takes two passes on at least $(1 - M^{-\alpha})$ fraction of all possible inputs and three passes on at most $M^{-\alpha}$ fraction of all possible inputs. If $M = 10^8$ and $\alpha = 2$, only on at most 10^{-14} % of all possible inputs, $\mathcal{A}$ will take more than two passes. Thus, algorithms of this kind will be of great practical importance.

All the algorithms given in Reference 34 use a block size of $\sqrt{M}$. Here is a list of some of the algorithms presented in Reference 34: (i) a three-pass algorithm for sorting $M\sqrt{M}$ keys. This algorithm is based on LMM sort and assumes that $B = \sqrt{M}$. In contrast, the algorithm of Chaudhry and Cormen [9] uses a block size of $M^{1/3}$ and sorts $M\sqrt{M}/\sqrt{2}$ keys in three passes; (ii) an expected two-pass algorithm that sorts nearly $M\sqrt{M}$ keys; (iii) an expected three-pass algorithm that sorts nearly $M^{1.75}$ keys; (iv) an expected six-pass algorithm that sorts nearly M^2 keys; and (v) a simple integer sorting algorithm that sorts integers in the range $[1, M^c]$ (for any constant c) in a constant number of passes (for any input size).

The integer sorting algorithm of Reference 34 has been employed in Reference 35 to design a simple randomized algorithm for sorting arbitrary keys. This algorithm is asymptotically optimal with high probability. Prior randomized algorithms have only been proven to be optimal on the average.

Notation. We say the amount of resource (like time, space, etc.) used by a randomized or probabilistic algorithm is $\widetilde{O}(f(n))$ if the amount of resource used is no more than $c\alpha f(n)$ with probability $\geq (1 - n^{-\alpha})$ for any $n \geq n_0$, where c and n_0 are constants and α is a constant ≥ 1. We could also define the asymptotic functions $\widetilde{\Theta}(\cdot)$, $\widetilde{o}(\cdot)$, and so forth in a similar manner.

Organization. The rest of this chapter is organized as follows. In Section 6.4, we provide details on the LMM sort algorithm of Reference 29. In Section 6.5, we show how to realize PDMs using mesh connected computers following the discussion in References 33 and 31. Section 6.6 is devoted to a discussion on sorting $M\sqrt{M}$ keys. Sorting algorithms with good expected performance are considered in Section 6.7. The integer sorting algorithm of Reference 34 is summarized in Section 6.8. Section 6.9 briefly explains how an integer sorting algorithm can be employed to devise a simple optimal randomized sorting algorithm for the PDM. Section 6.10 concludes the chapter.

6.4 The (ℓ, m)-Merge Sort (LMM)

Most of the sorting algorithms that have been developed for the PDS are based on merging. To begin with, these algorithms form N/M *runs* each of length M. A run refers to a sorted subsequence. These initial runs can be formed in one pass through the data (or equivalently N/DB parallel I/O operations). Thereafter, the algorithms merge R runs at a time. Let a *phase of merges* stand for the task of scanning through the input once and performing R-way merges. Note that each phase of merges will reduce the number of remaining runs by a factor of R. For instance, the DSM algorithm employs $R = M/DB$. The difference among the above sorting algorithms lies in how each phase of merges is done.

LMM of Reference 29 is also based on merging. It fixes $R = \ell$, for some appropriate ℓ. The LMM generalizes such algorithms as the odd-even merge sort, the s^2-way merge sort of Thompson and Kung [38], and the columnsort algorithm of Leighton [19].

6.4.1 A Merging Algorithm

The well-known odd–even mergesort algorithm has $R = 2$. It repeatedly merges two sequences at a time. There are n sorted runs each of length 1 to begin with. Thereafter, the number of runs decreases by a factor of 2 with each phase of merges. The odd–even merge algorithm is used to merge any two sequences. A description of odd–even merge follows:

1. Let $U = u_1, u_2, \ldots, u_q$ and $V = v_1, v_2, \ldots, v_q$ be the two sorted sequences to be merged. *Unshuffle* U into two: $U_{odd} = u_1, u_3, \ldots, u_{q-1}$ and $U_{even} = u_2, u_4, \ldots, u_q$. Similarly, unshuffle V into V_{odd} and V_{even}.
2. Recursively merge U_{odd} with V_{odd}. Let $X = x_1, x_2, \ldots, x_q$ be the resultant sequence. Also, merge U_{even} with V_{even} to get $Y = y_1, y_2, \ldots, y_q$.
3. Shuffle X and Y to form the sequence: $Z = x_1, y_1, x_2, y_2, \ldots, x_q, y_q$.
4. Do one step of *compare-exchange operation*, that is, sort successive subsequences of length two in Z. In particular, sort y_1, x_2; sort y_2, x_3; and so on. The resultant sequence is the merge of U and V.

The zero–one principle can be used to prove the correctness of this algorithm. Thompson and Kung's algorithm [38] is a generalization of the above algorithm. In this algorithm, R takes on the value s^2 for some appropriate function s of n. At any given time s^2 runs are merged using an algorithm similar to odd–even merge.

LMM generalizes s^2-way merge sort. LMM employs $R = \ell$. The number of runs is reduced by a factor of ℓ by each phase of merges. At any given time, ℓ runs are merged using the (ℓ, m)-merge algorithm. This merging algorithm is similar to the odd–even merge except that in Step 2, the runs are m-way unshuffled (instead of 2-way unshuffling). In Step 3, m sequences are shuffled and also in Step 4, the local sorting is done differently. A description of the merging algorithm follows:

Algorithm (l,m)-merge

1. Let $U_i = u_i^1, u_i^2, \ldots, u_i^r$, for $1 \leq i \leq l$, be the sequences to be merged. When r is *small* use a base case algorithm. Otherwise, unshuffle each U_i into m parts. That is, partition U_i into $U_i^1, U_i^2, \ldots, U_i^m$, where $U_i^1 = u_i^1, u_i^{1+m}, \ldots; U_i^2 = u_i^2, u_i^{2+m}, \ldots$; and so on.
2. Merge $U_1^j, U_2^j, \ldots, U_l^j$, for $1 \leq j \leq m$, recursively. Let the merged sequences be $X_j = x_j^1, x_j^2, \ldots, x_j^{lr/m}$, for $1 \leq j \leq m$.
3. Shuffle $X_1, X_2, \ldots, X_m$. In particular, form the sequence $Z = x_1^1, x_2^1, \ldots, x_m^1, x_1^2, x_2^2, \ldots, x_m^2, \ldots, x_1^{lr/m}, x_2^{lr/m}, \ldots, x_m^{lr/m}$.
4. At this point, the length of the "dirty sequence" (i.e., unsorted portion) can be shown to be no more than lm. We do not know where the dirty sequence is located. There are many ways to clean up the dirty sequences. One such way is given below.

    Let Z_1 denote the sequence of the first lm elements of Z; Let Z_2 denote the next lm elements as Z_2; and so on. Thus, Z is partitioned into $Z_1, Z_2, \ldots, Z_{r/m}$. Sort each one of the Z_i's. Then merge Z_1 and Z_2; merge Z_3 and Z_4; and so forth. Finally merge Z_2 and Z_3; merge Z_4 and Z_5; and so on.

Figure 6.1 illustrates this algorithm.

The above algorithm is not specific to any architecture. (The same can be said about any algorithm.) Rajasekaran [29] gives an implementation of LMM on PDS. The number of I/O operations used in this implementation is $\frac{N}{DB} \left[\frac{\log(N/M)}{\log\left(\min\{\sqrt{M}, M/B\}\right)} + 1 \right]^2$. This number is a constant when N is a polynomial in M, and M is a polynomial in B. In this case LMM is optimal. It has been demonstrated that LMM can be

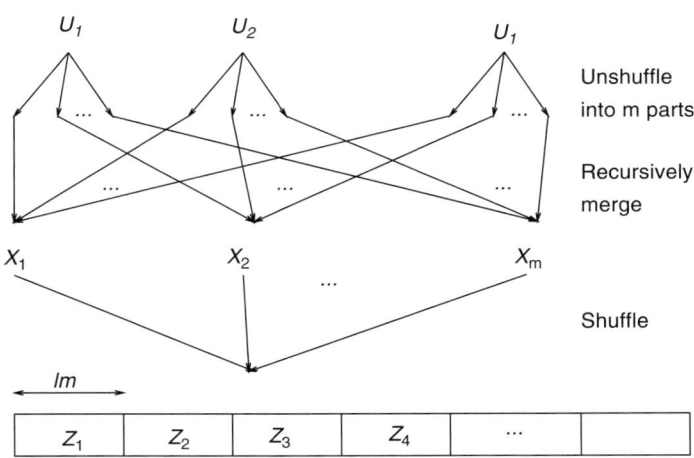

FIGURE 6.1 The (ℓ, m)-merge algorithm.

faster than the DSM when D is large [29]. Recent implementation results of Cormen and Pearson [13, 27] indicate that LMM is competitive in practice.

6.4.2 LMM on PDM

In this section we give details of the implementation of LMM on the PDS model. The implementation merges R runs at a time, for some appropriate R. We have to specify how the intermediate runs are stored across the D disks. Number the disks as well as the runs from zero. Each run will be striped across the disks. If $R \geq D$, the starting disk for the ith run is $i \bmod D$, that is, the zeroth block of the ith run will be in disk $i \bmod D$; its first block will be in disk $(i+1) \bmod D$; and so on. This strategy yields perfect disk parallelism since in one I/O read operation, one block each from D distinct runs can be accessed. If $R < D$, the starting disk for the ith run is iD/R. (Assume without loss of generality that D divides R.) Even now, we can obtain D/R blocks from each of the runs in one I/O operation and hence achieve perfect disk parallelism. See Figure 6.2.

The value of B will be much less than M in practice. For instance, when $M/B > \sqrt{M}$ the number of read passes made by LMM is no more than $\left(2 \frac{\log(N/M)}{\log M} + 1\right)^2$. However, the case $M/B \leq \sqrt{M}$ is also considered. The number of read passes made by LMM is upper bounded by $\left[\frac{\log(N/M)}{\log\left(\min\{\sqrt{M}, M/B\}\right)} + 1\right]^2$ in either case. LMM forms initial runs of length M each in one read pass through the data. After this, the runs will be merged R at a time. Throughout, we use $T(u, v)$ to denote the number of read passes needed to merge u sequences of length v each.

6.4.2.1 Some Special Cases

Some special cases will be considered first. The first case is the problem of merging $\sqrt{M}$ runs each of length M, when $M/B \geq \sqrt{M}$. In this case use $R = \sqrt{M}$. This merging can be done using **Algorithm** (l, m)-merge with $l = m = \sqrt{M}$.

Let the sequences to be merged be $U_1, U_2, \ldots, U_{\sqrt{M}}$. In Step 1, each U_i gets unshuffled into $\sqrt{M}$ parts. Each part is of length $\sqrt{M}$. This unshuffling takes one pass. In Step 2, $\sqrt{M}$ merges have to be done. Each merge involves $\sqrt{M}$ sequences of length $\sqrt{M}$ each. There are only M records in each merge and hence

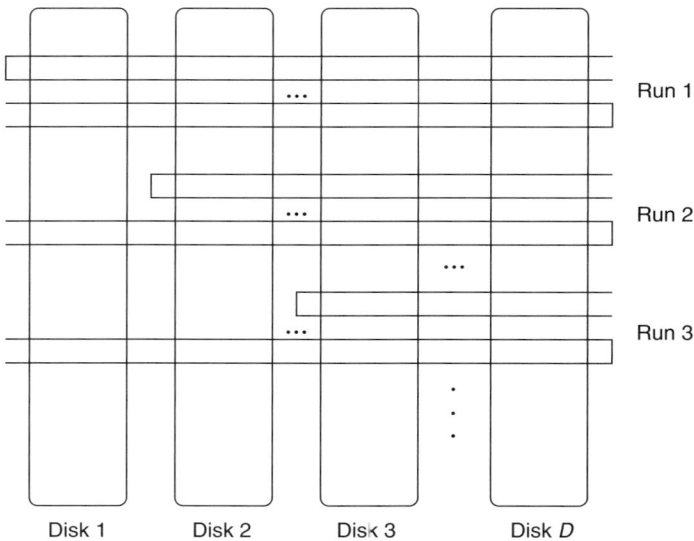

FIGURE 6.2 Striping of runs.

all the merges can be done in one pass through the data. Step 3 involves shuffling and Step 4 involves cleaning up. The length of the dirty sequence is $(\sqrt{M})^2 = M$. We can combine these two steps to complete them in one pass through the data. This can be done as follows. Have two successive Z_i's (c.f. **Algorithm** (l, m)-merge) (call these Z_i and Z_{i+1}) at any time in the main memory. Sort Z_i and Z_{i+1} and merge them. Ship Z_i to the disks. Bring in Z_{i+2}, sort it, and merge it with Z_{i+1}. Ship out Z_{i+1}; and so on.

Observe that perfect disk parallelism is maintained throughout. Thus we get:

Lemma 6.1 $T(\sqrt{M}, M) = 3$, if $\frac{M}{B} \geq \sqrt{M}$.

The second special case considered is that of merging M/B runs each of length M, when $M/B < \sqrt{M}$. Employ **Algorithm** (l, m)-merge with $l = m = M/B$. Along similar lines, we can prove the following Lemma.

Lemma 6.2 $T\left(\frac{M}{B}, M\right) = 3$, if $\frac{M}{B} < \sqrt{M}$.

6.4.2.2 The General Algorithm

The general algorithm utilizes the above special cases. The general algorithm is also presented in two cases, one for $M/B \geq \sqrt{M}$ and the other for $M/B < \sqrt{M}$. As usual, initial runs are formed in one pass. After this pass, N/M sorted sequences of length M each remain to be merged.

If $M/B \geq \sqrt{M}$, employ **Algorithm** (l, m)-merge with $l = m = \sqrt{M}$ and $R = \sqrt{M}$. Let K denote $\sqrt{M}$ and let $N/M = K^{2c}$. As a result, $c = \log(N/M)/\log M$. The following relation is easy to see.

$$T(K^{2c}, M) = T(K, M) + T(K, KM) + \cdots + T(K, K^{2c-1}M). \tag{6.1}$$

This relation means that we start with K^{2c} sequences of length M each; we merge K at a time to end up with K^{2c-1} sequences of length KM each; again merge K at a time to end up with K^{2c-2} sequences of length K^2M each; and so on. Finally there will remain K sequences of length $K^{2c-1}M$ each, which are merged. Each phase of merges is done using the **Algorithm** (l, m)-merge with $l = m = \sqrt{M}$.

$T(K, K^i M)$ for any i can be computed as follows. There are K sequences to be merged, each of length $K^i M$. Let these sequences be $U_1, U_2, \ldots, U_K$. In Step 1, each U_j is unshuffled into K parts each of size $K^{i-1}M$. This can be done in one pass. Now there are K merging problems, where each merging problem involves K sequences of length $K^{i-1}M$ each. The number of passes needed is $T(K, K^{i-1}M)$. The length of the dirty sequence in Steps 3 and 4 is $\leq K^2 = M$. Thus, Steps 3 and 4 can be completed in one pass. Therefore,

$$T(K, K^i M) = T(K, K^{i-1}M) + 2.$$

Expanding this, we see

$$T(K, K^i M) = 2i + T(K, M) = 2i + 3.$$

We have made use of the fact that $T(K, M) = 3$ (c.f. Lemma 6.1).

Upon substitution of this into Equation 6.1, we get

$$T(K^{2c}, M) = \sum_{i=0}^{2c-1} (2i + 3) = 4c^2 + 4c,$$

where $c = \log(N/M)/\log M$. If $N \leq M^3$, the above merging cost is ≤ 24 passes.

We have the following

Theorem 6.1 *The number of read passes needed to sort N records is $1 + 4\left(\frac{\log(N/M)}{\log M}\right)^2 + 4\frac{\log(N/M)}{\log M}$, if $\frac{M}{B} \geq \sqrt{M}$. This number of passes is no more than $\left[\frac{\log(N/M)}{\log\left(\min\{\sqrt{M}, M/B\}\right)} + 1\right]^2$.*

Now consider the case $M/B < \sqrt{M}$. **Algorithm** (l, m)-merge can be used with $l = m = M/B$ and $R = M/B$. The following theorem is proven in a similar fashion.

Theorem 6.2 *The number of read passes needed to sort N records is upper bounded by*
$$\left[\frac{\log(N/M)}{\log\left(\min\{\sqrt{M}, M/B\}\right)} + 1\right]^2, \text{ if } M/B < \sqrt{M}.$$

Theorems 6.1 and 6.2 yield the following.

Theorem 6.3 *We can sort N records in* $\leq \left[\frac{\log(N/M)}{\log\left(\min\{\sqrt{M}, M/B\}\right)} + 1\right]^2$ *read passes over the data, maintaining perfect disk parallelism. In other words, the total number of I/O read operations needed is* $\leq \frac{N}{DB} \left[\frac{\log(N/M)}{\log\left(\min\{\sqrt{M}, M/B\}\right)} + 1\right]^2$.

6.5 A Practical Realization of Parallel Disks

Though the existing models with multiple disks address the problem of data explosion, it is not clear how these models can be realized in practice. The assumption of bringing D blocks of data in one I/O operation may not be practical. A new model called a Parallel Machine with Disks (PMD) is proposed in Reference 33 that is a step in the direction of practical realization. A PMD is nothing but a parallel machine where each processor has an associated disk. The parallel machine can be structured or unstructured. The underlying topology of a structured parallel machine could be a mesh, a hypercube, a star graph, and so forth. Examples of unstructured parallel computers include SMP, NOW, a cluster of workstations (employing PVM or MPI), and so forth. The PMD is nothing but a parallel machine, where we study out-of-core algorithms. In the PMD model we not only count the I/O operations, but also the communication steps. One can think of a PMD as a realization of the PDM model.

Every processor in a PMD has an internal memory of size M. In one parallel I/O operation, a block of B records can be brought into the core memory of each processor from its own disk. There are a total of $D = P$ disks in the PMD, where P is the number of processors. Records from one disk can be sent to another through the communication mechanism available for the parallel machine after bringing the records into the main memory of the origin processor. The communication time could potentially be comparable to the time for I/O on the PMD. It is essential, therefore to not only account for the I/O operations but also for the communication steps, in analyzing any algorithm's run time on the PMD.

We can state the sorting problem on the PMD as follows. There are a total of N records to begin with. There are N/D records in each disk. The problem is to rearrange the records so that they are in either ascending order or descending order with N/D records ending up in each disk. We assume that the processors themselves are ordered so that the smallest N/D records will be output in the first processor's disk, the next smallest N/D records will be output in the second processor's disk, and so on.

We can apply LMM on a general PMD in which case the number of I/O operations will remain the same, that is,

$$\frac{N}{DB}\left[\frac{\log(N/M)}{\log(\min\{\sqrt{M}, M/B\})} + 1\right]^2.$$

In particular, the number of passes through the data will be

$$\left[\frac{\log(N/MD)}{\log(\min\{\sqrt{MD}, MD/B\})} + 1\right]^2.$$

Such an application needs mechanisms for $(k - k)$ routing and $(k - k)$ sorting.

The problem of $(k - k)$ routing in the context of a parallel machine is the problem of packet routing where there are at most k packets of information originating from each processor and at most k packets are destined for each processor. We are required to send all the packets to their correct destinations as quickly as possible. The problem of $(k - k)$ sorting is one where there are k keys to begin with at each processor and we are required to sort all the keys and send the smallest k keys to processor 1, the next smallest k keys to processor 2, and so on.

Let R_M and S_M denote the time needed for performing one $(M - M)$ routing and one $(M - M)$ sorting on the parallel machine, respectively. Then, in each pass through the data, the total communication time will be $\frac{N}{DM}(R_M + S_M)$, implying that the total communication time for the entire algorithm will be $\leq \frac{N}{DM}(R_M + S_M)\left[\frac{\log(N/MD)}{\log(\min\{\sqrt{MD},MD/B\})} + 1\right]^2$.

The following Theorem is proven in Reference 33.

Theorem 6.4 *Sorting on a PMD can be performed in* $\frac{N}{DB}\left[\frac{\log(N/MD)}{\log(\min\{\sqrt{MD},MD/B\})} + 1\right]^2$ *I/O operations. The total communication time is* $\leq \frac{N}{DM}(R_M + S_M)\left[\frac{\log(N/MD)}{\log(\min\{\sqrt{MD},MD/B\})} + 1\right]^2$.

6.6 Sorting $M\sqrt{M}$ Keys

6.6.1 A Lower Bound

The following lower bound result will help us judge the optimality of algorithms presented in this chapter.

Lemma 6.3 *At least two passes are needed to sort $M\sqrt{M}$ elements when the block size is $\sqrt{M}$. At least three passes are needed to sort M^2 elements when the block size is $\sqrt{M}$. These lower bounds hold on the average as well.*

Proof. The bounds stated above follow from the lower bound theorem proven in Reference 5. In particular, it has been shown in Reference 5 that $\log(N!) \leq N \log B + I \times (B \log((M - B)/B) + 3B)$, where I is the number of I/O operations taken by any algorithm that sorts N keys residing in a single disk. Substituting $N = M\sqrt{M}, B = \sqrt{M}$, we see that $I \geq \frac{2M\left(1 - \frac{1.45}{\log M}\right)}{\left(1 + \frac{6}{\log M}\right)}$. The right-hand side is very nearly equal to $2M$. In other words, to sort $M\sqrt{M}$ keys, at least two passes are needed. It is easy to see that the same is true for the PDM also. In a similar fashion, one could see that at least three passes are needed to sort M^2 elements. ∎

6.6.2 A Three-Pass Algorithm

In this section we present a three-pass algorithm for sorting on the PDM. This algorithm assumes a block size of $\sqrt{M}$.

We adapt the (l, m)-merge sort (LMM sort) algorithm of Rajasekaran [29]. The LMM sort partitions the input sequence of N keys into l subsequences, sorts them recursively, and merges the l sorted subsequences using the (l, m)-merge algorithm.

The (l, m)-merge algorithm takes as input l sorted sequences $X_1, X_2, \ldots, X_l$ and merges them as follows. Unshuffle each input sequence into m parts. In particular, X_i ($1 \leq i \leq l$) gets unshuffled into $X_i^1, X_i^2, \ldots, X_i^m$. Recursively merge $X_1^1, X_2^1, \ldots, X_l^1$ to get L_1; Recursively merge $X_1^2, X_2^2, \ldots, X_l^2$ to get L_2; $\cdots$; Recursively merge $X_1^m, X_2^m, \ldots, X_l^m$ to get L_m. Now shuffle $L_1, L_2, \ldots, L_m$. At this point, it can be shown that each key is at a distance of $\leq lm$ from its final sorted position. Perform local sorting to move each key to its sorted position.

Columnsort algorithm [19], odd–even merge sort [7], and the s^2-way merge sort algorithms are all special cases of LMM sort [29].

For the case of $B = \sqrt{M}$, and $N = M\sqrt{M}$, LMM sort can be specialized as follows to run in three passes.

Algorithm ThreePass

1. Form $l = \sqrt{M}$ runs each of length M. These runs have to be merged using (l, m)-merge. The steps involved are listed next. Let $X_1, X_2, \ldots, X_{\sqrt{M}}$ be the sequences to be merged.
2. Unshuffle each X_i into $\sqrt{M}$ parts so that each part is of length $\sqrt{M}$. This unshuffling can be combined with the initial runs formation task and hence can be completed in one pass.
3. In this step, we have $\sqrt{M}$ merges to do, where each merge involves $\sqrt{M}$ sequences of length $\sqrt{M}$ each. Observe that there are only M records in each merge and hence all the mergings can be done in one pass through the data.
4. This step involves shuffling and local sorting. The length of the dirty sequence is $(\sqrt{M})^2 = M$. Shuffling and local sorting can be combined and finished in one pass through the data as shown in Reference 29.

We get the following.

Lemma 6.4 *LMM sort sorts $M\sqrt{M}$ keys in three passes through the data when the block size is $\sqrt{M}$.*

Observation 6.6.1 *Chaudhry and Cormen [9] have shown that Leighton's columnsort algorithm [19] can be adapted for the PDM to sort $\sqrt{M^{1.5}/2}$ keys in three passes. In contrast, the three-pass algorithm of Lemma 6.4 (based on LMM sort) sorts $M^{1.5}$ keys in three passes.*

6.7 Algorithms with Good Expected Performance

6.7.1 A Useful Lemma

In this section, we prove a lemma that will be useful in the analysis of expected performance of sorting algorithms.

Consider a set $X = \{1, 2, \ldots, n\}$. Let $X_1, X_2, \ldots, X_m$ be a random partition of X into equal sized parts. Let $X_1 = x_1^1, x_1^2, \ldots, x_1^q$; $X_2 = x_2^1, x_2^2, \ldots, x_2^q$; $\ldots$; $X_m = x_m^1, x_m^2, \ldots, x_m^q$ in sorted order. Here $mq = n$.

We define the rank of any element y in a sequence of keys Y as $|\{z \in Y : z < y\}| + 1$. Let r be any element of X and let X_i be the part in which r is found. If $r = x_i^k$ (i.e., the rank of r in X_i is k) what can we say about the value of k?

Probability that r has a rank of k in X_i is given by

$$P = \frac{\binom{r-1}{k-1}\binom{n-r}{q-k}}{\binom{n-1}{q-1}}.$$

Using the fact that $\binom{a}{b} \leq \left(\frac{ae}{b}\right)^b$, we get

$$P \leq \frac{\left(\frac{r-1}{k-1}\right)^{k-1}\left(\frac{n-r}{q-k}\right)^{q-k}}{\left(\frac{n-1}{q-1}\right)^{q-1}}.$$

Ignoring the -1s and using the fact that $(1-u)^{1/u} \leq (1/e)$, we arrive at

$$P \leq \left(\frac{rq/n}{k}\right)^k e^{-(q-k)[r/n-k/q]}.$$

When $k = \frac{rq}{n} + \sqrt{(\alpha+2)q\log_e n + 1}$ (for any fixed α), we get, $P \leq n^{-\alpha-2}/e$. Thus, the probability that $k \geq \frac{rq}{n} + \sqrt{(\alpha+2)q\log_e n + 1}$ is $\leq n^{-\alpha-1}/e$.

In a similar fashion, we can show that the probability that $k \leq \frac{rq}{n} - \sqrt{(\alpha+2)q\log_e n + 1}$ is $\leq n^{-\alpha-1}/e$. This can be shown by proving that the number of elements in X_i that are greater than r cannot be higher than $\frac{(n-r)q}{n} + \sqrt{(\alpha+2)q\log_e n + 1}$ with the same probability.

Thus, the probability that k is not in the interval

$$\left[\frac{rq}{n} - \sqrt{(\alpha+2)q\log_e n + 1},\ \frac{rq}{n} + \sqrt{(\alpha+2)q\log_e n + 1}\right]$$

is $\leq n^{-\alpha-1}$.

As a consequence, probability that for any r the corresponding k will not be in the above interval is $\leq n^{-\alpha}$.

Now consider shuffling the sequences $X_1, X_2, \ldots, X_m$ to get the sequence Z. The position of r in Z will be $(k-1)m + i$. Thus, the position of r in Z will be in the interval:

$$\left[r - \frac{n}{\sqrt{q}}\sqrt{(\alpha+2)\log_e n + 1} - \frac{n}{q},\ r + \frac{n}{\sqrt{q}}\sqrt{(\alpha+2)\log_e n + 1}\right].$$

We get the following Lemma:

Lemma 6.5 *Let X be a set of n arbitrary keys. Partition X into $m = n/q$ equal sized parts randomly (or equivalently if X is a random permutation of n keys, the first part is the first q elements of X, the second part is the next q elements of X, and so on). Sort each part. Let $X_1, X_2, \ldots, X_m$ be the sorted parts. Shuffle the X_i's to get the sequence Z. At this time, each key in Z will be at most $\frac{n}{\sqrt{q}}\sqrt{(\alpha+2)\log_e n + 1} + \frac{n}{q} \leq \frac{n}{\sqrt{q}}\sqrt{(\alpha+2)\log_e n + 2}$ positions away from its final sorted position.*

Observation 6.7.1 *Let Z be a sequence of n keys in which every key is at a distance of at most d from its sorted position. Then one way of sorting Z is as follows: Partition Z into subsequences $Z_1, Z_2, \ldots, Z_{n/d}$ where $|Z_i| = d$, $1 \leq i \leq n/d$. Sort each Z_i ($1 \leq i \leq n/d$). Merge Z_1 with Z_2, merge Z_3 with Z_4, $\ldots$, merge $Z_{n/d-1}$ with $Z_{n/d}$ (assuming that n/d is even; the case of n/d being odd is handled similarly); Followed by this merge Z_2 with Z_3, merge Z_4 with Z_5, $\ldots$, and merge $Z_{n/d-2}$ with $Z_{n/d-1}$. Now it can be seen that Z is in sorted order.*

Observation 6.7.2 *The above discussion suggests a way of sorting n given keys. Assuming that the input permutation is random, one could employ Lemma 6.5 to analyze the expected performance of the algorithm. In fact, the above algorithm is very similar to the LMM sort [29].*

6.7.2 An Expected Two-Pass Algorithm

In this section we present an algorithm that sorts nearly $M\sqrt{M}$ keys when the block size is $\sqrt{M}$. The expectation is over the space of all possible inputs. In particular, this algorithm takes two passes for a large fraction of all possible inputs. Specifically, this algorithm sorts $N = \frac{M\sqrt{M}}{c\sqrt{\log M}}$ keys, where c is a constant to be fixed in the analysis. This algorithm is similar to the one in Section 6.7.1. Let $N_1 = N/M$.

Algorithm ExpectedTwoPass

1. Form N_1 runs each of length M. Let these runs be $L_1, L_2, \ldots, L_{N_1}$. This takes one pass.
2. In the second pass shuffle these N_1 runs to get the sequence Z (of length N). Perform local sorting as depicted in Section 6.7.1. Here are the details: Call the sequence of the first M elements of Z as Z_1; the next M elements as Z_2; and so on. In other words, Z is partitioned into $Z_1, Z_2, \ldots, Z_{N_1}$. Sort each one of the Z_i's. Followed by this merge Z_1 and Z_2; merge Z_3 and Z_4; and so forth. Finally merge Z_2 and Z_3; merge Z_4 and Z_5; and so on.

    Shuffling and the two steps of local sorting can be combined and finished in one pass through the data. The idea is to have two successive Z_is (call these Z_i and Z_{i+1}) at any time in the main memory. We can sort Z_i and Z_{i+1} and merge them. After this Z_i is ready to be shipped to the disks. Z_{i+2} will then be brought in, sorted, and merged with Z_{i+1}. At this point Z_{i+1} will be shipped out; and so on.

    It is easy to check if the output is correct or not (by keeping track of the largest key shipped out in the previous I/O). As soon as a problem is detected (i.e., when the smallest key currently being shipped out is smaller than the largest key shipped out in the previous I/O), the algorithm is aborted and the algorithm of Lemma 1.4 is used to sort the keys (in an additional three passes).

Theorem 6.5 *The expected number of passes made by Algorithm* ExpectedTwoPass *is very nearly two. The number of keys sorted is* $M\sqrt{\frac{M}{(\alpha+2)\log_e M+2}}$.

Proof. Using Lemma 6.5, every key in Z is at a distance of at most $N_1\sqrt{M}\sqrt{(\alpha+2)\log_e M+2}$ from its sorted position with probability $\geq (1 - M^{-\alpha})$. We want this distance to be $\leq M$. This happens when $N_1 \leq \sqrt{\frac{M}{(\alpha+2)\log_e M+2}}$.

For this choice of N_1, the expected number of passes made by ExpectedTwoPass is $2(1 - M^{-\alpha}) + 5M^{-\alpha}$ that is very nearly 2. ∎

As an example, when $M = 10^8$ and $\alpha = 2$, the expected number of passes is $2 + 3 \times 10^{-16}$. Only on at most 10^{-14} % of all possible inputs, ExpectedTwoPass will take more than two passes. Thus, this algorithm is of practical importance. Please also note that we match the lower bound of Lemma 6.3 closely.

Note: The columnsort algorithm [19] has eight steps. Steps 1, 3, 5, and 7 involve sorting the columns. In steps 2, 4, 6, and 8 some well-defined permutations are applied on the keys. Chaudhry and Cormen [9] show how to combine the steps appropriately, so that only three passes are needed to sort $M\sqrt{M/2}$ keys on a PDM (with $B = \Theta(M^{1/3})$). Here we point out that this variant of columnsort can be modified to run in an expected two passes. The idea is to skip Steps 1 and 2. Using Lemma 6.5, one can show that modified columnsort sorts $M\sqrt{\frac{M}{4(\alpha+2)\log_e M+2}}$ keys in an expected two passes. Contrast this number with the one given in Theorem 6.5.

6.7.3 An Expected Three-Pass Algorithm

In this section we show how to extend the ideas of the previous section to increase the number of keys to be sorted. In particular, we focus on an expected three-pass algorithm. Let N be the total number of keys to be sorted and let $N_2 = N\sqrt{(\alpha + 2)\log_e M + 2}/(M\sqrt{M})$.

Algorithm ExpectedThreePass

1. Using ExpectedTwoPass, form runs of length $M\sqrt{\frac{M}{(\alpha+2)\log_e M+2}}$ each. This will take an expected two passes. Now we have N_2 runs to be merged. Let these runs be $L_1, L_2, \ldots, L_{N_2}$.
2. This step is similar to Step 2 of ExpectedTwoPass. In this step, we shuffle the N_2 runs formed in Step 1 to get the sequence Z (of length N). Perform local sorting as depicted in ExpectedTwoPass.
    Shuffling and the two steps of local sorting can be combined and finished in one pass through the data (as described in ExpectedTwoPass).
    It is easy to check if the output is correct or not (by keeping track of the largest key shipped out in the previous I/O). As soon as a problem is detected (i.e., when the smallest key currently being shipped out is smaller than the largest key shipped out in the previous I/O), the algorithm is aborted and another algorithm is used to sort the keys.

Theorem 6.6 *The expected number of passes made by Algorithm* ExpectedThreePass *is very nearly three. The number of keys sorted is $\frac{M^{1.75}}{[(\alpha+2)\log_e M+2]^{3/4}}$.*

Proof. Here again we make use of Lemma 6.5.

In this case $q = M\sqrt{\frac{M}{(\alpha+2)\log_e M+2}}$. In the sequence Z, each key will be at a distance of at most $N_2 M^{3/4}[(\alpha + 2)\log_e M + 2]^{1/4}$ from its sorted position (with probability $\geq (1 - M^{-\alpha})$). We want this distance to be $\leq M$. This happens when $N_2 \leq \frac{M^{1/4}}{[(\alpha+2)\log_e M+2]^{1/4}}$.

For this choice of N_2, the expected number of passes made by ExpectedTwoPass is $3(1 - M^{-\alpha}) + 7M^{-\alpha}$ that is very nearly 3. ∎

Note: Chaudhry and Cormen [9] have recently developed a sophisticated variant of columnsort called *subblock columnsort* that can sort $M^{5/3}/4^{2/3}$ keys in four passes (when $B = \Theta(M^{1/3})$). This algorithm has been inspired by the Revsort of Schnorr and Shamir [37]. Subblock columnsort introduces the following step between Steps 3 and 4 of columnsort: Partition the $r \times s$ matrix into subblocks of size $\sqrt{s} \times \sqrt{s}$ each; Convert each subblock into a column; and sort the columns of the matrix. At the end of Step 3, there could be at most s dirty rows. With the absence of the new step, the value of s will be constrained by $s \leq \sqrt{r/2}$. At the end of the new step, the number of dirty rows is shown to be at most $2\sqrt{s}$. This is in turn because of the fact there could be at most $2\sqrt{s}$ dirty blocks.

The reason for this is that the boundary between the zeros and ones in the matrix is monotonous (see Figure 5 in Reference 9). The monotonicity is ensured by steps 1 through 3 of columnsort. With the new step in place, the constraint on s is given by $r \geq 4s^{3/2}$ and hence a total of $M^{5/3}/4^{2/3}$ keys can be sorted. If one attempts to convert subblock columnsort into a probabilistic algorithm by skipping Steps 1 and 2, it would not work since the monotonicity is not guaranteed. So, converting subblock columnsort into an expected three-pass algorithm (that sorts close to $M^{5/3}$ keys) is not feasible. In other words, the new step of forming subblocks (and the associated permutation and sorting) does not seem to help in expectation. On the other hand, ExpectedThreePass sorts $\Omega\left(\frac{M^{1.75}}{\log M}\right)$ keys in three passes with high probability.

Rajasekaran and Sen have also adapted LMM sort to sort M^2 keys on a PDM with $B = \sqrt{M}$. This adaptation runs in seven passes. They also present an algorithm that sorts almost M^2 keys in an expected six passes.

6.8 Optimal Integer Sorting

Often, the keys to be sorted are integers in some range $[1, R]$. Numerous sequential and parallel algorithms have been devised for sorting integers. Several efficient out-of-core algorithms have been devised by Arge et al. [4] for sorting strings. For instance, three of their algorithms have the I/O bounds of $O\left(\frac{N}{B} \log_{M/B} \frac{N}{B}\right)$, $O\left(\frac{N}{FB} \log_{M/B} \frac{N}{F} + \frac{N}{B}\right)$, and $O\left(\frac{K}{B} \log_{M/B} \frac{K}{B} + \frac{N}{B} \log_{M/B} |\Sigma|\right)$, respectively. These algorithms sort K strings with a total of N characters from the alphabet Σ. Here F is a positive integer such that $F|\Sigma|^F \leq M$ and $|\Sigma|^F \leq N$. These algorithms could be employed on the PDM to sort integers. For a suitable choice of F, the second algorithm (for example) is asymptotically optimal.

In this section, we analyze radix sort (see e.g., Reference 16) in the context of PDM sorting. This algorithm sorts an arbitrary number of keys. We assume that each key fits in one word of the computer. We believe that for applications of practical interest radix sort applies to run in no more than four passes for most of the inputs.

The range of interest in practice seems to be $[1, M^c]$ for some constant c. For example, weather data, market data, and so on are such that the key size is no more than 32 bits. The same is true for personal data kept by governments. For example, if the key is social security number, then 32 bits are enough. However, one of the algorithms given in this section applies for keys from an arbitrary range as long as each key fits in one word of the computer.

The first case we consider is one where the keys are integers in the range $[1, M/B]$. Also, assume that each key has a random value in this interval. If the internal memory of a computer is M, then it is reasonable to assume that the word size of the computer is $\Theta(\log M)$. Thus, each key of interest fits in one word of the computer. M and B are used to denote the internal memory size and the block size, respectively, in words.

The idea can be described as follows. We build M/B runs one for each possible value that the keys can take. From every I/O read operation, M keys are brought into the core memory. From out of all the keys in the memory, blocks are formed. These blocks are written to the disks in a striped manner. The striping method suggested in Reference 29 is used. Some of the blocks could be nonfull. All the blocks in the memory are written to the disks using as few parallel write steps as possible. We assume that $M = CDB$ for some constant C. Let $R = M/B$. More details follow.

Algorithm IntegerSort

for $i := 1$ **to** N/M **do**
1. In C parallel read operations, bring into the core memory $M = CDB$ keys.
2. Sort the keys in the internal memory and form blocks according to the values of keys. Keep a bucket for each possible value in the range $[1, R]$. Let the buckets be $\mathcal{B}_1, \mathcal{B}_2, \ldots, \mathcal{B}_R$. If there are N_i keys in $\mathcal{B}_i$, then $\lceil N_i/B \rceil$ blocks will be formed out of $\mathcal{B}_i$ (for $1 \leq i \leq R$).
3. Send all the blocks to the disks using as few parallel write steps as possible. The runs are striped across the disks (in the same manner as in Reference 29). The number of write steps needed is $\max_i\{\lceil N_i/B \rceil\}$.
4. Read the keys written to the disks and write them back so that the keys are placed contiguously across the disks.

Theorem 6.7 *Algorithm* IntegerSort *runs in $O(1)$ passes through the data for a large fraction ($\geq (1 - N^{-\alpha})$ for any fixed $\alpha \geq 1$) of all possible inputs. If step 4 is not needed, the number of passes is $(1 + \mu)$ and if step 4 is included, then the number of passes is $2(1 + \mu)$ for some fixed $\mu < 1$.*

Proof. Call each run of the **for** loop as a *phase* of the algorithm. The expected number of keys in any bucket is CB. Using Chernoff bounds, the number of keys in any bucket is in the interval $[(1 - \epsilon)CB, (1 + \epsilon)CB]$ with probability $\geq [1 - 2\exp(-\epsilon^2 CB/3)]$. Thus, the number of keys in every bucket is in this interval with probability $\geq \left(1 - \exp\left[\frac{-\epsilon^2 CB}{3} + \ln(2R)\right]\right)$. This probability will be $\geq (1 - N^{-(\alpha+1)})$ as long as $B \geq \frac{3}{C\epsilon^2}(\alpha + 1)\ln N$. This is readily satisfied in practice (since the typical assumption on B is that it is $\Omega(M^\delta)$ for some fixed $\delta > 1/3$.

As a result, each phase will take at most $\lceil (1+\epsilon)C \rceil$ write steps with high probability. This is equivalent to $\frac{\lceil (1+\epsilon)C \rceil}{C}$ passes through the data. This number of passes is $1 + \mu$ for some constant $\mu < 1$.

Thus, with probablity $\geq (1 - N^{-\alpha})$, IntegerSort takes $(1 + \mu)$ passes excluding step 4 and $2(1 + \mu)$ passes including step 4. ∎

As an example, if $\epsilon = 1/C$, the value of μ is $1/C$.

Observation 6.8.1 *The sorting algorithms of Reference 39 have been analyzed using asymptotic analysis. The bounds derived hold only in the limit. In comparison, our analysis is simpler and applies for any N.*

We extend the range of the keys using the following algorithm. This algorithm employs forward radix sorting. In each stage of sorting, the keys are sorted with respect to some number of their MSBs. Keys that have the same value with respect to all the bits that have been processed up to some stage are said to form a *bucket* in that stage. In the following algorithm, δ is any constant < 1.

Algorithm RadixSort

for $i := 1$ to $(1 + \delta)\frac{\log(N/M)}{\log(M/B)}$ do
    1. Employ IntegerSort to sort the keys with respect to their ith most significant $\log(M/B)$ bits.
    2. Now the size of each bucket is $\leq M$. Read and sort the buckets.

Theorem 6.8 *N random integers in the range $[1, R]$ (for any R) can be sorted in an expected $(1+\nu)\frac{\log(N/M)}{\log(M/B)} + 1$ passes through the data, where ν is a constant < 1. In fact, this bound holds for a large fraction $(\geq 1 - N^{-\alpha}$ for any fixed $\alpha \geq 1)$ of all possible inputs.*

Proof. In accordance with Theorem 6.7, each run of step 1 takes $(1 + \mu)$ passes. Thus, RadixSort takes $(1 + \mu)(1 + \delta)\frac{\log(N/M)}{\log(M/B)}$ passes. This number is $(1 + \nu)\frac{\log(N/M)}{\log(M/B)}$ for some fixed $\nu < 1$.

It remains to be shown that after $(1 + \delta)\frac{\log(N/M)}{\log(M/B)}$ runs of Step 1, the size of each bucket will be $\leq M$. At the end of the first run of step 1, the size of each bucket is expected to be $\frac{NB}{M}$. Using Chernoff bounds, this size is $\leq (1 + \epsilon)\frac{NB}{M}$ with high probability, for any fixed $\epsilon < 1$. After k (for any integer k) runs of Step 1, the size of each bucket is $\leq N(1 + \epsilon)^k (B/M)^k$ with high probability. This size will be $\leq M$ for $k \geq \frac{\log(N/M)}{\log\left[\frac{M}{(1+\epsilon)B}\right]}$. The right hand side is $\leq (1 + \delta)\frac{\log(N/M)}{\log(M/B)}$ for any fixed $\delta < 1$.

Step 2 takes one pass. ∎

Observation 6.8.2 *As an example, consider the case $N = M^2$, $B = \sqrt{M}$, and $C = 4$. In this case, RadixSort takes no more than 3.6 passes through the data.*

6.9 A Simple Optimal Randomized Sorting Algorithm

We show how to randomly permute N given keys such that each permutation is equally likely. We employ RadixSort for this purpose. The idea is to assign a random label with each key in the range $[1, N^{1+\beta}]$ (for any fixed $0 < \beta < 1$) and sort the keys with respect to their labels. This can be done in $(1+\mu)\frac{\log(N/M)}{\log(M/B)} + 1$ passes through the data with probability $\geq 1 - N^{-\alpha}$ for any fixed $\alpha \geq 1$. Here μ is a constant < 1. For many applications, this permutation may suffice. But we can ensure that each permutation is equally likely with one more pass through the data.

When each key gets a random label in the range $[1, N^{1+\beta}]$, the labels may not be unique. The maximum number of times any label is repeated is $\widetilde{O}(1)$ from the observation that the number of keys falling in a bucket is binomially distributed with mean $1/n$ and applying Chernoff bounds. We have to randomly permute keys with equal labels that can be done in one more pass through the data as follows. We think of the sequence of N input keys as $S_1, S_2, \ldots, S_{N/DB}$ where each S_i ($1 \leq i \leq N/(DB)$) is a subsequence of length DB. Note that keys with the same label can only span two such subsequences. We bring in DB keys at a time into the main memory. We assume a main memory of size $2DB$. There will be two subsequences at any time in the main memory. Required permutations of keys with equal labels are done and DB keys are shipped out to the disks. The above process is repeated until all the keys are processed.

With more care, we can eliminate this extra pass by combining it with the last stage of radix sort. Thus, we get the following:

Theorem 6.9 *We can permute N keys randomly in $O(\frac{\log(N/M)}{\log(M/B)})$ passes through the data with probability $\geq 1 - N^{-\alpha}$ for any fixed $\alpha \geq 1$, where μ is a constant < 1, provided $B = \Omega(\log N)$.*

Note: The above theorem has been presented in Reference 34.

Rajasekaran and Sen present a randomized sorting algorithm on the basis of the above permutation algorithm. This algorithm sorts N given keys in $\widetilde{O}\left(\frac{\log(N/M)}{\log(M/B)}\right)$ passes through the data. The basic idea is the following. The input keys are randomly permuted to begin with. After this, an R-way merge is employed for an appropriate value of R. If there are R runs to be merged, the algorithm brings a constant number of blocks from every run. A merging algorithm is applied on these blocks. The smallest key from these blocks will be sent to an output buffer. This key will be deleted from its block. The next smallest key will be identified from the remaining keys of the blocks. This key will be sent to the output buffer (and deleted from its block); and so on. When the output buffer is of size DB, these keys will be output to the disks in one parallel write step. The key to observe here is that the keys from the R runs will be "consumed" at nearly the same rate with high probability. This property is satisfied because of the random permutation done at the beginning. An advantage of this property is that it results in the avoidance of unnecessary read steps.

6.10 Conclusions

In this chapter, we have provided a survey of deterministic and randomized sorting algorithms that have been proposed for the PDM. Practical cases of interest such as sorting $\leq M^2$ keys and algorithms with good expected behavior have also been discussed.

Acknowledgment

This work has been supported in part by the NSF Grants CCR-9912395 and ITR-0326155.

References

[1] A. Aggarwal and C. G. Plaxton, Optimal Parallel Sorting in Multi-Level Storage, *Proc. 5th Annual ACM Symposium on Discrete Algorithms*, 1994, pp. 659–668.

[2] A. Aggarwal and J. S. Vitter, The Input/Output Complexity of Sorting and Related Problems, *Communications of the ACM*, 31(9), 1988, pp. 1116–1127.

[3] L. Arge, The Buffer Tree: A New Technique for Optimal I/O-Algorithms, *Proc. 4th International Workshop on Algorithms and Data Structures (WADS)*, 1995, pp. 334–345.

[4] L. Arge, P. Ferragina, R. Grossi, and J. S. Vitter, On Sorting Strings in External Memory, *Proc. ACM Symposium on Theory of Computing*, 1995.

[5] L. Arge, M. Knudsen, and K. Larsen, A General Lower Bound on the I/O-Complexity of Comparison-based Algorithms, *Proc. 3rd Workshop on Algorithms and Data Structures (WADS)*, 1993.

[6] R. Barve, E. F. Grove, and J. S. Vitter, Simple Randomized Mergesort on Parallel Disks, Technical Report CS-1996-15, Department of Computer Science, Duke University, October 1996.

[7] K. Batcher, Sorting Networks and Their Applications, *Proc. AFIPS Spring Joint Computing Conference* 32, 1968, pp. 307–314.

[8] M. Blum, R. W. Floyd, V. Pratt, R. L. Rivest, R. E. Tarjan, Time Bounds for Selection, *Journal of Computer and System Sciences* 7, 1973, pp. 448–461.

[9] G. Chaudhry and T. H. Cormen, Getting More from Out-of-Core Columnsort, *Proc. 4th Workshop on Algorithm Engineering and Experiments (ALENEX)*, 2002, pp. 143–154.

[10] G. Chaudhry, T. H. Cormen, and E. A. Hamon, Parallel Out-of-Core Sorting: The Third Way, to appear in *Cluster Computing*.

[11] G. Chaudhry, T. H. Cormen, and L. F. Wisniewski, Columnsort Lives! An Efficient Out-of-Core Sorting Program, *Proc. 13th Annual ACM Symposium on Parallel Algorithms and Architectures*, 2001, pp. 169–178.

[12] T. Cormen, Determining an Out-of-Core FFT Decomposition Strategy for Parallel Disks by Dynamic Programming, in *Algorithms for Parallel Processing*, IMA Volumes in Mathematics and Its Applications, Vol. 105, Springer-Verlag, 1999, pp. 307–320.

[13] T. Cormen and M. D. Pearson, *Personal Communication*.

[14] R. Dementiev and P. Sanders, Asynchronous Parallel Disk Sorting, *Proc. ACM Symposium on Parallel Algorithms and Architectures*, 2003, pp. 138–148.

[15] R. W. Floyd and R. L. Rivest, Expected Time Bounds for Selection, *Communications of the ACM* 18(3), 1975, pp. 165–172.

[16] E. Horowitz, S. Sahni, and S. Rajasekaran, *Computer Algorithms*, W. H. Freeman Press, 1998.

[17] M. Kaufmann, S. Rajasekaran, and J. F. Sibeyn, Matching the Bisection Bound for Routing and Sorting on the Mesh, *Proc. 4th Annual ACM Symposium on Parallel Algorithms and Architectures*, 1992, pp. 31–40.

[18] M. Kunde, Block Gossiping on Grids and Tori: Deterministic Sorting and Routing Match the Bisection Bound, *Proc. First Annual European Symposium on Algorithms*, Springer-Verlag Lecture Notes in Computer Science 726, 1993, pp. 272–283.

[19] T. Leighton, Tight Bounds on the Complexity of Parallel Sorting, *IEEE Transactions on Computers*, C34(4), 1985, pp. 344–354.

[20] Y. Ma, S. Sen, D. Scherson, The Distance Bound for Sorting on Mesh Connected Processor Arrays Is Tight, *Proc. 27th Symposium on Foundations of Computer Science*, 1986, pp. 255–263.

[21] G. S. Manku, S. Rajagopalan, and G. Lindsay, Approximate Medians and Other Quantiles in One Pass and with Limited Memory, *Proc. ACM SIGMOD Conference*, 1998.

[22] J. M. Marberg and E. Gafni. Sorting in Constant Number of Row and Column Phases on a Mesh. *Algorithmica*, 3(4), 1988, pp. 561–572.

[23] J. I. Munro and M. S. Paterson, Selection and Sorting with Limited Storage, *Theoretical Computer Science* 12, 1980, pp. 315–323.

[24] M. H. Nodine, J. S. Vitter, Large Scale Sorting in Parallel Memories, *Proc. 3rd Annual ACM Symposium on Parallel Algorithms and Architectures*, 1990, pp. 29–39.

[25] M. H. Nodine and J. S. Vitter, Greed Sort: Optimal Deterministic Sorting on Parallel Disks, *Journal of the ACM* 42(4), 1995, pp. 919–933.

[26] V. S. Pai, A. A. Schaffer, and P. J. Varman, Markov Analysis of Multiple-Disk Prefetching Strategies for External Merging, *Theoretical Computer Science*, 128(2), 1994, pp. 211–239.

[27] M. D. Pearson, Fast Out-of-Core Sorting on Parallel Disk Systems, Technical Report PCS-TR99-351, Dartmouth College, Computer Science, Hanover, NH, June 1999, ftp://ftp.cs.dartmouth.edu/TR/TR99-351.ps.Z.

[28] S. Rajasekaran, Sorting and Selection on Interconnection Networks, *DIMACS Series in Discrete Mathematics and Theoretical Computer Science* 21, Edited by D. F. Hsu, A. L. Rosenberg, and D. Sotteau, 1995, pp. 275–296.
[29] S. Rajasekaran, A Framework for Simple Sorting Algorithms on Parallel Disk Systems, *Theory of Computing Systems*, 34(2), 2001, pp. 101–114.
[30] S. Rajasekaran, Selection Algorithms for the Parallel Disk Systems, *Proc. International Conference on High Performance Computing*, 1998, pp. 54–61.
[31] S. Rajasekaran, Out-of-Core Computing on Mesh Connected Computers, *Journal of Parallel and Distributed Computing* 64(11), 2004, pp. 1311–1317.
[32] S. Rajasekaran and J.H. Reif, Derivation of Randomized Sorting and Selection Algorithms, in *Parallel Algorithm Derivation and Program Transformation*, Edited by R. Paige, J.H. Reif, and R. Wachter, Kluwer Academic Publishers, 1993, pp. 187–205.
[33] S. Rajasekaran and X. Jin, A Practical Model for Parallel Disks, *Proc. International Workshop on High Performance Scientific and Engineering Computing with Applications*, 2000.
[34] S. Rajasekaran and S. Sen, PDM Sorting Algorithms That Take a Small Number of Passes, *Proc. International Parallel and Distributed Processing Symposium (IPDPS)*, 2005.
[35] S. Rajasekaran and S. Sen, A Simple Optimal Randomized Sorting Algorithm for the PDM, *Proc. International Symposium on Algorithms and Computation (ISAAC)*, 2005, Springer-Verlag Lecture Notes in Computer Science 3827, 2005, pp. 543–552.
[36] I. D. Scherson, S. Sen, A. Shamir, Shear Sort: A True Two-Dimensional Sorting Technique for VLSI Networks, *Proc. International Conf. on Parallel Processing*, pp. 903–908, 1986.
[37] C. P. Schnorr and and A. Shamir, An optimal sorting algorithm for mesh connected computers, *Proc. 18th Annual ACM Symposium on Theory of Computing*, 1986, pp. 255–263.
[38] C. D. Thompson and H. T. Kung, Sorting on a Mesh Connected Parallel Computer, *Communications of the ACM* 20(4), 1977, pp. 263–271.
[39] J. S. Vitter and D. A. Hutchinson, Distribution Sort with Randomized Cycling, *Proc. 12th Annual SIAM/ACM Symposium on Discrete Algorithms*, 2001.
[40] J. S. Vitter and E. A. M. Shriver, Algorithms for Parallel Memory I: Two-Level Memories, *Algorithmica* 12(2-3), 1994, pp. 110–147.

7
A Programming Model and Architectural Extensions for Fine-Grain Parallelism

Alex Gontmakher
Assaf Schuster
Gregory Shklover
Technion–Israel Institute of Technology

Avi Mendelson
Intel Israel

7.1	Introduction...	7-1
7.2	Programming Model	7-2
	ISA Extension	
7.3	Code Generation..	7-3
	The Inthreads-C Language • Compilation of Explicitly Parallel Code • Internal Code Representation • Register Allocation	
7.4	Implementation..	7-8
	Misprediction Handling	
7.5	Performance Evaluation	7-11
	Parallelization Granularity • Benchmarks • Effect of Parallelization Degree • The Effect of Register Pressure	
7.6	Related Work...	7-17
	References	7-18

7.1 Introduction

Recent developments in semiconductor technology indicate that, unless fundamental breakthroughs are made, we can no longer expect a steady exponential growth in the processors' clock frequency. Thus, in order to sustain the rate of performance improvement of the last several decades, we must increase the average number of instructions that are executed in parallel.

The exploitable parallelism depends, on one hand, on the techniques that a processor employs to discover independent instructions, and on the other hand, on the actual amount of parallelism visible in the instruction stream. Current instruction level parallelism (ILP) exploitation techniques are rather efficient and approach the point of diminishing returns. Research indicates that for a reasonable out-of-order window, the parallelism of a single thread grows as log(*window_size*) [21].

As a characterization of the parallelism limit of a single-threaded processor, consider that the current state-of-the-art processors can hold at most several hundred instructions in flight [17,22], with processors supporting more than 1000 instructions being a target for long-term research [9]. Owing to these limitations, computer architectures naturally evolve in the direction of improving the parallelism properties of the programs. One direction, that of VLIW processors [12, 19, 37], relies on the compiler to produce instruction streams with explicitly parallel instructions. Techniques such as *predication* [25] and *speculative memory access* help reduce the effect of ILP-degrading events. Still, the parallelism available in single-threaded programs is fundamentally limited [23, 46].

An alternative way to increase the available parallelism is *multithreading*. Instructions of different threads are inherently less dependent on each other than instructions of the same thread. Not surprisingly, modern processors increasingly support multithreading in the form of Simultaneous Multithreading [43], Chip Multiprocessing [22], or both.

Multithreading is excellent for certain workloads such as server and HPTC applications, but is of limited utility to general purpose single-threaded code. Despite hardware support, the overhead of thread management limits the parallelization granularity to relatively coarse-grained tasks.

The magnitude of the overhead can be demonstrated with a simple experiment, parallelization of a simple nonrecursive quicksort into two threads. The benchmark was executed on a dual-CPU machine, and we plot the speedup of the program as a function of the sorted array size (Figure 7.1). We can see that even for the array size of 100, the overhead is so significant that the parallel version is drastically slower than the serial one. Only for an array size of about 1000 does the parallelization overcome the overhead!

We can see that there is a gap between the parallelization granularity of single-threaded processing and that of multithreading. To bridge the gap, we introduce *Inthreads*, a threading model that is sufficiently lightweight to use parallelization at very fine granularity. The *Inthreads* model is based on microthreads that operate within the context of a conventional thread. The processor supports a fixed number of microthreads that share the architectural registers, allowing for the thread management to be implemented entirely in hardware. In addition, the shared registers can be used as a communication medium, allowing extremely fast information transfer between threads.

7.2 Programming Model

7.2.1 ISA Extension

The Inthreads computational model is based on a fixed number of threads running over shared architectural registers in the context of a single SMT thread. As a result, the model provides an extremely lightweight threading mechanism: the fixed number of threads allows for thread management to be performed entirely in hardware. In addition, the shared registers provide a straightforward and efficient communication mechanism: a value can be transferred between threads by writing it into a register in

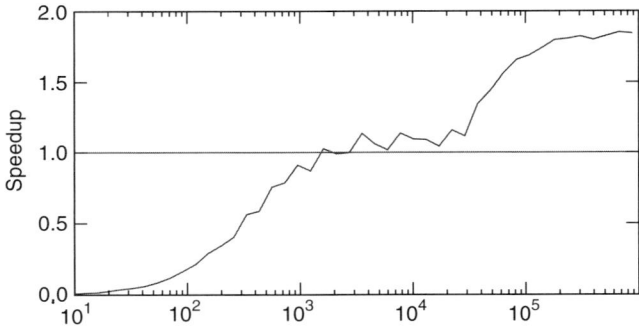

FIGURE 7.1 Speedup of quicksort parallelized into 2 threads.

one thread and reading it from the same register in another. All the general-purpose registers are shared, except for those that constitute the private state of each thread, such as the *Program Count* (PC).

Each thread has a thread ID (TID) that is determined at thread creation. The *main thread*, identified by TID = 0, is always active, whereas other threads can be started and terminated on demand. We provide three new instructions for starting and stopping threads: **inth.start**, **inth.halt** and **inth.kill**. **inth.start** creates a new thread with a given thread ID at a given address, specified by the **ID** and **OFFSET** parameters. To terminate itself, a thread issues an **inth.halt** instruction. **inth.halt** is executed synchronously, guaranteeing that all the preceding instructions complete. To kill another thread, a thread issues an **inth.kill** instruction, which receives the ID of the thread to be killed.

The synchronization mechanism consists of a set of binary semaphores stored in *condition registers* and three instructions to control them: **cond.wait**, **cond.set**, and **cond.clr**. All the synchronization instructions receive a single parameter—the condition register to access.

A **cond.wait** checks whether a given condition is set. If it is, the condition is cleared; otherwise the issuing thread is stalled until some other thread performs a **cond.set** to that condition. If several **cond.wait** instructions accessing the same condition are issued in parallel, only one of them will proceed. An **cond.set** sets the given condition. If there was a **cond.wait** suspended on the same condition, the **cond.wait** is awakened and the condition remains cleared. Finally, a **cond.clr** clears the given condition.

A fundamental requirement of a processor architecture is that the result of a computation for a compliant program must maintain the illusion of sequential and atomic instruction execution. A conventional processor uses the Reorder Buffer (ROB) to hide the out-of-order execution of instructions, whereas an Inthreads processor needs additional mechanisms because the register sharing exposes the out-of-order processing. The architecture achieves the illusion of sequential execution through collaboration between the hardware and the software by declaring *Release Consistency* as the memory model for interaction between threads and by requiring the programs to be free of data races.

Release Consistency distinguishes *strong* operations (*Release* and *Acquire*, which maintain the order with respect to all prior and subsequent operations, respectively) from regular *weak* operations. In our model, the *Release* semantics is required for **cond.set** and **inth.start** instructions, and the *Acquire* semantics is required for the **cond.wait** instruction. The implementation of the semantics is discussed in Section 7.4.

The data-race-free requirement, although requiring significant support from both the compiler and the microarchitecture, simplifies the implementation considerably. Conflicting accesses could either be handled by means of excessive serialization, or would require expensive rollback mechanisms to resolve violations of sequential execution. However, in absence of data races, it is enough for the processor to preserve the *Release* and *Acquire* semantics for the strong operations in order to ensure correct execution.

One consequence of the register sharing is the increase in register pressure. In general, threads in the Inthreads model have slightly lower register demand than SMT threads owing to the sharing of values, such as **SP**, **GP**, and common read-only variables, in shared registers. Still, parallelizing to a high degree may increase the register pressure to the point that extensive spilling is required, thus degrading the performance. This phenomenon effectively limits the practical number of threads to a constant. However, this limit is not inherently constraining because the requirement of hardware management for threads implies a fixed number of threads in the first place.

7.3 Code Generation

In order to facilitate experimentation and separate the concerns of parallelization from the technical details of low-level code generation, we split the process into two stages, the parallelization and the actual code generation, illustrated in Figure 7.2. As an interface between the stages, we define an intermediate language, *Inthreads-C*, which extends the C programming language with Inthreads-specific constructs. Code belonging to a thread is marked up by special **#pragma inthread** blocks. The Inthreads-C compiler

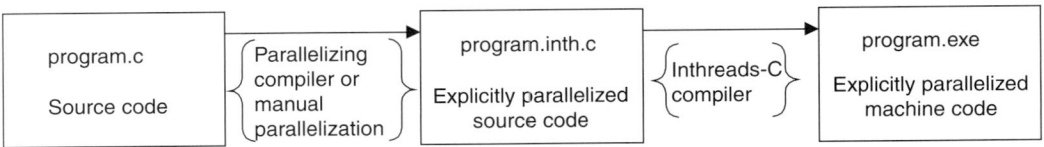

FIGURE 7.2 The compilation flow.

TABLE 7.1 Inthreads-C Commands Corresponding to the Architecture-Level Instructions

Instruction	Inthreads-C command	Language level semantics
inth.set	INTH_SET	The *Release* semantics of this operation imply that all the values produced before it must be visible by the other threads participating in the computation. The compiler must detect all such values and allocate them to shared variables. Moreover, the compiler must not move the instructions involved in the communication beyond the INTH_SET
inth.wait	INTH_WAIT	The *Acquire* semantics of this operation imply that all the values produced at the corresponding INTH_SET commands in the parallel threads will be visible by the code following the INTH_WAIT. The compiler considers only the instructions which operate on the same condition register as communicating
inth.clr	INTH_CLEAR	The operation can not participate in a communication and does not require special compiler support
inth.start	INTH_START	Both the fallback path and the address follow the instruction in the data flow graph. The *Release* semantics of INTH_START imply the same support of shared variable communication as for INTH_SET
inth.halt	INTH_HALT	Execution will not proceed beyond this point. All the values produced before the INTH_HALT and not communicated by explicit synchronization can be lost. Therefore, normally, an INTH_SET immediately precedes an INTH_HALT
inth.kill	INTH_KILL	A specified thread is killed asynchronously. Generally, it is unknown at which stage of computation will the other thread be killed. Instructions following the kill must not see any values provided by the thread except for those already communicated by synchronization. Therefore, the compiler is forbidden from reordering the instructions with respect to an INTH_KILL

analyzes the parallel code to identify the shared and private variables and performs optimizations according to the requirements of the Inthreads programming model.

The input for Inthreads-C can be generated either by a parallelizing compiler or manually. The results presented in this chapter were obtained by manual parallelization. In contrast to automatic parallelization in the general case, parallelization for Inthreads will require significantly simpler compiler analysis. The reason is that, owing to the low granularity, the Inthreads parallelization is performed in a local context, and is very technical in nature. Discussion of possible automatic parallelization techniques can be found in Reference 13.

7.3.1 The Inthreads-C Language

In the Inthreads-C language, the code for the threads is marked up syntactically and is designated by **pragma inthread** blocks. Threads that are potentially running together are parts of the same function.

Inthreads-C defines intrinsic commands that correspond to the thread management and synchronization instructions of the architecture. In addition to generation of the machine instructions, the commands require special compiler support to ensure correct execution. The commands and their compiler semantics are summarized in Table 7.1.

A Programming Model and Architectural Extensions for Fine-Grain Parallelism

```
next = 0;
for (i=2; i<=size; i++) {
   a = perm[i]->a;      /*1*/
   r = compute_r(a);    /*2*/
   if (cond(a,r)) {     /*3*/
      next++;           /*4*/
      perm[next]
         ->a = a;       /*5*/
      perm[next]
         ->cost = r;    /*6*/
   }
}
            (a)
```

```
worker1:
#pragma inthread
{
   int     i_p;
   double  a_p, r_p;
   for(i_p=2;i_p<=size;
        i_p+=2)
   {
      a_p = perm[i_p]->a;
      r_p = compute_r(a_p);
      res1 = cond(a_p, r_p);
      INTH_SET(1);
      INTH_WAIT(2);
      if (res1) {
         perm[d1]->a    =a_p;
         perm[d1]->cots =r_p;
      }
   }
   INTH_SET(1);
   INTH_HALT();
}
            (b)
```

```
INTH_START(1, worker1);
INTH_START(2, worker2);
#pragma inthread
{
   for (i=2, next=0; ;) {
      if (!(i <= size)) break;
      /*Interact with wkr 1*/
      INTH_WAIT(1);
      if (res1)
         d1 = ++next;
      INTH_SET(2);
      i++;
      /*Interact with wkr 2*/
      ..
   }
   /* wait for workers to
      complete */
   INTH_WAIT(1);
   ..
}
            (c)
```

FIGURE 7.3 Example of parallelization in Inthreads-C: (a) Original code, (b) Worker thread, and (c) Coordinating thread.

The assignment of code to the threads is determined statically at the compilation time: each inthread code block belongs to exactly one thread, which is identified by the corresponding INTH_START command that jumps to its beginning.

No valid control paths should lead from a thread's code to the surrounding code. To this end, the code of a thread must either terminate with an INTH_HALT instruction or contain an infinite loop from which it will be released by an INTH_KILL.

Figure 7.3 gives an example of Inthreads-C parallelized code, which is actually a part of the Mcf benchmark from SPEC2000. The original code is shown in Figure 7.3a. The condition computation in lines 1, 2, 3 and the updating in lines 5, 6 are independent in different iterations and can be performed in parallel. The only computation that needs to be serialized is **next++** in line 4.

To parallelize the code, we create several worker threads to carry out the corresponding invocations of lines 1, 2, 3, 5, and 6, and one coordinating thread to run the invocations of line 4 serially. The code for the first worker and the coordinating thread is shown in Figures 7.3b and 7.3c, respectively. For variables i, a, and r we create private copies (i_p, a_p, and r_p) inside each worker thread. Note that there are several instances of these variables, one for each thread; however, the instances are distinct both syntactically and semantically.

The result of the evaluation of **cond** is transferred to the coordinating thread through shared variable **res1**, and the value of **next** is transferred back through shared variable **d1**. Since **res1** and **d1** are shared, access to them is protected by synchronization through conditions 1 and 2, respectively.

7.3.2 Compilation of Explicitly Parallel Code

Explicitly parallel code compilation introduces unique challenges to each aspect of the compilation process. At the framework level, the internal code representation must adequately represent the semantics of the multithreaded code. Furthermore, the existing analysis and optimization algorithms, originally developed with single-threaded code in mind, might be incompatible with Inthreads. Such algorithms need to be adapted for explicitly parallel code.

To illustrate the problems, consider the example in Figure 7.4. Note that the value of a, defined in line 10, is read at line 5. However, traditional compilation methods, which are not aware of parallel execution, would analyze each thread independently and would not notice the communication. Applying dead code elimination [29] to the above code would therefore misidentify the line 10 as dead code–code that produces a value never used by any of the subsequently executing instructions. The compiler could then remove the

```
(1)     #pragma inthread
(2)     {
(3)         b = ...;
(4)         INTH_WAIT(1);
(5)         a += b;
(6)     }
(7)     return a;
(8)     #pragma inthread
(9)     {
(10)        a = ...;
(11)        INTH_SET(1);
(12)        INTH_HALT();
(13)    }
```

FIGURE 7.4 Parallel code example.

assignment to a, changing the program behavior. On the other hand, making conservative assumption about a would result in inefficient code if a was indeed private.

Applying traditional register allocation to the example above could also produce incorrect results: the allocator, ignorant of the fact that a is used for communication, could allocate it to different registers in the two threads, disrupting the program semantics. We can see that, in order to produce correct code, the compiler must process all the concurrent threads together, taking into account the interactions between the threads.

7.3.3 Internal Code Representation

In order to allow the compiler to handle multithreaded code, we must first extend the internal representation with the knowledge of parallel semantics.

Control flow graph (CFG) is the standard form of intermediate code representation. CFG models the program as a set of basic blocks that contain the program statements, connected by control edges that represent possible execution paths. Concurrent CFG is an extension of CFG for parallel code. CCFG adds parallel control flow edges at locations in code that spawn a thread to indicate that the execution follows both edges. In addition, synchronization operations between parallel threads are represented by *synchronization* edges, which connect **cond.set** instructions with corresponding **cond.wait** ones. Although synchronization edges do not represent real execution paths, they are used for parallel code analysis. The extensions imply additional basic block splitting: similar to branch instructions, an **inth.start** or an **cond.set** instruction terminates a basic block, whereas an **cond.wait** instruction starts a new one. Furthermore, an **inth.halt** instruction not only terminates a basic block, but also has no outgoing edges at all.

Figure 7.5 presents an example of a concurrent CFG. Solid arrows represent normal and parallel control flow edges, and dashed arrows denote synchronization edges.

Since the Inthreads architecture requires all the communication between threads to be protected by synchronization, all such communication necessarily proceeds along synchronization (or normal) edges. As a result, the mere presence of the additional edges in the Concurrent CFG is sufficient to ensure correctness of some of the analysis phases, including most of the phases based on data flow analysis (DFA) [29], such as dead code elimination and constant propagation.

7.3.4 Register Allocation

Our approach to register allocation is based on the graph coloring paradigm [7]. This method uses an interference graph to represent the basic register allocation principle—that a register can hold only one value at any given time. The nodes of the interference graph are *register ranges*, locations in the graph grouping each assignment to a variable with all its possible uses.

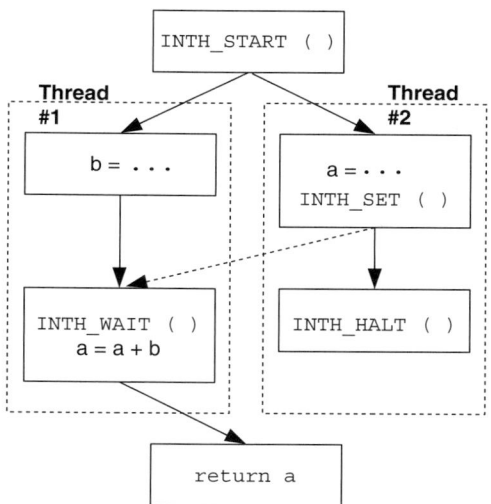

FIGURE 7.5 Parallel data flow graph example.

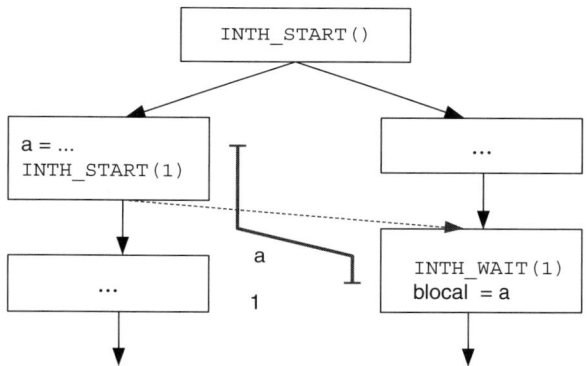

FIGURE 7.6 Using synchronization for finding register ranges.

Two register ranges are connected by an edge and are said to interfere if they cannot be allocated to the same machine register. In the case of parallel threads, communications through shared registers introduce additional connections, which span several threads and do not follow sequential control flow paths.

To address this change, we extend the register range building algorithms to consider the synchronization edges as well. Using this information, the compiler is able to construct register ranges that represent both sequential data flow and thread communication through shared variables. Since the architecture requires the concurrent accesses to be protected by synchronization, related accesses to a given variable will form a single register range.

In Figure 7.6, the definition of the shared variable a will be connected through the synchronization edges with the use of the variable in the parallel threads by a register range marked a_1.

The next step of building the parallel interference graph is to collect the interferences between the register ranges. Intuitively, for sequential case, two register ranges interfere if they are both live at some execution point. A variable v is live at an execution point e if there is a control flow path between e and some use of v, and v is not redefined on the path. The sequential version of the algorithm considers only execution points where new register ranges are defined. For each execution point e, the algorithm adds an interference edge between all register ranges defined at e and all the register ranges live at e.

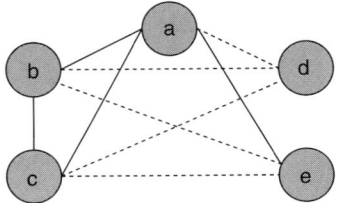

Thread 1	Thread 2
a = ...	d = ...
INTH_SET(1)	INTH_WAIT(1)
b = ...	e = d + a
c = ...	a = a + 1
INTH_WAIT(2)	INTH_SET(2)
c += b + a	... = e

FIGURE 7.7 Interference graph example.

In the parallel case, we should also consider the interferences between register ranges that are live in parallel threads. To this end, we modified the graph building algorithm to collect the interferences between the concurrently live register ranges. For each pair of statements (e_1, e_2) that may execute concurrently, the algorithm marks register ranges defined at e_1 as interfering with those live at e_2. We use a simple implementation that conservatively assumes that all statements of a thread may execute concurrently with all statements of a parallel thread. This is sufficient in most cases; however, a more accurate solution might improve the resulting interference graph.

Figure 7.7 shows a code example and the corresponding interference graph. Variable a is shared by both threads, whereas the rest of the variables are accessed only by one of the threads. The solid lines represent sequential interference, computed by traditional interference graph building algorithm. The dashed lines represent concurrent interference between register ranges accessed by parallel threads.

After constructing the parallel interference graph, the algorithm applies traditional simplification steps to find a coloring. When all the nodes in the graph are of a high degree, blocking the simplification, a heuristic step is taken to select a spill candidate. Some of the spill candidates may still be later allocated with a color, although others may be eventually spilled.

Spilling in presence of register sharing might cause races even for read-only values, therefore, it must be performed carefully. Detailed description of spilling and the rest of the compilation process can be found in Reference 38.

7.4 Implementation

In principle, the architecture of Inthreads is quite similar to SMT [43]—both architectures are based on multiple independent streams of instructions that are mixed in a shared execution core. Therefore, the microarchitecture reuses most of the mechanisms present in SMT processors, such as multiple fetch units, multiple Reorder Buffers (ROBs), and shared functional units.

The most notable difference with respect to SMT processors lies in the implementation of the register file. In an SMT processor, the register of the threads is distinct. To this end, such processors either have a completely private register file, including both the physical registers and the Register Allocation Table (RAT), or at least, have a per-thread copy of RAT. In contrast, in an Inthreads processor, the RAT must be shared between the threads.

The second difference in the implementation is that the Inthreads architecture requires mechanisms to support efficient execution of synchronization operations. Figure 7.8 presents these mechanisms (the Inthreads-related ones are shaded).

The *wait buffers (WB)* implement the delaying of **cond.wait** instructions for which the required condition is not ready. The delaying is performed independently for each thread, and therefore there is one WB per inthread. The WBs are placed before the renaming logic of the processor, therefore, the register dependencies, which are defined during renaming, follow the semantics of synchronization. To accommodate the overhead, a new pipeline stage is dedicated for the WBs.

The *thread control unit (TCU)* orchestrates the execution of the threads on the processor. The TCU receives all the thread-related instructions, and issues the necessary control signals to apply their side

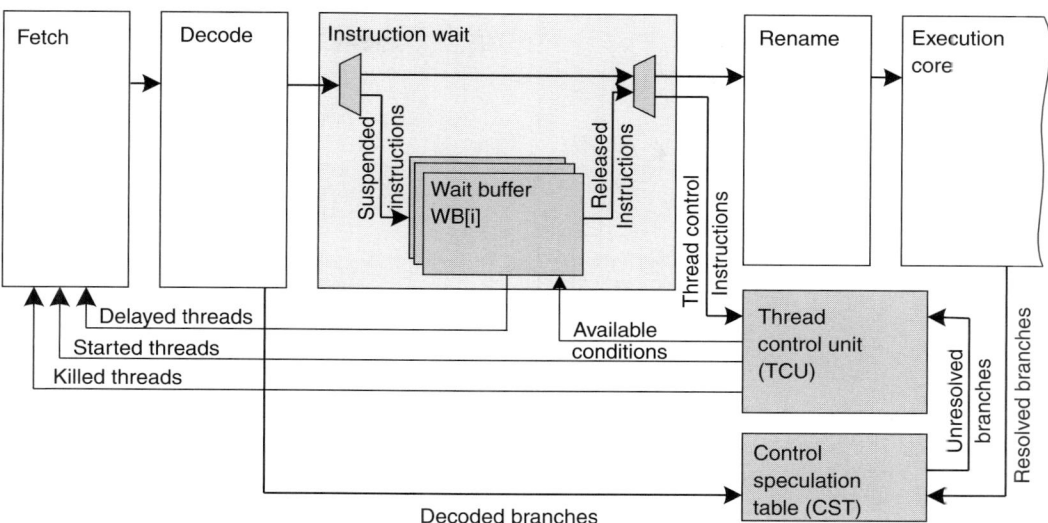

FIGURE 7.8 Inthreads implementation outline.

Sequential code	Thread t_k of T

`for(i=0;i<N;i++) {`
`  if(data[i%K].val > i )`
`  {`
`    data[i%K].count++;`
`  }`
`}`
	`for(i=k;i<N;i+=T) {`
	`  if(data[i%K].val > i )`
	`  {`
	`    MUTEX_ENTER`
	`    data[i%K].count++;`
	`    MUTEX_LEAVE`
	`  }`
	`}`

FIGURE 7.9 Low granularity parallelization with control speculation.

effects. First, the TCU notifies the WBs when conditions become available, allowing the corresponding cond.wait instructions to proceed. Second, it instructs the Fetch unit to start and stop fetching of instructions in response to inth.start, inth.kill, and inth.halt instructions.

The *Condition Speculation Table (CST)* keeps track of the unresolved branch instructions. This information is used by the TCU to determine the speculative status of instructions.

7.4.1 Misprediction Handling

One important aspect of efficient execution of Inthreads programs is the interaction of control flow speculation with multithreading. For an example, consider the program in Figure 7.9. When the sequential version of the program is executed, branch prediction may allow the processor to execute as many iterations in parallel as the hardware can accommodate. However, in the case of parallelized code, the presence of synchronization limits the number of iterations that can be active speculatively: since a mutex affects execution of other threads' instructions, it is dangerous to enter the mutex before the if has been resolved. As a result, in each thread, the if must be resolved before the next if can be issued. Therefore, the number

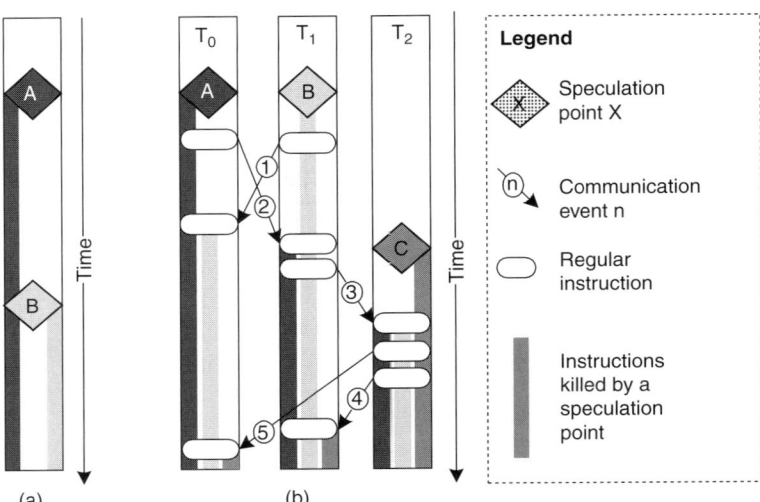

FIGURE 7.10 Speculation models for single-threaded and multithreaded execution. (a) single-threaded execution and (b) multi-threaded execution.

of iterations that can proceed in parallel is at most one per active thread, potentially leading to lower ILP than that of the serial, but speculative, code.

To realize the potential of both speculation and multithreading, we must enable speculative execution of instructions that involve communication between threads, such as interactions between instructions involved in a synchronization or between instructions accessing a shared variable. In order to recover from misspeculation, we must keep track of all the *communication events*, and take care of all the instructions, from all the threads, which have been affected by the misspeculated instruction.

Misprediction handling in case of multithreaded execution is inherently more complex than in the sequential case. Sequential speculation, as illustrated by Figure 7.10a is *linear*: misspeculation of A requires squashing of all the instructions following A, and all the instructions can be arranged in such a way that exactly the instructions following the misspeculated one are squashed.

On the other hand, misspeculation in the multithreaded case is nonlinear, as demonstrated in Figure 7.10b. The execution contains three threads T_0, T_1, and T_2 that contain speculation points A, B, and C, respectively. There are two communication events, 1 and 2, between T_0 and T_1 and two events, 3 and 4, between T_1 and T_2. Finally, there is a communication 5 from T_2 to T_0.

We first note that it is impossible to arrange instructions of all the threads in such an order that all the instructions following a speculation point are those affected by it. For example, A and B are independent, and neither of them should precede the other.

Another observation is that misspeculation recovery is timing sensitive. For example, consider the misspeculation recovery of B. The misspeculation propagates to thread T_2 along event 3, squashing the instructions following the consuming instruction of event 3. As a result, the producing instruction of event 4 is squashed, and therefore, the consuming instruction of 4 must be squashed as well. However, that instruction may have been already squashed since it follows B in the program order of T_1, and the pipeline may already contain the instructions for the correct execution path after B. In that case, the processor would squash instructions on the correct execution path. This situation may become even more complicated when misspeculation is propagated through longer sequences of events, such as the sequence of 2, 3, and 5 that would result from misspeculation of A.

Our solution to this problem is to process the misprediction atomically. To this end, we notice that a first squashed in each thread either follows a mispredicted branch or is an instruction receiving a communication. Therefore, a vector of timestamps can be used to determine the first squashed instruction in each thread. Upon computing the vector, we can squash in parallel the instructions in all the threads, without performing the propagation explicitly.

To compute the vector of squashed instructions, we use a table of all the currently active communication instructions. For each instruction i, we hold in the table its own timestamp vector (TSV_i), which determines the set of instructions in each thread that affect i. When a branch misprediction is detected, we compare the timestamp of the branch with the corresponding location in each one TSV_j for all the active communication instructions j to find which instructions need to be squashed.

The process of determining the set of misprediction instructions is efficient only if the number of the communication instructions that need to be handled is low. Fortunately, the Inthreads architecture helps reduce the number of communication instructions.

As stated above, Inthreads requires *Release Consistency* as the communication model for all the variable accesses, defining the synchronization and thread management instructions as the only *strong* instructions in the program. In its turn, the programs are required to avoid data races in the code, ordering all the potentially conflicting accesses with the synchronization instructions.

With regard to speculation, the avoidance of data races implies that every two regular instructions that can potentially participate in a communication must be separated by a chain of synchronizations. As a result, if the providing instruction is squashed, so is the synchronization instruction following it. Consequently, given a program that complies with the data-race-free requirement, handling misspeculation only for the synchronization instructions would automatically take care of all the instances of communication.

One additional effect to restricting the handled communication instructions to just the synchronization ones is that communication becomes easy to detect. Since in the Inthreads architecture each synchronization instruction accesses a specific condition register, the register number can be used as a tag for matching the pair of instructions participating in a communication. Matching the communication of regular instructions could be much more involved, both since the number of such instructions in flight is significantly higher, and since memory accesses, which could be involved in a communication, can be matched only after their target addresses are resolved.

In the Inthreads architecture we recognize several types of communication events.

A *Synchronization* event occurs between a cond.set instruction and a cond.wait instruction that accesses the same condition as the cond.set. Synchronization events are relatively easy to detect, as the number of synchronization instructions that are concurrently in flight is relatively low (see Section 7.5). Note that cond.wait and cond.clr instructions never take a producer part in communication, and therefore, can be executed speculatively with no need for any special recovery.

A *Variable value transfer* event occurs between any instruction writing a value to a variable (whether in a register or at some memory location) and a subsequent instruction reading the value from that variable. As described above, such transfers do not need to be handled by the processor since the handling of the other communication events would take care of the value transfers. Nevertheless, we measure the effect of the Release Consistency requirement by (optionally) disabling speculative execution of value transfers.

A *Thread Starting* event results from the communication between an inth.start instruction to the first instruction of the started thread. Thread starting events are handled similarly to the synchronization ones. The only difference is that the instruction receiving the communication does not belong to a specific instruction type, but is just the first instruction started by the thread. To handle this, we hold a *tsv* for every thread, as if the whole thread receives the communication.

A detailed description of the misprediction handling process can be found in Reference 14.

7.5 Performance Evaluation

We have extended the SimpleScalar-PISA model [6] to include the implementation of the new instructions and the microarchitectural extensions. The basic processor is modeled with a 8-stage pipeline. The Inthread-enabled version adds one stage after Decode to accommodate the synchronization

TABLE 7.2 Basic Processor Parameters

Parameter	No threads	Inthreads
Pipeline length	8	9
Supported threads	N/A	8
Fetch policy	8 per cycle	ICOUNT4.8
Branch predictor	bimodal	
L1I size	64 KB	
Logical registers	64GP + 64FP	
Physical registers	512GP,512FP	384GP,384FP
ROB size	512	128*8
Issue queue size	80	
Memory queue size	40	
Functional units	6 Int, 6FP, 4 Branch	
Maximum outstanding misses	16	
Memory ports	2	
L1D size	64 KB	
L2 size	1 MB	
L2 latency	20	
Memory latency	200	

functionality. Except where stated explicitly, the processor front-end was configured to fetch up to 8 instructions from up to 4 concurrent threads, managed by the ICOUNT policy. Table 7.2 lists the relevant processor parameters.

We assume that the instructions dispatched to the TCU execute in as few as two cycles: one to compute the *TSV*, and one cycle to issue the instruction if it is ready. We base our assumption on comparison of the TCU with the structure of the Issue Queue. The logic involved in the TCU is similar to that of the Issue Queue, although the number of instructions held in the TCU is significantly lower: in our experiments there was no measurable effect to increasing the TCU size over 16 instructions. Still, we measured the effect of increasing the TCU latency, shown in Figure 7.12.

The evaluation is based on three benchmarks from the SPEC2K suite [15]: 179.art, 181.mcf, and 300.twolf, and four programs from the MediaBench suite [24]: Adpcm encode, Adpcm decode, G721, and Mpeg2. In the benchmarks, we applied the parallelization only to those portions of the code that took a performance hit from high cache or branch miss rate (and constituted a significant part of the program's runtime). The programs were run until completion with reduced inputs. As a measure of the programs' runtime we used the number of clock cycles reported by the simulator.

We denote each value of speculation settings with three letters, each determining if the corresponding speculation mechanism from those described in Section 7.4.1 was turned on. For example, FFF means that the speculative execution of synchronization instructions, of value transfer between variables and of thread starting instructions have been turned off, and TTT means that all the mechanisms have been turned on.

7.5.1 Parallelization Granularity

One of the most important differences between Inthreads and conventional multithreading is the difference in achievable parallelization granularity. To evaluate the granularity, we measure the characteristic parameters of Inthreads-parallelized code.

Table 7.3 presents the results for our benchmarks. For each program, we counted the average number of instructions in the parallel and in the serial sections of the code (In the case of parallel code, we counted instructions of all the participating threads). In addition, we show the average number of serialization instructions performed by each thread (the number of committed **cond.wait**s).

We can see that the number of instructions in one parallelized section is very low—on the order of several hundred instructions. Compare this to Figure 7.1, which implies that the overhead of conventional

TABLE 7.3 Thread Parameters for the Benchmarks

	#KPars	#Insns/Par	#Synch/Par	#Insns/Ser
181.mcf	247	749	50.5	612
179.art	35	3926	0.0632	845
300.twolf	474	97.57	3.982	77.99
adpcm.enc	451	175.2	8.717	3.366
adpcm.dec	79	87.42	3.719	5.432
g721	1824	190.6	4.691	130.2
mpeg2	565	362.4	1.27	112.5
epic	88	559.5	0.3818	83.5

Note: #KPars = Number of parallel sections (in thousands);
#Insans/Par = Average instructions in a parallel section;
#Synch/Par = Average synchronizations per parallel section;
#Insns/Ser = Avegrage instructions per serial section.

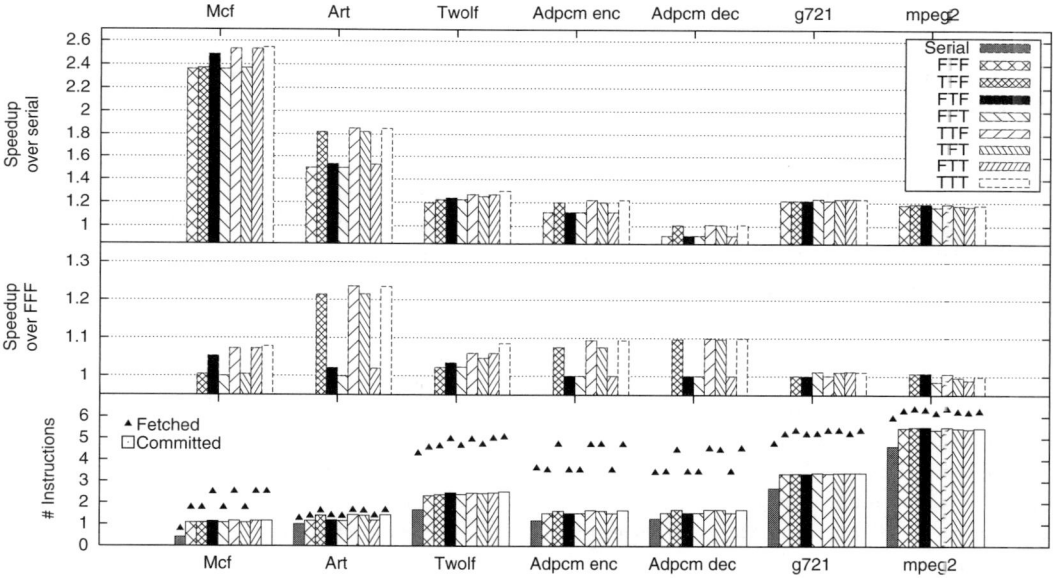

FIGURE 7.11 Performance of SPEC and Mediabench benchmarks.

multithreading is on the order of thousands of cycles. The largest parallelized sections are observed in 179.art, and are barely large enough to not suffer a slowdown when parallelized with conventional threads; however, as shown in Section 7.5.2, with Inthreads we have been able to speed up the benchmark considerably.

7.5.2 Benchmarks

Figure 7.11 presents the results of parallelization of the benchmarks with Inthreads. The largest speedup is achieved for Mcf and Art, which have problematic patterns of branches and memory accesses. The Adpcm code is highly sequential in nature, presenting little opportunity for Inthreads parallelization. On the other hand, the code in g721 and mpeg2 exhibits high level of instruction level parallelism, leaving little opportunity for further improvement in function unit utilization. Still, even in that case, a speedup of 20% is reached.

TABLE 7.4 Average Ages and Frequencies of Thread-Related Instructions

Benchmark	cond.set		inth.start	
	Age	Freq	Age	Freq
Mcf	41	0.018	7.0	0.003
Art	32	0.04	391	1.9×10^{-5}
Twolf	4.4	0.057	3.3	0.014
Adpcm enc.	6.3	0.059	2.0	2.9×10^{-5}
Adpcm dec.	2.4	0.061	3.0	3.1×10^{-5}
G721	3.7	0.05	3.1	0.012
Mpeg2	1.5	0.014	9.1	0.014

The benefit of speculative execution depends on the amount of time saved by earlier execution of thread-related instructions and the frequency of such instructions. Table 7.4 shows the frequencies and average age of the **cond.set** and **inth.start** instructions, measured from the time they enter the TCU and until they are issued. The age is measured for the processor running under FFF; when the speculation is turned on, the age becomes irrelevant as these instructions are executed almost immediately.

The results of applying speculative execution of thread-related instructions to the SPEC and Mediabench benchmarks are summarized in Figure 7.11. The first row shows the overall speedup achieved by parallelization, and the second one—the speedup increment caused by turning on the speculation of thread-related instructions. For each benchmark, we have exercised all the possible combinations of the speculation mechanisms in order to determine which of them have the most effect on performance.

We can see that Mcf benefits the most from speculative value transfer through variables, Art benefits from speculative execution of synchronization instructions and, to a lesser degree, from speculative variable transfer, and Twolf needs the combination of all the speculation mechanisms for maximal speedup. Both Art and Mcf perform thread starting relatively infrequently, as can be seen in Table 7.4, and therefore do not benefit from speculation in thread starting.

Mcf parallelizes consequent iterations that need to communicate small amount of information to each other. This communication is sped up when executed speculatively, and therefore, the most important speculation mechanism for Mcf is the speculation on variable transfer. On the other hand, Art performs synchronization under heavy speculation (which can be seen in the relatively high waiting time of the **cond.set** instructions), and thus receives most improvement from speculation on synchronization instructions. Twolf uses threads to parallelize small independent sections of code with heavy branching, and benefits from all the forms of speculative execution.

Both Adpcm programs split the iterations of their heaviest loops into portions executed by threads arranged in a virtual pipeline. Synchronization instructions are used to coordinate the execution of portions of the loop iterations by the threads, and thus speculation on such instructions speeds up these programs.

Both g721 and mpeg2 are barely affected by the speculation, albeit for opposite reasons. In g721, the hard-to-predict branches are executed just before the synchronization instructions, resulting in poor speculation success rate. In mpeg2, the synchronization is performed after long nonspeculative instruction sequences, obviating the need for speculative execution of synchronization instructions. Moreover, owing to the high resource utilization in mpeg2, in cases when synchronization speculation fails, the speculative thread starves the correct-path computation, resulting even a slight slowdown by the speculation.

The dependence of speedup on the TCU latency is shown in Figure 7.12. Mcf and Art are least sensitive to the latency owing to the relatively long age of the synchronization instructions. The speedup of adding speculation to thread-related operations, shown in Figure 7.12b, decreases with the latency when it grows

A Programming Model and Architectural Extensions for Fine-Grain Parallelism

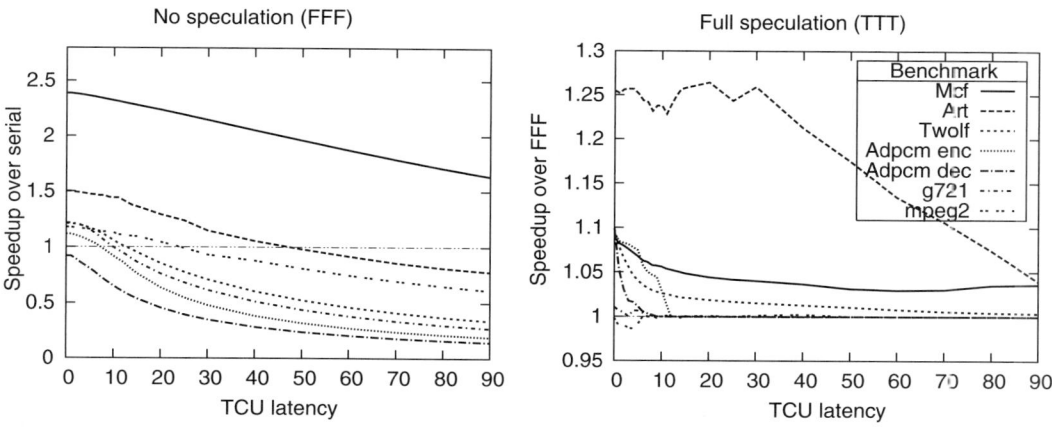

FIGURE 7.12 Performance of SPEC and Mediabench benchmarks under varying TCU latency.

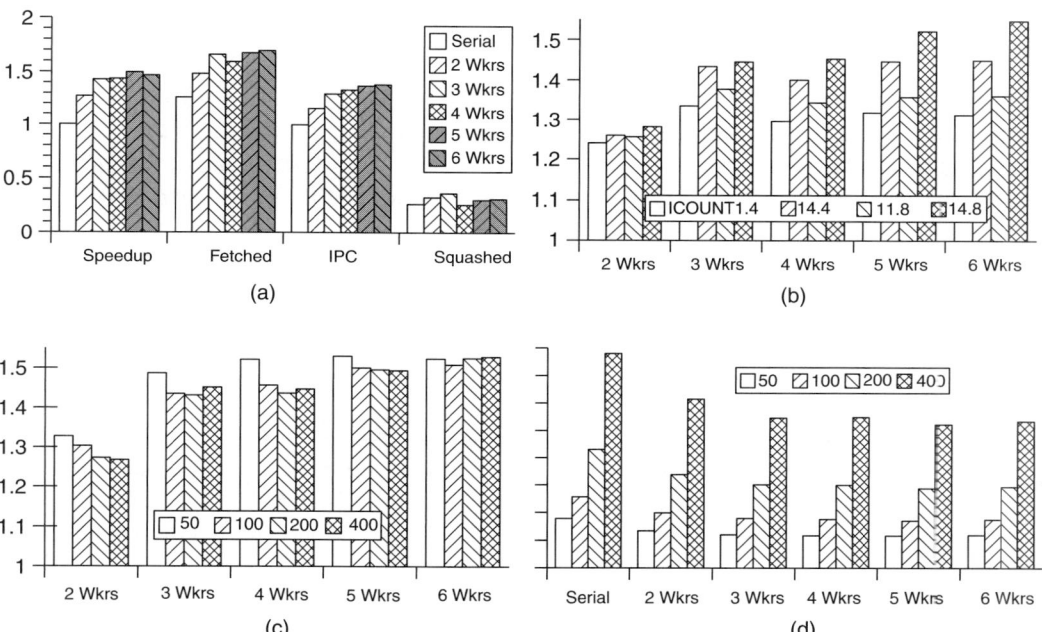

FIGURE 7.13 Execution results for Art for varied parallelization degree. In (a), Speedup is relative to the original program, while Fetched, IPC, and Squashed show the absolute number of instructions. (a) Execution result, (b) Speedup for fetch policy, (c) Speedup for memory latency, and (d) Execution time for memory latency.

larger than the benefit of speculation. It is interesting to note that the speculation speedup for Mcf almost does not decrease with latency. The reason is that when the latency grows, additional parallelism is achieved by an increase in the number of loop iterations that execute in parallel.

7.5.3 Effect of Parallelization Degree

For the Art and Mcf benchmarks, we have measured the effect of varying the number of threads used for the parallelization. The results are presented in Figures 7.13 and 7.14.

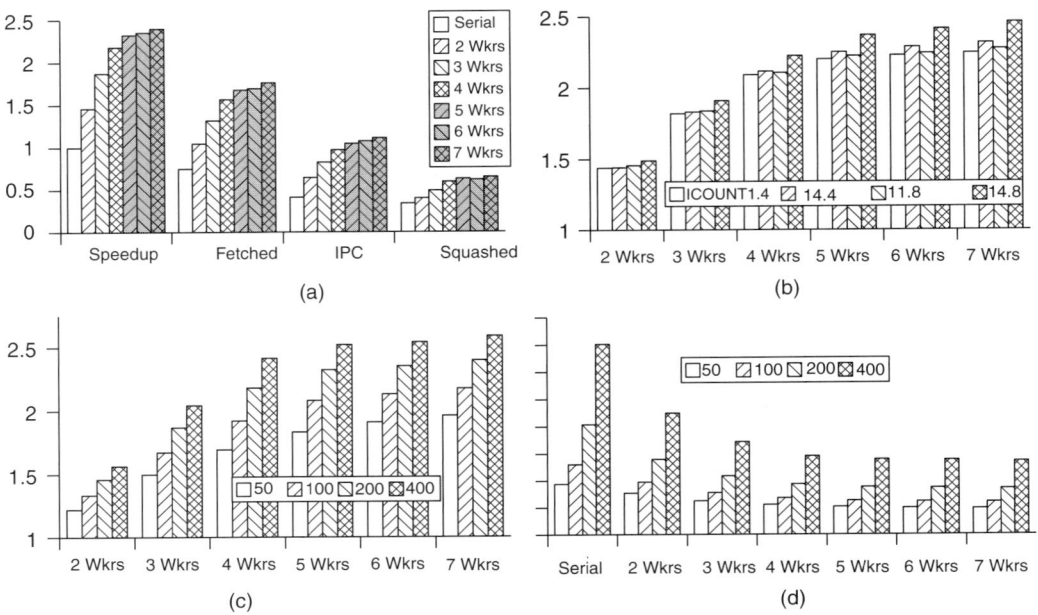

FIGURE 7.14 Execution results for Mcf for varied parallelization degree. In (a), Speedup is relative to the original program, while Fetched, IPC, and Squashed show the absolute number of instructions. (a) Execution results, (b) Speedup for fetch policy, (c) Speedup for memory latency, and (d) Execution time for memory latency.

In 179.art, threads are created to perform relatively small chunks of computation, on the order of several thousand instructions each. With conventional multithreading, the overhead of starting and terminating a thread this often would cancel any gains from parallelization, and parallelizing at larger granularity is impossible as the chunks are dependent.

Performance results for the complete execution of 179.art are presented in Figure 7.13a. We can see that the speedup is achieved by optimizing the fetch bandwidth while keeping number of squashed instructions constant. This implies that the fetch speed and speculative execution behavior are the dominant factors in the benchmark's performance. Indeed, as Figure 7.13b shows, the speedup improves with a more aggressive fetch policy.

Figures 7.13c and 7.13d show the speedup as a function of memory latency. We can see that the speedup is barely affected although the execution time depends strongly on the latency. This implies that 179.art contains little pointer chasing, and thus the behavior of the memory subsystem is not affected by parallelization.

The parallelization of 181.mcf is done at even finer granularity. The code contains small portions of serial code that require intensive synchronization: about 0.7 issued synchronization instructions, and related data transfers between threads, per cycle. Still the parallelization achieves a net speedup, owing to communication through registers and low-overhead synchronization.

The performance characterization of 181.mcf is presented in Figure 7.14a. Note that the percent of squashed instructions remains almost constant, indicating that there are no bottlenecks related to fetching or to speculation. Indeed, as Figure 7.14b shows, changing the fetch policy has little effect. On the other hand, the speedup does increase when the memory latency increases, as shown in Figure 7.14c. This is not surprising considering the amount of double pointer indirections in the code.

Figure 7.14d shows the execution time as a function of memory latency. It is interesting to note that the parallelized version is almost as fast at memory latency of 400 as the serial code at memory latency of 100.

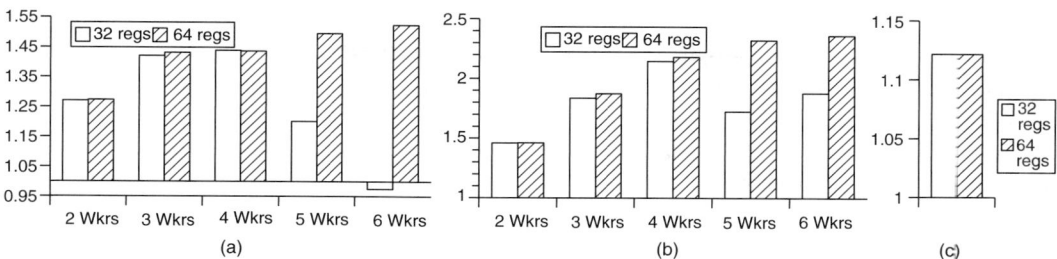

FIGURE 7.15 Effect of register pressure: speedup as function of register file size. (a) Speedup for 179.art, (b) Speedup for 181.mcf, (c) Speedup for 300.twolf.

7.5.4 The Effect of Register Pressure

To assess the sensitivity of the parallelization to the register file size, we compiled the benchmarks with 32 and 64 registers. The results are shown in Figure 7.15. For 179.art, the compiler finds enough registers for up to 4 threads, after which the performance deteriorates owing to spilling. Parallelization of 181.mcf suffers less from register pressure than that of art. With 5 threads, the compiler runs out of registers and spills several heavily used variables. With 6 threads, spilling increases only slightly, and the result improves again owing to increased parallelization. Finally, parallelization of 300.twolf with two threads does not cause a significant increase in register pressure.

7.6 Related Work

This work is based on the observation that programs often do not need many registers in order to reach top performance [5]. A natural way to take advantage of this fact is to divide the registers between several threads, gaining speedup by parallelization while hurting only slightly the performance of each thread.

Several researches have investigated methods for sharing registers between threads under compiler control. The most closely related work is the $_{mt}$SMT proposal by Redstone et al. [33], which adds multiple thread contexts for each SMT thread. The architectural register file is divided equally between registers; no communication is performed through registers. Redstone et al. examine several inherently parallelizable programs and observe an average 38% improvement in throughput with two contexts, with only 5% average slowdown of each thread. A comparable approach can be found in the work of Waldspurger and Weihl [45]. In contrast, in our work, the threads working in context of a single SMT thread may access all the registers. The threads use the shared registers as an extremely lightweight communication mechanism, which is used for parallelization by threads at the level of sequential code.

Several researchers have investigated parallelizing programs with microthreads in order to tolerate the memory latency. The Threadlets architecture by Rodrigues et al. [34] splits *basic blocks* into threads using a special trace-based compiler. The approach, applied to several scientific benchmarks, achieves 80% of the potential parallelism in most cases. In comparison, our model presents a more general threading mechanism, with control flow and generic communication capability, which is applicable to less uniform programs such as SPEC benchmarks.

Another method of tolerating long-latency accesses is to switch threads immediately when such an event occurs. Mowry and Ramkissoon [28] propose a system in which the processor traps into a thread-switch routine immediately after discovering a cache miss. To reduce the context switch overhead, the researchers use the compiler to statically partition the registers between the threads.

Fine-granularity threading requires a fast hardware-based synchronization mechanism. This mechanism usually takes a form of Full/Empty bit (FEB), which can be associated with memory locations [1, 3] or a register-like lock box [44]. The latter work achieves speedup for parallelizing programs that slow down when executed on a conventional architecture. Our approach of designating special instructions for synchronization allows us to implement communication through shared variables more efficiently.

Other works on fine-grain multithreading use threads that assist the main thread by warming up the cache and branch prediction or simply running forward speculatively to explore speculative paths to a considerable depth [4, 8, 10]. In these works, the communication between threads is either completely unnecessary, or is performed only in one direction, from the parent thread to the spawned one.

Synchronization operations are often redundant and can be sped up by ignoring them speculatively. Examples of this approach are *Speculative Synchronization* [27], *Transactional Memory* [16], and *Speculative Lock Elision* [32]. In contrast, we do not attempt to remove unnecessary synchronization but just speed it up by enabling speculative execution of the synchronization instructions. This avoids the need to monitor the execution for collisions, and makes the recovery simpler. As an additional benefit, our approach allows speculation in other thread interactions, like starting and killing threads.

Note that our use of the term *speculative execution* is different from the one used in the above works. Whereas *Speculative Synchronization* speculates on the fact that the lock is unnecessary, we perform the thread-related operations exactly as instructed by the program, but just allow them to proceed before the preceding control speculation instructions are resolved.

Thread Level Speculation issues threads derived from a serial program, speculating on a fact that the threads will turn out to be independent. In *Multiscalar processors* [39], the program is partitioned into speculative tasks that execute concurrently while observing data dependencies. *Speculative multithreaded processors* [26] and *Dynamic multithreading processors* [2] generate threads at control speculation points. In those works, the threads are usually very small, often with limited control flow. In contrast, the mechanisms in our work apply to general threads and provide a unified mechanism for speculation of both synchronization and thread management instructions.

Other works in this area are [18, 20, 30, 31, 40–42].

The high cost of branch misprediction has prompted several works that aim at reducing the penalty by retaining those instructions that would execute in the same way regardless of the branch outcome [11, 35, 36]. Our approach can be seen as achieving about the same in software by speculatively starting threads with dependencies marked up by communication instructions.

References

[1] A. Agarwal, J. Kubiatowicz, D. Kranz, B.-H. Lim, D. Yeung, G. D'Souza, and M. Parkin. Sparcle: An evolutionary processor design for large-scale multiprocessors. *IEEE Micro*, 13(3):48–61, 1993.

[2] H. Akkary and M. A. Driscoll. A dynamic multithreading processor. In *Proceedings of the 31st Annual ACM/IEEE International Symposium on Microarchitecture*, pp. 226–236. IEEE Computer Society Press, 1998.

[3] R. Alverson, D. Callahan, D. Cummings, B. Koblenz, A. Porterfield, and B. Smith. The Tera computer system. In *Proceedings of the 1990 International Conference on Supercomputing*, pp. 1–6, June 1990.

[4] R. Balasubramonian, S. Dwarkadas, and D. H. Albonesi. Dynamically allocating processor resources between nearby and distant ILP. Technical Report TR743, University of Rochester, April 2001.

[5] D. G. Bradlee, S. J. Eggers, and R. R. Henry. The effect on RISC performance of register set size and structure versus code generation strategy. In *Proceedings of the 18th Annual International Symposium on Computer Architecture*, pp. 330–339. ACM Press, 1991.

[6] D. Burger, T. M Austin, and S. Bennett. Evaluating future microprocessors: The SimpleScalar tool set. Technical Report CS-TR-1996-1308, University of Wisconsin-Madison, 1996.

[7] G. J. Chaitin. Register allocation and spilling via graph coloring. In *SIGPLAN Symposium on Compiler Construction*, 1982.

[8] R. Chappell, J. Stark, S. Kim, S. Reinhardt, and Y. Patt. Simultaneous subordinate multithreading. In *Proceedings of the International Symposium on Computer Achitecture*, May 1999.

[9] A. Cristal, O. J. Santana, M. Valero, and J. F. Martínez. Toward kilo-instruction processors. *ACM Transactions on Architecture and Code Optimizations*, 1(4):389–417, 2004.

[10] D. E. Culler, A. Sah, K. E. Schauser, T. von Eicken, and J. Wawrzynek. Fine-grain parallelism with minimal hardware support: A compiler-controlled threaded abstract machine. In *Proceedings of the 4th International Conference on Architectural Support for Programming Languages and Operating Systems*, pp. 164–175, 1991.

[11] R. Desikan, S. Sethumadhavan, D. Burger, and S. W. Keckler. Scalable selective re-execution for edge architectures. In *Proceedings of the 11th International Conference on Architectural Support for Programming Languages and Operating Systems*, pp. 120–132, New York, 2004. ACM Press.

[12] J. A. Fischer. Very long instruction word architectures and the ELI-12. In *Proceedings of the 10th Symposium on Computer Architectures*, pp. 140–150, June 1983.

[13] A. Gontmakher and A. Schuster. Intrathreads: Techniques for parallelizing sequential code. In *6th Workshop on Multithreaded Execution, Architecture, and Compilation* (in conjunction with Micro-35).

[14] A. Gontmakher, A. Schuster, Avi Mendelson, and G. Kovriga. Speculative synchronization and thread management for fine granularity threads. In *12th International Symposium on Computer Architecture*, 2006.

[15] J. L. Henning. SPEC CPU2000: Measuring CPU Performance in the New Millennium. *Computer*, 33(7): 28–35, 2000.

[16] M. Herlihy and J. E. B. Moss. Transactional memory: Architectural support for lock-free data structures. In *Proceedings of the 20th Annual International Symposium on Computer Architecture*, pp. 289–300, New York, 1993. ACM Press.

[17] G. Hinton, D. Sager, M. Upton, D. Boggs, D. Carmean, A. Kyker, and P. Roussel. The microarchitecture of the Pentium 4 processor. *Intel Technology Journal*, February 2001.

[18] K. Hiraki, J. Tamatsukuri, and T. Matsumoto. Speculative execution model with duplication. In *Proceedings of the 12th International Conference on Supercomputing*, pp. 321–328, New York, 1998. ACM Press.

[19] J. Huck, D. Morris, J. Ross, A. Knies, H. Mulder, and R. Zahir. Introducing the IA-64 architecture. *IEEE Micro*, 20(5):12–23, 2000.

[20] T. A. Johnson, Rudolf Eigenmann, and T. N. Vijaykumar. Min-cut program decomposition for thread-level speculation. In *Proceedings of the ACM SIGPLAN 2004 Conference on Programming Language Design and Implementation*, pp. 59–70, New York, 2004. ACM Press.

[21] S. Jourdan, P. Sainrat, and D. Litaize. Exploring configurations of functional units in an out-of-order superscalar processor. In *Proceedings of the 22nd Annual International Symposium on Computer Architecture*, pp. 117–125, 1995.

[22] R. Kalla, B. Sinharoy, and J. M. Tendler. IBM Power5 chip: A dual-core multithreaded processor. *IEEE Micro*, 24(2):40–47, March 2004.

[23] M. S. Lam and R. P. Wilson. Limits of control flow on parallelism. In *Proceedings of the 19th Annual International Symposium on Computer Architecture*, pp. 46–57, May 1992.

[24] C. Lee, M. Potkonjak, and W. H. Mangione-Smith. Mediabench: A tool for evaluating and synthesizing multimedia and communications systems. In *Proceedings of the 30th Annual ACM/IEEE International Symposium on Microarchitecture*, pp. 330–335, Washington, DC, 1997. IEEE Computer Society.

[25] S. Mahlke. Characterizing the impact of predicated execution on branch prediction. In *Proceedings of the 27th International Symposium on Microarchitecture*, December 1994.

[26] P. Marcuello, A. Gonzalez, and J. Tubella. Speculative multithreaded processors. In *Proceedings of the International Conference on Supercomputing (ICS-98)*, pp. 77–84, New York, July 13–17, 1998. ACM Press.

[27] J. F. Martínez and J. Torrellas. Speculative synchronization: Applying thread-level speculation to explicitly parallel applications. In *Proceedings of the 10th International Conference on Architectural Support for Programming Languages and Operating Systems*, pp. 18–29, New York, 2002. ACM Press.

[28] T. C. Mowry and S. R. Ramkissoon. Software-controlled multithreading using informing memory operations. In *HPCA*, pp. 121–132, 2000.

[29] S. Muchnik. *Advanced Compiler Design and Implementation*. Morgan Kaufmann Publishing, 1997.
[30] C.-L. Ooi, S. Wook Kim, Il. Park, R. Eigenmann, B. Falsafi, and T. N. Vijaykumar. Multiplex: unifying conventional and speculative thread-level parallelism on a chip multiprocessor. In *Proceedings of the 15th International Conference on Supercomputing*, pp. 368–380, New York, 2001. ACM Press.
[31] I. Park, B. Falsafi, and T. N. Vijaykumar. Implicitly-multithreaded processors. In *Proceedings of the 30th Annual International Symposium on Computer Architecture*, pp. 39–51, New York, 2003. ACM Press.
[32] R. Rajwar and J. R. Goodman. Speculative lock elision: Enabling highly concurrent multithreaded execution. In *Proceedings of the 34th Annual ACM/IEEE International Symposium on Microarchitecture*, pp. 294–305, Washington, DC, 2001. IEEE Computer Society.
[33] J. Redstone, S. Eggers, and H. Levy. Mini-threads: Increasing TLP on small-scale SMT processors. In *HPCA '03: Proceedings of the Ninth International Symposium on High-Performance Computer Architecture (HPCA'03)*, p. 19. IEEE Computer Society, 2003.
[34] A. Rodrigues, R. Murphy, P. Kogge, and K. Underwood. Characterizing a new class of threads in scientific applications for high end supercomputers. In *Proceedings of the 18th Annual International Conference on Supercomputing*, pp. 164–174. ACM Press, 2004.
[35] E. Rotenberg and J. Smith. Control independence in trace processors. In *Proceedings of the 32nd Annual ACM/IEEE International Symposium on Microarchitecture*, pp. 4–15, Washington, DC, 1999. IEEE Computer Society.
[36] A. Roth and G. S. Sohi. Register integration: A simple and efficient implementation of squash reuse. In *Proceedings of the 33rd Annual ACM/IEEE International Symposium on Microarchitecture*, pp. 223–234, New York, 2000. ACM Press.
[37] M. Schlansker and B. Rau. EPIC: An architecture for instruction-level parallel processors. Technical Report HPL-1999-111, Hewlett-Packard Laboratories, February 2000.
[38] G. Shklover, A. Gontmakher, A. Schuster, and A. Mendelson. Code generation for fine-granularity register sharing threads. In *INTERACT-10*, in conjunction with HPCA-12, 2006.
[39] G. S. Sohi, S. E. Breach, and T. N. Vijaykumar. Multiscalar processors. In *Proceedings of the 22nd Annual International Symposium on Computer Architecture*, pp. 414–425, New York, 1995. ACM Press.
[40] G. S. Sohi and A. Roth. Speculative multithreaded processors. *Computer*, 34(4):66–73, April 2001.
[41] Y. Solihin, J. Lee, and J. Torrellas. Using a user-level memory thread for correlation prefetching. In *Proceedings of the 29th Annual International Symposium on Computer Architecture*, pp. 171–182, Washington, DC, 2002. IEEE Computer Society.
[42] J. G. Steffan, C. B. Colohan, A. Zhai, and T. C. Mowry. A scalable approach to thread-level speculation. In *Proceedings of the 27th Annual International Symposium on Computer Architecture*, pp. 1–12, New York, 2000. ACM Press.
[43] D. M. Tullsen, S. Eggers, and H. M. Levy. Simultaneous multithreading: Maximizing on-chip parallelism. In *Proceedings of the 22th Annual International Symposium on Computer Architecture*, 1995.
[44] D. M. Tullsen, J. L. Lo, S. J. Eggers, and H. M. Levy. Supporting fine-grained synchronization on a simultaneous multithreading processor. In *Proceedings of the Fifth Annual International Symposium on High-Performance Computer Architecture*, pp. 54–58, Jan. 1999.
[45] C. A. Waldspurger and W.E. Weihl. Register relocation: Flexible contexts for multithreading. In *Proceedings of the 20th International Symposium on Computer Architecture*, May 1993.
[46] D. W. Wall. Limits of instruction-level parallelism. Technical Report TN-15, Digital Equipment Corporation, December 1990.

8
Computing with Mobile Agents in Distributed Networks

8.1	Introduction...	8-1
	What Is a Mobile Agent? • Outline of the Chapter	
8.2	Modeling Mobile Agents in Distributed Networks....	8-2
	Mobile Agents • Distributed Networks	
8.3	The Rendezvous Problem	8-3
	Solvability of Rendezvous • Asymmetric Rendezvous • Symmetric Rendezvous	
8.4	Graph Exploration	8-6
	The Problem • Efficiency Measures • Underlying Graph • Labeling of the Graph • Anonymity • Anonymous Tree Exploration	
8.5	Searching with Uncertainty	8-9
	Network and Search Models • Ring Network	
8.6	Game-Theoretic Approaches	8-12
	Search Games • Search and Rendezvous Models • Example of a Search Game • Example of a Rendezvous Search Game	
8.7	Network Decontamination and Black Holes...........	8-15
	Network Decontamination • Black Holes	
	Acknowledgments.....................................	8-17
	References	8-17

Evangelos Kranakis
Carleton University

Danny Krizanc
Wesleyan University

Sergio Rajsbaum
Universidad Nacional Autónoma de México

8.1 Introduction

Mobile agents are software entities with the capacity for motion that can act on behalf of their user with a certain degree of autonomy in order to accomplish a variety of computing tasks. Today they find applications in numerous computer environments such as operating system daemons, data mining, web crawlers, monitoring and surveillance, just to mention a few. Interest in mobile agents has been fueled by two overriding concerns. First, to simplify the complexities of distributed computing, and second to overcome the limitations of user interfaces.

8.1.1 What Is a Mobile Agent?

It is not easy to come up with a precise definition of mobile agent. A variety of formulations exist ranging from the naive to the sophisticated. According to Shoham [Sho97], in software engineering a mobile agent is *a software entity which functions continuously and autonomously in a particular environment, often inhabited by other agents and processes.* According to Bradshaw [Bra97] *an agent that inhabits an environment with other agents and processes is expected to be able to communicate and cooperate with them, and perhaps move from place to place in doing so.* Nwana and Ndumu [NN97] attempt to divide mobile agents in categories of either simple (cooperative, learning, autonomous) or composite (collaborative, collaborative learning, interface, smart) behavior.

An alternative approach is to define mobile agents by associating attributes and identifying properties that agents are supposed to be able to perform. Some of these properties may include (see [Woo02, Woo99]) *reactivity:* the ability to selectively sense and act, *autonomy:* goal-directedness, proactive and self-starting behavior, *collaborative behavior:* working in concert with other agents to achieve a common goal, *adaptivity:* being able to learn and improve with experience, and *mobility:* being able to migrate in a self-directed way from one host platform to another.

In this chapter, we examine mobile agents from the perspective of traditional research on distributed algorithms [Tel99]. We attempt to provide a framework for and a short survey of the recent research on the *theory of mobile agent computing*.

8.1.2 Outline of the Chapter

An outline of the chapter is as follows. Section 8.2 describes the mobile agent model being used and the distributed network they operate on. Section 8.3 focuses on the rendezvous problem and addresses its solvability in the symmetric and asymmetric cases of rendezvous. Section 8.4 concerns graph exploration. Section 8.5 surveys work on the problem of searching with uncertainty where agents are given faulty advice from the nodes during the search process. Section 8.6 is on game-theoretic approaches to search and rendezvous. Section 8.7 focuses on intrusion detection and avoidance by studying the decontamination problem and black hole search.

8.2 Modeling Mobile Agents in Distributed Networks

8.2.1 Mobile Agents

We are interested in modeling a set of software entities that act more or less autonomously from their originator and have the ability to move from node to node in a distributed network maintaining some sort of state with the nodes of the network providing some amount of (possibly long term) storage and computational support. Either explicitly or implicitly such a mobile (software) agent has most often been modeled using a finite automaton consisting of a set of states and a transition function. The transition function takes as input the agent's current state as well as possibly the state of the node it resides in and outputs a new agent state, possible modifications to the current node's state and a possible move to another node. In some instances we consider probabilistic automata that have available a source of randomness that is used as part of their input. Such agents are referred to as *randomized* agents.

An important property to consider is whether or not the agents are distinguishable, that is, if they have distinct labels or identities. Agents without identities are referred to as *anonymous* agents. Anonymous agents are limited to running precisely the same program, that is, they are identical finite automata. As the identity is assumed to be part of the starting state of the automaton, agents with identities have the potential to run different programs.

The knowledge the agent has about the network it is on and about the other agents can make a difference in the solvability and efficiency of various tasks. For example, knowledge of the size of the network or its topology or the number of and identities of the other agents may be used as part of an agent's program.

If available to the agents, this information is assumed to be part of its starting state. (One could imagine situations where the information is made available by the nodes of the network and not necessarily encoded in the agent.)

Other properties that may be considered in mobile agent computing include whether or not the agents have the ability to "clone" themselves, whether or not they have the ability to "merge" on meeting (sometimes referred to as "sticky" agents), or whether or not they can send self-generated messages. At this point, most of the theoretical research on mobile agents ignores these properties and they will not be discussed below.

8.2.2 Distributed Networks

The model of a distributed network is essentially inherited directly from the theory of distributed computing [Tel99]. We model the network by a graph whose vertices comprise the computing nodes and edges correspond to communication links.

The nodes of the network may or may not have distinct identities. In an *anonymous* network the nodes have no identities. In particular this means that an agent can not distinguish two nodes except perhaps by their degree. The outgoing edges of a node are usually thought of as distinguishable but an important distinction is made between a globally consistent edge-labeling versus a locally independent edge-labeling. A simple example is the case of a ring where clockwise and counterclockwise edges are marked consistently around the ring in one case, and the edges are arbitrarily—say by an adversary—marked 1 and 2 in the other case. If the labeling satisfies certain coding properties it is called a *sense of direction* [FMS98]. Sense of direction has turned out to greatly effect the solvability and efficiency of solution of a number of problems in distributed computing and has been shown to be important for mobile agent computing as well.

Networks are also classified by how they deal with time. In a synchronous network there exists a global clock available to all nodes. This global clock is inherited by the agents. In particular it is usually assumed that in a single step an agent arrives at a node, performs some calculation, and exits the node and that all agents are performing these tasks "in sync." In an asynchronous network such a global clock is not available. The speed with which an agent computes or moves between nodes, while guaranteed to be finite, is not a priori determined.

Finally, we have to consider the resources provided by the nodes to the agents. All nodes are assumed to provide enough space to store the agent temporarily and computing power for it to perform its tasks. (The case of malicious nodes refusing agents or even worse destroying agents—so-called *black holes*—is also sometimes considered.) Beyond these basic services one considers nodes that might provide some form of long-term storage, that is, state that is left behind when the agent leaves. In the rendezvous problem the idea of leaving an indistinguishable mark or *token* at a node (introduced in [BG01]) has been studied. In graph exploration a similar notion of a *pebble* is used [BS94]. More accommodating nodes might provide a *whiteboard* for agents to write messages to be left for themselves or for other agents.

8.3 The Rendezvous Problem

The mobile agent rendezvous problem is concerned with how mobile agents should move along the vertices of a given network in order to optimize the number of steps required for all of them to meet at the same node of the network. Requiring such agents to meet in order to synchronize, share information, divide up duties, and so forth would seem to be a natural fundamental operation useful as a subroutine in more complicated applications such as web-crawling, meeting scheduling, and so forth. For example, rendezvous is recognized as an effective paradigm for realizing the caching of popular information essential to the operation of a P2P network and thus reducing network traffic [ws].

In this section, we provide a short survey of recent work done on rendezvous within the distributed computing paradigm. We note that rendezvous has been studied in other settings such as robotics [RD01] and operations research [AG03]. This research is extensive and many of the solutions found can be applied here. But it is often the case that the models used and the concerns studied are sufficiently different as to require new approaches.

8.3.1 Solvability of Rendezvous

Given a particular agent model (e.g., deterministic, anonymous agents with knowledge they are on a ring of size n) and network model (e.g., anonymous, synchronous with tokens) a set of k agents distributed arbitrarily over the nodes of the network are said to *rendezvous* if after running their programs after some finite time they all occupy the same node of the network at the same time. It is generally assumed that two agents occupying the same node can recognize this fact (though in many instances this fact is not required for rendezvous to occur). As stated, rendezvous is assumed to occur at nodes. In some instances one considers the possibility of rendezvous on an edge, that is, if both agents use the same edge (in opposite directions) at the same time. (For physical robots this makes sense. For software agents this perhaps is not so realistic but sometimes necessary to allow for the possibility of rendezvous at all—especially in instances where the network lacks a sense of direction.)

The first question one asks for an instance of rendezvous is whether or not it is solvable. There are many situations where it is not possible to rendezvous at all. This will depend upon both the properties of the agents (deterministic or randomized, anonymous or with identities, knowledge of the size of the network or not, etc.) and the network (synchronous or asynchronous, anonymous or with identities, tokens available or not, etc.). The solvability is also a function of the starting positions chosen for the agents. For example, if the agents start at the same node and can recognize this fact, rendezvous is possible in this instance. Given a situation where some starting positions are not solvable (i.e., rendezvous is not possible) but others are, we distinguish between algorithms that are guaranteed to finish for all starting positions, with successful rendezvous when possible but otherwise recognizing that rendezvous is impossible, versus algorithms that are only guaranteed to halt when rendezvous is possible. Algorithms of the former type are said to solve *rendezvous with detection*. (The distinction is perhaps analogous to Turing machines deciding versus accepting a language.)

For solvable instances of rendezvous one is interested in comparing the efficiency of different solutions. Much of the research focuses on the time required to rendezvous. In the synchronous setting the time is measured through the global clock. (In some situations, it makes a difference if the agents begin their rendezvous procedure at the same time or there is possible delay between start times.) In the asynchronous setting we adapt the standard time measures from the distributed computing model. Also of interest is the size of the program required by the agents to solve the problem. This is referred to as the memory requirement of the agents and is considered to be proportional to the base two logarithm of the number of states required by the finite state machine encoding the agent.

As is often the case, researchers are interested in examining the extremes in order to get an understanding of the limits a problem imposes. Over time it has become clear that a rendezvous symmetry (of the agents and the network) plays a central role in determining its solvability and the efficiency of its solutions. As such we divide our discussion below into the asymmetric and symmetric cases. For simplicity we restrict ourselves to the case of just two agents in most of the discussions below.

8.3.2 Asymmetric Rendezvous

Asymmetry in a rendezvous problem may arise from either the network or the agents.

8.3.2.1 Network Asymmetry

A network is asymmetric if it has one or more uniquely distinguishable vertices. A simple example is the case of a network where all of the nodes have unique identities chosen from a subset of some totally ordered set such as the integers. In this case, the node labeled with the smallest identity (for example) is unique and may be used as a meeting point for a rendezvous algorithm. Uniqueness need not be conferred using node labels. For example, in a network where there is a unique node of degree one, it may be used as a focal point.

If a "map" of the graph with an agent's starting position marked on it is available to the agents then the problem of rendezvous is easily solved by just traversing the path to an agreed upon unique node. Algorithms that use an agreed upon meeting place are referred to by Alpern and Gal [AG03] as FOCAL strategies. In the case where the graph is not available in advance but the agents know that a focal point exists (e.g., they know the nodes are uniquely labeled and therefore there exists a unique minimum label node) this strategy reduces to the problem of graph traversal or graph exploration whereby all of the nodes (sometimes edges) of the graph are to be visited by an agent. This has been extensively studied and is surveyed in Section 8.4.

8.3.2.2 Agent Asymmetry

By agent asymmetry one generally means the agents have unique identities that allow them to act differently depending upon their values. In the simplest scenario of two agents, the agent with the smaller value could decide to wait at its starting position for the other agent to find it by exploring the graph as above. Alpern and Gal [AG03] refer to this as the wait for mommy (WFM) strategy and they show it to be optimal under certain conditions.

WFM depends upon the fact that the agents know in advance the identities associated with the other agents. In some situations this may be an unrealistic assumption. Yu and Yung [YY96] were the first to consider this problem. Under the assumption that the algorithm designer may assign the identities to the agents (as well as the existence of distinct whiteboards for each agent), they show that rendezvous may be achieved deterministically on a synchronous network in $O(nl)$ steps where n is the size of the network and l is the size of the identities assigned. Perhaps a more interesting case where an adversary assigns the labels was first considered in [DFP03]. Extensions to this work including showing rendezvous on an arbitrary graph is possible in time polynomial in n and l and that there exist graphs requiring $\Omega(n^2)$ time for rendezvous are described in [KP04, KM06]. The case of an asynchronous network is considered in [MGK+05] where a (nonpolynomial) upper bound is set for rendezvous in arbitrary graphs (assuming the agents have an upper bound on the size of the graph). Improvements (in some cases optimal) for the case of the ring network are discussed in each of the above papers.

8.3.3 Symmetric Rendezvous

In the case of symmetric rendezvous, both the (generally synchronous) network and the agents are assumed to be anonymous. Further, one considers classes of networks that in the worst case contain highly symmetric networks that do not submit to a FOCAL strategy. As might be expected some mechanism is required to break symmetry in order for rendezvous to be possible. The use of randomization and of tokens to break symmetry have both been studied extensively.

8.3.3.1 Randomized Rendezvous

Many authors have observed that rendezvous may be solved by anonymous agents on an anonymous network by having the agents perform a random walk. The expected time to rendezvous is then a (polynomial) function of the (size of the) network and is directly related to the cover time of the network. (See [MR95] for definitions relating to random walks.)

For example, it is straightforward to show that two agents performing a symmetric random walk on ring of size n will rendezvous within expected $O(n^2)$ time. This expected time can be improved by considering the following strategy (for a ring with sense of direction). Repeat the following until rendezvous is achieved: flip a (fair) coin and walk $n/2$ steps to the right if the result is heads, $n/2$ steps to the left if the result is tails. If the two agents choose different directions (which they do with probability 1/2), then they will rendezvous (at least on an edge if not at a node). It is easy to see that expected time until rendezvous is $O(n)$. Alpern refers to this strategy as Coin Half Tour and studies it in detail in [Alp95a]. Note that the agents are required to count up to n and thus seem to require $O(\log n)$ bits of memory to perform this algorithm (whereas the straightforward random walk requires only a constant number of states to implement). This can be reduced to $O(\log \log n)$ bits and this can be shown to be tight [Mor] for achieving linear expected rendezvous time.

8.3.3.2 Rendezvous Using Tokens

The idea of using tokens or marks to break symmetry for rendezvous was first suggested in [BG01] and expanded upon for the case of the ring in [Saw04]. The first observation to make is that rendezvous is impossible for deterministic agents with tokens (or whiteboards) on an even size ring when the agents start at distance $n/2$ as the agents will remain in symmetric positions indefinitely. However, this is the only starting position for the agents for which rendezvous is impossible. This leads one to consider algorithms for rendezvous with detection where rendezvous is achieved when possible and otherwise the agents detect they are impossible to rendezvous situation. In this case, a simple algorithm suffices (described here for the oriented case). Each agent marks their starting position with a token. They then travel once around the ring counting the distances between their tokens. If the two distances are the same, they halt declaring rendezvous impossible. If they are different they agree to meet (for example) in the middle of the shorter side.

Again, one observes that the algorithm as stated requires $O(\log n)$ bits of memory for each agent in order to keep track of the distances. Interestingly enough, this can be reduced to $O(\log \log n)$ bits and this can be shown to be tight for unidirectional algorithms [KKSS03]. If we are allowed two movable tokens (i.e., the indistinguishable marks can be erased and written with at most two marks total per agent at any time) then rendezvous with detection becomes possible with an agent with constant size memory [KKM06].

Multiagent rendezvous, that is, more than two agents, on the ring is considered in [FKK+04b,GKKZ06], the second reference establishing optimal memory bounds for the problem. Two agent rendezvous on the torus is studied in [KKM06] where tradeoffs between memory and the number of (movable) tokens used are given. Finally, [FKK+04a] considers a model in which tokens may disappear or fail over time.

8.3.3.3 Rendezvous and Leader Election

Consider k identical, asynchronous mobile agents on an arbitrary anonymous network of n nodes. Suppose that all the agents execute the same protocol. Each node has a whiteboard where the agents can write to and read from. *Leader election* refers to the problem of electing a leader among those agents.

In Barriere et al. [BFFS03] both the leader election and rendezvous problems are studied and shown to be equivalent for networks with appropriate sense of direction. For example, rendezvous and election are unsolvable (i.e., there are no deterministic generic solutions) if $\gcd(k, n) > 1$, regardless of whether or not the edge-labeling is a sense of direction. On the other hand, if $\gcd(k, n) = 1$ then the initial placement of the mobile agents in the network creates topological asymmetries that can be exploited to solve the problems.

8.4 Graph Exploration

8.4.1 The Problem

The main components of the graph exploration problem are a connected graph $G = (V, E)$ with V as its set of nodes and E as its set of links, as well as a unique start node s. The graph represents a distributed network, the nodes of which are autonomous processing elements with limited storage while s is the initial position of the mobile agent. The goal of graph exploration is for the agent (or team of agents) to visit all of the nodes (or possibly edges) of the graph. Apart from completing the exploration of the graph it is often required that the agent also draw a map of the graph. The core of graph exploration often involves the construction of a simpler underlying subgraph (typically some kind of spanner of the graph, like a tree) that spans the original network G. Some of the important constructions of such trees involve breadth-first search (BFS), depth-first search (DFS), and minimum spanning trees (MST). For a detailed discussion of distributed constructions of BFS, DFS, and MST, the reader is advised to consult Peleg [Pel00].

The performance of graph exploration algorithms is limited by the static and nonadaptive nature of the knowledge incorporated in the input data. Mobile agents can take sensory input from the environment, learn of potential alternatives, analyze information collected, possibly undertake local action consistent with design requirements, and report it to an authorized source for subsequent investigation and adaptation.

8.4.2 Efficiency Measures

There is a variety of efficiency measures that can be considered. Time is important when dealing with exploration and many of the papers mentioned below measure it by the number of edge traversals required by the agent in order to complete the task. On the other hand, they do not impose any restrictions on the memory of the agent, which can be an important consideration in applications. This of course gives rise to interesting time/memory tradeoffs. In Diks et al. [DFKP04] the problem of minimizing the memory of the agent is studied.

8.4.3 Underlying Graph

Exploration and navigation problems for agents in an unknown environment have been extensively studied in the literature (see the survey Rao et al. [RHSI93]). The underlying graph can be directed (e.g., in [AH00, BFR$^+$98, BS94], and DP99, the agent explores strongly connected directed graphs and it can move only in the direction from head to tail of an edge, not vice versa) or undirected (e.g., in References ABRS95, BRS95, DKK01, and PP99, the explored graph is undirected and the agent can traverse edges in both directions).

8.4.4 Labeling of the Graph

In graph exploration scenarios it can be assumed either that the nodes of the graph have unique labels recognizable by the agent or that they are anonymous. Exploration in directed graphs with such labels was investigated in [AH00, DP99]. Exploration of undirected labeled graphs was studied in [ABRS95, BRS95, DKK01, PP99]. In this case a simple exploration based on DFS can be completed in time $2e$, where e is the number of edges of the underlying graph. For an unrestricted agent an exploration algorithm working in time $e + O(n)$, with n being the number of nodes, was proposed in Panaite et al. [PP99]. Restricted agents were investigated in [ABRS95, BRS95, DKK01]. It was assumed that the agent has either a restricted tank [ABRS95, BRS95], forcing it to periodically return to the base for refueling, or that it is tethered, that is, attached to the base by a rope or cable of restricted length Duncan et al. [DKK01]. It was proved that exploration and mapping can be done in time $O(e)$ under both scenarios.

8.4.5 Anonymity

Suppose that a mobile agent has to explore an undirected graph by visiting all its nodes and traversing all edges, without any a priori knowledge either of the topology of the graph or of its size. This can be done easily by DFS if nodes and edges have unique labels. In some navigation problems in unknown environments such unique labeling either may not be available or simply the mobile agents cannot collect such labels because of limited sensory capabilities. Hence it is important to be able to program the mobile agent to explore graphs anonymously without unique labeling of nodes or edges. Arbitrary graphs cannot be explored under such weak assumptions. For example, an algorithm on a cycle of unknown size without any labels on nodes and without the possibility of putting marks on them cannot terminate after complete exploration.

Hence, [BFR+98, BS94] allow *pebbles* that the agent can drop on nodes to recognize already visited ones, and then remove them and drop in other places. They focus their attention on the minimum number of pebbles allowing for efficient exploration and mapping of arbitrary directed n-node graphs. (In the case of undirected graphs, one pebble suffices for efficient exploration.) In was shown in Bender et al. [BFR+98] that one pebble is enough if the agent knows an upper bound on the size of the graph, and $\Theta(\log \log n)$ pebbles are necessary and sufficient otherwise. Also, a comparison of the exploration power of one agent versus two cooperating agents with a constant number of pebbles is investigated in Bender et al. [BS94]. (Recall that a similar comparison for the rendezvous problem in a torus is investigated in Kranakis et al. [KKM06].)

8.4.6 Anonymous Tree Exploration

If we do not allow any marks it turns out that the class of graphs that can potentially be explored has to be restricted to trees, that is, connected graphs without cycles. Another important requirement is that the mobile agent be able to distinguish ports at a node, otherwise it would be impossible to explore even a star tree with three leaves (after visiting the second leaf the agent cannot distinguish the port leading to the first visited leaf from that leading to the unvisited one). Hence a natural assumption is that all ports at a node be locally labeled $1, \ldots, d$, where d is the degree of the node, although no coherence between those local labelings is required.

Exploration algorithms and solutions discussed in the sequel are from Diks et al. [DFKP04] where additional details can be found. Two exploration methodologies are considered. The results are summarized in Table 8.1. First, *exploration with stop*: starting at any node of the tree, the mobile agent has to traverse all edges and stop at some node. Second, *exploration with return*: starting at any node of the tree, the mobile agent has to traverse all edges and stop at the starting node. In both instances it is of interest to find algorithms for a mobile agent performing the given task using as little memory as possible. From the point of view of memory use, the most demanding task of exploration with stop is indeed stopping. This is confirmed by analyzing *perpetual exploration* whereby the mobile agent has to traverse all edges of the tree but is not required to stop. The following simple algorithm will traverse all edges of an n-node tree after at most $2(n-1)$ steps and will perform perpetual exploration using only $O(\log d)$ memory bits. The agent leaves the starting node by port 1. After entering any node of degree d by port i, the agent leaves it by port $i+1 \mod d$. If in addition we know an upper bound N on the size n of the tree, we can explore it with stop using $O(\log N)$ bits of memory.

For every agent there exists a tree of maximum degree 3 that this agent cannot explore with stop. Even more so, it can be shown that a agent that can explore with stop any n-node tree of maximum degree 3 must have $\Omega(\log \log \log n)$ bits of memory. Additional memory is essentially needed to decide when to stop in an anonymous environment. Exploration with stopping is even more demanding on memory. For exploration with return, a lower bound $\Omega(\log n)$ can be shown on the number of memory bits required for trees of size n. As for upper bounds, an efficient algorithm for exploration with return is given that uses only $O(\log^2 n)$ memory bits for all trees of size n.

TABLE 8.1 Upper and Lower Bounds on the Memory of a Mobile Agent for Various Types of Exploration in Trees on n Nodes and Maximum Degree d

Tree	Mobile agent	Memory bounds	
Exploration	Knowledge	Lower	Upper
Perpetual	∅	None	$O(\log d)$
w/Stopping	$n \leq N$	$\Omega(\log \log \log n)$	$O(\log N)$
w/Return	∅	$\Omega(\log n)$	$O(\log^2 n)$

8.5 Searching with Uncertainty

Another problem of interest in mobile agent computing is that of searching for an item in a distributed network in the presence of uncertainty. We assume that a mobile agent in a network is interested in locating some item (such as a piece of information or the location of a service) available at one node in the network. It is assumed that each of the nodes of the network maintains a database indicating the first edge on a shortest path to the items of interest. Uncertainty arises from the fact that inaccuracies may occur in the database for reasons such as the movement of items, out-of-date information, and so forth. The problem is how to use the inaccurate databases to find the desired item. The item occupies an unknown location but information about its whereabouts can be obtained by querying the nodes of the network. The goal of the agent is to find the requested item while traversing the minimum number of edges. In the sequel we discuss algorithms for searching in a distributed network under various models for (random) faults and different topologies.

8.5.1 Network and Search Models

The network has n nodes and is represented as a connected, undirected graph $G = (V, E)$ with V its set of nodes and E its set of links. The universe of items of interest is denoted by S. Let $Adj(x)$ be the set of links adjacent to node x in the network. For $x \in V$ and $v \in S$ define $S_v(x)$ to be the set of links adjacent to x which are on a shortest path from x to the node containing v. Clearly, $\emptyset \subset S_v(x) \subseteq Adj(x)$, for all x, v. Let $1 \geq p > 1/2$ be constant.

Among the several possible search models (see [KK99]) we consider only two. Under the *single shortest path* (SSP) model, we assume that for each item $v \in S$ a tree representing a SSP from each node in V to the node containing v has been defined. If $e \in S_v(x)$ is the edge chosen in this shortest path tree then e appears in the entry for v in the database of x with probability greater than or equal to p. The probability the entry is some other $e' \in Adj(x) \setminus \{e\}$ is less than or equal to $1 - p$. Furthermore, for each v and for each x and x' these events are independent, that is, an error occurs in x's entry for v with probability at most $1 - p$ independent of errors occurring in the v entry of any other nodes, where by an error we mean that the node reports an edge other than the shortest path tree edge when queried about v. Note the "error" may in fact be an edge on a different shortest path to v and therefore may not be an error at all. The *SSP with global information* is a sub-model of the SSP model. Here, some global information about how the shortest path trees were originally constructed is available to the search algorithm. For example, on a mesh we may assume that the original databases, before the introduction of errors, were constructed using the standard row column shortest path trees and this is used in the construction of the searching algorithm. Under the *all shortest paths* (ASP) model, the probability that the entry for v in the routing table of x is some $e \in S_v(x)$ is greater than or equal to p and the probability it is some $e \in Adj(x) \setminus S_v(x)$ is less or equal to $1 - p$. Again errors at distinct nodes (in this case reporting an edge in $Adj(x) \setminus S_v(x)$) occur independently.

Note that in both models, no restriction is placed upon what is reported by the database in the case of an error, that is, in the worst case the errors may be set by a malicious adversary. In the ASP model, no restriction is placed on which of the correct edges is reported, that is, again, the answers to queries may be set by a malicious adversary. Finally, we note that for any geodesic (unique shortest path) network the two models are the same, for example, trees, or rings of odd size.

8.5.2 Ring Network

For the n node ring we give below a searching algorithm in the SSP or ASP model which with high probability finds a given item in $d + O(\log n)$ steps where d is the distance from the starting node to the node containing the desired item. Note that for $d = \omega(\log n)$ this is optimal to within an additive term.

A searching algorithm on the ring could completely ignore the advice of the databases and arbitrarily choose a direction (say Left) and move around the ring in that direction until the item is found. In the worst case, this algorithm requires $n - d$ steps where $d \leq n/2$ is the actual distance to the item. Slightly better might be to consult the database of the initial node and use that direction to search for the item (ignoring all later advice). In this case, for any item the expected number of steps is $pd + (1 - p)(n - d)$. Better still, take the majority of the advice of more nodes. This leads directly to the following algorithm parameterized by k. Search in a given direction for k steps. Maintain a count of the advice given by the routing tables of the k processors along the way and make a decision on the basis of the principle of maximum likelihood either to stay the course or reverse direction until the desired item is found.

1. Let *dir* be the direction given by the initial node
2. **for** $j = 1$ **to** k or until item is found **do**
3. move in direction *dir*
4. **if** majority of processors agree with *dir*, continue until item is found **else** reverse direction and continue until item is found

Thus, the search proceeds for k steps and then the direction is chosen according to the principle of maximum likelihood. This means that it selects the direction agreed by the majority of nodes along this path of length k. The probability that this direction is correct is equal to p_k, where

$$p_k = \sum_{i \geq \lceil k/2 \rceil} \binom{k}{i} p^i (1-p)^{k-i}.$$

Let d be the distance of the initial node to the node containing the desired item, and let X the number of steps traversed by the algorithm. It is clear from the algorithm that if the direction decision made by the algorithm is correct then $X = d$ or $X = d + 2k$ depending on whether or not the algorithm reversed direction after k steps. Similarly, if the decision made was incorrect then $X = n - d$ or $X = n - d + 2k$ depending on whether or not the algorithm reversed direction after k steps. It follows that with probability at least p_k the number of steps traversed by the algorithm does not exceed $d + 2k$.

Observe that $p_k = \Pr[S_k \geq k/2]$ and using Chernoff–Hoeffding bounds (setting $\mu = kp$ and $\delta = 1 - 1/2p$) we derive that

$$p_k = = \Pr[S_k \geq (1-\delta)\mu] = 1 - \Pr[S_k < (1-\delta)\mu] > 1 - e^{-\mu\delta^2/2},$$

where S_k is the sum of k independent and identically distributed random variables $X_1, \ldots, X_k$ such that $\Pr[X_i = 1] = p$, for all i. Choosing $k \geq c2p/(p - 1/2)^2 \ln n$, we conclude the probability that the algorithm requires less than $d + 2k$ steps is $\geq 1 - n^{-c}$ for $c > 0$.

A similar approach can be followed for the torus network. For additional details the reader is advised to consult [KK99].

8.5.2.1 Complete Network

Next we consider the case of a complete network on n nodes. (The results of this section are mainly from Kirousis et al. [KKKS03].) The algorithms considered may use randomization or be deterministic, may be either *memoryless*, have *limited memory*, or have *unlimited memory* and can execute one of the following operations: (i) query a node of the complete network about the location of the information node (ii) follow the advice given by a queried node, and (iii) select the next node to visit using some probability distribution function on the network nodes (that may be, e.g., a function of the number of previously seen nodes or the number of steps up to now). A summary of the expected number of steps for various types of algorithms is summarized in Table 8.2. The algorithm given below, simply alternates between following the advice of the currently visited node and selecting a random node as the next node to visit. It only needs to remember if at the last step it followed the advice or not. Although one bit suffices, the

TABLE 8.2 Expected Number of Steps

	Memory less	Limited memory	Unlimited memory
Randomized	$n-1$	$2/p$	$1/p$
Deterministic	∞	∞	$1/p$

algorithm is stated as if it knew the number of steps it has taken. Then it simply checks if this number is odd or even. However, one bit would be sufficient.

Algorithm: Fixed Memory Search
Input : A clique (V, E) with a node designated as the information holder
Aim: Find the node containing the information
1. **begin**
2.   current = RANDOM(V)
3. $l \leftarrow 1$
4.   **while** current(information) $\neq$ **true**
5. $l \leftarrow l + 1$
6.   **if** $l \mod 2 = 0$
7.   current = current(advice)
8.   **else**
9.   current $\leftarrow$ RANDOM($V -$ current)
10.   **end while**
11. **end**

We will now give a randomized algorithm for locating the information node in a clique that has unlimited memory. More specifically, the algorithm can store the previously visited nodes and use this information in order to decide its next move. In this way, the algorithm avoids visiting again previously visited nodes. Such an algorithm always terminates within $n - 1$ steps in the worst case.

Algorithm: Nonoblivious Search
Input : A clique (V, E) with a node designated as the information holder
Aim: Find the node containing the information
1. **begin**
2. $l \leftarrow 1$
3.   current = RANDOM(V)
4. $M \leftarrow \{\text{current}\}$ // M holds the up to now visited nodes
5.   **while** current(information) $\neq$ **true**
6.   read(current(advice))
7.   **if** current $\notin M$
8. $M \leftarrow M \cup \{\text{current}\}$
9.   current = advice
10.   **end if**
11.   **else**
12.   current $\leftarrow$ RANDOM($V - M$)
13. $l \leftarrow l + 1$
14.   **end while**
15. **end**

For additional details, the reader is advised to consult Kirousis et al. [KKKS03] and Kaporis et al. [KKK+01]. For results in a model where the faults occurring worst-case deterministically, see Hanusse et al. [HKKK02].

8.6 Game-Theoretic Approaches

8.6.1 Search Games

In general, search theory deals with the problem of optimizing the time it takes in order to find a target in a graph. The P2P paradigm demands a fully distributed and cooperative network design, whereby nodes collectively form a system without any supervision. In such networks peers connect in an ad hoc manner and the location of resources is neither controlled by the system nor are there any guarantees for the success of a search offered to the users. This setup has its advantages as well, including anonymity, resource-sharing, self-organization, load balancing, and robustness to failures. Therefore, in this context, search games might provide a more flexible basis for offering realistic and effective solutions for searching. There is extensive literature on games and game theory; however, for our purposes, and Rubinstein it will be sufficient to refer the reader to Osborne [OR94] for a mathematical treatment and to the specialized but related multiagent perspective offered by Sandholm [San99].

8.6.2 Search and Rendezvous Models

As defined by Alpern and Gal [AG03], search games consist of a *search space* usually represented by a graph G (with vertex set V and edge set E) a *searcher* starting from a given vertex (usually called the origin) of the graph and a *hider* (mobile or otherwise) choosing its hiding point independently of the searcher and occupying vertices of the graph. It is assumed that neither the searcher nor the hider has any a priori knowledge of the movement or location of the other until they are some critical distance r apart, also called the *discovery radius*.

Search problems are defined as two-person, zero-sum games with associated strategies available to the searcher and receiver, respectively (see, e.g., Watson [Wat02], or Osborne and Rubinstein [OR94]). Strategies may be pure or mixed (i.e., probabilistic) and the set of pure strategies for the searcher (respectively, hider) is denoted by $\mathcal{S}$ (respectively, $\mathcal{H}$). A strategy $S \in \mathcal{S}$ (respectively, $H \in \mathcal{H}$) for the searcher (respectively, hider) is a trajectory in the graph G such that $S(t)$ (respectively, $H(t)$) is the point visited by the searcher (respectively, at which the hider is hiding) at time t. (If the hider is immobile then it assumed that $H(t) := H$ is a fixed vertex in the given graph.)

The cost of the game is specified by a cost function $c : \mathcal{S} \times \mathcal{H} \to R : (S, H) \to c(S, H)$ representing the *loss* (or effort) by searcher S when using trajectory S while the hider uses trajectory H. Since the game is zero-sum, $c(S, H)$ also represents the *gain* of the hider. Given the sets of available pure strategies $\mathcal{S}, \mathcal{H}$, and the cost function $c(\cdot, \cdot)$ we define the *value*, *minimax value*, and *minimax search trajectory* as follows:

$$\text{Value: } v(S) := \sup_{H \in \mathcal{H}} c(S, H),$$

$$\text{Minimax value: } \hat{V} := \inf_{S \in \mathcal{S}} v_c(S), \qquad (8.1)$$

$$\text{Minimax search trajectory: } \hat{S} \text{ s.t. } \hat{V} = v(\hat{S}),$$

where mention of the cost function c is suppressed in Identities 8.1. A natural cost function c is the time it takes until the searcher captures the hider (also called *capture time*) in which case V_c represents the minimal capture time. Formally, the capture time is defined as

$$c(S, H) := \min_{t}\{d(S(t), H(t)) \leq r\},$$

where r is the discovery radius.

In most interesting search games the searcher can do better by using random choices out of the pure strategies and these choices are called *mixed strategies*. In this case the capture time is a random variable and each player cannot guarantee a fixed cost but rather only an expected cost. Mixed strategies of a searcher and a hider are denoted by s, h, respectively, while $c(s, h)$ denotes the expected cost. More formally, we are given probability distributions $s = \{p_S : S \in \mathcal{S}\}, h = \{q_H : H \in \mathcal{H}\}$ on $\mathcal{S}$ and $\mathcal{H}$, respectively, and the expected cost $c(s, h)$ is defined by

$$c(s, h) = \sum_{S \in \mathcal{S}} \sum_{H \in \mathcal{H}} c(S, H) p_S q_H. \tag{8.2}$$

As with pure stategies, we define

$$\text{Value of } s: v(s) := \sup_h c(s, h) = \sup_{H \in \mathcal{H}} c(s, H),$$

$$\text{Value of } h: v(h) := \inf_s c(s, h) = \inf_{S \in \mathcal{S}} c(S, h),$$

$$\text{Minimax value: } v := \inf_s v(s) = \sup_h v(h). \tag{8.3}$$

When $\mathcal{S}$ and $\mathcal{H}$ are finite, the fundamental theorem of Von Neumann (see von Neumann and Morgoston [vNM53]) on two-person, zero-sum games applies and states that

$$\max_h \min_s c(s, h) = \min_s \max_h c(s, h).$$

(If only one of either $\mathcal{S}$ or $\mathcal{H}$ is finite then the minimax value can be proven to exist. If both are infinite the minimax value may not exist, but under certain conditions, like uniform convergence of trajectories, it can be proved to exist. See Alpern and Gal [AG03] for additional results.)

If the hider is immobile a pure hiding strategy in the search graph G is simply a vertex, whereas a mixed hiding strategy is a probability distribution on the vertices of G. A pure search strategy is a trajectory in G starting at the origin O of the graph. We denote by $X_S(t)$ the set of vertices of G which have been searched using strategy S by time t. Obviously, the set $X_S(0)$ of points discovered at time 0 does not depend on the strategy S but only on the discovery radius r. A mixed search strategy is a probability distribution over these pure search strategies.

The *rendezvous search problem* is also a search game but it differs from the search problem in that it concerns two searchers placed in a known graph that want to minimize the time required to rendezvous (usually) at the same node. At any given time the searchers may occupy a vertex of a graph and can either stay still or move from vertex to vertex. The searchers are interested in minimizing the time required to rendezvous. A detailed investigation of this problem has been carried out by Alpern and collaborators (see [Alp95b, Alp02]) and a full discussion can also be found in Alpern and Gal [AG03]. As in search games we can define pure and mixed rendezvous strategies as well as the expected rendezvous time.

8.6.3 Example of a Search Game

As an example, consider the *uniform* mixed hiding strategy whereby the hider is immobile and chooses the hiding vertex randomly with the uniform distribution among the vertices of a graph with n vertices. Further assume that the discovery radius satisfies $r = 0$, which means that the searcher discovers the hider only when it lands in the same vertex. Since the searcher discovers at most one new vertex per time unit we must have that $\Pr[T \leq t] \leq \min\{1, t/n\}$, where T is the capture time. It follows that the expected capture time satisfies

$$E[T] = \sum_t \Pr[T > t] \geq \sum_t \max\left\{0, 1 - \frac{t}{n}\right\} = \frac{n+1}{2}.$$

To show this is tight, consider the complete graph K_n on n vertices. Let the hider strategy be uniform as before. A pure search strategy is a permutation $S \in S_n$. For a given S, H the capture time $c(S, H)$ is equal to the smallest t such that $S_t = H$. Given t, H, there are exactly $(n-1)!$ permutations $S \in S_n$ such that $S_t = H$. Consider a mixed search strategy whereby the searcher selects a permutation $S \in S_n$ at random with the uniform distribution. We have that

$$\sum_{H=1}^{n} \sum_{S \in S_n} c(S, H) = \sum_{H=1}^{n} \sum_{t=1}^{n} \sum_{\substack{S \in S_n \\ S_t = H}} t = \frac{n(n+1)}{2} \times (n-1)!$$

Consequently, it follows that the expected capture time is

$$E[T] = \frac{1}{n!} \times \frac{1}{n} \times \sum_{H=1}^{n} \sum_{S \in S_n} c(S, H) = \frac{n+1}{2}.$$

The problem is more difficult on arbitrary networks. For example, on trees it can be shown that the optimal hiding strategy is to hide among leaves according to a certain probability distribution that is constructed recursively, while an optimal search strategy is a *Chinese postman tour* (i.e., a closed trajectory traversing all vertices of the graph that has minimum length). The value of this search game on trees is the minimum length of a Chinese postman tour. We refer the reader to Alpern and Gal [AG03] for a proof of this result as well as for search games on other networks.

8.6.4 Example of a Rendezvous Search Game

Consider again the complete graph K_n. If searchers select their initial position at random the probability they occupy the same vertex is $1/n$. In a rendezvous strategy the players select trajectories (i.e., paths) $S = s_0, s_1, \ldots$ and $S' = s'_0, s'_1, \ldots$, where s_i, s'_i are the nodes occupied by the two searchers at time i, respectively. For a given pair S, S' of trajectories the rendezvous time is equal to the smallest t such that $s_t = s'_t$. Hence, the probability that they meet within the first t steps is at most t/n. The expected rendezvous time is equal to

$$E[T] = \sum_{t=0}^{\infty} \Pr[T > t] = \sum_{t=0}^{\infty} (1 - \Pr[T \leq t]) \geq \sum_{t=0}^{n-1} \frac{n-t}{n} = \frac{n-1}{2}. \tag{8.4}$$

In general, the rendezvous problem is easy if there is a distinguished vertex since the searchers can rendezvous there. This possibility can be discounted by considering an appropriate symmetry assumption on the graph, for example, the group A of vertex automorphisms is a transitive group. As suggested by Anderson and Weber [AR90], it can be shown that the lower bound in Inequality 8.4 holds for any graph with a transitive group of automorphisms with essentially the same proof as above.

In the *asymmetric* version of the rendezvous game players are allowed to play different strategies. For example, assuming that the graph is hamiltonian, one of the searchers can stay stationary while the moving player can choose to traverse a hamiltonian path (this is also known as *wait for mommy* strategy), in which case it is not difficult to show that the expected meeting time is at most $(n-1)/2$, that is, the lower bound in Inequality 8.4 is also an upper bound.

In the *symmetric* rendezvous game the players do not have the option of playing different strategies. A natural strategy without searcher cooperation, for example, is a simultaneous random walk. If the searchers cooperate, they can build a spanning tree and then they can perform a preorder traversal on this tree that can search all the nodes exhaustively in $2n$ steps. Now the searchers in each time interval of length $2n - 1$ search exhaustively or wait each with probability $1/2$. This scheme has expected rendezvous time $2(2n - 1)$. For the complete graph there is an even better strategy proposed by Anderson and weber [AR90] that has expected rendezvous time $\sim 0.82 \cdot n$. Certainly, there are interesting questions concerning other topologies as well as memory and expected rendezvous time tradeoffs. We refer the reader to Alpern and Gal [AG03] for additional studies.

8.7 Network Decontamination and Black Holes

Mobile agents have been found to be useful in a number of network security applications. Intrusion is the act of (illegally) *invading* or *forcing* into the premises of a system without right or welcome in order to compromise the confidentiality, integrity, or availability of a resource. Computer systems need to be monitored for intrusions and their detection is a difficult problem in network and computer system security. Once detected it is important to remove the intruders or decontaminate the network. In this section, we look at two different problems related to these issues. First we discuss a model for decontaminating a network efficiently, and second we discuss algorithms for detecting malicious nodes called black holes.

8.7.1 Network Decontamination

In the graph searching problem a set of searchers want to catch a fugitive hidden in a graph. The fugitive moves with a speed that may be significantly higher than the speed of searchers, while the fugitive is assumed to have complete knowledge of the searchers' strategy. In an equivalent formulation, we are given a graph with *contaminated* edges and, through a sequence of steps using searchers, we want to arrive at a situation whereby all edges of the graph are simultaneously *clean*. The problem is to determine the minimum number of searchers sufficient to clean the graph and the goal is to obtain a state of the graph in which all edges are simultaneously clean. An edge is cleaned if a searcher traverses it from one of its vertices to the other. A clean edge is preserved from recontamination if either another searcher remains in a vertex adjacent to the edge, or all other edges incident to this vertex are clean. In other words, a clean edge e is recontaminated if there exists a path between e and a contaminated edge, with no searcher on any node of the path. The basic operations, also called *search steps*, are the following: (i) place a searcher on a node, and (ii) move a searcher along an edge, (iii) remove a searcher from a node. Graph searching is the problem of developing a search strategy, that is, a sequence of search steps resulting in all edges being simultaneously clean.

The main issues worth investigating include devising efficient search strategies and minimizing the number of searchers used by a strategy. In general, the smallest number of searchers for which a search strategy exists for a graph G is called the search number $s(G)$ of G. Determining the search number of a graph is NP-hard (see Meggido et al. [MHG+88]). Two properties of search strategies are of particular interest: first, absence of recontamination, and second, connectivity of the cleared area. A search strategy is *monotone* if no recontamination ever occurs. Monotone searching becomes more important when the cost of clearing an edge far exceeds the cost of traversing an edge. Hence each edge should be cleared only once. Lapaugh [Lap93] has proved that for every G there is always a monotone search strategy that uses $s(G)$ searchers.

Another type of search is *connected* search whereby clean edges always remain connected. Connectivity is an important safety consideration and can be important when searcher communication can only occur within clean areas of the network. Such strategies can be defined by not allowing operation searcher removals and allowing searcher insertions either only in the beginning of the search or when applied to vertices incident to an already cleared edge. Optimal connected strategies are not necessarily monotone. A family of graphs (depending on a parameter k) that can be contiguously searched by $280k + 1$ searchers but require at least $290k$ searchers for connected monotone search are constructed by Dyer [Dye05]. The problem of determining optimal search strategies under the connectivity constraint is NP-hard even on planar graphs, but it has been shown that optimal connected strategies can be computed in linear time for trees.

Let $G = (V, E)$ be a graph (with possible multiple edges and loops), with n vertices and m edges. For subsets $A, B \subseteq E$ such that $A \cap B = \emptyset$ we define $\delta(A, B)$ to be the set of vertices adjacent to at least one edge in A and to at least one edge in B. We also use $\delta(A)$ to denote $\delta(A, E - A)$. A *k-expansion* in G is a sequence $X_0, X_1, \ldots, X_r$, where $X_i \subseteq E$ for every $i = 0, \ldots, r$, $X_0 = \emptyset$, $X_r = E$, and satisfying the following:

- $|X_{i+1} - X_i| \leq 1$, for every $i = 0, \ldots, r-1$
- $|\delta(X_i)| \leq k$, for every $i = 0, \ldots, r$

The expansion number $x(G)$ of G is the minimum k for which there is a k-expansion in G. A graph with expansion number k can thus be obtained by adding one edge after the other, while at the same time preserving the property that no more than k nodes are on the boundary between the current and remaining set of edges. The following relation between expansion and searching holds: $x(G) \leq s(G) \leq x(G) + 1$, for any graph G.

Branch decomposition of a graph G is a tree T all of whose internal nodes have degree three, with a one-to-one correspondence between the leaves of the tree and the edges of G. Given an edge e of T, removing e from T results in two trees $T_1^{(e)}$ and $T_2^{(e)}$, and an e-cut is defined as the pair $\{E_1^{(e)}, E_2^{(e)}\}$, where $E_i^{(e)} \subset E$ is the set of leaves of $T_i^{(e)}$ for $i = 1, 2$. Note that $E_1^{(e)} \cap E_2^{(e)} = \emptyset$ and $E_1^{(e)} \cup E_2^{(e)} = E$. The width of T is defined as $\omega(T) = \max_e |\delta(E_1^{(e)})|$, where the maximum is taken over all e-cuts in T. The *branchwidth* $bw(G)$ of G is then $\min_T \omega(T)$, where the minimum is taken over all branch decompositions T of G. It follows easily that for any graph G, $bw(G) \leq \max\{x(G), 2\}$. Branchwidth is related to the ability of recursively splitting the graph into several components separated by few nodes. In particular, there is an edge of the optimal branch decomposition of a graph G whose removal corresponds to splitting G into components of size at most $m/2$ edges, and with at most $bw(G)$ nodes in common. For any graph G, we have that $bw(G) - 1 \leq s(G) = O(bw(G) \log n)$.

A k-expansion $X_0, X_1, \ldots, X_r$ of a graph G is connected if, for any $i = 1, \ldots, r$, the subgraph induced by X_i is connected. The *connected expansion number* $cx(G)$ of G is the minimum k for which there is a connected k-expansion in G. Similarly, a search strategy is connected if the set of clear edges induces a connected subgraph at every step of the search. The *connected search number* $cs(G)$ of a graph G is the minimum k for which there is a connected search strategy in G using at most k searchers. It can be proved that $cx(G) \leq cs(G) \leq cx(G) + 1$.

There is a polynomial-time algorithm that, given a branch decomposition T of a 2-edge-connected graph G of width k, returns a connected branch decomposition of G of width k (see Fomin et al. [FFT04]). Therefore, the connected branchwidth of any 2-edge-connected graph is equal to its branchwidth. In other words, there is no additional price for imposing the connectedness in branch decompositions. As a consequence, it is possible to partition the edges of any 2-edge-connected graph of branchwidth k into at most three connected subgraphs of size $m/2$ edges, sharing at most k nodes. The connected expansion cannot be too large in comparison with the expansion, and the same holds for graph searching. More specifically, it can be proved (see Fomin et al. [FFT04]) that for any connected graph G,

$$cx(G) \leq bw(G) \times (1 + \log m),$$
$$cx(G) \leq x(G) \times (1 + \log m),$$
$$cs(G) \leq s(G) \times (2 + \log m),$$

where m is the number of edges of the network.

8.7.2 Black Holes

Mobile code is designed to run on arbitrary computers and as such it may be vulnerable to "hijacking" and "brainwashing." In particular, when an agent is loaded onto a host, this host can gain full control over the agent and either change the agent code or alter, delete or read/copy (possibly confidential) information that the agent has recorded from previous hosts (see Schelderup and Ølnes [SØ99]). An important question is whether a mobile agent can shelter itself against an execution environment that tries to tamper and/or divert its intended execution (see Sander and Tschudin [ST98]).

Black holes were introduced by Dobrev et al. [DFPS01] in order to provide an abstract representation of a highly harmful object of unknown whereabouts but whose existence we are aware of. A black hole is a stationary process that destroys visiting mobile agents on their arrival without leaving any observable trace of such a destruction. The problem posed by Dobrev et al. [DFPS01] "is to unambiguously determine and report the location of the black hole." The black hole search (BHS) problem is to determine the location of the black hole using mobile agents. Thus, BHS is solved if at least one of the agent survives, while all surviving agents know the location of the black hole.

Consider a ring of n nodes and a set A of $k \geq 2$ mobile agents traversing the nodes of this ring. Assume that the size n of the ring is known to all the mobile agents but the number k of agents may not be a priori known to them. Suppose that the agents can move from node to (neighboring) node and have computing capabilities and bounded storage, follow the same protocol, and all their actions take a finite but otherwise unpredictable amount of time. Each node has a bounded amount of storage ($O(\log n)$ bits suffice), called a whiteboard, which can be used by the agents to communicate by reading from and writing on the whiteboards.

Let us consider the case of the ring network. At least two agents are needed to locate the black hole. Moreover, as a lower bound, it is shown in Dobrev et al. [DFPS01] that at least $(n-1)\log(n-1) + O(n)$ moves are needed in order to find a black hole, regardless of the number of agents being used. In addition, it is shown that two agents can find the black hole performing $2n \log n + O(n)$ moves (in time $2n \log n + O(n)$).

Questions of related interest include (i) how many agents are necessary and sufficient to locate a black hole, and (ii) time required to do so and with what a priori knowledge. For additional results and details on the BHS problem in arbitrary networks we refer the reader to Dobrev et al. [DFPS02]. For results on multiagent, BHS see [DFPS04].

Acknowledgments

Research of the first author was supported in part by Natural Sciences and Engineering Research Council of Canada (NSERC) and Mathematics of Information Technology and Complex Systems (MITACS) grants. Research of the third author was supported in part by PAPIIT-UNAM.

References

[ABRS95] Awerbuch, B., Betke, M., Rivest, R., and Singh, M. Piecemeal graph learning by a mobile robot. In *Proceedings of the 8th Conference on Computing Learning Theory*, pp. 321–328, 1995.

[AG03] Alpern, S. and Gal, S. *The Theory of Search Games and Rendezvous*. Kluwer Academic Publishers, Norwell, MA, 2003.

[AH00] Albers, S. and Henzinger, M. R. Exploring unknown environments. *SIAM Journal on Computing*, 29: 1164–1188, 2000.

[Alp95a] Alpern, S. The rendezvous search problem. *SIAM Journal of Control and Optimization*, 33:673–683, 1995.

[Alp95b] Alpern, S. The rendezvous search problem. *SIAM Journal of Control and Optimization*, 33:673–683, 1995. Earlier version: LSE CDAM Research Report, 53, 1993.

[Alp02] Alpern, S. Rendezvous search: A personal perspective. *Operations Research*, 50:772–795, 2002.

[AR90] Anderson, E. J. and Weber, R. R. The rendezvous problem on discrete locations. *Journal of Applied Probability*, 28:839–851, 1990.

[BFFS03]   Barriere, L., Flocchini, P., Fraigniaud, P., and Santoro, N. Election and rendezvous of anonymous mobile agents in anonymous networks with sense of direction. In *Proceedings of the 9th International Colloquium on Structural Information and Communication Complexity (SIROCCO)*, pp. 17–32, 2003.

[BFR+98]   Bender, M. A., Fernandez, A., Ron, D., Sahai, A., and Vadhan, S. The power of a pebble: Exploring and mapping directed graphs. In *Proc. 30th Ann. Symp. on Theory of Computing*, pp. 269–278, 1998.

[BG01]   Baston, V. and Gal, S. Rendezvous search when marks are left at the starting points. *Naval Research Logistics*, 47:722–731, 2001.

[Bra97]   Bradshaw, J. *Software Agents*. MIT Press, Cambridge, MA, 1997.

[BRS95]   Betke, M., Rivest, R., and Singh, M. Piecemeal learning of an unknown environment. *Machine Learning*, 18:231–254, 1995.

[BS94]   Bender, M. A. and Slonim, D. The power of team exploration: Two robots can learn unlabeled directed graphs. In *Proc. 35th Ann. Symp. on Foundations of Computer Science*, pp. 75–85, 1994.

[DFKP04]   Diks, K., Fraigniaud, P., Kranakis, E., and Pelc, A. Tree exploration with little memory. *Journal of Algorithms*, 51:38–63, 2004.

[DFP03]   Dessmark, A., Fraigniaud, P., and Pelc, A. Deterministic rendezvous in graphs. In *11th Annual European Symposium on Algorithms (ESA)*, pp. 184–195, 2003.

[DFPS01]   Dobrev, S., Flocchini, P., Prencipe, G., and Santoro, N. Mobile search for a black hole in an anonymous ring. In *DISC: International Symposium on Distributed Computing*. Springer-Verlag Lecture Notes in Computer Science, 2001.

[DFPS02]   Dobrev, Flocchini, Prencipe, and Santoro. Searching for a black hole in arbitrary networks: Optimal mobile agent protocols. In *PODC: 21th ACM SIGACT-SIGOPS Symposium on Principles of Distributed Computing*, 2002.

[DFPS04]   Dobrev, S., Flocchini, P., Prencipe, G., and Santoro, N. Multiple agents rendezvous in a ring in spite of a black hole. In *Symposium on Principles of Distributed Systems (OPODIS '03)*, pp. 34–46, 2004. Springer-Verlag Lecture Notes in Computer Science.

[DKK01]   Duncan, C. A., Kobourov, S. G., and Kumar, V.S.A. Optimal constrained graph exploration. In *Proc. 12th Ann. ACM-SIAM Symp. on Discrete Algorithms*, pp. 807–814, 2001.

[DP99]   Deng, X. and Papadimitriou, C. H. Exploring an unknown graph. *Journal of Graph Theory*, 32:265–297, 1999.

[Dye05]   Dyer, D. 2005. PhD Thesis, Simon Fraser University, Burnaby B.C., Canada.

[FFT04]   Fomin, F. V., Fraigniaud, P., and Thilikos, D. M. The price of connectedness in expansions. Technical Report LSI-04-28-R, Departament de Llenguatges i Sistemes Informaticas, Universitat Politecnica de Catalunya, Barcelona, Spain, 2004.

[FKK+04a]   Flocchini, P., Kranakis, E., Krizanc, D., Luccio, F., Santoro, N., and Sawchuk, C. Mobile agent rendezvous when tokens fail. In *Proceedings of 11th Sirocco*, 2004. Springer-Verlag Lecture Notes in Computer Science.

[FKK+04b]   Flocchini, P., Kranakis, E., Krizanc, D., Santoro, N., and Sawchuk, C. Multiple mobile agent rendezvous in the ring. In *LATIN*, pp. 599–608, 2004. Springer-Verlag Lecture Notes in Computer Science.

[FMS98]   Flocchini, P., Mans, B., and Santoro, N. Sense of direction: definition, properties and classes. *Networks*, 32:29–53, 1998.

[GKKZ06]   Gasieniec, L., Kranakis, E., Krizanc, D., and Zhang, X. Optimal memory rendezvous of anonymous mobile agents in a uni-directional ring. In *Proceedings of SOFSEM 2006, 32nd International Conference on Current Trends in Theory and Practice of Computer Science*, 2006. January 21–27, 2006 Merin, Czech Republic, to appear.

[HKKK02]   Hanusse, N., Kavvadias, D., Kranakis, E., and Krizanc, D. Memoryless search algorithms in a network with faulty advice. In *IFIP International Conference on Theoretical Computer Science*, pp. 206–216, 2002.

[KK99]   Kranakis, E. and Krizanc, D. Searching with uncertainty. In *Proceedings of SIROCCO'99*, pages 194–203, 1999. Gavoille, C., Bermond, J.-C., and Raspaud, A. eds., Carleton Scientific.
[KKK+01]   Kaporis, A., Kirousis, L. M., Kranakis, E., Krizanc, D., Stamatiou, Y., and Stavropulos, E. Locating information with uncertainty in fully interconnected networks with applications to world wide web retrieval. *Computer Journal*, 44:221–229, 2001.
[KKKS03]   Kirousis, L. M., Kranakis, E., Krizanc, D., and Stamatiou, Y. Locating information with uncertainty in fully interconnected networks: The case of non-distributed memory. *Networks*, 42:169–180, 2003.
[KKM06]   Kranakis, E., Krizanc, D., and Markou, E. Mobile agent rendezvous in a synchronous torus. In *Proceedings of LATIN 2006, March 20–24, Valdivia, Chile*, 2006. Springer-Verlag Lecture Notes in Computer Science.
[KKSS03]   Kranakis, E., Krizanc, D., Santoro, N., and Sawchuk, C. Mobile agent rendezvous search problem in the ring. In *International Conference on Distributed Computing Systems (ICDCS)*, pp. 592–599, 2003.
[KM06]   Kowalski, D. and Malinowski, A. How to meet in an anonymous network. In *Proceedings of the 13th Sirocco*, 2006. Springer-Verlag Lecture Notes in Computer Science.
[KP04]   Kowalski, D. and Pelc, A. Polynomial deterministic rendezvous in arbitrary graphs. In *Proceedings of the 15th ISAAC*, 2004. Springer-Verlag Lecture Notes in Computer Science.
[Lap93]   Lapaugh, A. Recontamination does not help to search a graph. *Journal of the ACM*, 40:224–245, 1993.
[MGK+05]   De Marco, G., L Gargano, Kranakis, E., Krizanc, D., Pelc, A., and Vacaro, U. Asynchronous deterministic rendezvous in graphs. In *Proceedings of the 30th MFCS*, pp. 271–282, 2005. Springer-Verlag Lecture Notes in Computer Science.
[MHG+88]   Megiddo, N., Hakimi, S., Garey, M., Johnson, D., and Papadimitriou, C. The complexity of searching a graph. *Journal of the ACM*, 35:18–44, 1988.
[Mor]   Morin, P. Personal Communication, 2006.
[MR95]   Motwani, R. and Raghavan, P. *Randomized Algorithms*. Cambridge University Press, New York, 1995.
[NN97]   Nwana, H. S. and Ndumu, D. T. An introduction to agent technology. In *Software Agents and Soft Computing*, pp. 3–26, 1997.
[OR94]   Osborne, M. J. and Rubinstein, A. *A Course in Game Theory*. MIT Press, Cambridge, MA, 1994.
[Pel00]   Peleg, D. *Distributed Computing: A Locality Sensitive Approach*. SIAM Monographs in Discrete Mathematics and Applications, Society for Industrial and Applied Mathematics, 2000.
[PP99]   Panaite, P. and Pelc, A. Exploring unknown undirected graphs. *Journal of Algorithms*, 33:281–295, 1999.
[RD01]   Roy, N. and Dudek, G. Collaborative robot exploration and rendezvous: Algorithms, performance bounds and observations. *Autonomous Robots*, 11:117–136, 2001.
[RHSI93]   Rao, N.S.V., Hareti, S., Shi, W., and Iyengar, S. S. Robot navigation in unknown terrains: Introductory survey of non-heuristic algorithms. Technical report, Oak Ridge National Laboratory, 1993. Tech. Report ORNL/TM-12410.
[San99]   Sandholm, T. W. Distributed rational decision making. In Weiss, G. ed., *Multiagent Systems: A Modern Approach to Distributed Artificial Intelligence*, pp. 201–258. MIT Press, Cambridge, MA, 1999.
[Saw04]   Sawchuk, C. *Mobile Agent Rendezvous in the Ring*. PhD thesis, Carleton University, School of Computer Science, Ottawa, Canada, 2004.
[Sho97]   Yoav Shoham. An overview of agent-oriented programming. In Bradshaw, J. M. ed., *Software Agents*, chapter 13, pp. 271–290. AAAI Press/The MIT Press, 1997.

[SØ99]   Schelderup, K. and J. Ølnes. Mobile agent security—issues and directions. In *6th International Conference on Intelligence and Services in Networks*, volume 1597, pp. 155–167, 1999. Springer-Verlag Lecture Notes in Computer Science.

[ST98]   Sander, T. and Tschudin, C. F. Protecting mobile agents against malicious hosts. In *Mobile Agents and Security*, pp. 44–60, 1998.

[Tel99]   Tel, G. Distributed control algorithms for ai. In Weiss, G., ed., *Multiagent Systems: A Modern Approach to Distributed Artificial Intelligence*, p. 539–580. MIT Press, Cambridge, MA, 1999.

[vNM53]   von Neumann, J. and Morgenstern, G. *Theory of Games and Economic Behavior*. Princeton University Press, Princeton, NJ, 1953.

[Wat02]   Watson, J. *Strategy: An Introduction to Game Theory*. W. W. Norton and Company, New York, NY, 2002.

[Woo99]   Wooldridge, M. Intelligent agents. In Weiss, G. ed., *Multiagent Systems: A Modern Approach to Distributed Artificial Intelligence*, pp. 27–77. MIT Press, Cambridge, MA, 1999.

[Woo02]   Wooldridge, M. *An Introduction to MultiAgent Systems*. Wiley, Hoboken, NJ, 2002.

[ws]   JXTA web site. http://www.jxta.org/. Accessed Jan 22, 2006.

[YY96]   Yu, X. and Yung, M. Agent rendezvous: A dynamic symmetry-breaking problem. In *Proceedings of ICALP '96*, pp. 610–621, 1996.

9
Transitional Issues: Fine-Grain to Coarse-Grain Multicomputers

9.1	Introduction...	9-1
9.2	The Main Issues..	9-4
9.3	The Model..	9-5
9.4	The Tools...	9-7
	Load Balancing • Merging • Sorting • Compaction • Library Structure	
9.5	Sample Problems.....................................	9-11
	Scalable Algorithms for Restricted-Domain Searching • Scalable Algorithms for List Ranking • Scalable Visibility Algorithms • Scalable Algorithms for Constrained Triangulation	
	References...	9-17

Stephan Olariu
Old Dominion University

9.1 Introduction

A great diversity of emergent architectures and paradigms promises to move the impact of parallel and distributed computing from the classroom and the laboratory to the marketplace. Many researchers around the world are now studying the design, analysis, and efficient implementation of algorithms for parallel and distributed platforms with the stated goal of making them available to the applications community.

In this context, one of the important technological trends that we witness today is the move away from fine-grain parallel processing and toward coarse-grain multiprocessor systems. This trend was motivated by the fact that, owing to a number of factors, fine-grain parallel machines could not keep pace with the increasing demands for high-performance computing arising in applications including scientific computing, multimedia, engineering simulations, 3D computer graphics, CAD, real-time transaction processing, and information-on-demand services that became more and more popular.

This trend was enabled and supported by the availability of high-speed networks and by improved microprocessor performance. What emerges, is a computing environment populated by platforms consisting of a large number of powerful processors connected by high-speed networks. In such a system the processors are executing their own program (which may be the same for all processors) and exchange data with other processors by sending and receiving appropriate messages over the network. At one end of the spectrum, one has loosely coupled multiprocessors, networks of workstations (NOW), and clusters of workstations (COW) relying solely on commodity hardware and software. Not surprisingly, networked or clustered workstations offer parallel processing at a relatively low cost, making them very appealing alternatives to multi-million-dollar supercomputers. In terms of performance, networked workstations are known to approach or even exceed supercomputer performance for certain application domains, especially when the communication and synchronization requirements are not exceedingly stringent [3, 47, 93].

However, these loosely coupled multiprocessors cannot compete with their more tightly coupled counterparts in applications where stringent synchronization and communications are the norm and cannot be avoided. At this end of the spectrum one encounters commercially available parallel multiprocessor machines including Intel Paragon, IBM SP2, Intel iPSC/860, MasPar, and the CM-5, all coarse-grain machines where, just as in the case of networked workstations, each individual processor has a considerable processing power and local memory [48–51, 92, 95, 99].

In the context of coarse-grained computing, the communicational aspects of the underlying parallel and/or distributed systems are typically modeled by message-passing systems. Many parallel computer systems, both loosely and tightly coupled, provide libraries of frequently used, and thus, standardized, communication operations. These include, on the one hand, *point-to-point* communication primitives such as send and receive, and on the other, *collective* communication primitives including broadcast, scatter, gather, combine, sort, among others. Among the better known communication libraries one can cite the PVM [9, 40], the CMMD [96, 97], the Intel iPSC/860 library [50], the IBM SP2 communication library [49, 94], and the MPI [41, 71, 91].

At the generic level, a coarse-grain parallel machine consists of p processors, each endowed with considerable computing power (of the same order of magnitude as that of a present-day workstation) and with its own sizable memory space. The processors are assumed to communicate through a high-speed network of an unspecified topology. When presented with a processing task involving n input items, each of the p processors is assumed to store, in its own memory space, roughly n/p of the input items. It is customary to say that the resulting computing system is *scalable* [47, 52, 56] if it has the capacity to increase speedup in proportion to the number of processors. In other words, the scalability reflects the system's ability to utilize increasing processor resources efficiently. Portability is yet another key issue: roughly speaking portability of a software package describes the property of the package to be adapted with a minimal amount of overhead to run on a large variety of parallel platforms. In this sense, portability can be equated with architecture independence [3, 47].

Until recently, the overwhelming majority of the existing algorithms have been designed either for the PRAM or for fine-grained models [5, 52]. The PRAM is an eminently theoretical model that is useful in evaluating the theoretical complexity of parallel algorithms. Unfortunately, the PRAM ignores communication issues among processors. It comes as no surprise, therefore, that there exists a rather wide discrepancy between the theoretical performance predicted by the PRAM model and the actual implementation on real-life machines. In fact, only a small number of real architectures (some bus-based multiprocessors like Encore and Sequent) can be considered conceptually similar in design with the PRAM model.

Although, in principle, any real-life machine can simulate the PRAM model with optimal speedup if enough slackness in the number of processors is allowed [98], it is nevertheless true that algorithms designed for network-based models will better match the architectures of existing parallel machines like Intel Paragon, IBM SP2, Intel iPSC/860, CM-5, MasPar MP-1, and so forth, where the processors have their own local memory and are interconnected through a high-speed network supporting message-based communication.

Historically, the realization that the PRAM does not and cannot serve as a working model for solving large-scale real-life problems, together with the technological reality of the mid-1980s has led to the proliferation of fine-grained interconnection-network-based multiprocessor systems including linear arrays, tree machines, meshes, hypercubes, various enhanced meshes, along with more exotic topologies [4, 27, 36, 43, 57, 61, 66, 66, 68, 70, 73, 100].

It is fair to say that among these, the mesh and its variants have acquired a well-deserved prominence owing to the fact that many problems feature data that maps naturally onto the mesh. In turn, there has been a lot of effort in designing algorithms for the fine-grained parallel machines, both with and without communication buses [28, 42, 43, 56–58, 61, 63, 66, 68, 73, 85, 92]. Some of these algorithms rely on very sophisticated data movement operations and paradigms. Among these, data reduction and sampling are extremely useful in divide-and-conquer algorithms where one has to combine solutions to subproblems in order to obtain the overall solution. Clearly, when combining solutions of subproblems, it is most desirable to work with small data sets. This reduction comes about as a result of well-known data reduction or sampling strategies. Our preliminary experimental results [42, 75] seem to indicate that a good starting point for solving problems in coarse-grain computer environments is to investigate data reduction and sampling strategies that result in reduced interprocessor communications. When combined with efficient data partitioning—which must be application dependent—and efficient sequential processing within each processor of the platform, this strategy results in efficient and scalable algorithms. As part of a larger-scale experiment we have applied successfully the above paradigm to the design of efficient fundamental algorithms on a number of coarse-grain platforms [42, 75–77].

This chapter is specifically targeted at the problem of understanding the transitional issues involved in migrating from fine-grain to coarse-grain platforms. Specifically, we will address the problem of designing scalable and portable parallel algorithms for a number of fundamental nonnumerical problems on coarse-grain machines. In the past, we have designed and implemented a large number of fine-grain algorithms motivated by applications to CAGD, CAD, scene analysis, VLSI, and robotics [10–16, 19, 20, 63, 76, 78–83]. Most of these algorithms were implemented on fine- and medium-grain commercially available parallel machines including the DAP [85] and the MasPar MP-1.

During the years, we have successfully ported some of these algorithms to coarse-grain platforms including the IBM SP2, the Intel Paragon, as well as to networks of SGI's running MPI. This experiment, reported in [42, 75], confirmed our intuition that, along with carefully reducing data dependencies, the key to designing algorithms for coarse-grain platforms is to use data reduction and sampling strategies that result in reduced interprocessor communications rounds. When combined with efficient data partitioning—which must be application dependent—and efficient sequential processing within each processor of the platform, this strategy has been shown [42, 75] to result in efficient and scalable coarse-grain algorithms. We have obtained results that impacted computing practice: we have obtained simpler, faster, and more robust scalable algorithms for fundamental problems in a number of application domains. At the same time, we have produced a portable library of fundamental algorithms that will be made available to the research and application community.

Specifically, we have selected a number of sample problems from various areas of computer science and engineering, including computer graphics, CAD, scene analysis, robotics, and VLSI. These problems are prototypal for a large class of other similar problems in their application domains and share the following features:

- They are representative problems from their application domains
- They are known to admit optimal solutions on fine-grain platforms
- Their solutions require sophisticated data movement operations and techniques
- Their solutions involve various data reduction/sampling techniques

What distinguishes these problems, are their different communicational patterns and requirements. Whereas in some of the problems the communication patterns between processors are quite regular and predictable, in some others these communication patterns are irregular, depending to a large extent on a number of dynamically varying factors.

This chapter proposes to identify these common threads in migrating these problems from the fine-grain scenario to a coarse-grain one. We are looking for common paradigms—lessons to learn—in an attempt to provide a unified approach to solving these problems in a generic coarse-grained, architecture-independent, parallel computational model. Our main challenge, and stated goal, is to develop architecture-independent solutions that can be viewed as *template algorithms*. From these template algorithms it will be possible to generate, with a minimum of additional effort, solutions for particular computational platforms including NOWs and a number of commercially available parallel machines including the SP2 and the Intel Paragon.

To set the stage for what follows, it is appropriate to introduce concepts concerning visibility problems. We begin with a brief survey of where and how visibility-related problems can be applied, which further lends emphasis to their significance across a wide variety of applications:

- In computer graphics, visibility from a point plays a crucial role in ray tracing, scene analysis, and hidden-line elimination [4, 27, 28, 58, 61, 89].
- Visibility relations among objects are of significance in path planning and collision avoidance problems in robotics [65] where a navigational course for a mobile robot is sought in the presence of various obstacles.
- In VLSI design, visibility plays a fundamental role in the compaction process of integrated circuit design [64, 67, 86, 90, 102]. It is customary to formulate the compaction problem as a visibility problem involving a collection of iso-oriented, nonoverlapping, rectangles in the plane.

Visibility-related problems have been widely studied in both sequential and parallel settings. As the challenge to solve large and complex problems has constantly increased, achieving high performance by using large-scale parallel machines became imperative. To effectively apply a high degree of parallelism to a single application, the problem data is spread across the processors. Each processor computes on behalf of one or a few data elements in the problem. This approach is called d—*data*—*level parallel* and is effective for a broad range of computation-intensive applications including problems in vision geometry and image processing.

9.2 The Main Issues

As already stated, the main goal of this chapter is to investigate important transitional issues involved in migrating optimal time algorithms from fine-grain multiprocessor platforms to coarse-grain ones. Given the unmistakable technological trend away from fine-grain machines and toward coarse-grain platforms, it is imperative to get a good handle on the issues involved in this transition. This is an especially important problem that is not yet well understood [60, 87]. From a theoretical perspective, we are seeking results in the form of computing paradigms that will apply to many particular instances of algorithm design for coarse-grain machines. Extensive experiments were conducted in the Parallel Research Lab at Old Dominion University and reported in References 42 and 75 and seem to indicate that a good heuristic for designing scalable algorithms in coarse-grain multicomputer environments is to investigate data reduction and sampling strategies that result in reduced interprocessor communications. When combined with efficient data partitioning—which must be application dependent—and efficient sequential processing within each processor of the platform, this paradigm results in efficient and scalable algorithms [75].

From a practical perspective, this investigation makes perfect sense, since a lot of effort has been devoted to algorithmic techniques and paradigms that work well in the arena of fine-grain computations. This has resulted in a large number of optimal or near-optimal algorithms on such platforms. From a usability perspective, it is important to be able to retain as many of the established tools—and lines of code—as possible, so as to make the transition as transparent as possible.

Specifically, we have selected a number of sample problems from various areas of computer science and engineering, including computer graphics, CAD, scene analysis, robotics, and VLSI. These problems are fundamental and share the following features:

- They are representative problems from their application domains
- They are known to admit optimal solutions on fine-grain platforms
- Their solutions require sophisticated data movement operations and techniques
- Their solutions involve various data reduction/sampling techniques

What distinguishes these problems, are their different communicational patterns and requirements. Whereas in some of the problems the communication patterns between processors are quite regular and predictable, in some others these communication patterns are irregular, depending to a large extent on a number of dynamically varying factors.

In this section we are identifying the dominating factors that affect the task of migrating these problems from the fine-grain scenario to a coarse-grain one. We are looking for common paradigms—lessons to learn—in an attempt to provide a unified approach to solving these problems in a generic coarse-grained, architecture-independent, parallel computational model. Our main challenge, and stated goal, is to develop architecture-independent solutions that can be viewed as *template algorithms*. From these template algorithms it will be possible to generate, with a minimum of additional effort, solutions for particular computational platforms including NOWs and a number of commercially available parallel machines including the SP2 and the Intel Paragon.

One of the first lessons that our experimentation has revealed in Reference 75 is that in order to minimize the overall running time of an algorithm in a coarse-grain model, the number of communication *rounds* between processors must be kept to a minimum. It is not hard to see that the amount of communication that takes place in the process of running an algorithm on a coarse-grain platform reveals, to a large extent, the data dependencies within the algorithm. One of the objectives is to look at the problem of partitioning the input data over processors in such a way that the data dependencies are as "local" as possible. Our experiments, conducted on a network of workstations, under MPI show that the number of communication phases is an important factor in the speedup that we measured [75]. The effect is even more dramatic in cases where the traffic on the network follows unpredictable patterns [42].

9.3 The Model

Until recently, the PRAM model [52] was the most widely used model for the design and analysis of parallel algorithms. It shared memory abstraction and the assumption that the processors operate synchronously made it relatively easy to use. Unfortunately, it was soon recognized that by ignoring the communication costs between processors, the PRAM is not well suited for predicting performance on actual parallel machines [1, 38, 54, 56, 98].

To address this shortcoming, the research community has turned to more realistic interconnection models that take into account, to some extent, interprocessor communications. In these models, communications at unit cost are allowed between directly connected processors [47, 52]. However, this does not seem to be a realistic approach either, because it encourages the design of algorithms specifically tailored to a particular interconnection topology and ultimately results in nonportable software [31, 47, 54, 69].

For these reasons, several alternative models including the BSP [98], the LogP [31,55], the LogGP [2,31], the Message-Passing Block PRAM [1], the Block Distributed Memory Model [53], the C^3 [44], and the GPPP [69] have been proposed in an attempt to capture in an architecture-independent fashion the interprocessor communication costs. However, with a few exceptions, no quantitative results comparing theoretical predictions within these models with experimental results are available in the literature [54] and the proliferation of these models shows that the underlying problem is a difficult one indeed.

The model of computation that we adopt in this chapter is a *coarse-grain multicomputer*, referred to, generically, as CGM(n, p), and meant to capture the intrinsic computational and communicational features of parallel systems ranging from networked/clustered workstations to commercially available parallel machines including Intel Paragon, IBM SP2, Intel iPSC/860, and CM-5.

The CGM(n, p) has p processors of the same class as those in present-day workstations, each endowed with a sizable amount of local memory in such a way that, with n standing for the size of the instance of the problem to be solved using this machine, each of the processors can store $O(n/p)$ of the input items. The p processors of the CGM(n, p) are enumerated as $P_1, P_2, \ldots, P_p$, each of them being assumed to know its own identity i, but not its position within the parallel machine. The processors are connected through an arbitrary interconnection fabric and exchange messages using various communication primitives. They are assumed to be operating in Single Program Multiple Data (SPMD) mode, where all of them are executing the same program but on different data items in their local memories. The CGM(n, p) represents an abstraction of the various commercially available parallel machines mentioned above.

To make the CGM(p, M) model practically relevant and to ensure that algorithms designed in this will be portable across various computational platforms, ranging from shared memory machines to NOWs and COWs, we assume a number of generic communication primitives compatible with, but distinct from, the *collective communication* primitives defined by the Message Passing Interface Standard, (MPI)* [71, 91]. More specifically, we assume the following communication primitives:

1. *Broadcast:* a processor, referred to as the root, sends a message to all the processors in the CGM(n, p); the time associated with a broadcast operation is denoted by $T_B(p)$.
2. *Gather* data from all processors to one processor. Every processor P_i stores data item a_i. After the gather operation, processor P_1 stores items $a_1, a_2, \ldots, a_p$.
3. *Scatter* data from one processor to all the processors. This is just the reverse of the gather operation. Processor P_i stores data items $a_1, a_2, \ldots, a_p$. After the scatter operation, every processor P_j stores item a_j.
4. *All-Gather* is a variation of gather where all the processors receive the result of the gather operation. Initially, every processor P_i has an item a_i; after the all-gather operation, every P_i has a copy of all the items $a_1, a_2, \ldots, a_p$.
5. *All-to-all* involves Scatter/Gather data from all processors. This is also called complete exchange operation. Initially, every processor stores p items, where the first item is to be sent to processor P_1, the second to processor P_2, and so on. After the all-to-all operation, every processor receives p items, one from each of the processors (including itself).
6. *Global Combine:* involves combining p elements, one from each processor, into a single result. The combining operating can be some arithmetic or logic operation (e.g., addition, maximum, multiplication, etc.).
7. *Partial Combine:* involves combining p' elements, one from each of a subset of P' processors, into a single result. As in the case of global combine, the operation is some arithmetic or logic operation. It is possible to have parallel partial reduction operations among pairwise disjoint subsets of processors of the CGM(n, p).

In the various algorithms designed on this model of computation, the time taken by some communication operation is denoted by $T_{\text{operation}}(n, p)$, where n is the number of data items involved in the communication operation, and p is the number of processors involved [42, 75].

One of the important features of our generic coarse-grain model is that a CGM(n, p) can be viewed as consisting of q independent CGM(n', p'), whenever identical subproblems are to be solved in each one of them in parallel. The running time of an algorithm in the CGM model is taken to be the sum of the total time spent on local computation within any of the p processors and the total time spent on

---

*The MPI standardization is an effort involving more than 40 organizations around the world, with the aim of providing a widely used standard for writing message-passing programs and thus establishing a practical, portable, efficient, and flexible standard for message passing.

interprocessor communication. Clearly, optimal solutions to various problems in this scenario require the designer to reduce the computational time, while keeping the number of communication rounds as low as possible.

9.4 The Tools

The purpose of this section is to discuss a number of simple coarse-grain algorithms in the CGM model that could serve as basic building blocks for more sophisticated ones. In particular, we use these, and similar tools, as stepping stones in our further investigations.

9.4.1 Load Balancing

Several problems on the CGM(n, p) require dynamic balancing of the load on the various processors, depending on the particular instance of the input. A generic situation in which this scheme is needed is described next. Given the following input:

- A sequence $S = \langle s_1, s_2, \ldots, s_n \rangle$ of n items stored n/p per processor in a CGM(n, p), where every processor P_i stores the subsequence of items $S_i = \langle s_{(i \times n/p)+1}, \ldots, s_{i \times n/p} \rangle$. Every item $s_i \in S$ is associated with a *solution*, as defined below. The goal is to determine the solution of every $s_j \in S$.
- A sequence $D = \langle d_1, d_2, \ldots, d_n \rangle$ of n elements stored n/p per processor in a CGM(n, p), where each processor P_i stores a subsequence of items $D_i = \langle d_{(i \times n/p)+1}, \ldots, d_{i \times n/p} \rangle$. Each D_i is referred to as a *pocket*. The solution of each $s_j \in S$ lies in exactly one of the pockets D_i, $1 \leq i \leq n/p$.
- A sequence $B = \langle b_1, b_2, \ldots, b_n \rangle$ of n elements stored n/p per processor in a CGM(n, p), where each processor P_i stores the subsequence of items $B_i = \langle b_{(i \times n/p)+1}, \ldots, b_{i \times n/p} \rangle$. Every element $b_j \in B$ is the subscript of the pocket D_{b_j} which determines the solution to the item $s_j \in S$.

Thus, every processor P_i is given B_i, the sequence corresponding to the pocket to which each $s_j \in S_i$ belongs, and has to determine the solution to every s_j. For every item $s_j \in S_i$ with $b_j = i$, the solution can be determined sequentially within the processor. However, if b_j is not equal to i, there is a need to send every such s_j to the processor storing the pocket D_{b_j}.

Let N_i be the number of items $s_j \in S$, such that $b_j = i$. In general, the value of N_i ($0 \leq i \leq p-1$) may vary from 0 to O(n) depending on the particular instance of the input. Since, a processor has at most O(n/p) memory, at most O(n/p) items with $b_j = i$ can be sent to the processor storing D_i, at one time. This motivates the need to schedule the movement of the every $s_j \in S$, in order to determine its solution. In this section, the dynamic load balancing scheme provides a solution to this scheduling problem. The various steps involved in obtaining the solution of every s_j, using the dynamic load balancing scheme, is discussed below.

Step 1: The purpose of this step is to determine N_i for every pocket D_i. Every processor P_l ($0 \leq l \leq p-1$) determines the number C_{lk} of items $s_j \in S_l$ such that $b_j = k$. This takes O(n/p) computation time. Next, every P_l obtains information about $C_{0l}, C_{1l}, \ldots, C_{(p-1)l}$ from processors $P_0, P_1, \ldots, P_{p-1}$, respectively. This step takes $T_{\text{Alltoall}}(p, p)$ time where each processor P_m sends the values $C_{m0}, C_{m1}, \ldots, C_{m(p-1)}$ to processors $P_0, P_1, \ldots, P_{p-1}$, respectively. On receiving $C_{0l}, C_{1l}, \ldots, C_{(p-1)l}$ from every processor, P_l determines their sum in O(p) time, to obtain the value N_l. The p items $N_0, N_1, \ldots, N_{p-1}$ are replicated in each of p processors using an all-gather operation. This step takes a communication time of $T_{\text{Allgather}}(p, p)$.

Let $c \times n/p$ (where c is an integer constant greater than or equal to 2) be a value that is known to every P_i. Now, a pocket D_k is said to be *sparse* if N_k is less than or equal to $c \times n/p$; otherwise D_k is said to be *dense*. In O(n/p) time, every P_i ($0 \leq i \leq p-1$) determines for every $b_j \in B_i$, whether D_{b_j} is a dense pocket or not.

Step 2: The aim of this step is to obtain the solution of every item $s_j \in S$, where pocket D_{b_j} is sparse.

Let every P_i send $s_j \in S_i$, to processor P_{b_j}, storing the pocket D_{b_j}, where pocket D_{b_j} is sparse. This can be accomplished by performing an all-to-all communication operation. Note that, any processor P_i would receive at most $O(n/p)$ items. This step would take $T_{\text{Alltoall}}(n, p)$ time for the communication operation. The solution to every item s_j that is sent to the processor storing the pocket containing its solution, can now be determined sequentially in each of processors P_i storing a sparse pocket. Let the time taken for this computation be $O(f(n/p))$. The solutions can be sent back by performing a reverse data movement to the one performed earlier in $T_{\text{Alltoall}}(n, p)$ time.

Step 3: Finally, let us determine the solution to every $s_j \in S$, where pocket D_{b_j} is dense. In order to ensure that at most $O(n/p)$ such s'_js are moved to any processor, there is a need to make copies of every dense pocket D_k. This is accomplished as follows.

Let n_d be the number of dense pockets. The number of copies that each dense pocket D_k should have is given by $\mathcal{N}_k = (N_k/)c \times n/p$. The total number of copies of all the dense pockets D_ks is given by $\mathcal{N}_0 + \mathcal{N}_1 + \cdots + \mathcal{N}_{n_d-1}$, which is not larger than $p/2$. Let the n_d dense pockets be enumerated, in increasing order of their subscripts, as $D_{m_1}, D_{m_2}, \ldots, D_{m_{n_d}}$. Similarly, let the $p - n_d$ sparse pockets be enumerated as $D_{q_1}, D_{q_2}, \ldots, D_{q_{p-n_d}}$. Since, the sparse pockets are already processed, the processors storing them are *marked* as available to hold copies of the dense pockets. Let the marked processors be enumerated as $P_{q_1}, P_{q_2}, \ldots, P_{q_{p-n_d}}$. Let every processor P_i, such that D_i is a dense pocket, retain a copy of pocket D_i. Now, the rest of the copies of each of the dense pockets are scheduled among the marked processors $P_{q_1}, P_{q_2}, \ldots, P_{q_{p-n_d}}$. The scheduling of the copies is done as follows. The copies of D_{m_1} are assigned to the first $\mathcal{N}_{m_1} - 1$ marked processors. The copies of D_{m_2} are assigned the next $\mathcal{N}_{m_2} - 1$ processors, and so on.

Now, each of the processors that should be storing the copy of the dense pocket D_k, including P_k, join a process group. Note that, there are exactly n_d process groups. Now, in a broadcast operation in each of the process groups, every processor P_l can obtain the copy of the dense pocket it is to store. Note that this operation can be performed using an all-to-all communication operation that takes $T_{\text{Alltoall}}(n, p)$ time.

Since there may be several copies of a dense pocket D_k, each processor P_i needs to determine to which copy it has to send its items s_j with $b_j = k$. This can be accomplished as follows: for each dense pocket D_k, the processor P_k is aware of $C_{0k}, C_{1k}, \ldots, C_{(p-1)k}$, and performs a prefix sum on this sequence giving the sequence $Q_{0k}, Q_{1k}, \ldots, Q_{(p-1)k}$. Every Q_{ik} is sent to processor P_i. This could also be performed in one all-to-all communication operation, in $T_{\text{Alltoall}}(p^2, p)$ time. Note that, at this stage, every processor P_i has information to determine to which processors each of the unsolved items $s_j \in S_i$ is to be sent.

Now, move the unsolved items $s_j \in S_i$ from every processor P_i to the processor containing the copy of dense pocket D_k determined in the previous step. The solution to each one of them is then determined in $O(f(n/p))$ time and sent back to the corresponding processor. Thus, the required dynamic load balancing operation is complete and the solution of every $s_j \in S$ has been determined.

Theorem 9.4.1.1 *An instance of size n of a problem applying the dynamic load balancing scheme can be solved in $O(n/p) + O(f(n/p))$ computational time, where function f depends on the particular problem, and a communication time of $O(T_{\text{Alltoall}}(n, p))$.*

9.4.2 Merging

The main goal of this subsection is to explore a possible solution to the classic merge problem on a CGM(n, p). Our solution uses the dynamic load balancing scheme discussed in Section 9.4.1. The computation time of the algorithm is $O(n/p)$, and since the sequential lower bound of the merge problem is $\Omega(n)$, this algorithm is computationally time-optimal.

Let $S_1 = \langle a_1, a_2, \ldots, a_{n/2}\rangle$ and $S_2 = \langle b_1, b_2, \ldots, b_{n/2}\rangle$, be two sorted sequences of $n/2$ items each. Let S_1 be stored in processors $P_0, P_1, \ldots, P_{(p/2)-1}$ of the CGM(n, p), n/p per processor. Similarly, let S_2 be stored in $P_{(p/2)}, P_{p/2+1}, \ldots, P_{p-1}$, n/p per processor. Any P_i ($0 \leq i \leq p/2 - 1$) stores items $S_{i1} = \langle a_{i \times n/p+1}, \ldots, a_{(i+1) \times n/p}\rangle$ belonging to S_1. Similarly, any P_i ($p/2 \leq i \leq p - 1$) stores items

$S_{i2} = \langle b_{(i-p/2) \times n/p + 1}, \ldots, b_{(i-p/2+1) \times n/p} \rangle$ belonging to S_2. The two sequences S_1 and S_2 are to be merged into a sorted sequence $S = \langle c_1, c_2, \ldots, c_n \rangle$, so that any processor P_i stores items $\langle c_{i \times n/p + 1}, \ldots, c_{(i+1) \times n^*p} \rangle$ in the sorted sequence. Define the *rank* of an item e in any sorted sequence $Q = \langle q_1, q_2, \ldots, q_r \rangle$ as the number of items in the sequence Q that are less than the item e, and is denoted as $rank(e, Q)$. In order to merge the sequences S_1 and S_2, determine $rank(a_i, S)$ for every $a_i \in S$ and $rank(b_j, S)$ for every $b_j \in S_2$. First, determine the $rank(a_i, S_2)$ for every $a_i \in S_1$. The sum of $rank(a_i, S_2)$ and $rank(a_i, S_1)$ given by i, gives the value of $rank(a_i, S)$. Similarly, $rank(b_j, S_1)$ and $rank(b_j, S_2)$ is are be determined for every $b_j \in S_2$, to obtain the value of $rank(b_j, S)$. This can be accomplished as described in the following steps:

Step 1: Let every processor P_m ($0 \leq m \leq p/2 - 1$) set the value of the $rank(a_i, S_1)$ to i, for every $a_i \in S_{m1}$. Similarly, let every processor P_m ($p/2 \leq m \leq (p-1)$) set the value of the $rank(b_j, S_2)$ to j, for every $b_j \in S_{m2}$. This can be accomplished in $O(n/p)$ time.

Step 2: Every processor P_m determines the *largest* item it holds, and that is referred to as the sample item l_m. Since the sequence of items stored by any P_m are already sorted, the value of l_m can be obtained in $O(1)$ time. Now, perform an all-gather operation so that every processor has a copy of the sequence of sample items $L = \langle l_0, l_1, \ldots, l_{p-1} \rangle$. This can be accomplished in $T_{\text{Allgather}}(p, p)$.

In every P_m ($0 \leq m \leq p/2-1$), perform the following computation in parallel. Determine the pocket for every $a_i \in S_{m1}$, where pocket for any a_i is determined as follows. Given the sequence of sample items $L = \langle l_0, l_1, \ldots, l_{p-1} \rangle$, a_i finds its rank in $L_2 = \langle l_{p/2}, \ldots, l_{(p-1)} \rangle$ (L_2 is determined from L). The value $rank(a_i, L_2)$ corresponds to the pocket of a_i. Similarly, in every P_m ($n/2 \leq m \leq (p-1)$), perform the following computation in parallel. Determine the pocket for every $b_j \in S_{m2}$, where pocket for any b_j is determined as follows. Given the sequence of sample items $L = \langle l_0, l_1, \ldots, l_{p-1} \rangle$, b_j finds its rank in $L_1 = \langle l_0, \ldots, l_{p/2-1} \rangle$ (L_1 is determined from L). The value $rank(b_j, L_1)$ gives the pocket of b_j.

Note that the value of $rank(a_i, S_{k2})$, where k is the pocket of a_i, gives the rank of c_i in the sorted list S_2 as $rank(a_i, S_2) = rank(a_i, S_{k2}) + (k - p/2) \times n/p$. Similarly, the value of $rank(b_j, S_{k1})$, where k is the pocket of b_j, gives the rank of b_j in the sorted list S_1 as $rank(b_j, S_1) = rank(b_j, S_{k1}) + (k \times n/p)$.

Now, each of the items $a_i \in S_1$ with pocket k, has to calculate $rank(a_i, S_{k2})$, in order to determine $rank(a_i, S)$. Also, each item $b_j \in S_2$ with pocket k, has to calculate $rank(b_j, S_{k1})$. In the worst case, it is possible that all the a_is have the same pocket and all the b_j's have the same pocket. Thus, there is a need to apply the dynamic load balancing scheme.

Step 3: The load balancing scheme is applied to determine the $rank(a_i, S_{k2})$ for every $a_i \in S_1$ and $rank(b_j, S_{k1})$ for every $b_j \in S_2$. This can be performed as described in Subsection 4.1 in $O(n/p)$ computational time and $O(T_{\text{Alltoall}}(n, p))$ communication time. Now, determine the rank of every $a_i \in S_1$, in the sorted sequence S as $rank(a_i, S_1) + rank(a_i, S_2)$. Equivalent computation is performed for every item $b_j \in S_2$.

Step 4: Once every item $a_i \in S_1$ and $b_j \in S_2$ determines its rank in S, denoted as $rank(c_i, S)$ and $rank(b_j, S)$, respectively, the destination processor for each item a_i is determined as $\lfloor rank(a_i, S)/(n/p) \rfloor$ and for b_j as $\lfloor rank(b_j, S)/(n/p) rfloor$, respectively. This is accomplished in $O(n/p)$ time. In one all-to-all communication operation, the items can be moved to their final positions giving the sorted sequence S. This step requires $T_{\text{Alltoall}}(n, p)$ communication time. Thus, the following result is obtained.

Theorem 9.4.2.1 *Consider two sorted sequences, $S_1 = \langle a_1, a_2, \ldots, a_{n/2} \rangle$, $S_2 = \langle b_1, b_2, \ldots, b_{n/2} \rangle$, stored n/p per processor, with S_1 stored in processors $P_0, P_1, \ldots, P_{p/2-1}$ and S_2 in processors $P_{p/2}, P_{p/2+1}, \ldots, P_{p-1}$, of a $CGM(n, p)$. The two sequences can be merged in $O(n/p)$ computational time, and $O(T_{\text{Alltoall}}(n, p))$ communication time.*

9.4.3 Sorting

Theorem 4.2.1 is the main stepping stone for the a coarse-grain sorting algorithm that we develop in this subsection. This algorithm implements the well-known strategy of sorting by merging. The computational time of the algorithm is $O(n \log n/p)$ and since the sequential lower bound for sorting is $\Omega(n \log n)$, this algorithm is computationally time-optimal.

Let $S = \langle a_1, a_2, \ldots, a_n \rangle$ be a sequence of n items from a totally ordered universe, stored $O(n/p)$ per processor on a CGM(n, p), where any processor P_i stores the items $a_{(i \times n/p)+1}, \ldots, a_{i \times n/p}$. The sorting problem requires the sequence S to be sorted in a specified order and the resulting sequence of items $\langle b_1, b_2, \ldots, b_n \rangle$, are stored n/p per processor so that any processor P_i stores the items, $\langle b_{(i \times n/p)+1}, \ldots, b_{i \times n/p} \rangle$ The details of the algorithm are as follows:

First, the input sequence is divided into a left subsequence containing the first $n/2$ items and a right subsequence containing the remaining $n/2$ items. Further, imagine dividing the original CGM(n, p) into two independent machines, CGM$(n/2, p/2)$. This can be accomplished by dividing the p processors into two process groups having $p/2$ processors each.

The algorithm proceeds to recursively sort the data in each of the two CGMs. The resulting sorted subsequences are merged using the algorithm described in Subsection 9.4.2. The recursion terminates when each of the CGM is a CGM$(n/p, 1)$, and the data items can be sorted using the sequential algorithm running in $O(n \log n/p)$ time. It is easy to see that the overall running time of this simple algorithm is $O(n \log n/p)$ computation time and $O(\log p T_{\text{Alltoall}}(n, p))$ communication time.

Theorem 9.4.3.1 *Given a sequence $S = \langle a_1, a_2, \ldots, a_n \rangle$ of n items from a totally ordered universe, stored $O(n/p)$ per processor on a CGM(n, p), sorting of the sequence can be accomplished in $O(n \log n/p)$ computation time and $O(\log p T_{\text{Alltoall}}(n, p))$ communication time.*

9.4.4 Compaction

The compaction operation involves a sequence of items $S = \langle a_1, a_2, \ldots, a_n \rangle$ stored n/p items per processor, in the p processors of an CGM(n, p), with r $(1 \leq r \leq n)$, items *marked*. The marked items are enumerated as $B = \langle b_1, b_2, \ldots, b_r \rangle$ and every a_i $(0 \leq i \leq n)$ knows its *rank* in the sequence B. The result of the compaction operation is to obtain the ordered sequence B, in order, in the first $O(\lceil \frac{r}{n/p} \rceil)$ processors storing S, so that any processor P_i $(0 \leq i \leq \lceil \frac{r}{n/p} \rceil)$ stores items $b_{i \times n/p+1}, \ldots, b_{(i+1) \times n/p}$. This data movement operation can be accomplished by determining the destination processors for each of the marked items as $\lfloor \frac{rank}{n/p} \rfloor$ in $O(n/p)$ computational time, followed by an all-to-all operation to move the marked items to their destination processors. This can be accomplished in $T_{\text{Alltoall}}(n, p)$ time. Thus the following result is obtained.

Theorem 9.4.4.1 *Consider a sequence $S = \langle a_1, a_2, \ldots, a_n \rangle$ of items stored n/p per processor in the p processors of a CGM(n, p), with r of the items marked. The marked items can be compacted to the first $\lceil \frac{r}{n/p} \rceil$ processors of the CGM(n, p) in $O(n/p)$ computation time and $O(T_{\text{Alltoall}}(n, p))$ communication time.*

9.4.5 Library Structure

In the previous sections, we have noted the variety of algorithms that we have already developed with the CGM model as the basis for communication. In addition, we have developed a library of CGM-based algorithms that proved of value to the parallel programming community in use on coarse-grained platforms.

Figure 9.1 illustrates the structure of an application built on this library. At the bottom level are the communications primitives that define the CGM communications model. Built on these are a variety of fundamental algorithms and other components that we believe have particularly wide utility to parallel applications. Among these are the component algorithms previously described.

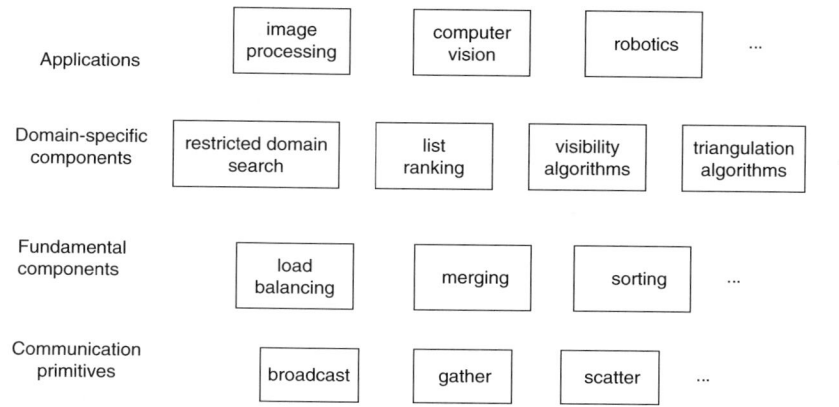

FIGURE 9.1 Structure of the CGM Library.

On these two layers rest the domain-specific components. As the name suggests, we cannot hope to provide a complete set of these for all application domains. But we provide a variety of useful components at this level by way of demonstrating the soundness and utility of the primitive and fundamental layers' components. Our contributions at this level are described subsequently in Section 9.5. In actual practice, we would expect the application developers to add additional reusable but domain-specific components at this level.

Of course, at the top of the application resides the code specific to that particular application. As will become apparent, some of the domain-specific components that we discuss in this proposal should be of immediate interest to such applications as image processing, computer vision, robotics, among others.

9.5 Sample Problems

The main goal of this section is to discuss the specific problems that the chapter is designed to address. These problems, as well as the solutions that we derive, will form a collection of "case studies" that are of import to both the algorithm-design community as well as to the application-oriented community. A number of issues involved in designing scalable parallel algorithms in various coarse-grain models have been reported in the literature [2, 4–6, 32–34, 37, 45, 54]. However, most of these are restricted to problems in computational geometry and, in this sense, fail to be representative of the entire spectrum.

The problems that we investigate are fundamental and, as already mentioned, have a number of common characteristics. Most important among these characteristics is that their solutions in the fine-grain model require a great amount of sophisticated nonlocal data movement in conjunction with data reduction and sampling. Therefore, it is expected that a naive implementation of these problems on coarse-grain machines will result in a correspondingly high degree of data dependency across processors and thus in a high amount of interprocessor communication [42, 75].

At the same time, these problems were selected because they feature a wide spectrum of intrinsic communicational requirements compounding the problem of data dependency noted above. Moreover, efficient and scalable solutions to these problems are especially important, as these problems are basic ingredients in the solution of more sophisticated computational tasks that one encounters in various application domains.

9.5.1 Scalable Algorithms for Restricted-Domain Searching

Query processing is a crucial transaction in various application domains including information retrieval, database design and management, and VLSI. Many of these applications involve data stored in a matrix

1	6	10	19
3	9	14	26
5	15	20	40
7	17	24	41

(a)

1	3	5	6
7	9	10	14
15	17	19	20
24	26	40	41

(b)

FIGURE 9.2 Illustrating sorted and fully sorted matrices.

satisfying a number of properties. One property that occurs time and again in applications specifies that the rows and the columns of the matrix are independently sorted [29, 30, 39]. It is customary to refer to such a matrix as *sorted*. A matrix is said to be *fully sorted* if its entries are sorted in row-major (resp. column-major) order. Figure 9.2a displays a sorted matrix; Figure 9.2b features a fully sorted version of the matrix in Figure 9.2a.

Sorted matrices provide a natural generalization of a number of real-life situations. Consider vectors $X = (x_1, x_2, \ldots, x_{\sqrt{n}})$ and $Y = (y_1, y_2, \ldots, y_{\sqrt{n}})$ with $x_i \leq x_j$ and $y_i \leq y_j$, whenever $i < j$. The Cartesian sum of X and Y, denoted $X + Y$ is the $\sqrt{n} \times \sqrt{n}$ matrix A with entries $a_{ij} = x_i + y_j$. It is clear that $X + Y$ is a sorted matrix. Moreover, $X + Y$ can be stored succinctly in $O(\sqrt{n})$ space [29, 37], since the entries a_{ij} can be computed, as needed, in constant time. Searching, ranking, and selecting in sorted matrices are key ingredients in fast algorithms in VLSI design, optimization, statistics, database design, and facility location problems and have received considerable attention in the literature [29, 30, 37, 39].

A generic instance of the BSR problem involves a sorted matrix A of size $\sqrt{n} \times \sqrt{n}$ and a collection $Q = \{q_1, q_2, \ldots, q_m\}$ $(1 \leq m \leq n)$, of queries; a query q_j can be of the following types:

1. *Search type:* determine the item of A that is closest to q_j.
2. *Rank type:* determine the *rank* of q_j in A, that is, the number of items in A that are strictly smaller than q_j.

We are interested in finding the solution to all the queries in parallel. It is important to note that search queries occur frequently in image processing, pattern recognition, computational learning, and artificial intelligence, where one is interested in returning the item in the database that best matches, in some sense, the query at hand [35]. On the other hand, rank queries are central to relational database design, histogramming, and pattern analysis [35]. Here, given a collection of items in a database along with a query, one is interested in computing the number of items in the database that have a lesser value than the query. In addition, rank queries finds applications to image processing, robotics, and pattern recognition [10, 35]. We note here that a variant of rank queries have also received attention in the literature. Specifically, a *range query* involves determining the number of items in a given database that fall in a certain range. It is not hard to see that range queries can be answered by specifying them as rank queries [35].

As it turns out, the BSR problem can be solved time optimally on meshes with multiple broadcasting [57, 85], that is, meshes enhanced by the addition of high-speed row and column buses. The optimal solution developed in Reference 18 relies on a novel and interesting cloning strategy for the queries. The scenario is the following. We partition the platform into a number of square blocks and clone the given queries in each of them. Having done that, we obtain the local solution of each query in each of the blocks. Finally, since the number of clones of each query is large—larger than the available bandwidth allowed to handle—we devise a strategy whereby we retrieve information gathered by *some* of the clones only. The interesting feature of this strategy is that there always exists a relatively small subset of the clones that, when retrieved, allows the resolution of all the queries. As a consequence, the algorithm devised in this chapter is completely different from that of Reference 12, showing the whole potential of meshes with multiple broadcasting.

In outline, our algorithm for the BSR problem proceeds in the following three stages:

Stage 0: The platform is subdivided into square blocks $R_{i,j}$ of the same size.

Stage 1: The set Q of queries is replicated in each block $R_{i,j}$, creating local instances of the BSR problem.

Stage 2: We determine in each block $R_{i,j}$, in parallel, the solution of the local instance of the BSR problem.

Stage 3: The solutions of the local instances of the BSR problem obtained is Stage 2 are combined into the solution of the original BSR problem.

The solution to the BSR problem in Reference 18 relies on a very intricate series of data movement operations in support of the cloning strategy. Somewhat surprisingly, the solution relies on creating a large number of clones of the given queries and then on a judicious sampling that eliminates the need to combine the solutions of all the clones. It should be clear at this point that the cloning strategy outline above is perfectly general. The fact that the domain was restricted allowed the sampling strategy to work in optimal time.

We plan to look at this problem in the coarse-grain scenario by relying in part on the tools that we have discussed in the previous section. We would also like to investigate domains other than sorted matrices for which our strategy yields optimal or close to optimal solutions.

We plan to begin by investigating possible partitioning strategies for the given matrix. As already mentioned, we have investigated the avenue of partitioning the matrix into square blocks, essentially, of size $\sqrt{n/p} \times \sqrt{n/p}$. This yields an optimal time algorithm for the fine-grain model. An alternative partitioning strategy would involve assigning each processor a *strip* of the given matrix, that is, a submatrix of size $\sqrt{n}\sqrt{n/p}$. This may work better in some cases that we would like to investigate, especially if the domain changes.

9.5.2 Scalable Algorithms for List Ranking

Finding a vast array of applications, the list ranking problem has emerged as one of the fundamental techniques in parallel algorithm design. Surprisingly, the best previously known algorithm to rank a list of n items on a reconfigurable mesh of size $n \times n$ was running in $O(\log n)$ time. It was open for more than eight years to obtain a faster algorithm for this important problem. Hayashi et al. [46] provided the first breakthrough: a deterministic list-ranking algorithm that runs in $O(\log^* n)$ time as well as a randomized one running in $O(1)$ expected time, both on a reconfigurable mesh of size $n \times n$. This result opens the door to an entire slew of efficient list-ranking-based algorithms on reconfigurable meshes.

The near-constant time solution developed in Reference 46 relies on sophisticated algorithms for computing a *good ruling set*, a concept that generalizes the notion of a ruling set [52]. In essence, the idea is the following. Given a linked list L we are interested in determining a sublist L' of L such that the items in L' are a certain "distance" apart in L. In Reference 62, the PIs have shown that the task of finding a good ruling set on an arbitrary reconfigurable linear array involving p processors can be computed in $O(\log p \log^* p)$ iterations.

Consider a linked list L of size n that we wish to rank. We assume a CGM(n,p) as discussed before. A priori, there is no good way of partitioning the n items among the p processors. The main idea of the algorithms in References 46 and 62 is that within each sublist determined by a good ruling set the ranking can be done fast. In the coarse-grain model, each processor proceeds to rank locally in $O(n/p)$ time the sublist that it has received.

As it turns out, each iteration in Reference 62 will translate into a communication round in the coarse-grain model. Therefore, it is not hard to obtain a list-ranking algorithm for the coarse-grain model that features $O(\log p \log^* /p)$ communication rounds and $O(n/p \log p \log^* /p)$ local computation time. We plan to further investigate this fundamental problem and reduce both the communication time as well as the local computation time.

9.5.3 Scalable Visibility Algorithms

To set the stage for what follows, it is appropriate to introduce some concepts concerning visibility problems. We begin with a brief survey of where and how visibility-related problems can be applied, which further lends emphasis to their significance across a wide variety of applications:

- In computer graphics, visibility from a point plays a crucial role in ray tracing, scene analysis, and hidden-line elimination.
- Visibility relations among objects are of significance in path planning and collision avoidance problems in robotics where a navigational course for a mobile robot is sought in the presence of various obstacles.
- In VLSI design, visibility plays a fundamental role in the compaction process of integrated circuit design. It is customary to formulate the compaction problem as a visibility problem involving a collection of iso-oriented, nonoverlapping, rectangles in the plane.

The input to the *object visibility* problem is a set S of n objects along with a viewpoint ω. The output is an ordered sequence of object portions that are visible to an observer positioned at ω. In Figure 9.3 we illustrate an instance of the problem, where the objects are restricted to line segments in the plane.

The *segment visibility* problem is fundamental since its solution turns out to be a key ingredient in determining the visibility relation among objects in the plane, such as a set of rectangles and disks, to name a few [64, 65]. Bhagavathi et al. have obtained a time-optimal solution to the segment visibility problem in a number of fine- and medium-grain computing models [14, 16]. In turn, these solutions were extended to a number of object visibility problems as part of our preliminary experiments [75]. The corresponding algorithms have been implemented on an SP2 and on a network of SGIs.

Our solution of the segment visibility problem (SV, for short) relies on a novel two-stage approach that we now outline. In essence, we use the first stage to determine, for each end point of a line segment, whether or not it is visible to an observer positioned at ω. The answer is "yes" or "no." Interestingly enough, this information is sufficient for the complete resolution of the problem in the second stage of the processing.

For an end point e of a line segment in S, let $e\omega$ denote the ray originating at e and directed toward ω. Similarly, let $e\overline{\omega}$ be the ray emanating from e, collinear with ω and away from ω. We now state a related problem, the *end point visibility* problem (EV, for short) that turns out to be intimately related to the segment visibility problem that we have already mentioned informally. Specifically, given a set S of line segments, the EV problem asks to determine, for every end point e of a segment in S, the closest segments (if any) intersected by the rays $e\omega$ and $e\overline{\omega}$. As an example, in Figure 9.3, the closest segments intersected by the rays $f_3\omega$ and $f_3\overline{\omega}$ are s_1 and s_6, respectively.

To state the SV problem formally, define the *contour* of S from ω to be the ordered sequence of segment portions that are visible to an observer positioned at ω. The SV problem asks to compute the contour of S from ω. For an illustration refer again to Figure 9.3 where the sequence of heavy lines, when traversed in increasing polar angle about ω, yields the contour of the set of segments.

Every line segment s_i in S has its end points denoted in increasing polar angle as f_i and l_i, standing for *first* and *last*, respectively. With a generic endpoint e_i of segment s_i, we associate the following variables:

- The identity of the segment to which it belongs (i.e., s_i)
- A bit indicating whether e_i is the first or last end point of s_i
- $t(e_i)$, the identity of the first segment, if any, that blocks the ray $e_i\omega$
- $a(e_i)$, the identity of the first segment, if any, that blocks the ray $e_i\overline{\omega}$

The notation $t(e_i)$ and $a(e_i)$ is meant to indicate directions "toward" and "away" from the viewpoint ω, respectively. Initially, $t(e_i)=a(e_i)=0$; when the algorithm terminates, $t(e_i)$ and $a(e_i)$ will contain the desired solutions. It is perhaps appropriate, before we get into details, to give a brief description of how the problem at hand is solved. The algorithm begins by computing an "approximate" solution to the EV problem: this involves determining for each of the rays $e_i\omega$ and $e_i\overline{\omega}$ whether it is blocked by some segment

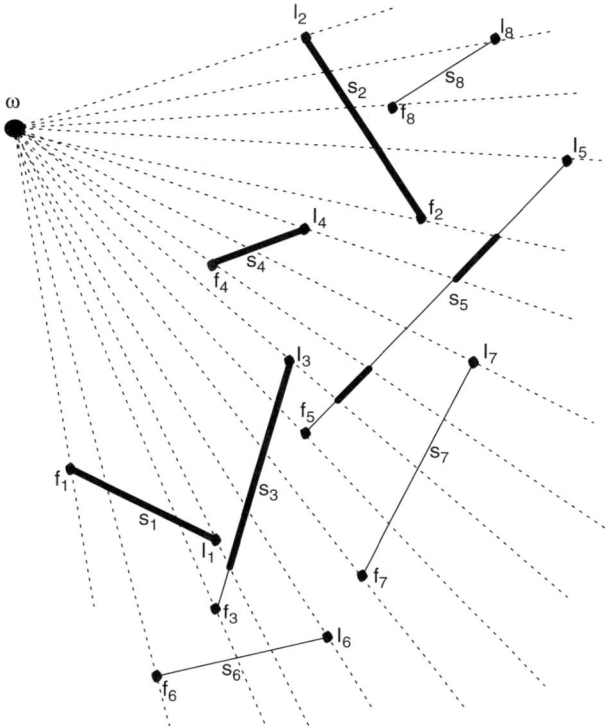

FIGURE 9.3 Illustrating the segment and endpoint visibility problem.

in S, without specifying the identity of the segment. This approximate solution is then refined into an exact solution.

Let us proceed with a high-level description of the algorithm. Imagine planting a complete binary tree T on S, with the leaves corresponding, in left-to-right order, to the segments in S. Given an arbitrary node v of T, we let $L(v)$ stand for the set of leaf-descendants of v. We further assume that the nodes in T are numbered level by level in left-to-right order. For a generic end point e_i of segment s_i, we let

- t-Blocked(e_i) stand for the identity of the first node in T on the path from the leaf storing the segment s_i to the root, at which it is known that the ray $e_i\omega$ is blocked by some segment in S.
- a-Blocked(e_i) stand for the identity of the first node in T on the path from the leaf storing s_i to the root, at which it is known that the ray $e_i\overline{\omega}$ is blocked by some segment in S.

Both t-blocked(e_i) and a-blocked(e_i) are initialized to 0.

Our algorithm proceeds in two stages. In the first stage, the tree T is traversed, in parallel, from the leaves to the root, computing for every endpoint e_i, t-blocked(e_i) and a-blocked(e_i). As we shall demonstrate, in case t-blocked(e_i) is not 0, we are guaranteed that some segment in S blocks the ray $e_i\omega$. However, the identity of the blocking segment is not known at this stage. Similarly, if a-blocked(e_i) is not 0, then we are guaranteed that some segment in S blocks the ray $e_i\overline{\omega}$. As before, the identity of the blocking segment is unknown. In the second stage of the algorithm, the tree T is traversed again, from the leaves to the root. In this process, the information in t-blocked(e_i) and a-blocked(e_i) is refined into t(e_i) and a(e_i).

Quite recently, Bhagavathi et al. [14] have shown that this approach leads to time-optimal solution not only to the segment visibility problems but also to visibility problems involving disks, rectangles, and convex polygons. These results were implemented on the DAP family of massively parallel machines. The algorithms rely heavily on very complicated data movement operations.

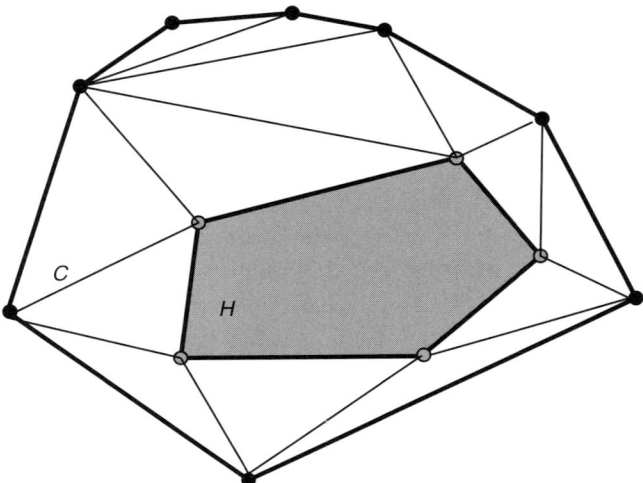

FIGURE 9.4 Triangulation of a convex region with a convex hole.

To design algorithms in the CGM model for visibility-related problems, we plan to rely on a judicious data partitioning along with the merge, sort, and compact operations that we have discussed in the previous section.

9.5.4 Scalable Algorithms for Constrained Triangulation

Triangulating a set of points in the plane is a central theme in computer-aided manufacturing, robotics, CAD, VLSI design, geographic data processing, and computer graphics. Even more challenging are constrained triangulations, where a triangulation is sought in the presence of a number of constraints such as prescribed edges and/or forbidden areas. Bokka et al. have obtained time-optimal for a number of constrained triangulation problems in the plane [22], solved on reconfigurable meshes. The algorithms are rather complex relying on a variety of data movement techniques.

A generic problem in this area asks to triangulate a set of points in the presence of a convex hole. Let $C = c_1, c_2, \ldots, c_n$ be a convex region of the plane and $H = h_1, h_2, \ldots, h_m$ be a convex hole within C. In many applications in computer graphics, computer-aided manufacturing and CAD, it is necessary to triangulate the region $C \setminus H$. The task at hand can be perceived as a constrained triangulation of C. For an illustration, refer to Figure 9.4.

The algorithm for triangulating a convex region with a convex hole will be a key ingredient in several of our subsequent algorithms for constrained triangulations.

For definiteness, let both C and H be stored one vertex per processor in the first row of a reconfigurable mesh $\mathcal{M}$ of size $n \times n$. Our triangulation algorithm proceeds as follows. Begin by choosing an arbitrary point interior to H and convert the vertices of C and H to polar coordinates having ω as pole and the positive x-direction as polar axis. Since ω is interior to C and H, convexity guarantees that the vertices of both C and H occur in sorted order about ω. Next, these two sorted sequences are merged into a sequence $b_1, b_2, \ldots, b_{n+m}$ sorted by polar angle. In the process of triangulating $C \setminus H$, several cases may occur.

Case 1: For some i $(1 \leq i \leq m)$ $h_i = b_j$ and $h_{i+1} = b_k$ with $j + 1 < k$.
We note that in this case, the line segment $b_{j+1}b_{k-1}$ lies in the wedge determined by h_i, h_{i+1}, and ω. Furthermore, the polygon $b_j, b_{j+1}, b_{j+2}, \ldots, b_{k-1}, b_k$ is convex. It is clear that this polygon can be triangulated in constant time by simply adding all the possible diagonals originating at b_{j+1}.
Case 2: For some i $(1 \leq i \leq n)$ $c_i = b_j$ and $c_{i+1} = b_k$ with $j + 1 < k$.

We now show that the polygon with vertices $c_i = b_j$, b_k, b_{k-1}, b_{k-2}, ..., b_{j+1} can be triangulated in constant time on a reconfigurable mesh. In preparation for this, we need the following simple observations:

- Let t $(j+1 \leq t \leq k-1)$ be such that b_t is visible from c_i. Then every vertex b_s with $j+1 \leq s \leq t$ is visible from c_i.
- Every vertex b_t $(j+1 \leq t \leq k-1)$ on H is visible from either c_i or c_{i+1}.

These two observations justify the following approach to triangulating the polygon $c_i = b_j$, b_k, b_{k-1}, b_{k-2}, ..., b_{j+1}. First, determine the vertex b_r. Add to the triangulation all the edges $c_i b_s$ with $j+1 \leq s \leq r$ and all the edges $c_{i+1} b_u$ with $t \leq u \leq k-1$.

To see how Case 1 and Case 2 are handled algorithmically, we assume that the sequence $b_1, b_2, \ldots, b_{n+m}$ is stored, in order, by the processors in the first row of the mesh, at most two vertices per processor. A simple data movement allows us to replicate the contents of the first row in all other rows of the mesh. We begin by handling the vertices in H first. Specifically, we determine all pairs h_i, h_{i+1} that are in Case 1 above. For each such pair, the vertices of C that lie between them are identified. Next, all pairs c_j, c_{j+1} as in Case 2 above are detected and all vertices of H lying between them are identified. Once this is done, triangulate them as specified in Case 1 and Case 2 above. It is important to note that in Case 2 the processors associated with the vertices of H can record in $O(1)$ time the edges added to the triangulation as well as the resulting triangles.

The above discussion along with the details that can be found in Bokka et al. [22] are sufficient to conclude that given an n-vertex convex region C with m-vertex hole H $(m \in O(n))$ within C, stored in one row or column of a reconfigurable mesh of size $n \times n$, the planar region $C \setminus H$ can be triangulated in $O(1)$ time.

Bokka et al. [22] went on to show that many other constrained triangulations can be constructed in optimal time on reconfigurable meshes and on meshes with multiple broadcasting. These include triangulating a region with rectangular holes as well as triangulating a region with prescribed line segments that must be part of the triangulation. All these algorithms have been implemented on fine-grain platforms and are good candidates for migration toward coarse-grain ones. The basic operations are again sorting, merging, and compaction. A very careful data partitioning and load balancing are required to minimize the interprocessor communication rounds required.

References

[1] A. Aggarwal, A. K. Chandra, and M. Snir, On communication latency in PRAM computations, *Proceedings of 2nd ACM Symposium on Parallel Algorithms and Architectures*, 1989, 11–21.

[2] A. Alexandrov, M. F. Ionescu, K. E. Schauser, and C. Scheiman, LogGP: Incorporating long messages into the LogP model, *Journal of Parallel and Distributed Computing*, 44 (1997), 71–79.

[3] C. Amza, A. L. Cox, S. Dwarkadas, P. Keleher, H. Lu, R. Rajamony, W. Yu, and W. Zwaenepoel, Treadmarks: shared memory computing on networks of workstations, *IEEE Computer* (1996), 18–28.

[4] M. J. Atallah, and D. Z. Chen, An optimal parallel algorithm for the visibility of a simple polygon from a point, *Proceedings of the 5th Annual ACM Symposium on Computational Geometry*, Saarbruchen, Germany, June 1989, 114–123.

[5] M. J. Atallah, F. Dehne, and S. E. Hambrusch. A coarse-grained, architecture-independent approach for connected component labeling, Manuscript, Purdue University, 1995.

[6] M. J. Atallah and J. J. Tsay, On the parallel-decomposability of geometric problems, *Proceedings Fifth Annual Symposium on Computational Geometry*, Saarbruchen, Germany, June 1989, 104–113.

[7] M. Barnett, R. Littlefield, D. G. Payne, and R. van de Geijn, Global combine on mesh architectures with wormhole routing, *Proceedings 6th International Parallel Processing Symposium*, April 1993, 156–162.

[8] K. E. Batcher, Design of massively parallel processor, *IEEE Transactions on Computers*, C-29 (1980), 836–840.

[9] A. Beguelin, J. Dongarra, A. Geist, R. Manchek, and V. Sunderam, A user's guide to the PVM Parallel Virtual Machine, Technical Report ORNL/TM-11826, Oak Ridge National Laboratory, May 1992.

[10] D. Bhagavathi, H. Gurla, R. Lin, S. Olariu, J. L. Schwing, and J. Zhang, Time- and VLSI-optimal sorting on meshes with multiple broadcasting, *Proceedings of International Conference on Parallel Processing*, St-Charles, IL, August 1993, III, 196–201.

[11] D. Bhagavathi, P. J. Looges, S. Olariu, J. L. Schwing, and J. Zhang, A fast selection algorithm on meshes with multiple broadcasting, *IEEE Transactions on Parallel and Distributed Systems*, 5 (1994), 772–778.

[12] D. Bhagavathi, S. Olariu, W. Shen, and L. Wilson, A time-optimal multiple search algorithm on enhanced meshes, with applications, *Journal of Parallel and Distributed Computing*, 22 (1994), 113–120.

[13] D. Bhagavathi, S. Olariu, J. L. Schwing, and J. Zhang, Convexity problems on meshes with multiple broadcasting, *Journal of Parallel and Distributed Computing*, 27 (1995), 142–156.

[14] D. Bhagavathi, V. Bokka, H. Gurla, S. Olariu, J. L. Schwing, and I. Stojmenović, Time-optimal visibility-related problems on meshes with multiple broadcasting, *IEEE Transactions on Parallel and Distributed Systems*, 6 (1995), 687–703.

[15] D. Bhagavathi, V. Bokka, H. Gurla, S. Olariu, J. L. Schwing, and J. Zhang, Square meshes are not optimal for convex hull computation, *IEEE Transactions on Parallel and Distributed Systems*, 7 (1996), 545–554.

[16] D. Bhagavathi, H. Gurla, S. Olariu, J. L. Schwing, and J. Zhang, Time-optimal parallel algorithms for dominance and visibility graphs, *Journal of VLSI Design*, 4 (1996), 33–40.

[17] V. Bokka, Doctoral Dissertation, Old Dominion University, May 1997.

[18] V. Bokka, H. Gurla, S. Olariu, J. L. Schwing, and L. Wilson, Time-optimal domain-specific querying on enhanced meshes, *IEEE Transactions on Parallel and Distributed Systems*, 8 (1997), 13–24.

[19] V. Bokka, H. Gurla, S. Olariu, I. Stojmenović, and J. L. Schwing, Time-optimal digital geometry algorithms on enhanced meshes, *Parallel Image Analysis: Theory and Applications*, L. S. Davis et al. Eds., Series in Machine Perception and Artificial Intelligence, 19, World Scientific, London, 1996.

[20] V. Bokka, H. Gurla, S. Olariu, and J. L. Schwing, Constant-time convexity problems on reconfigurable meshes, *Journal of Parallel and Distributed Computing*, 27 (1995), 86–99.

[21] V. Bokka, H. Gurla, S. Olariu, and J. L. Schwing, Time and VLSI-optimal convex hull computation on medium-grain enhanced meshes, *Proceedings Frontiers of Massively Parallel Computation*, Arlington, Virginia, February 1995, 506–513.

[22] V. Bokka, H. Gurla, S. Olariu, and J. L. Schwing, Podality-based time-optimal computations on enhanced meshes, *IEEE Transactions on Parallel and Distributed Systems*, 8 (1997), 1019–1035.

[23] V. Bokka, H. Gurla, S. Olariu, and J. L. Schwing, Time-optimal algorithms for constrained triangulations on meshes with multiple broadcasting, *IEEE Transactions on Parallel and Distributed Systems*, 9 (1998), 1057–1072.

[24] J. Bruck, L. de Coster, N. Dewulf, C.-T. Ho, and R. Lauwereins, On the design and implementation of broadcast and global combine operations using the postal model, *Proceedings 6th IEEE Symposium on Parallel and Distributed Systems*, October 1994, 594–602.

[25] J. Bruck, D. Dolev, C.-T. Ho, M.-C. Rosu, and R. Strong, Efficient message passing interface (MPI) for parallel computing on clusters of workstations, *Journal of Parallel and Distributed Computing*, 40 (1997), 19–34.

[26] J. Bruck and C.-T. Ho, Efficient global combine operations in multi-port message-passing systems, *Parallel Processing Letters*, 3 (1993), 335–346.

[27] I.-W. Chan, and D. K. Friesen, An optimal parallel algorithm for the vertical segment visibility reporting problem, *Proceedings ICCI'91*, Lecture Notes in Computer Science, 497, Springer-Verlag, 1991, 323–334.

[28] Y. C. Chen, W. T. Chen, G.-H. Chen, and J. P. Sheu, Designing efficient parallel algorithms on mesh connected computers with multiple broadcasting, *IEEE Transactions on Parallel and Distributed Systems*, 1, 1990.

[29] M. Cosnard, J. Dupras, and A. G. Ferreira, The complexity of searching in $X + Y$ and other multisets, *Information Processing Letters*, 34 (1990), 103–109.

[30] M. Cosnard and A. G. Ferreira, Parallel algorithms for searching in $X + Y$, *Proceedings of the International Conference on Parallel Processing*, St. Charles, IL, August 1989, III, 16–19.

[31] D. E. Culler, R. M. Karp, D. Patterson, A. Sahay, K, E. Schauser, E. E. Santos, R. Subromonian, and T. von Eicken, LogP: Towards a realistic model of parallel computation, *Proceedings 4th ACM Symposium on Principles of Programming Languages*, San Diego, May 1993, 1–12.

[32] D. E. Culler, A. Dusseau, R. Martin, and K. E. Schauser, Fast parallel sorting under LogP: from theory to practice, in *Portability and Performance in Parallel Computers*, T. Hey and J. Ferrante (Eds.) Wiley, 1994.

[33] F. Dehne, A. Fabri and A. Rau-Chaplin, Scalable parallel geometric algorithms for coarse grained multicomputers, *Proceedings of the ACM Symposium on Computational Geometry*, 1993, 298–307.

[34] O. Devillers, and A. Fabri, Scalable algorithms for bichromatic line segment intersection problems on coarse grained multicomputers, a manuscript, 1995.

[35] R. O. Duda and P. E. Hart, *Pattern Classification and Scene Analysis*, Wiley, 1973.

[36] H. ElGindy, An optimal speed-up parallel algorithm for triangulating simplicial point sets in space, *International Journal of Parallel Programming*, 15 (1986), 389–398.

[37] A. G. Ferreira, A. Rau-Chaplin, and S. Ubeda, Scalable convex hull and triangulation for coarse-grain multicomputers, *Proceedings Sixth IEEE Symposium on Parallel and Distributed Processing*, 1995, 561–569.

[38] H. Franke, P. Hochschild, P. Pattnaik, and M. Snir, An efficient implementation of MPI on IBM-SP1, *Proceedings International Conference on Parallel Processing*, 1994, III, 197–209.

[39] G. N. Frederickson and D. B. Johnson, Generalized selection and ranking: sorted matrices, *SIAM Journal on Computing*, 13 (1984), 14–30.

[40] A. Geist, A. Beguelin, J. Dongarra, W. Jiang, R. Manchek, and V. Sunderam, *PVM: Parallel Virtual Machinea User's Guide and Tutorial for Networked Parallel Computing*, MIT Press, 1994.

[41] W. Gropp, E. Lusk, and A. Skjellum, *Using MPI: Portable Parallel Programming with the Message-Passing Interface*, MIT Press, Cambridge, MA, 1994.

[42] H. Gurla, Time-optimal visibility algorithms on coarse-grain multiprocessors, PhD Thesis, Old Dominion University, August 1996.

[43] S. E. Hambrush, A. Hammed, and A. A. Khokhar, Communication operations on coarse-grained mesh architectures, *Parallel Computing*, 21 (1995), 731–751.

[44] S. E. Hambrush and A. A. Khokhar, C^3: an architecture-independent model for coarse-grained parallel machines, *Proceedings Fifth IEEE Symposium on Parallel and Distributed Systems*, 1994, 544–551.

[45] S. E. Hambrush, A. A. Khokhar, and Y. Liu, Scalable S-to-P broadcasting on message-passing MPPs, *Proceedings of International Conference on Parallel Processing*, 1996, 69–76.

[46] T. Hayashi, K. Nakano and S. Olariu, Efficient list ranking on the reconfigurable mesh, *Computing Systems Theory*, 31 (1998), 593–611.

[47] T. Hey and J. Ferrante, Eds., *Portability and Performance in Parallel Computers*, Wiley, 1994.

[48] D. Hillis, *The Connection Machine*, MIT Press, Cambridge, MA, 1985.

[49] International Business Machines Corporation, Scalable POWERparallel 2: Technical Review, GH23-2485-00, May 1994.

[50] Intel Corporation, Intel iPSC/860 Programmer's Reference Manual, 1991.

[51] Intel Corporation, Intel Paragon Systems Manual, Scalable Systems Division, 1993.

[52] J. JáJá, *An Introduction to Parallel Algorithms*, Addison-Wesley, Reading, MA, 1992.

[53] J. JáJá and K. W. Ryu, The Block Distributed Memory Model, Technical Report CS-TR-3207, UMIACS-TR-94-5, University of Maryland, 1994.

[54] B. H. H. Juurlink and H. A. G. Wijshoff, A quantitative comparison of parallel computational models, *Proceedings of the Eighth Annual ACM Symposium on Parallel Algorithms and Architectures*, June 1996, 13–24.

[55] R. M. Karp, A. Sahay, E. E. Santos, and K. Schauser, Optimal broadcast and summation in the LogP model, *Proceedings of the ACM 5th Annual Symposium on Parallel Algorithms and Architectures*, Velen, June 1993, 142–153.

[56] V. Kumar, A. Grama, A. Gupta, and G. Karypis, *Introduction to Parallel Computing: Design and Analysis of Algorithms*, Benjamin/Cummins, Redwood City, CA, 1994.

[57] V. Prasanna Kumar and C. S. Raghavendra, Array processor with multiple broadcasting, *Journal of Parallel and Distributed Computing*, 2 (1987), 173–190.

[58] V. Prasanna Kumar and D. I. Reisis, Image computations on meshes with multiple broadcast, *IEEE Transactions on Pattern Analysis and Machine Intelligence*, 11 (1989), 1194–1201.

[59] M. Lauria and A. Chien, MPI-FM: High performance MPI on workstation clusters, *Journal of Parallel and Distributed Computing*, 40 (1997), 4–18.

[60] W. G. Levelt, M. F. Kaashoek, and H. E. Ball, A comparison of two paradigms for distributed shared memory, *Software: Practice and Experience*, 22 (1992), 985–1010.

[61] H. Li and M. Maresca, Polymorphic-torus network for computer vision, *IEEE Transactions on Computers*, 38 (1989), 1345–1351.

[62] R. Lin, S. Olariu, and J. L. Schwing, Computing on reconfigurable buses—a new computational paradigm, *Parallel Processing Letters*, 4 (1994), 465–476.

[63] R. Lin, S. Olariu, J. L. Schwing, and J. Zhang, Simulating enhanced meshes, with applications, *Parallel Processing Letters*, 3 (1993), 59–70. Amsterdam, 1992, 16–27.

[64] E. Lodi and L. Pagli, A VLSI solution to the vertical segment visibility problem, *IEEE Transactions on Computers*, 35 (1986), 923–928.

[65] T. Lozano-Perez, Spatial planning: a configurational space approach, *IEEE Transactions on Computers*, 32 (1983), 108–119.

[66] M. Lu and P. Varman, Solving geometric proximity problems on mesh-connected computers, *Proceedings of the Workshop on Computer Architecture for Pattern Analysis and Image Database Management*, 1985, 248–255.

[67] A. A. Malik, An efficient algorithm for generation of constraint graph for compaction, *Proceedings of International Conference on CAD*, 1987, 130–133.

[68] M. Maresca and H. Li, Connection autonomy and SIMD computers: a VLSI implementation, *Journal of Parallel and Distributed Computing*, 7 (1989), 302–320.

[69] W. F. McColl, General-purpose parallel computing, in A. Gibbons and P. Spirakis, Eds., *Lectures on Parallel Computation*, Cambridge University Press, 1993.

[70] E. Merks, An optimal parallel algorithm for triangulating a set of points in the plane, *International Journal of Parallel Programming*, 15 (1986), 399–411.

[71] Message Passing Interface Forum, Document for a standard message-passing interface standard. Technical Report No. CS-93-214(revised), University of Tennessee, April 1994.

[72] K. Nakano and S. Olariu, An efficient algorithm for row minima computations on basic reconfigurable meshes, *IEEE Transactions on Parallel and Distributed Systems*, 9 (1998), 561–569.

[73] D. Nassimi and S. Sahni, Finding connected components and connected ones on a mesh-connected parallel computer, *SIAM Journal on Computing*, 9 (1980), 744–757.

[74] L. M. Ni, Should scalable parallel computers support efficient hardware multicast? *Proceedings of the ICPP Workshop on Challenges for Parallel Processing*, August 1995.

[75] S. Olariu and J. L. Schwing, An empirical study of the transition from fine- to coarse-grain multicomputers, Technical Report, Department of Computer Science, Old Dominion University, in preparation.

[76] S. Olariu and J. L. Schwing, A new deterministic sampling scheme with applications to broadcast-efficient sorting on the reconfigurable mesh, *Journal of Parallel and Distributed Computing*, 32 (1996), 215–222.

[77] S. Olariu and J. L. Schwing, A faster sorting algorithm in the Broadcast Communication Model, *Proc. International Parallel Processing Symposium*, Santa Barbara, April 1995, 319–323.
[78] S. Olariu, J. L. Schwing, and J. Zhang, Optimal convex hull algorithms on enhanced meshes, *BIT*, 33 (1993), 396–410.
[79] S. Olariu and I. Stojmenović, Time-optimal proximity algorithms on meshes with multiple broadcasting, *Journal of Parallel and Distributed Computing*, 36 (1996), 144–155.
[80] S. Olariu, J. L. Schwing, and J. Zhang, Fundamental data movement for reconfigurable meshes, *International Journal of High Speed Computing*, 6 (1994), 311–323.
[81] S. Olariu, J. L. Schwing, and J. Zhang, Fast computer vision algorithms on reconfigurable meshes, *Image and Vision Computing*, 10 (1992), 610–616.
[82] S. Olariu, J. L. Schwing, and J. Zhang, Computing the Hough Transform on reconfigurable meshes, *Image and Vision Computing*, 11 (1993), 623–628.
[83] S. Olariu, J. L. Schwing, and J. Zhang, Fast computer vision algorithms on reconfigurable meshes, *Image and Vision Computing*, 10 (1992), 610–616.
[84] S. Olariu, J. L. Schwing, and J. Zhang, Integer problems on reconfigurable meshes, with applications, *Journal of Computer and Software Engineering*, 1 (1993), 33–46.
[85] D. Parkinson, D. J. Hunt, and K. S. MacQueen, The AMT DAP 500, 33rd *IEEE Computer Society International Conference*, 1988, 196–199.
[86] B. T. Preas and M. J. Lorenzetti (Eds.) *Physical Design Automation of VLSI Systems*, Benjamin/Cummings, Menlo Park, 1988.
[87] U. Ramachandran and M. Khalidi, An implementation of distributed shared memory, *Software: Practice and Experience*, 21 (1991), 443–464.
[88] N. Rao, K. Maly, S. Olariu, S. Dharanikota, L. Zhang, and D. Game, Average waiting time profiles of the uniform DQDB, *IEEE Transactions on Parallel and Distributed Systems*, 6 (1995), 1068–1084.
[89] S. M. Rubin and T. Whitted, A 3-Dimensional representation for rendering of complex scenes, *ACM Computer Graphics*, (1988), 110–116.
[90] M. Schlag, F. Luccio, P. Maestrini, D. T. Lee, and C. K. Wong, A visibility problem in VLSI layout compaction, in F. P. Preparata (Ed.), *Advances in Computing Research*, Vol. 2, 1985, 259–282.
[91] M. Snir, S. W. Otto, S. Huss-Lederman, D. W. Walker, and J. Dongarra, *MPI: The Complete Reference*, MIT Press, Cambridge, MA, 1995.
[92] H. S. Stone, *High-Performance Computer Architecture*, 2nd ed., Addison-Wesley, 1990.
[93] V. Strumpen, Software-based communication latency hiding for commodity workstation networks, *Proceedings International Conference on Parallel Processing*, Indian Lakes Resort, IL, 1996, I, 146–153.
[94] C. B. Stunkel, G. D. Shea, B. Abali, M. Atkins, C. A. Bender, D. G. Grice, P. H. Hochschild, et al., The SP2 Communication Subsystem, URL:http://ibm.tc.cornell.edu/ibm/pps/doc/css/css.ps
[95] The Connection Machine CM-5, Technical Summary, Thinking Machines Corporation, 1991.
[96] CMMD Reference Manual 3.0, Thinking Machines Corporation, May 1993.
[97] L. W. Tucker and A. Mainwaring, CMMD: active messages on the Connection Machine CM-5, *Parallel Computing*, 20 (1994), 481–496.
[98] L. G. Valiant, A bridging model for parallel computation, *CACM*, 33 (1990), 103–111.
[99] D. L. Waltz, Application of the connection machine, *IEEE Computer* (1987), 85–97.
[100] C. A. Wang and Y. H. Tsin, An $O(\log n)$ time parallel algorithm for triangulating a set of points in the plane, *Information Processing Letters*, 25 (1988), 55–60.
[101] W. H. Wolf and A. E. Dunlop, Symbolic layout and compaction, in *Physical Design Automation of VLSI Systems*, B.T. Preas and M.J. Lorenzeth (Eds.), Benjamin/Cummings, Menlo Park, 1988, 211–281.
[102] D. A. Wood and M. D. Hill, Cost-effective parallel computing, *IEEE Computer* (1995), 69–72.
[103] X. Zhang, Z. Xu, and L. Sun, Performance predictions on implicit communication systems, *Proceedings of the Sixth IEEE Symposium on Parallel and Distributed Systems*, 1994, 560–568.

10
Distributed Computing in the Presence of Mobile Faults

Nicola Santoro
Carleton University

Peter Widmayer
ETH Zurich

10.1 Introduction... 10-1
   Distributed Computing and Communication Faults •
   Fault-Tolerant Computations and Mobile Faults
10.2 The Boundaries of Computability................... 10-3
   Agreement, Majority, and Unanimity • Impossibility •
   Possibility • Tightness of Bounds
10.3 Broadcast with Mobile Omissions.................. 10-10
   General Bounds • Specific Bounds
10.4 Function Evaluation in Complete Networks........ 10-12
   Problem and Basic Strategies • Omissions • Omissions and
   Corruptions • Applications of the Basic Techniques •
   Tightness of Bounds
10.5 Other Approaches 10-18
   Probabilistic • Fractional
10.6 Conclusions 10-19
References 10-19

10.1 Introduction

10.1.1 Distributed Computing and Communication Faults

A *distributed computing environment* is a collection of networked computational *entities* communicating with each other by means of *messages*, in order to achieve a common goal; for example, to perform a given task, to compute the solution to a problem, to satisfy a request either from the user (i.e., outside the environment) or from other entities. Although each entity is capable of performing computations it is the collection of all these entities that together will solve the problem or ensure that the task is performed. Such are for example distributed systems, grids of processors, data communication networks, distributed databases, transaction processing systems. More precisely, our universe is a system of $n \geq 2$ entities $x_1, \ldots, x_n$ connected through dedicated communication links. The connection structure is an arbitrary undirected, connected simple graph $G = (V, E)$. We will denote by $deg(G)$ the maximum node degree in G, by $con(G)$ the edge-connectivity of G, by $diam(G)$ the diameter of G, and by $m(G)$ the number of links of G.

In a distributed computing environment, the entities behave according to a specified set of rules, called *distributed algorithm* or *protocol*, that specify what each entity has to do. The collective but autonomous execution of those rules enables the entities to perform the desired task, to solve the problem. The process of designing a correct set of rules that will correctly solve the problem with a small cost is rather complex (e.g., [San06]). In all nontrivial computations, the entities need to communicate, and they do so by sending messages to (and receiving messages from) their neighbors in the underlying communication network G. The message sent by an entity need not be the same for all destination entities, that is separate links allow for different messages to different destinations at the same time.

A major obstacle to distributed computations is the possibility of *communication faults* in the system. A *communication* is a pair (α, β) of messages $\alpha, \beta \in M \cup \{\Omega\}$ for a pair (x, y) of neighboring entities called *source* and *destination*, where M is a fixed and possibly infinite message universe and Ω is the null message: α is the message sent by the source and β is the message received by the destination; by convention $\alpha = \Omega$ denotes that no message is sent, and $\beta = \Omega$ denotes that no message is received. A communication (α, β) is *faulty* if $\alpha \neq \beta$, nonfaulty otherwise. There are three possible types of *communication faults*:

1. An *omission*: the message sent by the source is never delivered to the destination ($\alpha \neq \beta = \Omega$).
2. An *addition*: a message is delivered to a destination although none was sent by the source ($\Omega = \alpha \neq \beta$).
3. A *corruption*: a message is sent by the source but one with different content is received by the destination ($\Omega \neq \alpha \neq \beta \neq \Omega$).

Although the nature of omissions and corruptions is quite obvious, that of additions is less so. Indeed, it describes a variety of situations. The most obvious one is when sudden noise in the transmission channel is mistaken for transmission of information by the neighbor at the other end of the link. The more important occurrence of an addition is rather subtle: it describes the reception of a "nonauthorized message," that is a message that appears to come from the source, but does not have its consent. In other words, additions represent messages surreptitiously inserted in the system by some outside, and possibly malicious, entity. Spam being transmitted from an unsuspecting network site clearly fits the description of an addition. Summarizing, additions do occur and can be very dangerous.

Note that other types of failures can be easily modeled by communication faults; for instance, omission of all messages sent by and to an entity can be used to describe the crash failure of that entity. Analogously, it is possible to model the intermittent failure of a link in terms of occasional omissions in the communication between the two entities connected by the link. In fact, most processor and link failure models can be seen as a special *localized* case of the communication failure model, where all the faults are restricted to the communications involving a fixed (though, a priori unknown) set of entities or of links.

10.1.2 Fault-Tolerant Computations and Mobile Faults

Not all communication faults lead (immediately) to computational errors (i.e., to incorrect results of the protocol), but some do. So the goal is to achieve *fault-tolerant* computations; that is, the aim is to design protocols that will proceed correctly in spite of the faults. Clearly, no protocol can be resilient to an arbitrary number of faults. In particular, if the entire system collapses, no protocol can be correct. Hence the goal is to design protocols that are able to withstand up to a certain amount of faults of a given type.

Another fact to consider is that not all faults are equally *dangerous*. The danger of a fault lies not necessarily in the severity of the fault itself but rather in the consequences that its occurrence might have on the correct functioning of the system. In particular, danger for the system is intrinsically related to the notion of detectability. In general, if a fault is easily detected, a remedial action can be taken to limit or circumvent the damage; if a fault is hard or impossible to detect, the effects of the initial fault may spread throughout the network creating possibly irreversible damage.

Distributed Computing in the Presence of Mobile Faults

An important distinction is whether the faults are localized or mobile.

- *Localized* faults occur always in the same region of the system; that is, only a fixed (although a priori unknown) set of entities/links will exhibit a faulty behavior.
- *Mobile* faults are not restricted to a fixed (but a priori unknown) set of links, but can occur between any two neighbors [SW89].

Note that, in general, the majority of failures are mostly transient and ubiquitous in nature; that is, faults can occur anywhere in the system and, following a failure, normal functioning can resume after a finite (although unpredictable) time. In particular, failures will occur on any communication link, almost every entity will experience at one time or another a send or receive failure, and so forth. Mobile faults are clearly more difficult to handle than the ones that occur always in the same places. In the latter case, once a fault is detected, we know that we can not trust that link; with mobile faults, detection will not help us with the future events.

With this in mind, when we talk about fault-tolerant protocols and fault-resilient computations we must always qualify the statements and clearly specify the type and number of faults that can be tolerated.

Unfortunately, the presence of communication faults renders the solution of problems difficult if not impossible. In particular, in *asynchronous* settings, the mere possibility of *omissions* renders all nontrivial tasks unsolvable, even if the faults are *localized* to (i.e., restricted to occur on the links of) a single entity [FLP85]. Owing to this impossibility result, the focus is on *synchronous* environments.

Since synchrony provides a perfect omission detection mechanism [CHT96], localized omissions are easily dealt with in these systems. Indeed, *localized* faults in synchronous environments have been extensively investigated in the *processor failure* model mostly for complete networks (e.g., see BG93, Dol82, DFF+82, DS82, GM98, and LSP82), as well as in the *link failure* model and in the *hybrid failure* models that consider both links and entities (e.g., CASD95, PT86 and SWR02).

The immediate question is then whether synchrony allows *mobile* communication faults, also to be tolerated—that is, faults that are not restricted to a fixed (but a priori unknown) set of links, but can occur between any two neighbors.

In this chapter, we will examine how we can deal with these communication failures, also called *dynamic faults* or *ubiquitous faults* in synchronous distributed computing systems. It is not surprising that the number of dynamic faults that can be tolerated at each time unit is by far less than that of the localized and permanent faults we can deal with. What is surprising is perhaps the fact that something *can* be done at all.

10.2 The Boundaries of Computability

The goal of this section is to examine the boundaries of computability in a synchronous distributed environment where communication faults are ubiquitous.

Defining what is computable in the presence of mobile faults seems to be a very arduous task. Similar to the case of localized faults, we will focus on a very basic problem, *agreement*, and study under what conditions this problem is solvable, if at all. The basic nature of the agreement problem is such that, if it cannot be solved under a set of conditions on the nature and amount of faults, then no nontrivial problem can be solved under those conditions. Determining under what conditions agreement can be achieved provides a threshold for the computability of all other nontrivial tasks and functions.

10.2.1 Agreement, Majority, and Unanimity

Each entity x_i has an input register with an initial value r_i, and an output register for which it must choose a value v_i as the result of its computation. For simplicity, we limit ourselves to Boolean inputs, that is $r_i \in \{0, 1\}$ for all i; all the results hold also for nonbinary values. In the *K-agreement problem*, at least K entities must terminally choose the same *decision* value $v \in \{0, 1\}$ within a finite amount of time, subject to the *validity* constraint that, if all input values r_i were the same, then v must be that value. Here, "terminally" means that, once made, the decision cannot be modified.

Depending on the value of parameter K, we have different types of agreement problems. Of particular interest are

- $K = \lceil n/2 \rceil + 1$ called *strong majority*
- $K = n$ called *unanimity* (or *consensus*), in which all entities must decide on the same value

Note that any agreement requiring *less* than a strong majority (i.e., $K \leq \lceil n/2 \rceil$) can be trivially reached without any communication; for example, each x_i chooses $v = r_i$. We are interested only in *nontrivial* agreements (i.e., $K > \lceil n/2 \rceil$).

A possible solution to the unanimity problem is, for example, having the entities compute a specified Boolean function (e.g., **AND** or **OR**) of the input values and choose its result as the decision value; another possible solution is for the entities to first elect a leader and then choose the input value of the leader as the decision value. In other words, unanimity (fault-tolerant or not) can be solved by solving any of a variety of other problems (e.g., function evaluation, leader election, etc.). For this reason, if the consensus problem cannot be solved, then none of those other problems can.

10.2.2 Impossibility

Let us examine how much more difficult it is to reach a nontrivial (i.e., $K > \lceil n/2 \rceil$) agreement in the presence of dynamic communication faults.

Consider for example a d-dimensional *hypercube*. From the results established in the case of entity failures, we know that if only one entity crashes, the other $n - 1$ can agree on the same value [Dol82]. Observe that with d omissions per clock cycle, we can simulate the "send failure" of a single entity: all messages sent from that entity are omitted at each time unit. This means that, if d omissions per clock cycle are localized to the messages sent by the same single entity all the time, then agreement among $n - 1$ entities is possible. What happens if those d omissions per clock cycle are *mobile* (i.e., not localized to the same entity all the time)?

Even in this case, at most a single entity will be isolated from the rest at any one time; thus, one might still reasonably expect that an agreement among $n - 1$ entities can be reached even if the faults are dynamic. Not only this expectation is false, but actually it is impossible to reach even strong majority (i.e., an agreement among $\lceil n/2 \rceil + 1$ entities).

If the communication faults are arbitrary (the *Byzantine* case), the gap between the localized and mobile cases is even stronger. In fact, in the hypercube, an agreement among $n - 1$ entities is possible in spite of d Byzantine communication faults if they are localized to the messages sent by a single entity [Dol82]; on the other hand, if the Byzantine faults are mobile then strong majority is impossible even if the number of faults is just $\lceil d/2 \rceil$.

These results for the hypercube are an instance of more general results by Santoro and Widmayer [SW05] that we are going to discuss next. Let **Omit**, **Add**, and **Corr** denote a system when the communication faults are only omissions, only additions, and only corruptions, respectively. We will denote by **X+Y** a system where the communication faults can be both of type **X** or **Y**, in an arbitrary mixture.

The first result is that, with $deg(G)$ omissions per clock cycle, strong majority cannot be reached:

Theorem 10.1 [SW05] *In **Omit**, no K-agreement protocol $\mathcal{P}$ is correct in spite of $deg(G)$ mobile communication faults per time unit if $K > \lceil n/2 \rceil$.*

If the failures are any mixture of corruptions and additions, the same bound $deg(G)$ holds for the impossibility of strong majority.

Theorem 10.2 [SW05] *In **Add+Corr**, no K-agreement protocol $\mathcal{P}$ is correct in spite of $deg(G)$ mobile communication faults per time unit if $K > \lceil n/2 \rceil$.*

In the case of arbitrary faults (omissions, additions, and corruptions: the Byzantine case), strong majority cannot be reached if just $\lceil deg(G)/2 \rceil$ communications may be faulty.

Distributed Computing in the Presence of Mobile Faults

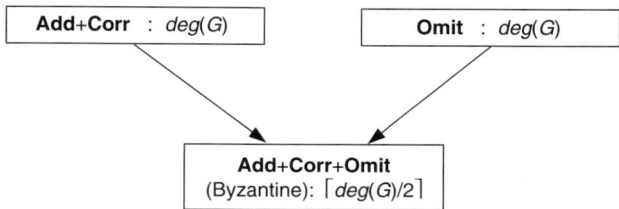

FIGURE 10.1 The minimum number of dynamic communication faults per clock cycle that may render strong majority impossible.

Theorem 10.3 [SW05] *In* **Omit+Add+Corr**, *no K-agreement protocol $\mathcal{P}$ is correct in spite of $\lceil deg(G)/2 \rceil$ mobile communication faults per time unit if $K > \lceil n/2 \rceil$.*

These results, summarized in Figure 10.1, are established using the proof structure for dynamic faults introduced in SW89. Although based on the bivalency argument of Fischer, Lynch, and Paterson [FLP85], the framework differs significantly from the ones for asynchronous systems since we are dealing with a *fully synchronous* system where time is a direct computational element, with all its consequences; for example, nondelivery of an expected message is detectable, unlike asynchronous systems where a "slow" message is indistinguishable from an omission. This framework has been first defined and used in SW89; more recently, similar frameworks have been used also in AT99, BJBO98, and MR02.

10.2.3 Possibility

Let us examine now when agreement among the entities is possible in spite of dynamic faults. We are interested in determining the maximum number F of dynamic communication faults of a given type that can be tolerated when achieving a nontrivial agreement.

Surprisingly, even *unanimity* can be achieved in several cases (a summary is shown in Figure 10.2). In these cases, we will reach unanimity, in spite of F communication faults per clock cycle, by computing the **OR** of the input values and deciding on that value. This is achieved by first constructing (if not already available) a mechanism for correctly broadcasting the value of a bit within a fixed amount of time T in spite of F communication faults per clock cycle. This reliable broadcast, once constructed, is then used to correctly compute the logical **OR** of the input values: all entities with input value 1 will reliably broadcast their value; if at least one of the input values is 1 (thus, the result of **OR** is 1), then everybody will be communicated this fact within time T; on the other hand, if all input values are 0 (thus, the result of **OR** is 0), there will be no broadcasts and everybody will be aware of this fact within time T. The variable T will be called *timeout*. The actual reliable broadcast mechanism as well as the value of F will obviously depend on the nature of the faults.

All the results of this section are from SW05.

10.2.3.1 Single Type Faults

Let us first consider systems with a single type of dynamic communication faults. Some of the results are sometimes counterintuitive.

10.2.3.1.1 Corruptions

If the faults are just *corruptions*, unanimity can be reached *regardless of the number of faults*.

Theorem 10.4 *In* **Corr**, *regardless of the number of mobile faults, unanimity can be reached in time $T = diam(G)$ transmitting at most $2\ m(G)$ bits.*

To understand this result, first consider that, since the only faults are corruptions, there are no omissions; thus, any transmitted message will arrive, although its content may be corrupted. This means that if an entity starts a broadcast protocol, every node will receive a message (although not necessarily the correct one). We can use this fact in computing the **OR** of the input values. Each entity with a input value 1 starts a broadcast. Regardless of its content, a message will always and only communicate the existence of an initial value 1. An entity receiving a message thus knows that the correct value is 1 regardless of the content of the message, and will forward it to all its neighbors (if it has not already done so). Each entity needs to participate in this computation (i.e., transmit to its neighbors) at most once. If there is an initial value 1, since there are no omissions, all entities will receive a message within time $T(G) = diam(G)$. If all initial values are 0, no broadcast is started and, since there are no additions, no messages are received; thus, all entities will detect this situation since they will not receive any message by time $T(G)$.

10.2.3.1.2 Additions

The same level of fault-tolerance holds also in systems where the faults are only *additions*, although the solution is more expensive. Indeed, if each entity transmits to its neighbors at each clock cycle, it leaves no room for additions. Hence, the entities can correctly compute the **OR** using a simple diffusion mechanism in which each entity transmits for the first $T(G) - 1$ time units: initially, an entity sends its value; if during this time it is aware of the existence of a 1 in the system, it will only send 1 from that moment on. The process clearly can terminate after $T(G) = diam(G)$ clock cycles. Hence,

Theorem 10.5 *In* **Add**, *regardless of the number of mobile faults, unanimity can be reached in time* $T = diam(G)$ *transmitting at most* $2m(G)diam(G)$ *bits.*

Observe that if a spanning-tree $\mathcal{T}(G)$ of G is available, it can be used for the entire computation. In this case, the number of bits is $2(n-1)diam(\mathcal{T}(G))$ while time is $diam(\mathcal{T}(G))$.

10.2.3.1.3 Omissions

Consider the case when the communication errors are just *omissions*. Let $F \leq con(G) - 1$. When broadcasting in this situation, it is rather easy to circumvent the loss of messages. In fact, it suffices that the initiator of the broadcast, starting at time 0, continues transmitting it to all its neighbors in each time step until time $T(G) - 1$ (the actual value of the timeout $T(G)$ will be determined later); an entity receiving the message at time $t < T(G)$ transmits the message to all its other neighbors in each time step until time $T(G) - 1$. Call this process *Bcast–Omit*.

It is not difficult to verify that for $T(G)$ large enough (e.g., $T(G) \geq con(G)(n-2) + 1$), the broadcast succeeds in spite of $f \leq con(G) - 1$ omissions per clock cycle. Let us denote by $T^*(G)$ the *minimum* timeout value ensuring that the broadcast is correctly performed in G in spite of $con(G) - 1$ omissions per clock cycle. Then, using *Bcast–Omit* to compute the **OR** we have

Theorem 10.6 *In* **Omit**, *unanimity can be reached in spite of* $F = con(G) - 1$ *mobile communication faults per clock cycle in time* $T = T^*(G)$ *transmitting at most* $2m(G) T^*(G)$ *bits.*

We will discuss the estimates on the actual value of $T^*(G)$ later in the chapter, when discussing broadcast.

10.2.3.2 Composite Types of Faults

Let us consider now systems where more than one type of mobile faults can occur. In each graph G there are at least $con(G)$ edge-disjoint paths between any pair of nodes. This fact has been used in the component failure model to show that, with enough redundant communications, information can be correctly propagated in spite of faults and that the entities can reach some form of agreement.

10.2.3.2.1 Omissions and Corruptions

If the system suffers from *omissions and corruptions*, the situation is fortunately no worse than that of systems with only omissions. Since there are no additions, no unintended message is received. Indeed, in the computation of the **OR**, the only intended messages are those originated by entities with initial value 1 and only those messages (possibly corrupted) will be transmitted along the network. Thus, an entity receiving a message knows that the correct value is 1, regardless of the content of the message. If we use the same mechanism we did for **Omit**, we are guaranteed that everybody will receive a message (regardless of its content) within $T = T^*(G)$ clock cycles in spite of $con(G) - 1$ or fewer omissions, if and only if at least one originated (i.e., if there is at least one entity with initial value 1). Hence,

Theorem 10.7 *In* **Corr+Omit**, *unanimity can be reached in spite of* $F = con(G) - 1$ *faults per clock cycle. The time to agreement is* $T = T^*(G)$ *and the number of bits is at most* $2\ m(G)\ T^*(G)$.

10.2.3.2.2 Omissions and Additions

In the case of systems with *omissions and additions*, consider the following strategy. To counter the negative effect of additions, each entity transmits to all its neighbors in every clock cycle. Initially, an entity sends its value; if at any time it is aware of the existence of a 1 in the system, it will send only 1 from that moment on. Since there are no corruptions, the content of a message can be trusted. Clearly, with such a strategy, no additions can ever take place. Thus, the only negative effects are due to omissions; however, if $F \leq con(G) - 1$, omissions cannot stop the nodes from receiving a 1 within $T = T^*(G)$ clock cycles if at least one entity has such an initial value. Hence,

Theorem 10.8 *In* **Add+Omit**, *unanimity can be reached in spite of* $F = con(G) - 1$ *mobile communication faults per clock cycle. The time to agreement is* $T = T^*(G)$ *and the number of bits is at most* $2\ m(G)\ T^*(G)$.

10.2.3.2.3 Additions and Corruptions

Consider the environment when faults can be both *additions and corruptions*. In this environment messages are not lost but none can be trusted; in fact the content could be incorrect (i.e., a corruption) or it could be a fake (i.e., an addition). This makes the computation of **OR** quite difficult. If we only transmit when we have 1 (as we did with only *corruptions*), how can we trust that a received message was really transmitted and not caused by an addition? If we always transmit the **OR** of what we have and receive (as we did with only *additions*), how can we trust that a received 1 was not really a 0 transformed by a corruption?

For this environment, indeed we need a more complex mechanism employing several techniques, as well as knowledge of the network G by the entities. The first technique we use is that of *time slicing* [SW90], in which entities are going to propagate 1 only at odd ticks and 0 at even ticks.

Technique *Time Slice*

1. Distinguish between *even* and *odd* clock ticks; an even clock tick and its successive odd tick constitute a *communication cycle*.
2. To broadcast 0 (respectively 1), x will send a message to all its neighbors only on *even* (respectively, *odd*) clock ticks.
3. When receiving a message at an *even* (respectively, *odd*) clock tick, entity y will forward it only on *even* (respectively, *odd*) clock ticks.

This technique, however, does not solve the problem created by additions; in fact, the arrival of a fake message created by an addition at an odd clock tick can generate an unwanted propagation of 1 in the system through the odd clock ticks. To cope with the presence of additions, we use another technique based on the edge-connectivity of the network. Consider an entity x and a neighbor y. Let $SP(x, y)$ be the set of the $con(G)$ shortest disjoint paths from x to y, including the direct link (x, y). To communicate a message from x to y, we use a technique in which the message is sent by x simultaneously on all the paths in $SP(x, y)$. This technique, called *Reliable Neighbor Transmission*, is as follows:

Technique *Reliable Neighbor Transmission*

1. For each pair of adjacent entities x, y and paths $SP(x, y)$, every entity determines in which of these paths it resides.
2. To communicate a message M to neighbor y, x will send along each of the $con(G)$ paths in $SP(x, y)$ a message, containing M and the information about the path, for t consecutive communication cycles (the value of t will be discussed later).

3. An entity z on one of those paths, upon receiving in communication cycle i a message for y with the correct path information, will forward it only along that path for $t - i$ communication cycles. A message with incorrect path information will be discarded.

Note that incorrect path information (due to corruptions and/or additions) in a message for y received by z is *detectable* and so is incorrect timing since (i) because of local orientation, z knows the neighbor w from which it receives the message; (ii) z can determine if w is really its predecessor in the claimed path to y; and (iii) z knows at what time such a message should arrive if really originated by x.

Let us now combine these two techniques. To compute the **OR**, all entities broadcast their input value using the *Time Slice* technique: the broadcast of 1s will take place at odd clock ticks, that of 0s at even ones. However, every step of the broadcast, in which every involved entity sends the bit to its neighbors, is done using the *Reliable Neighbor Transmission* technique. This means that each step of the broadcast now takes t communication cycles.

A corruption can now have one of two possible effects: Either it corrupts the path information in a message and causes the message not to reach its destination (regardless of whether the content of the message is correct or also corrupted), or it merely corrupts the content of the message, but leaves the path information correct. In the former case, the corruption will act like an omission (owing to the way messages travel along paths), whereas in the latter, the message will arrive, and the clock cycle in which it arrives at y will indicate the correct value of the bit (even cycles for 0, odd for 1). Therefore, if x transmits a bit and the bit is not lost, y will eventually receive one and be able to decide the correct bit value. This is however not sufficient. We need now to choose the appropriate value of t so that y will not mistakenly interpret the arrival of bits owing to additions, and will be able to decide if they were really originated by x. This value will depend on the length l of a longest one of the $con(G)$ shortest paths between any two entities. If no faults occur, then in each of the first $l - 1$ communication cycles, y will receive a message through its direct link to x. In each of the following $t - l + 1$ communication cycles, y will receive a message through each of its $con(G)$ incoming paths according to the reliable neighbor transmission protocol. Communication faults cause up to $f_c t$ messages to be lost during these t communication cycles (owing to corruptions of path information), and cause up to $f_a t$ incorrect messages to come in, owing to additions, with $f_c t + f_a t \leq (con(G) - 1)t$. Because the number of incoming correct messages must be larger than the number of incoming incorrect messages, we get

Lemma 10.1 *After $T \leq (con(G) - 1)(l - 1) + 1$ communication cycles, y can determine the bit that x has sent to y.*

Consider that broadcast requires $diam(G)$ steps, each requiring t communication cycles, each composed of two clock ticks. Hence

Lemma 10.2 *It is possible to compute the **OR** of the input value in spite of $con(G) - 1$ additions and corruptions in time at most $T \leq 2\, diam(G)\, (con(G) - 1)(l - 1)$.*

Hence, unanimity can be guaranteed if at most $con(G) - 1$ additions and corruptions occur in the system.

Theorem 10.9 *In* **Add+Corr**, *unanimity can be reached in spite of $F = con(G) - 1$ mobile communication faults per clock cycle; the time is $T \leq 2\, diam(G)\, (con(G) - 1)\, (l - 1)$ and the number of bits is at most $4m(G)(con(G) - 1)(l - 1)$.*

10.2.3.2.4 Byzantine Faults

In case of *Byzantine* faults, any type of faults can occur: omissions, additions, and corruptions. Still, using a simpler mechanism than that for additions and corruptions we are able to achieve consensus, albeit tolerating fewer ($F = \lceil con(G)/2 \rceil - 1$) faults per clock cycle. To broadcast, we use precisely the

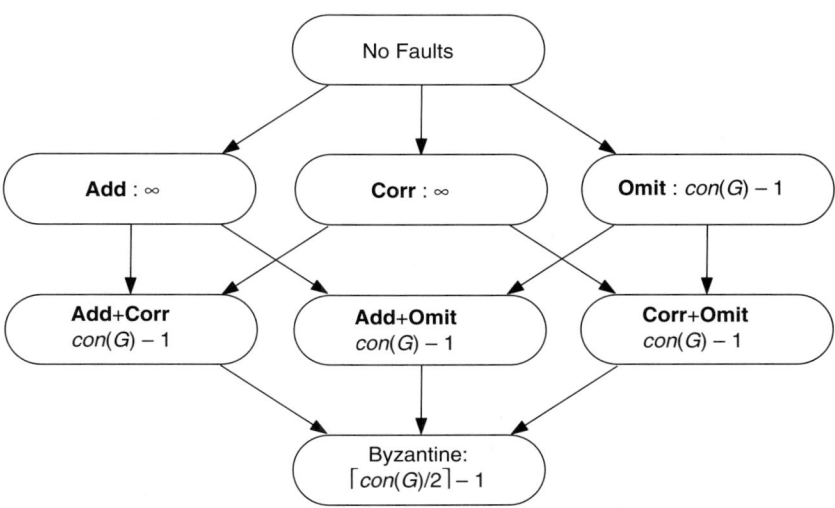

FIGURE 10.2 The maximum number of faults per clock cycle in spite of which unanimity is possible.

technique *Reliable Neighbor Transmission* introduced to deal with additions and corruptions; we do *not*, however, use time slicing: this time, a communication cycle lasts only one clock cycle; that is, any received message is forwarded along the path immediately. The decision process (i.e., how y, out of the possibly conflicting received messages, determines the correct content of the bit) is according to the following simple Acceptance Rule: *y selects as correct the bit value received most often during the t time units.*

The technique *Reliable Neighbor Transmission* with this *Acceptance Rule* will work. Indeed, by choosing $t > (con(G)-1)(l-1)$, communication to a neighbor tolerates $\lceil con(G)/2 \rceil - 1$ Byzantine communication faults per clock cycle, and uses $(con(G) - 1)(l - 1) + 1$ clock cycles. Since broadcasting requires $diam(G)$ rounds of Reliable Neighbor Transmission, we have

Theorem 10.10 *In* **Omit+Add+Corr**, *unanimity can be reached in spite of $\lceil con(G)/2 \rceil - 1$ mobile communication faults per clock cycle in time $T \leq diam(G) ((con(G) - 1) (l - 1) + 1)$.*

10.2.4 Tightness of Bounds

For all systems, except those where faults are just corruptions or just additions (and in which unanimity is possible regardless of the number of mobile faults), the bounds we have discussed are similar except that the possibility bounds are expressed in terms of the *connectivity con(G)* of the graph, while the impossibility bounds are in terms of the *degree deg(G)* of the graph.

This means that, in the case of *deg(G)-edge-connected graphs*, the impossibility bounds are indeed tight:

1. With the number of faults (or more) specified by the impossibility bound, even strong majority is impossible.
2. With one less fault than specified by the impossibility bound, even unanimity can be reached.
3. Any agreement among less than a strong majority of the entities can be reached without any communication.

In other words, in these networks, agreement is either trivial or complete or impossible. This large class of networks includes *hypercubes, toruses, rings, complete graphs*, and so forth. In these networks, there is a precise and complete "impossibility map" for the agreement problem in the presence of dynamic communication faults.

For those graphs where $con(G) < deg(G)$, there is a gap between possibility and impossibility. Closing this gap is clearly a goal of future research.

10.3 Broadcast with Mobile Omissions

Among the primitive operations of a distributed computing environment, a basic one is *broadcasting*: a single entity, the initiator, is in possession of some information, and the task is to inform all other entities. A generalization of this problem is the *wake up* in which one or more entities, the initiators, are *awake* while all others are *asleep*, and the task is to transform all entities into *awake*. Clearly, the broadcast is a wake-up with a single initiator; conversely, a wake-up is a broadcast when more than one entity initially has the information.

The solution to both problems is important because of the basic nature of these problems; it is also crucial, as we have seen, in the resolution of the agreement problem, and thus for the computability of non-trivial tasks. For these reasons, broadcast in the presence of mobile communication faults has been the object of extensive investigations, mostly in the case of **Omit**. Let us denote by $T_{\text{type}}(G)$ the *minimum* time ensuring that broadcast is possible in G in spite of a number of mobile communication faults of *type* not exceeding the bounds discussed before.

10.3.1 General Bounds

Consider the case when the communication errors are just *omissions*. We have just seen that broadcasting is possible if and only if $F \leq con(G) - 1$: the initiator, starting at time $t = 0$ transmits the information to all its neighbors for $T(G) - 1$ consecutive time units; an entity receiving the information for the first time at time $t < T(G)$ transmits the information to all its other neighbors at each time step until time $T(G) - 1$. Let us call BcastOmit this protocol; then the *minimum* timeout value $T^*(G)$ ensuring that the BcastOmit is correct in G in spite of $con(G) - 1$ omissions per clock cycle is precisely $T_{\text{omit}}(G)$:

$$T_{\text{omit}}(G) = T^*(G). \tag{10.1}$$

Clearly, $T_{\text{omit}}(G) \leq T_{\text{corr}+\text{omit}}(G)$ and $T_{\text{omit}}(G) \leq T_{\text{add}+\text{omit}}(G)$. On the other hand, by Theorems 10.7 and 10.8 we have $T_{\text{corr}+\text{omit}}(G) \leq T^*(G)$ and $T_{\text{add}+\text{omit}}(G) \leq T^*(G)$, respectively. Thus, by (10.1) we have

$$T_{\text{corr}+\text{omit}}(G) = T_{\text{add}+\text{omit}}(G) = T^*(G). \tag{10.2}$$

Hence, the key point is to determine (an upper bound on) the value $T^*(G)$. It is not difficult to verify that

Lemma 10.3 $T^*(G) \leq (n - 2) \, con(G) + 1$.

Proof. In G, there are at least $con(G)$ edge-disjoint paths between any two entities x and y; furthermore, each of these paths has length at most $n - 1$. In BcastOmit, an initiator x sends a message along all these $con(G)$ paths. At any time instant, there are $F < con(G)$ omissions; this means that at least one of these paths is free of faults. That is, at any time unit, the message from x will move one step further towards y along one of them. Since these paths have length at most $n - 1$, after at most $(n - 2)con(G) + 1 = n \, con(G) - 2 \, con(G) + 1$ time units the message from x would reach y. ∎

This upper bound is rather high, and depending on the graph G can be substantially reduced. Currently, the best available upper bounds are due to Chlebus et al. [CDP96].

Theorem 10.11 [CDP96] $T^*(G) = O(diam(G)^{con(G)})$.

Theorem 10.12 [CDP96] $T^*(G) = O(con(G)^{diam(G)/2-1})$.

Which general bound is better (i.e., smaller) depends on the graph G. Interestingly, in a hypercube H, as well as in other specific topologies, all these estimates are far from accurate.

10.3.2 Specific Bounds

Consider a d-dimensional *hypercube* H; in H, between any two nodes x and y there are d edge-disjoint paths of length at most d. That is, the paths in the proof of lemma 10.3 have length at most $\log n$ (and not n), yielding the obvious $\log^2 n$ upper bound to broadcast time. This upper bound on $T^*(G)$ has been first decreased to $O(\log n)$ by Ostromsky and Nedev [ON92], then to $\log n + O(\log \log n)$ by Fraigniaud and Peyrat [FP91], to $\log n + 7$ by De Marco and Vaccaro [MV98] who proved also a *lower bound* $T^*(H) \geq \log n + 2$, until finally Dobrev and Vrto [DV99] have shown that $\log n + 2$ clock cycles suffice! In other words, with only two time units more than in the fault-free case, broadcast can tolerate up to $\log n - 1$ message losses per time unit. Summarizing,

Theorem 10.13 [MV98, DV99] $T^*(H) = \log n + 2$.

The time to complete the broadcast in a d-dimensional hypercube has also been investigated when the number F of mobile faults is less than $d - 1$; Dobrev and Vrto [DV04] have shown that if $F \leq d - 3$ then the minimum broadcast time is $\log n$, while if $F = d - 2$ the broadcast time is always at most $\log n + 1$.

The case of d-dimensional general *tori* R has been investigated by Chlebus, et al. [CDP96] who showed that $T^*(R) = O(diam(R))$. In case of the d-dimensional k-ary even torus, De Marco and Rescigno [MR00] proved that $T^*(R) \leq diam(R) + 1$ provided k is limited by a polynomial in d. Without this assumption Dobrev and Vrto [DV02] showed that broadcast would be completed in at most one more time unit. Summarizing

Theorem 10.14 [DV02] $T^*(R) \leq diam(R) + 2$.

The case of a *complete graph* K is particularly interesting. Indeed, it is the first graph in which the effects of mobile faults were investigated [SW89, SW90]. The interest derives also from the fact that, while no nontrivial computation is possible with $F = con(K) = n - 1$ omissions per clock cycle, if the mobile omissions per clock cycle are $n - 2$ or fewer then any computable function can be computed in *constant* time. This is due to the fact that broadcast can be performed in four cycles:

Theorem 10.15 [SW89] $T^*(K) \leq 4$.

To see why this is true, consider the initiator x starting a broadcast at time $t = 0$; since at most $n - 2$ messages may be lost, at time $t = 1$ at least one more entity has the information; this means that these two entities send a total of $2(n-1)$ messages of which at most one in each direction is between the two sending entities (and thus wasted), and at most $n - 2$ are lost. This means that $2(n - 1) - 2 - (n - 2) = n - 2$ messages go towards the other entities. Since these messages come from two different sources (x and y), at least $(n - 2)/2 = n/2 - 1$ new entities will receive the information. That is, at time $t = 2$ at least $2 + n/2 - 1 = n/2 + 1$ entities have the information and forward it to all their neighbors. In the next step at least $n - 1$ entities have the information because $n/2 + 1$ omissions are required to not inform any new entity, and $\lfloor (n - 2)/(n/2 + 1) \rfloor = 1$, ensuring that at time $t = 4$ the broadcast is completed. A more detailed analysis depending on the number of initiators in each round has been done by Liptak and Nickelsen [LN00].

To achieve this time, the protocol *BcastOmit* requires all entities receiving the information to forward the information to all neighbors at each time unit of its execution. Thus, the total number of messages is in the worst case $(n - 1)(3n + 1)$. If the concern is on the amount of communication instead of time, a better broadcast protocol has been designed by Dobrev [Dob03] if the graph has weak sense of direction; in this case, for example, it is possible to broadcast in time $O(\log n)$ exchanging only $O(n \log n)$ messages.

10.4 Function Evaluation in Complete Networks

While in the previous sections we considered the problem of reaching unanimity and broadcasting in spite of faults, we now widen our view towards the computation of (Turing-computable) functions on the input.

10.4.1 Problem and Basic Strategies

The computation of (Turing-computable) functions on the input is a fundamental class of problems in distributed computing. The functions of interest include specific ones, such as counting the number of 1s among the input bits of the entities or reaching strong majority, as well as arbitrary, general ones. For initial value r_i of entity x_i, $i = 1, \ldots, n$, function computation requires that the result is known to all entities after a finite number of clock cycles, that is, the result value v_i must be the same, namely, the function value, for all i.

We limit, however, the type of network to the complete graph on n entities. That is, we consider n entities distributed over a network, where each entity is directly connected to each other entity through a distinct one-way link (often called *point-to-point* model of communication). Each entity has a unique identity, for example, one of the numbers 1 to n; that is, entities are not anonymous, and the size of the complete network is known to each entity. Now, the objective of a computation is to evaluate a nonconstant function on all entities' inputs.

Obviously, arbitrary function computation is a stronger computational requirement than mere unanimity (or any other form of agreement): whenever a distributed computing environment can compute a function, it can also agree unanimously on the function value. Therefore, the upper bounds on the number of tolerable faults for unanimity hold also for function computation. In particular, for omissions only, the upper bound of $F = con(G) - 1$ on the number of faults per clock cycle translates into $F = n - 2$, because there are $n - 1$ edge disjoint paths between any two entities in a complete graph.

We will consider two basic strategies of computing a function. In the first strategy, each entity is made to know enough information on input values to evaluate the desired function; in general, here each entity needs to know all input values. Then, each entity computes the function value locally.

In the second strategy, the function value is computed at some entity and made known to all others. Here we make use of the fact that for not too many communication faults, some entity will for sure know all the input values; then, the function value needs to be communicated to all entities. For both strategies, it is essential that we can propagate a value to all entities, that is, broadcast a value.

All results of this section are from SW90, unless otherwise specified.

10.4.2 Omissions

We have seen that for omissions only, broadcast works in 4 clock cycles when $F \leq n - 2$ (Theorem 10.15). The broadcast can be used to propagate all input values to all entities, where each message contains both the identifier and the input value of an entity. Note that the input value is not limited to be a single bit, but may just as well be a bitstring of arbitrary length (when we will look at omissions and corruptions next, the situation will no longer be this simple). After all input values have reached all entities, the desired function value can be computed at each entity locally:

Theorem 10.16 *In* **Omit**, *an arbitrary function of the input can be computed in 4 clock cycles, in spite of $F = n - 2$ mobile communication faults per clock cycle.*

This result is tight, since with $n - 1$ or more faults, not even strong majority can be achieved (see Theorem 10.1).

The problem of coping with omissions becomes far more difficult when we consider arbitrary anonymous networks. Owing to the anonymity, we cannot distinguish between inputs of different entities, and

Distributed Computing in the Presence of Mobile Faults   **10**-13

hence we are unable to propagate all input values to all entities and compute a function locally. Already the problem of determining the number of occurrences of all input values (their multiplicity) turns out to be very difficult and has been open for some time. A beautiful solution [Dob04] shows, however, that it is solvable whenever the number of faults is smaller than the edge connectivity and the network size is known.

Theorem 10.17 [Dob04] *In* **Omit**, *in an anonymous network of known size and of edge connectivity* $con(G)$, *the multiplicity of the input values can be computed in spite of* $F = con(G) - 1$ *mobile communication faults per clock cycle.*

As a consequence, any function on that multiplicity can also be computed, such as for instance the **XOR** of binary inputs.

10.4.3 Omissions and Corruptions

We show that the upper bound $n - 2$ on the number of tolerable omission and corruption faults for unanimity is tight for arbitrary function computation.

10.4.3.1 Basic Tools

Let us realize a communication of an arbitrary value by a series of communications of bits (we call this *bitwise communication*). Owing to the corruptions, we cannot distinguish for a received bit value between a 0 and a 1 that has been sent. We choose to limit ourselves to send either nothing or a 0.

Whenever we only need to broadcast a specific input value that is fixed before the algorithm starts to execute, such as a 0 for the Boolean **AND** of input bits or a 1 for the **OR**, this specific value is represented by the 0 that entities send. This leads immediately to the following:

Lemma 10.4 *In* **Omit+Corr**, *an arbitrary fixed value can be broadcast in 4 clock cycles, in spite of* $F = n - 2$ *mobile communication faults per clock cycle.*

Proof. To broadcast value v, each entity holding a v sends a 0 in the first cycle; all others are silent, that is, they send nothing. Each entity receiving a 0 or a 1 knows that the sent value represents v. In the following cycles, each entity knowing of a value v propagates a 0. In at most 4 cycles, all entities know of v, if there is an input value v in the system. ∎

Along the same lines of thought, we can broadcast a bitstring bit by bit with the *time slice* technique that we presented earlier: A 0 being sent represents a 0 in even numbered cycles and a 1 in odd clock cycles. Then, a bit can be sent in each *communication cycle* consisting of an even and an odd cycle. Now, various possibilities exist for *bitstring broadcasting*:

Techniques *Bitstring Broadcasting*

1. **Technique** *Bitstring Timeslicing* The most immediate approach is to broadcast bits in time slices. Since time slicing requires two cycles (one communication cycle) per bit to be communicated, in 4 communication cycles a one bit message will have been broadcast to every entity. In other words, if an entity has not received a message in these 4 communication cycles, nothing has been sent. We can use the freedom in the encoding of one message bit in two binary symbols to cope with the problem of terminating the bitstring, by encoding the bits as follows. In a communication cycle,

    - $0, 0$ stands for 0, and the message continues
    - $\Omega, 0$ stands for 1, and the message continues
    - $0, \Omega$ stands for 0, and the message ends
    - Ω, Ω stands for 1, and the message ends

For a message of size s (measured as its length in bits), in at most $4s$ communication cycles, that is, in $8s$ clock cycles, the bitstring reaches all entities.

2. **Technique** *Length Timeslicing* Instead of sacrificing an extra bit for each bit of the entire message, we might as well first send the length of the message (with time slicing), and then send the message by bitwise broadcast. In this case, each bit of the message can be encoded in one binary symbol of $\{0, \Omega\}$ to be broadcast. For a message of size s, we need to broadcast $\lceil \log s \rceil$ bits with time slicing, plus an additional s bits without time slicing, in $8\lceil \log s \rceil + 4s$ clock cycles altogether.

   Alternatively, the folklore concept of representing a bitstring in a *self-delimiting* way (see the section on binary strings in LV97), applied to our situation, would first send the length s of the bitstring in unary, by sending a 0 in each of s clock cycles, followed by one clock cycle with no communication to indicate that the unary sequence ends, followed by the bitwise broadcast of the message in the next s clock cycles. This would take $4(2s+1) = 8s+4$ clock cycles altogether and therefore not be attractive in our setting.

3. **Technique** *Iterated Length Timeslicing* The process of first sending the length of a message by time slicing, and then sending the message itself without time slicing may be applied repeatedly, in a fashion minimizing the number of clock cycles needed, if desired.

We summarize this discussion as follows.

Lemma 10.5 *In* **Omit+Corr**, *a string of size s can be broadcast in $8s$ clock cycles, in spite of $F = n-2$ mobile communication faults per clock cycle. It can alternatively be broadcast in $8\lceil \log s \rceil + 4s$ clock cycles.*

10.4.3.2 Input Bits

For the special case of input bits only (instead of input strings), Lemma 10.4 implies that the Boolean **AND** and **OR** can both be computed in 4 clock cycles, in spite of $n-2$ faults. At the same time with the same number of faults, it can also be determined whether the number of 1s in the input is at least k, for a fixed value k. To see whether at least k 1s occur in the input, each entity holding a 1 sends it in the first clock cycle; all other entities are silent. At least two entities will have received a bit (perhaps corrupted) for each of the 1s that occur in the input of other entities; we say such an entity knows all the 1s. However, no entity can detect at this stage whether it knows all the 1s. In the second cycle, each entity knowing of at least k 1s (including its own input value) starts broadcasting a 1 through the network. That is, the fixed value 1 is broadcast as outlined in the proof of Lemma 10.4. Since at least two entities start this broadcast in the second cycle, in at most 4 cycles in total each entity knows the result: if a 1 has reached the entity, the answer is affirmative. Similarly, one can solve the other three variants of the problem that ask for at least k 0s, at most k 1s, and at most k 0s (note that with a binary alphabet, at least k 1s is equivalent to at most $n-k$ 0s). We summarize as follows.

Theorem 10.18 *In* **Omit+Corr** *with bits as input, it can be determined in 4 clock cycles whether there are at least (at most) k 0s (1s) in the input, in spite of $F = n-2$ mobile communication faults per clock cycle.*

10.4.3.2.1 Functions on the Multiset of Inputs

We start to consider more general function computation by looking at functions on the multiset of inputs; after that, we will consider functions on the vector of inputs. For functions on the multiset of input bits, it is enough that each entity knows the exact number of 0s and 1s in the input. This can be determined by a binary search over the integers from 1 to n for the precise number of 1s in the input. To this end, each processor holding an input 1 sends it in the first cycle. Then, the three additional cycles to decide whether there are at least k 1s in the input are repeated exactly $\lceil \log n \rceil$ times, for values of k defined by binary search. At the end, each processor knows the number of 1s and 0s in the input, and can compute the function. Hence, we have the following:

Theorem 10.19 *In* **Omit+Corr** *with bits as input, any function on the multiset of inputs (as opposed to the input vector) can be determined in* $1 + 3\lceil \log n \rceil$ *clock cycles, in spite of* $F = n - 2$ *mobile communication faults per clock cycle.*

10.4.3.2.2 Functions of the Input Vector

Let us now turn to the computation of a function of the input vector. In the situation in which a function on a multiset of the inputs can be computed, as described above, all entities know the number of 1s in the input, say k_1, after $1 + 3\lceil \log n \rceil$ cycles. Now, each entity can check whether it has known of k_1 1s already after the first cycle (again, including its own input value). If so, an entity now also knows the input vector, and there are at least two of these entities in the system. These entities now compute an arbitrary function f on the input vector I, and then broadcast the result $f(I)$ by bitstring broadcast. Naturally, only entities having computed $f(I)$ will initiate the bitstring broadcasting; all other entities will only join in the broadcast.

Theorem 10.20 *In* **Omit+Corr** *with bits as input, any function f on the vector I of input bits can be computed in* $1 + 3\lceil \log n \rceil + 8\lceil \log |f(I)| \rceil + 4|f(I)|$ *clock cycles, where $|f(I)|$ is the length of the representation of $f(I)$, in spite of* $F = n - 2$ *mobile communication faults per clock cycle.*

10.4.3.3 Input Strings

For bitstrings as inputs, we still stick to communicating messages bitwise, owing to possible corruption. To compute an arbitrary function, we collect all input strings at all entities, and evaluate the function locally at each entity.

First, we perform a *length determination* step: If not known already, we first guess the length s of a longest input string by means of unbounded search for s. Recall that unbounded search goes through powers of two, starting at 2, until $s \leq 2^i$ for the first time for some i. It then performs binary search for s between 2^{i-1} and 2^i. A single one of these probes with value v in unbounded search works as follows. Each entity holding an input of length at least v initiates the broadcast of a message. In 4 cycles, the outcome of the probe is known to each entity: If no communication bit has been received in the last 4 cycles, no one of the other entities holds an input of length at least the probe value v, and otherwise at least one other entity has such an input. Since unbounded search terminates after at most $2\lceil \log s \rceil$ probes, $8\lceil \log s \rceil$ cycles suffice.

Then, in a *length adjustment* or *padding* step, all inputs are adjusted to maximum length locally at each entity. We assume this can be done, for example, by adding leading zeroes if the inputs represent binary numbers. (Otherwise, we apply the technique of doubling the length of each input first and marking its end, by applying the scheme described in the Bitstring Timeslicing Technique, and then we can pad the bitstring as much as needed beyond its designated end; this leads to an extra factor of 2 inside the time bound that we do not account for below.)

Finally, each entity in turn, as determined by a global order on entity identities, communicates its input string bit by bit according to the bitstring broadcasting scheme. This consumes no more than $8s$ cycles for each of the n inputs. Now, each entity knows all input strings and computes the function value locally. In total, this adds up as follows:

Theorem 10.21 *In* **Omit+Corr** *with bitstrings as inputs, any function can be computed in at most* $8\lceil \log s \rceil + 8ns$ *clock cycles, where s is the length of a longest input bitstring, in spite of* $F = n - 2$ *mobile communication faults per clock cycle.*

Specific functions, again, can be computed faster. For instance, the maximum (minimum) value in a set of input values represented as binary numbers can be found by unbounded search for that number. We get

Theorem 10.22 *In* **Omit+Corr** *with bitstrings as input, the maximum (minimum) value in the input can be computed in at most* $8\lceil \log v \rceil$ *clock cycles, where v is the result value, in spite of* $F = n - 2$ *mobile communication faults per clock cycle.*

Similarly, the rank r of a given query value v in the set of binary input numbers (i.e., the relative position that v would have in the sorted input sequence) can be determined by having each entity whose input value is at most v communicate in the first cycle; all other entities are silent. In the second cycle, each entity computes the number n_i of messages it has received, and increments it if its own input value is at most v. The largest value of any n_i now equals r and is known to at least two entities. Since $1 \leq n_i \leq n$ for all i, r can be determined by binary search in at most $4\lceil \log n \rceil$ clock cycles in total. Therefore, we get

Theorem 10.23 *In* **Omit+Corr** *with bitstrings as input, the rank among the input values of a given value can be computed in at most* $4\lceil \log n \rceil$ *clock cycles, in spite of* $F = n - 2$ *mobile communication faults per clock cycle.*

By using similar techniques, we can select the k-th largest input value, compute the sum of the input values (interpreted as binary numbers), and compute various statistical functions on the inputs, like mean, standard deviation, variance, within similar time bounds, that is, faster than arbitrary functions.

10.4.4 Applications of the Basic Techniques

10.4.4.1 Corruptions

We have seen that for the problem of reaching unanimous agreement, an arbitrary number of corruptions can be tolerated. The same is true for the computation of arbitrary functions, essentially for the same reason. Bits can be communicated without ambiguity, for example, by representing a 0 by the communication of a bit, and a 1 by the absence of a communication.

To compute a function, first all inputs are collected at all entities, and then the function is evaluated locally. The collection of all input strings at all entities starts by a *length determination* step for the length s of the longest input string by unbounded search in $2\lceil \log s \rceil$ probes, where each probe takes only one clock cycle: For probe value j, an entity sends a bit if its own input value is at least j bits long, and is otherwise silent. Hence, the length determination takes $2\lceil \log s \rceil$ clock cycles. Then, we perform a *padding* step locally at each entity so as to make all strings equally long. Finally, bit by bit, all input values are sent simultaneously, with the unambiguous encoding described above.

Because the communication of all inputs to all entities takes only s clock cycles, we get:

Theorem 10.24 *In* **Corr***, an arbitrary function of the input can be computed in* $2\lceil \log s \rceil + s$ *clock cycles, in spite of* $F = n(n-1)$ *mobile communication faults per clock cycle.*

Specific functions can, again, be computed faster in many cases. In one such case, a function asks for the number of input strings that have a certain, fixed property. Locally, each entity determines whether its input has the property, and sends a bit unambiguously telling the local result to all other entities. We therefore have

Theorem 10.25 *In* **Corr***, any function of the form* $f(I) = |\{ j \in I : P(j) \text{ holds}\}|$ *for a computable predicate P on the input I can be computed in 1 clock cycle, in spite of* $F = n(n-1)$ *mobile communication faults per clock cycle.*

10.4.4.2 Additions

If only spontaneous additions can occur, but no corruptions or loss of messages, all links may be faulty at all times, with no negative effect, because each entity simply sends its input string without ambiguity, and then each entity computes the desired function locally.

Theorem 10.26 *In* **Add***, any function can be computed in 1 clock cycle, in spite of* $F = n(n-1)$ *mobile communication faults per clock cycle.*

10.4.4.3 Additions and Corruptions

For the communication of single bits, additions and corruptions are isomorphic to omissions and corruptions, in the following sense. For omissions and corruptions, we made use of an unambiguous encoding that, for example, communicates a 0 to indicate a 0, and that sends nothing to indicate a 1. With an adapted behavior of the entities, this very encoding can also be used with additions and corruptions: Just like omissions were overcome by propagating a bit for 4 cycles, additions are overcome by propagating the absence of a message for 4 cycles. More precisely, we associate a phase of 4 cycles to the communication of 1s. If an entity has input bit 1, it is silent in the first cycle of the phase. If an entity does not receive $n - 1$ bits (one from each other entity) in a cycle, it knows that a 1 is being communicated, and it is silent in all future cycles of this phase, thus propagating the 1. This isomorphism can be used in the computations of arbitrary functions on input strings and input bits. We therefore get

Theorem 10.27 *In* **Add+Corr**, *arbitrary function computation tolerates the same number of mobile communication faults per clock cycle as in* **Omit+Corr**.

10.4.4.4 Additions and Omissions

For omissions only, our algorithms always communicate, for arbitrary function computation as well as for strong majority. Therefore, additions cannot occur, and hence, additions and omissions are not more difficult than omissions alone.

Theorem 10.28 *In* **Add+Omit**, *arbitrary function computation and strong majority tolerate the same number of mobile communication faults per clock cycle as in* **Omit**.

10.4.4.5 The Byzantine Case: Additions, Omissions, and Corruptions

Recall the *Acceptance Rule* that was used for unanimity in the Byzantine case (see Theorem 10.10), based on an application of *Reliable Neighbor Transmission*: in a communication of one bit from an entity x to a direct neighbor y of x, y selects the majority bit value that it has received from x on the $con(G)$ chosen paths within $t > (con(G) - 1)(l - 1)$ clock cycles, where l is the length of a longest one of all such paths. This rule guarantees correct communication of a bit from x to y if the number of Byzantine faults per clock cycle is limited to not more than $\lceil con(G)/2 \rceil - 1$. Applied to the Byzantine case in the complete network, this means that we can safely transmit a bit between any two entities in at most n (i.e., strictly more than $n - 1$) clock cycles, if there are not more than $\lceil (n-1)/2 \rceil - 1 = \lfloor (n-1)/2 \rfloor$ faults per cycle.

We can now build on the safe bitwise transmission by collecting all inputs of all entities at one distinguished entity, by then computing the function value there locally, and by then broadcasting the result to all other entities. To send one input bitstring between two entities, based on safe bitwise transmission, we can use the technique of doubling the length of the bitstring as in *Bitstring Timeslicing* to indicate its end, or the *Length Timeslicing* technique, or the *Iterated Length Timeslicing* technique. We therefore get

Lemma 10.6 *In* **Omit+Add+Corr**, *a string of size s can be communicated between two entities in a complete network in at most $T \leq 2sn$ clock cycles, in spite of $\lfloor (n-1)/2 \rfloor$ faults per clock cycle.*

Even if we do not care to overlap the bitstring communications in time, but simply collect and then distribute values sequentially, in $n - 1$ rounds of collecting and distributing, the Byzantine function evaluation is completed.

Theorem 10.29 *In* **Omit+Add+Corr**, *any function can be computed in at most $T \leq 2Sn$ clock cycles, where S is the total length of all input strings and the result string, in spite of $\lfloor (n-1)/2 \rfloor$ faults per clock cycle.*

10.4.5 Tightness of Bounds

Because the evaluation of an arbitrary function is a stronger requirement than mere unanimity, the lower bounds on the number of tolerable faults for unanimity carry over to function evaluation. That is, Figure 10.1 accurately describes the lower bound situation for the complete network, with $deg(G)$ replaced by $n-1$. The results in this section have shown that also Figure 10.2 accurately describes the situation in the complete network for the upper bounds, with $con(G)$ replaced by $n-1$. That is, for function evaluation in the complete network, the lower and the upper bounds on the numbers of faults are tight in all cases. Put differently, even the lower bound for unanimity in the complete network and the upper bound on arbitrary function evaluation are tight.

10.5 Other Approaches

The solutions we have seen so far are F-tolerant; that is, they tolerate up to F communication faults per time unit, where the value of F depends on the type of mobile faults in the system. The solution algorithms, in some cases, require for their correct functioning the transmission of a large number of messages (e.g., protocol *BcastOmit*). However, in some systems, as the number of message transmissions increases, so does the number of lost messages; hence those solutions would not behave well in those faulty environments.

10.5.1 Probabilistic

An approach that takes into account the interplay between the amount of transmissions and the number of losses is the *probabilistic* one. It does not assume an a priori upper bound F on the total number of dynamic faults per time unit; rather, it assumes that each communication has a known probability $p < 1$ to fail. The investigations using this approach have focused on designing broadcasting algorithms with low time complexity and high probability of success, and have been carried out by Berman et al. [BDP97] and by Pelc and Peleg [PP05]. The drawback of this approach is that obviously the solutions derived have no deterministic guarantee of correctness.

10.5.2 Fractional

The desire of a deterministic setting that explicitly takes into account the interaction between number of omissions and number of messages has been the motivation behind the *fractional* approach proposed by Královič et al. [KKR03]. In this approach, the amount of dynamic faults that can occur at time t is not fixed but rather a linear fraction $\alpha\ m_t$ of the total number m_t of messages sent at time t, where $0 \leq \alpha < 1$ is a (known) constant. The advantage of the fractional approach is that solutions designed for it tolerate the loss of up to a fraction of all transmitted messages [KKR03]. The anomaly of the fractional approach is that, in this setting, transmitting a single message per communication round ensures its delivery; thus, the model leads to very counterintuitive algorithms that might not behave well in real faulty environments.

To obviate this drawback, the fractional approach has been recently modified by Dobrev et al. [DKKS06] who introduced the *threshold fractional* approach. Similar to the fractional approach, the maximum number of faults occurring at time step t is a function of the amount m_t of messages sent in step t; the function is however

$$F(m_t) = \max\{T_h - 1, \lfloor \alpha\ m_t \rfloor\},$$

where $T_h \leq con(G)$ is a constant at most equal to the connectivity of the graph, and α is a constant $0 \leq \alpha < 1$. Note that both the traditional and the fractional approaches are particular, extreme instances of this model. In fact, $\alpha = 0$ yields the traditional setting: at most $T_h - 1$ faults occur at each time step. On the other hand, the case $T_h = 1$ results in the *fractional* setting. In between, it defines a spectrum of new settings never explored before.

The only new setting explored so far is the *simple threshold*, where to be guaranteed that at least one message is delivered in a time step, the total amount of transmitted messages in that time step must be above the threshold T_h; surprisingly, broadcast can be completed in (low) polynomial time for several networks including rings (with or without knowledge of n), complete graphs (with or without chordal sense of direction), hypercubes (with or without orientation), and constant-degree networks (with or without full topological knowledge) [DKKS06].

10.6 Conclusions

For graphs whose connectivity is the same as the degree, there exists a precise map of safe and unsafe computations in the presence of dynamic faults, generalizing the existing results for complete graphs. For those graphs where $con(G) < deg(G)$, the existing results leave a gap between possibility and impossibility. Closing this gap is the goal of future investigations. Preliminary results indicate that neither parameter provides the "true bound" for arbitrary graphs: there are graphs where $deg(G)$ is too large a parameter and there are networks where $con(G)$ is too low. An intriguing open question is whether there actually exists a single parameter for all graphs.

When the faults include additions and corruptions (with or without omissions) the current unanimity protocol assumes knowledge of the network. Is it possible to achieve unanimity without such a requirement?

With a few exceptions, the performance in terms of messages of the protocols have not been the main concern of the investigations. Designing more efficient reliable protocols would be of considerable interest, both practically and theoretically.

In the threshold fractional approach, the only new setting explored so far is the simple threshold. All the others are still unknown.

References

[AT99] Marcos Kawazoe Aguilera and Sam Toueg. A simple bivalency proof that t-resilient consensus requires t + 1 rounds. *Inf. Process. Lett.*, 71:155–158, 1999.

[BDP97] Piotr Berman, Krzysztof Diks, and Andrzej Pelc. Reliable broadcasting in logarithmic time with byzantine link failures. *J. Algorithms*, 22:199–211, 1997.

[BG93] Piotr Berman and Juan A. Garay. Cloture votes: n/4-resilient distributed consensus in t+1 rounds. *Math. Sys. Theory*, 26:3–19, 1993.

[BJBO98] Ziv Bar-Joseph and Michael Ben-Or. A tight lower bound for randomized synchronous consensus. In *PODC '98: Proceedings of the Seventeenth Annual ACM Symposium on Principles of Distributed Computing*, pp. 193–199, New York, NY, USA, 1998. ACM Press.

[CASD95] Flaviu Cristian, Houtan Aghili, Ray Strong, and Danny Dolev. Atomic broadcast: From simple message diffusion to byzantine agreement. *Inf. Comput.*, 118:158–179, 1995.

[CDP96] Bogdan Chlebus, Krzysztof Diks, and Andrzej Pelc. Broadcasting in synchronous networks with dynamic faults. *Networks*, 27:309–318, 1996.

[CHT96] Tushar Deepak Chandra, Vassos Hadzilacos, and Sam Toueg. The weakest failure detector for solving consensus. *J. ACM*, 43:685–722, 1996.

[DFF+82] Danny Dolev, Michael Fischer, Rob Fowler, Nancy Lynch, and Raymond Strong. Efficient byzantine agreement without authentication. *Inf. Control*, 52:256–274, 1982.

[DKKS06] Stefan Dobrev, Rastislav Královič, Richard Královič, and Nicola Santoro. On fractional dynamic faults with threshold. In *Proc. 13th Colloquium on Structural Information and Communication Complexity (SIROCCO'06)*, 2006.

[Dob03] Stefan Dobrev. Communication-efficient broadcasting in complete networks with dynamic faults. *Theory Comput. Syst.*, 36:695–709, 2003.

[Dob04] Stefan Dobrev. Computing input multiplicity in anonymous synchronous networks with dynamic faults. *J. Discrete Algorithms*, 2:425–438, 2004.

[Dol82] Danny Dolev. The byzantine generals strike again. *J. Algorithms*, 3:14–30, 1982.

[DS82] Danny Dolev and Raymond Strong. Polynomial algorithms for multiple processor agreement. In *STOC'82: Proceedings of the Fourteenth Annual ACM Symposium on the Theory of Computing*, pp. 401–407, New York, USA, 1982. ACM Press.

[DV99] Stefan Dobrev and Imrich Vrto. Optimal broadcasting in hypercubes with dynamic faults. *Inf. Process. Lett.*, 71:81–85, 1999.

[DV02] Stefan Dobrev and Imrich Vrto. Optimal broadcasting in tori with dynamic faults. *Parallel Process. Lett.*, 12:17–22, 2002.

[DV04] Stefan Dobrev and Imrich Vrto. Dynamic faults have small effect on broadcasting in hypercubes. *Discrete Appl. Math.*, 137:155–158, 2004.

[FLP85] Michael J. Fischer, Nancy A. Lynch, and Mike Paterson. Impossibility of distributed consensus with one faulty process. *J. ACM*, 32:374–382, 1985.

[FP91] Pierre Fraigniaud and Claudine Peyrat. Broadcasting in a hypercube when some calls fail. *Inf. Process. Lett.*, 39:115–119, 1991.

[GM98] Juan A. Garay and Yoram Moses. Fully polynomial byzantine agreement for $n > 3t$ processors in $t + 1$ rounds. *SIAM J. Comput.*, 27:247–290, 1998.

[KKR03] Rastislav Královic, Richard Královic, and Peter Ruzicka. Broadcasting with many faulty links. In *Proceedings of 10th Colloquium on Structural Information and Communication Complexity (SIROCCO'03)*, pp. 211–222, 2003.

[LN00] Z. Liptak and A. Nickelsen. Broadcasting in complete networks with dynamic edge faults. In *Proceedings of 4th International Conference on Principles of Distributed Systems (OPODIS 00)*, pp. 123–142, 2000.

[LSP82] Leslie Lamport, Robert E. Shostak, and Marshall C. Pease. The byzantine generals problem. *ACM Trans. Program. Lang. Syst.*, 4:382–401, 1982.

[LV97] Ming Li and Paul Vitanyi. *An Introduction to Kolmogorov Complexity and Its Applications*, 2nd ed., Springer-Verlag, Berlin, 1997.

[MR00] Gianluca De Marco and Adele A. Rescigno. Tighter time bounds on broadcasting in torus networks in presence of dynamic faults. *Parallel Process. Lett.*, 10:39–49, 2000.

[MR02] Yoram Moses and Sergio Rajsbaum. A layered analysis of consensus. *SIAM J. Comput.*, 31:989–1021, 2002.

[MV98] Gianluca De Marco and Ugo Vaccaro. Broadcasting in hypercubes and star graphs with dynamic faults. *Inf. Process. Lett.*, 66:321–326, 1998.

[ON92] Tz. Ostromsky and Z. Nedev. Broadcasting a message in a hypercube with possible link faults. In *Parallel and Distributed Processing '91* (K. Boyanov, ed.), pp. 231–240, Elsevier, 1992.

[PP05] Andrzej Pelc and David Peleg. Feasibility and complexity of broadcasting with random transmission failures. In *PODC '05: Proceedings of the 24th Annual ACM Symposium on Principles of Distributed Computing*, pp. 334–341, New York, 2005. ACM Press.

[PT86] Kenneth J. Perry and Sam Toueg. Distributed agreement in the presence of processor and communication faults. *IEEE Trans. Software Eng.*, 12:477–482, 1986.

[San06] Nicola Santoro. *The Design and Analysis of Distributed Algorithms*. John Wiley, Hoboken, NJ, 2006.

[SW89] Nicola Santoro and Peter Widmayer. Time is not a healer. In *Proceedings of 6th Annual Symposium on Theoretical Aspects of Computer Science (STACS 89)*, pp. 304–313, 1989.

[SW90] Nicola Santoro and Peter Widmayer. Distributed function evaluation in the presence of transmission faults. In *Proceedings International Symposium on Algorithms (SIGAL 90)*, pp. 358–367, 1990.

[SW05] Nicola Santoro and Peter Widmayer. Agreement in synchronous networks with ubiquitous faults (preliminary version; full version to appear in *Theoretical Computer Science*). In

Proceedings of 12th Colloquium on Structural Information and Communication Complexity (SIROCCO'05), 2005.

[SWR02] Ulrich Schmid, Bettina Weiss, and John M. Rushby. Formally verified byzantine agreement in presence of link faults. In *Proceedings of 22nd International Conference on Distributed Computing Systems (ICDCS 02)*, pp. 608–616, 2002.

11
A Hierarchical Performance Model for Reconfigurable Computers

11.1	Introduction...	11-1
	Reconfigurable Hardware	
11.2	Reconfigurable Computers............................	11-3
	Survey of Reconfigurable Computers • Element-Level Architectures • System- and Node-Level Architectures	
11.3	Performance Modeling	11-9
	Element-Level • Node-Level	
11.4	Example	11-13
	Molecular Dynamics Simulation • Performance Modeling • Experimental Results	
11.5	Related Work	11-18
11.6	Concluding Remarks	11-20
	Acknowledgments..	11-20
	References	11-20

Ronald Scrofano
Viktor K. Prasanna
University of Southern California

11.1 Introduction

For several years, reconfigurable hardware has been used successfully to provide application acceleration [5]. The state-of-the-art reconfigurable hardware is field-programmable gate arrays (FPGAs). These devices can be programmed and reprogrammed with application-specific hardware designs. Until recently, FPGAs did not contain enough logic resources to implement floating-point arithmetic and were, therefore, not applicable to many scientific computing applications. Instead, they were employed with great benefit in fields such as signal processing, cryptography, and embedded computing. Now, large FPGAs that also possess hardware features such as dedicated multipliers and on-chip memories have become attractive platforms for accelerating tasks in scientific applications. There has been a resulting development of *reconfigurable computers*—computers that have both general purpose processors (GPPs) and reconfigurable hardware, as well as memory and high-performance interconnection networks.

Our target application domain is scientific computing, although our work can be applied to many different areas. Scientific computing applications are very computationally demanding and often consist of many different tasks. Some of these tasks may be well-suited for execution in software while others may be well-suited for execution in hardware. This is where reconfigurable computers can be employed: the tasks well-suited for software execution can execute on the GPPs and the tasks that are well-suited for hardware acceleration can execute on the FPGAs.

Much of the early work in using reconfigurable hardware for scientific applications has focused on individual tasks [17,29,32,33,38,39]. Oftentimes, reconfigurable hardware can provide fantastic speed-ups for tasks in a scientific application but when the tasks are integrated back into the complete application, the speed-up for the complete application is not very impressive. Despite advances in high-level programming for reconfigurable hardware, it can be difficult to develop a working program for reconfigurable computers that utilizes both the GPP elements and the reconfigurable-hardware elements. A crucial step, then, is the determination of whether or not the approach considered will provide a significant enough speed-up to justify the implementation effort.

In this chapter, our goal is to develop a high-level performance model for reconfigurable computers in order to address these issues. To do so, we must first introduce reconfigurable computers and develop an abstract model for them. Thus, in Section 11.2, we introduce reconfigurable computers, survey currently available reconfigurable computers, and develop abstract architectural and programming models for reconfigurable computers. Section 11.3 develops an analytic performance model for reconfigurable computers. Section 11.4 shows an example of using the performance model in the porting of a scientific application to a reconfigurable computer. Section 11.5 presents some of the relevant work in the literature. Finally, Section 11.6 draws conclusions and presents areas for future work. We begin, though, with some background information to introduce the reader to reconfigurable hardware.

11.1.1 Reconfigurable Hardware

Reconfigurable hardware is used to bridge the gap between flexible but performance-constrained GPPs and very high-performance but inflexible application-specific integrated circuits (ASICs). Reconfigurable hardware is programmed and reprogrammed after fabrication. It thus combines the flexibility of software with the high performance of custom hardware [35]. Because reconfigurable hardware is programmed specifically for the problem to be solved rather than to be able to solve all problems, it can achieve higher performance and greater efficiency than GPPs, especially in applications with a regular structure and/or a great deal of parallelism. Because it is programmed and reprogrammed after fabrication, it is less expensive to develop designs for and much more flexible to use than ASICs.

The state-of-the-art reconfigurable-hardware devices are FPGAs [13]. The FPGAs in the reconfigurable computers considered here are SRAM-based. That is, they are configured by the writing of a configuration bitstream to the FPGA's SRAM-based configuration memory. FPGAs are implemented as a matrix of configurable logic blocks and programmable interconnect between and within the blocks. Figure 11.1 shows a generic FPGA architecture and its components. The large gray boxes represent the configurable logic blocks. The small gray boxes represent "switch boxes" that are used to program the routing. There are several routing architectures and we do not depict any particular one in the figure. Rather, we simply emphasize that each of the configurable logic blocks connects to the interconnection network and this network is configured such that the desired functionality is achieved on the FPGA.

The configurable logic blocks in FPGAs are made up of configurable combinational logic, flip-flops (FFs), multiplexers (MUXs), and high-speed intrablock connections (e.g., fast carry logic). The configurable combinational logic is implemented as look-up tables (LUTs). An l-input LUT can be configured to perform any logic function of l inputs. The LUT acts like a memory in which the inputs are address lines and the logic function that is implemented is determined by what is stored in the memory. The bits in a LUT memory are written as part of configuration. As an example, a 3-input LUT is shown on the right of Figure 11.1 (A, B, C are the inputs). The logic function implemented is a 3-input OR: if A, B, and C are each 0, the LUT outputs a 0, and it outputs a 1 otherwise. In practice, the most common type of LUT in FPGAs is the 4-input LUT.

The density of FPGAs has been steadily increasing, allowing more and more complex designs to be implemented in a single chip. In addition to increased density, FPGA vendors are increasing the amount

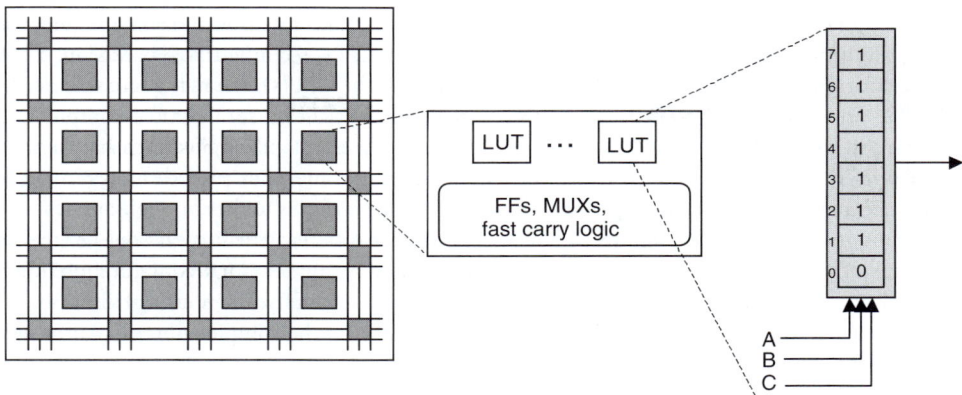

FIGURE 11.1 Generic FPGA and its components.

of embedded features within the FPGAs. These are dedicated (nonprogrammable) hardware features embedded into the FPGA fabric that increase the capability of the FPGAs by performing common functions without using programmable logic resources. Because they are dedicated, embedded hardware features take less space and can achieve higher clock rates than comparable features made in programmable logic. One embedded feature common in modern FPGAs is embedded RAM memory to provide local storage on the FPGAs [2, 40]. Large FPGAs have on the order of 1 MB of embedded memory. Embedded arithmetic units, such as multipliers and multiply-accumulate units, are also common [2, 40]. Even processors have been embedded into FPGAs. For example, the Xilinx Virtex-II Pro FPGAs have up to two PowerPC 405 cores embedded in the FPGA fabric.

There is no native support for floating-point arithmetic on FPGAs. That is, no currently available FPGAs provide embedded floating-point cores, though there has been some research in this direction [4]. Consequently, floating-point cores must be implemented using the programmable logic resources on the FPGA. Floating-point cores tend to require large amounts of area. In addition, in order to achieve high clock frequencies, they must be deeply pipelined. Nonetheless, it has been shown that floating-point performance on FPGAs is competitive with GPPs and that it is increasing at a faster rate than floating-point performance on GPPs [38].

There is a penalty for the hardware flexibility provided by FPGAs: reduced clock frequency. Especially because of the programmable interconnect, state-of-the-art FPGAs operate at clock frequencies about one-tenth those of state-of-the-art GPPs. Customization of the architecture to the problem is used to get around this large frequency discrepancy. As will be discussed more in Section 11.3.1, the main ways FPGAs obtain high performance is through the exploitation of pipelining and parallelism.

Designs for FPGAs have traditionally been written in hardware description languages such as VHDL and Verilog. Because this can be difficult, especially for large designs, there have also been many efforts aimed at facilitating higher-level programming [6, 9, 34]. These efforts focus on the development of compilers that input code written in a high-level language, such as C or a customized variant of C, and generate hardware designs. These efforts are becoming especially important with the development of reconfigurable computers because the community of users of FPGA technology is broadening beyond hardware designers to many users who are more comfortable with software programming.

11.2 Reconfigurable Computers

Reconfigurable computers bring together GPPs and reconfigurable hardware to exploit the benefits of each. With their high frequencies and von Neumann architecture, GPPs are well-suited for control-intensive tasks and tasks that are not too computationally intensive. Reconfigurable hardware, on the other hand, with its pipelining and fine-grained parallelism, is well-suited for data-intensive tasks that require a lot of processing. Reconfigurable computers consist of one or more GPPs, reconfigurable hardware, memory, and high-performance interconnect. Tasks can be executed on the GPP(s), the reconfigurable hardware, or both. The presence of high-performance interconnect distinguishes reconfigurable computers from other systems in which FPGAs are attached to a host processor over a standard bus, such as PCI or VME. For this reason, what we term reconfigurable computers here are sometimes referred to as *reconfigurable supercomputers* and *high-performance reconfigurable computing platforms* [13,31].

For reconfigurable computers, architecturally, there are three different areas to examine. At the *system level*, we consider the composition of the entire reconfigurable computer. That is, we are looking at all of the global properties of the system such as homogeneity/heterogeneity, memory distribution, and so forth. At the *node level*, we study the individual nodes that make up the complete system. At the *element level*, we look at the individual computing elements that make up nodes. Modeling at each of these levels is necessary in order to predict performance of the complete system.

As we proceed with our discussion of reconfigurable computers, we will make use of the following definitions for the computing elements:

RH element The reconfigurable hardware, likely in the form of one or more tightly coupled FPGAs, and off-chip (but onboard) memory that is spread over multiple banks, with each bank accessible independently.

GPP element One or more tightly coupled GPPs (e.g., Intel Pentium 4 processors) sharing a RAM memory.

We first present a survey of currently available reconfigurable computers. We then describe abstract architectures and develop abstract models that will facilitate performance modeling.

11.2.1 Survey of Reconfigurable Computers

Recently, reconfigurable computers have been developed and made available by several vendors. SRC Computers has developed several offerings [34]. The RH element in all of them is SRC's *MAP processor*. The MAP processor is depicted conceptually in Figure 11.2. In the Series C version of the MAP, there are 2 Xilinx Virtex-II XC2V6000 FPGAs available for the user to use for programmable logic. There are 6 banks of onboard memory. Each bank holds 4 MB worth of 64-bit words. The banks can be accessed independently, in parallel, leading to a maximum bandwidth of 4.8 GB/s. However, each bank is single-ported. That is, in a given cycle, a given memory bank may be read from or written to, but not both. There is also a penalty of two dead cycles for switching between read mode and write mode. Also, because the onboard memories are single-ported, both FPGAs cannot access the same onboard memory bank in the same cycle. The FPGAs can communicate and synchronize with one another, though. Data can also be streamed between the two FPGAs. In our work, we have been able to implement 53 single-precision floating-point cores

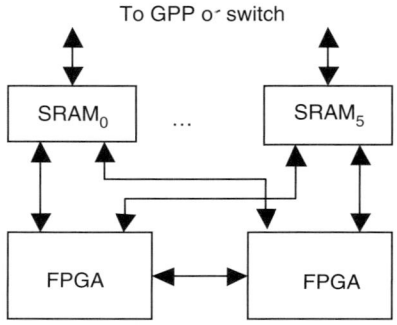

FIGURE 11.2 Conceptual drawing of the SRC MAP processor.

(11 adders, 12 subtracters, 19 multipliers, 5 squaring cores, 4 accumulators, 1 divider, and 1 square root core) in a single FPGA on the MAP. The newer Series H MAP promises to increase the available memory and memory bandwidth and will use larger, faster, FPGAs, boosting the performance even further.

The GPP element in all of SRC's reconfigurable computers is comprised of two 2.8 GHz Intel Xeon processors connected to a shared memory. The shared memory is 1 GB or more.

In SRC's basic MAPstation, there is one GPP element and one MAP processor as the RH element. Communication between the GPP element and the RH element is through DMA transfers. The transfer rate is 1.4 GB/s in each direction. It is also possible to stream data from/to the GPP element to/from the RH element. When data is streamed, the communication is overlapped with computation and the programmer can write the program as if new data elements arrive to the RH element every clock cycle. MAPstations can be connected over commodity networks to form clusters.

In SRC Hi-Bar systems, the GPP elements and RH elements are connected to SRC's Hi-Bar switch. The GPP and RH elements in this system are, then, more loosely coupled than in the basic MAPstation. There is no longer necessarily a one-to-one correspondence between GPP elements and RH elements. GPP elements can still communicate with RH elements through DMA transfers, but these transfers must go over the Hi-Bar switch, which has a bandwidth of 1.4 GB/s per port. There may also be memory connected to the switch that is shared by the GPP and RH elements. Each element reads/writes data from/to that shared memory through DMA transfers.

Silicon Graphics has developed the Reconfigurable Application Specific Computing Technology (RASC) RC100 Blade to provide hardware acceleration to their Altix systems [28]. This RH element consists of dual Xilinx Virtex 4 XC4VLX200 FPGAs, and four banks of onboard memory. The total amount of memory is anticipated to be 80 MB of SRAM or 20 GB of SDRAM. The onboard memory banks are 32 bits wide. The bandwidth between the onboard memory and the FPGA is 9.6 GB/s.

In this system, the RH elements are peers with the GPP elements on the interconnection network (SGI's NUMAlink 4) and all connect to the shared memory in the system. Through the use of NUMAlink, RH elements can communicate directly with one another and the entire system has access to the same coherency domain. The bandwidth between the FPGA and common memory over the NUMAlink is up to 12.4 GB/s. The GPP elements in this system have Intel Itanium 2s.

The Cray XD1 is another reconfigurable computer. An XD1 chassis consists of six blades. Each blade consists of one GPP element and one RH element. In the XD1, the RH element is an FPGA—a Xilinx Virtex-II Pro XC2VP50 or a Xilinx Virtex 4 XC4VLX160—and four onboard memory banks. The onboard memory banks each hold 4 or 8 MB and are 64 bits wide. The bandwidth from the onboard memory banks to the FPGA is 3.2 GB/s.

The GPP element in the XD1 consists of two AMD Opterons and shared RAM memory. The GPP element communicates with the RH element—and with other blades—through Cray's RapidArray Processor (RAP). In fact (see Figure 11.3), only one of the processors in the GPP element can communicate with the RAP that is connected to the RH element. The RH element can also use the RAP to communicate over the RapidArray interconnect with other processors and/or FPGAs. The bandwidth between the FPGA and the RAP and between the RAP and the GPP is 3.2 GB/s, whereas the bandwidth between the RAP and the rest of the interconnection network is 2 GB/s.

11.2.1.1 Massively Parallel Field-Programmable Gate Arrays

There have also been some efforts to develop reconfigurable computers consisting of only RH elements. Although these reconfigurable computers are beyond the scope of this work, we present them here for completeness. The reader is encouraged to read the cited works on these types of reconfigurable computers for more detail on their properties.

One example of a massively parallel FPGA array is the Berkeley Emulation Engine 2 [7]. In this system, an RH element consists of one Xilinx Virtex-II Pro FPGA and four onboard memories, each with a capacity of up to 1 GB. Four of these RH elements are grouped together and coupled with another RH element that is for control and has global communication interconnect interfaces that allow the RH elements

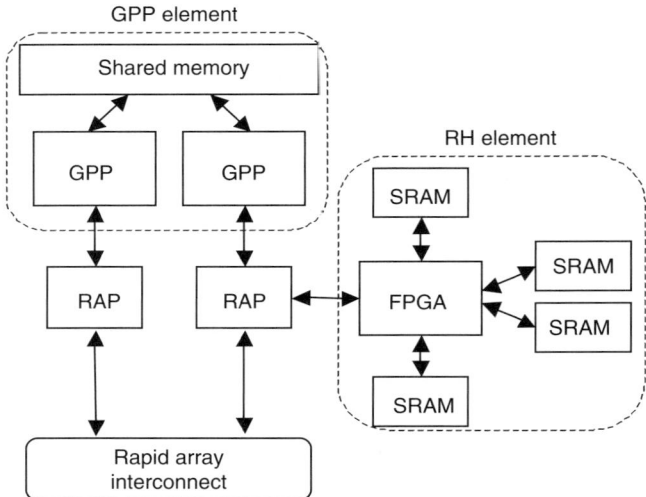

FIGURE 11.3 Conceptual drawing of a Cray XD1 blade.

to communicate with each other and any secondary devices in the system. At the system level, several interconnection schemes are possible, including Ethernet and Infiniband. For more details on BEE2 and its use in radio astronomy applications, see Reference 7.

Another massively parallel FPGA array is the TMD [23]. The RH elements in this system consist of Xilinx Virtex-II Pro FPGAs and four onboard memories, two with DRAM and two with SRAM, though in future versions the onboard memories may vary from RH element to RH element. The computing model for this reconfigurable computer is different from the others in that applications running on this reconfigurable computer are seen entirely as computing processes. The "compute engines" that do the processing are either custom hardware configured on the FPGA or embedded processors on the FPGA. Processes on the same FPGA communicate through FIFOs. Processes on different FPGAs that are on the same board use the high-performance I/O capabilities of the FPGAs that make up the system. Processes on different FPGAs on different boards use the I/O capabilities of the FPGAs that make up the system to emulate common interconnect standards like Ethernet and Infiniband to go through commodity switches. For more details on TMD, the limited MPI developed for it, and its application to molecular dynamics (MD) simulation, see Reference 23.

11.2.2 Element-Level Architectures

In our abstract model, the GPP element consists of one or more GPPs and RAM. If there are multiple GPPs within a GPP element, they are organized into a symmetric multiprocessor (SMP) system, sharing memory. This architecture is depicted in Figure 11.4. The GPP elements in each of the currently available reconfigurable computers conform to this abstract model (see Section 11.2.1). For instance, the GPP element in the SRC 6 MAPstation contains two Intel Xeon processors that share memory [34].

In our abstract model, the RH element consists of one or more reconfigurable-hardware devices, likely FPGAs, that have access to shared onboard memory. This is depicted in Figure 11.5. If there are multiple reconfigurable-hardware devices, we assume that they can locally communicate with one another and can thus be treated as one large reconfigurable-hardware device. The shared onboard memory in the RH element is organized into multiple banks that can be accessed in parallel. These banks provide single-cycle—possibly pipelined—access and are not implicitly cached. The onboard memory can be considered to be orders of magnitude smaller than the RAM in the GPP element. For example, in the Cray XD1, the

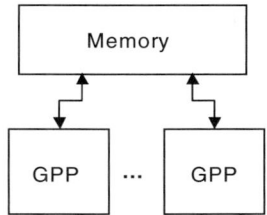

FIGURE 11.4 Abstract architecture of a GPP element.

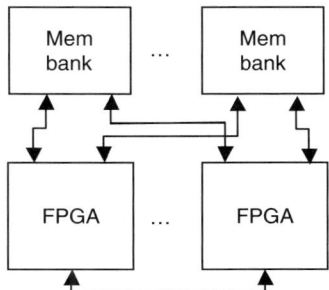

FIGURE 11.5 Abstract architecture of an RH element.

RH element has four banks of SRAM memory and each bank holds 4 MB (see Section 11.2.1). Each GPP element in the XD1, on the other hand, has access to up to 16 GB of DRAM.

11.2.3 System- and Node-Level Architectures

Reconfigurable computers can be considered as multiple instruction, multiple data (MIMD) machines by the Flynn classification [12]. That is, the GPP elements can act independently of one another and the RH elements. Similarly, the RH elements can act independently of one another and the GPP elements. Of course, it is more likely that the elements in the system will be used together to speed up the execution of a given application. Indeed, our example (see Section 11.4) application from scientific computing is accelerated by using the GPP element and RH element in a reconfigurable computer together.

The system-level architectures of reconfigurable computers, as described above, can be grouped mainly into two categories: *shared-memory* architectures and *cluster* architectures. We describe each below as well as the node-level architectures to support them.

11.2.3.1 Shared-Memory Architecture

As can be seen from the survey of available reconfigurable computers in Section 11.2.1, one common architecture for reconfigurable computers is the shared-memory architecture. In this architecture, the GPP elements and the RH elements are peers on the interconnection network and connect to a global shared memory. Each individual RH element and each individual GPP element is considered a node. Such a system naturally makes a heterogeneous reconfigurable computer because there are at least two types of nodes in the system: GPP elements and RH elements. If all of the GPP elements are identical to one another and all of the RH elements are identical to one another, then the system can be considered fairly homogeneous because there are only two different types of elements available. If, however, there are a variety of GPP elements and a variety of RH elements available, the system is then truly heterogeneous.

Silicon Graphics reconfigurable computers, as described in Section 11.2.1, fit the shared memory architectural model. The RASC RC100 blade is the RH element node and the Itanium processors are the GPP element nodes. This is a fairly homogeneous system. Similarly, the SRC Hi-Bar reconfigurable computers also fit this architectural model. The MAP processor(s) are the RH element node(s) and the dual-Pentiums and their memories make up the GPP element node(s).

11.2.3.2 Cluster Architecture

Another common architecture for reconfigurable computers is a cluster of accelerated nodes. We use this term to refer to architectures that are similar to both the traditional massively parallel processors (MPPs) and the traditional cluster of workstations architectures [18]. The cluster of accelerated nodes architecture is different from the shared-memory architecture in several ways.

One major way in which the cluster architecture is different from the shared-memory architecture is in what constitutes a node in the system. In the cluster architecture, a node consists of a GPP element, an RH element, and a high-performance intranode interconnect between them. The GPP element and the RH element are not peers on the interconnection network between nodes. Instead, the node communicates to other outside nodes as a unit. The RH element acts more as an accelerator for the GPP element.

Also, unlike in the shared-memory architecture, in the cluster architecture, there is no globally shared memory. The nodes must communicate with one another over the interconnection network using, for example, message passing. The interconnection network between the nodes can be either a commodity network (like the networks in traditional clusters of workstations) or a special purpose network (like the networks in traditional MPPs).

A cluster of SRC MAPstations is one example of this architecture. Each node in the cluster consists of one GPP element and one RH element. A commodity network would be used to connect the nodes. The Cray XD1 is another example of this architecture. In this reconfigurable computer, too, a node consists of one GPP element and one RH element. In this case, the nodes are connected through Cray's RapidArray Interconnect system.

11.2.3.3 Abstract Model and System-Level Computing Model

The abstract model of a reconfigurable computer that we proceed with in the rest of this chapter is based on the cluster architecture. We assume that each node in the system will consist of one GPP element, one RH element, and high-performance intranode interconnect. Memory is distributed to the nodes. That is, there is no central memory shared among the nodes.

Within a node, the GPP element can send blocks of data to the RH element or can *stream* data to the RH element. That is, one method of communication is for the GPP element to write the data necessary for the RH element to execute its computations into the RH element's memory before the computation starts. The other method of communication is for the communication of data between the GPP element and RH element to overlap with the computation on the RH element. Similarly, the RH element can write chunks of data to the GPP element or stream data to the GPP element. We assume that the elements can perform other computational tasks while data is streaming. In practice (e.g., in the XD1 [37]), there may be a performance difference between reading and writing between elements. We assume that this problem will be dealt with at implementation time and we thus consider all communication between the elements equal.

The computing model employed is that which we term the *accelerated single-program, multiple data* (ASPMD) model. Traditionally, the single-program, multiple data model of computation, means that the same program is run on all nodes in a system (where the nodes consist of one or more GPPs and memory), but on different data sets [18]. During a given cycle, the GPPs in the nodes may be at different points in the program. There may be synchronization commands that cause the nodes to block until all processors have reached a certain point in the program. If one node (A) requires data that resides on another node (B), that data must be communicated from node (B) to node (A). In practice, this communication is often achieved through *message-passing*.

The ASPMD model is basically the same, except now each node consists of one GPP element and one RH element. At the node level, the GPP element uses the RH element as an accelerator for intensive parts of the code. Either element can communicate with other nodes, but each node communicates as one single entity.

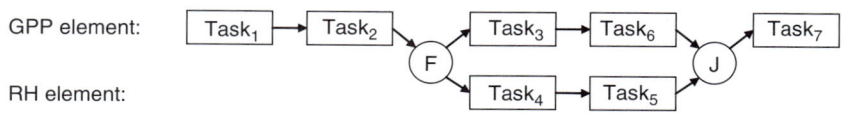

FIGURE 11.6 Example of task-level parallelism within a node (F represents forking and J represents joining).

We have chosen this model of computing for a two main reasons. First, the SPMD model is used commonly in many applications, including those in scientific computing, which are of the greatest interest to us. For example, the GAMESS computational chemistry code uses the SPMD model [21]. In the case of these applications, then, the tasks mapped to each node can be accelerated by the use of the RH element. Indeed, though not the focus of this work, even the massively parallel FPGA arrays can be programmed using this model.

The second reason that we have chosen the ASPMD model is that even though our architectural model and the ASPMD model match most closely to cluster architectures, programs written with the ASPMD computing model in mind can also run on reconfigurable computers that have shared-memory architectures. Indeed, this often happens with traditional shared-memory machines. For example, programs written in MPI can execute on SMPs. The messages are passed using shared variables. If there is not a one-to-one correspondence between RH elements and GPP elements, there may be a need for load balancing at the system level.

11.2.3.4 Node-Level Programming Model

With the ASPMD model used at the system-level, each node in the reconfigurable computer is executing the entire program. At the node level, then, it is possible to exploit task-level parallelism. One or more tasks can run on the RH element while one or more other tasks run on the GPP element. We assume, then, a fork-join programming model at the node level. An example is shown in Figure 11.6.

Note that the tasks in independent branches of the fork are independent of one another. For example, in Figure 11.6, task 5 does not depend on the output of task 3. If there were a dependency, there would need to be a join such that, for example, task 5 did not begin before task 3 completed. Since the tasks in each branch are independent of one another, there is also no data transfer between tasks in different branches.

The tasks need not take the same amount of time to execute. The program will wait until all branches of the fork have finished before proceeding with the next tasks. For better performance, load balancing between the GPP element and the RH element may also be possible, as in the *hybrid designs* in Reference 41.

11.3 Performance Modeling

The overall design flow for implementing an application on a reconfigurable computer is shown in Figure 11.7. The first step is to profile the application. In many cases, the application to be accelerated on a reconfigurable computer already exists as software. Thus, it can be profiled using existing commercial or academic tools such as TAU, VTune, gprof, or myriad others [16, 19, 27]. If the application does not already exist in software form, it can be analyzed using existing GPP modeling techniques. The results of the profiling step allow the user to determine the tasks in the application that are bottlenecks and are thus candidates for acceleration with reconfigurable hardware.

It is in the next two steps—developing designs and analyzing those designs—that the application is partitioned into hardware parts and software parts and that the hardware architectures for implementing tasks in the application are developed. During these two steps, a fairly accurate, high-level performance model is a necessity. One reason that a performance model is so important is that it may be the case that using a reconfigurable computer improves the execution time of a particular task or kernel, but the speed-up of the entire application is not significant enough to warrant the effort involved in developing a reconfigurable computer implementation. Thus, any hardware design must be evaluated in the context

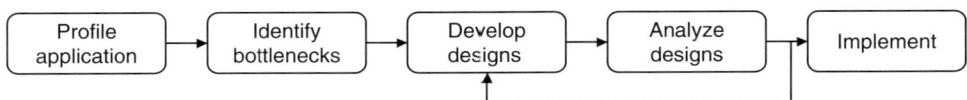

FIGURE 11.7 The design flow for mapping an application to a reconfigurable computer.

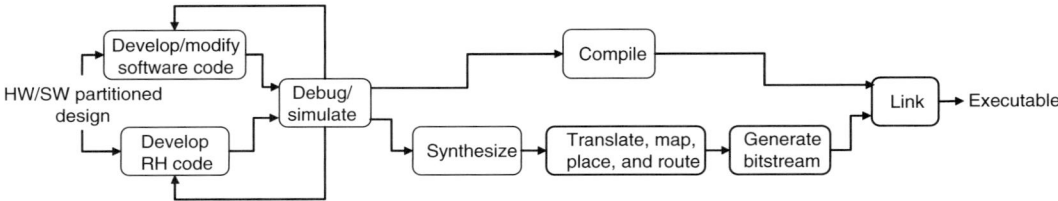

FIGURE 11.8 Steps in implementing an application on a reconfigurable computer.

of the entire application. In addition, despite advances in compiler tools and development environments, the final step in the design flow—implementation of an application on a reconfigurable computer—is still more difficult than implementation of an application on a GPP or a cluster of GPPs. This final step actually consists of several substeps, as illustrated in Figure 11.8. The steps in Figure 11.8 between synthesis and generate bitstream, inclusive, take the description of the reconfigurable-hardware design to be implemented on the RH element and convert that to a bitstream that can be used to configure the FPGA. These steps, especially place and route, take much longer than compiling software; collectively, they can take hours or even days to run.

We thus develop a performance model to address these issues. Our performance model is especially focused on the analysis step. The goal is to enable users to determine if moving tasks to the reconfigurable hardware in a reconfigurable computer will provide a significant enough speed-up to warrant the cost and effort required in the implementation step.

At the system level, the reconfigurable computer described by our abstract model behaves much like any other cluster or MPP. That is, each node executes a given program on a set of data and the nodes communicate with one another to exchange data, as necessary. For the most part, then, the system-level performance can be modeled using known techniques for modeling clusters and MPPs. See Section 11.5 for some references to existing techniques and a work specifically about system-level performance modeling of reconfigurable computers.

We thus focus our attention on node-level and element-level modeling. The element-level estimates will be used to make the node-level estimates.

11.3.1 Element-Level

11.3.1.1 General Purpose Processor Element

Modeling of GPP elements is a well-studied problem [3]. If the fork-join model of node-level programming is being used and code for the application exists already, a profiling tool can be used to get a good estimate for the time that the software tasks will take to execute. For example, in Section 11.4, we describe the use of profiling data in porting an MD application to the MAPstation. For the portions of the code that remain in software, we use the profiling results to estimate the performance of the GPP element. If profiling results do not exist, existing modeling techniques can be utilized [8].

Even if code and profiling results exist, the GPP element may incur additional costs if it must perform any *data remapping* in order to use the RH element. That is, the data in the original implementation of the application may be in a format that is well-suited for the GPP element but ill-suited for use by the RH element. In this case, the GPP element may need to reorganize the data such that it is in a suitable format for the RH element before it sends any data to the RH element. The remapping cost can be determined by implementing the remapping functions and profiling or by other, higher-level modeling techniques.

11.3.1.2 RH Element

To develop the RH element-level component of our performance model, we examine the factors that contribute to the high-performance, or lack thereof, of tasks mapped onto reconfigurable hardware. Typically, the reconfigurable hardware in RH elements is one or more FPGAs, so those devices are the focus of our examination.

The execution time of a task mapped to an FPGA is, of course, ultimately determined by the number of cycles required to finish the computation and the clock frequency at which the implementation executes. The two major factors affecting the number of cycles are pipelining and parallelism. Pipelining and parallelism are especially important in scientific applications that use floating-point arithmetic. FPGAs overcome their comparatively low clock frequencies by having the hardware specifically tailored to the application at hand—namely, by pipelining the execution and by exploiting both fine- and coarse-grained parallelism.

In scientific computing, floating-point arithmetic is often a necessity. Floating-point cores, especially those that conform to IEEE standard 754, are very complex and require a great deal of logic resources to implement. In order to achieve high clock frequencies, floating-point cores for FPGAs must be deeply pipelined: for example, double-precision adders and multipliers have about 10–20 pipeline stages while dividers and square rooters have about 40–60 stages [15]. The main difficulty with deeply pipelined units—for floating-point arithmetic or otherwise—is that their use can lead to data hazards, which may cause the pipeline to stall. Stalling the pipeline is often very detrimental to the performance of a task being executed on an FPGA. Some work addressing this problem is presented in Reference 26.

Customizing a design for a particular problem also involves exploiting parallelism, when possible. Fine-grained parallelism involves independent operations that are involved in producing the same result. This parallelism is similar to instruction-level parallelism in GPPs. For example, in the distance calculations that are part of force calculation in MD, additions and subtractions on each of the components of atomic position can occur in parallel, but the results of these parallel operations ultimately come together to provide a single result. Coarse-grained parallelism, on the other hand, involves independent operations that provide independent results. Extending the previous example, if there are two separate distance calculation pipelines, each providing independent results, then coarse-grained parallelism is being exploited.

Two factors constrain the amount of parallelism possible in a task being executed on an FPGA: limited logic resources available on the FPGA and limited bandwidth between the FPGA and memory. Logic resources available on the FPGA are still a limiting factor, especially in designs using floating-point arithmetic. Even though modern FPGAs are extremely large compared to those of even just a few years ago, floating-point cores, especially double-precision floating-point cores, still require large amounts of logic resources. This limits the amount of parallelism that can be achieved. If a design has a large number of floating-point cores, the floating-point cores will dominate the design's area requirement. So, a simple approximation to determine how many floating-point cores will fit on a chip, and thus how much parallelism is possible, is to add up the areas of the individual floating-point cores that are necessary in the design. Because the tools in the design flow perform optimizations, this approximation will likely provide an upper bound on the amount of area required by the floating-point cores. There are more detailed methods for area approximation as well (see Section 11.5).

The other limit on parallelism is bandwidth. If data cannot enter the FPGA (or leave it) fast enough in order for all the parallel computations to take place simultaneously, then the amount of parallelism in the design must be reduced. In streaming designs, it is the intranode bandwidth that can limit the parallelism. That is, the GPP element must be able to supply the RH element with enough new inputs each cycle to support the desired level of parallelism in the design mapped to the FPGAs in the RH element. In designs that do not involve streaming, it is the bandwidth between the FPGA and its onboard memory that can limit performance.

The onboard memory of the RH element in a reconfigurable computer is typically much smaller than the RAM in a GPP element, but it also has a lower access latency, in terms of cycles. In addition, this local memory is divided into b equally sized banks, where the banks can be accessed in parallel. Each access reads or writes one w-bit word. Clearly, then, a design cannot require more than bw bits worth of reading from and writing to onboard memory in a given cycle. This can limit parallelism and/or slow down a pipeline.

Another way in which accessing the onboard memory can affect performance is if there are costs associated with accessing it. In some cases, it may be that there is more than a single cycle of latency for a memory access. It is also possible that there are penalties (dead cycles) for switching a memory bank between read mode and write mode.

Because of the flexibility of FPGAs, it is difficult to develop one all-encompassing model for the hardware. Instead, all of the above factors should be taken into account, in order to estimate the number of cycles a design unit will require to produce a result. Then, knowing how many design units can be used in parallel, it is possible to estimate the total number of cycles to produce all the results.

Estimating frequency can be more challenging. The SRC 6 and 7 MAPstations require that all reconfigurable-hardware designs run at 100 and 150 MHz, respectively. Although that does not guarantee that a design will meet that frequency, it is the best that can be done for designs running on those machines. For other reconfigurable computers, there is no required frequency. The designer can use prior experience to make an estimate. For instance, in our prior work, we have seen up to a 28% decrease in achievable frequency between the lowest frequency of an individual floating-point core and a design in which floating-point cores use up most of the area in the FPGA [15]. There are other, more accurate methods, that require knowledge of the target FPGA. See Section 11.5 for details.

11.3.2 Node-Level

To determine the total node-level execution time of an application, we need to know the execution times of the individual tasks, the time spent in communication, and the time spent reconfiguring the hardware. The amount of time each task takes to execute is estimated using the models described in the last section. We need, however, to compensate for the fact that some tasks execute in parallel. For each collection of tasks within a fork-join section, the total execution time will be the time taken by the sequence of tasks that takes the longest to execute. That is, for fork-join section f, the time to execute is

$$T_f = \max(T_{\text{GPPtasks}_f}, T_{\text{RHtasks}_f}) \qquad (11.1)$$

where T_{GPPtasks_f} is the time taken by the tasks mapped to the GPP element to execute in fork-join section f and T_{RHtasks_f} is the time taken by the tasks mapped to the RH element to execute in fork-join section f. The total time for task execution is then

$$T = \sum_f T_f \qquad (11.2)$$

Take the tasks in Figure 11.6 as an example. Let x_i be the execution time of task i. Clearly, $T_1 = x_1 + x_2$ and $T_3 = x_7$ since in the first and third sections, only one element is used. In the second section, however, both elements are used. So, $T_2 = \max((x_3 + x_6), (x_4 + x_5))$. The total time for task execution is

$$\sum_{f=1}^{3} T_f = T_1 + T_2 + T_3 = (x_1 + x_2) + \max((x_3 + x_6), (x_4 + x_5)) + (x_7)$$

In order for task i to execute on a given element, the data required by task i must reside in the local memory of that element (we deal with the case of streaming shortly). Let m be the element on which the data resides and let n be the element on which task i will execute. That is, m is either "GPP element" or "RH element." If m is "GPP element," n is "RH element" and if m is "RH element," n is "GPP element." If, at the start of task i, the data task i needs reside in the local memory of m, it must be transferred to

A Hierarchical Performance Model for Reconfigurable Computers

n before task i can begin execution. This will incur a communication cost, $C_{i,m,n}$, that depends on the amount of data to be transferred, $d_{i,m,n}$, and the bandwidth, $B_{m,n}$, between m and n. That is,

$$C_{i,m,n} = \frac{d_{i,m,n}}{B_{m,n}} \quad (11.3)$$

The total time spent communicating data between elements is thus

$$C = \sum_i C_{i,m,n} \quad (11.4)$$

where the m and n are set on the basis of the node to which task i is mapped. An exceptional case is when it is possible to *stream* data from m to n. In that case, the computation and communication overlap. $C_{i,m,n}$ should be set equal to the startup cost for the streaming. Or, a coarser approximation can be made and $C_{i,m,n}$ can be set zero.

The last parameter needed for the node-level modeling is the amount of time that the reconfigurable computer must spend reconfiguring the reconfigurable hardware. Each time a different task is executed on the reconfigurable hardware, the reconfigurable hardware must be reconfigured. This takes time. How much time depends upon whether the reconfigurable hardware is fully or partially reconfigured. Also, note that it may be possible to overlap reconfiguration of the RH element with computation on the GPP element. We call the total amount of time spent in reconfiguration that cannot be overlapped with other computation R.

The total time for an application to execute on a given node is thus

$$L = T + C + R \quad (11.5)$$

11.4 Example

We now describe implementing an MD simulation application on the SRC 6 MAPstation. We will use one processor of the GPP element and one FPGA of the RH element. We model the performance and compare the predicted performance to the actual performance. We begin with a brief introduction to MD simulation, for which Reference 1 is the main reference.

11.4.1 Molecular Dynamics Simulation

Molecular dynamics is a simulation technique that is used in many different fields, from materials science to pharmaceuticals. It is a technique for simulating the movements of atoms in a system over time. The number of atoms in a system varies widely, from about 10,000 molecules in small simulations to over a billion molecules in the largest simulations.

In an MD simulation, each of the atoms i in the system has an initial position $\vec{r}_i(0)$ and an initial velocity $\vec{v}_i(0)$ at time $t = 0$. Given the initial properties of the system, such as the initial temperature, the initial density, and the volume of the system, the MD simulation determines the *trajectory* of the system from time $t = 0$ to some later time $t = t_f$. In the process, the simulation also keeps track of properties of the system such as total energy, potential energy, and kinetic energy.

In order to compute the system's trajectory, the positions of all the molecules at time $(t + \Delta t)$ are calculated on the basis of the positions of all the molecules at time t, where Δt is a small time interval, typically on the order of one femtosecond. There are many methods for calculating new positions; the most popular is the velocity Verlet algorithm. The steps in this algorithm are

1. Calculate $\vec{v}_i\left(t + \frac{\Delta t}{2}\right)$ based on $\vec{v}_i(t)$ and acceleration $\vec{a}_i(t)$
2. Calculate $\vec{r}_i(t + \Delta t)$ based on $\vec{v}_i\left(t + \frac{\Delta t}{2}\right)$

3. Calculate $\vec{a}_i(t + \Delta t)$ based on $\vec{r}_i(t + \Delta t)$ and $\vec{r}_j(t + \Delta t)$ for all atoms $j \neq i$
4. Calculate $\vec{v}_i(t + \Delta t)$ based on $\vec{a}_i(t + \Delta t)$

These steps are repeated for each iteration of the simulation. It is well known that finding the acceleration (Step 3), is the most time-consuming step in the velocity Verlet algorithm. Finding the acceleration requires finding the force acting upon each atom in the system. During force calculation, potential energy is also usually calculated.

There are two types of forces in MD simulations: bonded and nonbonded. Bonded forces are not usually the bottleneck. Nonbonded forces, which turn out to be the bottleneck in the simulation, can be calculated by various means. One of the most common is the *smooth particle mesh Ewald* (SPME) technique [11]. In this method, short-range forces are calculated using a cutoff technique. That is, for each atom i, all atoms j within the *cutoff radius*, r_c, of i exert forces on i and those atoms farther than r_c away from atom i do not. To find the atoms j that might be within r_c of atom i, we will utilize the Verlet neighbor list technique in which a list of atoms that might interact is built every x iterations, where x is usually 10–20.

Long-range electrostatic terms are broken up into a short-range (real-space) part and a long-range (reciprocal-space part). The short-range part is calculated with a cutoff method, while the long-range part is calculated with a gridded interpolation approach that makes use of the fast Fourier transform (FFT). Since they are calculated the same way, we will refer to the calculation of the real-space part and the short-range forces collectively as "the real-space part." The force atom j exerts on atom i is based upon the distance between atoms i and j as well as on the types of the atoms and their charges. Calculating a correction term is also necessary, but doing so is not a bottleneck in the simulation.

Therefore, the following tasks must be performed in each iteration of the simulation: velocity update (twice), position update, bonded force calculation, the real-space part of force calculation, the reciprocal-space part of force calculation, and correction term calculation. Once every 10 iterations, the neighbor list is rebuilt.

11.4.2 Performance Modeling

We begin with a preliminary baseline implementation of the tasks listed above, as described in Reference 25. We use that implementation to simulate two baseline simulations, one of palmitic acid in water and one of CheY protein in water. The profiles for these simulations, generated using the Oprofile profiling tool (see Reference 22), are shown in Tables 11.1 and 11.2 [25].

From the profile, we see that the real-space force calculation takes the most time and the reciprocal-space part takes the next-most time. Figure 11.9 shows the hardware/software partition that we will investigate for this application. In the figure, VU and PU represent the velocity and position update tasks, respectively; Recip and Real represent the reciprocal-space part and the real-space part of the force calculation, respectively; BF and Corr represent the bonded force calculation and force correction, respectively; and BNL represents building the neighbor list, with the dashed-line box indicating that this step is only executed on select iterations of the algorithm. With this partition, only one task is executing on the RH element while the rest execute on the GPP element.

TABLE 11.1 Profile of Software Implementation for Palmitic Acid Simulation

Task	Time (s/iteration)	% Computation time
Real-space part	1.87	56.23
Reciprocal space part	0.91	27.41
Building neighbor list	0.36	10.85
Other	0.06	2.31

TABLE 11.2 Profile of Software Implementation for CheY Protein Simulation

Task	Time (s/iteration)	% Computation time
Real-space part	1.08	65.28
Reciprocal space part	0.33	20.15
Building neighbor list	0.17	10.47
Other	0.04	2.05

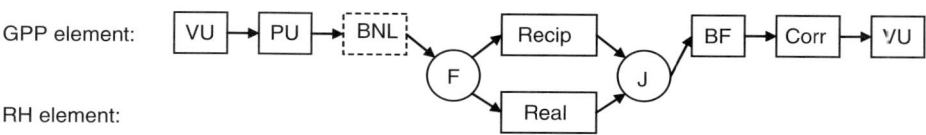

FIGURE 11.9 Assignment of MD application tasks to RH element and GPP element.

11.4.2.1 Element Level

For the GPP element, we will use the profiling results as our estimates. Very little data remapping is necessary for the RH design—described below—to function correctly, so we will not consider that cost. Thus, for the palmitic acid simulation, we estimate that the reciprocal-space part of the force calculation will take 0.91 s/iteration and the rest of the tasks mapped to the GPP element will take 0.42 s/iteration. Similarly, for the CheY protein simulation, we estimate that the reciprocal space part of the force calculation will take 0.33 s/iteration and the rest of the tasks mapped to the GPP element will take 0.21 s/iteration.

There are several possible designs for performing the real-space part of the force calculation in reconfigurable hardware. In Reference 24, it is shown that a *write-back* design is the best choice. Algorithm 11.1 is the algorithm for the write-back design. In the algorithm, positionOBM and forceOBM represent onboard memories, forceRAM represents an embedded RAM, and CALC_REAL represents applying the real-space force and potential calculation equations, as well as other necessary techniques, such as the minimum image convention [1]. The conceptual architecture for this design is shown in Figure 11.10. In this figure, the gray boxes are onboard memories, r_x represents the x component of the position vector $\vec{r}$, f_x represents the x component of the force vector $\vec{f}$, q represents atomic charge, and p represents the type of the atom. Distance Calculation represents all the calculations necessary to do the comparison in line 7 of the algorithm, Force Calculation represents the rest of the steps in the first inner loop, and Write Back represents the second inner loop.

In this design, the neighbor list is streamed in from the memory of the GPP element. The two inner loops are pipelined and the outer loop is not. Each atom i's neighbor list is traversed. If atom j is within r_c of i, the force that j exerts on i is calculated and accumulated and the equal and opposite force that i exerts on j is used to update the force on atom j. Intermediate force calculation results are stored on chip. Once all the neighbors of a given atom i have been checked, the pipeline is drained and the updated forces are written back to onboard memory in a pipelined fashion. Then, the next iteration of the outer loop can begin.

The biggest drawback of this architecture is the need to drain the pipeline after each iteration of the first inner loop. In Reference 24, we describe how this can be somewhat mitigated by using a FIFO to hold the neighbors that meet the cutoff condition in Line 7 of the algorithm. Nonetheless, in our performance modeling, we still must account for the draining of the pipeline. Since we do not know ahead of time how many neighbors will pass the condition in Line 7, we add a counter to the baseline software implementation and find the mean number of neighbors passing the condition for each atom. If we did not already have the software, we could make coarser estimates on the basis of physical properties of the atoms that would dictate how many neighbors would be within r_c of another atom at a given time.

```
 1  foreach atom i do
 2      $\vec{r}_i \leftarrow$ positionOBM[i]
 3      $\vec{f}_i \leftarrow$ forceOBM[i]
 4      $n \leftarrow 0$
 5      foreach neighbor j of i do
 6          $\vec{r}_j \leftarrow$ positionOBM[j]
 7          if $|\vec{r}_i - \vec{r}_j| < r_c$ then
 8              $\vec{f}_{ij} \leftarrow$ CALC_REAL($\vec{r}_i, \vec{r}_j, q_i, q_j, p_i, p_j$)
 9              $\vec{f}_i \leftarrow \vec{f}_i + \vec{f}_{ij}$
10              $\vec{f}_j \leftarrow$ forceOBM[j]
11              forceRAM[n] $\leftarrow \vec{f}_j - \vec{f}_{ij}$
12              $n \leftarrow n + 1$
13          end
14      end
15      forceOBM[i] $\leftarrow \vec{f}_i$
16      foreach $\vec{f}_j$ in forceRAM do
17          forceOBM[j] $\leftarrow \vec{f}_j$
18      end
19  end
```

Algorithm 11.1 Algorithm for write-back design.

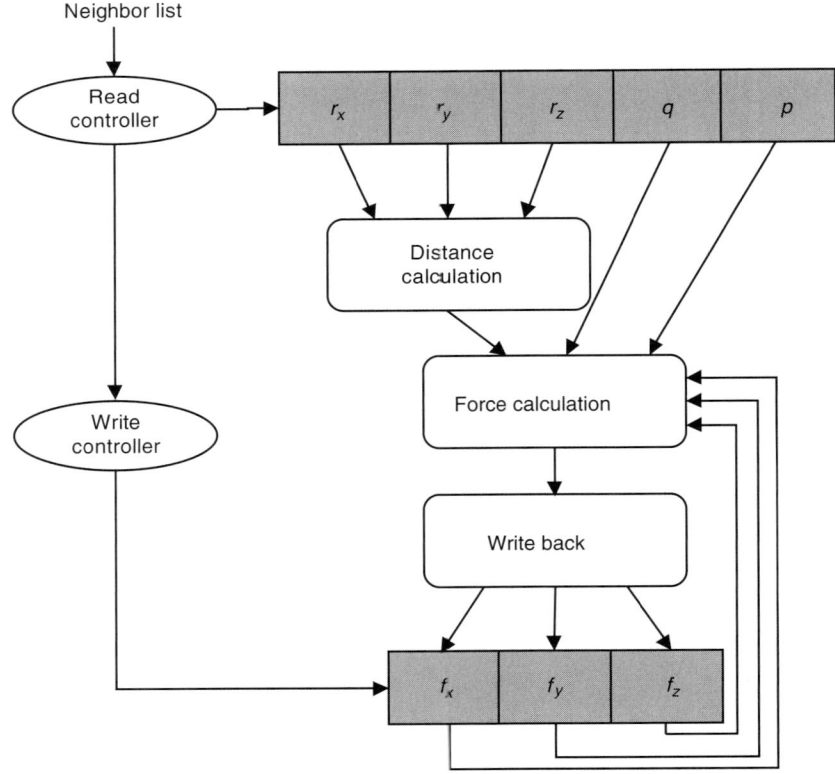

FIGURE 11.10 Conceptual architecture for the write-back design.

The equations for force calculation are quite complex, needing many floating-point operations. On the basis of estimates for the area of the floating-point cores, we determine that only one pipeline will fit in a single FPGA and, further, that the pipeline must use single-precision (32-bit) floating-point arithmetic. Even if there were sufficient area on the FPGA for double-precision (64-bit) floating-point arithmetic, we would be limited by the memory bandwidth from the onboard memories. The SRC 6 MAPstation has, at most, eight usable onboard memory banks, each 64 bits wide. Using double-precision arithmetic, there would not be enough banks to hold the three components of the position vector, three components of the force vector, atom type, and atom charge, as well as have an onboard memory bank available for streaming in the neighbor list, as is required by the MAPstation. To do the floating-point arithmetic, we use the floating-point cores provided by SRC [34].

Equation 11.6 is used to estimate the time required, in cycles, for this architecture to do the real-space part of the calculation, n is the number of atoms, N is the average number of neighbors within r_c of each atom, p is the number of pipeline stages that must be drained at the end of the first inner loop, and s is the number of cycles required to switch an onboard memory bank from read mode to write mode and vice versa. The first term gives the number of cycles spent in the pipeline of the first inner loop. The second term gives the time spent writing data back to memory. The third term gives the amount of time spent reading i atoms' data from memory in the outer loop. The last term gives the number of cycles needed to switch onboard memory modes.

$$T_{\text{RH}} = n(N + p) + n(N + 1) + n + sn \qquad (11.6)$$

For the palmitic acid simulation, $n = 52558$ and $N = 192$. On the MAPstation, $s = 2$ [34]. From the properties of SRC's floating-point cores and knowledge of the equations that need to be calculated, we estimate that $p = 211$ stages. All reconfigurable-hardware designs implemented on the MAPstation are required to run at 100 MHz. Using this information and Equation 11.6, we estimate that the real-space part of the simulation will take about 0.31 s/iteration.

For the CheY protein simulation, $n = 32932$ and $N = 176$; all the other parameters are the same. Thus, by Equation 11.6, we estimate that the real-space part of the simulation will take about 0.19 s/iteration.

11.4.2.2 Node Level

For the palmitic acid simulation, we estimate from element-level modeling that in total, the first and third collections of tasks in the task graph will take an average of 0.42 s/iteration to execute. We estimate that the real-space part of the force calculation will take 0.34 s/iteration and the reciprocal-space part of the force calculation will require 0.91 s/iteration. By Equation 11.1, the two tasks together will only need 0.91 s/iteration. Thus, by Equation 11.2, we estimate that the tasks will take 1.33 s/iteration to execute. Similarly, for the CheY protein simulation, we estimate that the tasks will take 0.54 s/iteration.

In each iteration of the simulation, we need to transfer the three components of each atom's position vector from the GPP element, where the new positions are calculated, to the RH element so that it can do its part of the force calculation. Similarly, at each iteration, the three components of the vector holding the real-space part of the force acting upon each atom must be transferred from the RH element back to the GPP element. We are using single-precision arithmetic, so it seems like each of these transfers will communicate $12n$ bytes, where n is the number of atoms. However, on the MAPstation, each transfer word is 8 bytes wide. So, we must actually transfer data in increments of 8 bytes, which means that the actual transfer must be $16n$ bytes.

For the palmitic acid simulation, $n = 52558$. The MAPstation can transfer data at a sustained rate of 1.4 GB/s. We have made the assumption that the neighbor list is streamed from the GPP element to the RH element, so we do not count it in the communication time. Thus, using Equations 11.3 and 11.4, we estimate that in each iteration, the combined communication time for both transfers will be

$$C = \frac{52558 \times 16}{1.4 \times 10^9} + \frac{52558 \times 16}{1.4 \times 10^9} = 1.2 \text{ ms/iteration}$$

Similarly, for the CheY protein simulation, $n = 32932$, so we estimate that the combined communication time in each iteration will be

$$C = \frac{32932 \times 16}{1.4 \times 10^9} + \frac{32932 \times 16}{1.4 \times 10^9} = 0.8 \text{ ms/iteration}$$

Because only one task is mapped to the RH element, the reconfigurable hardware is only configured once. This configuration happens in parallel with execution of tasks mapped to the GPP element, so we set $R = 0$.

Therefore, by Equation 11.5, we estimate that after implementation on the MAPstation, the palmitic acid simulation will take 1.33 s/iteration to execute and the CheY simulation will take 0.54 s/iteration to execute.

11.4.3 Experimental Results

We implemented the MD simulation on the MAPstation using the write-back design for the real-space part of the force calculation. During this process, we needed to make some modifications to the traditional techniques. See References 24 and 25 for details of the implementations.

Running the palmitic acid simulation, we found that it takes 1.51 s/iteration. This is a difference of about 12% from the time predicted by our model. One factor contributing to this discrepancy is that there was a threading conflict between the code generated by SRC's Carte compiler and the Intel MKL library used for performing the FFT needed by the reciprocal-space part of the force calculation [19]. This caused the FFTs on the GPP element to run more slowly in the hardware-accelerated version of the code than they do in the software-only version of the code. Accounting for such an implementation-time issue is beyond the scope of our model. Another factor that may contribute to the discrepancy between our model and the experimental results is the fact that we ignore start-up costs for data transfers, though these tend to be minor.

The CheY protein simulation takes 0.55 s/iteration, a difference of only 2% from the predicted time. For this simulation, the FFT threading conflict causes less of a discrepancy than in the palmitic acid simulation because the CheY protein simulation uses an FFT that is smaller and takes up a smaller percentage of the overall time than the FFT in the palmitic acid simulation.

From these two MD simulations, it is clear that the error in the performance model is relatively small given the high level of the modeling. This makes the performance model ideal for use early in the design process when determining if an application will get an appreciable benefit from acceleration with a reconfigurable computer.

11.5 Related Work

Reference 13 is an excellent introduction to the field of reconfigurable computing. It begins by describing low-level architectural details of FPGAs and various reconfigurable computing systems. It additionally covers languages for describing hardware designs and other compilation issues. This work also provides examples of the use of reconfigurable computing in several application fields, including signal and image processing, network security, bioinformatics, and supercomputing.

References 37 and 36 describe the implementation of a road traffic simulation application on the Cray XD1. This is one of the first implementations of a complete supercomputing application—both hardware and software—on a reconfigurable computer. While no performance model is proposed, some of the issues described in this chapter are also identified in Reference 36, such as the need to consider communication time between the RH element and the GPP element and that the performance of the whole application, not just a single accelerated kernel, should be considered when implementing an application on a reconfigurable computer.

Reference 14 presents a survey of available reconfigurable computers and then assesses how they would be used in several scientific computing applications. On the basis of the profiles of scientific computing applications from the fields of climate modeling and computational chemistry, among others, it concludes that an approach to using reconfigurable computers in scientific computing that is based solely on the use of libraries for common computational tasks will not be fruitful for large-scale scientific computing applications. Instead, it is crucial to carefully profile and partition the application—and possibly to re-write existing application code—in order to take advantage of reconfigurable computers.

In Reference 41, the hybrid design technique is presented. Hybrid designs are task designs that effectively utilize both the FPGAs and the processors in reconfigurable computers. They address key issues in the development of task designs for reconfigurable computers, most notably the exploitation of parallelism between the GPPs and FPGAs present in reconfigurable computers. At design time, load balancing is performed by spreading the floating-point operations over the GPP element and the RH element such that, based on frequency, both elements will take the same amount of time to finish their portion of the task. This is somewhat different than our approach in that tasks can be divided among elements whereas in the model presented here, each task is mapped to only one element on the node.

System-level modeling for reconfigurable computers is described in References 30 and 31. In this model, the reconfigurable computer consists of a number of distributed computing nodes connected by an interconnection network. Some or all of the nodes have reconfigurable hardware that provides acceleration. Hence, in this model, the reconfigurable computer is more heterogeneous than the reconfigurable computers considered in this chapter. The algorithms are modeled as synchronous iterative algorithms. The model considers background loading on the GPP elements in the reconfigurable computers. This work then applies the performance model to a scheduling problem. Unlike our model, it assumes that there are no conditional statements in the hardware. It also provides no element-level modeling techniques. Thus, References 30 and 31 and the work presented in this chapter are complimentary. References 30 and 31 could be used to fill in the system-level modeling that we have not addressed here.

GPP- and system-level modeling for traditional computing systems is discussed in Reference 3. The performance modeling methodology is based on application signatures, machine profiles, and convolutions. Application signatures are compact, machine-independent representations of the fundamental operations in an application. They are developed from running application code on some set of existing machines and merging the results into a machine-independent form. For example, one important part of the application signature is the memory access pattern. Machine profiles are generated from the results of low-level benchmarks. Most often, these benchmarks are focused on memory and communication performance. Convolution, as well as other techniques, are then used to combine application signatures with machine profiles in order to model the performance of the applications on the target machines.

References 10 and 20 describe estimating area and/or frequency of FPGA designs. In Reference 20, a data flow graph (DFG) is the input. The area of each operation in the data flow graph is estimated using parameterized models of common blocks, such as adders. This estimation technique is meant to be used by high-level language compilers. Similarly, Reference 10 begins with a DFG to estimate area and also frequency of applications mapped to FPGAs; the target applications are in the fields of telecommunications, multimedia, and cryptography. Algorithms are characterized by operations and their average word length and the degree of parallelism in the DFG. A "mapping model" of the target architecture is used to determine the area required by the operations in the DFG as well as the clock frequency at which the design will run. This mapping model requires knowledge of how the operations in the DFG are mapped to hardware. These techniques could be used to make the frequency and area estimates necessary in the element-level modeling phase.

11.6 Concluding Remarks

In this chapter, we have introduced reconfigurable computers and developed architectural models for reconfigurable computing. We then developed a hierarchical performance model for reconfigurable computers, describing key issues in the performance of reconfigurable-hardware designs in the process. As an example, we modeled and implemented an MD application on the SRC 6 MAPstation. The results show that the performance model described here is fairly accurate.

The emergence of reconfigurable computers makes hardware acceleration readily available to applications in scientific computing and many other fields. However, not all applications will greatly benefit from this new technology. Analytic performance modeling, as described in this chapter, is thus a necessary step that should take place early in the development process, especially given the cost and effort involved in porting applications to reconfigurable computers. Therefore, using the performance model and the techniques presented in this chapter will aid computer scientists and engineers as they develop applications that take advantage of this exciting new technology.

Acknowledgments

We wish to thank Maya Gokhale for her insights and observations that provided the initial inspiration for this work. We also wish to thank Frans Trouw for assisting us in understanding the MD application and for providing the input data for the example simulations.

This work is funded by Los Alamos National Laboratory under contract/award number 95976-001-04 3C. Access to the SRC 6 MAPstation was provided by the U.S. Army ERDC MSRC.

Xilinx, Virtex-II, Virtex-II Pro, and Virtex-4 are registered trademarks of Xilinx, Inc., San Jose, CA. MAP and Hi-Bar are registered trademarks of SRC Computers, Inc., Colorado Springs, CO. Intel, Pentium, Xeon, Itanium, and VTune are registered trademarks of Intel Corp., Santa Clara, CA. Cray is a registered trademark of Cray, Inc., Seattle, WA. Silicon Graphics, SGI, Altix, and NUMAlink are registered trademarks of Silicon Graphics, Inc., Mountain View, CA. AMD and Opteron are registered trademarks of Advanced Micro Devices, Inc., Sunnyvale, CA.

References

[1] Allen, M.P. and Tildeseley, D.J. *Computer Simulation of Liquids*. Oxford University Press, New York, 1987.
[2] Altera Corp. http://www.altera.com
[3] Bailey, D.H. and Snavely, A. Performance modeling: Understanding the present and predicting the future. In *Proc. Euro-Par 2005*, 2005.
[4] Beauchamp, M.J. et al. Embedded floating-point units in FPGAs. In *Proc. 14th ACM/SIGDA Int. Symp. on Field-Programmable Gate Arrays*, 2006, p. 12.
[5] Buell, D., Arnold, J., and Kleinfelder, W. *Splash2: FPGAs in a Custom Computing Machine*. IEEE Computer Society Press, 1996.
[6] Celoxica DK design suite. http://www.celoxica.com/products/dk/default.asp
[7] Chang, C. et al. The design and application of a high-end reconfigurable computing system. In Plaks, T., Ed., *Proc. Int. Conf. Engineering Reconfigurable Systems and Algorithms*, 2005.
[8] Chihaia, I. and Gross, T. Effectiveness of simple memory models for performance prediction. In *Proc. 2004 IEEE Int. Symp. Performance Analysis of Systems and Software*, 2004.
[9] Diniz, P. et al. Bridging the gap between compilation and synthesis in the DEFACTO system. In *Proc. Languages and Compilers for Parallel Computing Workshop*, 2001.
[10] Enzler, R. et al. High-level area and performance estimation of hardware building blocks on FPGAs. In *Proc. 10th Int. Conf. Field-Programmable Logic and Applications*, 2000, 525.
[11] Essmann, U. et al. A smooth particle mesh Ewald method. *J. Chem. Phys.*, 103, 8577, 1995.

[12] Flynn, M. Some computer organizations and their effectiveness. *IEEE Trans. Computers*, C-21, 948, 1972.

[13] Gokhale, M.B. and Graham, P.S. *Reconfigurable Computing: Accelerating Computation with Field-Programmable Gate Arrays.* Springer, Dordrecht, The Netherlands, 2005.

[14] Gokhale, M.B. et al. Promises and pitfalls of reconfigurable supercomputing. In *Proc. Int. Conf. Engineering Reconfigurable Systems and Algorithms*, 2006.

[15] Govindu, G., Scrofano, R., and Prasanna, V.K. A library of parameterizable floating-point cores for FPGAs and their application to scientific computing. In Plaks, T., Ed., *Proc. Int. Conf. on Engineering Reconfigurable Systems and Algorithms*, 2005.

[16] GNU gprof. http://www.gnu.org/software/binutils/manual/gprof-2.9.1/gprof.html

[17] Hemmert, K.S. and Underwood, K.D. An analysis of the double-precision floating-point FFT on FPGAs. In *Proc. IEEE Symp. Field-Programmable Custom Computing Machines*, 2005.

[18] Hwang, K. and Xu, Z. *Scalable Parallel Computing.* McGraw Hill, San Francisco, 1998.

[19] Intel Corp. http://www.intel.com

[20] Kulkarni, D. et al. Fast area estimation to support compiler optimizations in FPGA-based reconfigurable systems. In *Proc. 10th Annual IEEE Symp. Field-Programmable Custom Computing Machines*, 2002.

[21] Olson, R.M., Schmidt, M.W., and Gordon, M.S. Enabling the efficient use of SMP clusters: The GAMESS/DDI model. In *Proc. Supercomputing 2003*, 2003.

[22] OProfile. http://oprofile.sourceforge.net/about/

[23] Patel, A. et al. A scalable FPGA-based multiprocessor. In *Proc. 14th Annual IEEE Symp. Field-Programmable Custom Computing Machines*, 2006.

[24] Scrofano, R. et al. A hardware/software approach to molecular dynamics on reconfigurable computers. In *Proc. 14th Annual IEEE Symp. Field-Programmable Custom Computing Machines*, 2006.

[25] Scrofano, R. and Prasanna, V.K. Preliminary investigation of advanced electrostatics in molecular dynamics on reconfigurable computers. In *Proc. Supercomputing 2006*, 2006.

[26] Scrofano, R., Zhuo, L., and Prasanna, V.K. Area-efficient evaluation of arithmetic expressions using deeply pipelined floating-point cores. In Plaks, T., Ed., *Proc. Int. Conf. Engineering Reconfigurable Systems and Algorithms*, 2005.

[27] Shende, S. et al. Performance evaluation of adaptive scientific applications using TAU. In *Proc. Int. Conf. Parallel Computational Fluid Dynamics*, 2005.

[28] Silicon Graphics, Inc. http://www.sgi.com

[29] Singleterry, Jr., R.C., Sobieszczanski-Sobieski, J., and Brown, S. Field-programmable gate array computer in structural analysis: An initial exploration. In *Proc. 43rd AIAA/ASME/ASCE/AHS/ASC Structures, Structural Dynamics, and Materials Conf.*, 2002.

[30] Smith, M.C. *Analytical modeling of high performance reconfigurable computers: Prediction and analysis of system performance.* PhD thesis, University of Tennessee, 2003.

[31] Smith, M.C. and Peterson, G.D. Parallel application performance on shared high performance reconfigurable computing resources. *Performance Evaluation*, 60, 107, 2005.

[32] Smith, M.C., Vetter, J.S., and Liang, X. Accelerating scientific applications with the SRC-6 reconfigurable computer: Methodologies and analysis. In *Proc. 2005 Reconfigurable Architectures Workshop*, 2005.

[33] Smith, W.D. and Schnore, A.R. Towards an RCC-based accelerator for computational fluid dynamics applications. In Plaks, T., Ed., *Proc. Int. Conf. on Engineering Reconfigurable Systems and Algorithms*, 2003, p. 222.

[34] SRC Computers, Inc. http://www.srccomputers.com

[35] Tredennick, N. and Shimamoto, B. Reconfigurable systems emerge. In Becker, J., Platzner, M., and Vernalde, S., Eds., *Proc. 2004 Int. Conf. on Field Programmable Logic and Its Applications*, volume 3203 of Lecture Notes in Computer Science, Springer-Verlag, 2004, p. 2.

[36] Tripp, J.L. et al. Partitioning hardware and software for reconfigurable supercomputing applications: A case study. In *Proc. of Supercomputing 2005*, 2005.

[37] Tripp, J.L. et al. Metropolitan road traffic simulation on FPGAs. In *Proc. 2005 IEEE Symp. Field-Programmable Custom Computing Machines*, 2005.

[38] Underwood, K. FPGAs vs. CPUs: Trends in peak floating-point performance. In *Proc. 2004 ACM/SIGDA Twelfth Int. Symp. Field Programmable Gate Arrays*, 2004.

[39] Wolinski, C., Trouw, F., and Gokhale, M. A preliminary study of molecular dynamics on reconfigurable computers. In Plaks, T., Ed., *Proc. 2003 Int. Conf. Engineering Reconfigurable Systems and Algorithms*, 2003, p. 304.

[40] Xilinx, Inc. http://www.xilinx.com

[41] Zhuo, L. and Prasanna, V.K. Scalable hybrid designs for linear algebra on reconfigurable computing systems. In *Proc. 12th Int. Conf. Parallel and Distributed Systems*, 2006.

12
Hierarchical Performance Modeling and Analysis of Distributed Software Systems

Reda A. Ammar
University of Connecticut

12.1 Introduction 2-1
12.2 Related Work 2-2
12.3 The Hierarchical Performance Model 12-3
 Overview • Performance Modeling at System Level • Performance Modeling at Task Level
12.4 Case Study: Modeling and Evaluating Multilevel Caching in Distributed Database Systems 12-14
12.5 Multilevel Caching Algorithm 12-15
 Cache Searching • Cache Updating
12.6 Conclusions 12-19
References 12-20

12.1 Introduction

With the increased demand for electronic commerce and the continuing motivation for distributed systems to meet performance requirements when they are initially developed, a performance modeling technique is required for evaluating the communication and computation delays caused by the distributed system architecture and executing software for various hardware platforms. Performance evaluation and modeling techniques are essential when designing and implementing distributed software systems in application domains such as air-traffic control, medical systems, high-speed communications, and other real-time distributed systems. Unfortunately, many systems fail to satisfy performance requirements when they are initially developed [33–39].

Most often, software engineers design to satisfy the functional requirements. Performance does not become an issue until software implementation and integration are complete and critical paths within the system fail to meet the specified timing requirements. What if performance requirements fail to meet specified timing constraints? What if a bottleneck is known to exist in the system and the designer makes a conscious decision to modify the design with hopes of improving, but not diminishing the performance of the system? Ad hoc changes could impact the overall system design and may cause defects in functionality. If a performance requirement is not satisfied, designers should have a methodology to assist them in identifying software architecture or design modifications for improved performance.

In the absence of performance evaluation techniques, engineers must design and then implement a system before detecting critical performance defects. Waiting to identify performance defects until implementation/integration results in reduced productivity, increased project costs, slipped schedules, and the redesign of software, which may cause functional defects to be injected into the system. Software redesign may impact the understandability, maintainability, or reusability of the initial software structure. Systems that are designed for performance from the start of the software life cycle exhibit better performance than those employing a "fix-it-later" approach [34].

There are three principal methods for computer performance evaluation and analysis: direct measurement, simulation, and analytic modeling. Direct measurement is the most accurate of the methods, but requires the system to be implemented in order to collect specific performance information. Simulations are prototypes of the real system that provide a fairly accurate performance measurement, but are often time-consuming and difficult to construct. Analytic modeling, which exercises techniques such as queuing networks, Markov models, Petri-nets, state charts, and computation structure models (CSMs) [4,5,7,8,18,24,29,30], is the least expensive because hardware and software do not need to be implemented. Analytic modeling also provides insight into the variable dependencies and interactions that are difficult to determine using other methods. A combination of methods can be used to generate a performance model. This chapter focuses on an analytical software performance modeling technique, hierarchical performance modeling (HPM), used to support the evaluation of hardware platforms and software architectures for development of distributed real-time systems. The HPM includes analytical and direct measurement modeling techniques based on software performance engineering (SPE) [13,22,34,38,43,44], software performance modeling constructs [13,27,45], hardware and software codesign [28,30,40,42], queuing networks [2,3,16,17,41], software restructuring [30,42], time–cost evaluation criteria [4–6,18–20,24,25], and distributed computing concepts for conventional software systems. HPM evaluates the execution time (composed of both computation and communication costs) as a performance measure for a given distributed software system.

Related analytical performance modeling techniques are presented in the next section. A detailed discussion of the HPM framework is then presented. The final sections will present a case study of using HPM to model and analyze a distributed database system.

12.2 Related Work

In the 1970s, Sholl and Booth proposed performance-oriented development approaches [29], but most developers adopted the fix-it-later approach, which concentrated on software correctness. Performance considerations were deferred until software implementation and integration were completed and critical paths through the system failed to meet specified timing constraints. As the demands for computer resources increased, the number of developers with performance expertise decreased [31]. Thus, performance enhancements required major design changes and reimplementation of the software. Buzen presented queuing network models as a mechanism to model systems and published proficient algorithms for solving some significant models [9]. Analytical modeling techniques integrated with queuing networks were then developed to more accurately model the characteristics of software execution performance [7,8,33], thus providing the foundation for current software performance modeling techniques.

Queuing models have been used to model performance of software systems since the early 1970s [36]. Sensitivity analysis was introduced for performance models to provide a mechanism to identify performance bottlenecks and suggest which performance parameters have greatest influence on system performance. Software designs annotated with performance parameters provided the basis for sensitivity analysis of open queuing systems [22]. *Layered queuing models* provided a framework in which queuing models were utilized in a client/server fashion to model contention for both hardware and software layers [26]. *Angio traces* were recently introduced as performance traces that could be generated early in the software development life cycle to provide a mechanism for combining predictive and empirical performance modeling techniques [13]. Similar to HPM, these approaches utilize queuing networks to model contention delays of software architecture and hardware devices, but fail to generate performance models based on primitive operations for various hardware platforms for the evaluation of distributed system architecture alternatives.

Smith defined SPE and emphasized that quantitative methods should be used at the start of the software development life cycle to identify performance defects as early as possible and to eliminate designs that had unacceptable performance, thus reducing implementation and maintenance efforts [35]. Smith defined seven performance principles that can be used in conjunction with performance modeling techniques to improve the performance of systems at the start of the software development life cycle [34]. The criteria for constructing and evaluating software performance models are presented in Reference 37. The approach used by Smith consists of a software execution model and a system execution model [35]. The software execution model presents the flow of execution through the functional components. Graphical models are utilized for static analysis to derive the mean, best-, and worst-case scenarios. The system execution model represents a queuing network with the specification of environmental parameters. Smith and Williams include synchronization nodes and present an advanced model for performance evaluation of distributed software architecture [39]. These models closely relate to HPM, but like other models, do not address primitive operation costs for various hardware profiles or the generation of time–cost equations collaborated with flow value distributions for statistical performance evaluation and analysis.

Modeling software performance requires numerous parameters for each layer or model. Access and storage of this information within a layer or communication of information between layers is often complex. The maps, paths, and resources (MPR) framework used a Core model for management and storage of performance information needed by views of the framework [44]. The views presented for performance modeling provided designers with a methodology for evaluation of a variety of performance goals using the MPR framework. The author also motivated information management as an essential component for software performance modeling. Like the MPR framework, HPM also manages and distributes performance information at all levels of the HPM. Although the MPR framework supports mean execution time generation, HPM supports comprehensive statistical analysis and model-based simulation for the evaluation of software design and system architectures.

12.3 The Hierarchical Performance Model

12.3.1 Overview

Before a distributed system is designed, the performance of the system must be modeled to determine if the system can meet the user's needs based on the user demands and system architecture.

Performance models are abstractions of the functional and performance characteristics of a system. The HPM represents a vertical view of the performance modeling levels and is shown in Figure 12.1 [19].

Each of the following four levels of the HPM presents a different view of the system being specified:

1. *System level* is the highest level of abstraction. At this level queuing networks can be used to model the behavior of the system (hardware and software) and its connection with the external environment. A complete system can be modeled as a single entity, such as a queuing network. This level is concerned with input Arrivals and the interaction between software processes.
2. *Task level* is the second highest level of abstraction. The task level focuses on the interaction between software modules executing concurrently. Each module is an independent task (thread or process) that communicates with other software modules to complete a function. Thus, communication costs (between processors) and interrupt delays need to be estimated for task level analysis. Each

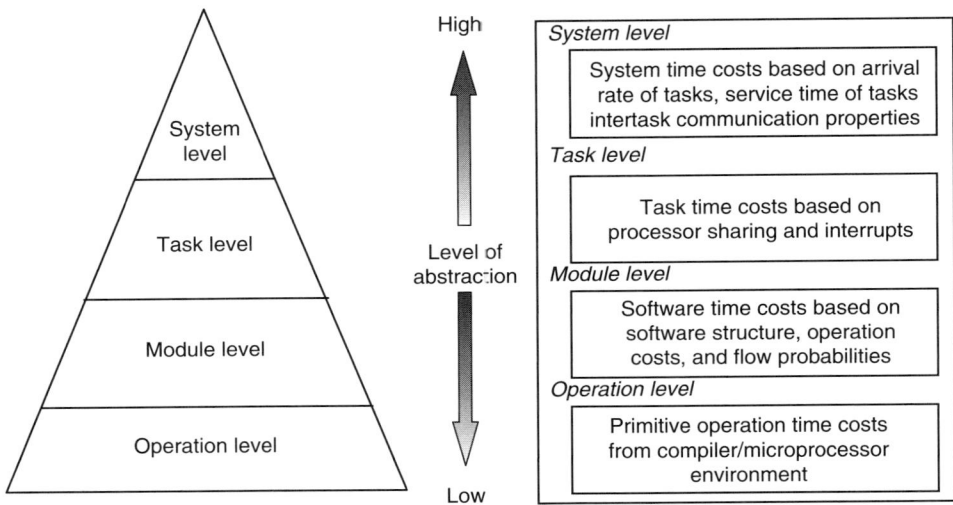

FIGURE 12.1 Overview of hierarchical performance model.

   software process at the System Level has a corresponding physical mapping at the Task Level for process to processor assignment and interprocessor communication definitions.
3. *Module level* is the level above the Operation Level. The module level calculates time–cost expressions for components of software (procedures and functions). The designer specifies software source code, a graphical design representation, or a mathematical model (equation) as the CSM. Graphical models use components to represent functional components, conditions, cases, and repetitions; links represent control flow. These models may be repeated or layered to accurately represent the software structure.
4. *Operation level* provides the lowest level of abstraction for performance modeling. The operation level provides time–cost measurements for primitive operations, built-in functions, function calls, and argument passing as specified in each statement of the specified software component. This level of analysis determines the interactions of the primitive computer instructions with the underlying processor architecture.

12.3.2 Performance Modeling at System Level

The system level is the highest level of abstraction. It represents a logical view of the application. At this level, queuing networks can be used to model the behavior of the system (hardware and software) and its connection with the external environment. A complete system can be modeled as a single entity, such as a queuing network. This level is concerned with input arrivals and the interaction between software processes.

The System Level is composed of two views, the Application View and the Node View; refer to Figure 12.2a. The Application View presents a global picture of the software system and represents communications and interactions between software processes. The Node View presents a more detailed view of the queuing properties associated with each software process. The arrival rates from external sources, software process service rates, message multipliers, coefficient of variation, number of classes, and flow probabilities for each class are performance parameters to specify for this level. A more detailed view of each "service center" (circle) is shown in the Node View at the System Level. Links represent flow of information from one software process to another. The C_{in} and C_{out} structures represent the flow probabilities of messages between processes for specified message class. The queue and servers represent the combined computation and communication waiting time and service delays, respectively. Multiple

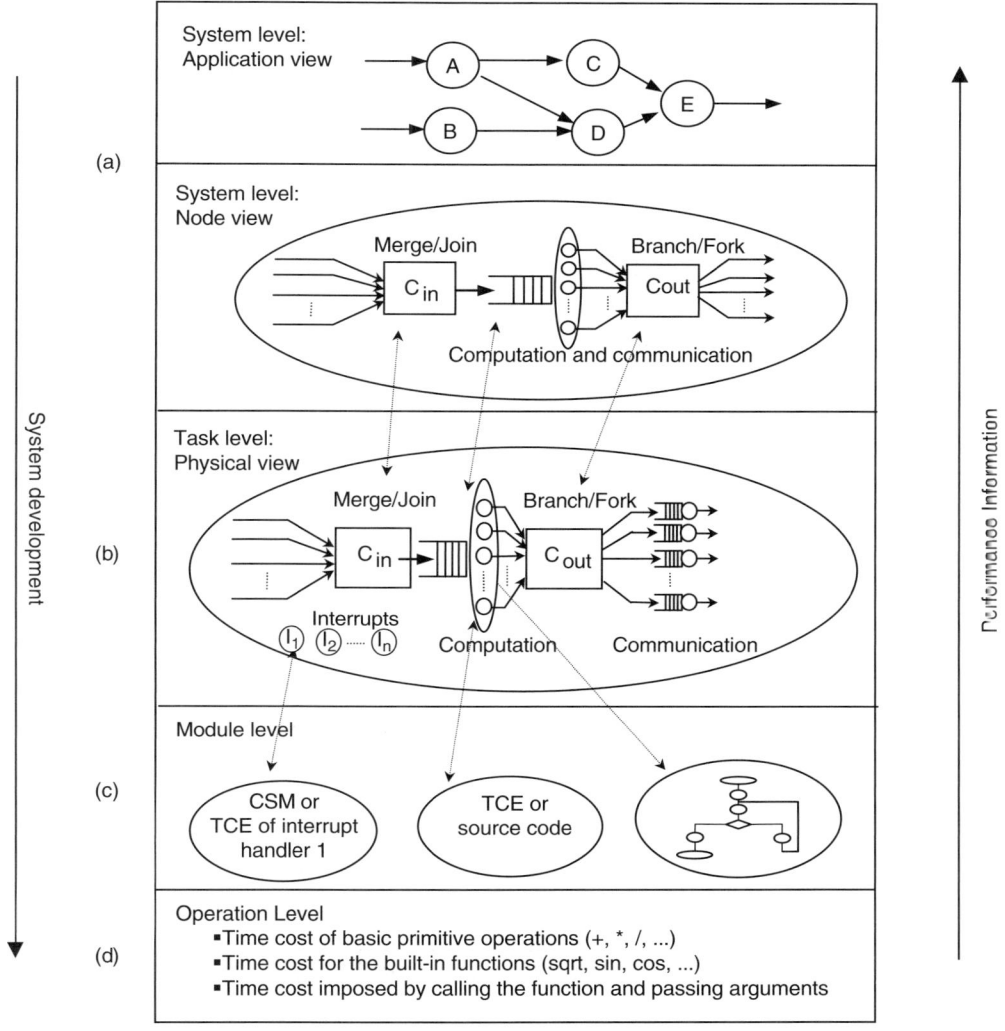

FIGURE 12.2 Detailed view of hierarchical performance model. (From Smarkusky, D., Ammar, R., and Antonios, I., *Proceedings of the 5th IEEE Symposium on Computers and Communication*, 2000.)

computation servers are represented, one for each message class. A performance equation is generated for the complete system or a user-specified subpath and can be expressed in terms of designer-specified estimated values or performance equations propagated from the Task Level.

To simplify the discussion, we assume a single class of user request and each queue is represented by an M/G/1 FCFS mode. Considering a system of n nodes, each node is represented as a two-stage tandem queue as in Figure 12.3. Each queue is a first-come first-serve, single server queue. The first queue is the execution queue that can be modeled as an M/G/1 queue. The second queue is the communication queue that can be modeled as a G/G/1 queue.

12.3.2.1 Execution Queue

Three different classes enter the execution queue:

1. The first class, class x, represents a global request received from a user with arrival rate λ_x. A portion of this class represents those subrequests that will be processed at other nodes with a probability α.

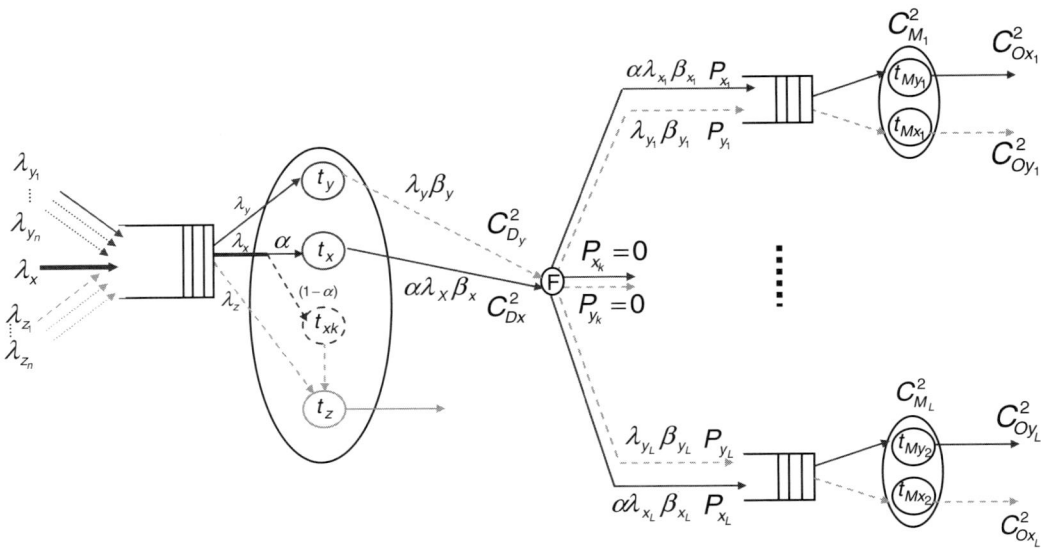

FIGURE 12.3 The distributed database model.

The other portion represents the share of node k from the request that will be processed locally with a probability $(1 - \alpha)$.

2. The second class, class y, represents subrequests received from other nodes to be evaluated at node k. The arrival rate of this class is λ_{y_i}, where i is a database node other than node k.
3. The third class, class z, represents responses from other nodes to subrequests sent by node k. The arrival rate of this class is λ_{z_i}, where i is a database node other than node k.

Hence, processing at the execution server can be divided into four main processes:

1. Processing a user's request, with expected execution time t_x and probability α.
2. Processing subrequests received from other nodes with expected execution time t_y.
3. $(1 - \alpha)$ is the probability that node k is involved in the global query; the site has to process its corresponding subrequests with expected execution time t_{x_k}.
4. Finally, the site formulates the final result of a user's request from partial results received from other nodes or it can hold its own partial result waiting for other results. The expected execution time of this process is t_z.

Calculations of the execution queue at node k can be represented as follows:

—The total arrival rate of class y is given by

$$\lambda_y = \sum_{i=1}^{n} \lambda_{y_i} \quad i \neq k \tag{12.1}$$

—The total arrival rate of class z is given by

$$\lambda_z = \sum_{i=1}^{n} \lambda_{z_i} \quad i \neq k \tag{12.2}$$

where node k can receive subrequests and responses to subrequests from all the database nodes except itself.

—The arrival rate of class x is λ_x as the node can receive only one request from a user at a time.
—The overall flow into the execution queue is given by

$$\lambda_S = \lambda_y + \lambda_x + \lambda_z \tag{12.3}$$

—The average execution time of processing subrequests received from other nodes is given by

$$\overline{t_y} = \frac{\lambda_y}{\lambda_S} t_y \tag{12.4}$$

—The average execution time of processing user requests received is given by

$$\overline{t_x} = \alpha \frac{\lambda_x}{\lambda_S} t_x \tag{12.5}$$

—The average execution time of processing part of the user request locally if site k is involved in it is given by

$$\overline{t_{x_k}} = (1 - \alpha) \frac{\lambda_x}{\lambda_S} t_{x_k} \tag{12.6}$$

—The average execution time of formalizing the final results or holding partial results received from other nodes is given by

$$\overline{t_z} = \frac{(1 - \alpha)\lambda_x + \lambda_z}{\lambda_S} t_z \tag{12.7}$$

—The total average execution time is given by

$$t_S = \overline{t_y} + \overline{t_x} + \overline{t_{x_k}} + \overline{t_z} \tag{12.8}$$

—The utilization of process y is given by

$$\rho_y = \lambda_y \overline{t_y} \tag{12.9}$$

—The utilization of the process x is given by

$$\rho_x = \alpha \lambda_x \overline{t_x} \tag{12.10}$$

—The utilization of the process x_k is given by

$$\rho_{x_k} = (1 - \alpha) \lambda_x \overline{t_{x_k}} \tag{12.11}$$

—The utilization of process z is given by

$$\rho_z = ((1 - \alpha)\lambda_x + \lambda_z) \overline{t_y} \tag{12.12}$$

—The total utilization of the execution queue over all classes is given by

$$\rho_S = \rho_y + \rho_x + \rho_{x_k} + \rho_z \tag{12.13}$$

— The service coefficient of variation for process x is given by

$$C_{\tau_x}^2 = \frac{\text{var}(t_x)}{t_x^2} \qquad (12.14)$$

— The service coefficient of variation for process x_k is given by

$$C_{\tau_{xk}}^2 = \frac{\text{var}(t_{x_k})}{t_{x_k}^2} \qquad (12.15)$$

— The service coefficient of variation for process y is given by

$$C_{\tau_y}^2 = \frac{\text{var}(t_y)}{t_y^2} \qquad (12.16)$$

— The service coefficient of variation for process y is given by

$$C_{\tau_z}^2 = \frac{\text{var}(t_z)}{t_z^2} \qquad (12.17)$$

— The overall service coefficient of variation of all classes is given in Reference 31:

$$C_\tau^2 = \frac{\sum_{i=x,y,z,x_k} p_i (1 + C_{\tau_i}^2) t_i^2}{t_S^2} - 1 \qquad (12.18)$$

$$C_\tau^2 = \frac{\left(\frac{\lambda_y}{\lambda_S}\right)\left(1 + C_{\tau_y}^2\right) t_y^2 + \left(\frac{\lambda_x}{\lambda_S}\right)\alpha \left(1 + C_{\tau_x}^2\right) t_x^2 \alpha + \left(\frac{\lambda_x}{\lambda_S}\right)(1-\alpha)\left(1 + C_{\tau_{xk}}^2\right) t_{x_k}^2 + \left(\frac{(1-\alpha)\lambda_x + \lambda_z}{\lambda_S}\right)\left(1 + C_{\tau_z}^2\right) t_z^2}{t_S^2} - 1 \qquad (12.19)$$

— As an M/G/1 queue, the total latency of the execution queue is given in Reference 31:

$$T_S = t_S \left(1 + \frac{\rho_S}{1-\rho_S}\left(\frac{1 + C_\tau^2}{2}\right)\right) \qquad (12.20)$$

— The interdeparture coefficient of variation for class x is given in Reference 31:

$$C_{D_x}^2 = 1 + \rho_x^2 \left(C_{\tau_x}^2 - 1\right) \qquad (12.21)$$

— The interdeparture coefficient of variation for class y is given by

$$C_{D_y}^2 = 1 + \rho_y^2 (C_{\tau_y}^2 - 1) \qquad (12.22)$$

12.3.2.2 Communication Queue

Two classes flow into the communication queue. Class y that represents responses to subrequests received from other nodes and Class x that represents subrequests sent to other nodes from node k as fragments from the global query submitted to node k. Having multiple channels and a point-to-point network, the communication queue calculations can be done as follows:

— Flow into channel j over all classes

$$\lambda_{Mj} = \beta_y \lambda_y P_{yj} + \alpha \beta_x \lambda_x P_{xj} \qquad (12.23)$$

where $\beta_y \lambda_y$ is the arrival rate from class y to channel j, P_{y_j} is the fraction of $\beta_y \lambda_y$ that is sent via channel to this channel, $\beta_x \lambda_x$ is the arrival rate from class x to channel j, and P_{x_j} is the fraction of $\beta_x \lambda_x$ going into the channel.

—Using the split approximations [3], the interarrival coefficient of variation of the flows going into channel j is given by

$$C_{Dj}^2 = C_{Dy_j}^2 P_{y_j} + C_{Dx_j}^2 P_{x_j} \tag{12.24}$$

—Probability of getting a message from each class to channel j

$$P_{Myj} = \frac{\beta_y \lambda_y P_{y_j}}{\lambda_{Mj}} \quad P_{Mxj} = \frac{\alpha \beta_x \lambda_x P_{x_j}}{\lambda_{Mj}} \tag{12.25}$$

—The overall probability of a message going to channel j is given by

$$P_{M_j} = \frac{\lambda_{M_j}}{\lambda_M} = \frac{\lambda_{M_j}}{\sum_{j=1}^{L} \lambda_{M_j}}, \tag{12.26}$$

where L is the number of channels.

—Expected communication service time for each class in channel j

$$t_{myj} = \frac{m_y}{R_j} \quad t_{mxj} = \frac{m_X}{R_j}, \tag{12.27}$$

where m is the message size and R is the communication channel capacity.

—Average communication service time for channel j over all classes is given by

$$t_{Mj} = P_{Myj} t_{myj} + P_{Mxj} t_{mxj} \tag{12.28}$$

—The utilization of channel j is given by

$$\rho_{Mj} = \lambda_{Mj} t_{Mj} \tag{12.29}$$

—The average communication service coefficient of variation for channel j over all classes is [31]:

$$C_{Mj}^2 \cong \frac{\left[P_{Myj} \left(1 + C_{Myj}^2\right) t_{myj}^2 \right] + \left[P_{Mxj} \left(1 + C_{Mxj}^2\right) t_{mxj}^2 \right]}{t_{Mj}^2} - 1 \tag{12.30}$$

—As a G/G/1 queue, the response time for channel j is given by

$$T_{Mj} = t_{Mj} \left[1 + \frac{\rho_{Mj} \left(C_{Dj}^2 + C_{Mj}^2\right)}{2(1 - \rho_{Mj})} \right] \tag{12.31}$$

—The overall response time for the communication queue over all channels is calculated as the weighted average of service time for each channel, and is given by

$$T_M = \sum_{j=1}^{L} P_{Mj} \cdot T_{Mj} \tag{12.32}$$

—The interdeparture coefficient of variation for channel j is given in Reference 30:

$$C_{Oj}^2 = \rho_{Mj}^2 C_{Mj}^2 + (1 - \rho_{Mj})\left(C_{Dj}^2 + \rho_{Mj}\right) \tag{12.33}$$

—The overall departure coefficient of variation for the communication queue can be estimated as [31]

$$C_O^2 = \sum_{j=1}^{L} P_{Mj} \cdot C_{Oj}^2 \tag{12.34}$$

In the aforementioned calculations, we assumed that a communication coprocessor is available. In this case, communication has no impact upon processor availability (to the execution task). We also assumed that there is no effect from the interrupts, and the available processing power is unity (e.g., $P = 1.0$). If, however, the communication is supported by the execution queue (no communications processor available), we assume that the communications task is given highest priority (e.g., over any other interrupts). In this case, we have to incorporate the processing power effect within the behavior of the service activities.

Having ρ_M that represents the processor utilization of the communications (message-delivery) queue, ρ_I, the interrupt processing power, and P_s, the fraction of the processor (utilization) that is available for the application, the effective processing power can be calculated as

$$P = P_S - \rho_M - \rho_I \tag{12.35}$$

In this case, the effective execution time for the execution queue is given by

$$t_S = \frac{\overline{t_y} + \overline{t_x} + \overline{t_{x_k}} + \overline{t_z}}{P_S} \tag{12.36}$$

where $\overline{t_y} + \overline{t_x} + \overline{t_{x_k}} + \overline{t_z}$ is the average execution time when a full processing power is available and P_S is the available processing power for the execution.

12.3.3 Performance Modeling at Task Level

The task level is the second highest level of abstraction. It focuses on the interaction between software modules executing concurrently. Each module is an independent task that communicates with other software modules to complete a function. Thus, communication costs (between processors) and interrupt delays need to be estimated for task level analysis. Each software process at the System Level has a corresponding physical mapping at the Task Level for process to processor assignment and interprocessor communication definitions.

The Task Level represents the physical views of the system under development and is also represented using a queuing model. Each Task Level model represents the hardware profile for a single software process. This level represents a single processor, unlike the System Level that represents multiple processors. Computation, communication, and interrupt costs are modeled at this level (Figure 12.2b). Multiple computation servers are represented; one for each message class. Multiple communication "service centers" are also modeled; one for each physical channel of the allocated processor. Interrupts, computation server, and communication servers each have a corresponding CSM (defined at the Module Level) that specifies the service to be performed. The C_{in} and C_{out} structure represents the arriving and departing flow probabilities for message classes on physical communication channels. Creation of this level requires that the software process defined at the System Level be allocated to a processor and assigned a processing power.

Message multipliers and flow probabilities at the Task Level are closely associated with those at the System Level. The number of messages on a per-class basis at the System Level must equal those at the Task Level. The service rates for both computation and communication, coefficient of variation, message sizes, communication rates, interrupt service times, interrupt cycle times, and communication coprocessor availability must be specified for this level dependent on one-moment or two-moment model specification. Although Task Level performance analysis is currently based on asynchronous point-to-point communications, the communication framework of HPM can be extended to support various communication protocols.

Performance parameters (service time, utilization, number of customers, etc.) are generated for computation and communication models. Interrupt utilization is also generated. Performance equation(s) are generated on a per-class basis and combined average over all classes. The tandem-queue computation and communication model is consolidated into a mathematically equivalent single-queue model for performance model generation and substitution at the System Level on a per-class basis.

12.3.3.1 Performance Modeling at Module Level

The module level is above the Operation Level. It calculates time–cost expressions for the software components (i.e., procedures and functions). The designer specifies software source code, a graphical design representation, or a mathematical model (equation) as the CSM, as shown in Figure 12.2c.

The CSM is a description of a detailed time-execution behavior of software application computations. The CSM contains two directed graphs, the control flow graph (CFG) and the data flow graph (DFG). The CFG indicates the order in which the operations and predicates are performed in order to carry out the computation, while the DFG shows the relations between operations and data. The two graphs model the time and space requirements of a computation (i.e., function, procedure, subroutine, or an entire program) (B. MacKay, 1996: #82).

The DFG is used to model the storage requirement of a computation. Hence, it is possible to use the CSM to develop both time– memory–cost expressions for a computation. In this chapter, we will focus only on the time–cost expressions.

A CFG is a complete representation of execution paths that a computation contains (B. MacKay, 1996 #82). Nodes represent actions that consume time while edges depict how the thread of execution reaches these nodes. This is represented by giving each node a time cost and each edge a flow count. The CFG is a directed, connected graph that has only one start node and only one end node. Each flow in the graph belongs to at least one cycle (B. MacKay, 1996: #82). When the computation structure is used to analyze the time–cost expressions of computations, only the CFG is considered (T. Pe. Carolyn, 1989: #73). From now on, the term CSM will be used to refer to the CFG.

12.3.3.2 Control Flow Graph Nodes

Table 12.1 (B. MacKay, 1996: #82) indicates the ten different types of nodes that can be used to characterize the execution time properties of an application algorithm. The START and END nodes have zero time cost, they serve the purpose of initiation and termination in the CFG, respectively. The START node will generate a unity flow on its outgoing edge, while the END node will absorb a unity flow on its incoming edge. The BASIC node denotes a sequential code segment, with an associated time cost, that is completely specified at the current level of refinement. The EXPAND node is used to represent an abstract operation such as a function call or procedure, which is not specified at the current level of refinement. The EXPAND node is normally used to represent an entire computation that has its own CFG specified elsewhere. It can also be used to specify other time delays that may require a task or a queue child model.

The BRANCH node denotes a data-dependent branch point of the flow control. The OR node serves as junction point for control flow through a decision. Together, the BRANCH and the OR nodes are used to construct any sequential control structure required such as IF statement, CASE statement, FOR loop, or WHILE loop. The time costs associated with a decision are normally embedded into adjacent BASIC nodes in the CFG. Thus, the BASIC and OR nodes do not, by themselves, contain a time cost.

TABLE 12.1 Control Flow Graph (CFG) Node Types

Name of node	Symbol	Input edges	Output edges	Represents	Time cost
START	Start	None	1	Activates computations at current level of refinement	None
END	END	1	None	Terminates computation at current level of refinement	None
BASIC	▭	1	1	Functional element at current level of refinement	Sequential code cost
EXPAND	▯▯	1	1	Module with CFG specified elsewhere	External CFG
BRANCH	⬡	1	2	Evaluates predicate at a branch point	None
OR	V	2	1	Serves a junction point for control flows	None

Source: MacKay, B.M., PhD dissertation, University of Connecticut, 1996.

The analysis of computations of time costs is a flow-based analysis, which relies on the general principle of conservation of flow. This principle is expressed with the following flow balance equation:

$$\sum (\text{input flows to node } v) = \sum (\text{output flows from node } v)$$

This equation holds for any BASIC, EXPAND, and DECISION nodes. START and END nodes generate and absorb a single flow unit, respectively. Given a CFG $G(V, E)$, $|V|$ number of flow balance equations can be derived. By solving the $|V|$ equations, E can be divided into two disjoint sets E' and $E - E'$ such that every flow in E' can be represented as a linear combination of the flows in $E - E'$. Flows in E' are defined as *dependent flows* and flows in $E - E'$ are defined as *independent flows*. For each cycle in the graph, which is a result of a branch point or a start node, there is a corresponding independent flow. From the basic electrical circuit theory, the following properties are known (B. MacKay, 1996: #82):

- The graph with only the dependent flows $G(V, E')$ forms a spanning tree of $G(V, E)$
- The graph $G(V, E)$ contains $|V| - 1$ dependent flows
- The graph $G(V, E)$ contains $|E| - |V| + 1$ independent flows

By directly solving the flow balance equation, we can find the flow counts for each dependent flow, given the flow counts of the independent flows.

The total time contribution from each node is given by

$$\text{Time}(v_i) = c_i \cdot b_i$$

where v_i is the flow of operation i.

The total time for the computation $G(V,E)$ is given by the sum of the time contributions of all the operation nodes:

$$\text{Time}(G(V, E)) = \sum_{i=1}^{|V|} \text{Time}(v_i) = \sum_{i=1}^{|V|} c_i \cdot b_i$$

HPM and Analysis of Distributed Software Systems

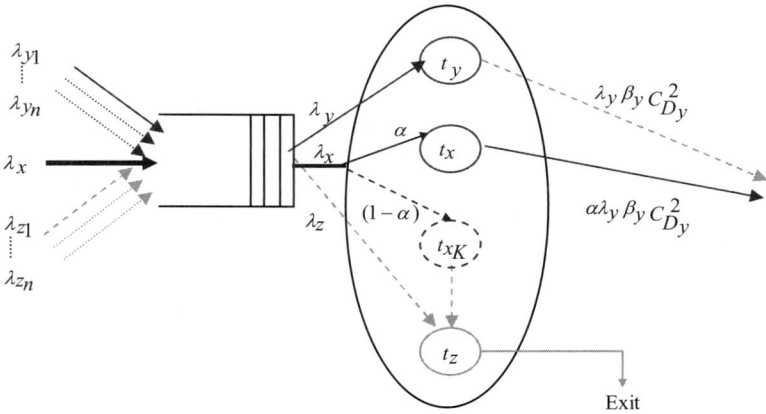

FIGURE 12.4 The execution queue.

TABLE 12.2 Flow Balance Equations for the CFG of the Distributed Query Processing

Node	Flow balance equation
v_1	$e_0 = e_1$
v_2	$e_1 = e_2 + e_3$
v_3	$e_2 + e_3 = e_4$
v_4	$e_4 = e_0$

Example: Figure 12.4 shows the CFG of distributed query processing. This graph consists of the seven basic EXPAND nodes that process the distributed query. Starting with this CFG, the flow balance equations that corresponds to each node in the graph are derived as follows (Table 12.2).

Flow e_0, which is the start flow, is always unity. The other independent flow is e_3, which corresponds to the only branching point in the graph. Solving the dependent flows in terms of the independent flows gives the following:

$$e_1 = 1, \quad e_2 = 1 - e_3, \quad e_4 = 1$$

The total time–cost equation of distributed query processing can be calculated as follows:

$$\begin{aligned}C_{\text{Decomposition}} = {} & C_{\text{Decomposition}} + C_{\text{Localization}} + C_{\text{Cache search}} + (1 - e_3)(C_{\text{Optimization}} \\ & + C_{\text{Cache update}}) + C_{\text{Exexution}} + C_{\text{Formulation}}\end{aligned} \quad (12.37)$$

Flow e_3 is a decision flow; it can be either 0 or 1 depending on the outcome of the decision node predicate. If $e_3 = 0$, this means that the query is not found in the cache and it has to be optimized and the cache should be updated with the query execution plan. If its value equals to 1, the query is found in the cache and there is no need to the optimization or cache update process.

Similarly, the time–cost equation for each EXPAND node in the aforementioned CFG will be calculated. In what follows, we will provide the CFG and the time–cost equation of each node excluding the calculation details. The time cost of each operation is evaluated from the next level.

12.3.3.3 Performance Modeling at Operation Level

This level provides the lowest level of abstraction for performance modeling (Figure 12.2d). The operation level provides time–cost measurements for primitive operations (addition, subtraction, multiplication, etc.), built-in functions (sqrt, sin, cos, etc.), function calls, and argument passing as specified in each statement of the specified software component. This level of analysis determines the interactions of the primitive computer instructions with the underlying processor architecture. Performance values are generated through actual experiments or gathered from hardware manufacturer's specifications. Primitive operations are dependent on the compiler, optimization settings, operating system, and platform profile parameters. Once the primitive time–cost values are measured, they are stored in a Primitive Assessment Table (PAT) for use by the Time–Cost Expression Deriver (TED) when time–cost equations are generated from source code or graphical models [18,19].

Using HPM to represent distributed systems provides a great flexibility in assessing application's performance attributes. This includes the computation time component for software processing, the communication delays of distributed processes, and the evaluation of hardware platform alternatives [32]. The information propagated from the operational and the model levels to the task and system levels of the hierarchy can be used to identify bottlenecks within critical paths of executing software [32]. The communication delays caused by distributed software architectures are modeled at the task level of HPM, where different alternatives of software process allocation can be evaluated. The model also supports the evaluation of interprocess communications based on point-to-point asynchronous communication channels [32]. Finally, HPM supports the assessment of hardware alternatives through the utilization of performance values of language primitives. These primitives are identified, measured, and selected on the basis of processor, compiler, and operating system characteristics [32].

12.4 Case Study: Modeling and Evaluating Multilevel Caching in Distributed Database Systems

Most query-caching techniques are confined to caching before query results in a single-level caching architecture to avoid accessing the underlying database each time a user submits the same query [10–12,15,21,23]. Although these techniques play a vital role in improving the performance of distributed queries, they do have some drawbacks. Caching query results need space and an appropriate updating strategy to always keep cached results consistent with the underlying databases. To overcome these drawbacks, we developed in References 11 and 12 a new query-caching approach that is based on caching query execution plans, subplans, and some results in a multilevel caching architecture [11]. By dividing the cache linearly into multiple levels, each level contains a subset of global queries subplans. Plans are cached in the form of interrelated but independent subplans. Subplans are interrelated as they all together represent a global execution plan. At the same time, they are independent as each subplan can be reused by itself to construct another execution plan with some other cached subplans. Each cache level will be further divided into two partitions: one for caching query execution plans (subplans) and the other for caching the results of those plans (subplans) that are frequently used, as depicted in Figure 12.1. New plans are cached on the top cache level. As the system is being used, subplans will be distributed among different cache levels according to their history.

Through this architecture, we extend the idea of query caching by doing the following:

— Instead of caching either the query results or the execution plans, we cache a combination of them. By caching some query results, the access time is reduced by avoiding accessing the underlying databases each time the same query is issued. By caching execution plans, we intend to make the cache independent of the database. Each time the data in a database are modified, we do not need to propagate these modifications to the cache. The plan will be reused whenever needed with the new values of its attributes.

—On the other hand, when caching global plans according to the proposed architecture, some of its subplans will be implicitly cached. Thus, cached plans can be reused to execute different queries instead of being used to execute only a sole general query. This will reduce the processing time by avoiding reoptimization and reconstruction of global execution plans (subplans).
—Furthermore, using multilevel cache architecture avoids searching the whole cache for a certain subplan. Hence, it reduces the search time.

12.5 Multilevel Caching Algorithm

Processing distributed queries usually go through four main layers. These layers perform functions of query decomposition, data localization, global optimization, and local queries optimization [14]. In the new approach, searching the cache is done directly after query decomposition and data localization layers. If the plan is found in the cache, the optimization layer will be avoided. If the plan (or part of it) is not found, the noncached queries (or subqueries) are sent to the optimizer to generate an execution plan. After optimization, the cache will be updated by the new plans. Three more phases are added to the generic distributed query-processing layers [14]: Searching the cache, updating it with new plans, and composing the global query plan from parts found in the cache and parts generated by the optimizer.

12.5.1 Cache Searching

As mentioned earlier, this phase starts directly after query decomposition and localization. Given a fragmented query, the algorithm tries to build an execution plan for it from the cache. We start searching the top cache level for fragment's results or subplans going down to subsequent cache levels to complete the global plan. The algorithm takes the fragmented query as an input and proceeds as follows:

—For each fragment, start searching the first cache level for its result or execution subplan. If the result is found, add it to the query result; or if the subplan itself is found, add it to the global execution plan.
—If the subplan is not found, continue searching the subsequent levels until the result or the subplan of this fragment is encountered. If the last cache level is searched without finding the result or the subplan, add it to the noncached query to be optimized.
—After finishing all fragments, start searching the cache for subplans that are used to combine the found fragments (intersite operations that join intermediate results).
—We do not have to search the cache for subplans to combine those fragments that are not found in the cache. Those subplans are added to the noncached query.
—Send the noncached query to the optimizer to get the optimized subplans and combine them with the cached subplans to generate the whole global execution plan.
—Update the cache with the new global plan.
—Send the global plan to subsequent query-processing layers.

12.5.2 Cache Updating

If the global query (or part of it) is not found in the cache, the second phase will start after the optimization layer to load the cache with the noncached optimized plans. When updating the cache, we have the following three cases:

1. *The whole query plan is not in the cache:* In this case, the whole plan will be cached at the top cache level. If this level is full, the cache will be reorganized.
2. *Part of the query plan is in the cache:* In this case, instead of caching only the noncached part, we will cache the whole plan after removing all its components that were previously found in the cache.

But if the new plan is too large to fit in one cache level, we can cache only the new part of the plan without removing the previously cached parts.

3. *The whole query plan can be constructed from the cache:* In this case, there is no need to update the cache as there are no new plans to be added to the cache.

We consider two cases: homogeneous distributed database and heterogeneous database.

12.5.2.1 Homogenous Distributed Database Systems

When implementing our model in a homogeneous distributed database system, we assume that all sites are identical in terms of hardware and software. They all use the same DBMS software with the same query optimizer. Besides, they all have the same database structure and the same multilevel caching structure and settings. The cache search time will be the same at each site. Similarly, the communication settings will also be the same at each site.

Using three different network topologies (ring, star, and bus), we measure the effect of having similar arrival rates on the query optimization time of a regular optimizer, single-level caching, and multilevel caching. Starting from 0.1, the arrival rate increments by 0.1 up to 1 query/s. We use the setting shown in Table 12.3.

TABLE 12.3 Performance Parameters of Multilevel Caching Case Study

Parameter	Symbol	Value
Number of nodes	N	5–10 nodes
Arrival rate at each node	λ_X	Incremental (0.1–1)
Probability of processing the query at other nodes	α_i	0.5
Number of query fragments	K	1
Regular optimization time for one fragment	P	1.05 s
Cache search time	t	0.001 s

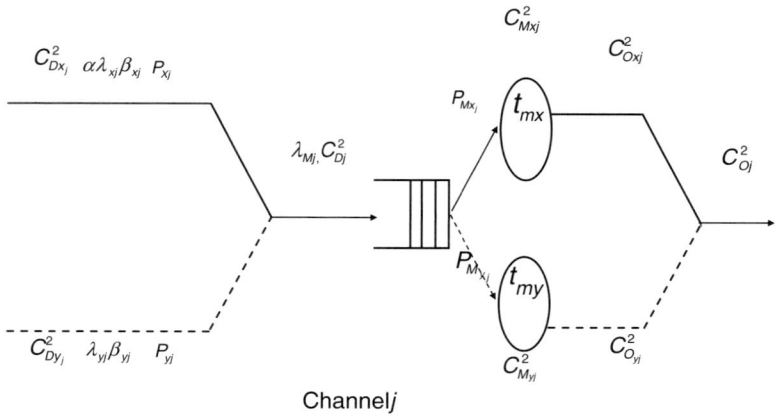

FIGURE 12.5 The communication queue for channel j.

TABLE 12.4 The Nodes Properties

Group	Number of nodes	CPU rate (GHz)	Clock cycle time (ns)
One	3	2	0.0005
Two	3	2.5	0.0004
Three	4	3	0.0006

TABLE 12.5 Hit Ration and Latency Delay for Different Cache Levels

Levels	L_1	L_2	L_3	L_4	L_5	L_6	L_7	L_8
Size (MB)	4	8	16	32	64	128	256	512
Hit ratio	17.22	20.14	23.2	26.55	29.08	29.61	29.96	30.6
Latency (cycles)	1	2	3	4	5	6	7	8

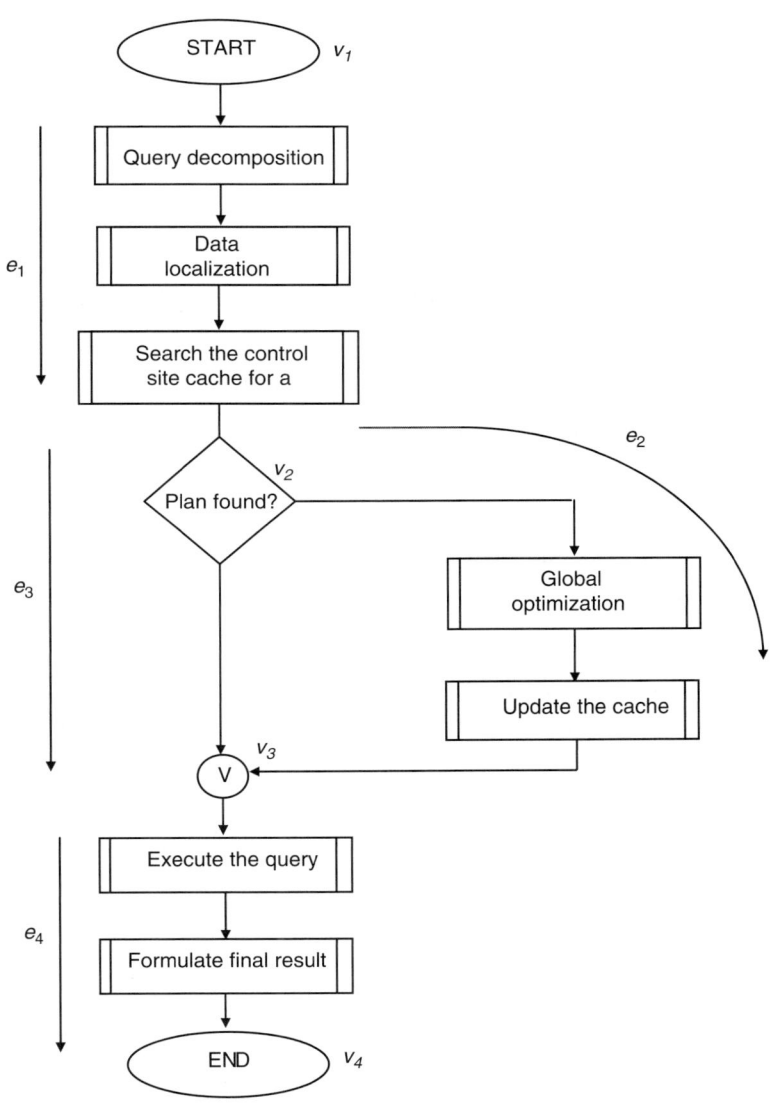

FIGURE 12.6 CFG for distributed query processing.

We used the HPM methodology described earlier to evaluate the average response time for each case of analysis. Figures 12.5 shows the effect of increasing the number of caching levels on the average optimization time of one-fragment query in a ring network formed of six nodes. We assume similar arrival rates for all the nodes that vary from 0.1 to 1 query/s. It can be seen that the average optimization time considerably decreases when the number of cache levels increases.

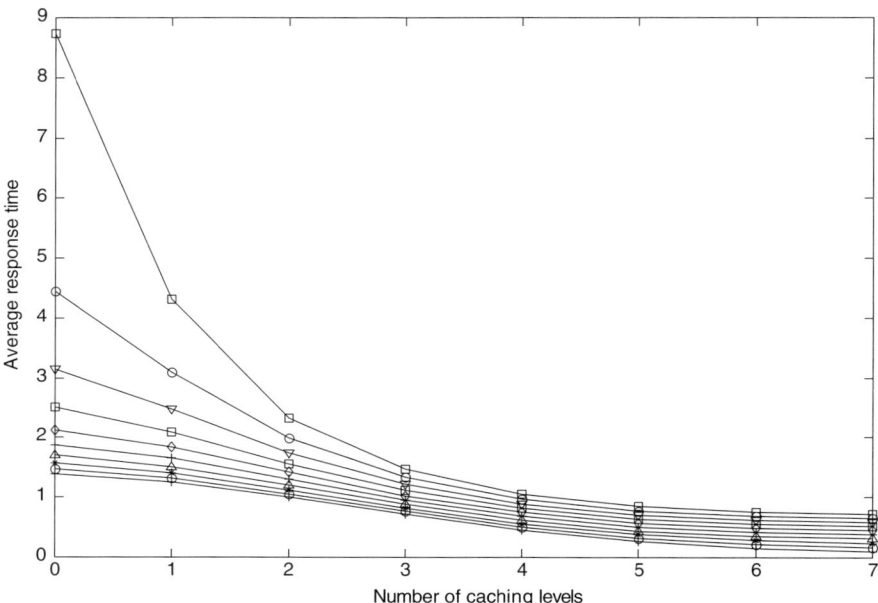

FIGURE 12.7 Optimization time proportional to different caching levels when varying the arrival rate in a 6-nodes ring network.

12.5.2.2 Heterogeneous Distributed Database System

In the case of heterogeneous distributed database systems, we implement our HPM model using the following system settings:

- The database is distributed among ten sites that are connected via a fully connected network. The ten sites are partitioned into three groups with characteristics as shown in Table 12.4.
- The cache of each site is divided linearly into eight levels. Each level is characterized by its search time and capacity. Assuming that we are using the Least Recently Used (LRU) replacement algorithm, Table 12.5 gives the hit ratio of each level, which is directly proportional to the capacity of its level.
- The arrival rate λ and the probability of processing the global query at other sites α of each node are randomly generated.
- The probability of sending part of the global query from node k to be processed at other sites is also generated randomly.

The HPM model is also used to derive the average response time as a function of different parameters (number of cache levels, etc.). Figures 12.6 and 12.7 used the derived equations [11,12] to demonstrate the efficiency of the multilevel caching technique using the aforementioned system settings when increasing the number of caching levels from 0 to 8 caching levels, varying the query size. At Figure 12.6, it can be seen that the percentage of time saving when using multilevel cache with respect to a regular optimizer significantly increases proportional to the number of caching levels. Figure 12.7 shows the percentage of time saving when using multilevel caching with respect to single-level cache when varying the query size. It is clear that increasing the number of cache levels significantly improves the optimization time no matter what query sizes we assume. For example, using six levels will reduce the optimization time 80% of the regular optimizer for any query size. It can be noticed that the improvements in the optimization time when using multilevel caching starts to approach 90% saving after the seventh and the eighth levels (Figure 12.8).

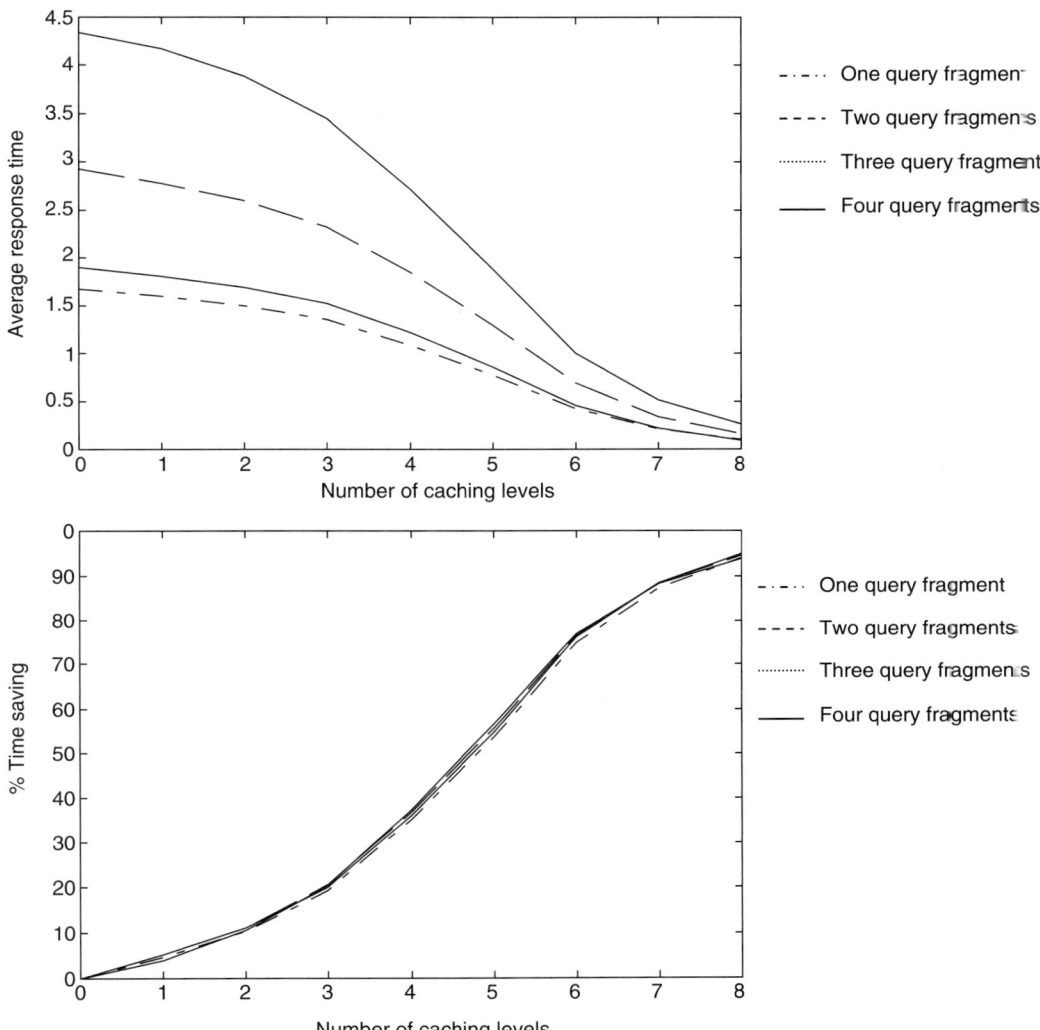

FIGURE 12.8 Percentage of time saving proportional to query size when using multilevel caching in a 10-nodes heterogeneous network.

12.6 Conclusions

In this chapter, we describe a performance analysis methodology called HPM. HPM provides an integrated scheme for interrelating the numerous variables in a composite modeling framework, encapsulating the layers of the model to focus upon specific functional regions of a computation. The layers of modeling involve the architectural level as a foundation for determining the starting behavior of a software routine's execution, progress upward through the multiple layers of the software subroutine calls, through the tightly multiple processor activities, and finally over the loosely coupled network that forms the communication network among different system modules. Each layer's model inherits the performance results of the lower, related layers as input properties for the model of the focus layer. This concept thus provides an integration of all performance variables while achievable a human-manageable focus at any view/level. The overall model then becomes a replica of the potential or real application, and can be tested, adjusted to achieve effective performance before the real application is put to work.

We also investigated a case study to evaluate the performance of a new distributed query-caching technique that is distinguished by caching query plans and results in a multilevel caching architecture. Unlike other distributed database analytical models, our model is not based on the homogeneity assumption. It can be applied to different distributed database architectures. Results of this case study show that caching query execution plans in a multilevel caching architecture in a distributed database systems significantly improves the query optimization time comparable to regular distributed query optimizers and single-level caching.

References

[1] Afrati, F.N., Li, C., and Ullman, J.D. Generating efficient plans for queries using views, *Proceedings ACM SIGMOD*, 2001.

[2] Allen, A. *Probability, Statistics, and Queueing Theory*, Academic Press, Inc., New York, 1978.

[3] Allen, A.O. Queuing models of computer systems, *IEEE Computer*, Vol. 13, April 1980.

[4] Ammar, R.A. and Qin, B. An approach to derive time costs of sequential computations, *Journal of Systems and Software*, Vol. 11, pp. 173–180, 1990.

[5] Bingman, T., MacKay, B., Schmitt, M., and Havira, M. ASAP: A tool for analytic performance prediction of software systems, *International Conference on Computer Applications in Industry and Engineering*, Orlando, FL, December 1996.

[6] Blasko, R. Hierarchical performance prediction for parallel programs, *Proceedings of the International Symposium and Workshop on Systems Engineering of Computer-Based Systems*, Tucson, AZ, March 1995.

[7] Booth, T.L. Performance optimization of software systems processing information sequences modeled by probabilistic languages, *IEEE Transactions on Software Engineering*, Vol. 5, No. 1, pp. 31–44, January 1979.

[8] Booth, T.L. Use of computation structure models to measure computation performance, *Proceedings of Conference on Simulation, Measurement, and Modeling of Computer Systems*, Boulder, CO, August 1979.

[9] Buzen, J.P. Queueing network models of multiprogramming, PhD dissertation, Harvard University, 1971.

[10] Dar, S., Franklin, M.J., and Jonsson, B. Semantic data caching and replacement, Proc. 22nd VLDB Conf., Bombay, India, 1996.

[11] Elzanfally, D.S., Eldean, A.S., and Ammar, R.A. Multilevel caching to speedup query processing in distributed databases, ISSPIT, 2003.

[12] Elzanfally, D.S., Ammar R.A., and Eldean, Ahmed S. Modeling and analysis of a multilevel caching database systems, *The 46th IEEE Midwest Symposium on Circuits and Systems*, Cairo, Egypt, December 2003.

[13] Hrischuk, C., Rolia, J., and Woodside, C. Automatic generation of a software performance model using an object-oriented prototype, *Proceedings of the IEEE International Workshop on Modeling, Analysis and Simulation of Computer and Telecommunication Systems*, Durham, NC, January 1995.

[14] Hwang, S.Y., Lee, K.S., and Chin, Y.H. Data replication in distributed system: A performance study, *The 7th International Conference on Data and Expert Systems*, 1996.

[15] Kossmann, D., Franklin, M., and Drasch, G. Cache investment: Integrating query optimization and dynamic data placement, *ACM Transaction on Database Systems*, 2000.

[16] Lazowska, E., Zahorjan, J., Graham, G., and Sevcik, K. *Quantitative System Performance*, Prentice Hall Inc., Upper Saddle River, NJ, USA, 1984.

[17] Lipsky, L. *Queueing Theory, A Linear Algebraic Approach*, Macmillan Publishing Co., New York, 1992.

[18] MacKay, B.M. and Sholl, H.A. Communication alternatives for a distributed real-time system, *Proceedings of the ISCA Computer Applications in Industry and Engineering Conference*, Honolulu, HI, November 1995.

[19] MacKay, B.M. Hierarchical modeling for parallel and distributed software applications, PhD dissertation, University of Connecticut, 1996.
[20] MacKay, B.M., Sholl, H.A., and Ammar, R.A. An overview of hierarchical modeling for parallel and distributed software applications, *International Conference on Parallel and Distributed Computing Systems*, Dijon, France, September 1996.
[21] Mitra, P. An algorithm for answering queries efficiency using views, *Proceedings in the 12th Australasian Database Conf (ADC'01)*, 2001.
[22] Opdahl, A. Sensitivity analysis of combined software and hardware performance models: Open queueing networks, *Performance Evaluation*, Vol. 22, No. 1, pp. 75–92, February 1995.
[23] Özsu, M.T. and Valduriez, P. *Principles of Distributed Database System*, 1999, Prentice Hall, USA.
[24] Qin, B., Sholl, H.A., and Ammar, R.A. Micro time cost analysis of parallel computations, *IEEE Transactions on Computers*, Vol. 40, No. 5, pp. 613–628, May 1991.
[25] Qin, B. and Ammar, R.A. An analytic approach to derive the time costs of parallel computations, *the International Journal on Computer Systems: Science & Engineering*, Vol. 8, No. 2, pp. 90–100, April 1993.
[26] Rolia, J.A. and Sevcik, K.C. The method of layers, *IEEE Transactions on Software Engineering*, Vol. 21, No. 8, pp. 689–700, August 1995.
[27] Rosiene, C.P., Ammar, R.A., and Demurjian, S.A. An evolvable and extensible modeling framework for performance, *Information and Systems Engineering*, Vol. 2, No. 3–4, December 1996, pp. 253–276.
[28] Shah, S. and Sholl, H.A. Task partitioning of incompletely specified real-time distributed systems, *ICSA 9th International Conference on Parallel and Distributed Systems*, Dijon, France, September 1996.
[29] Sholl, H.A. and Booth, T.L. Software performance modeling using computation structures, *IEEE Transactions on Software Engineering*, Vol. 1, No. 4, December 1975.
[30] Sholl, H. and Kim, S. An approach to performance modeling as an aid in structuring real-time distributed system software, *Proceedings of the 19th Hawaii International Conference on System Science*, January 1986.
[31] Sholl, H., Ammar, D., and Antonios, I. Two-moment hierarchical queueing model for the task level of the hierarchical performance modeling system, 2000, University of Connecticut, BRC-CSE TR-00-1.
[32] Smarkusky, D., Ammar, R., and Antonios, I. Hierarchical performance modeling for distributed system architecture, *Proceedings of the 5th IEEE Symposium on Computers and Communication*, 2000.
[33] Smith, C.U. The prediction and evaluation of the performance of software from extended design specifications, PhD dissertation, University of Texas, 1980.
[34] Smith, C.U. *Performance Engineering of Software Systems*. Addison-Wesley, 1990.
[35] Smith, C.U. Integrating new and "used" modeling tools for performance engineering, *Computer Performance Evaluation Modeling Techniques and Tools: Proceedings of the Fifth International Conference*, North-Holland, Amsterdam, The Netherlands, pp. 153–163, 1992.
[36] Smith, C.U. Software performance engineering, *Performance Evaluation of Computer and Communication Systems: Joint Tutorial Papers, Performance '93 and Sigmetrics '93*, Springer-Verlag, Berlin, Germany, pp. 509–536, 1993.
[37] Smith, C.U. and Williams, L.G. Software performance engineering: A case study including performance comparison with design alternatives, *IEEE Transactions on Software Engineering*, Vol. 19 No. 7, July 1993, pp. 720–741.
[38] Smith, C.U. Software performance engineering for object-oriented systems: A use case approach, Performance Engineering Services, Santa Fe, NM, and Software Engineering Research, Boulder, CO, 1998.
[39] Smith, C.U. and Williams, L.G. Performance evaluation of a distributed software architecture, *Proceedings of Computer Management Group*, Anaheim, December 1998.

[40] Suzuke, K. and Sangiovanni-Vincentelli, A. Efficient software performance estimation methods for hardware/software co-design, *Proceedings of the 33rd Annual Design Automation Conference*, Las Vegas, NV, June 1996.

[41] Trivedi, K.S. *Probability & Statistics with Reliability, Queuing and Computer Science Applications*, Prentice Hall, Inc., New Jersey, 1982.

[42] Wathne, S., Sholl, H.A., and Ammar, R.A., Task partitioning of multichannel, distributed, real-time systems, *ISCA 8th International Conference on Computer Applications in Industry*, November 1995.

[43] Williams, L.G. and Smith, C.U. Performance evaluation of software architectures, *Proceedings of the First International Workshop on Software and Performance*, Santa Fe, NM, pp. 164–177, October 1998.

[44] Woodside, C. Three-view model for performance engineering of concurrent software, *IEEE Transactions on Software Engineering*, Vol. 21, No. 9, September 1995.

[45] Zhang, P. A design and modeling environment to develop real-time, distributed software systems, PhD dissertation, University of Connecticut, 1993.

13
Randomized Packet Routing, Selection, and Sorting on the POPS Network

Jaime Davila
Sanguthevar Rajasekaran
University of Connecticut

13.1	Introduction	13-1
13.2	Preliminaries	13-3
13.3	Packet Routing	13-4
13.4	Selection	13-7
13.5	Integer Sorting	13-9
13.6	Sparse Enumeration Sorting	13-11
	Acknowledgment	13-13
	References	13-13

13.1 Introduction

A partitioned optical passive star network with parameters d and g, POPS(d, g), is a parallel computational model that was introduced in References 2, 8, 9, 12. In a POPS(d, g) there are $n = dg$ processors. These processors are divided onto g groups with d nodes in every group. We refer to these groups as $G_1, G_2, \ldots, G_g$. The j-th processor of the i-th group will be denoted as $G_i(j)$ or equivalently $P_{(i-1)d+j}$.

There is an optical coupler between every pair of groups, so that at any given time step any coupler can receive a message from a processor of the source group and broadcast it to the processors of the destination group. The coupler that connects processors in group j (the origin group) with processors in group i (the destination group) will be labeled $C(i, j)$. We use $C(*, i)$ to denote all the g couplers whose origin is G_i. Similarly, we use $C(j, *)$ to denote all the couplers whose destination is G_j.

In any time slot, a coupler (say $C(j, i)$) can receive only one message. This message gets broadcast to all the processors in G_j in the same slot. If more than one processor transmits a message to $C(j, i)$ in the same time slot, then no message is broadcast to G_j. In any time slot, a processor can receive a message from only one of the couplers that it is connected to. Also, in a time slot, a processor can send the same message to all (or a subset of) the (g) couplers that it is connected to.

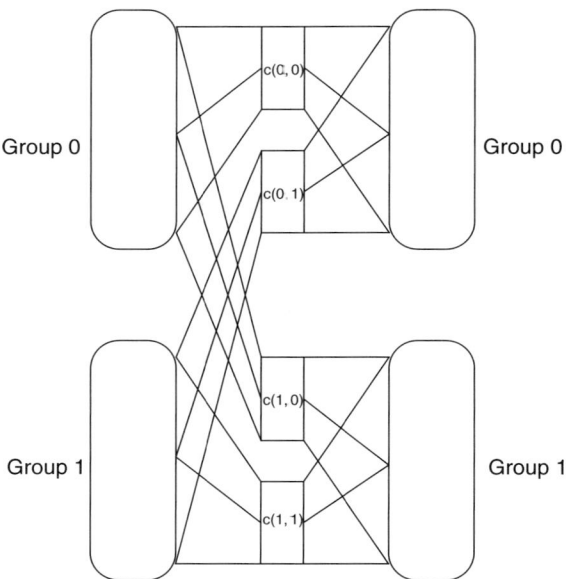

FIGURE 13.1 A POPS(3, 2) network.

Each processor has g transmitters and is associated with g couplers, one for each of the g couplers that its groups are connected to, so that there are ng transmitters and receivers in such a network. This implies that the diameter of such networks is 1, since a message can be sent from any source processor to a destination processor using the coupler that connects the two corresponding groups. Another important feature of a POPS(d, g) is that because g messages can be sent from each group, the network bandwith is g^2.

We assume that if more than one processor sends a message to the same coupler in the same time slot, then we assume that no message gets sent to any processor. This assumption is similar to the one made in the Optical Communication Parallel Computer (OCPC) [27].

Several embedding results are known for the POPS networks. For example, embedding results for rings and tori are given in Reference 8. Embeddings of rings and de Bruijn graphs into POPS networks have been presented in Reference 1. Emulations of the SIMD hypercube and the SIMD mesh on the POPS network are given in Reference 22.

Several fundamental algorithms have been devised for POPS networks including algorithms for problem such as prefix calculation [3], sorting [4], matrix multiplication [23], offline routing [11, 23], and online routing [3].

In this chapter, we present algorithms that address fundamental problems on the POPS(d, g) such as routing, selection, and particular kinds of sorting. The algorithms that we present use randomized POPS(d, g) for the case $d > g$. In general, the relationship between d and g is arbitrary, but the case $d > g$ seems more reasonable as has been pointed out in Reference 4. In addition, in some of the cases we will assume that $g = n^\epsilon$, for some fixed constant $0 < \epsilon < 1/2$. This assumption will be stated explicitly whenever it is made.

In Section 13.2, we introduce a few preliminaries and algorithms that will be used later on. In Section 13.3, we describe a near-optimal randomized algorithm that solves the online routing problem. In Section 13.4, we describe a near-optimal algorithm for the selection problem. In Section 13.5, we describe an algorithm for the integer sorting problem, and finally in Section 13.6, we present an algorithm for the case of sparse sorting.

Most of the results presented hear appear in References 5 and 20, but we improve on the presentation as well as some simplifications on the proofs and more general results were possible.

13.2 Preliminaries

The algorithms that will be presented in this chapter are of a randomized nature. In order to be able to talk about their time complexity one needs to introduce the following notation.

Definition 13.1 *We say that a function $g(x)$ (such as the time or space used by a randomized algorithm) is $\tilde{O}(f(x))$ if for all $c, \alpha > 0$ there exists a x_0 such that*

$$\Pr\left[g(x) \leq c\alpha f(x)\right] \geq 1 - x^{-\alpha} \qquad \text{for all } x \geq x_0.$$

In a similar way, we say that $g(x)$ is $\tilde{o}(f(x))$ if all $c > 0$ there exists $x_0 > 0$, such that for all $\alpha \geq 0$,

$$\Pr\left[g(x) \leq c\alpha f(x)\right] \geq 1 - x^{-\alpha} \qquad \text{for all } x \geq x_0.$$

By high probability *we mean a probability greater than or equal to $(1 - x^{-\alpha})$, for some fixed constant $\alpha \geq 0$.*

In the course of the paper we will be using the Chernoff bounds that could be stated in the following lemma.

Lemma 13.1 *If X is a binomial distribution with parameters (n, p), that is, $X \sim B(n, p)$, then*

$$\Pr(X \leq \lfloor (1-\epsilon)\bar{X} \rfloor) \leq e^{-\epsilon^2 \frac{\bar{X}}{2}}$$

$$\Pr(X \geq \lfloor (1+\epsilon)\bar{X} \rfloor) \leq e^{-\epsilon^2 \frac{\bar{X}}{3}}.$$

Note that there are g^2 couplers in a POPS(d, g) network. This means that at most g^2 different messages can be routed through the couplers in any time slot. This in turn means that at least $n/g^2 = d/g$ time slots are needed for routing a general permutation. This implies that a lower bound on problems that involve some sort of routing, will be $\lceil d/g \rceil$.

Algorithms for fundamental problems can be devised employing the emulation results known for the POPS networks. The following lemma is formulated by Sahni [22].

Lemma 13.2 *Each step of an n-node SIMD hypercube can be simulated in one time slot on a POPS(d, g) network (with $n = dg$), provided $d = 1$. If $d > 1$, then each step of the SIMD hypercube can be simulated in $2\lceil d/g \rceil$ time slots on the POPS(d, g) network.*

The above lemma can be used to obtain many algorithms for the POPS network. As an example, the following lemma is an immediate consequence of Lemma 13.2.

Lemma 13.3 *Two sorted sequences of length n each can be merged in $O(\lceil d/g \rceil \log n)$ time slots on a POPS(d, g) network (where $n = dg$). A sequence of length n can be sorted in $O(\lceil d/g \rceil \log^2 n)$ time slots on a POPS(d, g) network (using Batcher's algorithm). Also, a sequence of length n can be sorted in $O(\lceil d/g \rceil \log n \log \log n)$ time slots on a POPS(d, g) network (using Cypher and Plaxton's algorithm). Permutation routing problem can be solved using a randomized algorithm on the POPS network in $\tilde{O}(\lceil d/g \rceil \log n)$ time slots.*

The following lemma will be a nontrivial and interesting consequence of the simulation lemma and of known randomized algorithms on the Hypercube.

Lemma 13.4 *One can sort g^2 elements where g of them are in each group of the POPS(d, g) in $\tilde{O}(\log g)$ time.*

Proof. If there are g elements in every group we could think of the processors that have an element on them as a POPS(g, g). Using the randomized sorting on the Hypercube [26], we get a $\tilde{O}(\log g)$ time algorithm and using the simulation result in Lemma 13.3 we get the desired result. ∎

Lots of operations on parallel models can be seen as instances of a more general problem. In order to define this general problem, we introduce the following definition.

Definition 13.2 *Let $k_1, k_2, \ldots, k_n$ be a sequence of elements from some domain Σ. Let $\oplus$ be a binary, associative, and unit-time computable operation defined on Σ. Then, the* prefix computation problem *is defined to be that of computing the sequence $k_1, k_1 \oplus k_2, k_1 \oplus k_2 \oplus k_3, \ldots, k_1 \oplus k_2 \oplus \cdots \oplus k_n$.*

The following lemma has been proven in Reference 3 in the case of unit-time computable operations.

Lemma 13.5 *Prefix computation problem can be solved in $O(\lceil d/g \rceil + \log g)$ slots on a POPS(d, g) network.*

13.3 Packet Routing

The problem of permutation routing is defined as follows.

Definition 13.3 *There is a packet of information originating from every node in the network. The destinations of these packets form a permutation of the origins. The problem is to send all the packets to their destinations.*

The run time of any packet routing algorithm is defined to be the time taken by the last packet to reach its destination. Another criterion for measuring the performance of any routing algorithm is the queue length, which is defined to be the maximum number of packets that any node has to store while executing the algorithm.

Many packet routing algorithms have been proposed for the POPS network. Sahni [23] has presented optimal or near optimal algorithms for a class of permutations (called *bit-permute-complement permutations*). These algorithms are *offline*. A routing algorithm is said to be *offline* if the permutation to be routed is known in advance and certain computations are performed (e.g., to figure out the routes (i.e., paths) to be taken by each packet) before the actual routing algorithm starts. The preprocessing time is not included in the run time of the routing algorithm. Mei and Rizzi [11] present an offline routing algorithm that runs in one slot when $d = 1$ and $2\lceil d/g \rceil$ slots when $d > 1$.

Clearly, any sorting algorithm can be used to route permutations. Thus, as indicated in Lemma 13.3, permutation routing can be done in $O(\lceil d/g \rceil \log n \log \log n)$ slots. Also, the randomized algorithm of Valiant and Brebner [25] for routing on the hypercube in conjunction with the simulation Lemma 13.2 will imply that randomized routing on the POPS network can be accomplished in $\tilde{O}(\lceil d/g \rceil \log n)$ slots. A simple online permutation routing algorithm has been given by Datta and Soundaralakshmi [3] that takes $(8d)/g \log^2 g + (21d)/g + 3 \log g + 7$ slots.

We present a randomized algorithm that takes $\tilde{O}(\lceil d/g \rceil + \log g)$ time slots. The basic idea behind the algorithm is the two-phase routing scheme proposed by Valiant and Brebner [25]. This idea has been employed in the design of routing algorithms for numerous models of computing (e.g., References 16 and 19). In this scheme, each packet is sent to a random destination in the first phase. In the second phase, the packet is sent to its actual destination. Since the second phase is analogous to the first phase, it suffices to analyze just the first phase. The total run time of the algorithm can then be inferred to be twice the time it takes to complete the first phase.

Though the above general scheme has been employed for various networks, it is not straight forward to obtain optimal results using this scheme. Network-specific algorithmic innovations are needed to get optimal (or near optimal) run times.

In the following paragraphs, we present a summary of the first phase of the algorithm (the second phase is analogous to the first one and will not be presented). In this phase, each packet is sent to a random intermediate destination.

This phase will have two *subphases* (Phases I(a) and I(b)). These subphases will have several *stages* in which we route packets and which take $O(1)$ time slots.

We refer to any packet that has not yet been routed to a random destination as a *live packet* because the bandwidth of the network is g^2 we will route in every stage of Phase I(a) λg^2 of the live packets with high probability ($\lambda = 1/4e^2$). At the end of this phase (i.e., after $\lambda^{-1} d/g$ stages), there will be at most g^2 live packets (with high probability). Thus, the run time of Phase I(a) is $O(\lceil d/g \rceil)$ slots.

In Phase I(b), we route the remaining g^2 live packets by proceeding in several stages. In every stage of Phase I(b), a big fraction of the live packets will be routed with high probability. Thus, Phase I(b) takes $\widetilde{O}(\log \log g)$ slots.

In any stage of the algorithm there are two slots. In the first slot, each live packet π is transmitted to a random coupler C that the packet (i.e., its processor) is connected to. If more than one packet is transmitted to the same coupler (in the same slot), then C does not send any message to any of its receivers. In the same slot, each processor tunes to a random coupler it is connected to and receives the message from the coupler (if any). If the coupler $C(j, i)$ receives a single message π, then it is conceivable that more than one processor tunes to this message. Packets that have been successfully routed have to be informed so that they do not participate in future stages.

In the second slot of any stage, acknowledgments are sent to packets that have successfully reached their random destinations. If the processor $G_{i_2}^{j_2}$ receives a packet π originating from the processor $G_{i_1}^{j_1}$ through the coupler $C(i_2, i_1)$, then it sends an acknowledgment through the coupler $C(i_1, i_2)$, which is then read by $G_{i_1}^{j_1}$ in the same slot.

A processor $G_{i_1}^{j_1}$ gets a successful acknowledgment only if

- This processor chose some coupler $C(i_2, i_1)$ for transmitting its packet to and no other processor chose the same coupler.
- $G_{i_1}^{j_1}$'s message is received by some processor $G_{i_2}^{j_2}$ and no other processor receives this message.

Processors that receive acknowledgments in any stage do not participate in future stages. A message may be received by more than one processor in the same stage, but still an acknowledgment may not be sent to the packet's origin. (This happens when multiple processors in the same group try to send an acknowledgment through the same coupler in the same slot.) In this case, this message will participate in future stages. As a consequence, multiple copies of some packets may be sent to their destinations. But this is not a problem. More interesting is the fact that at least one copy of each packet will reach its destination.

In order to describe it more precisely, we write the following routine that is executed by a single processor $G_{i_1}^{i_2}$.

Routine SEND_AND_CONFIRM $G_{i_1}^{i_2}$.

1. First slot *(sending slot)*
    - Choose a random coupler $C(j_1, i_1)$ (from $C(*, i_1)$) and send the packet to this coupler.
    - Choose a random coupler $C(i_2, j_2)$ (from $C(i_2, *)$) and read the message from the coupler (if any).
2. Second slot *(confirmation slot)*
    - If a message from coupler $C(i_2, j_2)$ is read send an acknowledgment to the coupler $C(i_2, j_2)$.
    - Read coupler $C(j_1, i_1)$. If an acknowledgment is received set the processor to a *dead* state.

Using this routine we will describe our routing algorithm.

Algorithm POPS_ROUTE

Set the state of every processor to live.

Phase I(a)
for $k := 1$ to $\lceil d/g \rceil 4e^2$ do
\\\\ Each k corresponds to a stage.
Every processor $G_{i_1}^{i_2}$ with a live state decides to execute SEND_AND_CONFIRM with probability min $\left\{1, \frac{g}{d-(k-1)g/6e^2}\right\}$.

Phase I(b)
for $k := 1$ to $2 \log \log g$ do
Every processor $G_{i_1}^{i_2}$ with a live state decides to execute SEND_AND_CONFIRM

In order to prove the time complexity of this algorithm, we will need to introduce a couple of preliminary results. Note that we will assume in Lemmas 13.6 and 13.8 that $g \geq \log n$.

Lemma 13.6 *At the end of Phase I(a), the number of live packets is less than g^2 with high probability. Moreover, at the end of stage k ($0 \leq k < 4e^2 \lceil d/g \rceil$), the number of packets that are routed with acknowledgments is greater than $kg^2/4e^2$ with high probability.*

Proof. To illustrate the proof let us consider the first stage (the rest of the argument can be easily done by induction). All the n packets are live at the beginning. Let G_i be any group. Each of the d processors in G participates with probability g/d each. Thus, the expected number of participating processors from G_i is g. Since all the choices are independent we have that the number of participating processors g' has a binomial distribution $B(d, g/d)$ with mean g and by using Lemma 13.1 we obtain that

$$\Pr\left[g' - g \geq 2\alpha g^{\frac{3}{4}}\right] = \Pr\left[g' \geq g\left(1 + \frac{2\alpha}{g^{\frac{1}{4}}}\right)\right] \leq e^{-\sqrt{g}\alpha^2} < n^{-\alpha} \quad \text{for } g \geq \log n.$$

Hence, it is clear that $g' = g + \tilde{o}(g)$, for $g \geq \log n$.

Each such processor chooses a coupler at random from $(*, i)$. Consider a specific coupler $C(j, i)$. Probability that exactly one processor chooses $C(i, j)$ is $g' \times 1/g(1 - 1/g)^{g'-1} \geq 1/2e$. Hence, the number of couplers that get unique messages has a binomial distribution $B(g^2, p)$ where $p \geq 1/2e$. By using Lemma 13.1 again we get that the number of couplers that get unique messages is $\geq g^2/2e$ with high probability.

Let G_j any group. There are g couplers whose destination group is G_j. In these couplers there will be at least $g/2e$ messages with high probability. Probability that a specific coupler $C(i, j)$ with a message is picked by a unique destination processor is $g \times 1/g \times (1 - 1/g)^g \geq 1/2e$. Hence, the number of coupler whose message is picked by a unique destination has a binomial distribution $B(\tilde{g}, \tilde{p})$, where $\tilde{g} \geq g^2/2e$ with high probability and $\tilde{p} \geq 1/2e$. Thus, the number of processors that successfully get acknowledgments is $\geq g^2/4e^2$ with high probability. ∎

The statement of Lemma 13.7 was proven in Reference 21 and will be needed in the proof of Lemma 13.8.

Lemma 13.7 *Suppose that αm where $0 < \alpha < 1$ balls are thrown into m bins. With probability $1 - O(1/m)$ the number of balls that ended up in some bin with some other ball is at most $\alpha^2 m$.*

Lemma 13.8 *At the end of Phase I(b) we the number of live packets is 0 with high probability. Moreover, at the end of stage k, the number of packets that are routed with acknowledgments is greater than $(1 - 2\alpha^{2^k})g^2$ with high probability, for some $0 \leq \alpha < 1$.*

Proof. As a result of Lemma 13.6, the number of live packets after the end of Phase I(a) is less than g^2 with high probability, hence of live packets in any group is αg for some $0 \leq \alpha < 1$, with high probability.

Suppose we are at the end of the first stage of Phase I(b). By using an analysis similar to the one done in Lemma 13.6 we know that we have routed successfully $\geq g^2/4e^2$ messages with high probability.

Hence, in the second stage of Phase I(b) we have $(1 - \alpha)g$ live packets (with high probability) in every group, where $\alpha = 1/4e^2$. So, by applying Lemma 13.7 we have that

1. In the first slot $(1 - \alpha^2)g^2$ couplers get a unique message with high probability.
2. In the second slot the number of couplers that are chosen by a unique processor is $(1 - \alpha^2 - \alpha^4)g^2$ with high probability.

Hence, the number of packets that is successfully routed with acknowledgments in this stage is at least $(1 - \alpha^2 - \alpha^4)g^2 \geq (1 - 2\alpha^2)g^2$ with high probability.

Proceeding by induction we get that at stage i of Phase I (b), the number of alive packets is $(1 - 2\alpha^{2^i})g^2$ with high probability. ∎

Theorem 13.1 *Algorithm POPS_ROUTE runs in expected time $O(\lceil d/g \rceil + \log g)$. If $g \geq \log n$ we have the algorithm runs in $\tilde{O}(\lceil d/g \rceil + \log g)$ slots.*

Proof. By virtue of Lemma 13.6 we know that at the end of Phase I(a) the number of live packets is less than g^2 with high probability, and it is clear that this phase takes $(d)/g4e^2$ stages.

By using Lemma 13.8 we get that at stage i of Phase I (b) the number of alive packets is $(1 - 2\alpha^{2^i})g^2$ with high probability, hence the number of stages needed to route all the messages is less than $2 \log \log g$. Therefore, Phase I(b) takes $\tilde{O}(\log \log g)$ slots.

Finally, note that the prefix sum done at the end of Phase I(b) takes $\tilde{O}(\lceil d/g \rceil + \log g)$ by using Lemma 13.5, hence, we obtain the claimed complexity. ∎

Second phase of the algorithm. There are two ways of handling the second phase. The first way is to complete the first phase and then start the second phase. Since the second phase is analogous to the first phase, this phase will also take $\tilde{O}(\lceil d/g \rceil + \log g)$ slots.

Another way of handling the second phase is as follows. After the completion of each stage, packets that have been successfully routed to random destinations are routed to their correct destinations. Some packets may not reach their correct destinations (owing to conflicts). These packets will also participate in the next stage of the algorithm.

13.4 Selection

We start by defining explicitly the problem of selection.

Definition 13.4 *Given a sequence X of n numbers and an integer $1 \leq i \leq n$, the problem of selection is to identify the i-th smallest element of the sequence.*

Linear time sequential algorithms for selection have been devised (e.g., Reference 10). Optimal parallel algorithms are also known on various models (e.g., References 15 and 16). A common theme in deterministic and randomized selections has been to eliminate keys that cannot possibly be the element to be selected in stages (using sampling). Once the remaining keys are small in number, they could be sorted to identify the element of interest.

More precisely, a popular theme in randomized selection is the following scheme by Floyd and Rivest [6]. At any given time, the set of live keys is Y with $|Y| = N$ and at the beginning $Y = X$ and $N = n$.

1. Pick a random sample S of $o(N)$ keys at random.
2. Sort the set S. Let l be the key in S with rank $m = \lceil i(|S|/N) \rceil$. The expected rank of l in Y is i. We identify two keys l_1 and l_2 in S whose ranks in S are $m - \delta$ and $m + \delta$, respectively, δ being a "small" integer, such that the rank of l_1 in Y is $< i$, and the rank of l_2 in Y is $> i$, with high probability.
3. Next we eliminate all the keys in Y that are either $< l_1$ or $> l_2$ and also modify i appropriately.
4. Perform steps 1 through 3 of sampling and elimination until the remaining keys are small in number.
5. Finally sort the remaining keys and identify the key of interest.

In order to estimate the size of the keys in Y we need to state the following lemma, which is proven in Reference 17.

Lemma 13.9 *Let $S = \{k_1, k_2, \ldots, k_s\}$ be a random sample from a set X of size n. Let $k'_1, k'_2, \ldots, k'_s$ be the sorted order of this sample. If r_i is the rank of k'_i in X, we have that for every $\alpha \geq 1$,*

$$\text{Prob}\left(\left|r_i - i\frac{n}{s}\right| \geq c\alpha \frac{n}{\sqrt{s}} \sqrt{\log n}\right) < n^{-\alpha}.$$

Now, we are ready to present our selection algorithm. The algorithm selects the i-th smallest key from the input. N_j denotes the number of *live* keys at the beginning of the j-th iteration of the *repeat* loop. To begin with, the number of live keys, N_1, is the same as n. The number of live keys that do not get deleted in the j-th iteration is denoted by s_j. S_j is the set of sample keys employed in the j-th iteration.

Algorithm POPS_SELECT

$j = 1; N_j = n; i_j = i;$
repeat

1. Each live key decides to include itself in the sample S_j with probability $g/2N_j$.
2. Sort the keys using Lemma 13.4. Let q_j be the number of keys in the sample. Choose keys ℓ_j^1 and ℓ_j^2 from S_j with ranks $\lceil \frac{i_j q_j}{N_j} \rceil - d\sqrt{q_j \log N_j}$ and $\lceil \frac{i_j q_j}{N_j} \rceil + d\sqrt{q_j \log N_j}$, respectively, d being a constant $> \sqrt{3\alpha}$.
3. Broadcast ℓ_j^1 and ℓ_j^2 to all the processors and eliminate keys that fall outside the range $[\ell_j^1, \ell_j^2]$. Count the number, s_j, of surviving keys. By using Lemma 13.9. It can be shown that this number is $\widetilde{O}(\frac{N_j}{\sqrt{g}}\sqrt{\log N_j})$.
4. Count the number del_j of keys deleted that are $< \ell_j^1$. If the key to be selected is not one of the remaining keys (i.e., if $del_j \geq i_j$ or $i_j > del_j + s_j$), start all over again (i.e., go to step 1 with $j = 1$, $N_j = n$, and $i_j = i$).
Set $i_{j+1} = i_j - del_j$; $N_{j+1} = s_j$; and $j = j + 1$.

until $N_j \leq g$

Compact the surviving keys in G_1. This is done with a prefix computation followed by a partial permutation routing step just like in step 2. Sort the surviving keys and output the i_j-th smallest key from these.

In the following theorem, we assume that $g = n^\epsilon$ for some constant $\epsilon > 0$.

Theorem 13.2 *Algorithm POPS_SELECT takes $\widetilde{O}(\lceil d/g \rceil + \log g)$ slots on a POPS(d, g) network where $g = n^\epsilon, \epsilon > 0$.*

Proof. Step 1 takes $O(1)$ time. The distribution of the number of keys in the sample S_j is $B(N_j, g/2N_j)$. Using lemma 13.1, $|S_j|$ is $(g)/2 + \tilde{o}(g)$.

Step 2 takes $\tilde{O}(\lceil d/g \rceil + \log g)$ slots (cf. Lemma 13.5 and Theorem 13.1).

Step 3 takes $\tilde{O}(\log g)$ slots in accordance with Lemma 13.4.

Now, it suffices to show that the number of times the *repeat* loop will be executed is $\tilde{O}(1)$. The number of live keys at the end of j-th iteration is $\tilde{O}(\frac{N_j}{\sqrt{g}} \sqrt{\log N_j})$ by using Lemma 13.9. This in turn means that the number of live keys at the end of the first k iterations is $\tilde{O}\left(\frac{n}{g^{k/2}} (\log n)^{k/2}\right)$. If $g = n^\epsilon$ for some constant $\epsilon > 0$, then after $2/\epsilon$ iterations, the number of surviving keys is $\tilde{o}(g)$. Thus, with high probability, there will be only $\tilde{O}(1)$ iterations of the *repeat* loop. ∎

13.5 Integer Sorting

Definition 13.5 *Suppose we are given n numbers where each number is an integer in the range* $[1, n]$. *We will call the problem of sorting them,* integer sorting.

The problem of integer sorting is known to be solvable in linear sequential time by using Radix Sorting. Optimal parallel algorithms are also known for this problem on various parallel models (e.g., References 13, 14, 18, and 24).

In order to develop the algorithms of this section, we introduce the following definitions and lemmas that will be helpful.

Definition 13.6 *Let $a_1, a_2, \ldots, a_n$ be input keys for a sorting algorithm, and let $a_{\sigma(1)}, a_{\sigma(2)}, \ldots, a_{\sigma(n)}$ be the output of such an algorithm. We say that a sorting algorithm is* stable *if equal keys remain in the same relative order in the output as they were in the input. That is, if $a_i = a_{i+j}$ then $\sigma(i) < \sigma(i+j)$, for $i = 1, \ldots, n, j > 0$.*

Lemma 13.10 *If the sequence $a_1, a_2, \ldots, a_n$ (where each $k_i \in [0, R]$ for $i = 1, \ldots, n$) can be stable sorted using P processors in time T, then we can also stable sort n numbers in the range $[0, R^c]$ in $O(T)$ time.*

The following definition and lemma will allow us to state a principle according to which we can "share" the information among different subgroups of the POPS(d, g). If these subgroups are of size d/g we will be able to do such operation in optimal time d/g.

Definition 13.7 *Given a POPS(d, g) network, we divide every group (of size d) into subgroups of size d/g each so that there will be g such subgroups in a given group. We denote the i-th subgroup of the j-th group by S_j^i. The k-th element of S_j^i will be denoted by $S_j^i(k)$.*

Lemma 13.11 *Let $a_1, a_2, \ldots, a_n$ be an input sequence such that processor P_i contains element a_i and let $0 < h \leq d$. A POPS(d, g) can route these values so that processor P_{ih+q} has the values $(a_{ih+1}, \ldots, a_{(i+1)h})$, for $i = 1, \ldots, \lceil \frac{n}{h} \rceil$ and $q = 1, \ldots, h$ in $O\left(\frac{d}{g} + h\right)$ time.*

Proof. For simplicity of the proof we will assume h divides d and we will use the notation introduced in Definition 13.7.

The algorithm will proceed in d/g stages. In stage m of the algorithm we will be sending the value of $S_i^j(m)$ ($1 \leq i \leq g$ and $1 \leq j \leq g$) to its destination processors $G_i \left(\left\lfloor \frac{(j\frac{d}{g} + m)}{h} \right\rfloor + q \right)$ for $q = 1, \ldots, h$, in such a way that there are no collisions.

To do that we need to consider two cases. If $g \leq d/h$ we have that no two messages originating from different subgroups S_i^j and $S_{i'}^{j'}$ will have the same destination. Hence, we can route simultaneously g^2

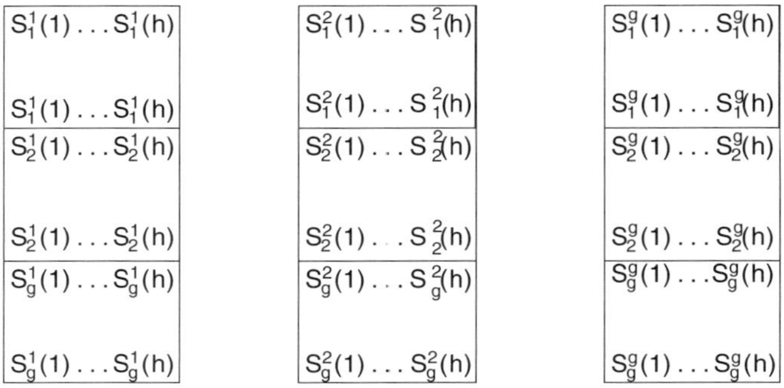

FIGURE 13.2 Illustration of Lemma 13.11 when $h = d/g$.

messages by the following two-step routing:

$$S_i^j(r) \to C(j,i) \to S_j^j(i) \to C(i,j) \to G_i\left(\left\lfloor\frac{jd + mg}{gh}\right\rfloor + q\right),$$

where $i, j \in \{1, \ldots, g\}$ and $q \in \{1, \ldots, h\}$.

So, we obtain that if $d/g \geq h$ we will expend $O(d/g)$ time.

When $g > d/h$ we have that all the messages originating from $S_i^j(r)$, $S_i^{j'}(r)$ will have the same destination if $\lfloor\frac{j}{d/h}\rfloor = \lfloor\frac{j'}{d/h}\rfloor$. In order to avoid these collisions, we will expend $\frac{g}{d/h}$ steps. In step l, we will send $g\frac{d}{h}$ messages simultaneously by using the following routing:

$$S_i^j(r) \to C(j,i) \to S_j^j(i) \to C(i,j) \to G_i\left(\left\lfloor\frac{jd + mg}{gh}\right\rfloor + q\right),$$

where $i, j \in \{1, \ldots, g\}, q \in \{1, \ldots, h\}$ and $j \equiv l \bmod d/h$.

So, we have that if $d/g < h$ the time we expend will be $O(\frac{d}{g} \times \frac{gh}{d}) = O(h)$. This means that the total time expent is $O(\max\{\frac{d}{g}, h\}) = O(\frac{d}{g} + h)$, hence the lemma follows. ∎

We are ready to describe an algorithm for sorting integers in the range $[1 \ldots h]$. We follow the approach used in **FineSort** of Reference 13. This strategy consists in dividing each group G_i into subgroups of size h. The j-th element of each one of these groups calculates the number of keys in its group that have the value j, as well as the relative order of its key with respect to its subgroup. Then, collectively all the processors count how many keys of a given value are. On the basis of this information, each processor sends its keys to its final destination.

For simplicity of the presentation and w.l.o.g we assume h divides n. Consider the following algorithm.

Algorithm POPS_INTEGER_SORT

We assume that $i \in \{0, \ldots, \frac{n}{h} - 1\}$ and $j \in \{1, \ldots, h\}$.

1. Send the values $(a_{ih+1}, \ldots, a_{(i+1)h})$ to processors $P_{ih+1}, \ldots, P_{(i+1)h}$. Note that by using Lemma 13.11 processor P_{ih+j} receives these values in h consecutive slots. In slot k it receives value a_{ih+k}.
2. Let us define $M_{i,j} := \{a_k : a_k = j \text{ and } ih + 1 \leq k \leq (i+1)h\}$.

    Processor P_{ih+j} calculates $L(i,j) := |M_{i,j}|$ as it receives the keys sent in step 1. It also calculates the relative order of its key $v := a_{ih+j}$ in $M_{i,v}$, which will be denoted by $d(i,j)$. Note that the processor does not need to store $M_{i,j}$ or $M_{i,v}$.

3. Processors P_{ih+j} send the value $L(i,j)$ to processors $P_{(j-1)\frac{n}{h}+i}$.
4. The processors collectively perform the prefix sum of

$$\begin{array}{cccc}
L(0,1), & L(1,1) & \ldots, & L(\frac{n}{h}-1,1), \\
L(0,2), & \ldots, & \ldots, & L(\frac{n}{h}-1,2), \\
\ldots, & \ldots, & \ldots, & \ldots, \\
L(0,h), & \ldots, & \ldots, & L(\frac{n}{h}-1,h).
\end{array}$$

We will call $S(i,j)$ the prefix value computed by processor $P_{(j-1)\frac{n}{h}+i}$.
5. Processors $P_{(j-1)\frac{n}{h}+i}$ send the value $S(i,j)$ to processor P_{ih+j}.
6. Send the values $(S(i,1), \ldots, S(i,h))$ to processors $P_{ih+1}, \ldots, P_{(i+1)h}$.
7. Processors P_{ih+j} reads value $S(i,v)$. It sends its key v to processor $P_{S(i,v)+d(i,j)}$.

Lemma 13.12 *Let $a_1, a_2, \ldots, a_n$ be a given input sequence, with $a_i \in [1,h]$ for $i=1, \ldots, n$ and $h \leq d$. A POPS(d,g) can stable sort these numbers in $\widetilde{O}(\lceil \frac{d}{g} \rceil + h + \log g)$ time. Furthermore, every processor needs constant memory.*

Proof. It is clear that the algorithm stable-sorts the given input. To see that it can do it within the claimed bounds we just have to notice that

1. Steps 1, 2, 6, and 7 can be done in $O(\frac{d}{g} + h)$ time by Lemma 13.11.
2. Steps 3, 5, and 7 take $\widetilde{O}(\frac{d}{g} + \log g)$ time in accordance with Lemma 13.1.
3. Step 4 takes time $O(\frac{d}{g} + \log g)$ as per Lemma 13.5. ∎

The following are simple corollaries of Lemma 13.12.

Corollary 13.1 *Let $a_1, \ldots, a_n$ be integers in the range $[1, \log n]$. A POPS(d,g) can sort these numbers in $\widetilde{O}(\lceil \frac{d}{g} \rceil + \log n)$ time.*

Corollary 13.2 *Let $a_1, \ldots, a_n$ be integers in the range $[1, d/g]$. A POPS(d,g) can sort these numbers in $\widetilde{O}(\lceil d/g \rceil + \log g)$ time.*

Theorem 13.3 *Let $a_1, \ldots, a_n$, be integers in the range $[1 \ldots n]$. A POPS(d,g) where $g = n^\epsilon$ and $0 < \epsilon < 1/2$ can stable sort these numbers in $\widetilde{O}((\lceil \frac{d}{g} \rceil + \log g))$ time.*

Proof. By using Corollary 13.2 and Lemma 13.10, the result follows immediately. Note that the constant associated with the analysis is $1/1 - 2\epsilon$. ∎

13.6 Sparse Enumeration Sorting

Definition 13.8 *The problem of sparse enumeration sorting is to sort n^α keys in a network of size n (for some fixed $\alpha < 1$).*

The basic idea to do sorting in this section is that of sampling as depicted by Frazer and McKellar [7]. This idea has been used extensively in the past to devise sequential and parallel randomized sorting algorithms. A brief description of the technique follows:

1. Randomly sample some keys from the input
2. Sort the sample (using an appropriate algorithm)
3. Partition the input using the sorted sample as splitter keys
4. Sort each part separately in parallel

One key observation regarding the size of each of the parts considered in step 3 is given in the following lemma [13]:

Lemma 13.13 *Let $X = x_1, \ldots, x_n$ be a given sequence of n keys and let $S = k_1, k_2, \ldots, k_s$ be a random sample of keys picked from X in sorted order. Define*

- $X_1 = \{k \in X : k \leq l_1\}$
- $X_j = \{k \in X : k_{j-1} < k \leq k_j\}$ for $2 \leq j \leq s$
- $X_{s+1} = \{k \in X : k > k_s\}$

The cardinality of each X_j (for $1 \leq j \leq (s+1)$) is $\widetilde{O}\left(\frac{n}{s} \log n\right)$.

Following this strategy, we can write the following algorithm:

Algorithm POPS_SPARSE_SORT

1. Each key decides to include itself in the sample with probability $g^2/2n^\alpha$.
2. Sort the sample keys and call the ordered keys $k_1, \ldots, k_s$, where s is the size of the sample. Send the first g keys to the first g processors in G_1, the next g keys to the first g processors in G_2, and so on.
3. Send the sample keys $k_g, k_{2g}, \ldots, k_{\lfloor s/g \rfloor}$ to every coupler of each group during $2 \log g$ iterations, using the following routing step where $i \in \{1, \ldots, g\}$.

$$G_i(g) \to C(i, i) \to G_i(j) \quad \text{for all } j \in \{1, \ldots, g\} \to C(j, i).$$

4. Each of the keys does a binary search over the sample keys that are in every coupler, and determines to which group it belongs.
5. If a key belongs to the j-th group, route the key to G_i.
6. Send the sample keys from each group to the couplers from the same group during $2 \log g$ iterations using

$$G_i(j) \to C(j, i) \to G_j(i) \to C(i, j) \quad \text{for all } i, j \in \{1, \ldots, g\}$$

7. The keys in group G_i, $i \in \{1, \ldots, g\}$ do a binary search over the couplers $C(*, i)$, deciding to which subgroup S_i^j they belong.
8. Route the keys from group G_i, $i \in \{1, \ldots, g\}$ to their corresponding subgroup S_i^j.
9. We broadcast the keys that are in S_i^j to every other processor in S_i^j.
10. Processor $S_i^j(k)$ calculates the k-th key in subgroup S_i^j.
11. We concentrate these elements by doing a prefix calculation, followed by a routing step.

Theorem 13.4 *A POPS(d, g) can solve sparse enumeration sorting in expected time $O(\lceil d/g \rceil + \log g)$ and in $\widetilde{O}(\lceil d/g \rceil + \log g)$ time when $g \geq \log n$, using $O(d/g)$ memory per processor.*

Proof. The number of sample keys s has a binomial distribution $B(n^\alpha, g^2/2n^\alpha)$ whose mean is $g^2/2$. By using Chernoff bounds we have that the number of splitter keys is $(g^2)/2 + \widetilde{o}(g)$, when $g \geq \log n$. This guarantees that step 2 can be done using constant space and in $\widetilde{O}((d/g) + \log n$ using Lemma 13.4 and Theorem 13.1.

When we consider $S = \{k_1, \ldots, k_s\}$ as the splitters we have that by using Lemma 13.13 the size of each X_i is $\widetilde{O}((n^\alpha)/g^2 \log n) = \widetilde{O}(d/g)$.

This implies that in step 5 the number of keys routed to each G_i is $\widetilde{O}(d)$ and that in step 8 the number of keys that are sent to each S_i^j is $\widetilde{O}(\frac{d}{g})$, so these routings can be done in $\widetilde{O}((d/g) + \log g)$ by using Theorem 13.1 and using constant space per processor.

Step 9 uses Lemma 13.11 that takes $O(d/g)$ time and $O(d/g)$ memory per processor.

Step 10 can be done in sequential time $O(d/g)$, and finally, step 11 can be done using Lemma 13.5 and Theorem 13.1 in $\widetilde{O}((d/g) + \log g)$ time. ∎

Acknowledgment

This research has been supported in part by the NSF Grants CCR-9912395 and ITR-0326155.

References

[1] P. Berthome and A. Ferreira, Improved embeddings in POPS networks through stack-graph models, *Proc. Third IEEE International Workshop on Massively Parallel Processing Using Optical Interconnections*, 1996, pp. 130–135.

[2] D. Chiarulli, S. Levitan, R. Melhem, J. Teza, and G. Gravenstreter, Multiprocessor interconnection networks using partitioned optical passive star (POPS) topologies and distributed control, *Proc. IEEE First International Workshop on Massively Parallel Processing Using Optical Interconnections*, 1994, pp. 70–80.

[3] A. Datta and S. Soundaralkshmi, Summation and routing on a partitioned optical passive stars network with large groups size, *IEEE Transactions on Parallel and Distributed Systems*, 14(12), 2003, pp. 306–315.

[4] A. Datta and S. Soundaralkshmi, Fast merging and sorting on a partitioned optical passive stars network, *Computer Science in Perspective*, Springer-Verlag, New York, 2003, pp. 115–127.

[5] J. Davila, S. Rajasekaran, Randomized sorting on the POPS network, *International Journal for Foundations of Computer Science* 16(1), 2005, pp. 105–116.

[6] Floyd, R.W., and Rivest, R.L., Expected time bounds for selection, *Communications of the ACM*, 18(3), 1975, pp. 165–172.

[7] W.D. Frazer and A. C. McKellar, Samplesort: A sampling approach to minimal storage tree sorting, *Journal of the ACM*, 17(3), 1970, pp. 496–507.

[8] G. Graventreter and R. Melhem, Realizing common communication patterns in partitioned optical passive star (POPS) networks, *IEEE Transactions on Computers*, 1998, pp. 998–1013.

[9] G. Graventreter, R. Melhem, D. Chiarulli, S. Levitan, and J. Teza, The partitioned optical passive stars (POPS) topology, *Proc. 9th IEEE International Parallel Processing Symposium*, 1995, pp. 4–10.

[10] E. Horowitz, S. Sahni, and S. Rajasekaran, *Computer Algorithms*, W. H. Freeman Press, 1998.

[11] A. Mei and R. Rizzi, Routing permutations in partitioned optical passive stars networks, *Proc. 16th International Parallel and Distributed Processing Symposium*, 2002.

[12] R. Melhem, G. Graventreter, D. Chiarulli, and S. Levitan, The communication capabilities of partitioned optical passive star networks, in *Parallel Computing Using Optical Interconnections*, Ed. K. Li, Y. Pan, and S. Zheng, Kluwer Academic Publishers, 1998, pp. 77–98.

[13] S. Rajasekaran and J.H. Reif, Optimal and sub-logarithmic time randomized parallel sorting algorithms, *SIAM Journal on Computing*, 18(3), 1989, pp. 594–607.

[14] S. Rajasekaran and S. Sen, On parallel integer sorting, *Acta Informatica*, 29, pp. 1–15, 1992.

[15] S. Rajasekaran, Sorting and selection on interconnection networks, *Workshop on Interconnection Networks and Mapping and Scheduling Parallel Computations*, eds. D.F. Hsu, A.L. Rosenberg, D. Sotteau, American Mathematical Society, Providence, RI, 1995, pp. 275–296.

[16] S. Rajasekaran, Computing on optical models, in *Randomization Methods in Algorithm Design*, Ed. P.M. Pardalos, S. Rajasekaran, and J. Rolim, AMS Press, 1999, pp. 239–249.

[17] S. Rajasekaran and J.H. Reif, Derivation of randomized sorting and selection algorithms, in *Parallel Algorithm Derivation and Program Transformation*, Ed. R. Paige, J.H. Reif, and R. Wachter, Kluwer Academic Publishers, 1993, pp. 187–205.

[18] S. Rajasekaran and S. Sahni, Sorting, selection, and routing on the array with reconfigurable optical buses, *IEEE Transactions on Parallel and Distributed Systems*, 8(11), 1997, pp. 1123–1131.

[19] S. Rajasekaran and S. Sahni, Fundamental algorithms for the array with reconfigurable optical buses, in *Parallel Computation Using Optical Interconnections*, Ed. K. Li, Y. Pan, and S. Q. Zheng, Kluwer Academic Publishers, 1998, pp. 185–204.

[20] S. Rajasekaran and J. Davila, Packet routing and selection on the POPS network, *Journal of Parallel Distributed Computing*, 65(8), 2005, pp. 927–933.

[21] S.B Rao and T. Tsantilas, Optical interprocessor communication protocols. *Proc. First International Workshop on Massively Parallel Processing Using Optical Interconnections*, 1994, pp. 266–274.

[22] S. Sahni, The partitioned optical passive stars network: Simulations and fundamental operations, *IEEE Transactions on Parallel and Distributed Systems*, 11(7), 2000, pp. 739–748.

[23] S. Sahni, Matrix multiplication and data routing using a partitioned optical stars network, *IEEE Transactions on Parallel and Distributed Systems*, 11(7), 2000, pp. 720–728.

[24] R. Vaidyanathan, C.R.P. Hartmann, and P.K. Varshney, Parallel integer sorting using small operations, *Acta Informatica*, 32, pp. 79–92, 1995.

[25] L.G. Valiant and G.J. Brebner, Universal schemes for parallel communication, *Proc. 13th Annual ACM Symposium on Theory of Computing*, 1981, pp. 263–277.

[26] L.G. Valiant and J.H. Reif, A logarithmic time sort for linear size networks. *Journal of the ACM(JACM)*, 34(1), January 1987, pp. 60–76.

[27] L.G. Valiant, General purpose parallel architectures, in *Handbook of Theoretical Computer Science: Vol. A*, Ed. J. van Leeuwen, North Holland, 1990.

14
Dynamic Reconfiguration on the R-Mesh

Ramachandran Vaidyanathan
Louisiana State University

Jerry L. Trahan
Louisiana State University

14.1	What is Dynamic Reconfiguration?	14-1
14.2	The R-Mesh	14-2
14.3	Basic Algorithms	14-5
	Permutation Routing • Neighbor Localization • Partitioning R-Meshes • Prefix Sums	
14.4	Priority Simulation	14-8
14.5	Sorting	14-11
	A Simple Suboptimal Algorithm • An Optimal Sorting Algorithm	
14.6	Graph Algorithms	14-14
	Connectivity • Minimum Spanning Tree • List Ranking	
14.7	Computational Complexity and Reconfiguration	14-20
	Relations to PRAMs • Relations to Circuits • Relations to TMs	
14.8	Concluding Remarks	14-25
	Acknowledgment	14-25
	References	14-25

14.1 What is Dynamic Reconfiguration?

Conventional computing handles different applications using either (i) a processor with fixed, general-purpose hardware to execute different software or (ii) different, specially designed, fixed hardware. The former approach has an advantage of flexibility with a disadvantage of a speed penalty owing to the use of general rather than special-purpose hardware. The latter approach swaps the advantage and disadvantage. With the advent of hardware that can change configurations relatively quickly, such as the introduction of field programmable gate arrays (FPGAs), an interest arose in exploiting the ability to reconfigure in order to combine speed and flexibility. One direction taken in this interest has been in using existing commercial devices, such as FPGAs, to solve problems with impressive speedups allowed by ever-faster reconfiguration times. A second direction has been investigating the possibilities of reconfiguration by capturing the ability to reconfigure in computational models, developing a body of algorithms on these models, and discovering the limits of the abilities of these models. This direction uncovered a rich collection of results and is the direction we recount in this chapter.

Researchers formulated many models to try to embody reconfigurability. These include models based on an underlying point-to-point topology such as the Reconfigurable Mesh (R-Mesh) [23], Polymorphic Torus [17], Processor Array with a Reconfigurable Bus System (PARBS) [51], and the, more general, Reconfigurable Network (RN) [2]. Other models that separate processors from the interconnection fabric (buses) include the Reconfigurable Multiple Bus Machine (RMBM) [44], Distributed Memory Bus Computer (DMBC) [35], and Reconfigurable Buses with Shift Switching (REBSIS) [19]. Models based on optical buses include the Linear Array with a Reconfigurable Pipelined Bus System (LARPBS) [30], Pipelined Reconfigurable Mesh (PR-Mesh) [40], Array Processors with Pipelined Buses (APPB) [21], and the Array with Reconfigurable Optical Buses (AROB) [31].

Most of these models comprise an array of simple processors, each with its own local memory, and a bus-based interconnection structure such that each processor can manipulate a separate portion of the interconnect. Typically, processors have the ability to selectively segment or fuse two or more pieces of the interconnection fabric. While each processor acts on the basis of local data, global bus configurations result from the collective local actions. Reconfiguration here involves changing the local (hence, global) configurations from step to step. The questions of interest become how (and how much) this ability to change the communication structure can benefit computation.

In this chapter we use the R-Mesh, a model that is both simple and versatile enough to capture features of most other models of dynamic reconfiguration. Section 14.2 describes the R-Mesh and a number of variants.

To demonstrate and explain the benefits of the ability to change configurations, we start in Section 14.3 with basic algorithms isolating key techniques. Permutation routing, neighbor localization, and prefix sums algorithms serve as building blocks for many of the results we report later. The ability to selectively embed a smaller R-Mesh in a larger one, in addition, allows the flexibility of structures dependent on input values. The need to select the highest priority item from a set of candidates arises often, and Section 14.4 explains how to take advantage of reconfiguration for this purpose.

Sections 14.5 and 14.6 demonstrate the versatility and speed obtainable with reconfiguration, using sorting and graph problems as the arena. The algorithms in these sections apply the building blocks established in earlier sections.

The limits on the advantage of reconfiguration form the final topic we address. Section 14.7 discusses the computational complexity of resource-bounded R-Meshes. Through simulations among models, it relates variants of the R-Mesh to each other and to well-known (nonreconfigurable) models. This aids defining which problems can most benefit from reconfiguration and gives a context for understanding algorithmic results.

While the results we present in this chapter cover a number of topics, they constitute but a small sample of the broad range of results developed on dynamically reconfigurable models. A more comprehensive coverage of the field, including models, algorithms, simulations, complexity classes, and implementation directions, appears in Vaidyanathan and Trahan [47].

14.2 The R-Mesh

An $x \times y$ R-Mesh [23] has xy processors arranged in an $x \times y$ array (Figure 14.1). In general, we will number rows and columns from 0 starting from the northwest and denote the processor in row i and column j by $p_{i,j}$.

As in a conventional mesh, each processor is connected to its four neighbors through its North (N), South (S), East (E), and West (W) ports. The R-Mesh also has connections internal to each processor. Each processor can independently connect its four ports in various ways. The processors in Figure 14.1 also show the 15 possible ways in which an R-Mesh processor can configure its ports. For example, processor $p_{0,0}$ connects its S and E ports, while keeping the remaining ports separate. On the other

Dynamic Reconfiguration on the R-Mesh

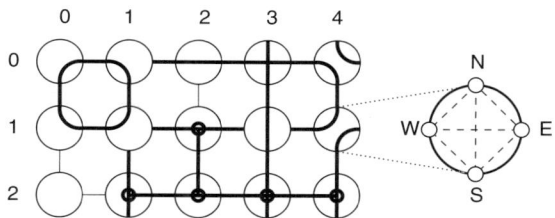

FIGURE 14.1 A 3×5 R-Mesh showing the 15 possible processor configurations.

hand, processor $p_{2,0}$ keeps all its ports separate, whereas processor $p_{2,3}$ connects all its ports together. A convenient representation of a processor's configuration is by a partition of the set $\{N, S, E, W\}$. In the above examples, processor $p_{0,0}$ is configured as the partition $\{\{N\}, \{S, E\}, \{W\}\}$, also written as $\{\overline{N}, \overline{SE}, \overline{W}\}$. Processors $p_{2,0}$ and $p_{2,3}$ correspond to partitions $\{\overline{N}, \overline{S}, \overline{E}, \overline{W}\}$ and $\{\overline{NSEW}\}$, respectively.

A set of processor configurations generates an *R-Mesh bus configuration* in which the ports of processors are connected together in different ways, forming buses. A *bus* is simply a maximally connected set of ports. For example, the bus formed by the ports of processors $p_{0,0}, p_{0,1}, p_{1,0}$, and $p_{1,1}$ is shown in bold in Figure 14.1. The figure also shows other buses in the R-Mesh. By independently configuring each processor, an R-Mesh can generate numerous bus configurations. This is crucial to the R-Mesh's ability to solve many problems extremely quickly.

An assumption key to the R-Mesh and other reconfigurable models is that information written anywhere on a bus is available to all ports on that bus within unit (constant) delay. An R-Mesh with this delay assumption is said to use the *unit-cost model*. Other cost models also exist (e.g., the log-cost and the bends-cost models). More details on these models are available in Vaidyanathan and Trahan [47] and El-Boghdadi [4]. We also assume a processor reading from a bus can ascertain whether a value has been written to the bus.

The R-Mesh is a synchronous model of computation in which a step can be divided into the following actions by each processor:

1. *Configure:* Each processor independently configures its ports according to one of the 15 possible port partitions.
2. *Communicate:* Each processor writes to and reads from a subset of its ports. We assume that a read operation reads the value written to the bus by a write operation of the same step.
3. *Compute:* Each processor performs a local computation. Typically, this includes commonly used arithmetic and logical operations (such as addition, subtraction, multiplication, division, comparison, and bitwise logical operations).

An R-Mesh can be classified according to different considerations. We describe some of these considerations below.

Variants Based on Bus Configuration
Here we categorize R-Meshes on the basis of the structure of their buses. This usually depends on the port partitions permitted to a processor.

- Each processor of the (*unconstrained*) R-Mesh described above can independently assume any of the 15 port partitions shown in Figure 14.1.
- In a *Fusing R-Mesh* (FR-Mesh), a processor is permitted to use only the partitions $\{\overline{NS}, \overline{EW}\}$ and $\{\overline{NSEW}\}$. One could think of the FR-Mesh as consisting of row and column buses, with each processor holding the ability to fuse together the horizontal and vertical buses traversing it. Despite its rather constrained appearance, the FR-Mesh is a powerful model of computation (Section 14.7).
- In a *Linear R-Mesh* (LR-Mesh), a processor cannot use the partitions $\{\overline{NSEW}\}$, $\{\overline{NSE}, \overline{W}\}$, $\{\overline{NSW}, \overline{E}\}$, $\{\overline{NEW}, \overline{S}\}$, and $\{\overline{SEW}, \overline{N}\}$ that connect more than two ports together. Consequently, all buses in an LR-Mesh are linear (nonbranching). The "cyclic bus" through processors $p_{0,0}, p_{0,1}, p_{1,0}$,

and $p_{1,1}$, of Figure 14.1 is a linear bus. The bus starting at the E port of processor $p_{0,1}$ and ending at the E port of $p_{1,3}$ (via processors $p_{0,2}$, $p_{0,3}$, $p_{0,4}$, and $p_{1,4}$) is an example of an "acyclic linear bus."

- The *Horizontal-Vertical R-Mesh* (*HVR-Mesh*) is a special case of the LR-Mesh in which a processor can use only the partitions $\{\overline{N}, \overline{S}, \overline{E}, \overline{W}\}$, $\{\overline{NS}, \overline{E}, \overline{W}\}$, $\{\overline{EW}, \overline{N}, \overline{S}\}$, and $\{\overline{NS}, \overline{EW}\}$, which prevent a bus from bending or branching. Consequently, all buses are segments of the horizontal and vertical buses spanning entire rows and columns.

The models become more powerful as we move from the HVR-Mesh to the unconstrained R-Mesh (as we show in Section 14.7). The models also become more difficult to implement as we go from the the HVR-Mesh to the unconstrained R-Mesh. Other variations of the R-Mesh that are based on its bus structure are discussed by Vaidyanathan and Trahan [47].

Variants Based on Bus Access
The R-Mesh can also be classified according to whether processors can access a bus concurrently (more than one processor at a time) or exclusively (at most one processor at a time). This classification is based on the well-known XRYW notation for the Parallel Random Access Machine (PRAM) [11], where $X, Y \in \{C, E\}$ for concurrent and exclusive.

- An EREW R-Mesh allows at most one processor to access (read from or write to) a bus at a time.
- A CREW R-Mesh permits at most one processor to write to a bus at a time, while allowing all processors to read simultaneously from a bus. This variant of the R-Mesh is one of the most commonly used because of its suitability for broadcasting information.
- The CRCW R-Mesh allows both reads and writes to be concurrent. As in PRAMs, a CRCW R-Mesh resolves concurrent writes to a bus by one of the following rules:
    — PRIORITY: The ports of processors are ordered by a fixed priority scheme (usually on the basis of the processor indices). Among those trying to write to a bus concurrently, the one with the highest priority succeeds in writing to the bus.
    — ARBITRARY: Any one of the processors (attempting to write to a bus concurrently) succeeds.
    — COMMON: All processors writing to a bus must write the same value.
    — COLLISION: Concurrent writes to a bus generate a special collision symbol on the bus.
    — COLLISION$^+$: This behaves as the COMMON rule when all concurrent writes are of the same value. Otherwise, it behaves as the COLLISION rule.

These rules present a gradation of concurrent write features and implementation obstacles on the R-Mesh. The COMMON, COLLISION, and COLLISION$^+$ rules are easy to implement on buses. On the other hand, PRIORITY and ARBITRARY are stronger rules that are not as simple to implement directly on buses. In Section 14.4, we will present simulations of the PRIORITY and ARBITRARY rules on COMMON, COLLISION, and COLLISION$^+$ R-Meshes.

As in the case of the PRAM, the Exclusive Read, Concurrent Write (ERCW) model is not used.

Variants Based on Processor Wordsize
Here we distinguish between R-Meshes on the basis of their processors' wordsize.

- *Word Model:* A processor of an $x \times y$ R-Mesh has a wordsize of $\Theta(\log xy)$, which allows a processor to address any other processor in the R-Mesh (including itself). Each processor has $O((xy)^{O(1)})$ bits of local memory, and the data size and bus width are assumed to be $\Theta(\log xy)$ bits.
- *Bit Model:* Each processor has a constant wordsize (regardless of the size of the underlying R-Mesh). The data size and bus width are constant, as is the size of the local storage. Consequently, processors do not possess unique processor ids. However, each processor may have a constant number of preloaded states that could, for example, indicate the position of a processor in the R-Mesh; this is similar to preloaded configurations in FPGAs [15].

Clearly, a word-model R-Mesh can simulate a bit-model R-Mesh of the same size without time overheads. The following result establishes that a bit-model R-Mesh of sufficient size can also simulate a word-model R-Mesh with significant time overheads [12].

Theorem 14.1 *Each step of a w-bit wordsize $x \times y$ word-model R-Mesh can be simulated on a $wx \times wy$ bit-model R-Mesh in constant time. Each word-model processor is assumed to be capable of addition, subtraction, multiplication, division, and logical operations on w-bit operands.*

Directed R-Mesh

So far, we have considered R-Meshes in which each port and bus is bidirectional. For example, in the partition $\{\overline{NS}, \overline{E}, \overline{W}\}$, data coming in at the N port exits the S port and vice versa.

A *directed R-Mesh* (*DR-Mesh*) permits the input and output directions of a port to be configured differently. For example, a processor may cause data coming into its N port to exit at the S port while allowing data entering at the S port to exit at the E port. In a sense, a DR-Mesh processor may be thought of as having four input ports and four output ports, with each configuration being a partition of these eight ports. Connecting multiple input ports together corresponds to a situation similar to a concurrent write.

Multidimensional R-Meshes

The R-Mesh described so far is a two-dimensional mesh with the added ability of each processor to partition its ports. This idea extends naturally to one- and multidimensional R-Meshes. A d-dimensional (nondirected) R-Mesh processor configures itself by partitioning its $2d$ ports. A two-dimensional R-Mesh can simulate any d-dimensional R-Mesh step in constant time [46] (see Theorem 14.2). A one-dimensional R-Mesh, however, cannot simulate a two-dimensional R-Mesh in constant time. Thus, a two-dimensional R-Mesh can be considered a universal R-Mesh model. We will consider higher dimensions (primarily 3-dimensions) as algorithmic conveniences. The following result will be used later.

Theorem 14.2 *For any $x, y, z > 1$, each step of an $x \times y \times z$ three-dimensional R-Mesh can be simulated on a $(7x + xy) \times 6yz$ two-dimensional R-Mesh in $O(1)$ time.*

It should be noted that the R-Mesh classifications discussed in this section are largely independent of each other. For example, a PRIORITY CRCW HVR-Mesh could be a word model or a bit model, nondirected or directed, and have one, two, or more dimensions. While we have described the most common variants of the R-Mesh in this chapter, other variants also exist [17, 22, 36, 50]. Vaidyanathan and Trahan [47, Chapter 3] also discuss these variants.

Unless mentioned otherwise, we will assume a word model, nondirected, CREW R-Mesh for the rest of this chapter. Where additional abilities are required, we will identify them and where more constrained models suffice, we will note these constraints.

14.3 Basic Algorithms

In this section, we introduce some basic algorithmic techniques on the R-Mesh that serve to introduce the computational model as well as construct stepping stones for the more advanced techniques presented in subsequent sections.

14.3.1 Permutation Routing

Let $\pi : \{0, 1, \ldots, n-1\} \longrightarrow \{0, 1, \ldots, n-1\}$ be a bijection (permutation). Consider an $n \times n$ R-Mesh in which each processor $p_{0,i}$ holds $\pi(i)$ and some data d_i. Permutation routing requires d_i to be conveyed to processor $p_{0,\pi(i)}$. The R-Mesh can accomplish this in three steps. First, the R-Mesh configures to construct a vertical bus in each column, and $p_{0,i}$ sends d_i to $p_{i,i}$. Next, the R-Mesh constructs a horizontal bus in each row, and $p_{i,i}$ sends d_i to $p_{i,\pi(i)}$. Finally, the R-Mesh once again constructs vertical buses, and $p_{i,\pi(i)}$ sends d_i to processor $p_{0,\pi(i)}$.

Theorem 14.3 *Permutation routing of n elements can be performed on an $n \times n$ LR-Mesh in $O(1)$ time. Initially, the elements to be routed are in a row of the R-Mesh.*

14.3.2 Neighbor Localization

Consider a one-dimensional, n-processor R-Mesh in which each processor p_i ($0 \leq i < n$) holds a flag f_i. The *neighbor localization problem* [45] is to generate a pointer α_i for each processor p_i with $f_i = 1$, such that $\alpha_i = j$ if and only if $j < i$, $f_j = 1$, and for all $j < k < i$, flag $f_k = 0$. If no such processor p_j exists, then $\alpha_i = \text{NIL}$. That is, α_i points to the nearest processor before p_i whose flag is set.

A one-dimensional R-Mesh has only two ports, the E and W ports, that it can either connect or keep separate. To solve the neighbor localization problem, processor p_i first connects its E and W ports if and only if $f_i = 0$. Next, each processor p_i with $f_i = 1$ writes index i to its E port and sets α_i to the value read from its W port. If no value is read, then $\alpha_i = \text{NIL}$. This solves the neighbor localization problem as a bus runs from the E port of a processor p_j with $f_j = 1$ to the W port of the next processor p_i with $f_i = 1$. Observe that if $\alpha_i = j$, then p_j is the only writer to, and p_j is the only reader from, the bus between p_j and p_i.

Theorem 14.4 *A one-dimensional, n-processor EREW R-Mesh can solve the neighbor localization problem in $O(1)$ time.*

Remarks By a symmetric algorithm, each processor can also obtain a pointer to the nearest flagged processor after itself. Neighbor localization can be performed on any linear acyclic bus of an R-Mesh, provided each processor can consistently tell one end of the bus from the other.

An important application of neighbor localization is the following result.

Corollary 14.5 *The OR or AND of n bits can be determined on a one-dimensional, n-processor EREW R-Mesh. Initially, the input bits are placed one to processor.*

Proof. By neighbor localization on the input bits, the first processor holding a 1 identifies itself (it is the only one with a NIL pointer). This processor declares the OR to be 1. If no such processor exists, then the result of the OR is a 0. Computing the AND is analogous as $x \text{ AND } y = (x' \text{ OR } y')'$ (DeMorgan's Law). ∎

14.3.3 Partitioning R-Meshes

As the underlying topology of an R-Mesh is the conventional mesh, any submesh of the underlying mesh induces a sub-R-Mesh. For example, in Figure 14.1, the submesh consisting of the six processors in the first two rows and first three columns also forms a sub-R-Mesh. However, while the processors, $p_{0,0}$, $p_{0,4}$, $p_{2,0}$, and $p_{2,4}$, in the four corners of the 3×5 mesh do not constitute a submesh, they form a sub-R-Mesh. They can function as one if the processors in row 0 (2) between $p_{0,0}$ and $p_{0,4}$ ($p_{2,0}$ and $p_{2,4}$) form a row bus and the processors in column 0 (4) between $p_{0,0}$ and $p_{2,0}$ ($p_{0,4}$ and $p_{2,4}$) form a column bus. Similarly, the portion of the topology shown in bold in Figure 14.2 can function as a sub-R-Mesh but not as a submesh. The following theorem formalizes this idea.

Theorem 14.6 *For any $A \subseteq \{0, 1, \ldots, x - 1\}$ and $B \subseteq \{0, 1, \ldots, y - 1\}$, an $x \times y$ R-Mesh can be treated as a $|A| \times |B|$ R-Mesh whose processors are those with row and column indices in A and B, respectively.*

Proof. For a row index $r \notin A$, each processor $p_{r,j}$ ($0 \leq j < y$) connects its N and S ports. For a column index $c \notin B$, each processor $p_{i,c}$ ($0 \leq i < x$) connects its E and W ports. This establishes row and column buses that connect neighbors in the "distributed sub-R-Mesh" defined by sets A and B (see Figure 14.2). The processors of this sub-R-Mesh can independently configure their port partitions. ∎

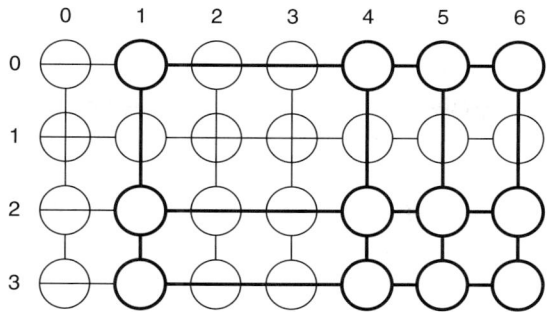

FIGURE 14.2 A 3 × 4 sub-R-Mesh of a 4 × 7 R-Mesh based on sets $A = \{0, 2, 3\}$ and $B = \{1, 4, 5, 6\}$.

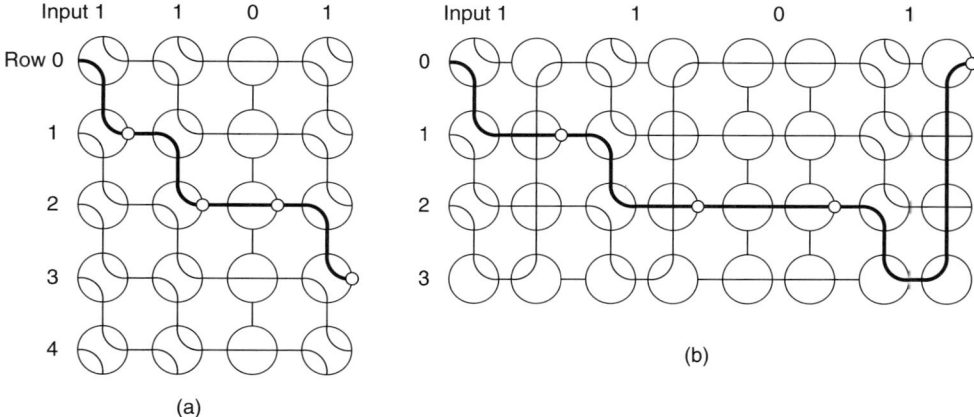

FIGURE 14.3 Prefix sums (a) and modulo-3 prefix sums (b) of four bits $(1, 1, 0, 1)$ on an R-Mesh. The bus carrying the signal is in bold. The ports that deduce the answer from the signal are shown as small circles.

14.3.4 Prefix Sums

Let $a_0, a_1, \ldots, a_{n-1}$ be a list of n numbers. Their *prefix sums* are $b_0, b_1, \ldots, b_{n-1}$, where $b_i = \sum_{j=0}^{i} a_j$, for $0 \leq i < n$. For any integer $q > 1$, the *modulo-q prefix sums* are $b'_0, b'_1, \ldots, b'_{n-1}$, where $b'_i = b_i \pmod{q}$.

The R-Mesh offers a simple method to find the prefix sums of n bits on an $(n + 1) \times n$ R-Mesh (see Figure 14.3a). The input bits are placed one to a column. The basic idea is to construct a pattern of buses that skip down by one row for each 1 in the input. This can be done by configuring processors of columns with a 0 as $\{\overline{N}, \overline{S}, \overline{EW}\}$ and of columns with a 1 as $\{\overline{NE}, \overline{SW}\}$. It is easy to prove that a signal originating at the W port of processor $p_{0,0}$ traverses the E port of processor $p_{i,j}$ if and only if the jth prefix sum b_j is i.

Lemma 14.7 *The prefix sums of n bits can be determined on an $(n + 1) \times n$ LR-Mesh in $O(1)$ time. Initially, the input bits are in a row of the R-Mesh.*

A variation of the above method is used to find the modulo-q prefix sums on a smaller R-Mesh of size $(q + 1) \times 2n$ (see Figure 14.3b). The only change is to wrap a bus back to row 0 whenever a 1 is encountered in row $q - 1$. An extra column is used per input bit to facilitate this bus wraparound.

Lemma 14.8 *For any integer $1 < q \leq n$, the modulo-q prefix sums of n bits can be determined on a $(q + 1) \times 2n$ LR-Mesh in $O(1)$ time. Initially, there is one input bit in every other column of the R-Mesh.*

Since the exclusive OR of n bits is their modulo-2 sum, we have the following corollary to Lemma 14.8.

Corollary 14.9 *The exclusive OR of n bits can be determined on a $3 \times n$ LR-Mesh in $O(1)$ time. Initially, the input bits are in a row of the R-Mesh.*

Remark A $2n$-column R-Mesh is not necessary as the computation can proceed in two rounds each handling at most $\lceil \frac{n}{2} \rceil$ inputs.

Coming back to prefix sums, a more efficient algorithm can be constructed using the modulo-q prefix sums algorithm of Lemma 14.8 by tracking the points where the signal wraps around on the bus. Let b_i and b'_i denote the ith prefix sum and modulo-q prefix sum, respectively. Let w_i be a flag indicating a wraparound of the signal; that is, $w_i = 1$ if and only if $a_i = 1$ and $b'_i = 0$. A $(q+1) \times 2n$ R-Mesh can compute b'_i and w_i using Lemma 14.8. Let $f_i = w_0 + w_1 + \cdots + w_i$ be the ith prefix sum of these flags. It can be shown that

$$\text{for all } 0 \leq i < n, \quad b_i = qf_i + b'_i.$$

One can now construct a procedure for computing the prefix sums b_i on a $(q+1) \times 2n$ R-Mesh that recursively computes f_i; as we noted earlier, the R-Mesh can compute b'_i and w_i. Each level of the recursion runs in constant time. Since a wraparound occurs once for every q ones in the input, there are at most n/q ones among the w_is. The recursion ends when there are no wraparounds. Since the number of 1s reduces by a factor of q for each level of recursion, there are at most $\log_q n = O\left(\frac{\log n}{\log q}\right)$ levels of recursion.

Theorem 14.10 *For any $1 \leq q < n$, the prefix sums of n bits can be performed on a $(q+1) \times 2n$ LR-Mesh in $O\left(\frac{\log n}{\log q}\right)$ time. The input bits are in every other row of the R-Mesh.*

Most of the results in this chapter have been gleaned from various sources, some adapted to the R-Mesh. Several of them were developed independently by different researchers. Neighbor localization, a generalization of the bus-splitting technique of Miller et al. [23], was applied to reconfigurable systems for sorting [45]. The exclusive OR algorithm was designed in the context of the computational powers of reconfigurable models [27, 44]. The basic prefix sums algorithms was developed by Wang et al. [51] and improved by Olariu et al. [29]. The distributed sub-R-Mesh structure was used by Trahan et al. [42] in a maximum finding algorithm.

14.4 Priority Simulation

As noted earlier, it is simple to implement the COMMON or the COLLISION rules for resolving concurrent writes on buses. Here we show that the PRIORITY (and therefore the ARBITRARY) rule can be simulated on a COMMON CRCW R-Mesh. This algorithm is an adaptation of a result for the Reconfigurable Multiple Bus Machine (RMBM) to the R-Mesh [44].

The basic idea of the simulation is to use COMMON, COLLISION, or COLLISION$^+$ writes to select the highest priority writer and then let that writer write exclusively.

The procedure to select the highest priority writer is best explained by an example. Consider an R-Mesh in which ports have been assigned unique indices according to their priority for concurrent writes. Consider a bus β on which seven ports (indexed 0, 8, 27, 33, 40, 46, 47) wish to write (see Figure 14.4a). We now describe a 6-iteration procedure that selects the highest indexed port attempting to write to bus β (port 47 in this example). Figure 14.4b shows the progress of this algorithm.

Initially, the seven writing ports are flagged as active as shown under the "before iteration 1" column of Figure 14.4b. In the first iteration, ports with a 1 in their most significant bit write a signal to the bus

Dynamic Reconfiguration on the R-Mesh

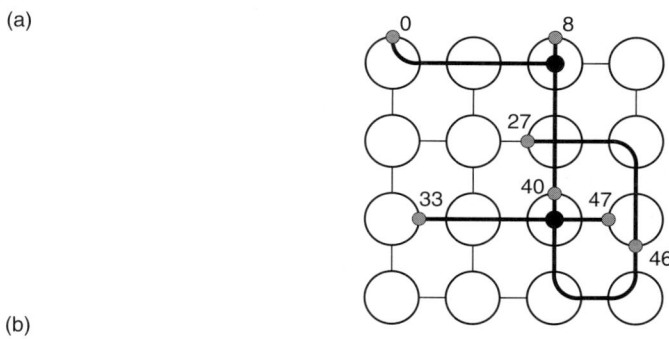

Port number	Active ports before iteration						
	1	2	3	4	5	6	
0	0 00000						
8	0 01000						
27	0 11011						
33	1 00001	1 0 0001	10 0 001				
40	1 01000	1 0 1000	10 1 000	101 0 00			
46	1 01110	1 0 1110	10 1 110	101 1 10	1011 1 0	10111 0	
47	1 01111	1 0 1111	10 1 111	101 1 11	1011 1 1	10111 1	101111

FIGURE 14.4 An illustration of the priority simulation algorithm with port indices in binary (a) Writing ports of bus β (in bold). (b) Active ports at different iterations of the algorithm.

and all active ports read from the bus. In the example, ports 33, 40, 46, and 47 write the signal and all seven ports detect it. Observe that all writes are of the same value, so a COMMON CRCW R-Mesh suffices. Ports 0, 8, and 27 (which did not write the signal) conclude that there is at least one other active port with a higher index than its own and flag themselves as inactive. At this point, ports 33, 40, 46, and 47 are the only active ports as shown under the column "before iteration 2" in Figure 14.4b. In iteration 2, active ports with a 1 in the next most significant bit write to the bus and the remaining active ports read from the bus. For our example, there is no write to the bus and no signal is detected; all active ports remain active for the next iteration. In iteration 3, ports 40, 46, and 47 write the signal to the bus, port 33 detects this signal and flags itself as inactive. Proceeding in the same manner, port 47 is the only one that remains active after all 6 bits have been considered (i.e., before iteration 7).

In general, in a π-port COMMON CRCW R-Mesh with a unique index from $\{0, 1, \cdots, \pi - 1\}$ for each port, the algorithm uses $\log \pi$ iterations to whittle down the initial set of active ports to the largest indexed writing port. While we describe the algorithm for a single bus, it works (simultaneously) on all buses in the R-Mesh. We describe all actions of the algorithm as if they were performed by individual ports. The action of a port σ is performed by the processor of port σ, reads and writes by the port are made from/to the bus incident on that port. The algorithm we describe below uses a three-dimensional R-Mesh and requires each port in the first two dimensions to use the two edges of a processor in the third dimension. We assume that each processor handles its four ports in the first two dimensions in some fixed order, so that the third dimension is available exclusively to each of them.

We now make two key observations that will lead to further generalization of the method and allow the algorithm to run in constant time.

- The algorithm proceeds in $\log \pi$ iterations because port addresses represented in binary require $\log \pi$ bits. By representing port addresses in base r, each address will only need $\log_r \pi$ digits, each from the set $\{0, 1, \ldots, r - 1\}$. Thus, with indices represented in base r, the algorithm can work in $\log_r \pi = O\left(\frac{\log \pi}{\log r}\right)$ iterations.

Port number	Active ports before iteration			
	1	2	3	4
0	0 | 00			
8	0 | 20			
27	1 | 23			
33	2 | 01	2 | 0 | 1		
40	2 | 20	2 | 2 | 0		
46	2 | 32	2 | 3 | 2	23 | 2	
47	2 | 33	2 | 3 | 3	23 | 3	233

FIGURE 14.5 An illustration of the priority simulation algorithm with port indices in base 4.

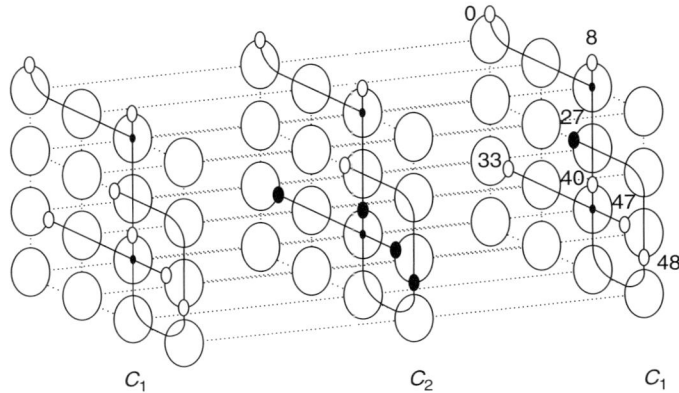

FIGURE 14.6 An illustration of the first iteration of the priority simulation algorithm example of Figure 14.5 with port indices in base 4. Ports writing the signal are shown as shaded ovals, the rest are unshaded.

- In iteration i of the algorithm so far, the R-Mesh selected active ports with the highest value in the ith bit position to continue to the next iteration. That is, all active ports with a 1 in the ith bit position went on to the next iteration; active ports with a 0 in the ith bit position continued to the next iteration if and only if there were no active ports with a 1 in the ith bit position at the start of the current iteration.

  Similarly, with the base-r representation, iteration i selects ports with the highest value in ith digit position.

Figure 14.5 illustrates the algorithm for the previous example with the port indices in base 4. The number of iterations needed now is $\log_4 64 = 3$.

To simulate an $x \times y$ PRIORITY CRCW R-Mesh, $\mathcal{P}$, with port indices in base r, we will use a three-dimensional $x \times y \times (r-1)$ COMMON CRCW R-Mesh, $\mathcal{C}$. This larger R-Mesh will allow each of the $\log_r \pi$ iterations to run in constant time; $\pi = 4xy$ is the number of ports in R-Mesh $\mathcal{P}$. As we noted earlier, we only need explain the functioning of iteration i (for $r - 1 \geq i \geq 1$).

View R-Mesh $\mathcal{C}$ as a stack of $r - 1$ two-dimensional COMMON CRCW sub-R-Meshes, $\mathcal{C}_1, \mathcal{C}_2, \ldots, \mathcal{C}_{r-1}$, each of size $x \times y$. This stack of R-Meshes is connected by edges in the third dimension as illustrated in Figure 14.6. Each port of $\mathcal{P}$ corresponds to a port in each sub-R-Mesh $\mathcal{C}_j$ (where $r - 1 \geq j \geq 1$). Each sub-R-Mesh $\mathcal{C}_j$ configures to the same bus configuration as $\mathcal{P}$. In iteration i (counting iterations from $\log_r \pi$ down to 1), only the active ports participate. Consider a port, σ, that is active at the start of iteration i (where $\log_r \pi \leq i \leq 1$). Let the radix-r digits of index σ be $s(\log_r \pi), s(\log_r \pi - 1), \ldots, s(i), \ldots, s(2), s(1)$. That is, the value of the ith digit is $s(i) \in \{0, 1, \ldots, r-1\}$.

If $s(i) > 0$, then port σ of sub-R-Mesh $C_{s(i)}$ writes a signal and ports σ of all sub-R-Meshes read from their buses. If for any $j > s(i)$, port σ detects a signal in C_j, then it also detects the presence of an active port on the bus, whose index is larger than its own. Here σ flags itself as inactive for iteration $i+1$. On the other hand, if for all $j > s(i)$, there is no signal in port σ of sub-R-Mesh C_j, then port σ continues to be active for the next iteration. Let port σ of sub-R-Mesh C_j set a flag $f_\sigma(j)$ to 1 if and only if it detects a signal.

To implement this method, use Corollary 14.5 to determine the AND of flags $f_\sigma(r-1)$, $f_\sigma(r-2), \ldots, f_\sigma(s(i)+2), f_\sigma(s(i)+1)$. Port σ continues to the next iteration if and only if the result of the above AND is a 0.

Observe that the function of the signal is simply to indicate the presence of an active port with a particular value in a particular digit in its index. On a COLLISION CRCW R-Mesh, the algorithm can proceed in a similar manner and interpret the collision symbol as the signal. Thus, the above algorithm works on the COLLISION and COLLISION$^+$ CRCW R-Meshes as well. These observations along with Theorem 14.2 give the following result.

Theorem 14.11 *For any $2 \leq r \leq 4xy$, each step of an $x \times y$ PRIORITY CRCW R-Mesh can be simulated on a three-dimensional $x \times y \times (r-1)$ or a two-dimensional $O(xr) \times O(yr)$ COMMON, COLLISION, or COLLISION$^+$ CRCW R-Mesh in $O\left(\frac{\log xy}{\log r}\right)$ time.*

A special case of this theorem occurs when r is arbitrarily small, but polynomial in xy.

Corollary 14.12 *For any constant $\epsilon > 0$, each step of an $x \times y$ PRIORITY CRCW R-Mesh can be simulated on a three-dimensional $x \times y \times (xy)^\epsilon$ or $(x^{1+\epsilon} y^\epsilon) \times (x^\epsilon y^{1+\epsilon})$ COMMON, COLLISION, or COLLISION$^+$ CRCW R-Mesh in $O(1)$ time.*

14.5 Sorting

Sorting is one of the most well-studied topics not only because of its use independently and in numerous computational settings but also because it provides valuable insight into solving several other problems. In this section we show that, for any $1 \leq t \leq \sqrt{n}$, an $\frac{n}{t} \times \frac{n}{t}$ R-Mesh can sort n elements in $O(t)$ time. This is the best possible, considering that sorting n elements on any word model of computation running in t time needs $\Omega\left(\left(\frac{n}{t}\right)^2\right)$ processors. This lower bound stems from the fact that for a word model of area A that sorts n elements in time t, $At^2 = \Omega(n^2)$ [16].

We begin with a suboptimal sorting algorithm and use it to derive the optimal result.

14.5.1 A Simple Suboptimal Algorithm

Consider the following sorting algorithm, whose correctness is obvious. Let $a_0, a_1, \ldots, a_{n-1}$ be the inputs to the algorithm. There is no loss of generality in assuming that the inputs are distinct.

- For each $0 \leq i, j < n$, set flag $g(i, j) = 1$ if and only if $a_i > a_j$.
- Compute the quantity $r(i) = \sum_{j=0}^{n-1} g(i, j)$. Clearly, $r(i)$ is the number of elements that are smaller than input a_i, that is, $r(i)$ is the rank of a_i in the final output. Since the inputs are distinct, so are their ranks.
- Relocate input a_i to the $(r(i))$th position in the output. Distinct ranks ensure that no two inputs are relocated to the same output position.

We now explain how this algorithm can be implemented to run in $O(t)$ time on a $2n^{1+\frac{1}{t}} \times n$ R-Mesh.

Let input a_j (where $0 \leq j < n$) be in processor $p_{0,j}$ of the $2n^{1+\frac{1}{t}} \times n$ R-Mesh. Configure the R-Mesh to form buses spanning the height of each column and broadcast input element a_j to each processor on column j. Next, partition the R-Mesh into n sub-R-Meshes, $\mathcal{S}_i$ (for $0 \leq i < n$), each of which is of size $2n^{\frac{1}{t}} \times n$. Sub-R-Mesh $\mathcal{S}_i$ is responsible for computing $r(i)$, the rank of a_i.

For $0 \leq \ell < 2n^{\frac{1}{t}}$ and $0 \leq j < n$, let $p_i(\ell, j)$ denote the processor in row ℓ and column j of $\mathcal{S}_i$. At this point in the algorithm, processor $p_i(0, j)$ holds input a_j. Using a row bus, processor $p_i(0, i)$ broadcasts a_i to all processors in row 0 of $\mathcal{S}_i$. Processor $p_i(0, j)$ now computes flag $g(i, j)$, which is 1 if and only if $a_i > a_j$. Finally, $\mathcal{S}_i$ uses Theorem 14.10 with $q = 2n^{\frac{1}{t}}$ to compute $r(i) = \sum_{j=0}^{n-1} g(i, j)$, the rank of a_i, in $O(t)$ time. We note that when $q \geq 3$, the algorithm of Theorem 14.10 can be modified to run on a $q \times n$ R-Mesh when the total sum (rather than the prefix sums) is required, as is the case here.

It is now straightforward to use Theorem 14.3 to relocate a_i to processor $p_{0,r(i)}$ of the original R-Mesh.

Lemma 14.13 *For any $1 \leq t \leq \log n$, a $2n^{1+\frac{1}{t}} \times n$ LR-Mesh can sort n elements, initially placed in a row, in $O(t)$ time.*

It is simple to prove that, for any $t > 3$, if $n \geq 2^{\frac{4t}{t-3}}$, then $n \geq 2n^{1+\frac{1}{t}}$. Thus we have the following corollary to Lemma 14.13 that will be used later in the main sorting algorithm of this section.

Corollary 14.14 *An $n^{\frac{3}{4}} \times n$ LR-Mesh can sort $n^{\frac{3}{4}}$ elements, initially placed in a row, in $O(1)$ time.*

14.5.2 An Optimal Sorting Algorithm

The optimal sorting algorithm for the R-Mesh is based on a technique called *columnsort* [16], which generalizes the well-known Batcher's odd-even merge sort.

14.5.2.1 Columnsort

Let $n = rs$ with $r \geq 2(s-1)^2$. Arrange the n elements to be sorted in an $r \times s$ array, henceforth called the "columnsort array." Columnsort is an 8-step algorithm that manipulates the elements of this array. Steps 1, 3, and 5 sort each column independently. Steps 2 and 4 permute the input elements. Step 2 transposes the columnsort array; that is, it picks up the elements in column-major order and places them back in row-major order in the same $r \times s$ array. Step 4 is the inverse of the permutation in Step 2. Steps 6–8 use an $r \times (s+1)$ columnsort array and require further elaboration.

Let L denote the n-element list of the elements of the columnsort array in column-major order. Let L_- be a list of $\lfloor \frac{r}{2} \rfloor$ elements, each of value $-\infty$ (a value smaller than all inputs). Let L_+ be a list of $\lceil \frac{r}{2} \rceil$ elements, each of value ∞ (a value larger than all inputs). Construct an $(r(s+1))$-element list L' by concatenating L_-, L, and L_+ in that order.

Step 6 constructs an $r \times (s+1)$ columnsort array by placing the elements of L' in column-major order. The action of this step is often called a *shift* (see Figure 14.7). Step 7 individually sorts the columns of the

0	11	22		$-\infty$	6	17	28
1	12	23		$-\infty$	7	18	29
2	13	24		$-\infty$	8	19	30
3	14	25		$-\infty$	9	20	31
4	15	26		$-\infty$	10	21	32
5	16	27		0	11	22	∞
6	17	28		1	12	23	∞
7	18	29		2	13	24	∞
8	19	30		3	14	25	∞
9	20	31		4	15	26	∞
10	21	32		5	16	27	∞
Before shift				After shift			

FIGURE 14.7 The shifting step of columnsort.

Dynamic Reconfiguration on the R-Mesh

$r \times (s+1)$ array. Observe that the $\pm\infty$ elements added in Step 6 retain their positions after the sort in Step 7. Step 8 removes the $\pm\infty$ elements and "unshifts" the $r \times (s+1)$ array back into an $r \times s$ array.

14.5.2.2 Implementing Columnsort on an $n \times n$ R-Mesh

To sort $n = rs$ elements, let $r = n^{\frac{3}{4}}$ and let $s = n^{\frac{1}{4}}$. For $n \geq 16$, the condition $r \geq 2(s-1)^2$ is satisfied. Let the columnsort array consisting of the n input elements be stored in column-major order in the top row of the $n \times n$ R-Mesh. Each column of the columnsort array corresponds to a row of $n^{\frac{3}{4}}$ contiguous elements of the top row of the R-Mesh. Thus, each column of the array can be viewed as being in the top row of an $n \times n^{\frac{3}{4}}$ sub-R-Mesh. Consequently, Steps 1, 3, and 5 of columnsort can be performed in constant time (Corollary 14.14). Steps 2 and 4 are permutation routing steps that also run in constant time (Theorem 14.3). Steps 6–8 can be performed by simply shifting sub-R-Mesh boundaries by $\lfloor \frac{r}{2} \rfloor$ columns in the $n \times n$ R-Mesh (see Figure 14.8). Observe that by virtue of the sort in Step 5 and because the values added in Step 6 are $\pm\infty$, the first $\lfloor \frac{r}{2} \rfloor$ and the last $\lceil \frac{r}{2} \rceil$ elements need not be sorted in Step 7.

Lemma 14.15 *An $n \times n$ LR-Mesh can sort n elements, initially placed in a row, in $O(1)$ time.*

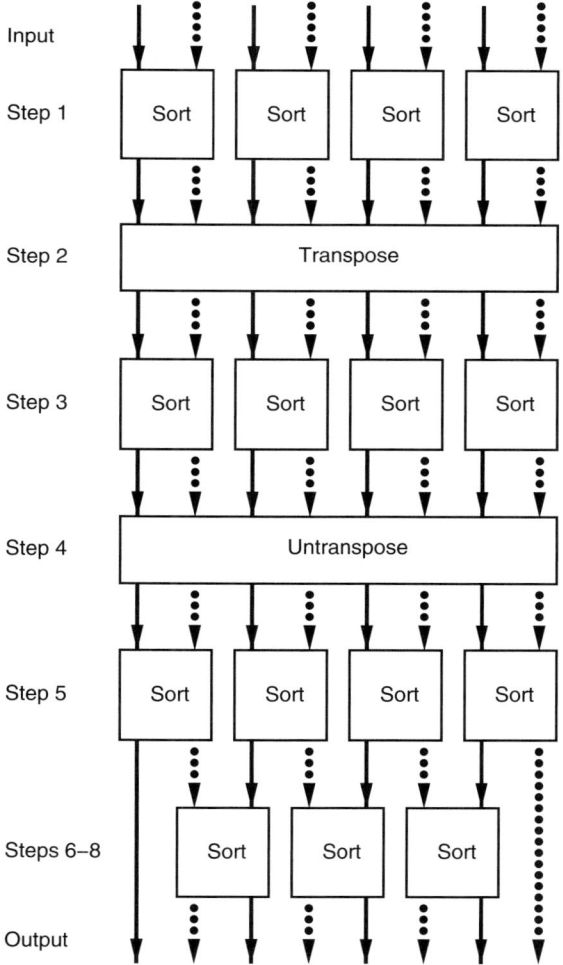

FIGURE 14.8 Columnsorting on the R-Mesh; the solid and dotted arrows represent $\lfloor \frac{r}{2} \rfloor$ and $\lceil \frac{r}{2} \rceil$ contiguous elements.

Scaling the Sorting Algorithm. Here we show that the sorting algorithm of Lemma 14.15 can be extended to run in $O(t)$ time on an $\frac{n}{t} \times \frac{n}{t}$ R-Mesh, for any $1 \leq t \leq \sqrt{n}$. This algorithm is important because it runs on an R-Mesh with n^2/t^2 area and therefore has an optimal AT^2 complexity of $\Theta(n^2)$ [16]. The result has added significance as scaling algorithms to run efficiently on smaller R-Meshes is neither automatic nor easy [47, Chapter 8].

Let the n inputs be in the first t rows of the $\frac{n}{t} \times \frac{n}{t}$ R-Mesh; observe that since $\frac{n}{t} \geq \sqrt{n}$, the R-Mesh has at least n processors. As before, we let $r = n^{3/4}$ and $s = n^{1/4}$. To perform a permutation routing of these elements, we will permute a row of $\frac{n}{t}$ elements at a time (using Theorem 14.3 with n replaced by n/t). Thus, Steps 2 and 4 run in $O(t)$ time on an $\frac{n}{t} \times \frac{n}{t}$ R-Mesh. All that remains is an explanation of how the R-Mesh sorts all s columns of the columnsort array and performs Steps 6–8 in $O(t)$ time. We consider two cases covering different ranges of values for t.

Sorting When $1 \leq t \leq n^{\frac{1}{4}}$. Observe that $\frac{n}{t} \geq n^{3/4} = r$ and $\frac{s}{t} \geq 1$. Also $\frac{n}{t} = \frac{rs}{t}$. Divide the first r rows of the $\frac{n}{t} \times \frac{n}{t}$ R-Mesh into s/t sub-R-Meshes, each of size $r \times r$. Each of these sub-R-Meshes can sort a column of the columnsort array in constant time (Lemma 14.15). Therefore, the R-Mesh can sort all s columns in $O(t)$ time.

The R-Mesh performs the shifting and unshifting in Steps 6 and 8 as before by realigning sub-R-Mesh boundaries. Unlike the previous case, however, some of the sub-R-Meshes will wraparound to include the last $\lfloor \frac{r}{2} \rfloor$ columns and first $\lceil \frac{r}{2} \rceil$ columns of the $\frac{n}{t} \times \frac{n}{t}$ R-Mesh. For Step 7, the R-Mesh performs the regular and "wraparound" sorts separately. Each of the at most $O(t)$ wraparound sorts will use the entire R-Mesh to run in constant time. By Theorem 14.6, the $\frac{n}{t} \times \frac{n}{t}$ R-Mesh can be treated as an $r \times r$ R-Mesh and by Theorem 14.3, the elements can be sorted so that the smallest elements are in the last $\lfloor \frac{r}{2} \rfloor$ columns and the largest elements are in the first $\lceil \frac{r}{2} \rceil$ columns.

Lemma 14.16 *For any $1 \leq t \leq n^{1/4}$, an $\frac{n}{t} \times \frac{n}{t}$ LR-Mesh can sort n elements, initially placed in the first t rows, in $O(t)$ time.*

Sorting When $n^{\frac{1}{4}} \leq t \leq n^{\frac{1}{2}}$. Let $\tau = \frac{t}{s}$. Then $\frac{n}{t} = \frac{rs}{t} = \frac{r}{\tau}$ and $1 \leq \tau \leq s \leq r^{\frac{1}{2}}$. The entire $\frac{r}{\tau} \times \frac{r}{\tau}$ R-Mesh can then sort one column of the columnsort array in $O(\tau)$ time (Lemma 14.16). Consequently, it can sort all s columns in $O(\tau s) = O(t)$ time. For Step 7, all sorts are wraparound sorts that can be handled as in the previous case.

Theorem 14.17 *For any $1 \leq t \leq \sqrt{n}$, an $\frac{n}{t} \times \frac{n}{t}$ LR-Mesh can sort n elements in $O(t)$ time. Initially, the n inputs are in a row of the R-Mesh.*

While the optimal algorithm of this section is based on the work of Jang and Prasanna [13], others [24, 26] have independently constructed algorithms for optimal, constant time sorting on the R-Mesh. Optimal sorting algorithms have also been developed for multidimensional R-Meshes [3, 24].

One result on multidimensional sorting [3] that we will use later is the one below.

Theorem 14.18 *A $\sqrt{n} \times \sqrt{n} \times \sqrt{n}$ LR-Mesh can sort n elements, in $O(1)$ time. Initially, the input elements are in some $\sqrt{n} \times \sqrt{n}$ slice of the R-Mesh.*

14.6 Graph Algorithms

The importance of graph algorithms stems from the ability of graphs to model a large variety of systems. In this section, we describe constant time solutions to a selection of important graph problems (connectivity-related problems, minimum spanning tree, and list ranking) that also serve to highlight the power of reconfigurable models. A common strategy followed by most of these algorithms is to somehow embed relevant information about the graph into the R-Mesh's bus structure and then use writes and reads to reveal the problem's solution. Most of these algorithms are based on the work of Wang

Dynamic Reconfiguration on the R-Mesh

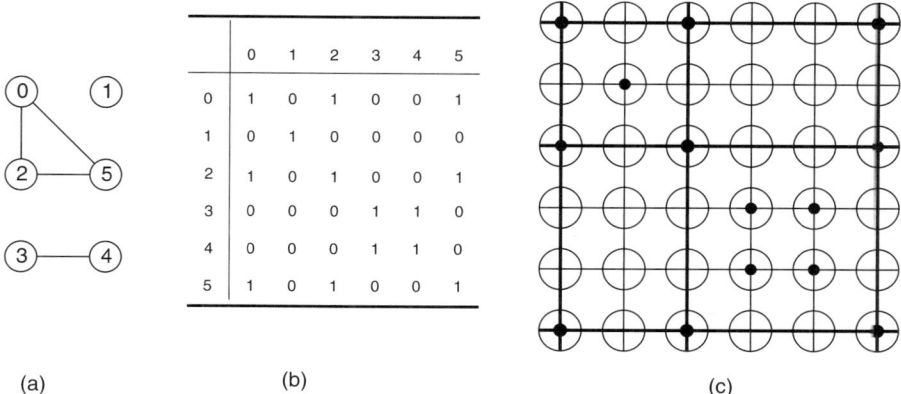

FIGURE 14.9 An illustration of the s–t connectivity algorithm. In this illustration, $s = 2$ and $t = 0$; the signal traverses the bus shown in bold. (a) $\mathcal{G}$, (b) A_r, and (c) FR-Mesh configuration.

and Chen [50]. There have been, however, numerous other important contributions in the area. Vaidyanathan and Trahan [47, Chapter 6] detail many more aspects of graph algorithms on reconfigurable models.

Throughout this section we will assume a graph $\mathcal{G} = (V, E)$ with n vertices and m edges as input to the problem. Without loss of generality, let $V = \{0, 1, 2, \ldots, n-1\}$. We will represent the graph as either a set of m edges or as an adjacency matrix A, for which $A[i, j] = 1$ if and only if there is an edge from vertex i to vertex j.

14.6.1 Connectivity

Here we present solutions to several problems that relate to the question of whether a pair of vertices of $\mathcal{G}$ is connected.

14.6.1.1 S–T Connectivity and Reachability

Given a pair of vertices s and t in an undirected graph $\mathcal{G}$, the s–t *connectivity* problem asks the question "Is there a path in $\mathcal{G}$ between s and t?" To solve this problem, we will embed the paths of $\mathcal{G}$ onto buses of the R-Mesh.

Let $\mathcal{G}$ be represented as an adjacency matrix. We will use an $n \times n$ R-Mesh to solve the problem. Let A_r denote the reflexive closure of the adjacency matrix A; that is, A_r is the same as A with all diagonal elements set to 1. If $A_r[i, j] = 0$, then configure processor $p_{i,j}$ using partition $\{\overline{\text{NS}}, \overline{\text{EW}}\}$, and with partition $\{\overline{\text{NSEW}}\}$ if $A_r[i, j] = 1$ (see Figure 14.9). Note that an FR-Mesh (see Section 14.2) suffices for this algorithm. It is easy to prove that (the ports of) processors $p_{s,s}$ and $p_{t,t}$ are on the same bus if and only if $\mathcal{G}$ has a path between s and t. Thus, once configured, processor $p_{s,s}$ writes a signal to its ports, and $p_{t,t}$ concludes that s and t are connected if and only if it receives the signal.

Lemma 14.19 *The s–t connectivity problem for an n-vertex graph can be solved on an $n \times n$ FR-Mesh in $O(1)$ time. Initially, the processors hold the corresponding bits of the adjacency matrix of the graph.*

The equivalent of s–t connectivity for directed graphs is *reachability*. The reachability problem asks the question "Given a source s and a terminus t in a directed graph $\mathcal{G}$, is there a directed path in $\mathcal{G}$ from s to t?" To solve this problem we use an approach similar to s–t connectivity, except that a DR-Mesh (see Section 14.2) replaces the (undirected) FR-Mesh to embed directed paths. Figure 14.10 illustrates the approach.

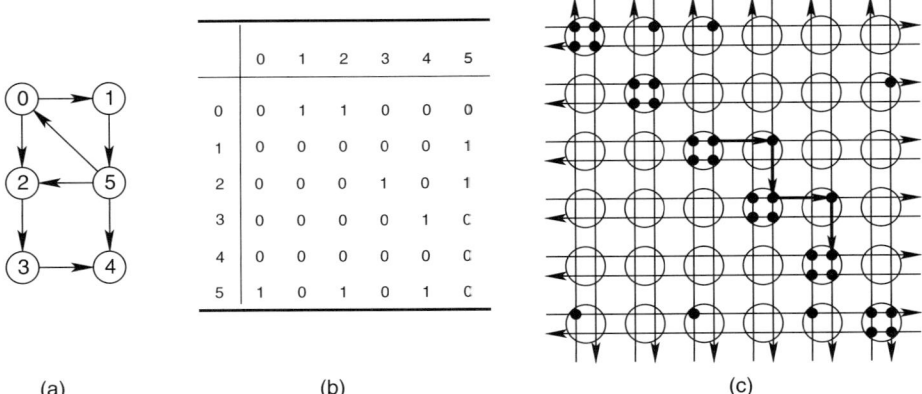

FIGURE 14.10 An illustration of the reachability algorithm. In this illustration, $s = 2$ and $t = 0$; the signal traverses the portion in bold and does not reach $p_{t,t}$. (a) $\mathcal{G}$, (b) A, and (c) DR-Mesh configuration.

Lemma 14.20 *The reachability problem for an n-vertex directed graph can be solved on an $n \times n$ DR-Mesh in $O(1)$ time. Initially, the processors hold the corresponding bits of the adjacency matrix of the graph.*

14.6.1.2 Connected Components

A *connected component* of graph $\mathcal{G}$ is a maximally connected subgraph of $\mathcal{G}$. Each pair of vertices in a connected component has a path between them. Finding the connected components of the graph involves assigning a unique component number to each vertex of a component. Typically, this component number is the index of one of the vertices of the component. The s–t connectivity algorithm (Lemma 14.19) readily converts to a connected components algorithm.

An $n \times n$ ARBITRARY CRCW FR-Mesh first configures as in the s–t connectivity algorithm. Next, each processor $p_{i,i}$ writes i to its bus and reads off the component number from the bus. This component number is the value written by one of the diagonal processors in the component (selected by the ARBITRARY rule). Using Theorem 14.11, the algorithm can be run on a three- or a two-dimensional COMMON, COLLISION, or COLLISION$^+$ CRCW R-Mesh. We now have the following result.

Theorem 14.21 *For any $1 \leq q \leq n$, the connected components of an n-vertex graph can be determined on a three-dimensional $n \times n \times (q - 1)$ or a two-dimensional $O(nq) \times O(nq)$ COMMON, COLLISION, or COLLISION$^+$ CRCW FR-Mesh in $O\left(\frac{\log n}{\log q}\right)$ time. The graph is input as an adjacency matrix, evenly distributed over an $n \times n$ sub-R-Mesh of the FR-Mesh.*

As in the case of Theorem 14.11, we have the following corollary.

Corollary 14.22 *For any constant $\epsilon > 0$, the connected components of an n-vertex graph can be determined on a three-dimensional $n \times n \times n^\epsilon$ or a two-dimensional $n^{1+\epsilon} \times n^{1+\epsilon}$ COMMON, COLLISION, or COLLISION$^+$ CRCW FR-Mesh in $O(1)$ time. The graph is input as an adjacency matrix, evenly distributed over an $n \times n$ sub-R-Mesh of the FR-Mesh.*

It should be noted that the reachability problem does not lend itself to a similar algorithm for finding the strongly connected components of a directed graph. We will present a slightly different method for strongly connected components later in this section.

14.6.1.3 Transitive Closure

The transitive closure $\mathcal{G}^*$ of a graph $\mathcal{G}$ has an edge from vertex u to vertex v if and only if $\mathcal{G}$ has a path from u to v. The problem we consider in this section is that of finding A^*, the adjacency matrix of $\mathcal{G}^*$.

Dynamic Reconfiguration on the R-Mesh

For an undirected graph, the connected components algorithm also determines the transitive closure. For any processor $p_{i,i}$ representing vertex i, let c_i be its connected component. Then $A^*[i, j] = 1$ if and only if the component number of vertex j is also c_i. Processor $p_{i,j}$ that ultimately holds $A^*[i, j]$ obtains the component numbers of vertices i and j from $p_{i,i}$ (using a row bus) and $p_{j,j}$ (using a column bus). Indeed, any connected components algorithm can be used to obtain the transitive closure.

Theorem 14.23 *For any $1 \leq q \leq n$, the transitive closure of an n-vertex graph can be determined on a three-dimensional $n \times n \times (q-1)$ or a two-dimensional $O(nq) \times O(nq)$ COMMON, COLLISION, or COLLISION$^+$ CRCW FR-Mesh in $O\left(\frac{\log n}{\log q}\right)$ time. The graph is input as an adjacency matrix evenly distributed over an $n \times n$ sub-R-Mesh of the FR-Mesh.*

Corollary 14.24 *For any constant $\epsilon > 0$, the transitive closure of an n-vertex graph can be determined on a three-dimensional $n \times n \times n^\epsilon$ or a two-dimensional $n^{1+\epsilon} \times n^{1+\epsilon}$ COMMON, COLLISION, or COLLISION$^+$ CRCW FR-Mesh in $O(1)$ time. The graph is input as an adjacency matrix evenly distributed over an $n \times n$ sub-R-Mesh of the FR-Mesh.*

For directed graphs, we will use an $n \times n \times n$ three-dimensional CREW DR-Mesh. Each $n \times n$ sub-R-Mesh determines the reachability (Lemma 14.20) from a different source node. That is, it determines the value in a particular row of A^*. The above algorithm can also be adapted to run on an $n^2 \times n^2$ DR-Mesh.

Theorem 14.25 *The transitive closure of an n-vertex directed graph can be determined on a three-dimensional $n \times n \times n$ or a two-dimensional $n^2 \times n^2$ CREW DR-Mesh in $O(1)$ time. The graph is input as an adjacency matrix in as few rows as possible.*

14.6.1.4 Strong Connectivity

Two vertices i and j of a directed graph $\mathcal{G}$ are strongly connected if i is reachable from j and j is reachable from i. A strongly connected component of $\mathcal{G}$ is a maximal subgraph of $\mathcal{G}$ in which each pair of vertices is strongly connected.

Let A^* be the adjacency matrix of the transitive closure of $\mathcal{G}$. Let A^t be the transpose of A^*, and let B be the bitwise AND of A^* and A^t. That is, $B[i, j] = 1$ if and only if $A^*[i, j] = 1$ and $A^t[i, j] = 1$ (or i is reachable from j and j is reachable from i). Thus, $B[i, j] = 1$ if and only if i and j are in the same strongly connected component. Neighbor localization (Theorem 14.4) can be used to find the position of the first such 1 in each row of B, and this position serves as the strongly connected component number.

From Theorem 14.25 we have the following result.

Theorem 14.26 *The strongly connected components of an n-vertex directed graph can be determined on a three-dimensional $n \times n \times n$ or a two-dimensional $n^2 \times n^2$ CREW DR-Mesh in $O(1)$ time. The graph is input as an adjacency matrix in as few rows as possible.*

14.6.2 Minimum Spanning Tree

A spanning tree $\mathcal{G}$ is a subtree of $\mathcal{G}$ that includes all vertices of $\mathcal{G}$. For a weighted graph $\mathcal{G}$, the minimum spanning tree is one with the least sum of edge weights. Assume the graph to be specified as a set of m edges and their weights. Without loss of generality, assume that the edge weights are distinct. Also, assume that the graph is connected and loopless and contains at least n edges. If $m < n$, then either $\mathcal{G}$ is disconnected or is already a tree.

Let $e_0, e_1, \ldots, e_{m-1}$ be the edges in increasing order of their weights. For each $0 \leq i < m$, let $\mathcal{G}_i$ be the subgraph of $\mathcal{G}$ with all n vertices but only edges $e_0, e_1, \ldots, e_{i-1}$. The minimum spanning tree algorithm is based on the (provable) fact that for each $0 \leq i < m$, edge $e_i = (u, v)$ belongs to the minimum spanning tree if and only if u and v are not connected in $\mathcal{G}_i$.

We have developed all the tools needed to perform the steps of the algorithm on an $n \times n \times m$ R-Mesh. The edges can be sorted by their weight using Theorem 14.18 on a $\sqrt{m} \times \sqrt{m} \times \sqrt{m}$ sub-R-Mesh. The ith

$n \times n$ slice of the R-Mesh constructs $\mathcal{G}_i$ independently and checks edge e_i for inclusion in the minimum spanning tree. It can be shown that this algorithm also runs on a nearly square $m\sqrt{n} \times m(2 + \sqrt{n})$ two-dimensional R-Mesh.

Theorem 14.27 *The minimum spanning tree of an n-vertex, m-edge connected graph can be determined on a three-dimensional $n \times n \times m$ or a two-dimensional $m\sqrt{n} \times m(2 + \sqrt{n})$ LR-Mesh in $O(1)$ time.*

14.6.3 List Ranking

While lists often provide a compact way of storing information, an array allows speedy access to the information. List ranking determines, for each list element ℓ, the number of list elements preceding ℓ. List ranking is key to converting a list into an array. Moreover, lists form naturally in bus-based dynamically reconfigurable models (as in neighbor localization (see Theorem 14.4) and Euler tour techniques [18,43]). Many applications rely on list ranking to further process this information.

In this section we start with a fast, but inefficient list ranking algorithm and then refine it to improve its efficiency. We will assume an n-element list L, represented as a set of n pointers $\alpha(i)$, where $0 \leq i < n$. The last element of the list has a NIL pointer. If $\ell_0, \ell_1, \cdots, \ell_{n-1}$ are the list elements, then the element following ℓ_i is $\ell_{\alpha(i)}$. The aim is to find the rank $r(i)$ of each element ℓ_i.

Let the input be in the "first n processors" of the R-Mesh. Since the algorithms use R-Meshes of size at least $n \times n$, the permutation routing algorithm of Theorem 14.3 allows us to assume that the inputs are available in any convenient arrangement over a row. It is also easy to determine the first element of list L.

Consider an $n^2 \times n^2$ R-Mesh. Assuming all processors to configure according to partition $\{\overline{\text{NS}}, \overline{\text{EW}}\}$, the R-Mesh can be viewed as having n^2 row buses. Group these buses into n groups, each with n buses and denote them by $b_{i,j}$, where $0 \leq i < n$ is the group number and $0 \leq j < n$ is the number within the group. Similarly, group column buses and denote them by $c_{i,j}$ for $0 \leq i,j < n$. We deviate from the standard notation for processors and let $p_{i,j}$ denote the processor on the top row through which column bus $c_{i,j}$ passes. Assume that each processor $p_{i,j}$ holds list element ℓ_i.

Each $p_{i,j}$ uses its column bus to connect row buses $b_{i,j}$ and $b_{\alpha(i),j+1}$, provided ℓ_i is not the last element of the list and $j < n - 1$. If a signal was to arrive at $b_{i,j}$ (indicating that rank $r(i) = j$), then it would also arrive at $b_{\alpha(i),j+1}$ indicating that $r(\alpha(i)) = j + 1$. Let ℓ_a be the first element of the list with $r(a) = 0$. By sending a signal at bus $b_{a,0}$, the algorithm will ensure that the signal arrives at bus $b_{i,r(i)}$. By detecting this signal, processor $p_{i,r(i)}$ would deduce the rank of ℓ_i to be $r(i)$. Figure 14.11 illustrates the algorithm for a 4-element list $\langle \ell_2, \ell_1, \ell_3, \ell_0 \rangle$; that is, $r(2) = 0$, $r(1) = 1$, $r(3) = 2$, and $r(0) = 3$.

It can be shown that the above algorithm can be adapted to run on an LR-Mesh.

Lemma 14.28 *An n-element list can be ranked by an $n^2 \times n^2$ LR-Mesh in $O(1)$ time. Initially, the input is in the top row of the R-Mesh.*

For any $2 \leq q \leq n$, the modulo-q rank of element ℓ_i of list L is $r(i) \pmod{q}$; recall that $r(i)$ is the (true) rank of ℓ_i. By restricting row and column bus groups' size to q, and by connecting buses $b_{i,j}$ and $b_{\alpha(i),(j+1) \pmod q}$, the algorithm of Lemma 14.28 will determine the modulo-q rank of each element of the list. This is similar to the method used in Lemma 14.8 to compute the modulo-q prefix sums of bits.

Lemma 14.29 *For any $2 \leq q \leq n$, an $nq \times nq$ LR-Mesh can find modulo-q ranks of elements of an n-element list in $O(1)$ time. Initially, the input is in the top row of the R-Mesh.*

By selecting the elements whose modulo-q ranks are 0, the original list can be contracted to a list of size n/q. If this contracted list can be ranked, then the modulo-q ranks can be used to rank the original list.

We now use Lemmas 14.28 and 14.29 to construct an efficient list ranking algorithm. The idea is to use an $nq \times nq$ R-Mesh, applying Lemma 14.29 several times to contract the list to a size of at most $\sqrt{nq}$, on which the available R-Mesh can then use Lemma 14.28. The total time taken will be the number of applications of Lemma 14.29 needed to reduce the list to size $\sqrt{nq}$.

Dynamic Reconfiguration on the R-Mesh

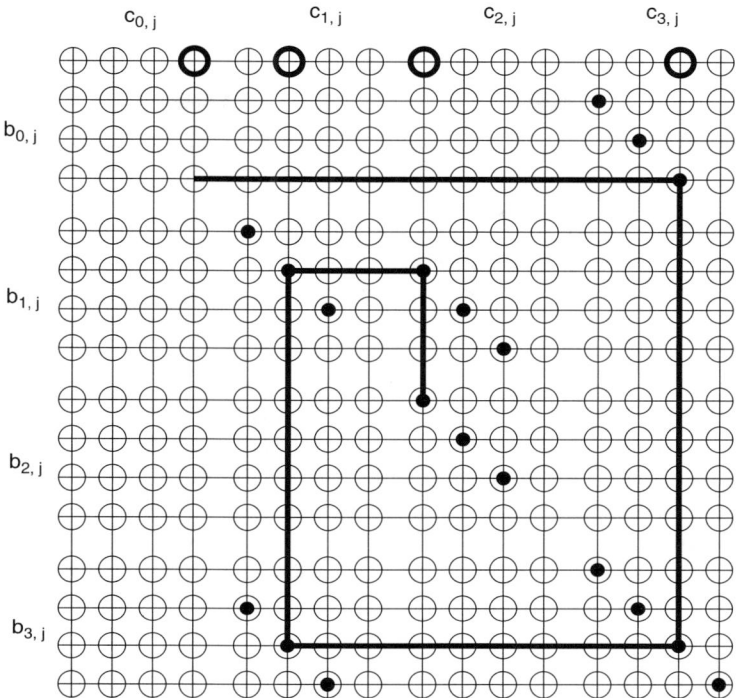

FIGURE 14.11 An illustration of the list ranking algorithm for a 4-element list $\langle \ell_2, \ell_1, \ell_3, \ell_0 \rangle$. The bus carrying the signal and the processors that detect it are shown in bold.

Let $n_0 = n$ and $q_0 = q$. In one application of Lemma 14.29, the list size is $n_1 = \frac{n_0}{q_0} = \frac{n}{q}$. Since an $nq \times nq = \left(\frac{n}{q}\right)q^2 \times \left(\frac{n}{q}\right)q^2$ R-Mesh is available, $q_1 = q^2$. After two applications of Lemma 14.29, the list reduces to size $n_2 = \frac{n_1}{q_1} = \frac{n}{q^3}$. Since $nq = n_2 q^4$, we have $q_2 = q^4$. Proceeding in the same way, after i applications of Lemma 14.29 it is easy to show that $n_i = \frac{n}{q^{2^i-1}}$ and $q_i = q^{2^i}$. If $n_i = \sqrt{nq}$, then $i = O\left(\log\left(\frac{\log n}{\log q}\right)\right)$.

Theorem 14.30 *For any $2 \leq q \leq n$, an $nq \times nq$ LR-Mesh can rank an n-element list in $O\left(\log\left(\frac{\log n}{\log q}\right)\right)$ time. Initially, a row holds the input list.*

With $q = n^\epsilon$ for some constant $0 < \epsilon \leq 1$, we have the following corollary.

Corollary 14.31 *For any constant $0 < \epsilon \leq 1$, an $n^{1+\epsilon} \times n^{1+\epsilon}$ LR-Mesh can rank an n-element list in $O(1)$ time. Initially, a row holds the input list.*

At the other extreme with $q = 2$, a $2n \times 2n$ R-Mesh ranks the list in $O(\log \log n)$ time. It can be shown that an $n \times n$ R-Mesh can also achieve this time.

Corollary 14.32 *An $n \times n$ LR-Mesh can rank an n-element list in $O(\log \log n)$ time. Initially, a row holds the input list.*

The inefficient list ranking algorithm and an independent derivation of Corollary 14.31 appear in Olariu et al. [28] and Trahan et al. [43]. The efficient list ranking algorithm of Theorem 14.30 is due to

Hayashi et al. [10]. They further improved the list ranking algorithm to run in $O(\log^* n)$ time* on an $n \times n$ R-Mesh.

14.7 Computational Complexity and Reconfiguration

A plethora of reconfigurable models exists. Models differ in allowed port configurations, in type of bus access, in number of dimensions, but what is the effect of these differences on the computational ability of the model? As noted in Section 14.2, a two-dimensional R-Mesh can simulate, in constant time, each step of an R-Mesh of any number of dimensions. Do any other variations have any more effect? This section addresses these questions by looking at simulations among models and resource-bounded complexity classes for these models.

Some relationships follow easily. A DR-Mesh can do anything an R-Mesh can do, which can do anything an LR-Mesh can do, which can do anything an HVR-Mesh can do. What happens, however, if we reverse the relations? Does a gap in computational power exist, and, if so, how much of a gap?

This section relates reconfigurable models to each other and to other more conventional models of computation. It also describes relationships among resource-bounded complexity classes of reconfigurable models and well-established complexity classes. Generally, the simulating model spends constant time to simulate each step of the simulated model. Some simulations precisely identify the number of processors employed by the simulating model, while others permit any polynomial number of processors.

We will relate R-Mesh variants first to the PRAM, then to circuits, then to the Turing machine (TM). For PRAMs, the relevant resources are processors, time, and memory. For circuits, the relevant resources are number of gates (size) and number of levels of gates between circuit inputs and outputs (depth). For TMs, the relevant resources are time and memory.

Notation: For a model ABC, let $(\text{ABC})^j$ denote the class of problems solvable by an ABC with polynomial number of processors (or size) in time (or depth) $O(\log^j n)$ for inputs of size n. Observe that $(\text{ABC})^0$ denotes the class of problems solvable in constant time.

14.7.1 Relations to PRAMs

A PRAM comprises a collection of synchronous processors executing a common program with a shared memory. Each processor also has a local memory. The shared memory is the only means for processors to communicate among themselves. The concurrent and exclusive write rules defined in Section 14.2 apply also to PRAMs. The PRAM is the standard model for theoretical investigation of the computational complexity of parallel computation [14]. As such, its complexity classes are long established. Its use of shared memory allows a focus on the maximum parallelism inherent in a problem.

The first simulation, of a PRAM by an HVR-Mesh, is straightforward. Wang and Chen [49] allocated a column of R-Mesh processors for each PRAM processor and a row of R-Mesh processors for each PRAM shared memory cell.

Theorem 14.33 *Each step of an n-processor CRCW PRAM with s shared memory cells can be simulated by an $s \times n$ CREW HVR-Mesh in $O(1)$ time.*

Proof. Let $\mathcal{M}$ denote a CRCW HVR-Mesh with n processors and s shared memory cells. Assume that the PRAM uses the PRIORITY concurrent write rule. Since an n-processor, s-memory, PRIORITY CRCW

---

* For any integer k, $\log^k n = (\log n)^k$, whereas $\log^{(k)} n = \underbrace{\log \log \cdots \log}_{k \text{ times}} n$. The quantity $\log^* n$ is the smallest integer k such that $\log^{(k)} n \leq 1$; it is an extremely slow growing function.

PRAM can simulate the ARBITRARY, COMMON, COLLISION, and COLLISION$^+$ rules in constant time, the simulation we present holds for all of the above concurrent write rules.

We construct an $s \times n$ CREW HVR-Mesh $\mathcal{R}$ that simulates $\mathcal{M}$. Each processor $p_{i,0}$ in the leftmost column of $\mathcal{R}$ simulates shared memory cell i of $\mathcal{M}$. Each processor $p_{0,j}$ in the top row of $\mathcal{R}$ simulates the processor with rank j in the priority order of $\mathcal{M}$. (Assume processor priority order to decrease with processor index.)

For each reading step of $\mathcal{M}$, each leftmost column processor of $\mathcal{R}$ broadcasts its data along its row. Each processor in the top row wishing to read broadcasts the address of the memory cell down its column. Each processor whose row index matches the address now forwards the data up its column.

On each writing step of $\mathcal{M}$, writing processors broadcast data and memory addresses down their columns and use neighbor localization (Theorem 14.4) to select the highest priority (rightmost) processor to perform the write exclusively. ∎

Next, we examine a simulation by a PRAM of a reconfigurable model, specifically, the HVR-Mesh. It melds together two simulations of Trahan et al. [44], eliminating an intermediate model.

Theorem 14.34 *Each step of an $x \times y$ CRCW HVR-Mesh can be simulated by a* PRIORITY *CRCW PRAM in $O(1)$ time using $O(xy^2 + yx^2)$ processors and $O(xy)$ shared memory cells.*

Proof. In an $x \times y$ HVR-Mesh, each processor has four ports and limited internal configurations. Each bus is either a horizontal bus connecting a contiguous subset of ports along a row or a vertical bus connecting a contiguous subset of ports along a column. Because each row and each column is independent of the others, we will describe the simulation of a one-dimensional HVR-Mesh (same as an R-Mesh for the one-dimensional case) then extend the bound to the two-dimensional case.

The ideas behind the simulation are to allot one shared memory cell to each port and, for each bus, to identify one port as representative of the bus. Communication via the corresponding shared memory cell takes the place of communication via the bus.

Let $\mathcal{R}$ denote an x-processor, one-dimensional, PRIORITY CRCW HVR-Mesh. We construct a PRIORITY CRCW PRAM $\mathcal{M}$ that simulates $\mathcal{R}$ using $O(x^2)$ processors and $O(x)$ shared memory cells. PRAM $\mathcal{M}$ can handle other write rules of $\mathcal{R}$ (ARBITRARY, COMMON, COLLISION, COLLISION$^+$) by conventional methods.

Designate the rightmost port of a bus as its *representative*. The E port of the rightmost processor is always a representative. Otherwise, all representatives are W ports. Denote the W port of p_j as W_j. For notational simplicity, denote the E port of p_{x-1} as W_x. Consequently, x possible representatives exist (dealing with rightmost port separately as a special case).

For each write to (or read from) an HVR-Mesh bus, the corresponding PRAM processor will determine the representative for the bus, then write to (read from) the corresponding shared memory cell. Let $s(W_j)$ denote the shared memory cell corresponding to W_j.

For any port W_j, $x-j$ potential representatives lie to its right in $\mathcal{R}$. PRAM $\mathcal{M}$ assigns $x-j+1$ processors $\pi_{j,k}$, where $j \leq k \leq x$, to W_j. For a write or read by p_j of $\mathcal{R}$ to W_j, processors $\pi_{j,k}$ of $\mathcal{M}$ test whether p_k has its E and W ports disconnected. If so, then $\pi_{j,k}$ writes k to a shared memory cell $s(j)$; note that $s(j)$ and $s(W_j)$ are different memory locations. By priority write resolution, $s(j)$ holds $\min\{k : j \leq k \leq x$ and p_k has disconnected its E and W ports$\}$, which is the index of the representative of p_j's bus. If $s(j)$ holds index g, then, for a write, $\pi_{j,0}$ writes to $s(W_g)$, and, for a read, $\pi_{j,0}$ reads from $s(W_g)$. For a write or read by p_j to its E port, $\pi_{j,0}$ uses the value in $s(j+1)$ as its representative.

To generalize the simulation to two dimensions, an $x \times y$ HVR-Mesh has x rows of length y and y columns of height x. For each row, the simulating PRAM employs y^2 processors and y shared memory cells, and, for each column, the PRAM employs x^2 processors and x shared memory cells. Overall, the PRAM uses $O(xy^2 + yx^2)$ processors and $O(xy)$ shared memory cells. ∎

Combining Theorems 14.33 and 14.34, we obtain the following.

Theorem 14.35 (CRCW PRAM)^j = (CREW HVR-Mesh)^j = (CRCW HVR-Mesh)^j, *for $j \geq 0$.*

14.7.2 Relations to Circuits

Strong relations hold between PRAMs and circuits. For instance, constant time on a PRAM with a polynomial number of processors is equivalent to constant depth on an unbounded fan-in circuit with polynomial number of gates. Theorem 14.35 implies then a relationship between HVR-Meshes and unbounded fan-in circuits. This section sketches a pair of results by Trahan et al. [40] weaving in relationships among other R-Mesh variants and other circuit types.

14.7.2.1 Circuit Definitions and Notation

Circuit definitions follow Venkateswaran [48], and complexity class definitions follow Allender et al. [1]. A *circuit* is a directed, acyclic structure comprising gates and wires. A gate with in-degree 0 is a circuit input labeled from the set $\{0, 1, x_0, x_1, \ldots, x_{n-1}, \overline{x_0}, \overline{x_1}, \ldots, \overline{x_{n-1}}\}$, where x_i is an input variable and $\overline{x_i}$ is its complement. All other gates have indegree of at least two and compute AND or OR functions of the values on their incoming wires. Gates whose outputs do not connect to the inputs of any other gate are circuit outputs.

A circuit has bounded fan-in if the fan-in of each gate is limited to a constant, unbounded fan-in if the fan-in is unlimited, and semi-unbounded fan-in if the fan-in of each AND gate is limited to a constant while the fan-in of each OR gate is unlimited. The *size* of a circuit is the number of wires in the circuit. The *depth* of a circuit is the length of the longest path from a circuit input to a circuit output. A circuit family $\mathcal{C}$ is a set $\{C_1, C_2, \ldots\}$ of circuits in which C_n has n inputs. A family $\mathcal{C}$ of size $S(n)$ is logspace uniform if, given an input of length n, a TM can construct a description of C_n using $O(\log S(n))$ workspace.

Class AC^j (respectively, SAC^j and NC^j) is the class of languages accepted by unbounded (respectively, semiunbounded and bounded) fan-in circuits of size polynomial in n and $O(\log^j n)$ depth. Let AC denote $\bigcup_{j=0}^{\infty} AC^j$; define SAC and NC analogously. To specify relations among these classes,

$$NC^j \subseteq SAC^j \subseteq AC^j = (CRCW\ PRAM)^j \subseteq NC^{j+1}.$$

These relations create a hierarchy of classes for increasing values of j.

14.7.2.2 Simulations with Circuits

By Theorem 14.35, HVR-Mesh^j = AC^j. The next two results, from Trahan et al. [40], insert the DR-Mesh and LR-Mesh (hence R-Mesh) into the hierarchy described above. A semiunbounded fan-in circuit simulates a DR-Mesh by adapting an unbounded fan-in circuit simulation of a PRAM in Theorem 14.36; an LR-Mesh simulates a bounded fan-in circuit by using list ranking to slice the circuit into layers and then simulates slices of layers in turn in Theorem 14.37.

Theorem 14.36 DR-Mesh$^j \subseteq SAC^{j+1}$.

Proof. Let $\mathcal{D}$ denote a DR-Mesh that, on inputs of size n, uses at most polynomial in n number of processors and $O(\log^j n)$ time. We construct a family of logspace-uniform, semiunbounded fan-in circuits $\mathcal{C}$ that simulates $\mathcal{D}$ with size polynomial in n and depth $O(\log^{j+1} n)$.

Each step of a CRCW PRAM can be simulated in constant depth by a polynomial size, unbounded fan-in circuit [39]. As in the proof of Theorem 14.34, one can map between reads and writes to (directed) buses and reads and writes to shared memory cells. In the earlier theorem, buses were undirected and confined to rows or columns. $\mathcal{D}$, however, has directed buses that need not be linear. This difference introduces the need for simulating circuits $\mathcal{C}$ to treat buses in $\mathcal{D}$ as subgraphs of a directed graph and to compute reachability in order to evaluate the value read from a bus.

Computing the transitive closure of the directed graph corresponding to $\mathcal{D}$'s bus connections reveals the desired reachability information. Class $\mathcal{C}$ squares the Boolean adjacency matrix corresponding to $\mathcal{D}$'s bus connections log n times to compute the transitive closure. Each squaring involves ANDing a pair of elements and ORing n elements, matching the bounded fan-in AND gates and unbounded fan-in OR gates of $\mathcal{C}$.

Consequently, $\mathcal{C}$ can simulate each step of $\mathcal{D}$ in $O(\log n)$ depth, and so the entire $O(\log^j n)$ time computation of $\mathcal{D}$ in $O(\log^{j+1} n)$ depth. Polynomial size follows from the preexisting PRAM simulation. ∎

The following theorem extends the base result $NC^1 \subseteq LR\text{-Mesh}^0$ owing to Ben-Asher et al. [2]. That result does not extend to circuit depth $\omega(\log n)$ because the number of processors is exponential in circuit depth.

Theorem 14.37 $NC^{j+1} \subseteq LR\text{-Mesh}^j$.

Proof. Let $\mathcal{C}$ denote a family of logspace-uniform, bounded fan-in circuits, where C_n is of size $O(n^c)$, for constant $c > 0$, and depth $O(\log^{j+1} n)$. We construct an LR-Mesh $\mathcal{R}$ with polynomial in n number of processors that will simulate C_n in $O(\log^j n)$ time.

LR-Mesh $\mathcal{R}$ will not evaluate C_n directly, but will work, rather, with a layered version of C_n with layers of gates such that each gate receives inputs only from outputs of the preceding layer. Let L_n denote the layered version of C_n. Circuit L_n is the same depth and only a constant factor of size larger than C_n [41].

LR-Mesh $\mathcal{R}$ generates a description of L_n in $O(1)$ time, as a polynomial size-bounded LR-Mesh can perform the output of a logspace TM in $O(1)$ time (see Theorem 14.38 in Section 14.7.3). For each wire in L_n from gate f to gate g, $\mathcal{R}$ creates a pointer from f to g. The resulting graph contains, for each circuit input h, a directed tree rooted at h. Run the list ranking algorithm (Theorem 14.30) to identify the layer to which each gate belongs.

LR-Mesh $\mathcal{R}$ now slices L_n into bands of depth $\log n$, where each band includes all gates at each included layer. Then $\mathcal{R}$ simulates these bands in sequence. Each simulation takes $O(1)$ time [2], and L_n has $O(\log^j n)$ bands, so the entire circuit simulation of C_n takes $O(\log^j n)$ time. ∎

Other simulations relate different R-Mesh variants: simulations establishing $FR\text{-Mesh}^j = R\text{-Mesh}^j$ [40], simulations of an R-Mesh on an LR-Mesh [9], of an acyclic DR-Mesh on an LR-Mesh [8], and of an R-Mesh on an HVR-Mesh [20] exist.

Generalizing the time bound to polylogarithmic produces equivalent classes for many models. Let ABC(polylog) denote the class of languages accepted by model ABC with polynomial number of processors in polylogarithmic time. We derive that the following classes are equivalent: DR-Mesh(polylog), R-Mesh(polylog), LR-Mesh(polylog), FR-Mesh(polylog), HVR-Mesh(polylog), CRCW PRAM(polylog), NC, and AC.

14.7.3 Relations to TMs

The past two sections related R-Mesh models to parallel models of computation. We next turn to a classical sequential model of computation, the Turing machine. As is common, the importance of these relationships is not because of the relationship to the TM itself but because of broad understanding of TM complexity classes and of how an enormous number of problems fit within these classes. Ben-Asher et al. [2] established the results that this section presents.

14.7.3.1 TM Definitions and Notation

A TM consists of a program with a finite number of states, an input tape, and a worktape. Each tape is partitioned into cells such that each cell holds a symbol from a finite alphabet. The TM has a tape head on the input tape that can read from one cell and a tape head on the worktape that can read from and write to one cell. In one step, the TM reads the symbols under the input and worktape heads, then, based on symbols read and its state, writes a symbol to the worktape, moves each tape head by one cell (or remains stationary), and changes state. The space used by a TM is the number of worktape cells accessed during the computation.

A deterministic TM (DTM) is one that, for each combination of symbols read and state, has at most one transition to next state, symbol to write to the worktape, and movements of the tape heads.

A nondeterministic TM (NTM) is one that has multiple possibilities for state transition, symbol written, and tape head movements. An NTM accepts its input if some possible computation path leads to an accepting state. A symmetric TM is a special case of an NTM in which, for each transition, the reverse transition is also possible.

Class L (respectively, NL and SL) is the class of languages accepted by DTMs (respectively, NTMs and symmetric TMs) in $\log n$ space. These classes relate to each other as

$$L = SL \subseteq NL.$$

14.7.3.2 Simulations with TMs

This section presents results equating constant-time bounded R-Mesh model complexity classes to TM classes. The simulation of an R-Mesh by a TM revolves around coping with the bus structure of the R-Mesh so that a TM can determine when a value written by one processor will be read by another. The simulation in the opposite direction revolves around the solution by an R-Mesh of a problem complete for the TM class. As the intricacies of the simulations are beyond space limitations of this chapter and as the simulations follow the same patterns, we sketch the first pair of simulations and then only indicate key changes for subsequent simulations.

Theorem 14.38 $\text{LR-Mesh}^0 = L$.

Proof outline. *TM simulation of LR-Mesh:* Let $\mathcal{R}$ denote an LR-Mesh that runs in constant time with polynomial in n processors for inputs of size n. We construct a TM $\mathcal{T}$ that simulates the computation of $\mathcal{R}$ in $\log n$ space. Suppose that $\mathcal{T}$ can determine the port configuration of a processor of $\mathcal{R}$ at time t. To determine whether a value written by processor p reaches processor q of $\mathcal{R}$, $\mathcal{T}$ traces the linear bus from p using processor configurations to determine whether the value that p writes reaches q. Observe that since $\mathcal{T}$ can determine the value read by q, then it can determine the port configuration taken by q for the next step. Consequently, $\mathcal{T}$ is able to determine port configurations at time t.

LR-Mesh simulation of TM: For a directed graph $\mathcal{G}$ with in-degree 1 and out-degree 1, the problem CYCLE is to determine for given vertices u and v whether u and v belong to a common cycle in $\mathcal{G}$. CYCLE is complete for L, so a constant-time LR-Mesh solution for CYCLE establishes $L \subseteq \text{LR-Mesh}^0$. To solve CYCLE, embed graph $\mathcal{G}$ in the buses of a COLLISION CRCW LR-Mesh, then let the processors corresponding to u and v write and read from their buses. A collision symbol indicates that u and v are on the same cycle of $\mathcal{G}$. ∎

Theorem 14.39 $\text{R-Mesh}^0 = L$.

Proof outline. For the TM simulation of an R-Mesh, the problem of determining whether a value written by processor p reaches processor q maps to the s–t connectivity problem on an undirected graph. A DTM can solve this problem in logarithmic space.

For the R-Mesh simulation of a TM, the R-Mesh solves the undirected s–t connectivity problem by Lemma 14.19. ∎

Note: At the time of Ben-Asher et al.'s solution, the s–t connectivity problem was known to be in the symmetric logspace oracle hierarchy, SLH, but was not known to be in L. Subsequently, Nisan and Ta-Shma [25] established that SLH = SL, and Reingold [32] established that s–t connectivity is in L, so SL = L, implying the result in Theorem 14.39. As a consequence of Reingold's result, $\text{LR-Mesh}^0 = \text{R-Mesh}^0$, which was not known previously.

Theorem 14.40 $\text{DR-Mesh}^0 = NL$.

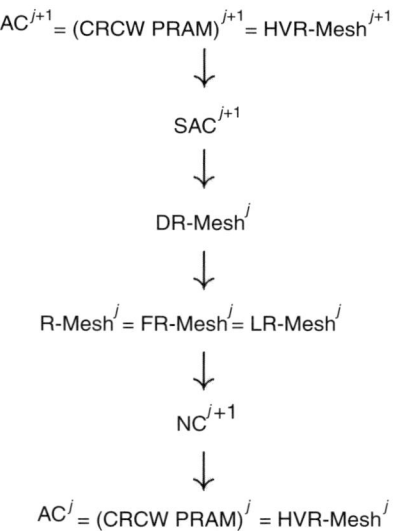

FIGURE 14.12 Hierarchy of complexity classes for reconfigurable models. An arrow from A to B denotes $B \subseteq A$.

Proof outline. For the TM simulation of a DR-Mesh, the problem of determining whether a value written by processor p reaches processor q now maps to a reachability problem on a directed graph, which an NTM can solve in log space.

This directed reachability problem is complete for NL, so the simulation of a TM by a DR-Mesh focuses on the solution of this problem by a DR-Mesh in constant time (Lemma 14.20). ∎

Figure 14.12 depicts the hierarchy of complexity classes obtained by the results described in this section.

14.8 Concluding Remarks

In this chapter, we have offered a glimpse of what dynamically reconfigurable architectures can do. Though the presentation has been with the R-Mesh as the backdrop, most results and algorithms have equivalents in other models of reconfiguration.

Though results on models like the R-Mesh hinge on the assumption of constant delay buses, they provide a useful algorithmic template for constructing more practical solutions. Some recent work has moved in this direction using concepts from R-Mesh-type models and FPGAs [4–7, 33, 34, 37, 38].

Acknowledgment

We are grateful to the U.S. National Science Foundation for its support of our research on dynamic reconfiguration (grants CCR-0310916, CCR-0073429, and CCR-9503882). Much of the insight that formed the basis for the presentation in this work was a result of this research.

References

[1] E. Allender, M. C. Loui, and K. W. Regan, "Complexity Classes," in *CRC Handbook of Algorithms and Theory of Computation*, M. Atallah, ed., CRC Press, Boca Raton, FL, 1998, pp. 27-1–27-23.

[2] Y. Ben-Asher, K.-J. Lange, D. Peleg, and A. Schuster, "The Complexity of Reconfiguring Network Models," *Info. and Comput.*, vol. 121, no. 1, pp. 41–58, 1995.

[3] Y.-C Chen and W.-T Chen, "Constant Time Sorting on Reconfigurable Meshes," *IEEE Trans. Computers*, vol. 43, no. 6, pp. 749–751, 1994.

[4] H. M. El-Boghdadi "On Implementing Dynamically Reconfigurable Architectures," PhD Thesis, Department of Electrical and Computer Engineering, Louisiana State University, 2003.

[5] H. M. El-Boghdadi, R. Vaidyanathan, J. L. Trahan, and S. Rai, "Implementing Prefix Sums and Multiple Addition Algorithms for the Reconfigurable Mesh on the Reconfigurable Tree Architecture," *Proc. Int. Conf. on Parallel and Distrib. Proc. Tech. and Appl.*, 2002, pp. 1068–1074.

[6] H. M. El-Boghdadi, R. Vaidyanathan, J. L. Trahan, and S. Rai, "On the Communication Capability of the Self-Reconfigurable Gate Array Architecture," *9th Reconfig. Arch. Workshop* in *Proc. Int. Parallel and Distrib. Proc. Symp.*, 2002.

[7] H. M. El-Boghdadi, R. Vaidyanathan, J. L. Trahan, and S. Rai, "On Designing Implementable Algorithms for the Linear Reconfigurable Mesh," *Proc. Int. Conf. on Parallel and Distrib. Proc. Tech. and Appl.*, 2002, pp. 241–246.

[8] J. A. Fernández-Zepeda, D. Fajardo-Delgado, J. A. Cárdenas-Haro, and A. G. Bourgeois, "Efficient Simulation of an Acyclic Directed Reconfigurable Model on an Undirected Reconfigurable Model," *Int. J. Found. Computer Sci.*, vol. 16, no. 1, pp. 55–70, 2005.

[9] J. A. Fernández-Zepeda, R. Vaidyanathan, and J. L. Trahan, "Using Bus Linearization to Scale the Reconfigurable Mesh," *J. Parallel Distrib. Comput.*, vol. 62, no. 4, pp. 495–516, 2002.

[10] T. Hayashi, K. Nakano, and S. Olariu, "Efficient List Ranking on the Reconfigurable Mesh, with Applications," *Theory of Comput. Syst.*, vol. 31, no. 5, pp. 593–611, 1999.

[11] J. JáJá, *An Introduction to Parallel Algorithms*, Addison-Wesley Publishing Co., Reading, MA, 1992.

[12] J. Jang, H. Park, and V. Prasanna, "A Bit Model of Reconfigurable Mesh," *Proc. 1st Reconfig. Arch. Workshop*, 1994.

[13] J.-w. Jang and V. K. Prasanna, "An Optimal Sorting Algorithm on Reconfigurable Mesh," *J. Parallel Distrib. Comput.*, vol. 25, no. 1, pp. 31–41, 1995.

[14] R. M. Karp and V. Ramachandran, "Parallel Algorithms for Shared-Memory Machines," in *Handbook of Theoretical Computer Science, Vol. A: Algorithms and Complexity*, J. van Leeuwen, ed., MIT Press, 1990, pp. 869–941.

[15] D. I. Lehn, K. Puttegowda, J. H. Park, P. Athanas, and M. Jones, "Evaluation of Rapid Context Switching on a CSRC Device," *Proc. Conf. on Eng. of Reconfigurable Systems and Algorithms*, 2002, pp. 154–160.

[16] T. Leighton, "Tight Bounds on the Complexity of Parallel Sorting," *IEEE Trans. Computers*, vol. 34, pp. 344–354, 1985.

[17] H. Li and M. Maresca, "Polymorphic Torus Network," *IEEE Trans. Computers*, vol. 38, pp. 1345–1351, 1989.

[18] R. Lin, "Fast Algorithms for Lowest Common Ancestors on a Processor Array with Reconfigurable Buses," *Info. Proc. Lett.*, vol. 40, pp. 223–230, 1991.

[19] R. Lin and S. Olariu, "Reconfigurable Buses with Shift Switching: Concepts and Applications," *IEEE Trans. Parallel and Distrib. Syst.*, vol. 6, no. 1, pp. 93–102, 1995.

[20] S. Matsumae and N. Tokura, "Simulation Algorithms among Enhanced Mesh Models," *IEICE Trans. Info. and Systems*, vol. E82-D, no. 10, pp. 1324–1337, 1999.

[21] M. Middendorf and H. ElGindy, "Matrix Multiplication on Processor Arrays with Optical Buses," *Informatica*, vol. 22, pp. 255–262, 1999.

[22] R. Miller, V. K. Prasanna-Kumar, D. I. Reisis, and Q. F. Stout, "Parallel Computations on Reconfigurable Meshes," Technical Report TR IRIS#229, Department of Computer Science, University of Southern California, 1987.

[23] R. Miller, V. N. Prasanna-Kumar, D. Reisis, and Q. Stout, "Parallel Computing on Reconfigurable Meshes," *IEEE Trans. Computers*, vol. 42, no. 6, pp. 678–692, 1993.

[24] M. Nigam and S. Sahni, "Sorting n Numbers on $n \times n$ Reconfigurable Meshes with Buses," *J. Parallel Distrib. Comput.*, vol. 23, pp. 37–48, 1994.

[25] N. Nisan and A. Ta-Shma, "Symmetric Logspace Is Closed under Complement," *Proc. 27th ACM Symp. Theory of Comput.*, 1995, pp. 140–146.

[26] S. Olariu and J. L. Schwing, "A Novel Deterministic Sampling Scheme with Applications to Broadcast-Efficient Sorting on the Reconfigurable Mesh," *J. Parallel Distrib. Comput.*, vol. 32, pp. 215–222, 1996.

[27] S. Olariu, J. L. Schwing, and J. Zhang, "On the Power of Two-Dimensional Processor Arrays with Reconfigurable Bus Systems," *Parallel Proc. Lett.*, vol. 1, no. 1, pp. 29–34, 1991.

[28] S. Olariu, J. L. Schwing, and J. Zhang, "Fundamental Data Movement Algorithms for Reconfigurable Meshes," *Proc. 11th Int. Phoenix Conf. on Comp. and Comm.*, 1992, pp. 472–479.

[29] S. Olariu, J. Schwing, and J. Zhang, "Applications of Reconfigurable Meshes to Constant Time Computations," *Parallel Comput.*, vol. 19, no. 2, pp. 229–237, 1993.

[30] Y. Pan and K. Li, "Linear Array with a Reconfigurable Pipelined Bus System: Concepts and Applications," *Info. Sc.*, vol. 106, nos. 3/4, pp. 237–258, 1998.

[31] S. Pavel and S. G. Akl, "Integer Sorting and Routing in Arrays with Reconfigurable Optical Buses," *Proc. Int. Conf. on Parallel Proc.*, 1996, pp. III-90–III-94.

[32] O. Reingold, "Undirected ST-Connectivity in Log-Space," *Proc. ACM Symp. on the Theory of Comput.*, 2005, pp. 376–385.

[33] K. Roy, J. L. Trahan, and R. Vaidyanathan, "Configuring the Circuit Switched Tree for Well-Nested and Multicast Communication," *Proc. IASTED Int. Conf. on Parallel and Distrib. Comput. Syst.*, 2006, pp. 271–285.

[34] K. Roy, R. Vaidyanathan, and J. L. Trahan, "Routing Multiple Width Communications on the Circuit Switched Tree," *Int. J. Foundations of Comp. Sc.*, vol. 17, no. 2, pp. 271–285, 2006.

[35] S. Sahni, "Data Manipulation on the Distributed Memory Bus Computer," *Parallel Proc. Lett.*, vol. 5, no. 1, pp. 3–14, 1995.

[36] A. Schuster "Dynamic Reconfiguring Networks for Parallel Computers: Algorithms and Complexity Bounds," PhD Thesis, Dept. of Computer Science, Hebrew University, Israel, 1991.

[37] R. P. S. Sidhu, A. Mei, and V. K. Prasanna, "String Matching on Multicontext FPGAs Using Self-Reconfiguration," *Proc. 1999 ACM/SIGDA 7th Int. Symp. on Field Programmable Gate Arrays*, 1999, pp. 217–226.

[38] R. Sidhu, S. Wadhwa, A. Mei, and V. K. Prasanna, "A Self-Reconfigurable Gate Array Architecture," *Proc. 10th Int. Workshop on Field-Programmable Logic and Appl.* (Springer-Verlag Lecture Notes in Comp. Sci. vol. 1896), 2000, pp. 106–120.

[39] L. Stockmeyer and U. Vishkin, "Simulation of Parallel Random Access Machines by Circuits," *SIAM J. Comput.*, vol. 13, no. 2, pp. 409–422, 1984.

[40] J. L. Trahan, A. G. Bourgeois, and R. Vaidyanathan, "Tighter and Broader Complexity Results for Reconfigurable Models," *Parallel Proc. Lett.*, vol. 8, no. 3, pp. 271–282, 1998.

[41] J. L. Trahan, M. C. Loui, and V. Ramachandran, "Multiplication, Division, and Shift Instructions in Parallel Random Access Machines," *Theoret. Comp. Sci.*, vol. 100, no. 1, pp. 1–44, 1992.

[42] J. L. Trahan, R. Vaidyanathan, and U. K. K. Manchikatla, "Maximum Finding on the Reconfigurable Mesh," manuscript, 1996.

[43] J. L. Trahan, R. Vaidyanathan, and C. P. Subbaraman, "Constant Time Graph Algorithms on the Reconfigurable Multiple Bus Machine," *J. Parallel Distrib. Comput.*, vol. 46, no. 1, pp. 1–14, 1997.

[44] J. L. Trahan, R. Vaidyanathan, and R. K. Thiruchelvan, "On the Power of Segmenting and Fusing Buses," *J. Parallel Distrib. Comput.*, vol. 34, no. 1, pp. 82–94, 1996.

[45] R. Vaidyanathan, "Sorting on PRAMs with Reconfigurable Buses," *Info. Proc. Lett.*, vol. 42, no. 4, pp. 203–208, 1992.

[46] R. Vaidyanathan and J. L. Trahan, "Optimal Simulation of Multidimensional Reconfigurable Meshes by Two-Dimensional Reconfigurable Meshes," *Info. Proc. Lett.*, vol. 47, no. 5, pp. 267–273, 1993.

[47] R. Vaidyanathan and J. L. Trahan, *Dynamic Reconfiguration: Architectures and Algorithms*, Kluwer Academic/Plenum Publishers, New York, 2003.

[48] H. Venkateswaran, "Properties That Characterize LOGCFL," *J. Comp. Syst. Sci.*, vol. 43, no. 2, pp. 380–404, 1991.

[49] B. F. Wang and G. H. Chen, "Two-Dimensional Processor Array with a Reconfigurable Bus System Is at Least as Powerful as CRCW Model," *Info. Proc. Lett.*, vol. 36, no. 1, pp. 31–36, 1990.

[50] B. F. Wang and G. H. Chen, "Constant Time Algorithms for the Transitive Closure and Some Related Graph Problems on Processor Arrays with Reconfigurable Bus Systems," *IEEE Trans. Parallel and Distrib. Syst.*, vol. 1, no. 4, pp. 500–507, 1990.

[51] B. F. Wang, G. H. Chen, and F. C. Lin, "Constant Time Sorting on a Processor Array with a Reconfigurable Bus System," *Info. Proc. Lett.*, vol. 34, pp. 187–192, 1990.

15
Fundamental Algorithms on the Reconfigurable Mesh

15.1	Introduction...	15-1
15.2	Basic Algorithms	15-3
	Leftmost Finding • Compression • Simple Prefix Sums • Prefix Remainders • Prime Numbers Selection • Logarithmic-Time Prefix Sums • Integer Summing Using the Carry Lookahead Generator	
15.3	Parallel Prefix-Remainders Technique..................	15-7
15.4	The Mixed Radix Representation Technique	15-9
15.5	Snake-Like Embedding Technique	15-11
15.6	Algorithm for Summing Integers	15-13
15.7	Conclusions	15-15
	References	15-16

Koji Nakano
Hiroshima University

15.1 Introduction

A *reconfigurable mesh* is a processor array that consists of processors arranged in a one- or two-dimensional grid with a reconfigurable bus system (Figure 15.1). The reconfigurable mesh of size $m \times n$ consists of mn processors arranged in a two-dimensional grid. For later reference, let PE(i, j) ($0 \leq i \leq m - 1$ and $0 \leq j \leq n - 1$) denote a processor at position (i, j). Also, one-dimensional reconfigurable mesh of size m is the reconfigurable mesh with a single row. Similarly, let PE(j) ($0 \leq j \leq m - 1$) denote a processor at position j. The control mechanism of the reconfigurable mesh is based on the SIMD principle. A single control unit dispatches instructions to each processor. Although all processors execute the same instructions, their behaviors may differ, because they work on different input and different coordinates. On the reconfigurable mesh, any two adjacent processors are connected with a single fixed link.

Each processor has locally controllable switches that can configure the connection patterns of its four ports denoted by N (*North*), E (*East*), W (*West*), and S (*South*). Figure 15.2 illustrates all possible 15 connection patterns of a processor. We can classify 15 connection patterns as follows:

Null No port is connected
Basic patterns Two ports are connected. There are six basic patterns, NS, WE, NW, NE, SE, and SW
Diagonal patterns A pair of two ports are connected such as NW·SE or NE·SW

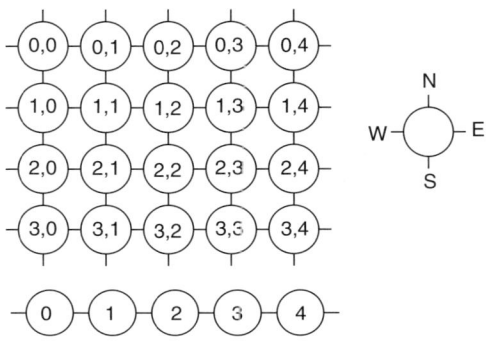

FIGURE 15.1 The reconfigurable mesh.

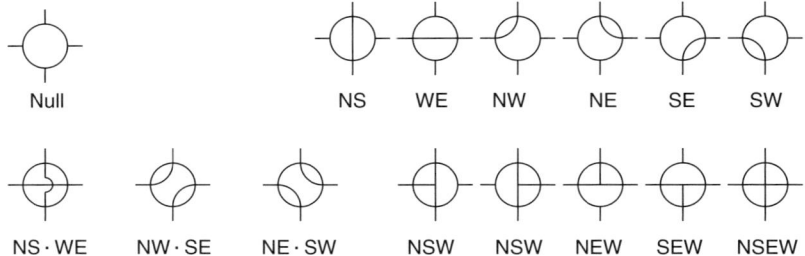

FIGURE 15.2 The local connection of processors.

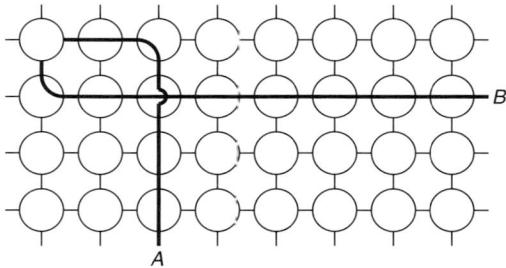

FIGURE 15.3 An example of subbuses.

Cross-over pattern A pair of two ports are connected such as NS·WE

Branch patterns Three or four ports are connected. There are five branch patterns, NSW, NSE, NEW, SEW, and NSEW

The computing power of the reconfigurable meshes depends on the connection patterns that are allowed. For example, if the *cross-over pattern* is not allowed, a $\sqrt{n} \times \sqrt{n}$ RM requires $\Omega(\log^* n)$ time to compute the parity of n binary values, whereas if all patterns are allowed it can compute the parity in constant time [21]. If the *branch patterns* are allowed, the connected components of an n-node graph can be labeled in constant time on an $n \times n$ RM [37], whereas we have not found an algorithm that can do this labeling in constant time when branch patterns are not allowed. From the practical point of view, the branch patterns are unrealistic, because it splits the energy and the signal decays. All parallel algorithms on the reconfigurable mesh presented in this chapter do not use the branch patterns.

The connected components formed by adjacent fixed links and internal connections constitute *subbuses*, through which the processors can communicate. Data sent to a port can be transferred through subbuses in one unit of time even if the length of the bus is very long. Figure 15.3 illustrates

Fundamental Algorithms on the Reconfigurable Mesh

an example of subbuses. The data sent to port E of PE(0, 0) is transferred through subbus A and it can be received by processors PE(0, 1), PE(0, 2), PE(1, 2), PE(2, 2), and PE(3, 2). Similarly, the data sent to port S of PE(0, 0) is transferred through subbus B and it can be received by processors PE(1, 0), PE(1, 1), PE(1, 2), ..., PE(1, 7).

The reconfigurable meshes have attracted considerable attention as theoretical models of parallel computation, and many studies have been devoted to developing efficient parallel algorithms on reconfigurable meshes. For example, they efficiently solve problems such as sorting [27, 33], selection [8, 10], arithmetic operations [2, 6, 20], graph problems [17, 22, 36], geometric problems [26, 30], and image processing [1, 3, 18, 19]. See Reference 34 for the comprehensive survey.

In this chapter, we focus on several fundamental techniques on the reconfigurable mesh. In particular, we show the parallel prefix-remainders technique that was first presented in Reference 25. This technique is very helpful for reducing the number of processors for solving various problems on the reconfigurable mesh. We also present the mixed radix representation technique and the snake-like embedding technique that are also used to reduce the number of processors. The remainder of this chapter is organized as follows: Section 15.2 shows basic algorithms on the reconfigurable mesh. In Section 15.3, we then go on to show the parallel-prefix remainders technique. Using this technique, we show that the sum of n binary values can be computed in $O(1)$ time on the reconfigurable mesh of size $(\log n)^2 / \log \log n \times n$. Section 15.4 presents the mixed radix representation technique on the reconfigurable mesh and shows that the prefix sums of n binary values can be computed in $O(\log \log n)$ time on the reconfigurable mesh of size $(\log n)^2/(\log \log n)^3 \times n$ using this technique. In Section 15.5, we present the snake-like embedding technique and show that the sum of n binary values can be computed in $O(\log^* n)$ time on the reconfigurable mesh of size $\sqrt{n} \times \sqrt{n}$. Finally, in Section 15.6, we show integer summing algorithms using these techniques.

15.2 Basic Algorithms

This section is devoted to show basic techniques on the bit and word reconfigurable mesh.

15.2.1 Leftmost Finding [25]

Suppose that n binary values $A = \langle a_0, a_1, \ldots, a_{n-1} \rangle$ given one for each processor on the one-dimensional reconfigurable mesh of size n. The leftmost 1 of A can be determined easily in constant time as follows. PE(i) connects WE if $a_i = 0$, and it sends 1 to E and tries to receive from W. After that, if PE(i) fails to receive 1, it is the leftmost processor. Figure 15.4 illustrates an example of the subbuses in the leftmost finding algorithms. Processors PE(2), PE(4), PE(5), and PE(7) send 1 to port E and PE(3), PE(4), ..., PE(7) receive 1 from W. Thus, PE(2) can learn that it has the leftmost 1. Therefore, it is easy to see that leftmost finding can be done in constant time.

Using the leftmost finding technique, the logical OR of n Boolean values can be computed in constant time. That is, using leftmost finding, the leftmost "true" can be found. If there exists a leftmost then the logical OR is true, and otherwise, it is false. Hence, the logical OR can be determined in constant time. Also, the logical AND can be determined in the same manner.

Lemma 15.1 *The leftmost 1, the logical OR, and the logical AND of n binary values can be computed in $O(1)$ time on the one-dimensional reconfigurable mesh of size n.*

FIGURE 15.4 Leftmost finding.

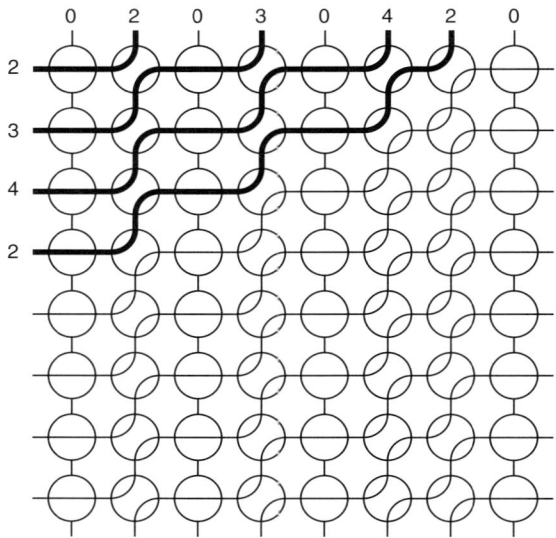

FIGURE 15.5 Compression.

15.2.2 Compression [25]

The compression is an algorithm on an $m \times n$ reconfigurable mesh that compresses a sequence of integers. That is, for a given sequence $A = \langle a_0, a_1, \ldots, a_{n-1} \rangle$ of integers, the compressed sequence $A' = \langle a_{i_0}, a_{i_1}, \ldots, a_{i_{m'-1}} \rangle$ of A $(i_0 < i_1 < \cdots < i_{m'-1})$ consists of the nonzero integers, where m' is the minimum of m and the number of nonzero integers. Hence, if m is smaller than the number of nonzero integers, some of them are lost by the compression. We assume that each a_j is given to PE$(0, j)$. At the end of compression, each a_{i_k} is stored in the local memory of PE$(k, 0)$. The compression can be done in constant time. Each column functions as a stack and each processor in the top row functions as the top of it. From the rightmost to the leftmost column, if an integer given to a column is not zero, then the processors on the column push up the integer on the stack by the connections of NW·SE. Otherwise, the processors hold the stack by connection of WE. Finally, each processor in the leftmost column knows a nonzero integer. See Figure 15.5 for an example of this bus configuration. Using this bus configuration, the compression can be done in constant time.

Lemma 15.2 *The compression of n integers can be done in $O(1)$ time on the reconfigurable mesh of size $m \times n$.*

15.2.3 Simple Prefix Sums [37]

The simple prefix sum is an algorithm on an $(n + 1) \times n$ reconfigurable mesh. Suppose that n binary values $A = \langle a_0, a_1, \ldots, a_{n-1} \rangle$ are given such that each PE$(0, j)$ is storing a_j $(0 \leq j \leq n - 1)$. Our goal is to compute all the prefix sums $s_j = a_0 + a_1 + \cdots + a_j$ $(0 \leq j \leq n - 1)$. If a binary value given to a column is 0, every processor in it connects ports E and W; otherwise, it sets buses just as for pushing down (Figure 15.6). Then, PE$(0, 0)$ sends the signal to port W, and every processor tries to receive it from E. The position of the processor that succeeds in receiving the signal corresponds to the prefix sums. It should have no difficulty to confirm that if PE(i, j) receives the signal from port E then, the prefix sum $s_j = i$. Therefore, the prefix sums problem can be solved in constant time on $(n + 1) \times n$, the reconfigurable mesh of size $(n + 1) \times n$.

Lemma 15.3 *The prefix sums of n binary values can be computed in $O(1)$ time on the reconfigurable mesh of size $(n + 1) \times n$.*

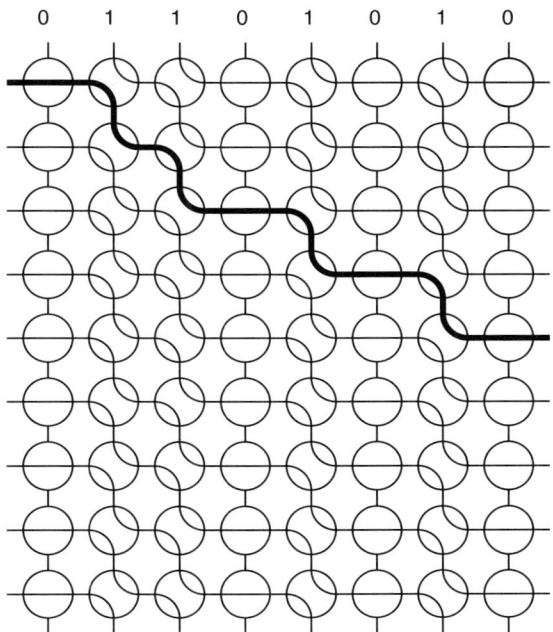

FIGURE 15.6 Simple prefix-sums algorithm.

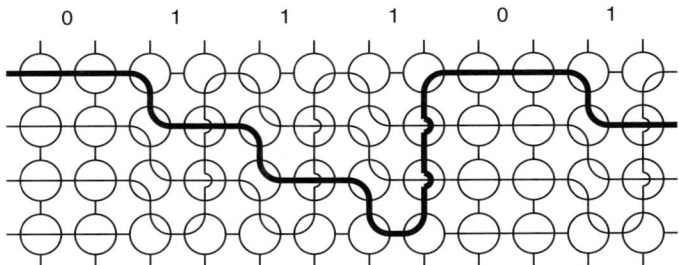

FIGURE 15.7 Prefix w-remainders.

15.2.4 Prefix Remainders [25]

The prefix w-remainders is an algorithm to compute a sequence $\langle s_0^w, s_1^w, \ldots, s_{n-1}^w \rangle$ of integers, where $s_i^w = (a_0 + a_1 + \cdots + a_i) \bmod w$ for given n binary values $A = \langle a_0, a_1, \ldots, a_{n-1} \rangle$. Each a_j ($0 \leq i \leq n-1$) is given to PE$(0, 2j)$. In the algorithm, each column functions as a cyclic shift register of size w. In the leftmost column, only the bottom element of the cyclic shift register is 1. In each column, if the element given to the column is 1, then the processors in the column shift the cyclic shift register, otherwise, they hold it. Then each prefix remainders is equal to the position where 1 is found on the cyclic shift register. More specifically, PE$(0, 0)$ sends the signal to port E. Every PE$(i, 2j + 1)$ tries to receive it from W. If PE$(i, 2j + 1)$ succeeds in receiving the signal, then $s_j^w = i$. See Figure 15.7 for an example of the subbuses for computing the prefix w-remainders. Hence, the prefix w-remainders can be computed in constant time on a $(w + 1) \times 2n$ reconfigurable mesh. Furthermore, by adding the prefix sums of even columns and those of odd columns, the prefix sums of all columns can be computed. Therefore, the prefix w-remainders of n binary values can also be computed in constant time on a $(w + 1) \times n$ reconfigurable mesh.

Lemma 15.4 *The prefix w-remainders of n binary values can be computed in O(1) time on the reconfigurable mesh of size $(w + 1) \times n$.*

15.2.5 Prime Numbers Selection

The goal of the prime numbers selection is to select prime numbers no more than n. The prime numbers selection can be done in constant time on the reconfigurable mesh of size $\sqrt{n} \times n$ as follows. First, each PE(i, j) $(2 \leq i \leq \sqrt{n}$ and $2 \leq j \leq n)$ computes $j \bmod i$ by the local computation. Using the logical OR in each j-th column, we can determine if there exists i $(2 \leq i \leq \sqrt{j})$ satisfying $j \bmod i = 0$. If such i does not exist, j is a prime number. Therefore, processors in each j-th column can learn whether j is a prime number in constant time.

Lemma 15.5 *The prime numbers no more than n can be selected in O(1) time on the reconfigurable mesh of size $\sqrt{n} \times n$.*

Also, by obvious application of the compression (Lemma 15.2), we can find first $\sqrt{n}$ prime numbers such that PE$(i, 0)$ knows the i-th smallest prime number.

15.2.6 Logarithmic-Time Prefix Sums

It is easy to compute the prefix sums for given n integers $A = \langle a_0, a_1, \ldots, a_{n-1}\rangle$ on a one-dimensional reconfigurable mesh. Each a_i is given to PE(i), and after executing the algorithm, each PE(i) knows the value of the i-th prefix sums, $s_i = a_0 + a_1 + \cdots + a_i$. The algorithm is based on the prefix-sums algorithm on the PRAM [11, 12]. If $n = 1$, then the prefix sum is a_0 and the algorithm is terminated. Otherwise, the respective prefix sums of $\langle a_0, a_1, \ldots, a_{n/2-1}\rangle$ on the processors PE(0), PE(1), $\ldots$, PE$(n/2 - 1)$, and of $\langle a_{n/2}, a_{n/2+1}, \ldots, a_{n-1}\rangle$ on the processors PE$(n/2)$, PE$(n/2 + 1)$, $\ldots$, PE$(n - 1)$ are computed independently. At that, each PE(i) $(0 \leq i \leq n/2 - 1)$ knows s_i and PE(i) $(n/2 \leq i \leq n - 1)$ knows $a_{n/2} + a_{n/2+1} + \cdots + a_n$. PE$(n/2 - 1)$ broadcasts $s_{n/2-1}$ to all processors. Each PE(i) $(n/2 \leq i \leq n - 1)$ receives it and adds it to the respective local prefix sums $a_{n/2} + a_{n/2+1} + \cdots + a_i$. Since $s_i = s_{n/2-1} + a_{n/2} + a_{n/2+1} + \cdots + a_i$, the values thus obtained are the prefix sums s_i of the input. It should be clear that the depth of the recursion is $\log n$ and each recursion can be done in $O(1)$ time. Therefore, the prefix sums can be computed in $O(\log n)$ time on a one-dimensional reconfigurable mesh of size n. Note that the binary operator "+" can be any associative binary operator. For example, all the prefix products $a_0 \cdot a_1 \cdots a_i$ can be computed in the same manner in $O(\log n)$ time.

Lemma 15.6 *The prefix sums of n integers can be computed in $O(\log n)$ time on the one-dimensional reconfigurable mesh of size n.*

15.2.7 Integer Summing Using the Carry Lookahead Generator [16]

Suppose that we need to compute the sum of n d-bit integers $A_0, A_1, \ldots, A_n$. Let the binary representation of each A_i be $a_{i,d-1} a_{i,d-2} \cdots, a_{i,0}$. Our goal is to compute the sum of the n d-bit integers in $O(1)$ time on the reconfigurable mesh. This can be done as follows. Let $B = b_{d-1} b_{d-2} \cdots b_0$ be the binary representation of the sum. It is easy to see that the binary representation can be computed by the following formulas:

$$b_0 = (a_{0,0} + a_{1,0} + \cdots + a_{n-1,0}) \bmod 2$$
$$c_0 = \lfloor (a_{0,0} + a_{1,0} + \cdots + a_{n-1,0})/2 \rfloor$$
$$b_j = (c_{j-1} + a_{0,j-1} + a_{1,j-1} + \cdots + a_{n-1,j-1}) \bmod 2 \quad (1 \leq j \leq d - 1)$$
$$c_j = \lfloor (c_{j-1} + a_{0,j-1} + a_{1,j-1} + \cdots + a_{n-1,j-1})/2 \rfloor \quad (1 \leq j \leq d - 1)$$

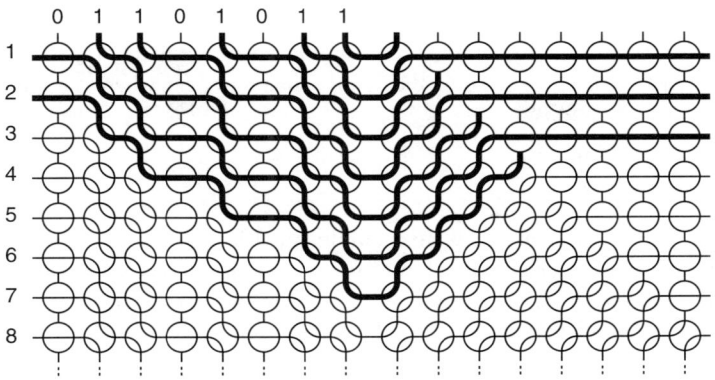

FIGURE 15.8 The sum of binary values and its division.

Note that $c_0, c_1, \ldots, c_{d-1}$ are the carry to the next bit. The readers should have no difficulty to confirm that the formula above is a straightforward method of the naive integer summing that computes the sum from the LSB. Using these formulas, the binary representation of the sum is computed. Figure 15.8 illustrates the subbuses of the reconfigurable mesh to compute the sum b_j and the carry c_j for $n = 8$. The left submesh computes $c_{j-1} + a_{0,j-1} + a_{1,j-1} + \cdots + a_{8,j-1}$ and the right submesh divides it by two. In the figure, the carry from the lower bit is $c_{j-1} = 2$. $0 + 1 + 1 + 0 + 1 + 0 + 1 + 1$ is added to it. Since in the rightmost column of the left submesh, the processors from 1st to 7th row receive the signal, the sum is 7. After that $c_j = \lfloor 7/2 \rfloor$ is computed in the right submesh. Since in the rightmost column of the right submesh, the processors from 1st to 3rd row receive the signal, $c_j = 3$. In this way, the sum and the carry can be computed. Since $c_j < n$ and $b_j < 2n$, b_j and c_j can be computed on the reconfigurable mesh of $2n \times 2n$. Therefore, $b_0, b_1, \ldots, b_{d-1}$ can be computed on the reconfigurable mesh of size $2n \times 2dn$. Consequently, we have the following.

Lemma 15.7 *The sum of n d-bit integers can be computed in $O(1)$ time on the reconfigurable mesh of size $2n \times 2dn$.*

The product of two n-bit integers can be computed by using the integer summing. Let $X = x_{n-1}x_{n-2}\cdots x_0$ and $Y = y_{n-1}y_{n-2}\cdots y_0$ be two n-bit integers. Let each X'_i ($0 \leq i \leq n-1$) be $(2n-1)$-bit integers such that

$$X'_i = x_{n-1}x_{n-2}\ldots x_0 0^i \quad \text{if } y_i = 1$$
$$X'_i = 0 \quad \text{if } y_i = 0$$

where 0^i denotes the consecutive i 0s. The product XY is equal to the sum $X'_0 + X'_1 + \cdots X'_{n-1}$. From Lemma 15.7, the product XY can be computed by computing the sum of n $(2n-1)$-bit integers. Therefore, we have the following.

Lemma 15.8 *The product of two n-bit integers can be computed in $O(1)$ time on the reconfigurable mesh of size $2n \times 4n^2$.*

15.3 Parallel Prefix-Remainders Technique

This section is devoted to show the parallel prefix-remainders technique on the reconfigurable mesh. Using this technique, we can compute the sum of binary values in a few number of processors on the reconfigurable mesh.

Suppose that n binary values $A = \langle a_0, a_1, \ldots, a_{n-1} \rangle$ are given to the reconfigurable mesh of size $m \times n$ such that each $PE(0, j)$ is storing a_j. Our goal is to compute the sum $x = a_0 + a_1 + \cdots + a_{n-1}$.

Let $p_1 (= 2), p_2 (= 3), p_3 (= 5), \ldots, p_q$ be the first q prime numbers, where q is the maximum integer satisfying the following conditions:

1. $q \leq \sqrt{m/\log m}$
2. $p_q \leq \sqrt{m \log m} - 1$

If m is too small (e.g., $m = 2$), then $q = 0$. To avoid this situation, we assume that $m \geq 16$ that ensures $q \geq 2$. If $m < 16$, the sum can be computed in $O(\log n)$ time using the algorithm for Lemma 15.6. We will show that $x \bmod p_1 \cdot p_2 \cdots p_q$ can be computed on the reconfigurable mesh of size $m \times n$.

The parallel prefix-remainders technique uses the Chinese remainder theorem as follows.

Theorem 15.9 (Chinese remainder theorem) *Suppose $n_1, n_2, \ldots, n_q$ are integers that are pairwise coprime. Then, for any given integers $a_1, a_2, \ldots a_q$, there exists an integer x solving the system of simultaneous congruences $x = a_i \pmod{n_i}$ $(1 \leq i \leq q)$.*

Using the Chinese remainder theorem, we can obtain the remainder for larger modulo. Suppose that we have $x \bmod p_1, x \bmod p_2, \ldots, x \bmod p_q$ for nonnegative integer x. From the Chinese Remainder Theorem, if $x \bmod p_i = y \bmod p_i$ for all i $(1 \leq i \leq q)$, then $x \bmod p_1 \cdot p_2 \cdots p_q = y \bmod p_1 \cdot p_2 \cdots p_q$ holds. Using this fact, the value of $x \bmod p_1 \cdot p_2 \cdots p_q$ can be computed.

The details of the parallel prefix-remainders algorithm are as follows:

Step 1: Select prime numbers $p_1, p_2, \ldots, p_q$ in $O(1)$ time by the prime number selection (Lemma 15.5).

Step 2: Suppose that an $m \times n$ reconfigurable mesh is divided into $\sqrt{m/\log m}$ submeshes, each of size $\sqrt{m \log m} \times n$ (Figure 15.9). Compute the prefix p_i-remainders of A on each i-th submesh $(1 \leq i \leq q)$. Each processor in the rightmost column of each i-th submesh $(1 \leq i \leq q)$ broadcasts the value of $x \bmod p_i$ to all processors on the same submesh.

Step 3: Processors in each j-th column $(0 \leq j \leq n-1)$ of each i-th submesh compute $j \bmod p_i$ by local computation.

Step 4: In each j-th column of each i-th submesh $(0 \leq j \leq n-1, 1 \leq i \leq q)$, check whether $x \bmod p_i = j \bmod p_i$ using the logical AND (Lemma 15.1). Let y_j $(0 \leq j \leq n-1)$ be the binary value such that $y_j = 1$ iff $x \bmod p_i = j \bmod p_i$ for all i $(1 \leq i \leq q)$.

Step 5: Find the minimum j such that $y_j = 1$ by the leftmost finding (Lemma 15.1). Clearly, for such j, $j = x \bmod p_1 \cdot p_2 \cdots p_q$ holds.

Since each step can be performed in constant time, we have the following.

Lemma 15.10 *For n binary values $A = \langle a_0, a_1, \ldots, a_{n-1} \rangle$, let $x = a_0 + a_1 + \cdots + a_{n-1}$. The remainder $x \bmod p_1 \cdot p_2 \cdots p_q$ can be computed in $O(1)$ time on the reconfigurable mesh of size $m \times n$.*

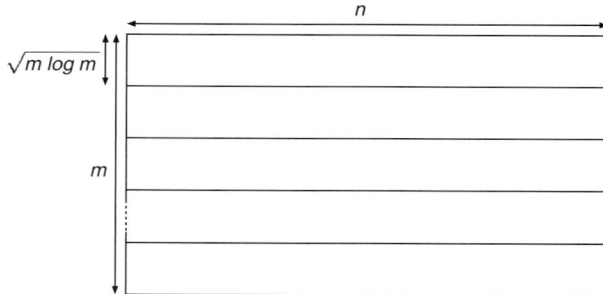

FIGURE 15.9 Division of reconfigurable mesh.

Next, to get the algorithm for computing the sum, we will estimate the value of $p_1 \cdot p_2 \cdots p_q$. From the prime number theorem,

$$\lim_{m \to \infty} \frac{q \ln p_q}{p_q} = 1$$

where ln is the natural logarithm. Hence, there exists a constant number c (> 0) and m_0 such that for all $m \geq m_0$,

1. $q/2 \geq c \times \sqrt{m/\log m}$
2. $p_{q/2} \geq c \times \sqrt{m \log m}$

Therefore, we have

$$\log(p_1 \cdot p_2 \cdots p_q) \geq \log p_{q/2} + \log p_{q/2+1} + \cdots + \log p_q$$
$$\geq c \times \sqrt{m/\log m} \times (\log c + \log m/2 + \log \log m/2)$$
$$\geq (c/2) \times \sqrt{m \log m}$$

Thus, there exists a constant number c (> 0) and m_0 such that for all $m \geq m_0$, $p_1 \cdot p_2 \cdot p_3 \cdots p_q \geq c^{\sqrt{m \log m}}$ holds.

We will show an algorithm to compute the sum of binary values by using the above remainder algorithm. Let $B = \langle b_0, b_1, \ldots, b_{n-1} \rangle$ be a binary sequence of length n determined as follows:

$$b_j = 1 \quad \text{if } a_j = 1, \text{ and } (a_0 + a_1 + \cdots + a_j) \bmod p_k = 0 \text{ for all } (1 \leq k \leq q)$$
$$= 0 \quad \text{otherwise}$$

The sequence B can be computed by the prefix-remainders technique (Lemma 15.10). It should be clear that $b_j = 1$ iff $a_j = 1$ and $(a_0 + a_1 + \cdots + a_j) \bmod p_1 p_2 \cdots p_q = 0$. Intuitively, a binary value in B is 1 for every $p_1 p_2 \cdots p_q$ 1s in A. Therefore, for such bs, the following equality holds:

$$x = (x \bmod p_1 \cdot p_2 \cdots p_q) + p_1 \cdot p_2 \cdots p_q \cdot (b_0 + b_1 + \cdots + b_{n-1})$$

Thus, recursively computing the sum of n binary values $B = \langle b_0, b_1, \ldots, b_{n-1} \rangle$ yields the sum of A. The recursion is terminated if all bs are zero. Let t be the depth of the recursion. Then t must be the minimum integer such that $(p_1 \cdot p_2 \cdots p_q)^t \geq n$. Since $p_1 \cdot p_2 \cdot p_3 \cdots p_q \geq c^{\sqrt{m \log m}}$, we have $t = O(\log n / \sqrt{m \log m})$. Therefore, we have

Theorem 15.11 *The sum of n binary values can be computed in $O(\log n / \sqrt{m \log m})$ time on the reconfigurable mesh of size $m \times n$.*

If $m = (\log n)^2 / \log \log n$, then $\log n / \sqrt{m \log m} = O(1)$. Thus, the sum of n binary values can be computed in $O(1)$ time on the reconfigurable mesh of size an $(\log n)^2 / \log \log n \times n$.

15.4 The Mixed Radix Representation Technique

This section shows the mixed radix representation technique that can be used to compute the prefix sums on the reconfigurable mesh. Again, let p_i be the i-th smallest prime number.

We usually use the fixed radix representation as follows: Let x be a nonnegative integer. The r-ary representation of x is $b_{k-1} b_{k-2} \cdots b_0$ such that $0 \leq b_i < r$ for all i ($0 \leq i \leq k-1$) and $x = b_0 r^0 + b_1 r^1 + \cdots + b_{k-1} r^{k-1}$. We use the following mixed radix representation as follows: The x-ary representation of x

is $v_{k-1}v_{k-2}\cdots v_0$ such that $0 \le v_i < p_{i+1}$ for all i and $x = v_0 + v_1 p_1 + v_2 p_1 p_2 + \cdots + v_{k-1} p_1 p_2 \cdots p_{k-1}$. If x is given, then the mixed radix representation can be computed by the following formula:

$$v_i = \lfloor \frac{x}{p_1 \cdot p_2 \cdots p_{i-1}} \rfloor \bmod p_i$$

where $p_1 \cdot p_2 \cdots p_{i-1} = 1$ if $i = 1$ for simplicity.

Using the mixed radix representation, all the prefix $p_1 p_2 \cdots p_q$-remainders can be computed efficiently on the reconfigurable mesh. Let $s_j = a_0 + a_1 + \cdots + a_j$ be the j-th prefix sum of $A = \langle a_0, a_1, \ldots, a_{n-1} \rangle$. Our idea is to compute the mixed radix representation $v_{j,k-1}v_{j,k-2}\cdots v_{j,0}$ of s_j for all j ($0 \le j \le n-1$). For this purpose, for each j and i ($0 \le j \le n-1$ and $1 \le i \le q$), $s_j \bmod p_i$ is computed by the prefix p_i-remainders. Then, let $b_{j,i}$ ($0 \le j \le n-1$ and $1 \le i \le q$) be binary value such that if $a_j = 1$ and $s_j \bmod p_1 = s_j \bmod p_2 = \cdots = s_j \bmod p_{i-1} = 0$ then $b_{j,i} = 1$, otherwise $b_{j,i} = 0$. Clearly, $b_{j,i} = 1$ if $a_j = 1$ and $s_j \bmod p_1 p_2 \cdots p_{i-1} = 0$ for all i. Thus, we have

$$b_{0,i} + b_{1,i} + \cdots + b_{j,i} = \lfloor \frac{s_j}{p_1 \cdot p_2 \cdots p_{i-1}} \rfloor.$$

Using this fact, we can obtain $v_{j,i} = \lfloor s_j / p_1 \cdot p_2 \cdots p_{i-1} \rfloor \bmod p_i$ by computing the prefix p_i-remainder of $\langle b_{0,i}, b_{1,i}, \ldots, b_{j,i} \rangle$. After that, each $p_1 p_2 \cdots p_q$-remainder $s_j = v_{j,0} + v_{j,1} p_1 + v_{j,2} p_1 p_2 + \cdots + v_{j,k-1} p_1 p_2 \cdots p_{q-1}$ is computed.

On the basis of the above idea, we have the prefix-remainders algorithm using the mixed radix representation as follows:

Step 1: Select prime numbers $p_1, p_2, \ldots, p_q$ in $O(1)$ time by the prime number selection (Lemma 15.5).

Step 2: Suppose that an $m \times n$ reconfigurable mesh is divided into $\sqrt{m/\log m}$ submeshes, each of size $\sqrt{m \log m} \times n$ (Figure 15.9). Compute the prefix p_i-remainders of A on each i-th submesh. At the end of Step 2, processors in the j-th column of the i-th submesh know the value of $s_j \bmod p_i$.

Step 3: In each j-th column such that $a_j = 1$, find minimum i such that $s_j \bmod p_i \ne 0$ using the leftmost finding (Lemma 15.1). Clearly, for such i, $s_j \bmod p_1 = s_j \bmod p_2 = \cdots = s_j \bmod p_{i-1} = 0$ and $s_j \bmod p_i \ne 0$ hold. For such i, let $b_{j,k} = 1$ if $a_j = 1$ and $k < i$, and $b_{j,k} = 0$ otherwise.

Step 4: In each i-th submesh, compute the prefix p_i-remainders of $B_i = \langle b_{0,i}, b_{1,i}, \ldots, b_{n-1,i} \rangle$. At the end of Step 4, processors in j-th column of i-th submesh know $v_{j,i} = (b_{0,i} + b_{1,i} + \cdots + b_{j,i}) \bmod p_i$.

Step 5: In each j-th column, compute $s_j = v_{j,0} + v_{j,1} p_1 + v_{j,2} p_1 p_2 + \cdots + v_{j,k-1} p_1 p_2 \cdots p_{q-1}$ by the logarithmic prefix-sums technique (Lemma 15.6).

Steps 1–4 can be done in $O(1)$ time. Step 5 needs $O(\log q) = O(\log m) = O(\log \log n)$ time. Thus, we have the following.

Lemma 15.12 *The prefix $p_1 p_2 \cdots p_q$-remainders of n binary values can be computed in $O(\log \log n)$ time on an $m \times n$ reconfigurable mesh.*

Similar to Theorem 15.11, we can prove that $O(\log n / \sqrt{m \log m})$ time iterations are sufficient to compute all the prefix sums. Consequently, we have the following.

Theorem 15.13 *The prefix sums of n binary values can be computed in $O(\log n / \sqrt{m \log m} + \log \log n)$ time on the reconfigurable mesh of size $n \times m$.*

If $m = (\log n)^2 / (\log \log n)^3$, then $\log n / \sqrt{m \log m} = O(\log \log n)$. Thus, the prefix sums of n binary values can be computed in $O(\log \log n)$ time on the reconfigurable mesh of size $(\log n)^2 / (\log \log n)^3 \times n$.

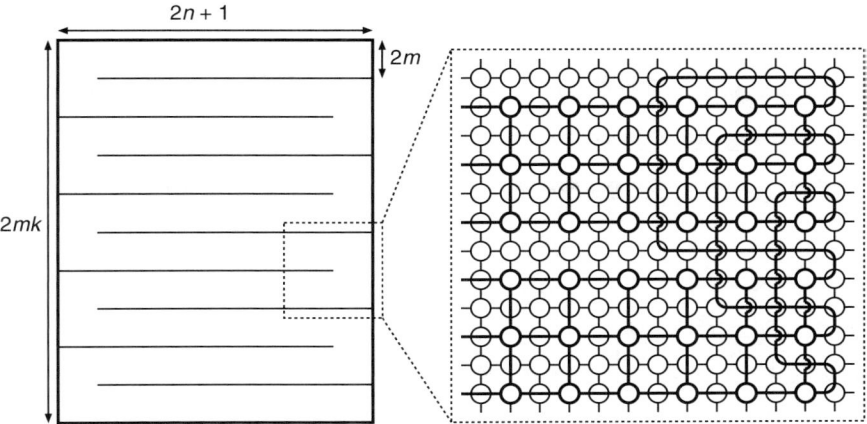

FIGURE 15.10 Snake-like embedding.

15.5 Snake-Like Embedding Technique

This section shows a snake-like embedding technique. Using this technique, we obtain an algorithm that computes the sum of n binary values on the reconfigurable mesh of size $\sqrt{n} \times \sqrt{n}$.

Figure 15.10 illustrates the *snake-like embedding* of an $m \times nk$ reconfigurable mesh in a $2mk \times (2n+1)$ reconfigurable mesh. In the snake-like embedding, a processor in the $m \times nk$ reconfigurable mesh corresponds to a processor whose position is in an even row and odd column of the $2mk \times (2n+1)$ reconfigurable mesh. In the figure, these processors are represented by thick circles, and the connections for the embedding are represented by thick lines. The $m \times nk$ reconfigurable mesh is bent $k-1$ times and is partitioned into k segments, each of which corresponds to consecutive $2m$ rows in the $2mk \times (2n+1)$ reconfigurable mesh. To complete the embedding, the connections between any two adjacent processors in the $m \times nk$ reconfigurable mesh is embedded in the $2mk \times (2n+1)$ reconfigurable mesh. Embedding of the connection between two adjacent processors in the same segment is trivial. To embed intersegment connections, m processors in the rightmost column of each segment should be connected one-to-one to m processors in the next segment. For one-to-one connections between the $2m$ processors, m odd rows in the $2mk \times (2n+1)$ reconfigurable mesh are used as shown in the figure. Similarly, m processors in the leftmost column can be connected to the corresponding processors in the previous segment.

Suppose that the algorithm for Lemma 15.10 is used to compute the remainder of nk binary values $A = \langle a_0, a_1, \ldots, a_{nk-1} \rangle$ on the snake-like embedding. The nk binary values are given to the reconfigurable mesh such that n binary values are given to each segment, one binary value for every two column (Figure 15.10). That is, the input is given to every $2m$ rows in the $2mk \times (2n+1)$ reconfigurable mesh. Let $x = a_0 + a_1 + \cdots + a_{nk-1}$. It should be clear that, using the algorithm for Lemma 15.10, the remainder $x \bmod p_1 p_2 \cdots p_q$ can be computed in $O(1)$ time.

As before, let $B = \langle b_0, b_1, \ldots, b_{nk-1} \rangle$ be the binary sequence such that $b_j = 1$ iff $a_j = 1$ and $s_j \bmod p_1 p_2 \cdots p_q = 0$, where $s_j = a_0 + a_1 + \cdots + a_j$. As shown in Theorem 15.11, by recursive computation of the remainder for B, the sum x of A can be computed in $O(\log(nk)/\sqrt{m \log m})$ time. We can reduce the computing time by compressing $B = \langle b_0, b_1, \ldots, b_{nk-1} \rangle$. Let $P_q = p_1 p_2 \cdots p_q$ and $C = \langle c_0, c_1, \ldots, c_{nk/P_q - 1} \rangle$ be the sequence such that $c_i = b_{i \cdot P_q} + b_{i \cdot P_q + 1} + \cdots + b_{i \cdot (P_q+1)-1}$ for all i ($0 \leq i \leq nk/P_q - 1$). Note that $b_j = 1$ iff $(a_0 + a_1 + \cdots + a_j) \bmod P_q = 0$ and $a_j = 1$. Hence, whenever $b_j = 1$, we have $b_{j+1} = b_{j+2} = \cdots = b_{j+P_q-1} = 0$. Since at most one of $b_{j \cdot P_q}, b_{j \cdot P_q + 1}, \ldots, b_{(j+1) \cdot P_q - 1}$ is 1 for each j, the value of each c_i is either 0 or 1 and can be determined by the logical OR of $b_{j \cdot P_q}, \ldots, b_{(j+1) \cdot P_q - 1}$. See Figure 15.11 for an example (in which a black circle denotes 1 and a white circle 0), where $n = 32$ and $P_q = 4$. Although 4 is an impossible value for P_q, the figure shows the essence of the algorithm. For c_is thus obtained, $x = (x \bmod P_q) + (c_0 + c_1 + \cdots + c_{nk/P_q - 1})P_q$ holds because each c_i with value 1

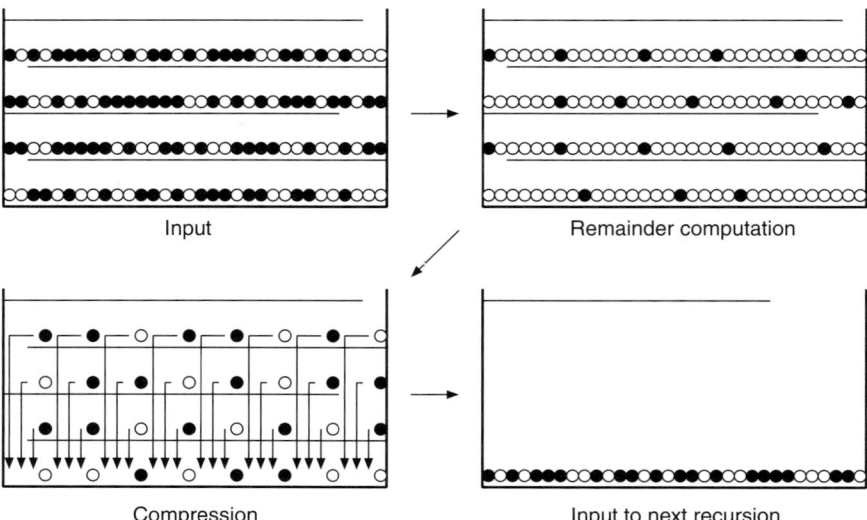

FIGURE 15.11 Remainder computation using snake-like embedding.

corresponds to P_q 1s in A. The sum x can therefore also be obtained by the recursive computation of the sum of $C = \langle c_0, c_1, \ldots, c_{nk/P_q-1} \rangle$. To accelerate the execution of the recursion, for each j, the values of $c_{jn}, \ldots, c_{(j+1)n-1}$ are transferred to a row, one for each processor. For example, $c_0, c_1, \ldots, c_{n-1}$ are transferred to a row as shown in Figure 15.11. Then all of the cs are distributed to every $2mP_q$ rows and the sum of $c_0, c_1, \ldots, c_{nk/P_q-1}$ is computed recursively. Therefore, in the following recursion, the computation of the sum of nk/P_q binary values is based on the embedding of an $mP_q \times nk/P_q$ reconfigurable mesh in the $m \times (2n+1)$ reconfigurable mesh. In each recursion, $(x \bmod P_q) + (c_0 + c_1 + \cdots + c_{nk/P_q-1})P_q$ can be computed in $O(1)$ time by using the algorithm for Lemma 1.8.

Finally, we will evaluate the depth of the recursion. Recall that, initially, $m \times nk$ reconfigurable mesh is embedded in the $2mk \times (2n+1)$ reconfigurable mesh. After the first iteration, $mP_q \times nk/P_q$ reconfigurable mesh is embedded. Let $R(t)$ be the number of rows of the reconfigurable mesh embedded in the $m \times (2n+1)$ reconfigurable mesh at the t-th iteration. It should be clear that $R(1) = m$ and $R(2) = mP_q$. We will compute the function $R(t)$ for all t. Again, let q be the maximum integer such that

1. $q \leq \sqrt{m/\log m}$
2. $P_q \leq \sqrt{m \log m} - 1$

For such q, let $Q(m)$ be the function such that

$$Q(m) = q_1 \cdot q_2 \cdots q_m$$

As we have proved before, there exists a constant $c > 0$ such that $Q(m) = c^{\sqrt{m \log m}}$. Using $Q(m)$, we have

$$R(t) = R(t-1)Q(R(t-1))$$
$$= R(t-1)c^{\sqrt{R(t-1) \log R(t-1)}}$$
$$> c^{R(t-2)}$$

The condition $R(t) \geq 2mk$ is sufficient for the termination of the recursion. Since $R(1) = m$, at most $t = O(\log^* k - \log^* m)$ is sufficient for $R(t) \geq 2mk$, where $\log^{(k+1)} n = \log(\log^{(k)} n)$ for all k, $\log^{(1)} n = \log n$, and $\log^* n$ is the minimum integer k such that $\log^{(k)} n \leq 1$. Replacing n, m, and k

by $\sqrt{n}$, $\sqrt{m}$, and $\sqrt{n}$, we can compute the sum of n binary values in $O(\log^* n - \log^* m)$ time on a $2\sqrt{nm} \times (2\sqrt{n} + 1)$ reconfigurable mesh.

Note that in this algorithm n binary values are given to a $2\sqrt{nm} \times (2\sqrt{n} + 1)$ reconfigurable mesh, one value for each $2\sqrt{m} \times 2$ submesh. Therefore, by executing this algorithm four times we can compute the sum of $4n$ binary values (given one for each $\sqrt{m} \times 1$ submesh). Using this technique, the size of the reconfigurable mesh can be reduced by a constant factor without asymptotically increasing the computing time. As a result, we have the following.

Theorem 15.14 *For n binary values given to every $\sqrt{m}$ rows of a $\sqrt{nm} \times \sqrt{n}$ reconfigurable mesh, the sum of these values can be computed in $O(\log^* n - \log^* m)$ ($1 \leq m \leq \log n$) time.*

Let $m = 1$. Then, we have the following.

Corollary 15.15 *The sum of n binary values given one for each processor on a $\sqrt{n} \times \sqrt{n}$ reconfigurable mesh can be computed in $O(\log^* n)$ time.*

15.6 Algorithm for Summing Integers

This section shows an algorithm that computes the sum of n d-bit integers in $O(\log^* n - \log^* m)$ ($1 \leq m \leq \log n$) time on a $\sqrt{nm} \times d\sqrt{n}$ reconfigurable mesh.

The algorithm is based on the *two-stage summing method* (Figure 15.12). For each i ($0 \leq i \leq n-1$), let A_i be a d-bit integer and $a_{i,d-1} a_{i,d-2} \cdots a_{i,0}$ its binary representation. The two-stage summing method computes the sum $x = A_0 + A_1 + \cdots + A_{n-1}$ as follows: In the first stage, for each j ($0 \leq j \leq d - 1$) the value of $B_j = a_{0,j} + a_{1,j} + \cdots + a_{n-1,j}$ is computed. In other words, the sum of each column of square A is computed and each B_j corresponds to each row of parallelogram B in the figure. For B_js thus obtained, computation of $B_0 \cdot 2^0 + B_1 \cdot 2^1 + \cdots + B_{d-1} \cdot 2^{d-1}$ in the second stage yields the sum x.

Figure 15.13 shows a reconfigurable mesh for computing the sum of n d-bit integers in a way based on the two-stage summing method. For each j, binary values $a_{0,j}, a_{1,j}, \ldots, a_{n-1,j}$ are given to a $\sqrt{nm} \times \sqrt{n}$ submesh corresponding to the j-th square A from the right in the figure. In each submesh A, $B_j = a_{0,j} + a_{1,j} + \cdots + a_{n-1,j}$ is computed in $O(\log^* n - \log^* m)$ time by the summing algorithm for Theorem 15.14.

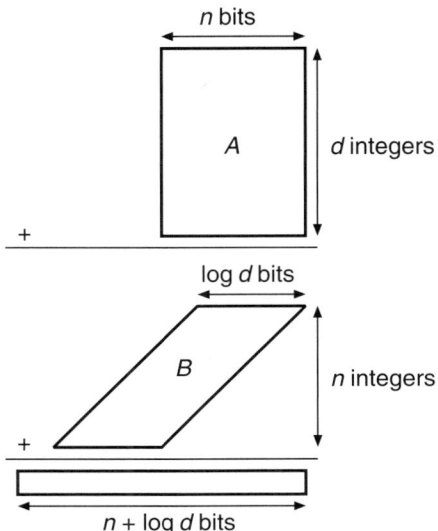

FIGURE 15.12 Two-stage summing method.

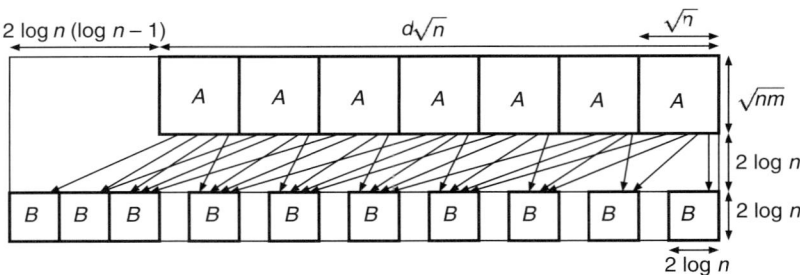

FIGURE 15.13 Integer summing.

Each sum thus obtained has $\log n$ bits, and each bit is transferred to a submesh B as follows: Since $B_j = b_{j,\log n} b_{j,\log n-1} \cdots b_{j,1}$ is computed in the j-th square A, $\log n$ bits are transferred to the j-th submesh B through the $(j + \log n - 1)$st subreconfigurable mesh B, one bit for each submesh. Hence, $2 \log n$ rows are enough to transfer every bit simultaneously. In the figure, these transfers are illustrated by arrows. After that, bits transferred to each submesh B in Figure 15.13 corresponds to a column of the parallelogram B in Figure 15.12. Therefore, to compute the sum of the parallelogram B in Figure 15.12, a $2 \log n \times 2 \log n$ carry lookahead generator is implemented in each submesh B. Each carry lookahead generator receives a carry from the previous carry generator, adds it to the sum of a column of the parallelogram B in Figure 15.12, and sends the carry to the following carry generator. Hence, in a way similar to the summing algorithm for Lemma 15.7, the sum of the parallelogram B can be computed in constant time. Therefore, the sum of n d-bit integers can be computed in $O(\log^* n - \log^* m)$ time on an $O(\sqrt{nm}) \times O(d\sqrt{n})$ reconfigurable mesh. Furthermore, big-'O' notation in the size of the reconfigurable mesh can be removed in the same way as Theorem 15.14. As a consequence, we have

Theorem 15.16 *The sum of n d-bit integers can be computed in $O(\log^* n - \log^* m)$ $(1 \le m \le \log n)$ time on a $\sqrt{nm} \times d\sqrt{n}$ reconfigurable mesh.*

Similar to Corollary 15.15, we have the following.

Corollary 15.17 *The sum of n d-bit integers can be computed in $O(\log^* n)$ time on a $\sqrt{n} \times d\sqrt{n}$ reconfigurable mesh.*

Next, we show an algorithm that computes the sum of n d-bit integers, given one for each processor, in $O(\log d + \log^* n)$ time on a $\sqrt{n} \times \sqrt{n}$ reconfigurable mesh.

If $d > n^{1/4}$, the sum can be computed in $O(\log n) = O(\log d)$ time from Lemma 15.6. Hence, we consider only the case where $d \le n^{1/4}$.

First, suppose that the reconfigurable mesh is partitioned into $\sqrt{n}/d \times \sqrt{n}/d$ submeshes each of size $d \times d$, as shown in Figure 15.14. The sum of each group can be computed in $O(\log d)$ time from Lemma 15.6. Since each sum has at most $d + \log d$ bits, we have to compute the sum of n/d^2 $(d+\log d)$-bit integers. To do this, we apply the two-stage summing method (Figure 15.12).

To apply the first stage of the two-stage summing method, the sum of each bit of n/d^2 $(d + \log d)$-bit integers is computed independently: Every processor that is not in the diagonal of each $d \times d$ submesh configures the cross-over pattern as shown in Figure 15.14. Then we can regard the reconfigurable mesh as $d\sqrt{n}/d \times \sqrt{n}/d$ submeshes. Using each $\sqrt{n}/d \times \sqrt{n}/d$ submesh, the algorithm for Corollary 15.15 can compute the sum of n/d^2 binary values in $O(\log^* n)$ time. Note that the sum of each bit has at most $\log(n/d^2)$ bits.

We next compute the sum of the parallelogram B in Figure 15.12. For this purpose, we use the method used in Lemma 15.7, which implements carry lookahead generators. In this case, the size of each carry lookahead generator is $2\log(n/d^2) \times 2\log(n/d^2)$ and the number of generators is $d + \log d + \log(n/d^2) -$

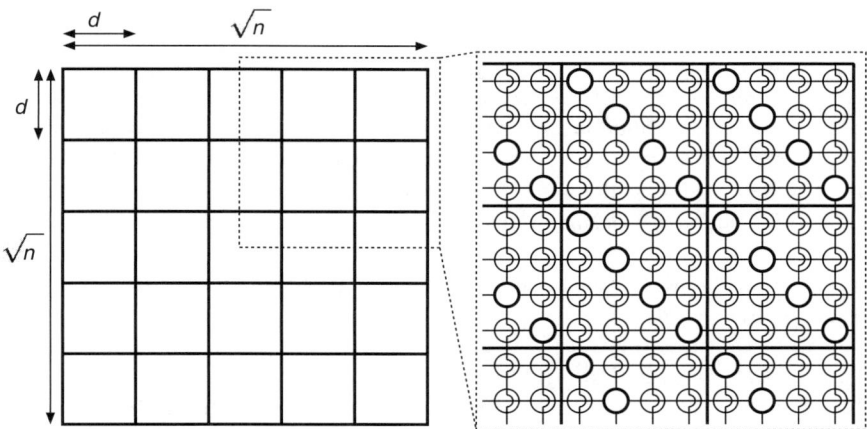

FIGURE 15.14 Integer summming algorithm.

$1 \leq d + \log n$. Therefore, $2 \log(n/d^2)(d + \log n) \times 2 \log(n/d^2) \leq 2n^{1/4} \log n \times 2 \log n$ is sufficient to compute all the bits in constant time. Note that the sum is thus obtained. Since computing the partial sum of each $d \times d$ requires $O(\log d)$ time and the two-stage summing method requires $O(\log^* n)$ time, we have the following.

Lemma 15.18 *The sum of n d-bit integers can be computed in $O(\log d + \log^* n)$ time on a $\sqrt{n} \times \sqrt{n}$ reconfigurable mesh.*

Furthermore, we can reduce the size of the reconfigurable mesh to $\sqrt{n/(\log d + \log^* n)} \times \sqrt{n/(\log d + \log^* n)}$. In this case $(\log d + \log^* n)$, d-bit integers are given to each processor. Each processor first independently computes the sum of $\log^* n + \log d$ d-bit integers in $O(\log d + \log^* n)$ time. The sum thus obtained has at most $d + \log(\log d + \log^* n)$ bits. Then, the sum of $n/(\log d + \log^* n)$ $(d + \log(\log d + \log^* n))$-bit integers is computed in $O(\log(d + \log(\log d + \log^* n)) + \log^*(n/(\log d + \log^* n))) = O(\log d + \log^* n)$ time on the $\sqrt{n/(\log d + \log^* n)} \times \sqrt{n/(\log d + \log^* n)}$ reconfigurable mesh by the algorithm for Lemma 15.18. Hence, the computing time is still $O(\log d + \log^* n)$. Therefore, we have the following.

Theorem 15.19 *The sum of n d-bit integers can be computed in $O(\log d + \log^* n)$ time on a $\sqrt{n/(\log d + \log^* n)} \times \sqrt{n/(\log d + \log^* n)}$ reconfigurable mesh.*

If $d = 1$, then the above algorithm computes the sum of binary values. Thus, we have the following.

Corollary 15.20 *The sum of n binary values can be computed in $O(\log^* n)$ time on a $\sqrt{n/\log^* n} \times \sqrt{n/\log^* n}$ reconfigurable mesh.*

Note that this corollary uses fewer number of processors than Corollary 15.17.

15.7 Conclusions

In this chapter, we have shown several useful techniques on the reconfigurable mesh that include the parallel prefix-remainders technique, the mixed radix representation technique, the snake-like embedding technique. These techniques are useful to reduce the number of processors on the reconfigurable mesh and to accelerate the computation.

References

[1] H. M. Alnuweiri. Fast algorithms for image labeling on a reconfigurable network of processors. In *Proc. 7th International Parallel Processing Symposium*, pp. 569–575. IEEE, April 1993.

[2] H. M. Alnuweiri, M. Alimuddin, and H. Aljunaidi. Switch models and reconfigurable networks: Tutorial and partial survey. In *Proc. Reconfigurable Architecture Workshop*. IEEE, April 1994.

[3] H. M. Alnuweiri and V. K. Prasanna. Parallel architectures and algorithms for image component labeling. *IEEE Transaction on Pattern Analysis and Machine Intelligence*, 14(10):1014–1034, October 1992.

[4] P. Beame and J. Hastad. Optimal bounds for decision problems on the CRCW PRAM. *Journal of the ACM*, 36(3):643–670, July 1989.

[5] Y. Ben-Asher, D. Peleg, R. Ramaswami, and A. Schuster. The power of reconfiguration. *Journal of Parallel and Distributed Computing*, 13:139–153, 1991.

[6] G.-H. Chen, B.-F. Wang, and H. Li. Deriving algorithms on reconfigurable networks based on function decomposition. *Theoretical Computer Science*, 120(2):215–227, November 1993.

[7] Y.-C. Chen and W.-T. Chen. Reconfigurable mesh algorithms for summing up binary values and its applications. In *Proc. 4th Symposium on Frontiers of Massively Parallel Computation*, pp. 427–433. IEEE, October 1992.

[8] H. ElGindy and P. Węgrowicz. Selection on the reconfigurable mesh. In *Proc. International Conference on Parallel Processing*, pp. III–26–33. CRC Press, 1991.

[9] P. Fragopoulou. On the efficient summation of n numbers on an n-processor reconfigurable mesh. *Parallel Processing Letters*, 3(1):71–78, March 1993.

[10] E. Hao, P. D. Mackenzie, and Q. F. Stout. Selection on the reconfigurable mesh. In *Proc. 4th Symposium on Frontiers of Massively Parallel Computation*, pp. 38–45. IEEE, October 1992.

[11] A. Gibbons and W. Rytter. *Efficient Parallel Algorithms*. Cambridge University Press, 1988.

[12] J. JáJá. *An Introduction to Parallel Algorithms*. Addison-Wesley, 1992.

[13] J.-W. Jang, H. Park, and V. K. Prasanna. A fast algorithm for computing histograms on a reconfigurable mesh. In *Proc. 4th Symposium on Frontiers of Massively Parallel Computation*, pp. 244–251. IEEE, October 1992.

[14] J.-W. Jang, H. Park, and V. K. Prasanna. An optimal multiplication algorithm for reconfigurable mesh. In *Proc. 4th Symposium on Parallel and Distributed Processing*, pp. 384–391. IEEE, 1992.

[15] J.-W. Jang, H. Park, and V. K. Prasanna. A bit model of reconfigurable mesh. In *Proc. Reconfigurable Architecture Workshop*. IEEE, April 1994.

[16] J.-W. Jang and V. K. Prasanna. An optimal sorting algorithm on reconfigurable mesh. In *Proc. 6th International Parallel Processing Symposium*, pp. 130–137. IEEE, 1992.

[17] T.-W. Kao, S.-J. Horng, and H.-R. Tsai. Computing connected components and some related applications on a RAP. In *Proc. International Conference on Parallel Processing*, pp. III–57–64. CRC Press, August 1993.

[18] T.-W. Kao, S.-J. Horng, Y.-L. Wang, and K.-L. Chung. A constant time algorithm for computing Hough transform. *Pattern Recognition*, 26(2):277–286, 1993.

[19] H. Li and M. Maresca. Polymorphic-torus architecture for computer vision. *IEEE Trans. on Pattern Analysis and Machine Intelligence*, 11(3):233–243, March 1989.

[20] R. Lin. Reconfigurable buses with shift switching—VLSI radix sort. In *Proc. International Conference on Parallel Processing*, pp. III–2–9. CRC Press, August 1992.

[21] P. D. MacKenzie. A separation between reconfigurable mesh models. In *Proc. 7th International Parallel Processing Symposium*, pp. 84–88. IEEE, April 1993.

[22] R. Miller, V. K. P. Kumar, D. I. Reisis, and Q. F. Stout. Parallel computations on reconfigurable meshes. *IEEE Transaction on Computers*, 42(6):678–692, June 1993.

[23] D. E. Muller and F. P. Preparata. Bounds to complexities of network for sorting and for switching. *Journal of the ACM*, 22(2):195–201, April 1975.

[24] K. Nakano. Efficient summing algorithm for a reconfigurable mesh. In *Proc. Reconfigurable Architecture Workshop*. IEEE, April 1994.

[25] K. Nakano, T. Masuzawa, and N. Tokura. A sub-logarithmic time sorting algorithm on a reconfigurable array. *IEICE Transaction (Japan)*, E-74(11):3894–3901, November 1991.

[26] M. Nigam and S. Sahni. Computational geometry on a reconfigurable mesh. In *Proc. 8th International Parallel Processing Symposium*, pp. 86–93. IEEE, April 1994.

[27] M. Nigam and S. Sahni. Sorting n numbers on $n \times n$ reconfigurable meshes with buses. *Journal of Parallel and Distributed Computing*, 23(1):37–48, October 1994.

[28] S. Olariu, J. L. Schwing, and J. Zhang. Fundamental algorithms on reconfigurable meshes. In *Proc. 29th Allerton Conference on Communications, Control, and Computing*, pp. 811–820, 1991.

[29] S. Olariu, J. L. Schwing, and J. Zhang. Fast computer vision algorithms for reconfigurable meshes. In *Proc. 6th International Parallel Processing Symposium*, pp. 258–261. IEEE, 1992.

[30] D. I. Reisis. An efficient convex hull computation on the reconfigurable mesh. In *Proc. 6th International Parallel Processing Symposium*, pp. 142–145. IEEE, 1992.

[31] P. Thangavel and V. P. Muthuswamy. Parallel algorithms for addition and multiplication on processor arrays with reconfigurable bus systems. *Information Processing Letters*, 46:89–94, May 1993.

[32] J. D. Ullman. *Computational Aspects of VLSI*. Computer Science Press, 1984.

[33] R. Vaidyanathan. Sorting on PRAMs with reconfigurable buses. *Information Processing Letters*, 42:203–208, June 1992.

[34] R. Vaidyanathan and J. L. Trahan *Dynamic Reconfiguration: Architectures and Algorithms*. Kluwer Academic/Plenum Publishers, 2003.

[35] K. Wada, K. Hagihara, and N. Tokura. Area and time complexities of VLSI computations. In *Proc. 7th IBM Symposium on Mathematical Foundations of Computer Science*, pp. 49–127, May 1982.

[36] B.-F. Wang and G.-H. Chen. Constant time algorithms for the transitive closure and some related graph problems on processor arrays with reconfigurable bus systems. *IEEE Transaction on Parallel and Distributed Systems*, 1(4):500–507, October 1990.

[37] B.-F. Wang, G.-H. Chen, and F.-C. Lin. Constant time sorting on a processor array with a reconfigurable bus system. *Information Processing Letters*, 34(4):187–192, April 1990.

16 Reconfigurable Computing with Optical Buses

Anu G. Bourgeois
Georgia State University

16.1 Introduction.. 16-1
16.2 Reconfiguration and Pipelining........................... 16-2
16.3 Model Descriptions.. 16-4
  Linear Array with a Reconfigurable Pipelined Bus System Structure • Addressing Techniques • Pipelined Reconfigurable Mesh Description • Other Optical Models • Relating Optical Models
16.4 Fundamental Algorithms................................... 16-14
  Binary Prefix Sums • Compression • Sorting and Selection Algorithms • PRAM Simulations
16.5 Conclusions.. 16-18
References... 16-18

16.1 Introduction

The main goals of telecommunication systems include maximizing capacity, throughput, and reliability, and minimizing distortion of messages. Many telecommunication networks have been employing optical fiber to achieve these goals [LPZ98b, VT04]. Optical technologies are also beneficial for interconnecting processors in parallel computers owing to the ability to pipeline multiple messages within a single bus cycle along an optical bus [QM93]. Such ability to pipeline multiple messages allows for efficient parallel computation within many applications [ER99, Guo92, PH96, RS97]. In such a model, many messages can be in transit simultaneously, pipelined in sequence on an optical bus, while the time delay between the furthest processors is only the end-to-end propagation delay of light over a waveguided bus. In parallel processing systems, communication efficiency determines the effectiveness of processor utilization, which, in turn, determines performance.

As a result, researchers have proposed several models based on pipelined optical buses as practical parallel computing platforms including the *Linear Array with a Reconfigurable Pipelined Bus System* (LARPBS) [LPZ98b, PL98], the *Linear Pipelined Bus* (LPB) [Pan95], the *Pipelined Optical Bus* (POB) [LPZ97b, ZL97], the *Linear Array with Pipelined Optical Buses* (LAPOB) [ER99], the *Pipelined Reconfigurable Mesh* (PR-Mesh) [TBV98], the *Array with Reconfigurable Optical Buses* (AROB) [PA96a, PA96b], the *Array Processors with Pipelined Buses* (APPB) [ME98], the *Array Processors with Pipelined Buses Using*

Switches (APPBS) [Guo94], the *Array Structure with Synchronous Optical Switches* (ASOS) [QM93], and the *Reconfigurable Array with Spanning Optical Buses* (RASOB) [Qia95].

Many parallel algorithms exist for arrays with POBs, such as sorting [ER99, Guo92, RS97], selection [HPS02, LZ96, Pan95, RS97], matrix operations [Li02, LPZ98a, PA96a], graph problems [Dat02, JL03, Raj05, WHW05], and string matching [SY05, XCH05]. There are also a number of algorithms for numerical problems [HP95], singular value decomposition [PH96], and image processing, such as Hough transform [CCPC04, Pan92] and Euclidean distance transform [CPX04], among others. This indicates that such systems are very efficient for parallel computation owing to the high bandwidth available by pipelining messages.

We will next describe some background specific to reconfigurable models with and without pipelined buses. Section 16.3 describes in detail the structure of the LARPBS and its addressing techniques. It extends this model to describe the PR-Mesh and other optical models as well as presenting relations among some of these models. Section 16.4 provides an overview of various fundamental algorithms developed for optical bus-based models and Section 16.5 concludes the chapter.

16.2 Reconfiguration and Pipelining

Recently, researchers have proposed many reconfigurable models such as the *Reconfigurable Mesh* (R-Mesh) [BALPS95, BSDE96, MS96], *Linear Reconfigurable Mesh* (LR-Mesh) [BALPS95], *Fusing Reconfigurable Mesh* (FR-Mesh) [FZTV97, FZVT98], *Processor Array with Reconfigurable Bus System* (PARBS) [WC90], *Reconfigurable Multiple Bus Machine* (RMBM) [TVT96], and *Reconfigurable Buses with Shift Switching* (RESBIS) [LO95]. Nakano presented a bibliography of published research on reconfigurable models [Nak95], and Vaidyanathan and Trahan have published a comprehensive book on dynamic reconfiguration [VT04].

Processors can fuse together the edges of a reconfigurable model to form buses (either electrical or optical buses) [BAPRS91]. The main characteristics of these models are as follows:

- Each processor can locally determine its internal port connections and/or switch settings at each step to create or segment buses
- The model assumes constant propagation delays on the buses
- The model uses the bus as a computational tool

The following examples demonstrate how reconfigurable models utilize these characteristics. Consider the OR operation on N bits, where each processor of an N-processor array holds an input. It is possible to perform this operation in constant time on an N-processor LR-Mesh. Assume $N = 8$ and that each processor holds one input bit and fuses its ports to form a single bus as shown in Figure 16.1a.

Each processor that holds a value of "1" internally disconnects the bus and writes on the bus through its left port. The leftmost processor, R_0, reads the value on the bus; this value corresponds to the result of the OR operation (Figure 16.1b). If one or more processors hold a "1," then R_0 reads a "1" from the leftmost processor (R_2 in Figure 16.1b) holding a "1." The processors between R_0 and R_2 all hold a "0," so they keep the bus intact and allow the value written by R_2 to reach R_0. If all processors hold a "0," then no value is written on the bus and the result is "0." All processors then fuse their ports to connect the bus, and processor R_0 broadcasts the result to all processors as in Figure 16.1c.

The time required to perform this computation on a parallel random access machine (PRAM) with exclusive writes is $O(\log N)$ steps for N input bits. The demonstrated example performs the computation in a constant number of steps using only exclusive writes on a one-dimensional R-Mesh. In the second step, although both R_2 and R_5 are writing simultaneously, the two processors are writing on separate buses, maintaining an exclusive write.

The example demonstrates some of the key features of reconfigurable models. First, processors determine their internal port configurations only on the basis of the local variable held; those with a "1" disconnect their ports and those with a "0" connect their ports. Second, broadcasting a value on a bus takes a single step owing to the assumption of constant propagation delay on a bus.

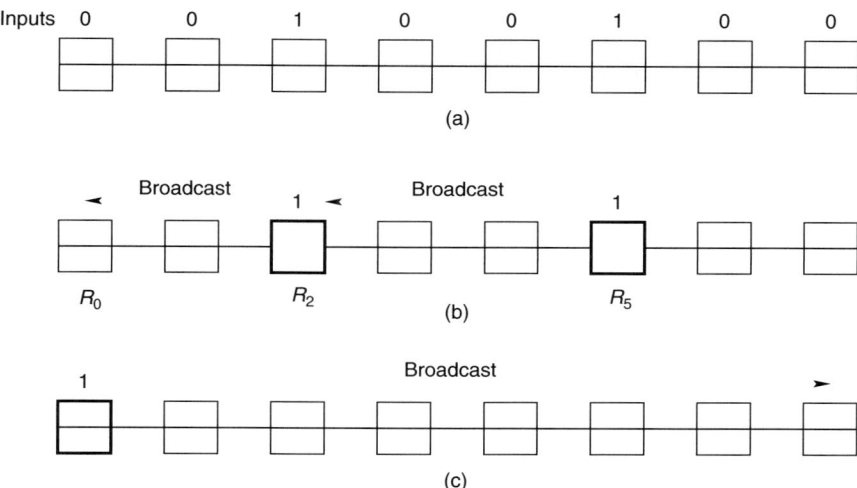

FIGURE 16.1 Computing the OR function on an LR-Mesh: (a) initial configuration; (b) disconnect bus and broadcast toward R_0; (c) broadcast result.

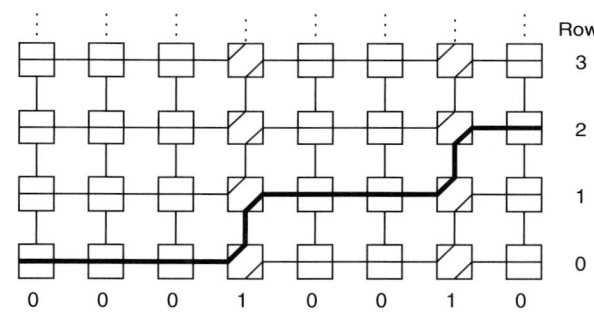

FIGURE 16.2 Summation of eight bits on an R-Mesh.

Next consider computing a binary sum on an R-Mesh. This is a two-dimensional model in which each processor has four ports (North, South, East, and West). The processors on the bottom row hold the input bit values.

First, all processors form vertical buses by fusing their North and South ports. Each processor on the bottom row broadcasts its input value to all processors on its vertical bus. A processor that reads a "0" on its vertical bus fuses its East and West ports together. A processor that reads a "1" on its vertical bus fuses its North and West ports together and its South and East ports. (Refer to Figure 16.2. The figure only shows the first four rows of the R-Mesh.)

The processor at the bottom left corner writes a signal at its West port. The internal port connections form staircase buses allowing a signal to step up a row for each "1" in the input. Figure 16.2 shows in bold the bus on which the signal propagates. The processors in the rightmost column read their East port. The processor that detects a signal determines the sum to be the same as its row value. This technique uses an $(N + 1) \times N$ R-Mesh to sum N bits in constant time. This example demonstrates the method of using the bus as a computational tool. In Section 16.4, we present a binary prefix-sums algorithm that runs on an N-processor LARPBS in constant time for N input bits.

The examples that we have considered thus far can all be executed on systems with either optical or electrical buses. Using optical waveguides provides us with the advantage of being able to pipeline messages on a bus. This is the ability of having multiple messages on a single bus concurrently. Section 16.3 provides more detail on how it is possible to pipeline messages on an optical bus.

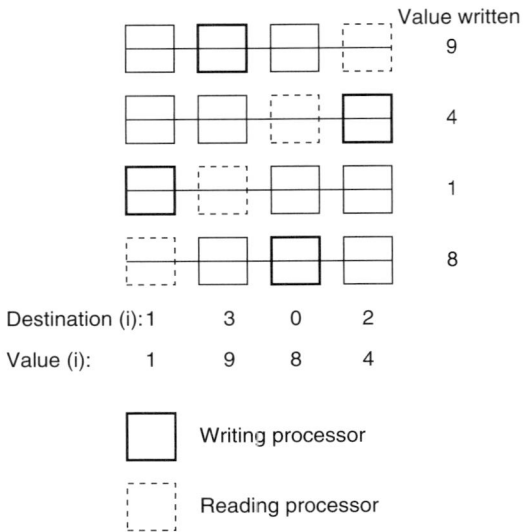

FIGURE 16.3 Permutation routing on an R-Mesh.

We will use a general permutation routing example to illustrate the benefit pipelining provides. Let $\mathcal{N} = \{0, 1, \ldots, N-1\}$ and let $\pi : \mathcal{N} \longrightarrow \mathcal{N}$ be a bijection. Permutation routing of N elements on an N-processor system refers to sending information from processor i to processor $\pi(i)$, for each $i \in \mathcal{N}$. We will first describe how to implement this on an R-Mesh, and then contrast this with how the LARPBS can perform a general permutation routing step more efficiently by using pipelining.

Consider a 4×4 R-Mesh in which each processor in column i on the bottom row holds $\pi(i)$ as shown in Figure 16.3. Assume each processor is to send value v_i to $\pi(i)$ on the bottom row.

First, each processor fuses its North and South ports forming vertical buses. Each processor on the bottom row broadcasts $\pi(i)$ and v_i along the vertical buses to all processors on the column. Next, all processors fuse their East and West ports forming horizontal buses. The processor with column index i and row index $\pi(i)$ writes v_i on the row bus as shown in Figure 16.3. Each processor with column index j and row index j reads from the bus with row index j. The processors then fuse their North and South ports forming vertical buses again. Each processor that reads a value in the previous step writes on the bus so that the processors in the bottom row can read the value from the permutation.

If there are N inputs, then an $N \times N$ R-Mesh is required to execute a permutation routing in $O(1)$ steps. If an N-processor one-dimensional R-Mesh is all that is available, then, by a simple bisection width argument, it would require N communication steps to route the permutation.

Pipelining enables an N-processor one-dimensional LARPBS to perform this general permutation routing in a single step. The properties of an optical waveguide support the propagation of multiple messages on a single bus during one communication step. (We discuss the details of pipelining messages in Section 16.3.) All processors of an LARPBS can concurrently select distinct destinations and each sends a message to its chosen destination in one bus cycle. To perform the permutation routing, each processor i selects $\pi(i)$ as its destination and sends its value v_i on the data waveguide. This ability of optical buses provides a saving in size and/or time.

16.3 Model Descriptions

A system with an optically pipelined bus uses optical waveguides instead of electrical buses to transfer information among processors. Signal (pulse) transmission on an optical bus possesses two advantageous properties: unidirectional propagation and predictable propagation delay per unit length. These two properties enable synchronized concurrent access to an optical bus in a pipelined fashion [GMH+91, MCL89, QM93, QMCL91]. Combined with the abilities of a bus structure to broadcast and multicast, this architecture suits many communication-intensive applications.

We adapt the following framework from Qiao and Melhem [QM93]. Organize data into fixed-length *data frames*, each comprising a train of optical pulses. The presence of an optical pulse represents a binary bit with value 1. The absence of an optical pulse represents a binary bit with value 0. Let ω denote the pulse duration. Define a unit pulse length Δ to be the spatial length of a single pulse; this is equivalent to the distance traveled by a pulse in ω units of time. The bus has the same length of fiber between consecutive processors, so propagation delays between consecutive processors are the same. Let τ denote the time for a signal to traverse the optical distance on the bus between two consecutive processors with spatial distance D_0; time τ is also referred to as a *petit cycle*.

As mentioned, the properties of an optical bus allow multiple processors to concurrently write on the bus by pipelining messages. This is possible provided that the following condition to assure no collisions is satisfied:

$$D_0 > b\omega c_g$$

where b is the number of bits in each message and c_g is the velocity of light in the waveguide [GMH+91]. The assurance that all processors start writing their messages on the bus at the same time is another condition that must be satisfied to guarantee that no two messages will collide. Let a *bus cycle* be the end-to-end propagation delay on the bus. We specify time complexity in terms of a step comprising one bus cycle and one local computation.

The next section describes the structure of the LARPBS. Section 16.3.3 extends the one-dimensional model to a multidimensional optical model, called the *Pipelined Reconfigurable Mesh* (PR-Mesh). Section 16.3.4 describes a myriad of other optical bus-based models, and Section 16.3.5 discusses some relations among these models.

16.3.1 Linear Array with a Reconfigurable Pipelined Bus System Structure

In the LARPBS, as described by Pan and Li [PL98], the optical bus is composed of three waveguides, one for carrying data (the *data waveguide*) and the other two (the *reference* and *select waveguides*) for carrying address information (see Figure 16.4). (For simplicity, the figure omits the data waveguide, as it resembles the reference waveguide.) Each processor connects to the bus through two directional couplers, one for transmitting and the other for receiving [GMH+91, QM93]. Note that optical signals propagate unidirectionally from left to right on the upper segment (transmitting segment) and from right to left on the lower segment (receiving segment), with a U-turn connecting the two segments. Referring to Figure 16.4, the processor furthest from the U-turn, R_0, is the *tail* of the bus, and the processor at the U-turn, R_4, is the *head*.

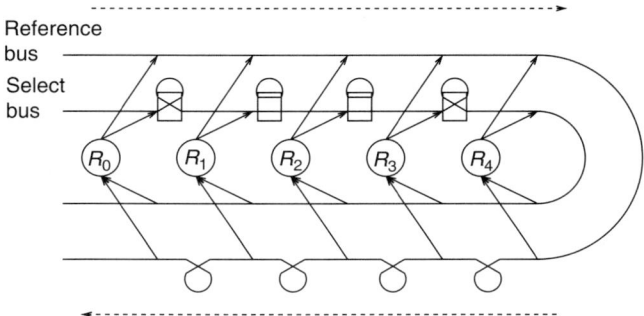

FIGURE 16.4 Structure of an LARPBS.

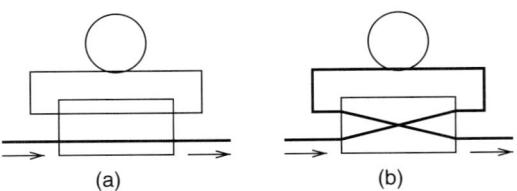

FIGURE 16.5 Conditional delay switch: (a) straight state; (b) cross state.

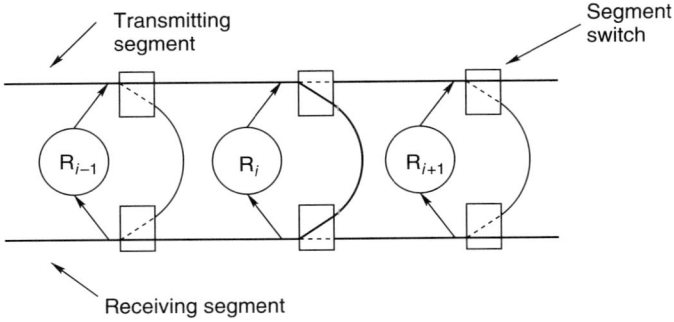

FIGURE 16.6 Segment switch.

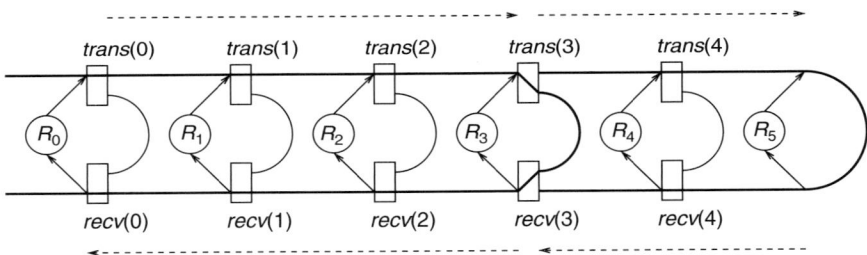

FIGURE 16.7 A six processor LARPBS model with two subarrays.

The receiving segments of the reference and data waveguides contain an extra segment of fiber of one unit pulse length, Δ, between each pair of consecutive processors (shown as a delay loop in Figure 16.4). The transmitting segment of the select waveguide has a switch-controlled conditional delay loop of length Δ between processors R_i and R_{i+1}, for each $0 \leq i \leq N-2$ (Figure 16.4). Processor $i+1$ controls the switch between processors i and $i+1$. A processor can set a switch to the *straight* or *cross* states, as shown in Figure 16.5. The length of a bus cycle for a system with N processors is $2N\tau + (N-1)\omega$.

To allow segmenting, the LARPBS has optical switches on the transmitting and receiving segments of each bus for each processor. Let $trans(i)$ and $recv(i)$ denote these sets of switches on the transmitting and receiving segments, respectively, on the three buses between processors R_i and R_{i+1}. Switches on the transmitting segment are 1×2 optical switches, and on the receiving segment are 2×1 optical switches as shown in Figure 16.6. With all switches set to *straight*, the bus system operates as a regular pipelined bus system. Setting $trans(i)$ and $recv(i)$ to *cross* segments the whole bus system into two separate pipelined bus systems, one consisting of processors $R_0, R_1, \ldots, R_i$ and the other consisting of $R_{i+1}, R_{i+2}, \ldots, R_{N-1}$. Figure 16.7 shows an LARPBS with six processors, in which switches in $trans(3)$ and $recv(3)$ are set to cross, splitting the array into two subarrays with the first having four processors and the second having two processors (for clarity, the figures show only one waveguide and omit conditional delay switches).

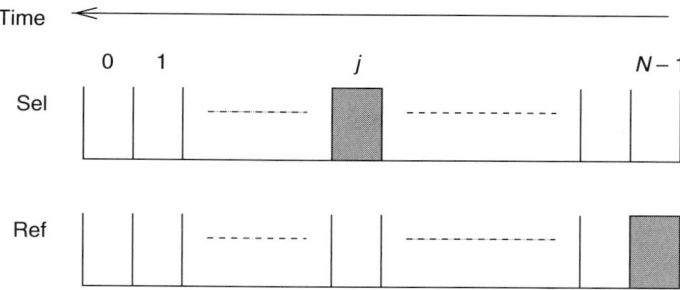

FIGURE 16.8 Select and reference frames.

16.3.2 Addressing Techniques

The LARPBS uses the *coincident pulse technique* [QM93] to route messages by manipulating the relative time delay of *select* and *reference* pulses on separate buses so that they will coincide only at the desired receiver. Each processor has a *select frame* of N bits (*slots*), of which it can inject a pulse into a subset of the N slots. For example, let all switches on the transmitting segment of the select waveguide be set straight to introduce no delay. Let source processor R_i send a reference pulse on the reference waveguide at time t_{ref} (the beginning of a bus cycle) and a select pulse on the select waveguide at time $t_{\text{sel}} = t_{\text{ref}} + (N - 1 - j)\omega$. Processor R_i also sends a data frame, on the data waveguide, that propagates synchronously with the reference pulse. After the reference pulse goes through $N - 1 - j$ fixed delay loops, the select pulse catches up to the reference pulse. As a result, processor R_j detects the double-height coincidence of reference and select pulses, then reads the data frame. Figure 16.8 shows a select frame relative to a reference pulse for addressing processor j. The coincident pulse technique admits broadcasting and multicasting of a single message by appropriately introducing multiple select pulses within a select frame.

The conditional delay switches on the transmitting segment introduce delays to the select pulses and can alter the location at which the select and reference pulses will coincide. These switches are useful as a computing tool to calculate binary prefix sums and perform compression, for example (Section 16.4). The length of the bus between two processors provides enough space for two frames of N slots to fit, although there is only one such frame on each waveguide for each processor. This prevents a pulse in the select frame of processor R_i from being shifted to overlap the reference frame of R_{i-1}.

When multiple messages arrive at the same processor in the same bus cycle, it receives only the first message and disregards subsequent messages that have coinciding pulses at the processor. This corresponds to the PRIORITY concurrent write rule. The PRIORITY write rule has the processor with the highest priority (in this case, the processor with the highest index or nearest the U-turn) win a write conflict when multiple processors are attempting to write to the same destination.

We will refer to the processor that has a select pulse injected in its slot in a select frame for a particular message as the *selected destination*. The *actual destination* will denote the processor that detects the coinciding reference and select pulses (the two may be different owing to conditional delay loops and segmenting). The *normal state of operation* is when the actual destinations of messages are the selected destinations. For the LARPBS, the normal state of operation is when all conditional delay switches and segment switches are set to straight.

Consider the LARPBS shown in Figure 16.4. Suppose processor R_1 injects a select pulse so that R_3 is its selected destination, and R_0 attempts to broadcast. The message sent by R_1 encounters one conditional delay switch set to cross, and the message sent by R_0 encounters two. As a result, the actual destination of R_1 is R_2 instead of R_3. The actual destinations of the message broadcast by R_0 are R_2, R_1, and R_0, rather than all five processors. Even though R_2 is the actual destination of the message sent by R_0, processor R_2 will receive only the message sent by R_1 because this message arrives prior to the one sent by R_0.

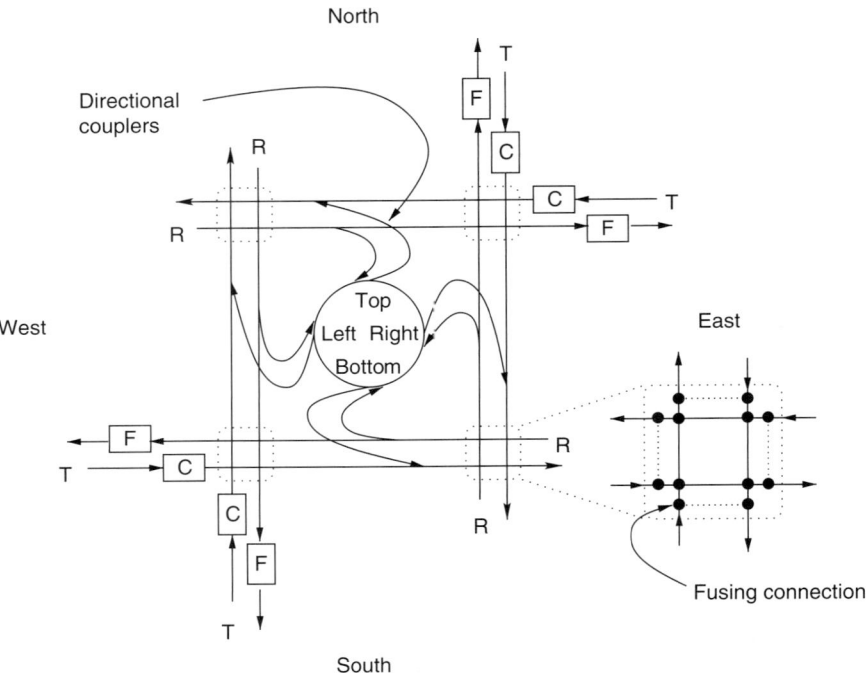

FIGURE 16.9 PR-Mesh processor connections. T: transmitting segment, R: receiving segment, C: conditional delay loop, and F: fixed delay loop.

16.3.3 Pipelined Reconfigurable Mesh Description

The *Pipelined Reconfigurable Mesh* (PR-Mesh) is a k-dimensional extension of the LARPBS [BT00, TBV98]. It is a mesh of processors in which each processor has $2k$ ports. Each processor can locally configure its ports by internally fusing pairs of ports or leaving ports as singletons, so all buses are linear. A two-dimensional PR-Mesh is an $R \times C$ mesh of processors in which each processor has four ports. The ports connect to eight segments of buses using directional couplers as shown in Figure 16.9. There are receiving and transmitting waveguides for the two dimensions, and within each dimension there are waveguides for both directions. Each processor locally controls a set of switches at each of the bus intersections that allow it to fuse bus segments together. The dashed boxes around each bus intersection contain these sets of switches. (The intersection for the lower right corner of the processor is shown larger to distinguish the connections.) Each fusing connection can be in one of ten possible settings. The dashed segments within the box are auxiliary segments that enable the processor to create U-turns. Figure 16.10 depicts the ten possible port partitions for each processor of a two-dimensional PR-Mesh. To implement these partitions, the switches can configure from within the same set of configurations at the switch level. Local fusing creates buses that run through fused switches to adjacent processors, then through their fused switches, and so on. Each such linear bus corresponds to an LARPBS. The switches may not be set, however, so that a cycle is formed. By allowing cycles, there would be no clear head or tail of a bus, therefore, it would be impossible to determine priority among the processors for concurrent write operations.

Each processor locally controls conditional delay switches on each of the transmitting segments. There are also fixed delay loops on each of the receiving segments. The switches at each bus intersection act as the segment switches. Refer to Figure 16.9 for the placement of these switches. A pair of receiving and transmitting buses that are traversed in opposite directions correspond to an LARPBS bus.

The following examples help to illustrate the processor and switch connections for different bus configurations. Consider a processor, R_i, that is connected to a segment of a horizontal bus, that is, it sets

Reconfigurable Computing with Optical Buses

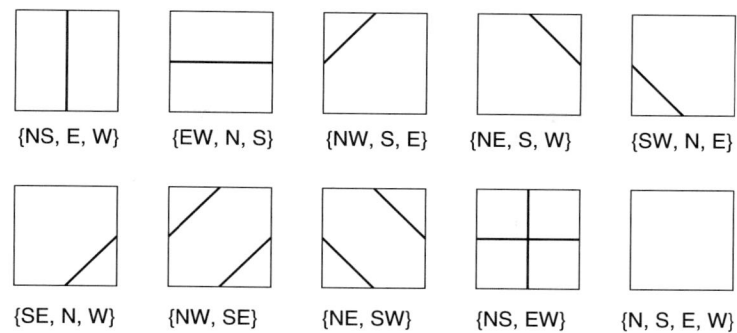

FIGURE 16.10 PR-Mesh switch connections.

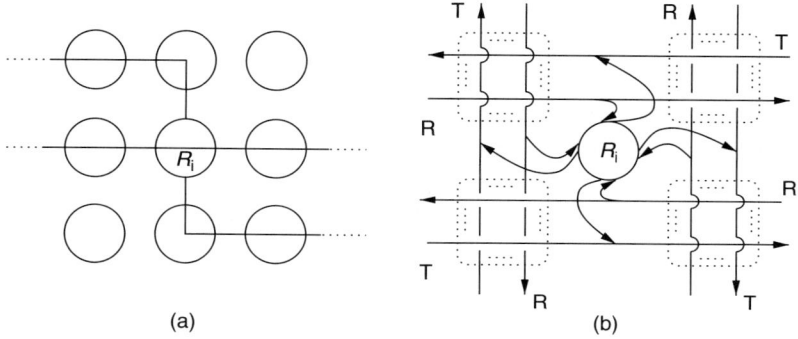

FIGURE 16.11 Example of PR-Mesh switch settings for {EW, N, S}.

its configuration so that the East and West ports are fused. Also, assume that the North and South ports are tails of separate buses, or open rather than fused. Figure 16.11a pictorially shows a possible set of bus formations at processor R_i. Processor R_i configures its switch settings so that the East and West ports are fused and the North and South ports are left open. Refer to Figure 16.11b to see the connections of each bus intersection. With this example, the left reading and writing connections do not necessarily correspond to the West port because of bus routing internal to the processors. For example, a read from the West port would be performed by either the Top or Bottom read connections. Read and write operations for the North port are performed by the Left connections, and read and write operations for the South port are performed by Right connections. The corresponding ports and connections are fixed for each bus configuration. Since there are only ten configurations, each processor can keep a table holding this information.

Once the bus is created, the orientation of the bus must be determined. To do this, the head of the bus broadcasts a message on the bus that corresponds to the correct direction and each processor connected waits for a message. For this example, if a message is sent on the upper horizontal segment, then R_i sends and receives messages using its Top port. If a message is sent on the lower horizontal segment, then R_i sends and receives messages using its Bottom port.

The next example illustrates the switch and port connections for creating U-turns. Consider a processor, R_j, that has each of its four ports at a U-turn of a bus, so that the processor is the head of four separate buses. Figure 16.12a pictorially shows a possible set of bus formations at processor R_j. Processor R_j configures its switch settings to create U-turns, utilizing the auxiliary segments, as shown in Figure 16.12b. For this example, the Right connections handle communications for the North port. Left connections handle communications for the South port, Top connections for the East port, and Bottom connections for the West port.

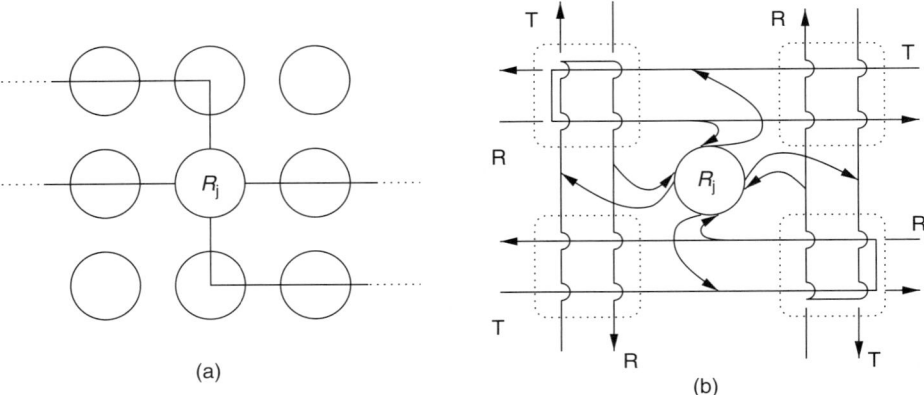

FIGURE 16.12 Example of PR-Mesh switch settings for {N, S, E, W}.

The PR-Mesh is similar to the LR-Mesh [BALPS95] in that both allow processors to dynamically change switch settings to construct different buses. The LR-Mesh, however, uses electrical buses rather than optical buses. The available internal port configurations are the same as those available to the PR-Mesh (Figure 16.10), thus forming only linear buses. The buses, however, can form cycles, unlike the PR-Mesh buses.

16.3.4 Other Optical Models

Most models based on optical buses similar to the LARPBS and PR-Mesh differ only by slight variations. For instance, they are all able to pipeline their messages. The differences among these models involve the switches used, the placement of the switches, and some other hardware features and capabilities.

The previous sections described the structure and addressing techniques of the LARPBS and PR-Mesh in detail. This section presents other optical models with the following section relating some of these models to each other.

The model most similar to the LARPBS is the LPB [Pan92]. This model is identical to the LARPBS with the exception that it does not have any segment switches. Therefore, the LPB is not able to segment its bus into smaller, independent subarrays.

The POB [LPZ97b, ZL97] is similar to the LARPBS and LPB as it also contains three waveguides. Conditional delay switches are on the receiving segment of the reference and data waveguides rather than the transmitting segment of the select waveguide, and like the LPB, the POB does not have segment switches. The POB contains no fixed delay loops, so the length of the bus cycle is actually shorter than that of the LARPBS and the LPB.

The POB also uses the coincident pulse technique to route messages. The effect of conditional delay switches on the POB is to delay the reference pulse relative to the select frame, so the POB is also able to perform one-to-one addressing, multicasting, and broadcasting. The location of the conditional delay switches on the receiving end enables the POB to multicast and broadcast without having to set multiple select pulses in a select frame, although multiple select pulses could be set as in the LARPBS and LPB. Consider the case when processor B_i is the selected destination, the delay switch between B_i and B_{i-1} is straight, and all remaining delay switches are set to cross. The select and reference pulses will coincide at B_i and again at B_{i-1}; therefore, both processors receive the message although only one select pulse was injected.

We now demonstrate the addressing of the POB by referring to the switch settings as shown in Figure 16.13. Suppose processor B_1 injects a select pulse so that B_3 is its selected destination, and B_0 injects a pulse so that B_2 is its selected destination. The settings of the straight switches will result in a multicast operation by B_0 to actual destinations B_2, B_1, and B_0. The actual destination of the message

Reconfigurable Computing with Optical Buses

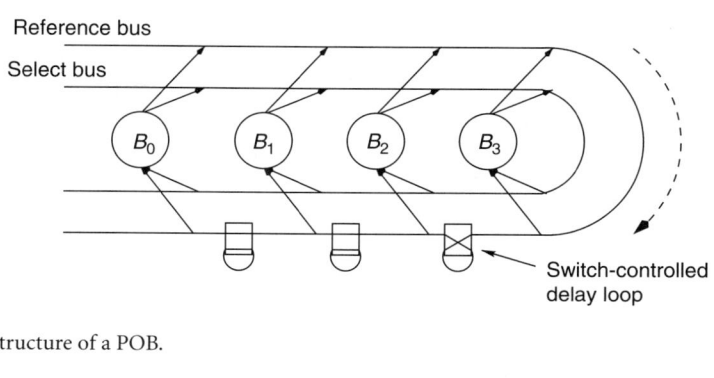

FIGURE 16.13 Structure of a POB.

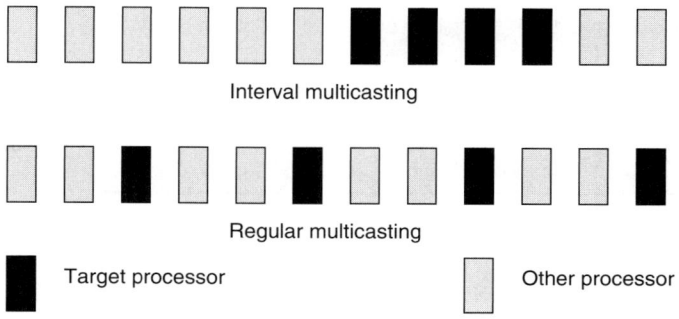

FIGURE 16.14 Multicasting patterns [ER99].

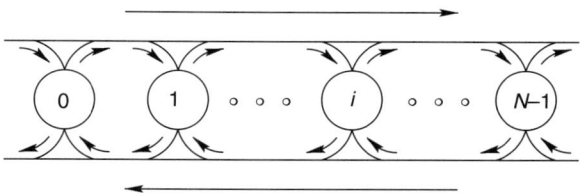

FIGURE 16.15 Linear array of processors with pipelined buses (APPB).

sent by B_1 is B_3. The normal state of operation for the POB is when all conditional delay switches are set to cross.

The LAPOB [ER99] is another model that uses directional couplers to connect to an optical bus. The model, however, does not possess either conditional delay or segment switches. Another restriction of the model is the method available to multicast. The LAPOB is able to address messages using either a *contiguous interval* or *regularly spaced* addressing pattern (refer to Figure 16.14).

A simpler optical model is the linear APPB [Guo94]. Each processor connects to two buses by two couplers, one for transmitting and the other for receiving (Figure 16.15). Unlike the LARPBS, processors transmit messages to and receive messages from the same bus segment. Extending this model to two dimensions, each processor connects to four buses.

The *Array Processors with Pipelined Buses using Switches* (APPBS) is a further extension [Guo94]. Unlike the structure of the PR-Mesh, the APPBS uses four switches at each processor to connect to each of the adjacent buses (Figure 16.16a). Four configurations are available to each switch. Figure 16.16b shows the configurations available to the top right switch at a processor. Each processor locally controls its switches, and can change its configuration once or twice at any petit cycle(s) within a bus cycle. (Recall that a petit cycle is the node-to-node propagation delay.) The available switch configurations form nonlinear buses that are not allowed in the PR-Mesh, though the model is restricted so that only one of two possible

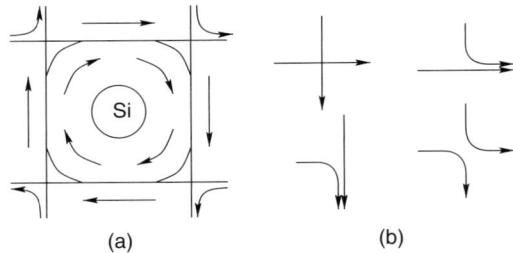

FIGURE 16.16 APPBS processor with switches: (a) switch connections at each APPBS processor; (b) switch configurations of top right switch at each APPBS processor.

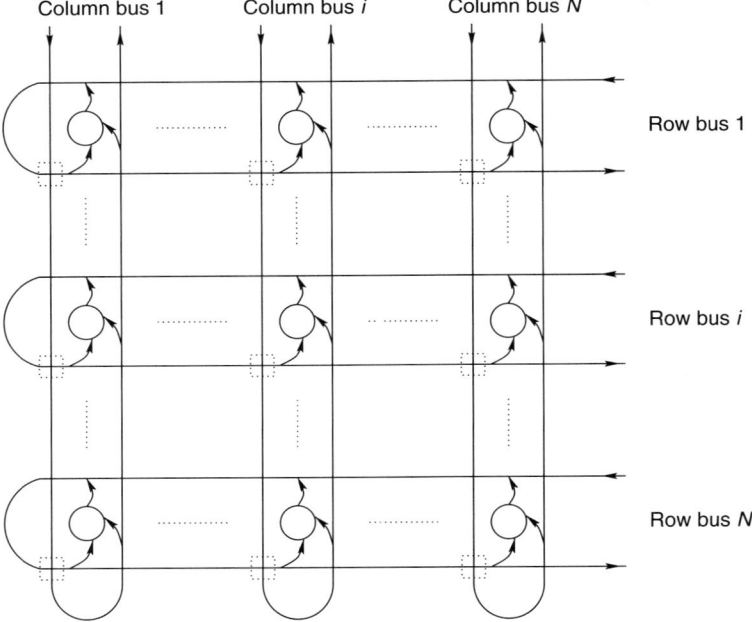

FIGURE 16.17 Array structure with synchronous optical switches (ASOS) [QM93].

converging paths can carry a message in any given petit cycle, so messages do not collide. This does allow messages to be interleaved from different buses. To overcome the obstacle of nonlinear buses or the "merged" switch configurations, we create copies of the buses for each message sent.

Another difference between the PR-Mesh and the APPBS is that the APPBS cannot end a bus in the middle of the mesh, so each bus must extend to the outer processors in the mesh. The APPBS can use either the coincident pulse technique or the control functions $send(m)$ and $wait(n)$ to send a message from processor m to processor n. These functions define the number of petit cycles processor m has to wait before sending a message and processor n must wait before reading a message. The ability of different switches to change their settings during different petit cycles could result in many different model configurations within a single bus cycle.

The ASOS [QM93] is another two-dimensional model that uses switches to connect row and column buses. Each processor is able to transmit on the upper segment of a row bus and receive from the lower segment of a row bus and the right segment of a column bus (Figure 16.17). The switches control the route a message takes. A switch set in the cross state causes messages to transfer from a row bus to a column bus.

The *Linear Array with Reconfigurable Optical Buses* (LAROB) [PA98, PA96a, PA96b] is similar to the LARPBS with extra hardware features. Each processor has switches that allow it to introduce up to N unit delays, unlike the one conditional delay of the LARPBS. Each processor also has a relative delay counter and an internal timing circuit to output a message during any petit cycle. An optical rotate-shift register and a counter are also present at each processor to assist in performing a bit polling operation within one step. This is the ability to select the kth bit of each of N messages and determine the number of these bits that are set to 1. Pavel and Akl [PA98] presented an extended version of the LAROB. The extended model allows online switch settings during a bus cycle and the transmission of up to N messages with arbitrary word size. They also presented a two-dimensional version of the LAROB called the *Array with Reconfigurable Optical Buses* (AROB). The AROB is also able to address processors using the control functions *send*(m) and *wait*(n) as the APPBS.

These extra features not possessed by the other optical models seem to suggest that the LAROB (AROB) has more "power." The next section presents results indicating that the AROB has the same complexity as the PR-Mesh, that is, both are able to solve the same problems in the same number of steps with a polynomial increase in the number of processors.

16.3.5 Relating Optical Models

In the previous sections, we have seen several similar models with "optically pipelined buses." Many of these models have different features, making it difficult to relate results from one model to another. To overcome these difficulties, there has been work to unify the models in order to increase understanding of which features are essential, and to be able to translate algorithms from one model to another. This section presents results that relate various optical models to each other.

In relating some of the one-dimensional optical models, three specific models are considered. Specifically, the LARPBS, LPB, and POB are related to each other. Of the three models, only the LARPBS has segment switches. It would seem likely that the ability to segment the bus would separate the LARPBS from the other two models in terms of computational power. However, through a cycle of simulations, it has been proven that the three optical models are equivalent [TBPV00]. That is, each model can simulate a step of either of the two other models in constant time, using the same number of processors. This implies an automatic translation of algorithms (without loss of speed or efficiency) among these models. In other words, any algorithm proposed for one of these models can be implemented on any of the others with the same number of processors, within a constant factor of the same time.

Theorem 16.1 *The LARPBS, LPB, and POB are equivalent models. Each one can simulate any step of one of the other models in $O(1)$ steps with the same number of processors.*

Recall that the only difference between the LARPBS and LPB is the segmenting ability of the former. The segmenting ability of the LARPBS simplifies algorithm design, yet, owing to the equivalence of these models, it is not necessary to include the segment switches. Moreover, this equivalence establishes dynamically selectable delay loops (conditional delay switches) as the key to the power of these models. This separation of the powers of segmentation and delays is similar to that established in the context of the RMBM [TVT96].

Now we will focus on results for two-dimensional optical models. Forming these relationships presents obstacles not present when analyzing linear arrays, such as the larger number of configurations possible owing to the multiple dimensions. To account for this, their equivalence was established in a slightly different context; complexity was considered by relating their time to within a constant factor and the number of processors to within a polynomial factor. Two models have the same complexity if either model can simulate any step of the other model in a constant number of steps, with up to a polynomial increase in the number of processors.

The models that were considered were the PR-Mesh, APPBS, and AROB. In particular, it was proven that the PR-Mesh, APPBS, and the AROB have equivalent complexity and can solve any problem of size N within class L^* in constant time using polynomial in N processors [BT00].

It was also established [TBV98] that the LR-Mesh and the Cycle-Free LR-Mesh (CF-LR-Mesh) have equivalent complexity as the three models mentioned above. (These two models are reconfigurable models that use electrical buses without the capability of pipelining.) The results lead to the conclusion that pipelining messages using optical buses provide us with better efficiency than electrical buses, as demonstrated in the next section. The PR-Mesh requires fewer buses than the CF-LR-Mesh; however, the PR-Mesh possesses the same limitations as the CF-LR-Mesh in solving graph problems since nonlinear connections are not allowed.

16.4 Fundamental Algorithms

Often, algorithms designed for pipelined optical models follow the approach of R-Mesh algorithms, but additionally exploit the ability to pipeline messages, multicast, and broadcast during a single step. This results in more efficient algorithms since multiple buses are not needed to transfer multiple messages concurrently. To demonstrate this, we present an array of fundamental algorithms for optical models, including those in the areas of sorting, selection, and PRAM simulations.

16.4.1 Binary Prefix Sums

Consider an LARPBS with N processors such that each one holds a binary value, v_i, for $0 \leq i < N$. The ith *binary prefix sum*, $psum_i$, is $v_0 + v_1 + \cdots + v_i$.

Theorem 16.2 *Binary prefix sums of N elements can be computed on an N-processor LARPBS in $O(1)$ steps.*

Proof. First, each processor R_i, $0 \leq i < N$ sets its conditional delay switch to straight if $v_i = 0$ and cross if $v_i = 1$. Referring to Figure 16.18c, R_1 and R_4 both hold a value of "1." Each processor sends a message containing its index addressed to processor R_{N-1}, that is, R_{N-1} is the selected destination for all messages. The conditional delay switches, however, will shift the pulses so that if $N - 1 - j$ is the number of switches set to cross after R_i, then the actual destination for processor R_i will be R_j. Processor R_j may receive multiple messages, however, it accepts only the first message to arrive in the bus cycle. Figure 16.18c shows the binary values held by processors that would induce switch settings as shown in Figure 16.4. On the basis of values, R_3 receives a message from R_1, R_2, and R_3, but accepts only the message from R_3, as shown in Figure 16.18a.

Next, processor R_j that received an index i then replies to R_i with a message containing its index. From the example, R_4 sends a message to itself, R_3 to itself, and R_2 to R_0 (Figure 16.18b). Since some messages may have been disregarded in the previous step, not all processors will receive a message in this step. To account for this, if R_i received a message from R_j during the second step, then it now segments the bus and broadcasts the index of j to its segment. The reason for this is that all processors within the same segment have the same prefix sums value. In our example, R_0, R_3, and R_4 segment the bus and broadcast the values 2, 3, and 4, respectively (Figure 16.18c). Each processor stores the value it receives as x_i.

Once R_0 receives the value x_0, it calculates the sum of all values in the array as $t = v_0 + (N - 1 - x_0) = 0 + (5 - 1 - 2) = 2$. Processor R_0 then broadcasts t to all processors, so that processor R_i can locally determine $psum_i = v_0 + v_1 + \cdots + v_i = t - (N - 1 - x_i)$. ∎

---

*The class L is the class of languages accepted by deterministic Turing machines with work space bounded by log N, where N denotes the input size. This class is contained inside P and the corresponding algorithms use less workspace than the size of their input [Joh90]. For example, a problem in L is one that can be solved in a reasonable amount of time by a polynomial number of computers.

Reconfigurable Computing with Optical Buses

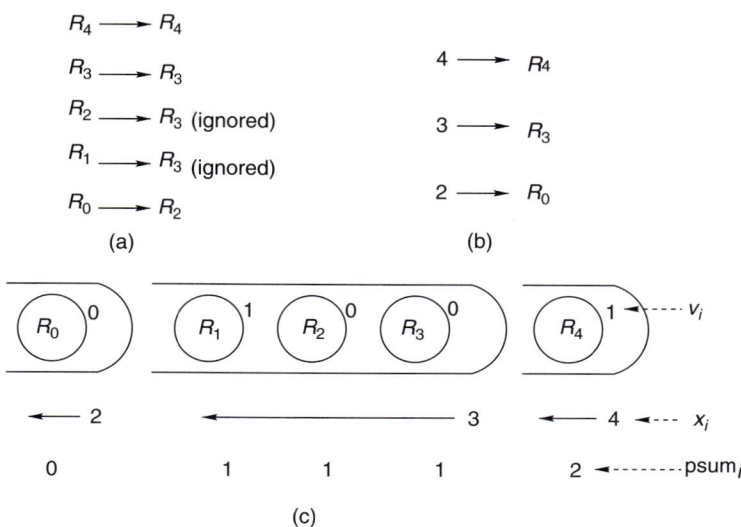

FIGURE 16.18 Binary prefix sums example: (a) actual destinations of first set of messages; (b) response to first message; (c) segmenting, broadcasting within segments, and computation steps.

The conditional delay switches are used to introduce unit delays, one unit delay for each input value of "1." The effect of this is that select and reference pulses of all processors with the same prefix sums value coincide at the same processor, however, only one message from this group of processors is received. The segment switches enable the highest indexed processor of such a group to segment the bus and broadcast data relaying information necessary for each processor to locally compute its prefix sum. The ability to pipeline messages allows each processor to compute its prefix sum simultaneously on a single bus.

16.4.2 Compression

Consider an LARPBS with N processors, such that each processor holds one element and some of the elements are marked. Let there be x such marked elements. The *compression* algorithm compacts all marked elements to the lower end of the array, namely processors R_0 through R_{x-1}, maintaining their relative order. The algorithm also compacts all unmarked elements to the upper end of the array, namely processors R_x through R_{N-1}, maintaining their relative order.

Theorem 16.3 *Compression of x elements, where $x \leq N$, can be performed on an N processor LARPBS in $O(1)$ steps.*

Proof. Consider processor R_i, where $0 \leq i < N$, holding a marked element v_i. Processor R_i sets its conditional delay switch to cross and sends a message with its index i addressed to processor R_{N-1}. All processors holding unmarked elements set their conditional delay switches to straight. If R_i holds the marked element with the kth largest index, then the actual destination for the message is R_{N-k}. Because of the conditional delays, each message written at this step arrives at a different destination processor.

Processor R_{N-k} that received an index i then replies to R_i with its index. Processor R_i stores k (i.e., N minus this index $N - k$) as $count_i$; this will contribute toward determining the final position for the marked element v_i. Next, each processor holding a marked element multicasts its index to all processors above it. The lowest indexed processor R_g holding a marked element will not receive a message, and will thus determine that it has the lowest index. Processor R_g then broadcasts $count_g$ to all processors so that each processor R_i with a marked element can then locally determine the final position for its element as $compress_i = count_g - count_i$.

Repeat the process for the unmarked elements; however, the position received in the second step is its final position. Once all processors have determined the final positions, route all elements to their proper destinations. ∎

16.4.3 Sorting and Selection Algorithms

Sorting and selection are basic operations finding use in many applications and have therefore been studied extensively. In this section, we sketch a variety of algorithms for sorting and selection.

ElGindy and Rajasekaran presented an $O(\log N \log \log N)$ step algorithm to sort N values on an N-processor LAPOB [ER99]. The algorithm uses a two-way merge sort in which there are $O(\log N)$ iterations of merges. Each successive merge is between larger pairs of sorted subsequences achieved by a multiway divide-and-conquer strategy. The merge procedure executes in $\log \log N$ recursive steps of partitioning the input sequences into subsequences that will then be merged in parallel on disjoint sets of processors. This algorithm can also be implemented on the LARPBS as well as some of the other one-dimensional optical arrays discussed.

The algorithm takes advantage of the pipelining ability of the LAPOB. This enables multiple merge operations to be executed in parallel on a single bus.

Theorem 16.4 *An N-processor LARPBS can sort N values in $O(\log N \log \log N)$ steps.*

Rajasekaran and Sahni designed an optimal algorithm to sort N elements in $O(1)$ steps using an $N^\epsilon \times N$ AROB, where ϵ is any constant greater than zero [RS97]. This algorithm is optimal owing to the lower bound of $\Omega(N^{1+\epsilon})$ processors for a comparison sort [AV87]. Rajasekaran and Sahni followed the column sorting algorithm of Leighton [Lei85], which assumes the elements are stored as a matrix of size $N^{2/3} \times N^{1/3}$. The algorithm consists of a constant number of column sorts and matrix transpositions. The transposition operations are basically permutation operations that the AROB can route in a single step by pipelining messages. The AROB performs column sort as follows.

First assume that an $N^{2/3} \times N$ AROB is available, then we will extend it for any $\epsilon > 0$. This provides an $N^{2/3} \times N^{2/3}$ subarray to sort each column of $N^{2/3}$ elements. Sort the elements of each subarray in $O(1)$ steps using the R-Mesh algorithm to sort N elements on an $N \times N$ R-Mesh in $O(1)$ steps [NS94]. This is possible owing to the ability to broadcast along a bus in a single step. In order to reduce the size of the AROB for any $\epsilon > 0$, recursively apply the sorting method for sorting columns for a total of $O(1)$ steps. This algorithm also runs on an $N^\epsilon \times N$ PR-Mesh.

Theorem 16.5 *An $N^\epsilon \times N$ AROB can sort N values in $O(1)$ steps, for constant $\epsilon > 0$.*

Integer sorting is a special case of sorting, and is usually performed by a series of radix sorts and compressions. This approach for sorting N k-bit integers takes $O(k)$ steps on an N-processor LARPBS [PL98]. Pavel and Akl presented an algorithm that runs in $O(k/\log \log N)$ steps on an N-processor LAROB [PA96a]. It takes advantage of the LAROB's bit polling operation and its ability to inject multiple delays onto the select waveguide. We will first describe the method for $k = O(\log \log N)$ bits and then extend it for $k = O(\log N)$ bits.

Each processor holds a value v_i, where $0 \leq i < \log N$. First, each processor p_i determines the number of processors p_j with $v_i = v_j$ and $i < j$ by using the bit polling operation. It then determines the total number of processors with the same value. The LAROB then uses the integer prefix-sums algorithm to rank the elements and determine the final destinations [PA96a]. The prefix-sums algorithm is similar to the binary prefix-sums algorithm of the LARPBS; however, a processor is able to introduce multiple delays to correspond to value v_i. Finally, route each element to its sorted position.

This algorithm stably sorts N integers with value $0 \leq v_i < \log N$ in $O(1)$ steps on an N-processor LAROB. To extend the range of values, divide the k bits in $k/\log \log N$ groups, each of $\log \log N$ bits. The LAROB performs the sorting algorithm in $k/\log \log N$ stages. During stage i, stably sort the values with respect to the ith least significant group of bits in $O(1)$ steps as above.

Theorem 16.6 *An N-processor LAROB can sort N k-bit values in $O(k/\log\log N)$ steps.*

The problem of *selection* is to select the kth smallest element out of N given elements. Li and Zheng designed a selection algorithm that runs in $O(\log N)$ time on an N-processor POB [LZ96]. The algorithm exploits the multicasting ability of the POB. It is recursive and proceeds as follows.

Let P denote the set of active processors; initially $|P| = N$. (The base case is when $|P| \leq 5$.) Partition P into groups of five contiguous processors each. In $O(1)$ steps, the tail of each group determines the median of its group. The POB compresses the $\lceil |P|/5 \rceil$ determined medians to the $\lceil |P|/5 \rceil$ leftmost processors. Recursively, find the median of these $\lceil |P|/5 \rceil$ values. Denote this value as m. The leftmost processor broadcasts m, and the POB computes prefix sums to count the number s of elements that are less than or equal to m. If $s = k$, then return m. If $s > k$ ($s < k$), then compress the elements less than or equal to (greater than) m and recursively call the select procedure on the s ($|P| - s$) elements. This algorithm also runs on an LARPBS.

Theorem 16.7 *The kth smallest element can be selected from N elements by an N-processor LARPBS in $O(\log N)$ steps.*

Rajasekaran and Sahni designed a randomized algorithm to perform selection on a $\sqrt{N} \times \sqrt{N}$ AROB in $O(1)$ steps with high probability (w.h.p.) [RS97]. (High probability is a probability $\geq (1 - n^{-\alpha})$ for any constant $\alpha \geq 1$.) The algorithm takes advantage of the constant time compression operation and sorting on an AROB. The algorithm first picks a random sample S of size $q = o(N)$. The AROB compresses the sample elements in the first row of the AROB and then sorts the sample. Next, choose two elements l_1 and l_2 from the sample whose ranks in S are $k(q/N) - \delta$ and $k(q/N) + \delta$ for some δ, where $\delta = f(N)$. These two elements bound the element to be selected w.h.p. Eliminate all elements outside of the range $[l_1, l_2]$. Repeat the process again for the remaining elements. The number of iterations required is less than four w.h.p.

Theorem 16.8 *The kth smallest element can be selected from N elements by a $\sqrt{N} \times \sqrt{N}$ AROB in $O(1)$ steps w.h.p.*

16.4.4 PRAM Simulations

An (N, M)-PRAM is a shared memory model that consists of N processors and M memory locations. The processors are able to read from and write to any of the shared memory locations. The read and write operations to a single memory location can either be concurrent or restricted to be exclusive to one processor at a time. Simulations of both *Exclusive Read Exclusive Write* (EREW) and *Concurrent Read Concurrent Write* (CRCW) PRAMs have been developed for some optical models.

The first result is a simulation of an N-processor CRCW PRAM with $O(N)$ memory locations by an N-processor LARPBS in $O(\log N)$ steps [LPZ97a]. The simulation takes advantage of an N-processor EREW PRAM with $O(N + M)$ memory locations being able to simulate an N-processor PRIORITY CRCW PRAM computation with M memory locations in $O(\log N)$ steps [J92]. Using this result, the LARPBS proceeds in simulating an N-processor EREW PRAM in $O(1)$ steps as follows.

First, assume that the EREW has $M = N$ shared memory locations. Let processor R_i of the LARPBS simulate PRAM processor P_i and hold memory location M_i. The LARPBS simulates a read step of the PRAM, where P_j reads from M_k, in two steps. In the first step, R_j sends its index to R_k, then in the second step, R_k sends the value of M_k to R_j. The LARPBS simulates a write step of the PRAM, where P_j writes value v_j into M_k, in a single step. Processor R_j sends v_j to R_k, and R_k stores this value. Since each step is an exclusive read or write step, the indices sent are all distinct and there are no conflicts. For the case when $M = O(N)$, there exists a constant c such that $M = cN$. In order to accommodate this, each processor of the LARPBS holds c memory locations and then simulates the read and write steps in c iterations. Combining the results provides the following theorem.

Theorem 16.9 *Each step of an N-processor* PRIORITY *CRCW PRAM with $O(N)$ shared memory locations can be simulated by an N-processor LARPBS in $O(\log N)$ steps.*

The simulation presented by Pavel and Akl [PA96b] is a randomized algorithm for the two-dimensional APPB model. They proceeded by first showing that a two-dimensional APPB with N processors can simulate any N-processor network, G, with constant degree in $O(1)$ steps. Map the processors of G to the APPB; however, the neighbors of a processor of G may not be neighbors in the APPB. To perform neighboring communications, construct a bipartite graph of G with k edges representing neighbor edges. From this, using k permutation routings, the APPB can simulate any communication step. This result implies that an N-processor APPB is able to simulate an N-processor butterfly network in $O(1)$ steps. Using Ranade's result [Ran91] that an N-processor butterfly network with $O(M)$ memory can simulate a step of a CRCW (N, M)-PRAM in $O(\log N)$ steps w.h.p. provides the following result.

Theorem 16.10 *Each step of an N-processor CRCW PRAM can be simulated by a $\sqrt{N} \times \sqrt{N}$ APPB in $O(\log N)$ steps w.h.p.*

16.5 Conclusions

The algorithms presented in this chapter are a small sample of the algorithms that have been developed for optically pipelined models. They demonstrate the key techniques used by most of these models and are building blocks for more complex algorithms. As mentioned in the Introduction, there have been many other algorithms developed that are more efficient on these models than on other reconfigurable models that do not use optical buses. This is achieved by exploiting key features of optical models, such as pipelining and constant propagation delays.

It is not always clear, however, which algorithms can run on which models, besides the one for which the algorithm was developed. The results relating many of the optical models alleviate this problem, enabling one to easily translate algorithms from one model to another.

Besides the work that has been cited in this chapter, including the array of algorithms and models, there has been much work considering the practical aspects of optical computing. For instance, we are often concerned with problem sizes that are too large to map directly to an available machine size. To address this issue, many papers have developed scalable algorithms for a variety of problems [CPX04, Li02, TBPV00]. Also, as it is not practical to allow an entire system to fail owing to the failure of a few nodes, fault tolerant algorithms have been developed for the LARPBS [BPP05, BT03]. Pan has presented a restricted LARPBS model to account for degradation of optical signals over lengthy bus spans [Pan03]. Roldán and d'Auriol have performed a detailed analysis of optical power and signal loss to determine bounds on the length of such optical buses [Rd03]. Many of these results provide the latitude of being able to develop algorithms without being concerned with the status or size of the available system.

References

[AV87] Y. Azar and U. Vishkin. Tight comparison bounds on the complexity of parallel sorting. *SIAM J. Comput.*, 16:458–464, 1987.

[BALPS95] Y. Ben-Asher, K. J. Lange, D. Peleg, and A. Schuster. The complexity of reconfiguring network models. *Inform. and Comput.*, 121:41–58, 1995.

[BAPRS91] Y. Ben-Asher, D. Peleg, R. Ramaswami, and A. Schuster. The power of reconfiguration. *J. Parallel Distrib. Comput.*, 13:139–153, 1991.

[BPP05] A. G. Bourgeois, Y. Pan, and S. K. Prasad. Constant time fault tolerant algorithms for a linear array with a reconfigurable pipelined bus system. *J. Parallel Distrib. Comput.*, 65:374–381, 2005.

[BSDE96] B. Beresford-Smith, O. Diessel, and H. ElGindy. Optimal algorithms for constrained reconfigurable meshes. *J. Parallel Distrib. Comput.*, 39:74–78, 1996.

[BT00]   A. G. Bourgeois and J. L. Trahan. Relating two-dimensional reconfigurable meshes with optically pipelined buses. *Int. J. Found Comp. Sci.*, 11:553–571, 2000.

[BT03]   A. G. Bourgeois and J. L. Trahan. Fault tolerant algorithms for a linear array with a reconfigurable pipelined optical bus system. *Parallel Algos. Appl.*, 18:139–153, 2003.

[CCPC04]   L. Chen, H. Chen, Y. Pan, and Y. Chen. A fast efficient parallel Hough transform algorithm on larpbs. *J. Supercomputing*, 29:185–195, 2004.

[CPX04]   L. Chen, Y. Pan, and X. H. Xu. Scalable and efficient parallel algorithms for euclidean distance transform on the larpbs model. *IEEE Trans. Parallel Distrib. Syst.*, 15:975–982, 2004.

[Dat02]   A. Datta. Efficient graph-theoretic algorithms on a linear array with a reconfigurable pipelined bus system. *J. Supercomputing*, 23:193–211, 2002.

[ER99]   H. ElGindy and S. Rajasekaran. Sorting and selection on a linear array with optical bus system. *Parallel Proc. Lett.*, 9:373–383, 1999.

[FZTV97]   J. A. Fernández-Zepeda, J. L. Trahan, and R. Vaidyanathan. Scaling the fr-mesh under different concurrent write rules. *Proc. World Multiconf. on Systemics, Cybernetics, and Informatics*, pp. 437–444, 1997.

[FZVT98]   J. A. Fernández-Zepeda, R. Vaidyanathan, and J. L. Trahan. Scaling simulation of the fusing-restricted reconfigurable mesh. *IEEE Trans. Parallel Distrib. Syst.*, 9:861–871, 1998.

[GMH+91]   Z. Guo, R. Melhem, R. Hall, D. Chiarulli, and S. Levitan. Array processors with pipelined optical buses. *J. Parallel Distrib. Comput.*, 12:269–282, 1991.

[Guo92]   Z. Guo. Sorting on array processors with pipelined buses. *Proc. Intl. Conf. Par. Process.*, pp. 289–292, 1992.

[Guo94]   Z. Guo. Optically interconnected processor arrays with switching capability. *J. Parallel Distrib. Comput.*, 23:314–329, 1994.

[HP95]   M. Hamdi and Y. Pan. Efficient parallel algorithms on optically interconnected arrays of processors. *IEE Proc.—Comput. Digital Techn.*, 142:87–92, 1995.

[HPS02]   Y. Han, Y. Pan, and H. Shen. Sublogarithmic deterministic selection on arrays with a reconfigurable bus. *IEEE Trans. Parallel Distrib Syst.*, 51:702–707, 2002.

[J92]   J. JáJá. *An Introduction to Parallel Algorithms*. Addison-Wesley Publishing Co., Reading, MA, 1992.

[JL03]   H. Shen, J. Li, and Y. Pan. More efficient topological sort using reconfigurable optical buses. *J. Supercomputing*, 24:251–258, 2003.

[Joh90]   D. S. Johnson. A catalog of complexity classes. In J. van Leeuwen, ed., *Handbook of Theoretical Computer Science, Volume A: Algorithms and Complexity*, pp. 67–162. MIT Press, Cambridge, MA, 1990.

[Lei85]   T. Leighton. Tight bounds on the complexity of parallel sorting. *IEEE Trans. Comput.*, 34:344–354, 1985.

[Li02]   K. Li. Fast and scalable parallel algorithms for matrix chain product and matrix powers on reconfigurable pipelined optical buses. *J. Inf. Sci. Eng.*, 18:713–727, 2002.

[LO95]   R. Lin and S. Olariu. Reconfigurable buses with shift switching: Concepts and applications. *IEEE Trans. Parallel Distrib. Syst.*, 6:93–102, 1995.

[LPZ97a]   K. Li, Y. Pan, and S. Q. Zheng. Simulation of parallel random access machines on linear arrays with reconfigurable pipelined bus systems. *Proc. Intl. Conf. Parallel Distrib. Proc. Tech. and App.*, pp. 590–599, 1997.

[LPZ97b]   Y. Li, Y. Pan, and S. Q. Zheng. Pipelined TDM optical bus with conditional delays. *Opt. Eng.*, 36:2417–2424, 1997.

[LPZ98a]   K. Li, Y. Pan, and S. Q. Zheng. Fast and efficient parallel matrix operations using a linear array with a reconfigurable pipelined bus system. In J. Schaeffer and R. Unrau, eds., *High Performance Computing Systems and Applications*. Kluwer Academic Publishers, Boston, MA, 1998.

[LPZ98b] K. Li, Y. Pan, and S. Q. Zheng. *Parallel Computing Using Optical Interconnections.* Kluwer Academic Publishers, Boston, MA, 1998.

[LZ96] Y. Li and S. Q. Zheng. Parallel selection on a pipelined TDM optical bus. *Proc. Intl. Conf. Parallel Distrib. Comput. Syst.*, pp. 69–73, 1996.

[MCL89] R. Melhem, D. Chiarulli, and S. Levitan. Space multiplexing of waveguides in optically interconnected multiprocessor systems. *Comput. J.*, 32:362–369, 1989.

[ME98] M. Middendorf and H. ElGindy. Matrix multiplication on processor arrays with optical buses. *Informatica*, 22, 1998.

[MS96] Y. Matias and A. Schuster. Fast, efficient mutual and self simulations for shared memory and reconfigurable mesh. *Parallel Algs. Appl.*, 8:195–221, 1996.

[Nak95] K. Nakano. A bibliography of published papers on dynamically reconfigurable architectures. *Parallel Proc. Lett.*, 5:111–124, 1995.

[NS94] M. Nigam and S. Sahni. Sorting n numbers on an $n \times n$ reconfigurable meshes with buses. *J. Parallel Distrib. Comput.*, 23:37–48, 1994.

[PA96a] S. Pavel and S. G. Akl. Matrix operations using arrays with reconfigurable optical buses. *Parallel Algs. Appl.*, 8:223–242, 1996.

[PA96b] S. Pavel and S. G. Akl. On the power of arrays with optical pipelined buses. *Proc. Intl. Conf. Par. Distr. Proc. Tech. and Appl.*, pp. 1443–1454, 1996.

[PA98] S. Pavel and S. G. Akl. Integer sorting and routing in arrays with reconfigurable optical buses. *Intl. J. Foundations of Comp. Sci.*, 9:99–120, 1998.

[Pan92] Y. Pan. Hough transform on arrays with an optical bus. *Proc. 5th Intl. Conf. Parallel Distrib. Comput. Syst.*, pp. 161–166, 1992.

[Pan95] Y. Pan. Order statistics on a linear array with a reconfigurable bus. *Future Gener. Comp. Syst.*, 11:321–328, 1995.

[Pan03] Y. Pan. Computing on the restricted LARPBS model. *Intl. Symp. Parallel Distrib. Proc. Appl. LNCS*, 2745:9–13, 2003.

[PH96] Y. Pan and M. Hamdi. Singular value decomposition on processor arrays with a pipelined bus system. *J. Network and Computer Appl.*, 19:235–248, 1996.

[PL98] Y. Pan and K. Li. Linear array with a reconfigurable pipelined bus system: Concepts and applications. *Inform. Sci.—An Int. J.*, 106:237–258, 1998.

[Qia95] C. Qiao. On designing communication-intensive algorithms for a spanning optical bus based array. *Parallel Proc. Lett.*, 5:499–511, 1995.

[QM93] C. Qiao and R. Melhem. Time-division optical communications in multiprocessor arrays. *IEEE Trans. Comput.*, 42:577–590, 1993.

[QMCL91] C. Qiao, R. Melhem, D. Chiarulli, and S. Levitan. Optical multicasting in linear arrays. *Int. J. Opt. Comput.*, 2:31–48, 1991.

[Raj05] S. Rajasekaran. Efficient parallel hierarchical clustering algorithms. *IEEE Trans. Parallel Distrib. Syst.*, 16:497–502, 2005.

[Ran91] A. G. Ranade. How to emulate shared memory. *J. Comput. Syst. Sci.*, 42:301–324, 1991.

[Rd03] R. Roldán and B. J. d'Auriol. A preliminary feasibility study of the LARPBS optical bus parallel model. *Proc. 17th Intl. Symp. High Performance Comput. Syst. and Appl.*, pp. 181–188, 2003.

[RS97] S. Rajasekaran and S. Sahni. Sorting, selection and routing on the arrays with reconfigurable optical buses. *IEEE Trans. Parallel Distrib. Syst.*, 8:1123–1132, 1997.

[SY05] D. Semé and S. Youlou. Computing the longest common subsequence on a linear array with reconfigurable pipelined bus system. *Proc. 18th Intl. Conf. Parallel Distrib. Comput. Syst.*, 2005.

[TBPV00] J. L. Trahan, A. G. Bourgeois, Y. Pan, and R. Vaidyanathan. An optimal and scalable algorithm for permutation routing on reconfigurable linear arrays with optically pipelined buses. *J. Parallel Distrib. Comput.*, 60:1125–1136, 2000.

[TBV98]   J. L. Trahan, A. G. Bourgeois, and R. Vaidyanathan. Tighter and broader complexity results for reconfigurable models. *Parallel Proc. Lett.*, 8:271–282, 1998.

[TVT96]   J. L. Trahan, R. Vaidyanathan, and R. K. Thiruchelvan. On the power of segmenting and fusing buses. *J. Parallel Distrib. Comput.*, 34:82–94, 1996.

[VT04]   R. Vaidyanathan and J. L. Trahan. *Dynamic Reconfiguration: Architectures and Algorithms.* Kluwer Academic/Plenum Publishers, New York, 2004.

[WC90]   B. F. Wang and G. H. Chen. Constant time algorithms for the transitive closure and some related graph problems on processor arrays with reconfigurable bus systems. *IEEE Trans. Parallel Distrib. Syst.*, 1:500–507, 1990.

[WHW05]   Y. R. Wang, S. J. Horng, and C. H. Wu. Efficient algorithms for the all nearest neighbor and closest pair problems on the linear array with a reconfigurable pipelined bus system. *IEEE Trans. Par. Distrib. Syst.*, 16:193–206, 2005.

[XCH05]   X. Xu, L. Chen, and P. He. Fast sequence similarity computing with LCS on LARPBS. *Proc. Intl. Workshop on Bioinformatics LNCS*, 3759:168–175, 2005.

[ZL97]   S. Q. Zheng and Y. Li. Pipelined asynchronous time-division multiplexing optical bus. *Opti. Eng.*, 36:3392–3400, 1997.

Algorithms

17
Distributed Peer-to-Peer Data Structures

17.1	Introduction...	17-1
	Model and Performance Measures • Chapter Overview	
17.2	Chord: A Distributed Hash Table......................	17-3
	Chord Structure • Adding and Removing Nodes Concurrently • Handling Node Failures	
17.3	Skip Graphs	17-7
	Skip Lists • Structure of Skip Graphs • Inserting and Deleting Concurrently • Congestion • Achieving $O(|K|)$ Space • Fault Tolerance • Hydra Components	
17.4	Further Information	17-15
	References	17-16

Michael T. Goodrich
Michael J. Nelson
University of California

17.1 Introduction

Peer-to-peer networks consist of a collection of machines connected to one another by some networking infrastructure, such as the Internet, that communicate and share resources directly rather than using an intermediary server. A fundamental component of such networks is a search protocol for determining the location of a desired resource. Generally, the resources are identified by an associated key. Queries on the keys specify which resource or resources the search function should return.

Some of the peer-to-peer systems currently in use, such as the Gnutella file-sharing system, implement searches by broadcasting the query to every peer in the network. This approach scales poorly as the number of users increases, since the number of messages generated by each search is directly proportional to the number of users. To circumvent the need to broadcast every query, other peer-to-peer networks maintain a record of which machine holds each key and its corresponding resource. Early peer-to-peer networks stored the record on a central server, or small collection of servers, that would receive and process queries from all of the peers. Centralized schemes have several drawbacks. Primary among them is that the scheme leads to a large volume of network traffic at the servers—the load on each server scales linearly with the number of queries. Moreover, the server must be notified whenever a new key is added or an old key is removed, which happens frequently in typical peer-to-peer systems. A second problem is that the functionality of the peer-to-peer network depends on a small set of machines; failure of the servers yields complete failure of the network. This dependency on a small collection of machines leaves the network vulnerable to denial of service attacks, in which the servers are flooded with messages from a malicious source until the servers become unable to process legitimate queries.

In light of these problems, researchers have begun to develop decentralized schemes in which a piece of the record is distributed to each of the peers in a highly structured fashion that allows for efficient location of the resources without burdening or relying upon one particular machine. These distributed peer-to-peer data structures are the focus of this chapter. We will look at structures that support some basic queries on a set of keys, including exact queries, range queries, and one-dimensional nearest-neighbor queries. We begin by outlining a typical networking model for peer-to-peer networks, and describe some of the fundamental performance properties sought when designing a distributed data structure.

17.1.1 Model and Performance Measures

Peer-to-peer networks are typically built upon some preexisting network, such as the Internet, that supports message passing from one machine to another if given the network address of the recipient. The data structures we discuss here are designed with the assumption that such a routing mechanism exists. The core of each distributed data structure will be a specification of which machines are connected to each other by knowledge of these addresses. The network addresses are often referred to as *pointers* because of their similarity to memory-addressing pointers. The structure of the pointers can be concisely represented by a graph where nodes correspond to machines and edges correspond to a pointer from one machine to another. Figure 17.1 gives an example of a graph for a centralized server, where every machine knows the address of the server.

We will measure the *space complexity* of a distributed data structure by the number of keys and pointers stored by each machine. If each node corresponds to a different machine, then the number of pointers is the degree of the corresponding node. In some cases, however, it will be more convenient to define a graph structure where nodes do not directly correspond to machines. For example, there may instead be one node corresponding to each key. In such cases, it is necessary to provide a function to map each of these nodes to a physical machine that will maintain pointers to adjacent machines and store any keys that are associated with the given node. In this chapter, we will outline an algorithm to map nodes to machines in a roughly balanced manner. More sophisticated algorithms might incorporate knowledge of each machine's storage and bandwidth resources to assign nodes to machines. In either case, a reasonable goal in the design of the data structure is to minimize the space used by each individual node.

Search operations in the data structure proceed by following pointers to nodes that are "closer" in some sense to the desired key, until eventually its precise location is found. As network bandwidth is typically the most limited network resource, the efficiency of a data structure operation is often evaluated according to its *message complexity*—the number of messages sent during the operation. Less attention is paid to the internal computation time; however, as the reader may wish to note, the internal computations associated with each message rarely exceed a logarithmic number of operations.

There is an inherent trade-off between message complexity and space complexity because message complexity is bounded by the dilation of the pointer graph—the maximum over all pairs of nodes of the

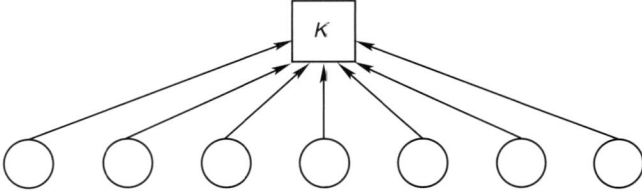

FIGURE 17.1 Graph representation for a client-server structure. Circular nodes represent client machines whereas the square node represents the server. Each client knows the network address of the server, where the set of keys is stored.

Distributed Peer-to-Peer Data Structures

shortest path length between that pair—whereas space complexity is a function of the maximum degree of the graph. The following theorem from graph theory captures this trade-off quantitatively.

Theorem 17.1 *A graph with maximum degree d has a dilation of* $\Omega(\log_d n) = \Omega\left(\frac{\log n}{\log d}\right)$.

One of the key properties of a distributed data structure is its *scalability*—the capability of the data structure to perform well as the number of machines increases. The space and message complexities are good measurements of scalability. A good rule of thumb for a data structure to be scalable is for the space and message complexity to be logarithmic or polylogarithmic in the number of machines and keys.

Another important property of a distributed data structure is its ability to tolerate failures. In peer-to-peer networks, machines may suddenly drop out of the network, either temporarily or permanently, for any number of reasons. For example, the machine may be shut down, or its network link may become overly congested. In such situations, there is no opportunity for this faulty machine to pass its keys or its structural information, such as pointers, to other machines. As this information is vital to the correct functioning of the data structure, a fault tolerant data structure should include mechanisms to provide some assurance that the functioning peers remain connected and that no data is permanently lost even if some number of machines cease operating.

The data structures in this chapter make use of randomization. As such, many of the performance bounds of the data structures are random variables. Therefore, we will often use a common convention of stating results as holding *with high probability* (*w.h.p.*), meaning the statement holds with probability $1 - O(n^{-k})$ for some constant $k \geq 1$, where, unless otherwise indicated, n is the number of nodes in the structure. In an actual implementation, it may be desirable to occasionally test for occurrences of the unlikely bad events.

For the remainder of the chapter, we denote the set of machines by N and the set of keys by K. The respective sizes are denoted with the standard set notation $|N|$ and $|K|$.

17.1.2 Chapter Overview

This chapter will survey representatives of recent distributed data structures. Section 17.2 looks at Chord, a distributed hash table that supports exact searches for keys. Section 17.3 looks at skip graphs, an ordered dictionary that supports queries such as one-dimensional range queries and nearest-neighbor queries. Section 17.4 provides pointers to additional information relating to the topics presented in this chapter.

17.2 Chord: A Distributed Hash Table

One of the most fundamental data structures for indexing is the hash table, which supports exact-match queries for keys by mapping them to memory locations on the basis of a hash function, which maps the set of keys to b-bit integers. The distributed hash table, or Distributed Hash Table (DHT), was one of the first distributed data structures to be developed. It extends the functionality of a hash table to the distributed setting by supporting the function *search(hash_value)*, which locates the node responsible for storing all keys that have a hash of *hash_value*.

Chord [20] is one of many distributed hash tables developed around the same time. Others include Pastry [17], Tapestry [22], CAN [17], and Viceroy [11]. Chord provides scalability and fault tolerance. It is also fairly simple, and hence is likely to be easily implementable. (In fact, an implementation already exists and may be found at http://pdos.csail.mit.edu/chord/)

17.2.1 Chord Structure

Distributed hash tables, including Chord [20], are built using a hash function that maps keys to a b-bit integer in the interval $[0, 2^b - 1]$. This interval is referred to as the *identifier space*. For the remainder of the

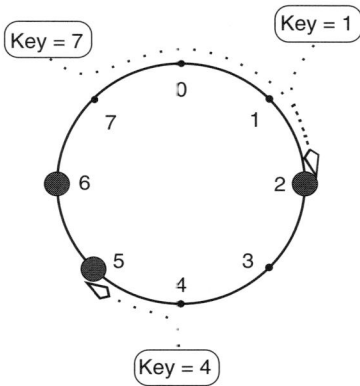

FIGURE 17.2 Identifier circle for $b = 3$ with node identifiers 2, 5, and 6 and keys 1, 4, and 7. Keys 1 and 7 are stored at node 2 and key 4 is stored at node 5.

section, we will use the term *key* interchangeably to refer to either a key or its hash value. The efficiency of Chord depends on assumptions of independence between the hash values of different keys. For simplicity of discussion, we will assume that the hash function behaves as an ideal random hash, that is that a given key hashes to a particular value in the identifier space with probability $1/2^b$.

Chord partitions the identifier space by assigning a random value from the identifier space to each node. We shall refer to this value as the *identifier* of the node, and shall denote the identifier of a node n by $n.id$. We again assume the value assigned is truly random, although a pseudo-random function would yield similar performance results.

The core of the Chord structure is a circularly linked list of the nodes ordered by their identifiers. Each node n with identifier $n.id$ keeps a pointer *successor* to the node with the next largest identifier, except if $n.id$ is the largest identifier, in which case the successor is the node with smallest identifier. We may think of the nodes as being arranged around an *identifier circle* labeled clockwise with values from 0 to $2^b - 1$. The successor of a node is then the node that follows it clockwise on the identifier circle. Each key and its associated data will be stored at the node that is nearest to it clockwise on the identifier circle, which we shall refer to as the *successor* of the key; the successor of a key k will be denoted by $successor(k)$. Similarly, we define the *predecessor* of a key to be the node preceding it clockwise on the identifier circle. Figure 17.2 gives an example of the identifier circle with $b = 3$ and three nodes with identifiers 2, 5, and 6 and keys that hash to 1, 4, and 7. The arrows point to the successor node of each key, at which that key will be stored.

Chord allows a node n to efficiently locate a key's predecessor by storing pointers to nodes on the identifier circle that are powers of 2 away from $n.id$. More precisely, a list of *finger* pointers is maintained at each node n such that $n.finger[i] = successor((n.id + 2^i) \bmod 2^b)$ for all i between 0 and $b - 1$. (Note that finger[0] corresponds to the successor of n.) The arithmetic is performed modulo 2^b so that the fingers will wrap around the identifier circle. For the remainder of this section, all arithmetic will be modulo 2^b and as such the modulus will not be stated explicitly. Utilizing these finger pointers, a binary search for predecessor(k) can be performed by moving to the finger node nearest but preceding k and then repeating the search from that node using a remote procedure call. Each such move reduces the distance from the current search node to the predecessor by at least a factor of $1/2$, for if this were not the case, then there would be a finger node closer to the predecessor than the one selected (at roughly twice the distance). Since the original distance to the predecessor cannot exceed 2^b, a predecessor search must terminate after at most b remote calls to finger nodes. Moreover, because the nodes are randomly distributed around the identifier circle, it will in fact terminate after $O(\log |N|)$ remote calls, regardless of how large b might be. Pseudocode for the function *find_predecessor(k)* is given in Figure 17.3. The successor of a key k is of course just the successor of predecessor(k); as such, we omit pseudocode for

Distributed Peer-to-Peer Data Structures

```
define find_predecessor(key):
    if successor.id not between self.id, key
        **Then current node is predecessor**
        return self.address
    else
        for i=b−1 downto 0
            if finger[i].id between self.id, key
                return finger[i].find_predecessor(key)
```

FIGURE 17.3 Pseudocode for the *find_predecessor* function. Returns the first node preceding the specified key in clockwise order on the identifier circle.

the function *find_successor*. Theorem 17.2 and its proof formalizes the argument regarding the message complexity of *find_predecessor*.

Theorem 17.2 *The message complexity of find_predecessor (and find_successor) in Chord is $O(\log |N|)$.*

Proof. We observe first that the distance (number of identifier space values) between predecessor (*key*) and the current node decreases by at least a factor of 1/2 each time a remote procedure call is made. Let d_p be the distance from the current node to the predecessor of the key, and let i be the index of the finger node to which the next remote procedure call is made. This finger node must be at a distance of at least 2^i from the current node by the definition of the finger nodes. Since this finger node was selected, $finger[i + 1].id$ must fall beyond the predecessor of the key. But then $d_p \leq 2^{i+1} = 2 * 2^i$, for otherwise predecessor(*key*) should take on the value of *finger* $[i + 1]$. Thus, the distance traveled, which was at least 2^i, must be greater than or equal to $d_p/2$.

The previous observation guarantees that after $\lceil \log |N| \rceil$ calls to *find_predecessor*, the distance from the current node to the predecessor cannot be more than $2^b/|N|$. Because the nodes are distributed randomly across the identifier space, the expected number of nodes in an interval of size $2^b/|N|$ is one, and with high probability it will not exceed $\log |N|$. This follows from the well-studied *balls-and-bins* problem in probability theory, from which it is known that if n balls are thrown into n bins, then with high probability no bin will contain more than $O(\log n)$ balls. In our case, the interval of size $2^b/|N|$ may be viewed as one of $|N|$ bins into which the nodes have been placed. Hence, once the distance is reduced to $2^b/|N|$, there will be only $O(\log |N|)$ more nodes traversed before reaching the predecessor. ∎

17.2.2 Adding and Removing Nodes Concurrently

For a new machine to join the system, it must have contact with some machine currently in the network. This machine will introduce it to the network by inserting a new node into the network that corresponds to the new machine. To maintain the Chord structure, several things must happen when a new node n is inserted into the Chord structure. The successor and predecessor of $n.id$—call them n_s and n_p—must be located and n inserted between them. The keys assigned to n_s that precede $n.id$ must also be transferred to n. In addition, the finger table for n must be computed by repeated calls to the *find_succesor* function. This step requires $O(b)$ calls in a straightforward implementation; however, by testing at each step whether finger $[i]$ and finger $[i + 1]$ are the same node, the number of calls will be bounded by $O(\log |N|)$. This follows from the same balls-and-bins observation made in the proof of Theorem 17.3—namely, the number of different finger nodes for distances smaller than $2^b/|N|$ is $O(\log |N|)$. Hence, a node insertion has $O(\log^2 |N|)$ message complexity. A final step that should be performed to maintain the Chord structure after an insertion is to update the finger table entries of nodes for which n is now the finger value. However, these updates may alternatively be performed by a periodic refreshing function described shortly.

Node removals are handled in a similar fashion. The keys stored by an exiting node must be transferred to its successor. The predecessor of the node must have its successor value updated. Lastly, the finger table entries that point to the exiting node should be updated. This functionality may also be left to the periodic refreshing function described subsequently.

Chord was designed with the peer-to-peer setting in mind; specifically, it was designed under the assumption that nodes frequently join and leave the system concurrently. In such a setting, it is unlikely that the finger tables will remain accurate at all times even if they are updated with each node join. In light of this, the designers of Chord suggest that only the successor pointers be maintained vigorously, in order to ensure the correctness of a query. Correctness of the search is guaranteed by correctness of the successor pointers, because of the nature of the search algorithm. When searching for the predecessor of a key, the search never passes beyond the identifier of that key. Thus, a step forward never increases the number of nodes between the current node and the predecessor being sought. The correctness of the successor pointers guarantees that after the loop in *find_predecessor* terminates, at least one step forward will be made until the predecessor is reached.

The problem of maintaining correctness of successor pointers with concurrent insertions and deletions is made easier by also maintaining predecessor pointers. Now the problem that must be addressed is the maintenance of a circular doubly linked list with concurrent insertions and deletions. An insertion into a list may be thought of as an update operation on the edge between the predecessor and successor. By associating a lock with each edge, we may effectively serialize the concurrent operations. We may arbitrarily assign control of the lock to either the predecessor or successor. A node that is to be inserted must acquire this lock before the edge can be changed. When multiple insertions occur at once, these nodes are placed in a queue. For insertions, it is clear that deadlocks will never occur. Each node being inserted changes only one edge—that which connects its successor and predecessor.

Deletions are more problematic, as a node being deleted must obtain the locks to both its left edge and right edge. In a worst case scenario, each node might be trying to delete itself and may obtain only one of the two edge locks. This is an instance of the problem better known as the Dining Philosophers Problem [10]. A randomized algorithm supplies a solution to the problem. Each node seeking the locks generates a random number and is granted the lock only if the neighboring node generated a smaller number, or did not request the lock at all. If it receives both locks, then the node performs the deletion. Otherwise, it gives up any lock it may have had and waits for another round of lock allocation. In this way, a node will delete itself in three rounds in expectation and $O(\log n)$ rounds with high probability.

If the successor pointers remain accurate, the finger entries may be refreshed periodically by invoking $successor(n.id + 2^i)$ to locate the correct current finger value. Under reasonable rates of joins and leaves, the finger tables can be updated well enough to support efficient queries.

17.2.3 Handling Node Failures

In any actual peer-to-peer network, machines will periodically fail and drop out of the network without providing notification in order to update the structure. The Chord structure may be augmented in order to provide a degree of tolerance to such node failures.

Before one may make formal statements about the fault tolerance of a data structure, some model of how nodes fail must be chosen. In the discussion below, we will assume that nodes fail independently of one another. In practice, of course, this is usually not the case. For example, a failure at one node will often be accompanied by the failure of other nodes that are physically nearby in the underlying routing network. Nevertheless, the independent-failure model provides useful insight into the fault tolerance of the data structure. In particular, because we assume the machines are assigned to independently random locations on the identifier circle, the position of failed nodes in the Chord network should be independent of one another even if there is some correlation between the physical locations of failed machines.

As noted in the previous section, the key to maintaining the Chord structure is to maintain correct successor pointers. The simplest method to maintain this when nodes may fail unexpectedly is to introduce

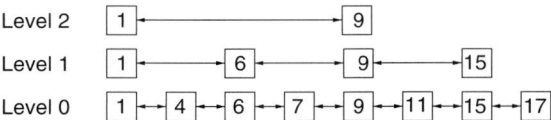

FIGURE 17.4 A perfect skip list on eight keys.

redundancy. Each node maintains a *successor list* of its nearest r successors and the keys assigned to those successors, where r is a tuneable parameter. If r is chosen to be $\log |N|$, then even if each node fails with probability $1/2$, there will be at least one live successor with probability $1 - 1/|N|$. Thus, if the successor of a node fails, it may use the successor list to locate the first live member of the successor list and update its successor value accordingly. The keys of the failed node should also be assigned to this first live successor. The refresh protocol will then eventually correct the finger table entries that point to failed nodes. This is guaranteed by the fact that a search for $n.finger[i]$—that is, $successor(n.id + 2^i)$—will always yield the correct finger value if the $finger[j]$ entries are correct at all nodes for $j < i$, since such a search uses only those finger entries. Hence, as long as the finger[0] entries, that is, the successor entries, are accurate, it will follow inductively that the other finger entries will eventually become accurate as well (in the absence of further faults).

This method of fault tolerance increases the space complexity by a factor of r. Later in the chapter, we will discuss how error-correcting codes can be used to reduce this factor to a constant.

17.3 Skip Graphs

The primary limitation of distributed hash tables is that they inherently provide support only for exact queries. Skip graphs support order-based queries such as nearest-neighbor and range queries. They were developed by Aspnes and Shah [3] and independently by Harvey et al. under the name of SkipNet [7]

Skip graphs also possess a property of *locality* not found in DHTs—keys that are close according to the ordering will be near each other in the skip graph. This may be useful, for example, in applications where there is some correlation between proximity of keys and physical proximity of the corresponding network resources.

Skip graphs present additional advantages over many distributed hash table implementations. For example, they do not require explicit knowledge of the number of machines in the network. In contrast, Chord requires this information in order to choose the parameter b that determines the number of bits in the hash function and the size of the corresponding identifier circle.

Skip graphs are an extension of skip lists. As such, we begin the discussion of skip graphs with a brief discussion of skip lists.

17.3.1 Skip Lists

The skip list structure was first presented by Pugh in Reference 16. A skip list is a collection of increasingly sparse doubly linked lists of keys, with each list sorted according to the ordering of the keys. The set of lists is ordered, and each list is labeled with a level on the basis of this ordering. The level 0 list contains every key. Each subsequent list at level i includes some subset of the list at level $i - 1$.

In a *perfect* skip list, every other element of level $i - 1$ is included in level i. Figure 17.4 gives an example of a perfect skip list. In this structure, moving from one node at level i to its successor at level i will travel over 2^i keys at level 0. This allows for the location of a key in logarithmic time much in the same way as Chord. Starting at the topmost level, if the next node to the right in the list is less than the key being sought, the search proceeds to that node. Otherwise, the search drops down to the same key in the next level and the process is repeated.

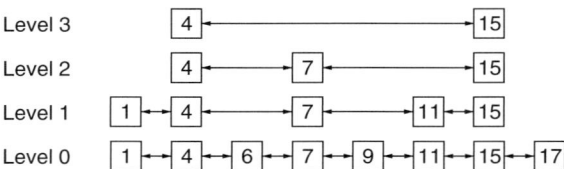

FIGURE 17.5 A randomized skip list on eight keys with a root size of two and a height of three.

A perfect skip list is difficult to maintain for a dynamic set of keys. *Randomized* skip lists were introduced to avoid this problem. In a randomized skip list, each key at level $i - 1$ is included independently of other keys in the list at level i with probability $1/2$. This process is repeated until some level contains at most some constant number of keys, say two. The index of this sparsest level is referred to as the *height* of the skip list, and the list at this uppermost level is referred to as the *root* list. The height of a randomized skip list is $O(\log |K|)$ with high probability. Figure 17.5 shows an example of a randomized skip graph with a height of three. Randomized skip lists retain the efficient search property of perfect skip lists—with high probability, a search takes $O(\log |K|)$ steps before reaching the search key, or its successor if the key is not present. Because the leftmost key is not necessarily present in the root list of a randomized skip list, the search algorithm must be amended slightly. When a search begins, a test should be performed to determine whether the search key is to the left or right of the initial root key. This will determine whether the lists are traversed leftward or rightward to find the key. A leftward search proceeds in the same manner as the rightward search described for perfect skip lists except that the opposite comparison operator is used to determine when to proceed to the left and when to move downward.

As already suggested, randomized skip lists allow for efficient insertion and deletion of keys. Because the presence of each element of each list is independent of other keys, the only operation needed for a new key is the insertion of the key into each list to which it belongs. This insertion mechanism is simple to implement. Once a key has been inserted at level i, it may be inserted into level $i + 1$ (if it is selected to by the randomized choice) by searching rightward in the level i list for the nearest key that is also present in level $i + 1$. This key corresponds to the successor of the new key in level $i + 1$. It is then easy to splice the new node into level $i + 1$ to the left of its successor. In the rare case that there is no successor of the key in the next list, its predecessor should be sought out instead. If neither are present, then the current list must be the root list. The process terminates here, unless the size of the root list exceeds whatever constant is chosen, in which case a new list is created at the next level by applying the random inclusion process. The insertion procedure as a whole takes $O(\log |K|)$ time in expectation, since, at each level, the expected distance to the new key's successor in the level above is two and the total number of levels, that is, the height of the skip list, is $O(\log |K|)$ with high probability.

Deleting a key is even simpler than inserting it. In each level to which it belongs, the deleted key's successor and predecessor are set to point to one another. In the rare case that the node being deleted is in the root list or the level below it, it may also be necessary to delete the root list if the level below now falls within the maximum root size. Since the height is $O(\log |K|)$ with high probability, the deletion process also requires $O(\log |K|)$ steps with high probability.

Any pointer-based data structure may be applied in a distributed setting simply by mapping some machine to each node of the pointer structure. A natural question to ask, then, is why a skip list does not provide a satisfactory solution to the distributed nearest neighbor problem. One answer is that it possesses some of the same problems as the centralized indexing scheme described in the beginning of the chapter. Assume that each key is mapped arbitrarily to some machine. Then, the nodes corresponding to keys in the root list serve as single points of failure; moreover, these nodes will likely have to process a large fraction of all queries. This can be seen by observing that any query originating from a key to the left of a root node with a search key located to the right of that root node must pass through the root node, either in the root list or in some lower level. (In fact, the search algorithm as described above must always use

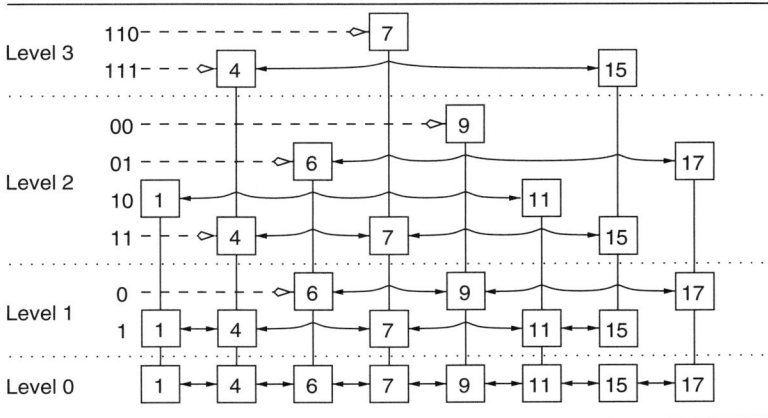

FIGURE 17.6 A skip graph on 8 keys. The lists at each level are partitioned randomly into two lists in the level above until each list has size two or less. Each binary string is associated with the list of elements whose membership vector agrees with that string.

the root list; however, a cleverer algorithm might be able to avoid this when the destination lies on the same side of the root key as the origin of the query.)

17.3.2 Structure of Skip Graphs

Skip graphs [3, 7] are designed in part to eliminate single points of failure and points of congestion. This is accomplished in essence by generating many different skip lists, any of which will be able to process a query. At the same time, these lists overlap with one another enough that the space complexity is not significantly increased.

Unlike Chord, which has one node for each machine, a skip graph has one node for each key. There are various methods of assigning machines to these nodes. One method is to assign a machine to the keys of all of the resources it hosts. Alternatively, one may use a distributed hash table to hash the keys onto the machines, yielding a more uniform distribution of nodes. In the discussion of fault tolerance, we will be assuming that the latter mechanism is used, as the former mechanism introduces added complexity to the failure model.

The structure of skip graphs resembles the structure of skip lists. A skip graph consists of a collection of doubly linked lists partitioned into levels. The level 0 list is identical to that of a skip list, consisting of one sorted list of all keys in the set of keys K. Level 1 and above, however, will in general contain more than one list. Level $i + 1$ is constructed by partitioning each list L at level i of size three or more into two lists, $L0$ and $L1$. Each element of L is randomly assigned to one of the two lists with equal probability. One way to view this in relation to skip lists is where the skip list throws out roughly half of the nodes at each level, making one sparser list, the skip graph takes the discarded half and creates a second list. Thus, level i may contain as many as 2^i lists, and in general every key will be present in one of the lists at each level.

To formulate this process, we associate a string of random bits with each key. The string in theory is infinite, but in practice need only be long enough to be different than the membership vectors of other keys. This string is referred to as the membership vector of the key, and for key k will be denoted by $m(k)$. For a binary string s we use the notation $s[i]$ to denote the ith bit of s and $s|i$ to denote the first i bits of s. Level i contains one linked list for each binary string of length i. Let s be such a string. Define the list associated with s as $L(s) = sort(\{k \mid m(k)|i = s\})$. A skip graph is then simply the union of all these lists. Figure 17.6 gives an example of a skip graph on 8 keys. A skip graph is in fact a collection of skip lists. Define the skip list associated with s to be $SL(s) = \{L(s|i) : \forall\, i \geq 0\}$.

```
define find_successor(key):
    if self.key < key
        return rightward_search(key, self.root_height)
    else if self.key > key
        return leftward_search(key, self.root_height)
    else
        return self.address

define rightward_search(key,height):
    while(height >=0)
        if right_neighbor[height].key < key
            return right_neighbor[height].rightward_search(key,height)
        else
            height = height - 1
    return right_neighbor[0].address
```

FIGURE 17.7 Pseudocode for the *find_successor* function and helper function *rightward_search*.

We must justify referring to such a set as a skip list. The definition of randomized skip list was an ordered set of lists satisfying the following two properties:

1. The list at level 0 contains all keys
2. A key in the list at level i is included in level $i+1$ with probability $1/2$

If we take $L(s|i)$ to be the level i list, then the two properties above will hold. Every key will be included at level 0 since the first 0 bits of every string is the empty string. If some key k is in level i, then $m(k)|i = s|i$. Thus, it will be included in level $i+1$ if and only if bit $i+1$ of $m(k)$ agrees with bit $i+1$ of s. This happens with probability $1/2$ since the bits of the string are taken to be random.

Thus, we see that a skip graph consists of many skip lists, each of which should support efficient search with high probability. We may extend the terminology used in the section on skip lists to apply to skip graphs. The root list of $SL(s)$ is the first list containing at most two keys. The height of $SL(s)$ is the level of the root list. Furthermore, the height of the skip graph is defined to be the maximum height of any root list. A minor modification to the analysis of the skip list height shows that the height of a skip graph is $3 \log |K|$ with probability $1 - O(1/|K|)$. Since each key occurs in $O(\log |K|)$ lists w.h.p., each node will require $O(\log |K|)$ pointers. Hence, the space complexity of a skip graph is $O(\log |K|)$ per key.

A natural choice for a search beginning from a node with key k_s is to search the skip list corresponding to $m(k_s)$, that is, $SL(m(k_s))$. This is a natural choice because k_s must be in the root list of $SL(m(k_s))$—the bits of $m(k_s)$ will always agree with themselves. A search then proceeds as it would in a normal skip list—searching either leftward or rightward depending on the direction of the search key. Figure 17.7 gives pseudocode for the functions *find_successor* and *rightward_search*. As noted in the previous subsection, leftward searches are symmetric to rightward searches.

Theorem 17.3 *The number of downward moves and sideways moves (leftward or rightward) for find_successor in a skip graph is $O(\log |K|)$ with probability $1 - O(1/|K|)$.*

Proof. The proof of this theorem is a direct adaptation of a similar theorem for skip lists. The number of downward moves is $O(\log |K|)$ with high probability because of the height bound of the skip graph. To determine the number of sideways moves, it is helpful to consider the path taken by a search operation in reverse, from the destination backward towards the root. In this reversed path, each step will be either a sideways move or an upward move. Suppose we are at some node in the path at level i. The next step will be an upward one if and only if the next membership bit of the current key agrees with the next bit of the root key's membership vector. This follows from the fact that whenever the current key is in the same level $i + 1$ list as the root key, it must be the case that it was encountered at level $i + 1$ before the search moved downward to level i. The membership vector bits agree with probability $1/2$. Thus each move will be an upward move with probability $1/2$, independent of any previous moves (in the reverse order). Let us consider an outcome causing an upward move to be a "heads," and an outcome for a sideways move a "tails." Suppose the height of the root is $k \log n$. Then, the number of sideways moves will be equal to the number of tails that occur before $k \log n$ heads are observed. This value is $k \log n$ in expectation.

To achieve the high probability bound, more sophisticated tools from probability theory are required. The total number of downward and sideways moves will be equal to the number of coin flips before $k \log n$ heads occur. This random variable follows a form of the negative binomial distribution [14]. The theorem follows from algebraic manipulation of the density function for this distribution, but the details are omitted here. ∎

17.3.3 Inserting and Deleting Concurrently

The basic insertion algorithm is identical to that of skip lists. A key must first be introduced to level 0 between its predecessor and successor. The key is then inserted into subsequent levels. To insert the key at level $i + 1$ once it is present in level i, its list at level i is traversed until another key is encountered that belongs to the same list in level $i + 1$. Since each key will be present in the list above with probability $1/2$, the search at any given level will take $O(1)$ messages in expectation and $O(\log |K|)$ messages with high probability. The total number of messages at all levels is $O(\log |K|)$ in expectation and with high probability as well, as it follows a negative binomial distribution. A key is deleted by linking its predecessor and successor at each level. This requires $O(1)$ messages at each level in the worst case, and hence $O(\log |K|)$ messages total.

Concurrency support comes almost directly from the algorithm described in Section 17.2 for concurrent insertion and deletion into a linked list. Insertions and deletions may be decomposed into the individual list insertions and list deletions. However, additional attention must be paid to the roots of the structure. Specifically, when a root lists grows beyond the maximum size bound, it is important that the root list be locked so that the correct split form of the root list is obtained in the level above. Pseudocode for key insertion in a skip graph is provided in Figure 17.8.

17.3.4 Congestion

Recall that one of the primary drawbacks to the centralized indexing scheme described at the beginning of the chapter was that the machine hosting the indexing data is subject to heavy amounts of network traffic owing to query handling. In certain settings, skip graphs provide a more balanced distribution of query-processing load across the machines.

In general, balancing query-handling load is a complicated problem because the distribution of queries may be heavily biased towards certain keys. Moreover, this query bias may change with time, making an adaptive load balancing approach difficult both to design and to analyze. As such, we will restrict ourselves to a uniform query distribution. That is, we consider the load balance when every pair of source and destination key is equally likely. Although this is a very restrictive model, it serves as a baseline for any distributed data structure. In other words, although it may not be sufficient for a data structure to perform well in this model, it is certainly necessary that a data structure performs well in the model for it to be of interest.

A useful metric for this model is the *congestion* of the data structure. The *congestion of a node* in a distributed data structure with n nodes is the probability that a query between a randomly chosen source and destination node passes through that node. Note that this probability is itself a random variable whose

```
define insert_key(k):
    height=0
    successor=find_successor(k)
    list_insert(successor,height)
    while current list size is greater than 2:
        successor=find_right_neighbor(height+1)
        if successor not NULL:
            list_insert_left(successor,height+1)
        else:
            predecessor=find_left_neighbor(height+1)
            if predecessor not NULL:
                list_insert_right(predecessor,height+1)
            else:
                create_new_list(height+1)

define find_right_neighbor(height):
    next = right_neighbor[height−1]
    while (m(k)[height] != m(next)) AND (next not NULL):
        next = next.right_neighbor[height−1]
    return next
```

FIGURE 17.8 Pseudocode for key insertion in a skip graph.

value depends on the random choices made in the construction of the data structure. The *congestion of a data structure* is the maximum congestion of any node.

The following theorem gives a probabilistic bound on the congestion of skip graphs.

Theorem 17.4 *The congestion of a skip graph is* $O\left(\frac{\log^2 |K|}{|K|}\right)$ *with high probability.*

17.3.5 Achieving $O(|K|)$ Space

A detriment of skip graphs in contrast to Chord is that there are $|K|$ nodes of degree $O(\log |K|)$ instead of $|N|$ nodes of degree $O(\log |N|)$. In some applications, $|K|$ might be significantly larger than $|N|$. In this section, we present an adaptation of skip graphs presented by Goodrich et al. in Reference 5. This structure still has $O(|K|)$ nodes total, but uses only $O(1)$ pointers per node. If we assume that one key requires at least as much space as one pointer, then the resulting structure uses only a constant factor more space per machine than an optimal solution, since any structure must use $|K|$ space for key storage. The resulting structure retains most of the other properties of skip graphs, including a logarithmic number of messages for searches and updates. Although these costs are logarithmic in $|K|$ rather than $|N|$, this amounts to only a constant factor more as long as $|K|$ is at most some polynomial function of $|N|$.

The basic structure of a skip graph uses $\Theta(\log |K|)$ space per node. This stems from the fact that every key occurs in $\Theta(\log |K|)$ levels with high probability. We reduce the space complexity by lowering the number of lists in which each key appears to a constant while still preserving the basic structure of a skip graph. We first partition the ordered list of keys into blocks of size $\Theta(\log |K|)$. Each such block is referred to as a *supernode*. A single key, called the supernode key, is selected from each supernode. The graph is defined by constructing a skip graph on only the supernode keys. This resulting structure is referred to as a *supernode skip graph*.

The space usage is reduced in the following manner: at each level, a different machine from the supernode is mapped to the list in which the supernode key appears. We refer to the machine assigned to the supernode key at level i as the representative of the supernode at level i. Since the height of a

Distributed Peer-to-Peer Data Structures

skip graph is $O(\log |K|)$, each machine will need to represent its supernode at only a constant number of levels, where the exact value of the constant depends on the exact size constraint imposed on the supernodes.

Searching a rainbow skip graph is nearly identical to searching a standard skip graph. The search begins at the root list in which the supernode key appears. Since only one machine is mapped to this level, other machines in the supernode begin a search by traversing a linked list of the supernode members until the root is reached. The successor of the search key is then located using a standard *find_successor* operation. However, the successor located is only the successor amongst the supernode keys. Hence, to locate the actual successor, the destination supernode is searched in a linear fashion to find the successor from amongst the entire set of keys. It may also be necessary to search the predecessor supernode as the real successor may lie anywhere between these two supernode key values.

Most key insertions will not change the structure of the supernode skip graph. Normally, the inserted key will simply be added to the supernode containing its successor. The exception comes when the supernode size exceeds the size restriction imposed on the supernode. The supernodes are maintained using a method similar to nodes in a B-tree. Each supernode should contain between $\log |K|$ and $2\log |K| - 1$ keys. If a supernode becomes too full, it may first try to pass some of its keys to its predecessor or successor in the supernode skip graph. Otherwise, the supernode and its neighbor must have at least $4\log |K| - 1$ keys between them. The two supernodes should then split into three supernodes. This can be implemented as an insertion of a third supernode to either side of the current two. The resulting supernodes should have roughly $4/3 \log |K|$ keys each, so that another supernode split will not occur until at least $2/3 \log |K|$ new insertions are performed.

Deletions are handled in a similar manner. One additional step necessary is to place a new representative key at each level where the key being removed was formerly a representative.

17.3.6 Fault Tolerance

Skip graphs may be augmented to handle node failures much in the same way as Chord. We must first, however, revisit the model of node failures used in analyzing the fault tolerance of the structure. The model used in Chord was a simple independent-failure model in which each node failed independently with some constant probability, say $1/2$. In Chord, this is a somewhat reasonable assumption because the nodes, which are in one-to-one correspondence with the machines, are randomly ordered in the identifier space. For skip graphs, we previously noted that there are multiple methods of assigning machines to the nodes. If machines are assigned keys through a distributed hash table, then each machine is equally likely to be assigned to any given node. Even so, the node failures are not truly independent because the machines and keys are not in one-to-one correspondence—thus, if it is given that one node has failed, then the probability that another node has failed is slightly larger than $1/2$. However, the analysis for this scenario is similar to the independent-failure model and yields similar results. We will present the results for the independent-failure model to keep the discussion simple.

Like Chord, the skip graph structure can be repaired easily, and searches will execute correctly, as long as the level 0 list remains connected. Thus, we may apply the simple solution of having each node store pointers to the next r nodes in the level 0 list. If the nodes fail independently with some constant probability, then every node will have a pointer to its next living neighbor with high probability when $r = \Theta(\log |K|)$. This choice of r does not change the asymptotic space complexity of the skip graph, as each node already uses $\Theta(\log |K|)$ space. However, if the goal is to achieve $O(1)$ space per node, as in the supernode skip graph, then this solution is inadequate.

Supernode skip graphs may be augmented with a form of error correcting codes known as erasure-resilient codes to provide fault tolerance while using only $O(1)$ space per key. The resulting structure is one called the rainbow skip graph, which was presented by Goodrich et al. in Reference [5].

A (c, r)-*erasure-resilient code* consists of an encoding algorithm and a decoding algorithm. The encoding algorithm takes a *message* of n bits and converts it into a sequence of constant-sized packets whose total size is cn bits. This sequence is called the coded message. The decoding algorithm is able to recover the original message from any set of packets whose total length is rn. In other words, the entire message can be computed from any r/c fraction of the coded message. Alon and Luby [1] provide a deterministic (c, r)-erasure-resilient code with linear-time encoding and decoding algorithms where r is a constant slightly larger than one. Although these codes are not generally the most practical, they give the most desirable theoretical results. It is also worth noting that since our model considers only the message complexity, we need not choose to use a linear-time code; we do so merely for the added comfort of maintaining efficient usage of computational power.

Erasure-resilient codes can be used in a distributed linked list to protect both the keys and pointer data from being lost when just a fraction of the nodes have failed. We do so by partitioning the list into blocks of contiguous keys of size $\Theta(\log |K|)$. Each block will encode all of its keys as well as the network addresses of the nodes in the block and the nodes in the two neighboring blocks. The code will be distributed evenly in pieces amongst the nodes in the block. In this way, as long as the code from a large enough fraction of the nodes is recovered, the decoding algorithm can recover all of the keys, as well as pointers to each member of the block. In addition, the pointers to adjacent blocks can be used to find the first nonfaulty successor and predecessor of the keys at the ends of the block.

Choosing the block size to be $\beta \log |K|$, for a large enough constant β, will ensure that nearly half of the nodes in the block will be nonfaulty with high probability. Thus, if the pieces of the coded message from these nodes could be collected together, the decoding will be possible by using a (c, r)-erasure-resilient code where r/c is slightly less than $1/2$. However, we have no assurance that the live nodes will be connected to one another after failures occur. Nevertheless, Goodrich et al. [5] provide some assurance that a large enough fraction of the nodes will be connected by augmenting each block with *expander graphs*—a special class of constant-degree graphs that are well connected. We call the resulting blocks *hydra components*, and describe them in detail in the following section.

17.3.7 Hydra Components

We now describe hydra components [5]—blocks of contiguous nodes organized in such a way that if each member fails independently with constant probability p, then with high probability the nodes that remain can collectively compute the critical structural information of all the nodes in that component, including those that have failed. This "critical information" will consist of all keys in the block as well as pointers to each node in the hydra component and to each node in the two neighboring components. Although in principal these failure-resilient blocks can be designed to handle any constant failure probability less than 1, for simplicity we define them to handle a failure probability of $1/2$.

To do so, we make use of a 2d-regular graph structure consisting of the union of d Hamiltonian cycles. The set of all such graphs on n vertices is denoted by $H_{n,d}$. A (μ, d, δ)-hydra-component consists of a sequence of μ nodes logically connected according to a random element of $H_{\mu,d}$, with each node storing an equal share of the critical information encoded by a suitably-chosen erasure-resilient code. The parameters of the erasure-resilient code should be chosen in such a way that the entire message can be reconstructed from the collective shares of information stored by any set of $\delta\mu$ nodes, that is, such that $r/c = \delta$. The message that is encoded will be the critical information of the component. If the critical information consists of M bits, then by evenly distributing the packets of the encoded message across all the nodes, $O(M/\mu)$ space is used per node, given that δ is a constant. In addition to this space, $O(\mu)$ space will be needed to store structures associated with the erasure-resilient code. However, in our applications μ will be no larger than the space needed for $O(1)$ pointers, that is, no more than $O(\log |K|)$ bits.

To achieve a high-probability bound on recovery, we rely upon the fact that random elements of $H_{\mu,d}$ are likely to be good expanders. In particular, we make use an adaptation of a theorem by Beigel et al. [4]. We state the theorem in the following.

Theorem 17.5 *Let V be a set of μ vertices, and let $0 < \gamma, \lambda < 1$. Let G be a member of $H_{\mu,d}$ defined by the union of d independent randomly chosen Hamiltonian cycles on V. Then, for all subsets W of V with $\lambda\mu$ vertices, G induces at least one connected component on W of size greater than $\gamma\lambda\mu$ with probability at least*

$$1 - e^{\mu[(1+\lambda)\ln 2 + d(\alpha \ln \alpha + \beta \ln \beta - (1-\lambda)\ln(1-\lambda))] + O(1)},$$

where $\alpha = 1 - \frac{1-\gamma}{2}\lambda$ and $\beta = 1 - \frac{1+\gamma}{2}\lambda$.

With suitably chosen μ, λ, and γ, Theorem 17.5 will guarantee that with high probability at least $\gamma\lambda\mu$ nodes are connected, conditioned on the event that $\lambda\mu$ nodes of the component have not failed. By applying Chernoff bounds, it can be shown that this event occurs with high probability for suitably chosen μ and λ. This fact is captured by Lemma 17.1, which follows.

Lemma 17.1 *If each node of a component of $\beta \log n$ nodes fails independently with probability $1/2$, then the number of nonfailing nodes is no less than $\lambda\mu$ with probability at least*

$$1 - \left(\frac{e^{\delta}}{(1+\delta)^{(1+\delta)}}\right)^{\beta \log n/2}$$

where $\delta = 1 - 2\lambda$.

These facts will directly yield the following theorem from Goodrich et al. [5].

Theorem 17.6 *For any constant k, there exist constants d, β, and δ such that with probability $1 - O(\frac{1}{n^k})$ the critical information of a $(\beta \log n, d, \delta)$ hydra component can be recovered with $O(\log n)$ messages when each node of the component is failed with probability $1/2$.*

Proof. Setting β to 40 and λ to $3/10$ provides a lower bound of $1 - O(1/n^2)$ on the probability that at least $3/10$ of the nodes do not fail. Theorem 17.5 can now be applied, with $\mu = \beta \log n$, $\gamma = 1/2$, and $d = 47$, to guarantee that there is a connected component amongst the $3/10$-fraction of unfailed nodes of size $3\mu/20$ with probability $1 - O(\frac{1}{n^2})$. Thus, the probability that the conditions of Theorem 17.5 and Lemma 17.1 both hold is $1 - O(1/n^2)$ for the given values of the parameters. This can be extended to any other value of k, for example by changing β, which appears in the exponent of both probability terms, by a factor of $k/2$.

By employing an intelligent flooding mechanism, the packets held by the $3\beta \log n/20$ connected nodes can be collected together in $O(\log n)$ messages. The erasure-resilient code can then be used to reconstruct the critical information by choosing parameters of the code so that $\delta = r/c = 3/20$. ∎

The integration of hydra components into supernode skip graphs thus provides a distributed dictionary using $O(|K|)$ pointers that tolerates significant random machine failures.

17.4 Further Information

The algorithms and analysis presented here are merely representatives of a large and growing literature on peer-to-peer data structures. Further theoretical analysis of Chord can be found in Reference 9. Applications of skip graphs are provided by Harvey et al. in their paper on SkipNet [7].

There is a considerable amount of work on variants of DHTs, including Koorde [8], Pastry [17], Scribe [18], Symphony [12], Tapestry [22], CAN [17], and Viceroy [11].

A number of extensions and additional applications of skip graphs can be found in recent computer science literature. Applications of skip graphs are provided by Harvey et al. in their paper on SkipNet [7]. Harvey and Munro [6] present a deterministic version of SkipNet, showing how to achieve worst-case $O(\log n)$ query times, albeit with increased update costs, which are $O(\log^2 n)$, and higher congestion, which is $O(\log n/n^{0.68})$. Zatloukal and Harvey [21] present an alternative method to obtain $O(1)$ space per key by modifying skip graphs to construct a structure they call family trees, achieving $O(\log n)$ expected time for search and update. Manku, et al. [13] show how to improve the expected query cost for searching skip graphs and SkipNet to $O(\log n/\log \log n)$ by having hosts store the pointers from their neighbors to their neighbor's neighbors (i.e., neighbors-of-neighbors (NoN) tables); see also Naor and Wieder [15]. Unfortunately, this improvement requires that the memory size and expected update time grow to be $O(\log^2 n)$. Arge et al. [2] lay out a framework for supporting a larger class of queries, including queries on multidimensional keys. The reader is also encouraged to consult the recent computer science literature such as the proceedings cited above to obtain additional and freshly emerging information.

References

[1] N. Alon and M. Luby. A linear time erasure-resilient code with nearly optimal recovery. *IEEE Transactions on Information Theory*, 42, 1996.

[2] L. Arge, D. Eppstein, and M. T. Goodrich. Skip-webs: Efficient distributed data structures for multi-dimensional data sets. In *24th ACM Symp. on Principles of Distributed Computing (PODC)*, 2005.

[3] J. Aspnes and G. Shah. Skip graphs. In *14th ACM-SIAM Symp. on Discrete Algorithms (SODA)*, pp. 384–393, 2003.

[4] R. Beigel, G. Margulis, and D. A. Spielman. Fault diagnosis in a small constant number of parallel testing rounds. In *5th Annual ACM Symp. on Parallel Algorithms and Architectures (SPAA)*, pp. 21–29, 1993.

[5] M. T. Goodrich, M. J. Nelson, and J. Z. Sun. The rainbow skip graph: A fault-tolerant constant-degree distributed data structure. In *17th ACM-SIAM Symp. on Discrete Algorithms (SODA)*, pp. 384–393, 2006.

[6] N. Harvey and J. Munro. Deterministic SkipNet. In *Twenty Second ACM Symp. on Principles of Distributed Computing (PODC)*, pp. 152–153, 2003.

[7] N. J. A. Harvey, M. B. Jones, S. Saroiu, M. Theimer, and A. Wolman. SkipNet: A scalable overlay network with practical locality properties. In *USENIX Symp. on Internet Technologies and Systems*, Lecture Notes in Computer Science, 2003.

[8] F. Kaashoek and D. R. Karger. Koorde: A simple degree-optimal distributed hash table. In *2nd Int. Workshop on Peer-to-Peer Systems*, 2003.

[9] D. Liben-Nowell, H. Balakrishnan, and D. R. Karger. Analysis of the evolution of peer-to-peer systems. In *21st ACM Symp. on Principles of Distributed Computing (PODC)*, pp. 233–242, 2002.

[10] N. A. Lynch. *Distributed Algorithms*. Morgan Kaufmann, San Francisco, CA, 1996.

[11] D. Malkhi, M. Naor, and D. Ratajczak. Viceroy: A scalable and dynamic emulation of the butterfly. In *21st ACM Symp. on Principles of Distributed Computing (PODC)*, pp. 183–192, 2002.

[12] G. S. Manku, M. Bawa, and P. Raghavan. Symphony: Distributed hashing in a small world. In *4th USENIX Symp. on Internet Technologies and Systems*, 2003.

[13] G. S. Manku, M. Naor, and U. Wieder. Know thy neighbor's neighbor: The power of lookahead in randomized P2P networks. In *Proceedings of the 36th ACM Symp. on Theory of Computing (STOC)*, pp. 54–63, 2004.

[14] R. Motwani and P. Raghavan. *Randomized Algorithms*. Cambridge University Press, Cambridge, NY, 1995.

[15] M. Naor and U. Wieder. Know thy neighbor's neighbor: Better routing in skip-graphs and small worlds. In *3rd Int. Workshop on Peer-to-Peer Systems*, 2004.

[16] W. Pugh. Skip lists: A probabilistic alternative to balanced trees. *Commun. ACM*, 33(6):668–676, 1990.

[17] S. Ratnasamy, P. Francis, M. Handley, R. M. Karp, and S. Shenker. A scalable content-addressable network. In *SIGCOMM*, pp. 161–172, 2001.

[18] A. Rowstron and P. Druschel. Pastry: Scalable, decentralized object location, and routing for large-scale peer-to-peer systems. *Lecture Notes in Computer Science*, 2218:329–350, 2001.

[19] A. I. T. Rowstron, A.-M. Kermarrec, M. Castro, and P. Druschel. SCRIBE: The design of a large-scale event notification infrastructure. In *Networked Group Communication*, pp. 30–43, 2001.

[20] I. Stoica, R. Morris, D. Karger, F. Kaashoek, and H. Balakrishnan. Chord: A scalable peer-to-peer lookup service for Internet applications. In *Proceedings of the 2001 ACM SIGCOMM Conference*, pp. 149–160, 2001.

[21] K. C. Zatloukal and N. J. A. Harvey. Family trees: An ordered dictionary with optimal congestion, locality, degree, and search time. In *15th ACM-SIAM Symp. on Discrete Algorithms (SODA)*, pp. 301–310, 2004.

[22] B. Y. Zhao, J. D. Kubiatowicz, and A. D. Joseph. Tapestry: An infrastructure for fault-tolerant wide-area location and routing. Technical Report UCB/CSD-01-1141, UC Berkeley, April 2001.

18
Parallel Algorithms via the Probabilistic Method

18.1	Introduction and Preliminaries	18-1
	Introduction • Preliminaries	
18.2	Limited Independence	18-6
	The Method of Conditional Probabilities • k-Wise Independence and Small Sample Spaces • Almost k-Wise Independence	
18.3	Parallel Algorithms Using k-Wise Independence	18-14
	Functions over $\mathbb{Z}_2^k$ • Multivalued Functions • Maximal Independent Sets in Graphs • Low Discrepancy Colorings • Lattice Approximation	
18.4	Fooling Automata	18-34
18.5	Parallel Algorithms for Packing Integer Programs	18-44
	Impact of k-Wise Independence: Multidimensional Bin Packing • Impact of Automata Fooling: Srinivasan's Parallel Packing	
18.6	Open Problems	18-53
	Acknowledgments	18-54
	References	18-54

Lasse Kliemann
Anand Srivastav
Christian Albrechts University

18.1 Introduction and Preliminaries

18.1.1 Introduction

We give an introduction to the design of parallel algorithms with the probabilistic method. Algorithms of this kind usually possess a randomized sequential counterpart. Parallelization of such algorithms is inherently linked with derandomization, either with the Erdős–Spencer method of conditional probabilities or exhaustive search in a polynomial-sized sample space.

The key notation is the treatment of random variables with various concepts of only limited independence, leading to polynomial-sized sample spaces with time and space efficient implementations of the conditional probability method.

Starting with the definition of limited independence in Section 18.2, we discuss in Section 18.3 algorithms using limited independence in a more or less direct way: the maximal independent set (MIS) algorithms of Luby (1986) [Lub86], which marks the beginning of the subject, and the algorithms formulated by Berger and Rompel (1989) [BR91] and Motwani et al. (1989, 1994) [MNN94] for parallel computation of low discrepancy colorings of hypergraphs and its extension to the lattice approximation problem. Although the lattice approximation problem can be viewed as a packing problem, the direct application of limited independence could not achieve parallel counterparts of well-known approximation algorithms for integer programs of packing type. An important step broadening the range of applicability has been the introduction of efficient parallelization by so-called fooling of automata by Karger and Koller [KK97],* which is presented in Section 18.4. The idea is to simulate a randomized algorithm by deterministic finite automata on a probability distribution for which the number of strings with nonzero probability is polynomial in the input size, while the original probability distributions for the randomized algorithm typically lead to an exponential number of such strings. A similar approach giving a little stronger results for the lattice approximation problem has been presented by Mahajan, Ramos, and Subrahmanyam [MRS97]. Finally, in Section 18.5, we first show the impact of limited independence to a packing type problem, the multidimensional bin packing put forth by Srivastav and Stangier [SS97a, SS97b] and then proceed to Srinivasan's [Sri01a] parallel algorithm for general packing integer problems, which is based on the automata fooling and a nice compression of the (exponential) state space to polynomial size. A comprehensive treatment of all applications of the probabilistic method in parallelizing algorithms is beyond the scope of this chapter. Though, we have tried to include many topics in the bibliography and remarks at the end of the specific sections. The intention of this chapter is to show the development of the method of k-wise independence in a straightforward way, starting with applications where constant independence is sufficient, passing applications of log n-wise independence like discrepancy problems and reaching the applications to integer programming. We give the proofs in the first three sections in full detail so that also nonexperts can find a self-contained introduction to the subject with detailed proofs of processor and time bounds without tracing a list of papers with sometimes varying notations. The examinations of so far presented proofs lead to the revision of some processor and time bounds. Sections 18.4 and 18.5 contain some state-of-the-art results on parallel algorithms through the probabilistic method. Our emphasis there is on methodical aspects rather than completeness. There are numerous open problems in the area. In Section 18.6, we give a selection of them that we find not only appealing but also important beyond the scope of parallelization.

18.1.2 Preliminaries

18.1.2.1 Vectors and Matrices

For every $n \in \mathbb{N}$ denote $[n] := \{1, \ldots, n\}$. The logarithm log is always taken to basis 2. Given a $(m \times n)$ matrix A over some field $\mathbb{F}$, denote by range A its range $\{Ax; \ x \in \mathbb{F}^n\} \subseteq \mathbb{F}^m$ and set rank $A :=$ dim range A. The kernel, that is, elements in $\mathbb{F}^n$ that are mapped to zero, is denoted by ker A. For a vector $x \in \mathbb{F}^n$ and an index set $I \subseteq [n]$ denote by x_I the vector of entries from x with indices in I. We choose the same notation, x_α for vectors of indices, that is, $\alpha \in [n]^k$ for some $k \in \mathbb{N}$. For a matrix A, the submatrix with rows from A according to I is A_I (according to a vector α, it is A_α). For two vectors $x, y \in \mathbb{F}^n$ denote the scalar (or inner) product $x \cdot y := \sum_{i \in [n]} x_i y_i$.

The ith component of a vector $v \in \mathbb{F}^n$ as usually is denoted by v_i, and if a family of vectors $v_1, \ldots, v_k$ is given, v_{ji} denotes the ith coordinate of vector v_j, $i \in [n], j \in [k]$.

18.1.2.2 Binary Representations

We use two different ways for the binary representation of natural and rational numbers. Given a set $\Omega := \{0, \ldots, d-1\}$ for some $d \in \mathbb{N}$ that is a power of 2, and a vector $x \in \Omega^n$ (or a sequence $x_1, \ldots, x_n \in \Omega$) we write $\mathbf{x} \in \{0, 1\}^{n \log d}$ for the binary representation of x. More precisely,

---

* We note that a close inspection and analysis of the lattice approximation algorithms of Reference MNN94 gives error bounds leading to weaker bounds for the support size in the basis crashing algorithm of Reference KK97.

x is chosen such that

$$x_i = \sum_{j=(i-1)\log(d)+1}^{i\log d} \mathbf{x}_j 2^{j-1-(i-1)\log d} \quad \text{for all } i \in [n]. \tag{18.1}$$

Given a number $x \in [0,1]$ with $L+1$ bits of precision, $\text{bit}_l(x)$ denotes the lth bit, that is, we have

$$x = \sum_{l=0}^{L} 2^{-l} \text{bit}_l(x).$$

Given a vector $p \in [0,1]^n$, where each of the n entries is given with $L+1$ bits of precision, we sometimes need the lth bit of each of the entries. We write $\vec{\text{bit}}_l(p) = (\text{bit}_l(p_1), \ldots, \text{bit}_l(p_n))$ for this.

18.1.2.3 Logarithmic Terms

Given parameters $a_1, \ldots, a_k$ and a function f depending on them, we write $f(a_1, \ldots, a_k) = O(\log^{(*)}(a_1, \ldots, a_k))$ if $f(a_1, \ldots, a_k) = O(\log(g(a_1, \ldots, a_k)))$, and g depends polynomially on $a_1, \ldots, a_k$. For example, $\log(a_1 \log^2(a_2 \log(a_3^5) + a_4)) = O(\log^{(*)}(a_1, \ldots, a_4))$. We drop the parameters $(a_1, \ldots, a_k)$ in some cases, if they are polynomially bounded in the input length of the given problem. This will be used to simplify the statement of some results.

18.1.2.4 Graphs and Hypergraphs

We use the standard notation of finite graphs and hypergraphs [Ber73]. A *graph* $G = (V, E)$ is a pair of a finite set V (the set of *vertices* or *nodes*) and a subset $E \subseteq \binom{V}{2}$, where $\binom{V}{2}$ denotes the set of all 2-element subsets of V. The elements of E are called *edges*. Denote the *neighbors* of a vertex $v \in V$ by $N(v) := \{w \in V;\ \exists \{v,w\} \in E\}$. For a subset $S \subseteq V$ of vertices denote their neighbors by $N(S) := \{w \in V \setminus S;\ \exists v \in S: \{v,w\} \in E\}$.

A *hypergraph* or set system $\mathcal{H} = (V, \mathcal{E})$ is a pair of a finite set V and a subset $\mathcal{E} \subseteq \mathcal{P}(V)$ of the powerset $\mathcal{P}(V)$. The elements of $\mathcal{E}$ are called *hyperedges*. The *degree* of a vertex $v \in V$ in $\mathcal{H}$, denoted by $\deg(v)$ or $d(v)$, is the number of hyperedges containing v, and $\deg(\mathcal{H}) = \max_{v \in V} d(v)$ is the *(vertex-)degree* of $\mathcal{H}$. For a pair of vertices $u, v \in V$, $\text{codeg}(u,v)$ is the *co-degree* of u and v, and is the number of edges containing both u and v, and $\text{codeg}(\mathcal{H})$ is the maximum over all $\text{codeg}(u,v)$. $\mathcal{H}$ is called r-*regular* if $\deg(v) = r$ for all $v \in V$. It is called s-*uniform* if $|E| = s$ for all $E \in \mathcal{E}$. It is convenient to order the vertices and hyperedges, $V = \{v_1, \ldots, v_n\}$ and $\mathcal{E} = \{E_1, \ldots, E_m\}$, and to identify vertices and edges with their indices. The *vertex-hyperedge incidence matrix* of a hypergraph $\mathcal{H} = (V, \mathcal{E})$, with $V = \{v_1, \ldots, v_n\}$ and $\mathcal{E} = \{E_1, \ldots, E_m\}$, is a matrix $A = (a_{ij}) \in \{0,1\}^{n \times m}$, where $a_{ij} = 1$ if $v_i \in E_j$, and 0 else.

For a modern treatment of graph theory, we refer to the books of Bollobás [Bol98], Diestel [Die97], and West [Wes96].

18.1.2.5 Probabilistic Tools

Throughout this chapter, we consider only finite probability spaces $(\Omega, \mathbb{P})$, where Ω is a finite set, usually a product of finite sets, and $\mathbb{P}$ is a probability measure with respect to the powerset $\mathcal{P}(\Omega)$ as the sigma field. Let $u_1, \ldots, u_n$ and $v_1, \ldots, v_n$ be integers, and let $X_1, \ldots, X_n$ be mutually independent (briefly independent) random variables, where X_j takes the values u_j and v_j, $j \in [n]$ and

$$\mathbb{P}[X_j = u_j] = p_j, \quad \mathbb{P}[X_j = v_j] = 1 - p_j$$

for $p_j \in [0,1]$ for all $j \in [n]$. For all $j \in [n]$ let w_j denote rational weights with

$$0 \le w_j \le 1 \quad \text{and} \quad \text{let } \psi := \sum_{j=1}^{n} w_j X_j$$

be the weighted sum. For $u_j = 1$, $v_j = 0$, $w_j = 1$, and $p_j = p$ for all $j \in [n]$, the sum $\psi = \sum_{j=1}^{n} X_j$ is the well-known binomially distributed random variable with mean np. The first large deviation inequality is implicitly given in Chernoff [Che52] in the binomial case. In explicit form it can be found in Okamoto [Oka58]. Its generalization to arbitrary weights is due to Hoeffding [Hoe63].

Theorem 18.1 (Hoeffding 1963) *Let $u_j = 1$, $v_j = 0$ for all $j \in [n]$ and let $\lambda > 0$. Then*

(i) $\mathbb{P}[\psi \geq \mathbb{E}[\psi] + \lambda] \leq \exp\left(-\frac{2\lambda^2}{n}\right)$

(ii) $\mathbb{P}[\psi \leq \mathbb{E}[\psi] - \lambda] \leq \exp\left(-\frac{2\lambda^2}{n}\right)$.

In the literature, Theorem 18.1 is well known as the Chernoff bound. A proof of a stronger version of it can be found in the Book of Habib et al. [HMRAR98]. The following theorem is also known as Chernoff bound. Its proof for the unweighted case can be found at several places (e.g., in the book of Motwani and Raghavan [MR95]). For convenience, we give a proof for the general case.

Theorem 18.2 *Let $u_j = 1$, $v_j = 0$. Then*

(i) $\mathbb{P}[\psi \geq (1+\beta)\mathbb{E}[\psi]] \leq \left(\frac{\exp(\beta)}{(1+\beta)^{(1+\beta)}}\right)^{\mathbb{E}[\psi]}$ *for all $0 \leq \beta$, and*

(ii) $\mathbb{P}[\psi \leq (1-\beta)\mathbb{E}[\psi]] \leq \left(\frac{\exp(-\beta)}{(1-\beta)^{(1-\beta)}}\right)^{\mathbb{E}[\psi]}$ *for all $0 \leq \beta \leq 1$.*

Proof. Fix some $t > 0$, which will be specified precisely later. By the Markov inequality and independence we have

$$\mathbb{P}[\psi \geq (1+\beta)\mathbb{E}[\psi]] = \mathbb{P}[\exp(t\psi) \geq \exp(t(1+\beta)\mathbb{E}[\psi])]$$

$$\leq \frac{\mathbb{E}[\exp(t\psi)]}{\exp(t(1+\beta)\mathbb{E}[\psi])}$$

$$= \frac{\prod_{j=1}^{n} \mathbb{E}[\exp(tw_j X_j)]}{\exp(t(1+\beta)\mathbb{E}[\psi])}. \tag{18.2}$$

For all $j \in [n]$, we can bound the expectation $\mathbb{E}[\exp(tw_j X_j)] = p_j \exp(tw_j) + (1-p_j) \leq \exp(p_j \exp(tw_j) - p_j)$, because $1 + x \leq \exp(x)$ holds for all $x \geq 0$. Plugging in $t = \ln(1+\beta)$ yields $\mathbb{E}[\exp(tw_j X_j)] \leq \exp(p_j \exp(\ln(1+\beta)w_j) - p_j) = \exp(p_j(1+\beta)^{w_j} - p_j)$. Because for all $x \geq 0$ and $y \in [0,1]$ we have $(1+x)^y \leq 1 + xy$, we find $\exp(p_j(1+\beta)^{w_j} - p_j) \leq \exp(p_j(1+\beta w_j) - p_j) = \exp(\beta p_j w_j)$. So we have

$$\prod_{j=1}^{n} \mathbb{E}[\exp(tw_j X_j)] \leq \prod_{j=1}^{n} \exp(\beta p_j w_j) = \exp\left(\sum_{j=1}^{n} \beta p_j w_j\right) = \exp(\beta \mathbb{E}[\psi]).$$

Using this in Equation 18.2 and further simplifying with $t = \ln(1+\beta)$ yields the assertion (i) of the theorem. The assertion (ii) can be proven along the same lines, using that $(1-x)^y \leq 1 - xy$ for all $x, y \in [0, 1]$. ∎

The right-hand sides of these bounds are known as functions G and H, respectively, in the literature:

$$G(\mu, \beta) := \left(\frac{\exp(\beta)}{(1+\beta)^{(1+\beta)}}\right)^{\mu}, \quad 0 \leq \mu, \beta,$$

$$H(\mu, \beta) := \left(\frac{\exp(-\beta)}{(1-\beta)^{(1-\beta)}}\right)^{\mu}, \quad 0 \leq \mu, \; 0 \leq \beta \leq 1.$$

By real analysis, one can show that for $0 \leq \beta \leq 1$ we have $G(\mu, \beta) \leq \exp\left((-\beta^2)\mu/3\right)$ and $H(\mu, \beta) \leq \exp\left(-(\beta^2 \mu)/2\right)$. This gives the following bound, originally formulated by Angluin and Valiant [AV79].

Theorem 18.3 (Angluin, Valiant 1979)
Let $u_j = 1$, $v_j = 0$ for all $j \in [n]$ and let $0 \leq \beta \leq 1$. Then

(i) $\mathbb{P}\left[\psi \geq (1+\beta)\mathbb{E}[\psi]\right] \leq \exp\left(-\frac{\beta^2 \mathbb{E}[\psi]}{3}\right)$,

(ii) $\mathbb{P}\left[\psi \leq (1-\beta)\mathbb{E}[\psi]\right] \leq \exp\left(-\frac{\beta^2 \mathbb{E}[\psi]}{2}\right)$.

For random variables with zero expectation, there are several inequalities that can be found in the book of Alon et al. [ASE92] and Alon and Spencer [AS00].

Theorem 18.4 (Hoeffding 1963)
Let $u_j = 1$, $v_j = -1$, $w_j = 1$ for all $j \in [n]$. For $\lambda > 0$ we have

(i) $\mathbb{P}[\psi \geq \lambda] \leq \exp\left(-\frac{\lambda^2}{2n}\right)$,

(ii) $\mathbb{P}[\psi \leq -\lambda] \leq \exp\left(-\frac{\lambda^2}{2n}\right)$.

The following, for the unweighted case (i.e., $w_j = 1$ for all $j \in [n]$), is [AS00, Th. A.1.6, A.1.7]. Their proof can be extended in a straightforward way to the general case.

Theorem 18.5 Let $u_j = 1 - p_j$, $v_j = -p_j$ for all $j \in [n]$ and let $\lambda > 0$. Then

(i) $\mathbb{P}[\psi \geq \lambda] \leq \exp\left(-\frac{2\lambda^2}{\sum_j w_j^2}\right)$,

(ii) $\mathbb{P}[\psi \leq -\lambda] \leq \exp\left(-\frac{2\lambda^2}{\sum_j w_j^2}\right)$.

Hence, $\mathbb{P}[|\psi| \geq \lambda] \leq 2\exp\left(-\frac{2\lambda^2}{\sum_j w_j^2}\right)$.

Proof. Assertion (ii) follows from (i) through symmetry (apply (i) to $-\psi$) and the assertion about $|\psi|$ clearly is a direct consequence of (i) and (ii).

We show (i). The idea is the same as in the proof of Theorem 18.2. Fix $t > 0$ to be specified later. By the Markov inequality and by independence we have

$$\mathbb{P}[\psi \geq \lambda] = \mathbb{P}[\exp(t\psi) \geq \exp(t\lambda)] \leq \frac{\mathbb{E}[\exp(t\psi)]}{\exp(t\lambda)} = \frac{\prod_{j=1}^n \mathbb{E}[\exp(tw_j X_j)]}{\exp(t\lambda)}. \quad (18.3)$$

For all $j \in [n]$ we have $\mathbb{E}[\exp(tw_j X_j)] = p_j \exp(tw_j(1-p_j)) + (1-p_j)\exp(-tw_j p_j) \leq \exp(t^2 w_j^2/8)$, because $x\exp(t(1-x)) + (1-x)\exp(-tx) \leq \exp(t^2/8)$ holds for all $x \in [0,1]$, see [AS00, Lem. A.1.6]. So we have

$$\prod_{j=1}^n \mathbb{E}[\exp(tw_j X_j)] \leq \prod_{j=1}^n \exp\left(\frac{t^2 w_j^2}{8}\right) = \exp\left(\frac{t^2}{8}\sum_{j=1}^n w_j^2\right).$$

With Equation 18.3, we see

$$\mathbb{P}[\psi \geq \lambda] \leq \exp\left(\frac{t^2}{8}\sum_{j=1}^n w_j^2 - t\lambda\right).$$

Choosing $t := 4\lambda/\sum_{j=1}^n w_j^2$ yields the bound in (i). Note that we may assume that not all weights are zero, since otherwise the statement of the theorem is trivial. ∎

Alon and Spencer improved the Hoeffding bound to $\exp\left(-\frac{2\lambda^2}{n}\right)$ replacing n by $pn = p_1 + \cdots + p_n$.

Theorem 18.6 (Alon, Spencer 1992)
Let $u_j = 1 - p_j$, $v_j = -p_j$, $w_j = 1$ for all $j \in [n]$ and let $\lambda > 0$. Set $p = \frac{1}{n}(p_1 + \ldots + p_n)$. Then

(i) $\mathbb{P}[\psi \geq \lambda] \leq \exp\left(-\frac{\lambda^2}{2pn} + \frac{\lambda^3}{2(pn)^2}\right)$,

(ii) $\mathbb{P}[\psi \leq -\lambda] \leq \exp(-\frac{\lambda^2}{2pn})$.

18.1.2.6 Computational Model

In this chapter, we consider the synchronous parallel random access machine model, the PRAM model. The various specifications for the PRAM are EREW, ERCW, CREW, and CRCW, indicating exclusive read/exclusive write, exclusive read/common write, common read/exclusive write, and common read/common write into the global memory, respectively. We consider in this chapter only the EREW PRAM model, thus do not specify in the statements of processor bounds and running times the parallel model explicitly. We sometimes write $\mathcal{NC}$ *algorithm* instead of *parallel algorithm*, refering to the class $\mathcal{NC}$ (for a definition see [Pap94, Chapter 15]). For model details, we refer the reader to standard text books (e.g., Papadimitriou [Pap94]). A quick introduction is given in the book of Motwani and Raghavan [MR95] in the chapter on parallel and distributed computing.

18.2 Limited Independence

18.2.1 The Method of Conditional Probabilities

The method of conditional probabilities has been invented by Erdős and Selfridge [ES73]. Further developments were given by Beck and Fiala [BF81], Beck and Spencer [BS83], Spencer [Spe87], and introduced as a derandomization technique in integer programming by Raghavan [Rag88]. With Raghavan's paper, the technique became a common and popular tool in the design of derandomized algorithms.

We consider the probability space $(\Omega, \mathbb{P})$ where $\Omega = \prod_{i=1}^n \Omega_i$, $\Omega_i = [N]$, the powerset $\mathcal{P}(\Omega)$ is the σ-field and $\mathbb{P}$ is a product (i.e., $\mathbb{P} = \bigotimes_{i=1}^n \mathbb{P}_i$), where $\mathbb{P}_i$ is a probability measure on Ω_i. Let $E \subseteq \Omega$ be an event with $\mathbb{P}[E] > 0$ and let E^c denote the complement of E. For $\omega_1, \ldots, \omega_l \in [N]$, $l \in [n]$ denote

$$\mathbb{P}[E^c \mid \omega_1, \ldots, \omega_l] := \frac{\mathbb{P}[E^c \cap \{y \in \Omega;\ y_1 = \omega_1, \ldots, y_l = \omega_l\}]}{\mathbb{P}_1[\{\omega_1\}] \cdot \ldots \cdot \mathbb{P}_l[\{\omega_l\}]}.$$

The following simple procedure constructs a vector in E:

Algorithm 1: CONDPROB.

Input: An event $E \subseteq \Omega$ with $\mathbb{P}[E] > 0$.
Output: A vector $x \in E$.

choose x_1 as a miminizer of the function $[N] \to [0, 1]$, $\omega \mapsto \mathbb{P}[E^c \mid \omega]$;
for $l \leftarrow 2$ **to** n **do**
    choose x_l as a miminizer of the function $[N] \to [0, 1]$, $\omega \mapsto \mathbb{P}[E^c \mid x_1, \ldots, x_{l-1}, \omega]$;
end

A similar algorithm works for conditional expectations.

Algorithm 2: CONDPROB.

Input: A function $F : \Omega \mapsto \mathbb{Q}$.
Output: A vector $x \in \Omega$ with $F(x) \geq \mathbb{E}[F]$.

choose x_1 as a miminizer of the function $[N] \to [0,1]$, $\omega \mapsto \mathbb{E}[F \mid \omega]$;
for $l \leftarrow 2$ **to** n **do**
 choose x_l as a miminizer of the function $[N] \to [0,1]$, $\omega \mapsto \mathbb{E}[F \mid x_1, \ldots, x_{l-1}, \omega]$;
end

Proposition 18.1 *The algorithms* CONDPROB *and* CONDEXP *are correct.*

Proof. The argument is put forth by Erdős and Selfridge [ES73]. Let $x_1, \ldots, x_{l-1} \in [N]$, $l \in [r]$. Conditional probabilities can be written as a convex combination:

$$\mathbb{P}[E^c \mid x_1, \ldots, x_{l-1}] = \sum_{\omega \in [N]} \mathbb{P}_l[\{\omega\}] \cdot \mathbb{P}[E^c \mid x_1, \ldots, x_{l-1}, \omega].$$

By the choice of the $x_1, \cdots, x_n$ and the assumption $\mathbb{P}[E] > 0$,

$$1 > \mathbb{P}[E^c] \geq \mathbb{P}[E^c \mid x_1] \geq \cdots \geq \mathbb{P}[E^c \mid x_1, \ldots, x_n] \in \{0, 1\},$$

so $\mathbb{P}[E^c \mid x_1, \ldots, x_n] = 0$ and $x \in E$. The proof for the algorithm CONDEXP is similar. ∎

Efficiency of these algorithms depends on the efficient computation of the conditional probabilities and expectations, respectively. In general, it seems to be hopeless to compute conditional probabilities directly. But for the purpose of derandomization it suffices to compute upper bounds for the conditional probabilities, which then play the role of the conditional probabilities. Such upper bounds have been introduced by Spencer [Spe87] in the hyperbolic cosine algorithm, and later defined in a rigorous way as so-called pessimistic estimators by Raghavan [Rag88]. Comprehensive treatment of sequential derandomization in combinatorial optimization can be found in *Handbook of Randomized Computing* [Sri01b].

18.2.2 k-Wise Independence and Small Sample Spaces

The conditional probability method can be viewed as a binary search for a "good" vector in the sample space Ω. If the size of Ω is polynomial in N and n, an exhaustive search is already a polynomial-time algorithm. Small sample spaces correspond to random variables with limited rather than full independence. Limited independence is applicable not only for derandomization, but also for parallelization.

Let $(\Omega, \Sigma, \mathbb{P})$ be a probability space. In this chapter, we consider only random variables $X_1, \ldots X_n$ with finite ranges $D_1, \ldots, D_n$.

Definition 18.1 (k-Wise Independence) $X_1, \ldots, X_n$ *are k-wise independent if for any $J \subseteq [n]$ with $|J| \leq k$ and any choice of $\alpha_j \in D_j, j \in J$, we have*

$$\mathbb{P}[X_j = \alpha_j \quad \text{for all } j \in J] = \prod_{j \in J} \mathbb{P}[X_j = \alpha_j]. \tag{18.4}$$

The definition usually given in papers from the theoretical computer science community is the following special case of Definition 18.1.

Remark 18.1 For uniformly distributed 0/1 random variables, k-wise independence reads as follows: $X_1, \ldots, X_n$ are k-wise independent if for any $r \in [k]$, $\alpha \in \{0,1\}^r$ and any choice of r variables

$X_{i_1}, \ldots, X_{i_r}$, $1 \le i_1 < i_2 < \cdots < i_r \le n$,

$$\mathbb{P}\,[(X_{i_1}, \ldots, X_{i_k}) = \alpha] = 2^{-r}.$$

A motivation for k-wise independence from the computational point of view is the following observation. If $X_1, \ldots, X_n$ are independent 0/1 random variables, then the generation of a vector $(X_1, \ldots, X_n) = (\omega_1, \ldots, \omega_n)$ requires n bits (n independent Bernoulli trials). Now suppose that there is a $(n \times l)$ matrix B such that

$$BY = X,$$

where $Y = (Y_1, \ldots, Y_l)$, $X = (X_1, \ldots, X_n)$, $Y_1, \ldots, Y_l$ are independent, and $X_1, \ldots, X_n$ are k-wise independent 0/1 random variables. If l is smaller than n, we need only $l < n$ random bits in order to generate $X_1, \ldots, X_n$. In other words, the sample space is $\{0, 1\}^l$ instead of $\{0, 1\}^n$ corresponding to the $X_1, \ldots, X_n$.

Alon et al. [ABI86] showed that k-wise independent 0/1 random variables can be constructed from mutually independent random variables using a matrix over $GF(2)$.

For simplicity, we make two assumptions.

1. We assume that $n = 2^{n'} - 1$ for some $n' \in \mathbb{N}$, $n' \ge 2$. If n is not of this form, we take the next larger number having this property. This enlarges the quantity n by at most a factor of 2. This will change the size of the sample space, see Theorem 18.7. It will not, however, change the bounds on the quantity l, which later will play the most important role. In all bounds on l (see Equations 18.7 and 18.11), there is only logarithmic dependence on n, and so bounds only change by a constant factor (hidden in the () notation).
2. We assume k to be odd, that is, $k = 2k' + 1$ for some $k' \in \mathbb{N}$. If k is even, take $k + 1$ instead of k. This does not change the bound on the size of the sample space in Theorem 18.7.

Let $b_1, \ldots, b_n$ be the n nonzero elements of $GF(2^{n'})$ and let B be the following $(n \times (1 + \frac{k-1}{2}))$ matrix over $GF(2^{n'})$:

$$B := \begin{bmatrix} 1 & b_1 & b_1^3 & \cdots & b_1^{k-2} \\ 1 & b_2 & b_2^3 & & b_2^{k-2} \\ 1 & b_3 & b_3^3 & & b_3^{k-2} \\ \vdots & \vdots & \vdots & \vdots & \vdots \\ 1 & b_n & b_n^3 & & b_n^{k-2} \end{bmatrix}. \quad (18.5)$$

In coding theory, B is well known as the parity check matrix of binary BCH codes [MS77].

Proposition 18.2 *Any k rows of B are linearly independent over $GF(2)$.*

Proof. Let $I \subseteq n$ be the set of indices of any k rows of B. The idea is to extend the submatrix defined by I to a Vandermonde matrix. This is done by inserting columns with even powers 2 to $k - 1$. Let $\lambda_i \in GF(2)$, $i \in I$ be such that

$$\sum_{i \in I} \lambda_i b_i^j = 0 \quad \text{for all } j \in \{1, 3, \ldots, k - 2\}. \quad (18.6)$$

Let $r \in \{2, 4, \ldots, k-1\}$. Then, we can write $r = 2^{r'} j$ with an odd $j \in \{1, 3, \ldots, k-2\}$. We have

$$0 = \Big(\sum_{i \in I} \lambda_i b_i^j\Big)^{2^{r'}} \quad \text{due to Equation 18.6}$$

$$= \sum_{i \in I} \lambda_i^{2^{r'}} b_i^{j \cdot 2^{r'}} \quad GF(2^{n'}) \text{ has characteristic 2}$$

$$= \sum_{i \in I} \lambda_i b_i^r \quad \text{since } \lambda_i \in \{0, 1\}.$$

Hence,

$$\sum_{i \in I} \lambda_i b_i^j = 0, \quad \text{for all } j \in [k-1].$$

As the Vandermonde matrix is nonsingular, it follows that $\lambda_i = 0$ for all $i \in I$. ∎

The additive group of $GF(2^{n'})$ is isomorphic to the vector space $\bigotimes_{i=1}^{n'} GF(2)$, which has dimension n' over $GF(2)$. Hence the elements of $GF(2^{n'})$ (and so the entries in B) can be represented by 0/1 vectors of length n', and their component-wise addition is the addition in $GF(2^{n'})$. Put

$$l := 1 + \frac{k-1}{2} n' = 1 + \frac{k-1}{2} \log(n+1) = O(k \log n). \quad (18.7)$$

In the following, we will regard B as a $(n \times l)$ matrix, where all columns except the first one are expanded to their 0/1 representation. It is clear that Proposition 18.2 still holds for this expanded version of B.

Let $Y_1, \ldots, Y_l$ be independent and uniformly distributed. Let Y be the vector $Y = (Y_1, \ldots, Y_l)$. Define $\Omega := \{0, 1\}$-valued random variables $X_1, \ldots, X_n$ by

$$X_i := (BY)_i, \quad \text{for all } i \in [n], \quad (18.3)$$

where the computation is done over $GF(2)$. The following theorem was proved by Alon et al. [ABI86].

Theorem 18.7 *The random variables $X_1, \ldots, X_n$ as defined in Equation 18.8 are k-wise independent and uniformly distributed. The size of the sample space is*

$$O((n+1)^{\lfloor k/2 \rfloor}).$$

Proof. Choose $I \subseteq [n]$ with $k_0 := |I| \le k$. Let $x \in \Omega^{k_0}$ be an arbitrarily chosen but fixed vector ($\Omega = \{0, 1\}$). Set $X_I = (X_i)_{i \in I}$. For k-wise independence we must show $\mathbb{P}[X_I = x] = 2^{-k_0}$. Let B_I be the submatrix of B with row indices from I. B_I is a $(k_0 \times l)$ matrix over $GF(2)$. Because of Proposition 18.2 we can extend B_I to an invertible $(l \times l)$ matrix C over $GF(2)$. Define

$$\Omega_x = \{x' \in \Omega^l; \; x'_i = x_i \quad \text{for all } i \in [k]\}. \quad (18.9)$$

Then $|\Omega_x| = 2^{l-k}$, and we have (all calculations done in $GF(2)$)

$$\begin{aligned}
\mathbb{P}[X_I = x] &= \mathbb{P}[B_I Y = x] \\
&= \mathbb{P}[CY \bmod d \in \Omega_x] \\
&= \sum_{x' \in \Omega_x} \mathbb{P}[CY - x' = 0] \\
&= \sum_{x' \in \Omega_x} \mathbb{P}[Y = C^{-1} x'] \\
&= \sum_{x' \in \Omega_x} 2^{-l} \quad (Y_1, \ldots, Y_l \text{ are independent}) \\
&= \frac{|\Omega_x|}{2^l} = \frac{2^{l-k_0}}{2^l} = 2^{-k_0}.
\end{aligned}$$

The random variables X_i constructed in this theorem can be viewed as mappings $X_i : \{0,1\}^l \to \{0,1\}$, $\omega \mapsto (B\omega)_i$ and the sample space corresponding to the $X_1, \ldots, X_n$ is $\{0,1\}^l$. According to Equation 18.7, its size is

$$2^l = O\left((n+1)^{\lfloor k/2 \rfloor}\right).$$ ∎

In view of the lower bound $\Omega(n^k)$ for sample spaces of certain k-wise independent random variables given by Chor et al. [CFG+85], this is best possible up to constants in the exponent. Note that the sample space constructed by Alon et al. is polynomial only if k is constant. In applications, k is often not constant, but $k = O(\log^c n)$ with some constant $c > 0$. Derandomization in such sample spaces can be done combining $O(\log^c n)$-wise random variables with the conditional probability method. A significant reduction of the size of the sample space was achieved by Naor and Naor [NN93]. The heart of their construction is the notation of almost k-wise independent random variables discussed in the next section.

The construction described above can be extended to multivalued random variables. Fix $d \in \mathbb{N}$ that is a power of 2, and set $\Omega := \{0, \ldots, d-1\}$. The aim is to construct k-wise independent uniformly distributed random variables $X_1, \ldots, X_n$ with values in Ω and a "small" sample space. To this end, use the above construction to find $N := n \log d$ uniformly distributed binary random variables $\mathbf{X}_1, \ldots, \mathbf{X}_N$ that are $(k \log d)$-wise independent.[*] These are interpreted as binary representations of the to construct random variables. For every $i \in [n]$ we then put

$$X_i := \sum_{j=(i-1)\log(d)+1}^{i \log d} \mathbf{X}_j 2^{j-1-(i-1)\log d}. \tag{18.10}$$

It can easily be verified that $X_1, \ldots, X_n$ are k-wise independent and uniformly distributed in Ω. This is summarized in the following corollary.

Corollary 18.1 *Let $n, d, k \in \mathbb{N}$ and $\Omega := \{0, \ldots, d-1\}$. Uniformly distributed Ω-valued and k-wise independent random variables $X_1, \ldots, X_n$ can be constructed from*

$$l := O(k \log d \cdot \log(n \log d)) \tag{18.11}$$

---

[*] We slightly abuse notation here, see Equation 18.1.

binary random variables $Y_1, \ldots, Y_l$ with a sample space of size

$$2^l = \left((1 + n \log d)^{\lfloor k \log(d)/2 \rfloor}\right).$$

Proof. We show that the random variables $X_1, \ldots, X_n$ defined in Equation 18.10 are k-wise independent and uniformly distributed. Let $I \subseteq [n]$ with $|I| \leq k$ and $x \in \{0, \ldots, d-1\}^n$. We only need the values x_i for $i \in I$, however, having x as an n dimensional vector allows easier notation of the binary representation of its entries, see Equation 18.1. We have

$$\mathbb{P}\left[\forall i \in I : X_i = x_i\right]$$
$$= \mathbb{P}\left[\forall i \in I \, \forall j \in [\log d] : \mathbf{X}_{j+(i-1)\log d} = \mathbf{x}_{j+(i-1)\log d}\right]$$
$$= \prod_{i \in I} \prod_{j \in [\log d]} \mathbb{P}\left[\mathbf{X}_{j+(i-1)\log d} = \mathbf{x}_{j+(i-1)\log d}\right] \quad \text{by } (k \log d)\text{-wise independence}$$
$$= \prod_{i \in I} \mathbb{P}\left[\forall j \in [\log d] : \mathbf{X}_{j+(i-1)\log d} = \mathbf{x}_{j+(i-1)\log d}\right] \quad \begin{array}{l}(k \log d)\text{-wise independence} \\ \text{implies } (\log d)\text{-wise independence}\end{array}$$
$$= \prod_{i \in I} \mathbb{P}\left[X_i = x_i\right].$$

To see that $X_1, \ldots, X_n$ are uniformly distributed, fix $i \in [n]$ and let $a \in \{0, \ldots, d-1\}$. We see easily that

$$\mathbb{P}\left[X_i = a\right] = \mathbb{P}\left[\sum_{j=(i-1)\log(d)+1}^{i \log d} \mathbf{X}_j 2^{j-1-(i-1)\log d} = a\right]$$
$$= \mathbb{P}\left[\forall j \in [\log d] : \mathbf{X}_{j+(i-1)\log d} = \mathbf{a}_j\right]$$
$$= \prod_{j \in [\log d]} \mathbb{P}\left[\mathbf{X}_{j+(i-1)\log d} = \mathbf{a}_j\right] \quad \begin{array}{l}(k \log d)\text{-wise independence} \\ \text{implies } (\log d)\text{-wise independence}\end{array}$$
$$= \prod_{j \in [\log d]} \frac{1}{2} = \frac{1}{2^{\log d}} = \frac{1}{d}. \qquad \blacksquare$$

In some applications, a uniform distribution is not enough. Instead, one is given probabilities $p_1, \ldots, p_n$ and the task is to construct random variables $X_1, \ldots, X_n$ such that $\mathbb{P}\left[X_j = 1\right] = p_j$ for all $j \in [n]$. This is, in fact, approximately possible.

Theorem 18.8 (Luby [Lub86]) *Given probabilities $p_1, \ldots, p_n$, one can construct 2-wise independent 0/1 random variables $X_1, \ldots, X_n$ such that*

$$\mathbb{P}\left[X_j = 1\right] = \frac{\lfloor p_i \cdot q \rfloor}{q} \quad \text{for all } j \in [n],$$

for a prime number $n \leq q \leq 2n$. The sample space has size $q^2 = O(n^2)$.

18.2.3 Almost k-Wise Independence

Almost k-wise independence is an approximate version of Equation 18.4. For uniformly distributed 0/1 random variables, it reads as follows.

Definition 18.2 (Almost k-Wise Independence) *Let $\varepsilon, \delta > 0$ and $k \in [n]$, $n \in \mathbb{N}$.*

(i) $X_1, \ldots, X_n$ *are* (ε, k)-*independent, if for any* $r \in [k]$, $\alpha \in \{0,1\}^r$ *and any choice of indices* $1 \leq i_1 < i_2 < \cdots < i_r \leq n$, *we have*

$$\left| \mathbb{P}\left[(X_{i_1}, \ldots, X_{i_r}) = \alpha \right] - 2^{-r} \right| \leq \varepsilon.$$

(ii) $X_1, \ldots, X_n$ *are δ-away from k-wise independence, if for any $r \in [k]$ and any choice of indices* $1 \leq i_1 < i_2 < \cdots < i_r \leq n$, *we have*

$$\sum_{\alpha \in \{0,1\}^r} \left| \mathbb{P}\left[(X_{i_1}, \ldots, X_{i_r}) = \alpha \right] - 2^{-r} \right| \leq \delta.$$

(ε, k)-independence measures the deviation from the uniform distribution in the maximum norm, while Definition 18.2 (ii) describes statistical closeness to the uniform distribution in the L^1-norm.

(ε, k)-independent random variables are at most $2^k \varepsilon$-away from k-wise independence, whereas if they are δ-away from k-wise independence, they are (δ, k)-independent. For a set $S \subseteq [n]$ let $X_S := \sum_{i \in S} X_i$ and put $X := \sum_{i=1}^n X_i$. The number $X_S \bmod 2$ is 0 if X_S is even and 1 else. It can be viewed as the parity of S. If the $X_1, \ldots, X_n$ are independent, and $\mathbb{P}[X_i = 1] = \mathbb{P}[X_i = 0]$ for all $i = 1, \ldots, n$, then

$$\mathbb{P}[X_s \bmod 2 = 1] = \mathbb{P}[X_s \bmod 2 = 0]. \tag{18.12}$$

An approximative version of Equation 18.12 leads to the concept of ε-biased random variables (See Vazirani [Vaz86], Naor and Naor [NN93], and Peralta [Per90]). The bias of S is

$$\text{bias}(S) := | \mathbb{P}[X_s \bmod 2 = 0] - \mathbb{P}[X_s \bmod 2 = 1] |. \tag{18.13}$$

So for independent uniform random variables, $\text{bias}(S) = 0$ according to Equation 18.12.

Definition 18.3 (ε-*Bias*)

(i) $X_1, \ldots, X_n$ *are ε-biased, if $\text{bias}(S) \leq \varepsilon$ for all $S \subseteq [n]$.*
(ii) $X_1, \ldots, X_n$ *are k-wise ε-biased, if $\text{bias}(S) \leq \varepsilon$ for all $S \subseteq [n]$, $|S| \leq k$.*

k-wise ε-biased random variables and almost k-wise independence are closely related. To see this, let $D : \Omega \longrightarrow [0,1]$ be the probability distribution induced by the measure $\mathbb{P}$, so $D(\omega) := \mathbb{P}[(X_1 \ldots, X_n) = \omega]$, and let U be the uniform distribution (i.e., $U(\omega) = 2^{-n}$ for all $\omega \in \Omega$). The variation distance $\|D - U\|$ between D and U is the L^1-Norm of $D - U$,

$$\|D - U\| := \sum_{\omega \in \Omega} |D(\omega) - U(\omega)| = \sum_{\omega \in \Omega} |D(\omega) - 2^{-n}|.$$

$\|D - U\|$ is a measure for the distance of D from the uniform distribution. Let $D(S)$ be the restriction of D and $U(S)$ the restriction of U to a subset $S \subseteq [n]$.

Definition 18.4 $X_1, \ldots, X_n$ *are k-wise δ-dependent, if for all subsets $S \subseteq [n]$ with $|S| \leq k$,*

$$\|D(S) - U(S)\| \leq \delta.$$

Note that if the $X_1, \ldots, X_n$ are k-wise δ-dependent, then they are δ-away from k-wise independence. Taking the Fourier transform $\widehat{D - U}$, Diaconis and Shahashahani [Dia88] proved

$$\|D - U\|^2 = \|\widehat{D - U}\|^2 \leq \sum_{S \subseteq [n]} \text{bias}_D(S).$$

This inequality immediately implies a corollary formulated by U. Vazirani (PhD thesis 1986 [Vaz86], see also the papers of Vazirani and Vazirani [VV84] and Chor et al. [CFG+85]).

Corollary 18.2 *If $X_1, \ldots, X_n$ are δ-biased, then they are k-wise $2^{k/2}\delta$-dependent and $(2^{k/2}\delta, k)$-independent.*

In conclusion, ε-biased random variables for small $\varepsilon > 0$ should behave as k-wise independent variables. For derandomization the hope is to replace k-wise independence by the weaker notion of almost k-wise independence and to obtain an even smaller sample space. Naor and Naor [NN93] proved that ε-biased random variables can be constructed with only "few" random bits.

Theorem 18.9 (Naor and Naor 1993) *Let $\varepsilon > 0$ and $n \in \mathbb{N}$, $k \leq n$.*

(i) *Uniformly distributed 0/1-valued random variables $X_1, \ldots, X_n$ that are ε-biased can be constructed using $O(\log n + \log 1/\varepsilon)$ random bits. The size of the corresponding sample space is $2^{O(\log n + \log 1/\varepsilon)} = \left(\frac{n}{\varepsilon}\right)^c$ for a constant $c > 0$.*

(ii) *Uniformly distributed 0/1-valued random variables $X_1, \ldots, X_n$ that are k-wise ε-dependent can be constructed using $O(\log \log n + k + \log 1/\varepsilon)$ random bits. The size of the corresponding sample space is $\left(\frac{2^k \log n}{\varepsilon}\right)^{O(1)}$.*

The constant c in Theorem 18.9 (i) depends on the expansion rate of an expander graph and is slightly larger than 4 for the asymptotically optimal expanders of Lubotzky et al. [LPS88]. For polynomially large $\frac{1}{\varepsilon}$, that is, $\frac{1}{\varepsilon} = O(\text{poly}(n))$, the size of the sample space is polynomial in n. In applications of the method of almost k-wise independence one would like to have k large and ε small while the size of the sample space should remain polynomial. The bound $\left((2^k \log n)/\varepsilon\right)^{O(1)}$ limits the growth of k, in fact for $1/\varepsilon = O(\text{poly}(n))$, k must be about $O \log(1/\varepsilon) = O(\log n)$. Seminal papers managed to reduce the size of sample spaces corresponding to ε-biased random variables (Azar et al. [AMN98], Even et al. [EGL+98] and Alon et al. [AGHP92]). Alon et al. achieved a size of roughly $(n/\varepsilon)^2$ for ε-biased random variables. In particular, they generate n random variables that are ε-away from k-wise independence with only $(2 + \sigma(1))(\log \log n + (k/2) + \log k + \log(1/\varepsilon))$ random bits. This beats the bound of Naor and Naor as long as $\varepsilon < 1/k \log n$. There are two critical aspects in all of these results:

1. The random variables $X_1, \ldots, X_n$ are 0/1-valued and uniform, that is, $\mathbb{P}[X_i = 1] = \mathbb{P}[X_i = 0] = 1/2$ for all i. In applications, this may not be the case.
2. The usual strategy for derandomization using k-wise independence is to construct a small sample space a priori and to simulate a randomized algorithm for a specific problem in such a space, if possible. Thus, the sample space is chosen independently of the problem!

Schulman [Sch92] gave an interesting, different approach for Problem 2: He observed that for concrete problems only certain sets of d random variables, so-called d-neighborhoods, among the n random variables need to be independent. Thus the choice of the magnitude of independence is *driven by the problem*. Koller and Megiddo [KM94] further developed Schulmans approach covering also multivalued, nonuniformly distributed random variables. Koller and Megiddo showed that the sample space of k-wise independent random variables $X_1, \ldots, X_n$ with nonuniform probabilities $\mathbb{P}[X_i = 1] = p_i$, $i \in [n]$ correspond to a sample space of size at most $m(n, k) = \binom{n}{k} + \binom{n}{k-1} + \cdots + \binom{n}{0}$. Karloff and Mansour [KM97] showed the existence of $p_1, \ldots, p_n$ such that the size of any k-wise independent 0/1 probability space over $p_1, \ldots, p_n$ is at least $m(n, k)$. An interesting connection between small hitting sets for combinatorial rectangles in high dimension and the construction of small sample spaces for general multivalued random variables is given in the paper of Linial et al. [LLSZ97].

A different approach to the construction of small sample spaces than k-wise independence was taken by Karger and Koller [KK97]. In Section 18.4, we describe their technique. It will be used later in Srinivasan's parallel algorithms for packing problems (Section 18.5.2).

Bibliography and Remarks. In probability theory, k-wise independence was already used by Joffe [Jof74]. This concept was adapted to combinatorics and the design of algorithms in the years 1985/86, where the fundamental papers of Karp and Widgerson [KW85] and Luby [Lub86] on 2-wise independence, the work of Alon et al. on k-wise independence [ABI86] and the lower bound proof for the size of k-wise independent sample space by Chor et al. [CFG+85] were published. Derandomization of space bounded computations is treated by Armoni [Arm98] and Saks [Sak96]. A survey on parallel derandomization techniques is given in the paper of Han [Han92].

In computational geometry, k-wise independence was applied by Berger et al. [BRS94] and later by Goodrich [Goo93b, Goo93a, Goo96], and by Amato et al. [AGR94]. For the parallelization of derandomized geometric algorithms Mulmuley [Mul96] extended earlier work of Karloff and Raghavan [KR93] on limiting random resources using bounded independence distributions and demonstrated that a polylogarithmetic number of random bits suffices for guaranteeing a good expected performance of many randomized incremental algorithms.

The method of k-wise independence has been applied successfully to reduce or remove randomness from probabilistic constructions and from algorithms in various fields, like hashing, pseudorandom generators, one-way functions, circuit and communication complexity, and Boolean matrix multiplication. Many of these aspects are treated in the lecture notes of Luby and Widgerson [LW95]. Applications to learning algorithms are discussed in Reference NSS95. Sitharam and Straney [SS01] applied derandomization to learn Boolean functions.

Yao [Yao82] introduced the concept of a pseudorandom generator (see also Reference Nie92). An excellent book on pseudorandom generators and applications to cryptography (and other areas) are the lectures of Luby [Lub96]. Blum and Micali [BM84] showed that the problem of constructing a pseudorandom generator in based on the concept of computational indistinguishability introduced by Goldwasser and Micali [GM84]. Pseudorandom generators have been designed by Karp et al. [KPS88], Ajtai et al. [AKS87], Chor and Goldreich [CG89], Nissan [Nis91, Nis92], Impagliazzo and Zuckerman [IZ90], Håstad et al. [HILL99], Sipser [Sip88], Håstad et al. [HILL97] and Impagliazzo and Widgerson [IW97], and Andreev et al. [ACR98], to name some of the researchers.

Santha [San87] showed how to sample with a small number of random points. Feder et al. [FKN95] applied almost k-wise independence to amortize the communication complexity. Application to hashing and Boolean matrix multiplication were discussed by Alon et al. [AGMN92], Alon and Galil [AG97], Alon and Naor [AN96], and Naor and Naor [NN93]. Approximation of DNF through derandomization was carried out by Luby and Velickovic [LV96] and approximate counting of depth-2 circuits is shown by Luby et al. [LWV93].

Blum et al. [BEG+94] applied hash functions to authenticate memories. An interesting technique for deterministic construction of small sample spaces for general multivalued random variables through the construction of point sequences with a discrepancy or ε-net property was introduced by Linial et al. [LLSZ97].

18.3 Parallel Algorithms Using k-Wise Independence

In this section, we show the algorithmic impact of k-wise independence, in particular its power for parallelizing algorithms.

18.3.1 Functions over $\mathbb{Z}_2^k$

For convenience, we formulate the derandomization problem in the following setting. Let $\mathbb{Z}_2^k$ denote the k-dimensional vector space over $GF(2)$. We describe a class of functions $F : \mathbb{Z}_2^n \to \mathbb{Q}$ for which the derandomization problem can be solved in parallel. Let $N = N(n, m, k)$ be an integer valued function.

At the moment we do not assume that N is polynomially bounded in m and n. Let F be of the form

$$F(x_1, \cdots, x_n) = \sum_{i=1}^{N} c_i f_i(x_{i_1}, \ldots, x_{i_k}), \tag{18.14}$$

with $c_i \in \mathbb{Q}$ and $f_i(x_{i_1}, \ldots, x_{i_k}) = (-1)^{\sum_{j=1}^{k} x_{i_j}}$. The so given function $[k] \to [n], j \mapsto i_j$ will occur in the following.

Theorem 18.10 *Let $k, n \in \mathbb{N}$, $k \leq n$, and $X_1, \ldots, X_n$ be k-wise independent uniformly distributed $0/1$ random variables as in Theorem 18.7. Let $F : \mathbb{Z}_2^n \to \mathbb{Q}$ be a function as in Equation 18.14. With $O(N)$ parallel processors we can construct $x_0 \in \{0, 1\}^n$ in time $O(k \log n \log N)$ such that $F(x_0) \leq \mathbb{E}[F(X_1, \ldots, X_n)]$. The same result holds for the construction of $x_0' \in \{0, 1\}$ such that $F(x_0') \geq \mathbb{E}[F(X_1, \ldots, X_n)]$.*

Proof. The proof is based on Reference BR91. The random variables $X_1, \ldots, X_n$ by definition have the form $X_i = (BY)_i$, where $Y = (Y_1, \ldots, Y_l)$, the $Y_1, \ldots, Y_l$ are uniformly distributed $0/1$ random variables, $l = 1 + \frac{k-1}{2} \log(n+1)$, B is the (expanded version of the) matrix in Equation 18.5, which is of dimension $(n \times l)$. Therefore, it suffices to give assignments for the $Y_1, \ldots, Y_l$. The conditional probability method then goes as follows:

Suppose that for some $t \in [l]$ we have computed the assignments

$$Y_1 = y_1, \ldots, Y_{t-1} = y_{t-1}.$$

Then we choose for Y_t the value $y_t \in \{0, 1\}$ that minimizes the function

$$w \mapsto \mathbb{E}[F(X_1, \ldots, X_n) \mid Y_1 = y_1, \ldots, Y_{t-1} = y_{t-1}, Y_t = w]. \tag{18.15}$$

Obviously, after l steps we get $(Y_1, \ldots, Y_l) = y$ for some $y \in \{0, 1\}^l$. Thus, $x_0 := By$ is a solution according to the correctness of the algorithm CONDEXP in Section 18.2.1. We are done, if the conditional expectations

$$\mathbb{E}[F(X_1, \ldots, X_n) \mid Y_1 = y_1, \ldots, Y_{t-1} = y_{t-1}, Y_t = y_t]$$

can be computed within the claimed processor and time bounds for every t. Fix now $t \in [l]$. By linearity of expectation, it is sufficient to compute for each $i \in [N]$

$$\mathbb{E}[f_i(X_{i_1}, \ldots, X_{i_k}) \mid Y_1 = y_1, \ldots, Y_{t-1} = y_{t-1}, Y_t = y_t]. \tag{18.16}$$

Let B_i be the ith row of B and put $b := \sum_{j=1}^{k} B_{i_j}$, computed over $GF(2)$. Note that b is a vector— a sum of rows of matrix B. Because in the exponent of f_i, only the parity is relevant, we have that $f_i(X_{i_1}, \ldots, X_{i_k}) = (-1)^{b \cdot Y}$. Let s be the last position in the vector b that contains a 1. To shorten notation put $\vec{Y}_t = (Y_1, \ldots, Y_t)$ and $\vec{y}_t = (y_1, \ldots, y_t)$. Now

$$\mathbb{E}\left[(-1)^{b \cdot Y} \mid \vec{Y}_t = \vec{y}_t\right] = \begin{cases} (-1)^{\sum_{j=1}^{s} b_j y_j} & \text{if } t \geq s \\ 0 & \text{if } t < s \end{cases} \tag{18.17}$$

Equation 18.17 follows from the following observations. If $t < s$ then

$$b \cdot Y = \underbrace{\sum_{j=1}^{t} b_j y_j}_{=:a_1} + \underbrace{\sum_{j=t+1}^{s} b_j Y_j}_{=:a_2}.$$

Hence $\mathbb{E}\left[(-1)^{a_1+a_2}\right] = (-1)^{a_1}\mathbb{E}\left[(-1)^{a_2}\right] = (-1)^{a_1} \cdot 0$, because the $Y_{t+1}, \ldots, Y_s$ are independent and uniformly distributed. If $t \geq s$, then obviously $b \cdot Y = \sum_{j=1}^{s} b_j y_j$, due to condition $\vec{Y}_t = \vec{y}_t$. Hence, for fix i and t we can compute $\mathbb{E}\left[f_i \mid \vec{Y}_t = y_t\right]$ in constant time. The total running time is computed as follows: we assign to each $f_i(X_{i_1}, \ldots, X_{i_k})$ one processor, so we have N processors in total. In the tth step, $t \in [l]$, we can compute the sum $\sum_{i=1}^{N} \mathbb{E}\left[f_i \mid \vec{Y}_t = y_t\right]$ in $O(\log N)$ time using N parallel processors. Summing up over the l steps we get a total running time of $O(l \log N)$. ∎

Theorem 18.10 gives a parallel derandomized algorithm, provided $N(n, m, k)$ is polynomially bounded in n and m. Its extension to uniformly distributed multivalued variables is presented in the next section.

18.3.2 Multivalued Functions

We will again make use of k-wise independent random variables $X_1, \ldots, X_n$ constructed through the matrix B, but this time with values in $\Omega := \{0, \ldots, d-1\}$. For the conditional probability method to apply, we need to compute conditional probabilities for subsets of the $X_1, \ldots, X_n$. We first show that this can be done efficiently.

Proposition 18.3 *Let $Y_1, \ldots, Y_l$, the matrix B, and $X_1, \ldots, X_n$ be as in Section 18.2.2. For every $t \in [l]$ denote $\vec{Y}_t := (Y_1, \ldots, Y_t)$. Now fix $t \in [l]$ and a vector $y \in \{0, 1\}^t$, as well as $s \in [n]$ and a vector $x \in \{0, 1\}^{|I|}, I \subseteq [n]$, and set $\hat{B} := B_I$. Let $\hat{B}_1$ be the first t columns of $\hat{B}$ and $\hat{B}_2$ the last $l-t$ ones. Then*

$$\mathbb{P}\left[X_I = x \mid \vec{Y}_t = y\right] = \begin{cases} 2^{-\operatorname{rank} \hat{B}_2} & \text{if } x - \hat{B}_1 y \in \operatorname{range} \hat{B}_2, \\ 0 & \text{else.} \end{cases}$$

This value can be computed using $O(n)$ parallel processors in $O(l^3 \log n)$ time. A similar result extends to $\{0, \ldots, d-1\}$-valued random variables, d a power of 2, with $l = O(k \log d \cdot \log(n \log d))$.

Proof. Because $Y_1, \ldots, Y_l$ are independent, we have

$$\mathbb{P}\left[X_I = x \mid \vec{Y}_t = y\right] = \mathbb{P}\left[\hat{B}Y = x \mid \vec{Y}_t = y\right]$$
$$= \mathbb{P}\left[\hat{B}_1 y + \hat{B}_2(Y_{t+1}, \ldots, Y_l) = x\right]$$
$$= \mathbb{P}\left[\hat{B}_2(Y_{t+1}, \ldots, Y_l) = x - \hat{B}_1 y\right].$$

We have $l - t = \dim \ker \hat{B}_2 + \dim \operatorname{range} \hat{B}_2$, hence $\dim \ker \hat{B}_2 = l - t - \dim \operatorname{range} \hat{B}_2$. The case of $x \notin \operatorname{range} \hat{B}_2$ is clear. If $x \in \operatorname{range} \hat{B}_2$, then the set of all preimages is a subspace of dimension $\dim \ker \hat{B}_2$. Because the field has two elements, there are $2^{\dim \ker \hat{B}_2}$ preimages of x. On the other hand, there are 2^{l-t} possible settings for the $Y_{t+1}, \ldots, Y_l$, and so the claim follows:

$$\mathbb{P}\left[\hat{B}_2(Y_{t+1}, \ldots, Y_l) = x - \hat{B}_1 y\right] = \frac{2^{\dim \ker \hat{B}_2}}{2^{l-t}} = 2^{l-t-\dim \operatorname{range} \hat{B}_2 - (l-t)} = 2^{-\dim \operatorname{range} \hat{B}_2}.$$

It is possible to compute $\dim \operatorname{range} \hat{B}_2 = \operatorname{rank} \hat{B}_2$ within the stated processor and time bounds by a simple devide-and-conquer application of Gauss elimination. The same can be used for testing $x - \hat{B}_1 y \in \operatorname{range} \hat{B}_2$.

For the extention to $\{0, \ldots, d-1\}$-valued random variables, use Corollary 18.1. ∎

Now we consider a more general class of functions than in the previous section. They are of a type later used in packing integer programs (Section 18.5.1). We give a $\mathcal{NC}$ algorithm for finding points below or above the expectation. Let $d, k, m, n \in \mathbb{N}$ and for each $(i, z, j) \in [m] \times \Omega \times [n]$ let $g_{izj} : \mathbb{R} \to \mathbb{R}$ such that for every $r \in \mathbb{R}$, the value $g_{izj}(r)$ can be computed on a single processor in constant time. For each $(i, z) \in [m] \times \Omega$ define the functions

$$f_{iz} : \Omega^n \to \mathbb{R}, \; (x_1, \ldots, x_n) \mapsto \sum_{j=1}^{n} g_{izj}(x_j),$$

and finally put them together

$$F : \Omega^n \to \mathbb{R}, \; (x_1, \ldots, x_n) \mapsto \sum_{(i,z) \in [m] \times \Omega} (f_{iz}(x_1, \ldots, x_n))^k.$$

Theorem 18.11 *Let $d, k, m, n \in \mathbb{N}$, $d \geq 2$ a power of 2, and $N := mdn^k$. Let $X_1, \ldots, X_n$ be k-wise independent random variables with values in $\Omega = \{0, \ldots, d-1\}$ as defined in Corollary 18.1 and let the g_{izj}, f_{iz}, and F as above. Then with $O(\max\{nd^k, N\})$ parallel processors we can construct $x_1, \ldots, x_n \in \Omega$ and $y_1, \ldots, y_n \in \Omega$ in time*

$$O(\log(n \log d) \cdot (k^4 \log^4 d \log^3(n \log d) \log n + k \log d \log N))$$

such that $F(x_1, \ldots, x_n) \geq \mathbb{E}[F(X_1, \ldots, X_n)]$, resp. $F(y_1, \ldots, y_n) \leq \mathbb{E}[F(X_1, \ldots, X_n)]$.

Proof. We prove the first assertion as the second follows from it. Fix $(i, z) \in [m] \times \Omega$ for a moment. We write the kth power of f_{iz} in a different way. For every multiindex $\alpha \in [n]^k$ and $(x_1, \ldots, x_k) \in \Omega^k$ define

$$f_\alpha^{(iz)}(x_1, \ldots, x_k) := \prod_{j=1}^{k} g_{iz\alpha_j}(x_j),$$

and observe that this function depends on k variables only. It can be evaluated with k processors in $O(\log k)$ time. It is easy to see that for all $x = (x_1, \ldots, x_n) \in \Omega^n$ we have, recalling that $x_\alpha = (x_{\alpha_1}, \ldots, x_{\alpha_k})$, $f_{iz}(x_1, \ldots, x_n) = \sum_{\alpha \in [n]^k} f_\alpha^{(iz)}(x_\alpha)$, and so

$$F(x_1, \ldots, x_n) = \sum_{(i,z) \in [m] \times \Omega} \sum_{\alpha \in [n]^k} f_\alpha^{(iz)}(x_\alpha).$$

By construction of the $X_1, \ldots, X_n$, it suffices to give an assignment to 0/1 random variables $Y_1, \ldots, Y_l$ with $l = O(k \log d \cdot \log(n \log d))$, see Corollary 18.1 and the discussion preceding it. The conditional probability method then goes as follows: Suppose that for some $t \in [l]$ we have computed the the values

$$Y_1 = y_1, \ldots, Y_{t-1} = y_{t-1},$$

then choose for Y_t a value $y_t \in \Omega$ that maximizes the function

$$w \mapsto \mathbb{E}[F(X_1, \ldots, X_n) \mid y_1, \ldots, y_{t-1}, Y_t = w]. \tag{18.18}$$

After l steps we have $Y = y$ with some $y \in \{0, 1\}^l$ and the solution can be computed from it by Equations 18.10 and 18.8. Let $\vec{Y}_t = (Y_1, \ldots, Y_t)$ and $\vec{y}_t = (y_1, \ldots, y_t)$. We are done, if we can compute the conditional expectations

$$\mathbb{E}[F(X_1, \ldots, X_n) \mid \vec{Y}_t = \vec{y}_t]$$

within the claimed processor and time bound. By linearity of expectation, it is sufficient to compute for each triple $(i, z, \alpha) \in [m] \times \Omega \times [n]^k$ the conditional expectation

$$\mathbb{E}\,[f_\alpha^{(iz)}(X_\alpha) \mid \vec{Y}_t = \vec{y}_t]\,.$$

To this end, write

$$\mathbb{E}\,[f_\alpha^{(iz)}(X_\alpha) \mid \vec{Y}_t = \vec{y}_t] = \sum_{x \in \Omega^k} f_\alpha^{(iz)}(x) \mathbb{P}\,[X_\alpha = x \mid \vec{Y}_t = \vec{y}_t]$$

$$= \sum_{x \in \Omega^k} f_\alpha^{(iz)}(x) \mathbb{P}\,[\mathbf{X}_\alpha = \mathbf{x} \mid \vec{Y}_t = \vec{y}_t]\,.$$

According to Proposition 18.3, for a fixed $x \in \Omega^k$, the conditional probability can be computed using $O(n)$ processors in time $O(l^3)$. Since we have d^k such vectors x, and each $f_\alpha^{(iz)}(x)$ uses k processors and $O(\log k)$ time, we can compute $\mathbb{E}\,[f_\alpha^{(iz)}(X_\alpha) \mid \vec{Y}_t = \vec{y}_t]$ for every $t \in [l]$ and $(i, z, \alpha) \in [m] \times \Omega \times [n]^k$ with $O(d^k(k+n)) = O(nd^k)$ parallel processors (recall $d \geq 2$) in $O(l^3 \log n + \log d^k + \log k) = O(l^3 \log n + k \log d)$ time. Then we compute for every y_t, $t \in [l]$, the expectation

$$\mathbb{E}\,[F(X_1, \ldots, X_n) \mid \vec{Y}_t = \vec{y}_t]$$

in $O(\log N)$ time using $O(N)$ parallel processors. Finally, a $y_t \in \Omega$ that maximizes Equation 18.18 can be computed finding the maximum of the d conditional expectations in $O(\log d)$ time with $O(d)$ processors. The maximum number of processors used is $O(\max\{nd^k, N\})$ and the total running time over all l steps is

$$O(l \cdot (l^3 \log n + k \log d + \log N + \log d))$$
$$= O(\log(n \log d) \cdot (k^4 \log^4 d \log^3(n \log d) \log n + k^2 \log^2 d + k \log d \log N + k \log^2 d))$$
$$= O(\log(n \log d) \cdot (k^4 \log^4 d \log^3(n \log d) \log n + k \log d \log N))\,. \blacksquare$$

18.3.3 Maximal Independent Sets in Graphs

We start with the celebrated parallel algorithm of Luby computing a maximal independent set in graphs. Historically, it was the first striking application of limited dependence to the design of a $\mathcal{NC}$ algorithm.

We consider a graph $G = (V, E)$ with $V = [n]$ and $|E| = m$. A subset I of vertices is called *independent*, if the induced subgraph on I contains no edges. A (MIS) has the additional property that adding an arbitrary vertex destroys independence. The following simple algorithm computes the lexicographically first MIS set in linear time:

Algorithm 3: LFMIS.

Input: A Graph $G = (V, E)$.
Output: A maximal independent set $I \subseteq V$.

$I \leftarrow \emptyset$;
for $i \leftarrow 1$ **to** n **do**
    $I \leftarrow I \cup \{i\}$;
    $V \leftarrow V \setminus (\{i\} \cup N(i))$;
end

Cook [Coo85] proved that deciding whether a vertex belongs to the lexicographically first MIS is logspace-complete in $\mathcal{P}$. Thus, there is no hope to parallelize LFMIS. Yet, the MIS problem is not inherently sequential. We will present a parallel randomized algorithm of Luby [Lub86] with expected running time $O(\log^2 n)$ on $O(n + m)$ processors. Removing the randomness by means of searching a small sample space will reveal that $O(n^2 m)$ processors can solve MIS in $O(\log^2 n)$ time. The key idea is to successively add to the MIS under construction not just a *single* vertex i but an independent set S that is computed in parallel.

Algorithm 4: PARALLELMIS (High Level).

Input: A Graph $G = (V, E)$.
Output: A maximal independent set $I \subseteq V$.

1   $I \leftarrow \emptyset$;
2   **while** $V \neq \emptyset$ **do**
3   select an independent set $S \subseteq V$;
4      $I \leftarrow I \cup S$;
5      $V \leftarrow V \setminus (S \cup N(S))$;
6   **end**

The high-level description makes clear that PARALLELMIS terminates with a MIS I. The parallelization is hidden in the selection of S. Luby analyzed the following randomized selection procedure:

Algorithm 5: PARALLELMIS (Details of the Selection Step).

Input: A Graph $G = (V, E)$.
Output: An independent set $S \subseteq V$.

1   **in parallel forall** $i \in V$ **do**
2   **if** $\deg(i) = 0$ **then** mark i;
3   **else** mark i with probability $\frac{1}{2 \cdot \deg(i)}$;
4   **end**
5   **in parallel forall** $\{i, j\} \in E$ **do**
6   **if** *both i and j are marked* **then** unmark the vertex with smaller degree;
7   **end**
8   $S \leftarrow \emptyset$;
9   **in parallel forall** $i \in V$ **do**
10   **if** *i is marked* **then** $S \leftarrow S \cup \{i\}$;
11   **end**

One can show that each iteration of PARALLELMIS can be implemented to run in $O(\log n)$ time on an EREW PRAM with $O(n + m)$ processors. We will prove that the algorithm terminates after an expected total number of $O(\log n)$ iterations, since a constant fraction of edges are expected to get removed in each update step in line 5 in Algorithm 4. We call a vertex $i \in V$ *good* if the degree of at least one-third of its neighbors is at most $\deg(i)$ and *bad* otherwise. Now consider an arbitrary iteration.

Lemma 18.1 *Let $i \in V$ be a good vertex. The probability that $i \in N(S)$ is at least*

  (a) $\frac{1-\exp(-(1/6))}{2}$ *if the random choices in line 3 in Algorithm 5 are completely independent*
  (b) $1/24$ *if the random choices are only pairwise independent.*

Proof outline. Observe that if a vertex j is marked, then it is selected into S with probability at least $1 - \mathbb{P}\left[\exists \text{ marked } k \in N(j) \text{ with } \deg(k) \geq \deg(j)\right] \geq 1 - \sum_{k \in N(j)} \frac{1}{2\deg(j)} = 1/2$.

(a) Since the vertex i is good, it has at least $\deg(i)/3$ neighbors with degree at most $\deg(i)$. Owing to complete independence, the probability that none of these is marked is at most $\left(1 - \frac{1}{2\deg(i)}\right)^{\deg(i)/3} \leq \exp(-1/6)$. Hence, with probability at least $\frac{1-\exp(-(1/6))}{2}$ a neighbor of i is selected into S, which means that i is in $N(S)$.

(b) Let X_j be the event that $j \in N(i)$ is marked in line 3, and let Y_j be the event that $j \in N(i)$ is selected into S. Since $\mathbb{P}[Y_j] \geq \mathbb{P}[X_j]/2$ and $Y_j \subseteq X_j$ we have by the principle of inclusion–exclusion that the probability for $i \in N(S)$ is

$$\mathbb{P}\left[\bigcup_{j \in N(i)} Y_j\right] \geq \sum_{j \in N(i)} \mathbb{P}[Y_j] - \sum_{j \neq k \in N(i)} \mathbb{P}[Y_j \cap Y_k]$$

$$\geq \frac{1}{2} \sum_{j \in N(i)} \mathbb{P}[X_j] - \sum_{j \neq k \in N(i)} \mathbb{P}[X_j \cap X_k]$$

$$\geq \frac{1}{2} \left(\sum_{j \in N(i)} \mathbb{P}[X_j] - \left(\sum_{j \in N(i)} \mathbb{P}[X_j] \right)^2 \right)$$

$$= \frac{1}{2} \left(\sum_{j \in N(i)} \mathbb{P}[X_j] \right) \left(1 - \sum_{j \in N(i)} \mathbb{P}[X_j] \right).$$

Since i is a good vertex,

$$\sum_{j \in N(i)} \mathbb{P}[X_j] \geq \frac{\deg(i)}{3} \cdot \frac{1}{2\deg(i)} = \frac{1}{6},$$

and assuming that $\sum_{j \in N(i)} \mathbb{P}[X_j] \leq \frac{1}{2}$* we get $\mathbb{P}\left[\bigcup_{j \in N(i)} Y_j\right] \geq 1/24$ as claimed. ∎

We call an edge *good*, if at least one of its endpoints is good.

Lemma 18.2 *In each iteration at least half of the edges are good.*

Proof outline. We direct the edges of the graph such that $(i, j) \in E$ implies $\deg(i) \leq \deg(j)$. Let V_g and V_b denote the sets of good and bad vertices, respectively, and let $E(X, Y)$ abbreviate the cardinality of the set of edges from $E \cap (X \times Y)$. Let $\deg^-(i)$ be the degree of a vertex i, (that is, the number of edges incident with i and pointing to i and $\deg^+(i)$ the out-degree of a vertex i (that is, the number of edges incident with i and pointing away from i). Using the fact that for every bad vertex i, $2\deg^-(i) \leq \deg^+(i)$ we get that

$$2E(V_b, V_b) + E(V_b, V_g) + E(V_g, V_b) = \sum_{i \in V_b} (\deg^+(i) + \deg^-(i))$$

$$\leq 3 \sum_{i \in V_b} (\deg^+(i) - \deg^-(i)) = 3 \sum_{i \in V_g} (\deg^-(i) - \deg^+(i))$$

$$= 3(E(V_b, V_g) - E(V_g, V_b)) \leq 3(E(V_b, V_g) + E(V_g, V_b)).$$

---

* Otherwise can choose an appropriate subset of neighbors s.t. $\sum \mathbb{P}[X_j] \approx 1/2$ and get $\mathbb{P}\left[\bigcup Y_j\right] \approx 1/8$.

This gives $E(V_b, V_b) \leq E(V_b, V_g) + E(V_g, V_b)$, implying that the number of bad edges is at most the number of good edges, so that at least half of the edges are good. ∎

Since an end point of a good edge is in $N(S)$ with probability at least $1/24$, we can conclude that a fraction of at least $1/48$ edges are expected to be deleted in each iteration. Hence the algorithm terminates after $O(\log n)$ iterations. Since only pairwise independence is needed, we can derandomize the algorithm by exhaustively searching a sample space of size $O(n^2)$, see Theorem 18.8. Thus, we get a deterministic algorithm for solving MIS in time $O(\log^2 n)$ with $O(n^2 m)$ processors. Goldberg and Spencer [GS89a] have shown that the number of processors can be reduced to $O(n + m/\log n)$ on cost of an increased running time of $O(\log^3 n)$.

Bibliography and Remarks. Karp et al. [KUW88] developed a randomized algorithm for finding a MIS in an independence system. Their algorithm can be adapted to compute a MIS in a general hypergraph in time $O(\sqrt{n}(\log n + \log m))$ on $O(m \cdot n)$ processors. Faster parallel algorithms have been described by Goldberg and Spencer [GS89b,GS89a]. Dahlhaus and Karpiński [DK89], and, independently, Kelsen [DKK92a] have given efficient $\mathcal{NC}$ algorithms for computing MIS in 3-uniform hypergraphs (journal version in [DKK92b]). Beame and Luby [BL90] studied the MIS problem for hypergraphs where the size of each edge is bounded by a fixed absolute constant c. They presented a generalized version of the randomized PARALLELMIS-algorithm running on $O(n+c)$ processors with running time polynomial in $\log(n+c)$, as verified by Kelsen [Kel92]. They also gave a similar algorithm for the general case of k-uniform hypergraphs, and on $\sigma(n \cdot m)$ processors conjectured a running time polynomial in $\log(n + m)$. Łuczak and Szymańska [ŁS97] observed that a linear hypergraph, that is, a hypergraph in which each pair of edges has at most one vertex in common, contains a large subhypergraph without vertices of large degree. They proved that the randomized algorithm of Beame and Luby finds a MIS in a linear hypergraph in polylogarithmic expected running time if a preprocessing step is added to make the algorithm perform on such an "equitable" subhypergraph. A derandomized version of their result has been given in the PhD thesis of Edita Szymańska [Szy98].

18.3.4 Low Discrepancy Colorings

One root for derandomization certainly is the application of the probabilistic method to combinatorial discrepancy theory. Combinatorial discrepancy theory deals with the problem of partitioning the vertices of a hypergraph such that all hyperedges are split into about equal parts by the partition classes. Discrepancy measures the deviation of an optimal partition from an ideal one, that is, one where all edges contain the same number of vertices in any partition class. An introduction to combinatorial discrepancy theory is given in the book of Alon et al. [ASE92] and the survey article of Beck and Sós [BS95]. Excellent sources covering many aspects (and proofs) of combinatorial, geometric as well as classical discrepancy theory are the books of Matoušek [Mat99] and Chazelle [Cha00] for the connection of derandomization and discrepancy theory. For an extension of 2-color discrepancy theory to c colors see Doerr and Srivastav [DS99, DS03].

Let $\mathcal{H} = (V, \mathcal{E})$ be a hypergraph, $\mathcal{E} = \{E_1, \ldots, E_m\}$, and let $\chi : V \to \{-1, +1\}$ be a function. Identifying -1 and $+1$ with colors, say red and blue, χ is a 2-*coloring* of V. The sets $\chi^{-1}(-1)$ and $\chi^{-1}(+1)$ build the partition of V induced by χ. The imbalance of a hyperedge $E \in \mathcal{E}$ with respect to χ can be expressed by

$$\chi(E) := \sum_{x \in E} \chi(x). \qquad (18.19)$$

The *discrepancy* of $\mathcal{H}$ with respect to χ is

$$\mathrm{disc}(\mathcal{H}, \chi) := \max_{E \in \mathcal{E}} |\chi(E)|, \qquad (18.20)$$

and the *discrepancy* of $\mathcal{H}$ is

$$\mathrm{disc}(\mathcal{H}) := \min_{\chi: V \to \{-1, +1\}} \mathrm{disc}(\mathcal{H}, \chi). \qquad (18.21)$$

Theorem 18.12 (Spencer 1987) *A 2-coloring of V with*

$$\mathrm{disc}(\chi) \leq \sqrt{2n \log(4m)}$$

can be constructed in $O(mn^2 \log(mn))$ time.

Proof. Put $\alpha := \sqrt{2n\log(4m)}$. Let $\chi_1, \ldots, \chi_n$ be independent random variables defined by $\mathbb{P}[\chi_j = +1] = \mathbb{P}[\chi_j = -1] = 1/2$ for all j. The vector $\chi = (\chi_1, \ldots, \chi_n)$ is a random 2-coloring of V. For each $i \in [m]$ let F_i be the event "$|\chi(E_i)| \leq \alpha$." From Hoeffding's inequality (Theorem 18.4) we infer $\mathbb{P}[F_i^c] \leq 1/(2m)$, thus,

$$\mathbb{P}\left[\bigcup_{i=1}^m F_i^c\right] \leq \frac{1}{2},$$

and a coloring with discrepancy at most α exists. With the algorithmic version of the Chernoff–Hoeffding inequality [SS96] it can be constructed in $O(mn^2 \log(mn))$ time. ∎

The original proof of Spencer is done with the hyperbolic cosine algorithm instead of the algorithmic Chernoff–Hoeffding bound. It is known that the discrepancy bound $O(\sqrt{n \ln m})$ is not optimal. The celebrated "six-standard-deviation" theorem of Spencer [Spe85] proves the existence of a 2-coloring with discrepancy at most $6\sqrt{n}$ (for $m = n$). In general, the bound $O(\sqrt{n})$ for discrepancy is optimal up to a constant factor, since the discrepancy of the set system induced by an $(n \times n)$ Hadamard matrix is at least $\frac{1}{2}\sqrt{n}$ (see Reference ASE92). It is a challenging open problem to give a randomized or deterministic polynomial-time algorithm that finds such a coloring. The main hinderance to the transformation of this existence result into an algorithm is the use of the pigeonhole principle in the proof.

We proceed to the parallelization of the discrepancy algorithm in Theorem 18.12. Berger and Rompel [BR91] and Motwani et al. [MNN94] in 1989 presented the first parallel algorithm for this problem. The key for the success of their approach is the computation of kth moments of the discrepancy function over a k-wise independent distribution.

Define $\Delta_i := \max\{|E_i|, 3\}$ for each $i \in [m]$. Let $A = (a_{ij})$ be the hyperedge-vertex incidence matrix of $\mathcal{H}$. Let $X_1, \ldots, X_n$ be $-1/1$ random variables with $\mathbb{P}[X_j = +1] = \mathbb{P}[X_j = -1] = 1/2$ and set

$$\psi_i := \sum_{j=1}^n a_{ij} X_j \text{ for each } i \in [m], \quad \text{and} \quad \Delta_0 := \min_{i \in [m]} \Delta_i \ (\geq 3), \quad \Delta := \max_{i \in [m]} \Delta_i.$$

In the following, we will always assume that all hyperedges have cardinality at least 3 (and so $|E_i| = \Delta_i$ and each ψ_i is the sum of Δ_i random variables). The established bounds will also hold in the general case, however, because they are always at least 2, which is an upper bound on the discrepancy of hyperedges of cardinality 1 and 2 regardless of the coloring.

For all $i \in [m]$ let k_i be a nonnegative integer and put

$$k := \max_{i \in [m]} k_i. \tag{18.22}$$

We assume that the $X_1, \ldots, X_n$ are k-wise independent. For every multiindex $\alpha \in [n]^{k_i}, \alpha = (\alpha_1, \ldots, \alpha_{k_i})$, we define

$$a_{i\alpha} := \prod_{l=1}^{k_i} a_{i\alpha_l} \quad \text{and} \quad \phi_\alpha := \prod_{l=1}^{k_i} X_{\alpha_l}.$$

We consider the k_ith powers $\psi_i^{k_i}$, $i \in [m]$. Let $\mathbb{E}_{\text{indep}}[\psi_i^k]$ be the expectation under the assumption of (completely) independent random variables $X_1, \ldots, X_n$.

Lemma 18.3 $\mathbb{E}[\psi_i^{k_i}] \leq 2(k_i \Delta_i)^{\frac{k_i}{2}}$ for all $i \in [m]$.

Proof. We have $\psi_i^{k_i} = \sum_{\alpha \in [n]^{k_i}} a_{i\alpha} \phi_\alpha$, hence $\mathbb{E}[\psi_i^{k_i}] = \mathbb{E}_{\text{indep}}[\psi_i^{k_i}]$, because there occur only $k_i \leq k$ random variables in each summand and the expectation is linear. So we may apply Theorem 18.4 to get

$$\mathbb{E}[\psi_i^{k_i}] = \mathbb{E}_{\text{indep}}[\psi_i^{k_i}] = \int_0^\infty \mathbb{P}_{\text{indep}}[|\psi_i|^{k_i} \geq x] \, dx$$

$$= \int_0^\infty \mathbb{P}_{\text{indep}}[|\psi_i| \geq x^{(1/k_i)}] \, dx$$

$$\leq \int_0^\infty 2 \cdot \exp\left(-\frac{x^{(2/k_i)}}{2\Delta_i}\right) dx \quad \text{by Theorem 18.4}$$

$$= 2\left(\frac{k_i}{2}\right)!(2\Delta_i)^{(k_i/2)}$$

$$\leq 2(k_i \Delta_i)^{(k_i/2)}. \qquad \blacksquare$$

Let $0 < \varepsilon < 1$ be a fixed parameter. For each $i \in [m]$ let $\beta_i > 1/\varepsilon$ be such that

$$k_i := \frac{\beta_i \log(2m)}{\log \Delta_i} \tag{18.23}$$

is the smallest even integer with $k_i \in \left\{ \left\lceil \frac{\log(2m)}{\varepsilon \log \Delta_i} \right\rceil, \left\lceil \frac{\log(2m)}{\varepsilon \log \Delta_i} \right\rceil + 1, \frac{\log(2m)}{\varepsilon \log \Delta_i} + 2 \right\}$. Define bounds for the discrepancies of each hyperedge E_i

$$\lambda_i := \Delta_i^{(1/2)+\varepsilon} \sqrt{\frac{\beta_i \log(2m)}{\log \Delta_i}}. \tag{18.24}$$

Note that because $\Delta_i \geq 3$, we have $\log \Delta_i > 1$.

Lemma 18.4 For all $i \in [m]$ we have

$$\mathbb{P}\left[|\psi_i| \geq \Delta_i^{(1/2)+\varepsilon} \sqrt{\frac{\log(2m)}{\varepsilon \log \Delta_i}} + 2\right] \leq \mathbb{P}[|\psi_i| \geq \lambda_i] \leq \frac{\mathbb{E}[\psi_i^{k_i}]}{\lambda_i^{k_i}} < \frac{1}{m}.$$

Hence, with positive probability $|\psi_i| \leq \lambda_i \leq \Delta_i^{(1/2)+\varepsilon} \sqrt{\frac{\log(2m)}{\varepsilon \log \Delta_i}} + 2$, for all $i \in [m]$.

Proof. We fix an arbitrary $i \in [m]$. The first inequality is trivial. The k_ith moment inequality [Fel67] gives

$$\mathbb{P}[|\psi_i| \geq \lambda_i] \leq \frac{\mathbb{E}[|\psi_i|^{k_i}]}{\lambda_i^{k_i}} = \frac{\mathbb{E}[\psi_i^{k_i}]}{\lambda_i^{k_i}}$$

$$\leq \frac{2(k_i \Delta_i)^{(k_i/2)}}{\lambda_i^{k_i}} \quad \text{by Lemma 18.3.} \tag{18.25}$$

The term in Equation 18.25 is

$$2\left(\frac{k_i \Delta_i}{\lambda_i^2}\right)^{k_i/2} = 2\left(\frac{1}{\Delta_i^{2\varepsilon}}\right)^{k_i/2}$$
$$= 2 \cdot 2^{-2\varepsilon \log \Delta_i \beta_i \log(2m)/(2 \log \Delta_i)}$$
$$= 2 \cdot 2^{-\beta_i \varepsilon \log(2m)}$$
$$< 2 \cdot 2^{-\log(2m)} = \frac{1}{m} \quad \text{since } \beta_i > \frac{1}{\varepsilon}.$$

The final assertion now follows with the union bound. ∎

Corollary 18.3 *If* $k := \max_{i \in [m]} k_i \leq \log(2m)$, *then for any* $i \in [m]$, *we have* $\varepsilon \geq \frac{1}{\beta_i} \geq \frac{1}{\log \Delta_i}$, *and with positive probability*

$$|\chi(E_i)| = |\psi_i| \leq \Delta_i^{(1/2)+\varepsilon} \sqrt{\log(2m) + 2} \leq \Delta^{(1/2)+\varepsilon} \sqrt{\log(2m) + 2},$$

holds for all $i \in [m]$, *hence* $\mathrm{disc}(\mathcal{H}, \chi) \leq \Delta^{(1/2)+\varepsilon} \sqrt{\log(2m) + 2}$.

Proof. The assumption $k \leq \log(2m)$ gives $\beta_i \leq \log \Delta_i$, so $\varepsilon \geq \frac{1}{\beta_i} \geq \frac{1}{\log \Delta_i}$, and further $(\varepsilon \log \Delta_i)^{-1} \leq 1$ and $\frac{\beta_i}{\log \Delta_i} \leq 1$, thus $\lambda_i \leq \Delta_i^{(1/2)+\varepsilon} \sqrt{\log(2m) + 2}$ for all $i \in [m]$. The union bound together with Lemma 18.4 implies the assertion of the corollary. ∎

Remark 18.2 Note that the discrepancy bound $\Delta^{(1/2)+\varepsilon} \sqrt{\log(2m) + 2}$ in Corollary 18.3 holds only for all i if ε is sufficiently large $\left(\text{i.e., } \varepsilon \geq \frac{1}{\log \Delta_c}\right)$. We can also take a different point of view: first we can assume $\Delta_i \geq \log(2m)$ for all i. Otherwise, the discrepancy bounds in Corollary 18.3 become $\Delta_i^{(1/2)+\varepsilon} \sqrt{\log(2m) + 2} < \Delta_i$, which is a trivial bound. So, the lower bound condition for ε now says $\varepsilon \geq \frac{1}{\log \log(2m)}$, which is certainly true for large enough m provided that ε is fix. Thus, Corollary 18.3 is true for any $\varepsilon > 0$ and hypergraphs where $2m \geq 2^{2^{1/\varepsilon}}$.

The following theorem in a similar version is formulated by Motwani et al. [MNN94]. Here we compute the exact processor and time bounds and give a compact proof using the derandomization stated in Theorem 18.10.

Theorem 18.13 *Let* $\mathcal{H} = (V, \mathcal{E})$ *be a hypergraph with* $|V| = n$, $|\mathcal{E}| = m$. *Let* $0 < \varepsilon < 1$ *fix and* k_i *for each* $i \in [m]$ *as in Equation 18.23,* $k = \max_{i \in [m]} k_i$.

(i) *Using* k-*wise independence, a 2-coloring* χ *of* $\mathcal{H}$ *with* $|\chi(E_i)| \leq \Delta_i^{(1/2)+\varepsilon} \sqrt{\frac{\log(2m)}{\varepsilon \log \Delta_i} + 2}$ *can be constructed with* $O(n^2 m^{1+\frac{1}{\varepsilon}})$ *processors in time* $O\left(\frac{1}{\varepsilon^2 \log \Delta_0} \log^2 n \log^2 m\right)$.

(ii) *If* $k \leq \log(2m)$, *then* $\varepsilon \geq \frac{1}{\beta_i} \geq \frac{1}{\log \Delta_i}$ *and the discrepancy bounds reduce to* $\Delta_i^{(1/2)+\varepsilon} \sqrt{\log(2m) + 2}$.

Proof. (ii) is an immediate consequence of (i) and Corollary 18.3. To prove (i), we invoke Theorem 18.10. Since we need 0/1 random variables, we introduce k-wise independent $Z_1, \ldots, Z_n$, think of each X_j as $X_j = (-1)^{Z_j}$ and regard ψ_i as a function of the $Z_1, \ldots, Z_n$, $\psi_i(Z_1, \ldots, Z_n) = \sum_{j=1}^n a_{ij} X_j(Z_j) = \sum_{j=1}^n a_{ij}(-1)^{Z_j}$.

For every multiindex $\alpha \in [n]^{k_i}$, $\alpha = (\alpha_1, \ldots, \alpha_{k_i})$, we have $\phi_\alpha(Z_1, \ldots, Z_n) = (-1)^{\sum_{j=1}^{k_i} Z_{\alpha_j}}$. Therefore

$$\sum_{i=1}^m \lambda_i^{-k_i} \psi_i^{k_i}(Z_1, \ldots, Z_n) = \sum_{i=1}^m \lambda_i^{-k_i} \sum_{\alpha \in [n]^{k_i}} a_{i\alpha} \phi_\alpha(Z_1, \ldots, Z_n)$$

$$= \sum_{i=1}^m \lambda_i^{-k_i} \sum_{\alpha \in [n]^{k_i}} a_{i\alpha} (-1)^{\sum_{j=1}^{k_i} Z_{\alpha_j}}$$

$$=: F(Z_1, \ldots, Z_n). \quad (18.26)$$

The function F has the form as in Theorem 18.10. For fixed i, the number of nonzero terms in the sum

$$\sum_{\alpha \in [n]^{k_i}} a_{i\alpha} (-1)^{\sum_{j=1}^{k_i} Z_{\alpha_j}}$$

is at most $\Delta_i^{k_i} \leq \Delta_i^{(\log(2m)/\varepsilon \log \Delta_i)+2} = (2m)^{\frac{1}{\varepsilon}} \Delta_i^2 = O(m^{\frac{1}{\varepsilon}} n^2)$. The number $N = N(n, m, k)$ of terms in Equation 18.26 thus is at most $N = m \cdot \max_{i \in [m]} \Delta_i^{k_i} = O(n^2 m^{1+\frac{1}{\varepsilon}})$. By Theorem 18.10, we can construct $y \in \{0, 1\}^n$ using $O(N) = O(n^2 m^{1+\frac{1}{\varepsilon}})$ parallel processors in $O(k \log n \log N) = O\left(k \log n \cdot (\log n + \frac{1}{\varepsilon} \log m)\right) = O\left(\frac{1}{\varepsilon^2 \log \Delta_0} \log^2 n \log^2 m\right)$ time such that $F(y) \leq \mathbb{E}[F(X_1, \ldots, X_n)]$. The vector $x = (x_1, \ldots, x_n)$, where $x_i = (-1)^{y_i}$, defines a 2-coloring χ.

Let A_i be the event that $|\psi_i| \geq \lambda_i$, for $i \in [m]$. Then $A := \bigcup_{i \in [m]} A_i$ is the event in which there exists some i such that A_i holds. We fix an arbitrary i. According to Lemma 18.4, the k_ith moment inequality, which in fact is the Markov inequality, states that

$$\mathbb{P}[A_i] = \mathbb{P}[|\psi_i| \geq \lambda_i] \leq \lambda_i^{-k_i} \mathbb{E}[\psi_i^{k_i}]. \quad (18.27)$$

Let $t \in [l]$, $\vec{y}_t \in \{0, 1\}^t$ and with $\vec{Y}_t = (Y_1, \ldots, Y_t)$ consider the event $\vec{Y}_t = \vec{y}_t$. It is easily checked that Equation 18.27 is also true conditioned on $\vec{Y}_t = \vec{y}_t$:

$$\mathbb{P}[A_i \mid \vec{Y}_t = \vec{y}_t] = \mathbb{P}[|\psi_i| \geq \lambda_i \mid \vec{Y}_t = \vec{y}_t] \leq \lambda_i^{-k_i} \mathbb{E}[\psi_i^{k_i} \mid \vec{Y}_t = \vec{y}_t]. \quad (18.28)$$

Then,

$$\mathbb{P}[A \mid \vec{Y}_t = \vec{y}_t] \leq \sum_{i=1}^m \lambda_i^{-k_i} \mathbb{E}[\psi_i^{k_i} \mid \vec{Y}_t = \vec{y}_t] = \mathbb{E}[F(Z_1, \ldots, Z_n) \mid \vec{Y}_t = \vec{y}_t].$$

Hence,

$$\mathbb{P}[A \mid Y = y] \leq \mathbb{E}[F(Z_1, \ldots, Z_n) \mid Y = y] = F(y). \quad (18.29)$$

Since by Lemma 18.4, $\mathbb{E}[F(Z_1, \ldots, Z_n)] < 1$, Equation 18.29 and $F(y) \leq \mathbb{E}[F(Z_1, \ldots, Z_n)]$ yield $\mathbb{P}[A \mid Y = y] < 1$, so $\mathbb{P}[A \mid Y = y] = 0$, because $\mathbb{P}[A \mid Y = y] \in \{0, 1\}$. In other words, the 2-coloring defined by y has the claimed discrepancy $|\chi(E_i)| \leq \lambda_i$ for all $i \in [m]$. ∎

Bibliography and Remarks. Beck and Fiala [BF81] showed for a general hypergraph $\mathcal{H}$ that $\text{disc}(\mathcal{H}) \leq 2 \deg(\mathcal{H})$. A long-standing conjecture of great interest is whether a bound of $O(\sqrt{\deg(\mathcal{H})})$ is valid. Beck [Bec81] showed for any hypergraph $\mathcal{H}$ of degree t, $\text{disc}(\mathcal{H}) = O(\sqrt{t} \ln t \ln n)$. Srinivasan [Sri97] improved the bound of Beck to $O(\sqrt{t} \ln n)$ and gave better bounds for the lattice approximation problem (see Beck and Fiala [BF81] and Raghavan [Rag88] for bounds derived with Chernoff–Hoeffding type inequalities). The best-known bound $O(\sqrt{t \log n})$ is put forward by Banaszczyk [Ban98].

Combinatorial discrepancies in more than two colors were introduced by Doerr and Srivastav [DS99, DS01, DS03]. They show that most of the important results for two colors hold also in arbitrary numbers of colors. The c-color discrepancy, $c \geq 2$, of a hypergraph on n points with n hyperedges is at most $\sigma((u/c)(uc))$, and there are examples of hypergraphs with c-color discrepancy at least $\Omega(\sqrt{n/c})$. Parallel algorithms for the computation of multicolorings with low discrepancy are not known, but we are confident that this should be possible. They also observed (see also Doerr [Doe02b]) that there are hypergraphs having very different discrepancies in different numbers of colors. A deep result of Doerr [Doe02a, Doe04] shows that this dichotomy is structurally inherent. Further insight was given by Doerr et al. [DGH06] and recently Doerr and Fouz [DF06]. One particular application of multicolor discrepancies is the declustering problem in computer science, which can be modeled as a discrepancy problem of the hypergraph of rectangles in the higher-dimensional grid. Doerr et al. [DHW04] give some of the current best results in this direction.

Among hypergraphs of special interest is the hypergraph of arithmetic progressions, where $[N]$ is the node set and all arithmetic progressions in $[N]$ are the hyperedges. For the hypergraph AP of arithmetic progressions in the first N integers Roth [Rot64] proved $\text{disk}(AP) \geq cN^{(1/4)}$. In 1994, Matoušek and Spencer [MS96] showed $\text{disk}(AP) \leq c'N^{(1/4)}$, resolving the problem ($c, c' > 0$ are constants). For the hypergraph of cartesian products of arithmetic progressions AP^d in $[N]^d$ Doerr et al. [DSW04] proved $\text{disc}(AP^d) = \Theta(N^{(d/4)})$, with constants depending only on d. For extension of these results to multicolor discrepancy we refer to Doerr and Srivastav [DS03]. For the hypergraph of arithmetic progressions over $[N]$ the c-color discrepancy is at most $O(c^{-0.16} \sqrt[4]{N})$, and is at least $O(c^{-0.5} \sqrt[4]{N})$. Another problem dealing with cartesian products was studied by Matoušek [Mat00], who proved—answering a question of Beck and Chen—that the discrepancy of the family of cartesian products of circular discs in the plane in $\mathbb{R}^4$ is $O(n^{(1/4)+\varepsilon})$ for an arbitrarily small constant $\varepsilon > 0$, thus it is essentially the same as for circular discs in the plane. *But all these results lack for efficient parallel algorithms!*

A breakthrough for 2-coloring uniform hypergraphs was achieved by Beck in 1991 [Bec91] through the Lovász-Local-Lemma. Alon [Alo91], using Beck's idea, gave a probabilistic algorithm and derandomized it with the method of almost k-wise independence for hypergraphs of degree at most $2^{n/500}$.

18.3.5 Lattice Approximation

The lattice approximation problem is stated as follows.

18.3.5.1 Lattice Approximation Problem

Input An $(m \times n)$ matrix $A = (a_{ij})$, $a_{ij} \in [0, 1]$ for all entries, vector $p \in [0, 1]^n$.
Task Find a vector $q \in \{0, 1\}^n$ such that $\|Ap - Aq\|_\infty$ is minimum.

We call it *half lattice approximation problem* or *vector balancing problem* if all entries of p are $1/2$. For convenience we define $c := Ap$. The construction of an $\mathcal{NC}$ algorithm for the lattice approximation problem in Reference MNN94 uses the $\mathcal{NC}$ algorithm for the discrepancy problem. In the first step we consider the half lattice approximation problem and then generalize it with a bit-by-bit rounding technique.

All our algorithms make use of the binary representation of the entries in A and p. It is necessary to limit the precision, so that we have a binary representation with a known number of bits. We assume that each a_{ij} and each p_j, $i \in [m]$, $j \in [n]$ has binary encoding length $L' = L + 1$. If $A' = (a'_{ij})$ is the matrix where each a'_{ij} denotes the entry a_{ij} truncated off to $L + 1$ bits of precision, and p' is the truncated version of p, it is straightforward to see that

$$\max\{\|Ap - A'p'\|_\infty, \|A'q - Aq\|_\infty\} \leq 2n2^{-L'} = n2^{-L}. \tag{18.30}$$

Hence, to push this error below some $\delta > 0$, it is sufficient to choose $L = \lceil \log(n(1/\delta)) \rceil$. As long as we are only interested in bounds of the kind $\|Ap - Aq\|_\infty = O(f(m, n, \varepsilon))$, it is sufficient to choose $L = \Theta(\log n)$, because then

$$\max\{\|Ap - A'p'\|_\infty, \|A'q - Aq\|_\infty\} = O(1).$$

Parallel Algorithms via the Probabilistic Method

Thus, in the following we may indeed assume that

$$a_{ij} = \sum_{k=0}^{L} 2^{-k} \operatorname{bit}_k(a_{ij}) \text{ and } p_j = \sum_{k=0}^{L} 2^{-k} \operatorname{bit}_k(p_j) \text{ for all } i \in [m] \text{ and } j \in [n],$$

knowing that L can be chosen logarithmic in n/δ or just in n, depending on the application, to achieve useful approximations.

(a) Half-Lattice Approximation

Let A, p be an instance of the half lattice approximation problem. We show its equivalence to a discrepancy problem. Let B be the $((L'm) \times n)$ matrix over $\{0, 1\}$ drawn from A by replacing each a_{ij} by the 0/1 vector* $(\operatorname{bit}_0(a_{ij}), \ldots, \operatorname{bit}_L(a_{ij}))$. We denote by $\mathcal{H}_A = (V, \mathcal{E})$ the hypergraph on $|V| = n$ nodes and $|\mathcal{E}| = L'm$ hyperedges $E_1, \ldots, E_{L'm}$ whose hyperedge-vertex incidence matrix is B. Let $\chi : V \to \{-1, 1\}$ be a 2-coloring of $\mathcal{H}_A$ and $q \in \{0, 1\}^n$ be the vector with entries $q_j = \frac{1+\chi(j)}{2}$, $j \in [n]$. (Hence, q is a 0/1 representation of the coloring χ.)

To shorten notations, define

$$h := h(m, L', \Delta_0, \varepsilon) := \frac{\log(2L'm)}{\varepsilon \log \Delta_0} + 2, \tag{18.31}$$

which is part of the discrepancy bound of Theorem 18.13 for the given hypergraph $\mathcal{H}_A$.

Theorem 18.14 *Let $0 < \varepsilon \leq 1/3$ and as before let $\Delta_0 = \min_{i \in [m]} \Delta_i$, where $\Delta_i = \max \{|E_i|, 3\}$.*

(i) *If for all $E \in \mathcal{E}$, $|\chi(E)| \leq |E|^{(1/2)+\varepsilon} \sqrt{\frac{\log(2L'm)}{\varepsilon \log |E|} + 2}$, then for all $i \in [m]$:*

$$|(Ap - Aq)_i| = O\left(c_i^{(1/2)+\varepsilon} \sqrt{h}\right).$$

(ii) *There is a parallel algorithm for the half lattice approximation problem using at most $O(n^2 (L'm)^{1+\frac{1}{\varepsilon}})$ processors and running in $O\left(\frac{1}{\varepsilon^2 \log \Delta_0} \log^2 n \log^2(L'm)\right)$ time, achieving the approximation guarantee stated above.*

Proof outline. (i) With $Q_1 := \{j; \ q_j = 1\}$ we have

$$|(Bq - Bp)_i| = \left||E_i \cap Q_1| - \frac{|E_i|}{2}\right|$$

$$= \frac{1}{2}|\chi(E_i)|$$

$$\leq \frac{1}{2}|E_i|^{(1/2)+\varepsilon} \sqrt{\frac{\log(2L'm)}{\varepsilon \log \Delta_i} + 2} \quad \text{by assumption}$$

$$\leq \frac{1}{2}|E_i|^{(1/2)+\varepsilon} \sqrt{h}. \tag{18.32}$$

---

* Note about our notation: This is *not* something like "$\vec{\operatorname{bit}}_l(a_{ij})$."

Now, for $i \in [m]$:

$$c_i = \sum_{j=1}^{n} \frac{1}{2} a_{ij} = \sum_{k=0}^{L} 2^{-k} \sum_{j=1}^{n} \frac{1}{2} \text{bit}_k(a_{ij})$$

$$= \sum_{k=0}^{L} 2^{-k} (Bp)_{(i-1)L+k}$$

$$= \sum_{k=0}^{L} 2^{-k} \frac{1}{2} |E_{(i-1)L+k}|, \quad (18.33)$$

and

$$|(Ap - Aq)_i| = \left| \sum_{j=1}^{n} \frac{1}{2} a_{ij} - \sum_{j=1}^{n} a_{ij} q_j \right|$$

$$= \left| \sum_{j=1}^{n} \frac{1}{2} \sum_{k=0}^{L} 2^{-k} \text{bit}_k(a_{ij}) - \sum_{j=1}^{n} \sum_{k=0}^{L} 2^{-k} \text{bit}_k(a_{ij}) q_j \right|$$

$$= \left| \sum_{k=0}^{L} 2^{-k} \left(\sum_{j=1}^{n} \frac{1}{2} \text{bit}_k(a_{ij}) - \sum_{j=1}^{n} \text{bit}_k(a_{ij}) q_j \right) \right|$$

$$= \left| \sum_{k=0}^{L} 2^{-k} (Bp - Bq)_{(i-1)L+k} \right|$$

$$= \sum_{k=0}^{L} 2^{-k-1} |E_{(i-1)L+k}|^{(1/2)+\varepsilon} \sqrt{h} \quad \text{by Equation 18.32}$$

$$= \sqrt{h} \sum_{k=0}^{L} 2^{(k+1)(\varepsilon-\frac{1}{2})} \underbrace{\left(2^{-k-1} |E_{(i-1)L+k}| \right)^{(1/2)+\varepsilon}}_{\leq c_i \text{ by Equation 18.33}}$$

$$= O\left(c_i^{(1/2)+\varepsilon} \sqrt{h} \right),$$

as $\varepsilon \leq 1/3$ and so the geometric sum $\sum_{k=0}^{L} 2^{(k+1)(\varepsilon-\frac{1}{2})}$ is $O(1)$.

(ii) The construction of B can be done with $O(L'mn)$ processors in constant time. The computation of a coloring χ with discrepancy

$$|\chi(E_i)| \leq |E_i|^{(1/2)+\varepsilon} \sqrt{\frac{\log(2L'm)}{\varepsilon \log |E_i|}} + 2 \quad \text{for all } i \in [L'm]$$

can be done according to Theorem 18.13 within the claimed processor and time bounds. ∎

(b) Lattice Approximation

In the general problem the input is given by a $(m \times n)$ matrix $A \in [0, 1]^{m \times n}$ and a vector $p \in [0, 1]^n$. We describe the bit-by-bit randomized rounding. In the following algorithm, after k iterations, each p_j, $j \in [n]$ is represented by at most $L + 1 - k$ bits of precision.

Algorithm 6: BIT-RR.

Input: $p = (p_1, \ldots, p_n) \in [0, 1]^n$, with $L + 1$ bits of precision in each component.
Output: Lattice point in $\{0, 1\}^n$.

for $l \leftarrow L$ **downto** 1 **do**
    forall $j \in [n]$ **do**
        if $\text{bit}_l(p_j) = 1$ **then**
            with probability $\frac{1}{2}$ **do** $p_j \leftarrow p_j - 2^{-l}$;   /* round down */
        else $p_j \leftarrow p_j + 2^{-l}$;   /* round up */
        end
    end
end
return $\vec{\text{bit}}_0(p)$

It is clear that the algorithm terminates after at most L iterations with the desired lattice point, which is the result of successively rounding the bits in p, until at most the most significant bit can be set to 1. Reviewing this rounding process, it is straightforward to prove that the probability of rounding the jth entry of p to 1 is exactly p_j. We proceed to the parallel derandomization of the algorithm BIT-RR:

Algorithm 7: DERAND-BIT.

Input: $p = (p_1, \ldots, p_n) \in [0, 1]^n$ with $L + 1$ bits of precision in each component and parameter $0 < \varepsilon \leq \frac{1}{4}$.
Output: Lattice point in $\{0, 1\}^n$.

$P^{(L)} \leftarrow p$;
for $l \leftarrow L$ **downto** 1 **do**
    $q^{(l)} \leftarrow$ solution to half LAP with input $\frac{1}{2}\vec{\text{bit}}_l(P^{(l)})$ and ε;
    $P^{(l-1)} \leftarrow P^{(l)} + 2^{-(l-1)}(q^{(l)} - \frac{1}{2}\vec{\text{bit}}_l(P^{(l)}))$;
end
return $P^{(0)}$

Recall the definition and the role of h; see Equation 18.31.

A similar version of the next theorem is a main result [MNN94]. In the following formulation we state the approximation as well as processor and time bounds with all necessary parameters explicitly.

Theorem 18.15 *Let $q \in \{0, 1\}^n$ be the lattice point computed by* DERAND-BIT *and $i \in [m]$. There is a constant[*] $\gamma \geq 1$ such that the following holds.*

(i) *If $c_i \geq \gamma h^{1/(1-2\varepsilon)}$, then*

$$|(Ap - Aq)_i| = O\big(c_i^{(1/2)+\varepsilon}\sqrt{h}\big).$$

(ii) *If $c_i < \gamma h^{1/(1-2\varepsilon)}$, then*

$$|(Ap - Aq)_i| = O\big(h^{(1/1-2\varepsilon)}\big).$$

The algorithm uses $O\big(n^2(L'm)^{1+\frac{1}{\varepsilon}}\big)$ processors and runs in $O\left(\frac{L'}{\varepsilon^2 \log \Delta_0} \log^2 n \log^2(L'm)\right)$ time. (Δ_0 is taken for the hypergraph induced by A, as before. We always have $\log \Delta_0 > 1$, because $\Delta_0 \geq 3$.)

---

[*] That is, independent of the LAP instance and of ε.

Proof. Denote $C^{(l)} := AP^{(l)}$ for all $l \in \{0, \ldots, L\}$ and

$$D_i^{(l)} := \left|C_i^{(l+1)} - C_i^{(l)}\right| \quad \text{for all } i \in [m] \quad \text{and} \quad l \in \{0, \ldots, L-1\}.$$

So, we have in particular $C^{(L)} = c$, and $D_i^{(l)}$ denotes the error in the ith component introduced when rounding to l bits of precision (from $l+1$ bits). We can prove a bound on these errors, regardless of in which of the two cases *(i)* or *(ii)* we are.

Claim. There is a constant $\alpha > 0$ such that

$$D_i^{(l)} \leq \alpha \frac{\left(C_i^{(l+1)}\right)^{(1/2)+\varepsilon} \sqrt{h}}{\left(2^{(1/2)-\varepsilon}\right)^l} \quad \text{for all } l \in \{0, \ldots, L-1\}. \tag{18.34}$$

For the proof, fix $l \in \{0, \ldots, L-1\}$. We have

$$2^l C_i^{(l+1)} = 2^l (AP^{(l+1)})_i = 2^l \left(A \sum_{k=0}^{l+1} 2^{-k} \vec{\text{bit}}_k(P^{(l+1)})\right)_i = \sum_{k=0}^{l+1} 2^{l-k} \left(A \vec{\text{bit}}_k(P^{(l+1)})\right)_i$$

$$\geq 2^{l-(l+1)} \left(A \vec{\text{bit}}_{l+1}(P^{(l+1)})\right)_i = \left(A \frac{1}{2} \vec{\text{bit}}_{l+1}(P^{(l+1)})\right)_i. \tag{18.35}$$

Let α be the constant from the first $O(.)$ in Theorem 18.14, *(i)*. We get

$$D_i^{(l)} = \left|C_i^{(l+1)} - C_i^{(l)}\right| = \left|(AP^{(l+1)})_i - (AP^{(l)})_i\right|$$

$$= \left|(AP^{(l+1)})_i - \left(A\left(P^{(l+1)} + 2^{-l}\left(q^{(l+1)} - \frac{1}{2}\vec{\text{bit}}_{l+1}(P^{(l+1)})\right)\right)\right)_i\right|$$

$$= 2^{-l}\left|\left(A q^{(l+1)} - A \frac{1}{2}\vec{\text{bit}}_{l+1}(P^{(l+1)})\right)_i\right|$$

$$\leq 2^{-l} \alpha \cdot \left(A \frac{1}{2}\vec{\text{bit}}_{l+1}(P^{(l+1)})\right)_i^{(1/2)+\varepsilon} \sqrt{h} \quad \text{by Theorem 18.14}$$

$$\leq 2^{-l} \alpha \cdot \left(2^l C_i^{(l+1)}\right)^{(1/2)+\varepsilon} \sqrt{h} \quad \text{by Equation 18.35}$$

$$= \alpha \frac{\left(C_i^{(l+1)}\right)^{(1/2)+\varepsilon} \sqrt{h}}{(2^{(1/2)-\varepsilon})^l}.$$

This proves the claim.

We need β such that

$$\beta \geq \sum_{k=1}^{L} \frac{2\alpha}{\left(2^{(1/2)-\varepsilon}\right)^k}.$$

Because $\varepsilon \leq 1/4$, we have

$$\frac{1}{1 - 2^{-\frac{1}{4}}} \geq \sum_{k=1}^{L} \frac{1}{\left(2^{(1/4)}\right)^k} \geq \sum_{k=1}^{L} \frac{1}{\left(2^{(1/2)-\varepsilon}\right)^k},$$

Parallel Algorithms via the Probabilistic Method

and hence choosing $\beta := 2\alpha \frac{1}{1-2^{-\frac{1}{4}}}$, which is a constant, is sufficient. Define then

$$\gamma := \max\left\{\left(\frac{\beta^2}{2^{\frac{4}{3}}-1}\right)^2, 1\right\}.$$

Consider first the case of (i), that is, $c_i \geq \gamma h^{1/(1-2\varepsilon)}$. We do a slightly complicated induction from L down to 0. Since $C^{(l)} = c$, the following inequality holds trivially for $l = L$:

$$C_i^{(l)} \leq c_i + \beta c_i^{(1/2)+\varepsilon} \sqrt{h}. \tag{18.36}$$

We will show that if it holds for some $l \in [L]$, then it also holds for $l-1$. At the same time, we will prove a better bound on the errors $D_i^{(l)}$. Now fix $l \in [L]$ and assume that Equation 18.36 holds. We have by Equation 18.34 that

$$D_i^{(l-1)} \leq \alpha \frac{\left(C_i^{(l)}\right)^{(1/2)+\varepsilon} \sqrt{h}}{\left(2^{(1/2)-\varepsilon}\right)^{l-1}}$$

$$\leq \alpha \frac{\left(c_i + \beta c_i^{(1/2)+\varepsilon} \sqrt{h}\right)^{(1/2)+\varepsilon} \sqrt{h}}{\left(2^{(1/2)-\varepsilon}\right)^{l-1}} \quad \text{by Equation 18.36}$$

$$\leq \frac{2\alpha c_i^{(1/2)+\varepsilon} \sqrt{h}}{\left(2^{(1/2)-\varepsilon}\right)^{l-1}} \quad \begin{array}{l}\text{because } c_i \geq \gamma h^{1/(1-2\varepsilon)} \\ \text{and the def. of } \gamma.\end{array}$$

To verify the last inequality, see that for it to be true it is sufficient that $c_i^{(1/2)-\varepsilon} \geq \frac{\beta\sqrt{h}}{2^{2/(1+2\varepsilon)}-1}$ and the fact that γ was chosen in a way that this holds when $c_i \geq \gamma h^{1/(1-2\varepsilon)}$.

Consider now for some $r \in \{0, \ldots, L\}$ the following statement, of which we will show that it holds for all r:

$$C_i^{(r)} \leq c_i + \sum_{k=1}^{L-r} D^{(L-k)}$$

$$\leq c_i + \sum_{k=1}^{L-r} \frac{2\alpha c_i^{(1/2)+\varepsilon} \sqrt{h}}{\left(2^{(1/2)-\varepsilon}\right)^{L-k}}$$

$$= c_i + c_i^{(1/2)+\varepsilon} \sqrt{h} \sum_{k=1}^{L-r} \frac{2\alpha}{\left(2^{(1/2)-\varepsilon}\right)^{L-k}}$$

$$\leq c_i + \beta c_i^{(1/2)+\varepsilon} \sqrt{h}. \tag{18.37}$$

This is obviously true for $r = L$. To see that it is true also for $r = L-1$, we use that Equation 18.36 holds for $l = L$ and the analysis following Equation 18.36. Then, Equation 18.37 establishes Equation 18.36 for $l = L-1$. Hence, we can prove the above bound on $D_i^{(l-1)}$ for this l and gain Equation 18.37 for

$r = L - 2$. Proceeding in this manner gives Equation 18.37 for all r, in particular for $r = 0$. This gives

$$|(Ap - Aq)_i| \leq \sum_{k=1}^{L} D_i^{(L-k)} \leq \beta c_i^{(1/2)+\varepsilon} \sqrt{h}.$$

The theorem is proved for the first case.

Consider now the case of *(ii)*, that is, $c_i < \gamma h^{1/(1-2\varepsilon)}$. As long as the values $C_i^{(l)}$ stay below $\gamma h^{1/(1-2\varepsilon)}$, the total error (so far) does so as well. If they always stay below this value, we are done. Otherwise, let l_0 be maximal such that $C_i^{(l_0)} \geq \gamma h^{1/(1-2\varepsilon)}$. We can now apply the analysis from case *(i)*, however* with $C_i^{(l_0)}$ instead of c_i. It is therefore necessary to upper-bound $C_i^{(l_0)}$. Fortunately, because $C_i^{(l_0+1)} < \gamma h^{1/(1-2\varepsilon)}$, Equation 18.34 gives us

$$\left|C_i^{(l_0+1)} - C_i^{(l_0)}\right| = D_i^{(l_0)} \leq \alpha \frac{\left(C_i^{(l_0+1)}\right)^{(1/2)+\varepsilon} \sqrt{h}}{\left(2^{(1/2)-\varepsilon}\right)^{l_0}} < \underbrace{\alpha \gamma^{\frac{3}{4}} \frac{h^{1/(1-2\varepsilon)}}{\left(2^{(1/2)-\varepsilon}\right)^{l_0}}}_{=:\alpha'} \leq \alpha' h^{1/(1-2\varepsilon)},$$

using $\varepsilon \leq 1/4$ and the identity

$$\frac{1}{1-2\varepsilon} \cdot \left(\frac{1}{2} + \varepsilon\right) + \frac{1}{2} = \frac{1}{1-2\varepsilon}. \tag{18.38}$$

Hence $C_i^{(l_0)} \leq (1+\alpha')h^{1/(1-2\varepsilon)}$. Using that $(1+\alpha')^{(1/2)+\varepsilon} \leq (1+\alpha')^{(3/4)}$, the result for case *(i)* with Equation 18.38 finally proves the assertion for case *(ii)*.

The statement about processors and running time follows from Theorem 18.14 because we have L' iterations. ∎

Remark 18.3 If we choose $L' = \lceil \log n \rceil$, the number of processors in Theorems 18.14 and 18.15 is $O(n^2(L'm)^{1+\frac{1}{\varepsilon}}) = O(n^2(m \lceil \log n \rceil)^{1+\frac{1}{\varepsilon}})$, which is close to the estimation implicitly given in Reference MNN94. The running times become $O\left(\frac{1}{\varepsilon^2 \log \Delta_0} \log^2 n \log^2(m \log n)\right)$ and $O\left(\frac{1}{\varepsilon^2 \log \Delta_0} \log^3 n \log^2(m \log n)\right)$, respectively.

If the entries in A are small compared to the right-hand side c, say $a_{ij} \leq \delta c_i$, for a sufficiently small δ, by a scaling technique one can construct a lattice point q with a *relative* error of $O(\delta^{(1/2)-\eta}\sqrt{h})$, and there is even a good upper bound on the number of positive entries in q. This will later be used in the application of basis crashing, which is an important building block for automata fooling.

We write $\text{supp}(q) := \{j;\ q_j > 0\}$ and call it the *support* of the vector q. The cardinality of the support is denoted by $|q| := |\text{supp}(q)|$. The sum of all entries in a vector p is denoted by $\bar{p}$. We have

Theorem 18.16 *Let $A \in [0,1]^{m \times n}$, $p \in [0,1]^n$ be a lattice approximation instance, $c := Ap$, $c \in \mathbb{R}^m_{>0}$. Put $L := \lceil \log n \rceil$, $h := \frac{\log(2L'm)}{\eta \log \Delta_0} + 2$, $0 < \eta \leq \frac{1}{4}$, γ as in Theorem 18.15. Assume that there exists $\delta > 0$ such that*

- $\delta \leq \left(\gamma h^{\frac{1}{1-2\eta}} + 1\right)^{-1}$
- $a_{ij} \leq \delta c_i$ for all i

---
* This point was not clearly stated in the original proof.

Then we can construct in $\mathcal{NC}$ a vector $q \in \{0,1\}^n$ such that

$$(Aq)_i \in \left(1 \pm O\bigl(\delta^{(1/2)-\eta}\sqrt{h}\bigr)\right)(Ap)_i$$

and $|q| = \bar{p} + O(\bar{p}^{1/2+\eta}\sqrt{h}) + 1$. The construction can be done with $O\!\left(n^2 O(m \log n)^{1+\frac{1}{\eta}}\right)$ parallel processors in $O\!\left(\frac{1}{\eta^2}\log^3 n \log^2(m \log n)\right)$ time. The entries in A and p may be of any precision.

Proof. Scale each row i of A by $1/(\delta c_i)$. The resulting matrix $\tilde{A}$ is in $[0,1]^{m \times n}$ and $\tilde{c} := \tilde{A}p$ has the property that $\tilde{c}_i = \frac{1}{\delta}$ for all i. In addition, we augment $\tilde{A}$ by an extra row of 1s at the bottom.

Truncate the entries in $\tilde{A}$ and p to $L+1$ bits of precision, yielding matrix A' and vector p'. We know from Equation 18.30 that for all vectors q and all $i \in [m]$ with $c' := A'p'$

$$|(\tilde{c} - c')_i| = |(\tilde{A}p - A'p)_i| \leq 1 \quad \text{and} \quad |(\tilde{A}p - \tilde{A}q)_i| \leq |(A'p' - A'q)_i| + 2.$$

We have $c'_i \geq \tilde{c}_i - 1 = \frac{1}{\delta} - 1 \geq \gamma h^{1/(1-2\eta)}$ for all $i \in [m]$. Hence, we can use case *(i)* of Theorem 18.15 to solve the LAP instance defined by A' and p' with parameter η yielding a vector $q \in \{0,1\}$ with

$$|(A'p' - A'q)_i| = O\bigl(c_i'^{(1/2)+\eta}\sqrt{h}\bigr) = O\!\left(\left(\frac{1}{\delta}\right)^{(1/2)+\eta}\sqrt{h}\right) \quad \text{for all } i \in [m], \tag{18.39}$$

using that trivially $c'_i \leq \tilde{c}_i = 1/\delta$. The relative error for the original problem so is

$$\frac{|(Ap - Aq)_i|}{(Ap)_i} = \frac{(\delta c_i)^{-1}|(Ap - Aq)_i|}{(\delta c_i)^{-1}(Ap)_i}$$

$$= \frac{|(\tilde{A}p - \tilde{A}q)_i|}{\delta^{-1}}$$

$$\leq \delta \left(O\!\left(\left(\frac{1}{\delta}\right)^{(1/2)+\eta}\sqrt{h}\right) + 2\right)$$

$$= O\bigl(\delta^{(1/2)-\eta}\sqrt{h}\bigr) + 2\delta$$

$$= O\bigl(\delta^{(1/2)-\eta}\sqrt{h}\bigr).$$

For the support of q, consider the last entry in c'. For this $c'_{m+1} = \sum_{j=1}^n p'_j = \bar{p}'$ we know because of $c_1 = \sum_{j=1}^n a_{1j}p_j \leq \delta c_1 \sum_{j=1}^n p_j = \delta c_1 \bar{p}$ that we have $c'_{m+1} = \bar{p}' \geq \bar{p} - 1 \geq \frac{1}{\delta} - 1 \geq \gamma h^{1/(1-2\eta)}$. Hence, we can also use case *(i)* of Theorem 18.15 for this row of the matrix. We get $|\bar{p} - |q|| \leq |\bar{p}' - |q|| + 1 = |(A'p' - A'q)_{m+1}| + 1 = O(\bar{p}^{1/2+\eta}\sqrt{h}) + 1$. The claimed bound follows.

The bounds for processors and running time follow from Theorem 18.15. ∎

Bibliography and Remarks. Srinivasan [Sri97] gave better bounds for the lattice approximation problem (see Beck and Fiala [BF81] and Raghavan [Rag88] for bounds derived with Chernoff–Hoeffding type inequalities). The quadratic lattice approximation problem, a generalization of the (linear) lattice approximation problem, has been studied by Srivastav and Stangier [SS93], and a derandomized algorithm using Azuma's martingale was obtained. Here, parallel algorithms are not known.

Closely related to the lattice approximation problem is the concept of linear discrepancy of hypergraphs. Solving a long-standing open problem, Doerr [Doe01] showed that linear discrepancy of a unimodular hypergraph is at most $n/(n+1)$. This problem is a particular case of another open problem, namely, the relation of hereditary and linear discrepancies of a hypergraph. The classical result put forth by Beck and Spencer [BS84] as well as Lovász et al. [LSV86] is $\text{disc}(\mathcal{H}) \leq 2\text{herdisc}(\mathcal{H})$. The latter paper also gives a class of hypergraphs fulfilling $\text{disc}(\mathcal{H}) = 2n/(n+1)\text{herdisc}(\mathcal{H})$ and conjectures that this should be the right factor. Doerr [Doe00] showed $\text{disc}(\mathcal{H}) \leq 2(2m-1)/(2m)\text{herdisc}(\mathcal{H})$. This is still the best upper bound known for this problem.

18.4 Fooling Automata

Many randomized algorithms can be modeled by a set of deterministic finite automata. If the input strings for these automata are drawn randomly according to a certain distribution ζ, the automata simulate the randomized algorithm as it works on some fixed input—the automata themselves may depend on the input to the algorithm. Such a representation of an algorithm allows a special kind of derandomization. The idea is to modify the distribution ζ to a second distribution ρ such that the size of the support, that is, the number of strings with nonzero probability, becomes polynomial, and that the automata approximately still simulate the algorithm. This is called *automata fooling*, because the automata are presented with a different distribution but still behave essentially as before. If the support of ρ is sufficiently small, all strings can be checked efficiently (in parallel), and hence we have a deterministic algorithm. We present the automata fooling technique from the paper of Karger and Koller [KK97].

Automata are represented by regular, directed multigraphs.

Definition 18.5 *An automaton $\mathcal{A}$ is a triple (V, E, s), where V is the set of states (the nodes or vertices of the graph) and E are labeled edges, which define possible transitions between two states, and $s \in V$ is a special state called the starting state of the automaton. Each state has outgoing edges labeled from 1 to r for some fixed number r. (In practice, automata will have ending states, from where there are no outgoing edges. To fit with this definition, for such a node v simply think of r loops (v, v).) Inputs for $\mathcal{A}$ are strings from $[r]^*$, that is, concatenations of the numbers from 1 to r.*

Giving to $\mathcal{A}$ a word $w = (w_1, w_2, \ldots, w_l) \in [r]^*$ as input means that starting from s, the automaton changes its state by traversing the edges as specified by w. In other words, w defines a walk in the graph (V, E) starting at node s.

Throughout this section, we fix a set $\mathcal{A}_1, \ldots, \mathcal{A}_m$ of m automata, write $\mathcal{A}_i = (V_i, E_i, s_i)$ for each $i \in [m]$, each of them taking inputs from the same set $[r]^*$.

Let $X_1, \ldots, X_n$ be independent random variables each taking values in $[r]$. They define a distribution on $W := [r]^n$. We denote this distribution by $\zeta : W \to [0, 1]$, that is, $\zeta(w)$ is the probability that $(X_1, \ldots, X_n)$ takes on the value $w \in W$. (This probability is equal to $\mathbb{P}[X_1 = w_1] \cdot \ldots \cdot \mathbb{P}[X_n = w_n]$.)

The automata $\mathcal{A}_1, \ldots, \mathcal{A}_m$ when provided with inputs drawn from W according to ζ are assumed to simulate some randomized algorithm **A** under consideration. The final states of the automata correspond to the output of the algorithm. The automata themselves may depend on the input for the algorithm **A**. We fix an input instance $\mathcal{I}$ for our presentation. It is assumed that the automata have polynomially many states, in the length of $\mathcal{I}$.

Definition 18.6 *The support of a distribution ξ on a set U is*

$$\text{supp}(\xi) := \{w \in U;\ \xi(w) > 0\}.$$

For the cardinality of the support, also called the size of the distribution, we write

$$|\xi| := |\text{supp}(\xi)|.$$

If we know that with positive probability (w.r.t. ζ), the algorithm (simulated by the automata) outputs a good solution,* it suffices to check the output of the automata for every $w \in \text{supp}(\zeta)$. However, this

---

* For example, a solution approximating some objective function within a certain margin.

Parallel Algorithms via the Probabilistic Method

set may be too large. If, on the other hand, we are able to replace ζ by some distribution ρ such that supp(ρ) is small and the automata still (approximately) simulate the algorithm, we can check every word in supp(ρ) instead. This can be done in parallel.

There are methods to reduce the support of a distribution while maintaining the probabilities of certain events. The challenge is that in general, the running times of these algorithms depend (polynomially) on the size of the original support. The solution to this is a divide-and-conquer approach. It will be described in the following. First, we look at the problem of reducing the support of a given distribution directly, and then consider intermediate transition probabilities, which help in the design of the divide-and-conquer technique.

Reducing Support Let ξ be a distribution and $B_1, \ldots, B_k$ events (e.g., transitions in the automata). The events are also called *constraints* in this context, and we denote the set of all constraints by $\mathcal{C}$.

For every $\varepsilon > 0$, by $(1 \pm \varepsilon)$ we denote the interval $(1 - \varepsilon, 1 + \varepsilon)$. Consequently, for a real number x, by $(1 \pm \varepsilon)x$ the interval $(x - \varepsilon x, x + \varepsilon x)$ is denoted.

Definition 18.7 *Let $1 > \varepsilon \geq 0$. Given a distribution ξ, a distribution ξ' is said to absolute ε-fool constraints $B_1, \ldots, B_k$, if*

$$|\mathbb{P}_{\xi'}[B_j] - \mathbb{P}_{\xi}[B_j]| < \varepsilon \quad \text{for all } j \in [k].$$

It is said to relative ε-fool constraints $B_1, \ldots, B_k$, if

$$\mathbb{P}_{\xi'}[B_j] \in (1 \pm \varepsilon)\mathbb{P}_{\xi}[B_j] \quad \text{for all } j \in [k].$$

Obviously, because we are dealing with probabilities, which are in $[0, 1]$, relative ε-fooling is stronger than absolute ε-fooling. In Reference KK97, only relative ε-fooling is considered. We include absolute ε-fooling in our analysis, because even in this case, we get interesting results, and there may be algorithms for reducing support with an absolute error that require fewer work (processors and time) than those known for relative fooling.

One technique for reducing the support of a distribution so that the resulting one absolute or relative ε-fools a given set of constraints is known as *basis crashing*. We speak of *absolute approximate basis crashing* or *relative approximate basis crashing* depending on the kind of error we intend to achieve. We speak of *basis crashing* if $\varepsilon = 0$ (in which case relative and absolute error coincides). Basis crashing generally works by expressing the constraints as a set of linear equations and then looking for a basic solution, which, by the properties of a basic solution, leads to a support of cardinality $\leq k$ (see [KK97, Thm. 2.2]). To our knowledge, the best algorithm for basis crashing ($\varepsilon = 0$) is given by Beling and Megiddo [BM98] and runs in $O(k^{1.62}|\xi|)$. Hence, if ξ has a large support (e.g., exponential in the input length), this technique alone does not yield a polynomial (or an $\mathcal{NC}$ algorithm) for the original problem.

As a building block for the devide-and-conquer approach, it suffices, however, to have an $\mathcal{NC}$ implementation for the reduction of the support, even if the number of processors and running time depend (polynomially and logarithmically, respectively) on the size of the original distribution, provided that the support is reduced strongly enough.

We will give an algorithm for relative approximate basis crashing. (This, of course, yields one for absolute basis crashing.) However, we will also show how the devide-and-conquer algorithm works if only an algorithm for absolute approximate basis crashing is available.

The nice idea presented by Reference KK97 is to formulate relative approximate basis crashing as a lattice approximation problem. The latter can then be solved using the parallel algorithm from Theorem 18.16. Consider some sample space Ω (e.g., the set of all words W). The given distribution is a mapping $\xi : \Omega \to [0, 1]$ and the constraints are subsets of Ω, that is, $B_i \subseteq \Omega$ for all $i \in [k]$. Denote the elements of the support supp(ξ) = $\{\omega_1, \ldots, \omega_S\}$ and consider the constraints as subsets of supp(ξ). If not already the case, we include the event B_0 consisting of all the elements from $\mathrm{supp}(\xi)$ in the set of constraints. This will help in ensuring that we really get a distribution out of our construction. So we may have in fact $k+1$ constraints. We have to pay attention here, because the term $O\left((k+1)^{1+\frac{1}{\eta}}\right)$, later occuring in the processor bounds, is not $O\left(k^{1+\frac{1}{\eta}}\right)$, unless η is appropriately lower bounded.

We use the algorithm from Theorem 18.16. Set $\varepsilon' := \frac{\varepsilon}{3}$, fix some $0 < \eta \leq 1/4$ and choose δ such that $\delta \leq (\varepsilon'^2 h^{-1})^{\frac{1}{1-2\eta}}$ and δ is small enough as required in the theorem. Reviewing the requirement of the theorem, we see that if ε is small enough compared to γ^{-1} (which we may assume), a setting of δ such that

$$\frac{1}{\delta} = (\varepsilon'^{-2} h)^{1/(1-2\eta)} = O\left(\left(\frac{h}{\varepsilon^2}\right)^{1/(1-2\eta)}\right) \tag{18.40}$$

suffices.

We construct the lattice approximation instance. The contraint-sample matrix associated to $(B_i)_{i=1,\ldots,k}$ is

$$\tilde{A} = (\tilde{a}_{ij})_{\substack{i=1,\ldots,k \\ j=1,\ldots,S}}, \quad \text{where } \tilde{a}_{ij} := \begin{cases} 1 & \text{if } \omega_j \in B_k \\ 0 & \text{else.} \end{cases}$$

Define a vector $b \in [0,1]^k$ by

$$b_i := \mathbb{P}_\xi[B_i] \quad \text{for all } i \in [k],$$

and vectors $r, p \in [0,1]^S$ by

$$r_j := \xi(\omega_j) \quad \text{for all } j \in [S],$$

$$p_j := \min\left\{\max_{i \in [k]} \frac{\tilde{a}_{ij} r_j}{\delta b_i}, 1\right\} \quad \text{for all } j \in [S],$$

and finally set

$$A = (a_{ij})_{\substack{i=1,\ldots,k \\ j=1,\ldots,S}}, \quad \text{where } a_{ij} := \frac{\tilde{a}_{ij} r_j}{p_j} \quad i \in [k], j \in [S].$$

The lattice approximation problem is defined by A and p, meaning that $c := Ap$ is the vector of constraint probabilities b. By the algorithm from Theorem 18.16 we get a vector $q \in \{0,1\}^S$ with a support bounded as in the theorem such that $|(Ap - Aq)_i| \leq \varepsilon'(Ap)_i$, that is,

$$(Aq)_i \in (1 \pm \varepsilon')(Ap)_i \quad \text{for all } i \in [k]. \tag{18.41}$$

Define $\tilde{\xi}'(\omega_j) := \frac{r_j}{p_j} q_j$ for all $j \in [S]$ and $\lambda := \sum_{j \in [S]} \tilde{\xi}'(\omega_j)$. The following setting yields a distribution:

$$\xi'(\omega) := \begin{cases} \frac{1}{\lambda}\tilde{\xi}'(\omega) & \text{if } \omega \in \{\omega_1, \ldots, \omega_S\} \\ 0 & \text{else} \end{cases} \quad \text{for all } \omega \in \Omega, \tag{18.42}$$

Parallel Algorithms via the Probabilistic Method

which is not far from the "pseudo" distribution $\tilde{\xi}'$, because, by the assumption that the event B_0 is among the constraints, we have

$$\lambda \in (1 \pm \varepsilon'). \tag{18.43}$$

Theorem 18.17 *Let ξ be a distribution, $0 < \varepsilon < 1$, and consider constraints (i.e., events) $B_1, \ldots, B_k$. A distribution ξ' can be constructed such that the given contraints are relative ε-fooled and*

$$|\xi'| = O\left(k \left(\frac{\log(k \log S)}{\eta \varepsilon^2}\right)^{1/(1-2\eta)}\right).$$

The construction can be done with $O\left(S^2 O(k \log S)^{1+\frac{1}{\eta}}\right)$ processors in time

$$O\left(\frac{1}{\eta^2} \log^3 S \log^2(k \log S)\right).$$

Proof. Let ξ' be as in Equation 18.42. By Equation 18.41, we see that the constraints are relative ε-fooled by ξ': for $i \in [k]$ we have $(Ap)_i = b_i = \mathbb{P}_\xi[B_i]$ and $(Aq)_i = \lambda \mathbb{P}_{\xi'}[B_i]$. Hence, $\lambda \mathbb{P}_{\xi'}[B_i] \in (1 \pm \varepsilon') \mathbb{P}_\xi[B_i]$. The choice of ε' and Equation 18.43 give the desired $\mathbb{P}_{\xi'}[B_i] \in (1 \pm \varepsilon) \mathbb{P}_\xi[B_i]$.

We proceed to bound the support size of ξ', which is the same as the support size of q. We know from Theorem 18.16 that $|q| \leq \bar{p} + O(\bar{p}^{(1/2)+\eta} \sqrt{h}) + 1$. Hence, we need an upper bound on $\bar{p}$. We say that $i \in [k]$ ownes $j \in [S]$ if the term $\frac{\tilde{a}_{ij} r_j}{\delta b_i}$ is maximal (over all possible i). Fix by own(j) an arbitrary i which owns j. We then have

$$\bar{p} = \sum_{j \in [S]} p_j = \sum_{i \in [k]} \sum_{\substack{j \in [S] \\ \text{s.t. } i = \text{own}(j)}} p_j \leq \sum_{i \in [k]} \sum_{\substack{j \in [S] \\ \text{s.t. } i = \text{own}(j)}} \frac{\tilde{a}_{ij} r_j}{\delta b_i}$$

$$= \sum_{i \in [k]} \frac{1}{\delta b_i} \sum_{\substack{j \in [S] \\ \text{s.t. } i = \text{own}(j)}} \tilde{a}_{ij} r_j \leq \sum_{i \in [k]} \frac{1}{\delta b_i} b_i = \sum_{i \in [k]} \frac{1}{\delta} = \frac{k}{\delta}.$$

The claimed bound for the support follows from a calculation using Equation 18.40 and the definition of h from Theorem 18.16.

The bounds on the number of processors and running time follow directly from Theorem 18.16. ∎

Remark 18.4 Karger and Koller [KK97, Th. 4.6] state a bound of $O\left(\frac{k \log k}{\varepsilon^2}\right)$ for the support size. We cannot see how to establish this bound for the following reasons. Our bound $O\left(k \left(\frac{\log(k \log S)}{\eta \varepsilon^2}\right)^{1/(1-2\eta)}\right)$ in the previous theorem relies on our rigorous analysis of the discrepancy algorithm, Theorem 18.13. The error parameter ε in Theorem 18.13, which is η here, can be found in the exponent as well as in the denominator (under the square root) in the discrepancy bound of our version of the theorem, while it only appears in the exponent in Reference MNN94. This factor of $1/\varepsilon$ persists also in the bounds for the lattice approximation, and thus also must appear in the bounds for the support size (at first hidden in the quantity h). Hence one cannot simply substitute η by $o(1)$, as it is done in Reference [KK97], where presumably the discrepancy result presented in Reference MNN94 was used, which did not show the factor of $1/\varepsilon$ (which is $\frac{1}{\eta}$ here). In any case putting $\eta = o(1)$ is questionable, because the number of processors depends exponentially on $1/\eta$, and the running time depends on it polynomially. Our analysis of the discrepancy algorithm also introduces a dependence on the number of variables, which in this application is the size of the support, since we cannot assume "$\Theta(\log n) = \Theta(\log m)$" in general. However, for $\eta = \frac{1}{16}$ we get the bound $O\left(k\frac{\log^{1.15}(k\log S)}{\varepsilon^{2.3}}\right)$. We have not optimized η here, so there might be room for some improvements, which we expect to be minor. Note that the constant hidden in the $O(.)$ notation of this bound goes to infinity as η goes to zero.

In the following, we will assume that some deterministic $\mathcal{NC}$ algorithm RED is given that can be used to reduce the support of a distribution while fooling a set of constraints. For example, we may use the basis crashing algorithm from Theorem 18.17.

Denote the output distribution of RED by $\text{RED}(\xi, \mathcal{C}, \varepsilon)$. The number of processors used is denoted by $\text{RED}_{\text{proc}}(|\xi|, k, \varepsilon)$ and the running time by $\text{RED}_{\text{time}}(|\xi|, k, \varepsilon)$. Define

$$\text{RED}_{\text{ratio}}(S, k, \varepsilon) := \sup_{\substack{\xi, \mathcal{C} \\ \text{s.t. } |\xi|=S \\ \text{and } |\mathcal{C}|=k}} \frac{|\text{RED}(\xi, \mathcal{C}, \varepsilon)|}{k}.$$

We assume RED_{proc}, RED_{time}, and $\text{RED}_{\text{ratio}}$ to be nondecreasing in the first two arguments, and nonincreasing in the third argument. All these quantities are dependent on the particular choice for RED. If we choose the algorithm from Theorem 18.17 as RED, we have $\text{RED}_{\text{ratio}}(S, k, \varepsilon) = O\left(\left(\frac{\log(k\log S)}{\eta\varepsilon^2}\right)^{1/(1-2\eta)}\right)$, while for the choice of the exact algorithm of Beling and Megiddo, it would be $\leq 1 + o(1)$.

For the algorithm RED to be efficient, $\text{RED}_{\text{ratio}}$ must be small enough and appropriately dependent on S, so that the size of the support behaves well under a certain iteration. To explain this, fix k and ε and define

$$f(S) := k \cdot \text{RED}_{\text{ratio}}(S^2, k, \varepsilon). \tag{18.44}$$

The following scheme will later be realized in Algorithm 18.4. Think of some distribution of size S_0. It is modified such that its support grows to at most S_0^2. Its support is then reduced to size S_1 using RED, and then it is modified again so that its support grows to at most S_1^2. Then it is treated with RED again, and so on. Doing so t times by definition never (meaning: in no step during the iteration) gives a distribution of size larger than

$$\text{RED}_{\text{iter}}^t(S_0, k, \varepsilon) := \max\{\underbrace{f(f(f...f(S_0)))}_{t' \text{ times}}; \ t' \leq t\}. \tag{18.45}$$

For fixed t, $\text{RED}_{\text{iter}}^t(S_0, k, \varepsilon)$ is nondecreasing in its first two arguments and nonincreasing in the third.

In Corollary 18.4, we will give an analysis of the iteration behavior of the algorithm from Theorem 18.17, showing that the support grows not too much.

Transition Probabilities

Let $i \in [m]$ and $a, b \in V_i$. Denote by $\mathbb{P}_\xi[a \to_i b]$ the probability that starting from state a, the automaton $\mathcal{A}_i$ will reach state b when it is given as input a word $w \in W$ generated according to distribution ξ. We call this the *transition probability* of i from a to b under ξ.

The transition probabilities where a is the initial state are the important properties of the automata. They are to be maintained approximately during the construction of a new distribution.

For $l \leq h \leq n$ we need the *intermediate transition probability* $\mathbb{P}_\xi[a \to_i b]^{l,h}$ defined as the probability for a transition from a to b when the input is generated according to the random variables $X_l, \ldots, X_h$ only. The extreme cases are $l = h$, where only one random variable is used, and $l = 1$ and $h = n$, where all random variables are used. We introduce the notation $W^{l,h} := [r]^{h-l+1}$.

Parallel Algorithms via the Probabilistic Method

A simple but important observation is that we can construct intermediate transition probabilities for larger words from smaller ones. If $l \leq g \leq h$, then

$$\mathbb{P}_\xi [a \to_i b]^{l,h} = \sum_{v \in V_i} \mathbb{P}_\xi [a \to_i v]^{l,g} \cdot \mathbb{P}_\xi [v \to_i b]^{g+1,h}. \tag{18.46}$$

Note that this extends to the case when ξ is *any* distribution with the property that

$$\xi(w_l, \ldots, w_h) = \xi(w_l, \ldots, w_g) \cdot \xi(w_{g+1}, \ldots, w_h), \quad \forall (w_l, \ldots, w_h) \in W^{l,h}. \tag{18.47}$$

Definition 18.8 *Let $\varepsilon \geq 0$. A distribution ρ is said to absolute ε-fool a set $\mathcal{A}_1, \ldots, \mathcal{A}_m$ of automata w.r.t. a distribution ζ, if*

$$|\mathbb{P}_\rho [s_i \to_i v] - \mathbb{P}_\zeta [s_i \to_i v]| < \varepsilon, \quad \forall v \in V_i, \quad \forall i \in [m].$$

It is said to relative ε-fool the automata if

$$\mathbb{P}_\rho [s_i \to_i v] \in (1 \pm \varepsilon) \mathbb{P}_\zeta [s_i \to_i v], \quad \forall v \in V_i, \quad \forall i \in [m].$$

In other words, it ε-fools the automata if it ε-fools all the relevant transition events. (The devide-and-conquer algorithm will, as a side-effect, ε-fool all transition events, not only those starting at s_i.)

Divide-and-Conquer Algorithm

We present an algorithm that constructs a distribution that ε-fools a given set of automata w.r.t. a given distribution. The algorithm works for absolute as well as for relative errors, provided that an adequate algorithm RED is given. The analysis for the case of relative error is taken from Reference KK97. We have added the analysis in the case of an absolute error, which requires some more work.

For each $1 \leq h \leq n$ we denote by $\mathcal{C}^{l,h}$ the set of constraints consisting of transition events of all automata under inputs generated according to $X_l, \ldots, X_h$. We have

$$\left|\mathcal{C}^{l,h}\right| \leq \sum_{i \in [m]} |V_i|^2 =: k. \tag{18.48}$$

Note that k is polynomial in the length of the input instance $\mathcal{I}$, because the automata by assumption have only polynomially many states.

Define $V_{\max} := \max_{i \in [m]} |V_i|$ and

$$\epsilon(\varepsilon) := \begin{cases} \min\left\{\frac{\varepsilon}{4}, \left(\frac{\varepsilon}{4V_{\max}}\right)^{\frac{1}{2}}\right\} & \text{for an absolute error} \\ \frac{\varepsilon}{4} & \text{for a relative error.} \end{cases}$$

The algorithm has to use smaller and smaller ε-values in each iteration. The function ϵ describes how these values decrease. It is crucial that the reciprocal of the ε-values stays polynomial. That is why we first take a look at the behavior of ϵ under iteration. Define for each $t \in \mathbb{N}$ the tth iterate by $\epsilon^{(t)}(\varepsilon) := \epsilon(\epsilon^{(t-1)}(\varepsilon))$ and $\epsilon^{(0)}(\varepsilon) := \varepsilon$.

Proposition 18.4 *Let $\varepsilon > 0$ and write $V := 4V_{\max}$. Then $\epsilon^{(\lceil \log \bar{n} \rceil)}(\varepsilon) \geq \frac{\varepsilon}{4n^2 V}$ in the case of an absolute error and $\epsilon^{(\lceil \log \bar{n} \rceil)}(\varepsilon) \geq \frac{\varepsilon}{4n^2}$ in the case of a relative error, hence the reciprocal is polynomial in the input length of $\mathcal{I}$.*

Proof. The case of relative error is easy; there we have $\epsilon^{(\lceil \log \bar{n} \rceil)}(\varepsilon) = \frac{\varepsilon}{4^{\lceil \log n \rceil}} \geq \frac{\varepsilon}{4n^2}$.

Now consider the absolute error. By the assumption that the automata have polynomially many states, V is polynomial. Assume that starting from some value ε_1, for t_1 iterations, the minimum is always attained by the first expression in the definition of ϵ. Then it is, for some t_2 iterations, attained by the second expression. For some t_3 iterations, it is again assumed by the first one, and so on. We have a sequence of numbers $t_1, \ldots, t_r$ such that $\sum_{s=1}^{r} t_s = \lceil \log n \rceil$, and the numbers with an odd index mark periods where the first expression matters and those with an even index mark periods where the second expression matters. A tedious calculation—the exponent of V is already significantly simplified in the second term of the following equation—shows that

$$\epsilon^{(\lceil \log n \rceil)}(\varepsilon) \geq \varepsilon^{(\frac{1}{2})^{\sum_{s \leq r} t_s}} 4^{-\sum_{s \leq r} t_s} V^{-\sum_{s=1}^{\lceil \log n \rceil} \frac{1}{2^s}} \geq \varepsilon 4^{-\lceil \log n \rceil} V^{-1} \geq \frac{\varepsilon}{4n^2 V}.$$ ∎

Algorithm 8: FOOL.

Input: Random variables $X_l, \ldots, X_h$ (with indices), parameter ε.
Output: Distribution ρ that absolute or relative ε-fools the automata, depending an ϵ and RED.

1 **if** $l = h$ **then**
2   **forall** $w \in W^{l,h}$ **do**   /* note that here $W^{l,h} = [r]$ */
3         $\tilde{\rho}(w) \leftarrow \mathbb{P}[X_l = w]$;
4   **end**
5 **end**
6 **else**
7     $g \leftarrow \lceil \frac{l+h}{2} \rceil$;
8     $\rho' \leftarrow \text{FOOL}(X_l, \ldots, X_g, \epsilon(\varepsilon))$ **in parallel with** $\rho'' \leftarrow \text{FOOL}(X_{g+1}, \ldots, X_h, \epsilon(\varepsilon))$;
9   **in parallel forall** $(w_l, \ldots, w_h) \in W^{l,h}$
10   such that $\rho'(w_l, \ldots, w_g) > 0$ and $\rho''(w_{g+1}, \ldots, w_h) > 0$ **do**
11         $\tilde{\rho}(w_l, \ldots, w_h) \leftarrow \rho'(w_l, \ldots, w_g) \cdot \rho''(w_{g-1}, \ldots, w_h)$;
12   **end**
13     $\rho \leftarrow \text{RED}(\tilde{\rho}, \mathcal{C}^{l,h}, \varepsilon)$;
14 **end**
15 **return** ρ;

The following tree illustrates how FOOL works for eight random variables $X_1, \ldots, X_8$. At the bottom, distributions are constructed for each random variable; this is what happens in lines 1 to 5. Moving upward in the tree means, for each branch, to "merge" two distributions; this is what happens in line 11.

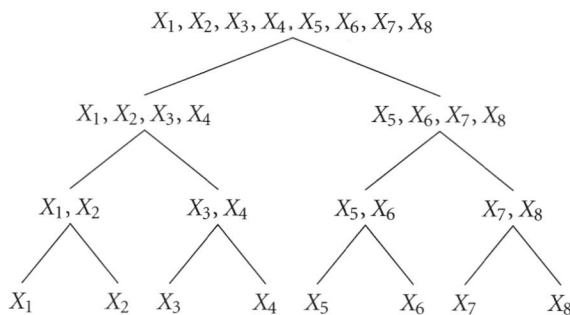

The depth of the tree is $\log n$. Note that without the calls to RED in line 13, the size of the support could be $r^{2^{\log n}} = r^n$ at the top, which is superpolynomial hence making it impractical to reduce it by just one final call to RED.

Theorem 18.18 *Let $1 > \varepsilon > 0$ and set $\varepsilon_{\min} := \epsilon^{(\lceil \log n \rceil)}(\varepsilon)$.*

(i) The algorithm FOOL *constructs a distribution ρ that ε-fools the automata and has size at most $S_{\max} := \text{RED}_{\text{iter}}^{\lceil \log n \rceil}(r, k, \varepsilon_{\min})$.*

(ii) It runs using

$$O(n \cdot \max\{O(S_{\max}^2), \text{RED}_{\text{proc}}(S_{\max}, k, \varepsilon_{\min})\})$$

parallel processors in time

$$O(\log n \cdot \text{RED}_{\text{time}}(S_{\max}, k, \varepsilon_{\min})).$$

Because RED *is an $\mathcal{NC}$ algorithm,* FOOL *is so as well – as long as $\text{RED}_{\text{iter}}^{\lceil \log n \rceil}$ is polynomial in its three arguments.*

Proof. *(i)* The recursion depth is $\lceil \log n \rceil$. The statement about the size of the final distribution is clear by Equation 18.45, noting that the treatment of the distribution in line 11 is expressed by taking the size of the support to the power of 2 in Equation 18.44, that the size of the starting distributions is bounded by r, and that $\text{RED}_{\text{iter}}^{\lceil \log n \rceil}$ is assumed to be nonincreasing in its third argument. So, we can argue with $\varepsilon_{\min}$ in $\text{RED}_{\text{iter}}^{\lceil \log n \rceil}(r, k, \varepsilon_{\min})$ and do not need to consider the intermediate values of "ε."

We now show by induction on $h - l$ that the algorithm has the required fooling property. If $h - l = 0$, this is easily seen. The distribution $\tilde{\rho}$ constructed in lines 2–4 exactly reflects the transition probabilities (of length one) for the random variable in question.

First consider an absolute error, let $h - l > 0$ and assume that the algorithm works correctly on smaller inputs. This means that ρ' and ρ'' have the property to absolute $\epsilon(\varepsilon)$-fool the (intermediate) transition events for $X_l, \ldots, X_g$ and $X_{g+1}, \ldots, X_h$, respectively. The distribution $\tilde{\rho}$ defined by the multiplications in line 11 by this very definition has the property given in Equation 18.47. Hence, we can use Equation 18.46. This yields for all $i \in [m]$ and $a, b \in V_i$

$$\mathbb{P}_{\tilde{\rho}}[a \to_i b] = \sum_{v \in V_i} \mathbb{P}_{\rho'}[a \to_i v] \cdot \mathbb{P}_{\rho''}[v \to_i b]$$

$$\leq \sum_{v \in V_i} (\mathbb{P}_\zeta[a \to_i v]^{l,g} + \epsilon(\varepsilon)) \cdot (\mathbb{P}_\zeta[v \to_i b]^{g+1,h} + \epsilon(\varepsilon))$$

$$\leq \left(\sum_{v \in V_i} \mathbb{P}_\zeta[a \to_i v]^{l,g} \cdot \mathbb{P}_\zeta[v \to_i b]^{g+1,h} \right)$$
$$+ \epsilon(\varepsilon) \underbrace{\sum_{v \in V_i} \mathbb{P}_\zeta[a \to_i v]^{l,g}}_{=1} + \epsilon(\varepsilon) \underbrace{\sum_{v \in V_i} \mathbb{P}_\zeta[v \to_i b]^{g+1,h}}_{=1} + |V_i| \epsilon(\varepsilon)^2$$

$$\leq \mathbb{P}_\zeta[a \to_i b]^{l,h} + 2\frac{\varepsilon}{4} + |V_i| \frac{\varepsilon}{4V_{\max}}$$

$$= \mathbb{P}_\zeta[a \to_i b]^{l,h} + \frac{3}{4}\varepsilon.$$

A corresponding lower bound can be established in the same way. The call to RED introduces another absolute error of at most $\varepsilon/4$, resulting in the distribution ρ to absolute ε-fool the (intermediate) transition events.

Consider now the case of a relative error. The distributions ρ' and ρ'' have the property to relative $\epsilon(\varepsilon)$-fool the (intermediate) transition events for $X_l, \ldots, X_g$ and $X_{g+1}, \ldots, X_h$, respectively, where $\epsilon(\varepsilon) = \varepsilon/4$.

The distribution $\tilde{\rho}$ defined by the multiplications in line 11 by this very definition has the property given in Equation 18.47. Hence, we can use Equation 18.46. This yields for all $i \in [m]$ and $a, b \in V_i$

$$\mathbb{P}_{\tilde{\rho}}[a \to_i b] = \sum_{v \in V_i} \underbrace{\mathbb{P}_{\rho'}[a \to_i v]}_{\in (1 \pm \frac{\varepsilon}{4})\mathbb{P}_\zeta[a \to_i v]^{l,g}} \cdot \underbrace{\mathbb{P}_{\rho''}[v \to_i b]}_{\in (1 \pm \frac{\varepsilon}{4})\mathbb{P}_\zeta[v \to_i b]^{g+1,h}}$$

$$\in \left(1 \pm \frac{\varepsilon}{4}\right)^2 \sum_{v \in V_i} \mathbb{P}_\zeta[a \to_i v]^{l,g} \cdot \mathbb{P}_\zeta[v \to_i b]^{g+1,h}$$

$$= \left(1 \pm \frac{\varepsilon}{4}\right)^2 \mathbb{P}_\zeta[a \to_i b]^{l,h}.$$

The call to RED introduces another relative error of $\varepsilon/4$, resulting in

$$\mathbb{P}_\rho[a \to_i b] \in \left(1 \pm \frac{\varepsilon}{4}\right)^3 \mathbb{P}_\zeta[a \to_i b]^{l,h}.$$

Because (calculations omitted)

$$1 - \varepsilon \leq 1 - \varepsilon + \frac{27}{64}\varepsilon^2 \leq \left(1 - \frac{\varepsilon}{4}\right)^3 \quad \text{and} \quad \left(1 + \frac{\varepsilon}{4}\right)^3 \leq 1 + \frac{61}{64}\varepsilon \leq 1 + \varepsilon,$$

the distribution ρ thus relative ε-fools the (intermediate) transition events.

(ii). First consider the time complexity. The statement follows from the fact that the recursion depth is $\lceil \log n \rceil$, the only expensive task in each recursion is the call to RED, and the definition of $\varepsilon_{\min}$. Keep in mind that we assumed RED$_{\text{time}}$ to be nondecreasing in the first two, and nonincreasing in the third argument.

Now consider the number of processors. There will only be n instances of FOOL running in parallel at a time. There are only $O(S_{\max}^2)$ multiplications to do in line 11, and with the same argument as above, a call to RED never uses more than RED$_{\text{proc}}(S_{\max}, k, \varepsilon_{\min})$ processors. ∎

Corollary 18.4 *Consider relative ε-fooling. If RED is chosen as the algorithm from Theorem 18.17, algorithm FOOL yields a distribution of size*

$$S_{\max} = O\left(k\left(n^4 \frac{\log(k \log^{(*)})}{\eta \varepsilon^2}\right)^{1/(1-2\eta)}\right), \tag{18.49}$$

using

$$O\left(nk^2 \left(n^4 \frac{\log(k \log^{(*)})}{\eta \varepsilon^2}\right)^{2/(1-2\eta)} (O(k) \log^{(*)})^{1+\frac{1}{\eta}}\right)$$

parallel processors in $O\left(\frac{1}{\eta^2} \log^{(*)}\right)$ time.

To prove this, we need an upper bound on RED$_{\text{iter}}$ for the case of the algorithm from Theorem 18.17. Recall Equations 18.44 and 18.45. The following technical proposition examines this iteration in a general setting.

Proposition 18.5 *Let a, b, c, p, S be large enough* such that $6p + 1 \leq b$, and define $\bar{f}(x) := a \log^p (b \log(cx^2))$. Then, for the tth iterate of f, for all $t \in \mathbb{N}$, we have*

$$f^{(t)}(S) \leq a \log^p \left(b^3 \left(\log \left(ca^2\right) + \log \left(cS^2\right)\right)\right) = a \log^p \left(2b^3 \left(\log \left(acS\right)\right)\right)$$
$$= O(a \log^p \left(b \log \left(acS\right)\right)). \tag{18.50}$$

Proof. Set $A := \log(ca^2) + \log(cS^2)$. We clearly have $f(S) \leq a \log^p(b^3 A)$ and furthermore

$$\begin{aligned}
f\left(a \log^p \left(b^3 A\right)\right) &= a \log^p \left(b \log \left(ca^2 \log^{2p} \left(b^3 A\right)\right)\right) \\
&= a \log^p \left(b \log \left(ca^2\right) + b \log \log^{2p} \left(b^3 A\right)\right) \\
&\leq a \log^p \left(bA + 2pb \log \log \left(b^3 A\right)\right) \\
&\leq a \log^p \left(bA + 2pb \left(\log \left(\log \left(b^3\right)\right) + \log (A)\right)\right) \\
&\leq a \log^p \left(bA + 4pb^2 + 2pb \log (A)\right) \\
&\leq a \log^p \left((b + 4pb^2 + 2pb) A\right) \\
&\leq a \log^p \left((1 + 6p) b^2 A\right) \\
&\leq a \log^p \left(b^3 A\right).
\end{aligned}$$

The second and third assertion is a simple calculation. ∎

The bound of the proposition can certainly be improved. We proceed to prove the corollary.

Proof. (of Corollary 18.4)
In terms of the previous proposition, we have $a = O\left(k(\eta \varepsilon_{\min}^2)^{-\frac{1}{1-2\eta}}\right)$, $b = k$, $c = 1$, and $p = \frac{1}{1-2\eta}$. The starting value is $S = r$, and we have $\varepsilon_{\min}^{-1} \leq \frac{4n^2}{\varepsilon}$. A calculation using the previous proposition leads to stated bound on $S_{\max}$. Using the equations for RED_{proc} and RED_{time} from Theorem 18.17 and $\frac{1}{1-2\eta} = O(1)$ we get the stated bounds for processors and running time. ∎

Remark 18.5 Fix one automaton A and let its final states give the values of some d-valued random variable Y. If we have a relative error, then

$$\mathbb{E}_\zeta [Y] = \sum_{y \in [d]} y \underbrace{\mathbb{P}_\zeta [Y = y]}_{\in (1 \pm \varepsilon) \mathbb{P}_\rho [Y = y]} \in (1 \pm \varepsilon) \sum_{y \in [d]} y \mathbb{P}_\rho [Y = y] = (1 \pm \varepsilon) \mathbb{E}_\rho [Y].$$

So, in other words, expectations are fooled just like probabilities. In the case of an absolute error, we have

$$\mathbb{E}_\zeta [Y] = \sum_{y \in [d]} y \mathbb{P}_\zeta [Y = y] \leq \sum_{y \in [d]} y (\mathbb{P}_\rho [Y = y] + \varepsilon)$$
$$= \sum_{y \in [d]} y \mathbb{P}_\rho [Y = y] + \sum_{y \in [d]} y \varepsilon \leq \mathbb{E}_\rho [Y] + d^2 \varepsilon,$$

and a similar lower bound. Hence, in case of absolute error, if expectations are to be ε-fooled, the fooling of the probabilities has to be done with an "ε" not larger than ε / d^2. If the automata have polynomially many states, d is polynomial, and hence such a requirement is reasonable.

---

* It is sufficient to choose $a \geq 2$, $c \geq 1$, $b \geq 12$, $S \geq 2$.

18.5 Parallel Algorithms for Packing Integer Programs

18.5.1 Impact of k-Wise Independence: Multidimensional Bin Packing

We choose the multidimensional bin packing problem as an example for a packing type problem for which a $\mathcal{NC}$ algorithm with the method of k-wise independence can be given. The material for this subsection is based on Srivastav and Stangier [SS97a]. In the next section, we will see the impact of the approach of Srinivasan [Sri01a] for general integer programs of packing type.

Definition 18.9 (*Bin Packing Problem* $\text{BIN}(\vec{b}, m)$)
Let $m, n \in \mathbb{N}$, and let $\vec{b} = (b_1, \ldots, b_m) \in \mathbb{N}^m$, called the bin size vector. Given vectors $\vec{v}_1, \ldots, \vec{v}_n \in [0, 1]^m$, pack* all vectors in a minimum number of bins such that in each bin B and for each coordinate $i \in [m]$ the condition $\sum_{\vec{v}_j \in B} v_{ji} \leq b_i$ holds. Define

$$C := \left\lceil \max_{i \in [m]} \frac{1}{b_i} \sum_{j=1}^n v_{ji} \right\rceil. \tag{18.51}$$

Observe that C is the minimum number of bins, if fractional packing is allowed. Let C_{opt} be the number of bins in an optimal integer solution.

$\text{BIN}(1, m)$ is the multidimensional bin packing problem, and $\text{BIN}(1, 1)$ is the classical bin packing problem. The intention behind the formulation with a bin size vector $\vec{b}$ is to analyze the relationship between bin sizes and polynomial-time approximability of the problem.

The known polynomial-time approximation algorithms for resource constrained scheduling applies also to multidimensional bin packing and are put forward by Garey et al. [GGJY76] and Röck and Schmidt [RS83]. Garey et al. constructed with the First-Fit-Decreasing heuristic a packing with C_{FFD} number of bins that asymptotically is a $(m + \frac{1}{3})$-factor approximation, that is, there is a nonnegative integer C_0 such that $C_{\text{FFD}} \leq C_{\text{opt}}(m + \frac{1}{3})$ for all instances with $C_{\text{opt}} \geq C_0$. De la Vega and Lueker [dlVL81] improved this result presenting for every $\varepsilon > 0$ a linear-time algorithm that achieves an asymptotic approximation factor of $m + \varepsilon$. For restricted bin packing and scheduling Błazewicz and Ecker gave a linear-time algorithm (see Reference BESW93). The first polynomial-time approximation algorithm for resource constrained scheduling and multidimensional bin packing with a constant factor approximation guarantee has been given in Reference SS97b. For every $\varepsilon > 0$, $1/\varepsilon \in \mathbb{N}$, a bin packing of size at most $\lceil (1 + \varepsilon) C_{\text{opt}} \rceil$ can be constructed in strongly polynomial time, provided that

$$b_i \geq \frac{3(1 + \varepsilon)}{\varepsilon^2} \log(8nd). \tag{18.52}$$

This approximation guarantee is independent of the dimension m. For $\varepsilon = 1$ this gives a sequential two-factor approximation algorithm for multidimensional bin packing under the above assumption on b.

There is no obvious way to parallelize this algorithm. In this section we will show that at least in some special cases there is an $\mathcal{NC}$ approximation algorithm. The algorithm is based on the method of $\log^c n$-wise independence. The parallel bin packing algorithm has four steps.

---

* Packing simply means to find a partitioning of vectors in a minimum number of sets so that the vectors in each partition (= bin) satisfy the upper bound conditions of the definition.

Step 1: *Finding an optimal fractional bin packing in parallel.* We wish to apply randomized rounding and therefore have to generate an optimal fractional solution. The following integer program is equivalent to the multidimensional bin packing problem.

$$\begin{aligned}
\min\ & D\\
& \sum_j v_{ji} x_{jz} \le b_i && \forall\, i \in [m],\, z \in [D]\\
& \sum_z x_{jz} = 1 && \forall\, j \in [n]\\
& x_{jz} = 0 && \forall\, j \text{ and } z > D\\
& x_{jz} \in \{0,1\}.
\end{aligned}$$

Note that this integer program is not an integer *linear* program, thus we cannot solve it with standard methods. But, fortunately, its optimal fractional solution is given by C, see Equation 18.51. This is easily checked by setting $x_{jz} = 1/C$ for all vectors j and all bins $z \in [C]$.

Step 2: *Enlarged fractional bin packing.* In order to define randomized rounding we enlarge the fractional bin packing. The reason for such an enlargement is that for the analysis of the rounding procedure we need some room. Fix $\alpha > 1$ and define

$$d := 2^{\lceil \log(\alpha C) \rceil}, \tag{18.53}$$

hence $d \le 2\alpha C$.

The new fractional assignments of vectors to bins are $\tilde{x}_{jz} = 1/d$ for all $j \in [n]$, $z \in [d]$. Note that this assignment defines a valid fractional bin packing, even with tighter bounds

$$\sum_{j=1}^n v_{ji} \tilde{x}_{jz} \le \frac{b_i}{\alpha} \tag{18.54}$$

for all $i \in [m]$, $z \in [d]$.

Step 3: *Randomized rounding.* The sequential randomized rounding procedure is to assign randomly and independently vector j to bin z with uniform probability $1/d$. With the method of limited independence we parallelize and derandomize this rounding scheme in the next step.

Step 4: *Derandomized rounding in $\mathcal{NC}$.* We apply the parallel conditional probability method for multivalued random variables (Theorem 18.11).

First, C can be computed in parallel with standard methods.

Lemma 18.5 *C can be computed with $O(nm)$ processors in $O(\log(nm))$ time.*

Recall the definitions preceding Theorem 18.11. For integers $r, z \in [d]$ define the indicator function

$$[r = z] := \begin{cases} 1 & \text{if } r = z \\ 0 & \text{else.} \end{cases}$$

For all $(i, z, j) \in [m] \times [d] \times [n]$ define the functions g_{izj} by

$$g_{izj} : \mathbb{R} \to \mathbb{R},\ r \mapsto v_{ji}\left([r = z] - \frac{1}{d}\right),$$

and, for an even k to be specified later, f_{iz} and F as in Theorem 18.11. We then have

$$F(x_1,\ldots,x_n) = \sum_{(i,z)\in[m]\times[d]} \underbrace{\left|\sum_{j=1}^n v_{ji}\left([x_j=z]-\frac{1}{d}\right)\right|^k}_{=f_{iz}(x_1,\ldots,x_n)}$$

for all $(x_1,\ldots,x_n) \in [d]^n$. Furthermore, we need estimates of the kth moments.

Lemma 18.6 *Let $\frac{1}{\log n} \leq \tau < 1/2$ and $k := 2\left\lceil \frac{\log(6\alpha nm)}{2\tau \log n} \right\rceil$. Then*

$$\mathbb{E}\left[F(X_1,\ldots,X_n)\right]^{1/k} \leq n^{(1/2)+\tau}(\log(6\alpha nm)+2)^{1/2}$$

for k-wise independent $[d]$-valued random variables $X_1,\ldots,X_n$.

Proof. Using Theorem 18.5, we can show as in the proof of Lemma 18.3

$$\mathbb{E}\left[|f_{iz}(X_1,\ldots,X_n)|^k\right] \leq 2\left(\frac{k}{2}\right)!\left(\frac{n}{2}\right)^{(k/2)}. \tag{18.55}$$

Robbins exact Stirling formula shows the existence of a constant γ_n with $1/(12n+1) \leq \gamma_n \leq 1/(12n)$ so that $n! = (\frac{n}{e})^n\sqrt{2\pi n}e^{\gamma_n}$ ([Bol98], p. 4), thus $n! \leq 3(\frac{n}{e})^n\sqrt{n}$. Furthermore, for all $k \geq 2$, we have $2\sqrt{k}(4e)^{-\frac{k}{2}} \leq 1$. With these bounds the right-hand side of inequality Equation 18.55 becomes $3(kn)^{k/2}$. The claimed bound now follows by summing over all $i \in [m]$, $z \in [d]$, using the trivial bound $C \leq n$ and taking the $1/k$th root. ∎

The main result is explained in Theorem 18.19.

Theorem 18.19 (Srivastav and Stangier 1997 [SS97a]) *Let $1/\log n \leq \tau < 1/2$ and $\alpha > 1$ and suppose that $b_i \geq \alpha(\alpha-1)^{-1}n^{(1/2)+\tau}(\log(6\alpha nm)+2)^{\frac{1}{2}}$ for all $i \in [m]$. Then there is an $\mathcal{NC}$ algorithm that runs on $O(n^3(nm)^{(1/\tau)+1})$ parallel processors and finds in $O\left(\left(\frac{\log^2 n \log(nm)}{\tau}\right)^4\right)$ time a bin packing of size at most $2^{\lceil \log(\alpha C) \rceil} \leq 2\alpha C$.*

Proof. Set $k = 2\left\lceil \frac{\log(6\alpha nm)}{2\tau \log n} \right\rceil$ and let d be as defined in Equation 18.53, so $d \leq 2\alpha C \leq 2\alpha n$. Thus, for fixed α we have $d^k = O(n^k)$ (we are not interested in large values of α). Also, because of $(1/\log n) \leq \tau$, we have $O(1)^{1/\tau} = O(1)$. By Theorem 18.11 we can construct integers $\widehat{x}_1,\ldots,\widehat{x}_n$, where $\widehat{x}_j \in [d]$ for all j such that

$$F(\widehat{x}_1,\ldots,\widehat{x}_n) \leq \mathbb{E}\left[F(X_1,\ldots,X_n)\right] \tag{18.56}$$

holds, using

$$O(n^{k+2}m) = O(n^3(nm)^{(1/\tau)+1})$$

parallel processors in time

$$O\Big(\underbrace{\log(n\log d)}_{=O(\log n)} \cdot (k^4 \underbrace{\log^4 d}_{=O(\log^4 n)}\ \underbrace{\log^3(n\log d)\log n}_{=O(\log^4 n)} + \underbrace{k\log d \log(mdn^k)}_{=O(k\log n \log(mn))})\Big)$$

$$= O\left(\log^4 n \cdot \left(\left(\frac{\log(6\alpha nm)}{\tau}\right)^4 \frac{\log^4 n}{\log^4 n}\log n + \left(\frac{\log(6\alpha nm)}{\tau}\right)^2 \log(nm)\right)\right)$$

$$= O\left(\left(\frac{\log^2 n \log(nm)}{\tau}\right)^4\right).$$

To complete the proof we must show that the vector $(\widehat{x}_1,\ldots,\widehat{x}_n)$ defines a valid bin packing. This is seen as follows. By definition of the function F we have for all coordinates i and all bins $z \in [d]$

$$\left|\sum_{j=1}^n v_{ji}\left(x_{jz}-\frac{1}{d}\right)\right| \le F(x_1,\ldots,x_n)^{(1/k)} \le \mathbb{E}\left[F(X_1,\ldots,X_n)\right]^{(1/k)}$$

$$\le n^{(1/2)+\tau}\left(\log(6\alpha nm)+2\right)^{(1/2)}. \qquad (18.57)$$

(the last inequality follows from Lemma 18.6). Hence, using Equations 18.57, 18.54, and the assumption $b_i \ge \alpha(\alpha-1)^{-1}n^{(1/2)+\tau}(\log(6\alpha nm)+2)^{1/2}$, we get for all coordinates i and all bins $z \in [d]$

$$\sum_{j=1}^n v_{ji}x_{jz} \le \left|\sum_{j=1}^n v_{ji}(x_{jz}-\frac{1}{d})\right| + \sum_{j=1}^n v_{ji}\frac{1}{d}$$

$$\le n^{(1/2)+\tau}(\log(6\alpha nm)+2)^{(1/2)} + \frac{b_i}{\alpha}$$

$$\le \left(1-\frac{1}{\alpha}\right)b_i + \frac{b_i}{\alpha} = b_i,$$

and the theorem is proved. ∎

Note that we cannot achieve a polynomial-time approximation better than a factor of $4/3$, even if $b_i = \Omega(\log(nm))$ for all bounds b_i and $\tau < 1/2$.

Theorem 18.20 (Srivastav and Stangier 1997 [SS97a]) *Let $0 < \tau < \frac{1}{2}$ and $3 \le \Delta$ be a fixed integer. Under the assumption $b_i = \Omega\big(n^{(1/2)+\tau}\sqrt{\log(nm)}\big)$ for all i, it is $\mathcal{NP}$-complete to decide whether or not there exists a multidimensional bin packing of size Δ.*

18.5.2 Impact of Automata Fooling: Srinivasan's Parallel Packing

Let $A \in ([0,1]\cap\mathbb{Q})^{m\times n}$, $b \in ((1,\infty)\cap\mathbb{Q})^m$, and $w \in ([0,1]\cap\mathbb{Q})^n$ such that $\max_{i\in[n]} w_i = 1$. We consider the *packing integer program*

$$\max w\cdot x$$
$$Ax \le b \qquad\qquad\qquad (\text{PIP})$$
$$x \in \{0,1\}^n$$

This is an $\mathcal{NP}$-hard problem, and hence good efficient approximation algorithms are sought. Let $x^* \in ([0,1] \cap \mathbb{Q})^n$ be an optimal solution to the LP relaxation of the PIP, that is, to the following linear program:

$$\max w \cdot x$$
$$Ax \leq b$$
$$x \in [0,1]^n$$

By OPT* we denote the value of x^* (i.e., $w \cdot x^*$). Set $B := \min\{b_i;\ i \in [m]\}$ (which is >1).

We briefly review previous results for PIPs. Let $0 < \varepsilon < 1$. If $B \geq 12\varepsilon^{-2}\ln(2m)$, then a vector $x \in \{0,1\}^n$ with $Ax \leq b$ and $w \cdot x \geq (1-\varepsilon)$OPT* can be constructed in deterministic polynomial time [SS96]. This result is based on an efficient derandomization of Chernoff–Hoeffding bounds. Srinivasan's work [Sri99] extends the range for the right-hand side b where a polynomial-time approximation algorithm does exist.

Theorem 18.21 (Srinivasan 1995, 1999)
There are constants K_0 and K_1 such that the following holds. A solution $x \in \{0,1\}^n$ for a given PIP can be constructed in polynomial time with

$$w \cdot x \geq K_0 \min\left\{\text{OPT}^*, \left(K_1 \frac{\text{OPT}^*}{m^{(1/B)}}\right)^{B/(B-1)}\right\}.$$

Remark 18.6 (Asymptotics of the approximation bounds)
Let us briefly compare Srinivasan's bound with the bounds of Reference SS96. If $B = B(m)$ is a function of m with $B(m) \longrightarrow \infty$, we may assume that $B/(B-1)$ is $1 + o(1)$. So

$$\left(K_1 \frac{\text{OPT}^*}{m^{(1/B)}}\right)^{B/(B-1)} = (K_1\text{OPT}^*)^{1+o(1)} e^{-\frac{\ln m}{B-1}} \approx K_1\text{OPT}^* e^{-\frac{\ln m}{B-1}}.$$

Then the approximation quality is governed by $e^{-\frac{\ln m}{B-1}}$. For $B = c\ln(m) + 1$, c a constant, we have $e^{-\frac{\ln m}{B-1}} = e^{-c}$, leading to a constant approximation factor $K_1 e^{-c}$. For any $c = \Omega(\varepsilon^{-2})$, $\varepsilon > 0$, e^{-c} is exponentially small, thus inferior to the $1 - \varepsilon$ factor presented in Reference SS96. But Srinivasan's bound gives a constant factor approximation for any constant $c > 0$, while in Reference [SS96], the "constant" is at least $12\varepsilon^{-2}$. On the other hand, if $\frac{\ln m}{B-1}$ tends to ∞, for example if $B = O(\ln\ln(m) + 1)$, we have that $e^{-\frac{\ln m}{B-1}} = e^{-\omega(m)} = o(1)$, leading to an asymptotically zero approximation factor. In this range of B, we have essentially no approximation guarantee.

In the rest of this section, we present Srinivasan's derandomized $\mathcal{NC}$ approximation algorithm [Sri01a]. We first describe a randomized approximation algorithm but will not try to make a statement about its success probability. Instead we will right away derandomize it using the automata fooling technique from Section 18.4. The randomized version merely serves as an illustration of the idea behind the to be constructed automata.

18.5.2.1 The Idea

To make the fractional optimum x^* an integral solution, consider the following rounding and alteration scheme. Fix a parameter $\lambda \geq 3$, which will be specified more precisely later. For each $j \in [n]$, set

$$y_j := \begin{cases} 1 & \text{with probability } \frac{x_j^*}{\lambda}, \\ 0 & \text{with probability } 1 - \frac{x_j^*}{\lambda}. \end{cases} \qquad (18.58)$$

Parallel Algorithms via the Probabilistic Method

This way we get an integral vector y with

$$\mathbb{E}[w \cdot y] = \frac{\text{OPT}^*}{\lambda}. \qquad (18.59)$$

However, y might be infeasible—some of the m constraints in "$Ay \le b$" might be violated. To repair this, y is modified in a greedy manner to fulfill all the constraints. Initialize $x := y$. For each row $i \in [m]$ of the matrix A sort the entries $(a_{ij})_{j=1,\ldots m}$ in decreasing order

$$a_{i,j_{1,i}} \ge a_{i,j_{2,i}} \ge \cdots a_{i,j_{m,i}}.$$

Then, starting with the first position $j = j_{1,i}$, set x_j to 0 until the constraint "$a_i \cdot x \le b_i$" is fulfilled.* This is done in parallel for all rows i, without any communication between the processes. Hence, there might in fact occur unnecessary alterations.

Let the random variables $Y_1, \ldots, Y_m$ denote the number of alterations for each row. (They are called this way because they are fully determined by the intermediate solution y.) Let I denote the input length of the PIP (clearly, $\max\{n, m\} \le I$). Let $C_0 \ge 1$ be a large enough constant and set

$$B' := \min\{C_0 \log I, B\} \quad (>1) \qquad (18.60)$$

and take

$$\lambda := K_0 \max\left\{1, \left(\frac{K_1 m}{\text{OPT}^*}\right)^{1/(B'-1)}\right\}, \qquad (18.61)$$

where $K_1 > 1$ and $K_0 \ge 1$ are some constants. Then, the following bounds can be proven using the Chernoff bound from Theorem 18.2.

$$\mathbb{E}[Y_i] = O\left(\left(\frac{e}{\lambda}\right)^{B'+1}\right) \quad \forall i \in [m]. \qquad (18.62)$$

This allows an estimation for the expected value of the objective function after the alteration process.

$$\mathbb{E}[w \cdot x] = \frac{\text{OPT}^*}{\lambda} - \sum_{i=1}^{m} \mathbb{E}[Y_i] = \frac{\text{OPT}^*}{\lambda} - m \cdot O\left(\left(\frac{e}{\lambda}\right)^{B'+1}\right).$$

18.5.2.2 Derandomization

Derandomization is done through the construction of finite deterministic automata $\mathcal{A}_1, \ldots, \mathcal{A}_{m+1}$ with polynomially many states that simulate the randomized algorithm described above. We need certain assumptions on the precision of the occuring numbers:

All entries in A, w, x^*, and x^*/λ are assumed to be rationals with denominator 2^d, where $d = \Theta(\log I)$, and all entries in A are assumed to be nonpositive powers of 2. $\qquad (18.63)$

This can be achieved by appropriate modifications of these numbers, and a simple observation how we can get a feasible solution for the original instance from a solution for the modified one.

---

* a_i denotes the ith row of A.

The automata (together with an appropriate distribution ζ) will simulate the randomzied rounding, where y is constructed from x^*, as follows.

- As inputs, all possible vectors from $[0,1]^n$ with rational entries, each with denominator 2^d are accepted. They are represented as elements from $\{0,\ldots,2^d\}^n$.
- Inputs are created randomly according to ζ, which is chosen to be the uniform distribution.
- Let $t_1,\ldots,t_n$ be such an input. For each l, the automata check whether $t_l \le 2^d \mathbb{E}[y_l] - 1$ and if so, behave as if $y_l = 1$. Otherwise, they behave as if $y_l = 0$.

The automata are layered with $n+1$ layers each, numbered from 0 to n. Transitions are allowed only from one layer to the next. Layer 0 has one state only—the starting state. The automata being in level l intuitively means that variables $y_1,\ldots,y_l$ (and so $x_1,\ldots,x_l$) have been fixed. This will become clearer in the following. Automata $\mathcal{A}_1,\ldots,\mathcal{A}_m$ correspond to the alteration process (where x is constructed from y) and $\mathcal{A}_{m+1}$ corresponds to the randomized rounding (where y is constructed from x^*).

We start by describing $\mathcal{A}_{m+1}$, which is the simplest. For each layer $l \in \{0,\ldots,n\}$, although there are exponentially many choices for a 0/1-vector $(y_1,\ldots,y_l)$, the following set Λ_l is of polynomial size only

$$\Lambda_l := \left\{ \left(\sum_{j=1}^l w_j y_j, l \right) ; (y_1,\ldots,y_l) \in \{0,1\}^l \right\} \subseteq \mathbb{Q}_{\ge 0} \times \{l\}.$$

This is true, because all these values are between 0 and n and all have denominator $2^d = 2^{\Theta(\log I)} = \Theta(I)$. Hence, only $O(nI)$ values are possible. The states in layer l are the elements from the set Λ_l.

For the transitions, assume that $l < n$ and consider a state $(\omega, l) \in \Lambda_l$. There are $2^d \mathbb{E}[y_{l+1}]$ edges going from (ω, l) to $(\omega + w_{l+1}, l+1) \in \Lambda_{l+1}$, with labels $0,\ldots,2^d \mathbb{E}[y_{l+1}] - 1$. This corresponds to the case "$y_{l+1} = 1$." Then, there are $2^d - (2^d \mathbb{E}[y_{l+1}] - 1)$ edges from (ω, l) to $(\omega, l+1) \in \Lambda_{l+1}$ with labels $2^d \mathbb{E}[y_{l+1}],\ldots,2^d$. This corresponds to the case "$y_{l+1} = 0$".

Note that by assumption, $\mathbb{E}[y_{l+1}] = x^*/\lambda$ is a fraction with denominator 2^d.

We now turn to automata $\mathcal{A}_1,\ldots,\mathcal{A}_m$. Their states have to express the number of variables that have to be altered (i.e., set to 0). $\mathcal{A}_i$ will essentially express Y_i for $i \in [m]$. Fix an $i \in [m]$ now. For given $y_1,\ldots,y_l$, we collect variables y_j that correspond to matrix entries a_{ij} of certain magnitudes and sum them up:

$$U_k(y_1,\ldots,y_l) := \sum_{\substack{j \le l \\ \text{s.t.} a_{ij} = 2^{-k}}} x_j \quad \text{for all } k \in \{0,\ldots,d\}. \tag{18.64}$$

Denote the vectors

$$T(y_1,\ldots,y_l) := (U_0(y_1,\ldots,y_l),\ldots,U_d(y_1,\ldots,y_l)). \tag{18.65}$$

From $T(y_1,\ldots,y_n)$ the value of Y_i for $y = (y_1,\ldots,y_n)$ can be inferred. A first idea is to define the states in layer l of $\mathcal{A}_i$ to be all possible values of $T(y_1,\ldots,y_l)$ with $(y_1,\ldots,y_l)$ ranging over all elements of $\{0,1\}^l$. This, however, can lead to superpolynomial state sets.

The solution is, intuitively, to summarize unimportant information so that the relevant information can be expressed more succinct. Before we start with that, one more assumption on the PIP has to be made. We need that $b_i \le C_0 \log I$ (for all i). This can be achieved by multiplying the offending constraints "$a_i \cdot x \le b_i$" with $C_0 \log I / b_i$. Note that this motivates Equation 18.60. If this reduces the value of B, that is, if $B' = C_0 \log I / b_i$, the value of λ increases (reducing the approximation guarantee) but is upper bounded by a constant. This can be seen by a calculation, using that we may assume* OPT* ≥ 1.

---

* A solution with value 1 can always be constructed by setting $x_j := 1$ for some j where $w_j = 1$ and all other entries to 0.

We now turn to the more succinct representation of information. This is done in two steps. First, values $U_k(.)$ for large k are put together, resulting in shorter vectors T', which take the role of the vectors T. Set

$$k_0 := \lceil \log(C_0 \log I) \rceil \quad \text{and} \quad U'(y_1, \ldots, y_l) := \sum_{\substack{j \leq l \\ \text{s.t.} a_{ij} \leq 2^{-k_0}}} x_j.$$

Then, define

$$T'(y_1, \ldots, y_l) := \left(U_0(y_1, \ldots, y_l), \ldots, U_{k_0-1}(y_1, \ldots, y_l), U'(y_1, \ldots, y_l) \right).$$

The second step is to characterize the interesting values the random variable T' can assume. This will be a polynomially sized set and hence our choice for the automata states (with one small addition of a "bad" state, see below). The following function will be useful as an upper bound on the interesting values:

$$g(k) := b_i 2^k + C_0 2^k \log I \log \log I.$$

Define the states V_l of layer l (for automata $\mathcal{A}_i$) as follows:

$$V_l := \left\{ (u_0, \ldots, u_{k_0-1}, u', l) \in \mathbb{N}_0^{k_0+1} \times \{l\}; \ \forall k \in \{0, \ldots, k_0 - 1\}: \ u_k \leq g(k) \text{ and } u' \leq b_i \right\} \qquad (18.66)$$
$$\cup \ \{\text{bad}_l\}$$

Transitions from layer l to $l+1$, similarly as in $\mathcal{A}_{m+1}$, roughly correspond to the cases of "$y_{l+1} = 0$" and "$y_{l+1} = 1$." However, more distinctions have to be made. First fix a state $u = (u_0, \ldots, u_{k_0-1}, u', l) \in V_l$ (which is not the bad state). Edges with labels $0, \ldots, 2^d \mathbb{E}[y_{l+1}] - 1$ go from u to $(u_0, \ldots, u_{k_0-1}, u', l+1) \in V_{l+1}$. This corresponds to the case "$y_{l+1} = 0$," in which obviously all values in u have to be maintained (except the indication for the level in the last entry).

For the case "$y_{l+1} = 1$," there are edges with labels $2^d \mathbb{E}[y_{l+1}], \ldots, 2^d$, which go from u to

- $(u_0, \ldots, u_k + 1, \ldots, u_{k_0-1}, u', l + 1) \in V_{l+1}$ if there exists $k \in \{0, \ldots, k_0 - 1\}$ s.t. $a_{il} = 2^{-k}$ and $u_k + 1 \leq g(k)$
- $(u_0, \ldots, u_{k_0-1}, u' + 1, l + 1) \in V_{l+1}$ if $a_{il} \leq 2^{-k_0}$ and $u' + 1 \leq b_i$
- bad$_l$ otherwise

From a bad state, all $2^d + 1$ transitions go only to the next bad state. This is consistent with the meaning of the states, because the entries in vectors used as the states are nondecreasing in l, and hence once one of the bounds ("$u_k \leq g(k)$" and "$u' \leq b_i$") is violated, it will so be further on. It is also clear that the value of Y_i can be read off the final states (in layer n), except from the bad state. If the automaton slips into the bad states, we have no information on Y_i, and hence to assume the worst, which is $Y_i = n$. Fortunately, we can bound the probability that this happens. The following lemma can be proven again using Theorem 18.2.

Lemma 18.7 *If the constant C_0 is chosen large enough, the probability that the automaton reaches a bad state is at most $1/I^3$.*

Let Y_i' be the random variable expressed by $\mathcal{A}_i$. Then, the expected value of Y_i' differs from that of Y_i by at most $1/I^2$ (because $n \leq I$). Hence the bound in Equation 18.62 is only weakened to

$$\mathbb{E}[Y_i'] = O\left(\left(\frac{e}{\lambda}\right)^{B'+1} + \frac{1}{I^2} \right).$$

Finally, we have to show that V_l has polynomial size. This can be checked by using the definition of the bounding function g and that $O(\log I)^{O(\log \log I)}$ can be bounded by a polynomial in I.

The construction is now ready for automata fooling. The approximation guarantee will be reduced depending on the ε used for the fooling. It suffices to do absolute fooling, so we give the analysis for this. A later remark will show a similar bound for the case of relative fooling.

Theorem 18.22 (Srinivasan 2001)

(i) *There is a derandomized $\mathcal{NC}$ algorithm for packing integer programs that fulfill Equation 18.63 that computes for every $1 > \varepsilon > 0$ a solution $x \in \{0,1\}^n$ for a PIP of input length I with value*

$$w \cdot x \geq \frac{\text{OPT}^*}{\lambda} - m \cdot O\left(\left(\frac{e}{\lambda}\right)^{B'+1} + \frac{1}{I^2}\right) - (m+1)\varepsilon.$$

(ii) *The following bound holds for the general case, when Equation 18.63 is not necessarily given*

$$w \cdot x \geq \frac{1}{2}\left(\frac{\text{OPT}^* - \frac{2n}{2^d}}{\lambda} - m \cdot O\left(\left(\frac{e}{\lambda}\right)^{B'+1} + \frac{1}{I^2}\right) - (m+1)\varepsilon\right).$$

Proof. (i) There are $\mathcal{NC}$ algorithms in the literature to construct a fractional solution with an arbitrary small relative error [LN93]. For simplicity, we have not noted this error in the given bounds, nor will do in the following.

Let X be the random variable denoting the final state of $\mathcal{A}_{m+1}$, which corresponds to the value of the constructed solution. Let $K_3 \geq 2$ be a constant such that X and $Y_1, \ldots, Y_m$ do not take on more than I^{K_3} different values. Such a constant—meaning: independent of the given PIP—exists, since already the automata expressing these random variables have polynomially many states. We use $\varepsilon' = \varepsilon/I^{K_3}$ for automata fooling with an absolute error, resulting in the expectations of X and $Y_1, \ldots, Y_m$ having absolute error ε (see Remark 18.5). Let ζ be the original distribution (which is the uniform distribution on $\{0, \ldots, 2^d\}^n$) and ρ the distribution created by the automata fooling algorithm. We have $\mathbb{E}_\rho[X] \geq \mathbb{E}_\zeta[X] - \varepsilon = \frac{\text{OPT}^*}{\lambda} - \varepsilon$, and also $\mathbb{E}_\rho[Y_i'] \leq \mathbb{E}_\zeta[Y_i'] + \varepsilon = O\left(\left(\frac{e}{\lambda}\right)^{B'+1} + \frac{1}{I^2}\right) + \varepsilon$.

(ii) If a given instance is not of the form in Equation 18.63, we have to modify it first. This can be done by rounding down all entries in w to the next real in $\left\{\frac{k}{2^d};\ k \in \{0, \ldots, 2^d\}\right\}$. And we only allow fractional solutions $x^* \in \left\{\frac{k}{2^d};\ k \in \{0, \ldots, 2^d\}\right\}^n$. For the fractional optimum of the new instance OPT_0^* it holds that $\text{OPT}_0^* \geq \text{OPT}^* - \frac{2n}{2^d}$. Next, we consider the impact of rounding the entries in A. They are rounded down to the next nonnegative power of 2. If x is feasible for this new instance, then $\frac{1}{2}x$ is feasible for the original one. ∎

Remark 18.7 If we use relative ε-fooling, we can prove in a similar manner

$$w \cdot x \geq (1-\varepsilon)\left(\frac{\text{OPT}^*}{\lambda} - m \cdot O\left(\left(\frac{e}{\lambda}\right)^{B'+1} + \frac{1}{I^2}\right)\right).$$

The number of parallel processors used by this algorithm and the running time depend on the algorithm used for automata fooling, which—for the case of the fooling algorithm presented in this chapter—in turn depends on the algorithm RED. Bounds for this are stated in Corollary 18.4 and Theorem 18.17. The polynomial size of the state space is essential, because the quantity k depends on it, see Equation 18.48.

Srinivasan claimes in Reference Sri01a that a result similar to Theorem 18.22 holds for covering integer programs and refers to the full verion. The full version had not appeared at the time this chapter was written.

Parallel Algorithms via the Probabilistic Method

Remark 18.8 Let $B' = \Omega(\log m)$, say $B' = C_1 \log m$ with $C_1 \geq 2$, and $K_0 \geq 2e$. Then the first bound of Theorem 18.22 can be simplified to

$$w \cdot x \geq \frac{\text{OPT}^*}{2K_0(2 + o(1))} - o(1) - (m+1)\varepsilon.$$

Proof. The assumption $B' = \Omega(\log m)$ leads to $K_1^{(1/B'-1)} = 1 + o(1)$ and $m^{1/(B'-1)} \leq 2 + o(1)$, thus

$$\lambda \leq K_0 \max\left\{1, (2+o(1))\text{OPT}^{* - \frac{1}{C_1 \log m - 1}}\right\} \leq K_0(2 + o(1)),$$

because $\text{OPT}^* \geq \frac{1}{2}$. Furthermore, $mO\left(\left(\frac{e}{\lambda}\right)^{B'+1}\right) = mO(m^{-C_1}) = o(1)$, and trivially $\frac{m}{l^2} = o(1)$. The stated bound follows. ∎

At the end of this section, a very legitimate question is whether Theorem 18.22 can help to improve the result for multidimensional bin packing (Theorem 18.19) in a way that the quite restrictive lower bound condition on the bin sizes $b_i = \left(n^{(1/2)+\tau}\sqrt{\log(nm)}\right)$ can be weakened to $b_i = \Omega(\log(nm))$ for all i. We think that this should be possible by adapting the proof of Theorem 18.22 to the scenario of randomized rounding with multivalued random variables *with* equality constraints.

Bibliography and Remarks. Earlier derandomized $\mathcal{NC}$ algorithms for packing integer programs were developed by Alon and Srinivasan [AS97]. For the case of $b_i = O(\log(m+n))$ for all $i \in [m]$ their algorithm gives results within a $1 + o(1)$ factor of the sequential bound $\Omega(\text{OPT}^*/m^{1/B})$ achieved before [RT87, Rag88]. By a scaling technique, for any $c > 1$, they can also solve general PIPs, but only within a factor $1/c$ of the sequential bound.

18.6 Open Problems

Parallel Computation of Low Discrepancy Colorings. The perhaps most challenging algorithmic problem in combinatorial discrepancy theory is to give an efficient algorithm that computes for a hypergraph $\mathcal{H} = (V, \mathcal{E})$, say, with $|V| = n = |\mathcal{E}|$, a 2-coloring with discrepancy at most $O(\sqrt{n})$, matching the celebrated six standard deviations bound of Spencer [Spe85]. It would be interesting to explore the impact of automata fooling and state space compression in this context.

Furthermore, 2-colorings for special hypergraphs with optimal or near-optimal discrepancy can be obtained by the *partial coloring* method of Beck [Bec81]. A $\mathcal{NC}$ algorithm here is not known. A very interesting example for research in this direction would be the hypergraph of all arithmetic progressions in $[n]$. Its discrepancy is $\Theta(\sqrt[4]{n})$ (Matoušek and Spencer [MS96], Roth [Rot64]). The probabilistic method gives only $O\left(\sqrt{n^2 \log n}\right)$ and the parallel algorithm from Section 18.3.4 computes colorings of discrepancy $O(n^{(1/2)+\varepsilon}\sqrt{\log n})$, $1/2 \leq \varepsilon \leq 1$. Any progress toward $O(\sqrt[4]{n})$, sequential or parallel, is of great interest.

Sum of Dependent Random Variables In the theoretical setting as well as applications of $\mathcal{NC}$ algorithm design with the probabilistic method, the underlying randomized sequential algorithm is usually analyzed with concentration inequalities for sums of independent random variables. But many combinatorial optimization problems show dependencies among the variables in a linear programming model in a natural way. An interesting example is the partial covering problem in graphs and hypergraphs (see Reference GKS04). Here $\mathcal{NC}$ algorithms are not known.

Parallelizing the Random Hyperplane Method The celebrated algorithms of Goemans and Williamson [GW95] for the max-cut problem can be efficiently derandomized (Mahajan and Ramesh [MR99]). An efficient parallelization ($\mathcal{RNC}$ or $\mathcal{NC}$ algorithm) for the random hyperplane method is *terra incognita* and a formidable challenge.

Practicability Today's research in the area of efficient algorithms is more and more challenged by the legitimate question of practicability. Algorithms based on *randomized rounding* are often practical, both because they are fast, easy to implement, and of very good performance of repeated several times [NRT87, BS04]. $\mathcal{NC}$ algorithms designed with the probabilistic method use a polynomial number of processors, and thus seem to be impractical in view of real-world applications, where we have access to only a constant number of parallel processors. Besides that, questions about communication overhead (e.g., synchronization, data communication between processors) or bottleneck situations when accessing a shared memory arise. Interdisciplinary work together with computer engineers will be necessary to get from theoretically efficient to practically efficient parallel algorithms.

This list is certainly incomplete. But we think that effort in these directions is sought to advance the design of parallel algorithms through an interaction of theoretical and practical issues.

Acknowledgments

We thank Andreas Baltz for his help in preparing the material for the paragraph on MIS (Section 18.3.3), and David Karger for clarifying proofs given in Reference KK97. We thank our students Sandro Esquivel and Carsten Krapp for setting parts of the chapter in LaTeX. The first author thanks the Deutsche Forschungsgemeinschaft, priority program 1126 "Algorithmics of Large and Complex Networks" for the support through the grant Sr7-3. Last but not least we sincerely thank the editors for the invitation to contribute to the handbook, their patience and cooperation.

References

[ABI86]   N. Alon, L. Babai, and A. Itai. A fast and simple randomized algorithm for the maximal independent set problem. *Journal of Algorithms*, 7:567–583, 1986.

[ACR98]   A.F. Andreev, A.E.F. Clementi, and J.D.P. Rolim. A new general derandomization method. *Journal of the ACM*, 45:179–213, 1998.

[AG97]   N. Alon and Z. Galil. On the exponent of all pairs shortest path problem. *Computer System Sciences*, 54:255–262, 1997. Preliminary version in *Proceedings of the 32nd IEEE Symposium on Foundations of Computer Science 1991 (FOCS '91)*, pp. 569–575.

[AGHP92] N. Alon, O. Goldreich, J. Håstad, and R. Peralta. Simple constructions for almost k-wise independent random variables. *Random Structures & Algorithms*, 3:289–304, 1992.

[AGMN92] N. Alon, Z. Galil, O. Margalit, and M. Naor. Witnesses for matrix multiplication and for shortest paths. In *Proceedings of the 33rd IEEE-Symposium on Foundations of Computer Science 1992 (FOCS '92)*, pp. 417–426, 1992.

[AGR94]   N.M. Amato, M.T. Goodrich, and E.A. Ramos. Parallel algorithms for higher dimensional convex hulls. In *Proceedings of the 35th IEEE Symposium on Foundation of Computer Science 1994 (FOCS '94)*, pp. 683–694, 1994.

[AKS87]   M. Ajtai, J. Komlós, and E. Szemeredi. Deterministic simulation in logspace. In *Proceedings of the 19th Annual ACM Symposium on the Theory of Computing 1987 (STOC '87)*, pp. 132–140, 1987.

[Alo91]   N. Alon. A parallel algorithmic version of the local lemma. *Random Structures & Algorithms*, 2:367–378, 1991.

[AMN98]   Y. Azar, R. Motwani, and J. Naor. Approximating probability distributions using small sample spaces. *Combinatorica*, 18:151–171, 1998.

[AN96]   N. Alon and M. Naor. Derandomization, witnesses for Boolean matrix multiplication and construction of perfect hash functions. *Algorithmica*, 16:434–449, 1996.

[Arm98]   R. Armoni. On the derandomization of space-bounded computations. In *Proceedings of the 2nd International Workshop on Randomization and Approximation Techniques in Computer Science (RANDOM '98)*, pp. 47–59, Barcelona, Spain, October 1998.

[AS97]   N. Alon and A. Srinivasan. Improved parallel approximation of a class of integer programming problems. *Algorithmica*, 17:449–462, 1997.

[AS00]   N. Alon and J. Spencer. *The Probabilistic Method*. John Wiley & Sons, Inc., New York, 2000.

[ASE92]   N. Alon, J. Spencer, and P. Erdős. *The Probabilistic Method*. John Wiley & Sons, Inc., New York, 1992.

[AV79]   D. Angluin and L.G. Valiant. Fast probabilistic algorithms for Hamiltonian circuits and matchings. *Computer System Sciences*, 18:155–193, 1979.

[Ban98]   W. Banaszczyk. Balancing vectors and Gaussian measure of n-dimensional convex bodies. *Random Structures & Algorithms*, 12:351–360, 1998.

[Bec81]   J. Beck. Roth's estimate of the discrepancy of integer sequences is nearly sharp. *Cominatorica*, 1:319–325, 1981.

[Bec91]   J. Beck. An algorithmic approach to the Lovász Local Lemma I. *Random Structures & Algorithms*, 2:343–365, 1991.

[BEG$^+$94] M. Blum, W. Evans, P. Gemmell, S. Kannan, and M. Naor. Checking the correctness of memories. *Algorithmica*, 12:225–244, 1994. Preliminary version in: *Proceedings of the 32nd IEEE Symposium on the Foundation of Computer Science 1991 (FOCS '91)*, pp. 90–99.

[Ber73]   C. Berge. *Graphs and Hypergraphs*. North Holland, Amsterdam, 1973.

[BESW93] J. Błazewicz, K. Ecker, G. Schmidt, and J. Węglarz. *Scheduling in Computer and Manufacturing Systems*. Springer-Verlag, Berlin, 1993.

[BF81]   J. Beck and T. Fiala. Integer-making theorems. *Discrete Applied Mathematics*, 3:1–8, 1981.

[BL90]   P. Beame and M. Luby. Parallel search for maximal independence given minimal dependence. In *Proceedings of the 1st Annual ACM-SIAM Symposium on Discrete Algorithms 1990 (SODA '90)*, pp. 212–218, 1990.

[BM84]   M. Blum and S. Micali. How to generate cryptographically strong sequences of pseudo-random bits. *SIAM Journal on Computing*, 13:850–864, 1984.

[BM98]   P.A. Beling and N. Megiddo. Using fast matrix multiplication to find basic solutions. *Theoretical Computer Science*, 205:307–316, 1998.

[Bol98]   B. Bollobás. *Modern Graph Theory*. Springer, New York, 1998.

[BR91]   B. Berger and J. Rompel. Simulating ($\log^c n$)-wise independence in $\mathcal{NC}$. *Journal of the ACM*, 38:1026–1046, 1991. Preliminary version in *Proceedings of the IEEE Symposium on the Foundation of Computer Science (FOCS) 1989*, pp. 2–7.

[BRS94]   B. Berger, J. Rompel, and P. Shor. Efficient $\mathcal{NC}$ algorithms for set cover with applications to learning and geometry. *Computer System Science*, 49:454–477, 1994. Preliminary Version in *Proceedings of the 30th Annual IEEE Symposium on the Foundations of Computer Science 1989 (FOCS '89)*, pp. 54–59.

[BS83]   J. Beck and J. Spencer. Balancing matrices with line shifts. *Combinatorica*, 3:299–304, 1983.

[BS84]   J. Beck and J. Spencer. Well distributed 2-colorings of integers relative to long arithmetic progressions. *Acta Arithmetica*, 43:287–298, 1984.

[BS95]   J. Beck and V. Sós. Discrepancy theory. *Handbook of Combinatorics*, Vol. II, pp. 1405–1446, 1995.

[BS04]   A. Baltz and A. Srivastav. Fast approximation of minimum multicast congestion–implementation versus theory. *RAIRO Operations Research*, 38:319–344, 2004.

[CFG$^+$85] B. Chor, J. Freidmann, O. Goldreich, J. Hastad, S. Rudich, and R. Smolensky. The bit extraction problem and t-resilient functions. In *Proceedings of the 26th IEEE Annual Symposium on the Foundations of Computer Science 1985 (FOCS '85)*, pp. 396–407, 1985.

[CG89]   B. Chor and O. Goldreich. On the power of two-point sampling. *Journal of Complexity*, 5:96–106, 1989.

[Cha00]   B. Chazelle. *The Discrepancy Method: Randomness and Complexity*. Cambridge University Press, UK, 2000.

[Che52]   H. Chernoff. A measure of asymptotic efficiency for test of a hypothesis based on the sum of observation. *Annals of Mathematical Statistics*, 23:493–509, 1952.

[Coo85]   S. Cook. A taxonomy of problems with fast parallel algorithms. *Information and Control*, 64: 2–21, 1985.

[DF06]   B. Doerr and M. Fouz. *Hereditary Discrepancies in Different Numbers of Colors II*. Preprint, 2006.

[DGH06]   B. Doerr, M. Gnewuch, and N. Hebbinghaus. Discrepancy of symmetric products of hypergraphs. *Electronic Journal of Combinatorics*, 13, 2006.

[DHW04]   B. Doerr, N. Hebbinghaus, and S. Werth. Improved bounds and schemes for the declustering problem. In Jirí Fiala, Václav Koubek, and Jan Kratochvíl, editors, *Proceedings of the 29th Symposium on Mathematical Foundations of Computer Science 2004 (MFCS '04)*, 2004. Lecture Notes in Computer Science, 3153 (2004), pp. 760–771, Berlin–Heidelberg, Springer-Verlag.

[Dia88]   P. Diaconis. *Group Representations in Probability and Statistics*, Institute of Mathematical Statistics, Hayward, CA, 1988.

[Die97]   R. Diestel. *Graph Theory*. Springer, New York, 1997.

[DK89]   E. Dahlhaus and M. Karpiński. An efficient algorithm for the 3MIS problem. Technical Report TR-89-052, International Computer Science Institute, Berkeley, CA, 1989.

[DKK92a]   E. Dahlhaus, M. Karpinski, and P. Kelsen. An efficient parallel algorithm for computing a maximal independent set in a hypergraph of dimension 3. *Information Processing Letters*, 42:309–314, 1992. Preliminary version: Manuscript, Department of Computer Sciences, University of Texas, 1990.

[DKK92b]   E. Dahlhaus, M. Karpiński, and P. Kelsen. An efficient parallel algorithm for computing a maximal independent set in a hypergraph of dimension 3. *Information Processing Letters*, 42:309–313, 1992.

[dlVL81]   W.F. de la Vega and C.S. Lueker. Bin packing can be solved within $1 + \varepsilon$ in linear time. *Combinatorica*, 1:349–355, 1981.

[Doe00]   B. Doerr. Linear and hereditary discrepancy. *Combinatorics, Probability and Computing*, 9:349–354, 2000.

[Doe01]   B. Doerr. Lattice approximation and linear discrepancy of totally unimodular matrices. In *Proceedings of the 12th Annual ACM-SIAM Symposium on Discrete Algorithms (SODA)*, pp. 119–125, 2001.

[Doe02a]   B. Doerr. Balanced coloring: Equally easy for all numbers of colors? In H. Alt and A. Ferreira, eds., *Proceedings of the 19th Annual Symposium on Theoretical Aspects of Computer Science (STACS '02)*, 2002. Lecture Notes in Computer Science, 2285 (2002), Berlin–Heidelberg, Springer-Verlag, pp. 112–120.

[Doe02b]   B. Doerr. Discrepancy in different numbers of colors. *Discrete Mathematics*, 250:63–70, 2002.

[Doe04]   B. Doerr. The hereditary discrepancy is nearly independent of the number of colors. *Proceedings of the American Mathematical Society*, 132:1905–1912, 2004.

[DS99]   B. Doerr and A. Srivastav. Approximation of multi-color discrepancy. In D. Hochbaum, K. Jansen, J. D. P. Rolim, and A. Sinclair, eds., *Randomization, Approximation and Combinatorial Optimization (Proceedings of APPROX-RANDOM 1999)*, 1999. Lecture Notes in Computer Science, 1671 (1999), Berlin–Heidelberg, Springer-Verlag, pp. 39–50.

[DS01]   B. Doerr and A. Srivastav. Recursive randomized coloring beats fair dice random colorings. In A. Ferreira and H. Reichel, editors, *Proceedings of the 18th Annual Symposium on*

	Theoretical Aspects of Computer Science (STACS '01), 2001. Lecture Notes in Computer Science, 2010 (2001), Berlin–Heidelberg, Springer-Verlag, pp. 183–194.
[DS03]	B. Doerr and A. Srivastav. Multicolour discrepancies. *Combinatorics, Probability and Computing*, 12:365–399, 2003.
[DSW04]	B. Doerr, A. Srivastav, and P. Wehr. Discrepancies of cartesian products of arithmetic progressions. *Electronic Journal of Combinatorics*, 11:12 pages, 2004.
[EGL+98]	G. Even, O. Goldreich, M. Luby, N. Nisan, and B. Velivković. Efficient approximation of product distributions. *Random Structures & Algorithms*, 13:1–16, 1998. Preliminary Version *Approximation of General Independent Distributions*. In *Proceedings of the 24th Annual ACM Symposium on Theory of Computing 1992 (STOC '92)*, pp. 10–16.
[ES73]	P. Erdős and J.L. Selfridge. On a combinatorial game. *Journal of Combinatorial Theory (A)*, 14:298–301, 1973.
[Fel67]	W. Feller. *An Introduction to the Theory of Probability and Its Applications*. John Wiley & Sons, 1967.
[FKN95]	T. Feder, E. Kushilevitz, and M. Naor. Amortized communication complexity. *SIAM Journal on Computing*, 24:736–750, 1995. Preliminary Version in *Proceedings of the 32nd IEEE Symposium on the Foundation of Computer Science 1991 (FOCS '91)*, pp. 239–248.
[GGJY76]	M.R. Garey, R.L. Graham, D.S. Johnson, and A.C.-C. Yao. Resource constrained scheduling as generalized bin packing. *Journal of Combinatorial Theory (A)*, 21:257–298, 1976.
[GKS04]	R. Gandhi, S. Khuller, and A. Srinivasan. Approximation algorithms for partial covering problems. *Journal of Algorithms*, 53:55–84, 2004.
[GM84]	S. Goldwasser and S. Micali. Probabilistic encryption. *Journal of Computer and System Sciences*, 28:270–299, 1984. Preliminary version in *Proceedings STOCS '82*, pp. 365–377.
[Goo93a]	M.T. Goodrich. Constructing arrangements optimally in parallel. *Discrete & Computational Geometry*, 9:371–385, 1993.
[Goo93b]	M.T. Goodrich. Geometric partitioning made easier, even in parallel. In *Proceedings of the 9th Annual ACM Symposium on Computational Geometry 1993*, pp. 73–82, 1993.
[Goo96]	M.T. Goodrich. Fixed-dimensional parallel linear programming via relative ϵ-approximations. In *Proceedings of the 7th Annual ACM-SIAM Symposium on Discrete Algorithms 1996 (SODA '96)*, pp. 132–141, 1996.
[GS89a]	M. Goldberg and T. Spencer. Constructing a maximal independent set in parallel. *SIAM Journal on Discrete Mathematics*, 2:322–328, 1989.
[GS89b]	M. Goldberg and T. Spencer. A new parallel algorithm for the maximal independent set problem. *SIAM Journal on Computing*, 18:419–427, 1989.
[GW95]	M.X. Goemans and D.P. Williamson. Improved approximation algorithms for maximum cut and satisfiability problems using semidefinite programming. *Journal of the ACM*, 42:1115–1145, 1995.
[Han92]	Y. Han. Parallel derandomization techniques. *Advances in Parallel Algorithms*, 368–395, 1992.
[HILL97]	J. Håstad, R. Impagliazzo, L. Levin, and M. Luby. Construction of a pseudo-random generator from any one-way function. Technical Report 91-068, ICSI, 1997.
[HILL99]	J. Håstad, R. Impagliazzo, L. Levin, and M. Luby. A pseudo-random generator from any one-way function. *SIAM Journal on Computing*, 28:1364–1396, 1999.
[HMRAR98]	M. Habib, C. McDiarmid, J. Ramirez-Alfonsin, and B. Reed. *Probabilistic Methods for Algorithmic Discrete Mathematics*. Springer Series Algorithms and Combinatorics, Vol. 16. Springer-Verlag, 1998.
[Hoe63]	W. Hoeffding. Probability inequalities for sums of bounded random variables. *American Statistical Association Journal*, 58:13–30, 1963.

[IW97] R. Impagliazzo and A. Wigderson. $P = BPP$ unless e has sub-exponential circuits, derandomizing the XOR lemma. In *Proceedings of the 29th Annual ACM Symposium on the Theory of Computing 1997 (STOC '97)*, pp. 220–229, 1997.

[IZ90] R. Impagliazzo and D. Zuckerman. How to recycle random bits. In *Proceedings of the 31st Annual IEEE Symposium on the Foundations of Computer Science 1990 (FOCS '89)*, pp. 248–253, 1990.

[Jof74] A. Joffe. On a sequence of almost deterministic k-independent random variables. *Annals of Probability*, 2:161–162, 1974. Preliminary version in *Proceedings of the AMS*, 29:381–382, 1971.

[Kel92] P. Kelsen. On the parallel complexity of computing a maximal independent set in a hypergraph. In *Proceedings of the 24th ACM Symposium on the Theory of Computing 1992 (STOC '92)*, pp. 339–350, 1992.

[KK97] D. Karger and D. Koller. (De)randomized construction of small sample spaces in $\mathcal{NC}$. *Computer System Sciences*, 55:402–413, 1997. Preliminary version in *Proceedings of the 35th IEEE Symposium on the Foundations of Computer Science 1994 (FOCS '94)*, pp. 252–263.

[KM94] D. Koller and N. Megiddo. Constructing small sample spaces satisfying given constraints. *SIAM Journal on Discrete Mathematics*, 7:260–274, 1994. Preliminary version in *Proceedings of the 25th ACM Symposium on Theory of Computing 1993 (STOC '93)*, pp. 268–277.

[KM97] H. Karloff and Y. Mansour. On construction of k-wise independent random variables. *Combinatorica*, 17:91–107, 1997. Preliminary version in *Proceedings of the 26th ACM Symposium on the Theory of Computing 1994 (STOC '94)*, pp. 564–573.

[KPS88] R. Karp, N. Pippenger, and M. Sipser. Expanders, randomness, or time versus space. *Journal of Computer and System Sciences*, 36:379–383, 1988. Preliminary version in *Proceedings of the First Annual Conference on Structure in Complexity Theory (1986)*, pp. 325–329.

[KR93] H. Karloff and P. Raghavan. Randomized algorithm and pseudorandom numbers. *Journal of the ACM*, 40:454–476, 1993.

[KUW88] R.M. Karp, E. Upfal, and A. Wigderson. The complexity of parallel search. *Computer and System Sciences*, 36:225–253, 1988.

[KW85] R.M. Karp and A. Wigderson. A fast parallel algorithm for the maximal independent set problem. *Journal of the ACM*, 32:762–773, 1985. Preliminary version in *Proceedings of the 16th ACM Symposium on Theory of Computing 1984 (STOC '84)*, pp. 266–272.

[LLSZ97] N. Linial, M. Luby, M. Saks, and D. Zuckerman. Efficient construction of a small hitting set for combinatorial rectangles in high dimension. *Combinatorica*, 17:215–234, 1997. Preliminary version in *Proceedings of the 25th ACM Symposium on the Theory of Computing 1993 (STOC '93)*, pp. 258–267.

[LN93] Michael Luby and Noam Nisan. A parallel approximation algorithm for positive linear programming. In *ACM Symposium on Theory of Computing*, pp. 448–457, 1993.

[LPS88] A. Lubotzky, R. Phillips, and P. Sarnak. Ramanujan graphs. *Combinatorica*, 8:261–277, 1988.

[ŁS97] T. Łuczak and E. Szymańska. A parallel randomized algorithm for finding a maximal independent set in a linear hypergraph. *Algorithms*, 25:311–320, 1997.

[LSV86] L. Lovász, J. Spencer, and K. Vesztergombi. Discrepancies of set-systems and matrices. *European Journal of Combinatorics*, 7:151–160, 1986.

[Lub86] M. Luby. A simple parallel algorithm for the maximal independent set problem. *SIAM Journal on Computing*, 15:1036–1053, 1986.

[Lub96] M. Luby. Pseudorandomness and cryptographic applications. *Princeton Computer Science Notes*, 1996.

[LV96]   M. Luby and B. Velickovic. On deterministic approximation of DNF. *Algorithmica*, 16:415–433, 1996.
[LW95]   M. Luby and A. Wigderson. Pairwise independence and derandomization. Technical Report TR-95-035, International Computer Science Institute, UC Berkeley, July 1995.
[LWV93]   M. Luby, A. Wigderson, and B. Velickovic. Deterministic approximate counting of depth−2 circuits. In *Proceedings of the Second Israel Symposium on Theory of Computing and Systems 1993*, pp. 18–24, 1993.
[Mat99]   J. Matoušek. *Geometric Discrepancy*. Springer-Verlag, Heidelberg, New York, 1999.
[Mat00]   J. Matoušek. On the discrepancy for Cartesian products. *Journal of the London Mathematical Society*, 61:737–747, 2000.
[MNN94]   R. Motwani, J. Naor, and M. Naor. The probabilistic method yields deterministic parallel algorithms. *Computer and System Sciences*, 49:478–516, 1994. Preliminary version in *Proceedings of the 30th Annual IEEE Symposium on Foundations of Computer Science 1989*, pp. 8–13.
[MR95]   R. Motwani and P. Raghavan. *Randomized Algorithms*. Cambridge University Press, 1995.
[MR99]   S. Mahajan and H. Ramesh. Derandomizing approximation algorithms based on semidefinite programming. *SIAM Journal on Computing*, 28:1641–1663, 1999. Preliminary version Derandomizing semidefinite programming based approximation algorithms. In *Proceedings of the 36th Annual IEEE Symposium on Foundations of Computer Science (1995)*, pp. 162–169.
[MRS97]   S. Mahajan, E.A. Ramos, and K.V. Subrahmanyam. Solving some discrepancy problems in *Proc. 17th Conference on Foundations of Software Technology and Theoretical Computer Science*, London, Springer-Verlag, 1997, pp. 22–36.
[MS77]   F.J. MacWilliams and N.J.A. Sloane. *The Theory of Error Correcting Codes*. North Holland, Amsterdam, 1977.
[MS96]   J. Matoušek and J. Spencer. Discrepancy of arithmetic progressions. *Journal of the American Mathematical Society*, 9:195–204, 1996.
[Mul96]   K. Mulmuley. Randomized geometric algorithms and pseudo-random generators. *Algorithmica*, 16:450–463, 1996. Preliminary version in *Proceedings of the 33rd Annual IEEE Symposium on the Foundations of Computer Science 1992 (FOCS '92)*, pp. 90–100.
[Nie92]   H. Niederreiter. *Random Number Generation and Quasi-Monte Carlo Methods*, volume 63 of *CBMS-NSF Regional Conference Series in Applied Mathematics*. SIAM, Philadelphia, PA, 1992.
[Nis91]   N. Nisan. Pseudorandom bits for constant depth circuits. *Combinatorica*, 1:63–70, 1991.
[Nis92]   N. Nisan. Pseudorandom generators for space-bounded computation. *Combinatorica*, 12:449–461, 1992.
[NN93]   J. Naor and M. Naor. Small bias probability spaces: Efficient constructions and applications. *SIAM Journal on Computing*, 22:838–856, 1993. Preliminary version in *Proceedings of the 22nd Annual ACM Symposium on the Theory of Computing 1990 (STOC '90)*, pp. 213–223.
[NRT87]   A.P.-C. Ng, P. Raghavan, and C.D. Thompson. Experimental results for a linear program global router. *Computers and Artificial Intelligence*, 6:229–242, 1987.
[NSS95]   M. Naor, L.J. Schulman, and A. Srinivasan. Splitters and near optimal derandomization. In *Proceedings of the 36th Annual IEEE Symposium on the Foundations of Computer Science 1995 (FOCS '95)*, pp. 182–191, 1995.
[Oka58]   M. Okamoto. Some inequalities relating to the partial sum of binomial probabilities. *Annals of the Institute of Statistical Mathematics*, 10:29–35, 1958.
[Pap94]   C.H. Papadimitriou. *Computational Complexity*. Addison-Wesley Publ. Company, Reading, MA, 1994.

[Per90] R. Peralta. On the randomness complexity of algorithm. Technical Report TR90-1, University of Wisconsin, Milwaukee, WI, 1990.

[Rag88] P. Raghavan. Probabilistic construction of deterministic algorithms: Approximating packing integer programs. *Computer System Sciences*, 37:130–143, 1988.

[Rot64] K.F. Roth. Remark concerning integer sequences. *Acta Arithmetica*, 9:257–260, 1964.

[RS83] H. Röck and G. Schmidt. Machine aggregation heuristics in shop scheduling. *Methods of Operations Research*, 45:303–314, 1983.

[RT87] P. Raghavan and C.D. Thompson. Randomized rounding: A technique for provably good algorithms and algorithmic proofs. *Combinatorica*, 7:365–374, 1987.

[Sak96] M. Saks. Randomization and derandomization in space-bounded computation. In *Eleventh Annual IEEE Conference on Computational Complexity*, pp. 128–149, Philadelphia, PA, May 1996.

[San87] M. Santha. On using deterministic functions to reduce randomness in probabilistic algorithms. *Information and Computation*, 74:241–249, 1987.

[Sch92] L. Schulman. Sample spaces uniform on neighborhoods. In *Proceedings of 24th Annual ACM Symposium on Theory of Computing 1992 (STOC '92)*, pp. 17–25, 1992.

[Sip88] M. Sipser. Expanders, randomness, or time versus space. *Computer and Systems Science*, 36:379–383, 1988.

[Spe85] J. Spencer. Six standard deviation suffice. *Transactions of the American Mathematical Society*, 289:679–706, 1985.

[Spe87] J. Spencer. *Ten Lectures on the Probabilistic Method*. SIAM, Philadelphia, 1987.

[Sri97] A. Srinivasan. Improving the discrepancy bound for sparse matrices: Better approximation for sparse lattice approximation problems. In *Proceedings of the 7th ACM/SIAM Symposium on Discrete Algorithms 1997 (SODA '97)*, pp. 692–701, 1997.

[Sri99] A. Srinivasan. Improved approximation guarantees for packing and covering integer programs. *SIAM Journal on Computing*, 29:648–670, 1999.

[Sri01a] A. Srinivasan. New approaches to covering and packing problems. In *Proceedings of the 12th Annual ACM-SIAM Symposium on Discrete Algorithms 2001 (SODA '01)*, pp. 567–576, 2001.

[Sri01b] A. Srivastav. Derandomization in combinatorial optimization. In Rajasekaran, Pardalos, Reif, and Rolim, editors, *Handbook of Randomized Computing*. Kluwer Academic Press, 2001, pp. 731–842.

[SS93] A. Srivastav and P. Stangier. On quadratic lattice approximations. In K.W. Ng, P. Raghavan, and N.V. Balasubramanian, editors, *Proceedings of the 4th International Symposium on Algorithms and Computation (ISAAC '93)*, Hong-Kong, 1993. Lecture Notes in Computer Science, 762 (1993), Springer-Verlag, pp. 176–184

[SS96] A. Srivastav and P. Stangier. Algorithmic Chernoff–Hoeffding inequalities in integer programming. *Random Structures & Algorithms*, 8:27–58, 1996.

[SS97a] A. Srivastav and P. Stangier. A parallel approximation algorithm for resource constrained scheduling and bin packing. In G. Bilardi, A. Ferreira, R. Lüling, and J. Rolim, editors, *Proceedings of the 4th International Symposium on Solving Irregularly Structured Problems in Parallel*, 1997. Lecture Notes in Computer Science, 1253 (1997), Springer-Verlag, pp. 147–159.

[SS97b] A. Srivastav and P. Stangier. Tight aproximations for resource constrained scheduling and bin packing. *Discrete Applied Mathematics*, 79:223–245, 1997.

[SS01] M. Sitharam and T. Straney. Derandomized learning of boolean functions. *International Journal of Foundations of Computer Science*, 12:491–516, 2001. Preliminary version in Lecture Notes in Computer Science, 1316 (1997), pp. 100–115.

[Szy98] E. Szymańska. PhD thesis. PhD thesis, Adam Mickiewicz University, Poznán, Poland, 1998.

[Vaz86]   U.V. Vazirani. Randomness, Adversaries and Computation. PhD thesis, University of California, Berkeley, CA, 1986.
[VV84]   U.V. Vazirani and V.V. Vazirani. Efficient and secure pseudo-random number generation. In *Proceedings of the 25th Annual IEEE Symposium on Foundations of Computer Science 1984 (FOCS '84)*, pp. 458–463, 1984.
[Wes96]   D.B. West. *Introduction to Graph Theory*. Prentice Hall, Upper Saddle River, NJ, 1996.
[Yao82]   A. Yao. Theory and applications of trapdoor functions. In *Proceedings of the 23th Annual IEEE Symposium on Foundations of Computer Science 1982 (FOCS '82)*, pp. 80–91, 1982.

19
Broadcasting on Networks of Workstations

19.1	Introduction...	19-1
19.2	Broadcasting in LogP Model	19-2
	Single-Item Broadcasting • All-to-All Broadcasting • Combining Broadcast Problem • Summation	
19.3	Broadcasting in Clusters of Workstations	19-5
	Problem Definition • Broadcasting • Multicasting • Bounding Global Transfers • Postal Model	
19.4	Broadcasting in Heterogenous Networks	19-19
	Problem Definition • Minimizing the Sum of Completion Times • 1.5-Approximation for Minimizing Broadcast Time • Polynomial Time Approximation Scheme • Multicast	
References	....	19-30

Samir Khuller
University of Maryland

Yoo-Ah Kim
University of Connecticut

Yung-Chun (Justin) Wan
Google Inc.

19.1 Introduction

Networks of workstations (NOWs) are a popular alternative to massively parallel machines and are widely used (e.g., the Condor project at Wisconsin [PL96] and the Berkeley NOW project [PCA95]). By simply using off-the-shelf PCs, a very powerful workstation cluster can be created, and this can provide a high amount of parallelism at relatively low cost. With the recent interest in grid computing [FK98] there is an increased interest to harness the computing power of these clusters to have them work together to solve large applications that involve intensive computation. Several projects such as Magpie [KHB+99, KBG00] are developing platforms to allow applications to run smoothly by providing primitives for performing basic operations such as broadcast, multicast, scatter, reduce, and so forth. Many of these primitives are implemented using simple heuristics. Our goal is to develop models, and an understanding of the difficult issues and challenges in implementing broadcast and multicast on such platforms. Several approximation algorithms have been developed in the theory literature, but they are for different models (typically an underlying communication graph exists that forbids any communication between nonadjacent nodes). The algorithms developed are computationally intensive and complex.

One fundamental operation that is used in such clusters is that of *broadcast* (this is a primitive in many message passing systems such as MPI [MPI94, BDH+97, GLDS96, HH98]). Some of this framework has been extended to clustered wide area systems (see References KHB+99 and KBG00) of the type we are addressing. In addition, it is used as a primitive in many parallel algorithms. The main objective of a broadcast operation is to quickly distribute data to the entire network for processing. Another situation is when the system is performing a parallel search, then the successful processor needs to inform all other processors that the search has concluded successfully. Various models for heterogenous environments have been proposed in the literature. One general model is the one proposed by Bar-Noy et al. [BNGNS98], where the communication costs between links are not uniform. In addition, the sender may engage in another communication before the current one is complete. An approximation factor with a guarantee of $O(\log k)$ is given for the operation of performing a multicast. Other popular models in the theory literature generally assume an underlying communication graph, with the property that only nodes adjacent in this graph may communicate. See References EK02 and EK03 for recent approximation algorithms on this model. However, this model is too restrictive and allows direct communication only between nodes adjacent in a certain communication graph. Broadcasting efficiently is an essential operation and many works are devoted to this (see References RL88, SHA88, KSSK93, BNK94, and BRP99 and references therein).

We consider three different models here. We first consider the LogP model proposed by Culller et al. [CKP+93b]. In the second model, we assume that it takes a long time to send a message to a machine in a different cluster, but sending a message to machine in the local cluster can be done quickly. Moreover, the time taken is the same, regardless of which machine in the cluster we send the message to. The third model we consider addresses a cluster that has several different machines (heterogenous cluster of workstations) and the time taken to send a message from different machines can vary.

19.2 Broadcasting in LogP Model

One of the popular models for parallel computing is the LogP model proposed by Culler et al. [CKP+93b]. The model extends Valiant's BSP model [Val90], and allows processors to work asynchronously and communicate by point-to-point messages. Four parameters are used to characterize a parallel system:

1. L: the latency (or maximum delay), associated with delivering a message
2. o: the overhead, representing the length of time for which a processor is busy during the transmission or reception of a message
3. g: the gap, a lower bound on the time between the transmission of successive messages, or the reception of successive messages, at the same processor
4. P: the number of processors

Figure 19.1b illustrates transmissions when $P = 8, L = 3, o = 2, g = 4$. The LogP model generalizes the Postal model introduced by Bar-Noy and Kipnis [BNK94]. In the postal model, a processor is busy for only one time unit when it sends a message and becomes free to send another message after that. It takes

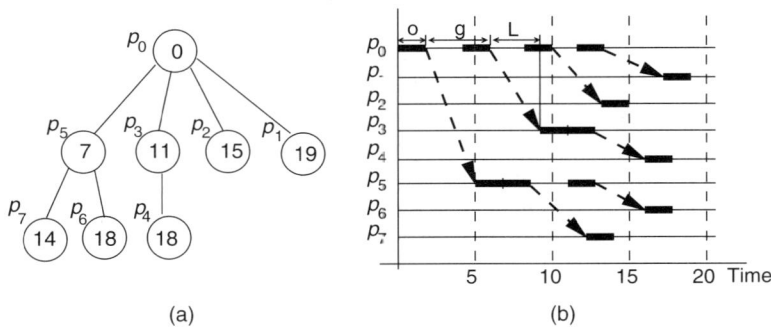

FIGURE 19.1 (a) An optimal broadcast tree $BT(8)$ when $P = 8, L = 3, o = 2, g = 4$. The numbers inside the nodes represent the completions times of processors. (b) Activity of each processor for the corresponding broadcast schedule.

L units of time (latency) for a message to arrive at the receiver. Therefore, Postal model is a special case of LogP model in which $g = 1$ and $o = 0$.

We describe the following algorithms for several variants of broadcast problem and a summation problem in LogP model [CKP+93b, RMKS92]:

- For the single-item broadcast problem, a simple greedy algorithm gives an optimal solution.
- For the all-to-all broadcast problem, where each of P processors has a data item to be sent to all other processors, there is an algorithm that gives an optimal schedule.
- In the combining broadcast problem, each processor holds a value and the P values has to be combined. We describe an optimal schedule to make the combined value available to all P processors.
- In the summation problem, we are given t time units and the objective is to add as many operands as possible for t time. We describe an optimal algorithm for the problem.

19.2.1 Single-Item Broadcasting

We assume that there is a message available at processor p_0 and want to find an optimal schedule to send the message to all other $P - 1$ processors. A simple greedy algorithm gives an optimal solution for this problem. That is, all informed processors send the message to uninformed processors as early and frequently as possible.

A broadcast operation can be represented as a broadcast tree BT as follows. The root of the tree is the source processor of the message (processor p_0). If a processor sends the message to a set of other processors, then the receivers become (ordered) children of the sender. We define the completion time c_i of processor p_i to be the time at which p_i receives the message. Given the completion time c_i of a parent node p_i, the completion times of k-th children are $c_i + L + 2o + (k-1) \times g$. Let $BT(P)$ be the subtree of BT with P nodes labeled with smallest completion times (ties broken arbitrarily). It is clear that $BT(P)$ gives an optimal schedule for the single-item broadcast problem. Figure 19.1 shows an example of a broadcast tree with $P = 8$ and the activity of each processor.

19.2.2 All-to-All Broadcasting

We next consider the all-to-all broadcast problem, in which each of P processors has a data item to be sent to all other processors. A trivial solution can be obtained by performing the single-item broadcast algorithm P times, but there exists a simple optimal solution for the problem. Consider the following algorithm:

All-to-All Broadcast

1 At each time $t \cdot g$, $t = 0, 1, \ldots, P - 2$
2   For each processor p_i, $i = 0, \ldots, P - 1$
3   p_i sends its message to processor $p_{i+t+1} \bmod P$.

It is easy to see that a processor can receive all $P - 1$ messages by time $L + 2o + (P - 2)g$ and this is an optimal solution for the all-to-all broadcast problem.

Theorem 19.1 *There is an optimal algorithm for all-to-all broadcast problem.*

Proof. The lowerbound on the all-to-all broadcast time is $L + 2o + (P - 2)g$ since a processor can receive the first message at time $L + 2o$ and it has to receive $P - 1$ messages. ∎

19.2.3 Combining Broadcast Problem

In this problem, each processor holds a value and we want to have a combined value of the P values (such as max or sum) available at every processor. A trivial solution would be to solve "all-to-one" broadcast (by reversing single-item broadcast) and then broadcast the combined value to all processors. The algorithm gives a 2-approximation. However, it is possible to devise an optimal algorithm for the problem.

We assume that the combining operation is commutative and associative. In addition, we assume that the operation takes 0 unit of time in each processor. Let x_i be a value that processor p_i has initially. $x[i:j]$ denotes the combined value of $x_i, x_{(i+1) \bmod P}, \ldots, x_{(j-1) \bmod P}, x_j$. For example, if $P = 5$, then $x[3:1]$ is the combined value of x_3, x_4, x_0, and x_1.

For simplicity, we describe the algorithm in the Postal model (i.e., $g = 1, o = 0$). Let $P(t)$ be the maximum number of processors that receive the item by time t in an optimal single-item broadcast algorithm. Then $P(t)$ in the Postal model can be computed recursively as follows:

$$P(t) = \begin{cases} 1 & \text{for } 0 \leq t < L, \\ P(t-1) + P(t-L) & \text{otherwise}. \end{cases}$$

Let $P = P(T)$. Then the following algorithm gives a schedule for combining broadcast problem and can be done in T time units:

All-to-All Broadcast with Combining

1   At each time $t, t = 0, 1, \ldots, T - L$
2   For each processor $p_i, i = 0, \ldots, P - 1$
3   p_i sends its current value to processor $p_{i+P(t+L-1) \bmod P}$.

Lemma 19.1 *At time t, processor p_i has the value $x[i - P(t) + 1 : i]$.*

Proof. We prove this by induction. At time 0, clearly it is true. Suppose that for any time $t' \leq t$ processor p_i has the value $x[i - P(t') + 1 : i]$. At time $t + 1$, processor p_i receives the value that processor $j = i - P(t)$ sent at time $t - L + 1$. Since the value sent by j is $x[i - P(t) - P(t-L+1) + 1 : i - P(t)]$ and i has the value $x[i - P(t) + 1 : i]$ at time t, the combined value that processor p_i has at time $t + 1$ is $x[i - P(t) - P(t-L+1) + 1 : i]$. By the definition of $P(t)$ we have the lemma. ∎

Theorem 19.2 *Using the above algorithm, every processor can receive value $x[0 : P - 1]$ at time T, where $P = P(T)$.*

Proof. By Lemma 19.1, at time T processor p_i has value $x[i - P(T) + 1 : i]$, which is the same as $x[0 : P - 1]$. ∎

19.2.4 Summation

We now consider the following summation problem. There are a set of operands $x_1, \ldots, x_n$ and we assume that an algorithm can choose how these are initially distributed among P processors. Let $X[i, j] = x_i + x_{i+1} + \cdots + x_j$. Given time t, the objective is to compute $X = X[1, n]$ for maximum n in t time units. Note that n may be bigger than P and it is possible that a processor has multiple operands at the beginning. Each addition takes one unit of time. We describe an algorithm to add the maximum number of operands in t time units. In the summation algorithm, a processor starts performing local additions and then performs a "reverse broadcast" schedule to add the partial sums computed in each processor.

It can be shown that there is an optimal solution that has the following properties:

1. No processor sends more than one message
2. No processor is idle until it sends a message
3. Suppose that a processor p receives a message at times $r_p^1, r_p^2, \ldots, r_p^k$ and sends at time s_p (except the last processor). Then $r_p^j = s_p - (o+1) + (k-j)g$

Lemma 19.2 *Given an optimal summation schedule, we can convert the schedule so that Properties (1)–(3) are satisfied and we can add as many operands as in the original schedule.*

Proof. (1) Suppose that a processor sends more than one partial sums $m_1, m_2, \ldots, m_k$ at time $t_1, t_2, \ldots, t_k$, respectively. We can modify the schedule so that the processor sends only one message without increasing the summation time. That is, instead of sending m_i, $i < k$, the processor adds m_i to m_{i+1} and only sends the sum of all partial sums $\sum_i m_i$ at time t_k.

(2) As we want to add as many operands as possible, processor can always add more operands without idle time.

(3) It is clear we have $s_p \geq r_p^j + o + 1 + (k-j)g$. If s_p is strictly greater than $r_p^j + o + 1 + (k-j)g$, then we can delay the reception of the message at p and add more operands. ∎

We consider a summation schedule that satisfies all three properties. Let p_i, $i > 0$ send its partial sum at time t_i and p_0 compute X at time t_0. The following lemma gives the total number of operands that can be added in the schedule.

Lemma 19.3 *Let p_i, $i > 0$ send its partial sum at time t_i and p_0 compute X at time t_0. Then the total number of operands added is $\sum_{i=0}^{P-1} t_i - o(P-1) + 1$.*

Proof. Let p_i receive k_i messages. Then the number of operands added by p_i is $t_i - k_i(o+1) + 1$. The total number of operands added is $\sum_i (t_i - k_i(o+1) + 1) = \sum_i t_i - (P-1)(o+1) + P = \sum_i t_i - (P-1)o + 1$. ∎

By Lemma 19.3, we need to maximize $\sum_i t_i$ in order to process the maximum number of operands. We can find sending time t_i for each processor p_i by reversing a single-item broadcast schedule. Consider an optimal single-item broadcast schedule A with parameters L, $o+1$, g, and P. We construct a summation schedule by reversing the broadcast schedule as follows: If processor p_i receives the message at time c_i in A, then in the summation algorithm A', p_i sends the partial sum at time $t_i = t - c_i$ (except the processor with $c_i = 0$). Processors perform additions until it sends its partial sum.

Lemma 19.4 *The optimal broadcast algorithm for the single-item broadcast problem in Section 19.2.1 minimizes $\sum_i c_i$, where c_i, is the receiving time of processor p_i.*

As an optimal single-item broadcast algorithm gives a schedule with maximum $\sum_i (t - c_i)$, our algorithm maximizes $\sum_i t_i$.

Theorem 19.3 *Given time t, there is a summation algorithm that maximizes the number of operands added by time t.*

19.3 Broadcasting in Clusters of Workstations

Consider several clusters of workstations. Each local cluster (sometimes this is also called a subnet) is connected on a fast local area network, and intercluster communication is through a wide area network. In such situations, the time taken for a pair of machines in the same cluster to communicate, can be significantly smaller than the communication time of a pair of machines in different clusters. In fact, in the work by Lowekamp and Beguelin [LB96] they also suggest methods for obtaining the subnets/clusters based on communication delays between pairs of processors.

Motivated by this, the *communication model* we consider is the following. There are k clusters of machines. Cluster i has size n_i. We will assume that in one time unit, a machine can send a message to any machine in its own cluster. However, sending a message from one machine to another machine in a different cluster takes C time units. Even if the machines in a cluster are heterogenous, their transmission times are usually much less than the communication time across clusters. We also assume that a machine can be sending or receiving a message from only one machine at any point of time. In this model, Lowekamp and Beguelin [LB96] propose some simple heuristics for performing broadcast and multicast. However, these heuristics may produce solutions arbitrarily far from optimal.

One potential concern with the above model is that it allows an arbitrary number of processors to communicate in every time step. This is of concern if the global network connecting the different clusters does not have enough capacity to permit such arbitrary communication patterns. There are several ways in which we can restrict the model. One model that we propose is the *bounded degree model* where each cluster i is associated with a parameter d_i that restricts the *number* of processors from this cluster that can communicate with processors outside this cluster in each time step. Another possible manner in which we may restrict global communication in each time step is to restrict the *total* number of simultaneous transfers that may be going on in each time step without restricting the number of transfers into/out of a single cluster. We call this model the *bounded size matching model*.

In addition, we consider a *postal model* [BNK94] where each message simply has a latency of C time units when the message is sent from one processor to another processor belonging to a different cluster. The sender is busy for only one time unit while the message is being injected into the network. The message takes C units of transit time and the receiver is busy for one unit of time when the message arrives. This model essentially captures the communication pattern as discussed in several papers that deal with implementations of systems to support such primitives (see References KHB$^+$99 and KBG00).

The results for these models can be described as follows:

- We develop algorithms for broadcasting and multicasting in the basic $1/C$ model, and show that these algorithms produce solutions where the time to perform the broadcast is not more than optimal by a factor of 2 (moreover, this bound is tight for both algorithms).
- For the *bounded degree model* we show how to reduce the problem to an instance of the basic model to develop a factor 3 approximation. The corresponding bound for multicasting is 3.
- For the *bounded size matching* model, we develop an algorithm for which we can prove a factor 2 approximation. The corresponding bound for multicasting is 2.
- For the *postal model*, our algorithm has a bound of 3. In addition, we present another algorithm, called *Interleaved* LCF, and show that the makespan is at most two times OPT$'$, where OPT$'$ is the minimum makespan among schedules that minimize the total number of global transfers.

In many of these cases, the algorithm we develop, called algorithm (largest cluster first) LCF, plays a central role. We first show that there is a simple analysis that shows that the worst case broadcast time can be bounded by a factor of 3 for this algorithm. We then improve the *lower bound* by introducing the concept of "experiencing" intercluster transfers to improve the approximation factor to 2. This lower bound in a sense combines the difficulty of propagating messages to a large number of clusters with the fact that some of the clusters may be very large.

The LogP model [CKP+93a] suggests an alternative framework when dealing with machines in a single cluster. Broadcasting algorithms [KSSK93] for the LogP model have been developed and shown to be optimal. An interesting generalized model would be to have a "two-level LogP" model with different parameters for the local networks (intracluster) and global networks (intercluster).

Finally, note that we do not know if the problem of minimizing the broadcast time is *NP*-hard or not (in any of these models). In addition, we are currently examining generalizations of this model when the communication time in different clusters may be different owing to different speed networks and different speed processors.

19.3.1 Problem Definition

We assume we have k clusters of processors. Cluster K_i has size $n_i, i = 0, \ldots, (k-1)$, the number of processors in the ith cluster. The total number of processors, denoted by N, is $\sum_{i=0}^{k-1} n_i$. We will assume that the broadcast/multicast originates at a processor in K_0. We order the *remaining* clusters in nonincreasing size order. Hence, $n_1 \geq n_2 \geq \cdots \geq n_{k-1}$. Clearly, n_0 could be smaller or larger than n_1; since it is simply the cluster that originates the broadcast/multicast.

A message may be sent from a processor, once it has received the message. If the message is sent to a processor in its cluster, the message arrives one time unit later. If the message is sent to a processor in a different cluster then the message arrives C time units later. Both the sending and receiving processors are busy during those C time units. In addition, we assume that each cluster advertises a single address to which messages are sent. Each cluster thus receives a message only once at this machine and then the message is propagated to different machines in the cluster. Thus, new machines may be added or dropped without having to inform other clusters of the exact set of new addresses (we only need to keep track of the sizes of the clusters). In some cases, the broadcast time can be reduced by having many messages arrive at the same cluster. *However, when we compare to the optimal solution we do not make any assumptions about the communication structure of the optimal solution.*

In Section 19.3.5, we consider a slightly different postal model where a processor is busy for only one time unit when it sends a message. The time a message arrives at a receiver depends on whether the sender and receiver are in the same cluster or not—it takes one time unit if it is a local transfer, and C time units otherwise.

19.3.2 Broadcasting

The high-level description of the algorithm is as follows: The source node first performs a local broadcast within its cluster. This takes $\lceil \log n_0 \rceil$ rounds. After all the nodes of K_0 have the message, we broadcast the message to the first n_0 clusters. Each node in K_0 sends a message to a distinct cluster. This takes exactly C rounds. Each cluster that receives a message does a local broadcast within its cluster. All nodes that have received the message then sends the message to distinct clusters. Again this takes C more rounds. While doing this, every node in K_0 keeps sending a message to a cluster that has not received a message as yet. Repeat this until all the processors receive the message. We call this algorithm LCF, as we always choose the largest cluster as a receiver among clusters that have not received the message.

Algorithm LCF

1. Broadcast locally in K_0.
2. Each cluster performs the following until all processors get informed.
   a. Cluster K_i in which all processors have messages, picks the first n_i clusters that have not received a message, and sends the message to them at every C time unit. Repeat until all clusters have at least one message.
   b. Each cluster that received a message does local broadcasting until all processors in the cluster have messages. (We interrupt all local broadcasting and do one global transfer, if the number of processors having the message is at least the number of clusters that have not received a message.)

Note that in our algorithm each cluster receives only one message from other clusters. That is, the total number of global transfers is minimized (we need $k-1$ global transfers). This property is important since we want to avoid wasting wide area bandwidth that is expensive.

19.3.2.1 Analysis

In this subsection, we prove that LCF gives a 2-approximation. For the purpose of analysis, we modify LCF slightly. The makespan of the schedule by the *modified* algorithm may be worse than the original algorithm (but no better) and it is at most two times the optimal.

In *modified* LCF, local and global phases take place in turn (see Figure 19.2). Let L_i be the set of clusters that receive the message at i-th global step. For example, L_0 includes K_0 and L_1 includes all clusters that receive the message from K_0 at the end of the first global phase. Let N_i be the total number of processors in clusters belonging to L_i. That is, $N_i = \sum_{K \in L_i} |K|$.

Algorithm *Modified* LCF

1. Broadcast locally in K_0. Then we have that $L_0 = K_0$ and $N_0 = n_0$.
2. At i-th step (repeat until all processors get informed).
   a. Global phase: Pick $\sum_{j=0,\ldots,i-1} N_j$ largest clusters that are not informed as yet. Each processor in $\bigcup_{j=0,\ldots,i-1} L_j$ sends one message to each of those clusters.
   b. Local phase: Clusters in L_i do local broadcasting.

Let p be the number of global transfer steps that *modified* LCF uses. Then, we have the following theorem.

Theorem 19.4 *The broadcast time of our algorithm is at most $2 \log N + pC + 3$.*

Define A_i (B_i) to be the biggest (smallest) cluster in L_i. We need the following two lemmas to prove this theorem.

Lemma 19.5 *For $i = 0 \ldots p - 1$, $n_0 \cdot |B_1| \cdots |B_i| \leq N_i$.*

Proof. We prove this by induction. For $i = 0$, it is true since $N_0 = n_0$. Suppose that for $i = k$ ($<p-1$) we have $n_0 \cdot |B_1| \cdots |B_k| \leq N_k$. Since at $(k+1)$-th global transfer step, every node in N_k will send the message to a cluster in L_{k+1}, $|L_{k+1}| \geq N_k$. Furthermore, the size of clusters in L_{k+1} is at least $|B_{k+1}|$ by

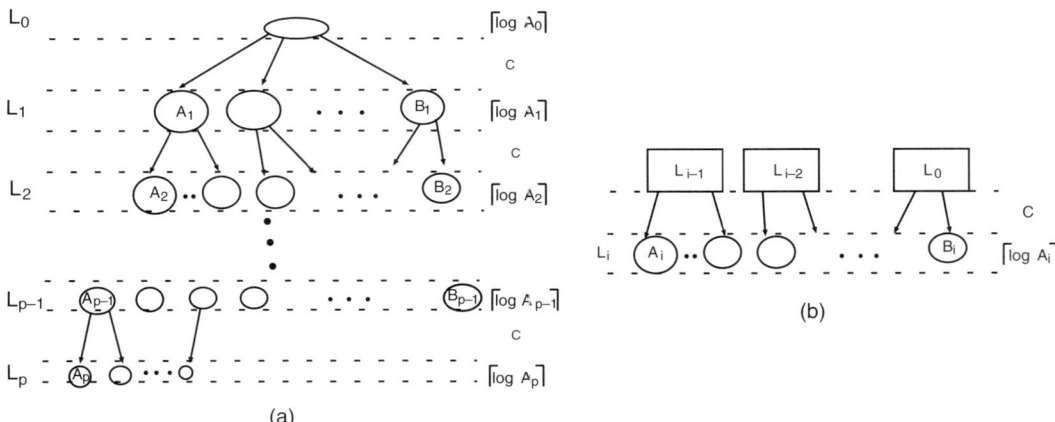

FIGURE 19.2 (a) It shows how *modified* LCF works (b) At i-th step, all processors in clusters $K \in L_j$ ($j = 0, \ldots, i-1$) send messages to L_i and then clusters in L_i perform local broadcasting. Therefore, i-th step takes $C + \lceil \log A_i \rceil$.

definition. Therefore,

$$N_{k+1} = \sum_{K \in L_{k+1}} |K| \geq \sum_{K \in L_{k+1}} |B_{k+1}| = |L_{k+1}| \cdot |B_{k+1}| \geq N_k \cdot |B_{k+1}| \geq n_0 \cdot |B_1| \cdots |B_k| \cdot |B_{k+1}|.$$

∎

Lemma 19.6 $\log |A_1| < \log N - (p - 2)$.

Proof. After we have all processors in A_1 receive the message, we need $p - 1$ more global transfer steps. With $|A_1|$ copies, we can make $|A_1| \cdot 2^i$ processors receive the message after i global transfer steps by doubling the number of copies in each global step. Therefore, $|A_1| \cdot 2^{p-2} < N$ (otherwise, we do not need p-th global broadcasting step). ∎

Proof of Theorem 19.4. The upper bound of the total broadcast time for local transfer phases is $\lceil \log n_0 \rceil + \lceil \log |A_1| \rceil + \cdots + \lceil \log |A_p| \rceil$. Since we have $|A_i| \leq |B_{i-1}|$ (for $2 \leq i \leq p$) in LCF, it is upper bounded by $\lceil \log n_0 \rceil + \lceil \log |A_1| \rceil + \lceil \log |B_1| \rceil + \cdots + \lceil \log |B_{p-1}| \rceil$. By Lemma 19.5 and Lemma 19.6, the total broadcast time only for local transfer steps is at most

$$\lceil \log n_0 \rceil + \lceil \log |A_1| \rceil + \lceil \log |B_1| \rceil + \cdots + \lceil \log |B_{p-1}| \rceil$$
$$\leq \log n_0 + \log |B_1| + \cdots + \log |B_{p-1}| + \log |A_1| + p + 1$$
$$< \log n_0 \cdot |B_1| \cdots |B_{p-1}| + \log N - (p - 2) + p + 1$$
$$\leq \log N_{p-1} + \log N + 3 \leq 2 \log N + 3.$$

Our schedule uses p global transfer steps, taking a total of pC additional rounds. ∎

In the next lemma, we prove that pC is a lower bound on the optimal broadcast time. To prove this we count the number of intercluster transfers a processor *experiences* as follows. Given a broadcast schedule, let the path from the source to the processor i, be $a_0, a_1, \ldots, a_l = i$. That is, the source a_0 sends the message to a_1 and a_1 sends to a_2, and so on. Finally a_{l-1} sends the message to processor i ($=a_l$). Let e_j ($j = 0, \ldots, l-1$) represent the number of processors that receive the message from a_j through intercluster transfers until a_{j+1} receives the message (including the transfer to a_{j+1} if they are in different clusters). In addition, let e_l be the number of processors that receive the message from processor i through intercluster transfers. That is, i sends the message to e_l processors in other clusters. Then, we say processor i experiences e intercluster transfers, where $e = \sum_{j=0}^{l} e_j$. Figure 19.3 shows an example of how to count the number of intercluster transfers that a processor experiences. In the example, processor i experiences 11 intercluster transfers. If there is any processor that experiences p intercluster transfers, then pC is a lower bound on the optimal solution.

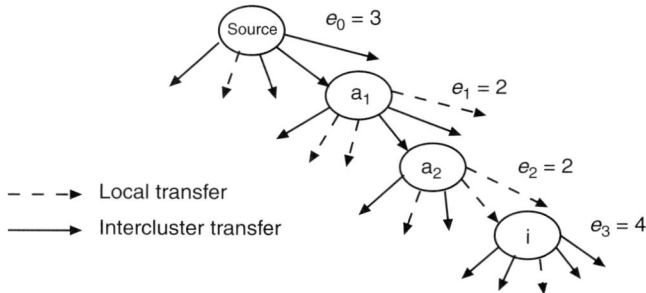

FIGURE 19.3 An example to show the intercluster transfers processor i experiences.

Lemma 19.7 *At least one node in the optimal solution experiences p intercluster transfers.*

Proof. Imagine a (more powerful) model in which once a node in a cluster receives the message, all nodes in the cluster receives the message instantly (i.e., local transfers take zero unit of time). In this model, the broadcast time is given by the maximum number of intercluster transfers that any processor experiences. We will prove that LCF gives an optimal solution for this model. Since LCF uses p global transfer steps, the optimal broadcast time is pC in this model. Since this lower bound is for a stronger model, it also works in our model.

Suppose that there is a pair of clusters K_i and K_j ($0 < i < j \leq k$) such that K_j receives the message earlier than K_i in the optimal solution. Let K_i receive the message at time t_i and K_j receive at time t_j ($t_i > t_j$). Modify the solution as follows. At time $t_j - C$ we send the message to K_i instead of K_j and K_i performs broadcasting as K_j does until t_i. (This can be done since the size of K_i is at least as big as K_j.) At time t_i, K_j receives the message and after that the same schedule can be done. This exchange does not increase the broadcast time, and therefore, LCF gives an optimal solution for the model with zero local transfer costs.

We now prove the lemma by contradiction. Suppose that there is an optimal solution (for the original model) in which all processors experience at most $p - 1$ intercluster transfers. Then, in the model with zero local transfer costs, we should be able to find a solution with broadcasting time $(p - 1)C$ by ignoring local transfers, which is a contradiction. ∎

Using the above lemma, we know that pC is a lower bound on the optimal broadcast time. Since $\log N$ is also a lower bound on the optimal solution, this gives us a 3-approximation (and an additive term of 3). However, the lower bound pC considers only the global communications of the optimal solution. On the other hand, the lower bound $\log N$ only counts the local transfers. To get a better bound, we prove the following theorem that combines the lower bounds developed above.

Theorem 19.5 *The optimal solution has to take at least $(p - 1)(C - 1) + \lceil \log N/2 \rceil$ rounds.*

Proof. Consider an optimal schedule, in the end all N nodes receive the message. We partition all nodes into two sets, S_l and S_s, where S_l contains all nodes which experienced at least $p - 1$ intercluster transfers, and S_s contains all nodes that experienced at most $p - 2$ intercluster transfers. We now show that $|S_s| < N/2$. Suppose this is not the case, it means that the optimal solution can satisfy at least $N/2$ nodes using at most $p - 2$ intercluster transfers. Using one more round of transfers, we can double the number of nodes having the message and satisfy all N nodes. This is a contradiction, since we use less than p intercluster transfers. Therefore, we have $|S_l| \geq N/2$. Since originally we have one copy of the message, satisfying nodes in S_l takes at least $\lceil \log N/2 \rceil$ transfers. So at least one node in S_l experienced $\lceil \log N/2 \rceil$ transfers (either intercluster or local transfers). We know that all nodes in S_l experienced at least $p - 1$ intercluster transfers. The node needs at least $(p - 1)C + (\lceil \log N/2 \rceil - (p - 1))$ rounds to finish. ∎

We now prove a central result about algorithm LCF, which will be used later.

Theorem 19.6 *Our algorithm takes at most $2OPT + 7$ rounds. Moreover, it takes at most $2OPT$ rounds when both p and C are not very small (i.e., when $(p - 2)(C - 2) \geq 7$).*

Proof. If C is less than 2, we can treat the nodes as one large cluster and do broadcasting. This takes at most $C\lceil \log N \rceil$ and is a 2-approximation algorithm. The problem is also trivial if p is 1, because in this case $n_0 \geq k$. Therefore, we consider the case where both values are at least 2. Here, we make use of Theorems 19.4 and 19.5.

$$(2OPT + 7) - (2\log N + pC + 3) \geq (2((p-1)(C-1) + \lceil \log \frac{N}{2} \rceil) + 7) - (2\log N + pC + 3)$$

$$\geq (p-2)(C-2) \geq 0 \quad \text{(when } p \geq 2 \text{ and } C \geq 2\text{)}. \quad \blacksquare$$

Remark 19.1 Our algorithm takes at most $2OPT$ rounds when both p and C are not very small (i.e., when $(p-2)(C-2) \geq 7$).

Corollary 19.1 *We have a polynomial-time 2-approximation algorithm for the broadcasting problem.*

19.3.2.2 Bad Example

There are instances for which the broadcast time of LCF is almost two times the optimal. Suppose that we have two clusters, K_0 and K_1, each of size n_0 and n_1 ($n_0 \leq n_1$), respectively. In addition, there are $n_0 - 1$ more clusters, each of size 1. A node in K_0 has a message to broadcast. It is easy to see that the makespan of our algorithm is $\lceil \log n_0 \rceil + C + \lceil \log n_1 \rceil$. However, the broadcasting can be made faster by sending a message to K_1 before local broadcast in K_0 is finished. A possible schedule is (i) make one local copy in K_0, (ii) one processor in K_0 sends a message to K_1 and another processor does local broadcast in K_0, (iii) after finishing local broadcast, processors in K_0 send messages to the remaining $n_0 - 1$ clusters, and (iv) clusters other than K_0 do local broadcasting as soon as they receive a message. The makespan of this solution is $1 + \max\{C + \lceil \log n_1 \rceil, \lceil \log(n_0 - 1) \rceil + C\}$. In the case where $n_0 \approx n_1$ and $\log n_0 \gg C$, the makespan of LCF is almost two times the optimal.

19.3.3 Multicasting

For multicasting, we need to have only a subset of processors receive the message. We may reduce the multicast time significantly by making use of large clusters that may not belong to the multicast group. Let n'_i denote the number of processors in K_i that belong to the multicast group. Let M denote the set of clusters (except K_0) in which some processor wants to receive the message and k' denote the size of set M. Formally, $M = \{K_i | n'_i > 0 \text{ and } i > 0\}$ and $k' = |M|$.

Let LCF(m) be algorithm LCF to make m copies. That is, LCF(m) runs in the same way as LCF but stops as soon as the total number of processors that received the message is at least m (we may generate up to $2(m-1)$ copies). For example, LCF for broadcasting is LCF(N). Here is the algorithm:

Algorithm LCF *Multicast*

1. Run LCF(k') by using any processor whether it belongs to the multicast group or not.
2. Send one copy to each cluster in M if it has not received any message yet.
3. Do local broadcast in clusters of M.

19.3.3.1 Analysis

Let p' be the number of global transfer steps LCF(k') uses. Suppose D be the number of rounds taken in the last local broadcast step in LCF(k') (after the p'-th global transfer steps). Note that some nodes in clusters performing the last local broadcast may not receive the message, since we stop as soon as the total number of nodes having the message is at least k', and hence $D \leq \lceil \log |A_{p'}| \rceil$. Nevertheless, D may be greater than $\lceil \log |B_{p'}| \rceil$ and thus some clusters may stop local broadcast before the D-th round. Let $A_{p'+1}$ be the biggest cluster among clusters that have not received a copy after LCF(k') has finished.

To get a 2-approximation algorithm, we need the following lemma, which bounds the sum of the number of rounds taken in the local phases in LCF(k') and in the last local broadcast in clusters of M.

The lemma still holds even when a cluster performing local broadcast in Phase 3 needs to broadcast to the whole cluster (i.e., $n'_i = n_i$).

Lemma 19.8 $(\log n_0 \cdot |B_1| \cdots |B_{p'-2}| + D) + \max(\lceil \log |A_{p'}| \rceil - D, \lceil \log |A_{p'+1}| \rceil) \leq \log k' + 2$.

Proof. *Case I:* $\lceil \log |A_{p'}| \rceil - D > \lceil \log |A_{p'+1}| \rceil$. It is easy to see that $(\log n_0 \cdot |B_1| \cdots |B_{p'-2}| + D) + (\lceil \log |A_{p'}| \rceil - D) \leq \log n_0 \cdot |B_1| \cdots |B_{p'-1}| + 1 \leq \log k' + 1$, and the lemma follows.

Case IIa: $\lceil \log |A_{p'}| \rceil - D \leq \lceil \log |A_{p'+1}| \rceil$ and $D > \lceil \log |B_{p'}| \rceil$. Note that $2^D \leq 2|A_{p'}| \leq 2|B_{p'-1}|$ and $|A_{p'+1}| \leq |B_{p'}|$; we have $(\log n_0 \cdot |B_1| \cdots |B_{p'-2}| + D) + (\lceil \log |A_{p'+1}| \rceil) < \log n_0 \cdot |B_1| \cdots |B_{p'}| + 2 \leq \log N_{p'-1} \cdot |B_{p'}| + 2$. After the p'-th global transfer step, one node in each of $N_{p'-1}$ clusters has just received the message. Each of these clusters will generate at least $|B_{p'}|$ copies (since $D > \lceil \log |B_{p'}| \rceil$), so $N_{p'-1} \cdot |B_{p'}| < k'$, and the lemma follows.

Case IIb: $\lceil \log |A_{p'}| \rceil - D \leq \lceil \log |A_{p'+1}| \rceil$ and $D \leq \lceil \log |B_{p'}| \rceil$. Note that $|A_{p'+1}| \leq |B_{p'-1}|$; we have $(\log n_0 \cdot |B_1| \cdots |B_{p'-2}| + D) + (\lceil \log |A_{p'+1}| \rceil) \leq \log n_0 \cdot |B_1| \cdots |B_{p'-1}| \cdot 2^D + 1 \leq \log N_{p'-1} \cdot 2^D + 1$. After the p'-th global transfer step, each cluster which has just received the message will generate at least 2^D copies, so $\log N_{p'-1} \cdot 2^{D-1} < k'$, and the lemma follows. ∎

Theorem 19.7 *Our multicast algorithm takes at most $2 \log k' + p'C + C + 4$ rounds.*

Proof. In a manner similar to the proof of Theorem 19.4, the broadcast time spent only in local transfer steps in LCF(k') is at most

$$\lceil \log n_0 \rceil + \lceil \log |A_1| \rceil + \cdots + \lceil \log |A_{p'-1}| \rceil + D$$
$$\leq \log n_0 + \log |B_1| + \cdots + \log |B_{p'-2}| + D + \log |A_1| + p'$$
$$< \log n_0 \cdot |B_1| \cdots |B_{p'-2}| + D + \log k' + 2.$$

The second inequality holds because $\log |A_1| < \log k' - (p' - 2)$ by Lemma 19.6. Moreover, the global transfer steps in LCF(k') and the second phase take $p'C$ and C rounds, respectively. Note that $n'_i \leq n_i$. In the third phase, all clusters that receive a message during the first phase need at most $\lceil \log |A_{p'}| \rceil - D$ rounds to do local broadcast. The remaining clusters, which receive a message during the second phase, are of size at most $|A_{p'+1}|$. Therefore, local broadcasting takes at most $\lceil \log |A_{p'+1}| \rceil$ rounds. Using Lemma 19.8, we have the theorem. ∎

Lemma 19.9 *At least one node in the optimal solution experiences p' intercluster transfers.*

Proof. The basic argument is the same as the one in Lemma 19.7. Note that LCF(k') uses any processor whether it belongs to the multicast group or not. If the optimal solution does not use any processor that LCF(k') uses, it cannot create new copies of the message faster than LCF(k'). ∎

Theorem 19.8 *The optimal solution takes at least $(p' - 1)(C - 1) + \lceil \log k'/2 \rceil$ rounds.*

Proof. The proof is very similar to the proof of Theorem 19.5. We partition all nodes into two sets, S_l and S_s. We now show that there are less than $k'/2$ distinct multicast clusters in S_s. Suppose this is not the case, it means that OPT can satisfy at least $k'/2$ distinct multicast clusters using at most $p' - 2$ intercluster transfers. Using one more round of transfers, all k' multicast clusters can receive the message, which is a contradiction. Therefore, we have at least $k'/2$ distinct multicast clusters in S_l and $|S_l| \geq k'/2$. ∎

Theorem 19.9 *Our algorithm takes at most $2\text{OPT} + 10$ rounds. Moreover, it takes at most 2OPT rounds when both p' and C are not very small (i.e., when $(p' - 3)(C - 2) \geq 10$).*

Proof. By making use of Theorems 19.7 and 19.8, and an analysis similar to that in Theorem 19.6, we can show that $(2OPT + 10) - (2\log k' + p'C + C + 4) \geq (p' - 3)(C - 2)$. The problem is trivial when C is less than 2 or $p' = 1$. When $p' = 2$, we can do an exhaustive search on the number of clusters in M which receives the message in the first global transfer step in LCF(k'). We can prove that it also takes at most $2OPT + 10$ rounds (details omitted). ∎

Remark 19.2 Our algorithm takes at most $2OPT$ rounds when both p' and C are not very small (i.e., when $(p' - 3)(C - 2) \geq 10$).

Corollary 19.2 *We have a polynomial-time 2-approximation algorithm for the multicast problem.*

19.3.4 Bounding Global Transfers

In the model we considered, we assume any node may communicate with any other node in other clusters, and the underlying network connecting clusters has unlimited capacity. A more practical model is to restrict the number of pairs of intercluster transfers that can happen simultaneously. In this section, we present two models to restrict the network capacity. The *bounded degree model* restricts the number of intercluster transfers associated with a particular cluster, while the *bounded size matching model* restricts the total number of intercluster transfers at any given time.

19.3.4.1 Bounded Degree Model

Associate an additional parameter d_i with each cluster i, which limits the number of intercluster transfers from or to nodes in cluster i in a time unit. We call this limitation a degree constraint. We denote an instance of this model be $I(n_i, d_i)$, meaning that there are n_i nodes in cluster K_i, and at most d_i of those may participate in intercluster transfers at any given time.

Algorithm *Bounded Degree Broadcast*

Given Instance $I(n_i, d_i)$, arbitrarily select a subset K'_i of d_i nodes in each cluster, and consider only the K'_i. We have a new instance $I(d_i, d_i)$. Note that $I(d_i, d_i)$ can be viewed as an instance of the general broadcast problem on the unrestricted model. In phase 1, run algorithm LCF in Section 19.3.2 on $I(d_i, d_i)$. In phase 2, since there are d_i informed nodes in each cluster, we do local broadcasting to send the message to the remaining $n_i - d_i$ nodes.

An important observation is that since there is only a unique message, it does not matter which subset of nodes in a cluster perform intercluster transfers. What matters is the number of informed nodes in the clusters at any given time. The following lemma compares the optimal number of rounds taken by instances using the two different models.

Lemma 19.10 *The optimal schedule of Instance $I(d_i, d_i)$ takes no more rounds than the optimal schedule of the corresponding instance $I(n_i, d_i)$.*

Proof. We argue that given an optimal schedule, which completes in OPT rounds, of Instance $I(n_i, d_i)$, we can create a schedule, which completes in at most OPT round, of the corresponding Instance $I(d_i, d_i)$.

Given an optimal schedule of Instance $I(n_i, d_i)$, let S_i be a set of the first d_i nodes in K_i that receive the message in the schedule. We can safely throw out all transfers (both intercluster and local transfers) in the schedule of which the receiving node is in $K_i \setminus S_i$, because we only need d_i nodes in $I(d_i, d_i)$. Let t_i be the time at which the last node in S_i receives the message. Consider any intercluster transfer that starts after t_i and is originated from a node in $K_i \setminus S_i$, we can modify the transfers so that it is originated from a node in S_i instead. It is not difficult to see that we can always find such a remapping, since there are at most d_i intercluster transfers at any given time, and nodes in S_i do not perform any local transfers after time t_i. Moreover, no node in $K_i \setminus S_i$ initiates a transfer on or before time t_i, because they have not received the message yet. Therefore, we have removed all transfers involving nodes in $K_i \setminus S_i$. We can safely remove all the nodes in $K_i \setminus S_i$, effectively making the schedule applicable to Instance $I(d_i, d_i)$. Since we do not add any new transfers, the total number of rounds used cannot go up. Therefore, the optimal number of rounds for $I(d_i, d_i)$ is at most that of $I(n_i, d_i)$. ∎

Theorem 19.10 *Our algorithm takes at most* $3OPT + 7$ *rounds.*

Proof. Using Theorem 19.6 and Lemma 19.10, the first phase takes at most $2OPT + 7$ rounds. Moreover, in phase 2, local broadcasting takes at most $\max_i \lceil \log n_i/d_i \rceil$ rounds, which is at most OPT. ∎

19.3.4.2 Bounded Degree Model: Multicasting

In this model, only a subset M_i (possibly empty) of nodes in cluster K_i needs the message. Nodes in $K_i \setminus M_i$ may help passing the message around. Let n'_i be $|M_i|$. Observe that although we may make use of nodes in $K_i \setminus M_i$, we never need more than d_i nodes in each cluster, because of the degree constraint in the number of intercluster transfers. Similarly, if $d_i \leq n'_i$, nodes in $K_i \setminus M_i$ are never needed.

Algorithm *Bounded Degree Multicast*

For all i, if $n'_i < d_i$ Then
  If $n_i > d_i$ Then $n_i \leftarrow d_i$ EndIf
Else (i.e., $n'_i \geq d_i$)
  If $n_i > n'_i$ Then $n_i \leftarrow n'_i$ EndIf
EndIf
Arbitrarily, select $\min(d_i, n_i)$ nodes for each cluster, with priority given to nodes in M_i. Run the LCF Multicast algorithm on the selected nodes. Now there are $\min(d_i, n'_i)$ nodes having the message in each cluster that belongs to the multicast group, so we can do local broadcasting to satisfy the remaining nodes.

After adjusting the n_i values, we have either $n'_i < n_i \leq d_i$ or $d_i \leq n'_i = n_i$. After selecting $\min(d_i, n_i)$ nodes for each cluster, we create a valid multicast instance as described in Section 19.3.3 (without the degree constraint). Thus, the algorithm is correct.

Theorem 19.11 *Our algorithm takes at most* $3OPT + 10$ *rounds.*

Proof. By Theorem 19.9, the multicast steps takes $2OPT + 10$ rounds. Moreover, the local broadcasting phase only needs to satisfy at most $n'_i - d_i$ nodes, which takes at most OPT rounds. ∎

19.3.4.3 Bounded Size Matching

In this model, we bound the number of intercluster transfers that can be performed simultaneously. Let us assume that we allow only B intercluster transfers at a time.

Note that we can assume $B \leq \lfloor N/2 \rfloor$, since this is the maximum number of simultaneous transfers allowed by our matching-based communication model.

Algorithm *Bounded Size Broadcast*

1. We run LCF(B) to make B copies of the message.
2. Every C time units we make B more copies by intercluster transfers until all clusters have at least one copy of the message.
3. Do local broadcast to inform all the processors in each cluster.

Let p_B be the number of global transfer steps LCF(B) uses, and p_L be the number of global transfer steps in the second stage of the algorithm.

Broadcasting on Networks of Workstations

Theorem 19.12 *We need $2 \log B + p_B C + p_L C + 4$ rounds for broadcasting when we allow only B intercluster transfers at a time.*

Proof. Note that the algorithm resembles the LCF multicast algorithm. In phase 1, we use LCF(B) instead of LCF(k'). In phase 2, we need p_L global transfer steps instead of 1 global transfer step to create at least one copy of the message in every cluster. In phase 3, the local broadcast always fills the entire cluster (i.e., we can treat $n'_i = n_i$.) Despite the differences, we can use the same techniques to prove the stated bound. ∎

Lemma 19.11 *In any schedule, there is a processor that experiences $p_B + p_L$ intercluster transfers.*

Proof. The proof is similar to the proof of Lemma 19.7. In the model where local transfers take zero unit of time, the same exchange argument will show that the optimal broadcast time is $(p_B + p_L)C$.

If there is a schedule in which all processors experience less that $p_B + p_L$ intercluster transfers (in the original model), it is a contradiction to the assumption that the optimal broadcast time is $(p_B + p_L)C$ in the model with zero local transfer costs. ∎

Theorem 19.13 *The optimal solution takes at least $(p_B + p_L - 1)(C - 1) + \lceil \log B \rceil$ rounds.*

Proof. The proof is very similar to the proof of Theorem 19.5. Note that in this case $|S_s| < N - B$, otherwise there is a way to satisfy all nodes using $p_B + p_L - 1$ rounds of intercluster transfers. Therefore, $|S_l| \geq B$ and the theorem follows. ∎

Theorem 19.14 *Our algorithm takes at most $2OPT + 6$ rounds.*

Proof. Using the proof technique in Theorem 19.6, the theorem follows from Theorems 19.12 and 19.13. ∎

Remark 19.3 Note that by setting $B = \lfloor N/2 \rfloor$, we can improve the makespan of the basic broadcasting (without any bound on the global transfers) by one round. This is because in this algorithm we stop performing local transfers when the number of copies is $\lfloor N/2 \rfloor$ (as more copies cannot contribute to global transfers) and start global transfers.

19.3.4.4 Bounded Size Matching: Multicasting

In this model, only a subset M_i of nodes in cluster K_i needs the message. Define $M = \{K_i | n'_i > 0 \text{ and } i > 0\}$ and $k' = |M|$. We assume $k' > B$ or otherwise we can use the LCF *multicast* algorithm. We run LCF(B) by using any processor available. Then, every C time units we make B more copies by intercluster transfers until all clusters in M have at least one copy of the message. Finally, do local broadcast to inform all the processors in each cluster in M.

Theorem 19.15 *The algorithm takes at most $2OPT + 10$ rounds.*

19.3.5 Postal Model

In this section, we consider a slightly different model. This model is motivated by the interest in Grid computing [FK98] and computing on clusters of machines. In fact, the work on the Magpie project [KHB+99, KBG00] specifically supports this communication model. In previous sections, we assumed that when processor p_i sends a message to processor p_j in another cluster, p_i is busy until p_j finishes receiving the message (this takes C time units). However, in some situations, it may not be realistic since p_i may become free after sending the message, and does not have to wait until p_j receives the message.

In this section, we assume that a processor is busy for only one time unit when it sends a message. The time a message arrives at the receiver depends on whether the sender and receiver are in the same cluster or not—it takes one time unit if it is a local transfer (within the cluster), and C time units if it is an intercluster transfer.

We first show that LCF gives a 3-approximation in this model. In addition, we present another algorithm, which we call *interleaved* LCF. Recall that we want to minimize the total number of global transfers as well as minimize the makespan. Let OPT′ denote the minimum makespan among all schedules that minimize the total number of global transfers. We can show that the makespan of the schedule generated by *interleaved* LCF is at most two times OPT′.

19.3.5.1 Analysis of Largest Cluster First

The analysis is similar to the one presented in Section 19.3.2. We modify the algorithm (for the analysis purpose) so that we have local and global phases in turn. However, in this model a processor can initiate more than one global transfer in a global phase since senders are busy for only one time unit per global transfer. We define A_i, B_i, L_i as follows. Let $A_0 = |K_0|$. After finishing local transfers in K_0, all processors in K_0 start global transfers. They can initiate a global transfer at every time unit. After C time units, n_0 clusters receive a message (denoted as L_1). Let $A_1 (B_1)$ be the biggest(smallest) cluster among them. Now K_0 stops global transfers for $\lceil \log A_1 \rceil$ rounds (global transfers already initiated continue to be done). For those $\lceil \log A_1 \rceil$ rounds, every cluster that received the message performs (only) local broadcasting. For all the clusters that receive the message in this step, they have exactly $\lceil \log A_1 \rceil$ rounds of local broadcasting. So for example, if a cluster receives t time unit later than A_1 then it can only start its global transfers t time units later than A_1 (it will be idle even if it finishes local broadcasting earlier). Note that A_1 is the biggest cluster in L_1, and therefore, $\lceil \log A_1 \rceil$ is enough for local broadcasting of those clusters. After $\lceil \log |A_1| \rceil$ time units, we have all clusters that have finished local broadcasting phase perform global transfers every time unit *for C rounds*. Clusters that have not finished local phase keep performing local broadcasting (or wait) and then start global transfers. Repeat this until all processors get informed. In general, we define L_i as clusters that receive messages in the first global transfers of i-th step (so they can participate in global phase of $(i+1)$-th step from the beginning) (see Figure 19.4a). A_i (B_i) is the biggest (smallest) cluster in L_i. Suppose that the schedule has p global phases. Then it is easy to see that Lemmas 19.5 and 19.6 hold. There is one subtle case where there are some clusters that receive the message later than A_p by the

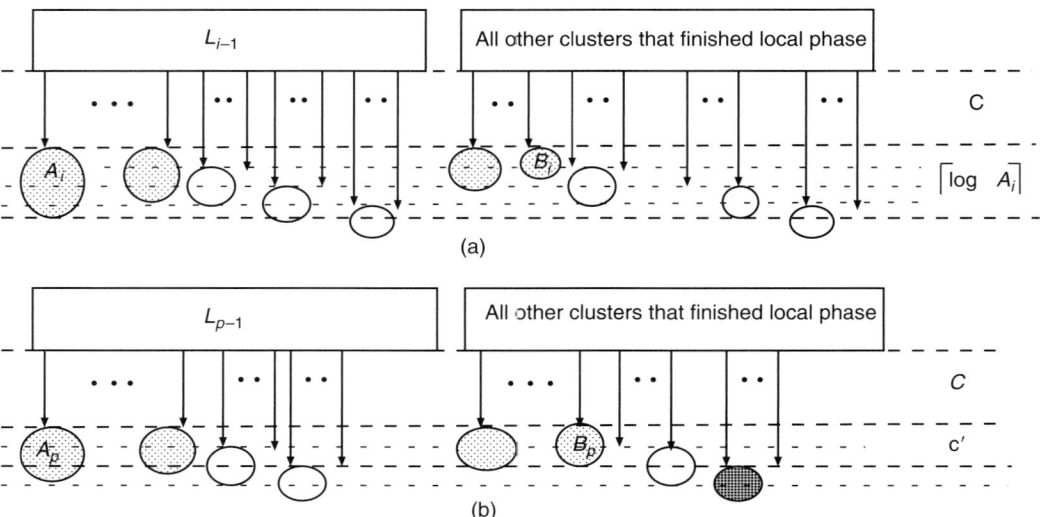

FIGURE 19.4 (a) i-th step of LCP in postal model. In this example $C = 4$, so a processor can initiate (at most) 4 global transfers. Dotted clusters belong to L_i. (b) In p-th step, there can be some processors that receive messages later than A_p. The dark circle is the last cluster that receives the message.

Broadcasting on Networks of Workstations

transfers initiated in p-th global phases (see Figure 19.4b). We first analyze the makespan of the simple case where A_p is one of the last clusters that receive the message.

Theorem 19.16 *The makespan of* modified LCF *is at most* $2 \log N + pC + 3$ *when A_p is one of the last clusters that receive the message.*

Proof. The total makespan taken for local transfers is

$$\lceil \log n_0 \rceil + \lceil \log |A_1| \rceil + \cdots + \log \lceil |A_p| \rceil \leq \log n_0 + \log |A_1| + \cdots + \log |A_p| + (p+1)$$
$$\leq \log n_0 + \log |A_1| + \log |B_1| \cdots + \log |B_{p-1}| + (p+1)$$
$$= \log n_0 \cdot |A_1| \cdot |B_1| \cdots |B_{p-1}| + (p+1)$$
$$\leq \log |A_1| + \log N_{p-1} + (p+1) \quad \text{(by Lemma 19.5)}$$
$$\leq 2 \log N + 3 \quad \text{(by Lemma 19.6)}.$$

Therefore, the makespan of the schedule is at most $2 \log N + pC + 3$. ∎

Lemma 19.12 *If we assume that local transfers take zero unit of time, the makespan is at least pC when A_p is one of the last clusters that receive the message. If there are some clusters that receive the message c' time units later than A_p, then the makespan is at least $pC + c'$.*

Proof. Consider the schedule given by LCF when local transfers take zero units of time. It is easy to see that LCF gives an optimal solution in this case. Moreover, it is the same as the schedule by the modified LCF, if we ignore local transfer phases. Since modified LCF, needs p global phases, pC is a lower bound for the optimal solution. ∎

We thus conclude

Theorem 19.17 *The makespan of the schedule generated by* LCF *is at most 3 times the optimal (with an additive term of 3) when A_p is one of the last clusters that receive the message.*

We now deal with the case when there are other clusters that receive the message later than A_p. Let A_{p+1} denote the biggest cluster that receives the message last, and it receives the message c' time units later than A_p ($c' < C$ since otherwise we would have another global phase). We need additional $c' + \lceil \log |A_{p+1}| \rceil$ time units (it is less than $c' + \lceil \log |B_p| \rceil$).

Lemma 19.13 $n_0 \cdot |B_1| \cdots |B_p| \leq N_p$.

Lemma 19.14 $pC + c'$ *is a lower bound on the optimal solution.*

Proof. Suppose that local transfers take zero unit of time. Then in LCF, A_{p+1} receives the message c' time units later than A_p (since the schedule can be obtained by ignoring local phases). Therefore, $pC + c'$ is a lower bound on the optimal solution. ∎

Theorem 19.18 *The makespan of the schedule generated by* LCF *is at most three times the optimal (with additive term of 4).*

19.3.5.2 Interleaved LCF

We present another algorithm called *Interleaved* LCF. We show that it is 2-approximation among schedules that use the minimum number of global transfers.

Algorithm *Interleaved* LCF

At every two rounds, a processor that has the message alternately performs the following two steps:

1. Local transfer: if there is any processor in the same cluster that has not received the message, then send the message to it.
2. Global transfer: if there is any cluster that has not received the message, choose the biggest cluster among them and send the message to a processor in the cluster.

We only consider a set of schedules (denoted as S) that minimize the total number of global transfers. Note that schedules in S have the property that each cluster receives only one message from outside ($k-1$ in total). Let OPT_S be the minimum makespan among all schedules in S.

Lemma 19.15 *There is a schedule in S with makespan OPT_S in which for any pair of clusters K_i, K_j ($n_i > n_j$), K_i receives a message no later than K_j.*

Proof. Given a schedule in S with makespan OPT_S, if there is a pair of clusters K_i, K_j ($n_i > n_j$) and K_i receives a message at time t_i and K_j receives a message at time t_j ($t_i > t_j$), then we can modify the schedule so that K_i receives the message no later than K_j without increasing the makespan.

At t_j K_i (instead of K_j) receives the message. K_i can do all transfers that K_j does till time t_i. At time t_i, K_j receives a message. Let x_t processors in K_i received the message just after time t in the *original* schedule. Similarly, let y_t processors in K_j receive the message just after time t. Then, $x_{t_i} = 1$ and $y_{t_i} \le n_j$. Note that we cannot swap the roles of two clusters just after t_i since K_i has y_{t_i} messages and K_j has only one message. Therefore, K_i should keep performing transfers as if it is K_j for some time. Let t' be the last time when $x_{t'} \le y_{t'}$. That is, $x_{t'+1} > y_{t'+1}$.

At time $t'+1$, we need to carefully choose which transfers we should do. Note that just before time $t'+1$, K_i has $y_{t'}$ messages and K_j has $x_{t'}$ messages. In K_i, we choose $x_{t'+1} - y_{t'}$ processors to make local transfers so that after $t'+1$, K_i has $x_{t'+1}$ copies of message. Since $x_{t'+1} \le 2x_{t'} \le 2y_{t'}$, $x_{t'+1} - y_{t'} \le y_{t'}$, and therefore, we have enough processors to choose. Similarly, in K_j $y_{t'+1} - x_{t'}$ processors do local transfers so that K_j has $y_{t'+1}$ after $t'+1$. The total number of global transfers coming from K_i and K_j in the original schedule is at most $x_{t'} + y_{t'} - (x_{t'+1} - x_{t'}) - (y_{t'+1} - y_{t'}) = x_{t'} + y_{t'} - (y_{t'+1} - x_{t'}) - (x_{t'+1} - y_{t'})$, and this is exactly the number of remaining processors. Therefore, we have the remaining processors enough to make global transfers. After $t'+1$, we can do transfers as in the original schedule. ∎

We can now consider schedules with the property in Lemma 19.15 only. Owing to the property, a processor knows the receiver to send a message when it performs a global transfer—the largest cluster that has not received any message. The only thing a processor needs to decide at each time is whether it will make a local transfer or global transfer. By performing local and global transfer alternatively, we can bound the makespan by a factor of two.

Theorem 19.19 *The makespan of* interleaved LCF *is at most two times OPT_S.*

Proof. Given an optimal schedule with the property in Lemma 19.15, modify the schedule so that each operation takes 2 units of time. That is, if a processor performs a local transfer then it is idle in the next time slot and if a processor performs a global transfer, it is idle in the previous time slot. The makespan of the modified schedule is at most two times the optimal. It is easy to see that the schedule by *interleaved* LCF should not be worse than the modified schedule since in *interleaved* LCF, the processors perform local and global transfers alternatively with no idle time. ∎

19.4 Broadcasting in Heterogenous Networks

Banikazemi et al. [BMP98] proposed a simple model where heterogeneity among processors is modeled by a nonuniform speed of the sending processor. A heterogenous cluster is defined as a collection of processors $p_1, p_2, \ldots, p_n$ in which each processor is capable of communicating with any other processor. Each processor has a transmission time that is the time required to send a message to any other processor in the cluster. Thus, the time required for the communication is a function of only the sender. Each processor may send messages to other processors in order, and each processor may be receiving only one message at a time.

Thus, a broadcast operation is implemented as a broadcast tree. Each node in the tree represents a processor of the cluster. The root of the tree is the source of the original message. The children of a node p_i are the processors that receive the message from p_i. The completion time of a node is the time at which it completes receiving the message from its parent. The completion time of the children of p_i is $c_i + j \times t_i$, where c_i is the completion time of p_i, t_i is the transmission time of p_i, and j is the child number. In other words, the first child of p_i has a completion time of $c_i + t_i$ ($j = 1$), the second child has a completion time of $c_i + 2t_i$ ($j = 2$), and so forth. See Figure 19.5 for an example.

A commonly used method to find a broadcast tree is referred to as the "fastest node first" (FNF) technique [BMP98]. This works as follows: In each iteration, the algorithm chooses a sender from the set of processors that have received the message (set S) and a receiver from the set of processors that have not yet received the message (set R). The algorithm then picks the sender from $s \in S$ so that s can finish the transmission as early as possible, and it chooses the receiver $r \in R$ as the processor with the minimum transmission time in R. Then r is moved from R to S and the algorithm continues. The intuition is that sending the message to fast processors first is a more effective way to propagate the message quickly. This technique is very effective and easy to implement. In practice, it works extremely well (using simulations) and in fact frequently finds optimal solutions as well [BMP98]. However, there are situations when this method also fails to find an optimal solution. A simple example is shown in Figure 19.5.

Despite several nontrivial advances in an understanding of the *fastest node first* method by Liu [Liu02] (see also work by Liu and Sheng [LS00]), it was still open as to what the complexity of this problem is. For example, is there a polynomial time algorithm to compute a solution that minimizes the broadcast time? Can we show that in all instances the FNF heuristic will find solutions close to optimal? The second question is of interest, regardless of the complexity of the problem; for example, if there was a complex polynomial time algorithm to find the optimal solution, it is unlikely that it would be used in practice and an understanding of the FNF heuristic is still very useful.

Liu [Liu02] (see also Reference LS00) shows that if there are only two classes of processors, then FNF produces an optimal solution. In addition, if the transmission time of every slower processor is a multiple of the transmission time of every faster processor, then again the FNF heuristic produces an optimal solution. So for example, if the transmission time of the fastest processor is 1 and the transmission times of all other processors are powers of 2, then the algorithm produces an optimal solution. It immediately

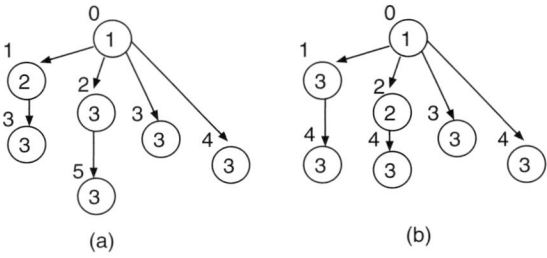

FIGURE 19.5 An example to show that FNF may not produce an optimal solution. Transmission times of processors are inside the circles. Times at which nodes receive a message are also shown (a) FNF and (b) optimal solution.

follows that by rounding all transmission times to powers of 2 we can obtain a solution using FNF whose cost is at most twice the cost of an optimal solution. However, this still does not explain the fact that this heuristic does much much better in practice. Is there a better worst case guarantee on this simple method?

We show several results:

- We show that the FNF heuristic actually produces an optimal solution for the problem of minimizing the sum of completion times (Section 19.4.3).
- We use this to show that the FNF method has a performance ratio of at most 1.5 when compared to the optimal solution for minimizing broadcast time (Section 19.4.3).
- We show that the performance ratio of FNF is at least $\frac{25}{22}$* (Section 19.4.3).
- As a corollary of the above approximation result, we are able to show that if the transmission times of the fastest $n/2$ processors are in the range $[1 \ldots C]$, then FNF produces a solution with makespan at most $T_{\text{OPT}} + C$ (Section 19.4.3).
- We prove that there is a polynomial time approximation scheme (PTAS) for minimizing the broadcast time (Section 19.4.4).
- We extend the above results to the problem of multicasting (Section 19.4.5).

The problem to minimize broadcast time was shown to be NP-hard when each processor has an arbitrary release time by Hall et al. [HLS98]. We establish that the problem is *NP*-hard even when all processors are available [KK04]. Hall et al. also proved that FNF minimizes the sum of completion times, and we provide a short proof for the problem. It was shown by Liu et al. [LW00] and Libeskind-Hadas et al. [LHHB+01] independently that FNF gives a 2-approximation for minimizing broadcast time. We prove that FNF gives a 1.5-approximation. Libeskind-Hadas et al. also showed that a schedule with the minimum broadcast time can be obtained in $O(n^{2k})$ using dynamic programming, where k is the number of different sending speeds of processors.

We also prove that there is a PTAS for this problem. However, this algorithm is not practical owing to its high running time, albeit polynomial.

19.4.1 Problem Definition

We are given a set of processors $p_0, p_1, \ldots, p_n$. There is one message to be broadcast from p_0 to all other processors $p_1, p_2, \ldots, p_n$. Each processor p_i can send a message to another processor with transmission time t_i once it has received the message. Each processor can be either sending a message or receiving a message at any point of time. Without loss of generality, we assume that $t_1 \leq t_2 \leq \cdots \leq t_n$ and $t_0 = 1$.

We define the completion time of processor p_i to be the time when p_i has received the message. Our objective is to find a schedule that minimizes $C_{\max} = \max_i c_i$, where c_i is the completion time of processor p_i. In other words, we want to find a schedule that minimizes the time required to send the message to all the processors.

Our proof makes use of the following results from Reference BMP98.

Theorem 19.20 *[BMP98] There exists an optimal broadcast tree in which all processors send messages without delay.*

Theorem 19.21 *[BMP98] There exists an optimal broadcast tree in which every processor has transmission time no less than its parent (unless the parent is the source node).*

19.4.2 Minimizing the Sum of Completion Times

In this section, we show that the FNF scheme finds an optimal schedule to minimize the sum of completion times of all the nodes, that is, it minimizes $C_{\text{sum}} = \sum_i c_i$.

---

*Observe that in the example shown in Figure 19.5 the performance ratio of FNF is 5/4, but this example does not appear to generalize easily to arbitrarily large instances.

Suppose that we are given an (optimal) schedule that minimizes the sum of completion times and the sum of completion times is C. If there is a processor p_i ($i \geq 1$) whose completion time is later than processor p_j ($i < j$), that is, $c_i > c_j$ in the schedule, then we can find a modified schedule such that processor p_i ($i \geq 1$) has completion time no later than processor p_j and the sum of completion times is no more than C.

Let us define a permutation $\pi : \{1, 2, \ldots, n\} \to \{1, 2, \ldots, n\}$. Then any schedule can be represented as an ordered list of processors $(p_{\pi(1)}, p_{\pi(2)}, \ldots, p_{\pi(n)})$ by sorting in nondecreasing order of completion times (ties broken in accordance with their indices).

We first prove a lemma similar to Theorem 19.21 for sum of completion times. The proof is the same as Theorem 19.21 [BMP98]. We include the proof for the completeness.

Lemma 19.16 *There exists a broadcast tree that minimizes the sum completion times, and for any processor p_i other than p_0, its children have transmission times no less than p_i.*

Proof. Suppose that in an optimal solution for sum of completion times, there are processors p_i and p_j ($i, j \neq 0$) with $t_i < t_j$ and p_i is a child of p_j. Then we can modify the solution and show that the sum of completion times for the modified solution is smaller than in the given optimal solution, which is a contradiction.

Let c_i and c_i' be the completion times of p_i in the given optimal solution and the modified solution, respectively. Let $P(c_i)$ denote a set of processors p_j such that $c_j \leq c_i$. We first modify the schedule *only for processors in $P(c_i)$*. The schedule is the same as the original schedule but p_i and p_j are exchanged. In other words, p_i receives the message in place of p_j and sends the message to children of p_j in $P(c_i)$. Clearly $c_j' < c_i$ and $c_i' = c_j$. For the rest of processors p_k in $P(c_i)$, c_k' is no later than c_k since p_i is faster than p_j.

For the processors not in $P(c_i)$, we can broadcast the message using the original schedule. Since $c_k' \leq c_k$ for all $p_k \in P(c_i) \setminus \{p_i, p_j\}$ and both p_i and p_j are avaiable to send a message at time c_i, the modified completion times are no later than the original schedule. Therefore, we have $\sum_k c_k' < \sum_k c_k$, which is a contradiction. ∎

Lemma 19.17 *Let $(p_{\pi(1)}, p_{\pi(2)}, \ldots, p_{\pi(n)})$ be an optimal schedule for the problem with sum of completion times C. Let s be the smallest index such that $\pi(s) \neq s$. Then we can find a schedule with sum of completion times no more than C and $i = \pi(i)$ for all $1 \leq i \leq s$.*

Proof. In the given optimal broadcast tree, let us call the node corresponding to processor p_i as node i. We denote the subtree rooted at node i as $T(i)$. Consider subtrees $T(s)$ and $T(s')$ where $s' = \pi(s)$. We know that s cannot be an ancestor of s' as $c_{s'} \leq c_s$. Also, s' cannot be an ancestor of s as $t_s < t_{s'}$ by Lemma 19.16. Therefore, $T(s)$ and $T(s')$ are disjoint.

Let node s have children $x_1, x_2, \ldots, x_k$ and node s' have $y_1, y_2, \ldots, y_{k'}$ as shown in Figure 19.6. We change the schedule as follows. First we exchange s and s'. In other words, the modified completion time of p_s becomes $c_{s'}$ and the completion time of $p_{s'}$ becomes c_s. Clearly, this does not increase the sum of completion times. For all i ($1 \leq i \leq \max(k, k')$), we compare the size of subtree $T(x_i)$ and $T(y_i)$ and attach the bigger one to s and the smaller one to s' as i-th child. (If there does not exist a child, simply consider the size of the subtree as zero.)

We can prove that this modification does not increase the sum of completion times. The difference of the sum of completion times for subtree $T(x_i)$ and $T(y_i)$ depends on which parent they are attached to. In case that $|T(x_i)| \geq |T(y_i)|$, the completion times of processors in $T(x_i)$ are decreased by $c_s - c_{s'}$ since the completion time of p_{x_i} is changed from $c_s + i \times t_s$ to $c_{s'} + i \times t_s$. The completion times of processors in $T(y_i)$ are increased by $c_s - c_{s'}$ since the completion time of p_{y_i} is changed from $c_{s'} + i \times t_{s'}$ to $c_s + i \times t_{s'}$. Therefore, the difference is

$$(c_s - c_{s'})(|T(y_i)| - |T(x_i)|) \leq 0.$$

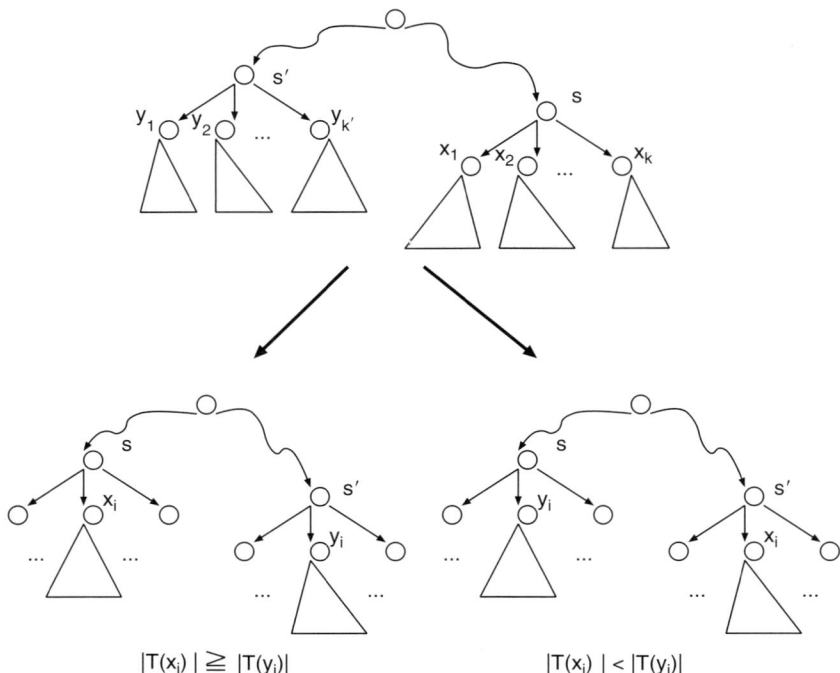

FIGURE 19.6 Figure shows how to modify the schedule of $T(s)$ and $T(s')$.

In case that $|T(x_i)| < |T(y_i)|$, the completion times of processors in $T(x_i)$ are increased by $i \times (t_{s'} - t_s)$ because the completion of x_i is modified from $c_s + i \times t_s$ to $c_s + i \times t_{s'}$, and the completion times of processors in $T(y_i)$ are decreased by $i \times (t_{s'} - t_s)$ because the completion of x_i is modified from $c_{s'} + i \times t_{s'}$ to $c_{s'} + i \times t_s$. Therefore, the difference is

$$i \cdot (t_{s'} - t_s)(|T(x_i)| - |T(y_i)|) \leq 0.$$

Therefore, we have the lemma. ∎

By repeating the procedure in Lemma 1.17, we can find a schedule such that processors receive the message in nondecreasing order of their transmission times and the sum of completion times is no more than the optimal. We thus conclude.

Theorem 19.22 *Algorithm* FNF *minimizes the sum of completion times.*

In fact, we can prove even stronger result by applying the same procedure only to a subset of processors p_i ($1 \leq i \leq k$) for any $k \leq n$.

Corollary 19.3 *Algorithm* FNF *minimizes the sum of completion times over all processors p_i ($1 \leq i \leq k$) for any $k \leq n$.*

Proof. We do the same procedure as in Lemma 19.17 except that we only count processors p_i ($1 \leq i \leq k$) when we compute the sizes of subtrees. ∎

We will use this corollary in the next section to prove that the FNF scheme gives a 1.5-approximation for minimizing broadcast time.

19.4.3 1.5-Approximation for Minimizing Broadcast Time

In this section, we prove that FNF scheme gives 1.5-approximation.

Let us consider the bar chart as shown in Figure 19.7a, where processors are listed in nondecreasing order of transmission time and each processor has a block whose height corresponds to its completion time in a schedule (these two charts correspond to the schedule created on the instance specified in Figure 19.5). We call these bars *gray blocks*. Each processor p_i can start sending messages as soon as it receives the message, and send to one processor in each t_i time units. A *white block* corresponds to each message sent by p_i. Therefore, the height of a white block is the same as the transmission times of the corresponding processors.

Definition 19.1 *We define the number of fractional blocks as the number of (fractional) messages a processor can send by the given time. In other words, given a time T, the number of fractional blocks of processor p is $(T - c_i)/t_i$.*

For example, the number of fractional blocks of p_1 by time T is 2 in the FNF schedule (see Figure 19.7a).

Definition 19.2 *We define $R(T)$ as the rectangular region bounded by time 0 and T and including processors from p_0 to $p_{\lfloor n/2 \rfloor}$ in the bar chart.*

An example is shown in Figure 19.7a.

We only include processors $p_0, p_1, \ldots, p_{\lfloor n/2 \rfloor}$ in $R(T)$ because of the following lemma.

Lemma 19.18 *There is an optimal schedule in which only the $\lfloor n/2 \rfloor$ fastest processors and the source processor send messages, that is, processor p_i ($\lfloor n/2 \rfloor + 1 \leq i \leq n$) need not send any messages.*

Proof. We prove this by showing that there is an optimal broadcast tree in which every internal node (except the source) has at least one child that is a leaf. Suppose an internal node $s(\neq p_0)$ does not have a child that is a leaf. Then we move the processor that receives the message last in subtree $T(s)$ to a child of s. It is easy to see that this does not increase the makespan of the schedule by Theorem 19.21. By repeatedly applying this modification to the given broadcast tree, we can find an optimal broadcast tree satisfying the property. ■

Lemma 19.19 *Algorithm FNF maximizes the number of fractional blocks in $R(T)$ for any T.*

To prove this lemma, we first prove the following proposition.

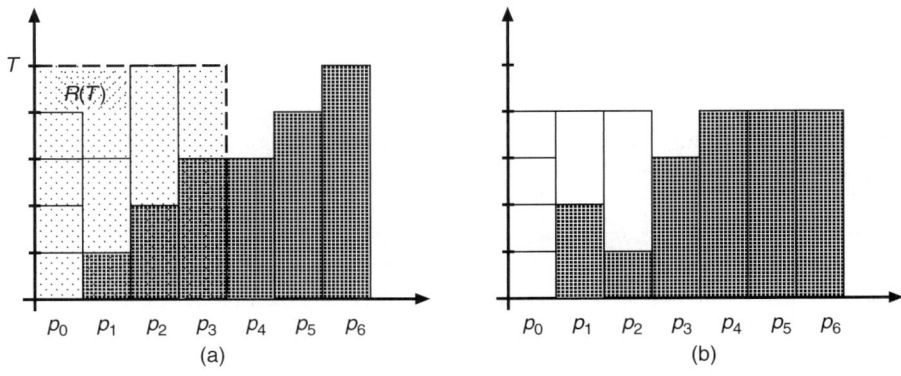

FIGURE 19.7 An example of bar charts corresponding to the schedules created by the instance specified in Figure 1.5 (a) FNF and (b) optimal solution.

Proposition 19.1 $\sum_{i=1}^{m} a_i/b_i \geq \sum_{i=1}^{m} a'_i/b_i$ if $\sum_{i=1}^{l} a_i \geq \sum_{i=1}^{l} a'_i$ for all $1 \leq l \leq m$ and $0 < b_i \leq b_{i+1}$ for all $1 \leq i \leq m-1$.

Proof. We will show that for all $1 \leq l \leq m$

$$\sum_{i=1}^{m} \frac{a_i}{b_i} \geq \sum_{i=1}^{l} \frac{a'_i}{b_i} + \sum_{i=l+1}^{m} \frac{a_i}{b_i} + \sum_{i=1}^{l} \frac{a_i - a'_i}{b_l}.$$

Then, if we set l as m, we have the proposition as $\sum_{i=1}^{m} a_i \geq \sum_{i=1}^{m} a'_i$.
We prove this by induction. For $l = 1$, it is clearly true since

$$\sum_{i=1}^{m} \frac{a_i}{b_i} = \frac{a'_1}{b_1} + \sum_{i=2}^{m} \frac{a_i}{b_i} + \frac{a_1 - a'_1}{b_1}.$$

Suppose that it is true when $l = k$. Then,

$$\sum_{i=1}^{m} \frac{a_i}{b_i} \geq \sum_{i=1}^{k} \frac{a'_i}{b_i} + \sum_{i=k+1}^{m} \frac{a_i}{b_i} + \sum_{i=1}^{k} \frac{a_i - a'_i}{b_k}$$

$$\geq \sum_{i=1}^{k} \frac{a'_i}{b_i} + \sum_{i=k+1}^{m} \frac{a_i}{b_i} + \sum_{i=1}^{k} \frac{a_i - a'_i}{b_{k+1}}$$

$$= \sum_{i=1}^{k+1} \frac{a'_i}{b_i} + \sum_{i=k+2}^{m} \frac{a_i}{b_i} + \sum_{i=1}^{k+1} \frac{a_i - a'_i}{b_{k+1}}. \blacksquare$$

Proof of Lemma 19.19. Let us denote the completion time of p_i in FNF and any other given schedule as c_i and c'_i, respectively. We need to show that $\sum_{i=0}^{\lfloor n/2 \rfloor} (T - c_i)/t_i \geq \sum_{i=0}^{\lfloor n/2 \rfloor} (T - c'_i)/t_i$. In fact, since $c_0 = c'_0 = 0$, it is enough to show that $\sum_{i=1}^{\lfloor n/2 \rfloor} (T - c_i)/t_i \geq \sum_{i=1}^{\lfloor n/2 \rfloor} (T - c'_i)/t_i$. Since we have Proposition 19.1 and $t_i \leq t_{i+1}$, it is enough to show that $\sum_{i=1}^{l} (T - c_i) \geq \sum_{i=1}^{l} (T - c'_i)$ for all $1 \leq l \leq \lfloor n/2 \rfloor$. This is true because $\sum_{i=1}^{l} c_i \leq \sum_{i=1}^{l} c'_i$ for all $1 \leq l \leq \lfloor n/2 \rfloor$ by Corollary 19.3.

Let the makespan of an optimal schedule and FNF be T_{OPT} and T_{FNF}, respectively. Then we have the following lemma. $\blacksquare$

Lemma 19.20 *The number of fractional blocks by FNF in $R(T_{\text{FNF}})$ is at most $3n/2$.*

Proof. If a processor receives the message from processor p_i, then it can be mapped to a white block of p_i. To finish the schedule, we should send the message to n processors, and therefore, there are n white blocks in $R(T_{\text{FNF}})$. In addition, each processor p_i ($0 \leq i \leq \lfloor n/2 \rfloor$) may have a fraction of block which is not finished by time T_{FNF}. But at least one processor should have no incomplete block since the makespan of FNF is T_{FNF}. Thus, the number of fractional blocks is at most $n + \lfloor n/2 \rfloor \leq 3n/2$. $\blacksquare$

Theorem 19.23 *There are at most n fractional blocks in $R(\frac{2}{3}T_{\text{FNF}})$ in any schedule.*

Proof. It is enough to show that FNF can have at most n fractional blocks in $R(\frac{2}{3}T_{\text{FNF}})$ since FNF maximizes the number of blocks by Lemma 19.19. By time $\frac{2}{3}T_{\text{FNF}}$, each processor can have only $2/3$ of the number of blocks it has in $R(T_{\text{FNF}})$. Let f_i be the number of fractional blocks of processor p_i in $R(T_{\text{FNF}})$ and f'_i be the number of fractional blocks in $R(\frac{2}{3}T_{\text{FNF}})$. Since $f_i = \frac{T_{\text{FNF}} - c_i}{t_i}$ and $f'_i = \frac{\frac{2}{3}T_{\text{FNF}} - c_i}{t_i}$, we have $f'_i \leq \frac{2}{3}(\frac{T_{\text{FNF}} - \frac{3}{2}c_i}{t_i}) \leq \frac{2}{3}f_i$. Therefore, we have at most $\frac{2}{3} \times 3n/2 = n$ fractional blocks in $R(\frac{2}{3}T_{\text{FNF}})$. $\blacksquare$

Corollary 19.4 *Algorithm FNF gives a 1.5-approximation.*

Proof. Since we need to send the message to n processors in the optimal schedule, we should have n blocks in $R(T_{\text{OPT}})$. It implies that $T_{\text{OPT}} \geq \frac{2}{3} T_{\text{FNF}}$. ∎

When transmission times are in a small range, the FNF heuristic has a better bound.

Theorem 19.24 *Suppose $C' = n/(2 \sum_{i=1}^{\lfloor n/2 \rfloor} 1/t_i)$, then the FNF heuristic finds a solution of cost at most $T_{\text{OPT}} + C'$.*

Proof. In the bar chart of an optimal solution, the number of fractional blocks we can have between height T_{FNF} and T_{OPT} is $t/t_0 + t/t_1 + \cdots + t/t_{\lfloor n/2 \rfloor}$, where $t = T_{\text{FNF}} - T_{\text{OPT}}$. Since we have at least n fractional blocks in $R(T_{\text{OPT}})$ and at most $3n/2$ fractional blocks in $R(T_{\text{FNF}})$, $\sum_{i=1}^{\lfloor n/2 \rfloor} t/t_i \leq n/2$. Therefore, $T_{\text{FNF}} - T_{\text{OPT}} \leq n/(2 \sum_{i=1}^{\lfloor n/2 \rfloor} 1/t_i)$. ∎

Theorem 19.25 *If the transmission times of the fastest $\lfloor n/2 \rfloor$ processors (except p_0) is at most C, then the FNF heuristic finds a solution of cost at most $T_{\text{OPT}} + C$.*

Proof. In the bar chart of an optimal schedule, let f_i be the number of fractional blocks of processor p_i by time T_{OPT} and g_i be the number of *complete* blocks of processor p_i by T_{OPT}. In the bar chart of FNF schedule, let f'_i be the number of fractional blocks of processor p_i by time T_{OPT}.

Define g'_i to be the number of complete blocks by time $T_{\text{OPT}} + C$ and we need to prove that $\sum_{i=0}^{i=\lfloor n/2 \rfloor} g'_i \geq n$. For $1 \leq i \leq \lfloor n/2 \rfloor$, since $t_i \leq C$, we have $g'_i \geq \lceil f'_i \rceil$. Therefore, the number of *complete* blocks in $R(T_{\text{OPT}} + C)$ is

$$\sum_{i=0}^{i=\lfloor n/2 \rfloor} g'_i \geq g'_0 + \sum_{i=1}^{i=\lfloor n/2 \rfloor} \lceil f'_i \rceil \geq g_0 + \sum_{i=1}^{i=\lfloor n/2 \rfloor} f'_i$$

$$\geq g_0 + \sum_{i=1}^{i=\lfloor n/2 \rfloor} f_i \geq g_0 + \sum_{i=1}^{i=\lfloor n/2 \rfloor} g_i.$$

The third inequality comes from the fact that FNF maximizes the number of fractional blocks in $R(T)$ for any T and $f_0 = f'_0$.

We also have $\sum_{i=0}^{i=\lfloor n/2 \rfloor} g_i \geq n$ by definition of T_{OPT}. Therefore, $T_{\text{FNF}} \leq T_{\text{OPT}} + C$. ∎

19.4.3.1 Bad Example

Bhat et al. [BRP99] gave an example and proved that the broadcast time by FNF can be 17/16 times the optimal in the example. In this section, we show that in fact, FNF gives the broadcast time of 25/22 times the optimal on the same example.

Consider the example shown in Figure 19.8. We have the source with transmission time 1 and $2n$ processors with a very large transmission time. Also, there are n processors with transmission times $n, n+1, \ldots, 2n-1$. In the optimal schedule, the source should send messages to processors with transmission time $2n-1, 2n-2, \ldots, n$, respectively. In other words, at time i the node with transmission time $2n-i$ receives the message from the source. Immediately after receiving the message, each of these processors sends a message to one of the slow processors. The schedule completes at time $2n$.

In the FNF schedule, the source sends messages to processors with transmission time $n, n+1, \ldots, 2n-1$, respectively, and again immediately after receiving the message, each of these processors sends a message to one of the slow processors. At time $2n$, $n/2$ of the slow processors have not yet received the message. After time $2n$, processor with transmission time $n+i-1$ will send another message to a slow processor

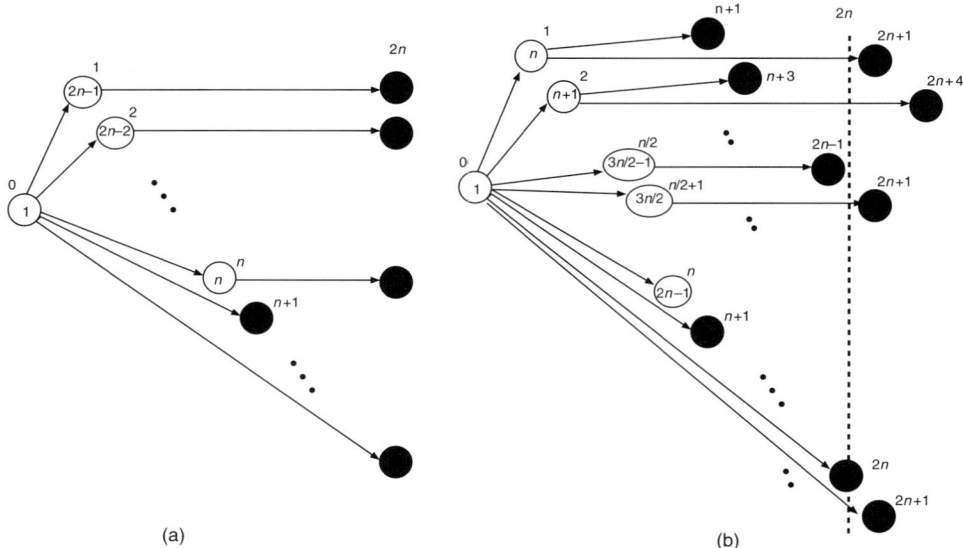

FIGURE 19.8 A bad example of FNF. The number inside a node is the transmission time of the processor and the number next to a node is the time it received the message. Black nodes are very slow processors. At time $2n$ (the dotted line), the optimal solution finishes broadcasting, but in the FNF schedule, $n/2$ processors have not received the messages as yet (a) an optimal solution and (b) FNF.

at time $2n + 3i - 2$. At the same time, processor with transmission time $(3n)/2 + i - 1$ sends a message at time $2n + 2i - 1$. The source sends a message every time unit. Therefore, if t is the time needed to send messages to the remaining $n/2$ processors, then we have

$$\left\lfloor \frac{t+2}{3} \right\rfloor + \left\lfloor \frac{t+1}{2} \right\rfloor + t \geq \frac{n}{2}.$$

Therefore, we need at least $3n/11 - 7/11$ additional time. That means that the broadcast time of FNF can be $25/22$ times the optimal for large values of n.

19.4.4 Polynomial Time Approximation Scheme

We now describe a polynomial time approximation scheme for the problem of performing broadcast in the minimum possible time. Unfortunately, the algorithm has a very high running time when compared to the *fastest node first* heuristic.

We will assume that we know the broadcast time T of the optimal solution. Since $t_0 = 1$, we know that the minimum broadcast time T is between 1 and n, and we can try all possible values of the form $(1 + \epsilon)^j$ for some fixed $\epsilon > 0$ and $j = 1, \ldots, \lceil \frac{\log n}{\log(1+\epsilon)} \rceil$. In this guessing process, we lose a factor of $(1 + \epsilon)$.

Let $\epsilon' > 0$ be a fixed constant. We define a set of fast processors F as all processors whose transmission time is at most $\epsilon' T$. Formally, $F = \{p_j | t_j \leq \epsilon' T, 1 \leq j \leq n\}$. Let S be the set of remaining (slow) processors. We partition S into collections of processors of similar transmissions speeds. For $i = 1, \ldots, k$, define $S_i = \{p_j | \epsilon' T (1 + \epsilon')^{i-1} < t_j \leq \epsilon' T (1 + \epsilon')^i, 1 \leq j \leq n\}$, where k is $\lceil \frac{\log(1/\epsilon')}{\log(1+\epsilon')} \rceil$. Since $t_1 \leq t_2 \leq \cdots \leq t_n$, $F = \{p_1, \ldots, p_{|F|}\}$ and $S = \{p_{|F|+1}, \ldots, p_n\}$.

Broadcasting on Networks of Workstations

We first send messages to processors in F using FNF. We prove that there is a schedule with broadcast time at most $(1 + O(\epsilon))T$ such that all processors in F receive the message first. We then find a schedule for slow processors using a dynamic programming approach.

Schedule for F: We use the FNF heuristic to send the message to processors in set F. Assume that the schedule for F has a broadcast time of T_{FNF}. In this schedule, every processor $p_j \in F$ becomes idle at some time between $T_{\text{FNF}} - t_j$ and T_{FNF}.

We will prove that there is a schedule with broadcast time at most $(1 + O(\epsilon))T$ such that all processors in F receive the message first, and then send it to the slow processors. In an optimal schedule, let P_F be the first $|F|$ processors (except the source) that finish receiving the message and T_F be the time when all processors in P_F finish receiving the message. The following lemma relates T_{FNF} with T_F.

Lemma 19.21 $T_F \geq T_{\text{FNF}} - \epsilon' T$.

Proof. We prove this by contradiction. Consider a FNF schedule with set P_F. If an optimal schedule is able to have all processors in P_F finish receiving the message before $T_{\text{FNF}} - \epsilon' T$, then FNF can finish sending the messages to P_F before time T_{FNF} (by Corollary 19.25). Since set F includes the fastest $|F|$ processors, this means that FNF can finish broadcasting for F before time T_{FNF}; it is a contradiction since T_{FNF} is the earliest time that processors in F receive the message in FNF. ∎

At time T_F, there can be some set of processors P'_F that have received the message *partially*. Note that $|P'_F| \leq |P_F|$ since every processor in P'_F should receive the message from a processor in P_F or from the source and at least one processor P_F completes receiving the message at time T_F.

Lemma 19.22 *There is a schedule in which all processors in F receive the message no later than any processor in S and the makespan of the schedule is at most $(1 + 3\epsilon')T$.*

Proof. The main idea behind the proof is to show that an optimal schedule can be modified to have a certain form. Note that by time T_F the optimal schedule can finish broadcasting for P_F and partially send the message to P'_F, and in additional time $T - T_F$ the optimal schedule can finish broadcasting for the remaining processors.

In FNF schedule, all processors in F have the message at time T_{FNF}. Since $|P_F \cup P'_F| \leq 2|F|$ and any processor in F has speed at most $\epsilon' T$, in additional $2\epsilon' T$ time we can finish broadcasting the message to $P_F \cup P'_F$. Once we broadcast the message to $P_F \cup P'_F$, we send the message to the remaining processors as in the optimal solution.

The broadcast time of this schedule is at most $T_{\text{FNF}} + 2\epsilon' T + T - T_F \leq T_{\text{FNF}} + 2\epsilon' T + T - (T_{\text{FNF}} - \epsilon' T) = (1 + 3\epsilon')T$. ∎

Create all possible trees of S: For the processors in S, we will produce a set $\mathcal{S}$ of labeled trees $\mathcal{T}$. A tree $\mathcal{T}$ is any possible tree with broadcast time at most T consisting of a subset of processors in S. Then we label a node in the tree as i if the corresponding processor belongs to S_i ($i = 1, \ldots, k$). We prove that the number of different trees is constant.

Lemma 19.23 *The size of $\mathcal{S}$ is constant for fixed $\epsilon' > 0$.*

Proof. First consider the size of a tree $\mathcal{T}$ (i.e., the number of processors in the tree). Let us denote it as $|\mathcal{T}|$. Since the transmission time of processors in S is greater than $\epsilon' T$, we need at least $\epsilon' T$ time units to double the number of processors that received the message. It means that given a processor as a root of the tree, within time T we can have at most $2^{1/\epsilon'}$ processors receive the message. Therefore, $|\mathcal{T}| \leq 2^{1/\epsilon'}$. Now each node in the tree can have different label $i = 1, \ldots, k$. To obtain an upperbound of the number of different trees, given a tree T we transform it to a complete binomial tree of size $2^{1/\epsilon'}$ by adding nodes labeled as 0. Then the number of different trees is at most $(k+1)^{2^{1/\epsilon'}}$. ∎

Attach T to F: Let the completion time of every processor $p_j \in F$ be c_j. Each processor p_j in F sends a message to a processor in S every t_j time unit. Therefore, a fast processor p_j can send messages to at most $X_j = \lfloor \frac{T-c_j}{t_j} \rfloor$ other processors. Let $X = \sum_{p_j \in F} X_j$. Let us consider the time x_i of each sending point in X. We sort those x_i in nondecreasing order and attach a tree from S to each point (see Figure 19.9). Note that we can attach at most $|X|$ trees of slow processors. Clearly, $|X| \leq n$.

We check if an attachment is feasible, using dynamic programming. Recall that we partition slow processors into a collection of processors $S_1, S_2, \ldots, S_k$ ($k = \log(1/\epsilon')/\log(1+\epsilon')$). Let s_i denote the number of processors in set S_i. We define a state $s[j, n_1, n_2, \ldots, n_k]$ ($0 \leq j \leq |X|, 0 \leq n_i \leq s_i$) to be true if there is a set of j trees in S that we can attach to first j sending points and the corresponding schedule satifies the following two conditions: (i) the schedule completes by time T and (ii) exactly n_i processors in S_i appear in j trees in total. Our goal is to find out whether $s[j, s_1, s_2, \ldots, s_k]$ is true for some j, which means that there is a feasible schedule with makespan at most T. The number of states is at most $O(n^{k+1})$ since we need at most n trees ($|X| \leq n$) and $s_i \leq n$.

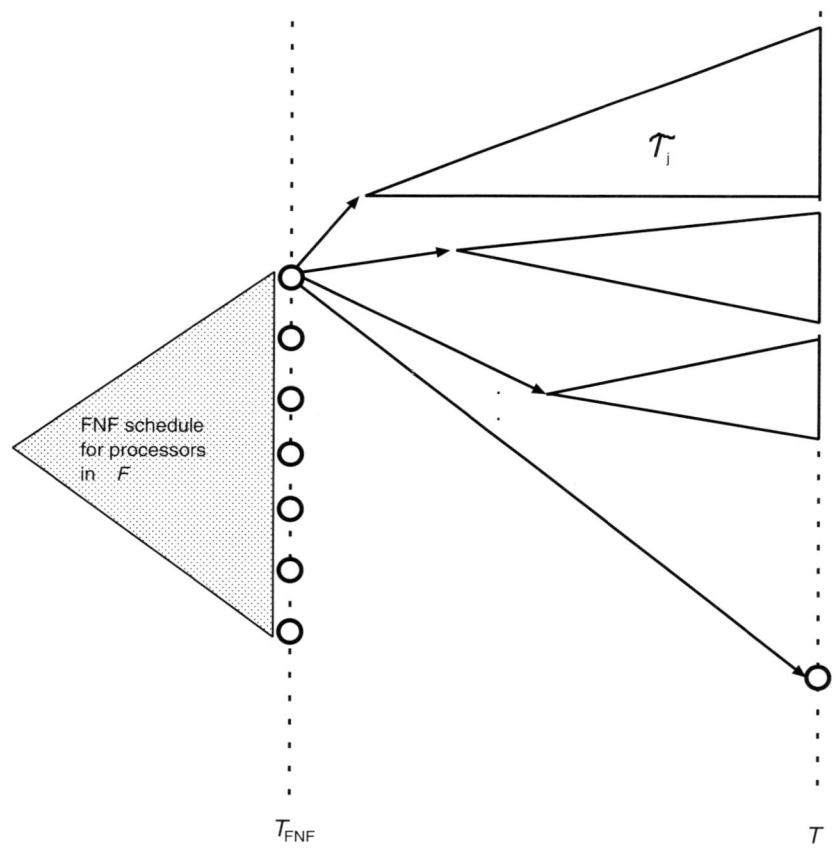

FIGURE 19.9 Attach trees for slow processors to fast processors.

Now we prove that each state can be computed in constant time. Given $s[j-1,\ldots]$, we compute $s[j, n_1, n_2, \ldots, n_k]$ as follows. We try to attach all possible trees in S to x_j. Then $s[j, n_1, n_2, \ldots, n_k]$ is true if there exists a tree T' such that the makespan of T' is at most $T - x_j$ and $s[j-1, n_1-m_1, n_2-m_1, \ldots, n_k-m_k]$ is true where T' has m_i slow processors belonging to set S_i. It can be checked in constant time since the size of S is constant (Lemma 19.23).

Theorem 19.26 *Given a value T, if a broadcast tree with broadcast time T exists, then the above algorithm will find a broadcast tree with broadcast time at most $(1+\epsilon')(1+3\epsilon')T$.*

Proof. Consider the best schedule among all schedules in which processors in F receive the message first. By Lemma 19.22, the broadcast time of this schedule is at most $(1+3\epsilon')T$. We round up the transmission time of p_j in S_i to $\epsilon'T(1+\epsilon')^i$ where i is the smallest integer such that $t_j \leq \epsilon'(1+\epsilon')^i T$. By this rounding, we increase the broadcast time by factor of at most $1+\epsilon'$. Therefore, the broadcast time of our schedule is at most $(1+\epsilon')(1+3\epsilon')T$. ∎

Theorem 19.27 *The algorithm takes as input the transmission times of the n processors, and constants $\epsilon, \epsilon' > 0$. The algorithm finds a broadcast tree with broadcast time at most $(1+\epsilon)(1+\epsilon')(1+3\epsilon')T$ in polynomial time.*

Proof. We try the above algorithm for all possible values of the form $T = (1+\epsilon)^j$ for $j = 1 \ldots \lceil \frac{\log n}{\log(1+\epsilon)} \rceil$. This will increase the broadcast time by factor of at most $1+\epsilon$. Therefore, the broadcast time of our schedule is at most $(1+\epsilon)(1+\epsilon')(1+3\epsilon')T$.

For each given value $(1+\epsilon)^j$, we find FNF schedule for processors in F (it takes at most $O(n \log n)$) and attach trees of slow processors to processors in F, using dynamic programming. As we discussed earlier, the number of states is $O(n^{k+1})$ and each state can be checked if it is feasible in $O((k+1)^{2^{1/\epsilon'}})$ time, which is constant. Thus, the running time of our algorithm is $O(\lceil \frac{\log n}{\log(1+\epsilon)} \rceil (n \log n + (k+1)^{2^{1/\epsilon'}+1} \times n^{k+1}))$, where k is $\lceil \frac{\log(1/\epsilon')}{\log(1+\epsilon')} \rceil$. ∎

19.4.5 Multicast

A multicast operation involves only a subset of processors. By utilizing fast processors that are not in the multicast group, we can reduce the multicasting time significantly. For example, suppose that we have n processors with transmission time t_1 and m more processors with transmission time t_2 where $t_1 < t_2$. Let we want to multicast a message to all processors with transmission time t_2. If we only use processors in the multicast group, it will take $t_2 \times \log m$ time. But if we utilize processors with transmission time t_1, we can finish the multicast in $t_1 \times (\log m + 1)$. Therefore, when $t_1 \ll t_2$, the speedup is significant.

Theorem 19.28 *Suppose that we have a ρ-approximation algorithm for broadcasting. Then, we can find a ρ-approximation algorithm for multicasting.*

Proof. Note that if an optimal solution utilizes k processors not in the multicast group, then those processors are the k fastest ones. Therefore, if we know how many processors participate in multicasting, we can use our ρ-approximation algorithm for broadcasting. By trying all possible k and taking the best one, we have ρ-approximation for multicasting. ∎

Theorem 19.29 *We have a polynomial time approximation scheme for multicasting.*

Proof. The proof is similar to Theorem 19.28. We find approximation schemes with k fastest processors not in the multicast group for all possible k, and take the best one. ∎

References

[BDH+97] J. Bruck, D. Dolev, C. Ho, M. Rosu, and R. Strong. Efficient message passing interface (MPI) for parallel computing on clusters of workstations. *Parallel Distributed Computing*, 40: 19–34, 1997.

[BMP98] M. Banikazemi, V. Moorthy, and D. K. Panda. Efficient collective communication on heterogeneous networks of workstations. In *International Conference on Parallel Processing*, 1998.

[BNGNS98] A. Bar-Noy, S. Guha, J. Naor, and B. Schieber. Multicasting in heterogeneous networks. In *Proceedings of the 13th Annual ACM Symposium on Theory of Computing*, pp. 448–453, 1998.

[BNK94] A. Bar-Noy and S. Kipnis. Designing broadcast algorithms in the postal model for message-passing systems. *Mathematical Systems Theory*, 27(5):431–452, 1994.

[BRP99] P. Bhat, C. Raghavendra, and V. Prasanna. Efficient collective communication in distributed heterogeneous systems. In *Proceedings of the International Conference on Distributed Computing Systems*, 1999.

[CKP+93a] D. E. Culler, R. M. Karp, D. A. Patterson, A. Sahay, K. E. Schauser, E. Santos, R. Subramonian, and T. von Eicken. Logp: Towards a realistic model of parallel computation. In *Proceedings of the 4th ACM SIGPLAN Symposium on Principles and Practice of Parallel Programming*, pp. 1–12, 1993.

[CKP+93b] David E. Culler, Richard M. Karp, David A. Patterson, Abhijit Sahay, Klaus E. Schauser, Eunice Santos, Ramesh Subramonian, and Thorsten von Eicken. LogP: Towards a realistic model of parallel computation. In *Principles Practice of Parallel Programming*, pp. 1–12, 1993.

[EK02] M. Elkin and G. Kortsarz. Combinatorial logarithmic approximation algorithm for the directed telephone broadcast problem. In *STOC 2002*, pp. 438–447, 2002.

[EK03] M. Elkin and G. Kortsarz. A sublogarithmic approximation algorithm for the undirected telephone broadcast problem: a path out of a jungle. In *SODA 2002*, pp. 76–85, 2003.

[FK98] Foster and K. Kesselman. *The Grid: Blueprint for a New Computing Infrastructure*. Morgan Kaufmann, 1998.

[GLDS96] W. Gropp, E. Lusk, N. Doss, and A. Skjellum. A high-performance, portable implementation of the MPI: a message passing interface standard. *Parallel Computing*, 22:789–828, 1996.

[HH98] P. Husbands and J. Hoe. A survey of gossiping and broadcasting in communication networks. In *Supercomputing*, 1998.

[HLS98] N. Hall, W.-P. Liu, and J. Sidney. Scheduling in broadcast networks. *Networks*, 32:233–253, 1998.

[KBG00] T. Kielmann, H. Bal, and S. Gorlatch. Bandwidth-efficient collective communication for clustered wide area systems. In *International Parallel and Distributed Processing Symposium*, 2000.

[KHB+99] T. Kielmann, R. Horfman, H. Bal, A. Plaat, and R. Bhoedjang. Magpie: Mpi's collective communication operations for clustered wide area systems. In *Symposium on Principles and Practice of Parallel Programming*, 1999.

[KK04] S. Khuller and Y.-A. Kim. On broadcasting in heterogenous networks. In *Proceedings of the 15th Annual ACM/SIAM Symposium on Discrete Algorithms*, pp. 1004–1013, 2004.

[KSSK93] R. Karp, A. Sahay, E. Santos, and K. E. Schauser. Optimal broadcast and summation in the logp model. In *Proceedings of 5th Annual Symposium on Parallel Algorithms and Architectures*, pp. 142–153, 1993.

[LB96] B. B. Lowekamp and A. Beguelin. Eco: Efficient collective operations for communication on heterogeneous networks. In *International Parallel Processing Symposium (IPPS)*, pp. 399–405, Honolulu, HI, 1996.

[LHHB+01] R. Libeskind-Hadas, J. R. K. Hartline, P. Boothe, G. Rae, and J. Swisher. On multicast algorithms for heterogeneous networks of workstations. *Journal of Parallel and Distributed Computing*, 61(11):1665–1679, 2001.

[Liu02] P. Liu. Broadcasting scheduling optimization for heterogeneous cluster systems. *Journal of Algorithms*, 42:135–152, 2002.

[LS00] P. Liu and T.-H. Sheng. Broadcasting scheduling optimization for heterogeneous cluster systems. In *ACM Symposium on Parallel Algorithms and Architectures (SPAA)*, pp. 129–136, 2000.

[LW00] P. Liu and D. Wang. Reduction optimization in heterogeneous cluster environments. In *Proceedings of the International Parallel and Distributed Processing Symposium*, pp. 477–482, 2000.

[MPI94] Message passing interface form, March 1994.

[PCA95] D. A. Patterson, D. E. Culler, and T. E. Anderson. A case for nows (networks of workstations). *IEEE Micro*, 15(1):54–64, 1995.

[PL96] J. Pruyne and M. Livny. Interfacing condor and pvm to harness the cycles of workstation clusters. *Journal on Future Generations of Computer Systems*, 12:67–85, 1996.

[RL88] D. Richards and A. L. Liestman. Generalization of broadcasting and gossiping. *Networks*, 18:125–138, 1988.

[RMKS92] A. Sahay, R.M. Karp, and E. Santos. Optimal broadcast and summation in the logp model. Technical Report UCB/CSD-92-721, EECS Department, University of California, Berkeley, 1992.

[SHA88] S. M. Hedetniemi, S. T. Hedetniemi, and A. L. Liestman. A survey of broadcasting and gossiping in communication networks. *Networks*, 18:319–349, 1988.

[Val90] L. G. Valiant. A bridging model for parallel computation. *Communications of the ACM*, 33(8):103–111, August 1990.

20
Atomic Selfish Routing in Networks: A Survey

Spyros Kontogiannis
University of Ioannina

Paul G. Spirakis
University of Patras

20.1	Introduction...	20-2
	Roadmap	
20.2	The Model...	20-2
	Dealing with Selfish Behavior • Potential Games • Configuration Paths and Discrete Dynamics Graph • Isomorphism of Strategic Games • Layered Networks	
20.3	Related Work...	20-5
	Existence and Tractability of Pure Nash Equilibrium • Price of Anarchy in Congestion Games	
20.4	Unweighted Congestion Games	20-7
20.5	Existence and Complexity of Constructing PNE	20-10
	Efficient Construction of Pure Nash Equilibrium in Unweighted Congestion Games • Existence and Construction of Pure Nash Equilibrium in Weighted Congestion Games	
20.6	Parallel Links and Player-Specific Payoffs	20-16
	Players with Distinct Weights	
20.7	The Price of Anarchy of Weighted Congestion Games......................................	20-20
	Flows and Mixed Strategies Profiles • Flows at Nash Equilibrium • Maximum Latency versus Total Latency • An Upper Bound on the Social Cost • Bounding the Price of Anarchy	
20.8	The Pure Price of Anarchy in Congestion Games	20-27
20.9	Conclusions	20-27
References		20-28

In this survey, we present some recent advances in the *atomic congestion games* literature. Our main focus is on a special case of congestion games, called *network congestion games*, which is of particular interest for the networking community. The algorithmic questions that we are interested in have to do with the *existence* of pure Nash equilibria, the *efficiency* of their construction when they exist, as well as the *gap* of the best/worst (mixed in general) Nash equilibria from the social optima in such games, typically called the *Price of Anarchy* and the *Price of Stability*, respectively.

20.1 Introduction

Consider a model where selfish individuals (henceforth called **players**) in a communication network having varying service demands compete for some shared resources. The quality of service provided by a resource decreases with its *congestion*, that is, the amount of demands of the players willing to be served by it. Each player may reveal his actual, unique choice of a subset of resources (called a *pure strategy*) that satisfies his service demand, or he may reveal a probability distribution for choosing (independently of other players' choices) one of the possible (satisfactory for him) subsets of resources (called a *mixed strategy*). The players determine their actual behavior on the basis of other players' behaviors, but they do not cooperate. We are interested in situations where the players have reached some kind of stable state, that is, an equilibrium. The most popular notion of equilibrium in noncooperative game theory is the *Nash equilibrium*: a "stable point" among the players, from which no player is willing to deviate unilaterally. In Reference 23, the notion of the *coordination ratio* or *price of anarchy* was introduced as a means for measuring the performance degradation due to lack of players' coordination when sharing common goods. A more recent measure of performance is the *price of stability* [2], capturing the gap between the best possible Nash equilibrium and the globally optimal solution. This measure is crucial for the network designer's perspective, who would like to propose (rather than let the players end up in) a Nash equilibrium (from which no player would like to defect unilaterally) that is as close to the optimum as possible.

A realistic scenario for the above model is when *unsplittable* traffic demands are routed selfishly in general networks with load-dependent edge delays. When the underlying network consists of two nodes and parallel links between them, there has been an extensive study on the existence and computability of equilibria, as well as on the price of anarchy. In this survey, we study the recent advances in the more general case of arbitrary *congestion games*. When the players have identical traffic demands, the congestion game is indeed isomorphic to an *exact potential game* ([29], see also Theorem 20.1 of this survey) and thus always possesses a *pure Nash equilibrium*, that is, an equilibrium where each player adopts a pure strategy. We shall see that varying demands of the players crucially affect the nature of these games, which are no longer isomorphic to exact potential games. We also present some results in a variant of congestion games, where the players' payoffs are not resource-dependent (as is typically the case in congestion games) but *player-specific*.

20.1.1 Roadmap

In Section 20.2, we formally define the congestion games and their variants considered in this survey. We also give some game-theoretic definitions. In Section 20.3, we present most of the related work in the literature, before presenting in detail some of the most significant advances in the area. In Section 20.4, we present some of the most important results concerning unweighted congestion games and their connection to the *potential games*. In Section 20.5, we study some complexity issues of unweighted congestion games. In Section 20.6, we present Milchtaich's extension of congestion games to allow *player-specific* payoffs, whereas in Section 20.5.2, we study some existence and computability issues of Pure Nash Equilibrium (PNE) in weighted congestion games. Finally, in Section 20.7, we study the price of anarchy of weighted congestion games. We close this survey with some concluding remarks and unresolved questions.

20.2 The Model

Consider having a set of resources E in a system. For each $e \in E$, let $d_e(\cdot)$ be the **delay** per player that requests his service, as a function of the total usage (that is, the *congestion*) of this resource by all the players. Each such function is considered to be *nondecreasing* in the total usage of the corresponding resource. Each resource may be represented by a pair of points: an entry point to the resource and an exit point from it. So, we represent each resource by an arc from its entry point to its exit point, and we associate with this arc the **charging cost** (e.g., the delay as a function of the load of this resource) that each player has to pay if he is served by this resource. The entry/exit points of the resources need not be unique; they may coincide in order to express the possibility of offering a *joint service* to players, which consists of a sequence of resources. We denote by V the set of all entry/exit points of the resources in the system. Any nonempty collection of resources corresponding to a directed path in $G \equiv (V, E)$ comprises an **action** in the system.

Let $N \equiv [n]^*$ be the set of players, each willing to adopt some action in the system. $\forall i \in N$, let w_i denote player i's **traffic demand** (e.g., the flow rate from a source node to a destination node), while $\mathcal{P}^i \equiv \{a_1^i, \ldots, a_{m_i}^i\} \subseteq 2^E \setminus \emptyset$ (for some $m_i \geq 2$) is the collection of actions, any of which would satisfy player i (e.g., alternative routes from a source to a destination node, if G represents a communication network). The collection $\mathcal{P}^i$ is called the *action set* of player i, and each of its elements contains at least one resource. Any n-tuple $\varpi \in \mathcal{P} \equiv \times_{i=1}^{n} \mathcal{P}^i$ is a **pure strategies profile**, or a **configuration** of the players. Any real vector $\mathbf{p} = (\mathbf{p}^1, \mathbf{p}^2, \ldots, \mathbf{p}^n)$ s.t. $\forall i \in N$, $\mathbf{p}^i \in \Delta(\mathcal{P}^i) \equiv \{\mathbf{z} \in [0,1]^{m_i}: \sum_{k=1}^{m_i} z_k = 1\}$ is a probability distribution over the set of allowable actions for player i, and is called a **mixed strategies profile** for the n players.

A **congestion model** $((\mathcal{P}^i)_{i \in N}, (d_e)_{e \in E})$ typically deals with players of identical demands, and thus, the resource delay functions depend only on the *number* of players adopting each action [12, 29, 31]. In the more general case, that is, a **weighted congestion model** is the tuple $((w_i)_{i \in N}, (\mathcal{P}^i)_{i \in N}, (d_e)_{e \in E})$. That is, we allow the players to have different (but fixed) demands for service (denoted by their weights) from the whole system, and thus affect the resource delay functions in a different way, depending on their own weights. We denote by $W_{\text{tot}} \equiv \sum_{i \in N} w_i$ and $w_{\max} \equiv \max_{i \in N}\{w_i\}$.

The **weighted congestion game** $\Gamma \equiv (N, E, (w_i)_{i \in N}, (\mathcal{P}^i)_{i \in N}, (d_e)_{e \in E})$ associated with this model is the game in strategic form with the set of players N and players' demands $(w_i)_{i \in N}$, the set of shared resources E, the action sets $(\mathcal{P}^i)_{i \in N}$, and players' cost functions $(\lambda_{\varpi^i}^i)_{i \in N, \varpi^i \in \mathcal{P}^i}$ defined as follows: For any configuration $\varpi \in \mathcal{P}$ and $\forall e \in E$, let $\Lambda_e(\varpi) = \{i \in N : e \in \varpi^i\}$ be the set of players wishing to exploit resource e according to ϖ (called the **view** of resource e with respect to configuration ϖ). We also denote by $x_e(\varpi) \equiv |\Lambda_e(\varpi)|$ the *number* of players using resource e with respect to ϖ, whereas $\theta_e(\varpi) \equiv \sum_{i \in \Lambda_e(\varpi)} w_i$ is the **load** of e with respect to to ϖ. The **cost** $\lambda^i(\varpi)$ **of player** i **for adopting strategy** $\varpi^i \in \mathcal{P}^i$ in a given configuration ϖ is equal to the cumulative delay $\lambda_{\varpi^i}(\varpi)$ of all the resources comprising this action:

$$\lambda^i(\varpi) = \lambda_{\varpi^i}(\varpi) = \sum_{e \in \varpi^i} d_e(\theta_e(\varpi)). \quad (20.1)$$

On the other hand, for a mixed strategies profile $\mathbf{p}$, the (**expected**) **cost of player** i **for adopting strategy** $\varpi^i \in \mathcal{P}^i$ with respect to $\mathbf{p}$ is

$$\lambda_{\varpi^i}^i(\mathbf{p}) = \sum_{\varpi^{-i} \in \mathcal{P}^{-i}} P(\mathbf{p}^{-i}, \varpi^{-i}) \cdot \sum_{e \in \varpi^i} d_e\left(\theta_e(\varpi^{-i} \oplus \varpi^i)\right), \quad (20.2)$$

where $\varpi^{-i} \in \mathcal{P}^{-i} \equiv \times_{j \neq i} \mathcal{P}^j$ is a configuration of all the players except for i, $\mathbf{p}^{-i} \in \times_{j \neq i} \Delta(\mathcal{P}^j)$ is the mixed strategies profile of all players except for i, $\varpi^{-i} \oplus a$ is the new configuration with i definitely choosing the action $a \in \mathcal{P}^i$, and $P(\mathbf{p}^{-i}, \varpi^{-i}) \equiv \prod_{j \neq i} p_{\varpi^j}^j$ is the occurrence probability of ϖ^{-i} according to $\mathbf{p}^{-i}$.

Remark 20.1 We abuse notation a little bit and consider the player costs $\lambda_{\varpi^i}^i$ as functions whose exact definition depends on the other players' strategies: In the general case of a mixed strategies profile $\mathbf{p}$, Equation (20.2) is valid and expresses the expected cost of player i with respect to $\mathbf{p}$, conditioned on the event that i chooses path ϖ^i. If the other players adopt a pure strategies profile ϖ^{-i}, we get the special form of Equation (20.1) that expresses the exact cost of player i choosing action ϖ^i.

---

* $\forall k \in \mathbf{N}$, $[k] \equiv \{1, 2, \ldots, k\}$.

Remark 20.2 Concerning the players' private cost functions, instead of charging them for the *sum* of the expected costs of the resources that each of them chooses to use (call it the SUMCOST objective), we could also consider the maximum expected cost over all the resources in the strategy that each player adopts (call it the MAXCOST objective). This is also a valid objective, especially in scenarios dealing with bandwidth allocation in networks. Nevertheless, in the present survey we focus our interest on the SUMCOST objective, unless stated explicitly that we use some other objective.

A congestion game in which all players are indistinguishable (i.e., they have the traffic demands and the same action set) is called **symmetric**. When each player's action set $\mathcal{P}^i$ consists of sets of resources that comprise (simple) paths between a unique origin–destination pair of nodes (s_i, t_i) in (V, E), we refer to a **(multicommodity) network congestion game**. If additionally all origin–destination pairs of the players coincide with a unique pair (s, t), we have a **single-commodity network congestion game** and then all players share exactly the same action set. Observe that in general a single-commodity network congestion game is not necessarily symmetric because the players may have different demands and thus their cost functions will also differ.

20.2.1 Dealing with Selfish Behavior

Fix an arbitrary (mixed in general) strategies profile **p** for a congestion game that is described by the tuple $((w_i)_{i \in N}, (\mathcal{P}^i)_{i \in N}, (d_e)_{e \in E})$. We say that **p** is a **Nash equilibrium (NE)** if and only if

$$\forall i \in N, \forall \alpha, \beta \in \mathcal{P}^i, \ p_\alpha^i > 0 \ \Rightarrow \ \lambda_\alpha^i(\mathbf{p}) \leq \lambda_\beta^i(\mathbf{p}).$$

A configuration $\varpi \in \mathcal{P}$ is a **PNE** if and only if

$$\forall i \in N, \forall \alpha \in \mathcal{P}^i, \ \lambda^i(\varpi) = \lambda_{\varpi^i}(\varpi) \leq \lambda_\alpha(\varpi^{-i} \oplus \alpha) = \lambda^i(\varpi^{-i} \oplus \alpha).$$

The **social cost** SC(**p**) in this congestion game is

$$\mathrm{SC}(\mathbf{p}) = \sum_{\varpi \in \mathcal{P}} P(\mathbf{p}, \varpi) \cdot \max_{i \in N}\{\lambda_{\varpi^i}(\varpi)\}, \quad (20.3)$$

where $P(\mathbf{p}, \varpi) \equiv \prod_{i=1}^n p_{\varpi^i}^i$ is the probability of configuration ϖ occurring with respect to the mixed strategies profile **p**. The **social optimum** of this game is defined as

$$\mathrm{OPT} = \min_{\varpi \in \mathcal{P}} \left\{ \max_{i \in N}[\lambda_{\varpi^i}(\varpi)] \right\}. \quad (20.4)$$

The **price of anarchy** for this game is then defined as

$$\mathcal{R} = \max_{\mathbf{p} \text{ is a NE}} \left\{ \frac{\mathrm{SC}(\mathbf{p})}{\mathrm{OPT}} \right\}. \quad (20.5)$$

20.2.2 Potential Games

Fix an arbitrary game in strategic form $\Gamma = (N, (\mathcal{P}^i)_{i \in N}, (U^i)_{i \in N})$ and some vector $\mathbf{b} \in \mathbf{R}_{>0}^n$. A function $\Phi : \mathcal{P} \to \mathbf{R}$ is called

- An **ordinal potential** for Γ, if $\forall \varpi \in \mathcal{P}, \forall i \in N, \forall \alpha \in \mathcal{P}^i$,

$$\mathrm{sign}[]\, \lambda^i(\varpi) - \lambda^i(\varpi^{-i} \oplus \alpha) = \mathrm{sign}[]\, \Phi(\varpi) - \Phi(\varpi^{-i} \oplus \alpha).$$

- A **b-potential** for Γ, if $\forall \varpi \in \mathcal{P}$, $\forall i \in N, \forall \alpha \in \mathcal{P}^i$,

$$\lambda^i(\varpi) - \lambda^i(\varpi^{-i} \oplus \alpha) = b_i \cdot [\Phi(\varpi) - \Phi(\varpi^{-i} \oplus \alpha)].$$

- An **exact potential** for Γ, if it is a **1**-potential for Γ.

20.2.3 Configuration Paths and Discrete Dynamics Graph

For a congestion game $\Gamma = (N, E, (w_i)_{i \in N}, (\mathcal{P}^i)_{i \in N}, (d_e)_{e \in E})$, a **path** in $\mathcal{P}$ is a sequence of configurations $\gamma = (\varpi(0), \varpi(1), \ldots, \varpi(k))$ s.t. $\forall j \in [k]$, $\varpi(j) = (\varpi(j-1))^{-i} \oplus \pi_i$, for some $i \in N$ and $\pi_i \in \mathcal{P}^i$. γ is a **closed path** if $\varpi(0) = \varpi(k)$. It is a **simple path** if no configuration is contained in it more than once. γ is an **improvement path** with respect to Γ, if $\forall j \in [k]$, $\lambda^{i_j}(\varpi(j)) < \lambda^{i_j}(\varpi(j-1))$, where i_j is the unique player differing in his strategy between $\varpi(j)$ and $\varpi(j-1)$. That is, the unique defector of the jth move in γ is actually willing to make this move because it improves his own cost. The **Nash dynamics graph** of Γ is a directed graph whose vertices are configurations, and there is an arc from a configuration ϖ to a configuration $\varpi^{-i} \oplus \alpha$ for some $\alpha \in \mathcal{P}^i$ if and only if $\lambda^i(\varpi) > \lambda^i(\varpi^{-i} \oplus \alpha)$. The set of best replies of a player i against a configuration $\varpi^{-i} \in \mathcal{P}^{-i}$ is defined as $BR_i(\varpi^{-i}) = \arg\max_{\alpha \in \mathcal{P}^i}\{\lambda^i(\varpi^{-i} \oplus \alpha)\}$. Similarly, the set of best replies against a mixed profile $\mathbf{p}^{-i}$ is $BR_i(\mathbf{p}^{-i}) = \arg\max_{\alpha \in \mathcal{P}^i}\{\lambda^i_\alpha(\mathbf{p}^{-i} \oplus \alpha)\}$. A path γ is a **best-reply improvement path** if each defector jumps to a best-reply pure strategy. The **best response dynamics graph** is a directed graph whose vertices are configurations, and there is an arc from a configuration ϖ to a configuration $\varpi^{-i} \oplus \alpha$ for some $\alpha \in \mathcal{P}^i \setminus \{\varpi^i\}$ if and only if $\alpha \in BR_i(\varpi^{-i})$ and $\varpi^i \notin BR_i(\varpi^{-i})$.

A (finite) strategic game Γ possesses the **finite improvement property** (FIP) if any improvement path of Γ has finite length. Γ possesses the **finite best-reply property** (FBRP) if every best-reply improvement path is of finite length.

20.2.4 Isomorphism of Strategic Games

Two games in strategic form $\Gamma = (N, (\mathcal{P}^i)_{i \in N}, (U^i)_{i \in N})$ and $\tilde{\Gamma} = (N, (\tilde{\mathcal{P}}^i)_{i \in N}, (\tilde{U}^i)_{i \in N})$ are called **isomorphic** if there exist bijections $g : \times_{i \in N} \mathcal{P}^i \mapsto \times_{i \in N} \tilde{\mathcal{P}}^i$ and $\tilde{g} : \times_{i \in N} \tilde{\mathcal{P}}^i \mapsto \times_{i \in N} \mathcal{P}^i$ s.t $\forall \varpi \in \times_{i \in N} \mathcal{P}^i, \forall i \in N$, $U^i(\varpi) = \tilde{U}^i(g(\varpi))$ and $\forall \tilde{\varpi} \in \times_{i \in N} \tilde{\mathcal{P}}_i, \forall i \in N$, $\tilde{U}^i(\tilde{\varpi}) = U^i(\tilde{g}(\tilde{\varpi}))$.

20.2.5 Layered Networks

We consider a special family of networks (an example is provided in Figure 20.1) whose behavior with respect to the price of anarchy, as we shall see, is asymptotically equivalent to that of the parallel links model of Reference 23 (which is actually a 1-layered network): Let $\ell \geq 1$ be an integer. A directed network $G = (V, E)$ with a distinguished source–destination pair (s, t), $s, t \in V$, is an ℓ-**layered network** if every (simple) directed $s - t$ path has length exactly ℓ and each node lies on a directed $s - t$ path. In a layered network, there are no directed cycles and all directed paths are simple. In the following, we always use $m = |E|$ to denote the number of edges in an ℓ-layered network $G = (V, E)$.

20.3 Related Work

20.3.1 Existence and Tractability of Pure Nash Equilibrium

It is already known that the class of unweighted (atomic) congestion games (i.e., players have the same demands and, thus, the same affection on the resource delay functions) is guaranteed to have at least one PNE: actually, Rosenthal [31] proved that any potential game has at least one PNE and it is easy to write any unweighted congestion game as an exact potential game using Rosenthal's potential function[*]

---

[*] For more details on potential games, see Reference 29.

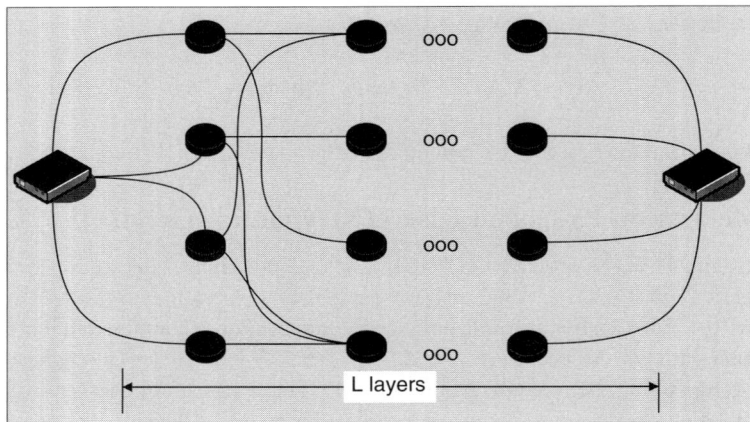

FIGURE 20.1 An example of a layered network.

(e.g., [12, Theorem 1]). In Reference 12, it is proved that a PNE for any unweighted single-commodity network congestion game* (no matter what resource delay functions are considered, so long as they are nondecreasing with loads) can be constructed in polynomial time, by computing the optimum of Rosenthal's potential function, through a nice reduction to min-cost flow. On the other hand, it is shown that even for a symmetric congestion game or an unweighted multicommodity network congestion game, it is PLS-complete to find a PNE (though it certainly exists).

The special case of single-commodity, parallel-edges network congestion game, where the resources are considered to behave as parallel machines, has been extensively studied in recent literature. In Reference 14, it was shown that for the case of players with varying demands and uniformly related parallel machines, there is always a PNE that can be constructed in polynomial time. It was also shown that it is NP-hard to construct the best or the worst PNE. In Reference 17, it was proved that the fully mixed NE (FMNE), introduced and thoroughly studied in Reference 27, is worse than any PNE, and any NE is at most $(6 + \varepsilon)$ times worse than the FMNE, for varying players and identical parallel machines. In Reference 26, it was shown that the FMNE is the worst possible for the case of two related machines and tasks of the same size. In Reference 25, it was proved that the FMNE is the worst possible when the global objective is the sum of squares of loads.

Reference 13 studies the problem of constructing a PNE from any initial configuration, of social cost at most equal to that of the initial configuration. This immediately implies the existence of a PTAS for computing a PNE of minimum social cost: first compute a configuration of social cost at most $(1 + \varepsilon)$ times the social optimum [18], and consequently transform it into a PNE of at most the same social cost. In Reference 11, it is also shown that even for the unrelated parallel machines case a PNE always exists, and a potential-based argument proves a convergence time (in case of integer demands) from arbitrary initial configuration to a PNE in time $\mathcal{O}\left(mW_{\text{tot}} + 4^{W_{\text{tot}}/m + w_{\max}}\right)$.

Reference 28 studies the problem of weighted parallel-edges network congestion games with player-specific costs: each allowable action of a player consists of a single resource and each player has his own private cost function for each resource. It is shown that (i) weighted (parallel-edges network) congestion games involving only two players, or only two possible actions for all the players, or equal delay functions (and thus, equal weights), always possess a PNE; (ii) even a single-commodity, 3-players, 3-actions, weighted (parallel-edges network) congestion game may not possess a PNE (using 3-wise linear delay functions).

---

*Since Reference 12 only considers unit-demand players, this is also a symmetric network congestion game.

20.3.2 Price of Anarchy in Congestion Games

In the seminal paper [23], the notion of coordination ratio, or price of anarchy, was introduced as a means for measuring the performance degradation owing to lack of players' coordination when sharing common resources. In this work it was proved that the price of anarchy is 3/2 for two related parallel machines, while for m machines and players of varying demands, $\mathcal{R} = \Omega\left(\frac{\log m}{\log \log m}\right)$ and $\mathcal{R} = \mathcal{O}(\sqrt{m \log m})$. For m identical parallel machines, Reference 27 proved that $\mathcal{R} = \Theta\left(\frac{\log m}{\log \log m}\right)$ for the FMNE, while for the case of m identical parallel machines and players of varying demands it was shown in Reference 22 that $\mathcal{R} = \Theta\left(\frac{\log m}{\log \log m}\right)$. In Reference 9, it was finally shown that $\mathcal{R} = \Theta\left(\frac{\log m}{\log \log \log m}\right)$ for the general case of related machines and players of varying demands. Reference 8 presents a thorough study of the case of general, monotone delay functions on parallel machines, with emphasis on delay functions from queuing theory. Unlike the case of linear cost functions, they show that the price of anarchy for nonlinear delay functions in general is far worse and often even unbounded.

In Reference 32, the price of anarchy in a multicommodity network congestion game among infinitely many players, each of negligible demand, is studied. The social cost in this case is expressed by the total delay paid by the whole flow in the system. For linear resource delays, the price of anarchy is at most 4/3. For general, continuous, nondecreasing resource delay functions, the total delay of any Nash flow is at most equal to the total delay of an optimal flow for double flow demands. Reference 33 proves that for this setting, it is actually the class of allowable latency functions and not the specific topology of a network that determines the price of anarchy.

20.4 Unweighted Congestion Games

In this section, we present some fundamental results connecting the classes of unweighted congestion games and (exact) potential games [29, 31]. Since we refer to players of identical (say, unit) weights, the players' cost functions are

$$\lambda^i(\varpi) \equiv \sum_{\varepsilon \in \varpi^i} d_e(x_e(\varpi)),$$

where $x_e(\varpi)$ indicates the *number* of players willing to use resource e with respect to configuration $\varpi \in \mathcal{P}$. The following theorem proves the strong connection of congestion games with the exact potential games.

Theorem 20.1 [29,31] *Every (unweighted) congestion game is an exact potential game.*

Proof. Fix an arbitrary (unweighted) congestion game $\Gamma = (N, E, (\mathcal{P}^i)_{i \in N}, (d_e)_{e \in E})$. For any configuration $\varpi \in \mathcal{P}$, consider the function $\Phi(\varpi) = \sum_{e \in \cup_{i \in N} \varpi^i} \sum_{k=1}^{x_e(\varpi)} d_e(k)$, which we shall call **Rosenthal's potential**. We can easily show that Φ is an *exact potential* for Γ: For this, consider arbitrary configuration $\varpi \in \mathcal{P}$, an arbitrary player $i \in N$, and an alternative (pure) strategy $\alpha \in \mathcal{P}^i \setminus \{\varpi^i\}$ for this player. Let also $\hat{\varpi} = \varpi^{-i} \oplus \alpha$. Then,

$$\Phi(\hat{\varpi}) - \Phi(\varpi) = \sum_{e \in \cup_j \hat{\varpi}^j} \sum_{k=1}^{x_e(\hat{\varpi})} d_e(k) - \sum_{e \in \cup_j \varpi^j} \sum_{k=1}^{x_e(\varpi)} d_e(k)$$

$$= \sum_{e \in \cup_i \hat{\varpi}^i \setminus \varpi^i} \left[\sum_{k=1}^{x_e(\varpi)+1} d_e(k) - \sum_{k=1}^{x_e(\varpi)} d_e(k)\right]$$

$$+ \sum_{e \in \cup_i \varpi^i \setminus \hat{\varpi}^i} \left[\sum_{k=1}^{x_e(\varpi)-1} d_e(k) - \sum_{k=1}^{x_e(\varpi)} d_e(k)\right]$$

$$= \sum_{e \in \hat{\varpi}^i \setminus \varpi^i} d_e(x_e(\varpi) + 1) - \sum_{e \in \varpi^i \setminus \hat{\varpi}^i} d_e(x_e(\varpi))$$

$$= \sum_{e \in \hat{\varpi}^i} d_e(x_e(\hat{\varpi})) - \sum_{e \in \varpi^i} d_e(x_e(\varpi)) = \lambda^i(\hat{\varpi}) - \lambda^i(\varpi),$$

where we have exploited the fact that $\forall e \in E \setminus (\varpi^i \cup \hat{\varpi}^i)$ and $\forall e \in \varpi^i \cap \hat{\varpi}^i$ the load of each of these resources (i.e., the number of players using them) remains the same. In addition, $\forall e \in \hat{\varpi}^i \setminus \varpi^i$, $x_e(\hat{\varpi}) = x_e(\varpi) + 1$ and $\forall e \in \varpi^i \setminus \hat{\varpi}^i$, $x_e(\hat{\varpi}) = x_e(\varpi) - 1$. ∎

Remark 20.3 The existence of a (not necessarily exact) potential for any game in strategic form directly implies the existence of a PNE for this game. The existence of an exact potential may help (as we shall see later) the *efficient* construction of a PNE, but this is not true in general.

More interestingly, Monderer and Shapley [29] proved that every (finite) potential game is isomorphic to an unweighted congestion game. The proof presented here is a new one provided by the authors of the survey. The main idea is based on the proof of Monderer and Shapley, yet it is much simpler and easier to follow.

Theorem 20.2 [29] *Every finite (exact) potential game is isomorphic to an unweighted congestion game.*

Proof. Consider an arbitrary (finite) strategic game $\Gamma = (N, (Y^i)_{i \in N}, (U^i)_{i \in N})$ among $n = |N|$ players, which admits an exact potential $\Phi : \times_{i \in N} Y^i \mapsto \mathbf{R}$. Suppose that $\forall i \in N$, $Y^i = \{1, 2, \ldots, m_i\} \equiv [m_i]$ for some finite integer $m_i \geq 2$ (players having a unique allowable action can be safely removed from the game), and let $Y \equiv \times_{i \in N} Y^i$ be the action space of the game. We want to construct a proper unweighted congestion game $C = (N, E, (\mathcal{P}^i)_{i \in N}, (d_e)_{e \in E})$ and a bijection $\varpi : Y \mapsto \mathcal{P}$ (recall that $\mathcal{P} \equiv \times_{i \in N} \mathcal{P}^i$ is the action space of C) such that

$$\forall i \in N, \forall \mathbf{s} \in Y, \quad \lambda^i(\varpi(\mathbf{s})) = U^i(\mathbf{s}). \tag{20.6}$$

First of all, observe that for any (unweighted) congestion game C, we can express the players' costs as follows: $\forall i \in N, \forall \varpi \in \mathcal{P}$,

$$\lambda^i(\varpi) = \sum_{e \in \varpi^i} d_e(x_e(\varpi))$$

$$= \sum_{e \in \varpi^i \cap (E \setminus \cup_{j \neq i} \varpi^j)} d_e(1) + \sum_{e \in \cup_{k \neq i} \varpi^i \cap \varpi^k \cap (E \setminus \cup_{j \neq i,k} \varpi^j)} d_e(2)$$

$$+ \cdots + \sum_{e \in \cap_{k \in N} \varpi^k} d_e(n). \tag{20.7}$$

We proceed by determining the set of shared resources $E = E_1 \cup E_2$ for our congestion game. We construct two sets of resources: The first set E_1 contains resources that determine exactly what the players in the potential game actually do. The second set E_2 contains resources that represent what all other players (except for the one considered by the resource) should not do. More specifically, fix an arbitrary configuration $\mathbf{s} \in Y$ of the players with respect to the game Γ. We construct the following vector $\mathbf{e}(\mathbf{s}) \in E_1$ that is nothing more than the binary representation of this configuration (see Figure 20.2):

$$\forall k \in N, \forall j \in [m_k], \quad e_j^k(\mathbf{s}) = \begin{cases} 1 & \text{if } s^k = j \\ 0 & \text{otherwise} \end{cases}$$

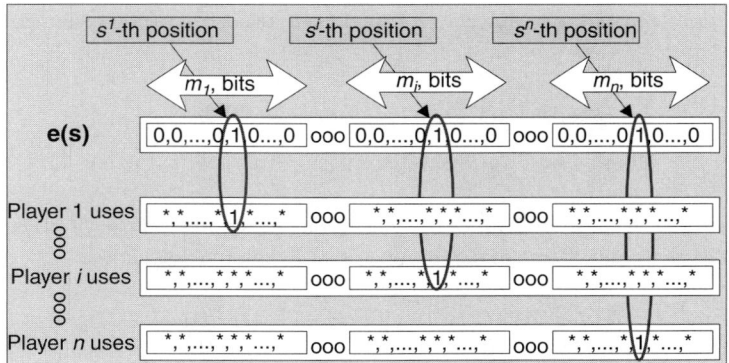

FIGURE 20.2 The set of resources representing the binary encodings of the configurations in the potential game.

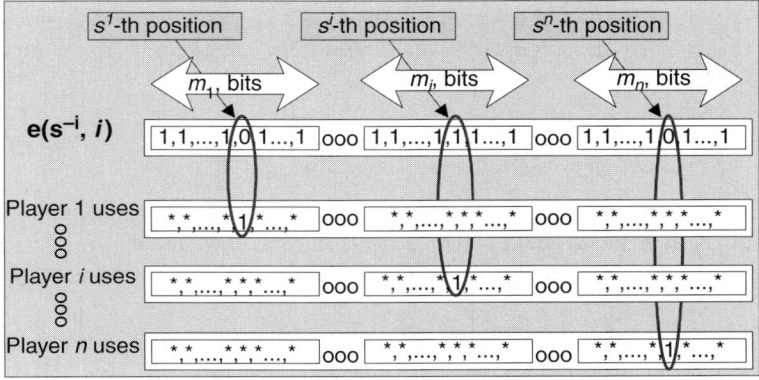

FIGURE 20.3 The set of resources representing the binary encodings of the forbidden configuration for the other players in the potential game.

Fix now an arbitrary player $i \in N$. We define n more resources whose binary vectors represent what the other players (i.e., other than player i) in the potential game *should not* do (see Figure 20.3):

$$\forall k \in N, \forall j \in [m_k], \quad e_j^k(\mathbf{s}, i) = \begin{cases} 0 & \text{if } k \neq i \text{ and } s^k = j \\ 1 & \text{otherwise} \end{cases}$$

It is easy to see that $\mathbf{e}(\mathbf{s}, i)$ actually says that none of the players $k \neq i$ agrees with the profile $\mathbf{s}^{-i} \in Y^{-i}$. So, we force this binary vector $\mathbf{e}(\mathbf{s}, i)$ to have definitely 0s exactly at the positions where $\mathbf{s}^{-i}$ has 1s for any of the players except for player i, and we place 1s anywhere else. For $m = m_1 + m_2 + \cdots + m_n$, let $E_1 = \{\mathbf{e}(\mathbf{s}) \in \{0,1\}^m : \mathbf{s} \in Y\}$ and $E_2 = \{\mathbf{e}(\mathbf{s}, i) \in \{0,1\}^m : (i \in N) \wedge (\mathbf{s}^{-i} \in Y^{-i})\}$. The set of resources in our congestion game is then $E = E_1 \cup E_2$.

We now proceed with the definition of the action sets of the players with respect to C. Indeed, $\forall i \in N$, $\mathcal{P}^i = \{\pi_1^i, \ldots, \pi_{m_i}^i\}$, where $\forall j \in [m_i]$, $\pi_j^i = \{\mathbf{e} \in E : e_j^i = 1\}$. Now the bijective map that we assume is almost straightforward: $\forall \mathbf{s} = (s^i)_{i \in N} \in Y$, $\varpi(\mathbf{s}) = (\pi_{s^i}^i)_{i \in N}$.

Some crucial observations are the following: First, $\forall \mathbf{s} \in Y$, the resource $\mathbf{e}(\mathbf{s})$ is the only resource in E_1 that is used by *all* the players when the configuration $\varpi(\mathbf{s})$ is considered. For any other configuration $\mathbf{z} \in Y \setminus \{\mathbf{s}\}$, the resource $\mathbf{e}(\mathbf{s})$ is *not* used by at least one player, assuming $\varpi(\mathbf{z})$. Similarly, $\forall \mathbf{s} \in Y, \forall i \in N$, the resource $\mathbf{e}(\mathbf{s}, i)$ is the only resource in E_2 that is *exclusively used* by player i, assuming the configuration $\varpi(\mathbf{s})$.

Now, it is very simple to set the resource delay functions d_e in such a way that we assure equality (20.6). Indeed, observe that $\forall i \in N, \forall \mathbf{s} \in Y, \forall \alpha \in Y^i \setminus \{s^i\}$,

$$U^i(\mathbf{s}) - U^i(\mathbf{s}^{-i} \oplus \alpha) = \Phi(\mathbf{s}) - \Phi(\mathbf{s}^{-i} \oplus \alpha) \Leftrightarrow$$
$$Q^i(\mathbf{s}^{-i}) \equiv U^i(\mathbf{s}) - \Phi(\mathbf{s}) = U^i(\mathbf{s}^{-i} \oplus \alpha) - \Phi(\mathbf{s}^{-i} \oplus \alpha).$$

That is, the quantity $Q^i(\mathbf{s}^{-i})$ is invariant of player i's strategy. Recall the form of player i's cost given in equality (20.7). The charging scheme of the resources that we adopt is as follows:

$$\forall e \in E, \ d_e(k) = \begin{cases} Q^i(\mathbf{s}^{-i}) & \text{if } (k=1) \wedge (\mathbf{e} = \mathbf{e}(\mathbf{s}, i) \in E_2) \\ 0 & \text{if } (1 < k < n) \vee (k = 1 \wedge \mathbf{e} \in E_1) \\ \Phi(\mathbf{s}) & \text{if } (k = n) \wedge (\mathbf{e} = \mathbf{e}(\mathbf{s}) \in E_1) \end{cases}$$

Now, it is easy to see that for any player $i \in N$ and any configuration $\mathbf{s} \in Y$, the only resource used exclusively by i with respect to $\mathbf{s}$ that has nonzero delay is $\mathbf{e}(\mathbf{s}, i)$. Similarly, the only resource used by all the players that has nonzero delay is $\mathbf{e}(\mathbf{s})$. All other resources charge zero delays, no matter how many players use them. Thus,

$$\forall i \in N, \forall \mathbf{s} \in Y, (20.7) \Rightarrow \lambda^i(\varpi(\mathbf{s})) = d_{\mathbf{e}(\mathbf{s},i)}(1) + d_{\mathbf{e}(\mathbf{s})}(n) = Q^i(\mathbf{s}^{-i}) + \Phi(\mathbf{s}) = U^i(\mathbf{s}). \ \blacksquare$$

Remark 20.4 The size of the congestion game that we use to represent a potential game is at most $(|N|+1)$ times larger than the size of the potential game. Since an unweighted congestion game is itself an exact potential game, this implies an essential equivalence of exact potential and unweighted congestion games.

20.5 Existence and Complexity of Constructing PNE

In this section, we deal with issues concerning the existence and complexity of constructing PNE in weighted congestion games. Our main references for this section are References 12, 15, 24. We start with some complexity issues concerning the construction of PNE in unweighted congestion games (in which a PNE always exists), and consequently, we study existence and complexity issues in weighted congestion games in general. Fix an arbitrary (weighted, in general) congestion game $\Gamma = (N, E, (\mathcal{P}^i)_{i \in N}, (w_i)_{i \in N}, (d_e)_{e \in E})$, where the w_is denote the (positive) weights of the players.

A crucial class of problems containing the weighted network congestion games is Polynomial Local Search (PLS) [21] (stands for *Polynomial Local Search*). This is the subclass of total functions in NP that are guaranteed to have a solution because of the fact that "*every finite directed acyclic graph has a sink.*"

A **local search** problem Π has a set of instances D_Π that are strings. To each instance $x \in D_\Pi$, there corresponds a set $S_\Pi(x)$ of solutions, and a standard solution $s_0 \in S_\Pi(x)$. Each solution $s \in S_\Pi(x)$ has a cost $f_\Pi(s, x)$ and a neighborhood $N_\Pi(s, x)$. The search problem is, given an instance $x \in D_{P^i}$, to find a locally optimal solution $s \in S_\Pi(x)$, that is, a solution $s^* \in S_\Pi(x)$ s.t. $f_\Pi(s^*, x) = \max_{s \in N_\Pi(s^*, x)}\{f_\Pi(s, x)\}$ (assuming that we consider as our objective the minimization of the cost function).

Definition 20.1 (**PLS**) *A local search problem Π is in PLS if its instances D_Π and solutions $S_\Pi(x: x \in D_\Pi)$ are binary strings, there is a polynomial p such that the length of solutions $S_\Pi(x)$ is bounded by $p(|x|)$, and there are three polynomial-time algorithms A_Π, B_Π, C_Π with the following properties:*

1. *Given a string $x \in \{0,1\}^*$, A_Π determines whether $x \in D_\Pi$, and if so, returns some initial solution $s_0 \in S_\Pi(x)$.*
2. *Given an instance $x \in D_\Pi$ and a string s, B_Π determines whether $s \in S_\Pi(x)$, and if so, computes the value $f_\Pi(s, x)$ of the cost function at s.*

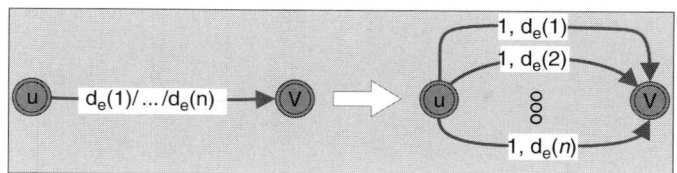

FIGURE 20.4 The reduction to a min-cost flow problem [12].

3. *Given an instance $x \in D_\Pi$ and a solution $s \in S_\Pi(x)$, C_Π determines whether s is a local optimum of $f_\Pi(\cdot, x)$ in its neighborhood $N_\Pi(s, x)$, and if not, returns a neighbor $s' \in N_\Pi(s, x)$ having a better value (i.e., $f_\Pi(s', x) < f_\Pi(s, x)$ for a minimization problem and $f_\Pi(s', x) > f_\Pi(s, x)$ for a minimization problem).*

The problem of constructing a PNE for a weighted congestion game is in PLS, in the following cases: (i) for any unweighted congestion game, since it is an exact potential game (see Theorem 20.1), and (ii) for any weighted network congestion game with linear resource delays, which admits (as we shall prove in Theorem 20.6) a **b**-potential with $b_i = 1/2w_i, \forall i \in N$, and thus, finding PNE is equivalent to finding local optima for the optimization problem with state space, the action space of the game, and objective, the potential of the game. On the other hand, this does not imply a polynomial-time algorithm for constructing a PNE, since (as we shall see more clearly in the weighted case) the improvements in the potential can be very small and too many. In addition, although problems in PLS admit a PTAS, this does not imply also a PTAS for finding ε-approximate PNE (approximation of the potential does not imply also approximation of each player's payoff).

20.5.1 Efficient Construction of Pure Nash Equilibrium in Unweighted Congestion Games

In this subsection, we shall prove that for unweighted single-commodity network congestion games a PNE can be constructed in polynomial time. On the other hand, even for multicommodity network congestion games, it is PLS complete to construct a PNE. The main source of this subsection is the work of Fabrikant et al. [12].

Theorem 20.3 [12] *There is a polynomial time algorithm for finding a PNE in any unweighted single-commodity network congestion game.*

Proof. Fix an arbitrary unweighted, single-commodity network congestion game $\Gamma = (N, E, (\mathcal{P}_{s-t})_{i \in N}, (d_e)_{e \in E})$ and let $G = (V, E)$ be the underlying network. Recall that this game admits Rosenthal's exact potential $\Phi(\varpi) = \sum_{e \in E} \sum_{k=1}^{x_e(\varpi)} d_e(k)$. The algorithm computes the optimum value of $\Phi : \mathcal{P} \mapsto \mathbf{R}$ in the action space of the game. The corresponding configuration is thus a PNE.

The algorithm exploits a reduction to a min-cost flow problem. This reduction is done as follows: We construct a new network $G' = (V', E')$ with the same set of vertices $V' = V$, while the set of arcs (i.e., resources) is defined as follows (see also Figure 20.4): $\forall e = (u, v) \in E$, we add n parallel arcs from u to v, $e'(1), \ldots, e'(n) \in E$, where these arcs have a *capacity* of 1 (i.e., they allow at most one player to use each of them) and fixed delays (when used) $\forall k \in [d_e(n)]$, $d'_{e'(k)} = d_e(k)$. Let $\mathbf{f}^* \in \mathbf{R}_{\geq 0}^{|E'|}$ be a minimizer of the following min-cost flow problem.

$$\min \quad \sum_{e' \in E'} d'_{e'} \cdot f_{e'}$$

$$\text{s.t.} \quad \sum_{e'=(s,u) \in E'} f_{e'} - \sum_{e'=(u,s) \in E'} f_{e'} = n$$

$$\sum_{e'=(u,t) \in E'} f_{e'} - \sum_{e'=(t,u) \in E'} f_{e'} = n$$

$$\forall v \in V \setminus \{s,t\}, \quad \sum_{e'=(v,u) \in E'} f_{e'} = \sum_{e'=(u,v) \in E'} f_{e'}$$

$$\forall e' \in E', \quad 0 \leq f_{e'} \leq 1.$$

This problem can be solved in polynomial time [1] (e.g., in time $\mathcal{O}(r \log k (r + k \log k))$ using scaling algorithms) where $r = |E'| = |E| \times n$ and $k = |V|$ are the number of arcs and the number of nodes in the underlying (augmented) network, respectively.

It is straightforward to see that any min-cost flow in G' is integral and it corresponds to a configuration in Γ that minimizes the potential of the game (which is then a PNE). ∎

On the other hand, the following theorem proves that it is not that easy to construct a PNE, even in an unweighted multicommodity network congestion game. We give this theorem with a sketch of its proof.

Theorem 20.4 [12] *It is PLS-complete to find a PNE in unweighted congestion games of the following types: (i) General congestion games, (ii) symmetric congestion games, and (iii) multicommodity network congestion games.*

Proof. We only give a sketch of the first two cases. The PLS completeness proof of case (iii) is rather complicated and therefore is not presented in this survey. The interested reader may find it in Reference 12.

We first prove the PLS completeness of general congestion games, and consequently, we show that an arbitrary congestion game can be transformed into an equivalent symmetric congestion game. In order to prove the PLS completeness of a congestion game, we shall use the following problems:

NOTALLEQUAL3SAT: Given a set N of $\{0,1\}$-variables and a collection C of clauses s.t. $\forall c \in C, |c| \leq 3$, is there an assignment of values to the variables so that no clause has all its literals assigned the same value?

POSNAE3FLIP: Given an instance (N, C) of NOTALLEQUAL3SAT with positive literals only and a weight function $w : C \mapsto \mathbf{R}$, find an assignment for the variables of N, s.t. The total weight of the unsatisfied clauses and the totally satisfied (i.e., with all their literals set to 1) clauses cannot be decreased by a unilateral flip of the value of any variable.

It is known that POSNAE3FLIP is PLS complete [34]. Given an instance of POSNAE3FLIP, we shall construct a congestion game $\Gamma = (N, E, (\mathcal{P}^v)_{v \in N}, (d_e)_{e \in E})$ as follows: The player set of the game is exactly the set of variables N. $\forall c \in C$, we construct two resources $e_c, e'_c \in E$ whose delay functions are $d_e(k) = w(c) \times \mathbf{I}_{\{k=3\}}$ and $d_{e'}(k) = w(c) \times \mathbf{I}_{\{k=3\}}$. That is, resource e_c (or e'_c) has delay $w(c)$ only when all the 3 players it contains actually use it. Each player $v \in N$ has exactly two allowable actions indicating the possible *true* (i.e., 1) or *false* (i.e., 0) values of the corresponding variable: $\mathcal{P}^v = \{\{e_c \in E : v \in c\}, \{e'_c \in E : v \in c\}\}$. Smaller clauses (i.e., of two literals) are implemented similarly. Clearly, a flip of a variable corresponds to the change in the strategy of the corresponding player. The changes in the total weight due to a flip equal the changes in the cumulative delay over all the resources. Thus, any PNE of the congestion game Γ is a local optimum (i.e., a solution) of the POSNAE3FLIP problem, and vice versa.

We now proceed to show that any unweighted congestion game can be transformed into a *symmetric* congestion game. Let $\Gamma = (N, E, (\mathcal{P}^v)_{v \in N}, (d_e)_{e \in E})$ be an arbitrary congestion game. We construct an equivalent symmetric congestion game $\hat{\Gamma} = (N, \hat{E}, \hat{\mathcal{P}}, (\hat{d}_e)_{e \in \hat{E}})$ as follows: First of all we add to the set of resources n new distinct resources: $\hat{E} = E \cup \{e_i\}_{i \in N}$. The delays of these resources are

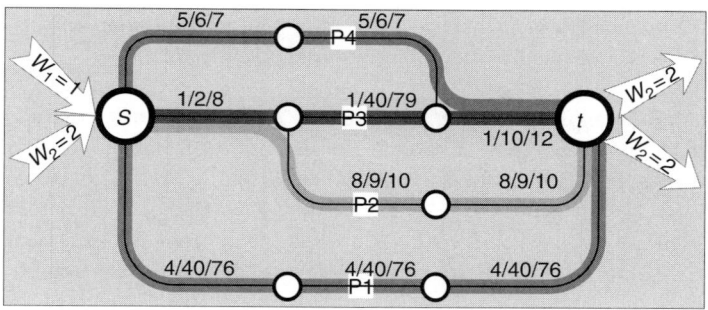

FIGURE 20.5 A weighted single-commodity network congestion game that may have no PNE. Consider two players with demands $w_1 = 1$ and $w_2 = 2$. The notation $a/b/c$ means that a load of 1 has delay a, a load of 2 has delay b, and a load of 3 has delay c.

$\forall i \in N$, $\hat{d}_{e_i}(k) = M \times \mathbf{I}_{\{k \geq 2\}}$ for some sufficiently large constant M. The old resources maintain the same delay functions: $\forall e \in E, \forall k \geq 0$, $\hat{d}_e(k) = d_e(k)$. Now each player has the same action set $\hat{\mathcal{P}} = \times_{i \in N} \{a \cup \{e_i\} : a \in \mathcal{P}^i\}$. Observe that (by setting the constant M sufficiently large) in any PNE of $\hat{\Gamma}$ each of the distinct resources $\{e_i\}_{i \in N}$ is used by exactly one player (these resources act as if they have capacity 1 and there are only n of them). So, for any PNE in $\hat{\Gamma}$ we can easily get a PNE in Γ by simply dropping the unique new resource used by each of the players. This is done by identifying the "anonymous" players of $\hat{\Gamma}$ according to the unique resource they use, and match them with the corresponding players of Γ. ∎

20.5.2 Existence and Construction of Pure Nash Equilibrium in Weighted Congestion Games

In this subsection, we deal with the existence and tractability of PNE in weighted network congestion games. First, we show that it is not always the case that a PNE exists, even for a weighted single-commodity network congestion game with only linear and 2-wise linear (i.e., the maximum of two linear functions) resource delays. Recall that, as discussed previously, any unweighted congestion game has a PNE, for any kind of nondecreasing delays, owing to the existence of an exact potential for these games. This result was independently proved by References 15 and 24, on the basis of similar constructions. In this survey, we present the version of Reference 15 owing to its clarity and simplicity.

Lemma 20.1 [15] *There exist instances of weighted single-commodity network congestion games with resource delays being either linear or 2-wise linear functions of the loads, for which there is no PNE.*

Proof. We demonstrate this by the example shown in Figure 20.5. In this example, there are exactly two players of demands $w_1 = 1$ and $w_2 = 2$, from node s to node t. The possible paths that the two players may follow are labeled in the figure. The resource delay functions are indicated by the 3 possible values they may take given the two players. Observe now that this example has no PNE: there is a simple closed path $\gamma = ((P3, P2), (P3, P4), (P1, P4), (P1, P2), (P3, P2))$ of length 4 that is an improvement path (actually, each defecting player moves to his new best choice, so this is a best-reply improvement path) and in addition, any other configuration not belonging to γ is either one or two best-reply moves away from some of these nodes. Therefore, there is no sink in the Nash dynamics graph of the game, and thus, there exists no PNE. Observe that the delay functions are *not* player-specific in our example, as was the case in Reference 28. ∎

Consequently, we show that there may exist no exact potential function for a weighted single-commodity network congestion game, even when the resource delays are identical to their loads. The next argument shows that Theorem 20.1 does not hold anymore even in this simplest case of weighted congestion games.

Lemma 20.2 [15] *There exist weighted single-commodity network congestion games that are not exact potential games, even when the resource delays are identical to their loads.*

Proof. Let $\Gamma = (N, E, (w_i)_{i \in N}, (\mathcal{P}^i)_{i \in N}, (d_e)_{e \in E})$ denote a weighted single-commodity network congestion game with $d_e(x) = x$, $\forall e \in E$. Recall the definition of players' costs for a configuration (Equation 20.1). Let us now define, for any path $\gamma = (\varpi(0), \ldots, \varpi(r))$ in the configuration space $\mathcal{P}$ of Γ, the quantity $I(\gamma, \lambda) = \sum_{k=1}^{r} \left[\lambda^{i_k}(\varpi(k)) - \lambda^{i_k}(\varpi(k-1)) \right]$, where i_k is the unique player in which the configurations $\varpi(k)$ and $\varpi(k-1)$ differ. Our proof is based on the fact that Γ is an (exact) potential game if and only if every simple closed path γ of length 4 has $I(\gamma, \lambda) = 0$ [29, Theorem 2.8].

For the sake of contradiction, assume that every closed simple path γ of length 4 for Γ has $I(\gamma, \lambda) = 0$, fix arbitrary configuration ϖ and consider the path $\gamma = (\varpi, x = \varpi^{-1} \oplus \pi_1, y = \varpi^{-(1,2)} \oplus (\pi_1, \pi_2), z = \varpi^{-2} \oplus \pi_2, \varpi)$ for some actions $\pi_1 \neq \varpi^1$ and $\pi_2 \neq \varpi^2$. We shall demonstrate that $I(\gamma, \lambda)$ cannot be identically 0 when there are at least two players of different demands. So, consider that the first two players have different demands: $w_1 \neq w_2$. We observe that

$$\lambda^1(x) - \lambda^1(\varpi) = \sum_{e \in \pi_1} \theta_e(x) - \sum_{e \in \varpi^1} \theta_e(\varpi)$$

$$= |\pi_1 \setminus \varpi^1| \times w_1 + \sum_{e \in \pi_1 \setminus \varpi^1} \theta_e(\varpi) - \sum_{e \in \varpi^1 \setminus \pi_1} \theta_e(\varpi),$$

since the resources in $\varpi^1 \cap \pi_1$ retain their initial loads. Similarly, we have

$$\lambda^2(y) - \lambda^2(x) = \sum_{e \in \pi_2 \setminus \varpi^2} [\theta_e(x) + w_2] - \sum_{e \in \varpi^2 \setminus \pi_2} \theta_e(x)$$

$$= |\pi_2 \setminus \varpi^2| \times w_2 + \sum_{e \in \pi_2 \setminus \varpi^2} \theta_e(x) - \sum_{e \in \varpi^2 \setminus \pi_2} \theta_e(x)$$

$$\lambda^1(z) - \lambda^1(y) = \sum_{e \in \varpi^1 \setminus \pi_1} \theta_e(z) - \sum_{e \in \pi_1 \setminus \varpi^1} [\theta_e(z) + w_1]$$

$$= \sum_{e \in \varpi^1 \setminus \pi_1} \theta_e(z) - \sum_{e \in \pi_1 \setminus \varpi^1} \theta_e(z) - |\pi_1 \setminus \varpi^1| \times w_1$$

$$\lambda^2(\varpi) - \lambda^2(z) = \sum_{e \in \varpi^2 \setminus \pi_2} \theta_e(\varpi) - \sum_{e \in \pi_2 \setminus \varpi^2} \theta_e(\varpi) - |\pi_2 \setminus \varpi^2| \times w_2.$$

Thus, since $I \equiv I(\gamma, \lambda) = \lambda^1(x) - \lambda^1(\varpi) + \lambda^2(y) - \lambda^2(x) + \lambda^1(z) - \lambda^1(y) + \lambda^2(\varpi) - \lambda^2(z)$, we get

$$I = \sum_{e \in \pi_1 \setminus \varpi^1} [\theta_e(\varpi) - \theta_e(z)] + \sum_{e \in \pi_2 \setminus \varpi^2} [\theta_e(x) - \theta_e(\varpi)] +$$

$$+ \sum_{e \in \varpi^1 \setminus \pi_1} [\theta_e(z) - \theta_e(\varpi)] + \sum_{e \in \varpi^2 \setminus \pi_2} [\theta_e(\varpi) - \theta_e(x)].$$

Observe now that

$$\forall e \in \pi_1 \setminus \varpi^1,$$
$$\theta_e(\varpi) - \theta_e(z) = \theta_e(\varpi) - \theta_e(\varpi^{-2} \oplus \pi_2) = w_2 \times \left(\mathbf{I}_{\{e \in \varpi^2 \setminus \pi_2\}} - \mathbf{I}_{\{e \in \pi_2 \setminus \varpi^2\}}\right)$$

$$\forall e \in \pi_2 \setminus \varpi^2,$$
$$\theta_e(x) - \theta_e(\varpi) = \theta_e(\varpi^{-1} \oplus \pi_1) - \theta_e(\varpi) = w_1 \times t \left(\mathbf{I}_{\{e \in \pi_1 \setminus \varpi^1\}} - \mathbf{I}_{\{e \in \varpi^1 \setminus \pi_1\}}\right)$$

$$\forall e \in \varpi^1 \setminus \pi_1,$$
$$\theta_e(z) - \theta_e(\varpi) = \theta_e(\varpi^{-2} \oplus \pi_2) - \theta_e(\varpi) = w_2 \times \left(\mathbf{I}_{\{e \in \pi_2 \setminus \varpi^2\}} - \mathbf{I}_{\{e \in \varpi^2 \setminus \pi_2\}}\right)$$

$$\forall e \in \varpi^2 \setminus \pi_2,$$
$$\theta_e(\varpi) - \theta_e(x) = \theta_e(\varpi) - \theta_e(\varpi^{-1} \oplus \pi_1) = w_1 \times \left(\mathbf{I}_{\{e \in \varpi^1 \setminus \pi_1\}} - \mathbf{I}_{\{e \in \pi_1 \setminus \varpi^1\}}\right).$$

Then,

$$I = (w_1 - w_2) \times \left[|(\pi_1 \setminus \varpi^1) \cap (\pi_2 \setminus \varpi^2)| + |(\varpi^1 \setminus \pi_1) \cap (\varpi^2 \setminus \pi_2)| \right.$$
$$\left. - |(\varpi^1 \setminus \pi_1) \cap (\pi_2 \setminus \varpi^2)| - |(\pi_1 \setminus \varpi^1) \cap (\varpi^2 \setminus \pi_2)|\right],$$

which is typically not equal to zero for a single-commodity network. It should be noted that the second parameter, which is network dependent, can be nonzero even for some cycle of a very simple network. For example, in the network of Figure 20.5 (which is a simple 2-layered network), the simple closed path $\gamma = (\varpi(0) = (P1, P3), \varpi(1) = (P2, P3), \varpi(2) = (P2, P1), \varpi(3) = (P1, P1), \varpi(4) = (P1, P3))$ has this quantity equal to -4, and thus, no weighted single-commodity network congestion game on this network can admit an exact potential. ∎

Our next step is to focus our interest on the ℓ-layered networks with resource delays identical to their loads. We shall prove that any weighted ℓ-layered network congestion game with these delays admits at least one PNE, which can be computed in pseudopolynomial time. Although we already know that even the case of weighted ℓ-layered network congestion games with delays equal to the loads cannot have any exact potential,[*] we will next show that $\Phi(\varpi) \equiv \sum_{e \in E}[\theta_e(\varpi)]^2$ is a **b**-potential for such a game and some positive n-vector **b**, assuring the existence of a PNE.

Theorem 20.5 [15] *For any weighted ℓ-layered network congestion game with resource delays equal to their loads, at least one PNE exists and can be computed in pseudopolynomial time.*

Proof. Fix an arbitrary ℓ-layered network (V, E) and denote by L all the $s - t$ paths in it from the unique source s to the unique destination t. Let $\varpi \in \mathcal{P} \equiv L^n$ be an arbitrary configuration of the players for the corresponding congestion game on (V, E). Also, let i be a user of demand w_i and fix some path $\alpha \in L$.

---

[*] The example at the end of the proof of Lemma 20.2 involves the 2-layered network of Figure 20.5.

Denote $\hat{\varpi} \equiv \varpi^{-i} \oplus \alpha$. Observe that

$$\Phi(\varpi) - \Phi(\hat{\varpi}) = \sum_{e \in E} \left(\theta_e^2(\varpi) - \theta_e^2(\hat{\varpi})\right)$$

$$= \sum_{e \in \varpi^i \setminus \alpha} \left(\theta_e^2(\varpi) - \theta_e^2(\hat{\varpi})\right) + \sum_{e \in \alpha \setminus \varpi^i} \left(\theta_e^2(\varpi) - \theta_e^2(\hat{\varpi})\right)$$

$$= \sum_{e \in \varpi^i \setminus \alpha} \left([\theta_e(\varpi^{-i}) + w_i]^2 - \theta_e^2(\varpi^{-i})\right) + \sum_{e \in \alpha \setminus \varpi^i} \left(\theta_e^2(\varpi^{-i}) - [\theta_e(\varpi^{-i}) + w_i]^2\right)$$

$$= \sum_{e \in \varpi^i \setminus \alpha} \left(w_i^2 + 2w_i \theta_e(\varpi^{-i})\right) - \sum_{e \in \alpha \setminus \varpi^i} \left(w_i^2 + 2w_i \theta_e(\varpi^{-i})\right)$$

$$= w_i^2 \times \left(|\varpi^i \setminus \alpha| - |\alpha \setminus \varpi^i|\right) + 2w_i \times \left(\sum_{e \in \varpi^i \setminus \alpha} \theta_e(\varpi^{-i}) - \sum_{e \in \alpha \setminus \varpi^i} \theta_e(\varpi^{-i})\right)$$

$$= 2w_i \times \left(\sum_{e \in \varpi^i \setminus \alpha} \theta_e(\varpi^{-i}) - \sum_{e \in \alpha \setminus \varpi^i} \theta_e(\varpi^{-i})\right) = 2w_i \times \left[\lambda^i(\varpi) - \lambda^i(\hat{\varpi})\right],$$

since, $\forall e \in \varpi^i \cap \alpha$, $\theta_e(\varpi) = \theta_e(\hat{\varpi})$, in ℓ-layered networks $|\varpi^i \setminus \alpha| = |\alpha \setminus \varpi^i|$, $\lambda^i(\varpi) = \sum_{e \in \varpi^i} \theta_e(\varpi) = \sum_{e \in \varpi^i \setminus \alpha} \theta_e(\varpi^{-i}) + w_i |\varpi^i \setminus \alpha| + \sum_{e \in \varpi^i \cap \alpha} \theta_e(\varpi)$ and $\lambda^i(\hat{\varpi}) = \sum_{e \in \alpha} \theta_e(\hat{\varpi}) = \sum_{e \in \alpha \setminus \varpi^i} \theta_e(\varpi^{-i}) + w_i |\alpha \setminus \varpi^i| + \sum_{e \in \varpi^i \cap \alpha} \theta_e(\varpi)$. Thus, Φ is a **b**-potential for our game, where $\mathbf{b} = (1/(2w_i))_{i \in N} \in \mathbf{R}_{>0}^n$, assuring the existence of at least one PNE.

We proceed with the construction of a PNE in pseudopolynomial time. Wlog assume now that the players have integer weights. Then, each player performing any improving defection must reduce his cost by at least 1, and thus, the potential function decreases by at least $2w_{\min} \geq 2$ along each arc of the Nash dynamics graph of the game. Consequently, the naive algorithm that, starting from an arbitrary initial configuration $\varpi \in \mathcal{P}$, follows any improvement path that leads to a sink (i.e., a PNE) of the Nash dynamics graph cannot contain more than $1/2|E|W_{\text{tot}}^2$ defections, since $\forall \varpi \in \mathcal{P}$, $\Phi(\varpi) \leq |E|W_{\text{tot}}^2$. ∎

A recent improvement, based essentially on the same technique as above, generalizes the last result to the case of arbitrary multicommodity network congestion games with linear resource delays (we state the result here without its proof):

Theorem 20.6 [16] *For any weighted multicommodity network congestion game with linear resource delays, at least one PNE exists and can be computed in pseudopolynomial time.*

20.6 Parallel Links and Player-Specific Payoffs

In Reference 28, Milchtaich studies a variant of the classical (unweighted) congestion games, where the resource delay functions are not universal, but player-specific. In particular, there is again a set E of shared resources and a set N of (unit demand) players with their action sets $(\mathcal{P}_i)_{i \in N}$, but rather than having a single delay $d_e : \mathbf{R}_{\geq 0} \mapsto \mathbf{R}$ per resource $e \in E$, there is actually a different utility function* per player $i \in N$ and resource $e \in E$, $U_e^i : \mathbf{R}_{\geq 0} \mapsto \mathbf{R}$ determining the payoff of player i for using resource e, given a configuration $\varpi \in \mathcal{P}$ and the load $\theta_e(\varpi)$ induced by it on that resource. On the other hand, Milchtaich

---

* In this case, we refer to the **utility** of a player, which each player wishes to *maximize* rather than *minimize*. For example, this is the negative of a player's private cost function on a resource.

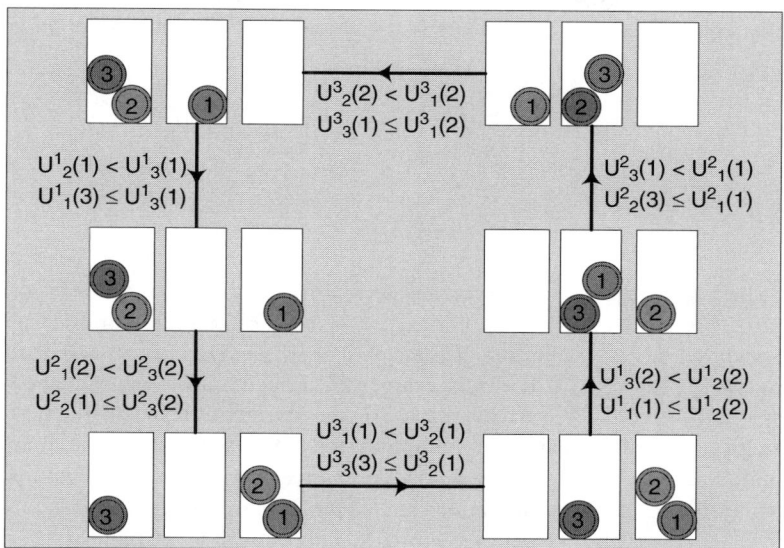

FIGURE 20.6 An example of a 3-players, 3-actions unweighted congestion game with a best-reply cycle.

makes two simplifying (yet crucial) assumptions:

1. Each player may choose only one resource from a pool E of resources (shared to all the players) for his service (i.e., this is modeled as the *parallel-links* model of Koutsoupias and Papadimitriou [23]).
2. The received payoff is *monotonically nonincreasing* with the number of players selecting it. Although they do not always admit a potential, these games always possess a PNE.

In Reference 28, Milchtaich proved that unweighted congestion games on parallel links with player-specific payoffs, involving only two strategies, possess the FIP. It is also rather straightforward that any 2-players unweighted congestion game on parallel links with player-specific payoffs possesses the FBRP.

Milchtaich also gave an example of an unweighted congestion game on three parallel links with three players, for which there is a best-reply cycle (although there is a PNE). The example is shown in Figure 20.6. In this example, we only determine the necessary conditions on the (player-specific) payoff functions of the players for the existence of the best-response cycle (see figure). It is easily verified that this system of inequalities is feasible, and that configurations $(3, 1, 2)$ and $(2, 3, 1)$ are PNE for the game.

Theorem 20.7 [28] *Every unweighted congestion game on parallel links with player-specific, nonincreasing payoffs of the resources possesses a PNE.*

Proof. First of all, we need to prove the following lemma that bounds the lengths of best-reply paths of a specific kind.

Lemma 20.3 *Let $(e(0), e(1), \ldots, e(M))$ be a sequence of (possibly repeated) resources from E.*

type (a) path: *Let $(\varpi(1), \varpi(2), \ldots, \varpi(M))$ be a best-reply improvement path of the game, s.t. $\forall t \geq 1, \forall e \in E$,*

$$x_e(t) = \begin{cases} x_e(t-1), & \text{if } e \notin \{e(t-1), e(t)\} \\ x_e(t-1) + 1, & \text{if } e = e(t) \\ x_e(t-1) - 1, & \text{if } e = e(t-1) \end{cases}$$

Then $M \leq |N|$.

type (b) path: Let $(\varpi(1), \varpi(2), \ldots, \varpi(M))$ be a best-reply improvement path of the game, s.t. $\forall t \geq 1, \forall e \in E$,

$$x_e(t) = \begin{cases} x_e(t-1), & \text{if } e \notin \{e(t-1), e(t)\} \\ x_e(t-1) + 1, & \text{if } e = e(t-1) \\ x_e(t-1) - 1, & \text{if } e = e(t) \end{cases}$$

Then, $M \leq |N| \cdot (|E| - 1)$.

Proof. (a) Fix a best-reply, type (a) improvement path $\gamma = (\varpi(1), \varpi(2), \ldots, \varpi(M))$ (see Figure 20.7a). For each resource $e \in E$, let $x_e^{\min} \equiv \min_{0 \leq t \leq M}\{x_e(t)\}$ denote the minimum load of e among the configurations involved in γ. By definition of γ as a best-reply sequence, we observe that $\forall e \in E, \forall t \geq 1, x_{e(t)}(t) = x_{e(t)}(t-1) + 1 = x_e^{\min} + 1$. That is, a resource reaches its maximum load exactly when it is the unique node in γ that currently receives a new player, reaches its minimum load in the very next time step (because of losing a player), and remains at its minimum load until before it appears again in this sequence. But then, the unique deviator in each move goes to a resource that reaches its maximum load, whereas all the other resources are already at their minimum loads. This implies that there is no chance that the specific player will move away from this new resource that he selfishly chose, till the end of the best-reply path under consideration. More formally, $\forall t \geq 1$, let $i(t)$ be the unique deviator that moves from $e(t-1)$ to $e(t)$. Then, $\forall t \in [M]$, we can verify (by induction on t') that $\forall t + 1 \leq t' \leq M, \forall e \in E \setminus \{e(t)\}$,

$$U_{e(t)}^{i(t)}\left(x_e(t')\right) \geq U_{e(t)}^{i(t)}\left(x_{e(t)}^{\min} + 1\right) \geq U_e^{i(t)}\left(x_e^{\min} + 1\right) \geq U_e^{i(t)}\left(x_e(t') + 1\right),$$

and so, $i(t)$ will not selfishly move away from $e(t)$ till the end of the best-reply improvement path. The first and the third inequality are true because of the monotonicity of the payoff functions. The second inequality is owing to the selfish choice (by player $i(t)$) of resource $e(t)$ at time t, despite the fact that $e(t)$ reaches its maximum load and all the other resources have their minimum loads after this move. Since this holds for all players, this path may have length at most $|N|$.

(b) Fix now a best-reply, type (b) improvement path $\gamma = (\varpi(1), \varpi(2), \ldots, \varpi(M))$ (see Figure 20.7b). For each resource $e \in E$, let again $x_e^{\min} \equiv \min_{0 \leq t \leq M} x_e(t)$. We observe that $\forall 1 \leq t \leq M, x_{e(t)}(t) = x_{e(t)}^{\min}$ and $x_{e(t-1)}(t) = x_{e(t-1)}^{\min} + 1$. Owing to the best-reply moves considered in the path, if $i(t)$ is again the unique deviator from $e(t)$ toward $e(t-1)$ at time t, then $\forall 1 \leq t \leq M$,

$$U_{e(t)}^{i(t)}\left(x_{e(t)}(t-1)\right) = U_{e(t)}^{i(t)}\left(x_{e(t)}^{\min} + 1\right)$$
$$< U_{e(t-1)}^{i(t)}\left(x_{e(t-1)}(t-1) + 1\right) = U_{e(t-1)}^{i(t)}\left(x_{e(t-1)}^{\min} + 1\right),$$

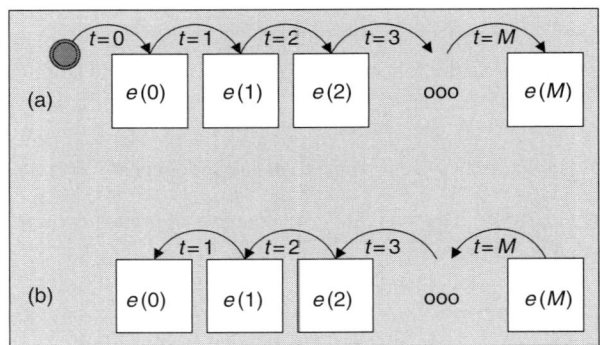

FIGURE 20.7 The two types of best-reply improvement paths considered by Reference 28.

which implies that $\forall t' > t$ player $i(t)$ residing at a resource $e \neq e(t)$ could never prefer to deviate to $e(t)$ rather than deviate to (or, stay at, if already there) resource $e(t-1)$. That is, player $i(t)$ can never go back to the resource from which it defected once, till the end of the best-reply path. Thus, each player can make at most $|E|-1$ moves and so, $M \leq |N| \times (|E|-1)$. ∎

The proof of the theorem proceeds by induction on the number of players. Trivially, for a single player we know that as soon as he is at a best reply, this is also a PNE (there is nothing else to move). We assume now that any unweighted congestion game on parallel links with player-specific payoffs and $|\tilde{N}| < n$, $\tilde{\Gamma} = \left(\tilde{N}, E, (U_e^i)_{i \in \tilde{N}, e \in E}\right)$ possesses a PNE. We want to prove that this is also the case for any such game with n players. Let $N = [n]$ and $\Gamma = \left(N, E, (U_e^i)_{i \in N, e \in E}\right)$ be such a game. We temporarily pull player n out of the game and let the remaining $n-1$ players continue playing, until they reach a PNE (without player n) $\mathbf{s} \in E^{n-1}$. That the PNE $\mathbf{s}$ exists for the pruned game $\tilde{\Gamma} = \left([n-1], E, (U_e^i)_{i \in [n-1]}, e \in E\right)$ holds by inductive hypothesis. Now we let player n be assigned to a best-reply resource $e \in BR_n(\mathbf{s})$, and we thus construct a configuration $\varpi(0)$ as follows: $\varpi^n(0) = e$; $\forall i \in [n-1], \varpi^i(0) = s^i$.

Consequently, starting from $\varpi(0)$, we construct a best-reply *maximal* improvement path of type (a) (see previous lemma) $\Pi(a) = (\varpi(0), \varpi(1), \ldots, \varpi(M))$, which we already know that is of length at most n. We claim that the terminal configuration of this path is a PNE for Γ. Clearly, any player that deviated to a best-reply resource in $\Pi(a)$ does not move again and is also at a best-reply resource in $\varpi(M)$ (see the proof of the lemma). So, we only have to consider players that have not defected during $\Pi(a)$. Fix any such player i, residing at resource $e = \varpi^i(0) = s^i$. Observe that in $\varpi(M)$ the following holds:

$$\forall e \in E, x_e(M) = \begin{cases} x_e(0), & \text{if } e \neq e(M), \\ x_e(0) + 1, & \text{if } e = e(M) \end{cases}$$

Observe that if some player $i : s^i(M) = e(M)$ that has not moved during $\Pi(a)$ is not at his best reply in $\varpi(M)$, then we can set $e(M+1) \in BR_i(\varpi^{-i}(M))$ and thus augment the best-reply improvement path $\Pi(a)$, which contradicts the maximality assumption. Consider now any player $i \in e \neq e(M)$ that has not moved during $\Pi(a)$. This player is certainly at a best-reply resource $e \in BR_i(\varpi^{-i}(M))$, since this resource has exactly the same load as in $\mathbf{s}$ and any other resource has at least the load it had in $\mathbf{s}$. So, $\varpi(M)$ is a PNE for Γ since every player is at a best-reply resource with respect to $\varpi(M)$. This completes the proof of the theorem. ∎

Remark 20.5 Observe that the proof of this theorem is constructive, and thus also implies a path of length at most $|N|$ that leads to a PNE. But this is not necessarily an improvement path, when all players are considered to coexist all the time, and therefore, there is no justification of the adoption of such a path by the (selfish) players. Milchtaich [28], using an argument of the same flavor as in the above theorem, proves that from an arbitrary initial configuration and allowing only best-reply defections, there is a best-reply improvement path of length at most $|E| \times \binom{|N|+1}{2}$. The idea is, starting from an arbitrary configuration $\varpi(0)$, to let the players construct a best-reply improvement path of type (a), and then (if necessary) construct a best-reply improvement path of type (b), starting from the terminal configuration of the previous path. It is then easily shown that the terminal configuration of the second path is a PNE of the game.

The unweighted congestion games on parallel links and with player-specific payoffs are **weakly acyclic** games, in the sense that from any initial configuration $\varpi(0)$ of the players, there is at least one best-reply improvement path connecting it to a PNE. Of course, this does not exclude the existence of best-reply cycles (see example of Figure 20.6). But, it is easily shown that when the deviations of the players occur sequentially and in each step the next deviator is chosen randomly (among the potential deviators) to a randomly chosen best-reply resource, then this path will converge almost surely to a PNE in finite time.

20.6.1 Players with Distinct Weights

Milchtaich proposed a generalization of his variant of congestion games, by allowing the players to have distinct weights, denoted by a weight vector $\mathbf{w} = (w_1, w_2, \ldots, w_n) \in \mathbf{R}^n_{>0}$. In that case, the (player-specific) payoff of each player on a resource $e \in E$ depends on the load $\theta_e(\varpi) \equiv \sum_{i: e \in \varpi^i} w_i$, rather than the number of players willing to use it.

For the case of weighted congestion games on parallel links with player-specific payoffs, it is easy to verify (in a similar fashion as for the unweighted case) that

- If there are only two available strategies, then FIP holds
- If there are only two players, then FBRP holds
- For the special case of resource-specific payoffs, FIP holds

On the other hand, there exists a 3-players, 3-actions game that possesses no PNE. For example, see the instance shown in Figure 20.8, where the three players have essentially two strategies each (a "LEFT" and a "RIGHT" strategy) to choose from, while their third strategy gives them strictly minimal payoffs and can never be chosen by selfish moves. The rationale of this game is that, in principle, player 1 would like to avoid using the same link as player 3, which in turn would like to avoid using the same link as player 2, which would finally want to avoid player 1.

The inequalities shown in Figure 20.8b demonstrate the necessary conditions for the existence of a best-reply cycle among six configurations of the players. It is easy to verify also that any other configuration has either at least one MIN strategy for some player (and in that case this player wants to defect) or is one of (2, 2, 1), (3, 3, 3). The only thing that remains to assure is that (2, 3, 1) is strictly better for player 2 than (2, 2, 1) (i.e., player 2 would like to avoid player 1) and that (3, 3, 1) is better for player 3 than (3, 3, 3) (i.e., player 3 would like to avoid player 2). The feasibility of the whole system of inequalities can be trivially checked to hold, and thus this game cannot have any PNE since there is no sink in its dynamics graph.

20.7 The Price of Anarchy of Weighted Congestion Games

In this section, we focus our interest on weighted ℓ-layered network congestion games where the resource delays are identical to their loads. Our source for this section is Reference 15. This case comprises a highly

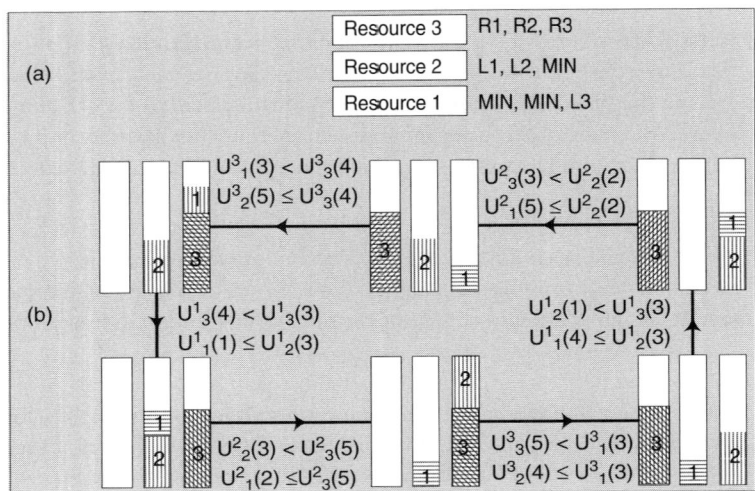

FIGURE 20.8 A 3-players weighted congestion game (from Reference 28) on three parallel links with player-specific payoffs, which does not possess a PNE. (a) The LEFT–RIGHT strategies of the players. (b) The best-reply cycle.

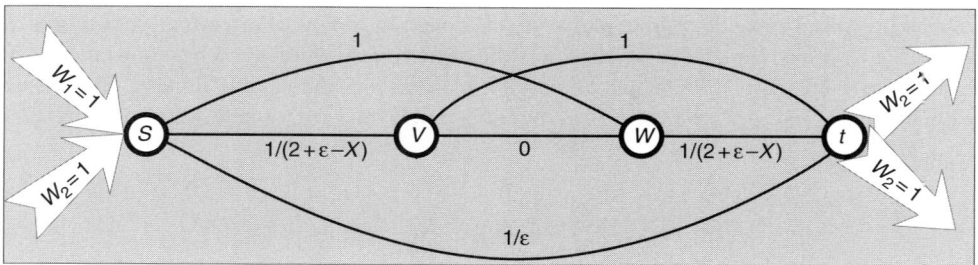

FIGURE 20.9 An example of a single-source network congestion game without a PNE [32].

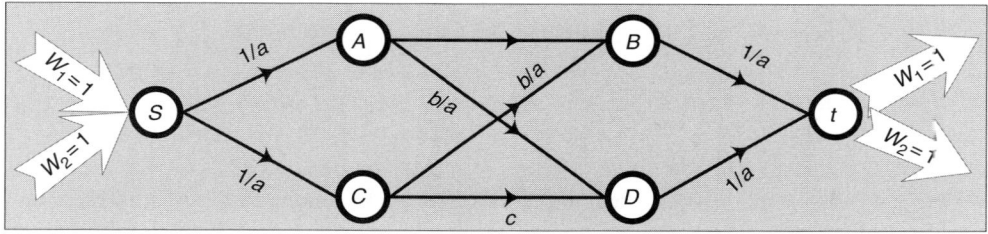

FIGURE 20.10 Example of an ℓ-layered network with linear resource delays and unbounded price of anarchy.

nontrivial generalization of the well-known model of selfish routing of atomic (i.e., indivisible) traffic demands via identical parallel channels [23]. The main reason why we focus on this specific category of resource delays is that there exist instances of (even unweighted) congestion games on layered networks that have unbounded price of anarchy even if we only allow linear resource delays. In page 256 of Reference 32 an example is given where the price of anarchy is indeed unbounded (see Figure 20.9). This example is easily converted into an ℓ-layered network. The resource delay functions used are either constant or M/M/1-like (i.e., of the form $1/c - x$) delay functions. However, we can be equally bad even with linear resource delay functions: Observe the following example depicted in Figure 20.10. Two players, each having a unit of traffic demand from s to t, choose their selected paths selfishly. The edge delays are shown above them in the usual way. We assume that $a \gg b \gg 1 \geq c$. It is easy to see that the configuration (sCBt,sADt) is a PNE of social cost $2 + b$ while the optimum configuration is (sABt,sCDt), the (optimum) social cost of which is $2 + c$. Thus, $\mathcal{R} = (b + 2/c + 2)$.

In the following, we restrict our attention to ℓ-layered networks whose resource delays are equal to their loads. Our main tool is to interpret a strategies profile as a flow in the underlying network.

20.7.1 Flows and Mixed Strategies Profiles

Fix an arbitrary ℓ-layered network $G = (V, E)$ and a set $N = [n]$ of distinct players willing to satisfy their own traffic demands from the unique source $s \in V$ to the unique destination $t \in V$. Again, $\mathbf{w} = (w_i)_{i \in [n]}$ denotes the varying demands of the players. Fix an arbitrary mixed strategies profile $\mathbf{p} = (p_1, p_2, \ldots, p_n)$ where, for sake of simplicity, we consider that $\forall i \in [n], p_i : \mathcal{P}_{s-t} \mapsto [0, 1]$ is a real function (rather than a vector) assigning nonnegative probabilities to the $s - t$ paths of G (which are the allowable actions for player i).

A **feasible flow** for the n players is a function $\rho : \mathcal{P}_{s-t} \mapsto \mathbf{R}_{\geq 0}$ mapping amounts of nonnegative traffic (on behalf of all the players) to the $s-t$ paths of G, in such a way that $\sum_{\pi \in \mathcal{P}_{s-t}} \rho(\pi) = W_{\text{tot}} \equiv \sum_{i \in [n]} w_i$. That is, all players' demands are actually satisfied. We distinguish between unsplittable and splittable (feasible) flows. A feasible flow is **unsplittable** if each player's traffic demand is satisfied by a unique path of $\mathcal{P}_{s-t}$. In the general case, any feasible flow is **splittable**, in the sense that the traffic demand of each player is possibly routed over several paths of $\mathcal{P}_{s-t}$.

We map the mixed strategies profile **p** to a feasible flow $\rho_\mathbf{p}$ as follows: For each $s-t$ path $\pi \in \mathcal{P}_{s-t}$, $\rho_\mathbf{p}(\pi) \equiv \sum_{i \in [n]} w_i \times p_i(\pi)$. That is, we handle the *expected load traveling along π according to* **p** as a splittable flow, where player i routes a fraction of $p_i(\pi)$ of his total demand w_i along π. Observe that, if **p** is actually a pure strategies profile, the corresponding flow is then unsplittable. Recall now that for each edge $e \in E$,

$$\theta_e(\mathbf{p}) = \sum_{i=1}^{n} \sum_{\pi: e \in \pi} w_i p_i(\pi) = \sum_{\pi: e \in \pi} \rho_\mathbf{p}(\pi) \equiv \theta_e(\rho_\mathbf{p})$$

denotes the expected load (and in our case, also the expected delay) of e with respect to **p**, and can be expressed either as a function $\theta_e(\mathbf{p})$ of the mixed profile **p** or as a function $\theta_e(\rho_\mathbf{p})$ of its associated feasible flow $\rho_\mathbf{p}$. As for the expected delay along a path $\pi \in \mathcal{P}_{s-t}$ according to **p**, this is

$$\theta_\pi(\mathbf{p}) = \sum_{e \in \pi} \theta_e(\mathbf{p}) = \sum_{e \in \pi} \sum_{\pi' \ni e} \rho_\mathbf{p}(\pi') = \sum_{\pi' \in \mathcal{P}_{s-t}} |\pi \cap \pi'| \rho_\mathbf{p}(\pi') \equiv \theta_\pi(\rho_\mathbf{p}).$$

Let $\theta^{\min}(\rho) \equiv \min_{\pi \in \mathcal{P}_{s-t}} \{\theta_\pi(\rho)\}$ be the minimum expected delay among all $s-t$ paths. From now on for simplicity we drop the subscript of **p** from its corresponding flow $\rho_\mathbf{p}$, when this is clear by the context. When we compare network flows, two typical measures are those of total latency and maximum latency. For a feasible flow ρ the **maximum latency** is defined as

$$L(\rho) \equiv \max_{\pi: \rho(\pi) > 0} \{\theta_\pi(\rho)\} = \max_{\pi: \exists i, \, p_i(\pi) > 0} \{\theta_\pi(\mathbf{p})\} \equiv L(\mathbf{p}). \qquad (20.8)$$

$L(\rho)$ is nothing but the *maximum expected delay paid by the players*, with respect to **p**. From now on, we use ρ^* and ρ_f^* to denote the optimal unsplittable and splittable flows, respectively. The objective of **total latency** is defined as follows:

$$C(\rho) \equiv \sum_{\pi \in \mathcal{P}} \rho(\pi) \theta_\pi(\rho) = \sum_{e \in E} \theta_e^2(\rho) = \sum_{e \in E} \theta_e^2(\mathbf{p}) \equiv C(\mathbf{p}). \qquad (20.9)$$

The second equality is obtained by summing over the edges of π and reversing the order of the summation. We have no direct interpretation of the total latency of a flow to the corresponding mixed profile. Nevertheless, observe that $C(\mathbf{p})$ was used as the **b**-potential function of the corresponding game that proves the existence of a PNE.

20.7.2 Flows at Nash Equilibrium

Let **p** be a mixed strategies profile and let ρ be the corresponding flow. For an ℓ-layered network with resource delays equal to the loads, the cost of player i on path π is $\lambda_\pi^i(\mathbf{p}) = \ell w_i + \theta_\pi^{-i}(\mathbf{p})$, where $\theta_\pi^{-i}(\mathbf{p})$ is the expected delay along path π if the demand of player i was removed from the system:

$$\theta_\pi^{-i}(\mathbf{p}) = \sum_{\pi' \in \mathcal{P}} |\pi \cap \pi'| \sum_{j \neq i} w_j p_j(\pi') = \theta_\pi(\mathbf{p}) - w_i \sum_{\pi' \in \mathcal{P}} |\pi \cap \pi'| p_i(\pi'). \qquad (20.10)$$

Thus, $\lambda_\pi^i(\mathbf{p}) = \theta_\pi(\mathbf{p}) + \left[\ell - \sum_{\pi' \in \mathcal{P}} |\pi \cap \pi'| p_i(\pi')\right] w_i$. Observe now that if **p** is a NE, then $L(\mathbf{p}) = L(\rho) \leq \theta^{\min}(\rho) + \ell w_{\max}$. Otherwise, the players routing their traffic on a path of expected delay greater than $\theta^{\min}(\rho) + \ell w_{\max}$ could improve their delay by defecting to a path of expected delay $\theta^{\min}(\rho)$. We sometimes say that a flow ρ corresponding to a mixed strategies profile **p** is a NE, with the understanding that it is actually **p** that is a NE.

20.7.3 Maximum Latency versus Total Latency

We show that if the resource delays are equal to their loads, a splittable flow is optimal with respect to the objective of maximum latency if and only if it is optimal with respect to the objective of total latency. As a corollary, we obtain that the optimal splittable flow defines a NE where all players adopt the same mixed strategy.

Lemma 20.4 [15] *There is a unique feasible flow ρ that minimizes both $L(\rho)$ and $C(\rho)$.*

Proof. For every feasible flow ρ, the average path latency $\frac{1}{W_{\text{tot}}} C(\rho)$ of ρ cannot exceed its maximum latency among the used paths $L(\rho)$:

$$C(\rho) = \sum_{\pi \in \mathcal{P}} \rho(\pi) \theta_\pi(\rho) = \sum_{\pi: \rho(\pi) > 0} \rho(\pi) \theta_\pi(\rho) \leq L(\rho) W_{\text{tot}}. \qquad (20.11)$$

A feasible flow ρ minimizes $C(\rho)$ if and only if for every $\pi_1, \pi_2 \in \mathcal{P}$ with $\rho(\pi_1) > 0$, $\theta_{\pi_1}(\rho) \leq \theta_{\pi_2}(\rho)$ (e.g., [6], [30, Section 7.2], [32, Corollary 4.2]). Hence, if ρ is optimal with respect to the objective of total latency, for all paths $\pi \in \mathcal{P}$, $\theta_\pi(\rho) \geq L(\rho)$. Moreover, if $\rho(\pi) > 0$, then $\theta_\pi(\rho) = L(\rho)$. Therefore, if ρ minimizes $C(\rho)$, then the average latency is indeed *equal* to the maximum latency:

$$C(\rho) = \sum_{\pi \in \mathcal{P}: \rho(\pi) > 0} \rho(\pi) \theta_\pi(\rho) = L(\rho) W_{\text{tot}}. \qquad (20.12)$$

Let ρ be the feasible flow that minimizes the total latency and let ρ' be the feasible flow that minimizes the maximum latency. We prove the lemma by establishing that the two flows are identical. Observe that $L(\rho') \geq \frac{C(\rho')}{W_{\text{tot}}} \geq \frac{C(\rho)}{W_{\text{tot}}} = L(\rho)$. The first inequality follows from Inequality 20.11, the second from the assumption that ρ minimizes the total latency, and the last equality from Equation 20.12. On the other hand, it must be $L(\rho') \leq L(\rho)$ because of the assumption that the flow ρ' minimizes the maximum latency. Hence, it must be $L(\rho') = L(\rho)$ and $C(\rho') = C(\rho)$. In addition, since the function $C(\rho)$ is strictly convex and the set of feasible flows forms a convex polytope, there is a unique flow that minimizes the total latency. Thus, ρ and ρ' must be identical. ∎

The following corollary is an immediate consequence of Lemma 20.4 and the characterization of the flow minimizing the total latency.

Corollary 20.1 *A flow ρ minimizes the maximum latency if and only if for every $\pi_1, \pi_2 \in \mathcal{P}$ with $\rho(\pi_1) > 0$, $\theta_{\pi_1}(\rho) \leq \theta_{\pi_2}(\rho)$.*

Proof. By Lemma 20.4, the flow ρ minimizes the maximum latency if and only if it minimizes the total latency. Then, the corollary follows from the fact that ρ minimizes the total latency if and only if for every $\pi_1, \pi_2 \in \mathcal{P}$ with $\rho(\pi_1) > 0$, $\theta_{\pi_1}(\rho) \leq \theta_{\pi_2}(\rho)$ (e.g., [30, Section 7.2], [32, Corollary 4.2]). ∎

The following corollary states that the optimal splittable flow defines a mixed NE where all players adopt exactly the same strategy.

Corollary 20.2 *Let ρ_f^* be the optimal splittable flow and let $\mathbf{p}$ be the mixed strategies profile where every player routes his traffic on each path π with probability $\rho_f^*(\pi)/W_{\text{tot}}$. Then, $\mathbf{p}$ is a NE.*

Proof. By construction, the expected path loads corresponding to $\mathbf{p}$ are equal to the values of ρ_f^* on these paths. Since all players follow exactly the same strategy and route their demand on each path π with probability ρ_f^*/W_{tot}, for each player i,

$$\theta_\pi^{-i}(\mathbf{p}) = \theta_\pi(\mathbf{p}) - w_i \sum_{\pi' \in \mathcal{P}} |\pi \cap \pi'| \frac{\rho_f^*(\pi')}{W_{\text{tot}}} = \left(1 - \frac{w_i}{W_{\text{tot}}}\right) \theta_\pi(\mathbf{p}).$$

Since the flow ρ_f^* also minimizes the total latency, for every $\pi_1, \pi_2 \in \mathcal{P}$ with $\rho_f^*(\pi_1) > 0$, $\theta_{\pi_1}(\mathbf{p}) \leq \theta_{\pi_2}(\mathbf{p})$ (e.g., [6], [30, Section 7.2], [32, Corollary 4.2]), which also implies that $\theta_{\pi_1}^{-i}(\mathbf{p}) \leq \theta_{\pi_2}^{-i}(\mathbf{p})$. Therefore, for every player i and every $\pi_1, \pi_2 \in \mathcal{P}$ such that player i routes his traffic demand on π_1 with positive probability, $\lambda_{\pi_1}^i(\mathbf{p}) = \ell w_i + \theta_{\pi_1}^{-i}(\mathbf{p}) \leq \ell w_i + \theta_{\pi_2}^{-i}(\mathbf{p}) = \lambda_{\pi_2}^i(\mathbf{p})$. Consequently, $\mathbf{p}$ is a NE. ∎

20.7.4 An Upper Bound on the Social Cost

Next, we derive an upper bound on the social cost of every strategy profile whose maximum expected delay (i.e., the maximum latency of its associated flow) is within a constant factor from the maximum latency of the optimal unsplittable flow.

Lemma 20.5 *Let ρ^* be the optimal unsplittable flow, and let $\mathbf{p}$ be a mixed strategies profile and ρ its corresponding flow. If $L(\mathbf{p}) = L(\rho) \leq \alpha L(\rho^*)$, for some $\alpha \geq 1$, then*

$$SC(\mathbf{p}) \leq 2e(\alpha + 1) \left(\frac{\log m}{\log \log m} + 1\right) L(\rho^*),$$

where $m = |E|$ denotes the number of edges in the network.

Proof. For each edge $e \in E$ and each player i, let $X_{e,i}$ be the random variable describing the actual load routed through e by i. The random variable $X_{e,i}$ is equal to w_i if i routes his demand on a path π including e and 0 otherwise. Consequently, the expectation of $X_{e,i}$ is equal to $\mathbf{E}\{X_{e,i}\} = \sum_{\pi: e \in \pi} w_i p_i(\pi)$. Since each player selects his path independently, for every fixed edge e, the random variables in $\{X_{e,i}\}_{i \in [n]}$ are independent from each other.

For each edge $e \in E$, let $X_e = \sum_{i=1}^n X_{e,i}$ be the random variable that describes the actual load routed through e, and thus, also the actual delay paid by any player traversing e. X_e is the sum of n independent random variables with values in $[0, w_{\max}]$. By linearity of expectation,

$$\mathbf{E}\{X_e\} = \sum_{i=1}^n \mathbf{E}\{X_{e,i}\} = \sum_{i=1}^n w_i \sum_{\pi \ni e} p_i(\pi) = \theta_e(\rho).$$

By applying the standard Hoeffding bound* with $w = w_{\max}$ and $t = e\kappa \max\{\theta_e(\rho), w_{\max}\}$, we obtain that for every $\kappa \geq 1$,

$$\mathbf{P}\{X_e \geq e\kappa \max\{\theta_e(\rho), w_{\max}\}\} \leq \kappa^{-e\kappa}.$$

For $m \equiv |E|$, by applying the union bound we conclude that

$$\mathbf{P}\{\exists e \in E : X_e \geq e\kappa \max\{\theta_e(\rho), w_{\max}\}\} \leq m\kappa^{-e\kappa}. \quad (20.13)$$

---

* We use the standard version of Hoeffding bound [19]: Let $X_1, X_2, \ldots, X_n$ be independent random variables with values in the interval $[0, w]$. Let $X = \sum_{i=1}^n X_i$ and let $\mathbf{E}\{X\}$ denote its expectation. Then, $\forall t > 0$, $\mathbf{P}\{X \geq t\} \leq (e\,\mathbf{E}\{X\}/t)^{t/w}$.

For each path $\pi \in \mathcal{P}$ with $\rho(\pi) > 0$, we define the random variable $X_\pi = \sum_{e \in \pi} X_e$ describing the actual delay along π. The social cost of **p**, which is equal to the expected maximum delay experienced by some player, cannot exceed the expected maximum delay among paths π with $\rho(\pi) > 0$. Formally,

$$SC(\mathbf{p}) \leq \mathbf{E}\left\{\max_{\pi:\rho(\pi)>0}\{X_\pi\}\right\}.$$

If for all $e \in E$, $X_e \leq e\kappa \max\{\theta_e(\rho), w_{\max}\}$, then for every path $\pi \in \mathcal{P}$ with $\rho(\pi) > 0$,

$$\begin{aligned}
X_\pi = \sum_{e \in \pi} X_e &\leq e\kappa \sum_{e \in \pi} \max\{\theta_e(\rho), w_{\max}\} \\
&\leq e\kappa \sum_{e \in \pi}(\theta_e(\rho) + w_{\max}) \\
&= e\kappa \left(\theta_\pi(\rho) + \ell w_{\max}\right) \\
&\leq e\kappa \left(L(\rho) + \ell w_{\max}\right) \\
&\leq e(\alpha + 1)\kappa \, L(\rho^*).
\end{aligned}$$

The third equality follows from $\theta_\pi(\rho) = \sum_{e \in \pi} \theta_e(\rho)$, the fourth inequality from $\theta_\pi(\rho) \leq L(\rho)$ since $\rho(\pi) > 0$, and the last inequality from the hypothesis that $L(\rho) \leq \alpha L(\rho^*)$ and the fact that $\ell w_{\max} \leq L(\rho^*)$ because ρ^* is an unsplittable flow. Therefore, using Equation 20.13, we conclude that

$$\mathbf{P}\left\{\max_{\pi:\rho(\pi)>0}\{X_\pi\} \geq e(\alpha + 1)\kappa \, L(\rho^*)\right\} \leq m\kappa^{-e\kappa}.$$

In other words, the probability that the actual maximum delay caused by **p** exceeds the optimal maximum delay by a factor greater than $2e(\alpha + 1)\kappa$ is at most $m\kappa^{-e\kappa}$. Therefore, for every $\kappa_0 \geq 2$,

$$\begin{aligned}
SC(\mathbf{p}) \leq \mathbf{E}\left\{\max_{\pi:\rho(\pi)>0}\{X_\pi\}\right\} &\leq e(\alpha + 1)L(\rho^*)\left(\kappa_0 + \sum_{k=\kappa_0}^{\infty} km k^{-ek}\right) \\
&\leq e(\alpha + 1)L(\rho^*)\left(\kappa_0 + 2m\kappa_0^{-e\kappa_0+1}\right).
\end{aligned}$$

If $\kappa_0 = \frac{2 \log m}{\log \log m}$, then $\kappa_0^{-e\kappa_0+1} \leq m^{-1}$, $\forall m \geq 4$. Thus, we conclude that

$$SC(\mathbf{p}) \leq 2e(\alpha + 1)\left(\frac{\log m}{\log \log m} + 1\right)L(\rho^*).$$

∎

20.7.5 Bounding the Price of Anarchy

Our final step is to show that the maximum expected delay of every NE is a good approximation to the optimal maximum latency. Then, we can apply Lemma 20.5 to bound the price of anarchy for our selfish routing game.

Lemma 20.6 *For every flow ρ corresponding to a mixed strategies profile* **p** *at NE, $L(\rho) \leq 3L(\rho^*)$.*

Proof. The proof is based on Dorn's theorem [10] that establishes strong duality in quadratic programming.* We use quadratic programming duality to prove that for any flow ρ at Nash equilibrium, the minimum expected delay $\theta^{\min}(\rho)$ cannot exceed $L(\rho_f^*) + \ell\, w_{\max}$. This implies the lemma because $L(\rho) \leq \theta^{\min}(\rho) + \ell\, w_{\max}$, since ρ is at Nash equilibrium, and $L(\rho^*) \geq \max\{L(\rho_f^*), \ell\, w_{\max}\}$, since ρ^* is an unsplittable flow.

Let Q be the square matrix describing the number of edges shared by each pair of paths. Formally, Q is a $|\mathcal{P}| \times |\mathcal{P}|$ matrix and for every $\pi, \pi' \in \mathcal{P}$, $Q[\pi, \pi'] = |\pi \cap \pi'|$. By definition, Q is symmetric. Next, we prove that Q is positive semidefinite.†

$$\begin{aligned}
x^T Q x &= \sum_{\pi \in \mathcal{P}} x(\pi) \sum_{\pi' \in \mathcal{P}} Q[\pi, \pi'] x(\pi') \\
&= \sum_{\pi \in \mathcal{P}} x(\pi) \sum_{\pi' \in \mathcal{P}} |\pi \cap \pi'| x(\pi') \\
&= \sum_{\pi \in \mathcal{P}} x(\pi) \sum_{e \in \pi} \sum_{\pi' : e \in \pi'} x(\pi') \\
&= \sum_{\pi \in \mathcal{P}} x(\pi) \sum_{e \in \pi} \theta_e(x) \\
&= \sum_{e \in E} \theta_e(x) \sum_{\pi : e \in \pi} x(\pi) \\
&= \sum_{e \in E} \theta_e^2(x) \geq 0.
\end{aligned}$$

First, recall that for each edge e, $\theta_e(x) \equiv \sum_{\pi : e \in \pi} x(\pi)$. The third and the fifth equalities follow by reversing the order of summation. In particular, in the third equality, instead of considering the edges shared by π and π', for all $\pi' \in \mathcal{P}$, we consider all the paths π' using each edge $e \in \pi$. On both sides of the fifth inequality, for every edge $e \in E$, $\theta_e(x)$ is multiplied by the sum of $x(\pi)$ over all the paths π using e.

Let ρ also denote the $|\mathcal{P}|$-dimensional vector corresponding to the flow ρ. Then, the π-th coordinate of $Q\rho$ is equal to the expected delay $\theta_\pi(\rho)$ on the path π, and the total latency of ρ is $C(\rho) = \rho^T Q \rho$.

Therefore, the problem of computing a feasible splittable flow of minimum total latency is equivalent to computing the optimal solution to the following quadratic program: $\min\{\rho^T Q \rho : \mathbf{1}^T \rho \geq W_{\text{tot}}, \rho \geq \mathbf{0}\}$, where $\mathbf{1/0}$ denotes the $|\mathcal{P}|$-dimensional vector having $1/0$ in each coordinate. Also, note that no flow of value strictly greater than W_{tot} can be optimal for this program. This quadratic program is clearly feasible and its optimal solution is ρ_f^* (Lemma 20.4).

The Dorn's dual of this quadratic program is $\max\{z W_{\text{tot}} - \rho^T Q \rho : 2Q\rho \geq \mathbf{1}z, z \geq 0\}$ (e.g., [10], [5, Chapter 6]). We observe that any flow ρ can be regarded as a feasible solution to the dual program by setting $z = 2\theta^{\min}(\rho)$. Hence, both the primal and the dual programs are feasible. By Dorn's theorem [10], the objective value of the optimal dual solution is exactly $C(\rho_f^*)$. More specifically, the optimal dual solution is obtained from ρ_f^* by setting $z = 2\theta^{\min}(\rho_f^*)$. Since $L(\rho_f^*) = \theta^{\min}(\rho_f^*)$ and $C(\rho_f^*) = L(\rho_f^*) W_{\text{tot}}$, the objective value of this solution is $2\theta^{\min}(\rho_f^*) W_{\text{tot}} - C(\rho_f^*) = C(\rho_f^*)$.

Let ρ be any feasible flow at Nash equilibrium. Setting $z = 2\theta^{\min}(\rho)$, we obtain a dual feasible solution. By the discussion above, the objective value of the feasible dual solution $(\rho, 2\theta^{\min}(\rho))$ cannot

---

* Let $\min\{x^T Q x + c^T x : Ax \geq b, x \geq \mathbf{0}\}$ be the primal quadratic program. The Dorn's dual of this program is $\max\{-y^T Q y + b^T u : A^T u - 2Q y \leq c, u \geq \mathbf{0}\}$. Dorn [10] proved strong duality when the matrix Q is symmetric and positive semidefinite. Thus, if Q is symmetric and positive semidefinite and both the primal and the dual programs are feasible, their optimal solutions have the same objective value.

† An $n \times n$ matrix Q is positive semidefinite if for every vector $x \in \mathbf{R}^n$, $x^T Q x \geq 0$.

exceed $C(\rho_f^*)$. In other words,

$$2\,\theta^{\min}(\rho)\,W_{\text{tot}} - C(\rho) \leq C(\rho_f^*). \tag{20.14}$$

Since ρ is at Nash equilibrium, $L(\rho) \leq \theta^{\min}(\rho) + \ell\,w_{\max}$. In addition, by Equation 20.11, the average latency of ρ cannot exceed its maximum latency. Thus,

$$C(\rho) \leq L(\rho)\,W_{\text{tot}} \leq \theta^{\min}(\rho)\,W_{\text{tot}} + \ell\,w_{\max}\,W_{\text{tot}}.$$

Combining the inequality above with Inequality 20.14, we obtain that $\theta^{\min}(\rho)\,W_{\text{tot}} \leq C(\rho_f^*) + \ell\,w_{\max}\,W_{\text{tot}}$. Using $C(\rho_f^*) = L(\rho_f^*)\,W_{\text{tot}}$, we conclude that $\theta^{\min}(\rho) \leq L(\rho_f^*) + \ell\,w_{\max}$. ∎

The following theorem is an immediate consequence of Lemmas 20.6 and 20.5.

Theorem 20.8 [15] *The price of anarchy of any ℓ-layered network congestion game with resource delays equal to their loads is at most* $8\mathrm{e}\left(\frac{\log m}{\log\log m} + 1\right)$.

A recent development that is complementary to the last theorem is the following, which we state without a proof:

Theorem 20.9 [16] *The price of anarchy of any unweighted, single-commodity network congestion game with resource delays* $(d_e(x) = a_e \cdot x, a_e \geq 0)_{e \in E}$ *is at most* $24\mathrm{e}\left(\frac{\log m}{\log\log m} + 1\right)$.

20.8 The Pure Price of Anarchy in Congestion Games

In this last section, we overview some recent advances in the *Pure Price of Anarchy* (PPoA) of congestion games, that is, the worst-case ratio of the social cost of a PNE over the social optimum of the game.

The case of linear resource delays has been extensively studied in the literature. The PPoA with respect to the total latency objective has been proved that it is $(3+\sqrt{5})/2$, even for weighted multicommodity network congestion games [4,7]. This result is also extended to the case of mixed equilibria. For the special case of identical players, it has been proved (independently by the papers [4,7]) that the PPoA drops down to 5/2. When considering identical users and single-commodity network congestion games, the PPoA is again 5/2 with respect to the maximum latency objective, but explodes to $\Theta(\sqrt{n})$ for multicommodity network congestion games [7]. Earlier, it was implicitly proved by Reference 15 that the PPoA of any weighted congestion game on a layered network with resource delays identical to the congestion is at most 3.

20.9 Conclusions

In this survey, we have presented some of the most significant advances concerning the atomic (mainly network) congestion games literature. We have focused on issues dealing with existence of PNE, construction of an arbitrary PNE when such equilibria exist, as well as the price of anarchy for many broad subclasses of network congestion games.

We highlighted the significance of allowing distinguishable players (i.e., players with different action sets, or with different traffic demands, or both) and established some kind of "equivalence" between games with unit-demand players on arbitrary networks with delays equal to the loads and games with players of varying demands on layered networks.

Still, there remain many unresolved questions. The most important question is a complete characterization of the subclasses of games having PNE and admitting polynomial time algorithms for constructing them, in the case of general networks.

In addition, a rather recent trend deals with the network-design perspective of congestion games. For these games [2,3,20], the measure of performance is the **price of stability**, that is, the ratio of the *best* NE over the social optimum of the game, trying to capture the notion of the gap between solutions proposed by the network designer to the players and the social optimum of the game (which may be an unstable state for the players). This seems to be a rather intriguing and very interesting (complementary to the one presented here) approach of congestion games, in which there are still many open questions.

References

[1] Ahuja R., Magnanti T., Orlin J. *Network Flows: Theory, Algorithms and Applications.* Prentice Hall, 1993.

[2] Anshelevich E., Dasgupta A., Kleinberg J., Tardos E., Wexler T., Roughgarden T. The price of stability for network design with fair cost allocation. In *Proc. of the 45th IEEE Symp. on Foundations of Computer Science (FOCS '04)*, pp. 295–304, 2004.

[3] Anshelevich E., Dasgupta A., Tardos E., Wexler T. Near-optimal network design with selfish agents. In *Proc. of the 35th ACM Symp. on Theory of Computing (STOC '03)*, pp. 511–520, 2004.

[4] Awerbuch B., Azar Y., Epstein A. The price of routing unsplittable flow. In *Proc. of the 37th ACM Symp. on Theory of Computing (STOC '05)*, pp. 57–66, 2005.

[5] Bazaraa M.S., Sherali H.D., Shetty C.M. *Nonlinear Programming: Theory and Algorithms*, 2nd edn. John Wiley & Sons, Inc., 1993.

[6] Beckmann M., McGuire C.B., Winsten C.B. *Studies in the Economics of Transportation.* Yale University Press, 1956.

[7] G. Christodoulou and E. Koutsoupias. The Price of Anarchy of Finite Congestion Games. In *Proc. of the 37th ACM Symp. on Theory of Computing (STOC '05)*, pp. 67–73, 2005.

[8] Czumaj A., Krysta P., Voecking B. Selfish traffic allocation for server farms. In *Proc. of the 34th ACM Symp. on Theory of Computing (STOC '02)*, pp. 287–296, 2002.

[9] Czumaj A., Voecking B. Tight bounds for worst-case equilibria. In *Proc. of the 13th ACM-SIAM Symp. on Discrete Algorithms (SODA '02)*, pp. 413–420, 2002.

[10] Dorn W.S. Duality in quadratic programming. *Quarterly of Applied Mathematics*, 18(2):155–162, 1960.

[11] Even-Dar E., Kesselman A., Mansour Y. Convergence time to Nash equilibria. In *Proc. of the 30th International Colloquium on Automata, Languages and Programming (ICALP '03)*, pp. 502–513. Springer-Verlag, 2003.

[12] Fabrikant A., Papadimitriou C., Talwar K. The complexity of pure Nash equilibria. In *Proc. of the 36th ACM Symp. on Theory of Computing (STOC '04)*, 2004.

[13] Feldmann R., Gairing M., Luecking T., Monien B., Rode M. Nashification and the coordination ratio for a selfish routing game. In *Proc. of the 30th International Colloquium on Automata, Languages and Programming (ICALP '03)*, pp. 514–526. Springer-Verlag, 2003.

[14] Fotakis D., Kontogiannis S., Koutsoupias E., Mavronicolas M., Spirakis P. The structure and complexity of Nash equilibria for a selfish routing game. In *Proc. of the 29th International Colloquium on Automata, Languages and Programming (ICALP '02)*, pp. 123–134. Springer-Verlag, 2002.

[15] Fotakis D., Kontogiannis S., Spirakis P. Selfish unsplittable flows. In *Proc. of the 31st International Colloquium on Automata, Languages and Programming (ICALP '04)*, pp. 593–605. Springer-Verlag, 2004.

[16] Fotakis D., Kontogiannis S., Spirakis P. Symmetry in network congestion games: Pure equilibria and anarchy cost. In *Proc. of the 3rd Workshop on Approximation and Online Algorithms (WAOA '05)*. Springer-Verlag, 2005.

[17] Gairing M., Luecking T., Mavronicolas M., Monien B., Spirakis P. Extreme Nash equilibria. In *Proc. of the 8th Italian Conference on Theoretical Computer Science (ICTCS '03)*. Springer-Verlag, 2003.

[18] Hochbaum D., Shmoys D. A polynomial approximation scheme for scheduling on uniform processors: Using the dual approximation approach. *SIAM J. Comput.*, 17(3):539–551, 1988.

[19] Hoeffding W. Probability inequalities for sums of bounded random variables. *J. Am. Stat. Assoc.*, 58(301):13–30, 1963.

[20] Johari R., Tsitsiklis J. Efficiency loss in a network resource allocation game. To appear in the *Mathematics of Operations Research*, 2004.

[21] Johnson D., Papadimitriou C., Yannakakis M. How easy is local search? *J. Comput. Sys. Sci.*, 37:79–100, 1988.

[22] Koutsoupias E., Mavronicolas M., Spirakis P. Approximate equilibria and ball fusion. *ACM Trans. Comput. Sys.*, 36:683–693, 2003.

[23] Koutsoupias E., Papadimitriou C. Worst-case equilibria. In *Proc. of the 16th Annual Symposium on Theoretical Aspects of Computer Science (STACS '99)*, pp. 404–413. Springer-Verlag, 1999.

[24] Libman L., Orda A. Atomic resource sharing in noncooperative networks. *Telecomm. Sys.*, 17(4):385–409, 2001.

[25] Lücking T., Mavronicolas M., Monien B., Rode M. A new model for selfish routing. In *Proc. of the 21st Annual Symposium on Theoretical Aspects of Computer Science (STACS '04)*, pp. 547–558. Springer-Verlag, 2004.

[26] Lücking T., Mavronicolas M., Monien B., Rode M., Spirakis P., Vrto I. Which is the worst-case Nash equilibrium? In *26th International Symposium on Mathematical Foundations of Computer Science (MFCS'03)*, pp. 551–561. Springer-Verlag, 2003.

[27] Mavronicolas M., Spirakis P. The price of selfish routing. In *Proc. of the 33rd ACM Symp. on Theory of Computing (STOC '01)*, pp. 510–519, 2001.

[28] Milchtaich I. Congestion games with player-specific payoff functions. *Games and Econom. Behav.*, 13:111–124, 1996.

[29] Monderer D., Shapley L. Potential games. *Games and Economic Behavior*, 14:124–143, 1996.

[30] Papadimitriou C., Steiglitz K. *Combinatorial Optimization: Algorithms and Complexity*. Prentice-Hall, Inc., 1982.

[31] Rosenthal R.W. A class of games possessing pure-strategy Nash equilibria. *Int. J. Game Theory*, 2:65–67, 1973.

[32] Roughdarden T., Tardos E. How bad is selfish routing? *J. Assoc. Comput. Mach.*, 49(2):236–259, 2002.

[33] Roughgarden T. The price of anarchy is independent of the network topology. In *Proc. of the 34th ACM Symp. on Theory of Computing (STOC '02)*, pp. 428–437, 2002.

[34] Schäffer A., Yannakakis M. Simple local search problems that are hard to solve. *SIAM J. Comput.*, 20(1):36–87, 1991.

21

Scheduling in Grid Environments

21.1	The Scheduling Problem	21-2
	Traditional Multiprocessor Scheduling • Grid Scheduling	
21.2	Models	21-4
	Application Model • Grid Model • Grid Scheduler Model	
21.3	Scheduling Procedure in Grid Environments	21-6
	Job Description and Submission • Job Analysis • Preselection of Resources • Resource Allocation • Execution Preparation • Monitoring • Postexecution	
21.4	Grid Scheduling Systems	21-9
	AppLeS • Condor • Moab Grid Scheduler (Silver) • NetSolve and GridSolve • Nimrod/G	
21.5	Scheduling Algorithms for Grids	21-12
	List Scheduling with Round-Robin Order Replication • Storage Affinity • XSufferage	
21.6	Grid Simulation Tools	21-13
	Bricks • GridSim • MicroGrid • OptorSim • SimGrid	
21.7	Conclusions	21-16
References		21-16

Young Choon Lee
Albert Y. Zomaya
The University of Sydney

As the demand for more powerful computing resources continually increases, especially in areas of science, engineering, and commerce, a myriad of high-performance computing systems, such as supercomputers and computer clusters, have been built with various different architectures. In general, the use of these specialist computing systems is confined to specific groups of people. Moreover, each of these systems is generally restricted to independent use; that is, it is highly unlikely that a user of one system can access other organizations' systems. A solution to this is grid computing. A grid enables a virtual computing system interconnecting these geographically distributed heterogeneous computing systems with a variety of resources to be constituted. Here, resources refer not only to physical computers, networks, and storage systems but also to much broader entities, such as databases, data transfer, and simulation [1]. The grid creates the illusion that its users are accessing a single, very powerful, and reliable computing system.

A vast number of researchers have been putting in a lot of effort to facilitate building and efficiently utilizing grids. Some significant results for grid computing include the Globus toolkit [2], Legion [3], and GrADS [4]. These tools, especially the Globus toolkit, which has become the de facto standard, have been used to build many grids [5–9].

Many issues that can be relatively easily handled in conventional computing environments become seriously challenging problems in grids, mainly because a grid consists of multiple administrative domains. Two very crucial issues among them are security and scheduling. They have been investigated and researched over time. The grid security infrastructure (GSI) in the Globus toolkit addresses and effectively deals with the security issue. However, there are still a considerable number of difficulties to overcome for efficiently scheduling jobs in grids.

Grid scheduling requires a series of challenging tasks. These include searching for resources in collections of geographically distributed, heterogeneous computing systems and making scheduling decisions, taking into consideration the quality of service. A grid scheduler differs from a scheduler for conventional computing systems in several respects. One of the primary differences is that the grid scheduler does not have full control over the grid. More specifically, the local resources are, in general, not controlled by the grid scheduler, but by the local scheduler [10]. Another difference is that the grid scheduler cannot assume that it has a global view of the grid.

Despite efforts that current grid schedulers with various scheduling algorithms have made to provide comprehensive and sophisticated functionalities, they have difficulty guaranteeing the quality of schedules they produce. The single most challenging issue they encounter is the dynamicity of resources, that is, the availability and capability of a grid resource change dynamically. In addition, applications deployed in grids are diverse in their models. As attempts to overcome hurdles arising from this dynamicity issue, several noticeable approaches such as resource provisioning, accurate state estimation, and job migration have been proposed and studied [11–16].

21.1 The Scheduling Problem

The efficient use of resources is one of the most critical and studied issues in many different fields including computing. This is known as scheduling. Traditional scheduling is the process of allocating tasks to resources over time on the basis of certain performance measurement criteria, in order to meet constraints [17]. Scheduling can be divided into two forms: static and dynamic [18]. Since dynamic scheduling makes scheduling decisions at runtime to adapt to the dynamically varying resource conditions, it is a preferable approach for grids [19].

Scheduling in high-performance computing systems, such as massively parallel processor (MPP) systems and computer clusters, in particular, plays a very crucial role and has been thoroughly studied. Many of the scheduling techniques used for these systems can also be applied to grid scheduling. However, owing to the dynamic nature of the grid, there are a series of issues that complicate scheduling in the grid.

21.1.1 Traditional Multiprocessor Scheduling

In parallel processing, the problem of scheduling parallelizable tasks onto multiple processing units has been extensively studied for quite some time. However, owing to the NP-completeness of the task scheduling problem in most cases [18], the myriad of existing scheduling algorithms is based on heuristic algorithms. In this traditional multiprocessor scheduling, homogeneous computing systems are the most predominant target platforms, even though some scheduling and work load management studies in the past couple of decades have been conducted for heterogeneous multiprocessor systems, such as clusters of computers and networks of workstations (NOWs).

A typical scenario in traditional multiprocessor scheduling is running a message-passing parallel application on a homogeneous multiprocessor system, such as shared-memory supercomputers and clusters of workstations. Two main characteristics of traditional multiprocessor scheduling that can be derived from this scenario are that scheduling is centralized and that it is performed in a single administrative domain. In other words, a traditional multiprocessor scheduling system (i) has complete control over the system, (ii) is able to obtain all necessary scheduling information, and (iii) operates over a high-speed interconnection network.

21.1.2 Grid Scheduling

Although the grid can provide powerful processing capability, users may not be able to utilize it effectively if their jobs are not appropriately scheduled. Owing to unique characteristics of the grid, scheduling on grids becomes far more complex than traditional multiprocessor scheduling. Grid scheduling differs from traditional multiprocessor scheduling primarily in the following three respects:

1. *Lack of control*: Resources in the grid are dispersed across multiple administrative domains that are often in different geographical locations. This implies that resources belonging to a particular administrative domain are more dedicated to the local users of the domain than users in other administrative domains; that is, the alien users have less control over those resources compared to the local users. Moreover, resources in a grid may not be equally accessible to all the users of the grid. More specifically, several levels of access privilege apply to different users. For example, system administrators probably have more privilege than ordinary users. Note that, owing to this factor, a grid scheduler may not be able to deliver the best possible schedule even if the resources on which the job can run most efficiently are available.
2. *Dynamicity of resource*: The availability and capability of resources can change dynamically. This is one of the major hurdles to overcome when scheduling on grids. Resources join and disjoin the grid at any time. Grid resources, especially network resources, fail more frequently compared to those in conventional parallel computing systems. This is the primary reason that makes grid resources unavailable. Since resources in the grid are used by both grid and local users, their capabilities fluctuate dynamically.
3. *Application diversity*: It is often the case that a particular conventional parallel computing system is used by a certain group of users for applications in specific fields such as bioinformatics and high-energy physics; that is, the applications run in the system are of a similar model, if not the same. However, applications deployed in grids are typically from various disciplines, and thus, they are of different application models.

While scheduling decisions in traditional scheduling are made primarily on the basis of static information, scheduling decisions in grid scheduling are more heavily influenced by dynamic information. In general, an information service such as the Globus Monitoring and Discovering Service (MDS) [20] is provided in a grid. Both static and dynamic information about grid resources can be obtained from this information service.

When a job is running in a controlled local environment, for example, a departmental cluster, one can easily expect to get a certain level of quality of service. Conversely, it is highly unlikely to get such quality of service in a grid environment. One way to overcome this difficulty is to adopt resource provisioning (i.e., advance reservation) in grid scheduling. A critical issue in advance reservation is cooperation between different local schedulers who are operating with their own policies. Another approach being studied and practiced is performance prediction on the application and the resource. Accurate performance prediction can substantially help to make better scheduling decisions. Monitoring the job during its execution is also an effective technique to ensure quality of service, since abnormalities in the progress of the job can be effectively handled at the appropriate time. Note that efficient grid scheduling involves developing complex and advanced mechanisms and techniques to deliver a high quality of performance.

To our best knowledge, none of the current grid schedulers is providing a comprehensive set of services that enables the grid scheduler to support scheduling various types of applications, effectively interacting with different local schedulers. A major obstacle for a grid scheduler to implement advanced techniques is lack of control and ownership of resources. At present, one way to provide a broad range of grid scheduling services, such as advance reservation and migration, is by using a prepackaged grid/local scheduler pair. For example, the Moab grid scheduler, Sliver [21], enables such services with use of the Maui scheduler [22] as prepackaged local scheduler.

An abstract model of grid scheduling is shown in Figure 21.1.

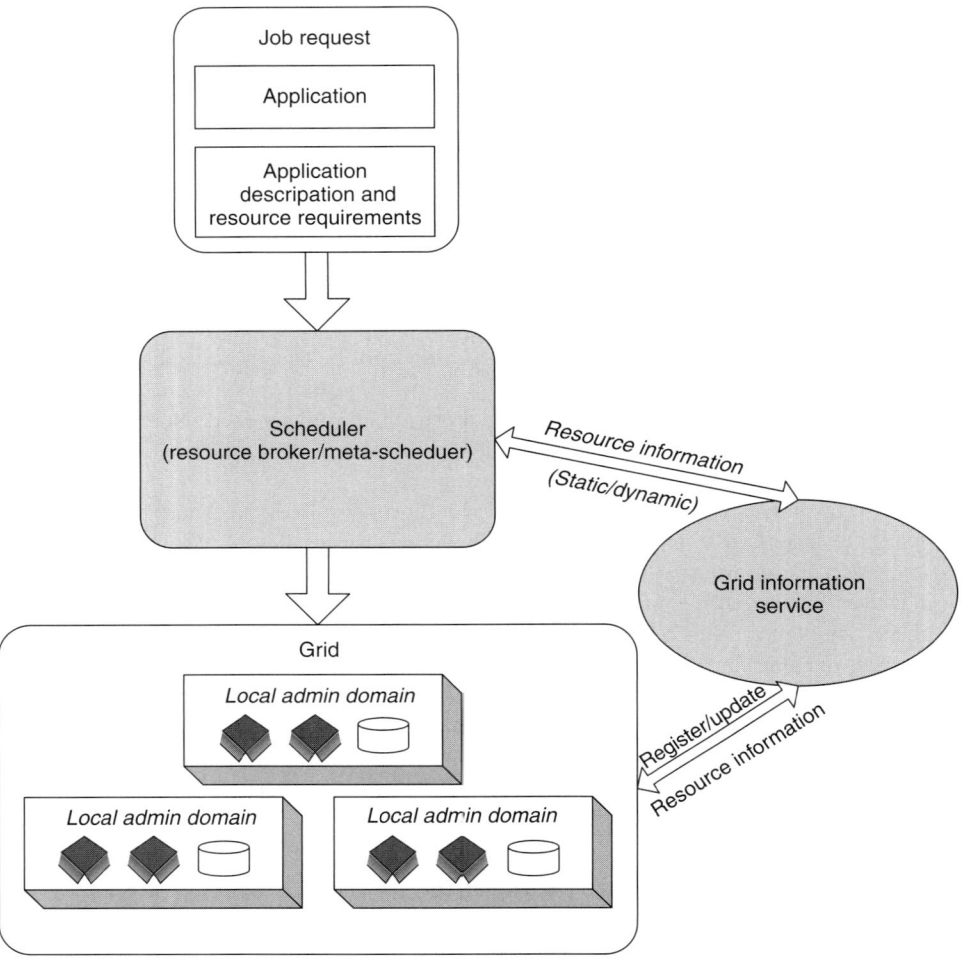

FIGURE 21.1 An abstract model of grid scheduling.

21.2 Models

21.2.1 Application Model

It is obvious that large-scale and coarse-grained applications are most suitable for grids. As the grid has emerged as a promising platform to tackle large-scale problems, an increasing number of applications in various areas, including bioinformatics, high-energy physics, image processing, and data mining, have been developed and ported for grid environments. In general, these applications are designed with parallel and/or distributed processing in mind. Two typical application models found among them are independent and interdependent (i.e., workflow). Some examples of the former model are bag-of-tasks and synchronous iterative applications. An application in the independent model consists of tasks without any dependencies, and thus, no specific order of task execution. BLAST [23], MCell [24], INS2D [25], and many data mining applications fall into this model. On the other hand, an application in the workflow model is composed of interdependent tasks. Therefore, the execution of the tasks is subject to their precedence constraints. Applications in this model impose a great strain on grid scheduling. LIGO [26], image processing on the grid [27], and Montage [28] are examples in the workflow model.

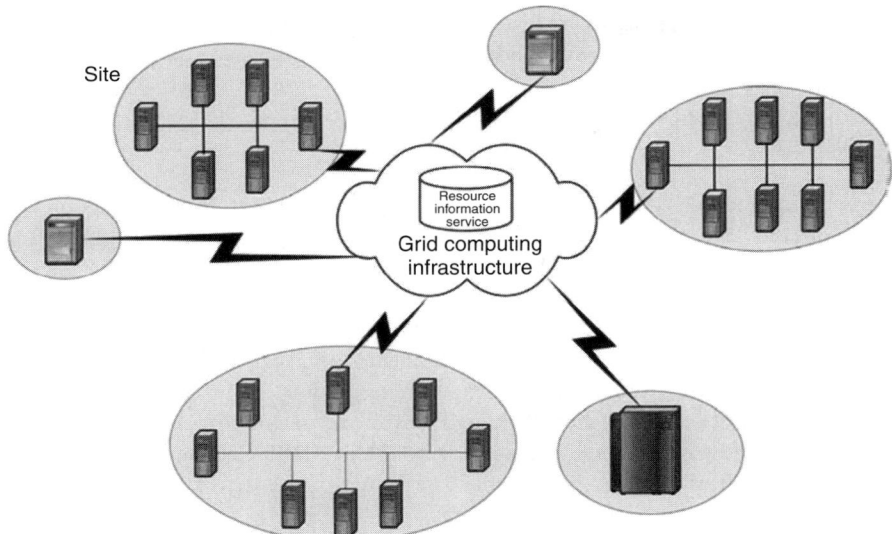

FIGURE 21.2 Grid model.

Applications deployed in grids can also be classified into compute-intensive and data-intensive. In the case of running applications in the former category, the performance of computing resources is the most influential factor. However, the management of data transfers plays a crucial role with applications in the latter category.

21.2.2 Grid Model

A grid G consists of a number of sites/virtual organizations, in each of which a set of m computational hosts is participating in a grid (Figure 21.2). More formally,

$$G = \{S_1, S_2, \ldots, S_r\}, \quad \text{and} \quad S_i, 1 \leq i \leq r, = \{H_{i,1}, H_{i,2}, \ldots, H_{i,m}\},$$

where S_i is the ith site participating in G, and H_i is a set of host at S_i, respectively.

Each site is an autonomous administrative domain that has its own local users who use the resources in it. In general, these sites are connected with each other through WAN, whereas hosts in the same site are connected through a high-bandwidth LAN. Hosts are composed of both space- and time-shared machines with various processing speeds. These resources are not entirely dedicated to the grid. In other words, they are used for both local and grid jobs. Each of these hosts has one or more processors, memory, disk, and so forth.

21.2.3 Grid Scheduler Model

Local schedulers and local resource managers have autonomy over resources, and thus can apply domain-specific scheduling strategies to improve the utilization of resources as well as the performance of jobs. In addition, resources in conventional computing environments, for example, supercomputers, clusters, and NOWs, are static in general. However, these two factors are highly unlikely in grid computing environments. Therefore, a grid scheduler must take into consideration the following four fundamental characteristics:

1. The scalability of grid schedulers is a crucial issue since a grid can grow and shrink dynamically [29,30].

2. Complex security requirements are involved in grid scheduling.
3. A grid scheduler should perform application/resource matching according to application quality of service constraints.
4. In a grid environment, failure is not the exception, but the rule. Hence, fault tolerance must be considered.

A centralized structure is the single dominating model with most conventional multiprocessor schedulers. However, this is a costly solution for grid schedulers due to the scalability issue in particular. In general, a grid scheduler can be classified as centralized, decentralized, or hierarchical in structure [31,32].

In a *centralized* model, a single grid scheduler has the capability of orchestrating the entire scheduling activity in the grid. Note that a centralized grid scheduler makes scheduling decisions with global scheduling information. This results in delivering better schedules compared to those produced by schedulers in the other two approaches. Although this model seems to be quite effective, it is not wholly desirable for grid environments because it is not easily scalable as the grid expands over time. One can also note that this scheme is vulnerable to a failure of the system in which the scheduler is operating.

In a *decentralized* model, multiple grid schedulers independently operate in a grid. More specifically, each grid scheduler queries a resource information service, in order to gather resource information and to make scheduling decisions, without interacting with each other. Two primary advantages of this scheme are scalability and fault tolerance.

Since each scheduler can only get a limited amount of scheduling information, the quality of schedules produced by the decentralized scheduler appears to be less competitive than that of schedules delivered by the centralized scheduler. Therefore, the coordination between schedulers should be addressed, so that more efficient scheduling with better resource information can be achieved.

A *hierarchical* scheme is a hybrid model. At the highest level, this model can be seen as centralized. However, it also contains the essential properties of the decentralized model, that is, scalability and fault tolerance. These factors make the hierarchy model desirable for grid environments.

21.3 Scheduling Procedure in Grid Environments

Scheduling a job in a grid environment consists of a series of complex steps as shown in Figure 21.3. The majority of these steps heavily rely on facilities provided by grid middleware (e.g., Globus toolkit). All the steps of grid scheduling presented in Figure 21.3 can also be found in traditional local scheduling. However, the dynamic nature of the grid considerably complicates many of these steps, especially the resource allocation and the monitoring steps. A three-phase architecture for grid scheduling that consists of similar steps to the ones presented in this section can be found in Reference 33.

21.3.1 Job Description and Submission

The first step in grid scheduling involves two tasks—job description and job submission. In the job description task, the job submission entity, either the user or some application, identifies the job configuration, for example, directory, executable arguments, and so forth, and determines a set of resources that the job requires. Once the job description is complete, a job request is submitted to the grid scheduler with the job description.

A grid job is generally submitted with a set of specific resource requirements, such as machine type, number of nodes, memory, operating system, and so forth. These requirements can vary significantly between jobs. Although the job request is believed to be accurately specified, it cannot be assumed that the expected performance of the job will be obtained. This is mainly because the required resources and the actual resources allocated to the job are different.

Scheduling in Grid Environments 21-7

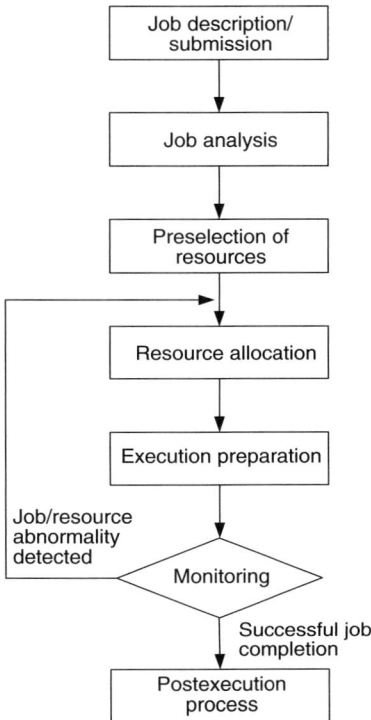

FIGURE 21.3 Grid scheduling procedure.

The job description normally contains the following:

1. *Resource requirements*: processor architecture, number of central processing units (CPUs), memory, a minimum bandwidth, operating system, and so forth
2. *Application properties*: type (e.g., independent, mpi, parameter sweep [34]), location of executables, arguments, input/output files, specific software packages, dependencies, and so forth
3. *Time requirements*: start time and finish time

In addition to these core requirements, some extra requirements can be specified (e.g., cost). A job is submitted with this data as part of the command line or submission script.

In current grid environments, two well-known and widely used job/resource description languages are an extensible markup language (XML)-based job description language in the Globus Toolkit and condor classified advertisements (ClassAds) [35].

21.3.2 Job Analysis

To date, jobs are typically described and submitted by the user manually. For this reason, the user has to either know or be informed about details of the application, such as the detailed application structure and execution time, in order to accurately describe the job. It is true that the more application information the scheduler has the better performance it delivers.

Job analysis is a step in which the job is scrutinized and optimized automatically by the grid scheduler or an automatic tool integrated in the grid scheduler. This process consists of two subtasks: (i) Additional application characteristics are identified; for example, the execution time of the application is predicted using existing techniques (e.g., References 36 and 37). (ii) The job request is optimized on the basis of the information obtained from (i). It should be noted that the job analysis step is not implemented in most current grid schedulers.

21.3.3 Preselection of Resources

In a computing system governed by a single administrative domain, comprehensive resource information is available to users. However, this is not true in the grid where users have very limited knowledge of the participating resources. In general, grid information services (GIS) are implemented in grids in order to resolve this matter. Both static and dynamic information are accumulated and maintained by the GIS. The Globus MDS is a widely implemented and deployed GIS.

In this step, a collection of resources that satisfy the resource requirements is identified on the basis of the static information gathered. In general, static information is obtained from the GIS by sending a query. Some examples of static information include the number of nodes, memory, specialized hardware devices, bandwidth, operating systems, software packages, and so forth. Three major categories of static information are services, machines, and networks. The collection of resources identified earlier may contain a large number of resources. It is, therefore, necessary to select more preferable resources on the basis of certain criteria, such as preferences specified by the user and/or application characteristics identified from the job analysis step. Heuristics can also be used to reduce the number of matched resources. For example, a grid scheduler filters out those resources that have provided poor quality of service according to the historical information maintained by the grid scheduler.

21.3.4 Resource Allocation

Once the best set of resources is determined from the preselection of resources step, the grid scheduler allocates the job to these selected resources using a certain mapping scheme. Note that a grid scheduler equipped with an efficient job/resource mapping scheme can significantly reduce the application completion time and improve the system utilization. This has popularized the research in scheduling algorithms for grid environments.

The resource allocation decision is affected primarily by dynamic information along with user-specified criteria. More specifically, the preselected resources are evaluated as to whether they are available and will be available, and can deliver a certain quality of service that meets the minimum job requirements, such as the user-specified time and cost constraints. Dynamic information about the preselected resources is critical in grid scheduling, since the current states of these resources, such as availabilities and capabilities, cannot be precisely known from static information.

To help better scheduling, performance prediction information is also gathered from performance analysis/prediction tools [e.g., the Network Weather Service (NWS)] [38–40]. The NWS provides performance information on the basis of historical performance measurements obtained from periodic monitoring. Performance information contains short-term performance forecasts of various network and computational resources. In addition, a grid scheduler may gather more dynamic information from local schedulers. However, this is only possible if these two parties have agreed to cooperate.

The resource allocation process may involve advance resource reservation, especially for applications of the workflow model. Some serious efforts [22,41–45] in advance reservation have been made in the recent past. However, a number of issues that hinder reserving resources in advance are still floating. These issues include the cooperation between grid schedulers and local schedulers, the variance of the job requirements during runtime, and so forth.

The Condor ClassAd/Matchmaker is a well-known resource selector for distributed systems. The resource selection is conducted by a designated matchmaker using ClassAds described by both resource providers and consumers containing resource characteristics and requirements, respectively.

21.3.5 Execution Preparation

In most cases, jobs require some preparation tasks (e.g., file staging, setup, etc.) before execution. Since the resource set on which a job is running is typically different from where the job is submitted, any necessary input files and settings in order to run the job have to be prepared. The input files are, in general, transferred to the target resource set using file transferring services, such as GridFTP [46], an enhanced file transfer protocol. In addition to file staging, some environment variables are set and temporary directories are created, which are used during job execution. These job-specific tasks must be undone after the job is run.

21.3.6 Monitoring

Once the job starts running, a monitoring service gets initiated and monitors the progress of the job. Monitoring jobs in a grid environment has special significance, in that, job completion cannot be guaranteed owing to unforeseen factors, such as failure of resources or interruption by high-priority jobs. Therefore, the progress of a job should be constantly monitored during the runtime to appropriately react against exceptions occurring to the job. The grid scheduler may take actions on the basis of the progress information provided by the monitoring service. Some examples of these actions include migrating, rescheduling, or terminating the job and renegotiating resources. Rescheduling takes place not only owing to resource failures but also because of some other reasons such as balancing load or increasing throughput. For example, while the job is running, if some other powerful resources that are previously occupied by other jobs become available and the grid scheduler estimates they will give better performance for the job, a decision of rescheduling may be made. Although rescheduling is an attractive alternative to improve performance, there are some challenging issues to be addressed, such as checkpointing, if migration is involved in rescheduling. It is said that migrating a job may be too costly because of time-consuming checkpointing [47].

21.3.7 Postexecution

Once the job is complete, some postexecution tasks are generally needed. For example, the output files have to be transferred to the user and any temporary files and settings have to be cleaned up [33].

21.4 Grid Scheduling Systems

In the past decade, the advent of grid computing has resulted in the development of various grid scheduling approaches with a number of significant advances. Although these approaches are substantially different, they all have the same objective, namely, efficient resource utilization and improvement of job performance. Most of these approaches have been implemented and evaluated in real grids. In addition to grid schedulers, an increasing number of local schedulers, such as the Portable Batch System (PBS) [44] and Platform Load Sharing Facility (LSF) [45], have been enabled to support grid environments. Despite these efforts, a broad range of challenging issues in grid scheduling still remains. Many of these issues are discussed in different research and working groups within the Global Grid Forum (GGF) Scheduling and Resource Management Area [48]. The primary goal of this area is to address various issues relating to resource scheduling and resource management in grid environments.

The grid scheduling systems presented in this section are merely a selected subset among many existing ones. In this section, the main characteristics and techniques of their approaches are described. At the highest level, grid scheduling can be classified into metascheduling and resource brokering. This classification along with some examples of each category is shown in Figure 21.4.

21.4.1 AppLeS

The core objective of the Application Level Scheduler (AppLeS) [49], as the name implies, is to perform application-centric scheduling in order to improve the performance of the application. A customized schedule for the application is obtained on the basis of the concept that the performance improvement of the application is the sole concern in making the scheduling decision.

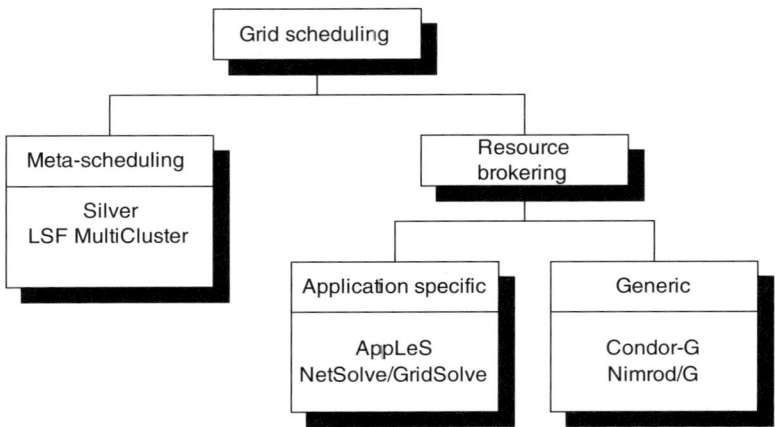

FIGURE 21.4 A taxonomy of the grid scheduling approach.

With the AppLeS, each application is tied to its own AppLeS agent that performs scheduling for the tied application. An AppLeS agent consists of four subsystems and one central coordinator. The four subsystems are the resource selector, planner, performance estimator, and actuator. The coordinator plays the central role in coordinating these four subsystems. The coordinator performs scheduling with static and dynamic application and resource information, directing different phases of scheduling to these four subsystems in turn.

Three sources of information used in AppLeS to conduct scheduling are (i) Static information, the description of the application, the preferences having been provided by the user. (ii) Dynamic information about the resources including prediction of resources states gathered with the NWS. (iii) A collection of grid application class models and application-specific models, which is used to extract general characteristics from the model to which the application belongs.

A spin-off from the AppLeS project is the development of AppLeS templates, such as the AppLeS Parameter Sweep Template (APST) [50] and Master Slave Template. The rationale behind this is to develop templates that can be used for different classes of applications; that is, each template can be used for a certain class of application, not just for a single application as is the case with the traditional AppLeS agent.

21.4.2 Condor

The Condor project [51] has made a series of efforts in developing an extensive set of mechanisms to facilitate high-throughput computing. Condor [52] is a specialized workload management system that attempts to achieve better utilization of resources, in particular, workstations owned by different users. More specifically, Condor enables poorly utilized powerful workstations owned by different individuals and small groups to be efficiently managed. The primary types of jobs that Condor targets are compute-intensive.

In Condor, a designated machine that performs scheduling is called the Central Manager (CM) that can be viewed as a matchmaker. The two types of information that the CM receives are the properties of a machine and the resource requirements of a job. The former is periodically advertised to the CM by the starter daemon, *Startd*, of the machine. The scheduler daemon, *Schedd*, of the machine from which the job is submitted, advertises the latter. These two daemons reside in each machine that Condor manages. The matchmaking that CM conducts is based on the information provided by these two daemons. Once a match is found, the rest of the steps necessary to actually run the job are taken care by the two daemons. These include establishing a connection between the resource consumer and provider, and sending the job to the resource provider.

Condor provides a number of distinctive features, such as job checkpoint and migration, remote system calls, support for multiple job models, and so forth. Condor-G [53], the job management part of Condor, is developed using the Globus toolkit to enable access to grid resources.

21.4.3 Moab Grid Scheduler (Silver)

Silver is a centralized grid scheduler that provides a rich set of sophisticated features incorporating local schedulers (e.g., Maui, PBSPro, Loadleveler, etc.). These features include a gridwide common user interface, advanced coallocation, advance reservation, intelligent load balancing, and supporting jobs spanning multiple computing systems. The core technique that Silver adopts to enable most of these functionalities is advance reservation. A centralized scheduling model of Silver also makes many scheduling tasks such as load balancing and job monitoring easier.

Silver has been designed on the basis of the following three mottos:

1. Optimal resource utilization
2. Ease of use and management
3. Maximum flexibility in global and local policies and facilities

A job is submitted using an extended version of the PBS job language (xPBS). Silver is able to guarantee the start time of the job by using advance reservations.

Note that the target system of Silver is a group of multiple computer clusters, which is similar to an intragrid.

21.4.4 NetSolve and GridSolve

Two major user groups of powerful computational resources are science and engineering professionals. The typical applications they run require specialist software packages or libraries such as Matlab [54] and Mathematica [55], and high-performance computing systems. The key to the NetSolve project [56] is to allow these users, more specifically their applications, to remotely access various software resources available on powerful machines in a network. Users in the NetSolve system simply launch their applications in the traditional manner without knowing anything about computer networking or distributed computing. NetSolve takes care of the technical details of running applications over diverse resources in a network.

NetSolve is a client–server application that enables remote access to dispersed computational resources using the TCP/IP protocol. The primary focus of the NetSolve system is to enable NetSolve clients to access scientific packages or libraries in NetSolve servers. These servers are normally more sophisticated and powerful machines. The NetSolve client accesses these software resources by sending a remote procedure call to the server that can provide the best service for the client's request.

The NetSolve system is composed of three major components: the client, the server, and the agent. A NetSolve client is a user's application written in C or FORTRAN, or using Matlab or the Web with the NetSolve client library linked in. It sends a request to the NetSolve agent that attempts to find the best NetSolve servers for the request. The NetSolve agent consists of a series of functionalities, such as resource discovery, load balancing, resource allocation, and fault tolerance. The agent returns to the client with a list of servers that are capable of servicing the request. The client then contacts a server and sends input parameters consisting of a problem name (e.g., dgesv()), the name of a LAPACK [57] subroutine, and input data. Once the server completes the request, it returns either the result or error status to the client.

The GridSolve project [58] is another effort under way from the NetSolve project. It can be seen as the next generation of NetSolve that will allow the broad scientific community to benefit from grid computing. GridSolve uses a remote procedure call mechanism redesigned for grid computing, GridRPC [59].

21.4.5 Nimrod/G

Nimrod/G [60] is a grid-enabled version of Nimrod [61] that performs resource brokering for parametric modeling experiments based on a model of computational economy. In Nimrod/G, a parametric experiment, that is, a parameter sweep application, is described by using a declarative parametric modeling language or the Clustor GUI [62]. Nimrod/G uses a computational economy, meaning that an application is submitted with a deadline and a price, and is allocated to a resource or a resource set that meets the deadline with minimum cost. The deadline denotes the latest completion time. The deadline of the application and the price are specified by the user. In addition to these user-specified parameters, another key parameter in the computational economy is resource cost set by the resource provider. Hence, three most influential factors on resource scheduling in Nimrod/G are deadline, budget, and resource cost.

Nimrod/G consists of five main components:

1. The client or user station is a user interface that is capable of initiating, controlling, and monitoring experiments. The user is also allowed to control an experiment at different locations by running multiple instances of the same client.
2. The second component is the parametric engine, the core part of the Nimrod/G system that controls the whole experiment coordinating the other four components.
3. The actual scheduling processes, such as resource discovery and resource allocation, are performed by the scheduler using the Globus middleware services, such as MDS.
4. The dispatcher delegates a task to the job wrapper on the selected resource for the execution.
5. A job wrapper is a program that is responsible for the preparation, execution, and postexecution of the task.

21.5 Scheduling Algorithms for Grids

As illustrated in the previous section, a grid scheduling system typically consists of a series of components, such as resource discovery and allocation, and monitoring. Although all these components are playing crucial roles for delivering high quality schedules, the greatest attention drawn in grid scheduling is probably scheduling algorithms or task/resource mapping schemes. Owing to the diversity of applications deployed in grids, it is quite challenging for a single scheduling algorithm to deal with various types of applications. Owing to the NP-complete nature of the task scheduling problem, in most cases [63], heuristic algorithms are usually employed.

In this section, three well-known scheduling algorithms for grids are presented. They are selected, among many other previously proposed scheduling algorithms, on the basis of both their simplicity and practicality.

21.5.1 List Scheduling with Round-Robin Order Replication

Round-robin order replication (RR) [64] is a grid scheduling algorithm for independent coarse-grained tasks. As the name implies, its distinctiveness comes from the RR scheme that makes replicas of running tasks in a round-robin fashion after conducting list scheduling for all the unscheduled tasks. RR first randomly assigns a task to each host in the grid and then waits until one or more of those assigned hosts complete their tasks. On the completion of a task, the next unscheduled task is dispatched to the host on which the completed task has run. This tends to result in fast resources getting more tasks. Once all the tasks are dispatched, RR starts replicating running tasks hoping that some or all of these replicas finish earlier than their originals. Note that RR performs scheduling without any dynamic information on resources and tasks. Nevertheless, the algorithm is compelling and comparable to other scheduling heuristics that require such performance information.

21.5.2 Storage Affinity

The storage affinity (SA) algorithm [65] primarily aims at the minimization of data transfer by making scheduling decisions incorporating the location of data previously transferred. In addition, it considers task replication as soon as a host becomes available between the time the last unscheduled task gets assigned and the time the last running task completes its execution.

SA determines task/host assignments on the basis of "the storage affinity metric." The SA of a task to a host is the amount of the task's input data already stored in the site to which the host belongs. Although the scheduling decision SA makes is between task and host, SA is calculated between task and site. This is because in the grid model used for SA, each site in the grid uses a single data repository that can be fairly accessible by the hosts in the site.

For each scheduling decision, SA calculates SA values of all unscheduled tasks and dispatches the task with the largest SA value. If none of the tasks has a positive SA value, one of them is scheduled at random. By the time this initial scheduling gets completed, there would be a running task on each host in the grid. On the completion of any of these running tasks, SA starts task replication. Now each of the remaining running tasks is considered for replication and the best one is selected. The selection decision is based on the SA value and the number of replicas.

21.5.3 XSufferage

XSufferage extends the Sufferage scheduling heuristic [34] by taking site-level data sharing into account. It makes scheduling decisions on the basis of the sufferage value of tasks. The sufferage value of a task in XSufferage is defined as the difference between its earliest site-level completion time and its second earliest site-level completion time. While sufferage values used in Sufferage are host level, those adopted by XSufferage are site level. The sufferage value of a task is used as a measure of possible increase on makespan; that is, a task with a large sufferage value implies that the completion time of the task seriously increases, causing a possible increase of makespan, if it is not assigned to the host on which the earliest site-level completion time is attainable. Therefore, the larger the sufferage value of a task the higher scheduling priority the task gets.

At each scheduling decision, it computes sufferage values of all the unscheduled tasks and schedules the task whose sufferage value is the largest. This approach can be effective in that a serious increase of makespan is reduced. However, this does not guarantee that the overall makespan is shortened. It should be noted that XSufferage makes an arguable assumption that it has access to 100% accurate performance information on resources and tasks.

21.6 Grid Simulation Tools

It should be noted that when new grid scheduling schemes are designed they are typically implemented and deployed in real grid environments for performance evaluation. This intuitively implies their evaluation studies are subject to the underlying grid environments. Moreover, states of resources in a grid fluctuate dynamically over time. Therefore, results obtained from an evaluation study are not repeatable. This fact also makes comparison between different grid scheduling approaches difficult. Simulations are, therefore, an attractive alternative in that the simulation of repeatable and controllable environments is feasible. Simulations enable different grid scheduling schemes to be tested over diverse virtual grid environments with a broad range of scenarios.

As grid scheduling has been actively studied, some tools that enable the simulation of grids have been developed. Five well-known grid simulation tools include Bricks [66], GridSim [67], MicroGrid [68,69], OptorSim [70], and SimGrid [71]. These simulation tools have the common objective of developing and evaluating scheduling algorithms for grid environments.

21.6.1 Bricks

Bricks is a client–server paradigm-based performance evaluation system for grid scheduling algorithms developed at the Tokyo Institute of Technology. Bricks aims to facilitate analysis and evaluation of various grid scheduling schemes.

The two major components of Bricks are the simulated grid environment and the scheduling unit. In the evaluation of a scheduling approach, the former is in charge of simulating a grid environment and actual scheduling-related processes are taken care of by the latter. Both components can be easily customized by the Bricks script or user-supplied modules.

The simulated grid environment consists of three entities: client, network, and server. The characteristics and behaviors of these entities can be dynamically specified using the Bricks script.

The five subcomponents in the Scheduling Unit are the network monitor, the server monitor, the resourceDB, the predictor, and the scheduler. Their functions are implicitly indicated by their names. The main task performed by the first two of these components is gathering dynamic information by monitoring network and servers. More specifically, they monitor network and servers, gather dynamic information such as network bandwidth and load of servers, and send the gathered information to the resourceDB. The resourceDB is a GIS in the Bricks system that is queried by the predictor and scheduler to make scheduling decisions. Any of these subcomponents is replaceable with a user-supplied Java module.

21.6.2 GridSim

GridSim is a Java-based discrete-event grid simulation toolkit that provides facilities for modeling and simulation of grid entities such as resources, users, and application models, to design and evaluate scheduling schemes for grid environments. GridSim is implemented on the basis of SimJava [72], a discrete-event simulation package implemented in Java. Simulations using GridSim are enabled by coordinating entities (e.g., resources, brokers, network based I/O, information service, and statistics). These entities run in parallel in their own threads.

GridSim is composed of five layers: (i) The bottom layer is Java Virtual Machine (JVM) that is responsible for actually running Java bytecode. (ii) SimJava sits on top of JVM providing discrete-event simulation functionalities. (iii) The GridSim toolkit as the third layer simulates Grid environments modeling resources and fundamental grid services (e.g., GIS). (iv) In the fourth layer, grid schedulers or resource brokers are simulated. (v) Modeling basic simulation entities such as applications and users are placed at the top layer. The fourth and fifth layers are enabled using the GridSim toolkit.

21.6.3 MicroGrid

MicroGrid is a set of simulation tools that enable virtualized grid environments to be created. It aims to facilitate the accurate simulation of controlled and repeatable grid environments in order to conduct various studies on grid-related issues such as grid topologies, middleware, and application performance. Unlike other grid simulation tools presented in this chapter, MicroGrid is an emulator, meaning that it creates a virtual grid on which real applications, including Globus applications, actually run. Therefore, one can expect to get more precise results than those produced with other grid simulation tools.

The simulation tools provided from MicroGrid support core elements of the grid such as resources and information service being virtualized. The two main components that are in charge of virtualizing grid resources are the MicroGrid network online simulator (MaSSF) and CPU controller. Grid resources modeled and simulated by these components are network resources and compute resources, respectively. The interception and/or manipulation of live network streams and processes play crucial roles for the simulation of the behaviors of these two resource types. More specifically, the socket and other I/O related function calls (e.g., send, recv, fread, fwrite, etc.) are intercepted and redirected appropriately to effectively simulate network behaviors. In addition, processes of a virtual machine may be suspended to ensure the correct behaviors of the machine.

The number of virtual machines that MicroGrid emulates is not limited by the number of physical machines. For example, one can virtualize a grid with hundreds of machines on ten physical machines. This is made possible by using the virtual Internet Protocol (IP) mapping service. In other words, each physical machine is mapped with one or more virtual machines.

One obvious shortcoming of MicroGrid is the amount of time needed for the MicroGrid simulation due to running actual applications. It should also be noted that MicroGrid does not support multiprocessor virtual machines.

21.6.4 OptorSim

An increasing number of applications in many fields of science and engineering are required to deal with a massive amount of data. These large-scale data-intensive applications are normally deployed in data grids. When running this type of application, an efficient data access and replication strategy significantly alleviates the burden of data transfer, which makes a major impact on the completion time of the application.

OptorSim is a simulation package that provides a series of grid simulation features including simulating background network traffic. The primary intent of the development of OptorSim is the evaluation of data replication strategies in grid environments. The simulation model it adopts is time based instead of event driven. Similar to GridSim and SimGrid resources, computing resources and the scheduling entity (i.e., resource broker) are simulated using threads.

A simulation using OptorSim is configured by three major input files and additional simulation properties specified in a parameters file. The three main configuration files characterize the grid and its resources, the job, and the background network traffic. Other simulation details that can be dictated in the parameters file include the number of jobs, the interarrival time of the job, the scheduling strategy of the resource broker, the replication optimization algorithm, and the data access pattern. While the scheduling strategy determines which job is assigned onto which resource, the replication optimization algorithm makes decisions on when and where a particular data file is replicated and/or deleted.

Currently, OptorSim only supports large sequential jobs and does not consider resource failures.

21.6.5 SimGrid

SimGrid is a simulation toolkit that facilitates the development and evaluation of scheduling algorithms for heterogeneous, distributed computing environments (e.g., computational grids). Two versions of SimGrid have been implemented. The first version of SimGrid [73], SimGrid v1 provides a low-level interface called SG that requires the user to specify fine details to conduct the simulation. The second and latest version of SimGrid, SimGrid v2, introduces a high-level interface, MSG, that is implemented on the basis of SG. Although SG gives more flexibility, explicit description and configuration work for the simulation are required. Moreover, its use is more focused on the simulation of specific application domains (e.g., precedence-constraint applications and/or computing environment topologies). These drawbacks are overcome in SimGrid v2 with the abstraction of foundational entities such as agents, locations, tasks, paths, and channels. Agents are responsible for scheduling. Locations, and paths and channels are compute and network entities, respectively. Hereafter, SimGrid denotes the second version.

SimGrid allows one to model and simulate grid environments with synthetic data and/or data collected from a real environment. It should be noted that the latter certainly enables more realistic simulations. The two primary sources of data obtained from a real environment used in SimGrid are platform descriptions and network traffic traces. They are obtained with Effective Network View (ENV) [74] and NWS. The implementation of bandwidth-sharing models for simulating contention among data transfers also makes realistic simulations possible. Two bandwidth-sharing models that are supported in SimGrid are the packet-level network simulation model and the macroscopic model.

21.7 Conclusions

This chapter briefly reviewed the problem of scheduling for grid computing environments. It was pointed out that grid scheduling requires a series of challenging tasks. These include searching for resources in collections of geographically distributed heterogeneous computing systems and making scheduling decisions taking into consideration quality of service. A grid scheduler differs from a scheduler for conventional computing systems in several respects that were highlighted in the chapter. Also, a number of simulation tools and development environments were examined.

References

[1] I. Foster, C. Kesselman (eds.), *The Grid: Blueprint for a Future Computing Infrastructure*, Morgan Kaufmann, USA, 1999.

[2] I. Foster, C. Kesselman, Globus: A metacomputing infrastructure toolkit, *International Journal of Supercomputer Applications*, 11:115–128, 1997.

[3] A. Grimshaw, W. Wulf, The legion vision of a worldwide virtual computer, *Communications of the ACM*, 40:39–45, 1997.

[4] F. Berman et al. [Authors are Berman, Chien, Cooper, Dongarra, Foster, Gannon, Johnsson, Kennedy, Kesselman, Mellor-Crummey, Reed, Torczon, and Wolski], TheGrADS project: Software support for high-level grid application development, *International Journal of High Performance Computing Applications*, 15:327–344, Winter 2001.

[5] R. Stevens, P. Woodward, T. DeFanti, C. Catlett, From the I-WAY to the National Technology Grid, *Communications of the ACM*, 40:50–61, 1997.

[6] W. E. Johnston, D. Gannon, B. Nitzberg, Grids as production computing environments: The engineering aspects of NASA's Information Power Grid, *Proceedings of the 8th International Symposium on High Performance Distributed Computing*, August 3–6, 1999, pp. 197–204.

[7] D. A. Reed, Grids, the TeraGrid and beyond, *IEEE Computer*, 36:62–68, January 2003.

[8] P. Eerola, et al., The Nordugrid production grid infrastructure, status and plans, *Proceedings of the 4th International Workshop on Grid Computing*, November 17, 2003, pp. 158–165.

[9] A. Ghiselli, DataGrid Prototype 1, *Proceedings of the TERENA Networking Conference*, June 3–6, 2002.

[10] I. Foster, C. Kesselman, S. Tuecke, The anatomy of the grid: Enabling scalable virtual organizations, *International Journal of Supercomputer Applications*, 15:200–222, 2001.

[11] W. Smith, I. Foster, V. Taylor, Scheduling with advanced reservations, *Proceedings of the 14th International Parallel and Distributed Processing Symposium*, IEEE Computer Society, Cancun, May 1–5, 2000, pp. 127–132.

[12] R. M. Badia, J. Labarta, J. Gimenez, F. Escale, DIMEMAS: Predicting MPI applications behavior in Grid environments, Workshop on Grid Applications and Programming Tools (GGF8), 2003.

[13] G. R. Nudd, D. J. Kerbyson, E. Papaefstathiou, S. C. Perry, J. S. Harper, D. V. Wilcox, Pace: A toolset for the performance prediction of parallel and distributed systems, *International Journal of High Performance Computing Applications*, 14:228–251, 1999.

[14] I. Foster, C. Kesselman (eds.), *The Grid 2: Blueprint for a Future Computing Infrastructure*, Elsevier Inc., USA, 2004.

[15] K. Cooper et al. [Authors are Cooper, Dasgupta, Kennedy, Koelbel, Mandal, Marin, Mazina, Mellor-Crummey, Berman, Casanova, Chien, Dail, Liu, Olugbile, Sievert, Xia, Johnsson, Liu, Patel, Reed, Deng, Mendes, Shi, YarKhan, and Dongarra], New Grid Scheduling and Rescheduling Methods in the GrADS Project, *Proceedings of the International Parallel and Distributed Processing Symposium*, IEEE Computer Society, Santa Fe, 2004, pp. 199–206.

[16] D. Zhou, V. Lo, Wave scheduler: Scheduling for faster turnaround time in peer-based desktop grid systems, *Proceedings of the Job Scheduling Strategies for Parallel ProcessingWorkshop*, Feitelson, D., Frachtenberg, E., Rudolph, L., and Schwiegelshohn, U., Eds., Springer-Verlag GmbH, Cambridge, MA, 2005, pp. 194–218.

[17] M. Pinedo, *Scheduling Theory, Algorithms and Systems*, First Edition, Prentice Hall, 1995.
[18] A. Grama, A. Gupta, G. Karypis, V. Kumar, *Introduction to Parallel Computing*, Second Edition, Addison Wesley, 2003.
[19] H. Dail, H. Casanova, F. Berman, A decoupled scheduling approach for the grads program development environment, *Proceedings of the 2002 ACM/IEEE Conference on Supercomputing*, IEEE Computer Society, Baltimore, 2002, pp. 1–14.
[20] Globus Monitoring and Discovering Service (MDS), http://www.globus.org/toolkit/mds/
[21] Moab Grid Scheduler (Sliver), http://www.clusterresources.com/products/mgs/docs/index.shtml
[22] Maui Cluster Scheduler, http://www.clusterresources.com/pages/products/maui-cluster-scheduler.php
[23] S. F. Altschul, W. Gish, W. Miller, E. W. Myers, D. J. Lipman, Basic local alignment search tool, *Journal of Molecular Biology*, 1:403–410, 1990.
[24] J. Stiles, T. Bartol, E. Salpeter, M. Salpeter, Monte Carlo simulation of neuromuscular transmitter release using MCell, a general simulator of cellular physiological processes, *Computational Neuroscience*, 279–284, 1998.
[25] S. Rogers, D. Ywak, Steady and unsteady solutions of the incompressible Navier–Stokes equations, *AIAA Journal*, 29:603–610, 1991.
[26] B. Barish, R.Weiss, LIGO and detection of gravitational waves, *Physics Today*, 50, 1999.
[27] S. Hastings, et al. [Authors are Hastings, Kurç, Langella, Çatalyürek, Pan, and Saltz], Image processing on the Grid: A toolkit or building grid-enabled image processing applications, *Proceedings of the International Symposium of Cluster Computing and the Grid*, IEEE Computer Society, Tokyo, 2003, pp. 36–43.
[28] G. B. Berriman, et al. [Authors are Berriman, Good, Laity, Bergou, Jacob, Katz, Deelman, Kesselman, Singh, Su, and Williams], Montage a grid enabled image mosaic service for the national virtual observatory, *Proceedings of the conference on Astronomical Data Analysis Software and Systems XIII*, Astronomical Society of the Pacific, Strasbourg, Ochsenbein, F., Allen, M., and Egret, D., Eds., 2003, pp. 593–596.
[29] Q. Snell, K. Tew, J. Ekstrom, M. Clement, An enterprise-based grid resource management system, *Proceedings of the 11th IEEE International Symposium on High Performance Distributed Computing*, IEEE Computer Society, Edinburgh, July 23–26, 2002, pp. 83–90.
[30] J. Cao, D. J. Kerbyson, G. R. Nudd, Performance evaluation of an agent-based resource management infrastructure for grid computing, *Proceedings of the 1st IEEE/ACM International Symposium on Cluster Computing and the Grid*, IEEE Computer Society, Brisbane, May 15–18, 2001, pp. 311–318.
[31] K. Krauter, R. Buyya, M. Maheswaran, Ataxonomy and survey of grid resource management systems for distributed computing, *International Journal of Software: Practice and Experience*, 32:135–164, 2002.
[32] L. Zhang, Scheduling algorithm for real-time applications in grid environment, *Proceedings of the IEEE International Conference on Systems, Man and Cybernetics*, Volume 5, October 6–9, 2002.
[33] J. M. Schopf, Ten Actions When Grid Scheduling, in *Grid ResourceManagement: State of the Art and Future Trends*, J. Nabrzyski, J. M. Schopf, J. Weglarz, Ed., Kluwer Academic Publishers, 2003, chap. 2.
[34] H. Casanova, A. Legrand, D. Zagorodnov, F. Berman, Heuristics for scheduling parameter sweep applications in grid environments, *Proceedings of the 9th Heterogeneous Computing Workshop (HCW)*, IEEE Computer Society, Cancun, 2000, pp. 349–363.
[35] R. Raman, M. Livny, M. Solomon, Matchmaking: distributed resource management for high throughput computing, *Proceedings of the 7th International Symposium on High Performance Distributed Computing*, IEEE Computer Society, Chicago July 28–31, 1998, pp. 140–146.
[36] W. Smith, I. Foster, V. Taylor, Predicting application run times using historical information, *Proceedings of the Workshop on Job Scheduling Strategies for Parallel Processing*, Springer-Verlag, Orlando, 1998, pp. 122–142.

[37] S. Krishnaswamy, A. Zaslasvky, S. W. Loke, Predicting application run times using rough sets, *Proceedings of the 9th International Conference on Information Processing and Management of Uncertainty in Knowledge-based Systems (IPMU 2002)*, Annecy, France, July, IEEE Press, 2002, pp. 455–462.

[38] R. Wolski, Dynamically forecasting network performance using the network weather service, *Cluster Computing*, 1:119–132, January 1998.

[39] R. Wolski, N. T. Spring, J. Hayes, The network weather service: A distributed resource performance forecasting system, *Journal of Future Generation Computing Systems*, 15:757–768, 1999.

[40] The network weather service home page, http://nws.cs.ucsb.edu

[41] I. Foster et al. [Authors are Foster, Kesselman, Lee, Lindell, Nahrstedt, and Roy], A distributed resource management architecture that supports advance reservations and co-allocation, *Proceedings of the International Workshop on Quality of Service*, IEEE Computer Society, June 1999, pp. 27–36.

[42] I. Foster et al. [Authors are Foster, Roy, Sander, and Winkler], End-to-end quality of service for high-end applications, tech. rep., Argonne National Laboratory, 1999. http://www.mcs.anl.gov/qos/qos-papers.htm

[43] I. Foster, A. Roy, V. Sander, A quality of service architecture that combines resource reservation and application adaptation, *Proceedings of the International Workshop on Quality of Service*, IEEE Computer Society, Pittsburgh, 2000, pp. 181–188.

[44] PBSPro, http://www.pbspro.com/features.html

[45] Platform LSF, http://www.platform.com/products/LSF/

[46] W. Allcock et al. [Authors are Allcock, Bester, Bresnahan, Chervenak, Liming, and Tuecke], GridFTP: Protocol Extension to FTP for the Grid, Grid Forum Internet-Draft, March 2001.

[47] J. MacLaren et al. [Authors are MacLaren, Sakellariou, Krishnakumar, Garibaldi, and Ouelhadj], Towards service level agreement based scheduling on the Grid, position paper in the Workshop on Planning and Scheduling for Web and Grid Services, 2004, www.isi.edu/ikcap/icaps04-workshop/final/maclaren.pdf

[48] Global Grid Forum Scheduling and Resource Management Area, http://www-unix.mcs.anl.gov/~schopf/ggf-sched/

[49] F. Berman, R. Wolski, The AppLeS project: A status report, *8th NEC Research Symposium*, Berlin, Germany, May 1997.

[50] H. Casanova et al. [Authors are Casanova, Obertelli, Bermand, and Wolski], The AppLeS Parameter sweep template: User-level middleware for the Grid, *Scientific Programming*, 8:111–126, August 2000.

[51] The Condor Project, University of Wisconsin Condor: High throughput computing, http://www.cs.wisc.edu/condor/

[52] M. J. Litzkow, M. Livny, M. W. Mutka, Condor—A hunter of idle workstations, *Proceedings of the 8th International Conference on Distributed Computer Systems*, IEEE Computer Society, San Jose, 1988, pp. 104–111.

[53] J. Frey et al. [Authors are Frey, Tannenbaum, Livny, Foster, and Tuecke], Condor-G: A computation management agent for multi-institutional grids, *Cluster Computing*, 5:237–246, July, 2002.

[54] The MathWorks, Inc., *MATLAB Reference Guide*, 1992.

[55] *The Mathematica Book*, Third Edition, Wolfram Media, Inc. and Cambridge University Press, 1996.

[56] H. Casanova, J. Dongarra, Netsolve: A network-enabled server for solving computational science problems, *International Journal of Supercomputer Applications and High Performance Computing*, 11:212–223, Fall 1997.

[57] E. Anderson et al. [Authors are Anderson, Bai, Bischof, Blackford, Demmel, Dongarra, Du Croz, Greenbaum, Hammarling, McKenney, and Sorensen], *LAPACK Users' Guide*, Third Edition, SIAM, 1999.

[58] Project Description—GridSolve: A system for Grid-enabling general purpose scientific computing environments, http://icl.cs.utk.edu/netsolvedev/files/gridsolve/GridSolve-description.pdf

HANDBOOK OF PARALLEL COMPUTING: MODELS,
ALGORITHMS AND APPLICATIONS. ED. BY SANGUTHEVAR
RAJASEKARAN.
BOCA RATON: CHAPMAN & HALL/CRC, 2008   Cloth   1224 P.
SER: CHAPMAN & HALL/CRC COMPUTER AND INFORMATION
SCIENCE SERIES
ED: UNIVERSITY OF CONNECTICUT. NEW COLLECTION.

LCCN 2007-025276
ISBN 1584886234

181 UNIV OF MAINE/ORONO   APPROVALS   QA76.58
DATE 2/20/08 OSTA   5718-11   139.95
                                                                                LIST 139.95
                                                                                DISC 0%
                                                                                NET 139.95

YBP - CONTOOCOOK, N.H. 03229

SUBJ: PARALLEL PROCESSING (ELECTRONIC COMPUTERS)--
HANDBOOKS, MANUALS, ETC.

CLASS QA76.58   DEWEY# 005.1   LEVEL ADV-AC

2587

[59] K. Seymour et al. [Authors are Seymour, Nakada, Matsuoka, Dongarra, Lee, and Casanova], Overview of GridRPC: A remote procedure call API for Grid computing, *Proceedings of the Third International Workshop on Grid Computing*, Springer-Verlag, Baltimore, MD, USA, November 18, 2002, pp. 274–278.

[60] R. Buyya, D. Abramson, J. Giddy, Nimrod/G: An architecture for a resource management and scheduling system in a global computational grid, *Proceedings of HPC ASIA '2000*, China, IEEE Computer Society, 2000, pp. 283–289.

[61] D. Abramson, R. Sosic, J. Giddy, B. Hall, Nimrod: A tool for performing parameterized simulations using distributed workstations, *Fourth IEEE International Symposium on High Performance Distributed Computing*, IEEE Computer Society, Washington, August 1995, pp. 112–121.

[62] Clustor Manual, *Writing Job Plans*, Chapter 4, 1999, http://www.activetools.com/manhtml20/plans.htm

[63] M. R. Garey, D. S. Johnson, *Computers and Intractability: A Guide to the Theory of NP-Completeness*, W. H. Freeman and Co., 1979.

[64] N. Fujimoto, K. Hagihara, Near-optimal dynamic task scheduling of independent coarse-grained tasks onto a computational grid, *Proceedings of the International Conference on Parallel Processing*, IEEE Computer Society, Taiwan, 2003, pp. 391–398.

[65] E. Santos-Neto et al. [Authors are Santos-Neto, Cirne, Brasileiro, and Lima.], Exploiting replication and data reuse to efficiently schedule data-intensive applications on Grids, *Proceedings of the 10th Workshop on Job Scheduling Strategies for Parallel Processing*, Springer Berlin, New York, 2004, pp. 210–232.

[66] A. Takefusa [Authors are Takefusa, Matsuoka, Nakada, Aida, and Nagashima], Overview of a performance evaluation system for global computing scheduling algorithms, *Proceedings of the 8th IEEE International Symposium on High Performance Distributing Computing*, IEEE Computer Society, Redondo Beach, 1999, pp. 97–104.

[67] R. Buyya, M. Murshed, GridSim: A toolkit for the modeling and simulation of distributed resource management and scheduling for Grid computing, *Journal of Concurrency and Computation: Practice and Experience*, 14:1507–1542, 2002.

[68] H. J. Song et al. [Authors are Song, Liu, Jakobsen, Bhagwan, Zhang, Taura, and Chien], The MicroGrid: A scientific tool for modeling computational grids, *Scientific Programming*, 8:127–141, 2000.

[69] H. Xia, H. Dail, H. Casanova, A. Chien, The MicroGrid: Using emulation to predict application performance in diverse grid network environments, *Proceedings of the Workshop on Challenges of Large Applications in Distributed Environments*, IEEE Press, Honolulu, Hawaii, June 2004, pp. 52–61.

[70] W. H. Bell, et al. [Authors are Bell, Cameron, Capozza, Millar, Stockinger, Zini.], Optorsim—A grid simulator for studying dynamic data replication strategies, *International Journal of High Performance Computing Applications*, 17:403–416, 2003.

[71] A. Legrand, L. Marchal, H. Casanova, Scheduling distributed applications: The SimGrid simulation framework, *Proceedings of the 3rd IEEE/ACM International Symposium on Cluster Computing and the Grid*, Tokyo, Japan, May 12–15, 2003, pp. 138–145.

[72] F. Howell R. McNab, Simjava, http://www.dcs.ed.ac.uk/home/hase/simjava/

[73] H. Casanova, Simgrid: A toolkit for the simulation of application scheduling, *Proceedings of the 1st IEEE/ACM International Symposium on Cluster Computing and the Grid*, IEEE Computer Society, Brisbane, Australia, May 15–18, 2001, pp. 430–437.

[74] G. Shao, F. Berman, R. Wolski, Using effective network views to promote distributed application performance, *Proceedings of the International Conference on Parallel and Distributed Processing Techniques and Applications*, CSREA Press, Las Vegas, June 1999, pp. 2649–2656.

22
QoS Scheduling in Network and Storage Systems

22.1	Introduction...	22-1
22.2	Queuing Model	22-3
22.3	Fluid Flow Server Model	22-4
	Fair Queuing in the Fluid Flow Model • Weighted Fair Queuing in the Fluid Flow Model	
22.4	Discrete Server Model	22-8
	Service Allocation • Latency in a Discrete Server Model • Weighted Fair Queuing • Performance Analysis of Discrete Model Schedulers • Advanced Schedulers	
22.5	Parallel Server Models	22-17
	Fair Lexicographic Scheduling • Multiple Uniform Servers	
22.6	Bibliographic References	22-20
	Acknowledgment...	22-21
	References	22-21

Peter J. Varman
Ajay Gulati
Rice University

22.1 Introduction

Resource sharing is a ubiquitous facet of computer systems. Time-sharing systems, virtual memory, and I/O device virtualization were among the early innovations by which physical resources such as the processor, primary memory, mass storage devices, and input/output units like card readers and printers were shared among concurrent applications. The primary motivation for these solutions was to increase the utilization of expensive hardware and amortize their cost over many concurrent users. Continuing evolution in technology, computing environments, and applications have dramatically changed the cost, requirements, and performance profiles of computer systems. Nevertheless, resource sharing remains a fundamental characteristic of modern computing. Cluster computing systems, for instance, must manage hundreds of processors, storage devices, and network connections on behalf of several concurrent applications, while giving each the abstraction of executing in isolation on a dedicated supercomputer. A router in a communication network is accessed by multiple traffic streams and must multiplex both its internal resources and the bandwidth of the shared communication link equitably among the streams to provide all users with acceptable rates of service. In data management applications, huge repositories of shared data are stored on thousands of storage devices, and accessed by hundreds of concurrent applications that query or update portions of the data. The storage manager needs to coordinate the traffic to and from the devices to ensure data integrity and provide performance guarantees.

Large consolidated data centers and supercomputer installations provide a shared base of hardware and software for use by multiple clients. The advantage of this approach is the amortization of both the capital costs and the recurring costs of system maintenance and administration. However, the cost advantages need to be traded off against the greater security and performance predictability inherent in a dedicated system. In terms of performance, users expect application execution times and query response times to be consistent across different invocations. A large drop in performance due to increased system load or due to aberrant behavior of other concurrent applications is a strong deterrent to adopting a consolidated server approach. A fundamental requirement for effective resource sharing is therefore one of *performance isolation*. The service provider should guarantee a certain contractually agreed-upon service to a client at all times, independent of the behavior of other clients that may violate their end of the service agreement. The server must be able to guarantee this Quality-of-Service (QoS) to each client using appropriate runtime system management policies. The abstract interface presented to a client in terms of the bandwidth and response time guarantees is referred to as *performance virtualization*, and is enforced by admission control and resource scheduling policies at the server.

In certain environments, resource sharing can be achieved by statically partitioning the resources physically among the concurrent clients. Multiprocessor systems, for instance, are often space multiplexed, with each task receiving a dedicated subset of the processors and system memory, and, to the extent possible, dedicated storage and inter-processor network bandwidth. However, static space multiplexing has several limitations: (i) it may not be possible to physically partition the resource like a single, high-performance processor, or a communication link, (ii) static physical partitioning requires accurate a priori estimates of resource requirements, (iii) the allocation must provision for the worst-case and does not exploit statistical multiplexing due to temporal variations in resource usage, (iv) the need to share huge amounts of dynamically changing information encourages the use of shared devices rather than replicated, dedicated data storage. A viable solution must be able to dynamically schedule resources so as to obtain high system utilization while simultaneously meeting all QoS guarantees.

To illustrate the issues arising in performance isolation consider sharing a single resource using two different scheduling policies: First Come First Served (FCFS) and Time-Division Multiplexing (TDM). We can assume that all requests are of the same size. In the case of FCFS resource scheduling, a client can make a burst of requests at some instant, and thereby lock out other clients during the period that its burst is being served. Performance isolation implies that this sort of negative influence on the performance of well-behaved clients by unexpected actions of others should be avoided. In TDM, time is divided into fixed-length slots and statically assigned to clients in a round-robin manner. Such a schedule automatically spreads out the requests that come in a burst, and prevents it from starving other clients. However, because the slots are statically assigned, a slot is wasted if a client does not have a request to schedule at that time. While TDM can isolate the clients effectively it does so at the price of under utilizing the resources.

The twin concerns of performance guarantees and resource utilization appear in several contexts. In networking systems, for instance, circuit switching can provide very tight performance guarantees but can cause poor resource utilization. On the other hand, packet switching uses statistical multiplexing leading to very high utilization but the performance guarantees are quite loose. The problem gets aggravated when a particular traffic stream goes through multiple routers connected in some arbitrary topology. IO workloads typically have unknown correlated service times and often exhibit skewed patterns of access that cause temporal hot spot activity on a subset of the resources, aggravating the scheduling problem. In addition, while the performance benefits of caching makes it ubiquitous in data management systems, defining fair service in the presence of caching is a difficult issue. All these can cause unpredictable delays and increase the complexity of the scheduling algorithms.

In this chapter we will survey a set of scheduling algorithms that work for different models of sharing, resources, and request requirements. First we describe an idealized fluid flow model of fair service, followed by a discussion of practical schedulers using discrete service models such as WFQ, SC, and

QoS Scheduling in Network and Storage Systems

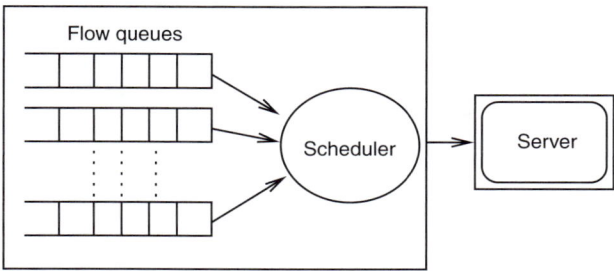

FIGURE 22.1 Queuing system model.

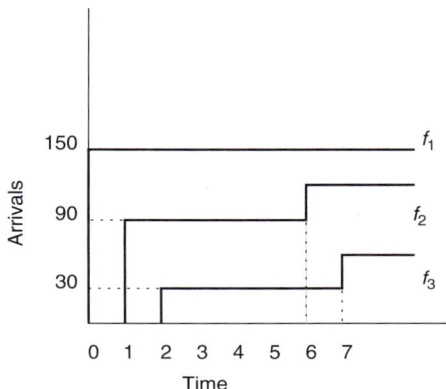

FIGURE 22.2 Cumulative arrival functions.

WF^2Q. Finally, we consider algorithms for parallel storage servers where additional resource constraints need to be considered. The references to the original works are presented in Section 22.6.

22.2 Queuing Model

A generic model of the queuing system is shown in Figure 22.1. Service is provided by a server of **capacity** C units/sec. The actual service units depend on the application: typical units are CPU instructions executed, bytes transmitted, or I/O operations completed. Applications are modeled by **flows** that make requests for service. Requests from a flow are held in a **flow queue** associated with the flow, till they are dispatched to the server. The requests of a flow are served in FCFS order. However, the order in which requests from different flows are serviced depends on the scheduling algorithm. In general, requests can be of arbitrary size up to some maximum limit, and may arrive at arbitrary time instants. A flow is called **active** if it has at least one request in the system; the request may be either in service or in the flow queue. An active flow will also be called a **backlogged** flow. A flow that is not active is **idle**. At any time instant some flows will be active whereas the others will be idle. The set of active flows at time t is denoted by $\mathcal{A}(t)$, and the number of such flows, $|\mathcal{A}(t)|$, by $n(t)$. Flow f_i is said to be backlogged in the interval $[t_1, t_2]$ if it is backlogged at all time instants $t \in [t_1, t_2]$. The system is said to be backlogged at time t if there is at least one backlogged flow at that time. A server is said to be **work-conserving** if its capacity is completely utilized whenever the system is backlogged. We will assume work-conserving servers unless specified otherwise. The maximum value of $|\mathcal{A}|$ which is the maximum number of flows that may be simultaneously active is denoted by N.

A flow f_i is characterized by its cumulative arrival and service functions. The **cumulative arrival function** $CA_i(t)$ is the total amount of service requested by f_i in the interval $[0, t]$, and the **cumulative service function** $CS_i(t)$ is the total amount of service provided to the flow in $[0, t]$. Figure 22.2 shows

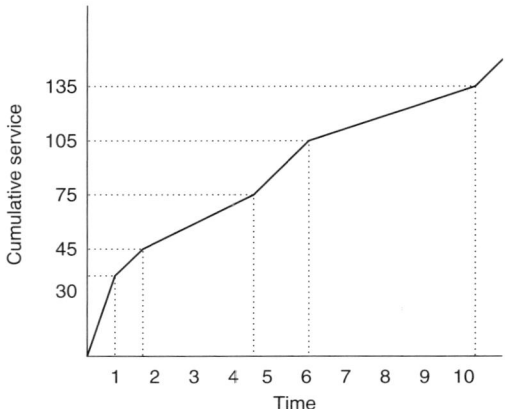

FIGURE 22.3 Cumulative service function for flow f_1.

example arrival functions for three flows f_1, f_2, and f_3. f_1 has a request of size 150 units arriving at $t = 0$. f_2 has two requests of size 90 and 30 units arriving at $t = 1$ and $t = 6$, respectively. Similarly, f_3 has requests of size 30 at times $t = 2$ and $t = 7$. Figure 22.3 shows a hypothetical cumulative service function for f_1. The instantaneous service rate of f_1 (represented by the slope of the cumulative service curve at any time) varies considerably as requests from other flows share the server's capacity. Note that if f_i is backlogged at t, then $CS_i(t) < CA_i(t)$. The service received by f_i in a time interval $[t_1, t_2]$ is denoted by $S_i(t_1, t_2) = CS_i(t_2) - CS_i(t_1)$.

In a **fair queuing** model the server aims to provide each backlogged flow with an equal share of the server capacity by dividing it equally among the active flows. In an ideal fair queuing schedule, the service provided to any two flows that are continuously backlogged in an interval should be equal. That is, for any interval $[t_1, t_2]$ in which flows f_i and f_j are both backlogged, $S_i(t_1, t_2) = S_j(t_1, t_2)$.

In a more general setting, each flow f_i demands a different amount of service from the system. Let ρ_i denote the service rate requested by f_i. The admission control mechanism must ensures that the aggregate service demand of the flows does not exceed the server capacity. Hence, for a server of capacity C the system will admit at most N flows $f_1, f_2, \ldots, f_N$, such that $\sum_{i=1}^{N} \rho_i < C$. Define the **weight** of f_i to be $w_i = \rho_i / \sum_{j=1}^{N} \rho_j$, for $1 \leq i \leq N$. From the admission constraint $Cw_i > \rho_i$. The **normalized service** received by f_i in $[t_1, t_2]$ is defined as $S_i(t_1, t_2)/w_i$. In an ideal **weighted fair queuing** (WFQ) schedule the normalized service provided to any two flows that are continuously backlogged in any interval should be equal. That is, for any interval $[t_1, t_2]$ in which both flows f_i and f_j are backlogged, $S_i(t_1, t_2)/w_i = S_j(t_1, t_2)/w_j$. When the time interval over which the service is measured tends to zero, the fairness criterion implies that the instantaneous normalized service rates of any two backlogged flows should be equal.

In a WFQ schedule, the server will provide f_i a service rate of at least $Cw_i > \rho_i$. To see this let $\mathcal{A}$ denote the set of flows that are continuously backlogged in the interval $[t, t + \delta t]$. Let $K = S_j(t, t + \delta t)/w_j$ be the normalized service received by any of the active flows in the interval $[t, t + \delta t]$. Then $\sum_{f_j \in \mathcal{A}} S_j(t, t + \delta t) = K \sum_{f_j \in \mathcal{A}} w_j = C\delta t$, where the last equality follows because the server is work conserving. Hence, $S_i(t, t + \delta t) = Kw_i = Cw_i \delta t / \sum_{f_j \in \mathcal{A}} w_j \geq Cw_i \delta t$. Hence the instantaneous service rate given to f_i, $S_i(t, t + \delta t)/\delta t \geq Cw_i > \rho_i$. Thus, ρ_i is a lower bound on the service rate given to an active flow f_i in a WFQ schedule.

22.3 Fluid Flow Server Model

The **fluid flow model** (abbreviated FF) is an idealized server in which all active flows simultaneously receive service at any time instant. This means that the server is multiplexed at infinitesimally small

QoS Scheduling in Network and Storage Systems

TABLE 22.1 Example Arrival Pattern

Arrival time	Flow Id	Size of request
0	f_1	150
1	f_2	90
2	f_3	30
6	f_2	30
7	f_3	30

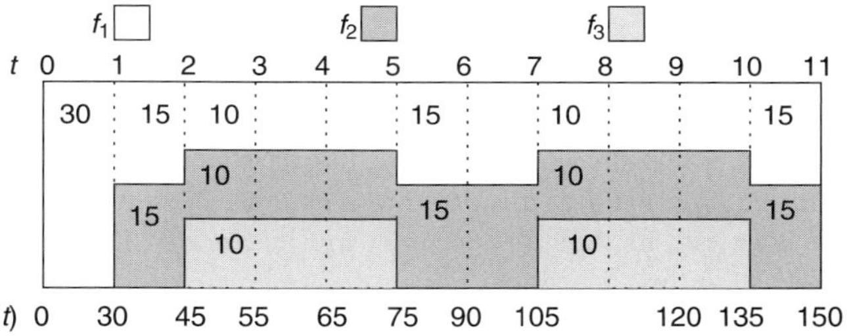

FIGURE 22.4 System evolution: Fluid flow model.

time intervals, so that all $|\mathcal{A}(t)|$ flows that are active at time t receive nonzero service. We will refer to such a server as a fluid server. When a flow is activated it begins service immediately. The service allocation for the flows already in service are reduced, so that the capacity is reapportioned among the new set of active flows. Similarly, when a flow becomes idle, the service rates of each of the remaining active flows immediately increases to use up the share of the capacity freed up by the newly idled flow.

Example 22.1 *Assume a server capacity $C = 30$. Table 22.1 shows a sequence of service requests indicating the arrival time, flow id, and service requirements of the requests. The example corresponds to the cumulative arrival function shown in Figure 22.2.*

22.3.1 Fair Queuing in the Fluid Flow Model

In a **fair queuing fluid flow server**, the instantaneous service rate at time t of any flow $f_i \in \mathcal{A}(t)$ is $C/n(t)$. Figure 22.4 shows stylistically the state of the system during the time instants t, $0 \leq t \leq 11$. In interval $[0, 1)$, f_1 is the only flow using the server and utilizes the entire server capacity, completing 30 of its required 150 service units. In the interval $[1, 2)$ flows f_1 and f_2 share the server equally, receiving 15 units of service each. Hence, when f_3 enters into service at time 2, the residual service required for each of the three flows are 105, 75, and 30 units, respectively. Each of the flows will now receive a service rate of 10 until the next event: this will, in general, be either the completion of one of the current requests or the arrival of a new request. f_3 has the smallest amount of residual service (30 units) at $t = 2$, and will complete in an additional 3 seconds at $t = 5$, provided no new flow arrives in the time interval $[2, 5)$. Since the next arrival is only at time 6, this will be the case. Following the departure of the request at time 5, flows f_1 and f_2 share the capacity equally. Since f_2 has 45 units of service remaining at $t = 5$, it will complete at $t = 8$ if its service rate does not change in this interval. However, there are arrivals at times 6 and 7, which need to be considered. At $t = 6$, the arrival is a packet for f_2 whose earlier packet is still in service. Hence, this packet will be held in its flow queue until the current request of f_2 finishes service; consequently, there is no change in the service rate allocations to f_2 and f_3 at this time. At $t = 7$ there is an arrival of a request from an idle flow f_3. This newly activated flow will immediately begin service, whence the service rates of f_1 and f_2 will also fall from 15 to 10 at this time. Since f_2 with 15 units has the smallest amount of residual service remaining, it completes service in an additional 1.5 seconds at time $t = 8.5$.

The next packet from f_2 immediately begins service, and the residual service for the flows at this time are 30, 30, and 15, respectively. f_3 completes after 1.5 seconds at $t = 10$, after which f_1 and f_2 complete in an additional one second at $t = 11$. Note that the cumulative service curve of Figure 22.3 shows the service obtained by f_1 in the fluid flow model as described above.

As we will see later, practical scheduling schemes will need to simulate the behavior of the idealized fluid flow model in order to match its service profile. A straightforward simulation of a fluid flow system given a sequence of arrivals and their service requirements, will emulate the events as described in the example. At any instant t we maintain the residual service $R_i(t)$ required for each flow $f_i \in \mathcal{A}(t)$. The next event will be either an arrival or the completion of the requests in service. Assuming the next event is a service completion, it must be from the flow f_j with the smallest residual service $R_j(t)$; and the time at which it completes will be $t' = t + R_j(t)/n(t)$. If t' is less than the arrival time of the next request, then the next event is the departure from flow f_j. We update the residual service of each active flow i to $R_i(t') = R_i(t) - (t' - t)C/n(t)$, and if the flow with the service completion becomes idle, we update $n(t') = n(t) - 1$. On the other hand, if this flow has a pending request it begins service immediately after setting its residual time to its total service time. In the case that the next event is the arrival of a request for some flow f_k at time $\hat{t}$, we first update the residual service $R_i(\hat{t})$ for each $f_i \in \mathcal{A}(t)$ as before. If the arrival activates flow f_k (i.e., it was idle just prior to $\hat{t}$), then we add f_k to $\mathcal{A}(t)$ and update $n(\hat{t}) = n(t) + 1$. However, if f_k is backlogged at $\hat{t}$ then the new request is simply added to the end of its flow queue. Although the simulation described above is straightforward to implement it has the drawback of being computationally expensive. The algorithm requires $O(n(t))$ operations for every event (arrival or departure) at time t, since the residual time for each active flow needs to be updated.

An efficient method of simulating a fluid flow system is based on the concept of **virtual time**. A virtual clock is associated with the system. The virtual clock ticks at a *variable rate* depending on the number of flows that are currently active. One tick of the virtual clock is defined as the interval of time required to complete one unit of service from each of the currently active flows. Hence, the more flows that are active the slower the virtual clock ticks. A request of size L will require exactly L virtual clock ticks to complete, *independent of the number of active flows*. Since a fluid flow system allocates each flow that is active at t a service rate of $C/n(t)$, one tick of the virtual clock at time t corresponds to $n(t)/C$ units of real time. The virtual time at physical time t will be denoted by $V(t)$. The bottom line of Figure 22.4 shows the evolution of virtual time $V(t)$ for the example system shown.

A fundamental property of virtual time is that if a request of size L begins service in a fluid server at virtual time $V(t)$ it will complete service at virtual time $V(t) + L$, *independent of the future arrival or departures of the other flows*. Note, however, that the real time at which the service completes does not share this property, since the service rate obtained by the flow depends on the future arrivals and departures. A request begins service either the instant it arrives (if the flow is idle at this time) or when the last request in its flow queue completes service. Hence, the virtual start time of a request is the later of the the virtual arrival time of the request and the virtual finish time of the request (from the same flow) immediately before it.

By introducing the artifact of virtual time it becomes much more efficient to track the dynamics of a fluid server. The behavior is completely captured by the **virtual time function**, $V(t)$, that tracks the value of virtual time as a function of real time. The function $V(t)$ is a piecewise linear function defined as follows. Let t_1 and t_2 be the real time instants corresponding to two successive events. In the interval $[t_1, t_2)$, $V(t)$ is a straight line of slope $C/n(t_1)$ through the point $(t_1, V(t_1))$. For any real time $t \in [t_1, t_2)$, the virtual time corresponding to t is given by $V(t) = V(t_1) + (t - t_1)C/n(t_1)$. Note that the set of active flows does not change between two successive events, hence the number of active flows during the entire time interval $[t_1, t_2)$ is $n(t_1)$. Using $V(t)$, the dynamics of the fluid flow system can be simulated efficiently in virtual time. Initially both real time and virtual time are initialized to zero when the system is idle. At the first arrival, set $t = 0$, $V(0) = 0$, and $n(0) = 1$. The virtual start and finish times of the request are set to 0 and L(the request size) respectively. At any time t we maintain a priority queue of $n(t)$ departure events sorted by the virtual finish times of the requests. Arrivals are also maintained in a FCFS queue in order of their real arrival times. To determine the next event we compare the virtual time of the earliest departure with the virtual time of the next arrival. Suppose the next arrival is at real time s. The virtual time $V(s)$, assuming there are no other events in the interval $[t, s)$, is given by $V(s) = t + (s - t)C/n(t)$. If $V(s)$ is less than the earliest virtual finish time, then the next event is the arrival at virtual time $V(s)$. If not, then the next event is the service completion of the request with the smallest virtual finish time.

If the next event is the arrival of a request of size L at real time s, the virtual time is advanced to $V(s)$ and $n(s)$ is updated if necessary. If the arrival is to a newly activated flow, $n(s) = n(t) + 1$; else $n(s) = n(t)$, and the request is merely placed at the end of its flow queue. In the former case, the virtual finish time of the new request is computed as $V(s) + L$, and the request is inserted into the priority queue in the order of its virtual finish time. If the next event is a departure at time s, we update $V(s)$ as before, and remove the event from the priority queue. If this flow queue is not empty, then the next request from the queue begins service; its virtual finish time is set to $V(s) + L$ and the request is inserted into the priority queue. If the flow queue is empty at time s, this flow now becomes idle, and we update $n(s) = n(t) - 1$. In this way the $O(n(t))$ updates of the residual service time at every event required by the straightforward implementation, is replaced by an $O(1)$ update to the virtual time. In both cases there is an additional $O(\log(n(t)))$ time required to maintain the priority queue of finish times of the active flows. In one case, the entries are sorted on the basis of real time and in the other the virtual times are used as the key.

22.3.2 Weighted Fair Queuing in the Fluid Flow Model

A useful generalization of the fair queuing policy described in the previous section is to prioritize the flows so that each receives a service rate that reflects its relative priority. The weight assigned to a flow represents this priority. Rather than dividing the server capacity equally among all concurrent flows, the allocation is made in proportion to the weight w_i assigned to flow f_i. In particular, at time t every flow $f_i \in \mathcal{A}(t)$, is allocated a service rate of $Cw_i / \sum_{j \in \mathcal{A}(t)} w_j$. In the special case when all flows have the same weight this is the same as Fair Queuing where C is divided equally among $|\mathcal{A}(t)|$ concurrent flows.

Example 22.2 *Using the arrivals in Table 22.1 and a system capacity of $C = 30$ as before, we describe the system dynamics when the weights w_1, w_2, and w_3 are assumed to be $1/2$, $1/3$, and $1/6$, respectively.*

Figure 22.5 shows a trace of the system for the server of Example 22.2. In the interval $[1, 2)$ the server capacity is divided in the ratio $3 : 2$ between f_1 and f_2. Hence, f_1 receives a service rate of 18 and f_2 a service rate of 12. In the interval $[2, 3)$ all three flows are active and will receive service rates of 15, 10, and 5, respectively, reflecting their relative weights $3 : 2 : 1$. Also, note that all flows finish at time 11, exactly the same time as in Example 1. This is not coincidental. The fluid server is work conserving. In a work conserving system the capacity is completely utilized whenever there is pending work. Consequently, for the same inputs the total amount of service done by the two systems at any time must be identical; all that can change is the distribution of the service among different flows.

The concept of virtual time extends naturally to the case of weighted flows. By definition, in one unit of virtual time every active flow f_i receives one unit of *normalized service* or w_i units of actual service. Note

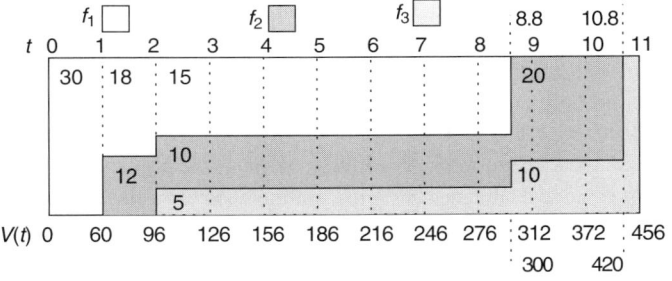

FIGURE 22.5 System evolution: WFQ in the fluid flow model.

that f_i receives service at the rate $Cw_i/\sum_{j\in\mathcal{A}(t)} w_j$ at time t, and so each unit of virtual time corresponds to $\sum_{j\in\mathcal{A}(t)} w_j/C$ units of physical time. The virtual time function $V(t)$ will still be a piecewise linear function whose slope at any time instant depends on the weights of the active flows at that time. In particular, the slope of the linear segment of $V(t)$ in the real time interval $[t_1, t_2)$ between two successive events is $C/\sum_{j\in\mathcal{A}(t_1)} w_j$. The virtual time $V(t)$ in Example 22.2 is shown in the bottom line of Figure 22.5.

The dynamics of a weighted fluid flow server are captured by the following formulas that specify the virtual times at which a request starts and finishes service. It uses the virtual time function $V(t)$ to map the arrival time of a request to virtual time. Note that $V(t)$ depends on the history of arrivals and requests up to real time t. Let $A_i(k)$ denote the real arrival time of the kth request from f_i with service requirement of $L_i(k)$. Let $S_i(k)$ denote the virtual time at which the request begins service and $F_i(k)$ the virtual time at which it finishes its service.

$$S_i(k) = \max(F_i(k-1), V(A_i(k))) \tag{22.1}$$

$$F_i(k) = S_i(k) + L_i(k)/w_i \tag{22.2}$$

$V(t)$ is a continuous, piecewise linear function whose derivative at time t is given by: $dV(t)/dt = C/\sum_{j\in\mathcal{A}(t)} w_j$.

Note that the virtual start time of a request is either the finish time of the previous request from the flow (if the flow is active when the request arrives) or is the virtual time of its arrival (in which case it begins service immediately). Its virtual finish time is obtained by adding to the virtual start time, the amount of virtual time required to service the request. Since f_i is serviced at a rate of w_i per unit of virtual time, the virtual service time is obtained by scaling the service requirement by the weight of the flow. Using these equations it is possible to efficiently simulate the fluid flow system exactly as described for Fair Queuing, using a priority queue of departure instants keyed on virtual time, and by updating the virtual time at each event.

22.4 Discrete Server Model

The fluid server serves as a *benchmark* for the behavior of a fair queuing system. In systems where the server is a single resource like a CPU, communication link, or a disk, it is not possible to have more than one request in service at any time instant. In this case the resource needs to be time-multiplexed among the requests in order to mimic the behavior of the fluid server. If the resource is such that service can be efficiently preempted and resumed (like CPU cycles) then it is possible to develop a quantized approximation to a fluid server. Within a time quantum, the resource must be switched rapidly among all the active flows to present the illusion of simultaneous service at the granularity of a time quanta.

In networking and storage systems the resource cannot be readily preempted. In the former case, variable-length packets are time multiplexed on a single communication link. A packet is a logical unit of transmission that contains a single header containing metadata and routing information, and must be transmitted in its entirety as a single unit. A request that arrives during the transmission of one packet must wait until the current packet is completely sent before it can begin its transmission; in contrast, in a fluid server both packets will be simultaneously transmitted using a fraction of the channel bandwidth. In a storage system preemption of a request in service is not a viable option. An I/O operation reads or writes a block of data. It is not practical to read only part of a block and then resume reading the remainder after servicing another request that preempted it. In addition to the implementation complexities, the overhead of incurring repeated mechanical latencies to reposition the head for a single request is prohibitive.

The **discrete server model** deals with providing weighted fair service in the situation where the service cannot be preempted. A request that begins service must complete before another request can be scheduled on the server. There are three main issues that need to be considered: (i) approximating instantaneous weighted service allocation in a nonpreemptible system, (ii) latency of requests due to the serial use of the server, and (iii) the computational complexity of the scheduling algorithm. We will consider the case of multiple parallel servers in Section 22.5.

We first present a general overview of the scheduling issues involved in the discrete server model, before describing specific algorithms that have been proposed. To focus on the essential aspects, we will consider a simple example system that nevertheless illustrates the issues that arise in a more general setting.

Example 22.3 *Flows f_a and f_b equally share a discrete server. f_a sends a batch of 12 requests at time 0 and f_b sends two batches of 4 requests each at times 0 and 12, respectively. All requests are of size 1 and the capacity of the server $C = 1$. The discrete server scheduler will attempt to emulate the behavior of a fair fluid server with respect to the service received by a flow in any time interval.*

We denote the i-th, $i \geq 1$, request of f_a and f_b by a_i and b_i, respectively. A simple algorithm to schedule the requests is derived by tagging each request with its *sequence number*; the i-th request of a flow is tagged with the sequence number i. Whenever the server becomes free the scheduler chooses the request with the smallest sequence number among all the flow queues, and dispatches that request to the server. Ties are broken using an arbitrary but consistent policy. In this example we assume that flow f_a is always given priority when the sequence numbers are tied.

22.4.1 Service Allocation

Figure 22.6a shows the cumulative service function for the discrete server system described above using the data of Example 22.3. It is divided into three phases to demonstrate different aspects of the system behavior. First we consider the time interval $[0, 8]$ in which both the flows are always backlogged. The server will alternate between serving requests of f_a and f_b. Requests a_i and b_i complete service at odd and even time steps, $2i - 1$ and $2i$, respectively, $1 \leq i \leq 4$. At any time instant either f_a or f_b will receive service at the full capacity of the server. Hence the instantaneous service rate of a flow alternates between 0 and C. The mean service rates of the flows averaged over a window of two time units are both equal to $C/2$. Note that in a fluid server the cumulative service function would be a straight line of slope 1/2 through the origin for both the flows, and $S_a[0, t] = S_b[0, t]$ for all $t \in [0, 8]$. The discrete server will, in contrast, satisfy the cumulative service equality only at the discrete instants $t \in \{0, 2, 4, 6, 8\}$.

In the next phase of the example in the interval $[8, 12]$, we note that f_b has no outstanding requests at time 8. It becomes idle at this time, and remains idle till it is activated at time 12 by the arrival of fresh requests. Consequently, it does not receive any service in the interval $[8, 12]$ as shown by its flat cumulative service function in this interval. Since the server is work conserving, after completing a_5 at time 9, the scheduler will pick the next request from flow f_a to immediately begin service. Consequently, the algorithm will schedule the requests a_5, a_6, a_7, and a_8 back-to-back, completing all four requests at time 12. The algorithm automatically allocates the service capacity freed up by f_b to flow f_a, so that its service rate in the interval $[8, 12)$ is double that in $[0, 8]$. The behavior of the fluid server during this period is exactly the same as shown for the discrete server.

The third phase of the scheduler is the most interesting, and begins at time 12 when f_b again becomes backlogged. Suppose the algorithm continues as usual without being sensitive to the idle period that f_b has just completed. The newly arrived requests of f_b, b_5, b_6, b_7, and b_8 will be assigned sequence numbers 5, 6, 7, and 8, respectively, while the remaining requests of f_a, a_9, a_{10}, a_{11}, and a_{12}, will have sequence numbers 9, 10, 11, and 12, respectively. Note that all four requests of f_b have smaller tags than any pending request of f_a. Consequently, the scheduler will service the next four requests of f_b followed by the remaining four requests of f_a. During the interval $[12, 16]$ flow f_b has exclusive use of the server and f_a is starved, as the scheduler favors f_b to make up for the excess service that f_a received in the interval $[8, 12]$ while f_b was idle.

Is this behavior of the scheduler in the third phase reasonable? The answer depends on the higher-level goals of the system. If the aim is to emulate the fluid server and provide instantaneous (or fine-grained) fairness to the flows, then the algorithm as stated fails to achieve that goal. In the fluid server, as shown in Figure 22.6b, from time 12 onwards both f_a and f_b will equally share service and finish their requests

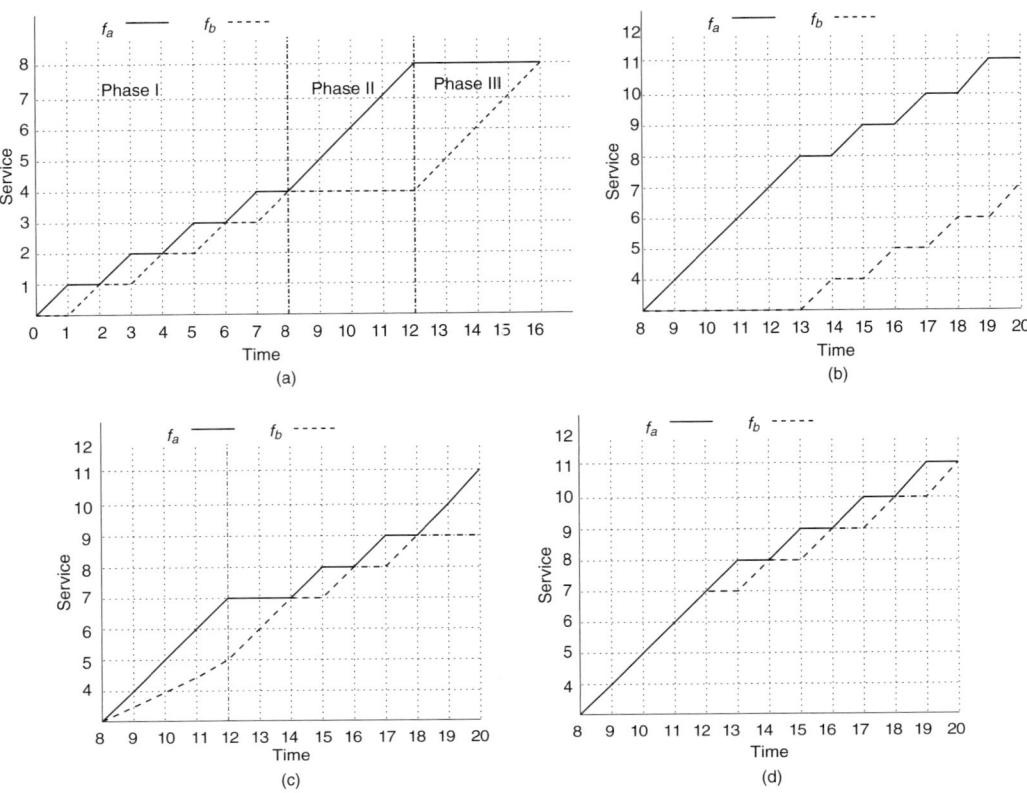

FIGURE 22.6 Cumulative service allocation in the discrete server (a) Basic algorithm, (b) Emulating fluid flow, (c) Partial solution, (d) Desired behavior.

simultaneously at time 20. In the discrete server the scheduler attempts to equalize the average service received by the flows over the entire interval [0, 16].

The fluid flow model of instantaneous fairness implicitly defines a history-independent fairness policy. Under this policy the past behavior of a flow does not influence how it is viewed by the system in the future. One consequence of this definition is that a flow cannot save up its share of the server resources for use in the future by under utilizing its designated share in the past. Under this policy if a flow does not utilize its share of the resources fully it loses them permanently, and cannot stake a claim in the future for preferential treatment based on its past frugality. Otherwise, it may be possible for a flow to starve the other flows as it monopolizes the server after a long break of inactivity. A corollary is that a flow should not be penalized for using spare system capacity. By utilizing spare capacity the system utilization is improved and the flows receive better service during that period. Under this policy the additional service is treated as a bonus and the flow receiving the extra service should not be penalized by having its service reduced in the future.

A general strategy to emulate the instantaneous fairness policy of a fluid server is to adjust the starting sequence number of a newly activated flow so as to artificially catch up with flows that have been active during the interval. To negate the effect of credit gained during an idle period, the starting sequence number of a freshly activated flow can be increased to account for the service it would have received at its fair share had it remained backlogged during its idle period. In the example, during the period [8, 12), two requests for f_b would have been serviced at its fair service rate of 1/2. Hence, the starting sequence number for flow f_b at time 12 would be increased from 5 to to 7. The evolution of the system under this policy is shown in Figure 22.6c; note that the requests b_5, b_6, b_7, and b_8 that arrived at time 12 are assigned tags of 7, 8, 9, and 10, respectively. The order of scheduling would be: b_5, b_6, a_9, b_7, a_{10}, b_8, a_{11}, a_{12}. This results in a schedule that is closer to that of a fluid server since it halves the time for which f_a is locked out of service. Nevertheless, depending on the length of time for which f_b was idle idle, the length of the period during which f_a is starved can be considerable.

In order to prevent f_a from paying a penalty for its use of the excess capacity while f_b was idle, the difference between the sequence numbers of f_a and f_b can be adjusted to account for the extra service received by f_a during that interval. Since the excess is exactly the service that f_b did not use, the adjustment of the sequence number is the same as before. The sequence number of f_b is increased to 7, and the subsequent evolution is the same as shown in Figure 22.6c. Clearly, adjusting the sequence numbers to enforce just one of these policies is insufficient; both need to be simultaneously enforced to emulate the behavior of the fluid server. A correct solution would increase the sequence number of flow f_b to 9, so that on activation both f_a and f_b will compete fairly from time 8 onwards as shown in Figure 22.6d.

Several different schemes using the basic framework described above have been described in the literature, and are detailed in subsequent sections. Arriving requests are tagged appropriately and served in order of their tags, so that during continuously backlogged periods the service allocations are proportional to the weights associated with the flows. The major differences among the schemes are in their handling of the activation of a flow, and result in different service allocations, latency, and scheduling complexities.

22.4.2 Latency in a Discrete Server Model

In a general queuing system, the latency experienced by a request depends on the arrival and service pattern of all the flows sharing the system. In a system based on fair queuing the goal is to logically isolate the flows from each other so that the latency experienced by requests of one flow is relatively independent of the behavior of the remaining flows. The latency is made up of two components: the *queuing delay* is the time the request waits in the flow queue before it is dispatched to the server, and the *service time* is the actual time it spends in the server. We first consider the service time component. In a fluid flow server the request of size L_i from f_i requires L_i/w_i units of virtual service time; the physical service time depends on the rate of the virtual clock, which is determined by set of flows that are active during the interval that the request is in service. Note that this set may change as flows become active and idle during this interval. However, since ρ_i is a lower-bound on the service rate guaranteed to f_i, the maximum physical service time is bounded by L_i/ρ_i. In the discrete server the physical service time is a fixed amount L_i/C since the entire server capacity is dedicated to the request once it is dispatched to the server.

We next consider the queuing delay component of the latency. In the fluid server an arriving request from f_i that sees requests with total service demand σ_i in its flow queue ahead of it, will experience a virtual queuing delay of σ_i/w_i units, and a *worst-case* physical queuing time of σ_i/ρ_i. If the request arrives at an empty flow queue then in the fluid server it begins service *immediately* and has a queuing delay of zero. In a discrete server estimating the queuing delay is more involved. When a request arrives to an empty flow queue in a discrete server it will not, in general, begin service immediately. It must at least wait till the request currently undergoing service completes, and then wait for its turn at the server. The former delay of L_{max}/C, where L_{max} is an upper bound on the request size, cannot be avoided in general since service in the discrete server is not preemptible. The latter delay depends on the requests from other flows present in the system and the tag assignment policy used by the scheduling algorithm. The different scheduling algorithms vary in this regard. The ideal is to approach the latency behavior experienced in a fluid server. The worst-case latency (difference between the finish and arrival times) of a request arriving to an empty flow queue will be referred to as the **flow intrinsic delay**. Hence, it is desirable that the worst-case latency of a request arriving at an empty flow queue should be $L_{max}/C + L_i/\rho_i$, and a request arriving at a flow queue with service demand σ_i ahead of it should experience a worst-case delay of $L_{max}/C + \sigma_i/\rho_i + L_i/\rho_i$.

22.4.3 Weighted Fair Queuing

The WFQ scheduling algorithm for a discrete server uses the concept of virtual time described for a fluid flow server in Section 22.3.2 as its basis. The idea is to schedule requests so that they finish in the same order as they would in a fluid server. The scheduler exactly simulates a fluid flow server using Equations 22.1 and 22.2 to determine the virtual start time and virtual finish time of each arriving request. However, unlike the situation in a fluid flow model where the inverse virtual time function precisely determines the physical start and finish times of the request, in the discrete model the virtual times are used to guide the relative order of scheduling. Each request is assigned a tag on arrival. Whenever the server becomes free the request with the *smallest tag* is dispatched. Two variants of WFQ scheduling are defined based on the choice of tags. In **Start Time Weighted Fair Queuing** (S-WFQ) the virtual start time of a request defined by Equation 22.1 is used to tag a request. In **Finish Time Weighted Fair Queuing** (F-WFQ) the virtual finish time of a request, defined by Equation 22.2 is used as the tag. We consider each of these scheduling strategies in turn below.

22.4.3.1 Start Time Weighted Fair Queuing

The operation of the system using the data of Example 22.1 is used to illustrate the algorithm. Each of the flows has a weight of $1/3$. The tags assigned to the requests equals their virtual start times in the fluid flow server. Table 22.2 shows the virtual start and finish times of the requests using the results from Figure 22.4. The virtual times in the table have been scaled by 3 to reflect the weight of $1/3$ of the flows. Note that the request from f_3 arriving at 7 has a smaller virtual start time than the request from f_2 arriving at 6. This reflects the fact that an earlier request from f_2 is in service in the fluid server at time 6, and will only complete at virtual time 360; hence, $S_2(2) = F_2(1) = 360$. However, there is no request from f_3 in the system at 7 in the fluid server and so its virtual start time is the value of the current virtual time; hence $S_3(2) = V(7) = 315$.

Figure 22.7a illustrates the actual order in which requests are serviced in a S-WFQ schedule. When the first request finishes service at real time 5, the pending requests from f_2 and f_3 with tags 90 and 135 are compared. The request of f_2 with size 90 begins service and will complete at time 8. Meanwhile, additional requests from f_2 and f_3 arrive at times 6 and 7, and are tagged with virtual start times 360 and 315, respectively. When the server becomes free at time 8, the pending requests have tags of 135, 315, and 360, and the request with tag 135 (the first request from f_3) begins service. This is followed by the second request of f_3 that begins service at time 9, and finally the second request of f_2 that begins service at time 10.

While it may be appear that S-WFQ is similar to a FCFS scheduler, there are fundamental differences between the two. In particular, under FCFS scheduling a flow could send a burst of requests in close temporal proximity, and *all* of them would receive service before a request from a different flow that came after the burst. However, in S-WFQ the arrival time alone does not determine the scheduling priority (tag) of the request. The requests in a burst would receive tags spaced by the their sizes, so that competing flows can receive service between the requests of the burst. In the example the request from f_2 at time 6 has a tag larger than the tag of the arrival from f_3 at time 7, reflecting the fact that f_2 has pending requests.

TABLE 22.2 Start and Finish Tags for WFQ and SC Schedules

Arrival time	Flow Id	Size of request	Virtual start time (WFQ)	Virtual finish time (WFQ)	Virtual start time (SC)	Virtual finish time (SC)
p	f_1	150	0	450	0	450
1	f_2	90	90	360	450	720
2	f_3	30	135	225	450	540
6	f_2	30	360	450	720	810
7	f_3	30	315	405	720	810

QoS Scheduling in Network and Storage Systems

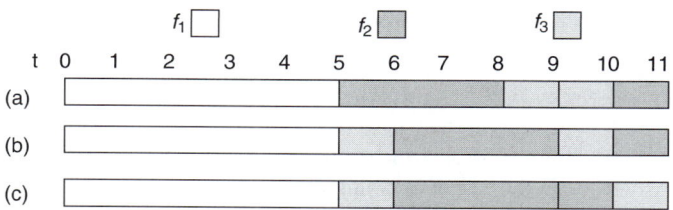

FIGURE 22.7 Comparison of packet order in (a) Start WFQ (b) Finish WFQ (c) SCFQ.

Observe that in S-WFQ the virtual time that a request begins service is at least as large as its start tag. Also note that the real times at which requests begins service in S-WFQ are consistent with their start tag values, and hence their service order in the fluid flow system. That is, request m begins service before request n in S-WFQ, if and only if the start tag of m is no more than the start tag of n. To show this we need only consider the case of m and n belonging to different flows. If n has already arrived at the time m is scheduled then its tag could not be smaller than m's tag, since m was served in preference to n. If n had not arrived at the time m was scheduled, its start tag must be at least the virtual time of its arrival, which must be greater than the virtual start time at which m begins service, which, from the observation above, must be no less than the start tag of m. Note, however, there is no relation between the times at which a request finishes in the fluid flow and S-WFQ models. A request in the fluid server can finish much later than the request under S-WFQ even though it began service earlier. For instance, the request from f_1 and the first request from f_2 finish at times 11 and 8.5 in the fluid server, but at time 5 and 8 in S-WFQ. Finally, note that the times at which a request is active in the two systems may be completely disjoint. For instance, the request from f_3 is serviced in the interval [2, 5) in the fluid server and [8, 9) in S-WFQ.

22.4.3.2 Finish Time Weighted Fair Queuing

In F-WFQ requests are tagged with their virtual finish times in the equivalent fluid flow system. As usual, among the pending requests the one with the smallest tag is chosen to service next. Using the example of Table 22.2 the schedule is shown in Figure 22.7b. The request from f_1 at time 0 is tagged with the virtual finish time 450, and immediately begins service since there are no other pending requests. Requests arriving at times 1 and 2 are tagged with 360 and 225, respectively, their virtual finish times in the fluid server. When the first request finishes service at time 5, the pending request with the smaller tag (225) begins service, and completes at time 6. Requests arriving at times 6 and 7 are tagged with their virtual finish times of 450 and 405, respectively. At time 6, the smallest tag (360) belongs to the first request from f_2; this is followed by the second request of f_3 that begins service at time 9, and finally the second request of f_2 that begins service at time 10.

In contrast to S-WFQ in which all requests are serviced in the order of their tags, in F-WFQ it is possible for requests to serviced out-of-order. This happens when a request n arrives after some request m is scheduled. While the start tag assigned to n must be larger than the virtual time at which m begins service, its finish tag depends on the size of n. In particular, if m has a large normalized size, then its finishing tag can be larger than the tag of n. However, n will be scheduled *after* m in the F-WFQ server. For instance, requests arriving at times 1 and 2 in the above example, have smaller finish tags than the first request that begins service at time 0.

Also, as can be seen from the example, the schedule created by the F-WFQ policy is different from the Shortest Job First (SJF) policy. While a burst of short requests from a flow would all garner immediate service in SJF, under F-WFQ requests from other flows can be interleaved. For instance, SJF would schedule the arrivals at time 6 and 7 before the arrival at time 1. Finally, note that the times at which a request is active in the F-WFQ and fluid flow systems may be completely disjoint. For instance, the request from f_3 is serviced in the interval [2, 5) in the fluid flow system and [5, 6) in F-WFQ.

22.4.3.3 Self-Clocked Weighted Fair Queuing

The S-WFQ and F-WFQ schedulers described in the previous section provide an excellent discrete approximation to the ideal fluid server. However, this scheme requires a real-time simulation of the fluid server in order to track the virtual time accurately. In applications like high-speed networking where there are real-time constraints on the time available to schedule requests, the computation required for this simulation can be onerous. Although the cost of updating virtual time at any event is a constant, the possibility of a large number of events (flow activations or deactivations) occurring in a small time interval (a service time), would require many updates in that interval. To ensure that such bursts do not overwhelm the system, several approximations to the WFQ schemes have been proposed that use simpler notions of virtual time to synchronize the flows. The trade off is a possible reduction in the fairness of the resulting schedule or increased latencies for some flows.

The **Self-Clocked Weighted Fair Queuing** (abbreviated SC) scheduling algorithm approximates the current virtual time by the virtual finish time of the request currently in service. Packets are tagged with virtual start and virtual finish times defined by Equations 22.3 and 22.4 below, similar to that for WFQ. The virtual time function $v(t)$ for SC is defined as the *virtual finish time* of the request that is *currently in service*. If the system is idle then $v(t)$ can be reset to zero. On a service completion, the pending request with the *smallest* value of *virtual finish time* is selected to be scheduled next. This is similar to the virtual finish time based scheduling policy of F-WFQ where the finish tag is used for scheduling.

$$S_i(k) = \max(F_i(k-1), v(A_i(k))) \tag{22.3}$$

$$F_i(k) = S_i(k) + L_i(k)/w_i \tag{22.4}$$

$v(t)$ is a discontinuous function whose value equals the virtual finish time of the request currently in service.

The function $v(t)$ differs appreciably from $V(t)$, the virtual time function that governs the evolution of the fluid flow system. First notice that $v(t)$ is defined with respect to the parameters of the actual packet based system rather than the idealized fluid flow system used in WFQ. Hence there is no need to simulate an ideal system during scheduling. Also while $V(t)$ is a piecewise linear function whose slope that changes whenever a flow is activated or deactivated, $v(t)$ is a piecewise discontinuous function with constant value in the real time interval during which a request is in service. It can be shown that the differences between $V(t)$ and $v(t)$ can grow unbounded with time, and the scheduling order for WFQ and SC can differ.

We use the data of Example 1 with weights of 1/3 for the flows to illustrate the SC algorithm. To avoid ambiguity when an arrival and request both occur at the same time t, we assume that the arrival occurs first at t^-; a different but consistent schedule can be constructed by assuming the requests arrive later, at t^+. Figure 22.7c shows the schedule created by the SC algorithm. Table 22.2 shows the start and finish tags assigned to each request. When the first request arrives at time 0, $v(t) = 0$, and the finish tag for the request, $F_1(1)$ is set to 450. The real departure time of the request will be at $t = 5$. For the interval $t \in (0, 5]$, $v(t) = 450$. During this interval requests arrive at times 1 and 2, and are respectively time stamped with $S_2(1) = v(1) = 450$, and $S_3(1) = v(2) = 450$, and $F_2(1) = 720$ and $F_3(1) = 540$. When the first request completes service at $t = 5$, the request with the smaller finish tag (from f_3) begins service and the virtual time $v(t)$ is updated to $v(t) = 540$ for $t \in (5, 6]$. At $t = 6^-$, there is an arrival of a second request from f_2, which will be time stamped with $S_2(2) = \max(720, 540) = 720$, the virtual finish time of the earlier pending request from this flow. The finish tag, $F_2(2) = 810$. When the request from f_3 finishes service at time 6, the first request from f_2 begins service till $t = 9$. During the interval $t \in (6, 9], v(t) = 720$. The arrival at $t = 7^-$ is tagged $S_3(2) = v(7) = 720$ and $F_3(2) = 810$. The remaining two requests have equal finish tags and the tie can be arbitrarily broken.

22.4.4 Performance Analysis of Discrete Model Schedulers

Let $S_i^{\mathcal{P}}(t_1, t_2)$ denote the amount of service received by a flow f_i in the real time interval $[t_1, t_2^-)$, and let $D_i^{\mathcal{P}}$ denote the intrinsic delay of f_i using a scheduling policy $\mathcal{P}$. Let $L_i^{\max}$ denote the size of largest request from flow f_i. Recall that ρ_i is the minimum service rate guaranteed to flow f_i.

Lemma 22.1 *For any two flows f_i and f_j that are continuously backlogged in an interval $[t_1, t_2]$ let $\mathcal{D}_{i,j}^{\mathcal{P}}(t_1, t_2) = S_i^{\mathcal{P}}(t_1, t_2)/w_i - S_j^{\mathcal{P}}(t_1, t_2)/w_j$ be the difference in normalized service received by the two flows.*

$$\mathcal{D}_{i,j}^{FF}(t_1, t_2) = 0 \tag{22.5}$$

$$\mathcal{D}_{i,j}^{SC}(t_1, t_2) \leq L_i^{\max}/w_i + L_j^{\max}/w_j \tag{22.6}$$

Proof. By definition, the FF server provides equal instantaneous normalized service to all active flows. Hence, the cumulative service in any period in which the flows f_i and f_j are both continuously active must be equal.

For the SC scheduler we argue as follows (see Figure 22.8a). First, observe that in any backlogged interval in SC, the total normalized service provided for k consecutive requests of a flow equals the difference between the finish tag of the last request and the start tag of the first request of the sequence. We can assume that t_1 and t_2 are time instants when either a request from f_i or f_j is in service, since the end intervals where this is not true do not change either $S_i(t_1, t_2)$ or $S_j(t_1, t_2)$. Without loss of generality one can assume that requests a_1 and a_s of f_i are being serviced at t_1 and t_2 respectively; otherwise, the service that f_i receives relative to f_j will only reduce. Let the requests of f_j that are serviced during $[t_1, t_2]$ be $b_1, b_2, \ldots, b_m$. Since f_j is backlogged at t_2 there must be a request b_{m+1} that is serviced after a_s. Therefore, the finish tags must satisfy $F_i(a_s) \leq F_j(b_{m+1})$. Also $F_j(b_{m+1}) = F_j(b_m) + L_j(b_{m+1})/w_j$. Hence $F_i(a_s) - F_j(b_m) \leq L_j(b_{m+1})/w_j$.

The service received by f_i and f_j in the interval $[t_1, t_2]$ satisfies: $S_j^{SC}(t_1, t_2) = F_j(b_m) - S_j(b_1)$ and $S_i^{SC}(t_1, t_2) \leq F_i(a_s) - S_i(a_1)$. Subtracting, $\mathcal{D}_{i,j}^{SC}(t_1, t_2) \leq F_i(a_s) - F_j(b_m) + S_j(b_1) - S_i(a_1) \leq L_j(b_{m+1})/w_j + S_j(b_1) - F_i(a_1) + L_i(a_1)/w_i$. To complete the proof we show that $S_j(b_1) \leq F_i(a_1)$. First consider the case when b_1 arrives to an empty flow at time t. Since $t < t_1$, $S_j(b_1) = v(t) < v(t_1) = F_i(a_1)$. The remaining alternative is for a request b' of f_j to be active when b_1 arrives, so that $S_j(b_1) = F_j(b')$. Since b_1 is the first request of f_j to get service after t_1, b' must have finished at some time $t' < t_1$. Hence, $S_j(b_1) = F_j(b') = v(t') < v(t_1) = F_i(a_1)$.

To see that the bound is tight consider requests of maximum length from flows other than f_i arriving at time 0, and a request from f_i of length $L_i^{\max}$ arriving at time 0^+. The request from f_i will be assigned a finish tag $F_i(1)$. Since a flow $f_j, j \neq i$ will receive service before f_i till its virtual finish tag exceeds $F_i(1)$, the

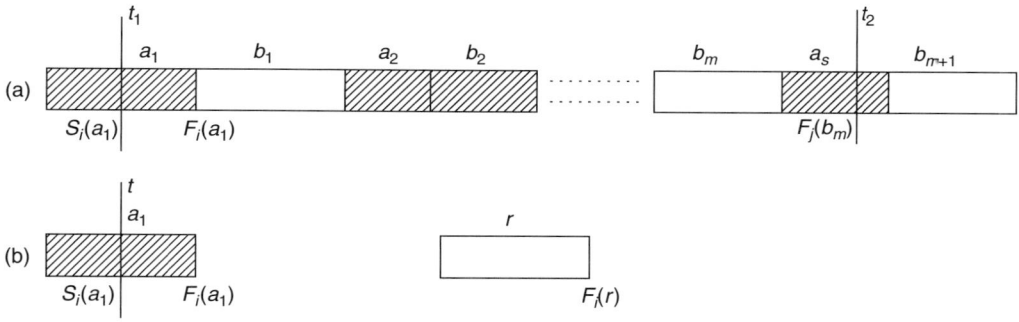

FIGURE 22.8 Service done in time interval $[t_1, t_2]$ for flows f_i and f_j.

difference in normalized service received by f_j and f_i just before f_i begins service, can be as large as $F_i(1)$. In SC, $F_i(1) = L_j^{\max}/w_j + L_i^{\max}/w_i$, where the request from f_j is in service at ϵ. ∎

Lemma 22.2 *Let D_i^P denote the intrinsic delay for a flow f_i using policy P.*

$$D_i^{\text{FF}} \leq L_i^{\max}/\rho_i \tag{22.7}$$

$$D_i^{\text{SC}} \leq L_i^{\max}/\rho_i + \sum_{j \neq i} L_j^{\max}/C \tag{22.8}$$

Proof. A request from f_i arriving to an empty flow queue in the FF system begins service immediately and has service time bounded by its maximum size divided by its minimum guaranteed service rate.

To prove Equation 22.8 we proceed as follows (see Figure 22.8b). Let the arrival time of a request, r, from f_i to an empty flow queue be at real time t. Assume that the request in service at t is a_1 from some flow f_a, and the real start and finish times of this request be s_a and f_a, respectively. The finish tag of r will be $F_i(r) = v(t) + L_i(r)/w_i$, where $L_i(r)$ is the size of r. Now $v(t) = F_a(a_1)$, the virtual finish time of a_1, since r arrives to an empty flow queue while a_1 is in service.

We first show that the only requests of a flow $f_j, j \neq i$ that can delay r are those whose finish tags lie between $F_a(a_1)$ and $F_i(r)$. Any request that is in the system after s_a with finish tag greater than $F_i(r)$ will be serviced after r and so does not delay it. Any request arriving after s_a will have a virtual start time, and hence finish tag, greater than $F_a(a_1)$. Any request arriving before s_a that can delay r must not have finished before s_a and must have a finish tag greater than $F_a(a_1)$. Let $\Delta F = F_i(r) - F_a(a_1)$. Of the set of requests that can delay r consider those from some flow f_j whose virtual start time is at least $F_a(a_1)$. The maximum amount of service that these requests can incur equals $\Delta F w_j$, and hence delay r by at most $\Delta F w_j/C$. The remaining requests of f_j that could delay r must have virtual start time less than $F_a(a_1)$. Since the request must also have virtual finish time no less than $F_a(a_1)$ there can be *at most one* such request. The delay caused by this request of f_j is bounded by $L_j^{\max}/C$. Adding the queuing delay to the service time of r (bounded by $L_i^{\max}/C$) we get $\sum_{j \neq i}(\Delta F w_j)/C + L_i(r)/C + \sum_{j \neq i}(L_j^{\max}/C)$. Noting that $\Delta F = L_i(r)/w_i$ and that $\sum_j w_j \leq 1$, the above bound is $L_i(r)/Cw_i - \sum_{j \neq i} L_j^{\max}/C \leq L_i^{\max}/\rho_i + \sum_{j \neq i} L_j^{\max}/C$.

To see that the bounds are tight consider flows $f_j, j \neq i$ that are continuously backlogged from $t = 0$ onwards. Assume that a request of size $L_k^{\max}$ of f_k is in service when a request r from f_i arrives at time $t = 0^+$. To match the bound we will assume that f_k is the worst-case flow in that $L^{\max} = L_k^{\max}$. Request r will be assigned a finish tag $F_i(r) = L_k^{\max}/w_k + L_i(r)/w_i$. All flows $f_j, j \neq i$ will receive service before r until the virtual finish tag of the request of f_j exceeds $F_i(r)$, which will lead to a delay that will match the upper bound.

It is difficult to provide similar bounds for WFQ scheduling. Equations 22.9 and 22.10 from Reference 18 bound the performance of WFQ in comparison with the FF policy. The equations indicate that WFQ cannot get too far behind a fluid flow server in terms of the service provided to a backlogged flow, nor can the finish time in an FF server lag by more than the maximum service time of a request. ∎

Lemma 22.3 *For request r of flow f_i, let $C_i(r)^P$ denote the time the request completes service using scheduling policy $\mathcal{P}$.*

$$S_i^{\text{FF}}(0, t) - S_i^{F-\text{WFQ}}(0, t) \leq L_i^{\max} \tag{22.9}$$

$$C_i^{F-\text{WFQ}}(0, t) - C_i^{\text{FF}}(0, t) \leq L_i^{\max}/C \tag{22.10}$$

22.4.5 Advanced Schedulers

In this section, we briefly review some additional issues that have been raised concerning the design of discrete schedulers and the proposed solutions. The reader is referred to the original sources for further details.

We begin with motivating **Worst-Case Fair Weighted Fair Queuing** (WF^2Q), a scheduler used to reduce the variability in flow behavior. In certain circumstances, the WFQ scheduler may provide significantly greater service to a flow in a time interval than it would in an equivalent fluid flow server. To illustrate the issues consider a unit capacity system with 6 flows where f_1 through f_5 have weight $1/10$ and f_6 has weight $1/2$. All requests are assumed to be of size one. Assume that one request for each of f_1 to f_5 and 5 requests for f_6 arrive at at $t = 0$ and the sixth request of f_6 arrives at $t = 5$. The requests of f_1 to f_5 will get a finish tag of 10; the five requests of f_6 will gets finish tags of 2, 4, 6, 8, and 10. The first 5 requests of f_6 might get service back-to-back finishing at time 5. The sixth request will get a finish tag of 12 and will wait in the flow queue until the 5 requests from f_1 to f_5 complete at time 10. This pattern of f_6 getting a burst of service of 5 seconds followed by an idle period of 5 seconds can continue indefinitely. The WF^2Q scheduling algorithm aims to avoid this bursty behavior that occurs because the service provided by WFQ to a flow runs far ahead of the service provided by the equivalent fluid server. For instance, in this example, at $t = 1$ when the first request of f_6 completes, the virtual time in the fluid flow server would have only advanced to $V(1) = 1$, although the finish tag of the request is 2. Therefore, WF^2Q works by throttling such a flow that has run ahead of its virtual finish time in the FF schedule. In particular, at any scheduling instant, WF^2Q will only consider those requests that have started service in the FF server. In other words, the only requests eligible for scheduling at time t are those whose start tags are at least as high as $V(t)$, the current virtual time of the FF server. Of the eligible requests, the one with the smallest virtual finish time is chosen for service. In the example, at time 1, the requests of f_6 are ineligible and a request from one of the other flows will be serviced. When this request completes at time 2, the second request of f_6 becomes eligible, and having the smallest finish tag will be served next. In this way, all the requests of f_6 are served at the uniform rate of one request every two time units.

WF^2Q helps smooth out the variance in service times caused by flows that have widely differing weights. Other schemes that address the issue of variability while trading off computation complexity, and service fairness have been invented. These include **Leap Forward Virtual Clock**, **Rate Proportional** and **Latency Rate Servers**, and **Minimum-Delay Self-Clocked Fair Queuing**. Other approaches to fair scheduling have been proposed that do not use the notion of virtual time, such as **Deficit Round Robin** and **Lottery Scheduling**. References to these works are discussed in Section 22.6.

22.5 Parallel Server Models

A generalization of server scheduling addressed so far is to consider the use of multiple concurrent servers in the system. The simplest model in this regard is a system made up of multiple uniform servers, that are indistinguishable in terms of the service afforded to the flows. An example of such a system is a multiprocessor cluster performing database or web transactions for clients, and where any free processor may be allocated a request. A more complex situation arises when the allocation of servers to requests are constrained because only some server or subset of servers can be assigned to a request. An example of such a system is a storage server managing a farm of storage devices that hold the data. Requests for service are directed to the particular device that holds the required data and the resource scheduler needs to account for these constraints when scheduling the flows. Since the uniform server case may be considered, in some sense, a special case of the resource-constrained model we will concentrate on the latter.

A model of a parallel I/O server that consists of N independent disks, $D_1, D_2, \ldots, D_N$ is shown in Figure 22.9. There are m flows f_1 through f_m that must access a specific disk for service in each request. Requests are held in the flow queues and scheduled in batches. In a scheduling event one request is selected for service at each disk, unless there are no requests at all directed to that disk. The dispatched requests are buffered in *disk queues* associated with each disk. When a disk completes servicing its current request

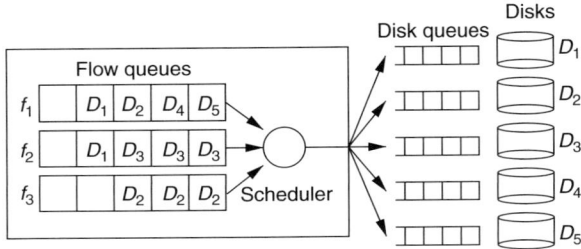

FIGURE 22.9 Parallel server model.

it chooses the next request from its disk queue to serve next. To avoid idling a disk, the scheduler must dispatch requests before any disk queue becomes empty; other than that there is considerable flexibility in the invocation of the scheduler. In an idealized model in which the disk service times are uniform for all disks, one can dispense with the disk queues and dispatch requests directly to the disks at each scheduling step.

22.5.1 Fair Lexicographic Scheduling

We present an algorithm for fairly multiplexing the server among concurrent flows. The simple generalization to handle weighted fair allocation will be described later. We use the following definitions to describe the algorithm. The **allocation vector** at scheduling step t, $A(t) = [b_1, b_2, \ldots, b_m]$, where b_j denotes the number of disks assigned to flow j at step t, and $\sum_{1 \leq i \leq m} b_i \leq N$. The **cumulative allocation vector** at I/O step t, $CA(t) = [B_1, B_2, \ldots, B_m]$, where B_j is the total number of I/O requests dispatched for flow j up to step t. That is, $CA(t) = \sum_{k=1}^{t} A(k)$. The **weight** of a vector is the sum of its components. If f and g are n-vectors of the same weight whose components are sorted in nonincreasing order, then f is *lexicographically smaller* than g, if there is some $1 \leq k \leq n$ such that f and g agree in the first $k-1$ components and $f_k < g_k$.

The scheduling algorithm known as **LexAS** is based on the *lexicographic minimization* of the *cumulative allocation vector* that tracks the service received by a flow in the current system backlog period. At any time a flow has a certain cumulative allocation equal to the number of its requests that have been dispatched to the disk queues. Given the current cumulative allocation and the requests in the flow queues, the scheduler will create a schedule for this step that is work conserving and allocates disks to flows so as to make the cumulative allocation at the end of this step is as even as possible. This is captured by the following notion of **lexicographic fairness**: given $CA(t-1)$ and a set of requests, find a feasible allocation vector $A(t)$ of maximal weight such that $CA(t) = CA(t-1) + A(t)$ and for all feasible values of $A(t)$, $CA(t)$ is lexicographically minimum.

For instance, in the example of Figure 22.10, suppose that each flow has the same cumulative allocation prior to the start of this step. Since there are five disks with pending requests, a work-conserving schedule will construct an allocation vector of weight 5. For instance, $[4, 1, 0]$ is a feasible allocation where disks $\{D_1, D_2, D_4, D_5\}$ are allocated to f_1, and $\{D_3\}$ is allocated to f_2. Some other feasible allocation vectors in this case are: $[3, 2, 0]$, $[3, 1, 1]$, $[2, 2, 1]$. The fairest allocation in this case is $[2, 2, 1]$, and the schedule will dispatch requests from $\{D_4, D_5\}$ for f_1, $\{D_1, D_3\}$ for f_2 and $\{D_2\}$ for f_3. On the other hand, suppose the initial cumulative vector was $[10, 12, 12]$; then the fairest allocation vector at this step would be $[3, 1, 1]$ leading to a new cumulative vector of $[13, 13, 13]$. The algorithm that we present will compute the allocation vector that results in the fairest cumulative allocation vector after the current step.

LexAS works by finding a set of paths in an augmented bipartite resource graph $G = (V \cup \{\alpha, \omega\}, E \cup E_\omega \cup E_\alpha)$ defined as follows (see Figure 22.10): $V = \{f_1, f_2, \ldots, f_m\} \cup \{D_1, D_2, \ldots, D_N\}$ is a set of $m + N$ nodes, one for each flow and disk in the system. E is the set of directed edges between nodes representing flows and nodes representing disks: there is an edge (f_i, D_j) whenever there is a request for disk D_j in the queue for flow f_i. Distinguished vertices α and ω will serve as the source and sink of

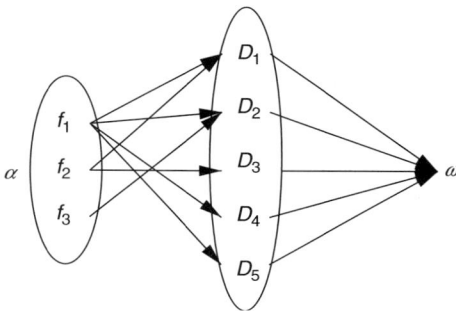

FIGURE 22.10 Resource allocation graph.

paths through the graph. $E_\omega = \{(D_j, \omega), 1 \leq j \leq N\}$, is the set of directed edges from each disk to ω and $E_\alpha \subseteq \{(\alpha, f_i), 1 \leq i \leq m\}$, is a subset of the directed edges from α to flow nodes f_i. The algorithm, formally stated in Algorithm LexAS, maintains a *priority vector*, $\mathcal{P} = [p_1, p_2, \ldots, p_m] : p_i$ is the priority for flow node f_i. At each step the node f_i with highest-priority is selected, and an edge from α to f_i is added to G. The algorithm then attempts to find a path in the current resource graph from α to ω. If a path cannot be found, then the currently selected node f_i is marked as *saturated*, and the algorithm will adjust the priority of f_i so that it will not be selected again. If the search for a path is successful this means that one additional request made by flow f_i can be satisfied, without decreasing the number of disks already allocated to other flows by the algorithm. The actual assignment of disks to flows might get changed to make this possible, but the *number of disks* assigned to any other flow will *not change* by this reassignment. In preparation for the next iteration, the resource graph is modified to reflect the new assignments using the step referred to as *path conditioning* in the algorithm. This step replaces every edge in the newly identified path by its reverse edge, and removes the first edge (α, f_i) from the graph.

The algorithm maintains the following *invariant*: an edge (D_r, f_t) in the resource graph at the start of an iteration means that currently disk D_r is assigned to flow f_t. Suppose the algorithm chooses to allocate a disk to flow f_u at some iteration, and finds a path $(\alpha, f_u, D_{j_1}, f_{j_1}, D_{j_2}, f_{j_2}, \ldots, D_{j_k}, f_{j_k}, D_{j_{k+1}}, \omega)$ through f_u. This implies that at the start of the iteration there were at least k assignments: disk D_{j_i} was assigned to flow $f_{j_i}, 1 \leq i \leq k$. When the edges between flows and disk nodes are reversed by the path conditioning step at the end of the iteration, the $k + 1$ new assignments will be: D_{j_1} to f_u, D_{j_2} to f_{j_1}, and so on ending with $D_{j_{k+1}}$ assigned to f_{j_k}. Note that for each of the flow vertices f_{j_i} only the identity of the assigned disk was changed.

To handle weighted fair allocation we define a **normalized cumulative allocation vector** $\mathcal{N}(t)$, consisting of weight-scaled values of the cumulative allocation vector $CA(t)$. That is, $\mathcal{N}(t) = [\eta_1, \eta_2, \ldots \eta_m]$, where $\eta_i = B_i/w_i, i = 1, \ldots, m$. We modify algorithm LexAS as follows. The *priority vector* $\mathcal{P}$ in step 4 of the algorithm is initialized to $\mathcal{N}(t)$ at the start. Step 7 that chooses the flow to be allocated a disk is modified as follows: among all nonsaturated flow nodes, we select the one with the minimum value of $\eta_i + (1/w_i)$. Finally in Step 12, we increment p_i by $1/w_i$.

22.5.2 Multiple Uniform Servers

In this section, we will present an adaptation of LexAS for the uniform server model, where each of the requests can be handled by any of the available servers.

When there are no resource constraints, the lexicographic minimization based scheduling becomes greatly simplified. When a request belonging to flow f completes service at one of the servers, its service demand c_f is known; the scheduler updates the value of the normalized service allocation for flow f, η_f, by adding c_f/w_f to its current value. The scheduler next attempts to schedule a new request on the freed server. From among the nonempty queues it chooses the request belonging to the flow with the smallest value of normalized service allocation, η_i, to schedule next (ties are broken randomly or based

```
1  allocations = 0;
2  d = number of disks with at least one request pending;
3  G is augmented resource graph with $E_\alpha = \phi$;
4  Let priority vector $\mathcal{P} = [p_1, p_2, \cdots, p_m]$, where $p_i = B_i$, at the time of scheduling;
5  Mark all $p_i$'s as nonsaturated;
6  while (allocations < d) do
7      Choose the minimum nonsaturated element $p_i$ of $\mathcal{P}$ with ties broken arbitrarily. Add edge
       $(\alpha, f_i)$ to G;
8      Search for a path from $\alpha$ to $\omega$ that includes $(\alpha, f_i)$ ;
9      if (no path is found) then
10         Mark $p_i$ as saturated in $\mathcal{P}$;
11     else
12         $p_i = p_i + 1$;
13         allocations = allocations + 1;
14     end
15     Update graph G by path conditioning;
16 end
```

Algorithm 1: Algorithm LexAS.

on weights/priority). The above simple algorithm achieves the same goal as the algorithm described for the constrained server case. In this case, since only one server is scheduled at a time, the allocation that results in the lexicographically minimum vector $N + A$ will assign the server to the flow with the current smallest value of N.

It may be seen that the amortized scheduling overhead for each arrival is $O(1)$ and that for scheduling a new request is $O(\log m)$ by use of an appropriate priority queue data structure.

22.6 Bibliographic References

The discussions in this chapter are based directly on the material in the original papers published in the literature. The fluid flow model, also referred to as Generalized Processor Sharing [15], has been proposed as an ideal algorithm for sharing communication links among contending flows. The discrete models using virtual start and finish time tagging are based on the seminal papers on Virtual Clock [28], WFQ [7, 10], and PGPS [18, 19]. Self-clocked approximations of WFQ are presented in the works of Reference 6, SCFQ [8], and STFQ [9]. WF^2Q was introduced in References 1 and 2. Efficient scheduling algorithms were proposed in Leap Forward Virtual Clock [24] and MD-SCFQ [4]. New algorithms with better performance guarantees and a framework for comparing various fair queuing mechanisms were proposed in References 22 and 23. Other algorithms such as deficit round robin (DRR) [21] have also been proposed and analyzed. Bounds in terms of worst case delay and bandwidth allocation using rate based flow control and the PGPS algorithm for weighted allocation were shown in References 18 and 19. Delay analysis of Virtual Clock servers was presented in Reference 26. Service curve-based scheduling for guaranteeing delay bounds was presented in References 5, 17 and 20.

A lucid overview of the issues in QoS scheduling for storage systems appears in Reference 25. The SFQ(D) model [13] generalizes STFQ for servers with internal concurrency. Stonehenge [12] is a storage virtualization system system based on a two-level scheme using virtual time tags and a low-level real-time scheduler. Avatar [27] is a recent two-level scheme for meeting throughput and latency guarantees. Façade [16] is a system to provide latency guarantees by controlling the size of the disk queues, to trade off throughput and response time. Lexicographic Scheduling [11] provides weighted bandwidth allocation in a parallel server with resource preferences. Other approaches proposed on the basis of control system modeling include Triage [14] and SLEDS [3].

Acknowledgment

Support by the National Science Foundation under grant CNS-0541369 is gratefully acknowledged.

References

[1] Jon C. R. Bennett and H. Zhang. WF^2Q: Worst-case fair weighted fair queueing. In *INFOCOM (1)*, pages 120–128, 1996.

[2] Jon C. R. Bennett and H. Zhang. Hierarchical packet fair queueing algorithms. *IEEE/ACM Trans. on Networking*, 5(5):675–689, 1997.

[3] D. D. Chambliss, G. A. Alvarez, P. Pandey, D. Jadav, J. Xu, R. Menon, and T. P. Lee. Performance virtualization for large-scale storage systems. In *Symposium on Reliable Distributed Systems*, 109–118, Oct 2003.

[4] F. Chiussi and A. Francini. Minimum-delay self clocked fair queueing algorithm for packet-switched networks. In *INFOCOMM'98*, 1998.

[5] R. L. Cruz. Quality of service guarantees in virtual circuit switched networks. *IEEE J. Selected Areas in Comm.*, 13(6):1048–1056, 1995.

[6] J. Davin and A. Heybey. A simulation study of fair queuing and policy enforcement. *Computer Communications Review*, 20(5):23–29, October 1990.

[7] A. Demers, S. Keshav, and S. Shenker. Analysis and simulation of a fair queuing algorithm. *J. Internetworking Res. Exp.*, 1(1):3–26, September 1990.

[8] S. Golestani. A self-clocked fair queueing scheme for broadband applications. In *INFOCOMM'94*, pages 636–646, April 1994.

[9] P. Goyal, H. M. Vin, and H. Cheng. Start-time fair queuing: A scheduling algorithm for integrated services packet switching networks. Technical Report CS-TR-96-02, UT Austin, January 1996.

[10] A. G. Greenberg and N. Madras. How fair is fair queuing? *J. ACM*, 39(3):568–598, 1992.

[11] A. Gulati and P. Varman. Lexicographic QoS scheduling for parallel I/O. In *Proceedings of the ACM Symposium on Parallelism in Algorithms and Architectures*, pages 29–38, Las Vegas, Nevada, 2005. ACM Press.

[12] L. Huang, G. Peng, and T. Chiueh. Multi-dimensional storage virtualization. *SIGMETRICS Perform. Eval. Rev.*, 32(1):14–24, 2004.

[13] W. Jin, J. S. Chase, and J. Kaur. Interposed proportional sharing for a storage service utility. *SIGMETRICS Perform. Eval. Rev.*, 32(1):37–48, 2004.

[14] M. Karlsson, C. Karamanolis, and X. Zhu. Triage: Performance isolation and differentiation for storage systems. In *IWQoS*, June 2004.

[15] L. Kleinrock. *Queuing Systems, Volume 2: Computer Applications*. Wiley Interscience, 1976.

[16] C. Lumb, A. Merchant, and G. Alvarez. Façade: Virtual storage devices with performance guarantees. *File and Storage technologies (FAST'03)*, pages 131–144, March 2003.

[17] T. S. Eugene Ng, D. C. Stephens, I. Stoica, and H. Zhang. Supporting best-effort traffic with fair service curve. In *Measurement and Modeling of Computer Systems*, pages 218–219, 1999.

[18] A. K. Parekh and R. G. Gallager. A generalized processor sharing approach to flow control in integrated services networks: The single-node case. *IEEE/ACM Trans. Netw.*, 1(3):344–357, 1993.

[19] A. K. Parekh and R. G. Gallagher. A generalized processor sharing approach to flow control in integrated services networks: The multiple node case. *IEEE/ACM Trans. Netw.*, 2(2):137–150, 1994.

[20] H. Sariowan, R. L. Cruz, and G. C. Polyzos. Scheduling for quality of service guarantees via service curves. In *Proceedings of the International Conference on Computer Communications and Networks*, pages 512–520, 1995.

[21] M. Shreedhar and G. Varghese. Efficient fair queueing using deficit round robin. In *SIGCOMM '95: Proceedings of the conference on Applications, technologies, architectures, and protocols for computer communication*, pages 231–242, New York, 1995. ACM Press.

[22] D. Stiliadis and A. Varma. Efficient fair queueing algorithms for packet-switched networks. *IEEE/ACM Transactions on Networking*, 6(2):175–185, 1998.

[23] D. Stiliadis and A. Varma. Latency-rate servers: A general model for analysis of traffic scheduling algorithms. *IEEE/ACM Transactions on Networking*, 6(5):611–624, 1998.

[24] S. Suri, G. Varghese, and G. Chandramenon. Leap forward virtual clock: A new fair queueing scheme with guaranteed delay and throughput fairness. In *INFOCOMM'97*, April 1997.

[25] J. Wilkes. Traveling to Rome: QoS specifications for automated storage system management. In *International Workshop on QoS*, pages 75–91, June 2001.

[26] G. G. Xie and S. S. Lam. Delay guarantees of virtual clock server. *IEEE/ACM Transactions On Networking*, 3(6):683–689, December 1995.

[27] J. Zhang, A. Sivasubramaniam, Q. Wang, A. Riska, and E. Riedel. Storage performance virtualization via throughput and latency control. In *MASCOTS*, pages 135–142, 2005.

[28] L. Zhang. VirtualClock: A new traffic control algorithm for packet-switched networks. *ACM Trans. Comput. Syst.*, 9(2):101–124, 1991.

23
Optimal Parallel Scheduling Algorithms in WDM Packet Interconnects

23.1	Introduction and Background	23-1
23.2	Wavelength Conversion	23-3
23.3	Formalization of the Packet Scheduling Problem	23-4
23.4	Outline of the Parallel Segment Expanding Algorithm	23-5
23.5	Phase One: Matching Output Vertices	23-6
23.6	Phase Two: Matching Buffer Vertices	23-9
	Direct Insertion • An Outline for Augmenting Path Search • Identifying Forward and Backward Segments • Expanding the Reachable Set in Parallel • Details of Reachable Set Expanding • Proof for the Optimality of Reachable Set Expanding	
23.7	Implementation Issues and Complexity Analysis	23-15
23.8	Performance Evaluation	23-17
23.9	Summary	23-18
	References	23-18

Zhenghao Zhang
Yuanyuan Yang
State University of New York

In this chapter, we consider packet scheduling in wavelength-division-multiplexing (WDM) optical interconnects with limited range wavelength conversion and shared buffer. We will describe how an optimal packet schedule, by which packet loss and packet delay can be minimized, can be found by a parallel algorithm. We will formalize the problem as a matching problem in a bipartite graph and show step-by-step how the algorithm is derived on the basis of properties of the special bipartite graph.

23.1 Introduction and Background

All-optical networking is now widely regarded as a promising candidate for future high-speed networks [4,6,8] because of the huge bandwidth of optics. The bandwidth can be divided into a number of independent channels, which is referred to as WDM. In this chapter, we consider the packet scheduling problem in WDM packet interconnects.

In a communication network, an interconnect (or a switch) forwards the arriving packets to their destinations. In a WDM interconnect, *output contention* arises when more than one packet on the same wavelength is destined to the same output fiber at the same time. To resolve the contention, we can either temporarily store some of the packets in a buffer or convert the wavelengths of the packets to some idle wavelengths by *wavelength converters* [8]. However, these methods are expensive, since at present, optical buffers can only be implemented with optical delay lines (ODL) that are very expensive and bulky, and optical wavelength converters, if *full range* (i.e., capable of converting a wavelength to any other wavelengths), are very expensive and difficult to implement. Therefore, to reduce the cost, we can let all output fibers share a common ODL buffer pool [24,25], thus reducing the overall size of the buffer. We can also use *limited range* wavelength converters instead of *full range* wavelength converters, where a limited range wavelength converter can only convert a wavelength to a limited number of wavelengths. Recent results show that limited range wavelength converters will give a performance close to that of full range wavelength converters even when the conversion range is small [5,9,10].

Therefore, in this chapter, we consider a WDM interconnect with shared buffer and limited range wavelength conversion. As in References 4, 6, and 8, we assume that (i) the network is time slotted, (ii) the packets arrive at the interconnect at the beginning of time slots, (iii) the duration of an optical packet is one time slot, and (iv) each packet is destined to only one output fiber. Whenever possible, an arriving packet will be sent, by the switching fabric, to its destined output fiber. However, if it cannot be sent to the output fiber, it will be sent to one of the delay lines. After being delayed for one time slot, it will come out of the delay line and compete with other delayed packets as well as the newly arrived packets for access to the output fiber again. If it fails, it will be sent to a delay line again to wait for the next time slot. Figure 23.1 shows such an interconnect with N input fibers, N output fibers, and B delay lines. On each fiber, including the input fiber, the output fiber, and the delay line, there are k wavelengths. Each wavelength may carry one packet, and packets on different wavelengths are separated by *demultiplexers* following the input fibers or the delay lines. At one time slot, there can be as many as $(N + B)k$ packets, Nk from the input fibers and Bk from the delay lines, that are waiting to be switched by the switching fabric. To help resolve contention, each input *to the switching fabric* has a limited range wavelength converter that will convert the wavelength of a packet to a wavelength determined by the packet scheduling algorithm.

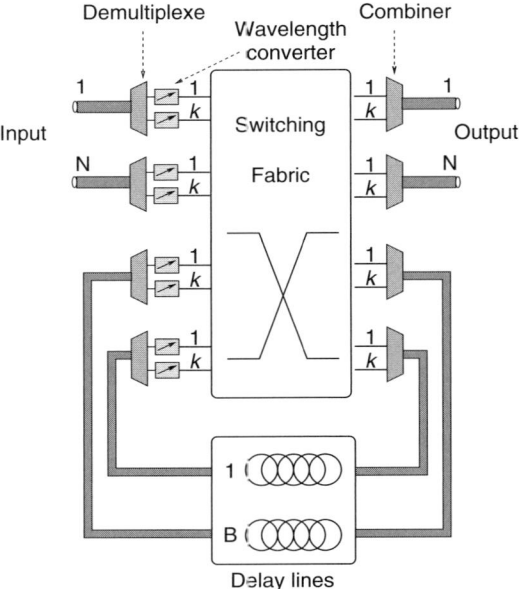

FIGURE 23.1 Optical packet interconnect with recirculating buffering and wavelength conversion.

The switching fabric is capable of connecting any of the $(N + B)k$ packets to any of the N output fibers or B delay lines. In front of the output fibers and the delay lines, *combiners* are used to combine packets on different wavelengths into one composite signal that can be sent to the output fibers or the delay lines.

The best packet scheduling algorithms for such interconnects should maximize throughput and minimize packet delay. To maximize throughput, we should send as many packets to the output fibers and ODLs as possible such that the fewest possible number of packets are dropped. In the mean time, to minimize packet delay, we should send a packet to the output fiber instead of sending it to the ODLs, whenever possible. As will be seen, this problem can be formalized as a weighted matching problem in a bipartite graph, and an optimal schedule is an optimal matching in the bipartite graph. However, if directly applying existing matching algorithms for general bipartite graphs, it will take $O((N-B)^3 k^3)$ time to find the matching, which is too slow for an optical interconnect. We will show that, owing to the limited range wavelength conversion, the bipartite graph will have some nice properties, and we can design a parallel algorithm called parallel segment expanding algorithm that runs in $O(Bk^2)$ time to find the optimal schedule.

23.2 Wavelength Conversion

All-optical wavelength conversion is usually achieved by conveying information from the input light signal to a probe signal [7,19]. The probe signal is generated by a tunable laser tuned to the desired output wavelength. The tuning range of the laser is continuous, but under limited range wavelength conversion, it is only part of the whole spectrum owing to constraints such as tuning speed, loss, and so forth.

We can see that a wavelength that can be converted to an interval of wavelengths, because the tuning range of the laser is continuous and covers an interval of wavelengths. Also, note that if the laser for the conversion of λ_1 can be tuned to λ_3, then the laser for the conversion of λ_2 should also be able to be tuned to λ_3, since λ_2 is closer to λ_3 than λ_1 is. These two observations lead to the following two assumptions of wavelength conversion:

Assumption 23.1 *The wavelengths that can be converted from λ_i for $i \in [1, k]$ can be represented by interval $[Begin(i), End(i)]$, where $Begin(i)$ and $End(i)$ are positive integers in $[1, k]$. Wavelengths that belong to this interval are called the* adjacency set *of λ_i.*

Assumption 23.2 *For two wavelengths λ_i and λ_j, if $i < j$, then $Begin(i) \leq Begin(j)$ and $End(i) \leq End(j)$.*

This type of wavelength conversion is called "ordered interval" because the adjacency set of a wavelength can be represented by an interval of integers, and the intervals for different wavelengths are "ordered." The cardinality of the adjacency set is called the *conversion degree* of the wavelength. Different wavelengths may have different conversion degrees. The *conversion distance* of a wavelength, denoted as d, is defined as the largest difference between a wavelength and a wavelength that can be converted from it.

A bipartite graph can be used to visualize the wavelength conversion. Let the left side vertices represent input wavelengths and the right side vertices represent output wavelengths. λ_i on the left is adjacent to λ_j on the right if λ_i can be converted to λ_j. Figure 23.2 shows such a conversion graph for $k = 8$. The adjacency set of λ_3, for example, can be represented as in References 2 and 4. The conversion degree and conversion distance of λ_3 are 3 and 1, respectively.

Note that the above assumptions on wavelength conversion are very general, only relying on the two facts observed at the beginning of this section. It is allowed for different wavelengths to have different conversion degrees and different conversion distances. This type of wavelength conversion is also used in other research works, for example, References 17, 20, and 21. Full range wavelength conversion can also be considered as a special case, by letting the conversion degrees for all wavelengths be k.

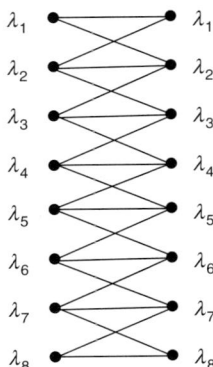

FIGURE 23.2 Wavelength conversion of an 8-wavelength system.

23.3 Formalization of the Packet Scheduling Problem

As mentioned earlier, the goal of the packet scheduling algorithm should be (i) To minimize the packet loss, drop as few packets as possible. (ii) To minimize the packet delay, send as many packets directly to the output fibers as possible.

This is a typical resource allocation problem and can be formalized as a matching problem in a bipartite graph. In this bipartite graph, left side vertices represent the packets and right side vertices represent the wavelength channels. For example, Figure 23.3b shows such a graph for a simple interconnect where $N = 2$, $B = 2$, and $k = 4$, when wavelength conversion is defined as in Figure 23.3a. Each vertex in Figure 23.3b is represented by a box. Vertices are arranged according to their wavelengths, lower wavelength first. Note that since there are two input fibers and two delay lines, there are a total of four boxes for each wavelength. On the left side, number "1" or "2" was put in the box as the index of the destination fiber of the packet. If a box is empty, there is no packet on this wavelength channel. On the right side, there are two types of vertices: the *output vertices* and the *buffer vertices*. The output vertices represent wavelength channels on the output fibers. The buffer vertices represent wavelength channels on the delay lines. In the figure, an output vertex is labeled as "1" or "2" according to the output fiber it is in, and a buffer vertex is labeled by a cross. A left side vertex, say, a, is adjacent to a right side vertex, say, b, if and only if the wavelength channel represented by b can be assigned to the packet represented by a. A necessary condition for a left side vertex to be adjacent to b is that its wavelength must be able to be converted to the wavelength of b. If b is a buffer vertex, all such left side vertices are adjacent to b, regardless of their destinations. However, if b is an output vertex, b is only adjacent to vertices representing packets destined to the output fiber where b is in.

In this bipartite graph, let E denote the set of edges. Any packet schedule can be represented by M that is a subset of E, where edge $ab \in M$ if wavelength channel b is assigned to packet a. Since any packet needs only one wavelength channel and a wavelength channel can be assigned to only one packet, the edges in M are vertex disjoint, since if two edges share a vertex, either one packet is assigned to two wavelength channels or one wavelength channel is assigned to two packets. Thus, M is a *matching* in G. If $ab \in M$, it is said that M "covers" a and b. If a is not covered by M, it is said that a is "free," "unmatched," or "unsaturated."

To maximize network throughput, a maximum cardinality matching should be found, because in such a matching the maximum number of packets are assigned to wavelength channels. To minimize the total delay, this matching should cover the maximum number of output vertices, because in such a matching the minimum number of packets will experience delay. To do this, we can assign weight 1 to the output vertices and weight 0 to the buffer vertices, then find an optimal matching, that is, a matching with

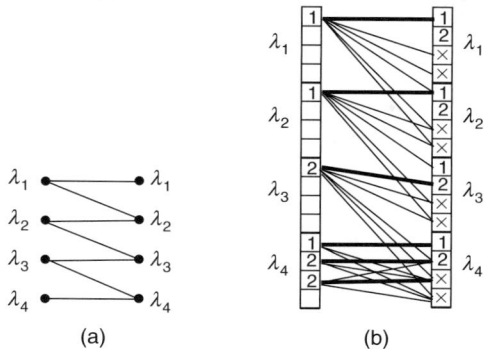

FIGURE 23.3 (a) Wavelength conversion of a 4-wavelength interconnect. (b) Packets and wavelength channels in an interconnect where $N = 2$ and $B = 2$.

maximum cardinality and also maximum total weight, where the weight of a matching is defined as the total weight of the vertices it covers. In the example in Figure 23.3b, an optimal matching is shown in heavy lines.

23.4 Outline of the Parallel Segment Expanding Algorithm

In this section, we give an outline of the parallel segment expanding algorithm that finds an optimal matching.

Optimal matching in a bipartite graph can be found by the following genetic algorithm, which can be called the matroid greedy algorithm [22,23]. The algorithm starts with an empty set Π and checks the weighted vertices one by one. In general, in step s it will check the vertex with the s_{th} largest weight. When checking vertex b, if there is a matching that covers b and all the vertices in Π, it will add b to Π; otherwise, b will not be added to Π. Then $s \leftarrow s + 1$ and repeat until all vertices have been checked. When finished, Π stores weighted vertices that can be covered by an optimal matching.

In an arbitrary bipartite graph with n vertices, to check whether a vertex can be covered along with vertices in Π needs $O(n^2)$ time; thus, the time complexity of the matroid greedy algorithm is $O(n^3)$. For our application, it would be as high as $O((N + B)^3 k^3)$, where N is the number of input/output fibers, B is the number of delay lines, and k is the number of wavelength channels, which is apparently too slow since the scheduling must be carried out in real time. In the following, we will give a fast optimal scheduling algorithm called the parallel segment expanding algorithm that solves the problem in $O(Bk^2)$ time. The algorithm is based on the matroid greedy algorithm; however, the running time is greatly reduced owing to the following two reasons.

First, in our bipartite graph, vertices have only two types of weight: output vertices have weight 1 and buffer vertices have weight 0. According to the matroid greedy algorithm, the output vertices should be checked first since they have a larger weight. After that, the buffer vertices should be checked. As a result, our algorithm runs in only two phases. In the first phase, it will find a matching that covers the maximum number of output vertices, such that the resulting matching will have the largest weight; in the second phase it will augment the matching until it covers as many buffer vertices as possible, such that the resulting matching will also have the maximum cardinality.

Second, our algorithm is run in parallel. We will not use a centralized scheduler that works on the bipartite graph introduced in Section 23.3, instead, we will "divide" the bipartite graph into N subgraphs, one for each output fiber, and use N processing units to find the matching in parallel.

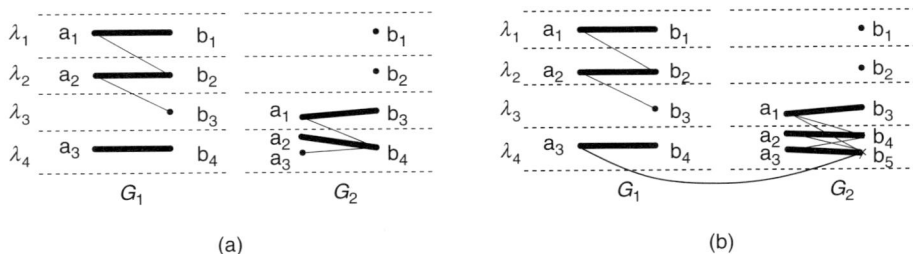

FIGURE 23.4 (a) Figure 23.3 is decomposed into two subgraphs when only considering the output vertices. (b) Assigning a buffer vertex on λ_4 to G_2.

The subgraphs will be denoted as G_i for $1 \leq i \leq N$. In G_i, left side vertices represent packets destined for output fiber i. Right side vertices include output vertices on output fiber i and buffer vertices that are matched to left side vertices in G_i. These buffer vertices are said to be "assigned" to G_i. Buffer vertices that cannot be matched are not shown in any subgraphs, because they cannot be used to augment the matching, as can be seen in the following sections. Note that this is an equivalent way of representing the input/output relationship as using the "whole" bipartite graph, nevertheless, this enables us to develop an algorithm that runs in parallel. For example, Figure 23.3b can be shown equivalently as in Figure 23.4b. It is important to note that output vertices are only adjacent to left side vertices in one particular subgraph while buffer vertices can be adjacent to left side vertices in different subgraphs.

For any subgraph G_i, left side vertices are denoted as a_1 to a_m and right side vertices are denoted as b_1 to b_n according to their wavelengths, where vertices on lower wavelengths have smaller indices, and vertices on the same wavelength are in an arbitrary order.

Note that in phase one, the parallelism of our algorithm is quite natural. Recall that in this phase only the output vertices need to be considered. Since an output vertex in G_i is only adjacent to left side vertices in G_i, G_i has no connections to other subgraphs, and the N subgraphs will be isolated from each other. For example, if we only consider the output vertices in Figure 23.3b, the whole bipartite graph is decomposed into two *isolated* subgraphs shown in Figure 23.4a. Therefore, to match the maximum number of output vertices, we can find a maximum matching for each of the subgraphs *in parallel* and then combine them together.

The problem becomes more complex in phase two, since after assigning buffer vertices to the subgraphs, the subgraphs will not be isolated. For example, a buffer vertex on λ_4 in the example of Figure 23.4b can be matched to a_3 in G_2, therefore it is assigned to G_2, and is denoted as b_5. Note that it is not only adjacent to vertices in G_2 but also to vertices in G_1. However, we can still have an algorithm that runs in all subgraphs in parallel by taking advantage of the properties on the subgraphs.

In phase two, according to the matroid greedy algorithm, check the buffer vertices one by one to see whether they can be matched along with all the previously matched vertices. Buffer vertices on lower wavelengths are checked first. The buffer vertex that is being checked is denoted as b_x. When checking b_x, the matching in G_i is denoted as M_i, and the union of M_i for $i = 1, 2, \ldots, N$ is denoted as M. If b_x can be matched, M is updated to cover b_x; otherwise, M is not changed and we proceed to the next buffer vertex. The details of the method for matching buffer vertices will be described in later sections. Some of the most frequently used notations are listed in Table 23.1.

23.5 Phase One: Matching Output Vertices

As explained earlier, in this phase we need only to consider N *isolated* subgraphs and find a maximum matching for each of them in parallel. Since the same method can be used for finding a maximum matching in each of the subgraphs, we will only explain it for one subgraph G_i.

Optimal Parallel Scheduling Algorithms in WDM Packet Interconnects

TABLE 23.1 List of Symbols

b_x	The buffer vertex being checked
G_i	Subgraph for output i
M_i	Current matching in G_i
M	Union of M_i for $1 \leq i \leq N$
$mat[a]$	The vertex matched to a
R	Reachable set of b_x
FS_p^i	The p_{th} forward segment in G_i
BS_q^i	The q_{th} backward segment in G_i
$breg_i$	Wavelengths of buffer vertices discovered in G_i
$lreg$	Wavelengths of newly discovered left side vertices

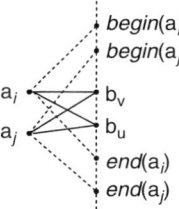

FIGURE 23.5 Illustration of Proposition 23.3. If $a_i b_u \in E$ and $a_j b_v \in E$, then $a_i b_v \in E$ and $a_j b_u \in E$.

Before presenting the algorithm, we first describe the properties of G_i. Owing to the two properties of ordered interval wavelength conversion, G_i has the following two properties:

Proposition 23.1 *The adjacency set of any left side vertex, say, a_i, is an interval and can be represented as $[begin(a_i), end(a_i)]$.*

Proposition 23.2 *If $i < j$, then $begin(a_i) \leq begin(a_j)$ and $end(a_i) \leq end(a_j)$.*

A bipartite graph with these two properties is called *request graph*. A request graph has the following property [27].

Proposition 23.3 *If edge $a_i b_u \in E$, $a_j b_v \in E$ and $i < j$, $u > v$, then $a_i b_v \in E$, $a_j b_u \in E$.*

This property can be called *crossing edge property*, because if $i < j$ and $u > v$, $a_i b_u$ and $a_j b_v$ will appear crossing each other in the request graph. A visual illustration of this property is shown in Figure 23.5.

Proof. By Proposition 23.1, to show $a_i b_v \in E$ is to show that $begin(a_i) \leq v \leq end(a_i)$. First, since $i < j$, by Proposition 23.2, $begin(a_i) \leq begin(a_j)$. Also, since $a_j b_v \in E$, $begin(a_j) \leq v$. Therefore, $begin(a_i) \leq v$. Next, since $a_i b_u \in E$, $u \leq end(a_i)$. Therefore, $v \leq u \leq end(a_i)$. Similarly, we can show $a_j b_u \in E$. ∎

It is a good exercise to prove that in a request graph, right side vertices have the same properties as the left side vertices, that is,

Proposition 23.4 *The adjacency set of any right side vertex, say, b_u, is an interval and can be represented as $[begin(b_u), end(b_u)]$.*

Proposition 23.5 *If $u < v$, then $begin(b_u) \leq begin(b_v)$ and $end(b_u) \leq end(b_v)$.*

Maximum matching in request graphs can be found by the first available algorithm described in Table 23.2 [27]. This algorithm checks the right side vertices from the top to the bottom. A right side

TABLE 23.2 First Available Algorithm

---
for $u := 1$ **to** n **do**
    let a_j be the free vertex adjacent to b_u
    with the smallest index
    if no such a_j exists
        b_u is not matched
    else
        match b_u to a_j
    end if
end for

---

vertex b_u will be matched to its *first available neighbor*, which is the free left side vertex adjacent to it with the smallest index. The time complexity of this algorithm is $O(n)$, where n is the number of right side vertices, since the loop is executed n times, and the work within the loop can be done in constant time. For example, after running the first available algorithm, the matchings in Figure 23.4a are shown in heavy lines.

Theorem 23.1 *The first available algorithm finds a maximum matching in a request graph.*

Proof. We first define the "top edge" of a request graph as the edge connecting the first non-isolated left side vertex to the first nonisolated right side vertex. For example, edge a_1b_1 in G_1 of Figure 23.4a is the top edge. We claim that the top edge must belong to some maximum matching. To see this, let a_ib_u be the top edge of request graph G_i. Given any maximum matching of G_i, if it contains edge a_ib_u, such a maximum matching has been found. Otherwise, we show that it can be transformed into a maximum matching that contains edge a_ib_u. Note that any maximum matching must cover at least one of a_i and b_u, since otherwise we can add in edge a_ib_u to obtain a matching with a larger cardinality. If exactly one of a_i and b_u is matched, the edge covering a_i or b_u can be removed and edge a_ib_u will be added in, and the resulting matching will still be of maximum cardinality. For example, if in G_1 of Figure 23.4a a matching matches a_1 to b_2 and b_1 is unmatched, we can match a_1 to b_1 and leave b_2 unmatched, since this new matching will be of the same cardinality as the old matching. Hence, we only need to consider the case when a_i and b_u are both matched, but not to each other. Let a_i be matched to b_v and b_u matched to a_j. Since a_i and b_u are the first nonisolated vertices, $i < j$ and $u < v$. Thus, by Proposition 3, we have $a_ib_u \in E$ and $a_jb_v \in E$. Therefore, we can match a_i to b_u and a_j to b_v, and the resulting matching is still maximum and also contains edge a_ib_u.

Having seen that a_ib_u must belong to a maximum matching, we can remove a_i and b_u from G_i, because it can be easily verified that a maximum matching in the residual graph, plus edge a_ib_u, is a maximum matching in G_i. Recursively applying this fact, we can see that a maximum matching can be found by repeatedly taking the top edge. Note that this is precisely what the first available algorithm does by matching a right side vertex to its first available neighbor in each step. ∎

Before ending this section, we give some more properties of the request graph that will be used in the remaining sections.

Proposition 23.6 *There must be a maximum matching in a request graph with no crossing edges.*

This is because given any matching, if there is a pair of crossing edges, say, a_ib_u and a_jb_v, we can replace them with a_ib_v and a_jb_u that do not cross each other. Such a matching is called a noncrossing matching, in which the i-th matched left side vertex is matched to the i-th matched right side vertex.

It can be easily verified that

Proposition 23.7 *The matching found by first available algorithm is noncrossing.*

Another property of the matching found by the first available algorithm, owing to the fact that a vertex is always matched to its first available neighbor, is

Proposition 23.8 *If b_u is matched to a_i, all the left side vertices adjacent to b_u with smaller indices than a_i must be matched to right side vertices with smaller indices than b_u.*

Here we introduce a more sophisticated example shown in Figure 23.6, which will be used throughout this chapter to help understand our algorithm. In this example, $N = 4$, $B = 2$, and $k = 8$, and the wavelength conversion is as defined in Figure 23.2. For notational convenience, we use a 1×8 vector called the arrival vector to represent the number of packets destined to an output fiber, in which the ith element is the number of packets destined to this output fiber on λ_i. The arrival vector is [2, 1, 1, 0, 3, 0, 0, 2] for output fiber 1, [2, 5, 3, 0, 0, 0, 0, 0] for output fiber 2, [2, 0, 2, 5, 0, 0, 0, 0] for output fiber 3, and [0, 0, 0, 0, 0, 0, 0, 0] for output fiber 4. There is no packet destined to output fiber 4, therefore, only 3 subgraphs, G_1, G_2, and G_3 are shown. Figure 23.6a shows the matching after running first available algorithm on each of the subgraphs (to make the figures readable, not all edges in the graphs are shown).

23.6 Phase Two: Matching Buffer Vertices

After phase one, we need to augment the matching to match buffer vertices. The goal is to match as many buffer vertices as possible while keeping the previously matched vertices matched.

As mentioned earlier, we will check the buffer vertices one by one. We will first try to use a simple method called "direct insertion" to match b_x, which is the buffer vertex being checked. Direct insertion, roughly speaking, is to match b_x in each subgraph by running the first available algorithm in these subgraphs in parallel. b_x can be matched if it can be matched in one of the subgraphs. However, if direct insertion fails, it may still be able to be matched. In this case, we will try a more complex method, which is to find an M-augmenting path with one end being b_x, where M denotes the current matching. By graph theory, b_x can be matched if and only if there exists such a path. An M-augmenting path is an M-alternating path with both ends being unmatched vertices, where an M-alternating path is a path that alternates between edges in M and not in M. If such an augmenting path is found, a "flip" operation can be performed on the edges in the path, that is, the edges that were in M can be removed and the edges that were not in M can be added in. The new matching will contain one more edge than the old matching, and will cover all previously covered vertices plus the two vertices at the ends of the path.

23.6.1 Direct Insertion

To match buffer vertex b_x by direct insertion in subgraph G_i, a vertex on the same wavelength as b_x can be added to G_i, then the first available algorithm is run. If all the right side vertices in G_i can be matched, including the added vertex, b_x can be matched in G_i and will be assigned to it. Note that this can be done in parallel in all subgraphs, and if b_x can be directly inserted to more than one subgraphs, we can arbitrarily pick one.

After assigning a buffer vertex to a subgraph, an index is given to the vertex and the indices of some right side vertices are also updated. For example, in Figure 23.6b, buffer vertex b_x on wavelength λ_1 can be matched in G_1. After assigning it to G_1, give an index 2 to it, that is, refer to it as b_2, and increment the indices of the right side vertices following it by 1.

23.6.2 An Outline for Augmenting Path Search

When b_x cannot be directly inserted to any of the N subgraphs, b_x may still be able to be matched. We can try our second method, which is to find an M-augmenting path.

An example is shown in Figure 23.6c. In this figure, on the right side of the subgraphs, a dot represents an output vertex and a cross represents a buffer vertex. Edges in M are represented by solid lines and edges

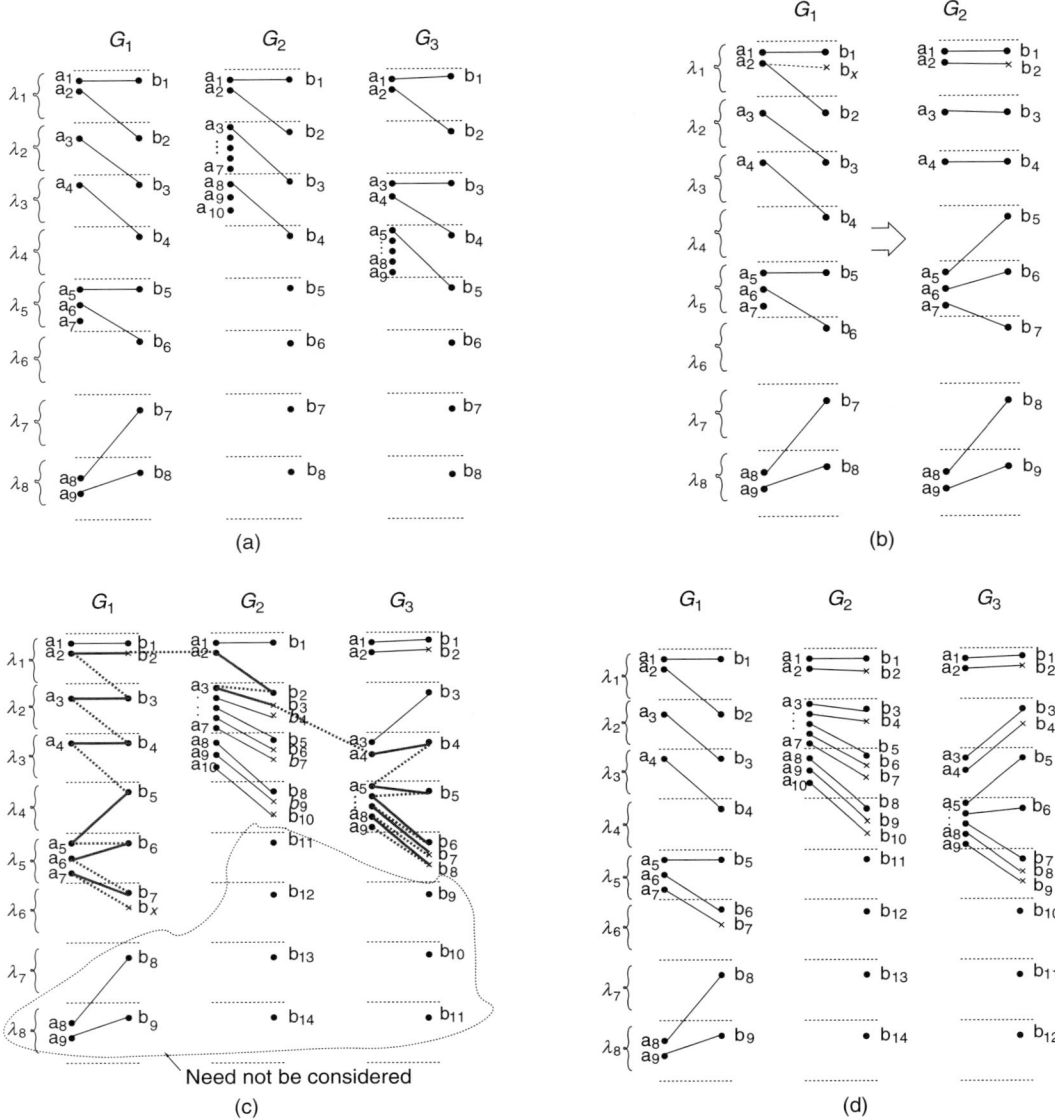

FIGURE 23.6 Matchings in an interconnect where $N = 4$, $B = 2$, and $k = 8$, and the wavelength conversion is as defined in Figure 23.2. G_4 is not shown here because there is no packet destined to output fiber 4. On the right side, a dot is an output vertex and a cross is a buffer vertex. (a) After matching output vertices by the first available algorithm. (b) Direct insertion of a buffer vertex on λ_1 to G_1. (c) An augmenting path for b_x on λ_6. (d) b_x is matched in G_1 and is given index 7.

not in M are represented by dashed lines. Let b_x be a buffer vertex on wavelength λ_6. It is not hard to see that b_x cannot be directly inserted in any of the subgraphs. However, it can still be matched, because as shown in the figure, there is an M-augmenting path starting at b_x, traversing G_1, G_2, and G_3, and ending at an unmatched left side vertex a_9 in G_3. After the flip operation, the new matching is shown in Figure 23.6d. Note that we have moved the buffer vertex on λ_1 from G_1 to G_2 and moved the buffer vertex on λ_2 from G_2 to G_3, because according to the new matching they should be assigned to G_2 and G_3, respectively.

It is important to note that in the augmenting path, some buffer vertices serve as "bridges" to extend the path from one subgraph to another. Also, note that to update the matching according to the augmenting path, we are actually moving some buffer vertices from one subgraph to another. The purpose of this moving can be considered as "making room" for b_x in one of the subgraphs. The optimal scheduling can also be considered as optimally assigning the buffer vertices to the subgraphs such that maximum number of them can be matched.

To find the augmenting path, as algorithms for general bipartite graphs, our algorithm also searches for the *reachable set* of b_x. The reachable set, denoted as R, is defined as the set of vertices that can be reached from b_x via M-alternating paths. If there is an unmatched left side vertex in this set, the augmenting path is found. Note that the differences between our algorithm and algorithms for more general bipartite graphs are profound. First, our algorithm is executed in all subgraphs in parallel, as a result, many vertices in different subgraphs can be added to R simultaneously. Second, in the general method, a vertex can be added to R only if it is adjacent to some vertices already in R. In our algorithm, many vertices can be added to R even if they are not adjacent to any vertices currently in R, as long as these vertices are in the same *forward segment* or *backward segment* that will be defined soon. More importantly, this can be done in *constant* time regardless of the number of vertices in the segment, which makes a parallel algorithm possible.

Before presenting the algorithm, owing to the importance of the forward and backward segments, we will first give formal definitions of them and describe their properties in the next subsection.

23.6.3 Identifying Forward and Backward Segments

First, assume that M_i is the matching found by running the first available algorithm in G_i. Recall that owing to the properties of the first available algorithm, M_i is noncrossing.

The forward and backward segments are defined within a subgraph, hence in this subsection we consider only one subgraph, say, G_i. Note that unmatched right side vertices in G_i need not be considered because they cannot be used to expand R. Also, note that we need only to work on a subgraph of G_i with vertex set of a_1 to a_i and b_1 to b_u, where b_u is the right side vertex on the same wavelength as b_x with the largest index and a_i is the vertex matched to b_u. This is because if b_x cannot be directly inserted in G_i, left side vertices with larger indices than a_i have to be matched to right side vertices with indices larger than b_u. For example, in Figure 23.6c, a_i is vertex a_7 and b_u is vertex b_7, and the subgraph of G_1 we need to work on has vertex set a_1 to a_7 and b_1 to b_7. However, to avoid introducing too many notations, this subgraph will be referred to as G_i with the understanding that when searching for the augmenting path G_i means a subgraph of itself.

First, we define *forward segment*. For notational simplicity, for any vertex b, we use $mat[b]$ to denote the vertex matched to it. Note that the index of $mat[b_p]$ is larger than $mat[b_q]$ if $p > q$, since M_i is noncrossing. Now, imagine scanning the right side vertices from $b_1, b_2, \ldots$, until b_v when one of the following conditions is satisfied: (i) b_v is not adjacent to $mat[b_{v+1}]$; or (ii) there are some unmatched left side vertices adjacent to b_v. If such b_v is found, b_1 to b_v and all the left side vertices matched to them, plus the unmatched left side vertices adjacent to b_v with smaller indices than $mat[b_{v+1}]$ are called a *forward segment*. After finding the first forward segment, scan from b_{v+1} to find the second forward segment, then the third until reaching b_u, the last right side vertex in G_i.

As an example, in Figure 23.6c, G_1 and G_2 both have only one forward segment. However, when scanning right side vertices in G_3, we found that although b_1 is adjacent to a_2 ($mat[b_2]$), b_2 is not adjacent to a_3 ($mat[b_3]$). Therefore, a_1 to a_2 and b_1 to b_2 should be a forward segment. The second forward segment in G_3 is a_3 to a_9 and b_3 to b_8.

We now describe the properties of a forward segment.

Proposition 23.9 *In a forward segment, the index of an unmatched left side vertex is larger than any matched left side vertex.*

Proof. Without loss of generality, consider the first forward segment. Let a_j be the unmatched left side vertex and let b_v be the last right side vertex in this segment. It suffices to show that the index of $mat[b_v]$ must be smaller than a_j. However, this is immediate owing to Proposition 23.8. ∎

The following is the most important property of a forward segment. It enables us to add many vertices to the reachable set at the same time.

Proposition 23.10 *Suppose left side vertex a_s is matched to b_p. If a_s is in R, then in the same forward segment as a_s, left side vertices with indices no less than a_s and right side vertices with indices no less than b_p can all be added to R.*

Proof. Let b_v be the last right side vertex in this forward segment and suppose b_v is matched to a_t. Since the matching is noncrossing, left side vertices in the same forward segment with larger indices than a_s are either matched to $b_{p+1}, b_{p+2}, \ldots, b_v$ or unmatched (if there are any). By the definition of the forward segment, a possible alternating path is a_s to b_p, b_p to $mat[b_{p+1}]$, $mat[b_{p+1}]$ to $b_{p+1}, \ldots, b_{v-1}$ to a_t, a_t to b_v, then to any of the unmatched vertices. ∎

For example, in Figure 23.6c, once the alternating path reaches a_4 in G_3, left side vertices with indices no less than a_4 and right side vertices with indices no less than b_4 can all be added to R. Note that this is a much efficient way to expand the reachable set, since many vertices, including those not adjacent to a_4, are added to R in one single operation.

The following properties can also be easily verified.

Proposition 23.11 *Any nonisolated vertex is in exactly one forward segment.*

Proposition 23.12 *Nonisolated vertices on the same wavelength are in the same forward segment.*

We will denote the p-th forward segment in G_i as FS_p^i. For a forward segment, the following information is needed in our algorithm: (i) Whether there is an unmatched left side vertex in this segment, (ii) The wavelength indices of buffer vertices in the segment. The first item can be stored in a single bit. The second item is stored in a k-bit register, denoted as $fsreg_p^i$ for FS_p^i, in which bit l is one if there is a buffer vertex on λ_l in FS_p^i. For example, the information about the second forward segment in G_3 is $\{yes; \text{"000010XX"}\}$, which means that in this segment there is an unmatched left side vertex, and there are buffer vertices on λ_5. "X" means "do not care," since there cannot be right side vertices on wavelengths higher than λ_6 assigned to any subgraphs when checking b_x because vertices on lower wavelengths are checked first.

The *backward segment* is defined in a similar way, only reversing the scanning direction to backwards. Imagine starting at b_u where b_u is the last right side vertex in G_i, and scan back to $b_{u-1}, b_{u-2}, \ldots$, till b_w when b_w is not adjacent to $mat[b_{w-1}]$. b_w to b_u and all the left side vertices matched to them is called a *backward segment*. After finding the first backward segment, we can start scanning from b_{w-1} to find other backward segments. Note that unlike the forward segments that are numbered from top to bottom, the backward segments are numbered in the reverse direction, or from bottom to top, in accordance to the scanning direction. For example, in Figure 23.6c, both G_1 and G_3 have only one backward segment. G_2 has two backward segments. The first one has vertex set a_8 to a_{10} and b_8 to b_{10}, and the second one has vertex set a_1 to a_7 and b_1 to b_7.

There are also similar properties regarding backward segments.

Proposition 23.13 *For a backward segment, let b_u and b_w be the right side vertices with the largest and the smallest indices, respectively. Suppose b_u is matched to a_i and b_w is matched to a_j. Then left side vertices between a_i and a_j are all matched.*

Proof. By contradiction. If this is not true, suppose a_l is not matched where $i > l > j$. Let a_s and a_t be matched left side vertices in this segment where $s < l$ and $t > l$. Furthermore, let a_s and a_t be such vertices that are closest to a_l. Note that if a_t is matched to b_z, then a_s must be matched to b_{z-1} and b_z must be adjacent to a_s. By Proposition 23.1 of the request graph, b_z is also adjacent to a_l. But since $l < t$, First Available Algorithm would not have matched b_z to a_t. This is a contradiction. ∎

Similar to forward segments, we can show that the following.

Proposition 23.14 *Suppose left side vertex a_s is matched to b_w. If a_s is in R, then in the same backward segment as a_s, left side vertices with indices no more than a_s and right side vertices with indices no more than b_w can all be added to R.*

Proposition 23.15 *Any matched vertex is in exactly one backward segment.*

Proposition 23.16 *Matched vertices on the same wavelength are in the same backward segment.*

We will denote the q_{th} backward segment in G_i as BS_q^i. Note that unmatched vertices are not in any backward segment, and the information of a backward segment needed in our algorithm is only the wavelength indices of buffer vertices, stored in register $bsreg_q^i$ for BS_q^i. For example, in Figure 23.6c, $bsreg_1^1$ is {"100000XX"}, which means that there is a buffer vertex on λ_1.

23.6.4 Expanding the Reachable Set in Parallel

We are now ready to present the core of the algorithm, which is to expand the reachable set R in parallel. In the expanding process, we say a buffer vertex in R *discovers* a forward or backward segment if it is adjacent to some left side vertex in the segment. Left side vertices and right side vertices satisfying conditions described in Propositions 23.10 and 23.14 are also said to be discovered by this buffer vertex, and the buffer vertex is called the "discoverer" of these vertices. All the discovered vertices can be added to R, hence in the rest of the chapter we will interchangeably say a vertex is "discovered" or a vertex is "added to R."

Despite the technical details, the idea is quite simple, which is explained in the following. At the beginning, b_x itself is added to R. Then *in parallel*, each subgraph checks for vertices in itself that can be discovered by b_x. These vertices, if there are any, must be in the first backward segment of each subgraph. Note that no unmatched left side vertices are discovered at this time since there are no unmatched vertices in backward segments. However, at this time some, buffer vertices may have been discovered and they can be used to reach from one subgraph to other subgraphs. Each subgraph will announce buffer vertices discovered in itself. The union of the buffer vertices announced by all N subgraphs can be used to further discover "new" vertices, which is also done *in parallel* in each subgraph, just as how b_x was used to discover vertices at the beginning. If unmatched left side vertices are discovered, the searching is done. Otherwise, as before, each subgraph can announce the newly discovered buffer vertices, which can be used to discover new vertices. This process is repeated until an unmatched left side vertex is discovered or until after a round no new buffer vertex is discovered.

In short, the algorithm runs in "rounds." Before each round, the buffer vertices discovered in the previous round were made known to every subgraph. Then *in parallel*, each subgraph uses these buffer vertices to discover vertices in it. If no unmatched left side vertices were discovered, the subgraphs will announce the newly discovered buffer vertices and then start the next round.

We can now explain how the augmenting path was found in the example in Figure 23.6c.

Round 1. Since b_x is on λ_6, it is only adjacent to a_5, a_6, and a_7 in G_1. It can discover the first backward segment in G_1, and vertices a_1 to a_7 and b_1 to b_7. Only one buffer vertex, b_2 in G_1, can be discovered. Thus, G_1 announces that there is a buffer vertex on λ_1, while G_2 and G_3 do not make any announcement.

Round 2. Each of the subgraphs checks which forward and backward segments can be discovered by a buffer vertex on λ_1. G_1 finds out that these segments have been discovered before. G_2 finds that the buffer vertex can discover the first forward and the second backward segment. G_3 finds that the buffer vertex can discover the first forward and the first backward segment. There are no unmatched left side vertices in these segments, however, buffer vertices on λ_2, λ_3, and λ_4 in the first forward segment of G_2 can be discovered.

Round 3. G_3 finds that a buffer vertex on λ_2 can discover the second forward segment in it, and there is an unmatched left side vertex a_9 in this segment, thus an augmenting path is found.

23.6.5 Details of Reachable Set Expanding

After explaining the general idea, in this subsection we give the details of parallel reachable set expanding.

First, note that after a round, to announce the newly discovered buffer vertices, subgraph G_i needs only to write to a k-bit register, denoted as $breg_i$, in which bit i is "1" if a buffer vertex on λ_i in G_i has been discovered. The union of the N announcements can thus be obtained by doing an "or" operation to $breg_i$ for all $1 \leq i \leq N$, and suppose the result is stored in register $breg$. With $breg$ we can obtain $lreg$, in which bit i is "1" if left side vertices on λ_i can be discovered by the newly discovered buffer vertices indicated by $breg$, and if these left side vertices have not been discovered before.

In the next round, each subgraph will examine $lreg$ by scanning through it. If subgraph G_i finds that bit l of $lreg$ is "1," it will check whether there are left side vertices on λ_l, and if yes, these left side vertices have been discovered. It can then find the forward and backward segment to which these left side vertices belong, and suppose they are FS_p^i and BS_q^i, respectively. From previous discussions, we know that left side vertices and right side vertices in the same forward and backward satisfying conditions described in Propositions 23.10 and 23.14 are also discovered. To be specific, left side vertices in FS_p^i on wavelengths no lower than λ_l and left side vertices in BS_q^i on wavelengths no higher than λ_l are discovered. If there are unmatched left side vertices in FS_p^i, the search is over. Otherwise, suppose a right side vertex on λ_h is matched to one of the left side vertices on λ_l. Then right side vertices in FS_p^i on wavelengths no lower than λ_h and right side vertices in BS_q^i on wavelengths no higher than λ_h are discovered. This right side vertex is referred to as the "pivot" vertex. Any right side vertex matched to left side vertex on λ_l can be a pivot vertex, and in practice we can choose the one with the smallest index. The subgraph needs to write a few "1"s to $breg_i$ according to the wavelengths of discovered buffer vertices. Let κ_h^f be a k-bit register where bit t is "1" if $t \geq h$. Let κ_h^b be a k-bit register where bit t is "1" if $t \leq h$. To update $breg_i$ we can "or" $breg_i$ with $fsreg_p^i \& \kappa_h^f$ and $bsreg_q^i \& \kappa_h^b$. Note that all these operations take *constant* time in all subgraphs; therefore, all subgraphs will finish checking $lreg$ at the same time in each round, which is the desired behavior of a parallel algorithm.

When an unmatched left side vertex is found, the algorithm needs to establish the M-augmenting path. Note that if a vertex was discovered in a certain round, its discoverer must be discovered in the previous round, and there must be an M-alternating path from a vertex to its discoverer. When the unmatched left side vertex, say, a_t, is found, we will have a sequence of vertices, a_t, b_u, b_v, b_w, ..., b_z, b_x, where each vertex is discovered by the vertex next to it. In case a vertex has multiple discoverers, we can arbitrarily choose one. Therefore, we can start with a_t, and first find the M-alternating path to its discoverer b_u, then extend the M-alternating path from b_u to b_v, and so on, until the path extends to b_x.

However, in implementation, the augmenting path need not be explicitly established. As mentioned earlier, to update the matching is actually to move some buffer vertices from one subgraph to another. We can first move b_u to the subgraph where a_t is in, then move b_v into the subgraph where b_u used to be in, then move b_w into the subgraph where b_v used to be in, ..., until b_x is moved into the subgraph where b_z used to be in. Then we can run the first available algorithm in the subgraphs in parallel to obtain the new matching.

23.6.6 Proof for the Optimality of Reachable Set Expanding

We now prove the optimality of the algorithm.

Theorem 23.2 *The algorithm is capable of finding an M-augmenting path with one end being b_x if there exists one.*

Proof. It suffices to show that if the algorithm terminates without reporting an augmenting path, it has discovered all the left side vertices reachable from b_x. We show this by contradiction. If this is not true, then there will be a left side vertex a_t, either matched or unmatched, that can be reached by an alternating path starting from b_x but was not discovered by the algorithm. Suppose the last three vertices in this hypothetical alternating path are a_s, b_u, and a_t, in this order. We claim that if a_t was not discovered, neither would a_s.

Suppose the claim is not true, that is, the algorithm has discovered a_s. Note that since b_u must be matched to a_s, if the algorithm has discovered a_s, it should also have discovered b_u. If b_u is a buffer vertex, the algorithm would have discovered a_t, since all vertices adjacent to buffer vertices were scanned. Thus, b_u must be an output vertex. Note that if b_u is an output vertex, a_t must be in the same subgraph as a_s.

First, consider when $s < t$. In this case, if a_s and a_t are in the same forward segment, the algorithm would have discovered a_t, thus a_s and a_t must be in different forward segments. However, if this is true, there must be an unmatched left side vertex between a_s and a_t since b_u is adjacent to a_t, by the definition of forward segments. Thus, the algorithm would have terminated by discovering an unmatched left side vertex, which is a contradiction. Thus, s cannot be smaller than t.

However, if $s > t$, according to Proposition 23.8 in Section 23.5, a_t must be matched. It is not hard to see that a_s and a_t must be in the same backward segment since b_u is adjacent to a_t. But, in this case, the algorithm must also have discovered a_t, which is also a contradiction. Therefore, the claim is true, that is, the algorithm did not discover a_s.

Following the same argument, we can show that the algorithm also did not discover the left side vertex before a_s in the alternating path, and this can be carried on until the conclusion that the algorithm did not discover any left side vertex in this alternating path. But, this contradicts with the fact that the algorithm at least discovers all the left side vertices directly adjacent to b_x and any alternating path starting from b_x must visit one of these vertices first. ∎

We summarize all the previous discussions and present the parallel segment expanding algorithm in Table 23.3 for finding optimal schedules in the WDM interconnect with shared buffer.

23.7 Implementation Issues and Complexity Analysis

In this section, we give the complexity analysis of the algorithm. We show that by using N processing units or decision-making units, the algorithm runs in $O(Bk^2)$ time. It should be mentioned that it is not difficult to implement the N processing units in hardware to reduce the cost and further speedup the scheduling process.

The key to reducing the running time and storage space is that the algorithm works on *wavelengths* of vertices rather than the indices, because vertices on the same wavelength have the same adjacency sets and are in the same forward and backward segments. For each subgraph, both left side and right side vertices can be stored in an array with k entries, denoted by $LS[]$ and $RS[]$, respectively, where each entry stores (i) the number of vertices on this wavelength; and (ii) the index of forward and backward segments these vertices belong to. The forward and backward segments can be described by two intervals of integers, because in a segment, left side vertices and right side vertices are all consecutive. For example, the second forward segment in G_3 in Figure 23.6 can be represented by References 3 and 4 and References 2 and 4, corresponding to left and right side vertices, respectively, which means that vertices with wavelengths in this range are in this segment. Each segment also needs a k-bit register to store the wavelengths of buffer vertices. In addition, the forward segment needs one bit to store the information of whether there is an unmatched left side vertex. The total storage space needed for all N subgraphs is $O(Nk(k + \log Nk))$ bits.

The first task, covering maximum number of output vertices, is to run first available algorithm on each subgraph with only output vertices. It can be done in parallel and takes $O(k)$ time.

TABLE 23.3 Parallel Segment Expanding Algorithm

Run First Available Algorithm in all subgraphs in parallel
to cover maximum number of output vertices.
for every buffer vertex b_x **do**
   Try direct insertion in all subgraphs in parallel.
   If successful, **continue** to the next buffer vertex.
   for $i := 1$ **to** N **do in parallel**
     if b_x can discover BS_1^i
       $breg_i \leftarrow bsreg_1^i$
     end if
   end for
   Generate $lreg$ according to $breg_i$, $i \in [1, N]$.
   while $lreg$ is not all zero
     for $i := 1$ **to** N **do in parallel**
       Clear $breg_i$.
       for each '1' bit in $lreg$, say, bit l, **do**
         if no left side vertex on λ_l
           continue to next '1' bit;
         end if
         Suppose left side vertices on λ_l are in FS_p^i and BS_q^i.
         if there are unmatched left side vertices in FS_p
           break from the while loop;
         end if
         Update $breg_i$ according to $fsreg_p^i$ and $bsreg_q^i$.
       end for
     end for
     Generate $lreg$ according to $breg_i$, $i \in [1, N]$. Reset a bit in
     $lreg$ to '0' if it has been checked in previous rounds.
   end while
   Update the matching if an augmenting path is found
end for

After this, we need to match buffer vertices one by one. We will show that to match a buffer vertex, in total, $O(k)$ time is needed. Suppose buffer vertex b_x is on wavelength λ_j. As mentioned earlier, first we will try direct insertion in parallel. For each subgraph, this is to increment the number of vertices of $RS[j]$ by one, then run the first available algorithm, which takes $O(k)$ time.

If direct insertion fails, we will try to find the augmenting path. We can first use $O(k)$ time to find the forward and backward segments in each subgraph by a linear search. Then, at each round of reachable set expansion, each subgraph checks the content of register $lreg$. While scanning through $lreg$, we can use simple hardware to skip the "0" bits and scan only the "1" bits. As explained earlier, for each "1" bit in $lreg$, the work can be done in constant time for all subgraphs. Also, note that in each round the subgraph only checks the new wavelengths that have not been checked before. Thus, the total time spent on this task in the entire expansion process is $O(k)$, since there are k wavelengths.

Between two expansions, the content of $breg_i$ for $1 \leq i \leq N$ should be "or"ed, and then be used to generate $lreg$. We can implement this function in hardware that finishes this job in constant time. Also, note that the number of expansion rounds can never exceed k, because after each round, there must be some "1" bit in $lreg$, that is, there must be some newly discovered left side vertices. Therefore, the total time spent in generating $lreg$ is also $O(k)$.

If one unmatched left side vertex is found, the matching should be updated. As explained earlier, this is to move some buffer vertices from one subgraph to another, which is to increment or decrement the corresponding entries in $RS[]$ of the subgraphs. Since the number of expansion rounds cannot exceed k, this will also take $O(k)$ time. Then the new matching can be obtained in $O(k)$ time by running the first available algorithm on each subgraph.

Therefore, overall, it needs $O(k)$ time to match a buffer vertex. Thus, the total running time of the algorithm is $O(Bk^2)$ since there are Bk buffer vertices.

23.8 Performance Evaluation

We implemented parallel segment expanding algorithm in software and evaluated the performance of the WDM interconnect. The interconnect simulated has 8 input fibers and 8 output fibers with 8 wavelengths on each fiber. We assume that the arrivals of the packets at the input channels are bursty: an input channel alternates between two states, the "busy" state and the "idle" state. When in the "busy" state it continuously receives packets and all the packets go to the same destination. When in the "idle" state it does not receive any packets. The length of the busy and idle periods follows geometric distribution. The durations of the packets are all one time slot and for each experiment the simulation program was run for 100,000 time slots.

In Figure 23.7a, we show the packet loss probability (PLP) of the interconnect as a function of the number of fiber delay lines. The traffic load is 0.8. The average burst length and the average idle period are

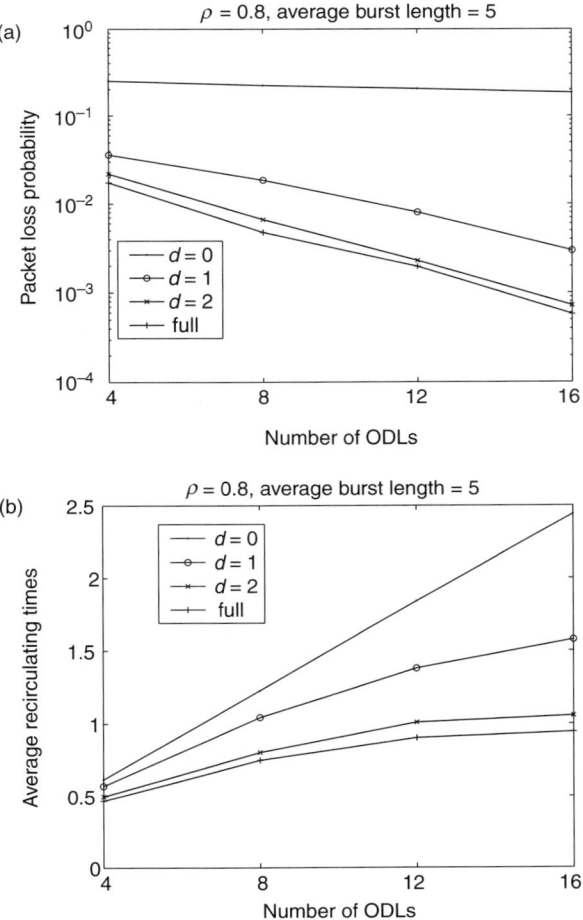

FIGURE 23.7 Performance of a WDM interconnect with 8 input/output fibers and 8 wavelengths per fiber under bursty traffic. (a) Packet loss probability. (b) Average delay.

5 and 1.25 time slots, respectively. Two wavelength conversion distances, $d = 1$ and $d = 2$, were tested, along with the no conversion case ($d = 0$) and the full conversion case.

As expected, packet loss probability decreases as the number of delay lines increases. However, more significant improvements seem to come from the increase of the conversion ability. For example, when $d = 0$, the PLP curve is almost flat, which means that without wavelength conversion, there is scarcely any benefit by adding buffer to the interconnect. But when $d = 1$, the PLP drops by a great amount as compared to $d = 0$, and the slope of the PLP curve is also much larger. We can also see that the PLP for $d = 2$ is already very close to the full wavelength conversion, therefore, there is no need to further increase the conversion ability. All this suggests that in a WDM interconnect, wavelength conversion is crucial, but the conversion need not be full range, and with a proper conversion distance, adding buffer will significantly improve the performance.

In Figure 23.7b, we show the average packet delay as a function of the number of fiber delay lines. The average delay is actually the average rounds a packet needs to be recirculated before being sent to the output fiber. We can see that a larger conversion distance results in a shorter delay, and the delay for $d = 2$ is very close to the delay for full conversion. Fixing the conversion distance, the delay is longer for larger buffer sizes, since when buffer is larger more packet are directed to buffers rather than being dropped.

23.9 Summary

In this chapter, we studied WDM optical interconnects with limited range wavelength conversion and shared buffer. We focused on optimal packet scheduling to achieve minimum packet loss and packet delay. We defined the request graph on the basis of the properties of ordered interval wavelength conversion. The parallel segment expanding algorithm was given for finding the optimal schedule. The algorithm is derived after defining the forward and backward segments in the request graphs and runs in N decision-making units in parallel, where N is the number of input/output fibers. The time complexity of the algorithm is $O(Bk^2)$, as compared to $O((N + B)^3 k^3)$ time if directly applying other existing algorithms to find the optimal schedule, where B is the number of optical delay lines and k is the number of wavelengths per fiber.

References

[1] B. Mukherjee, WDM optical communication networks: progress and challenges, *IEEE Journal on Selected Areas in Communications*, vol. 18, no. 10, pp. 1810–1824, 2000.

[2] D. K. Hunter, M. C. Chia, and I. Andonovic buffering in optical packet switches, *Journal of Lightwave Technology*, vol. 16 no. 12, pp. 2081–2094, 1998.

[3] M. Kovacevic and A. Acampora, Benefits of wavelength translation in all-optical clear-channel networks, *IEEE Journal on Selected Areas in Communications*, vol. 14, no. 5, pp. 868–880, 1996.

[4] S.L. Danielsen, C. Joergensen, B. Mikkelsen, and K.E. Stubkjaer Analysis of a WDM packet switch with improved performance under bursty traffic conditions due to tunable wavelength converters, *Journal of Lightwave Technology*, vol. 16, no. 5, pp. 729–735, 1998.

[5] T. Tripathi and K. N. Sivarajan, Computing approximate blocking probabilities in wavelength routed all-optical networks with limited-range wavelength conversion, *IEEE Journal on Selected Areas in Communications*, vol. 18, pp. 2123–2129, 2000.

[6] G. Shen, S.K. Bose, T.H. Cheng, C. Lu, and T.Y. Chai, Performance study on a WDM packet switch with limited-range wavelength converters, *IEEE Communications Letters*, vol. 5, no. 10, pp. 432–434, 2001.

[7] R. Ramaswami and K. N. Sivarajan, *Optical Networks: A Practical Perspective*, Morgan Kaufmann Publishers, San Fransisco, USA, 1998.

[8] L. Xu, H. G. Perros and G. Rouskas, Techniques for optical packet switching and optical burst switching, *IEEE Communications Magazine*, pp. 136–142, 2001.

[9] R. Ramaswami and G. Sasaki, Multiwavelength optical networks with limited wavelength conversion, *IEEE/ACM Transactions on Networking*, vol. 6, pp. 744–754, 1998.

[10] X. Qin and Y. Yang, Nonblocking WDM switching networks with full and limited wavelength conversion, *IEEE Transactions on Communications*, vol. 50, no. 12, pp. 2032–2041, 2002.

[11] Y. Yang, J. Wang and C. Qiao Nonblocking WDM multicast switching networks, *IEEE Transactions on Parallel and Distributed Systems*, vol. 11, no. 12, pp. 1274–1287, 2000.

[12] S.L. Danielsen, et. al, WDM packet switch architectures and analysis of the influence of tunable wavelength converters on the performance, *Journal of Lightwave Technology*, vol. 15, no. 2, pp. 219–227, 1998.

[13] S. Nakamura and G.M. Masson, Lower bounds on crosspoint in concentrators, *IEEE Transactions on Computers*, pp. 1173–1178, 1982.

[14] A. Yavuz Oruc and H. M. Huang crosspoint complexity of sparse crossbar concentrators, *IEEE Transactions on Information Theory*, pp. 1466–1471, 1996.

[15] N. McKeown, The iSLIP scheduling algorithm input-queued switch, *IEEE/ACM Transactions on Networking*, vol. 7, pp. 188–201, 1999.

[16] M. Karpinski and Wojciech Rytter, *Fast Parallel Algorithms for Graph Matching Problems*, Oxford University Press, Oxford, UK, 1998.

[17] V. Eramo, M. Listanti and M. DiDonato; Performance evaluation of a bufferless optical packet switch with limited-range wavelength converters, *IEEE Photonics Technology Letters*, vol. 16, no. 2, pp. 644–646, 2004.

[18] Z. Zhang and Y. Yang, Optimal scheduling in buffered WDM packet switching networks with arbitrary wavelength conversion capability, *IEEE Transactions on Computers*, vol. 55, no. 1, pp. 71–82, January 2006.

[19] W.J. Goralski, *Optical Networking and WDM*, 1st Edition, McGraw-Hill, Columbus, OH, USA, 2001.

[20] H. Qin, S. Zhang and Z. Liu, Dynamic routing and wavelength assignment for limited-range wavelength conversion, *IEEE Communications Letters*, vol. 5, no. 3, pp. 136–138, 2003.

[21] X. Masip-Bruin, et al., Routing and wavelength assignment under inaccurate routing information in networks with sparse and limited wavelength conversion, *Proc. IEEE GLOBECOM '03.*, vol. 5, pp. 2575–2579, 2003.

[22] E.L. Lawler, *Combinatorial Optimization: Networks and Matroids*, Holt, Rinehart and Winston, 1976.

[23] W. Lipski Jr and F.P. Preparata Algorithms for maximum matchings in bipartite graphs, *Naval Research Logistics Quarterly*, vol. 14, pp. 313–316, 1981.

[24] D. K. Hunter and I. Andronovic, Approaches to optical Internet packet switching, *IEEE Communications Magazine*, vol. 38, no. 9, pp. 116–122, 2000.

[25] C. Develder, M. Pickavet and P. Demeester, Assessment of packet loss for an optical packet router with recirculating buffer, *Optical Network Design and Modeling 2002*, pp. 247–261, Torino, Italy, 2002.

[26] Z. Zhang and Y. Yang, WDM optical interconnects with recirculating buffering and limited range wavelength conversion, *IEEE Transactions on Parallel and Distributed Systems*, vol. 17, no. 5, pp. 466–480, May 2006.

[27] Z. Zhang and Y. Yang, Optimal scheduling in WDM optical interconnects with arbitrary wavelength conversion capability, *IEEE Transactions on Parallel and Distributed Systems*, vol. 15, no. 1, pp. 1012–1026, November 2004.

[28] D. B. West, *Introduction to Graph Theory*, Prentice-Hall, Upper Saddle River, NJ, USA, 1996.

[29] G. Bendeli et al., Performance assessment of a photonic ATM switch based on a wavelength controlled fiber loop buffer, *OFC'96 Technical Digest*, pp. 106–107, OFC, 1996.

[30] D.K. Hunter, et al., WASPNET: A wavelength switched packet network, *IEEE Communications Magazine*, vol. 37, no. 3, pp. 120–129, March 1999.

24
Real-Time Scheduling Algorithms for Multiprocessor Systems

Michael A. Palis
Rutgers University

24.1	Introduction...	24-1
24.2	Background on Real-Time Systems.....................	24-2
24.3	Scope and Map...	24-3
24.4	Some Results in Uniprocessor Real-Time Scheduling...	24-4
	Static-Priority Scheduling • Dynamic-Priority Scheduling	
24.5	Multiprocessor Scheduling Algorithms	24-6
24.6	Partitioned Scheduling on Multiprocessors...........	24-7
	Partitioned Scheduling Using Rate Monotonic • Partitioned Scheduling Using Earliest Deadline First	
24.7	Global Scheduling on Multiprocessors	24-12
	Global Static-Priority Scheduling • Global Dynamic-Priority Scheduling	
24.8	Proportionate-Fair Scheduling	24-15
24.9	Online Multiprocessor Scheduling	24-18
24.10	Concluding Remarks	24-22
	References	24-23

24.1 Introduction

Multiprocessor systems have become commonplace, providing increased levels of performance to a broad range of software applications that span the desktop to the data center. At the same time, applications have become increasingly sophisticated, imposing new demands on the operating systems that harness the power of these machines. One such class of applications are *real-time* applications, which do not necessarily demand speed, as much as they demand the *timeliness* and *predictability* of the execution of the underlying computational tasks. Examples of such applications abound. They include applications requiring dedicated hardware systems such as manufacturing process controls, aircraft avionics, and robotics. They also include various multimedia applications—such as videoconferencing, collaborative work, and online gaming—that are now proliferating on desktops and servers and that coexist, and share system resources, with traditional nonreal-time applications.

A *real-time operating system* (or *real-time system* for short) is one that is specifically designed to handle applications whose resource requirements are time-constrained. The resource requirement most often considered is the amount of processor time that a given computational task would need in order to complete execution. The timing constraint on this resource is typically specified as a *deadline*, which is the time by which a given computational task must complete. An integral component of the real-time system is the *scheduler*, which is responsible for selecting and dispatching tasks for execution and ensuring that their deadlines are met. Real-time systems are broadly classified as *hard* or *soft* depending on the criticality of deadlines. In a *hard real-time system*, all deadlines must be met or the system is considered to have failed. In a *soft real-time system*, deadlines may be occasionally missed with no catastrophic effect, except possibly for a degradation in performance.

This chapter presents a summary of past and current research on *multiprocessor scheduling algorithms* for *hard* real-time systems. It highlights some of the most important theoretical results on this topic over a 30-year period, beginning with the seminal paper of Liu and Layland in 1973 that provided impetus to much of the subsequent research in real-time scheduling theory. This chapter also identifies some open problems and future directions for research in multiprocessor scheduling. Clearly, a book chapter such as this could cover only a small area of the vast and fertile field of research in real-time system scheduling. The reader is encouraged to read Reference 55 for an excellent comprehensive survey of this topic, past to present.

24.2 Background on Real-Time Systems

In a real-time system, the basic unit of execution is a *job*, which is characterized by three parameters: an *arrival time* a, which is the time when the job becomes available for execution; an *execution requirement* e, which is the amount of processor time that the job needs to complete execution; and a *relative deadline* $d \geq e$, which is the amount of time after its arrival within which the job must complete. That is, the job must be executed for e time units within the time interval $[a, a + d]$. The quantity $a + d$ is called the job's *absolute deadline*. A *real-time task* is a (finite or infinite) sequence of jobs $T = \{J_1, J_2, \ldots\}$. A *task system* is a finite set of real-time tasks $\tau = \{T_1, T_2, \ldots, T_n\}$.

The most studied real-time task model is the *periodic task* model of Liu and Layland [42]. A periodic task is an infinite sequence of jobs in which the arrival times of successive jobs are separated by a fixed amount p, called the task's *period*. Formally, a periodic task T is characterized by a 4-tuple (a, e, p, d), where a is the task arrival time, which coincides with the arrival time of T's first job; e is the execution requirement of each job; $p \geq e$ is the period; and $d \geq e$ is the relative deadline of each job. In other words, for $i \geq 1$, job J_i of T arrives at time $a + (i - 1)p$ and must be executed for e time units by its absolute deadline $a + (i - 1)p + d$. It is assumed that all task parameters are nonnegative integers. A *sporadic task* [43] is similar to a periodic task except that the parameter p denotes the *minimum*, rather than exact, separation between job arrival times of a task. An *aperiodic* task is a task that consists of only one job. An aperiodic task corresponds to the concept of a "job" in classical job shop scheduling theory. The *utilization* of a task T is the ratio between its execution requirement and its period, that is, $u = e/p$. The utilization of a task is also sometimes referred to as its *weight* [15]. The utilization of a task system $\tau = \{T_1, T_2, \ldots, T_n\}$ is the sum of the utilizations of its constituent tasks, that is, $U(\tau) = \sum_{i=1}^{n} u_i$.

A task system is called *synchronous* if all tasks have the same arrival time (typically assumed to be time 0); otherwise it is called *asynchronous*. In an *implicit-deadline* task system, every task has a relative deadline that is equal to its period. In a *constrained-deadline* system, every task has a relative deadline not larger than its period. In an *arbitrary-deadline* system, relative deadlines are not related to periods.

The basis periodic and sporadic task models have been extended in various ways by other researchers. Some of these extended models are discussed later in Section 24.10.

Scheduling algorithms may be nonpreemptive or preemptive. In *nonpreemptive* scheduling, a job, once it begins execution, runs to completion without interruption. In *preemptive scheduling*, a job executing on a processor may be preempted before it is completed and resumed at a later time.

Much of the research in real-time scheduling has focussed on *priority-based* scheduling algorithms. In such algorithms, tasks are assigned different *priority levels* (or *priorities*, for short) to determine the relative order in which they are to be executed on the processors of the system. Priorities can either be static or dynamic. In *static-priority* scheduling, a task is assigned a fixed priority that does not change during its lifetime. For a periodic task, this means that the jobs comprising the task all have the same priority—that is, the static priority assigned to the entire task. In *dynamic-priority* scheduling, a task's priority may change (perhaps several times) during its lifetime. One can further distinguish between two schemes for assigning dynamic priorities [21]. In *job-level dynamic-priority* assignment, a task's priority changes only at the job boundaries—that is, the jobs generated by the task may have different priorities but a job's priority remains fixed during its lifetime. In *fully dynamic-priority* assignment, no restrictions are placed on the priorities that may be assigned to jobs; in particular, a job's priority may change at any time during its lifetime.

A task system τ is *schedulable* by scheduling algorithm A if A produces a schedule that meets the absolute deadline of every job of every task of τ. τ is *feasible* if there exists *some* schedule for τ that meets all deadlines. Hence, schedulablity is a property of a task system with respect to a scheduling algorithm, while feasibility is an inherent property of a task system, independent of any scheduling algorithm. Algorithm A is *optimal* for a class of task systems if every feasible task system in the class is schedulable by A.

24.3 Scope and Map

A fundamental problem in real-time scheduling is the *feasibility analysis problem*: given a task system and a scheduling environment, determine whether there exists a schedule that will meet all deadlines. This problem turns out to be difficult even for uniprocessor systems. In Reference 37 Leung and Merrill showed that the problem of determining whether an arbitrary periodic task system is feasible on a single processor is co-NP-hard. This result was subsequently tightened by Baruah et al. [19] who showed that the problem is, in fact, co-NP-complete in the strong sense, even if the utilization of the task system is known to be bounded from above by some fixed positive constant. A comprehensive summary of complexity results on uniprocessor real-time scheduling can be found in Reference 17.

Because of the intractability of the feasibility analysis problem for arbitrary periodic task systems, much of the work on real-time scheduling has focussed on restricted task systems for which efficient scheduling algorithms can be found. The most studied is the class of **implicit-deadline synchronous periodic task systems**—that is, task systems in which every task has a period equal to its relative deadline and all tasks have the same arrival time, which is assumed to be zero. In this chapter, we will focus primarily on scheduling algorithms for this class of task systems. Henceforth, unless otherwise specified, an "implicit-deadline synchronous periodic task system" will simply be referred to as a "periodic task system." In addition, we shall use τ to denote a periodic task system with n tasks $\{T_1, T_2, \ldots, T_n\}$, such that the i-th task $T_i = (e_i, p_i)$ has execution requirement e_i and period p_i. The utilization of task T_i is $u_i = e_i/p_i$. The utilization of the entire task system τ is $U(\tau)$.

With regards to scheduling, we shall limit the discussion to **preemptive** scheduling algorithms for **identical** multiprocessor systems. In an identical multiprocessor, all processors are the same. In addition, we shall only consider multiprocessor scheduling algorithms in which **job parallelism is forbidden;** that is, each job may execute on at most one processor at any given instant of time. Although this limits the space of multiprocessor scheduling algorithms that we consider, it is within this space that some of the most fertile research in real-time scheduling has been conducted over the last three decades.

The remainder of this chapter is organized as follows. In the next section, we set the stage by first presenting some important results in uniprocessor real-time scheduling that are relevant to the study of multiprocessor scheduling algorithms. Section 24.5 gives an overview of multiprocessor scheduling. Subsequent sections discuss, in turn, the three main approaches to multiprocessor scheduling: *partitioned* scheduling, *global* scheduling, and *proportionate-fair* (*Pfair*) scheduling.

As noted earlier, much of the research on multiprocessor scheduling has focussed on *synchronous* task systems in which tasks have the same arrival time, which is often assumed to be time 0. For this class of task systems, a scheduling algorithm can, for all practical purposes, make scheduling decisions *offline*—that is, with full knowledge of all task parameters. In practice, however, tasks can arrive at any time and hence the scheduling algorithm must necessarily make scheduling decisions *online*—that is, only on the basis of the information about tasks that have already arrived, but not on information about future tasks. Section 24.9 discusses some recent results in online real-time scheduling. Finally, Section 24.10 ends the chapter with some concluding remarks.

24.4 Some Results in Uniprocessor Real-Time Scheduling

Research in multiprocessor scheduling owes much to previous work in scheduling algorithms for uniprocessor systems. This section briefly describes some important results in uniprocessor scheduling that are relevant to the study of the multiprocessor case.

24.4.1 Static-Priority Scheduling

We first consider static-priority uniprocessor scheduling algorithms. In such algorithms, each task (and all of its jobs) is assigned a fixed priority that does not change during its lifetime. At each time step, the *active* job (i.e., a job that has arrived but not yet completed) with the highest priority is executed, preempting the currently running job, if that job has a lower priority.

The most studied static-priority scheduling algorithm is the *rate monotonic* (RM) algorithm [42, 54]. RM assigns task priorities in inverse proportion to their periods—that is, the task with the smallest period is assigned the highest priority (with ties broken arbitrarily). Equivalently, if we define the *request rate* (or simple *rate*) of a task T_i as $r_i = 1/p_i$, where p_i is the task's period, then RM prioritizes the tasks according to monotonically decreasing rates (hence, the name).

In their ground-breaking paper [42], Liu and Layland proved that RM is *optimal* in the following sense: if a periodic task system is feasible under static-priority assignment (i.e., is schedulable by some static-priority scheduling algorithm) then the task system is schedulable by RM. Another important result on RM, which was shown by both Serlin [54] and Liu and Layland [42], is the following:

Theorem 24.1 (Condition RM-LL) [42,54] *A periodic task system τ is schedulable by* RM *if*

$$U(\tau) \leq n(2^{1/n} - 1). \tag{24.1}$$

For example, consider the task system

$$\tau = \{T_1 = (2, 10), T_2 = (4, 15), T_3 = (10, 35)\}.$$

For this task system, $u_1 = 0.2$, $u_2 \approx 0.267$, $u_3 \approx 0.286$. Since the total utilization $U(\tau) \approx 0.753 \leq 3(2^{1/3} - 1) \approx 0.779$, then according to Condition RM-LL this task system is schedulable by RM.

Condition RM-LL is a *sufficient* condition for schedulability, but it is not a *necessary* condition because there are task systems whose utilizations may exceed the bound of $n(2^{1/n} - 1)$ and yet are schedulable by RM. For example, the task system

$$\tau = \{T_1 = (4, 10), T_2 = (5, 10)\}$$

has a utilization $U(\tau) = 0.4 + 0.5 = 0.9 > 2(2^{1/2} - 1) \approx 0.828$ and hence fails the schedulability test of Condition RM-LL. However, one can easily show that this task system is schedulable by RM.

As n approaches infinity, the value of $n(2^{1/n} - 1)$ approaches $\ln 2 \approx 0.693$. Hence any periodic task system, regardless of the number of tasks, is schedulable by RM as long as its utilization is at most 69.3%.

Unfortunately, this is a pessimistic estimate of RM's actual performance; a statistical study conducted by Lehoczky et al. [36] showed that for randomly generated task sets, the average real schedulable utilization of RM is approximately 88%.

In Reference 36 Lehoczky et al. presented an *exact* (i.e., necessary and sufficient) schedulability test for RM, given below:

Theorem 24.2 (Condition RM-Exact) [36] *Let τ be a periodic task system such that $p_1 \leq p_2 \leq \cdots \leq p_n$. Let $S_i = \{kp_j | j = 1, \ldots, i; k = 1, \ldots, \lfloor p_i/p_j \rfloor\}$ and $W_i(t) = \sum_{j=1}^{i} e_j \lceil t/p_j \rceil$. Then T_i is schedulable by RM if and only if*

$$L_i = \min_{t \in S_i}(W_i(t)/t) \leq 1. \tag{24.2}$$

The entire task system τ is schedulable by RM if and only if

$$L = \max_{1 \leq i \leq n} L_i \leq 1. \tag{24.3}$$

Condition RM-LL can be tested in $O(n)$ time, where n is the number of tasks. On the other hand, Condition RM-Exact has *pseudopolynomial* time complexity because it requires testing Equation 24.2 for all scheduling points in S_i, whose cardinality is a function of the (relative) values of the task periods. Hence, even though it is an exact test, it is not often used in practice because of its high complexity. Nonetheless, the exactness of Condition RM-Exact, combined with the fact that RM is an optimal static-priority uniprocessor scheduling algorithm, implies that the feasibility analysis problem for implicit-deadline synchronous periodic task systems on a single processor under static-priority assignment can be solved in pseudopolynomial time.

There exists a large body of research literature on other schedulability tests for RM—see Reference 51 for an excellent survey.

It should be pointed out that rate monotonic scheduling is applicable only to periodic task systems with implicit deadlines, that is, for tasks whose period and relative deadline are the same. As described in Section 24.2 a more general task model is the constrained-deadline periodic task system in which every task has a relative deadline than can be different from, but no larger than, its period. For such task systems, the *deadline monotonic* algorithm (DM) of Leung and Whitehead [41] is an optimal static-priority scheduling algorithm. In DM, tasks are assigned priorities in inverse proportion to their relative deadlines: the task with the shortest relative deadline is assigned the highest priority, with ties broken arbitrarily. (Observe that when the period and relative deadline of each task are the same, DM priority assignment becomes RM priority assignment.) Despite the optimality of DM for constrained-deadline periodic task systems, it has not received as much attention as RM, probably because of the relative scarcity of efficient schedulability tests for DM. The few that have been developed include those of Audsley et al. [9, 10] and Baker [11, 12].

It turns out that RM and DM are optimal, not only for synchronous periodic task systems, but also for sporadic task systems [17]. However, they are no longer optimal for asynchronous task systems (i.e., in which tasks may not have the same arrival times) [17, 30, 35]. Finally, DM, while optimal for constrained-deadline systems, is no longer optimal for systems with arbitrarily relative deadlines (i.e., in which some tasks may have relative deadlines greater than their periods), even if the system is synchronous [17, 35].

24.4.2 Dynamic-Priority Scheduling

We next consider dynamic-priority uniprocessor scheduling algorithms. In such algorithms, the priority assigned to a task may change (perhaps several times) during its lifetime. For periodic task systems, this means that the jobs generated by an individual task may be assigned different priorities as they become active during the course of the task's execution. At each time step, the active job with the highest priority is executed, preempting the currently running job, if that job has a lower priority.

The most popular dynamic-priority scheduling algorithm is the earliest deadline first algorithm (EDF) [25,42]. EDF assigns priorities to jobs in inverse proportion to their absolute deadlines, where the absolute deadline is the job's arrival time plus its relative deadline. The job with the earliest absolute deadline is given the highest priority (with ties broken arbitrarily).

In the same paper [42] in which they proved the optimality of RM under static-priority assignment, Liu and Layland proved that EDF is optimal under dynamic-priority assignment for uniprocessor systems—that is, any (implicit-deadline synchronous) periodic task system that is schedulable on a single processor by some dynamic-priority scheduling algorithm is also schedulable by EDF. They also proved the following *exact* schedulability condition for EDF:

Theorem 24.3 (Condition EDF-LL) [42] *A periodic task system τ is schedulable by* EDF *if and only if*

$$U(\tau) = \sum_{i=1}^{n} u_i \leq 1. \tag{24.4}$$

Note that the utilization bound of 1 of Condition EDF-LL is the best possible since periodic task systems with utilizations greater than 1 are clearly not feasible. Indeed, Dertouzos [25] showed that EDF is optimal among all preemptive scheduling algorithms for scheduling arbitrary task systems (including periodic, sporadic, and aperiodic task systems) on uniprocessors.

The optimality of EDF implies that the feasibility analysis problem for arbitrary periodic task systems on a single processor under dynamic-priority assignment can be solved in exponential time by essentially simulating the behavior of EDF on the periodic task system for a sufficiently long interval [17]. For implicit-deadline synchronous periodic task systems, the feasibility analysis problem can be solved in $O(n)$ time by virtue of Condition EDF-LL.

In Reference 43, Mok presented another dynamic-priority scheduling algorithm, *least laxity first* (LLF). The *laxity* of an active job (of some periodic task) at time t is the absolute deadline of the job minus the remaining execution time needed to complete the job. At each time instant t, LLF executes the active job whose laxity is least, preempting the currently running job, if that job has a greater laxity. Mok [43] showed that, like EDF, LLF is an optimal scheduling algorithm under dynamic-priority assignment for uniprocessor systems. LLF is a *fully dynamic-priority* scheduling algorithm because an individual job's priority may change at any time (relative to other jobs). In contrast, EDF is a *job-level dynamic-priority* algorithm because a job's priority remains fixed during its lifetime. As a result, LLF generally has a larger overhead than EDF owing to the greater number of context switches caused by laxity changes at runtime. For this reason, LLF has not received as much attention as EDF among real-time scheduling researchers.

In Reference 15, Baruah et al. introduced another fully dynamic-priority scheduling algorithm called *Proportionate-fair* (Pfair). Under Pfair scheduling, each periodic task is executed at a uniform rate (corresponding to its utilization) by breaking it into quantum-length subtasks. The subtasks are assigned *pseudodeadlines* and are scheduled in an earliest-pseudodeadline-first manner, with ties resolved by special tie-breaking rules. Being a fully dynamic-priority algorithm, Pfair suffers from the large overhead of frequent context switches as does LLF. However, it is of important theoretical interest because it is currently the only known method for optimally scheduling periodic tasks on multiprocessor systems. Pfair scheduling is discussed in Section 24.8.

24.5 Multiprocessor Scheduling Algorithms

Just as in the uniprocessor case, multiprocessor scheduling algorithms can be classified according to the scheme used in assigning task priorities: *static-priority* (such as RM) and *dynamic-priority* algorithms (such as EDF). In addition, multiprocessor scheduling algorithms can be classified according to how the tasks are allocated to processors. In *partitioned* scheduling, all jobs generated by a task are required to execute on the same processor. In *nonpartitioned* or *global* scheduling, task migration is allowed—that is, different jobs of the same task may execute on different processors. Job migration is also allowed—a job that has been preempted on a particular processor may resume execution on the same or a different processor.

The two classification schemes described above—static versus dynamic and partitioned versus global—give rise to the following four general classes of multiprocessor scheduling algorithms:

1. Partitioned static-priority scheduling
2. Partitioned dynamic-priority scheduling
3. Global static-priority scheduling
4. Global dynamic-priority scheduling

As we shall soon see, much of the work on multiprocessor real-time scheduling has focussed on the development of scheduling algorithms that achieve provably good *utilization bounds*. An example is Liu and Layland's $n(2^{1/n} - 1)$ utilization bound for RM scheduling of n periodic tasks on uniprocessors (see Condition RM-LL of Theorem 24.1). In general, this value represents the *minimum schedulable utilization* $U_A(n)$ achievable by a scheduling algorithm A, in the sense that any periodic task system τ with n tasks is guaranteed to be schedulable by A if $U(\tau) \leq U_A(n)$. Similar utilization bounds have been obtained for multiprocessor scheduling algorithms; the difference is that these bounds are in general a function of both the number of tasks, n, and the number of processors, m.

Before discussing individually the four classes of multiprocessor scheduling algorithms described above, we state some general results that will be helpful in putting into context the specific results for each of the four classes that are presented later in this chapter.

One important result, which was shown by Leung and Whitehead in Reference 41, is that the partitioned and global approaches to static-priority scheduling on multiprocessors are *incomparable* in that there are task systems that are feasible under static-priority assignment using partitioned scheduling on m processors but which are not feasible under static-priority assignment using global scheduling on m processors, and vice versa. This result has provided strong motivation to study both partitioned and global approaches to static-priority multiprocessor scheduling, since neither approach is strictly better than the other.

Another important result, reported by Andersson et al. [8] and Srinivasan and Baruah [57], gives the following upper bound on the schedulable utilization of *any* static-priority or job-level dynamic-priority multiprocessor scheduling algorithm, partitioned or global.

Theorem 24.4 [8,57] *No static-priority or job-level dynamic-priority multiprocessor scheduling algorithm—partitioned or global—can feasibly schedule all periodic task systems with utilization at most U on m processors if $U > (m + 1)/2$.*

Note that the above bound holds for both RM (a static-priority algorithm) and EDF (a job-level dynamic-priority algorithm). Thus, unless one employs a *fully dynamic* multiprocessor scheduling algorithm, the best one could hope for is a utilization of $(m + 1)/2$ on m processors.

In Section 24.8, we show that Pfair scheduling [15], a fully dynamic-priority scheduling algorithm, breaks the $(m + 1)/2$ bound given by Theorem 24.4 and achieves an optimal schedulable utilization of m on m processors.

24.6 Partitioned Scheduling on Multiprocessors

Partitioned scheduling is generally viewed as consisting of two distinct algorithms: a *task partitioning* (or *task assignment* or *task allocation*) algorithm that finds a partition of the set of tasks among the processors, and a *scheduling* algorithm that determines the order of execution of the tasks assigned to the same processor. Task migration is not allowed, that is, a task assigned to a particular processor is only executed on that processor.

Partitioned scheduling has several advantages over global scheduling. First, it incurs less runtime overhead than global scheduling because task partitioning can be performed before runtime, and at runtime tasks do not incur migration costs simply because migration is not allowed. Second, once tasks are assigned to processors, well-known scheduling algorithms with efficient uniprocessor schedulability tests (such as RM and EDF) can be used for each processor. For these reasons, the partitioned approach has received greater attention than the global approach (until recently).

The task partitioning problem is analogous to the *bin-packing problem* [22], in which the processors are regarded as bins of a given capacity and the tasks are regarded as items with different weights to be packed into the bins (processors). The "capacity" of a processor is often determined by a schedulability condition that specifies when a processor is considered "full." For example, using Liu and Layland's schedulability condition for RM scheduling (Condition RM-LL of Theorem 24.1), the capacity of a processor is $n(2^{1/n} - 1)$, where n is the number of tasks allocated to the processor. Note that the processor capacity is not constant, as it depends on n, and so the problem of finding an optimal task partition is at least as hard as the bin-packing problem, which is known to be NP-hard in the strong sense [29]. Consequently, bin-packing heuristics are often used to approximately solve the task partitioning problem.

Among bin-packing heuristics (see Reference 22 for details), the following four algorithms are the most popular ones used for task allocation in partitioned multiprocessor scheduling:

- **First Fit (FF)**: FF allocates a new object to the nonempty bin with the lowest index, such that the weight of the new object along with the weights of the objects already in the bin, do not exceed the capacity of the bin. If the object cannot fit into any nonempty bin, it is allocated to the next empty bin (if available).
- **Next Fit (NF)**: NF maintains a pointer to the "current" bin, which is the bin to which the last object was allocated. A new object is allocated to the current bin if it fits into that bin; otherwise, it is allocated to the next empty bin. Note that NF does not revisit previous bins.
- **Best Fit (BF)**: BF allocates a new object to the nonempty bin with the smallest remaining capacity into which it can fit. If an object cannot be allocated to a nonempty bin, it is allocated to the next empty bin.
- **Worst Fit (WF)**: WF is similar to BF, except that it allocates a new object to the bin with the largest remaining capacity into which it can fit.

The common approach to partitioned multiprocessor scheduling is to first perform task partitioning using some bin-packing heuristic (e.g., FF), then schedule the tasks allocated to the same processor using some uniprocessor scheduling algorithm (e.g., RM). A schedulability condition associated with the scheduling algorithm (e.g., Condition RM-LL for RM) is used to determine processor capacity. Thus, such an approach offers three degrees of freedom in designing a partitioned multiprocessor scheduling algorithm: the bin-packing heuristic, the uniprocessor scheduling algorithm, and the schedulability condition associated with the scheduling algorithm.

In most of the research in partitioned multiprocessor scheduling, two algorithms—RM and EDF—have been the uniprocessor scheduling algorithms of choice. This is not surprising in light of the fact that RM and EDF are, respectively, the most studied static-priority and dynamic-priority scheduling algorithms.

24.6.1 Partitioned Scheduling Using Rate Monotonic

We first consider partitioned mulitprocessor scheduling using RM. To illustrate this approach, consider the scheduling algorithm that uses the FF bin-packing heuristic to partition the tasks among the processors; RM to (locally) schedule the tasks allocated to the same processor; and Liu and Layland's utilization bound for RM on a single processor (i.e., Condition RM-LL of Theorem 24.1) to determine processor capacity. Let us call this algorithm, RM-FF. On a m-processor system, RM-FF schedules the tasks of a task system $\tau = \{T_1, T_2, \ldots, T_n\}$ as follows. For each task T_i, $i = 1, 2, \ldots, n$, RM-FF allocates T_i to the first (i.e., lowest-indexed) processor P_j with enough remaining capacity to accommodate the task—that is,

$(u_i + U_j) \leq (n_j + 1)(2^{1/(n_j+1)} - 1)$, where n_j is the number of tasks previously allocated to P_j and U_j is the sum of the utilizations of these tasks. Tasks allocated to the same processor are executed according to the rate-monotonic policy (i.e., in nondecreasing order of their periods). RM-BF, RM-NF, and RM-WF can be similarly defined. In addition, one can define variants of these algorithms that first sort the tasks in increasing or decreasing order of their utilizations before the tasks are allocated to processors. For example, RM-FFI (RM-FFD) is just like RM-FF except that the tasks are first sorted so that $u_i \leq u_{i+1}$ (respectively, $u_i \geq u_{i+1}$) for all i.

Oh and Baker [45] investigated algorithm RM-FF and proved the following:

Theorem 24.5 [45] *A periodic task system τ is schedulable by RM-FF on $m \geq 2$ processors if*

$$U(\tau) \leq m(2^{1/2} - 1) \approx 0.414m. \tag{24.5}$$

Note that the utilization bound of $0.414m$ for RM-FF is reasonably close to the $(m+1)/2$ upper bound given by Theorem 24.4 on the schedulable utilization achievable by any static-priority multiprocessor scheduling algorithm. In fact, Oh and Baker proved a slightly better upper bound than $(m+1)/2$ for the case of *partitioned* (but not global) static-priority scheduling:

Theorem 24.6 [45] *There exist periodic task systems τ with utilization*

$$U(\tau) \geq \frac{m+1}{1 + 2^{1/(m+1)}} + \varepsilon, \tag{24.6}$$

for arbitrarily small $\varepsilon > 0$, that are not schedulable on $m \geq 2$ processors by any partitioned static-priority scheduling algorithm.

López et al. [39] refined and generalized the results of Oh and Baker by considering periodic task systems consisting of tasks with utilization at most α, for some fixed α, $0 < \alpha \leq 1$. Their main results are summarized in the following theorem:

Theorem 24.7 [39] *Let $T(n, \alpha)$ be the class of periodic task systems with n tasks, in which every task has utilization at most α. Let*

$$\beta = \lfloor 1/(\lg(\alpha + 1)) \rfloor \tag{24.7}$$

and

$$U_{LL}^{RM}(n, m, \alpha) = \begin{cases} n(2^{1/n} - 1) & \text{if } m = 1 \\ (m\beta + 1)(2^{1/(\beta+1)} - 1) & \text{if } m > 1. \end{cases} \tag{24.8}$$

Suppose $n > m\beta$.

1. *There exist periodic task systems $\tau \in T(n, \alpha)$ with utilization*

$$U(\tau) \geq U_{LL}^{RM} + \varepsilon, \tag{24.9}$$

for arbitrarily small $\varepsilon > 0$, that are not schedulable on m processors by any partitioned RM scheduling algorithm that uses Condition RM-LL to determine processor capacity.

2. *Any partitioned RM scheduling algorithm that is reasonable allocation decreasing (RAD) can feasibly schedulable every periodic task system $\tau \in T(n, \alpha)$ provided that*

$$U(\tau) \leq U_{LL}^{RM}. \tag{24.10}$$

The condition $n > m\beta$ in the above theorem is imposed because if $n \leq m\beta$ the scheduling problem becomes trivial. When Condition RM-LL is used to determine processor capacity, a processor can accommodate up to β tasks, each with utilization at most α. Hence, an m-processor system can trivially schedule up to $m\beta$ such tasks.

Part 1 of the theorem states that if we restrict ourselves to partitioned scheduling algorithms that use RM to locally schedule tasks and Condition RM-LL to determine processor capacity, then regardless of the task partitioning scheme used, a schedulable utilization of at most U_{LL}^{RM} is the best that we can hope for.

Part 2 states that there are, in fact, partitioning schemes that achieve this bound, namely, RAD algorithms that have the following properties:

- Tasks are ordered by decreasing utilization before making allocation (i.e., $u_1 \geq u_2 \geq \cdots \geq u_n$), then allocated to processors in that order.
- The algorithm fails to allocate a task only when there is no processor with sufficient capacity to accommodate the task.

For example, RM-FFD, RM-BFD, and RM-WFD are RAD algorithms. Note that RM-NFD is not a RAD algorithm because it disallows revisiting processors previously used for allocation, even though they could accommodate more tasks.

For the unrestricted case, where $\alpha = 1$ (and hence $\beta = 1$), Theorem 24.7 gives a utilization bound is $(m+1)(2^{1/2} - 1) = 0.414(m+1)$ for RAD algorithms. Thus, there remains a gap, though small, between this bound and the Oh and Baker upper bound of $(m+1)/(1 + 2^{1/(m+1)})$ on the schedulable utilization of any partitioned static-priority scheduling algorithm. It is an open question whether this gap can be closed by using RM in conjunction with a better schedulability test (to determine processor capacity) or by using some other static-priority assignment to locally schedule tasks on processors.

Experimental results conducted by Lauzac et al. [34] indicate that a better schedulability test can improve the utilization achievable by partitioned RM scheduling. They compared (among others) two versions of RM-FF, one using the Liu and Layland test (Condition RM-LL) and the other using a new schedulability test they developed called R-BOUND. They showed that for randomly generated task sets, RM-FF with R-BOUND achieved an average utilization of 96% compared to just 72% for RM-FF with Condition RM-LL. The theoretical performance of R-BOUND, however, has yet to be determined.

It is interesting to ask whether Oh and Baker's utilization bound of $m(2^{1/2} - 1)$ for RM-FF can be improved, perhaps even matching that of RAD algorithms. Note that, unlike RAD algorithms, RM-FF does not presort the tasks by decreasing utilization before allocating them to processors. In another paper [40], López et al. derived a *tight* utilization bound for RM-FF that is higher than Oh and Baker's but slightly lower than U_{LL}^{RM}. They also showed that the same tight bound holds for RM-BF. Thus, presorting tasks by decreasing utilization does help, albeit in a small way. Interestingly, while RM-WFD is as good as RM-FFD and RM-BFD in terms of achievable utilization, RM-WF behaves much worse than RM-FF and RM-BF [39].

For the partitioned scheduling algorithms described above, the number of processors in the system is assumed to be a fixed constant m. In Reference 24, Dhall and Liu studied the same problem but under the assumption that the system has an *infinite* supply of processors. They asked the following question: how many more processors would a (nonoptimal) partitioned RM scheduling algorithm A need than an optimal partitioned scheduling algorithm OPT in order to feasibly schedule a given periodic task system? In particular, they were interested in finding the worst-case bound (over all periodic task systems) for $\Re_A = \lim_{N_{opt} \to \infty} N_A / N_{OPT}$, where N_A (N_{OPT}) is the minimum number of processors required by A (respectively, OPT) to feasibly schedule a given task system. Dhall and Liu showed that for RM-NFD, $2.4 \leq \Re_{RM-NFD} \leq 2.67$ and for RM-FF, $2 \leq \Re_{RM-FF} \leq (4 \times 2^{1/3})/(1 + 2^{1/3}) \approx 2.23$. These results were subsequently improved by other researchers and extended to other partitioned RM scheduling algorithms [20, 46–48]. Because of space limitations, we do not discuss these other works here; a comprehensive survey can be found in Reference 52.

24.6.2 Partitioned Scheduling Using Earliest Deadline First

For partitioned multiprocessor scheduling under dynamic-priority assignment, EDF has been the local scheduling algorithm of choice for the simple reason that on a single processor, EDF achieves a schedulable utilization of 1 (see Condition EDF-LL of Theorem 24.3), which is the best possible. Just like RM, one can design a partitioned EDF scheduling algorithm in conjunction with a bin packing heuristic (e.g., FF) to partition the tasks among the processors. Thus, for example, EDF-FF is the partitioned EDF scheduling algorithm that uses FF for task allocation.

In Reference 38, López et al. carried out a systematic study of partitioned EDF scheduling algorithms. Just as in Reference [39], they considered periodic task systems consisting of tasks with utilization at most α, for some fixed α, $0 < \alpha \leq 1$. Their main results are summarized in the following theorem:

Theorem 24.8 [38] *Let $T(n,\alpha)$ be the class of periodic task systems with n tasks, in which every task has utilization at most α. Let*

$$\beta = \lfloor 1/\alpha \rfloor \qquad (24.11)$$

and

$$U_{LL}^{EDF}(n, m, \alpha) = \frac{\beta m + 1}{\beta + 1}. \qquad (24.12)$$

Suppose $n > m\beta$.

1. *There exist periodic task systems $\tau \in T(n,\alpha)$ with utilization*

$$U(\tau) \geq U_{LL}^{EDF} + \varepsilon, \qquad (24.13)$$

   for arbtrarily small $\varepsilon > 0$, that are not schedulable on m processors by any partitioned EDF scheduling algorithm.
2. *EDF-FF and EDF-BF each achieve a schedulable utilization of U_{LL}^{EDF}; that is, each can feasibly schedule every periodic task system $\tau \in T(n,\alpha)$ provided that*

$$U(\tau) \leq U_{LL}^{EDF}. \qquad (24.14)$$

The above theorem is the EDF analog of Theorem 24.7 for RM described in the previous subsection. The condition $n > \beta m$ is imposed because if the number of tasks in the task set is $n \leq \beta m$, the task set can be trivially scheduled on m processors using EDF.

Part 1 of the theorem states that, regardless of the task partitioning scheme used, a schedulable utilization of at most U_{LL}^{EDF} is the best that we can hope for.

Part 2 states that the First Fit and Best Fit partitioning schemes achieve this bound. Clearly, the same is true for the increasing-utilization (EDF-FFI, EDF-BFI) and decreasing-utilization (EDF-FFD, EDF-BFD) versions of these partitioning schemes. Just as in the case of partitioned RM scheduling, López et al. showed that partitioned EDF scheduling using Worst Fit (i.e., EDF-WF) does not achieve the optimal bound, but instead achieves a lower schedulable utilization of $m - (m-1)\alpha$. However, if the tasks are presorted according to increasing or decreasing utilization (i.e., EDF-WFI or EDF-WFD), then Worst Fit achieves the optimal bound on schedulable utilization.

For the unrestricted case, where $\alpha = 1$ (and hence $\beta = 1$), Theorem 24.8 gives a utilization bound is $(m+1)/2$, which is tight in light of Theorem 24.4.

As in the case of RM, other researchers have studied the performance of partitioned EDF scheduling assuming an infinite number of processors. The reader is referred to Reference 52 for a survey of results on this problem.

24.7 Global Scheduling on Multiprocessors

In global scheduling, all active jobs are stored in a single queue regardless of the tasks that generated these jobs. A single system-wide priority space is assumed; the highest-priority job in the queue is selected for execution whenever a processor becomes idle. Moreover, if a new active job is added to the queue and this job has a higher priority than some currently running job, it preempts the running job with the lowest priority, and the preempted job is returned to the queue to be resumed at a later time. Note that task migration may occur, that is, different jobs of the same task may be executed on different processors. Job migration may also occur, that is, a job that has been preempted on a particular processor may be resumed on a different processor. All the global scheduling algorithms we shall discuss are *work-conserving* in the sense that no processor is left idle while there remain active jobs awaiting execution.

As in the partitioned method, static-priority or dynamic-priority assignment may be used with global scheduling. Thus, a global RM (EDF) scheduling algorithm assigns to each job a global priority equal to the reciprocal of its task's period (respectively, equal to its absolute deadline) and the highest-priority job is the one with the smallest period (respectively, with the earliest absolute deadline). All jobs are considered for execution in accordance with their global priorities, as described in the previous paragraph.

Because global scheduling schemes have the flexibility to migrate tasks and jobs, one would suspect that global scheduling algorithms achieve higher utilization bounds that partitioned scheduling algorithms. However, this is not the case. In Reference 24, Dhall and Liu showed that global RM and EDF scheduling suffer from an anomaly, called the *Dhall effect*, in that they cannot schedule certain feasible task systems with very low utilizations. For example, consider the periodic task system τ consisting of the following $n = m + 1$ tasks:

$$\tau = \{T_1 = (2\epsilon, 1), T_2 = (2\epsilon, 1), \ldots, T_m = (2\epsilon, 1), T_{m+1} = (1, 1 + \epsilon)\}$$

If this task set is scheduled on m processors using either RM or EDF, the first job of task T_{m+1} will have the lowest global priority and hence will only be executed after the first jobs of the m other tasks have been executed in parallel. Clearly, this job of T_{m+1} will miss its deadline. Therefore, the task set cannot be scheduled by either RM or EDF on m processors. On the other hand, the task set is feasible because it is possible to meet all task deadlines by assigning T_{m+1} the highest priority and letting it execute on its own processor and letting the other m tasks share the remaining $m - 1$ processors. Note that as $\epsilon \to 0$, the utilization of the task set, and hence the schedulable utilization of global RM or EDF scheduling, approaches 1 no matter how many processors are used. In contrast, as was shown in Section 24.6, *partitioned* RM and EDF scheduling using Best Fit (RM-FF and EDF-FF) can successfully schedule periodic task systems with utilizations of up $0.414(m + 1)$ and $0.5(m + 1)$ on m processors, respectively.

Looking at the above example, it appears that the difficulty in global scheduling arises when there is a mixture of low-utilization (i.e., "light") tasks, such as $T_1, \ldots, T_m$, and high-utilization (i.e., "heavy") tasks, such as T_{m+1}. Intuitively, heavy tasks are harder to schedule because they have less "slack" than light tasks and hence they should be scheduled as soon as possible to avoid missing a deadline. In contrast, light tasks are easier to schedule because they have more slack and hence their execution can be delayed without missing their deadlines. Unfortunately, neither RM nor EDF priority assignment is able to guarantee that heavy tasks receive higher priority than light ones.

24.7.1 Global Static-Priority Scheduling

Andersson and Jonsson [7] used the above observation to develop a global static-priority scheduling algorithm, called AdaptiveTkC, that circumvents the Dhall effect by assigning higher priorities to tasks with less slack. More precisely, each periodic task $T_i = (e_i, p_i)$ is assigned a priority equal to $p_i - ke_i$, where k is a fixed nonnegative real number called the global *slack factor*. The task with the least value of

$p_i - ke_i$ obtains the highest priority. Note that when $k = 0$ the algorithm is equivalent to RM. Andersson and Jonsson proved the following:

Theorem 24.9 [7] *Let τ be a periodic task system consisting of $n = m + 1$ tasks. If the tasks are assigned priorities according to nondecreasing values of $p_i - ke_i$ and the m highest-priority tasks have the same period and execution requirement, then the following value of k maximizes the utilization of the task system when scheduled on m processors:*

$$k = \frac{1}{2} \times \frac{m - 1 + \sqrt{5m^2 - 6m + 1}}{m}. \quad (24.15)$$

The corrresponding utilization of the task system τ is

$$U(\tau) = 2 \frac{m^2}{3m - 1 + \sqrt{5m^2 - 6m + 1}}. \quad (24.16)$$

Note that $\lim_{m \to \infty} U(\tau) > 0.38m$, which shows that the Dhall effect is circumvented by the scheduling algorithm. It should be pointed out, however, that the above theorem assumes a constrained task system in which all but one of the $m+1$ tasks have the same period and execution requirement. (The task system used earlier to illustrate the Dhall effect has this property.) The authors conjecture that this constrained task system represents a worst-case scenario and that AdaptiveTkC has a minimum schedulable utilization given by Equation 24.16. However, this remains to be proven.

Motivated by the same observation that heavy tasks should be given higher priority than light tasks in order the avoid the Dhall effect, Andersson et al. [8] developed a variant of RM, called RM-US[m/(3m−2)], which assigns static priorities to periodic tasks according to the following rule:

- If $u_i > \frac{m}{3m-2}$ then T_i has the highest priority (ties broken arbitrarily).
- If $u_i \leq \frac{m}{3m-2}$ then T_i has rate-monotonic priority.

(RM-US[m/(3m−2)] stands for rate monotonic utilization-separation with separator $m/(3m - 2)$.) As an example of the priorities assigned by algorithm RM-US[m/(3m−2)], consider the periodic task system

$$\tau = \{T_1 = (1, 7), T_2 = (2, 10), T_3 = (9, 20), T_4 = (11, 22), T_5 = (2, 25)\}$$

to be scheduled on a 3-processor system. The utilizations of these five tasks are $\approx 0.143, 0.2, 0.45, 0.5,$ and 0.08, respectively. For $m = 3$, $m/(3m - 2)$ equals $3/7 \approx 0.4286$; hence, tasks T_3 and T_4 will each be assigned the highest priority, and the remaining three tasks will be assigned rate-monotonic priorities. Thus, the possible priority assignments are therefore as follows (highest-priority task listed first):

$$T_3, T_4, T_1, T_2, T_5$$

or

$$T_4, T_3, T_1, T_2, T_5.$$

Andersson et al. proved the following utilization bound for RM-US[m/(3m−2)]:

Theorem 24.10 [8] *Any periodic task system τ with utilization $U(\tau) \leq m^2/(3m - 2)$ is schedulable on m processors by* RM-US[m/(3m−2)].

In AdaptiveTkC and RM-US[m/(3m−2)], the Dhall effect is circumvented by modifying the priority scheme so that it is no longer strictly rate-monotonic. Another way to avoid the Dhall effect, while retaining the rate-monotonic priority scheme, is to require that the task system only have light tasks, as the following result by Baruah and Goossens shows:

Theorem 24.11 [16] *Any periodic task system τ in which each task's utilization is no more than $1/3$ is schedulable on m processors by the global RM scheduling algorithm provided that $U(\tau) \leq m/3$.*

In Reference 11 and 12, Baker studied the global scheduling of *constrained-deadline* periodic task systems. In such a task system, each periodic task $T_i = (e_i, p_i, d_i)$ has a relative deadline d_i different from, but no larger than, its period p_i. As discussed in Subsection 24.4.1, uniprocessor scheduling for this class of task systems can be done optimally by the DM algorithm. In DM, tasks are assigned priorities in inverse proportion to their relative deadlines: the task with the shortest relative deadline is assigned the highest priority, with ties broken arbitrarily. Baker developed two sufficient schedulability tests for global DM scheduling on m processors; the simpler of these two tests is given below:

Theorem 24.12 [11, 12] *A constrained-deadline periodic task system $\tau = \{T_1 = (e_1, d_1, p_1), \ldots, T_n = (e_n, d_n, p_n)\}$ is schedulable on m processors by the global DM scheduling algorithm if*

$$\sum_{i=1}^{n} \frac{e_i}{p_i}\left(1 + \frac{p_i - \delta_i}{d_k}\right) \leq m(1-\alpha) + \alpha, \qquad (24.17)$$

where $\alpha = \max\{e_i/d_i | 1 \leq i \leq n\}$, $\delta_i = e_i$ for $i < k$, and $\delta_k = d_k$.

The special case when $d_i = p_i$ for all i corresponds to an implicit-deadline periodic task system, in which case DM priority assignment becomes RM priority assignment. Thus, as a corollary to the above theorem, we get

Corollary 24.1 [11, 12] *An (implicit-deadline) periodic task system τ is schedulable on m processors by the global RM scheduling algorithm if*

$$U(\tau) \leq \frac{m}{2}(1-\alpha) + \alpha, \qquad (24.18)$$

where $\alpha = \max\{u_i | 1 \leq i \leq n\}$.

The above corollary generalizes the result of Baruah and Goossens stated in Theorem 24.11 to periodic task systems with tasks having utilizations at most α, for some constant α, $0 < \alpha \leq 1$. In particular, when $\alpha = 1/3$, the corollary gives a schedulable utilization of $m/3 + 1/3$, which is slightly better than the bound given by Theorem 24.11.

24.7.2 Global Dynamic-Priority Scheduling

Just like RM, global EDF scheduling suffers from the Dhall effect in that it is not able to schedule certain feasible task systems with very low utilizations, particularly those with a mixture of heavy tasks and and light tasks. In Reference 28, Goossens, Funk, and Baruah proved the following result:

Theorem 24.13 [28] *A periodic task system τ is schedulable on m processors by the global EDF scheduling algorithm if*

$$U(\tau) \leq m - \alpha(m-1), \qquad (24.19)$$

where $\alpha = \max\{u_i | 1 \leq i \leq n\}$.

Note as the maximum task utilization become larger, the schedulable utilization that can be achieved by global EDF becomes smaller, indicating the increasing influence of the Dhall effect.

Baker [11,12] generalized the above result to the case of *constrained-deadline* periodic task systems, in which each task $T_i = (e_i, p_i, d_i)$ has a relative deadline d_i different from, but no larger than, its period p_i. He presented two sufficient schedulability tests for global EDF scheduling on m processors; the simpler of these two tests is given below:

Theorem 24.14 [11,12] *A constrained-deadline periodic task system* $\tau = \{T_1 = (e_1, d_1, p_1), \ldots, T_n = (e_n, d_n, p_n)\}$ *is schedulable on m processors by the global EDF scheduling algorithm if*

$$\sum_{i=1}^{n} \min\left\{1, \frac{e_i}{p_i}\left(1 + \frac{p_i - d_i}{d_{\min}}\right)\right\} \leq m(1-\alpha) + \alpha, \quad (24.20)$$

where $\alpha = \max\{e_i/d_i | 1 \leq i \leq n\}$, $d_{\min} = \min\{d_i | 1 \leq i \leq n\}$.

Theorem 24.13 becomes a special case of the above theorem by setting $d_i = p_i$ for all i.

One can circumvent the Dhall effect by using a modified dynamic-priority assignment. In the previous subsection, we described RM-US[m/(3m−2)] [8], a global static-priority scheduling algorithm that assigns *modified* rate-monotonic priorities to tasks to ensure that heavy tasks can be scheduled, thereby guaranteeing that the minimum schedulable utilization is not too low. In Reference 57, Srinivasan and Baruah adapted this algorithm to develop a global dynamic-priority scheduling algorithm, called EDF-US[m/(2m−1)], which assigns priorities to *jobs* of tasks T_i in a periodic task system according to the following rule:

- If $u_i > \frac{m}{2m-1}$ then the jobs of T_i are assigned the highest priority (ties broken arbitrarily).
- If $u_i \leq \frac{m}{2m-1}$ then T_i's jobs are assigned priorities according to EDF.

Srinivasan and Baruah proved the following utilization bound for EDF-US[m/(2m−1)]:

Theorem 24.15 [57] *Any periodic task system τ with utilization $U(\tau) \leq m^2/(2m-1)$ is schedulable on m processors by* EDF-US[m/(2m-1)].

Note that $\lim_{m\to\infty} m^2/(2m-1) > m/2$. Therefore, the schedulable utilization of EDF-US[m/(2m-1)] nearly matches the upper bound of $(m+1)/2$ given by Theorem 24.4.

In Reference 13, Baruah developed an improved algorithm, called fpEDF, that assigns priorities to jobs of tasks in the following way. First, the tasks are sorted in nonincreasing order of their utilizations; that is, $u_i \geq u_{i+1}$ for $1 \leq i < n$. Next, the first $(m-1)$ tasks in the sorted list are considered and from among these tasks, those with utilizations $> 1/2$ have their jobs assigned the highest priority. The remaining tasks (i.e., those from the first $m-1$ tasks that have utilizations $\leq 1/2$ as well as the $(n-m+1)$ tasks at the tail end of the list) have their jobs assigned priorities according to EDF. Baruah proved the following:

Theorem 24.16 [13] *Any periodic task system τ with utilization $U(\tau) \leq (m+1)/2$ is schedulable on m processors by* fpEDF.

Thus, fpEDF is optimal—its schedulable utilization matches the upper bound of $(m+1)/2$ given by Theorem 24.4 for any job-level dynamic-priority scheduling algorithm.

24.8 Proportionate-Fair Scheduling

At this point, it is useful to summarize the known results for the four classes of multiprocessor scheduling algorithms that we have discussed.

1. *Partitioned static-priority scheduling:* The best known algorithms are RM scheduling algorithms that use reasonably allocation decreasing (RAD) bin-packing heuristics, such as RM-FFD and

RM-BFD, all of which achieve a schedulable utilization of $0.414(m + 1)$ on m processors. No optimal algorithm for this class is currently known.

2. *Partitioned dynamic-priority scheduling:* The best known algorithms are EDF-FF and EDF-BF, both of which achieve a schedulable utilization of $(m + 1)/2$ on m processors, which is optimal for the class of *job-level* dynamic-priority scheduling algorithms.
3. *Global static-priority scheduling:* The best known algorithm is RM-US[m/(3m-2)], which achieves a schedulable utilization of $m^2/(3m - 2)$ on m processors. No optimal algorithm for this class is currently known.
4. *Global dynamic-priority scheduling:* The best known algorithm is fpEDF that achieves a schedulable utilization of $(m + 1)/2$, which is optimal for the class of *job-level* dynamic priority scheduling algorithms.

It is also known that no static-priority or job-level dynamic-priority multiprocessor scheduling algorithm—partitioned or global—can achieve a schedulable utilization greater than $(m + 1)/2$ for all periodic task systems (Theorem 24.4). Therefore, only a *fully dynamic-priority* scheduling algorithm can possibly surpass this utilization bound.

In the remainder of this section, we describe a global, fully dynamic-priority, scheduling algorithm, called *Pfair* scheduling, that achieves an optimal utilization of m on m processors. First introduced by Baruah et al. in Reference 15, Pfair scheduling is presently the only known optimal method for scheduling periodic tasks on a multiprocessor system.

Under Pfair scheduling, each periodic task $T_i = (e_i, p_i)$ is assigned a *weight* $w(T_i) = e_i/p_i$. (Note that the weight of T_i is identical to its utilization u_i; we use the term weight instead of utilization in keeping with the notation used in Reference 15.) Processor time is allocated in discrete time units called *quanta*. Time interval $[t, t + 1)$, where t is a nonnegative integer, is called *slot t*. (Hence, time t refers to the beginning of slot t.) It is assumed that all task parameters are integer multiples of the quantum size.

In each slot, each processor can be allocated at most one task. A task may be allocated time on different processors, but not within the same slot (i.e., migration is allowed but parallelism is not). The sequence of allocation decisions over time defines a *schedule S*. Formally, $S : \tau \times \mathcal{N} \mapsto \{0, 1\}$, where τ is the set of n tasks to be scheduled and $\mathcal{N}$ is the set of nonnegative integers. $S(T_i, t) = 1$ if and only if T_i is scheduled in slot t. In any m-processor schedule, $\sum_{i=1}^{n} S(T_i, t) \leq m$ holds for all t.

A task's weight essentially defines the rate at which it is to be scheduled. In an ideal "fluid" schedule in which the quantum size can be arbitrarily small, every task T_i should be allocated $w(T_i) \times t$ units of processor time over the interval $[0, t)$ for all t. For any fixed (integer) quantum size, this is clearly impossible to guarantee. Instead, Pfair scheduling attempts to follow closely the ideal schedule by making sure that the task is never allocated too little or too much processor time than the ideal schedule would in any time interval. The deviation from the the fluid schedule is formalized in the notion of *lag*. Formally, the *lag* of a task T_i at time t is given by

$$lag(T_i, t) = w(T_i) \times t - \sum_{u=0}^{t-1} S(T_i, u). \qquad (24.21)$$

That is, the lag of a task is the difference between a task's allocation in the Pfair schedule and the allocation it would receive in an ideal schedule. A schedule is *Pfair* if and only if

$$(\forall T_i \in \tau, t :: -1 < lag(T_i, t) < 1). \qquad (24.22)$$

That is, the allocation error for each task is always less than one quantum. It follows that in a Pfair schedule, each task T_i receives either $\lfloor w(T_i) \times t \rfloor$ or $\lceil w(T_i) \cdot t \rceil$ units of processor time over the interval $[0, t)$ [15]. Note that if t is a multiple of task's period, that is, $t = kp_i$ for some $k > 0$, $\lfloor w(T_i) \times t \rfloor$ and $\lceil w(T_i) \times t \rceil$ both reduce to ke_i, which is exactly the amount of processor time that the task should receive over the interval $[0, kp_i)$ in a *periodic* schedule. Thus, every Pfair schedule is also a periodic schedule, although the converse is not always true.

Baruah et al. proved the following schedulability condition for Pfair scheduling:

Theorem 24.17 [15] *A periodic task system τ has a Pfair schedule on m processors if*

$$\sum_{i=1}^{n} w(T_i) \leq m. \qquad (24.23)$$

Since the weight of a task is the same as its utilization, and since a periodic task system τ with utilization $U(\tau) > m$ has no feasible schedule on m processors, it follows from the above theorem that Pfair scheduling achieves *optimal* schedulable utilization.

The proof of Theorem 24.17 makes use of a reduction to maximum flow in a network to prove the existence of a Pfair schedule. However, the size of the network generated by the reduction is exponential in the size of the problem instance, and hence, the reduction argument does not by itself provide an efficient algorithm. In the same paper [15], Baruah et al. presented a Pfair scheduling algorithm, called PF, that runs in polynomial time.

In PF, each task T_i is conceptually divided into an infinite sequence of quantum-length *subtasks*. Let $T_{i,j}$ denote the j-th subtask of task T_i. Subtask $T_{i,j+1}$ is called the *successor* of $T_{i,j}$.

Each subtask $T_{i,j}$ is assigned a *pseudorelease* $r(T_{i,j})$ and a *pseudodeadline* $d(T_{i,j})$, defined as follows:

$$r(T_{i,j}) = \left\lfloor \frac{j-1}{w(T_i)} \right\rfloor \quad d(T_{i,j}) = \left\lceil \frac{j}{w(T_i)} \right\rceil. \qquad (24.24)$$

Intuitively, $r(T_{i,j})$ and $d(T_{i,j}) - 1$ are, respectively, the earliest and latest time slots that subtask $T_{i,j}$ must be scheduled or else Condition 24.22 will be violated. The interval $W(T_{i,j} = [r(T_{i,j}), d(T_{i,j}))$ is called $T_{i,j}$'s *window*. Note that $r(T_{i,j})$ is either $d(T_{i,j}) - 1$ or $d(T_{i,j})$. Thus, consecutive windows of the same task either overlap by one slot or are disjoint.

PF schedules the subtasks (of all tasks) on an earliest-pseudodeadline-first basis—that is, in each new slot, the m subtasks with the earliest pseudodeadlines are allocated to the m processors. If multiple subtasks have the same pseudodeadline, then a *tie-breaking* rule is used to prioritize the subtasks. It turns out that it is *necessary* to resolve tie-breaks among subtasks with the same pseudodeadline: simply breaking ties arbitrarily does not necessarily produce a Pfair schedule for the given task system, even there is one.

The PF tie-breaking rule involves a bit, denoted $b(T_{i,j})$, which is defined to be 1 if $T_{i,j}$'s window overlaps with its $T_{i,j+1}$'s, and 0 otherwise. If two subtasks have the same pseudo-deadline, the subtask with a b-bit of 1 is given higher priority than the subtask with a b-bit of 0. Informally, it is better to execute $T_{i,j}$ "early" if its window overlaps that of $T_{i,j+1}$'s, because this potentially leaves more slots available to $T_{i,j+1}$. If two subtasks have the same pseudodeadline and have b-bits of 1, then the tie-breaking rule just described is applied repeatedly to the successors of these subtasks until the tie is broken.

The aforementioned tie-breaking procedure, as such, does not run in polynomial time because the number of successor subtasks to be examined may in the worst case be proportional to the execution requirements of the corresponding tasks (hence, it is pseudopolynomial). Baruah et al. obtained an equivalent, but faster, GCD-like tie-breaking procedure that runs in $O(n)$ time per slot, but the constant of proportionality can be arbitrarily high.

Subsequently, Baruah et al. [18] developed a more efficient algorithm, called PD, that has a per-slot running time of $O(\min\{m \lg n, n\})$. Briefly, PD limits the tie-breaking rule to look only at a constant number of pseudodeadlines into the future. Instead of a recursive procedure, PD uses four constant-time tie-breaking rules. Anderson and Srinivasan [2] improved upon the PD algorithm by showing that only two of the four tie-breaking rules are necessary to produce a Pfair schedule. Their new algorithm, called PD^2, is currently the most efficient Pfair scheduling algorithm.

Under Pfair scheduling, if some subtask of a task T_i executes "early" within its window, then T_i will not be eligible to execute until the beginning of the window of its next subtask. Thus, Pfair scheduling algorithms are not necessarily work-conserving. (Recall that a work-conserving scheduling algorithm is one that never leaves a processor idle when there are uncompleted jobs in the system that could be executed on that processor.) In Reference 2, Anderson and Srinivasan introduced a work-conserving variant of Pfair scheduling, called *Early-Release fair* (*ERfair*) scheduling. Under ERfair scheduling, if two subtasks are part of the same job, then the second subtask may be released "early," that is, before the beginning of its Pfair window, as soon as the preceding subtask of the same job has completed. The notion of ERfair scheduling is obtained by simply dropping the -1 lag constraint from Condition 24.22 That is, a schedule is *ERfair* if and only if

$$(\forall T_i \in \tau, t :: lag(T_i, t) < 1). \qquad (24.25)$$

Anderson and Srinivasan developed an ERfair version of PD^2 and showed that this algorithm can (optimally) schedule all periodic task systems with utilization at most m on m processors.

In a more recent work [3, 56], Anderson and Srinivasan extended the early-release task model to also allow subtasks to be released "late", that is, there may be a temporal separation between the windows of consecutive subtasks of the same task. The resulting model, called the *intrasporadic* (*IS*) model, is a generalization of the sporadic task model. (In the sporadic task model, there can be a separation between consecutive *jobs* of the same task. The IS model allows separation between consecutive *subtasks* of the same task.) Anderson and Srinivasan showed that PD^2 can also optimally schedule IS task systems with utilization at most m on m processors [56].

24.9 Online Multiprocessor Scheduling

As we have seen from previous sections, much of the research on real-time scheduling has focussed on offline algorithms that have *a priori* knowledge of the tasks to be scheduled. However, in many applications (e.g., web servers), the real-time system does not have knowledge of future task arrivals. That is, tasks may arrive at any time and the task parameters (e.g., execution requirement, deadline, period, etc.) become known to the system only at the time of the task's arrival. Thus, the system has to make scheduling decisions *online*—that is, using only information about tasks that have already arrived, but not on information about future tasks. Moreover, a surge in the task arrival rate could easily bring the system to a state of overload that makes it impossible to meet all task deadlines. Thus, the system has to perform some form of *admission control* wherein it selectively admits some tasks and rejects others to ensure that all admitted tasks can run to completion without missing their deadlines.

From the above discussion, it is clear that designing good online scheduling algorithms is a challenging problem. The problem is exacerbated by the fact that *no optimal online scheduler can exist* for $m > 1$ processors, even for simple task systems (such as aperiodic tasks). This is implied by a result derived by Dertouzos and Mok in Reference 26 who proved that any algorithm that does not have *a priori* knowledge of at least one of the following task parameters—arrival time, execution requirement, and deadline—cannot schedule all feasible task sets. In Reference 31, Hong and Leung proved a similar result and, in fact, showed that there is no optimal online scheduler even if the tasks are known to just have two distinct deadlines. In contrast, there is an optimal online scheduler for uniprocessor systems: EDF 25.

Despite this discouraging result, it is interesting to compare the performance of online scheduling algorithms with their offline counterparts. As discussed in previous sections, provably good utilization bounds are known for offline scheduling algorithms. Therefore, a natural question to ask is whether similar utilization bounds can be obtained for online scheduling algorithms. In References 4 and 5 Andersson et al. considered this question for the case of *aperiodic task* scheduling using the partitioned and global approaches.

In Reference 4, Andersson et al. developed an online version of the partitioned scheduling algorithm, EDF-FF (see Subsection 24.6.2), as follows. Let $\tau = \{T_1, T_2, \ldots, T_n\}$ be an aperiodic task system. For any time instant $t \geq 0$, let to be $V(t) = \{T_i | a_i \leq t < a_i + d_i\}$, where a_i and d_i are, respectively, the arrival time and relative deadline of task T_i. Let the utilization at time t be defined as $U(t) = \sum_{T_i \in V(t)} e_i/d_i$. Then, τ is schedulable on a single processor by EDF if $\forall t : U(t) \leq 1$ [4].

Call a processor *occupied* at time t if there is at least one task T_k that has been assigned to the processor and for this task, $a_k \leq t < a_k + d_k$. If a processor is not occupied, it is called *empty*. For each new task arrival, the online EDF-FF algorithm assigns the task to the first processor that makes the earliest transition from being empty to occupied and for which the task passes the uniprocessor schedulability test for that processor, as described in the preceding paragraph. Tasks allocated to the same processor are scheduled according to EDF. If the task cannot be assigned to any occupied processor, it is assigned to the next empty processor. If no empty processor is available, then the task cannot be scheduled, and the algorithm declares failure.

Andersson et al. proved the following:

Theorem 24.18 [4] *Any asynchronous aperiodic task system τ is schedulable by the online EDF-FF algorithm on $m \geq 3$ processors if $U(\tau) \leq 0.31m$.*

This utilization bound should be compared to the upper bound of $(m+1)/2$ of Theorem 24.4, which also applies to aperiodic tasks.

In another paper [5], Andersson et al. considered online global scheduling and developed an online version of algorithm EDF-US[m/(2m-1)] (see Subsection 24.7.2) with a suitably modified schedulability test. They proved the following:

Theorem 24.19 [5] *Any asynchronous aperiodic task system τ is schedulable online by EDF-US[m/(2m−1)] on m processors if $U(\tau) \leq m^2/(2m-1)$.*

Note that this is the same utilization bound that the *offline* algorithm achieves for synchronous periodic task systems. For large numbers of processors, this bound approaches $m/2$, which nearly matches the upper bound of $(m+1)/2$ of Theorem 24.4.

The online algorithms of References 4 and 5 do not perform admission control per se, as the objective is to find utilization bounds for these algorithms assuming that the input task system is feasible. (Thus, the algorithms simply declare failure and halt upon encountering a task that is unschedulable.) In practice, however, the online algorithm does not know in advance whether the sequence of incoming tasks constitutes a feasible set. Consequently, it must work under the assumption that the task sequence could be *infeasible*, and hence it would need to perform admission control to admit (possibly) only a subset of the tasks that can be guaranteed to be schedulable.

In Reference 23, Dasgupta and Palis considered the problem of online scheduling with admission control of aperiodic task systems. Unlike References 4 and 5, they measured the performance of the online algorithm in terms of its *cumulative load* (or simply *load*), which is the sum of the execution requirements of all admitted tasks. They used *competitive analysis*, a standard analysis technique for online algorithms, to compare the performance of the online scheduling algorithm with an *optimal offline* scheduling algorithm that knows the entire task system in advance. In particular, the optimal offline algorithm is able to construct a feasible schedule for the given task system if one exists; if the task system is infeasible, it is able to construct a schedule for the feasible subset with the largest possible load.

More formally, let A be an online scheduling algorithm and let OPT be an optimal offline scheduling algorithm. Let $\mathcal{L}_A(\tau)$ and $\mathcal{L}_{OPT}(\tau)$ be the loads of A and OPT, respectively, for a task system τ. We say that A achieves a *competitive ratio* of c (or is c-competitive) if

$$\frac{\mathcal{L}_{OPT}(\tau)}{\mathcal{L}_A(\tau)} \leq c \qquad (24.26)$$

for every task system τ.

Clearly, $\mathcal{L}_{OPT} \geq \mathcal{L}_A$; hence the goal is to make $\mathcal{L}_A$ as close as possible to $\mathcal{L}_{OPT}$, that is, to make the competitive ratio as close to 1 as possible.

Dasgupta and Palis considered an online version of EDF-FF, which is similar to that of Reference 4, but which performs admission control, as follows: For each new task, the algorithm admits the task if there is at least one processor in which it can be feasibly scheduled and allocates it to the first such processor. Otherwise, the algorithm rejects the task. Call this algorithm EDF-FF-AC (EDF-FF with Admission Control).

For an aperiodic task $T_i = (e_i, d_i)$ let its weight be $w_i = e_i/d_i$. Dasgupta and Palis proved the following:

Theorem 24.20 [23] *Let $\mathcal{T}(n, \alpha)$ be the class of asynchronous aperiodic task systems with n tasks, in which every task has weight at most α, $0 < \alpha \leq 1$. On an m-processor system, EDF-FF-AC achieves a competitive ratio of at most $1/(1 - \alpha)$ for every $\tau \in \mathcal{T}(n, \alpha)$.*

Thus, EDF-FF-AC attains a better (i.e., smaller) competitive ratio when the tasks are lighter (i.e., have less weight). This result is analogous to results for the offline case wherein better utilization bounds can be achieved when the tasks are lighter.

Dasgupta and Palis also proved the following lower bound on the competitive ratio.

Theorem 24.21 [23] *Let $\mathcal{T}(n, \alpha)$ be the class of asynchronous aperiodic task systems with n tasks, in which every task has weight at most α, $0 < \alpha \leq 1$. For task systems in $\mathcal{T}(n, \alpha)$ scheduled on m processors, any online scheduling algorithm has a competitive ratio $\rho(n, m, \alpha)$, where*

$$\rho(n, m, \alpha) \geq \begin{cases} \dfrac{1}{1 - \alpha + \varepsilon} & \text{if } m = 1 \\ \dfrac{1}{1 - 1/(m\lceil 1/\alpha \rceil)} & \text{if } m > 1, \end{cases} \quad (24.27)$$

for arbitrarily small $\varepsilon > 0$.

Thus, EDF-FF-AC achieves optimal competitive ratio for uniprocessor systems ($m = 1$). For $m > 1$, there is a gap between the above lower bound and the upper bound given by Theorem 24.20; closing this gap remains an open problem.

EDF-FF-AC uses a *greedy* admission control strategy: a new task is *always* admitted if it can be feasibly scheduled along with previously admitted tasks. This seems to be a natural and sensible policy for admission control—without any knowledge of future task arrivals (including the possibility that no further tasks will arrive), the scheduling algorithm would be better off admitting every schedulable task that it encounters.

In References 49 and 50, Palis asked the question whether the greedy policy is the best admission control strategy for online scheduling. He considered a refinement of the aperiodic task model, called the (r, g) *task model*. In this model, an aperiodic task T is characterized by four parameters (a, e, r, p), where a is the arrival time and e is the execution requirement. The last two parameters specify a *rate of progress* requirement for executing the task. Specifically, upon its arrival at time a, task T must be executed for *at least krp time units in the time interval $[a, a + kp)$ for every in integer $k > 1$.*

Note that the rate of progress requirement specifies the *minimum*, rather than exact, amount of processor time that the task would need over a given time interval. Thus, on a lightly loaded system, the task could potentially run at a faster rate and hence complete sooner. Moreover, if the system load fluctuates (e.g., from lightly loaded to heavily loaded), a task that has been running "ahead" could be temporarily blocked, or be run at slower rate, to allow the processor to execute other tasks or newly arrived tasks. However, the system has to guarantee that the task never lags behind its "reserved" rate of krp time units over kp time units for every k. Note that, although the parameter p is called the period, a task in the (r, g) task model is not a periodic task since it does not consist of distinct "jobs" that are individually "released" every p time units. It is intended to model real-time tasks (such as in continuous media applications) that can continuously run on an unloaded system, but that nonetheless require a minimum execution rate when the system becomes heavily loaded.

Define $g \equiv rp/e$ as the *granularity* of task T. We assume $e \geq rp$; hence $0 < g \leq 1$. Intuitively, the granularity is a measure of the "smoothness" of the task's execution profile: the "finer" the granularity (i.e., the smaller g is), the "smoother" the task's rate of progress will be during its execution. For example, consider two tasks $T_1 = (0, 12, 0.5, 2)$ and $T_2 = (0, 12, 0.5, 8)$. Both tasks have the same execution time (12) and the same rate (0.5). However, the tasks have different periods and hence different granularities: T_1 has granularity $g_1 = 0.5 \times 2/12 = 1/12$ and T_2 has granularity $g_2 = 0.5 \times 8/12 = 1/3$. When executed, T_1, which has the smaller granularity, will have a "smoother" rate of progress than T_2. Specifically, T_1 would run for at least k time units over the first $2k$ time units—that is, at least one out of every two time units would be allocated to T_1. On the other hand, T_2 would only need to run for at least 4 time units over the first 8 time units, 8 over the first 16, and 12 over the first 24. In particular, it is possible that T_2 is not executed during the first 4 time units, then is executed continuously during the next 4 time units, yet its rate of progress requirement for the first period would be satisfied. That is, T_2's execution profile would be more "jittery" than T_1's. As this example illustrates, finer-grain tasks impose more stringent rate of progress requirements than coarser-grain tasks and hence, intuitively, they would be more difficult to schedule. The interesting problem is to quantify the effect of task granularity on scheduling performance.

A task system $\tau = \{T_1, T_2, \ldots, T_n\}$, where $T_i = (a_i, e_i, r_i, p_i)$, is an (r, g) task system if $r \geq \max\{r_i | 1 \leq i \leq n\}$ and $g = \min\{g_i | 1 \leq i \leq n\}$. That is, in an (r, g) task system, every task has rate at most r and granularity at least g. Note that a task system consisting of ordinary aperiodic tasks T_i with execution requirement e_i and relative deadline d_i is a special case of an (r, g) task system in which task T_i has execution requirement e_i, period d_i, and rate e_i/d_i. The granularity of every task is 1.

In References 49 and 50, Palis investigated the impact of admission control on online scheduling of (r, g) task systems on a *single* processor, and proved the following:

Theorem 24.22 [49, 50]

1. For (r, g) task systems satisfying $r \lg(1/g) \leq 1$, there is an online scheduling algorithm with a nongreedy admission control policy, called **NonGreedy-EDF**, that achieves a competitive ratio of $O(\lg(1/g))$. Moreover, this algorithm is optimal: the competitive ratio achievable by any online algorithm for this class of task systems in $\Omega(\lg(1/g))$.
2. The rate bound $r \lg(1/g)$ is necessary: without this bound, any online scheduling algorithm has a competitive ratio of $\Omega((1/g)^r)$.
3. **Greedy-EDF**, the EDF scheduling algorithm with greedy admission control, has a competitive ratio of $\Theta(1/g(1-r))$. In particular, for rate-bounded task systems satisfying $r \lg(1/g) \leq 1$, this algorithm has a competitive ratio of $\Omega(1/g)$.

The above theorem presents two important results. First, the product $r \lg(1/g)$ delineates the boundary between task systems that are difficult to schedule and those for which efficient scheduling algorithms exist. The latter consists of (r, g) task systems that are rate-bounded, that is, satisfying $r \lg(1/g) \leq 1$. Second, a greedy admission control policy does *not* yield an efficient scheduling algorithm for rate-bounded task systems; a *nongreedy* policy does and the resulting algorithm achieves optimal competitive ratio.

Greedy-EDF always admits a new task if it passes a schedulability test indicating that it can be feasibly scheduled along with previously admitted tasks. In contrast, NonGreedy-EDF employs a *throttle test* that places a threshold on the "profitability" of admitting the new task, given the possibility that there may be more "profitable" tasks that may arrive in the future. In other words, a new task that passes the schedulability test may not be admitted because it is not profitable enough. Doing so allows the scheduling algorithm to admit future, more profitable tasks, which would otherwise be rejected if the new task were admitted.

Briefly, for each new task T_j, NonGreedy-EDF performs the following test:

NonGreedy-EDF Throttle Test. Admit task T_j if

$$r_j \int_{t \geq 0} \left(\mu^{R_j(t)} - 1 \right) dt \leq e_j, \qquad (24.28)$$

where $\mu = (2/g) + 2$; $R_j(t) = \sum_{T_i \in S_j} \bar{r}_i(t)$; $S_j(t)$ is the set of tasks admitted before T_j; and $\bar{r}_i(t) = r_i(t)$ if $a_i \leq t < d_i$, else $\bar{r}_i = 0$.

That is, task T_j is admitted if its rate, multiplied by a *weight factor* $\int_{t \geq 0} \left(\mu^{R_j(t)} - 1 \right) dt$, is less than or equal to its execution requirement, which is the "profit" accrued by the algorithm for admitting the task. The weight factor is a measure of how much "loaded" the processor already is at the time of T_j's arrival; this "load" is expressed as an exponential function of the sum of the rates of previously admitted tasks that have not yet completed. Note that the base of the exponential function, μ, is inversely proportional to the granularity; thus, for the same set of admitted tasks, the load on the processor owing to these tasks is considered to be higher when the tasks are fine-grain than when they are coarse-grain. Therefore, the "load" properly takes into account, not only the processing requirements of these tasks (which depend on their rates) but also the degree of difficulty in meeting their rate of progress requirements (which is inversely proportional to their granularity).

The results given in Theorem 24.22 are applicable only to uniprocessor systems. Although the algorithms Greedy-EDF and NonGreedy-EDF are straightforward to extend to the multiprocessor case, an analysis of their competitive ratio for $m > 1$ processors remains an interesting open problem.

24.10 Concluding Remarks

The increasing prevalence of multiprocessor systems and the growing need to support applications with real-time characteristics have made multiprocessor scheduling an important and challenging problem in real-time systems design. This chapter presented some of the most important theoretical results in multiprocessor real-time scheduling over the past three decades. However, despite significant research progress, a host of interesting and important open problems remain, some of which are discussed below:

- **Other real-time task models.** Research in real-time multiprocessor scheduling algorithms has focussed primarily on the periodic and sporadic task models. It would be interesting to extend these results to more general task models. Examples include the *multiframe task model* [44], which generalizes the sporadic task model by allowing jobs to have different execution requirements, and the *generalized multiframe task model* [14], which, in addition, allows different job relative deadlines and different minimum separation times between job arrivals. Also of interest are real-time task models, such as the *rate-based execution (RBE)* model [32] and *CPU Reservations* [33], which specify timing constraints in terms of an average execution rate for each task (rather than an explicit period or deadline).
- **Resource augmentation.** The fact that no optimal online multiprocessor scheduling algorithms can exist [26, 31] has motivated some researchers to ask whether optimality can be achieved by equipping the system with faster processors or extra processsors. Some promising results in this direction have been obtained by Phillips et al. in Reference 53, where they showed that EDF and LLF can feasibly schedule every aperiodic task system on m processors if given processors are $(2 - 1/m)$ times faster than those of an optimal offline algorithm. Interestingly, they also showed that adding *extra* processors does not help EDF: there are feasible task systems that cannot be scheduled by EDF even if given $m\Delta$ processors, where Δ is the ratio between the longest task duration and the shortest task duration. On the other hand, LLF becomes optimal when given $m \lg \Delta$ processors. It would be interesting to extend these results to static-priority (e.g., RM) and Pfair scheduling algorithms.

- **General multiprocessor systems.** This chapter focussed primarily on scheduling algorithms for *identical multiprocessor* systems—that is, systems in which all processors have the same computing speed. It would be interesting to extend these results to more general multiprocessor models. One such model is the *uniform multiprocessor* model in which each processor has its own computing speed: a task that executes on a processor with speed s for t time units completes $s \times t$ units of execution. This can be further generalized to the *unrelated multiprocessor* model, in which each task-processor pair (T_i, P_j) has an associated speed $s_{i,j}$: task T_i, when executed on processor P_j for t time units, completes $s_{i,j} \times t$ units of execution. Among the few results known for more general models are those derived by Baruah and his colleagues, who have developed schedulability tests for global scheduling using RM [16] and EDF [28] on uniform multiprocessor systems.
- **Admission control.** Until References 49 and 50, admission control in real-time scheduling has largely been considered as a byproduct of schedulability and/or feasibility testing. The default implementation of admission control is a *greedy* policy that always admits a task whenever it passes the schedulability or feasibility test used by the scheduling algorithm. In References 49 and 50, Palis showed that the greedy policy is, in theory, not the best strategy for admission control and that a nongreedy policy achieves better theoretical performance. This result opens up an important new direction of research in online real-time scheduling. The natural next step is to extend the results in References 49 and 50, which are only applicable to uniprocessor scheduling of aperiodic task systems, to multiprocessor scheduling of periodic, sporadic, and more general task systems.
- **Allowing job parallelism.** In designing multiprocessor scheduling algorithms for real-time systems, researchers have often made the assumption that real-time tasks consist of *sequential* jobs that cannot be parallelized, that is, a job cannot be sped up by executing it on multiple processors. Ironically, there is a vast research literature on *parallel job scheduling* that posits exactly the opposite—that is, many operations, including those used in real-time applications, are efficiently parallelizable (e.g., References 1 and 27). To take full advantage of the computational power provided by multiprocessor systems, operating system designers must develop schedulers that embody the characteristics of both real-time scheduling (namely, meeting all task deadlines) and parallel job scheduling (namely, executing jobs and/or tasks in parallel).

References

[1] Alves Barbosa da Silva, F. and Scherson, I.D., Efficient parallel job scheduling using gang service, *International Journal of Foundations of Computer Science*, 12:3, pp. 265–284, 2001.

[2] Anderson, J. and Srinivasan, A., Early-release fair scheduling, *Proc. 12th Euromicro Conference on Real-Time Systems*, pp. 35–43, June 2000.

[3] Anderson, J. and Srinivasan, A., Pfair scheduling: beyond periodic task systems, *Proc. 7th International Conference on Real-Time Computing Systems and Applications*, pp. 297–306, December 2000.

[4] Andersson, B., Abdelzaher, T., and Jonsson, J., Partitioned aperiodic scheduling on multiprocessors, *Proc. 17th International Parallel and Distributed Processing Symposium*, 2003.

[5] Andersson, B., Abdelzaher, T., and Jonsson, J., Global priority-driven aperiodic scheduling on multiprocessors, *Proc. 17th International Parallel and Distributed Processing Symposium*, 2003.

[6] Andersson B. and Jonsson J., Fixed priority preemptive multiprocessor scheduling: to partition or not to partition, in *IEEE International Conference on Real-Time Computing Systems and Applications*, December 2000.

[7] Andersson B. and Jonsson J., Some insights on fixed-priority preemptive non-partitioned multiprocessor scheduling, in *IEEE Real-Time Systems Symposium*, Work-in-Progress Session, Orlando, Florida, November 27–30, pp. 53–56, December 2000.

[8] Andersson, B., Baruah, S., and Jonsson, J., Static priority scheduling on multiprocessors, *Proc. 22nd IEEE Real-Time Systems Symposium*, London, UK, pp. 193–202, 2001.

[9] Audsley, N.C., Deadline monotonic scheduling, Tech. Report YCS 146, Dept. of Computer Science, University of York, September 1990.

[10] Audsley, N.C., Burns, A., Richardson, M.F., and Wellings, A.J., Hard real-time scheduling: the deadline monotonic approach, *Proc. 8th IEEE Workshop on Real-Time Operating Systems*, pp. 133–137, 1991.

[11] Baker. T.P., An analysis of deadline-monotonic scheduling on a multiprocessor, Tech. Rept. TR-030301, Florida State University, Dept. of Computer Science, February 2003.

[12] Baker, T.P., Multiprocessor EDF and deadline monotonic schedulability analysis, *Proc. 24th IEEE Real-Time Systems Symposium*, pp. 120–129, December 2003.

[13] Baruah, S.K., Optimal utilization bounds for the fixed-priority scheduling of periodic task systems on identical multiprocessors, *IEEE Transactions on Computers*, 53:6, pp. 781–784, 2004.

[14] Baruah, S., Chen, D., Gorinsky, S., and Mok, A., Generalized multiframe tasks, *International Journal of Time-Critical Computing Systems*, 17, pp. 5–22, 1999.

[15] Baruah, S., Cohen, N., Plaxton, C.G., and Varvel, D., Proportionate progress: a notion of fairness in resource allocation, *Algorithmica*, 15, pp. 600–625, 1996.

[16] Baruah, S.K. and Goossens, J., Rate-monotonic scheduling on uniform multiprocessors, *IEEE Transactions on Computers*, 52:77, pp. 966–970, 2003.

[17] Baruah, S. and Goossens, J., Scheduling real-time tasks: algorithms and complexity, in *Handbook of Scheduling: Algorithms, Models, and Performance Analysis*, Leung, J.Y., Ed., Chapman and Hall/CRC, Boca Raton, FL, 2004.

[18] Baruah, S.K., Gehrke, J.E., and Plaxton, C.G., Fast scheduling of periodic tasks on multiple resources, *Proc. 9th International Parallel Processing Symposium*, pp. 280–288, April 1995.

[19] Baruah, S., Howell, R., and Rosier, L., Algorithms and complexity concerning the preemptive scheduling of periodic, real-time tasks on one processor, *Real-Time Systems: The International Journal of Time-Critical Computing*, 2, pp. 301–324, 1990.

[20] Burchard, A., Liebeherr, J., Oh, Y., and Son, S.H., Assigning real-time tasks to homogeneous multiprocessor systems, *IEEE Transactions on Computers*, 44:12, pp. 1429–1442, 1995.

[21] Carpenter, J., Funk, S., Holman, P., Srinivasan, A., Anderson, J., and Baruah, S., A categorization of real-time multiprocessor scheduling problems and algorithms, in *Handbook of Scheduling: Algorithms, Models, and Performance Analysis*, Leung, J.Y., Ed., Chapman and Hall/CRC, Boca Raton, FL, 2004, pp. 30-1–30-19.

[22] Coffman, E.G., Galamos, G., Martello, S., and Vigo, D., Bin packing approximation algorithms: combinatorial analysis, in *Handbook of Combinatorial Optimization* (Supplement Volume A), Du, D.-Z. and Pardalos, P.M., Eds., Kluwer Academic Publishers, 1999, pp. 151–208.

[23] Dasgupta, B. and Palis. M.A., Online real-time preemptive scheduling of jobs with deadlines on multiple machines, *Journal of Scheduling*, 4, pp. 297–312, 2001.

[24] Dhall, S.K. and Liu, C.L., On a real-time scheduling problem, *Operations Research*, 26:1, pp. 127–140, 1978.

[25] Dertouzos, M.L., Control robotics: the procedural control of physical processes. *Proc. IFIP Congress*, pp. 807–813, 1974.

[26] Dertouzos, M.L. and Mok, A. Multiprocessor on-line scheduling of hard-real-time tasks, *IEEE Transactions on Software Engineering*, 15:12, pp. 1497–1506, 1989.

[27] Edmonds, J., Chinn, D.D., Brecht. T., and Deng, X. Non-clairvoyant multiprocessor scheduling of jobs with changing execution characteristics, *Journal of Scheduling*, 6:3, pp. 231–250, 2003.

[28] Funk, S., Goossens, J., and Baruah, S., On-line scheduling on uniform multiprocessors, *Real-Time Systems*, 25, pp. 197–205, 2003.

[29] Garey, M. and Johnson, D. *Computers and Intractability*, W.H. Freeman, New York, 1979.

[30] Goossens, J. and Devillers, R., The non-optimality of the monotonic priority assignments for hard real-time offset free systems, *Real-Time Systems: The International Journal of Time-Critical Computing*, 13:2, pp. 107–126, 1997.

[31] Hong, K.S. and Leung, J.Y.-T., On-line scheduling of real-time tasks, *IEEE Transactions on Computers*, 41:10, pp. 1326–1330, 1992.

[32] Jeffay, J. and Goddard, S., A theory of rate-based execution, *Proc. 20th IEEE Real-Time Systems Symposium*, pp. 304–314, December 1999.

[33] Jones, M., Rosu, D., and Rosu, M.-C., CPU reservations and time constraints: efficient, predictable scheduling of independent activities, *Proc. 16th ACM Symposium on Operating Systems Principles*, pp. 198–211, October 1997.

[34] Lauzac, S., Melhem, R., and Mossé, D., An efficient RMS admission control algorithm and its application to multiprocessor scheduling, *Proc. 1998 International Parallel Processing Symposium*, pp. 511–518, April 1998.

[35] Lehoczky, J.P., Fixed priority scheduling of periodic tasks with arbitrary deadlines, *Proc. 11th IEEE Real-Time Systems Symposium*, pp. 201–209, 1990.

[36] Lehoczky, J.P., Sha, L., and Ding, Y., The rate monotonic scheduling algorithm: exact characterization and average case behavior, *Proc. 10th IEEE Real-Time Systems Symposium*, pp. 166–171, 1989.

[37] Leung, J. and Merrill, M., A note on the preemptive scheduling of periodic, real-time tasks, *Information Processing Letters*, 11, pp. 115–118, 1980.

[38] López, J.M., García, M., Díaz, J.L., and García, D.F., Utilization bounds for EDF scheduling on real-time multiprocessor systems, *Real-Time Systems*, 28, pp. 39–68, 2004.

[39] López, J.M., Díaz, J.L., and García, D.F., Minimum and maximum utilization bounds for multiprocessor rate monotonic scheduling, *IEEE Transactions on Parallel and Distributed Systems*, 15:7, pp. 642–653, 2004.

[40] López, J.M., García, M., Díaz, J.L., and García, D.F., Utilization bounds for multiprocessor rate-monotonic scheduling, *Real-Time Systems*, 24, pp. 5–28, 2003.

[41] Leung, J.Y.-T. and Whitehead, J., On the complexity of fixed priority scheduling of periodic real-time tasks, *Performance Evaluation*, 2:4, pp. 237–250, 1982.

[42] Liu, C.L. and Layland, W., Scheduling algorithms for multiprogramming in a hard real-time environment, *Journal of the ACM*, 20:1, pp. 46–61, 1973.

[43] Mok, A.K., Fundamental design problems of distributed systems for the hard real-time environment, PhD Thesis, Massachusetts Institute of Technology, Dept. of Electrical Engineering and Computer Science, May 1983.

[44] Mok, A. and Chen, D., A multiframe model for real-time tasks, *IEEE Transactions on Software Engineering*, 23:10, pp. 635–645, 1997.

[45] Oh. D.-I. and Baker, T., Utilization bounds for N-processor rate monotone scheduling with static processor assignment, *Real-Time Systems*, 15:2, pp. 183–193, 1998.

[46] Oh, Y. and Son, S.H., Tight performance bounds of heuristics for a real-time scheduling problem, Tech. Report CS-93-24, Univ. of Virginia, Dept. of Computer Science, May 1993.

[47] Oh, Y. and Son, S.H., Fixed priority scheduling of periodic tasks on multiprocessor systems, Tech. Report CS-95-16, Univ. of Virginia, Dept. of Computer Science, March 1995.

[48] Oh, Y. and Son, S.H., Allocating fixed priority periodic tasks on multiprocessor systems, *Real-Time Systems*, 9:3, pp. 207-239, 1995.

[49] Palis, M. A., Competitive algorithms for fine-grain real-time scheduling, *Proc. 25th IEEE International Real-Time Systems Symposium*, pp. 129–138, Dec. 2004.

[50] Palis, M.A., The granularity metric for fine-grain real-time scheduling, *IEEE Transactions on Computers*, 54:12, pp. 1572–1583, 2005.

[51] Pereira Zapata, O.U., Alvarez, P.M., and Leyva del Foyo, L.E., Comparative analysis of real-time scheduling algorithms on one processor under rate monotonic, Rept. No. CINVESTAV-CS-RTG-01, CINVESTAV-IPN, Departamento de Ingenieria Electrica, Seccíon de Computacíon, 2005.

[52] Pereira Zapata, O.U. and Alvarez, P.M., EDF and RM multiprocessor scheduling algorithms: survey and performance evaluation, Rept. No. CINVESTAV-CS-RTG-02, CINVESTAV-IPN, Departamento de Ingenieria Electrica, Seccíon de Computacíon, 2005.

[53] Phillips, C.A., Stein, C., Torng, E., and Wein, J., Optimal time-critical scheduling via resource augmentation, *Proc. 29th Annual Symposium on the Theory of Computing*, pp. 140–149, May 1997.

[54] Serlin, O., Scheduling of time critical processes, *Proc. Spring Joint Computer Conference*, Atlantic City, NJ, pp. 925–932, 1972.

[55] Sha, L., Abdelzaher, T. Arzen, K.-E., Cervin, A., Baker, T., Burns, A., Buttazzo, G., Caccamo, M., Lehoczky, J., and Mok, A.K., Real-time scheduling theory: a historical perspective, *Real-Time Systems*, 28:2ô3, pp. 101–155, November 2004.

[56] Srinivasan, A., and Anderson, J., Optimal rate-based scheduling on multiprocessors, *Proc. 34th Annual ACM Symposium on Theory of Computing*, pp. 189–198, May 2002.

[57] Srinivasan, A. and Baruah, S., Deadline-based scheduling of periodic task systems on multiprocessors, *Information Processing Letters*, 84, pp. 93–98, 2002.

25
Parallel Algorithms for Maximal Independent Set and Maximal Matching

Yijie Han
University of Missouri-Kansas City

25.1 Introduction.. 25-1
25.2 The Bit Pairs PROFIT/COST Problem 25-2
25.3 The General Pairs PROFIT/COST Problem 25-4
25.4 Maximal Independent Set 25-7
 Overview of the Algorithm • The Initial Stage •
 The Speedup Stage
25.5 Maximal Matching....................................... 25-17
References 25-18

25.1 Introduction

Maximal independent set and maximal matching are fundamental problems studied in computer science. In the sequential case, a greedy linear time algorithm can be used to solve these problems. However, in parallel computation these problems are not trivial. Currently fastest parallel algorithms for these problems are obtained through derandomization [7–9,11]. Owing to the complications of the application of derandomization technique, the parallel algorithms obtained [9] were not well understood. We will explain the details of the derandomization technique and its applications to the maximal independent set and maximal matching problem in this chapter. Instead of using an $O(2^n)$ sized sample space for the bit pairs PROFIT/COST problem as was done in Reference 9, we follow Luby's algorithm [18,19], and use an $O(n)$ sized sample space. This simplifies the approach presented in Reference 9.

The basic idea of derandomization is to start with a randomized algorithm and then obtain a deterministic algorithm by applying the technique of derandomization.

Derandomization is a powerful technique because with the aid of randomization the design of algorithms, especially the design of parallel algorithms, for many difficult problems becomes manageable. The technique of derandomization offers us the chance of obtaining a deterministic algorithm that would be difficult to obtain otherwise. For some problems the derandomization technique enables us to obtain better algorithms than those obtained through other techniques.

Although the derandomization technique has been applied to the design of sequential algorithms [20, 21], these applications are sequential in nature and cannot be used directly to derandomize parallel algorithms. To apply derandomization techniques to the design of parallel algorithms, we have to study how to preserve or exploit parallelism in the process of derandomization. Thus, we put emphasis on derandomization techniques that allow us to obtain fast and efficient parallel algorithms.

Every technique has its limit, so does the derandomization technique. In order to apply derandomization techniques, a randomized algorithm must be first designed or be available. Although every randomized algorithm with a finite sample space can be derandomized, it does not imply that the derandomization approach is always the right approach to take. Other algorithm design techniques might yield much better algorithms than those obtained through derandomization. Thus, it is important to classify situations where derandomization techniques have a large potential to succeed. In the design of parallel algorithms, we need to identify situations where derandomization techniques could yield good parallel algorithms.

Since derandomization techniques are applied to randomized algorithms to yield deterministic algorithms, the deterministic algorithms are usually derived at the expense of a loss of efficiency (time and processor complexity) from the original randomized algorithms. Thus, we have to study how to obtain randomized algorithms that are easy to derandomize and have small time and processor complexities.

The parallel derandomization technique that results in efficient parallel algorithms for maximal independent set and maximal matching originates from Luby's results [18]. Luby formulated maximal independent set and maximal matching problems as PROFIT/COST problems that we will study in detail in the following sections.

We have succeeded in applying PROFIT/COST algorithms in the derandomization of Luby's randomized algorithms for the $\Delta + 1$ vertex coloring problem, the maximal independent set problem, and the maximal matching problem, and obtained more efficient deterministic algorithms for the three problems [9].

25.2 The Bit Pairs PROFIT/COST Problem

The *bit pairs PROFIT/COST* problem (BPC for short) as formulated by Luby [18] can be described as follows:

Let $\vec{x} = \langle x_i \in \{0, 1\}: i = 0, \ldots, n-1\rangle$. Each point $\vec{x}$ out of the 2^n points is assigned probability $1/2^n$. Given function $B(\vec{x}) = \sum_{i,j} f_{i,j}(x_i, x_j)$, where $f_{i,j}$ is defined as a function $\{0, 1\}^2 \to \mathcal{R}$. The PROFIT/COST problem is to find a good point $\vec{y}$ such that $B(\vec{y}) \geq E[B(\vec{x})]$. B is called the BENEFIT function and $f_{i,j}$s are called the PROFIT/COST functions.

The size m of the problem is the number of PROFIT/COST functions present in the input. The input is dense if $m = \theta(n^2)$ and is sparse if $m = o(n^2)$.

The vertex partition problem is a basic problem that can be modeled by the BPC problem [18]. The vertex partition problem is to partition the vertices of a graph into two sets such that the number of edges incident with vertices in both sets is at least half of the number of edges in the graph. Let $G = (V, E)$ be the input graph. $|V|$ 0/1-valued uniformly distributed mutually independent random variables are used, one for each vertex. The problem of partitioning vertices into two sets is now represented by the 0/1 labeling of the vertices. Let x_i be the random variable associated with vertex i. For each edge $(i, j) \in E$ a function $f(x_i, x_j) = x_i \oplus x_j$ is defined, where $\oplus$ is the exclusive-or operation. $f(x_i, x_j)$ is 1 iff edge (i, j) is incident with vertices in both sets. The expectation of f is $E[f(x_i, x_j)] = (f(0,0) + f(0,1) + f(1,0) + f(1,1))/4 = 1/2$. Thus, the BENEFIT function $B(x_0, x_1, \ldots, x_{|V|-1}) = \sum_{(i,j) \in E} f(x_i, x_j)$ has expectation $E[B] = \sum_{(i,j) \in E} E[f(x_i, x_j)] = |E|/2$. If we find a good point p in the sample space such that $B(p) \geq E[B] = |E|/2$, this point p determines the partition of vertices such that the number of edges incident with vertices in both sets is at least $|E|/2$.

The BPC problem is a basic problem in the study of derandomization, that is, converting a randomized algorithm to a deterministic algorithm. The importance of the BPC problem lies in the fact that it can be used as a key building block for the derandomization of more complicated randomized algorithms [8,9,19].

Luby [18] gave a parallel algorithm for the BPC problem with time complexity $O(\log^2 n)$, using a linear number of processors on the EREW PRAM model [4]. He used a sample space with $O(n)$ sample points and designed $O(n)$ uniformly distributed pairwise independent random variables on the sample space. His algorithm was obtained through a derandomization process in which a good sample point is found by a binary search of the sample space.

A set of n 0/1-valued uniformly distributed pairwise independent random variables can be designed on a sample space with $O(n)$ points [18]. Let $k = \log n$ (w.l.g. assuming it is an integer). The sample space is $\Omega = \{0, 1\}^{k+1}$. For each $a = a_0 a_1 \ldots a_k \in \Omega$, $Pr(a) = 2^{-(k+1)}$. The value of random variables x_i, $0 \leq i < n$, on point a is $x_i(a) = (\sum_{j=0}^{k-1}(i_j \cdot a_j) + a_k) \bmod 2$, where i_j is the j-th bit of i starting with the least significant bit. It is not difficult to verify that x_is are the desired random variables. Because $B(x_0, x_1, \ldots, x_{n-1}) = \sum_{i,j} f_{i,j}(x_i, x_j)$, where $f_{i,j}$ depends on two random variables, pairwise independent random variables can be used in place of the mutual independent random variables. A good point can be found by searching the sample space. Luby's scheme [18] uses binary search that fixes one bit of a at a time (therefore partitioning the sample space into two subspaces) and evaluates the conditional expectations on the subspaces. His algorithm [19] is shown below:

Algorithm Convert1:
for $l := 0$ **to** k
    begin
        $B_0 := E[B(x_0, x_1, \ldots, x_{n-1}) a_0 = r_0, \ldots, a_{l-1} = r_{l-1}, a_l = 0]$;
        $B_1 := E[B(x_0, x_1, \ldots, x_{n-1}) a_0 = r_0, \ldots, a_{l-1} = r_{l-1}, a_l = 1]$;
        if $B_0 \geq B_1$ **then** $a_l := 0$ **else** $a_l := 1$;
        /*The value for a_l decided above is denoted by r_l. */
    end
output($a_0, a_1, \ldots, a_k$);

Since each time the sample space is partitioned into two subspaces the subspace with larger expectation is preserved while the other subspace is discarded, the sample point $(a_0, a_1, \ldots, a_k)$ found must be a good point, that is, the value of B evaluated at $(a_0, a_1, \ldots, a_k)$ is $\geq E[B]$.

By the linearity of expectation, the conditional expectation evaluated in the above algorithm can be written as $E[B(x_0, x_1, \ldots, x_{n-1}) \mid a_0 = r_0, \ldots, a_l = r_l] = \sum_{i,j} E[f_{i,j}(x_i, x_j) \mid a_0 = r_0, \ldots, a_l = r_l]$. It is assumed [18] that constant operations(instructions) are required for a single processor to evaluate $E[f_{i,j}(x_i, x_j) \mid a_0 = r_0, \ldots, a_l = r_l]$. Algorithm Convert1 uses a linear number of processors and $O(\log^2 n)$ time on the EREW PRAM model.

Han and Igarashi gave a CREW PRAM algorithm for the PROFIT/COST problem with time complexity $O(\log n)$, using a linear number of processors [7,11,12]. They used a sample space of $O(2^n)$ points. The problem is also solved by locating a good point in the sample space. They obtained time complexity $O(\log n)$ by exploiting the redundancy of a shrinking sample space and the mutual independence of random variables.

If we use a random variable tree T_{chain} as shown in Figure 25.1 to form the sample space, then we can choose to fix one random bit on the chain in one round as we did in Luby's binary search [18], or we could choose to fix more than one random bit in one round. For example, we may choose to fix t random bits on the chain in one round. If we choose to do so then in one round we have to enumerate all 2^t possible situations (subsample spaces). The total number of rounds we need is $(k + 1)/t$. The algorithm wherein one round t random bits are fixed is as follows:

Algorithm Convert2:
for $l := 1$ **to** $(k + 1)/t$
    begin

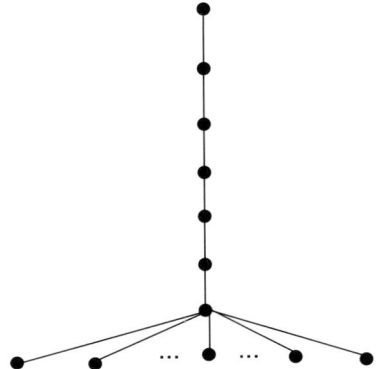

FIGURE 25.1 A random variable tree.

$$B_i := E[B(x_0, x_1, \ldots, x_{n-1}) \mid$$
$$a_0 = r_0, \ldots, a_{(l-1)t} = r_{(l-1)t}, a_{(l-1)t+1}a_{(l-1)t+2}\cdots a_{lt} = i], 0 \le i < 2^t;$$
$$B_j = \max_{0 \le i < 2^t} B_i.$$
$$a_{(l-1)t+1}a_{(l-1)t+2}\cdots a_{lt} = j;$$
/*The value for a_i decided above is denoted by r_i, $(l-1)t+1 \le i \le lt$. */

end
output($a_0, a_1, \ldots, a_k$);

Algorithm Convert2 involves more operations but has advantage in time. In the case of vertex partitioning problem, it can be done in $O((\log^2 n)/t)$ time with $O(2^t m)$ processors.

Let $n = 2^k$ and A be an $n \times n$ array. Elements $A[i, j]$, $A[i, j\#0]$, $A[i\#0, j]$, and $A[i\#0, j\#0]$ form a gang of level 0, which is denoted by $g_A^{(0)}[\lfloor i/2 \rfloor, \lfloor j/2 \rfloor]$. Here, $i\#a$ is obtained by complementing the a-th bit of i. All gangs of level 0 in A form array $g_A^{(0)}$. Elements $A[\lfloor i/2^t \rfloor 2^t + a, \lfloor j/2^t \rfloor 2^t + b]$, $0 \le a, b < 2^t$, form a gang of level t, which is denoted by $g_A^{(t)}[\lfloor i/2^t \rfloor, \lfloor j/2^t \rfloor]$. All gangs of level t in A form array $g_A^{(t)}$.

When visualized on a two-dimensional array A, a step of algorithm Convert1 can be interpreted as follows. Let function $f_{i,j}$ be stored at $A[i,j]$. Setting the random bit at level 0 of the random variable tree is done by examining the PROFIT/COST functions in the diagonal gang of level 0 of A. Setting the random bit at level t of the random variable tree is done by examining the PROFIT/COST functions in the diagonal gang of level t.

A derandomization tree D that reflects the way the BPC functions are derandomized can be built. D is of the following form. The input BPC functions are stored at the leaves, $f_{i,j}$ is stored in $A_0[i,j]$. A node $A_t[i,j]$ at level $t > 0$ is defined if there exist input functions in the range $A_0[i2^t + u, j2^t + v]$, $0 \le u, v < 2^t$. A derandomization tree is shown in Figure 25.2. A derandomization tree can be built by sorting the input BPC functions and therefore take $O(\log n)$ time with $m + n$ processors.

25.3 The General Pairs PROFIT/COST Problem

Luby [18] formulated the general pairs PROFIT/COST problem (GPC for short) as follows:

Let $\vec{x} = \langle x_i \in \{0,1\}^q : i = 0, \ldots, n-1 \rangle$. Each point $\vec{x}$ out of the 2^{nq} points is assigned probability $1/2^{nq}$. Given function $B(\vec{x}) = \sum_i f_i(x_i) + \sum_{i,j} f_{i,j}(x_i, x_j)$, where f_i is defined as a function $\{0,1\}^q \to \mathcal{R}$ and $f_{i,j}$ is defined as a function $\{0,1\}^q \times \{0,1\}^q \to \mathcal{R}$. The general pairs PROFIT/COST problem is to find a good point $\vec{y}$ such that $B(\vec{y}) \ge E[B(\vec{x})]$. B is called the general pairs BENEFIT function and $f_{i,j}$s are called the general pairs PROFIT/COST functions.

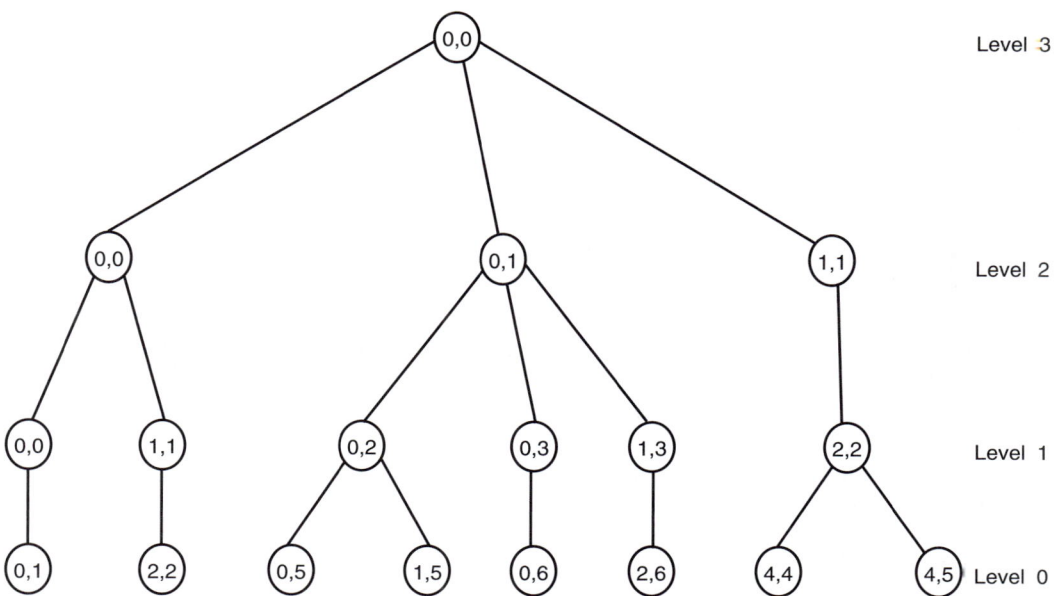

FIGURE 25.2 A derandomization tree. Pairs in circles at the leaves are the subscripts of the BPC functions.

We shall present a scheme where the GPC problem is solved by pipelining the BPC algorithm to solve BPC problems in the GPC problem.

First, we give a sketch of our approach. The incompleteness of the description in this paragraph will be fulfilled later. Let P be the GPC problem we are to solve. P can be decomposed into q BPC problems to be solved sequentially [18]. Let P_u be the u-th BPC problem. Imagine that we are to solve P_u, $0 \le u < k$, in one pass, that is, we are to fix $\vec{x}_0, \vec{x}_1, \ldots, \vec{x}_{k-1}$ in one pass, with the help of enough processors. For the moment, we can have a random variable tree T_u and a derandomization tree D_u for P_u, $0 \le u < k$. In step j, our algorithm will work on fixing the bits at level $j - u$ in T_u, $0 \le u \le \min\{k-1, j\}$. The computation in each tree D_u proceeds as we have described in the last section. Note that BPC functions $f_{i_v, j_v}(x_{i_v}, x_{j_v})$ depends on the setting of bits $x_{i_u}, x_{j_u}, 0 \le u < v$. The main difficulty with our scheme is that when we are working on fixing $\vec{x}_v$, $\vec{x}_u$, $0 \le u < v$, have not been fixed yet. The only information we can use when we are fixing the random bits at level l of T_u is that random bits at levels 0 to $l+c-1$ are fixed in $T_{u-c}, 0 \le c \le u$. This information can be accumulated in the pipeline of our algorithm and transmitted in the pipeline. Fortunately, this information is sufficient for us to speed up the derandomization process without resorting to too many processors.

Suppose we have $c \sum_{i=0}^{k}(m \times 4^i)$ processors available, where c is a constant. Assign $cm \times 4^u$ processors to work on P_u for $\vec{x}_u$. We shall work on $\vec{x}_u$, $0 \le u \le k$, simultaneously in a pipeline. The random variable tree for P_u (except that for P_0) is not constructed before the derandomization process begins, rather it is constructed as the derandomization process proceeds. We use F_u to denote the random variable tree for P_u. We are to fix the random bits on the l-th level of F_v (for $\vec{x}_v$) under the condition that random bits from level 0 to level $l+c-1$, $0 \le c \le v$, in F_{v-c} have already been fixed. We are to perform this fixing in $O(\log n)$ time. The random variable trees are built bottom up as the derandomization process proceeds. Immediately before the step we are to fix the random bits on the l-th level of F_u, the random variable trees F_{u-i} have been constructed up to the $(l+i)$-th level. The details of the algorithm for constructing the random variable trees will be given later in this section.

Consider a GPC function $f_{i,j}(x_i, x_j)$ under the condition stated in the last paragraph. When we start working on $\vec{x}_v$, we should have the BPC functions $f_{i_v, j_v}(x_{i_v}, x_{j_v})$ evaluated and stored in a table. However, because $\vec{x_u}$, $0 \leq u < v$, have not been fixed yet, we have to try out all possible situations. There are a total of 4^v patterns for bits x_{i_u}, x_{j_u}, $0 \leq u < v$, we use 4^v BPC functions for each pair (i,j). By $f_{i_v, j_v}(x_{i_v}, x_{j_v})(y_{v-1}y_{v-2} \cdots y_0, z_{v-1}z_{v-2} \cdots z_0)$, we denote the function $f_{i_v, j_v}(x_{i_v}, x_{j_v})$ obtained under the condition that $(x_{i_{v-1}} x_{i_{v-2}} \cdots x_{i_0}, x_{j_{v-1}} x_{j_{v-2}} \cdots x_{j_0})$ is set to $(y_{v-1}y_{v-2} \cdots y_0, z_{v-1}z_{v-2} \cdots z_0)$.

Let $a(u:v)$ be the v-th bit to the u-th bit of a. Let $max\_diff(a,b) = t$ mean that t-th bit is the most significant bit that $a(t:t) \neq b(t:t)$.

Lemma 25.1 *In P_d at step $d + t$ there are exactly 2^d conditional bit patterns that remain for each BPC function $f_{a,b}(x,y)$ satisfying $max\_diff(a,b) = t$.*

Proof. At step $d + t$, the random bits at t-th level in F_v (random variable tree for P_v), $0 \leq v < d$, have been fixed. This limits $x_v y_v$ to two patterns (either 00, 11 or 01, 10). Thus, there are 2^d conditional bit patterns $(x_{d-1}, x_{d-2} \ldots x_0, y_{d-1} y_{d-2} \ldots y_0)$ for P_d.

The set of remaining conditional bit patterns for $f_{a,b}$ is called the surviving set.

Let a BPC function $f_{a,b}$ in P_d have conditional bit pattern $p_{a,b}(y_{d-1}y_{d-2} \cdots y_0, z_{d-1}z_{d-2} \cdots z_0)$. Let $S_i = \{r | r(\log n : l+1) = i\}$. Let $L_i = \{r | r \in S_i, r(l:l) = 0\}$ and $R_i = \{r | r \in S_i, r(l:l) = 1\}$. We compare

$$F_0^{(l)} = \sum_{i \in \{0,1\}^{\log n - l + 1}} \sum_{a \in L_i, b \in R_i}$$
$$F_{a,b}(\Psi(a(l:0), r^{(l-1:0)}0), \Psi(b(l:0), r^{(l-1:0)}0)) p_{a,b}(y_{d-1}y_{d-2} \cdots y_0, z_{d-1}z_{d-2} \cdots z_0) +$$
$$F_{b,a}(\Psi(b(l:0), r^{(l-1:0)}0), \Psi(a(l:0), r^{(l-1:0)}0)) p_{b,a}(z_{d-1}z_{d-2} \cdots z_0, y_{d-1}y_{d-2} \cdots y_0)$$

with

$$F_1^{(l)} = \sum_{i \in \{0,1\}^{\log n - l + 1}} \sum_{a \in L_i, b \in R_i}$$
$$F_{a,b}(\Psi(a(l:0), r^{(l-1:0)}1), \Psi(b(l:0), r^{(l-1:0)}1)) p_{a,b}(y_{d-1}y_{d-2} \cdots y_0, z_{d-1}z_{d-2} \cdots z_0) +$$
$$F_{b,a}(\Psi(b(l:0), r^{(l-1:0)}1), \Psi(a(l:0), r^{(l-1:0)}1)) p_{a,b}(y_{d-1}y_{d-2} \cdots y_0, z_{d-1}z_{d-2} \cdots z_0)$$

and select 0 if former is no less than the latter and select 1 otherwise. Note that $p_{a,b}(y_{d-1}y_{d-2} \cdots y_0, z_{d-1}z_{d-2} \cdots z_0)$ indicates the conditional bit pattern, not a multiplicand. $r^{(l-1:0)}$ is the fixed random bits at levels 0 to $l-1$ in the random variable tree. $\Psi(a,b) = (\sum_i a_i \cdot b_i) \mod 2$, where a_i (b_i) is the i-th bit of a (b). ∎

We have completed a preliminary description of our derandomization scheme for the GPC problem. The algorithm for processors working on $\vec{x_d}$, $0 \leq d < k$ can be summarized as follows:

Step t ($0 \leq t < d$): Wait for the pipeline to be filled.

Step $d + t$ ($0 \leq t < \log n$): Fix random variables at level t for all conditional bit patterns in the surviving set. (* There are 2^d such patterns in the surviving set. At the same time the bit setting information is transmitted to P_{d+1}. *)

Step $d + \log n$: Fix the only remaining random variable at level $\log n$ for the only bit pattern in the surviving set. Output the good point for $\vec{x_d}$. (* At the same time the bit setting information is transmitted to p_{d+1}.*)

Theorem 25.1 *The GPC problem can be solved on the CREW PRAM in time $O((q/k+1)(\log n + \tau) \log n)$ with $O(4^k m)$ processors, where τ is the time for computing the BPC functions $f_{i_d, j_d}(x_{i_d}, x_{j_d})(\alpha, \beta)$.*

Proof. The correctness of the scheme comes from the fact that as random bits are fixed a smaller space with higher expectation is obtained, and thus when all random bits are fixed a good point is found. Since k $\vec{x_u}$s are fixed in one pass that takes $O((\log n + \tau) \log n)$ time, the time complexity for solving the GPC problem is $O((q/k + 1)(\log n + \tau) \log n)$. The processor complexity is obvious from the description of the scheme. ∎

We now give the algorithm for constructing the random variable trees. This algorithm will help understand better the whole scheme.

The random bit at the l-th level of the random variable trees for P_u is stored in $T_u^{(l)}$. The leaves are stored in $T_u^{(-1)}$. The algorithm for constructing the random variable trees for P_u is given below:

Procedure **RV-Tree**

begin

Step t ($0 \le t < u$): Wait for the pipeline to be filled.

Step $u + t$ ($0 \le t < \log n$):

(* In this step we are to build $T_u^{(t)}$. At the beginning of this step $T_{u-1}^{(t)}$ has already been constructed. Let $T_{u-1}^{(t-1)}$ be the child of $T_{u-1}^{(t)}$ in the random variable tree. The setting of the random bit r at level t for P_{u-1}, i.e. the random bit in $T_{u-1}^{(t)}$, is known. *)

make $T_u^{(t-1)}$ as the child of
$T_u^{(t)}$ in the random variable tree for P_u;

Fix the random variables in $T_u^{(t)}$;

Step $u + \log n$:

(* At the beginning of this step the random variable trees have been built for T_i, $0 \le i < u$. Let $T_{u-1}^{(\log n)}$ be the root of T_{u-1}. The random bit r in $T_{u-1}^{(\log n)}$ has been fixed.*)

make $T_u^{(\log n - 1)}$ as the child of
$T_u^{(\log n)}$ in the random variable tree;

fix the random variable in $T_u^{(\log n)}$;

output $T_u^{(\log n)}$ as the root of T_u;

end

25.4 Maximal Independent Set

Let $G = (V, E)$ be an undirected graph. For $W \subseteq V$ let $N(W) = \{i \in V \mid \exists j \in W, (i, j) \in E\}$. Known parallel algorithms [2, 5, 6, 14, 17, 19] for computing a maximal independent set have the following form.

Procedure General Independent
begin
    $I := \phi$;
    $V' := V$;
    while $V' \ne \phi$ do
        begin
            Find an independent set $I' \subseteq V'$;
            $I := I \cup I'$;
            $V' := V' - (I' \cup N(I'))$;
        end
end

Luby's work [19] formulated each iteration of the while loop in general independent as a GPC problem. We now adapt Luby's formulation [17, 19].

Let k_i be such that $2^{k_i-1} < 4d(i) \leq 2^{k_i}$. Let $q = \max\{k_i | i \in V\}$. Let $\vec{x} = <x_i \in \{0,1\}^q, i \in V>$. The length $|x_i|$ of x_i is defined to be k_i. Define*

$$Y_i(x_i) = \begin{cases} 1 & \text{if } x_i(|x_i|-1) \cdots x_i(0) = 0^{|x_i|} \\ 0 & \text{otherwise} \end{cases}$$

$$Y_{i,j}(x_i, x_j) = -Y_i(x_i)Y_j(x_j)$$

$$B(\vec{x}) = \sum_{i \in V} \frac{d(i)}{2} \sum_{j \in adj(i)} \left(Y_j(x_j) + \sum_{k \in adj(j), d(k) \geq d(j)} Y_{j,k}(x_j, x_k) + \sum_{k \in adj(i) - \{j\}} Y_{j,k}(x_j, x_k) \right)$$

where $x_i(p)$ is the p-th bit of x_i.

Function B sets a lower bound on the number of edges deleted from the graph [17, 19] should vertex i be tentatively labeled as an independent vertex if $x_i = (0 \cup 1)^{q-|x_i|} 0^{|x_i|}$. The following lemma was proven in Reference 17 (Theorem 1 in Reference 17).

Lemma 25.2 [17] $E[B] \geq |E|/c$ for a constant $c > 0$.

Function B can be written as

$$B(\vec{x}) = \sum_{j \in V} \left(\sum_{i \in adj(j)} \frac{d(i)}{2} \right) Y(x_j) + \sum_{(j,k) \in E, d(k) \geq d(j)} \left(\sum_{i \in adj(j)} \frac{d(i)}{2} \right) Y_{j,k}(x_j, x_k)$$

$$+ \sum_{i \in V} \frac{d(i)}{2} \sum_{j,k \in adj(i), j \neq k} Y_{j,k}(x_j, x_k)$$

$$= \sum_i f_i(x_i) + \sum_{(i,j)} f_{i,j}(x_i, x_j)$$

where

$$f_i(x_i) = \left(\sum_{j \in adj(i)} \frac{d(j)}{2} \right) Y(x_i)$$

and

$$f_{i,j}(x_i, x_j) = \delta(i,j) \left(\sum_{k \in adj(i)} \frac{d(k)}{2} \right) Y_{i,j}(x_i, x_j) + \left(\sum_{k \in V \text{ and } i,j \in adj(k)} \frac{d(k)}{2} \right) Y_{i,j}(x_i, x_j)$$

---

* In Luby's formulation [19] $Y_i(x_i)$ is zero unless the first $|x_i|$ bits of x_i are 1s. In order to be consistent with the notations in our algorithm, we let $Y_i(x_i)$ be zero unless the first $|x_i|$ bits of x_i are 0s.

$$\delta(i,j) = \begin{cases} 1 & \text{if } (i,j) \in E \text{ and } d(j) \geq d(i) \\ 0 & \text{otherwise} \end{cases}$$

Thus, each execution of a GPC procedure eliminates a constant fraction of the edges from the graph.

We take advantage of the special properties of the GPC functions to reduce the number of processors to $O(m + n)$. The structure of our algorithm is complicated. We first give an overview of the algorithm.

25.4.1 Overview of the Algorithm

Because we can reduce the number of edges by a constant fraction after solving a GPC problem, a maximal independent set will be computed after $O(\log n)$ GPC problems are solved. Our algorithm has two stages, the initial stage and the speedup stage. The initial stage consists of the first $O(\log^{0.5} n)$ GPC problems. Each GPC problem is solved in $O(\log^2 n)$ time. The time complexity for the initial stage is thus $O(\log^{2.5} n)$. When the first stage finishes the remaining graph has size $O((m+n)/2^{\sqrt{\log n}})$. There are $O(\log n)$ GPC problems in the speedup stage. A GPC problem of size s in the speedup stage is solved in time $O(\log^2 n/\sqrt{k})$ with $O(c^k s \log n)$ processors. Therefore, the time complexity of the speedup stage is $O(\sum_{i=O(\sqrt{\log n})}^{O(\log n)} (\log^2 n/\sqrt{i})) = O(\log^{2.5} n)$. The initial stage is mainly to reduce the processor complexity while the speedup stage is mainly to reduce the time complexity.

We shall call the term $\sum_{i \in V} \frac{d(i)}{2} \sum_{j,k \in adj(i), j \neq k} Y_{j,k}(x_j, x_k)$ in function B the vertex cluster term. There is a cluster $C(v) = \{x_w | (v, w) \in E\}$ for each vertex v. We may use $O(\sum_{v \in V} d^2(v))$ processors, $d^2(v)$ processors for cluster $C(v)$, to evaluate all GPC functions and to apply our derandomization scheme given in Section 25.3. However, to reduce the number of processors to $O(m + n)$, we have to use a modified version of our derandomization scheme in Section 25.3.

Consider the problem of fixing a random variable r in the random variable tree. The GPC function $f(x, y)$, where x and y are the leaves in the subtree rooted at r, is in fact the sum of several functions scattered in the second term of function B and in several clusters. As we have explained in Section 25.2, setting r requires $O(\log n)$ time because of the summation of function values. (Note that the summation of n items can be done in time $O(n/p + \log n)$ time with p processors.) A BPC problem takes $O(\log^2 n)$ time to solve. We pipeline all BPC problems in a GPC problem and get time complexity $O(\log^2 n)$ for solving a GPC problem.

The functions in B have a special property that we will exploit in our algorithm. Each variable x_i has a length $|x_i| \leq q = O(\log n)$. $Y_{i,j}(x_i, x_j)$ is zero unless the first $|x_i|$ bits of x_i are 0s and the first $|x_j|$ bits of x_j are 0s. When we apply our scheme there is no need to keep BPC functions $Y_{i_u, j_u}(x_{i_u}, x_{j_u})$ for all conditional bit patterns because many of these patterns will yield zero BPC functions. In our algorithm, we keep one copy of $Y_{i_u, j_u}(x_{i_u}, x_{j_u})$ with conditional bits set to 0s. This of course helps reduce the number of processors.

There are $d^2(i)$ BPC functions in cluster $C(i)$ while we can allocate at most $d(i)$ processors in the very first GPC problem because we have at most $O(m + n)$ processors for the GPC problem. What we do is use an *evaluation tree* for each cluster. The evaluation tree $TC(i)$ for cluster $C(i)$ is a "subtree" of the random variable tree. The leaves of $TC(i)$ are the variables in $C(i)$. When we are fixing j-th random bit the contribution of $C(i)$ can be obtained by evaluating the function $f(x, y)$, where $max\_diff(x, y) = j$. Here, $max\_diff(x, y)$ is the most significant bit that x and y differs. If there are a leaves l with $l(j : j) = 0$ and b leaves r with $r(j : j) = 1$, then the contribution from $TC(i)$ for fixing j-th random bit is the sum of ab function values. We will give the details of evaluating this sum using a constant number of operations per cluster.

Let us summarize the main ideas. We achieve time $O(\log^2 n)$ for solving a BPC problem; we put all BPC problem in a GPC problem as one batch into a pipeline to get $O(\log^2 n)$ time for solving a GPC problem; we use a special property of functions in B to maintain one copy for each BPC function for only conditional bits of all 0s; and we use evaluation trees to take care of the vertex cluster term.

We now sketch the speedup stage. Since we have to solve $O(\log n)$ GPC problems in this stage, we have to reduce the time complexity for a GPC problem to $o(\log^2 n)$ in order to obtain $o(\log^3 n)$ time. We view the random variable tree as containing blocks with each block having a levels and a random bits. We fix a block in a step instead of fixing a level in a step. Each step takes $O(\log n)$ time and a BPC problem takes $O(\log^2 n/a)$ time. If we have as many processors as we want, we could solve all BPC problems in a GPC problem by enumerating all possible cases instead of putting them through a pipeline; that is, in solving P_u we could guess all possible settings of random bits for P_v, $0 \leq v < u$. We have explained this approach in algorithm Convert2 in Section 2. In doing so, we would speed up the GPC problem. In reality, we have extra processors, but they are not enough for us to enumerate all possible situations. We therefore put a BPC problems of a GPC problem in a team. All BPC problems in a team are solved by enumeration. Thus, they are solved in time $O(\log^2 n/a)$. Let b be the number of teams we have. We put all these teams into a pipeline and solve them in time $O((b + \log n/a) \log n)$. The approach of the speedup stage can be viewed as that of the initial stage with added parallelism that comes with the help of extra processors.

25.4.2 The Initial Stage

We first show how to solve a GPC problem for function B in time $O(\log^2 n)$ using $O((m+n)\log n)$ processors.

$O(m+n)$ processors will be allocated to each BPC problem. The algorithm for processors working on F_u has the following form:

Step t $(0 \leq t < u)$: Wait for the pipeline to be filled
Step $u + t$ $(0 \leq t < \log n)$: Fix random variables at level t
Step $u + \log n$: Fix the only remaining random variable at level $\log n$. Output the good point for $\vec{x}_u$

We will allow $O(\log n)$ time for each step and $O(\log^2 n)$ time for the whole algorithm.

We can view algorithm RV-Tree as one that distributes random variables x_i into different sets. Each set is indexed by (u, t, i, j). We call these sets BD sets because they are obtained from conditional bit transmission and the derandomization trees. x is in $BD(u, t, i, j)$ if x is a leaf in $T_u^{(t)}$, $x(\log n : t) = i$, and $x(t - 1 : 0) = j$. When u and t are fixed, $BD(u, t, i, j)$ sets are disjoint. Because we allow $O(\log n)$ time for each step in RV-Tree, the time complexity for constructing the random variable trees is $O(\log^2 n)$.

Example 25.1 *See Figure 25.3 for an execution of RV-Tree. Variables are distributed into the BD sets as shown below:*

Step 0:
$x_0, x_1 \in BD(0, 0, 0, \epsilon);$
$x_2, x_3 \in BD(0, 0, 1, \epsilon);$
$x_4, x_5 \in BD(0, 0, 2, \epsilon);$
$x_6, x_7 \in BD(0, 0, 3, \epsilon).$

Step 1:
$x_0, x_1, x_2, x_3 \in BD(0, 1, 0, \epsilon);$
$x_4, x_5, x_6, x_7 \in BD(0, 1, 1, \epsilon);$
$x_0, x_1 \in BD(1, 0, 0, 0);$
$x_2, x_3 \in BD(1, 0, 1, 0);$
$x_4 \in BD(1, 0, 2, 0);$
$x_5 \in BD(1, 0, 2, 1);$
$x_6, x_7 \in BD(1, 0, 3, 0).$

Parallel Algorithms for Maximal Independent Set and Maximal Matching

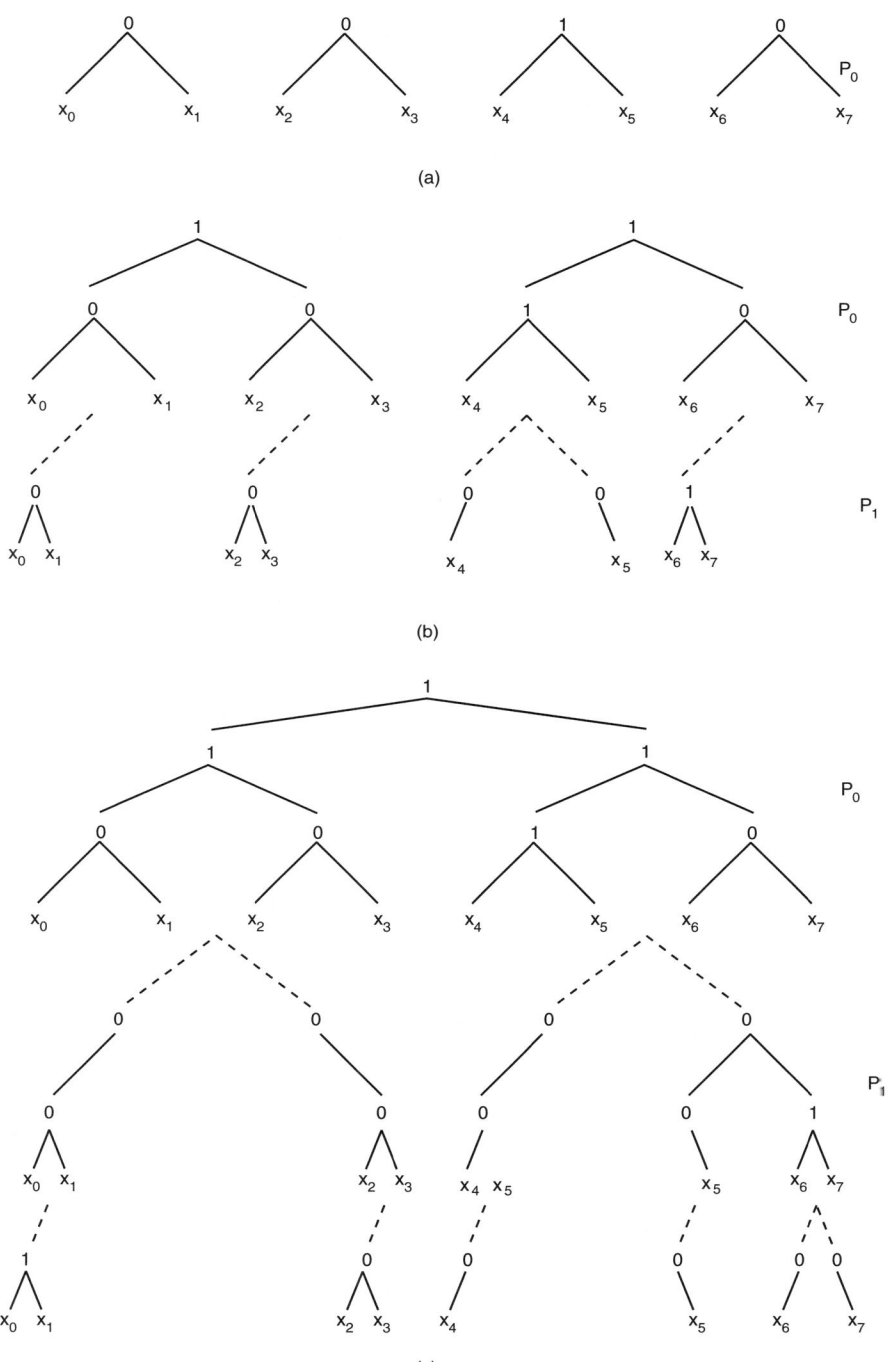

FIGURE 25.3 An execution of RV-Tree. Solid lines depicts the distribution of random variables into BD sets. Dotted lines depicts conditional bit transmission: (a) step 0, (b) step 1, (c) step 2, (d) step 3, (e) step 4, and (f) step 5.

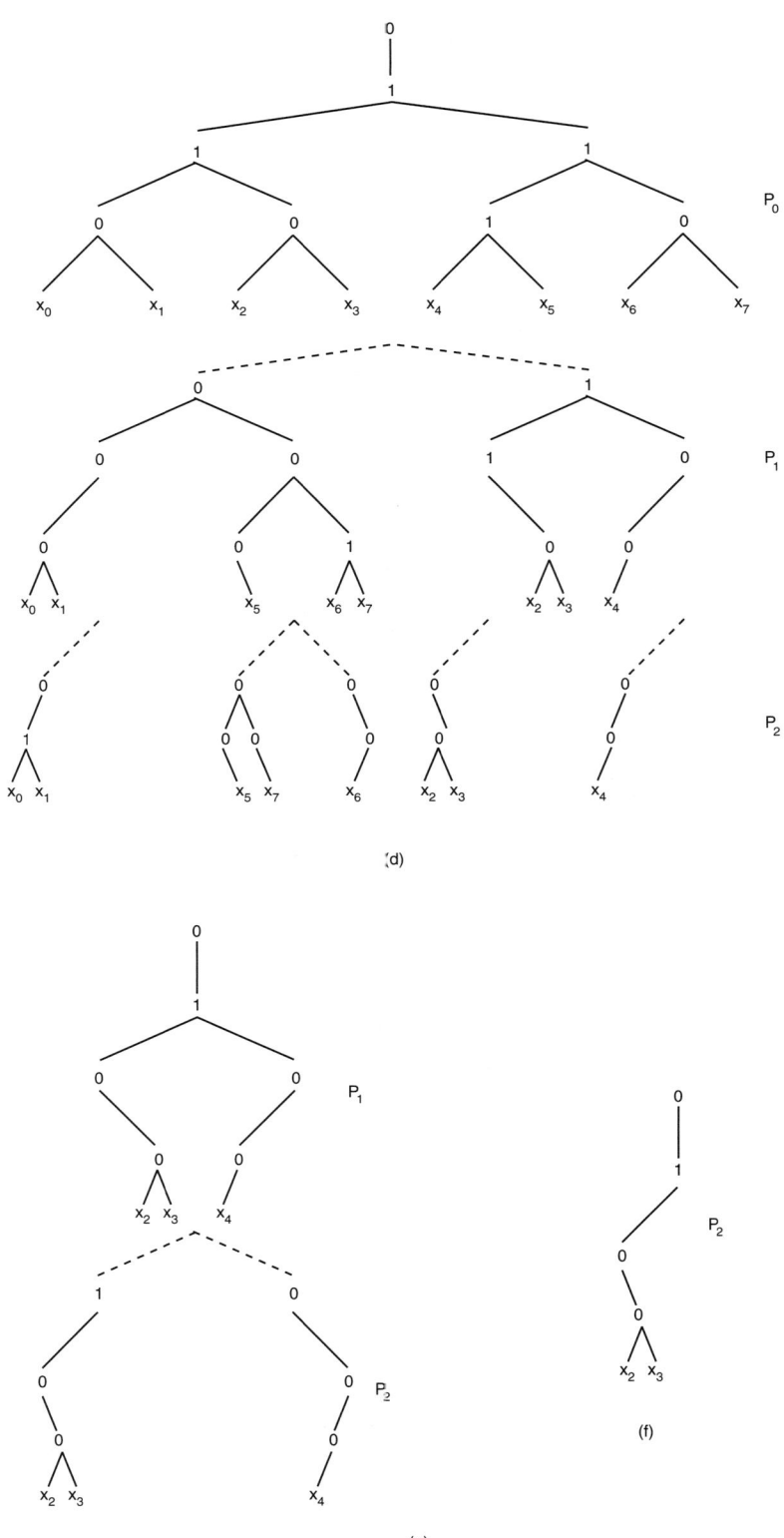

FIGURE 25.3 Continued.

Step 2:

$$x_0, x_1, x_2, x_3,$$
$$x_4, x_5, x_6, x_7 \in BD(0, 2, 0, \epsilon);$$
$$x_0, x_1 \in BD(1, 1, 0, 0);$$
$$x_2, x_3 \in BD(1, 1, 0, 1);$$
$$x_4 \in BD(1, 1, 1, 0);$$
$$x_5, x_6, x_7 \in BD(1, 1, 1, 1);$$
$$x_0, x_1 \in BD(2, 0, 0, 00);$$
$$x_2, x_3 \in BD(2, 0, 1, 00);$$
$$x_4 \in BD(2, 0, 2, 00);$$
$$x_5 \in BD(2, 0, 2, 10);$$
$$x_6 \in BD(2, 0, 3, 00);$$
$$x_7 \in BD(2, 0, 3, 01).$$

Step 3:

$$x_0, x_1, x_2, x_3,$$
$$x_4, x_5, x_6, x_7 \in BD(0, 3, 0, \epsilon);$$
$$x_0, x_1, x_5, x_6, x_7 \in BD(1, 2, 0, 0);$$
$$x_2, x_3, x_4 \in BD(1, 2, 0, 1);$$
$$x_0, x_1 \in BD(2, 1, 0, 00);$$
$$x_5, x_7 \in BD(2, 1, 0, 10);$$
$$x_6 \in BD(2, 1, 0, 11);$$
$$x_2, x_3 \in BD(2, 1, 1, 00);$$
$$x_4 \in BD(2, 1, 1, 10).$$

Step 4:

$$x_2, x_3, x_4 \in BD(1, 3, 0, 0);$$
$$x_2, x_3 \in BD(2, 2, 0, 00);$$
$$x_4 \in BD(2, 2, 0, 01).$$

Step 5:

$$x_2, x_3 \in BD(2, 3, 0, 00).$$

Now, consider GPC functions of the form $Y_i(x_i)$ and $Y_{i,j}(x_i, x_j)$ except the functions in the vertex cluster term. Our algorithm will distribute these functions into sets $BDF(u, t, i', j')$ by the execution of RV-Tree, where $BDF(u, t, i', j')$ is essentially the BD set except it is for functions. $Y_{i,j}$ is in $BDF(u, t, i', j')$ iff both x_i and x_j are in $BD(u, t, i', j')$, $\max\{k|$ (the k-th bit of i XOR j) $= 1\} = t$, $|x_i| > u$ and $|x_j| > u$, where XOR is the bitwise exclusive-or operation, with the exception that all functions belong to $BDF(u, \log n, 0, j')$ for some j'. The condition $\max\{k|$ (the k-th bit of i XOR j) $= 1\} = t$ ensures that x_i and x_j are in different "subtrees." The conditions $|x_i| > u$ and $|x_j| > u$ ensure that x_i and x_j are still valid. The algorithm for the GPC functions for P_u is shown below:

Procedure **FUNCTIONS**
   begin
      Step t $(0 \le t < u)$:
         (* Functions in $BDF(0, t, i', \Lambda)$ reach depth 0 of P_t. *)
         Wait for the pipeline to be filled;
      Step $u + t$ $(0 \le t < \log n)$:
         (* Let $S = BDF(u, t, i', j')$. *)
         if S is not empty **then**
            begin
               for each GPC function $Y_{i,j}(x_i, x_j) \in S$
                   compute the BPC function $Y_{i_u, j_u}(x_{i_u}, x_{j_u})$ with conditional bits set to all 0s;

               (* To fix the random bit in $T_u^{(t)}$. *)
               $T_u^{(t)} := 0;$
               $F_0 := \sum_{Y_{i,j} \in S} Y_{i_u, j_u}(\Psi(i(t:0), T_u^{(t:0)}), \Psi(j(t:0), T_u^{(t:0)}))$
                   $+ \sum_{Y_{i,j} \in S} Y_{i_u, j_u}(\Psi(i(t:0), T_u^{(t:0)}) \oplus 1, \Psi(j(t:0), T_u^{(t:0)}) \oplus 1) + VC;$
               (* VC is the function value obtained for functions in the vertex cluster term. We shall explain how to compute it later. $\oplus$ is the exclusive-or operation. $T_u^{(t:0)}$ is the random bits in level 0 to t in T_u.*)

               $T_u^{(t)} := 1;$
               $F_1 := \sum_{Y_{i,j} \in S} Y_{i_u, j_u}(\Psi(i(t:0), T_u^{(t:0)}), \Psi(j(t:0), T_u^{(t:0)}))$

$$+ \sum_{Y_{i,j} \in S} Y_{i_u, j_u}(\Psi(i(t:0), T_u^{(t:0)}) \oplus 1, \Psi(j(t:0), T_u^{(t:0)}) \oplus 1) + VC;$$

if $F_0 \geq F_1$ **then** $T_u^{(t)} := 0$
else $T_u^{(t)} := 1;$
(* The random bit is fixed. *)

(* To decide whether $Y_{i,j}$ should remain in the pipeline. *)
for each $Y_{i,j} \in S$
  begin
    if $\Psi(i(t:0), T_u^{(t:0)}) \neq \Psi(j(t:0), T_u^{(t:0)})$ **then** remove $Y_{i,j}$;
    (* $Y_{i,j}$ is a zero function in the remaining computation of P_u and also a zero function
    in P_v, $v > u$. *)
    if $(\Psi(i(t:0), T_u^{(t:0)}) = \Psi(j(t:0), T_u^{(t:0)})) \wedge (|x_i| \geq u+1) \wedge (|x_j| \geq u+1)$
then
      (* Let $b = T_u^{(t)}$. *)
      put $Y_{i,j}$ into $BDF(u+1, t, i', j'b)$; (* $Y_{i,j}$ to be processed in Y_{u+1}. *)

    end
  end
Step $u + \log n$:
  if $S = BDF(u, \log n, 0, j')$ is not empty **then**
  (* S is the only set left for this step. *)
    begin
      for each GPC function $Y_{i,j}(x_i, x_j)$ $(Y_i(x_i)) \in S$
        compute the BPC function $Y_{i_u, j_u}(x_{i_u}, x_{j_u})$ $(Y_{i_u}(x_{i_u}))$ with conditional bits set to all 0s;

      (* To fix the random bit in $T_u^{(\log n)}$. *)
      $T_u^{(\log n)} := 0;$
      $F_0 := \sum_{Y_{i,j} \in S} Y_{i_u, j_u}(\Psi(i(\log n : 0), T_u^{(\log n:0)}), \Psi(x_j(\log n : 0), T_u^{(\log n:0)}))$
      $+ \sum_{Y_i \in S} Y_{i_u}(\Psi(i(\log n : 0), T_u^{(\log n:0)})) + VC;$

      $T_u^{(\log n)} := 1;$
      $F_1 := \sum_{Y_{i,j} \in S} Y_{i_u, j_u}(\Psi(i(\log n : 0), T_u^{(\log n:0)}), \Psi(x_j(\log n : 0), T_u^{(\log n:0)}))$
      $+ \sum_{Y_i \in S} Y_{i_u}(\Psi(x_i(\log n : 0), T_u^{(\log n:0)})) + VC;$

      if $F_0 \geq F_1$ **then** $T_u^{(\log n)} := 0$
      else $T_u^{(\log n)} := 1;$
      (* The random bit is fixed. *)

      (* To decide whether $T_{i,j}$ should remain in the pipeline. *)
      for each $Y_{i,j} \in S$
        begin
          (* Let $b = T_u^{(\log n)}$). $\Psi(x_i(\log n : 0), T_u^{(\log n:0)})$ and $\Psi(x_j(\log n : 0), T_u^{(\log n:0)})$
          must be equal
          here. *)
          if $(\Psi(i(\log n : 0), T_u^{(\log n:0)}) = \Psi(x_j(\log n : 0), T_u^{(\log n:0)}) = 0) \wedge ((|x_i| \geq$

$u+1) \vee (|x_j| \geq u+1))$
   then
      put $Y_{i,j}$ into $BDF(u+1, \log n, 0, j'b)$;
   else remove $Y_{i,j}$;
end

(* To decide whether Y_i should remain in the pipeline. *)
for each $Y_i \in S$
begin
   (* Let $b = T_u^{(\log n)}$. *)
   if $(\Psi(i(\log n:0), T_u^{(\log n:0)}) = 0) \wedge (|x_i| \geq u+1)$ **then**
      put Y_i into $BDF(u+1, \log n, 0, j'b)$;
   else remove Y_i;
end
end
end

There are $O(\log n)$ steps in RV-Tree and FUNCTIONS, each step takes $O(\log n)$ time and $O((m+n)\log n)$ processors.

Now, we describe how the functions in the vertex cluster term are evaluated. Each function $Y_{i,j}(x_i, x_j)$ in the vertex cluster term is defined as $Y_{i,j}(x_i, x_j) = -1$ if the first $|x_i|$ bits of x_i are 0s and the first $|x_j|$ bits of x_j are 0s, and otherwise as $Y_{i,j}(x_i, x_j) = 0$. Let $l(i) = |x_i| - u$. Then, $Y_{i_u, j_u}(\Lambda, \Lambda)(0^u, 0^u) = -1/2^{l(i)+l(j)}$ and $Y_{i_u, j_u}(0, 0)(0^u, 0^u) = -1/2^{l(i)+l(j)-2}$ if $|x_i| > u$ and $|x_j| > u$. Procedure RV-Tree is executed in parallel for each cluster $C(v)$ to build an *evaluation tree* $TC(v)$ for $C(v)$. An evaluation tree is similar to the random variable tree. The difference between the random variable tree and $TC(v)$ is that the leaves of $TC(v)$ consist of variables from $C(v)$. Let $r_j = T_{u,v}^{(j)}$ be the root of a subtree T' in $TC(v)$, which is to be constructed in the current step. Let r_j be at the j-th level. Let $S_i = \{r|r(\log n : j+1) = i\}$. Let $L_i = \{a|a \in S_i, a(j:j) = 0\}$ and $R_i = \{a|a \in S_i, a(j:j) = 1\}$. At the beginning of the current step, L_i and R_i have already been constructed. Random bits at levels less than j have been fixed. Define $M(v, i, v_j, b) = \sum_{\Psi(k(j:0), r(j:0))=b} \frac{1}{2^{l(k)}}$, where ks are leaves in the subevaluation tree rooted at i (i.e., $k(\log n : j+1) = i$). At the beginning of the current step $M(v, i0, r_{j-1}, b)$ and $M(v, i1, r_{j-1}, b)$, $b = 0, 1$, have already been computed. During the current step, $i0$s are at the left side of r_j and $i1$s are at the right side of r_j. Now, r_j is tentatively set to 0 and 1 to obtain the value VC in procedure FUNCTIONS. We first compute $VC(v, r_j)$ for each v. $VC(v, r_j) = 2 \sum_{i \in \{0,1\}^{\log n-j}} \sum_{b=0}^{1} M(v, i0, r_{j-1}, b) M(v, i1, r_{j-1}, b \oplus r)$, where $\oplus$ is the exclusive-or operation. The VC value used in procedure FUNCTIONS is $-\sum_{\{v|L_i \text{ or } R_i \text{ is not empty}\}} \frac{d(v)}{2} VC(v, T_{u,v}^{(t)})$. After setting r_j, we obtain an updated value for $M(v, i, r_j, b)$ as $M(v, i, r_j, b) = M(v, i0, r_{j-1}, b) + M(v, i1, r_{j-1}, b \oplus r_j)$.

The above paragraph shows that we need only spend $O(T_{VC})$ operations for evaluating VC for all vertex clusters in a BPC problem, where T_{VC} is the total number of tree nodes of all evaluation trees. T_{VC} is $O(m \log n)$ because there are a total of $O(m)$ leaves and some nodes in the evaluation trees have one child.

We briefly describe the data structure for the algorithm. We build the random variable tree and evaluation trees for P_0. Nodes $T_0^{(t)}$ in the random variable tree and nodes $T_{0,v}^{(t)}$ in the evaluation trees and functions in $BDF(0, t, i, \Lambda)$ are sorted by the pair (t, i). This is done only once and takes $O(\log n)$ time with $O(m + n)$ processors [1, 3]. As the computation proceeds, each BDF set will split into several sets, one for each distinct conditional bit pattern. A BDF set in P_u can split into at most two in P_{u+1}. Since we allow $O(\log n)$ time for each step, we can allocate memory for the new level to be built in the evaluation trees. We use pointers to keep track of the conditional bit transmission from P_u to P_{u+1} and the evaluation trees. The nodes and functions in the same BD and BDF sets (indexed by the same (u, t, i', j')) should be arranged to occupy consecutive memory cells to facilitate the computation of F_0 and F_1 in FUNCTIONS. These operations can be done in $O(\log n)$ time using $O((m + n)/\log n)$ processors.

It is now straightforward to verify that our algorithm for solving a GPC problem takes $O(\log^2 n)$ time, $O(\log n)$ time for each of the $O(\log n)$ steps. We note that in each step for each BPC problem we have used $O(m+n)$ processors. This can be reduced to $O((m+n)/\log n)$ processors because in each step $O(m+n)$ operations are performed for each BPC problem. They can be done in $O(\log n)$ time using $O((m+n)/\log n)$ processors. Since we have $O(\log n)$ BPC problems, we need only $O(m+n)$ processors to achieve time complexity $O(\log^2 n)$ for solving one GPC problem.

We use $O((m+n)/\log^{0.5} n)$ processors to solve the first $O(\log^{0.5} n)$ GPC problems in the maximal independent set problem. Recall that the execution of a GPC algorithm will reduce the size of the graph by a constant fraction. For the first $O(\log \log n)$ GPC problems the time complexity is $O(\sum_{i=1}^{O(\log \log n)} \log^{2.5} n / c^i) = O(\log^{2.5} n)$, where $c > 1$ is a constant. In the i-th GPC problem we solve $O(c^i \log^{0.5} n)$ BPC problems in a batch, incurring $O(\log^2 n)$ time for one batch and $O(\log^{2.5} n / c^i)$ time for the $O(\log^{0.5} n / c^i)$ batches. The time complexity for the remaining GPC problems is $O(\sum_{i=O(\log \log n)}^{\log^{0.5} n} \log^2 n) = O(\log^{2.5} n)$.

25.4.3 The Speedup Stage

The input graph here is the output graph from the initial stage. The speedup stage consists of the rest of the GPC problems.

We have to reduce the time complexity for solving one GPC problem to under $O(\log^2 n)$ in order to obtain an $o(\log^3 n)$ algorithm for the maximal independent set problem. After the initial stage, we have a small size problem and we have extra processor power to help us speed up the algorithm.

We redesign the random variable tree T for a BPC problem. We use $S = (\log n + 1)/a$ blocks in T, where a is a parameter.

Lemma 25.3 *If all random bits up to level j are fixed, then random variables in $S_i = \{a | a(\log n : j) = i\}$ can assume only two different patterns.*

Proof. This is because the random bits from the root to the $(j+1)$st level are common to all random variables in S_i.

In fact, we have implicitly used this lemma in transmitting conditional bits from P_u to P_{u+1} in the design of our GPC algorithm.

The q BPC problems in a GPC problem are divided into $b = q/a$ teams (w.l.g. assuming it is an integer). Team i, $0 \leq i < b$, has a BPC problems. Let J_w be w-th team. The algorithm for fixing the random variables for J_w can be expressed as follows:

Step t ($0 \leq t < w$): Wait for the pipeline to be filled
Step $t + w$ ($0 \leq t < S$): Fix random variables in block t in random variable trees for J_w

Each step will be executed in $O(\log n)$ time. Since there are $O(b + S)$ steps, the time complexity is $O(\log^2 n/a)$ for the above algorithm since $q = O(\log n)$.

For a graph with m edges and n vertices, to fix random bits in block 0 for P_0 we need $2^a(m+n)$ processors to enumerate all possible 2^a bit patterns for the a bits in block 0. To fix the bits in block 0 for P_v, $v < a$, we need $2^{a(v+1)}$ patterns to enumerate all possible $a(v+1)$ bits in block 0 for P_u, $u \leq v$. For each of the $2^{a(v+1)}$ patterns, there are 2^v conditional bit patterns. Thus, we need $c^{a^2}(m+n)$ processors for team 0 for a suitable constant c. Although the input to each team may have many conditional bit patterns, it contains at most $O(m+n)$ BD and BDF sets. We need keep working for only those conditional bit patterns that are not associated with empty BD or BDF sets. Thus, the number of processors needed for each team is the same because when team J_w is working on block i the bits in block i have already been fixed for teams J_u, $u < w$, and because we keep only nonzero functions. The situation here is similar to the situation in the initial stage. Thus, the total number of processors we need for solving one GPC problem in time $O(\log^2 n/a)$ is $c^{a^2}(m+n) \log n/a = O(c^{a^2}(m+n) \log n)$. We conclude that one GPC problem can be solved in time $O(\log^2 n/\sqrt{k})$ with $O(c^k(m+n)\log n)$ processors. Therefore, the time complexity for the speedup stage is $O(\sum_{k=1}^{\log n}(\log^2 n/\sqrt{k})) = O(\log^{2.5} n)$. ∎

Theorem 25.2 *There is an EREW PRAM algorithm for the maximal independent set problem with time complexity $O(\log^{2.5} n)$ using $O((m+n)/\log^{0.5} n)$ processors.*

25.5 Maximal Matching

Let $N(M) = \{(i,k) \in E, (k,j) \in E \mid \exists (i,j) \in M\}$. A maximal matching can be found by repeatedly finding a matching M and removing $M \cup N(M)$ from the graph. The maximal matching problem has been studied by several researchers [13, 15, 16, 18, 19].

We adapt Luby's work [19] to show that after an execution of the GPC procedure a constant fraction of the edges will be reduced.

Let k_i be such that $2^{k_i-1} < 4d(i) \leq 2^{k_i}$. Let $q = \max\{k_i \mid i \in V\}$. Let $\vec{x} = \{x_{ij} \in \{0,1\}^q, (i,j) \in E\}$. The length $|x_{ij}|$ of x_{ij} is defined to be $\max\{k_i, k_j\}$. Define

$$Y_{ij}(x_{ij}) = \begin{cases} 1 & \text{if } x_{ij}(|x_{ij}|-1) \cdots x_{ij}(0) = 0^{|x_{ij}|} \\ 0 & \text{otherwise} \end{cases}$$

$$Y_{ij,i'j'}(x_{ij}, x_{i'j'}) = -Y_{ij}(x_{ij})Y_{i'j'}(x_{i'j'})$$

$$B(\vec{x}) = \sum_{i \in V} \frac{d(i)}{2} \left(\sum_{j \in adj(i)} \left(Y_{ij}(x_{ij}) + \sum_{k \in adj(j), k \neq i} Y_{ij,jk}(x_{ij}, x_{jk}) \right) \right.$$
$$\left. + \sum_{j,k \in adj(i), j \neq k} Y_{ij,ik}(x_{ij}, x_{ik}) \right) \quad (25.1)$$

where $x_{ij}(p)$ is the p-th bit of x_{ij}.

Function B sets a lower bound on the number of edges deleted from the graph [19] should edge (i,j) be tentatively labeled as an edge in the matching set if $x_{ij} = (0 \cup 1)^{q-|x_{ij}|}0^{|x_{ij}|}$. The following lemma can be proven by following Luby's proof for Theorem 1 in Reference 17.

Lemma 25.4 $E[B] \geq |E|/c$ *for a constant $c > 0$.*

Function B can be written as

$$B(\vec{x}) = \sum_{(i,j) \in E} \frac{d(i) + d(j)}{2} Y_{ij}(x_{ij}) + \sum_{j \in V} \sum_{i,k \in adj(j), i \neq k} \frac{d(i)}{2} Y_{ij,jk}(x_{ij}, x_{jk})$$
$$+ \sum_{i \in V} \frac{d(i)}{2} \sum_{j,k \in adj(i), j \neq k} Y_{ij,ik}(x_{ij}, x_{ik}) \quad (25.2)$$

There are two cluster terms in function B. We only need explain how to evaluate the cluster term $\sum_{j \in V} \sum_{i,k \in adj(j), i \neq k} \frac{d(i)}{2} Y_{ij,jk}(x_{ij}, x_{jk})$. The rest of the functions can be computed as we have done for the maximal independent set problem in Section 5.

Again, we build an evaluation tree $TC(v)$ for each cluster $C(v)$ in the cluster term. Let $l(ij) = |x_{ij}| - u$. Let $r_j = T_{u,v}^{(t)}$ be the root of a subtree T' in $TC(v)$ at the j-th level, which is to be constructed in the current step. Let $S_i = \{a | a(\log n : j+1) = i\}$. Let $L_i = \{a | a \in S_i, a(j:j) = 0\}$ and $R_i = \{a | a \in S_i, a(j:j) = 1\}$. Let $r_{L_i} = i0$ and $r_{R_i} = i1$ be the roots of L_i and R_i, respectively. At the beginning of the current step, L_is and R_is have already been constructed. Random bits at level 0 to $j-1$ have been fixed. Define $M(v, i, v_j, b) = \sum_{\Psi(ij(j:0), r(j:0)) = b} \frac{1}{2^{l(ij)}}$. Define $N(v, i, v_j, b) = \sum_{\Psi(ij(j:0), r(j:0)) = b} \frac{d(v)}{2} \frac{1}{2^{l(ij)}}$. At the beginning of the current step, $M(v, i0, v_{j-1}, b)$, $M(v, i1, r_{j-1}, b)$, $N(v, i0, v_{j-1}, b)$, and $N(v, i1, v_{j-1}, b)$, $b = 0, 1$, have already been computed and associated with $i0$ and $i1$, respectively. During the current step, $i0$s are at the left side of r_j and $i1$s are at the right side of r_j. Now r_j is tentatively set to 0 and 1 to obtain value VC for fixing r. We first compute $VC(v, r_j)$ for each v. $VC(v, r_j) = \sum_{i \in \{0,1\}^{\log n - j}} \sum_{b=0}^{1} (N(v, i0, v_{j-1}, b) M(v, i1, v_{j-1}, b \oplus r_j) + M(v, i0, v_{j-1}, b) N(v, i1, v_{j-1}, b \oplus r_j))$. The VC value is $-\sum_{i | i \in \{0,1\}^{\log n - j}, L_i \text{ or } R_i \text{ is not empty}\}} VC(v, r_j)$. After setting r_j, we obtain updated value $M(v, i, r_j, b)$ and $N(v, i, r_j, b)$ as $M(v, i, r_j, b) = \sum_i M(v, i0, r_{j-1}, b) + M(v, i1, r_{j-1}, b \oplus r_j)$, $N(v, i, r_j, b) = \sum_i N(v, i0, r_{j-1}, b) + N(v, i1, v_{j-1}, b \oplus r_j)$.

Since this computation does not require more processors, we have the following theorem.

Theorem 25.3 *There is an EREW PRAM algorithm for the maximal matching problem with time complexity $O(\log^{2.5} n)$, using $O((m+n)/\log^{0.5} n)$ processors.*

Later development improved the results presented here [10, 15, 16]. In particular, $O(\log^{2.5} n)$ time with optimal number of processors ($(m+n)/\log^{2.5} n$ processors) has been achieved [16].

References

[1] M. Ajtai, J. Komlós, and E. Szemerédi. An $O(N \log N)$ sorting network. *Proc. 15th ACM Symp. on Theory of Computing*, pp. 1–9 (1983).

[2] N. Alon, L. Babai, and A. Itai. A fast and simple randomized parallel algorithm for the maximal independent set problem. *J. of Algorithms*, 7, 567–583 (1986).

[3] R. Cole. Parallel merge sort. *Proc. 27th Symp. on Foundations of Computer Science*, IEEE, pp. 511–516 (1986).

[4] S. Fortune and J. Wyllie. Parallelism in random access machines. *Proc. 10th Annual ACM Symp. on Theory of Computing*, San Diego, CA, pp. 114–118 (1978).

[5] M. Goldberg and T. Spencer. A new parallel algorithm for the maximal independent set problem. *SIAM J. Comput.* 18(2), 419–427 (April 1989).

[6] M. Goldberg, and T. Spencer. Constructing a maximal independent set in parallel. *SIAM J. Dis. Math.* 2(3), 322–328 (August 1989).

[7] Y. Han. A parallel algorithm for the PROFIT/COST problem. *Proc. of 1991 Int. Conf. on Parallel Processing*, Vol. III, pp. 107–114, St. Charles, IL.

[8] Y. Han. A fast derandomization scheme and its applications. *Proc. 1991 Workshop on Algorithms and Data Structures*, Ottawa, Canada, Lecture Notes in Computer Science 519, pp. 177–188 (August 1991).

[9] Y. Han. A fast derandomization scheme and its applications. *SIAM J. Comput.* 25(1), 52–82 (February 1996).

[10] Y. Han. An improvement on parallel computation of a maximal matching. *Inf. Process. Lett.* 56, 343–348 (1995).

[11] Y. Han and Y. Igarashi. Derandomization by exploiting redundancy and mutual independence. *Proc. Int. Symp. SIGAL'90*, Tokyo, Japan, LNCS 450, 328–337 (1990).

[12] Y. Han and Y. Igarashi. Fast PROFIT/COST algorithms through fast derandomization. *Acta Informatica* 36(3), 215–232 (1999).

[13] A. Israeli and Y. Shiloach. An improved parallel algorithm for maximal matching. *Inf. Process. Lett.* 22, 57–60 (1986).

[14] R. Karp and A. Wigderson. A fast parallel algorithm for the maximal independent set problem. *JACM* 32(4), 762–773 (October 1985).

[15] P. Kelson. An optimal parallel algorithm for maximal matching. *Inf. Process. Lett.* 52, 223–228 (1994).

[16] KVRCN. Kishore, S. Saxena. An optimal parallel algorithm for general maximal matchings is as easy as for bipartite graphs. *Inf. Process. Lett.*, 75(4), 145–151 (2000).

[17] M. Luby. A simple parallel algorithm for the maximal independent set problem. *SIAM J. Comput.*, 15(4), 1036–1053 (November 1986).

[18] M. Luby. Removing randomness in parallel computation without a processor penalty. *Proc. 29th Symp. on Foundations of Computer Science, IEEE*, pp. 162–173 (1988).

[19] M. Luby. Removing randomness in parallel computation without a processor penalty. TR-89-044, *Int. Comp. Sci. Institute*, Berkeley, CA.

[20] P. Raghavan. Probabilistic construction of deterministic algorithms: approximating packing integer programs. *JCSS* 37(4), 130–143, (October 1988).

[21] J. Spencer. *Ten Lectures on the Probabilistic Method.* SIAM, Philadelphia (1987).

26
Efficient Parallel Graph Algorithms for Multicore and Multiprocessors

26.1	Designing Efficient Parallel Algorithms on Multiprocessors	26-2
	Limited Parallelism • Synchronization • Cache and Memory Access Patterns • Algorithmic Optimizations	
26.2	A Fast, Parallel Spanning Tree Algorithm	26-6
	Analysis of the Symmetric Multiprocessor Spanning Tree Algorithms • Experimental Results • Discussion	
26.3	Fast Shared-Memory Algorithms for Computing Minimum Spanning Tree...............................	26-13
	Data Structures for Parallel Borůvka's Algorithms • Analysis • A New Parallel Minimum Spanning Tree Algorithm • Analysis • Experimental Results	
26.4	Biconnected Components...............................	26-21
	The Tarjan–Vishkin Algorithm • Implementation of the Tarjan–Vishkin Algorithm • A New Algorithm to Reduce Overheads • Performance Results and Analysis • Discussion	
26.5	Parallel Algorithms with Mutual Exclusion and Lock-Free Protocols.......................................	26-30
	Lock-Free Parallel Algorithms • Lock-Free Protocols for Resolving Races among Processors • Parallel Algorithms with Fine-Grained Mutual Exclusion Locks • Experimental Results • Discussion	
26.6	Conclusions and Future Work	26-39
References		26-40

David A. Bader
Georgia Institute of Technology

Guojing Cong
IBM T.J. Watson Research Center

Graph abstractions are used in many computationally challenging science and engineering problems. For instance, the minimum spanning tree (MST) problem finds a spanning tree of a connected graph G with the minimum sum of edge weights. MST is one of the most studied combinatorial problems with practical applications in VLSI layout, wireless communication, and distributed networks [67,87,95], recent problems in biology and medicine such as cancer detection [12,55,56,66], medical imaging [3], and proteomics [28,73], and national security and bioterrorism such as detecting the spread of toxins through populations in the case of biological/chemical warfare [13], and is often a key step in other graph problems [69,70,86,93]. Graph abstractions are also used in data mining, determining gene function, clustering in semantic webs, and security applications. For example, studies (e.g., References 21 and 59) have shown that certain activities are often suspicious not because of the characteristics of a single actor, but because of the interactions among a group of actors. Interactions are modeled through a graph abstraction where the entities are represented by vertices, and their interactions are the directed edges in the graph. This graph may contain billions of vertices with degrees ranging from small constants to thousands.

There are plenty of theoretically fast parallel algorithms for graph problems in the literature, for example, work-time optimal parallel random access machine (PRAM) algorithms and communication-optimal bulk synchronous parallel (BSP) algorithms; however, in practice there are seldom any parallel implementations that beat the best sequential implementations for arbitrary, sparse graphs. Moderate parallel speedups are achieved for either dense graphs that can be represented as matrices (e.g., MST [27] and biconnected components [94]) or sparse graphs with regular structures that can be easily partitioned onto the processors (e.g., connected components [60], MST [20]). The gap between theory and practice suggests nonnegligible gaps between parallel models assumed by the algorithms and the current parallel computer architectures. Take the PRAM model [51,80] as an example. Great efforts have been invested to design PRAM algorithms for various problems. PRAM is a simple, ideal model to use when the algorithm designer tries to expose the maximum parallelism in the problem because it abstracts away details of the machines, for example, number of processors, synchronization and communication cost, and memory hierarchy. However, when a PRAM algorithm is implemented on modern architectures, these factors turn out to be critical for performance, and deserve special attention from algorithm designers. For example, communication cost for distributed-memory architecture and cache misses for shared-memory architecture can dominate the execution time, there are algorithms (communication-optimal algorithms and cache-aware/oblivious algorithms) designed to reduce the communication or noncontiguous memory access complexities.

The gap between model and architecture for parallel algorithms is masked by the regularity of computation, communication, and memory access pattern so that communication is rare and locality is abundant. For graph problems, however, especially the arbitrary, sparse instances, irregularity exposes the mismatch between the architecture and the problem domain. With most current parallel computers, arbitrary, sparse instances turn out to be very challenging to deal with. First, no efficient techniques are currently available to partition the graphs evenly onto the processors. This constitutes a big problem for load-balancing with distributed-memory architectures. Second, most algorithms rely on the adjacency information of the graph, with arbitrary, sparse instances, a vertex's neighbor can be any other vertex. Poor locality hinders the cache performance of the algorithm on modern architectures, where optimizing for cache performance has a significant impact on the performance.

To bridge the gap between theory and practice, improvement can be made from both sides, architecture and algorithm design. We focus on algorithm design and engineering techniques that better fit current (or near future) architectures and help achieve good performance for solving arbitrary, sparse graph instances. Our target architecture is shared-memory machines, including symmetric multiprocessors (SMPs) and the multithreaded architecture. We choose shared-memory architecture because it alleviates the difficulty to partition the irregular inputs (a big problem for distributed-memory machines), and the communication among processors through shared memory reads/writes is generally fast.

26.1 Designing Efficient Parallel Algorithms on Multiprocessors

Symmetric multiprocessor architectures, in which several processors operate in a true, hardware-based, shared-memory environment and are packaged as a single machine, are becoming commonplace. Indeed, most of the new high-performance computers are clusters of SMPs having from 2 to over 100 processors per node. Moreover, as supercomputers increasingly use SMP clusters, SMP computations will play a significant role in supercomputing and computational science.

The significant features of SMPs are that the input can be held in the shared memory without having to be partitioned and they provide much faster access to their shared memory than an equivalent message-based architecture. Of course, there remains a large difference between the access time for an element in the local processor cache and that for an element that must be obtained from memory—and that difference increases as the number of processors increases, so that cache-aware implementations are even more important on large SMPs than on single workstations.

Recently, increased available space on the processor owing to refined manufacturing technology and the demand for higher performance that will not be sustained by increasing clock frequency led to the logical creation of multicore CPUs. Many current processors adopt the multicore design that can utilize thread-level parallelism (TLP) in one chip. This form of TLP is known as chip-level multiprocessing, or CMP. As both SMPs and CMPs employ TLP, algorithm designs share many common features on these architectures. For the rest of chapter we focus on SMPs, however, the techniques presented should also work on CMPs.

While an SMP is a shared-memory architecture, it is by no means the PRAM used in theoretical work—synchronization cannot be taken for granted, memory bandwidth is limited, and performance requires a high degree of locality. Emulation of "fast" PRAM algorithms by scaling down the parallelism to the number of processors available from an SMP often times does not yield fast parallel implementations.

For example, in our study of the spanning tree and biconnected-components problems, the implementations (with various optimizations) of PRAM algorithms do not beat the sequential implementation on modern SMPs. (Experimental results for the two problems are presented in Section 26.2 and 26.4, respectively.) Spanning tree and biconnected components are representative of series graph problems that have fast theoretic parallel algorithms but no fast parallel implementations. The sequential implementations are usually very efficient with small constant factors. It calls for new algorithm design and optimization techniques to achieve parallel speedups over fast sequential implementations.

We present general techniques for designing efficient parallel algorithms for graph problems on SMPs. These techniques are summarized after our study of the spanning tree, MST, Euler-tour construction and rooted spanning tree, biconnected components problems, parallel random permutation generation, and lock-free parallel algorithms. Here we discuss the impact of the techniques that target at features of SMPs transparent to a parallel model but crucial to performance. Our goal is good performance on current and emerging systems. Although lower asymptotic complexity usually means faster execution with large enough problem size, for many of the "fast" parallel algorithms for graph problems with polylog running times, we do not expect to see parallel speedups with any realistic input sizes because of the tremendous constant factors involved. Instead of striving to develop work-time optimal parallel algorithms on an idealistic model, we design algorithms that are implementation-friendly, and are able to achieve good performance and scalability.

The rest of the chapter is organized as follows. In the remaining of Section 26.1, that is, Sections 26.1.1–26.1.4, we discuss features of SMPs that have great impact on the performance of an algorithm and present techniques for designing fast parallel algorithms; Sections 26.2, 26.3, and 26.4 present fast algorithms for the spanning tree, MST, biconnected components problems that demonstrate the techniques presented in Section 26.1; and Section 26.5 is devoted to fine-grained mutual exclusion locks and lock-free protocols in parallel algorithms; Section 26.6 is our conclusions and future work.

26.1.1 Limited Parallelism

To solve a problem on a parallel computer, a fundamental question to ask is whether the problem is in *NC*, that is, whether it can be solved on PRAM in polylogarithmic time with a polynomial number of processors. PRAM encourages effort to explore the maximum parallelism of a problem, and PRAM algorithms may utilize a large number of processors. Although there are now machines with thousands of processors, the number of processors available to a parallel program is still nowhere near the size of the problem. The "lesser" question then is how to simulate an *NC* algorithm on a target architecture that has a realistic number of p processors. This can be trivially done through *Brent's scheduling principle*. However, exploring the maximum degree of parallelism does not always yield fast implementations because often times the *NC* algorithms take drastically different approaches than the simple sequential algorithms and bring along large overhead.

In our studies with many of the graph problems, we initially attempted to use various algorithm engineering techniques to optimize the simulations of PRAM algorithms on SMPs, hoping that the similarity between PRAM and SMPs could bring break-through speedups for irregular problems. The fact is that simulations of PRAM algorithms very rarely give reasonable speedups. This brought us to look for alternative algorithm designs.

Acknowledging the fact that parallel computers can only utilize limited levels of parallelism of the algorithm, Kruskal et al. [61] argued that nonpolylogarithmic time algorithms (e.g., sublinear time algorithms) could be more suitable than polylog algorithms for implementation on real machines with practically large input size. The design focus is shifted from lowering complexities for an idealistic model to designing algorithms that solve problems of realistic sizes efficiently with a realistic number of processors.

The shifting of design focus brings performance advantages coming from limited parallelism. Usually larger granularity of parallelism can be exploited in the algorithm design, which in turn results in fewer synchronizations, less overhead, and similarity to the sequential algorithm. Algorithms with coarse-grained parallelism also tend to work well with distributed-memory environments (e.g., many practical LogP and BSP algorithms [26, 90]).

Together with performance advantages, designing for $p \ll n$ processors also brings complications that are otherwise transparent to PRAM algorithms. One is load-balancing, the other is the various memory consistency models that come with different architectures. Load-balancing is usually not an issue for PRAM algorithms because of fine-grained parallelism and the plenty number of processors. For real computers and efficient algorithms that exploit coarse-grained parallelism, however, load-balancing is crucial to performance. The well-studied technique of randomization can be used for load-balancing. In Section 26.2, a "work-stealing" load-balancing technique for SMPs developed by Blumofe and Leiserson [11] is applied for solving the spanning tree problem. Memory consistency model is generally transparent to parallel algorithm design. However, it affects the correctness of an algorithm on real computers. Different protocols about how a processor's behavior is observed by its peers may produce different results for the same program. Algorithms that are aware of memory consistency models may have performance advantages. The spanning tree algorithm presented in Section 26.2 is one such example. Memory consistency issue is also related to synchronization that we will discuss in the next section.

26.1.2 Synchronization

Synchronization among processors is implicit in each step of the PRAM algorithm. With most real parallel computers, however, each processor has its own clock and works asynchronously. Synchronization is crucial to the correctness and performance of the algorithms, and is usually achieved by using barriers or locks to preserve the desired program semantics.

In the extreme case, barriers can be placed after each statement of the program to ensure almost PRAM-like synchronous execution. For many algorithms, however, rigorous PRAM-like synchronizations are not necessary to guarantee correctness and they may seriously degrade the performance. Barriers are only necessary to guarantee that certain events are observed by all processors. It is desirable to have a minimum number of barriers in an algorithm. When adapting a PRAM algorithm to SMPs, a thorough understanding of the algorithm suffices to eliminate unnecessary barriers. In this case, synchronization is a pure implementation issue. Alternatively, with asynchronous algorithms, greater granularity of parallelism are explored and the number of barriers can be further reduced. The spanning tree algorithm presented in Section 26.2 employs only a constant number of barriers.

Mutual exclusion locks are widely used for interprocess synchronization owing to their simple programming abstractions. For certain problems the use of locks can simplify algorithm designs, however, there are performance complications to consider. Locking cost is dependent on architecture, the implementation, and contention pattern associated with inputs. Theoretic algorithm designs usually do not employ locks, yet algorithms with fast locking can have potential performance advantages for the graph problems we study. Fine-grained mutual exclusion is further discussed in Section 26.5.

Lock-free shared data structures in the setting of distributed computing have received a fair amount of attention. Major motivations of lock-free data structures include increasing fault tolerance of a (possibly heterogeneous) system and alleviating the problems associated with critical sections such as *priority inversion* and *deadlock*. For parallel computers with tightly coupled processors and shared memory, these issues are no longer major concerns. Although many of the results are applicable especially when the model used is shared memory multiprocessors, no earlier studies have considered improving the performance of a parallel implementation by way of lock-free programming. As a matter of fact, often times in practice lock-free data structures in a distributed setting do not perform as well as those that use locks. As the data structures and algorithms for parallel computing are often drastically different from those in distributed computing, it is possible that lock-free programs perform better. In Section 26.5, we compare the similarity and difference of lock-free programming in both distributed and parallel computing environments and explore the possibility of adapting lock-free programming to parallel computing to improve performance. Lock-free programming also provides a new way of simulating PRAM and asynchronous PRAM algorithms on current parallel machines.

26.1.3 Cache and Memory Access Patterns

Modern processors have multiple levels of memory hierarchy, a fact that is generally not taken into consideration by most of the parallel models. The increasing speed difference between processor and main memory makes cache and memory access patterns of a program important factors to consider for performance. The number (and pattern) of memory accesses could be the dominating factor of performance instead of computational complexities. Quite a number of cache-aware algorithms emerged. For parallel computing, *the SMP model* proposed by Helman and JáJá is the first effort to model the impact of memory access and cache over an algorithm's performance [41]. Under this model, there are two parts to an algorithm's complexity, M_E the memory access complexity and T_C the computation complexity. The M_E term is the number of noncontiguous memory accesses, and the T_C term is the running time. Parameters of the model include the problem size n and the number of processors p. Having the M_E term recognizes the effect that memory accesses have over an algorithm's performance. It is not directly related to cache, for no parameters of cache size, associativity or replacement scheme are included, instead, it reflects that most cache-friendly algorithms make memory accesses in a regular way with a higher degree of locality.

Standard techniques for improving cache performances attempt to lay out data structures and schedule computations for better cache utilization. For graph problems, as the memory access pattern is largely determined by the irregular structure of the graph, no mature techniques are available to design cache aware algorithms. Our experimental studies show that on current SMPs often times memory writes have larger impact than reads on the performance of the algorithm. This is because with the snoopy-cache consistency protocol for current SMPs, memory writes tend to generate more protocol transactions than memory reads and concurrent writes also create consistency issues of memory management for the operating system and the corresponding memory management is of higher cost than reads. We could have incorporated this fact into a parallel model by separating the number of memory reads and writes as different parameters, but then other questions like how to weigh their different importance would emerge. Here the message is that for graph problems that are generally irregular in nature, when standard techniques of optimizing the algorithm for cache performance does not apply, trading memory reads for writes might be an option. This technique is used in Section 26.3.

26.1.4 Algorithmic Optimizations

For many problems, parallel algorithms are more complicated than the sequential counterparts, incurring large overheads with many algorithm steps. Instead of lowering the asymptotic complexities, we can reduce the constant factors and improve performance. Algorithmic optimizations are problem specific. We demonstrate the benefit of such optimizations with our biconnected components algorithm in Section 26.4.

26.2 A Fast, Parallel Spanning Tree Algorithm

Finding a spanning tree of a graph is an important building block for many parallel algorithms, for example, biconnected components and ear decomposition [69], and can be used in graph planarity testing [58]. The best sequential algorithm for finding a spanning tree of a graph $G = (V, E)$ where $n = |V|$ and $m = |E|$ uses depth- or breadth-first graph traversal that runs in $O(m + n)$. The implementation of the sequential algorithms is very efficient (linear time with a very small hidden constant), and the only data structure used is a stack or queue which has good locality features. Previous approaches for parallel spanning tree algorithms use techniques other than traversal that are conducive to parallelism and have polylogarithmic time complexities. In practice, none of these parallel algorithms has shown significant parallel speedup over the best sequential algorithm for irregular graphs, because the theoretic models do not realistically capture the cost for communication on current parallel machines (e.g., References 6, 15, 18, 24, 34, 39, 40, 47, 50, 52–54, 61, 72, 77, and 83), the algorithm is too complex for implementation (e.g., References 18 and 39), or there are large constants hidden in the asymptotic notation that could not be overcome by a parallel implementation (e.g., References 20, 27, 38, 49, and 60).

Earlier experimental studies [38, 49, 60] implement algorithms on the basis of the "graft-and-shortcut" approach, as represented by the Shiloach-Vishkin (SV) Algorithm. None of the studies achieve parallel speedup for arbitrary, sparse inputs. SV is originally a connected components algorithm, and can be extended for solving the spanning tree problem. SV takes an edge list as input and starts with n isolated vertices and m processors. Each processor P_i $(1 \le i \le m)$ inspects edge $e_i = (v_{i_1}, v_{i_2})$ and tries to graft vertex v_{i_1} to v_{i_2} under the constraint that $i_1 < i_2$. Grafting creates $k \ge 1$ connected components in the graph, and each of the k components is then shortcut to a single supervertex. Grafting and shortcutting are iteratively applied to the reduced graphs $G' = (V', E')$ (where V' is the set of supervertices and E' is the set of edges among supervertices) until only one supervertex is left. For a certain vertex v with multiple adjacent edges, there can be multiple processors attempting to graft v to other smaller-labeled vertices. Only one grafting is allowed, and the corresponding edge that causes the grafting is labeled as a spanning tree edge.

We present a new spanning tree algorithm for SMPs that exhibits parallel speedups on graphs with regular and irregular topologies. Our experimental study of parallel spanning tree algorithms reveals the superior performance of our new approach compared with the previous algorithms. For realistic problem sizes ($n \gg p^2$), the expected running time for our new SMP spanning tree algorithm on a graph with n vertices and m edges is given by $T(n, p) = \langle M_E; T_C \rangle \le \left\langle O\left(\frac{n+m}{p}\right); O\left(\frac{n+m}{p}\right)\right\rangle$ where p is the number of processors, using the SMP complexity model.

Our new parallel spanning tree algorithm for shared-memory multiprocessors has two main steps: (i) stub spanning tree, and (ii) work-stealing graph traversal. The overall strategy is first to generate a small stub spanning tree with one processor, and then let each processor start from vertices in the stub tree and traverse the graph simultaneously, where each processor follows a DFS-order. When all the processors are done, the subtrees grown by graph traversal are connected by the stub tree into a spanning tree. Work-stealing is a randomized technique used for load balancing the graph traversals and yields an expected running time that scales linearly with the number of processors for suitably large inputs. Unlike the SV approach, the labeling of vertices does not affect the performance of our new algorithm.

Stub spanning tree: In the first step, one processor generates a stub spanning tree, that is, a small portion of the spanning tree by randomly walking the graph for $O(p)$ steps. The vertices of the stub spanning tree are evenly distributed into each processor's stack, and each processor in the next step traverses from the first element in its stack. After the traversals in Step 2, the spanning subtrees are connected to each other by this stub spanning tree.

Parallel Graph Algorithms for Multiprocessors

Work-stealing graph traversal: The basic idea of this step is to let each processor traverse the graph similar to the sequential algorithm in such a way that each processor finds a subgraph of the final spanning tree. In order for this step (see Algorithm 1) to perform correctly and efficiently, we need to address the following two issues: (i) coloring the same vertex simultaneously by multiple processors; that is, a vertex may appear in two or more subtrees of different processors, and (ii) balancing the load among the processors.

Data : (1) An adjacency list representation of graph $G = (V, E)$ with n vertices, (2) a starting vertex *root* for each processor, (3) *color*: an array of size n with each element initialized to 0, and (4) *parent*: an array of size n.
Result: p pieces of spanning subtrees, except for the starting vertices, each vertex v has *parent*[v] as its parent
begin
    1. color my starting vertex with my label i and place it into my stack S
        color[*root*] = i
        Push(S, *root*)
    2. start depth-first search from *root*, color the vertices that have not been visited with my label i until the stack is empty.
    2.1 **while** Not-Empty(S) **do**
        2.2 v = Pop(S)
        2.3 **for** each neighbor w of v **do**
            2.4 **if** (*color*[w] = 0) **then**
                2.5 *color*[w] = i
                2.6 *parent*[w] = v
                2.7 Push(S, w)
end

Algorithm 1: Graph traversal step for our SMP algorithm for processor i ($1 \leq i \leq p$).

As we will show the algorithm runs correctly even when two or more processors color the same vertex. In this situation, each processor will color the vertex and set as its parent the vertex it has just colored. Only one processor succeeds at setting the vertex's parent to a final value. For example, using Figure 26.1, processor P_1 colored vertex u, and processor P_2 colored vertex v, at a certain time they both find w

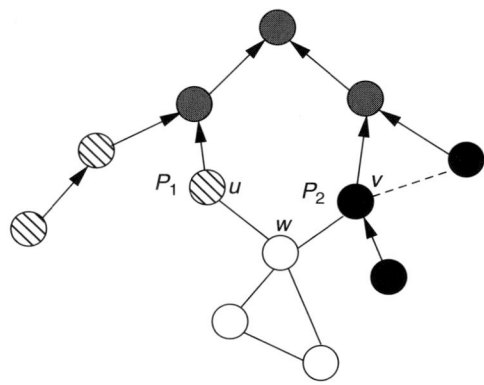

FIGURE 26.1 Two processors P_1 and P_2 work on vertex u and v, respectively. They both see vertex w as unvisited, so each is in a race to color w and set w's parent pointer. The gray vertices are in the stub spanning tree; the shaded vertices are colored by P_1; the black vertices are marked by P_2; and the white vertices are unvisited. Directed solid edges are the selected spanning tree edges; dashed edges are nontree edges; and undirected solid edges are not yet visited.

unvisited and are now in a race to color vertex w. It makes no difference which processor colored w last because w's parent will be set to either u or v (and it is legal to set w's parent to either of them; this will not change the validity of the spanning tree, only its shape). Further, this event does not create cycles in the spanning tree. Both P_1 and P_2 record that w is connected to each processor's own tree. When various processors visit each of w's unvisited children, its parent will be set to w, independent of w's parent.

Lemma 26.2.1 *On an SMP with sequential memory consistency, Algorithm 1 does not create any cycles in the spanning tree.*

Proof. (by contradiction) Suppose in Algorithm 1 processors $P_1, P_2, \ldots, P_j$ create a cycle sequence $< s_1, s_2, \ldots, s_k, s_1 >$, that is, P_i sets s_i's parent to s_{i+1}, and P_j sets s_k's parent to s_1. Here any P_i and P_j with $1 \leq i, j \leq p$ and $1 \leq k \leq n$ could be the same or different processors. According to the algorithm, s_i's parent is set to s_{i+1} only when P_i finds s_{i+1} at the top of its stack (and s_{i+1} was colored before and put into the stack), and s_i is s_{i+1}'s unvisited (uncolored) neighbor. This implies that for P_i the coloring of s_{i+1} happens before the coloring of s_i. In other words, processor P_i observes the memory write to location $color[s_{i+1}]$ happen before the write to location $color[s_i]$. On an SMP with sequential memory consistency, this means each processor should see the sequence in this order. Let t_i be the time at which s_i is colored; we have $t_i > t_{i+1}$, that is, $t_1 > t_2 > t_3 > \cdots > t_k > t_1$, which is a contradiction. Thus, the SMP graph traversal step creates no cycles. ∎

Lemma 26.2.2 *For connected graph G, Algorithm 1 will set parent[v] for each vertex $v \in V$ that is colored 0 before the start of the algorithm.*

Proof. First, we prove (by contradiction) that each vertex with color 0 before the start of the algorithm will be colored from the set $\{1, 2, \ldots, p\}$ after the algorithm terminates. Suppose there exists a vertex $v \in V$ that still has color 0 after Algorithm 1 terminates. This implies that each neighbor w of v is never placed into the stack, otherwise Step 2.3 in Algorithm 1 would have found that v is w's neighbor, and would have colored v as one of $1, 2, \ldots, p$. If w is never placed in the stack, then w has color 0, which in turn means that all w's neighbors have color 0. By induction, and because G is connected, we find all of the vertices in G are colored 0 after the algorithm terminates, which is clearly a contradiction. Further, since each vertex is colored, Step 2.6 in Algorithm 1 guarantees that each vertex's *parent* is set. ∎

For certain shapes of graphs or ordering of traversals, some processors may have little work to do whereas others are overloaded. For example, using Figure 26.2, after generating a stub spanning tree (black vertices), processors P_1, P_2, P_3, and P_4, start a traversal from designated starting points. In this case P_1, P_2, and P_3, color no other vertices than u, v, and w, while processor P_4, starting from vertex x, has significant work to do. In this example for instance, this results in all but one processor sitting idle while

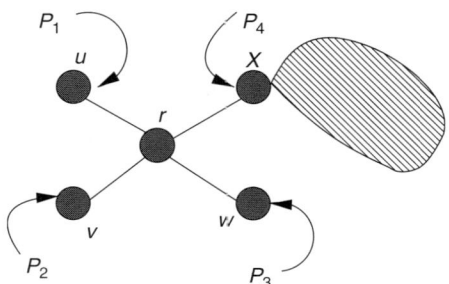

FIGURE 26.2 Unbalanced load: processors P_1, P_2, and P_3, each color only one vertex while processor P_4 colors the remaining $n - 5$ vertices.

a single processor performs almost all the work, and obviously no speedup will be achieved. We remedy this situation as follows.

To achieve better load-balancing across the processors, we add the technique of work-stealing to our algorithm. Whenever any processor finishes with its own work (i.e., it cannot reach any other unvisited vertex), it randomly checks other processors' stacks. If it finds a nonempty stack, the processor steals part of the stack. Work-stealing does not affect the correctness of the algorithm, because when a processor takes elements from a stack, all of the elements are already colored and their *parents* have already been set, and no matter which processor inspects their unvisited children, they are going to be set as these children's *parents*. As we show later in our experimental results, we find that this technique keeps all processors equally busy performing useful work, and hence, evenly balances the workload for most classes of input graphs.

Arguably there are still pathological cases where work-stealing could fail to balance the load among the processors. For example, when connectivity of a graph (or portions of a graph) is very low (the diameter of the graph is large), stacks of the busy processors may only contain a few vertices. In this case work awaits busy processors while idle processors starve for something to do. Obviously this is the worst case for the SMP traversal algorithm. We argue that this case is very rare (see Section 26.2.1); however, we next propose a detection mechanism that can detect the situation and invoke a different spanning tree algorithm that is robust to this case.

The detection mechanism uses condition variables to coordinate the state of processing. Whenever a processor becomes idle and finds no work to steal, it will go to sleep for a duration on a condition variable. Once the number of sleeping processors reaches a certain threshold, we halt the SMP traversal algorithm, merge the grown spanning subtree into a supervertex, and start a different algorithm, for instance, the SV approach. In theoretic terms, the performance of our algorithm could be similar to that of SV in the worst case, but in practical terms this mechanism will almost never be triggered; for instance, in our experimental studies with a collection of different types of graphs, we never encountered such a case.

When an input graph contains vertices of degree two, these vertices along with a corresponding tree edge can be eliminated as a simple preprocessing step. Clearly, this optimization does not affect correctness of the algorithm, and we can assume that this procedure has been run before the analysis in the next section.

Theorem 26.2.1 *For connected graph G, suppose we generate a stub spanning tree and store the vertices into each processor's stack. Let each processor start the traversal from the first vertex stored in its stack. Then after the work-stealing graph traversal step terminates, we have a spanning tree of G.*

Proof. Theorem 26.2.1 follows from Lemmas 26.2.1 and 26.2.2. ∎

26.2.1 Analysis of the Symmetric Multiprocessor Spanning Tree Algorithms

We compare our new SMP algorithm with the implementation of SV both in terms of complexity and actual performance (Section 26.2.2). Our analysis are based on the SMP complexity model.

Our spanning tree algorithm takes advantage of the shared-memory environment in several ways. First, the input graph's data structure can be shared by the processors without the need for the difficult task of partitioning the input data often required by distributed-memory algorithms. Second, load balancing can be performed asynchronously using the lightweight work-stealing protocol.

The first step that generates a stub spanning tree is executed by one processor in

$$T(n, p) = \langle M_E; T_C \rangle = \langle O(p); O(p) \rangle.$$

In the second step, the work-stealing graph traversal step needs one noncontiguous memory access to visit each vertex, and two noncontiguous accesses per edge to find the adjacent vertices, check their colors, and set the parent. For almost all graphs, the expected number of vertices processed per processor is $O(n/p)$ with the work-stealing technique; and hence, we expect the load to be evenly balanced. (Palmer [74] proved that almost all random graphs have diameter two.) During the tree-growing process, a small number of vertices may appear in more than one stack because of the races among the processors. Analytically, we could model this as a random process that depends on parameters related to system and problem characteristics. However, this number will not be significant. Our experiments show that the number of vertices that appear in multiple processors' stacks at the same time are a minuscule percentage (e.g., less than ten vertices for a graph with millions of vertices).

We expect each processor to visit $O(n/p)$ vertices; hence, the expected complexity of the second step is

$$T(n, p) = \langle M_E; T_C \rangle = \left\langle \frac{n}{p} + 2\frac{m}{p}; O\left(\frac{n+m}{p}\right) \right\rangle.$$

Thus, the expected running time for our SMP spanning tree algorithm is given as

$$T(n, p) = \langle M_E; T_C \rangle \leq \left\langle \frac{n}{p} + 2\frac{m}{p} + O(p); O\left(\frac{n+m}{p}\right) \right\rangle, \quad (26.1)$$

with high probability. For realistic problem sizes ($n \gg p^2$), this simplifies to

$$T(n, p) = \langle M_E; T_C \rangle \leq \left\langle O\left(\frac{n+m}{p}\right); O\left(\frac{n+m}{p}\right) \right\rangle. \quad (26.2)$$

The algorithm scales linearly with the problem size and number of processors, and we use only a constant number of barrier synchronizations.

In comparison, assuming the worst-case of $\log n$ iterations, the total complexity for emulating SV on SMPs is

$$T(n, p) = \langle M_E; T_C \rangle \leq \left\langle \frac{n \log^2 n}{p} + \left(4\frac{m}{p} + 2\right) \log n; O\left(\frac{n \log^2 n + m \log n}{p}\right) \right\rangle. \quad (26.3)$$

Comparing the analysis, we predict that our approach has less computation $\left(O\left(\frac{n+m}{p}\right)\right)$ than the SV approach that has worst-case computational requirements of $O\left(\frac{n \log^2 n + m \log n}{p}\right)$. Even if SV iterates only once, there is still approximately $\log n$ times more work per iteration. Looking at memory accesses, our SMP algorithm is more cache friendly, having small number of noncontiguous memory access per the input size. On the other hand, SV has a multiplicative factor of approximately $\log^2 n$ more non-contiguous accesses per vertex assigned to each processor. Our SMP approach also uses less synchronization ($O(1)$) than the SV approach that requires $O(\log n)$.

26.2.2 Experimental Results

As we focus on practical implementation, we present experimental results to demonstrate the performance advantage of our algorithms. Throughout the chapter unless otherwise stated, experiments are conducted on the Sun E4500, a uniform-memory-access (UMA) shared memory parallel machine with 14 UltraSPARC II 400MHz processors and 14 GB of memory. Each processor has 16 kb of direct-mapped data (L1) cache and 4 MB of external (L2) cache. The algorithms are implemented using POSIX threads and software-based barriers. Although there are more powerful SMPs available today, there is no drastic change in the architecture and the SUN E4500 is still a representative. We also show results on other systems when necessary.

For inputs, a collection of sparse graph generators are used to compare the performance of various parallel implementations. These generators include several employed in previous experimental studies of parallel graph algorithms for related problems [20, 36, 38, 49, 60].

- **Regular and Irregular Meshes** Computational science applications for physics-based simulations and computer vision commonly use mesh-based graphs.
  — **2D Torus** The vertices of the graph are placed on a 2D mesh, with each vertex connected to its four neighbors.
  — **2D60** 2D mesh with the probability of 60% for each edge to be present.
  — **3D40** 3D mesh with the probability of 40% for each edge to be present.
- **Random Graph** We create a random graph of n vertices and m edges by randomly adding m unique edges to the vertex set. Several software packages generate random graphs this way, including LEDA [68].
- **Geometric Graphs and AD3** In these k-regular graphs, n points are chosen uniformly and at random in a unit square in the Cartesian plane, and each vertex is connected to its k nearest neighbors. Moret and Shapiro [71] use these in their empirical study of sequential MST algorithms. **AD3** is a geometric graph with $k = 3$.
- **Structured Graphs** These graphs are used by Chung and Condon (see Reference 20 for detailed descriptions) to study the performance of parallel Borůvka's algorithm. They have recursive structures that correspond to the iteration of Borůvka's algorithm and are degenerate (the input is already a tree).
  — **str0** At each iteration with n vertices, two vertices form a pair. So with Borůvka's algorithm, the number of vertices decrease exactly by a half in each iteration.
  — **str1** At each iteration with n vertices, $\sqrt{n}$ vertices form a linear chain.
  — **str2** At each iteration with n vertices, $n/2$ vertices form linear chain, and the other $n/2$ form pairs.
  — **str3** At each iteration with n vertices, $\sqrt{n}$ vertices form a complete binary tree.

For weighted graphs, we assign random weights to the edges. Owing to limited space, we present experimental results for random graphs and meshes. Results for other graph types are also presented if with such inputs the algorithms reveal interesting features.

We offer a collection of our performance results that demonstrate for the first time a parallel spanning tree algorithm that exhibits speedup when compared with the best sequential approach over a wide range of input graphs. The performance plots in Figure 26.3 are for the regular and irregular meshes (torus, **2D60** and **3D40**), in Figure 26.4 are for the random, geometric and **AD3**, and geographic classes of graphs. Note that only the meshes are regular; all of the remaining graphs used are irregular. In these plots, the horizontal dashed line represents the time taken for the best sequential spanning tree algorithm to find a solution on the same input graph using a single processor.

In the case of the torus inputs, we observe that the initial labeling of vertices greatly affects the performance of the SV algorithm, but the labeling has little impact on our algorithm. In all of these graphs, we note that the SV approach runs faster as we employ more processors. However, in many cases, the SV parallel approach is slower than the best sequential algorithm. For $p > 2$ processors, in our testing with a variety of classes of large graphs, our new spanning tree algorithm is always faster than the sequential algorithm, and executes faster as more processors are available. This is remarkable, given that the sequential algorithm runs in linear time with a very small hidden constant in the asymptotic complexity.

26.2.3 Discussion

The design of the new spanning tree algorithm is efficient with small parallel overheads, which is crucial in competing with the sequential implementation. Also the algorithm is of a "chaotic" nature,

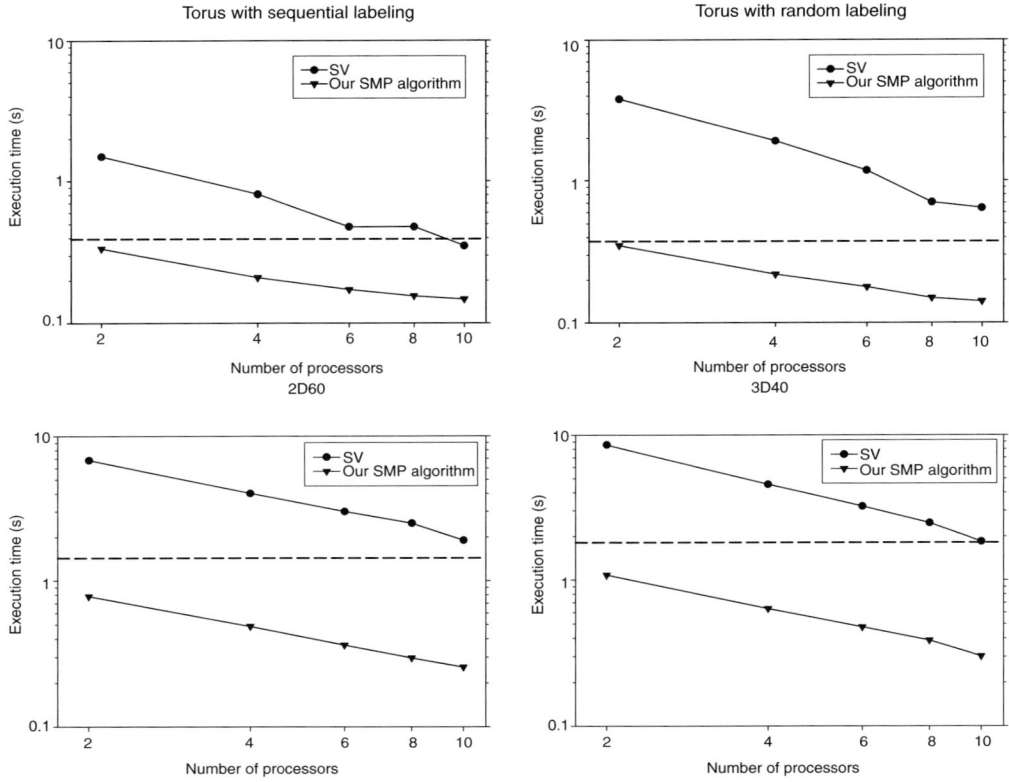

FIGURE 26.3 Comparison of parallel spanning tree algorithms for regular and irregular meshes with $n = 1M$ vertices. The top-left plot uses a row-major order labeling of the vertices in the torus, while the top-right plot uses a random labeling. The bottom-left and -right plots are for irregular torus graphs **2D60** and **3D40**, respectively. The dashed line corresponds to the best sequential time for solving the input instance. Note that these performance charts are log–log plots.

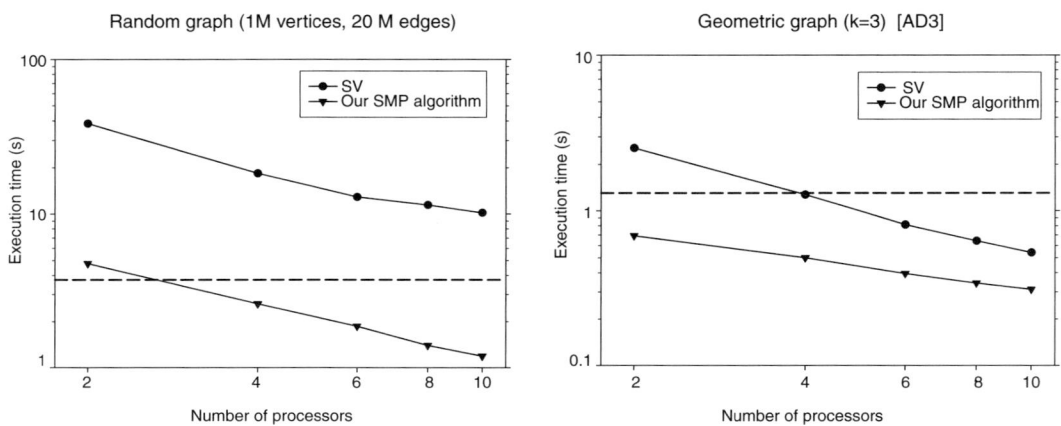

FIGURE 26.4 Comparison of parallel spanning tree algorithms for graphs with $n = 1M$ vertices. The left plot uses a random graph with $m = 20M \approx n \log n$ edges. The right plot uses **AD3**, a geometric graph with $k = 3$. The dashed line corresponds to the best sequential time for solving the input instance. Note that these performance charts are log–log plots.

that is, each run over the same input may yield different spanning trees, and there are races among processors. One implication of running the algorithm on current architectures is that many shared-memory multiprocessors provide relaxed consistency models other than the sequential consistency model. In such cases, memory access ordering instructions such as "membar" should be used to preserve the desired orderings.

26.3 Fast Shared-Memory Algorithms for Computing Minimum Spanning Tree

Given an undirected connected graph G with n vertices and m edges, the MST problem finds a spanning tree with the minimum sum of edge weights. Pettie and Ramachandran [76] designed a randomized, time-work optimal MST algorithm for the EREW PRAM, and using EREW to QSM and QSM to BSP emulations from Reference 35, mapped the performance onto QSM and BSP models. Cole et al. [22, 23] and Poon and Ramachandran [78] earlier had randomized linear-work algorithms on CRCW and EREW PRAM. Chong et al. [16] gave a deterministic EREW PRAM algorithm that runs in logarithmic time with a linear number of processors. On the BSP model, Adler et al. [1] presented a communication-optimal MST algorithm. While these algorithms are fast in theory, many are considered impractical because they are too complicated and have large constant factors hidden in the asymptotic complexity. Earlier experimental studies (e.g., References 20 and 27) either do not achieve speedup on or are unsuitable for the more challenging arbitrary, sparse instances.

We present our implementations of MST algorithms on shared-memory multiprocessors that achieve for the first time in practice reasonable speedups over a wide range of input graphs, including arbitrary sparse graphs. In fact, if G is not connected, our algorithms find the MST of each connected component; hence, solving the minimum spanning forest problem.

We start with the design and implementation of a parallel Borůvka's algorithm. Borůvka's algorithm is one of the earliest MST approaches, and the Borůvka iteration (or its variants) serves as a basis for several of the more complicated parallel MST algorithms, hence its efficient implementation is critical for parallel MST. Three steps characterize a Borůvka iteration: *find-min*, *connect-components*, and *compact-graph*. *Find-min* and *connect-components* are simple and straightforward to implement, and the *compact-graph* step performs bookkeeping that is often left as a trivial exercise to the reader. JáJá [51] describes a compact-graph algorithm for dense inputs. For sparse graphs, though, the compact-graph step often is the most expensive step in the Borůvka iteration. Section 26.3.1 explores different ways to implement the compact-graph step, then proposes a new data structure for representing sparse graphs that can dramatically reduce the running time of the compact-graph step with a small cost to the find-min step. The analysis of these approaches is given in Section 26.3.2.

In Section 26.3.3, we present a new parallel MST algorithm for SMPs that marries the Prim and Borůvka approaches. In fact, the algorithm when run on one processor behaves as Prim's, and on n processors becomes Borůvka's, and runs as a hybrid combination for $1 < p < n$, where p is the number of processors.

26.3.1 Data Structures for Parallel Borůvka's Algorithms

Borůvka's MST algorithm lends itself more naturally to parallelization. Three steps comprise each iteration of parallel Borůvka's algorithm:

1. *Find-min*: for each vertex v label the incident edge with the smallest weight to be in the MST.
2. *Connect-components*: identify connected components of the induced graph with edges found in Step 1.
3. *Compact-graph*: compact each connected component into a single supervertex, remove self-loops and multiple edges; and relabel the vertices for consistency.

Steps 1 and 2 (find-min and connect-components) are relatively simple and straightforward; Step 3 (compact-graph) shrinks the connected components and relabels the vertices. For dense graphs that can be represented by an adjacency matrix, JáJá [51] describes a simple and efficient implementation for this step. For sparse graphs this step often consumes the most time yet no detailed discussion appears in the literature. In the following we describe our design of three Borůvka approaches that use different data structures, and compare the performance of each implementation.

We use the edge list representation of graphs, in one implementation of Borůvka's algorithm (designated Bor-EL), with each edge (u, v) appearing twice in the list for both directions (u, v) and (v, u). An elegant implementation of the compact-graph step sorts the edge list (using an efficient parallel sample sort [41]) with the supervertex of the first end point as the primary key, the supervertex of the second end point as the secondary key, and the edge weight as the tertiary key. When sorting completes, all of the self-loops and multiple edges between two supervertices appear in consecutive locations, and can be merged efficiently using parallel prefix sums.

With the adjacency list representation (but using the more cache-friendly adjacency arrays [75]) each entry of an index array of vertices points to a list of its incident edges. The compact-graph step first sorts the vertex array according to the supervertex label, then concurrently sorts each vertex's adjacency list using the supervertex of the other end point of the edge as the key. After sorting, the set of vertices with the same supervertex label are contiguous in the array, and can be merged efficiently. We call this approach Bor-AL.

Both Bor- and Bor-AL achieve the same goal that self-loops and multiple edges are moved to consecutive locations to be merged. Bor-EL uses one call to sample sort whereas Bor-AL calls a smaller parallel sort and then a number of concurrent sequential sorts. We make the following algorithm engineering choices for the sequential sorts used in this approach. The $O(n^2)$ insertion sort is generally considered a bad choice for sequential sort, yet for small inputs, it outperforms $O(n \log n)$ sorts. Profiling shows that there could be many short lists to be sorted for very sparse graphs. For example, for one of our input random graphs with 1M vertices, 6M edges, 80% of all 311,535 lists to be sorted have between 1 and 100 elements. We use insertion sort for these short lists. For longer lists we use a nonrecursive $O(n \log n)$ merge sort.

For the previous two approaches, conceivably the compact-graph step could be the most expensive step for a parallel Borůvka's algorithm. Next, we propose an alternative approach with a new graph representation data structure (that we call *flexible adjacency list*) that significantly reduces the cost for compacting the graph.

The flexible adjacency list augments the traditional adjacency list representation by allowing each vertex to hold multiple adjacency lists instead of just a single one; in fact it is a linked list of adjacency lists (and similar to Bor-AL, we use the more cache-friendly adjacency array for each list). During initialization, each vertex points to only one adjacency list. After the connect-components step, each vertex appends its adjacency list to its supervertex's adjacency list by sorting together the vertices that are labeled with the same supervertex. We simplify the compact-graph step, allowing each supervertex to have self-loops and multiple edges inside its adjacency list. Thus, the compact-graph step now uses a smaller parallel sort plus several pointer operations instead of costly sortings and memory copies, whereas the find-min step gets the added responsibility of filtering out the self-loops and multiple edges. Note that for this new approach (designated Bor-FAL) there are potentially fewer memory write operations compared with the previous two approaches. This is important for an implementation on SMPs because memory writes typically generate more cache coherency transactions than do reads.

In Figure 26.5, we illustrate the use of the flexible adjacency list for a 6-vertex input graph. After one Borůvka iteration, vertices 1, 2, and 3, form one supervertex and vertices 4, 5, and 6, form a second supervertex. Vertex labels 1 and 4 represent the supervertices and receive the adjacency lists of vertices 2 and 3, and vertices 5 and 6, respectively. Vertices 1 and 4 are relabeled as 1 and 2. Note that most of the original data structure is kept intact so that we might save memory copies. Instead of relabeling vertices in the adjacency list, we maintain a separate lookup table that holds the supervertex label for each vertex. We easily obtain this table from the connect-components step. The find-min step uses this table to filter out self-loops and multiple edges.

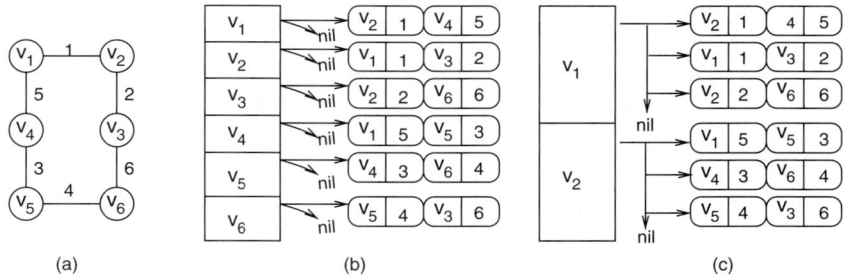

FIGURE 26.5 Example of flexible adjacency list representation (a) Input Graph, (b) Initialized flexible adjacency list, (c) Flexible adjacency list after one iteration.

26.3.2 Analysis

Here we analyze the complexities of the different Borůvka variants under the SMP model. For a sparse graph G with n vertices and m edges, as the algorithm iterates, the number of vertices decreases by at least half in each iteration, so there are at most $\log n$ iterations for all of the Borůvka variants.

First we consider the complexity of *Bor-EL*. The find-min and connect-component steps are straightforward, and their aggregate complexity in one iteration (assuming balanced load among processors) is characterized by

$$T(n,p) = \langle M_E; T_C \rangle = \left\langle \frac{n + n \log n}{p}; O\left(\frac{m + n \log n}{p}\right) \right\rangle.$$

The parallel sample sort that we use in *Bor-EL* for compact-graph has the complexity of

$$T(n,p) = \left\langle \left(4 + 2\frac{c \log \frac{l}{p}}{\log z}\right) \frac{l}{p}; O\left(\frac{l}{p} \log l\right) \right\rangle$$

with high probability, where l is the length of the list and c and z are constants related to cache size and sampling ratio [41]. The cost of the compact-graph step, by aggregating the cost for sorting and for manipulating the data structure, is

$$T(n,p) = \left\langle \left(4 + 2\frac{c \log(2m/p)}{\log z}\right) \frac{2m}{p}; O\left(\frac{2m}{p} \log 2m\right) \right\rangle.$$

The value of m decreases with each successive iteration dependent on the topology and edge weight assignment of the input graph. Since the number of vertices is reduced by at least half each iteration, m decreases by at least $n/2$ edges each iteration. For the sake of simplifying the analysis, though, we use m unchanged as the number of edges during each iteration; clearly an upper bound of the worst case. Hence, the complexity of *Bor-EL* is given as

$$T(n,p) = \left\langle \left(\frac{8m + n + n \log n}{p} + \frac{4mc \log(2m/p)}{p \log z}\right) \log n; O\left(\frac{m}{p} \log m \log n\right) \right\rangle.$$

Without going into the input-dependent details of how vertex degrees change as the Borůvka variants progress, we compare the complexity of the first iteration of *Bor-AL* with *Bor-EL* because in each iteration these approaches compute similar results in different ways. For *Bor-AL* the complexity of the first iteration is

$$T(n,p) = \left\langle \left(\frac{8n + 5m + n \log n}{p} + 2c \frac{n \log \frac{n}{p} + m \log \frac{m}{n}}{p \log z} \right); O\left(\frac{n}{p} \log m + \frac{m}{p} \log \frac{m}{n} \right) \right\rangle.$$

Whereas for *Bor-EL*, the complexity of the first iteration is

$$T(n,p) = \left\langle \left(\frac{8m + n + n \log n}{p} + \frac{4mc \log \frac{2m}{p}}{p \log z} \right); O\left(\frac{m}{p} \log m \right) \right\rangle.$$

We see that *Bor-AL* is a faster algorithm than *Bor-EL*, as expected, since the input for *Bor-AL* is "bucketed" into adjacency lists, versus *Bor-EL* that is an unordered list of edges, and sorting each bucket first in *Bor-AL* saves unnecessary comparisons between edges that have no vertices in common. We can consider the complexity of *Bor-EL* then to be an upper bound of *Bor-AL*.

In *Bor-FAL* n reduces at least by half while m stays the same. Compact-graph first sorts the n vertices, then assigns $O(n)$ pointers to append each vertex's adjacency list to its supervertex's. For each processor, sorting takes $O\left(\frac{n}{p} \log n\right)$ time, and assigning pointers takes $O(n/p)$ time assuming each processor gets to assign roughly the same amount of pointers. Updating the lookup table costs each processor $O(n/p)$ time. As n decreases at least by half, the aggregate running time for compact-graph is

$$T_C(n,p)_{cg} = \frac{1}{p} \sum_{i=0}^{\log n} \frac{n}{2^i} \log \frac{n}{2^i} + \frac{2}{p} \sum_{i=0}^{\log n} \frac{n}{2^i} = O\left(\frac{n \log n}{p}\right),$$

$$M_E(n,p)_{cg} \leq \frac{8n}{p} + \frac{4cn \log \frac{n}{p}}{p \log z}.$$

With *Bor-FAL*, to find the smallest weight edge for the supervertices, all the m edges will be checked, with each processor covering $O(m/p)$ edges. The aggregate running time is $T_C(n,p)_{fm} = O(m \log n/p)$ and the memory access complexity is $M_E(n,p)_{fm} = m/p$. For the finding connected component step, each processor takes $T_{cc} = O(n \log n/p)$ time, and $M_E(n,p)_{cc} \leq 2n \log n$. The complexity for the whole Borůvka's algorithm is

$$T(n,p) = T(n,p)_{fm} + T(n,p)_{cc} + T(n,p)_{cg}$$

$$\leq \left\langle \frac{8n + 2n \log n + m \log n}{p} + \frac{4cn \log \frac{n}{p}}{p \log z}; O\left(\frac{m+n}{p} \log n\right) \right\rangle.$$

It would be interesting and important to check how well our analysis and claim fit with the actual experiments. Detailed performance results are presented in Section 26.3.5. Here we show that *Bor-AL* in practice runs faster than *Bor-EL*, and *Bor-FAL* greatly reduces the compact-graph time. Figure 26.6 shows for the three approaches the breakdown of the running time for the three steps.

Immediately we can see that for *Bor-* and *Bor-AL* the compact-graph step dominates the running time. *Bor-EL* takes much more time than *Bor-AL*, and only gets worse when the graphs get denser. In contrast the execution time of compact-graph step of *Bor-FAL* is greatly reduced: in the experimental section with a random graph of 1M vertices and 10M edges, it is over 50 times faster than *Bor-EL*, and over 7 times faster than *Bor-AL*. Actually the execution time of the compact-graph step of *Bor-FAL* is almost the same for the three input graphs because it only depends on the number of vertices. As predicted, the execution time of the find-min step of *Bor-FAL* increases. And the connect-components step only takes a small fraction of the execution time for all approaches.

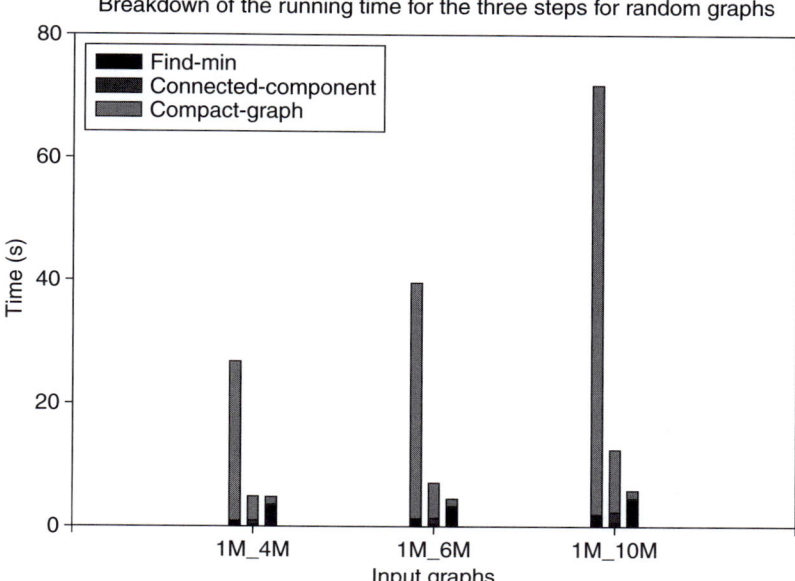

FIGURE 26.6 Running times for the find-min, connect-component, and compact-graph steps of the *Bor-EL*, *Bor-AL*, and *Bor-ALM* approaches (the three groups from left to right, respectively) of the parallel MST implementations using random graphs with $n = 1M$ vertices and $m = 4M$, $6M$, and $10M$ edges (the bars from left to right, respectively, in each group).

26.3.3 A New Parallel Minimum Spanning Tree Algorithm

In this section, we present a new shared-memory algorithm for finding a MST/forest that is quite different from Borůvka's approach in that it uses multiple, coordinated instances of Prim's sequential algorithm running on the graph's shared data structure. In fact, the new approach marries Prim's algorithm (known as an efficient sequential algorithm for MST) with that of the naturally parallel Borůvka approach. In our new algorithm essentially we let each processor simultaneously run Prim's algorithm from different starting vertices. We say a tree is *growing* when there exists a lightweight edge that connects the tree to a vertex not yet in another tree, and *mature* otherwise. When all of the vertices have been incorporated into mature subtrees, we contract each subtree into a supervertex and call the approach recursively until only one supervertex remains. When the problem size is small enough, one processor solves the remaining problem using the best sequential MST algorithm. If no edges remain between supervertices, we halt the algorithm and return the minimum spanning forest. The detailed algorithm is given in Algorithm 2.

In Algorithm 2, Step 1 initializes each vertex as uncolored and unvisited. A processor *colors* a vertex if it is the first processor to insert it into a heap, and labels a vertex as *visited* when it is extracted from the heap; that is, the edge associated with the vertex has been selected to be in the MST. In Step 2 (Algorithm 3) each processor first searches its own portion of the list for uncolored vertices from which to start Prim's algorithm. In each iteration a processor chooses a unique color (different from other processors' colors or the colors it has used before) as its own color. After extracting the minimum element from the heap, the processor checks whether the element is colored by itself, and if not, a collision with another processor occurs (meaning multiple processors try to color this element in a race), and the processor stops growing the current sub-MST. Otherwise it continues.

We next prove that Algorithm 2 finds a MST of the graph. Note that we assume without loss of generality that all the edges have distinct weights.

> **Input:** Graph $G = (V, E)$ represented by adjacency list A with $n = |V|$
> n_b: the base problem size to be solved sequentially.
> **Output:** MSF for graph G
> **begin**
>   **while** $n > n_b$ **do**
>   1. Initialize the *color* and *visited* arrays
>   **for** $v \leftarrow i\frac{n}{p}$ to $(i+1)\frac{n}{p} - 1$ **do**
> $color[v] = 0, visited[v] = 0$
>   2. Run Algorithm 3.
>   3. **for** $v \leftarrow i\frac{n}{p}$ to $(i+1)\frac{n}{p} - 1$ **do**
>   **if** $visited[v] = 0$ **then** find the lightest incident edge e to v, and label e to be in MST
>   4. With the found MST edges, run connected components on the induced graph, and shrink each component into a supervertex
>   5. Remove each supervertex with degree 0 (a connected component)
>   6. Set $n \leftarrow$ the number of remaining supervertices; and $m \leftarrow$ the number of edges between the supervertices
>   7. Solve the remaining problem on one processor
> **end**

Algorithm 2: Parallel algorithm for new MSF approach, for processor p_i, for $(0 \le i \le p-1)$. Assume w.l.o.g. that p divides n evenly.

Lemma 26.3.1 *On an SMP with sequential memory consistency, subtrees grown by Algorithm 3 do not touch each other, in other words, no two subtrees share a vertex.*

Proof. Step 1.4 of Algorithm 3 grows subtrees following the fashion of Prim's algorithm. Suppose two subtrees T_1 and T_2 share one vertex v. We have two cases:

1. *Case 1:* T_1 and T_2 could be two trees grown by one processor
2. *Case 2:* each tree is grown by a different processor

v will be included in a processor's subtree only if when it is extracted from the heap and found to be colored as the processor's current color (Step 1.4 of Algorithm 3).

Case 1 If T_1 and T_2 are grown by the same processor p_i (also assume without loss of generality T_1 is grown before T_2 in iterations k_1 and k_2, respectively, with $k_1 < k_2$), and processor p_i chooses a unique color to color the vertices (Step 1.2 of Algorithm 3), then v is colored $k_1 p + i$ in T_1, and later colored again in T_2 with a different color $k_2 p + i$. Before coloring a vertex, each processor will first check whether it has already been colored (Step 1.1 of Algorithm 3); this means when processor p_i checks whether v has been colored, it does not see the previous coloring. This is a clear contradiction of sequential memory consistency.

Case 2 Assume that v is a vertex found in two trees T_1 and T_2 grown by two processors p_1 and p_2, respectively. We denote t_v as the time that v is colored. Suppose when v is added to T_1, it is connected to vertex v_1, and when it is added to T_2, it is connected to v_2. Since v is connected to v_1 and v_2, we have that $t_{v_1} < t_v$ and $t_{v_2} < t_v$. Also $t_v < t_{v_1}$ and $t_v < t_{v_2}$ since after adding v to T_1 we have not seen the coloring of v_2 yet, and similarly after adding v to T_2 we have not seen the coloring of v_1 yet. This is a contradiction of Step 1.4 in Algorithm 3, and hence, a vertex will not be included in more than one growing tree. ∎

Lemma 26.3.2 *No cycles are formed by the edges found in Algorithm 2.*

Proof. In Step 2 of Algorithm 2, each processor grows subtrees. Following Lemma 26.3.1, no cycles are formed among these trees. Step 5 of Algorithm 2 is the only other step that labels edges, and the edges found in this step do not form cycles among themselves (otherwise it is a direct contradiction of the

> **Input:** (1) p processors, each with processor ID p_i, (2) a partition of adjacency list for each processor
> (3) array *color* and *visited*
> **Output:** A spanning forest that is part of graph G's MST
> **begin**
>   1. **for** $v \leftarrow i\frac{n}{p}$ to $(i+1)\frac{n}{p} - 1$ **do**
>   1.1 **if** $color[v] \neq 0$ **then** v is already colored, continue
>   1.2 $my\_color = color[v] = v + 1$
>   1.3 insert v into heap H
>   1.4 **while** H *is not empty* **do**
> $w = heap\_extract\_min(H)$
>   **if** $(color[w] \neq my\_color)$ OR *(any neighbor u of w has* color *other than 0 or my_color)*
>   **then** break
>   **if** $visited[w] = 0$ **then**
> $visited[w] = 1$, and label the corresponding edge e as in MST
>   **for** *each neighbor u of w* **do**
>   **if** $color[u] = 0$ **then** $color[u] = my\_color$
>   **if** u *in heap H* **then** $heap\_decrease\_key(u, H)$
>   **else** $heap\_insert(u, H)$
> **end**

Algorithm 3: Parallel algorithm for new MST approach based on Prim's that finds parts of MST, for processor p_i, for $(0 \leq i \leq p-1)$. Assume w.l.o.g. that p divides n evenly.

correctness of Borůvka's algorithm). Also these edges do not form any cycles with the subtrees grown in Step 2. To see this, note that each of these edges has at least one end point that is not shared by any of the subtrees, so the subtrees can be treated as "vertices." Suppose l such edges and m subtrees form a cycle, we have l edges and $l + m$ vertices, which means $m = 0$. Similarly, edges found in Step 5 do not connect two subtrees together, but may increase the sizes of subtrees. ∎

Lemma 26.3.3 *Edges found in Algorithm 2 are in the MST.*

Proof. Consider a single iteration of Algorithm 2 on graph G. Assume after Step 5, we run parallel Borůvka's algorithm to get the MST for the reduced graph. Now we prove that for the spanning tree T we get from G, every edge e of G that is not in T is a T-heavy edge. Let us consider the following cases:

- Two end points of e are in two different subtrees. Obviously e is T-heavy because we run Borůvka's algorithm to get the MST of the reduced graph (in which each subtree is now a vertex).
- Two end points u, v of e are in the same subtree that is generated by Step 1.4. According to Prim's algorithm e is T-heavy.
- Two end points u, v of e are in the same subtree, u is in the part grown by Step 1.4 and v is in part grown by step 3. It is easy to prove that e has larger weight than all the weights of the edges along the path from u to v in T.
- Two end points u, v are in the the same subtree, both u and v are in parts generated by Step 5. Again e is T-heavy.

In summary, we have a spanning tree T, yet all the edges of G that are not in T are T-heavy, so T is a MST. ∎

Theorem 26.3.1 *For connected graph G, Algorithm 2 finds the MST of G.*

Proof. Theorem 26.3.1 follows by repeatedly applying Lemma 26.3.3. ∎

The algorithm as given may not keep all of the processors equally busy, since each may visit a different number of vertices during an iteration. We balance the load simply by using the work-stealing technique as follows. When a processor completes its partition of n/p vertices, an unfinished partition is randomly selected, and processing begins from a decreasing pointer that marks the end of the unprocessed list. It is theoretically possible that no edges are selected for the growing trees, and hence, no progress made during an iteration of the algorithm (although this case is highly unlikely in practice). For example, if the input contains n/p cycles, with cycle i defined as vertices $\left\{i\frac{n}{p}, (i+1)\frac{n}{p}, \ldots, (i+p-1)\frac{n}{p}\right\}$, for $0 \leq i < n/p$, and if the processors are perfectly synchronized, each vertex would be a singleton in its own mature tree. A practical solution that guarantees progress with high probability is to randomly reorder the vertex set, which can be done simply in parallel and without added asymptotic complexity [81].

26.3.4 Analysis

Our new parallel MST algorithm possesses an interesting feature: when run on one processor the algorithm behaves as Prim's, and on n processors becomes Borůvka's, and runs as a hybrid combination for $1 < p < n$, where p is the number of processors. In addition, our new algorithm is novel when compared with Borůvka's approach in the following ways:

1. Each of p processors in our algorithm finds for its starting vertex the smallest-weight edge, contracts that edge, and then finds the smallest-weight edge again for the contracted supervertex. We do not find all the smallest-weight edges for all vertices, synchronize, and then compact as in the parallel Borůvka's algorithm.
2. Our algorithm adapts for any number p of processors in a practical way for SMPs, where p is often much less than n, rather than in parallel implementations of Borůvka's approach that appear as PRAM emulations with p coarse-grained processors that emulate n virtual processors.

The performance of our new algorithm is dependent on its granularity n/p, for $1 \leq p \leq n$. The worst case is when the granularity is small, that is, a granularity of 1 when $p = n$ and the approach turns to Borůvka. Hence, the worst case complexities are similar to that of the parallel Borůvka variants analyzed previously. Yet, in practice, we expect our algorithm to perform better than parallel Borůvka's algorithm on sparse graphs because their lower connectivity implies that our algorithm behaves like p simultaneous copies of Prim's algorithm with some synchronization overhead.

26.3.5 Experimental Results

This section summarizes the experimental results of our implementations and compares our results with previous experimental results.

We implemented three sequential algorithms: Prim's algorithm with binary heap, Kruskal's algorithm with nonrecursive merge sort (which in our experiments has superior performance over qsort, GNU quicksort, and recursive merge sort for large inputs), and the $m \log m$ Borůvka's algorithm.

Previous studies such as Reference 20 compare their parallel implementations with sequential Borůvka (even though they report that sequential Borůvka is several times slower than other MST algorithms) and Kruskal's algorithm. We observe Prim's algorithm can be three times faster than Kruskal's algorithm for some inputs.

In our performance results we specify which sequential algorithm achieves the best result for the input and use this algorithm when determining parallel speedup. In our experimental studies, *Bor-EL*, *Bor-AL*, *Bor-ALM*, and *Bor-FAL*, are the parallel Borůvka variants using edge lists, adjacency lists, adjacency lists and our memory management, and flexible adjacency lists, respectively. **BC-MSF** is our new minimum spanning forest parallel algorithm.

Parallel Graph Algorithms for Multiprocessors

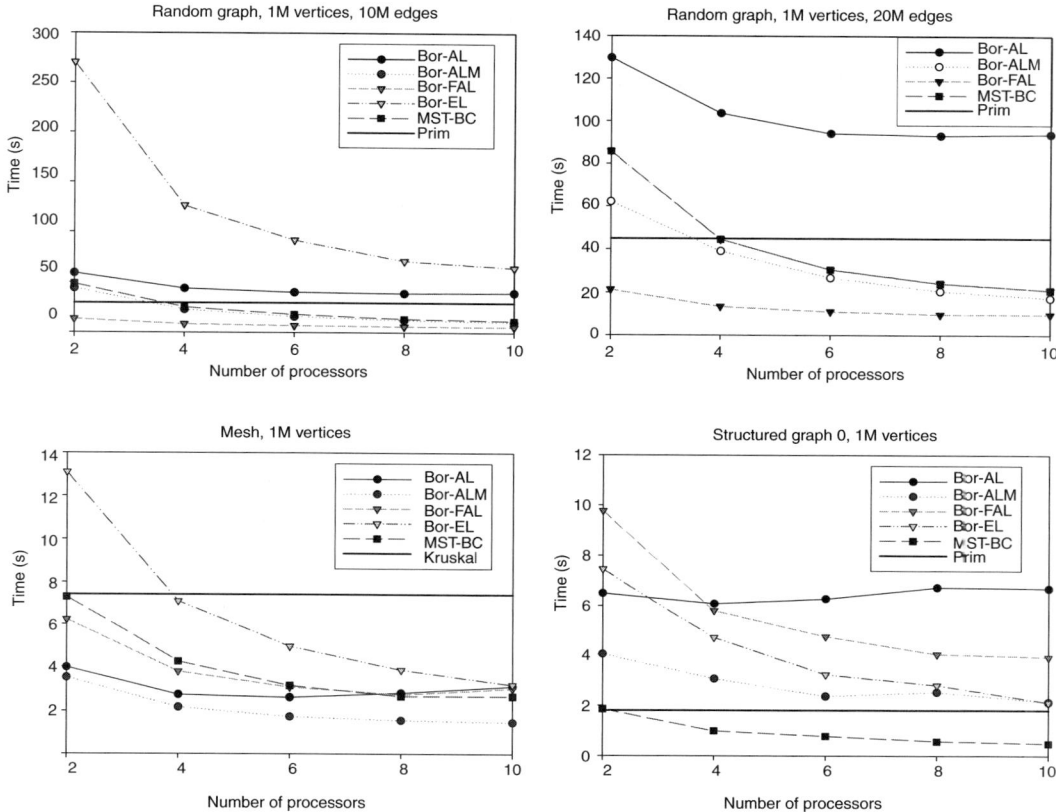

FIGURE 26.7 Comparison of parallel MST algorithms for random graph with $n = 1M$ vertices and $m = 10M$, and 20M, edges; for a regular mesh with $n = 1M$ vertices; and for a structured graph with 1M vertices.

The performance plots in Figure 26.7 are for the random graphs, a mesh and a structured graph. In these plots, the thick horizontal line represents the time taken for the best sequential MST algorithm (named in each legend) to find a solution on the same input graph using a single processor.

For the random, sparse graphs, we find that our Borůvka variant with flexible adjacency lists often has superior performance, with a speedup of approximately five using eight processors over the best sequential algorithm (Prim's in this case).

For the structured graph that is worst-cases for Borůvka algorithms, our new MST algorithm is the only approach that runs faster than the sequential algorithm.

26.4 Biconnected Components

A connected graph is said to be *separable* if there exists a vertex v such that removal of v results in two or more connected components of the graph. Given a connected, undirected graph $G = (V, E)$ with $|V| = n$ and $|E| = m$, the biconnected components problem finds the maximal induced subgraphs of G that are not *separable*. Tarjan [84] presents an optimal $O(n + m)$ algorithm that finds the biconnected components of a graph based on depth-first search (DFS). Eckstein [31] gave the first parallel algorithm that takes $O(d \log^2 n)$ time with $O((n + m)/d)$ processors on CREW PRAM, where d is the diameter of the graph. Savage and JáJá [82] designed two parallel algorithms on CREW PRAM. The first one takes $O(\log^2 n)$ time with $O(n^3 / \log n)$ processors. The second one is suitable for sparse graphs, and requires $O(\log^2 n \log k)$ time and $O(mn + n^2 \log n)$ processors where k is the number of biconnected components in the graph. Tsin and Chin [89] developed an algorithm on CREW PRAM that takes $O(\log^2 n)$ time with $O(n^2/\log^2 n)$ processors and is optimal for dense graphs. Tarjan and Vishkin [85] present an $O(\log n)$ time algorithm on CRCW PRAM that uses $O(n+m)$ processors. The fundamental Euler-tour technique is also introduced in Reference 85. Liang et al. [64] studied the biconnected components problems for graphs with special properties, for example, interval graphs, circular-arc graphs, and permutation graphs, and achieved better complexity bounds. There are also other biconnected components related studies, for example, finding the smallest augmentation to make a graph biconnected [48], and finding the smallest biconnected spanning subgraph (an NP-hard problem) [17,57].

Woo and Sahni [94] presented an experimental study of computing biconnected components on a hypercube for Tarjan and Vishkin's algorithm and Read's algorithm [79]. Their test cases are graphs that retain 70 and 90% edges of the complete graphs, and they achieved parallel efficiencies up to 0.7 for these dense inputs. The implementation uses adjacency matrix as input representation, and the size of the input graphs is limited to less than 2K vertices.

The Tarjan–Vishkin biconnected components algorithm builds upon many other parallel algorithms such as prefix sum, pointer jumping, list ranking, sorting, connected components, spanning tree, Euler-tour construction, and tree computations. Studies have demonstrated reasonable parallel speedups for these parallel primitives on SMPs [7–9,25,42]. It is not clear whether an implementation using these techniques achieves good speedup compared with the best sequential implementation because of the cost of parallel overheads. Here we focus on algorithmic overhead instead of communication and synchronization overhead. For example, Tarjan's sequential biconnected components algorithm [84] uses DFS with an auxiliary stack, while the Tarjan–Vishkin parallel algorithm (denoted as TV) employs all the parallel techniques mentioned earlier. Another factor that makes it hard to achieve good parallel speedups is the discrepancies among the input representations assumed by different primitives. TV finds a spanning tree, roots the tree, and performs various tree computations. Algorithms for finding spanning trees take edge list or adjacency list data structures as input representations, while rooting a tree and tree computations assume an Eulerian circuit for the tree that is derived from a circular adjacency list representation. Converting representations is not trivial, and incurs a real cost in implementations. In our studies, direct implementation of TV on SMPs does not outperform the sequential implementation even at 12 processors. In our optimized adaptation of TV onto SMPs, we follow the major steps of TV, yet we use different approaches for several of the steps. For example, we use a different spanning tree algorithm, a new approach to root the tree, construct the Euler-tour, and perform the tree computations. With new algorithm design and engineering techniques, our optimized adaptation of TV is able to achieve speedups upto 2.5 when employing 12 processors.

We also present a new algorithm that eliminates edges that are not essential in computing the biconnected components. For any input graph, edges are first eliminated before the computation of biconnected components is done so that at most $\min(m, 2n)$ edges are considered. Although applying the filtering algorithm does not improve the asymptotic complexity, in practice, the performance of the biconnected components algorithm can be significantly improved. In fact we achieve speedups upto 4 with 12 processors using the filtering technique. This is remarkable, given that the sequential algorithm runs in linear time with a very small hidden constant in the asymptotic complexity.

The rest of the section is organized as follows. Section 26.4.1 introduces TV; Section 26.4.2 discusses the implementation and optimization of TV on SMPs; Section 26.4.3 presents our new edge-filtering algorithm; Section 26.4.4 is analysis and performance results; and in Section 26.4.5 we give our conclusions.

26.4.1 The Tarjan–Vishkin Algorithm

First, we give a brief review of the Tarjan–Vishkin biconnected components algorithm. For an undirected, connected graph $G = (V, E)$ and a spanning tree T of G, each nontree edge introduces a simple cycle that itself is biconnected. If two cycles C_1 and C_2 share an edge, then $C_1 \cup C_2$ are biconnected. Let R_c be the relation that two cycles share an edge, then the transitive closure of R_c (denoted as R_c^*) partitions the

Parallel Graph Algorithms for Multiprocessors

graph into equivalence classes of biconnected components. If we are able to compute R_c, we can find all the biconnected components of graph G.

The size of R_c is too large ($O(n^2)$) even for sparse graphs where $m = O(n)$) to be usable in fast parallel algorithms. Tarjan and Vishkin defined a smaller relation R'_c with $|R'_c| = O(m)$ and proved that the transitive closure of R'_c is the same as that of R_c [51,85]. For any pair (e, g) of edges, $(e, g) \in R'$ (or simply denoted as $eR'_c g$) if and only if one of the following three conditions holds (denote the parent of a vertex u as $p(u)$, and the root of T as r):

1. $e = (u, p(u))$ and $g = (u, v)$ in $G - T$, and $v < u$ in preorder numbering
2. $e = (u, p(u))$ and $g = (v, p(v))$, and (u, v) in $G - T$ such that u and v are not related (having no ancestral relationships)
3. $e = (u, p(u))$, $v \neq r$, and $g = (v, p(v))$, and some nontree edge of G joins a descendant of u to a nondescendant of v

Once R'_c is computed, TV builds an auxiliary graph $G' = (V', E')$ where V' is the set E of edges of G, and $(e, g) \in E'$ if $eR'_c g$. The connected components of G' correspond to the equivalence classes of R'^*_c and identify the biconnected components of G.

TV has six steps:

1. *Spanning-tree* computes a spanning tree T for the input graph G. A spanning tree algorithm derived from the Shiloach–Vishkin's connected components algorithm [83] is used.
2. *Euler-tour* constructs an Eulerian circuit for T.
3. *Root-tree* roots T at an arbitrary vertex by applying the Euler-tour technique on the circuit obtained in the previous step.
4. *Low-high* computes two values low(v) and high(v) for each vertex v. The value low(v) denotes the smallest vertex (in preorder numbering), that is, either a descendant of v or adjacent to a descendant of v by a nontree edge. Similarly, high(v) denotes the largest vertex (in preorder numbering), that is, either a descendant of v or adjacent to a descendant of v by a nontree edge.
5. *Label-edge* tests the appropriate conditions of R'_c and builds the auxiliary graph G' using the low, high values.
6. *Connected-components* finds the connected components of G' with the Shiloach–Vishkin connected components algorithm.

TV takes an edge list as input. The parallel implementations of the six steps within the complexity bound of $O(\log n)$ time and $O(m)$ processors on CRCW PRAM are straightforward except for the *Label-edge* step. Tarjan and Vishkin claim that the *Label-edge* step takes constant time with $O(m)$ processors because the conditions for R'_c can be tested within these bounds. Note that if two edges e and g satisfy one of the conditions for R'_c, mapping $(e, g) \in R'_c$ into an edge in $G' = (V', E')$ is not straightforward because no information is available about which vertices e and g are mapped to in V'. Take condition 1 for example. For each nontree edge $g_1 = (u, v) \in E$, if $u < v$ and let $g_2 = (u, p(u))$, (g_1, g_2) maps to an edge in E'. If we map each edge $e \in E$ into a vertex $v' \in V'$ whose number is the location of e in the edge list, we need to search for the location of g_2 in the edge list.

Here, we present an algorithm for this missing step in TV that builds the auxiliary graph in $O(\log m)$ time with $O(m)$ processors, which does not violate the claimed overall complexity bounds of TV. The basic idea of the algorithm is as follows. Assume, w.l.o.g., $V = [1, n]$ (we use $[a, b]$ denote the integer interval between a and b) and $V' = [1, m]$. We map each tree edge $(u, p(u)) \in E$ to vertex $u \in V'$. For each nontree edge e, we assign a distinct integer n_e between $[0, m - n]$, and map e to vertex $n_e + n \in V'$. Assigning numbers to nontree edges can be done by a prefix sum. The formal description of the algorithm is shown in Algorithm 4.

We prove Algorithm 4 builds an auxiliary graph within the complexity bound of TV.

Theorem 26.4.1 *Algorithm 4 builds an auxiliary graph $G' = (V', E')$ in $O(\log m)$ time with $O(m)$ processors and $O(m)$ space on EREW PRAM.*

```
Input: L: an edge list representation for graph G = (V, E) where |V| = n and |E| = m
Preorder: preorder numbering for the vertices
Output: X: an edge list representation of the auxiliary graph
begin
    for 0 ≤ i ≤ m − 1 parallel do
        if L[i].is_tree_edge=true then N[i] ← 1;
        else N[i] ← 0;

    prefix-sum(N,m);
    for 0 ≤ i ≤ m − 1 parallel do
        u=L[i].v1; v=L[i].v2;
        if L[i].is_tree_edge=true then
            if Preorder[v]<Preorder[u] then L'[i] ← (u,N[i] + n);
            if u and v not related then L'[m + i] ← (u,v);
        else
            if u ≠ root and v ≠ root then L'[2m − i] ← (u,v);
    compact L' into X using prefix-sum;
end
```

Algorithm 4: Building the auxiliary graph.

Proof. According to the mapping scheme, $V' = [1, m + n]$. Each tree edge $L[i] = (u, p(u))$ is uniquely mapped to $u \in V'$. For each nontree edge $L[j]$, a unique number $N[j] \in [1, m]$ is assigned. Nontree edge $L[j]$ is mapped to $N[j]+n \in V'$ so that it is not mapped to a vertex number assigned to a tree edge and no two nontree edges share the same vertex number. It is easy to verify that this is a one-to-one mapping from E to V' and can be done in $O(\log m)$ time with $O(m)$ processors. As for E', testing the conditions, that is, finding all the edges in E', can be done in constant time with $O(m)$ processors.

The tricky part is to determine where to store the edge information each time we add a new edge e' (image of (e, g) where $e, g \in E$) to E'. The easy way is to use an $(n+m) \times (n+m)$ matrix so that each edge of E' has a unique location to go. If we inspect the conditions for R'_c closely, we see that for each condition we add at most m edges to the edge list. L' is a temporary structure that has $3m$ locations. Locations $[0, m − 1]$, $[m, 2m − 1]$, and $[2m, 3m − 1]$ are allocated for condition 1, 2, and 3, respectively. After all the edges are discovered, L' is compacted into X using prefix sum. Prefix sums dominate the running time of Algorithm 4, and no concurrent reads or writes are required. So Algorithm 4 builds X (the auxiliary graph) in $O(\log m)$ time with $O(m)$ processors and $O(m)$ space on EREW PRAM. ∎

26.4.2 Implementation of the Tarjan–Vishkin Algorithm

We adapt TV on SMPs (*TV-SMP*) to serve as a baseline implementation for comparison with our new algorithm. *TV-SMP* emulates TV in a coarse-grained fashion by scaling down the parallelism of TV to the number of processors available from an SMP. The emulation of each step is straightforward except for the *Euler-tour* step. In the literature, the Euler-tour technique usually assumes a circular adjacency list as input where there are cross pointers between the two antiparallel arcs (u, v) and (v, u) of an edge $e = (u, v)$ in the edge list. For the tree edges found in the *Spanning-tree* step, such a circular adjacency list has to be constructed on the fly. The major task is to find for an arc (u, v) the location of its antiparallel twin, (v, u). After selecting the spanning tree edges, we sort all the arcs (u, v) with $\min(u, v)$ as the primary key, and $\max(u, v)$ as the secondary key. The arcs (u, v) and (v, u) are then next to each other in the resulting list so that the cross pointers can be easily set. We use the efficient parallel sample sorting routine designed by Helman and JáJá [41]. Our experimental study shows that the parallel overhead of *TV-SMP* is too much for the implementation to achieve parallel speedup with a moderate number of processors. The performance results are given in Section 26.4.4.

We apply algorithm engineering techniques to reduce the parallel overhead of TV to run on SMPs. The major optimization is to merge *Spanning-tree* and *Root-tree* into one single step owing to the observation that when a spanning tree is being computed, we can also root the tree on the fly. For example, the new spanning tree algorithm presented in Section 26.2 also computes a rooted tree with minimal overhead. We denote this implementation as *TV-opt*.

26.4.3 A New Algorithm to Reduce Overheads

The motivation to further improve TV comes from the following observation for many graphs: not all nontree edges are necessary for maintaining the biconnectivity of the biconnected components. We say an edge e is *nonessential* for biconnectivity if removing e does not change the biconnectivity of the component it belongs to. Filtering out *nonessential* edges when computing biconnected components (of course we will place these edges back in later) may produce performance advantages. Recall that TV is all about finding $R_c'^*$. Of the three conditions for R_c', it is trivial to check for condition 1 which is for a tree edge and a nontree edge. Conditions 2 and 3, however, are for two tree edges and checking involves the computation of *high* and *low* values. To compute *high* and *low*, we need to inspect every nontree edge of the graph, which is very time consuming when the graph is not extremely sparse. The fewer edges the graph has, the faster the *low-high* step. Also when we build the auxiliary graph, the fewer edges in the original graph means the smaller the auxiliary graph and the faster the *Label-edge* and *Connected-components* steps.

Take Figure 26.8 for example. On the left in Figure 26.8 is a biconnected graph G_1. After we remove nontree edges e_1 and e_2, we get a graph G_2 shown on the right in Figure 26.8, which is still biconnected. G_1 has a R_c' relation of size 11 (4, 4, and 3 for conditions 1, 2, and 3, respectively), while graph G_2 has a R_c' relation of size 7 (2, 2, and 3 for conditions 1, 2, and 3, respectively). So the auxiliary graph of G_1 has 10 vertices and 11 edges, while the auxiliary graph for G_2 has only 8 vertices and 7 edges. When there are many *nonessential* edges, filtering can greatly speed up the computation.

Now the questions are how many edges can be filtered out and how to identify the *nonessential* edges for a graph $G = (V, E)$ with a spanning tree T. We postpone the discussion of the first question until later in this section because it is dependent on how filtering is done. First, we present an algorithm for identifying *nonessential* edges. The basic idea is to compute a spanning forest F for $G-T$. It turns out that if T is a breadth-first search (BFS) tree, the nontree edges of G that are not in F can be filtered out.

Assuming T is a BFS tree, next we prove several lemmas.

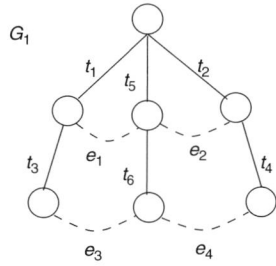
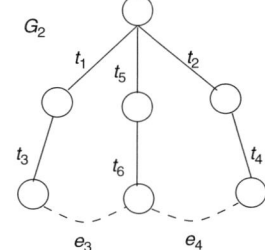

FIGURE 26.8 Two graphs G_1 and G_2. The solid edges are tree edges and the dashed edges are nontree edges. G_2 is derived from G_1 by removing *nonessential* nontree edges e_1 and e_2. Below the graphs are the corresponding R_c' relations defined by the three conditions.

Lemma 26.4.1 *For any edge $e = (u, v)$ in F, there is no ancestral relationship between u and v in T.*

Proof. Clearly u and v cannot be the parent of each other as e is not in T. Suppose w.l.o.g. that u is an ancestor of v, and w is the parent of v ($w \neq u$), considering the fact that T is a BFS tree, v is at most one level away from u and w is at least one level away from u. So w cannot be v's parent, and we get a contradiction. ∎

Lemma 26.4.2 *Each connected component of F is in some biconnected component of graph G.*

Proof. Let C be a connected component of F. Note that C itself is also a tree. Each edge in C is a nontree edge to T, and is in a simple cycle, hence some biconnected component, of G. We show by induction that the simple cycles determined by the edges of C form one biconnected component.

Starting with an empty set of edges, we consider the process of growing C by adding one edge at each step and keeping C connected. Suppose there are k edges in C, and the sequence in which they are added is $e_1, e_2, \ldots, e_k$.

As e_1 is a nontree edge to T, e_1 and the paths from its two end points to the lowest common ancestor (lca) of the two end points form a simple cycle. And e_1 is in a biconnected component of G.

Suppose the first l edges in the sequence are in one biconnected component Bc. We now consider adding the $(l + 1)$th edge. As C is connected, $e_{l+1} = (u, w)$ is adjacent to some edge, say, $e_s = (v, w)$ (where $1 \leq s \leq l$) in the tree we have grown so far at vertex w. By Lemma 26.4.1 there are no ancestral relationships between u and w, and v and w in tree T. If there is also no ancestral relationship between u and w as illustrated in part (a) of Figure 26.9, then the paths in T from u to $lca(u, v)$ and from v to $lca(u, v)$ plus the edges (u, w) and (v, w) in C form a simple cycle S. As (v, w) is in Bc and (u, w) is in S, and Bc shares with S the edge (v, w), so (u, w) and (v, w) are both in the biconnected component that contains $Bc \cup S$. If there is some ancestral relationship between u and v, then there are two cases: either

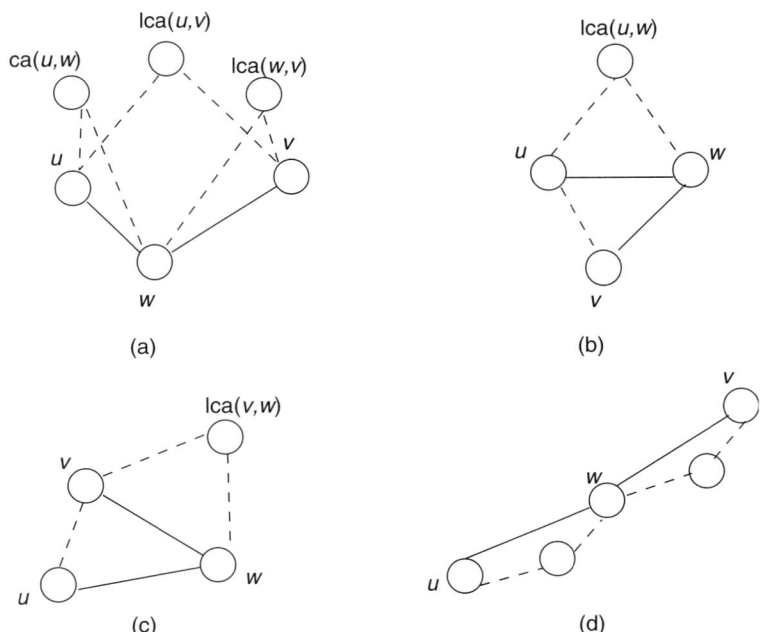

FIGURE 26.9 Illustration of the proof of Theorem 26.4.2. $e_{l+1} = (u, w)$, and $e_s = (v, w)$. The dotted lines are the paths in T while the solid lines are edges in C.

u is the ancestor of v or v is the ancestor of u. These two cases are illustrated respectively by parts (b) and (c) in Figure 26.9. Let's first consider the case that u is the ancestor of v. The paths in T from u to $lca(u, w)$, from w to $lca(u, w)$, and from v to u, and edge (v, w) form a simple cycle S. S shares with B_c edge (v, w), again (u, w) and (v, w) are both in the biconnected component that includes $B_c \cup S$. Similarly, we can prove (u, w) and (v, w) are in one biconnected component for the case that v is the ancestor of u. By induction, it follows that all edges of C are in one biconnected component. ∎

Figure 26.9d shows an example that (u, w) and (v, w) are not in one biconnected component if T is not a BFS tree and there are ancestral relationships between u, v, and w.

Theorem 26.4.2 *The edges of each connected component of $G-T$ are in one biconnected component.*

Proof. Let C be a connected component of $G-T$. If C is a tree, by Lemma 26.4.2, all edges of C are in a biconnected component. If C is not a tree, then there exits a spanning tree T_C of C. All edges of T_C are in a biconnected component by Lemma 26.4.2. Each nontree edge e (relative to T_C) in C forms a simple cycle with paths in T_C, and the cycle shares the paths with the biconnected component that T_C is in, so e is also in the biconnected component. ∎

The immediate corollary to Theorem 26.4.2 is that we can compute the number of biconnected components in a graph using breadth-first traversal. The first run of BFS computes a rooted spanning tree T. The second run of BFS computes a spanning forest F for $G-T$, and the number of components in F is the number of biconnected components in G.

Next, we apply the idea of filtering out *nonessential* edges to the parallel biconnected components problem. First T and F for G are computed. Then biconnected components for $T \cup F$ are computed using a suitable biconnected components algorithm, for example, the Tarjan-Vishkin algorithm. We are yet to find the biconnected components to which they belong for all the edges that are filtered out, that is, edges in $G - (T \cup F)$. According to condition 1 (which holds for arbitrary rooted spanning tree), edge $e = (u, v) \in G - (T \cup F)$ is in the same biconnected component of $(u, p(u))$ if $v < u$. A new algorithm using the filtering technique is shown in Algorithm 5.

Input: A connected graph $G = (V, E)$
Output: Biconnected components of G
begin
    1. compute a breadth-first search tree T for G;
    2. for $G-T$, compute a spanning forest F;
    3. invoke TV on $F \cup T$;
    4. **for** *each edge* $e = (u, v) \in G - (F \cup T)$ **do**
        label e to be in the biconnected component that contains $(u, p(u))$;
end

Algorithm 5: An improved algorithm for biconnected components.

In Algorithm 5, Step 1 takes $O(d)$ time with $O(n)$ processors on arbitrary CRCW PRAM where d is the diameter of the graph; Step 2 can be done in $O(\log n)$ time with $O(n)$ processors on arbitrary CRCW PRAM [85]; Step 3 is the Tarjan–Vishkin algorithm that can be done in $O(\log n)$ time with $O(n)$ processors; finally, Step 4 can be implemented in $O(1)$ time with $O(n)$ processors. So Algorithm 5 runs in $max(O(d), O(\log n))$ time with $O(n)$ processors on CRCW PRAM.

Asymptotically, the new algorithm does not improve the complexity bound of TV. In practice, however, Step 2 filters out at least $max(m - 2(n - 1), 0)$ edges. The denser the graph becomes, the more edges are filtered out. This can greatly speed up the execution of Step 3. Recall that TV inspects each nontree edge to compute the *low* and *high* values for the vertices, and builds an auxiliary graph with the number of vertices equal to the number of edges in G. In Section 26.4.4, we demonstrate the efficiency of this edge filtering technique.

For very sparse graphs, d can be greater than $O(\log n)$ and becomes the dominating factor in the running time of the algorithm. One extreme pathological case is that G is a chain ($d = O(n)$), and computing the BFS tree takes $O(n)$ time. However, pathological cases are rare. Palmer [74] proved that almost all random graphs have diameter two. And even if $d > \log n$, in many cases, as long as the number of vertices in the BFS frontier is greater than the number of processors employed, the algorithm will perform well on a machine with p processors ($p \ll n$) with expected running time of $O\left(\frac{n+m}{p}\right)$. Finally, if $m \leq 4n$, we can always fall back to *TV-opt*.

26.4.4 Performance Results and Analysis

This section summarizes the experimental results of our implementation. We test our implementation on arbitrary, sparse inputs that are the most challenging instances for previous experimental studies. We create a random graph of n vertices and m edges by randomly adding m unique edges to the vertex set. The sequential implementation implements Tarjan's algorithm.

Figure 26.10 shows the performance of *TV-SMP*, *TV-opt*, and *TV-filter* on random graphs of $1M$ vertices with various edge densities. For all the instances, TV does not beat the best sequential implementation

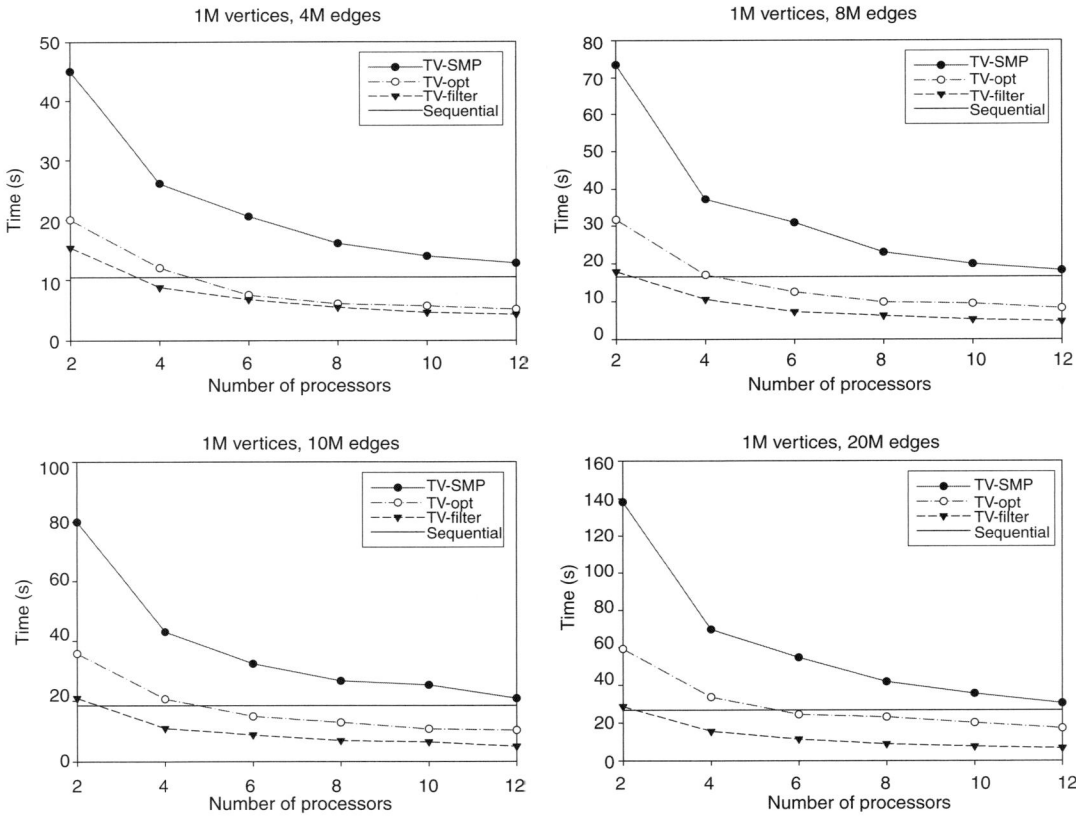

FIGURE 26.10 Comparison of the performance of *TV-SMP*, *TV-opt*, and *TV-filter* for graphs with $n = 1M$ vertices and various edge densities. "Sequential" is the time taken for the implementation of Tarjan's biconnected components algorithm for the same problem instance.

even at 12 processors. *TV-opt* takes roughly half the execution time of TV. As predicted by our analysis earlier, the denser the graph, the better the performance of *TV-filter* compared with *TV-opt*. For the instance with 1M vertices, 20M edges ($m = n \log n$), *TV-filter* is twice as fast as *TV-opt* and achieves speedups up to 4 compared with the best sequential implementation.

Figure 26.11 shows the breakdown of execution time for different parts of the algorithm for *TV-SMP*, *TV-opt*, and *TV-filter*. Comparing *TV-SMP* and *TV-opt*, we see that *TV-SMP* takes much more time to compute a spanning tree and constructing the Euler-tour than *TV-opt* does. Also for tree computations, *TV-opt* is much faster than TV because in *TV-opt* prefix sum is used whereas in *TV-SMP* list ranking is used. For the rest of the computations, *TV-SMP* and *TV-opt* take roughly the same amount of time. Compared with *TV-opt*, *TV-filter* has an extra step, that is, filtering out *nonessential* edges. The extra cost of filtering out edges is worthwhile if the graph is not extremely sparse. As our analysis predicted in Section 26.4.3, we expect reduced execution time for *TV-filter* in computing low-high values, labeling and computing connected components. Figure 26.11 confirms our analysis.

26.4.5 Discussion

Although in theory *TV-filter* is not faster than TV, we see the effect of algorithmic optimizations on performance. As quite a few fundamental parallel primitives and routines such as prefix sum, list ranking, Euler-tour construction, tree computation, connectivity, and spanning tree are employed as building blocks, our study shows promising results for parallel algorithms that take a drastically different approach than the straightforward sequential approach.

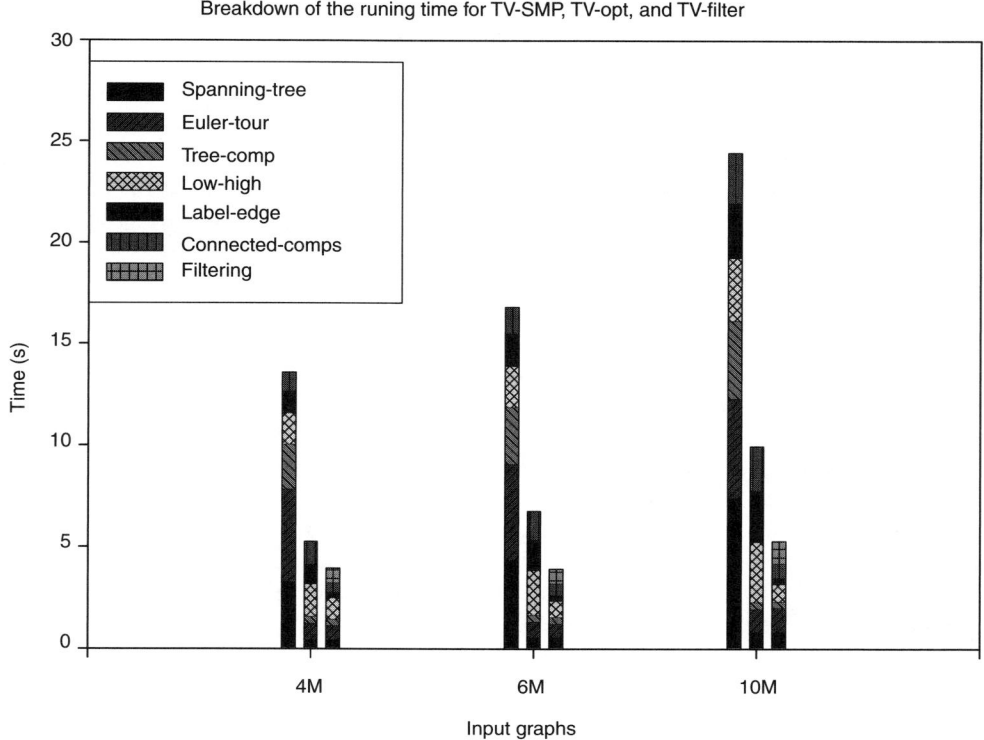

FIGURE 26.11 Breakdown of execution time at 12 processors for the *Spanning-tree*, *Euler-tour*, *Root-tree*, *Low-high*, *Label-edge*, *Connected-components*, and *Filtering* steps. All graphs have 1M vertices with different number of edges shown on the x-axis. The three columns for each input graph, from left to right, are the execution times for *TV-SMP*, *TV-opt*, and *TV-filter*, respectively.

26.5 Parallel Algorithms with Mutual Exclusion and Lock-Free Protocols

Irregular problems, such as graph problems with arbitrary, sparse instances, are challenging to parallelize and to achieve high performance because typically their memory access patterns or execution paths are not predictable a priori, and straightforward data decompositions or load balancing techniques, such as those used for regular problems, often are not efficient for these applications. Fine-grained synchronization, a technique for managing the coordination of work within an irregular application, can be implemented through lock-free protocols, system mutex locks, and spinlocks. However, fine-grained locks and lock-free protocols are seldom employed in implementations of parallel algorithms.

Most theoretic parallel algorithmic models are either synchronous (e.g., PRAM [51]) or for network-based systems (e.g., LogP [26] and BSP [90]) with no explicit support for fine-grained synchronization. In these models, coarse synchronization is performed through a variety of mechanisms such as lock-step operation (as in PRAM), algorithm supersteps (as in BSP), and collective synchronization primitives such as barriers (as in LogP), rather than fine-grained coordination of accesses to shared data structures. In practice, the performance of parallel algorithms that use locks and lock-free protocols are highly dependent on the parallel computer architecture and the contention among processors to shared regions.

System mutex locks, widely used for interprocess synchronization due to their simple programming abstractions, provide a common interface for synchronization, and the performance depends on the implementation and the application scenario. User-defined spinlocks are customizable; however, the disadvantages are that the user is exposed to low-level hardware details, and portability can be an issue. For large-scale application of locks in a high-performance computing environment, spinlocks have the advantage of economic memory usage and simple instruction sequences. Each spinlock can be implemented using one memory word, while a system mutex lock can take multiple words for its auxiliary data structures, which exacerbates the problem with accessing memory.

Mutual exclusion locks have an inherent weakness in a (possibly heterogeneous and faulty) distributed computing environment; that is, the crashing or delay of a process in a critical section can cause deadlock or serious performance degradation of the system [43,65]. Lock-free data structures (sometimes called concurrent objects) were proposed to allow concurrent accesses of parallel processes (or threads) while avoiding the problems of locks. In theory we can coordinate any number of processors through lock-free protocols. In practice, however, lock-free data structures are primarily used for fault-tolerance.

In this section, we investigate the performance of fine-grained synchronization on irregular parallel graph algorithms using shared-memory multiprocessors. In Sections 26.2 and 26.3, we presented fast parallel algorithms for the spanning tree and MST problems, and demonstrated speedups compared with the best sequential implementation. The algorithms achieve good speedups due to algorithmic techniques for efficient design and better cache performance. We have excluded certain design choices that involve fine-grained synchronizations. This section investigates these design choices with lock-free protocols and mutual exclusion. Our main results include novel applications of fine-grained synchronization where the performance beats the best previously known parallel implementations.

The rest of the section is organized as follows: Section 26.5.1 presents lock-free parallel algorithms with an example of lock-free spanning tree algorithm; Section 26.5.3 presents parallel algorithms with fine-grained locks; and Section 26.5.4 compares the performance of algorithms with fine-grained synchronizations with earlier implementations.

26.5.1 Lock-Free Parallel Algorithms

Early work on lock-free data structures focused on theoretical issues of the synchronization protocols, for example, the power of various atomic primitives and impossibility results [5,19,29,30,32,37], by considering the simple *consensus problem* where n processes with independent inputs communicate through a set of shared variables and eventually agree on a common value. Herlihy [44] unified much of the earlier theoretic results by introducing the notion of *consensus number* of an object and defining a hierarchy on the concurrent objects according to their consensus numbers. Consensus number measures the relative power of an object to reach distributed consensus, and is the maximum number of processes for which the object can solve the consensus problem. It is impossible to construct lock-free implementations of many simple and useful data types using any combination of atomic *read, write, test&set, fetch&add,* and *memory-to-register swap,* because these primitives have consensus numbers either one or two. On the other hand, *compare&swap* and *load-linked, store-conditional* have consensus numbers of infinity, and hence are *universal* meaning that they can be used to solve the consensus problem of any number of processes. Lock-free algorithms and protocols are proposed for many commonly used data structures, for example, linked lists [92], queues [46,62,88], set [63], union-find sets [4], heaps [10], and binary search trees [33,91]; and also for the performance improvement of lock-free protocols [2, 10]. While lock-free data structures and algorithms are highly resilient to failures, unfortunately, they seem to come at a cost of degraded performance. Herlihy et al. studied practical issues and architectural support of implementing lock-free data structures [43, 45], and their experiments with small priority queues show that lock-free implementations do not perform as well as lock-based implementations.

26.5.2 Lock-Free Protocols for Resolving Races among Processors

A parallel algorithm often divides into phases and in each phase certain operations are applied to the input with each processor working on portions of the data structure. For irregular problems there usually are overlaps among the portions of data structures partitioned onto different processors. Locks provide a mechanism for ensuring mutually exclusive access to critical sections by multiple working processors. Fine-grained locking on the data structure using system mutex locks can bring large memory overhead. What is worse is that many of the locks are never acquired by more than one processor. Most of the time each processor is working on distinct elements of the data structure owing to the large problem size and relatively small number of processors. Yet still extra work of locking and unlocking is performed for each operation applied to the data structure, which may result in a large execution overhead depending on the implementation of locks.

Here, we show that lock-free protocols through atomic machine operations are an elegant solution to the problem. When there is work partition overlap among processors, it suffices that the overlap is taken care of by one processor. If other processors can detect that the overlap portion is already taken care of, they no longer need to apply the operations and can abort. Atomic operations can be used to implement this "test-and-work" operation. As the contention among processors is low, we expect the overhead of using atomic operations to be small. Note that this is very different from the access patterns to the shared data structures in distributed computing; for example, two producers attempting to put more work into the shared queues. Both producers must complete their operations, and when there is conflict they will retry until success.

To illustrate this point in a concrete manner, we consider the application of lock-free protocols to the Shiloach–Vishkin parallel spanning tree algorithm (see Section 26.2).

Recall that with SV, grafting and shortcutting are iteratively applied to the graph until only one supervertex is left. For a certain vertex v with multiple adjacent edges, there can be multiple processors attempting to graft v to other smaller-labeled vertices. Yet only one grafting is allowed, and we label the corresponding edge that causes the grafting as a spanning tree edge. This is a partition conflict problem.

Two-phase election is one method that can be used to resolve the conflicts. The strategy is to run a race among processors, where each processor that attempts to work on a vertex v writes its processor id into a tag associated with v. After all the processors are done, each processor checks the tag to see whether it is the winning processor. If so, the processor continues with its operation, otherwise it aborts. A global barrier synchronization among processors is used instead of a possibly large number of fine-grained locks. The disadvantage is that two runs are involved.

Another more natural solution to the work partition problem is to use lock-free atomic instructions. When a processor attempts to graft vertex v, it invokes the atomic *compare&swap* operation to check on whether v has been inspected. If not, the atomic nature of the operation also ensures that other processors will not work on v again. The detailed description of the algorithm is shown in Algorithm 6, and the implementations of inline assembly functions for *compare&swap* are architecture dependent.

Data : (1) EdgeList[1...2m]: edge list representation for graph $G = (V, E), |V| = n, |E| = m$; each element of EdgeList has two field, v_1 and v_2 for the two end points of an edge
(2) integer array D[1...n] with D[i] = i
(3) integer array Flag[1...n] with Flag[i] = 0
Result: a sequence of edges that are in the spanning tree
begin
   $n' = n$
   while $n' \neq 1$ **do**
      for $k \leftarrow 1$ *to* n' **in parallel do**
         $i =$ EdgeList[k].v_1
         $j =$ EdgeList[k].v_2
         if D[j] < D[i] *and* D[i] = D[D[i]] *and compare&swap*(&Flag[D[i]], 0, *PID*) = 0 **then**
            label edge EdgeList[k] to be in the spanning tree
            D[D[i]] = D[j]

      for $i \leftarrow 1$ *to* n' **in parallel do**
         while D[i] $\neq$ D[D[i]] **do**
            D[i] = D[D[i]]

      $n' =$ the number of super-vertices
end

Algorithm 6: Parallel lock-free spanning tree algorithm (**span-lockfree**).

26.5.3 Parallel Algorithms with Fine-Grained Mutual Exclusion Locks

Mutual exclusion provides an intuitive way for coordinating synchronization in a parallel program. For example, in the spanning algorithm in Section 26.5.2, we can also employ mutual exclusion locks to resolve races among processors. Before a processor grafts a subtree that is protected by critical sections, it first gains access to the data structure by acquiring locks, which guarantees that a subtree is only grafted once.

To illustrate the use of mutex locks, in this section we present a new implementation of the MST problem based on parallel Borůvka's algorithm that outperforms all previous implementations. We next introduce parallel Borůvka's algorithm and previous experimental results.

In Section 26.3, we studied the performance of different variations of parallel Borůvka's algorithm. The major difference among them is the input data structure and the implementation of *compact-graph*. Bor-FAL takes our *flexible adjacency list* as input and runs parallel sample sort on the vertices to compact the graph. For most inputs, Bor-FAL is the fastest implementation. In the *compact-graph* step, Bor-FAL merges each connected components into a single supervertex that combines the adjacency list of all the vertices in the component. Bor-FAL does not attempt to remove self-loops and multiple edges, and avoids runs of extensive sortings. Instead, self-loops and multiple edges are filtered out in the *find-min* step. Bor-FAL greatly reduces the number of shared memory writes at the relatively small cost of an increased number of reads, and proves to be efficient as predicted on current SMPs.

Now we present an implementation with fine-grained locks that further reduces the number of memory writes. In fact the input edge list is not modified at all in the new implementation, and the *compact-graph* step is completely eliminated. The main idea is that instead of compacting connected components, for each vertex there is now an associated label *supervertex* showing to which supervertex it belongs. In each

Parallel Graph Algorithms for Multiprocessors

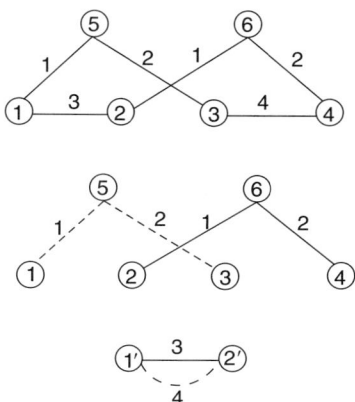

FIGURE 26.12 Example of the race condition between two processors when Borůvka's algorithm is used to solve the MST problem.

iteration all the vertices are partitioned as evenly as possible among the processors. For each vertex v of its assigned partition, processor p finds the adjacent edge e with the smallest weight. If we compact connected components, e would belong to the supervertex v' of v in the new graph. Essentially, processor p finds the adjacent edge with smallest weight for v'. As we do not compact graphs, the adjacent edges for v' are scattered among the adjacent edges of all vertices that share the same supervertex v', and different processors may work on these edges simultaneously. Now the problem is that these processors need to synchronize properly in order to find the edge with the minimum weight. Again, this is an example of the irregular work-partition problem. Figure 26.12 illustrates the specific problem for the MST case.

On the top in Figure 26.12 is an input graph with six vertices. Suppose we have two processors P_1 and P_2. Vertices 1, 2, and 3, are partitioned on to processor P_1 and vertices 4, 5, and 6 are partitioned on to processor P_2. It takes two iterations for Borůvka's algorithm to find the MST. In the first iteration, the *find-min* step labels $\langle 1, 5 \rangle$, $\langle 5, 3 \rangle$, $\langle 2, 6 \rangle$, and $\langle 6, 4 \rangle$, to be in the MST. *connected-components* finds vertices 1, 3, and 5, in one component, and vertices 2, 4, and 6, in another component. The MST edges and components are shown in the middle of Figure 26.12. Vertices connected by dashed lines are in one component, and vertices connected by solid lines are in the other component. At this time, vertices 1, 3, and 5, belong to supervertex $1'$, and vertices 2, 4, and 6, belong to supervertex $2'$. In the second iteration, processor P_1 again inspects vertices 1, 2, and 3, and processor P_2 inspects vertices 4, 5, and 6. Previous MST edges $\langle 1, 5 \rangle$, $\langle 5, 3 \rangle$, $\langle 2, 6 \rangle$, and $\langle 6, 4 \rangle$ are found to be edges inside supervertices and are ignored. On the bottom of Figure 26.12 are the two supervertices with two edges between them. Edges $\langle 1, 2 \rangle$ and $\langle 3, 4 \rangle$ are found by P_1 to be the edges between supervertices $1'$ and $2'$, edge $\langle 3, 4 \rangle$ is found by P_2 to be the edge between the two supervertices. For supervertex $2'$, P_1 tries to label $\langle 1, 2 \rangle$ as the MST edge while P_2 tries to label $\langle 3, 4 \rangle$. This is a race condition between the two processors, and locks are used in to ensure correctness. The formal description of the algorithm is given in Algorithm 7. Note that Algorithm 7 describes the parallel MST algorithm with generic locks. The locks in the algorithm can be either replaced by system mutex locks or spinlocks.

Depending on which types of locks are used, we have two implementations, *Bor-spinlock* with spinlocks and *Bor-lock* with system mutex locks. We compare their performance with the best previous parallel implementations in Section 26.5.4.

26.5.4 Experimental Results

We ran our shared-memory implementations on two platforms as implementations of fine-grained synchronization are architecture dependent. In addition to the Sun Enterprise 4500 (E4500), we also consider IBM pSeries 570 (p570). The IBM p570 has 16 IBM Power5 processors and 32 GB of memory, with 32

Data : (1) graph $G = (V, E)$ with adjacency list representation, $|V| = n$
  (2) array $D[1 \ldots n]$ with $D[i] = i$
  (3) array $Min[1 \ldots n]$ with $Min[i] =$ **MAXINT**
  (4) array $Graft[1 \ldots n]$ with $Graft[i] = 0$
Result: array E_{MST} of size $n - 1$ with each element being a MST tree edge
begin
  while *not all D[i] have the same value* **do**
    for $i \leftarrow 1$ *to n* **in parallel do**
      for *each neighbor j of vertex i* **do**
        if $D[i] \neq D[j]$ **then**
          lock($D[i]$)
          if $Min[D[i]] < w(i, j)$ **then**
            $Min[D[i]] \leftarrow w(i, j)$
            $Graft[D[i]] \leftarrow D[j]$
            Record/update edge $e = \langle i, j \rangle$ with the minimum weight
          unlock($D[i]$)

    for $i \leftarrow 1$ *to n* **in parallel do**
      if $Graft[i] \neq 0$ **then**
        $D[i] \leftarrow Graft[i]$
        $Graft[i] \leftarrow 0$
        $Min[i] \leftarrow$ **MAXINT**
        Retrieve the edge e that caused the grafting
        Append e to the array E_{MST}

    for $i \leftarrow 1$ *to n* **in parallel do**
      while $D[i] \neq D[D[i]]$ **do**
        $D[i] \leftarrow D[D[i]]$

end

Algorithm 7: Parallel Borůvka minimum spanning tree algorithm.

Kbytes L1 data cache, 1.92 MB L2 cache. There is a L3 cache with 36 MB per two processors. The processor clock speed is 1.9 GHz.

Before discussing experimental results for spanning tree and MST algorithms in Sections 26.2.2 and 26.2.2, we show that for large data inputs, algorithms with fine-grained synchronizations do not incur serious contention among processors.

With fine-grained parallelism, contention may occur for access to critical sections or to memory locations in shared data structures. The amount of contention is dependent on the problem size, number of processors, memory access patterns, and execution times for regions of the code. In this section, we investigate contention for our fine-grained synchronization methods and quantify the amount of contention in our graph theoretic example codes.

To measure contention, we record the number of times a spinlock spins before it gains access to the shared data structure. For lock-free protocols it is difficult to measure the actual contention. For example, if *compare&swap* is used to partition the workload, it is impossible to tell whether the failure is due to contention from another contending processor or due to the fact that the location has already been claimed before. However, inspecting how spinlocks behave can give a good indication of the contention for lock-free implementations as in both cases processors contend for the same shared data structures.

Figure 26.13 shows the contention among processors for the spanning tree and MST algorithms with different number of processors and sizes of inputs. The level of contention is represented by *success rate*, which is calculated as the total number of locks acquired divided by the total number of times the locks spin. The larger the *success rate*, the lower the contention level. We see that contention level increases for a certain problem size with the number of processors. This effect is more obvious when the input size is small, for example, with hundreds of vertices. For large problem size, for example, millions of vertices, there is no clear difference in contention for 2 and 16 processors. In our experiments, *success rate* is above 97% for input sizes with more than 4096 vertices, and is above 99.98% for 1M vertices, regardless of the number of processors (between 2 and 16).

26.5.4.1 Spanning Tree Results

We compare the performance of the lock-free Shiloach–Vishkin spanning tree implementation with four other implementations that differ only in how the conflicts are resolved. In Table 26.1, we briefly describe the four implementations.

Among the four implementations, *span-race* is not a correct implementation and does not guarantee correct results. It is included as a baseline to show how much overhead is involved with using lock-free protocols and spinlocks.

In Figures 26.14 and 26.15, we plot the performance of our spanning tree algorithms using several graph instances on Sun E4500 and IBM p570. Note that we use larger instances on the IBM p570 than on the Sun E4500 because of the IBM's larger main memory. In these performance results, we see that *span-2phase*, *span-lockfree*, and *span-spinlock* scale well with the number of processors, and the execution time of *span-lockfree* and *span-spinlock* is roughly half of that of *span-2phase*. It is interesting to note that *span-lockfree*, *span-spinlock*, and *span-race* are almost as fast as each other for various inputs, which suggests similar overhead for spinlocks and lock-free protocols, and the overhead is negligible on both systems although the implementation of lock-free protocols and spinlocks use different hardware atomic operations on the two systems. The performance differences in these approaches are primarily due to the nondeterminism inherent in the algorithm.

There is some difference in the performance of *span-lock* on the two platforms. The scaling of *span-lock* on IBM p570 is better than on the Sun E4500. This may be due to the different implementations of mutex locks on the two systems. The implementation of system mutex locks usually adopts a hybrid approach, that is, the lock busy waits for a while before yielding control to the operating system. Depending on the processor speed, the cost of context switch, and the application scenario, the implementation of system mutex lock chooses a judicious amount of time to busy wait. On the Sun E4500, the mutex lock implementation is not particularly friendly for the access pattern to shared objects generated by our algorithms.

26.5.4.2 Minimum Spanning Tree Results

Performance results on Sun E4500 are shown in Figure 26.16. These empirical results demonstrate that *Bor-FAL* is the fastest implementation for sparse random graphs, and *Bor-ALM* is the fastest implementation for meshes. From our results we see that with 12 processors *Bor-spinlock* beats both *Bor-FAL* and

TABLE 26.1 Five Implementations of Shiloach–Vishkin's Parallel Spanning Tree Algorithm

Implementation	Description
span-2phase	Conflicts are resolved by two-phase election
span-lock	Conflicts are resolved using system mutex locks
span-lockfree	No mutual exclusion, races are prevented by atomic updates
span-spinlock	Mutual exclusion by spinlocks using atomic operations
span-race	No mutual exclusion, no attempt to prevent races

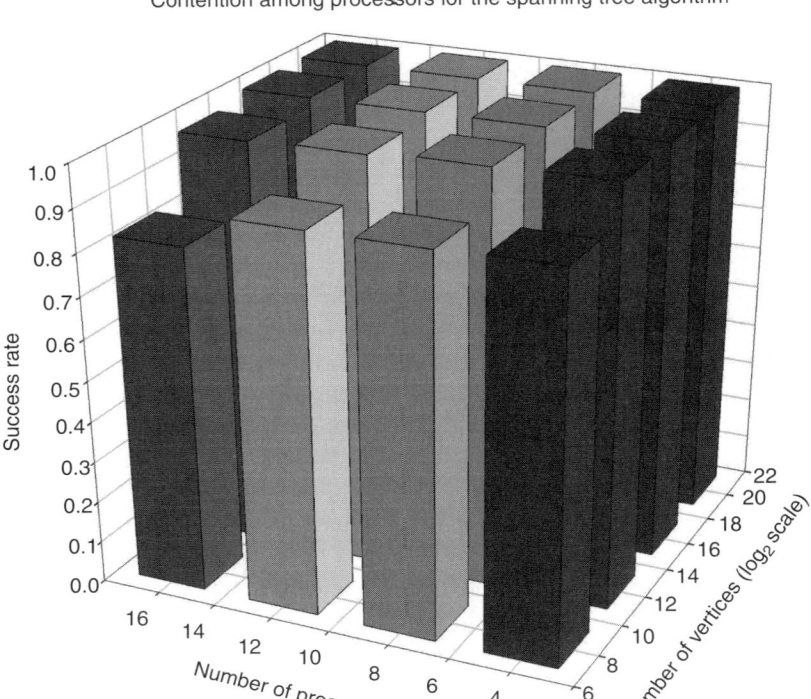
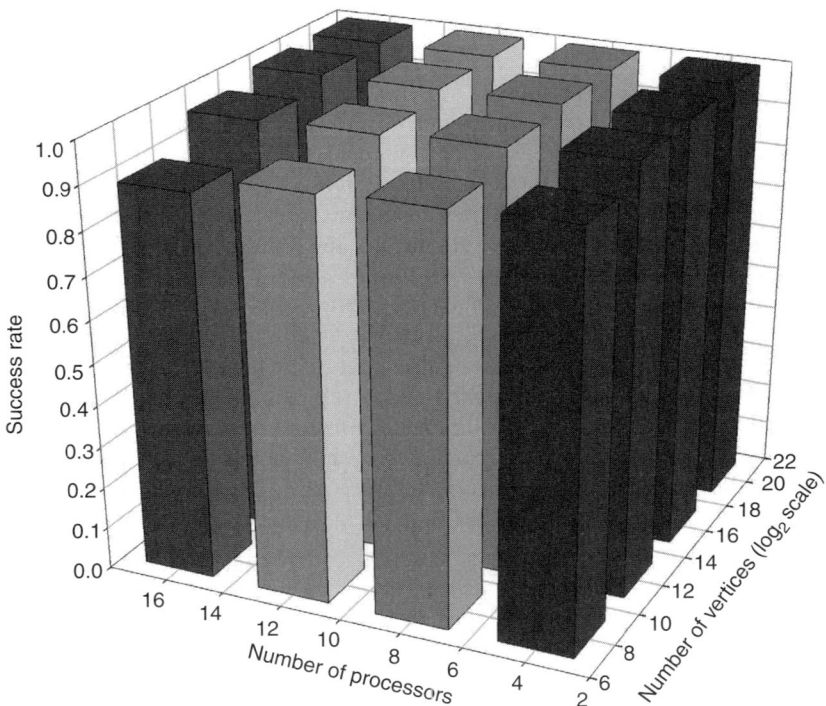

FIGURE 26.13 Contention among processors for *span-* and *Bor-spinlock*. The input graphs are random graphs with n vertices and $4n$ edges.

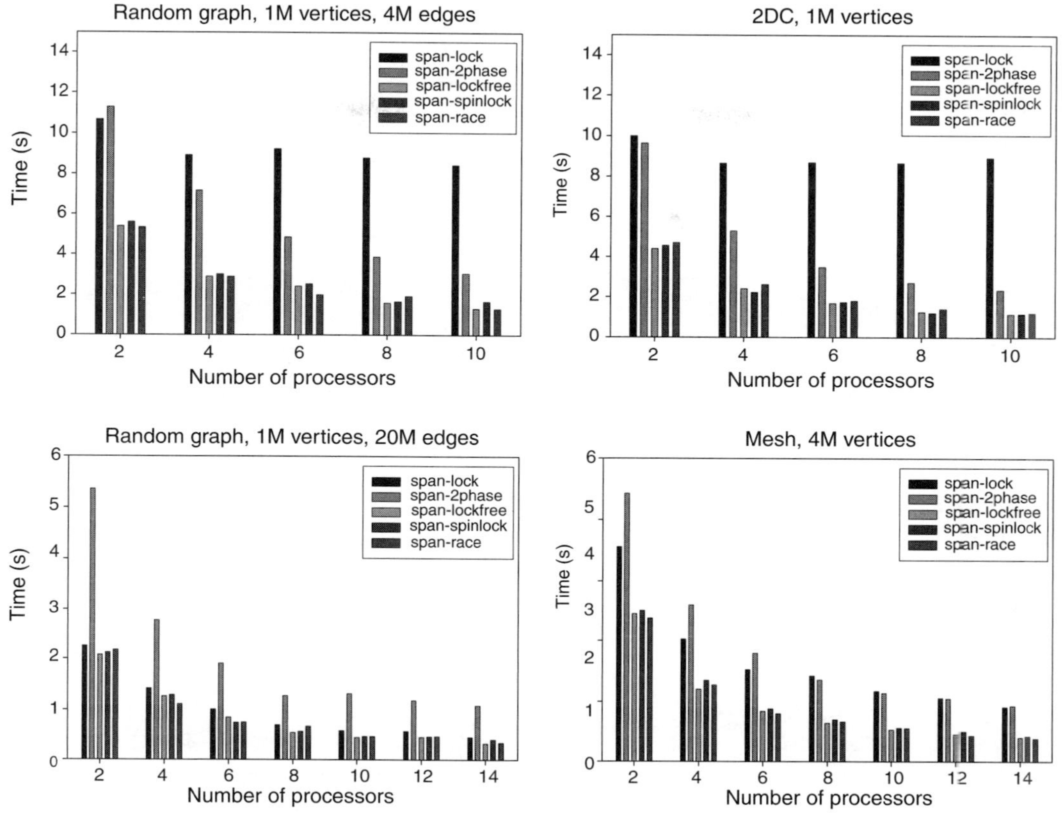

Parallel Graph Algorithms for Multiprocessors

FIGURE 26.14 The performance of the spanning tree implementations on the Sun E4500. The vertical bars from left to right are *span-lock*, *span-2phase*, *span-lockfree*, *span-spinlock*, and *span-race*, respectively.

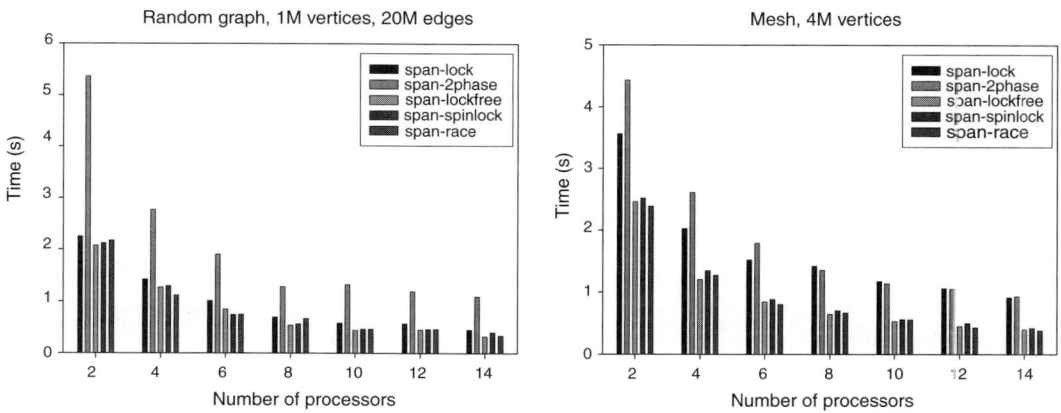

FIGURE 26.15 The performance of the spanning tree implementations on the IBM p570. The vertical bars from left to right are *span-lock*, *span-2phase*, *span-lockfree*, *span-spinlock*, and *span-race*, respectively.

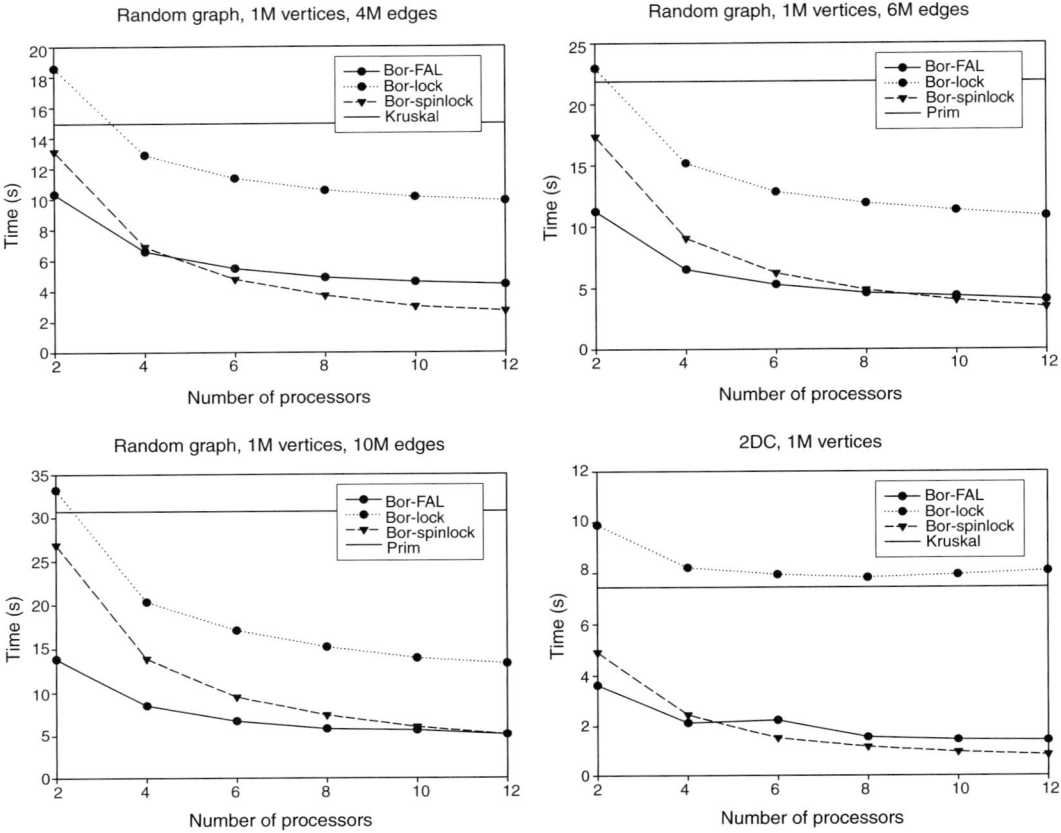

FIGURE 26.16 Comparison of the performance of *Bor-spinlock* on the Sun E4500 against the previous implementations on random graphs with 1M vertices and 4M, 6M, and 10M edges, and a mesh with 1M vertices, respectively. The horizontal line in each graph shows the execution time of the best sequential implementation.

Bor-ALM, and performance of *Bor-spinlock* scales well with the number of processors. Performance of *Bor-lock* is also plotted. *Bor-lock* is the same as *Bor-spinlock* except that system mutex locks are used. *Bor-lock* does not scale with the number of processors. The performance of the best sequential algorithms among the three candidates, Kruskal, Prim, and Borůvka, is plotted as a horizontal line for each input graph. For all the input graphs shown in Figure 26.14, *Bor-spinlock* tends to perform better than the previous best implementations when more processors are used. These performance results demonstrate the potential advantage of spinlock-based implementations for large and irregular problems. Aside from good performance, *Bor-spinlock* is also the simplest approach as it does not involve sorting required by the other approaches.

Performance results on p570 are shown in Figure 26.17. Compared with results on the Sun E4500, again *Bor-lock* scales better on the IBM p570, yet there is still a big gap between *Bor-lock* and *Bor-spinlock* due to the economic memory usage of spinlock and its simple implementation.

26.5.5 Discussion

We present novel applications of lock-free protocols and fine-grained mutual exclusion locks to parallel algorithms and show that these protocols can greatly improve the performance of parallel algorithms for large, irregular problems. As there is currently no direct support for invoking atomic instructions from

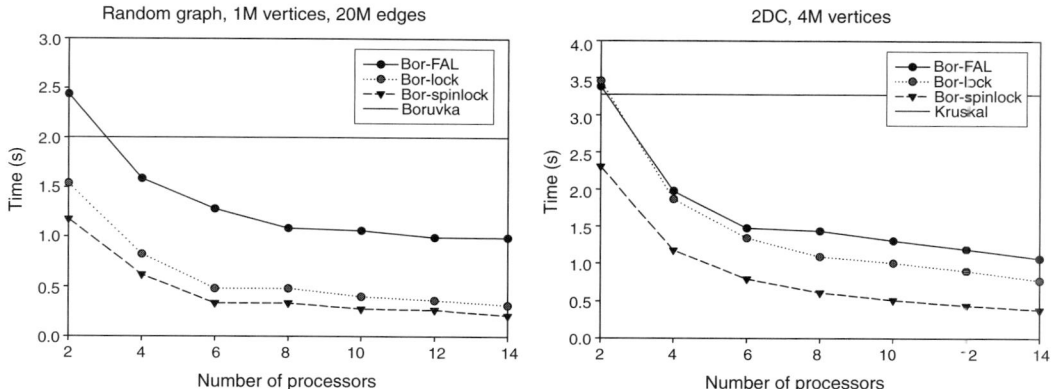

FIGURE 26.17 Comparison of the performance of *Bor-spinlock* on the IBM p570 against the previous implementations on a random graph with 1M vertices, 20M edges, and a mesh with 4M vertices. The horizontal line in each graph shows the execution time of the best sequential implementation.

most programming languages, our results suggest it necessary that there be orchestrated support for high performance algorithms from the hardware architecture, operating system, and programming languages.

26.6 Conclusions and Future Work

We have shown promising results for solving irregular graph problems on current architectures. For the spanning tree, MST, and biconnected components, our algorithms achieve for the first time parallel speedups for arbitrary, sparse instances on SMPs. We propose techniques for designing efficient parallel algorithms on SMPs (and CMPs) and expect the techniques applicable in solving more complicated problems.

Experimental studies play a very important role in the case of parallel computing to evaluate various algorithms. The gap between theory and practice is especially outstanding for parallel algorithms solving irregular problems, which results from the mismatch between the algorithmic model and current architectures. Although many researchers are aware of this problem, there still is not yet a model that is both easy-to-use and more realistic. Before the advent of a simple and omnipotent model (which is certainly closely related to architecture research), we find it helpful to take a two-step approach in solving a problem with parallel computers. First consider the problem under PRAM and gain insights into the inherent structure, then design algorithms for the target architecture.

Quite a few new architecture designs are currently being studied and evaluated, which will certainly boost the study of parallel algorithms, both theoretic and experimental. Different techniques are proposed to remove the bottleneck of memory access speed by using better caching scheme or by bringing the processing elements closer to memory (e.g., processing-in-memory and memory in processor). There are also processors targeted at specific workloads, for example, streaming processors that are designed for streaming data and multimedia workloads. Some designs (e.g., the Cell Broadband Engine [14]) offer very high memory bandwidth, but as a compromise crude cache coherence protocol. As computer architectures are becoming more sophisticated than ever, it is unlikely that any parallel model with a reasonable number of parameters can capture all features crucial to performance; experimental studies are becoming more important. And we believe our study of parallel graph algorithms on SMPs should provide valuable insight in solving problems efficiently on future architectures.

In future, we plan to continue and extend our study in three directions. The first is to solve more complicated problems with higher-level algorithms on current architectures. The second is to study algorithm designs for emerging new architectures. And finally, we expect to apply our techniques to solve problems coming from application domains.

References

[1] M. Adler, W. Dittrich, B. Juurlink, M. Kutylowski, and I. Rieping. Communication-optimal parallel minimum spanning tree algorithms. In *Proc. 10th Ann. Symp. Parallel Algorithms and Architectures (SPAA-98)*, pp. 27–36, Newport, RI, June 1998. ACM.

[2] J. Allemany and E.W. Felton. Performance issues in non-blocking synchronization on shared-memory multiprocessors. In *Proc. 11th ACM Symposium on Principles of Distributed Computing*, pp. 125–134, August 1992.

[3] L. An, Q.S. Xiang, and S. Chavez. A fast implementation of the minimum spanning tree method for phase unwrapping. *IEEE Trans. Med. Imaging*, 19(8):805–808, 2000.

[4] R.J. Anderson and H. Woll. Wait-free parallel algorithms for the union-find problem. In *Proc. 23rd Annual ACM Symposium on Theory of Computing*, pp. 370–380, May 1991.

[5] J. Aspnes and M.P. Herlihy. Wait-free data structures in the asynchronous PRAM model. In *Proc. 2nd Ann. Symp. Parallel Algorithms and Architectures (SPAA-90)*, pp. 340–349, July 1990.

[6] B. Awerbuch and Y. Shiloach. New connectivity and MSF algorithms for shuffle-exchange network and PRAM. *IEEE Trans. Computers*, C-36(10):1258–1263, 1987.

[7] D.A. Bader and G. Cong. A fast, parallel spanning tree algorithm for symmetric multiprocessors (SMPs). In *Proc. Int'l Parallel and Distributed Processing Symp. (IPDPS 2004)*, pp. 38–47, Santa Fe, NM, April 2004.

[8] D.A. Bader and G. Cong. Fast shared-memory algorithms for computing the minimum spanning forest of sparse graphs. In *Proc. Int'l Parallel and Distributed Processing Symp. (IPDPS 2004)*, Santa Fe, NM, April 2004.

[9] D.A. Bader, S. Sreshta, and N. Weisse-Bernstein. Evaluating arithmetic expressions using tree contraction: A fast and scalable parallel implementation for symmetric multiprocessors (SMPs). In S. Sahni, V.K. Prasanna, and U. Shukla, ed., *Proc. 9th Int'l Conf. on High Performance Computing (HiPC 2002)*, volume 2552 of Lecture Notes in Computer Science, pp. 63–75, Bangalore, India, December 2002. Springer-Verlag.

[10] G. Barnes. Wait-free algorithms for heaps. Technical Report TR-94-12-07, University of Washington, Seattle, WA, 1994.

[11] R. Blumofe and C. Leiserson. Scheduling multithreaded computations by work stealing. In *Proc. 35th Ann. Sym. Foundations of Computer Science*, Santa Fe, NM, pp. 356–368, November 1994.

[12] M. Brinkhuis, G.A. Meijer, P.J. van Diest, L.T. Schuurmans, and J.P. Baak. Minimum spanning tree analysis in advanced ovarian carcinoma. *Anal. Quant. Cytol. Histol.*, 19(3):194–201, 1997.

[13] C. Chen and S. Morris. Visualizing evolving networks: Minimum spanning trees versus pathfinder networks. In *IEEE Symp. on Information Visualization*, Seattle, WA, pp. 69–74, October 2003.

[14] T. Chen, R. Raghavan, J. Dale, and E. Iwata. Cell broadband engine architecture and its first implementation, November 2005. http://www-128.ibm.com/developerworks/power/library/pa-cellperf/.

[15] F.Y. Chin, J. Lam, and I.-N. Chen. Efficient parallel algorithms for some graph problems. *Commun. ACM*, 25(9):659–665, 1982.

[16] K.W. Chong, Y. Han, and T.W. Lam. Concurrent threads and optimal parallel minimum spanning tree algorithm. *J. ACM*, 48:297–323, 2001.

[17] K.W. Chong and T.W. Lam. Approximating biconnectivity in parallel. *Algorithmica*, 21:395–410, 1998.

[18] K.W. Chong and T.W. Lam. Finding connected components in $O(\log n \log \log n)$ time on the EREW PRAM. *J. Algorithms*, 18:378–402, 1995.

[19] B. Chor, A. Israeli, and M. Li. On processor coordination using asynchronous hardware. In *Proc. 6th ACM Symposium on Principles of Distributed Computing*, pp. 86–97, Vancouver, British Columbia, Canada, 1987.

[20] S. Chung and A. Condon. Parallel implementation of Borůvka's minimum spanning tree algorithm. In *Proc. 10th Intl Parallel Processing Symp. (IPPS'96)*, pp. 302–315, April 1996.

[21] T. Coffman, S. Greenblatt, and S. Marcus. Graph-based technologies for intelligence analysis. *Commun. ACM*, 47(3):45–47, 2004.

[22] R. Cole, P.N. Klein, and R.E. Tarjan. A linear-work parallel algorithm for finding minimum spanning trees. In *Proc. 6th Ann. Symp. Parallel Algorithms and Architectures (SPAA-94)*, pp. 11–15, Newport, RI, June 1994. ACM.

[23] R. Cole, P.N. Klein, and R. E. Tarjan. Finding minimum spanning forests in logarithmic time and linear work using random sampling. In *Proc. 8th Ann. Symp. Parallel Algorithms and Architectures (SPAA-96)*, pp. 243–250, Newport, RI, June 1996. ACM.

[24] R. Cole and U. Vishkin. Approximate parallel scheduling. Part II: applications to logarithmic-time optimal graph algorithms. *Information and Computation*, 92:1–47, 1991.

[25] G. Cong and D.A. Bader. The Euler tour technique and parallel rooted spanning tree. In *Proc. Int'l Conference on Parallel Processing (ICPP 2004)*, Montréal, Canada, August 2004.

[26] D.E. Culler, R.M. Karp, D.A. Patterson, A.Sahay, K. E. Schauser, E. Santos, R. Subramonian, and T. von Eicken. LogP: Towards a realistic model of parallel computation. In *4th Symp. Principles and Practice of Parallel Programming*, pp. 1–12. ACM SIGPLAN, May 1993.

[27] F. Dehne and S. Götz. Practical parallel algorithms for minimum spanning trees. In *Workshop on Advances in Parallel and Distributed Systems*, pp. 366–371, West Lafayette, IN, October 1998. co-located with the 17th IEEE Symp. on Reliable Distributed Systems.

[28] J.C. Dore, J. Gilbert, E. Bignon, A. Crastes de Paulet, T. Ojasoo, M. Pons, J.P. Raynaud, and J.F. Miquel. Multivariate analysis by the minimum spanning tree method of the structural determinants of diphenylethylenes and triphenylacrylonitriles implicated in estrogen receptor binding, protein kinase C activity, and MCF7 cell proliferation. *J. Med. Chem.*, 35(3):573–583, 1992.

[29] C. Dwork, D. Dwork, and L. Stockmeyer. On the minimal synchronism needed for distributed consensus. *J. ACM*, 34(1):77–97, 1987.

[30] C. Dwork, N. Lynch, and L. Stockmeyer. Consensus in the presence of partial synchrony. *J. ACM*, 35(2):288–323, 1988.

[31] D.M. Eckstein. BFS and biconnectivity. Technical Report 79-11, Dept of Computer Science, Iowa State Univ. of Science and Technology, Ames, IW, 1979.

[32] M. Fischer, N.A. Lynch, and M.S. Paterson. Impossibility of distributed commit with one faulty process. *J. ACM*, 32(2):374–382, 1985.

[33] K. Fraser. Practical lock-freedom. PhD thesis, King's College, University of Cambridge, United Kingdom, September 2003.

[34] H. Gazit. An optimal randomized parallel algorithm for finding connected components in a graph. *SIAM J. Comput.*, 20(6):1046–1067, 1991.

[35] P.B. Gibbons, Y. Matias, and V. Ramachandran. Can shared-memory model serve as a bridging model for parallel computation? In *Proc. 9th Ann. Symp. Parallel Algorithms and Architectures (SPAA-97)*, pp. 72–83, Newport, RI, June 1997. ACM.

[36] S. Goddard, S. Kumar, and J.F. Prins. Connected components algorithms for mesh-connected parallel computers. In S. N. Bhatt, editor, *Parallel Algorithms: 3rd DIMACS Implementation Challenge October 17–19, 1994*, volume 30 of DIMACS Series in Discrete Mathematics and Theoretical Computer Science, pp. 43–58. American Mathematical Society, 1997.

[37] A. Gottlieb, R. Grishman, C.P. Kruskal, K.P. McAuliffe, L. Rudolph, and M. Snir. The NYU ultracomputer—designing a MIMD, shared-memory parallel machine. *IEEE Trans. Computers*, C-32(2):175–189, 1984.

[38] J. Greiner. A comparison of data-parallel algorithms for connected components. In *Proc. 6th Ann. Symp. Parallel Algorithms and Architectures (SPAA-94)*, pp. 16–25, Cape May, NJ, June 1994.

[39] S. Halperin and U. Zwick. An optimal randomised logarithmic time connectivity algorithm for the EREW PRAM. In *Proc. 7th Ann. Symp. Discrete Algorithms (SODA-96)*, pp. 438–447, 1996. Also published in *J. Comput. Syst. Sci.*, 53(3):395–416, 1996.

[40] Y. Han and R.A. Wagner. An efficient and fast parallel-connected component algorithm. *J. ACM*, 37(3):626–642, 1990.

[41] D.R. Helman and J. JáJá. Designing practical efficient algorithms for symmetric multiprocessors. In *Algorithm Engineering and Experimentation (ALENEX'99)*, volume 1619 of Lecture Notes in Computer Science, pp. 37–56, Baltimore, MD, January 1999. Springer-Verlag.

[42] D.R. Helman and J. JáJá. Prefix computations on symmetric multiprocessors. *J. Parallel Distrib. Comput.*, 61(2):265–278, 2001.

[43] M. Herlihy and J.E.B. Moss. Transactional memory: Architectural support for lock-free data structures. In *Proc. 20th Int'l Symposium in Computer Architecture*, pp. 289–300, May 1993.

[44] M.P. Herlihy. Wait-free synchronization. *ACM Trans. Program. Lang. Sys.*, 13(1):124–149, 1991.

[45] M.P. Herlihy. A methodology for implementing highly concurrent data objects. *ACM Trans. Program. Lang. Syst.*, 15(5):745–770, 1993.

[46] M.P. Herlihy and J.M. Wing. Axioms for concurrent objects. In *Proc. 14th ACM SIGACT-SIGPLAN Symposium on Principles of Programming Languages*, pp. 13–26, January 1987.

[47] D.S. Hirschberg, A.K. Chandra, and D.V. Sarwate. Computing connected components on parallel computers. *Commun. ACM*, 22(8):461–464, 1979.

[48] T. Hsu and V. Ramachandran. On finding a smallest augmentation to biconnect a graph. *SIAM J. Computing*, 22(5):889–912, 1993.

[49] T.-S. Hsu, V. Ramachandran, and N. Dean. Parallel implementation of algorithms for finding connected components in graphs. In S. N. Bhatt, editor, *Parallel Algorithms: 3rd DIMACS Implementation Challenge October 17–19, 1994*, volume 30 of DIMACS Series in Discrete Mathematics and Theoretical Computer Science, pp. 23–41. American Mathematical Society, 1997.

[50] K. Iwama and Y. Kambayashi. A simpler parallel algorithm for graph connectivity. *J. Algorithms*, 16(2):190–217, 1994.

[51] J. JáJá. *An Introduction to Parallel Algorithms*. Addison-Wesley Publishing Company, New York, 1992.

[52] D.B. Johnson and P. Metaxas. Connected components in $O(\log^{3/2} |v|)$ parallel time for the CREW PRAM. In *Proc. of the 32nd Annual IEEE Symp. on Foundations of Computer Science*, pp. 688–697, San Juan, Puerto Rico, 1991.

[53] D.B. Johnson and P. Metaxas. A parallel algorithm for computing minimum spanning trees. In *Proc. 4th Ann. Symp. Parallel Algorithms and Architectures (SPAA-92)*, pp. 363–372, San Diego, CA, 1992.

[54] D.R. Karger, P.N. Klein, and R.E. Tarjan. A randomized linear-time algorithm to find minimum spanning trees. *J. ACM*, 42(2):321–328, 1995.

[55] K. Kayser, S.D. Jacinto, G. Bohm, P. Frits, W.P. Kunze, A. Nehrlich, and H.J. Gabius. Application of computer-assisted morphometry to the analysis of prenatal development of human lung. *Anat. Histol. Embryol.*, 26(2):135–139, 1997.

[56] K. Kayser, H. Stute, and M. Tacke. Minimum spanning tree, integrated optical density and lymph node metastasis in bronchial carcinoma. *Anal. Cell Pathol.*, 5(4):225–234, 1993.

[57] S. Khuller and U. Vishkin. Biconnectivity approximations and graph carvings. *J. ACM*, 41(2):214–235, March 1994.

[58] P.N. Klein and J.H. Reif. An efficient parallel algorithm for planarity. *J. Computer and Syst. Sci.*, 37(2):190–246, 1988.

[59] V.E. Krebs. Mapping networks of terrorist cells. *Connections*, 24(3):43–52, 2002.

[60] A. Krishnamurthy, S.S. Lumetta, D.E. Culler, and K. Yelick. Connected components on distributed memory machines. In S. N. Bhatt, editor, *Parallel Algorithms: 3rd DIMACS Implementation Challenge October 17–19, 1994*, volume 30 of DIMACS Series in Discrete Mathematics and Theoretical Computer Science, pp. 1–21. American Mathematical Society, 1997.

[61] C.P. Kruskal, L. Rudolph, and M. Snir. Efficient parallel algorithms for graph problems. *Algorithmica*, 5(1):43–64, 1990.

[62] L. Lamport. Specifying concurrent program modules. *ACM Trans. Program. Lang. Syst.*, 5(2):190–222, 1983.

[63] V. Lanin and D. Shaha. Concurrent set manipulation without locking. In *Proc. 7th ACM SIGACT-SIGMOD-SIGART Symposium on Principles of Database Systems*, pp. 211–220, March 1988.

[64] Y. Liang, S.K. Dhall, and S. Lakshmivarahan. Efficient parallel algorithms for finding biconnected components of some intersection graphs. In *Proceedings of the 19th Annual Conference on Computer Science*, pp. 48–52, San Antonio, TX, United States, 1991.

[65] H. Massalin and C. Pu. Threads and input/output in the synthesis kernel. In *Proc. 12th ACM Symposium on Operating Systems Principles*, pp. 191–201, December 1989.

[66] M. Matos, B.N. Raby, J.M. Zahm, M. Polette, P. Birembaut, and N. Bonnet. Cell migration and proliferation are not discriminatory factors in the in vitro sociologic behavior of bronchial epithelial cell lines. *Cell Motil. Cytoskel.*, 53(1):53–65, 2002.

[67] S. Meguerdichian, F. Koushanfar, M. Potkonjak, and M. Srivastava. Coverage problems in wireless ad-hoc sensor networks. In *Proc. INFOCOM '01*, pp. 1380–1387, Anchorage, AK, April 2001. Proc. IEEE.

[68] K. Mehlhorn and S. Näher. *The LEDA Platform of Combinatorial and Geometric Computing*. Cambridge University Press, 1999.

[69] G.L. Miller and V. Ramachandran. Efficient parallel ear decomposition with applications. Manuscript, UC Berkeley, MSRI, January 1986.

[70] Y. Moan, B. Schieber, and U. Vishkin. Parallel ear decomposition search (EDS) and st-numbering in graphs. *Theor. Comput. Sci.*, 47(3):277–296, 1986.

[71] B.M.E. Moret and H.D. Shapiro. An empirical assessment of algorithms for constructing a minimal spanning tree. In *Computational Support for Discrete Mathematics*, DIMACS Monographs in Discrete Mathematics and Theoretical Computer Science, 15, pp. 99–117. American Mathematical Society, 1994.

[72] D. Nash and S.N. Maheshwari. Parallel algorithms for the connected components and minimal spanning trees. *Inf. Process. Lett.*, 14(1):7–11, 1982.

[73] V. Olman, D. Xu, and Y. Xu. Identification of regulatory binding sites using minimum spanning trees. In *Proc. 8th Pacific Symp. Biocomputing (PSB 2003)*, pp. 327–338, Hawaii, 2003. World Scientific Pub.

[74] E. Palmer. *Graphical Evolution: An Introduction to the Theory of Random Graphs*. John Wiley & Sons, New York, 1985.

[75] J. Park, M. Penner, and V.K. Prasanna. Optimizing graph algorithms for improved cache performance. In *Proc. Int'l Parallel and Distributed Processing Symp. (IPDPS 2002)*, Fort Lauderdale, FL, April 2002.

[76] S. Pettie and V. Ramachandran. A randomized time-work optimal parallel algorithm for finding a minimum spanning forest. *SIAM J. Comput.*, 31(6):1879–1895, 2002.

[77] C.A. Phillips. Parallel graph contraction. In *Proc. 1st Ann. Symp. Parallel Algorithms and Architectures (SPAA-89)*, pp. 148–157. ACM, 1989.

[78] C.K. Poon and V. Ramachandran. A randomized linear work EREW PRAM algorithm to find a minimum spanning forest. In *Proc. 8th Intl Symp. Algorithms and Computation (ISAAC'97)*, volume 1350 of Lecture Notes in Computer Science, pp. 212–222. Springer-Verlag, 1997.

[79] R. Read. Teaching graph theory to a computer. In W. Tutte, ed., *Proc. 3rd Waterloo Conf. on Combinatorics*, pp. 161–173, Waterloo, Canada, May 1969.

[80] J.H. Reif, editor. *Synthesis of Parallel Algorithms*. Morgan Kaufmann Publishers, 1993.

[81] P. Sanders. Random permutations on distributed, external and hierarchical memory. *Inf. Process. Lett.*, 67(6):305–309, 1998.

[82] C. Savage and J. JáJá. Fast, efficient parallel algorithms for some graph problems. *SIAM J. Computing*, 10(4):682–691, 1981.

[83] Y. Shiloach and U. Vishkin. An $O(\log n)$ parallel connectivity algorithm. *J. Algs.*, 3(1):57–67, 1982.

[84] R.E. Tarjan. Depth-first search and linear graph algorithms. *SIAM J. Comput.*, 1(2):146–160, 1972.

[85] R.E. Tarjan and J. Van Leeuwen. Worst-case analysis of set union algorithms. *J. ACM*, 31(2):245–281, 1984.

[86] R.E. Tarjan and U. Vishkin. An efficient parallel biconnectivity algorithm. *SIAM J. Comput.*, 14(4):862–874, 1985.

[87] Y.-C. Tseng, T.T.-Y. Juang, and M.-C. Du. Building a multicasting tree in a high-speed network. *IEEE Concurr.*, 6(4):57–67, 1998.

[88] P. Tsigas and Y. Zhang. A simple, fast and scalable non-blocking concurrent FIFO queue for shared memory multiprocessor systems. In *Proc. 13th Ann. Symp. Parallel Algorithms and Architectures (SPAA-01)*, pp. 134–143, September 2001.

[89] Y. H. Tsin and F. Y. Chin. Efficient parallel algorithms for a class of graph theoretic problems. *SIAM J. Comput.*, 13(3):580–599, 1984.

[90] L. G. Valiant. A bridging model for parallel computation. *Commun. ACM*, 33(8):103–111, 1990.

[91] J. Valois. Lock-free data structures. PhD thesis, Rensselaer Polytechnic Institute, Troy, NY, May 1995.

[92] J.D. Valois. Lock-free linked lists using compare-and-swap. In *Proc. 14th Ann. ACM Symp. on Principles of Distributed Computing*, pp. 214–222, Ottowa, Canada, August 1995.

[93] U. Vishkin. On efficient parallel strong orientation. *Information Processing Letters*, 20(5):235–240, 1985.

[94] J. Woo and S. Sahni. Load balancing on a hypercube. In *Proc. 5th Intl Parallel Processing Symp.*, pp. 525–530, Anaheim, CA, April 1991. IEEE Computer Society Press.

[95] S.Q. Zheng, J.S. Lim, and S.S. Iyengar. Routing using implicit connection graphs. In *9th Intl Conf. on VLSI Design: VLSI in Mobile Communication*, Bangalore, India, January 1996. IEEE Computer Society Press.

27

Parallel Algorithms for Volumetric Surface Construction

27.1	Introduction	27-1
27.2	Isosurface Extraction	27-3
27.3	A Simple Optimal Out-of-Core Isosurface Extraction Algorithm	27-5
27.4	Strategies for Parallel Isosurface Extraction	27-7
27.5	A Scalable and Efficient Parallel Algorithm for Isosurface Extraction	27-9
27.6	View-Dependent Isosurface Extraction and Rendering	27-11
	A Sequential Algorithm for Structured Grids • The Multipass Occlusion Culling Algorithm • A Single Pass Occlusion Culling Algorithm with Random Data Partitioning • View-Dependent Isosurface Extraction by Ray Tracing	
27.7	Conclusion	27-15
	References	27-15

Joseph JaJa
Amitabh Varshney
Qingmin Shi
University of Maryland

27.1 Introduction

Isosurfaces have long been used as a meaningful way to represent feature boundaries in volumetric datasets. Surface representations of volumetric datasets facilitate visual comprehension, computational analysis, and manipulation. The earliest beginnings in this field can be traced to microscopic examination of tissues. For viewing opaque specimen, it was considered desirable to slice the sample at regular intervals and to view each slice under a microscope. The outlines of key structures in such cross-sectional slices were traced on to transparent sheets that were then sequentially stacked with transparent spacers in between them [WC71, LW72]. A user holding such a translucent stack would then be able to mentally reconstruct the location of the surface interpolating the contours on successive slices. It is, therefore, only natural that one of the first algorithms for surface reconstruction from volumes proceeded by interpolation between planar contours [FKU77]. Fuchs et al.'s method proceeded by casting the problem of optimal surface reconstruction between planar contours as one of finding minimum cost cycles in a directed toroidal graph. Their method computed the minimum-area interpolating surface between the given planar contours. The advent of x-ray computed tomography (CT) in 1973 through theoretical advances

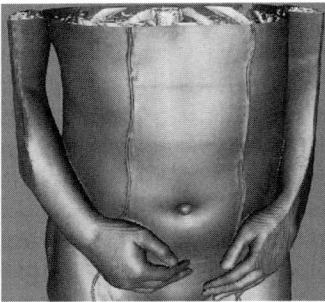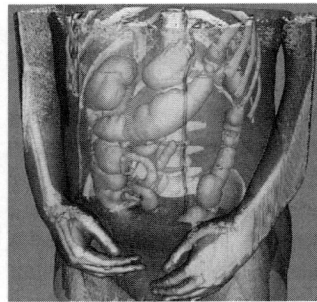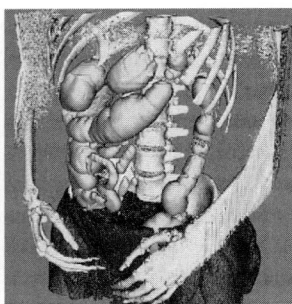

FIGURE 27.1 These images show various isosurfaces corresponding to different anatomical organs extracted and rendered using the Marching–Cubes algorithm. These images are reproduced from [Lorensen01].

by Cormack [Cormack63, Cormack64] and their practical realization by Hounsfield [Hounsfield73] made it significantly easier to acquire highly detailed cross-sectional images. The use of magnetic fields and radio waves to create images of physical objects by Lauterbur [Lauterbur73] and Mansfield [GGM74] led to the development of magnetic resonance imaging (MRI). The obvious implications of such volumetric imaging technologies in medical diagnosis and therapy sparked a lively interest among radiologists and scientific visualization researchers and greatly accelerated research in a variety of volume visualization and analysis tools and techniques [Levoy88, Kaufman91]. The Visible Human Project (VHP) of the National Library of Medicine (NLM) has greatly spurred the development of fast volume visualization techniques, including fast techniques for isosurface extraction by making available 1971 digital anatomical images (15 GB) for the VHP male dataset and 5189 digital images (39 GB) for the VHP female dataset [Yoo04]. Some isosurfaces extracted from the VHP male dataset appear in the following text [Lorensen01] (Figure 27.1).

Isosurface extraction also became important in another area—modeling of three-dimensional (3D) objects with implicit functions. The basic idea in this research is to model 3D objects as zero crossings of a scalar field induced by an implicit function. The basis functions used in such implicit functions could be Gaussians (blobby models) [Blinn82], quadratic polynomials (metaballs) [NHK+86], or higher-degree polynomials (soft objects) [WMW86]. Subsequent work on level-set methods by Osher and Sethian [OS88] added a temporal dimension to implicit surfaces and enabled them to evolve through time using numerical solutions of a time-dependent convection equation. Techniques based on level-set methods have been extensively used in modeling a variety of time-evolving phenomena in scientific computing, computer-aided design/manufacturing (CAD/CAM), computational fluid dynamics, and computational physics [OF03].

Simplicity of shape control, smooth blending, and smooth dynamics characteristics have led to the rapid adoption of isosurfaces for modeling a number of objects and phenomenon in digital entertainment media. For instance, metaballs have been used in modeling of musculature and skin as isosurfaces of implicit radial basis functions (Figure 27.2).

In dynamic settings, morphing and fluid flow simulations have used isosurfaces computed as level sets of a scalar field. For instance, level-set-based morphing techniques developed by Whitaker and Breen [BW01] have been used in movies such as *Scooby-Doo 2* and have been extended to work directly from a compressed format in a hierarchical run-length-encoded scheme [HNB+06] (Figure 27.3).

Although fairly sophisticated volume visualization techniques have evolved over the last decade, isosurface-based visualization of volumetric features continues to occupy an important cornerstone of modern-day volume visualization. The main reasons behind the continued interest in isosurface-based representations for volumes include:

- Simple algorithms for extraction of the isosurface.
- Triangle-mesh-based isosurfaces map well to the graphics hardware that is optimized for rendering triangles.
- Isosurface-based depiction of volumetric features is intuitive and greatly facilitates their visual comprehension.

FIGURE 27.2 An animation still that depicts battle between two battling Tyrannosauruses. The mesh was taken from New Riders IPAS Plugin Reference CD. The muscular structure is created using the Metareyes plug-in. This involves creating hundreds of metaballs to gradually build up a solid mesh. This image was created by Spencer Arts and is reproduced here from the website: http://www.siggraph.org/education/materials/HyperGraph/modeling/metaballs/trex.htm

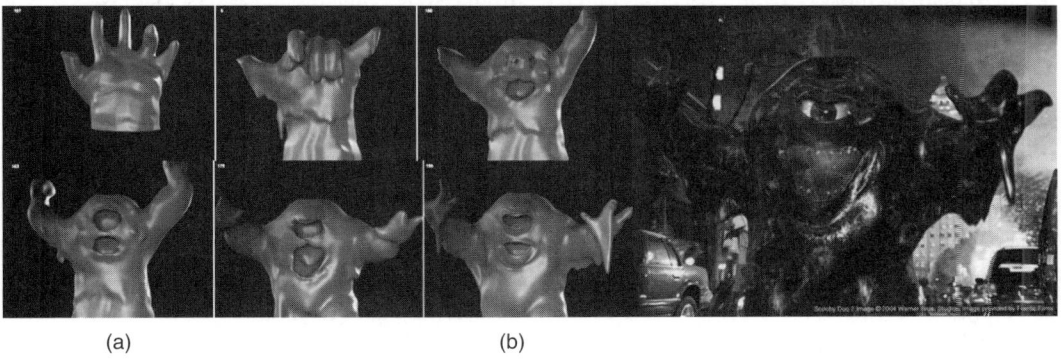

(a)   (b)

FIGURE 27.3 Tar Monster simulation in *Scooby-Doo 2* (Warner Brothers) (a) shows an early test sequence of the Tar Monster morphing and (b) shows the final rendered scene (images copyright Frantic Films).

27.2 Isosurface Extraction

We start by introducing a general framework for addressing the isosurface extraction problem. A scalar field function f defined over a domain $D \subset R^3$ is sampled at a set of points, $\{v_i \in D\}_{i=1}^n$, where D has been decomposed into a set Σ of polyhedral cells $\{C_j\}$ whose vertices are the points $\{v_i\}$. The decomposition will be referred to as *structured* grid if the cells in Σ form a rectilinear or curvilinear mesh. Otherwise, we have an *unstructured* grid. In a structured grid, the connectivity of the cells is implicitly defined and Σ can be represented as a 3D array of scalar field values of its vertices. In particular, cells in a rectilinear grid are axis aligned but their vertices can be spaced arbitrarily in each of the dimensions. On the other hand, the connectivity of the cells in an unstructured grid has to be defined explicitly and such a grid will typically be represented by a list of all the cells (where each cell is specified by a list of its vertices) and another list consisting of the coordinates of the vertices. Given a scalar value λ, the isosurface corresponding to λ consists of the set of all points $p \in D$, such that $f(p) = \lambda$. Under fairly general conditions on f, it can be shown that the isosurface is a manifold that separates space into the surface itself and two connected open sets: "above" (or "outside") and "below" (or "inside") the isosurface. In particular, a point $p \in D$ is above the isosurface if $f(p) > \lambda$ and below the isosurface if $f(p) < \lambda$. The isosurface extraction problem is to determine, for any given isovalue, a piecewise linear approximation of the isosurface from the grid sampling of the scalar field.

The popular Marching–Cubes (MC) algorithm, developed by Lorensen and Cline in 1987 [LC87], computes a triangular mesh approximation of the isosurface on the basis of the following strategy. A cell C is cut by the isosurface for λ if, and only if, $f_{min} \leq \lambda \leq f_{max}$, where f_{min} and f_{max} are, respectively, the minimum and the maximum values of the scalar field over the vertices of C. Such a cell is called *active*; otherwise the cell is called *inactive*. The MC algorithm scans the entire cell set, one cell at a time, determines whether a cell is active and, in the affirmative, generates a local triangulation of the isosurface intersection with the cell. More specifically, each vertex of an active cell is marked as "above" or "below" the isosurface depending on whether its scalar field value is larger or smaller than the isovalue, followed by a triangulation based on linear interpolation performed through a table look-up. This algorithm provides an effective solution to the isosurface extraction problem but can be extremely slow for moderate to large size grids. Since its introduction, many improvements have appeared that attempt to reduce the exhaustive search through the grid by indexing the data during a preprocessing phase. There are two general approaches to index the data and speed up the search. The first approach, primarily applicable to structured grids, amounts to developing a hierarchical spatial decomposition of the grid using an *octree* as the indexing structure. Each node of the octree is tagged by the minimum and maximum values of the scalar field over the vertices lying within the spatial region represented by the node. Such a structure restricts, in general, the search to smaller subsets of the data, and hence lead to a much faster search process than the MC algorithm. Moreover, it can be effective in dealing with large-scale data that does not fit in main memory. This approach was initially introduced by Wilhelms and Van Gelder [WG92]. We should note that it is possible that, for some datasets, the octree search could be quite inefficient as many subtrees may need to be explored to determine relatively very few active cells.

The second approach focuses on the value ranges $[f_{min}, f_{max}]$ of all the cells, and how they can be organized to speed up the search. These ranges can be either viewed (i) as points of the *span space* defined by a coordinate system in which one axis corresponds to the minimum value and the other axis corresponds to the maximum value or (ii) as intervals over the real line. The problem of determining the active cells amounts to determining the points in the span space lying in the upper rectangle as indicated in Figure 27.4 or equivalently determining all the intervals that contain the isovalue λ. Algorithms based on the span space involve partitioning the span space into tiles and the encoding of the partition by some indexing structure. For example, we can partition the span space using equal-size or equal-frequency rectangular tiles [SHL+96] or using a *kd*-tree [LSJ96]. Theoretically, optimal algorithms organize the intervals $[f_{min}, f_{max}]$ into an *interval tree* defined recursively as follows. The root of the tree stores the median of the endpoints of the intervals as the *splitting value*, and is associated with two lists of the intervals $[f_{min}, f_{max}]$ that contain the splitting value, one list in increasing order of f_{min} while the second is in decreasing order of f_{max}. The size of this indexing structure is proportional to the number N of intervals and the search complexity is $O(\log N + k)$, where N is the total number of intervals and k is the number of intervals that contain the isovalue. It is easy to see that this search complexity is asymptotically the best possible for main memory algorithms.

It is worth noting that a hybrid technique was developed in [BPT+99] and involves the construction of a set of *seed cells*, such that, for any given isovalue, this set contains at least one cell from each connected component of the corresponding isosurface. The search for active cells involves contour propagation starting from some seed cells pertaining to the isosurface. Note that such a technique requires a structured grid to effectively explore neighboring cells as the contour is being propagated. A main issue with this technique is how to determine and organize the set of seed cells whose size may be quite significant.

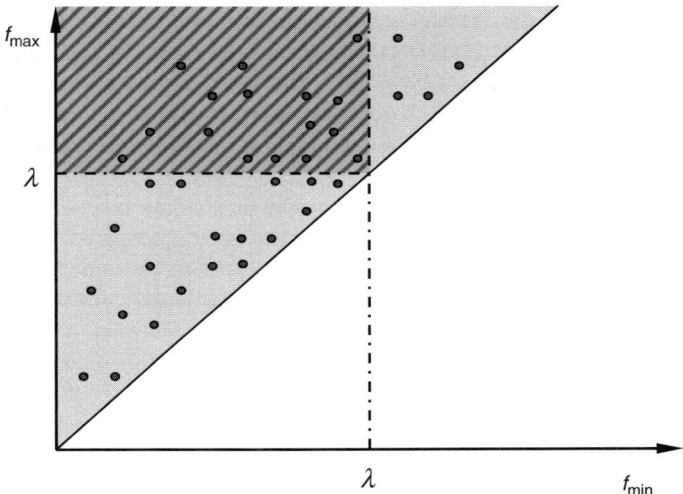

FIGURE 27.4 Span space and the subset corresponding to the isovalue.

In this chapter, we will primarily focus on isosurface extraction for large-scale data, that is, the corresponding datasets are at least an order of magnitude larger than the size of the main memory of a processor and hence have to reside on disk storage. Owing to their electromechanical components, disks have two to three orders of magnitude longer access time than random access main memory. A single disk access reads or writes a block of contiguous data at once. The performance of an external memory (or out-of-core) algorithm is typically dominated by the number of input–output (I/O) operations, each such operation involving the reading or writing of a single disk block. Chiang and Silva [CS97] present the first out-of-core isosurface extraction algorithm based on an I/O optimal interval tree, which was later improved in [CSS98] using a variation of the interval tree called the binary-blocked I/O interval tree and the metacell notion to be introduced next. A *metacell* consists of a cluster of neighboring cells that are stored contiguously on a disk, such that each metacell is about the same size that is a small multiple of the disk block size. For structured grids, a metacell consists of a subcube that can be represented by a sequence of the scalar values appearing in a predefined order. The metacell technique turns out to be quite useful when dealing with out-of-core algorithms.

Next, we introduce a very efficient out-of-core isosurface extraction algorithm that will be easy to implement on a multiprocessor system with optimal performance as we will see later.

27.3 A Simple Optimal Out-of-Core Isosurface Extraction Algorithm

Introduced in [WJV06] the compact interval tree is a simple indexing structure that enables the extraction of isosurfaces by accessing only active metacells. It is a variation of the interval tree but makes use of the concepts of span space and metacells, and can be used for both structured and unstructured grids. We start by associating with each metacell an interval $[f_{min}, f_{max}]$ corresponding respectively to the minimum and maximum values of the scalar field over the metacell. Each node of the indexing tree contains a splitting value as in the case of the standard interval tree, except that no sorted lists of intervals are stored at each node. Instead, we store the distinct values of the f_{max} endpoints of these intervals in decreasing order, and associate with each such value a list of the left endpoints of the corresponding intervals sorted in increasing order. We present the details next.

Consider the span space consisting of all possible combinations of the (f_{min}, f_{max}) values of the scalar field. With each such pair, we associate a list that contains all the metacells whose minimum scalar field value is f_{min} and whose maximum scalar field value is f_{max}. The essence of the scheme for our compact interval tree is illustrated through Figure 27.5 representing the span space, and Figure 27.6 representing the compact interval tree built upon the n distinct values of the endpoints of the intervals corresponding to the metacells.

Let f_0 be the median of all the endpoints. The root of the interval tree corresponds to all the intervals containing f_0. Such intervals are represented as points in the square of Figure 27.5, whose bottom right corner is located at (f_0, f_0). We group together all the metacells having the same f_{max} value in this square, and store them consecutively on disk from left to right in increasing order of their f_{min} values. We will refer to this contiguous arrangement of all the metacells having the same f_{max} value within a square as a *brick*. The bricks within the square are, in turn, stored consecutively on disk in decreasing order of the f_{max} values. The root will contain the value f_0, the number of nonempty bricks in the corresponding square, and an index list of the corresponding bricks. This index list consists of at most $n/2$ entries corresponding

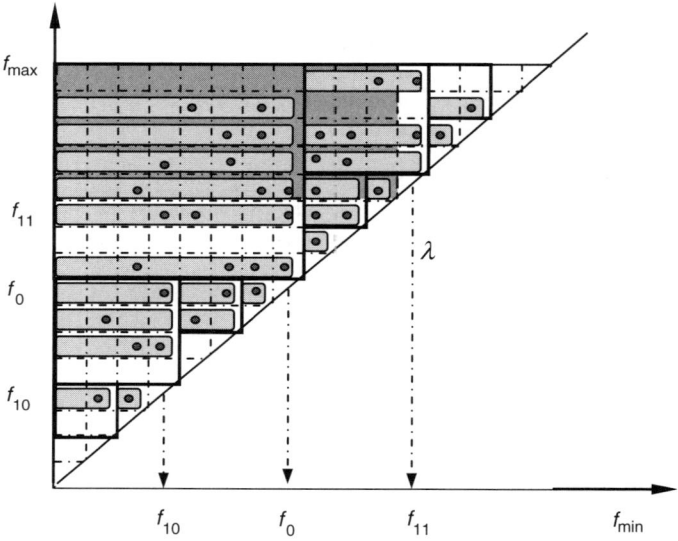

FIGURE 27.5 Span space and its use in the construction of the compact interval tree.

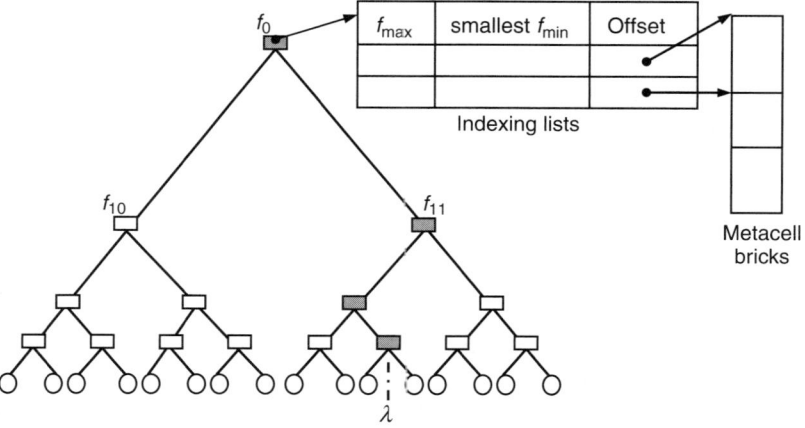

FIGURE 27.6 Structure of the compact interval tree.

TABLE 27.1 Size Comparison between the Standard Interval Tree and Our Indexing Structure

Data set	Scalar field size (bytes)	N	n	Standard interval tree (MB)	Compact interval tree (KB)
Bunny	2	2,502,103	5,469	38.2	42.7
MRBrain	2	756,982	2,894	11.6	22.6
CTHead	2	817,642	3,238	12.5	25.3
Pressure	4	24,507,104	20,748,433	374	158.3
Velocity	4	27,723,399	22,108,501	423	168.7

to the nonempty bricks, each entry containing three fields: the f_{max} value of the brick, the smallest f_{min} value of the metacells in the brick, and a pointer that indicates the start position of the brick on the disk (see Figure 27.6). Each brick contains contiguous metacells in increasing order of f_{min} values. We recursively repeat the process for the left and right children of the root. We will then obtain two smaller squares whose bottom right corners are located, respectively, at (f_{10}, f_{10}) and (f_{11}, f_{11}) in the span space, where f_{10} and f_{11} are the median values of the endpoints of the intervals associated, respectively, with the left and right subtrees of the root. In this case, each child will have at most $n/4$ nonempty index entries associated with its corresponding bricks on the disk. This recursive process is continued until all the intervals are exhausted. At this point, we have captured all possible pairs (f_{min}, f_{max}) and their associated metacell lists.

Note that the size of the standard interval tree is, in general, much larger than the compact interval tree. We can upper bound the size of the compact interval tree as follows. There are at most $n/2$ index entries at each level of the compact interval tree and the height of the tree is no more than $\log_2 n$. Hence, our compact interval tree is of size $O(n \log n)$, while the size of the standard interval tree is $\Omega(N)$, where N is the total number of intervals and hence can be as large as $\Omega(n^2)$. Note that the size of the preprocessed dataset will be about the same size as the original input.

In Table 27.1, we compare the sizes of the two indexing structures for some well-known datasets, including the datasets Bunny, MRBrain, and CTHead from the Stanford Volume Data Archive [Stanford]. As can be seen from the table, our indexing structure is substantially smaller than the standard interval tree, even when $N \approx n$.

We now describe the algorithm that uses the compact interval tree to extract the isosurface. Given a query isovalue λ, consider the unique path from the leaf node labeled with the largest value $\leq \lambda$ to the root. Each internal node on this path contains an index list with pointers to some bricks. For each such node, two cases can happen depending on whether λ belongs to the right or left subtree of the node:

Case 1: λ falls within the range covered by the node's right subtree. In this case, the active metacells associated with this node can be retrieved from the disk sequentially starting from the first brick until we reach the brick with the smallest value f_{max} larger than λ.

Case 2: falls within the range covered by the node's left subtree. The active metacells are those whose f_{min} values satisfy $f_{min} \leq \lambda$ from each of the bricks on the index list of the node. These metacells can be retrieved from the disk starting from the first metacell on each brick until a metacell is encountered with a $f_{min} > \lambda$.

Note that since each entry of the index list contains the f_{min} of the corresponding brick, no I/O access will be performed if the brick contains no active metacells. It is easy to verify that the performance of our algorithm is I/O optimal, in the sense that the active metacells are the only blocks of data that are moved into main memory for processing.

27.4 Strategies for Parallel Isosurface Extraction

To describe the basic strategies used for extracting isosurfaces on multiprocessor systems, we need to establish some general framework for discussing and evaluating parallel algorithms. We start by noting that a number of multiprocessor systems with different architectures are in use today. Unfortunately, the different architectures may require different implementations of a given algorithm in order to achieve *effective and scalable performance*. By effective and scalable performance, we mean a performance time that scales linearly relative to the best-known sequential algorithm, without increasing the size of the problem. In particular, such parallel algorithm should match the best sequential algorithm when run on a single processor. The development of such an algorithm is, in general, a challenging task unless the problem is embarrassingly parallel (i.e., the input can be partitioned easily among the processors, and each processor will run the sequential algorithm almost unchanged).

A major architectural difference between current multiprocessors relates to the interconnection of the main memory to the processors. There are two basic models. The first is the *shared-memory model* in which all the processors share a large common main memory and communicate implicitly by reading from or writing into the shared memory. The second model is a *distributed memory* architecture in which each processor has its own local memory, and processors communicate through an interconnection network using message passing. In general, distributed memory systems tend to have a better cost/performance ratio and tend to be more scalable, while shared-memory systems can have an edge in terms of ease of programmability. Many current systems use a cluster of symmetric multiprocessors (SMPs), for which all processors within an SMP share the main memory, and communication between the SMPs is handled by message passing through an interconnection network.

Another architectural difference, which is quite important for us, is the interconnection between the disk storage system and the processors. For a shared storage system, the processors are connected to a large shared pool of disks and a processor either has very small or no local disk. Another possibility is to assume that each processor has its own local disk storage, and processors can exchange the related data through an interconnection network using message passing. These two models are the most basic and other variations have been suggested in the literature.

In this chapter, we will primarily focus on distributed memory algorithms (not involving SMPs) such that each processor has its own disk storage. Given that we are primarily interested in isosurface extraction for large-scale data of size ranging from hundreds of gigabytes to several terabytes, we will make the assumption that our data cannot fit in main memory, and hence it resides on the local disks of the different processors. Under these assumptions, a parallel isosurface extraction algorithm has to deal with the following issues:

1. What is the input data layout on the disks of the processors performed during the preprocessing step? And what is the overhead in arranging the data in this format from the initial input format?
2. How is the overall dataset indexed and where is the index stored? What is the size of the index set?
3. What is the role of each processor in identifying the active cells and generating the triangles?

Critical factors that influence the performance include the amount of work required to index and organize the data during the initial preprocessing phase, the relative computational loads on the different processors to identify active cells and generate corresponding triangles for any isovalue query, and the amount of work required for coordination and communication among the processors. More specifically, the performance of a parallel isosurface extraction algorithm depends on the following factors:

1. *Preprocessing Overhead*, which include the time required to layout the dataset on the processors' disks and to generate the indexing structure.
2. *Query Execution Time*, which depends on the following three measures:
    a. *Total Disk Access Time*, which can be captured by the maximum number of disk blocks accessed by any processor from its local disk.
    b. *Processor Computation Time*, which amounts to the maximum time spent by any processor to search through the local metacells (when in memory) and generate the corresponding triangles.
    c. *Interprocessor Communication*, which can be captured by the number of communication steps and the maximum amount of data exchanged by any processor during each communication step.

We focus here on the query execution time while taking note of the complexities of the preprocessing phase. Regarding the (parallel) execution time, the relative importance of the three measures depend on the multiprocessor system used and on their relative magnitudes. For example, interprocessor communication tends to be significantly slower for clusters assembled using off-the-shelf interconnects than for vendor-supplied multiprocessors with a proprietary interconnection network, and hence, minimizing interprocessor communication should be given a high priority in this case.

Having set a context for evaluating parallel isosurface extraction algorithms, we group the recently published parallel algorithms under the following two general categories.

Tree-Based Algorithms
Under this category, we group the algorithms that make use of a single tree structure for indexing the whole dataset using either an octree or an external version of the interval tree. For example, using a shared-memory model, a parallel construction of an octree is described in Reference [BGE+98], after which the isosurface extraction algorithm amounts to a sequential traversal of the octree until the appropriate active blocks (at the leaves) are identified. These blocks are assigned in a round-robin fashion to the available processors (or different threads on a shared memory). Each thread then executes the MC algorithm on its local data. The preprocessing step described in [CS99, CFS+01] and CSS98 involves partitioning the dataset into metacells, followed by building a B-tree like interval tree, called Binary-Blocked I/O interval tree (BBIO tree). The computational cost of this step is similar to an external sort, which is not insignificant. The isosurface generation requires that a host traverses the BBIO tree to determine the indices of the active metacells, after which jobs are dispatched on demand to the available processors. This class of algorithms suffers from the sequential access to the global indexing structure, the amount of coordination and communication required to dynamically allocate the corresponding jobs to the available processors, and the overhead in accessing the data required by each processor.

Range Partitioning-Based Algorithms
Under this category, we group all the algorithms that partition the range of scalar field values and then allocate the metacells pertaining to each partition to a different processor. In this case, the strategy is an attempt to achieve load balancing among the processors that use the local data to extract active cells. For example, the algorithm described in [ZBB01] follows this strategy. Blocks are assigned to different processors on the basis of the partitions a block spans (in the hope that for any isovalue, the loads among the processors will be even). An external interval tree (BBIO tree) is then built separately for the local data available on each processor. The preprocessing algorithm described in [ZN03] is based on partitioning the range of scalar values into equal-sized subranges, creating afterward a file of subcubes (assuming a structured grid) for each subrange. The blocks in each range file are then distributed across the different processors, on the basis of a work estimate of each block. In both cases, the preprocessing is computationally demanding and there is no guarantee that the loads on the different processors will be about the same in general. In fact, one can easily come up with cases in which the processors will have very different loads, and hence the performance will degrade significantly in such cases.

Next, we describe an effective and scalable algorithm based on the compact interval tree indexing structure reported in [WJV06].

27.5 A Scalable and Efficient Parallel Algorithm for Isosurface Extraction

The isosurface extraction algorithm based on the compact interval tree, described in Section 27.3, can be mapped onto a distributed memory multiprocessor system in such a way as to provably achieve load balancing and a linear scaling of the I/O complexity, without increasing the total amount of work relative to the sequential algorithm. Recall that the indexing structure is a tree such that each node contains a list of f_{max} values and pointers to bricks, each brick consisting of a contiguous arrangement of metacells with increasing f_{min} (but with the same f_{max} value). Let p be the number of processors available on our

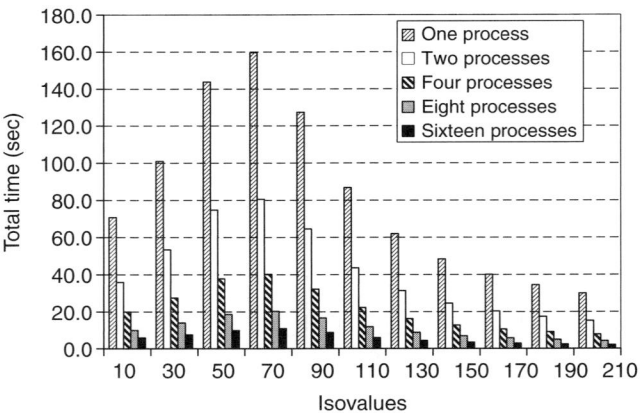

FIGURE 27.7

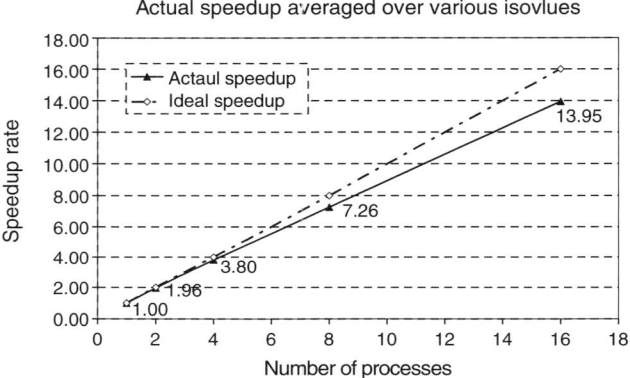

FIGURE 27.8

multiprocessor system, each with its own local disk. We now show how to distribute the metacells among the local disks in such a way that the active metacells corresponding to any isovalue are spread evenly among the processors.

For each index in the structure, we stripe the metacells stored in a brick across the p disks, that is, the first metacell on the first brick is stored on the disk of the first processor, the second on the disk on the second processor, and so on, wrapping around as necessary. For each processor, the indexing structure will be the same except that each entry contains the f_{min} of the metacells in the corresponding local brick and a pointer to the local brick. It is clear that, *for any given isovalue, the active metacells are spread evenly among the local disks of the processors.*

Given a query, we start by broadcasting the isovalue to all the processors. Each processor will then proceed to use its own version of the indexing structure to determine the active metacells and extract the triangles as before. Given that each processor will have roughly the same number of active metacells, approximately the same number of triangles will be generated by each processor simultaneously. Note that it is easy to verify that we have split the work almost equally among the processors, without incurring any significant overhead relative to the sequential algorithm.

We have performed extensive experimental tests on this algorithm, all of which confirmed the scalability and efficiency of our parallel implementation. Here we only cite experimental results achieved on a 16-node cluster, each node is a 2-way Dual-CPU running at 3.0 GHz with 8 GB of memory and 60 GB local disk, interconnected with a 10 Gbps Topspin InfiniBand network. We used the Richtmyer–Meshkov instability dataset produced by the ASCI team at Lawrence Livermore National Labs (LLNL). This dataset represents a simulation in which two gases, initially separated by a membrane, are pushed against a wire mesh, and then perturbed with a superposition of long wavelength and short wavelength disturbances and a strong shock wave. The simulation data is broken up over 270 time steps, each consisting of a $2048 \times 2048 \times 1920$ grid of one-byte scalar field. Figures 27.7 and 27.8 illustrate the performance and scalability of the algorithm described in this section using 1, 2, 4, 8, and 16 processors. Note that the one-processor algorithm is exactly the same as the one described in Section 27.3 and hence the scalability is relative to the fastest sequential algorithm.

27.6 View-Dependent Isosurface Extraction and Rendering

The isosurface extraction techniques described thus far aim at extracting a triangular mesh that closely approximates the complete isosurface. However, as the sizes of the datasets increase, the corresponding isosurfaces often become too complex to be computed and rendered at interactive speed. There are several ways to handle this growth in complexity. If the focus is on interactive rendering alone, one solution is to extract the isosurface as discussed in previous sections and then render it in a view-dependent manner using ideas developed by Xia and Varshney [XV96], Hoppe [Hoppe97], and Luebke and Erikson [LE97]. However, full-resolution isosurface extraction is very time-consuming and such a solution can result in a very expensive preprocessing and an unnecessarily large collection of isosurface triangles. An alternative is to extract view-dependent isosurfaces from volumes that have been preprocessed into an octree [FGH+03] or a tetrahedral hierarchy of detail [DDL+02, GDL+02]. This is a better solution, but it still involves traversing the multiresolution volume hierarchy at runtime and does not directly consider visibility-based simplification. The most aggressive approaches compute and render only the portion of the isosurface that is visible from a given arbitrary viewpoint. As illustrated in Figure 27.9, a 2D isocontour consists of three sections: A, B, and C. Only the visible portion, shown as bold curves, will be computed and rendered, given the indicated viewpoint and screen.

Experiments have shown that the view-dependent approach may reduce the complexity of the rendered surfaces significantly [LH98]. In this section, we will describe a typical sequential view-dependent isosurface extraction algorithm and then introduce two existing parallel algorithms that more or less are based on this approach. We will also briefly discuss another view-dependent approach based on *ray tracing*.

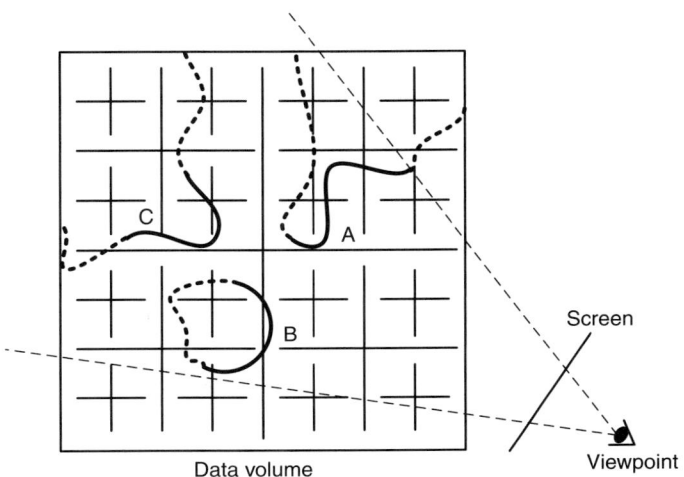

FIGURE 27.9 Illustration of the visibility of isosurface patches with respect to a viewpoint.

27.6.1 A Sequential Algorithm for Structured Grids

The view-dependent isosurface extraction algorithm requires that cells be examined in a front-to-back order with respect to the viewpoint in order to guarantee that the isosurface patches on the front are generated first and are used to test the visibility of the remaining cells. Because the visibility test is in the volume space, an indexing structure based on volume space partitioning rather than value space partitioning is more suitable. One such data structure is the octree mentioned earlier. An octree recursively subdivides the 3D rectangular volume into eight subvolumes. For the purpose of isosurface generation, each node v is augmented with a pair of *extreme values*, namely, the minimum and maximum scalar values $f_{\min}(v)$ and $f_{\max}(v)$ of the points within the corresponding subvolume. When traversing the octree, we can check at each node v visited whether $f_{\min}(v) \leq \lambda \leq f_{\max}(v)$ for a given isovalue λ. If this is the case, then the subtree rooted at v is recursively traversed. Otherwise, the subvolume corresponding to v does not intersect the isosurface and the entire subtree can be skipped.

A particular version of the octree called *Branch-On-Need Octree (BONO)* [WG92] has been shown to be space efficient when the resolution of the volume is not a power of two in all three dimensions. Unlike the standard octree, where the subvolume corresponding to an internal node is partitioned evenly, a BONO recursively partitions the volume as if its resolution is a power of two in each dimension. However, it allocates a node only if its corresponding spatial region actually "covers" some portion of the data. An illustration of a standard octree and a BONO is given in Figure 27.10 for a 1D case.

The visibility test of a region is performed using the *coverage mask*. A coverage mask consists of one bit for each pixel on the screen. This mask is incrementally "covered" as we extract the visible pieces of the isosurface. A bit is set to 1 if it is covered by the part of the isosurface that has already been drawn and 0 otherwise. The visibility of a subvolume is determined as follows.

The subvolume under consideration is first projected onto the screen. If every pixel in the projection is set to one in the coverage mask, then the corresponding subvolume is not visible and can therefore be safely pruned. Otherwise, that subvolume needs to be recursively examined. Typically, to reduce the cost of identifying the bits to be tested, the visibility of the *bounding box* of the projection instead of the projection itself is tested. This is a conservative test but guarantees that no visible portion of the isosurface will be falsely classified as invisible. Once an active cell is determined to be visible, a triangulation of the isosurface patches inside it is generated as before. The triangles thus generated are used to update the coverage mask. Figure 27.11 gives the pseudocode of the overall algorithm.

Mapping a sequential view-dependent algorithm to a distributed memory multiprocessor system presents several challenges. First, the set of visible active cells are not known beforehand, which makes it difficult to partition the triangulation workload almost equally among the available processors. Second, once the processors have finished triangulating a batch of active cells, the coverage mask needs to be updated using the triangles just computed, a process that requires significant interprocessor communication. Finally, in the case where dataset cannot be duplicated at each node, careful data placement is required so that each processor is able to access most of the data it needs locally. We next describe two parallel algorithms that attempt to address these problems.

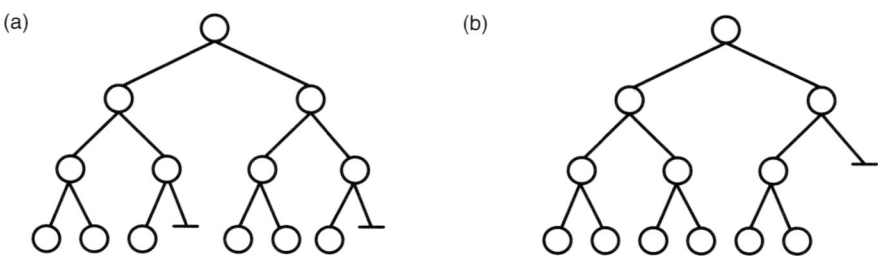

FIGURE 27.10 A 1D illustration of the (a) standard octree and (b) BONO, both having six leaf nodes.

```
procedure traverse()
begin
    Set all bits in the coverage mask to 0;
    traverse(root);
end


procedure traverse(node)
begin
    if f_min(node) > λ or f_max(node) < λ then
        return;
    else
        Compute the bounding box b of the subvolume corresponding to node.
        if b is invisible then
            return;
        else
            if node is a leaf then
                Compute the triangulation of node;
                Update the coverage mask using the triangles just generated;
                return;
            else
                for each child c of node do
                    traverse(c, λ);
                endfor
            endif
        endif
    endif
end
```

FIGURE 27.11 The framework of a sequential view-dependent isosurface extraction algorithm.

27.6.2 The Multipass Occlusion Culling Algorithm

The *multipass occlusion culling* algorithm was proposed by Gao and Shen [GS01]. In this algorithm, one of the processors is chosen as the *host* and the other processors serve as the *slaves*. Blocks, each consisting of $m \times n \times l$ cells, are the units of the visibility test. The host is responsible for identifying active and visible blocks and for distributing them to the slaves on the fly. The slaves are responsible for generating the isosurface patches within the active blocks they receive, and for updating the corresponding portions of the coverage mask.

To improve the workload balance of the slave nodes, the entire set of active blocks is extracted at once at the host using a front-to-back traversal of an octree and is stored in an *active block list*. This list is then processed in multiple passes. Each pass picks some active blocks for triangulation and discards those determined as invisible. This process continues until the active cell list becomes empty.

A modified coverage mask is used to perform the visibility test. Each bit of the coverage mask has three states: *covered*, *pending*, and *vacant*. A covered bit indicates that the corresponding pixel has been drawn already; a vacant bit indicates a pixel that has not been drawn; and a pending bit represents a pixel that *may be* drawn in the current pass.

In each pass, the host first scans the active block list and classifies each block as *visible*, *invisible*, or *pending*. An active block is visible if its projection covers at least one vacant bit in the coverage mask. An active block is determined to be invisible if every bit in it is covered. Otherwise, this active block is classified as pending. A visible block is put into the *visible block list* and will be rendered in the current pass. Its projection is also used to update the corresponding invisible bits to pending. An invisible block is discarded. A pending block may be visible depending on the triangulation result of the current pass and, therefore, will remain in the active blocks list and will be processed in the next pass.

Once the visible blocks are identified, they are distributed to the slaves for triangulation. To keep the workloads at the slaves as balanced as possible, the number of active cells in each active and visible block is estimated. A recursive image space partitioning algorithm is used to partition the image space into k tiles, where k is the number of slaves, such that each tile has approximately the same number of estimated active cells. The active blocks, whose projected bounding boxes have centers that lie in the same tile, are sent to the same slave. Each slave maintains a local coverage mask. Once the current batch of active blocks is processed, the slave updates its local coverage mask. The local coverage masks are then merged to update the global coverage mask, which resides at the host. A technique called *binary swap* [MPH+94] is used to reduce the amount of communication needed to update the global coverage mask.

27.6.3 A Single Pass Occlusion Culling Algorithm with Random Data Partitioning

Another parallel view-dependent isosurface extraction algorithm is due to Zhang et al. [ZBR02]. Unlike the algorithm of Reference GS01, which assumes that each node has a complete copy of the original dataset, this algorithm statically partitions the data and distributes the corresponding partitions across all the processors. Deterministic data partitioning techniques based on scalar value ranges of the cells such as the one discussed in the previous section and the one in Reference ZBB01 can be used for this partitioning task. Zhang et al. show that random data distribution could be equally effective in terms of maintaining workload balance. In their scheme, the dataset is partitioned into blocks of equal size and each block is randomly assigned to a processor with equal probability. They show that, in such a scheme, it is very unlikely that the workload of a particular processor will greatly exceed the expected average workload. The blocks assigned to a processor are stored on its local disk and an external interval tree is built to facilitate the identification of active blocks among them.

To address the problem of the high cost in updating the coverage mask, they opt to create a high-quality coverage mask at the beginning and use it throughout the entire extraction process. This coverage mask is constructed using ray casting. A set of rays are shot from the viewpoint toward the centers of each boundary face of the boundary blocks. Some additional randomly shot rays are also used. For each ray, the first active block b it intersects is added to the *visible block set* V. To further improve the quality of the coverage mask, the nearest neighboring blocks of b is also added to V. The ray casting step can be done in parallel with each processor responsible for the same number of rays. The range information of all the blocks is made available at each processor for this purpose. The occluding blocks computed during the ray casting step are collected from all the processors and broadcast so that every processor gets the complete set of occluding blocks.

Using the local external interval tree, each processor obtains the set of local active blocks. Those local active blocks that are also occluding blocks are processed immediately and the isosurface patches generated are used to create the local coverage mask. The local coverage mask produced by each processor is then merged into a global coverage mask, which is broadcast to all the processors. Notice that, unlike the multipass algorithm, the computation of the global coverage mask in this case happens only once.

Once the global coverage mask is available to all the processors, each processor uses it to decide which of the remaining active blocks are visible and processes only those blocks. The images created by different processors are then combined to yield the final image.

27.6.4 View-Dependent Isosurface Extraction by Ray Tracing

Finally, we describe another parallel scheme for extracting and rendering isosurface in a view-dependent fashion, namely, *distributed ray tracing*. It was first proposed by Parker et al. [PSL+98], and was implemented on a shared-memory multiprocessor machine and later ported to distributed memory systems [DPH+03].

Unlike the view-dependent isosurface extraction algorithm we have just discussed, the ray tracing method does not actually compute the linear approximation of the isosurface. Instead, it directly renders the image pixel-by-pixel. For each pixel p, a ray r_p is created that shoots from the viewpoint through p toward the volume. The first active cell, r_p intersects, can be identified using an indexing structure such as the k-D tree [WFM+05] or the octree [WG92]. The location and orientation of the isosurface at its intersection point with r_p is then analytically computed and is used to draw the pixel p. To find the intersection point, we need to be able to determine the function value $f(p)$ of any point p within the active cell. This can be done using trilinear interpolation of the function values sampled at the vertices of the active cell.

One benefit of this scheme is that the tracing of a ray is totally independent of the tracing of other rays. Thus, the parallelization of this scheme is quite straightforward. However, one needs to be aware that tracing thousands of rays can be very expensive compared to finding and rendering a linear approximation of the isosurface. Also, the tracing of each ray may involve any portion of the volume. Therefore, a shared-memory system is more suitable for this scheme. Nevertheless, as demonstrated in Reference DPH+03, by carefully exploiting the locality of the search and by caching data blocks residing at remote processors, the same scheme can still achieve good performance in a distributed memory system.

To avoid tracing thousands of rays in the aforementioned scheme, Liu et al. [LFL02] have used ray casting to only identify the active cells for a given view. These active cells then serve as seeds from which the visible isosurfaces are computed by outward propagation as suggested by Itoh and Koyamada [IK95].

27.7 Conclusion

We presented in this chapter an overview of the best-known techniques for extracting and rendering isosurfaces. For large-scale datasets, the interactive rendering of isosurfaces is quite challenging and requires some form of extra hardware support either through parallel processing on a multiprocessor or through effective programming of graphics processors. We have focused on the best-known parallel techniques to handle isosurface extraction and rendering, and examined their performance in terms of scalability and efficiency. We believe that the rendering techniques do not achieve the desired performance and more research is required to improve on these techniques.

References

[Stanford] http://graphics.stanford.edu/data/voldata/
[LLNL] http://www.llnl.gov/CASC/asciturb/
[BPT+99] C. L. Bajaj, V. Pascucci, D. Thompson, and X. Zhang. Parallel accelerated isocontouring for out-of-core visualization. *Proceedings of 1999 IEEE Parallel Visual and Graphics Symposium*, pp. 97–104, 1999.
[BGE+98] D. Bartz, R. Grosso, T. Ertl, and W. Straßer. Parallel construction and isosurface extraction of recursive tree structures. *Proceedings of WSCG '98, Volume III*, pp. 479–486, 1998.
[Blinn82] J. Blinn. A generalization of algebraic surface drawing. *ACM Transactions on Graphics*, 1, 235–256, July, 1982.
[BW01] D. Breen and R. Whitaker. A level-set approach for the metamorphosis of solid models. *IEEE Transactions on Visualization and Computer Graphics*, 7, 173–192, 2001.

[CFS+01] Y.-J. Chiang, R. Farias, C. Silva, and B. Wei. A unified infrastructure for parallel out-of-core isosurface and volume rendering of unstructured grids. *Proceedings of IEEE Symposium on Parallel and Large-Data Visualization and Graphics*, pp. 59–66, 2001.

[CS97] Y.-J. Chiang and C. T. Silva. I/O optimal isosurface extraction. *Proceedings IEEE Visualization*, pp. 293–300, 1997.

[CS99] Y. Chiang and C. Silva. External memory techniques for isosurface extraction in scientific visualization. *External Memory Algorithms and Visualization*, Vol. 50, DIMACS Book Series, American Mathematical Society, 1999, pp. 247–277.

[CSS98] Y.-J. Chiang, C. T. Silva, and W. J. Schroeder. Interactive out-of-core isosurface extraction. *Proceedings of IEEE Visualization*, pp. 167–174, 1998.

[Cormack63] A. M. Cormack. Representation of a function by its line integrals, with some radiological applications. *Journal of Applied Physics*, 34, 2722–2727, 1963.

[Cormack64] A. M. Cormack. Representation of a function by its line integrals, with some radiological applications. II. *Journal of Applied Physics*, 35, 2908–2913, 1964.

[DDL+02] E. Danovaro, L. De Floriani, M. Lee, and H. Samet. Multiresolution tetrahedral meshes: An analysis and a comparison. *Proceedings of the International Conference on Shape Modeling*, Banff, Canada, pp. 83–91, 2002.

[DPH+03] D. DeMarle, S. Parker, M. Hartner, C. Gribble, and Charles Hansen. Distributed interactive ray tracing for large volume visualization. *Proceedings of the 2003 IEEE Symposium on Parallel and Large-Data Visualization and Graphics (PVG'03)*, pp. 87–94, 2003.

[FGH+03] D. Fang, J. Gray, B. Hamann, and K. I. Joy. Real-time view-dependent extraction of isosurfaces from adaptively refined octrees and tetrahedral meshes. In *SPIE Visualization and Data Analysis*, 2003.

[FKU77] H. Fuchs, Z. Kedem, and S. P. Uselton. Optimal surface reconstruction from planar contours. *Communications of the ACM*, 20, 10, 693–702, October 1977.

[GS01] Jinzhu Gao and Han-Wei Shen. Parallel view-dependent isosurface extraction using multi-pass occlusion culling. In *Parallel and Large Data Visualization and Graphics*, IEEE Computer Society Press, October 2001, pp. 67–74.

[GGM74] A. N. Garroway, P. K. Grannell, and P. Mansfield. Image formation in NMR by a selective irradiative process. *Journal of Physics C: Solid State Physics*, 7, L457–L462, 1974.

[GDL+02] B. Gregorski, M. Duchaineau, P. Lindstrom, V. Pascucci, and K. I. Joy. Interactive view-dependent rendering of large isosurfaces. *Proceedings of IEEE Visualization 2002 (VIS'02)*, pp. 475–482, 2002.

[Hounsfield73] G. N. Hounsfield. Computerized transverse axial scanning (tomography): part 1. Description of system. *The British Journal of Radiology*, 46, 1016–1022, 1973.

[Hoppe97] H. Hoppe. View-dependent refinement of progressive meshes. *Proceedings of SIGGRAPH 97*, pp. 189–198, 1997.

[HNB+06] B. Houston, M. Nielsen, C. Batty, O. Nilsson and K. Museth. Hierarchical RLE level set: A compact and versatile deformable surface representation. *ACM Transactions on Graphics*, 25, 1–24, 2006.

[IK95] T. Itoh and K. Koyamada. Automatic isosurface propagation using an extreme graph and sorted boundary cell lists. *IEEE Transactions on Visualization and Computer Graphics*, 1, 319–327, December 1995.

[Kaufman91] A. Kaufman. *Volume Visualization*, IEEE Computer Society Press, ISBN 0-8186-9020-8, 1991.

[Lauterbur73] P. Lauterbur. Image formation by induced local interactions: Examples employing nuclear magnetic resonance. *Nature*, 242, 190–191, 1973.

[Levoy88] M. Levoy. Display of surfaces from volume data. *IEEE Computer Graphics and Applications*, 8, 29–37, May 1988.

[LW72] C. Levinthal and R. Ware. Three-dimensional reconstruction from serial sections. *Nature*, 236, 207–210, March 1972.

[LFL02] Z. Liu, A. Finkelstein, and K. Li. Improving progressive view-dependent isosurface propagation. *Computers and Graphics*, 26, 209–218, 2002.

[LH98] Y. Livnat and C. Hansen. View dependent isosurface extraction. In *Visualization '98*, pp. 175–180, October 1998.

[LSJ96] Y. Livnat, H. Shen, and C. R. Johnson. A near optimal isosurface extraction algorithm using the span space. *IEEE Transactions on Visualization and Computer Graphics*, 2, 73–84, 1996.

[Lorensen01] W. Lorensen. Marching through the visible man, at http://www.crd.ge.com/esl/cgsp/projects/vm/, August 2001.

[LC87] W. E. Lorensen and H. E. Cline. Marching cubes: A high resolution 3D surface construction algorithm. In Maureen C. Stone, editor. *Computer Graphics (SIGGRAPH'87 Proceedings)*, 21, pp. 161–169, July 1987.

[LE97] D. Luebke and C. Erikson. View-dependent simplification of arbitrary polygonal environments. *Proceedings of SIGGRAPH 97*, pp. 199–208, 1997.

[MPH+94] K. Ma, J. S. Painter, C. D. Hansen, and M. F. Krogh. Parallel volume rendering using binary-swap composition. *IEEE Computer Graphics and Applications*, 14, pp. 59–68, 1994.

[NHK+86] H. Nishimura, M. Hirai, T. Kawai, T. Kawata, I. Shirakawa, and K. Omura. Object modeling by distribution function and a method of image generation. *Electronics Communication Conference 85 J68-D(4)*, pp. 718–725, 1986 [in Japanese].

[OF03] S. Osher and R. Fedkiw. *Level Set Methods and Dynamic Implicit Surfaces*, Springer, New York, 2003.

[OS88] S. Osher and J. Sethian. Fronts propagating with curvature-dependent speed: algorithms based on Hamilton-Jacobi formulations. *Journal of Computational Physics*, 79, 12–49, 1988.

[PSL+98] S. Parker, P. Shirley, Y. Livnat, C. Hansen, and P.-P. Sloan. Interactive ray tracing for isosurface rendering. In *Visualization 98*. IEEE Computer Society Press, October 1998, pp. 233–238.

[SHL+96] H. W. Shen, C. D. Hansen, Y. Livnat, and C. R. Johnson. Isosurfacing in span space with utmost efficiency (ISSUE). *IEEE Visualization '96*. October 1996, pp. 281–294.

[WC71] M. Weinstein and K. R. Castleman. Reconstructing 3-D specimens from 2-D section images. *Proc. SPIE 26*, pp. 131–138, May 1971.

[WFM+05] I. Wald, H. Friedrich, G. Marmitt, P. Slusallek, and H.-P. Seidel. Faster Isosurface ray tracing using implicit KD-Trees. *IEEE Transactions on Visualization and Computer Graphics*, 11, 562–572, 2005.

[WJV06] Q. Wang, J. JaJa, and A. Varshney. An efficient and scalable parallel algorithm for out-of-core isosurface extraction and rendering. *IEEE International Parallel and Distributed Processing Symposium*, 67(5): 592–603, 2007.

[WG92] J. Wilhelms and A. Van Gelder. Octrees for faster isosurface generation. *ACM Transactions on Graphics*, 11, 201–227, July 1992.

[WMW86] G. Wyvill, C. McPhetters, and B. Wyvill. Data structure for soft objects, *The Visual Computer*, 2, 227–234, 1986.

[XV96] J. C. Xia and A. Varshney. Dynamic view-dependent simplification of polygonal models. *Proceedings of IEEE Visualization 96*, pp. 327–334, 1996.

[Yoo04] T. Yoo. *Insight into Images: Principles and Practice for Segmentation, Registration, and Image Analysis*, A. K. Peters Ltd., MA, 2004.

[ZBB01] X. Zhang, C. L. Bajaj, and W. Blanke. Scalable isosurface visualization of massive datasets on cots clusters. *Proceedings of IEEE Symposium on Parallel and Large-Data Visualization and Graphics*, pp. 51–58, 2001.

[ZBR02] X. Zhang, C. L. Bajaj, and V. Ramachandran. Parallel and out-of-core view-dependent isocontour visualization using random data distribution. *Proceedings Joint Eurographics-IEEE TCVG Symposium on Visualization and Graphics*, pp. 9–18, 2002.

[ZN03] H. Zhang and T. S. Newman. Efficient parallel out-of-core isosurface extraction. *Proceedings of IEEE Symposium on parallel and large-data visualization and graphics (PVG) '03*, pp. 9–16, October 2003.

28

Mesh-Based Parallel Algorithms for Ultra Fast Computer Vision

28.1	Introduction...	28-1
28.2	The Parallel Model of Computation.....................	28-3
28.3	Basic Building Blocks: Ultra Fast Algorithms..........	28-5
28.4	A New Paradigm: Computing on Buses...............	28-6
28.5	The Main Topics	28-9
28.6	Some Low- and Mid-Level Vision Tasks of Interest ...	28-10
	Computing the Hough Transform • Connected Component Labeling Algorithms • Image Segmentation • Region Representation and Description • Moment Computations	
28.7	Concluding Remarks	28-18
	References	28-18

Stephan Olariu
Old Dominion University

28.1 Introduction

In the area of computing, the only thing that seems to increase faster than processing speed is the need for ever more processing speed. The last few decades have seen cycle times move from milliseconds to nanoseconds, yet applications ranging from medical research to graphics to military applications, to the design of the processing chips themselves have resulted in a demand for faster processing than is available.

It was known for some time that this phenomenal growth could not continue forever. Long before electronic processing speeds began to approach limitations imposed by the laws of physics, experts expressed the opinion that the future of computer science was in parallel processing. The technological difficulties of this fundamental change to the von Neumann model of computing were considerable, but recently the many years of research have finally begun to pay off and the technology needed for parallel processing is finally a reality.

One major difference between increasing the speed of a single processor and moving to parallel computing is that parallel computing also requires the development of new algorithms to take advantage of the new computing model. Graham Carey introduced his edited book titled *Parallel Supercomputing: Methods, Algorithms, and Applications* [6] with the apt comment that the study of parallel methods and algorithms lags behind the hardware development.

While some computations are readily adapted to parallel processing, many others, including algorithms in image processing and robot and computer vision require fundamental changes in the approach to the problem. Computer vision tasks are among the most computationally intensive known today. A typical machine and robot vision system integrates algorithms from a large number of disciplines including statistics, numerical analysis, artificial intelligence, optics, computer architecture, image processing, pattern recognition, graph theory, and many others. Computer vision is one of the "Grand Challenges" identified by the federal Government's High-Performance Computing and Communication (HPCC) Initiative.

Computer vision tasks are traditionally partitioned into three distinct categories, depending on the types of objects they operate on. It is fairly standard to refer to the computational tasks involving two-dimensional arrays of pixels as *low-level* tasks. At the other end of the spectrum, *high-level* vision deals with more refined structures and are closely related to scene analysis and image understanding. *Mid-level* vision tasks fall between these extremes. As with low-level vision tasks, they operate on two-dimensional arrays of pixels. Their primary role, however, is to create symbolic representations of the image under investigation.

Given the wide spectrum of life and safety-critical applications in medical applications, computer vision tasks often require fast solutions. It is common knowledge that these tasks involve an enormous amount of computation, especially in the low- and mid-level stages of the processing. One way to achieve fast solutions for computer vision problems is to increase the performance of present-day machines. Although single-processor computer performance has grown rapidly in the past, it is becoming increasingly clear that this rate of growth cannot be sustained: the speed of light and the laws of quantum mechanics act as formidable barriers that cannot be breached.

In the last decade, it has been realized that a different and more cost-effective way to increase the processing power is to turn to parallel and massively parallel computing devices. But parallel machines by themselves will not necessarily bring about a dramatic improvement in the running time of computer vision tasks, unless more efficient solutions to fundamental vision problems are obtained. Indeed, in many contexts, inefficient solutions to low- and mid-level vision tasks are *the* bottleneck hampering any increase in computational power from translating into increased performance of the same order of magnitude. To illustrate the importance of fast algorithms for fundamental vision problems consider the task of computing the Fast Fourier Transform (FFT) of an 512×512 image. The direct implementation of the naive algorithm would take a PC performing 40,000 real multiplications per second about 2 months. Even a supercomputer with a computational power of 1000 MFLOPS would take about 3 min, which is unacceptable. The reader is referred to References 8, 37, 77, 78, 93 where a strong case is made for *fast* algorithms for fundamental vision tasks.

This chapter looks at a number of fundamental low- and mid-level vision problems in the context of reconfigurable architectures. Specifically, our stated goal is to obtain fast and ultra fast (to be made precise in Section 28.2.2) algorithms for these problems on reconfigurable meshes, that is, on mesh-connected computers enhanced by the addition of a reconfigurable bus system.

We review a powerful algorithmic paradigm on reconfigurable architectures that holds the promise of achieving very high speed for a large number of computer vision tasks. Our work should have a direct impact on such fundamental issues as discriminating texture features for detecting man-made objects in a given terrain and the automatic target recognition and acquisition.

Common wisdom has it that buses are entities created to support fast communication and data transfer needs within a parallel machine. Our main point is that buses can, and indeed must, also be used as computational devices or as topological descriptors of relevant objects within the image at hand. In fact, in some cases, confining the buses to serve as communication paths only, leads to inefficiencies in the algorithm. For example, in a number of applications once buses are established in response to communication needs of the algorithm, it becomes necessary to carry out computations involving items stored by the processors belonging to various buses in the bus system. It is important to note that the dynamic nature and shape of these buses do not allow the algorithm designer to efficiently use the computational power of processors outside these buses. Thus, processors on buses have to be able to cooperate to solve the problem at hand, in spite of (indeed, because of) the dynamic nature of the bus system.

We also survey methodologies that make it possible to better harness the computational power of bus systems in those cases where it is impractical to call upon processors outside the buses to carry relevant computation.

Along similar lines, in some contexts (image segmentation being a prime example) it is of interest to mandate the construction of buses that topologically characterize a region of the image, by separating it from other regions. Up to now, nothing has been done in this direction. Recently this powerful paradigm has been applied for the purpose of obtaining a fast contour tracing and component labeling algorithm for reconfigurable meshes. But this is just one application, with many more to be discovered.

Th remainder of the chapter is organized as follows: Section 28.1 describes the reconfigurable mesh, the parallel computational model adopted, and the rationale behind adopting it. Section 28.3 briefly surveys relevant building blocks—ultra fast algorithms devised to solve a number of low-level vision tasks. Section 28.4 outlines our new computational paradigm that we call *computing on buses*. It is this new paradigm that is at the heart of the chapter. Section 28.5 outlines the main research topics we survey. Section 28.6 discusses a number of fundamental low- and mid-level image tasks. Finally, Section 28.7 offers concluding remarks.

28.2 The Parallel Model of Computation

Recent advances in VLSI have made it possible to build massively parallel machines featuring many thousands of cooperating processors. Among these, the mesh-connected computer architecture has emerged as one of the most natural choices for solving image processing and computer vision tasks. Its regular structure and simple interconnection topology makes the mesh particularly well suited for VLSI implementation [30, 33, 45]. In addition, matrices and digitized images map naturally onto the mesh. It comes as no surprise, therefore, that many image processing and computer vision algorithms have been reported on mesh-connected computers [4, 8, 11, 16, 24, 36, 41, 54, 59, 72, 79, 88, 95].

However, owing to their large communication diameter, meshes tend to be slow when it comes to handling data transfer operations over long distances. In an attempt to overcome this problem, mesh-connected computers have recently been augmented by the addition of various types of bus systems [5, 28, 32, 33, 75]. A common feature of these bus structures is that they are static in nature, which means that the communication patterns among processors cannot be modified during the execution of the algorithm.

Typical computer and robot vision tasks found today in industrial, medical, and military applications involve digitized images featuring millions of pixels [2, 8, 11, 19, 27, 34, 41]. The huge amount of data contained in these images, combined with real-time processing requirements have motivated researchers to consider adding reconfigurable features to high-performance computers. Along this line of thought, a number of bus systems whose configuration can change, under program control, have been proposed in the literature: such a bus system is referred to as *reconfigurable*. Examples include the *bus automaton* [83], the *reconfigurable mesh* [57], and the *polymorphic torus* [49, 55].

A reconfigurable mesh of size $M \times N$ consists of MN identical processors positioned on a rectangular array with M rows and N columns. As usual, it is assumed that every processor knows its own coordinates within the mesh: we let $P(i, j)$ denote the processor placed in row i and column j, with $P(1, 1)$ in the north-west corner of the mesh. Every processor $P(i, j)$ is connected to its four neighbors $P(i - 1, j)$, $P(i + 1, j)$, $P(i, j - 1)$, and $P(i, j + 1)$, provided they exist.

Each processor has four ports denoted by N, S, E, and W, as illustrated in Figure 28.1. Local connections between these ports can be established, under program control, creating a powerful bus system that changes dynamically to accommodate various computational needs. Our computational model allows at most two connections to be set in each processor at any one time. Furthermore, these two connections must involve disjoint pairs of ports, as shown in Figure 28.2.

It is worth noting that in the absence of local connections, the reconfigurable mesh is functionally equivalent to the mesh-connected computer.

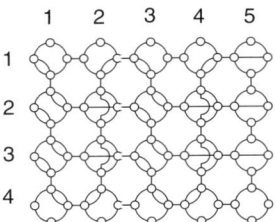

FIGURE 28.1 A reconfigurable mesh of size 4×5.

FIGURE 28.2 Examples of allowed connections.

By adjusting the local connections within each processor several subbuses can be dynamically established. We assume that the setting of local connection is destructive in the sense that setting a new pattern of connections destroys the previous one.

At any given time, only one processor can broadcast a value onto a bus. Processors, if instructed to do so, read the bus: if no value is being transmitted on the bus, the read operation has no result. We assume that the processing elements have a constant number of registers of $O(\log MN)$ bits and a very basic instruction set. Each instruction can consist of setting local connections, performing a simple arithmetic or boolean operation, broadcasting a value on a bus, or reading a value from a specified bus. More sophisticated operations including floor and modulo computation as specifically excluded. As it turns out, these operations can, in fact, be added to the repertoire with very little overhead, thus setting the stage for a very powerful computing environment [62]. We assume a SIMD model: in each time unit the same instruction is broadcast to all processors, which execute it and wait for the next instruction. Its low wiring cost and regular structure make the reconfigurable mesh suitable for VLSI implementation [50]. In addition, it is not hard to see that the reconfigurable mesh can be used as a universal chip capable of simulating any equivalent-area architecture without loss of time.

It is worth mentioning that at least two VLSI implementations have been performed to demonstrate the feasibility and benefits of the two-dimensional reconfigurable mesh: one is the Yorktown Ultra-Parallel Polymorphic Image Engine (YUPPIE) chip [50, 55] and the other is the Gated-Connection Network (GCN) chip [85,86]. These two implementations suggested that the broadcast delay, although not constant, is very small. For example, only 16 machine cycles are required to broadcast on a 10^6-processor YUPPIE. The GCN has further shortened the delay by adopting precharged circuits. Recently, it has been shown in [84] that the broadcast delay is even further reduced if the reconfigurable bus system is implemented using Optical Fibers [20] as the underlying global bus system and Electrically Controlled Directional Coupler Switches (ECS) [20] for connecting or disconnecting two fibers. Owing to these new developments, the reconfigurable mesh will become a feasible architecture in the near future. Consequently, algorithms developed on the reconfigurable mesh will become of a practical relevance.

To express the complexity of our algorithms in a realistic way, it is assumed that broadcasting information along buses does not come for free: we let $\delta(N)$ stand for the time needed to broadcast a piece of information along a bus of length $O(N)$.

One of the important goals of this chapter is to identify bottlenecks that prevent progress in the design of fast parallel algorithms for fundamental vision tasks of immediate relevance to medical applications.

28.3 Basic Building Blocks: Ultra Fast Algorithms

Throughout this chapter, we assume that a digitized image I of size $N \times N$ is stored in the natural way, one pixel per processor, in a reconfigurable mesh of size $N \times N$. Let A be an algorithm. We shall say that A is *fast* if its running time is bounded by $O(\delta(N) \log N)$; algorithm A is said to be *ultra fast* if its running time is bounded by $O(\delta(N) \log \log N)$. Indeed, obtaining ultra fast algorithms for computer vision is one of the most ambitious goals of the machine vision community these days [8, 41, 78]. Our belief is that such algorithms for a number of fundamental low- and mid-level vision tasks are possible to obtain, if judicious use is made of the buses in a reconfigurable architecture.

The purpose of this section is to present some ultra fast algorithms recently devised. These results have been successfully used to devise a number of algorithms for low and mid-level vision tasks on reconfigurable meshes.

As it turns out, in many important applications in pattern recognition, image processing, computer and robot vision, the problem at hand can be reduced to a computational problem involving only integer values. These problems are collectively referred to as integer problems. It is often possible to design specialized algorithms to handle these integer problems more efficiently than would standard, general-purpose algorithms.

With this observation in mind, Olariu et al. [66,67] have obtained a $O(\delta(N))$ time algorithm to compute the prefix sums of an N-element integer sequence on a reconfigurable mesh of size $N \times N$ [66,67]. As an interesting consequence of this algorithm, we have obtained $O(\delta(N) \log \log N)$ time algorithms to solve a number of matrix-related tasks that have direct applications to image processing and machine vision. Specifically, we have shown that our technique affords ultra fast algorithms to compute low-level descriptors including the perimeter, area, histogram, and median row of a binary image on a reconfigurable mesh [67–69].

The problem of sorting n items in constant time on a reconfigurable mesh of size $n \times n$ was open for almost 10 years. Quite surprisingly, the problem was solved, quasi independently, by a number of research groups at roughly the same time. For example, Lin et al. [52] have shown that N items from an ordered universe stored in one row of a reconfigurable mesh of size $N \times N$ can be sorted in $O(\delta(N))$ time, where $\delta(N)$ is the time delay incurred as a message is broadcast along a bus of length N. As mentioned before, for all practical values of N, this translates as an $O(1)$ time algorithm, with a very small proportionality constant. Although this algorithm has been devised to handle the "sparse" case, it can be used as a key ingredient in a number of dense situations by reducing, in stages, the dense case to a sparse one. To see how this paradigm works, assume that a certain computation involves all the pixels in an $N \times N$ image. Begin by subdividing the image into subimages of a suitable size (problem dependent) and compute a certain value in each subimage; as a second step, several values are grouped together and a further intermediate result is evaluated; this process is continued until sufficiently few items remain to fit in one row of the mesh. At this moment the fast sparse algorithm can be used. For example, this algorithmic paradigm has been used by Olariu et al. [66–68] to obtain a number of other fast algorithms on reconfigurable meshes.

These algorithms are basic tools that can be used to devise other fast and ultra fast algorithms for low- and mid-level vision tasks of immediate relevance to medical applications. But more and more tools are needed. Past experience shows that existing tools often lead to better understanding of several algorithmic issues as well as to more sophisticated algorithmic paradigms.

On the methodological side, one of our key contributions thus far was to show that all the computational tasks mentioned above belong to the same class, in the sense that they can be solved using the same technique. All the algorithms mentioned in this section vastly improve on the state of the art. Indeed, the fastest known algorithms [39,50] compute the area and the perimeter of an image in $O(\delta(N) \log N)$ time. For small values of $\delta(N)$, our algorithms discussed in this subsection perform *exponentially* faster, while using the *same* number of processors.

28.4 A New Paradigm: Computing on Buses

Throughout this chapter, we shall take the view that buses are dynamic entities that are created under program control to fulfill a number of algorithmic needs. Surprisingly, up to now, buses within a bus system have been used *exclusively* to ferry data around. To the best of our knowledge, no attempts have been made to use the processors belonging to buses to perform any other role within the overall computational effort.

Olariu et al. [70,71] has proposed a new way of looking at a reconfigurable bus system, that has resulted in a new and potentially powerful computational paradigm. Specifically, they have shown that

1. The processors on buses can be used to carry out meaningful computation.
2. The bus can be configured to reflect essential aspects of the topology of the image at hand and bus processors are then used in a way supportive of the computational goals of the algorithm.

The purpose of this section is to briefly discuss this new point of view and the advantages are that it confers algorithms for computer vision. It is often the case that to suit various communicational needs among groups of processors, several distinct buses are created within a reconfigurable mesh. Moreover, the number of the buses created can be quite large and, as a rule, their topology cannot be predicted before the algorithm is run. This being the case, it is obvious that we cannot rely on processors outside of these buses to carry out meaningful computations involving items stored by the processors belonging to buses.

To illustrate this last point, consider an example from medical image processing. The algorithm has to locate and count various foci of infection within the image at hand. This task is best performed by having processors that detect a certain condition set their local connection in a prescribed way. What may result is a set of disjoint closed buses. For computational purposes, we need to elect a *leader* on every such bus, that is one processor of the bus that will be responsible for reporting the results of the computation. It is easy to see that the dynamic reconfigurability of the bus system precludes us from having processors outside of a given bus elect the leader or count the number of foci detected. Therefore, unless we elect the leaders of various buses sequentially, we have to rely on the processors on the bus only.

Yet another example arises in the context of connected component labeling. Here, we "wrap" a bus around every connected component in the image at hand and exploit the topology of the resulting buses to inform all the pixels of the identity of the connected component of which they are a part [70].

A bus is best thought of as a doubly linked list of processors, with every processor on the bus being aware of its immediate neighbors, if any, in the positive and negative direction. When restricted to a given bus, the processors will be assumed to have two ports: one is the positive port, the other being the negative one. The *positive rank* of a processor on a bus is taken to be one larger than the number of processors preceding the given processor when the bus is traversed in the positive direction. A bus is said to be *ranked* when every processor on the bus knows its positive and negative rank. The *length* of a bus coincides with the highest rank of a processor on that bus.

Typically, processors on ranked buses will be identified by their positive rank on the bus. Thus, $P(i)$ will denote the i-th processor on the bus in the positive direction. Let B be a ranked bus of length N. In many algorithms it is important to solve the following problem:

Problem 28.1 For every processor $P(i)$ on the bus B determine the largest j for which i is a multiple of 2^j.

As it turns out, even if the processors have a very restricted instruction repertoire, specifically excluding floor and modulo computations, Problem 28.1 can be solved in time proportional to $\log N$. The details of a simple algorithm that provides a solution to Problem 28.1 follow. During the course of the algorithm certain processors are marked. We refer to all the processors that are not marked as *unmarked*. In particular, initially all processors are unmarked. Every processor $P(i)$ on the bus uses three registers to store intermediate results. These registers are denoted by R1(i), R2(i), and R3(i). For all i, R1(i) is initialized to 0.

Step 1: All processors connect their positive and negative ports. In $\lfloor \log N \rfloor$ time processor $P(1)$ generates successive powers of 2, from 2^0 to $2^{\lfloor \log N \rfloor}$, and broadcasts on the bus, in the positive direction, ordered pairs of the form $(k, 2^k)$.

Upon receiving such an ordered pair, the unique processor $P(i)$ for which $i = 2^k$ performs the following:

R1(i) ← R2(i) ← $i = 2^k$;
R3(i) ← k;
$P(i)$ marks itself.

Note that at the end of Step 1, all processors whose rank is a power of 2 are marked.

Step 2: Every processor on the bus, except for the marked processors, connects its positive and negative ports. In $\lfloor \log N \rfloor + 1$ iterations, with k starting out at $\lfloor \log N \rfloor$ and ending at 0, in steps of -1, the marked processor whose register R3 contains k broadcasts the ordered pair $(k, 2^k)$ in the positive direction of the bus, and then connects its positive and negative ports.

Upon receiving the ordered pair $(k, 2^k)$, every unmarked processor $P(i)$ does the following:

if R1(i) + $2^k \leq i$ **then**
    R1(i) ← R1(i) + 2^k;
    R2(i) ← 2^k;
    R3(i) ← k;
endif.

An easy inductive argument establishes the following result:

Lemma 28.1 *When Step 2 terminates, for every i ($1 \leq i \leq N$), R3(i) contains the largest j for which i is a multiple of 2^j. Furthermore, R2(i) contains 2^j.*

Consider an arbitrary unranked bus B of length N. Every processor on the bus is assumed to store an item from a totally ordered universe. We are interested to solve the following problem:

Problem 28.2 *Solve the prefix maxima problem on the bus B.*

To make the problem precise, we need to introduce some terminology. Let P be an arbitrary processor on the bus. We let Neg(P) stand for the set of processors preceding P in the positive direction of B, and we let Pos(P) stand for the set of processors following P in the positive direction of B. Let $v(P)$ be the item stored by processor P.* Now the prefix maxima problem involves computing for every processor P on the bus value

$$\max_{Q \in \text{Neg}(P) \cup \{P\}} v(Q).$$

Initially, every processor on the bus marks itself "active." During the course of the algorithm certain processors will become inactive, as we are about to explain. For a processor P on the bus, its (current) active neighbors are the closest active processors, say P' in Neg(P) and P'' in Pos(P), provided they exist. An active processor P is said to be a *local maximum* if

$$v(P) > v(P') \quad \text{and} \quad v(P) > v(P'').$$

The algorithm begins by having all processors disconnect their positive and negative ports. We equate active processor with those processors that have set no connections (thus, initially, all processors are

---
*For simplicity we assume that all values on the bus are distinct.

active). The algorithm comprises at most $\lfloor \log N \rfloor$ iterations. Specifically, every iteration involves the following sequence of steps:

Step 1: Every active processor checks its two active neighbors to detect whether or not it is a local maximum.

Step 2: Every processor that is not a local maximum connects its positive and negative ports (and, thus, becomes "inactive").

Step 3: Every processor that is a local maximum in the current iteration broadcasts the value it holds in the positive direction on the bus.

Step 4: Every inactive processor updates the prefix maximum by comparing the value received with the current maximum it stores.

Since no two active processors on the bus can be local maxima in a given iteration, every iteration eliminates half of the active processors. From this it follows that this simple algorithm terminates in at most $\lfloor \log N \rfloor$ iterations. To summarize our findings we state the following result.

Lemma 28.2 *The prefix maxima problem on an unranked bus of length N can be solved in $O(\log N)$ time.*

Quite often, the dynamic setting of connections within a reconfigurable mesh will result in establishing a number of buses, some of them being closed. Such is the case in some computer vision applications where buses are "wrapped" around regions of interest in the image at hand [70]. Subsequent computation calls for selecting an ID for every region (object) under consideration. This can be done by labeling every region with the identity of one of its pixels.

Put differently, it is necessary to elect a *leader* on the bus that was created around the region and broadcast the identity of this leader to all the pixels concerned. Therefore, we shall formulate the following general problem.

Problem 28.3 Consider an unranked closed bus of N processors, with every processor holding an item from a totally ordered universe. Identify the processor that holds the largest of these items.

Clearly, the first two steps of the previous algorithm will also solve this problem. Specifically, every iteration involves the following sequence of steps:

Step 1: Every active processor checks its two active neighbors and determines whether or not it is a local maximum.

Step 2: Every processor that is not a local maximum connects its positive and negative ports, thus becoming inactive.

As for the previous problem, at most $\lfloor \log N \rfloor$ iterations are performed. Since electing a leader is equivalent (in our formulation) to computing the maximum of the items on the bus we have proved the following result.

Lemma 28.3 *A leader can be elected on an unranked bus of length N in $O(\log N)$ time.*

Next, consider an arbitrary bus B. Applying the divide-and-conquer paradigm on B assumes that every processor on the bus knows its (positive) rank. Therefore, an important task is to determine the rank of every processor on B. Specifically, we state this as the following fundamental problem.

Problem 28.4 Let B be an arbitrary bus. Compute the rank of every processor on B.

Obviously, in case the bus B is closed, our first task is to elect a leader on this bus and to transform B into a "linear" bus. We will assume that the number N of processors on the bus is known. Our algorithm for ranking a bus is an adaptation of an elegant algorithm of Cole and Vishkin [14]. They define the *r*-ruling set problem as follows: Let $G = (V, E)$ be a directed graph such that the in-degree and the out-degree of every vertex is exactly one. For obvious reasons such a graph is termed a *ring*. A subset U of V is an *r*-ruling set if (i) no two vertices in U are adjacent, and (ii) for each vertex v in V there is a directed path from v to some vertex in U whose edge length is at most r. Cole and Vishkin [14] prove the following result.

Proposition 28.1 *A 2-ruling set of a ring with N vertices can be obtained in $O(\log^* N)$ time using N processors in the EREW-PRAM.*

As noted in Reference 14, the same algorithm applies to models of computations where only local communications between successive nodes in the ring are allowed. This is precisely the case of an unranked bus. To rank an arbitrary bus B of length N we let every processor except for the first one in the positive direction store a 1 (the first processor stores a 0). Now repeatedly find a 2-ruling set in B and eliminate the nodes in the current ruling set after having added the value they contain to the next remaining element. Clearly, this process terminates in $O(\log N)$ iterations. Consequently, we have the following result.

Lemma 28.4 *An arbitrary bus of length N can be ranked in $O(\log N \log^* N)$ time using the processors on the bus only.*

It is important to note that once a bus is ranked it becomes computationally equivalent to a one-dimensional reconfigurable array. Therefore, all the results derived for one-dimensional reconfigurable arrays apply directly to ranked buses. In particular, several machine architectures can be simulated by buses. One such important architecture is the *tree machine* consisting of a complete binary trees of processors [45]. Now a well-known result of Lengauer [46] concerning embedding trees into linear arrays has the following consequence.

Lemma 28.5 *Every computational step of an N-node tree machine can be simulated on a ranked bus with N processors with a slowdown of $O(\log N)$. Furthermore, this is optimal.*

Note that Lemma 28.5 guarantees that any ranked bus independent of "shape" can be used to solve problems for which efficient tree solutions exist. These problems include parenthesis matching, expression evaluation, language recognition, to name just a few.

One of the main goals of this chapter is to investigate other uses for buses within a reconfigurable bus system, for the purpose of designing fast and ultra fast vision algorithms that are simpler and faster than the best algorithms to date. The central focus is the development of algorithms of immediate relevance in the detection of man-made objects in natural surroundings and target identification and acquisition.

28.5 The Main Topics

The study of parallel algorithms for computer vision applications is important from both practical and theoretical points of view. On the practical side, it provides a context in which one may identify difficult computational problems; on the theoretical side it provides a framework within which to understand inherent similarities and differences between computational tasks specific to low- and mid-level vision:

This chapter focuses on designing fast and ultra fast algorithms for a number of low- and mid-level vision problems that are fundamental building blocks for more sophisticated machine and robot vision tasks. Specifically, we survey fast and ultra fast algorithms on reconfigurable meshes for the following problems that have been recognized as bottlenecks:

1. Computing the Hough transform for detecting lines and curves in digitized images.
2. Connected component labeling.
3. Image segmentation strategies including thresholding and split and merge region growing.

4. Region representation and description.
5. Moment computations. This is motivated by the fact that both geometric and Legendre moments have been recognized as attractive low-level descriptors that can also be effective in image reconstruction.

In addition, we investigate the advantage that the new paradigm of computing on buses offers for image processing and computer vision on a reconfigurable architecture. The problems we look at share the following features:

1. All are fundamental building blocks for high-level machine vision tasks and are key ingredients in image understanding efforts.
2. None of the vision tasks is known to have fast, let alone ultra fast, solutions on reconfigurable meshes.
3. Fast solutions to these problems will certainly translate into correspondingly fast solution to many other problems in machine and robot vision.

28.6 Some Low- and Mid-Level Vision Tasks of Interest

28.6.1 Computing the Hough Transform

One of the fundamental problems in computer vision and image processing is the detection of shape. An important subproblem involves detecting straight lines and curves in binary or gray-level images. The task of detecting lines is often accomplished by a computational method referred to as the Hough transform [1–4,7,12,15,17,21,22,36,38,40,50,72–74,76,91,98,99]. The basic idea of the Hough transform involves converting lines and curves in the two-dimensional image space into points in a certain parameter space. For example, when considering line segments, a straight line L can be uniquely represented by two parameters θ and ρ, where θ is the angle determined by the normal to L and the positive direction of the x axis, and ρ is the signed distance from the origin to the line L. For each angle, the number of edge pixels at a distance ρ can be computed and recorded. A large incidence of edge pixels for a given ρ and θ are evidence for the existence of a line with those characteristics.

Recently, a number of Hough transform algorithms have been designed for different parallel architectures including the shared-memory [12], the Butterfly [7], tree [35,36], pyramid [40], hypercube [7,76], and mesh-connected machines [15,21,24,45,80].

To put things in perspective, we now discuss published results on reconfigurable architectures, namely, the results of Pan and Chuang [72], Jenq and Sahni [38], Maresca and Li [50], and Shu et al. [87]. Pan and Chuang [72] assume that all the M edge pixels have been compacted into an $N \times M/N$ array and that every processing element has $O(N)$ memory. With these assumptions they propose an $O\left(N \frac{\log M}{\log N}\right)$ algorithm to compute the Hough transform. However, there are a number of missing details in Reference 72. First, it is not clear how, starting with a square image as input, the edge pixels can be compacted in the way assumed by the algorithm. Second, they assume that every processor of the mesh has $O(N)$ memory, which is not always the case.

More recently, Jenq and Sahni [38] proposed an algorithm for the Hough transform running on reconfigurable meshes and featuring a running time proportional to $p \log(N/p)$ time units, where p is the number of different angles in the quantization. Although the algorithm in Reference 38 runs on the reconfigurable mesh, it does not use the buses as computational devises, their algorithm being token-based as in the standard mesh algorithm. An algorithm for the Hough transform on the polymorphic torus has been proposed by Maresca and Li [50]. We note that the polymorphic torus is a much more powerful model that the reconfigurable mesh as *any* number of local connections are allowed within processors. The algorithm in Reference 50 is simply a standard mesh algorithm for the Hough transform that is run on the polymorphic torus by embedding the mesh into this latter architecture.

Finally, Shu et al. [87] have proposed an algorithm for the Hough transform on the Gated-Connection Network (GCN, for short). The GCN architecture is more powerful than the reconfigurable mesh used in this work as multiple write conflicts are allowed and handled by special-purpose hardware. The algorithm in Reference 87 involves two stages taking advantage of the flexible communication capabilities of the GCN and proceeds in two stages. In the first stage, the algorithm detects consecutive connected subsegments through projections in coarse orientation quantization and then joins them by line segments. In the second stage, the segments are merged in fine quantization to compute the beginning and end of the detected lines.

The purpose of this subsection is to illustrate how the proposed paradigm of using buses to carry out meaningful computation applies to the problem in hand. Let L be a straight line in the plane, and consider the two parameters ρ and θ described above. It is obvious that if we restrict θ to the range $0\ldots\pi$, the ordered pair (θ, ρ) uniquely determines the line L. In addition, a point (x, y) of the plane belongs to L whenever

$$x \cos \theta + y \sin \theta = \rho. \tag{28.1}$$

Assume that all the edge pixels (i.e., pixels in the image space that may belong to a straight line) are 1s and all the remaining pixels are 0s. Next, let the θ-space be quantized with $\theta_1, \theta_2, \ldots, \theta_p$ standing for the angles in the quantization. For an image of size $N \times N$, the Hough transform involves building a matrix $H[1\ldots p, -N\sqrt{2}\ldots N\sqrt{2}]$ such that for every i ($1 \leq i \leq p$), $H[i, \rho]$ contains the number of the edge pixels (x, y) for which $\lfloor x \cos \theta + y \sin \theta \rfloor = \rho$.

We are now in a position to describe how our Hough transform algorithm on reconfigurable meshes works. For this purpose, a reconfigurable mesh of size $N \times N$ is assumed, with every processor $P(i, j)$ storing pixel(i, j) whose x and y coordinates are i and j, respectively.

For simplicity, consider an angle θ satisfying $0 \leq \theta < \frac{\pi}{4}$. Now for every pixel$(i, j)$ in the image space $\rho_{i,j} = \lfloor i \cos \theta + j \sin \theta \rfloor$ is computed. Let R_ρ be the collection of all pixels that have $\rho_{i,j} = \rho$. Olariu et al. [71] have shown that R_ρ is the collection of vertical intervals in the image where the bottom of one interval is immediately adjacent to the top of the next.

A bus can thus be passed through each of the pixels in R_ρ in a natural way in constant time. Note that all pixels in the image will belong to some of the R_ρ's. Thus, the processors used to add the number of edge pixels in each R_ρ must be restricted to the bus connecting the pixels/processors making up R_ρ. We now note that every bus has length $O(N)$ and that processors along these buses can be ranked by their "city block" distance from the processor at the beginning of the bus. At this point we can apply the technique in Reference 57 to compute the number of edge pixels in $O(\delta(N) \log N)$ time.

Since the sums for all ρ at a given θ are computed in two steps, and each step takes $O(\delta(N) \log N)$ time, the row of values in H corresponding θ can be computed in $O(\delta(N) \log N)$ time. Thus, the complete set of values for H can be computed in $O(p\delta(N) \log N)$ time.

Consequently, the Hough transform matrix H of a $N \times N$ digitized image stored one pixel per processor by a reconfigurable mesh of size $N \times N$ can be computed in $O(p\delta(N) \log N)$ time, where p is the number of angles in the quantization of the θ-space.

In fact, we can do better. The idea of the improved algorithm is to pipeline the p angles $\theta_1, \theta_2, \ldots, \theta_p$ through p submeshes $S_1, S_2, \ldots, S_p$ of size $(N/p) \times N$. Initially, the mesh S_1 processes angle θ_1 as described above for $\delta(N) \log(N/p)$ time units. In the next stage, the submesh S_2 receives the partial result computed by S_1 and continues processing θ_1; at the same time S_1 starts processing the angle θ_2, and so on. It is easy to confirm that the whole computation is finished in $(2p - 1)\delta(N) \log(N/p)$ time units.

In spite of its simplicity and intuitive appeal, the result we just described is certainly not optimal, although it matches the fastest known algorithm to date [38]. We propose to further look into alternate approaches for using buses for this important problem and to obtain a much faster algorithm for computing the Hough Transform.

A number of situations occurring in medical, weather forecast, and metallurgical applications call for extremely fast identification and processing of circular shapes, occurring as projections of various kinds of droplets: rain, blood, molten metal. Duda and Hart [17] have suggested the Hough transform as a natural and stable method for circle detection. A circle C is parameterized by its center of coordinates (a, b) and by its radius r. Clearly, an edge point (x_1, y_1) belongs to C if it satisfies the equation

$$(x - a)^2 + (y - b)^2 = r^2. \tag{28.2}$$

It is easy to see that, in the context of circle detection, the parameter space is three-dimensional making the task computationally intensive. Thus, the choice of quantization along the three parameter axes is very important and reflects the required accuracy and amount of computational resources available. We note that if the quantization is 1 pixel along each of the axes chosen, then to find all circles of radius r ($1 \leq r \leq N$) in an image of size $N \times N$ requires a three-dimensional matrix of size N^3.

However, in many practical applications we do not have to contend with the full generality of the problem. For instance, in medical image processing applications, the radii of the droplets (blood, for example) are known to be much smaller than the size of the image. The same observation holds in the case of rain drops and metal drops. Consequently, a natural way of restricting the circle detection task is to assume that the possible radii involved occur in a certain range only.

With this observation in mind, we shall now assume that the task at hand involves identifying and counting circles of radii $r_1, r_2, \ldots, r_k$ where k is problem-dependent. Otherwise, the setup is standard: we assume a gray image of size $N \times N$ pretiled onto a reconfigurable mesh of size $N \times N$, one pixel per processor, in the natural way. We now point out how the paradigm of computing on buses can be used to solve the problem in two special scenarios that we describe next.

Scenario 1 The circles do not occlude each other.

This scenario is typical of a quality control image processing environment where one is interested in inspecting and counting steel balls produced in a manufacturing context. If the steel balls, as they are being produced, are directed onto a planar conveyor belt then occlusions are not expected to occur.

We now show that the task of counting the number of balls of each radius of interest can be performed in $O(\delta(N) \log N)$ time. The algorithm proceeds in the following sequence of steps:

Step 1: Solve the connected component labeling problem on the image at hand; if the algorithm of Olariu et al. [70] is used, then this step runs in $O(\delta(N) \log N)$ time. As a byproduct of the algorithm in [ ], a bus will be wrapped around every component in the image. Note that since the circles are not occluding each other, the buses created in this step are disjoint.

Step 2: For each edge pixel on each bus perform the center detection as described above.

Step 3: Compute the number of votes received by each of the potential centers on every bus and choose the largest as the center of the corresponding circle. Note that using the bus operations detailed previously, the task specific to this step can be performed in $O(\delta(N) \log N)$ time.

We note that we do not really have to restrict the range of radii in the above algorithm. It is easy to see that the nonocclusion requirement is sufficient to guarantee that the entire computation will take place in $O(\delta(N) \log N)$ time, with all the circles processed in parallel.

Scenario 2 No occlusion involves more than p circles.

This scenario occurs routinely in the detection of fuel depots from aerial images as well as in medical image processing involving the detection of certain types of tumors in x-rays as well as in processing/counting blood cells.

As in the previous case, we show that the task of counting the number of "circles" of each radius of interest can be performed in $O(p\delta(N) \log N)$ time. The algorithm proceeds in the following sequence of steps:

Step 1: Solve the connected component labeling problem on the image at hand; if the algorithm of Olariu et al. [70] is used, then this step runs in $O(\delta(N) \log N)$ time. As noted above, a by-product of the algorithm in Reference 70 is that a bus will be wrapped around every component in the image. Note that since in this scenario the circles are allowed to occlude each other, the components obtained at the end of Step 1 are not circles. However, it is the case that the buses created in this step are disjoint.

Step 2: For each edge pixel on each bus perform the center detection as described above.

Step 3: On every bus in parallel, subbuses consisting of those processors for which the same center has been voted for are delimited. Since, by assumption, at most p circles occlude each other we can sum up votes on each bus in p sequential steps. Note that using the bus operations detailed previously, the task specific to this step can be performed in $O(p\delta(N)\log N)$ time.

We plan to further investigate and generalize the algorithms presented above my adopting less and less restrictive assumptions about the number of occlusions that occur. In addition, we plan to look at the problem of detecting and processing occluding ellipses, since as pointed out in Reference 98 most of the interesting applications of the Hough transform involve reasoning about these geometric figures. Our stated goal is to provide fast and ultra fast algorithms to solve these problems on reconfigurable meshes.

28.6.2 Connected Component Labeling Algorithms

A fundamental mid-level vision task involves detecting and labeling the various connected components present in the image. The task at hand is commonly referred to as *component labeling*, and has been extensively studied both in the sequential and parallel settings [2, 16, 18, 28, 34, 47, 48, 53, 55, 59, 90].

The purpose of this subsection is to show how our paradigm of computing on buses leads to a fast algorithm to label all the components in a given image. We assume that the input is a binary image I of size $N \times N$; we let pixel(i, j) denote the pixel in row i and column j of I, with pixel$(1,1)$ situated in the north-west corner of the image. The pixels of the binary image I will be referred to as 1-pixels and 0-pixels, in the obvious way.

Connectivity in images is traditionally defined in terms of neighborhoods. For the purpose of this paper, the neighbors of pixel(i, j) are pixel$(i, j-1)$, pixel$(i, j+1)$, pixel$(i-1, j)$, and pixel$(i+1, j)$. In this context, two pixels pixel(i, j) and pixel(i', j') are said to be *connected* if there exists a sequence of pixels originating at pixel(i, j) and ending at pixel(i', j'), such that consecutive pixels in the sequence are neighbors. A maximal region of pairwise connected 1-pixels is referred to as a component. The task of *labeling* the components of an image involves assigning a unique "name" to every component, and informing every 1-pixel in the image of the name of the component it belongs to.

The bulk of the component labeling algorithms in the literature are divide-and-conquer based, or else proceed along lines originating in Levialdi [47]. Our approach is different: we exploit the dynamic reconfigurability of the bus system of the mesh to "wrap" a bus around every component of the image. As it turns out, once these buses have been created, the subtasks of uniquely identifying every component and of informing every 1-pixel of the name of the component it belongs to can be performed fast.

For simplicity, we assume that the image I has been preprocessed resulting in a new image I' of size $2N \times 2N$: for all i, j ($1 \leq i, j \leq N$), pixel(i, j) is mapped to the group of pixels consisting of pixel$(2i-1, 2j-1)$, pixel$(2i-1, 2j)$, pixel$(2i, 2j-1)$, and pixel$(2i, 2j)$. We further assume that the new image I' has been pretiled in a reconfigurable mesh of size $2N \times 2N$ such that for every i, j, processor $P(i, j)$ stores pixel(i, j) of I'.

A 1-pixel pixel(i, j) is termed a *boundary* pixel if the 3×3 window centered at pixel(i, j) contains both 0 and 1-pixels. By convention, the portion of the 3×3 window that exceeds the borders of the image is assumed to contain 0-pixels. A processor storing a boundary pixel will be termed a boundary processor. On every boundary we elect a *leader*: this is a boundary processor $P(i, j)$ for which j is minimized and i is maximized, among all the processors on the same boundary as $P(i, j)$. A high level description of our component labeling algorithm follows:

Step 1: Begin by identifying boundary pixels; once this is done, the corresponding boundary processors connect a certain pair of ports, thus establishing buses where each boundary processor is connected to its neighboring boundary processors.

Step 2: The goal of this step is to determine the leader of every boundary bus created in Step 1. Arbitrarily, define the leader to be the pixel on the boundary with minimum x and maximum y coordinate. Note that the corresponding processor will have its N and E ports connected. The collection of all processors on the bus that have their N and E ports connected form the candidates for leader. Candidates pass their coordinates on the bus, first in one direction and then in the other. Processors can eliminate themselves from the candidate set by comparing their own coordinates with those of their neighbor candidates read from the bus.

This process continues until only one candidate remains active. By our previous discussion the leader election process ends in at most $O(\log N)$ steps. Now our assumption that broadcasting along a bus of length $O(N)$ takes $O(\delta(N))$ time, implies that the leader election process runs in time $O(\delta(N) \log N)$. The leader can then look locally to determine whether the corresponding bus represents an internal or external boundary.

Step 3: Every internal bus determines the identity of its corresponding external bus. Internal leaders "look" left. There are four possible results. In each case it is possible to relink the bus so that a given external bus and all its corresponding internal buses form a single bus.

Next, the identity of the external bus is broadcast along the entire bus. The total time needed for this step is $O((\delta(N))^2)$.

Step 4: The goal of this last step is to inform every processor holding a 1-pixel of the identity of the leader of the corresponding external bus. To achieve the goal of this step, we let every nonboundary processor storing a 1-pixel set its local connection to EW; processors storing 0-pixels set no connection. Next, every boundary processor broadcasts westbound the identity of the leader of the external bus of the component. This, obviously, achieves the desired result, and Step 4 is complete. Since this step depends upon communication along buses it is performed in $O(\delta(N))$ time.

Consequently, the component labeling problem of a binary image of size $N \times N$ stored one pixel per processor by a reconfigurable mesh of size $N \times N$ can be solved in $O((\delta(N))^2 + \delta(N) \log N)$ time.

One of our main objectives is to obtain a faster, possibly even an ultra fast, algorithm for this extremely important machine vision task. For this, we need to further investigate basic operations on buses. Note that except for electing the leader, the dominant cost in the remaining steps is bounded by $O((\delta(N))^2)$. It is not known whether this can be done faster. Any improvement in this algorithm will impact other vision tasks, including image segmentation and clustering techniques.

28.6.3 Image Segmentation

Image segmentation is one of the most important computational steps in automated image analysis because it is at this moment that objects of interest are identified and extracted from the image for further processing.

The purpose of image segmentation is to partition the image at hand into disjoint regions (or objects) that are meaningful with respect to some application. The partition generated by image segmentation typically features the following properties:

- Every region is homogeneous with respect to some classifier such as gray-level or texture
- Adjacent regions are significantly different with respect to the chosen classifier
- Regions are topologically simple with as few disparities (e.g., holes) as possible

Image segmentation has been well studied and numerous image segmentation techniques have been proposed in the literature [2, 4, 9, 10, 31, 43, 60, 61, 81, 82, 95, 97]. A basic ingredient in image segmentation is *clustering* [2, 51, 81], that is, the process of grouping similar patterns into larger and larger aggregates called clusters. It has been recognized that iterative clustering schemes are computationally inefficient [8, 41, 78] since they have to go through the image many times. We are therefore focusing on parallel segmentation and clustering schemes only. Thresholding is one of the most natural approaches to image segmentation, and is motivated by the observation that objects in the image tend to appear as clusters in some measurement space.

Suppose that the image at hand consists of several light objects on a black background, with the gray levels grouped into k dominant modes of the histogram (this is the case in the presence of $k-1$ types of light objects). Computationally, the thresholding technique involves identifying and ranking the local minima (i.e., "valleys") of the histogram. Further, all the gray-level values between consecutive valleys are grouped together into a cluster, labeled by the rank of the first valley to its left. Finally, the desired image segmentation is accomplished by mapping the clusters back to the image domain and computing the components of the resulting assignment.

Note how well our technique discussed in Section 28.2.2 works in this context: as noted in Reference 66 we can compute the histogram of an $N \times N$ image stored in a natural way by an $N \times N$ reconfigurable mesh in $O(\delta(N) \log \log N)$ time. Once the histogram is computed, the corresponding valleys are computed in $O(\delta(N))$ time. Finally, to complete the histogram-based image segmentation, we "wrap" buses around pixel regions that correspond to gray-levels identified as one cluster and apply the connected component algorithm discussed in the preceding subsection. The running time of the segmentation algorithm just described is dominated by the running time of the component labeling algorithm, which is $O(\delta(N) \log N)$.

Yet another important image segmentation technique goes by the name of *region growing* [2, 31, 37, 53, 81, 89]. The basic scheme involves combining pixels with adjacent pixels to form regions. Further, regions are combined with adjacent regions to form larger regions. The association of neighboring pixels and neighboring regions in the region "growing" process is typically governed by a homogeneity criterion that must be satisfied in order for regions to be merged. It is well known that the homogeneity criterion used for merging is application dependent; sometimes it varies dynamically even within the same application [37, 81].

A variety of homogeneity criteria have been investigated for region growing, ranging from statistical techniques to state-space approaches that represent regions by their boundaries [9, 11, 22, 41]. One of the segmentation techniques often associated with region growing is referred to as *split and merge* [2]. The idea is as follows: once a relevant homogeneity criterion χ has been chosen the image at hand is tested. If χ is satisfied, then there is nothing else to do. Otherwise, the image is partitioned into four quadrants and χ is tested for each of them in parallel. Quadrants that satisfy the criterion are left alone; those that do not are further subdivided recursively.

In the merge stage, adjacent homogeneous regions are combined to form larger and larger regions. There are a number of problems, however. First, the result of the segmentation depends on the order on which the individual regions are combined. In extreme cases, this leads to undesired effects like image fragmentation.

The single most useful data structure to handle the merge stage phase of this algorithm is the *quadtree* [89]. Quadtrees are simple and can be created directly in the split phase of the region growing algorithm. Many operations on quadtrees are simple and elegant. To the best of our knowledge, however, none of these operations have been implemented on reconfigurable architectures.

As a goal of this research effort we plan to investigate fast implementation of quadtree operations on reconfigurable meshes from which fast image segmentation algorithms will be developed.

There are a number of points about the split and merge approach to segmentation that we now point out. First, there is nothing magic about partitioning the mesh into quadrants. In fact, partitioning the image into submeshes of size $\sqrt{N} \times \sqrt{N}$ seems to be more promising in the context of reconfigurable meshes. Note that by applying the techniques mentioned in Section 28.2.2, we can compute the maximum and minimum gray levels in the whole image in $O(\delta(N) \log \log N)$ time. Using a partition as described above, and proceeding recursively will result into an algorithm for the split phase running in $O((\delta(N) \log \log N)^2)$ time, which is much better than the state of the art.

At this time, it is not clear how the merge phase should proceed. In particular, it is not obvious that the merge phase will run as fast as the split phase, but this has to be further elucidated. One possible approach is to use buses as topological descriptors in the split phase. Specifically, at the end of the split stage of the algorithm, every homogeneous region is characterized by a bus that follows its contour. The merge operation should be conducted in a way that makes the operation a simple bus reconfiguration operation, meaning that the buses surrounding adjacent regions to be merged will be made into a single bus wrapped around the grown region. This would lead to an $O(\delta(N) \log N)$ algorithm for the merge stage, which by our definition is fast but not ultra fast. We want to further investigate the possibility of obtaining ultra fast algorithms for the merge stage of this natural segmentation scheme. An ultra fast image segmentation algorithm would be an invaluable tool in image understanding.

Yet another important issue that we are interested in investigating is how to take advantage of the reconfigurability of the bus system to speed up other clustering and image segmentation schemes. Of particular interest here are hierarchical and agglomerative clustering [37, 51].

28.6.4 Region Representation and Description

A fundamental operation that follows image segmentation is region representation and description for further processing. Traditionally, there are two ways of representing a region:

1. By topologically describing its boundary
2. By giving its internal characteristics such as gray-levels or texture

The next task is to provide a description of each region of interest. It has been recognized that good descriptors must have a number of desirable properties such as invariance with respect to size, rotation, and translation [2, 19, 48, 81].

An important goal is to look into ways of performing the representation and description tasks ultra fast on reconfigurable meshes. The purpose of this subsection is to investigate some ideas related to these goals. A natural first step in describing the topology of some region R obtained by segmentation is to establish a bus around the region. This bus is either created as a result of the segmentation process itself, or it can be created by some other ad hoc method. For this purpose, every processor that stores a boundary pixel for R sets its local connection in a way dictated by a 3×3 window centered at the boundary pixel. What results is a closed bus $B(R)$ with the desired property. As it turns out, $B(R)$ can be used to obtain quite a few topological descriptors for R.

Probably, the most basic descriptor of region R is the length of its boundary [2]. Note that once $B(R)$ is available, this involves computing the length of the bus itself that can be done in $O(\delta(N) \log N)$ time. A second important descriptor is the convex hull of R whose computation we discuss next.

The first operation that has to be done is identifying pixels that belong to the convex hull of R. It is easy to see that all these pixels must belong to $B(R)$. Note that one of the difficulties of this approach is that the buses are unranked, that is, the processors on $B(R)$ do not know their ranks on the bus itself. We next argue that out of every boundary bus we can extract a meaningful subbus that can be ranked in $O(\delta(N) \log N)$ time. For this purpose, we begin by computing the leader of $B(R)$ in $O(\delta(N) \log N)$ time. Note that establishing a positive and negative direction on this bus is direct and can be computed locally by looking at a 3×3 window. To be more specific, we can impose the counterclockwise direction (with respect to the interior of the object) as positive. Next, we compute the prefix minima of all the row numbers of processors on $B(R)$.

Now every processor on the bus can determine by comparing its row number with the prefix minimum whether or not is can hold an extreme pixel of the object. All the processors that do not, set their connection along the bus. For all practical purposes, we have now constructed a monotone subbus B' of $B(R)$ what can be ranked in $O(\delta(N))$ time.

It should be clear that the key ingredients needed to perform divide and conquer on the bus B' are available at every processor of B' in $O(\delta(N) \log N)$ time. Finally, the convex hull of R can be easily constructed in $O((\delta(N) \log N)^2)$ time by divide and conquer. Note that the only processors used in this process are those belonging to $B(R)$. Note further that the convex hull of every region in the image at hand is computed in parallel. Since, obviously the buses involved are disjoint, no conflict will ever occur.

A further descriptor of a region R is the diameter of its boundary, defined as $\max_{x,y \in B(R)} d(z, y)$, that is the largest distance between any two pixels on the boundary of R [81]. Surprisingly, computing the diameter of a region seems to be hard. It is not known how to compute this in less than $O((\delta(N) \log N^2)$ time. This is in contrast with the well-known fact in computational geometry [81] that computing the diameter of a planar object once the convex hull is known is a linear-time operation.

We plan to look at a number of possible alternatives for computing the diameter and other natural topological descriptors of regions. In particular, the *shape* of the boundary of R can be quantified by using moments [13, 81, 92]. Our goal is to obtain fast and possibly ultra fast algorithms for these important tasks.

28.6.5 Moment Computations

An important class of low-level descriptors, namely, the *moments* find wide applications in pattern recognition and low-level computer and robot vision [2, 8, 37, 81]. Moments have the desirable property of capturing many invariant features of digitized images. In it well-known that in addition to their attractive properties as descriptors, moments feature the interesting property that by using a sufficiently large number of moments, most of the original image information can be restored [81]. This makes moments an invaluable tool in image analysis and restoration.

Less informally, the geometric (p, q)-th moment of an image $I[1...N, 1...N]$ is given by

$$M_{pq} = \sum_{i=1}^{n} \sum_{j=1}^{n} i^p j^q I[i, j]. \tag{28.3}$$

As it turns out, M_{00} is precisely the area of the image. As pointed out in Reference 81 this is a simple and robust measure for locating the approximate position of an object in the field of view. The *center* of an image is a point of coordinates x_c, y_c defined as

$$x_c = \frac{M_{10}}{M_{00}} \quad \text{and} \quad y_c = \frac{M_{01}}{M_{00}}. \tag{28.4}$$

The usual practice is to choose the center of area (also referred to as *centroid*) as a standard location descriptor. In addition, the center of an image has many applications in matching the shapes of two closed contours (see Reference 81 for details).

Finally, M_{02} and M_{20} are the moments of inertia around the i and j axes, respectively. Furthermore, $M_{02} + M_{20}$ yields the moment of inertia around the origin (assumed to be at $I[1,1]$. It is easy to verify that $M_{02} + M_{20}$ is invariant under rotation and thus an important descriptor in many applications [2, 19, 81, 92]

In addition to the geometric moments, several other moments turn out to be of interest as image descriptors. For example, the Legendre moments of order $(p + q)$ [92] are defined as follows:

$$\lambda(p, q) = \frac{(2p + 1)(2q + 1)}{4} \sum_{j=0}^{p} \sum_{k=0}^{q} a_{pj} a_{qk} M_{jk}. \tag{28.5}$$

Here, the a_{pj}s (respectively a_{qk}s) are the coefficients of the p order (respectively q order) Legendre polynomials defined as follows:

$$P_p(x) = \sum_{j=0}^{p} a_{pj} x^j = \frac{1}{2^p p!} \frac{d^p}{dx^p} (x^2 - 1)^p. \tag{28.6}$$

It is important to note that for any fixed p and q, the corresponding Legendre moment $\lambda(p, q)$ depends on the same or lower order geometric moments only. It is argued in Reference 92 that Legendre moments are less sensitive to noise than the usual geometric moments and can be used successfully in the process of image reconstruction, even in the presence of considerable noise.

Note that the techniques in References 66 and 67 that allowed to compute the area and perimeter of a binary image of size $N \times N$ in $O(\log \log N)$ time do not seem to extend to allow ultra fast computation of higher moments. It would be of interest to know whether the techniques presented in Reference 66 can be extended to handle the computation of general moments. In fact, one of the subgoals of this research effort is to devise fast and ultra fast algorithms for computing generalized moments on reconfigurable meshes. As it stands now, all the $(p+q)$-moments of the objects present in an image of size $N \times N$ can be computed in $O(\log N)$ time. At present it is not known whether $O(\log \log N)$ algorithms are achievable.

28.7 Concluding Remarks

Faced with real-time processing requirement for applications in industry, medicine, and the military, the machine vision community needs faster and faster algorithms. It has been recognized that to achieve this goal it is necessary to devise ultra fast algorithms for a number of fundamental low- and mid-level vision tasks that are key ingredients in many machine understanding activities. A possible first step in this direction is to provide a kernel of problems that are well understood and for which ultra fast solutions are possible.

Massively parallel machines have been shown to be very effective for highly regular algorithms such as image filtering and computing the FFT. For nonuniform tasks, that is tasks that exhibit a nonuniform and unpredictable load distribution, effective vision algorithms are very hard to design.

This chapter surveyed a number of fundamental low- and mid-level vision problems on a novel and very versatile architecture—the reconfigurable mesh. All the problems are motivated by, and find applications to, basic image understanding tasks. Indeed, these problems share the following characteristics:

- From a theoretical point of view, solving these problems require techniques and insight into parallel algorithmics on reconfigurable meshes that are likely to contribute to a general knowledge-base of paradigms. At the same time, revealing new computational features of the problems provide one with much needed lower-bound results of import to subsequent algorithmic efforts.
- From the algorithmic point of view, fast solutions to these problems impact image processing and machine vision at large. They will result in simpler and faster algorithms that can be applied to larger and larger classes of problems with a wide spectrum of application.

References

[1] W. J. Austin, A. M. Wallace, and V. Fraitot, Parallel algorithms for plane detection using an adaptive Hough transform, *Image and Vision Computing*, 9 (1991), 372–384.

[2] D. H. Ballard and C. M. Brown, *Computer Vision*, Prentice-Hall, Englewood Cliffs, NJ, 1982.

[3] D. Ben-Tzvi, A. ey, and M. Sandler, Synchronous multiprocessor implementation of the Hough transform, *Computer Vision, Graphics, and Image Processing*, 52 (1990), 437–446.

[4] W.-E. Blanz, D. Petkovic, and J. L. C. Sanz, Algorithms and architectures for machine vision, in C. H. Chen, Ed., *Signal Processing Handbook*, M. Dekker, New York, 1989, 276–316.

[5] S. H. Bokhari, Finding maximum on an array processor with a global bus, *IEEE Transactions on Computers*, C-33, 2 (1984), 133–139.

[6] G. F. Carey, Ed., *Parallel Supercomputing: Methods, Algorithms and Applications*, Wiley Interscience, 1989.

[7] S. Chandran and L. Davis, The Hough transform on the butterfly and the NCUBE, Technical Report, Univ. of Maryland, 1989.

[8] R. Chellappa, *Digital Image Processing*, IEEE Computer Society Press, 1992.

[9] L. T. Chen, L. S. Davis, and C. P. Kruskal, Efficient parallel processing of image contours, *IEEE Transactions on PAMI*, 15 (1993), 69–81.

[10] A. P. Choo, A. J. Maeder, and B. Pham, Image segmentation for complex natural scenes, *Image and Vision Computing*, 8 (1990), 155–163.

[11] A. N. Choudhary and J. H. Patel, *Parallel Architectures and Parallel Algorithms for Integrated Vision Systems*, Kluwer Academic Publishers, 1990.

[12] A. N. Choudhary and R. Ponnusami, Implementation and evaluation of Hough transform algorithms on a shared-memory multiprocessor, *Journal of Parallel and Distributed Computing*, 12 (1991), 178–188.

[13] L. P. Cordella and S. Sanniti Di Baja, Geometric properties of the union of maximal neighborhoods, *IEEE Trans. Pattern Analysis and Machine Intelligence*, 11 (1989), 214–223.

[14] R. Cole and U. Vishkin, Deterministic coin tossing with applications to optimal parallel list ranking, *Information and Control*, 70 (1986), 32–53.

[15] R. E. Cypher, J. L. C. Sanz, and Y. Hung, Hough transform has $O(N)$ complexity on a SIMD $N \times N$ mesh array architecture, *Proc. IEEE Workshop on Comp. Architectures for Pattern Analysis and Machine Intelligence*, 1987.

[16] R. E. Cypher, J. L. C. Sanz, and L. Snyder, Practical algorithms for image component labeling on SIMD mesh-connected computers, *Proc. Internat. Conf. on Parallel Processing*, 1987, 772–779.

[17] R. O. Duda and P. E. Hart, Use of the Hough transformation to detect lines and curves in pictures, *Communications of the ACM*, 15 (1972), 11–15.

[18] H. Embrechts, D. Roose, and P. Wambacq, Component labelling on a MIMD multiprocessor, *CVGIP: Image Understanding*, 57 (1993), 155–165.

[19] M. Fairhurst, *Computer Vision and Robotic Systems*, Prentice-Hall, Englewood Cliffs, NJ, 1988.

[20] D. G. Feitelson, *Optical Computing*, MIT Press, 1988.

[21] A. Fishburn and P. Highnam, Computing the Hough transform on a scan line of processors, *Proc. IEEE Workshop on Comp. Architectures for Pattern Analysis and Machine Intelligence*, 1987, 83–87.

[22] G. Gerig and F. Klein, Fast contour identification through efficient Hough transform and simplified interpretation strategy, *Proc. 8th Joint Conf. Pattern Recognition*, Paris, 1986, 498–500.

[23] M. Gokmen and R. W. Hall, Parallel shrinking algorithms using 2-subfield approaches, *Computer, Vision, Graphics, and Image Processing*, 52 (1990), 191–209.

[24] C. Guerra and S. Hambrush, Parallel algorithms for line detection on a mesh, *Proc. IEEE Workshop on Comp. Architectures for Pattern Analysis and Machine Intelligence*, 1987, 99–106.

[25] Z. Guo and R. W. Hall, Fast fully parallel thinning algorithms, *CVGIP: Image Understanding*, 55 (1992), 317–328.

[26] S. E. Hambrush and M. Luby, Parallel asynchronous connected components in a mesh, *Information Processing Letters*, 38 (1991), 257–263.

[27] R. M. Haralick and L. G. Shapiro, *Computer and Robot Vision*, Vol. 1, 2nd edn., Addison-Wesley, 1992.

[28] S. E. Hambrush and L. TeWinkel, A study of connected component algorithms on the MPP, *Proc. Third Internat. Conf on Supercomputing*, 1988, 477–483.

[29] L. Hayat, A. Naqvi, and M. B. Sandler, Comparative evaluation of fast thinning algorithms on a multiprocessor architecture, *Image and Vision Computing*, 10 (1992), 210–219.

[30] J. L. Hennessy and D. A. Patterson, *Computer Architecture, A Quantitative Approach*, Morgan Kaufmann Publishers, San Manteo, CA, 1990.

[31] S. Hopkins, G. J. Michaelson, and A. M. Wallace, Parallel imperative and functional approaches to visual scene labelling, *Image and Vision Computing*, 7 (1990), 178–193.

[32] K. Hwang, H. M. Alnumeiri, V. K. P. Kumar, and D. Kim, orthogonal multiprocessor sharing memory with an enhanced mesh for integrated image understanding, *CVGIP: Image Understanding*, 53 (1991), 31–45.

[33] K. Hwang and F. A. Briggs, *Computer Architecture and Parallel Processing*, McGraw-Hill, New York, 1984.

[34] S.-Y. Hwang and S. L. Tanimoto, Parallel coordination of image operators: model, algorithm and performance, *Image and Vision Computing*, 11 (1993), 129–138.

[35] H. Ibrahim, J. Kender, and D. E. Shaw, On the application of the massively parallel SIMD tree machines to certain intermediate vision tasks, *Computer Vision, Graphics, and Image Processing*, 36 (1986), 53–75.

[36] J. Illingworth and J. Kittler, A survey of the Hough transform, *Computer Vision, Graphics, and Image Processing*, 44 (1988) 87–116.

[37] B. Jagde, *Digital Image Processing*, Springer-Verlag, 1991.

[38] J.-F. Jenq and S. Sahni, Reconfigurable mesh algorithms for the Hough transform, *Proc. International Conference on Parallel Processing*, 1991, III, 34–41.

[39] J.-F. Jenq and S. Sahni, Serial and parallel algorithms for the medial axis transform, *IEEE Transactions on PAMI*, 14 (1992), 1218–1224.

[40] J. Jolion and A. Rosenfeld, An $O(\log n)$ pyramid Hough transform, *Pattern Recognition Letters*, 9 (1989), 343–349.

[41] R. Kasturi and R. C. Jain, *Computer Vision: Principles*, IEEE Computer Society Press, 1991.

[42] A. Kundu and J. Zhou, Combination median filter, *IEEE Transactions on Image Processing*, 1 (1992), 422–429.

[43] P. C. K. Kwok, A thinning algorithm by contour generation, *Communications of the ACM*, 31 (1988), 1314–1324.

[44] V. F. Leavers, Use of the Radon transform as a method of extracting information about shape in two dimensions, *Image and Vision Computing*, 10 (1992), 99–107.

[45] F. Thomson Leighton, *Introduction to Parallel Algorithms and Architectures: Arrays, Trees, Hypercubes*, Morgan Kaufmann Publishers, San Mateo, CA, 1992.

[46] T. Lengauer, Upper and lower bounds on the complexity of the min-cut linear arrangement problem on trees, *SIAM Journal on Algebraic and Discrete Methods*, 3 (1982), 99–113.

[47] S. Levialdi, On shrinking binary picture patterns, *Communications of the ACM*, 15 (1972), 7–10.

[48] M. D. Levine, *Vision in Man and Machine*, McGraw-Hill, New York, 1985.

[49] H. Li and M. Maresca, Polymorphic-torus network, *IEEE Transactions on Computers*, C-38, 9 (1989), 1345–1351.

[50] H. Li and M. Maresca, Polymorphic-torus architecture for computer vision, *IEEE Transaction on Pattern Analysis and Machine Intelligence*, 11, 3 (1989), 233–243.

[51] X. Li, Parallel algorithms for hierarchical clustering and cluster validity, *IEEE Transactions on Pattern Analysis and Machine Intelligence*, 12 (1990), 1088–1092.

[52] R. Lin, S. Olariu, J. L. Schwing, and J. Zhang, Sorting in O(1) Time on a Reconfigurable Mesh of Size $N \times N$, *Parallel Computing: From Theory to Sound Practice, Proceedings of EWPC'92*, Plenary Address, IOS Press, 1992, 16–27.

[53] H.-H. Liu, T. Y. Young, and A. Das, A multilevel parallel processing approach to scene labeling problems, *IEEE Transaction on Pattern Analysis and Machine Intelligence*, 10 (1988), 586–590.

[54] M. Mahonar and H. R. Ramapriyan, Connected component labeling on binary images on a mesh connected massively parallel processor, *Computer Vision, Graphics, and Image Processing*, 48 (1989), 133–149.

[55] M. Maresca and H. Li, Connection autonomy and SIMD computers: A VLSI implementation, *Journal of Parallel and Distributed Computing*, 7 (1989), 302–320.

[56] M. Milgram and T. de Saint-Pierre, Boundary detection and skeletonization with a massively parallel architecture, *IEEE Transaction on Pattern Analysis and Machine Intelligence*, 12 (1990), 74–78.

[57] R. Miller, V. K. P. Kumar, D. Reisis, and Q. F. Stout, Meshes with reconfigurable buses, *Proc. Fifth MIT Conf. on Advanced Research in VLSI*, (1988), 163–178.

[58] P. J. Narayanan and L. D. Davis, Replicated data algorithms in image processing, *CVGIP: Image Understanding*, 56 (1992), 351–365.

[59] D. Nassimi and S. Sahni, Finding connected components and connected ones on a mesh-connected parallel computer, *SIAM Journal on Computing*, 9 (1980), 744–757.

[60] D. Nichol and M. Fiebig, Image segmentation and matching using the binary object forest, *Image and Vision Computing*, 9 (1991), 139–149.

[61] J. A. Noble, Finding half boundaries and junctions in images, *Image and Vision Computing* 10 (1992), 219–232.

[62] S. Olariu, J. L. Schwing, and J. Zhang, Fundamental algorithms on reconfigurable meshes, *Proc. 29th Annual Allerton Conf. on Communication, Control, and Computing*, 1991, 811–820.

[63] S. Olariu, J. L. Schwing, and J. Zhang, Fundamental data movement for reconfigurable meshes, *Int'l J. High Speed Computing*, 6, 311–323, 1994. A preliminary version has appeared in *Proc. of the International Phoenix Conf. on Computers and Communications*, Scottsdale, Arizona, 1992, 472–479.

[64] S. Olariu, J. L. Schwing, and J. Zhang, integer problems on reconfigurable meshes, with applications, *Proc. 29-th Annual Allerton Conference on Communications, Control, and Computing*, 1991, 821–830.

[65] S. Olariu, J. L. Schwing, and J. Zhang, A fast adaptive convex hull algorithm on processor arrays with a reconfigurable bus system, *Proc. Third Annual NASA Symposium on VLSI Design*, 1991, 13, 2.1–2.9.

[66] S. Olariu, J. L. Schwing, and J. Zhang, Fast computer vision algorithms on reconfigurable meshes, *Image and Vision Computing Journal*, 1992. A preliminary version of this work has appeared in *Proc. 6th International Parallel Processing Symposium*, Beverly Hills, 1992.

[67] S. Olariu, J. L. Schwing, and J. Zhang, Constant time integer sorting on an $n \times n$ reconfigurable mesh, *Proc. of the International Phoenix Conf. on Computers and Communications*, Scottsdale, Arizona, 1992, 480–484.

[68] S. Olariu, J. L. Schwing, and J. Zhang, Fast mid-level vision algorithms on reconfigurable meshes, *Parallel Computing: From Theory to Sound Practice, Proceedings of EWPC'92*, IOS Press, 1992, 188–191.

[69] S. Olariu, J. L. Schwing, and J. Zhang, Efficient image processing algorithms for reconfigurable meshes, *Proc. of Vision Interface'1992*, Vancouver, British Columbia, May 1992.

[70] S. Olariu, J. L. Schwing, and J. Zhang, Fast component labeling on reconfigurable meshes, *Image and Vision Computing*, to appear. A preliminary version has appeared in *Proc. International Conference on Computing and Information*, Toronto, May 1992.

[71] S. Olariu, J. L. Schwing, and J. Zhang, Computing the Hough transform on reconfigurable meshes, *Proc. of Vision Interface'1992*, Vancouver, British Columbia, May 1992.

[72] Y. Pan and H. Y. H. Chuang, Improved mesh algorithm for straight line detection, *Proc 3rd IEEE Symposium on the Frontiers of Massively Parallel Computation*, 1990, 30–33.

[73] D. C. W. Pao, H. F. Li, and R. Jayakumar, Shape recognition using the straight line Hough transform: Theory and generalization, *IEEE Transactions on PAMI*, 14 (1992), 1076–1089.

[74] J. Princen, J. Illingworth, and J. Kittler, A hierarchical approach to line extraction based on the Hough transform, *Computer Vision, Graphics, and Image Processing*, 52 (1990), 57–77.

[75] V. K. P. Kumar and C. S. Raghavendra, Array processor with multiple broadcasting, *Journal of Parallel and Distributed Computing*, 4 (1987), 173–190.

[76] S. Ranka and S. Sahni, Computing Hough transform on hypercube multicomputers, *Journal of Supercomputing*, 4 (1990), 169–190.

[77] A. P. Reeves, Parallel computer architecture for image processing, *Computer Vision, Graphics, and Image Processing*, 25 (1984), 68–88.

[78] A. Rosenfeld, The impact of massively parallel computers on image processing, *Proc. 2nd Symposium on the Frontiers of Parallel Computing*, 1988, 21–27.

[79] A. Rosenfeld, Parallel image processing using cellular arrays, *IEEE Computer*, 16 (1983), 14–20.

[80] A. Rosenfeld, Image analysis and computer vision: 1991 (Survey), *CVGIP: Image Understanding*, 55 (1992), 349–380.

[81] A. Rosenfeld and A. Kak, *Digital Picture Processing*, Vols. 1–2, Academic Press, 1982.

[82] P. L. Rosin and G. A. W. West, Segmentation of edges into lines and arcs, *Image and Vision Computing*, 7 (1990), 109–114.

[83] J. Rothstein, Bus automata, brains, and mental models, *IEEE Transaction on Systems Man Cybernetics* 18 (1988), 423–435.

[84] A. Schuster and Y. Ben-Asher, Algorithms and optic implementation for reconfigurable networks, *Proc. 5th Jerusalem Conference on Information Technology*, October 1990.

[85] D. B. Shu, L. W. Chow, and J. G. Nash, A content addressable, bit serial associate processor, *Proc. IEEE Workshop on VLSI Signal Processing*, Monterey, CA, November 1988.

[86] D. B. Shu and J. G. Nash, The gated interconnection network for dynamic programming, in S. K. Tewsburg et al. Eds., *Concurrent Computations*, Plenum Publishing, 1988.

[87] D. B. Shu, J. G. Nash, M. M. Eshaghian, and K. Kim, Implementation and application of a gated-connection network in image understanding, in H. Li and Q. F. Stout, Eds., Reconfigurable massively parallel computers, Prentice-Hall, Englewood Cliffs, NJ, 1991.

[88] T. de Saint Pierre and M. Milgram, New and efficient cellular algorithms for image processing, *CVGIP: Image Understanding*, 55 (1992), 261–274.

[89] H. Samet, The design and analysis of spatial data structures, Addison-Wesley, 1990.

[90] C. Allen Sher and A. Rosenfeld, Detecting and extracting compact textured regions using pyramids, *Image and Vision Computing*, 7 (1990), 129–134.

[91] I. D. Svalbe, Natural representations for straight lines and the Hough transform on discrete arrays, *IEEE Transactions on Pattern Analysis and Machine Intelligence*, 11 (1989), 941–953.

[92] C.-H. Teh and R. T. Chin, On image analysis by the method of moments, *IEEE Transactions on Pattern Analysis and Machine Intelligence*, 10 (1988), 496–513.

[93] D. Vernon, *Machine Vision, Automated Visual Inspection and Robot Vision*, Prentice-Hall, 1991.

[94] L. Westberg, Hierarchical contour-based segmentation of dynamic scenes, *IEEE Transactions on PAMI*, 14 (1992), 946–952.

[95] M. Willebeek-Lemair and A. P. Reeves, Region growing on a highly parallel mesh-connected SIMD computer, *Proc. Frontiers of Massively Parallel Computing*, 1988, 93–100.

[96] A. Y. Wu, S. K. Bhaskar, and A. Rosenfeld, Parallel computation of geometric properties from the medial axis transform, *Computer Vision, Graphics, and Image Processing*, 41 (1988), 323–332.

[97] B. Yu, X. lin, and Y. Wu, Isothetic polygon representation for contours, *CVGIP: Image Understanding*, 56 (1992), 264–268.

[98] H. K. Yuen, J. Illingworth, and J. Kittler, Detecting partially occluded ellipses using the Hough transform, *Image and Vision Computing*, 7 (1989), 31–37.

[99] H. K. Yuen, J. Prince, J. Illingworth, and J. Kittler, Comparative study of Hough transform methods for circle finding, *Image and Vision Computing*, 8 (1990), 71–77.

29
Prospectus for a Dense Linear Algebra Software Library

James Demmel
Beresford Parlett
William Kahan
Ming Gu
David Bindel
Yozo Hida
E. Jason Riedy
Christof Voemel
University of California

Jack Dongarra
University of Tennessee and Oak Ridge National Laboratory

Jakub Kurzak
Alfredo Buttari
Julie Langou
Stanimire Tomov
University of Tennessee

Xiaoye Li, Osni Marques
Lawrence Berkeley National Laboratory

Julien Langou
University of Colorado

Piotr Luszczek
Math Works

29.1	Introduction	29-1
29.2	Motivation	29-2
29.3	Challenges of Future Architectures	29-3
29.4	Better Algorithms	29-4
	Algorithmic Improvements for the Solution of Linear Systems • Algorithmic Improvements for the Solution of Eigenvalue Problems	
29.5	Added Functionality	29-7
	Putting More of LAPACK into ScaLAPACK • Extending Current Functionality	
29.6	Software	29-10
	Improving Ease of Use • Improved Software Engineering • Multithreading	
29.7	Performance	29-15
	References	29-16

29.1 Introduction

Dense linear algebra (DLA) forms the core of many scientific computing applications. Consequently, there is continuous interest and demand for the development of increasingly *better* algorithms in the field. Here "better" has a broad meaning, and includes improved reliability, accuracy, robustness, ease of use, and most importantly new or improved algorithms that would more efficiently use the available computational resources to speed up the computation. The rapidly evolving high end computing systems and the close dependence of DLA algorithms on the computational environment is what makes the field particularly dynamic.

A typical example of the importance and impact of this dependence is the development of LAPACK [4] (and later ScaLAPACK [19]) as a successor to the well-known and formerly widely used LINPACK [39] and EISPACK [39] libraries. Both LINPACK and EISPACK were based, and their efficiency depended on optimized Level 1 BLAS [21]. Hardware development trends though, and in particular an increasing Processor-to-Memory speed gap of approximately 50% per year, started to increasingly show the inefficiency of Level 1 BLAS versus Level 2 and 3 BLAS, which prompted efforts to reorganize DLA algorithms to use block matrix operations in their innermost loops. This formed LAPACK's design philosophy. Later ScaLAPACK extended the LAPACK library to run scalably on distributed memory parallel computers.

There are several current trends and associated challenges that influence the development of DLA software libraries. The main purpose of this work is to identify these trends, address the new challenges, and consequently, outline a prospectus for new releases of the LAPACK and ScaLAPACK libraries.

29.2 Motivation

LAPACK and ScaLAPACK are widely used software libraries for numerical linear algebra. LAPACK provides routines for solving systems of simultaneous linear equations, least-squares solutions of linear systems of equations, eigenvalue problems, and singular value problems. The ScaLAPACK (Scalable LAPACK) library includes a subset of LAPACK routines redesigned for distributed memory MIMD parallel computers. There have been over 58 million web hits at www.netlib.org (for the associated libraries LAPACK, ScaLAPACK, CLAPACK, and LAPACK95). LAPACK and ScaLAPACK are used to solve leading edge science problems and they have been adopted by many vendors and software providers as the basis for their own libraries, including AMD, Apple (under Mac OS X), Cray, Fujitsu, HP, IBM, Intel, NEC, SGI, several Linux distributions (such as Debian), NAG, IMSL, the MathWorks (producers of MATLAB), InteractiveSupercomputing.com, and PGI. Future improvements in these libraries will therefore have a large impact on the user communities and their ability to advance scientific and technological work.

The ScaLAPACK and LAPACK development is mostly driven by

- Algorithm research
- The result of the user/vendor survey
- The demands and opportunities of new architectures and programming languages
- The enthusiastic participation of the research community in developing and offering improved versions of existing Sca/LAPACK codes [70]

The user base is both large and diverse, ranging from users solving the largest and most challenging problems on leading edge architectures to the much larger class of users solving smaller problems of greater variety.

The rest of this chapter is organized as follows: Section 29.3 discusses challenges in making current algorithms run efficiently, scalably, and reliably on future architectures. Section 29.4 discusses two kinds of improved algorithms: faster ones and more accurate ones. Since it is hard to improve both simultaneously, we choose to include a new faster algorithm if it is about as accurate as previous algorithms, and we include a new more accurate algorithm if it is at least about as fast as the previous algorithms. Section 29.5 describes new linear algebra functionality that will be included in the next Sca/LAPACK release. Section 29.6 describes our proposed software structure for Sca/LAPACK. Section 29.7 describes a few initial performance results.

29.3 Challenges of Future Architectures

Parallel computing is becoming ubiquitous at all scales of computation: It is no longer just exemplified by the TOP 500 list of the fastest computers in the world, where over 400 have 513 or more processors, and over 100 have 1025 or more processors. In a few years, typical laptops are predicted to have 64 cores per multicore processor chip, and up to 256 hardware threads per chip. So unless all algorithms (not just numerical linear algebra!) can exploit this parallelism, they will not only cease to speed up, but in fact slow down compared to machine peak.

Furthermore, the gap between processor speed and memory speed continues to grow exponentially: processor speeds are improving at 59% per year, main memory bandwidth at only 23%, and main memory latency at a mere 5.5% [54]. This means that an algorithm that is efficient today, because it does enough floating point operations per memory reference to mask slow memory speed, may not be efficient in the near future. The same story holds for parallelism, with communication network bandwidth improving at just 26%, and network latency unimproved since the Cray T3E in 1996 until recently.

The large-scale target architectures of importance for LAPACK and ScaLAPACK include platforms like the Cray X1 at Oak Ridge National Laboratory, the Cray XT3 (Red Storm) at Sandia National Laboratory, the IBM Blue Gene/L system at Lawrence Livermore National Laboratory, a large Opteron cluster at Lawrence Berkeley National Laboratory (LBNL), a large Itanium-2 cluster at Pacific Northwest National Laboratory, the IBM SP3 at LBNL, and near term procurements underway at various DOE and NSF sites.

Longer term the High Productivity Computing Systems (HPCS) program [64]. At the current time three vendors are designing hardware and software platforms that should scale to a petaflop by 2010: Cray's Cascade system (with the Chapel programming language), IBM's PERCS system (with X10), and Sun Microsystems' Hero system (with Fortress). Other interesting platforms include IBM's Blue Planet [87], and vector extensions to IBM's Power systems called ViVA and ViVA-2.

As these examples illustrate, LAPACK and ScaLAPACK will have to run efficiently and correctly on a much wider array of platforms than in the past. It will be a challenge to map LAPACK's and ScaLAPACK's current software hierarchy of BLAS/BLACS/PBLAS/LAPACK/ScaLAPACK efficiently to all these platforms. For example, on a platform with multiple levels of parallelism (multicores, SMPs, distributed memory), would it be better to treat each SMP node as a ScaLAPACK process, calling BLAS that are themselves parallel, or should each processor within the SMP be mapped to a process, or something else? At a minimum, the many tuning parameters (currently hard-coded within ILAENV) will have to be tuned for each platform; see Section 29.6.

A more radical departure from current practice would be to make our algorithms asynchronous. Currently our algorithms are block synchronous, with phases of computation followed by communication with (implicit) barriers. But on networks that can overlap communication and computation, or on multithreaded shared memory machines, block synchrony can leave a significant fraction of the platform idle at any time. For example, the LINPACK benchmark version of LU decomposition exploits such asynchrony and can runs 2x faster than its block synchronous ScaLAPACK counterpart; see Section 29.6.3.

Future platforms are also likely to be heterogeneous, in performance and possibly floating point semantics. One example of heterogeneous performance is a cluster purchased over time, with both old, slow processors and new, fast processors. Varying user load can also cause performance heterogeneity. But there is even performance heterogeneity within a single processor. The best-known examples are the x86/x86-64 family of processors with SSE and 3DNow! instruction sets, PowerPC family chips with Altivec extensions such as IBM Cell, scalar versus vector units on the Cray X1, X1E, and XD1 platforms, and platforms with both GPUs and CPUs. In the case of Cell, single precision runs at 10 times the speed of double precision. Speed differences like this motivate us to consider using rather different algorithms than currently in use; see Section 29.4.

Heterogeneities can also arise in floating point semantics. This is obvious in case of clusters of different platforms, but also within platforms. To illustrate one challenge imagine that one processor runs fastest when handling denormalized numbers according to the IEEE 754 floating point standard [5, 65] and another runs fastest when flushing them to zero. On such a platform sending a message containing a denormalized number from one processor to another can change its value (to zero) or even lead to a trap. Either way, correctness is a challenge, and not just for linear algebra.

29.4 Better Algorithms

Three categories of routines are going to be addressed in Sca/LAPACK: (i) improved algorithms for functions in LAPACK, which also need to be put in ScaLAPACK (discussed here) (ii) functions now in LAPACK but not ScaLAPACK (discussed in Section 29.5), and (iii) functions in neither LAPACK nor ScaLAPACK (also discussed in Section 29.5).

There are a number of faster and/or more accurate algorithms that are going to be incorporated in Sca/LAPACK. The following is a list of each set of future improvements.

29.4.1 Algorithmic Improvements for the Solution of Linear Systems

1. The recent developments of extended precision arithmetic [11, 21, 22, 71] in the framework of the new BLAS standard allow the use of higher precision iterative refinement to improve computed solutions. Recently, it has been shown how to modify the classical algorithm of Wilkinson [62, 101] to compute not just an error bound measured by the infinity (or max) norm, but also a componentwise, relative error bound, that is, a bound on the number of correct digits in each component. Both error bounds can be computed for a tiny $O(n^2)$ extra cost after the initial $O(n^3)$ factorization [36].
2. As mentioned in Section 29.3, there can be a large speed difference between different floating point units on the same processor, with single precision running 10x faster than double precision on an IBM Cell, and 2 times faster on the SSE unit than the x86 unit on some Intel platforms. We can exploit iterative refinement to run faster in these situations. Suppose we want to solve $Ax = b$ where all the data types are double. The idea is to round A to single, perform its LU factorization in single, and then do iterative refinement with the computed solution $\hat{x}$ and residual $r = A\hat{x} - b$ in double, stopping when the residual is as small as it would be had the LU decomposition been done in double. This means that the $O(n^3)$ work of LU decomposition will run at the higher speed of single, and only the extra $O(n^2)$ of refinement will run in double. This approach can get a factor of 2 times speedup on a laptop with x86 and SSE units, and promises more on Cell. Of course it only converges if A's condition number is less than about $1/\sqrt{\epsilon}$, where ϵ is machine precision, but in this common case it is worthwhile. For algorithms beyond solving $Ax = b$ and least squares, such as eigenvalue problems, current iterative refinement algorithms increase the $O(n^3)$ work by a small constant factor. But if there is a factor of 10x to be gained by working in single, then they are still likely to be worthwhile.
3. Gustavson, Kågström and others have recently proposed a new set of *recursive data structures* for dense matrices [48, 57, 59]. These data structures represent a matrix as a collection of small rectangular blocks (chosen to fit inside the L1 cache), which are stored using one of several "space filling curve" orderings. The idea is that the data structure and associated recursive matrix algorithms are *cache oblivious* [49], that is, they optimize cache locality without any explicit blocking such as conventionally done in LAPACK and ScaLAPACK, or any of the tuning parameters (beyond the L1 cache size).

    The reported benefits of these data structures and associated algorithms to which they apply are usually slightly higher peak performance on large matrices and a faster increase towards peak performance as the dimension grows. Sometimes, slightly modified, tuned BLAS are used for operations on matrices assumed to be in L1 cache. The biggest payoff by far is for factoring symmetric matrices stored in packed format, where the current LAPACK routines are limited to the performance of Level 2 BLAS, which do $O(1)$ flops per memory reference. Whereas the recursive algorithms can use the faster Level 3 BLAS, which do $O(n)$ flops per memory reference and can be optimized to hide slower memory bandwidth and latencies.

The drawback of these algorithms is their use of a completely different and rather complicated data structure, which only a few expert users could be expected to use. That leaves the possibility of copying the input matrices in conventional column-major (or row-major) format into the recursive data structure. Furthermore, they are only of benefit for "one-sided factorizations" (LU, LDL^T, Cholesky, QR), but none of the "two-sided factorizations" needed for the EVD or SVD (there is a possibility they might be useful when no eigenvectors or singular vectors are desired).

The factorization of symmetric packed matrices will be incorporated into LAPACK using the recursive data structures, copying the usual data structure in-place to the recursive data structure. The copying costs $O(n^2)$ contrast to the overall $O(n^3)$ operation count, so the asymptotic speeds should be the same. Even if the recursive data structures will be used for other parts of LAPACK, the same column-major interface data structures will be kept for the purpose of ease of use.

4. Ashcraft et al. [9] proposed a variation of Bunch–Kaufman factorization for solving symmetric indefinite systems $Ax = b$ by factoring $A = LDL^T$ with different pivoting. The current Bunch–Kaufman factorization is backward stable for the solution of $Ax = b$ but can produce unbounded L factors. Better pivoting provides better accuracy for applications requiring bounded L factors, like optimization and the construction of preconditioners [31, 62].

5. A Cholesky factorization with diagonal pivoting [61, 62] that avoids a breakdown if the matrix is nearly indefinite/rank-deficient is valuable for optimization problems (and has been requested by users) and is also useful for the high accuracy solution of the symmetric positive definite EVD, see below. For both this pivoting strategy and the one proposed above by Ashcraft/Grimes/Lewis, published results indicate that, on uniprocessors (LAPACK), the extra search required (compared to Bunch–Kaufman) has a small impact on performance. This may not be the case for distributed memory (ScaLAPACK) in which the extra searching and pivoting may involve nonnegligible communication costs.

6. Progress has been made in the development of new algorithms for computing or estimating the condition number of tridiagonal [38, 60] or triangular matrices [47]. These algorithms play an important role in obtaining error bounds in matrix factorizations, and the most promising algorithms should be evaluated and incorporated in the future release.

29.4.2 Algorithmic Improvements for the Solution of Eigenvalue Problems

Algorithmic improvements to the current LAPACK eigensolvers concern both accuracy and performance.

1. Braman, Byers, and Mathias proposed in their SIAM Linear Algebra Prize winning work [23, 24] an up to 10x faster Hessenberg QR-algorithm for the nonsymmetric EVD. This is the bottleneck of the overall nonsymmetric EVD, for which significant speedups should be expected. Byers recently spent a sabbatical with James Demmel where he did much of the software engineering required to convert his prototype into LAPACK format. An extension of this work with similar benefits will be extended to the QZ algorithm for Hessenberg-triangular pencils, with collaboration from Mehrmann. Similar techniques will be exploited to accelerate the routines for (block) companion matrices; see Section 29.5.2.

2. An early version of an algorithm based on Multiple Relatively Robust Representations (MRRR) [80–83] for the tridiagonal symmetric eigenvalue problem (STEGR) was incorporated into LAPACK version 3. This algorithm promised to replace the prior $O(n^3)$ QR algorithm (STEQR) or $\approx O(n^{2.3})$ divide and conquer (STEDC) algorithm with an $O(n^2)$ algorithm. It should have cost $O(nk)$ operations to compute the nk entries of k n-dimensional eigenvectors, the minimum possible work, in a highly parallel way. In fact, the algorithm in LAPACK v.3 did not cover all possible eigenvalue distributions and resorted to a slower and less accurate algorithm based on classical inverse iteration for "difficult" (highly clustered) eigenvalue distributions. The inventors of MRRR, Parlett and Dhillon, have continued to work on improving this algorithm. Very recently,

Parlett and Vömel have proposed a solution for the last hurdle [84] and now pass the Sca/LAPACK tests for the most extreme examples of highly multiple eigenvalues. (These arose in some tests from very many large "glued Wilkinson matrices" constructed so that large numbers of mathematically distinct eigenvalues agreed to very high accuracy, much more than double precision. The proposed solution involves randomization, making small random perturbations to an intermediate representation of the matrix to force all eigenvalues to disagree in at least 1 or 2 bits.) Given the solution to this last hurdle, this algorithm may be propagated to all the variants of the symmetric EVD (e.g., banded, generalized, packed, etc.) in LAPACK. A parallelized version of this algorithm will be available for the corresponding ScaLAPACK symmetric EVD routines. Currently, ScaLAPACK only has parallel versions of the oldest, least efficient (or least accurate) LAPACK routines. This final MRRR algorithm requires some care at load balancing because the Multiple Representations used in MRRR represent subsets of the spectrum based on how clustered they are, which may or may not correspond to a good load balance. Initial work in this area is very promising [16].

3. The MRRR algorithm can and should also be applied to the SVD, replacing the current $O(n^3)$ or $\approx O(n^{2.3})$ bidiagonal SVD algorithms with an $O(n^2)$ algorithm. The necessary theory and a preliminary prototype implementation have been developed [55].

4. There are three phases in the EVD (or SVD) of a dense or band matrix: (i) reduction to tridiagonal (or bidiagonal) form, (ii) the subsequent tridiagonal EVD (or bidiagonal SVD), and (iii) backtransforming the eigenvectors (or singular vectors) of the tridiagonal (or bidiagonal) to correspond to the input matrix. If many (or all) eigenvectors (or singular vectors) are desired, the bottleneck will be phase 2. But now the MRRR algorithm promises to make phase 2 cost just $O(n^2)$ in contrast to the $O(n^3)$ costs of phases 1 and 3. In particular, Howell and Fulton [51] recently devised a new variant of reduction to bidiagonal form for the SVD that has the potential to eliminate half the memory references by reordering the floating point operations (flops). Howell and Fulton fortunately discovered this algorithm during the deliberations of the recent BLAS standardization committee, because its implementation required new BLAS routines (routines GEMVT and GEMVER [21]) that were then added to the standard. These routines are called "Level 2.5 BLAS" because they do many more than $O(1)$ but fewer than $O(n)$ flops per memory reference. Preliminary tests indicate speedups of up to nearly 2x.

5. For the SVD, when only left or only right singular vectors are desired, there are other variations on phase 1 to consider that reduce both floating point operations and memory references [13,85]. Initial results indicate reduced operation counts by a ratio of up to .75, but at the possible cost of numerical stability for some singular vectors.

6. When few or no vectors are desired, the bottleneck shifts entirely to phase 1. Bischof and Lang [18] have proposed a successive band reduction (SBR) algorithm that will asymptotically (for large dimension n) change most of the Level 2 BLAS operations in phase 1 to Level 3 BLAS operations (see the attached letter from Lang). They report speedups of almost 2.4x. This approach is not suitable when a large number of vectors are desired because the cost of phase 3 is much larger per vector. In other words, depending on how many vectors are desired, either the SBR approach will be used or the one-step reduction (the Howell/Fulton variant for the SVD and the current LAPACK code for the symmetric EVD). And if only left or only right singular vectors are desired, the algorithms described in bullet 5 might be used. This introduces a machine-dependent tuning parameter to choose the right algorithm; see tuning of this and other parameters in Section 29.6.2. It may also be possible to use Gustavson's recursive data structures to accelerate SBR.

7. Drmač and Veselić have made significant progress on the performance of the one-sided Jacobi algorithm for computing singular values with high relative accuracy [35, 46]. In contrast to the algorithms described above, their's can compute most or all of the significant digits in tiny singular values when these digits are determined accurately by the input data and when the above algorithms return only roundoff noise. The early version of this algorithm introduced by James Demmel in [35,37] was quite slower than the conventional QR-iteration-based algorithms and much slower than the MRRR algorithms discussed above. But recent results reported by Drmač at [66] show that a combination of clever optimizations have finally led to an accurate algorithm that is *faster* than

the original QR-iteration-based algorithm. Innovations include preprocessing by QR factorizations with pivoting, block application of Jacobi rotations, and early termination. Two immediate applications include the (full matrix) SVD and the symmetric positive-definite EVD, by first reducing to the SVD using the Cholesky-with-pivoting algorithm discussed earlier.

8. Analogous, high accuracy algorithms for the symmetric indefinite EVD have also been designed. One approach by Slapničar [88,89] uses a J-symmetric Jacobi algorithm with hyperbolic rotations, and another one by Dopico/Molera/Moro [45] does an SVD, which "forgets" the signs of the eigenvalues and then reconstructs the signs. The latter can directly benefit by the Drmač/Veselić algorithm above.

29.5 Added Functionality

29.5.1 Putting More of LAPACK into ScaLAPACK

Table 29.1 compares the available data types in the latest releases of LAPACK and ScaLAPACK. After the data type description, the prefixes used in the respective libraries are listed. A blank entry indicates that the corresponding type is not supported. The most important omissions in ScaLAPACK are as follows: (i) There is no support for packed storage of symmetric (SP,PP) or Hermitian (HP,PP) matrices, nor the triangular packed matrices (TP) resulting from their factorizations (using $\approx n^2/2$ instead of n^2 storage); these have been requested by users. The interesting question is what data structure to support. One possibility is recursive storage as discussed in Section 29.5.2 [2, 48, 57, 59]. Alternatively, the packed storage may be partially expanded into a 2D array in order to apply Level 3 BLAS (GEMM) efficiently. Some preliminary ScaLAPACK prototypes support packed storage for the Cholesky factorization and the

TABLE 29.1 Data Types Supported in LAPACK and ScaLAPACK

	LAPACK	ScaLAPACK
General band	GB	GB, DB
General (i.e., unsymmetric, in some cases rectangular)	GE	GE
General matrices, generalized problem	GG	GG
General tridiagonal	GT	DT
(Complex) Hermitian band	HB	
(Complex) Hermitian	HE	HE
Upper Hessenberg matrix, generalized problem	HG	
(Complex) Hermitian, packed storage	HP	
Upper Hessenberg	HS	LAHQR only
(Real) orthogonal, packed storage	OP	
(Real) orthogonal	OR	OR
Positive definite band	PB	PB
General positive definite	PO	PO
Positive definite, packed storage	PP	
Positive definite tridiagonal	PT	PT
(Real) symmetric band	SB	
Symmetric, packed storage	SP	
(Real) symmetric tridiagonal	ST	ST
Symmetric	SY	SY
Triangular band	TB	
Generalized problem, triangular	TG	
Triangular, packed storage	TP	
Triangular (or in some cases quasitriangular)	TR	TR
Trapezoidal	TZ	TZ
(Complex) unitary	UN	UN
(Complex) unitary, packed storage	UP	

A blank entry indicates that the corresponding format is not supported in ScaLAPACK.

TABLE 29.2 LAPACK Codes and the Corresponding Parallel Version in ScaLAPACK

	LAPACK	ScaLAPACK
Linear equations	GESV (LU)	PxGESV
	POSV (Cholesky)	PxPOSV
	SYSV (LDL^T)	Missing
Least squares (LS)	GELS (QR)	PxGELS
	GELSY (QR w/pivoting)	Missing
	GELSS (SVD w/QR)	Missing
	GELSD (SVD w/D&C)	Missing
Generalized LS	GGLSE (GRQ)	Missing
	GGGLM (GQR)	Missing
Symmetric EVD	SYEV (QR)	PxSYEV
	SYEVD (D&C)	PxSYEVD
	SYEVR (RRR)	Missing
Nonsymmetric EVD	GEES (HQR)	Missing driver
	GEEV (HQR + vectors)	Missing driver
SVD	GESVD (QR)	PxGESVD (missing complex C/Z)
	GESDD (D&C)	Missing
Generalized symmetric EVD	SYGV (inverse iteration)	PxSYGVX
	SYGVD (D&C)	Missing
Generalized nonsymmetric EVD	GGES (HQZ)	Missing
	GGEV (HQZ + vectors)	Missing
Generalized SVD	GGSVD (Jacobi)	Missing

The underlying LAPACK algorithm is shown in parentheses. "Missing" means both drivers and computational routines are missing. "Missing driver" means that the underlying computational routines are present.

symmetric eigenvalue problem [20]. (ii) ScaLAPACK only offers limited support of band matrix storage and does not specifically take advantage of symmetry or triangular form (SB, HB, TB). (iii) ScaLAPACK does not support data types for the standard (HS) or generalized (HG, TG) nonsymmetric EVDs; see further below.

Table 29.2 compares the available functions in LAPACK and ScaLAPACK. The relevant user interfaces ("drivers") are listed by subject, and acronyms are used for the software in the respective libraries. Table 29.2 also shows that, in the ScaLAPACK library, the implementation of some driver routines and their specialized computational routines are currently missing.

For inclusion in ScaLAPACK:

1. The solution of symmetric linear systems (SYSV), combined with the use of symmetric packed storage (SPSV), will be a significant improvement with respect to both memory and computational complexity over the currently available LU factorization. It has been requested by users and is expected to be used widely. In addition to solving systems it is used to compute the inertia (number of positive, zero and negative eigenvalues) of symmetric matrices.

2. EVD and SVD routines of all kinds (standard for one matrix and generalized for two matrices) are missing from ScaLAPACK. For SYEV, ScaLAPACK has `p_syev` (QR algorithm), `p_syevd` (divide and conquer), `p_syevx` (bisection and inverse iteration); a prototype code is available for MRRR (`p_syevr`). For SVD, ScaLAPACK has `p_gesvd` (QR algorithm). And for NEV, ScaLAPACK has `p_gehrd` (reduction to Hessenberg form), `p_lahqr` (reduction of a Hessenberg matrix to Schur form). These two routines enable users to get eigenvalues of a nonsymmetric matrix but not (easily) the eigenvectors. The MRRR algorithm is expected to be exploited for the SVD and symmetric EVD as are new algorithms of Braman/Byers/Mathias for the nonsymmetric EVD (see Section 29.4); work is under way to get the QZ algorithm for the nonsymmetric generalized eigenvalue problem.

3. LAPACK provides software for the linearly constrained (generalized) least squares problem, and users in the optimization community will benefit from a parallel version. In addition, algorithms

for rank deficient, standard least squares problems based on the SVD are missing from ScaLAPACK; it may be that a completely different algorithm based on the MRRR algorithm (see Section 29.4) may be more suitable for parallelism instead of the divide and conquer (D&C) algorithm that is fastest for LAPACK.

4. Expert drivers that provide error bounds, or other more detailed structural information about eigenvalue problems, should be provided.

29.5.2 Extending Current Functionality

This subsection outlines possible extensions of the functionalities available in LAPACK and ScaLAPACK. These extensions are mostly motivated not only by users but also by research progress:

1. Several updating facilities are planned to be included in a new release. Whereas updating matrix factorizations like Cholesky, LDL^T, LU, QR [53] have a well-established theory and unblocked (i.e., noncache optimized) implementations exist, for example, in LINPACK [39], the efficient update of the SVD is a current research topic [56]. Furthermore, divide and conquer-based techniques are promising for a general framework of updating eigendecompositions of submatrices.

2. Semiseparable matrices are generalizations of the inverses of banded matrices, with the property that any rectangular submatrix lying strictly above or strictly below the diagonal has a rank bounded by a small constant. Recent research has focused on methods exploiting semiseparability, or being a sum of a banded matrix and a semiseparable matrix, for better efficiency [30,97]. The development of such algorithms is being considered in a future release. Most exciting are the recent observations of Gu, Bini and others [17] that a companion matrix is banded plus semiseparable, and that this structure is preserved under QR iteration to find its eigenvalues. This observation lets us accelerate the standard method used in MATLAB and other libraries for finding roots of polynomials from $O(n^3)$ to $O(n^2)$. An initial rough prototype of this code becomes faster than the highly tuned LAPACK eigensolver for n between 100 and 200 and then becomes arbitrarily faster for larger n. Whereas the current algorithm has been numerically stable on all examples tested so far, more work needs to be done to guarantee stability in all cases. The same technique should apply to finding eigenvalues of block companion matrices, that is, matrix polynomials, yielding speedups proportional to the degree of the matrix polynomial.

3. Eigenvalue problems for matrix polynomials [52] are common in science and engineering. The most common case is the quadratic eigenvalue problem $(\lambda^2 M + \lambda D + K)x = 0$, where typically M is a mass matrix, D a damping matrix, K a stiffness matrix, λ a resonant frequency, and x a mode shape. The classical solution is to linearize this eigenproblem, asking instead for the eigenvalues of a system of twice the size:

$$\lambda \cdot \begin{bmatrix} 0 & I \\ M & D \end{bmatrix} \cdot \begin{bmatrix} y_1 \\ y_2 \end{bmatrix} + \begin{bmatrix} I & 0 \\ 0 & K \end{bmatrix} \cdot \begin{bmatrix} y_1 \\ y_2 \end{bmatrix} = 0,$$

where $y_2 = x$ and $y_1 = \lambda x$. But there are a number of ways to linearize, and some are better at preserving symmetries in the solution of the original problem or saving more time than others. There has been a great deal of recent work on picking the right linearization and subsequent algorithm for its EVD to preserve desired structures. In particular, for the general problem $\sum_{i=0}^{k} \lambda^i \cdot A_i \cdot x = 0$, the requested cases are symmetric ($A_i = A_i^T$, arising in mechanical vibrations without gyroscopic terms), its even ($A_i = (-1)^i A_i^T$) and odd ($A_i = (-1)^{i+1} A_i^T$) variations (used with gyroscopic terms), and palindromic ($A_i = A_{k-i}^T$, arising in discrete time periodic and continuous time control). Recent references include [6–8, 14, 72–74, 92, 93, 95].

4. Matrix functions (square root, exponential, sign function) play an important role in the solution of differential equations in both science and engineering, and have been requested by users. Recent research progress has led to the development of several new algorithms [10, 25–27, 34, 63, 68, 69, 79, 90] that could be included in a future release.

5. The eigenvalue and singular value decomposition of products and quotients of matrices play an important role in control theory. Such functionalities, incorporated from the software library SLICOT [15] are being considered, using the improved underlying EVD algorithms. Efficient solvers for Sylvester and Lyapunov equations that are also currently in SLICOT could be incorporated.
6. Multiple user requests concern the development of out-of-core versions of matrix factorizations. ScaLAPACK prototypes [20] are under development that implement out-of-core data management for the LU, QR, and Cholesky factorizations [40,41]. Users have asked for two kinds of parallel I/O: to a single file from a sequential LAPACK program (possible with sequential I/O in the reference implementation), and to a single file from MPI-based parallel I/O in ScaLAPACK.

29.6 Software

29.6.1 Improving Ease of Use

"Ease of use" can be classified as follows: ease of programming (which includes the possibility to easily convert from serial to parallel, from LAPACK to ScaLAPACK, and the possiblity to use high level interfaces), ease of obtaining predictable results in dynamic environments (for debugging and performance), and ease of installation (including performance tuning). Each will be discussed in turn.

There are tradeoffs involved in each of these subgoals. In particular, ultimate ease of programming, exemplified by typing $x = A \backslash b$ in order to solve $Ax = b$ (paying no attention to the data type, data structure, memory management, or algorithm choice) requires an infrastructure and user interface best left to the builders of systems like MATLAB, and may come at a significant performance and reliability penalty. In particular, many users now exercise, and want to continue to exercise, detailed control over data types, data structures, memory management, and algorithm choice, to attain both peak performance and reliability (e.g., not running out of memory unexpectedly). But, some users would also like Sca/LAPACK to handle work space allocation automatically, make it possible to call ScaLAPACK on a greater variety of user-defined data structures, and pick the best algorithm when there is a choice.

To accomodate these "ease of programming" requests as well as requests to make the Sca/LAPACK code accessible from other languages than Fortran, the following steps are considered:

1. Produce new F95 modules for the LAPACK drivers, for work-space allocation and algorithm selection.
2. Produce new F95 modules for the ScaLAPACK drivers, which convert, if necessary, the user input format (e.g., a simple block row layout across processors) to the optimal one for ScaLAPACK (which may be a 2D block cyclic layout with block sizes that depend on the matrix size, algorithm, and architecture). Allocate memory as needed.
3. Produce LAPACK and ScaLAPACK wrappers in other languages. On the basis of current user surveys, these languages will tentatively be C, C++, Python, and MATLAB.

See Section 29.6.2 for details on the software engineering approach to these tasks.

Ease of conversion from serial code (LAPACK) to parallel code (ScaLAPACK) is done by making the interfaces (at least at the driver level) as similar as possible. This includes expanding ScaLAPACK's functionality to include as much of LAPACK as possible (see Section 29.5).

Another ease-of-programming request is improved documentation. The Sca/LAPACK websites are continously developed to enable ongoing user feedback and support and the websites employ tools like bugzilla to track reported bugs.

Obtaining predictable results in a dynamic environment is important for debugging (to get the same answer when the code is rerun), for reproducibility, auditability (for scientific or legal purposes), and for performance (so that runtimes do not vary widely and unpredictably).

First consider getting the same answer when the problem is rerun. To establish reasonable expectations, consider three cases: (i) Rerunning on different computers with different compilers. Reproducibility here is more than one can expect. (ii) Rerunning on the same platform (or cluster) but with different data layouts or blocking parameters. Reproducibility here is also more than the user can expect, because roundoff errors will differ as, say, matrix products are computed in different orders. (iii) Just typing "a.out" again with the same inputs. Reproducibility here should be the user goal, but it is not guaranteed because asynchronous communication can result in, for example, dot products of distributed vectors being computed in different orders with different roundoff errors. Using an existing switch in the BLACS to insist on the same order of communication (with an unavoidable performance penalty) will address this particular problem, but investigations are underway to know whether this is the only source of nonreproducibility (asynchrony in tuned SMP BLAS is another possible source, even for "sequential" LAPACK). As more fully asynchronous algorithms are explored to get higher performance (see Section 29.4), this problem becomes more interesting. Reproducibility will obviously come with a performance penalty but is important for debugging of codes that call Sca/LAPACK and for situations where auditability is critical.

Now consider obtaining predictable performance, in the face of running on a cluster where the processor subset that actually performs the computation may be chosen dynamically, and have a dynamically varying load of other jobs. This difficult problem is discussed further in Section 29.6.2.

Ease of installation, which may include platform-specific performance tuning, depends on the multiple modes of access of the Sca/LAPACK libraries: (i) Some users may use vendor-supplied libraries prebuilt (and pretuned) on their platforms, so their installation needs are already well addressed. (ii) Some users may use netlib to download individual routines and the subroutines they call (but not the BLAS, which must be supplied separately). These users have decided to perform the installations on their own (perhaps for educational purposes). For these users, it can be made possible to select an architecture from a list, so that the downloaded code has parameter values that are quite likely to optimize the user's performance. (iii) Other users may download the entire package from netlib and install and tune it on their machines. The use of autoconf and automatic performance tuning should be expected.

Different users will want different levels of installation effort since complete testing and performance tuning can take a long time. For performance tuning, a database of pretuned parameters for various computing platforms will be built, and if the user indicates that he is happy with the pretuned parameters, performance tuning can be sidestepped. As discussed in Section 29.6.2, there is a spectrum of tuning effort possible, depending on what fraction of peak performance one seeks and how long one is willing to take to tune. Similarly, testing the installation for correctness can be done at at least four levels (and differently for each part of the library): (i) No test cases need be run at all if the user does not want to. (ii) A few small test cases can be run to detect a flawed installation but not failures on difficult numerical cases. (iii) The current test set can be run, which is designed for a fairly thorough numerical testing, and in fact not infrequently indicates error bounds exceeding tight thresholds on some problems and platforms. Some of these test cases are known failure modes for certain algorithms. For example, inverse iteration is expected to sometimes fail to get orthogonal eigenvectors on sufficiently clustered eigenvalues, but it depends on the vagaries of roundoff. (iv) More extensive "torture test" sets are used internally for development, which the most demanding users (or competing algorithm developers) should use for testing. If only for the sake of future algorithm developers, these extensive test sets should be easily available.

29.6.2 Improved Software Engineering

The following is a description of the Sca/LAPACK software engineering (SWE) approach. The main goals are to keep the substantial code base maintainable, testable, and evolvable into the future as architectures and languages change. Maintaining compatibility with other software efforts and encouraging third party contributions to the efforts of the Sca/LAPACK team are also goals [70].

These goals involve tradeoffs. One could explore starting "from scratch," using higher level ways to express the algorithms from which specific implementations could be generated. This approach yields high flexibility allowing the generation of a code that is optimized for future computing platforms with different layers of parallelism, different memory hierarchies, different ratios of computation rate to bandwidth to latency, different programming languages, compilers, and so forth. Indeed, one can think of the problem as implementing the following *metaprogram*:

```
(1) for all linear algebra problems (linear systems, eigenproblems,...
(2)    for all matrix types (general, symmetric, banded, ...)
(3)       for all data types (real, complex, single, double, higher precision)
(4)          for all machine architectures and communication topologies
(5)             for all programming interfaces
(6)                provide the best algorithm(s) available in terms of performance and accuracy
                      (''algorithms'' is plural because sometimes no single one is always best)
```

The potential scope can appear quite large, requiring a judicious mixture of prioritization and automation. Indeed, there is prior work in automation [58], but so far this work has addressed only part of the range of algorithmic techniques Sca/LAPACK needs (e.g., not eigenproblems), it may not easily extend to more asynchronous algorithms and still needs to be coupled to automatic performance tuning techniques. Still, some phases of the meta-program are at least partly automatable now, namely, steps (3) through (5) (see below).

Note that line (5) of the metaprogram is "programming interfaces" not "programming languages," because the question of the best implementation language is separate from providing ways to call it (from multiple languages). Currently, Sca/LAPACK is written in F77. Over the years, the Sca/LAPACK team and others have built on this to provide interfaces or versions in other languages: LAPACK95 [12] and LAPACK3E [3] for F95 (LAPACK3E providing a straightforward wrapper, and LAPACK95 using F95 arrays to simplify the interfaces at some memory and performance costs), CLAPACK in C [32] (translated, mostly automatically, using f2c [50]), LAPACK++ [42], TNT [94] in C++, and JLAPACK in Java [67] (translated using f2j).

First is the SWE development plan and then the SWE research plan:

1. The core of Sca/LAPACK will be maintained in Fortran, adopting those features of F95 that most improve ease-of-use and ease-of-development, but do not prevent the most demanding users from attaining the highest performance and reliable control over the runtime environment. Keeping Fortran is justified for cost and continuity reasons, as well as the fact that the most effective optimizing compilers still work best on Fortran (even when they share "back ends" with the C compiler, because of the added difficulty of discerning the absence of aliasing in C) [33].

    The F95 features adopted include recursion (to support new matrix data structures and associated algorithms discussed in Section 29.4), modules (to support production of versions for different precisions, beyond single and double), and environmental enquiries (to replace xLAMCH), but not automatic workspace allocation (see the next bullet).

2. F95 versions of the Sca/LAPACK drivers will be provided (which will usually be wrappers) to improve ease-of-use, possibly at some performance and reliability costs. For example, automatically allocating workspace using assumed-size arrays (as in the more heavily used LAPACK95, as opposed to LAPACK3E), and performing algorithm selection when appropriate (based on performance models described below) may be done.

3. ScaLAPACK drivers will be provided that take a variety of parallel matrix layouts and automatically identify and convert to the optimal layout for the problem to be solved. Many users have requested accomodation of simpler or more general input formats, which may be quite different from the more complicated performance-optimized 2D block-cycle (and possibly recursive) layouts used internally. Using the performance models described below, these drivers will determine the optimal layout for the input problem, estimate the cost of solving "in-place" versus converting to the optimal layout, solving, and converting back, and choose the fastest solution. Separate layout conversion routines will be provided to help the user identify and convert to better layouts.

4. Wrappers for the Sca/LAPACK drivers in other languages will be provided, with interfaces that are "natural" for those languages, chosen on the basis of importance and demand from the user survey. Currently, this list includes C, Python, and MATLAB.

In particular, no native implementations in these languages will be provided, depending instead on interoperability of these languages with the F95 subset used in (1).

5. The new Sca/LAPACK routines will be converted to use the latest BLAS standard [11, 21, 22], which also provides new high precision functionality necessary for new routines required in Section 29.4 [11, 36], systematically ensure thread-safety, and deprecate superseded routines.

6. Appropriate tools will be used (e.g., autoconf, bugzilla, svn, automatic overnight build and test, etc.) to streamline installation and development and encourage third party contributions. Appropriate source control and module techniques will be used to minimize the number of versions of each code, reducing the number of versions in the cross product {real,complex} × {single, double, quad, ...} to one or two.

7. Performance tuning will be done systematically. Initial experiments are showing up to 10 times speedups using different communication schemes that can be applied in the BLACS [76, 96]. In addition to tuning the BLAS [99, 100] and BLACS, there are over 1300 calls to the ILAENV routine in LAPACK, many of which return tuning parameters that have never been systematically tuned. Some of these parameters are block sizes for blocked algorithms, some are problem size thresholds for choosing between different algorithms, and yet others are numerical convergence thresholds. ScaLAPACK has yet more parameters associated with parallel data layouts and communication algorithms.

   As described in Section 29.6.1, there are different levels of effort possible for performance tuning, and it may also be done at different times. For example, as part of the development process, a database of pretuned parameters and performance models will be built for various computing platforms, which will be good enough for many users [78]. Still, a tool is needed that at user-install time systematically searches a very large parameter space of tuning parameters to pick the best values, where users can "dial" the search effort from quick to exhaustive and then choose different levels of search effort for different routines depending on which are more important. Sophisticated data modeling techniques may be used to minimize search time [98].

Beyond these development activities, the following research tasks are being performed, which should influence and improve the development:

1. As discussed in Section 29.3 emerging architectures offer parallelism at many different levels, from multicore chips to vector processors to SMPs to distributed memory machines to clusters, many of which will appear simultaneously in one computing system. Alongside these layers is the current software hierarchy in ScaLAPACK: BLAS, BLACS, PBLAS, LAPACK, and ScaLAPACK. An investigation needs to be done to decide how to best map these software layers to the hardware layers. For example, should a single BLAS call use all the parallelism available in an SMP and its multicore chips, or should the BLAS call just use the multicore chips with individual ScaLAPACK processes mapping to different SMP nodes? The number of possible combinations is large and growing. These mappings have to be explored to identify the most effective ones, using performance modeling to both measure progress (what fraction of peak is reached?) and limit the search space.

2. There is extensive research in other high performance programming languages funded by DOE, NSA, and the DARPA HPCS program. UPC [29], Titanium [102], CAF [77], Fortress [1], X10 [86], and Cascade [28] are all languages under active development, and are being designed for high productivity SWE on high peformance machines. It is natural to ask if these are appropriate programming languages for ScaLAPACK, especially for the more asynchronous algorithms that may be fastest and for more fine-grained architectures like Blue Gene. Prototype selected ScaLAPACK codes will be provided in some of these languages to assess their utility.

3. Further automating the production of the Sca/LAPACK software is a worthy goal. One can envision expressing each algorithm once in a sufficiently high level language and then having a compiler (or source-to-source translator) automatically produce versions for any architecture or programming interface. Work in this direction includes [58]. The hope is that by limiting the scope to dense

linear algebra, the translation problem will be so simplified that writing the translator pays off in being able to create the many needed versions of the code. But there are a number of open research problems that need to be solved for such an approach to work. First, it has been demonstrated for one-sided factorizations (e.g., LU, Cholesky, and QR), but not on more complex algorithms like some two-sided ones needed by eigenproblems, including the successive-band-reduction (SBR) schemes discussed in Section 29.4. Second, much if not most of Sca/LAPACK involves iterative algorithms for eigenproblems with different styles of parallelism, and it is not clear how to extend to these cases. Third, it is not clear how to best express the more asynchronous algorithms that can achieve the highest performance; this appears to involve either more sophisticated compiler analysis of more synchronous code (which this approach hoped to avoid) or a different way of expressing dependencies. In particular, one may want to use one of the programming languages mentioned above. Finally, it is not clear how to best exploit the multiple levels of parallelism discussed above, that is, which should be handled by the high-level algorithm, the programming language, the compiler, the various library levels (BLAS, PBLAS), and so on. These are worthy research goals to solve before using this technique for development.
4. Performance tuning may be very time consuming if implemented in a brute force manner, by running and timing many algorithms for all parameter values in a large search space. A number of users have already expressed concern about installation time and difficulty, when this additional tuning may occur. On the basis of earlier experience [98], it is possible to use statistical models to both limit the search space and more accurately determine the optimal parameters at run time.
5. In a dynamically changing cluster environment, a call to Sca/LAPACK might be run locally and sequentially, or remotely on one or more processors chosen at call-time. The use of performance models will be available with dynamically updated parameters (based on system load) to choose the right subset of the cluster to use, including the cost of moving the problem to a different subset and moving the answer back. This will help achieve performance predictability.

29.6.3 Multithreading

With the number of cores on multicore chips expected to reach tens in a few years, efficient implementations of numerical libraries using shared memory programming models is of high interest. The current message passing paradigm used in ScaLAPACK and elsewhere introduces unnecessary memory overhead and memory copy operations, which degrade performance, along with making it harder to schedule operations that could be done in parallel. Limiting the use of shared memory to fork-join parallelism (perhaps with OpenMP) or to its use within the BLAS does not address all these issues.

On the other hand, a number of goals can be achieved much more easily in shared memory than on distributed memory systems. The most striking simplification is no need for data partitioning, which can be replaced by work partitioning. Still, cache locality has to be preserved and, in this aspect, the locality of the two-dimensional block cyclic (or perhaps recursive) layout cannot be eliminated.

There are several established programming models for shared memory systems appropriate for different levels of parallelism in the applications with the the client/server, work-crew, and pipeline being the most established ones. Since matrix factorizations are rich in data dependencies, the pipeline seems to be the most applicable model. Another advantage of pipelining is its match to hardware for streaming data processing, like the IBM Cell Broadband Engine. To achieve efficiency we must avoid pipeline stalls (also called bubbles) when data dependencies block execution. The next paragraph illustrates this approach for LU factorization.

LU and other matrix factorization have left- and right-looking formulations [43]. It has even been observed that transition between the two can be done by automatic code transformations [75], although more powerful methods than simple dependency analysis is necessary. It is known that lookahead can be used to improve performance, by performing panel factorizations in parallel with the update to the trailing matrix from the previous step of the algorithm [91]. The lookahead can be of arbitrary depth; this fact is exploiting by the LINPACK benchmark [44].

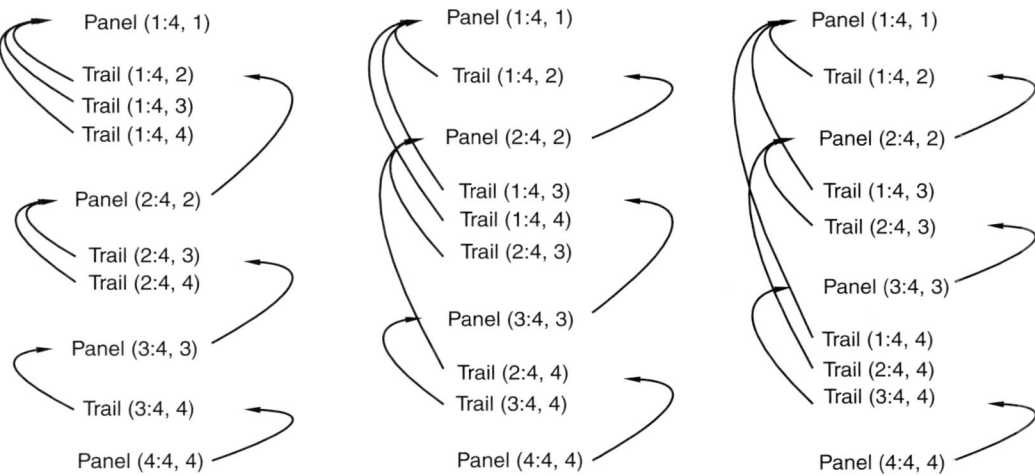

FIGURE 29.1 Different possible factorization schemes: right-looking with no lookahead, right-looking with a lookahead of 1, left-looking (right-looking with maximum lookahead). Arrows show data dependencies.

We observe that the right- and the left-looking formulations are two extremes of a spectrum of possible execution paths, with the lookahead providing a smooth transition between them. We regard the right-looking formulation as having zero lookahead, and the left-looking formulation as having maximum lookahead; see Figure 29.1.

The deeper the lookahead, the faster panels can be factorized and the more matrix multiplications are accumulated at the end of a factorization. Shallow lookaheads introduce pipeline stalls, or bubbles, at the end of a factorization. On the other hand, deep lookaheads introduce bubbles at the beginning of a factorization. Any fixed depth lookahead may stall the pipeline at both the beginning and at the end of execution.

Recent experiments show that pipeline stalls can be greatly reduced if unlimited lookahead is allowed and the lookahead panel factorizations are dynamically scheduled in such a way that their issues do not stall the pipeline. Dynamic work scheduling can easily and elegantly be implemented on shared memory, whereas it is a much more complex undertaking in distributed memory arrangements. It is also worth observing that distributed memory implementations do not favor deep lookaheads owing to storage overhead, which is not a problem in shared memory environments.

29.7 Performance

We give a few recent performance results for ScaLAPACK driver routines on recent architectures. We discuss strong scalability, that is, we keep the problem size constant while increasing the number of processors.

Figure 29.2 gives the time to solution for factoring a linear system of equations of size $n = 8000$ on a cluster of dual processor 64 bit AMD Opterons interconnected with a Gigabit ethernet. As the number of processors increases from 1 to 64, the time decreases from 110 Section (3.1 GFlops) to 9 Section (37.0 GFlops*).

Figure 29.3 gives the time to solution to find the eigenvalue and eigenvectors of two symmetric matrices, one of size $n = 4000$, the other of size 12000. The number of processors grows from 1 to 16 in the $n = 4000$ case and from 4 to 64 in the $n = 12000$ case. The machine used a cluster of dual processor 64 bit Intel Xeon EMTs interconnected with a Myrinet MX interconnect. The matrices are generated randomly using

---

*Thanks to Emmanuel Jeannot for sharing the result.

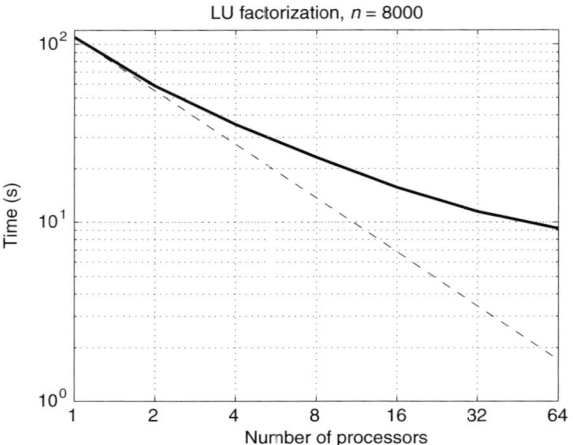

FIGURE 29.2 Scalability of the LU factorization for ScaLAPACK (pdgetrf) for a matrix of size $n = 8000$.

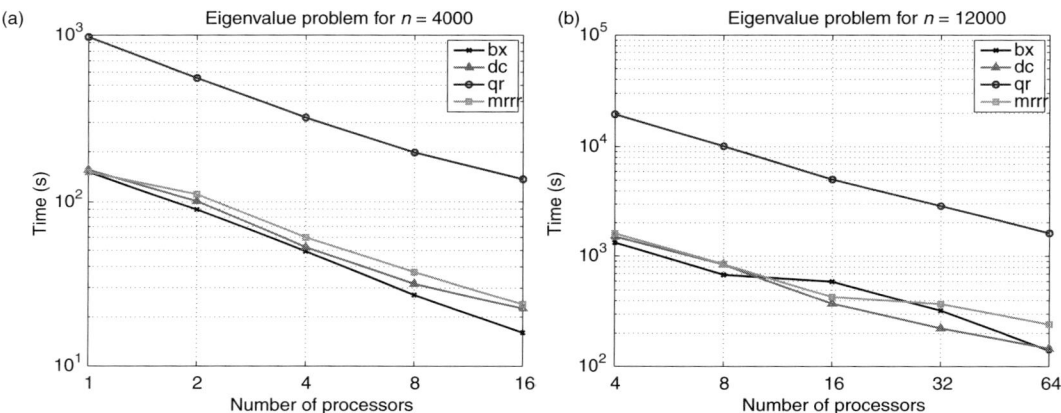

FIGURE 29.3 Scalability of the Symmetric Eigenvalue Solver routines in ScaLAPACK (pdsyev) with a matrix of size 4000 (a) and one of size 12000 (b). Four different methods for the tridiagonal eigensolve have been tested: the BX (pdsyevx), QR (pdsyev), DC (pdsyevd), and MRRR (pdsyevr).

the same generator as in the Linpack Benchmark, and so have random eigenvalue distributions with no tight clusters.

References

[1] E. Allen, D. Chase, V. Luchangco, J.-W. Maessen, S. Ryu, G. L. Steele Jr., and S. Tobin Hochstadt. The Fortress language specification, version 0.707. research.sun.com/projects/plrg/fortress0707.pdf

[2] B. S. Andersen, J. Wazniewski, and F. G. Gustavson. A recursive formulation of Cholesky factorization of a matrix in packed storage. *Trans. Math. Soft.*, 27(2):214–244, 2001.

[3] E. Anderson. LAPACK3E. www.netlib.org/lapack3e, 2003.

[4] E. Anderson, Z. Bai, C. Bischof, J. Demmel, J. Dongarra, J. Du Croz, A. Greenbaum, S. Hammarling, A. McKenney, S. Ostrouchov, and D. Sorensen. *LAPACK Users' Guide, Release 1.0*. SIAM, Philadelphia, 1992, 235 pages.

[5] ANSI/IEEE, New York. *IEEE Standard for Binary Floating Point Arithmetic*, Std 754-1985 edition, 1985.

[6] T. Apel, V. Mehrmann, and D. Watkins. Structured eigenvalue methods for the computation of corner singularities in 3D anisotropic elastic structures. Preprint 01–25, SFB393, Sonderforschungsbereich 393, Fakultät für Mathematik, Chemnitz University of Technology, 2001.

[7] T. Apel, V. Mehrmann, and D. Watkins. Structured eigenvalue methods for the computation of corner singularities in 3D anisotropic elastic structures. *Comp. Meth. App. Mech. Eng.*, 191:4459–4473, 2002.

[8] T. Apel, V. Mehrmann, and D. Watkins. Numerical solution of large scale structured polynomial eigenvalue problems. In *Foundations of Computational Mathematics*. Springer-Verlag, 2003.

[9] C. Ashcraft, R. G. Grimes, and J. G. Lewis. Accurate symmetric indefinite linear equation solvers. *SIAM J. Mat. Anal. Appl.*, 20(2):513–561, 1998.

[10] Z. Bai, J. Demmel, and M. Gu. Inverse free parallel spectral divide and conquer algorithms for nonsymmetric eigenproblems. *Num. Math.*, 76:279–308, 1997. UC Berkeley CS Division Report UCB//CSD-94-793, February 94.

[11] D. Bailey, J. Demmel, G. Henry, Y. Hida, J. Iskandar, W. Kahan, S. Kang, A. Kapur, X. Li, M. Martin, B. Thompson, T. Tung, and D. Yoo. Design, implementation and testing of extended and mixed precision BLAS. *ACM Trans. Math. Soft.*, 28(2):152–205, June 2002.

[12] V. Barker, S. Blackford, J. Dongarra, J. Du Croz, S. Hammarling, M. Marinova, J. Wasniewski, and P. Yalamov. *LAPACK95 Users' Guide*. SIAM, 2001. www.netlib.org/lapack95

[13] J. Barlow, N. Bosner, and Z. Drmač. A new stable bidiagonal reduction algorithm. www.cse.psu.edu/~barlow/fastbidiag3.ps, 2004.

[14] P. Benner, R. Byers, V. Mehrmann, and H. Xu. Numerical computation of deflating subspaces of embedded Hamiltonian pencils. Technical Report SFB393/99–15, Fakultät für Mathematik, TU Chemnitz, 09107 Chemnitz, FRG, 1999.

[15] P. Benner, V. Mehrmann, V. Sima, S. Van Huffel, and A. Varga. SLICOT—a subroutine library in systems and control theory. *Appl. Comput. Contr. Sig. Cir.*, 1:499–539, 1999.

[16] P. Bientinisi, I. S. Dhillon, and R. van de Geijn. A parallel eigensolver for dense symmetric matrices based on multiple relatively robust representations. Technical Report TR-03-26, Computer Science Dept., University of Texas, 2003.

[17] D. Bini, Y. Eidelman, L. Gemignani, and I. Gohberg. Fast QR algorithms for Hessenberg matrices which are rank-1 perturbations of unitary matrices. Dept. of Mathematics report 1587, University of Pisa, Italy, 2005. www.dm.unipi.it/gemignani/papers/begg.ps

[18] C. H. Bischof, B. Lang, and X. Sun. A framework for symmetric band reduction. *Trans. Math. Soft.*, 26(4):581–601, 2000.

[19] L. S. Blackford, J. Choi, A. Cleary, E. D'Azevedo, J. Demmel, I. Dhillon, J. Dongarra, et al. *ScaLAPACK Users' Guide*. SIAM, Philadelphia, 1997.

[20] L. S. Blackford, J. Choi, A. Cleary, J. Demmel, I. Dhillon, J. J. Dongarra, S. Hammarling, et al. Scalapack prototype software. Netlib, Oak Ridge National Laboratory, 1997.

[21] L. S. Blackford, J. Demmel, J. Dongarra, I. Duff, S. Hammarling, G. Henry, M. Heroux, et al. An updated set of Basic Linear Algebra Subroutines (BLAS). *ACM Trans. Math. Soft.*, 28(2), June 2002.

[22] L. S. Blackford, J. Demmel, J. Dongarra, I. Duff, S. Hammarling, G. Henry, M. Heroux, et al. Basic Linear Algebra Subprograms Technical (BLAST) Forum Standard. *Intern. J. High Performance Comput.*, 15(3–4):1–181, 2001.

[23] K. Braman, R. Byers, and R. Mathias. The multishift QR algorithm. Part I: Maintaining well-focused shifts and Level 3 performance. *SIAM J. Mat. Anal. Appl.*, 23(4):929–947, 2001.

[24] K. Braman, R. Byers, and R. Mathias. The multishift QR algorithm. Part II: Aggressive early deflation. *SIAM J. Mat. Anal. Appl.*, 23(4):948–973, 2001.

[25] R. Byers. Numerical stability and instability in matrix sign function based algorithms. In C. Byrnes and A. Lindquist, eds, *Computational and Combinatorial Methods in Systems Theory*, pp. 185–200. North-Holland, 1986.
[26] R. Byers. Solving the algebraic Riccati equation with the matrix sign function. *Lin. Alg. Appl.*, 85:267–279, 1987.
[27] R. Byers, C. He, and V. Mehrmann. The matrix sign function method and the computation of invariant subspaces, *SIAM J. Mat. Anal. Appl.*, 18:615–632, 1997.
[28] D. Callahan, B. Chamberlain, and H. Zima. The Cascade high-productivity language. In *9th International Workshop on High-Level Parallel Programming Models and Supportive Environments (HIPS 2004)*, pp. 52–60. IEEE Computer Society, April 2004. www.gwu.edu/ upc/publications/ productivity.pdf
[29] F. Cantonnet, Y. Yao, M. Zahran, and T. El-Ghazawi. Productivity analysis of the UPC language. In *IPDPS 2004 PMEO workshop*, 2004. www.gwu.edu/ upc/publications/productivity.pdf
[30] S. Chandrasekaran and M. Gu. Fast and stable algorithms for banded plus semiseparable systems of linear equations. *SIAM J. Mat. Anal. Appl.*, 25(2):373–384, 2003.
[31] S. H. Cheng and N. Higham. A modified Cholesky algorithm based on a symmetric indefinite factorization. *SIAM J. Mat. Anal. Appl.*, 19(4):1097–1110, 1998.
[32] CLAPACK: LAPACK in C. http://www.netlib.org/clapack/
[33] C. Coarfa, Y. Dotsenko, J. Mellor-Crummey, D. Chavarria-Miranda, F. Contonnet, T. El-Ghazawi, A. Mohanti, and Y. Yao. An evaluation of global address space languages: Co-Array Fortran and Unified Parallel C. In *Proc. 10th ACM SIGPLAN Symp. on Principles and Practice and Parallel Programming (PPoPP 2005)*, 2005. www.hipersoft.rice.edu/caf/publications/index.html
[34] P. Davies and N. J. Higham. A Schur–Parlett algorithm for computing matrix functions. *SIAM J. Mat. Anal. Appl.*, 25(2):464–485, 2003.
[35] J. Demmel, M. Gu, S. Eisenstat, I. Slapničar, K. Veselić, and Z. Drmač. Computing the singular value decomposition with high relative accuracy. *Lin. Alg. Appl.*, 299(1–3):21–80, 1999.
[36] J. Demmel, Y. Hida, W. Kahan, X. S. Li, S. Mokherjee, and E. J. Riedy. Error bounds from extra precise iterative refinement, *ACM Trans. Math. Soft.*, 32(2):325–351, 2006.
[37] J. Demmel and K. Veselić. Jacobi's method is more accurate than QR. *SIAM J. Mat. Anal. Appl.*, 13(4):1204–1246, 1992.
[38] I. S. Dhillon. Reliable computation of the condition number of a tridiagonal matrix in $O(n)$ time. *SIAM J. Mat. Anal. Appl.*, 19(3):776–796, 1998.
[39] J. Dongarra, J. Bunch, C. Moler, and G. W. Stewart. *LINPACK User's Guide*. SIAM, Philadelphia, PA, 1979.
[40] J. Dongarra and E. D'Azevedo. The design and implementation of the parallel out-of-core ScaLAPACK LU, QR, and Cholesky factorization routines. Computer Science Dept. Technical Report CS-97-347, University of Tennessee, Knoxville, TN, January 1997. www.netlib.org/ lapack/lawns/lawn118.ps
[41] J. Dongarra, S. Hammarling, and D. Walker. Key concepts for parallel out-of-core LU factorization. Computer Science Dept. Technical Report CS-96-324, University of Tennessee, Knoxville, TN, April 1996. www.netlib.org/lapack/lawns/lawn110.ps
[42] J. Dongarra, R. Pozo, and D. Walker. Lapack++: A design overview of object-oriented extensions for high performance linear algebra. In *Supercomputing 93*. IEEE, 1993. math.nist.gov/lapack++
[43] J. J. Dongarra, I. S. Duff, D. C. Sorensen, and H. A. van der Vorst. *Numerical Linear Algebra for High-Performance Computers*. SIAM, Philadelphia, PA, 1998.
[44] J. J. Dongarra, P. Luszczek, and A. Petitet. The LINPACK benchmark: Past, present and future. *Concurrency Computat.: Pract. Exper.*, 15:803–820, 2003.
[45] F. M. Dopico, J. M. Molera, and J. Moro. An orthogonal high relative accuracy algorithm for the symmetric eigenproblem. *SIAM. J. Mat. Anal. Appl.*, 25(2):301–351, 2003.
[46] Z. Drmač and K. Veselić. New fast and accurate Jacobi SVD algorithm. Technical Report, Dept. of Mathematics, University of Zagreb, 2004.

[47] I. S. Duff and C. Vömel. Incremental norm estimation for dense and sparse matrices. *BIT*, 42(2):300–322, 2002.

[48] E. Elmroth, F. Gustavson, I. Jonsson, and B. Kågström. Recursive blocked algorithms and hybrid data structures for dense matrix library software. *SIAM Review*, 46(1):3–45, 2004.

[49] L. Arge, et al. Cache-oblivious and cache-aware algorithms. Schloss Dagstuhl International Conference, www.dagstuhl.de/04301, 2004.

[50] f2c: Fortran-to-C translator. http://www.netlib.org/f2c

[51] G. W. Howell, J. W. Demmel, C. T. Fulton, S. Hammarling, and K. Marmol. Cache efficient bidiagonalization using BLAS 2.5 operators. University of Tennessee, CS Tech Report, LAPACK Working Note 174, Netlib, 2005. http://www.netlib.org/lapack/lawns/downloads/

[52] I. Gohberg, P. Lancaster, and L. Rodman. *Matrix Polynomials*. Academic Press, New York, 1982.

[53] G. Golub and C. Van Loan. *Matrix Computations*. Johns Hopkins University Press, Baltimore, MD, 3rd ed., 1996.

[54] S. Graham, M. Snir, and eds. C. Patterson. *Getting Up to Speed: The Future of Supercomputing*. National Research Council, 2005.

[55] B. Grosser. Ein paralleler und hochgenauer $O(n^2)$ Algorithmus für die bidiagonale Singulärwertzerlegung. PhD thesis, University of Wuppertal, Wuppertal, Germany, 2001.

[56] M. Gu and S. C. Eisenstat. A stable and fast algorithm for updating the singular value decomposition. Yale University, Research Report YALE DCR/RR-966, 1994.

[57] J. A. Gunnels and F. G. Gustavson. Representing a symmetric or triangular matrix as two rectangular matrices. Poster presentation. SIAM Conference on Parallel Processing, San Francisco, 2004.

[58] John A. Gunnels, Fred G. Gustavson, Greg M. Henry, and Robert A. van de Geijn. FLAME: Formal linear algebra methods environment. *ACM Trans. Math. Soft.*, 27(4):422–455, 2001.

[59] F. Gustavson. Recursion leads to automatic variable blocking for dense linear-algebra algorithms. *IBM J. of Res. and Dev.*, 41(6):737–755, 1997. www.research.ibm.com/journal/rd/416/gustavson.html

[60] G. I. Hargreaves. Computing the condition number of tridiagonal and diagonal-plus-semiseparable matrices in linear time. Technical Report submitted, Department of Mathematics, University of Manchester, Manchester, England, 2004.

[61] N. J. Higham. Analysis of the Cholesky decomposition of a semi-definite matrix. In M. G. Cox and S. Hammarling, eds, *Reliable Numerical Computation*, Chapter 9, pp. 161–186. Clarendon Press, Oxford, 1990.

[62] N. J. Higham. *Accuracy and Stability of Numerical Algorithms*. 2nd edn, SIAM, Philadelphia, PA, 2002.

[63] N. J. Higham and M. I. Smith. Computing the matrix cosine. *Numerical Algorithms*, 34:13–26, 2003.

[64] High productivity computing systems (hpcs). www.highproductivity.org

[65] IEEE Standard for Binary Floating Point Arithmetic Revision. grouper.ieee.org/ groups/754, 2002.

[66] Fifth International Workshop on Accurate Solution of Eigenvalue Problems. www.fernuni-hagen.de/MATHPHYS/iwasep5, 2004.

[67] JLAPACK: LAPACK in Java. http://icl.cs.utk.edu/f2j

[68] C. Kenney and A. Laub. Rational iteration methods for the matrix sign function. *SIAM J. Mat. Anal. Appl.*, 21:487–494, 1991.

[69] C. Kenney and A. Laub. On scaling Newton's method for polar decomposition and the matrix sign function. *SIAM J. Mat. Anal. Appl.*, 13(3):688–706, 1992.

[70] LAPACK Contributor Webpage. http://www.netlib.org/lapack-dev/contributions.html

[71] X. S. Li, J. W. Demmel, D. H. Bailey, G. Henry, Y. Hida, J. Iskandar, W. Kahan, et al. Design, implementation and testing of extended and mixed precision BLAS. *Trans. Math. Soft.*, 28(2):152–205, 2002.

[72] K. Meerbergen. Locking and restarting quadratic eigenvalue solvers. Report RAL-TR-1999-011, CLRC, Rutherford Appleton Laboratory, Dept. of Comp. and Inf., Atlas Centre, Oxon OX11 0QX, GB, 1999.
[73] V. Mehrmann and D. Watkins. Structure-preserving methods for computing eigenpairs of large sparse skew-Hamiltonian/Hamiltonian pencils. *SIAM J. Sci. Comput.*, 22:1905–1925, 2001.
[74] V. Mehrmann and D. Watkins. Polynomial eigenvalue problems with Hamiltonian structure. *Elec. Trans. Num. Anal.*, 13:106–113, 2002.
[75] V. Menon and K. Pingali. Look left, look right, look left again: An application of fractal symbolic analysis to linear algebra code restructuring. *Int. J. Parallel Comput.*, 32(6):501–523, 2004.
[76] R. Nishtala, K. Chakrabarti, N. Patel, K. Sanghavi, J. Demmel, K. Yelick, and E. Brewer. Automatic tuning of collective communications in MPI. poster at SIAM Conf. on Parallel Proc., San Francisco, www.cs.berkeley.edu/ rajeshn/poster_draft_6.ppt, 2004.
[77] R. Numrich and J. Reid. Co-array Fortran for parallel programming. Fortran Forum, 17, 1998.
[78] OSKI: Optimized Sparse Kernel Interface. http://bebop.cs.berkeley.edu/oski/
[79] M. Parks. *A Study of Algorithms to Compute the Matrix Exponential.* PhD thesis, University of California, Berkeley, CA, 1994.
[80] B. N. Parlett and I. S. Dhillon. Fernando's solution to Wilkinson's problem: An application of double factorization. *Lin. Alg. Appl.*, 267:247–279, 1997.
[81] B. N. Parlett and I. S. Dhillon. Relatively robust representations of symmetric tridiagonals. *Lin. Alg. Appl.*, 309(1–3):121–151, 2000.
[82] B. N. Parlett and I. S. Dhillon. Multiple representations to compute orthogonal eigenvectors of symmetric tridiagonal matrices, 2004.
[83] B. N. Parlett and I. S. Dhillon. Orthogonal eigenvectors and relative gaps. *SIAM J. Mat. Anal. Appl.*, 25(3):858–899, 2004.
[84] B. N. Parlett and C. Vömel. Tight clusters of glued matrices and the shortcomings of computing orthogonal eigenvectors by multiple relatively robust representations. *Lin. Alg. Appl.*, 387:1–28, 2004.
[85] R. Ralha. One-sided reduction to bidiagonal form. *Lin. Alg. Appl.*, 358:219–238, 2003.
[86] V. Saraswat. Report on the experimental language X10, v0.41. IBM Research Technical Report, 2005.
[87] H. Simon, W. Kramer, W. Saphir, J. Shalf, D. Bailey, L. Oliker, M. Banda, et al. Science-driven system architecture: A new process for leadership class computing. *J. Earth Sim.*, 2:2–10, March 2005.
[88] I. Slapničar. Highly accurate symmetric eigenvalue decomposition and hyperbolic SVD. *Lin. Alg. Appl.*, 358:387–424, 2002.
[89] I. Slapničar and N. Truhar. Relative perturbation theory for hyperbolic singular value problem. *Lin. Alg. Appl.*, 358:367–386, 2002.
[90] M. I. Smith. A Schur algorithm for computing matrix pth roots. *SIAM J. Mat. Anal. Appl.*, 24(4):971–989, 2003.
[91] P. E. Strazdins. A comparison of lookahead and algorithmic blocking techniques for parallel matrix factorization. *Int. J. Parallel Distrib. Systems Networks*, 4(1):26–35, 2001.
[92] F. Tisseur. Backward error analysis of polynomial eigenvalue problems. *Lin. Alg. App.*, 309:339–361, 2000.
[93] F. Tisseur and K. Meerbergen. A survey of the quadratic eigenvalue problem. *SIAM Rev.*, 43:234–286, 2001.
[94] TNT: Template Numerical Toolkit. http://math.nist.gov/tnt
[95] F. Triebsch. Eigenwertalgorithmen für symmetrische λ-Matrizen. PhD thesis, Fakultät für Mathematik, Technische Universität Chemnitz, 1995.
[96] S. S. Vadhiyar, G. E. Fagg, and J. Dongarra. Towards an accurate model for collective communications. *Intern. J. High Perf. Comp. Appl.*, special issue on Performance Tuning, 18(1):159–167, 2004.

[97] R. Vandebril, M. Van Barel, and M. Mastronardi. An implicit QR algorithm for semiseparable matrices to compute the eigendecomposition of symmetric matrices. Report TW 367, Department of Computer Science, K.U.Leuven, Leuven, Belgium, 2003.

[98] R. Vuduc, J. Demmel, and J. Bilmes. Statistical models for automatic performance tuning. In *Intern. Conf. Comput. Science*, May 2001.

[99] R. C. Whaley and J. Dongarra. The ATLAS WWW home page. http://www.netlib.org/atlas/

[100] R. C. Whaley, A. Petitet, and J. Dongarra. Automated empirical optimization of software and the ATLAS project. *Parallel Comput.*, 27(1–2):3–25, 2001.

[101] J. H. Wilkinson. *Rounding Errors in Algebraic Processes*. Prentice Hall, Englewood Cliffs, 1963.

[102] K. Yelick, L. Semenzato, G. Pike, C. Miyamoto, B. Liblit, A. Krishnamurthy, P. Hilfinger, et al. Titanium: A high-performance Java dialect. *Concurrency: Practice and Experience*, 10:825–836, 1998.

30
Parallel Algorithms on Strings

30.1	Introduction...	**30**-1
30.2	Parallel String Matching................................	**30**-2
	Reduction of Periodic Case to Aperiodic • A Simple Algorithm for Strongly Aperiodic Case • Nonperiodic Patterns: Witnesses and Duels • Preprocessing the Pattern • Searching Phase	
30.3	Naming Technique..	**30**-9
30.4	Parallel Construction of Suffix Arrays.................	**30**-12
30.5	Transformation of Suffix Arrays into Suffix Trees	**30**-13
	Algorithm BUILD1 • Algorithm BUILD2	
30.6	Parallel Construction of Suffix Trees by Refining.......	**30**-19
	Bibliographic Notes..	**30**-21
	References	**30**-21

Wojciech Rytter
Warsaw University

30.1 Introduction

We present several *polylogarithmic*-time parallel string algorithms using a high level description of the *parallel random access machine* (PRAM). We select only basic problems: *string matching*, construction of the *dictionary of basic subwords*, *suffix arrays*, and *suffix trees*. Especially, the suffix tree is a very useful data structure in string algorithmics, once it is constructed it can be used to solve easily in parallel many other problems expressed in terms of suffix trees. Most problems related to trees are easily computable in parallel.

A very general model is assumed, since we are interested mainly in exposing the parallel nature of some problems without going into the details of the parallel hardware. The PRAM, a parallel version of the random access machine, is used as a standard model for presentation of parallel algorithms. The PRAM consists of a number of processors working synchronously and communicating through a common random access memory. Each processor is a random access machine with the usual operations. The processors are indexed by consecutive natural numbers, and synchronously execute the same central program; but, the action of a given processor also depends on its number (known to the processor). In one step, a processor can access one memory location. The models differ with respect to simultaneous access to the same memory location by more than one processor. For the CREW (concurrent read, exclusive write) variety of PRAM machine, any number of processors can read from the same memory location simultaneously, but write conflicts are not allowed: no two processors can attempt to write simultaneously into the same location. If we allow many processors to write in a single step into a same location, provided that they all write the same, then such a model is called CRCW PRAM. In this chapter we assume the

CREW model of the PRAM. Parallelism will be expressed by the following type of parallel statement: **for all** $i \in X$ **do in parallel** action(i). The execution of this statement consists in

- Assigning a processor to each element of X
- Executing in parallel by assigned processors the operations specified by action(i)

Usually the part "$x \in X$" looks like "$1 \leq i \leq n$" if X is an interval of integers.

A typical basic problem computed on PRAM is known as *prefix computation*. Given a vector x of n values the problem is to compute all prefix products:

$$y[1] = x[1], \quad y[2] = x[1] \otimes x[2], \quad y[3] = x[1] \otimes x[2] \otimes x[3], \quad \ldots.$$

The prefix computation problem can be computed in logarithmic time with $O(n/\log(n))$ processors.

The total work, measured as the product of time and number of processors is linear. Such parallel algorithms are called *optimal*.

There is another measure of optimality, the total number of operations. Some processors are idle at some stages and they do not contribute in this stages to the total number of operations. There is a general result that *translates* the number of operations into the total work.

Lemma 30.1 [Brent's Lemma]. *Assume we have a PRAM algorithm working in parallel time $t(n)$ and performing $W(n)$ operations, assume also that we can easily identify processors that are not active in a given parallel step. Then we can simulate this algorithm in $O(t(n) + \lceil W(n)/t(n) \rceil)$ parallel time with $O(W(n)/t(n))$ processors.*

The lemma essentially says that in most situations the total work and total number of operations are asymptotically equal. This happens in situations that appear in this chapter.

30.2 Parallel String Matching

We say that a word is *basic* iff its length is a power of two. The **string-matching** problem is to find all occurrences of a pattern-string $x[1..m]$ in a text y of size n, where $m \leq n$.

Denote by *period*(x) the size of the smallest period of text x, for example,

$$period(abaababa) = 5.$$

A string x is called iff $period(x) \leq |x|/2$. Otherwise it is called *aperiodic*. We consider three cases:

- **Periodic patterns**: $period(x) \leq |x|/2$
- **Aperiodic patterns**: $period(x) > |x|/2$
- **Strongly aperiodic patterns**: all basic prefixes of x are aperiodic

30.2.1 Reduction of Periodic Case to Aperiodic

Assume the pattern is periodic. Suppose that v is the shortest prefix of the pattern that is a period of the pattern. Then vv^- is called the *nonperiodic part of the pattern* (v^- denotes the word v with the last symbol removed). We omit the proof of the following lemma, which justifies the name "nonperiodic part" of the pattern.

Lemma 30.2 *If the pattern is periodic then its nonperiodic part is nonperiodic.*

Lemma 30.3 *Assume that the pattern is periodic, and that all occurrences (in the text) of its nonperiodic part are known. Then, we can find all occurrences of the whole pattern in the text in $O(\log m)$ time with $n/\log m$ processors.*

Parallel Algorithms on Strings   30-3

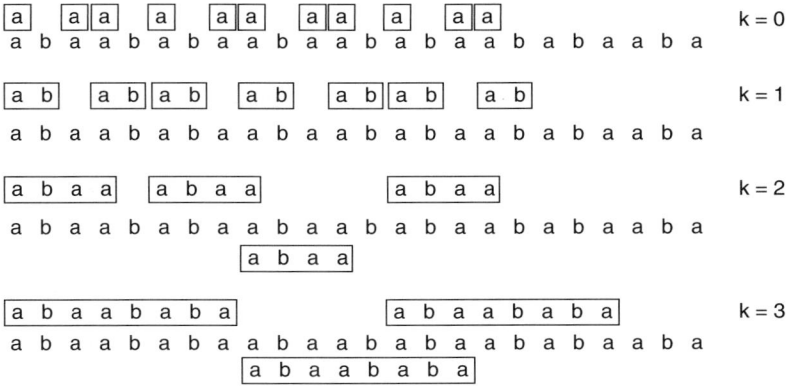

FIGURE 30.1 Performance of the algorithm *Strongly Aperiodic String Matching* for the strongly aperiodic pattern $x = abaababa$.

Proof. We reduce the general problem to unary string matching. Let $w = vv^-$ be the nonperiodic part of the pattern. Assume that w starts at position i on the text. By a segment containing position i we mean the largest segment of the text containing position i and having a period of size $|v|$. We assign a processor to each position. All these processors simultaneously write 1 into their positions if the symbol at distance $|v|$ to the left contains the same symbol. The last position containing 1 to the right of i (all positions between them also contain ones) is the end of the segment containing i. Similarly, we can compute the first position of the segment containing i. It is easy to compute it optimally for all positions i in $O(\log n)$ time by a parallel prefix computation. The string matching is now reduced to a search for a unary pattern, which is an easy task. ■

30.2.2 A Simple Algorithm for Strongly Aperiodic Case

Assume the pattern x is strongly aperiodic and assume that its length is a power of two (otherwise a straightforward modification is needed). Let *TestPartial*(i, r) be the test for a match of $x[1..r]$ starting at position i in text y. The value of *TestPartial*(i, r) is *true* iff $y[i..i + r - 1] = x[1..r]$. It can be checked in $O(\log r)$ time with $O(r/\log r)$ processors.

The algorithm is using an *elimination strategy*.

Figure 30.1 illustrates performance of this algorithm for

$$x = abaababa, \quad y = abaababaabaababa.$$

Initially, all positions are candidates for a match.

Then, we perform the following algorithm:

Algorithm *Strongly Aperiodic String Matching*
    for $k = 0$ to $\log |x|$ **do**
        for each candidate position i **do in parallel**
        if *TestPartial*$(i, 2^k)$=false **then**
            remove i as a candidate for a match;
    return the set of remaining candidates as occurrences of x.

The algorithm *Strongly Aperiodic String Matching* works in $O(\log^2 n)$ time with $O(n)$ processors. The total number of operations is $O(n \log n)$, so the number of processors can be reduced to $O(n/\log n)$ owing to Brent's Lemma.

30.2.3 Nonperiodic Patterns: Witnesses and Duels

The first *optimal* polylogarithmic time parallel algorithm for string matching has been given by Uzi Vishkin, who introduced the crucial notion of a duel, an operation that eliminates in a constant time one of two candidates for a match. The total work of Vishkin's algorithm is $O(n)$.

We assume later in this section that the pattern x is aperiodic, which means that its smallest period is larger than $|x|/2$.

This assumption implies that two consecutive occurrences of the pattern in a text (if any) are at a distance greater than $|x|/2$. However, it is not clear how to use this property for searching the pattern. We proceed as follows: after a suitable preprocessing phase, given too close positions in the text, we eliminate one of them as a candidate for a match. In a window of size $m/2$ of the text we can eliminate all but one candidate using an elimination strategy, the time is logarithmic and the work is linear with respect to m. Hence, the total work is linear with respect to n, since there are only $O(n/m)$.

We define the following *witness table WIT*: for $0 < i < |m|$,

$$WIT[i] = \text{any } k \text{ such that } x[i+k-1] \neq x[k], \quad \text{or}$$

$$WIT[i] = 0, \quad \text{if there is no such } k.$$

This definition is illustrated in Figures 30.2 and 30.3.

A position i_1 on y is said to be *in the range of* a position i_2 if $|i_1 - i_2| < m$. We also say that two positions $i_1 < i_2$ on the text are *consistent* if i_2 is not in the range of i_1, or if $WIT[i_2 - i_1] = 0$.

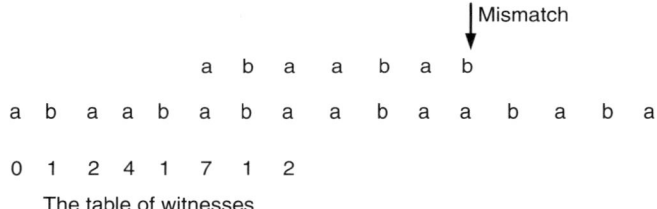

FIGURE 30.2 The witness table $WIT = [0, 1, 2, 4, 7, 1, 2]$ for the text *abaababaabaababa*, $WIT[7] = 7$, owing to the mismatch shown in the figure by the arrow.

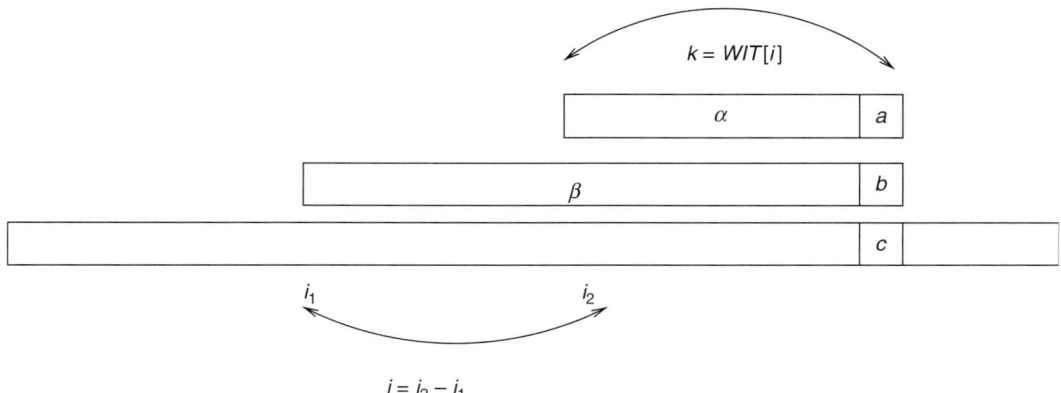

FIGURE 30.3 The duel between positions $i1$, $i2$ and computation of the witness for the loser. We know that $a \neq b$, so $a \neq c$ or $b \neq c$. The strings αa, βb are prefixes of the pattern. If $i2$ is a loser then we know a possible value $WIT[i2] = k$, otherwise a possible value $WIT[i1] = i + k - 1$ is known.

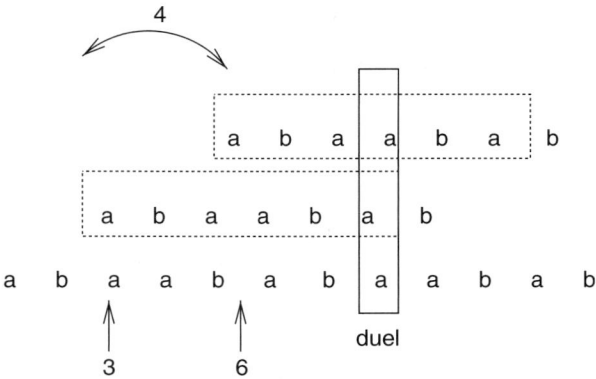

FIGURE 30.4 A duel between two inconsistent positions 3 and 6. The first of them is eliminated since the prefix of the pattern of size 7 placed at starting position 3 does not match the pattern at position $6 + WIT[6 - 3 + 1] - 1$. By the way, we know that possible value of $WIT[3]$ is 7.

If the positions are not consistent, then we can remove one of them as a candidate for the starting position of an occurrence of the pattern just by considering position $i_2 + WIT[i_2 - i_1 + 1]$ on the text. This is the operation called a **duel** (see Figure 30.4).

Let $i = i_2 - i_1 + 1$, and $k = WIT[i]$. Assume that we have $k > 0$, that is, positions i_1, i_2 are not consistent. Let $a = x[k]$ and $b = x[i + k - 1]$, then $a \neq b$. Let c be the symbol in the text y at position $i_2 + k - 1$. We can eliminate at least one of the positions i_1 or i_2 as a candidate for a match by comparing c with a and b.

In some situations, both positions can be eliminated, but, for simplicity, the algorithm below always removes exactly one position.

Let us define, with $a = x[WIT[i_2 - i_1 + 1]]$:

$$duel(i_1, i_2) = (\text{if } a = c \text{ then } i_1 \text{ else } i_1).$$

The position that "survives" is the value of *duel*, the other position is eliminated. We also say that the survival position is a **winner** and the other one is a **loser**.

The *duel* operation is illustrated in Figures 30.4 and 30.3.

In the preprocessing phase we have $y = x$, the pattern makes duels on itself.

30.2.4 Preprocessing the Pattern

In the preprocessing phase we assume that $y = x$ and each duel makes additional operation of computing the witness value for the losing position. This means that if $k = duel(i_1, i_2)$ is the *winning position* and $j = \{i_1, i_2\} - \{k\}$, then j is a *loser* and as a side-effect of $duel(i_1, i_2)$ the value of $WIT[j]$ is computed.

We use a straightforward function $ComputeWit(i)$ which computes the first witness j for i in time $O(\log n \cdot \log\lceil (i+j)/i \rceil)$ using $O(i + j)$ processors. If there is no witness then define $ComputeWit(i) = 0$. The function uses a kind of prefix sum computation to find the first mismatch for $j \in [1 \ldots i]$, then for $j \in [1 \ldots 2i]$ and so forth, doubling the "distance" until the first witness is computed or the end of x is reached.

The preprocessing algorithm operates on a logarithmic sequence of *prefix windows* that has sizes growing at least as powers of two.

The i-th **window** $\mathcal{W}_i$ is an interval $[1 \ldots j]$. The algorithm maintains the following **invariant**:

after the i-th iteration the witnesses for positions in $\mathcal{W}_i$ are computed and for each $k \in \mathcal{W}_i$, $WIT[k] \in \mathcal{W}_i$ or x is periodic with period k,
the size of the window $\mathcal{W}_i$ is at least 2^i.

If $\mathcal{W}_i$ is an interval $[1 \ldots j]$ then define the i-th **working interval** as $\mathcal{I}_i = [j+1 \ldots 2j]$.

The preprocessing algorithm works in iterations, in the $(i+1)$-st **iteration** we perform as follows: witnesses are computed for the positions in the *working interval* $\mathcal{I}_i$ using already computed part of the witness table for positions in $\mathcal{W}_i$ and performing in parallel the tree of duels; see Figures 30.5 and 30.6. The witnesses for losers are computed. Only one position j in $[k+1 \ldots 2k]$, the final winner, has no witness value. We compute its witness value k in parallel using *ComputeWit(j)*. There are three cases:

Case 1: aperiodic case: $k \in \mathcal{W}_i$, $x[1 \ldots j-1]$ does not start a periodic prefix, see Figure 30.6. The iteration is finished. New window $calW_{i+1}$ is twice larger than $\mathcal{W}_i$.

Case 2: periodic case: $k > 0$ and $k \notin \mathcal{W}_i$, $x[1 \ldots j-1]$ starts a periodic prefix that ends at k. Then the witness table for positions in $[j \ldots k-j]$ are computed using periodicity copying from the witness values in $\mathcal{W}_i$, see Figure 30.7.

Case 3: $k = 0$, the whole pattern is periodic. We have already computed witness table for aperiodic part. This is sufficient for the searching phase, owing to Lemma 30.3.

Our construction implies the following fact.

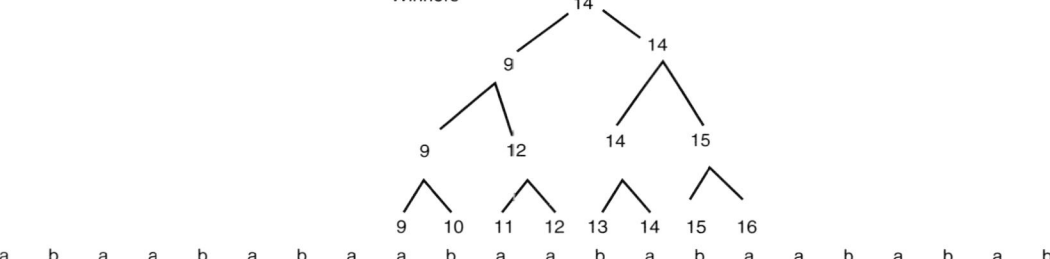

FIGURE 30.5 Preprocessing the pattern *abaababaabaababaababab*..., the iteration corresponds to the *working interval* in x of positions $[9 \ldots 16]$: the tree of winners in the duels between positions in the interval $[9 \ldots 16]$.

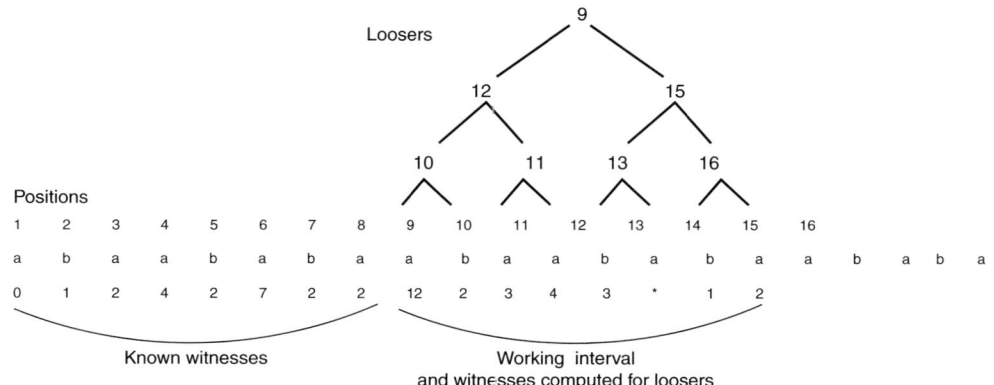

FIGURE 30.6 Processing the working interval $[9 \ldots 16]$ and computing $WIT[9 \ldots 16]$: the tree of losers in the duels between positions in the interval $[9 \ldots 16]$. For each loser, as a side-effect of the corresponding duel, the value of its witness is computed. The witness of position 14 is not computed when making duels since it has not lost any duel and it should be computed by applying the procedure *ComputeWit*(14). It computes $WIT[14] = 9$.

Parallel Algorithms on Strings

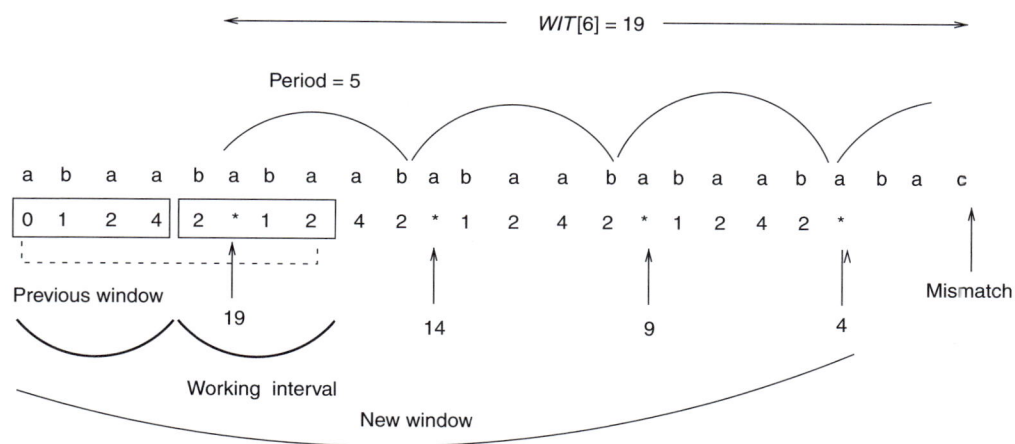

FIGURE 30.7 Periodic case, one iteration of preprocessing the pattern $(abaab)^4 aba\#$. During computation of witnesses in the working interval [5..8] we found that the witness for position 6 is very far away: $ComputeWit(6) = WIT[6] = 19$. This implies a long periodic segment and witnesses are copied to positions to the right of the working interval in a periodic way, except positions 6, 6+5, 6+10, 6+15 denoted by $*$. For these positions, the witnesses are computed by the formula $WIT[6+5k] = 19 - 5k$ for $k = 0, 1, 2, 3$.

Lemma 30.4 *We can locate nonperiodic part of x (possibly the whole x) and compute its witness table in $O(\log^2 n)$ parallel time with $O(n)$ totalwork.*

30.2.5 Searching Phase

Assume that the pattern is aperiodic and the witness table has been computed. Define the operation $\otimes$ by

$$i \otimes j = duel(i, j).$$

The operation $\otimes$ is "practically" associative. This means that the value of $i_1 \otimes i_2 \otimes i_3 \otimes \ldots \otimes i_{m/2}$ depends on the order of multiplications, but all values (for all possible orders) are good for our purpose. We need any of the possible values.

Once the witness table is computed, the string-matching problem reduces to instances of the parallel prefix computation problem. We have the following algorithm. Its behavior on an example string is shown in Figure 30.8 for searching the pattern *abaababa*, for which the witness table is precomputed:

> **Algorithm** *Vishkin-string-matching-by-duels*;
>   consider windows of size $m/2$ on y;
>   { sieve phase }
>   **for** each window **do in parallel**
>   { $\otimes$ can be treated as if it were associative }
>   compute the surviving position $i_1 \otimes i_2 \otimes i_3 \otimes \cdots \otimes i_{m/2}$
>   where $i_1, i_2, i_3, \ldots, i_{m/2}$ are consecutive positions
>   in the window;
>   { naive phase }
>   **for** each surviving position i **do in parallel**
>   check naively an occurrence of x at position i
>   using m processors;

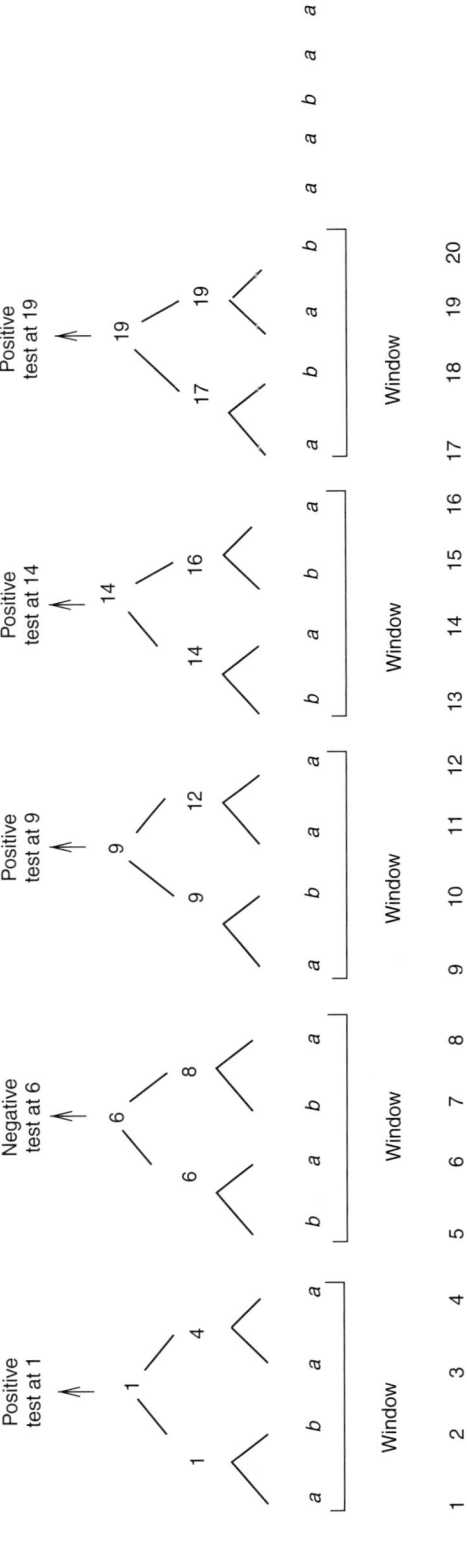

FIGURE 30.8 Searching for *abaababa* in *abaababaabaababaabababaabaa*, the witness values $WIT[1..4]$ for the first *half* of the pattern are used to eliminate in each window all candidates but one, for a possible match of *abaababa*.

Parallel Algorithms on Strings

Theorem 30.1 *Assume we know the witness table and the period of the pattern. Then, the string-matching problem can be solved optimally in $O(\log m)$ time with $O(n/\log m)$ processors of a CREW PRAM.*

Proof. Let $i_1, i_2, i_3, \ldots, i_{m/2}$ be the sequence of positions in a given window. We can compute $i_1 \otimes i_2 \otimes i_3 \otimes \cdots \otimes i_{m/2}$ using an optimal parallel algorithm for the parallel prefix computation. Then, in a given window, only one position survives; this position is the value of $i_1 \otimes i_2 \otimes i_3 \otimes \cdots \otimes i_{m/2}$. This operation can be executed simultaneously for all windows of size $m/2$.

For all windows, this takes $O(\log m)$ time with $O(n/\log m)$ processors of a CREW PRAM.

Afterward, we have $O(n/m)$ surviving positions altogether. For each of them we can check the match using $m/\log m$ processors. Again, a parallel prefix computation is used to collect the result, that is, to compute the conjunction of m Boolean values (match or mismatch, for a given position). This takes again $O(\log m)$ time with $O(n/\log m)$ processors.

Finally, we collect the $O(n/m)$ Boolean values using a similar process. ∎

Corollary 30.1 *There is an $O(\log^2 n)$ time parallel algorithm that solves the string-matching problem (including preprocessing) with $O(n/\log^2 n)$ processors of a CREW PRAM.*

30.3 Naming Technique

The numbering or naming of subwords of a text, corresponding to the sorted order of these subwords, can be used to build a useful data structure. We assign names to certain subwords, or pairs of subwords. Assume we have a sequence

$$S = (s_1, s_2, \ldots, s_t)$$

of at most n different objects. The *naming* of S is a table

$$X[1], X[2], \ldots, X[t]$$

that satisfies conditions (1–2). If, in addition, it satisfies the third condition then the naming is called a *sorted naming*.

1. $s_i = s_j \Leftrightarrow X[i] = X[j]$, for $1 \leq i, j \leq t$
2. $X[i] \in [1..n]$ for each position i, $1 \leq i \leq t$
3. $X[i]$ is the rank of s_i in the ordered list of the different elements of S

Given the string y of length n, we say that two positions are k-equivalent if the subwords of length k starting at these positions are equal. Such an equivalence is best represented by assigning to each position i a name or a number to the subword of length k starting at this position. The name is denoted by $Name_k[i]$ and called a k-name. We assume that the table $Name_k$ is a good and sorted numbering of all subwords of a given length k.

We consider only those subwords of the text whose length k is a power of two. Such subwords are called *basic subwords*. The *name of a subword* is denoted by its rank in the lexicographic ordering of subwords of a given length. For each k-name r we also require (for further applications) a link $Pos_k[r]$ to any one position at which an occurrence of the k-name r starts. Symbol # is a special end marker that has the highest rank in the alphabet. The text is padded with enough end markers to let the k-name at position n defined.

The tables *Name* and *Pos* for a given text w are together called its **dictionary of basic subwords** and is denoted by $DBS(w)$.

Example The tables below show the dictionary of basic subwords for an example text: tables of k-names and of their positions. The k-name at position i corresponds to the subword $y[i..i+k-1]$; its name is its rank according to lexicographic order of all subwords of length k (order of symbols is $a < b < \#$). Indices k are powers of two. The tables can be stored in $O(n \log n)$ space.

Positions	=	1	2	3	4	5	6	7	8			
y	=	a	b	a	a	b	b	a	a	#	#	#
$Name_1$	=	1	2	1	1	2	2	1	1			
$Name_2$	=	2	4	1	2	5	4	1	3			
$Name_4$	=	3	6	1	4	8	7	2	5			
Name of subword	=	1	2	3	4	5	6	7	8			
Pos_1	=	1	2									
Pos_2	=	3	1	8	2	5						
Pos_4	=	3	7	1	4	8	2	6	5			

Remark String matching for y and pattern x of length m can be easily reduced to the computation of a table $Name_m$. Consider the string $w = x\&y$, where $\&$ is a special symbol not in the alphabet. Let $Name_m$ be the array which is part of $DBS(x\&y)$. If $q = Name_m[1]$ then the pattern x starts at all positions i on y such that $Name_m[i + m + 1] = q$.

The tables above display $DBS(abaabbaa\#)$. Three additional #'s are appended to guarantee that all subwords of length 4 starting at positions 1, 2, ..., 8 are well defined. The figure presents tables *Name* and *Pos*. In particular, the entries of Pos_4 give the lexicographically sorted sequence of subwords of length 4. This is the sequence of subwords of length 4 starting at positions 3, 7, 1, 4, 8, 2, 6, 5. The ordered sequence is

$$aabb, aa\#\#, abaa, abba, a\#\#\#, baab, baa\#, bbaa.$$

We introduce the procedure *Sort-Rename* that is defined now. Let S be a sequence of total size $t \leq n$ containing elements of some linearly ordered set. The output of *Sort-Rename*(S) is an array of size t, which is a good and sorted naming of S.

Example Let $S = (ab, aa, ab, ba, ab, ba, aa)$. Then,

$$Sort\text{-}Rename(S) = (2, 1, 2, 3, 2, 3, 1).$$

For a given k, define
$$Composite\text{-}Name_k[i] = (Name_k[i], Name_k[i+k]).$$

KMR algorithm is based on the following simple property of naming tables.

Lemma 30.5 [Key-Lemma], $Name_{2k} = Sort\text{-}Rename(Composite\text{-}Name_k)$.

The main part of algorithm *Sort-Rename* is the lexicographic sort. We explain the action of *Sort-Rename* on the following example:

$$x = ((1, 2), (3, 1), (2, 2), (1, 1), (2, 3), (1, 2)).$$

The method to compute a vector X of names satisfying conditions (1–3) is as follows. We first create the vector y of composite entries $y[i] = (x[i], i)$. Then, the entries of y are lexicographically sorted. Therefore, we get the ordered sequence

$$((1, 1), 4), ((1, 2), 1), ((1, 2), 6), ((2, 2), 3), ((2, 3), 5), ((3, 1), 2).$$

Parallel Algorithms on Strings

Next, we partition this sequence into groups of elements having the same first component. These groups are consecutively numbered starting from 1. The last component i of each element is used to define the output element $X[i]$ as the number associated with the group. Therefore, in the example, we get

$$X[4] = 1,\ X[1] = 2,\ X[6] = 2,\ X[3] = 3,\ X[5] = 4,\ X[2] = 5.$$

Doing so, the procedure *Sort-Rename* has the same complexity as sorting n elements.

Lemma 30.6 *If the vector x has size n, and its components are pairs of integers in the range $(1, 2, \ldots, n$, Sort-Rename(x) can be computed in parallel time $O(\log n)$ with $O(n)$ processors.*

The dictionary of basic subwords is computed by the algorithm Compute-DBS. Its correctness results from fact (∗) below. The number of iterations is logarithmic, and the dominating operation is the call to procedure *Sort-Rename*.

Once all vectors $Name_p$, for all powers of two smaller than r, are computed, we easily compute the vector $Name_q$ in linear time for each integer $q < r$. Let t be the greatest power of two not greater than r. We can compute $Name_q$ by using the following fact:

(∗) $Name_q[i] = Name_q[j]$ iff $Name_t[i] = Name_t[j]$ and $Name_t[i + q - t] = Name_t[j + q - t]$

Algorithm Compute-DBS;
{ a parallel version of the Karp-Miller-Rosenberg algorithm }
  $K :=$ largest power of 2 not exceeding n;

  $Name_1 :=$ Sort-Rename(x);
  for $k := 2, 4, \ldots, K$ **do**
    $Name_{2k} :=$ Sort-Rename(Composite-Name$_k$);

  for each $k = 1, 2, 4, \ldots, K$ **do in parallel**
    for each $1 \leq i \leq n$ **do in parallel**
      $Pos_k[Name_k[i]] := i$;

This construction proves the following theorem.

Theorem 30.2 *The dictionary of basic subwords of a text of length n can be constructed in $\log^2(n)$ time with $O(n)$ processors of a CREW PRAM.*

We show a straightforward application of the dictionary of basic subwords. Denote by $LongestRepFactor(x)$ the length of the *longest repeated subword* of y. It is the longest word occurring at least twice in y (occurrences are not necessarily consecutive). When there are several such longest repeated subwords, we consider any of them. Let also $LongestRepFactor_k(x)$ be the maximal length of the subword that occurs at least k times in y. Let us denote by $REP_k(r, y)$ the function that tests is there is a k-repeating subword of size r. Such function can be easily implemented to run in $\log n$ parallel time if we have $DBS(w)$.

Theorem 30.3 *The function $LongestRepFactor_k(x)$ can be computed in $O(\log^2 n)$ time with $O(n)$ processors.*

Proof. We can assume that the length n of the text is a power of two; otherwise a suitable number of "dummy" symbols are appended to the text. The algorithm *KMR* is used to compute $DBS(x)$. We then apply a kind of binary search using function $REP_k(r, y)$: the binary search looks for the maximum r such that $REP_k(r, y) \neq$ nil. If the search is successful then we return the longest (k times) repeated subword. Otherwise, we report that there is no such repetition. The sequence of values of $REP_k(r, y)$ (for $r = 1, 2, \ldots, n-1$) is "monotonic" in the following sense: if $r1 < r2$ and $REP_k(r2, y) \neq$ nil, then $REP_k(r1, y) \neq$ nil. The binary search behaves similarly to searching an element in a monotonic sequence. It has $\log n$ stages; at each stage the value $REP_k(r, y)$ is computed in logarithmic time. Altogether the computation takes $O(\log n)$ parallel time. ∎

30.4 Parallel Construction of Suffix Arrays

Assume we are given a string x of length n, assume also that we append to x a special end marker symbol $x_{n+1} = \#$, which is smaller than any other symbol of x. We assume in the paper that the alphabet is a subset of $[1 \ldots n]$.

The **suffix array** $SUF = [i_1, i_2, \ldots, i_n]$ is the permutation that gives the sorted list of all suffixes of the string x. This means that

$$suf(i_1) < suf(i_2) < suf(i_3) < \cdots < suf(i_n),$$

where $<$ means here lexicographic order, and $suf(i_k) = x[i_k \ldots n] \#$.

In other words $suf(i_k)$ is the k-th suffix of x in sense of lexicographic order (Figure 30.9).

There is another useful table LCP of the lengths of adjacent common prefixes:

For $1 < i < n$, $LCP[k]$ is the length of the longest common prefix of $suf(i_k)$ and $suf(i_{k-1})$. In addition, define $LCP[1] = 0$.

Example Let us consider an example string: $x =$ b a b a a b a b a b b a #.

For this string we have:

$$SUF = [4, 2, 5, 7, 9, 12, 3, 1, 6, 8, 11, 10]$$

$$LCP = [0, 1, 3, 4, 2, 1, 0, 2, 4, 3, 2, 1]$$

p_4 a a b a b a b b a #
p_2 a b a a b a b a b b a #
p_5 a b a b a b a b #
p_7 a b a b a b #
p_9 a b b a #
p_{12} a #
p_3 b a a b a b a b b a #
p_1 b a b a a b a b a b b a #
p_6 b a b a b b a #
p_8 b a b b a #
p_{11} b a #
p_{10} b b a #

FIGURE 30.9 The sorted sequence (top-down) of 12 suffixes of the string $x =$ babaabababba# with the LCP-values between suffixes (length of shaded straps). The special suffix # and the empty suffix are not considered.

The data structure consisting of both tables SUF, LCP is called the *suffix array*. The dictionary of basic subwords can be used to show directly the following fact.

Theorem 30.4 [12] *The suffix array can be constructed in $O(\log^2 n)$ time with $O(n)$ processors.*

The work of the algorithm from this theorem is $O(n \log^2 n)$.

Once $DBS(x)$ and suffix arrays of x has been computed the table LCP can be computed in a straightforward way in $O(\log n)$ time with $O(n)$ processors. We perform similarly as in the computation of the longest repeating subword. We assign a single processor to each i that computes the length of the longest common prefix between two consecutive (in lexicographic order) suffixes of x in logarithmic sequential time using a kind of binary search.

The work of constructing suffix arrays has been improved in Reference 12 to $O(n \log n)$ using the *Skew Algorithm* which computed only a part of the dictionary DBS of basic subwords of linear size. Observe that the whole dictionary DBS has $O(n \log n)$ size.

Theorem 30.5 [12] *The arrays SUF, LCP can be constructed in $O(\log^2 n)$ time and $O(n \log n)$ work.*

We shortly describe the construction behind the proof of this theorem. The construction is especially interesting in sequential setting where a linear time algorithm is given.

Let $Z \subseteq \{1, 2, 3, \ldots\}$. Denote by SUF_Z the part of the suffix array corresponding only to positions in Z. The set Z is *good* iff for each x of length n it satisfied the following conditions: (1) We can reduce the computation of SUF_Z of x to the computation of SUF for a string x' of size at most cn, where $c < 1$ is a constant; (2) If we know the table SUF_Z then we can compare lexicographically any two suffixes of x in constant work.

We take later $Z = \{i \geq 1 : i \text{ modulo } 3 \in \{0, 2\}\}$. We leave to the reader the proof that our set Z satisfies the second condition for a *good* set. We show on an example how the first condition is satisfied.

Example   For our example string $x = babaabababba\#$ we have

$$SUF_Z = [2, 5, 9, 12, 3, 6, 8, 11].$$

Sorting suffixes starting at positions 2, 5, 8, 11 correpsonds to sorting suffixes of the string [aba][aba][bab][ba#], where 3-letter blocks can be grouped together. Similarly sorting suffixes starting at positions 3, 6, 9, 12 corresponds to sorting suffixes of the string [baa][bab][bab][abb][a##]. We can encode, preserving lexicographic order, the blocks into single letters A, B, C, D, E, F; see Figure 30.10, where it is shown how the computation of SUF_Z is reduced to the computation of the suffix array for $x' = BBFD\$EFCA$. The symbol $\$$ is lexicographically smallest.

The *Skew Algorithm* recursively computes the suffix array for x'. Then we sort all suffixes using the property (2) of *good* sets. We can reduce it to merging, since the sort of suffixes at unclassified positions (outside Z) can be done by referring to the sorting sequence of positons in Z. Then we can merge two sorted sequence, and this can be done in logarithmic time with $O(n)$ work.

The table LCP can be computed in the process of parallel computation of the suffix array in the *Skew Algorithm*. This means that we recursively compute the suffix array and LCP table for the smaller string x', then we reconstruct the suffix array together with LCP for the initial string x. We refer the reader to Reference 12.

30.5 Transformation of Suffix Arrays into Suffix Trees

A simple algorithm for suffix arrays suggests that the simplest way to compute in parallel suffix trees could be to transform in parallel suffix arrays into suffix trees. Assume that the tables SUF and LCP for an input string are already computed. We show two simple transformations. Especially, the first one is very simple. If we compute suffix arrays using $DES(x)$, then descibed below algorithm BUILD1 is a very unsophisticated construction of the suffix tree in polylogarithmic time with the work only slightly larger than for other algorithms. We have simplicity against sophisticated, losing a polylogarithmic factor.

30.5.1 Algorithm BUILD1

We use a parallel *divide-and-conquer* approach. Its correctness is based on the following fact.

Observation Assume $i < j < k$, then the lowest common ancestor in the suffix tree of leaves $SUF[i]$ and $SUF[j]$ is on the branch from the root to the leaf $SUF[k]$. Hence, for a fixed k all lowest common ancestors of $SUF[i]$ and $SUF[j]$ for all $i < k < j$ are on a same branch.

Let $SUF = [i_1, i_2, \ldots, i_n]$. The input to the function is a subsegment α of the sequence $[i_1, i_2, \ldots, i_n]$. The suffix trees are built for the subsegments of of $suf(i_1), suf(i_2), \ldots, suf(i_n)$. α is split into two parts α_1, α_2 of approximately same size. The suffix trees $T1$ and $T2$ are constructed in parallel for the left and the right parts. The last element of α_1 is the same as of α_2. Hence the resulting suffix tree results by merging together the rightmost branch of $T1$ with the leftmost branch of $T2$. The algorithm is written as a recursive function $BUILD1$.

Example In our example, see Figure 30.11b, the sequence corresponding to the suffix array is $\alpha = [i_1, i_2, \ldots, i_n] = [12, 11, 3, 6, 9, 1, 4, 7, 10, 2, 5, 8]$. Let us consider $SUF[6] = 1$, and the branch from the root to 1 is a separator of the suffix tree, it separates the tree into two subtrees with a common branch: the first subtree is for the first half $\alpha_1 = [12, 11, 3, 6, 9, 1]$ and the second subtree for $\alpha_2 = [1, 4, 7, 10, 2, 5, 8]$. The sizes of "halves" differ at most by one.

Theorem 30.6 *Assume that the arrays SUF, LCP are given for an input string of size n. Then the algorithm BUILD1 computes the suffix tree for x in $O(\log^2 n)$ time and $O(n)$ space with $O(n \log n)$ total work.*

Proof. If $|\alpha| = 2$ and we have only two suffixes $suf(i_{j-1}), suf(i_j)$, for some j, then we can create for them a partial suffix tree in $O(1)$ time as follows. The tree has a root, two leaves corresponding to whole suffixes $suf(i_{j-1}), suf(i_j)$, and one internal node corresponding to the longest common prefix of $suf(i_{j-1}), suf(i_j)$. The length of this prefix is given by $LCP[j]$.

The main operation is that of *merging* two branches corresponding to a same suffix. The *total depth* of a node is the length of a string corresponding to this node in the suffix tree. The nodes on the branches

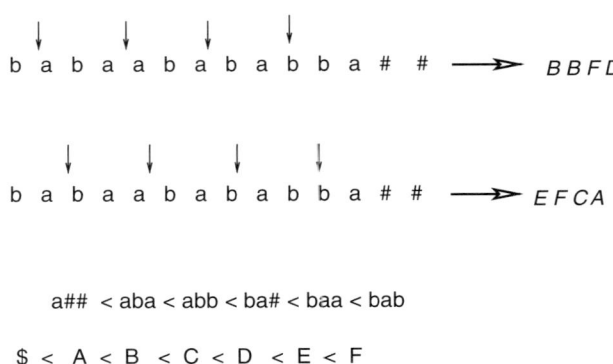

FIGURE 30.10 Sorting suffixes starting at positions pointed by arrow in the text $x = babaabababba\#$ corresponds to sorting suffixes of a "compressed" representation $x' = BBFD\$EFCA$. The set $Z \cap \{1, 2, \ldots, n\}$ corresponds to positions pointed by arrows.

Parallel Algorithms on Strings

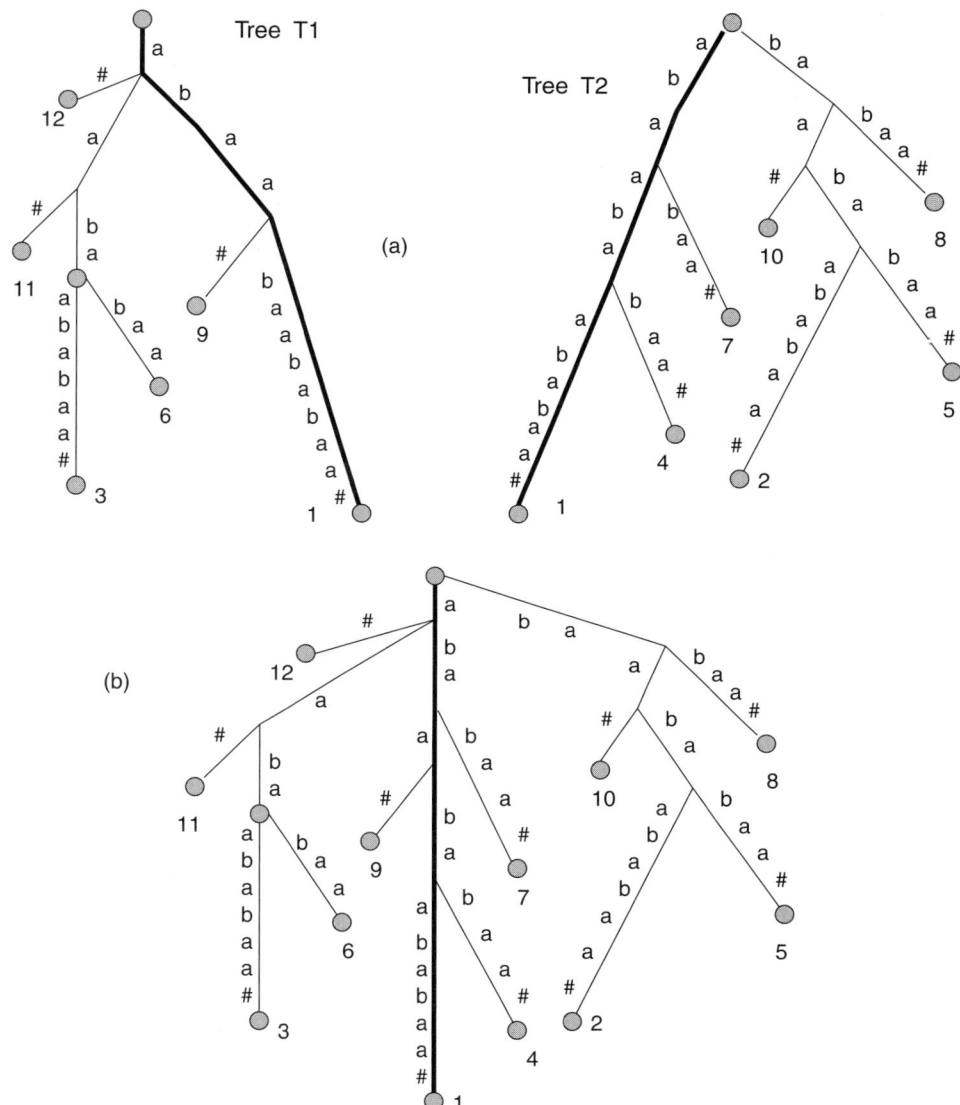

FIGURE 30.11 (a) The suffix trees for two "halves" of the sorted sequence of suffixes: (12, 11, 3, 6, 9, 1) and (1, 4, 7, 10, 2, 5, 8). The "middle" suffix (in bold) belongs to both parts. (b) The suffix tree resulting after merging the nodes corresponding to the common suffix branch corresponding to $suf(1)$.

are sorted in sense of their total depth. Hence to merge two branches we can do a parallel merge of two sorted arrays of size n.

This can be done in logarithmic time with linear work, see Reference 45. The total time is $O(\log^2 n)$ since the depth of the recursion of the function $BUILD1(\alpha)$ is logarithmic. ∎

The total work satisfies similar recurrence as the complexity of the merge-sort.

$$Work(n) = Work(n/2) + Work(n/2 + 1) + O(n).$$

The solution is a function of order $O(n \log n)$. This completes the proof.

Algorithm BUILD1(α); { Given SUF = $[i_1, i_2, \ldots i_n]$}
{ $\alpha = (s_1, s_2, \ldots s_k)$ is a subsegment of $[i_1, i_2, \ldots i_n]$}
if $|\alpha| \leq 2$ **then**

compute the suffix tree T sequentially in $O(1)$ time;
{Comment: the table LCP is needed for $O(1)$ time computation in this step}

else

in parallel do
{$T1 := \text{BUILD1}(s_1, s_2, \ldots s_{k/2})$; $T2 := \text{BUILD1}(s_{k/2}, s_{k/2+1}, \ldots s_k)$};
merge in parallel rightmost branch of $T1$ with leftmost branch of $T2$;

return the resulting suffix tree T;

The history of the algorithm for our example string is shown in Figure 30.1.

30.5.2 Algorithm BUILD2

We design now the algorithm BUILD2, the preprocessing phase for this algorithm consists in the computation of two tables: the *Nearest Smaller Neighbor* table NS and the table $Leftmost[1 \ldots n]$. We show that both tables can be computed in $O(\log n)$ time with $O(n)$ processors. For all $i > 1$ such that $LCP[i] > 0$ define:

$$SN[i] = \begin{cases} L[i], & \text{if } LCP[L[i]] \geq LCP[R[i]], \\ R[i], & \text{otherwise,} \end{cases}$$

where $L[i] = \max\{j < i: LCP[j] < LCP[i]\}$ $R[i] = \min\{j > i: LCP[j] < LCP[i]\}$.

$L[i]$ is the first position to the left of i with smaller value, symmetrically $R[i]$ is the first smaller neighbor on the right side. $Leftmost[i]$ is defined for each $1 \leq i \leq n$ as:

$$\min\{1 \leq j \leq i: LCP[j] = LCP[i] \text{ and } LCP[k] \geq LCP[i] \text{ for each } j \leq k \leq i\}.$$

In other words, $Leftmost[i]$ is the leftmost position to the left of i with the same value as in position i and such that intermediate values are not smaller (Figure 30.12).

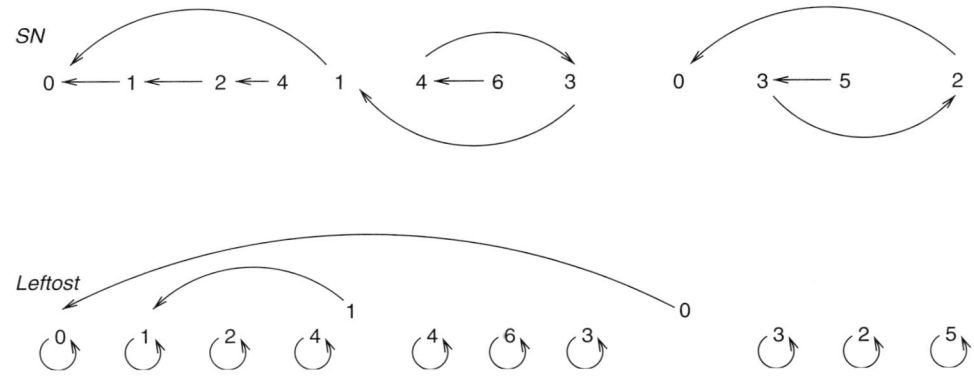

FIGURE 30.12 The tables SN and $Leftmost$.

Parallel Algorithms on Strings

Lemma 30.7 [Preprocessing-Lemma]. *The tables SN and Leftmost can be computed on a CREW PRAM in $O(\log n)$ time with n processors.*

Proof. We show how to compute the table L within claimed complexities. Assume w.l.o.g. that n is a power of 2 and construct complete regular binary tree R of height $\log n$ over positions $1 \ldots n$. Using a classical algorithm for parallel prefix computation we compute for each node v of R the minimum of LCP over the interval corresponding to leaf-positions of the subtree rooted at v. Then for each position i we compute $L[i]$ by first traversing up in R from the i-th leaf until we hit a node v whose value is smaller then $LCP[i]$, and than we go down to the position $L[i]$. ∎

The second algorithm is nonrecursive. Assume that the sequence of leaves $(i_1, i_2, \ldots, i_n)$ read from left to right corresponds to the lexicographically ordered sequence of suffixes. The leaf nodes are represented by starting positions i_k of suffixes. Denote by LCA the lowest common ancestor function.

Fact 30.7 *Each internal node v equals $LCA(i_{k-1}, i_k)$ for some k. As the representative of v we choose i_k such that in the sequence $(i_1, i_2, \ldots, i_n)$ i_k is leftmost among all r_{i_r} such that $LCA(i_{k-1}, i_k) = v$, see Figure 30.13. The internal node $i_k = LCA(i_{k-1}, i_k)$ is the node on the branch from the root to i_k, the string corresponding to the path from root to v is of length $LCP[k]$.*

Denote by $string\_repr(v)$ the string corresponding to labels from the root to v. We have:

If i_k is an internal node then $string\_repr(i_k) = x[i_k \ldots i_k + LCP[k] - 1]$;
If i_k is a leaf node then $string\_repr(i_k) = x[i_k \ldots n]$.

The following fact follows directly from these properties.

Lemma 30.8 [Edge-Labels-Lemma]. *Assume i_s is the father of i_k and $r = LCP[k]$, then the label of the edge from i_s to i_k is*
(a.) If i_k is an internal node then $x[i_k + p \ldots i_k + r - 1]$, where $p = LCP[s]$.
(b.) $x[i_k + r \ldots n]$, otherwise.

If i_k is an internal node then we denote its father by $InFather[i_k]$, otherwise we denote it by $LeafFather[i_k]$. information whether it is a leaf or internal node.

Lemma 30.9 [Fathers-Lemma]. *Assume $(i_1, i_2, \ldots, i_n)$ is the suffix array, then*
(a.) $InFather[i_k] = i_j$, where $j = Leftmost(SN(k))$.
(b.) If $LCP[t] \leq LCP[t+1]$ or $t = n$ then $LeafFather[i_t] = i_j$, where $j = Leftmost[t]$, otherwise $LeafFather[i_t] = i_{t+1}$.

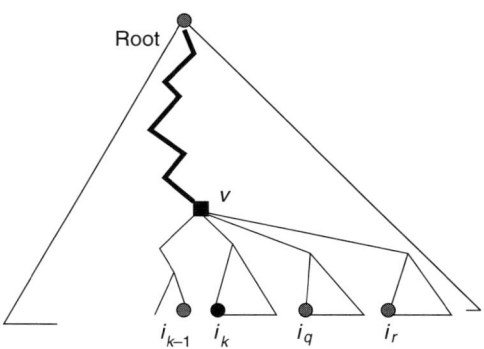

FIGURE 30.13 The node v is identified with i_k, the length of the string $string\_repr(v)$ corresponding to the path from *root* to v equals $LCP[k]$. $Leftmost[r] = Leftmost[q] = Leftmost[k] = k$.

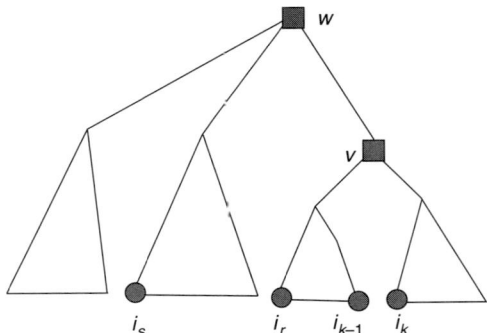

FIGURE 30.14 Computation of the father $w = i_s$ of $v = i_k$. $SN[k] = r$, $Leftmost[r] = s$.

Proof. Define the height of a node v as the length of $string\_repr(v)$. For an internal node i_k its height equals $LCP[k]$. Assume $w = i_s$ is the father of an internal node v, then all nodes between w and v should have height larger or equal to $height(v)$ and $height(w) < height(v)$, see Figure 30.14. Owing to our representation of internal nodes i_s is the leftmost element to the left of i_k in the suffix array with the same LCP as $SN[k]$. Hence, $s = Leftmost[SN[k]]$. Similar argument works for leaf nodes. ∎

Both leaves and internal nodes are identified with integers in the range $[1 \ldots n]$, however, to distinguish between internal node and a leaf with a same number in the implementation we add additional one-bit information.

Algorithm BUILD2(x);

{ The names of internal nods and leaves are integers in $[1 \ldots n]$
{ one-bit information is kept to distinguish internal nodes from leaves}

compute in parallel the tables SN and $Leftmost$;

for each $k \in [1 \ldots n]$

Create leaf node i_k; Create internal node i_k if $Leftmost[k] = k$;

for each node v **do in parallel**

Compute the father w of v according to the formula from Lemma 30.9;
Compute the label of $w \Rightarrow v$ according to formula from Lemma 30.8;

Theorem 30.8 *Assume that the arrays SUF, LCP are given for an input string of size n. The algorithm BUILD2 computes the suffix tree for x on a CREW PRAM in $O(\log n)$ time with $O(n \log n)$ work and $O(n)$ space.*

Proof. The algorithm BUILD2 is presented informally above. Once the table SN is computed all other computations are computed independently *locally* in $O(\log n)$ time with n processors. ∎

The main idea in the construction is to identify the internal nodes of the suffix tree with leaves i_k, such that $v = LCA(i_{k-1}, i_k)$.

The correctness follows from Lemmas 30.7 and 30.9.

30.6 Parallel Construction of Suffix Trees by Refining

The dictionary of basic subwords leads to an efficient alternative construction of suffix trees in $\log^2 n$) time with $O(n)$ processors.

To build the suffix tree of a text, a coarse approximation of it is first built. Afterward the tree is refined step by step, see Figure 30.15. We build a series of a logarithmic number of trees $T_n, T_{n/2}, \ldots, T_1$: each successive tree is an approximation of the suffix tree of the text; the key invariant is:

$inv(k)$: for each internal node v of T_k there are no two distinct outgoing edges for which the labels have the same prefix of the length k; the label of the path from the root to leaf i is $y[i..i+n]$; there is no internal node of outdegree one.

Remark If $inv(1)$ holds, then, the tree T_1 is essentially the suffix tree for x. Just a trivial modification may be needed to delete all #'s padded for technical reasons, but one. Note that the parameter k is always a power of two. This gives the logarithmic number of iterations.

The core of the construction is the procedure $REFINE(k)$ that transforms T_{2k} into T_k. The procedure maintains the invariant: if $inv(2k)$ is satisfied for T_{2k}, then $inv(k)$ holds for T_k after running $REFINE(k)$ on T_{2k}. The correctness (preservation of invariant) of the construction is based on the trivial fact expressed graphically in Figure 30.15. The procedure $REFINE(k)$ consists of two stages:

1. Insertion of new nodes, one per each nonsingleton k-equivalence class
2. Deletion of nodes of outdegree one (*chain nodes*)

We need the following procedure of **local refining**:

> **procedure** $REFINE(k)$;
>   **for** each internal node v of T **do** $LocalRefine(k, v)$;
>   delete all nodes of outdegree one;

The informal description of the construction of the suffix tree is summarized by the algorithm below:

Algorithm *Suffix-Tree-by-Refining*;
    let T be the tree of height 1 which leaves are $1, 2, \ldots, n$,
    the label of the i-th edge is $y[i..n]$ encoded as $[i, *]$;
    $k := n$;
    repeat { T satisfies $inv(k)$ }
        $k := k/2$;
        $REFINE(k)$;
        {insert in parallel new nodes,
        then remove nodes of outdegree 1;}
    until $k = 1$;

In the first stage the operation $LocalRefine(k, v)$ is applied to all internal nodes v of the current tree.

This local operation is graphically presented in Figure 30.15. The k-equivalence classes, labels of edges outgoing a given node, are computed. For each nonsingleton class, we insert a new (internal) node.

The algorithm is informally presented on the example text *abaabbaa#*, see Figures 30.16 and 30.17.

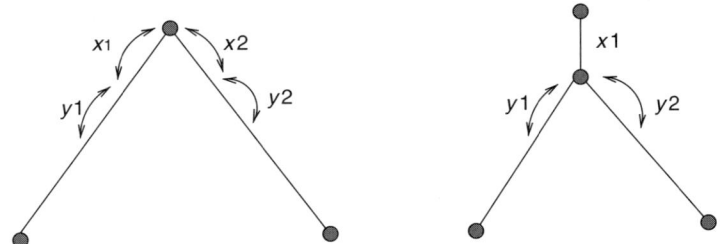

FIGURE 30.15 Insertion of a new node to the temporary suffix tree. We have $|x1| + |y1| = |x2| + |y2| = 2k$ and $|x1| = |x2| = |y1| = |y2|$. We have $|x1| = |x2|, y1 \neq |y2|$. After insertion of a new node, if $inv(2k)$ is locally satisfied on the left, $inv(k)$ holds locally on the right.

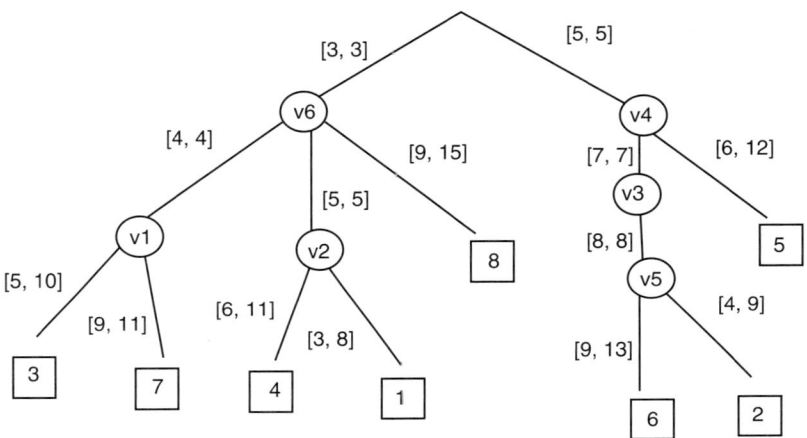

FIGURE 30.16 Tree T_1 after the first stage of $REFINE(1)$: insertion of new nodes v_4, v_5, v_6.

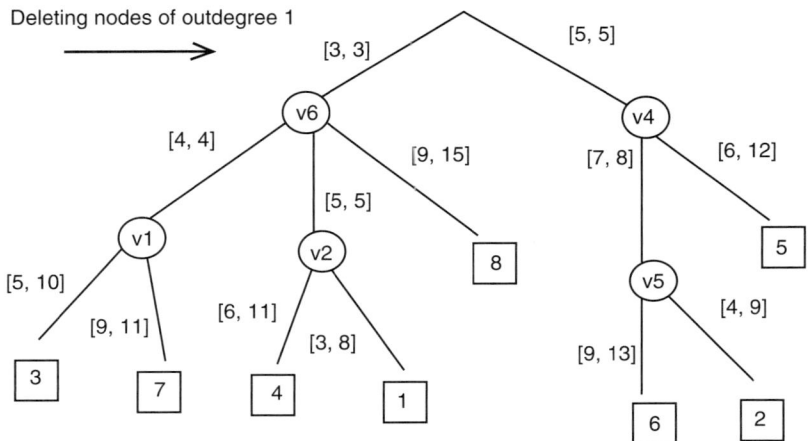

FIGURE 30.17 Tree T_1 after the second stage of $REFINE(1)$: deletion of chain nodes.

We start with the tree T_8 of all subwords of length 8 starting at positions $1, 2, \ldots, 8$. The tree is almost a suffix tree: the only condition that is violated is that the root has two distinct outgoing edges in which labels have a common nonempty prefix. We attempt to satisfy the condition by successive refinements: the prefixes violating the condition become smaller and smaller, divided by two at each stage, until the final tree is obtained.

Bibliographic Notes

The discussion of general models and techniques related to PRAM's can be found in References 11 and 45. The basic books on string algorithms are References 2, 3, 7, and 18.

The first optimal parallel algorithm for string matching was presented by Galil in Reference 41 for constant size alphabets. Vishkin improved on the notion of the duels of Galil, and described the more powerful concept of (fast) duels that leads to an optimal algorithm independently of the size of the alphabet [60]. The idea of witnesses and duels is used also by Vishkin in Reference 61 in the string matching by sampling. The concept of deterministic sampling is very powerful. It has been used by Galil to design a constant-time optimal parallel searching algorithm (the preprocessing is not included). This result was an improvement upon the $O(\log^* n)$ result of Vishkin, though $O(\log^* n)$ time can also be treated practically as a constant time. There are other parallel-string matching algorithms on different models of computation (CRCW) or randomized ones [30, 31, 40, 42].

Two-dimensional string matching was considered in References 21, 22, 34, and 38.

Approximate string matching has been considered in References 24, 52, and 55.

The Karp–Miller–Rosenberg naming technique is from Reference 48. It has been extended in Reference 59. The dictionary of basic subwords and application of Karp–Miller–Rosenberg naming technique in parallel setting is from References 37 and 48.

Suffix arrays were considered in sequential setting in References 4, 5, 12–14, and 16. The parallel sorting algorithms on CREW PRAM working in logarithmic time and $O(n \log n)$ work is from Reference 33.

The fundamental algorithms for suffix trees are in References 1, 8, 17, 19, and 20. The construction of suffix tree by refining was presented in Reference 29. Other parallel algorithms for suffix trees are in References 9, 10, 15, 35, and 46.

The parallel algorithm computing the edit distance is from Reference 24. It was also observed independently in Reference 55. There are other problems for strings considered in parallel settings, see References 23, 25, 36, 42, 54, 56, 57, and 58.

References

[1] Apostolico, A., The myriad virtues of suffix trees, in: Apostolico, A., and Galil, Z., eds., *Combinatorial Algorithms on Words*, NATO Advanced Science Institutes, Series F, Vol. 12, Springer-Verlag, Berlin, 1985, 85–96.

[2] Charras, C. and Lecroq, T., *Handbook of Exact String Matching Algorithms*, King's College Publications, London, February 2004.

[3] Crochemore, M., and Rytter, W., *Jewels of Stringology: Text Algorithms*, World Scientific, New York, 2003.

[4] Farach, M., Optimal Suffix Tree Construction with Large Alphabets, FOCS, IEEE Computer Society, 1997, 137–143.

[5] Farach, M., Ferragina, P., and Muthukrishnan, S., On the sorting complexity of suffix tree construction, *Journal of the ACM*, 47(6): 987–1011, 2000.

[6] Gibbons, A., and Rytter, W., *Efficient Parallel Algorithms*, Cambridge University Press, 1938.

[7] Gusfield, D., *Algorithms on Strings, Trees, and Sequences: Computer Science and Computational Biology*, Cambridge University Press, Cambridge, 1997.

[8] Grossi, R., and Italiano, G., Suffix trees and their applications in string algorithms, Techn. Report CS-96-14, Universita di Venezia, 1996.

[9] Hariharan, R., Optimal parallel suffix tree construction. *STOC* 1994: 290–299.

[10] Iliopoulos, C., Landau, G.M., Schieber, B., and Vishkin, U., Parallel construction of a suffix tree with applications, *Algorithmica* 3 (1988): 347–365.

[11] JaJa, J., *An Introduction to Parallel Algorithms*, Addison Wesley, New York, 1992.

[12] Karkkainen, J., and Sanders, P., *Simple Linear Work Suffix Array Construction*, ICALP, Springer Berlin/Heidelberg, 2003, 943–955.

[13] Kasai, T., Lee, G., Arimura, H., Arikawa, S., and Park, K., Linear-time longest-common-prefix computation in suffix arrays and its applications, in *Proc. 12th Combinatorial Pattern Matching*, 181–192.

[14] Ko, P., and Aluru, S., Space efficient linear time construction of suffix arrays, *14th Annual Symposium, Combinatorial Pattern Matching*, LNCS 2676, 2003, 200–210.

[15] Landau, G.M., Schieber, B., and Vishkin, U., Parallel construction of a suffix tree, in *Automata, Languages and Programming*, Lecture Notes in Computer Science 267, Springer-Verlag, Berlin, 1987, 314–325.

[16] Manber, U. and Myers, E., Suffix arrays: A new method for on-line string searches, in: *Proc. of Ist ACM-SIAM Symposium on Discrete Algorithms*, American Mathematical Society, Providence, R.I., 1990: 319–327.

[17] McCreight, E.M., A space-economical suffix tree construction algorithm, *J. ACM* 23(2) (1976): 262–272.

[18] Smyth, B., *Computing Patterns in Strings*, Pearson Addison Wesley, New York, 2003.

[19] Ukkonen, E., Constructing suffix trees on-line in linear time, *Proc. Information Processing 92*, Vol. 1, IFIP Transactions A-12, Elsevier, 1992, 484–492.

[20] Weiner, P., Linear pattern matching algorithms, in *Proc. 14th IEEE Annual Symposium on Switching and Automata Theory*, Washington, DC, 1973, 1–11.

[21] Amir, A., Benson, G., and Farach, M., Optimal parallel two-dimensional pattern matching, in *5th Annual ACM Symposium on Parallel Algorithms and Architectures*, ACM Press, 1993, 79–85.

[22] Amir, A., and Landau, G.M., Fast parallel and serial multidimensional approximate array matching, *Theoret. Comput. Sci.*, 81 (1991): 97–115.

[23] Apostolico, A., Fast parallel detection of squares in strings, *Algorithmica* 8 (1992): 285–319.

[24] Apostolico, A., Atallah, M.J., Larmore, L.L., and McFaddin, H.S., Efficient parallel algorithms for string editing and related problems, *SIAM J. Comput.*, 19(5) (1990): 968–988.

[25] Apostolico, A., Breslauer, D., and Galil, Z., Optimal parallel algorithms for periods, palindromes and squares, in *3rd Proceedings of SWAT, Lecture Notes in Computer Science 621*, Springer-Verlag, 1992, 296–307.

[26] Berkman, O., Breslauer, D., Galil, Z., Schieber, B., and Vishkin, U., Highly parallelizable problems, in *Proc. 21st ACM Symposium on Theory of Computing*, Association for Computing Machinery, New York, 1989, 309–319.

[27] Apostolico, A., and Crochemore, M., Fast parallel Lyndon factorization and applications, *Math. Syst. Theory* 28(2) (1995): 89–108.

[28] Apostolico, A., and Guerra, C., The longest common subsequence problem revisited, *J. Algorithms* 2 (1987): 315–336.

[29] Apostolico, A., Iliopoulos, C., Landau, G.M., Schieber, B., and Vishkin, U., Parallel construction of a suffix tree with applications, *Algorithmica* 3 (1988): 347–365.

[30] Breslauer, D., and Gall, Z., An optimal $O(\log \log n)$-time parallel string-matching, *SIAM J. Comput.*, 19(6) (1990): 1051–1058.

[31] Chlebus, B.S., Leszek Gasieniec: Optimal pattern matching on meshes. *STACS* 1994: 213–224.

[32] Capocelli, R., ed., *Sequences: Combinatorics, Compression, Security and Transmission*, Springer-Verlag, New York, 1990.

[33] Cole, R., Parallel merge sort, *SIAM J.Comput.*, 17 (1988): 770–785.

[34] Cole, R., Crochemore, M., Galil, Z., Gasieniec, L., Hariharan, R., Muthakrishnan, S., Park, K., and Rytter, W., Optimally fast parallel algorithms for preprocessing and pattern matching in one and two dimensions, in *FOCS'93*, 1993, 248–258.

[35] Crochemore, M., and Rytter, W., Parallel construction of minimal suffix and factor automata, *Inf. Process. Lett.*, 35 (1990): 121–128.
[36] Crochemore, M., and Rytter, W., Efficient parallel algorithms to test square-freeness and factorize strings, *Inf. Process. Lett.*, 38 (1991): 57–60.
[37] Crochemore, M. and Rytter, W., Usefulness of the Karp–Miller–Rosenberg algorithm in parallel computations on strings and arrays, *Theoret. Comput. Sci.*, 88 (1991): 59–82.
[38] Crochemore, M. and Rytter, W., Note on two-dimensional pattern matching by optimal parallel algorithms, in *Parallel Image Analysis*, Lecture Notes in Computer Science 654, Springer-Verlag, 1992, 100–112.
[39] Crochemore, M., Galil, Z., Gasieniec, L., Park, K., and Rytter, W., Constant-time randomized parallel string matching. *SIAM J. Comput.*, 26(4): (1997): 950–960.
[40] Czumaj, A., Galil, Z., Gasieniec, L., Park, K., and Plandowski, W., Work-time-optimal parallel algorithms for string problems. *STOC* 1995: 713–722.
[41] Galil, Z., Optimal parallel algorithm for string matching, *Information and Control* 67 (1985): 144–157.
[42] Czumaj, A., Gasieniec, L., Piotrw, M., and Rytter, W., Parallel and sequential approximations of shortest superstrings. *SWAT* 1994: 95–106.
[43] Galil, Z., and Giancarlo, R., Parallel string matching with k mismatches, *Theoret Comput. Sci.*, 51 (1987) 341–348.
[44] Gasieniec, L. and Park, K., Work-time optimal parallel prefix matching (Extended Abstract). *ESA* 1994: 471–482.
[45] Gibbons, A. and Rytter, W., *Efficient Parallel Algorithms*, Cambridge University Press, Cambridge, U.K., 1988.
[46] Hariharan, R., Optimal parallel suffix tree construction. *J. Comput. Syst. Sci.*, 55(1) (1997): 44–69.
[47] Jàjà, J., *An Introduction to Parallel Algorithms*, Addison-Wesley, Reading, MA, 1992.
[48] Karp, R.M., Miller, R.E., and Rosenberg, A.L., Rapid identification of repeated patterns in strings, arrays and trces, in *Proc. 4th ACM Symposium on Theory of Computing, Association for Computing Machinery*, New York, 1972, 125–136.
[49] Kedem, Z.M., Landau, G.M., and Palem, K.V., Optimal parallel suffix-prefix matching algorithm and applications, in *Proc. ACM Symposium on Parallel Algorithms, Association for Computing Machinery*, New York, 1989, 388–398.
[50] Kedem, Z.M. and Palem, K.V., Optimal parallel algorithms for forest and term matching. *Theoret. Comput. Sci.*, 93(2) (1989): 245–264.
[51] Landau, G.M., Schieber, B., and Vishkin, U., Parallel construction of a suffix tree, in *Automata, Languages and Programming*, Lecture Notes in Computer Science 267, Springer-Verlag, Berlin, 1987, 314–325.
[52] Landau, G.M. and Vishkin, U., Introducing efficient parallelism into approximate string matching, in (STOC, 1986): 220–230.
[53] Landau, G.M. and Vishkin, U., Fast parallel and serial approximate string matching, *J. Algorithms*, 10 (1989): 158–169.
[54] Robert, Y. and Tchuente, M., A systolic array for the longest common subsequence problem, *Inf. Process. Lett.*, 21 (1885): 191–198.
[55] Rytter, W., On efficient computations of costs of paths of a grid graph, *Inf. Process. Lett.*, 29 (1988): 71–74.
[56] Rytter, W., On the parallel transformations of regular expressions to non-deterministic finite automata, *Inf. Process. Lett.*, 31 (1989): 103–109.
[57] Rytter, W., and Diks, K., On optimal parallel computations for sequences of brackets, in: [32]: 92–105.
[58] Schieber, B. and Vishkin, U., On finding lowest common ancestors: simplification and parallelization, *SIAM J. Comput.*, 17 (1988): 1253–1262.

[59] Sahinalp, S. C. and Vishkin, U., Efficient approximate and dynamic matching of patterns using a labeling paradigm (extended abstract). *FOCS* 1996: 320–328.
[60] Vishkin, U., Optimal parallel pattern matching in strings, *Inf. and Control* 67 (1985): 91–113.
[61] Vishkin, U., Deterministic sampling, a new technique for fast pattern matching, *SIAM J. Comput.*, 20(1) (1991): 22–40.

31
Design of Multithreaded Algorithms for Combinatorial Problems

David A. Bader
Kamesh Madduri
Georgia Institute of Technology

Guojing Cong
IBM T.J. Watson Research Center

John Feo
Microsoft Corporation

31.1	Introduction..	31-1
31.2	The Cray MTA-2	31-2
31.3	Designing Multithreaded Algorithms for the MTA-2	31-3
	Expressing Parallelism • Synchronization • Analyzing Complexity	
31.4	List Ranking	31-4
	Multithreaded Implementation • Performance Results	
31.5	Graph Traversal	31-8
	Related Work • A Multithreaded Approach to Breadth-First Search • *st*-Connectivity and Shortest Paths • Performance Results	
31.6	Algorithms in Social Network Analysis	31-16
	Centrality Metrics • Algorithms for Computing Betweenness Centrality • Parallel Algorithms • Performance Results	
31.7	Conclusions	31-24
	Acknowledgments...	31-25
	References	31-25

31.1 Introduction

Graph theoretic and combinatorial problems arise in several traditional and emerging scientific disciplines such as VLSI design, optimization, databases, and computational biology. Some examples include phylogeny reconstruction [65,66], protein–protein interaction networks [89], placement and layout in VLSI chips [59], data mining [52,55], and clustering in semantic webs. Graph abstractions are also finding increasing relevance in the domain of large-scale network analysis [28,58]. Empirical studies show that many social and economic interactions tend to organize themselves in complex network structures. These networks may contain billions of vertices with degrees ranging from small constants to thousands [14,42].

31-1

The Internet and other communication networks, transportation, and power distribution networks also share this property. The two key characteristics studied in these networks are *centrality* (which nodes in the graph are best connected to others, or have the most influence) and *connectivity* (how nodes are connected to one another). Popular metrics for analyzing these networks, like betweenness centrality [18, 43], are computed using fundamental graph algorithms such as breadth-first search (BFS) and shortest paths.

In recognition of the importance of graph abstractions for solving large-scale problems on high performance computing (HPC) systems, several communities have proposed graph theoretic computational challenges. For instance, the 9th DIMACS Implementation Challenge [38] is targeted at finding shortest paths in graphs. The DARPA High Productivity Computer Systems (HPCS) [35] program has developed a synthetic graph theory benchmark called SSCA#2 [9, 11], which is composed of four kernels operating on a large-scale, directed multigraph.

Graph theoretic problems are typically memory intensive, and the memory accesses are fine-grained and highly irregular. This leads to poor performance on cache-based systems. On distributed memory clusters, few parallel graph algorithms outperform their best sequential implementations due to the high memory latency and synchronization costs. Parallel shared memory systems are a more supportive platform. They offer higher memory bandwidth and lower latency than clusters, as the global shared memory avoids the overhead of message passing. However, parallelism is dependent on the cache performance of the algorithm and scalability is limited in most cases. Whereas it may be possible to improve the cache performance to a certain degree for some classes of graphs, there are no known general techniques for cache optimization, because the memory access pattern is largely dependent on the structure of the graph.

In this chapter, we present fast parallel algorithms for several fundamental graph theoretic problems, optimized for multithreaded architectures such as the Cray MTA-2. The architectural features of the MTA-2 aid the design of simple, scalable, and high-performance graph algorithms. We test our implementations on large scale-free and sparse random graph instances, and report impressive results, both for algorithm execution time and parallel performance. For instance, BFS on a scale-free graph of 200 million vertices and 1 billion edges takes less than 5 seconds on a 40-processor MTA-2 system with an absolute speedup of close to 30. This is a significant result in parallel computing, as prior implementations of parallel graph algorithms report very limited or no speedup on irregular and sparse graphs, when compared to the best sequential implementation.

This chapter is organized as follows. In Section 31.2, we detail the unique architectural features of the Cray MTA-2 that aid irregular application design. Section 31.3 introduces the programming constructs available to express parallelism on the MTA-2. In Section 31.4, we compare the performance of the MTA-2 with shared memory systems such as symmetric multiprocessors (SMP) by implementing an efficient shared memory algorithm for list ranking, an important combinatorial problem. In the subsequent sections, we present multithreaded algorithms and performance results for BFS, *st*-connectivity and shortest paths, and apply them to evaluate real-world metrics in Social Network Analysis (SNA).

31.2 The Cray MTA-2

The Cray MTA-2 [34] is a high-end shared memory system offering two unique features that aid considerably in the design of irregular algorithms: fine-grained parallelism and zero-overhead synchronization. The MTA-2 has no data cache; rather than using a memory hierarchy to hide latency, the MTA-2 processors use hardware multithreading to tolerate the latency. The low-overhead synchronization support complements multithreading and makes performance primarily a function of parallelism. Since combinatorial problems often have an abundance of parallelism, these architectural features lead to superior performance and scalability.

The computational model for the MTA-2 is *thread-centric*, not processor-centric. A thread is a logical entity comprised of a sequence of instructions that are issued in order. An MTA processor consists of 128 hardware *streams* and one instruction pipeline. A stream is a physical resource (a set of 32 registers, a status word, and space in the instruction cache) that holds the state of one thread. An instruction is three-wide: a memory operation, a fused multiply add, and a floating point add or control operation. Each stream can have up to eight outstanding memory operations. Threads from the same or different programs are mapped to the streams by the runtime system. A processor switches among its streams every cycle, executing instructions from nonblocked streams. As long as one stream has a ready instruction, the processor remains fully utilized. No thread is bound to any particular processor. System memory size and the inherent degree of parallelism within the program are the only limits on the number of threads used by a program.

The interconnection network is a partially connected 3-D torus capable of delivering one word per processor per cycle. The system has 4 GB of memory per processor. Logical memory addresses are hashed across physical memory to avoid stride-induced hot spots. Each memory word is 68 bits: 64 data bits and 4 tag bits. One tag bit (the full-empty bit) is used to implement synchronous load and store operations. A thread that issues a synchronous load or store remains blocked until the operation completes; but the processor that issued the operation continues to issue instructions from nonblocked streams.

Cray's XMT [94,95] system, formerly called the Eldorado, is a follow-on to the MTA-2 that showcases the massive multithreading paradigm. The XMT is anticipated to scale from 24 to over 8000 processors, providing over one million simultaneous threads and 128 terabytes of globally shared memory. The basic building block of the XMT, the Threadstorm processor, is very similar to the thread-centric MTA processor.

31.3 Designing Multithreaded Algorithms for the MTA-2

The MTA-2 is closer to a theoretical PRAM machine than a shared memory SMP system. Since the MTA-2 uses parallelism to tolerate latency, algorithms must often be parallelized at very fine levels to expose sufficient parallelism. However, it is not necessary that all parallelism in the program be expressed such that the system can exploit it; the goal is simply to saturate the processors. The programs that make the most effective use of the MTA-2 are those that express the parallelism of the problem in a way that allows the compiler to best exploit it.

31.3.1 Expressing Parallelism

The MTA-2 compiler automatically parallelizes *inductive* loops of three types: parallel loops, linear recurrences, and reductions. A loop is inductive if it is controlled by a variable that is incremented by a loop-invariant stride during each iteration, and the loop-exit test compares this variable with a loop-invariant expression. An inductive loop has only one exit test and can only be entered from the top. If each iteration of an inductive loop can be executed completely independently of the others, then the loop is termed parallel. The compiler does a dependence analysis to determine loop-carried dependencies, and then automatically spawns threads to parallelize the loop. To attain the best performance, we need to write code (and thus design algorithms) such that most of the loops are implicitly parallelized.

There are several compiler directives that can be used to parallelize various sections of a program. The three major types of parallelization schemes available are

1. *Single-processor (fray) parallelism*: The code is parallelized in such a way that just the 128 streams on the processor are utilized.
2. *Multiprocessor (crew) parallelism*: This has higher overhead than single-processor parallelism. However, the number of streams available is much larger, bounded by the size of the whole machine rather than the size of a single processor. Iterations can be statically or dynamically scheduled.
3. *Future parallelism*: The *future* construct (detailed below) is used in this form of parallelism. This does not require that all processor resources used during the loop be available at the beginning of the loop. The runtime growth manager increases the number of physical processors as needed. Iterations are always dynamically scheduled. We illustrate the use of the future directive to handle nested parallelism in the subsequent sections.

A *future* is a powerful construct to express user-specified explicit parallelism. It packages a sequence of code that can be executed by a newly created thread running concurrently with other threads in the program. Futures include efficient mechanisms for delaying the execution of code that depends on the computation within the future, until the future completes. The thread that spawns the future can pass information to the thread that executes the future through parameters. Futures are best used to implement task-level parallelism and the parallelism in recursive computations.

31.3.2 Synchronization

Synchronization is a major limiting factor to scalability in the case of practical shared memory implementations. The software mechanisms commonly available on conventional architectures for achieving synchronization are often inefficient. However, the MTA-2 provides hardware support for fine-grained synchronization through the full-empty bit associated with every memory word. The compiler provides a number of generic routines that operate atomically on scalar variables. We list a few useful constructs that appear in the algorithm pseudocodes in subsequent sections:

- The `int_fetch_add` routine (`int_fetch_add(&v, i)`) atomically adds integer i to the value at address v, stores the sum at v, and returns the original value at v (setting the full-empty bit to full). If v is an empty sync or future variable, the operation blocks until v becomes full.
- `readfe(&v)` returns the value of variable v when v is full and sets v empty. This allows threads waiting for v to become empty to resume execution. If v is empty, the read blocks until v becomes full.
- `writeef(&v, i)` writes the value i to v when v is empty, and sets v back to full. The thread waits until v is set empty.
- `purge(&v)` sets the state of the full-empty bit of v to empty.

31.3.3 Analyzing Complexity

To analyze algorithm performance, we use a complexity model similar to the one proposed by Helman and JáJá [50], which has been shown to provide a good cost model for shared-memory algorithms on current SMP systems [10, 13, 49, 50]. The model uses two parameters: the problem's input size n, and the number p of processors. Running time $T(n, p)$ is measured by the triplet $\langle T_M(n,p); T_C(n,p); B(n,p)\rangle$, where $T_M(n, p)$ is the maximum number of noncontiguous main memory accesses required by any processor, $T_C(n, p)$ is an upper bound on the maximum local computational complexity of any of the processors, and $B(n, p)$ is the number of barrier synchronizations. This model, unlike the idealistic PRAM, is more realistic in that it penalizes algorithms with noncontiguous memory accesses that often result in cache misses, and also considers synchronization events in algorithms.

Since the MTA-2 is a shared memory system with no data cache and no local memory, it is comparable to an SMP where all memory reference are remote. Thus, the Helman–JáJá model can be applied to the MTA with the difference that the magnitudes of $T_M(n, p)$ and $B(n, p)$ are reduced through multithreading. In fact, if sufficient parallelism exists, these costs are reduced to zero and performance is a function of only $T_C(n, p)$. Execution time is then a product of the number of instructions and the cycle time.

The number of threads needed to reduce $T_M(n, p)$ to zero is a function of the memory latency of the machine, about 100 cycles. Usually a thread can issue two or three instructions before it must wait for a previous memory operation to complete; thus, 40–80 threads per processor are usually sufficient to reduce $T_M(n, p)$ to zero. The number of threads needed to reduce $B(n, p)$ to zero is a function of intrathread synchronization. Typically, it is zero and no additional threads are needed; however, hotspots can occur. Usually these can be worked around in software, but they do occasionally impact performance.

31.4 List Ranking

List ranking [30,53,77,78] is a key technique often needed in efficient parallel algorithms for solving many graph-theoretic problems; for example, computing the centroid of a tree, expression evaluation, minimum spanning forest, connected components, and planarity testing. Helman and JáJá [49, 50] present an efficient list ranking algorithm with implementation on SMP servers that achieves significant parallel speedup. Using this implementation of list ranking, Bader et al. [10] have designed fast parallel algorithms and demonstrated speedups compared with the best sequential implementation for graph-theoretic problems such as ear decomposition, tree contraction and expression evaluation [13], spanning tree [6], rooted spanning tree [31], and minimum spanning forest [7].

31.4.1 Multithreaded Implementation

List ranking is an instance of the more general prefix problem. Let X be an array of n elements stored in arbitrary order. For each element i, let $X(i).value$ be its value and $X(i).next$ be the index of its successor. Then for any binary associative operator $\oplus$, compute $X(i).prefix$ such that $X(head).prefix = X(head).value$ and $X(i).prefix = X(i).value \oplus X(predecessor).prefix$, where $head$ is the first element of the list, i is not equal to $head$, and $predecessor$ is the node preceding i in the list. If all values are 1 and the associative operation is addition, then prefix reduces to list ranking.

The operations are difficult to parallelize because of the noncontiguous structure of lists and asynchronous access of shared data by concurrent tasks. Unlike arrays, there is no obvious way to divide the list into even, disjoint, continuous sublists without first computing the rank of each node. Moreover, concurrent tasks may visit or pass through the same node by different paths, requiring synchronization to ensure correctness.

The MTA implementation (described in high-level in the following four steps and also given in detail in Algorithm 1) is similar to the Helman and JáJá algorithm:

1. Choose NWALK nodes (including the head node) and mark them. This step divides the list into NWALK sublists.
2. Traverse each sublist computing the prefix sum of each node within the sublist.
3. Compute the rank of each marked node.
4. Retraverse the sublists incrementing the local rank of each node by the rank of the marked node at the head of the sublist.

The first and third steps are $O(n)$. They consist of an outer loop of $O(NWALK)$ and an inner loop of $O(length\ of\ the\ sublist)$. Since the lengths of the local walks can vary, the work done by each thread will vary. The second step is also $O(NWALKS)$ and can be parallelized using any one of the many parallel array prefix methods. In summary, the MTA algorithm has three parallel steps with NWALKS parallelism. Our studies show that by using 100 streams per processor and approximately 10 list nodes per walk, we achieve almost 100% utilization—so a linked list of length $1000p$ fully utilizes an MTA system with p processors.

Since the lengths of the walks are different, the amount of work done by each thread is different. If threads are assigned to streams in blocks, the work per stream will not be balanced. Since the MTA is a shared memory machine, any stream can access any memory location in equal time; thus, it is irrelevant which stream executes which walk. To avoid load imbalances, we instruct the compiler through a pragma to dynamically schedule the iterations of the outer loop. Each stream gets one walk at a time; when it finishes its current walk, it increments the loop counter and executes the next walk. The `int_fetch_add` instruction is used to increment the shared loop counter.

31.4.2 Performance Results

This section summarizes the performance results of our list ranking implementations on SMP and the MTA-2 shared-memory systems. Our SMP implementation is also based on the Helman–JáJá algorithm [49, 50] and is detailed in [8]. We tested our implementation on the Sun E4500, a uniform-memory-access (UMA) shared memory parallel machine with 14 UltraSPARC II 400 MHz processors and 14 GB of memory. Each processor has 16 KB of direct-mapped data (L1) cache and 4 MB of external (L2) cache. We implement the algorithms using POSIX threads and software-based barriers.

```
int list[NLIST+1], rank[NLIST+1];
void RankList(list, rank)
int *list, *rank;
{
  int i, first;
  int tmp1[NWALK+1], tmp2[NWALK+1];
  int head[NWALK+1], tail[NWALK+1], lnth[NWALK+1], next[NWALK+1];
#pragma mta assert noalias *rank, head, tail, lnth, next, tmp1, tmp2

  first = 0;
#pragma mta use 100 streams
  for (i = 1; i <= NLIST; i++) first += list[i];

  first = ((NLIST * NLIST + NLIST) / 2) - first;
  head[0] = 0; head[1]     = first;
  tail[0] = 0; tail[1]     = 0;
  lnth[0] = 0; lnth[1]     = 0;
  rank[0] = 0; rank[first] = 1;

  for (i = 2; i <= NWALK; i++) {
      int node = i * (NLIST / NWALK);
      head[i]    = node;
      tail[i]    = 0;
      lnth[i]    = 0;
      rank[node] = i;
  }

#pragma mta use 100 streams
#pragma mta assert no dependence lnth
  for (i = 1; i <= NWALK; i++) {
    int j, count, next_walk;

    count = 0;
    j     = head[i];
    do {count++; j = list[j];} while (rank[j] == -1);

    next_walk = rank[j];

    tail[i]          = j;
    lnth[next_walk]  = count;
    next[i]          = next_walk;
  }

  while (next[1] != 0) {
#pragma mta assert no dependence tmp1
    for (i = 1; i <= NWALK; i++) {
      int n    = next[i];
      tmp1[n]  = lnth[i];
      tmp2[i]  = next[n];
    }

    for (i = 1; i <= NWALK; i++) {
      lnth[i] += tmp1[i];
      next[i]  = tmp2[i];
      tmp1[i]  = 0;
    }
  }
```

Design of Multithreaded Algorithms for Combinatorial Problems

```
#pragma mta use 100 streams
#pragma mta assert no dependence *rank
  for (i = 1; i <= NWALK; i++) {
    int j, k, count;
    j = head[i];
    k = tail[i];
    count = NLIST - lnth[i];
    while (j != k) {
      rank[j] = count; count--; j = list[j];
    }
  }
```

Algorithm 1: The MTA list ranking code.

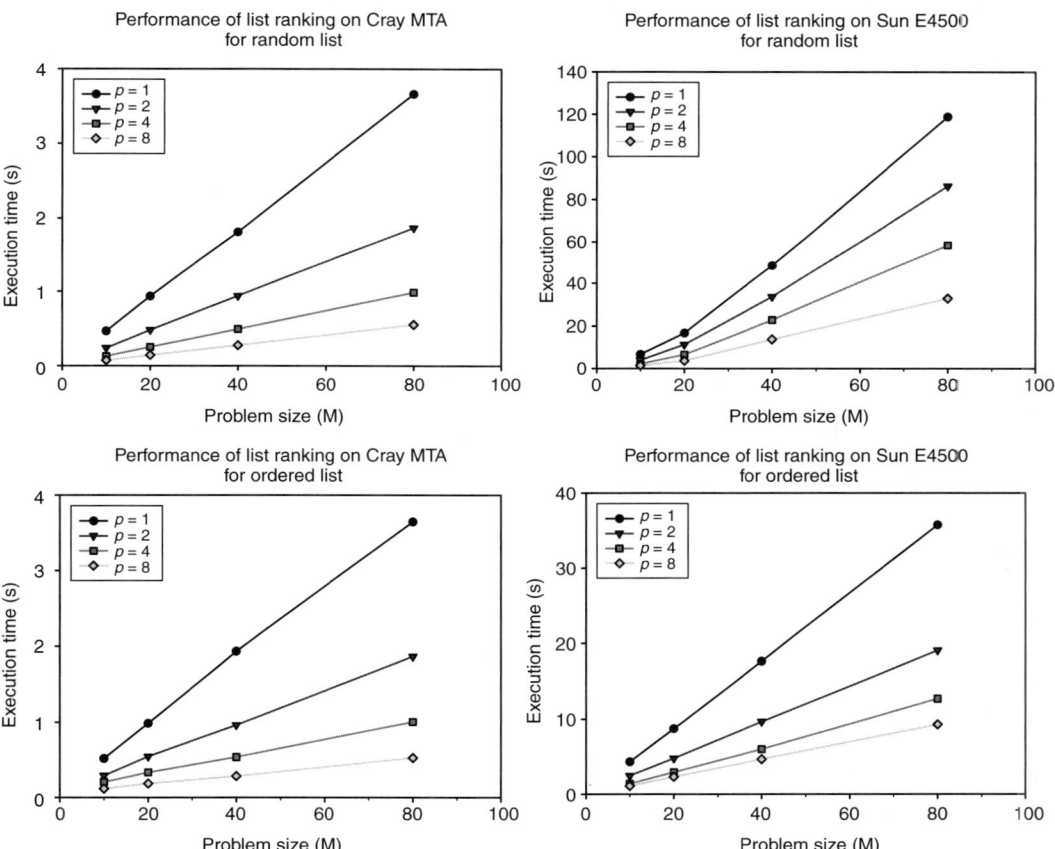

FIGURE 31.1 Running times for list ranking on the Cray MTA (left) and Sun SMP (right) for $p = 1, 2, 4$, and 8 processors.

For list ranking, we use two classes of lists to test our algorithms: *Ordered* and *Random*. Ordered places each element in the array according to its rank; thus, node i is the i-th position of the array and its successor is the node at position $(i + 1)$. Random places successive elements randomly in the array. Since the MTA maps contiguous logical addresses to random physical addresses the layout in physical memory for both classes is similar. We expect, and in fact see, that performance on the MTA is independent of order. This is in sharp contrast to SMP machines that rank Ordered lists much faster than Random lists. The running times for list ranking on the SMP and MTA are given in Figure 31.1. First, all of the

TABLE 31.1 Processor Utilization for List Ranking on the Cray MTA-2

	List Ranking (%)	
Number of processors	Random list	Ordered list
1	98	97
4	90	85
8	82	80

implementations scaled well with problem size and number of processors. In all cases, the running times decreased proportionally with the number of processors, quite a remarkable result on a problem such as list ranking whose efficient implementation has been considered a "holy grail" of parallel computing. On the Cray MTA, the performance is nearly identical for random or ordered lists, demonstrating that locality of memory accesses is a nonissue; first, since memory latency is tolerated, and second, since the logical addresses are randomly assigned to the physical memory. On the SMP, there is a factor of 3 to 4 difference in performance between the best case (an ordered list) and the worst case (a random list). On the ordered lists, the MTA is an order of magnitude faster than this SMP, while on the random list, the MTA is approximately 35 times faster. We also attain a high percentage of processor utilization (see Table 31.1).

Thus, we confirm the results of previous SMP studies and present the first results for combinatorial problems on multithreaded architectures [8]. These results highlight the benefits of latency-tolerant processors and hardware support for synchronization. The MTA-2, because of its randomization between logical and physical memory addresses, and its multithreaded execution techniques for latency hiding, performed extremely well on the list ranking problem, oblivious to the spatial locality of the list. The MTA-2 also allows the programmer to focus on the concurrency in the problem, while an SMP system forces the programmer to optimize for locality and cache. We find the latter results in longer, more complex programs that embody both parallelism and locality.

31.5 Graph Traversal

Breadth-First Search [32] is one of the basic paradigms for the design of efficient graph algorithms. Given a graph $G = (V, E)$ (m edges and n vertices) and a distinguished source vertex s, BFS systematically explores the edges of G to *discover* every vertex that is reachable from s. It computes the *distance* (smallest number of edges) from s to each reachable vertex. It also produces a *breadth-first tree* with root s that contains all the reachable vertices. All vertices at a distance k (or *level k*) are first visited, before discovering any vertices at distance $k + 1$. The *BFS frontier* is defined as the set of vertices in the current level. BFS works on both undirected and directed graphs. A queue-based sequential algorithm runs in optimal $O(m + n)$ time.

In this section, we present fast parallel algorithms for BFS and st-connectivity, for directed and undirected graphs, on the MTA-2. We extend these algorithms to compute single-source shortest paths, assuming unit-weight edges. The implementations are tested on four different classes of graphs—random graphs generated on the basis of the Erdős–Rényi model, scale-free graphs, synthetic sparse random graphs that are hard cases for parallelization, and SSCA#2 benchmark graphs. We also outline a parallel implementation of BFS for handling high-diameter graphs.

31.5.1 Related Work

Distributed BFS [4,92] and st-connectivity [15,45] are both well-studied problems, with related work on graph partitioning and load balancing schemes [5,86] to facilitate efficient implementations. Other problems and algorithms of interest include shortest path variants [26,33,39,63,83,85] and external memory algorithms and data structures [1,22,62] for BFS. Several PRAM and BSP [36] algorithms have been proposed to solve this problem. However, there are very few parallel implementations that achieve significant parallel speedup on sparse, irregular graphs when compared against the best sequential implementations.

31.5.2 A Multithreaded Approach to Breadth-First Search

> **Input**: $G(V, E)$, source vertex s
> **Output**: Array $d[1..n]$ with $d[v]$ holding the length of the shortest path from s to $v \in V$, assuming unit-weight edges
>
> 1 **for** all $v \in V$ **in parallel do**
> 2 $d[v] \leftarrow -1$;
> 3 $d[s] \leftarrow 0$;
> 4 $Q \leftarrow \phi$;
> 5 *Enqueue* $s \leftarrow Q$;
> 6 **while** $Q \neq \phi$ **do**
> 7   **for** all $u \in Q$ **in parallel do**
> 8   *Delete* $u \leftarrow Q$;
> 9   **for** each v adjacent to u **in parallel do**
> 10   **if** $d[v] = -1$ **then**
> 11 $d[v] \leftarrow d[u] + 1$;
> 12   *Enqueue* $v \leftarrow Q$;

Algorithm 2: Level-synchronized parallel BFS.

Unlike earlier parallel approaches to BFS, on the MTA-2 we do not consider load balancing or the use of distributed queues for parallelizing BFS. We employ a simple level-synchronized parallel algorithm (Algorithm 2) that exploits concurrency at two key steps in BFS:

1. All vertices at a given *level* in the graph can be processed simultaneously, instead of just picking the vertex at the head of the queue (step 7 in Algorithm 2).
2. The adjacencies of each vertex can be inspected in parallel (step 9 in Algorithm 2).

We maintain an array d to indicate the level (or distance) of each visited vertex and process the global queue Q accordingly. Algorithm 2 is, however, a very high-level representation, and hides the fact that thread-safe parallel insertions to the queue and atomic updates of the distance array d are needed to ensure correctness. Algorithm 3 details the MTA-2 code required to achieve this (for the critical steps 7–12), which is simple and very concise. The loops will not be automatically parallelized as there are dependencies involved. The compiler can be forced to parallelize them using the *assert parallel* directive on both the loops. We then note that we have to handle and exploit the nested parallelism in this case. We can explicitly indicate that the iterations of the outer loop can be handled concurrently, and the compiler will dynamically schedule threads for the inner loop. We do this using the compiler directive *loop future* (see Algorithm 3) to indicate that the iterations of the outer loop can be concurrently processed. We use the low-overhead instructions `int_fetch_add`, `readfe()`, and `writeef()` to atomically update the value of d, and insert elements to the queue in parallel.

Once correctness is assured, we optimize the code further. Note that we used the *loop future* directive on the outer loop in Algorithm 3 to concurrently schedule the loop iterations. Using this directive incurs an overhead of about 200 instructions. So we do not use it when the number of vertices to be visited at a given level is <50 (experimentally determined figure for this particular loop). Clearly, the time taken to spawn threads must be considerably less than the time spent in the outer loop.

```
/* While the Queue is not empty */
#pragma mta assert parallel
#pragma mta loop future
for (i = startIndex; i < endIndex; i++) {
  u = Q[i];

  /* Inspect all vertices adjacent to u */
  #pragma mta assert parallel
    for (j = 0; j < degree[u]; j++) {
      v = neighbor[u][j];

      /* Check if v has been visited yet? */
      dist = readfe(&d[v]);

      if (dist == -1) {
        writeef(&d[v], d[u] + 1);
        /*Enqueue v */
        Q[int_fetch_add(&count, 1)] = v;
      } else {
        writeef(&d[v], dist);
      }
    }
}
```

Algorithm 3: MTA-2 parallel C code for steps 7–12 in Algorithm 2.

High-degree vertices pose a major problem. Consider the case when the majority of vertices at a particular level are of low-degree (<100), but a few vertices are of very high-degree (order of thousands). If Algorithm 2 is applied, most of the threads will be done processing the low-degree vertices quickly, but only a few threads will be assigned to inspect the adjacencies of the high-degree nodes. The system will be heavily underutilized then, until the loop finishes. To prevent this, we first need to identify high-degree nodes at each level and work on them sequentially, but inspect their adjacencies in parallel. This ensures that work is balanced among the processors. We can choose the low-degree cutoff value appropriately so that parallelization of adjacency visits would be sufficient to saturate the system. We take this approach for BFS on Scale-free graphs. In general, given an arbitrary graph instance, we can determine which algorithm to apply based on a quick evaluation of the degree distribution. This can be done either during graph generation stage (when reading from an uncharacterized data set and internally representing it as a graph) or in a pre-processing phase before running the actual BFS algorithm.

We observe that the above parallelization schemes will not work well for high-diameter graphs (for instance, consider a chain of vertices with bounded degree). For arbitrary sparse graphs, Ullman and Yannakakis offer high-probability PRAM algorithms for transitive closure and BFS [87] that take $\tilde{O}(n^\epsilon)$ time with $\tilde{O}(mn^{1-2\epsilon})$ processors, provided $m \geq n^{2-3\epsilon}$. The key idea here is as follows. Instead of starting the search from the source vertex s, we expand the frontier up to a distance d in parallel from a set of randomly chosen n/d *distinguished* vertices (that includes the source vertex s also) in the graph. We then construct a new graph whose vertices are the distinguished vertices, and we have edges between these vertices if they were pair-wise reachable in the previous step. Now a set of n/d^2 *superdistinguished* vertices are selected among them and the graph is explored to a depth d^2. After this step, the resulting graph would be dense and we can determine the shortest path of the source vertex s to each of the vertices. Using this information, we can determine the shortest paths from s to all vertices.

31.5.3 *st*-Connectivity and Shortest Paths

st-connectivity is a related problem, also applicable to both directed and undirected graphs. Given two vertices *s* and *t*, the problem is to decide whether or not they are connected, and determine the shortest path between them, if one exists. It is a basic building block for more complex graph algorithms, has linear time complexity, and is complete for the class SL of problems solvable by symmetric, nondeterministic, log-space computations [60].

Input: $G(V, E)$, vertex pair (s, t)
Output: The smallest number of edges *dist* between *s* and *t*, if they are connected

1 **for** *all* $v \in V$ **in parallel do**
2      $color[v] \leftarrow$ WHITE;
3      $d[v] \leftarrow 0$;
4 $color[s] \leftarrow$ RED; $color[t] \leftarrow$ GREEN; $Q \leftarrow \phi$; *done* $\leftarrow$ FALSE; *dist* $\leftarrow \infty$;
5 Enqueue $s \leftarrow Q$; Enqueue $t \leftarrow Q$;
6 **while** $Q \neq \phi$ and *done* = FALSE **do**
7   **for** *all* $u \in Q$ **in parallel do**
8   Delete $u \leftarrow Q$;
9   **for** *each v adjacent to u* **in parallel do**
10             $color \leftarrow$ readfe($\&color[v]$);
11   **if** $color =$ WHITE **then**
12                 $d[v] \leftarrow d[u] + 1$;
13   Enqueue $v \leftarrow Q$;
14   writeef($\&color[v], color[u]$);
15   **else**
16   **if** $color \neq color[u]$ **then**
17   *done* $\leftarrow$ TRUE;
18                     $tmp \leftarrow$ readfe($\&dist$);
19   **if** $tmp > d[u] + d[v] + 1$ **then**
20   writeef($\&dist, d[u] + d[v] + 1$);
21   **else**
22   writeef($\&dist, tmp$);
23   writeef($\&color[v], color$);

Algorithm 4: *st*-connectivity (STCONN-FB): concurrent BFSes from *s* and *t*.

We can easily extend the BFS algorithm for solving the *st*-connectivity problem too. A naïve implementation would be to start a BFS from *s*, and stop when *t* is visited. However, we note that we could run BFS concurrently both from *s* and to *t*, and if we keep track of the vertices visited and the expanded frontiers on both sides, we can correctly determine the shortest path between *s* and *t*. The key steps are outlined in Algorithm 4 (termed STCONN-FB), which has both high-level details as well as MTA-specific synchronization constructs. Both *s* and *t* are added to the queue initially, and newly discovered vertices are either colored RED (for vertices reachable from *s*) or GREEN (for vertices that can reach *t*). When a *back edge* is found in the graph, the algorithm terminates and the shortest path is evaluated. As in the previous case, we encounter nested parallelism here and apply the same optimizations.

The pseudocode for STCONN-FB is elegant and concise, but it is also very easy to introduce race conditions and potential deadlocks. Figure 31.2 illustrates a subtle race condition if we do not update *dist* atomically in Algorithm 4. Consider the directed graph presented in the figure and the problem

Input: $G(V, E)$, vertex pair (s, t)
Output: The smallest number of edges *dist* between s and t, if they are connected

1 **for** *all $v \in V$* **in parallel do**
2     $color[v] \leftarrow WHITE$;
3     $d[v] \leftarrow 0$;
4 $color[s] \leftarrow GRAY$; $color[t] \leftarrow GRAY$; $Qs \leftarrow \phi$; $Qt \leftarrow \phi$;
5 $done \leftarrow FALSE$; $dist \leftarrow -1$;
6 Enqueue $s \leftarrow Qs$; Enqueue $t \leftarrow Qt$; $extentS \leftarrow 1$; $extentT \leftarrow 1$;
7 **while** $(Qs \neq \phi$ *or* $Qt \neq \phi)$ *and done = FALSE* **do**
8   Set Q appropriately;
9   **for** *all $u \in Q$* **in parallel do**
10   Delete $u \leftarrow Q$;
11   **for** *each v adjacent to u* **in parallel do**
12         $color \leftarrow$ `readfe(&`$color[v]$`)`;
13   **if** $color = WHITE$ **then**
14           $d[v] \leftarrow d[u] + 1$;
15   Enqueue $v \leftarrow Q$;
16   `writeef(&`$color[v]$`, `$color[u]$`)`;
17   **else**
18   **if** $color \neq color[v]$ **then**
19             $dist \leftarrow d[u] + d[v] + 1$;
20             $done \leftarrow TRUE$;
21   `writeef(&`$color[v]$`, `$color$`)`;
22     $extentS \leftarrow |Qs|$; $extentT \leftarrow |Qt|$;

Algorithm 5: *st*-connectivity (STCONN-MF): alternate BFSes from s and t.

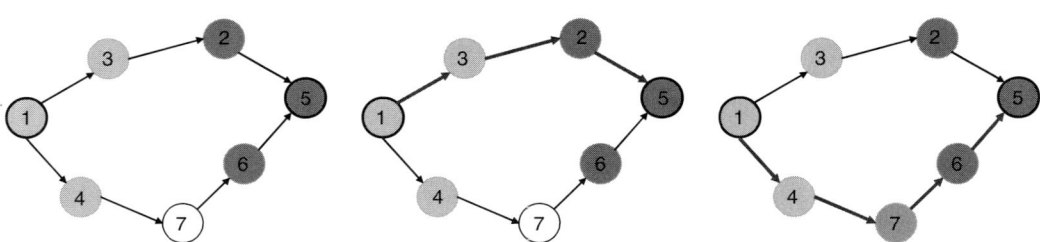

FIGURE 31.2 An illustration of a possible race condition to be avoided in *st*-connectivity Alg. STCONN-FB.

of determining whether vertices 1 and 5 are connected. After one iteration of the loop and concurrent expansion of frontiers from both 1 and 5, we have the vertices colored as shown in the leftmost diagram. The shortest path is clearly $(1, 3, 2, 5)$ and of length 3. Let us assume that the search from 2 finds 3 (or vice-versa), updates the *dist* value to 3, and mark *done* TRUE. Now vertex 4 will find 7 and color it, and then vertex 6 will encounter 7 colored differently. It may then update the value of *dist* to 4 if we do not have the conditional statement at line 19.

We also implement an improved algorithm for *st*-connectivity (STCONN-MF, denoting *minimum frontier*) applicable to graphs with a large percentage of high degree nodes, detailed in Algorithm 5. In this case, we maintain two different queues Q_s and Q_t and expand the smaller frontier (Q in Algorithm 5 is either Q_s or Q_t, depending on the values of *extentS* and *extentT*) on each iteration. This algorithm

Design of Multithreaded Algorithms for Combinatorial Problems

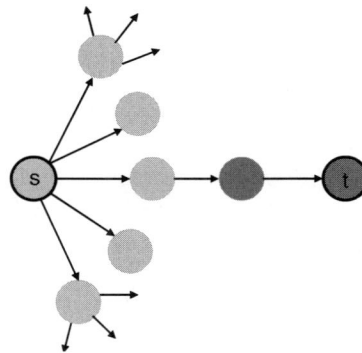

FIGURE 31.3 A case when st-connectivity Alg. STCONN-MF will be faster.

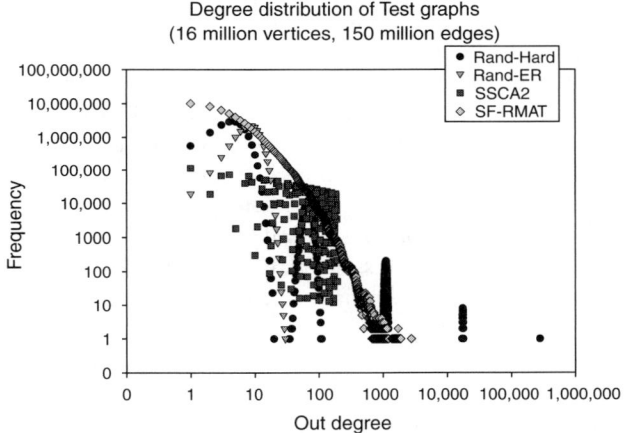

FIGURE 31.4 Degree distributions of the four test graph classes.

would be faster for some graph instances (see Figure 31.3) for an illustration). Intuitively, we try to avoid inspecting the adjacencies of high-degree vertices, thus doing less work than STCONN-FB. Both Algorithms 4 and 5 are discussed in more detail in [12].

31.5.4 Performance Results

This section summarizes the experimental results of our BFS and st-connectivity implementations on the Cray MTA-2. We report results on a 40-processor MTA-2, with each processor having a clock speed of 220 MHz and 4 GB of RAM. From the programmer's viewpoint, the MTA-2 is, however, a global shared memory machine with 160 GB memory.

We test our algorithms on four different classes of graphs (see Figure 31.4):

1. Random graphs generated based on the Erdős–Rényi $G(n, p)$ model (Rand-ER): A random graph of m edges is generated with $p = m/n^2$ and has very little structure and locality.
2. Scale-free graphs (SF-RMAT), used to model real-world large-scale networks: these graphs are generated using the R-MAT graph model [25]. They have a significant number of vertices of very high degree, although the majority of vertices are low-degree ones. The degree distribution plot on a log–log scale is a straight line with a heavy tail, as seen in Figure 31.4.

3. Synthetic sparse random graphs that are hard cases for parallelization (Rand-Hard): As in scale-free graphs, a considerable percentage of vertices are high-degree ones, but the degree distribution is different.
4. DARPA SSCA#2 benchmark (SSCA2) graphs: A typical SSCA#2 graph consists of a large number of highly interconnected clusters of vertices. The clusters are sparsely connected, and these intercluster edges are randomly generated. The cluster sizes are uniformly distributed and the maximum cluster size is a user-defined parameter. For the graph used in the performance studies in Figure 31.4, we assume a maximum cluster size of 10.

We generate directed graphs in all four cases. Our algorithms work for both directed and undirected graphs, as each vertex stores all its neighbors, and the edges in both directions. In this section, we report results for the undirected case. By making minor changes to our code, we can analyze directed graphs also.

Figure 31.5(a) plots the execution time and speedup attained by the BFS algorithm on a random graph of 134 million vertices and 940 million edges (average degree 7). The plot in the inset shows the scaling when the number of processors is varied from 1 to 10, and the main plot for 10–40 processors. We define the *Speedup* on p processors of the MTA-2 as the ratio of the execution time on p processors to that on one processor. Since the computation on the MTA is thread-centric, system utilization is also an important metric to study. We observed utilization of close to 97% for single processor runs. We also note that the system utilization was consistently high (around 80% for 40 processor runs) across all runs. We achieve a speedup of nearly 10 on 10 processors for random graphs, 17 on 20 processors, and 28 on 40 processors. This is a significant result, as random graphs have no locality and such instances would offer very limited on no speedup on cache-based SMPs and other shared memory systems. The decrease in efficiency as the number of processors increases to 40 can be attributed to two factors: hot spots in the BFS queue, and a performance penalty due to the use of the future directive for handling nested parallelism.

Figure 31.5(b) gives the BFS execution time for a Scale-free graph of 134 million vertices and 940 million edges, as the number of processors is varied from 1 to 40. The speedups are slightly lower than the previous case, owing to the variation in the degree distribution. We have a preprocessing step for high-degree nodes as discussed in the previous sections; this leads to an additional overhead in execution time (when compared to random graphs), as well as insufficient work to saturate the system in some cases. Figures 31.5(c) and (d) summarize the BFS performance for SSCA#2 graphs. The execution time and speedup (Figure 31.5(c)) are comparable to random graphs. We also varied the user-defined cluster size parameter to see how BFS performs for dense graphs. Figure 31.5(d) shows that the dense SSCA#2 graphs are also handled well by our BFS algorithm.

Figure 31.5(e) and (f) show the performance of BFS as the edge density is varied for Rand-ER and Rand-Hard graphs. We consider a graph of 2.147 billion edges and vary the number of vertices from 16 to 536 million. In case of Rand-ER graphs, the execution times are comparable as expected, since the dominating term in the computational complexity is the number of edges, 2.147 billion in this case. However, in case of the Rand-Hard graphs, we note an anomaly: the execution time for the graph with 16 million vertices is comparatively more than the other graphs. This is because this graph has a significant number of vertices of very large degree. Even though it scales with the number of processors, since we avoid the use of nested parallelism in this case, the execution times are higher.

Figure 31.6 summarizes the performance of st-connectivity. Note that both the st-connectivity algorithms are based on BFS, and if BFS is implemented efficiently, we would expect st-connectivity also to perform well. Figure 31.6(a) shows the performance of STCONN-MF on random graphs as the number of processors is varied from 1 to 10. Note that the execution times are highly dependent on (s, t) pair we choose. In this particular case, just 45,000 vertices were visited in a graph of 134 million vertices. The st-connectivity algorithm shows near-linear scaling with the number of processors. The actual execution time is bounded by the BFS time, and is dependent on the shortest path length and the degree distribution of the vertices in the graph. In Figure 31.6(b), we compare the performance of

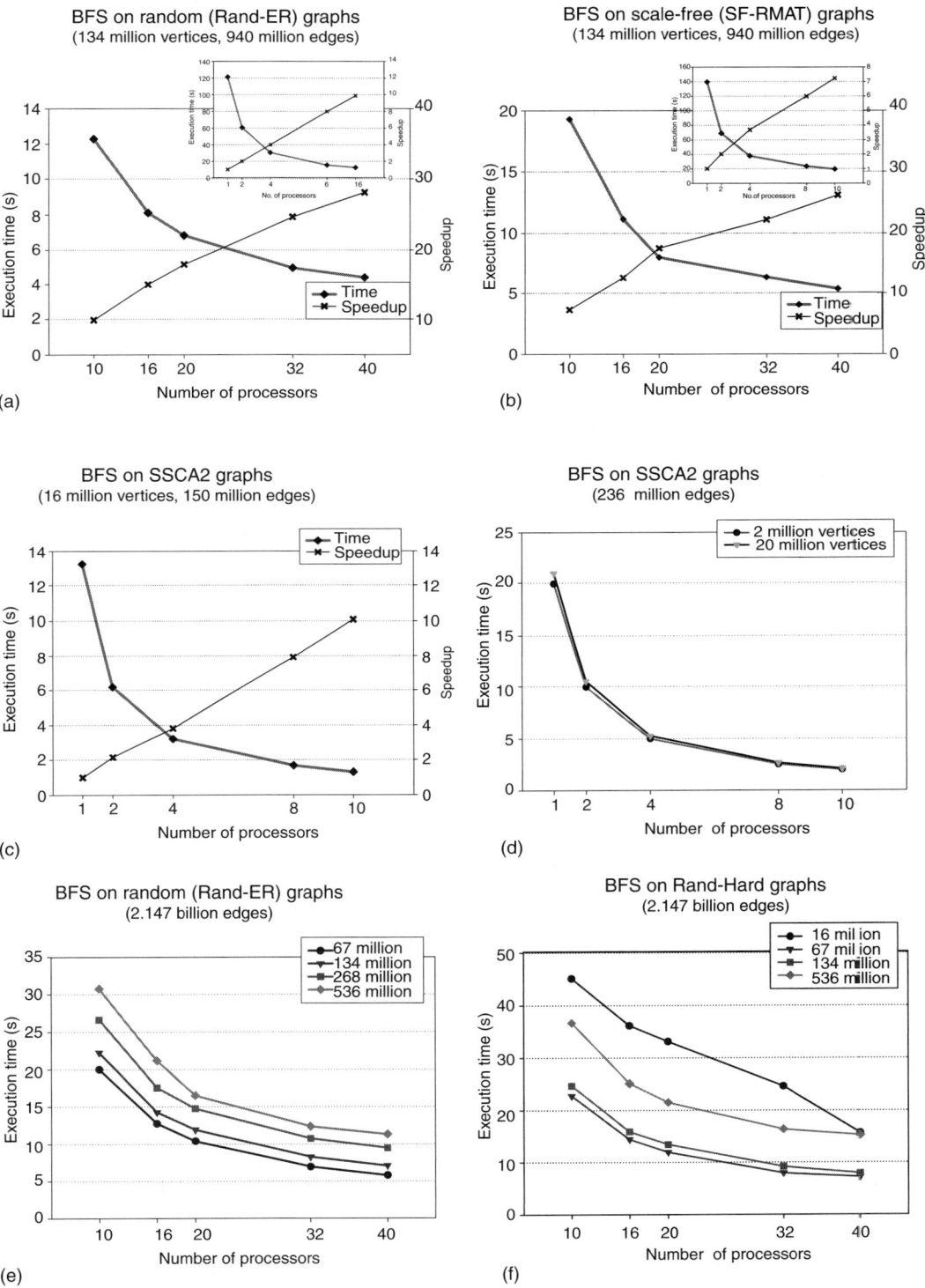

FIGURE 31.5 Breadth First Search performance results (a) Execution time and speedup for random graphs: 1–10 processors (inset), and 10–40 processors (b) Execution time and speedup for scale-free (SF-RMAT) graphs: 1–10 processors (inset), and 10–40 processors (c) Execution time and speedup for SSCA2 graphs (d) Execution time variation as a function of average degree for SSCA2 graphs (e) Execution time variation as a function of average degree for Rand-ER graphs (f) Execution time variation as a function of average degree for Rand-Hard graphs.

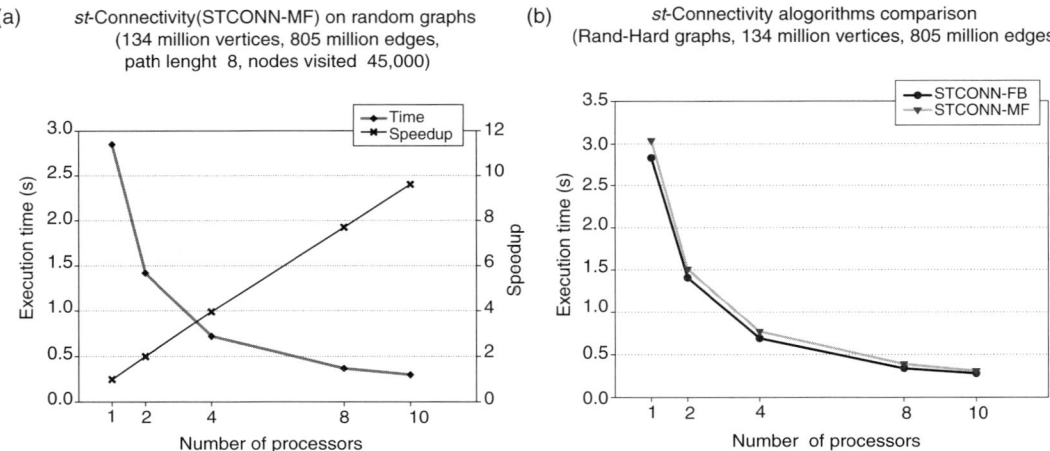

FIGURE 31.6 *st*-connectivity results: Execution time and speedup for Rand-ER graphs (a) and comparison of the two *st*-connectivity algorithms (b).

the two algorithms, concurrent BFS from *s* and *t* (STCONN-FB), and expanding the smaller frontier in each iteration (STCONN-MF). Both of them scale linearly with the number of processors for a problem size of 134 million vertices and 805 million edges. STCONN-FB performs slightly better for this graph instance. They were found to perform comparably in other experiments with random and SSCA#2 graphs.

In summary, we present fast multithreaded algorithms for fundamental graph traversal problems in this section. Our implementations show strong scaling for irregular and sparse graphs chosen from four different graph classes, and also achieve high system utilization. The absolute execution time values are significant; problems involving large graphs of billions of vertices and edges can be solved in seconds to minutes.

31.6 Algorithms in Social Network Analysis

Large-scale network analysis is an exciting field of research with applications in a variety of domains such as social networks (friendship circles, organizational networks), the Internet (network topologies, web graphs, peer-to-peer networks), transportation networks, electrical circuits, genealogical research and bioinformatics (protein-interaction networks, food webs). These networks seem to be entirely unrelated and indeed represent quite diverse relations, but experimental studies [14, 21, 42, 68, 69] have shown that they share some common traits such as a low average distance between the vertices, heavy-tailed degree distributions modeled by power laws, and high local densities. Modeling these networks on the basis of experiments and measurements, and the study of interesting phenomena and observations [24, 29, 75, 93], continue to be active areas of research. Several models [25, 47, 70, 73, 91] have been proposed to generate synthetic graph instances with scale-free properties.

Network analysis is currently a very active area of research in the social sciences [44, 79, 82, 90], and seminal contibutions to this field date back to more than 60 years. There are several analytical tools [16, 57, 80] for visualizing social networks, determining empirical quantitative indices, and clustering. In most applications, graph abstractions and algorithms are frequently used to help capture the salient features. Thus, network analysis from a graph theoretic perspective is about extracting interesting information, given a large graph constructed from a real-world dataset.

Network analysis and modeling have received considerable attention in recent times, but algorithms are relatively less studied. Real-world networks are often very large in size, ranging from several hundreds of thousands to billions of vertices and edges. A space-efficient memory representation of such graphs is itself a big challenge, and dedicated algorithms have to be designed exploiting the unique characteristics of these networks. On single processor workstations, it is not possible to do exact in-core computations on large graphs owing to the limited physical memory. Current high-end parallel computers have sufficient physical memory to handle large graphs, and a naïve in-core implementation of a graph theory problem is typically two orders of magnitude faster the best external memory implementation [2]. Algorithm design is further simplified on parallel shared memory systems; owing to the global address memory space, there is no need to partition the graph, and we can avoid the the overhead of message passing. However, attaining good performance is still a challenge, as a large class of graph algorithms are combinatorial in nature, and involve a significant number of noncontiguous, concurrent accesses to global data structures with low degrees of locality.

In this section, we present fast multithreaded algorithms for evaluating several centrality indices frequently used in complex network analysis. These algorithms have been optimized to exploit properties typically observed in real-world large scale networks, such as the low average distance, high local density, and heavy-tailed power law degree distributions. We test our implementations on real datasets such as the web graph, protein-interaction networks, movie-actor and citation networks, and report impressive parallel performance for evaluation of the computationally intensive centrality metrics (betweenness and closeness centrality) on the Cray MTA-2. We demonstrate that it is possible to rigorously analyze networks in the order of millions of vertices and edges, which is three orders of magnitude larger than instances that can be handled by existing SNA software packages.

We give a brief overview of the various centrality metrics in Section 31.6.1, followed by the parallel algorithms in Section 31.6.3 and performance results in Section 31.6.4 on a variety of real-world datasets and synthetic scale-free graphs. We have also designed algorithms for these centrality metrics optimized for SMP and compare the performance of the MTA-2 and a high-end SMP (IBM p5 570) in Section 31.6.4.

31.6.1 Centrality Metrics

One of the fundamental problems in network analysis is to determine the *importance* of a particular vertex or an edge in a network. Quantifying *centrality* and *connectivity* helps us identify portions of the network that may play interesting roles. Researchers have been proposing metrics for centrality for the past 50 years, and there is no single accepted definition. The metric of choice is dependent on the application and the network topology. Almost all metrics are empirical, and can be applied to element-level [20], group-level [40], or network-level [88] analyses. We present a few commonly used indices in this section.

31.6.1.1 Preliminaries

Consider a graph $G = (V, E)$, where V is the set of vertices representing *actors* or *nodes* in the social network, and E, the set of edges representing the relationships between the actors. The number of vertices and edges are denoted by n and m, respectively. The graphs can be directed or undirected. Let us assume that each edge $e \in E$ has a positive integer weight $w(e)$. For unweighted graphs, we use $w(e) = 1$. A *path* from vertex s to t is defined as a sequence of edges $\langle u_i, u_{i+1}\rangle$, $0 \leq i \leq l$, where $u_0 = s$ and $u_l = t$. The *length* of a path is the sum of the weights of edges. We use $d(s, t)$ to denote the distance between vertices s and t (the minimum length of any path connecting s and t in G). Let us denote the total number of shortest paths between vertices s and t by σ_{st}, and the number passing through vertex v by $\sigma_{st}(v)$.

31.6.1.2 Degree Centrality

The degree centrality DC of a vertex v is simply the degree $deg(v)$ for undirected graphs. For directed graphs, we can define two variants: in-degree centrality and out-degree centrality. This is a simple local measure, based on the notion of neighborhood. This index is useful in case of static graphs, for situations when we are interested in finding vertices that have the most direct connections to other vertices.

31.6.1.3 Closeness Centrality

This index measures the closeness, in terms of *distance*, of an actor to all other actors in the network. Vertices with a smaller total distance are considered more important. Several closeness-based metrics [17,71,81] have been developed by the SNA community. A commonly used definition is the reciprocal of the total distance from a particular vertex to all other vertices:

$$CC(v) = \frac{1}{\sum_{u \in V} d(v,u)}$$

Unlike degree centrality, this is a global metric. To calculate the closeness centrality of a vertex v, we may apply BFS (for unweighted graphs) or a single-source shortest paths (SSSP, for weighted graphs) algorithm from v.

31.6.1.4 Stress Centrality

Stress centrality is a metric based on shortest paths counts, first presented in [84]. It is defined as

$$SC(v) = \sum_{s \neq v \neq t \in V} \sigma_{st}(v)$$

Intuitively, this metric deals with the *work* done by each vertex in a communications network. The number of shortest paths that contain an element v will give an estimate of the amount of stress a vertex v is under, assuming communication will be carried out through shortest paths all the time. This index can be calculated using a variant of the all-pairs shortest-paths algorithm, that calculates and stores all shortest paths between any pair of vertices.

31.6.1.5 Betweenness Centrality

Betweenness centrality is another shortest paths enumeration-based metric, introduced by Freeman in [43]. Let $\delta_{st}(v)$ denote the *pairwise dependency*, or the fraction of shortest paths between s and t that pass through v:

$$\delta_{st}(v) = \frac{\sigma_{st}(v)}{\sigma_{st}}$$

Betweenness centrality of a vertex v is defined as

$$BC(v) = \sum_{s \neq v \neq t \in V} \delta_{st}(v)$$

This metric can be thought of as *normalized* stress centrality. Betweenness centrality of a vertex measures the control a vertex has over communication in the network, and can be used to identify key actors in the network. High centrality indices indicate that a vertex can reach other vertices on relatively short paths, or that a vertex lies on a considerable fraction of shortest paths connecting pairs of other vertices. We discuss algorithms to compute this metric in detail in the next section.

This index has been extensively used in recent years for analysis of social as well as other large-scale complex networks. Some applications include biological networks [37,54,76], study of sexual networks and AIDS [61], identifying key actors in terrorist networks [28,58], organizational behavior [23], supply chain management [27], and transportation networks [48].

There are a number of commercial and research software packages for SNA (e.g., Pajek [16], InFlow [57], UCINET [3]), which can also be used to determine these centrality metrics. However, they can only be used to study comparatively small networks (in most cases, sparse graphs with less than 40,000 vertices). Our goal is to develop fast, high-performance implementations of these metrics so that we can analyze large-scale real-world graphs of millions to billions of vertices.

31.6.2 Algorithms for Computing Betweenness Centrality

A straightforward way of computing betweenness centrality for each vertex would be as follows:

1. Compute the length and number of shortest paths between all pairs (s, t)
2. For each vertex v, calculate every possible pair-dependency $\delta_{st}(v)$ and sum them up

The complexity is dominated by Step 2, which requires $\Theta(n^3)$ time summation and $\Theta(n^2)$ storage of pair-dependencies. Popular SNA tools like UCINET use an adjacency matrix to store and update the pair-dependencies. This yields an $\Theta(n^3)$ algorithm for betweenness by augmenting the Floyd–Warshall algorithm for the all-pairs shortest-paths problem with path counting [18].

Alternately, we can modify Dijkstra's single-source shortest paths algorithm to compute the pair-wise dependencies. Observe that a vertex $v \in V$ is on the shortest path between two vertices $s, t \in V$, iff $d(s, t) = d(s, v) + d(v, t)$. Define a set of *predecessors* of a vertex v on shortest paths from s as $pred(s, v)$. Now each time an edge $\langle u, v \rangle$ is scanned for which $d(s, v) = d(s, u) + d(u, v)$, that vertex is added to the predecessor set $pred(s, v)$. Then, the following relation would hold:

$$\sigma_{sv} = \sum_{u \in pred(s,v)} \sigma_{su}$$

Setting the initial condition of $pred(s, v) = s$ for all neighbors v of s, we can proceed to compute the number of shortest paths between s and all other vertices. The computation of $pred(s, v)$ can be easily integrated into Dijkstra's SSSP algorithm for weighted graphs, or BFS Search for unweighted graphs. But even in this case, determining the fraction of shortest paths using v, or the pair-wise dependencies $\delta_{st}(v)$, proves to be the dominant cost. The number of shortest $s - t$ paths using v is given by $\sigma_{st}(v) = \sigma_{sv} \cdot \sigma_{vt}$. Thus computing $BC(v)$ requires $O(n^2)$ time per vertex v, and $O(n^3)$ time in all. This algorithm is the most commonly used one for evaluating betweenness centrality.

To exploit the sparse nature of typical real-world graphs, Brandes [18] came up with an algorithm that computes the betweenness centrality score for all vertices in the graph in $O(mn + n^2 \log n)$ time for weighted graphs, and $O(mn)$ time for unweighted graphs. The main idea is as follows. We define the *dependency* of a source vertex $s \in V$ on a vertex $v \in V$ as

$$\delta_s(v) = \sum_{t \in V} \delta_{st}(v)$$

The betweenness centrality of a vertex v can be then expressed as $BC(v) = \sum_{s \neq v \in V} \delta_s(v)$. The dependency $\delta_s(v)$ satisfies the following recursive relation:

$$\delta_s(v) = \sum_{w: v \in pred(s,w)} \frac{\sigma_{sv}}{\sigma_{sw}} (1 + \delta_s(w))$$

The algorithm is now stated as follows. First, n SSSP computations are done, one for each $s \in V$. The predecessor sets $pred(s, v)$ are maintained during these computations. Next, for every $s \in V$, using the information from the shortest paths tree and predecessor sets along the paths, compute the dependencies $\delta_s(v)$ for all other $v \in V$. To compute the centrality value of a vertex v, we finally compute the sum of all dependency values. The $O(n^2)$ space requirements can be reduced to $O(m + n)$ by maintaining a *running centrality score*.

31.6.3 Parallel Algorithms

In this section, we present our new parallel algorithms and implementation details of the centrality metrics on the Cray MTA-2. Since there is abundant coarse-grained parallelism to be exploited in the centrality metrics algorithms, we have also designed algorithms for SMPs, and compare the performance of the

MTA-2 and the IBM p5 570 SMP system for large graph instances. We also evaluated the performance of connected components on SMPs and the MTA-2 in one of our previous works [8].

31.6.3.1 Degree Centrality

We use a cache-friendly adjacency array representation [74] for internally storing the graph, storing both the in- and out-degree of each vertex in contiguous arrays. This makes the computation of degree centrality straight-forward, with a constant time look-up on the MTA-2 and the p5 570. As noted earlier, degree centrality is a useful metric for determining the graph structure, and for a first pass at identifying important vertices.

31.6.3.2 Closeness Centrality

Recall the definition of closeness centrality

$$CC(v) = \frac{1}{\sum_{u \in V} d(v, u)}$$

We need to calculate the distance from v to all other vertices, and then sum over all distances. One possible solution to this problem would be to precompute a *distance matrix* using the $O(n^3)$ Floyd–Warshall All-Pairs Shortest Paths algorithm. Evaluating a specific closeness centrality value then costs $O(n)$ on a single processor ($O(n/p + \log p)$ using p processors) to sum up an entire row of distance values. However, since real-world graphs are typically sparse, we have $m \ll n^2$ and this algorithm would be very inefficient in terms of actual running time and memory utilization. Instead, we can just compute n shortest path trees, one for each vertex $v \in V$, with v as the source vertex for BFS or Dijkstra's algorithm. On p processors, this would yield $T_C = O\left(\frac{nm+n^2}{p}\right)$ and $T_M = nm/p$ for unweighted graphs. For weighted graphs, using a naïve queue-based representation for the expanded frontier, we can compute all the centrality metrics in $T_C = O\left(\frac{nm+n^3}{p}\right)$ and $T_M = 2nm/p$. The bounds can be further improved with the use of efficient priority queue representations.

Since the evaluation of closeness centrality is computationally intensive, it is valuable to investigate approximate algorithms. Using a random sampling technique, Eppstein and Wang [41] show that the closeness centrality of all vertices in a weighted, undirected graph can be approximated with high probability in $O\left(\frac{\log n}{\epsilon^2}(n \log n + m)\right)$ time, and an additive error of at most $\epsilon \Delta_G$ (ϵ is a fixed constant, and Δ_G is the diameter of the graph). The algorithm proceeds as follows. Let k be the number of iterations needed to obtain the desired error bound. In iteration i, pick vertex v_i uniformly at random from V and solve the SSSP problem with v_i as the source. The estimated centrality is given by

$$CC_a(v) = \frac{k}{n \sum_{i=1}^{k} d(v_i, u)}.$$

The error bounds follow from a result by Hoeffding [51] on probability bounds for sums of independent random variables. We parallelize this algorithm as follows. Each processor can run SSSP computations from k/p vertices and store the evaluated distance values. The cost of this step is given by $T_C = O\left(\frac{k(m+n)}{p}\right)$ and $T_M = km/p$ for unweighted graphs. For real-world graphs, the number of sample vertices k can be set to $\Theta\left(\frac{\log n}{\epsilon^2}\right)$ to obtain the error bounds given above. The approximate closeness centrality value of each vertex can then be calculated in $O(k) = O\left(\frac{\log n}{\epsilon^2}\right)$ time, and the summation for all n vertices would require $T_C = O\left(\frac{n \log n}{p\epsilon^2}\right)$ and constant T_M.

31.6.3.3 Stress and Betweenness Centrality

These two metrics require shortest paths enumeration and we design our parallel algorithms based on Brandes' [18] sequential algorithm for sparse graphs. Algorithm 6 outlines the general approach for the

Input: $G(V, E)$
Output: Array $BC[1..n]$, where $BC[v]$ gives the centrality metric for vertex v

1 **for** *all $v \in V$ **in parallel*** **do**
2     $BC[v] \leftarrow 0$;
  for *all $s \in V$ **in parallel*** **do**
3     $S \leftarrow$ empty stack;
4     $P[w] \leftarrow$ empty list, $w \in V$;
5     $\sigma[t] \leftarrow 0, t \in V; \sigma[s] \leftarrow 1$;
6     $d[t] \leftarrow -1, t \in V; d[s] \leftarrow 0$;
7     $Q \rightarrow$ empty queue;
8   enqueue $s \leftarrow Q$;
9   **while** *Q not empty* **do**
10   dequeue $v \leftarrow Q$;
11   push $v \rightarrow S$;
12   **for** *each neighbor w of v **in parallel*** **do**
13   **if** $d[w] < 0$ **then**
14   enqueue $w \rightarrow Q$;
15                 $d[w] \leftarrow d[v] + 1$;
16   **if** $d[w] = d[v] + 1$ **then**
17                 $\sigma[w] \leftarrow \sigma[w] + \sigma[v]$;
18   append $v \rightarrow P[w]$;
19     $\delta[v] \leftarrow 0, v \in V$;
20   **while** *S not empty* **do**
21   pop $w \leftarrow S$;
22   **for** $v \in P[w]$ **do**
23             $\delta[v] \leftarrow \delta[v] + \frac{\sigma[v]}{\sigma[w]}(1 + \delta[w])$;
24   **if** $w \neq s$ **then**
25             $BC[w] \leftarrow BC[w] + \delta[w]$;

Algorithm 6: Parallel betweenness centrality for unweighted graphs.

case of unweighted graphs. On each BFS computation from s, the queue Q stores the current set of vertices to be visited, S contains all the vertices reachable from s, and $P(v)$ is the predecessor set associated with each vertex $v \in V$. The arrays d and σ store the distance from s, and shortest path counts, respectively. The centrality values are computed in Steps 22–25, by summing the dependencies $\delta(v), v \in V$. The final scores need to be divided by two if the graph is undirected, as all shortest paths are counted twice.

We observe that parallelism can be exploited at two levels:

1. The BFS/SSSP computations from each vertex can be done concurrently, provided the centrality running sums are updated atomically.
2. The actual BFS/SSSP can be also be parallelized. When visiting the neighbors of a vertex, edge relaxation can be done concurrently.

It is theoretically possible to do all the SSSP computations concurrently (Steps 3–25 in Algorithm 6). However, the memory requirements scale as $O(p(m + n))$, and we only have a modest number of processors on current SMP systems. So for the SMP implementation, we do a coarse-grained partitioning of work and assign each processor a fraction of the vertices from which to initiate SSSP computations. The loop iterations are scheduled dynamically so that work is distributed as evenly as possible. There are no synchronization costs involved, as a processor can compute its own partial sum of the centrality value for each vertex, and all the sums can be merged in the end using a efficient reduction operation.

Thus, the stack S, list of predecessors P and the BFS queue Q, are replicated on each processor. Even for a graph of 100 million edges, with a conservative estimate of 10 GB memory usage by each processor, we can employ all 16 processors on our target SMP system, the IBM p570. We further optimize our implementation for scale-free graphs. Observe that in Algorithm 6, Q is not needed as the BFS frontier vertices are also added to S. To make the implementation cache-friendly, we use dynamic adjacency arrays instead of linked lists for storing elements of the predecessor set S. Note that $|\Sigma_{v \in V} P_s(v)| = O(m)$, and in most cases the predecessor sets of the few high-degree vertices in the graph are of comparatively greater size. Memory allocation is done as a preprocessing step before the actual BFS computations. Also, the indices computation time and memory requirements can be significantly reduced by decomposing the undirected graph into its biconnected components.

However, for a multithreaded system like the MTA-2, this simple approach will not be efficient. We need to exploit parallelism at a much finer granularity to saturate all the hardware threads. So we parallelize the actual BFS computation, and also have a coarse grained partition on the outer loop. In the previous section, we show that our parallel BFS implementation attains scalable performance for up to 40 processors on low-diameter, scale-free graphs. We use the loop *future* parallelism so that the outer loop iterations are dynamically scheduled, thus exploiting parallelism at a coarser granularity.

For the case of weighted graphs, Algorithm 6 must be modified to consider edge weights. The relaxation condition changes, and using a Fibonacci heap or pairing heap priority queue representation, it is possible to compute betweenness centrality for all the n vertices in $T_C = O(mn + n^2 \log n/p)$. We can easily adapt our SMP implementation to consider weighted graphs also. We intend to work on a parallel multithreaded implementation of SSSP on the MTA-2 for scale-free graphs in future.

An approximate algorithm for betweenness centrality is detailed in [19], which is again derived from the Eppstein–Wang algorithm [41]. As in the case of closeness centrality, the sequential running time can be reduced to $O(\log n/\epsilon^2 (n + m))$ by setting the value of k appropriately. The parallel algorithms can also be derived similarly, by computing only k dependencies ($\delta[v]$ in Algorithm 6) instead of v, and taking a normalized average.

31.6.4 Performance Results

This section summarizes the experimental results of our centrality implementations on the Cray MTA-2 and the IBM p5 570. We report results on a 40-processor MTA-2, with each processor having a clock speed of 220 MHz and 4 GB of RAM. The IBM p5 570 is a 16-way SMP with 16 1.9 GHz Power5 cores with simultaneous multithreading (SMT), and 256 GB shared memory.

We test our centrality metric implementations on a variety of real-world graphs, summarized in Table 31.2. Our implementations have been extended to read input files in both PAJEK and UCINET graph formats. We also use a synthetic graph generator [25] to generate graphs obeying small-world

TABLE 31.2 Test Dataset Characteristics

Dataset	Source	Network description
ND-actor	[67]	An undirected graph of 392,400 vertices (movie actors) and 31,788,592 edges. An edge corresponds to a link between two actors, if they have acted together in a movie. The dataset includes actor listings from 127,823 movies
ND-web	[67]	A directed network with 325,729 vertices and 1,497,135 arcs (27,455 loops). Each vertex represents a web page within the Univ. of Notredame *nd.edu* domain, and the arcs represent from $\rightarrow$ to links
ND-yeast	[67]	Undirected network with 2114 vertices and 2277 edges (74 loops). Vertices represent proteins, and the edges interactions between them in the yeast network
PAJ-patent	[72]	A network of about 3 million U.S. patents granted between January 1963 and December 1999, and 16 million citations made among them between 1975 and 1999
PAJ-cite	[72]	The *Lederberg* citation dataset, produced using HistCite, in PAJEK graph format with 8843 vertices and 41,609 edges

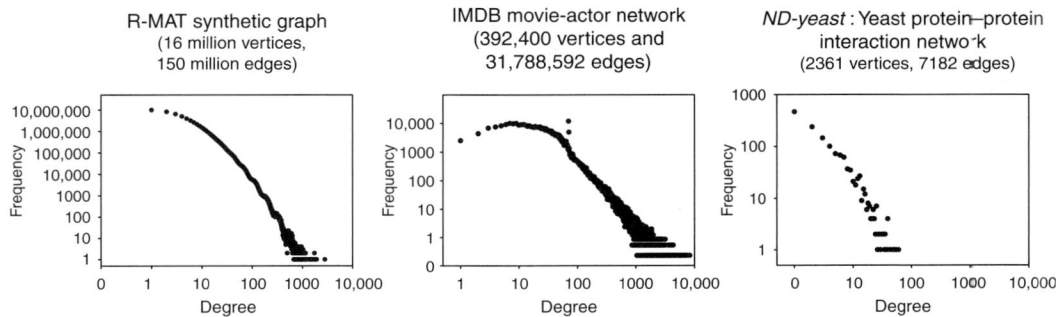

FIGURE 31.7 Degree distributions of some test graph instances.

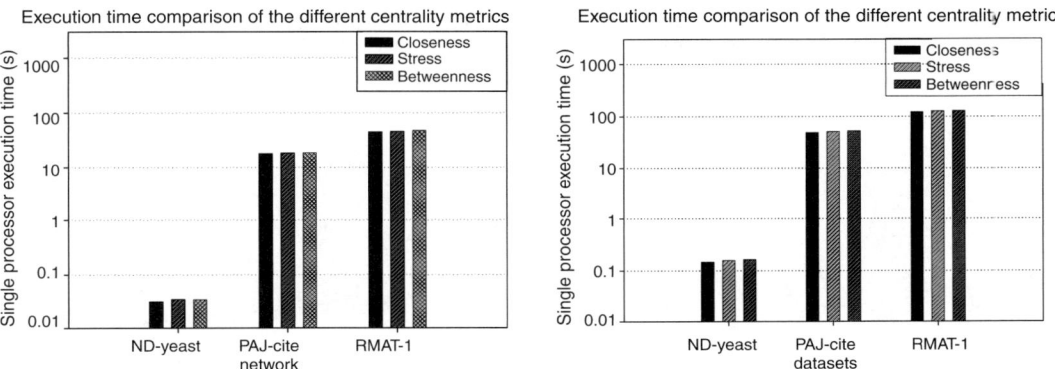

FIGURE 31.8 Single processor execution time comparison of the centrality metric implementations on the IBM p5 570 (left) and the Cray MTA-2 (right).

characteristics. The degree distributions of some test graph instances are shown in Figure 31.7. We can observe that on a log–log plot, the out-degree distribution is a straight line with a heavy tail, which is in agreement with prior experimental studies on scale-free graphs.

Figure 31.8 compares the single processor execution time of the three centrality metrics for three graph instances, on the MTA-2 and the p5 570. All three metrics are of the same computational complexity and show nearly similar running times in practice.

Figure 31.9 summarizes multiprocessor execution times for computing betweenness centrality, on the p5 570 and the MTA-2. Figure 31.9(a) gives the running time for the ND-actor graph on the p570. As expected, the execution time falls nearly linearly with the number of processors. It is possible to evaluate the centrality metric for the entire graph in around 42 minutes, on 16 processors. We observe similar performance for the patents citation graph. However, note that the execution time is highly dependent on the size of the largest nontrivial connected component in these real graphs. The patents network, for instance, is composed of several disconnected subgraphs, representing patents and citations in unrelated areas. However, it did not take significantly more time to compute the centrality indices for this graph compared to the ND-actor graph, even though this is a much bigger graph instance.

Figures 31.9(c) and (d) summarize the performance of the MTA-2 on ND-web, and a synthetic graph instance of the same size generated using the R-MAT algorithm. Again, note that the actual execution time is dependent on the graph structure; for the same problem size, the synthetic graph instance takes much longer than the ND-web graph. Compared to a four processor run, the execution time reduces significantly for 40 processors, but we do not attain performance comparable to our prior graph algorithm implementations [8, 12]. We need to optimize our implementation to attain better system utilization, and

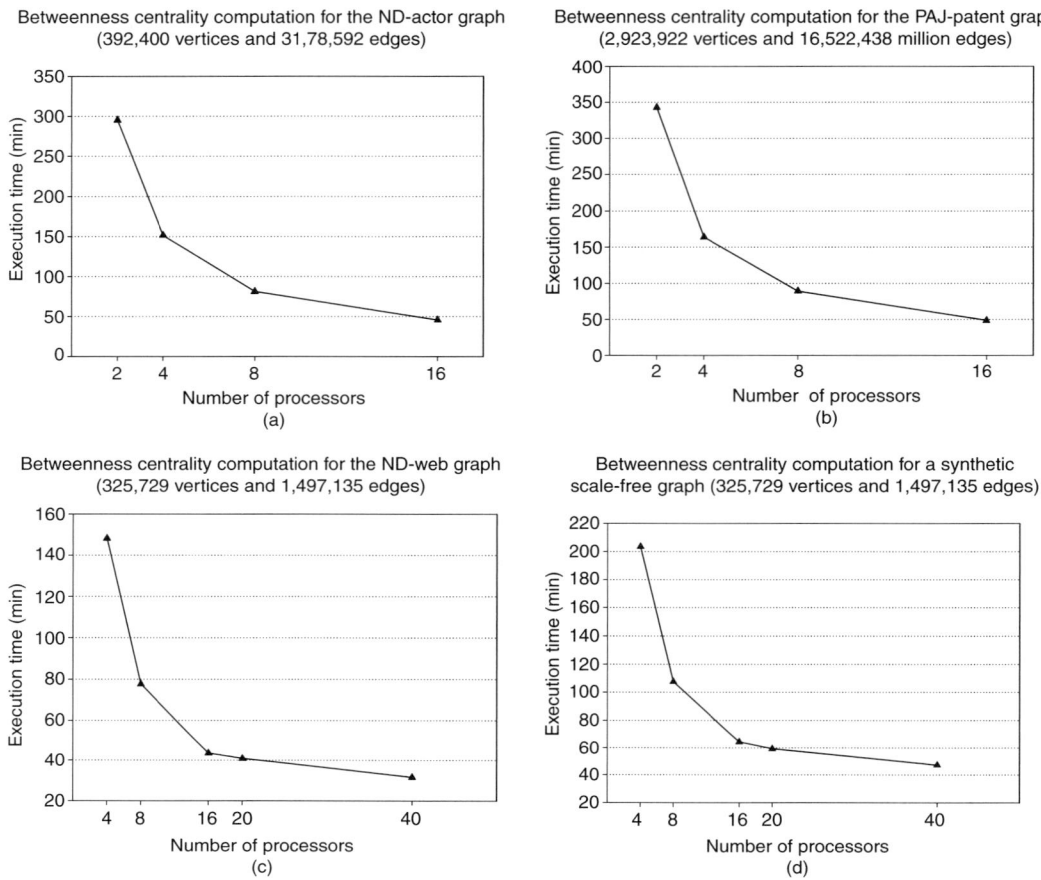

FIGURE 31.9 Multiprocessor performance of betweenness centrality for various graph instances on the IBM p5 570 [(a), (b)] and the Cray MTA-2 [(c), (d)].

the current bottleneck is due to automatic handling of nested parallelism. For better load balancing, we need to further preprocess the graph and manually assign teams of threads to the independent BFS computations. Our memory management routines for the MTA-2 implementation are not as optimized p570 routines, and this is another reason for drop in performance and scaling.

In summary, we present parallel algorithms for evaluating several network indices, including betweenness centrality, optimized for multithreaded architectures and SMP. To our knowledge, this is the first effort at parallelizing these widely used SNA tools. Our implementations are designed to handle problem sizes in the order of billions of edges in the network, and these are three orders of magnitude larger than instances that can be handled by current SNA tools. We are currently working on improving the betweenness centrality implementation on the MTA-2, and also extend it to efficiently compute scores for weighted graphs. In future, we plan to implement and analyze performance of approximate algorithms for closeness and betweenness centrality, and also apply betweenness centrality values to solve harder problems like graph clustering.

31.7 Conclusions

In this chapter, we discuss fast parallel algorithms for several fundamental combinatorial problems, optimized for multithreaded architectures. We implement the algorithms on the Cray MTA-2, a massively multithreaded parallel system. One of the objectives of this chapter is to make the case for the multithreading paradigm for solving large-scale combinatorial problems. We demonstrate that the MTA-2 outperforms conventional parallel systems for applications where fine-grained parallelism and synchronization are critical. For instance, breadth-first search on a scale-free graph of 200 million vertices and 1 billion edges takes less than 5 seconds on a 40-processor MTA-2 system with an absolute speedup of close to 30. This is a significant result in parallel computing, as prior implementations of parallel graph algorithms report very limited or no speedup on irregular and sparse graphs, when compared to the best sequential implementation. It may now be possible to tackle several key PRAM algorithms [46,53,56,64] that have eluded practical implementations so far.

Acknowledgments

This work was supported in part by NSF Grants CAREER CCF-0611589, NSF DBI-0420513, ITR EF/BIO 03-31654, IBM Faculty Fellowship and Microsoft Research grants, NASA grant NP-2005-07-375-HQ, and DARPA Contract NBCH30390004. We thank Bruce Hendrickson and Jon Berry of Sandia National Laboratories for discussions on large-scale graph problems. We thank Richard Russell for sponsoring our Cray MTA-2 accounts, and Simon Kahan and Petr Konecny for their advice and suggestions on MTA-2 code optimization. John Feo (jofeo@microsoft.com), currently employed by Microsoft, performed this work while he was with his former employer Cray Inc.

References

[1] J. M. Abello and J. S. Vitter, editors. *External Memory Algorithms.* American Mathematical Society, Boston, MA, 1999.

[2] D. Ajwani, R. Dementiev, and U. Meyer. A computational study of external-memory bfs algorithms. In *Proc. 17th Ann. Symp. Discrete Algorithms (SODA-06)*, pp. 601–610. ACM-SIAM, 2006.

[3] Analytic Technologies. UCINET 6 social network analysis software. http://www.analytictech.com/ucinet.htm

[4] B. Awerbuch and R. G. Gallager. A new distributed algorithm to find Breadth First search trees. *IEEE Trans. Inf. Theor.*, 33(3):315–322, 1987.

[5] B. Awerbuch, A. V. Goldberg, M. Luby, and S. A. Plotkin. Network decomposition and locality in distributed computation. In *IEEE Symp. on Foundations of Computer Science*, pp. 364–369, 1989.

[6] D. A. Bader and G. Cong. A fast, parallel spanning tree algorithm for symmetric multiprocessors (SMPs). In *Proc. Int'l Parallel and Distributed Processing Symp. (IPDPS 2004)*, Santa Fe, NM, April 2004.

[7] D. A. Bader and G. Cong. Fast shared-memory algorithms for computing the minimum spanning forest of sparse graphs. In *Proc. Int'l Parallel and Distributed Processing Symp. (IPDPS 2004)*, Santa Fe, NM, April 2004.

[8] D.A. Bader, G. Cong, and J. Feo. On the architectural requirements for efficient execution of graph algorithms. In *Proc. 34th Int'l Conf. on Parallel Processing (ICPP)*, Oslo, Norway, June 2005.

[9] D.A. Bader, J. Feo, and *et al.* HPCS SSCA#2 Graph Analysis Benchmark Specifications v1.1, January 2006.

[10] D.A. Bader, A.K. Illendula, B.M.E. Moret, and N. Weisse-Bernstein. Using PRAM algorithms on a uniform-memory-access shared-memory architecture. In G.S. Brodal, D. Frigioni, and A. Marchetti-Spaccamela, ed., *Proc. 5th Int'l Workshop on Algorithm Engineering (WAE 2001)*, volume 2141 of *Lecture Notes in Computer Science*, pp. 129–144, Århus, Denmark, Springer-Verlag, 2001.

[11] D.A. Bader and K. Madduri. Design and implementation of the HPCS graph analysis benchmark on symmetric multiprocessors. In *Proc. 12th Int'l Conf. on High Performance Computing*, Goa, India, Springer-Verlag, December 2005.

[12] D. A. Bader and K. Madduri. Designing multithreaded algorithms for breadth-first search and st-connectivity on the Cray MTA-2. In *Proc. 35th Int'l Conf. on Parallel Processing (ICPP)*, Columbus, OH, IEEE Computer Society, August 2006.

[13] D.A. Bader, S. Sreshta, and N. Weisse-Bernstein. Evaluating arithmetic expressions using tree contraction: A fast and scalable parallel implementation for symmetric multiprocessors (SMPs). In S. Sahni, V.K. Prasanna, and U. Shukla, editors, *Proc. 9th Int'l Conf. on High Performance Computing (HiPC 2002)*, volume 2552 of Lecture Notes in Computer Science, pp. 63–75, Bangalore, India, Springer-Verlag, December 2002.

[14] A-L. Barabasi and R. Albert. Emergence of scaling in random networks. *Science*, 286(5439):509–512, October 1999.

[15] G. Barnes and W. L. Ruzzo. Deterministic algorithms for undirected s-t connectivity using polynomial time and sublinear space. In *Proc. 23rd Annual ACM Symp. on Theory of Computing*, pp. 43–53, New York, ACM Press, 1991.

[16] V. Batagelj and A. Mrvar. Pajek—Program for large network analysis. *Connections*, 21(2):47–57, 1998.

[17] A. Bavelas. Communication patterns in task oriented groups. *J. Acous. Soc. Am.*, 22:271–282, 1950.

[18] U. Brandes. A faster algorithm for betweenness centrality. *J. Math. Soc.*, 25(2):163–177, 2001.

[19] U. Brandes and T. Erlebach, editors. *Network Analysis: Methodological Foundations*, volume 3418 of Lecture Notes in Computer Science. Springer-Verlag, 2005.

[20] S. Brin and L. Page. The anatomy of a large-scale hypertextual web search engine. *Computer Networks and ISDN Systems*, 30(1–7):107–117, 1998.

[21] A. Broder, R. Kumar, F. Maghoul, P. Raghavan, S. Rajagopalan, R. Stata, A. Tomkins, and J. Wiener. Graph structure in the web. *Computer Networks*, 33(1–6):309–320, 2000.

[22] A. L. Buchsbaum, M. Goldwasser, S. Venkatasubramanian, and J.R. Westbrook. On external memory graph traversal. In *Proc. 11th Annual ACM-SIAM Symp. on Discrete Algorithms*, pp. 859–860, Philadelphia, PA, 2000.

[23] N. Buckley and M. van Alstyne. Does email make white collar workers more productive? Technical report, University of Michigan, 2004.

[24] D. S. Callaway, M. E. J. Newman, S. H. Strogatz, and D. J. Watts. Network robustness and fragility: percolation on random graphs. *Physics Review Letters*, 85:5468–5471, 2000.

[25] D. Chakrabarti, Y. Zhan, and C. Faloutsos. R-MAT: A recursive model for graph mining. In *Proc. 4th SIAM Intl. Conf. on Data Mining*, Florida, April 2004.

[26] B. V. Cherkassky, A.V. Goldberg, and T. Radzik. Shortest paths algorithms: theory and experimental evaluation. *Math. Programm.*, 73:129–174, 1996.

[27] D. Cisic, B. Kesic, and L. Jakomin. Research of the power in the supply chain. International Trade, Economics Working Paper Archive EconWPA, April 2000.

[28] T. Coffman, S. Greenblatt, and S. Marcus. Graph-based technologies for intelligence analysis. *Comm. ACM*, 47(3):45–47, 2004.

[29] R. Cohen, K. Erez, D. Ben-Avraham, and S. Havlin. Breakdown of the internet under intentional attack. *Phys. Rev. Lett.*, 86:3682–3685, 2001.

[30] R. Cole and U. Vishkin. Faster optimal prefix sums and list ranking. *Inf. Comput.*, 81(3):344–352, 1989.

[31] G. Cong and D. A. Bader. The Euler tour technique and parallel rooted spanning tree. In *Proc. Int'l Conf. on Parallel Processing (ICPP)*, pp. 448–457, Montreal, Canada, August 2004.

[32] T. H. Cormen, C. E. Leiserson, and R. L. Rivest. *Introduction to Algorithms*. MIT Press, Inc., Cambridge, MA, 1990.

[33] A. Crauser, K. Mehlhorn, U. Meyer, and P. Sanders. A parallelization of Dijkstra's shortest path algorithm. In *Proc. 23rd Int'l Symp. on Mathematical Foundations of Computer Science*, pp. 722–731, London, UK, Springer-Verlag, 1998.

[34] Cray Inc. The MTA-2 multithreaded architecture. http://www.cray.com/products/systems/mta

[35] DARPA Information Processing Technology Office. High productivity computing systems (HPCS) project, 2004. http://www.darpa.mil/ipto/programs/hpcs/
[36] F. Dehne, A. Ferreira, E. Cáceres, S. W. Song, and A. Roncato. Efficient parallel graph algorithms for coarse-grained multicomputers and BSP. *Algorithmica*, 33:183–200, 2002.
[37] A. del Sol, H. Fujihashi, and P. O'Meara. Topology of small-world networks of protein–protein complex structures. *Bioinformatics*, 21(8):1311–1315, 2005.
[38] C. Demetrescu, A. Goldberg, and D. Johnson. 9th DIMACS implementation challenge—shortest paths. http://www.dis.uniroma1.it/~challenge9
[39] E. W. Dijkstra. A note on two problems in connexion with graphs. *Numer. Math.*, 1:269–271, 1959.
[40] P. Doreian and L.H. Albert. Partitioning political actor networks: some quantitative tools for analyzing qualitative networks. *Quant. Anthropol.*, 161:279–291, 1989.
[41] D. Eppstein and J. Wang. Fast approximation of centrality. In *Proc. 12th Ann. Symp. Discrete Algorithms (SODA-01)*, Washington, DC, 2001.
[42] M. Faloutsos, P. Faloutsos, and C. Faloutsos. On power–law relationships of the Internet topology. In *Proc. ACM SIGCOMM*, pp. 251–262, 1999.
[43] L. C. Freeman. A set of measures of centrality based on betweenness. *Sociometry*, 40(1):35–41, 1977.
[44] L. C. Freeman. *The Development of Social Network Analysis: A Study in the Sociology of Science.* Booksurge Pub., 2004.
[45] H. Gazit and G. L. Miller. An improved parallel algorithm that computes the BFS numbering of a directed graph. *Inf. Process. Lett.*, 28(2):61–65, 1988.
[46] J. Greiner. A comparison of data-parallel algorithms for connected components. In *Proc. 6th Ann. Symp. Parallel Algorithms and Architectures (SPAA-94)*, pp. 16–25, Cape May, NJ, June 1994.
[47] J-L. Guillaume and M. Latapy. Bipartite graphs as models of complex networks. In *Proc. 1st Int'l Workshop on Combinatorial and Algorithmic Aspects of Networking*, 2004.
[48] R. Guimera, S. Mossa, A. Turtschi, and L.A.N. Amaral. The worldwide air transportation network: Anomalous centrality, community structure, and cities' global roles. *Proc. Natl. Acad. Sci. USA*, 102(22):7794–7799, 2005.
[49] D. R. Helman and J. JáJá. Designing practical efficient algorithms for symmetric multiprocessors. In *Algorithm Engineering and Experimentation (ALENEX'99)*, volume 1619 of Lecture Notes in Computer Science, pp. 37–56, Baltimore, MD, Springer-Verlag, January 1999.
[50] D. R. Helman and J. JáJá. Prefix computations on symmetric multiprocessors. *J. Parallel Distrib. Comput.*, 61(2):265–278, 2001.
[51] W. Hoeffding. Probability inequalities for sums of bounded random variables. *J. Am. Stat. Assoc.*, 58:713–721, 1963.
[52] A. Inokuchi, T. Washio, and H. Motoda. An apriori-based algorithm for mining frequent substructures from graph data. In *Proc. 4th European Conf. on Principles of Data Mining and Knowledge Discovery*, pp. 13–23, London, UK, 2000.
[53] J. JáJá. *An Introduction to Parallel Algorithms.* Addison-Wesley Publishing Company, New York, 1992.
[54] H. Jeong, S. P. Mason, A.-L. Barabasi, and Z. N. Oltvai. Lethality and centrality in protein networks. *Nature*, 411:41, 2001.
[55] G. Karypis, E. Han, and V. Kumar. Chameleon: Hierarchical clustering using dynamic modeling. *Computer*, 32(8):68–75, 1999.
[56] P. N. Klein and J. H. Reif. An efficient parallel algorithm for planarity. *J. Comp. Syst. Sci.*, 37(2):190–246, 1988.
[57] V. Krebs. InFlow 3.1—Social network mapping software, 2005. http://www.orgnet.com
[58] V. E. Krebs. Mapping networks of terrorist cells. *Connections*, 24(3):43–52, 2002.
[59] T. Lengauer. *Combinatorial Algorithms for Integrated Circuit Layout.* John Wiley & Sons, Inc., New York, 1990.

[60] H. R. Lewis and C. H. Papadimitriou. Symmetric space-bounded computation (extended abstract). In *Proc. 7th Colloquium on Automata, Languages and Programming*, pp. 374–384, London, UK, 1980.

[61] F. Liljeros, C. R. Edling, L. A. N. Amaral, H. E. Stanley, and Y. Aberg. The web of human sexual contacts. *Nature*, 411:907, 2001.

[62] U. Meyer. External memory BFS on undirected graphs with bounded degree. In *Proc. 12th Annu. ACM-SIAM Symp. on Discrete Algorithms*, pp. 87–88, Philadelphia, PA, 2001.

[63] U. Meyer and P. Sanders. Δ-stepping: a parallelizable shortest path algorithm. *J. Algorithms*, 49(1):114–152, 2003.

[64] G. L. Miller and J. H. Reif. Parallel tree contraction and its application. In *Proc. 26th Ann. IEEE Symp. Foundations of Computer Science*, pp. 478–489, Portland, OR, October 1985.

[65] B. M. E. Moret, D. A. Bader, and T. Warnow. High-performance algorithm engineering for computational phylogenetics. In *Proc. Int'l Conf. on Computational Science*, volume 2073–2074 of Lecture Notes in Computer Science, San Francisco, CA, Springer-Verlag, 2001.

[66] B. M. E. Moret, D. A. Bader, T. Warnow, S.K. Wyman, and M. Yan. GRAPPA: a high-performance computational tool for phylogeny reconstruction from gene-order data. In *Proc. Botany*, Albuquerque, NM, August 2001.

[67] Notredame CNet resources. http://www.nd.edu/~networks

[68] M. Newman. The structure and function of complex networks. *SIAM Review*, 45(2):167–256, 2003.

[69] M. E. J. Newman. Scientific collaboration networks: shortest paths, weighted networks and centrality. *Phys. Rev. E*, 64, 2001.

[70] M. E. J. Newman, S. H. Strogatz, and D. J. Watts. Random graph models of social networks. *Proc. Natl. Acad. of Sci.*, USA, 99:2566–2572, 2002.

[71] U. J. Nieminen. On the centrality in a directed graph. *Soc. Sci. Res.*, 2:371–378, 1973.

[72] PAJEK datasets. http://www.vlado.fmf.uni-lj.si/pub/networks/data/

[73] C. R. Palmer and J. G. Steffan. Generating network topologies that obey power laws. In *Proc. ACM GLOBECOM*, November 2000.

[74] J. Park, M. Penner, and V. K. Prasanna. Optimizing graph algorithms for improved cache performance. In *Proc. Int'l Parallel and Distributed Processing Symp. (IPDPS 2002)*, Fort Lauderdale, FL, April 2002.

[75] R. Pastor-Satorras and A. Vespignani. Epidemic spreading in scale-free networks. *Phys. Rev. Lett.*, 86:3200–3203, 2001.

[76] J. W. Pinney, G. A. McConkey, and D. R. Westhead. Decomposition of biological networks using betweenness centrality. In *Proc. Poster Session of the 9th Ann. Int'l Conf. on Research in Computational Molecular Biology (RECOMB 2004)*, Cambridge, MA, May 2005.

[77] M. Reid-Miller. List ranking and list scan on the Cray C-90. In *Proc. 6th Ann. Symp. Parallel Algorithms and Architectures (SPAA-94)*, pp. 104–113, Cape May, NJ, June 1994.

[78] M. Reid-Miller. List ranking and list scan on the Cray C-90. *J. Comput. Syst. Sci.*, 53(3):344–356, December 1996.

[79] W. Richards. International network for social network analysis, 2005. http://www.insna.org

[80] W. Richards. Social network analysis software links, 2005. http://www.insna.org/INSNA/soft_inf.html

[81] G. Sabidussi. The centrality index of a graph. *Psychometrika*, 31:581–603, 1966.

[82] J. P. Scott. *Social Network Analysis: A Handbook*. SAGE Publications, 2000.

[83] R. Seidel. On the all-pairs-shortest-path problem in unweighted undirected graphs. *J. Comput. Syst. Sci.*, 51(3):400–403, 1995.

[84] A. Shimbel. Structural parameters of communication networks. *Math. Biophy.*, 15:501–507, 1953.

[85] M. Thorup. Undirected single-source shortest paths with positive integer weights in linear time. *J. ACM*, 46(3):362–394, 1999.

[86] J. L. Träff. An experimental comparison of two distributed single-source shortest path algorithms. *Parallel Comput.*, 21(9):1505–1532, 1995.

[87] J. Ullman and M. Yannakakis. High-probability parallel transitive closure algorithms. In *Proc. 2nd Ann. Symp. Parallel Algorithms and Architectures (SPAA-90)*, pp. 200–209, New York, 1990. ACM Press.

[88] University of Virginia. Oracle of Bacon. http://www.oracleofbacon.org

[89] A. Vazquez, A. Flammini, A. Maritan, and A. Vespignani. Global protein function prediction in protein–protein interaction networks. *Nat. Biotechnol.*, 21(6):697–700, June 2003.

[90] S. Wasserman and K. Faust. *Social Network Analysis: Methods and Applications.* Cambridge University Press, 1994.

[91] D. J. Watts and S. H. Strogatz. Collective dynamics of small world networks. *Nature*, 393:440–442, 1998.

[92] A. Yoo, E. Chow, K. Henderson, W. McLendon, B. Hendrickson, and Ü. V. Çatalyürek. A scalable distributed parallel breadth-first search algorithm on Blue Gene/L. In *Proc. Supercomputing (SC 2005)*, Seattle, WA, November 2005.

[93] D. H. Zanette. Critical behavior of propagation on small-world networks. *Phys. Rev. E*, 64, 2001.

[94] Cray Inc. The Cray XMT platform. http://www.cray.com/products/xmt/

[95] J. Feo, D. Harper, S. Kahan, and P. Konecny. ELDORADO. In N. Bagherzadeh, M. Valero, and A. Ramírez, eds., *Proc. 2nd Conf. on Computing Frontiers*, pp. 28–34, Ischia, Italy, ACM Press, 2005.

32
Parallel Data Mining Algorithms for Association Rules and Clustering

Jianwei Li
Wei-keng Liao
Alok Choudhary
Northwestern University

Ying Liu
Graduate University of the Chinese Academy of Sciences

32.1	Introduction...	32-1
32.2	Parallel Association Rule Mining	32-2
	A Priori-Based Algorithms • Vertical Mining • Pattern-Growth Method • Mining by Bitmaps • Comparison	
32.3	Parallel Clustering Algorithms	32-12
	Parallel *k*-Means • Parallel Hierarchical Clustering • Parallel *HOP*: Clustering Spatial Data • Clustering High-Dimensional Data	
32.4	Summary	32-19
References	....	32-19

32.1 Introduction

Volumes of data are exploding in both scientific and commercial domains. Data mining techniques that extract information from huge amount of data have become popular in many applications. Algorithms are designed to analyze those volumes of data automatically in efficient ways, so that users can grasp the intrinsic knowledge latent in the data without the need to manually look through the massive data itself. However, the performance of computer systems is improving at a slower rate compared to the increase in the demand for data mining applications. Recent trends suggest that the system performance has been improving at a rate of 10–15% per year, whereas the volume of data collected nearly doubles every year. As the data sizes increase, from gigabytes to terabytes or even larger, sequential data mining algorithms may not deliver results in a reasonable amount of time. Even worse, as a single processor alone may not have enough main memory to hold all the data, a lot of sequential algorithms could not handle large-scale problems or have to process data out of core, further slowing down the process.

In recent years, there is an increasing interest in the research of parallel data mining algorithms. In parallel environment, by exploiting the vast aggregate main memory and processing power of parallel processors, parallel algorithms can have both the execution time and memory requirement issues well addressed. However, it is not trivial to parallelize existing algorithms to achieve good performance as well as

Database			Frequent itemsets (minsup = 33%)	
TID	Items		k	Frequent itemsets : support
1	f d b e		1	a:67% b:50% c:33% d:50% e:83% f:67%
2	f e b		2	ac:33% ad:33% ae:50% af:33% bd:33% be:33% bf:33% ce:33% cf:33% de:33% ef:67%
3	a d b			
4	a e f c		3	ace:33% acf:33% aef:33% bef:33% cef:33%
5	a d e			
6	a c f e		4	acef:33%

FIGURE 32.1 Example database and frequent itemsets.

scalability to massive data sets. First, it is crucial to design a good data organization and decomposition strategy so that workload can be evenly partitioned among all processes with minimal data dependence across them. Second, minimizing synchronization and/or communication overhead is important in order for the parallel algorithm to scale well as the number of processes increases. Workload balancing also needs to be carefully designed. Last, disk I/O cost must be minimized.

In this chapter, parallel algorithms for association rule mining (ARM) and clustering are presented to demonstrate how parallel techniques can be efficiently applied to data mining applications.

32.2 Parallel Association Rule Mining

Association rule mining is an important core data mining technique to discover patterns/rules among items in a large database of variable-length transactions. The goal of ARM is to identify groups of items that most often occur together. It is widely used in market-basket transaction data analysis, graph mining applications like substructure discovery in chemical compounds, pattern finding in web browsing, word occurrence analysis in text documents, and so on. The formal description of ARM can be found in References AIS93 and AS94. And most of the research focuses on the frequent itemset mining subproblem, that is, finding all frequent itemsets each occurring at more than a minimum frequency (*minsup*) among all transactions. Figure 32.1 gives an example of mining all frequent itemsets with $minsup = 33\%$ from a given database. Well-known sequential algorithms include *A priori* [AS94], *Eclat* [ZPOL97a], *FP-growth* [HPY00], and *D-CLUB* [LCJL06]. Parallelizations of these algorithms are discussed in this section, with many other algorithms surveyed in Reference Zak99.

32.2.1 *A Priori*-Based Algorithms

Most of the parallel ARM algorithms are based on parallelization of *A priori* that iteratively generates and tests candidate itemsets from length 1 to k until no more frequent itemsets are found. These algorithms can be categorized into *Count Distribution*, *Data Distribution*, and *Candidate Distribution* methods [AS96, HKK00]. The *Count Distribution* method follows a data-parallel strategy and statically partitions the database into horizontal partitions that are independently scanned for the local counts of all candidate itemsets on each process. At the end of each iteration, the local counts will be summed up across all processes into the global counts so that frequent itemsets can be found. The *Data Distribution* method attempts to utilize the aggregate main memory of parallel machines by partitioning both the database and the candidate itemsets. Since each candidate itemset is counted by only one process, all processes have to exchange database partitions during each iteration in order for each process to get the global counts of the assigned candidate itemsets. The *Candidate Distribution* method also partitions candidate itemsets but selectively replicates instead of partition-and-exchanging the database transactions, so that each process can proceed independently. Experiments show that the *Count Distribution* method exhibits better performance and scalability than the other two methods. The steps for the *Count Distribution* method are generalized as follows for distributed-memory multiprocessors.

Parallel Data Mining Algorithms for Association Rules and Clustering

(a) Database partitioning

TID	Items	Process number
1	f d b e	P0
2	f e b	
3	a d b	P1
4	a e f c	
5	a d e	P2
6	a c f e	

(b) Mining frequent 1-itemsets

Item	Local counts			Global count
	P0	P1	P2	
a	0	2	2	4
b	2	1	0	3
c	0	1	1	2
d	1	1	1	3
e	2	1	2	5
f	2	1	1	4

(c) Mining frequent 2-itemsets

2-itemset	Local counts			Global count
	P0	P1	P2	
ab	0	1	0	1
ac	0	1	1	2
ad	0	1	1	2
ae	0	1	2	3
af	0	1	1	2
bc	0	0	0	0
bd	1	1	0	2
be	2	0	0	2
bf	2	0	0	2
cd	0	0	0	0
ce	0	1	1	2
cf	0	1	1	2
de	1	0	1	2
df	1	0	0	1
ef	2	1	1	4

(d) Mining frequent 3-itemsets

3-itemset	Local counts			Global count
	P0	P1	P2	
ace	0	1	1	2
acf	0	1	1	2
ade	0	0	1	1
aef	0	1	1	2
bde	1	0	0	1
bef	2	0	0	2
cef	0	1	1	2

(e) Mining frequent 4-itemsets

4-itemset	Local counts			Global count
	P0	P1	P2	
acef	0	1	1	2

FIGURE 32.2 Mining frequent itemsets in parallel using the *Count Distribution* algorithm with three processes. The itemset columns in (b), (c), (d), and (e) list the candidate itemsets. Itemsets that are found infrequent are grayed out.

1. Divide the database evenly into horizontal partitions among all processes.
2. Each process scans its local database partition to collect the local count of each item.
3. All processes exchange and sum up the local counts to get the global counts of all items and find frequent 1-itemsets.
4. Set level $k = 2$.
5. All processes generate candidate k-itemsets from the mined frequent $(k-1)$-itemsets.
6. Each process scans its local database partition to collect the local count of each candidate k-itemset.
7. All processes exchange and sum up the local counts into the global counts of all candidate k-itemsets and find frequent k-itemsets among them.
8. Repeat steps 5–8 with $k = k + 1$ until no more frequent itemsets are found.

As an example, to mine all frequent itemsets in Figure 32.1, the *Count Distribution* algorithm needs to scan the database four times to count the occurrences of candidate 1-itemsets, 2-itemsets, 3-itemsets, and 4-itemsets, respectively. As illustrated in Figure 32.2, the counting workload in each scan is distributed over three processes in such a way that each process scans only an assigned partition (1/3) of the whole database. The three processes proceed in parallel and each one counts the candidate itemsets locally from its assigned transactions. Summation of the local counts for one itemset generates the global count that is used to determine the support of that itemset. The generation and counting of candidate itemsets is based on the same procedures as in *A priori*.

In the *Count Distribution* algorithm, communication is minimized since only the counts are exchanged among the processes in each iteration, that is, in Steps 3 and 7. In all other steps, each process works independently, relying only on its local database partition. However, since candidate and frequent itemsets are replicated on all processes, the aggregate memory is not utilized efficiently. Also, the replicated work of generating those candidate itemsets, and selecting frequent ones among them on all processes can be very costly if there are too many such itemsets. In that case, the scalability will be greatly impaired when the number of processes increases. If on a shared-memory machine, since candidate/frequent itemsets and their global counts can be shared among all processes, only one copy of them needs to be kept. So the tasks of getting global counts, generating candidate itemsets, and finding frequent ones among them, as in Steps 3, 5, and 7, can be subdivided instead of being repeated among all processes. This actually leads to another algorithm, *CCPD* [ZOPL96], which works the same way as the *Count Distribution* algorithm in other steps. Nevertheless, both algorithms cannot avoid the expensive cost of database scan and interprocess synchronization per iteration.

32.2.2 Vertical Mining

To better utilize the aggregate computing resources of parallel machines, a localized algorithm [ZPL97] based on parallelization of *Eclat* was proposed and exhibited excellent scalability. It makes use of a vertical data layout by transforming the horizontal database transactions into vertical tid-lists of itemsets. By name, the tid-list of an itemset is a sorted list of IDs for all transactions that contain the itemset. Frequent k-itemsets are organized into disjoint equivalence classes by common $(k-1)$-prefixes, so that candidate $(k+1)$-itemsets can be generated by joining pairs of frequent k-itemsets from the same classes. The support of a candidate itemset can then be computed simply by intersecting the tid-lists of the two component subsets. Task parallelism is employed by dividing the mining tasks for different classes of itemsets among the available processes. The equivalence classes of all frequent 2-itemsets are assigned to processes and the associated tid-lists are distributed accordingly. Each process then mines frequent itemsets generated from its assigned equivalence classes independently, by scanning and intersecting the local tid-lists. The steps for the parallel *Eclat* algorithm are presented below for distributed-memory multiprocessors:

1. Divide the database evenly into horizontal partitions among all processes.
2. Each process scans its local database partition to collect the counts for all 1-itemsets and 2-itemsets.
3. All processes exchange and sum up the local counts to get the global counts of all 1-itemsets and 2-itemsets, and find frequent ones among them.
4. Partition frequent 2-itemsets into equivalence classes by prefixes.
5. Assign the equivalence classes to processes.
6. Each process transforms its local database partition into vertical tid-lists for all frequent 2-itemsets.
7. Each process exchanges the local tid-lists with other processes to get the global ones for the assigned equivalence classes.
8. For each assigned equivalence class on each process, recursively mine all frequent itemsets by joining pairs of itemsets from the same equivalence class and intersecting their corresponding tid-lists.

Steps 1 through 3 work in a similar way as in the *Count Distribution* algorithm. In Step 5, the scheduling of the equivalence classes on different processes needs to be carefully designed in a manner of minimizing the workload imbalance. One simple approach would be to estimate the workload for each class and assign the classes in turn in descending workload order to the least loaded process. Since all pairs of itemsets from one equivalence class will be computed to mine deeper level itemsets, $\binom{s}{2}$ can be used as the estimated workload for an equivalence class of s itemsets. Other task scheduling mechanisms can also be applied once available. Steps 6 and 7 construct the tid-lists for all frequent 2-itemsets in parallel. As each process scans only one horizontal partition of the database, it gets a partial list of transaction IDs for each itemset. Concatenating the partial lists of an itemset from all processes will generate the global tid-list covering all the transactions. In many cases, the number of frequent 2-itemsets can be so large that assembling all their tid-lists may be very costly in both processing time and memory usage. As an alternative, tid-lists of frequent items can be constructed instead and selectively replicated on all processes so that each process has the tid-lists of all the member items in the assigned equivalence classes. However, this requires generating the tid-list of a frequent 2-itemset on the fly in the later mining process, by intersecting the tid-lists of the two element items. Step 8 is the asynchronous phase where each process mines frequent itemsets independently from each of the assigned equivalence classes, relying only on the local tid-lists. Computing on each equivalence class usually generates a number of child equivalence classes that will be computed recursively.

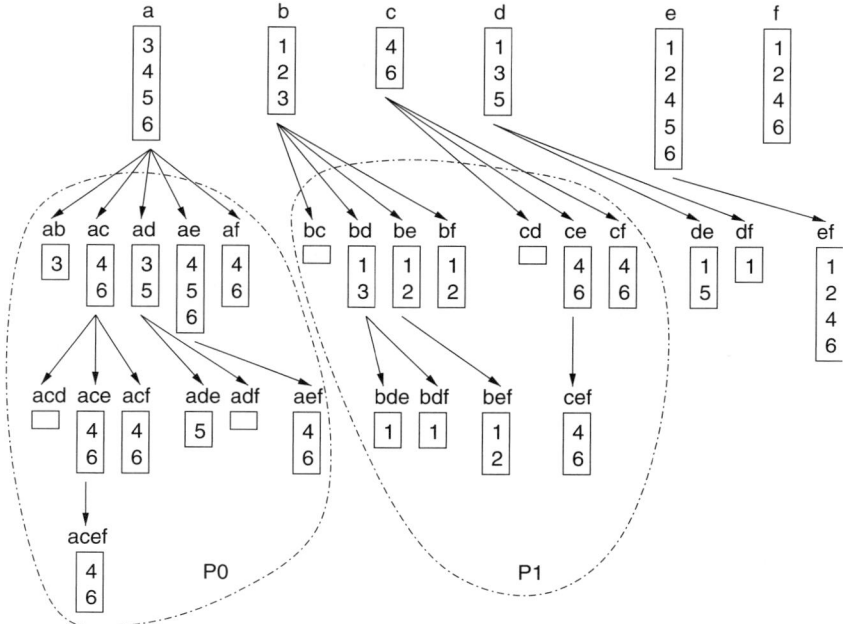

FIGURE 32.3 Mining frequent itemsets using the parallel *Eclat* algorithm. Each itemset is associated with its tid-list. Itemsets that are found infrequent are grayed out.

Taking Figure 32.1 as an example, Figure 32.3 illustrates how the algorithm mines all frequent itemsets from one class to the next using the intersection of tid-lists. The frequent 2-itemsets are organized into five equivalence classes that are assigned to two processes. Process P0 will be in charge of the further mining task for one equivalence class, $\{ac, ad, ae, af\}$, while process P1 will be in charge of two, $\{bd, be, bf\}$ and $\{ce, cf\}$. The rightmost classes, $\{de\}$ and $\{ef\}$, do not have any further mining task associated. The two processes then proceed in parallel, without any data dependence across them. For example, to mine the itemset "*ace*," process P0 only needs to join the two itemsets, "*ac*" and "*ae*," and intersect their tid-lists, "46" and "456," to get the result tid-list "46." At the same time, process P1 can be independently mining the itemset "*bef*" from "*be*," "*bf*" and their associated tid-lists that are locally available.

There are four variations of parallel *Eclat*—*ParEclat, ParMaxEclat, ParClique,* and *ParMaxClique*—as discussed in Reference ZPOL97b. All of them are similar in parallelization and only differ in the itemset clustering techniques and itemset lattice traversing strategies. *ParEclat* and *ParMaxEclat* use prefix-based classes to cluster itemsets, and adopt bottom-up and hybrid search strategies, respectively, to traverse the itemset lattice. *ParClique* and *ParMaxClique* use smaller clique-based itemset clusters, with bottom-up and hybrid lattice search, respectively.

Unlike the *A priori*-based algorithms that need to scan the database as many times as the maximum length of frequent itemsets, the *Eclat*-based algorithms scan the database only three times and significantly reduce the disk I/O cost. Most importantly, the dependence among processes is decoupled right in the beginning so that no communication or synchronization is required in the major asynchronous phase. The major communication cost comes from the exchange of local tid-lists across all processes when the global tid-lists are set up. This one time cost can be amortized by later iterations. For better parallelism, however, the number of processes should be much less than that of equivalence classes (or cliques) for frequent 2-itemsets, so that task assignment granularity can be relatively fine to avoid workload imbalance. Also, more effective workload estimation functions and better task scheduling or workload balancing strategies are needed in order to guarantee balanced workload for various cases.

32.2.3 Pattern-Growth Method

In contrast to the previous itemset generation-and-test approaches, the pattern-growth method derives frequent itemsets directly from the database without the costly generation and test of a large number of candidate itemsets. The detailed design is explained in the *FP-growth* algorithm. Basically, it makes use of a novel frequent-pattern tree (FP-tree) structure where the repetitive transactions are compacted. Transaction itemsets are organized in that frequency-ordered prefix tree such that they share common prefix part as much as possible, and reccurrences of items/itemsets are automatically counted. Then the FP-tree is traversed to mine all frequent patterns (itemsets). A partitioning-based, divide-and-conquer strategy is used to decompose the mining task into a set of smaller subtasks for mining confined patterns in the so-called conditional pattern bases. The conditional pattern base for each item is simply a small database of counted patterns that cooccur with the item. That small database is transformed into a conditional FP-tree that can be processed recursively.

Owing to the complicated and dynamic structure of the FP-tree, it may not be practical to construct a single FP-tree in parallel for the whole database. However, multiple FP-trees can be easily built in parallel for different partitions of transactions. And conditional pattern bases can still be collected and transformed into conditional FP-trees for all frequent items. Thereafter, since each conditional FP-tree can be processed independently, task parallelism can be achieved by assigning the conditional FP-trees of all frequent items to different processes as in References ZEHL01 and PK03. In general, the *FP-growth* algorithm can be parallelized in the following steps, assuming distributed-memory multiprocessors:

1. Divide the database evenly into horizontal partitions among all processes.
2. Scan the database in parallel by partitions and mine all frequent items.
3. Each process constructs a local FP-tree from its local database partition with respect to the global frequent items (items are sorted by frequencies in descending order within each scanned transaction).
4. From the local FP-tree, each process generates local conditional pattern bases for all frequent items.
5. Assign frequent items (hence their associated conditional FP-trees) to processes.
6. For each frequent item, all its local conditional pattern bases are accumulated and transformed into the conditional FP-tree on the designated process.
7. Each process recursively traverses each of the assigned conditional FP-trees to mine frequent itemsets in the presence of the given item.

Like its sequential version, the parallel algorithm also proceeds in two stages. Steps 1 through 3 is the first stage to construct the multiple local FP-trees from the database transactions, Using the transactions in the local database partition, each process can build its own FP-tree independently. For each transaction, global frequent items are selected and sorted by frequency in descending order, and then fed to the local FP-tree as follows. Starting from the root of the tree, check if the first item exists as one of the children of the root. If it exists then increase the counter for this node, or else add a new child node under root for this item with 1 count. Then, taking the current item node as the new temporary root, repeat the same procedure for the next item in the sorted transaction. The nodes of each item are linked together with the head in the header table. Figure 32.4 shows the parallel construction of the multiple local FP-trees on two processes for the example database in Figure 32.1.

The second stage is to mine all frequent itemsets from the FP-trees, as in Steps 4 through 7. The mining process starts with a bottom-up traversal of the local FP-trees to generate the conditional pattern bases starting from their respective items in the header tables. Each entry of a conditional pattern base is a list of items that precede a certain item on a path of a FP-tree up to the root, with the count of each item set to be that of the considered item node along that path. The assignment of items among processes will be based on some workload estimation heuristic, which is part of ongoing research. For simplicity, the size of the conditional pattern base can be used to estimate the workload associated with an item, and items in the header table can be assigned to processes consecutively. Transforming the conditional pattern bases into conditional FP-trees is no different than constructing FP-trees from database transactions, except

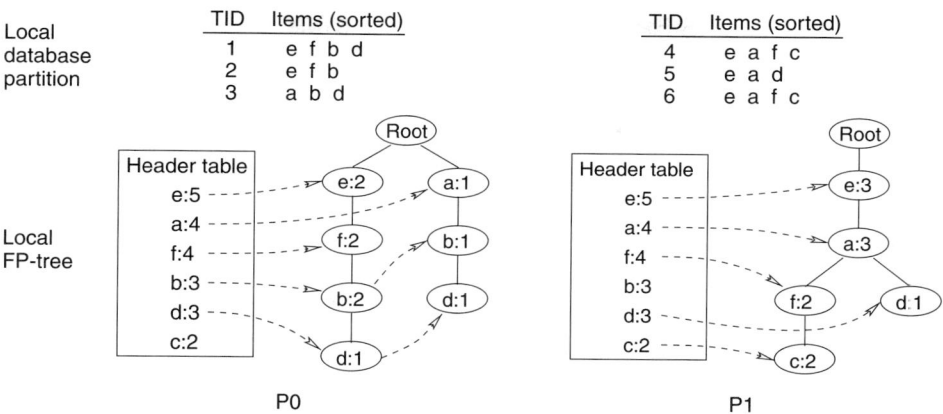

FIGURE 32.4 Construction of local FP-trees from local database partitions over two processes. In the transactions and header tables, items are sorted by frequencies in descending order.

Process number	P0				P1	
Item	e:5	a:4	f:4	b:3	d:3	c:2
Conditional pattern base		(e:3)	(e:2) (e:2, a:2)	(e:2, f:2) (a:1)	(e:1, f:1, b:1) (a:1, b:1) (e:1, a:1)	(e:2, a:2, f:2)
Conditional FP-tree		(tree with e:3)	(tree with e:4, a:2)	(tree with e:2, f:2)	(tree with a:2, b:2, e:2, b:1, a:2, e:1, b:1, e:1)	(tree with a:2, e:2, f:2)
Mined frequent itemsets	e:5	a:4 ea:3	f:4 ef:4 af:2 eaf:2	b:3 eb:2 fb:2 efb:2	d:3 ed:2 ad:2 bd:2	c:2 ac:2 ec:2 fc:2 afc:2 efc:2 aec:2 aefc:2

FIGURE 32.5 Mining frequent itemsets from conditional pattern bases and conditional FP-trees in parallel.

that the counter increment is the exact count collected for each item instead of 1. For each frequent item, as each local FP-tree will derive only part of the conditional pattern base, building the global conditional FP-tree needs the accumulation of local conditional pattern bases from all processes. Then a call to the recursive *FP-growth* procedure on each conditional FP-tree will generate all the conditional patterns on the designated process independently. If on a shared-memory machine, since the multiple FP-trees can be made accessible to all processes, the conditional pattern base and conditional FP-tree can be generated on the fly for the designated process to mine the conditional patterns, for one frequent item after another. This can largely reduce memory usage by not generating all conditional pattern bases and conditional FP-trees for all frequent items at one time. Figure 32.5 gives the conditional pattern bases and conditional FP-trees for all frequent items. They are assigned to the two processes. Process P0 computes on the conditional FP-trees for items *a*, *f*, and *b*, while process P1 does those for *d* and *c*. Item *e* has an empty conditional pattern base and does not have any further mining task associated. Frequent itemsets derived from the conditional FP-tree of each item are listed as the conditional patterns mined for the item.

In parallel *FP-growth*, since all the transaction information is compacted in the FP-trees, no more database scan is needed once the trees are built. So the disk I/O is minimized by scanning the original database only twice. The major communication/synchronization overhead lies in the exchange of local conditional pattern bases across all processes. Since the repetitive patterns are already merged, the total size of the conditional pattern bases is usually much smaller than the original database, resulting in relatively low communication/synchronization cost.

32.2.4 Mining by Bitmaps

All previous algorithms work on ID-based data that is either organized as variable-length records or linked by complicated structures. The tedious one-by-one search/match operations and the irregular layout of data easily become the hurdle for higher performance that can otherwise be achieved as in fast scientific computing over well-organized matrices or multidimensional arrays.

On the basis of *D-CLUB*, a parallel bitmap-based algorithm *PD-CLUB* can be developed to mine all frequent itemsets by parallel adaptive refinement of clustered bitmaps using a differential mining technique. It clusters the database into distributed association bitmaps, applies a differential technique to digest and remove common patterns, and then independently mines each remaining tiny bitmaps directly through fast aggregate bit operations in parallel. The bitmaps are well organized into rectangular two-dimensional matrices and adaptively refined in regions that necessitate further computation.

The basic idea of parallelization behind *PD-CLUB* is to dynamically cluster the itemsets with their associated bit vectors, and divide the task of mining all frequent itemsets into smaller ones, each to mine a cluster of frequent itemsets. Then the subtasks are assigned to different processes and accomplished independently in parallel. A dynamic load balancing strategy can be used to reassign clusters from overloaded processes to free ones. The detailed explanation of the algorithm will be based on a number of new definitions listed below:

- *FI-cluster:* A FI-cluster is an ordered set of frequent itemsets. Starting from the FI-cluster (C_0) of all frequent 1-itemsets (sorted by supports in ascending order), other FI-clusters can be defined recursively as follows: from an existing FI-cluster, joining one itemset with each of the succeeding ones also generates a FI-cluster if only frequent itemsets are collected from the results. Itemsets can be reordered in the generated FI-cluster.
- *Bit vector:* For a given database of d transactions, each itemset is associated with a bit vector, where one bit corresponds to each transaction and is set to 1 iff the itemset is contained in that transaction.
- *Clustered Bitmap:* For each FI-cluster, the bit vectors of the frequent itemsets are also clustered. Laying out these vertical bit vectors side by side along their itemsets in the FI-cluster will generate a two-dimensional bitmap, called the clustered bitmap of the FI-cluster.
- *D-CLUB:* In the clustered bitmap of a FI-cluster, the following patterned bit rows are to be removed: *e-rows* (each with all 0s), *a-rows* (each with only one 1), *p-rows* (each with zero or more leading 0s followed by trailing 1s), *o-rows* (each with only one 0), and *c-rows* (each with zero or more leading 1s followed by trailing 0s). The remaining rows with different bits mixed disorderly form the differential clustered bitmap (*D-CLUB*) of the FI-cluster.

Recursively, generating FI-clusters from the initial C_0 results in a cluster tree that covers all frequent itemsets exactly once. The root of the tree is C_0, each node is a FI-cluster (or simply a cluster), and the connection between two FI-clusters denotes the generation relationship. Taking the frequent itemsets in Figure 32.1 as an example, Figure 32.6a shows the cluster tree of all frequent itemsets. For instance, FI-cluster {*caf*, *cae*} is generated from {*ca*, *cf*, *ce*} by joining itemset "*ca*" with "*cf*" and "*ce*," respectively.

Given the initial FI-cluster C_0, all frequent itemsets can be generated by traversing the cluster tree top down. Bit vectors are used to compute the supports of the generated itemsets. The count of an itemset in the database is equal to the number of 1s contained in its bit vector. Since the bitwise *AND* of the bit vectors for two itemsets results in the bit vector of the joined itemset, the clustered bitmap of a given FI-cluster sufficiently contains the count information for all the itemsets and their combinations. So a subtree of FI-clusters can be independently mined from the root FI-cluster and its clustered bitmap, by generating the

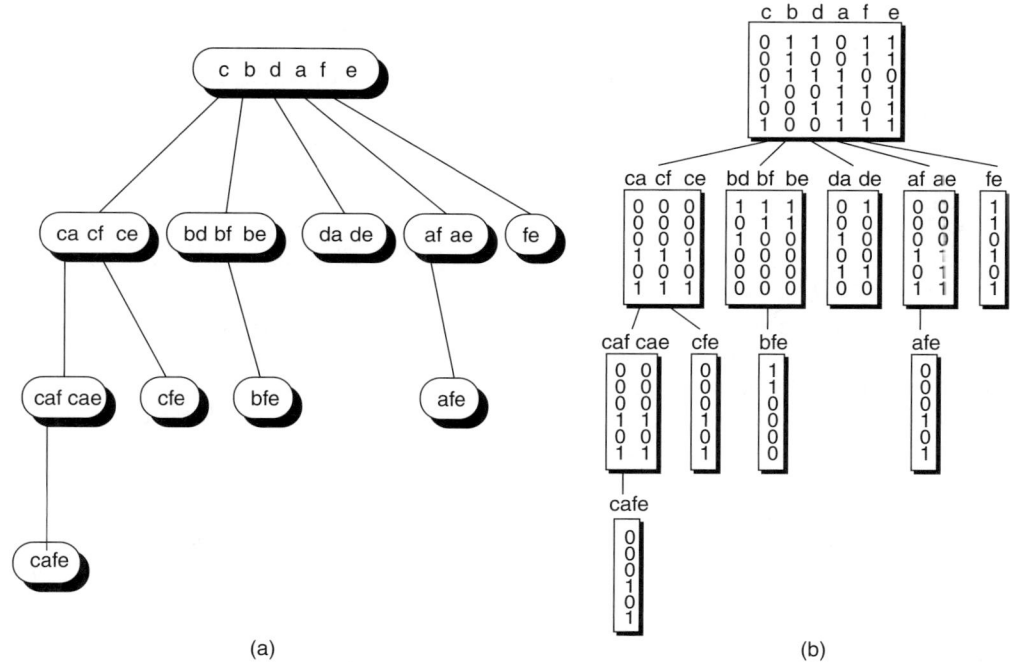

FIGURE 32.6 Mining frequent itemsets by clustered bitmaps. (a) Cluster tree of frequent itemsets. (b) Clustered bitmaps.

progeny FI-clusters and their associated clustered bitmaps as follows. When pairs of itemsets in a parent FI-cluster are joined to generate a child FI-cluster as described in the definition, the corresponding bit vectors from the parent's clustered bitmap are also operated via bitwise AND to form the child's clustered bitmap. Figure 32.6b shows how the clustered bitmaps are bound with their FI-clusters so that all frequent itemsets can be mined with supports computed along the cluster tree hierarchy.

In most cases, the cluster bitmaps may be too big to be processed efficiently, as they could be very sparse and contain too many obvious patterns as in e-, a-, p-, o-, and c-rows. So D-CLUBs are used in place of clustered bitmaps. To make e- and p-rows take the majority in a clustered bitmap, the itemsets in the FI-cluster can be reordered by ascending counts of 1s in their bit columns. With most of the sparse as well as dense rows removed, the size of the bitmap can be cut down by several orders of magnitude. Generating D-CLUBs along the cluster tree hierarchy is basically the same as doing clustered bitmaps, except that those patterned rows are to be removed. Removed rows are digested and turned into partial supports of the itemsets to be mined, through a number of propagation counters that can carry over from parent to children clusters. The support of an itemset is then computed by combining the count obtained from the D-CLUB and that from the propagation counters.

In practice, the algorithm starts by building the D-CLUBs for the level-2 clusters (FI-clusters of frequent 2-itemsets) from the original database. It then recursively refines each of the D-CLUBs along the cluster tree hierarchy, via bitwise AND of the corresponding bit columns. After each refinement, only selected rows and columns of the result bitmap are to be further refined. Selection of rows is achieved by the removal of those patterned rows, while selection of columns is by retaining only the columns for frequent itemsets. That gives the result D-CLUB so that it can be recursively refined. Propagation counters are incrementally accumulated along the traversing paths of the cluster tree when the patterned rows are digested and removed. The D-CLUBs are organized into a two-dimensional matrix of integers, where each bit column is grouped into a number of 32-bit integers. So, generating children D-CLUBs from their parents is performed by fast aggregate bit operations in arrays. Since the FI-clusters are closely bound to their *D-CLUB*s and the associated propagation counters, refinement of the *D-CLUB*s will directly generate the frequent itemsets with exact supports. The refinement stops where the *D-CLUB*s become empty, and all frequent itemsets in the subtree rooted at the corresponding FI-cluster can then be inferred, with supports calculated directly from the propagation counters.

The following steps outline the spirit of the *PD-CLUB* algorithm for shared-memory multiprocessors:

1. Scan the database in parallel by horizontal partitions and mine frequent 1-itemsets and 2-itemsets by clusters.
2. For clusters of frequent 2-itemsets, build their partial *dCLUB*s over the local database partition on each process, recording local propagation counters at the same time.
3. Combine the partial *D-CLUB*s into global ones for the level-2 clusters, and sum up the local propagation counters to get the global ones for each of the itemsets.
4. Sort the level-2 clusters by estimated workloads in descending order.
5. Assign each level-2 cluster in turn to one of the free processes and recursively refine its *D-CLUB* to mine all frequent itemsets in the FI-cluster subtree rooted at that cluster.

In Step 1, the count distribution method is used in mining frequent 1- and 2-itemsets. The *D-CLUB*s and propagation counters for the level-2 clusters are initially set up in Steps 2 and 3. The subsequent mining tasks are dynamically scheduled on multiple processes, as in Steps 4 and 5. The granularity of task assignment is based on the workload associated with mining the whole FI-cluster subtree rooted at each of the level-2 clusters. The workload for each cluster can be estimated as the size of the *D-CLUB* that is to be refined. Owing to the mining independence between subtrees belonging to different branches of the cluster tree, each of the assigned tasks can be independently performed on the designated process. Each of the subtrees is mined recursively in a depth-first way on a single process to better utilize the cache locality. Since tasks are assigned in a manner from coarse to fine grain, workload balance can be fairly well kept among all processes. However, it may happen that there are not enough clusters to be assigned to some free processes while others are busy with some extra work. To address this issue, a fine-tune measure can be taken in Step 5 to reassign branches of clusters to be mined on a busy process to a free one. It works as follows. The initial tasks from Step 4 are added to an assignment queue and assigned to processes as usual. Whenever a process finds the queue empty after it generates a new cluster, it goes on to mine only the leftmost subtree of the cluster. The further mining task for the right part of that cluster is added to the queue so that it can be reassigned for some free process to mine all other subtrees. After the queue becomes filled again, all processes settle back to independent mining of their assigned cluster subtrees in full in a depth-first way. Figure 32.7 shows such an example of cluster reassignment among two processes. At some moment when the assignment queue is empty, process P1 becomes free while process P0 is mining cluster 1. After generating cluster 1, P0 adds the right part of it to the queue and continues to mine only its left subtree, that is, clusters 3, 5, and 6. At the same time, P1 gets the new assignment to mine clusters 2 and 4, and the queue becomes empty again. Then clusters 2 and 3 are generated successively on P1 and P0. Cluster 2 does not have subtrees while cluster 3 has some. So P1 continues mining cluster 4, and P0 moves forward to mine cluster 5, adding right part of cluster 3 to the queue for P1 to mine cluster 6 later. P0 and P1 do approximately equal amounts of work and finish roughly at the same time.

For distributed-memory multiprocessors, owing to the expensive synchronization and communication cost, the algorithm can use a static task scheduling strategy instead to assign tasks as equally as possible at the very beginning. Then each process can perform their tasks independently without communicating or being synchronized with other processes. The basic steps can be expressed as follows:

1. Scan the database in parallel by horizontal partitions and mine frequent 1- and 2-itemsets by clusters.
2. For clusters of frequent 2-itemsets, build their partial *D-CLUB*s over the local partition of database on each process.
3. Initially assign the level-2 clusters to processes.

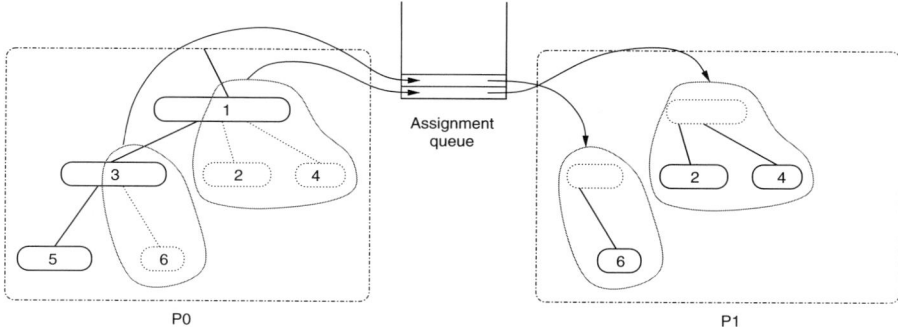

FIGURE 32.7 Dynamic cluster reassignment. Extra cluster subtrees originally to be mined by process P0 (which is overloaded) are reassigned for the free process P1 to mine, via a global assignment queue. Clusters in solid lines are mined locally, while those in dotted lines are mined on remote processes. Clusters are numbered by the order in which they are generated in full.

4. For the assigned clusters on each process, combine the partial *D-CLUB*s from all processes to get the global ones.
5. Each process recursively refines the *D-CLUB* to mine all frequent itemsets for each assigned cluster.

The first two steps work the same way as previously. In Step 3, the mining tasks for the cluster subtrees rooted at the level-2 clusters are prescheduled on all processes. A greedy scheduling algorithm can be used to sort the level-2 clusters by estimated workloads in descending order and then assign the clusters in turn to the least loaded process. Communication is needed in Step 4 for all processes to exchange partial *D-CLUB*s. The global *D-CLUB*s for the level-2 clusters are constructed only on the designated processes so that the refinement of each *D-CLUB* in Step 5 can be performed independently. For a workload balancing purpose, cluster reassignment is possible by sending a FI-cluster, its *D-CLUB* and the associated propagation counters as a unit from one process to another. Then the cluster subtree rooted at that FI-cluster can be mined on the second process instead of on the originally designated one. However, since the tasks are already prescheduled, the ongoing processes need a synchronization mechanism to detect the moment when cluster reassignment is needed and release some tasks for rescheduling. A communication mechanism is also needed for the source process to send data to the destination process. A dedicated communication thread can be added to each process for such purposes.

32.2.5 Comparison

In general performance, experiments show that *D-CLUB* performs the best, followed by *FP-growth* and *Eclat*, with *A priori* doing the worst. Similar performance ranking holds for their parallel versions. However, each of them has its own advantages and disadvantages.

Among all of the parallel ARM algorithms, the *A priori*-based algorithms are the most widely used because of the simplicity and easy implementation. Also, association rules can be directly generated on the way of itemset mining, because all the subset information is already computed when candidate itemsets are generated. These algorithms scale well with the number of transactions, but may have trouble handling too many items and/or numerous patterns as in dense databases. For example, in the *Court Distribution* method, if the number of candidate/frequent itemsets grows beyond what can be held in the main memory of each processor, the algorithm cannot work well no matter how many processors are added. The performance of these algorithms is dragged behind mainly by the slow itemset counting procedure that repeatedly searches the profuse itemsets against the large amount of transactions.

The *Eclat*-based algorithms have the advantage of fast support computing through tid-list intersection. By independent task parallelism, they gain very good speedups on distributed-memory multiprocessors. The main drawback of these algorithms is that they need to generate and redistribute the vertical tid-lists of which the total size is comparable to that of the original database. Also, for a long frequent itemset, the major common parts of the tid-lists are repeatedly intersected for all its subsets. To alleviate this situation, diffset optimization [ZG03] has been proposed to track only the changes in tid-lists instead of keeping the entire tid-lists through iterations so that it can significantly reduce the amount of data to be computed.

Parallel *FP-growth* handles dense databases very efficiently and scales particularly well with the number of transactions, benefiting from the fact that repeated or partially repeated transactions will be merged into paths of the FP-trees any way. However, this benefit does not increase accordingly with the number of processes, because multiple FP-trees for different sets of transactions are purely redundant. The benefit is also very limited for sparse databases with a small number of patterns scattered. The algorithm can handle a large number of items by just assigning them to multiple processes, without worrying about the exponentially large space of item/itemset combinations.

The *PD-CLUB* algorithm is self-adaptive to the database properties, and can handle both dense and sparse databases very efficiently. With the data size and representation fundamentally improved in the differential clustered bitmaps, the mining computation is also substantially reduced and simplified into fast aggregate bit operations in arrays. Compared to parallel *Eclat*, the D-CLUBs used in *PD-CLUB* have much smaller sizes than the tid-lists or even the diffsets, which results in much less communication cost when the D-CLUBs need to be exchanged among processes. The independent task parallelism plus the dynamic workload balancing mechanism gives the algorithm near-linear speedups on multiple processes.

32.3 Parallel Clustering Algorithms

Clustering is to group data objects into classes of similar objects on the basis of their attributes. Each class, called a cluster, consists of objects that are similar between themselves and dissimilar to objects in other classes. The dissimilarity or distance between objects is measured by the given attributes that describe each of the objects. As an unsupervised learning method, clustering is widely used in many applications, such as pattern recognition, image processing, gene expression data analysis, market research, and so on.

Existing clustering algorithms can be categorized into partitioning, hierarchical, density-based, grid-based and model-based methods [HK00], each generating very different clusters for various applications. Representative algorithms are introduced and their parallelizations are studied in this section.

32.3.1 Parallel *k-Means*

As a partitioning method, the *k-means* algorithm [Mac67] takes the input parameter, k, and partitions a set of n objects into k clusters with high intracluster similarity and low intercluster similarity. It starts by randomly selecting k objects as the initial cluster centroids. Each object is assigned to its nearest cluster on the basis of the distance between the object and the cluster centroid. It then computes the new centroid (or mean) for each cluster. This process is repeated until the sum of squared error (*SSE*) for all objects converges. The *SSE* is computed by summing up all the squared distances, one between each object and its nearest cluster centroid.

In parallel *k-means* [DM00], data parallelism is used to divide the workload evenly among all processes. Data objects are statically partitioned into blocks of equal sizes, one for each process. Since the main computation is to compute and compare the distances between each object and the cluster centroids, each process can compute on its own partition of data objects independently if the k-cluster centroids are maintained on all processes. The algorithm is summarized in the following steps:

1. Partition the data objects evenly among all processes.
2. Select k objects as the initial cluster centroids.
3. Each process assigns each object in its local partition to the nearest cluster, computes the *SSE* for all local objects, and sums up local objects belonging to each cluster.

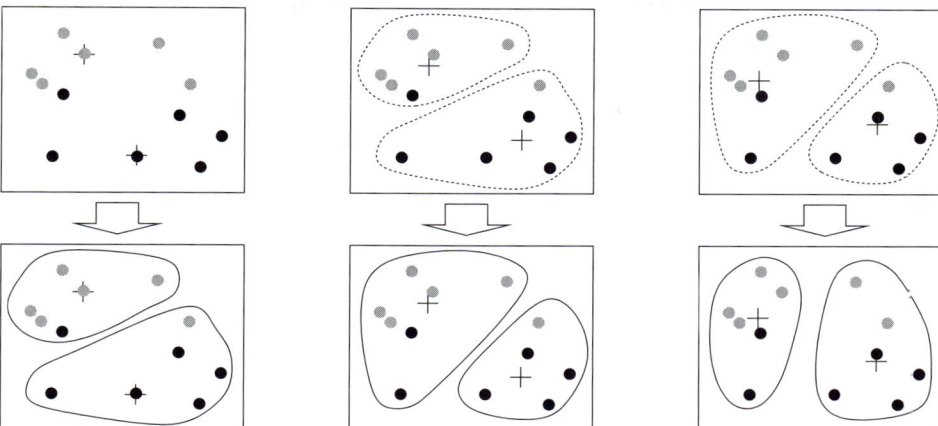

FIGURE 32.8 *k-means* clustering ($k = 2$) on two processes. One process computes on the objects in gray color, and the other is in charge of those in black. Cluster centroids, marked by "+," are maintained on both processes. Circles in solid lines denote the cluster formation in each iteration, based on the current cluster centroids. Dashed circles mark the previous cluster formation that is used to compute new cluster centroids.

4. All processes exchange and sum up the local *SSE*s to get the global *SSE* for all objects and compute the new cluster centroids.
5. Repeat Steps 3–5 until there is no change in the global *SSE*.

The algorithm is proposed on distributed-memory multiprocessors but works similarly on shared-memory systems as well. Step 3 is the major computation step where clusters are formed. Objects belonging to one cluster may be distributed over multiple processes. In Step 4, each of the new cluster centroid is computed in the same way as the global *SSE* for all objects, except that the summational result will be further divided by the count of objects in the cluster in order to get the mean. Figure 32.8 shows an example of *k-means* clustering for $k = 2$. The data objects are partitioned among two processes and the two clusters are identified through three iterations.

In parallel *k-means*, the workloads per iteration are fairly well balanced between processes, which results in linear speedups when the number of data objects is large enough. Between iterations, there is a small communication/synchronization overhead for all processes to exchange the local *SSE*s and the local member object summations for each cluster. The algorithm needs a full scan of the data objects in each iteration. For large disk-resident data sets, having more processors could result in superlinear speedups as each data partition then may be small enough to fit in the main memory, avoiding disk I/O except in the first iteration.

32.3.2 Parallel Hierarchical Clustering

Hierarchical clustering algorithms are usually applied to bioinformatics procedures such as grouping of genes and proteins with similar structure, reconstruction of evolutionary trees, gene expression analysis, and so forth. An agglomerative approach is commonly used to recursively merge pairs of closest objects or clusters into new clusters until all objects are merged into one cluster or until a termination condition is satisfied. The distance between two clusters can be determined by single link, average link, complete link, or centroid-based metrics. The single link metric uses the minimum distance between each pair of intercluster objects, average link uses the average distance, and complete link uses the maximum. The centroid-based metric uses the distance between the cluster centroids. Figure 32.9 gives an example of

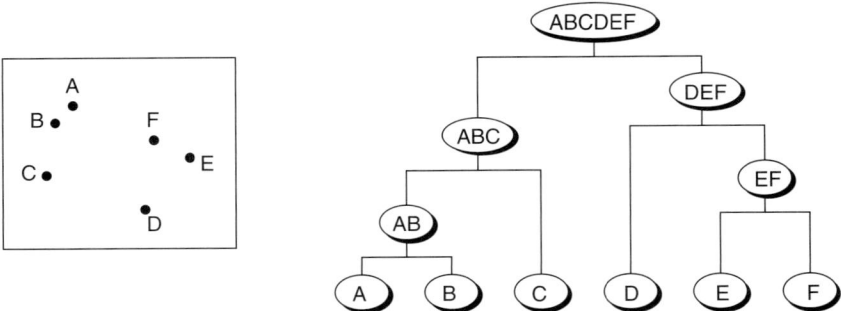

FIGURE 32.9 Example of hierarchical clustering.

agglomerative hierarchical clustering using the single link metric. The dendrogram on the right shows which clusters are merged at each step.

The hierarchical mergence of clusters can be parallelized by assigning clusters to different processes so that each process will be in charge of a disjoint subset of clusters, as in Reference Ols95. When two clusters are merged, they are released from the owner processes, and the new cluster will be assigned to the least loaded process. The basic parallel algorithm for agglomerative hierarchical clustering takes the following steps:

1. Initially treat each data object as a cluster and assign clusters among all processes evenly so that each process is responsible for a disjoint subset of clusters.
2. Calculate the distance between each pair of clusters in parallel, maintaining a nearest neighbor for each cluster.
3. All processes synchronously find the closest pair of clusters, agglomerate them into a new cluster, and assign it to the least loaded process.
4. Calculate the distance between the new cluster and each of the remaining old clusters, and update the nearest neighbor for each cluster on the responsible process.
5. Repeat Steps 3–5 until there is only one cluster remaining or certain termination conditions are satisfied.

The algorithm works for both shared and distributed memory multiprocessors, though the latter case requires all processes to exchange data objects in the initialization Step of 2. It is crucial to design an appropriate data structure for the parallel computation of the all-pair distances between clusters. For Steps 2, 3, and 4, a two-dimensional array can be used to store distances between any two clusters. It is distributed over all processes such that each one is responsible for those rows corresponding to its assigned clusters. Similarly, the nearest neighbor of each cluster and the associated distance are maintained in a one-dimensional array, with the responsibility divided among processes accordingly. The closest pair of clusters can then be easily determined on the basis of the nearest neighbor distances. Step 3 is a synchronous phase so that all processes know which two clusters are merged. After two clusters are merged into a new cluster, each process only updates the assigned rows to get the distances from the assigned clusters to the new cluster. Specifically, the distance from cluster i to the new cluster is computed from the distances between cluster i and the two old clusters that are merged, for example, by taking the less value when the single link metric is used. The nearest neighbor information is then recomputed, taking into account the newly computed distances. The assignment of the new cluster requires estimation of the workload on each process. For simplicity, the number of assigned clusters can be used as the estimated workload, which is pretty effective. Taking Figure 32.9 as an example, Figure 32.10 illustrates how the all-pair distances between clusters and the neighbor information are divided for two processes to compute in parallel, before and after the first merge of clusters. The new cluster "AB" is assigned to process P0.

	Intercluster distances					Nearest
	A	B	C	D	E	F	neighbor
A		10	38	60	60	41	B: 10
P0 B	10		26	60	62	44	A: 10
C	38	26		50	66	50	B: 26
D	60	60	50		38	39	E: 38
P1 E	60	62	66	38		20	F: 20
F	41	44	50	39	20		E: 20

(a) Before merge

	Intercluster distances					Nearest
	AB	C	D	E	F	neighbor
AB		26	60	60	41	C: 26
P0 C	26		50	66	50	AB: 26
D	60	50		38	39	E: 38
P1 E	60	66	38		20	F: 20
F	41	50	39	20		E: 20

(b) After merge

FIGURE 32.10 Compute/update intercluster distances and nearest neighbors in parallel over two processes. From (a) to (b), clusters "A" and "B" are merged into a new cluster "AB." In (b), only entries in bold font require new computation or update.

By task parallelism and employing a dynamic workload balancing strategy, this algorithm can achieve good speedups when the number of processes is much less than that of clusters. However, there is some communication and/or synchronization overhead in every step of cluster agglomeration, because all processes have to obtain the global information to find the minimum distance between clusters and also have to keep every process informed after the new cluster has been formed. The algorithm assumes the data structures being able to be kept in main memory so that it scans the data set only once.

32.3.3 Parallel *HOP*: Clustering Spatial Data

HOP [EH98], a density-based clustering algorithm proposed in astrophysics, identifies groups of particles in N-body simulations. It first constructs a KD tree by recursively bisecting the particles along the longest axis so that nearby particles reside in the same subdomain. Then it estimates the density of each particle by its N_{dens} nearest neighbors that can be efficiently found by traverse of the KD tree. Each particle is associated to its densest neighbor within its N_{hop} nearest neighbors. A particle can hop to the next densest particle and continue hopping until it reaches a particle that is its own densest neighbor. Finally, *HOP* derives clusters from groups that consist of particles associated to the same densest neighbor. Groups are merged if they share a sufficiently dense boundary, according to some given density threshold. Particles whose densities are less than the density threshold are excluded from groups.

Besides the cosmological N-body problem, *HOP* may find its application in other fields, such as molecular biology, geology, and astronomy, where large spatial data sets are to be processed with similar clustering or neighbor-finding procedures:

To parallelize the *HOP* clustering process, the key idea is to distribute the data particles across all processes evenly with proper data placement so that the workloads are balanced and communication cost for remote data access is minimized. The following steps explain the parallel *HOP* algorithm [LLC03] in detail, assuming distributed-memory multiprocessors:

1. *Constructing a Distributed KD Tree:* The particles are initially distributed among all processes in blocks of equal sizes. Starting from the root node of the KD tree, the algorithm first determines the longest axis d and then finds the median value m of all particles' d coordinates in parallel. The whole spatial domain is bisected into two subdomains by m. Particles are exchanged between processes such that the particles whose d coordinates are greater than m go to one subdomain and the rest of the particles to the other one. Therefore, an equal number of particles are maintained in each subdomain after the bisection. This procedure is repeated recursively in every subdomain till the number of subdomains is equal to the number of processes. Then, each process continues to build its own local tree within its domain until the desired bucket size (number of particles in each leaf) is reached. Note that interprocess communication is not required in the construction of the local trees.

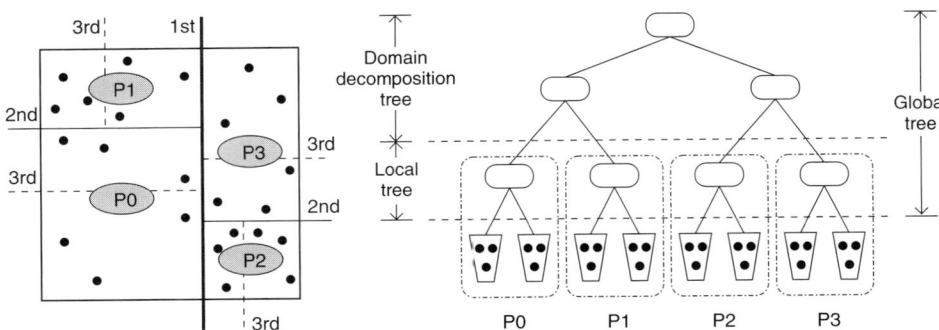

FIGURE 32.11 Two-dimensional KD tree distributed over four processes. Each process contains 6 particles. Bucket size is 3 and the global tree has 3 levels. Local tree can be built concurrently without communication. Every process maintains the same copy of the global tree.

A copy of the whole tree is maintained on every process so that the communication overhead incurred at performing search domain intersection test with the remote local trees at the stages 2 and 3 can be reduced. Therefore, at the end of this stage, local trees are broadcasted to all processes. As shown in Figure 32.11, the root node of the KD tree represents the entire simulation domain while each of the rest tree nodes represents a rectangular subdomain of its parent node. The information contained in a nonleaf tree node includes the aggregated mass, center of mass, number of particles, and domain boundaries. When the KD tree is completed, particles are divided into spatially closed regions of approximately equal number. The advantage of using a KD tree is not only its simplicity but also the balanced data distribution.

2. *Generating Density:* The density of a particle is estimated by its N_{dens} nearest neighbors, where N_{dens} is a user-specified parameter. Since it is possible that some of the N_{dens} neighbors of a particle are owned by remote processes, communication is required to access nonlocal neighbor particles at this stage. One effective approach is, for each particle, to perform an intersection test by traversing the global tree with a given initial search radius r, while keeping track of the nonlocal intersected buckets, as shown in Figure 32.12. If the total number of particles in all intersected buckets is less than N_{dens}, the intersection test is reperformed with a larger radius. Once tree walking is completed for all local particles, all the remote buckets containing the potential neighbors are obtained through communication. Note that there is only one communication request to each remote process to gather the intersected buckets. No further communication is necessary when searching for its N_{dens} nearest neighbors. Since the KD tree displays the value of spatial locality, particle neighbors are most likely located in the same or nearby buckets. According to the experimental results, the communication volume is only 10–20% of the total number of particles. However, with highly irregular particle distribution, communication costs may increase.

    To calculate the density for particle p, the algorithm uses a PQ tree (priority queue) [CLR90] to maintain a sorted list of particles that are currently the N_{dens} nearest neighbors. The root of the PQ tree contains the neighbor farthest from p. If a new neighbor's distance to p is shorter than the root, replace the root with the second farthest one and update the PQ tree. Finally, the particles that remain in the PQ tree are the N_{dens} nearest neighbors of p.

3. *Hopping:* This stage first associates each particle to its highest density neighbor among its N_{hop} nearest neighbors that are already stored in the PQ tree generated at the previous stage. Each particle, then, hops to the highest density neighbor of its associated neighbor. Hopping to remote particles is performed by first keeping track of all the remote particles and then by making a communication request to the owner processes. This procedure may repeat several times until all the needed nonlocal particles are already stored locally. Since the hopping is in density increasing order, the convergence is guaranteed.

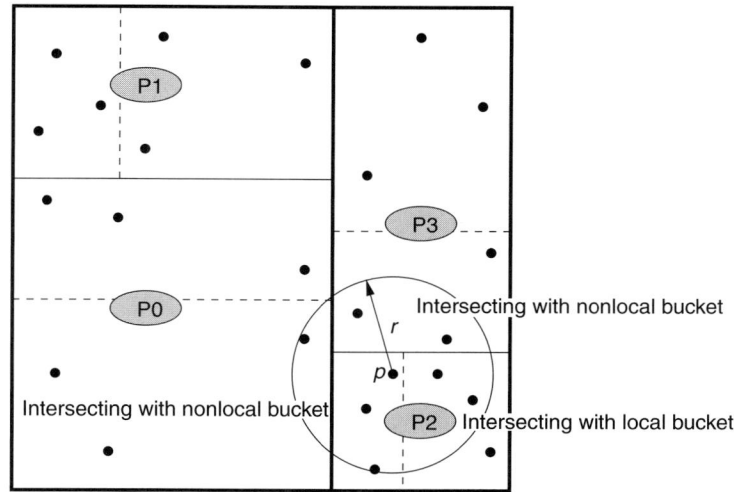

FIGURE 32.12 Intersection test for particle p on process P2 for $N_{dens} = 7$. The neighborhood is a spherical region with a search radius r. The neighborhood search domain of p intersects with the subdomains of P0 and P3.

4. *Grouping*: Particles linked to the same densest particle are defined as a group. However, some groups should be merged or refined according to the chosen density thresholds. Thus, every process first builds a boundary matrix for the groups constructed from its local particles and then exchanges the boundary matrix among all processes. Particles whose densities are less than a given threshold are excluded from groups, and two groups are merged if their boundary satisfies some given thresholds.

The parallel *HOP* clustering algorithm distributes particle data evenly among all processes to guarantee balanced workload. It scans the data set only once and stores all the particle data in the distributed KD tree over multiple processes. This data structure helps to minimize interprocess communication as well as improves the neighbor search efficiency, as the spatially closed particles are usually located in the same buckets or a very few neighbor buckets. The communication cost comes from the particle redistribution during the KD tree construction and the remote particle access in the neighbor-based density generation. Experiments showed that it gained good speedups on a number of different parallel machines.

32.3.4 Clustering High-Dimensional Data

For high-dimensional data, grid-based clustering algorithms are usually used. *CLIQUE* [AGGR98] is one of such algorithms. It partitions the n-dimensional data space into nonoverlapping rectangular units of uniform size, identifying the dense units among these. On the basis of a user-specified global density threshold, it iteratively searches dense units in the subspaces from 1-dimension through k-dimension until no more dense units are found. The generation of candidate subspaces is based on the *A priori property* used in ARM. The dense units are then examined to form clusters.

The *pMAFIA* algorithm [NGC01] improves *CLIQUE* by using adaptive grid sizes. The domain of each dimension is partitioned into variable sized adaptive grid bins that capture the data distribution. Also, variable density thresholds are used, one for each bin. Adaptive dense units are then found in all possible subspaces. A unit is identified as dense if its population is greater than the density thresholds of all the bins that form the unit. Each dense unit of dimension d can be specified by the d dimensions and their corresponding d bin indices. Figure 32.13 illustrates the dense unit identification for both uniform grid size and adaptive grid sizes.

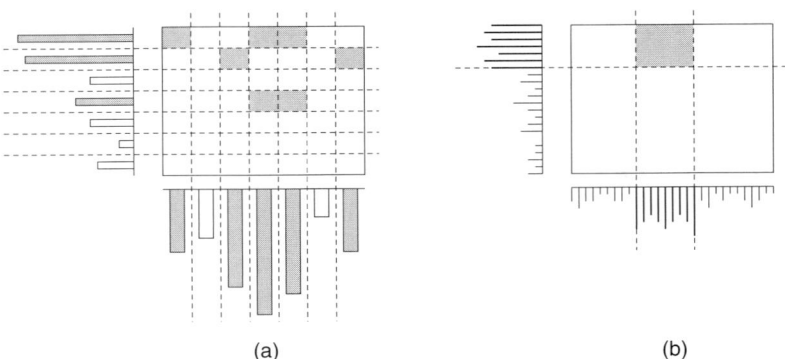

FIGURE 32.13 Identify dense units in two-dimensional grids: (a) uses a uniform grid size, while (b) uses adaptive grid sizes by merging fine bins into adaptive grid bins in each dimension. Histograms are built for all dimensions.

PMAFIA is one of the first algorithms that demonstrate a parallelization of subspace clustering for high-dimensional large-scale data sets. Targeting distributed-memory multiprocessors, it makes use of both data and task parallelism. The major steps are listed below:

1. Partition the data objects evenly among all processes.
2. For each dimension, by dividing the domain into fine bins, each process scans the data objects in its local partition and builds a local histogram independently.
3. All processes exchange and sum up the local histograms to get the global one for each dimension;
4. Determine adaptive intervals using the global histogram in each dimension and set the density threshold for each interval.
5. Each process finds candidate dense units of current dimensionality (initially 1) and scans data objects in its local partition to populate the candidate dense units.
6. All processes exchange and sum up the local populations to get the global one for each candidate dense unit.
7. Identify dense units and build their data structures.
8. Increase the dimensionality and repeat Steps 5 through 8 until no more dense units are found.
9. Generate clusters from identified dense units.

The algorithm spends most of its time in making repeated passes over the data objects and finding out the dense units among the candidate dense units formed from 1- to k-dimensional subspaces until no more dense units are found. Building the histograms in Step 2 and populating the candidate dense units in Step 5 are performed by data parallelism such that each process scans only a partition of data objects to compute the local histograms and local populations independently. After each of the independent steps, the local values are collected from all processes and they add up to the global values, as in Steps 3 and 6, respectively. In Step 4, all processes perform the same adaptive grid computation from the global histograms they gathered. The adaptive grid bins in each dimension are then generated, each considered to be a candidate dense unit of dimension 1. Candidate dense units from subspaces of higher dimensions need to be generated in Step 5. Similar to the candidate subspace generation procedure in *CLIQUE*, *pMAFIA* generates candidate dense units in any dimension k by combining dense units of dimension $k-1$ such that they share any $k-2$ dimensions. Since each pair of the dense units needs to be checked for possible intersection into a candidate dense unit, the amount of work is known and the task can be easily subdivided among all processes such that each one intersects an equal number of pairs. Task parallelism is also used in Step 7 to identify dense units from the populated candidates such that each process scans an equal number of candidate dense units.

pMAFIA is designed for disk-resident data sets and scans the data as many times as the maximum dimension of the dense units. By data parallelism, each partition of data is stored in the local disk once it is retrieved from the remote shared disk by a process. That way, subsequent data accesses can see a much larger I/O bandwidth. It uses fine bins to build the histogram in each dimension, which results in large sizes of histograms and could add to the communication cost for all processes to exchange local histograms. Compared to the major time spent in populating the candidate dense units, this communication overhead can be negligible. Also, by using adaptive grids that automatically capture the data distribution, the number of candidate dense units is minimized, so is the communication cost in the exchange of their local populations among all processes. Experiments showed that it could achieve near-linear speedups.

32.4 Summary

As data accumulates in bulk volumes and goes beyond the processing power of single-processor machines, parallel data mining techniques become more and more important for scientists as well as business decision makers to extract concise, insightful knowledge from the collected information in an acceptable time. In this chapter, various algorithms for parallel ARM and clustering are studied, spanning distributed and shared memory systems, data and task parallelism, static and dynamic workload balancing strategies, and so on. Scalability, interprocess communication/synchronization, workload balance, and I/O issues are discussed for these algorithms. The key factors that affect the parallelism varies for different problems or even different algorithms of the same problem. For example, owing to the dynamic nature of ARM, the workload balance is a big issue for many algorithms that use static task scheduling mechanisms. And for the density-based parallel *HOP* clustering algorithms, the focus of effort should be put on minimizing the data dependence across processes. By efficiently utilizing the aggregate computing resources of parallel processors and minimizing the interprocess communication/synchronization overhead, high-performance parallel data mining algorithms can be designed to handle massive data sets.

While extensive research has been done in this field, a lot of new exciting work is being explored for future development. With the rapid growth of distributed computing systems, the research on distributed data mining has become very active. For example, the emerging pervasive computing environment is becoming more and more popular, where each ubiquitous device is a resource-constrained distributed computing device. Data mining research in such systems is still in its infancy but most algorithm designs can be theoretically based on existing parallel data mining algorithms. This chapter can serve as a reference for the state of the art for both researchers and practitioners who are interested in building parallel and distributed data mining systems.

References

[AGGR98] Rakesh Agrawal, Johannes Gehrke, Dimitrios Gunopulos, and Prabhakar Raghavan. Automatic subspace clustering of high dimensional data for data mining applications. In *Proc. of the ACM SIGMOD Int'l Conf. on Management of Data*, pp. 94–105, June 1998.

[AIS93] Rakesh Agrawal, Tomasz Imielinski, and Arun N. Swami. Mining association rules between sets of items in large databases. In *Proc. of the ACM SIGMOD Int'l Conf. on Management of Data*, pp. 207–216, May 1993.

[AS94] Rakesh Agrawal and Ramakrishnan Srikant. Fast algorithms for mining association rules. In *Proc. of the 20th Int'l Conf. on Very Large Databases*, pp. 487–499, September 1994.

[AS96] Rakesh Agrawal and John C. Shafer. Parallel mining of association rules. *IEEE Trans. on Knowledge and Data Engineering*, 8(6):962–969, December 1996.

[CLR90] Thomas T. Cormen, Charles E. Leiserson, and Ronald L. Rivest. *Introduction to Algorithms*. MIT Press, Cambridge, MA, June 1990.

[DM00] Inderjit S. Dhillon and Dharmendra S. Modha. A data-clustering algorithm on distributed memory multiprocessors. *Large-Scale Parallel Data Mining*, Lecture Notes in Artificial Intelligence, 1759:245–260, 2000.

[EH98]   Daniel J. Eisenstein and Piet Hut. Hop: A new group-finding algorithm for n-body simulations. *Journal of Astrophysics*, 498:137–142, 1998.

[HK00]   Jiawei Han and Micheline Kamber. *Data mining: concepts and techniques*. Morgan Kaufmann Publishers Inc., San Francisco, CA, August 2000.

[HKK00]   Eui-Hong Han, George Karypis, and Vipin Kumar. Scalable parallel data mining for association rules. *IEEE Transactions on Knowledge and Data Engineering*, 12(3):337–352, 2000.

[HPY00]   Jiawei Han, Jian Pei, and Yiwen Yin. Mining frequent patterns without candidate generation. In *Proc. of the ACM SIGMOD Int'l Conf. on Management of Data*, pp. 1–12, May 2000.

[LCJL06] Jianwei Li, Alok Choudhary, Nan Jiang, and Wei-keng Liao. Mining frequent patterns by differential refinement of clustered bitmaps. In *Proc. of the SIAM Int'l Conf. on Data Mining*, April 2006.

[LLC03]   Ying Liu, Wei-keng Liao, and Alok Choudhary. Design and evaluation of a parallel HOP clustering algorithm for cosmological simulation. In *Proc. of the 17th Int'l Parallel and Distributed Processing Symposium*, April 2003.

[Mac67]   James B. MacQueen. Some methods for classification and analysis of multivariate observations. In *Proc. of the 5th Berkeley Symposium on Mathematical Statistics and Probability*, Vol. 1, pp. 281–297, 1967.

[NGC01]   Harsha Nagesh, Sanjay Goil, and Alok Choudhary. Parallel algorithms for clustering high-dimensional large-scale datasets. In Robert Grossman, Chandrika Kamath, Philip Kegelmeyer, Vipin Kumar, and Raju Namburu, eds., *Data Mining for Scientific and Engineering Applications*, pp. 335–356. Kluwer Academic Publishers, 2001.

[Ols95]   Clark F. Olson. Parallel algorithms for hierarchical clustering. *Parallel Computing*, 21:1313–1325, 1995.

[PK03]   Iko Pramudiono and Masaru Kitsuregawa. Tree structure based parallel frequent pattern mining on PC cluster. In *Proc. of the 14th Int'l Conf. on Database and Expert Systems Applications*, pp. 537–547, September 2003.

[Zak99]   Mohammed J. Zaki. Parallel and distributed association mining: A survey. *IEEE Concurrency*, 7(4):14–25, 1999.

[ZEHL01] Osmar R. Zaiane, Mohammad El-Hajj, and Paul Lu. Fast parallel association rule mining without candidacy generation. In *Proc. of the IEEE Int'l Conf. on Data Mining*, November 2001.

[ZG03]   Mohammed J. Zaki and Karam Gouda. Fast vertical mining using diffsets. In *Proc. of the ACM SIGKDD Int'l Conf. on Knowledge Discovery and Data Mining*, pp. 326–335, August 2003.

[ZOPL96] Mohammed J. Zaki, Mitsunori Ogihara, Srinivasan Parthasarathy, and Wei Li. Parallel data mining for association rules on shared-memory multi-processors. In *Proc. of the ACM/IEEE Conf. on Supercomputing*, November 1996.

[ZPL97]   Mohammed J. Zaki, Srinivasan Parthasarathy, and Wei Li. A localized algorithm for parallel association mining. In *Proc. of the 9th ACM Symposium on Parallel Algorithms and Architectures*, pp. 321–330, June 1997.

[ZPOL97a] Mohammed J. Zaki, Srinivasan Parthasarathy, Mitsunori Ogihara, and Wei Li. New algorithms for fast discovery of association rules. Technical Report TR651, University of Rochester, July 1997.

[ZPOL97b] Mohammed J. Zaki, Srinivasan Parthasarathy, Mitsunori Ogihara, and Wei Li. Parallel algorithms for discovery of association rules. *Data Mining and Knowledge Discovery: An International Journal*, special issue on Scalable High-Performance Computing for KDD, 1(4):343–373, December 1997.

33
An Overview of Mobile Computing Algorithmics

33.1	Introduction...	33-1
33.2	Cellular Networks..	33-2
	Bandwidth Allocation • Location Management	
33.3	Low Earth Orbiting Satellite Networks.................	33-10
	Related Work • A Survey of Basic Medium Access Control Protocols • Mobility Model and Traffic Parameters	
33.4	Resource Management and Call Admission Control Algorithm	33-17
	Resource Management Strategy • Connection Admission Control Algorithm • Performance Evaluation	
33.5	Mobile Ad Hoc Networks	33-21
	Routing	
33.6	Wireless Sensor Networks	33-25
	The Network Model • Network Organization and Clustering • The Work Model	
33.7	Looking into the Crystal Ball.................................	33-29
33.8	Concluding Remarks	33-29
	References	33-30

Stephan Olariu
Old Dominion University

Albert Y. Zomaya
The University of Sydney

33.1 Introduction

The widening availability of mobile information systems is being driven by the increasing demand to have information available for users *anytime, anywhere*. As the availability of wireless devices and mobile services increases, so will the load on available radio frequency resources [13]. In fact, wireless and mobile hosts (MHs) are the fastest growing segment of the personal computer market but are not well supported by current network protocols including those in use in the Internet. Wireless networks have fundamentally different properties than typical wired networks, including higher error rates, lower bandwidths, nonuniform transmission propagation, increased usage costs, and increased susceptibility to interference and eavesdropping.

Similarly, MHs behave differently and have fundamentally different limitations than stationary hosts. For example, a MH may move and become disconnected from or change its point of connection to the

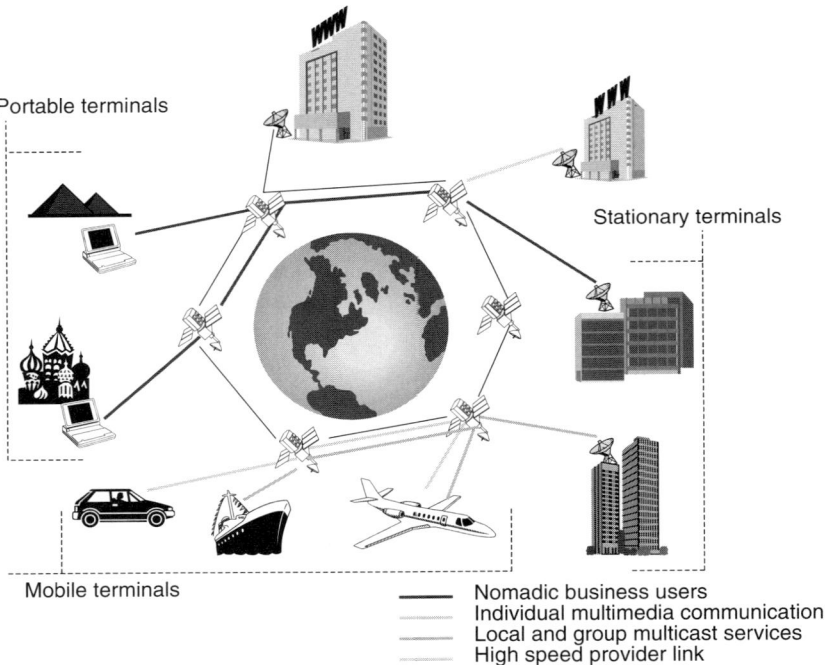

FIGURE 33.1 A global integrated terrestrial and LEO satellite system.

network, and MHs generally operate on limited battery power. These differences in network and host properties produce many new challenges for network protocols.

It is increasingly clear that the use of m-commerce is becoming an important new tool for the manufacturing and retail industries and an imperative new development for all industries striving to maintain a competitive edge. Companies can use the mechanisms and technologies offered by m-commerce to put their stores online. At the same time, customer mobility has emerged an important catalyst triggering a new paradigm shift that is redefining the way we conduct business [5,6,8,12,14,15,18,29,50,51]. An important feature of mobile commerce is that the online store is accessible 24 h per day to a potentially huge customer base scattered around the world. By their very nature, mobile commerce applications will rely increasingly on wireless communications. To be cost effective and, thus economically viable, these applications will have to reach a wide consumer base. In turn, this suggests that a wide coverage—indeed, a global one—is essential in mobile commerce. As illustrated in Figure 33.1, such a global coverage is only achieved by a combination of terrestrial and satellite networks.

33.2 Cellular Networks

A cellular network is a mobile network in which radio resources are managed in cells. A cell is a two-dimensional area bounded logically by an interference threshold. By exploiting the ability to borrow radio bandwidth, one can ensure that areas of high load are assigned sufficient resources. Further, resources can be assigned in a way that maximizes their availability at a future time. Overall, one can divide mobile computing subjects that have great potential for parallel algorithmics research into three main categories: bandwidth allocation, location management, and routing protocols. Of course, other areas are of equal importance, such as, security, but in this paper we will confine ourselves to these three areas.

33.2.1 Bandwidth Allocation

Adding bandwidth to a mobile network is an expensive operation. When developing a mobile network's infrastructure, wasting bandwidth should be avoided because it is very costly. Thus, the objective is to try and get the most out of the minimum infrastructure. The same problem applies to a mobile network that is already installed, where it is cheaper to utilize the available resources more effectively than to add more bandwidth.

The channel allocation problem involves how to allocate borrrowable channels in such a way as to maximize the long- and/or short-term performance of the network [22]. The performance metrics that can be used to evaluate the solutions proposed will be primarily the number of hosts blocked and the number of borrowings. A host is blocked when it enters a cell and cannot be allocated a channel. Obviously, the more hosts that are blocked, the worse will be the performance of the network. The other major metric is the number of channel borrowings. This should be minimized because channel borrow requests generate network traffic. There are other metrics that can be used to evaluate the performance of the solution such as the number of "hot cells" that appear in a cellular environment.

There are, in fact, a number of ways to deal with excess load in mobile networks in addition to channel borrowing, such as channel sharing and cell splitting [25]. In particular, cell splitting is commonly used in many real cellular networks [13]. Cell splitting works by breaking down cells into smaller cells. This is accomplished by having several different levels of cell coverage. These different levels are often called macrocells and microcells. A macrocell is essentially an umbrella over a set of microcells. When traffic becomes too great for a cell to handle and there is a microcell structure in place, the cell can be switched out and the microcells switched in. This enables the original cell site to handle more loads. There are obvious drawbacks to this scheme that prevent it from being implemented throughout the network. The obvious problem is, of course, cost. A secondary concern is the extra network traffic introduced by the additional cells.

33.2.1.1 Cells

A general graph model can represent arrangement of cells in real cellular network as featured in Figure 33.2. In the graph model, each node represents a base station (center of cell), and neighboring cells are represented by edges connecting the nodes. Other simpler models have also been used for simulation purposes, which include one-dimensional (1D) and structured two-dimensional (2D) models.

In the 1D model, each user has two possible opposing moves (e.g., left or right)—see Figure 33.3. Common 2D models include the mesh configuration (Figure 33.4), and the hexagonal configuration (Figure 33.5). In the hexagonal configuration, each cell can have a maximum of six neighboring cells. In the mesh configuration, each cell can have a maximum of either four or eight neighbors, depending on whether diagonal movements are allowed.

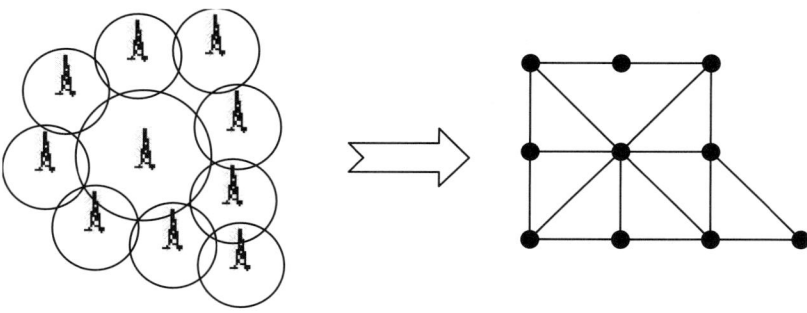

FIGURE 33.2 Cellular network and corresponding graph model.

FIGURE 33.3 One-dimensional network.

FIGURE 33.4 Two-dimensional, mesh-configuration network.

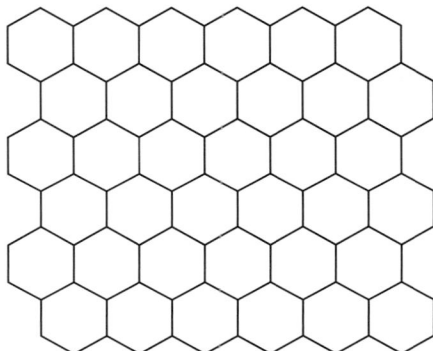

FIGURE 33.5 Two-dimensional, hexagonal-configuration network.

33.2.1.2 Frequency Reuse

Frequency reuse is a technique used in mobile networks to make more effective use of the available radio resources. Put simply, frequency reuse is a way of assigning the same channel to more than one base station. By assigning a channel to more than one cell, we get two major benefits: lower transmitter power and increased capacity. Lower transmission power can result in longer battery life in mobile devices.

Channels that use the same frequencies in a cellular system are called cochannels. The distance between cochannels is determined primarily by the transmitter power used at each cell site. A quantity that is related to this distance is called the Carrier-to-Interference Ratio (CIR). The CIR is used to measure to what extent two cells that use the same frequencies interfere with each other.

Owing to cochannel interference, the same radio frequencies are generally not used in neighboring cells. In both the aforementioned networks, all of the available radio spectrum is divided into sevenths and assigned to each cell in a group of seven cells. This scheme demonstrates a classic trade-off. If the network has smaller cells, the channel reuse distance is also smaller. The advantage of using smaller cells is such that they will support more hosts, as the channel density is higher. It will, however, cost more to use smaller cells because more of them are required to cover a service area. There is also an advantage to using larger cells because fewer are needed to cover a given area, and hence, the cost of implementing the network is less. However, to support larger cells, more transmitter power is required and hence the mobile battery's life is lower.

33.2.1.3 Mobile Hosts

In a wireless environment, a cell is a bounded 2D area that is created by a mobile service station (MSS). A MH is an entity that exists in a mobile network and is free to move around. The definition of MH is always changing as mobile devices become more advanced. A new paradigm is emerging whereby MHs are containing more local intelligence. Examples of this are Personal Digital Assistants (PDAs) and many new mobile telephony devices. With increasing inbuilt computational power, a mobile network may yet emerge as a dynamic parallel computing platform.

In some parts of the literature, MHs are treated as "dumb" entities that simply consume radio resources and follow some path. In other papers [25], mobile agents are used to convey load information around the network. These agents could potentially be executed on MHs. This may reduce the load on the MSS in the future.

33.2.1.4 Load Coping Schemes

Limited radio bandwidth is a major issue in mobile networking [16]. When a cellular layout plan is designed, each cell is assigned a fixed number of channels that is able to handle ambient and predicted system loads. Situations can occur, however, where the load at points of the system increases beyond which the system can handle. Specifically, when all channels in a cell are allocated to hosts, the cell is considered "hot." Hot cells must be avoided because MHs will be denied a service upon entering the cell and handoff requests will also be denied. There are a number of schemes available that can be used to cope with dynamic load. Some popular schemes are described next.

33.2.1.5 Channel Borrowing with Locking

Channel borrowing with locking, or simple borrowing (SB), is a scheme that allows neighbors to use each other's channels. This scheme uses cochannel locking to eliminate cochannel interference that occurs when a neighboring cell borrows a channel. A cell's neighbor can borrow a channel if it is not already allocated to a local host. There are a number of decisions to be made when borrowing a channel, including where to borrow from and when to borrow. A dynamic load-balancing algorithm must make these decisions.

Another more complex scheme exists, known as channel borrowing without locking (CBWL). This scheme uses power control to avoid locking the cochannels of the lending cell and is described in Reference 25. This technique uses the borrowed channel under less power.

The constrained power problem has also been researched as a method of improving channel utilization [2,16,52]. These particular research efforts suggest a dynamic resource allocation algorithm based on constrained power control. The goal is to minimize all transmitter power subject to a finite frequency spectrum and finite transmitter power.

There are some other models for mobile networks that differ from the classical fixed base station configuration. There is a great deal of research occurring in the area of robotics where each MH or robot is actually a MSS as well. This creates a new set of problems, particularly to do with routing algorithms. Since there may be no underlying fixed network structure, message routing tables will be changing constantly. This requires an advanced dynamic mobile routing algorithm that can cope with the changing structure of the virtual network links. Networks of this class have much tighter resource requirements and require efficient resource allocation schemes.

33.2.2 Location Management

One of the challenges facing mobile computing is the tracking of the current location of users—the area of location management. In order to route incoming calls to appropriate mobile terminals, the network must from time-to-time keep track of the location of each mobile terminals [43].

Mobility tracking expends the limited resources of the wireless network. Besides, the bandwidth used for registration and paging between the mobile terminal and base stations, power is also consumed from the portable devices. Furthermore, frequent signaling may result in degradation of Quality of Service

(QoS) due to interferences. On the other hand, a miss on the location of a mobile terminal will necessitate a search operation on the network when a call comes in. Such an operation, again, requires the expenditure of limited wireless resources. The goal of mobility tracking, or location management, is to balance the registration (location update) and search (paging) operation, so as to minimize the cost of mobile terminal location tracking.

Two simple location management strategies are the *Always-Update* strategy, and the *Never-Update* strategy. In the *Always-Update* strategy, each mobile terminal performs a location update whenever it enters a new cell. As such, the resources used (overhead) for location update would be high. However, no search operation would be required for incoming calls. In the *Never-Update* strategy, no location update is ever performed. Instead, when a call comes in, a search operation is conducted to find the intended user. Clearly, the overhead for the search operation would be high, but no resources would be used for the location update. These two simple strategies represent the two extremes of location management strategies, whereby one cost is minimized and the other maximized. Most existing cellular systems use a combination of the two strategies mentioned previously.

One of the common location management strategy used in existing systems today is the location area (LA) scheme. In this scheme, the network is partitioned into *regions* or LA, with each region consisting of one or more cells (Figure 33.6). The never-update strategy can then be used within each region, with location update performed only when a user moves out to another region/LA. When a call arrives, only cells within the LA for which the user is in needs to be searched. For example, in Figure 33.6, if a call arrives for user X, then search is confined to the 16 cells of that LA. It is recognized that optimal LA partitioning (one that gives the minimum location management cost) is an NP-complete problem [2].

Another location management scheme similar to the LA scheme is suggested in References 2 and 9. In this strategy, a subset of cells in the network is designated as the *reporting cells* (Figure 33.7). Each mobile terminal performs a location update only when it enters one of these reporting cells. When a call arrives, search is confined to the reporting cell the user last reported, and the neighboring bounded nonreporting cells. For example, in Figure 33.7, if a call arrives for user X, then search is confined to the reporting cell the user last reported in, and the nonreporting cells marked **P**. Obviously, certain reporting cells configuration leads to unbounded nonreporting cells. It was shown in Reference 44 that finding an optimal set of reporting cells, such that the location management cost is minimized, is an NP-complete problem. In Reference 18, a heuristic method to find near optimal solutions is described.

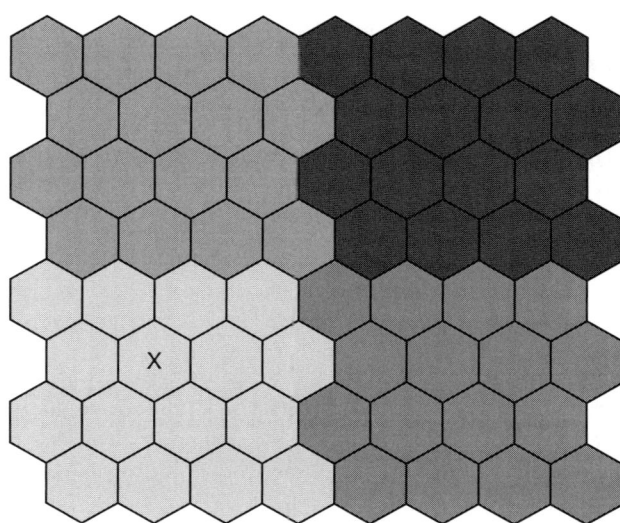

FIGURE 33.6 A 64-cell network with four LAs.

An Overview of Mobile Computing Algorithmics

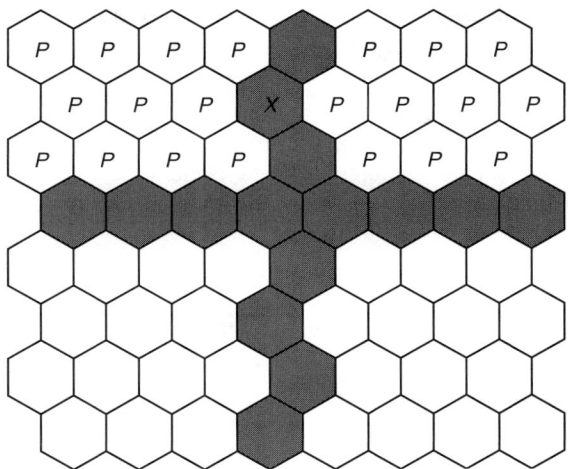

FIGURE 33.7 A 64-cell network with reporting cells (shaded areas represent reporting cells).

33.2.2.1 Location Management Cost

To be able to effectively compare the different location management techniques available, one needs to associate with each location management technique a value, or cost.

As previously noted, location management involves two elementary operations of *location update* and *location inquiry*, as well as *network interrogation* operations. Clearly, a good location update strategy would reduce the *overhead* for location inquiry. At the same time, location updates should not be performed excessively, as it expends on the limited wireless resources.

To determine the *average cost* of a location management strategy, one can associate a cost component to each location update performed, as well as to each polling/paging of a cell. The most common cost component is the wireless bandwidth used (wireless traffic load imposed on the network). That is, the wireless traffic from mobile terminals to base stations (and vice versa) during location updates and location inquiry.

The total cost of the aforementioned two cost components (location update and cell paging) over a period of time T, as determined by simulations (or analytically) can then be averaged to give the average cost of a location management strategy [45].

For example, the following simple equation can be used to calculate the total cost of a location management strategy:

$$\text{Total cost} = C \cdot N_{\text{LU}} + N_{\text{P}} \tag{33.1}$$

where N_{LU} denotes the number of location updates performed during time T, N_{P} denotes the number of paging performed during time T, and C is a constant representing the cost ratio of location update and paging. The aforementioned cost formula can be used to compare the efficiency of different location management techniques. Several things, however, should be noted:

- The more complex location management strategy will almost always require more computational power at the mobile terminal, the system, or both. It may also require greater database cost (record size, etc.). This part of the location management cost is usually ignored, as it is hard to quantify.
- The cost of location update is usually much higher than the cost of paging—up to several times higher [44], mainly owing to the need to set up a "signaling" channel.
- Depending on the update strategy used, the total number of location updates performed over a period T may depend on the mobile users' *mobility* and *call arrival* patterns.

The total cost of location inquiry over a period T would undoubtedly depend on the number of calls received by the users, that is, on the users' call arrival patterns. It may also be influenced by the users' mobility pattern.

Noting the previously mentioned issues, it is clear that any simulation result would be strongly influenced by users' mobility and call arrival patterns chosen for the simulation [44]. Several users' mobility and call arrival patterns commonly used to determine the effectiveness of a location management strategy is discussed in the following text.

33.2.2.2 User Mobility Patterns

In the area of location management, two mobile user's patterns are of interest: their mobility pattern and their call arrival pattern. Several approaches have been proposed in the literature to model a mobile user's movement pattern. These include

Memoryless movement model: The user's next cell location does not depend on his or her previous cell location. That is, the next cell location is selected with equal probability from the neighboring cells.

Markovian model: In this model, neighboring cells have different probability of being the user's next cell location, which depends on the set of cell(s) the user has visited. Other probability models have also been proposed [44].

Gauss-Markov model: The model, used in Reference 44, captures some essential characteristics of real mobile users' behavior, including the correlation of users' velocity in time. In the extreme cases, the Gauss-Markov model simplifies to the Memoryless movement model, and constant velocity fluid model (described in the following text).

Shortest distance model: In this model [44], users are assumed to follow a shortest path from source to destination. At each intersection, a user chooses a path that maintains the shortest distance assumption. The model is particularly suited for *vehicular* traffic, whereby each user has a source and destination. Under such condition, the shortest distance assumption is certainly reasonable.

Activity-based model: This relatively new model is developed to try to reflect the actual observed users' mobility patterns as closely as possible.

With portable communication devices becoming more and more common in everyday life, simple probability model no longer accurately reflects reality, as there is now a whole range of users with very different mobility patterns.

The central concept of an activity-based model is that of *activity*. Each activity is associated with several parameters, including time of day, duration, as well as location. New activities are then selected on the basis of previous activities and time of day. When a new activity is selected, it is once again assigned a duration and location. All of these parameters can be obtained from a population's survey. Activity-based model is discussed in more detail in Reference 44.

Actual mobility trace of users in a cellular network can also be used for simulation, or a trace generator based on actual mobility traces can be created. Output from such a generator can be obtained from Reference 44.

While the aforementioned models describe individual user's mobility, there are also models that describe system-wide (macroscopic) movement behavior. These include

Fluid model: Here, mobile users' traffic flow is modeled as fluid flow, describing the macroscopic movement pattern of the system. The model is suitable for vehicular traffic, where users do not make regular stops and interruptions.

Gravity model: In this model, movement between two sites i, j is a function of each site's gravity P_i, P_j (e.g., population), and an adjustable parameter $K(i, j)$.

While there have been many approaches suggested to model "human movement," there is only one model commonly used for call arrivals. The commonly implemented model comes from queuing theory, that of Poisson arrival rate, or exponential interarrival rate. Data suggests that macroscopically (in aggregate), the Poisson call arrival rate accurately reflects call arrivals in existing cellular networks. However, individual user may not have Poisson call arrival rate. For individual user, other factors come into play, such as time of day.

In order to overcome the problem, actual call arrival trace of an existing cellular network can be used for simulation. As in the case of mobility pattern, a trace generator based on actual call arrival trace can also be created [44].

33.2.2.3 Location Queries

Several location inquiry strategies have been proposed in the literature. Note that some of these location inquiry strategies are not applicable to the location update strategies discussed earlier. These are detailed as follows:

In its simplest form, locating a mobile terminal can be done by simultaneously paging all the cells within the network. This technique will also take the least time to locate a mobile terminal. However, this technique will result in enormous signaling traffic, particularly for moderate to large networks [44].

Note, though, that in the case of LA strategies, the simultaneous polling needs to be done only on the cells within a LA, as the required mobile terminal would be in a cell within its LA.

One simple improvement to simultaneously paging all the cells within the network is to group cells into *paging areas* (PA). Cells within each paging area are polled simultaneously. Each paging area can then be polled sequentially, until the required user is found. The order of PA's polling can be done in several ways:

Arbitrarily: This is undoubtedly the easiest to implement, but may not be necessarily efficient.
Geographical knowledge: For example, if a user lives on an island, may start by polling all PA within the island.

For the probability of locating a mobile user within each paging area—start polling paging area with the highest probability. This method can be classified under the *intelligent paging method*, discussed later in this chapter.

It should be noted, however, that this method would result in time delay in locating a mobile user, especially in large networks. However, the size of the PA can be easily adjusted to accommodate any delay constraint, or QoS requirements of the network. Paging under delay constraints is discussed in Reference 44.

In the case of a network with LA, each paging area can correspond to a LA. Alternatively, one LA may contain several PA. The number of PA within a LA can be determined, among other things, by the maximum delay constraint. One example of paging area partitioning within each LA is described in Reference 1.

There is also a variant discussed in Reference 8, whereby each mobile terminal performs a location update whenever it crosses a paging area boundary. However, the signaling is only limited to the Mobile Switching Centre (MSC), and therefore does not "affect" the system's traffic. Still, this introduces extra signaling traffic from the location update operations.

This technique tries to improve on the sequential paging of PA described earlier. Using this technique, time-consuming page area planning can be avoided, and the high time delay that may be incurred using the paging area technique previously described can be avoided.

In this technique, the last known cell location (or other cell deemed to have the highest location probability) of a mobile terminal is paged first. On a miss, all the cells surrounding the last known cell are paged. This "ring of cell" paging continues until the required mobile terminal is found. In the case of networks with LA, the "ring of cell" can be confided to within the LA. Note that the complexity for this technique is also greater than the simple paging area technique described earlier, as the "ring" must be computed, or predefined for each cell.

Other than the simultaneous networkwide search, the other paging methods described earlier sequentially pages group of cells, until the required user is found. Which cells are to be paged, and in which order—so as to minimize the paging cost, however, remains an open problem. Ideally, we want to page the cell(s) in the order of location probability (with cells that have the highest probability of being paged first). In a sense, we would need to *predict* the current location of the user. In order to achieve this, many factors should be taken into account, such as geographical, user's mobility pattern, and time.

This extended paging strategy has been studied extensively. Successful find in the first paging operation ensures limited time delay in finding a mobile terminal, thus maintaining a required QoS. In order to minimize such delay, the probability of success in the first paging step should be sufficiently high—the whole aim of intelligent paging.

Undoubtedly, the added "intelligence" leads to a rather complex system with considerable computing and memory requirement. However, very high paging cost savings have been reported for this paging method [44], as evident by the high success in the first paging operation. It was also suggested in Reference 44 to use intelligent paging on specific classes of mobile users, instead of applying it to all users. That is, since some users are simply unpredictable, then to reduce the computational power required, it is suggested to use the more conventional methods of paging on these users.

This new, so-called intelligent paging method is very general in nature, and can certainly be used in any location update strategies. In particular, its use in LA planning allows for larger-size LA implementation, resulting in fewer location update operations.

33.3 Low Earth Orbiting Satellite Networks

In response to the increasing demand for truly global coverage needed by personal communication services (PCS), a new generation of mobile satellite networks intended to provide *anytime–anywhere* communication services was proposed in the literature [3,10,27]. Low Earth Orbiting (LEO) mobile satellite networks, deployed at altitudes ranging from 500 to 2000 km, are well suited to handle bursty Internet and multimedia traffic and to offer anytime–anywhere connectivity to mobile users. LEO satellite networks offer numerous advantages over terrestrial networks including global coverage and low cost-per-minute access to end users equipped with handheld devices. Since LEO satellite networks are expected to support real-time interactive multimedia traffic, they must be able to provide their users with QoS guarantees including bandwidth, delay, jitter, call dropping, and call blocking probability (CBP).

While providing significant advantages over their terrestrial counterparts, LEO satellite networks present protocol designers with an array of daunting challenges, including handoff, mobility, and location management [8,32]. Because LEO satellites are deployed at low-altitude, Kepler's third law implies that these satellites must traverse their orbits at a very high speed. The coverage area of a satellite—a circular area of the surface of the Earth—is referred to as its *footprint*. For spectral efficiency reasons, the satellite footprint is partitioned into slightly overlapping cells, called *spotbeams*. Refer to Figure 33.8 for an illustration. As their coverage area changes continuously, in order to maintain connectivity, end users must switch from spotbeam to spotbeam and from satellite to satellite, resulting in frequent intra- and inter-satellite handoffs. Owing to the large number of handoffs experienced by a typical connection during its lifetime, resource management and connection admission control are very important tasks if the system is to provide fair bandwidth sharing and QoS guarantees. In particular, a reliable handoff mechanism is needed to maintain connectivity and to minimize service interruption to on-going connections, as end users roam about the system. In fact, one of the most important QoS parameters for LEO satellite networks is the *call dropping probability* (CDP), quantifying the likelihood that an on-going connection will be force-terminated owing to an unsuccessful handoff attempt.

33.3.1 Related Work

In wireless mobile networks, the radio bandwidth is shared by a large number of users. An important property of the network is that a user would change its access points several times. This fact causes

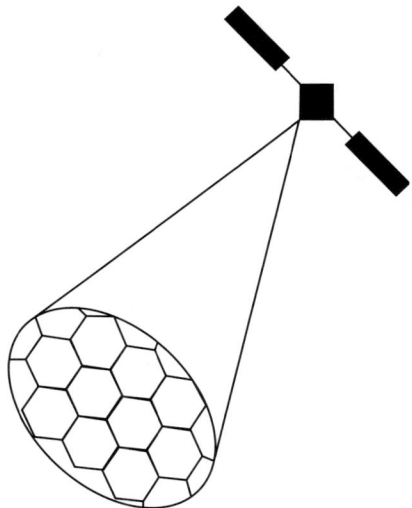

FIGURE 33.8 Illustrating the partition of the footprint into cell-like spotbeams.

TABLE 33.1 Recent Admission Control Strategies for LEO Systems

Papers	Multiservices	Reservation strategy	Resource allocation	Necessary information	CAC criteria	QoS issues
[9]	No	FCA	DCA	Residual time	Local	No
[27]	Yes	Fixed	FCA	Residual time	Local	No
[47]	No	No	FCA	User location	Threshold	Yes
[7]	No	Adaptive	DCA	User location	Local	No
[11]	Yes	Probabilistic	FCA	Residual time	Local and nonlocal	Yes
[46]	Yes	Predictive	DCA	Residual time	Local and nonlocal	Yes
[33]	Yes	Adaptive	DCA	User location	Local and nonlocal	Yes

technical problems in which fair sharing of bandwidth between handoff connections and new connections is required. Resource allocation and call admission control (CAC, for short) are very important tasks that need to be performed efficiently in order to achieve high bandwidth utilization and QoS provisioning in terms of low dropping probability for handoffs and reasonably low blocking probability for new connections.

In the remainder of this section, we survey a number of call admission algorithms proposed in the literature. We refer to Table 33.1 for an at-a-glance comparison between several resource allocation strategies and CAC algorithms for LEO satellite networks found in the literature.

Del Re et al. [9] proposed a mobility model and different resource allocation strategies for LEO satellite networks. The queuing of handoff request is proposed aiming to reduce the handoff blocking. The dynamic channel allocation scheme is expected to be the best scheme to optimize system performance.

Later, Mertzanis et al. [28] introduced two different mobility models for satellite networks. In the first model, only the motion of satellite is taken into account, whereas in the second model, other motion components like earth rotation and user movements are considered. To design a CAC algorithm for mobile satellite systems, the authors introduced a new metric called "mobility reservation status," which provides the information about the current bandwidth requirements of all active connections in a specific spotbeam in addition to the "possible" bandwidth requirements of mobile terminals currently connected to the neighboring spotbeams. A new call request is accepted in the spotbeam where it originated, say m, if there is sufficient available bandwidth in the spotbeam and the mobility reservation status of particular neighboring spotbeams have not exceeded a predetermined threshold T_{NewCall}. If a new call is accepted, the mobility reservation status of a particular number S of spotbeams will be updated. A handoff request is accepted if bandwidth is available in the new spotbeam and also the handoff threshold (T_{HO}) is not exceeded. The key idea of the algorithm is to prevent handoff dropping during a call by reserving the bandwidth in a particular number S of spotbeams that the call would handoff into. The balance between new call blocking and handoff call blocking depends on the selection of predetermined threshold parameters for new and handoff calls. However, during simulation implementation, we found that the scheme has a problem of determining threshold points for the case of LEO satellite networks.

Uzunalioglu [47] proposed a call admission strategy based on the user location. In his scheme, a new call is accepted only if the handoff CBP of the system is below the target blocking rate at all times. Thus, this strategy ensures that the handoff blocking probability averaged over the contention area is lower than a target handoff blocking probability P_{QoS} (QoS of the contention area). The system always traces the location of all the users in each spotbeam and updates the user's handoff blocking parameters. The algorithm involves high processing overhead to be handled by the satellite, unsuitable for high capacity systems where a satellite footprint consists of many small-sized spotbeams, each having many active users.

Cho [7] employs user location information as the basis for adaptive bandwidth allocation for handoff resource reservation. In a spotbeam, bandwidth reservation for handoff is allocated adaptively by calculating the possible handoffs from neighboring beams. A new call request is accepted if the beam where it originated has enough available bandwidth for new calls. The reservation mechanism provides a low handoff blocking probability compared to fixed guard channel strategy. However, the use of location information in handoff management suffers from the disadvantage of updating locations, which results in high processing load to onboard handoff controller, thereby increasing the complexity of terminals. The method seems suitable only for fixed users.

El-Kadi et al. [11] proposed a probabilistic resource reservation strategy for real-time services. They also introduced a novel call admission algorithm where real-time and nonreal-time service classes are treated differently. The concept of *sliding window* is proposed in order to predict the necessary amount of reserved bandwidth for a new call in its future handoff spotbeams. For real-time services, a new call request is accepted if the spotbeam where it originated has available bandwidth and resource reservation is successful in future handoff spotbeams. For nonreal-time services, a new call request is accepted if the spotbeam where it originated satisfies its maximum required bandwidth. Handoff requests for real-time traffic are accepted if the minimum bandwidth requirement is satisfied. Nonreal-time traffic handoff requests are honored if there is some residual bandwidth available in the cell.

Todorova et al. [46] proposed a selective look-ahead strategy, which they call SILK, specifically tailored to meet the QoS needs of multimedia connections where real- and nonreal-time service classes are treated differently. The handoff admission policies introduced distinguish between both types of connections. Bandwidth allocation only pertains to real-time connection handoffs. To each accepted connection, bandwidth is allocated in a look-ahead horizon of k cells along its trajectory, where k is referred to as the depth of the look-ahead horizon. The algorithm offers low CDP, providing for reliable handoff of on-going calls and acceptable CDP for new calls.

More recently, Olariu et al. [33] proposed Q-Win, a novel admission control and handoff management strategy for multimedia LEO satellite networks. A key ingredient in Q-Win is a novel predictive resource allocation protocol. Extensive simulation results have confirmed that Q-Win offers low CDP, acceptable CBP for new call attempts, and high bandwidth utilization. Q-Win involves some processing overhead. However, as the authors show, this overhead is transparent to the MHs, being absorbed by the onboard processing capabilities of the satellite. Their simulation results show that Q-Win scales to a large number of users.

33.3.2 A Survey of Basic Medium Access Control Protocols

Medium Access Control (MAC) protocols have been developed for different environments and applications. A survey of MAC protocols for wireless ATM networks is given in Reference 3. For a survey of MAC protocols for satellite networks, we refer the reader to Peyravi [35,36].

MAC protocols classification proposed in Reference 35 is based on the static or dynamic nature of the channel, the centralized or distributed control mechanism for channel assignments, and the adaptive behavior of the control algorithm.

According to this classification, one can define five groups of MAC protocols:

1. Fixed assignment protocols
2. Demand assignment protocols
3. Random access (RA) protocols
4. Hybrid of RA and reservation protocols
5. Adaptive protocol

Fixed and demand assignment protocols are contention-free protocols using static allocation (fixed assignment) or dynamic allocation (demand assignment).

The last three groups of protocols are contention-oriented protocols, where collision takes place on either the data channel (RA) or signaling channel (hybrid of RA and reservation).

In *Fixed Bandwidth Assignment* (FBA) schemes, an earth terminal is given a fixed number of slots per frame for the duration of the connection. The technique has the advantage of simplicity, but lacks flexibility and reconfigurability.

The main disadvantage of this solution is that during idle periods of a connection, slots go unused. Fixed allocation of channel bandwidth leads to inefficient use of transponder capacity. Because of the constant assignment of capacity, such techniques are not suitable for VBR applications.

In *Demand Assignment Multiple Access* (DAMA) schemes, the capacity of the uplink channel is dynamically allocated on demand in response to stations' request on the basis of their queue occupancies. Therefore, the time varying bandwidth requirements of the stations can be accommodated and no bandwidth will be wasted. Dynamic allocation using reservation (implicit or explicit) increases transmission throughput. Real-time VBR applications can use this scheme. DAMA does not presuppose any particular physical layer transmission format. It can be implemented using TDMA, FDMA, and CDMA.

DAMA protocols consist of a phase for bandwidth request and a phase of data transmission. The main disadvantage of this solution is the long delay experienced by the data in the queues from the time the stations send bandwidth requests to the satellites until the time the data is received in the destination. These protocols also have limited dynamical characteristics by QoS control and adaptation to bursty Internet and multimedia traffic.

In RA schemes [3,5], all the slots of the frame are available to all the users. Each station can send its traffic over the randomly selected slot(s) without making any request. There is no attempt to coordinate the ready stations to avoid collision. In case of collision, data will be corrupted and have to be retransmitted.

The technique is simple to implement and adaptive to varying demands, but very inefficient in bandwidth utilization. The resulting delays are, in most of the cases, not acceptable for real-time traffic.

Hybrid MAC schemes take the good features of different techniques to improve QoS. Hybrid protocols derive their efficiency from the fact that reservation periods are shorter than transmission periods. Some examples of Hybrid MAC schemes are

> *Combined Random/Reservation* schemes [6] combine DAMA with RA. Main advantage is the very good bandwidth utilization at high loads and low delay. The disadvantage of the technique proposed is the necessity to monitor the RA part of the channel for possible collisions, which leads to large processing time.

In the *Combined Fixed/Demand Assignment* schemes [7], a fixed amount of uplink bandwidth is always guaranteed to the stations. The remaining bandwidth is assigned by DAMA. The efficiency of the technique depends on the amount of fixed bandwidth allocated to the stations.

In *Combined Free/Demand Assignment* (CFDAMA) scheme, the reserved slots are assigned to the stations by DAMA based on their demands. The unused slots are distributed in a round-robin manner on the basis of weighting algorithms [6].

In the *Adaptive Protocol*, the number of the contenders is controlled to reduce the conflicts, or the channel switches between RA mode and reservation mode depending on the traffic load. Adaptive protocols are a wide area with many possible solutions having different advantages and disadvantages. In respect of the previously described design aspects for MAC protocols for satellite communication, these protocols could not be suitable for satellite communication because of the high implementation complexity [9–13].

33.3.3 Mobility Model and Traffic Parameters

A mobility model provides statistical estimates for residency time as well as for the location and time of the next handoff. Several mobility models for LEO satellite networks were proposed in the recent literature [9,27]. As a rule, these models consider only satellite movement neglecting the effects of Earth rotation and user mobility.

33.3.3.1 The Linear Mobility Model

Although several mobility models exist for LEO satellites [20,31] it is customary to assume a one-dimensional mobility model where the end users move in straight lines and at a constant speed, essentially the same as the orbital speed of the satellite [9,10]. We assume an orbital speed of 26,000 km/h. All the spotbeams (also referred to as cells) are identical in shape and size. Although each spotbeam is circular, we use squares to approximate spotbeams (we note that some authors use regular hexagons instead of squares). The diameter of a cell is taken to be 425 km [9,10]. Further, MHs traverse each cell along its maximum diameter. Thus, the time t_s it takes a MH to cross a cell is roughly 65 s.

The diameter of a cell is taken to be 425 km [9,10]. Further, end users traverse each cell along its maximum diameter. Thus, the time t_s it takes an end user to cross a cell is, roughly, 65 s. The end user remains in the cell where the connection was initiated for t_f time, where t_f is uniformly distributed between 0 and t_s. Thus, t_f is the time until the first handoff request, assuming that the call does not end in the original cell. After the first handoff, a constant time t_s is assumed between subsequent handoff requests until call termination.

As illustrated in Figure 33.9, when a new connection is requested in cell N, it is associated with a *trajectory*, consisting of a list $N, N+1, N+2, \ldots, N+k$, of cells that the connection may visit during its lifetime. The holding time of connections is assumed to be exponentially distributed with mean $1/\mu$. Assume that connection C was accepted in cell N. After t_f time units, C is about to cross into cell $N+1$. Let p_f be the probability of this first handoff request. El-Kadi et al. [11] have shown that

$$p_f = \frac{1 - e^{-\mu t_s}}{\mu t_s}.$$

Moreover, the probability of the $(k+1)$-th handoff request is $e^{-\mu t_s}$, which, as expected, is independent of k [11]. Consequently, we will let $p_s = e^{-\mu t_s}$ denote the probability of a *subsequent* handoff request. It is important to note that t_f, t_s, p_f, and p_s are mobility parameters that can be easily evaluated by the satellite using its onboard processing capabilities.

33.3.3.2 A Two-Dimensional Mobility Model

While providing a simple model to work with, this mobility model is not realistic. We now review the 2D mobility model proposed by Nguyen et al. [31]. In this mobility model, the

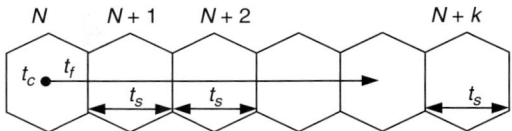

FIGURE 33.9 Illustrating the linear mobility model.

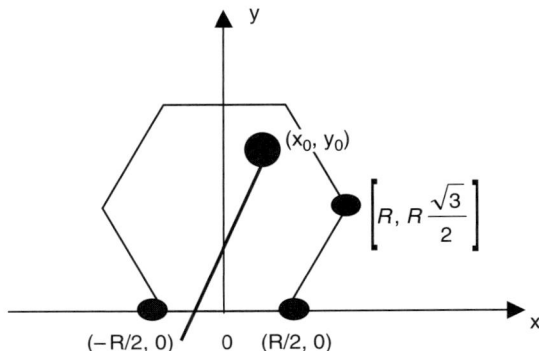

FIGURE 33.10 User mobility parameters.

combined effect of two factors is considered: satellite motion and the Earth rotation. Specifically, we assume that

- The satellite coverage areas (footprints) move on the Earth surface with a speed of $V_{sat} = 26,000$ km/h.
- The footprints are partitioned into spotbeams, each approximated by a regular hexagon.
- "Y-axis" handoffs occur due to the very high speed of LEO satellites.
- The rotation of the Earth and user motion cause "x-axis" handoffs. Because the Earth rotation speed (approximately $V_{earth} = 1600$ km/h at the Equator) is much higher than the velocity of ground users (about 100 km/h for cars or trains), the effect of user mobility is ignored in our model too.

Referring to Figure 33.10, user mobility has two components: V_y and V_x, where V_x is the Earth rotation speed (V_{earth}) while V_y is the satellite rotation speed (V_{sat}). As illustrated in Figure 33.11, the speed of the mobile user is

$$V_{user} = \sqrt{V_x^2 + V_y^2}$$

Consider an arbitrary hexagonal spotbeam and the coordinate system illustrated in Figure 33.12. Let R be the radius of the enclosing circle. Let (x_0, y_0) be the position of the user. We are interested in determining *where* and *when* the next handoff will take place. This involves determining the identity of the cell into which the user will migrate, as well as the moment in time when the handoff occurs.

Let λ be the (slope of the) actual direction of movement of our user. Clearly,

$$\lambda = tg\theta = \frac{V_Y}{V_X}.$$

The equation of the line describing the movement of the user is

$$y - y_0 = \lambda(x - x_0). \tag{33.2}$$

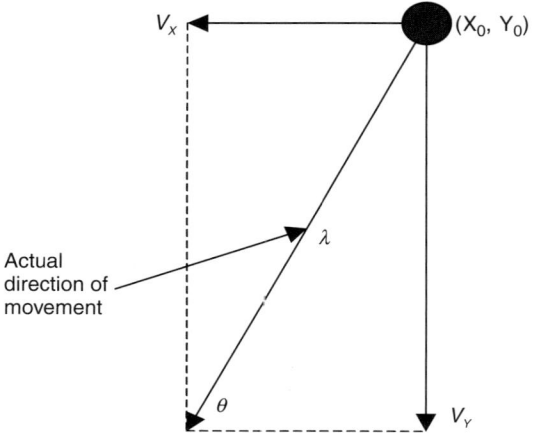

FIGURE 33.11 Illustrating mobility parameters.

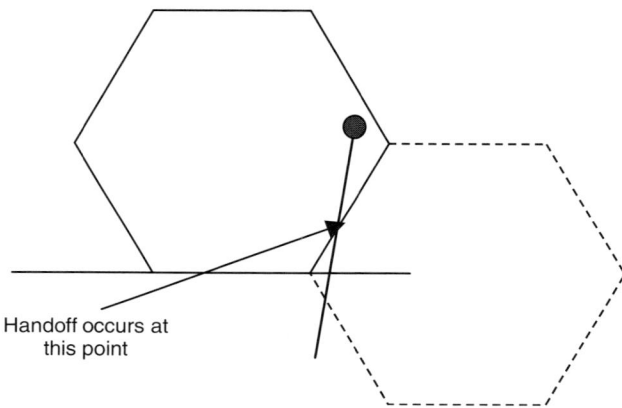

FIGURE 33.12 Illustrating the location of handoff in case (B).

To see where the next handoff takes place, we compute the intersection of this line with the x-axis by setting

$$-y_0 = \lambda(x - x_0).$$

In turn, this yields

$$x = x_0 - \frac{y_0}{\lambda}. \tag{33.3}$$

Equation 33.3 has the following interpretation:

If $-R/2 \leq x \leq R/2$, handoff will occur into the cell *below* the given cell. The time of the next handoff is

$$t_{\text{next}} = \frac{y_0}{V_Y}.$$

If $x > R/2$ then, as illustrated in Figure 33.11, the actual place where the handoff takes place is the intersection of the line in Equation 33.1 with the line

$$y = \sqrt{3}\left(x - \frac{R}{2}\right)$$

More precisely, the coordinates of the handoff point are

$$x = \frac{\lambda x_0 - y_0 - R\frac{\sqrt{3}}{2}}{\lambda - \sqrt{3}}$$

$$y = \frac{\sqrt{3}(\lambda x_0 - y_0 - \frac{R\lambda}{2})}{\lambda - \sqrt{3}}$$

The time to the next handoff can be computed easily as

$$t_{\text{next}} = \frac{x_0\sqrt{3} - y_0 - \frac{R\sqrt{3}}{2}}{V_X(\lambda - \sqrt{3})}.$$

If $x < -R/2$, then the actual place of handoff is symmetric to (B) and is not repeated here. Similarly, the time until the next handoff can be derived in a straightforward manner.

33.4 Resource Management and Call Admission Control Algorithm

We will now show how the 2D mobility model discussed earlier [31] can be used to derive an efficient resource management and CAC algorithm for LEO satellite networks.

33.4.1 Resource Management Strategy

We distinguish two different service classes: real-time (Class 1) and nonreal-time (Class 2) connections. A Class 1 connection is characterized by minimum bandwidth ($B_{\min}$) and maximum bandwidth ($B_{\max}$) requirement. Class 2 traffic only has a required initial bandwidth (B_{init}).

1. During their lifetime, Class 1 connections are guaranteed at least the QoS provided when they entered the network. In the case of radio resource management, the QoS is a value between $B_{\min}$ and $B_{\max}$.
2. Class 2 connections can take free bandwidth not used by Class 1 connections. In fact, Class 2 traffic is not time-critical and can easily tolerate fluctuations in the bandwidth they receive. At setup time, each Class 2 connection is allocated at least B_{init} bandwidth. However, a Class 2 handoff is accepted as long as there is some available bandwidth in the new spotbeam.

Probabilistic bandwidth reservation strategies for handoff traffic have been reported in the literature [25,30]. Figure 33.13 illustrates the details of our Sequential Probabilistic resource Reservation (SPR).

Assume that a Class 1 handoff request is accepted in a spotbeam i at time t_i and might handoff in the future to the spotbeam j. An amount of the resource B_R is reserved for a Class 1 connection in the

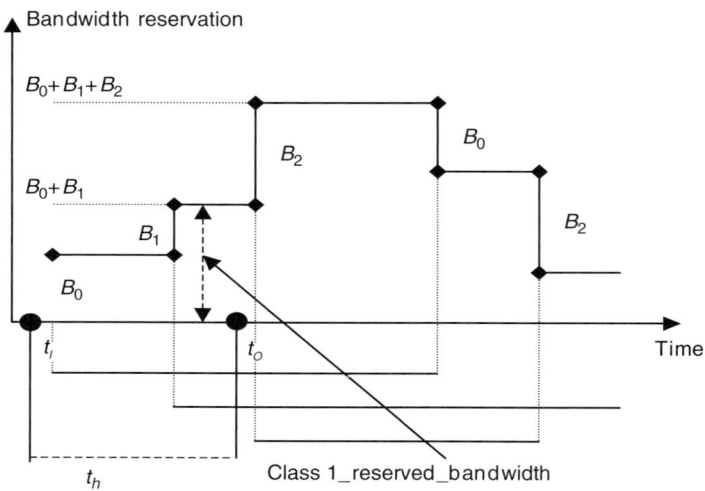

FIGURE 33.13 Illustrating the resource reservation strategy.

spotbeam j during the time $[t_i, +\tau_i, t_i, +\tau_i + \tau_j]$, where τ_i and τ_j are the dwelling times of the connections in the spotbeam i and j, respectively, estimated according to proposed mobility model:

$$B_R = P_R \times B_C$$

P_R is the probability that the user is active in spotbeam i until the next handoff to spotbeam j and B_C is the initial bandwidth assigned when a new call is accepted.

We denote t_C as the lifetime of the connection since it was accepted until it enters the spotbeam i. Assuming that the distribution of the call holding time to be exponential with mean $1/\mu$, then we have

$$P_R = P[T > (t_C + \tau_i)] = e^{-\mu(t_C + \tau_i)}$$

33.4.2 Connection Admission Control Algorithm

The following pseudocode captures the essence of our CAC algorithm:

IF *new-connection-request* **THEN**
    IF Class 1 **THEN**
      1st criterion:
      check if:
      available_bandwidth >= *min_required_bandwidth*
      2nd criterion:
      Try to reserve bandwidth in the next spotbeam.
      Accept the connection if reservation is successful
      Allocate *initial_bandwidth* **B**$_C$
    ELSE /* Class 2*/
      IF *available_bandwidth* >= *initial_required_bandwidth* **THEN**
      Accept the request
      Allocate *initial_required_bandwidth*
ELSE /* handoff request*/
    IF Class 1 **THEN**
      IF *free_bandwidth* > *initial_bandwidth* **THEN**

```
            Accept the handoff request
            Allocate:
            min{max_required_bandwidth, free_bandwidth}
            Try to reserve bandwidth in the next spotbeam
    ELSE
            Reject the request
    ELSE /* Class 2*/
        IF available_bandwidth >0 THEN
            Allocate
        min{available_bandwidth,initial_required_bandwidth}
ELSE
            Reject the handoff request
```

Our CAC algorithm decides whether to accept a new connection request or not depending on different criteria for different service classes. The algorithm decides an admission depending on spotbeam bandwidth status. The amount of *available_bandwidth* in a spotbeam is computed as the difference *free_bandwidth-reserved_bandwidth*.

The amount of *reserved_bandwidth* is calculated as shown in Figure 33.13. Let t_I be the time when a request (new connection or handoff) arrives to a spotbeam of the satellite. The satellite can estimate the interval t_h, that is, how long the connection will reside in the spotbeam. The departure time t_O is obtained and thus the *reserved_bandwidth* can be estimated:

- For a Class 1 new connection request, accept it if available bandwidth is larger than the minimum requirement and a *successful reservation* is done in the next spotbeam.
- *Successful reservation* means that reservation amount in the next handoff spotbeam at the time (currently) does not exceed a predetermined percentage of the free bandwidth in the spotbeam (in our simulations, we chose 80%).
- For the Class 2 connection, the *initial* amount of bandwidth is at least guaranteed in the originating spotbeam. However, after a handoff request the amount of bandwidth received by the connection could be drastically reduced. This, in turn, translated into increased delays, which are easily tolerated by Class 2 connections.

33.4.3 Performance Evaluation

The performance of the proposed resource reservation strategy and new call admission algorithm was obtained by discrete-event simulation under studied traffic and system parameters. In our simulation experiments, there are 36 spotbeams, divided to three arrays. Homogeneous traffic is assumed and new call request is Poisson distribution. The connection duration is exponentially distributed. System parameters are described in Table 33.2, on the basis of the well-known Iridium satellite system [10]. In Figures 33.14 and 33.15, simulation results of two different resource management strategies are presented: fixed reservation (FRES) [34] and SPR.

TABLE 33.2 Simulation Parameters

System parameters				
Spotbeam	Radius	Capacity	Speed	
	212 km	5000 Kbit	26000 km/h	
	Maximum bandwidth	Minimum bandwidth	Initial bandwidth	Mean duration
Service parameters				
Class 1	30 Kbits/s	20 Kbit/s	10 Kbits/s	180s
Class 2				60s

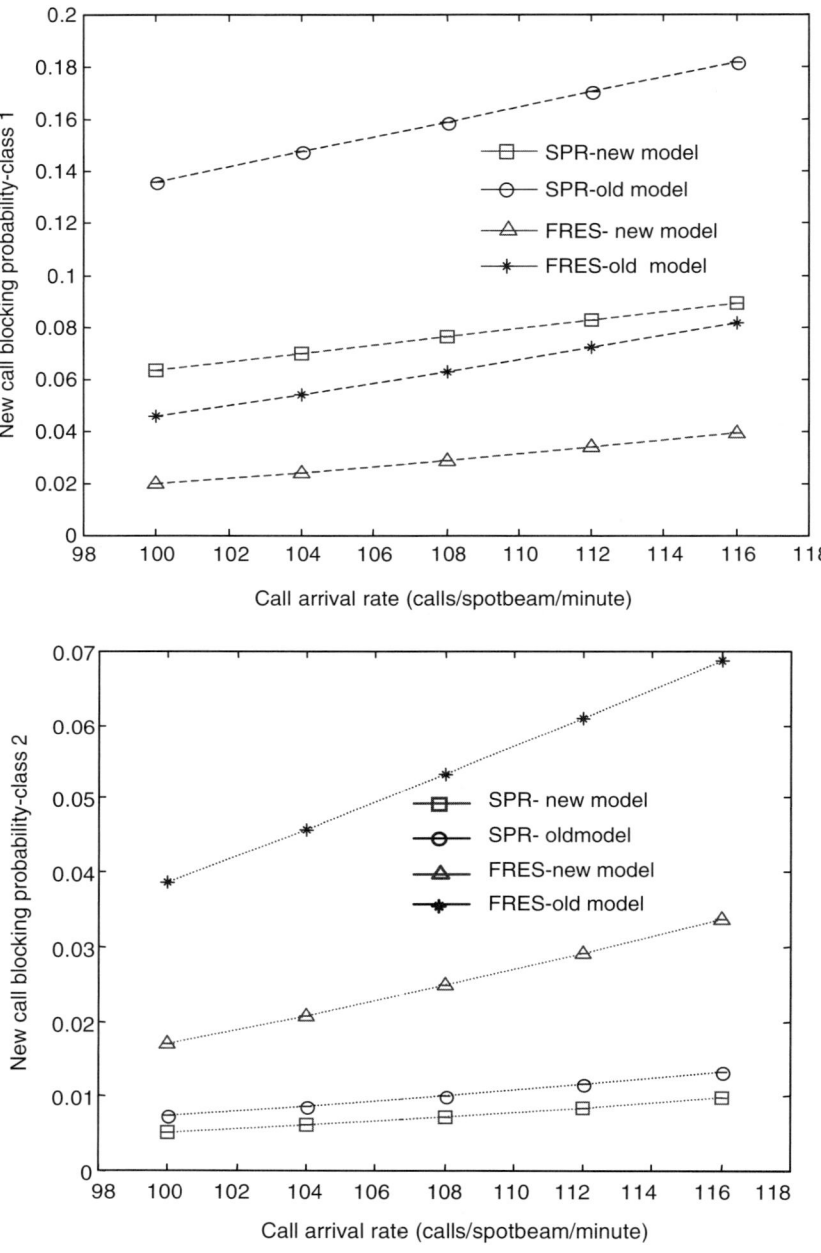

FIGURE 33.14 New connection blocking probability.

We assume that each spotbeam has a fixed amount of radio capacity (FCA). In the FRES strategy, 3% of spotbeam capacity is reserved only for Class 1 handoff traffic. We also compare the impact of mobility models on network performance. The well-known one-dimensional mobility model and our novel mobility model are used in the simulation experiments.

Our simulation results show that there are significant differences between the performance evaluated according to the novel mobility (*new*) model and the existing mobility (*old*) model, showing that the rotation of the Earth should not be ignored when modeling mobility and evaluating network performance.

An Overview of Mobile Computing Algorithmics

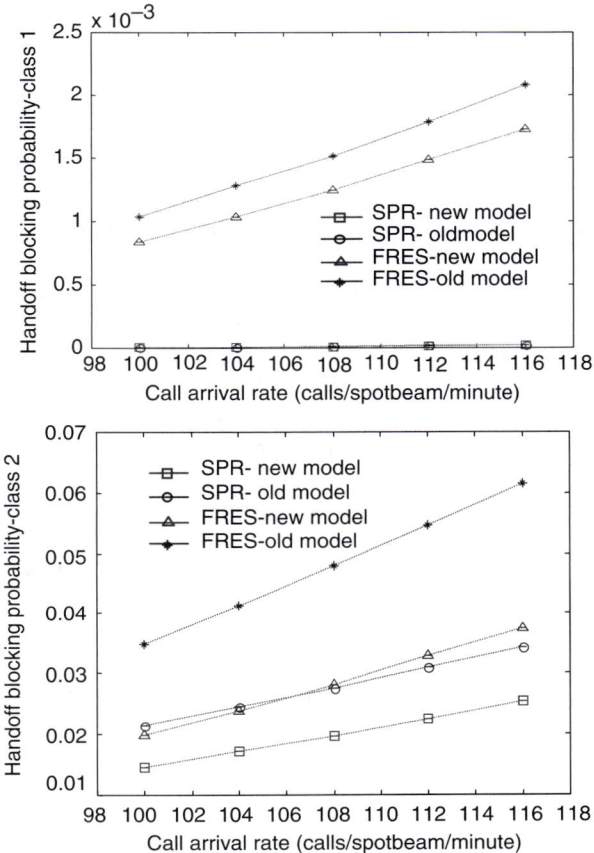

FIGURE 33.15 Handoff blocking probability.

When compared with the FRES strategy, our SPR strategy features a higher new CBP for Class 1 service. However, the new CBP of Class 2 service obtained by SPR is much lower. Our reservation strategy gains also extremely high performances in terms of CBP for both service classes, and especially for Class 1 CBP (10^{-4}). This is because the SPR reserves bandwidth adaptively and dynamically for Class 1 traffic so that Class 2 services can use the available bandwidth that is not used by Class 1 traffic.

33.5 Mobile Ad Hoc Networks

A mobile ad hoc network (MANET) is a collection of wireless MHs forming a temporary network without the aid of any centralized administration or standard support services that are regularly available on the wide-area network to which the hosts are connected [39,42,43]. As the name suggests, there is no formal structure or topology in the network. It is possible for hosts to move in any manner, change their availability and transmission range at any time or in any situation. This differs from regular mobile networks that have either designated access points, such as a wireless local area network (WLAN), or a base station, such as cellular networks.

A MANET is a packet-based network where short messages are sent between hosts. These networks are capable of both unicast and multicast message transmissions [42]. Unicast messages are messages sent from one host to one other host within the network, and multicast messages are messages sent from one host to many hosts within the network.

A MANET may operate in isolation from other networks or have an interface with a fixed wired network, acting as a "stub" network [42]. This connection to the wired network usually occurs because MHs in a MANET often operate within a limited bandwidth, at most 11 Mbit/s with newer technologies but usually 2–3 Mbit/s, and are not suited to transferring large quantities of data [3,25]. "Stub" networks allow messages to pass if they were originated at or are destined to a host in the MANET, but will not allow other messages into the MANET. This means that "stub" networks will not allow messages from one wired network through the "stub" network to the same or another wired network. "Stub" networks are a wireless local area network (LAN) extension and could possibly have multiple connections to the wired network.

MANETs have industrial and commercial applications that require data exchange between MHs, such as in a seminar, meeting, conference, or lecture. Simply moving the hosts within range of each other can form the network. Emergency operations and other scenarios that require rapidly deployable dynamic data communications can also benefit from using a MANET. As the location of the hosts cannot be trivially determined owing to the lack of infrastructure, MANETs can be particularly useful for some military applications. The location of each host can be determined with the newer forms of MANETs that include global positioning system (GPS) technologies.

A MANET has a few features that differentiate it from more conventional network topologies:

All hosts in a MANET are mobile and act as routers [25]. When each host receives a message it has the ability to determine where it should be retransmitted.

MANETs act as an autonomous system of MHs. This means that MANETs can operate without any specific administration built into the infrastructure.

MANETs may operate in isolation or have an interface to a fixed network, acting as a "stub" or WLAN extensions.

MANET topologies are dynamic and may change with time, as hosts move and the transmission and reception environment changes. MANET topologies may be assumed by the protocol to be flat or hierarchical.

Communication links between hosts may be bi- or unidirectional. A unidirectional communication link is where one host can transmit to a second host, but the second host cannot transmit back. A bidirectional transmission link is where the two hosts can transmit to each other [17]. This occurs due to having heterogeneous hosts, with different transmission properties, and the external environment interfering with the transmission and reception of each host.

Available bandwidth for a mobile network is significantly lower than for wired networks. In addition, there are overhead and physical conditions that reduce the data throughput within the network.

MHs often operate off a limited power supply, such as batteries or unreliable power devices.

MANETs have limited physical security and are prone to unauthorized access. Hosts encrypting data before transmission and using authorization techniques are rectifying this.

33.5.1 Routing

Routing is the transferring information from a source host to a destination host across a network. In most networks, intermediate hosts are required to move the information between the hosts. Hence, the task of moving the information is not trivial and algorithms have been and are continually being researched to improve how information is moved through the network [39,42]. Most routing algorithms operate in the network layer of the Open System Interconnection (OSI) reference model.

As the MHs have limited transmission range, hosts that wish to communicate may be out of range of each other, so intermediate hosts may be required to relay, or retransmit, the message so that it can reach the destination host [38]. Figure 33.16 shows a simple mobile computing network consisting of three hosts (A, B, and C). For example, in the network, where this transmission range of each host is indicated by the corresponding circles, host A cannot send a message directly to host C, as host C is outside of host A's transmission range. The message will need to be sent to host B and then relayed to host C. This is often referred to as multihop networking. As sending of messages may require using multiple hosts, a nontrivial routing problem is introduced [38].

In both wired and wireless networks, there is an ambition to find a route where a message will be successfully sent in the shortest period. In wireless networks, a large portion of the time used when transmitting a message from a source host to a destination host is spent while the message is being analyzed at each intermediate host. For this reason, the fastest route, between a source and a destination, is usually the route that contains the least number of hops (or intermediate hosts). Hence, using the route with the least number of hops will improve network performance.

As the bandwidth of the mobile network is limited, congestion in a network is more often the norm rather than the exception. When this is the case, it is important to maintain a high average ratio of bits transmitted to bits delivered.

As hosts in mobile networks are commonly radio-based, they, in simple terms, transmit data over a specific range or band of frequencies. Interference occurs when a host is simultaneously receiving two messages from two hosts using the same frequency band. The message may be lost at the host that is receiving the two messages simultaneously due to the interference. In the network displayed previously in Figure 33.1, host A and host C can both transmit to host B, but not to each other. Host B can transmit to both host A and host C. Interference occurs when both host A and host C transmit to host B simultaneously. Host B can distinguish that messages have been sent but it cannot understand the messages. Selecting routes that can avoid interference can significantly increase the ratio of bits transmitted to bits delivered.

No matter what protocol or technique is used, it must be able to have distributed operation and provide loop-free routes. In this circumstance, distributed operation means that no one host administers the network, and that any authorized host has the ability to operate in the network, regardless of the other hosts in the network. A loop-free route is a route through the network that does not use any host more than once. It is desirable that the techniques used by the protocol have unidirectional link support, manage bandwidth and power requirements as well as high error rates, and provide a degree of security.

33.5.1.1 Power-Aware Routing

There has been substantial effort to design protocols that deal with power- and cost-aware routing. However, most of these protocols related to power- and cost-aware routing are theoretical models, which have mostly been tested in simulated environments. The development of actual protocols in this area has only recently been looked at in a serious and in-depth manner [41,43].

The aim of power- and cost-aware routing is to increase the network and node life. Recently, several power-aware metrics based on power consumption by routing algorithms were devised and verified. These metrics focused on the adjustment of transmission powers and packet redirection, in order to quantify the usage of power in a mobile network and propose ways of minimizing the power used.

Most of the older routing protocols assumed that every transmission from one node to another would use the same amount of power. On the basis of this assumption, they deduced that a protocol based on the shortest path from one node to another would also be the one that minimizes power usage. However, after

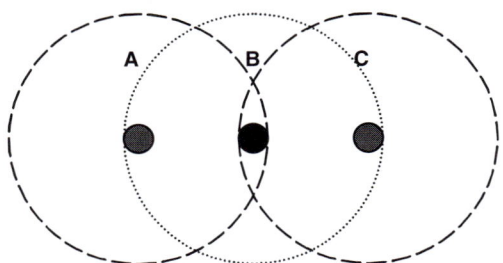

FIGURE 33.16 A simple mobile ad hoc network.

looking at the physics of transmitting radio and other radiating signals in space, in order to communicate between mobile units, it becomes clear that there is a great difference between transmitting over a distance of 10 and 100 m. When this new variability in power usage is taken into account, it becomes clear that using the mobile unit's ability to adjust its transmission power could lead to power being saved by the network and individual nodes. This insight was used in Reference 43 to minimize power usage in their proposed protocol for efficient power-aware routing. In this case, instead of each unit broadcasting at the maximum transmission power when sending packets to another unit, the mobile unit should only transmit with enough energy for the receiving node to correctly read the transmission.

Attenuation of the power used by broadcasting units would also lead to a reduction in the occurrences of collisions. Collisions occur when two or more mobile units attempt to broadcast at the same time in the same area. When this occurs, the receiving node will be unable to detect which node is sending the data; indeed, it will be unable to read any of the transmissions properly. Stojmenovic and Lin [43] propose that if the mobile units only transmit with enough energy to reach the destination unit then the probability of a collision occurring will also be greatly reduced. This will lead to a further saving of power, because in current protocols, when a collision occurs the sending nodes have to retransmit the packet at a later time.

In order for a node to attenuate its broadcasting signal to the appropriate amount of power, it must know how far the receiving node is from it and the minimum amount of energy that it must receive in order to read the broadcast. The work in Reference 43 assumes that each of the mobile units has a GPS available to it and that through this each unit will know exactly where all the other units in the network are located. While this may appear to be a simple solution to the problem, when looked at carefully, it becomes clear that using such GPS equipment would itself cost power. This is clearly counterproductive to the power minimization efforts of the algorithm.

Other researches on the topic have suggested that the protocol should be limited to networks in which every node can directly communicate with every other node in the network directly, without having to route through a number of other nodes [41]. Utilizing this property, every node can communicate with any other node initially at maximum broadcast and determine how far away each node is. While this strategy does require less power usage, it greatly limits the topology allowed for the network and especially the spanning capability of the network. Such networks could only be the size of the maximum broadcasting range of a mobile unit.

Another method of getting around the problem is to have a periodic update of the whole network. This way, the new positions of the nodes can be found and the power attenuated accordingly. Using this method would require that the mobile units not move too rapidly because if the units moved too quickly, the periodic update would have to occur quite frequently. From experimental results, it is clear that updating the network using a flooding technique uses a large amount of power; therefore, frequent updates would not be helpful toward the goal of minimizing power loss.

Clearly, a different strategy must be used in order to minimize energy use and keep the flexibility of having spanning networks. The solution proposed by this paper is to use a limited form of the flooding algorithm, in conjunction with additional information about the broadcasting power in the packets, to allow every node to find out where every other node in the network is. While the flooding algorithm does use more power, it is only employed when the network structure has to be detected. It would also generally use less power than the addition of a GPS unit. This strategy also allows for a spanning network, where every node does not have to be able to directly connect to another node but needs to be transitively connected to all other nodes (i.e., it must be able to find a route using other nodes to get to any node in the network).

33.5.1.2 Packet Redirection

Packet redirection is a crucial idea in the area of power- and cost-aware routing protocols. Reducing power through redirection relies on the fact that routing a packet in a direct path from a node to another does not always lead to paths that minimize power. There are two scenarios in which redirection can be used to minimize power. First, if a broadcasting node is weak, it can use another node that has abundant amounts of power in order to minimize its own power usage. The second situation occurs when a direct broadcast from one node to another uses less power than using multiple hops through other nodes to the destination node. The second scenario is linked to the way the signal power of a radiating type of broadcasting unit dissipates as it spreads in space.

The initial results of the experimental protocols appear to be positive. They have all shown to increase the life of individual nodes and the network in general. Initially, there was some criticism that algorithms using the redirection methods of minimizing power would reduce throughput and speed. However, experimental results have shown that using power- and cost-aware algorithms can reduce congestion in certain nodes in a network. If one node is overused in a network for communication, then clearly the throughput of that node will be reduced. If power- and cost-aware metrics are used, then the node with excessive network traffic will be avoided and the congestion will be reduced, thus increasing throughput.

33.6 Wireless Sensor Networks

Recent advances in nanotechnology made it technologically feasible and economically viable to develop low-power, battery-operated devices that integrate general-purpose computing with multiple sensing and wireless communications capabilities. It is expected that these small devices, referred to as *sensor nodes*, will be mass-produced, making production costs negligible. Individual sensor nodes have a nonrenewable power supply and, once deployed, must work unattended. We envision a massive random deployment of sensor nodes, numbering in the thousands or tens of thousands. Aggregating sensor nodes into sophisticated computation and communication infrastructures, called sensor networks, will have a significant impact on a wide array of applications including military, scientific, industrial, health, and domestic. The fundamental goal of a wireless sensor network is to produce, over an extended period of time, meaningful global information from local data obtained by individual sensor nodes [2,21,22,40].

However, a wireless sensor network is only as good as the information it produces. In this respect, perhaps the most important concern is *information* security. Indeed, in most application domains, sensor networks will constitute a mission critical component requiring commensurate security protection. Sensor network communications must prevent disclosure and undetected modification of exchanged messages. Owing to the fact that individual sensor nodes are anonymous and that communication among sensors is via wireless links, sensor networks are highly vulnerable to security attacks. If an adversary can thwart the work of the network by perturbing the information produced, stopping production, or pilfering information, then the perceived usefulness of sensor networks will be drastically curtailed. Thus, security is a major issue that must be resolved in order for the potential of wireless sensor networks to be fully exploited. The task of securing wireless sensor networks is complicated by the fact that the sensors are mass-produced anonymous devices with a severely limited energy budget and initially unaware of their location [2,21,48,49,53].

In many application domains, safeguarding *output data assets*, data produced by the sensor network and consumed by the end user (application), against loss or corruption is a major security concern. In these application domains, a sensor network is deployed into a hostile target environment for a relatively extended amount of time. The network self-organizes and works to generate output data that is of importance to the application. For example, a sensor network may be deployed across a vast expanse of enemy territory ahead of a planned attack; the network system monitors the environment and produces reconnaissance data that is key to a mission-planning application. Periodically, during the network lifetime, a mobile gateway that is mounted on a person, land, or airborne vehicle or a satellite collects the output data assets from the network system, to maintain an up-to-date state. This means the network system must store the output data assets from the time it is produced until it is collected. Therefore, securing the output data assets in the network is an important problem in this class of applications.

We view an attack on the output data assets in the sensor network as a type of denial of service attacks. This is based on the abstraction that output data is stored in a logical *repository* and that *access* to this output data repository constitutes, in effect, a service provided by the network system to the application; corruption or loss of output data denies the application access to that service. Many wireless sensor networks are mission-oriented, must work unattended, and espouse data-centric processing. Consequently, they are significantly different in their characteristics from ad hoc networks. Security solutions designed specifically for wireless sensor networks are, therefore, required.

33.6.1 The Network Model

The network model used in this chapter is a derivative of the model introduced in References 48 and 49. Specifically, we assume a class of wireless sensor networks consisting of a large number of sensors nodes randomly deployed in the environment of interest. A training process, as explained later, establishes a coordinate system and defines a clustering of all nodes. Posttraining, the network undergoes multiple operation cycles during its lifetime. The training process also endows the role of *sink* upon one or more of the defined clusters. The sink role is transient, however, since new sink clusters are designated at the beginning of each operation cycle. Each sink cluster, henceforth called sink, acts as a repository for a portion of the sensory data, generated in the network during an operation cycle. At the end of an operation cycle, each sink transfers data stored in its repository to a *gateway*. In the following text, we describe the three primary entities in our network model in more detail.

33.6.1.1 The Sensor Node

We assume a sensor to be a device that possesses three basic capabilities; sensing, computing, and communication. At any point in time, a sensor node is either performing one of a fixed set of *sensor primitive operations* or is idle (asleep).

We assume that individual sensor nodes have four fundamental constraints:

1. Sensors are anonymous; initially a sensor node has no unique identifier.
2. Each sensor has a limited nonrenewable energy budget.
3. Each sensor attempts to maximize the time it is in sleep mode; a sensor wakes up at specific (possibly random) points in time for short intervals under the control of a timer.
4. Each sensor has a modest transmission range, perhaps a few meters with the ability to send and receive over a wide range of frequencies. In particular, communication among sensor nodes in the sensor network must be multihop.

33.6.1.2 The Sink

The sink granularity in our network model is a cluster in a trained network. For each operation cycle, a number of clusters are designated to serve as sinks. This means that all nodes in a sink cluster serve as sink nodes. Coarser sink granularity supports higher sink storage capacity, and thus potentially longer operation cycles. However, coarser sink granularity, as envisioned here, comes at a potentially higher risk of anonymity attacks due to the space correlation among sink nodes. If it is discovered that a node x is a sink node, then it immediately follows that there exists at least one other sink node (typically more) in the vicinity of x. Thus, the success of an anonymity attack in our model is sensitive to the probability of identifying the *first* node in a sink. One approach to decrease the latter probability is to have only a subset of the nodes in a sink cluster serve as sink nodes. There is a trade-off between sink granularity, and hence its capacity and operation longevity, and the amount of security a sink has against anonymity attacks.

A notable advantage of our sink model is that sinks are dynamic configurations of regular nodes in the sensor network, as opposed to being, for example, special supernodes. This makes the system less complex to design and operate, and eliminates a would-be attractive focus for anonymity attacks. Another major advantage is that the sink, in our model, is not a single point of failure, in two distinct ways. First, the network uses multiple sink entities, as opposed to a single sink entity. Second, a sink entity is not a single node; rather, it is a set of nodes.

The solution proposed in this paper for the anonymity problem uses whole cluster granularity for the sink. It encompasses techniques for randomly and securely choosing sinks for an operation cycle, and maintaining the anonymity of these sinks.

33.6.1.3 The Gateway

The gateway is an entity that connects the sensor network system to the outside world. The gateway is not constrained in mobility, energy, computation, or communication capabilities. There are two basic functions for the gateway in our network model:

Training: The gateway performs the network training process, postdeployment. For training purposes, the gateway is assumed to be able to send long-range, possibly, directional broadcasts to all sensors. It should be noted that our training process does not involve any transmission from the sensor nodes deployed in the network. Thus, in principle, the gateway does not have to be geographically colocated with the sensor nodes during the training process.

Harvesting (data collection): At the end of an operation period, the gateway must collect sensory data stored in each sink. In a simple collection scenario, the gateway traverses the deployment environment to collect data from all the sinks. If we assume that nodes comprising a sink perform collaborative data fusion, then the harvesting process requires a relatively short period of time. Specific harvesting protocols are beyond the scope of this paper.

33.6.2 Network Organization and Clustering

Figure 33.17a features an untrained sensor network immediately after deployment in an environment that is represented here as a 2D plane. For simplicity, we assume that the trainer is centrally located relative to all deployed nodes. Namely, considering the nodes to be points in the plane, we assume, in this paper, that the trainer is located at the center of the smallest circle in the plane that contains all deployed nodes. It should be noted that this is not a necessary condition. To be specific, our proposed training scheme is applicable if the trainer is located either at the center of or at a point outside of the smallest circle that contains all deployed nodes.

The primary goal of training is to establish a *coordinate system*, to provide the nodes *location awareness* in that system, and to organize the nodes into *clusters*. The coordinate system and clustering are briefly explained next. We refer the interested reader to References 48 and 49 for an in-depth description of the training process.

33.6.2.1 The Coordinate System

The training process establishes a polar coordinate system as exemplified by Figure 33.17b. The coordinate system divides the sensor network area into equiangular *wedges*. In turn, these wedges are divided into

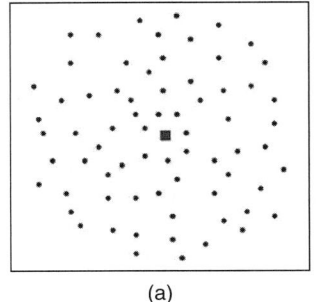
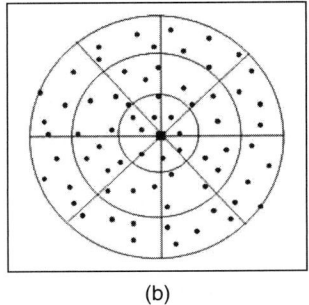

FIGURE 33.17 (a) An untrained network with a centrally located trainer; (b) a trained network.

sectors by means of concentric circles or *coronas* centered at the trainer location. Corona radii can be determined on the basis of several criteria, for example, in References 48 and 49, they are designed to maximize the efficiency of *sensors-to-sink* multihop communication. The intersection of every wedge and corona defines a unique sector; each sector is uniquely identifiable by the combination of its unique wedge identifier and unique corona identifier. The training process guarantees that each node belongs to one and only one sector in the coordinate system, and that each node knows the identity of its sector [48,49].

Let c and w be, respectively, the set of coronas, and the set of wedges defined by the training process. The resulting coordinate system can thus be formally represented by $\{(r_0, r_1, \ldots, r_{|c|-1}), \theta\}$, where r_i is the radius of corona i, $0 \leq i \leq |c| - 1$, θ is the wedge angle, and $|w| = 2\pi/\theta$. A fundamental assumption here is that any coordinate system is designed such that all nodes located in the same sector can communicate using direct (single hop) transmission.

33.6.2.2 Clustering

A major advantage of Wadaa et al.'s coordinate system [48,49] is that sectors implement the concept of clustering (at no additional cost). A sector effectively constitutes a cluster; clusters are disjoint and are uniquely identifiable. All nodes located in the same sector are members of the same cluster and have the same location coordinates, namely, the corona and wedge identifiers corresponding to that sector. This clustering scheme is ideally suited for sensor nodes that are intrinsically anonymous. Wadaa et al. [48,49] proposed a scalable training protocol where each untrained node incurs a communication cost equal to $\log |w| + \log |c|$, and the nodes do not transmit any messages during the training process.

33.6.3 The Work Model

The work model defines how sensor nodes work collaboratively to generate and store sensory data during an operation cycle. We propose a work model that is based on two principles:

1. *Intracluster* activity generates sensory data of interest to the application.
2. *Intercluster* activity transports each granule of sensory data to a sink, randomly chosen from among those available, for storage.

33.6.3.1 Intracluster Activity

In our model, the sensory data resulting from intracluster activity encodes states of a process of interest. Namely, we assume that the goal of intracluster activity is to monitor a process (or phenomenon), and report on its *local state* at any point in time. The state space of the phenomenon is given by $\{s_0, s_1, s_2, \ldots, s_z\}$. State s_0 denotes the *normal state*, and each of s_i, $1 \leq i \leq z$ denotes an *exception state*. The assumption here is that each state s_i, $1 \leq i \leq z$ corresponds to an application-defined exception of a particular type. The normal state corresponds to the fact that no exception of any type is detected.

We propose a transaction-based model for managing the computation and reporting of target process states. The model is a specialization of a transaction-based management model for sensor networks introduced in References 48 and 49. In this model, intracluster activity proceeds as follows. For a given cluster, subsets of nodes located in the cluster dynamically band together forming *workforces*.

Periodically, members of each workforce collaborate to perform an instance of a state computation transaction preloaded into each node. The transaction computes and reports the local process state. Note that the system allows for a fresh transaction to be downloaded to the nodes at the beginning of each operation cycle. In the simplest case, performing an instance of the state computation transaction entails that each member in the corresponding workforce performs a sensing operation and formulate a node report. A specific member of the workforce, designated as a transaction instance manager, then receives all node reports, and formulates a *Transaction Instance Report* (TIR). The TIR is the encoding of the local process state of interest at the time. This TIR is subsequently transported to a sink for storage. In principle, after transmitting the TIR, the corresponding workforce disbands. For simplicity, we assume that at most one transaction instance is in progress in a given cluster at any given point in time. Two important design parameters in the aforementioned transaction-based model are *workforce size* and *workforce setup*. These are discussed in the following text:

Workforce size: This is application dependent. Applications negotiate QoS requirements, for example, the "confidence level" associated with TIRs. These QoS parameters are, in turn, mapped to constraints at the transaction level. A constraint we assume in the anonymity solution proposed in this paper is that the average workforce size must be larger than or equal to λ, where λ is derived from application-level QoS parameters.

Workforce setup: This is governed by criteria such as load balancing or energy conservation, and thus represents a system concern. We distinguish two basic approaches to workforce setup, *static* and *dynamic*. The dynamic approach trades dynamic load balancing at the node level, and robustness to node failures for added delay time and energy overhead of the setup protocol. The static approach trades savings in energy and time of static setup for dynamic load balancing at the node level and robustness to node failures. In our proposed anonymity solution, we organize the node population of a cluster, before each operation cycle, into a fixed number of static workforces that have the same size on the average. Workforces are tasked to work (do an instance of the state computation transaction) in a round-robin fashion. The goal of our approach is to eliminate the energy and time overhead of dynamic setup while achieving load balancing at the coarser granularity of a workforce.

33.6.3.2 Intercluster Activity

As indicated earlier, the goal of intercluster activity in our work model is to route TIRs from their clusters of origin to the sinks, by means of multihop communication. We define a hop in a route as a direct transmission from one cluster to a *neighbor* cluster. A cluster u_j is a neighbor of a cluster u_i if and only if both u_j and u_i are located in the same corona, or the same wedge, for all i and j, $i \neq j$. It follows that in any instance of the coordinate system defined in Section 33.3, each cluster has either three or four neighbors. The set of all neighbors of a cluster u_i is called the *neighborhood* of u_i. In the proposed anonymity solution, we define a distributed intercluster routing protocol that yields optimal routes in terms of the number of hops from source to sink, and is highly scalable in the number of clusters in the network. Scalability can be attributed to two characteristics of the protocol. First, the protocol uses *no* dynamic global or regional state information. This eliminates the need for control messages to support routing. Second, the protocol uses distributed (incremental) route computation; the destination of hop i, computes, in turn, the destination of hop $i + 1, i \geq 1$.

33.7 Looking into the Crystal Ball

It is quite clear from the previous brief description of some of the active areas of research in the field of mobile computing that there are a lot of ideas and open research problems. These open problems present ample opportunity for researchers from the fields of parallel and distributed computing (e.g., computational geometry and graph algorithms) to borrow ideas and solution methodologies from their respective disciplines and extend them to solve mobile computing problems. The time is ripe for such a concerted effort. Some researchers have begun to take advantage of these opportunities and only the future will tell how successful these efforts will be.

33.8 Concluding Remarks

This chapter presented a brief overview of some of the research areas in mobile computing that could benefit from the lessons learned by parallel and distributed computing researchers. It is a well-recognized fact that there is a great deal of overlap between the underlying fundamentals of mobile environments and parallel and distributed environments. This fact can be used to influence research and open new avenues for the design of new algorithms that can address problems in mobile computing.

References

[1] A. Abutaleb and V. O. K. Li, Paging strategy optimization in personal communication systems, *Wireless Networks*, 3, 195–204, 1997.

[2] F. Akyildiz, J. S. M. Ho, and B. L. Yi, Movement-based location update and selective paging for PCS networks, *IEEE/ACM Transactions on Networking*, 4, 629–638, 1996.

[3] F. Akiyldiz and S. H. Jeong, Satellite ATM networks: A survey, *IEEE Communications*, 35(7), 569–587, 1997.

[4] F. Akyildiz, W. Su, Y. Sankarasubramanian, and E. Cayirci, Wireless sensor networks: A survey, *Computer Networks*, 38(4), 393–422, 2002.

[5] S. J. Barnes and S. L. Huff, Rising sun: iMode and the wireless Internet, *Communications of the ACM*, 46(11), 79–84, 2003.

[6] H. C. B. Chan and V. C. M. Leung, Dynamic bandwidth assignment multiple access for efficient ATM-based service integration over LEO satellite systems, *Proc. Northcon'98*, Seattle, WA, pp. 15–20, October 1998.

[7] S. Cho, Adaptive dynamic channel allocation scheme for spotbeam handover in LEO satellite networks, *Proc. IEEE VTC 2000*, pp. 1925–1929, 2000.

[8] G. Comparetto and R. Ramirez, Trends in mobile satellite technology, *IEEE Computer*, 30(2), 44–52, 1997.

[9] E. Del Re, R. Fantacci, and G. Giambene, Efficient dynamic channel allocation techniques with handover queuing for mobile satellite networks, *IEEE Journal of Selected Areas in Communications*, 13(2), 397–405, 1995.

[10] E. Del Re, R. Fantacci, and G. Giambene, Characterization of user mobility in Low Earth Orbiting mobile satellite systems, *Wireless Networks*, 6, 165–179, 2000.

[11] M. El-Kadi, S. Olariu, and P. Todorova, Predictive resource allocation in multimedia satellite networks, *Proc. IEEE, GLOBECOM 2001*, San Antonio, TX, November 25–29, 2001.

[12] J. Farserotu and R. Prasad, A survey of future broadband multimedia satellite systems, issues and trends, *IEEE Communications*, 6, 128–133, 2000.

[13] K. Feher. *Wireless Digital Communications*, New York: McGraw-Hill, 1995.

[14] I. Foster and R. G. Grossman, Data integration in a bandwidth-rich world, *Communications of the ACM*, 46(11), 51–57, 2003.

[15] E. C. Foudriat, K. Maly, and S. Olariu, H3M—A rapidly deployable architecture with QoS provisioning for wireless networks, *Proc. 6th IFIP Conference on Intelligence in Networks (SmartNet 2000)*, Vienna, Austria, September 2000.

[16] S. Grandhi, R. Yates, and D. Goodman, Resource allocation for cellular radio systems, *IEEE Transactions on Vehicular Technology*, 46, 581–586, 1997.

[17] P. Gupta and P. R. Kumar, The capacity of wireless networks, *IEEE Transactions on Information Theory*, 46(2), 388–407, 2000.

[18] A. Hac and X. Zhou, Locating strategies for personal communication networks, a novel tracking strategy, *IEEE Journal on Selected Areas in Communications*, 15, pp. 1425–1436, 1997.

[19] I. Huang, M.-J. Montpetit, and G. Kesidis, ATM via satellite: A framework and implementation, *Wireless Networks*, 4, 1998, 141–153.

[20] I. Jamalipour and T. Tung, The role of satellites in global IT: trends and implications, *IEEE Personal Communications*, 8(3), 2001, 5–11.

[21] K. Jones, A. Wadaa, S. Olariu, L. Wilson, and M. Eltoweissy, Towards a new paradigm for securing wireless sensor networks, *Proc. New Security Paradigms Workshop* (NSPW'2003), Ascona, Switzerland, August 2003.

[22] J. M. Kahn, R. H. Katz, and K. S. J. Pister, Mobile networking for Smart Dust, *Proc. 5th Annual International Conference on Mobile Computing and Networking* (MOBICOM'99), Seattle, WA, August 1999.

[23] I. Katzela and M. Naghshineh, Channel assignment schemes for cellular mobile telecommunication systems: A comprehensive survey, *IEEE Personal Communications*, June 1996, pp. 10–31.

[24] T. Le-Ngoc and I. M. Jahangir, Performance analysis of CFDAMA-PB protocol for packet satellite communications, *IEEE Transactions on Communications*, 46(9), 1206–1214, 1998.

[25] D. Levine, I. F. Akyildiz, and M. Naghshineh, A resource estimation and call admission algorithm for wireless multimedia networks using the shadow cluster concept, *IEEE/ACM Transactions on Networking*, 5, 1–12, 1997.

[26] K. Y. Lim, M. Kumar, and S. K. Das, Message ring-based channel reallocation scheme for cellular networks, *Proc. International Symposium on Parallel Architectures, Algorithms, and Networks*, 1999.

[27] M. Luglio, Mobile multimedia satellite communications, *IEEE Multimedia*, 6, 10–14, 1999.

[28] I. Mertzanis, R. Tafazolli, and B. G. Evans, Connection admission control strategy and routing considerations in multimedia (Non-GEO) satellite networks, *Proc. IEEE VTC*, 1997, pp. 431–436.

[29] Mobile Media Japan, Japanese mobile Internet users: http://www.mobilemediajapan.com/, January 2004.

[30] M. Naghshineh and M. Schwartz, Distributed call admission control in mobile/wireless networks, *IEEE Journal of Selected Areas in Communications*, 14, 711–717, 1996.

[31] H. N. Nguyen, S. Olariu, and P. Todorova, A novel mobility model and resource allocation strategy for multimedia LEO satellite networks, *Proc. IEEE WCNC*, Orlando, FL, March 2002.

[32] H. N. Nguyen, J. Schuringa, and H. R. van As, Handover schemes for QoS guarantees in LEO satellite networks, *Proc. ICON'2000*.

[33] S. Olariu, S. A. Rizvi, R. Shirhatti, and P. Todorova, Q-Win—A new admission and handoff management scheme for multimedia LEO satellite networks, *Telecommunication Systems*, 22(1–4), 151–168, 2003.

[34] I. Oliviera, J. B. Kim, and T. Suda, An adaptive bandwidth reservation scheme for high-speed multimedia wireless networks, *IEEE Journal of Selected Areas in Communications*, 16, 858–874, 1998.

[35] H. Peyravi, Multiple Access Control protocols for the Mars regional networks: A survey and assessments, Technical Report, Department of Math and Computer Science, Kent State University, September 1995.

[36] H. Peyravi, Medium access control protocol performance in satellite communications, *IEEE Communications Magazine*, 37(3), 62–71, 1999.

[37] H. Peyravi and D. Wieser, Simulation modeling and performance evaluation of multiple access protocols, *Simulation*, 72(4), 221–237, 1999.

[38] X. Qiu and V. O. K. Li, Dynamic reservation multiple access (DRMA): A new multiple access scheme for Personal Communication Systems (PCS), *Wireless Networks*, 2, 117–128, 1996.

[39] E. M. Royer and C.-K. Toh, A review of current routing protocols for ad-hoc mobile wireless networks, *IEEE Personal Communications Magazine*, 23, 46–55, 1999.

[40] P. Saffo, Sensors, the next wave of innovation, *Communications of the ACM*, 40(2), 93–97, 1997.

[41] S. Singh, M. Woo, and C. S. Raghavendra, Power-aware routing in mobile ad hoc networks, *Proc. 4th Annual ACM/IEEE International Conference on Mobile Computing and Networking*, 1998, pp. 181–190.

[42] I. Stojmenovic. *Handbook of Wireless Networks and Mobile Computing*, New York: Wiley, 2002

[43] I. Stojmenovic and X. Lin, Power-aware localized routing in wireless networks, in *Proc. of Parallel and Distributed Computing Symposium*, pp. 371–376, 2000.

[44] R. Subrata and A. Y. Zomaya, Location management in mobile computing, in the *Proc. of the ACS/IEEE International Conference on Computer Systems and Applications*, pp. 287–289, Beirut, Lebanon, June 26–29, 2001.

[45] R. Subrata and A. Y. Zomaya, Evolving cellular automata for location management in mobile computing networks, *IEEE Transactions on Parallel and Distributed Systems*, 13(11), 2002.

[46] P. Todorova, S. Olariu, and H. N. Nguyen, A selective look-ahead bandwidth allocation scheme for reliable handoff in multimedia LEO satellite networks, *Proc. ECUMN'2002*, Colmar, France, April 2002.

[47] H. Uzunalioglu, A connection admission control algorithm for LEO satellite networks, *Proc. IEEE ICC*, 1999, pp. 1074–1078.

[48] A. Wadaa, S. Olariu, L. Wilson, K. Jones, and Q. Xu, On training wireless sensor networks, *Proc. 3rd International Workshop on Wireless, Mobile and Ad Hoc Networks* (WMAN'03), Nice, France, April 2003.

[49] A. Wadaa, S. Olariu, L. Wilson, K. Jones, and Q. Xu, A virtual infrastructure for wireless sensor networks, *Journal of Mobile Networks and Applications (MONET)*, 10, 151–167, 2005.

[50] M. Werner, J. Bostic, T. Ors, H. Bischl, and B. Evans, Multiple access for ATM-based satellite networks, *European Conference on Network & Optical Communications*, NOC'98, Manchester, UK, June 1998.

[51] M. Werner, C. Delucchi, H.-J. Vogel, G. Maral, and J.-J. De Ridder, ATM-Based Routing in LEO/MEO Satellite networks with intersatellite links, *IEEE Journal on Selected Areas in Communications*, 15(2), 69–82, 1997.

[52] J. E. Wieselthier and A. Ephremides, Fixed- and movable-boundary channel-access schemes for integrated voice/data wireless networks, *IEEE Transactions on Communications*, 43(1), 64–74, 1995.

[53] Wood and J. A. Stankovic, Denial of service in sensor networks, *IEEE Computer*, 35(4), 54–62, 2002.

[54] A. Y. Zomaya and M. Wright, Observations on using genetic algorithms for channel allocation in mobile computing, *IEEE Transactions on Parallel and Distributed Systems*, 13(9), 948–962, 2002.

[55] A. Y. Zomaya, F. Ercal, and S. Olariu (Eds.). *Solutions to Parallel and Distributed Computing Problems: Lessons from Biological Sciences*, New York: Wiley, 2001.

Applications

34
Using FG to Reduce the Effect of Latency in Parallel Programs Running on Clusters

34.1	Latency in Parallel Computing	34-1
34.2	Pipeline-Structured Algorithms	34-2
34.3	FG Overview	34-4
34.4	Multiple Pipelines with FG	34-5
34.5	Intersecting Pipelines	34-6
34.6	Additional FG Features	34-7
	Macros • Hard Barriers • Soft Barriers • Fork/Join • Stage Replication • Directed Acyclic Graphs	
34.7	How FG Speeds Up Programs	34-11
34.8	How FG Speeds Up Programming	34-13
34.9	Conclusion	34-15
	References	34-15

Thomas H. Cormen
Elena Riccio Davidson
Dartmouth College

34.1 Latency in Parallel Computing

We use parallel computing to get high performance, and one of the most significant obstacles to high-performance computing is latency. We also use parallel computing—particularly distributed-memory clusters—to operate on massive datasets, taking advantage of the memory and disk space across the multiple nodes of a cluster. But working with massive data on a cluster induces latency in two forms: accessing data on disk and interprocessor communication.

In order to mitigate latency, we rely on two algorithmic techniques that date back to at least 1988 [1]. The first is to access data in relatively large blocks in order to amortize the fixed costs of I/O and communication operations. That is, we access disk-resident data by reading and writing blocks rather than individual words, and we organize interprocessor communication into fewer, longer messages. The second technique is to design algorithms in order to minimize the number of high-latency operations. Both the I/O model of Aggarwal and Vitter [1] and the Parallel Disk Model of Vitter and Shriver [21] were developed to promote algorithms that access disk-resident data in blocks while minimizing the number of blocks read and written.

In addition to these algorithmic techniques, we can also mitigate latency by overlapping I/O and communication with other operations. While we access a disk, the CPU is available for other work, and so we can perform an I/O operation concurrently with computation on memory-resident data. Likewise, interprocessor communication uses the network, again leaving the CPU available for other work. When we overlap operations, we try to hide lower-latency operations behind higher-latency ones to reduce the overall running time of the program.

The remainder of this chapter is organized as follows. Section 34.2 discusses how pipelines can mitigate latency. Section 34.3 describes FG, a programming environment that improves pipeline-structured programs. Sections 34.4 and 34.5 expand upon FG's basic capabilities, showing how it supports multiple pipelines that may even intersect. Section 34.6 outlines several additional features of FG. Sections 34.7 and 34.8 document how FG speeds up programs and the programmers who write them. Finally, Section 34.9 offers some concluding remarks.

34.2 Pipeline-Structured Algorithms

Pipelines are a proven technique for overlapping operations in both hardware and software systems. We find that software pipelines are well suited to certain algorithms that work with massive datasets. This section provides two examples of such software pipelines.

By "massive datasets," we mean datasets that do not fit in main memory. We call algorithms designed to work with these datasets *out-of-core* algorithms. The data reside on one or more disks, and many out-of-core algorithms make multiple *passes* over the data. In a pass, each datum moves from disk to main memory, participates in one or more operations, and finally moves from memory back to disk. In many out-of-core algorithms in the literature (e.g., References 2, 8, 9, 20, and 21), each pass falls into a common framework: a pipeline.

A pipeline is a serial arrangement of *stages*. In the high-latency environment of a cluster, each processor runs a process that has its own copy of the pipeline. Each pipeline stage receives a buffer of data from its predecessor stage in the pipeline, operates on the data in that buffer, and then passes the buffer on to its successor stage. Buffers often are configured to be the size of the I/O or communication blocks. When a buffer has traversed the entire pipeline, we say that the pipeline has completed a *round*. Because buffers are smaller than the total amount of data, a pipeline must execute many rounds in each pass.

We can structure some out-of-core permutations as pipelines, such as the pipelines shown in Figure 34.1, where each processor runs its own copy. Consider an out-of-core permutation that comprises a number of passes, in which each pass reads data from disk into a buffer, permutes the data across the cluster among buffers, and then writes the data from a buffer back to disk. We can further break permuting data across the cluster into three stages, each of which operates on a buffer. First, each processor permutes locally within the buffer to collect data destined for the same processor into consecutive memory locations. Second, a collective communication operation transfers data among the processors. Third, each processor permutes locally the data it has received into the correct order for writing to disk. Thus, a buffer completes a round of the out-of-core permutation by traversing the five-stage pipeline of Figure 34.1.

In order for a single buffer to serve for both sending and receiving within the collective communication operation, each processor must send and receive the same amount of data in each round. Not all permutations have this characteristic, but several useful ones do (e.g., cyclic rotations and BMMC permutations [8]).

Even for something as seemingly data dependent as sorting, we can use a similar pipeline structure. Figure 34.2 shows the pipelines for a single pass of out-of-core columnsort [3, 4, 7]. The specifics of each stage differ from those of the permutation pipeline above, and a sorting stage replaces the first permutation stage. Two characteristics of out-of-core columnsort make it suitable for a pipeline structure. First, as in the permutation algorithms above, the collective communication operation has the property that each processor sends and receives the same amount of data in each round. Second, out-of-core columnsort has the unusual property that the sequence of I/O and communication operations is independent of the

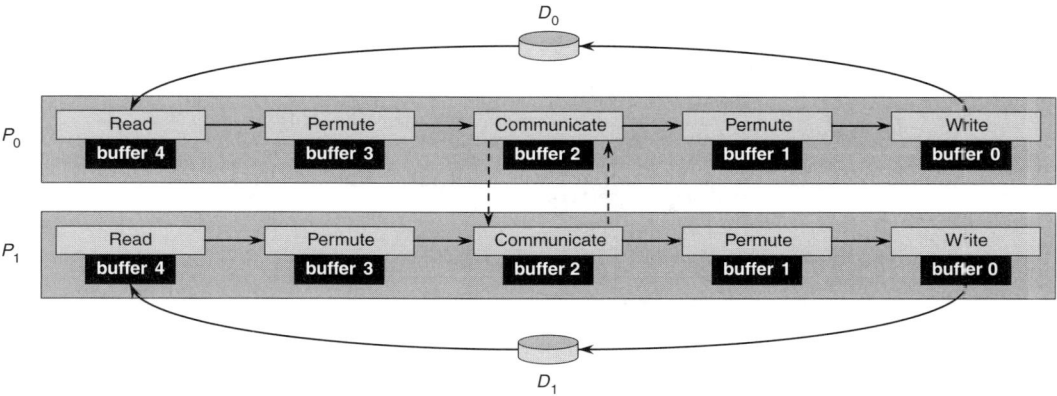

FIGURE 34.1 The pipelines for one pass of an out-of-core permutation on two processors, P_0 and P_1. Each processor owns one disk. The pipelines run concurrently on P_0 and P_1. At any moment, each of the five stages on a processor may be working on a distinct buffer. The pipeline on P_0 reads and writes disk D_0, and the pipeline on P_1 reads and writes disk D_1. During the communicate stage, P_0 sends a message to P_1 and P_1 sends a message to P_0.

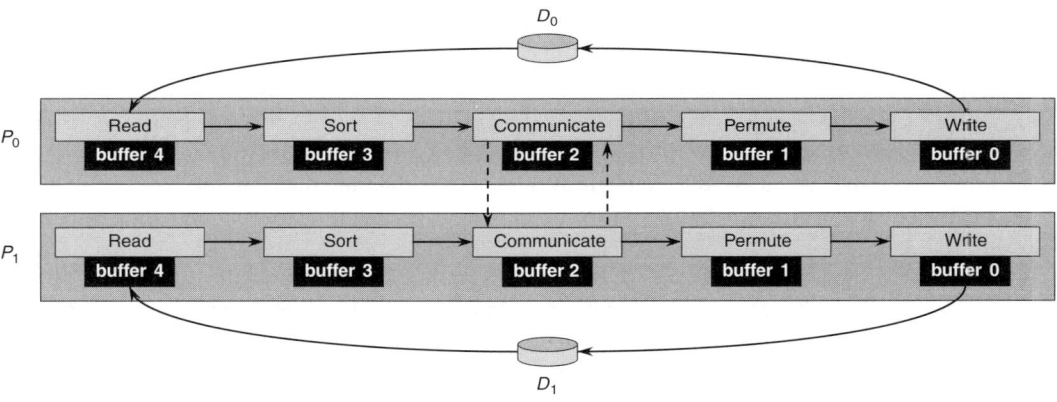

FIGURE 34.2 The pipelines for one pass of out-of-core columnsort on two processors, P_0 and P_1. These pipelines are similar to those in Figure 34.1, except that here, the second stage is a local sort.

data being sorted. That is, the exact sequence of disk reads, disk writes, and communication operations depends only on the amount of data to sort, the buffer size, and the cluster configuration, rather than on the input data itself.

We could organize a pipeline as a sequence of function calls, where each function implements one stage. Observe, however, that we can instead organize the pipeline as a sequence of asynchronous stages. Each stage receives a buffer from its predecessor stage, operates on the buffer, and then passes the buffer on to its successor stage. When stages are asynchronous, a stage can yield the CPU to other stages while it is in the midst of a high-latency operation. At any moment, each stage may be working on a distinct buffer (assuming that there are sufficient buffers).

There are multiple approaches for overlapping the operations of pipeline stages. For example, we could use asynchronous I/O and communication calls to allow these high-latency operations to yield. The problem with this approach is that programming with asynchronous calls is messy. For example, Figure 34.3 shows two program schemas for out-of-core permuting on a single processor. Disk reads and writes are asynchronous in Figure 34.3a, but they are synchronous in Figure 34.3b. Observe how much

```
(a)     b = 0
        start read into buffer [b, 0]
        while some read has not been started do
            wait for read into buffer [b, 0]
            if not first or second time through
                then wait for write to complete from target buffer [b, 1]
            if not working on the final buffer
                then start read into buffer [1 − b, 0]
            permute (in-core) from read buffer [b, 0] into target buffer [b, 1]
            start write of target buffer [b, 1]
            b = 1 − b
        wait for writes to complete from target buffers [0, 1] and [1, 1]
```

```
(b)     while some read has not been started do
            read into buffer 0
            permute (in-core) from read buffer 0 into target buffer 1
            write target buffer 1
```

FIGURE 34.3 An out-of-core permutation using asynchronous and synchronous I/O operations. (a) Using asynchronous I/O. While waiting for a read or write to complete, we can begin the in-core permutation, but we must schedule it statically. (b) Using synchronous I/O. It is much simpler, but much less efficient, than its asynchronous counterpart.

simpler the schema of Figure 34.3b is, and imagine how much more complicated the schema of Figure 34.3a would have been if it had been designed for a cluster by means of asynchronous communication calls. Moreover, the schema of Figure 34.3a suffers from a larger problem: it violates the pipeline abstraction by rolling all stages into a single loop.

Instead of making asynchronous I/O and communication calls, we can respect the pipeline abstraction by mapping each pipeline stage to its own thread. This approach yields at least two advantages. One is that we can implement each stage with simpler, synchronous calls, as in Figure 34.3b. Another advantage is that with a threaded implementation, scheduling is dynamic rather than the static approach of Figure 34.3a. Section 34.7 will demonstrate that programs run faster with dynamic thread scheduling than with static scheduling.

Of course, a programmer always has the option of creating and working with threads and buffers directly. This code, which we call *glue*, is cumbersome to write and difficult to debug. We find that it also can be voluminous—it accounts for approximately 25% of the source code for out-of-core columnsort [7]. A software package, FG, provides a much simpler and more efficient alternative to directly working with threads and buffers.

34.3 FG Overview

FG's goal is to lighten the burden on the programmer by reducing the amount and complexity of glue without degrading performance. In order to overlap operations, we need code that manages buffers and creates asynchrony, but it is not necessary for the programmer to write such code. FG,* a software framework for creating asynchronous pipelines, relieves the programmer of this responsibility; in other words, it provides the glue. This section presents an overview of FG's fundamental features.

The central paradigm of FG is a pipeline of asynchronous stages. Each processor runs its own copy of the pipeline. The programmer writes a straightforward function for each stage, containing only synchronous calls. FG assumes the jobs of managing the buffers of data, moving the buffers through the pipeline, and running the stages asynchronously.

FG assumes all aspects of buffer management. It allocates buffers at the start of execution and deallocates them at the end. The programmer need specify only the number and size of buffers. FG creates and manages

---

*The name "FG" is short for "ABCDEFG," which is an acronym for "Asynchronous Buffered Computation Design and Engineering Framework Generator." We pronounce "FG" as the word "effigy."

buffer queues between consecutive stages, and it recycles buffers as they reach the end of the pipeline so that their memory can be reused for additional rounds.

For each stage, the programmer provides a function that implements the stage, and FG maps the stage to its own thread. FG runs on any threads package that implements the POSIX standard pthreads interface [14]. The programmer does not write any code associated with pthreads but instead has the simpler task of creating FG-defined objects. FG spawns all the threads, creates the buffer queues and the semaphores that guard them, and kills the threads after the pipeline has completed executing. In addition to the thread for each stage, FG creates two more threads for a source stage and a sink stage. The source stage emits buffers into the pipeline, and the sink stage recycles buffers back to the source once they have traversed the entire pipeline.

FG associates with each buffer a small data structure called a *thumbnail*. A thumbnail contains a pointer to the buffer, the number of the current round for that buffer, a *caboose* flag indicating whether this buffer is the final one through the pipeline, and other fields including programmer-defined ones. In reality, it is the thumbnails that traverse the pipeline, with their buffers accessed via the pointers in the thumbnails. The source stage sets the round number each time it emits a thumbnail. When a buffer flagged as the caboose reaches the sink stage, FG shuts down the entire pipeline and frees up its resources.

Some stages might require services before and after the pipeline runs. For example, a stage that reads from disk might need to open a file before the pipeline runs and to close the file once the pipeline has completed execution. Although the programmer could write code for tests that determine whether a stage is executing for the first time (e.g., test whether the round number is 0) or the last time (e.g., test whether the caboose flag is set), FG allows the programmer to attach initialization and cleanup functions to any stage. An initialization function runs before the pipeline executes, and a cleanup function runs after the pipeline completes.

A stage might need an extra buffer to do its job. For example, the permute stages of the example pipelines in Section 34.2 copy data from the buffers they receive in the pipeline (we call these *pipeline buffers*) into another buffer. Where does this other buffer come from? FG supplies *auxiliary buffers* for this purpose. FG allocates auxiliary buffers and associated thumbnails when it creates the pipeline, and they remain available for any stage to request. In a stage such as a permuting stage, once all the data have been copied into the auxiliary buffer, it is the auxiliary buffer that we want to send on to the next stage. (The inefficient alternative is to copy the entire auxiliary buffer back into the pipeline buffer.) FG allows the programmer to simply swap the buffer pointers in the pipeline and auxiliary thumbnails, so that the pipeline thumbnail can continue through the pipeline, but now pointing to the appropriate buffer. Because the original pipeline buffer is now referenced by the auxiliary thumbnail, it becomes available for future use as an auxiliary buffer.

34.4 Multiple Pipelines with FG

A single pipeline per processor, as described in Section 34.3, suffices when data enter and exit each stage at the same rate. The two out-of-core computations described in Section 34.2 have this property, because within each collective communication operation, every processor sends and receives the same amount of data.

Not every cluster computation has this property, however. Consider a computation in which during collective communication operations, each processor might receive more or less data than it sends. For example, in the distribution pass of an out-of-core, distribution-based, sorting algorithm, each round would entail an all-to-all communication step in which the amount of data that one processor sends to another depends on the data. Consequently, the amount of data that a processor receives usually differs from the amount that it sends. If the all-to-all communication was confined to a single pipeline stage, then the rate of data entering this stage (i.e., the data sent) would differ from the rate of data exiting this stage (i.e., the data received).

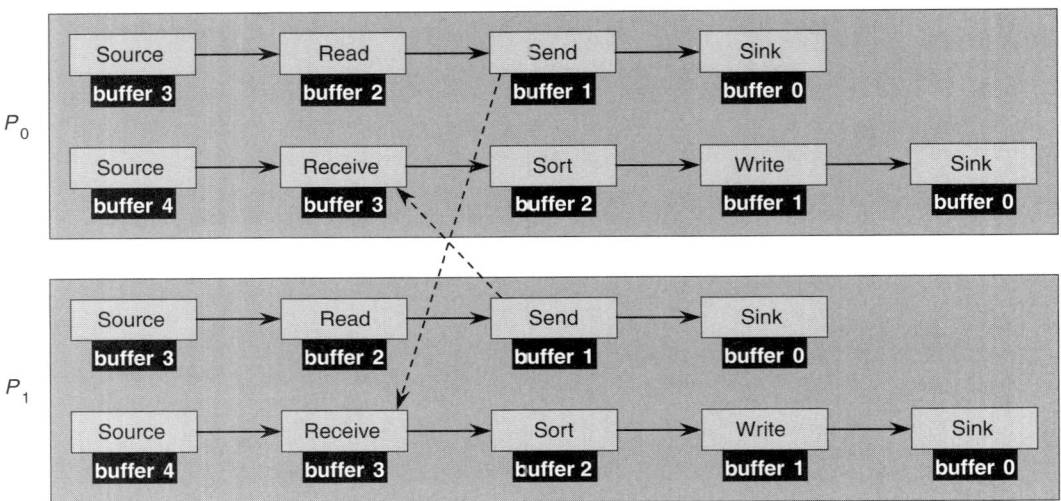

FIGURE 34.4 The pipelines for the first pass of out-of-core distribution sort, running on two processors. Each processor runs a send pipeline and a receive pipeline concurrently. On processor P_0, the send pipeline reads data from disk and sends a message to P_1, and the receive pipeline receives a message from P_1, sorts locally, and writes to disk. On processor P_1, the send pipeline reads data from disk and sends a message to P_0, and the receive pipeline receives a message from P_0, sorts locally, and writes to disk.

FG solves this problem by allowing the programmer to create multiple pipelines running concurrently on the same processor. Each pipeline may have its own rate at which data move through it. For example, Figure 34.4 shows the pipelines for the distribution pass in a cluster-based, out-of-core, distribution sort. In each processor, the top pipeline, which we call the "send pipeline," has two stages other than the FG-supplied source and sink. The read stage reads data into a buffer. The buffer travels to the send stage, which scatters the data among all the processors, depending on the data values and the splitter values (i.e., predetermined values that demarcate the data values destined for each processor). The bottom pipeline, which we call the "receive pipeline," has three stages other than the source and sink. The receive stage receives the data sent by all the processors in the send stage of the send pipeline. From there, each buffer goes to a sort stage, which creates a sorted run the size of a buffer, and finally the write stage writes each sorted run out to disk. A further pass, which we do not describe here, completes the job of sorting.

Buffers circulate through these two pipelines at different rates. Each send stage always sends a full buffer's worth of data. But we do not know in advance how much data a given receive stage will receive. Therefore, each receive stage accepts empty buffers from the source stage in its pipeline and conveys full buffers to the sort stage at a rate that depends on how much data it receives in each round. Using separate pipelines for sending and receiving allows each one to operate at its own rate.

FG allows any number of pipelines to run on each processor, subject to system-imposed limits on how many threads may run concurrently.

34.5 Intersecting Pipelines

FG allows multiple pipelines to intersect at a particular stage. Figure 34.5 shows pipelines that merge several sorted runs of data on a single processor. The pipelines intersect at the merge stage. Each of the vertical pipelines reads a section of a sorted run into a buffer and conveys the buffer to the single merge stage. The merge stage merges all the sorted runs together into a single sorted sequence. It accepts empty buffers from the source stage in the horizontal pipeline and fills them by merging data from the buffers

Using FG to Reduce the Effect of Latency in Parallel Programs Running on Clusters

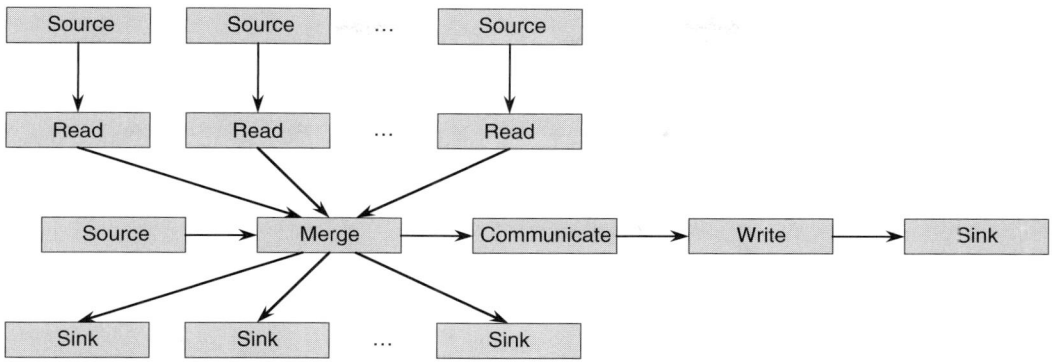

FIGURE 34.5 The pipelines for one pass of a program that merges several runs of sorted data. All the pipelines run concurrently on one processor. Buffers are no longer shown. Small buffers traverse the vertical pipelines, and larger buffers traverse the horizontal pipeline. The merge stage is common to all the pipelines and merges the data in the small buffers into longer, sorted sequences, which the merge stage moves into the larger buffers.

conveyed to it by the various read stages. The merge stage then conveys these full, sorted buffers to its successor along the horizontal pipeline.

Each buffer is assigned to a particular pipeline and always traverses only that pipeline. If the merge stage accepts a buffer from a particular pipeline, when it conveys that buffer, it conveys the buffer along that same pipeline. Different pipelines may have different buffer configurations. That is, both the sizes and numbers of buffers may vary among pipelines. Each thumbnail contains a pipeline identifier, and so when a stage conveys a buffer, FG automatically keeps the buffer in the correct pipeline.

Let us examine a buffer as it travels through a vertical pipeline. The source stage emits the buffer to the read stage, which reads part of a sorted run into the buffer. Next, the buffer travels to the merge stage, which picks off items from the buffer until the buffer is exhausted. The buffer then travels to the sink stage for its pipeline, where it is recycled back to the source stage.

Now, we consider how the merge stage operates. Each buffer that the merge stage accepts from a vertical pipeline contains a section of a sorted run. Whereas the vertical pipeline provides input buffers to the merge stage, the horizontal pipeline manages its output buffers. The merge stage repeatedly examines all of its input buffers to find the smallest item not yet taken. It then copies this item into its output buffer, which it conveys along the horizontal pipeline once this output buffer fills. When the merge stage exhausts an input buffer, it conveys that buffer along the appropriate vertical pipeline for reuse. Then it accepts another input buffer from the same vertical pipeline; this buffer contains the next section of that pipeline's sorted run.

The buffers in the vertical pipelines are considerably smaller than the buffers in the horizontal pipeline. That is because, in practice, there are many sorted runs and their union would not fit in the processor's memory. Instead, each buffer in the vertical pipelines contains a relatively small section of a sorted run so that the data being considered at any one moment fits in memory. In practice, we should not make these section sizes too small; we choose them to be the minimum underlying size of a disk-read operation. Because there is only one horizontal pipeline per processor, we can choose its buffer size to be much larger than the size of the vertical buffers.

34.6 Additional FG Features

An FG programmer has options beyond pipeline structures. FG allows buffers to travel among stages in various ways. The programmer may specify alternative ways for stages to accept and convey buffers, as well as nonlinear arrangements of stages. This section outlines specific FG features that provide these

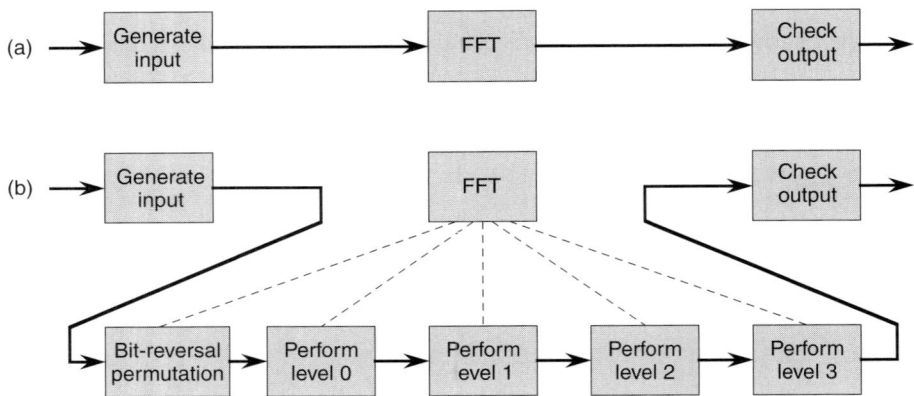

FIGURE 34.6 (a) A three-stage pipeline, where the middle stage is a macro that computes the FFT of each buffer. (b) The same pipeline for $N = 16$ values per buffer, with the macro expanded into $\lg N + 1 = 5$ stages. Dashed lines indicate the macro expansion. The first expanded stage performs the bit-reversal permutation, and the other four perform the levels of butterfly computations. The solid line represents the actual pipeline executed.

capabilities. In particular, we examine macros, which enable the programmer to treat a section of the pipeline much like a subroutine; soft and hard barriers, which change how stages accept and convey buffers; fork/join constructs, which split the pipeline into multiple paths that eventually rejoin; stage replication, in which multiple copies of a stage operate on distinct buffers; and a generalization of linear pipelines to directed acyclic graphs (DAGs).

34.6.1 Macros

A *macro* stage allows the programmer to collect several stages into one stage, which then plugs into another pipeline. A macro is like a subroutine, in which many lines of code are invoked by a single call elsewhere. With macros, a stage is no longer restricted to being a single function. A stage within a macro may itself be another macro, and so macros have a tree structure. The root represents the macro invocation, and the frontier of the tree (i.e., the leaves, taken from left to right) represents the actual stages executed and their order of execution.

For example, consider a pipeline that, at its highest level, has three stages, as shown in Figure 34.6a:

1. Generate N complex input values in each buffer.
2. Perform a fast Fourier transform (FFT) on the N complex numbers in each buffer. To perform an FFT, we perform a bit-reversal permutation, followed by $\lg N$ levels* of butterfly computations.
3. Check the result of the FFT on each buffer.

Although we could implement the middle stage of this pipeline as a single function, we can pipeline the FFT computation by placing the bit-reversal permutation and each of the $\lg N$ levels of the butterfly graph in its own stage. If we organize the FFT stage as a macro with $\lg N + 1$ stages, we get the structure shown in Figure 34.6b. With $N = 16$, the macro expands into a frontier of five stages. The resulting pipeline has seven stages: one to generate values in each buffer, the five from the FFT macro, and one to check the result in each buffer.

34.6.2 Hard Barriers

Suppose we wish to perform a particular permutation on a two-dimensional image and then run a two-dimensional FFT on it. The permutation must be fully completed before running the FFT. In other

---

*We use $\lg N$ to denote $\log_2 N$.

words, as long as the permutation is complicated enough that the FFT code cannot simply adjust where it reads from, all the data has to be written by the permutation code before the FFT code can read any of the data. An example of a sufficiently complicated permutation would be a torus permutation, in which the value at each position (i, j) maps to position $((i + a) \bmod r, (j + b) \bmod s)$, where the shift is by (a, b) and the dimensions of the image are $r \times s$.

If the image is large enough that the permutation and two-dimensional FFT are out-of-core operations, then we need to ensure that the permutation code writes out every buffer before the FFT code reads its first buffer. We could structure this sequence as two distinct pipelines. Alternatively, we could place a *hard barrier* between the permutation and FFT pipeline sections of a single pipeline to effectively split the pipeline into two. No buffers progress beyond a hard barrier until a caboose arrives.

A pipeline may contain any number of hard barriers. By introducing h hard barriers, we effectively split the pipeline into $h + 1$ pipelines, each of which runs in turn.

34.6.3 Soft Barriers

The merge stage that we saw in Section 34.5 takes as input a number of sorted runs. Depending on the problem size and machine configuration, there may be a very large number of sorted runs to merge. Although it is easy to find the next input item to copy to the merged output when there are few input runs, it becomes more time consuming when there are several. Therefore, we might alter the pipelines that produce the sorted runs to instead create fewer, longer, sorted runs.

For example, suppose that we wish to double the size of each sorted run. At some point, a stage would need to have accepted two buffers, which it would then sort together and convey to a write stage. But this stage cannot start sorting until both buffers arrive. We call such a stage a *soft barrier*. A stage that is designated as a soft barrier accepts a programmer-specified number of buffers at once so that it can operate on them together.

34.6.4 Fork/Join

FG supports pipelines in which different buffers take different paths. Consider a pipeline that implements the shear sort algorithm [17,18], which sorts a square matrix in "snake-like" ordering. That is, the output matrix is sorted row by row, but with the top row going from left to right, the next row from right to left, then left to right, and so on. One phase of the shear sort algorithm sorts every other row in ascending order and the remaining rows in descending order. In the FG context, each buffer contains one row of the input matrix, and so half the buffers must visit an ascending-sort stage and half must visit a descending-sort stage. FG provides a mechanism to allow an ascending sort and a descending sort to be stages in the same pipeline. An FG *fork* stage, shown in Figure 34.7, splits a linear pipeline into two paths: one for the ascending sort and one for the descending sort. The fork stage has two successor stages instead of the usual single, linear successor. An FG *join* stage regathers the split pipeline into one single path. The join stage has two predecessor stages instead of the usual single, linear predecessor. In reality, FG does not introduce a fork stage and a join stage into the pipeline. Rather, it uses the fork and join placement as cues from the user to reassign how it orders stages in the pipeline. FG alters the way buffers move from queue to queue, but it does not add new stages or threads to the pipeline.

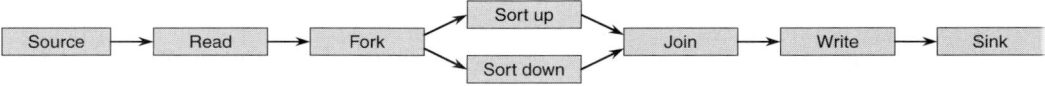

FIGURE 34.7 The pipeline for shear sort. The fork-join construct causes every other row to be sorted into ascending order and the remaining rows to be sorted into descending order. The fork and join elements are conceptual only: FG moves buffers from the read stage directly to one of the sort stages, and it moves buffers from the sort stages directly to the write stage.

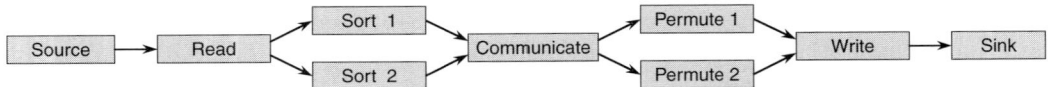

FIGURE 34.8 A pass of columnsort with two replicated stages. We replicate the sort stage with sort 2, and we replicate the permute stage with permute 2. Each coupling of a stage and its copy can work on different buffers concurrently. Stage replication is effective when each node contains multiple processors.

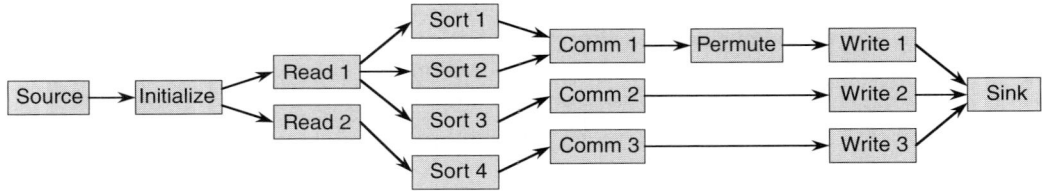

FIGURE 34.9 All passes of out-of-core columnsort, configured in one DAG structure. There are two distinct read stages, four sort stages, three communicate stages, one permute stage, and three write stages. Different buffers follow different paths according to the pass number. The initialize stage determines the pass number on the fly and assigns the corresponding path to each buffer. (Elena Riccio Davidson and Thomas H. Cormen, Proceedings of the 19th International Parallel and Distributed Processing Symposium (IPDPS 2005), © 2005 IEEE.)

34.6.5 Stage Replication

In general, we expect to see the best performance from a balanced pipeline—one in which all the stages make progress at about the same rate. FG provides the ability to compensate for a stage that is processing buffers more slowly than the other stages in a pipeline. The programmer can replicate such a stage, creating another copy of it to run in a new thread. Figure 34.8 shows our out-of-core columnsort pipeline with the sort stage and permute stage replicated. The two copies of each stage work on different buffers concurrently, thus passing buffers through the entire pipeline more quickly. Stage replication can be done statically by the user or dynamically by FG.

FG does not limit how many replicas a stage may have. There is a practical limit, however. To exploit parallelism, each copy of a replicated stage should run on a separate processor of the same node. Therefore, after a certain point, which depends on how many processors each node of the cluster has, replicating stages yields no further performance improvement.

34.6.6 Directed Acyclic Graphs

An FG programmer can stray even further from the linear pipeline model by creating programs that fit a DAG structure. A stage may convey its buffers to any other stage, rather than to only its linear successor, as long as the structure of the stages is acyclic. Moreover, stages can make these decisions on the fly, so that FG need not know in advance the paths that buffers take through the pipeline.

Recall the pipeline for one pass of out-of-core columnsort shown in Figure 34.2. Columnsort requires four separate passes to successfully sort an input dataset. For each pass, the programmer builds a new pipeline that resembles the one shown in Figure 34.2, but pipelines for the passes are not quite identical to one another. Although each pass has the same basic structure, the four passes together contain 13 distinct stages: two read stages, four sort stages, three communicate stages, one permute stage, and three write stages. The particular combination of read, sort, communicate, permute, and write stages changes in every pass.

With only a linear pipeline at our disposal, we would build the four pipelines for columnsort one at a time: we run the pipeline for the first pass, dismantle it, build an entirely new pipeline for the second pass, dismantle that pipeline, and so on. Figure 34.9 illustrates how we can instead build only one pipeline that is able to run all four passes of columnsort. This pipeline has a DAG structure and contains all 13 columnsort stages, as well as a new stage, initialize, which assigns each buffer to a particular path through the pipeline according to the buffer's round number. (The first several buffers traverse the path for pass 1, the next several buffers traverse the path for pass 2, and so on.) We use the same DAG, with the same 13 stages, throughout the execution of columnsort, instead of building a new pipeline, with different stages, in each pass.

34.7 How FG Speeds Up Programs

We have seen that FG uses two methods for creating high-performance implementations. It manages the buffers that amortize the cost of transferring data from disk to memory, and it overlaps work by means of dynamic thread scheduling. In this section, we will see how these techniques enable FG to improve performance for some parallel programs by up to approximately 26%.

We use as benchmarks three applications from the High-Performance Computing Challenge (HPCC) benchmark suite [13]. The HPCC benchmark codes were designed to measure how parallel machines perform, and we use them here to measure the overall wall-clock time of the original implementations against code rewritten for FG. The original HPCC benchmark programs are written in Fortran. To obtain a fair comparison with FG, which uses C++, we ported the original Fortran code to C.

Because the HPCC benchmark applications generate data in memory, the main performance bottleneck turns out to be interprocessor communication, rather than disk I/O. The original Fortran and equivalent C versions of the HPCC benchmarks do not use threads because it is assumed that the cost of a context switch is too great to make multithreading worthwhile. Yet, FG's performance improvements come from the use of dynamic, thread-based scheduling.

Table 34.1 summarizes running times, in seconds, for the benchmark applications in their original Fortran implementations, equivalent C implementations, and converted FG implementations. Input dataset sizes range from 256 to 4096 MB. The improvements are the result of experiments run on a 16-node Beowulf cluster with two 2.8-GHz processors and 4 GB of RAM per node.

The first benchmark application, matrix transpose, is based on the algorithm described in Reference 5. The pipeline for matrix transpose works as follows. Operations are performed on submatrices, called blocks, rather than on the individual matrix elements. First, we generate a matrix A that is distributed across the cluster so that each processor has some portion of A in its local memory. Second, the processors perform communication operations: every block $A(I, J)$ moves to the appropriate destination processor—specifically, the processor that originally held the block $A(J, I)$. Finally, after the communication operations, each block is transposed so that the final transposed matrix A^T is distributed across all the processors. Table 34.1 shows that, compared to the C implementation, FG reduces running time significantly in all cases, but we see the most improvement from the smallest input dataset of 256 MB. In that case, FG reduces running time by 26.88% over the C implementation. As the problem size increases, the communication time dominates the overall running time of the application, and so there is less opportunity for FG to overlap operations. Consequently, we see a drop in improvement as the problem size increases up to 4096 MB, where the benefit from FG is 16.20%.

The second benchmark application, matrix multiplication, is based on the algorithm described in Reference 6. The pipeline for matrix multiplication operates as follows. Each processor owns some columns of input matrix A and some rows of input matrix B. Each processor also is a destination for some portion of the result matrix C. In the first stage of the pipeline, each processor makes some initial summations into its portion of C using the columns of A and rows of B in its local memory. The next stage is a communication stage: the processors transfer some rows of B to one another, and each processor sums into C with its new row data. Then, communication operations transfer some columns of A, and again each processor sums into C. Additional pipeline stages perform similar communication and local computation operations until each processor has received enough columns of A and rows of B to compute its portion of C. As shown in Table 34.1, the greatest reduction in running time comes from the smallest input data size of 256 MB, for which FG yields an improvement of 24.70% over the C implementation. As with matrix

TABLE 34.1 Running Times in Seconds for Parallel Matrix Transpose, Matrix Multiplication, and FFT

Data size (MB)	Fortran	C	FG	Improvement (%)
Matrix Transpose				
256	60.13	60.31	44.10	26.88
512	118.74	121.63	91.00	25.18
1024	199.46	202.76	157.29	22.43
2048	365.74	387.75	316.05	18.49
4096	726.89	758.68	635.76	16.20
Matrix Multiplication				
256	54.19	54.94	41.37	24.70
512	159.37	161.80	124.92	22.79
1024	572.64	589.55	475.70	19.31
2048	1188.64	1275.95	1059.07	17.00
4096	2810.36	2820.06	2416.48	14.31
FFT				
256	4.61	4.73	3.74	20.80
512	12.91	13.25	10.25	22.64
1024	26.12	26.66	20.80	21.98
2048	56.52	58.12	45.45	21.80
4096	121.70	124.54	95.73	23.13

Times shown are for the original Fortran implementations, converted C implementations, and FG implementations, each running on 16 processors. The improvement shown is the FG implementation over the C implementation. Each time shown is the average of three runs.

transpose, this application becomes more communication bound as the problem size increases, and so we see a corresponding decrease in the improvements brought by FG.

The third benchmark application, a three-dimensional FFT implementation, is based on the algorithm described in Reference 19. The input dataset is three-dimensional matrix X of dimensions $m \times n \times r$. We can think of the FFT algorithm as computing $F_n X(F_m \otimes F_r)$, where F_j denotes the j-dimensional Fourier transpose matrix and $A \otimes B$ is the Kronecker product of A and B. First, the input matrix X is packed into a two-dimensional array, and each processor owns some columns of this matrix. If we let I_j be the j-dimensional identity matrix and use auxiliary matrix Y, then we can apply the split-transpose algorithm to decompose FFT into the following five steps [12]:

1. $X \leftarrow X(I_m \otimes F_r)$
2. $X \leftarrow X(F_m \otimes I_r)$
3. $Y \leftarrow X^T$
4. $Y \leftarrow Y F_n$
5. $X \leftarrow Y^T$

In the distributed-memory case, most of these steps occur locally in memory. In the first two steps, each processor performs the Kronecker and multiplication operations only on the columns of X that it has in its local memory. In the third step, the processors engage in one all-to-all communication operation to transpose the columns of X across all the processors. The final two steps are done in memory: a local multiplication and a local transpose.

Table 34.1 shows that FG reduces running time for FFT by as much as 23.13% compared to the C implementation. We see the smallest improvement, 20.80%, from the smallest dataset, and we see the greatest improvement, 23.13%, from the largest dataset. The running times for FFT are far smaller than the times for matrix multiplication and matrix transpose, owing to the communication patterns of the algorithms. FFT requires only one communication operation per round, whereas the other algorithms use communication patterns that are tied to the input size. We see a performance improvement nevertheless, because our experimental testbed comprises 2-processor nodes, and so two computation-based stages can overlap.

34.8 How FG Speeds Up Programming

In addition to reducing running time for some parallel programs, FG reduces the source code size of some programs, and it reduces the time required to design, write, and debug them as well. We call the time a programmer devotes to produce a working program the *code-development time*. A set of usability studies confirms that FG reduces lines of source code and code-development time for an asynchronous, pipelined program. Indeed, we will see in this section that for a set of programmers implementing this program, FG reduces source code size by an average of 36.6% and code-development time by an average of 32.5%.

The usability studies involved 18 graduate and undergraduate computer science students from three universities. The subjects were split into two groups: FG programmers and threads programmers. Both groups were asked to implement in-core columnsort. The FG programmers implemented an asynchronous version of in-core columnsort by using the FG programming environment. The threads programmers implemented an asynchronous version of in-core columnsort by writing the code that used standard pthreads and managed 10 buffers. All the subjects wrote code in C/C++. The subjects were initially supplied with a synchronous version of columnsort containing functions implementing each stage and a function that called these stage functions in order. The synchronous version passed just one buffer from function to function. The entire task was to convert this synchronous code into a program that runs asynchronously, passing ten buffers along a pipeline. Each stage of the pipeline runs one of the supplied stage functions.

The FG programmers required far less time to produce working code and used far fewer lines of source code than the threads programmers, as shown in Table 34.2. In total, 23 subjects participated in the experiment, but results shown are from the 18 subjects who completed the assignment. On average, the eight threads programmers required 256.87 min to convert the synchronous implementation of columnsort to an asynchronous implementation. The 10 FG programmers required an average of only 173.40 min, an improvement of 32.5% over the threads programmers. Moreover, the threads programmers produced 586 lines of source code on average, whereas the FG programmers averaged only 371.78 lines, a reduction

TABLE 34.2 Code-Development Times in Minutes for the Threads Programmers and FG Programmers

	Threads programmers	FG programmers
	185	93
	215	140
	220	147
	225	160
	235	170
	265	175
	305	185
	405	100
		214
		250
Average	256.87	173.40

The threads programmers required between 185 and 405 min to convert the synchronous program into an asynchronous one, and they averaged 256.87 min. The FG programmers required between 93 and 250 min, and they averaged 173.40 min.

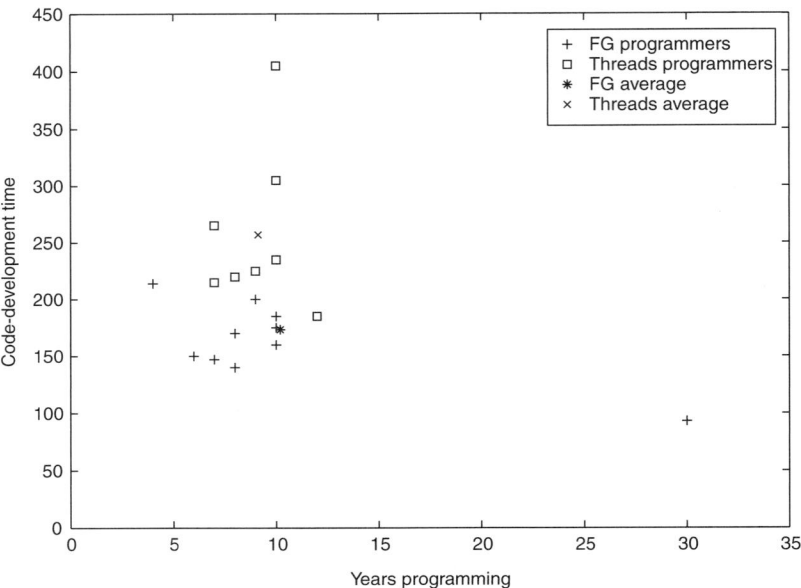

FIGURE 34.10 A scatter plot showing years of experience and code-development time for each subject who participated in the usability experiments, along with averages for both groups of programmers.

of 36.6%. The FG programmers were more efficient even though none of them was familiar with FG prior to the experiment, whereas all of the threads programmers had some experience coding with threads.

The FG programmers had more experience programming, on average, than the threads programmers. The FG programmers averaged 10.20 years of experience coding, compared with 9.13 years for the threads programmers. There is not a direct correlation between time and experience, however, as illustrated in Figure 34.10. Indeed, the programmer with the longest code-development time is a threads programmer with 10 years of programming experience—more than the threads programmers' average of 9.13 years. In addition, several FG and threads programmers have identical years of experience, and so we can compare them directly. For example, there are three FG programmers and three threads programmers with 10 years of coding experience. All of the FG programmers have shorter code-development times than any of the three equivalent threads programmers. The average time required for the FG programmers with 10 years' experience was 185 min, whereas the average time required for the threads programmers with 10 years' experience was 315 min. In other words, for these programmers, the FG group required 41.3% less time. We see similar trends for the FG and threads programmers with 7, 8, and 9 years of experience.

Although the FG programmers are slightly more experienced than the threads programmers, the results are not heavily biased in favor of FG. Researchers such as Curtis et al. [10] have shown that a programmer's skill and experience have a direct impact on the time required to write a program from scratch. They have also shown, however, that these variables matter little for the time required to comprehend and modify an existing program, which was the task asked of the subjects in the FG study.

Without accounting for individual factors such as experience, we can see that FG reduces code-development time by using a statistical hypothesis test [11]. This tool applies because, in addition to experience, other individual factors, such as effort exerted, may directly impact a subject's code-development time but are difficult to quantify [10, 15, 16]. Thus, it is useful to visit the experimental results without considering even years of experience as a mitigating factor [11]. We can create a statistical hypothesis by assuming that the subjects in the study are equal in intrinsic ability. The probability that a programmer randomly chosen from a pool completes a given task faster than another randomly chosen programmer is 1/2. That is, we can liken the experiment to a series of coin tosses. The statistical hypothesis test works by choosing a null hypothesis and showing it to be likely or unlikely enough that we can accept or reject it.

From the 18 subjects in the FG studies, there are 80 pairwise comparisons of FG programmers to threads programmers. In 72 of these pairs, the FG programmer was faster than the threads programmer. Our statistical hypothesis test shows that this result would be unlikely to occur by chance. We define our null hypothesis, H_0, as follows: FG does not reduce code-development time. That is, for a given pair of programmers, one drawn from the FG group and the other from the threads group, H_0 states that the FG programmer exhibits a faster code-development time than the threads programmer with probability $1/2$. Because each comparison is between an FG programmer and a threads programmer, all comparisons are independent. With our null hypothesis and independence, the binomial formula gives the probability that the FG programmer is faster in 72 out of 80 pairs: 1.199×10^{-14}, which is the same as the probability of 72 out of 80 coin flips turning up heads. Therefore, it is highly unlikely that 72 out of 80 pairs would favor the FG programmer if FG itself were not a contributing factor to reducing code-development time.

As shown in Figure 34.10, there is one outlier subject in the experiments, with by far the most experience (30 years) and shortest code-development time (93 min). He was designated an FG programmer; if we discard his data, the average experience of the remaining FG programmers is 8 years, less than the inclusive FG experience of 10.20 years and also less than the 9.13 average for all the threads programmers. By excluding this outlier, the FG programmers have less experience, on average, than the threads programmers, but still exhibit far faster code-development times. Indeed, after discarding the outlier, the average code-development time for FG programmers is 182.33 min: 29% faster than the threads programmers' average time of 256.87 min.

34.9 Conclusion

Latency has proven to be a significant obstacle to high-performance computing. Parallel programming on a distributed-memory cluster introduces latency that arises from reading and writing data on disk and from interprocessor communication. We have seen that FG, a programming environment meant to mitigate latency in parallel programs, improves the program-development process in two ways. First, it improves performance. Compared to C implementations of three HPCC benchmark applications, FG reduced running time by up to approximately 26%. Second, FG improves usability by making programs smaller and less complex. For an asynchronous, pipelined program, usability studies show that FG reduces lines of source code by an average of 36.6% and code-development time by an average of 32.5%. Furthermore, FG applies to several program structures, including linear pipelines, multiple pipelines, intersecting pipelines, and acyclic variations on pipelines.

References

[1] Alok Aggarwal and Jeffrey Scott Vitter. The input/output complexity of sorting and related problems. *Communications of the ACM*, 31:1116–1127, September 1988.

[2] Lauren M. Baptist and Thomas H. Cormen. Multidimensional, multiprocessor, out-of-core FFTs with distributed memory and parallel disks. In *Proceedings of the Eleventh Annual ACM Symposium on Parallel Algorithms and Architectures*, pp. 242–250, June 1999.

[3] Geeta Chaudhry, Thomas H. Cormen, and Leonard F. Wisniewski. Columnsort lives! An efficient out-of-core sorting program. In *Proceedings of the Thirteenth Annual ACM Symposium on Parallel Algorithms and Architectures*, pp. 169–178, July 2001.

[4] Geeta Chaudhry and Thomas H. Cormen. Getting more from out-of-core columnsort. In *4th Workshop on Algorithm Engineering and Experiments (ALENEX 02)*, pp. 143–154, January 2002.

[5] Jaeyoung Choi, Jack J. Dongarra, and David W. Walker. Parallel matrix transpose algorithms on distributed memory concurrent computers. Technical Report UT-CS-93-215, University of Tennessee, 1993.

[6] Jaeyoung Choi, Jack J. Dongarra, and David W. Walker. PUMMA: Parallel universal matrix multiplication algorithms on distributed memory concurrent computers. *Concurrency: Practice and Experience*, 6:543–570, 1994.

[7] Thomas H. Cormen and Elena Riccio Davidson. FG: A framework generator for hiding latency in parallel programs running on clusters. In *Proceedings of the 17th International Conference on Parallel and Distributed Computing Systems (PDCS-2004)*, pp. 137–144, September 2004.

[8] Thomas H. Cormen, Thomas Sundquist, and Leonard F. Wisniewski. Asymptotically tight bounds for performing BMMC permutations on parallel disk systems. *SIAM Journal on Computing*, 28: 105–136, 1999.

[9] Thomas H. Cormen, Jake Wegmann, and David M. Nicol. Multiprocessor out-of-core FFTs with distributed memory and parallel disks. In *Proceedings of the Fifth Workshop on I/O in Parallel and Distributed Systems (IOPADS '97)*, pp. 68–78, November 1997.

[10] Bill Curtis, Sylvia B. Sheppard, and Phil Milliman. Third time charm: Stronger prediction of programmer performance by software complexity metrics. In *ICSE '79: Proceedings of the 4th International Conference on Software Engineering*, pp. 356–360, Piscataway, NJ, USA, IEEE Press, 1979.

[11] Oren Etzioni and Ruth Etzioni. Statistical methods for analyzing speedup learning experiences. *Machine Learning*, 14:333–347, 1993.

[12] Markus Hegland. Real and complex fast Fourier transforms on the Fujitsu VPP 500. *Parallel Computing*, 22:539–553, 1996.

[13] HPC Challenge. http://icl.cs.utk.edu/hpcc, May 2005.

[14] IEEE. Standard 1003.1-2001, Portable operating system interface, 2001.

[15] Michael G. Morris, Cheri Speier, and Jeffrey A. Hoffer. The impact of experience on individual performance and workload differences using object-oriented and process-oriented systems analysis techniques. In *Proceedings of the 29th Hawaii International Conference on System Sciences*, pp. 232–241, 1996.

[16] Ronald H. Rasch and Henry L. Tosi. Factors affecting software developers' performance: An integrated approach. *Management Information Systems Quarterly*, 16:395–413, 1992.

[17] Kazuhiro Sado and Yoshihide Igarashi. Some parallel sorts on a mesh-connected processor array and their time efficiency. *Journal of Parallel and Distributed Computing*, 3:398–410, September 1986.

[18] Isaac D. Scherson, Sandeep Sen, and Adi Shamir. Shear sort: A true two-dimensional sorting technique for VLSI networks. In *Proceedings of the 1986 International Conference on Parallel Processing*, pp. 903–908, August 1986.

[19] Daisuke Takahashi. Efficient implementation of three-dimensional FFTs on clusters of PCs. *Computer Physics Communications*, 152:144–150, 2003.

[20] Jeffrey Scott Vitter. External memory algorithms and data structures: Dealing with MASSIVE DATA. *ACM Computing Surveys*, 33:209–271, June 2001.

[21] Jeffrey Scott Vitter and Elizabeth A. M. Shriver. Algorithms for parallel memory I: Two-level memories. *Algorithmica*, 12:110–147, August and September 1994.

35
High-Performance Techniques for Parallel I/O

Avery Ching
Kenin Coloma
Jianwei Li
Wei-keng Liao
Alok Choudhary
Northwestern University

35.1	Introduction...	35-1
35.2	Portable File Formats and Data Libraries	35-1
	File Access in Parallel Applications • NetCDF and Parallel NetCDF • HDF5	
35.3	General MPI–IO Usage and Optimizations............	35-6
	MPI–IO Interface • Significant Optimizations in ROMIO • Current Areas of Research in MPI–IO	
35.4	Parallel File Systems	35-15
	Summary of Current Parallel File Systems • Noncontiguous I/O Methods for Parallel File Systems • I/O Suggestions for Application Developers	
	References	35-22

35.1 Introduction

An important aspect of any large-scale scientific application is data storage and retrieval. I/O technology lags other computing components by several orders of magnitude with a performance gap that is still growing. In short, much of I/O research is dedicated to narrowing this gap.

Applications that utilize high-performance I/O do so at a specific level in the parallel I/O software stack depicted in Figure 35.1. In the upper levels, file formats and libraries such as netCDF and HDF5 provide certain advantages for particular application groups. MPI–IO applications can leverage optimizations in the MPI specification [Mes] for various operations in the MPI–IO and file system layers. This chapter explains many powerful I/O techniques applied to each stratum of the parallel I/O software stack.

35.2 Portable File Formats and Data Libraries

Low level I/O interfaces, like UNIX I/O, treat files as sequences of bytes. Scientific applications manage data at a higher level of abstraction where users can directly read/write data as complex structures instead of byte streams and have all type information and other useful metadata automatically handled. Applications commonly run on multiple platforms also require portability of data so that the data generated from one platform can be used on another without transformation. As most scientific applications are programmed

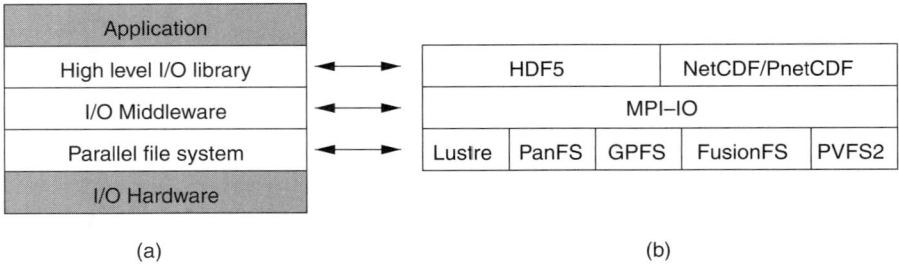

FIGURE 35.1 (a) Abstract I/O software stack for scientific computing. (b) Current components of the commonly used I/O software stack.

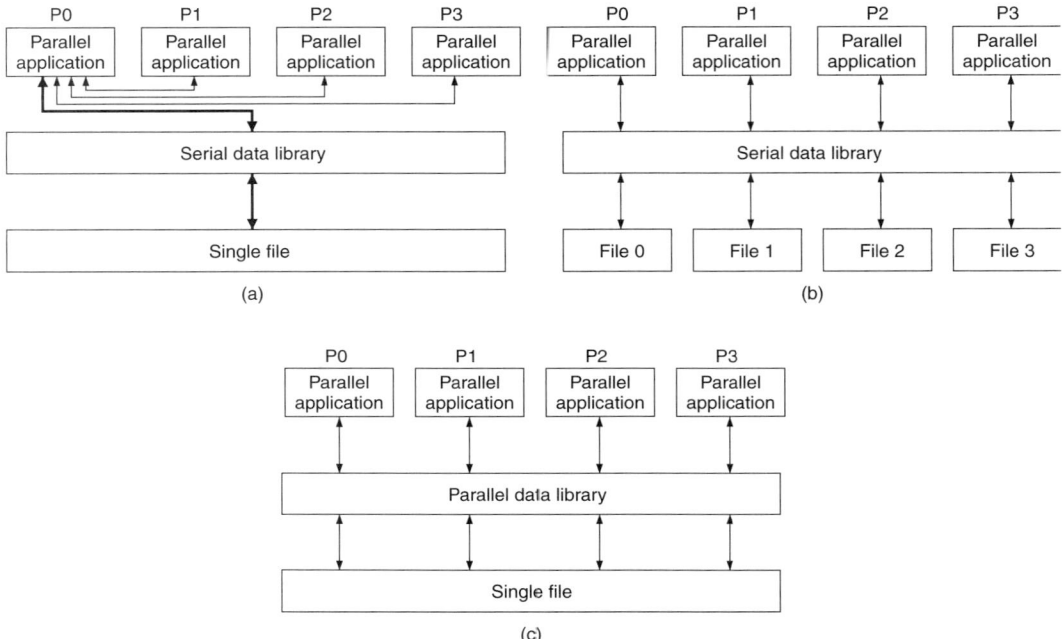

FIGURE 35.2 Using data libraries in parallel applications: (a) using a serial API to access single files through a single process; (b) using a serial API to access multiple files concurrently and independently; (c) using a parallel API to access single files cooperatively or collectively.

to run in parallel environments, parallel access to the data is desired. This section describes two popular scientific data libraries and their portable file formats, netCDF and HDF5.

35.2.1 File Access in Parallel Applications

Before presenting a detailed description of library design, general approaches for accessing portable files in parallel applications (in a message-passing environment) are analyzed. The first and most straightforward approach is described in the scenario of Figure 35.2a where one process is in charge of collecting/distributing data and performing I/O to a single file using a serial API. The I/O requests from other processes are carried out by shipping all the data through this single process. The drawback of this approach is that collecting all I/O data on a single process can easily create an I/O performance bottleneck and also overwhelm its memory capacity.

In order to avoid unnecessary data shipping, an alternative approach has all processes perform their I/O independently using the serial API, as shown in Figure 35.2b. In this way, all I/O operations can proceed concurrently, but over separate files (one for each process). Managing a dataset is more difficult, however, when it is spread across multiple files. This approach undermines the library design goal of easy data integration and management.

A third approach introduces a parallel API with parallel access semantics and an optimized parallel I/O implementation where all processes perform I/O operations to access a single file. This approach, as shown in Figure 35.2c, both frees the users from dealing with parallel I/O intricacies and provides more opportunities for various parallel I/O optimizations. As a result, this design principle is prevalent among modern scientific data libraries.

35.2.2 NetCDF and Parallel NetCDF

NetCDF [RD90], developed at the Unidata Program Center, provides applications with a common data access method for the storage of structured datasets. Atmospheric science applications, for example, use netCDF to store a variety of data types that include single-point observations, time series, regularly spaced grids, and satellite or radar images. Many organizations, such as much of the climate modeling community, rely on the netCDF data access standard for data storage.

NetCDF stores data in an array-oriented dataset that contains dimensions, variables, and attributes. Physically, the dataset file is divided into two parts: file header and array data. The header contains all information (metadata) about dimensions, attributes, and variables except for the variable data itself, while the data section contains arrays of variable values (raw data). Fix-sized arrays are stored contiguously starting from given file offsets, while variable-sized arrays are stored at the end of the file as interleaved records that grow together along a shared unlimited dimension.

The netCDF operations can be divided into the five categories as summarized in Table 35.1. A typical sequence of operations to write a new netCDF dataset is to create the dataset; define the dimensions, variables, and attributes; write variable data; and close the dataset. Reading an existing netCDF dataset involves first opening the dataset; inquiring about dimensions, variables, and attributes; then reading variable data; and finally, closing the dataset.

The original netCDF API was designed for serial data access, lacking parallel semantics and performance. Parallel netCDF (PnetCDF) [LLC[+]03], developed jointly between Northwestern University and Argonne National Laboratory (ANL), provides a parallel API to access netCDF files with significantly better performance. It is built on top of MPI–IO, allowing users to benefit from several well-known optimizations already used in existing MPI–IO implementations, namely, the data sieving and two phase I/O strategies in ROMIO. MPI–IO is explained in further detail in Section 35.3. Figure 35.3 describes the overall architecture for PnetCDF design.

In PnetCDF, a file is opened, operated, and closed by the participating processes in an MPI communication group. Internally, the header is read/written only by a single process, although a copy is cached in local memory on each process. The *root* process fetches the file header, broadcasts it to all processes when opening a file, and writes the file header at the end of the define mode if any modifications occur

TABLE 35.1 NetCDF Library Functions

Function type	Description
Dataset functions	Create/open/close a dataset, set the dataset to define/data mode, and synchronize dataset
Define mode functions	Define dataset dimensions and variables
Attribute functions	Manage adding, changing, and reading attributes of datasets
Inquiry functions	Return dataset metadata: dim(id, name, len), var(name, ndims, shape, id)
Data access functions	Provide the ability to read/write variable data in one of the five access methods: single value, whole array, subarray, subsampled array (strided subarray), and mapped strided subarray

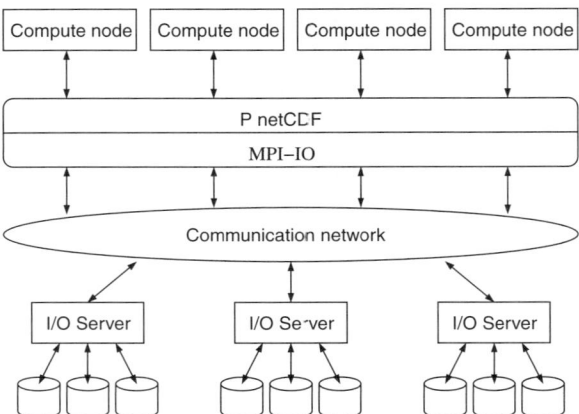

FIGURE 35.3 Design of PnetCDF on a parallel I/O architecture. PnetCDF runs as a library between the user application and file system. It processes parallel netCDF requests from user compute nodes and, after optimization, passes the parallel I/O requests down to MPI–IO library. The I/O servers receive the MPI–IO requests and perform I/O over the end storage on behalf of the user.

in the header. The define mode functions, attribute functions, and inquiry functions all work on the local copy of the file header. All define mode and attribute functions are made collectively and require all the processes to provide the same arguments when adding, removing, or changing definitions so the local copies of the file header are guaranteed to be the same across all processes from the time the file is collectively opened until it is closed.

The parallelization of the data access functions is achieved with two subset APIs, the *high-level API* and the *flexible API*. The high-level API closely follows the original netCDF data access functions and serves as an easy path for original netCDF users to migrate to the parallel interface. These calls take a single pointer for a contiguous region in memory, just as the original netCDF calls did, and allow for the description of single elements (`var1`), whole arrays (`var`), subarrays (`vara`), strided subarrays (`vars`), and multiple noncontiguous regions (`varm`) in a file. The flexible API provides a more MPI-like style of access by providing the user with the ability to describe noncontiguous regions in memory. These regions are described using MPI datatypes. For application programmers that are already using MPI for message passing, this approach should be natural. The file regions are still described using the original parameters. For each of the five data access methods in the flexible data access functions, the corresponding data access pattern is presented as an MPI file view (a set of data visible and accessible from an open file) constructed from the variable metadata (shape, size, offset, etc.) in the netCDF file header and user provided starts, counts, strides, and MPI datatype arguments. For parallel access, each process has a different file view. All processes can collectively make a single MPI–IO request to transfer large contiguous data as a whole, thereby preserving useful semantic information that would otherwise be lost if the transfer were expressed as per process noncontiguous requests.

35.2.3 HDF5

Hierarchical Data Format (HDF) is a portable file format and software, developed at the National Center for Supercomputing Applications (NCSA). It is designed for storing, retrieving, analyzing, visualizing, and converting scientific data. The current and most popular version is HDF5 [HDF], which stores multi-dimensional arrays together with ancillary data in a portable, self-describing file format. It uses a hierarchical structure that provides application programmers with a host of options for organizing how data is stored in HDF5 files. Parallel I/O is also supported.

HDF5 files are organized in a hierarchical structure, similar to a UNIX file system. Two types of primary objects, groups and datasets, are stored in this structure, respectively resembling directories and files in the UNIX file system. A group contains instances of zero or more groups or datasets while a dataset stores a multi-dimensional array of data elements. Both are accompanied by supporting metadata. Each group or dataset can have an associated attribute list to provide extra information related to the object.

A dataset is physically stored in two parts: a header and a data array. The header contains miscellaneous metadata describing the dataset as well as information that is needed to interpret the array portion of the dataset. Essentially, it includes the name, datatype, dataspace, and storage layout of the dataset. The name is a text string identifying the dataset. The datatype describes the type of the data array elements and can be a basic (atomic) type or a compound type (similar to a *struct* in C language). The dataspace defines the dimensionality of the dataset, that is, the size and shape of the multi-dimensional array. The dimensions of a dataset can be either fixed or unlimited (extensible). Unlike netCDF, HDF5 supports more than one unlimited dimension in a dataspace. The storage layout specifies how the data arrays are arranged in the file.

The data array contains the values of the array elements and can be either stored together in contiguous file space or split into smaller *chunks* stored at any allocated location. Chunks are defined as equally sized multi-dimensional subarrays (blocks) of the whole data array and each chunk is stored in a separate contiguous file space. The chunked layout is intended to allow performance optimizations for certain access patterns, as well as for storage flexibility. Using the chunked layout requires complicated metadata management to keep track of how the chunks fit together to form the whole array. Extensible datasets whose dimensions can grow are required to be stored in chunks. One dimension is increased by allocating new chunks at the end of the file to cover the extension.

The HDF5 library provides several interfaces that are categorized according to the type of information or operation the interface manages. Table 35.2 summarizes these interfaces.

To write a new HDF5 file, one needs to first create the file, adding groups if needed; create and define the datasets (including their datatypes, dataspaces, and lists of properties like the storage layout) under the desired groups; write the data along with attributes; and finally close the file. The general steps in reading an existing HDF5 file include opening the file; opening the dataset under certain groups; querying the dimensions to allocate enough memory to a read buffer; reading the data and attributes; and closing the file.

HDF5 also supports access to portions (or selections) of a dataset by *hyperslabs*, their unions, and lists of independent points. Basically, a hyperslab is a subarray or strided subarray of the multi-dimensional dataset. The selection is performed in the file dataspace for the dataset. Similar selections can be done in the memory dataspace so that data in one file pattern can be mapped to memory in another pattern as long as the total number of data elements is equal.

TABLE 35.2 HDF5 Interfaces

Interface		Function name prefix and functionality
Library functions	H5:	General HDF5 library management
Attribute interface	H5A:	Read/write attributes
Dataset interface	H5D:	Create/open/close and read/write datasets
Error interface	H5E:	Handle HDF5 errors
File interface	H5F:	Control HDF5 file access
Group interface	H5G:	Manage the hierarchical group information
Identifier interface	H5I:	Work with object identifiers
Property list interface	H5P:	Manipulate various object properties
Reference interface	H5R:	Create references to objects or data regions
Dataspace interface	H5S:	Defining dataset dataspace
Datatype interface	H5T:	Manage type information for dataset elements
Filters and compression interface	H5Z:	Inline data filters and data compression

HDF5 supports both sequential and parallel I/O. Parallel access, supported in the MPI programming environment, is enabled by setting the file access property to use MPI–IO when the file is created or opened. The file and datasets are collectively created/opened by all participating processes. Each process accesses part of a dataset by defining its own file dataspace for that dataset. When accessing data, the data transfer property specifies whether each process will perform independent I/O or all processes will perform collective I/O.

35.3 General MPI–IO Usage and Optimizations

Before MPI, there were proprietary message passing libraries available on several computing platforms. Portability was a major issue for application designers and thus more than 80 people from 40 organizations representing universities, parallel system vendors, and both industrial and national research laboratories formed the Message Passing Interface (MPI) Forum. MPI-1 was established by the forum in 1994. A number of important topics (including parallel I/O) had been intentionally left out of the MPI-1 specification and were to be addressed by the MPI Forum in the coming years. In 1997, the MPI-2 standard was released by the MPI Forum that addressed parallel I/O among a number of other useful new features for portable parallel computing (remote memory operations and dynamic process management). The I/O goals of the MPI-2 standard were to provide developers with a portable parallel I/O interface that could richly describe even the most complex of access patterns. ROMIO [ROM] is the reference implementation distributed with ANL's MPICH library. ROMIO is included in other distributions and is often the basis for other MPI–IO implementations. Frequently, higher level libraries are built on top of MPI–IO, which leverage its portability across different I/O systems while providing features specific to a particular user community. Examples such as netCDF and HDF5 were discussed in Section 35.2.

35.3.1 MPI–IO Interface

The purposely rich MPI–IO interface has proven daunting to many. This is the main obstacle to developers using MPI–IO directly, and also one of the reasons most developers subsequently end up using MPI–IO through higher level interfaces like netCDF and HDF5. It is, however, worth learning a bit of advanced MPI–IO, if not to encourage more direct MPI–IO programming, then to at least increase general understanding of what goes on in the MPI–IO level beneath the other high level interfaces. A very simple execution order of the functions described in this section is as follows:

1. MPI_Info_create/MPI_Info_set (optional)
2. datatype creation (optional)
3. MPI_File_open
4. MPI_File_set_view (optional)
5. MPI_File_read/MPI_File_write
6. MPI_File_sync (optional)
7. MPI_File_close
8. datatype deletion (optional)
9. MPI_Info_free (optional)

35.3.1.1 Open, Close, and Hints

```
MPI_File_open(comm, filename, amode, info, fh)
MPI_File_close(fh)
MPI_Info_create(info)
MPI_Info_set(info, key, value)
MPI_Info_free(info)
```

While definitely far from "advanced" MPI–IO, `MPI_File_open` and `MPI_File_close` still warrant some examination. The `MPI_File_open` call is the typical point at which to pass optimization information to an MPI–IO implementation. `MPI_Info_create` should be used to instantiate and initialize an `MPI_Info` object, and then `MPI_Info_set` is used to set specific hints (*key*) in the info object. The info object should then be passed to `MPI_File_open` and later freed with `MPI_Info_free` after the file is closed. If an info object is not needed, `MPI_INFO_NULL` can be passed to open. The hints in the info object are used to either control optimizations directly in an MPI–IO implementation or to provide additional access information to the MPI–IO implementation so it can make better decisions on optimizations. Some specific hints are described in Section 35.3.2. To get the hints of the info object back from the MPI–IO library the user should call `MPI_File_get_info` and be sure to free the info object after use.

35.3.1.2 Derived Datatypes

Before delving into the rest of the I/O interface and capabilities of MPI–IO, it is essential to have a sound understanding of derived datatypes. Datatypes are what distinguish the MPI–IO interface from the more familiar standard POSIX I/O interface.

One of the most powerful features of the MPI specification is user defined derived datatypes. MPI's derived datatypes allow a user to describe an arbitrary pattern in a memory space. This access pattern, possibly noncontiguous, can then be logically iterated over the memory space. Users may define derived datatypes based on elementary MPI predefined datatypes (`MPI_INT`, `MPI_CHAR`, etc.) as well as previously defined derived datatypes. A common and simple use of derived datatypes is to single out values for a specific subset of variables in multi-dimensional arrays.

After using one or more of the basic datatype creation functions in Table 35.3, `MPI_Type_commit` is used to finalize the datatype and must be called before use in any MPI–IO calls. After the file is closed, the datatype can then be freed with `MPI_Type_free`.

Seeing as a derived datatype simply maps an access pattern in a logical space, while the discussion above has focused on memory space, it could also apply to file space.

35.3.1.3 File Views

`MPI_File_set_view(fh, disp, etype, filetype, datarep, info)`

File views specify accessible file regions using derived datatypes. This function should be called after the file is opened, if at all. Not setting a file view allows the entire file to be accessed. The defining datatype is referred to as the *filetype*, and the *etype* is a datatype used as an elementary unit for positioning. Figure 35.4 illustrates how the parameters in `MPI_File_set_view` are used to describe a "window" revealing only certain bytes in the file. The displacement (*disp*) dictates the start location of the initial filetype in terms of etypes. The file view is defined by both the displacement and filetype together. While this function is collective, it is important that each process defines its own individual file view. All processes in the same communicator must use the same etype. The *datarep* argument is typically set to "native," and has to do with file interoperability. If compatibility between MPI environments is needed or the environment is

TABLE 35.3 Commonly Used MPI Datatype Constructor Functions.
Internal Offsets Can Be Described in Terms of the Base Datatype or in Bytes

Function	Internal offsets	Base types
MPI_Type_contiguous	None	Single
MPI_Type_vector	Regular (old types)	Single
MPI_Type_hvector	Regular (bytes)	Single
MPI_Type_index	Arbitrary (old types)	Single
MPI_Type_hindex	Arbitrary (bytes)	Single
MPI_Type_struct	Arbitrary (old types)	Mixed

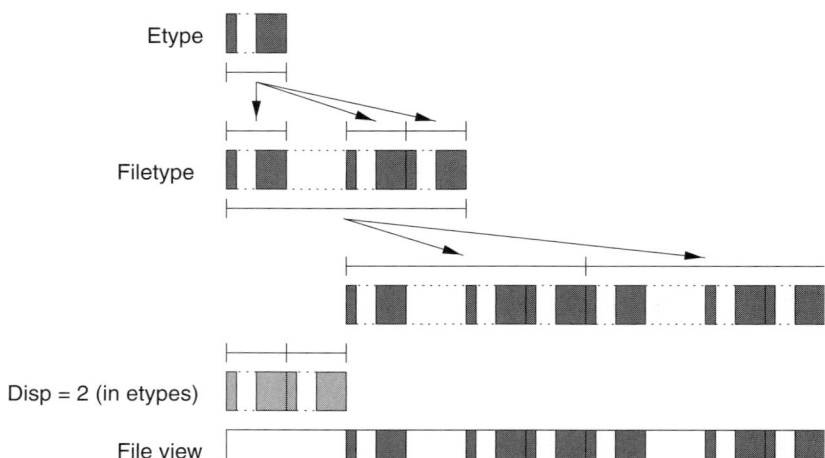

FIGURE 35.4 File views illustrated: filetypes are built from etypes. The filetype access pattern is implicitly iterated forward starting from the disp. An actual count for the filetype is not required as it conceptually repeats forever, and the amount of I/O done is dependent on the buffer datatype and count.

heterogeneous, then "external32" or "internal" should be used. File views allow an MPI–IO read or write to access complex noncontiguous regions in a single call. This is the first major departure from the POSIX I/O interface, and one of the most important features of MPI–IO.

35.3.1.4 Read and Write

```
MPI_File_read(fh, buf, count, datatype, status)
MPI_File_write(fh, buf, count, datatype, status)
MPI_File_read_at(fh, offset, buf, count, datatype, status)
MPI_File_write_at(fh, offset, buf, count, datatype, status)
MPI_File_sync(fh)
```

In addition to the typical MPI specific arguments like the MPI communicator, the datatype argument in these calls is the second important distinction of MPI–IO. Just as the file view allows one MPI–IO call to access multiple noncontiguous regions in file, the datatype argument allows a single MPI–IO call to access multiple memory regions in the user buffer with a single call. The *count* is the number of datatypes in memory being used.

The functions MPI_File_read and MPI_File_write use MPI_File_seek to set the position of the file pointer in terms of etypes. It is important to note that the file pointer position respects the file view, skipping over inaccessible regions in the file. Setting the file view resets the individual file pointer back to the first accessible byte.

The MPI_File_read_at and MPI_File_write_at, "_at" variations of the read and write functions, explicitly set out a starting position in the additional *offset* argument. Just as in the seek function, the offset is in terms of etypes and respects the file view.

Similar to MPI non-blocking communication, non-blocking versions of the I/O functions exist and simply prefix read and write with "i" so the calls look like MPI_File_iread. The I/O need not be completed before these functions return. Completion can be checked just as in non-blocking communication with completion functions like MPI_Wait.

The `MPI_File_sync` function is a collective operation used to ensure written data is pushed all the way to the storage device. Open and close also implicitly guarantee data for the associated file handle is on the storage device.

35.3.1.5 Collective Read and Write

```
MPI_File_read_all(fh, buf, count, datatype, status)
MPI_File_write_all(fh, buf, count, datatype, status)
```

The collective I/O functions are prototyped the same as the independent `MPI_File_read` and `MPI_File_write` functions and have "_at" equivalents as well. The difference is that the collective I/O functions must be called collectively among all the processes in the communicator associated with the particular file at open time. This explicit synchronization allows processes to actively communicate and coordinate their I/O efforts for the call. One major optimization for collective I/O is disk directed I/O [Kot97]. Disk directed I/O allows I/O servers to optimize the order in which local blocks are accessed. Another optimization for collective I/O is the two phase method described in further detail in the next section.

35.3.2 Significant Optimizations in ROMIO

The ROMIO implementation of MPI–IO contains several optimizations based on the POSIX I/O interface, making them portable across many file systems. It is possible, however, to implement a ROMIO driver with optimizations specific to a given file system. In fact, the current version of ROMIO as of this writing (2005-06-09) already includes optimizations for PVFS2 [The] and other file systems. The most convenient means for controlling these optimizations is through the MPI–IO hints infrastructure mentioned briefly above.

35.3.2.1 POSIX I/O

All parallel file systems support what is called the POSIX I/O interface, which relies on an offset and a length in both memory and file to service an I/O request. This method can service noncontiguous I/O access patterns by dividing them up into contiguous regions and then individually accessing these regions with corresponding POSIX I/O operations. While such use of POSIX I/O can fulfill any noncontiguous I/O request with this technique, it does incur several expensive overheads. The division of the I/O access pattern into smaller contiguous regions significantly increases the number of I/O requests processed by the underlying file system. Also, the division often forces more I/O requests than the actual number of noncontiguous regions in the access pattern as shown in Figure 35.5a. The serious overhead sustained from servicing so many individual I/O requests limits performance for noncontiguous I/O when using the POSIX interface. Fortunately for users which have access to file systems supporting only the POSIX interface, two important optimizations exist to more efficiently perform noncontiguous I/O while using only the POSIX I/O interface: data sieving I/O and two phase I/O.

35.3.2.2 Data Sieving

Since hard disk drives are inherently better at accessing large amounts of sequential data, the *data sieving* technique [TGL99a] tries to satisfy multiple small I/O requests with a larger contiguous I/O access and later "sifting" the requested data in or out of a temporary buffer. In the read case, a large contiguous region of file data is first read into a temporary data sieving buffer and then the requested data is copied out of the temporary buffer into the user buffer. For practical reasons, ROMIO uses a maximum data sieving buffer size so multiple data sieving I/O requests may be required to service an access patterns. ROMIO will always try to fill the entire data sieving buffer each time in order to maximize the number of file regions encompassed. In the write case, file data must first be read into the data sieving buffer unless the user define regions in that contiguous file region cover the entire data sieving region. User data can then be copied into the data sieving buffer and then the entire data sieving buffer is written to the file in a single I/O call. Data sieving writes require some concurrency control since data that one process does

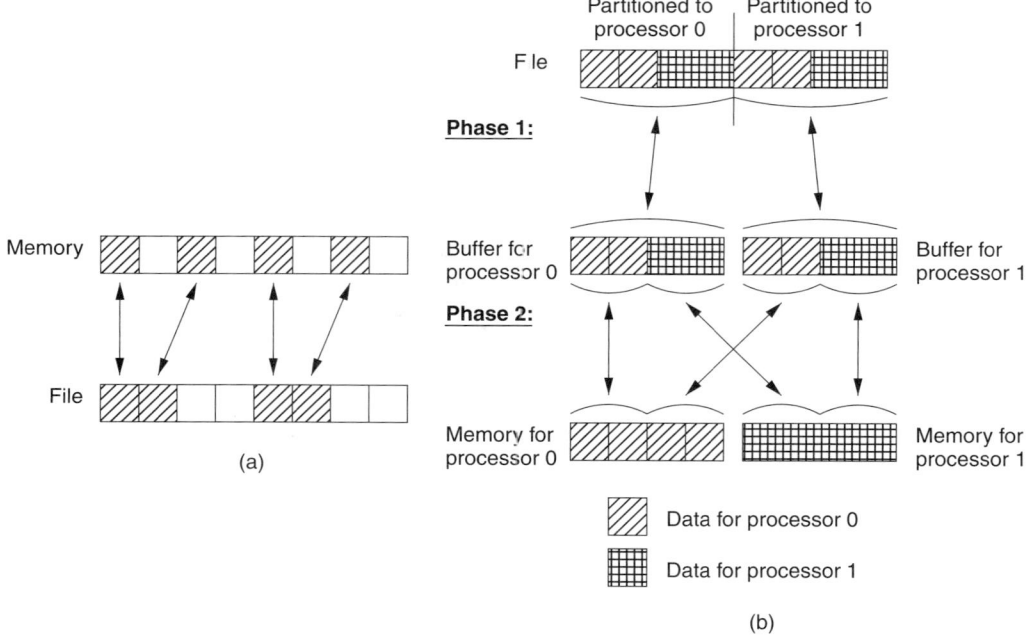

FIGURE 35.5 (a) Example POSIX I/O request. Using traditional POSIX interfaces for this access pattern cost four I/O requests, one per contiguous region. (b) Example two phase I/O request. Interleaved file access patterns can be effectively accessed in larger file I/O operations with the two phase method.

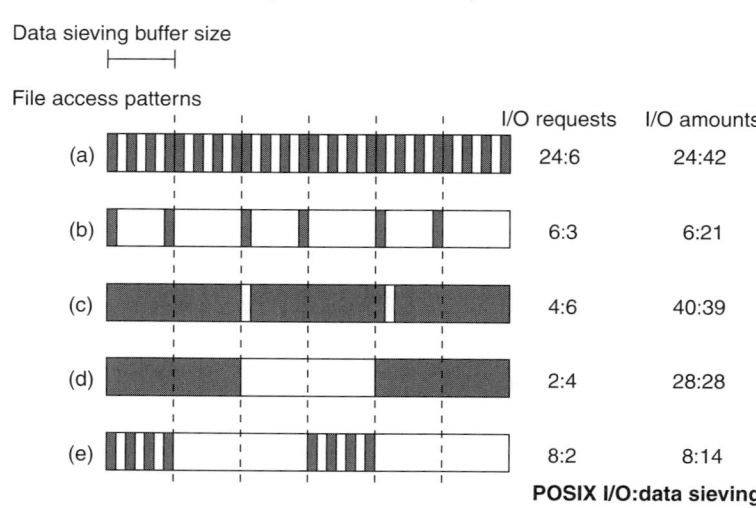

FIGURE 35.6 (a) Probably data sieve: Data sieving reduces I/O requests by a factor of 4, but almost doubles the I/O amount (b) Do not data sieve: Data sieving I/O requests are reduced by half, but almost 4 (8 if write) times more data is accessed (c) Do not data sieve: Data sieving increases I/O requests and only marginally reduces I/O amount. (d) Do not data sieve (Pareto optimal): Data sieving doubles I/O requests, but has no effect on I/O amount. (e) Probably data sieve: Data sieving reduces I/O requests by a factor of 4, but almost doubles I/O.

not intend to modify is still read and then written back with the potential of overwriting changes made by other processes.

Data sieving performance benefits come from reducing the number of head seeks on the disk, the cut in the accrued overhead of individual I/O requests, and large I/O accesses. Figure 35.6a,e illustrates specific cases where data sieving may do well. Data sieving is less efficient when data is either sparsely distributed or the access pattern consists of contiguous regions much larger than the data sieving buffer size (end case would be a completely contiguous access pattern). In the sparse case, as in Figure 35.6b,d, the large data sieving I/O request may only satisfy a few user requests, and in even worse, may be accessing much more data than will actually be used (Figure 35.6b). The number of I/O accesses may not be reduced by much, and the extra time spent accessing useless data may be more than the time taken to make more small I/O requests. In the case where the user's access pattern is made up of contiguous regions nearly the size of or greater than the data sieving buffer size, shown in Figure 35.6c,d, the number of I/O requests generated may actually be greater than the number of I/O requests generated had the user's I/O requests been passed directly to the file system. Additionally, data sieving will have been double buffering, and paid an extra memory copy penalty for each time the data sieve buffer was filled and emptied.

One factor not yet considered is the user memory buffer. If the user memory buffer is noncontiguous with small regions (relative to the data sieving buffer), it will have the effect of breaking up, but not separating what might have been large contiguous regions in file, thus creating an numerous I/O requests for POSIX I/O. This effect is illustrated in Figure 35.7, and presents an ideal opportunity for data sieving to reduce the overall number of I/O calls, as well as making efficient use of the data sieving buffer. Even if the original filetype consisted of large sparsely distributed regions, data sieving would still likely prove to be very beneficial.

So while data sieving could conceivably result in worse performance (the point at which would be sooner in the case of read-modify-write data sieving writes), some simple considerations can be kept in mind to determine whether data sieving will be a benefit or detriment. Assuming data is fairly uniformly spaced (no locally dense, overall sparse distributions), and the user access pattern is indeed noncontiguous, Figure 35.8 provides a quick table for determining when data sieving is most appropriate. Small, big, sparse, and dense metrics are all relative to the data sieving buffer size. An MPI–IO implementation ought to preprocess the user's access pattern at least to some degree to determine the appropriateness of data sieving on its own. As mentioned earlier, however, less uniform access patterns may require some user intervention as an automated runtime determination may not catch certain cases. In the previous example (Figure 35.6e), an access pattern that consists of clusters of densely packed data will likely benefit from data sieving. Using only the data sieving technique for I/O will be referred to as *data sieving* I/O.

35.3.2.3 Two Phase I/O

Figure 35.5b illustrates the two phase method for collective I/O [TGL99b], which uses both PCSIX I/O and data sieving. This method is referred to as *two phase* I/O throughout this chapter. The two phase method identifies a subset of the application processes that will actually do I/O; these processes are called *aggregators*. Each aggregator is responsible for I/O to a specific and disjoint portion of the file.

In an effort to heuristically balance I/O load on each aggregator, ROMIO calculates these *file realms* dynamically on the basis of the aggregate size and location of the accesses in the collective operation. When performing a read operation, aggregators first read a contiguous region containing desired data from storage and put this data in a local temporary buffer. Next, data is redistributed from these temporary buffers to the final destination processes. Write operations are performed in a similar manner. First, data is gathered from all processes into temporary buffers on aggregators. Aggregators read data from storage to fill in the holes in the temporary buffers to make contiguous data regions. Next, this temporary buffer is written back to storage using POSIX I/O operations. An approach similar to data sieving is used to optimize this write back to storage when there are still gaps in the data. As mentioned earlier, data sieving is also used in the read case. Alternatively, other noncontiguous access methods, such as the ones described in Section 35.4.2, can be leveraged for further optimization.

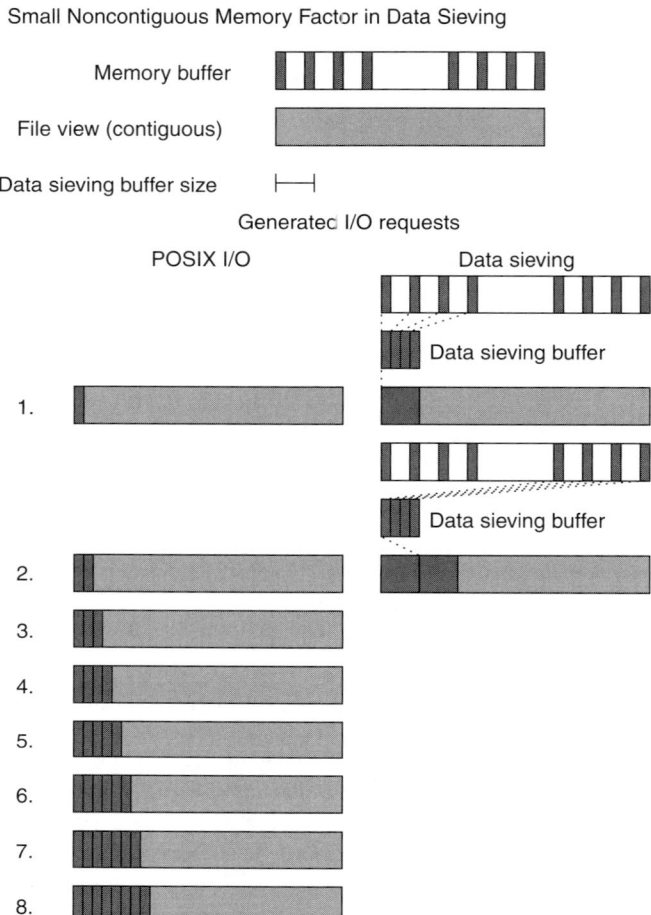

FIGURE 35.7 Evaluating the file access pattern alone in this case does not paint the entire I/O picture. The small noncontiguous memory pieces break up the large contiguous file access pattern into many small I/O requests. Since these small I/O requests end up next to each other, data sieving can reduce the number of I/O requests by a factor of 4 without accessing any extraneous data, making data sieving Pareto optimal, assuming it takes longer to read/write 1 unit of data 4 times than to copy 4 units of data into or out of the buffer and to read/write 4 units of data.

The big advantage of two phase I/O is the consolidation by aggregators of the noncontiguous file accesses from all processes into only a few large I/O operations. One significant disadvantage of two phase I/O is that all processes must synchronize on the open, set view, read, and write calls. Synchronizing across large numbers of processes with different sized workloads can be a large overhead. Two phase I/O performance relies heavily on the particular MPI implementation's data movement performance. If the MPI implementation is not significantly faster than the aggregate I/O bandwidth in the system, the overhead of the additional data movement in two phase I/O will likely prevent two phase I/O from outperforming direct access optimizations like list I/O and datatype I/O discussed later.

35.3.2.4 Common ROMIO Hints

There are a few reserved hints in the MPI–IO specification and are therefore universal across MPI–IO implementations, but for the most part, MPI–IO implementers are free to make up hints. The MPI–IO specification also dictates that any unrecognized hints should just be ignored, leaving data unaffected.

General Guide to Using Data Sieving vs POSIX I/O

Small noncontiguous memory regions

Noncontig file regions	Noncontig file region distribution	
	Sparse	Dense
Small	No	Yes
Large	Yes	Yes

Other memory regions and sizes

Noncontig file regions		
	Sparse	Dense
Small	No	Yes
Large	No	No

FIGURE 35.8 The three main factors to consider in determining whether to use data sieving are whether the user buffer is noncontiguous with small pieces, the size of the noncontiguous file regions, and the distribution of the file accesses all with respect to the data sieving buffer size. If both memory and file descriptions are contiguous, do not use data sieving.

TABLE 35.4 The *Same* Column Indicates Whether the Hint Passed Needs to Be the Same across All Processes in the Communicator

Hint	Value	Same	Std.	Basic description
romio_cb_read	E/D/A	Yes	No	Ctrl two phase collective reads
romio_cb_write	E/D/A	Yes	No	Ctrl two phase collective writes
cb_buffer_size	Integer	Yes	Yes	two phase collective buffer size
cb_nodes	Integer	yes	Yes	Number of collective I/O aggregators
cb_config_list	String list	Yes	Yes	List of collective aggregator hosts
romio_ds_read	E/D/A	No	No	Ctrl data sieving for indep. reads
romio_ds_write	E/D/A	No	No	Ctrl data sieving for indep. writes
romio_no_indep_rw	Bool	Yes	No	No subsequent indep. I/O
ind_rd_buffer_size	Integer	No	no	Read data-sieve buffer sz.
ind_wr_buffer_size	Integer	No	No	Write data-sieve buffer sz.
striping_factor	Integer	Yes	Yes	Number of I/O servers to stripe file across
striping_unit	Integer	Yes	Yes	Stripe sz distributed on I/O servers

The *Std.* column indicates official MPI reserved hints.
E/D/A = Enable/Disable/Auto.

In this way user applications that specify hints relevant to either a certain file system or MPI–IO implementation should still be portable, though the hints may be disregarded.

Table 35.4 lists some hints that are typically used. The exact data sieving hints are specific to ROMIO, but the collective I/O hints are respected across MPI–IO implementations. While an MPI–IO developer may choose not to implement two phase collective I/O, if they do decide to, they should use the hints in the table for user configuration. The *striping_factor* and *striping_unit* are standardized MPI–IO hints used to dictate file distribution parameters to the underlying parallel file system.

Although hints are an important means for applications and their developers to communicate with MPI implementations, it is usually more desirable for the MPI–IO implementation to automate the use and configuration of any optimizations.

35.3.3 Current Areas of Research in MPI–IO

Although data sieving I/O and two phase I/O are both significant optimizations, I/O remains a serious bottleneck in high-performance computing systems. MPI–IO remains an important level in the software stack for optimizing I/O.

35.3.3.1 Persistent File Realms

The Persistent File Realm (PFR) [CCL+04] technique modifies the two phase I/O behavior to ensure valid data in an incoherent client-side file system cache. Following MPI consistency semantics, non-overlapping file writes should be immediately visible to all processes within the I/O communicator. An underlying file system must provide coherent client-side caching if any at all.

Maintaining cache coherency over a distributed or parallel file system is no easy task, and the overhead introduced by the coherence mechanisms sometimes outweigh the performance benefits of providing a cache. This is the exact reason that PVFS [CLRT00] does not provide a client-side cache.

If an application can use all collective MPI I/O functions, PFRs can carefully manage the actual I/O fed to the file system in order to ensure access to only valid data in an incoherent client-side cache. As mentioned earlier, the ROMIO two phase I/O implementation heuristically load balances the I/O responsibilities of the I/O aggregators. Instead of rebalancing and reassigning the *file realms* according to the accesses of each collective I/O call, the file realms "persist" between collective I/O calls. The key to PFRs is recognizing that the data cached on a node is not based on its own MPI–IO request, but the combined I/O accesses of the communicator group. Two phase I/O adds a layer of I/O indirection. As long as each I/O aggregator is responsible for the same file realm in each collective I/O call, the data it accesses will always be the most recent version. With PFRs, the file system no longer needs to worry about the expensive task of cache coherency, and the cache can still safely be used. The alternative is to give up on client-side caches completely as well as the performance boost they offer.

35.3.3.2 Portable File Locking

The MPI–IO specification provides an *atomic mode*, which means file data should be sequentially consistent. As it is, even with a fully POSIX compliant file system, some extra work is required to implement atomicity in MPI–IO because of the potential for noncontiguous access patterns. MPI–IO functions must be sequentially consistent across noncontiguous file regions in atomic mode. In high-performance computing, it is typical for a file system to relax consistency semantics or for a file system that supports strict consistency to also support less strict consistency semantics. File locking is the easiest way to implement MPI's atomic mode. It can be done in three different ways based on traditional contiguous byte-range locks. The first is to lock the entire file being accessed during each MPI–IO call, the down side being potentially unneeded access serialization for all access to the file. The second is to lock a contiguous region starting from the first byte accessed ending at the last byte access. Again, since irrelevant bytes between a noncontiguous access pattern are locked, there is still potential for false sharing lock contention. The last locking method is two phase locking where byte range locks for the entire access (possibly noncontiguous) must be acquired before performing I/O [ACTC06]. While file locking is the most convenient way to enforce MPI's atomic mode, it is not always available.

Portable file locking at the MPI level [LGR+05] provides the necessary locks to implement the MPI atomic mode on any file system. This is accomplished by using MPI-2's Remote Memory Access (RMA) interface. The "lock" is a remote accessible boolean array the size of the number of processes N. Ideally, the array is a bit array, but it may depend on the granularity of a particular system. To obtain the lock, the process puts a *true* value to its element in the remote array and gets the rest of the array in one MPI-2

High-Performance Techniques for Parallel I/O

RMA epoch (in other words both the put and get happen simultaneously and atomically). If the array obtained is clear, the lock is obtained, otherwise the lock is already possessed by another process, and the waiting process sets up a `MPI_Recv` to receive the lock from another process. To release the lock, the locking process writes a *false* value to its element in the array and gets the rest of the array. If the array is clear, then no other process is waiting for the lock. If not, then should pass the lock with a `MPI_Send` call to the next waiting process in the array. Once lock contention manifests itself, the lock is passed around in the array sequentially (and circularly), and thus this algorithm does not provide fairness.

35.3.3.3 MPI–IO File Caching

It is important to remember that caching is a double-edged sword that can some times help and other times impede. Caching is not always desirable, and these situations should be recognized during either compile or run time [VSK+03]. Caching is ideally suited to applications performing relatively small writes that can be gathered into larger more efficient writes. Caching systems need to provide run time infrastructure for identifying patterns and monitoring cache statistics to make decisions such as whether to keep caching or bypass the cache. Ideally, the cache could self-tune other parameters like page sizes, cache sizes, and eviction thresholds as well.

Active buffering [MWLY02] gathers writes on the client and uses a separate I/O thread on the client to actually write the data out to the I/O system. I/O is aggressively interleaved with computation, allowing computation to resume quickly.

DAChe [CCL+05] is a coherent cache system implemented using MPI-2 for communication. DAChe makes local client-side file caches remotely available to other processes in the communicator using MPI-2 RMA functions. The same effect can be achieved using threads on the clients to handle remote cache operations [LCC+05a, LCC+05b]. A threaded version of coherent caching provides additional functionality over DAChe to intermittently write out the cache. Though some large-scale systems lack thread support, most clusters do support threads, and multi-core microprocessors are starting to become commonplace in high-performance computing. Modern high-performance computers also commonly use low latency networks with native support for one-sided RMA, which provides DAChe with optimized low level transfers. Cache coherency is achieved by allowing any given data to be cached on a single client at a time. Any peer accessing the same data passively accesses the data directly on the client caching the page, assuring a uniform view of the data by all processes.

35.4 Parallel File Systems

35.4.1 Summary of Current Parallel File Systems

Currently there are numerous parallel I/O solutions available. Some of the current major commercial efforts include Lustre [Lus], Panasas [Pan], GPFS [SH02], and IBRIX Fusion [IBR]. Some current and older research parallel file systems include PVFS [CLRT00, The], Clusterfile [IT01], Galley [NK96], PPFS [EKHM93], Scotch [GSC+95], and Vesta [CF96].

This section begins by describing some of the parallel file systems in use today. Next, various I/O methods for noncontiguous data access and how I/O access patterns and file layouts can significantly affect performance are discussed with a particular emphasis on structured scientific I/O parameters (region size, region spacing, and region count).

35.4.1.1 Lustre

Lustre is widely used at the U.S. National Laboratories, including Lawrence Livermore National Laboratory (LLNL), Pacific Northwest National Laboratory (PNNL), Sandia National Laboratories (SNL), the National Nuclear Security Administration (NNSA), Los Alamos National Laboratory (LANL), and NCSA. It is an open source parallel file system for Linux developed by Cluster File Systems, Inc. and HP.

Lustre is built on the concept of objects that encapsulate user data as well as attributes of that data. Lustre keeps unique inodes for every file, directory, symbolic link, and special file which holds references to objects on object storage targets (OSTs). Metadata and storage resources are split into metadata servers (MDSs) and OSTs, respectively. MDSs are replicated to handle failover and are responsible for keeping track of the transactional record of high-level file system changes. OSTs handle actual file I/O directly to and from clients once a client has obtained knowledge of which OSTs contain the objects necessary for I/O. Since Lustre meets strong file system semantics through file locking, each OST handles locks for the objects that it stores. OSTs handle the interaction of client I/O requests and the underlying storage, which are called object-based disks (OBDs). While OBD drivers for accessing journaling file systems such as ext3, ReiserFS, and XFS are currently used in Lustre, manufacturers are working on putting OBD support directly into disk drive hardware.

35.4.1.2 PanFS

Panasas ActiveScale File System (PanFS) has customers in the areas of government sciences, life sciences, media, energy, and many others. It is a commercial product which, like Lustre, is also based on an object storage architecture.

The PanFS architecture consists of both MDSs and object-based storage devices (OSDs). MDSs have numerous responsibilities in PanFS including authentication, file and directory access management, cache coherency, maintaining cache consistency among clients, and capacity management. The OSD is very similar to Lustre's OST in that it is also a network-attached device smart enough to handle object storage, intelligent data layout, management of the metadata associated with objects it stores, and security. PanFS supports the POSIX file system interface, permissions, and ACLs. Caching is handled at multiple locations in PanFS. Caching is performed at the compute nodes and is managed with callbacks from the MDSs. The OBDs have a write data cache for efficient storage and a third cache is used for metadata and security tokens to allow secure commands to access objects on the OSDs.

35.4.1.3 General Parallel File System

The General Parallel File System (GPFS) from IBM is a shared disk file system. It runs on both AIX and Linux and has been installed on numerous high-performance clusters such as ASCI Purple. In GPFS, compute nodes connect to file system nodes. The file system nodes are connected to shared disks through a switching fabric (such as fibre channel or iSCSI). The GPFS architecture uses distributed locking to guarantee POSIX semantics. Locks are acquired on a byte-range granularity (limited to the smallest granularity of a disk sector). A *data shipping* mode is used for fine-grain sharing for applications, where GPFS forwards read/write operations originating from other nodes to nodes responsible for a particular data block. Data shipping is mainly used in the MPI–IO library optimized for GPFS [PTH[+]01].

35.4.1.4 FusionFS

IBRIX, founded in 2000, has developed a commercial parallel file system called FusionFS. It was designed to have a variety of high-performance I/O needs in scientific computing and commercial spaces. Some of its customers include NCSA, the Texas Advanced Computing Center at the University of Austin at Texas, Purdue University, and Electromagnetic Geoservices.

FusionFS is a file system that is a collection of *segments*. Segments are simply a repository for files and directories with no implicit namespace relationships (e.g., not necessarily a directory tree). Segments can be of variable sizes and not necessarily the same size. In order to get parallel I/O access, files can be spread over a group of segments. Segments are managed by segment servers in FusionFS, where a segment server may "own" one or more segments. Segment servers maintain the metadata and lock the files stored in their segments. Since the file system is composed of segments, additional segments may be added for increasing capacity without adding more servers. Segment servers can be configured to handle failover responsibilities, where multiple segment servers have access to shared storage. A standby segment server would automatically take control of another server's segments if it were to fail. Segments may be taken offline for maintenance without disturbing the rest of the file system.

High-Performance Techniques for Parallel I/O

35.4.1.5 Parallel Virtual File System

The Parallel Virtual File System 1 (PVFS1) is a parallel file system for Linux clusters developed at Clemson University. It has been completely redesigned as PVFS2, a joint project between ANL and Clemson University. Whereas PVFS1 was a research parallel file system, PVFS2 was designed as a production parallel file system made easy for adding/removing research modules. A typical PVFS2 system is composed of server processes that can handle metadata and/or file data responsibilities.

PVFS2 features several major distinctions over PVFS1. First of all, it has a modular design that makes it easy to change the storage subsystem or network interface. Clients and servers are stateless, allowing the file system to cleanly progress if a connected client crashes. Most important to this chapter, PVFS2 has strong built-in support for noncontiguous I/O and an optimized MPI–IO interface. PVFS2 also uses the concept of objects that are referred to by handles in the file system. Data objects in PVFS2 are stored in the servers with metadata information about the group of objects that make up a file as well as attributes local to a particular object. It is best to access PVFS2 through the MPI–IO interface, but a kernel driver provides access to PVFS2 though the typical UNIX I/O interface.

35.4.2 Noncontiguous I/O Methods for Parallel File Systems

Numerous scientific simulations compute on large, structured multi-dimensional datasets that must be stored at regular time steps. Data storage is necessary for visualization, snapshots, checkpointing, out-of-core computation, postprocessing [NSLD99], and numerous other reasons. Many studies have shown that the noncontiguous I/O access patterns evident in applications as IPARS [IPA] and FLASH [FOR+00] are common to most scientific applications [BW95, CACR95]. This section begins by describing the important noncontiguous I/O methods that can be leveraged through MPI–IO, which require specific file system support (list I/O and datatype I/O). POSIX I/O and two phase I/O were discussed in depth in Section 35.3.2. In this section, noncontiguous I/O methods are described and compared.

35.4.2.1 List I/O

The list I/O interface is an enhanced file system interface designed to support noncontiguous accesses and is illustrated in Figure 35.9a. The list I/O interface describes accesses that are both noncontiguous in memory and file in a single I/O request by using offset-length pairs. Using the list I/O interface, an

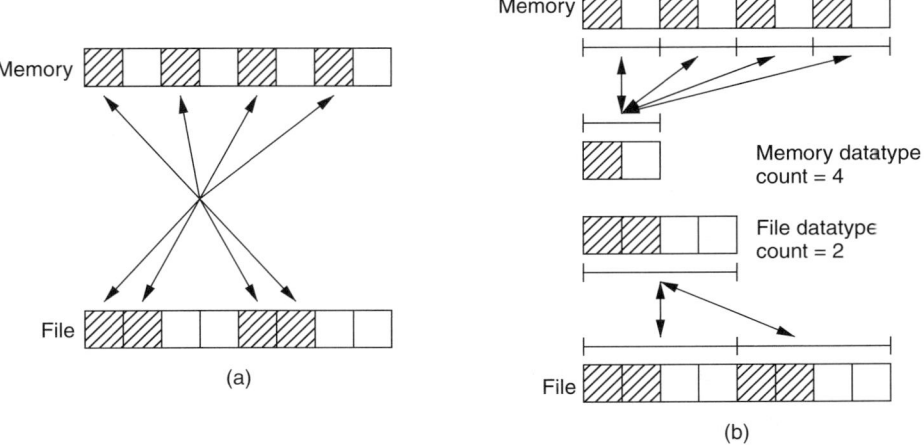

FIGURE 35.9 (a) Example list I/O request. All offsets and lengths are specified for both memory and file using offset-length pairs. (b) Example datatype I/O request. Memory and file datatypes are built for expressing structured I/O patterns and passed along with an I/O request.

MPI–IO implementation can flatten the memory and file datatypes (convert them into lists of contiguous regions) and then describe an MPI–IO operation with a single list I/O request. Given an efficient implementation of this interface in a file system, list I/O improves noncontiguous I/O performance by significantly reducing the number of I/O requests needed to service a noncontiguous I/O access pattern. Previous work [CCC+03] describes an implementation of list I/O in PVFS and support for list I/O under the ROMIO MPI–IO implementation. Drawbacks of list I/O are the creation and processing of these large lists and the transmission of the file offset-length pairs from client to server in the parallel file system. Addition, list I/O request sizes should be limited when going over the network; only a fixed number of file regions can be described in one request. So while list I/O significantly reduces the number of I/O operations (in [CCC+03], by a factor of 64), a linear relationship still exists between the number of noncontiguous regions and the number of actual list I/O requests (within the file system layer). In the rest of this section, this maximum number of offset-length pairs allowed per list I/O request is addressed as *ol-max*.

Using POSIX I/O for noncontiguous I/O access patterns will often generate the same number of I/O requests as noncontiguous file regions. In that case, previous results have shown that list I/O performance runs parallel to the POSIX I/O bandwidth curves and shifted upward due to a constant reduction of total I/O requests by a factor of *ol-max*. List I/O is an important addition to the optimizations available in MPI–IO. It is most effective when the I/O access pattern is noncontiguous and irregular, since datatype I/O is more efficient for structured data access.

35.4.2.2 Datatype I/O

While list I/O provides a way for noncontiguous access patterns to be described in a single I/O request, it uses offset-length pairs. For structured access patterns, more concise solutions exist for describing the memory and file regions. Hence, datatype I/O (Figure 35.9b) borrows from the derived datatype concept used in both message passing and I/O for MPI applications. MPI-derived datatype constructors allow for concise descriptions of the regular, noncontiguous data patterns seen in many scientific applications (such as extracting a row from a two-dimensional dataset). The datatype I/O interface replaces the lists of I/O regions in the list I/O interface with an address, count, and datatype for memory, and a displacement, datatype, and offset into the datatype for file. These parameters correspond directly to the address, count, datatype, and offset into the file view passed into an MPI–IO call and the displacement and file view datatype previously defined for the file. While the datatype I/O interface could be directly used by programmers, it is best used as an optimization under the MPI–IO interface. The file system must provide its own support for understanding and handling datatypes. A datatype I/O implementation in PVFS1 [CCL+03] demonstrates its usefulness in several I/O benchmarks.

Since datatype I/O can convert a MPI–IO operation directly into a file system request with a one-to-one correspondence, datatype I/O greatly reduces the amount of I/O requests necessary to service a structured noncontiguous request when compared to the other noncontiguous access methods. Datatype I/O is unique with respect to other methods in that increasing the number of noncontiguous regions that are regularly occurring does not require additional I/O access pattern description data traffic over the network. List I/O, for example, would have to pass more file offset-length pairs in such a case. When presented with an access pattern of no regularity, datatype I/O regresses into list I/O behavior.

35.4.3 I/O Suggestions for Application Developers

Application developers often create datasets using I/O calls in a simple to program manner. Although convenient, the logical layout of a dataset in a file can have a significant impact on I/O performance. When creating I/O access patterns, application developers should consider three major I/O access pattern characteristics that seriously affect I/O performance. These suggestions focus on structured, interleaved, and noncontiguous I/O access, all of which are common for scientific computing (as mentioned earlier in

Section 35.4.2). The following discussion addresses the effect of changing each parameter while holding the others constant:

- *Region count*: Changing the region count (whether in memory or file) will cause some I/O methods to increase the amount of data sent from the clients to the I/O system over the network. For example, increasing the region count when using POSIX I/O will increase the number of I/O requests necessary to service a noncontiguous I/O call. Datatype I/O may be less affected by this parameter since changing the region count does not change the size of the access pattern representation in structured data access. List I/O could be affected by the region count since it can only handle so many offset-length pairs before splitting into multiple list I/O requests. Depending on the access pattern, two phase I/O may also be less affected by the region count, since aggregators only make large contiguous I/O calls.
- *Region Size*: Memory region size should not make a significant difference in overall performance if all other factors are constant. POSIX I/O and list I/O could be affected since smaller region sizes could create more noncontiguous regions, therefore requiring more I/O requests to service. File region size makes a large performance difference since hard disk drive technology provides better bandwidth to larger I/O operations. Since two phase I/O already uses large I/O operations, it will be less affected by file region size. The other I/O methods will see better bandwidth (up to the hard drive disk bandwidth limit) for larger file region sizes.
- *Region spacing*: Memory region spacing should not make an impact on performance. However, file regions spacing changes the disk seek time. Even though this section considers region spacing as the logical distance to the next region, there is a some correlation with respect to actual disk distance due to the physical data layout many file systems choose. If the distance between file regions is small, two phase I/O will improve performance owing to internal data sieving. Also, when the file region spacing is small enough to fit multiple regions into a file system block, file system block operations may help with caching. Spacing between regions is usually different in memory and in file owing to the interleaved data operation that is commonly seen in scientific datasets that are accessed by multiple processes. For example, in the FLASH code [FOR+00], the memory structure of the block is different from that of the file structure, since the file dataset structure takes into account multiple processes.

An I/O benchmark, *Noncontiguous I/O Test* (NCIO), was designed in [CCL+06] for studying I/O performance using various I/O methods, I/O characteristics, and noncontiguous I/O cases. This work validated many of the I/O suggestions listed here. Tables 35.5 and 35.6 show a summary of how I/O parameters affect memory and file descriptions. If application developers understand how I/O access

TABLE 35.5 The Effect of Changing the Memory Description Parameters of an I/O Access Pattern Assuming That All Others Stay the Same

I/O method	Increase region count from 1 to c	Increase region size from 1 to s	Increase region spacing from 1 to p
POSIX	Increase I/O ops from 1 to c	No change	No change
List	Increase I/O ops from 1 to $c/ol\text{-}max$	No change	No change
Datatype	Increment datatype count from 1 to c	No change	No change
Data sieving	Minor increase of local memory movement	Surpassing buffer size requires more I/O ops	Surpassing buffer size requires more I/O ops
Two phase	Minor increase of memory movement across network	Surpassing aggregate buffer requires more I/O ops	Surpassing aggregate buffer requires more I/O ops

TABLE 35.6 The Effect of Changing the File Description Parameters of an I/O Access Pattern Assuming That All Others Stay the Same

I/O method	Increase region count from 1 to c	Increase region size from 1 to s	Increase region spacing from 1 to p
POSIX	Increase I/O ops from 1 to c	Improves disk bandwidth	Increases disk seek time
List	Increase I/O ops from 1 to c/ol-max	Improves disk bandwidth	Increases disk seek time
Datatype	Increment datatype count from 1 to c	Improves disk bandwidth	Increase disk seek time
Data sieving	Minor increase of local memory movement	Lessens buffered I/O advantages and surpassing buffer size requires more I/O ops	Lessens buffered I/O advantages and surpassing buffer requires more I/O ops
Two phase	Minor increase of memory movement across network	Lessens buffered I/O advantages and surpassing aggregate buffer size requires more I/O ops	Lessens buffered I/O advantages and surpassing aggregate buffer size requires more I/O ops

```
read(fd, buf, 4)                    MPI_create_vector(3, 4, 12 MPI_BYTE, filetype)
lseek(fd, 8, SEEK_CUR)              MPI_File_set_view(fh, 0, MPI_BYTE, filetype, "native," info)
read(fd, buf, 4)                    MPI_File_read(fd, buf, 12, MPI_BYTE, status)
lseek(fd, 8, SEEK_CUR)
read(fd, buf, 4)
```

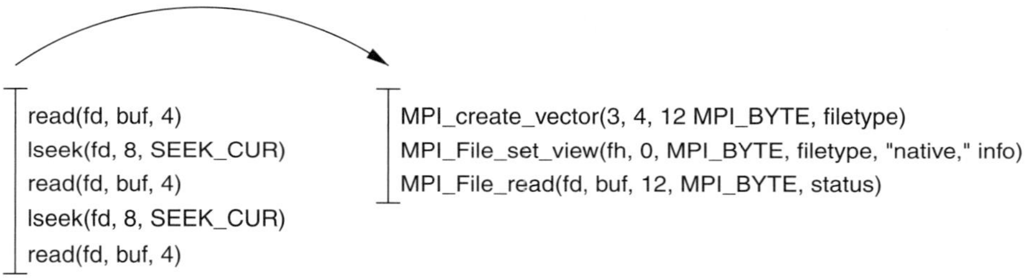

FIGURE 35.10 Example code conversion from the POSIX interface to the MPI–IO interface.

patterns affect overall performance, they can create optimized I/O access patterns that will both attain good performance as well as suit the needs of the application.

35.4.3.1 Possible I/O Improvements

- *All large-scale scientific applications should use the MPI–IO interface (either natively or through higher level I/O libraries).* MPI–IO is a portable parallel I/O interface that provides more performance and functionality over the POSIX I/O interface. Whether using MPI–IO directly or through a higher-level I/O library that uses MPI–IO (such as PnetCDF or HDF5), applications can use numerous I/O optimizations such as collective I/O and data sieving I/O. MPI–IO provides a rich interface to build descriptive access patterns for noncontiguous I/O access. Most programmers will benefit from the relaxed semantics in MPI–IO when compared to the POSIX I/O interface. If a programmer chooses to use a particular file system's custom I/O interface (i.e., not POSIX or MPI–IO), portability will suffer.
- *Group individual I/O access to make large MPI–IO calls.* Even if an application programmer uses the MPI–IO interface, they need to group their read/write accesses together into larger MPI datatypes and then do a single MPI–IO read/write. Larger MPI–IO calls allow the file system to use optimizations such as data sieving I/O, list I/O, and datatype I/O. It also provides the file system with more information about what the application is trying to do, allowing it to take advantage of data locality on the server side. A simple code conversion example in Figure 35.10 changes three POSIX read() calls into a single MPI_File_read() call, allowing it to use data sieving I/O, list I/O, or datatype I/O to improve performance.

High-Performance Techniques for Parallel I/O

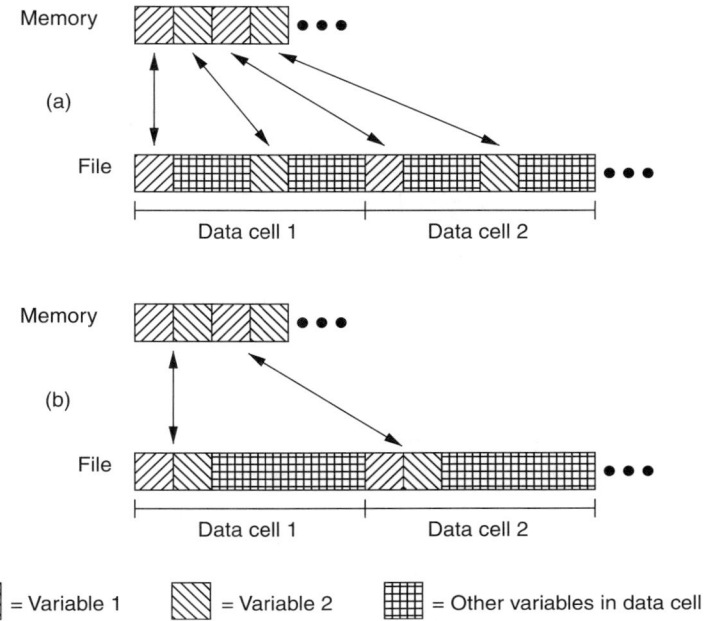

FIGURE 35.11 (a) Original layout of variables in data cells. (b) Reorganization of data to combine file regions during write operations increases I/O bandwidth.

- *Whenever possible, increase the file region size in an I/O access pattern.* After creating large MPI–IO calls that service noncontiguous I/O access patterns, try to manipulate the I/O access pattern such that the file regions are larger. One way to do this is data reorganization. Figure 35.11 shows how moving variables around in a data cell combined file regions for a better performance. While not always possible, if a noncontiguous file access pattern can be made fully contiguous, performance can improve by up to two orders of magnitude [CCL+06]. When storing data cells, some programmers write one variable at a time. Making a complex memory datatype to write his data contiguously in file in a single MPI-IO I/O call will be worth the effort.
- *Reduce the file region spacing in an I/O access pattern.* When using data sieving I/O and two phase I/O, this will improve buffered I/O performance by accessing less unused data. POSIX I/O, list I/O, and datatype I/O will suffer less disk seek penalties. Again, a couple of easy ways to do this is to reorganize the data layout or combine multiple I/O calls to make fewer, but larger I/O calls.
- *Consider individual versus collective (two phase I/O).* Two phase I/O provides good performance over the other I/O methods when the file regions are small (bytes or tens of bytes) and near-by since it can make large I/O calls, while the individual I/O methods (excluding data sieving I/O) have to make numerous small I/O accesses and disk seeks. The advantages of larger I/O calls outweigh the cost of passing network data around in that case. Similarly, the file system can process accesses in increasing order across all the clients with two phase I/O. If the clients are using individual I/O methods, the file system must process the interleaved I/O requests one at a time, which might require a lot of disk seeking. However, in many other cases, list I/O and datatype I/O outperform two phase I/O [CCL+06]. More importantly, two phase I/O has an implicit synchronization cost. All processes must synchronize before any I/O can be done. Depending on the application, this synchronization cost can be minimal or dominant. For instance, if the application is doing a checkpoint, since the processes will likely synchronize after the checkpoint is written, the synchronization cost is minimal. However, if the application is continually processing and writing results in an embarrassingly parallel manner, the implicit synchronization costs of two phase I/O can dominate the overall application running time as shown in Figure 35.12.

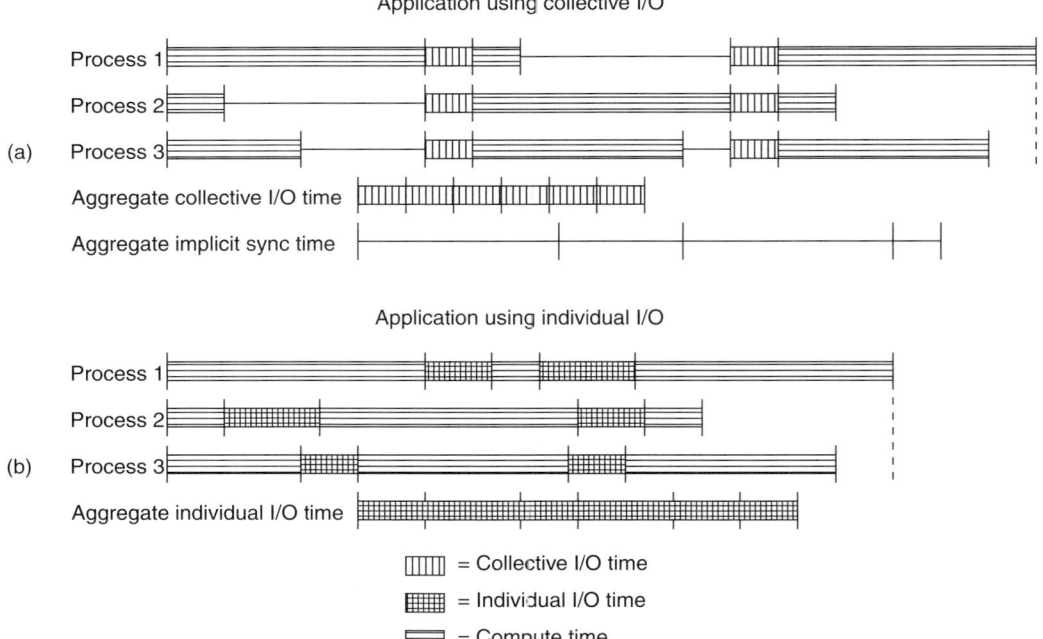

FIGURE 35.12 Cost of collective I/O synchronization. Even if collective I/O (a) can reduce the overall I/O times, individual I/O (b) outperforms it in this case because of no implicit synchronization costs.

- *When using individual I/O methods, choose datatype I/O.* In nearly all cases datatype I/O exceeds the performance of the other individual I/O methods. The biggest advantage of datatype I/O is that it can compress the regularity of an I/O access pattern into datatypes, keeping a one-to-one mapping from MPI–IO calls to file system calls. In the worst case (unstructured I/O), datatype I/O breaks down to list I/O, which is still much better than POSIX I/O.
- *Do not use data sieving I/O for interleaved writes.* Interleaved writes will have to be processed one at a time by the file system because the read-modify-write behavior in the write case requires concurrency control. Using data sieving I/O for writes is only supported by file systems that have concurrency control. Data sieving I/O is much more competitive with the other I/O methods when performing reads, but should still be used in limited cases.
- *Generally, there is no need to reorganize the noncontiguous memory description if file description is noncontiguous.* Some programmers might be tempted to copy noncontiguous memory data into a contiguous buffer before doing I/O, but recent results suggest that it will not make any difference in performance [CCL+06]. It would most likely just incur additional programming complexity and memory overhead.

References

[ACTC06] Peter Aarestad, Avery Ching, George Thiruvathukal, and Alok Choudhary. Scalable approaches for supporting MPI–IO atomicity. In *Proceedings of the IEEE/ACM International Symposium on Cluster Computing and the Grid*, Singapore, May 2006. IEEE Computer Society Press.

[BW95] Sandra Johnson Baylor and C. Eric Wu. Parallel I/O workload characteristics using Vesta. In *Proceedings of the IPPS '95 Workshop on Input/Output in Parallel and Distributed Systems*, pp. 16–29, Santa Barbara, CA, April 1995. IEEE Computer Society Press.

[CACR95] Phyllis E. Crandall, Ruth A. Aydt, Andrew A. Chien, and Daniel A. Reed. Input/output characteristics of scalable parallel applications. In *Proceedings of Supercomputing '95*, San Diego, CA, IEEE Computer Society Press, December 1995.

[CCC+03] Avery Ching, Alok Choudhary, Kenin Coloma, Wei Keng Liao, Robert Ross, and William Gropp. Noncontiguous access through MPI–IO. In *Proceedings of the IEEE/ACM International Symposium on Cluster Computing and the Grid*, Tokyo, Japan, May 2003. IEEE Computer Society Press.

[CCL+03] Avery Ching, Alok Choudhary, Wei Keng Liao, Robert Ross, and William Gropp. Efficient structured data access in parallel file systems. In *Proceedings of the IEEE International Conference on Cluster Computing*, Hong Kong, December 2003. IEEE Computer Society Press.

[CCL+04] Kenin Coloma, Alok Choudhary, Wei Keng Liao, Lee Ward, Eric Russell, and Neil Pundit. Scalable high-level caching for parallel I/O. In *Proceedings of the IEEE International Parallel and Distributed Processing Symposium*, Sante Fe, NM, April 2004. IEEE Computer Society Press.

[CCL+05] Kenin Coloma, Alok Choudhary, Wei Keng Liao, Lee Ward, and Sonja Tideman. DACHe: Direct access cache system for parallel I/O. In *Proceedings of the International Supercomputer Conference*, Heidelberg, June 2005. Prometeus GmbH.

[CCL+06] Avery Ching, Alok Choudhary, Wei Keng Liao, Lee Ward, and Neil Pundit. Evaluating I/O characteristics and methods for storing structured scientific data. In *Proceedings of the International Conference of Parallel Processing*, Rhodes Island, Greece, April 2006. IEEE Computer Society Press.

[CF96] Peter F. Corbett and Dror G. Feitelson. The Vesta parallel file system. *ACM Transactions on Computer Systems*, 14(3):225–264, August 1996.

[CLRT00] Philip H. Carns, Walter B. Ligon III, Robert B. Ross, and Rajeev Thakur. PVFS: A parallel file system for Linux clusters. In *Proceedings of the 4th Annual Linux Showcase and Conference*, pp. 317–327, Atlanta, GA, October 2000. USENIX Association.

[EKHM93] Chris Elford, Chris Kuszmaul, Jay Huber, and Tara Madhyastha. Portable parallel file system detailed design. Technical report, University of Illinois at Urbana-Champaign, November 1993.

[FOR+00] Bruee Fryxell, Kevin Olson, Paul Ricker, Frank Timmes, Michael Zingale, Donald Lamb, Peter MacNeice, Robert Rosner, and Henry Tufo. FLASH: An adaptive mesh hydrodynamics code for modelling astrophysical thermonuclear flashes. *Astrophysical Journal Supplement*, 131:273, 2000.

[GSC+95] Garth A. Gibson, Daniel Stodolsky, Pay W. Chang, William V. Courtright II, Chris G. Demetriou, Eka Ginting, Mark Holland, Qingming Ma, LeAnn Neal, R. Hugo Patterson, Jiawen Su, Rachad Youssef, and Jim Zelenka. The Scotch parallel storage systems. In *Proceedings of 40th IEEE Computer Society International Conference (COMPCON 95)*, pp. 403–410, San Francisco, CA, Spring 1995. IEEE Computer Society Press.

[HDF] HDF5 home page. http://hdf.ncsa.uiuc.edu/HDF5/

[IBR] IBRIX FusionFS. http://www.ibrix.com/

[IPA] IPARS: integrated parallel accurate reservoir simulation. http://www.ticam.utexas.edu/CSM/ACTI/ipars.html

[IT01] Florin Isaila and Walter Tichy. Clusterfile: A flexible physical layout parallel file system. In *Proceedings of the IEEE International Conference on Cluster Computing*, Newport Beach, CA, October 2001. IEEE Computer Society Press.

[Kot94] David Kotz. Disk-directed I/O for MIMD multiprocessors. In *Proceedings of the 1994 Symposium on Operating Systems Design and Implementation*, pp. 61–74, November 1994. USENIX Association. Updated as Dartmouth TR PCS-TR94-226 on November 8, 1994.

[Kot97] David Kotz. Disk-directed I/O for MIMD multiprocessors. *ACM Transactions on Computer Systems*, 15(1):41–74, February 1997.

[LCC+05a] Wei Keng Liao, Alok Choudhary, Kenin Coloma, Lee Ward, and Sonja Tideman. Cooperative write-behind data buffering for MPI I/O. In *Proceedings of the Euro PVM/MPI Conference*, Sorrento, Italy, September 2005. IEEE Computer Society Press.

[LCC+05b] Wei Keng Liao, Kenin Coloma, Alok Choudhary, Lee Ward, Eric Russell, and Sonja Tideman. Collective caching: Application-aware client-side file caching. In *Proceedings of the 14th IEEE International Symposium on High Performance Distributed Computing*, Research Triangle Park, NC, July 2005. IEEE Computer Society Press.

[LGR+05] Robert Latham, William Gropp, Robert Ross, Rajeev Thakur, and Brian Toonen. Implementing MPI–IO atomic mode without file system support. In *Proceedings of the IEEE Conference on Cluster Computing Conference*, Boston, MA, September 2005. IEEE Computer Society Press.

[LLC+03] Jianwei Li, Wei Keng Liao, Alok Choudhary, Robert Ross, Rajeev Thakur, William Gropp, Rob Latham, Andrew Sigel, Brad Gallagher, and Michael Zingale. Parallel netCDF: A high-performance scientific I/O interface. In *Proceedings of Supercomputing*, Phoenix, AZ, November 2003. ACM Press.

[Lus] Lustre. http://www.lustre.org

[Mes] Message passing interface forum. http://www.mpi-forum.org

[MWLY02] Xiaosong Ma, Marianne Winslett, Jonghyun Lee, and Shengke Yu. Faster collective output through active buffering. In *Proceedings of the IEEE International Parallel and Distributed Processing Symposium*, Fort Lauderdale, FL, April 2002. IEEE Computer Society Press.

[NK96] Nils Nieuwejaar and David Kotz. The Galley parallel file system. In *Proceedings of the 10th ACM International Conference on Supercomputing*, pp. 374–381, Philadelphia, PA, May 1996. ACM Press.

[NSLD99] Michael L. Norman, John Shalf, Stuart Levy, and Greg Daues. Diving deep: Data-management and visualization strategies for adaptive mesh refinement simulations. *Computing in Science and Engineering*, 14(4):36–47, 1999.

[Pan] Panasas. http://www.panasas.com

[PTH+01] Jean-Pierre Prost, Richard Treumann, Richard Hedges, Bin Jia, and Alice Koniges. MPI–IO/GPFS, an Optimized Implementation of MPI–IO on top of GPFS. In *Proceedings of Supercomputing*, Denver, CO, November 2001. ACM Press.

[RD90] Russ Rew and Glenn Davis. The Unidata netCDF: Software for scientific data access. In *Proceedings of the 6th International Conference on Interactive Information and Processing Systems for Meterology, Oceanography and Hydrology*, Anaheim, CA, February 1990. American Meterology Society.

[ROM] ROMIO: A high-performance, portable MPI–IO implementation. http://www.mcs.anl.gov/romio

[SH02] Frank Schmuck and Roger Haskin. GPFS: A shared-disk file system for large computing clusters. In *Proceedings of the Conference on File and Storage Technologies*, San Jose, CA, January 2002.

[TGL99a] Rajeev Thakur, William Gropp, and Ewing Lusk. Data sieving and collective I/O in ROMIO. In *Proceedings of the Seventh Symposium on the Frontiers of Massively Parallel Computation*, pp. 182–189, Annapolis, MD, February 1999. IEEE Computer Society Press.

[TGL99b] Rajeev Thakur, William Gropp, and Ewing Lusk. On implementing MPI–IO portably and with high performance. In *Proceedings of the Sixth Workshop on Input/Output in Parallel and Distributed Systems*, pp. 23–32, Atlanta, GA, May 1999. ACM Press.

[The] The parallel virtual file system 2 (PVFS2). http://www.pvfs.org/pvfs2/

[VSK+03] Murali Vilayannur, Anand Sivasubramaniam, Mahmut Kandemir, Rajeev Thakur, and Robert Ross. Discretionary caching for I/O on clusters. In *Proceedings of the Third IEEE/ACM International Symposium on Cluster Computing and the Grid*, pp. 96–103, Tokyo, Japan, May 2003. IEEE Computer Society Press.

36
Message Dissemination Using Modern Communication Primitives

36.1	Introduction...	36-1
36.2	Definitions and Related Work	36-5
36.3	Multimessage Multicasting..........................	36-9
	NP-completeness of the MM_C • Upper and Lower Bounds for the MM_C • Receiving Buffers • Messages with a Fixed Number of Destinations	
36.4	Multimessage Multicasting with Forwarding	36-15
36.5	Distributed or On-Line Multimessage Multicasting ..	36-17
36.6	Message Dissemination over Arbitrary Networks	36-18
36.7	Multimessage Multicasting in Optical Networks	36-19
36.8	Discussion	36-20
	References	36-20

Teofilo F. Gonzalez
University of California

36.1 Introduction

In this chapter, we discuss algorithms, complexity issues, and applications for message dissemination problems defined over networks based on modern communication primitives. More specifically, we discuss the problem of disseminating messages in parallel and distributed systems under the multicasting communication mode. These problems arise while executing in a parallel computing environment iterative methods for solving scientific computation applications, dynamic programming procedures, sorting algorithms, sparse matrix multiplication, discrete Fourier transform, and so forth. These problems also arise when disseminating information over sensor and ad hoc wireless networks.

The goal of this chapter is to collect scattered research results developed during the last decade to establish that the multicasting communication environment is a powerful communication primitive that allows for solutions that are considerable better than those achievable under the telephone (or one-to-one) communication environment. The multicasting communication environment has been available for quite some time in parallel computing systems. We also establish that within the multicasting communication mode forwarding plays an important role by allowing solutions that are considerable better than when restricting to direct communications, even when the communication load is balanced and the network is completely connected. Offline communication scheduling allows for considerably better solutions over online scheduling. Though online scheduling provides added flexibility and it is applicable to a larger set of scenarios.

The goal of parallel and distributed systems is to speedup the processing of programs by a factor proportional to the number of processing elements. To accomplish this goal a program must be partitioned into tasks, and the communications that must take place between these tasks must be identified to ensure a correct execution of the program. To achieve high performance one must assign each task to a processing unit (statically or dynamically) and develop communication programs to perform all the intertask communications efficiently. Efficiency depends on the algorithms used to route messages to their destinations, which is a function of the communication environment. Given a communication environment and a set of messages that need to be exchanged, the message dissemination problem is to find a schedule to transmit all the messages in the least total number of communication rounds. Generating an optimal communication schedule, that is, one with the least total number of communication rounds (or minimum total communication time), for our message dissemination problems over a wide range of communication environments is an NP-hard problem. To cope with intractability efficient message dissemination approximation algorithms have been developed for different types of communication environments. Let us begin by discussing the different components of our message dissemination problems: message communication patterns (the communications that must take place), and communication environment. The communication environment consists of the communication network (the direct communications allowed for each processor), primitive operations (the basic communication operations allowed by the system), and the communication model (possible operations during each communication round or step).

Message Communication Patterns: We refer to one of the most general message communication problem as the *multimessage multicasting* problem. In this problem every processor needs to communicate a set of messages, each to a set of processors. We use *multimessage* to mean that each processor sends a set of messages, and *multicasting* means that each message is sent to a set of processors. The extreme cases of multicasting is when messages are sent to only one destination (called *unicasting*) and to all possible destinations (called *broadcasting*). A more general version of the problem is when the messages have different lengths, but it is "equivalent" to the multimessage multicasting problem when the messages are partitioned into packets. A restricted version of the multimessage multicasting is the multimessage unicasting problem, in which all the messages have exactly one destination. Another restricted form of the problem is the *gossiping* problem where each processor sends only one message, but each message is sent to all the other processors. Further restrictions include the single message broadcasting, and single message multicasting problems. For some communication environments even these simple versions of the message dissemination problem are NP-complete.

Network: The complete static communication network (there are bidirectional links between every pair of processors) is the simplest and most flexible when compared to other static networks (or simply *networks*) with restricted structure like rings, mesh, star, binary trees, hypercube, cube connected cycles, shuffle exchange, and so forth, and dynamic networks (or multistage interconnection networks), like Omega Networks, Benes Networks, Fat Trees, and so forth. The minimum number of communications rounds needed to carry out any communication pattern over a complete network is an obvious lower bound for the corresponding problem over any communication network.

Primitive Operations: Under the classical *telephone* communication primitive when a processor sends a message it is sent to just one of its neighbor processors at a time. In order to send a message to k destinations one needs to send a copy of the message to one destination at a time. This takes k communication rounds. When *message forwarding* is allowed the processor sending the message may send fewer than k messages, but one or more of the intermediate processors that receives the message must forward it to other destinations.

The *multicasting* communication primitive allows a processor to send a message during each communication round to a subset of its adjacent processors. This communication mode allows multidestination messages to be transmitted faster, and it has been available for quite some time in parallel computing systems. It is a natural communication mode in wireless networks. In our communication problems one may use multicasting efficiently by identifying messages with disjoint origins and destinations so that they may be transmitted concurrently. This is not always possible; however, to deal with problem instances that lack concurrency one may conveniently partition message destinations into groups and messages will be transmitted to each group of destinations during different communication rounds. We refer to this operation as *splitting message destinations*. Forwarding is also available when using the multicasting communication primitive. In this chapter when forwarding is allowed it is specified explicitly. There are situations when forwarding is not allowed. For example, forwarding is expensive in all optical networks because every forwarding operation implies that an optical signal is transformed into an electrical one and then to an optical one. These are expensive operations. For many years there have been discussions about the multicasting communication mode over the Internet, but it is only partially operational and the part that is operational is not fully utilized.

Communication Model: The *single port communication mode* allows every processor to send at most one message and receive at most one message during each communication round. The *multiport communication mode* is a more general model that allows a processor to send/receive more than one message per communication round provided that only one message is transmitted per communication link. The advantage of this more general model is that all the communications may be carried out faster, but there is an additional cost associated with the hardware equipment needed to achieve this communication concurrency. In what follows we discuss only the single port communication mode. A relatively inexpensive way to reduce the total communication time (which is just the total number of communication rounds) is to add buffers at the receiving end of each processor. Additional hardware is needed for the buffers and to control their behavior. However, the buffers make a single port communication mode behave more like a multiport communication at a fraction of the cost.

The problem is to develop algorithms that generate near-optimal solutions to message dissemination problems under the multicasting communication environment. These problems have wide applicability in all sorts of systems. The algorithms we have developed provide the initial solution that may then be fine tuned as new technological innovations take place.

Executing in a parallel processing environment iterative methods for large scientific computations give rise to multimessage multicasting problems. A simple example is solving large sparse system of linear equations through stationary iterative methods [1]. Other applications arise when dynamic programming procedures are being executed in a parallel or distributed computing environment. Sorting, matrix multiplication, discrete Fourier transform, and so forth [2,3] are applications with significant message dissemination operations. Multicasting information over a b-channel ad hoc wireless communication network is another important application in information systems. The multicasting communication environment arises naturally in sensor and ad hoc wireless networks. Because the cost of deployment is decreasing rapidly, these types of networks are finding new applications. Ad hoc wireless networks are suited for many different scenarios including situations where wired Internet or Intranet is not practical or possible.

The solution of sparse systems of linear equations through an iterative method in a parallel computing environment gives rise to one of the simplest forms of multimessage multicasting. Given vector $X[0] = x_1[0], x_2[0], \ldots, x_n[0]$, the vectors $X[t]$, for $t = 1, 2, \ldots$, need to be computed as $x_i[t+1] = f_i(X[t])$. However, f_i includes very few terms, because the system is sparse. In a parallel computing environment, the x_is are assigned to the processors where the function f_i computes the new values. The x_is computed at each iteration need to be transmitted to the processors that require them to evaluate the functions f_i at the next iteration. The computation starts by transmitting the $X[0]$ elements to the appropriate processors and then use them to compute $X[1]$. Then $X[1]$ is transmitted and used to compute $X[2]$, and so on. This process repeats until the difference between the old and new values are smaller than some given tolerance, or a certain number of iterations have taken place. The same communication schedule is used at each iteration. The schedule is computed offline once the placement of the x_is has been determined. When the computation and communication loads are balanced, full speedups are possible provided the computation and communication time can be overlapped. The solution to the linear equations problem is more dynamic and converges faster when one uses the newly computed values as soon as they become available. A very important problem in this whole process is the placement of the x_is, but for brevity it is not discussed in this chapter.

The offline multimessage multicasting problem is fully justified by the above application. The "distributed" or "nearly online" multimessage multicasting problem is an interesting variation. Message destinations are not known until a portion of a task has been executed and this information is available only locally at the processor executing the task. At a given synchronization point all the processors start performing their communication steps. The communication schedule needs to be constructed online. Globalization of the information and the time required to compute the communication schedule need to be taken into consideration when the performance of the algorithm is being evaluated. Message dissemination problems are solved offline, unless we mention explicitly that they are distributed or online versions of the problem.

A class of parallel computers use dynamic networks for communications. These computers include the now extinct Meiko CS-2 and the IBM GF11 machines [4] that use a variation of the Benes communication network. The computer chips implementing the basic Meiko CS-2 switches are still operational connecting processors in distributed environments. A *pr-dynamic network* [1] is a dynamic network that performs in one communication round any single permutation communication step, that is, every processor sends and receives one message, and the "switches" can replicate its input data to any subset of its output ports. The most interesting property of pr-dynamic networks is that any communication pattern over a complete network can be automatically converted into an equivalent communication pattern for any pr-dynamic network that carries out all the communications in just two communication rounds. In the first round, data is replicated and in the second one, data is distributed to the appropriate location [5–7]. Communication networks that are about 50% larger can reduce the two communication rounds to one.

In our analysis we concentrate on the multimessage multicasting over complete undirected networks. This version of the problem has a simple structure that is simple to analyze. Furthermore, any communication schedule for a complete undirected network can be automatically transformed into one for any pr-dynamic network. This transformation doubles the total communication time. However, for the best of our algorithms, this communication step does not introduce any additional communication rounds. Specifically, when multicasting operations have adjacent processors as destinations, or the destinations of any two messages transmitted currently are not interleaving (if a message has destination processors i and j for $i < j$, then no other message may have destination k, for $i < k < j$ during the communication round), can be implementation in a pr-dynamic network in a single communication round. The schedules constructed by the best of our forwarding algorithms satisfy the above property. Another reason for concentrating our study on the multimessage multicasting over completely connected networks is that it also models a fixed set of processors over an optical communication ring.

In an n processor system connected through a Benes network messages may need to traverse $\Omega(\log n)$ switches to reach their destination. Our communication model refers to this path as a single communication step that takes one communication round. This simplification is acceptable because the amount of time needed to prepare a message for transmission (moving the data, and setting up the routing table) is more than the message transmission time through the switches (which are implemented in silicon chips). So the problem is simply minimizing the total number of communication rounds required for the transmission of all messages, even when n is not too large. For large n, messages may interfere with the transmission of other messages and cause the total communication time to be much larger than the time to set up all the messages. Strictly speaking, our analysis is not scalable, but our assumptions are acceptable for systems with up to several hundreds of millions of processors. We do not expect to have networks with more than this number of processors in the near future. Processors interconnected through wrap-around pr-networks allow for some communication patterns to traverse only a constant number of switches to reach their destination. Algorithms may be designed to take advantage of this, but most of the time it is impossible to gain from the implementation of this strategy. Since there is a gain only in relatively few cases, we do not take into consideration this level of detail in our model. Our assumptions are reasonable ones that simplify the analysis and allow us to concentrate on the most important communication issues.

As we have discussed earlier, our analysis has direct applications from sensor networks to parallel and distributed computing. Other applications in high-performance communication systems include voice and video conferencing, operations on massive distributed data, scientific applications and visualization, high-performance supercomputing, medical imaging, and so forth. The delivery of multidestination messages will continue to increase in the future and so our problems will find more applications with time. This work is laying the foundations over which applications can run smoothly.

When the communications are restricted to the telephone mode, the best possible schedules have total communication time that is not bounded by any function in terms of d. For example, when there is only one message to be delivered to n destinations, any schedule requires $\log n$ communication rounds even when the network is fully connected. However, under the multicasting communication primitive it requires only one communication round. The multicasting communication primitive delivers performance that is not attainable under the telephone communication mode. This is an important consequence of the research accomplishments reported in this chapter.

The distributed version of the multimessage multicasting problem is much more general than the offline versions, but their communication schedules have $\Omega(d + \log n)$ expected number of communication rounds. One can always construct schedules for the multimessage multicasting problem with a number of communication rounds of at most d^2, and just $2d$ when forwarding is allowed. Therefore, there is a clear advantage when we have knowledge ahead of time of all the communication requirements. Forwarding is another important factor in reducing the total communication time. These are important consequences of the research accomplishments reported in this chapter.

In Section 36.2 we formally define our problems and discuss related work. Then in Section 36.3 we discuss the multimessage multicasting problem. We establish NP-completeness results, provide upper and lowers bounds for the communication time, and discuss restricted versions as well as generalizations of the problem. Section 36.4 discusses similar issues, but for the multimessage multicasting problem with forwarding. Communication schedules with significantly better communication time are achievable when forwarding is allowed. The distributed or online version of the multimessage multicasting problem with forwarding is discussed in Section 36.5.

Section 36.6 discusses the *gossiping* communication problem under the multicasting communication mode. From our definitions we know gossiping is a restricted version of multimessage multicasting; however, we treat these two problems separately because the work on multimessage multicasting has been limited to a class of interconnection architectures, whereas the research results for the gossiping problems apply to all kinds of networks. That is, by restricting the communication patterns provable good solutions can be generated for arbitrary networks.

Section 36.7 considers the multiport communication model in the form of multimessage multicasting for an all-optical multifiber star networks. In this case the messages are carried on different wavelengths in one of several fibers. There is limited switching that permits messages on the same wavelength but different fibers to be interchanged. The effects of limited switching on the total communication time are discussed.

36.2 Definitions and Related Work

The multimessage multicasting communication problem under the single port fully connected network with the multicasting communication primitive is denoted by MM_C. Formally, there is a set of n processors, $P = \{P_1, P_2, \ldots, P_n\}$, interconnected through a complete network (all possible bidirectional links are available). Processors alternate synchronously between computation and communication. During each computation phase, the processors generate a set of messages each of which is to be transmitted to a subset

TABLE 36.1 Source and Destination Processors for the Messages in Example 36.1

Message	A	B	C	D	E	F
Source	1	1	2	2	3	3
Destination	{4, 5, 6}	{7, 8, 9}	{4, 5, 7, 8}	{6, 9}	{5, 6, 8, 9}	{4, 7}

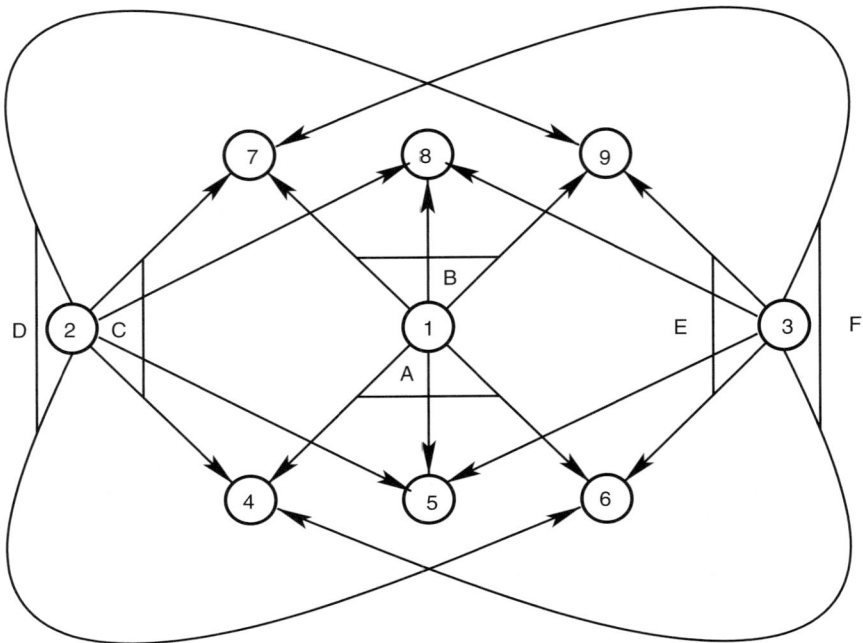

FIGURE 36.1 Directed Multigraph representation for Example 36.1.

of processors (*destinations*). The processor that generates a message is called the *source or originating* processor for the message. All the source messages for P_i form the *hold* set (h_i). By using the message destination information one may compute the messages each processor P_i needs to receive (n_i).* The maximum number of messages that any processor needs to send or receive is denoted by $d = \max\{|h_i|, |n_i|\}$. Example 36.1 gives an instance of the MM_C problem.

Example 36.1 *The number of processors, n, is nine. Six messages need to be disseminated. The source processor and destination processors for the messages are given in Table 36.1. In general all processors need to send and receive messages. For simplicity in this example the first three processors (P_1, P_2, and P_3) just sent messages and processors P_4 through P_9 just receive messages. Every processor sends and receives at most three messages. Therefore, the degree of the instance, denoted by d, is 3.*

There are several ways one may visualize an MM_C problem instance. In the directed multigraph representation, processor P_i is represented by a vertex labeled i and a message is represented by a set of edges or branches originating at the source processor with an edge ending at each destination processor. All the edges associated with a message are *bundled* together. This is represented by drawing a line joining all the directed edges bundled together (see Figure 36.1). Figure 36.1 depicts the directed multigraph

---

*The n_is do not need to be computed explicitly, but the communication process must deliver to every processor all the messages it needs.

representation for the problem instance given in Example 36.1. Another way to represent the edges associated with a message is by using a directed hyperedge.

The message communications in our complete network must satisfy the restrictions imposed by our communication model given below:

1. During each communication round each processor P_i may multicast a message in its initial hold set h_i (or the current hold set when forwarding is allowed) to a set of processors. The message transmitted remains in the hold set h_i.
2. During each communication round each processor may receive at most one message. A message that processor P_i receives (if any) is added to its hold set h_i and it is available for the next communication round. When two or more messages are sent during the same communication round to the same processor, the processor does not receive any of the messages.* The processors sending the messages will be informed that their message it sent did not reach all its destination. The processors will not know the destinations that were not reached.

When every processor receives all the messages it needs, $n_i \subseteq h_i$ for all i, the communication phase terminates. The total number of communication rounds is referred to as the *total communication time* or simply TCT.

By using the directed graph representation the MM_C problem is an edge coloring problem where the coloring is such that no two edges emanating from the same processor belonging to different bundles are colored identically, and no two edges incident to the same vertex are colored identically. The colors correspond to the communication rounds and the coloring restrictions enforce the communication rules. Our notation is not consistent so we use interchangeably colors, communication rounds or steps, and time; vertices and processors; bundle, branches or edges, and messages. Note that when message forwarding is allowed, the above representation breaks down. One may use edge multicolorings to describe classes of forwarding algorithms [8], but describing an MM_C problem as an edge coloring problem is awkward since forwarding means that an edge is replaced by a sequence of edges whose coloring must satisfy some additional constraints.

From the communication rules it is clear that any message may be transmitted during one or more communication rounds. In other words, a processor may send a message to a subset of its destinations at a given time and to another subset of destinations at another time. We refer to this as *splitting message destinations*. Table 36.2 gives a communication schedule (with message splitting) for Example 36.1 that takes four communication rounds. The i-th row in the table specifies the multicasts to be performed during the i-th communication round. A stricter version of the problem requires every message to be sent to all its destinations during one communication round. We refer to this case as the *unsplittable message destinations*. The problem instance given in Example 36.1 requires five communication rounds when message destinations are unsplittable. The reason for this is that messages originating at the same processor or having at least one common destination cannot be transmitted during the same communication round. Therefore, the only two messages that may be transmitted concurrently are messages D and F, and any schedule for this version of the problem must have TCT at least five. This is not too different from the TCTs of the schedule with splittable message destinations given in Table 36.2. However, there are unsplittable message destination problem instances where all schedules have TCT that cannot be bounded above by any function in terms of d, but, as we shall see later on, all the instances with splittable message destinations have a schedule with TCT at most d^2. Our model allows splittable message destinations because it can significantly improve the TCT and it can be implemented in practice.

We claim that all the communication schedules for the problem instance given in Example 36.1 require at least four communication steps (when splitting message destinations is allowed). The proof of this fact is by contradiction. Suppose there is a schedule with total communication time equal to three. Then either message A or B is transmitted to all their destinations in one time unit. Without loss of generality let's

---

*This situation does not normally arise when constructing schedules offline, however, it is common in "distributed" or "online" versions of the problem.

TABLE 36.2 Message Transmissions for Schedule S

Communication round	Concurrent communications		
1	$A: 1 \to \{5,6\}$	$C: 2 \to \{7\}$	$E: 3 \to \{8,9\}$
2	$B: 1 \to \{7,8,9\}$	$C: P_2 \to \{4\}$	$E: 3 \to \{5,6\}$
3	$D: 2 \to \{6,9\}$	$F: 3 \to \{4,7\}$	—
4	$C: 2 \to \{5,8\}$	—	—

TABLE 36.3 Message Transmissions (with Forwarding) for Schedule T

Communication round	Concurrent communications			
1	$C: 2 \to \{4,7\}$	$E: 3 \to \{5,6,8,9\}$	—	—
2	$A: 1 \to \{4,5,6\}$	$D: 2 \to \{9\}$	$F: 3 \to \{7\}$	$C: 7 \to \{8\}$
3	$B: 1 \to \{7,8,9\}$	$D: P_2 \to \{6\}$	$F: 3 \to \{4\}$	$D: 4 \to \{5\}$

say that it is message B and that the message is transmitted during communication round 1. Then each of the messages C, D, E, and F emanating from processors P_2 and P_3 with destinations P_7, P_8, and P_9 must be transmitted during communication rounds 2 and 3 in one time unit each. But message E cannot be transmitted concurrently with C, D, or F. Also, messages C, D, or F cannot be transmitted concurrently. Therefore, a schedule with TCT at most three does not exist for this problem instance. For this version of the problem all schedules must require least four communication rounds. However, forwarding allows for communications schedules with TCT equal to three. Table 36.3 gives one such schedule. The only message that is forwarded is message D that is transmitted to processor P_5 through processor P_4. This is possible because the network is fully connected. The difference in the TCT is not significant for the instance given in Example 36.1. However, as we shall see later on there are problem instances that require d^2 TCT without forwarding, but all problem instances have a communication schedule with TCT at most $2d$ when forwarding is allowed.

The maximum number of destinations for messages in a problem instance is called the *fan-out k*. When $k = 1$ each message has at most one destination so the problem is referred to the *multimessage unicasting MU_C* problem. Gonzalez [1] established that this problem corresponds to the minimum makespan openshop preemptive scheduling problem that can be solved in polynomial time. Since all the nonzero tasks have the same length, the schedule constructed by the algorithm in Reference 9 does not have preemptions and, every problem instance of degree d has a schedule with makespan equal to d. Therefore, there is always a schedule with TCT equal to d for the MU_C problem. The makespan openshop preemptive scheduling problem may be viewed as a generalization of the problem of coloring the edges of bipartite multigraphs. There are well-known algorithms that generate optimal colorings for these graphs [10–14]. The fastest of these algorithms is the one by Cole et al. [13]. This algorithm is the best one to solve the MU_C problem when the edge multiplicities are small [14]. However, Gonzalez and Sahni's [9] openshop algorithm is currently the fastest method for the general case. The algorithm given in Reference 13 is for multigraphs so there are multiple edges between pair of nodes. The algorithm in Reference 9, for the openshop problem, represents multiple edges between vertices by a weighted edge. In both of these papers m (or $|E|$) is the total number of edges, but in Reference 13 m is not $O(n^2)$ and may be very large compared to n^2, whereas in Reference 9 it is always less than n^2. One should keep this in mind when comparing these algorithms.

The MU_C problem and its variants have been studied. The basic results include: heuristics, approximation algorithms, polynomial time algorithms for restricted versions of the problem, and NP-completeness results. Coffman et al. [15] present approximation algorithms for a generalization of the MU_C problem where messages have different lengths, but need to be communicated without interruption (preemptions are not allowed), and there are $\gamma(P_i)$ sending or receiving ports per processor. The power of message forwarding was reported by Whitehead [16]. When preemptions are allowed, messages may be transmitted with interruptions, generalizations of the MU_C problem have been considered by

Choi and Hakimi [17, 18], Hajek and Sasaki [19], Gopal et al. [20]. The n-port MU_C problem over complete networks, for transferring files, was studied by Rivera-Vega et al. [21]. A variation of the MU_C problem, called *the message exchange problem*, has been studied by Goldman et al. [22]. The communication model in this version of the problem is asynchronous and it closely corresponds to actual distributed memory machines. Another restricted version of the MU_C problem, called *data migration*, was studied by Hall et al. [23].

The main difference between the above research and the one we discuss in this chapter is that we concentrate on the multicasting communication mode, rather than on the *telephone communication mode*. Previous research on the basis of the multicasting communication mode has concentrated on single messages. Multimessage multicasting results are reported in References 1, 8, and 24–35. The initial research by Shen [34] on multimessage multicasting was for n-cube processor systems. The objective function was very general and attempted to minimize the maximum number of hops, amount of traffic, and degree of message multiplexing, so only heuristics could be developed. More recently, different strategies for solving multimessage multicasting problems over all-optical networks were surveyed by Thaker and Rouskas [35]. The multimessage multicasting problem based on the telephone communication mode has been studied under the name of *data migration with cloning* by Khuller et al. [36]. The total communication time for optimal schedules for the telephone communication mode is considerably larger to the one over the multicasting communication mode. A generalization of this message dissemination problem allowing for limited broadcasting or multicasting was considered by Khuller et al. [37]. These problems are used to model networks of workstations and grid computing. These data migration problems have also been studied under the sum of weighted completion time objective function by Gandhi et al. [38]. This version of the message dissemination problem is based on the telephone communication mode.

The edge coloring problem has also been studied under the "sums of colors" objective function [39]. It is interesting to note that even though the edge coloring and open shop problems are equivalent when the objective is to minimize the number of colors used and the makespan of the schedule, the same relationship does not hold when the objective is to minimize the "sum of the colors" (when colors are represented by integers) and the sum of completion time for the jobs. The correspondence is between the average of the colors and the average completion time for the tasks (which is different from the one for the jobs). The weighted versions of these problems have also been studied.

36.3 Multimessage Multicasting

We begin in Section 36.3.1 by presenting a reduction that shows that the decision version of the multimessage multicasting problem defined over complete networks is an NP-complete problem. Then we present in Section 36.3.2 upper and lower bounds for the TCT of the MM_C problem. Specifically we discuss a simple algorithm that shows that for every problem instance a communication schedule with TCT at most d^2 can be constructed in linear time. Then we show that there are problem instances for which all communication schedules have TCT at least d^2. Since these bounds are large we seek ways to improve on these results. In Section 36.3.3, we show that by adding buffers at the receiving end of every processor one may decrease the TCT required to solve MM_C problems. Specifically, we show that if there are l buffers at the receiving end of each processor it is possible to construct communication schedules with TCT at most $d^2/l + l - 1$. The algorithm corresponds to the one given in Section 36.3.2 when $l = 1$, and it is a generalization of that algorithm for $l > 1$. When $l = d$, the TCT reduces to $2l - 1$. The lower bounds in Section 36.3.2 apply when $l = 1$, but the instances involved in the lower bound have huge number of processors and messages have very large fan-out. In Section 36.3.4, we discuss algorithms for problems where the messages have restricted fan-out.

36.3.1 NP-Completeness of the MM_C

Gonzalez [1] established that the decision version of the multimessage multicasting problem is NP-complete even when every message is sent to at most two destinations. The reduction is from the edge

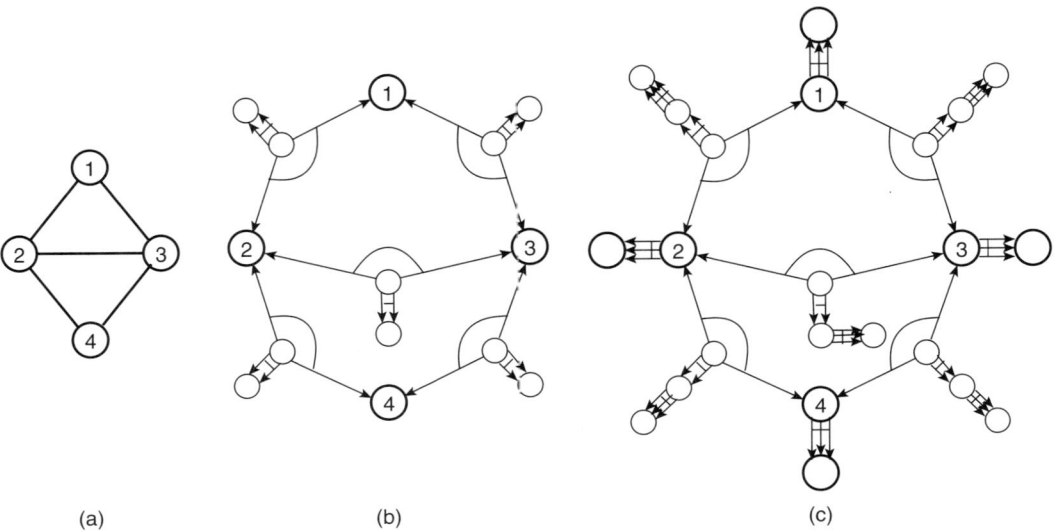

FIGURE 36.2 Graph edge coloring instance (a), corresponding MM_C instance (b), and corresponding instance of the MM_C problem with forwarding (c).

coloring problem. Given an undirected graph $G = (V, E)$ in which each vertex is of (graph) degree d, the edge coloring problem is to decide whether or not the edges of the graph can be colored with at most d colors so that every two edges incident to the same vertex are colored differently is an NP-complete problem (Holyer [40]).

The idea behind the reduction from the edge coloration problem to the MM_C is not too complex. Given an instance (G, d) of the edge coloration problem, one constructs an instance of the MM_C problem with fan-out $k = d$ as follows. Every vertex in G is represented by a processor, and for every edge there are two processors called *edge* and *extra-edge*. An edge in G is represented by d messages emanating from the edge processor representing the edge. One message has as destination the two processors representing the vertices in G that the edge joins. The other $d - 1$ messages have the same destination which is the corresponding extra-edge processor. Figure 36.2a gives a graph G with $d = 3$ and Figure 36.2b gives the instance of MM_C constructed from it. The claim is that the instance of the MM_C problem has a schedule with TCT equal to d iff the edges of graph G can be colored with exactly d colors. The equivalence between the problems can be established by observing that in a schedule with total communication d each of the messages emanating from an edge processor must be transmitted during a single communication round. Therefore, the message with two destinations must be multicasted to its two destinations during a single communication round. So it corresponds to the edge in G and thus both problems are equivalent. Clearly, the reduction takes polynomial time with respect to the number of vertices and edges in G. A formal proof for Theorem 36.1 is given in Reference 24.

Theorem 36.1 *The decision version of the MM_C problem is NP-complete even when $k = 2$.*

36.3.2 Upper and Lower Bounds for the MM_C

We show that every instance of the MM_C of degree d has a communication schedule with total communication time at most d^2. Furthermore, one such schedule can be constructed in linear time with respect to the total number of message destinations. Then we present a problem instance, which can be generalized to be of degree $d > 2$, such that all its communication schedules have TCT at least d^2.

Gonzalez algorithm [1] constructs a communication schedule with TCT at most d^2 for every instance of the MM_C problem of degree d. The algorithm begins by defining d^2 colors with a pair of integers i and

j with values between 1 and d. First all the messages emanating out of each processor are arranged in some order (any order) and the edges incident to every processor are also arranged in some order (any order is fine). We select one edge at a time. When considering edge $e = (p, q)$ it is colored with the color (i, j) if the edge e represents a message-destination pair corresponding to the i-th message originating at processor P_p and it is the j-th edge incident upon processor P_q. No two edges incident upon a vertex have the same second color component, and not two edges originating at the same processor that represent different messages are assigned the same first color component. Therefore the coloring just defined is valid. From this coloring one can easily construct a schedule for the MM_C problem with TCT equal to d^2. Furthermore, this schedule can be constructed in linear time with respect to the input length. These claims are summarized in the following theorem whose proof follows the arguments defined above and is given in Reference 1.

Theorem 36.2 *The above algorithm generates a communication schedule with TCT at most d^2 for every instance of the MM_C problem of degree d. The time complexity for the algorithm is linear with respect to the total number of message destinations.*

Now let us establish a lower bound for the TCT for problem instances of degree d. We show that there are problem instances such that all their schedules have TCT at least d^2. Consider the instance of the MM_C problem with $d = 2$ given in Figure 36.3. The problem instance consists of 28 processors. Each of the four rectangles represents a different processor and the solid black dots represent 24 additional processors. Each of the rectangle processors sends two messages which are labeled with the two letters. We claim that the problem instance does not have a schedule with TCT less than four. We prove this by contradiction. Suppose there is a schedule with total communication time three. Then each of the four rectangle processors must send one of its messages during exactly one communication round. But since there are four of those messages and only three communications rounds, we know that two of such messages must be transmitted during the same communication round. Any two of the messages originating at different processors have a destination in common. Therefore, there does not exist a communication schedule with TCT at most three. So all communication schedules have TCT at least $d^2 = 4$.

Gonzalez [1] defined instances of the MM_C problem for all $d > 2$ such that all their communication schedules have TCT at least d^2. The proof is a generalization for the one discussed above. The following theorem established in Reference 1 summarizes the discussion.

Theorem 36.3 *There are MM_C problem instances of degree d, for all $d > 1$, such that all their communication schedules have TCT at least d^2.*

The instance given above for $d = 2$ has 28 processors, but the one for $d = 4$ has more than one million processors. These problem instances have messages with a very large fan-out in order to achieve the d^2 bound. These are scenarios that are not likely to arise in commercial systems in the near future. There are different ways to construct schedules with smaller TCT. In Section 36.3.3, we discuss the case when there are buffers and in Section 36.3.4 we discuss algorithms for problem instances with fixed fan-out. This is a more likely scenario for the near future.

36.3.3 Receiving Buffers

One relatively inexpensive way to achieve high performance is by adding l buffers [41] at the receiving end of each processor and developing controlling hardware so the buffering behaves as follows: (i) if at least one buffer has a message during a communication round, then one such message (probably in a FIFO order) will be transferred to the processor and at the next communication round the buffer will be said to be free; and (ii) a processor may receive as many messages as the number of free buffers it has. If more messages are sent to a processor than the number of free buffers, then none of the buffers will receive the new messages and the processors that sent the message will be informed that their message did not

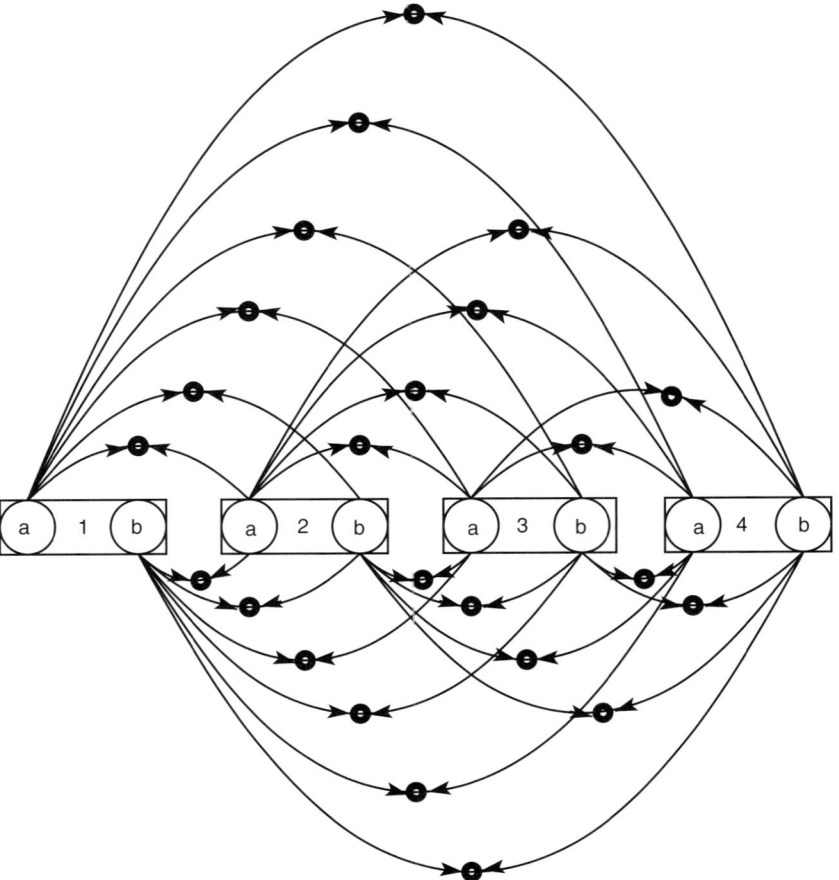

FIGURE 36.3 Problem instance of degree 2 that requires 4 colors. The triangles and solid circles represent processors.

reach all its destinations, but will not know the destinations that received the message. For this message dissemination problem, Gonzalez has recently developed [32] an efficient approximation algorithms that generates communication schedules with TCT at most $d^2/l + l - 1$. When there is one buffer, $l = 1$, the TCT is just d^2 and the algorithm reduces to the algorithm discussed in the previous subsection. For $l = d$, the TCT is $2d - 1$, which is slightly better than the one we will discuss in the next section where forwarding is allowed. The main advantage of this algorithm is that its time complexity is linear. The disadvantage is that the buffers and the hardware have an additional cost, but forwarding does not normally have any hidden costs. In what follows we outline the algorithm for the case when there are l buffers per processor. The algorithm, which is a generalization of the one discussed in the previous subsection, was initially reported in Reference 32.

Assume that d is a multiple of l (the number of buffers). The set $\{(i,j) | 1 \le i \le d \text{ and } 1 \le j \le d/l\}$ defines a set of d^2/l colors. Number (with the integers $[1, d]$) in any order all the messages originating at each vertex and label all the edges for each message with the message number. The incoming edges to each processor are arranged in nondecreasing order of their labels. The first l incoming edges to a processor, with respect to the ordering just defined, are given the value of 1, the next l edges are given the value of 2, and so on. The color (i,j) is assigned to edge $e = \{p, q\}$, when e represents the i-th message originating at processor P_p, and e was assigned the value j as an edge incident to vertex P_q.

A communication schedule with TCT at most $d^2/l+l-1$ is constructed from the above edge coloration as follows. The messages multicasted at time i are the ones represented by the edges colored $(i, 1)$, for

$1 \leq i \leq d$. Each processor has at most l incoming messages colored $(i, 1)$, so it follows that the processor buffers will not overflow. But it may be that the l messages for a processor fill up the buffers. The messages colored $(i, 2)$ will be sent at time $i + d$, for $1 \leq i \leq d$. Then next d time slots are for the messages colored $(i, 3)$, and so forth. The schedule overlaps the process of freeing the buffers for the messages colored (i, j) with the arrival of the messages $(i, j + 1)$, for all $1 \leq j < d/l$. All the $(i, 1)$ messages are received from the buffers by the processor by time $d + l - 1$, hence the last $l - 1$ time units overlap with the reception of the first $(i, 2)$ messages. We need to show that there will not be an overflow of the buffers on any of the processors. Let P_r be a processor that receives a messages colored $(1, 2)$. This message must be the first message originating at some processor. From the way we assigned colors to the incoming edges to processor P_r, we know that all the messages colored $(i, 1)$ that were received by P_r were colored $(1, 1)$. Therefore, the buffers are empty by the time the messages $(1, 2)$ are received. So even if l messages are received concurrently, there will be buffers to store them. Essentially the same argument can be used to show that there will be room in the buffers for all the messages colored $(i, 2)$. The proof that there are buffers available when the messages $(i, 3), (i, 4), \ldots, (i, d/l)$ arrive is similar. After all the messages are received one needs $l - 1$ time units to empty all the buffers. Therefore our procedure constructs a communication schedule with TCT at most $d * (d/l) + l - 1$. This algorithm and the proof of correctness appears in Reference 32, and it is summarized in the following theorem which we state without a proof. Readers interested in additional details are referred to Reference 32.

Theorem 36.4 *The above algorithm constructs a communication schedule with TCT at most $d^2/l - l - 1$ for every instance of the MM_C problem of degree d, where l is the number of buffers available at each processor. The time complexity for the algorithm is linear with respect to the total number of message destinations.*

For the case when $l = 1$, the lower bound constructed earlier in this chapter automatically provides the lower bound d^2. In other words, there are problem instances that require TCT of d^2, which is exactly the same as the upper bound provided by the schedules generated by the above algorithm. Gonzalez [32] has established the lower bound of $2d - 1$ for the case when there are $l = d$ buffers. This lower bound matches the upper bound for the TCT of the schedules generated by the above algorithm. However, for other values of l, the current lower and upper bounds do not match. We conjecture that the upper bounds are best possible. This remains as an interesting open problem.

36.3.4 Messages with a Fixed Number of Destinations

First we present approximation algorithms for the case when every message has at most two destinations, and then we discuss the case when every message is sent to at most k destinations.

36.3.4.1 Two-Destination Messages

As we established earlier, the MM_C problem where every message has at most two destinations is an NP-hard problem. In this subsection we present two approximation algorithms for this restricted version of the problem. The idea for the first algorithm is simple. We take every message with two destinations and break it into two messages, each with one destination. The resulting instance is an instance of the multimessage unicasting problem and of degree $2d$. As mentioned, before the multimessage unicasting problem, can be solved in polynomial time [1] and it has a schedule with TCT equal to $2d$. The time complexity to construct such schedule is $O(r(\min\{r, n^2\} + n \log n))$ time, where $r \leq dn$.

Another algorithm to construct a communication schedule with TCT equal to $2d - 1$ is presented in Reference 1. The algorithm is faster than the one just described. At each iteration the idea is to color all the messages originating at one of the processors. A maximal set of messages is selected such that each message can be colored with exactly one color. The remaining messages are colored with two colors. The two colors used by each message are obtained from a complete matching in a bipartite graph constructed from the current coloration. Hall's theorem is used to show that the bipartite graph has a complete matching. This result was initially established in Reference 1 and it is summarized in the following theorem.

Theorem 36.5 *Algorithm GM given in Reference 1 constructs for every instance of the MM_C problem with fan-out $k = 2$ and degree d a communication schedule with TCT at most $2d - 1$. The time complexity for the procedure is $O(nd^{2.5})$, where n is the number of processors.*

36.3.4.2 Multidestination Messages

As in the previous subsection, we can split every message with k destinations into k messages with one destination. Then by solving the MU_C problem we generate a schedule with TCT equal to kd. In what follows we present a set of algorithms all of which, except for the first one, generate schedules with TCT smaller than kd.

The following algorithms send every message during no more than q communication rounds, where q is an input value. The algorithms color the messages originating at P_1, then P_2, and so on. We describe first the algorithm for $q = 1$. When considering P_j each message destination pair originating at P_j has a potential conflict with at most $d - 1$ other messages with destination P_r. If all of these edges have been colored, there are at most $d - 1$ colors that cannot be used to color the current edge. The k edges for the message will have at most $k(d-1)$ colors, none of which can be used to color all the edges representing the message. These colors are called *destination-forbidden* colors. When every message is to be colored with exactly one color, then there are at most $d - 1$ *source-forbidden* colors. Therefore this coloring procedure may require at most $d + k(d - 1)$ different colors. This means that there is a schedule with TCT at most $d + k(d - 1)$. This bound is larger than the previous kd bound when $d > k$; however, the schedule can be constructed faster than the previous one.

To decrease the number of colors required by the above coloring strategy, one can increase the number of different colors used to color every message to two, that is, $q = 2$ which means the messages are split into at most two groups and each group is to be colored uniquely. This increases the number of source-forbidden colors and does not reduce the total number of destination-forbidden colors for each edge. But one may use any of the two colors for a message provided both of these colors are not destination-forbidden colors in any of the edges representing the message. For example, any message with two destinations cannot be colored when there are only $d - 1$ colors and we only take into consideration the destination-forbidden colors. The reason for this is that the overlap of destination-forbidden colors between the two edges representing the message can be $d - 1$, and therefore none of the $d - 1$ colors may be used to color the message. When there are d colors available it is possible to color the two edges for the message with one or two colors as follows. If one color is not a destination-forbidden color in both of the edges, then such color can be used to color the message. On the other hand when all the d colors are destination-forbidden colors in at least one of the edges, then at most $d - 2$ colors are destination-forbidden colors for both edges and the two remaining colors can be used to color the two edges for the message. What is the maximum number of colors, as a function to d and k, such that the k edges representing a message cannot be colored with any two colors when considering only destination-forbidden conflicts? As we just established, this value is $d - 1$ for $k = 2$. When $k = 3$ and $d = 9$ the maximum number is 12. The sets of destination-forbidden colors $\{1, 2, 3, 4, 5, 6, 7, 8\}$, $\{1, 2, 3, 4, 9, 10, 11, 12\}$, and $\{5, 6, 7, 8, 9, 10, 11, 12\}$ for each of the three edges can be used to show that no pair of the 12 colors can be used to color the three edges representing the message. The reason is that every pair of colors appears as a pair of destination-forbidden colors in at least one of the sets, so the pair of colors cannot be used to color the three edges. We claim that if there are 13 colors available it is possible to color the three edges using two colors. To prove this we show that there is always at least one color that is destination-forbidden in at most one of the three edges. This is because if the 13 colors are in the two lists, then the lists must contain at least 26 elements. But each of the three lists contain only eight elements. So there is at least one color that is destination-forbidden in at most one list. This color can therefore be used to color two of the edges, and the edge where it is destination-forbidden can be colored with another color since the list contains at most eight destination-forbidden colors. Therefore, it is possible to color all messages when 13 colors are available. Gonzalez [1] has established that the maximum number of colors such that no two of them can color completely a message is $d - 1$ for $k = 2$, about $1.5(d - 1)$ for $k = 3$, and so forth. Gonzalez [1] also derived asymptotic bounds for this case $q = 2$, but for brevity we do not include this result.

The above analysis can be used for the following greedy algorithm initially proposed in Reference 1. Given any instance of the MM_C problem and a value of q, the algorithm colors the messages originating at P_1 then the ones originating at P_2, and so on. When coloring the messages originating at P_j it colors one message at a time. Each message is colored with at most q colors by using a greedy strategy. It selects first a color that can be assigned to the largest number of edges, then it selects a second color that can be assigned to the largest number of uncolored edges, and so on. Gonzalez [1] shows that all the messages can be colored, each with at most q colors, by using at most $qd + k^{1/q}(d-1)$ colors. The procedure takes $O(q \cdot d \cdot e)$ time, where $e \leq nd$ is total number of message destinations. A proof of this result is given in Reference 1 and it is summarized in the following theorem.

Theorem 36.6 *The above algorithm constructs in $O(q \cdot d \cdot e)$ time a communication schedule with TCT $qd + k^{1/q}(d-1)$ for the MM_C problem with fan-out k of degree d using at most q colors per message.*

Gonzalez [27] has developed improved approximation algorithms for the MM_C problem. These approximation algorithms are more complex and the proofs for their approximation bounds is complex because they involve the manipulation of lengthy inequalities. But the proof technique is similar to the ones in this subsection. For $q = 3$, the approximation ratio of the improved algorithm is about 10% smaller for $k = 15$, about 40% smaller for $k = 100$, than when $q = 2$.

36.4 Multimessage Multicasting with Forwarding

The MMF_C problem is the MM_C when *forwarding* is allowed. Forwarding means that messages may be sent indirectly through other processors even when there is a direct link from the source to the destination processor. Can forwarding allow significantly better solutions for our problem? Remember that in the MM_C problem the communication load is balanced, that is, every processor sends and receives at most d messages, and all possible direct links between processors exist. It seems then that forwarding will not generate significantly better solutions, that is, communication schedules with significantly smaller TCT. However, as we shall see in this section there is significant difference between the TCT of schedules for the same instance of the MM_C and MMF_C problems. It is important to point out that message forwarding consumes additional resources (links). The least amount of link capacity is utilized when forwarding is not permitted.

NP-completeness: The reduction in the previous section for the MM_C cannot be used to show that the decision version of the MMF_C problem is NP-complete. The reason is that the extra-edge processor may be used for forwarding in schedules with TCT d (see Figure 36.2b). This destroys the equivalence with the edge coloration problem. However, the reduction can be easily modified for the MMF_C problem. The idea is to introduce another processor, called *additional-edge*, for each edge in G (see Figure 36 2c). This additional-edge processor will receive d messages from the corresponding extra-edge processor. In any schedule with TCT equal to d the extra-edge processors will be busy all the time sending its own messages and will not be able to forward any of the messages originating at an edge processors. This observation can be used to establish the equivalence with the edge coloration problem. Readers interested in additional details are referred to References 1 and 8 for a formal proof for all cases, rather than just the informal arguments given above.

The most interesting aspect of the MMF_C problem is that for every problem instance one can construct in polynomial time a schedule with TCT at most $2d$ [8, 26]. The two known algorithms to construct such schedules [8, 26] are slower than the ones we discussed for the MM_C problem in the previous section. However, the communication schedules have TCT that is significantly lower than that for the MM_C problem. The only exception is when every processor has $l = d$ buffers, but this requires additional hardware to store the data and control the buffer operations.

Both of the algorithms developed by Gonzalez [8, 26] have a two-phase structure. The first phase forwards messages in such a way that the resulting problem to be solved is a multimessage unicasting problem of degree d. As we have seen, this problem is an instance of the makespan openshop preemptive scheduling problem in which all tasks have unit execution time requirements. The second phase generates a communication schedule for this openshop problem from which a solution to the MMF_C can be obtained. The time complexity of the second phase is $O(r(\min\{r, n^2\} + n \log n))$, where r is the total number of message destinations ($r \leq dn$). The difference between the two algorithms is the first phase. The algorithm in Reference 26 tries to forward all the messages in the least number of communication rounds (no more than d), but the procedure is complex. The algorithm in Reference 8 is very simple, but performs all the message forwarding in d communication rounds. In what follows we discuss the simpler of the two algorithms through an example. Readers interested in additional details are referred to Reference 8.

The algorithm forwards all the messages. This is accomplished in d communication rounds. At each communication round every processor forwards at most one message and receives at most one message that it must deliver in the second phase to a set of at most d destinations. We explain this algorithm using the following example.

Example 36.2 *There are 7 processors and 15 messages. Table 36.4 lists the messages originating at each process and their destinations.*

The messages are ordered with respect to their originating processors. For Example 36.2 they are ordered in alphabetic order. Define message–destination pairs for all the messages and their destinations. Order the tuple first with respect to the ordering of the messages $A, B, \ldots, O$, and next with respect to their destinations (in increasing order). The messages in the first three pairs are forwarded from their originating processor to the first processor. The messages in the second three pairs are forwarded from their originating processor to the second processor, and so on. Clearly, each processor sends at most three messages, but the messages will be forwarded to one or more destinations. To avoid sending more than one message at a time to the same processors, the first three messages (A, B, C) must be multicast at time 1, 2, and 3, respectively; the next three messages (D, E, F) must be multicast at time 1, 2, and 3, respectively; and so on. Table 36.5 lists all the multicasting operations performed in phase 1 for Example 36.2. In our example, some messages are sent to multiple processors (G, M), while others are sent to single destinations. It is important to note that problem instances, like the one given above, some of the messages will end up being sent to and from the same processor. Obviously, these messages do not need to be transmitted. In some problem instances the elimination of these transmissions decreases the TCT, but in general it does not.

The resulting multimessage unicasting problem instance is given in Table 36.6. In Reference 8, a formal procedure is presented to perform the transformation outlined above. The algorithm is a generalization of

TABLE 36.4 Source Processor and Message–Destination Pairs for Example 36.2

Source	1	2	3	4	5	6	7
Message destination	—	A: {7}	C: {1}	E: {2, 3}	H: {1, 6}	J: {4}	M: {4, 5}
Message destination	—	B: {4}	D: {2}	F: {1, 7}	I: {6, 7}	K: {2}	N: {3}
Message destination	—	—	—	G: {5, 6}	—	L: {5}	O: {3}

TABLE 36.5 Message Forwarding Operations in Phase 1 for Example 36.2

Source	1	2	3	4	5	6	7
Round 1	—	A → {1}	D → {2}	G → {3, 4}	—	J → {5}	M → {6, 7}
Round 2	—	B → {1}	—	E → {2}	H → {4}	K → {6}	N → {7}
Round 3	—	—	C → {1}	F → {3}	I → {5}	L → {6}	O → {7}

TABLE 36.6 Resulting Multimessage Unicasting Problem Constructed for Example 36.2

Source	1	2	3	4	5	6	7
Message destination	A: {7}	D: {2}	F: {1}	G: {6}	I: {6}	K: {2}	M: {5}
Message destination	B: {4}	E: {2}	F: {7}	H: {1}	I: {7}	L: {5}	N: {3}
Message destination	C: {1}	E: {3}	G: {5}	H: {6}	J: {4}	M: {4}	O: {3}

the above procedure and it is shown that it constructs an instance of the multimessage unicasting problem whose solution provides a solution to the original problem instance [8].

The above construction is such that all the multicasting operations have as destination adjacent processors. This property is important because every communication round can also be performed by a pr-dynamic network. In other words, the schedule for a fully connected network can be used directly by any pr-dynamic network. Gonzalez [8] established the following theorem on the basis of a generalization of the above arguments.

Theorem 36.7 *Given any instance of the MMF_C of degree d, the algorithm presented in Reference [8] constructs a communication schedule with TCT at most $2d$. The procedure takes $O(r(\min\{r, n^2\} + n \log n))$, where r is the total number of messages destinations ($r \le dn$) and the communication schedule generated can be used for any pr-dynamic network.*

It is important to note that the algorithms in the previous section are faster than the one discussed in this section. However, the schedules generated by the algorithm in this section have considerably smaller TCT, and are also valid for pr-dynamic networks [8].

For restricted versions of the MMF_C it is possible to generate schedules with smaller TCT. Let us discuss briefly these versions of the problem. The $l - MMF_C$, for $2 \le l \le d$, is the MMF_C where every processor sends messages to at most ld destinations. Gonzalez [8] discusses an algorithm to construct communication schedule with TCT at most $\lfloor (2 - 1/l)d \rfloor + 1$ for the $l - MMF_C$ problem. For $l = 2$ the approximation ratio of the algorithm reduces from 2 to 1.5, but the difference becomes smaller as l increases. The algorithm is similar in nature to the one in this subsection. The lower TCT is a result of exploiting the limitation on the total number of message destinations.

A generalization of the MMF_C problem is called the *Multisource MMF_C*. The difference is that messages may originate in several processors, rather than just in one. Gonzalez [8] presents an algorithm for this problem. The algorithm constructs an instance of the MMF_C problem of least possible degree by selecting an origin for each message. This origin for each message can be determined by finding matchings in several bipartite graphs.

36.5 Distributed or On-Line Multimessage Multicasting

The $DMMF_C$ problem is the *distributed* or *online* version of the MMF_C problem. In this generalization of the MMF_C problem processors know local information and almost no global information. That is, each processor knows the destination of all the messages that it must send, and the values for d and n. Every processor in the MMF_C problem knows the messages it will receive, but in the $DMMF_C$ this information needs to be computed and transmitted to every processor. The schedule must be constructed online, and that is a time-consuming process. To speed-up the process, the above information and the schedule are not computed explicitly, but the operations performed by the algorithm form an implicit schedule.

The algorithm in Reference 28 for the $DMMF_C$ problem is based on the following idea. First it computes some global information and uses it to perform the first phase of the forwarding algorithm discussed in the previous section. Then it solves the resulting problem, which is the distributed multimessage unicasting problem, with forwarding, $DMUF_C$, problem.

The DMUF$_C$ problem is a fundamental problem for optical-communication parallel computers [10,42–44]. Valiant's [44] distributed algorithm constructs a communication schedule with $O(d + \log n)$ expected TCT. The expected TCT is optimal, to within a constant, for $d = \Omega(\log n)$. For $d = o(\log n)$, Goldberg et al. [43] showed that no more than $O(d + \log \log n)$ communication rounds are needed with certain probability. When forwarding is not allowed, the expected total communication time is larger, $\Theta(d + \log n \log \log n)$.

Gonzalez's [28] approach for the DMMF$_C$ problem begins by computing some global information through the well-known parallel prefix algorithm. Then the information is used to execute the forwarding phase of the algorithm discussed in the previous section. As in the case of the forwarding algorithm discussed in the previous section, the resulting problem is simply a multimessage unicasting problem. This problem can be solved by using Valiant's [44] distributed algorithm mentioned above. For every instance of the DMMF$_C$ problem the distributed algorithm performs all the communications in $O(d + \log n)$ expected total communication time [28]. The following theorem was established in Reference 28 by using a generalization of the analysis discussed in this section.

Theorem 36.8 *[28] For every instance of the DMMF$_C$ problem the algorithm given in [28] constructs a schedule with $O(d + \log n)$ expected TCT.*

Clearly, the DMMF$_C$ problem is much more general than the MMF$_C$ problem, but communication schedules for the DMMF$_C$ problem have expected total communication time $\Omega(d + \log n)$, but the TCT for the MM$_C$ problem is d^2, and just $2d$ for the MMF$_C$ problem. The knowledge ahead of time of all the communication patterns permits us to construct schedules with significantly smaller TCT. Forwarding is a very important factor in reducing the TCT. This is the most important consequence of the results discussed in this chapter. If the communication is restricted to the telephone mode, the best possible schedules will have TCT that is not bounded by any function in terms of d. In fact when there is only one message (to be sent from one processor to all other processors) the TCT is about $\log n$ and that is best possible. So the multicasting communication primitive delivers performance that is not attainable under the telephone communication mode. Even though the scheduling problems are NP-hard, schedules with suboptimal TCT can be constructed in polynomial time. In other words, the approximation problem is tractable.

36.6 Message Dissemination over Arbitrary Networks

Generating schedules with near optimal TCT is a complex problem for message dissemination problems over arbitrary graph. One problem is that the degree of the problem is no longer a good lower bound for the optimal TCT. Gonzalez [30,31] has had some success with restricted message dissemination problems. More specifically, with the gossiping problem. In what follows we discuss the gossiping problem when the multicasting communication primitive is available.

As we defined in the first section, the *broadcasting* problem defined over an n processor network N consists of sending a message from one processor in the network to all the remaining processors. The *gossiping* problem over a network N consists of broadcasting n messages each originating at a different processor. The network in this case is an arbitrary one, so messages need to be forwarded and the messages are routed through paths that are not necessarily the ones with the fewest number of links.

Gossiping problems can be solved optimally when the network has a Hamiltonian circuit, but finding a Hamiltonian circuit in an arbitrary network is a computationally intractable problem. Fortunately, it is not necessary for a network to have a Hamiltonian circuit for the gossiping problem to be solvable in $n-1$ steps, where n is the number of nodes in the network. There are networks that do not have a Hamiltonian circuit in which gossiping can be performed in $n-1$ communication steps under the multicasting communication mode, but not under the unicasting mode [30,31]. This shows again the power of the multicasting communication mode. Not all networks allow for communication schedules with total communication time $n-1$. For example, the straight line network does not have one such schedule because it is impossible to deliver a new message to each end of the line of processors at each communication round.

The broadcasting and gossiping problems have been studied for the past three decades [45]. However, most of the work is for the telephone type of communication. Most of the previous algorithms are not scalable because the communication mode allows for the transmission of packets of size 1 and n in one communication round. Under the traditional communication modes these problems are computationally difficult, that is, NP-hard. However, there are efficient algorithms to construct optimal communication schedules for restricted networks under some given communication modes [46–48]. There is no known polynomial time approximation algorithm with an approximation ratio bounded by a constant for the broadcasting problem defined over arbitrary graphs. The best approximation algorithms appear in References 48 and 49. A randomized algorithm is proposed in Reference 50. Broadcasting under the multicasting communication mode is a trivial problem to solve. That is, it is possible to construct a schedule with optimal TCT in linear time, for every problem instance.

A variation of the gossiping problem in which there are costs associated with the edges and there is a bound on the maximum number of packets that can be transmitted through a link at each time unit has been studied in Reference 51. Approximation algorithms for several versions of this problem with nonconstant approximation ratio are given in References 51–53.

We discuss now the algorithms for gossiping under the multicasting mode given in References 30 and 31. These algorithms take polynomial time. The algorithms begin by constructing a spanning tree of minimum radius, a problem that can be solved in polynomial time [30,31]. Then all the communications are performed in the resulting tree network. A vertex whose closest leaf is the farthest is called the *root* of the tree. The number of nodes in the network is n, and r is the radius of the network.

The simplest of algorithms [30], which has been used to solve other message routing problems, performs the gossiping in $2n + r$ time units. The idea is to send to the root all the messages and then send them down using the multicasting operation at each node. The best of the algorithms [31] is based on the observation that all the other operations can be carried out in a single stage by maximizing the concurrent operations performed at each step and finding appropriate times to transmit all messages down the tree. The total communication time of the schedules generated by this algorithm is just $n + r$. A lower bound of $n + r - 1$ for the number of communication rounds has been established for a large number of problem instances. Therefore, the algorithm given in Reference 31 generates schedules with TCT that is as close to optimal as possible for this set of problem instances. The *weighted gossiping* problem is a generalization of the gossiping problem where each processor needs to broadcast a nonempty set of messages, rather than just one. The currently best algorithm for the unrestricted version of the problem can be easily adapted to solve the weighted gossiping problem [31]. Furthermore, the algorithm can be easily transformed into a distributed algorithm. An interesting open problem is to design an efficient algorithm for the case when some of the weights could be zero.

36.7 Multimessage Multicasting in Optical Networks

Let us now consider the multimessage multicasting under the multiport communication model that allows multicasting. The architecture studied in this case is the star network with multiple fibers between nodes that allows optical switching between fibers along the same wavelength. The specific problem we consider is given any $(n + 1)$-node star network, a predetermined number of fibers that connect its nodes, and a set of multicasts (or multidestination messages) to be delivered in one communication round, find a conflict free message transmission schedule that uses the least number of wavelengths per fiber [33]. When the number of wavelengths, r, needed exceeds the number available, λ_{min}, one may transform the schedule into one with $\lceil r/\lambda_{min} \rceil$ communication phases or rounds over the same network, but restricted to λ_{min} wavelengths per fiber.

High transmission speeds are needed for many applications [54]. It is very likely that future communication networks will include a large amount of multidestination traffic [54,55]. Furthermore, since one of the biggest costs when building an optical network is the actual physical laying of the optical fibers, often many fibers may be installed at the same time, for about the same cost, resulting in multifiber networks [56]. Wide area testbeds are currently being developed, to transmit data over numerous wavelengths in realtime [57].

For the multimessage unicasting problem over a g fiber star network Li and Simha [58] developed a polynomial time algorithm to route all messages using $\lceil d/g \rceil$ wavelengths per fiber. Clearly, multifibers allow for solutions using fewer wavelengths, but the complexity of the switches needed for switching between fibers increases. Upper and lower bounds have been developed for the number of wavelengths needed to transmit all messages in one communication round (which translates directly to minimizing the number of communication rounds when the number of wavelengths per fiber is a fixed constant) [33]. Brandt and Gonzalez [33] have established an upper bound for m, the number of wavelengths per fiber, equal to $\lceil \frac{d^2}{i(g-1)} \rceil$, where i is any positive integer such that $1 \leq i \leq g-1$. The idea is to partition the fibers incoming into every processor into two groups: *receiving* and *sending* fibers. As their name indicates, the i receiving fibers are used to receive messages and the $g-i$ sending fibers are used for sending multicasts and the switching is at the center node. The optical switching can be set to deliver the messages from the sending fiber to the receiving fiber provided that they are sent and received along the same wavelength. To guarantee that this is always possible, we assign multiple wavelengths for the reception of each message on one of the receiving fibers, and we send along multiple wavelengths each message on one of the fibers. Specifically, the set of wavelengths is partitioned into d/i sets each with $m/(d/i)$ wavelengths. Each of these sets is called a *receiving set* of wavelengths. Now each multicast is sent on one wavelength from each of these sets. This guarantees that a switching to send the multicast to all its destinations always exists. The product $m \cdot g$ is minimum when $i = g/2$. In this case $m = \lceil 4d^2/g^2 \rceil$, which is about four times the lower bound $m = \lceil \frac{d^2}{g^2} + \frac{d}{g} \rceil$ we have established in Reference 33. For restricted values of g and d, Brandt and Gonzalez [33] have proved tight bounds for m, some of which are quite elaborate to establish. The reader is referred to Reference 33 for further information.

36.8 Discussion

It is simple to see that the $DMMF_C$ problem is more general than the MMF_C and the MM_C problems, but the best communication schedule for the $DMMF_C$ problem has TCT $\Omega(d + \log n)$ whereas for the MMF_C problem is just $2d$, and the MM_C problem is d^2. Therefore, knowing all the communication information ahead of time allows one to construct significantly better communication schedules. Also, forwarding plays a very important role in reducing the total communication time for our scheduling problems.

The most important open problem is to develop distributed algorithms with similar performance guarantees for processors connected through pr-dynamic networks. Algorithms exist for the nondistributed version of this problem [8]. The main difficulty in extending that work to the distributed case is the construction of the routing tables with only local information.

Another very challenging open problem is to develop efficient approximation algorithms for the MMF_C problem that generate schedules with communication time significantly smaller than $2d$. There are several variations of the MM_C problem that are worth studying. For example the case when there are precedence constraints between the messages seems to be one that arises in several applications.

References

[1] Gonzalez, T.F., Complexity and approximations for multimessage multicasting, *Journal of Parallel and Distributed Computing*, 55, 215, 1998.

[2] Bertsekas, D.P., and Tsitsiklis, J.N., *Parallel and Distributed Computation: Numerical Methods*, Prentice Hall, Englewood Cliffs, NJ, 1989.

[3] Krumme, D. W., Venkataraman, K.N., and Cybenko, G., Gossiping in minimal time, *SIAM Journal on Computing* 21(2), 111, 1992.

[4] Almasi, G.S. and Gottlieb, A., *Highly Parallel Computing*, Benjamin/Cummings Publishing, New York, 1994.
[5] Lee, T.T., Non-blocking copy networks for multicast packet switching, *IEEE Journal of Selected Areas of Communication*, 6(9), 1455, 1988.
[6] Liew, S.C., A general packet replication scheme for multicasting in interconnection networks, *Proceedings of the IEEE INFOCOM*, 1, 1995, 394.
[7] Turner, J.S., A practical version of Lee's multicast switch architecture, *IEEE Transactions on Communications*, 41(8), 1993, 1166.
[8] Gonzalez, T.F., Simple multimessage multicasting approximation algorithms with forwarding, *Algorithmica*, 29, 511, 2001.
[9] Gonzalez, T.F. and Sahni, S., Open shop scheduling to minimize finish time, *JACM*, 23(4), 665, 1976.
[10] Anderson, R.J. and G. L. Miller, G.L., Optical communications for pointer based algorithms, TRCS CRI 88 – 14, USC, 1988.
[11] Cole, R. and Hopcroft, J., On edge coloring bipartite graphs, *SIAM Journal on Computing*, 11(3), 540, 1982.
[12] Gabow, H. and Kariv, O., Algorithms for edge coloring bipartite graphs and multigraphs, *SIAM Journal Computing*, 11, 117, 1982.
[13] Cole, R., Ost, K., and Schirra, S., Edge-coloring bipartite multigraphs in $O(E \log D)$, *Combinatorica*, 21, 5, 2001.
[14] Gonzalez, T.F., Open shop scheduling, in *Handbook of Scheduling: Algorithms, Models, and Performance Analysis*, Leung, Y.J.-T., (ed) Chapman & Hall, CRC, 2004, Chap. 6.
[15] Coffman, Jr, E.G., Garey, M.R., Johnson, D.S., and LaPaugh, A.S., Scheduling file transfers in distributed networks, *SIAM Journal on Computing*, 14(3), 744, 1985.
[16] Whitehead, J., The complexity of file transfer scheduling with forwarding, *SIAM Journal on Computing* 19(2), 222, 1990.
[17] Choi, H.-A. and Hakimi, S.L., Data transfers in networks with transceivers, *Networks*, 17, 393, 1987.
[18] Choi, H.-A. and Hakimi, S.L., Data transfers in networks, *Algorithmica*, 3, 223, 1988.
[19] Hajek, B., and Sasaki, G., Link scheduling in polynomial time, *IEEE Transactions on Information Theory*, 34(5), 910, 1988.
[20] Gopal, I.S., Bongiovanni, G., Bonuccelli, M.A., Tang, D.T., and Wong, C.K., An optimal switching algorithm for multibeam satellite systems with variable bandwidth beams, *IEEE Transactions on Communications*, 30(11), 2475, 1982.
[21] Rivera-Vega, P.I., Varadarajan, R., and Navathe, S.B., Scheduling file transfers in fully connected networks, *Networks*, 22, 563, 1992.
[22] Goldman, A., Peters, J. G., and Trystram, D., Exchanging messages of different sizes, *Journal of Parallel and Distributed Computing*, 66, 18, 2006.
[23] Hall, J., Hartline, J., Karlin, A.R., Saia, J., and Wilkes, J., *Proc. of SODA*, 2001, 620.
[24] Gonzalez, T. F., MultiMessage Multicasting, *Proc. of the Third International Workshop on Parallel Algorithms for Irregularly Structured Problems*, LNCS 1117, Springer, 1996, 217.
[25] Gonzalez, T.F., Proofs for improved approximation algorithms for multimessage multicasting, UCSB TRCS-96-17, 1996.
[26] Gonzalez, T.F., Algorithms for multimessage multicasting with forwarding, *Proc. of the 10th PDCS*, 1997, 372.
[27] Gonzalez, T.F., Improved approximation algorithms for multimessage multicasting, *Nordic Journal on Computing*, 5, 196, 1998.
[28] Gonzalez, T.F., Distributed multimessage multicasting, *Journal of Interconnection Networks*, 1(4), 303, 2000.
[29] Gonzalez, T. F., On solving multimessage multicasting problems, *International Journal of Foundations of Computer Science*, Special Issue on Scheduling—Theory and Applications, 12(6), 791, 2001.

[30] Gonzalez, T. F., Gossiping in the multicasting communication environment, *Proc. of IPDPS*, 2001.
[31] Gonzalez, T. F., An efficient algorithm for gossiping in the multicasting communication environment, *IEEE Transactions on Parallel and Distributed Systems*, 14(7), 701, 2003.
[32] Gonzalez, T. F., Improving the computation and communication time with buffers, *Proc. 17th IASTED PDCS*, 2006, 336.
[33] Brandt, R. and Gonzalez, T.F., Multicasting using WDM in multifiber optical star networks, *Journal of Interconnection Networks*, 6(4), 383, 2005.
[34] Shen, H., Efficient multiple multicasting in hypercubes, *Journal of Systems Architecture*, 43(9), 1997.
[35] Thaker, D. and Rouskas, G., Multi-destination communication in broadcast WDM networks: A survey, *Optical Networks*, 3(1), 34, 2002.
[36] Khuller, S., Kim, Y.-A., and Wan, Y-C., Algorithms for data migration with cloning, *SIAM Journal on Computing*, 33(2), 448, 2004
[37] Khuller, S., Kim, Y.-A., and Wan, Y-C., Broadcasting on networks of workstations, *Proc. of SPAA*, 2005.
[38] Gandhi, R., Halldorsson, M.M., Kortsarz, M., and Shachnai, H., Improved results for data migration and open shop scheduling, *ACM Transactions on Algorithms*, 2(1), 116, 2006.
[39] Bar-Noy, A., Bellare, M., Halldorsson, M.M., Shachnai, H., and Tamir, T., On chromatic sums and distributed resource allocation, *Inf. Comput.*, 140, 183, 1998.
[40] Holyer, I., The NP-completeness of edge-coloring, *SIAM Journal of Computing*, 11, 117, 1982.
[41] Bruno, J., Personal communication, 1995.
[42] Gereb-Graus, M. and Tsantilas, T., Efficient optical communication in parallel computers, *Proc. of 4th SPAA*, 1992, 41.
[43] Goldberg, L.A., Jerrum, M., Leighton, T., and Rao., S., Doubly logarithmic communication algorithms for optical-communication parallel computers, *SIAM J. Comp.*, 26(4), 1100, 1997.
[44] Valiant, L.G., General purpose parallel architectures, in *Handbook of Theoretical Computer Science*, van Leeuwen, J., ed., Elsevier, 1990, Chap. 18.
[45] Hedetniemi, S., Hedetniemi, S., and Liestman, A survey of gossiping and broadcasting in communication networks, *NETWORKS*, 18, 129, 1988.
[46] Even, S. and Monien, B., On the number of rounds necessary to disseminate information, *Proc. SPAA*, 1989, 318.
[47] Fujita, S. and Yamashita, M., Optimal group gossiping in hypercube under circuit switching model, *SIAM Journal on Computing* 25(5), 1045, 1996.
[48] Ravi, R., Rapid rumor ramification, *Proc. FOCS*, 1994, 202.
[49] Hromkovic, J., Klasing, R., Monien, B., and Peine, R., *Dissemination of Information in Interconnection Networks (Broadcasting and Gossiping)*, Du D.Z. and Hsu, D.F., eds., Kluwer Academic, 1995, 273.
[50] Feige, U., Peleg, D., Raghavan, P., and Upfal, E., Randomized broadcast in networks, *Proc. of SIGAL*, LNCS, Springer-Verlag, 1990, 128.
[51] Fraigniaud, P. and Vial, S., Approximation algorithms for broadcasting and gossiping, *Journal of Parallel and Distributed Computing*, 43, 47, 1997.
[52] Bermond, J. C., Gargano, L., Rescigno, C.C., and Vaccaro, U., Fast gossiping by short messages, *SIAM Journal on Computing* 27(4), 917, 1998.
[53] Gargano, L., Rescigno, A.A., and Vaccaro, U., Communication complexity of gossiping by packets, *Journal of Parallel and Distributed Computing*, 45, 73, 1997.
[54] Rouskas, G. and Ammar, M., Multi-destination communication over single-hop lightwave WDM networks, *Proc. of INFOCOM*, 1994, 1520.
[55] Rouskas, G. and Ammar, M., Multi-destination communication over tunable-receiver single-hop WDM networks, *IEEE Journal on Selected Areas in Communications*, 15(3), 501, 1997.

[56] Ferreira, A., Perennes, S., Richa, A., Rivano, H., and Stier, N., On the design of multifiber WDM networks, *Proc. of AlgoTel*, 2002, 25.

[57] Chien, A., OptIPuter software: System software for high bandwidth network applications. http://www-csag.ucsd.edu/projects/Optiputer.html, 2002.

[58] Li, G. and Simha, R., On the wavelength assignment problem in multifiber WDM star and ring networks, *IEEE/ACM Transactions on Networking*, 9(1), 60, 2001.

37
Online Computation in Large Networks

37.1	Introduction...	37-1
37.2	Packet Buffering...	37-2
	Single Buffer Problems • Multibuffer Problems with Uniform Packet Values • Multibuffer Problems with Arbitrary Packet Values • Managing CIOQ Switches	
37.3	Online Routing..	37-9
	Model and Known Results • A Simple Oblivious Routing Scheme	
37.4	Transmission Control Protocol and Multicast Acknowledgment	37-14
	TCP and Multicast Acknowledgment • Multicast Acknowledgment	
37.5	Caching Problems in Networks	37-17
	Document Caching • Connection Caching	
37.6	Conclusions	37-21
	References	37-21

Susanne Albers
University of Freiburg

37.1 Introduction

With the advent of the Internet, computational problems arising in large networks have received tremendous research interest. The considerable body of work also addresses the design and analysis of algorithms for fundamental optimization problems. These include, for instance, routing and scheduling as well as advanced resource management problems. The general goal is to devise strategies having a *provably* good performance.

Many important networking problems are inherently *online*, that is, at any time decisions must be made without knowledge of future inputs or future states of the system. In this context, competitive analysis [41] has proven to be a powerful tool. Here an online algorithm ALG is compared to an optimal offline algorithm OPT that knows the entire input in advance. ALG is called c-competitive if, for any input, the solution computed by ALG is at most a factor of c away from that of OPT. We will make this notion more precise when studying specific problems.

In this chapter, we will study fundamental network problems that have attracted considerable attention in the algorithms community over the past 5–10 years. Most of the problems we will investigate are related to communication and data transfer, which are premier issues in high-performance networks today.

In computer networks and telecommunications, packet switching is the now-dominant communications paradigm. The data to be transmitted is partitioned into packets, which are then sent over the network. At the hardware level router and switches are essential components of a network infrastructure, ensuring that packets reach their correct destination. In Section 37.2, we will consider various basic buffer management problems that arise in current switch architectures and survey known results. We will present numerous algorithms with an excellent competitive performance.

At the transport layer of a network routing protocols determine the routing paths along which data packets are transferred. Initiated by a breakthrough result, the past 4 years have witnessed extensive research activities on online oblivious routing schemes. In Section 37.3, we review recent results and also present the most simple oblivious routing algorithm with a polylogarithmic competitiveness that has been developed so far. Also, at the transport layer of a network, the Transmission Control Protocol (TCP) is the most common protocol for exchanging data. As networks are to some extent unreliable and packets may get lost, data sent over a TCP connection must be acknowledged by the receiving node. In order to keep the network congestion low, acknowledgments of data packets may be aggregated. In Section 37.4, we study TCP acknowledgment problems and address related multicast acknowledgment issues.

In Section 37.5, we investigate two basic caching problems that occur in large networks. We first consider document caching, that is, the problem of maintaining a local cache with frequently accessed web documents, and review proposed document replacement strategies. We then address connection caching, which consists in handling a limited number of open TCP connections, and show that concepts from document caching can be used to solve this problem efficiently.

The chapter concludes with some final remarks in Section 37.6.

37.2 Packet Buffering

The performance of high-speed networks critically depends on switches that route data packets arriving at the input ports to the appropriate output ports so that the packets can reach their correct destination in the network. To reduce packet loss when the traffic is bursty, ports are equipped with buffers where packets can be stored temporarily. However, the buffers are of limited capacity so that effective buffer management strategies are important to maximize the throughput at a switch. Although packet buffering strategies have been investigated in the applied computer science and, in particular, networking communities for many years, only a seminal paper by Kesselman et al. [32] in 2001 has initiated profound algorithmic studies.

Input-queued (IQ) switches, output-queued (OQ) switches, and switches, with a combination of input and output queuing (CIOQ) represent the dominant switch architectures today. In an IQ switch with m input and m output ports, each input i maintains for each output j a separate queue Q_{ij}, $1 \leq i, j \leq m$, storing those packets that arrive at input i and have to be routed to output j. In each time step, for any output j, one packet from the queues Q_{ij}, $1 \leq i \leq m$, can be sent to that output. Figure 37.1 depicts the basic architecture of an IQ switch with three input and three output ports. Each input port is equipped with buffers for each of the three output ports. The small rectangles represent data packets.

In an OQ switch, queues are located at the output ports. Each output port j maintains a single queue buffering those packets that have to be sent through that output.

Finally, an CIOQ switch is equipped with buffers at the input and output ports. In this case, the internal speed of the switch is larger than the frequency with which the switch can forward packets.

The switch architectures described above give rise to various algorithmic buffering problems. The goal is to maintain the packet buffers so as to minimize the packet loss or, equivalently, to maximize the packet throughput. The problems are online in that at any point in time future packet arrivals are unknown. In the following, we will first study single buffer problems and then settings with multiple buffers. We will use the terms *buffer* and *queue* interchangeably.

Online Computation in Large Networks

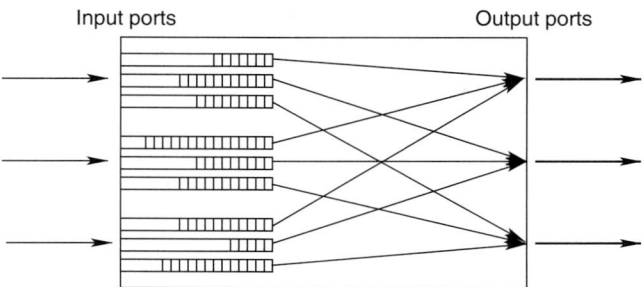

FIGURE 37.1 An IQ switch.

37.2.1 Single Buffer Problems

Single buffer problems arise when maintaining a fixed queue Q_{ij} of an IQ switch or the queue located at an output port of an OQ switch. From an algorithmic point of view the problems are interesting when packets have *values* representing QoS parameters of the respective data streams. We wish to always forward those packets having the highest values.

Formally, consider a buffer that can simultaneously store up to B data packets. Packets arrive online and can be buffered if space permits. Time is assumed to be discrete. In any time step t, let $Q(t)$ be the set of packets currently stored in the buffer and let $A(t)$ be the set of newly arriving packets. Each packet p has a value $v(p)$, which is a nonnegative real number. If $|Q(t)| + |A(t)| \leq B$, then all new packets can be admitted to the buffer; otherwise, $|Q(t)| + |A(t)| - B$ packets from $Q(t) \cup A(t)$ must be dropped. In the time step, we can select one packet from the buffer and transmit it to the output port. We assume that the packet arrival step precedes the transmission step. The goal is to maximize the throughput, that is, the total value of the transmitted packets.

Given a packet arrival sequence σ, let $T_{\text{ALG}}(\sigma)$ be the throughput achieved by an online algorithm ALG and let $T_{\text{OPT}}(\sigma)$ be the throughput of an optimal offline algorithm OPT that knows the entire sequence σ in advance. Since we are dealing with a maximization problem, ALG is called c-competitive if there exists a constant a such that $T_{\text{ALG}}(\sigma) \geq \frac{1}{c} T_{\text{OPT}}(\sigma) + a$, for all sequences σ. The constant a must be independent of σ.

Several problem variants are of interest. In a FIFO model, packets must be transmitted in the order they arrive. If packet p is transmitted before p', then p must not have arrived later than p'. In a non-FIFO model, there is no such restriction. In a preemptive model we may drop packets from the buffer, while in a nonpreemptive model this is not allowed.

Kesselman et al. [32] analyzed a natural *Greedy* algorithm in the preemptive FIFO model and proved that it is 2-competitive.

Algorithm Greedy: In the event of a buffer overflow, drop the packets with the smallest values.

In the following, let α be the ratio of the largest to smallest packet value.

Theorem 37.1 [32] *Greedy achieves a competitive ratio of* $\min\left\{2 - \frac{1}{B+1}, 2 - \frac{2}{\alpha+1}\right\}$.

Kesselman et al. [33] gave an algorithm that is 1.98-competitive, thus beating the bound of 2 for any B and α. In a subsequent paper, Bansal et al. [13] modified that strategy and proved an upper bound of 1.75. The algorithm works as follows.

Algorithm β-Preemptive Greedy: When a packet of value $v(p)$ arrives, execute the following steps (i) Find the first packet p' in FIFO order such that p' has value less than $v(p)/\beta$ and less than the value of the packet following p' in the buffer (if any). If such a packet exists, preempt it. (ii) If there is free space in the buffer, accept p. (iii) Otherwise (the buffer is full), evict a packet p' with the smallest value among p and those in the buffer, while keeping the rest of the buffer.

Theorem 37.2 [13] *Setting $\beta = 4$, β-Preemptive Greedy is 1.75-competitive.*

Theorem 37.3 [33] *The competitive ratio of any deterministic online algorithm in the preemptive FIFO model is at least 1.419.*

The above lower bound shows that there is room for improvement.

Aiello et al. [1] investigated nonpreemptive single buffer problems. In this case, the buffer can simply be maintained as a FIFO queue. Andelman et al. [5] gave asymptotically tight bounds for this scenario. They analyzed the following algorithm. Suppose that the packet values are in the range $[1, \alpha]$.

Algorithm Exponential Interval Round Robin: Divide the buffer into k partitions of size B/k, where $k = \lceil \ln \alpha \rceil$. Split the interval $[1, \alpha]$ into k subintervals $[\alpha_0, \alpha_1), [\alpha_1, \alpha_2), \ldots, [\alpha_{k-1}, \alpha_k)$, where $\alpha_j = \alpha^{j/k}$. Each partition of the buffer is associated with one of the subintervals, accepting packets from that subinterval if space permits. The partitions take turn in sending packets. If a partition is empty, its turn is passed to the next partition.

Theorem 37.4 [5] *Exponential Interval Round Robin achieves a competitive ratio of $e\lceil \ln \alpha \rceil$.*

Theorem 37.5 [5] *No online algorithm can achieve a competitive ratio smaller than $1 + \ln \alpha$ in the nonpreemptive model.*

Kesselman et al. [32] also introduced a *bounded delay model* where packets have deadlines. A packet that has not been transmitted by its deadline is lost. There is no bound on the buffer size and packets may be reordered. Kesselman et al. analyzed a *Greedy* strategy that at any time transmits the packet of highest value among those with unexpired deadlines. This strategy is 2-competitive. Chrobak et al. [19] gave an improved algorithm that is 1.939-competitive. A lower bound equal to the Golden Ratio $\Phi = (1 + \sqrt{5})/2 \approx 1.618$ was shown by Andelman et al. [5].

37.2.2 Multibuffer Problems with Uniform Packet Values

We next investigate settings with multiple buffers and start by studying a very basic scenario in which all data packets are equally important, that is, they all have the same value. Most current networks, in particular IP networks, treat packets from different data streams equally at intermediate switches. As we shall see, even problems with uniform packet values are highly nontrivial when several buffers have to be maintained.

Formally, we are given m buffers, each of which can simultaneously store up to B data packets. Again, the buffers are organized as queues. In any time step, in a packet arrival phase, new packets may arrive at the buffers and can be appended to the queues if space permits. Suppose that buffer i currently stores b_i packets and that a_i new packets arrive there. If $b_i + a_i \leq B$, then all new packets can be accepted; otherwise, $a_i + b_i - B$ packets must be dropped. Furthermore, in any time step, in a transmission phase, an algorithm can select one nonempty buffer and transmit the packet at the head of the queue to the output. We assume w.l.o.g. that the packet arrival phase precedes the transmission phase. The goal is to maximize the throughput, which is the total number of successfully transmitted packets. Obviously, the m buffers correspond to the queues Q_{ij}, $1 \leq i \leq m$, of an IQ switch competing for service at output port j.

For this multibuffer problem, tight or nearly tight upper and lower bounds on the competitiveness of deterministic and randomized policies have been developed. As we consider a scenario where all packets have the same value the algorithms apply a greedy admission policy: At any time t and for any of the m buffers, whenever new data packets arrive, an algorithm accepts as many packets as possible subject to the constraint that a buffer can only store up to B data packets simultaneously. Thus, the algorithms we present here only specify which buffer to serve in each time step.

A simple observation shows that any *reasonable* algorithm ALG, which just serves any nonempty queue, is 2-competitive: Partition σ into subsequences σ_ℓ such that ALG's buffers are empty at the end of each σ_ℓ. W.l.o.g. we postpone the beginning of $\sigma_{\ell+1}$ until OPT has emptied its buffers, too. Let T be the length of σ_ℓ, that is, the number of time steps until ALG's buffers are empty. If OPT buffers b_i packets in queue i at the end of σ_ℓ, then at least b_i packets must have arrived there in σ_ℓ. ALG has transmitted T packets, where $T \geq \sum_{i=1}^{m} b_i$, while OPT delivers at most $T + \sum_{i=1}^{m} b_i$ packets.

Considering concrete strategies, the most simple and natural algorithm is *Greedy*. In the following, let the *load* of a queue be the number of packets currently stored in it.

Algorithm Greedy: In each time step serve a queue currently having the maximum load.

From a practical point of view, *Greedy* is very interesting because it is fast and uses little extra memory. More specifically, the algorithm just has to determine the most loaded queue and, regarding memory, only has to store its index. Serving the longest queue is a very reasonable strategy to avoid packet loss if future packet arrival patterns are unknown. Implementations of *Greedy* can differ in the way how ties are broken when several queues store a maximum number of packets. It turns out that this does not affect the competitive performance. As *Greedy*, obviously, is a reasonable algorithm, the following theorem is immediate.

Theorem 37.6 *Greedy is 2-competitive, no matter how ties are broken.*

Unfortunately, *Greedy* is not better than 2-competitive as the lower bound of $2 - 1/B$ stated in the next theorem can be arbitrarily close to 2.

Theorem 37.7 [4] *For any B, the competitive ratio of Greedy is not smaller than $2 - 1/B$, no matter how ties are broken.*

Thus, *Greedy* is exactly 2-competitive, for arbitrary buffer sizes B. On the basis of the fact that *Greedy*'s competitive performance is not better than that of arbitrary reasonable strategies, researchers have worked on finding improved policies. The first deterministic online algorithm beating the bound of 2 was devised in [4] and is called *Semi Greedy*. The strategy deviates from standard *Greedy* when all the queues are lightly populated, that is, they store at most $B/2$ packets, and the current risk of packet loss is low. In this situation, the algorithm preferably serves queues that have never experienced packet loss in the past.

Algorithm Semi Greedy: In each time step execute the first of the following three rules that applies to the current buffer configuration. (i) If there is a queue buffering more than $\lfloor B/2 \rfloor$ packets, serve the queue currently having the maximum load. (ii) If there is a queue the hitherto maximum load of which is less than B, then among these queues serve the one currently having the maximum load. (iii) Serve the queue currently having the maximum load. In each of the three rules, ties are broken by choosing the queue with the smallest index. Furthermore, whenever all queues become empty, the hitherto maximum load is reset to 0 for all queues.

Theorem 37.8 [4] *Semi Greedy achieves a competitive ratio of 17/9.*

The ratio of 17/9 is approximately 1.89.

A deterministic online algorithm that has an almost optimal competitiveness is the *Waterlevel* strategy [9]. We give a condensed presentation of the algorithm; in the original paper, the description was more general. We use q_i to refer to the i-th buffer/queue. *Waterlevel* consists of three algorithms that simulate each other. At the bottom level there is a fractional *Waterlevel* algorithm, denoted by FWL, that allows us to process fractional amounts of packets. FWL is based on the fact that a packet switching schedule can be viewed as a matching that maps any time step t to the packet p transmitted during that step. For any packet arrival sequence σ, consider the following bipartite graph $G_\sigma = (U, V, E)$ in which vertex sets U and V represent time steps and packets, respectively. If T is the last point in time at which packets arrive in σ, then packets may be transmitted up to time $T + mB$. Thus, for any time t, $1 \leq t \leq T + mB$, set U contains a vertex u_t. For any packet p that ever arrives, V contains a vertex v_p. Let P_t^i be the set of the last B packets that arrive at queue q_i until (and including) time t and let $P_t = \bigcup_{i=1}^{m} P_t^i$. The set of edges is defined as $E = \{(u_t, v_p) \mid p \in P_t\}$.

In a standard matching, each edge is either part of the matching or not, that is, for any edge $(u_t, v_p) \in E$ we can define a variable x_t^p that takes the value 1 if the edge is part of the matching and 0 otherwise. In a fractional matching we relax this constraint and allow $x_t^p \in [0, 1]$. Intuitively, x_t^p is the extent to which packet p is transmitted during time t. Of course, at any time t, a total extent of at most 1 can be transmitted and, over all time steps, any packet can be transmitted at most once. Formally

$$\sum_{p \in P_t} x_t^p \leq 1, \quad \text{for any } t,$$

$$\sum_{t=1}^{T+mB} x_t^p \leq 1, \quad \text{for any } p.$$

The graph G_σ evolves over time. At any time t, $1 \leq t \leq T + mB$, a new node u_t is added to U and new packet nodes, depending on the packet arrivals, may be added to V. A switching algorithm has to construct an online matching, mapping time steps (fractionally) to data packets that have arrived to far. The idea of the *Waterlevel* algorithm is to serve the available data packets as evenly as possible. For any time t and any packet p, let $s_t^p = \sum_{t' < t} x_{t'}^p$ be the extent to which p has been served to far. The fractional *Waterlevel* algorithm FWL works as follows.

Algorithm FWL: At any time t, for any packet $p \in P_t$, match an extent of $x_t^p = \max\{h - s_t^p, 0\}$, where h is the maximum number such that $\sum_{p \in P_t} x_t^p \leq 1$.

The goal of the following steps is to discretize FWL. This is done by first using larger buffers.

Algorithm D(FWL): Work with queues of size $B + 1 + \lfloor H_m \rfloor$. Run a simulation of FWL. At any time t and for any queue q_i, let S_i be the total number of packets transmitted from queue q_i by D(FWL) before time t and let S_i' be the total extent to which packets from queue q_i have been served by FWL up to (and including) time t. Transmit a packet from the nonempty queue for which the residual service extent $S_i' - S_i$ is largest.

Next, we take care of the large buffer sizes and present the actual *Waterlevel* strategy.

Algorithm Waterlevel: Work with queues of size B. Run a simulation of D(FWL). In each time step, accept a packet if D(FWL) accepts it and the corresponding queue is not full. Transmit packets as D(FWL) if the corresponding queue is not empty.

Although *Waterlevel* is quite involved, it has the best possible competitiveness, for large B, that deterministic strategies can achieve. The ratio of $e/(e-1)$ is ~ 1.58.

Theorem 37.9 [9] *Waterlevel achieves a competitive ratio of* $\frac{e}{e-1}\left(1 + \frac{\lfloor H_m + 1 \rfloor}{B}\right)$.

Theorem 37.10 [4] *The competitive ratio of any deterministic online algorithm is at least* $e/(e-1)$.

We next discuss randomized strategies. The competitive performance ratios we mention below hold against oblivious adversaries [14]. The first randomized online algorithm proposed was *Random Schedule* [10] attaining a competitiveness of $e/(e-1)$. We omit the presentation here and instead give a description of *Random Permutation* [40], the policy that has the smallest competitiveness known.

The basic approach of *Random Permutation* is to reduce the packet switching problem with m buffers of size B to one with mB buffers of size 1. To this end, a packet buffer q_i of size B is associated with a set $Q_i = \{q_{i,0}, \ldots, q_{i,B-1}\}$ of B buffers of size 1. A packet arrival sequence σ for the problem with size B buffers is transformed into a sequence $\tilde{\sigma}$ for unit-size buffers by applying a round robin strategy. More specifically, the j-th packet ever arriving at q_i is mapped to $q_{i,j \bmod B}$ in Q_i. *Random Permutation* at any time runs a simulation of the following algorithm *SimRP* for $m' = mB$ buffers of size 1.

Algorithm SimRP(m'): The algorithm is specified for m' buffers for size 1. Initially, choose a permutation π uniformly at random from the permutations on $\{1, \ldots, m'\}$. In each step transmit the packet from the nonempty queue whose index occurs first in π.

The algorithm for buffers of arbitrary size then works as follows.

Algorithm Random Permutation: Given a packet arrival sequence σ that arrives online, run a simulation of *SimRP(mB)* on $\tilde{\sigma}$. At any time, if *SimRP(mB)* serves a buffer from Q_i, transmit a packet from q_i. If the buffers of *SimRP(mB)* are all empty, transmit a packet from an arbitrary nonempty queue if there is one.

The algorithm achieves a nearly optimal competitive ratio.

Theorem 37.11 [40] *Random Permutation is 1.5-competitive.*

Theorem 37.12 [4] *The competitive ratio of any randomized online algorithm is not smaller than 1.4659, for any buffer size B.*

37.2.3 Multibuffer Problems with Arbitrary Packet Values

In this section, we investigate a generalized scenario in which packets may have values. Again, consider m packet buffers, each of which can simultaneously store up to B packets. Each packet p has a nonnegative value $v(p)$. At any time t, let $Q_i(t)$ be the set of packets stored in buffer i and let $A_i(t)$ be the set of packets arriving at that buffer. If $|Q_i(t)| + |A_i(t)| \leq B$, then all arriving packets can be admitted to buffer i; otherwise, $|Q_i(t)| + |A_i(t)| - B$ packets must be dropped. At any time, the switch can select one nonempty buffer and transmit the packet at the head to the output port. The goal is to maximize the total value of the transmitted packets.

Azar and Richter [10] presented a general technique that transforms a buffer management strategy for a single queue (for both the preemptive and nonpreemptive models) into an algorithm for m queues. The technique is based on the algorithm *Transmit-Largest* that works in the preemptive non-FIFO model.

Algorithm Transmit-Largest (TL)
1. Admission control: Use a greedy approach for admission control in any of the m buffers. More precisely, enqueue a packet arriving at buffer i if buffer i is not full or if the packet with the smallest value in the buffer has a lower value than the new packet. In the latter case, the packet with the smallest value is dropped.
2. Transmission: In each time step, transmit the packet with the largest value among all packets in the m queues.

Using this algorithm, Azar and Richter [10] designed a technique *Generic Switch* that takes a single buffer management algorithm *ALG* as input parameter. We are interested in the preemptive FIFO and the nonpreemptive models. Here, packets are always transmitted in the order they arrive (this holds w.l.o.g. in the nonpreemptive model) and only *ALG*'s admission control strategy is relevant to us.

Algorithm Generic Switch

1. Admission control: Apply admission control strategy *ALG* to any of the m buffers.
2. Transmission: Run a simulation of *TL* (in the preemptive non-FIFO model) with online paket arrival sequence σ. In each time step, transmit the packet from the head of the queue served by *TL*.

Theorem 37.13 [10] *If ALG is a c-competitive algorithm, then Generic Switch is 2c-competitive.*

Using this statement, applying upper bounds from Section 37.2.1, one can derive a number of results for multiqueue problems. In the preemptive FIFO model *Greedy* gives a competitiveness of $\min\{4 - \frac{2}{B+1}, 4 - \frac{4}{\alpha+1}\}$. The improved algorithm by Bansal et al. [13] yields a 3.5-competitive strategy. In the nonpreemptive setting we obtain a $2e\lceil \ln \alpha \rceil$-competitive algorithm.

For the preemptive scenario, Azar and Richter [11] gave an improved 3-competitive algorithm.

Algorithm Transmit-Largest Head

1. Admission control: Each new packet p arriving at buffer i is handled as follows. Admit p to the buffer if it is not full or if it is full but $v(p)$ is larger than the smallest value $v(p')$ currently stored in the buffer. In the latter case, discard p' before admitting p.
2. Transmission: In each time step, transmit the packet with the largest value among all packets stored at the head of the queues.

Azar and Richter [11] introduced an interesting 0/1-principle for switching networks, which allowed them to analyze the above algorithm on packet sequences consisting of 0/1-values only. They established the following result.

Theorem 37.14 [11] *Algorithm Transmit-Largest Head is 3-competitive.*

37.2.4 Managing CIOQ Switches

We finally study the most general scenario of managing all the buffers of a CIOQ switch with m input and m output ports. At the input ports we have queues Q_{ij} storing those packets that arrive at input port i and have to be routed to output port j, $1 \leq i, j \leq m$. At any output port j we have a queue Q'_j keeping those packets that have to be transmitted through that port, $1 \leq j \leq m$. All the queues are organized as preemptive FIFO buffers. Each time step t consists of three phases. As usual, in a first packet arrival phase, packets arrive at the input queues Q_{ij} and can be admitted subject to capacity constraints. A second *scheduling phase* consists of S rounds, where S represents the speed of the switch. In each round, at most one packet can be removed at any input port i and at most one packet can be appended at any output port j. More precisely, for any i, an algorithm may remove only one packet stored in $Q_{i1}, \ldots, Q_{im}$ and, for any j, may add only one packet to Q'_j. In a third transmission phase, each output queue Q'_j may send out the packet at the head. We will consider settings with uniform as well as arbitrary packet values $v(p)$. The goal is to maximize the total number/value of packets transmitted by the output ports.

For any time t and any queue Q, let $Q(t)$ be the set of packets currently stored in Q and let $h(Q(t))$ be the packet currently located at the head. Furthermore, let $\min(Q(t))$ be the minimum value of any packets in $Q(t)$. Every scheduling round can be viewed as a matching mapping packets at the heads of the Q_{ij}, $1 \leq i, j \leq m$, to the respective output ports. For any time step t, let t_s denote the time of scheduling round s, $1 \leq s \leq S$. Any subset $H(t_s) \subseteq \{h(Q_{ij}(t_s)) \mid 1 \leq i, j \leq m\}$ of eligible packets induces a bipartite graph $G(t_s) = (U, V, E(t_s))$. For any input port i there is a vertex in U, and for any output port j there is a vertex in V. Edge set $E(t_s)$ contains edge (i, j) if and only if $H(t_s)$ contains a packet p from Q_{ij}. This edge has weight $v(p)$. A scheduling step amounts to finding a maximum (weighted) matching in $G(t_s)$.

We present two constant competitive algorithms, one for unit value packets and a second one for arbitrary valued packets. The algorithms specify how buffers are managed when data packets have to be appended to queues and how the scheduling rounds proceed. Clearly, in the transmission phase of each time step, each nonempty output queue transmits the packet at the head. Both algorithms are greedy-based strategies. The algorithms also work when the buffers Q_{ij} and Q'_j have different sizes. We start by considering the scenario with unit value packets. The following algorithm was proposed by Kesselman and Rosen [34].

Algorithm Greedy: Each time step t is specified as follows:
1. *Input/output buffer management:* Admit a new packet arriving at a buffer if space permits; otherwise drop the packet.
2. *Scheduling:* In each round t_s, compute a maximum matching in the graph $G(t_s)$ induced by $H(t_s) = \{h(Q_{ij}(t_s)) \mid 1 \leq i,j \leq m \text{ and } Q'_j(t_s) \text{ is not full}\}$.

Theorem 37.15 [34] *For any speed S, Greedy is 3-competitive.*

Azar and Richter [12] developed an algorithm for packets with arbitrary values. The strategy employs a buffer admission policy $Admin(\beta)$ that, in the case of a full buffer, admits a new packet if its value is β times larger than the minimum value in the queue. Here, $\beta \geq 1$ is a real parameter specified later.

Algorithm Admin(β): At any time t, admit a packet p arriving at a buffer Q if the buffer is not full or if it is full and $v(p) > \beta \min(Q(t))$. In the latter case, discard a packet p' with $v(p') = \min(Q(t))$ from the buffer.

Algorithm Greedy(β): Each time step t is specified as follows:
- Buffer management: At the input queues use *Greedy(1)*. At the output queues use *Greedy(β)*.
- Scheduling: In each scheduling round t_s, compute a maximum weighted matching in the graph $G(t_s)$ induced by $H(t_s) = \{h(Q_{ij}(t_s)) \mid 1 \leq i,j \leq m \text{ and } Q'_j(t_s) \text{ is not full or } v(h(Q_{ij}(t_s))) > \beta \min(O'_j(t_s))\}$.

Theorem 37.16 [12] *For any speed S, setting $\beta = 1 + \sqrt{5}$, Greedy(β) is 9.47-competitive.*

37.3 Online Routing

Efficient routing protocols represent a key technological ingredient in large networks. There exists a vast body of literature investigating many problem variants and developing lower and upper bounds for the respective settings. Recently, the problem of online virtual circuit routing has received considerable research interest, initiated by a breakthrough of Räcke [37]. In this section, we survey the recent results and also present the most simple online oblivious routing scheme that has been developed so far.

37.3.1 Model and Known Results

Consider a network modeled by a weighted undirected graph $G = (V, E)$ with $|V| = n$ vertices and $|E|$ edges. Any edge $e \in E$ has a capacity $c(e)$, which is a nonnegative real value. A sequence of routing requests, each defined by a source node s and a target node t, arrives online. In response to each request, a routing algorithm has to specify a path from the source to the target in the network. This increases the load of each edge on the path by 1. When the entire sequence of routing requests is scheduled, let the *absolute load* of a link e be the number of routing paths using that link. The *relative load* of e is the absolute load divided by the capacity $c(e)$. The goal is to minimize the *congestion*, that is, the maximum of the relative loads occurring on the links of the network. An online algorithm is c-competitive if, for any sequence of routing requests, its congestion is not larger than c times that of an optimal solution.

Aspnes et al. [6] presented an algorithm that achieves a competitive ratio of $O(\log n)$ and proved that this is the best asymptotic bound possible. However, the algorithm is centralized and thus hard to implement in a real network. Awerbuch and Azar [7] gave a less centralized algorithm with the same competitive performance. However, this algorithm also has to repeatedly scan the entire network when making routing decisions, which is not feasible in practice. From a practical point of view *oblivious* routing algorithms are important. A routing scheme is oblivious if the path selected for a routing request is independent of other requests; the path just depends on the source and destination of the request. Valiant and Brebner [42] considered oblivious routing schemes for specific network topologies and gave an efficient randomized algorithm for hypercubes. Borodin and Hopcroft [16] and Kaklamanis et al. [30] showed that randomization is essential in the sense that deterministic oblivious routing algorithms cannot well approximate the congestion in nontrivial networks. It was a long-standing open problem whether or not efficient randomized oblivious routing algorithms exist for arbitrary network topologies. In 2002, Räcke [37] achieved a breakthrough by proving the following result.

Theorem 37.17 [37] *For arbitrary network topologies, there exists a randomized oblivious routing algorithm that achieves a competitive ratio of $O(\log^3 n)$.*

One of the main concepts of Räcke's result is a hierarchical decomposition of the given network. The decomposition is defined by a series of nested edge cuts of subgraphs into smaller and smaller pieces until every piece eventually consists of only one network node. A hierarchical decomposition can be associated with a tree in which each node represents a subgraph and the children of a node represent the connected components when the subgraph is split into smaller pieces. The tree is used as a hint how to route in the original graph. Räcke proves that there exists a good hierarchical decomposition such that one can route in the original graph almost as well as in the tree. To this end, he shows that routing in the tree can be simulated by a hierarchical set of multicommodity flows.

Unfortunately, Räcke's result is nonconstructive and the actual oblivious routing scheme cannot be found in polynomial time. In a subsequent paper, Azar et al. [8] used linear programming to show that an optimally competitive oblivious routing scheme can be found in polynomial time in arbitrary networks. The result applies to undirected as well as directed networks. Azar et al. [8] also show that there are directed networks for which the competitiveness is $\Omega(\sqrt{n})$. Unfortunately, the algorithm in [8] is based on linear programming with an infinite number of constraints and uses the Ellipsoid method with separation oracle to compute the solution. This approach cannot be applied for large networks.

To overcome this problem, Harrelson et al. [28] and Bienkowski et al. [15] presented explicit polynomial time algorithms for constructing good tree decompositions and associated routing schemes. The solution of Harrelson et al. [28] achieves an even improved competitive ratio of $O(\log^2 n \log \log n)$. The scheme by Bienkowski et al. [15] has a slightly higher competitiveness of $O(\log^4 n)$. However, their algorithm is the simplest one among the proposed approaches and we will therefore present it in the next section. We state again the result with the smallest competitiveness.

Theorem 37.18 [28] *For arbitrary undirected network topologies, there exists a randomized oblivious routing algorithm that achieves a competitive ratio of $O(\log^2 n \log \log n)$.*

As mentioned above, there exist simple *directed* graphs for which every oblivious routing algorithm has a competitiveness of $\Omega(\sqrt{n})$ [8]. This lower bound holds even if all routing requests specify the same single sink [26]. Motivated by these high lower bounds, Hajiaghayi et al. [25] study oblivious routing in directed graphs with *random demands*. Here, each routing request has an associated demand that is chosen randomly from a known demand distribution. Hajiaghayi et al. [25] show that if an oblivious routing algorithm is *demand-independent*, that is, it ignores the demands when making routing decisions, it cannot be better than $\Omega(\sqrt{n}/\log n)$-competitive, with high probability. Therefore, the authors study *demand-dependent* oblivious routing where algorithms take into account the demands of requests and prove the following result.

Theorem 37.19 [25] *For general directed graphs and for any c, there exists a demand-dependent oblivious routing algorithm that achieves a competitive ratio of $c \cdot O(\log^2 n)$ with probability at least $1 - 1/n^c$.*

As for lower bounds, there exist networks for which no (demand dependent) oblivious routing algorithm can be better than $\Omega(\log n / \log \log n)$-competitive [25]. This bound also holds for undirected graphs [27].

37.3.2 A Simple Oblivious Routing Scheme

We describe the routing scheme of Bienkowski et al. [15] achieving a polylogarithmic competitiveness of $O(\log^4 n)$. Recall that the network is modeled by a weighted undirected graph $G = (V, E)$, where edge $(u, v) \in E$ has a nonnegative real capacity $c(u, v)$. For simplicity, for any pair of nodes $u, v \in V$ that is not connected by an edge, we set $c(u, v) = 0$. For any two sets $X, Y \subseteq V$, let $cap(X, Y) = \sum_{x \in X, y \in Y} c(x, y)$ be the total edge capacity between nodes in X and nodes in Y. Furthermore, for any $X \subseteq V$, let $out(X) = cap(X, V \setminus X)$ be the total capacity of links leaving X.

Bienkowski et al. construct a randomized oblivious routing scheme that consists of a probability distribution over (s, t)-paths, for each source-target pair $s, t \in V$. Such a probability distribution can be viewed as a unit flow between s and t. A result by Raghavan and Thompson [38] implies that probabilistically choosing a fixed routing path, where the probability of choosing a path is equal to its flow value w.r.t. the unit flow, yields the same expected loads on the links, with high probability. Given a sequence of routing requests, let D be the nonnegative $n \times n$ demand matrix specifying for each source-target pair s, t the volume of routing requests from s to t. The diagonal entries are 0. Given a unit flow from s to t, let $f_{s,t}(e)$ be the flow across edge $e \in E$. Then the total load on edge e is given by $load(e) = \sum_{s,t} d_{s,t} f_{s,t}(e)$, where $d_{s,t}$ is the demand of routing requests from s to t as specified in D. The congestion of the edge is $load(e)/c(e)$, and this value is to be minimized over all edges.

As already sketched in Section 37.3.1, the core of the oblivious routing scheme developed by Räcke [37] and also by Bienkowsi et al. [15] is a hierarchical decomposition of the graph $G = (V, E)$. A hierarchical decomposition $\mathcal{H}$ is a set system over the node set V with the following properties:

- For any two subsets $X, Y \in \mathcal{H}$, either $X \setminus Y$, $Y \setminus X$ or $X \cap Y$ is empty, that is, the two sets X and Y are either properly contained in each other or disjoint.
- $\mathcal{H}$ contains V and all singleton sets $\{v\}$, $v \in V$.

The sets $X \in \mathcal{H}$ are also called clusters.

A hierarchical decomposition $\mathcal{H}$ of G is associated with a decomposition tree $T_\mathcal{H} = (V_t, E_t)$. For the cluster V, the node set V_t contains a red node. For all other clusters $H \in \mathcal{H}$, $H \neq V$, set V_t contains a red node r_t as well as a blue node b_t. For a node $v_t \in T_t$, let H_{v_t} be the cluster represented by v_t. The tree is rooted at a red node corresponding to V and consists of the following edges. For any $H \in \mathcal{H}$, $H \neq V$, the corresponding blue and red nodes b_t and r_t are connected by an edge. A blue node b_t corresponding to H_{b_t} is connected with a red node r_t corresponding to H_{r_t} if $H_{b_t} \subseteq H_{r_t}$ and there is no $H \in \mathcal{H}$, $H \neq H_{b_t}$ and $H \neq H_{r_t}$, with $H_{b_t} \subseteq H \subseteq H_{r_t}$. Thus, starting with a red root, $T_\mathcal{H}$ consists of blue and red node layers, where leaves are again colored red. We finally define layers on the nodes and clusters. A node $v_t \in V_t$ belongs to layer ℓ if the path from v_t to the root contains ℓ red nodes, not counting v_t. A cluster belongs to layer ℓ if its corresponding nodes are in layer ℓ. Note that for any cluster $H \neq V$, the blue and red nodes are in the same layer. Figure 37.2 depicts a hierarchical decomposition of a graph and the associated decomposition tree. In the tree, large circles represent red nodes and small circles represent blue nodes.

The oblivious routing scheme that specifies for each source-target pair in the graph a unit flow is based on solutions to concurrent multicommodity flow problems (CMCF-problems) that are specified for any cluster $H \in \mathcal{H}$. We need one more definition. For any edge $e \in E$ in the original graph we define a *level* with respect to $\mathcal{H}$ and $T_\mathcal{H}$. If both end points of e belong to the same level $\ell - 1$ cluster but to different level ℓ clusters, then e belongs to level ℓ, that is, $level(e) = \ell$. For any node set $X \subseteq V$ and $\ell \in \{0, \ldots, \text{height}(T_\mathcal{H})\}$, let

$$w_\ell(X) = \sum_{\substack{e \in X \times V \\ level(e) \leq \ell}} c(e)$$

be the total capacity of edges having an endpoint in X and a level of at most ℓ.

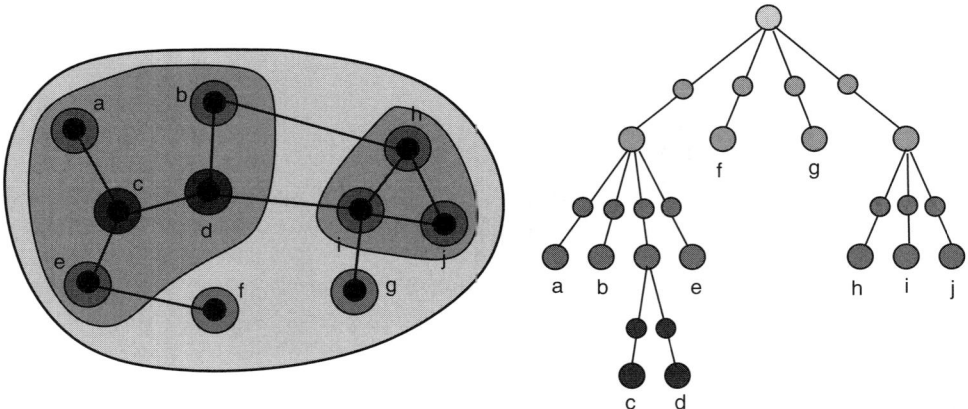

FIGURE 37.2 A graph decomposition and the corresponding decomposition tree.

For any level ℓ cluster H of the decomposition tree, a CMCF-problem is defined as follows. For any ordered pair $u, v \in H$ of nodes there is a commodity $d_{u,v}$ with source u, sink v and demand

$$\text{dem}(u, v) = \frac{w_{\ell+1}(u) w_{\ell+1}(v)}{w_{\ell+1}(H)}$$

Using known multicommodity flow algorithms, such a CMCF-problem will be solved *completely inside* the cluster H, that is, no edges outside H may be used. For each edge e inside the cluster, the original capacity $c(e)$ must be observed. Because of these capacity constraints, the demands might not be met. The *throughput fraction* of a solution to a CMCF-problem is the minimum, over all commodities, of the fraction of the commodity's demand that is met by the solution. We are interested in solutions that maximize the minimum throughput fraction $q_{\min}$ that is achieved for all the clusters $H \in \mathcal{H}$.

Suppose that we are given good solutions to the CMCF-problems for clusters $H \in \mathcal{H}$. Then an oblivious routing scheme is constructed as follows.

Flow paths for source–target pairs $s, t \in V$.

1. In $T_\mathcal{H}$ consider the path from s to t. Let $v_1, \ldots, v_r$ be the nodes on this path and let $\ell(i)$ be the level of v_i.
2. A unit flow is transferred by the network nodes of the clusters corresponding to nodes $v_1, \ldots, v_r$. If v_i is a blue node, then any network node $u \in H_{v_i}$ receives a fraction of $w_{\ell(i)}(u)/w_{\ell(i)}(H_{v_i})$ of the flow. If v_i is a red node, then $u \in H_{v_i}$ receives a fraction $w_{\ell(i)+1}(u)/w_{\ell(i)+1}(H_{v_i})$.
3. The flow transformation from nodes in H_{v_i} to nodes in $H_{v_{i+1}}$ is done as follows. Let $f_i(u)$ be the amount of flow received by $u \in H_{v_i}$, and let r_t be the red node from $\{v_i, v_{i+1}\}$. Node $u \in H_{v_i}$ sends a fraction $f_i(u) f_{i+1}(v)$ to node $v \in H_{v_{i+1}}$ using flow paths of commodity $d_{u,v}$ a given by the solution to the CMCF-problem for cluster H_{r_t}.

In Step 2, intuitively, the flow for blue nodes v_i is distributed according to the weight of edges entering or leaving H_{v_i} because the flow has just entered H_{v_i} or will leave this cluster in the next step. For red nodes v_i, the flow is distributed according to the weight of edges leaving or entering the subclusters of H_{v_i}, because the flow has to enter the subclusters of H_{v_i} or leave H_{v_i} in the next step. Bienkowski et al. [15] proved the following lemma.

Lemma 37.1 [15] *The above routing scheme achieves a competitive ratio of $O(h/q_{\min})$, where h is the height of $T_\mathcal{H}$.*

It remains to show that there exists a hierarchical decomposition $\mathcal{H}$ such the height of $T_{\mathcal{H}}$ is $O(\log n)$ and the minimum throughput fraction $q_{\min}$ of any CMCF-problem for a cluster $H \in \mathcal{H}$ is at $\Omega(1/\log^3 n)$. Starting with the universe V, clusters H are partitioned into smaller and smaller subclusters. To ensure that for each cluster a minimum throughput fraction can be guaranteed, a certain *precondition* must be fulfilled. For a subgraph induced by a cluster $H \in \mathcal{H}$, consider a cut $(A, H \setminus A)$ and let $cap(A, H \setminus A)/dem(A, H \setminus A)$ be the sparsity of the cut, where $dem(A, H \setminus A)$ is the demand of the CMCF-problem separated by the cut. Let σ be the minimum possible ratio between the throughput fraction of a CMCF-problem and the sparsity of an approximate sparsest cut. For general graphs there exist algorithms attaining $\sigma = O(\log n)$. Let $\lambda = 64\sigma \log n$ and $q_{\min} = 1/(24\sigma\lambda)$. We note that $q_{\min} = \Omega(1/\log^3 n)$.

We are interested in constructing clusters H that *fulfill the throughput property*, that is, the solution to the CMCF-problem in H has a throughput of at least $q_{\min}$. Bienkowski et al. show that this can be achieved recursively for all levels of the decomposition if a certain precondition is satisfied. A level ℓ cluster is said to *fulfill the precondition* if, for all sets $U \subseteq H$ with $|U| \leq \frac{3}{4}|H|$ inequality $\lambda cap(U, H \setminus U) \geq w_\ell(U)$ holds.

Bienkowski et al. developed an algorithm that, given a level ℓ cluster H consisting of at least two nodes and fulfilling the precondition, partitions H into disjoint subclusters H_i having the following properties:

(P1) H fulfills the throughput condition.
(P2) For each subcluster H_i, inequality $|H_i| \leq 2/3|H|$ is satisfied.
(P3) Each H_i fulfills the precondition.

The algorithm runs in polynomial time with respect to $|H|$ and the maximum capacity of a network link. Note that property (P2) ensures that the height of the decomposition tree is $O(\log n)$.

The actual partitioning algorithm described below makes use of a subroutine *Assure Precondition*. As shown by Bienkowski et al., the algorithm returns for any node set $R \subseteq V$ a partition in which every subset R_i fulfills the precondition. The running time is polynomial in $|R|$. Starting with a partition consisting of only R, the algorithm in each iteration takes every set R_i of the current partition $\mathcal{P}_R$. For each R_i, it computes an approximate sparsest cut for concurrent multicommodity flow problem $\mathcal{G}(R_i)$ with demands $dem(u, v) = w_\ell(u)/|R_i|$, for each ordered pair $u, v \in R_i$. Let $\psi = cap(A, B)/(|B|w_\ell(A)/|R_i| + |A|w_\ell(B)/|R_i|)$ be the sparsity of the cut. If ψ is small, then R_i is replaced by A and B in the current partition.

Algorithm Assure Precondition (R):
$\mathcal{P}_R := \{R\}$;
repeat
    for each $R_i \in \mathcal{P}_R$ **do**
        Compute an approximate sparsest cut (A, B) for $\mathcal{G}(R_i)$;
        if $\psi \leq 4\sigma/\lambda$ **then** $\mathcal{P}_R := (\mathcal{P}_R \setminus \{R_i\}) \cup \{A, B\}$;
until no changes are made to $\mathcal{P}_R$;
return $\mathcal{P}_R$;

We finally describe the algorithm for partitioning a cluster H into subclusters H_i such that properties (P1–3) are satisfied. Initially, the algorithm starts with singleton sets $\{v\}$, $v \in H$. As long as H does not fulfill the throughput property, the algorithm computes an approximate sparsest cut (A, B). W.l.o.g. let $|A| \leq |B|$. The algorithm identifies all the subclusters H_i having a large intersection with A and removes them from the current partition. For these subclusters *Assure Precondition* is called, which returns a new partitioning fulfilling the precondition. Bienkowski et al. show that by choosing subclusters H_i having a large intersection with A, they identify an index set I of subclusters such that $out(\cup_{i \in I} H_i) << w_{l+1}(\cup_{i \in I} H_i)$. In the next iteration of the algorithm the throughput property is then easier to satisfy.

Algorithm Partition (H):
$\mathcal{P}_H := \{\{v\} \mid v \in H\}$;
while H does not fulfill the throughput property **do**

Compute an approximate sparsest cut (A, B) of H (w.l.o.g. $|A| \leq |B|$);
$U^* := \{H_i \in H \mid |A \cap H_i| \geq \frac{3}{4}|H_i|\}$;
for each $H_i \in U^*$ **do** $\mathcal{P}_H := \mathcal{P}_H \setminus H_i$;
$\mathcal{P}_H := \mathcal{P}_H \cup (\text{Assure Precondition}(\cup_{H_i \in U^*} H_i))$;
return $\mathcal{P}_H$;

In summary, Bienkowski et al. prove the following result.

Theorem 37.20 [15] *For general undirected graphs, the oblivious routing algorithm described above achieves a competitive ratio of $O(\log^4 n)$.*

37.4 Transmission Control Protocol and Multicast Acknowledgment

The most common protocol for data transmission in large networks is TCP. If two network nodes wish to exchange data, then there has to exist an open TCP connection between these two nodes. The data is partitioned into packets, which are then sent across the connection. A node receiving data must acknowledge the receipt of each incoming packet so that the sending node is aware that the transmission was successful. In most TCP implementations today data packets do not have to be acknowledged individually. Instead, there is some delay mechanism that allows the TCP to acknowledge multiple packets with a single acknowledgment and, possibly, to piggyback the acknowledgment on an outgoing data packet. This reduces the number of acknowledgments sent and hence the network congestion as well as the overhead at the network nodes for sending and receiving acknowledgments. On the other hand, by reducing the number of acknowledgments, we add latency to the TCP connection, which is not desirable. Thus, the goal is to balance the reduction in the number of acknowledgments with the increase in latency.

The TCP acknowledgment problem just described occurs in generalized form in multicast protocols as well. Suppose that a sender wishes to transmit data to a group of receivers in a communication network. Instead of sending separate messages to the individual receivers, the sender performs a *multicast*, that is, it transmits a single message along the tree spanned by the entire group of receivers. Again, receivers must acknowledge each arriving packet by sending a corresponding control message to the root of the multicast tree. At the nodes of the tree, acknowledgments may be delayed and aggregated to reduce the communication load. As before, the goal is to balance the reduction in the communication overhead with the increase in acknowledgment delays.

37.4.1 TCP and Multicast Acknowledgment

Formally, TCP acknowledgment can be modeled as follows [20]. A network node receives a sequence of m data packets. Let a_i denote the arrival time of packet i, $1 \leq i \leq m$. At time a_i, the arrival times a_j, $j > i$, are not known. We have to partition the sequence $\sigma = (a_1, \ldots, a_m)$ of packet arrival times into n subsequences $\sigma_1, \ldots, \sigma_n$, for some $n \geq 1$, such that each subsequence ends with an acknowledgment. We use σ_i to denote the set of arrivals in the partition. Let t_i be the time when the acknowledgment for σ_i is sent. We require $t_i \geq a_j$, for all $a_j \in \sigma_i$. If data packets are not acknowledged immediately, there are acknowledgment delays. Dooley et al. [20] considered the objective function that minimizes the number of acknowledgments and the sum of the delays incurred for all of the packets, that is, we wish to minimize $f = n + \sum_{i=1}^{n} \sum_{a_j \in \sigma_i} (t_i - a_j)$. It turns out that a simple greedy strategy is optimal for this problem.

Algorithm Greedy: Send an acknowledgment whenever the total delay of the unacknowledged packets is equal to 1, that is, equal to the cost of an acknowledgment.

Theorem 37.21 [20] *The Greedy algorithm is 2-competitive and no deterministic online algorithm can achieve a smaller competitive ratio.*

Noga [36] and independently Seiden [39] showed that no randomized algorithm can achieve a competitive ratio smaller than $e/(e-1) \approx 1.58$ against oblivious adversaries. Karlin et al. [31] presented a randomized strategy that achieves this factor. Let $P(t, t')$ be the set of packets that arrive after time t but up to (and including) time t'. The following algorithm works for positive real numbers between 0 and 1. It sends an acknowledgment when, in hindsight, z time units of latency could have been saved by sending an earlier acknowledgment.

Algorithm Save(z): Let t be the time when the last acknowledgment was sent. Send the next acknowledgment at the first time $t' > t$ such that there is a time τ with $t \leq \tau \leq t'$ and $P(t, t')(t' - \tau) = z$.

Theorem 37.22 [31] *If z is chosen according to the probability density function $p(z) = e^z/(e-1)$, Save(z) achieves a competitive ratio of $e/(e-1)$.*

Albers and Bals [3] investigated another family of objective functions that penalize long acknowledgment delays of individual data packets more heavily. When TCP is used for interactive data transfer, long delays are not desirable as they are noticeable to a user. Hence, we wish to minimize the function $g = n + \max_{1 \leq i \leq n} d_i$, where $d_i = \max_{a_j \in \sigma_i}(t_i - a_j)$ is the maximum delay of any packet in σ_i. The following family of algorithms is defined for any positive real z.

Algorithm Linear Delay(z): Initially, set $d = z$ and send the first acknowledgment at time $a_1 + d$. In general, suppose that the i-th acknowledgment has just been sent and that j packets have been processed so far. Set $d = (i+1)z$ and send the $(i+1)$-st acknowledgment at time $a_{j+1} + d$.

Theorem 37.23 [3] *Setting $z = \pi^2/6 - 1$, Linear Delay(z) achieves a competitive ratio of $\pi^2/6 \approx 1.644$ and no deterministic strategy can achieve a smaller competitiveness.*

It is well known that $\pi^2/6 = \sum_{i=1}^{\infty} 1/i^2$. In addition, Albers and Bals [3] investigate a generalization of the objective function g where delays are taken to the p-th power and hence are penalized even more heavily. They proved that the best competitive ratio is an alternating sum of Riemann's zeta function. The ratio is decreasing in p and tends to 1.5 as $p \to \infty$. Frederiksen and Larsen [24] studied a variant of the TCP acknowledgment problem, where it is required that there is some minimum delay between sending two acknowledgments to reflect the physical properties of the network.

37.4.2 Multicast Acknowledgment

In multicast acknowledgment, we are given a weighted tree T rooted at a node r. All edges of T are directed toward this root node r. Packets arrive at the nodes of T and must be sent to the root. These packets represent the acknowledgments of broadcast messages. The cost incurred in this process consists of two components. First, each packet p incurs a delay cost equal to its waiting time in the system. More specifically, if p arrives at time a_p and reaches the root at time r_p, then the delay is $r_p - a_p$. Second, there is a communication cost. When a set of packets is delivered along an edge e, a delivery cost equal to the weight $w(e)$ of the edge is incurred. This cost is independent of the size of the set of packets sent. We make the simplifying assumption that there is no propagation delay along the links, that is, sending a packet along a link does not increase the packet's delay. The delay of a packet is given by the total waiting time at the nodes. We wish to minimize the total cost consisting of the entire communication cost and the sum of the delay costs of all the packets.

Brito et al. [17] introduced three models for studying multicast acknowledgment:

- *Asynchronous Model*: All the network nodes act in isolation and there is no way for them to coordinate.
- *Synchronous Model*: There is a global clock allowing the network nodes to synchronize their actions.
- *Full Information Model*: At any time each node has complete knowledge of the state of the network. However, future packet arrivals are unknown.

Brito et al. [17] settled the competitiveness in the Asynchronous Model. The competitiveness is expressed in terms of a parameter $\chi(T)$ that is defined as follows. Let $\mathcal{S}_T$ be the set of subtrees T' of T that are rooted at r. For any $T' \in \mathcal{S}_T$, let $\mathcal{P}(T)$ be the set of paths π that start at a node in T' and end at r. Let $w(\pi)$ be the sum of the weights associated with edges in π and let $w(T')$ be the total weight of edges in T'. Define

$$\chi(T) = \max_{T' \in \mathcal{S}_T} \frac{\sum_{\pi \in \mathcal{P}(T')} w(\pi)}{w(T')}$$

Brito et al. [17] showed that if all edge costs are 1, then $\chi(T)$ is Θ(height of T). In general, $\chi(T)$ can take values between 1 and the cost of the most expensive path in T. The authors proposed the following online algorithm, which is nearly optimal. For a network node v, let π_v be the path from v to the root.

Algorithm Wait: Node v in the network delays the delivery of packets arriving from the external source until the accumulated delay is at least $w(\pi_v)/\chi(T)$. Then it delivers the packets to the root. At any time, packets arriving from other nodes in the network are forwarded immediately upon arrival, generating no delay.

Theorem 37.24 [17] *In the Asynchronous Model, Wait is $(\chi(T) + 1)$-competitive.*

Theorem 37.25 [17] *In the Asynchronous Model, any deterministic online algorithm has a competitive ratio of at least $\chi(T)$.*

Khanna et al. [35] studied the Synchronous and Full Information Models and proposed greedy-like strategies. They assume that each new packet arriving in the network must wait at least one time unit before it can be forwarded. At any time t and for any network node v, let $P(v, t)$ be the set of packets waiting at v at time t.

Algorithm Greedy1: At any time t, network node v delivers $P(t, v)$ along the outgoing edge e to the parent node if the total delay associated with $P(v, t)$, denoted by $\text{delay}(P(v, t))$ is at least $w(e)$. Upon arrival at the parent node, the cost associated with $P(v, t)$ is reduced to $\text{delay}(P(v, t)) - w(e)$, that is, intuitively, the packets have "paid" for crossing e.

Theorem 37.26 [35] *In the Synchronous Model, Greedy1 is $O(h \log(w(T)))$-competitive, where h is the height of T.*

Khanna et al. [35] also developed lower bounds for *oblivious* algorithms. A strategy is oblivious if every node makes decisions solely on the basis of local information.

Theorem 37.27 [35] *In the Synchronous Model, any oblivious online algorithm has a competitive ratio of at least $\Omega(\sqrt{h})$.*

Finally, consider the Full Information Model. Khanna et al. [35] proposed the following policy that represents a generalized rent-or-buy strategy.

Algorithm Greedy2: At any time t identify the maximal subtree $T' \in \mathcal{S}_T$ for which the total delay of packets waiting at nodes in T' is at least $w(T')$. Deliver the packets in T' to the root.

Theorem 37.28 [35] *In the Full Information Model, Greedy2 is $O(\log(w(T)))$-competitive.*

Brito et al. [17] gave constant competitive algorithms for line networks and so-called shallow networks.

37.5 Caching Problems in Networks

In this section, we investigate two basic caching problems in large networks. We will first present algorithms for managing caches with web documents. We will then show that similar strategies can be used in connection caching where one has to maintain a set of open TCP connections.

37.5.1 Document Caching

Web caching is a fundamental resource management problem that arises when users, working on computers, surf the web and download documents from other network sites. Downloaded documents can be stored in local caches so that they do not have to be retransmitted when users wish to access these documents again. Caches can be maintained by web clients or servers. Storing frequently accessed documents in local caches can substantially reduce user response times as well as the network congestion. Web caching differs from standard paging and caching in operating systems in that documents here have varying sizes and incur varying costs when being downloaded. The loading cost depends, for instance, on the size of the document and on the current congestion in the network. The goal is to maintain the local caches in such a way that a sequence of document accesses generated in the network can be served with low total loading cost. It turns out that this distributed problem can be decomposed into local ones, one for each cache in the network. Note that the cache configurations are essentially independent of each other. Each cache may store an arbitrary subset of the documents available in the network. Thus, the problem of minimizing the loading cost globally in the network is equivalent to minimizing the loading cost incurred locally at each cache. Therefore, the following problem abstraction has been studied extensively.

We are given a two-level memory system consisting of a small fast memory and a large slow memory. The fast memory represents a local cache we have to maintain; the slow memory represents the remaining network, that is, the universe of all documents accessible in the network. We assume that the fast memory has a capacity of K, counted in bits. For any document d, let $size(d)$ be the size and $cost(d)$ be the cost of d. The size is again measured in bits. A sequence of requests for documents arrives online at the fast memory. At any time, if the referenced document is in cache, the request can be served with 0 cost. If the requested document is not in cache, it must be retrieved over the network at a cost of $cost(d)$. Immediately after this download operation the document may be brought at no extra cost into cache. The goal is to serve a sequence of requests so that the total loading cost is as small as possible. Various cost models have been proposed in the literature.

1. *The Bit Model*: For each document d, we have $cost(d) = size(d)$. (The delay in bringing the document into fast memory depends only upon its size.)
2. *The Fault Model*: For each document d, we have $cost(d) = 1$ while the sizes can be arbitrary.
3. *The General Model*: For each document d, both the cost and size can be arbitrary.

In the following, let k be the ratio of K to the size of the smallest document ever requested.

A first natural idea for solving the web caching problem is to extend caching algorithms known from operating systems. A classical strategy with an excellent performance in practice is *Least Recently Used (LRU)*. When a referenced memory page is not in cache, LRU evicts the page from cache whose last access is longest ago and then loads the missing object. For the Bit and the Fault Models, the LRU strategy is $(k + 1)$-competitive [21]. The analysis assumes that a missing document does not necessarily have to be brought into fast memory. For the General Model, Young [43] gave a k-competitive online algorithm called *Landlord*. He assumes that referenced documents must be brought into cache if not already present.

Algorithm Landlord: For each document d in fast memory, the algorithm maintains a variable $credit(d)$ that takes values between 0 and $cost(d)$. If a requested document d is already in fast memory, then $credit(d)$ is reset to any value between its current value and $cost(d)$. If the requested page is not in fast memory, then the following two steps are executed until there is enough room to load d: (i) For each document d' in fast memory, decrease $credit(d')$ by $\Delta \cdot size(d')$, where $\Delta = \min_{d' \in F} credit(d')/size(d')$ and F is the set of documents in fast memory. (ii) Evict any document d' from fast memory with $credit(d') = 0$. When there is enough room, load d and set $credit(d)$ to $cost(d)$.

Theorem 37.29 [43] *Landlord is k-competitive in the General Model.*

A result by Sleator and Tarjan [41] implies that, in any cost model, no deterministic online algorithm can be better than k-competitive. An interesting question is if the bound of k can be improved using randomization. For the Bit and the Fault Models, Irani presented randomized online algorithms that achieve a competitiveness of $O(\log^2 k)$. In the following, we present these algorithms.

We first normalize all document sizes such that the smallest document has size 1. The algorithms divide the documents into $\lfloor \log k \rfloor + 1$ classes, where class ℓ documents have a size of at least 2^ℓ and less than $2^{\ell+1}$. The documents in class ℓ are also referred to as ℓ-documents. Let $cs(e)$ denote the class of documents of size e. The randomized algorithms developed by Irani [29] make use of randomized marking strategies developed by Fiat et al. [22] for the standard paging problem. In standard paging documents or pages, have uniform sizes and incur a cost of 1 when being referenced but not residing in cache. A randomized *Marking* algorithm operates in phases. At the beginning of each phase all pages are unmarked. Whenever a page is requested, it is marked. If a referenced page is not in cache, that is, a page fault occurs, the algorithm evicts a page that is chosen uniformly at random from the unmarked pages in cache and then loads the missing object. A phase ends when a page fault occurs and all pages in cache are marked. At that time, all marks are erased and a new phase is started. Fiat et al. [22] showed that the randomized *Marking* algorithm is $2H_k$-competitive, where $H_k \approx \ln k$ is the k-th Harmonic number.

The randomized algorithms by Irani apply marking policies to each document class. For each class, independent phase partitionings are generated. The algorithms are described in a relaxed model that allows one to evict documents partially. The real algorithms will evict the entire documents once a portion has been discarded. At any time, let $size(d)_{in}$ and $size(d)_{out}$ be the number of bits of d that reside inside or outside the cache, respectively. When evicting documents, the algorithms will call the following subroutine $Evict(\ell, C)$ that evicts C bits from class ℓ documents. After first evicting documents that reside only partially in cache, the subroutine applies the marking policy.

Algorithm Evict (ℓ, C):
if ∃ partially evicted document d in cache **then**
    $x := \min\{size(d)_{in}, C\}$;
    Evict x bits from d;
    $C := C - x$;
while $C \neq 0$ **do**
    if ∃ ℓ-documents in cache **then**
        if ∄ unmarked ℓ-documents **then** unmark all ℓ-documents;
        Choose an unmarked ℓ-document d uniformly at random;
        $x := \min\{size(d), C\}$;
        Evict x bits of d from the cache and set $C := C - x$;
if there are less than 2^ℓ bits belonging to unmarked ℓ-documents in cache **then**
    Evict them;

The randomized algorithm *Marking (FM)* for the Fault Model is now easy to state. For ease of exposition, it is assumed that a referenced missing document d may be brought into cache while exceeding temporarily the cache capacity of K. The capacity constraint will be established again by evicting documents. The crucial question is from which document class to evict. Irani resolves this problem by evicting from every class. However, this causes an extra factor of $O(\log k)$ in the competitiveness of the algorithm, compared to the classical randomized *Marking* algorithm.

> **Algorithm Marking (FM):** A request to an ℓ-document d is processed a follows:
> Bring d into cache, if not already there, and mark it;
> **if** d was requested in the previous phase **then**
> $\quad$ Let u_ℓ be number of bits belonging to ℓ-documents in cache;
> $\quad$ $Evict(\ell, \min\{u_\ell, \text{size}(d)_{\text{out}}\})$;
> **if** cache capacity is exceeded **then**
> $\quad$ Execute $Evict(j, 2^{j+1})$ twice for all j;

Theorem 37.30 [29] *Algorithm Marking (FM) is $O(\log^2 k)$-competitive.*

The algorithm *Marking (BM)* for the Bit Model is very similar in structure. An additional feature is that the algorithm maintains a counter $c(\ell)$ for each class ℓ. These counters are initialized to 0 at the beginning of the request sequence. The purpose of these counters is the following. When a missing document is brought into cache and the cache capacity is exceeded by only a small amount of e, in the Bit Model one cannot afford to evict a large document. Thus, for any document class j with $j > cs(e)$ one adds to the class's counter. When the counter is large enough, one cashes the counter by evicting a document.

> **Algorithm Marking (BM):** A request to an ℓ-document d is processed as follows:
> Bring d into cache, if not already there, and mark it;
> **if** d was requested in the previous phase **then**
> $\quad$ Let u_ℓ be number of bits belonging to ℓ-documents in cache;
> $\quad$ $Evict(\ell, \min\{u_\ell, \text{size}(d)_{\text{out}}\})$;
> **if** cache capacity is exceeded by e bits **then**
> $\quad$ **if** total volume of ℓ'-documents, $\ell' \leq cs(e)$, in cache is less than e **then**
> $\quad\quad$ Let $\ell_0, \ell_0 > cs(e)$, be smallest class having an ℓ_0-document in cache;
> $\quad\quad$ $m := 2^{\ell_0}$;
> $\quad$ **else** $m := e$;
> $\quad$ **for all** $j \leq cs(e)$ **do** $Evict(j, m)$;
> $\quad$ **for all** $j > cs(e)$ **do**
> $\quad\quad$ $c(j) := c(j) + m$;
> $\quad\quad$ **while** $c(j) \geq 2^j$ and $\exists$ j-documents in cache **do**
> $\quad\quad\quad$ $Evict(j, 2^j)$;
> $\quad\quad\quad$ $c(j) := c(j) - 2^j$;

Theorem 37.31 [29] *Algorithm Marking (BM) is $O(\log^2 k)$-competitive.*

37.5.2 Connection Caching

In this last technical section, we investigate a connection caching problem that can be solved using techniques similar to those in document caching. Here, we have to maintain a limited set of open TCP connections not knowing which connections will be needed next. Communication between clients and servers in the web is performed using HTTP (Hyper Text Transfer Protocol), which in turn uses TCP to transmit data. The current protocol HTTP/1.1 works with *persistent connections,* that is, once a TCP connection is established, it may be kept open and used for transmission until the connection is explicitly closed by one of the endpoints. Of course, each network node can simultaneously maintain only a limited number of open TCP connections. If a connection is closed, there is a mechanism by which one endpoint can signal the close to the other endpoint [23].

Formally, in connection caching, we are given a network modeled as an undirected graph G. The nodes of the graph represent the nodes in the network. The edges represent the possible connections. Each node has a cache in which it can maintain information on *open* connections. A connection $c = (u, v)$ is open if information on c is stored in the caches of both u and v. For a node v, let $k(v)$ denote the number of open connections that v can maintain simultaneously. Let k be the size of the largest cache in the network. For a connection $c = (u, v)$, let $cost(c)$ be the *establishment cost* of c that is incurred when c is opened. An algorithm for connection caching is presented with a request sequence $\sigma = \sigma(1), \sigma(2), \ldots, \sigma(m)$, where each request $\sigma(t)$ specifies a connection $c_t = (u_t, v_t)$, $1 \leq t \leq m$. If the requested connection c_t is already open, then the request can be served at cost 0; otherwise, the connection has to be opened at a cost of $cost(c_t)$. The goal is to serve the request sequence σ so that the total cost is as small as possible.

An important feature of this problem is that local cache configurations are not independent of each other. When one endpoint of an open connection decides to close the connection, then the other endpoint also cannot use that connection anymore.

Cohen et al. [18], who initiated the theoretical study of connection caching in the world wide web, investigated *uniform connection caching* where the connection establishment cost is uniform for all the connections. They considered various models of communication among the network nodes. Cohen et al. showed that any c-competitive algorithm for standard paging can be transformed into a $2c$-competitive algorithm for uniform connection caching. Each local node simply executes a paging strategy ignoring notifications of connections that were closed by other nodes. Using the k-competitive paging algorithm LRU [41], one obtains a $2k$-competitive algorithm. Cohen et al. [18] also considered deterministic *Marking* strategies, which work in the same way as their randomized counterparts except that on a fault an *arbitrary* unmarked page may be evicted.

Theorem 37.32 [18] *Deterministic Marking strategies can be implemented in uniform connection caching such that a competitive ratio of k is achieved. For each request, at most 1 bit of extra communication is exchanged between the two corresponding network nodes.*

Obviously, the above performance is optimal since the lower bound of k for deterministic standard paging [41] carries over to uniform connection caching. Cohen et al. [18] also investigated randomized *Marking* strategies and showed that they are $4H_k$-competitive against oblivious adversaries.

In [2], Albers investigated *generalized connection caching* where the connection establishment cost can be different for the various connections. She showed that the *Landlord* algorithm known for document caching can be adapted so that it achieves an optimal competitiveness. The implementation is as follows.

Algorithm Landlord: For each cached connection c, the algorithm maintains a credit value $credit(c)$ that takes values between 0 and $cost(c)$. Whenever a connection is opened, $credit(c)$ is set to $cost(c)$. If a requested connection (u, v) is not already open, then each node $w \in \{u, v\}$ that currently has $k(w)$ open connections executes the following steps. Let $\delta = \min_c$ open at w $credit(c)$. Close a connection c_w at w with $credit(c_w) = \delta$ and decrease the credit of all the other open connections at w by δ.

Theorem 37.33 [2] *Landlord is k-competitive for generalized connection caching.*

Ideally, we implement *Landlord* in a distributed fashion such that, for each open connection $c = (u, v)$, both end points u and v keep their copies of $credit(c)$. If one end point, say u, reduces the credit by δ, then this change has to be communicated to v so that v can update its $credit(c)$ value accordingly. The amount of extra communication for an open connection can be large if the repeated δ reductions are small. It is possible to reduce the amount of extra communication at the expense of increasing slightly the competitiveness of the algorithm. For any $0 < \epsilon \leq 1$, *Landlord* can be modified so that it is $(1 + \epsilon)k$-competitive and uses at most $\lceil 1/\epsilon \rceil - 1$ bits of extra communication for each open connection [2]. Setting

$\epsilon = 1$, we obtain a $2k$-competitive algorithm that does not use any extra communication. For $\epsilon = 1/2$, the resulting algorithm is $3k/2$-competitive and uses only one bit of extra communication.

37.6 Conclusions

In this chapter, we have studied various algorithmic networking problems. There are of course interesting topics that we have not touched upon here. One such subject is *broadcast*, that is, the problem of disseminating information in a network so as to optimize a given objective function that is usually defined with respect to the issued requests. A full chapter of this book is dedicated to broadcast so that we did not address it here. Another fundamental topic is network design, that is, the problem of setting up a network so that, for instance, certain cost or connectivity requirements are met. However, network design is typically studied as an offline problem as layout decisions have long-term impacts. Another important research direction in the area of networking is that of mechanism design and algorithmic game theory. A separate chapter of this book studies this issue.

References

[1] W. Aiello, Y. Mansour, S. Rajagopolan, and A. Rosén. Competitive queue policies for differentiated services. *Journal of Algorithms*, 55:113–141, 2005.
[2] S. Albers. Generalized connection caching. *Theory of Computing Systems*, 35:251–267, 2002.
[3] S. Albers and H. Bals. Dynamic TCP acknowledgement: Penalizing long delays. *Proc. 14th ACM-SIAM Symposium on Theory of Computing*, 47–55, 2003.
[4] S. Albers and M. Schmidt. On the performance of greedy algorithms in packet buffering. *SIAM Journal on Computing*, 35:278–304, 2006.
[5] N. Andelman, Y. Mansour, and A. Zhu. Competitive queueing policies in QoS switches. *Proc. 14th ACM-SIAM Symposium on Discrete Algorithms*, 761–770, 2003.
[6] J. Aspnes, Y. Azar, A. Fiat, S.A. Plotkin, and O. Waarts. On-line routing of virtual circuits with applications to load balancing and machine scheduling. *Journal of the ACM*, 44:486–504, 1997.
[7] B. Awerbuch and Y. Azar. Local optimization of global objectives: Competitive distributed deadlock resolution and resource allocation. *Proc. 35th Annual IEEE Symposium on Foundations of Computer Science*, 240–249, 1994.
[8] Y. Azar, E. Cohen, A. Fiat, H. Kaplan, and H. Räcke. Optimal oblivious routing in polynomial time. *Proc. 35th Annual ACM Symposium on Theory of Computing*, 383–388, 2003.
[9] Y. Azar and A. Litichevskey. Maximizing throughput in multi-queue switches. *Proc. 12th Annual European Symposium on Algorithms (ESA)*, Springer LNCS 3221, 53–64, 2004.
[10] Y. Azar and Y. Richter. Management of multi-queue switches in QoS networks. *Algorithmica*, 43:81–96, 2005.
[11] Y. Azar and Y. Richter. The zero-one principle for switching networks. *Proc. 36th Annual ACM Symposium on Theory of Computing*, 64–71, 2004.
[12] Y. Azar and Y. Richter. An improved algorithm for CIOQ switches. *Proc. 12th Annual European Symposium on Algorithms (ESA)*, Springer LNCS 3221, 65–76, 2004.
[13] N. Bansal, L. Fleischer, T. Kimbrel, M. Mahdian, B. Schieber, and M. Sviridenko. Further improvements in competitive guarantees for QoS buffering. *Proc. 31st International Colloquium on Automata, Languages and Programming*, Springer LNCS 3142, 196–207, 2004.
[14] S. Ben-David, A. Borodin, R.M. Karp, G. Tardos, and A. Wigderson. On the power of randomization in on-line algorithms. *Algorithmica*, 11:2–14, 1994.
[15] M. Bienkowski, M. Korzeniowski, and H. Räcke. A practical algorithm for constructing oblivious routing schemes. *Proc. 15th Annual ACM Symposium on Parallel Algorithms*, 24–33, 2003.

[16] A. Borodin and J.E. Hopcroft. Routing, merging, and sorting on parallel models of computation. *Journal of Computer and System Sciences*, 30:130–145, 1985.

[17] C. Brito, E. Koutsoupias, and S. Vaya. Competitive analysis of organization networks or multicast acknowledgement: How much to wait? *Proc. 15th Annual ACM-SIAM Symposium on Discrete Algorithms*, 627–635, 2004.

[18] E. Cohen, H. Kaplan, and U. Zwick. Connection caching: Model and algorithms. *Journal of Computer and System Sciences*, 67:92–126, 2003.

[19] M. Chrobak, W. Jawor, J. Sgall, and T. Tichy. Improved online algorithms for buffer management in QoS switches. *Proc. 12th Annual European Symposium on Algorithms (ESA)*, Springer LNCS 3221, 204–215, 2004.

[20] D.R. Dooly, S.A. Goldman, and D.S. Scott. On-line analysis of the TCP acknowledgment delay problem. *Journal of the ACM*, 48:243–273, 2001.

[21] A. Feldmann, A. Karlin, S. Irani, and S. Phillips. Private communication cited in [29].

[22] A. Fiat, R.M. Karp, L.A. McGeoch, D.D. Sleator, and N.E. Young. Competitive paging algorithms. *Journal of Algorithms*, 12:685–699, 1991.

[23] R. Fielding, J. Getty, J. Mogul, H. Frystyk, and T. Berners-Lee. Hypertext transfer protocol – HTTP/1.1. http://www.cis.ohio-state.edu/htbin/rfc/rfc2068.html

[24] J.S. Frederiksen and K.S. Larsen. Packet bundling. *Proc. 8th Scandinavian Workshop on Algorithm Theory*, Springer LNCS 2368, 328–337, 2002.

[25] M.T. Hajiaghayi, J.H. Kim, F.T. Leighton, and Harald Räcke. Oblivious routing in directed graphs with random demands. *Proc. 37th Annual ACM Symposium on Theory of Computing*, 193–201, 2005.

[26] M.T. Hajiaghayi, R.D. Kleinberg, F.T. Leighton, and H. Räcke. Oblivious routing on node-capacitated and directed graphs. *Proc. 16th Annual ACM-SIAM Symposium on Discrete Algorithms*, 782–790, 2005.

[27] M.T. Hajiaghayi, R.D. Kleinberg, F.T. Leighton, and H. Räcke. New lower bounds for oblivious routing in undirected graphs. *Proc. 17th Annual ACM-SIAM Symposium on Discrete Algorithms*, 918–927, 2006.

[28] C. Harrelson, K. Hildrum, and S. Rao. A polynomial-time tree decomposition to minimize congestion. *Proc. 15th Annual ACM Symposium on Parallel Algorithms*, 34–43, 2003.

[29] S. Irani. Page replacement with multi-size pages and applications to Web caching. *Algorithmica*, 33:384–409, 2002.

[30] C. Kaklamanis, D. Krizanc, and T. Tsantilas. Tight bounds for oblivious routing in the hypercube. *Proc. 2nd Annual ACM Symposium on Parallel Algorithms*, 31–36, 1990.

[31] A.R. Karlin, C. Kenyon, and D. Randall. Dynamic TCP acknowledgement and other stories about $e/(e-1)$. *Algorithmica*, 36:209–224, 2003.

[32] A. Kesselman, Z. Lotker, Y. Mansour, B. Patt-Shamir, B. Schieber, and M. Sviridenko. Buffer overflow management in QoS switches. *Proc. 33rd Annual ACM Symposium on Theory of Computing*, 520–529, 2001.

[33] A. Kesselman, Y. Mansour, and R. van Stee. Improved competitive guarantees for QoS buffering. *Proc. 11th European Symposium on Algorithms*, Springer LNCS 2832, 361–372, 2003.

[34] A. Kesselman and A. Rosen. Scheduling policies for CIOQ switches. *Proc. 15th Annual ACM Symposium on Parallel Algorithms*, 353–362, 2003.

[35] S. Khanna, J. Naor, and D. Raz. Control message aggregation in group communication protocols. *Proc. 29th International Colloquium on Automata, Languages and Programming*, Springer LNCS 2380, 135–146, 2002.

[36] J. Noga. Private communication, 2001.

[37] H. Räcke. Minimizing congestion in general networks. *Proc. 43rd Annual IEEE Symposium on Foundations of Computer Science*, 43–52, 2002.

[38] P. Raghavan and C.D. Thompson. Randomized rounding: A technique for provably good algorithms and algorithmic proofs. *Combinatorica*, 7:365–374, 1987.

[39] S.S. Seiden. A guessing game and randomized online algorithms. *Proc. 32nd Annual ACM Symposium on Theory of Computing*, 592–601, 2000.

[40] M. Schmidt. Packet buffering: Randomization beats deterministic algorithms. *Proc. 22nd Annual Symposium on Theoretical Aspects of Computer Science (STACS)*, Springer LNCS 3404, 293–304, 2005.

[41] D.D. Sleator and R.E. Tarjan. Amortized efficiency of list update and paging rules, *Communications of the ACM*, 28:202–208, 1985.

[42] L.G. Valiant and G.J. Brebner. Universal schemes for parallel communication. *Proc. 13th Annual ACM Symposium on Theory of Computing*, 263–277, 1981.

[43] N.E. Young. Online file caching. *Algorithmica*, 33:371–383, 2002.

38

Online Call Admission Control in Wireless Cellular Networks

Ioannis Caragiannis
Christos Kaklamanis
Evi Papaioannou
University of Patras

38.1	Introduction	38-1
38.2	Two Simple Algorithms	38-3
38.3	The Algorithm p-Random	38-5
38.4	"Classify and Randomly Select"-Based Algorithms	38-11
38.5	A Lower Bound	38-15
38.6	Extensions and Open Problems	38-18
	References	38-19

38.1 Introduction

We study frequency spectrum management issues in wireless cellular networks where base stations connected through a high-speed network are used to build the required communication infrastructure. A geographical area in which communication takes place is divided into regions. Each region is the calling area of a base station. In such systems, communication is established in the following way. When a user A wishes to communicate with some other user B, a path must be established between the base stations of the regions where users A and B are located. Then communication is performed in three steps: (i) wireless communication between A and its base station, (ii) communication between the base stations, and (iii) wireless communication between B and its base station. At least one base station is involved in the communication even if both users are located in the same region or only one of the two users is part of the cellular network (and the other uses for example the PSTN). Improving the access of users to base stations is the aim of this work.

The network topology usually adopted [9–11, 21] is the one shown in the left part of Figure 38.1. All regions are regular hexagons (cells) of the same size. This shape results from the assumption of uniform distribution of identical base stations within the network, as well as from the fact that the calling area of a base station is a circle which, for simplicity reasons, is idealized as a regular hexagon. Owing to the shape of the cells, we call these networks wireless cellular networks.

Many users of the same region can communicate simultaneously with their base station through Frequency Division Multiplexing (FDM). The base station is responsible for allocating distinct frequencies from the available spectrum to users so that signal interference is avoided. Signal interference manifests

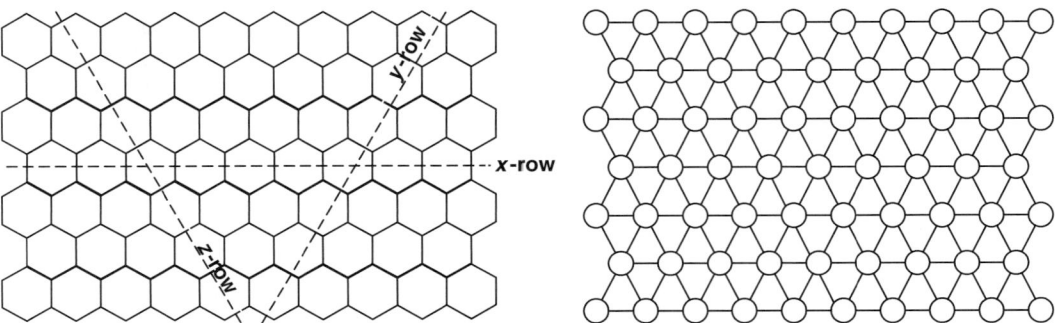

FIGURE 38.1 A cellular network and its interference graph.

itself when the same frequency is assigned to users located in the same or adjacent cells. It can be represented by an *interference graph* G whose vertices correspond to cells, and an edge (u, v) indicates that the assignment of the same frequency to two users lying at the cells corresponding to nodes u and v will cause signal interference. The interference graph of a cellular network is depicted in the right part of Figure 38.1. If the assumption of uniform distribution of identical base stations does not hold, arbitrary interference graphs can be used to model the underlying network.

Since the spectrum of available frequencies is limited, important engineering problems related to the efficient reuse of frequencies arise [12, 13, 17, 18]. We study the *call admission control* (or, simply, call control) problem that is defined as follows: Given users that wish to communicate with their base station, the *call control* problem in a network that supports a spectrum of w available frequencies is to assign frequencies to users so that at most w frequencies are used in total, signal interference is avoided, and the number of users served is maximized.

We assume that calls corresponding to users that wish to communicate with their base station appear in the cells of the network in an online manner. When a call arrives, a call control algorithm decides either to accept the call (assigning a frequency to it) or to reject it. Once a call is accepted, it cannot be rejected (preempted). Furthermore, the frequency assigned to the call cannot be changed in the future. We assume that all calls have infinite duration; this assumption is equivalent to considering calls of similar duration.

We use competitive analysis [3] to evaluate the performance of online call control algorithms with the competitive ratio being the performance measure. In this setting, given a sequence of calls, the performance of an online call control algorithm A is compared to the performance of the optimal offline algorithm OPT. Let $B_A(\sigma)$ be the benefit of the online algorithm A on the sequence of calls σ, that is, the number of calls of σ accepted by A, and $B_{OPT}(\sigma)$ the benefit of the optimal algorithm OPT. If A is a deterministic algorithm, we define its competitive ratio ρ as

$$\rho = \max_{\sigma} \frac{B_{OPT}(\sigma)}{B_A(\sigma)},$$

where the maximum is taken over all possible sequences of calls. If A is a randomized algorithm, we define its competitive ratio ρ as

$$\rho = \max_{\sigma} \frac{B_{OPT}(\sigma)}{\mathcal{E}[B_A(\sigma)]},$$

where $\mathcal{E}[B_A(\sigma)]$ is the expectation of the number of calls accepted by A, and the maximum is taken over all possible sequences of calls.

We compare the performance of deterministic algorithms against *offline adversaries*, that is, adversaries that have knowledge of the behavior of the deterministic algorithm in advance. In the case of randomized algorithms, we consider *oblivious adversaries* whose knowledge is limited to the probability distribution of the random choices of the randomized algorithm.

In the next sections, we survey related work on the call control problem. We adapt to wireless cellular networks ideas that have been proposed in similar context (such as call control in optical networks [1,2,22]) and present recent work of the authors [5–8]. Presenting very briefly the related results, deterministic algorithms do not provide efficient online solutions to the problem. The greedy algorithm (studied in References 19 and 20) is probably the simplest such algorithm. Simple arguments can show that its competitive ratio is at least 3 in cellular networks and, furthermore, no deterministic algorithm has better competitive ratio. So, better solutions are possible only through randomized algorithms. Adapting the classify and randomly select idea from Reference 2, a simple 3-competitive algorithm can be achieved. Although not beating the lower bound for deterministic algorithms, unlike the greedy algorithm, this algorithm achieves competitiveness of 3 even in networks with arbitrarily many frequencies.

Improving the bound of 3 was the aim of previous work of the authors. As a first step, Reference 6 studies the randomized algorithm p-RANDOM, which marginally deviates from the behavior of the greedy algorithm. Intuitively, p-RANDOM accepts a call with probability p whenever this is possible. Although simple, this idea is powerful enough to cross the barrier of 3 in single-frequency cellular networks. The analysis is interesting, as well. In order to account for the expected benefit of the algorithm, the benefit is amortized to optimal calls and lower bounds on the expectation of the amortized benefit per optimal call are enough to prove upper bounds on the competitive ratio. For obtaining good bounds, detailed case analysis is required.

In another step, in order to investigate the power of randomization, the authors in References 7 and 8 develop simple randomized algorithms extending the "classify and randomly select" paradigm. These algorithms are based on colorings of the interference graph satisfying certain properties. Besides improving further the upper bounds on the competitiveness of randomized call control, these algorithms are particularly simple since they use only a small constant number of random bits or comparably weak random sources. On the other hand, p-RANDOM requires as many bits as the number of calls in a sequence. Furthermore, the "classify and randomly select" based algorithms work for cellular networks supporting many frequencies without any overhead in performance.

In the rest of the chapter, we survey the above results. We first present in Section 38.2 the greedy algorithm and the "classify and randomly select" algorithm produced by naively applying the ideas of Reference 2 in a cellular network and present the simple argument that proves the lower bound on the competitive ratio of deterministic algorithms. As part of the analysis of the greedy algorithm, we present an interesting technique from Reference 1 for transforming call control algorithms for one frequency to algorithms for many frequencies with a small sacrifice on performance. In Section 38.3, we present the amortized benefit analysis of algorithm p-RANDOM by presenting the related case analysis. The "classify and randomly select"-based algorithms are presented in Section 38.4. In Section 38.5, we demonstrate how Yao's principle can be used in order to prove lower bounds for randomized algorithms against oblivious adversaries. We conclude by briefly discussing interesting extensions of our model and open problems.

38.2 Two Simple Algorithms

In this section, we describe two well-known online algorithms for call control in wireless networks: the greedy algorithm and a randomized algorithm based on the "classify and randomly select" paradigm. Also, we present a lower bound on the competitiveness of deterministic online call control algorithms. These results will be the starting point for the improvements we will present in the next sections.

Assume that a sequence of calls σ appears in a network that support w frequencies $1, 2, \ldots, w$. The greedy algorithm is an intuitive deterministic algorithm. For any new call c at a cell v, the greedy algorithm searches for the minimum available frequency, that is, for the minimum frequency among frequencies $1, 2, \ldots, w$ that has not been assigned to calls in cell v or its adjacent cells. If such a frequency exists, the call c is accepted and is assigned this frequency; otherwise, the call is rejected.

Pantziou et al. [20] have proved that this algorithm is at most $(\Delta + 1)$-competitive against offline adversaries for networks supporting many frequency, where Δ is the degree of the network. The following statement slightly extends this result.

Theorem 38.1 *Let $G = (V, E)$ be an interference graph, v a vertex of G, and Γ_v the maximum independent set in the neighborhood of v. The greedy algorithm is $\frac{1}{1-\exp^{-1/\gamma}}$-competitive against an offline adversary, where $\gamma = \max_{v \in V} |\Gamma_v|$.*

To prove this statement, we will first show that the greedy algorithm is γ-competitive if the network supports only one frequency. Given a sequence of calls, denote by B_A be the set of calls accepted by the greedy algorithm and B_{OPT} the set of calls accepted by the optimal algorithm. Observe that for each optimal call c not accepted by the algorithm, there are at most γ calls in $B_{\text{OPT}} \setminus B_A$ that are rejected because of the acceptance of c. Thus,

$$|B_{\text{OPT}}| = |B_{\text{OPT}} \setminus B_A| + |B_{\text{OPT}} \cap B_A|$$
$$\leq \gamma |B_A \setminus B_{\text{OPT}}| + |B_{\text{OPT}} \cap B_A|$$
$$\leq \gamma |B_A|.$$

To prove Theorem 38.1, we will use a technique of Awerbuch et al. [1] who present a simple way for transforming call control algorithms designed for networks with one frequency to call control algorithms for networks with arbitrarily many frequencies, with a small sacrifice in competitiveness. Consider a wireless cellular network and a (deterministic or randomized) online call control algorithm ALG-1 for one frequency. A call control algorithm ALG for w frequencies can be constructed in the following way. For each call c, we execute the algorithm ALG-1 for each of the w frequencies until either c is accepted or the frequency spectrum is exhausted (and the call c is rejected), that is,

1. for any new call c
2.   for $i = 1$ to w
3.   run ALG-1(c) for frequency i
4.   if c was accepted then
5.   assign frequency i to c
6.   stop
7.   reject c.

Lemma 38.1 *(Awerbuch et al. [1]) If ALG-1 is ρ-competitive, then ALG has competitive ratio*

$$\frac{1}{1 - \left(1 - \frac{1}{\rho w}\right)^w} \leq \frac{1}{1 - \exp(-1/\rho)}.$$

Hence, Theorem 38.1 immediately follows by Lemma 38.1 and the fact that the greedy algorithm is γ-competitive for one frequency. For cellular networks, where the interference graph is a hexagon graph, it is $\gamma = 3$, and Theorem 38.1 yields the following corollary.

Corollary 38.1 *The greedy algorithm is 3.53-competitive in cellular networks with arbitrarily many frequencies.*

Note that γ is a lower bound for the competitive ratio of every deterministic algorithm. Consider a network that supports one frequency and consists of a cell v and γ mutually non-adjacent cells $v_1, v_2, \ldots, v_\gamma$ that are adjacent to v. Consider, now, the following sequence of calls produced by an adversary that has knowledge of the way that the algorithm makes its decisions. First, a call c is presented in cell v. If the algorithm rejects c, then the adversary stops the sequence. In this case, the algorithm has no benefit from its execution. If the algorithm accepts the call c, the adversary presents γ calls $c_1, c_2, \ldots, c_\gamma$ in cells v_1, $v_2, \ldots, v_\gamma$, respectively. The benefit of the algorithm is then 1 while the optimal algorithm would obtain benefit γ by rejecting call c and accepting the calls $c_1, \ldots, c_\gamma$. Adapting this argument to cellular networks, we obtain the following statement.

Theorem 38.2 *No deterministic algorithm can be better than 3-competitive against an offline adversary.*

Obviously, the best the algorithm A can do is to accept all calls presented in cells that are nonadjacent to cells where previously accepted calls are located. But this is exactly what the greedy algorithm does for networks that support one frequency.

The "classify and randomly select" paradigm (introduced in a different context in Reference 2; see also Reference 20) uses a coloring of the cells of the network (coloring of the interference graph) with positive integer (colors) $1, 2, \ldots$ in such way that adjacent cells are assigned different colors. The randomized algorithm classifies the calls of the sequence into a number of classes; class i contains calls appeared in cells colored with color i. It then selects uniformly at random one of the classes, and considers only calls that belong to the selected class, rejecting all other calls. Once a call of the selected class appears, the greedy algorithm is used.

Using simple arguments, we can prove that the "classify and randomly select" algorithm CRS is χ-competitive against oblivious adversaries, where χ is the number of colors used in the coloring of the cells of the network. This may lead to algorithms with competitive ratio equal to the chromatic number (and no better, in general) of the corresponding interference graph, given that an optimal coloring (i.e., with the minimum number of colors) is available. In cellular networks, the interference graph is 3-colorable. Hence, we obtain the following statement.

Theorem 38.3 *Algorithm CRS is 3-competitive against oblivious adversaries in cellular networks supporting arbitrarily many frequencies.*

38.3 The Algorithm p-Random

As a first attempt in order to prove that randomization indeed helps in order to beat the lower bound of 3 on the competitiveness of deterministic algorithms, we present and analyze the algorithm p-RANDOM, a randomized call control algorithm for cellular networks that supports one frequency. Algorithm p-RANDOM receives as input a sequence of calls in an online manner and works as follows:

1. Initially, all cells are unmarked.
2. for any new call c in a cell v
3.   if v is marked then reject c.
4.   if v has an accepted call or is adjacent to a cell
       with an accepted call, then reject c
5.   else
6.   with probability p accept c.
7.   with probability $1 - p$ reject c and mark v.

The algorithm uses a parameter $p \in [1/3, 1]$. Obviously, if it is $p < 1/3$, the competitive ratio will be greater than 3, since the expected benefit of the algorithm on a sequence of a single call will be p. The algorithm is simple and can be easily implemented with small communication overhead (exchange of messages) between the base stations of the network.

Marking cells on rejection guarantees that algorithm p-RANDOM does not simulate the greedy deterministic one. Assume otherwise, that marking is not used. Then, consider an adversary that presents t calls in a cell v and one call in 3 (mutually not adjacent) cells adjacent to v. The probability that the randomized algorithm does not accept a call in cell v drops exponentially as t increases, and the benefit approaches 1, while the optimal benefit is 3.

Note that algorithm p-RANDOM may accept at most one call in each cell, but this is also the case for any algorithm running in networks that support one frequency (including the optimal one). Thus, for the competitive analysis of algorithm p-RANDOM, we will only consider sequences of calls with at most one call per cell. Also, there is no need for taking into account the procedure of marking cells during the analysis.

We now prove the upper bound on the competitive ratio of algorithm p-RANDOM as a function of p. Our main statement is the following.

Theorem 38.4 *For $p \in [1/3, 1]$, algorithm p-RANDOM has competitive ratio at most*

$$\frac{3}{5p - 7p^2 + 3p^3}$$

against oblivious adversaries.

Proof. Let σ be a sequence of calls. We assume that σ has been fixed in advance and will be revealed to the algorithm in an online manner. We make this assumption because we are interested in the competitiveness of the algorithm against oblivious adversaries whose knowledge is limited to the probability distribution of the random choices of the algorithm (i.e., the parameter p).

Consider the execution of algorithm p-RANDOM on σ. For any call $c \in \sigma$, we denote by $X(c)$ the random variable that indicates whether the algorithm accepted c. Clearly, the benefit of algorithm p-RANDOM on σ can be expressed as

$$B(\sigma) = \sum_{c \in \sigma} X(c).$$

Let $A(\sigma)$ be the set of calls in σ accepted by the optimal algorithm. For each call $c \in A(\sigma)$, we define the amortized benefit $\bar{b}(c)$ as

$$\bar{b}(c) = X(c) + \sum_{c' \in \gamma(c)} \frac{X(c')}{d(c')},$$

where $\gamma(c)$ denotes the set of calls of the sequence in cells adjacent to c. For each call $c' \notin A(\sigma)$, $d(c')$ is the number of calls in $A(\sigma)$ that are in cells adjacent to the cell of c. By the two equalities above, it is clear that

$$B(\sigma) = \sum_{c \in A(\sigma)} \bar{b}(c).$$

Furthermore, note that for any call $c' \notin A(\sigma)$, $d(c') \leq 3$. We obtain that

$$\bar{b}(c) \geq X(c) + \frac{\sum_{c' \in \gamma(c)} X(c')}{3}$$

and, by linearity of expectation,

$$\mathcal{E}[B(\sigma)] \geq \sum_{c \in A(\sigma)} \left(\mathcal{E}\left[X(c) + \frac{\sum_{c' \in \gamma(c)} X(c')}{3} \right] \right) \tag{38.1}$$

Let $\gamma'(c)$ be the set of calls in cells adjacent to the cell of c that appear prior to c in the sequence σ. Clearly, $\gamma'(c) \subseteq \gamma(c)$, which implies that

$$\sum_{c' \in \gamma(c)} X(c') \geq \sum_{c' \in \gamma'(c)} X(c').$$

Thus, Equation 38.1 yields

$$\mathcal{E}[B(\sigma)] \geq \sum_{c \in A(\sigma)} \left(\mathcal{E}\left[X(c) + \frac{\sum_{c' \in \gamma'(c)} X(c')}{3} \right] \right) \tag{38.2}$$

In what follows, we will try to bound from below the expectation of the random variable $X(c) + \frac{\sum_{c' \in \gamma'(c)} X(c')}{3}$, for each call $c \in A(\sigma)$.

We concentrate on a call $c \in A(\sigma)$. Let $\Omega = 2^{\gamma'(c)}$ be the set that contains all possible subsets of $\gamma'(c)$. We define the *effective neighborhood* of c, denoted by $\Gamma(c)$, to be the subset of $\gamma'(c)$ that contains the calls of $\gamma'(c)$ which, when it appears, is unconstrained by calls of σ at distance 2 from c. Clearly, $\Gamma(c)$ is a random variable taking its values from the sample space Ω. Intuitively, whether an optimal call c is accepted by the algorithm depends on its effective neighborhood $\Gamma(c)$. We have

$$\mathcal{E}\left[X(c) + \frac{\sum_{c' \in \gamma'(c)} X(c')}{3} \right] =$$

$$\sum_{\gamma \in \Omega} \mathcal{E}\left[X(c) + \frac{\sum_{c' \in \gamma} X(c')}{3} \middle| \Gamma(c) = \gamma \right] \cdot \Pr[\Gamma(c) = \gamma] \geq$$

$$\min_{\gamma \in \Omega} \left\{ \mathcal{E}\left[X(c) + \frac{\sum_{c' \in \gamma} X(c')}{3} \middle| \Gamma(c) = \gamma \right] \right\} =$$

$$\min_{\gamma \in \Omega} \left\{ \mathcal{E}[X(c) | \Gamma(c) = \gamma] + \frac{\mathcal{E}\left[\sum_{c' \in \gamma} X(c') | \Gamma(c) = \gamma \right]}{3} \right\}. \tag{38.3}$$

To compute $\mathcal{E}[X(c)|\Gamma(c) = \gamma]$, we observe that algorithm p-RANDOM may accept c only if it has rejected all calls in its effective neighborhood γ. The probability that all calls of γ are rejected given that $\Gamma(c) = \gamma$ is $(1-p)^{|\gamma|}$, and then c is accepted with probability p. Thus,

$$\mathcal{E}[X(c)|\Gamma(c) = \gamma] = p(1-p)^{|\gamma|}. \tag{38.4}$$

We now bound from below $\mathcal{E}\left[\sum_{c' \in \gamma} X(c')|\Gamma(c) = \gamma\right]$ by distinguishing between cases according to the size of the effective neighborhood $|\gamma|$.

Claim 38.5 For all $p \in [1/3, 1]$,

$$\mathcal{E}\left[\sum_{c' \in \gamma} X(c')|\Gamma(c) = \gamma\right] \geq \begin{cases} 0 & \text{if } |\gamma| = 0 \\ p & \text{if } |\gamma| = 1 \\ 2p - p^2 & \text{if } |\gamma| = 2 \\ 3p - 2p^2 & \text{if } |\gamma| = 3 \\ 4p - 3p^2 + p^3 & \text{if } |\gamma| = 4 \\ 5p - 4p^2 + p^3 & \text{if } |\gamma| = 5 \\ 6p - 5p^2 + p^3 & \text{if } |\gamma| = 6 \end{cases}$$

Proof. In Figures 38.2 through 38.6, we give all possible cases for the effective neighborhood of an optimal call c in a sequence of calls σ. In each figure the optimal call is denoted by the black circle in the middle cell while black circles in the outer cells denote calls in the effective neighborhood γ of c. An arrow from a call c_1 to another call c_2 indicates that c_1 appears in σ prior to c_2. In the figures, we have eliminated the symmetric cases.

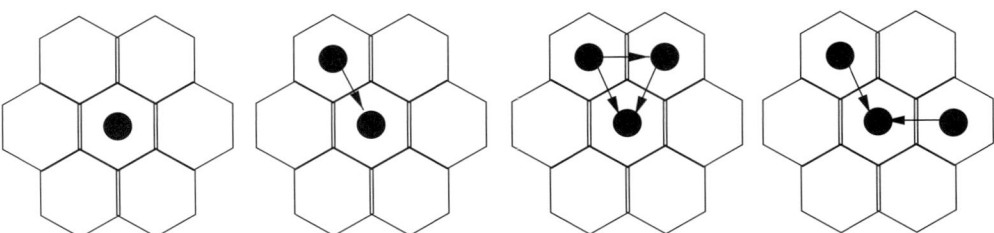

FIGURE 38.2 The cases $|\gamma| = 0, 1, 2$.

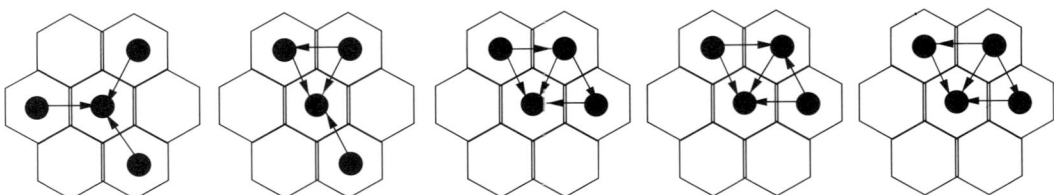

FIGURE 38.3 The case $|\gamma| = 3$.

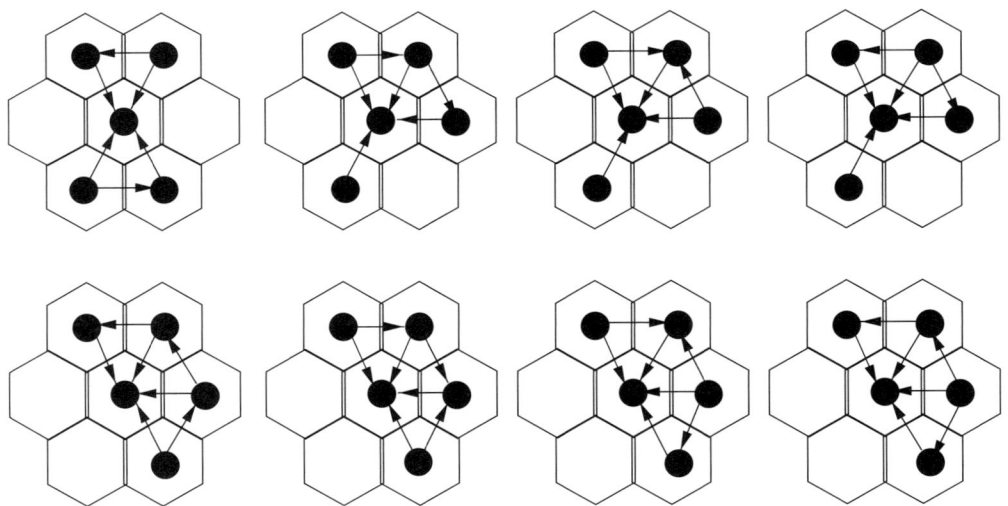

FIGURE 38.4 The case $|\gamma| = 4$.

The proof is trivial for the cases $|\gamma| = 0$ and $|\gamma| = 1$ (the two leftmost cases in Figure 38.2). In the third case of Figure 38.2 (where $|\gamma| = 2$), we observe that the algorithm accepts the first call in γ with probability p and the second one with probability $p(1-p)$. In total, the expectation of the number of accepted calls in γ is $2p - p^2$. In the rightmost case of Figure 38.2, the expectation of the number of accepted calls in γ is $2p < 2p - p^2$.

Similarly, we can compute the desired lower bounds on $\mathcal{E}\left[\sum_{c' \in \gamma} X(c') | \Gamma(c) = \gamma\right]$ for the cases $|\gamma| = 3, 4, 5, 6$. ∎

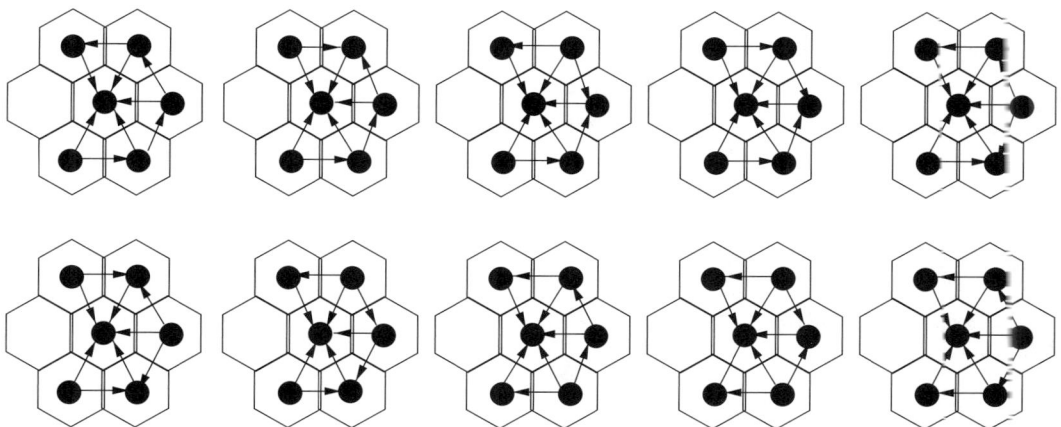

FIGURE 38.5 The case $|\gamma| = 5$.

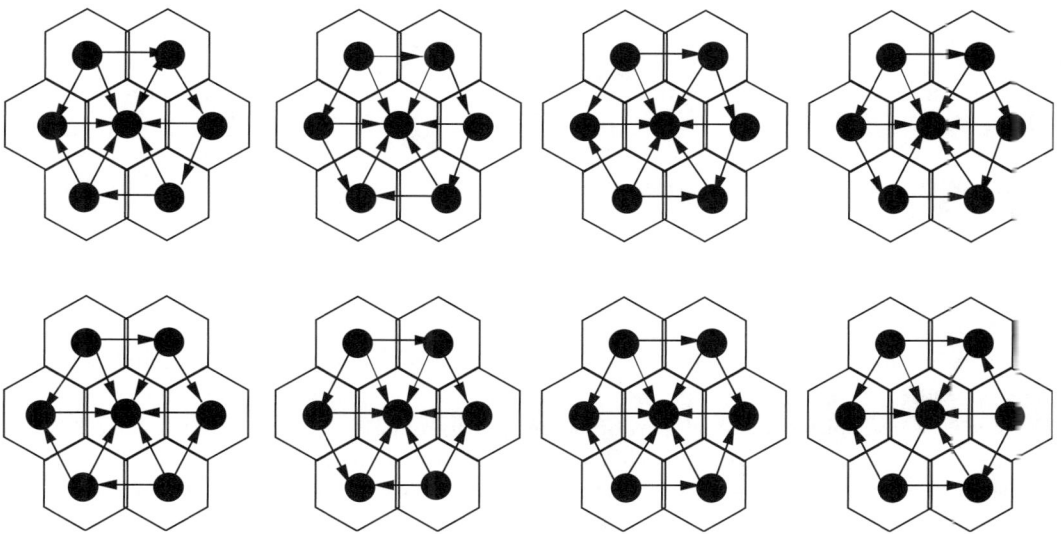

FIGURE 38.6 The case $|\gamma| = 6$.

By making calculations with Equations 38.3, 38.4, and Claim 38.5, we obtain that

$$\mathcal{E}\left[X(c) + \frac{\sum_{c' \in \gamma'(c)} X(c')}{3}\right] \geq$$

$$\min_{\gamma \in \Omega} \left\{ \mathcal{E}[X(c)|\Gamma(c) = \gamma] + \frac{\mathcal{E}\left[\sum_{c' \in \gamma} X(c')|\Gamma(c) = \gamma\right]}{3} \right\} \geq$$

$$\min_{\gamma \in \Omega: |\gamma|=2} \left\{ \mathcal{E}[X(c)|\Gamma(c) = \gamma] + \frac{\mathcal{E}\left[\sum_{c' \in \gamma} X(c')|\Gamma(c) = \gamma\right]}{3} \right\} \geq$$

$$p(1-p)^2 + \frac{2p - p^2}{3} = \frac{5p - 7p^2 + 3p^3}{3}.$$

Now, using Equation 38.2 we obtain that

$$\mathcal{E}[B(\sigma)] \geq \sum_{c \in A(\sigma)} \frac{5p - 7p^2 + 3p^3}{3}$$

$$= \frac{5p - 7p^2 + 3p^3}{3} \cdot B_{\text{OPT}}(\sigma).$$

This completes the proof of Theorem 38.4. ■

The expression in Theorem 38.4 is minimized to $729/265 = 2.651$ for $p = 5/9$. Thus, we obtain the following result.

Corollary 38.2 *There exists an online randomized call control algorithm for cellular networks with one frequency that is at most 2.651-competitive against oblivious adversaries.*

In the following, we show that our analysis is not far from being tight. In particular, we prove the following.

Theorem 38.6 *For any $p \in (1/3, 1)$, algorithm p-RANDOM is at least 2.469-competitive against oblivious adversaries.*

Proof. We will show that the competitive ratio of algorithm p-RANDOM against oblivious adversaries is at least

$$\max\left\{\frac{3}{4p - 3p^2}, \frac{3}{5p - 7p^2 + 4p^3 - p^4}\right\} \tag{38.5}$$

by constructing two sequences σ_1 and σ_2 of calls for which the competitive ratio of algorithm p-RANDOM is $\frac{3}{4p-3p^2}$ and $\frac{3}{5p-7p^2+4p^3-p^4}$, respectively.

Sequence σ_1 is depicted in the left part of Figure 38.7. In round 1, a call appears at some cell c, and in round 2 one call appears in each one of the three mutually adjacent cells in the neighborhood of c. Clearly, the benefit of the optimal algorithm is 3. To compute the expectation of the benefit of algorithm p-RANDOM, we observe that with probability p, the call presented in round 1 is accepted, and with probability $1 - p$, the call presented in round 1 is rejected and each of the three calls presented in round 2 is accepted with probability p. Thus, the expectation of the benefit of the algorithm on sequence σ_1 is $p + (1 - p)3p = 4p - 3p^2$.

Sequence σ_2 is depicted in the right part of Figure 38.7. Calls appear in four rounds. The labels on the calls denote the round in which the calls appear. Clearly, the benefit of the optimal algorithm is 18 since the optimal algorithm would accept the calls that appear in rounds 3 and 4. To compute the expectation of the benefit of algorithm p-RANDOM on sequence σ_2, we first compute the probability that each call is accepted.

- A call that appears in round 1 is accepted with probability p.
- A call that appears in round 2 can be accepted if its adjacent call that appeared in round 1 has been rejected; thus, the probability that a call that appears in round 2 is accepted is $p(1 - p)$.
- A call that appears in round 3 can be accepted if its adjacent calls that appeared in rounds 1 and 2 have been rejected; thus, the probability that a call that appears in round 3 is accepted is $p(1 - p)^2$.
- A call that appears in round 4 can be accepted if its adjacent calls that appeared in rounds 1 and 2 have been rejected. The probability that a call that appears in round 1 is rejected is $1 - p$ while the probability that a call that appears in round 2 is rejected is $1 - p - (1 - p)^2$. Thus, the probability that a call that appears in round 4 is accepted is $p(1 - p)\left(1 - p - (1 - p)^2\right)$.

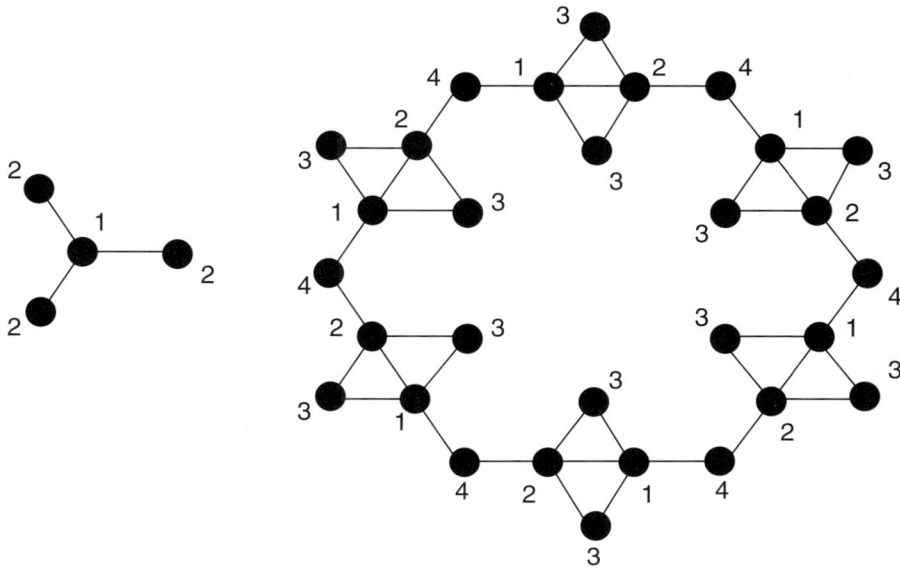

FIGURE 38.7 The lower bound on the performance of algorithm p-Random.

Note the number of calls that appear in rounds 1, 2, 3, and 4 is 6, 6, 12, and 6, respectively. Thus, we obtain that the expectation of the benefit of the algorithm is

$$6p + 6p(1-p) + 12p(1-p)^2 + 6p(1-p)\left(1 - p - (1-p)^2\right) = 30p - 42p^2 + 24p^3 - 6p^4.$$

By the above constructions, it is clear that the competitive ratio of algorithm p-Random is lower bounded by Equation 38.5. This expression is minimized for $p \approx 0.6145$ to 2.469. This completes the proof of the theorem. ∎

Using algorithm p-Random, a call control algorithm for w frequencies can be constructed using the technique of Awerbuch et al. [1] presented in Section 38.2. For each call c, we execute the algorithm p-Random for each of the w frequencies until either c is accepted or the frequency spectrum is exhausted (and the call c is rejected). Using Lemma 38.1, we obtain that the competitive ratio we achieve in this way for two frequencies is 2.927. Unfortunately, for larger numbers of frequencies, the competitive ratio becomes larger than 3.

38.4 "Classify and Randomly Select"-Based Algorithms

In this section, we present "classify and randomly select"-based algorithms originally presented in References 7 and 8. We start with algorithm CRS-A that works in networks with one frequency and achieves a competitive ratio against oblivious adversaries similar (but slightly inferior) to that proved for algorithm p-Random.

Algorithm CRS-A uses a coloring of the cells with four colors 0, 1, 2, and 3, such that only two colors are used in the cells belonging to the same axis. This can be done by coloring the cells in the same x-row with either the colors 0 and 1 or the colors 2 and 3, coloring the cells in the same y-row with either the colors 0 and 2 or the colors 1 and 3, and coloring the cells in the same z-row with either the colors 0 and 3 or the colors 1 and 2. Such a coloring is depicted in the left part of Figure 38.8.

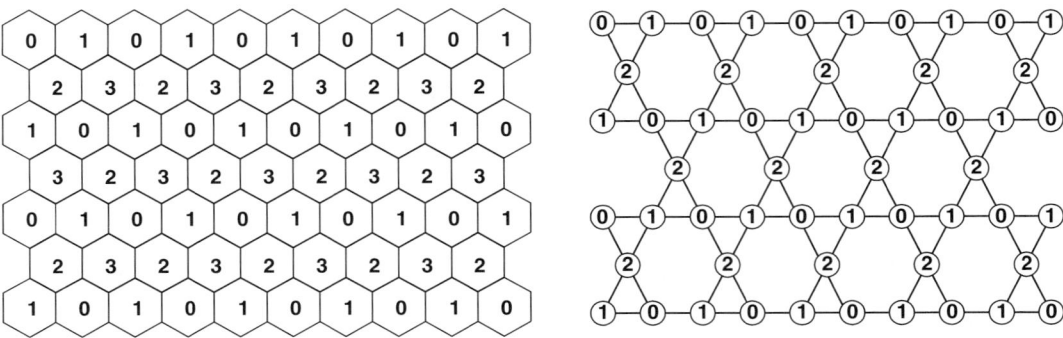

FIGURE 38.8 The 4-coloring used by algorithm CRS-A and the corresponding subgraph of the interference graph induced by the nodes not colored with color 3.

Algorithm CRS-A randomly selects one out of the four colors and executes the greedy algorithm on the cells colored with the other three colors, ignoring (i.e., rejecting) all calls in cells colored with the selected color.

Theorem 38.7 *Algorithm CRS-A in cellular networks supporting one frequency is 8/3-competitive against oblivious adversaries.*

Proof. Let σ be a sequence of calls and denote by O the set of calls accepted by the optimal algorithm. Denote by σ' the set of calls in cells that are not colored with the color selected and by O' the set of calls the optimal algorithm would have accepted on input σ'. Clearly, $|O'|$ will be at least as large as the subset of O that belongs to σ'. Since the probability that the cell of a call in O is not colored with the color selected is $3/4$, it is $\mathcal{E}[|O'|] \geq 3/4|O|$.

Now let B be the set of calls accepted by algorithm CRS-A, that is, the set of calls accepted by the greedy algorithm when executed on sequence σ'. Observe that each call in O' either belongs in B or it is rejected because some other call is accepted. Furthermore, a call in $B \setminus O'$ can cause the rejection of at most two calls of O'. This implies that $|B| \geq |O'|/2$, which yields that the competitive ratio of algorithm CRS-A is

$$\frac{|O|}{\mathcal{E}[|B|]} \leq \frac{2|O|}{\mathcal{E}[|O'|]} \leq \frac{8}{3}.$$ ∎

The main advantage of algorithm CRS-A is that it uses only two random bits. In the next section, we present simple online algorithms with improved competitive ratios that use slightly stronger random sources and work on networks with arbitrarily many frequencies.

Algorithm CRS-A can be seen as an algorithm based on the "classify and randomly select" paradigm. It uses a coloring of the interference graph (not necessarily using the minimum possible number of colors) and a classification of the colors. It starts by randomly selecting a color class (i.e., a set of colors) and then run the greedy algorithm in the cells colored with colors from this color class, ignoring (i.e., rejecting) calls in cells colored with colors not belonging to this class. Algorithm CRS-A uses a coloring of the interference graph with four colors 0, 1, 2, and 3, and the four color classes $\{0, 1, 2\}, \{0, 1, 3\}, \{0, 2, 3\}$, and $\{1, 2, 3\}$. Note that, in the previously known algorithms based on the "classify and randomly select" paradigm, color classes are singletons (e.g., References 1 and 20).

The following simple lemma gives a sufficient condition for obtaining efficient online algorithms based on the "classify and randomly select" paradigm.

Lemma 38.2 *Consider a network with interference graph $G = (V, E)$ that supports w frequencies and let χ be a coloring of the nodes of V with the colors of a set X. If there exist ν sets of colors $s_0, s_1, \ldots, s_{\nu-1} \subseteq X$ and an integer λ such that*

- *Each color of X belongs to at least λ different sets of the sets $s_0, s_1, \ldots, s_{\nu-1}$.*
- *For $i = 0, 1, \ldots, \nu - 1$, each connected component of the subgraph of G induced by the nodes colored with colors in s_i is a clique.*

then there exists an online randomized call control algorithm for the network G, which has competitive ratio ν/λ against oblivious adversaries.

Proof. Consider a network with interference graph G that supports w frequencies and the randomized online algorithm working as follows. The algorithm randomly selects one out of the ν color classes $s_0, \ldots, s_{\nu-1}$ and executes the greedy algorithm on the cells colored with colors of the selected class, rejecting all calls in cells colored with colors not in the selected class.

Let σ be a sequence of calls and let O be the set of calls accepted by the optimal algorithm on input σ. Assume that the algorithm selects the color class s_i. Let σ' be the sequence of calls in cells colored with colors in s_i, and O' be the set of calls accepted by the optimal algorithm on input σ'. Also, we denote by B the set of calls accepted by the algorithm.

First, we can easily show that $|B| = |O'|$. Let G_j be a connected component of the subgraph of G induced by the nodes of G colored with colors in s_i. Let σ_j be the subsequence of σ' in cells corresponding to nodes of G_j. Clearly, any algorithm (including the optimal one) will accept at most one call of σ_j at each frequency. If the optimal algorithm accepts w calls, this means that the sequence σ_j has at least w calls and the greedy algorithm, when executed on σ', will accept w calls from σ_j (one call in each one of the available frequencies). If the optimal algorithm accepts $w' < w$ calls from σ_j, this means that σ_j contains exactly $w' < w$ calls and the greedy algorithm will accept them all in w' different frequencies. Since a call of σ_j is not constrained by a call in $\sigma_{j'}$ for $j \neq j'$, we obtain that $|B| = |O'|$.

The proof is completed by observing that the expected benefit of the optimal algorithm on input σ' over all possible sequences σ' defined by the random selection of the algorithm is $\mathcal{E}[|O'|] \geq \frac{1}{\lambda}|O|$, since, for each call in O, the probability that the color of its cell belongs to the color class selected is at least ν/λ. Hence, the competitive ratio of the algorithm against oblivious adversaries is

$$\frac{|O|}{\mathcal{E}[|B|]} = \frac{|O|}{\mathcal{E}[|O'|]} \leq \lambda/\nu.$$

∎

Next, we present simple randomized online algorithms for call control in cellular networks, namely, CRS-B, CRS-C, and CRS-D, which are also based on the "classify and randomly select" paradigm and achieve even better competitive ratios.

Consider a coloring of the cells with five colors 0, 1, 2, 3, and 4 such that for each $i \in \{0, 1, 2, 3, 4\}$, and for each cell colored with color i, the two adjacent cells in the same x-row are colored with colors $(i-1) \bmod 5$ and $(i+1) \bmod 5$, while the remaining four of its adjacent cells are colored with colors $(i+2) \bmod 5$ and $(i+3) \bmod 5$. Such a coloring is depicted in the left part of Figure 38.9. Also, define $s_i = \{i, (i+1) \bmod 5\}$, for $i = 0, 1, \ldots, 4$. Observe that, for each $i = 0, 1, \ldots, 4$, each pair of adjacent cells colored with the colors i and $(i+1) \bmod 5$ is adjacent to cells colored with colors $(i+2) \bmod 5$, $(i+3) \bmod 5$, and $(i+4) \bmod 5$, that is, colors not belonging to s_i. Thus, the coloring together with the color classes s_i satisfy the conditions of Lemma 38.2 with $\nu = 5$ and $\lambda = 2$. We call CRS-B the algorithm that uses this coloring and works according to the "classify and randomly select" paradigm as in the proof of Lemma 38.2. We obtain the following:

Theorem 38.8 *Algorithm CRS-B in cellular networks is 5/2-competitive against oblivious adversaries.*

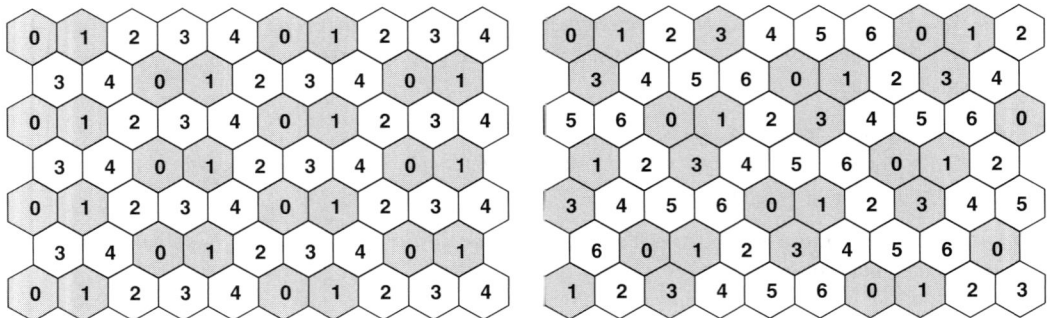

FIGURE 38.9 The 5-coloring used by algorithm CRS-B and the 7-coloring used by algorithm CRS-C. The gray cells are those colored with the colors in set s_0.

Now consider a coloring of the cells with seven colors $0, 1, \ldots, 6$ such that for each cell colored with color i (for $i = 0, \ldots, 6$), its two adjacent cells in the same x-row are colored with the colors $(i - 1) \bmod 7$ and $(i + 1) \bmod 7$, while its two adjacent cells in the same z-row are colored with colors $(i - 3) \bmod 7$ and $(i + 3) \bmod 7$. Such a coloring is depicted in the right part of Figure 38.9. Also, define $s_i = \{i, (i + 1) \bmod 7, (i + 3) \bmod 7\}$, for $i = 0, 1, \ldots, 6$. Observe that, for each $i = 0, 1, \ldots, 6$, each triangle of cells colored with the colors i, $(i + 1) \bmod 7$, and $(i + 3) \bmod 7$ is adjacent to cells colored with colors $(i+2) \bmod 7$, $(i+4) \bmod 7$, $(i+5) \bmod 7$, and $(i+6) \bmod 7$, that is, colors not belonging to s_i. Thus, the coloring together with the color classes s_i satisfy the conditions of Lemma 38.2 with $\nu = 7$ and $\lambda = 3$. We call CRS-C the algorithm that uses this coloring and works according to the "classify and randomly select" paradigm as in the proof of Lemma 38.2. We obtain the following.

Theorem 38.9 *Algorithm CRS-C in cellular networks is $7/3$-competitive against oblivious adversaries.*

Algorithm CRS-D uses a coloring of the cells with 16 colors $0, \ldots, 15$ defined as follows. The cell with coordinates $(x, y, x + y)$ is colored with color $4(x \bmod 4) + y \bmod 4$. The color classes are defined as s_{4i+j} for $0 \le i, j \le 3$ as follows:

$$s_{4i+j} = \{4i + j, 4i + (j + 1) \bmod 4, 4((i + 1) \bmod 4) + j,$$

$$4((i + 1) \bmod 4) + (j + 2) \bmod 4, 4((i + 2) \bmod 4) + (j + 1) \bmod 4,$$

$$4((i + 2) \bmod 4) + (j + 2) \bmod 4, 4((i + 3) \bmod 4) + (j + 3) \bmod 4\}.$$

An example of this coloring is depicted in Figure 38.10.

We now show that the coloring and the color classes used by algorithm CRS-D satisfy the conditions of Lemma 38.2. Each color $k = 0, 1, \ldots, 15$ belongs to 7 of the 16 color classes $s_0, s_1, \ldots, s_{15}$; color $4i + j$ belongs to the color classes $4i + j$, $4(i + 1) + (j + 1)$, $4(i + 2) + (j + 2)$, $4(i + 2) + (j + 3)$, $4(i + 3) + j$, $4(i+3) + (j+2)$, and $4i + (j+3) \bmod 4$ for $0 \le i, j \le 3$. Now, consider the cells colored with colors from the color class s_{4i+j} and the corresponding nodes of the interference graph. The connected components of the subgraph of the interference graph defined by these nodes are of the following types:

- Cliques of three nodes corresponding to cells colored with colors $4i + j$, $4i + (j + 1) \bmod 4$, and $4((i+1) \bmod 4) + j$, respectively. Indeed, the neighborhood of such nodes contains nodes colored with colors $4i+(j+2) \bmod 4$, $4i+(j+3) \bmod 4$, $4(i+1) \bmod 4+(j+1) \bmod 4$, $4(i+1) \bmod 4 + (j+3) \bmod 4$, $4(i+3) \bmod 4+j$, $4(i+3) \bmod 4+(j+1) \bmod 4$, $4(i+3) \bmod 4+(j+2) \bmod 4$, $4(i + 2) \bmod 4 + j$, and $4(i + 2) \bmod 4 + (j + 3) \bmod 4$ that do not belong to class s_{4i+j}.
- Cliques of three nodes corresponding to cells colored with colors $4((i + 1) \bmod 4) + (j + 2) \bmod 4$, $4((i + 2) \bmod 4) + (j + 1) \bmod 4$, and $4((i + 2) \bmod 4) + (j + 2) \bmod 4$,

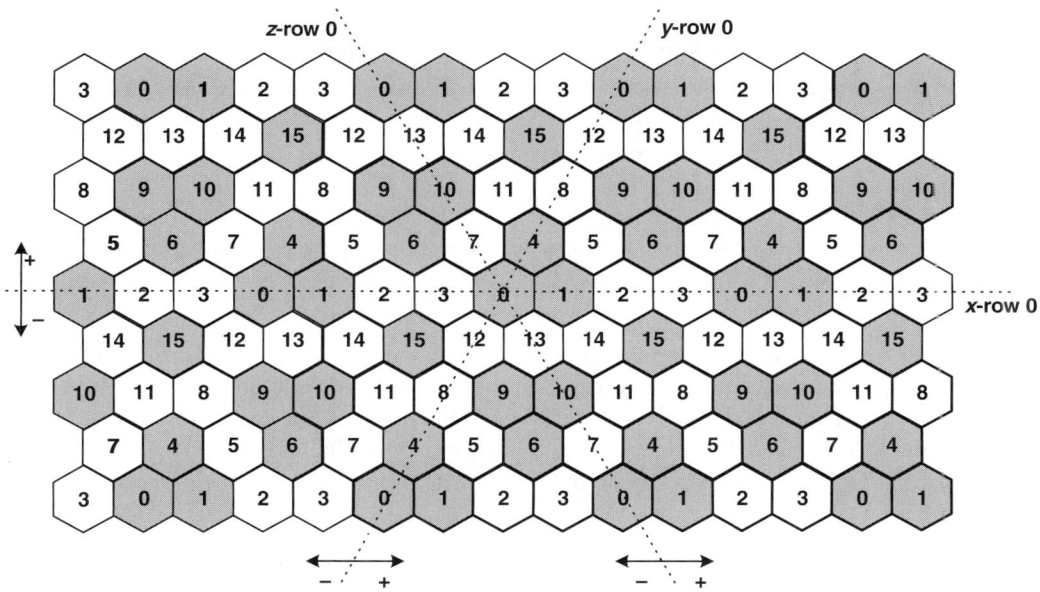

FIGURE 38.10 The 16-coloring used by algorithm CRS-D. The grey cells are those colored with colors in the class s_0.

respectively. Again, the neighborhood of such nodes contains nodes colored with colors $4i + (j + 2) \bmod 4$, $4i + (j + 3) \bmod 4$, $4(i + 1) \bmod 4 + (j + 1) \bmod 4$, $4(i + 1) \bmod 4 + (j+3) \bmod 4$, $4(i+3) \bmod 4 + j$, $4(i+3) \bmod 4 + (j+1) \bmod 4$, $4(i+3) \bmod 4 + (j+2) \bmod 4$, $4(i + 2) \bmod 4 + j$, and $4(i + 2) \bmod 4 + (j + 3) \bmod 4$ that do not belong to class s_{4i+j}.

- Isolated nodes corresponding to cells colored with color $4((i + 3) \bmod 4) + (j + 3) \bmod 4$. The neighborhood of such a cell consists of cells colored with colors $4i+(j+2) \bmod 4$, $4i+(j+3) \bmod 4$, $4(i + 2) \bmod 4 + j$, $4(i + 2) \bmod 4 + (j + 3) \bmod 4$, $4(i + 3) \bmod 4 + j$, and $4(i + 3) \bmod 4 + (j + 2) \bmod 4$ that do not belong to class s_{4i+j}.

Hence, the coloring and the color classes used by algorithm CRS-D satisfy the conditions of Lemma 38.2 for $\lambda = 7$ and $\nu = 16$. This yields the following:

Theorem 38.10 *Algorithm CRS-D for call control in cellular networks is 16/7-competitive against oblivious adversaries.*

Obviously, the algorithm uses only 4 random bits for selecting equiprobably one out of the 16 color classes.

38.5 A Lower Bound

The randomized algorithms presented in the previous section significantly beat the lower bound on the competitiveness of deterministic algorithms. In what follows, using the Minimax Principle [23] (see also Reference 16), we prove a lower bound on the competitive ratio, against oblivious adversaries, of any randomized algorithm for call control in cellular networks. We consider networks that support one frequency; our lower bounds can be easily extended to networks that support multiple frequencies. In our proof, we use the following lemma.

Lemma 38.3 *(Minimax Principle, Yao [16])* Given a probability distribution $\mathcal{P}$ over sequences of calls σ, denote by $\mathcal{E}_\mathcal{P}[B_A(\sigma)]$ and $\mathcal{E}_\mathcal{P}[B_{\text{OPT}}(\sigma)]$ the expected benefit of a deterministic algorithm A and the optimal offline algorithm on sequences of calls generated according to $\mathcal{P}$. Define the competitiveness of A under $\mathcal{P}$, $c_A^\mathcal{P}$ to be such that

$$c_A^\mathcal{P} = \frac{\mathcal{E}_\mathcal{P}[B_{\text{OPT}}(\sigma)]}{\mathcal{E}_\mathcal{P}[B_A(\sigma)]}.$$

Let A_R be a randomized algorithm. Then, the competitiveness of A under $\mathcal{P}$ is a lower bound on the competitive ratio of A_R against an oblivious adversary, that is, $c_A^\mathcal{P} \leq c_{A_R}$.

So, in order to prove a lower bound for any randomized algorithm, it suffices to define an adversary that produces sequences of calls according to a probability distribution and prove that the ratio of the expected optimal benefit over the expected benefit of any deterministic algorithm (which may know the probability distribution in advance) is above some value; by Lemma 38.3, this value will also be a lower bound for any randomized algorithm against oblivious adversaries.

Theorem 38.11 *No randomized online call control algorithm can be better than 2-competitive against oblivious adversaries in cellular networks.*

Proof. We present an adversary $\mathcal{ADV}$-2 that produces sequences of calls according to a probability distribution $\mathcal{P}_2$ that yields the lower bound. We show that the expected benefit of every deterministic algorithm (which may know $\mathcal{P}_2$ in advance) for such sequences of calls is at most 2, whereas the expected optimal benefit is at least 4. Then, Theorem 38.11 follows by Lemma 38.3. First, we describe the sequences of calls produced by $\mathcal{ADV}$-2 without explicitly giving the cells where they appear; then, we show how to construct them in a cellular network.

We start by defining a simpler adversary $\mathcal{ADV}$-1 that works as follows: It first produces two calls in two nonadjacent cells v_0 and v_1. Then it tosses a fair coin.

- On HEADS, it produces two calls in cells v_{00} and v_{01}, which are mutually nonadjacent, adjacent to v_0 and nonadjacent to v_1. Then, it stops.
- On TAILS, it produces two calls in cells v_{10} and v_{11}, which are mutually nonadjacent, adjacent to v_1 and nonadjacent to v_0. Then, it stops.

Now, consider the set of all possible deterministic algorithms $\mathcal{A}_1$ working on the sequences produced by $\mathcal{ADV}$-1. Such an algorithm $A_1 \in \mathcal{A}_1$ may follow one of the following strategies:

- It may accept both calls in cells v_0 and v_1, presented at the first step. This means that the calls presented in the second step cannot be accepted.
- It may reject both calls in cells v_0 and v_1, and then either accept one or both calls presented in the second step or reject them both.
- It may accept only one of the two calls in cells v_0 and v_1, and, if the calls produced at the second step by $\mathcal{ADV}-1$ are nonadjacent to the accepted call, either accept one or both calls presented in the second step or reject them both.

In the first two cases, the expected benefit of the algorithm A_1 is at most 2. In the third case, the expected benefit is 1 (in the first step) plus the expected benefit in the second step. The latter is either zero with probability $1/2$ (this is the case where the cells of the calls produced by the adversary in the second step are adjacent the cell of the call accepted by the algorithm in the first step) or at most 2 with probability $1/2$. Overall, the expected benefit of the algorithm is at most 2.

The adversary $\mathcal{ADV}$-2 works as follows: It first produces two calls in two nonadjacent cells v_0 and v_1. Then it tosses a fair coin.

- On HEADS, it produces two calls in cells v_{00} and v_{01}, which are mutually nonadjacent, adjacent to v_0, and nonadjacent to v_1. Then, it tosses a fair coin.
    - On HEADS, it produces two calls in cells v_{000} and v_{001}, which are mutually nonadjacent, adjacent to v_0 and v_{00}, and nonadjacent to v_1 and v_{01}. Then, it stops.
    - On TAILS, it produces two calls in cells v_{010} and v_{011}, which are mutually nonadjacent, adjacent to v_0 and v_{01}, and nonadjacent to v_1 and v_{00}. Then, it stops.
- On TAILS, it produces two calls in cells v_{10} and v_{11}, which are mutually nonadjacent, adjacent to v_1, and nonadjacent to v_0. Then, it tosses a fair coin.
    - On HEADS, it produces two calls in cells v_{100} and v_{101}, which are mutually nonadjacent, adjacent to v_1 and v_{10}, and nonadjacent to v_0 and v_{11}. Then, it stops.
    - On TAILS, it produces two calls in cells v_{110} and v_{111}, which are mutually nonadjacent, adjacent to v_1 and v_{11}, and nonadjacent to v_0 and v_{10}. Then, it stops.

Observe that, the subsequence of the last four calls produced by $\mathcal{ADV}$-2 essentially belongs to the set of sequences of calls produced by $\mathcal{ADV}$-1.

Now, consider the set of all possible deterministic algorithms $\mathcal{A}_2$ working on the sequences produced by $\mathcal{ADV}$-2. Such an algorithm $A_2 \in \mathcal{A}_2$ may follow one of the following strategies:

- It may accept both calls in cells v_0 and v_1 presented at the first step. This means that the calls presented in the next steps cannot be accepted.
- It may reject both calls in cells v_0 and v_1 and then apply a deterministic algorithm A_1 on the subsequence presented after the first step.
- It may accept only one of the two calls in cells v_0 and v_1, and, then, if the calls produced at the next steps by $\mathcal{ADV}$-2 are nonadjacent to the accepted call, apply a deterministic algorithm A_1 on the subsequence presented after the first step.

In the first case, the expected benefit of the algorithm A_2 is at most 2. In the second case, the expected benefit of A_2 is the expected benefit of A_1 on the sequence of calls presented after the first step, that is, at most 2. In the third case, the expected benefit is 1 (in the first step) plus the expected benefit in the next steps. The benefit of the algorithm in the next steps is either zero with probability $1/2$ (this is the case where the cells of the calls produced by the adversary in the next steps are nonadjacent to the cell of the call accepted by the algorithm in the first step) or the expected benefit of A_1 on the sequence of calls presented after the first step, that is, at most 2 with probability $1/2$. Overall, the expected benefit of the algorithm is at most 2.

Furthermore, the expected optimal benefit on sequences produced by $\mathcal{ADV}$-2 is at least 4. Indeed, in each of the possible sequences

$$\sigma_2^{00} = \langle v_0, v_1, v_{00}, v_{01}, v_{000}, v_{001}\rangle, \quad \sigma_2^{01} = \langle v_0, v_1, v_{00}, v_{01}, v_{010}, v_{011}\rangle,$$
$$\sigma_2^{10} = \langle v_0, v_1, v_{10}, v_{11}, v_{100}, v_{101}\rangle, \quad \sigma_2^{11} = \langle v_0, v_1, v_{10}, v_{11}, v_{110}, v_{111}\rangle$$

generated by $\mathcal{ADV}$-2, the calls in cells $\langle v_1, v_{01}, v_{000}, v_{001}\rangle$, $\langle v_1, v_{00}, v_{010}, v_{011}\rangle$, $\langle v_0, v_{11}, v_{100}, v_{101}\rangle$, and $\langle v_0, v_{10}, v_{110}, v_{111}\rangle$ can be accepted, respectively. Overall, the ratio of the expected optimal benefit over the expected benefit of algorithm A_2 on the sequences generated by the adversary $\mathcal{ADV}$-2 is at least 2, which (by Lemma 38.3) is a lower bound on the competitive ratio of any randomized algorithm for call control.

Figure 38.11 shows how to locate the cells used by $\mathcal{ADV}$-2 so that all the above restrictions hold. This completes the proof of the theorem. ∎

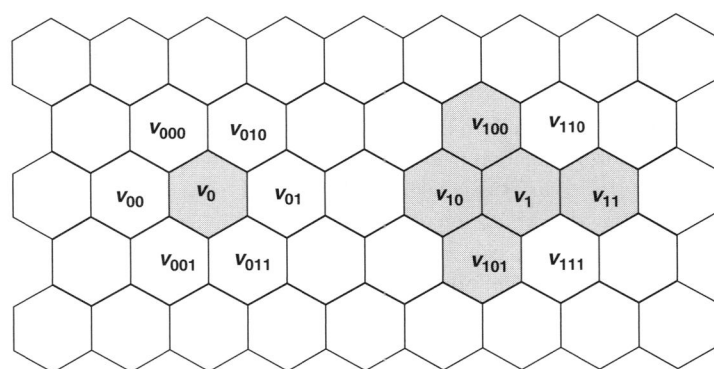

FIGURE 38.11 The calls that may be produced by the adversary $\mathcal{ADV}$-2. The gray cells host calls of the sequence σ_2^{10}.

38.6 Extensions and Open Problems

We conclude this chapter by briefly discussing extensions of our model and open problems. The cellular networks described so far have the property that the same frequency can be safely assigned to calls in cells at distance of at least 2 from each other so that no signal interference arises. These networks are said to be of *reuse distance* 2. We can generalize this constraint and consider cellular networks of reuse distance $k > 2$. Using counterexamples similar to the one described in Section 38.2, we can show that the competitive ratio of deterministic algorithms is at least 4 and 5 in cellular networks of reuse distance $k \in \{3, 4, 5\}$ and $k \geq 6$, respectively. So, investigating randomized online algorithms for these networks is interesting as well. The techniques of Section 38.4 can be used to beat the deterministic lower bounds. Note that Lemma 38.2 is general enough to be useful also in this case. In Reference 7, we present colorings of the interference graphs of these networks that satisfy the requirements of Lemma 38.2. These constructions can be thought of as generalizations of the coloring used by algorithm CRS-C presented in Section 38.4. The interested reader may refer to Reference 7 for a description of these colorings. Here, we only state the corresponding result.

Theorem 38.12 *There exists an online randomized call control algorithm that uses at most $O(\log k)$ random bits and has competitive ratio $4 - \epsilon(k)$ against oblivious adversaries in cellular networks of reuse distance $k \geq 2$.*

The term $\epsilon(k)$ is quite large for small values of k. For any value of reuse distance k, the exact bound implied by Theorem 38.12 is significantly smaller than the corresponding lower bound of deterministic algorithms.

Suitable colorings could be used in wireless networks with different interference graphs. However, for more general interference graphs, irregularities may be difficult to handle. An important class of wireless networks that also contains the cellular networks of reuse distance 2 is the class of planar graphs. These graphs are known to be 4-colorable. Using such a coloring and applying the "classify and randomly select" paradigm, we obtain a 4-competitive online algorithm [20]. No better bounds are known for this case.

However, planarity is a property broad enough to accurately model the wireless networks deployed in practice. In Reference 5, we study wireless networks in which the interference graph has small degree. This case can model the irregularity in the distribution and shape of cells in areas where establishing a regular infrastructure is difficult owing to environmental factors (e.g., mountains, etc.) or just economically infeasible to achieve. In Reference 5, we present the analysis of p-RANDOM in networks with interference graphs of maximum degree 3 or 4. For suitable values of parameter p, the algorithm achieves competitive ratios that cannot be attained either by deterministic algorithms or by "classify and randomly select" algorithms, and, furthermore, this result extends to wireless networks that support many frequencies. The analysis based on the amortized benefit can be further generalized to show bounds on the competitiveness of p-Random on arbitrary interference graphs beating the lower bounds of deterministic algorithms.

Concerning lower bounds on the competitiveness of randomized algorithms, Yao's principle is the main tool to extend the result presented in Section 38.5 to the networks discussed above. By extending the construction of Theorem 38.11 to cellular networks of reuse distance $k \geq 6$, we obtain a lower bound of 2.5 [8]. A different construction in Reference 6 yields a 2.086 lower bound for randomized call control in planar networks. The results of Bartal et al. [4] show limitations on the power of randomized call control algorithms for networks with more general interference graphs.

Definitely, there are still interesting issues to investigate. First, the upper and lower bounds discussed in this chapter still have small gaps. Determining the best possible competitiveness of call control that can be achieved in cellular networks is a challenging task (especially for cellular networks of small reuse distance). Furthermore, our study is mostly of a combinatorial nature. The assumption that calls have equal duration is not always realistic in practice. But, under the classical competitive analysis model, time seems to be a hard opponent to beat. Introducing durations to the calls would imply meaningless (i.e., extremely high) bounds on the competitiveness of call control, even in very simple scenario; this is a rather disappointing result that mainly shows the limitations of competitive analysis. Considering statistical adversaries [3] that construct sequences of calls according to probability distributions known to the algorithm seems to provide a realistic model in order to handle calls with different duration.

Finally, we point out that we have focused on the optimization of communication between users and base stations. Extending the notion of calls to direct connections between pairs of users is worth investigating. By applying similar arguments to those presented in this chapter, we obtain lower bounds of 6 and 3.5 for deterministic and randomized online call control algorithms, respectively. Randomization is definitely helpful in this setting as well, since our preliminary results demonstrate that the competitiveness of p-Random is significantly smaller than 5 for single-frequency wireless cellular networks. Unfortunately, the techniques of Reference 1 for extending the single-frequency case to the case of many frequencies are not applicable to this model. Here, new analysis techniques deserve some attention.

References

[1] B. Awerbuch, Y. Azar, A. Fiat, S. Leonardi, and A. Rosen. On-line competitive algorithms for call admission in optical networks. *Algorithmica*, Vol. 31(1), pp. 29–43, 2001.

[2] B. Awerbuch, Y. Bartal, A. Fiat, and A. Rosen. Competitive non-preemptive call control. In *Proc. of the 5th Annual ACM-SIAM Symposium on Discrete Algorithms (SODA '94)*, pp. 312–320, 1994.

[3] A. Borodin and R. El-Yaniv. *Online Computation and Competitive Analysis*. Cambridge University Press, Cambridge, UK, 1998.

[4] Y. Bartal, A. Fiat, and S. Leonardi. Lower bounds for on-line graph problems with applications to on-line circuit and optical routing. In *Proc. of the 28th Annual ACM Symposium on Theory of Computing (STOC '96)*, pp. 531–540, 1996.

[5] I. Caragiannis, C. Kaklamanis, and E. Papaioannou. Randomized call control in sparse wireless cellular networks. In *Proc. of the 8th International Conference on Advances in Communications and Control (COMCON 01)*, pp. 73–82, 2001.

[6] I. Caragiannis, C. Kaklamanis, and E. Papaioannou. Efficient on-line frequency allocation and call control in cellular networks. *Theory of Computing Systems*, Vol. 35, pp. 521–543, 2002. Preliminary results in *ACM SPAA '00* and *IPDPS '01*.

[7] I. Caragiannis, C. Kaklamanis, and E. Papaioannou. Simple on-line algorithms for call control in cellular networks. In *Proc. of the 1st Workshop on Approximation and On-line Algorithms (WAOA '03)*, LNCS 2909, Springer, pp. 67–80, 2003.

[8] I. Caragiannis, C. Kaklamanis, and E. Papaioannou. New bounds on the competitiveness of randomized online call control in cellular networks. In *Proc. of the Euro-Par '05*, LNCS 3648, Springer, pp. 1089–1099, 2005.

[9] D. Dimitrijević and J. Vučetić. Design and performance analysis of algorithms for channel allocation in cellular networks. *IEEE Transactions on Vehicular Technology*, Vol. 42(4), pp. 526–534, 1993.

[10] W. K. Hale. Frequency assignment: Theory and applications. *Proceedings of the IEEE*, 68(12), pp. 1497–1514, 1980.

[11] J. Janssen, K. Kilakos, and O. Marcotte. Fixed preference frequency allocation for cellular telephone systems. *IEEE Transanctions on Vehicular Technology*, Vol. 48(2), pp. 533–541, 1999.

[12] J. Janssen, D. Krizanc, L. Narayanan, and S. Shende. Distributed on-line frequency assignment in cellular networks. *Journal of Algorithms*, Vol. 36(2), pp. 119–151, 2000.

[13] S. Jordan and E. Schwabe. Worst-case performance of cellular channel allocation policies. *ACM Baltzer Journal on Wireless Networks*, Vol. 2(4), pp. 265–275, 1996.

[14] S. Kim and S. L. Kim. A Two-phase algorithm for frequency assignment in cellular mobile systems. *IEEE Transactions on Vehicular Technology*, Vol. 43(3), pp. 542–548, 1994.

[15] S. Leonardi, A. Marchetti-Spaccamela, A. Prescuitti, and A. Rosen. On-line randomized call-control revisited. In *Proccedings of the 9th Annual ACM-SIAM Symposium on Discrete Algorithms (SODA '98)*, pp. 323–332, 1998.

[16] R. Motwani and B. Raghavan. *Randomized Algorithms*. Cambridge University Press, Cambridge, UK, 1995.

[17] C. McDiarmid and B. Reed. Channel assignment and weighted coloring. *Networks*, Vol. 36, pp. 114–117, 2000.

[18] L. Narayanan and S. Shende. Static frequency assignment in cellular networks. In *Algorithmica*, Vol. 29(3), pp. 396–409, 2001.

[19] L. Narayanan and Y. Tang. Worst-case analysis of a dynamic channel assignment strategy. *Discrete Applied Mathematics*, Vol. 140(1-3), pp. 115–141, 2004.

[20] G. Pantziou, G. Pentaris, and P. Spirakis. Competitive call control in mobile networks. *Theory of Computing Systems*, Vol. 35(6), pp. 625–639, 2002.

[21] P. Raymond. Performance analysis of cellular networks. *IEEE Transactions on Communications*, Vol. 39(12), pp. 1787–1793, 1991.

[22] P.-J. Wan and L. Liu. Maximal throughput in wavelength-routed optical networks. Multichannel optical networks: Theory and practice, *DIMACS Series in Discrete Mathematics and Theoretical Computer Science*, AMS, Vol. 46, pp. 15–26, 1998.

[23] A. C. Yao. Probabilistic computations: Towards a unified measure of complexity. In *Proc. of the 17th Annual Symposium on Foundations of Computer Science (FOCS '77)*, pp. 222–227, 1977.

39
Minimum Energy Communication in Ad Hoc Wireless Networks

39.1	Introduction...	39-1
39.2	Symmetric Wireless Networks	39-4
	Symmetric Connectivity Requirements • Multicasting and Broadcasting	
39.3	Asymmetric Wireless Networks.....................	39-7
	Multicasting, Broadcasting, and Group Communication • Bidirected Connectivity Requirements	
39.4	The Geometric Model	39-13
	The Linear Case • Multidimensional Wireless Networks	
39.5	Extensions	39-16
References		39-18

Ioannis Caragiannis
Christos Kaklamanis
Panagiotis Kanellopoulos
University of Patras

39.1 Introduction

Wireless networks have received significant attention during the recent years. Especially, *ad hoc wireless networks* emerged owing to their potential applications in environmental monitoring, sensing, specialized ad hoc distributed computations, emergency disaster relief, battlefield situations, and so forth [37, 41]. Unlike traditional wired networks or cellular wireless networks, no wired backbone infrastructure is installed for ad hoc networks.

A node (or station) in these networks is equipped with an omnidirectional antenna that is responsible for sending and receiving signals. Communication is established by assigning to each station a transmitting power. In the most common power attenuation model [37], the signal power falls proportionally to $1/r^\alpha$, where r is the distance from the transmitter and α is a constant that depends on the wireless environment (typical values of α are between 1 and 6). So, a transmitter can send a signal to a receiver if $P_s/(d(s,t)^\alpha) \geq \gamma$, where P_s is the power of the transmitting signal, $d(s,t)$ is the Euclidean distance between the transmitter and the receiver, and γ is the receiver's power threshold for signal detection, which is usually normalized to 1.

Communication from a node s to another node t may be established either directly if the two nodes are close enough and s uses adequate transmitting power or by using intermediate nodes. Observe that owing to the nonlinear power attenuation, relaying the signal between intermediate nodes may result in energy conservation.

A crucial issue in ad hoc networks is to support communication patterns that are typical in traditional networks. These include broadcasting, multicasting, and gossiping (all-to-all communication). Since establishing a communication pattern strongly depends on the use of energy, the important engineering question to be solved is to guarantee a desired communication pattern, minimizing the total energy consumption. In this chapter, we consider a series of *minimum energy communication* problems in ad hoc wireless networks, which we formulate below.

We model an ad hoc wireless network by a complete directed graph $G = (V, E)$, where $|V| = n$, with a nonnegative edge cost function $c : E \rightarrow R^+$. Given a nonnegative node weight assignment $w : V \rightarrow R^+$, the *transmission graph* G_w is the directed graph defined as follows. It has the same set of nodes as G and a directed edge (u, v) belongs to G_w if the weight assigned to node u is at least the cost of the edge (u, v) (i.e., $w(u) \geq c(u, v)$). Intuitively, the weight assignment corresponds to the energy levels at which each node operates (i.e., transmits messages), while the cost between two nodes indicates the minimum energy level necessary to send messages from one node to the other. Usually, the edge cost function is symmetric (i.e., $c(u, v) = c(v, u)$). An important special case, which usually reflects the real-world situation, henceforth called geometric case, is when nodes of G are points in a Euclidean space and the cost of an edge (u, v) is defined as the Euclidean distance between u and v raised to a fixed power α ranging from 1 to 6 (i.e., $c(u, v) = d(u, v)^\alpha$). Asymmetric edge cost functions can be used to model medium abnormalities or batteries with different energy levels [32].

The problems we study in this work can be stated as follows. Given a complete directed graph $G = (V, E)$, where $|V| = n$, with nonnegative edge costs $c : E \rightarrow R^+$, find a nonnegative node weight assignment $w : V \rightarrow R^+$ such that the transmission graph G_w maintains a connectivity property and the sum of weights is minimized. Such a property is defined by a requirement matrix $R = (r_{ij}) \in \{0, 1\}$, where r_{ij} is the number of directed paths required in the transmission graph from node v_i to node v_j. Depending on the connectivity property for the transmission graph, we may define a variety of problems.

Several communication requirements are of interest. In minimum energy steiner subgraph (MESS), the requirement matrix is symmetric. Alternatively, we may define the problem by a set of nodes $D \subseteq V$ partitioned into p disjoint subsets $D_1, D_2, ..., D_p$. The entries of the requirement matrix are now defined as $r_{ij} = 1$ if $v_i, v_j \in D_k$ for some k and $r_{ij} = 0$, otherwise. The minimum energy subset strongly connected subgraph (MESSCS) is the special case of MESS with $p = 1$ while the minimum energy strongly connected subgraph (MESCS) is the special case of MESSCS with $D = V$ (i.e., the transmission graph is required to span all nodes of V and to be strongly connected). Althaus et al. [1] and Călinescu et al. [11] study MESCS under the extra requirement that the transmission graph contains a bidirected subgraph (i.e., a directed graph in which the existence of a directed edge implies that its opposite directed edge also exists in the graph), which maintains the connectivity requirements of MESCS. By adding this extra requirement to MESS and MESSCS, we obtain the bidirected MESS and bidirected MESSCS, respectively, that is, the requirement for the transmission graph in bidirected MESS (resp., bidirected MESSCS) is to contain as a subgraph a bidirected graph satisfying the connectivity requirements of MESS (resp., MESSCS). Minimum energy communication problems with symmetric requirement matrices are usually referred to as *group communication* problems.

In minimum energy multicast tree (MEMT), the connectivity property is defined by a root node v_0 and a set of nodes $D \subseteq V - \{v_0\}$ such that $r_{ij} = 1$ if $i = 0$ and $v_j \in D$ and $r_{ij} = 0$, otherwise. The minimum energy broadcast tree (MEBT) is the special case of MEMT with $D = V - \{v_0\}$. By inverting the connectivity requirements, we obtain the following two problems: the minimum energy inverse multicast tree (MEIMT) where the connectivity property is defined by a root node v_0 and a set of nodes $D \subseteq V - \{v_0\}$ such that $r_{ij} = 1$ if $v_i \in D$ and $j = 0$ and $r_{ij} = 0$, otherwise, and the minimum energy inverse broadcast tree (MEIBT), which is the special case of MEIMT with $D = V - \{v_0\}$.

TABLE 39.1 Abbreviations for Problems Used in This Chapter

Abbreviation	Problem
MESS	Minimum Energy Steiner Subgraph
MESSCS	Minimum Energy Subset Strongly Connected Subgraph
MESCS	Minimum Energy Strongly Connected Subgraph
MEMT	Minimum Energy Multicast Tree
MEBT	Minimum Energy Broadcast Tree
MEIMT	Minimum Energy Inverse Multicast Tree
MEIBT	Minimum Energy Inverse Broadcast Tree
SF	Steiner Forest
ST	Steiner Tree
DST	Directed Steiner Tree
MSA	Minimum Spanning Arborescence
NWSF	Node-Weighted Steiner Forest
NWST	Node-Weighted Steiner Tree

The terminology used in this chapter is the same as the one in Reference 14. Table 39.1 summarizes the abbreviations used for the problems studied, as well as for other combinatorial problems used several times in the rest of this chapter.

In the following sections, we usually refer to classical combinatorial optimization problems and show the relations of minimum energy communication problems to them. For completeness, we present their definitions here. The steiner forest (SF) problem is defined as follows. Given an undirected graph $G = (V, E)$ with an edge cost function $c : E \to R^+$ and a set of nodes $D \subseteq V$ partitioned into p disjoint sets $D_1, \ldots, D_p$, compute a subgraph H of G of minimum total edge cost such that any two nodes v_i, v_j belonging to the same set D_k for some k are connected through a path in H. Steiner Tree (ST) is the special case of SF with $p = 1$. An instance of the Directed Steiner Tree (DST) is defined by a directed graph $G = (V, E)$ with an edge cost function $c : E \to R^+$, a root node $v_0 \in V$, and a set of terminals $D \subseteq V - \{v_0\}$. Its objective is to compute a tree of minimum edge cost that is directed out of v_0 and spans all nodes of D. The special case of DST, with $D = V - \{v_0\}$ is called Minimum Spanning Arborescence (MSA). The Node-Weighted Steiner Forest (NWSF) problem is defined as follows. Given an undirected graph $G = (V, E)$ with a node cost function $c : V \to R^+$ and a set of nodes $D \subseteq V$ partitioned into p disjoint sets $D_1, \ldots, D_p$, compute a subgraph H of G of minimum total node cost such that any two nodes v_i, v_j belonging to the same set D_k for some k are connected through a path in H. Node-Weighted Steiner Tree (NWST) is the special case of NWSF with $p = 1$.

We study symmetric wireless networks in Section 39.2. In this setting, communication problems with symmetric connectivity requirements admit algorithms with constant approximation ratio. Multicasting and broadcasting are inherently more difficult, admitting only logarithmic approximations. Almost all communication problems become even harder in the more general case of asymmetric wireless networks. For example, MEMT is equivalent to DST in terms of hardness of approximation; this implies similar complexity for other communication problems as well. Surprisingly, broadcasting still admits logarithmic approximation in asymmetric wireless networks. Results on the asymmetric model are surveyed in Section 39.3. Table 39.2 summarizes the known results in symmetric and asymmetric wireless networks.

Better results exist for minimum energy communication problems in geometric wireless networks. The linear case where nodes correspond to points on a line has been proved to be tractable while most problems become hard in higher dimensions. We discuss the related results in Section 39.4. A summary of them is presented in Table 39.3.

We conclude this chapter by presenting extensions of the network model and briefly discussing results on related communication problems in Section 39.5.

TABLE 39.2 The Best Known Results for the Problems Discussed in This Chapter in Symmetric and Asymmetric Wireless Networks

	Approximability in asymmetric networks		Approximability in symmetric networks	
Problem	Lower bound	Upper bound	Lower bound	Upper bound
MESS	$\Omega(\log^{2-\epsilon} n)$ [14]		313/312 [21]	4 [14]
MESSCS	$\Omega(\log^{2-\epsilon} n)$ [14]		313/312 [21]	3.1 [14]
MESCS	$\Omega(\log n)$ [9]	$O(\log n)$ [9, 14]	313/312 [21]	2 [30]
Bidirected MESS	$\Omega(\log |D|)$ [1]	$O(\log |D|)$ [14]	96/95 [14]	4 [14]
Bidirected MESSCS	$\Omega(\log |D|)$ [1]	$O(\log |D|)$ [14]	96/95 [14]	3.1 [14]
Bidirected MESCS	$\Omega(\log n)$ [1]	$O(\log n)$ [9, 14]	313/312 [21]	5/3 [1]
MEMT	$\Omega(\log^{2-\epsilon} n)$ [9, 14]	$O(|D|^{\epsilon})$ [32]	$\Omega(\log n)$ [18]	$O(\log n)$ [14]
MEBT	$\Omega(\log n)$ [18]	$O(\log n)$ [9, 14]	$\Omega(\log n)$ [18]	$O(\log n)$ [6, 13]
MEIMT	$\Omega(\log^{2-\epsilon} n)$ [14]	$O(|D|^{\epsilon})$ [14]	96/95 [14]	1.55 [14]
MEIBT	1	1	1	1

TABLE 39.3 The Best Known Results under the Geometric Model

	Linear networks	Multidimensional networks	
	Complexity	Complexity	Approximability
MESCS	P [30]	NP-hard [21, 30]	2 [30]
		APX-hard ($d \geq 3$) [21]	
MEMT	P [13, 19]	NP-hard [18]	9.3 ($\alpha = d = 2$) [1]
MEBT	P [13, 19]	NP-hard [18]	6 ($\alpha = d = 2$) [2]
			$3^d - 1$ [23]

39.2 Symmetric Wireless Networks

In this section, we consider ad hoc wireless networks with symmetric edge cost functions in the underlying subgraph. Almost all the minimum energy communication problems discussed in the previous section are NP-hard in these networks. Observe that the geometric model is a special case of symmetric networks, hence, the hardness results for the geometric case hold in this case as well. We postpone the discussion on the particular hardness results until Section 39.4 where we discuss the results on geometric networks in more detail. Here, we only state the stronger inapproximability bounds that hold owing to the generality of symmetric wireless networks. On the positive side, we present approximation algorithms for each problem.

39.2.1 Symmetric Connectivity Requirements

We start by presenting a constant approximation algorithm for MESS; this is the most general problem falling in this category. The algorithm constructs a solution to MESS by exploiting the solution of a corresponding instance for problem SF. The reduction presented in the following appeared in Reference 14. It can be thought of as a generalization of an algorithm in Reference 30, which uses minimum spanning trees (MST) to approximate instances of MESCS.

Consider an instance I_{MESS} of MESS that consists of a complete directed graph $G = (V, E)$, a symmetric edge cost function $c : E \to R^+$, and a set of terminals $D \subseteq V$ partitioned into p disjoint subsets $D_1, \ldots, D_p$. Construct the instance I_{SF} of SF that consists of the complete undirected graph $H = (V, E')$, the edge cost function $c' : E' \to R^+$, defined as $c'(u, v) = c(u, v) = c(v, u)$ on the undirected edges of E', and the set of terminals D together with its partition into the sets $D_1, \ldots, D_p$. Consider a solution for I_{SF} that consists of a subgraph $F = (V, A)$ of H. Construct the weight assignment w to the nodes of V

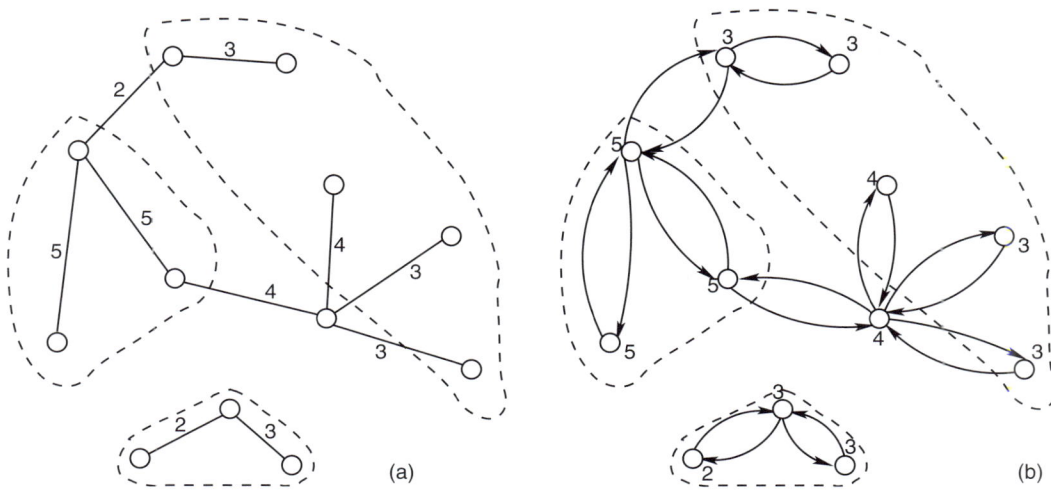

FIGURE 39.1 Transforming a solution for I_{SF} (a) to a solution for I_{MESS} (b). In (a), numbers on the edges of the solution denote their cost. In (b), numbers are associated with the nodes and denote their weight. In both cases, the dashed closed lines indicate the subsets in which the set of terminals is partitioned.

by setting $w(u) = 0$, if there is no edge touching u in A and $w(u) = \max_{v:(u,v) \in A}\{c'(u,v)\}$, otherwise. An example of this construction is presented in Figure 39.1. The following statement has been proved in Reference 14 (see also Reference 30).

Lemma 39.1 *If F is a ρ-approximate solution for I_{SF}, then w is a 2ρ-approximate solution for I_{MESS}.*

We can solve I_{SF} using the 2-approximation algorithm of Goemans and Williamson [25] for SF. When $p = 1$ (i.e., when I_{MESS} is actually an instance of MESSCS), the instance I_{SF} is actually an instance of ST that can be approximated within $1 + \frac{1}{2}\ln 3 \approx 1.55$, using an algorithm of Robins and Zelikovsky [38]. We obtain the following result.

Theorem 39.1 *There exist a 4-, a 3.1-, and a 2-approximation algorithm for MESS, MESSCS, and MESCS in symmetric wireless networks, respectively.*

Clearly, the transmission graph constructed by the above technique contains a bidirected subgraph that maintains the connectivity requirements of MESS, and thus, the algorithms for MESS, MESSCS, and MESCS actually provide solutions to bidirected MESS, bidirected MESSCS, and bidirected MESCS, respectively. The analysis presented in Reference 14 still holds; thus, the approximation guarantees of Theorem 39.1 hold for bidirected MESS and bidirected MESSCS in symmetric wireless networks as well.

Next, we present a simple approximation-preserving reduction from ST to bidirected MESSCS in symmetric wireless networks.

Given an instance I_{ST} of ST that consists of an undirected graph $G = (V, E)$ with edge cost function $c : E \to R^+$, and a set of terminals $D \subseteq V$, construct the instance $I_{bMESSCS}$ as follows. $I_{bMESSCS}$ consists of a complete directed graph $H = (U, A)$ with symmetric edge cost function $c' : A \to R^+$, and a set of terminals $D' \subseteq U$. The set of nodes U contains a node h_v for each node v of V and two nodes $h_{(u,v)}$ and $h_{(v,u)}$ for each edge (u, v) of E. The edge cost function c' is defined as $c'(h_u, h_{(u,v)}) = c'(h_{(u,v)}, h_u) = c'(h_{(v,u)}, h_v) = c'(h_v, h_{(v,u)}) = 0$ and $c'(h_{(u,v)}, h_{(v,u)}) = c'(h_{(v,u)}, h_{(u,v)}) = c(u,v)$, for each edge (u, v) of E, while all other directed edges of A have infinite cost. The construction is presented in Figure 39.2. The set of terminals is defined as $D' = \{h_u \in U | u \in D\}$. It is not difficult to see that a ρ-approximate solution for instance $I_{bMESSCS}$ reduces in polynomial time to a ρ-approximate solution

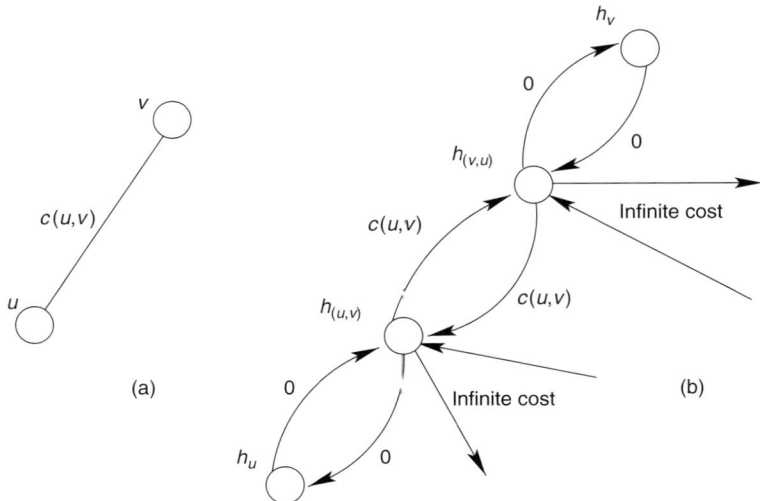

FIGURE 39.2 An edge in I_{ST} (a) and the corresponding structure in $I_{bMESSCS}$ (b). In (b), all edges that are incident to $h_{(u,v)}$ but are not incident to either $h_{(v,u)}$ or h_u, as well as all edges that are incident to $h_{(v,u)}$ but are not incident to either $h_{(u,v)}$ or h_v have infinite cost.

for instance I_{ST}. Thus, using an inapproximability result for ST presented in Reference 17, we obtain the following:

Theorem 39.2 *For any $\epsilon > 0$, bidirected MESSCS in symmetric wireless networks is not approximable within $96/95 - \epsilon$, unless $P = NP$.*

Clearly, this result also applies to bidirected MESS. For MESCS, a weaker inapproximability result of 313/312 follows by adapting a reduction of Reference 20 to Vertex Cover in bounded-degree graphs and using a known inapproximability result for the latter problem [5].

39.2.2 Multicasting and Broadcasting

In this section, we present logarithmic approximation algorithms for MEMT and MEBT in symmetric wireless networks. Such results have been obtained independently in References 6, 7, 13, and 14. The algorithms in References 6 and 7 use set covering techniques, while the algorithm in References 13 reduces the problem to Node Weighted Connected Dominating Set. We present the algorithm from Reference 14 here, which is probably the simplest one; it reduces the problem to NWST.

Consider an instance I_{MEMT} of MEMT, which consists of a complete directed graph $G = (V, E)$, a symmetric edge cost function $c : E \to R^+$, a root node $v_0 \in V$, and a set of terminals $D \subseteq V - \{v_0\}$.

Construct an instance I_{NWST} of NWST, which consists of an undirected graph $H = (U, A)$, a node weight function $c' : U \to R^+$, and a set of terminals $D' \subseteq U$. For a node $v \in V$, we denote by n_v the number of different edge costs in the edges directed out of v, and, for $i = 1, \ldots, n_v$, we denote by $X_i(v)$ the ith smallest edge cost among the edges directed out of v. The set of nodes U consists of n disjoint sets of nodes called *supernodes*. Each supernode corresponds to a node of V. The supernode Z_v corresponding to node $v \in V$ has the following $n_v + 1$ nodes: an *input node* $Z_{v,0}$ and n_v *output nodes* $Z_{v,1}, \ldots, Z_{v,n_v}$. For each pair of nodes $u, v \in V$, the set of edges A contains an edge between the output node $Z_{u,i}$ and the input node $Z_{v,0}$ such that $X_i(u) \geq c(u,v)$. Also, for each node $v \in V$, A contains an edge between the input node $Z_{v,0}$ and each output node $Z_{v,i}$, for $i = 1, \ldots, n_v$. The cost function c' is defined as $c'(Z_{v,0}) = 0$ for the input nodes and as $c'(Z_{v,i}) = X_i(v)$ for $i = 1, \ldots, n_v$, for the output nodes. The set of terminals D' is defined as $D' = \{Z_{v,0} \in U | v \in D \cup \{v_0\}\}$. An example of this reduction is depicted in Figure 39.3.

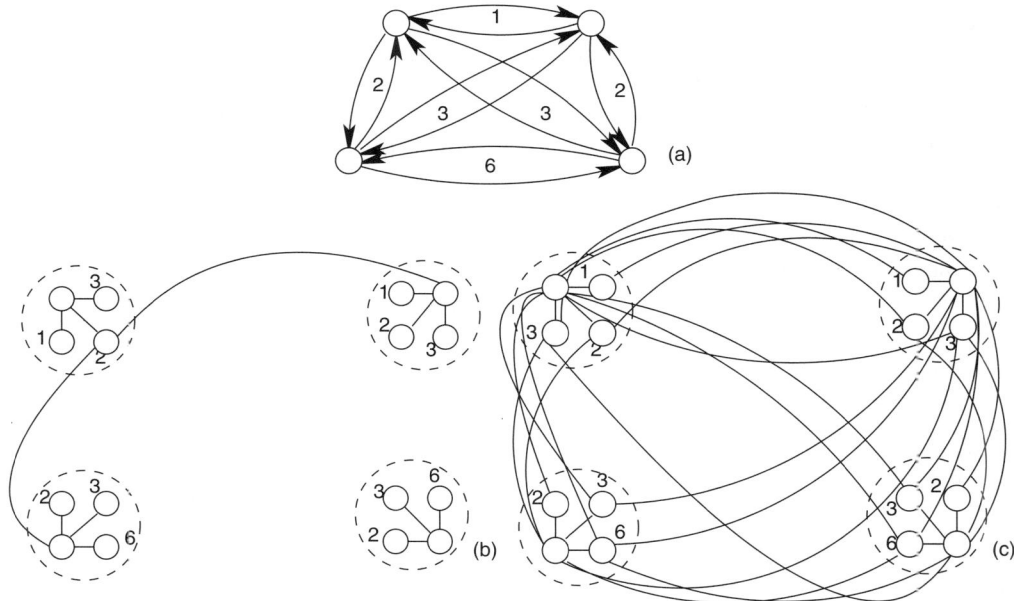

FIGURE 39.3 The reduction to Node-Weighted Steiner Tree. (a) The graph G of an instance of MEMT. (b) The graph H of the corresponding instance of NWST. Each large cycle indicates a supernode. Only the edges incident to the node of weight 2 of the upper left supernode are shown. These edges are those that correspond to edges in (a) of cost at most 2, directed out of the left upper node. (c) The graph H of the corresponding instance of NWST.

Consider a subgraph $F = (S, A')$ of H, which is a solution for I_{NWST}. We compute a spanning tree $T' = (S, A'')$ of F and, starting from $Z_{v_0,0}$, we compute a breadth-first search (BFS) numbering of the nodes of T'. For each $v \in S$, we denote by $m(v)$ the BFS number of v. We construct a tree $T = (V, E')$ that, for each edge of F between a node $Z_{u,i}$ of supernode Z_u and a node $Z_{v,j}$ of another supernode Z_v such that $m(Z_{u,i}) < m(Z_{v,j})$, contains a directed edge from u to v. The output of our algorithm is the weight assignment w defined as $w(u) = \max_{(u,v) \in T} c(u, v)$ if u has at least one outgoing edge in T, and $w(u) = 0$, otherwise. The following lemma is proved in Reference 14 and relates the quality of the two solutions.

Lemma 39.2 *If F is a ρ-approximate solution to I_{NWST}, then w is a 2ρ-approximate solution to I_{MEMT}.*

Guha and Khuller [26] present a $1.35 \ln k$-approximation algorithm for NWST, where k is the number of terminals in the instance of NWST. Given an instance I_{MEMT} of MEMT with a set of terminals D, the corresponding instance I_{NWST} has $|D| + 1$ terminals. Thus, the cost of the solution of I_{MEMT} is within $2 \times 1.35 \ln(|D| + 1) = 2.7 \ln(|D| + 1)$ of the optimal solution. We remind that MEBT is the special case of MEMT with $D = V - \{v_0\}$.

Theorem 39.3 *There exist a $2.7 \ln(|D|+1)$- and a $2.7 \ln n$-approximation algorithm for MEMT and MEBT in symmetric wireless networks, respectively.*

39.3 Asymmetric Wireless Networks

In general, minimum energy communication problems are more difficult in the asymmetric model. For example, MEIMT is equivalent to DST. This is a rather disappointing result since DST has polylogarithmic inapproximability while the best known algorithm has polynomial approximation ratio. The following simple reduction from References 9 and 14 demonstrates the equivalence of MEIMT and DST.

Assume that we have an instance I_{MEIMT} of MEIMT defined by a complete directed graph $G = (V, E)$, an edge cost function $c : E \to R^+$, a root node $v_0 \in V$, and a set of terminals $D \subseteq V - \{v_0\}$. Consider the instance I_{DST} of DST that consists of G, the edge cost function $c' : E \to R^+$ defined as $c'(u, v) = c(v, u)$ for any edge $(u, v) \in E$, the set of terminals D, and the root node v_0. Also, we may start by an instance I_{DST} of DST and construct I_{MEIMT} in the same way. Then, it is not difficult to see that any ρ-approximate solution for I_{DST} reduces (in polynomial time) to a ρ-approximate solution for I_{MEIMT} while any ρ-approximate solution for I_{MEIMT} also reduces to a ρ-approximate solution for I_{DST}.

As corollaries, using the approximability and inapproximability results presented in References 16 and 29, we obtain that MEIMT is approximable within $O(|D|^\epsilon)$ and inapproximable within $O(\ln^{2-\epsilon} n)$, for any constant $\epsilon > 0$. Note that DST in symmetric wireless networks is equivalent to ST. Going back to symmetric networks, using the approximability and inapproximability results presented in References 38 and 17, we obtain that MEIMT in symmetric wireless networks is approximable within 1.55 and inapproximable within $96/95 - \epsilon$, for any $\epsilon > 0$. Also, instances of DST having all nonroot nodes as terminals are actually instances of MSA that is known to be computable in polynomial time [22]. Thus, MEIBT can be solved in polynomial time (even in asymmetric wireless networks).

39.3.1 Multicasting, Broadcasting, and Group Communication

MEMT and MESSCS are as hard to approximate as DST. Consider an instance I_{DST} of DST that consists of a directed graph $G = (V, E)$ with an edge cost function $c : E \to R^+$, a root node v_0, and a set of terminals $D \subseteq V - \{v_0\}$. Without loss of generality, we may assume that G is a complete directed graph with some of its edges having infinite cost.

We construct the instance I_{MEMT} of MEMT that consists of a complete directed graph $H = (U, A)$ with edge cost function $c' : A \to R^+$, a root node $v'_0 \in U$, and a set of terminals $D' \subseteq U - \{v'_0\}$. The set of nodes U has a node h_v for each node $v \in V$ and a node $h_{(u,v)}$ for each directed edge (u, v) of E. For each directed edge (u, v) of E, the directed edge $(h_u, h_{(u,v)})$ of A has zero cost and the directed edge $(h_{(u,v)}, h_v)$ of A has cost $c'(h_{(u,v)}, h_v) = c(u, v)$, while all other edges of A have infinite cost. This construction is presented in Figure 39.4. The set of terminals is defined as $D' = \{h_u \in U | u \in D\}$, while $v'_0 = h_{v_0}$. It is not difficult to see that a ρ-approximate solution to I_{MEMT} reduces in polynomial time to a ρ-approximate solution to I_{DST} [14].

A similar reduction can be used to show inapproximability of MESSCS. We construct the instance I_{MESSCS} of MESSCS, which consists of the graph G, the set of terminals $D \cup v_0$, and an edge cost function $c'' : E \to R^+$ defined as follows. For each directed edge (u, v) of E such that $u \neq v_0$, it is $c''(u, v) = c(v, u)$, while all edges of E directed out of v_0 have zero cost. An example of this construction

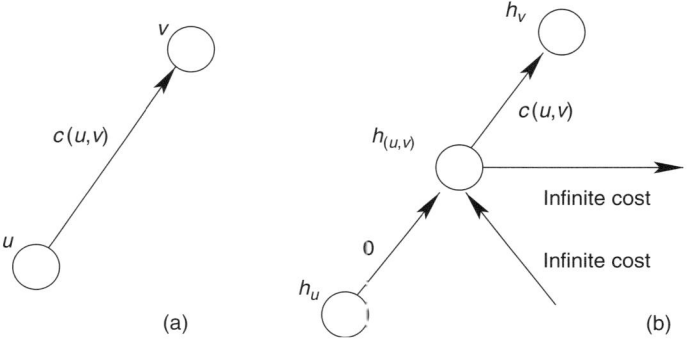

FIGURE 39.4 An edge in I_{DST} (a) and the corresponding structure in I_{MEMT} (b). In (b), edges directed out of $h_{(u,v)}$ that are not incident to h_v, as well as edges that are not incident to h_u and are destined for $h_{(u,v)}$ have infinite cost.

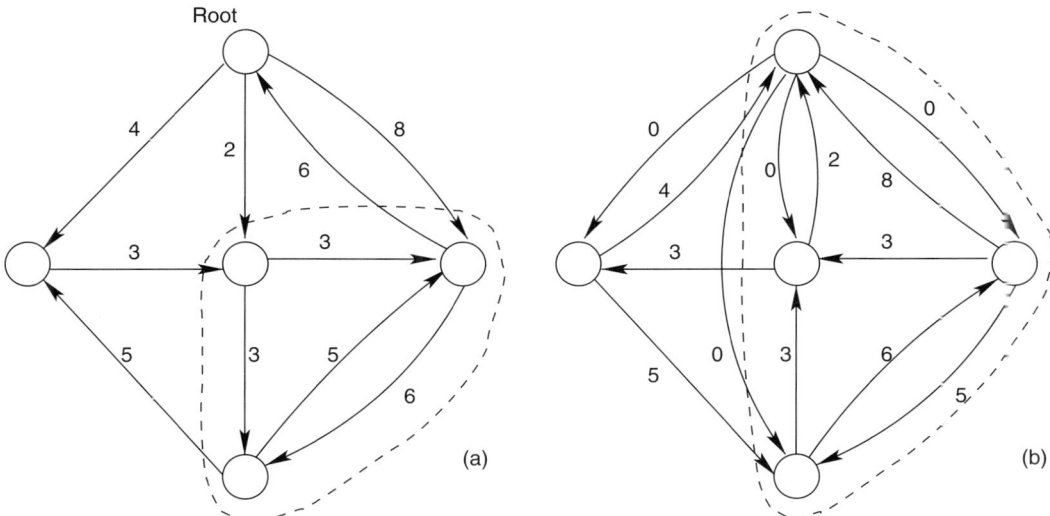

FIGURE 39.5 Transforming an instance of I_{DST} (a) to an instance of I_{MESSCS} (b). Dashed closed lines indicate the sets of terminals.

is presented in Figure 39.5. Again, any ρ-approximate solution to I_{MESSCS} reduces in polynomial time to a ρ-approximate solution to I_{DST} [14].

Using the inapproximability result for DST [29], we obtain the following:

Theorem 39.4 *For any $\epsilon > 0$, MEMT and MESSCS in asymmetric wireless networks are not approximable within $O(\ln^{2-\epsilon} n)$, unless $NP \subseteq ZTIME(n^{polylog(n)})$.*

On the positive side, Liang [32] presents an intuitive reduction for transforming an instance I_{MEMT} of MEMT into an instance I_{DST} of DST in such a way that a ρ-approximate solution for I_{DST} implies a ρ-approximate solution for I_{MEMT}.

We describe this reduction here. Assume that I_{MEMT} consists of a complete directed graph $G = (V, E)$ with an edge cost function $c : E \rightarrow R^+$ and a root node $r \in V$. Then, the instance I_{DST} consists of a directed graph $H = (U, A)$ with an edge cost function $c' : A \rightarrow R^+$, a root node $r' \in U$, and a set of terminals $D \subseteq U - \{r'\}$. For a node $v \in V$, we denote by n_v the number of different edge costs in the edges directed out of v, and, for $i = 1, \ldots, n_v$, we denote by $X_i(v)$ the ith smallest edge cost among the edges directed out of v. For each node $v \in V$, the set of nodes U contains $n_v + 1$ nodes $Z_{v,0}, Z_{v,1}, \ldots, Z_{v,n_v}$. For each directed edge $(v, u) \in E$ and for $i = 1, \ldots, n_v$, the set of edges A contains a directed edge of zero cost from $Z_{v,i}$ to $Z_{u,0}$ if $X_i(v) \geq c(v, u)$. Also, for each node $v \in V$, and $i = 1, \ldots, n_v$, the set of edges A contains a directed edge from $Z_{v,0}$ to $Z_{v,i}$ of cost $c'(Z_{v,0}, Z_{v,i}) = X_i(v)$. An example of this construction is presented in Figure 39.6. The set of terminals is defined by $D = \{Z_{v,0} | v \in V - \{r\}\}$ and $r' = Z_{r,0}$.

Now, a solution for the original instance I_{MEMT} of MEMT is obtained by assigning energy to each node v equal to the cost of the most costly outgoing edge of $Z_{v,0}$ that is used in the solution of I_{DST}. If the solution of I_{DST} is ρ-approximate, the solution obtained for I_{MEMT} in this way is ρ-approximate as well. Using the approximation algorithm for DST presented in Reference 16, we obtain the following:

Theorem 39.5 *For any $\epsilon > 0$, there exists an $O(|D|^\epsilon)$ approximation algorithm for MEMT in asymmetric wireless networks.*

Note that the algorithm of Liang for approximating MEMT actually computes a solution to an instance of DST with $O(n^2)$ nodes. This means that a polylogarithmic approximation algorithm for DST would immediately yield polylogarithmic approximation algorithms for MEMT.

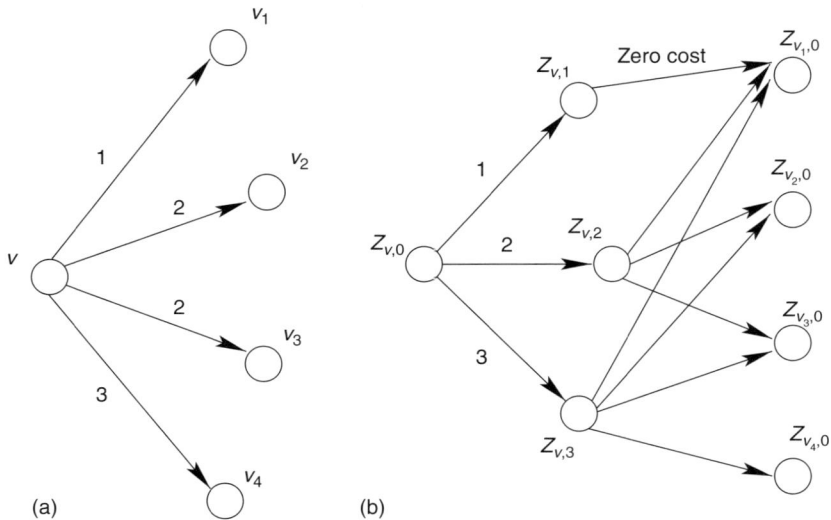

FIGURE 39.6 Liang's reduction of MEMT to DST. A node v and its outgoing edges in I_{MEMT} (a) and the corresponding structure in I_{DST} (b). All edges in (b) directed out of $Z_{v,1}, Z_{v,2}$, and $Z_{v,3}$ have zero cost.

Surprisingly, MEBT admits logarithmic approximations. This was independently proved in References 9 and 14. The algorithm presented in Reference 9 uses sophisticated set covering arguments. We briefly discuss the algorithm in Reference 14 that uses Liang's reduction and an algorithm of Zosin and Khuller [43] that efficiently approximates special instances of DST.

The algorithm in Reference 43 approximates I_{DST} by repeatedly solving instances of the minimum density directed tree (MDDT) problem. An instance of MDDT is defined in the same way as instances of DST, and the objective is to compute a tree directed out of the root node such that the ratio of the cost of the tree over the number of terminals it spans is minimized. The algorithm in Reference 43 repeatedly solves instances I_{MDDT}^i of MDDT derived by the instance I_{DST}. The instance I_{MDDT}^1 is defined by the graph H with edge cost function c, the set of terminals $D_1 = D$, and the root node $r_1 = r'$. Initially, the algorithm sets $i = 1$. While $D_i \neq \emptyset$, it repeats the following. It finds a solution T to I_{MDDT}^i that consists of a tree $T_i = (V(T_i), E(T_i))$, defines the instance I_{MDDT}^{i+1} by contracting the nodes of T_i into the root node r_{i+1} and by setting $D_{i+1} = D_i \setminus V(T_i)$, and increments i by 1.

Using standard arguments in the analysis of set covering problems, Reference 43 shows that if the solution T_i is a ρ-approximate solution for I_{MDDT}^i in each iteration i, then the union of the trees T_i computed in all iterations is an $O(\rho \ln n)$-approximate solution for I_{DST}. They also show how to find a $(d + 1)$-approximate solution for I_{MDDT}^i if the graph obtained when removing the terminals from G has depth d. Observe that, given an instance I_{MEBT} of MEBT, the graph H obtained by applying the reduction of Liang is bipartite, since there is no edge between nodes of $D \cup \{r'\}$ and between nodes of $V - (D \cup \{r'\})$. Thus, the graph obtained by removing the terminals of D from H has depth 1. Following the reasoning presented in Reference 43 and the reduction of Liang, we obtain a logarithmic approximation algorithm for MEBT.

Theorem 39.6 *There exists an $O(\ln n)$-approximation algorithm for MEBT in asymmetric wireless networks.*

We now present a method for approximating MESSCS. Let I_{MESSCS} be an instance of I_{MESSCS} that consists of a complete directed graph $G = (V, E)$ with edge cost function $c : E \to R^+$ and a set of terminals $D \subseteq V$. Pick an arbitrary node $v_0 \in D$ and let I_{MEMT} and I_{MEIMT} be the instances of MEMT and MEIMT, respectively, consisting of the graph G with edge cost function c, the root node v_0, and the set of terminals $D - \{v_0\}$.

Assume that we have weight assignments w_1 and w_2 to the nodes of V which are solutions for I_{MEMT} and I_{MEIMT}, respectively. Construct the weight assignment w_3 defined as $w_3(u) = \max\{w_1(u), w_2(u)\}$, for every $u \in V$. An example of this construction is presented in Figure 39.7. The following statement is proved in References 9 and 14.

Lemma 39.3 *If the weight assignments w_1 and w_2 are ρ_1 and ρ_2 approximate solutions for I_{MEMT} and I_{MEIMT}, respectively, then the weight assignment w_3 is a $(\rho_1 + \rho_2)$-approximate solution to I_{MESSCS}.*

Hence, we can solve I_{MEMT} and I_{MEIMT} using the reduction of Liang and the $O(|D|^\epsilon)$-approximation algorithm in Reference 16 for DST, in order to obtain the following result.

Theorem 39.7 *For any $\epsilon > 0$, there exists an $O(|D|^\epsilon)$-approximation algorithm for MESSCS in asymmetric wireless networks.*

Similarly, we can solve any instance of MESCS by solving an instance of MEBT (using the $O(\ln n)$-approximation algorithm described above) and an instance of MEIBT (this can be done optimally in polynomial time), and then merging the two solutions. In this way, we obtain the following result.

Theorem 39.8 *There exists an $O(\ln n)$-approximation algorithm for MESCS in asymmetric wireless networks.*

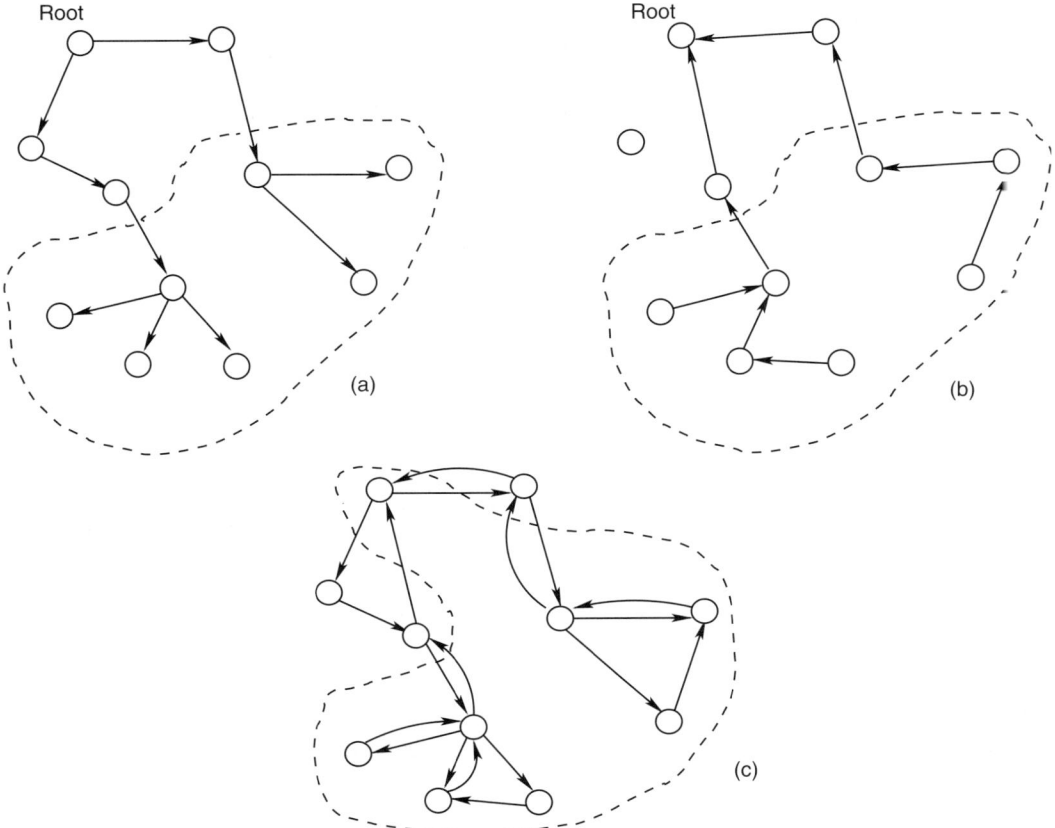

FIGURE 39.7 An example of combining solutions for I_{MEMT} (a) and I_{MEIMT} (b) in order to construct a solution for I_{MESSCS} (c). Dashed closed lines indicate the sets of terminals.

Intuitive reductions of problems MEBT and MESCS to SET COVER show that the results in Theorems 39.6 and 39.8 are tight [18].

39.3.2 Bidirected Connectivity Requirements

In this section, we present a logarithmic approximation algorithm for bidirected MESS in asymmetric wireless networks from Reference 14. We point out that a logarithmic approximation algorithm for the special case of bidirected MESCS based on set covering techniques was obtained independently in Reference 9.

The algorithm in Reference 14 is substantially simpler and uses a reduction of instances of bidirected MESS to instances of NWSF. The main idea behind this reduction is similar to the one used to approximate MEMT in symmetric wireless networks in Section 39.2.2. However, both the construction and the analysis have subtle differences.

Consider an instance $I_{b\text{MESS}}$ of bidirected MESS that consists of a complete directed graph $G = (V, E)$, an edge cost function $c : E \rightarrow R^+$, and a set of terminals $D \subseteq V$ partitioned into p disjoint subsets $D_1, D_2, ..., D_p$. We construct an instance I_{NWSF} of NWSF consisting of an undirected graph $H = (U, A)$, a node weight function $c' : U \rightarrow R^+$, and a set of terminals $D' \subseteq U$ partitioned into p disjoint sets $D'_1, D'_2, \ldots, D'_p$. For a node $v \in V$, we denote by n_v the number of different edge costs in the edges directed out of v, and, for $i = 1, \ldots, n_v$, we denote by $X_i(v)$ the ith smallest edge cost among the edges directed out of v. The set of nodes U consists of n disjoint sets of nodes called *supernodes*. Each supernode corresponds to a node of V. The supernode Z_v corresponding to node $v \in V$ has the following $n_v + 1$ nodes: a *hub node* $Z_{v,0}$ and n_v *bridge nodes* $Z_{v,1}, \ldots, Z_{v,n_v}$. For each pair of nodes $u, v \in V$, the set of edges A contains an edge between the bridge nodes $Z_{v,i}$ and $Z_{u,j}$ such that $X_i(u) \geq c(u, v)$ and $X_j(v) \geq c(v, u)$. Also, for each node $v \in V$, A contains an edge between the hub node $Z_{v,0}$ and each bridge node $Z_{v,i}$, for $i = 1, \ldots, n_v$. The cost function c' is defined as $c'(Z_{v,0}) = 0$ for the hub nodes and as $c'(Z_{v,i}) = X_i(v)$ for $i = 1, \ldots, n_v$, for the bridge nodes. The set of terminals D' is defined as $D' = \cup_i D'_i$, where $D'_i = \{Z_{v,0} \in U | v \in D_i\}$. An example of this reduction is depicted in Figure 39.8.

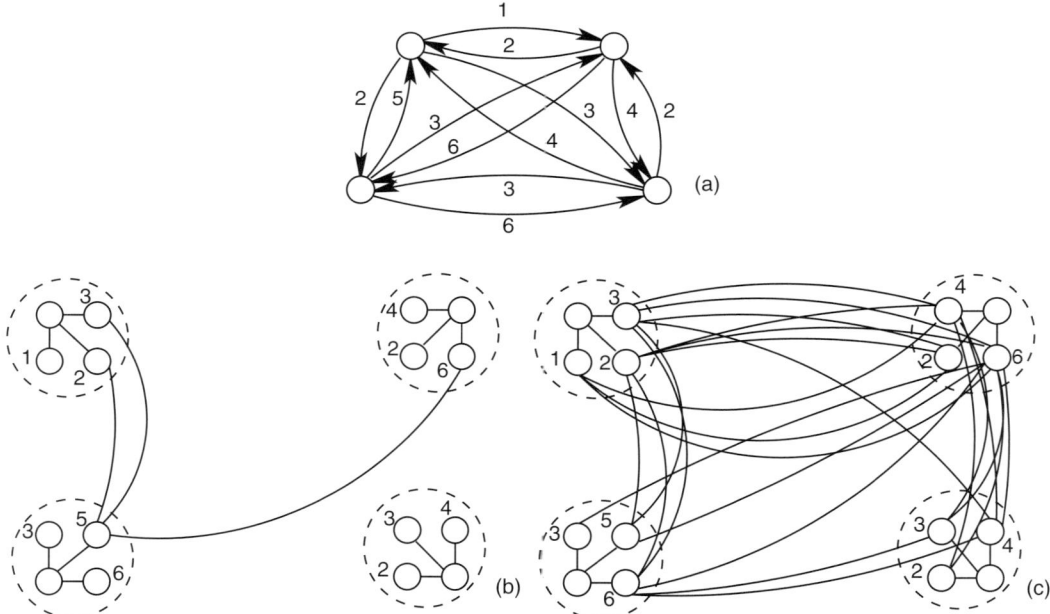

FIGURE 39.8 The reduction to Node-Weighted Steiner Forest. (a) The graph G of an instance of bMESS. (b) The graph H of the corresponding instance of NWSF. Each large cycle indicates a supernode. Only the edges incident to the node of weight 5 of the lower left supernode are shown. (c) The graph H of the corresponding instance of NWSF.

Consider a subgraph $F = (S, A')$ of H, which is a solution for I_{NWSF}. We construct a weight assignment w on the nodes of G by setting $w(v) = 0$, if S contains no node from supernode Z_v, and $w(v) = \max_{u \in (Z_v \cap S)} c'(u)$, otherwise. The next statement is proved in Reference 14 and relates the quality of the two solutions.

Lemma 39.4 *If F is a ρ-approximate solution to I_{NWSF}, then w is a ρ-approximate solution to I_{bMESS}.*

Guha and Khuller [26] present a 1.61 ln k-approximation algorithm for NWSF, where k is the number of terminals in the graph. Using this algorithm to solve I_{NWSF}, we obtain a solution of I_{bMESS} that is within 1.61 ln $|D|$ of optimal. Moreover, when $p = 1$ (i.e., when I_{bMESS} is actually an instance of bidirected MESSCS), the instance I_{NWSF} is actually an instance of NWST that can be approximated within 1.35 ln k, where k is the number of terminals in the graph [26]. The next theorem summarizes the discussion of this section.

Theorem 39.9 *There exist an 1.61 ln $|D|$-, an 1.35 ln $|D|$-, and an 1.35 ln n-approximation algorithm for bidirected MESS, bidirected MESSCS, and bidirected MESCS in asymmetric wireless networks, respectively.*

39.4 The Geometric Model

In this section, we survey results in geometric wireless networks. Recall that, in these networks, nodes of G correspond to points in a Euclidean space, and the cost of an edge (u, v) is defined as the Euclidean distance between u and v raised to a fixed power α ranging from 1 to 6, (i.e., $c(u, v) = d(u, v)^\alpha$). As we will show in the following, finding an optimal solution becomes easier if nodes are restricted to be placed on a line, while for higher dimensions almost all connectivity requirements lead to hard problems. The most important cases and the best-examined ones are those of MEBT and MESCS.

39.4.1 The Linear Case

A linear wireless network consists of n points on a line having coordinates $x_1 \leq x_2 \leq \cdots \leq x_n$. For MEBT, when we are also given a special node x_r that is the root, the following theorem has been proved independently in References 13 and 19.

Theorem 39.10 *MEBT can be solved in polynomial time in linear wireless networks for any $\alpha \geq 1$.*

The main idea of the corresponding algorithms is the exploitation of structural properties of the optimal solution in order to drastically reduce the search space. More specifically, if we partition the set of nodes into two sets (called *left* and *right*) depending on their position on the line with respect to the root, the transmission graph corresponding to the optimal solution has at most one node that reaches nodes both from the left and the right set. Such a node is called *root-crossing*. The main idea in the proof of this fact is that if there exist $k \geq 2$ root-crossing nodes, then a solution having no greater cost with $k-1$ root-crossing nodes also exists. Examples of situations with two root-crossing nodes are depicted in Figure 39.9. Thus, we can always reduce the number of root-crossing nodes to one without increasing the total cost. So, all possible transmission graphs that are candidates to be the optimal solution can be examined in polynomial time.

For MESCS, an optimal algorithm running in time $O(n^4)$ is presented in Reference 30. This algorithm relies on the use of dynamic programming.

Theorem 39.11 *MESCS can be solved in polynomial time in linear wireless networks for any $\alpha \geq 1$.*

The main idea of the algorithm is that the optimal solution can be computed in a recursive way. Unfortunately, the assumption that starting from an optimal solution for k points $x_1, \ldots, x_k$ we can

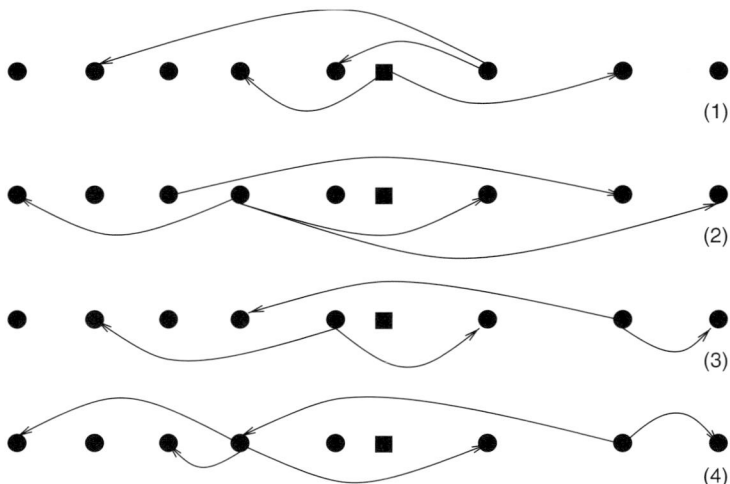

FIGURE 39.9 The four possible cases we have to consider when there are two root-crossing nodes. Only edges directed out of root-crossing nodes are shown. The root is the squared node. In all cases, we can reduce the number of root-crossing nodes to 1.

extend it to include point x_{k+1} does not work, since the energy assigned to point x_{k+1} may lead to a situation where the energy previously assigned to some points in $\{x_1, \ldots, x_k\}$ should be reduced. On the positive side, the following stronger recursive statements are proved in Reference 30. For any $l \geq k$ and any $i \leq k$, there is an assignment that has minimum cost among the assignments with the following properties:

1. There is a path between any pair in $\{x_1, \ldots, x_k\}$ in the transmission graph.
2. x_l is within the reach of a node in $\{x_1, \ldots, x_k\}$.
3. In the transmission graph, any backward edge from x_k to x_i is free of cost. These edges enable connectivity without adding to the cost.

Using the above statement, starting from an empty solution we gradually extend the assignment until covering all nodes, thus obtaining an optimal solution.

39.4.2 Multidimensional Wireless Networks

Almost all connectivity requirements besides MEIBT lead to NP-hard problems when graph G consists of points on a d-dimensional space for $d \geq 2$ and $\alpha \geq d$. Most of the theoretical and experimental work in this framework is for MEMT and MEBT, and especially for the case where the points are located on a Euclidean plane (i.e., $d = 2$). Note that in the case $\alpha = 1$ an optimal solution that consists of assigning sufficient energy to the root so that it reaches the node that is the furthest away can be computed for any d.

The following theorem was presented in Reference 18.

Theorem 39.12 *For any $d \geq 2$ and any $\alpha \geq d$, MEBT is NP-hard.*

Naturally, this result led to the design and analysis of approximation algorithms. The majority of research has focused on the case $\alpha = d = 2$ and, unless stated otherwise, the bounds discussed in the following correspond to this setting. The first algorithms were proposed in the seminal work of Wieselthier et al. [41]. These algorithms are based on the construction of MST and shortest path trees (SPT) on the graph representing the network. The energy assigned to each node is then the minimum required in order to be able to reach its neighbors in the tree. The approach followed in Reference 41 for computing solutions of MEMT was to prune the trees obtained in solutions of MEBT. Another algorithm presented in Reference 41, called *Broadcast Incremental Algorithm* (BIP), constructs a tree starting from the root and gradually augmenting it by adding the node that can be reached with the minimum additional energy from some node already in the tree. BIP can be seen as a node version of Dijkstra's algorithm for computing SPT, with one fundamental difference on the operation whenever a new node is added. Whereas Dijkstra's algorithm updates the node weights (representing distances), BIP updates the edge costs (representing the additional energy required in order to reach a node out of the tree). This update is performed by subtracting the cost of the added edge from the cost of every edge from the source node of the added edge to any node that is not yet included in the tree.

Experimental results presented in Reference 41 showed that BIP outperforms algorithms MST and SPT. In subsequent work, Wan et al. [39] study the algorithms presented in Reference 41 in terms of efficiency in approximating the optimal solution. Their main result is an upper bound of 12 on the approximation ratio of algorithm MST. Slightly weaker approximation bounds for MST have been presented in Reference 18. In Reference 39, it is also proved that the approximation ratio of BIP is not worse than that of MST, and that other intuitive algorithms have very poor approximation ratio. The upper bound in Reference 39 for MST was improved in References 23 and 35. Recently, Ambühl [2] proved an upper bound of 6. This result is tight since there exists a corresponding lower bound [39] presented in Figure 39.10. We should also note that there exists a 13/3 lower bound on the approximation ratio of BIP.

Theorem 39.13 *For* $\alpha = d = 2$, MST *and* BIP *are 6-approximation algorithms for MEBT.*

This result implies a constant approximation algorithm of 6ρ for MEMT as well [40], where ρ is the approximation ratio for ST (currently, $\rho \leq 1 + \ln 3/2$ [38]). For the more general case of arbitrary d and $\alpha \geq d$, the authors of Reference 23 prove a $3^d - 1$ upper bound on the performance of MST that holds independently of α. Very recently, for the case $d = 3$, Navarra [36] presented an 18.8-approximation algorithm that is an improvement over the 26 bound that stems from the aforementioned formula. In this case, there is still a significant gap since the best lower bound is 12.

On the other side, several intuitive algorithms have been experimentally proved to work very well on random instances of MEBT and MEMT. In References 33 and 42, algorithms based on shortest paths are enhanced with the *potential power saving* idea. These algorithms examine whether establishing a new path could also include nodes that had been included in the multicast tree in previous phases, and could now be connected to the multicast tree as children of some node in the path. In this way, the energy of some nodes could be decreased. Čagalj et al. [8] introduced a heuristic called *embedded wireless multicast advantage*

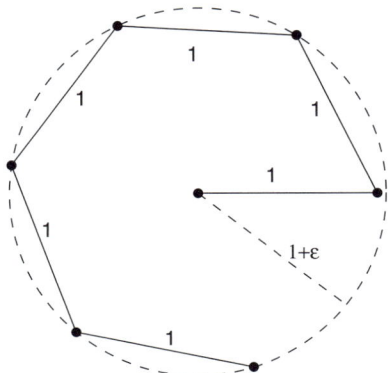

FIGURE 39.10 The lower bound on the performance of MST when $\alpha = d = 2$. The node in the center is the root; an optimal solution would assign energy $(1 + \epsilon)^2$ to it. Solid lines represent the minimum spanning tree that leads to a solution of total energy 6.

(EWMA) for computing efficient solutions to MEBT instances. Starting with a broadcast tree, EWMA "walks" on broadcast trees by performing the following two types of changes in each step: (i) outgoing edges are added to a single node v; this node is said to be *extended* and (ii) all outgoing edges are removed from some descendants of v; in this case we say that the particular descendants of v are *excluded*. Define the *gain* of a node v as the decrease in the energy of the broadcast tree obtained by excluding some of the nodes of the tree in exchange for the increase in node v's energy in order to establish edges to all excluded nodes and their children. Intuitively, EWMA repeatedly examines the nodes of the tree and tries to make use of nodes with maximum gain, so as to make local modifications on the structure of the tree. As it was observed in Reference 4, EWMA can be easily converted to work for MEMT as well. Another heuristic called Sweep was proposed in Reference 41; this also takes as input a tree and transforms it to an energy-efficient tree by performing local improvements. Sweep works as follows. Starting from the broadcast tree, it proceeds in steps; in the ith step it examines node v_i. If for some nodes $v_{i_1}, v_{i_2}, \ldots$ that are not ancestors of v_i the energy of v_i in the broadcast tree is not smaller than the cost of all the edges from v_i to $v_{i_1}, v_{i_2}, \ldots$, Sweep removes the incoming edges of $v_{i_1}, v_{i_2}, \ldots$ and adds edges from v_i to $v_{i_1}, v_{i_2}, \ldots$ in the broadcast tree. The algorithm terminates when all nodes have been examined. Clearly, it can be used on any multicast tree as well.

Another issue of apparent importance is to design algorithms for MEMT that are amenable to implement in a distributed environment (e.g., References 8, 15, and 42). In Reference 4, a characterization of experimental algorithms is presented, while the authors introduced some algorithms that establish *dense shortest paths*, that is, they add to the solution the shortest path that has the lowest ratio of additional energy over the number of newly added nodes. Experimental comparison between these algorithms and already existing ones suggests that density is a useful property.

MESCS has received less attention. A first proof that MESCS is NP-hard for d-dimensional Euclidean spaces appeared in Reference 30 for the case $d \geq 3$. This negative result was strengthened to APX-hardness while the problem was proved to be NP-hard also for $d = 2$ in Reference 21. The above proofs assume that $\alpha \geq 2$, while in Reference 24 MESCS is also proved to be NP-hard for $\alpha = 1$.

Theorem 39.14 *MESCS is NP-hard for any $d \geq 2$ and any $\alpha \geq 1$, and APX-hard for $d \geq 3$ and $\alpha \geq 2$.*

A simple 2-approximation algorithm based on MST was presented in Reference 30. Essentially, this is the algorithm we discussed in Section 39.2.1 in a more general setting. Again, it is not hard to show that the total energy is at most twice the cost of the MST, the latter being a lower bound on cost of the optimal solution.

When the transmission graph is required to contain a bidirected subgraph satisfying the connectivity requirements, Althaus et al. [1] present a 5/3-approximation algorithm, by establishing a connection between bidirected MESCS and k-restricted Steiner trees, and using a 5/3-approximation algorithm for the latter problem. This bound cannot be obtained by the algorithm in Reference 30. As the authors in Reference 1 note, the cost of an optimal solution for MESCS can be half the cost of an optimal solution for bidirected MESCS. Consider a set of $n^2 + n$ nodes, consisting of n groups of $n + 1$ nodes each, that are located on the sides of a regular $2n$-gon. Each group has 2 "thick" nodes in distance 1 of each other and $n - 1$ equally spaced nodes the line segment between them. It is easy to see that an optimal solution for MESCS assigns energy 1 to the one thick node in each group and an amount of energy equal to $\epsilon^2 = (1/n)^2$ to all other nodes in the group. The total energy then equals $n + 1$. For bidirected MESCS it is necessary to assign energy equal to 1 to all but two of the thick nodes, and of ϵ^2 to the remaining nodes, which results in a total energy of $2n - 1 - 1/n + 2/n^2$. An example when $\alpha = 2$ and $n = 3$ is depicted in Figure 39.11.

39.5 Extensions

The connectivity requirements we have considered in this work can be defined by $0 - 1$ requirement matrices. A natural extension is to consider matrices with nonnegative integer entries r_{ij} denoting that

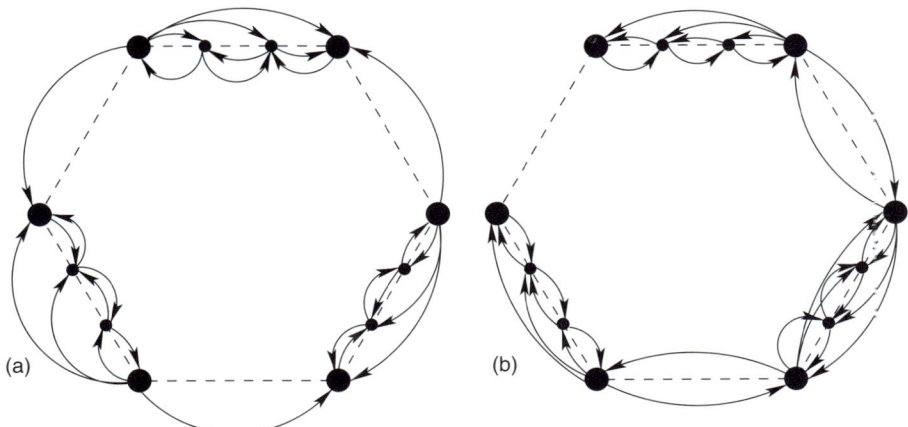

FIGURE 39.11 The transmission graphs of the optimal solution for MESCS (a) and of the optimal solution for bidirected MESCS (b) in the counterexample presented in Reference 1 for $n = 3$. Extending this construction to a regular $2n$-gon proves that the cost of (b) can be twice the cost of (a).

r_{ij} node/edge-disjoint paths are required from node v_i to node v_j. This extension leads to combinatorial problems that capture important engineering questions related to the design of *fault-tolerant* ad hoc wireless networks. In this direction, some results on symmetric wireless networks are presented in References 12, 28, and 31. More specifically, Lloyd et al. [31] examined the problem of establishing a transmission graph containing a bidirected subgraph that is biconnected with respect to both the nodes and the edges. For both cases, they presented 8-approximation algorithms by using algorithms for network design. These results were improved by Călinescu and Wan [12] and were extended to hold also for the case when the transmission graph is not required to contain a bidirected subgraph. In addition, Călinescu and Wan [12] extended the problem to the case of k-connectivity. For the latter case, they presented corresponding $2k$-approximation algorithms, both with respect to nodeconnectivity and edgeconnectivity. Hajiaghayi et al. [28] focus on bidirected MESCS and prove APX-hardness for k-node connectivity and k-edge connectivity and present an $O(\log^4 n)$-approximation algorithm for the first case and an $O(\sqrt{n})$-approximation algorithm for the second case, improving on the previous results for large values of k.

Another direction of research is when the transmission graph is required to have a bounded diameter. Thus, *bounded-hop* versions of the problems studied in this chapter arise. Călinescu et al. [10] examine bounded-hops MEBT and MESCS both for the general and the geometric model. For MEBT, they present an $(O(\log n), O(\log n))$-bicriteria approximation algorithm, that is, the transmission graph has depth at most $O(h \log n)$ and total energy at most $O(\log n)$ times the optimal solution, where h is the bound on the number of hops. For the geometric model, their algorithm can be modified and achieve an approximation ratio of $O(\log^\alpha n)$. Ambühl et al. [3] focus on bounded-hop MEBT in the Euclidean plane and present a polynomial time algorithm when $h = 2$, using dynamic programming and a polynomial time approximation scheme (PTAS) for any fixed bound h on the number of hops. Similarly, for MESCS in the general model, Călinescu et al. [10] presents an $(O(\log n), O(\log n))$-bicriteria approximation algorithm, which can be modified to achieve an $O(\log^{\alpha+1} n)$ approximation algorithm for the geometric model. For the geometric model and specifically for linear networks, Clementi et al. [20] present a polynomial algorithm that returns the optimal solution in time $O(n^3)$ when $h = 2$ and a 2-approximation algorithm for arbitrary values of h.

Several other extensions of the model described in this chapter are also interesting. In the *network lifetime* problem [9], the objective is to establish a communication pattern so that the time until the first node exhausts its available energy is maximized. When we drop the assumption that each node is equipped with an omnidirectional antenna, we obtain *directional* equivalents of the discussed problems. Finally, all the cases considered in this work make the assumption that interference [34] is not a concern and no transmission is lost owing to collisions. Excluding this assumption is also worth investigating.

References

[1] E. Althaus, G. Călinescu, I. Măndoiu, S. Prasad, N. Tchervenski, and A. Zelikovsky. Power Efficient Range Assignment in Ad Hoc Wireless Networks. *Wireless Networks*, 2006, to appear. Preliminary version in *Proc. of the IEEE Wireless Communications and Networking Conference (WCNC '03)*, IEEE Computer Society Press, pp. 1889–1894, 2003.

[2] C. Ambühl. An Optimal Bound for the MST Algorithm to Compute Energy Efficient Broadcast Trees in Wireless Networks. In *Proc. of 32nd International Colloquium on Automata, Languages and Programming (ICALP '05)*, pp. 1130–1150, 2005.

[3] C. Ambühl, A. E. F. Clementi, M. Di Ianni, N. Lev-Tov, A. Monti, D. Peleg, G. Rossi, R. Silvestri. Efficient Algorithms for Low-Energy Bounded-Hop Broadcast in Ad Hoc Wireless Networks. In *Proc. of the 21st Annual Symposium on Theoretical Aspects of Computer Science (STACS '04)*, LNCS 2996, Springer, pp. 418–427, 2004.

[4] S. Athanassopoulos, I. Caragiannis, C. Kaklamanis, and P. Kanellopoulos. Experimental Comparison of Algorithms for Energy-Efficient Multicasting in Ad Hoc Networks. In *Proc. of the 3rd International Conference on AD-HOC Networks & Wireless (ADHOC NOW '04)*, LNCS 3158, Springer, pp. 183–196, 2004.

[5] P. Berman and M. Karpinski. On Some Tighter Inapproximability Results. In *Proc. of the 26th International Colloquium on Automata, Languages, and Programming (ICALP '99)*, LNCS 1644, Springer, pp. 200–209, 1999.

[6] F. Bian, A. Goel, C. Raghavendra, and X. Li. Energy-Efficient Broadcast in Wireless Ad Hoc Networks: Lower Bounds and Algorithms. *Journal of Interconnection Networks*, 149–166, 2002.

[7] V. Biló and G. Melideo. An Improved Approximation Algorithm for the Minimum Energy Consumption Broadcast Subgraph. In *Proc. of Euro-Par 2004—Parallel Processing*, LNCS 3149, Springer, pp. 949–956, 2004.

[8] M. Čagalj, J.-P. Hubaux, and C. Enz. Energy-Efficient Broadcasting in All-Wireless Networks. 11 *Wireless Networks*, pp. 177–188, 2005.

[9] G. Călinescu, S. Kapoor, A. Olshevsky, and A. Zelikovsky. Network Lifetime and Power Assignment in Ad Hoc Wireless Networks. In *Proc. of the 11th Annual European Symposium on Algorithms (ESA '03)*, LNCS 2832, Springer, pp. 114–126, 2003.

[10] G. Călinescu, S. Kapoor, and M. Sarwat. Bounded Hops Power Assignment in Ad Hoc Wireless Networks. *Discrete Applied Mathematics*, 154(9), 1358–1371, 2006.

[11] G. Călinescu, I. Măndoiu, and A. Zelikovsky. Symmetric Connectivity with Minimum Power Consumption in Radio Networks. In *Proc. of the 2nd IFIP International Conference on Theoretical Computer Science*, pp. 119–130, 2002.

[12] G. Călinescu and P.-J. Wan. Range Assignment for High Connectivity in Wireless Ad Hoc Networks. In *Proc. of the 2nd International Conference on Ad Hoc, Mobile, and Wireless Networks (ADHOC-NOW '03)*, LNCS 2865, Springer, pp. 235–246, 2003.

[13] I. Caragiannis, C. Kaklamanis, and P. Kanellopoulos. New Results for Energy-Efficient Broadcasting in Wireless Networks. In *Proc. of the 13th Annual International Symposium on Algorithms and Computation (ISAAC '02)*, LNCS 2518, Springer, pp. 332–343, 2002.

[14] I. Caragiannis, C. Kaklamanis, and P. Kanellopoulos. Energy Efficient Wireless Network Design. *Theory of Computing Systems*, 39(5), 593–617, 2006.

[15] J. Cartigny, D. Simplot-Ryl, and I. Stojmenovic. Localized Minimum-Energy Broadcasting in Ad Hoc Networks. In *Proc. of the 22nd Annual Joint Conference of the IEEE Computer and Communications Societies (INFOCOM '03)*, 2003.

[16] M. Charikar, C. Chekuri, T.-Y. Cheung, Z. Dai, A. Goel, S. Guha, and M. Li. Approximation Algorithms for Directed Steiner Problems. *Journal of Algorithms*, 33, 73–91, 1999.

[17] M. Chlebík and J. Chlebíková. Approximation Hardness of the Steiner Tree Problem on Graphs. In *Proc. of the 8th Scandinavian Workshop on Algorithm Theory (SWAT '02)*, LNCS 2368, Springer, pp. 170–179, 2002.

[18] A. E. F. Clementi, P. Crescenzi, P. Penna, G. Rossi, and P. Vocca. On the Complexity of Computing Minimum Energy Consumption Broadcast Subgraphs. In *Proc. of the 18th Annual Symposium on Theoretical Aspects of Computer Science (STACS '01)*, LNCS 2010, Springer, pp. 121–131, 2001.

[19] A. E. F. Clementi, M. Di Ianni, and R. Silvestri. The Minimum Broadcast Range Assignment Problem on Linear Multi-hop Wireless networks. *Theoretical Computer Science*, 299, 751–761, 2003.

[20] A. E. F. Clementi, A. Ferreira, P. Penna, S. Perennes, and R. Silvestri. The Minimum Range Assignment Problem on Linear Radio Networks. *Algorithmica*, 35, 95–110, 2003.

[21] A. E. F. Clementi, P. Penna, and R. Silvestri. On the Power Assignment Problem in Radio Networks. *Mobile Network and Applications*, 9, 125–140, 2004.

[22] J. Edmonds. Optimum Branchings. *Journal of Research of the National Bureau of Standards*, 71B, 233–240, 1967.

[23] M. Flammini, A. Navarra, R. Klasing, and S. Perennes. Improved Approximation Results for the Minimum Energy Broadcasting Problem. In *Proc. of the DIALM-POMC Joint Workshop on Foundations of Mobile Computing (DIALM-POMC '04)*, pp. 85–91, 2004.

[24] B. Fuchs. On the Hardness of Range Assignment Problems. Unpublished manuscript. 2005.

[25] M. X. Goemans and D. P. Williamson. A General Approximation Technique for Constrained Forest Problems. *SIAM Journal on Computing*, 24, 296–317, 1995.

[26] S. Guha and S. Khuller. Improved Methods for Approximating Node Weighted Steiner Trees and Connected Dominating Sets. *Information and Computation*, 150, pp. 57–74, 1999.

[27] M. Hajiaghayi, N. Immorlica, and V. Mirrokni. Power Optimization in Fault-Tolerant Topology Control Algorithms for Wireless Multi-hop Networks. In *Proc. of the 9th ACM International Conference on Mobile Computing and Networking (MOBICOM '03)*, pp. 300–312, 2003.

[28] M. Hajiaghayi, G. Kortsarz, V. Mirrokni, and Z. Nutov. Power Optimization for Connectivity Problems. In *Proc. of the 11th International Conference on Integer Programming and Combinatorial Optimization (IPCO '05)*, pp. 349–361, 2005.

[29] E. Halperin and R. Krauthgamer. Polylogarithmic Inapproximability. In *Proc. of the 35th Annual ACM Symposium on Theory of Computing (STOC '03)*, pp. 585–594, 2003.

[30] L. M. Kirousis, E. Kranakis, D. Krizanc, and A. Pelc. Power Consumption in Packet Radio Networks. *Theoretical Computer Science*, 243, pp. 289–305, 2000.

[31] E. Lloyd, R. Liu, M. Marathe, R. Ramanathan and S. S. Ravi. Algorithmic Aspects of Topology Control Problems for Ad Hoc Networks. In *Proc. of 3rd ACM International Symposium on Mobile Ad Hoc Networking and Computing (MOBIHOC '02)*, pp. 123–134, 2002.

[32] W. Liang. Constructing Minimum-Energy Broadcast Trees in Wireless Ad Hoc Networks. In *Proc. of 3rd ACM International Symposium on Mobile Ad Hoc Networking and Computing (MOBIHOC '02)*, pp. 112–122, 2002.

[33] P. Mavinkurve, H. Ngo, and H. Mehra. MIP3S: Algorithms for Power-Conserving Multicasting in Wireless Ad Hoc Networks. In *Proc. of the 11th IEEE International Conference on Networks (ICON '03)*, 2003.

[34] T. Moscibroda and R. Wattenhofer. Minimizing Interference in Ad Hoc and Sensor Networks. In *Proc. of the 3rd ACM Join Workshop on Foundations of Mobile Computing (DIALM – POMC)*, pp. 24–33, 2005.

[35] A. Navarra. Tighter Bounds for the Minimum Energy Broadcasting Problem. In *Proc. of the 3rd International Symposium on Modeling and Optimization in Mobile, Ad Hoc, and Wireless Networks (WiOpt '05)*, pp. 313–322, 2005.

[36] A. Navarra. 3-D Minimum Energy Broadcasting. In *Proc. of the 13th Colloquium on Structural Information and Communication Complexity (SIROCCO '06)*, LNCS 4056, Springer, pp. 240–252, 2006.

[37] T. S. Rappaport. *Wireless Communications: Principles and Practices*. Prentice Hall, 1996.

[38] G. Robins and A. Zelikovsky. Improved Steiner Tree Approximations in Graphs. In *Proc. of the 11th Annual ACM-SIAM Symposium on Discrete Algorithms (SODA '00)*, pp. 770–779, 2000.

[39] P.-J. Wan, G. Călinescu, X.-Y. Li, and O. Frieder. Minimum-Energy Broadcasting in Static Ad Hoc Wireless Networks. *Wireless Networks*, 8, pp. 607–617, 2002.

[40] P.-J. Wan, G. Călinescu, and C.-W. Yi. Minimum-Power Multicast Routing in Static Ad Hoc Wireless Networks. *IEEE/ACM Transactions on Networking*, 12, 507–514, 2004.

[41] J. E. Wieselthier, G. D. Nguyen, and A. Ephremides. On the Construction of Energy-Efficient Broadcast and Multicast Trees in Wireless Networks. In *Proc. of IEEE INFOCOM 2000*, pp. 585–594.

[42] V. Verma, A. Chandak, and H. Q. Ngo. DIP3S: A Distributive Routing Algorithm for Power-Conserving Broadcasting in Wireless Ad Hoc Networks. In *Proc. of the Fifth IFIP-TC6 International Conference on Mobile and Wireless Communications Networks (MWCN '03)*, pp. 159–162, 2003.

[43] L. Zosin and S. Khuller. On Directed Steiner Trees. In *Proc. of the 13th Annual ACM/SIAM Symposium on Discrete Algorithms (SODA '02)*, pp. 59–63, 2002.

40
Power Aware Mapping of Real-Time Tasks to Multiprocessors

	40.1	Introduction...	**40**-1
		Overview of Power Aware Scheduling for Parallel Systems	
	40.2	Variable Speed Processors and Power Management ..	**40**-4
	40.3	Frame-Based RT-Applications and Canonical Schedule	**40**-4
		Frame-Based Real-Time Applications • Task Scheduling for Parallel Systems • Canonical Schedule and Static Power Management	
	40.4	Power Aware Mapping with Slack Sharing............	**40**-8
		Greedy Slack Reclamation • Slack Sharing among Processors	
	40.5	Fixed Priority Slack Sharing for Dependent Tasks.....	**40**-12
	40.6	Slack Shifting for Applications with Different Execution Paths	**40**-13
Dakai Zhu		AND/OR Application Model • Shifting Schedule Sections in Canonical Schedule • Speculation for Better Performance	
University of Texas			
Bruce R. Childers	40.7	Practical Considerations	**40**-16
Daniel Mossé		Slack Reservation for Speed Adjustment Overhead	
Rami Melhem	40.8	Summary	**40**-18
University of Pittsburgh		References	**40**-19

40.1 Introduction

The performance of modern processors has increased at the expense of drastically increased power consumption. The increased power consumption not only reduces the operation time for battery-powered embedded systems (such as PDAs and laptops), but also increases the sophistication/cost of the cooling infrastructures for dense clusters (such as web servers). Therefore, energy has been promoted to be a first-class resource in a system [45] and power aware computing has emerged to be a major research area, which aims at "using the right amount of energy in the right place at the right time" [23]. Various studies have been conducted to manage power consumption of different components in a system (such as CPU, memory, disk, and network interfaces), interested readers are referred to References 16 and 20 for more comprehensive information. In this chapter, we focus on CPU power consumption when executing a set of real-time tasks on multiprocessor platforms.

As the simplest scheme, a system can finish the required computation as fast as possible and turn off the processors when the system is (and likely to stay) idle to save energy. However, this shutdown technique is suboptimal even with the perfect knowledge of idle intervals in a schedule. The reason comes from the convex relation between CPU speeds and its power consumption [7, 10], which implies that a better approach should uniformly scale down the processing speed during the computation across all processors [4]. *Voltage scaling (VS)*, as one powerful and energy efficient technique, scales down the CPU speed and supply voltage simultaneously for lower performance requirements to get more energy savings [40, 43]. However, with scaled processing speeds, an application takes more time to complete. For real-time tasks, which normally have timing constraints and have to finish the execution before their deadlines, special cares are needed when scaling down their processing speed for energy savings. On the basis of the VS technique, various power aware scheduling schemes have been studied for different task models on uniprocessor systems [2, 21, 36]. However, much less work has been done for power awareness in multiprocessor systems.

Multiprocessor real-time scheduling is one of the most extensively studied areas. There are two major approaches for scheduling/mapping real-time tasks to processors: *partition* and *global* scheduling [9, 11]. In partition scheduling, each task is assigned to a specific processor and, each processor has its own scheduler and fetches tasks for execution from its own queue. In global scheduling, all tasks are put in a global queue and processors select from the queue the task with the highest priority for execution.

For real-time systems, to ensure that tasks can meet their deadlines in the worst case scenario, scheduling decisions are normally based on the worst case execution time (WCET) of tasks. For different partitions or different priority assignments to tasks (therefore different orders of tasks in the global queue), the mappings of tasks to processors are different, which result in different schedule length (i.e., the total time needed for executing the tasks is different). It is well known that the optimal partition as well as optimal priority assignment in global scheduling for minimizing the schedule length are NP-hard in the strong sense [11]. Therefore, many heuristics (such as First Fit, Best Fit [6], and Longest Task First (LTF) [35]) have been studied to efficiently obtain close-to-optimal schedules. Moreover, considering the runtime behaviors (which normally vary substantially [13]) of real-time tasks, optimal priority assignment in terms of schedule length does not necessarily lead to minimum energy consumption [47].

Despite the simplicity of exploiting the well-studied uniprocessor energy management schemes on individual processors, the fixed mapping of tasks to processors in partition scheduling limits the power management opportunity across different processors. Hence, the power aware mapping introduced in this chapter will base on global scheduling. Moreover, instead of designing new heuristics on priority assignment, we adopt LTF heuristic when assigning priorities to tasks. First, more slack may be expected from the execution of longer tasks and thus can be reclaimed by remaining tasks for more energy savings. Second, the ratio of the schedule length under LTF heuristic over the one under optimal priority assignment is bounded by a constant [35].

For the LTF priority assignment, if the resulting schedule completes all tasks well before the deadline in the worst case, *static slack* exists. Following the idea of static power management (SPM) for uniprocessors, we can uniformly stretch the schedule and scale down the processing speed to finish all tasks *just-in-time* and obtain energy savings [17]. However, for parallel systems, owing to the possible dependencies among tasks, gaps exist in the schedule and result in different parallelism for different sections in a schedule. A better approach should take such parallelism into consideration when exploiting the static slack for more energy savings [32].

In addition, *dynamic slack* can be expected online owing to the runtime behaviors of real-time tasks, which normally only use a small fraction of their WCETs [13]. The efficient power aware mapping should also take such slack into consideration and adjust the mapping of tasks to processors at runtime for better performance on energy savings.

In this chapter, we first introduce an efficient static scheme for parallel applications, especially the ones with dependent tasks. The scheme explores different degrees of parallelism in a schedule and generates an energy efficient *canonical schedule*, which is defined as the one corresponding to the case where all tasks take their WCETs. Then, on the basis of global scheduling, we focus on the online schemes that exploit the runtime behaviors of tasks and dynamic slack for more energy savings. Specifically, for independent and dependent real-time tasks, the *slack sharing* power aware scheduling dynamically adjusts the mapping of tasks to processors at runtime. By sharing the dynamic slack across processors, the scheme intends to scale down the processors more uniformly with balanced *actual* workload and thus achieves better energy savings. For any given task priority assignment heuristic (e.g., LTF), if the *canonical schedule* can finish all tasks in time, the power aware mapping with slack sharing will meet tasks' deadlines as well. For applications with multiple execution paths consisting of different real-time tasks, *slack shifting* further explores slack time from the execution paths other than the longest one. Speculation schemes that exploit statistical characteristics about applications are discussed with the intention to scale down all tasks more evenly for better energy savings. Practical issues related to power management are addressed in the end, and the idea of *slack reservation* is introduced to incorporate CPU speed adjustment overhead.

40.1.1 Overview of Power Aware Scheduling for Parallel Systems

The research on power aware scheduling in real-time system can be traced back to the important seminar work of Weiser et al. [40] and Yao et al. [43], where the authors introduced the concept of VS for interval-based adjustment as well as task-based speed adjustment. For systems with processors running at different fixed speeds (and thus with different power profiles), several task assignment and scheduling schemes have been proposed to minimize system energy consumption while still meeting applications' deadlines, where applications are usually represented by directed acyclic graphs (DAG) [18, 27, 42].

For systems with variable speed processors, SPM can be accomplished by deciding beforehand the best speed for each processor given the fixed task sets and predictable execution times [17]. Gruian et al. [19] proposed a priority-based energy sensitive list scheduling heuristic to determine the amount of time allocated to each task, considering energy consumption and critical path timing requirement in the priority function. In Reference 46, Zhang et al. proposed a mapping heuristic for fixed task graphs to maximize the opportunities for VS algorithms, where the VS problem was formulated as an integer programming problem.

For independent periodic hard real-time tasks to be executed on identical multiprocessors, Aydin et al. address the problem of energy minimization for partition-based scheduling. The authors investigate the joint effects of partitioning heuristics on energy consumption and feasibility based on EDF scheduling [4] and rate-monotonic scheduling (RMS) [1], respectively. The system synthesis problem using global EDF is addressed by Baruah et al. [5], where utilization-based approach is used to determine the speed for each processor. When considering heterogeneous multiprocessors, the problem of power aware resource allocation is formulated as a Generalized Assignment Problem (GAP) by Yu et al. [44]. On the basis of integer linear programming (ILP), the authors propose a LR-heuristic-based solution with earliest deadline first (EDF) scheduling.

For distributed systems where communication cost is significant and task migration is prohibitive, Luo et al. proposed a static optimization algorithm by shifting the static schedule to redistribute static slack according to the average slack ratio on each processor element for periodic task graphs and aperiodic tasks. The scheme reduces energy consumption and response time for soft aperiodic tasks at run time [29]. The authors improved the static optimization by using critical path analysis and task execution order refinement to get the maximal static slow-down factor for each task [30], where a dynamic VS scheme is also proposed.

For web servers, where static power from other components may be significant, Elnozahy et al. [12] presented cluster-wide power management and evaluated policies that combine VS and turning on/off individual server nodes. The policies are further extended and evaluated by Xu et al. [41]. Sharma et al. investigated adaptive algorithms for dynamic VS in QoS-enabled web servers to minimize the energy consumption subject to service delay constraints [38]. For surveillance systems based on sensor network, Maniezzo et al. further explore the tradeoff between the energy consumption for computation and communication components [31].

40.2 Variable Speed Processors and Power Management

In this chapter, we consider a frame-based application to be executed on a shared memory multiprocessor system. The processors in the system are assumed to be homogeneous with identical performance-power characteristics and tasks in the application can be executed on any processor.

The power consumption of a processor when executing a task is dominated by its dynamic power dissipation P_d, which is quadratically related to the processor supply voltage and linearly related to its processing speed [7, 10]. With the knowledge of almost linear relation between processing speed and supply voltage, VS techniques reduce processor supply voltage for lower processing speeds [40,43]. From now on, we refer to *speed adjustment* as changing both CPU supply voltage and frequency. Therefore, the dynamic power dissipation of a processor can be simply modeled as

$$P_d = kf^3 \tag{40.1}$$

where k is a system dependent constant and f is the processor clock frequency (i.e., the processor speed). The maximum speed of processors is denoted by f_{max} and, without loss of generality, the processing speeds are normalized with respect to f_{max} (e.g., $f_{max} = 1.0$).

Note that energy is the integration of power over time. For a task with fixed amount of computation c, the energy consumption of executing the task at speed f is $E(f) = P_d \times (c/f) = k \times c \times f^2$. That is, although scaling down the processing speed linearly increases the processing time (i.e., c/f) of the task, it saves power as well as energy. The lower the speed is, the more energy will be saved. However, with lower speeds, a task takes more time to finish. For real-time tasks, the extended execution may cause deadline misses, and utmost care is needed to select the appropriate processing speed to meet the timing constraints and obtain the maximum energy savings.

For example, consider a task that, with the maximum speed f_{max}, needs 20 time units to complete its work. If we have 40 time units allocated to this task (e.g., the task has a deadline 40 time units away), we can reduce the processor speed and supply voltage by half while still finishing the task on time. The new power when executing the task would be: $P'_d = k \times (f_{max}/2)^3 = (1/8) \times P_d$ and the new energy consumption would be: $E' = P'_d \times 40 = \frac{1}{4} \times k \times f_{max}^3 \times 20 = \frac{1}{4} \times E$, where P_d is the power and E is the energy consumption of the task with normal execution at the maximum processor speed f_{max}.

When no task is available for execution, processors become idle and are put into a power savings sleep state. For simplicity, it is assumed that no power is consumed when processors are in sleep state. That is, we consider only dynamic power dissipation in this chapter. However, in addition to the dominated dynamic power, static leakage power consumption becomes more significant as the technology size shrinks [39]. Recent research shows that, the effect of static leakage power on energy management is equivalent to imposing a *minimum energy efficient speed* limitation. That is, if the processing speed is not scaled down below this limit, a task will consume less energy at lower processing speeds. Interested readers are referred to References 14, 26, 37, and 48 for more details.

40.3 Frame-Based RT-Applications and Canonical Schedule

40.3.1 Frame-Based Real-Time Applications

A frame-based real-time application consists of a set of tasks $\Gamma = \{T_1, \ldots, T_n\}$ and has a frame length D. Within each frame, the tasks are mapped to processors and the execution of all tasks needs to complete before the end of the frame (i.e., after the tasks are released at the beginning of one frame, they have a relative deadline D). The mapping of tasks to processors and the schedule within a frame are repeated for the following frames [28]. Because of the periodicity, we consider only the problem of scheduling the tasks in a single frame with the deadline D.

For each task $T_i \in \Gamma$, the estimated WCET is denoted by c_i. Note that the execution time of a task depends on the processing speed of the processor that executes the task. As variable speed processors are considered in the systems, we assume that the WCETs of tasks are based on the maximum processor speed f_{max}. When a task is executed at a scaled speed (e.g., f), its execution time (e.g., $c_i(f_{max}/f)$) will increase linearly. Moreover, we consider only nonpreemptive scheduling schemes; that is, a task will *run-to-completion* whenever it begins to execute.

The execution of one task may or may not depend on the results of other tasks, which corresponds to *dependent* and *independent* tasks, respectively. For independent tasks, all tasks are ready for execution at the beginning of each frame. Whereas the readiness of dependent tasks relies on the completion time of their predecessor(s).

In general, the dependencies among tasks are represented by a directed acyclic graphic (DAG), where the nodes represent tasks and the edges represent dependencies. There is an edge, $T_i \to T_j$, if and only if T_i is a *direct* predecessor of T_j, which means that T_j needs the data produced by T_i. The tasks without predecessors are referred to as *root tasks* and are ready for execution at the beginning of each frame. The communication between different tasks is achieved through accessing shared memory. Therefore, it is assumed that there is no communication cost even if tasks T_i and T_j are mapped onto different processors. After T_i finishes execution, the produced data is available *immediately* and T_j is *ready* to execute.

Consider an example with three dependent tasks as shown in Figure 40.1a, where tasks T_1 and T_2 are root tasks that will be ready for execution at the beginning of each frame. Task T_3 has two direct predecessors and depends on both task T_1 and task T_2. Only after T_1 and T_2 finish their execution, can task T_3 begin to run. For the two numbers associated with each node, the first one is the WCET of the corresponding task, and the second one is the task's *average case execution time (ACET)*, which will be used in the speculation schemes as discussed in Section 40.6.3.

40.3.2 Task Scheduling for Parallel Systems

Two approaches have been used traditionally to schedule a set of tasks on multiple processors: partition scheduling and global scheduling [9,11]. The mapping of tasks to processors is fixed in partition scheduling and there is a task queue for each processor. In contrast, global scheduling only has a global queue and the mapping of tasks to processors will depend on runtime information.

For partition-based scheduling, the well-developed uniprocessor power management schemes may be applied directly on each processor after assigning tasks to processors. However, it has been shown that, due to the convex relation between CPU speed and its power consumption, the optimal energy efficient schedule can *only* be obtained with perfect load balancing among all processors [4]. Considering the runtime behaviors of the tasks [13], the *actual* workload on each processor under partition scheduling may vary substantially. Therefore, the fixed assignment of tasks to processors under partition scheduling limits the opportunities of managing power consumption across different processors, which may lead to suboptimal energy efficient schemes. On the contrary, global scheduling has the merit of *automatically* balancing the workload among processors at runtime. Moreover, the nonconstant mapping of tasks to

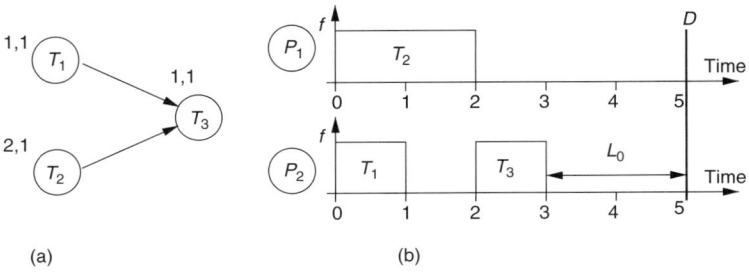

FIGURE 40.1 Dependent tasks and canonical schedule. (a) example DAG and (b) canonical schedule

processors makes the problem of global queue-based power aware scheduling more interesting and we will focus on global scheduling in this chapter.

When tasks are available at the same time, the order of tasks entering the global queue (i.e., the priority assignment of tasks) affects *which* task goes *where* (i.e., task's mapping) as well as the workload on each processor. It is well known that the optimal priority assignment to balance the workload and minimize the schedule length is NP-hard [11]. Therefore, many heuristics (e.g., LTF [35]) have been studied to efficiently obtain close-to-optimal schedules.

Our research has shown that, the optimal priority assignment for tasks that minimizes static schedule length based on tasks' WCETs may not result in the minimum energy consumption due to the runtime behaviors of tasks [47]. Instead of seeking another priority assignment heuristic, in this chapter we focus on exploring the available slack time for energy savings. For illustration purpose, we use the LTF heuristic (based on tasks' WCETs) when assigning task's priority. Intuitively, large tasks are likely to generate more slack at runtime and thus more energy savings may be expected. Moreover, the ratio of schedule length with LTF over the one with optimal priority assignment is bounded [35] and it is likely to obtain a feasible schedule with LTF. However, the power management schemes discussed next do not rely on any specific heuristic and can work with any priority assignment.

40.3.3 Canonical Schedule and Static Power Management

In order to simplify the discussion, a *canonical schedule* is defined as the mapping/schedule of tasks to processors when all tasks use their WCETs. To focus on power management, for the application considered, it is assumed that the canonical schedule under the LTF priority assignment could finish all tasks in time. Otherwise, a better priority assignment heuristic may be used to obtain a feasible canonical schedule.

For the example shown in Figure 40.1a, the canonical schedule for the tasks executing on a dual-processor system is illustrated in Figure 40.1b. In the figure, the X-axis represents time, the Y-axis represents processing speed, and the area of a task box defines the amount of work needed for executing the task.

Initially, both task T_1 and task T_2 are ready for execution and, under LTF heuristic, T_2 has higher priority. After processor P_1 fetches and executes T_2, the second processor P_2 gets task T_1 for execution. However, when processor P_2 finishes the execution of task T_1 at time 1, no task is ready and it becomes idle. When processor P_1 finishes task T_2, task T_3 becomes ready. Figure 40.1b shows that processor P_2 obtains task T_3 and finishes at time 3.

Suppose that the frame length of the application considered is $D = 5$. There are 2 units of slack available. Following the idea of SPM for uniprocessor systems, where the slack is proportionally distributed to tasks according to their WCETs [25], static slack can be uniformly distributed over the length of a schedule for energy savings [17]. Figure 40.2 shows the case of uniformly scaling down the processing of all tasks, where every task runs at speed $\frac{3}{5}f_{max}$. Compared to the case where no power management (NPM) is applied (i.e., tasks are executed at speed f_{max} and finish before the deadline), the simple approach of scaling down task's processing uniformly could save $0.64E$, where E is the energy consumed under NPM.

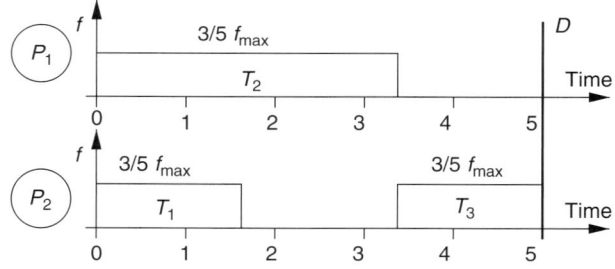

FIGURE 40.2 Static power management with uniformly scaling down.

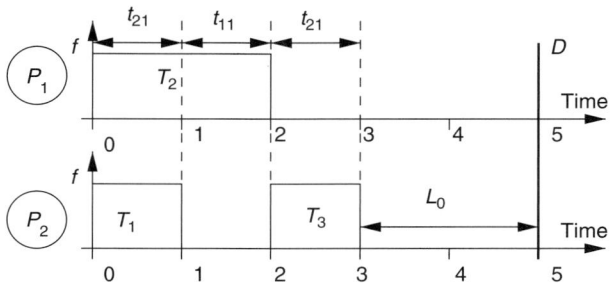

FIGURE 40.3 Parallelism in the schedule.

However, despite the simplicity and its optimality for uniprocessor systems, SPM is not optimal in terms of energy savings for multiprocessor systems. The reason comes from the fact that the degree of parallelism in a schedule is different for different schedule sections. For example, as shown in Figure 40.3, due to the dependency among tasks, processor P_2 is idle from time 1 to time 2 and a gap exists in the schedule. If we partition the schedule along the time-line at the time points when processors change their states from busy/idle to idle/busy, we will get a sequence of schedule sections with different *degrees of parallelism* (i.e., the number of processors that are active and execute tasks).

Define t_{ij} as the length of the j-th section of a schedule with parallelism of i. The aggregated length for the schedule sections with degree of parallelism as i is denoted as $t_i = \sum_j t_{ij}$. For the example, the schedule section from time 0 to time 1 is referred to as $t_{21} = 1$ and has the degree of parallelism 2. The schedule sections from time 1 to 2 and from time 2 to time 3 have the degree of parallelism 1 and are referred to as $t_{11} = 1$ and $t_{12} = 1$, respectively. Hence, there are $t_1 = t_{11} + t_{12} = 2$ and $t_2 = t_{21} = 1$.

Instead of distributing the static slack over the schedule and scaling down the processing of tasks uniformly, considering the different degrees of parallelism in a schedule, it will be more energy efficient to allocate more slack for schedule sections with higher degrees of parallelism. The intuition is that, the same amount of slack can be used to scale down more computation for the schedule sections with higher degrees of parallelism and thus obtain more energy savings.

For a system with N processors, there are at most N processors that can concurrently execute tasks for any schedule sections and the degree of parallelism in a schedule ranges from 1 to N. Suppose that the static slack in the system is L_0. Moreover, the sections in a schedule with degree of parallelism i has total length of t_i (which may consist of several sections $t_{ij}, j = 1, \ldots, u_i$, where u_i is the total number of sections with parallelism i) and the amount of static slack allocated to them is denoted by l_i ($i = 1, \ldots, N$). After scaling down the schedule sections accordingly, the total energy consumption E_{total} will be

$$E_{\text{total}} = \sum_{i=1}^{N} E_i = \sum_{i=1}^{N} (k \times i \times f_i^3 \times (t_i + l_i))$$

$$= k \times \sum_{i=1}^{N} \left(i \times \left(\frac{t_i}{t_i + l_i} \times f_{\max} \right)^3 \times (t_i + l_i) \right)$$

$$= k \times f_{\max}^3 \times \sum_{i=1}^{N} \left(i \times \frac{t_i^3}{(t_i + l_i)^2} \right) \quad (40.2)$$

where k is the system dependent constant and f_i is the speed for sections with parallelism i. As illustrated, different allocation for the slack will result in different amount of energy savings. The problem of finding an optimal allocation of L_0 in terms of energy savings will be to find $l_1, \ldots, l_N$ so as to

$$\min(E_{\text{total}})$$

subject to

$$l_i \geq 0, \quad i = 1, \ldots, N$$

$$\sum_{i=1}^{N} l_i \leq L_0$$

The constraints put limitations on how to allocate the static slack. Solving the above problem is similar to solving the constrained optimization problem presented in Reference 3. Interested readers are referred to Reference 3 for more details.

For the special case of dual-processor systems, by differentiating Equation 40.2 with respect to l_1 and l_2, we can get the following optimal solutions:

$$l_1 = \frac{T_1 \times (L_0 - (2^{1/3} - 1) \times T_2)}{T_1 + 2^{1/3} \times T_2} \tag{40.3}$$

$$l_2 = \frac{T_2 \times (2^{1/3} \times L_0 + (2^{1/3} - 1) \times T_1)}{T_1 + 2^{1/3} \times T_2} \tag{40.4}$$

From the above equations, if $L_0 \leq (2^{1/3} - 1)T_2$ then $l_1 = 0$ (since there is a constraint that $l_i \geq 0$) and $l_2 = L_0$, that is, all the static slack will be allocated to the sections of schedule with parallelism 2. For the example in Figure 40.1, there will be $l_1 = 1.0676$ and $l_2 = 0.9324$. The minimum energy consumption is $0.3464E$ and the amount of energy saving is $0.6536E$, which is better than the case of SPM (where the energy savings is $0.64E$).

Note that the above optimal solution has an assumption that the schedule sections with the same degree of parallelism are executed at the same speed. However, the scheduling unit in our system is a task. After the static slack is distributed to schedule sections according to their parallelism, tasks should collect the slack from different sections that involve them and run at a uniform speed. For the example, task T_2 gets $l_2 = 0.9324$ from the section t_{21} and $(1/2)l_1 = 0.5338$ from the section t_{12}. Therefore, task T_2 will obtain 1.4662 units of slack in total and could run at a reduced speed of $0.577 f_{\max}$. In fact, with each task collecting its slack and running at a uniform scaled speed, more energy savings could be obtained. As an exercise, interested readers are encouraged to find out the amount of energy savings for the example after T_2 collects its slack and run at a single speed.

When a task finishes early, the difference between the actual time taken and the one allocated to it is referred to as *dynamic slack* that can be used by the remaining task(s) to scale down their execution and save energy. In what follows, we will focus on online power aware mapping schemes that exploit dynamic slack for energy savings while guaranteeing to meet all timing constraints. To simplify the discussion, we assume that the canonical schedule of the applications considered finish all tasks *just-in-time* (i.e., no static slack exists). Otherwise, the schedule with scaled execution of tasks that finishes just-in-time is referred to as canonical schedule and used as a base for the online schemes.

40.4 Power Aware Mapping with Slack Sharing

As a first step, we assume that tasks are independent. That is, there is no data dependency between tasks and all tasks are *ready* for execution at the beginning of each frame. Sections 40.5 and 40.6 further discuss the cases for dependent tasks.

With LTF heuristic, larger tasks have higher priorities and the order of tasks in the global queue follows the decreasing order of tasks' WCETs, where larger tasks are in the front of the queue. When a task uses less time than its WCET and finishes early at run time, the processor on which the task just finished will fetch the next ready tasks from the head of the global queue for execution. The task may have been supposed

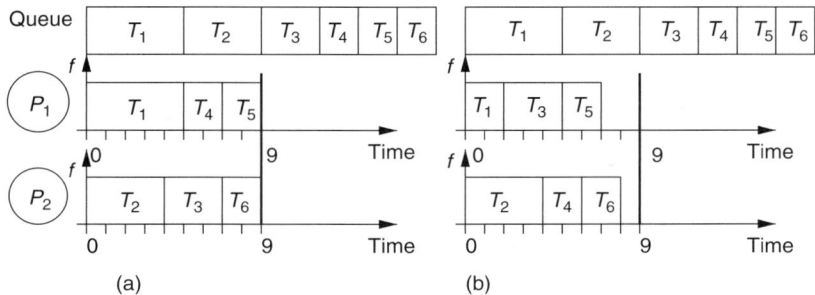

FIGURE 40.4 Canonical schedule and actual execution for independent tasks. (a) Canonical schedule and (b) actual execution with NPM

to run on a different processor, and thus, the mapping/schedule of tasks to processors could be different from the canonical schedule.

For example, consider an application consisting of six tasks $\Gamma = \{T_1(5), T_2(4), T_3(3), T_4(2), T_5(2), T_6(2)\}$ to be executed on a dual-processor system, where the number associated with each task is its WCET. With LTF priority assignment, Figure 40.4a shows the order of tasks in the global queue and the canonical schedule. Here, the canonical schedule finishes all tasks just-in-time with the frame length of $D = 9$. Suppose that, during one execution, tasks T_1 to T_6 take 2, 4, 3, 2, 2, 2 time units, respectively, Figure 40.4b shows that the actual schedule also meets the deadline. However, it can also be seen that the mapping of tasks to processors, where task T_3 is mapped to processor P_1 and task T_4 is mapped to processor P_2, is different from the canonical schedule. Without slack reclamation for power management, it can be proven that, if tasks use less than or equal to their WCETs, the actual schedule will take less time than the canonical schedule. That is, all tasks can finish their executions before the deadline.

40.4.1 Greedy Slack Reclamation

The dynamic slack generated from the early completion of tasks can be reclaimed by the remaining tasks to scale down their execution and save energy. In *greedy slack reclamation*, all the slack generated from the early completion of one task on a processor will be allocated to the next task (if any) to be executed on this processor. It has been shown that the greedy scheme is an effective power management technique in uniprocessor systems [2,33]. However, as the mapping of tasks to processors may change in multiprocessor systems due to the runtime behaviors of tasks, such slack reclamation scheme is not guaranteed to meet the timing constraints of tasks as illustrated below.

For the above example, Figure 40.5a shows that task T_1 finishes early at time 2, and 3 units of slack time are available on processor P_1. Under global scheduling, the next task to be run on processor P_1 is T_3. With greedy slack reclamation, all the available slack on P_1 will be given to task T_3. Therefore, including its WCET, T_3 will have 6 units of time and the processor speed can be reduced to $3/6 \times f_{max}$ accordingly. When T_3 uses up all its allocated time, task T_6 will miss the deadline D as illustrated in Figure 40.5b. Hence, even if the canonical schedule can finish before D, greedy slack reclamation under global scheduling cannot guarantee that all tasks will finish in time.

40.4.2 Slack Sharing among Processors

Compared with the canonical schedule, we can see that task T_3 should finish its execution no later than time 7. However, when processor P_1 finishes task T_1 early, with greedy slack reclamation, task T_3 obtains all the 3 units of slack on processor P_1, which makes it possible for task T_3 to finish after time 7 with the scaled processing speed, thus leading to a deadline miss. Note that, only 2 units of the slack on processor P_1 are available before time 4, which is the time point when task T_3 is supposed to start its execution following the completion of task T_2 on processor P_2 in the canonical schedule. Therefore, we may divide

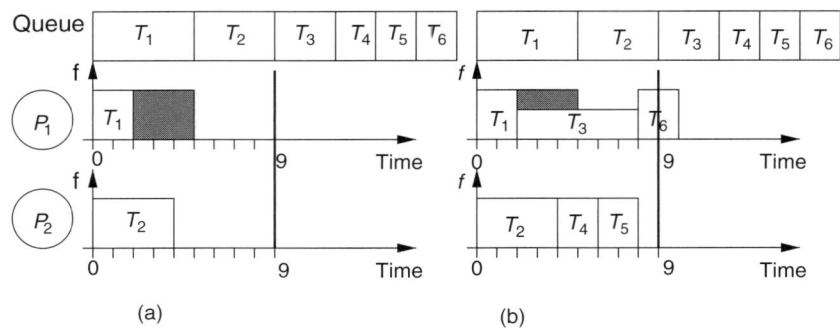

FIGURE 40.5 Greedy slack reclamation. (a) T_1 finished early and (b) T_6 missed deadline.

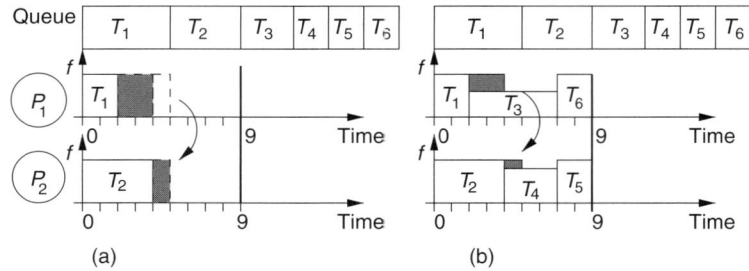

FIGURE 40.6 Slack sharing reclamation. (a) Share slack among processors and (b) all tasks finish in time.

the slack on processor P_1 into two parts: part one with 2 units is available before time 4 that can be reclaimed by task T_3; part two with 1 unit that will be *shared* with processor P_2 (see Figure 40.6a).

With *slack sharing*, task T_3 reclaims the 2 units of slack on processor P_1 and emulates the timing as in the canonical schedule. That is, without considering these 2 units of slack, it looks like T_3 starts at time 4 and finishes at time 7, the same as in the canonical schedule. With the slack, T_3 starts at time 2, runs for 5 time units at the speed of $(3/5) \times f_{max}$ and also finishes its execution at time 7. Task T_4 starts at time 4 on processor P_2, reclaims the shared 1 unit of slack, executes for 3 time units at the speed of $(2/3) \times f_{max}$ and ends at time 7. As illustrated in Figures 40.6b, all tasks complete before the deadline. Therefore, if a processor is supposed to finish executing its task later than other processors in the worst case, but finishes early and obtains some dynamic slack, such slack should be shared with other processors appropriately before reclaiming it for energy savings.

From a different perspective, sharing the slack emulates the mapping of tasks to processors as in the canonical schedule. For the example, it can be looked at as task T_1 being allocated 4 time units on processor P_1 instead of 5, while task T_2 is allocated 5 time units on processor P_2 instead of 4. The next task T_3 will run on the processor P_1 where the current running task on it supposes to complete early. Hence, when task T_1 finishes early, processor P_1 only gets 2 units of slack and P_2 gets 1 unit. So, for the sake of the remaining tasks, the situation looks like task T_1 is assigned to processor P_2 that is supposed to finish at time 5 and task T_2 is assigned to processor P_1 that is supposed to finish at time 4. All the remaining tasks that are assigned to P_2 in the canonical schedule will now be assigned to P_1 and vice versa.

Following the slack sharing idea illustrated in the example, the online slack sharing mapping for power management can be generalized for systems with N (≥ 2) processors. Before formally presenting the algorithm, we first define two notations for ease of discussion. The *estimated finish time (EFT)* for a task executing on a processor is defined as the time at which the task is expected to finish its execution if it consumes all the time allocated to it. The *start time of the next task (STNT)* for a processor is defined as

```
 1: if ((T_k = Dequeue(Ready-Q)) != NULL) then
 2:    Find P_r such that:
       STNT_r = min{STNT_1, ..., STNT_n};
 3:    if (STNT_id > STNT_r) then
 4:       STNT_id ↔ STNT_r;
 5:    end if
 6:    EFT_k = STNT_id + c_k;
 7:    STNT_id = EFT_k;
 8:    f_k = f_max · c_k/(EFT_k − t);
 9:    Execute T_k at speed f_k;
10: else
11:    wait();
12: end if
```

Algorithm 1: Slack sharing mapping algorithm.

the time at which the next task can begin its execution on that processor, which is actually the EFT of the task currently running on the processor.

The slack sharing mapping algorithm is presented in Algorithm 1. It is assumed that all tasks are put into the global queue *Ready-Q* at the beginning of each frame following the LTF priority assignment. Each processor invokes the algorithm individually. The shared memory holds the control information, such as the *Ready-Q* and *STNT* values, which must be updated within a critical section (not shown in the algorithm for simplicity). At the beginning of each frame, the value of *STNTs* for processors are initialized to 0. The current processor invoking the algorithm is P_{id}, t is the current time and f_k is the processing speed of P_{id} to execute task T_k.

When processor P_{id} finishes a task at time t, it will try to fetch the next ready task from *Ready-Q* for execution. If there are no more tasks in *Ready-Q*, processor P_{id} will stall and sleep until the next frame. Here, we use the function *wait()* to put one processor to sleep (line 11). Otherwise, the processor gets the next ready task T_k from *Ready-Q*. To emulate the mapping of tasks to processors as in the canonical schedule, where task T_k supposes to run on the processor with the smallest *STNT*, we exchange the value of $STNT_{id}$ with that of the minimum $STNT_r$ (lines 2, 3, and 4). By exchanging $STNT_{id}$ with $STNT_r$, processor P_{id} shares part of its slack (specifically, $STNT_{id} - STNT_r$) with processor P_r and task T_k reclaims other part (with amount of $STNT_r - t$) to scale down its execution for energy savings (lines 6, 7, and 8). Therefore, the slack sharing scheme is still a *greedy-oriented* approach. That is, after sharing the slack appropriately, all the remaining slack is allocated to the next task to be executed.

From the algorithm, we notice that at any time (except when *Ready-Q* is empty) the values of *STNT* for processors are always equal to the N biggest values of *EFT* of the tasks running on the processors. One of these tasks is the most recently started task (from line 6 to 7). The task that starts next will be executed by the processor with the smallest *STNT* (after exchanging *STNT* values between processors). Therefore, by sharing the slack among the processors, the slack sharing mapping algorithm emulates the mapping of tasks to processors as in the canonical schedule and uses no more time than the canonical schedule while reclaiming the slack for energy savings. The result is further summarized in the following theorem. Interested readers are referred to Reference 47 for the proof of this theorem.

Theorem 40.1 *For any given heuristic of assigning task's priority, if the canonical schedule for the independent tasks of a frame-based real-time application finishes within the frame length D, any execution with the same priority assignment under the slack sharing mapping algorithm will complete all tasks in time.*

40.5 Fixed Priority Slack Sharing for Dependent Tasks

List scheduling is a standard scheduling technique for dependent tasks [11], where a task becomes *ready* for execution when all of its predecessors finish execution. The *root* tasks that have no predecessors are ready at the beginning of each frame. List scheduling puts tasks into a global ready queue as soon as they become ready and processors fetch tasks from the front of the ready queue for execution. When more than one task is ready at the same time (e.g., have the same predecessors), finding the optimal order (i.e., priority) of tasks being added to the global queue that minimizes execution time is NP-hard [11]. Here, we use the same LTF heuristic as for independent tasks and put into the ready queue first the longest task (based on WCET) among the tasks that become ready simultaneously.

Hence, in list scheduling, the order of tasks entering the global queue (i.e., the order of tasks being fetched and executed by processors) depends on both tasks' ready time and the priority assignment heuristics. Note that, owing to the dependencies among tasks, the ready time of a task depends on the actual execution time of its predecessor(s). Therefore, the runtime behaviors of tasks will result in different orders of tasks in the global queue, which in turn leads to different mapping of tasks to processors. The different mapping may cause the application to take more time than the canonical schedule, even if tasks use no more time than their WCETs as shown below.

For example, consider a dependent task set with six tasks ($\Gamma = \{T_1, T_2, T_3, T_4, T_5, T_6\}$), where the data dependencies are represented by the DAG as shown in Figure 40.7a. The canonical schedule is shown in Figure 40.7b and the frame length of the application is $D = 12$. Task nodes are labeled with the tuple (c_i, a_i), where c_i is the WCET for task T_i and a_i is the ACET. From the canonical schedule, we can see that T_1 and T_2 are root tasks and ready at time 0. T_3 and T_4 are ready at time 2 when their predecessor T_1 finishes execution. T_5 is ready at time 3, and T_6 is ready at time 6.

For the execution where tasks take their ACET a_i, Figure 40.8 shows the order of tasks in the global queue and the schedule for the execution. It is clear that the actual schedule takes more time than the canonical schedule and misses the deadline. The reason is that the ready time of task T_5 becomes earlier than that of T_3 and T_4 due to early completion of task T_1. Thus, the order of tasks in the global queue changes, which in turn leads to a different mapping of tasks to processors and thus to the deadline miss.

To guarantee that the actual execution takes no more time than the canonical schedule, we need to prevent T_5 from executing before T_3 and T_4 and keep the order of tasks in the global queue the same as in the canonical schedule. In the beginning of each frame, regardless of whether tasks are ready or not, all tasks are put into the global queue in the same order as in the canonical schedule. Processors fetch tasks for execution from the head of the global queue. Note that, owing to the runtime behavior of tasks, it is possible that ready tasks exist in the global queue while the header task is not ready. In this case, the processor that tries to fetch the header task for execution cannot fetch the ready tasks from the middle of the global queue and is blocked until the header task becomes ready.

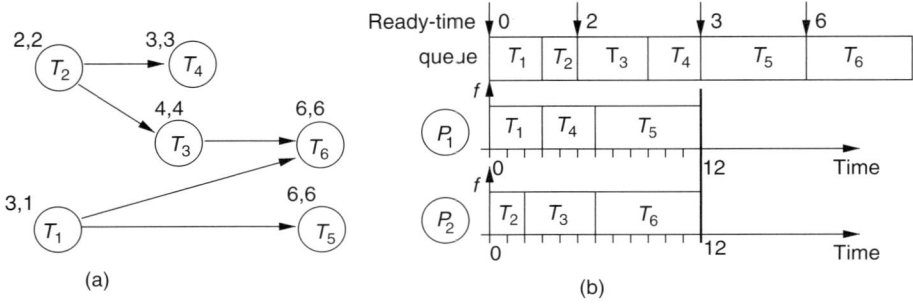

FIGURE 40.7 List scheduling with a dual-processor system: (a) precedence graph and (b) canonical execution, finish at $D = 12$.

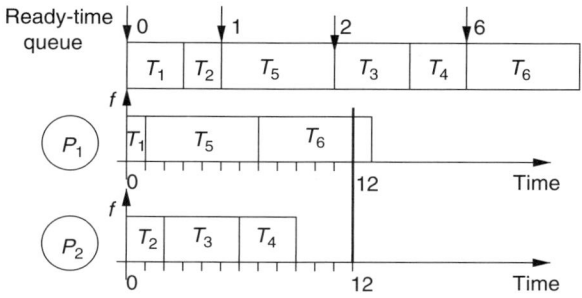

FIGURE 40.8 The execution where tasks take their ACET.

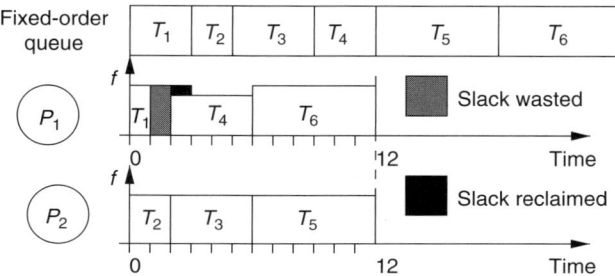

FIGURE 40.9 Fixed priority list scheduling with slack sharing.

By keeping the same execution order of tasks as in the canonical schedule, the fixed priority list scheduling ensures that any execution will take no more time than the canonical schedule. Moreover, the power aware mapping with slack sharing can work with the fixed priority list scheduling for energy savings. The scheme is illustrated by the above example as in Figure 40.9. As the header task may block processors due to not being ready, slack time may be wasted as shown in the figure. However, by enforcing the execution order, the same as for independent tasks, the slack sharing with fixed priority list scheduling ensures the same timing constraints as in the canonical schedule while achieving energy savings.

40.6 Slack Shifting for Applications with Different Execution Paths

We have presented the slack sharing mapping algorithm for applications where all tasks will be executed within every frame. However, for applications that have different execution paths following different branches, only a subset of the tasks on a specific execution path will be executed within one frame. For these applications, in addition to task's less than WCET execution, dynamic slack is also expected at the task set level (e.g., shorter execution paths other than the longest one).

40.6.1 AND/OR Application Model

The AND/OR model that extends the one in Reference 15 is used to represent branches as well as parallelism within such applications. In addition to computation nodes that represent normal tasks, two other different kinds of nodes are defined in the model: AND nodes and OR nodes. For the vertices in a DAG, computation nodes are represented by a *circle*, AND nodes are represented by *diamond* and OR synchronization nodes are represented by *dual-circle*. The computation requirement for AND/OR is assumed to be 0. That is, they are considered as *dummy* tasks. For a synchronization node with nonzero computation requirement, it is easy to transform it to a synchronization node and a computation node.

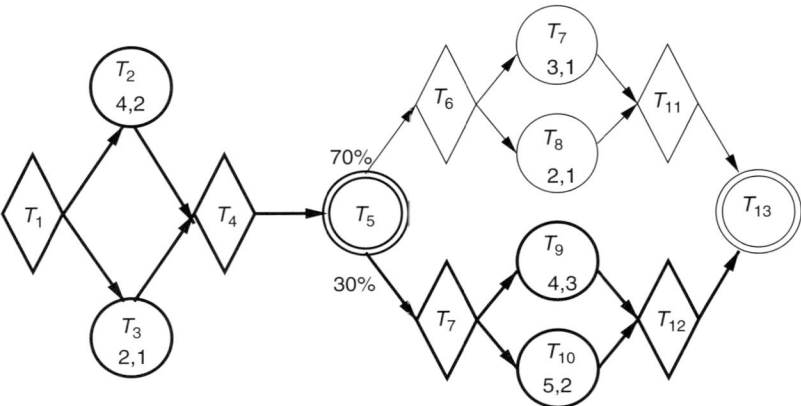

FIGURE 40.10 An example AND/OR graph.

For an OR node, it depends on only one of its predecessors. In other words, it is ready for execution when any one of its predecessors completes. Moreover, only one of its successors depends on it. That is, exactly one of its successors will be executed after its execution. The OR nodes are used to model branches and explore the different execution paths in the applications. To represent the probability of taking each branch after an OR synchronization node, a number is associated with each successor of an OR node, which will be used by speculation schemes for more energy savings as discussed in Section 40.6.3.

An AND node depends on *all* its predecessors. Thus, only after all its predecessors finish execution can it be executed. All successors of an AND node depend on it, which implies that all its successors are executed after the completion of an AND node. The AND nodes are used to explore the parallelism in the applications.

For simplicity, we only consider the case where an OR node cannot be processed concurrently with other paths. All processors will synchronize at an OR node. We define *application segment* as a set of tasks that are separated by two adjacent OR nodes. There will be several application segments, one for each of the branches, between two corresponding OR nodes.

For example, Figure 40.10 shows one AND/OR application. Recall that the AND/OR synchronization nodes are dummy tasks with computation time as 0. In this example, there are two execution paths after the OR node T_5. The upper branch has the computation nodes T_7 and T_8, and the lower branch has the computation nodes T_9 and T_{10}. The longest path of the application is shown bold.

40.6.2 Shifting Schedule Sections in Canonical Schedule

To find out the amount of slack from different paths after a branch point, we need to generate the canonical schedule for the application segments between adjacent OR nodes. For the above example running on a dual-processor system, the canonical schedule is shown in Figure 40.11a.

For the synchronization nodes, which are considered as dummy tasks with 0 computation requirement, they are shown as bold vertical lines. The dotted rectangles represent the application segments that, as a whole, will or will not be executed (*integrated segments*). For this example, if execution follows the upper branch after task T_5, it takes 2 time units less compared with the path along the lower branch. To utilize such possible slack online, we can *shift* the canonical schedule for each application segment toward the deadline as late as possible. The shifted canonical schedule for the example is shown in Figure 40.11b.

Here L_1 comprises the possible slack that can be reclaimed when the execution follows the upper branch. Note that the tasks in one application segment are shifted together for simplicity. We do not consider shifting individual tasks in a segment. For example, for the segment consisting tasks T_7 and T_8, we may shift task T_8 one time unit more without violating the timing requirement. But shifting single

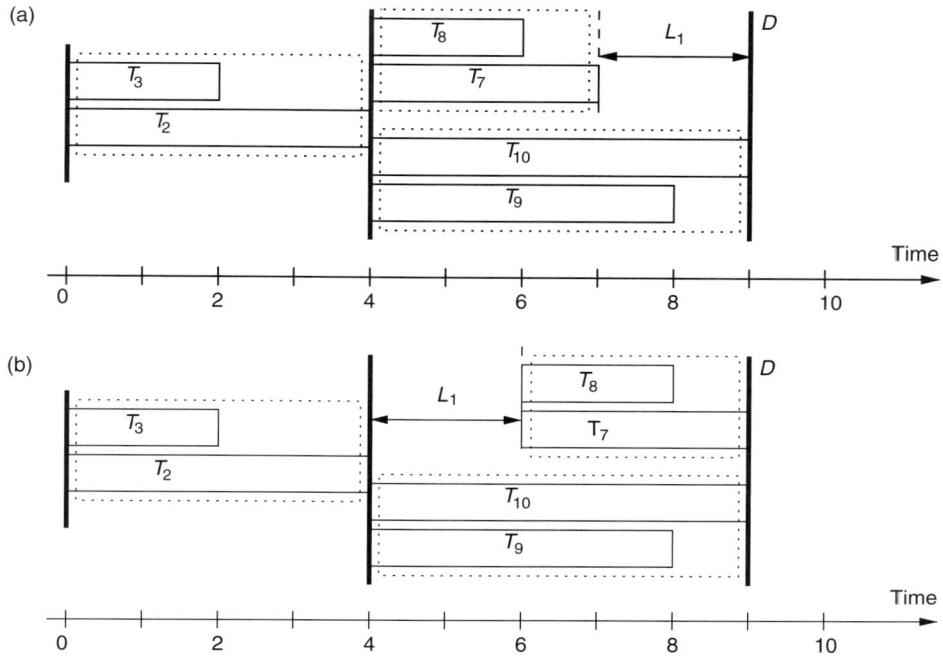

FIGURE 40.11 The canonical schedules for the example in Figure 40.10: (a) canonical schedule and (b) shifted canonical schedule.

tasks increases the complexity of the algorithm and hopefully that one unit of slack could be claimed by the subsequent tasks at runtime.

When there are nested OR nodes, different segments may obtain different amounts of slack from shifting. For the section of canonical schedule corresponding to each application segment, the power aware mapping with slack sharing can be applied at runtime, which will guarantee the timing constraints as explained in previous sections. For reclaiming slack from different execution paths at runtime, additional timing information regarding the tasks in the shifted canonical schedule need to be collected offline. Please refer to Reference 49 for detailed algorithms and explanations.

40.6.3 Speculation for Better Performance

Note that, the above discussed *slack sharing/shifting* schemes are *greedy-oriented*. That is, all available slack will be allocated to the next ready task after sharing/shifting the slack appropriately. Each task would run at a different speed after reclaiming the slack, and many speed adjustments are expected.

As mentioned earlier, owing to the convex relation between CPU speed and its power consumption, the energy consumption will be minimized when all tasks run at the same scaled speed across all processors [4]. Considering the speed adjustment overhead as discussed in the next Section, the single speed setting is even more attractive. From this intuition, we can speculate one processing speed at which tasks should, on average, be executed by exploiting the statistical information (e.g., the ACET of tasks and the probability for each branch to be taken, which can be obtained from profiling) about an application.

As the simplest approach, *one* processing speed will be speculated at the beginning of each frame by considering the average behaviors of all tasks and paths. More specifically, for a given application with a set of tasks to be executed on a multiprocessor system, a static schedule may be generated assuming that each task takes its ACET. Thus, the average schedule length Π_a could be obtained from the average schedule length for the application segments between corresponding branch points and the probability of the segments being executed. The single speculated speed will be

$$f_{ss} = \frac{\Pi_a}{D} f_{max} \qquad (40.5)$$

For tasks that may receive excessive amount of dynamic slack, the processing speed calculated from the greedy-oriented slack sharing scheme (line 8 in Algorithm 1) could be less than f_{ss}. In this case, we would rather execute the task at speed f_{ss} and save some slack for future tasks with the expectation of running all tasks at the same speed. However, the speculative speed f_{ss} is optimistic and does not take into consideration the worst-case behaviors of tasks. That is, it is possible that there is no enough slack for a task and the speed calculated from Algorithm 1 is higher than f_{ss}. In this case, the task will be executed at the higher speed calculated from the slack sharing scheme to guarantee that all future tasks meet their timing constraints.

Therefore, the speed at which a task is executed will be the higher speed between the speculation speed f_{ss} and the one calculated from the greedy-oriented slack sharing schemes. By executing tasks at the speed no lower than the one from the greedy slack sharing schemes, the speculation guarantees that all tasks will finish no later than their completion times under slack sharing schemes. Thus, if all tasks could finish in time in canonical schedule, slack sharing schemes can meet all timing constraints, so does the speculation scheme.

If the statistical characteristics of tasks in an application vary substantially (e.g., the tasks at the beginning of one application have the average/maximum execution time ratio as 0.9 while for tasks at the end of the application the ratio is 0.1), it may be better to respeculate the speed while the application execution progresses based on the statistical information about the remaining tasks. Considering the difficulty of speculating a speed after each task's completion for multiprocessor systems and significant amount of slack may be expected after an OR node, the adaptive scheme speculates one new speed after each OR node. That is, the speculative speed would be

$$f_{as} = \frac{\Pi_a^i}{D - t} \qquad (40.6)$$

where t is the current time (when the OR node is processed) and Π_a^i is the average remaining schedule length when branch b_i is taken after that OR node. Again, to guarantee the timing constraints to be met, the speed for executing a task will be the higher speed between f_{as} and the one from the slack sharing/shifting scheme. Please refer to Reference 49 for more detailed discussion regarding how the speculative speeds are calculated and used.

40.7 Practical Considerations

In the previous discussion, we have assumed that tasks can run at any arbitrary speed and there is no overhead for speed adjustment. However, for the available commodity variable voltage/speed processors, such as Transmeta [24] and Intel XScale [22], only a few speed levels are available and changing speed takes time and energy. We will address the issues with discrete speed levels and present the idea of *slack reservation* to incorporate the overhead of speed adjustment.

For the processors that only have several working speeds, the power aware mapping algorithms can be easily adapted. Specifically, after reclaiming the slack and obtaining the new processing speed f for the next task, if f falls between two speed levels ($f_k < f \leq f_{k+1}$), setting f to the next higher speed level f_{k+1} will always guarantee that the task meets its deadline.

With the higher speed, the task does not need all the slack initially allocated to it and part of the slack is saved for future use. By sharing the slack with future tasks, the processing speed for all tasks may get more close to each other, which results in more energy savings. The simulation results show that the power aware mapping with discrete speed levels may achieve better performance, in terms of energy savings, than the case where processors have continuous speeds [47].

40.7.1 Slack Reservation for Speed Adjustment Overhead

Depending on different techniques, the timing overhead of speed adjustment varies substantially, ranging from tens of microseconds [8] to a few milliseconds [34]. Moreover, such overhead may also depend on the scope of speed adjustment. For a general speed adjustment time overhead model, we assume that the overhead consists of two parts: a constant part that imitates the set-up time and a variable part that is proportional to the scope of speed adjustment. Hence,

$$O_{\text{adjustment}} = O_c + \lambda \cdot |f_1 - f_2| \tag{40.7}$$

where O_c and λ are constants; f_1 and f_2 are the speeds before and after speed adjustment. Here, the choice of $\lambda = 0$ results in a constant time overhead, which could be the maximum amount of time needed to adjust the speed with the full scope from/to the highest speed to/from the lowest one.

As the power aware mapping with slack sharing is a nonpreemptive scheme, processors will at most change speed twice for each task. Therefore, as the simple approach, the time overhead can be incorporated by adding twice the maximum overhead to the WCET for all tasks. This ensures that the processor has enough time to scale down the processing speed for each task and speed up afterward (which may be necessary for the remaining tasks finishing in time). However, this approach is conservative and the inflated WCET of tasks could lead to *false* deadline misses in the canonical schedule, which is used as a baseline.

Instead of statically adding the maximum speed adjustment overhead to task's WCET, we could reserve slack for speed adjustment whenever needed. Specifically, when there is available slack, before using it to scale down the processing speed for the next task, we set aside enough slack for scaling down and speeding up the processing speed, which make sure that future tasks could run at the appropriate speed level and meet the timing constraints. The idea is further illustrated in Figure 40.12.

When task T_i finishes early and leaves the amount of slack S_i, part of S_i will be used to change the processing speed for task T_{i+1}. Moreover, enough slack is also reserved for changing the processing speed back to $f_{\max}$ in the future. In case task T_{i+1} uses up its allocated time, the reserved slack makes sure that the remaining tasks could run at the maximum speed $f_{\max}$ and meet their deadlines. The rest of the slack is used as additional time for task T_{i+1} to scale down its processing speed.

Suppose that the current speed for task T_i is f_i and assume that the speed for task T_{i+1} is f_{i+1} (to be computed). The overhead, O_i, to change speed from f_i to f_{i+1}, and the overhead, R_i, to change speed from f_{i+1} back to $f_{\max}$ are

$$O_i = O_c + \lambda \times |f_{i+1} - f_i| \tag{40.8}$$

$$R_i = O_c + \lambda \times (f_{\max} - f_{i+1}) \tag{40.9}$$

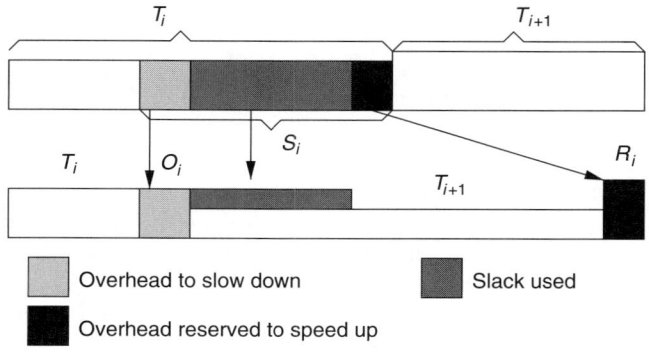

FIGURE 40.12 Slack reservation for speed adjustment overhead.

Hence, f_{i+1} can be calculated by giving additional time, $(S_i - O_i - R_i)$, to task T_{i+1} as

$$f_{i+1} = f_{\max} \times \frac{c_{i+1}}{c_{i+1} + S_i - O_i - R_i} \tag{40.10}$$

Assuming that $f_{i+1} < f_i$, the above equation is a quadratic equation in f_{i+1}

$$2 \times \lambda \times f_{i+1}^2 + [c_{i+1} + S_i - 2 \times O_c - \lambda \times (f_{\max} + f_i)] \times f_{i+1} - f_{\max} \times c_{i+1} = 0 \tag{40.11}$$

If no solution is obtained with $f_{i+1} < f_i$ from the above equation, the assumption is wrong; that is, we should have $f_{i+1} \geq f_i$. It is possible to set $f_{i+1} = f_i$ if the slack $S_i - R_i$ is enough for task T_{i+1} to scale down the speed from $f_{\max}$ to f_i, that is, if $f_{\max} \times \frac{c_{i+1}}{c_{i+1} + S_i - R_i} \leq f_i$, we can set $f_{i+1} = f_i$. If it is not possible to set $f_{i+1} \leq f_i$, we have $f_{i+1} > f_i$ and f_{i+1} can be solved as

$$f_{i+1} = f_{\max} \times \frac{c_{i+1}}{c_{i+1} + S_i - 2 \times O_c - \lambda \times (f_{\max} - f_i)} \tag{40.12}$$

Finally, if f_{i+1} computed from the above equation is larger than the maximum speed $f_{\max}$, the available slack is not enough to make two speed adjustments and we will set $f_{i+1} = f_{\max}$.

In most cases, the reserved slack, R_i, will not be used and becomes part of the slack S_{i+1} from the early completion of task T_{i+1}. However, in some cases, the useful slack, $S_{i+1} - R_i$, is not enough for task T_{i+2} to use it. In this case, R_i will be used to change the speed back to $f_{\max}$ and task T_{i+2} will run at $f_{\max}$. Moreover, because of the speed adjustment overhead, special care is needed before sharing the slack among processors. That is, if there is not enough time for the current processor to change its speed back to the maximum speed $f_{\max}$ after sharing the slack, the slack sharing process should not be invoked. Instead, the current processor needs to change its speed back to $f_{\max}$ first and share the slack after that (if possible). Please refer to Reference 47 for more detailed discussions.

40.8 Summary

Power awareness has become an important system requirement not only for portable devices, which are normally battery powered, but also for systems with extremely high computation density (such as server clusters). Although many efforts have been devoted to uniprocessor systems, the research for parallel system is quite limited.

In this chapter, we addressed the problem of power aware mapping of a set of real-time tasks on a multiprocessor system. VS, which scales down the processing speed for tasks by exploring slack time in the system, is explored for power management. Despite the simplicity of partition scheduling, the fixed mapping of tasks to processors limits the runtime mapping flexibility and thus the opportunity of managing power across processors. In global scheduling, the optimal priority assignment for tasks to minimize the schedule length is known to be NP-hard, and many heuristics have been proposed based on tasks' WCET (e.g., LTF). Instead of working on another heuristic to get the close-to-optimal static schedule, we focus on the mapping of tasks to processors and explore the slack (static and dynamic) for energy savings.

Considering the different degrees of parallelism in a schedule due to dependencies among tasks, one scheme that aims at more efficiently allocating static slack is first discussed. As a reference, the *canonical schedule* is defined as the schedule/mapping of tasks to processors in which all tasks take their WCETs. It is assumed that the canonical schedule meets the timing constraints under a given priority assignment heuristic (e.g., LTF).

On the basis of global scheduling, for a set of independent tasks, power aware mapping with slack sharing shares slack among processors at runtime, scales down the processing speed of tasks for energy savings while ensuring that the actual schedule takes no more time than the canonical schedule. Therefore, for the same priority assignment heuristic, if the tasks can finish in time in the canonical schedule, the power aware mapping with slack sharing will meet all tasks' timing constraints.

For dependent tasks, the readiness of a task depends on the runtime behaviors of its predecessors, which may result in different execution order of tasks and different mapping of tasks to processors as the ones in the canonical schedule. To ensure the timeliness of tasks, the fixed order list scheduling executes all tasks in the same order as in the canonical schedule. The slack sharing is further combined with the fixed order list scheduling for energy savings.

For applications with different execution paths, the AND/OR application model is extended to represent the branches within an application. For such applications, only a subset of tasks along a certain execution path will be executed. In addition to the slack sharing that is applied between two adjacent branch points for reclaiming the slack resulting from task's less than WCET execution, slack shifting is used to reclaim the additional slack resulting from executing paths other than the longest one.

Motivated by the fact that minimum energy consumption is obtained by executing all tasks with a single speed across all processors, one speculation scheme is discussed. The scheme speculates one speed at which tasks should be executed by exploiting the statistical information of an application (e.g., the ACET of tasks and the probability for each branch to be taken). Considering the possible uneven distribution of dynamic slack in an application, an adaptive scheme is further addressed, which considers only remaining tasks for speculation.

Finally, some practical issues are discussed. For discrete speed levels, after obtaining an arbitrary speed, running at the next higher discrete speed always guarantees the task finish before its deadline. Slack reservation sets aside part of the available slack whenever needed for incorporating the speed adjustment overhead.

References

[1] T. A. AlEnawy and H. Aydin. Energy-aware task allocation for rate monotonic scheduling. In *Proc. of the 11th IEEE Real-Time and Embedded Technology and Applications Symposium (RTAS)*, San Francisco, CA, March 2005.

[2] H. Aydin, R. Melhem, D. Mossé, and P. Mejia-Alvarez. Dynamic and aggressive scheduling techniques for power-aware real-time systems. In *Proc. of the 22nd IEEE Real-Time Systems Symposium*, London, UK, December 2001.

[3] H. Aydin, R. Melhem, D. Mossé, and P. Mejia-Alvarez. Optimal reward-based scheduling for periodic real-time systems. *IEEE Transaction on Computers*, 50(2):111–130, 2001.

[4] H. Aydin and Q. Yang. Energy-aware partitioning for multiprocessor real-time systems. In *Proc. of the 17th International Parallel and Distributed Processing Symposium (IPDPS), Workshop on Parallel and Distributed Real-Time Systems (WPDRTS)*, Nice, France, April 2003.

[5] S. K. Baruah and J. Anderson. Energy-efficient synthesis of periodic task systems upon identical multiprocessor platforms. In *Proc. of The 24th IEEE International Conference on Distributed Computing Systems*, Hachioji, Tokyo, Japan, March 2004.

[6] A. Burchard, J. Liebeherr, Y.Oh, and S. H. Son. New strategies for assigning real-time tasks to multiprocessor systems. *IEEE Transaction on Computers*, 44(12): 1429–1442, 1995.

[7] T. D. Burd and R. W. Brodersen. Energy efficient CMOS microprocessor design. In *Proc. of the HICSS Conference*, Maui, UW, January 1995.

[8] T. D. Burd, T. A. Pering, A. J. Stratakos, and R. W. Brodersen. A dynamic voltage scaled microprocessor system. *IEEE Journal of Solid-State Circuits*, 35(11):1571–1580, 2000.

[9] J. Carpenter, S. Funk, P. Holman, A. Srinivasan, J. Anderson, and S. Baruah. A categorization of real-time multiprocessor scheduling problems and algorithms. *Handbook of Scheduling: Algorithms, Models and Performance Analysis*. Joseph Y. Leung (ed.) Chapman & Hall/CRC, 2004.

[10] A. Chandrakasan, S. Sheng, and R. Brodersen. Low-power CMOS digital design. *IEEE Journal of Solid-State Circuit*, 27(4):473–484, 1992.

[11] M. L. Dertouzos and A. K. Mok. Multiprocessor on-line scheduling of hard-real-time tasks. *IEEE Trans. on Software Engineering*, 15(12):1497–1505, 1989.

[12] E. (Mootaz) Elnozahy, M. Kistler, and R. Rajamony. Energy-efficient server clusters. In *Proc. of Power Aware Computing Systems*, Cambridge, MA, February 2002.

[13] R. Ernst and W. Ye. Embedded program timing analysis based on path clustering and architecture classification. In *Proc. of The International Conference on Computer-Aided Design*, San Jose, CA, pp. 598–604, November 1997.

[14] X. Fan, C. Ellis, and A. Lebeck. The synergy between power-aware memory systems and processor voltage. In *Proc. of the Workshop on Power-Aware Computing Systems*, San Diego, CA, December 2003.

[15] D. W. Gillies and J. W.-S. Liu. Scheduling tasks with and/or precedence constraints. *SIAM J. Compu.*, 24(4):797–810, 1995.

[16] R. Graybill and R. Melhem (eds). *Power Aware Computing*. Kluwer Academic/Plenum Publishers, 2002.

[17] F. Gruian. System-level design methods for low-energy architectures containing variable voltage processors. In *Proc. of the Workshop on Power-Aware Computing Systems*, Cambridge, MA, November 2000.

[18] F. Gruian and K. Kuchcinski. Low-energy directed architecture selection and task scheduling for system-level design. In *Proc. of 25th IEEE Euromicro Conference*, pp. 296–302, Milan, Italy, September 1999.

[19] F. Gruian and K. Kuchcinski. Lenes: Task scheduling for low-energy systems using variable supply voltage processors. In *Proc. of Asia and South Pacific Design Automation Conference (ASP-DAC)*, Yokohama, Japan, January 2001.

[20] P.J.M. Havinga and G. J. M. Smith. Design techniques for low-power systems. *Journal of Systems Architecture*, 46(1):1–21, 2000.

[21] I. Hong, G. Qu, M. Potkonjak, and M. Srivastava. Synthesis techniques for low-power hard real-time tasks on variable voltage processors. In *Proc. of the 19th IEEE Real-Time Systems Symposium*, Madrid, Spain, December 1998.

[22] http://developer.intel.com/design/intelxscale/.

[23] http://www.darpa.mil/IPTO/Programs/pacc/index.htm, 2006.

[24] http://www.transmeta.com

[25] T. Ishihara and H. Yauura. Voltage scheduling problem for dynamically variable voltage processors. In *Proc. of The 1998 International Symposium on Low Power Electronics and Design*, Monterey, CA, August 1998.

[26] R. Jejurikar, C. Pereira, and R. Gupta. Leakage aware dynamic voltage scaling for real-time embedded systems. In *Proc. of the 41st Annual Design Automation Conference (DAC)*, San Diego, CA, January 2004.

[27] D. Kirovski and M. Potkonjak. System-level synthesis of low-power hard real-time systems. In *Proc. of the Design Automation Conference*, Anaheim, CA, June 1997.

[28] F. Liberato, S. Lauzac, R. Melhem, and D. Mossé. Fault-tolerant real-time global scheduling on multiprocessors. In *Proc. of the 10th IEEE Euromicro Workshop in Real-Time Systems*, York, England, June 1999.

[29] J. Luo and N. K. Jha. Power-conscious joint scheduling of periodic task graphs and aperiodic tasks in distributed real-time embedded systems. In *Proc. of International Conference on Computer Aided Design*, San Jose, CA, November 2000.

[30] J. Luo and N. K. Jha. Static and dynamic variable voltage scheduling algorithms for real-time heterogeneous distributed embedded systems. In *Proc. of 15th International Conference on VLSI Design*, Bangalore, India, January 2002.

[31] D. Maniezzo, K. Yao, and G. Mazzini. Energetic trade-off between computing and communication resource in multimedia surveillance sensor networks. In *Proc. of the 4th IEEE Conference on Mobile and Wireless Communications Networks*, Stockholm, Sweden, September 2002.

[32] R. Mishra, N. Rastogi, D. Zhu, D. Mossé, and R. Melhem. Energy aware scheduling for distributed real-time systems. In *Proc. of International Parallel and Distributed Processing Symposium (IPDPS)*, Nice, France, April 2003.

[33] D. Mossé, H. Aydin, B. R. Childers, and R. Melhem. Compiler-assisted dynamic power-aware scheduling for real-time applications. In *Proc. of Workshop on Compiler and OS for Low Power*, Philadelphia, PA, October 2000.

[34] W. Namgoong, M. Yu, and T. Meg. A high efficiency variable voltage CMOS dynamic DC-DC switching regulator. In *Proc. of IEEE International Solid-State Circuits Conference*, San Francisco, CA, February 1996.

[35] D.-T. Peng. Performance bounds in list scheduling of redundant tasks on multi-processors. In *Proc. of the 22nd Annual International Symposium on Fault-Tolerant Computing*, Boston, MA, July 1992.

[36] P. Pillai and K. G. Shin. Real-time dynamic voltage scaling for low-power embedded operating systems. In *Proc. of 18th ACM Symposium on Operating Systems Principles (SOSP'01)*, Chateau Lake Louise, Banff, Canada, October 2001.

[37] S. Saewong and R. Rajkumar. Practical voltage scaling for fixed-priority RT-systems. In *Proc. of the 9th IEEE Real-Time and Embedded Technology and Applications Symposium*, Washington, DC, May 2003.

[38] V. Sharma, A. Thomas, T. Abdelzaher, K. Skadron, and Z. Lu. Power-aware QOS management in web servers. In *Proc. of the 24th IEEE Real-Time System Symposium*, Cancun, Mexico, December 2003.

[39] A. Sinha and A. P. Chandrakasan. Jouletrack—a web based tool for software energy profiling. In *Proc. of Design Automation Conference*, Los Vegas, NV, June 2001.

[40] M. Weiser, B. Welch, A. Demers, and S. Shenker. Scheduling for reduced CPU energy. In *Proc. of the First USENIX Symposium on Operating Systems Design and Implementation*, Monterey, CA, November 1994.

[41] R. Xu, D. Zhu, C. Rusu, R. Melhem, and D. Mossé. Energy efficient policies for embedded clusters. In *Proc. of the Conference on Language, Compilers, and Tools for Embedded Systems (LCTES)*, Chicago, IL, June 2005.

[42] P. Yang, C. Wong, P. Marchal, F. Catthoor, D. Desmet, D. Kerkest, and R. Lauwereins. Energy-aware runtime scheduling for embedded-multiprocessor SOCs. *IEEE Design & Test of Computers*, 18(5):46–58, 2001.

[43] F. Yao, A. Demers, and S. Shenker. A scheduling model for reduced CPU energy. In *Proc. of The 36th Annual Symposium on Foundations of Computer Science*, Milwaukee, WN, October 1995.

[44] Y. Yu and V. K. Prasanna. Power-aware resource allocation for independent tasks in heterogeneous real-time systems. In *Proc. of the 9th International Conference on Parallel and Distributed Systems*, Taiwan, ROC, December 2002.

[45] H. Zeng, X. Fan, C. Ellis, A. Lebeck, and A. Vahdat. Ecosystem: Managing energy as a first class operating system resource. In 10th *International Conference on Architectural Support for Programming Languages and Operating Systems (ASPLOS X)*, San Jose, CA, October 2002.

[46] Y. Zhang, X. Hu, and D. Z. Chen. Task scheduling and voltage selection for energy minimization. In *Proc. of the 39th Design Automation Conference*, New Orleans, LA, June 2002.

[47] D. Zhu, R. Melhem, and B. R. Childers. Scheduling with dynamic voltage/speed adjustment using slack reclamation in multi-processor real-time systems. *IEEE Transactions on Parallel and Distributed Systems*, 14(7):686–700, 2003.

[48] D. Zhu, R. Melhem, and D. Mossé. The effects of energy management on reliability in real-time embedded systems. In *Proc. of the International Conference on Computer Aidded Design (ICCAD)*, San Jose, CA, November 2004.

[49] D. Zhu, D. Mossé, and R. Melhem. Power aware scheduling for and/or graphs in real-time systems. *IEEE Transactions on Parallel and Distributed Systems*, 15(9):849–864, 2004.

41
Perspectives on Robust Resource Allocation for Heterogeneous Parallel and Distributed Systems

41.1	Introduction...	41-2
41.2	Generalized Robustness Metric	41-3
41.3	Derivations of Robustness Metric for Example Systems	41-6
	Independent Application Allocation System • The HiPer-D System • A System in Which Machine Failures Require Reallocation	
41.4	Robustness against Multiple Perturbation Parameters...	41-12
41.5	Computational Complexity	41-13
41.6	Demonstrating the Utility of the Proposed Robustness Metric	41-14
	Overview • Independent Application Allocation System • The HiPer-D System	
41.7	Designing Static Heuristics to Optimize Robustness for the HiPer-D System	41-18
	Overview • Computation and Communication Models • Heuristic Descriptions	
41.8	Simulation Experiments and Results	41-24
41.9	Related Work	41-25
41.10	Future Work...	41-27
41.11	Conclusions	41-27
Acknowledgments...		41-28
References		41-28

Shoukat Ali
Intel Corporation

Howard Jay Siegel
Anthony A. Maciejewski
Colorado State University

41.1 Introduction

This research focuses on the robustness of a resource allocation in parallel and distributed computing systems. What does robustness mean? Some dictionary definitions of robustness are (i) strong and healthy, as in "a robust person" or "a robust mind," (ii) sturdy or strongly formed, as in "a robust plastic," (iii) suited to or requiring strength as in "a robust exercise" or "robust work," (iv) firm in purpose or outlook as in "robust faith," (v) full-bodied as in "robust coffee," and (vi) rough or rude as in "stories laden with robust humor." In the context of resource allocation in parallel and distributed computing systems, how is the concept of robustness defined?

The allocation of resources to computational applications in heterogeneous parallel and distributed computer systems should maximize some system performance measure. Allocation decisions and associated performance prediction are often based on estimated values of application parameters, whose actual values may differ; for example, the estimates may represent only average values, or the models used to generate the estimates may have limited accuracy. Furthermore, parallel and distributed systems may operate in an environment where certain system performance features degrade due to unpredictable circumstances, such as sudden machine failures, higher than expected system load, or inaccuracies in the estimation of system parameters (e.g., References 2–4, 9, 10, 26, 31, 32, 34, and 38). Thus, an important research problem is the development of resource management strategies that can guarantee a particular system performance given bounds on such uncertainties. A resource allocation is defined to be robust with respect to specified system performance features against perturbations (uncertainties) in specified system parameters if degradation in these features is constrained when limited perturbations occur. An important question then arises: given a resource allocation, what extent of departure from the assumed circumstances will cause a performance feature to be unacceptably degraded? That is, how robust is the system?

Any claim of robustness for a given system must answer these three questions: (i) What behavior of the system makes it robust? (ii) What uncertainties is the system robust against? (iii) Quantitatively, exactly how robust is the system? To address these questions, we have designed a model for deriving the degree of robustness of a resource allocation, that is, the maximum amount of collective uncertainty in system parameters within which a user-specified level of system performance can be guaranteed. The model will be presented and we will demonstrate its ability to select the most robust resource allocation from among those that otherwise perform similarly (based on the primary performance criterion). The model's use in static (offline) resource allocation heuristics also will be demonstrated. In particular, we will describe a static heuristic designed to determine a robust resource allocation for one of the example distributed systems. In general, this work is applicable to different types of computing and communication environments, including parallel, distributed, cluster, grid, Internet, embedded, and wireless.

The rest of the chapter is organized as follows. Section 41.2 defines a generalized robustness metric. Derivations of this metric for three example parallel and distributed systems are given in Section 41.3. Section 41.4 extends the definition of the robustness metric given in Section 41.2 to multiple specified

TABLE 41.1 Glossary of Notation

Φ	The set of all performance features
ϕ_i	The i-th element in Φ
$\langle \beta_i^{\min}, \beta_i^{\max} \rangle$	A tuple that gives the bounds of the tolerable variation in ϕ_i
Π	The set of all perturbation parameters
π_j	The j-th element in Π
n_{π_j}	The dimension of vector π_j
μ	A resource allocation
$r_\mu(\phi_i, \pi_j)$	The robustness radius of resource allocation μ with respect to ϕ_i against π_j
$\rho_\mu(\Phi, \pi_j)$	The robustness of resource allocation μ with respect to set Φ against π_j
$\mathcal{A}$	The set of applications
$\mathcal{M}$	The set of machines
P	A weighted concatenation of the vectors $\pi_1, \pi_2, \ldots, \pi_{|\Pi|}$

perturbation parameters. The computational complexity of the robustness metric calculation is addressed in Section 41.5. Section 41.6 presents some experiments that highlight the usefulness of the robustness metric. Three static heuristics to derive a robust resource allocation for an example distributed system are described in Section 41.7, and are evaluated through simulations in Section 41.8. A sampling of the related work is given in Section 41.9. Section 41.10 gives some possibilities of future work in this area, and Section 41.11 concludes the chapter. A glossary of the notation used in this chapter is given in Table 41.1. Note that, throughout this chapter, new symbols are underlined when they are introduced. Such underlining is not a part of the symbology.

41.2 Generalized Robustness Metric

This section proposes a general procedure, called FePIA, for deriving a general robustness metric for any desired computing environment. The name for the above procedure stands for identifying (i) the performance features, (ii) the perturbation parameters, (iii) the impact of perturbation parameters on performance features, and (iv) the analysis to determine the robustness. Specific examples illustrating the application of the FePIA procedure to sample systems are given in the next section. Each step of the FePIA procedure is now described.

1. Describe quantitatively the requirement that makes the system robust. On the basis of this *robustness requirement*, determine the quality of service, (QoS), performance features that should be limited in variation to ensure that the robustness requirement is met. Identify the acceptable variation for these feature values as a result of uncertainties in system parameters. Consider an example where (i) the QoS performance feature is makespan (the total time it takes to complete the execution of a set of applications) for a given resource allocation, (ii) the acceptable variation is up to 120% of the makespan that was calculated for the given resource allocation using estimated execution times of applications on the machines they are assigned, and (iii) the uncertainties in system parameters are inaccuracies in the estimates of these execution times.
   Mathematically, let Φ be the set of system performance features that should be limited in variation. For each element $\phi_i \in \Phi$, quantitatively describe the tolerable variation in ϕ_i. Let $\langle \beta_i^{min}, \beta_i^{max} \rangle$ be a tuple that gives the bounds of the tolerable variation in the system feature ϕ_i. For the makespan example, ϕ_i is the time the i-th machine finishes its assigned applications, and its corresponding $\langle \beta_i^{min}, \beta_i^{max} \rangle$ could be $\langle 0, 1.2 \times \text{(estimated makespan value)} \rangle$.
2. Identify all of the system and environment uncertainty parameters whose values may impact the QoS performance features selected in step 1. These are called the perturbation parameters (these are similar to hazards in Reference 10), and the performance features are required to be robust with respect to these perturbation parameters. For the makespan example above, the resource allocation (and its associated estimated makespan) was based on the estimated application execution times. It is desired that the makespan be robust (stay within 120% of its estimated value) with respect to uncertainties in these estimated execution times.
   Mathematically, let Π be the set of perturbation parameters. It is assumed that the elements of Π are vectors. Let π_j be the j-th element of Π. For the makespan example, π_j could be the vector composed of the actual application execution times, that is, the i-th element of π_j is the actual execution time of the i-th application on the machine it was assigned. In general, representation of the perturbation parameters as separate elements of Π would be based on their nature or kind (e.g., message length variables in π_1 and computation time variables in π_2).
3. Identify the impact of the perturbation parameters in step 2 on the system performance features in step 1. For the makespan example, the sum of the actual execution times for all of the applications assigned a given machine is the time when that machine completes its applications. Note that 1 (i) implies that the actual time each machine finishes its applications must be within the acceptable variation.

Mathematically, for every $\phi_i \in \Phi$, determine the relationship $\phi_i = f_{ij}(\boldsymbol{\pi}_j)$, if any, that relates ϕ_i to $\boldsymbol{\pi}_j$. In this expression, f_{ij} is a function that maps $\boldsymbol{\pi}_j$ to ϕ_i. For the makespan example, ϕ_i is the finishing time for machine m_i, and f_{ij} would be the sum of execution times for applications assigned to machine m_i. The rest of this discussion will be developed assuming only one element in Π. The case where multiple perturbation parameters can affect a given ϕ_i simultaneously will be examined in Section 41.4.

4. The last step is to determine the smallest collective variation in the values of perturbation parameters identified in step 2 that will cause any of the performance features identified in step 1 to violate its acceptable variation. This will be the degree of robustness of the given resource allocation. For the makespan example, this will be some quantification of the total amount of inaccuracy in the execution times estimates allowable before the actual makespan exceeds 120% of its estimated value.

Mathematically, for every $\phi_i \in \Phi$, determine the *boundary values of* $\boldsymbol{\pi}_j$, that is, the values satisfying the boundary relationships $f_{ij}(\boldsymbol{\pi}_j) = \beta_i^{min}$ and $f_{ij}(\boldsymbol{\pi}_j) = \beta_i^{max}$. (If $\boldsymbol{\pi}_j$ is a discrete variable then the boundary values correspond to the closest values that bracket each boundary relationship. See Subsection 41.3.3 for an example.) These relationships separate the region of robust operation from that of nonrobust operation. Find the smallest perturbation in $\boldsymbol{\pi}_j$ that causes any $\phi_i \in \Phi$ to exceed the bounds $\langle \beta_i^{min}, \beta_i^{max} \rangle$ imposed on it by the robustness requirement.

Specifically, let $\boldsymbol{\pi}_j^{orig}$ be the value of $\boldsymbol{\pi}_j$ at which the system is originally assumed to operate. However, owing to inaccuracies in the estimated parameters or changes in the environment, the value of the variable $\boldsymbol{\pi}_j$ might differ from its assumed value. This change in $\boldsymbol{\pi}_j$ can occur in different "directions" depending on the relative differences in its individual components. Assuming that no information is available about the relative differences, all values of $\boldsymbol{\pi}_j$ are possible. Figure 41.1 illustrates this concept for a single feature, ϕ_i, and a two-element perturbation vector $\boldsymbol{\pi}_j \in \mathbf{R}^2$. The curve shown in Figure 41.1 plots the set of boundary points $\{\boldsymbol{\pi}_j | f_{ij}(\boldsymbol{\pi}_j) = \beta_i^{max}\}$ for a resource allocation μ. For this figure, the set of boundary points $\{\boldsymbol{\pi}_j | f_{ij}(\boldsymbol{\pi}_j) = \beta_i^{min}\}$ is given by the points on the π_{j1}-axis and π_{j2}-axis.

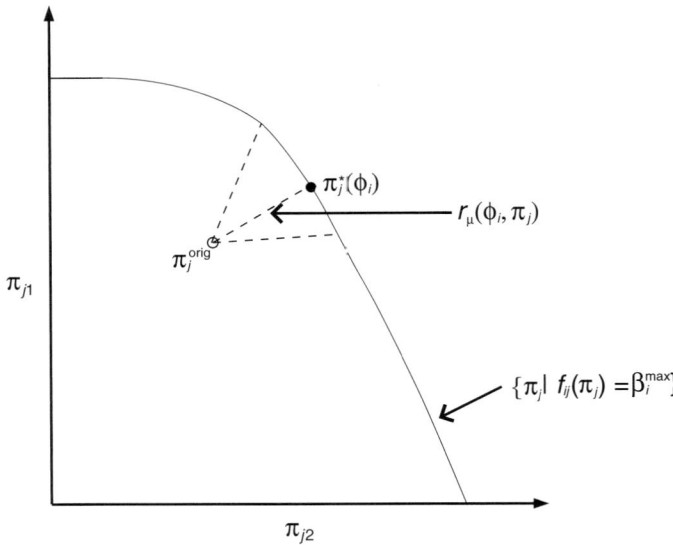

FIGURE 41.1 Some possible directions of increase of the perturbation parameter $\boldsymbol{\pi}_j$, and the direction of the smallest increase. The curve plots the set of points, $\{\boldsymbol{\pi}_j | f_{ij}(\boldsymbol{\pi}_j) = \beta_i^{max}\}$. The set of boundary points, $\{\boldsymbol{\pi}_j | f_{ij}(\boldsymbol{\pi}_j) = \beta_i^{min}\}$ is given by the points on the π_{j1}-axis and π_{j2}-axis.

The region enclosed by the axes and the curve gives the values of π_j for which the system is robust with respect to ϕ_i. For a vector $\mathbf{x} = [x_1\ x_2\ \cdots\ x_n]^T$, let $\|\mathbf{x}\|_2$ be the ℓ_2-norm (Euclidean norm) of the vector, defined by $\sqrt{\sum_{r=1}^{n} x_r^2}$. The point on the curve marked as $\pi_j^\star(\phi_i)$ has the property that the Euclidean distance from π_j^{orig} to $\pi_j^\star(\phi_i)$, $\|\pi_j^\star(\phi_i) - \pi_j^{\text{orig}}\|_2$, is the smallest over all such distances from π_j^{orig} to a point on the curve. An important interpretation of $\pi_j^\star(\phi_i)$ is that the value $\|\pi_j^\star(\phi_i) - \pi_j^{\text{orig}}\|_2$ gives the largest Euclidean distance that the variable π_j can change in *any* direction from the assumed value of π_j^{orig} without the performance feature ϕ_i exceeding the tolerable variation. Let the distance $\|\pi_j^\star(\phi_i) - \pi_j^{\text{orig}}\|_2$ be called the robustness radius, $r_\mu(\phi_i, \pi_j)$, of ϕ_i against π_j. Mathematically,

$$r_\mu(\phi_i, \pi_j) = \min_{\pi_j:\, (f_{ij}(\pi_j) = \beta_i^{\max}) \vee (f_{ij}(\pi_j) = \beta_i^{\min})} \|\pi_j - \pi_j^{\text{orig}}\|_2. \qquad (41.1)$$

This work defines $r_\mu(\phi_i, \pi_j)$ to be the *robustness of resource allocation μ with respect to performance feature ϕ_i against the perturbation parameter π_j*.

The robustness definition can be extended easily for all $\phi_i \in \Phi$ (see Figure 41.2). The robustness metric is simply the minimum of all robustness radii. Mathematically, let

$$\rho_\mu(\Phi, \pi_j) = \min_{\phi_i \in \Phi} \left(r_\mu(\phi_i, \pi_j) \right). \qquad (41.2)$$

Then, $\rho_\mu(\Phi, \pi_j)$ is the *robustness metric of resource allocation μ with respect to the performance feature set Φ against the perturbation parameter π_j*. Figure 41.2 shows a small system with a single two-element perturbation parameter, and only two performance features.

Even though the ℓ_2-norm has been used for the robustness radius in this general formulation, in practice, the choice of a norm should depend on the particular environment for which a robustness

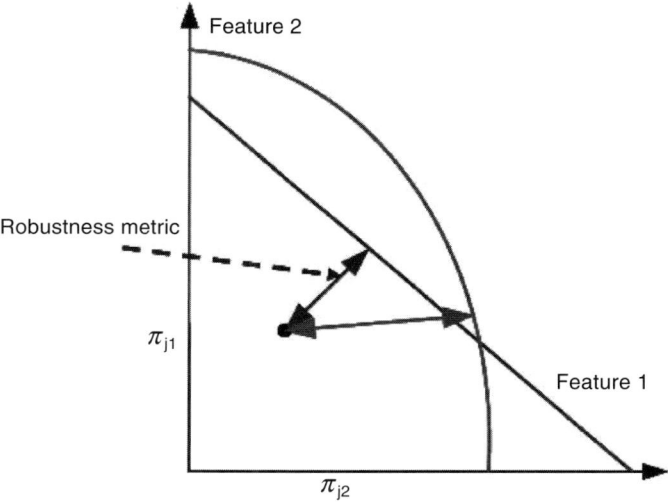

FIGURE 41.2 This figure shows a small system with a single perturbation parameter with elements π_{j1} and π_{j2}, and only two performance features. There is one robustness radius for each performance feature. The robustness metric is the smallest of the two robustness radii.

measure is being sought. Subsection 41.3.3 gives an example situation where the ℓ_1-norm is preferred over the ℓ_2-norm.

In addition, in some situations, changes in some elements of π_j may be more probable than changes in other elements. In such cases, one may be able to modify the distance calculation so that the contribution from an element with a larger probability to change has a proportionally larger weight. This is a subject for future study.

41.3 Derivations of Robustness Metric for Example Systems

41.3.1 Independent Application Allocation System

The first example derivation of the robustness metric is for a system that allocates a set of independent applications to a set of machines [12]. In this system, it is required that the makespan be robust against errors in application execution time estimates. Specifically, the actual makespan under the perturbed execution times must be no more than a certain factor (>1) times the estimated makespan calculated using the assumed execution times. It is obvious that the larger the "factor," the larger the robustness. Assuming that ℓ_2-norm is used, one might also reason that as the number of applications assigned to a given machine increases, the change in the finishing time for that machine will increase due to errors in the application computation times. As will be seen shortly, the instantiation of the general framework for this system does reflect this intuition.

A brief description of the system model is now given. The applications are assumed to be independent, that is, no communications between the applications are needed. The set $\mathcal{A}$ of applications is to be allocated to a set $\mathcal{M}$ of machines so as to minimize the makespan. Each machine executes a single application at a time (i.e., no multitasking), in the order in which the applications are assigned. Let C_{ij} be the estimated time to compute (ETC) for application a_i on machine m_j. It is assumed that C_{ij} values are known for all i, j, and a resource allocation μ is determined using the ETC values. In addition, let F_j be the time at which m_j finishes executing all of the applications allocated to it.

Assume that unknown inaccuracies in the ETC values are expected, requiring that the resource allocation μ be robust against them. More specifically, it is required that, for a given resource allocation, its actual makespan value M (calculated considering the effects of ETC errors) may be no more than τ (>1) times its estimated value, M^{orig}. The estimated value of the makespan is the value calculated assuming the ETC values are accurate (see Figure 41.3). Following step 1 of the FePIA procedure in Section 41.2, the system performance features that should be limited in variation to ensure the makespan robustness are the finishing times of the machines. That is, $\Phi = \{F_j | 1 \leq j \leq |\mathcal{M}|\}$.

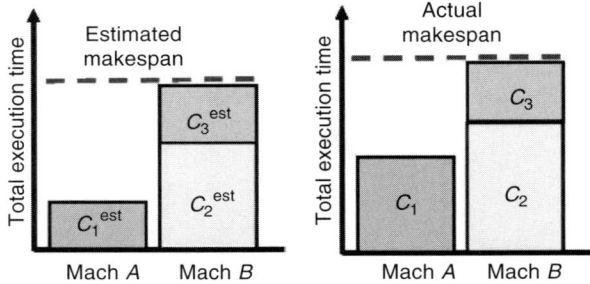

FIGURE 41.3 Three applications are executing in this 2-machine cluster. The estimated value of the makespan, calculated using the estimated computation times, is equal to the estimated finishing time of machine B in the graph on left. The actual makespan value, calculated using the actual computation times, is equal to the actual finishing time of machine B in the graph on right, and is more than the estimated value.

According to step 2 of the FePIA procedure, the perturbation parameter needs to be defined. Let C_i^{orig} be the ETC value for application a_i on the machine where it is allocated by resource allocation μ. Let C_i be equal to the actual computation time value (C_i^{orig} plus the estimation error). In addition, let $\boldsymbol{C}$ be the vector of the C_i values, such that $\boldsymbol{C} = [C_1\ C_2 \cdots C_{|\mathcal{A}|}]$. Similarly, $\boldsymbol{C}^{\text{orig}} = [C_1^{\text{orig}}\ C_2^{\text{orig}} \cdots C_{|\mathcal{A}|}^{\text{orig}}]$. The vector $\boldsymbol{C}$ is the perturbation parameter for this analysis.

In accordance with step 3 of the FePIA procedure, F_j has to be expressed as a function of $\boldsymbol{C}$. To that end,

$$F_j(\boldsymbol{C}) = \sum_{i:\, a_i \text{ is allocated to } m_j} C_i. \tag{41.3}$$

Note that the finishing time of a given machine depends only on the actual execution times of the applications allocated to that machine, and is independent of the finishing times of the other machines. Following step 4 of the FePIA procedure, the set of boundary relationships corresponding to the set of performance features is given by $\{F_j(\boldsymbol{C}) = \tau M^{\text{orig}} |\ 1 \leq j \leq |\mathcal{M}|\}$.

For a two-application system, $\boldsymbol{C}$ corresponds to $\boldsymbol{\pi}_j$ in Figure 41.1. Similarly, C_1 and C_2 correspond to π_{j1} and π_{j2}, respectively. The terms $\boldsymbol{C}^{\text{orig}}$, $F_j(\boldsymbol{C})$, and τM^{orig} correspond to $\boldsymbol{\pi}_j^{\text{orig}}$, $f_{ij}(\boldsymbol{\pi}_j)$, and $\beta_i^{\max}$, respectively. The boundary relationship "$F_j(\boldsymbol{C}) = \tau M^{\text{orig}}$" corresponds to the boundary relationship "$f_{ij}(\boldsymbol{\pi}_j) = \beta_i^{\max}$."

From Equation 41.1, the robustness radius of F_j against $\boldsymbol{C}$ is given by

$$r_\mu(F_j,\ \boldsymbol{C}) = \min_{\boldsymbol{C}:\ F_j(\boldsymbol{C})=\tau M^{\text{orig}}} \|\boldsymbol{C} - \boldsymbol{C}^{\text{orig}}\|_2. \tag{41.4}$$

That is, if the Euclidean distance between any vector of the actual execution times and the vector of the estimated execution times is no larger than $r_\mu(F_j,\ \boldsymbol{C})$, then the finishing time of machine m_j will be at most τ times the estimated makespan value.

For example, assume only applications a_1 and a_2 have been assigned to machine j (depicted in Figure 41.4), and $\boldsymbol{C}$ has two components C_1 and C_2 that correspond to execution times of a_1 and a_2 on machine j, respectively. The term $F_j(\boldsymbol{C}^{\text{orig}})$ is the finishing time for machine j computed on the basis of the ETC values of applications a_1 and a_2. Note that the right hand side in Equation 41.4 can be interpreted as the perpendicular distance from the point $\boldsymbol{C}^{\text{orig}}$ to the hyperplane described by the

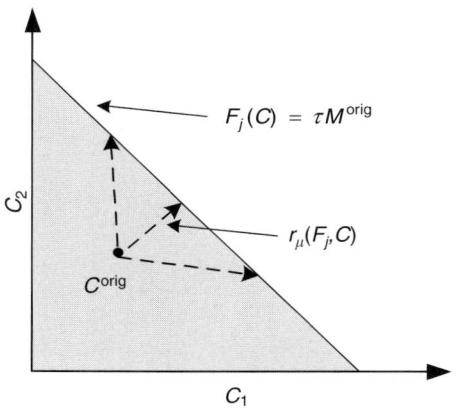

FIGURE 41.4 Some possible directions of increase of the perturbation parameter $\boldsymbol{C}$. The set of boundary points is given by $F_j(\boldsymbol{C}) = \tau M^{\text{orig}}$. The robustness radius $r_\mu(F_j,\ \boldsymbol{C})$ corresponds to the smallest increase that can reach the boundary. The shaded region represents the area of robust operation.

equation $\tau M^{orig} - F_j(C) = 0$. Let $n(m_j)$ be the number of applications allocated to machine m_j. Using the point-to-plane distance formula [42], Equation 41.4 reduces to

$$r_\mu(F_j, C) = \frac{\tau M^{orig} - F_j(C^{orig})}{\sqrt{n(m_j)}}. \tag{41.5}$$

The robustness metric, from Equation 41.2, is $\rho_\mu(\Phi, C) = \min_{F_j \in \Phi} r_\mu(F_j, C)$. That is, if the Euclidean distance between any vector of the actual execution times and the vector of the estimated execution times is no larger than $\rho_\mu(\Phi, C)$, then the actual makespan will be at most τ times the estimated makespan value. The value of $\rho_\mu(\Phi, C)$ has the units of C, namely, time.

41.3.2 The HiPer-D System

The second example derivation of the robustness metric is for a HiPer-D [28] like system that allocates a set of continuously executing, communicating applications to a set of machines. It is required that the system be robust with respect to certain QoS attributes against unforeseen increases in the "system load."

The HiPer-D system model used here was developed in Reference 2, and is summarized here for reference. The system consists of heterogeneous sets of sensors, applications, machines, and actuators. Each machine is capable of multitasking, executing the applications allocated to it in a round-robin fashion. Similarly, a given network link is multitasked among all data transfers using that link. Each sensor produces data periodically at a certain rate, and the resulting data streams are input into applications. The applications process the data and send the output to other applications or to actuators. The applications and the data transfers between them are modeled with a directed acyclic graph, shown in Figure 41.5. The figure also shows a number of *paths* (enclosed by dashed lines) formed by the applications. A path is a chain of producer–consumer pairs that starts at a sensor (the driving sensor) and ends at an actuator (if it is a trigger path) or at a multiple-input application (if it is an update path). In the context of Figure 41.5, path 1 is a trigger path, and path 2 is an update path. In a real system, application d could be a missile firing program that produces an order to fire. It needs target coordinates from application b in path 1, and an updated map of the terrain from application c in path 2. Naturally, application d must respond to any output from b, but must not issue fire orders if it receives an output from c alone; such an output is used only to update an internal database. So while d is a multiple input application, the rate at which it produces data is equal to the rate at which the trigger application b produces data (in the HiPer-D model). That rate,

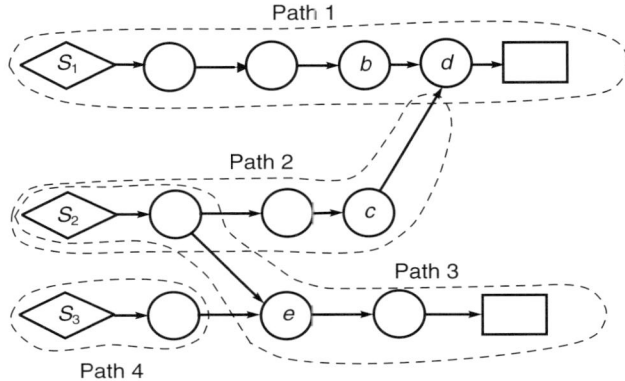

FIGURE 41.5 The DAG model for the applications (circles) and data transfers (arrows). The diamonds and rectangles denote sensors and actuators, respectively. The dashed lines enclose each path formed by the applications.

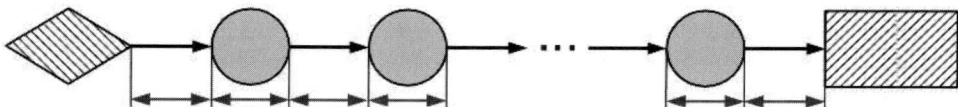

FIGURE 41.6 There is one throughput constraint for each application and one for each data transfer. It imposes a limit on each application's computation time and its data transfer time.

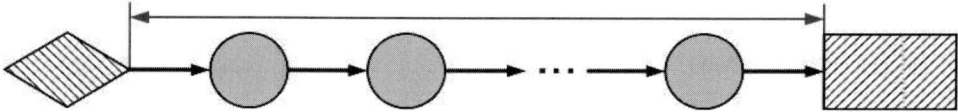

FIGURE 41.7 There is one latency constraint for each path. It imposes a limit on the time taken from the sensor output to the path "end point," which could be an actuator or a multiple-input application.

in turn, equals the rate at which the driving sensor, S_1, produces data. The problem specification indicates the path to which each application belongs, and the corresponding driving sensor. Example uses of this path model include defense, monitoring vital signs medical patients, recording scientific experiments, and surveillance for homeland security.

Let $\mathcal{P}$ be the set of all paths, and $\mathcal{P}_k$ be the list of applications that belong to the k-th path. Note that an application may be present in multiple paths. As in Subsection 41.3.1, $\mathcal{A}$ is the set of applications.

The sensors constitute the interface of the system to the external world. Let the maximum periodic data output rate from a given sensor be called its output data rate. The minimum throughput constraint states that the computation or communication time of any application in $\mathcal{P}_k$ is required to be no larger than the reciprocal of the output data rate of the driving sensor for $\mathcal{P}_k$ (see Figure 41.6). For application $a_i \in \mathcal{P}_k$, let $R(a_i)$ be set to the output data rate of the driving sensor for $\mathcal{P}_k$. In addition, let T_{ij}^c be the computation time for application a_i allocated to machine m_j. Also, let T_{ip}^n be the time to send data from application a_i to application a_p. Because this analysis is being carried out for a specific resource allocation, the machine where a given application is allocated is known. Let $\mathcal{D}(a_i)$ be the set of successor applications of a_i. It is assumed that a_i is allocated to m_j, and the machine subscript for T_{ij}^c is omitted in the ensuing analysis for clarity unless the intent is to show the relationship between execution times of a_i at various possible machines. For the throughput constraint, $T_i^c \leq 1/R(a_i)$ and $T_{ip}^n \leq 1/R(a_i)$ for all $a_p \in \mathcal{D}(a_i)$.

The maximum end-to-end latency constraint states that, for a given path $\mathcal{P}_k$, the time taken between the instant the driving sensor outputs a data set until the instant the actuator or the multiple-input application fed by the path receives the result of the computation on that data set must be no greater than a given value, $L_k^{\max}$ (see Figure 41.7). Let L_k be the actual (as opposed to the maximum allowed) value of the end-to-end latency for $\mathcal{P}_k$. The quantity L_k can be found by adding the computation and communication times for all applications in $\mathcal{P}_k$ (including any sensor or actuator communications)

$$L_k = \sum_{\substack{i:\, a_i \in \mathcal{P}_k \\ p:\, (a_p \in \mathcal{P}_k) \wedge (a_p \in \mathcal{D}(a_i))}} \left[T_i^c + T_{ip}^n \right]. \tag{41.6}$$

It is desired that a given resource allocation μ of the system be robust with respect to the satisfaction of two QoS attributes: the latency and throughput constraints. Following step 1 of the FePIA procedure in Section 41.2, the system performance features that should be limited in variation are the latency values for the paths and the computation and communication time values for the applications. The set Φ is given by

$$\Phi = \{T_i^c | \ 1 \leq i \leq |\mathcal{A}|\} \bigcup \{T_{ip}^n | \ (1 \leq i \leq |\mathcal{A}|) \land (\text{for } p \text{ where } a_p \in \mathcal{D}(a_i))\} \bigcup \{L_k | \ 1 \leq k \leq |\mathcal{P}|\} \tag{41.7}$$

This system is expected to operate under uncertain outputs from the sensors, requiring that the resource allocation μ be robust against unpredictable increases in the sensor outputs. Let λ_z be the output from the z-th sensor in the set of sensors, and be defined as the number of objects present in the most recent data set from that sensor. The system workload, λ, is the vector composed of the load values from all sensors. Let $\boldsymbol{\lambda}^{\text{orig}}$ be the initial value of $\boldsymbol{\lambda}$, and λ_i^{orig} be the initial value of the i-th member of $\boldsymbol{\lambda}^{\text{orig}}$. Following step 2 of FePIA, the perturbation parameter π_j is identified to be $\boldsymbol{\lambda}$.

Step 3 of the FePIA procedure requires that the impact of $\boldsymbol{\lambda}$ on each of the system performance features be identified. The computation times of different applications (and the communication times of different data transfers) are likely to be of different complexities with respect to $\boldsymbol{\lambda}$. Assume that the dependence of T_i^c and T_{ip}^n on $\boldsymbol{\lambda}$ is known (or can be estimated) for all i, p. Given that, T_i^c and T_{ip}^n can be reexpressed as functions of $\boldsymbol{\lambda}$ as $T_i^c(\boldsymbol{\lambda})$ and $T_{ip}^n(\boldsymbol{\lambda})$, respectively. Even though d is triggered only by b, its computation time depends on the outputs from both b and c. In general, $T_i^c(\boldsymbol{\lambda})$ and $T_{ip}^n(\boldsymbol{\lambda})$ will be functions of the loads from all those sensors that can be traced back from a_i. For example, the computation time for application d in Figure 41.5 is a function of the loads from sensors S_1 and S_2, but that for application e is a function of S_2 and S_3 loads (but each application has just one driving sensor: S_1 for d and S_2 for e). Then Equation 41.6 can be used to express L_k as a function of $\boldsymbol{\lambda}$.

Following step 4 of the FePIA procedure, the set of boundary relationships corresponding to Equation 41.7 is given by

$$\{T_i^c(\boldsymbol{\lambda}) = 1/R(a_i) | \ 1 \leq i \leq |\mathcal{A}|\} \bigcup$$
$$\{T_{ip}^n(\boldsymbol{\lambda}) = 1/R(a_i) | \ (1 \leq i \leq |\mathcal{A}|) \land (\text{for } p \text{ where } a_p \in \mathcal{D}(a_i))\} \bigcup$$
$$\{L_k(\boldsymbol{\lambda}) = L_k^{\max} | \ 1 \leq k \leq |\mathcal{P}|\}.$$

Then, using Equation 41.1, one can find, for each $\phi_i \in \Phi$, the robustness radius, $r_\mu(\phi_i, \boldsymbol{\lambda})$. Specifically,

$$r_\mu(\phi_i, \boldsymbol{\lambda}) = \begin{cases} \min_{\boldsymbol{\lambda}: T_x^c(\boldsymbol{\lambda})=1/R(a_x)} \|\boldsymbol{\lambda} - \boldsymbol{\lambda}^{\text{orig}}\|_2 & \text{if } \phi_i = T_x^c \quad (41.8a) \\ \min_{\boldsymbol{\lambda}: T_{xy}^n(\boldsymbol{\lambda})=1/R(a_x)} \|\boldsymbol{\lambda} - \boldsymbol{\lambda}^{\text{orig}}\|_2 & \text{if } \phi_i = T_{xy}^n \quad (41.8b) \\ \min_{\boldsymbol{\lambda}: L_k(\boldsymbol{\lambda})=L_k^{\max}} \|\boldsymbol{\lambda} - \boldsymbol{\lambda}^{\text{orig}}\|_2 & \text{if } \phi_i = L_k \quad (41.8c) \end{cases}$$

The robustness radius in Equation 41.8a is the largest increase (Euclidean distance) in load in any direction (i.e., for any combination of sensor load values) from the assumed value that does not cause a throughput violation for the computation of application a_x. This is because it corresponds to the value of $\boldsymbol{\lambda}$ for which the computation time of a_x will be at the allowed limit of $1/R(a_x)$. The robustness radii in Equations 41.8b and 41.8c are the similar values for the communications of application a_x and the latency of path $\mathcal{P}_k$, respectively. The robustness metric, from Equation 41.2, is given by $\rho_\mu(\Phi, \boldsymbol{\lambda}) = \min_{\phi_i \in \Phi} (r_\mu(\phi_i, \boldsymbol{\lambda}))$. For this system, $\rho_\mu(\Phi, \boldsymbol{\lambda})$ is the largest increase in load in any direction from the assumed value that does not cause a latency or throughput violation for any application or path. Note that $\rho_\mu(\Phi, \boldsymbol{\lambda})$ has the units of $\boldsymbol{\lambda}$, namely, objects per data set. In addition, note that although $\boldsymbol{\lambda}$ is a discrete variable, it has been treated as a continuous variable in Equation 41.8 for the purpose of simplifying the illustration. A method for handling a discrete perturbation parameter is discussed in Subsection 41.3.3.

41.3.3 A System in Which Machine Failures Require Reallocation

In many research efforts (e.g., References 27, 31, and 38), flexibility of a resource allocation has been closely tied to its robustness, and is described as the quality of the resource allocation that can allow it to be changed easily into another allocation of comparable performance when system failures occur. This section briefly sketches the use of the FePIA procedure to derive a robustness metric for systems where resource reallocation becomes necessary owing to dynamic machine failures. In the example of derivation analysis given below, it is assumed that resource reallocation is invoked because of permanent simultaneous failure of a number of machines in the system (e.g., owing to a power failure in a section of a building).

In our example, for the system to be robust, it is required that (i) the total number of applications that need to be reassigned, $N^{\text{re-asgn}}$, has to be less than $\tau_1\%$ of the total number of applications, and (ii) the value of a given objective function (e.g., average application response time), J, should not be any more than τ_2 times its value, J^{orig}, for the original resource allocation. It is assumed that there is a specific resource reallocation algorithm, which may not be the same as the original resource allocation algorithm. The resource reallocation algorithm will reassign the applications originally allocated to the failed machines to other machines, as well as reassign some other applications if necessary (e.g., as done in Reference 40). As in Subsection 41.3.1, $\mathcal{A}$ and $\mathcal{M}$ are the sets of applications and machines, respectively.

Following step 1 of the FePIA procedure, $\Phi = \{N^{\text{re-asgn}}, J\}$. Step 2 requires that the perturbation parameter π_j be identified. Let that be F, a vector that indicates the identities of the machines that have failed. Specifically, $F = [f_1 \, f_2 \, \cdots \, f_{|\mathcal{M}|}]^T$ such that f_j is 1 if m_j fails, and is 0 otherwise. The vector F^{orig} corresponds to the original value of F, which is $[0 \, 0 \, \cdots \, 0]^T$.

Step 3 asks for identifying the impact of F on $N^{\text{re-asgn}}$ and J. The impact depends on the resource reallocation algorithm, as well as F, and can be determined from the resource allocation produced by the resource reallocation algorithm. Then $N^{\text{re-asgn}}$ and J can be reexpressed as functions of F as $N^{\text{re-asgn}}(F)$ and $J(F)$, respectively.

Following step 4, the set of boundary values of F needs to be identified. However, F is a discrete variable. The boundary relationships developed for a continuous π_j, that is, $f_{ij}(\pi_j) = \beta_i^{\min}$ and $f_{ij}(\pi_j) = \beta_i^{\max}$, will not apply because it is possible that no value of π_j will lie on the boundaries $\beta_i^{\min}$ and $\beta_i^{\max}$. Therefore, one needs to determine all those pairs of the values of F such that the values in a given pair bracket a given boundary ($\beta_i^{\min}$ or $\beta_i^{\max}$). For a given pair, the "boundary value" is taken to be the value that falls in the robust region. Let $F^{(+1)}$ be a perturbation parameter value such that the machines that fail in the scenario represented by $F^{(+1)}$ include the machines that fail in the scenario represented by F, and exactly one other machine. Then, for $\phi_1 = N^{\text{re-asgn}}$, the set of "boundary values" for F is the set of all those "inner bracket" values of F for which the number of applications that need to be reassigned is less than the maximum tolerable number. Mathematically,

$$\left\{ F | \left(N^{\text{re-asgn}}(F) \leq \tau_1 |\mathcal{A}|\right) \wedge \left(\exists F^{(+1)} \quad N^{\text{re-asgn}}(F^{(+1)}) > \tau_1 |\mathcal{A}|\right) \right\}.$$

For $\phi_2 = J$, the set of "boundary values" for F can be written as

$$\left\{ F | \left(J(F) \leq \tau_2 J^{\text{orig}}\right) \wedge \left(\exists F^{(+1)} \quad J(F^{(+1)}) > \tau_2 J^{\text{orig}}\right) \right\}.$$

Then, using Equation 41.1, one can find, the robustness radii for the set of constraints given above. However, for this system, it is more intuitive to use the ℓ_1-norm (defined as $\sum_{r=1}^{n} |x_r|$ for a vector $\mathbf{x} = [x_1 \, x_2 \, \cdots \, x_n]^T$) for use in the robustness metric. This is because, with the ℓ_2-norm, the term $\|F - F^{\text{orig}}\|_2$ equals the square root of the number of the machines that fail, rather than the (more natural) number of machines that fail. Specifically, using the ℓ_1-norm,

$$r_\mu(N^{\text{re-asgn}}, F) = \min_{F:\ (N^{\text{re-asgn}}(F) \leq \tau_1|\mathcal{A}|) \wedge (\exists F^{(+1)}\ N^{\text{re-asgn}}(F^{(+1)}) > \tau_1|\mathcal{A}|)} \|F - F^{\text{orig}}\|_1, \quad (41.9)$$

and

$$r_\mu(J, F) = \min_{F:\ (J(F) \leq \tau_2 J^{\text{orig}}) \wedge (\exists F^{(+1)}\ J(F^{(+1)}) > \tau_2 J^{\text{orig}})} \|F - F^{\text{orig}}\|_1. \quad (41.10)$$

The robustness radius in Equation 41.9 is the largest number of machines that can fail in any combination without causing the number of applications that have to be reassigned to exceed $\tau_1|\mathcal{A}|$. Similarly, the robustness radius in Equation 41.10 is the largest number of machines that can fail in any combination without causing the objective function in the reallocated system to degrade beyond $\tau_2 J(F)$. The robustness metric, from Equation 41.2, is given by $\rho_\mu(\Phi, F) = \min(r_\mu(N^{\text{re-asgn}}, F), r_\mu(J, F))$. As in Subsection 41.3.2, it is assumed here that the discrete optimization problems posed in Equations 41.9 and 41.10 can be solved for optimal or near-optimal solutions using combinatorial optimization techniques [37].

To determine if the robustness metric value is k, the reallocation algorithm must be run for all combinations of k machines failures out of a total of $|\mathcal{M}|$ machines. Assuming that the robustness value is small enough, for example five machine failures in a set of 100 machines, then the number of combinations would be small enough to be computed offline in a reasonable time. If one is using a fast greedy heuristic for reallocation (e.g., those presented in Reference 2), the complexity would be $O(|\mathcal{A}||\mathcal{M}|)$ for each combination of failures considered. For a Min-min-like greedy heuristic (shown to be effective for many heterogeneous computing systems, see Reference 2 and the references provided at the end of Reference 2), the complexity would be $O(|\mathcal{A}|^2|\mathcal{M}|)$ for each combination of failures considered.

41.4 Robustness against Multiple Perturbation Parameters

Section 41.2 developed the analysis for determining the robustness metric for a system with a single perturbation parameter. In this section, that analysis is extended to include multiple perturbation parameters.

Our approach for handling multiple perturbation parameters is to concatenate them into one parameter, which is then used as a single parameter as discussed in Section 41.2. Specifically, this section develops an expression for the robustness radius for a single performance feature, ϕ_i, and multiple perturbation parameters. Then the robustness metric is determined by taking the minimum over the robustness radii of all $\phi_i \in \Phi$.

Let the vector π_j have n_{π_j} elements, and let $\diamond$ be the vector concatenation operator, so that $\pi_1 \diamond \pi_2 = [\pi_{11}\ \pi_{12}\ \ldots\ \pi_{1n_{\pi_1}}\ \pi_{21}\ \pi_{22}\ \ldots\ \pi_{2n_{\pi_2}}]^T$. Let $\mathbf{P}$ be a weighted concatenation of the vectors $\pi_1, \pi_2, \ldots, \pi_{|\Pi|}$. That is, $\mathbf{P} = (\alpha_1 \times \pi_1) \diamond (\alpha_2 \times \pi_2) \diamond \ldots \diamond (\alpha_{|\Pi|} \times \pi_{|\Pi|})$, where α_j ($1 \leq j \leq |\Pi|$) is a weighting constant that may be assigned by a system administrator or be based on the sensitivity of the system performance feature ϕ_i toward π_j (explained in detail later).

The vector $\mathbf{P}$ is analogous to the vector π_j discussed in Section 41.2. Parallel to the discussion in Section 41.2, one needs to identify the set of boundary values of $\mathbf{P}$. Let f_i be a function that maps $\mathbf{P}$ to ϕ_i. (Note that f_i could be independent of some π_j.) For the single system feature ϕ_i being considered, such a set is given by $\{\mathbf{P} | (f_i(\mathbf{P}) = \beta_i^{\min}) \bigvee (f_i(\mathbf{P}) = \beta_i^{\max})\}$.

Let $\mathbf{P}^{\text{orig}}$ be the assumed value of $\mathbf{P}$. In addition, let $\mathbf{P}^\star(\phi_i)$ be, analogous to $\pi_j^\star(\phi_i)$, the element in the set of boundary values such that the Euclidean distance from $\mathbf{P}^{\text{orig}}$ to $\mathbf{P}^\star(\phi_i)$, $\|\mathbf{P}^\star(\phi_i) - \mathbf{P}^{\text{orig}}\|_2$, is the smallest over all such distances from $\mathbf{P}^{\text{orig}}$ to a point in the boundary set. Alternatively, the value $\|\mathbf{P}^\star(\phi_i) - \mathbf{P}^{\text{orig}}\|_2$ gives the largest Euclidean distance that the variable $\mathbf{P}$ can move in *any* direction from an assumed value

Perspectives on Robust Resource Allocation

of $\mathbf{P}^{\text{orig}}$ without exceeding the tolerable limits on ϕ_i. Parallel to the discussion in Section 41.2, let the distance $\|\mathbf{P}^\star(\phi_i) - \mathbf{P}^{\text{orig}}\|_2$ be called the robustness radius, $r_\mu(\phi_i, \mathbf{P})$, of ϕ_i against $\mathbf{P}$. Mathematically,

$$r_\mu(\phi_i, \mathbf{P}) = \min_{\mathbf{P}:\, (f_i(\mathbf{P})=\beta_i^{\min})\,\vee\,(f_i(\mathbf{P})=\beta_i^{\max})} \|\mathbf{P} - \mathbf{P}^{\text{orig}}\|_2. \tag{41.11}$$

Extending for all $\phi_i \in \Phi$, the robustness of resource allocation μ with respect to the performance feature set Φ against the perturbation parameter set Π is given by $\rho_\mu(\Phi, \mathbf{P}) = \min_{\phi_i \in \Phi} \left(r_\mu(\phi_i, \mathbf{P})\right)$.

The sensitivity-based weighting procedure for the calculation of α_js is now discussed. Typically, π_1, $\pi_2, \ldots, \pi_{|\Pi|}$ will have different dimensions, that is, will be measured in different units, for example, seconds, objects per data set, bytes. Before the concatenation of these vectors into $\mathbf{P}$, they should be converted into a single dimension. In addition, for a given ϕ_i, the magnitudes of α_j should indicate the relative sensitivities of ϕ_i to different π_js. One way to accomplish the above goals is to set $\alpha_j = 1/r_\mu(\phi_i, \pi_j)$. With this definition of α_j,

$$\mathbf{P} = \frac{\pi_1}{r_\mu(\phi_i, \pi_1)} \diamond \frac{\pi_2}{r_\mu(\phi_i, \pi_2)} \diamond \cdots \diamond \frac{\pi_{|\Pi|}}{r_\mu(\phi_i, \pi_{|\Pi|})}. \tag{41.12}$$

Note that a smaller value of $r_\mu(\phi_i, \pi_j)$ makes α_j larger. This is desirable because a small value of the robustness against π_j indicates that ϕ_i has a big sensitivity to changes in π_j, and therefore the relative weight of π_j should be large. Also, note that the units of $r_\mu(\phi_i, \pi_j)$ are the units of π_j. This fact renders $\mathbf{P}$ dimensionless.

41.5 Computational Complexity

To calculate the robustness radius, one needs to solve the optimization problem posed in Equation 41.1. Such a computation could potentially be very expensive. However, one can exploit the structure of this problem, along with some assumptions, to make this problem somewhat easier to solve. An optimization problem of the form $\min_{x:\, l(x)=0} f(x)$ or $\min_{x:\, c(x)\geq 0} f(x)$ could be solved very efficiently to find the global minimum if $f(x)$, $l(x)$, and $c(x)$ are convex, linear, and concave functions, respectively. Some solution approaches, including the well-known interior-point methods, for such *convex optimization* problems are presented in Reference 11.

Because all norms are convex functions [11], the optimization problem posed in Equation 41.1 reduces to a convex optimization problem if $f_{ij}(\pi_j)$ is linear. One interesting problem with linear $f_{ij}(\pi_j)$ is given in Subsection 41.3.1.

If $f_{ij}(\pi_j)$ is concave, and the constraint "$f_{ij}(\pi_j) = \beta_i^{\min}$" is irrelevant for some scenario (as it is for the system in Subsection 41.3.2 where the latency of a path must be no larger than a certain limit, but can be arbitrarily small), then once again the problem reduces to a convex optimization problem. Because the distance from a point to the boundary of a region is the same as the distance from the point to the region itself, $\min_{\pi_j:\, (f_{ij}(\pi_j)=\beta_i^{\max})} \|\pi_j - \pi_j^{\text{orig}}\|_2$ is equivalent to $\min_{\pi_j:\, (f_{ij}(\pi_j)\geq \beta_i^{\max})} \|\pi_j - \pi_j^{\text{orig}}\|_2$. In such a case, the optimization problem would still be convex (and efficiently solvable) even if $f_{ij}(\pi_j)$ were concave [11].

Similarly, if $f_{ij}(\pi_j)$ is convex, and the constraint "$f_{ij}(\pi_j) = \beta_i^{\max}$" is irrelevant for some scenario (e.g., for a network, the throughput must be no smaller than a certain value, but can be arbitrarily large), then the optimization problem reduces to a convex optimization problem.

However, if the above conditions are not met, the optimization problem posed in Equation 41.1 could still be solved for near-optimal solutions using heuristic approaches (some examples are given in Reference 15).

41.6 Demonstrating the Utility of the Proposed Robustness Metric

41.6.1 Overview

The experiments in this section seek to establish the utility of the robustness metric in distinguishing between resource allocations that perform similarly in terms of a commonly used metric, such as makespan. Two different systems were considered: the independent task allocation system discussed in Subsection 41.3.1 and the HiPer-D system outlined in Subsection 41.3.2. Experiments were performed for a system with five machines and 20 applications. A total of 1000 resource allocations were generated by assigning a randomly chosen machine to each application, and then each resource allocation was evaluated with the robustness metric and the commonly used metric.

41.6.2 Independent Application Allocation System

For the system in Subsection 41.3.1, the ETC values were generated by sampling a Gamma distribution. The mean was arbitrarily set to 10, the task heterogeneity was set to 0.7, and the machine heterogeneity was also set to 0.7 (the heterogeneity of a set of numbers is the standard deviation divided by the mean). See Reference 5 for a description of a method for generating random numbers with given mean and heterogeneity values.

The resource allocations were evaluated for robustness, makespan, and load balance index (defined as the ratio of the finishing time of the machine that finishes first to the makespan). The larger the value of the load balance index, the more balanced the load (the largest value being 1). The tolerance, τ, was set to 120% (i.e., the actual makespan could be no more than 1.2 times the estimated value). In this context, a robustness value of x for a given resource allocation means that the resource allocation can endure any combination of ETC errors without the makespan increasing beyond 1.2 times its estimated value as long as the Euclidean norm of the errors is no larger than x seconds.

Figure 41.8a shows the "normalized robustness" of a resource allocation against its makespan. The normalized robustness equals the absolute robustness divided by the estimated makespan. A similar graph for the normalized robustness against the load balance index is shown in Figure 41.8b. It can be seen in Figure 41.8 that some resource allocations are clustered into groups, such that for all resource allocations within a group, the normalized robustness remains constant as the estimated makespan (or load balance index) increases.

The cluster of the resource allocations with the highest robustness has the feature that the machine with the largest finishing time has the smallest number of applications allocated to it (which is two for the experiments in Figure 41.8). The cluster with the smallest robustness has the largest number, 11, of applications allocated to the machine with the largest finishing time. The intuitive explanation for this behavior is that the larger the number of applications allocated to a machine, the more the degrees of freedom for the finishing time of that machine. A larger degree of freedom then results in a shorter path to constraint violation in the parameter space. That is, the robustness is then smaller (using the ℓ_2-norm).

If one agrees with the utility of the observations made above, one can still question if the same information could be gleaned from some traditional metrics (even if they are not traditionally used to measure robustness). In an attempt to answer that question, note that sharp differences exist in the robustness of some resource allocations that have very similar values of makespan. A similar observation could be made from the robustness against load balance index plot (Figure 41.8b). In fact, it is possible to find a set of resource allocations that have very similar values of the makespan, and very similar values of the load balance index, but with very different values of the robustness. These observations highlight the fact that the information given by the robustness metric could not be obtained from two popular performance metrics. A typical way of using this robustness measure in a resource allocation algorithm would be to have a biobjective optimization criterion where one would attempt to optimize makespan while trying to maximize robustness.

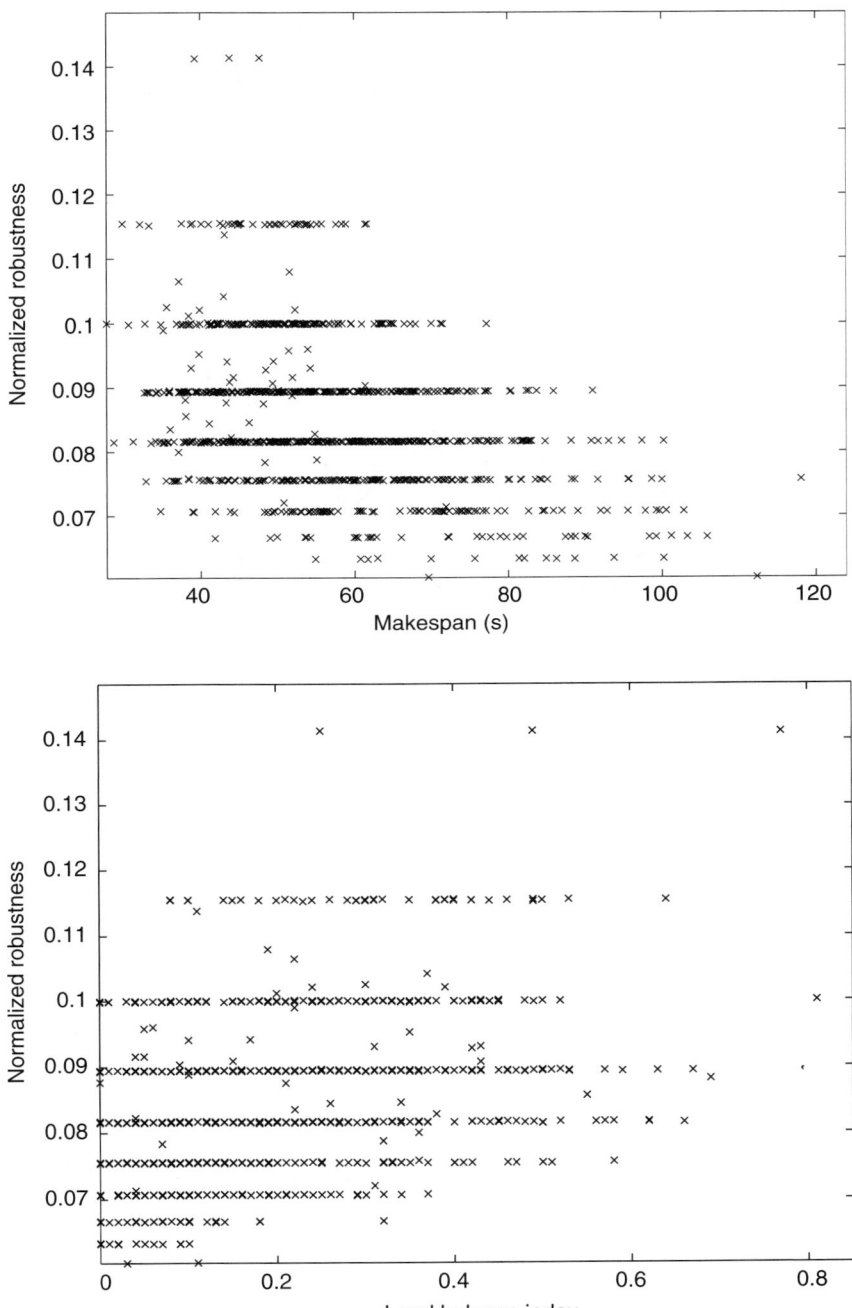

FIGURE 41.8 The plots of normalized robustness against (a) makespan, and (b) load balance index for 1000 randomly generated resource allocations.

The clustering seen in Figure 41.8 can be explained using Equation 41.5. Consider a machine m_g that finishes last for a given resource allocation. Then the finishing time for m_g is equal to the makespan, M^{orig}, of that resource allocation. Call m_g the makespan machine. Now $F_g(\mathbf{C}^{\text{orig}}) = M^{\text{orig}}$. From Equation 41.5 for m_g, we have

$$r_\mu(F_g, \mathbf{C}) = \frac{\tau M^{\text{orig}} - M^{\text{orig}}}{\sqrt{n(m_g)}}$$

$$\frac{r_\mu(F_g, \mathbf{C})}{M^{\text{orig}}} = \frac{\tau - 1}{\sqrt{n(m_g)}}. \tag{41.13}$$

The LHS in Equation 41.13 is the normalized robustness radius for m_g. If the makespan machine is also equal to the robustness machine, that is, m_g is such that $g = \operatorname{argmin}_j(r_\mu(F_j, \mathbf{C}))$, then the LHS equals to the overall normalized robustness plotted in Figure 41.8. Now consider a set of resource allocations that have different makespans but all share one common feature: for all, the makespan machine is the same as the robustness machine and the number of applications on the makespan machine is 2. These resource allocations will give rise to the top cluster in Figure 41.8. The lower clusters are for higher values of $n(m_g)$. Note that "inter-cluster" distance decreases as $n(m_g)$ increases, as indicated by Equation 41.13. The outlying points belong to the resource allocations for which the makespan and robustness machines are different.

41.6.3 The HiPer-D System

For the model in Subsection 41.3.2, the experiments were performed for a system that consisted of 19 paths, where the end-to-end latency constraints of the paths were uniformly sampled from the range [750, 1250]. The system had three sensors (with rates 4×10^{-5}, 3×10^{-5}, and 8×10^{-6}), and three actuators. The experiments made the following simplifying assumptions. The computation time function, $T_i^c(\lambda)$, was assumed to be linear in λ. Specifically, for real number b_{iz}, $T_i^c(\lambda)$ was assumed to be of the form $\sum_{1 \le z \le 3} b_{iz} \lambda_z$. In case there was no route from the z-th sensor to application a_i, we set b_{iz} to 0 to ensure $T_i^c(\lambda)$ would not depend on the load from the z-th sensor. If there was a route from the z-th sensor to application a_i, b_{iz} was sampled from a Gamma distribution with a mean of 10 and a task heterogeneity of 0.7. For simplicity in the presentation of the results, the communication times were all set to zero. These assumptions were made only to simplify the experiments, and are *not* a part of the formulation of the robustness metric. The salient point in this example is that the utility of the robustness metric can be seen even when simple complexity functions are used.

The resource allocations were evaluated for robustness and "slack." In this context, a robustness value of x for a given resource allocation means that the resource allocation can endure any combination of sensor loads without a latency or throughput violation as long as the Euclidean norm of the increases in sensor loads (from the assumed values) is no larger than x. Slack has been used in many studies as a performance measure (e.g. References 18 and 31) for resource allocation in parallel and distributed systems, where a resource allocation with a larger slack is considered to be more "robust" in a sense that it can better tolerate additional load. In this study, slack is defined mathematically as follows. Let the fractional value of a given performance feature be the value of the feature as a percentage of the maximum allowed value. Then the percentage slack for a given feature is the fractional value subtracted from 1. Intuitively, this is the percentage increase in this feature possible before reaching the maximum allowed value (see Figure 41.9). The system-wide percentage slack is the minimum value of percentage slack taken over all performance features, and can be expressed mathematically as

$$\min\left(\min_{k:\mathcal{P}_k \in \mathbb{P}}\left(1 - \frac{L_k(\lambda)}{L_k^{\max}}\right),\ \min_{i:\,a_i \in \mathcal{A}}\left(1 - \frac{\max\left(T_i^c(\lambda),\ \max_{a_p \in \mathcal{D}(a_i)} T_{ip}^n(\lambda)\right)}{1/R(a_i)}\right)\right). \tag{41.14}$$

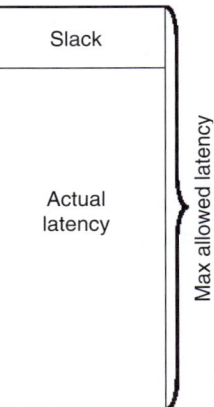

FIGURE 41.9 Slack in latency is the percentage increase in the actual latency possible before reaching the maximum allowed latency value.

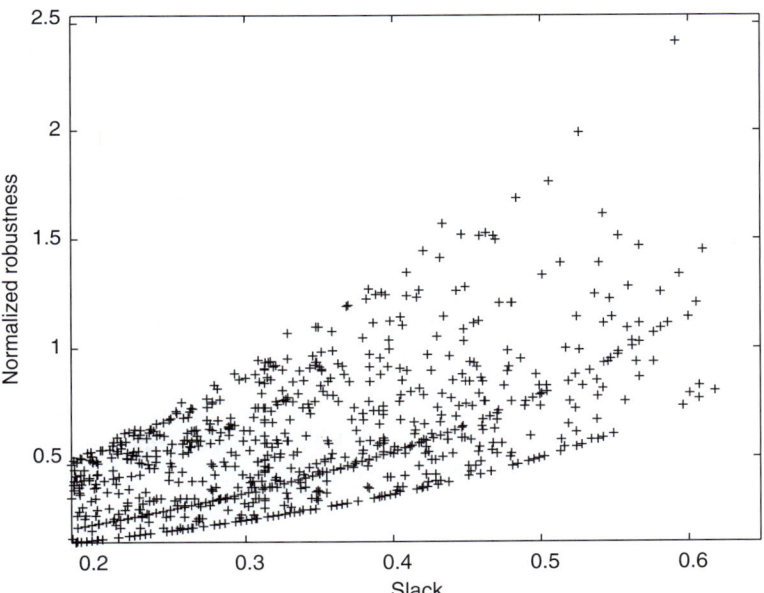

FIGURE 41.10 The plot of normalized robustness against slack for 1000 randomly generated resource allocations.

Recall that all resource allocations meet the primary performance criteria of obeying all of the throughput and latency constraints. Thus, we can select a resource allocation that, in addition to meeting primary criteria, is most robust.

Figure 41.10 shows the normalized robustness of a resource allocation against its slack. For this system, the normalized robustness equals the absolute robustness divided by $\|\lambda^{orig}\|_2$. It can be seen that the normalized robustness and slack are not correlated. *If*, in some research study, the purpose of using slack is to measure a system's ability to tolerate additional load, *then* our measure of robustness is a better indicator of that ability than slack. This is because the expression for slack, Equation 41.14, does not directly take into account how the sensor loads affect the computation and communication times. It could be conjectured that for a system where all sensors affected the computation and communication times of all applications in exactly the same way, the slack and this research's measure of robustness would be tightly correlated. This, in fact, is true. Other experiments performed in this study show that for a system with small heterogeneity, the robustness and slack are tightly correlated, thereby suggesting that robustness measurements are not needed if slack is known. As the system heterogeneity increases, the robustness and slack become less correlated, indicating that the robustness measurements can be used to distinguish between resource allocations that are similar in terms of the slack. As the system size increases, the correlation between the slack and the robustness decreases even further. In summary, for heterogeneous systems, using slack as a measure of how much increase in sensor load a system can tolerate may cause system designers to grossly misjudge the system's capability.

41.7 Designing Static Heuristics to Optimize Robustness for the HiPer-D System

41.7.1 Overview

The rest of this chapter describes the design and development of several static resource allocations heuristics for the HiPer-D system described in Subsection 41.3.2. The resource allocation problem has been shown, in general, to be NP-complete [16, 21, 29]. Thus, the development of heuristic techniques to find near-optimal resource allocation is an active area of research, for example, References 1, 7, 8, 12, 13, 20, 22, 24, 33, 35, 36, 39, 41, 44, 45 and 46.

Static resource allocation is performed when the applications are mapped in an offline planning phase such as in a production environment. Static resource allocation techniques take a set of applications, a set of machines, and generate a resource allocation. These heuristics determine a resource allocation offline, and must use estimated values of application computation times and interapplication communication times.

Recall that for a HiPer-D system there are usually a number of QoS constraints that must be satisfied. A heuristic failure occurs if the heuristic cannot find a resource allocation that allows the system to meet its throughput and latency constraints. The system is expected to operate in an uncertain environment where the workload, that is, the load presented by the set of sensors is likely to change unpredictably, possibly invalidating a resource allocation that was based on the initial workload estimate. The focus here is on designing a static heuristic that (i) determines an optimally *robust* resource allocation, that is, a resource allocation that maximizes the allowable increase in workload until a runtime reallocation of resources is required to avoid a QoS violation, and (ii) has a very low *failure rate*.

We propose a heuristic that performs well with respect to the failure rates and robustness to unpredictable workload increases. This heuristic is, therefore, very desirable for systems where low failure rates can be a critical requirement and where unpredictable circumstances can lead to unknown increases in the system workload.

41.7.2 Computation and Communication Models

41.7.2.1 Computation Model

This subsection summarizes the computation model for this system as originally described in Reference 2. For an application, the ETC a given data set depends on the load presented by the data set, and the machine executing the application. Let $C_{ij}(\lambda)$ be the ETC for application a_i on m_j for a given workload (generated by λ) when a_i is the only application executing on m_j. This research assumes that $C_{ij}(\lambda)$ is a function known for all i, j, and λ.

The effect of multitasking on the computation time of an application is accounted for by assuming that all applications mapped to a machine are processing data continuously. Let N_j be the number of applications executing on machine m_j. Let $T^c_{ij}(\lambda)$ be the computation time for a_i on machine m_j when a_i

shares this machine with other applications.* If the overhead due to context switching is ignored, $T_{ij}^c(\lambda)$ will be $N_j \times C_{ij}(\lambda)$.

However, the overhead in computation time introduced by context switching may not be trivial in a round-robin scheduler. Such an overhead depends on the estimated-time-to-compute value of the application on the machine, and the number of applications executing on the machine. Let $O_{ij}^{cs}(\lambda)$ be the context switching overhead incurred when application a_i executes a given workload on machine m_j. Let T_j^{cs} be the time m_j needs in switching the execution from one application to another. Let T_j^q be the size (quantum) of the time slice given by m_j to each application in the round-robin scheduling policy used on m_j. Then,

$$O_{ij}^{cs}(\lambda) = \begin{cases} 0 & \text{if } N_j = 1 \\ \dfrac{C_{ij}(\lambda) \times N_j}{T_j^q} \times T_j^{cs} & \text{if } N_j > 1 \end{cases}$$

$T_{ij}^c(\lambda)$ can now be stated as,

$$T_{ij}^c(\lambda) = \begin{cases} C_{ij}(\lambda) & \text{if } N_j = 1 \\ C_{ij}(\lambda) \times N_j \left(1 + \dfrac{T_j^{cs}}{T_j^q}\right) & \text{if } N_j > 1 \end{cases} \quad (41.15)$$

41.7.2.2 Communication Model

This section develops an expression for the time needed to transfer the output data from a source application to a destination application at a given load. This formulation was originally given in Reference 2. Let $M_{ip}(\lambda)$ be the size of the message data sent from application a_i to a destination application a_p at the given load. Let $m(a_i)$ be the machine on which a_i is mapped. Let $T_{ip}^n(\lambda)$ be the transfer time, that is, the time to send data from application a_i to application a_p at the given load.

A model for calculating the communication times should identify the various steps involved in effecting a data transfer. The two steps identified in Reference 23 for communication in a similar domain are the communication setup time and the queuing delay. These steps contribute to the overall communication time to different extents depending on the intended communications environment.

Communication Setup Time. Before data can be transferred from one machine to another, a communication setup time is required for setting up a *logical* communication channel between the sender application and the destination application. Once established, a logical communication channel is torn down only when the sending or receiving application is finished executing. Because this study considers continuously executing applications (because the sensors continually produce data), the setup time will be incurred only once, and will, therefore, be amortized over the course of the application execution. Hence, the communication setup time is ignored in this study.

Queuing Delay. A data packet is queued twice enroute from the source machine to the destination machine. First, the data packet is queued in the output buffer of the source machine, where it waits to use the communication link from the source machine to the switch. The second time, the data packet is queued in the output port of the switch, where it waits to use the communication link from the switch to the destination machine. The switch has a separate output port for each machine in the system.

The queuing delay at the sending machine is modeled by assuming that the bandwidth of the link from the sending machine to the switch is shared equally among all data transfers originating at the sending machine. This will underestimate the bandwidth available for each transfer because it assumes

---

*The function $T_i^c(\lambda)$ used in Subsection 41.3.2 is essentially the same as the function $T_{ij}^c(\lambda)$. The subscript j was absent in the former because the identity of the machine where application a_i was mapped was not relevant to the discussion in Subsection 41.3.2.

that all of the other transfers are always being performed. Similarly, the queuing delay at the switch is modeled by assuming that the bandwidth of the link from the switch to the destination machine is equally divided among all data transfers originating at the switch and destined for the destination machine. Let $B(m(a_i), \text{swt})$ be the bandwidth (in bytes per unit time) of the communication link between $m(a_i)$ and the switch, and $B(\text{swt}, m(a_p))$ be the bandwidth (in bytes per unit time) of the communication link between the switch and $m(a_p)$. The abbreviation "swt" stands for "switch." Let $N^{\text{ct}}(m(a_i), \text{swt})$ be the number of data transfers using the communication link from $m(a_i)$ to the switch. The superscript "ct" stands for "contention." Let $N^{\text{ct}}(\text{swt}, m(a_p))$ be the number of data transfers using the communication link from the switch to $m(a_p)$. Then, $T_{ip}^n(\lambda)$, the time to transfer the output data from application a_i to a_p, is given by

$$T_{ip}^n(\lambda) = M_{ip}(\lambda) \times \left(\frac{N^{\text{ct}}(m(a_i), \text{swt})}{B(m(a_i), \text{swt})} + \frac{N^{\text{ct}}(\text{swt}, m(a_p))}{B(\text{swt}, m(a_p))} \right). \quad (41.16)$$

The above expression can also accommodate the situations when a sensor communicates with the first application in a path, or when the last application in a path communicates with an actuator. The driving sensor for $\mathcal{P}_k$ can be treated as a "pseudoapplication" that has a zero computation time and is already mapped to an imaginary machine, and, as such, can be denoted by $\alpha_{k,0}$. Similarly, the actuator receiving data from $\mathcal{P}_k$ can also be treated as a pseudoapplication with a zero computation time, and will be denoted by $\alpha_{k,|\mathcal{P}_k|+1}$. Accordingly, for these cases: $M_{ip}(\lambda)$ corresponds to the size of the data set being sent from the sensor; $B(m(\alpha_{k,0}), \text{swt})$ is the bandwidth of the link between the sensor and the switch; and $B(\text{swt}, m(\alpha_{k,|\mathcal{P}_k|+1}))$ corresponds to the bandwidth of the link between the switch and the actuator. Similarly, $N^{\text{ct}}(m(\alpha_{k,0}), \text{swt})$ and $N^{\text{ct}}(\text{swt}, \alpha_{k,|\mathcal{P}_k|+1})$ both are 1. In the situation where $m(a_i) = m(a_p)$, one can interpret $N^{\text{ct}}(m(a_i), \text{swt})$ and $N^{\text{ct}}(\text{swt}, m(a_f))$ both as 0; $T_{ip}^n(\lambda) = 0$ in that case.

41.7.3 Heuristic Descriptions

41.7.3.1 Overview

This section develops three greedy heuristics for the problem of finding an initial static allocation of applications onto machines to maximize $\rho_\mu(\Phi, \lambda)$, where Φ is as defined in Equation 41.7. From this point on, for the sake of simplicity we will denote $\rho_\mu(\Phi, \lambda)$ by $\Delta\Lambda$. Greedy techniques perform well in many situations, and have been well studied (e.g., Reference 29). One of the heuristics, most critical task first (MCTF), is designed to work well in heterogeneous systems where the throughput constraints are more stringent than the latency constraints. Another heuristic, the most critical path first (MCPF) heuristic, is designed to work well in heterogeneous systems where the latency constraints are more stringent than the throughput constraints.

It is important to note that these heuristics use the $\Delta\Lambda$ value to guide the heuristic search; however, the procedure given in Section 41.3.2 (Equation 41.8) for calculating $\Delta\Lambda$ assumes that a complete resource allocation of all applications is known. During the course of the execution of the heuristics, not all applications are mapped. In these cases, for calculating $\Delta\Lambda$, the heuristics assume that each such application a_i is mapped to the machine where its computation time is smallest over all machines, and that a_i is using 100% of that machine. Similarly, for communications where either the source or the destination application (or both) are unmapped, it is assumed that the data transfer between the source and destination occurs over the highest speed communication link available in the network, and that the link is 100% utilized by the data transfer. With these assumptions, $\Delta\Lambda$ is calculated and used in any step of a given heuristic.

Before discussing the heuristics, some additional terms are now defined. Let $\Delta\Lambda^T$ be the robustness of the resource allocation when only throughput constraints are considered, that is, all latency constraints are ignored. Similarly, let $\Delta\Lambda^L$ be the robustness of the resource allocation when only latency constraints are considered. In addition, let $\Delta\Lambda_{ij}^T$ be the robustness of the assignment of a_i to machine m_j with respect to the throughput constraint, that is, it is the largest increase in load in any direction from the initial value that does not cause a throughput violation for application a_i, either for the computation of a_i on machine m_j or for the communications from a_i to any of its successor applications. Similarly, let $\Delta\Lambda_k^L$ be the robustness of the assignment of applications in $\mathcal{P}_k$ with respect to the latency constraint, that is, it is the largest increase in load in any direction from the initial value that does not cause a latency violation for the path $\mathcal{P}_k$.

41.7.3.2 Most Critical Task First Heuristic

The MCTF Heuristic makes one application to machine assignment in each iteration. Each iteration can be split into two phases. Let $\mathcal{M}$ be the set of machines in the system. Let $\Delta\Lambda^*(a_i, m_j)$ be the value of $\Delta\Lambda$ if application a_i is mapped on m_j. Similarly, let $\Delta\Lambda^{T*}(a_i, m_j)$ be the value of $\Delta\Lambda_{ij}^T$ if application a_i is mapped on m_j. In the first phase, each unmapped application a_i is paired with its "best" machine m_j such that

$$m_j = \underset{m_k \in \mathcal{M}}{\operatorname{argmax}}(\Delta\Lambda^*(a_i, m_k)). \tag{41.17}$$

(Note that $\operatorname{argmax}_x f(x)$ returns the value of x that maximizes the function $f(x)$. If there are multiple values of x that maximize $f(x)$, then $\operatorname{argmax}_x f(x)$ returns the set of all those values.) If the RHS in Equation 41.17 returns a set of machines, $G(a_i)$, instead of a unique machine, then $m_j = \operatorname{argmax}_{m_k \in G(a_i)}(\Delta\Lambda^{T*}(a_i, m_k))$, that is, the individual throughput constraints are used to break ties in the overall system-wide measure. If $\Delta\Lambda^*(a_i, m_j) < 0$, this heuristic cannot find a resource allocation. The first phase does not make an application to machine assignment; it only establishes application-machine pairs (a_i, m_j) for all unmapped applications a_i.

The second phase makes an application to machine assignment by selecting one of the (a_i, m_j) pairs produced by the first phase. This selection is made by determining the most "critical" application (the criterion for this is explained later). The method used to determine this assignment in the first iteration is totally different from that used in the subsequent iterations.

Consider the motivation of the heuristic for the special first iteration. Before the first iteration of the heuristic, all applications are unmapped, and the system resources are entirely unused. With the system in this state, the heuristic selects the pair (a_x, m_y) such that

$$(a_x, m_y) = \underset{\substack{(a_i, m_j) \text{ pairs from} \\ \text{the first phase}}}{\operatorname{argmin}} (\Delta\Lambda^*(a_i, m_j)).$$

That is, from all of the (application, machine) pairs chosen in the first step, the heuristic chooses the pair (a_x, m_y) that leads to the smallest robustness. The application a_x is then assigned to the machine m_y. It is likely that if the assignment of this application is postponed, it might have to be assigned to a machine where its maximum allowable increase in the system load is even smaller. (The discussion above does not imply that an optimal resource allocation must contain the assignment of a_x on m_y.) Experiments conducted in this study have shown that the special first iteration significantly improves the performance.

The criterion used to make the second phase application to machine assignment for iterations 2 to $|\mathcal{A}|$ is different from that used in iteration 1, and is now explained. The intuitive goal is to determine the (a_i, m_j) pair, which if not selected, may cause the most future "damage," that is, decrease in $\Delta\Lambda$. Let $\mathcal{M}^{a_i}$ be the ordered list, $\langle m_1^{a_i}, m_2^{a_i}, \cdots, m_{|\mathcal{M}|}^{a_i}\rangle$, of machines such that $\Delta\Lambda^*(a_i, m_x^{a_i}) \geq \Delta\Lambda^*(a_i, m_y^{a_i})$ if $x < y$. Note that $m_1^{a_i}$ is the same as a_i's "best" machine. Let v be an integer such that $2 \leq v \leq |\mathcal{M}|$, and let $r(a_i, v)$ be the percentage decrease in $\Delta\Lambda^*(a_i, m_j)$ if a_i is mapped on $m_v^{a_i}$ (its v-th best machine) instead of $m_1^{a_i}$, that is,

$$r(a_i, v) = \frac{\Delta\Lambda^*(a_i, m_1^{a_i}) - \Delta\Lambda^*(a_i, m_v^{a_i})}{\Delta\Lambda^*(a_i, m_1^{a_i})}.$$

```
 1: initialize: v = 2; F = the set of (a_i, m_j) pairs from the first phase
 2: for v = 2 to |M| do
 3:     if argmax_{(a_i,m_j) ∈ F} (r(a_i, v)) is a unique pair (a_x, m_y) then
 4:         return (a_x, m_y)
 5:     else
 6:         F = the set of pairs returned by argmax_{(a_i,m_j) ∈ F} (r(a_i, v))
 7:     end if
 8: end for
    /*program control reaches here only if no application, machine pair has been */
    /*selected in Lines 1 to 8 above. F is now the set of (a_i, m_j) pairs from the last */
    /*execution of Line 6 */
 9: if argmax_{(a_i,m_j) ∈ F} (T(a_i, 2)) is a unique pair (a_x, m_y) then
10:     return (a_x, m_y)
11: else
12:     arbitrarily select and return an application, machine pair from the set of pairs
        given by argmax_{(a_i,m_j) ∈ F} (T(a_i, 2))
13: end if
```

FIGURE 41.11 Selecting the most critical application to map next given the set of (a_i, m_j) pairs from the first phase of MCTF.

In addition, let $T(a_i, 2)$ be defined such that

$$T(a_i, 2) = \frac{\Delta \mathbf{\Lambda}^{T^*}(a_i, m_1^{a_i}) - \Delta \mathbf{\Lambda}^{T^*}(a_i, m_2^{a_i})}{\Delta \mathbf{\Lambda}^{T^*}(a_i, m_1^{a_i})}.$$

Then, in all iterations other than the first iteration, MCTF maps the most critical application, where the most critical application is found using the pseudocode in Figure 41.11. The program in Figure 41.11 takes the set of (a_i, m_j) pairs from the first phase of MCTF as its input. Then for each pair (a_i, m_j) it determines the percentage decrease in $\Delta \mathbf{\Lambda}^*(a_i, m_j)$ if a_i is mapped on its second best machine instead of its best machine. Then, once this information is calculated for all of the pairs obtained from the first phase, the program in Figure 41.11 chooses the pair for which the percentage decrease in $\Delta \mathbf{\Lambda}^*(a_i, m_j)$ is the largest. It may very well be the case that all pairs are alike, that is, all have the same percentage decrease in $\Delta \mathbf{\Lambda}^*(a_i, m_j)$ when the program considers their second best and the best machines. In such a case, the program makes the comparisons again, however, using the third best and the best machines this time. This continues until a pair is found that has the largest percentage decrease in $\Delta \mathbf{\Lambda}^*(a_i, m_j)$ or until all possible comparisons have been made and still there is no unique "winner." If we do not have a unique winner even after having compared the $|M|$-th best machine with the best machine, we use $T(a_i, 2)$ to break the ties, that is, we choose the pair that maximizes $T(a_i, 2)$. If even that is not unique, we arbitrarily select one pair. This method is explained in detail in Figure 41.11. The technique shown in Figure 41.11 builds on the idea of the Sufferage heuristic given in Reference 35.

41.7.3.3 Two-Phase Greedy Heuristic

This research also proposes a modified version of the Min-min heuristic. Variants of the Min-min heuristic (first presented in Reference 29) have been studied, for example, References 6, 12, 35, and 46, and have been seen to perform well in the environments for which they were proposed. Two-phase greedy (TPG), a Min-min style heuristic for the environment discussed in this research, is shown in Figure 41.12.

Like MCTF, the TPG heuristic makes one application to machine assignment in each iteration. Each iteration can be split into two phases. In the first phase, each unmapped application a_i is paired with its "best" machine m_j such that $m_j = \operatorname{argmax}_{m_k \in \mathcal{M}}(\Delta \mathbf{\Lambda}^*(a_i, m_k))$. The first phase does not make an application to machine assignment; it only establishes application-machine pairs (a_i, m_j) for all unmapped applications a_i. The second phase makes an application to machine assignment by selecting one of the

```
 1: while all applications are not mapped do
 2:      for each unmapped application a_i do
 3:          find the machine m_j such that m_j = argmax_{m_k ∈ M} (ΔΛ*(a_i, m_k))
 4:          resolve ties arbitrarily
 5:          if ΔΛ*(a_i, m_j) < 0 then
 6:              exit (this heuristic cannot find a resource allocation)
 7:          end if
 8:      end for
 9:      from the (a_i, m_j) pairs found above, select the pair(s) (a_x, m_y) such that (a_x, m_y) =
            argmax_{(a_i, m_j) pairs} (ΔΛ*(a_i, m_j))
10:      resolve ties arbitrarily
11:      map a_x on m_y
12: end while
```

FIGURE 41.12 The TPG heuristic.

(a_i, m_j) pairs produced by the first phase. It chooses the pair that maximizes $\Delta\Lambda^*(a_i, m_j)$ over all first phase pairs. Ties are resolved arbitrarily.

41.7.3.4 Most Critical Path First Heuristic

The MCPF heuristic explicitly considers the latency constraints of the paths in the system. It begins by ranking the paths in the order of the most "critical" path first (defined below). Then it uses a modified form of the MCTF heuristic to map applications on a path-by-path basis, iterating through the paths in a ranked order. The modified form of MCTF differs from MCTF in that the first iteration has been changed to be the same as the subsequent iterations.

The ranking procedure used by MCPF is now explained in detail. Let $\hat{\Lambda}^L(\mathcal{P}_k)$ be the value of $\Delta\Lambda_k^L$ assuming that each application a_i in $\mathcal{P}_k$ is mapped to the machine m_j, where it has the smallest computation time, and that a_i can use 100% of m_j. Similarly, for the communications between the consecutive applications in $\mathcal{P}_k$, it is assumed that the data transfer between the applications occurs over the highest speed communication link in the system, and that the link is 100% utilized by the data transfer. Note that the entire ranking procedure is done before any application is mapped.

The heuristic ranks the paths in an ordered list $\langle \mathcal{P}_1^{crit}, \mathcal{P}_2^{crit}, \cdots, \mathcal{P}_{|\mathcal{P}|}^{crit} \rangle$ such that $\hat{\Lambda}^L(\mathcal{P}_x^{crit}) \leq \hat{\Lambda}^L(\mathcal{P}_y^{crit})$ if $x < y$. Once the ranking of the paths has been done, the MCPF heuristic uses MCTF to map each application in a path, starting from the highest ranked path first.

41.7.3.5 Duplex

For an arbitrary HC system, one is not expected to know if the system is more stringent with respect to latency constraints or throughput constraints. In that case, this research proposes running both MCTF and MCPF, and taking the better of the two mappings. The Duplex heuristic executes both MCTF and MCPF, and then chooses the resource allocation that gives a higher $\Delta\Lambda$.

41.7.3.6 Other Heuristics

To compare the performance of the heuristics proposed in this research (MCTF, MCPF, and TPG), three other greedy heuristics were also implemented. These included two-phase greedy X (TPG-X) and two fast greedy heuristics. TPG-X is an implementation of the Max-min heuristic [29] for the environment discussed in this research. TPG-X is similar to the TPG heuristic except that in Line 9 of Figure 41.12, "argmax" is replaced with "argmin." The first fast greedy heuristic, denoted FGH-L, iterates through the unmapped applications in an arbitrary order, assigning an application a_i to the machine m_j such that (i) $\Delta\Lambda^*(a_i, m_j) \geq 0$, and (ii) $\Delta\Lambda^L$ is maximized (ties are resolved arbitrarily). The second fast greedy heuristic, FGH-T, is similar to FGH-L except that FGH-T attempts to maximize $\Delta \Lambda^T$. For a given application a_i, if FGH-L or FGH-T cannot find a machine m_j such that $\Delta \Lambda^*(a_i, m_j) \geq 0$, then the heuristic fails.

41.7.3.7 An Upper Bound

An upper bound, UB, on the $\Delta \Lambda$ value also is calculated for comparing the absolute performance of a given heuristic. The UB is equal to the $\Delta \Lambda$ for a system where the following assumptions hold: (i) the communication times are zero for all applications, (ii) each application a_i is mapped on the machine m_j where $\Delta \Lambda_{ij}^T$ is maximum over all machines, and (iii) each application can use 100% of the machine where it is mapped. These assumptions are, in general, not physically realistic.

41.8 Simulation Experiments and Results

In this study, several sets of simulation experiments were conducted to evaluate and compare the heuristics. Experiments were performed for different values of $|\mathcal{A}|$ and $|\mathcal{M}|$, and for different types of HC environments. For all experiments, it was assumed that an application could execute on any machine.

The following simplifying assumptions were made for performing the experiments. Let n_s be the total number of sensors. The ETC function $C_{ij}(\lambda)$ for application a_i on m_j was assumed to be of the form $\sum_{1 \leq z \leq n_s} b_{ijz} \lambda_z$, where $b_{ijz} = 0$ if there is no route from the z-th sensor to application a_i. Otherwise, b_{ijz} was sampled from a Gamma distribution with a given mean and given values of task heterogeneity and machine heterogeneity. The $T_{ij}^c(\lambda)$ value would depend on the actual resource allocation as well as $C_{ij}(\lambda)$, and can be calculated using the computation model given in Subsection 41.7.2.1. Similarly, the $M_{ip}(\lambda)$ functions for the size of the message data sent from application a_i to a destination application a_p at a given load were similarly generated, except that machine heterogeneity was not involved. The communication time functions, $T_{ip}^n(\lambda)$, would depend on the actual resource allocation as well as $M_{ip}(\lambda)$, and can be calculated using the communication model given in Subsection 41.7.2.2. For a given set of computation and communication time functions, the experimental set up allowed the user to change the values of sensor output rates and end-to-end latency constraints so as to change the "tightness" of the throughput and latency constraints. The reader is directed to Reference 6 for details.

An experiment is characterized by the set of system parameters (e.g., $|\mathcal{A}|$, $|\mathcal{M}|$, application and machine heterogeneities) it investigates. Each experiment was repeated 90 times to obtain good estimates of the mean and standard deviation of $\Delta \Lambda$. Each repetition of a given experiment will be referred to as a trial. For each new trial, a DAG with $|\mathcal{A}|$ nodes was randomly regenerated, and the values of $C_{ij}(\lambda)$ and $M_{ip}(\lambda)$ were regenerated from their respective distributions.

Results from a typical set of experiments are shown in Figure 41.13. The first bar for each heuristic, titled "$\Delta \Lambda^N$," shows the normalized $\Delta \Lambda$ value averaged for all those trials in which the given heuristic successfully found a resource allocation. The normalized $\Delta \Lambda$ for a given heuristic is equal to $\Delta \Lambda$ for the resource allocation found by that heuristic divided by $\Delta \Lambda$ for the upper bound defined in Subsection 41.7.3.7. The second bar, titled "$\delta \lambda^N$," shows the normalized $\Delta \Lambda$ averaged only for those trials in which every heuristic successfully found a resource allocation. This figure also shows, in the third bar, the value of the failure rate for each heuristic. The failure rate or FR is the ratio of the number of trials in which the heuristic could not find a resource allocation to the total number of trials. The interval shown at the tops of the first two bars is the 95% confidence interval [30].

Figure 41.13 shows the relative performance of the heuristics for the given system parameters. In this figure, FGH-T and FGH-L are not shown because of their poor failure rate and $\Delta \Lambda^N$, respectively. It can be seen that the $\Delta \Lambda^N$ performance difference between MCTF and MCPF is statistically insignificant. The traditional Min-min and Max-min like heuristics, that is, TPG and TPG-X, achieve $\Delta \Lambda^N$ values significantly lower than those for MCTF or MCPF (i.e., much poorer robustness). To make matters worse, the FR values for TPG and TPG-X are significantly higher than those for MCTF or MCPF. Even though

Perspectives on Robust Resource Allocation

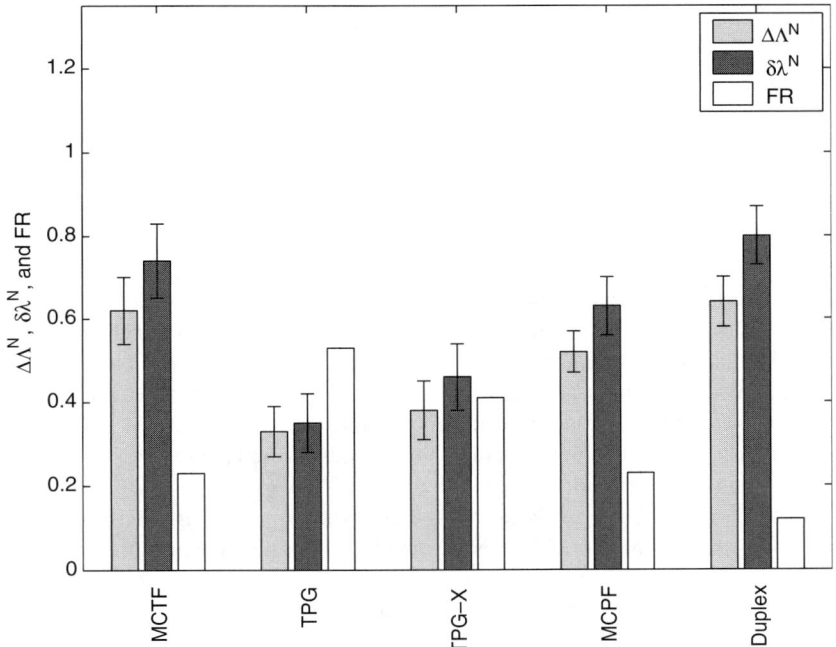

FIGURE 41.13 The relative performance of heuristics for a system where $|\mathcal{M}| = 6$, $|\mathcal{A}| = 50$. Number of sensors = number of actuators = 7. Task heterogeneity = machine heterogeneity = 0.7. All communication times were set to zero. A total of 90 trials were performed.

Duplex's $\Delta\Lambda^N$ value is statistically no better than that of MCTF, its FR value, 12%, is about half that of MCTF (23%).

Additional experiments were performed for various other combinations of $|\mathcal{A}|$, $|\mathcal{M}|$, and tightness of QoS constraints, and the relative behavior of the heuristics was similar to that in Figure 41.13. Note that all communication times were set to zero in Figure 41.13 (but not in all experiments). Given the formulation of UB, it is expected that if the communication times are all zero in a given environment, then UB will be closer to the optimal value, and will make it easier to evaluate the performance of the heuristics with respect to the upper bound.

41.9 Related Work

Although a number of robustness measures have been studied in the literature (e.g., Reference 10, 14, 17, 18, 19, 25, 31, 34, 38, and 43), those measures were developed for specific systems. The focus of the research in this chapter is a general mathematical formulation of a robustness metric that could be applied to a variety of parallel and distributed systems by following the FePIA procedure presented in this chapter.

Given an allocation of a set of communicating applications to a set of machines, the work in Reference 10 develops a metric for the robustness of the makespan against uncertainties in the estimated execution times of the applications. The paper discusses in detail the effect of these uncertainties on the value of makespan, and how the robustness metric could be used to find more robust resource allocations. On the basis of the model and assumptions in Reference 10, several theorems about the properties of robustness are proven. The robustness metric in Reference 10 was formulated for errors in the estimation of application execution times, and was not intended for general use (in contrast to our work). In addition, the formulation in Reference 10 assumes that the execution time for any application is at most k times the estimated value, where $k \geq 1$ is the same for all applications. In our work, no such bound is assumed.

In Reference 14, the authors address the issue of probabilistic guarantees for fault-tolerant real-time systems. As a first step towards determining such a probabilistic guarantee, the authors determine the maximum frequency of software or hardware faults that the system can tolerate without violating any hard real-time constraint. In the second step, the authors derive a value for the probability that the system will not experience faults at a frequency larger than that determined in the first step. The output of the first step is what our work would identify as the robustness of the system, with the satisfaction of the real-time constraints being the robustness requirement, and the occurrence of faults being the perturbation parameter.

The research in Reference 17 considers a single-machine scheduling environment where the processing times of individual jobs are uncertain. The system performance is measured by the total flow time (i.e., the sum of *completion* times of all jobs). Given the probabilistic information about the processing time for each job, the authors determine the normal distribution that approximates the flow time associated with a given schedule. A given schedule's robustness is then given by 1 minus the risk of achieving substandard flow time performance. The risk value is calculated by using the approximate distribution of flow time. As in Reference 10, the robustness metric in Reference 17 was formulated for errors in the estimation of processing times, and was not intended for general use.

The study in Reference 18 explores slack-based techniques for producing robust resource allocations in a job-shop environment. The central idea is to provide each task with extra time (defined as slack) to execute so that some level of uncertainty can be absorbed without having to reallocate. The study uses slack as its measure of robustness. The study does not develop a robustness metric; instead, it implicitly uses slack to achieve robustness.

The Ballista project [19] explores the robustness of commercial off-the-shelf software against failures resulting from invalid inputs to various software procedure calls. A failure causes the software package to crash when unexpected parameters are used for the procedure calls. The research quantifies the robustness of a software procedure in terms of its failure rate—the percentage of test input cases that cause failures to occur. The Ballista project extensively explores the robustness of different operating systems (including experimental work with IBM, FreeBSD, Linux, AT&T, and Cisco). However, the robustness metric developed for that project is specific to software systems.

The research in Reference 25 introduces techniques to incorporate fault tolerance in scheduling approaches for real-time systems by the use of additional time to perform the system functions (e.g., to reexecute, or to execute a different version of, a faulty task). Their method guarantees that the real-time tasks will meet the deadlines under transient faults, by reserving sufficient additional time, or slack. Given a certain system slack and task model, the paper defines its measure of robustness to be the "fault tolerance capability" of a system (i.e., the number and frequency of faults it can tolerate). This measure of robustness is similar, in principle, to ours.

In Reference 31, a "neighborhood-based" measure of robustness is defined for a job-shop environment. Given a schedule s and a performance metric $P(s)$, the robustness of the schedule s is defined to be a weighted sum of all $P(s')$ values such that s' is in the set of schedules that can be obtained from s by interchanging two consecutive operations on the same machine.

The work in Reference 34 develops a mathematical definition for the robustness of makespan against machine breakdowns in a job-shop environment. The authors assume a certain random distribution of the machine breakdowns and a certain rescheduling policy in the event of a breakdown. Given these assumptions, the robustness of a schedule s is defined to be a weighted sum of the expected value of the makespan of the rescheduled system, M, and the expected value of the schedule delay (the difference between M and the original value of the makespan). Because the analytical determination of the schedule delay becomes very hard when more than one disruption is considered, the authors propose surrogate measures of robustness that are claimed to be strongly correlated with the expected value of M and the expected schedule delay.

The research in Reference 38 uses a genetic algorithm to produce robust schedules in a job-shop environment. Given a schedule s and a performance metric $P(s)$, the "robust fitness value" of the schedule s is a weighted average of all $P(s')$ values such that s' is in a set of schedules obtained from s by adding a small "noise" to it. The size of this set of schedules is determined arbitrarily. The "noise" modifies s by randomly changing the ready times of a fraction of the tasks. Similar to Reference 31, Reference 38 does not explicitly state the perturbations under which the system is robust.

Our work is perhaps closest in philosophy to Reference 43, which attempts to calculate the stability radius of an optimal schedule in a job-shop environment. The stability radius of an optimal schedule, s, is defined to be the radius of a closed ball in the space of the numerical input data such that within that ball the schedule s remains optimal. Outside this ball, which is centered at the assumed input, some other schedule would outperform the schedule that is optimal at the assumed input. From our viewpoint, for a given optimal schedule, the robustness requirement could be the persistence of optimality in the face of perturbations in the input data. Our work differs and is more general because we consider the given system requirements to generate a robustness requirement, and then determine the robustness. In addition, our work considers the possibility of multiple perturbations in different dimensions.

Our earlier study in Reference 2 is related to the one in Sections 41.7 and 41.8 of this chapter. However, the robustness measure we used in Reference 2 makes a simplifying assumption about the way changes in λ can occur. Specifically, it is assumed that λ changes so that all components of λ increase in proportion to their initial values. That is, if the output from a given sensor increases by $x\%$, then the output from all sensors increases by $x\%$. Given this assumption, for any two sensors σ_p and σ_q, $(\lambda_p - \lambda_p^{\text{init}})/\lambda_p^{\text{init}} = (\lambda_q - \lambda_q^{\text{init}})/\lambda_q^{\text{init}} = \Delta\lambda$.

41.10 Future Work

There are many directions in which the robustness research presented in the chapter can be extended. Examples include the following:

1. Deriving the boundary surfaces for different problem domains.
2. Incorporating multiple types of perturbation parameters (e.g., uncertainties in input sensor loads and uncertainties in estimated execution times). Challenges here are how to define the collective impact to find each robust radius and how to state the combined bound on multiple perturbation parameters to maintain the promised performance.
3. Incorporating probabilistic information about uncertainties. In this case, a perturbation parameter can be represented as a vector of random variables. Then, one might have probabilistic information about random variables in the vector (e.g., probability density functions) or probabilistic information describing the relationship between different random variables in the vector or between different vectors (e.g., a set of correlation coefficients).
4. Determining when to use Euclidean distance versus other distance measures when calculating the collective impact of changes in the perturbation parameter elements.

41.11 Conclusions

This chapter has presented a mathematical description of a metric for the robustness of a resource allocation with respect to desired system performance features against multiple perturbations in various system and environmental conditions. In addition, the research describes a procedure, called FePIA, to methodically derive the robustness metric for a variety of parallel and distributed computing resource allocation systems. For illustration, the FePIA procedure is employed to derive robustness metrics for three example distributed systems. The experiments conducted in this research for two example parallel and distributed systems illustrate the utility of the robustness metric in distinguishing between the resource allocations that perform similarly otherwise based on the primary performance measure (e.g., no constraint violation for HiPer-D environment and minimizing makespan for the cluster environments). It was shown that the robustness metric was more useful than approaches such as slack or load balancing.

Also, this chapter described several static resource allocation heuristics for one example distributed computing system. The focus was on designing a static heuristic that will (i) determine a maximally robust resource allocation, that is, a resource allocation that maximizes the allowable increase in workload until a runtime reallocation of resources is required to avoid a QoS violation, and (ii) have a very low failure rate. This study proposes a heuristic, called Duplex, that performs well with respect to the failure rate and the robustness towards unpredictable workload increases. Duplex was compared under a variety of simulated heterogeneous computing environments, and with a number of other heuristics taken from the literature. For all of the cases considered, Duplex gave the lowest failure rate, and a robustness value better than those of other evaluated heuristics. Duplex is, therefore, very desirable for systems where low failure rates can be a critical requirement and where unpredictable circumstances can lead to unknown increases in the system workload.

Acknowledgments

This research was supported by the DARPA/ITO Quorum Program through the Office of Naval Research under Grant No. N00014-00-1-0599, by the Colorado State University Center for Robustness in Computer Systems (funded by the Colorado Commission on Higher Education Technology Advancement Group through the Colorado Institute of Technology), and by the Colorado State University George T. Abell Endowment.

References

[1] S. Ali, T. D. Braun, H. J. Siegel, N. Beck, L. Bölöni, M. Maheswaran, A. I. Reuther, J. P. Robertson, M. D. Theys, and B. Yao, Characterizing resource allocation heuristics for heterogeneous computing systems, *Parallel, Distributed, and Pervasive Computing*, A. R. Hurson, ed., Vol. 63 of *Advances in Computers*, Elsevier Academic Press, San Diego, CA, 2005, pp. 91–128.

[2] S. Ali, J.-K. Kim, Y. Yu, S. B. Gundala, S. Gertphol, H. J. Siegel, A. A. Maciejewski, and V. Prasanna, Greedy heuristics for resource allocation in dynamic distributed real-time heterogeneous computing systems, *2002 International Conference on Parallel and Distributed Processing Techniques and Applications*, Vol. II, June 2002, pp. 519–530.

[3] S. Ali, A. A. Maciejewski, H. J. Siegel, and J.-K. Kim, Measuring the robustness of a resource allocation, *IEEE Transactions on Parallel and Distributed Systems*, Vol. 15, No. 7, July 2004, pp. 630–641.

[4] S. Ali, A. A. Maciejewski, H. J. Siegel, and J.-K. Kim, Robust resource allocation for sensor-actuator distributed computing systems, *2004 International Conference on Parallel Processing (ICPP 2004)*, Las Vegas, August 2004, pp. 178–185.

[5] S. Ali, H. J. Siegel, M. Maheswaran, D. Hensgen, and S. Sedigh-Ali, Representing task and machine heterogeneities for heterogeneous computing systems, *Tamkang Journal of Science and Engineering*, Vol. 3, No. 3, November 2000, pp. 195–207, invited.

[6] S. Ali, Robust resource allocation in dynamic distributed heterogeneous computing systems. PhD thesis, School of Electrical and Computer Engineering, Purdue University, August 2003.

[7] H. Barada, S. M. Sait, and N. Baig, Task matching and scheduling in heterogeneous systems using simulated evolution, *10th IEEE Heterogeneous Computing Workshop (HCW 2001) in the Proceedings of the 15th International Parallel and Distributed Processing Symposium (IPDPS 2001)*, April 2001.

[8] I. Banicescu and V. Velusamy, Performance of scheduling scientific applications with adaptive weighted factoring, *10th IEEE Heterogeneous Computing Workshop (HCW 2001) in the proceedings of the 15th International Parallel and Distributed Processing Symposium (IPDPS 2001)*, April 2001.

[9] P. M. Berry. Uncertainty in scheduling: Probability, problem reduction, abstractions and the user, *IEE Computing and Control Division Colloquium on Advanced Software Technologies for Scheduling*, Digest No. 1993/163, April 26, 1993.

[10] L. Bölöni and D. C. Marinescu, Robust scheduling of metaprograms, *Journal of Scheduling*, Vol. 5, No. 5, September 2002, pp. 395–412.

[11] S. Boyd and L. Vandenberghe, *Convex Optimization*, available at http://www.stanford.edu/class/ee364/index.html

[12] T. D. Braun, H. J. Siegel, N. Beck, L. L. Bölöni, M. Maheswaran, A. I. Reuther, J. P. Robertson, et al., A comparison of eleven static heuristics for mapping a class of independent tasks onto heterogeneous distributed computing systems, *Journal of Parallel and Distributed Computing*, Vol. 61, No. 6, June 2001, pp. 810–837.

[13] T. D. Braun, H. J. Siegel, and A. A. Maciejewski, Heterogeneous computing: Goals, methods, and open problems (invited keynote presentation for the 2001 International Multiconference that included PDPTA 2001), *2001 International Conference on Parallel and Distributed Processing Techniques and Applications (PDPTA 2001)*, Vol. I, pp. 1–12, June 2001.

[14] A. Burns, S. Punnekkat, B. Littlewood, and D. Wright, Probabilistic guarantees for fault-tolerant real-time systems, Technical report, Design for Validation (DeVa) TR No. 44, Esprit Long Term Research Project No. 20072, Dept. of Computer Science, University of Newcastle upon Tyne, UK, 1997.

[15] Y. X. Chen, Optimal anytime search for constrained nonlinear programming, Master's thesis, Dept. of Computer Science, Univ. of Illinois, Urbana, IL, May 2001.

[16] E. G. Coffman, Jr., ed., *Computer and Job-Shop Scheduling Theory*, John Wiley & Sons, New York, NY, 1976.

[17] R. L. Daniels and J. E. Carrillo, β-Robust scheduling for single-machine systems with uncertain processing times, *IIE Transactions*, Vol. 29, No. 11, 1997, pp. 977–985.

[18] A. J. Davenport, C. Gefflot, and J. C. Beck, Slack-based techniques for robust schedules, *6th European Conference on Planning (ECP-2001)*, September 2001, pp. 7–18.

[19] J. DeVale and P. Koopman, Robust software—no more excuses, *IEEE International Conference on Dependable Systems and Networks (DSN 2002)*, June 2002, pp. 145–154.

[20] M. M. Eshaghian, ed., *Heterogeneous Computing*, Artech House, Norwood, MA, 1996.

[21] D. Fernandez-Baca, Allocating modules to processors in a distributed system, *IEEE Transaction on Software Engineering*, Vol. SE-15, No. 11, November 1989, pp. 1427–1436.

[22] I. Foster and C. Kesselman, eds., *The Grid 2: Blueprint for a New Computing Infrastructure*, Morgan Kaufmann, San Fransisco, CA, 2004.

[23] I. Foster, *Designing and Building Parallel Programs*, Addison-Wesley, Reading, MA, 1995.

[24] R. F. Freund and H. J. Siegel, Heterogeneous processing, *IEEE Computer*, Vol. 26, No. 6, June 1993, pp. 13–17.

[25] S. Ghosh, Guaranteeing fault tolerance through scheduling in real-time systems. PhD thesis, Faculty of Arts and Sciences, University of Pittsburgh, 1996.

[26] S. D. Gribble, Robustness in complex systems, *8th Workshop on Hot Topics in Operating Systems (HotOS-VIII)*, May 2001, pp. 21–26.

[27] E. Hart, P. M. Ross, and J. Nelson, Producing robust schedules via an artificial immune system, *1998 International Conference on Evolutionary Computing*, May 1998, pp. 464–469.

[28] R. Harrison, L. Zitzman, and G. Yoritomo, High performance distributed computing program (HiPer-D)—engineering testbed one (T1) report, Technical report, Naval Surface Warfare Center, Dahlgren, VA, November 1995.

[29] O. H. Ibarra and C. E. Kim, Heuristic algorithms for scheduling independent tasks on nonidentical processors, *Journal of the ACM*, Vol. 24, No. 2, April 1977, pp. 280–289.

[30] R. Jain, *The Art of Computer Systems Performance Analysis*, John Wiley & Sons, Inc., New York, 1991.

[31] M. Jensen, Improving robustness and flexibility of tardiness and total flowtime job shops using robustness measures, *Journal of Applied Soft Computing*, Vol. 1, No. 1, June 2001, pp. 35–52.

[32] E. Jen, Stable or robust? What is the difference?, *Santa Fe Institute Working Paper No. 02-12-069*, 2002.

[33] A. Khokhar, V. K. Prasanna, M. Shaaban, and C. L. Wang, Heterogeneous computing: Challenges and opportunities, *IEEE Computer*, Vol. 26, No. 6, June 1993, pp. 18–27.
[34] V. J. Leon, S. D. Wu, and R. H. Storer, Robustness measures and robust scheduling for job shops, *IEE Transactions*, Vol. 26, No. 5, September 1994, pp. 32–43.
[35] M. Maheswaran, S. Ali, H. J. Siegel, D. Hensgen, and R. F. Freund, Dynamic mapping of a class of independent tasks onto heterogeneous computing systems, *Journal of Parallel and Distributed Computing*, Vol. 59, No. 2, November 1999, pp. 107–131.
[36] Z. Michalewicz and D. B. Fogel, *How to Solve It: Modern Heuristics*, Springer-Verlag, New York, 2000.
[37] G. L. Nemhauser and L. A. Wolsey, *Integer and Combinatorial Optimization*, John Wiley & Sons, New York, 1988.
[38] M. Sevaux and K. Sörensen, Genetic algorithm for robust schedules, *8th International Workshop on Project Management and Scheduling (PMS 2002)*, April 2002, pp. 330–333.
[39] V. Shestak, H. J. Siegel, A. A. Maciejewski, and S. Ali, Robust resource allocations in parallel computing systems: Model and heuristics, *The 2005 IEEE International Symposium on Parallel Architectures, Algorithms, and Networks*, December 2005.
[40] S. Shivle, P. Sugavanam, H. J. Siegel, A. A. Maciejewski, T. Banka, K. Chindam, S. Dussinger et al., Mapping subtasks with multiple versions on an ad hoc grid, *Parallel Computing*, Special Issue on Heterogeneous Computing, Vol. 31, No. 7, July 2005, pp. 671–690.
[41] V. Shestak, H. J. Siegel, A. A. Maciejewski, and S. Ali, The robustness of resource allocations in parallel and distributed computing systems, *19th IEEE International Conference on Architecture of Computing Systems: System Aspects in Organic Computing (ARCS 2006)*, March 2006.
[42] G. F. Simmons, *Calculus with Analytic Geometry*, Second Edition, McGraw-Hill, New York, 1995.
[43] Y. N. Sotskov, V. S. Tanaev, and F. Werner, Stability radius of an optimal schedule: A survey and recent developments, *Industrial Applications of Combinatorial Optimization*, G. Yu, ed., Kluwer Academic Publishers, Norwell, MA, 1998, pp. 72–108.
[44] P. Sugavanam, H. J. Siegel, A. A. Maciejewski, M. Oltikar, A. Mehta, R. Pichel, A. Horiuchi et al., Robust static allocation of resources for independent tasks under makespan and dollar cost constraints, *Journal of Parallel and Distributed Computing*, Vol. 67, No. 4, April 2007, pp. 400–416.
[45] L. Wang, H. J. Siegel, V. P. Roychowdhury, and A. A. Maciejewski, Task matching and scheduling in heterogeneous computing environments using a genetic-algorithm-based approach, *Journal of Parallel and Distributed Computing*, Vol. 47, No. 1, November 1997, pp. 8–22.
[46] M.-Y. Wu, W. Shu, and H. Zhang, Segmented min-min: A static mapping algorithm for meta-tasks on heterogeneous computing systems, *9th IEEE Heterogeneous Computing Workshop (HCW 2000)*, May 2000, pp. 375–385.

42
A Transparent Distributed Runtime for Java

Michael Factor
IBM Research Laboratory

Assaf Schuster
Konstantin Shagin
Technion–Israel Institute of Technology

42.1	Introduction	42-1
42.2	System Model	42-3
42.3	JavaSplit Overview	42-3
	Class File Instrumentation • Shared Memory Management • Distributed Synchronization	
42.4	Fault Tolerance	42-10
	Java Thread Checkpointing • Replication Groups • Distributed Shared Memory Persistence • Checkpointing • Fault-Tolerant Distributed Queueing	
42.5	Performance Evaluation	42-15
42.6	Related Work	42-17
	Java-Based Distributed Computing • Fault-Tolerant DSMs	
42.7	Conclusions	42-21
References		42-22

42.1 Introduction

Collections of interconnected workstations are widely considered a cost-efficient alternative to supercomputers. Despite this consensus and high demand, there is no general parallel processing framework that is accepted and used by all. It is hard to adapt existing solutions because they often target a specific platform, require special networking hardware, or enforce complicated programming conventions. As a result, many organizations choose to implement custom frameworks that, for the same reasons, cannot be used by others. The effort invested in devising such frameworks compromises the cost benefit of using off-the-shelf computing resources.

Another difficulty in using a network of workstations for parallel processing is that a participating workstation may suddenly become unable to continue its part of the computation. This can be caused by a failure, for example, a power outage, or by user intervention, for example, shutdown or restart. These events result in the immediate loss of the data and processes located on the node. Unless the application or the runtime that executes it are fault-tolerant, the above actions will surely violate the integrity of the computation. This problem is especially acute in nondedicated environments, where the runtime system has little control over the participating machines.

In this chapter, we address the aforementioned issues. We provide a runtime environment for a heterogenous collection of failure-prone workstations which transparently supports a well-known high-level parallel programming paradigm. The proposed system, which we call *JavaSplit-FT*, executes standard, possibly preexisting, multithreaded Java programs written for a single standard Java Virtual Machine (JVM). It makes the distributed and unreliable nature of the underlying hardware completely transparent to the programmer. Consequently, any Java programmer can write applications for our framework without any additional training.

Each runtime node executes the application threads assigned to it using nothing but its local standard unmodified JVM. A node can use any JVM implementation, as long as it complies with the JVM specification [LY99]. Unlike systems that utilize specialized networking hardware [ABH$^+$01, VBB00], JavaSplit-FT employs IP-based communication. It accesses the network through the Java socket interface. The use of standard JVMs in conjunction with IP-based communication enables JavaSplit-FT to perform computation on any given heterogenous set of machines interconnected by an IP network, such as the Internet or the intranet of a large organization. In addition, each node can locally optimize the performance of its JVM, for example, via a *just-in-time* compiler.

JavaSplit-FT is resilient to multiple node failures. Fault tolerance is accomplished by checkpointing and replication. At certain execution points each application thread independently checkpoints itself. A snapshot of a thread does not include the application's shared objects, that is, objects that can be accessed by more than one thread. Shared objects and thread states are made persistent by replication in the volatile memory of other nodes. Unlike many checkpointing algorithms, ours keeps only a single checkpoint per thread.

Scalability considerations played an important role in the design of our fault-tolerance scheme. As a result, neither failure-free execution nor recovery from a failure require global cooperation of nodes. Moreover, recovery does not roll back nonfailing nodes. During recovery, a nonfailing node continues its normal execution, unless it wishes to acquire a lock held by a failed thread. In that case, it must wait until the thread is restored.

The shared memory abstraction is supported by a distributed shared memory (DSM) subsystem. Our fault-tolerance scheme is tightly integrated into the DSM management protocol that is based on *home-based lazy release consistency* (HLRC) [ZIL96]. We exploit the fact that in the employed DSM protocol a thread may affect the data observed by remote threads only during lock ownership transfer. By checkpointing a thread when it is about to transfer ownership, we ensure that the most recent thread snapshots are interconsistent.

The proposed fault-tolerance protocol has a minor effect on DSM management. It prevents optimizations that make a thread's modifications visible on other nodes at times other than lock transfer. Nevertheless, it does not prohibit prefetching of shared data and therefore allows the number of data misses to be reduced

Transparency, portability, and the ability to capture and restore a thread state are all achieved by means of bytecode instrumentation. JavaSplit-FT instruments the program to intercept events that are interesting in the context of distributed execution, for example, synchronization, creation of new threads, and accesses to shared objects. In addition, the JavaSplit-FT bytecode rewriter augments the program and a subset of runtime classes with thread checkpointing capabilities. Combined with the runtime modules, which are also written in Java, the resulting bytecode can be executed on any standard JVM.

JavaSplit-FT requires minimal management efforts. A Java-enabled workstation that wishes to participate in the computation does not need to install any software. It merely needs to run a thin Java client that notifies the system core that the current workstation is available. The client does not include all the runtime modules. The application and runtime classes are loaded from the network using the customizable Java class loading mechanism. Moreover, the client program can be incorporated in a web page applet. Consequently, a workstation completely unaware of JavaSplit-FT can join the runtime simply by pointing its Java-enabled browser to that page.

The proposed system is suitable for utilization of idle interconnected computing resources. The portable client program mentioned above can be incorporated into a screen saver. Since the system is resilient to node failures, it will remain consistent even if a machine's user reboots it or disconnects it from the network. Moreover, fault tolerance allows us to minimize a workstation's response time to its native user by treating node reclamations as node failures. When a user's presence is detected (e.g., by monitoring input devices), we can immediately kill all local runtime processes and release all the memory they occupy.

JavaSplit-FT is a fault-tolerant extension of *JavaSplit* [FSS]. To fully grasp our fault-tolerance scheme, it is essential to understand the memory management and distributed synchronization protocols employed by JavaSplit. For completeness, we will describe these mechanisms in Section 42.3.

42.2 System Model

We assume a collection of interconnected processor nodes. Each node consists of a uniprocessor or a symmetric multiprocessor (SMP) and a volatile memory. The nodes do not share any physical memory and communicate by message passing. The messages may be lost, duplicated, or delayed (and therefore may arrive out of order). However, there are no undetected transmission errors in delivered messages.

We assume a fail-stop model [19], that is, when a node fails, it stops communicating. Failures may occur simultaneously and even during recovery. We assume the existence of a mechanism that allows verification of node liveness.

Node failures may be permanent or transient. However, since the system is designed to treat a node recovered from a transient failure as an entirely new node, we can assume that all failures are permanent. Transient failures are modeled as permanent by assigning a new ID to every recovering node and ignoring messages destined to their previous incarnations. Note that messages sent from the prior incarnations of the recovering nodes should not be a problem: the assumption of unreliable channels means that the system must be able to ignore outdated messages.

42.3 JavaSplit Overview

The JavaSplit runtime administers a pool of worker nodes. An external resource management system may be employed to detect available machines. Each worker can execute one or more application threads. The computation begins by starting the application's *main* method on an available worker. Each newly created thread is placed for execution on one of the worker nodes, according to a plug-in load balancing function.

JavaSplit instruments all classes used by the application submitted for distributed execution and combines them with the runtime modules, which are also written in Java. Thus, the bytecode becomes aware of the distributed nature of the underlying environment. JavaSplit uses twin class hierarchy (TCH) [FSS04], a novel instrumentation technique that allows transformation of both user-defined classes and Java *standard library classes*, including the library classes with native methods. The instrumentation may be performed at run time, when the class is loaded into the JVM, or before the execution begins. In any case, all participating JVMs use the instrumented classes instead of the originals.

To support the shared memory abstraction, JavaSplit incorporates a DSM. The design choice of using a DSM rather than the master-proxy model is orthogonal to our bytecode instrumentation techniques. The instrumentation allows interception of accesses to shared data and calls to methods of remote objects. Therefore, both *data shipping* and *method shipping* paradigms are possible. We consider the former more suitable for scalable high-performance computing.

Our DSM dynamically classifies objects into *local* and *shared*, managing only the latter. An object is considered *local* until the system detects that more than one thread accesses it. In this case, the object receives a globally unique ID and is registered with the DSM. Dynamic detection of shared objects contributes to the system's performance because the maintenance cost of a local object is lower. This performance gain can be significant because, in many Java applications, there are only a few shared objects.

To support the synchronization primitives of Java, for example, lock acquire and wait operations, JavaSplit incorporates a distributed synchronization mechanism. This mechanism is based on the distributed queue (DQ) algorithm [WS02]. (Note that an earlier version of JavaSplit [FSS] employed a different protocol.) Our protocol differs from the classic DQ in two aspects. First, it does not assume reliable channels and therefore is designed to tolerate message loss. Second, it supports Java-specific synchronization operations, such as `wait` and `notify`.

42.3.1 Class File Instrumentation

In JavaSplit, the main goals of instrumentation are (i) distributing the threads and objects of an application among the worker nodes and (ii) preserving consistency of data accesses and lock operations. In order to achieve these goals, JavaSplit must instrument any class used by the original application, including the Java *standard library classes*, also known as *system classes*.

Owing to their special status within the JVM, system classes are difficult to instrument. Most JVMs make assumptions regarding the structure and loading order of system classes. If these assumptions do not hold, a JVM may terminate abnormally. For example, if the instrumentation process augments one of the classes `java.lang.Object`, `java.lang.Class` or `java.lang.String` with a field, the JVM crashes. Moreover, dynamic instrumentation using custom class loaders is hindered by the fact that a subset of system classes (approximately 200 in Sun JDK 1.4.2) is already loaded by the JVM, before the class loading mechanism can be modified to enable rewriting. Finally, some system classes have *native methods*, that is, methods whose implementation is not expressed in bytecode, but rather in a machine dependent assembler. To enable sound instrumentation of system classes, we employ a bytecode instrumentation strategy, which we call the TCH approach [FSS04].

42.3.1.1 Twin Class Hierarchy

At the heart of the TCH approach lies the idea of renaming all the instrumented classes. In JavaSplit, for each original class `mypackage.MyClass` we produce a rewritten version called `javasplit.mypackage.MyClass`. Thus, we create a hierarchy of classes, parallel to the original one, encapsulated in a package called `javasplit`. Figure 42.1 illustrates this transformation.

In a rewritten class, all referenced class names are replaced with new *javasplit* names. For example, in the bytecode, the renaming affects such instructions as `instanceof`, `invokevirtual`, `new`, and `getfield`. During the execution, the runtime uses the *javasplit* classes instead of the originals. Figure 42.2 demonstrates this change at the source code level. In practice, however, the transformation is performed in the bytecode. As a result of the TCH class name transformation, the rewritten system classes become user classes and thus no longer have special status within the JVM. This eliminates most of the instrumentation difficulties, except the problem of classes with native methods, which cannot be automatically transformed.

Implementation of a native method is always bound to a particular method in a particular class. It can be accessed only by calling that method, but it cannot be reused in the implementation of another method (or in another class). Therefore, after TCH class renaming, native methods in an instrumented class cannot use their original implementation. Fortunately, the TCH renaming allows the original and the instrumented versions of a class to coexist. Thus, an instrumented class can use the original class and thus regain access to the native functionality. (See Reference FSS04 for a more detailed discussion.)

42.3.1.2 JavaSplit-pecific Transformations

The bytecode instrumentation intercepts events that are important in the context of a distributed runtime. First, the bytecodes that start execution of new threads are replaced by calls to a handler that ships the thread to a node chosen by the load balancing function. Second, the lock-related operations, identified by the instructions `monitorenter` and `monitorexit` or by *synchronized methods*, are replaced by synchronization handlers. Third, in order to preserve memory consistency, the rewriter inserts access checks before accesses to fields and array elements, for example, `getfield`, `putstatic`, `iaload`, and `lastore` (see Figure 42.3). If an access check fails, a newer version of the object is obtained from another node.

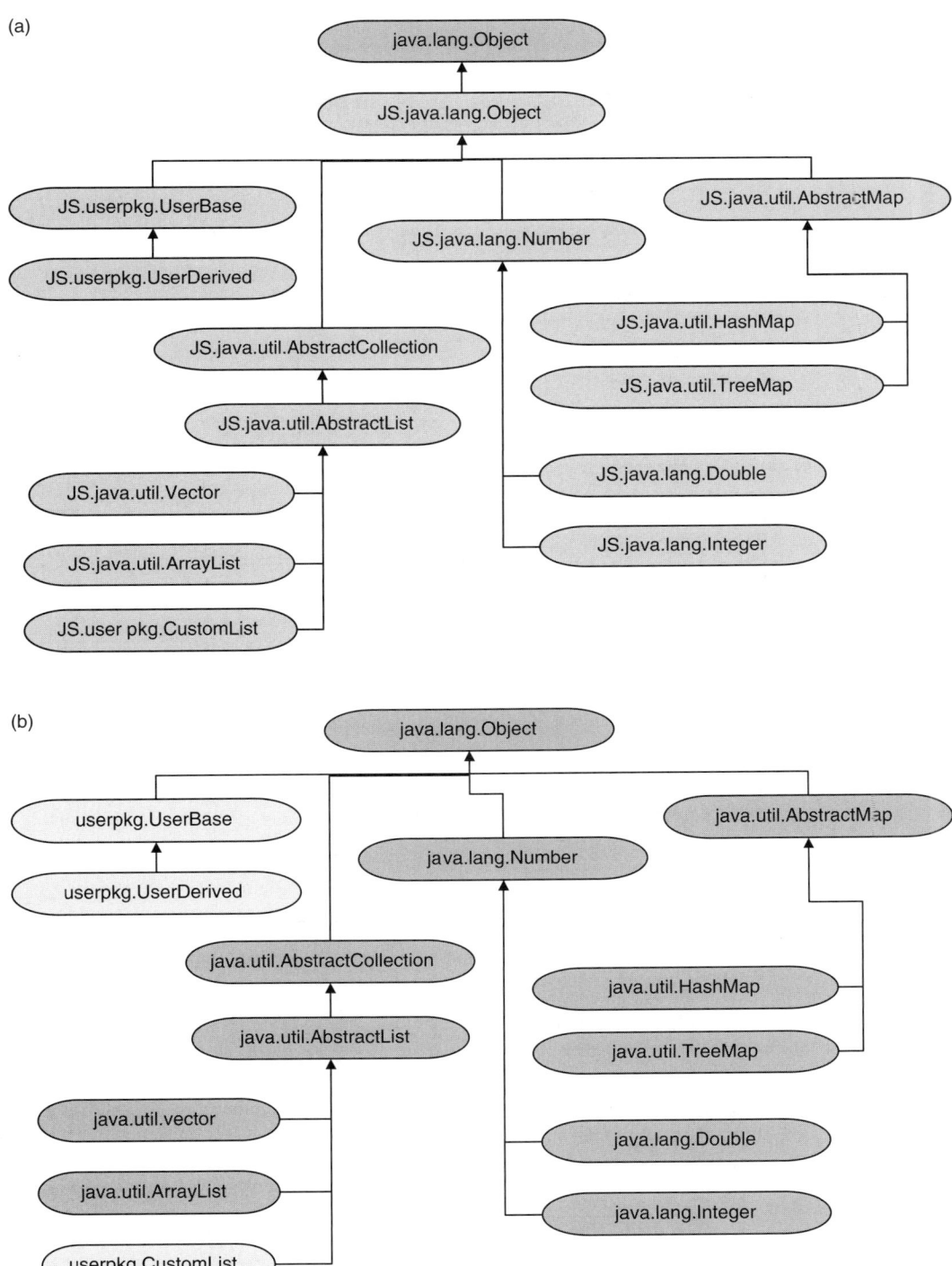

FIGURE 42.1 A fragment of the class hierarchy, before and after the class renaming transformation. The instrumented versions of system classes become user classes: (a) original class hierarchy and (b) instrumented class hierarchy. (Java system classes are designated by dark gray.)

(a)
```java
class A extends somepackage.C {
    // fields
    private int myIntField;
    public B myRefField;
    public java.util.Vector myVectorField;
    // methods
    protected void doSomething(B b, int n) {
        if(b instanceof java.util.List){ ... }
        java.lang.Class vecClass
            = java.lang.Class.forName("java.util.Vector");
        ...
    }
    public B doSomethingElse(java.lang.String str) {
        java.lang.System.out.println(str);
        java.io.File f = new java.io.File(str);
        ...
    }
}
```

(b)
```java
class JS.A extends JS.somepackage.C {
    // fields
    private int myIntField;
    public JS.B myRefField;
    public JS.java.util.Vector myVectorField;
    // methods
    protected void doSomething(JS.B b, int n) {
        if(b instanceof JS.java.util.List){ ... }
        // The string parameter of forName is not augmented with the prex  "JS."
        // Instead, the implementation of JS.java.lang.Class.forName is modied
        // to produce an instance of JS.java.lang.Class representing JS.java.util.Vector
        JS.java.lang.Class vecClass
            = JS.java.lang.Class.forName("java.util.Vector");
        ...
    }
    public JS.B doSomethingElse(JS.java.lang.String str) {
        JS.java.lang.System.cut.println(str);
        JS.java.io.File f = new JS.java.io.File(str);
        ...
    }
}
```

FIGURE 42.2 A Java class before and after JavaSplit instrumentation: (a) original class and (b) instrumented class.

....	....
ALOAD 1	// load the instance of class A
DUP	
GETFIELD	A::byte __javasplit__state__
IFNE	// jump to the next GETFIELD
DUP	
INVOKESTATIC	Handler::readMiss
GETFIELD	A::myIntField
....	....

FIGURE 42.3 Instrumented read access of the field *myIntField* in class A. The instructions in bold are added by the instrumentation. If the object is valid for read, that is, the value of the state field is not 0, only the first 3 added instructions are executed.

In addition to the above modifications, the classes are augmented with utility fields and methods. The class at the top of the inheritance tree is augmented with fields indicating the state of the object during the execution, for example, access permission, version, and whether it is locked. This approach enables quick retrieval of the state information and allows the garbage collector to discard it together with the object. Each class is augmented with several utility methods, which are generated on the basis of the fields of the specific class. The most important utility methods are for serializing, deserializing, and comparing (*diffing*).

JavaSplit provides distribution-aware versions for a subset of most commonly used native classes. The new classes are implemented in pure Java and do not have native methods.

In Java applications there are a lot of unnecessary lock operations [CGS+99]. Often, especially in the system classes, locks protect accesses to objects that are accessed by no more than one thread during the entire execution. The overhead of the unnecessary locking may be negligible in Java. However, when rewriting bytecodes for distributed execution, one must be extra careful to avoid the performance degradation that may result from the increased cost of the transformed lock operations, which were unnecessary to begin with.

JavaSplit reduces the cost of lock operations of local objects by avoiding the invocation of lock handlers when locking a local object. Instead, a counter is associated with a local object, counting the number of times the current owner of the object has locked it since becoming its owner. (In Java, a thread can acquire the same lock several times without releasing it, and the object is considered locked until the owner performs the same number of releases.) Acquire (lock) and release (lock) operations on a local object simply increase and decrease the counter. Thus, the object is locked only when the counter is positive. If the object becomes shared when another thread wishes to acquire it, the lock counter is used to determine whether the object is locked. The cost of lock counter operations is not only low in comparison to the invocation of lock handlers for shared objects, but is also cheaper than the original Java-acquire (`monitorenter`) operation (see Section 42.5).

42.3.2 Shared Memory Management

The DSM incorporated in JavaSplit is object-based. The object-based approach fits well on top of the object-based memory management of a JVM. By contrast, the *page-based* approach would be quite unnatural in the current context, since the hardware paging support for detection of memory accesses cannot be attained without modifying the JVM. In addition, allocation of multiple objects on the same page would result in *false sharing*.

Any shared memory, distributed or not, is subject to a set of constraints, which constitute the *memory model*. For example, the *Java Memory Model* (JMM) sets constraints on the implementation of shared memory in the context of a single JVM. The new JMM [JSR] is based on the principles of Lazy Release Consistency (LRC) [KCZ92]. Therefore, we implement a DSM protocol that complies with LRC. Since there is a natural mapping for Java volatile variables to the release-acquire semantics of LRC in the revised JMM, we encapsulate accesses to volatile variables with acquire-release blocks. Our DSM protocol is inspired by the *HLRC* [ZIL96] protocol.

42.3.2.1 Lazy Release Consistency

Lazy Release Consistency uses locks for synchronization. There are two synchronization operations, `acquire` and `release`, which are equivalent to the JMM primitives `lock` and `unlock`, respectively [GJSB00]. The *release-acquire* pairs define a partial order among operations performed by the participating processes. This partial order, called *happened-before-1* [KCZ92] and denoted by $\rightarrow$, is defined as follows:

- If a_1 and a_2 are accesses by the same process, and a_1 occurs before a_2 in program order, then $a_1 \rightarrow a_2$.

- If a_1 is a release by process p_1, and a_2 is a subsequent acquire of the same lock by process p_2, then $a_1 \to a_2$.
- If $a_1 \to a_2$ and $a_2 \to a_3$, then $a_1 \to a_3$.

LRC requires that when a process p performs an operation op (which can be a read, write, acquire, or release), all operations that precede op in the *happened-before-1* partial order appear completed to p.

42.3.2.2 Home-Based Implementation of lazy Release Consistency

The DSM protocol of JavaSplit employs the *home-based* approach to implement the LRC memory model. In home-based protocols, each shared memory unit (page or object) has a *home node* to which all writes are propagated and from which all copies are derived.

Our protocol refines the scalability of HLRC. In HLRC, a *vector timestamp* proportional to the number of processes in the system is attached to each copy of a shared memory unit. In a large-scale environment, vector timestamps can occupy a lot of space and induce nonnegligible communication overhead. We reduce the timestamp to a fixed-size value, which consists of the ID of the most recently (locally) released lock and a counter of release operations associated with that lock. In object-based systems like ours, the gain may be significant because the number of objects, and hence the number of timestamps, may be large.

The employed protocol is a *multiple-writer* protocol. In multiple-writer protocols more than one process can write into the same shared memory unit simultaneously. The multiple-writer approach is advantageous because it does not require that the writer have exclusive ownership of the shared memory unit.

In multiple-writer schemes, there must be a way for a writer to identify its modifications in order to be able to merge them with modifications of other writers. Like most multiple writer schemes, ours also uses the *twinning and diffing* technique. Before modifying a cached copy of a shared memory unit that has not yet been modified, the writer makes another copy of it, called a *twin*. At a later point of execution, the modifications can be calculated by XOR-ing (diffing) the cached copy with its twin.

The *happened-before-1* order is preserved by passing consistency data during lock transfer from process to process. During lock transfer, the releasing process R sends to the acquiring process A the list of shared memory units that were modified before R's current release, according to *happened-before-1*. The sent list does not include shared memory units whose modifications A is aware of. An entry of this list is called a *write notice*. On the basis of the received write notices, the acquiring process may invalidate the local copies of corresponding shared memory units. Consequently, when accessing an invalidated copy, the process may need to obtain a newer version of the shared memory unit.

In JavaSplit, a thread that wishes to access a shared object but does not have a valid cached copy fetches it from the object's home. A timestamp that indicates the required version of the object is attached to the fetch request message. If the requested version is not available, the home delays the reply until the necessary updates are received (from another thread).

When a thread obtains the required version of the object from the object's home, it creates a local cached copy. The thread uses the cached copy for both reading and writing, until a write notice with a more recent timestamp is received. When this happens, the thread invalidates the cached copy and remembers the received timestamp. The next time the thread accesses the cached copy, it discovers that the copy is invalid and sends a fetch request containing the received timestamp.

Before transferring ownership of some lock to another thread, the current thread calculates the recent object modifications using the *diffing* technique. Then it sends the modifications to the homes of the modified objects. A timestamp that identifies the current lock transfer operation is attached to each such outgoing modification request. After sending the updates, the current thread sends the requester of the lock the write notices for the objects it updated. The aforementioned timestamp is attached to this message as well. Note that in HLRC each write notice has its own vector timestamp, whereas our protocol uses a single fixed-size timestamp.

Figure 42.4 illustrates *happened-before-1*DSM protocol with a simple example. Note that the lock request message is not sent directly from B to A, but rather through the home of L, as described in Section 42.3.3.

A Transparent Distributed Runtime for Java

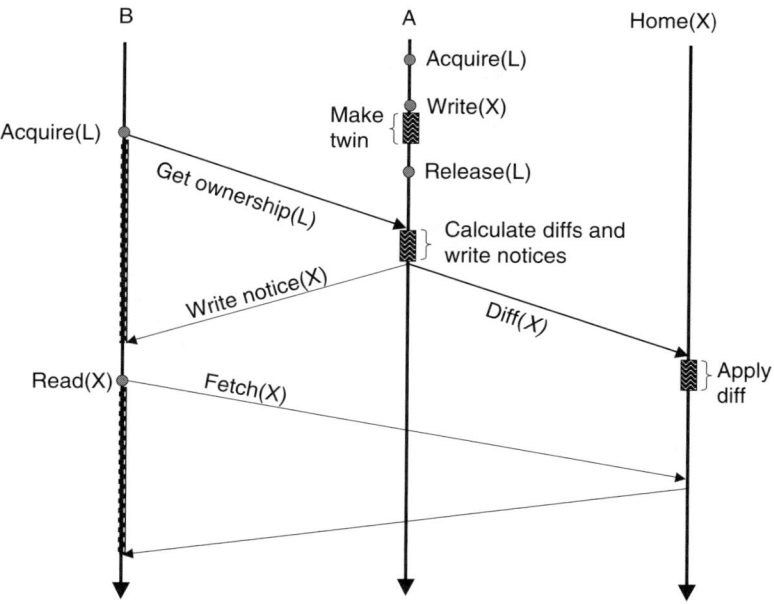

FIGURE 42.4 Example of DSM activity.

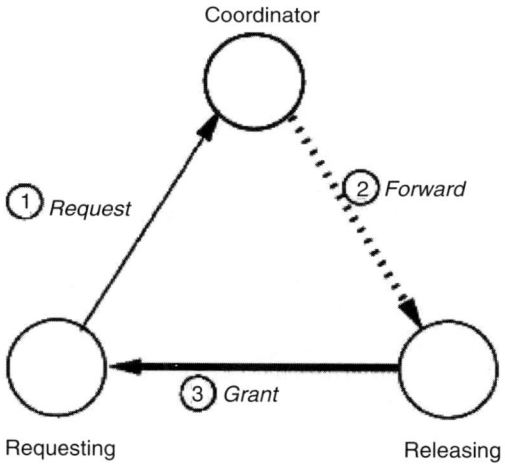

FIGURE 42.5 Distributed queue basics.

42.3.3 Distributed Synchronization

In Java, each object has a lock associated with it. An object's lock does not necessarily synchronize accesses to the object.

The distributed synchronization protocol employed in JavaSplit is based on the DQ protocol [WS02]. In DQ, a lock is assigned to a coordinator node. A process wishing to acquire the lock sends a request message to the coordinator. The coordinator forwards each new request to the previous requester, thus forming a DQ of requests. Upon exiting the critical section, the releasing node passes the ownership directly to the next requester or keeps it if no forwarded requests have been received from the coordinator. Figure 42.5 illustrates the basics of DQ.

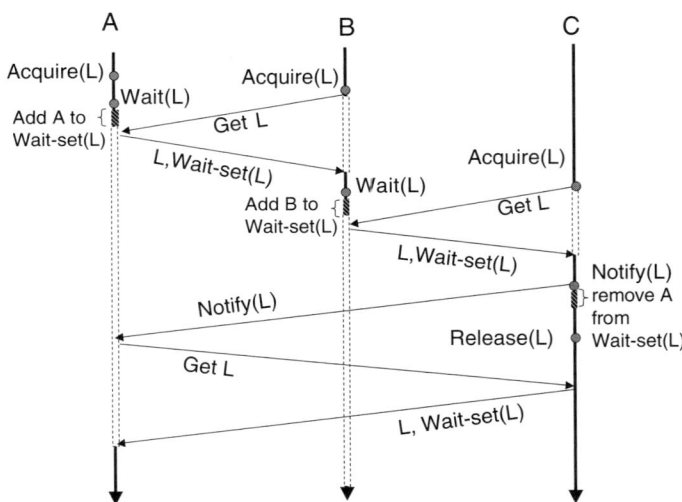

FIGURE 42.6 Conditional waiting for distributed token-based synchronization algorithms.

Our protocol extends DQ by adding support for unreliable channels and Java-specific conditional waiting primitives, such as wait and notify [GJSB00]. An object's home node plays the role of the coordinator of the lock associated with that object.

42.3.3.1 Distributed Conditional Waiting

In Java, a *wait set* is associated with each object. A thread that invokes the wait method of an object is inserted into the wait set. Invocation of an object's notify method removes an arbitrary thread from the object's wait set. Invocation of the method notifyAll removes all threads from the object's wait set. In order to perform the above operations, a thread must be the owner of the object's lock. A thread that invokes the wait method releases the lock and blocks until a notification is received. When a notification is received, the thread must reacquire the lock in order to proceed.

In our implementation of conditional waiting, the main idea is to pass the wait set along with the lock ownership. Thus, the owner of the lock can always access the wait set locally.

The wait operation is translated to the following sequence of operations: (i) unlock, (ii) add the current thread to the wait set, (iii) wait for notification, and (iv) lock. The notify operation removes the required number of threads from the wait set and sends a notification message to each of them. Figure 42.6 shows an example of the protocol execution. Note that lock request messages between the nodes are actually sent through the home of the lock.

Our implementation of conditional waiting is suitable not only for DQ but for all the algorithms that transfer lock ownership by sending a message from the current owner to the next one. In the literature, these algorithms are classified as *token based*.

42.4 Fault Tolerance

To enable employment of JavaSplit-FT in large-scale environments, designing a scalable fault-tolerant scheme has been one of our primary goals. Consequently, the resulting protocol never requires global cooperation of nodes—not during failure-free execution nor during recovery. During failure-free execution, the application threads independently checkpoint their state. Unlike most *independent checkpointing* algorithms, our scheme does not require rollback of nonfailing threads and allows us to keep a single (most recent) snapshot of each thread.

Our fault-tolerance protocol combines thread checkpointing with replication. A thread snapshot includes local application objects, and certain system data associated with a thread. The shared objects stored in the DSM are made persistent independently. The shared objects as well as the thread snapshots are replicated in the volatile memory of a group of nodes.

The key idea of our fault tolerance scheme is to capture the states of the application threads in such a way as to guarantee that rollback of any thread to its latest saved checkpoint will not violate system consistency. If this requirement is fulfilled, a failed thread will recover by simply restarting from its most recent saved checkpoint.

In JavaSplit-FT, each thread executes in a separate JVM process. For simplicity, the reader can assume that each application thread runs on a separate node. In reality, failure of a node that executes a number of threads is equivalent to simultaneous failure of several virtual nodes.

42.4.1 Java Thread Checkpointing

The Java platform does not support capturing and reestablishment of a thread state. Fortunately, there are several frameworks that use bytecode instrumentation to augment Java with this functionality [SSY00, Tao01, TRV+00]. In our implementation we employ the *Brakes* project [TRV+00].

The execution state of a Java thread is a stack of *method frames*. A method frame contains the method's stack and its local variables. (Each method has its own stack.) When a method is invoked, a new frame is created and is put on top of the previous frames. When a method returns, its frame is removed from the stack. The main problem in capturing the thread state in a standard JVM is that the contents of a method frame are not accessible outside the method.

To enable capturing of a thread state, the Brakes transformer inserts code after each method invocation. This code is executed only when a certain Boolean variable indicates that the thread wishes to capture its state. This inserted code saves the contents of the frame of the method that contains the invocation. It also saves the program counter of the invocation and returns from the current method. Thus, when capturing the execution state, the stack of the method frames is unrolled and its contents are saved.

To enable reestablishment of a thread state, the Brakes transformer inserts code at the beginning of each method. The inserted code consists of several state reestablishment code blocks, one for each method invocation in the current method. It is executed only when the thread is restoring its state, which is also indicated by a Boolean flag. The inserted code restores the contents of the current frame and jumps to the saved program counter, thus skipping the already executed instructions. The state reestablishment procedure terminates when the topmost execution stack is restored.

The instrumentation overhead of a normal execution is 4 bytecode instructions per method call. The time overhead of state capturing and reestablishment greatly depends on the application.

Only the methods that might be involved in the stack manipulations described above must be instrumented. Currently, JavaSplit-FT does not analyze the application to determine these methods and transforms all user code. Such an analysis will improve the speed of local execution. The methods of runtime classes, however, are instrumented selectively.

Brakes bytecode transformation is applied after the basic JavaSplit transformation described in Section 42.3.1.2. Performing transformation in the opposite order would make the code inserted by Brakes distribution-aware and therefore incorrect. Similar to the JavaSplit transformation, the Brakes transformation can be performed statically, before the execution begins, or dynamically, during class loading.

42.4.2 Replication Groups

The participating nodes are organized into roughly equal-sized groups. The minimal size of a group is a parameter of the runtime. The group members cooperate in order to ensure the safety and consistency of the data, locks, and threads mapped to them. Thus, the group provides the aforementioned entities with the illusion that they are situated on a single infallible node.

An existing pure-Java group communication library such as JavaGroups [Ban] can be used to implement a group. Such a module must support group membership services, intra-group broadcast, and state transfer. However, since JavaSplit-FT requires a reduced set of functions we have, for greater efficiency, created our own group communication module. This module maintains open TCP connections between every two group members. Failure detection is implemented by monitoring the validity of a connection. A group member who closes a connection is suspected of a failure, which is verified by additional attempts to reestablish the connection.

In order to guarantee that the runtime is resilient to F simultaneous node failures, the size of a group should be at least F+1. In practice, the runtime can handle any number of failures within a group as long as the members do not all fail simultaneously. To improve fault tolerance, the system should add new nodes to groups whose size drops below a predefined minimum. Note that a node joining a group must copy the group state from some member.

The policy of partitioning nodes into groups is important. On the one hand, geographic dispersal of group members can improve fault tolerance because it prevents dependencies between failures of members of the same group. For instance, in the context of idle cycle utilization, workstations situated in the same room might be reclaimed simultaneously, when their users return from a coffee break. On the other hand, since members of the same group frequently exchange messages (to keep the replicas consistent), their geographical proximity might increase overall system efficiency. The tradeoff between efficiency and fault tolerance is inherent to our work.

42.4.3 Distributed Shared Memory Persistence

As mentioned in Section 42.3, the application objects are dynamically classified into *local* and *shared*. Each shared object has a home node that manages the object's *master copy*. A node keeps the master copies of all the shared objects mapped to it in a special repository, which we call the shared object repository (SOR). The SORs of the participating nodes constitute the DSM of our system.

A node's SOR is not included in a thread snapshot. This significantly reduces both the time required to capture a thread state and the size of the resulting snapshot. SOR persistence is ensured by replication on the members of the node's group.

Each modification received by an SOR is propagated to the replicas and applied to them as well. Only after all the replicas have been updated is the modification applied to the primary replica of SOR and a modification request acknowledged. Thus, a modification to an object is not revealed to the threads that wish to fetch it until all the replicas have received the modification and applied it. Figure 42.7 illustrates the basics of this protocol.

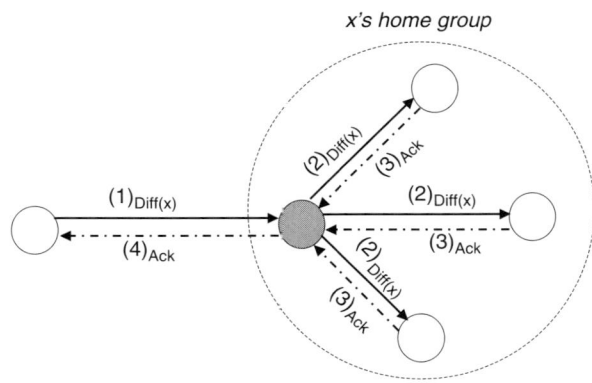

FIGURE 42.7 Shared data replication.

An SOR's node may fail before all its replicas are updated, which leaves the replicas inconsistent. In this case, an acknowledgment of the modification request will not be sent. Consequently, when the SOR is recovered, possibly from a less updated replica, it will receive a retransmitted modification request and propagate it to the other replicas. Since update messages are identified by versions, only the less updated replicas will apply the modifications. The recovered primary SOR will acknowledge the receipt of the update message only after all the replicas receive the modification and apply it, if necessary. Thus, the consistency of SOR replicas is restored.

Note that the thread checkpointing policy described in Section 42.4.4 guarantees that a modification request will be retransmitted, despite failures of its source, until acknowledgment of the request is received.

42.4.4 Checkpointing

To preserve the integrity of the system despite failures, we must ensure that the states of every two threads are consistent. In JavaSplit, there are two ways a thread can affect other threads: by updating the shared application data (stored in some SOR) and through lock-related operations. In this section, we describe a thread checkpointing policy that ensures interthread consistency with respect to shared data. The consistency of lock operations is guaranteed by an additional mechanism that is described in Section 42.4.5.

Recall that in the original DSM protocol described in Section 42.3.2.2, a thread may affect the application data observed by other threads only by updating the SORs. The modifications are sent to SORs when the current thread transfers lock ownership to another thread. Therefore, a thread affects the data observed by other threads only when it transfers a lock. Consequently, if a thread rolls back past the latest lock transfer operation, the system state becomes inconsistent. Figure 42.8 illustrates this point. If we roll back the thread R to q_2, this will have the same effect as stopping R at q_2 and then resuming it after a while. However, if we roll back R to q_1, the system state becomes inconsistent, because: (i) R returns to a state in which it has not sent the modifications, although they have already been applied to the SORs, and (ii) there is no guarantee R will recreate these modifications after the rollback. After the rollback, the execution of R may differ from the original execution, because R may fetch different (more recent) copies of shared objects or be requested to transfer a lock at a different time. Moreover, R may read its own "future" modifications to SORs, which is illegal.

To avoid inconsistency between the state of a rolled back thread and the rest of the system, each thread captures and saves its state just before sending out the modifications. In Figure 42.8, the checkpoints are denoted by c_i. When a thread fails, it restarts from its latest snapshot and retransmits the modifications. This retransmission is the only effect the failure has on the SORs and consequently on the other application threads. However, since the modifications can be applied on SORs only once, the repeated modification requests will be ignored. The retransmission following thread recovery is necessary because the previously transmitted modification requests may have failed to arrive at their target as a result of channel unreliability or target node failure.

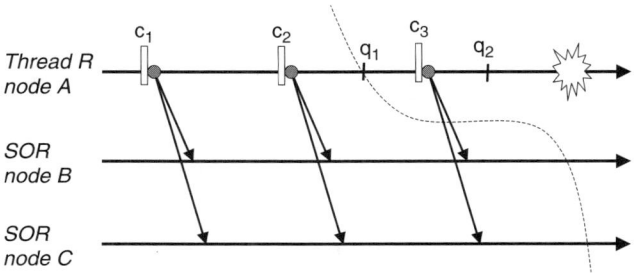

FIGURE 42.8 Checkpointing.

The thread's snapshot is made persistent by replication in the group to which thread's node belongs. When a group member receives the newer snapshot, it discards the previous one. If a node fails, its thread is restored on the least loaded member of its group from the most recent checkpoint found in the group. To avoid execution of several threads on a single processor, a failed thread can be restored on a new node added to the group especially for that purpose. It is also possible to have a backup member that does not execute any threads unless a group member fails.

If the node of a thread fails during snapshot transmission to the group members, some replicas might not receive the latest snapshot. In this situation, the thread is recovered from the most recent snapshot. Before restarting the thread, the less updated group members must receive the most recent snapshot in order to ensure the snapshot's persistence.

42.4.5 Fault-Tolerant Distributed Queueing

In JavaSplit-FT, we augment JavaSplit's synchronization protocol, described in Section 42.3.3, with resilience to node failures. This results in a protocol, which we call fault-tolerant distributed queue (*FT-DQ*). Below we present *FT-DQ* and show how it complies with the checkpointing policy described in Section 42.4.4.

In *FT-DQ*, there are several mechanisms that preserve the integrity of the DQ. First, each queue node, including the current owner and excluding the queue tail, periodically sends a *your-request-registered* message to the next node in the queue. If a queue node stops receiving this message, it will suspect that something is wrong and retransmit its request to the coordinator. Second, the coordinator of a lock not only remembers the tail but also keeps record of the entire queue. This allows a repeated lock request to be forwarded to the same node it was forwarded to in the past. Finally, request versions are attached to protocol messages in order to identify repeated and outdated packets.

42.4.5.1 Recovery from Queue Member Failures

The coordinator detects failures of queue members by monitoring repeated requests. If a queue member M_{i+1} sends several repeated requests, the coordinator suspects that its predecessor in the queue, M_i, has failed. This suspicion is verified by trying to contact M_i. If indeed M_i has failed, the coordinator restores the consistency of the queue according to the following procedure.

First, the coordinator contacts M_{i-1}, the predecessor of M_i in the DQ. If M_{i-1} has not already passed the lock ownership to M_i, the coordinator notifies it that the next node in the queue is M_{i+1}. Otherwise, the coordinator waits until M_i is restored on another node and queries it whether it considers itself the current lock owner. If M_i considers itself the lock owner, the coordinator reminds it that the next node in the queue is M_{i+1}. (A queue considers itself the owner despite a failure if the checkpoint from which it is restored was taken after it received the ownership.) Otherwise, the coordinator links between M_{i-1} and M_{i+1}. In any case, if the recovery procedure determines that M_i does not hold the lock, the coordinator removes M_i from its local queue record.

The current owner may detect that its successor failed while trying to pass ownership to it. If that happens, it may inform the coordinator. The coordinator will then instruct the owner to pass the lock to the successor of the failed thread, if it exists. This is merely an optimization, since the coordinator would eventually find out about the failure by receiving repeated requests from the successor of the failed queue member.

If a thread checkpoints itself while waiting for lock ownership and then fails, it is restored in a state in which it expects to receive the ownership. However, it will not receive the ownership because its location has changed. Therefore, a thread keeps track of its pending lock requests. During recovery it initiates a new request for each recorded pending request. Note that it is not enough to merely retransmit the requests, because the coordinator may consider them outdated and ignore them.

42.4.5.2 Recovery from Coordinator Failures

If the coordinator fails, its knowledge about the queue is lost. It knows neither which node is the current owner nor which node is the queue's tail. To recover, the new coordinator initiates a current owner detection procedure. When the current owner is detected, it is instructed not to pass the ownership to the next node in the outdated queue, but rather wait for a new lock transfer command from the coordinator. Then the coordinator sets the detected current owner to be the first node in its local queue and resumes normal operation. (Consequently, the next lock request is forwarded to the current owner.)

In order that the coordinator can determine the current owner, each node keeps record of the last node to which it transferred the lock. Since, according to Section 42.4.4, a thread captures its state just before the lock transfer, a thread always remembers to whom it has passed the lock despite node failures. Thus, the current owner may be tracked by following these records, starting from the coordinator. To speed up the discovery of the current owner, the coordinator may save the identity of the most recent owner known to it in a thread snapshot. To keep the coordinator updated, queue members should periodically notify the coordinator after they pass ownership to another node.

42.4.5.3 Support for Conditional Waiting

Conditional waiting introduces additional interthread dependencies, which in the presence of node failures may violate consistency. A thread that performs a `notify` operation may fail and recover from a point prior to the notification. This compromises consistency because after the rollback it may choose to notify another thread or not to perform notify at all. A waiting thread may fail after receiving a notification and consequently "forget" the fact that it has received it. This may result in a deadlock.

To prevent the above inconsistencies, the notifying thread checkpoints itself before sending a notification and the waiting thread checkpoints itself upon receiving a notification. The notifier keeps retransmitting the notification until it receives an acknowledgment from the waiting thread. The waiting thread transmits the acknowledgment only after it has checkpointed itself.

The additional checkpoint of the notifier may be avoided by delaying the notification transmission until the lock's `release` operation. If at the time of lock release the notifier discovers that it needs to transfer the lock to another thread, it will have to checkpoint itself anyway. This optimization does not violate the semantics of the Java `notify` operation, because a notification has no effect until the lock is released.

42.5 Performance Evaluation

We have evaluated the performance of JavaSplit-FT using a collection of Intel Xeon dual processor (2×1.7 MGHz) machines running on Windows XP and interconnected by a Gigabit Ethernet. In our experiments we employ SUN JRE 1.4.2, one of the most popular JVMs. In our performance evaluation we use five applications, the first two of which are from the *Java Grande Forum Benchmark Suite* [SB01].

1. **Series**: Computes the first N Fourier coefficients of the function $f(x)=(x+1)x$. The calculation is distributed between threads in a block manner. We run Series for $N = 250,000$.
2. **RayTracer**: Renders a scene containing 64 spheres at resolution of $N \times N$ pixels. The worker threads of this application render different rows of the scene. We run Ray Tracer for $N = 1600$.
3. **TSP**: The TSP application searches for the shortest path through all N vertices of a given graph. The threads eliminate some permutations using the length of the minimal path known so far. A thread discovering a new minimal path propagates its length to the rest of the threads. The threads also cooperate to ensure that no permutation is processed by more than one thread by managing a global queue of jobs. We run TSP for $N = 14$.
4. **KeySearch**: Implements a known plain-text attack on DES. The inputs of the application are an original message and the result of its encryption. The threads cooperate to find the encryption key in the range 2^{24} of keys. Coordination among threads is needed to allow load balancing and in order to ensure the threads terminate as soon as the key is found.
5. **PrimeFinder**: Searches for prime numbers in the range [2,N]. The detected primes are written to the standard output. For better load balancing, the range is dynamically distributed among threads, which cooperate to avoid checking the same number more than once. In our experiments, $N = 3,000,000$.

TABLE 42.1 Heap Access Latency (*Nanoseconds*)

	Original	Rewritten	Slowdown
Field read	0.84	1.82	2.2
Field write	0.97	2.48	2.6
Static read	0.84	1.84	2.2
Static write	0.97	2.97	3.1
Array read	0.98	5.45	5.6
Array write	1.23	5.05	4.1

TABLE 42.2 Local Acquire Cost (*Nanoseconds*)

Original	Local object	Shared object
90.6	19.6	281

We execute our benchmark applications in the following settings:

1. *Original Java:* An original (not instrumented) Java program running on a standard JVM.
2. *JavaSplit:* A program instrumented to be distribution-aware running on JavaSplit.
3. *JavaSplit-FT:* A program instrumented for distribution-awareness and checkpointing, running on JavaSplit-FT.

Since JavaSplit and JavaSplit-FT are distinguished mainly by their fault-tolerance, we can learn about the overhead of our fault-tolerance scheme by comparing their performance.

The most significant performance issues introduced by the JavaSplit bytecode rewriter are (i) the addition of access checks before accesses to heap data (i.e., accesses to object fields and array elements), (ii) replacing lock operations (e.g., `monitorenter` and `monitorexit`) with distributed synchronization code. Table 42.1 shows the cost of heap data accesses in the rewritten code in comparison with their cost in the original Java program. Table 42.2 shows the cost of a local acquire operation, that is, an acquire that does not result in communication. Although there is considerable overhead in acquiring a shared object, acquiring local objects costs less than in original Java. This is owing to the lock optimization described in Section 42.3.1.2.

To explore how instrumentation affects local execution time, we ran each benchmark application with a single worker thread. Since JavaSplit and JavaSplit-FT do not perform any networking when running on a single node, the observed overhead is mostly owing to instrumentation. The results of this experiment are presented in Figure 42.9.

In PrimeFinder, the instrumentation overhead is negligible. This is because most of its runtime PrimerFinder spends printing prime numbers on screen, rather than in computation, which is instrumented more heavily. In Series, JavaSplit execution time is smaller than the original. The reason for this phenomena is the aforementioned optimization of a local lock acquisition. Since JavaSplit executes a single application thread, all lock acquisitions are local, which, according to Table 42.2, give JavaSplit an advantage over the original application. The overhead of Brakes transformation is not very significant, because there are few method invocations in Series. TSP and RayTracer exhibit overhead of approximately 130% on JavaSplit. These applications frequently access arrays and object fields. Since JavaSplit instruments heap accesses to make them distribution-aware, it introduces nonnegligible overhead into these benchmarks. The overhead of Brakes transformation in RayTracer is quite significant, because of the frequent use of method invocations in the original application. Brakes instrumentation introduces time overhead of 4 bytecode instructions per each method invocation.

It is important to understand that in its current state, the system does not perform any optimizations on the bytecode rewriting process. The existing optimizations [ABH[+]01, VHB[+]01] based on flow analysis

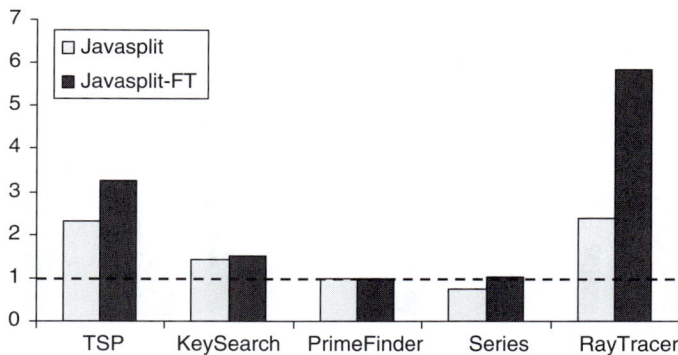

FIGURE 42.9 Normalized execution times of single-threaded runs. The value 1 on the vertical axis corresponds to the execution time of the original application on a single JVM.

of the original bytecodes, for example, access check elimination and batching, can reduce most of the JavaSplit instrumentation overhead. The Brakes instrumentation can also be optimized. Although Brakes transformation needs to be applied only to the methods that are involved in stack capturing and reestablishment (see Section 42.4.1), it is currently applied to all application methods. Bytecode flow analysis can be employed in order to determine the exact set of methods that need to be rewritten. The above optimizations have not been implemented yet. In contrast to scalability, they are not among our main research goals.

Figures 42.10 and 42.11 present the results of our scalability-related experiments. In these experiments, two threads are executed on each of the dual-processor nodes. Each thread runs in a separate JVM process. The bytecode instrumentation is performed statically, before the execution begins. In general, both systems exhibit good scalability. Naturally, the scalability of the former is better, however, not by far. In the latter, the increased overhead is mostly owing to the checkpointing and instrumentation. The overall checkpointing time is not significant. However, it has great impact on the average lock acquire time. A delay in lock transfer as a result of checkpointing affects all threads that wish to acquire the lock.

42.6 Related Work

Our work is related to two main areas: Java-based distributed computing and fault tolerance.

42.6.1 Java-Based Distributed Computing

Several works have devised a distributed runtime for Java [ABH+01, AFT99, PZ97, SNM01, TSCI01, TS02, VBB00, YC97, ZWL02]. None of these systems are fault tolerant and therefore, in the case of node failure, a distributed computation must be restarted.

The above works can be classified into three main categories: (i) cluster-aware VMs, (ii) compiler-based DSM systems, and (iii) systems using standard JVMs. JavaSplit and JavaSplit-FT belong to the last category. JavaSplit's distinguishing feature in comparison to previous work is its combination of transparency and portability. JavaSplit-FT adds fault tolerance to the list of JavaSplit features.

42.6.1.1 Cluster-Aware Virtual Machines

Java/DSM [YC97], Cluster VM for Java [AFT99], and JESSICA2 [ZWL02] implement distributed JVMs. These systems require that each node contain a custom JVM.

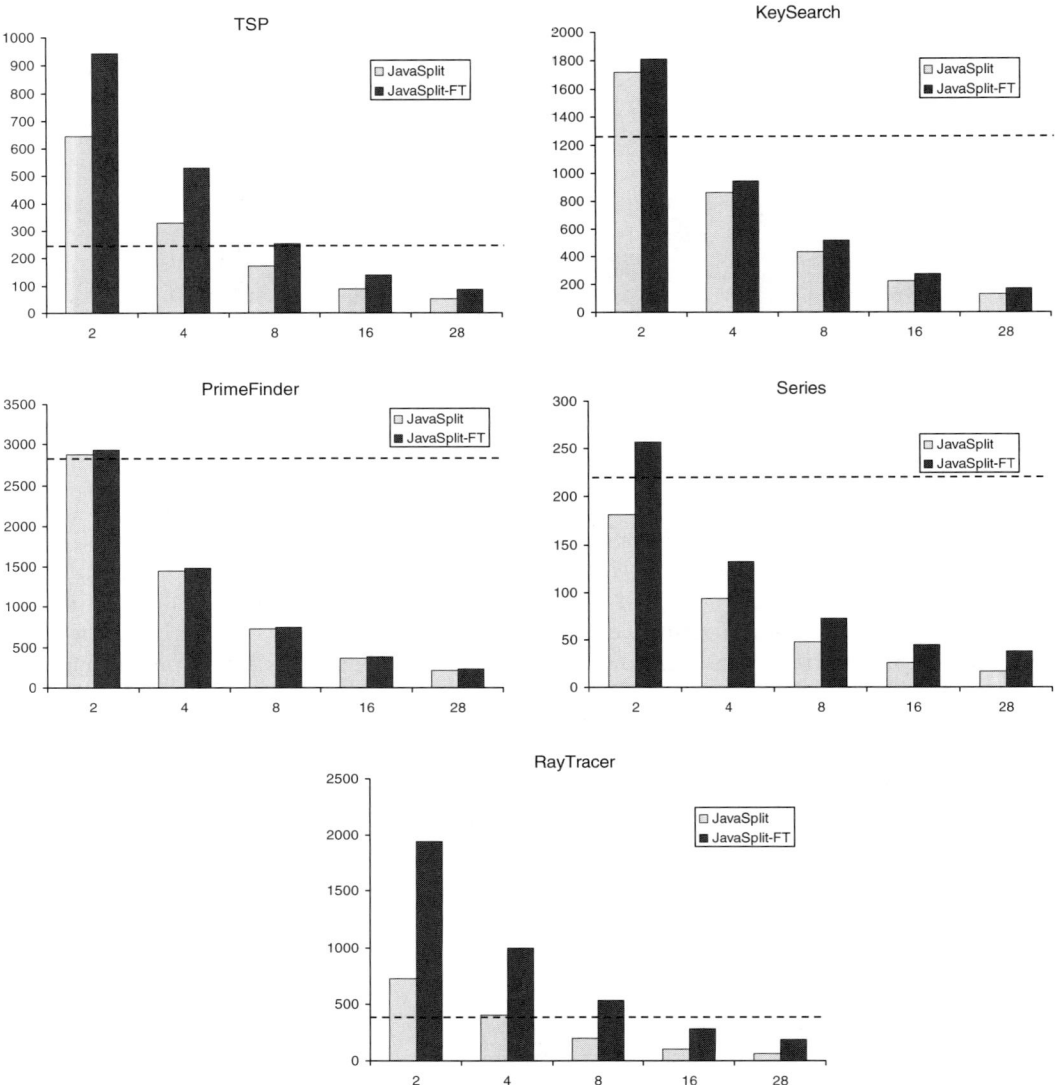

FIGURE 42.10 Benchmark execution times (in seconds). The dashed line indicates the execution time of the original benchmark using two threads in a single JVM running on a dual-processor machine. The horizontal axis of each graph represents the number of processors.

In Java/DSM the local VM is similar to a standard JVM, except that all objects are allocated on an existing C-based software DSM, called TreadMarks [KDCZ94]. Similar to our work, TreadMarks implements LRC. The single system image provided by Java/DSM is incomplete: a thread's location is not transparent to the programmer, and the threads cannot migrate between machines. In contrast, Cluster VM for Java and JESSICA2 provide a complete single system image of a traditional JVM.

Instead of using a DSM, Cluster VM for Java uses a proxy design pattern with various caching and object migration optimizations. JESSICA2 uses a home-based global object space (GOS) to implement a distributed Java heap. JESSICA2 has many desirable capabilities, for example, support for load balancing through thread migration, an adaptive migrating-home protocol for the GOS, and a dedicated JIT compiler. The latter feature distinguishes JESSICA2 from the majority of similar systems. For instance, Cluster VM for Java and Java/DSM are unable to use a standard JIT and do not implement a dedicated one. In contrast, JavaSplit-FT is able to utilize any standard JIT supplied with the local JVM.

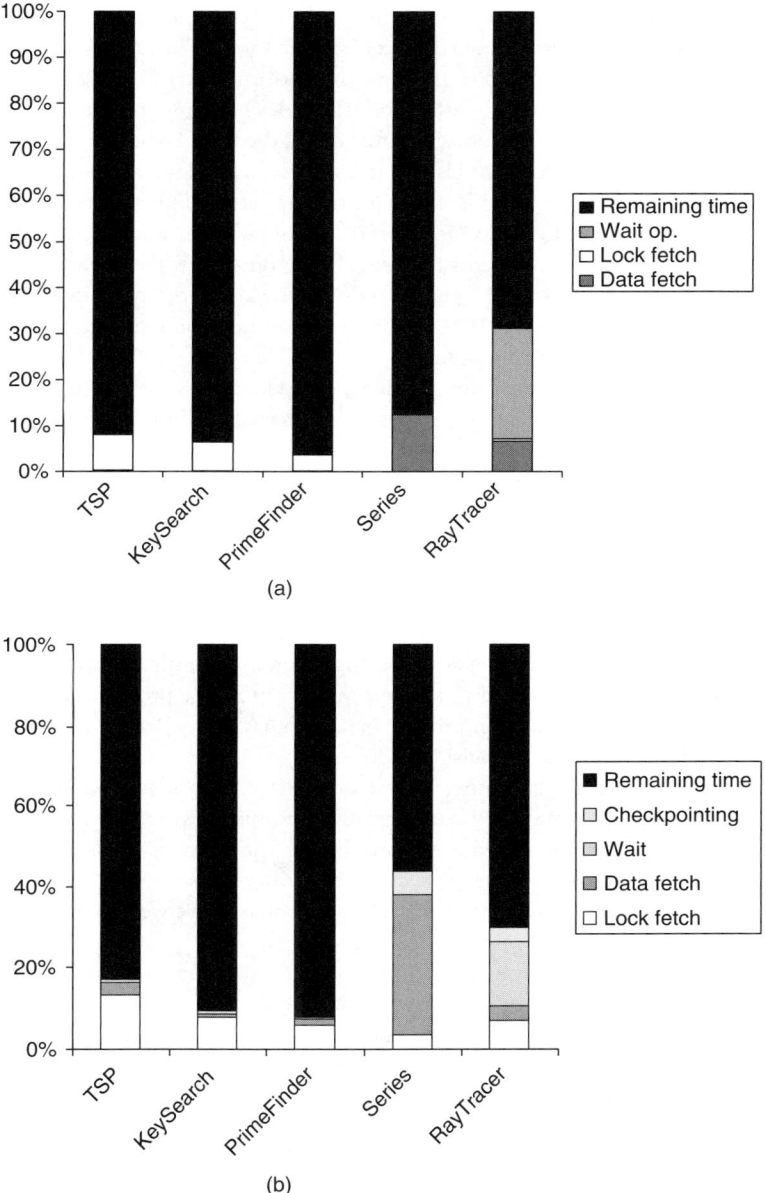

FIGURE 42.11 Time breakdown of JavaSplit and JavaSplit-FT on 28 processors: (a) Java Split and (b) Java Split-FT.

The cluster-aware VMs are potentially more efficient than systems using standard JVMs: they are able to access machine resources, for example, memory and network interface, directly rather than through the JVM. On the downside, because a node's local JVM is modified in these systems, none of them have true cross-platform portability.

42.6.1.2 Compiler-Based Distributed Shared Memory Systems

Compiler-based DSM systems compile Java programs into native machine code, while adding DSM capabilities. There are two compiler-based systems known to us: Hyperion [ABH+01] and Jackal [VBB00]. Both systems support standard Java and do not require changes in the programming paradigm.

Hyperion translates Java bytecodes to *C* source code and then compiles the *C* source using a native *C* compiler. The DSM handlers are inserted during the translation to *C*. The Java-bytecode-to-C translator performs various optimizations in order to improve the performance of the DSM. For example, if a shared object is referenced in each iteration of a loop (that does not contain synchronization), the code for obtaining a locally cached copy of the object is lifted out of the loop. Hyperion employs existing DSM libraries to implement its DSM protocol and is able to use various low-level communication layers.

Jackal combines an extended Java compiler and runtime support to implement a fine-grain DSM. The compiler translates Java sources into *Intel x86* code rather than Java bytecode. The Jackal compiler stores Java objects in shared regions and augments the program it compiles with access checks that drive the memory consistency protocol. Similar to Hyperion, it performs various optimizations, striving to achieve a more efficient distributed execution [VHB$^+$01]. Jackal incorporates a distributed garbage collector and provides thread and object location transparency.

In the compiler-based systems, the use of a dedicated compiler allows various compiler optimizations to be performed. These might significantly improve performance. In addition, since the application is compiled to machine code, the speed of a local execution is increased without requiring a just-in-time compiler. On the downside, use of native machine code sacrifices portability. Similar to the cluster-aware VMs, the compiler-based systems have direct access to the node resources.

42.6.1.3 Systems That Use Standard Java Virtual Machines

The systems that use a collection of standard JVMs [PZ97, SNM01, TSCI01, TS02, CKV98, Her99, CLNR97] are not transparent. They either introduce unorthodox programming constructs and style or require user intervention to enable distributed execution of an existing program. In contrast, JavaSplit-FT is completely transparent, both to the programmer and to the person submitting the program for execution. Among the systems that deviate from the Java programming paradigm, JavaParty [PZ97] and JSDM [SNM01] are close enough to pure Java in order to be considered here.

JavaParty supports parallel programming by extending Java with a preprocessor and a runtime. JavaParty modifies the Java programming paradigm by introducing a new reserved word, *remote*, to indicate classes that should be distributed across machines. The source code is transformed into regular Java code plus RMI hooks that are passed to the RMI compiler. The single system image is further flawed by the fact that the programmer must also distinguish between remote and local method invocations.

In JSDM, access checks, in the form of method invocation to memory consistency operations, are inserted manually by the user (or possibly by instrumentation) for field read/write accesses. JSDM requires that an input program be an SMPD-style multithreaded program. Moreover, the programmer must use special classes provided by JSDM when writing the program and mark the shared objects.

Similar to JavaSplit-FT, Addistant [TSCI01] and jOrchestra [TS02] instrument Java bytecode for distributed execution (but not necessarily to achieve better performance). Both systems require nontrivial user intervention to transform the classes used by the application. This intervention demands knowledge of the application structure, further compromising the transparency of these systems. Addistant and jOrchestra employ the master-proxy pattern to access remote objects. In contrast, JavaSplit uses an object-based DSM. Unlike JavaSplit, Addistant and jOrchestra are unable to instrument system classes and therefore treat them as *unmodifiable code*. This results in several limitations, mostly related to the placement of data. In contrast to jOrchestra and Addistant, JavaSplit supports arbitrary distribution of objects and does not require user intervention in the instrumentation process.

Addistant provides only class-based distribution: all the instances of a class must be allocated on the same host. The user has to explicitly specify whether instances of an unmodifiable class are created only by modifiable code, whether an unmodifiable class is accessed by modifiable code, and whether instances of a class can be safely passed by-copy. This information is application-specific and getting it wrong results in a partitioning that violates the original application semantics.

In jOrchestra, user involvement in the instrumentation process is less significant than in Addistant. The former provides a *profiler* tool that determines class interdependencies and a *classifier* tool that ensures the correctness of the partition chosen by the user. jOrchestra cannot support remote accesses from system classes to any remote object. To solve this problem, jOrchestra statically partitions the objects among the nodes, placing all (user-defined and system) objects that can be referenced from a certain system class on the same node with that class.

Owing to the strong class dependencies within Java packages, this usually results in partitions that coincide with package boundaries. The transparency of jOrchestra is further flawed by the incomplete support for Java synchronization: *synchronized* blocks and `wait/notify` calls do not affect remote nodes.

There are several advantages to using standard JVMs. First, the system can use heterogenous collections of nodes. Second, each node can locally optimize its performance, through a JIT, for example. Third, local garbage collection can be utilized to collect local objects that are not referenced from other nodes. The main drawback of these systems is their indirect access to the node resources. Since JavaSplit-FT uses only standard JVMs, it has all the advantages mentioned above.

42.6.2 Fault-Tolerant DSMs

There have been several works that augmented a LRC-based DSM with a fault tolerance mechanism. Most of these works, for example, [CGS$^+$96, SJF95, YY98], use a homeless implementation of LRC as their base memory consistency protocol.

We know of only three other works that devise a recoverable home-based LRC-compliant DSM. Sultan et al. [SIN00] follow a log-based approach that can tolerate single node failures in a home-based, lazy release consistent DSM. They use volatile logging of protocol data combined with independent checkpointing to stable storage. Because their scheme is log based, their work focuses on how to dynamically optimize log trimming and checkpoint garbage collection to efficiently control the size of the logs and the number of checkpoints kept. Kongmunvattana et al. [KT99] also use a log-based mechanism to tackle the problem. The main disadvantage of both schemes is that, during recovery, the number of nodes that need to cooperate with the failing one is unbounded. Moreover, in the latter scheme, recovery may require some of the nonfaulty processes to roll back to an earlier checkpoint. In contrast, recovery in our scheme does not require cooperation of other nodes. A thread is simply restarted from its latest checkpoint. Another important difference is that in our system the shared data is not a part of a checkpoint, which significantly reduces the snapshot time and size.

Christodoulopoulou et al. [CAB03] extend a home-based LRC protocol designed for clusters of SMP nodes. As in our system, data persistence is achieved through replication in volatile memory of nodes and checkpointing is integrated into synchronization operations. Despite the similarity of the basic ideas, their system can tolerate only a single node failure.

42.7 Conclusions

In this chapter, we have presented and evaluated a fault-tolerant distributed runtime for standard Java applications. Our work combines elements of several well-established fields in the core of distributed systems. First, we have extended a line of works in the area of OSM. Second, we have refined several techniques in fault-tolerant computing. Finally, our research is motivated by ideas in works on harnessing free cycles.

We view this work as a first step in providing a convenient computing infrastructure for nondedicated environments. The underlying question is whether (mostly idle) enterprise interconnects are sufficiently wide to efficiently support high-level programming paradigms, such as DSM. If this is the case, then, contrary to current community wisdom, it is not mandatory to restrict the communication pattern of a program that utilizes idle resources. Java, as a popular multithreaded programming language, is best suited for this experiment.

References

[ABH+01] G. Antoniu, L. Bougé, P. Hatcher, M. MacBeth, K. McGuigan, and R. Namyst. The Hyperion system: Compiling multithreaded Java bytecode for distributed execution. *Parallel Computing*, 27(10):1279–1297, 2001.

[AFT99] Y. Aridor, M. Factor, and A. Teperman. cJVM: A single system image of a JVM on a cluster. *International Conference on Parallel Processing*, pp. 4–11, 1999.

[Ban] B. Ban. Design and implementation of a reliable group communication toolkit for Java. Cornell University.

[CAB03] R. Christodoulopoulou, R. Azimi, and A. Bilas. Dynamic data replication: An approach to providing fault-tolerant shared memory clusters. *Proceedings of the Ninth Annual Symposium on High Performance Computer Architecture*, February 2003.

[CGS+96] M. Costa, P. Guedes, M. Sequeira, N. Neves, and M. Castro. Lightweight logging for lazy release consistent distributed shared memory. *Operating Systems Design and Implementation*, pp. 59–73, 1996.

[CGS+99] J.-D. Choi, M. Gupta, M. J. Serrano, V. C. Sreedhar, and S. P. Midkiff. Escape analysis for Java. *Proceedings of the Conference on Object-Oriented Programming Systems, Languages, and Applications (OOPSLA)*, pp. 1–19, 1999.

[CKV98] D. Caromel, W. Klauser, and J. Vayssière. Towards seamless computing and metacomputing in Java. *Concurrency: Practice and Experience*, 10(11–13):1043–1061, 1998.

[CLNR97] N. Camiel, S. London, N. Nisan, and O. Regev. The POPCORN project: Distributed computation over the Internet in Java. *6th Int'l World Wide Web Conference*, 1997.

[FSS] M. Factor, A. Schuster, and K. Shagin. JavaSplit: A runtime for execution of monolithic Java programs on heterogeneous collections of commodity workstations. *IEEE Fifth Int'l Conference on Cluster Computing*.

[FSS04] M. Factor, A. Schuster, and K. Shagin. Instrumentation of standard libraries in object-oriented languages: The twin class hierarchy approach. *OOPSLA '04: Proceedings of the 19th Annual ACM SIGPLAN Conference on Object-Oriented Programming, Systems, Languages, and Applications*, pp. 288–300. ACM Press, 2004.

[GJSB00] J. Gosling, B. Joy, G. Steele, and G. Bracha. *The Java Language Specification*, Second Edition. Addison-Wesley, Boston, MA, 2000.

[Her99] M. Herlihy. The Aleph toolkit: Support for scalable distributed shared objects. *Workshop on Communication, Architecture, and Applications for Network-Based Parallel Computing*, Orlando, January 1999.

[JSR] JSR 133. Java Memory Model and thread specification revision. http://jcp.org/jsr/detail/133.jsp.

[KCZ92] P. Keleher, A. L. Cox, and W. Zwaenepoel. Lazy release consistency for software distributed shared memory. *Proceedings of the 19th Annual Int'l Symposium on Computer Architecture (ISCA'92)*, pp. 13–21, 1992.

[KDCZ94] P. Keleher, S. Dwarkadas, A. L. Cox, and W. Zwaenepoel. Treadmarks: Distributed shared memory on standard workstations and operating systems. *Proceedings of the Winter 1994 USENIX Conference*, pp. 115–131, 1994.

[KT99] A. Kongmunvattana and N.-F. Tzeng. Coherence-centric logging and recovery for home-based software distributed shared memory. *ICPP '99: Proceedings of the 1999 International Conference on Parallel Processing*, page 274, Washington, DC, 1999. IEEE Computer Society.

[LY99] T. Lindholm and F. Yellin. *The Java(TM) Virtual Machine Specification*, 2nd Edition. 1999.

[PZ97] M. Philippsen and M. Zenger. JavaParty—Transparent remote objects in Java. *Concurrency: Practice and Experience*, 9(11):1225–1242, 1997.

[SB01] L. A. Smith and J. M. Bull. A multithreaded Java Grande benchmark suite. *Proceedings of the Third Workshop on Java for High Performance Computing*, 2001.

[SIN00] F. Sultan, L. Iftode, and T. Nguyen. Scalable fault-tolerant distributed shared memory. *Supercomputing '00: Proceedings of the 2000 ACM/IEEE Conference on Supercomputing (CDROM)*, p. 20, Washington, DC, 2000. IEEE Computer Society.

[SJF95]　G. Suri, R. Janssens, and W. K. Fuchs. Reduced overhead logging for rollback recovery in distributed shared memory. *FTCS-25: 25th International Symposium on Fault Tolerant Computing Digest of Papers*, pp. 279–288, Pasadena, California, 1995.

[SNM01]　Y. Sohda, H. Nakada, and S. Matsuoka. Implementation of a portable software DSM in Java. *Java Grande*, 2001.

[SSY00]　T. Sakamoto, T. Sekiguchi, and A. Yonezawa. Bytecode transformation for portable thread migration in Java. *ASA/MA*, pp. 16–28, 2000.

[Tao01]　W. Tao. A portable mechanism for thread persistence and migration. PhD thesis, University of Utah, 2001.

[TRV+00]　E. Truyen, B. Robben, B. Vanhaute, T. Coninx, W. Joosen, and P. Verbaeten. Portable support for transparent thread migration in Java. *ASA/MA*, pp. 29–43, 2000.

[TS02]　E. Tilevich and Y. Smaragdakis. J-Orchestra: Automatic Java application partitioning. *European Conference on Object-Oriented Programming (ECOOP)*, Malaga, Spain, June 2002.

[TSCI01]　M. Tatsubori, T. Sasaki, S. Chiba, and K. Itano. A bytecode translator for distributed execution of "legacy" Java software. *ECOOP 2001 Object-Oriented Programming*, LNCS, 2072, 2001.

[VBB00]　R. Veldema, R. A. F. Bhoedjang, and H. E. Bal. Distributed shared memory management for Java. *Proceedings of the Sixth Annual Conference of the Advanced School for Computing and Imaging (ASCI 2000)*, pp. 256–264, 2000.

[VHB+01]　R. Veldema, R. F. H. Hofman, R. A. F. Bhoedjang, C. J. H. Jacobs, and H. E. Bal. Source-level global optimizations for fine-grain distributed shared memory systems. *ACM SIGPLAN Notices*, 36(7):83–92, 2001.

[WS02]　M.-Y. Wu and Wei Shu. Note: An efficient distributed token-based mutual exclusion algorithm with central coordinator. *Journal of Parallel Distributed Computing*, 62(10):1602–1613, 2002.

[YC97]　W. Yu and A. L. Cox. Java/DSM: A platform for heterogeneous computing. *Concurrency—Practice and Experience*, 9(11):1213–1224, 1997.

[YY98]　T. Park Y. Yi and H. Y. Yeom. A causal logging scheme for lazy release consistent distributed shared memory systems. *Proceedings of the 1998 Int'l Conference on Parallel and Distributed Systems (ICPADS'98)*, 1998.

[ZIL96]　Y. Zhou, L. Iftode, and K. Li. Performance evaluation of two home-based lazy release consistency protocols for shared memory virtual memory systems. *Proceedings of the Second Symposium on Operating Systems Design and Implementation (OSDI'96)*, pp. 75–88, 1996.

[ZWL02]　W. Zhu, C.-L. Wang, and F. C. M. Lau. JESSICA2: A distributed Java Virtual Machine with transparent thread migration support. *IEEE Fourth International Conference on Cluster Computing*, Chicago, September 2002.

43
Scalability of Parallel Programs

43.1	Introduction...	43-1
43.2	Metrics of Parallel Performance	43-3
43.3	Metrics for Scalability Analysis	43-4

Scaling Characteristics of Parallel Programs • The Isoefficiency Metric of Scalability • Cost-Optimality and the Isoefficiency Function • A Lower Bound on the Isoefficiency Function • The Degree of Concurrency and the Isoefficiency Function • Scaling Properties and Parallel Benchmarks • Other Scalability Analysis Metrics

43.4	Heterogeneous Composition of Applications	43-14
43.5	Limitations of Analytical Modeling	43-14
	Acknowledgments...	43-15
	References	43-15

Ananth Grama
Purdue University

Vipin Kumar
University of Minnesota

43.1 Introduction

Faced with the challenge of reducing the power consumption of conventional microprocessors while delivering increased performance, microprocessor vendors have leveraged increasing transistor counts to deliver multicore processors. Packaging multiple, possibly simpler cores, operating at lower voltage results in lower power consumption, a primary motivation for multicore processors. A byproduct of this is that parallelism becomes the primary driver for performance. Indeed, multicore designs with up to 16 cores are currently available, and some with up to 80 cores have been prototyped.

Multicore processors bring tremendous raw performance to the desktop. Harnessing this raw compute power in real programs, ranging from desktop applications to scientific codes is critical to their commercial and technological viability. Beyond the desktop, the packaging and power characteristics of multicore processors make it possible to integrate a large number of such processors into a parallel ensemble. This has the potential to realize the long-held vision of truly scalable parallel platforms. Indeed, systems with close to hundred thousand cores are likely to become available in the foreseeable future. The peak performance of such systems is anticipated to be several petaFLOPS.

The ability to build scalable parallel systems with very large number of processing cores puts forth a number of technical challenges. With large component counts, the reliability of such systems is a major issue. System software support must provide a reliable logical machine view to the programmer, while supporting efficient program execution. Programming models and languages must allow a programmer to express concurrency and interactions in a manner that maximizes concurrency, minimizes overheads, while at the same time providing high-level abstractions of the underlying hardware. Runtime systems must continually monitor faults and application performance, while providing diagnostic, prognostic, and remedial services. While all of these are important issues, in our belief, there is an even more fundamental question—*are there algorithms that can efficiently scale to such configurations, given hardware parameters, limitations, and overheads?* The answer to this question and the methodology that answers this question are critical.

Estimating the performance of a program on a larger hardware configuration from known performance on smaller configurations and smaller problem instances is generally referred to as scalability analysis. Indeed, this is precisely the problem we are faced with, when we try to assess applications and algorithms capable of efficiently utilizing emerging large-scale parallel platforms. Scalability analysis has been well-studied in literature—with automated, analytical, and experimental methods proposed and investigated. Each class of methods has its limitations and advantages. Automated techniques [CMW+05, GC01] work on realizations of algorithms (programs), as opposed to algorithms themselves. From this point of view, their role in guiding algorithm development is limited—since they are implementation based. Furthermore, automated techniques generally rely on parameterizing quanta of communication and computation, and using or simulations (Monte Carlo, discrete event simulation). For this reason, these methods have limited accuracy, or may themselves be computationally expensive.

Experimental approaches [BDL98, ABE+06, (Ed01, ABB+00, FP05] to scalability analysis typically perform detailed quantification of the program instance and underlying platform, in order to generate a prediction model. These models can be used to predict performance beyond the observed envelope. As is the case with automated techniques, since these methods use realizations of algorithms, their ability to guide the algorithm development process is limited. Furthermore, because of the sensitive interaction between programs and underlying systems, the prediction envelope of such methods is often limited.

Analytical methods for scalability analysis rely on asymptotic estimates of basic metrics, such as parallel time and efficiency. Where available, such analytical estimates hold the potential for accurate scalability predictions, while providing guidance for algorithm and architecture development. For example, scalability analysis predicts the necessary network bandwidth/compute power in a system to efficiently scale Fast Fourier Transforms (FFTs) to large numbers of processors. The major drawback of these methods, of course, is that analytical quantification of parallel performance is not always easy. Even for simple algorithms, the presence of system (as opposed to programmer) controlled performance optimizations (caches and associated replacement policies are the simple examples) make accurate analytical quantification difficult.

In this chapter, we focus on analytical approaches to scalability analysis. Our choice is motivated by the fact that in relatively early stages of hardware development, such studies can effectively guide hardware design. Furthermore, we believe that for many applications, parallel algorithms scaling to very large configurations may not be the same as those that yield high efficiencies on smaller configurations. For these desired predictive properties, analytical modeling is critical.

There are a number of scaling scenarios, which may be suited to various application scenarios. In a large class of applications, the available memory limits the size of problem instance being solved. The question here is, given constraints on memory size, how does an algorithm perform? In other applications, for example, weather forecasting, there are definite deadlines on task completion. The question here becomes one of, what is the largest problem one can solve, given constraints on execution time, on a specific machine configuration. In yet other applications, the emphasis may be on efficiency. Here, one asks the question, what is the smallest problem instance I must solve on a machine so as to achieve desired execution efficiency. Each of these scenarios is quantified by a different scalability metric. We describe these metrics and their application to specific algorithms.

The rest of this chapter is organized as follows: in Section 43.2, we describe basic metrics of parallel performance, in Section 43.3, we describe various scalability metrics and how they can be applied to emerging parallel platforms. In Section 43.4, we discuss alternate parallel paradigms of task mapping

Scalability of Parallel Programs 43-3

for composite algorithms and how our scalability analysis applies to these scenarios. We conclude in Section 43.5 with a discussion of limitations of our analysis framework.

43.2 Metrics of Parallel Performance

Throughout this chapter, we refer to a single processing core as a processor. Where we talk of a chip with multiple cores, we refer to it as a chip multiprocessor, or a multicore processor. We assume that a parallel processor (or ensemble) consists of p identical processing cores. The cores are connected through an interconnect. Exchanging a message of size m between cores incurs a time of $t_s + t_w m$. Here, t_s corresponds to the network latency and t_w the per-word transfer time determined by the network bandwidth. The communication model used here is a simplification, since it does not account for network congestion, communication patterns, overlapped access, and interconnect topology, among others. We define a parallel system as a combination of a parallel program and a parallel machine. We assume that the serial time of execution of an algorithm T_s is composed of W basic operations.

Perhaps the simplest and most intuitive metric of parallel performance is the parallel runtime, T_p. This is the time elapsed between initiation of the parallel program at the first processor to the completion of the program at the last processor. Indeed, an analytical expression for parallel time captures all the performance parameters of a parallel algorithm. The problem with parallel runtime is that it does not account for the resources used to achieve the execution time. Specifically, if one were to indicate that the parallel runtime of a program, which took 10s on a serial processor, is 2s, we would have no way of knowing whether the parallel program (and associated algorithm) performs well or not. The obvious question w.r.t. parallel time is, how much lower it is compared to its serial counterpart. This speedup, S is defined as the ratio of the serial time T_s and the parallel time T_p. Mathematically, $S = T_s/T_p$. For the example above, the speedup is $10/2 = 5$.

While we have a sense of how much faster our program is compared to its serial counterpart, we still cannot estimate its performance, since we do not know how many resources were consumed to achieve this speedup. Specifically, if one were told that a parallel program achieved a speedup of 5, could we say anything about the performance of the program? For this purpose, the speedup can be normalized with respect to the number of processors p to compute efficiency. Formally, efficiency $E = S/p$. In the example above, if the speedup of 5 is achieved using 8 processors, the efficiency is 5/8 or 0.625 (or 62.5%). If the same speedup is achieved using 16 processors, the corresponding efficiency is 0.312 (or 31.2%). Clearly, the algorithm can be said to perform better in the former case.

Efficiency, by itself, is not a complete indicator of parallel performance either. A given problem can be solved using several possible serial algorithms, which may be more or less amenable to parallel execution. For example, given a sparse linear system, we can solve it using simple Jacobi iterations. These are easily parallelizable and mostly involve near-neighbor communication (other than global dot products). On the other hand, the same system can be solved using a more powerful solver, say GMRES with an approximate inverse preconditioner. This method solves the problem much faster in the serial context, but is less amenable to parallelism, compared to the first algorithm. In such cases, one must be cautious relying simply on efficiency as a metric for performance.

To account for these issues, vendors and users sometimes rely on total aggregate FLOPS as a metric. Indeed, when we refer to a platform as being capable of ten petaFLOPS, it is in the context of some program and input, or the absolute peak performance, which may, or may not be achievable by any program. The most popular choice of a benchmark in this context is the Linpack benchmark. The Linpack benchmark has favorable data reuse and computation/communication characteristics, and therefore yields parallel performance closer to peak for typical parallel platforms. However, this begs the obvious question of what, if any, this implies for other applications that may not have the same data reuse characteristics. Indeed, Linpack numbers, referring to dense matrix operations are rarely indicative of machine performance on typical applications that have sparse kernels (PDE solvers), molecular dynamics type sparse interaction potentials, access workloads of large-scale data analysis kernels, or commercial server workloads. It is important to note that the oft-used sparse kernels often yield as low as 5–10% of peak performance on conventional processors because of limited data reuse.

In addition to the shortcomings mentioned above, basic metrics do not explicitly target scaling. Specifically, if the parallel time of program (algorithm) A for solving a problem is lower than that of algorithm B for solving the same problem, what does it say if the problem size is changed, the number of processors is increased, or if any of the computation/ communication parameters are varied. It turns out that a single sample point in the multidimensional performance space of parallel programs says nothing about the performance at other points. This provides strong motivation for the development of scalability metrics.

43.3 Metrics for Scalability Analysis

We motivate scalability analysis through two examples at two diverse ends of the spectrum. The first example relates to emerging large-scale platforms, and the second, to emerging scalable multicore desktop processors.

An important challenge to the parallel computing community currently is to develop a core set of algorithms and applications that can utilize emerging petascale computers. The number of cores in these platforms will be in the range of 100,000, in an energy efficient configuration. A number of important questions are posed in this context—(i) which, if any, of the current applications and algorithms are likely to be able to scale to such configurations, (ii) what fundamentally new algorithms will be required to scale to very large machine configurations, (iii) what are realizable architectural features that will enhance scaling characteristics of wide classes of algorithms, (iv) what are the memory and I/O requirements of such platforms, and (v) how can time-critical applications, such as weather forecasting, effectively utilize these platforms. Scalability analysis can answer many of these questions, while guiding algorithm and architecture design.

At the other end of the spectrum, desktop software vendors are grappling with questions of how to develop efficient software with meaningful development and deployment cycles. Specifically, will software, ranging from operating systems (Windows Vista, Linux, etc.) to desktop applications (word processing, media applications), be able to use 64 cores or beyond, which is likely to become available in high-end desktops and engineering workstations in a 5-year time frame. The difficulty associated with this is that such platforms do not exist currently, therefore, one must design algorithms and software that can scale to future platforms. Again, scalability analysis can guide algorithm and software development in fundamental ways.

The general theme of these two motivating examples is that often, programs are designed and tested for smaller problems on fewer processing elements. However, the real problems these programs are intended to solve are much larger, and the machines contain larger number of processing elements. Whereas code development is simplified by using scaled-down versions of the machine and the problem, their performance and correctness is much more difficult to establish based on scaled-down systems.

Example 43.1 Why is performance extrapolation so difficult? *Consider three parallel algorithms for computing an n-point Fast Fourier Transform (FFT) on 64 processing cores. Figure 43.1 illustrates speedup as the value of n is increased to 18K. Keeping the number of processing cores constant, at smaller values of n, one would infer from observed speedups that binary exchange and 3-D transpose algorithms are the best. However, as the problem is scaled up to 18 K points and beyond, it is evident from Figure 43.1 that the 2-D transpose algorithm yields best speedup [GGKK03].*

Similar results can be shown relating to the variation in a number of processing cores as the problem size is held constant. Unfortunately, such parallel performance traces are the norm as opposed to the exception, making performance prediction based on limited observed data very difficult.

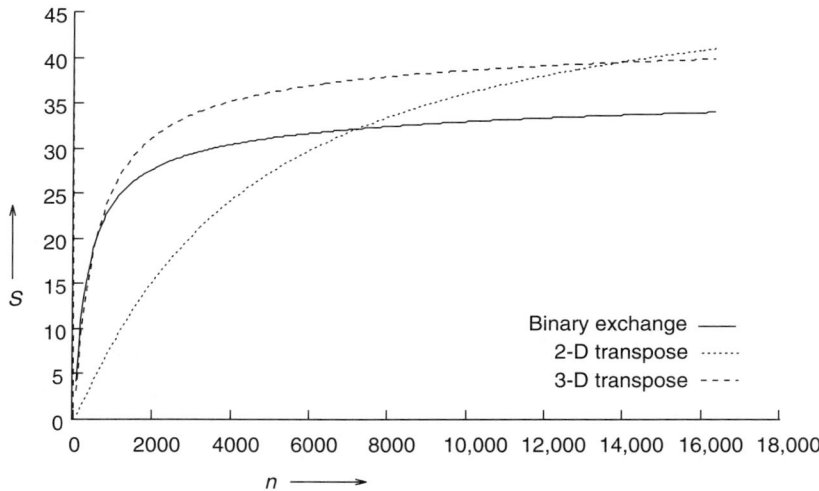

FIGURE 43.1 A comparison of the speedups obtained by the binary-exchange, 2D transpose, and 3D transpose algorithms on 64 cores with $t_c = 2ns$, $t_w = 4ns$, $t_s = 25ns$, and $t_h = 2ns$.

43.3.1 Scaling Characteristics of Parallel Programs

The parallel runtime of a program, summed across all processing cores, (pT_p), consists of essential computation that would be performed by its serial counterpart (T_s), and overhead incurred in parallelization (T_o). Note that we use T_o to denote the total (or parallel) overhead summed across all processors. Formally, we write this as

$$pT_p = T_s + T_o \qquad (43.1)$$

We also know that the efficiency of a parallel program can be written as

$$E = \frac{S}{p} = \frac{T_s}{pT_p}$$

Using the expression for parallel overhead (Equation 43.1), we can rewrite this expression as

$$E = \frac{1}{1 + \frac{T_o}{T_s}}. \qquad (43.2)$$

The total overhead function T_o is an increasing function of p [FK89, Fla90]. This is because every program must contain some serial component. If this serial component of the program takes time t_{serial}, then during this time all the other processing elements must be idle. This corresponds to a total overhead function of $(p-1) \times t_{\text{serial}}$. Therefore, the total overhead function T_o grows at least linearly with p. Furthermore, owing to communication, idling, and excess computation, this function may grow superlinearly in the number of processing cores. Equation 43.2 gives us several interesting insights into the scaling of parallel programs. First, for a given problem size (i.e., the value of T_s remains constant), as we increase the number of processing elements, T_o increases. In this scenario, it is clear from Equation 43.2 that the overall efficiency of the parallel program goes down. This characteristic of decreasing efficiency with increasing number of processing cores for a given problem size is common to all parallel programs. Often, in our quest for ever more powerful machines, this fundamental insight is lost—namely, the same problem instances that we solve on today's computers are unlikely to yield meaningful performance on emerging highly parallel platforms.

Let us investigate the effect of increasing the problem size keeping the number of processing cores constant. We know that the total overhead function T_o is a function of both problem size T_s and the number of processing elements p. In many cases, T_o grows sub-linearly with respect to T_s. In such cases, we can see that efficiency increases if the problem size is increased keeping the number of processing elements constant. Indeed, when demonstrating parallel performance on large machine configurations, many researchers assume that the problem is scaled in proportion to the number of processors. This notion of experimentally scaling problem size to maintain *constant computation per processor* is referred to as *scaled speedup* [GMB88, Gus88, Gus92], and is discussed in Section 43.3.7. For such algorithms, it should be possible to keep the efficiency fixed by increasing both the size of the problem and the number of processing elements simultaneously. This ability to maintain efficiency constant by simultaneously increasing the number of processing cores and the size of the problem is essential when utilizing scalable parallel platforms. Indeed, a number of parallel systems exhibit such characteristics. We call such systems *scalable* parallel systems. The *scalability* of a parallel system is a measure of its capacity to increase performance (speedup) in proportion to the number of processing cores. It reflects a parallel system's ability to utilize increasing processing resources effectively.

43.3.2 The Isoefficiency Metric of Scalability

We summarize our discussion in the section above in the following two observations:

1. For a given problem size, as we increase the number of processing cores, the overall efficiency of the parallel system goes down. This phenomenon is common to all parallel systems.
2. In many cases, the efficiency of a parallel system increases if the problem size is increased while keeping the number of processing cores constant.

These two phenomena are illustrated in Figure 43.2a and b, respectively. Following from these two observations, we define a scalable parallel system as one in which the efficiency can be kept constant as the number of processing elements is increased, provided that the problem size is also increased.

An important question that follows naturally is, what is the rate at which the problem size must be increased in order to keep efficiency constant. This rate is critical for a number of reasons—primarily, this rate determines the rate at which the total memory in the system must be increased. If problem size is linear in memory size, as is the case for a number of algorithms (applications), and the rate of increase

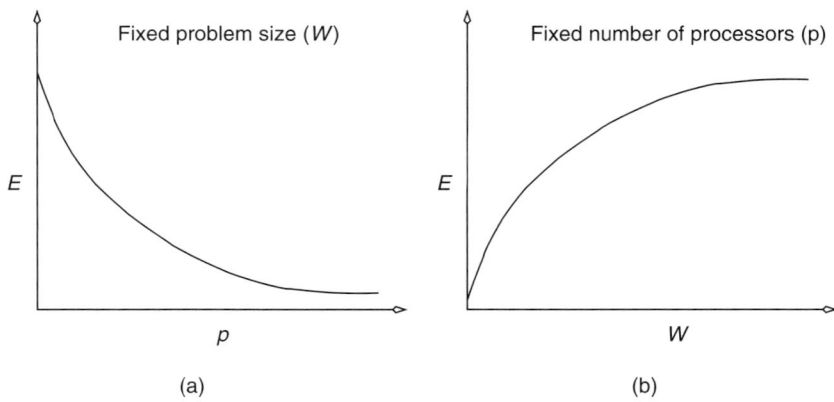

FIGURE 43.2 Variation of efficiency: (a) as the number of processing elements is increased for a given problem size; and (b) as the problem size is increased for a given number of processing elements. The phenomenon illustrated in graph (b) is not common to all parallel systems.

in problem size is superlinear, this implies that the total memory in the parallel machine must increase superlinearly. This is a critical observation. Conversely, if the increase in system memory is sublinear with respect to the number of cores, one can trivially show that the parallel efficiency will decline. The relation between problem size, memory size, and efficiency, is a critical determinant of the scaling characteristics of parallel systems. A second implication of increasing problem size is that a superlinear increase in problem size w.r.t p results in an increasing (parallel) time to solution since speedup is limited by p. For applications with constraints on time to solution (a classic example is in short-term weather forecasting), this may eventually limit growth in problem size. We discuss these three issues, namely, increase in problem size to maintain efficiency, impact of memory constraints, and impact of parallel solution time constraints separately.

For different parallel systems, the problem size must increase at different rates in order to maintain a fixed efficiency as the number of processing elements is increased. This rate is determined by the overheads incurred in parallelization. A lower rate is more desirable than a higher growth rate in problem size. This is because it allows us to extract good performance on smaller problems, and consequently, improved performance on even larger problem instances. We now investigate metrics for quantitatively determining the degree of scalability of a parallel system. We start by formally defining Problem Size.

43.3.2.1 Problem Size

When analyzing parallel systems, we frequently encounter the notion of the size of the problem being solved. Thus far, we have used the term *problem size* informally, without giving a precise definition. A naive way to express problem size is as a parameter of the input size; for instance, n, in case of a matrix operation involving $n \times n$ matrices. A drawback of this definition is that the interpretation of problem size changes from one problem to another. For example, doubling the input size results in an eight-fold increase in the execution time for matrix multiplication and a four-fold increase for matrix addition (assuming that the conventional $\Theta(n^3)$ algorithm is the best matrix multiplication algorithm, and disregarding more complicated algorithms with better asymptotic complexities). Using memory size as a measure of problem size has a similar drawback. In this case, the memory size associated with the dense matrices is $\Theta(n^2)$, which is also the time for, say, matrix addition and matrix-vector multiplication, but not for matrix–matrix multiplication.

A consistent definition of the size or the magnitude of the problem should be such that, regardless of the problem, doubling the problem size always means performing twice the amount of computation. Therefore, we choose to express problem size in terms of the total number of basic operations required to solve the problem. By this definition, the problem size is $\Theta(n^3)$ for $n \times n$ matrix multiplication (assuming the conventional algorithm) and $\Theta(n^2)$ for $n \times n$ matrix addition. In order to keep it unique for a given problem, we define *problem size* as the number of basic computation steps in the best sequential algorithm to solve the problem on a single processing core. Since it is defined in terms of sequential time complexity, problem size W is a function of the size of the input.

In general, we often assume that it takes unit time to perform one basic computation step of an algorithm. This assumption does not impact the analysis of any parallel system because the other hardware-related constants, such as message startup time, per-word transfer time, and per-hop time, can be normalized with respect to the time taken by a basic computation step. With this assumption, the problem size W is equal to the serial runtime T_s of the fastest known algorithm to solve the problem on a sequential computer.

43.3.2.2 The Isoefficiency Function

Parallel execution time can be expressed as a function of problem size, overhead function, and the number of processing elements. We can write parallel runtime as

$$T_\text{p} = \frac{W + T_\text{o}(W, p)}{p} \tag{43.3}$$

The resulting expression for speedup is

$$S = \frac{W}{T_p}$$

$$= \frac{Wp}{W + T_o(W,p)}. \qquad (43.4)$$

Finally, we write the expression for efficiency as

$$E = \frac{S}{p}$$

$$= \frac{W}{W + T_o(W,p)}$$

$$= \frac{1}{1 + T_o(W,p)/W}. \qquad (43.5)$$

In Equation 43.5, if the problem size is kept constant and p is increased, the efficiency decreases because the total overhead T_o increases with p. If W is increased keeping the number of processing elements fixed, then for scalable parallel systems, the efficiency increases. This is because T_o grows slower than $\Theta(W)$ for a fixed p. For these parallel systems, efficiency can be maintained at a desired value (between 0 and 1) for increasing p, provided W is also increased.

For different parallel systems, W must be increased at different rates with respect to p in order to maintain a fixed efficiency. For instance, in some cases, W might need to grow as an exponential function of p to keep the efficiency from dropping as p increases. Such parallel systems are poorly scalable. The reason is that on these parallel systems it is difficult to obtain good speedups for a large number of processing elements unless the problem size is enormous. On the other hand, if W needs to grow only linearly with respect to p, then the parallel system is highly scalable. That is because it can easily deliver speedups proportional to the number of processing elements for reasonable problem sizes.

For scalable parallel systems, efficiency can be maintained at a fixed value (between 0 and 1) if the ratio T_o/W in Equation 43.5 is maintained at a constant value. For a desired value E of efficiency,

$$E = \frac{1}{1 + T_o(W,p)/W},$$

$$\frac{T_o(W,p)}{W} = \frac{1-E}{E},$$

$$W = \frac{E}{1-E} T_o(W,p). \qquad (43.6)$$

Let $K = E/(1-E)$ be a constant depending on the efficiency to be maintained. Since T_o is a function of W and p, Equation 43.6 can be rewritten as

$$W = KT_o(W,p). \qquad (43.7)$$

From Equation 43.7, the problem size W can usually be obtained as a function of p by algebraic manipulations. This function dictates the growth rate of W required to keep the efficiency fixed as p increases. We call this function the *isoefficiency function* of the parallel system. The isoefficiency function determines the ease with which a parallel system can maintain constant efficiency, and hence achieve speedups increasing in proportion to the number of processing elements. A small isoefficiency function means that small increments in the problem size are sufficient for the efficient utilization of an increasing number of processing elements, indicating that the parallel system is highly scalable. However, a large

Scalability of Parallel Programs

isoefficiency function indicates a poorly scalable parallel system. The isoefficiency function does not exist for unscalable parallel systems, because in such systems the efficiency cannot be kept at any constant value as p increases, no matter how fast the problem size is increased.

Example 43.2 Isoefficiency function of computing an n-point FFT. *One parallel formulation of FFT, called the Binary Exchange Algorithm, computes an n-point FFT on a p processor machine (with $O(p)$, or full bisection bandwidth) in time*

$$T_p = t_c \frac{n}{p} \log n + t_s \log p + t_w \frac{n}{p} \log p$$

Recall here that t_c refers to unit computation, t_s to the message startup time, and t_w to the perword message transfer time. Recall also that the problem size, W, for an n-point FFT is given by

$$W = n \log n$$

The efficiency, E, of this computation can be written as

$$E = \frac{T_s}{pT_p} = \frac{t_c n \log n}{t_c \frac{n}{p} \log n + t_s \log p + t_w \frac{n}{p} \log p}$$

Through simple algebraic manipulations, we can rewrite the above expression as

$$\frac{t_s p \log p}{t_c n \log n} + \frac{t_w \log p}{t_c \log n} = \frac{1-E}{E}$$

If efficiency E is held constant, the right hand side is constant. The equality can be forced in the asymptotic sense by forcing each of the terms in the left hand side to be constant, individually. For the first term, we have,

$$\frac{t_s p \log p}{t_c n \log n} = \frac{1}{K},$$

for constant K. This implies that the corresponding isoefficiency term is given by $W \approx n \log n \approx p \log p$. For the second term, we have,

$$\frac{t_w \log p}{t_c \log n} = \frac{1}{K},$$

$$\log n = K \frac{t_w}{t_c} \log p,$$

$$n = p^{K t_w / t_c},$$

$$W = n \log n = K \frac{t_w}{t_c} p^{K t_w / t_c}.$$

In addition to these two isoefficiency terms resulting from communication overhead, since the Binary Exchange algorithm can only use $p = n$ processors, there is an isoefficiency term resulting from concurrency. This term is given by $W = n \log n = p \log p$.

The combined isoefficiency from the three terms is the dominant of the three. If $K t_w < t_c$ (which happens when we have very high bandwidth across the processing cores), the concurrency and startup terms dominate, and we have an isoefficiency of $p \log p$. Otherwise, the isoefficiency is exponential, and given by $W \approx p^{K t_w / t_c}$. This observation demonstrates the power of isoefficiency analysis in quantifying both scalability of the algorithm, as well as desirable features of the underlying platform. Specifically, if we would like to use the Binary Exchange algorithm for FFT on large platforms, the balance of the machine should be such that $K t_w < t_c$ (i.e., it must have sufficient perprocessor bisection bandwidth).

The isoefficiency function [GGK93] captures, in a single expression, the characteristics of a parallel algorithm as well as the parallel platform on which it is implemented. Isoefficiency analysis enables us to predict performance on a larger number of processing cores from observed performance on a smaller number of cores. Detailed isoefficiency analysis can also be used to study the behavior of a parallel system with respect to changes in hardware parameters such as the speed of processing cores and communication channels. In many cases, isoefficiency analysis can be used even for parallel algorithms for which we cannot derive a value of parallel runtime, by directly relating the total overhead to the problem size and number of processors [GGKK03].

43.3.3 Cost-Optimality and the Isoefficiency Function

A parallel system is cost-optimal if the product of the number of processing cores and the parallel execution time is proportional to the execution time of the fastest known sequential algorithm on a single processing element. In other words, a parallel system is cost-optimal if and only if

$$pT_p = \Theta(W). \tag{43.8}$$

Substituting the expression for T_p from the right-hand side of Equation 43.3, we get the following:

$$W + T_o(W, p) = \Theta(W)$$

$$T_o(W, p) = O(W) \tag{43.9}$$

$$W = \Omega(T_o(W, p)) \tag{43.10}$$

Equations 43.9 and 43.10 suggest that a parallel system is cost-optimal if and only if its overhead function does not asymptotically exceed the problem size. This is very similar to the condition specified by Equation 43.7 for maintaining a fixed efficiency while increasing the number of processing elements in a parallel system. If Equation 43.7 yields an isoefficiency function $f(p)$, then it follows from Equation 43.10 that the relation $W = \Omega(f(p))$ must be satisfied to ensure the cost-optimality of a parallel system as it is scaled up.

43.3.4 A Lower Bound on the Isoefficiency Function

We discussed earlier that a smaller isoefficiency function indicates higher scalability. Accordingly, an ideally scalable parallel system must have the lowest possible isoefficiency function. For a problem consisting of W units of work, no more than W processing elements can be used cost-optimally; additional processing cores will be idle. If the problem size grows at a rate slower than $\Theta(p)$ as the number of processing elements increases, the number of processing cores will eventually exceed W. Even for an ideal parallel system with no communication, or other overhead, the efficiency will drop because processing elements added beyond $p = W$ will be idle. Thus, asymptotically, the problem size must increase at least as fast as $\Theta(p)$ to maintain fixed efficiency; hence, $\Omega(p)$ is the asymptotic lower bound on the isoefficiency function. It follows that the isoefficiency function of an ideally scalable parallel system is $\Theta(p)$.

This trivial lower bound implies that the problem size must grow atleast linearly, or the problem size per processor must be atleast constant. For problems where memory size and problem size W are linearly related, the memory per processing core must increase linearly to maintain constant efficiency.

43.3.5 The Degree of Concurrency and the Isoefficiency Function

A lower bound of $\Omega(p)$ is imposed on the isoefficiency function of a parallel system by the number of operations that can be performed concurrently. The maximum number of tasks that can be executed simultaneously at any time in a parallel algorithm is called its *degree of concurrency*. The degree of concurrency is a measure of the number of operations that an algorithm can perform in parallel for a problem of size W; it is independent of the parallel architecture. If $C(W)$ is the degree of concurrency of a parallel algorithm, then for a problem of size W, no more than $C(W)$ processing elements can be employed effectively.

Example 43.3 Effect of concurrency on isoefficiency function. *Consider solving a system of n equations in n variables by using Gaussian elimination. The total amount of computation is $\Theta(n^3)$. But the n variables must be eliminated one after the other (assuming pivoting), and eliminating each variable requires $\Theta(n^2)$ computations. Thus, at most $\Theta(n^2)$ processing elements can be kept busy at any time. Since $W = \Theta(n^3)$ for this problem, the degree of concurrency $C(W)$ is $\Theta(W^{2/3})$ and at most $\Theta(W^{2/3})$ processing elements can be used efficiently. On the other hand, given p processing cores, the problem size should be at least $\Omega(p^{3/2})$ to use them all. Thus, the isoefficiency function of this computation due to concurrency is $\Theta(p^{3/2})$.*

The isoefficiency function due to concurrency is optimal (i.e., $\Theta(p)$) only if the degree of concurrency of the parallel algorithm is $\Theta(W)$. If the degree of concurrency of an algorithm is less than $\Theta(W)$, then the isoefficiency function due to concurrency is worse (i.e., greater) than $\Theta(p)$. In such cases, the overall isoefficiency function of a parallel system is given by the maximum of the isoefficiency functions due to concurrency, communication, and other overheads.

43.3.6 Scaling Properties and Parallel Benchmarks

At this point, we can further investigate why dense linear algebra kernels are frequently used as benchmarks for parallel systems. Consider the example of matrix multiplication—the first thing to note is that it has significant data reuse, even in the serial sense ($\Theta(n^3)$ operations on $\Theta(n^2)$ data). The parallel runtime of a commonly used matrix multiplication algorithm (Cannon's algorithm) is given by

$$T_p = \frac{n^3}{p} + 2\sqrt{p}\, t_s + 2 t_w \frac{n^2}{\sqrt{p}}$$

The corresponding isoefficiency is given by $W = p^{1.5}$. The favorable memory to FLOPS ratio of matrix multiplication, combined with the relatively favorable isoefficiency function makes this an ideal parallel benchmark. Specifically, if $M \approx n^2$ represents the memory requirement, since $W \approx n^3$, we have $W \approx M^{1.5}$. Substituting in the expression for isoefficiency, we have, $W \approx M^{1.5} \approx p^{1.5}$. From this, we can see that we can keep efficiency constant with $M \approx p$. In other words, the increase in problem size to hold efficiency constant requires only a linear increase in total memory, or constant memory per processor.

The relatively benign memory requirement and their high processor utilization makes dense kernels, such as matrix–matrix multiplication and linear direct solvers ideal benchmarks for large-scale parallel platforms. It is important to recognize the favorable characteristics of these benchmarks, since a number of real applications, for example, sparse solvers and molecular dynamics methods, do not have significant data reuse and the underlying computation is linear in memory requirement.

43.3.7 Other Scalability Analysis Metrics

We have alluded to constraints on runtime and memory as being important determinants of scalability. These constraints can be incorporated directly into scalability metrics. One such metric, called Scaled Speedup, increases the problem size linearly with the number of processing cores [GMB88, Gus88, Gus92, SN93, Wor91]. If the scaled-speedup curve is close to linear with respect to the number of processing cores, then the parallel system is considered scalable. This metric is related to isoefficiency if the parallel algorithm under consideration has linear or near-linear isoefficiency function. In this case the scaled-speedup metric provides results very close to those of isoefficiency analysis, and the scaled-speedup is linear or near-linear with respect to the number of processing cores. For parallel systems with much worse isoefficiencies, the results provided by the two metrics may be quite different. In this case, the scaled-speedup versus number of processing cores curve is sublinear.

Two generalized notions of scaled speedup have been investigated. They differ in the methods by which the problem size is scaled up with the number of processing elements. In one method, the size of the problem is increased to fill the available memory on the parallel computer. The assumption here is that aggregate memory of the system increases with the number of processing cores. In the other method, the size of the problem grows with p subject to a bound on parallel execution time. We illustrate these using an example.

Example 43.4 Memory and time-constrained scaled speedup for matrix–vector products. *The serial runtime of multiplying a matrix of dimension $n \times n$ with a vector is n^2 (with normalized unit execution time). The corresponding parallel runtime using a simple parallel algorithm is given by:*

$$T_P = \frac{n^2}{p} + \log p + n$$

and the speedup S is given by

$$S = \frac{n^2}{\frac{n^2}{p} + \log p + n} \tag{43.11}$$

The total memory requirement of the algorithm is $\Theta(n^2)$. Let us consider the two cases of problem scaling. In the case of memory-constrained scaling, we assume that the memory of the parallel system grows linearly with the number of processing cores, that is, $m = \Theta(p)$. This is a reasonable assumption for most current parallel platforms. Since $m = \Theta(n^2)$, we have $n^2 = c \times p$, for some constant c. Therefore, the scaled speedup S' is given by

$$S' = \frac{c \times p}{\frac{c \times p}{p} + \log p + \sqrt{c \times p}}$$

or

$$S' = \frac{c_1 p}{c_2 + c_3 \log p + c_4 \sqrt{p}}.$$

In the limiting case, $S' = O(\sqrt{p})$.

In the case of time-constrained scaling, we have $T_P = O(n^2/p)$. Since this is constrained to be constant, $n^2 = O(p)$. We notice that this case is identical to the memory-constrained case. This happened because the memory and runtime of the algorithm are asymptotically identical.

Example 43.5 Memory and time-constrained scaled speedup for matrix–matrix products. *The serial runtime of multiplying two matrices of dimension $n \times n$ is n^3. The corresponding parallel runtime using a simple parallel algorithm is given by*

$$T_P = \frac{n^3}{p} + \log p + \frac{n^2}{\sqrt{p}}$$

and the speedup S is given by

$$S = \frac{n^3}{\frac{n^3}{p} + \log p + \frac{n^2}{\sqrt{p}}} \tag{43.12}$$

The total memory requirement of the algorithm is $\Theta(n^2)$. Let us consider the two cases of problem scaling. In the case of memory-constrained scaling, as before, we assume that the memory of the parallel system grows

Scalability of Parallel Programs

linearly with the number of processing elements, that is, $m = \Theta(p)$. Since $m = \Theta(n^2)$, we have $n^2 = c \times p$, for some constant c. Therefore, the scaled speedup S' is given by

$$S' = \frac{(c \times p)^{1.5}}{\frac{(c \times p)^{1.5}}{p} + \log p + \frac{c \times p}{\sqrt{p}}} = O(p)$$

In the case of time-constrained scaling, we have $T_p = O(n^3/p)$. Since this is constrained to be constant, $n^3 = O(p)$, or $n^3 = c \times p$ (for some constant c). Therefore, the time-constrained speedup S'' is given by

$$S'' = \frac{c \times p}{\frac{c \times p}{p} + \log p + \frac{(c \times p)^{2/3}}{\sqrt{p}}} = O(p^{5/6})$$

This example illustrates that memory-constrained scaling yields linear speedup, whereas time-constrained speedup yields sublinear speedup in the case of matrix multiplication.

Serial Fraction f. The experimentally determined serial fraction f can be used to quantify the performance of a parallel system on a fixed-size problem. Consider a case when the serial runtime of a computation can be divided into a totally parallel and a totally serial component, that is,

$$W = T_{ser} + T_{par}.$$

Here, T_{ser} and T_{par} correspond to totally serial and totally parallel components. From this, we can write

$$T_p = T_{ser} + \frac{T_{par}}{p}.$$

Here, we have assumed that all of the other parallel overheads such as excess computation and communication are captured in the serial component T_{ser}. From these equations, it follows that

$$T_p = T_{ser} + \frac{W - T_{ser}}{p} \tag{43.13}$$

The serial fraction f of a parallel program is defined as

$$f = \frac{T_{ser}}{W}.$$

Therefore, from Equation 43.13, we have

$$T_p = f \times W + \frac{W - f \times W}{p}$$

$$\frac{T_p}{W} = f + \frac{1-f}{p}$$

Since $S = W/T_p$, we have

$$\frac{1}{S} = f + \frac{1-f}{p}.$$

Solving for f, we get

$$f = \frac{1/S - 1/p}{1 - 1/p}. \tag{43.14}$$

It is easy to see that smaller values of f are better since they result in higher efficiencies. If f increases with the number of processing elements, then it is an indicator of rising communication overhead, and thus an indicator of poor scalability.

Example 43.6 Serial component of the matrix–vector product. *From Equations 43.14 and 43.11, we have*

$$f = \frac{\frac{\frac{n^2}{p} + \log p + n}{n^2}}{1 - 1/p} \tag{43.15}$$

Simplifying the above expression, we get

$$f = \frac{p \log p + np}{n^2} \times \frac{1}{p-1}$$

$$f \approx \frac{\log p + n}{n^2}$$

It is useful to note that the denominator of this equation is the serial runtime of the algorithm and the numerator corresponds to the overhead in parallel execution.

A comprehensive discussion of various scalability and performance measures can be found in the survey by Kumar and Gupta [KG94]. Although over 10 years old, the results cited in this survey and their applications are particularly relevant now, as scalable parallel platforms are finally being realized.

43.4 Heterogeneous Composition of Applications

Many applications are composed of multiple steps, with differing scaling characteristics. For example, in commonly used molecular dynamics techniques, the potentials on particles are evaluated as sums of near- and far-field potentials. Near-field potentials consider particles in a localized neighborhood, typically resulting in a linear algorithmic complexity. Far-field potentials, along with periodic boundary conditions are evaluated using Ewald summations, which involve computation of an FFT. Short-range potentials can be evaluated in a scalable manner on most platforms, since they have favorable communication to computation characteristics (communication is required only for particles close to the periphery of a processing core's subdomain). FFTs, on the other hand have more exacting requirement on network bandwidth, as demonstrated earlier in this chapter.

Traditional parallel paradigms attempt to distribute computation associated with both phases across all processing cores. However, since FFTs generally scale worse than the short-range potentials, it is possible to use a heterogeneous partition of the platform—with one partition working on short-range potentials, and the other on computing the FFT. The results from the two are combined to determine total potential (and force) at each particle.

Scalability analysis is a critical tool in understanding and developing heterogeneous parallel formulations. The tradeoffs of sizes of heterogeneous partitions, algorithm scaling, and serial complexities are critical determinants of overall performance.

43.5 Limitations of Analytical Modeling

We have, in this chapter, demonstrated the power of scalability analysis, and more generally, analytical modeling. However, it is not always possible to do such modeling under tight bounds. Consider the simple case of a quad-core processor with partitioned L1 caches, shared L2 caches, and a shared bus to memory. This is a typical configuration for such processors. If the bus bandwidth is high enough, the bus does not cause contention and each processing core can be assumed to have contention-free paths to memory and to the other cores, independent of communication patterns. If, on the other hand, the bus bandwidth is limited, effective access time to memory (in effect interprocessor communication) must account for contention on the bus and coherence overheads. Even if this were possible, finite cache sizes impact contention and access time significantly. Specifically, if the working set of each core exceeds local cache, local accesses may leak on to the bus as well. In this case, it is impossible to disambiguate local and remote (communication) accesses. Analytical modeling in such scenarios must account for finite resources and parameters such as line sizes, replacement policies, and other related intricacies. Clearly such analysis, if performed precisely, becomes specific to problem instance and platform, and does not generalize.

While aforementioned limitations apply specifically to small-to-moderate-sized systems, larger aggregates (terascale and beyond) are generally programmed through explicit message transfers. This is accomplished either through MPI-style messaging, or explicit (put-get) or implicit (partitioned global arrays) one-way primitives. In such cases, even with loose bounds on performance of individual multicore processors, we can establish meaningful bounds on scalability of the complete parallel system. Analytical modeling and scalability analysis can be reliably applied to such systems as well.

Acknowledgments

The work presented in this chapter is supported by DARPA's High Productivity Computing Systems (HPCS) program, and by the National Science Foundation.

References

[ABB+00]   Vikram S. Adve, Rajive Bagrodia, James C. Browne, Ewa Deelman, Aditya Duteb, Elias N. Houstis, John R. Rice et al. POEMS: End-to-end performance design of large parallel adaptive computational systems. *Software Engineering*, 26(11):1027–1048, 2000

[ABE+06]   Brian Armstrong, Hansang Bae, Rudolf Eigenmann, Faisal Saied, Mohamed Sayeed, and Yili Zheng. HPC benchmarking and performance evaluation with realistic applications. In *Proceedings of Benchmarking Workshop*, 2006.

[BDL98]   Shirley Browne, Jack Dongarra, and Kevin London. Review of performance analysis tools for MPI parallel programs. *NHSE Review*, 1998.

[CMW+05]   Nilesh Choudhury, Yogesh Mehta, Terry L. Wilmarth, Eric J. Bohm, and Laxmikant V. Kale. Scaling an optimistic parallel simulation of large-scale interconnection networks. *WSC '05: Proceedings of the 37th conference on Winter simulation*, pages 591–600. Winter Simulation Conference, 2005.

[(Ed01)]   Rudolf Eigenmann (Editor). *Performance Evaluation and Benchmarking with Realistic Applications*. MIT Press, Cambridge, MA, 2001.

[FK89]   Horace P. Flatt and Ken Kennedy. Performance of parallel processors. *Parallel Computing*, 12:1–20, 1989.

[Fla90]   Horace P. Flatt. Further applications of the overhead model for parallel systems. Technical Report G320-3540, IBM Corporation, Palo Alto Scientific Center, Palo Alto, CA, 1990.

[FP05]   T. Fahringer and S. Pllana. Performance prophet: A performance modeling and prediction tool for parallel and distributed programs. Technical Report 2005-08, AURORA Technical Report, 2005.

[GC01]   D.A. Grove and P.D. Coddington. A performance modeling system for message-passing parallel programs. Technical Report DHPC-105, Department of Computer Science, Adelaide University, Adelaide, SA 5005, 2001.

[GGK93]   Ananth Grama, Anshul Gupta, and Vipin Kumar. Isoefficiency function: A scalability metric for parallel algorithms and architectures. *IEEE Parallel and Distributed Technology, Special Issue on Parallel and Distributed Systems: From Theory to Practice*, 1(3):12–21, 1993.

[GGKK03] Ananth Grama, Anshul Gupta, George Karypis, and Vipin Kumar. *Introduction to Parallel Computing*. Addison Wesley, 2003.

[GMB88] John L. Gustafson, Gary R. Montry, and Robert E. Benner. Development of parallel methods for a 1024-processor hypercube. *SIAM Journal on Scientific and Statistical Computing*, 9(4):609–638, 1988.

[Gus88] John L. Gustafson. Reevaluating Amdahl's law. *Communications of the ACM*, 31(5):532–533, 1988.

[Gus92] John L. Gustafson. The consequences of fixed time performance measurement. In *Proceedings of the 25th Hawaii International Conference on System Sciences: Volume III*, pages 113–124, 1992.

[KG94] Vipin Kumar and Anshul Gupta. Analyzing scalability of parallel algorithms and architectures. *Journal of Parallel and Distributed Computing*, 22(3):379–391, 1994. Also available as Technical Report TR 91-18, Department of Computer Science Department, University of Minnesota, Minneapolis, MN.

[SN93] X.-H. Sun and L. M. Ni. Scalable problems and memory-bounded speedup. *Journal of Parallel and Distributed Computing*, 19:27–37, September 1993.

[Wor91] Patrick H. Worley. Limits on parallelism in the numerical solution of linear PDEs. *SIAM Journal on Scientific and Statistical Computing*, 12:1–35, January 1991.

44
Spatial Domain Decomposition Methods in Parallel Scientific Computing

44.1	Introduction...	44-1
44.2	Spatial Locality-Based Parallel Domain Decomposition...	44-3
	Motivating Applications • Analyzing Parallel Domain Decompositions	
44.3	Orthogonal Recursive Bisection	44-6
	Orthogonal Recursive Bisection Construction • Data Ownership • Quality of Partitioning	
44.4	Space Filling Curves	44-9
	Space Filling Curves Construction • Data Ownership • Quality of Partitioning	
44.5	Octrees and Compressed Octrees......................	44-13
	Octrees and Space Filling Curves • Constructing Parallel Compressed Octrees • Data Ownership • Quality of Partitioning	
44.6	Application to the N-Body Problem...................	44-19
	Acknowledgment...	44-22
	References	44-22

Sudip Seal
Srinivas Aluru
Iowa State University

44.1 Introduction

Increasing availability of more powerful and cost-effective hardware in terms of both processing as well as networking elements has allowed us to harness the power of large parallel computers to numerically solve larger and larger problems in a wide variety of scientific areas like electromagnetics, fluid dynamics, molecular dynamics, and quantum chromodynamics, to mention only a few. In most cases, a given problem is usually defined over a domain that can either be a predefined interval of space-time coordinates or an interval in some other abstract space. In the context of scientific computing, the term *domain decomposition* is often used to refer to the process of partitioning the underlying domain of the governing equation(s) such that it provides a more accurate result and/or takes fewer numerical steps to achieve a predetermined degree of accuracy of the result. In parallel scientific computing, the same term is used to refer to the process of partitioning the underlying domain of the problem across processors in a manner that attempts to balance the work performed by each processor while minimizing the number and sizes of communications between them. In this chapter, we adopt the latter definition. As such, the techniques that are described in this chapter do not attempt to improve the accuracy or efficiency (in terms of number of steps) of numerical techniques that are employed to solve the governing equations of the scientific application. Instead, we will focus on those techniques and algorithms that are widely used in scientific computing to partition the problem domain across multiple processors with the objective of optimizing certain metrics that will be introduced shortly.

Domain decomposition is the first step in many scientific computing applications. Computations within any subregion often require information from other, mostly adjacent, subregions of the domain. For example, in the finite element method, computations at a particular element need information available in the elements adjacent to it. Whenever information from neighboring elements is not locally available, processors need to communicate to access information. As communication is significantly slower than computation, domain decomposition methods attempt to minimize interprocessor communications and, in fact, try to overlap computation and communication for even better performance.

Another important design goal is to achieve load balance. The load on a processor refers to the amount of computation that it is responsible for. Achieving load balance while simultaneously minimizing communication is often nontrivial. This stems from the fact that the input data need not necessarily be uniformly embedded in the underlying domain of the problem. Moreover, the type of accesses required vary widely from application to application, although as a general rule of thumb accesses typically involve spatial neighborhoods. Therefore, optimal domain decomposition may vary from application to application. Furthermore, some applications may require multiple types of accesses, and designing a domain decomposition method that simultaneously optimizes these can be challenging.

A natural way to model any domain decomposition problem is through balanced graph partitioning. An interaction graph can be constructed in which units of computation are represented by nodes and data dependencies by edges. Weights may or may not be associated with the nodes and/or the edges depending on the particular model. Once such a graph is obtained, the task of parallel domain decomposition is reduced to partitioning the node set into p (where p denotes the number of processors) roughly equal sized/weighted subsets such that the total number/weight of the cross-edges (which represent interprocessor communications) between the p subsets is minimized. Graph partitioning has, however, been shown to be NP-complete [14]. Consequently, widely used tools such as Chaco [26] and METIS [30] developed for such purposes rely on heuristics to yield good quality partitions. Though they work well in practice, the above algorithms are based on a simplified model as indicated in References [24] and [25]. For example, the number of cross-edges between the p subsets of nodes is considered to be proportional to the total communication volume in the simplified models underlying both the above tools. However, such a metric overcounts the communication volume since multiple edges from one node in a processor to nodes in a different processor represent communicating the same information. Further advances have been made in recent years in graph partitioning models; for example, see research on hypergraph partitioning [9, 10].

While graph partitioning is a general way to model spatial domain decomposition, the nodes of the graph have spatial coordinates and graph edges typically exist between nodes that are proximate. Taking advantage of this, several methods for spatial domain decomposition have been developed that simply strive to create partitions that preserve spatial locality without explicitly considering edge information at all. The chief advantage of such methods in comparison to explicit graph partitioning is that they are much simpler, relatively easy to parallelize, and result in much faster algorithms. Their chief disadvantage is that they lack the generality of graph partitioning methods. Nevertheless, they provide good quality partitions for many applications and in some cases provably so. Moreover, in certain applications, graph edges can be described in terms of spatial geometry: as an example, a node may require information from all nodes that are within a sphere of radius r centered at the node.

Such geometric constraints can easily be incorporated into spatial locality-based domain decomposition techniques.

This chapter will focus on spatial locality-based parallel domain decomposition methods. We present an overview of some of the most widely used techniques such as orthogonal recursive bisection (ORB), space filling curves (SFCs), and parallel octrees. We present parallel algorithms for computing the specified domain decompositions and examine the partitioning quality in the context of certain motivating applications. Finally, we will look at a case study of the N-body problem, to demonstrate how spatial locality-based parallel domain decomposition and the execution of the underlying numerical method can sometimes be integrated to provide a provably good parallel algorithm for the application.

44.2 Spatial Locality-Based Parallel Domain Decomposition

In this section, we describe some of the relevant issues in designing and evaluating spatial locality-based parallel domain decomposition methods. The decomposition methods themselves will be described in greater detail in subsequent sections. To illustrate what is expected of these techniques, we will briefly review certain representative applications in scientific computing.

44.2.1 Motivating Applications

Several applications in scientific computing require simulating physical phenomena whose underlying behavior can be described by a set of partial differential equations (PDEs). Owing to lack of closed form solutions for the governing equations, these applications are solved using numerical techniques that require the underlying domain to be discretized into triangular or hexagonal meshes, rectangular grids, and so forth. For example, triangular meshes arise in finite element methods (FEMs) and rectangular grids are used in several numerical methods including finite difference, adaptive mesh refinement, and multigrid methods. Several applications use such numerical methods—heat transfer, shock-wave propagation, stress analysis, structural integrity analysis, climatic simulations, and so forth. These are broadly classified as grid-based methods.

The accuracy of the solutions computed often depends upon the resolution of discretization that is used and the computational work increases with increase in resolution. For computational efficiency, different parts of the domain are discretized to different resolutions depending upon the rate at which the solution of the underlying PDEs changes with space and/or time. The objective is to have higher resolution in rapidly changing areas and lower resolution in relatively stable areas of the domain so as to achieve balanced numerical accuracy throughout the domain. For example, in studying the propagation of a shock wave, a higher resolution is required at the frontier of the shock wave as compared to areas that are farther from the wave. Also, as the shock wave propagates through the domain over time, the discretization itself needs to be varied with time. Note that a variable resolution is the best way to achieve highest numerical accuracy possible for a given computational cost. On the other hand, there are a few applications, such as in plasma physics, where a uniform discretization of the domain is sufficient. Clearly, designing and proving the efficiency of domain decomposition methods for such applications, known as regularly structured problems, is substantially easier. Most applications fall under the category of irregularly structured problems, for which proving the optimality of a domain decomposition method is often difficult and sometimes not known.

Applications can be classified as static or dynamic, depending upon whether the simulation being carried out is to find the steady state or to study the evolution of a system with time. Finding the heat gradient in a domain whose boundaries are held at constant temperatures, or finding the electromagnetic field radiated by a perfect electrically conducting surface, fall under the former category. Studying shock-wave propagation and performing a crash analysis, where the crash causes deformation to bring into contact previously nontouching surfaces that require further analysis iteratively, are examples of dynamic applications. The primary difference with respect to parallel domain decomposition is that static problems require decomposition once which is held constant through the iterations, while dynamic problems require domain decomposition or incremental load balancing at frequent time intervals, or at every time step. Another differentiating factor in grid-based methods is whether the domain discretization is hierarchical or not. For example, multigrid methods use a hierarchy of grids over the domain and computation is carried out across multiple grids in a particular sequence. Therefore, parallel domain decomposition methods for these applications should optimize decomposition across multiple grids simultaneously.

Another important class of scientific applications are particle-based methods. Although gird-based methods attempt to sample the otherwise continuous solution at discrete points in the domain, particle-based methods study the evolution of a system of particles or atoms. An example is the gravitational N-body problem, in which the initial positions and velocities of a set of particles are given and their final positions and velocities after a specified time period are sought under the influence of mutual gravitational attraction. Another example is molecular dynamics, in which atomic interactions based on van der Waal forces, Lennard–Jones potential, and Coulomb's forces are calculated. Molecular dynamics is used to study molecular conformations and to compute energetically stable conformations of protein molecules. There are other methods, such as mesh less or mesh free methods, where a simulation of virtual particles is carried out. Other applications of particle-based methods include fluid dynamics and smoothed particle hydrodynamics.

Spatial locality-based domain decomposition methods make particular sense for particle-based methods. In these methods, particles interact with other particles, often based on spatial locality, providing a direct justification for such parallel domain decomposition methods. For example, the rapidly decaying Lennard–Jones potential is evaluated in molecular dynamics simulations by taking into account only those atoms that lie within a specified cutoff distance. Such description of interactions using geometric constraints particularly suits spatial locality-based parallel domain decomposition methods.

In grid-based methods, interactions are often specified in terms of adjacency. In a FEM with triangular decomposition, interactions between a triangular element with other triangular elements sharing an edge are computed. Multigrid methods involve interactions between adjacent grid cells at a particular level of the multigrid hierarchy, based on cell containment at two successive levels of the hierarchy. Domain decomposition for particle-based methods is equivalent to computing a partition of multidimensional point data. For the purposes of unified treatment, grid-based methods can be viewed in the same way by associating points with geometric shapes. For instance, a triangular element can be represented by its centroid and a square or cubic grid cell can be represented using its center. Because of this, one can focus on partitioning multidimensional point data without loss of generality, and we take this approach in this chapter.

44.2.2 Analyzing Parallel Domain Decompositions

In this section, we outline some issues that must be considered in evaluating the merits of a parallel domain decomposition and in comparing one decomposition technique to another. Although quality of the partitions generated by the decomposition is an obvious factor, other factors such as runtime of the parallel domain decomposition algorithm itself and its amenability to incremental load balancing and so on are important. Even for judging the quality of parallel domain decomposition, there is no single metric that can be uniformly used as a benchmark. This is because application needs vary and performance on each type of access pattern needs to be evaluated both theoretically and experimentally.

44.2.2.1 Runtime of the Domain Decomposition Algorithm

Because parallel domain decomposition is a necessary overhead that does not have a counterpart when solving the problem sequentially, this overhead must be kept as small as possible as a fraction of the overall execution time. This is particularly important for scaling to large systems, as the overhead can potentially become a limiting factor. For dynamic problems, this is even more important as parallel domain decomposition or load balancing is needed frequently. Spatial locality-based decomposition methods have a particular advantage here because of the simplicity of the underlying model of partitioning multidimensional point data. It is also easy to parallelize the decomposition algorithm itself, which is useful in reducing the runtime overhead along with the remaining application runtime, when scaling to larger and larger systems.

44.2.2.2 Computation of Data Ownership

When an application is run after parallel domain decomposition, it is necessary for each processor to be able to compute the identity of processors containing nonlocal data on the basis of spatial coordinates. Two types of accesses may be involved: retrieving the data associated with a point given its spatial coordinates, or retrieving data associated with all the points that are specified by geometric constraints. Here, the word *point* refers to the location of a particle, or atom, or the center of a grid cell, or whatever form in which the input data or discretization is interpreted as a multidimensional point set. Note that while geometric constraints typically specify a region of continuous space, the points we are interested in are those from the input or current state of the system that lie in the region. It is essential that computation of data ownership be possible directly from the spatial coordinates. It should also be as efficient as possible.

44.2.2.3 Quality of Partitioning

Although quality of partitioning is of paramount importance, there is no straightforward way to define it and what is required can vary from application to application. Balancing the computational load per processor is important. Many decomposition methods achieve this by partitioning the data equally among processors. What is needed is to arrange this in a way that minimizes the number of remote accesses as well as the number of processors that contain the remote data needed by each processor amongst other considerations. Intuitively, it makes sense to allocate a continuous region as opposed to multiple disjoint regions to a processor to reduce the total number of remote accesses. For a uniform decomposition, it is easy to argue that computational work is proportional to the volume of the subregion and communication cost is proportional to its surface area. Most applications are irregular and do not lend themselves to such simplistic analysis.

Some general guides on the quality of partitioning are easy to observe. Long, tube-like regions are less preferable to cube-like partitions. A domain decomposition that places many processor domains adjacent to a few others will generally result in unbalanced communication. Beyond such general rules of thumb, a rigorous analysis may vary from application to application.

The access patterns required by some scientific applications can be abstracted as queries on multidimensional point data. Thus, developing efficient parallel domain decomposition methods that are optimized for such queries can simultaneously be beneficial to several applications. Some examples of such queries are given below:

- *Spherical region query:* In a spherical region query, each point requires information from all other points that lie within a specified radius, say r, from it. Spherical region queries are common to molecular dynamics simulations [12, 21, 34, 37].
- *Nearest neighbor queries:* In this query, each point needs information from its nearest neighbor(s). These arise in almost all finite element and finite difference methods [5, 50].
- *Annular region query:* Given two distances d_1 and d_2 ($d_1 < d_2$), each point in an annular region query needs information from all points that lie at a distance d such that $d_1 \leq d \leq d_2$. These queries, sometimes also referred to as doughnut region queries, are used in algorithms for N-body problems [20, 46]. In practice, the region for this kind of query is most often constituted by the volume outside a cuboid C_1 but within a larger cuboid C_2 that completely encloses C_1.

In the following sections, we describe several domain decomposition techniques that are widely used in parallel scientific computing and analyze their relative merits on the basis of the above measures of quality. Typical scientific computing problems are defined in three dimensions, though some two-dimensional applications exist. For simplicity, we will present supporting illustrations only in two dimensions.

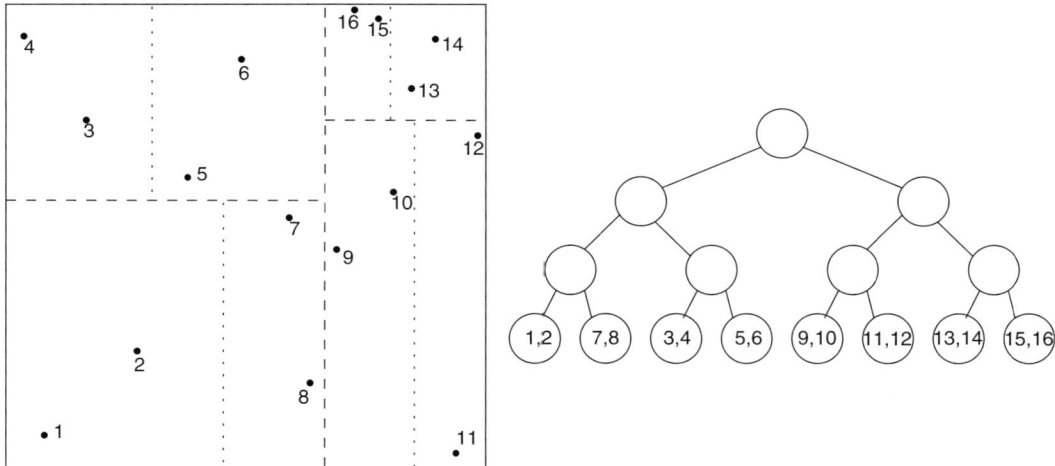

FIGURE 44.1 Orthogonal recursive bisection is used to create 8 partitions of a 2D domain that encloses 16 points. The associated ORB tree is shown on the right.

44.3 Orthogonal Recursive Bisection

Orthogonal recursive bisection (ORB) as a tool for parallel domain decomposition was first introduced in 1987 [8] and has been widely used in several implementations of partial differential equation solvers since then. Consider a cube enclosing the given n multidimensional points. The cube is split along a chosen dimension x, using the median value of the x coordinates of all the points (see Figure 44.1). This creates two subdomains each containing approximately equal number of points. The process is recursively repeated within each subdomain, alternately choosing between the dimensions, until the total number of partitions is equal to the number of processors p. This works when p is a power of 2 and each point represents equal computational load. Points with differing computational loads can be accommodated by associating points with weights and computing a weighted median. Similarly, if p is not a power of 2, the following strategy can be recursively used: If p is even, split using the median. If p is odd, split using the element with rank $n(p-1)/2p$. The subdomain with smaller number of points is recursively subdivided among $(p-1)/2$ processors and the other subdomain is recursively subdivided among $(p+1)/2$ processors.

Since each split is a bifurcation, ORB can be represented as a binary tree (see Figure 44.1), with the root node representing the bounding cube and the leaf nodes representing the final partitions.

44.3.1 Orthogonal Recursive Bisection Construction

For convenience of presentation, we assume that n is a multiple of p and p is a power of 2. As described earlier, it is easy to relax these restrictions. Initially, the n points are distributed such that each processor has n/p points. The first step is to find the median element along a chosen dimension. This is done by running a parallel median finding algorithm. Several parallel median finding algorithms have been presented, and theoretically and experimentally evaluated in Reference [3]. Median finding algorithms differ in the number of iterations they take to converge to the median, which has implications on the communication complexity. Randomized median finding algorithms perform better because they use fewer number of iterations. Two good choices are parallelization of Floyd's classic quicksort type median finding algorithm [13] that takes $O(\log p)$ expected number of iterations and the randomized algorithm by Rajasekaran et al. [38] that takes $O(\log \log p)$ expected number of iterations. In Floyd's algorithm, the rank of a randomly chosen element, say x, is found. If this rank is less than the median rank, all elements smaller than or equal to x are discarded. If the rank is greater than the median, all elements greater than

ORB (*lev, leftProc, rightProc*)
1: **if** *lev* < log *p* **then**
2:    *midProc* = ⌊(*leftProc* + *rightProc*)/2⌋
3:    *m* = FindMedian (*lev* mod *d*) {*d* is the number of dimensions}
4:    ExchangePoints (*lev*, *m*)
5:    **if** *myRank* < *midProc* **then**
6:      ORB (*lev* + 1, *leftProc*, *midProc* − 1)
7:    **else**
8:      ORB (*lev* + 1, *midProc*, *rightProc*)
9:    **end if**
10: **end if**

Algorithm 1: Orthogonal recursive bisection.

x are discarded. This reduction of the number of elements is recursively carried out until the median of the original set of points is found or the number of elements reduced to a small number that can be directly examined. The algorithm by Rajasekaran et al. achieves the much smaller $O(\log \log p)$ expected number of iterations by finding, with high probability, two elements whose ranks are likely to straddle the median so that all elements whose ranks do not lie in between the ranks of the two chosen elements can be discarded.

One issue with median finding algorithms is that even though the initial data of n elements is distributed evenly across processors, the surviving data in subsequent iterations may not be evenly distributed. In the best case, where the distribution is even throughout the iterations, the computational work throughout is bounded by $O(n/p)$. In the worst case, where there is a processor that has all its data elements surviving through the iterations, the computational work is worse by a factor equal to the number of iterations. Experimental results show the former is more likely [3].

Once the median element is identified, it is used to split the domain into subdomains with equal number of points. The data on each processor is split into two subsets, according to the respective subdomains they fall in. An all-to-all communication is necessary to move the points in the first subdomain onto the first half of the processors and similarly for the second half. This process is recursively applied on each half of the processors until subdomains spanning one processor and containing n/p points are obtained. Pseudocode for the ORB construction algorithm is given in Algorithm 1. The algorithm is initially called with *lev* = 0, *leftProc* = 0 and *rightProc* = $p − 1$. This method has a best case parallel computational time of $O\left(\frac{n}{p} \log p\right)$ (assuming data is always evenly distributed during median finding, an effect that can be obtained by prerandomization of data if needed), and requires $O(\log p)$ communication rounds involving all-to-all communication.

Another approach for constructing ORB decomposition relies on presorting the input points along each of the dimensions so that medians can be identified in $O(1)$ time [2]. When partitioning data according to subdomains, the sorted order should be maintained within each subdomain. This requires a data redistribution strategy in which the order of the data as given by processor rank and index within a processor must be maintained. Median-based methods allow less communication intensive data redistribution as data only needs to be shifted from processors containing excess points to processors containing fewer.

Sorting requires $O\left(\frac{n}{p} \log n\right)$ parallel computational time when compared to the $O\left(\frac{n}{p} \log p\right)$ time of median-based approach; however, the chief advantage of sorting is that efficient parallel sorting algorithms such as sample sort use just one all-to-all communication whereas a median-finding method requires $O(\log p)$ or $O(\log \log p)$ iterations in which all-to-all communication is used. Moreover, some applications require constructing a multidimensional binary search tree, also known as k-d tree, on the input set of points [7]. A k-d tree can be viewed as an extension of the ORB tree where decomposition is continued until each leaf contains one point. If such a tree is needed, the cost of sorting can be effectively amortized over the log n levels of the tree and the sorting-based method turns out to be superior [2]. Another crucial factor is the number of dimensions. As sorting-based methods require sorting along each dimension, media-based approaches are favored for problems with higher dimensional data. This is, however, not an issue for most scientific computing applications, which are three dimensional.

44.3.2 Data Ownership

In ORB decomposition, each processor is responsible for a rectangular region and the points associated with it. To determine ownership, a copy of the ORB tree is stored on every processor, whose size is $O(p)$. At each internal node in the tree, the value of the coordinate that is used to split the region at the node into its two child regions is stored. The dimension itself may be implicit (such as when they are used in a sequence), or explicitly stored (e.g., when using the dimension with the largest span). Given a query point, the processor that owns the region containing the point can be found using a binary search on the ORB tree starting from the root in $O(\log p)$ time.

Suppose we are asked to retrieve all the points within a spatial query region Q. This requires determining the processors whose regions overlap with Q. Let D denote the domain and D' and D'' denote its left and right child in the ORB tree, respectively. If $Q \cap D' = Q' \neq \phi$ (similarly, $Q \cap D'' = Q'' \neq \phi$), recursively search the tree rooted at D' with query region Q' (similarly, search D'' with query region Q''). This identifies a set of leaf nodes in the ORB tree, which reveals the processors whose regions overlap with the query region.

44.3.3 Quality of Partitioning

Orthogonal recursive bisection is a balanced partitioning of multidimensional point data to processors. Another advantage of ORB is its ability to accommodate local rebalancing without having to reconstruct the entire partitioning from scratch. The need for rebalancing can occur in two ways: if the locations of the points change slightly from iteration to iteration, such as when simulating a particle-based system, the new ORB tree is likely to be closer to the previous one. Another type of rebalancing need arises in applications such as AMR, where the level of refinement within a subregion of the domain may change from one iteration to the next. In this case, each subdomain in the current ORB decomposition may end up with different numbers of points. Note that exact load balance may not be necessary or desirable in view of the cost of load balancing itself. Incremental load balancing can be carried out at any subtree of the ORB tree that is deemed sufficient to achieve approximate balance across all processors. Even within that, the older ORB partitioning can be used as a guide and partitions readjusted.

Using ORB in the straightforward manner as described above may create long and thin partitions that may lead to higher and unbalanced communication costs. One of the ways in which such skewed partitions are discouraged is by carefully choosing the dimension used for bisection. Instead of alternating the dimension of bisection at each level of recursion, one of the strategies that is often employed is to partition each subdomain along the dimension in which the subdomain has the largest span (see Figure 44.2).

Another modification that has been suggested to make the ORB algorithm more communication efficient is the *unbalanced recursive bisection (URB)* as has been done in SUMAA3D [29]. Instead of dividing the unknowns in half, the cut that minimizes the partition aspect ratio and divides the unknowns into kn/p and $(p-k)n/p$ (where $k = 1, 2, \ldots, p$) sized groups is chosen. This algorithm leads to an even distribution of points with more balanced partition aspect ratios and, consequently, tends to decrease the communication to computation ratio. Other variants of the ORB method have been attempted, for example, hierarchical ORB (ORB-H) [48] or rectilinear ORB [33]. The basic disadvantage of these algorithms is that all the cuts are, by definition, straight, which forces a $d-1$ dimensional hyperplane to carry the load imbalance introduced by a cut in d-dimensions. The ORB-Median of Medians (ORB-MM) [44] variant addresses this issue at the expense of introducing highly irregular partitions that do not have a compact mathematical representation thereby incurring large book-keeping overheads in practical

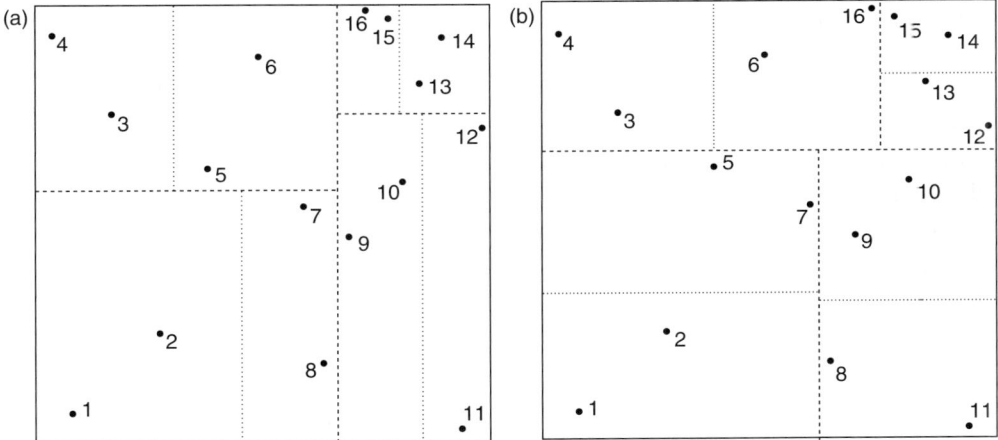

FIGURE 44.2 (a) The shape of partitions obtained using an ORB can have very skewed aspect ratios. (b) The strategy which bisects each domain along the dimension with longest span can yield better aspect ratios.

implementations. In addition, the absence of a robust mathematical representation of the partitions makes it more difficult to compute data ownership and repartitioning of the domain as desired. To avoid these drawbacks, one would like to be able to decompose the domain in a manner that is not only easy to implement in practice, but also possesses a robust mathematical representation that enables fast computation of data ownership and be devoid of the load balancing limitations that are imposed by straight sectioning cuts. In the next section, we will describe a space filling curves-based technique that overcomes some of the above drawbacks.

44.4 Space Filling Curves

In 1890, to demonstrate an earlier counterintuitive result that the number of points in a unit interval has the same cardinality as the number of points in a unit square, Giuseppe Peano introduced SFCs as curves that pass through every point of a closed unit square. Since then SFCs have been generalized to higher dimensions. Though in their original form SFCs are continuous, in this chapter we are primarily interested in discrete SFCs. The reader is referred to Reference [39] for an in-depth exposition of continuous SFCs. For our purpose, we will adopt the following definition of a SFC: Consider a d dimensional hypercube. Bisecting this hypercube k times recursively along each dimension results in a d dimensional matrix of $2^k \times 2^k \times \cdots \times 2^k = 2^{dk}$ nonoverlapping hypercells of equal size. A SFC is a mapping of these hypercells, the location of each of which in the cell space is given by d coordinates, to a one-dimensional linear ordering. The process of mapping discrete multidimensional data into a one-dimensional ordering is often referred to as *linearization*. There are many different mappings that yield a linearization of multidimensional data. Three common examples of linearization of two-dimensional data are shown in Figure 44.3 for $k = 1, 2,$ and 3. As can be easily observed from the figure, each curve is recursively drawn in that a $2^{k+1} \times 2^{k+1}$ SFC contains four $2^k \times 2^k$ SFCs. For the Hilbert curve [27], appropriate rotations need to be performed as well. The above curves can be easily generalized to higher dimensions.

In general, a SFC-based decomposition of a domain that contains n points is carried out by first enclosing the points in a hypercube of side length, say L. Consider the decomposition of this hypercube into 2^{kd} hypercells each of side length $L/2^k$, such that each cell is occupied by at most one point. The *resolution* of the resulting decomposition is defined by k. A SFC is then used to map the nonempty cells into a one-dimensional ordering. An example with $k = 3$ in two dimensions is shown in Figure 44.4(a).

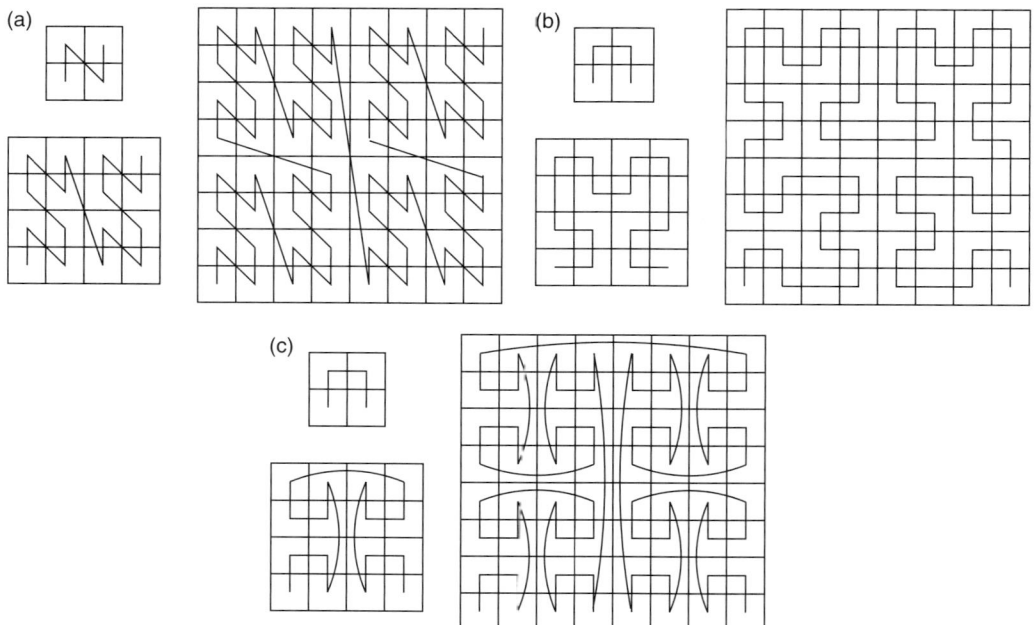

FIGURE 44.3 The (a) z-curve, (b) Hilbert curve, and (c) gray code curve for $k = 1, 2$, and 3.

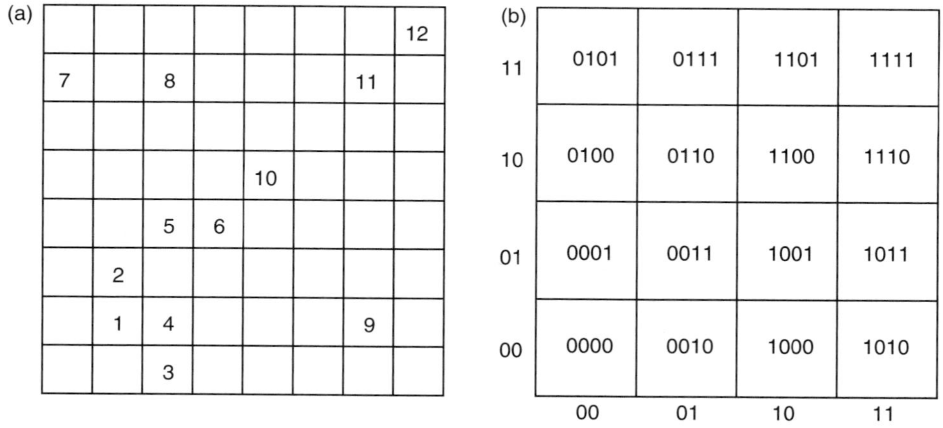

FIGURE 44.4 (a) A 8×8 decomposition of two-dimensional space containing 12 points. The points are labeled in the order in which the cells containing them are visited using a Z-SFC. (b) Bit interleaving scheme for generating index of a cell as per Z-SFC.

In this figure, the cells that are numbered are considered to be occupied by one point. The rest of the cells are empty. A Z-SFC is used to visit each of the cells. The order in which the occupied cells are visited along the Z-SFC is indicated by the numbers in the cells. Note that these numbers do not indicate the index (along the SFC) of the cells that enclose the points; only the order in which the nonempty cells are visited along the SFC. The resulting one-dimensional ordering is divided into p equal partitions that are assigned to processors. In practice, the value of k is fixed and multiple points that reside in a cell can be listed in an arbitrary order in the SFC ordering.

44.4.1 Space Filling Curves Construction

Consider a d-dimensional cube containing n points and a resolution k that separates the points into different cells. Note that a SFC specifies a linear ordering of the 2^{kd} cells and the SFC order of the n points is defined as the order of the n cells containing the points as per the SFC linearization of cells. However, directly implementing this definition results in a poorly performing algorithm. The runtime to order the 2^{kd} cells, $\Theta(2^{kd})$, is expensive because typically $n \ll 2^{kd}$. To overcome this, a method to directly order cells containing the points is needed.

Let D denote the side length of the domain whose corner is anchored at the origin of a d-dimensional coordinate system. The position of a cell in the cell space can be described by d integer coordinates, each in the range $[0 \ldots 2^k - 1]$. For example, the tuple (i, j) can be used to describe the two-dimensional cell whose lower left corner is located at $\left(i\frac{D}{2^k}, j\frac{D}{2^k}\right)$. It turns out that the position or index of a cell in SFC linearization can be directly computed from its coordinates and this mapping function is the characteristic defining function of the SFC. For example, consider the Z-SFC, also known as Morton ordering [32]. The index of a cell in this order can be computed by representing the integer coordinates of the cell using k bits and then interleaving the bits starting from the first dimension to form a dk bit integer. In Figure 44.4(b), the index of the cell with coordinates $(2, 3) = (10, 11)$ is given by $1101 = 13$. Similar, but more complex, characteristic functions exist for other SFCs including the gray code and Hilbert curves. For the gray code, carrying out bit interleaving as done for the Z-SFC but treating the result as a binary reflected gray code is sufficient to generate the index; that is, the index is obtained by converting the bit interleaved result from gray to binary. Generating indices for the Hilbert curve is more complex and can be found in Reference [36]. Generating indices for the Z-SFC is particularly simple and elegant and as a result, Z-SFC has been used in many large-scale implementations [20, 49]. The Hilbert SFC is also often used, because it avoids sudden jumps and any two consecutive cells in Hilbert order are neighbors in the d-dimensional space. Hilbert curve also has the property that any partitioning of its linearization induces a partitioning of the domain into contiguous subdomains.

The characteristic function of a SFC can be used to linearize a set of multidimensional points. Typically, an appropriate resolution k is chosen either based on knowledge of the domain, the resolution required, or based on limiting cell indices to 32 or 64 bit integers (e.g., a resolution of 21 for three-dimensional data generates 63 bit numbers). The first step in SFC linearization is to find the coordinates of the cell containing each of the input points. Given a point in d-dimensional space with coordinate x_i along dimension i, the integer coordinate along dimension i of the cell containing the point is given by $\left\lfloor \frac{2^k x_i}{D} \right\rfloor$. From the integer coordinates of the cell, the index of the cell can be generated. Thus, the index of the cell containing a given point can be computed in $O(kd)$ time in d dimensions, or $O(k)$ time in three dimensions. Once the indices corresponding to all the points are generated, SFC decomposition is achieved by a parallel integer sort [22, 23]. Parallel sort also has the desirable side-effect of partitioning the SFC linearization to processors.

If needed, one can also find the lowest resolution such that each point lies in a separate cell. To find this, compute the smallest distance between any pair of points along each of the dimensions. Let the smallest of these distances be l. Then, the resolution is given by $k = \lceil \log_2 \frac{D}{l} \rceil$.

44.4.2 Data Ownership

Since the linearized mapping is stored across the processors in a sorted array on the basis of the integer keys of each point, computation of the processor identity given a query point can be accomplished if each

SFC Linearization
1: Choose a resolution k.
2: For each point, compute the index of the cell containing the point.
3: Parallel sort the resulting set of integer keys.

Algorithm 2: SFC linearization.

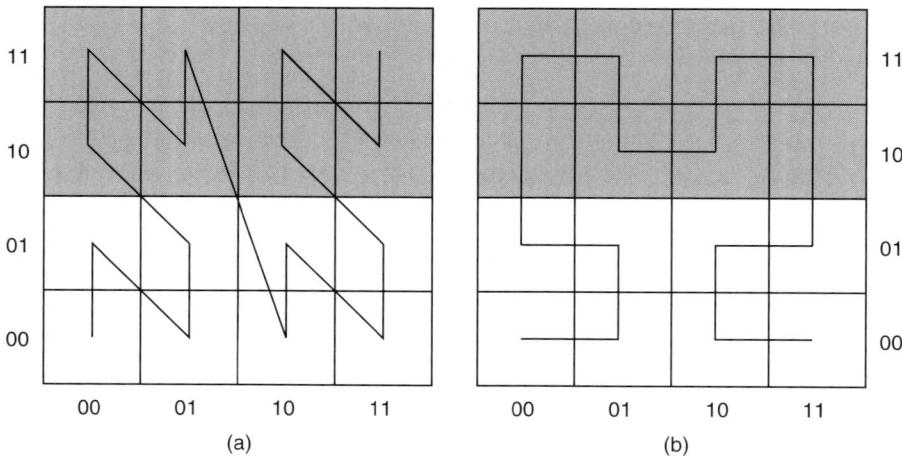

FIGURE 44.5 Example of a range query in a domain that is mapped using (a) Z-SFC and (b) Hilbert curve.

processor is aware of the boundary elements of all other processors. These boundary elements can be stored in an array of length p in each processor. A binary search in this array can identify the processor that owns a given integer key in $O(\log p)$ runtime and hence the point that generated the integer key.

44.4.3 Quality of Partitioning

Partitioning the SFC linearization of the points equally to processors ensures load balancing. In addition, SFC-based domain decomposition avoids the adverse effects of straight sectional cuts through the domain as described at the end of Section 44.3. If the computational load per point is not identical but known, the SFC linearization can be partitioned such that the total load per processor is about the same. If the load is not known, the first iteration can be run with an equally partitioned SFC linearization, and load on each point estimated. This can be used to readjust the load balance, by simply moving the boundaries of the processors on this one-dimensional mapping. If the point locations are dynamic, then SFC linearization can be recomputed as necessary, or incremental load balancing performed if a few points change in the order. In general, computing the SFC linearization is equivalent to parallel sorting and maintaining the SFC linearization is equivalent to maintaining the sorted order in the presence of local changes to it (the sorted order).

In any mapping from a higher dimension to a lower dimension, proximity information will be inherently lost. For example, in Figure 44.4(a), both points 2 and 4 are nearest neighbors of point 1 in the two-dimensional space. But, along the linear array, point 2 continues to remain as an immediate (nearest) neighbor of point 1 whereas point 4 does not. As such, if the 12 points in the example are partitioned across four processors, then both points 1 and 2 will be local to a processor whereas point 4 will be in a different one. If a nearest neighbor query is generated by point 1, then it can only be satisfied through a communication with another processor to retrieve point 4. In fact, it can be analytically shown [41] that for the case where each cell is occupied, the average distance to a nearest neighbor in two dimensions is $\Theta(\sqrt{n})$ where $n = 2^k \times 2^k$. It is easy to see that a spherical region query generated by point 1 with a radius greater than or equal to the side length of the smallest subsquare will also result in a similar communication. Typical database applications and geographic information systems often generate yet another kind of queries, called *range queries*, in which all points within a specified hyperrectangular region of the underlying domain are to be reported. In Figure 44.5(a) and (b), the shaded regions represent such a range query in two dimensions. Clearly, the space filling curve on the left (Z-SFC) generates two noncontiguous portions of the linearized array compared to the one on the right (Hilbert curve), which contains the query region in one contiguous subarray.

Spatial Domain Decomposition Methods in Parallel Scientific Computing

Therefore, it comes as no surprise that different space filling curves exhibit different locality preserving properties. Over the years, several results have been reported in this context. Different measures of locality have been proposed and locality properties of different SFCs based on such definitions have been studied. Gotsman and Lindenbaum [15] bounded the locality of multidimensional SFCs based on a measure that reflects the extent to which two points that are close on the linearized data can be far apart in the original multidimensional space under a Euclidean norm. They showed that the Hilbert curve comes closest to achieving optimal locality based on the above measure. Clustering properties have been analyzed by Abel and Mark [1] in two dimensions and in higher dimensions by Jagadish [28]. Closed form formulas for the number of clusters in a query region of arbitrary shape have been derived for the Hilbert curve in Reference [31].

Note that SFC-based partitions can result in noncontiguous regions of the domain to be mapped to the same processor. For example, in Figure 44.4(a), partitioning the 12 points across four processors results in points 7, 8, and 9 to be mapped to a single processor. Unlike the other three partitions, the region of space that maps to the processor owning the above three points is disjoint. Even if the mapped domain is continuous, it may be of a complicated shape. Complicated shapes of partitions naturally raises the question of how good the computation to communication ratio is for a SFC-based parallel domain decomposition. Parallel applications [11, 18, 35, 36, 45] that use SFCs to partition the problem domain exhibit good scaling and runtime results. Empirical justification remains the mainstay of the popularity of SFCs. More recently, it has been shown [47] that if a three-dimensional cube is decomposed into $n = 2^k \times 2^k \times 2^k$ cells such that the probability of occupancy is uniform across all cells, then partitioning any SFC linearization across $p = n^\alpha (0 < \alpha \leq 1)$ processors exhibits the following properties: (i) for nearest neighbor queries, the expected total number of points requiring remote accesses is $\Theta\left(n^{3/4+\alpha/4}\right)$ and (ii) For spherical region queries with radius r, the expected total number of points requiring remote accesses is $\Theta\left(n^{3/4+\alpha/4}\right)$. These results show that for any sublinear number of processors ($p = n^\alpha, \alpha < 1$), the total number of remote accesses remains sublinear. The number of remote accesses is $\Omega\left(n^{3/4}\right)$ and increases only as $\sqrt[4]{p}$ as the number of processors is increased. For n^α processors, the ratio of computation to communication is $\Theta\left(n^{1/4(1-\alpha)}\right)$, showing that communication costs can be contained for sufficiently large values of n. However, this analysis is only for the simpler case of a uniform distribution. Such analysis to study the quality of SFC-based domain decomposition for irregular data is not known.

Comparative works on SFC- and ORB-based decompositions in specific areas of scientific computing such as vortex dynamics and smoothed particle hydrodynamics [36] have consistently validated the greater efficiency of the former. Because of its ease of repartitioning, SFCs have been widely used as a tool for domain decomposition in parallel implementations of AMR [18, 45]. In general, the above applications follow a two-step procedure. In the first step, ORB or SFC is used to preprocess the input data such that the computational domain is partitioned across multiple processors. In the second step, other data structures, usually tree based [40, 49], are subsequently built locally on each processor on the part of the domain that it is mapped to. The numerical computations that drive the scientific application are then carried out using these latter data structures that are built using the locally available data. The next section describes a more sophisticated approach of using the same data structure for both parallel domain decomposition and subsequent numerical computations.

44.5 Octrees and Compressed Octrees

Octrees are hierarchical tree data structures that organize multidimensional points using a recursive decomposition of the space containing them. Such a tree is called a *quadtree* in two dimensions, *octree* in three dimensions, and *hyperoctree* in higher dimensions. In this chapter, we will simply use the term octree irrespective of the dimensionality of the domain space. The dimensionality will be clear from the context. Consider a hypercube enclosing the given n multidimensional points. This domain is recursively bisected in a manner that is similar, though not identical, to a SFC decomposition. The domain enclosing all the points forms the root of the octree. This is subdivided into 2^d subregions of equal size by bisecting

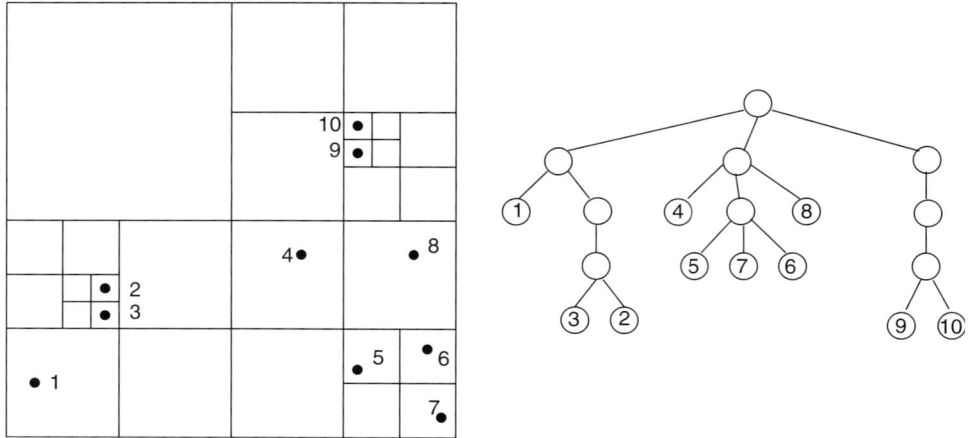

FIGURE 44.6 A quadtree built on a set of 10 points in two dimensions.

along each dimension. Each of these regions that contain at least one point is represented as a child of the root node. The same procedure is recursively applied to each child of the root node terminating when a subregion contains at most one point. The resulting tree is called a region octree to reflect the fact that each node of the tree corresponds to a nonempty subdomain. An example is shown in Figure 44.6. In practice, the recursive subdivision is stopped when a predetermined resolution level is reached, or when the number of points in a subregion falls below a preestablished constant threshold.

In octrees, the manner in which any subregion is bisected is independent of the specific location of the points within it. The size of an octree is, therefore, sensitive to the spatial distribution of the points. This is because chains may be formed when many points lie within a small volume of space. An example is shown in the tree in Figure 44.6. The chain below the rightmost child of the root is caused by points labeled 9 and 10 because they are very close to each other and can be separated only after several recursive subdivisions. Points labeled 2 and 3 also result in such a chain. Though nodes in these chains represent different volumes of the underlying space, they do not contain any extra information and only make the size of the resulting tree dependent upon the distribution of the points. An octree built on a uniform distribution of points is height balanced and has a size linear in the number of input points. But, for input points with a nonuniform distribution, the tree can be imbalanced.

Though multiple nodes on a chain represent different regions of space, they all contain the same points. Thus, in any application where points are associated with elements of interest (particles, grid cells, finite elements, etc.) and region subdivision is for purposes of convenience or enabling faster algorithms, different nodes on a chain essentially contain the same information. As such, no information is lost if the chains are compressed. However, such a compressed node should still encapsulate the fact that it represents multiple regions of space unlike the nodes that are not compressed. These observations motivate the development of *compressed octree*. A compressed octree is simply an octree with each of its chains compressed into a single node. Thus, each node in a compressed octree is either a leaf or has at least two children. This ensures that the size of the resulting compressed octree is $O(n)$ and is independent of the spatial distribution of the points. The compressed octree corresponding to the octree in Figure 44.6 is shown in Figure 44.7.

To understand how a compressed octree encapsulates the spatial information in the compressed nodes, we need to establish some terminology. Given n points in d dimensions, the *root cell* refers to the smallest hypercubic region that encloses all the points. Bisecting the root cell along each dimension results in 2^d hypercubic regions, each having half the side length as the root cell. Each of these smaller disjoint hypercubic regions are called *cells*. All hypercubic regions obtained by a recursive subdivision of any cell are also called cells. The term *subcell* is used to refer to a cell that is completely contained in another cell

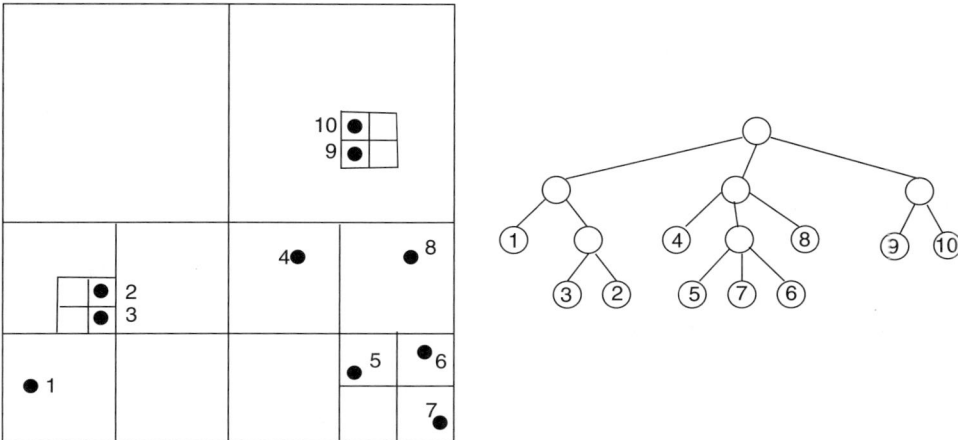

FIGURE 44.7 A compressed quadtree corresponding to the quadtree of Figure 44.6.

and the cell containing a subcell is called its *supercell*. *Immediate subcell* refers to those cells obtained after bisecting a cell along each dimension. A cell is the *immediate supercell* of any of its immediate subcells. Note that for any two cells, either one is completely contained in the other or they are disjoint. Cells are considered disjoint if they share a boundary and are adjacent.

To encapsulate the spatial information otherwise lost in the compression, two cells are stored in each node v of a compressed octree, *large cell of v* and *small cell of v*, denoted by $L(v)$ and $S(v)$, respectively. The large cell is defined as the largest cell that encloses all the points the node represents. Similarly, the small cell is the smallest cell that encloses all the points that the node represents. If a node is not a result of compression of a chain, then the large cell and the small cell of that node are the same; otherwise, they are different. Observe that the large cell of a node is an immediate subcell of the small cell of its parent. Since a leaf contains a single point, its small cell is defined to be the hypothetical cell with zero length containing the point. Or, when the maximum resolution is specified, the small cell of a point is defined to be the cell at the highest resolution containing the point.

Note that the side length of a cell decreases exponentially in an octree as one walks down a path from the root to a leaf. This reason, coupled with the fact that computers can only handle finite precision, implies that the height of an octree that can be represented and that is sufficient for the application at hand, is rather small. For example, a tree of height 20 is sufficient to capture a length scale of $2^{20}:1 \geq 10^6:1$. This prompted some researchers to treat the height of an octree as a constant for practical purposes, even though it is $\Omega(\log n)$ as is the case for any tree. Note that a tree of height 20 can have as many as 2^{60} uniformly distributed points, even at 1 point per leaf. Because of this practical rationale, chains in octrees pose no serious performance degradation in runtime or storage in practice, and the appeal of compressed octrees might appear largely theoretical. However, the properties of compressed octrees enable design of elegant algorithms that can be extended to octrees, if necessary. For example, a compressed octree can be constructed, and later expanded to create the octree, if desired. Furthermore, since the size of compressed octrees are distribution independent, they can be used to prove rigorous runtime bounds, which in turn can explain why octree-based methods perform well in practice.

44.5.1 Octrees and Space Filling Curves

Given the similarity in decomposition used by octrees and SFCs, it is hardly surprising that they can be related. Here, we establish this relation and show how it can be exploited to unify ideas on SFC-based parallel domain decomposition and parallel octrees, and derive good algorithms and tree implementation strategies.

Given a cell, the corresponding *cell space* is defined as the set of all cells obtained by dividing the root cell into cells of the same size. Octree nodes contain cells that are from cell spaces at different resolutions. SFCs order the cell space at a particular level of resolution. Thus, octrees can be viewed as multiple SFCs at various resolutions. For convenience, suppose we pick the Z-SFC. When drawing an octree, we can draw the children on a node in the order in which Z-SFC visits the subcells represented by the children. By doing so, we ensure that the order of octree nodes at the same level is the same as the SFC order of the corresponding subcells. The membership test is easy: A cell at a particular resolution is present in an octree if the cell is not empty (i.e., it has one or more points). The same concepts apply for a compressed octree except that a cell is present as a small cell at a node in the compressed octree only if the cell is not empty and none of its immediate subcells contains all the points in the cell. Although the Z-SFC is used here, it is possible to use any SFC order while applying the same concepts.

Apart from establishing a linearization of cells at a particular resolution, it is also beneficial to define a linearization that cuts across multiple levels. Note that given any two cells, they are either disjoint or one is contained in the other. This observation can be exploited to establish a total order on the cells of an octree [42]: Given two cells, if one is contained in the other, the subcell is taken to precede the supercell; if they are disjoint, they are ordered according to the order of the immediate subcells of the smallest supercell enclosing them. A nice property that follows from these rules is the resulting linearization of all cells in an octree (or compressed octree) is identical to its postorder traversal.

We use the term *cell* to designate the cell at a node in an octree, or the small cell at a node in a compressed octree. Note that large cells can be easily derived from the corresponding small cells at parent nodes. As each node in an octree is uniquely described by the corresponding cell, it can be represented by its index in SFC linearization, as shown in Figure 44.8. However, ambiguity may arise when distinguishing indices of cells at different levels of resolution. For example, it is not possible to distinguish between 00 (cell of length $D/2$ with coordinates (0,0)), and 0000 (cell of length $D/4$ with coordinates (00,00)), when both are stored in, say, standard 32-bit integer variables. A simple mechanism suggested to overcome this is to prepend the bit representation of an index with a "1" bit [49]. With this, the root cell is 1, the cells with Z-SFC indices 00 and 0000 are now 100 and 10,000, respectively.

The process of assigning indices to cells can also be viewed hierarchically. A cell at resolution i can be described using i-bit integer coordinates. The first $i-1$ of these bits are the same as the coordinates of its immediate supercell. Thus, the index of a cell can be obtained by taking the least significant bit of each of its coordinates, concatenating them into a d-bit string, and appending this to the index of its immediate supercell. Note that bit representations of cells is meaningful under the assumption that the resolution of an octree is small and fixed, which is valid in practice. The advantage of such bit representation is that it allows primitive operations on cells using fast bit operations:

- Check if a cell C_1 is contained in another cell C_2: If C_2 is a prefix of C_1, then C_1 is contained in C_2; otherwise, not.
- Find the smallest cell containing two cells C_1 and C_2: This is obtained by finding the longest common prefix of C_1 and C_2 that is a multiple of d. If node w is the lowest common ancestor of nodes u and v in a compressed octree, then $S(w)$ is the smallest cell containing $S(u)$ and $S(v)$. Thus, lowest common ancestors in compressed octrees can be computed using fast bit operations on cells.
- Find the immediate subcell of C_1 that contains a given cell C_2: If $3k+1$ is the number of bits representing C_1, the required immediate subcell is given by the first $3k+4$ bits of C_2. This operation is useful in computing the large cell $L(v)$ of a node v, as the immediate subcell of its parent u's small cell $S(u)$ that contains $S(v)$.

44.5.2 Constructing Parallel Compressed Octrees

Consider n points equally distributed across p processors, and let k denote the prespecified maximum resolution. For each point, generate the index of the leaf cell containing it, which is the cell at resolution k containing the point. In the next step, the leaf cells are sorted in parallel using any optimal parallel integer

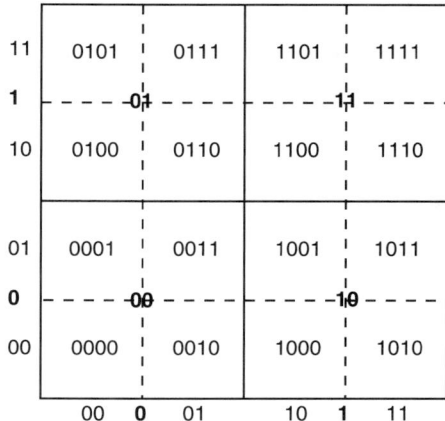

FIGURE 44.8 Bit interleaving scheme for a hierarchy of cells.

Parallel Octree Construction
1: For each point, compute the index of the leaf cell containing it.
2: Parallel sort the leaf indices to compute their SFC linearization.
3: Each processor obtains the leftmost leaf cell of the next processor.
4: On each processor, construct a local compressed octree for the leaf cells within it and the borrowed leaf cell.
5: Send the out of order nodes to appropriate processors.
6: Insert the received out of order nodes in the already existing sorted order of nodes.

Algorithm 3: Parallel compressed octree.

sorting algorithm [22, 23]. This creates the SFC-linearization of the leaf cells, or the left to right order of leaves in the compressed octree. If multiple points fall in a leaf cell, then duplication can be eliminated during parallel sort and the points falling within a leaf cell can be recorded. The subsequent tree building will take place on the leaf cells alone. For convenience and without loss of generality, we assume each point falls in a different leaf cell, giving rise to a tree with n leaves.

The procedure so far is identical to parallel domain decomposition using SFCs. Subsequently, parts of the compressed octree are generated on each processor, such that each node and edge of the tree is generated on some processor. To do this, each processor first borrows the leftmost leaf cell from the next processor and runs a sequential algorithm to construct the compressed octree for its leaf cells together with the borrowed leaf cell in $O(n/p)$ time [4] as follows: initially, the tree has a single node that represents the first leaf cell in SFC-order. The remaining leaf cells are inserted one at a time as per the SFC-order. During the insertion process, keep track of the most recently inserted leaf. If C is the next leaf cell to be inserted, then starting from the most recently inserted leaf, walk up the path toward the root until the first node v such that $C \subseteq L(v)$ is encountered. Two possibilities arise:

- *Case I:* If C is not contained in $S(v)$, then C is in the region $L(v) - S(v)$, which was empty previously. The smallest cell containing C and $S(v)$ is a subcell of $L(v)$ and contains C and $S(v)$ in different immediate subcells. Create a new node u between v and its parent and insert a new child of u with C as small cell.
- *Case II:* If C is contained in $S(v)$, v is not a leaf node. The compressed octree presently does not contain a node that corresponds to the immediate subcell of $S(v)$ that contains C, that is, this immediate subcell does not contain any of the points previously inserted. Therefore, it is enough to insert C as a child of v corresponding to this subcell.

The number of nodes visited per insertion is not bounded by a constant but the total number of nodes visited over all insertions is $O(n/p)$ [4]. The local tree is stored in its postorder traversal order in an array. For each node, indices of its parent and children in the array are stored. The following lemma shows that each node in the compressed octree is generated as part of the local tree on at least one processor [19].

Lemma 44.1 *Consider the compressed octree for the n leaf cells, which we term the global compressed octree. Each node in this tree can be found in the local tree of at least one processor.*

Proof. Every internal node is the lowest common ancestor of at least one consecutive pair of leaves. More specifically, let u be a node in the global compressed octree and let $v_1, v_2, \ldots, v_k$ be its children ($k \geq 2$). Consider any two consecutive children v_i and v_{i+1} ($1 \leq i \leq k - 1$). Then, u is the lowest common ancestor of the rightmost leaf in v_i's subtree and the leftmost leaf in v_{i+1}'s subtree. If we generate the lowest common ancestor of every consecutive pair of leaf nodes, we are guaranteed to generate every internal node in the compressed octree. This is the reason the first leaf in the next processor is borrowed before constructing each local tree. However, a node may be generated multiple times—each internal node is generated at least once but at most $2^d - 1$ times. ∎

We need to generate the postorder traversal order of the global compressed octree. To do so, nodes that should actually appear later in the postorder traversal of the global tree, should be sent to the appropriate processors. Such nodes, termed *out of order nodes*, appear consecutively after the borrowed leaf in the postorder traversal of the local tree and hence can be easily identified. This is owing to the linearization defined earlier on cells across multiple levels. The first leaf cell in each processor is gathered into an array of size p. Using a binary search in this array, the destination processor for each out of order node can be found. Nodes that should be routed to the same processor are collected together and sent using an all-to-all communication.

Lemma 44.2 *The number of out of order nodes in a processor is bounded by k, where k is the maximum resolution.*

Proof. There can be at most one node per level in the local tree that is out of order. Suppose this is not true. Consider any two out of order nodes in a level, say v_1 and v_2. Suppose v_2 occurs to the right of v_1 (v_2 comes after v_1 in cell order). Since v_1 and v_2 are generated, some leaves in their subtrees belong to the processor. Because the leaves allocated to a processor are consecutive, the rightmost leaf in the subtree of v_1 belongs to the processor. This means v_1 is not an out of order node, a contradiction. ∎

Lemma 44.3 *The total number of nodes received by a processor is at most $k(2^d - 1)$.*

Proof. We first show that a processor cannot receive more than one distinct node per level. Suppose this is not true. Let v_1 and v_2 be two distinct nodes received at the same level. Without loss of generality, let v_2 be to the right of v_1. A node is received by a processor only if it contains the rightmost leaf in the subtree of the node. Therefore, all the leaf nodes between the rightmost leaf in the subtree of v_1 and the rightmost leaf in the subtree of v_2 must be contained in the same processor. In that case, the entire subtree under v_2 is generated on the processor itself and the processor could not have received v_2 from another processor, a contradiction. Hence, a processor can receive at most k distinct nodes from other processors. As there can be at most $2^d - 1$ duplicate copies of each node, a processor may receive no more than $k(2^d - 1)$ nodes. ∎

The received nodes are merged with the local postorder traversal array, and their positions are communicated back to the sending processors. The net result is the postorder traversal of the global octree distributed across processors. Each node contains the position of its parent and each of its children in this array. The storage required per processor is $O\left(\frac{n}{p} + k\right)$.

44.5.3 Data Ownership

Because of the total order established across all cells (nodes) in the tree, data ownership can be computed in much the same way as in the case of a SFC-based domain decomposition. The leftmost (or rightmost) boundary cells on each processor are stored in an array of length p in each processor. A binary search in this array can identify the processor that owns a given cell in $O(\log p)$ time.

Data ownership can also be computed by simply using the linearization of all the leaf cells. For any cell C of a bigger size, consider a decomposition of C into cells at the leaf level. Consider the SFC linearization of all of these cells and let C' denote the last cell in that order. Then, C is owned by the same processor containing C'. Computing the index of C' is easy. If C is at resolution l, then the index of C' is obtained by appending to the index of C a bit string of $d(k - l)$ consecutive 1s. Thus, data ownership on compressed octrees can be computed using the same information as stored for SFCs.

44.5.4 Quality of Partitioning

The parallel octree decomposition described here retains all the features of the SFC-based parallel domain decomposition described in Section 44.4.3. At the leaf level, the decomposition is identical to SFC-based parallel domain decomposition. There are two ways to view the distributed representation of the octree: One can view it as a parallel domain decomposition for cells at every level represented by the octree. This view is beneficial to applications such as AMR and multigrid, where numerical computations at multiple levels are expected. Another view is to look at the whole structure as a load balanced distributed representation of the octree data structure. In several applications such as the N-body problem, molecular dynamics, and so forth, octrees are used for solving the applications. In such cases, one can parallelize the application at hand using the parallel octree data structure. As for load balancing, each processor has a local tree of size $O\left(\frac{n}{p} + k\right) \approx O\left(\frac{n}{p}\right)$.

Note that similar to parallel octrees, parallel multidimensional binary search trees or k-d trees [2] are also used for parallel domain decomposition and application parallelization. The reason they are not described in detail here is because parallel ORB decomposition, described in Section 44.3, already shows how to generate the top $\log p$ levels of the k-d tree. All that remains is to construct one subtree per processor locally using a serial k-d tree construction algorithm. The contrast is that while ORB decomposition computes the top $\log p$ levels of the k-d tree, SFC decomposition computes the leaf order of octree nodes.

44.6 Application to the N-Body Problem

In this section, we focus on a well-studied application to illustrate how parallel domain decomposition techniques are used in scientific computing. The N-body problem is to simulate the evolution of n particles under the influence of mutual pairwise interactions, such as gravitational or electrostatic forces. Since a closed-form solution is not possible, iterative methods are used to solve the N-body problem. At each discrete time interval, the total force acting on each particle is computed and this information is used to update its position and velocity. A straightforward computation by considering all pairwise interactions requires $\Theta(n^2)$ work per iteration. The rapid growth with n effectively limits the number of particles that can be simulated by this method.

We confine our discussion to two classes of methods that have been developed to reduce the work per iteration: methods that compute particle–cell interactions, and those that compute cell–cell interactions. The general idea behind both the schemes is that the force or potential due to a cluster of particles can be approximated at sufficiently faraway distances by treating the entire cluster as a single mass located at the center of mass of the cluster. The former methods use this idea to compute the potential by a cluster on a particle. The latter methods further reduce computation by approximating the potential due to a cluster on another cluster. To carry out such approximations at various levels, both methods use octrees to hierarchically decompose the domain containing the particles.

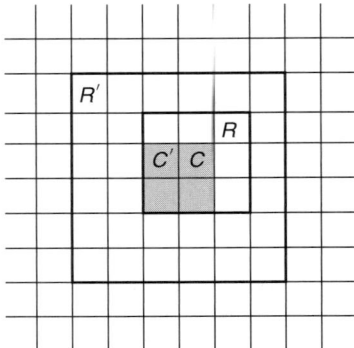

FIGURE 44.9 Illustration of partial local expansion calculation. Cells inside R' but outside R are required for C's partial local expansion.

The particle–cell interaction method, popularized by Barnes and Hut [6], works by carrying out the following procedure for each particle: the octree is traversed starting from the root cell. Consider a cell C encountered at some point during the traversal. If C is sufficiently far away from the particle as measured by the ratio of its side length to its distance to the particle, the potential due to the cell on the particle is directly computed. Otherwise, each of C's child cells are evaluated recursively in the same manner. For a particle, one can mark the nodes in the tree that are encountered in this traversal. This will include all the nodes from the root to the leaf containing the particle and cells that are spatial neighbors to the cells on this path. The portion of the octree induced by these nodes is called the *locally essential tree* for the particle.

ORB-based parallel domain decomposition forms the basis of one of the early large scale n-body simulation reported [40]. In this work, ORB is used to partition the n particles to p processors in a load balanced manner. Once this is accomplished, each processor is responsible for computing the potential on the set of particles assigned to it. To do this, a combined locally essential tree that consists of all nodes in the octree needed by any particle assigned to the processor through ORB decomposition is created. As there is no global octree to draw from and only an ORB decomposition is computed, all the particles that lie in the geometric region needed to build the combined locally essential tree are brought to the processor, followed by which the locally essential tree is built serially. The locally essential trees are built simultaneously in parallel on all processors. Note that this requires an undetermined number of nodes in the octree to be replicated to processors. In addition, replication is high at the top of the tree but reduces at lower levels. Because of the locality preserved by ORB, it is expected that many of the interior points have all the required points locally.

Note that the work required per particle need not be the same. Therefore, the actual computation per particle in an iteration is measured (such as number of nodes in its locally essential tree), and is assumed to reflect the computation load in the next iteration. Assigning these estimates as weights to points, ORB performance can be improved in the next iteration. Also, as the particles move during each time step, the ORB is either recomputed from scratch or incrementally adjusted. A particle–cell interaction-based N-body code with the same ideas, but using SFC decomposition instead of ORB, is reported in Reference [49].

One complication in the above described approaches is that while ORB or SFC is used for parallel domain decomposition purposes, the application itself relies on octrees. Apart from the complexities in reconciling the two different schemes both in methodology and software complexity, there is no way to prove one method minimizes communication to bring nonlocal data essential to build the other. These problems can potentially be overcome by directly using the same scheme for both needs, such as the parallel octree. We describe such a method in the context of methods that directly compute cell–cell interactions.

The fast multipole method (FMM) is a robust mathematical method with guaranteed error bounds [16, 17] that compute cell–cell interactions directly. In FMM, each cell in the octree is associated with a multipole expansion and a local expansion. The multipole expansion at a cell C is a truncated infinite series that describes the effect of the particles within the cell C at a distant point. It converges at points that are sufficiently far off from cell C. The local expansion at a cell C is a truncated infinite series that describes the effect of all distant particles on the points within the cell C.

The multipole expansions are computed by a bottom-up traversal of the octree. At the leaf cells, they are directly computed. For an internal cell, its multipole expansion is computed by aggregating the multipole expansions of its immediate subcells. The total time for computing the multipole expansions is proportional to the number of nodes in the octree.

The local expansion at a cell C is obtained by appropriately combining the multipole expansions of the cells that converge at every point in cell C. In Figure 44.9, the local expansion at cell C should include the effect of all particles outside region R. However, the effect of all particles outside region R' are included in the local expansion of C's parent C'. When considering cell C, we only compute the effect of particles that are outside region R but within region R'. We call this the *partial local expansion* at cell C. This is added to the local expansion at cell C' to compute the local expansion at cell C. Once the partial local expansions are calculated for each node in the tree, the local expansions can be calculated in a top-down traversal, in a reversal of the procedure used for computing multipole expansions. Finally, the local expansions are evaluated at each leaf. For further details on FMM, the reader is referred to the original sources [16, 17]. Some careful analysis by adapting FMM to compressed octrees results in $O(n)$ runtime per iteration [4].

The entire FMM computation, including parallel domain decomposition, can be effectively achieved by using parallel compressed octrees [20, 42]. To begin with, the parallel compressed octree of the particles is computed as described in Section 44.4. The tree is stored in an array in its postorder traversal order, split across processors such that each processor has n/p leaves. The next step of computing multipole expansions is achieved as follows: each processor scans its local array from left to right. When a leaf node is reached, its multipole expansion is directly computed from the particles within the leaf cell. If the node's multipole expansion is known, it is shifted to its parent and added to the parent's multipole expansion, provided the parent is local to the processor. As the tree is stored in postorder traversal order, if all the children of a node are present in the same processor, it is encountered only after all its children are. This ensures that the multipole expansion at a cell is known when the scan reaches it. This computation takes $O(n/p + k)$ time, where k is the highest resolution. During the scan, some nodes are labeled *residual nodes* on the basis of the following rules:

- If the multipole expansion owing to a cell is known but its parent lies in a different processor, it is labeled a *residual leaf node*.
- If the multipole expansion at a node is not yet computed when it is visited, it is labeled a *residual internal node*.

Each processor copies its residual nodes into an array. It is easy to see that the residual nodes form a tree (termed the *residual tree*) and the tree is present in its postorder traversal order, distributed across processors.

Lemma 44.4 *The number of residual internal nodes in a processor is at most $k - 1$.*

Proof. We show that there can be at most one residual internal node per level in a processor. Suppose this is not true. Consider any two residual internal nodes in a level, say v_1 and v_2. Suppose v_2 occurs to the right of v_1 (v_2 comes after v_1 in cell order). Because of postorder property, the right most leaf in v_1's subtree and the rightmost leaf in v_2's subtree should belong to this processor. Because the leaves allocated to a processor are consecutive, all the leaves in the subtree of v_2 belong to the processor. This means v_2 is not a residual internal node, a contradiction. Therefore, there can be only one residual internal node at each level in a processor. Furthermore, no leaf can be a residual internal node. Hence, the claim. ∎

Corollary 44.1 *The size of the residual tree is at most $8pk$.*

Proof. From the previous lemma, the total number of residual internal nodes is at most $p(k-1)$. It follows that the maximum number of residual leaf nodes can at most be $7p(k-1)+1$, for a total tree size of at most $8p(k-1)+1 < 8pk$. ∎

Multipole expansion calculation has the associative property. Because of this, multipole expansions on the residual tree can be computed using an efficient parallel upward tree accumulation algorithm [43]. The main advantage of using the residual tree is that its size is independent of the number of particles, and is rather small. Local expansions can similarly be computed in a reverse of the algorithm used to compute multipole expansions in parallel. The intermediate step of computing partial local expansion calculation at each node is carried out by identifying the geometric region containing the relevant cells, and using the SFC or octree data ownership computations described earlier to locate the processors containing them. This is carried out simultaneously for all nodes, such that one all-to-all communication can be used to receive relevant nonlocal data. Once again, because of the underlying SFC-based decomposition, many of the cells required for partial local expansion calculation are expected to be available locally.

The chief advantage of the parallel octree-based approach are that (i) the same scheme is used for domain decomposition and application solution, and (ii) as is evident from the above discussion, it leads to provably good and communication efficient algorithms. As demonstrated here, parallel domain decomposition is just a first but important step in parallelizing scientific computing applications. However, it may well determine how effectively the application can be solved in parallel. To gain further knowledge of parallel domain decomposition methods, it is important to study the various applications in parallel scientific computing and how domain decomposition is used in the context of solving them. The references listed at the end of this chapter are a good starting point.

Acknowledgment

This work is supported in part by the National Science Foundation under CCF-0306512 and CCF-0431140.

References

[1] Abel, D. J. and Mark, D. M. A comparative analysis of some two-dimensional orderings. *International Journal of Geographical Information Systems* 4, 1 (1990), 21–31.

[2] Al-furaih, I., Aluru, S., Goil, S., and Ranka, S. Parallel construction of multidimensional binary search trees. *IEEE Transactions of Parallel and Distributed Systems* 11, 2 (2000).

[3] Al-furiah, I., Aluru, S., Goil, S., and Ranka, S. Practical algorithm for selection on coarse grained parallel computers. *IEEE Transactions on Parallel and Distributed Systems* 8, 8 (1997), 813–824.

[4] Aluru, S. and Sevilgen, F. Dynamic compressed hyperoctrees with applications to n-body problem. *Proc. of Foundations of Software Technology and Theoretical Computer Science* (1999), pp. 21–33.

[5] Bank, R. E. and Jimack, P. K. A new parallel domain decomposition method for the adaptive finite element solution of elliptic partial differential equations. *Concurrency and Computation: Practice and Experience* 13, 5 (2001), 327–350.

[6] Barnes, J. and Hut, P. A hierarchical $O(n \log n)$ force-calculation algorithm. *Nature* 324, 4 (1986), 446–449.

[7] Bentley, J. Multidimensional binary search trees used for associative searching. *Communications of the ACM* 18, 9 (1975), 509–517.

[8] Berger, M. J. and Bokhari, S. H. A partitioning strategy for nonuniform problems on multiprocessors. *IEEE Transaction Computers* 36, 5 (1987), 570–580.

[9] Çatalyürek, Ü. V. and Aykanat, C. A fine-grain hypergraph model for 2D decomposition of sparse matrices. *Intl. Parallel and Distributed Processing Symposium* (2001), p. 118.

[10] Çatalyürek, Ü. V. and Aykanat, C. A hypergraph-partitioning approach for coarse-grain decomposition. *Proc. Supercomputing* (2001), p. 28.

[11] Dennis, J. M. Partitioning with space-filling curves on the cubed-sphere. *Intl. Parallel and Distributed Processing Symposium* (2003).

[12] Eisenhauer, G. and Schwan, K. Design and analysis of a parallel molecular dynamics application. *Journal of Parallel and Distributed Computing* 35, 1 (1996), 76–90.

[13] Floyd, R. W. and Rivest, R. L. Expected time bounds for selection. *Communications of the ACM* 18 (1975).

[14] Garey, M., Johnson, D., and Stockmeyer, L. Some simplified NP-complete graph problems. *Theoretical Computer Science* 1 (1976), 237–267.

[15] Gotsman, C. and Lindenbaum, M. On the metric properties of discrete space filling curves. *IEEE Transactions on Image Processing* 5, 5 (1996), 794–797.

[16] Greengard, L. and Rokhlin, V. A fast algorithm for particle simulations. *Journal of Computational Physics* 73 (1987), 325–348.

[17] Greengard, L. F. *The Rapid Evaluation of Potential Fields in Particle Systems*. MIT Press, 1988.

[18] Griebel, M. and Zumbusch, G. Hash based adaptive parallel multilevel methods with space filling curves. *Proc. Neumann Institute for Computing Symposium* (2002), pp. 479–492.

[19] Hariharan, B. and Aluru, S. Efficient parallel algorithms and software for compressed octrees with applications to hierarchical methods. *Parallel Computing* 31 (2005), 311–331.

[20] Hariharan, B., Aluru, S., and Shanker, B. A scalable parallel fast multipole method for analysis of scattering from perfect electrically conducting surfaces. *Proc. Supercomputing* (2002), p. 42.

[21] Hayashi, R. and Horiguchi, S. Efficiency of dynamic load balancing based on permanent cells for parallel molecular dynamics simulation. *Int'l Parallel and Distributed Processing Symposium* (2000), p. 85.

[22] Helman, D. R., Bader, D. A., and JaJa, J. Parallel algorithms for personalized communication and sorting with an experimental study. *ACM Symposium on Parallel Algorithms and Architectures* (1996), pp. 211–222.

[23] Helman, D. R., JáJá, J., and Bader, D. A. A new deterministic parallel sorting algorithm with an experimental evaluation. *ACM Journal of Experimental Algorithms* 3 (1998), 4.

[24] Hendrickson, B. Graph partitioning and parallel solvers: Has the emperor no clothes? *Solving Irregularly Structured Problems in Parallel*, Irregular'98, LNCS 1457, Springer-Verlag (1998), 218.

[25] Hendrickson, B. and Kolda, T. G. Graph partitioning models for parallel computing. *Parallel Computing* 26, 12 (2000), 1519–1534.

[26] Hendrickson, B. and Leland, R. The Chaco user's guide, Version 2.0. Tech. Rep. SAND95-2344, Sandia National Laboratory, Albuquerque, NM, 1995.

[27] Hilbert, D. Uber die stegie abbildung einer linie auf flachenstuck, *Mathematische Annalen*, 38(1891), 459–460.

[28] Jagadish, H. V. Linear clustering of objects with multiple attributes. *Proc. ACM SIGMOD* (1990), pp. 332–342.

[29] Jones, T. M. and Plassmann, P. E. Parallel algorithms for the adaptive refinement and partitioning of unstructured meshes. *Proc. of Scalable High-Performance Computing Conference* (1994), pp. 478–485.

[30] Karypis, G. and Kumar, V. Parallel multilevel k-way partitioning scheme for irregular graphs. *Proc. Supercomputing* (1996), p. 35.

[31] Moon, B., Jagadish, H. V., Faloutsos, C., and Saltz, J. H. Analysis of the clustering properties of hilbert space-filling curve. *IEEE Transactions on Knowledge and Data Engineering* 13, 1 (2001), 124–141.

[32] Morton, G. A computer oriented geodetic data base and a new technique in file sequencing. Tech. rep., IBM, Ottawa, Canada, 1966.

[33] Nicol, D. M. Rectilinear partitioning of irregular data parallel computations. *Journal of Parallel and Distributed Computing 23*, 2 (1994), 119–134.
[34] Nyland, L., Prins, J., Yun, R. H., Hermans, J., Kum, H., and Wang, L. Achieving scalable parallel molecular dynamics using dynamic spatial domain decomposition techniques. *Journal of Parallel and Distributed Computing 47*, 2 (1997), 125–138.
[35] Parashar, M. and Browne, J. C. On partitioning dynamic adaptive grid hierarchies. *Hawaii Intl. Conf. on System Sciences* (1996), pp. 604–613.
[36] Pilkington, J. and Baden, S. Dynamic partitioning of non-uniform structured workloads with spacefilling curves. *IEEE Transaction on Parallel and Distributed Systems 7*, 3 (1996), 288–300.
[37] Plimpton, S. Fast parallel algorithms for short-range molecular dynamics. *Journal of Computational Physics 117*, 1 (1995), 1–19.
[38] Rajasekaran, S., Chen, W., and Yooseph, S. Unifying themes for parallel selection. *Proc. 4th International Symposium on Algorithms and Computation*, Springer-Verlag LCNS 834 (1994), pp. 92–100.
[39] Sagan, H. *Space Filling Curves*. Springer-Verlag, 1994.
[40] Salmon, J. K. Parallel hierarchical N-body methods. PhD thesis, California Institute of Technology, 1990.
[41] Seal, S. and Aluru, S. Communication-aware parallel domain decomposition using space filling curves. *Proc. Parallel and Distributed Computing Symposium (ISCA)* (2006), pp. 159–164.
[42] Sevilgen, F., Aluru, S., and Futamura, N. A provably optimal, distribution-independent parallel fast multipole method. *Intl. Parallel and Distributed Processing Symposium* (2000), pp. 77–84.
[43] Sevilgen, F., Aluru, S., and Futamura, N. Parallel algorithms for tree accumulations. *Journal of Parallel and Distributed Computing* (2005), 85–93.
[44] Singh, J. P., Holt, C., Hennessy, J. L., and Gupta, A. A parallel adaptive fast multipole method. *Proc. Supercomputing* (1993), pp. 54–65.
[45] Steensland, J., Chandra, S., and Parashar, M. An application-centric characterization of domain-based SFC partitioners for parallel SAMR. *IEEE Transactions on Parallel and Distributed Systems 13*, 12 (2002), 1275–1289.
[46] Teng, S.-H. Provably good partitioning and load balancing algorithms for parallel adaptive n-body simulation. *SIAM Journal of Scientific Computing 19*, 2 (1998), 635–656.
[47] Tirthapura, S., Seal, S., and Aluru, S. A formal analysis of space filling curves for parallel domain decomposition. *Proc. of Intl. Conf. of Parallel Processing* (2006), pp. 505–512.
[48] Walker, D. W. The hierarchical spatial decomposition of three-dimensional p-in-cell plasma simulations on mind distributed memoru multiprocessors. Tech. Rep. ORNL/TM-12071, Oak Ridge National Laboratory, July 1992.
[49] Warren, M. S. and Salmon, J. K. A parallel hashed oct-tree n-body algorithm. *Proc. Supercomputing* (1993), pp. 12–21.
[50] Zhuang, Y. and Sun, X. A highly parallel algorithm for the numerical simulation of unsteady diffusion processes. *Int'l Parallel and Distributed Processing Symposium* (2005), p. 16a.

45
Game Theoretical Solutions for Data Replication in Distributed Computing Systems

Samee Ullah Khan
Colorado State University

Ishfaq Ahmad
University of Texas

45.1	Introduction..	45-1
45.2	Related Work...	45-2
45.3	Problem Formulation	45-4
45.4	Some Essential Background Material..................	45-6
45.5	Casting DRP into an Incentive Compatible Game Theoretical Auction.......................................	45-7
	The Ingredients • The Casting • Further Discussion on the Casting Process	
45.6	Experimental Setup and Discussion of Results	45-11
	Network Topologies • Access Patterns • Further Clarifications on the Experimental Setup • The Determination of the Relocation Period • Comparative Techniques • Comparative Analysis	
45.7	Concluding Remarks and Future Directions...........	45-20
References		45-22

45.1 Introduction

A major objective of distributed computing systems is to make data available to the geographically dispersed users in a fast, seamless fashion, while keeping the contents up-to-date [9]. Caching and replication are the two popular techniques to cater to these needs [49]. Both of these techniques play complementary roles, that is, caching attempts to store the most commonly accessed data objects as close to the users as possible, while replication distributes contents across the system. Their comparison, however, reveals some interesting analogies, for instance, cache replacement algorithms are examples of online, distributed, locally greedy algorithms for data allocation in replicated systems. Furthermore, caches do not have full server capabilities and thus can be viewed as a replicated system that sends requests for specific object types (e.g., dynamic pages) to a single server. Essentially, every major aspect of a caching scheme has its equivalent in replicated systems, but not vice versa [48]. Replication also leads to load balancing and increases client-server proximity [10].

One can find numerous techniques that propose solutions toward the data replication problem (DRP) in distributed computing systems. These range from the traditional linear programming techniques to bio-inspired genetic and meta-heuristic [49]. All of the techniques contribute in their own way with some overlap in terms of system architecture, performance evaluation, workload, and so forth. There are two popular models [65] to tackle the DRP: (i) centralized replication model and (ii) distributed replication model. In the first model, a central body is used to make the decisions about when, where, and what to replicate. In the second model, geographically distributed entities (servers, program modules, etc.) are used to make the decisions. Both the techniques have several pros and cons, for instance, the centralized model is a potential point of failure and overloaded by all the computations involved in resolving to a decision. On the other hand, the distributed model suffers from the possibility of mediocre optimization due to the localized view of the distributed entities [33]. A natural way to counteract both the extremities is to view the decision-making process as a "semi-distributed" [1] procedure, where all the data-intensive computing is performed at the geographically distributed entities, while the final decision on replication is taken by a single entity. This would lessen the burden on the decision-making entity, make it more fault tolerant since, in case of a failure, it can easily be replaced. It would also improve the overall solution quality since the distributed entities would leverage on the central body's ability to provide a global snapshot of the system [34].

Game theory has the natural ability to absorb a distributed optimization scenario into its realm [63]. Within the context of the DRP, the geographically distributed entities would be termed as the players, and the central decision-making body as the referee, where the players compete to replicate data objects onto their servers so that the users accessing their servers would experience reduced access time. A closer look at this process (competing for data objects and refereeing) reveals a close resemblance between the DRP and auctions. When an object is brought for auction, the bidders in a distributed fashion propose a bid for that object, without knowing what the other bidders are bidding, and the object is allocated to the bidder only when the auctioneer approves it. Of course, there is more detail to this process, which relies explicitly on the environment, situation, players involved, objects that are up for auction, purpose of the auction, and so forth. The theory that deals with these details is called auction theory, which is a special branch of game theory [62].

Using game theoretical techniques, we can tailor-make an auction procedure for a given scenario (problem at hand) and guarantee certain performance criteria, for instance, we can make sure that the players always project the correct worth of an object. This is a difficult problem when the players have to rely on local data, but with the help of game theory, this can be achieved without an extra overhead [36,62]. To put things into perspective, we will describe how game theory can be used to create techniques for the DRP in distributed computing systems.

The rest of the text is organized as follows. Section 45.2 summarizes the related work on the DRP. Section 45.3 formulates the DRP. Section 45.4 encapsulates the background material on game theory, and identifies specific structural properties concerning auction theory. Section 45.5 concentrates on modeling the game theory-based techniques for the DRP. The experimental results and concluding remarks are provided in Sections 45.6 and 45.7, respectively.

45.2 Related Work

The DRP is an extension of the classical file allocation problem (FAP). Chu [11] studied the FAP with respect to multiple files in a multiprocessor system. Casey [9] extended this work by distinguishing between updates and read file requests. Eswaran [18] proved that Casey's formulation was NP-complete. Mahmoud and Riordon [51] provide an iterative approach that achieves good solution quality when solving the FAP for infinite server capacities. A complete, although old, survey on the FAP can be found in [17]. Apers [2] considered the data allocation problem (DAP) in distributed databases where the query execution

Game Theoretical Solutions for Data Replication in Distributed Computing Systems

TABLE 45.1 Summary of Related Work

Category	Number of objects	Storage constraints	References
Category 1 No object access	Single object	No storage constraint	[24]
		Storage constraint	—
		No storage constraint	[22], [25]
	Multiple objects	Storage constraint	—
Category 2 Read accesses only	Single object	No storage constraint	[17], [22], [25], [30], [39], [42]
		Storage constraint	[8], [12], [29], [39], [44]
		No storage constraint	[7], [28], [39], [46], [52], [70]
	Multiple objects	Storage constraint	[12], [57]
Category 3 Read and write accesses	Single object	No storage constraint	[2], [11], [15], [52], [74], [75]
		Storage constraint	[10], [27], [40], [50], [52], [69]
		No storage constraint	[23], [42], [65], [66]
	Multiple objects	Storage constraint	[5], [6], [51], [55]

strategy influences allocation decisions. In [45], the authors proposed several algorithms to solve the DAP in distributed multimedia databases (without replication), also called as video allocation problem (VAP). Replication algorithms fall into the following three categories:

1. The problem definition does not cater for the user accesses.
2. The problem definition only accounts for read access.
3. The problem definition considers both read and write access including consistency requirements.

These categories are further classified into four categories according to whether a problem definition takes into account single or multiple objects, and whether it considers storage costs. Table 45.1 shows the categorized outline of the previous work reported.

The main drawback of the problem definition in category 1 is that they place the replicas of every object in the same node. Clearly, this is not practical when many objects are placed in the system. However, they are useful as a substitute of the problem definition of category 2, if the objects are accessed uniformly by all the clients in the system and utilization of all nodes in the system is not a requirement. In this case, category 1 algorithms can be orders of magnitude faster than the ones for category 2, because the placement is decided once and it applies to all objects.

Most of the research papers tackle the problem definition of category 2. They are applicable to read-only and read-mostly workloads. In particular, this category fits well in the context of content distribution networks (CDNs). Problem definitions [22,29,52,69] have all been used in CDNs. The two main differences between them are whether they consider single or multiple objects, and whether they consider storage costs or not. The cost function in [51] also captures the impact of allocating large objects and could possibly be used when the object size is highly variable. In [17] the authors tackled a similar problem—the proxy cache placement problem. The performance metric used there was the distance parameter, which consisted of the distance between the client and the cache, plus the distance between the client and the node for all cache misses. It is to be noted that in CDN, the distance is measured between the cache and the closest node that has a copy of the object.

The storage constraint is important since it can be used in order to minimize the amount of changes to the previous replica placements. As far as we know only the works reported in [31] and [48] have evaluated the benefits of taking storage costs into consideration. Although there are research papers that consider storage constraints in their problem definition, yet they never evaluate this constraint (e.g., see [18], [25], [47], and [64]).

Considering the impact of writes, in addition to that of reads, is important, if content providers and applications are able to modify documents. This is the main characteristic of category 3. Some research papers in this category also incorporate consistency protocols—in many different ways. For most of them, the cost is the number of writes times the distance between the client and the closest node that has the object, in addition to the cost of distributing these updates to the other replicas of the object. In [24], [28], [30], and [47], the updates are distributed in the system using a minimum spanning tree. In [25] and [64], one update message is sent from the writer to each copy, while in [31] and [48], a generalized update mechanism is employed. There a broadcast model is proposed in which any user can update a copy. Next, a message is sent to the primary (original) copy holder server that broadcasts it to the rest of the replicas. This approach is shown to have lower complexity than any of the aforementioned techniques. In [42] and [66], it is not specified how updates are propagated. The other main difference among the above definitions is that [28,30,31,47,48,70] minimize the maximum link congestion, while the rest minimize the average client access latency or other client perceived costs. Minimizing the link congestion would be useful, if bandwidth is scare.

Some ongoing work is related to dynamic replication of objects in distributed systems when the read–write patterns are not known a priori. Awerbuch et al.'s work [5] is significant from a theoretical point of view, but the adopted strategy for commuting updates (object replicas are first deleted), can prove difficult to implement in a real-life environment. Wolfson et al. [75] proposed an algorithm that leads to optimal single file replication in the case of a tree network. The performance of the scheme for general network topologies is not clear though. Dynamic replication protocols were also considered under the Internet environment. Heddaya et al. [23] proposed protocols that load balance the workload among replicas. Rabinovich et al. [65] proposed a protocol for dynamically replicating the contents of an Internet service provider in order to improve client–server proximity without overloading any of the servers. However, updates were not considered.

45.3 Problem Formulation

Consider a distributed computing system comprising M servers, with each server having its own processing power, memory (primary storage), and media (secondary storage). Let S^i and s^i be the name and the total storage capacity (in simple data units, for example, blocks), respectively, of server i, where $1 \leq i \leq M$. The M servers of the system are connected by a communication network. A link between two servers S^i and s^j (if it exists) has a positive integer $c(i,j)$ associated with it, giving the communication cost for transferring a data unit between servers S^i and s^j. If the two servers are not directly connected by a communication link then the aforementioned cost is given by the sum of the costs of all the links in a chosen path from server S^i to the server s^j. Without the loss of generality, we assume that $c(i,j) = c(j,i)$. This is a common assumption (e.g., see [24], [31], [48], [64], etc.). Let there be N objects, each identifiable by a unique name O_k and size in simple data unites O_k, where $1 \leq k \leq N$. Let r_k^i and w_k^i be the total number of reads and writes, respectively, initiated from S^i for O_k during a certain time period. Our replication policy assumes the existence of one primary copy for each object in the network. Let P_k be the server that holds the primary copy of O_k, that is, the only copy in the network that cannot be deallocated, hence referred to as primary server of the k-th object. Each primary server P_k, contains information about the whole replication scheme R_k of O_k. This can be done by maintaining a list of the servers where the k-th object is replicated at, called from now on the *replicators* of O_k. Moreover, every server S^i stores a two-field record for each object. The first field is its primary server P_k and the second is the nearest neighborhood server NN_k^i of server S^i, which holds a replica of object k. In other words, NN_k^i is the server for which the reads from S^i for O_k, if served there, would incur the minimum possible communication cost. It is possible that $NN_k^i = S^i$, if S^i is a *replicator* or the primary server of O_k. Another possibility is that $NN_k^i = P_k$, if the primary server is the closest one holding a replica of O_k. When a server S^i reads an object, it does so by addressing the request to the corresponding NN_k^i. For the updates, we assume that every server can update every object. Updates of an object O_k are performed by sending the updated version to its primary server P_k, which afterward broadcasts it to every server in its replication scheme R_k.

For the DRP under consideration, we are interested in minimizing the total network transfer cost due to object movement [i.e., the object transfer cost (OTC)]. The communication cost of the control messages has minor impact on the overall performance of the system, therefore, we do not consider it in the transfer cost model, but it is to be noted that incorporation of such a cost would be a trivial exercise. There are two components affecting OTC. The first component of OTC is due to the read requests. Let R_k^i denote the total OTC, due to S^is' reading requests for object O_k, addressed to the nearest server NN_k^i. This cost is given by the following equation:

$$R_k^i = r_k^i O_k c(i, NN_k^i) \tag{45.1}$$

where $NN_k^i = \{Server\ j | j \in R_k \wedge \min c(I, j)\}$. The second component of OTC is the cost arising due to the writes. Let W_k^i be the total OTC, due to S^is' writing requests for object O_k, addressed to the primary server P_k. This cost is given by the following equation:

$$W_k^i = w_k^i O_k (c(i, P_k) + \sum_{\forall (j \in R_k), j \neq i} c(P_k j)). \tag{45.2}$$

Here, we made the indirect assumption that in order to perform a write we need to ship the whole updated version of the object. This of course is not always the case, as we can move only the updated parts of it (modeling such policies can also be done using our framework). The cumulative OTC, denoted as C_{overall}, due to reads and writes is given by:

$$C_{\text{overall}} = \sum_{i=1}^{M} \sum_{k=1}^{N} (R_k^i + W_k^i). \tag{45.3}$$

Let $X_{ik} = 1$ if S^i holds a replica of object O_k, and 0 otherwise. X_{ik}s define an $M \times N$ replication matrix, named X, with boolean elements. Equation 45.3 is now refined to:

$$X = \sum_{i=1}^{M} \sum_{k=1}^{N} (1 - X_{ik}) \left[r_k^i O_k \min\{c(i,j) | X_{jk} = 1\} + w_k^i O_k c(i, P_k) \right] + X_{ik} \left(\sum_{x=1}^{M} w_k^x \right) O_k c(i, P_k). \tag{45.4}$$

Servers that are not the *replicators* of object O_k create OTC equal to the communication cost of their reads from the nearest *replicator*, in addition to that of sending their writes to the primary server of O_k. Servers belonging to the replication scheme of O_k, are associated with the cost of sending/receiving all the updated versions of it. Using the aforementioned formulation, the DRP can be defined as:

Find the assignment of 0, 1 values in the X matrix that minimizes C_{overall}, subject to the storage capacity constraint:

$$\sum_{k=1}^{N} X_{ik} O_k \leq s^i \forall (1 \leq i \leq M), \text{ and subject to the primary copies policy:}$$

$$X_{P_k^k} = 1 \forall (1 \leq k \leq N).$$

The minimization of C_{overall} has the following two impacts on the distributed computing system under consideration. First, it ensures that the object replication is done in such a way that it minimizes the maximum distance between the replicas and their respective primary objects. Second, it ensures that the maximum distance between an object k and the user(s) accessing that object is also minimized. Thus, the solution aims for reducing the overall OTC of the system. In the generalized case, the DRP is NP complete [48].

45.4 Some Essential Background Material

Game theory is widely thought to have originated in the early twentieth century, when von Neumann gave a concrete proof of the min–max theorem [72]. Although, it was the first formally stated major work in this field, the roots of game theory can be traced back to the ancient Babylonian Talmud. The Talmud is a compilation of the ancient laws set forth in the first five centuries AD. Its traces can be found in various religions and the modern civil and criminal laws. One related problem discussed in the Talmud is the marriage contract problem: A man has three wives. Their marriage contracts specify that in the case of the husband's death, the wives receive 1:2:3 of his property. The Talmud in all mystery gives self-contradictory recommendations. It states that: If the man dies leaving an estate of only 100, there should be an equal division. If the estate is worth 300, it recommends proportional division (50,100,150), while for an estate of 200, it recommends (50,75,75). In 1985, Aumann and Maschler [4] reported that the marriage contract problem and its weird solution discussed in the Talmud are only justifiable via cooperative game theoretical analysis. The foundation of the famous min–max theorem is credited to Waldegrave, who on November 13, 1713 wrote a letter to de Montmort describing a card game *le Her* and his solution [43]. It would take two centuries for Waldegrave's result to be formally acknowledged [72].

Some of the most pioneering results were reported within a year, when Nobel Laureate John Nash made seminal contributions to both cooperative and noncooperative games. Nash [58,59] proved the existence of a strategic equilibrium for noncooperative games (Nash Equilibrium). He also proposed that cooperative games were reducible to noncooperative games. In the next two papers [60,61], he eventually accomplished that and founded the axiomatic bargaining theory and proved the existence of the Nash Bargaining Solution for cooperative games (a notion similar to the Nash Equilibrium). The beauty of game theory is in its abstractly defined mathematics and notions of optimality. In no other branch of sciences do we find so many understandable definitions and levels of optimality [63].

Auction theory is a special branch of game theory that deals with biddings and auctions, which have long been an important part of the market mechanisms. Auctions allow buyers to opt for prices often lesser than the original market prices, but they have to compete and in doing so they have to realize their needs and constraints. Analysis of such games began with the pioneering work of Vickrey [71]. An auction is a market institution with an explicit set of rules determining resource allocation and prices on the basis of bids from the market participants [41]. For instance, we can formulate an auction as

1. Bidders send bids to indicate their willingness to exchange goods.
2. The auction may post-price quotes to provide summarized information about the status of the price-determination process. (Steps 1 and 2 may be iterated.)
3. The auction determines an allocation and notifies the bidders as to who purchases what from whom at what price. (The aforementioned sequence may be performed once or be repeated any number of times.)

There are four standard types of auctions [54]:

1. The English auction (also called the oral, open, or ascending-bid auction).
2. The Dutch auction (or descending-bid auction).
3. The first-price sealed-bid auction.
4. The second-price sealed-bid (or Vickrey) auction.

It is to be noted that the English and the Dutch auctions are collectively called progressive auctions; similarly the first- and second-price sealed-bid auctions are collectively know as sealed-bid auctions.

The English auction is the auction form most commonly used for the selling of goods. In the English auction, the price is successfully raised until only one bidder remains. This can be done by having an auctioneer announce prices, or by having bidders call the bids themselves, or by having bids submitted electronically with the current best bid posted. The essential feature of the English auction is that, at any point in time, each bidder knows the level of the current best bid. Antiques and artwork, for example, are often sold by English auction.

The Dutch auction is the converse of the English auction [76]. The auctioneer calls an initial high price and then lowers the price until one bidder accepts the current price. The Dutch auction is used, for instance, for selling cut flowers in Netherlands, fish in Israel, and tobacco in Canada.

With the first-price sealed-bid auction, potential buyers submit sealed bids and the highest bidders are awarded items for the price they bid [54]. The basic difference between the first-price sealed-bid auction and the English auction is that, with the English auction, bidders are able to observe their rival's bids and accordingly, if they choose, revise their own bids; with the sealed-bid auction, each bidder can submit only one bid. First-price sealed-bid auctions are used in the auctioning of mineral rights to U.S. government-owned land; they are also sometimes used in the sales of artwork and real estate [41]. Of greater quantitative significance is the use, already noted, of sealed-bid tendering for government procurement contracts [71].

Under the second-price sealed-bid auction, bidders submit sealed bids having been told that the highest bidder wins the item but pays a price equal not to his own bid but to the second-highest bid. While this auction has useful theoretical properties, it is seldom used in practice. The most significant application of this type of auction is found in the selling of FCC bandwidths [41].

45.5 Casting DRP into an Incentive Compatible Game Theoretical Auction

To begin, we first describe when in the lifespan of the distributed computing system a replication algorithm (an incentive compatible auction if we want to use the correct term) is to be invoked. The answer to that question depends specifically on the system at hand, but usually the replication algorithms are invoked when the system experiences the least amount of queries to access the data objects. This is to ensure that the least amount of users would be affected from the movement of data objects in the system; furthermore, it reduces the workload on the entities to compute their preferences toward the objects they prefer to host—remember they are already busy answering all the queries directed to them.

In the subsequent text, we will first extract the necessary ingredients from the discussion on auction theory (Section 45.4), and use them to cast the DRP into an incentive compatible game theoretical auction.

45.5.1 The Ingredients

The Basics: The auction mechanism contains M players. Each player i has some private information $t^i \in \Re$. This data is termed as the player's *type*. Only player i has knowledge of t^i. Everything else in the auction mechanism is public knowledge. Let t denote the vector of all the true types $t = (t^1, t^2, \ldots, t^M)$.

Communications: Since the players are self-interested (selfish) in nature, they do not communicate the value t^i. The only information that is relayed is the corresponding bid b^i. Let b denote the vector of all the bids ($b = (b^1, b^2, \ldots, b^M)$, and let b^{-i} denote the vector of bids not including player i, that is, $b^{-i} = (b^1, b^2, \ldots, b^{i-1}, b^{i+1}, \ldots, b^M)$. It is to be understood that we can also write $b = (b^{-i}, b^i)$.

Components: The auction mechanism has two components: (i) the algorithmic output $o(\cdot)$ and (ii) the payment mapping function $p(\cdot)$.

Algorithmic output: The auction mechanism allows a set of outputs O, based on the output function that takes in as the argument, the bidding vector, that is, $o(b) = \{o^1(b), o^2(b), \ldots, o^M(b)\}$, where $o(v) \in O$. This output function relays a unique output given a vector b. That is, when $o(\cdot)$ receives b, it generates an output, which is of the form of allocations $o^i(b)$. Intuitively it would mean that the algorithm takes in the vector bid b and then relays to each player its allocation.

Monetary cost: Each player i incurs some monetary cost $c^i(t^i, o)$, that is, the cost to accommodate the allocation $o^i(b)$. This cost is dependent on the output and the player's private information.

Payments: To offset c^i, the auction mechanism makes a payment $p^i(b)$ to player i. A player i always attempts to maximize its profit (utility) $u^i(t^i, b) = p^i(b) - c^i(t^i, o)$. Each player i cares about the other players' bid only insofar as they influence the outcome and the payment. While t^i is only known to player i, the function c^i is public. (Note that when we were previously describing the properties of auctions, the payments were made by the players and not the auction mechanism. That is fine in that context since the incentive for the players there was to acquire the object. In the context of the DRP, since the players have to conform to the global optimization criteria and host the objects, an incentive for the players would be to receive payments for hosting objects rather than making payments.)

Bids: Each player i is interested in reporting a bid b^i such that it maximizes its profit, regardless of what the other players bid (dominant strategy), that is, $u^i(t^i, (b^{-i}, t^i)) \geq u^i(t^i, (b^{-i}, b^i))$ for all b^{-i} and b^i.

The incentive compatible auction mechanism: We now put all the pieces together. An incentive compatible auction mechanism (I-CAM) consists of a pair $(o(b), p(b))$, where $o(\cdot)$ is the output function and $p(\cdot)$ is the payment mapping function. The objective of the auction mechanism is to select an output o, that optimizes a given objective function $f(\cdot)$.

Desirable properties: In conjunction to the discussion in Section 45.4, we desire our auction to exhibit the utility maximization property. In economic theory, a utility maximization property is also known as the efficient outcome of the auction mechanism [16]. In the subsequent text we will consider other properties, but for the time being let us limit our discussion to the utility maximization property. Remember that a utility maximization property is only attainable when an auction is an incentive compatible auction [41].

45.5.2 The Casting

We follow the same pattern as discussed in Section 45.5.1.

The Basics: The distributed system described in Section 45.3 is considered, where each server is represented by a player, that is, the auction mechanism contains M players. In the context of the DRP, a player holds two key elements of information (i) the available server capacity ac^i and (ii) the access frequencies (both read r_k^i and write w_k^i). Let us consider what possible cases for information holding there can be

1. DRP $[\pi]$: Each player i holds the access frequencies $\{r_k^i, w_k^i\} = t^i$ associated with each object k as private information, where as the available server capacity ac^i and everything else (this includes all the auction-related functions, network, and system parameters) is public knowledge.
2. DRP $[\sigma]$: Each player i holds the available server capacity $ac^i = t^i$ as private information, where as the access frequencies $\{r_k^i, w_k^i\}$ and everything else is public knowledge.
3. DRP $[\pi, \sigma]$: Each player i holds both the access frequencies $\{r_k^i, w_k^i\}$ and the server capacity ac^i as private information $\{ac^i, \{r_k^i, w_k^i\}\} = t^i$, whereas everything else is public knowledge.

Intuitively, if players know the available server capacities of other players, that gives them no advantage whatsoever. However, if they come to know about their access frequencies, then they can modify their valuations and alter the algorithmic output. Everything else such as the network topology, latency on communication lines, and even the server capacities can be public knowledge. Therefore, DRP $[\pi]$ is the only natural choice.

Communications: The players in the auction mechanism are assumed to be selfish and therefore, they project a bid b^i to the auction mechanism.

Components: The auction mechanism has two components (i) the algorithmic output $o(\cdot)$ and (ii) the payment mapping function $p(\cdot)$.

Algorithmic output: In the context of the DRP, the replication algorithm accepts bids from all the players, and outputs the maximum beneficial bid, that is, the bid that incurs the minimum replication cost overall (Equation 45.3). We will give a detailed description of the algorithm in the later text.

Monetary cost: When an object is allocated (for replication) to a player i, the player becomes responsible to entertain (read and write) requests to that object. For example, assume object k is replicated by player i. Then the amount of traffic that the player's server has to entertain due to the replication of object k is exactly equivalent to the replication cost (i.e., $c^i = R_k^i + W_k^i$). This fact is easily deducible from Equation 45.4.

Payments: To offset c^i, the auction mechanism makes a payment $p^i(b)$ to player i. This payment is chosen by the auction mechanism such that it eliminates incentives for misreporting by imposing on each player the cost of any distortion it causes. The payment for player i is set so that i's report cannot effect the total payoff to the set of other players (excluding player i), $M - i$. With this principle in mind, let us derive a formula for the payments. To capture the effect of i's report on the outcome, we introduce a hypothetical *null report*, which corresponds to player i reporting that it is indifferent among the possible decisions and cares only about payments. When player i makes the null report, the auction optimally chooses the allocation $o(t^{-i})$. The resulting total value of the decision for the set of players $M - i$ would be $V(M - i)$, and the auction might also "collect" a payment $h^i(t^{-i})$ from player i. Thus, if i makes a null report, the total payoff to the players in set $M - i$ is $V(M - i) - h^i(t^{-i})$.

The auction is constructed so that this, $V(M - i) - h^i(t^{-i})$, amount is the total payoff to those players regardless of i's report. Thus, suppose when the reported type is t, i's payment is $p^i(t) + h^i(t^{-i})$, so that $p^i(t)$ is i's additional payment over what i would pay if it made the null report. The decision $o(t)$ generally depends on i's report, and the total payoff to members of $M - i$ is then $\sum_{i \in M - i} v^i(o(t), t^i) + p^i(t) - h^i(t^{-i})$. We equate this total value with the corresponding total value when player i makes the null report:

$$\sum_{i \in M-i} v^i\left(o(t), t^i\right) + p^i(t) + h^i\left(t^{-i}\right) = V(M - i) + h^i\left(t^{-i}\right). \tag{45.5}$$

Using Equation 45.6, we solve for the extra payment as

$$p^i(t) = V(M - i) - \sum_{i \in M-i} v^i\left(o(t), t^i\right). \tag{45.6}$$

According to Equation 45.6, if player i's report leads to a change in the decision o, then i's extra payment $p^i(t)$ is specified to compensate the members of $M - i$ for the total losses they suffer on the account [33].

The derived payment procedure is in its most general form. A careful observation would reveal that its special cases include every possible payment procedure. The most famous of them all is the Vickrey payment. To say the least, we will show the derived payment procedure is equivalent to Vickrey payments. A player's value for any decision depends only on the objects that the player acquires, and not on the objects acquired by other players. That is, $v^i(t^i) = 1$ if the player acquires the object and $v^i(t^i) = 0$ otherwise. Since the losing players are not pivotal [41] (because their presence does not affect the allocation o), they obtain zero payments in our mechanism. According to Equation 45.7, the price a winning player pays in the (derived) payment procedure is equal to the difference between the two numbers. The first number is the maximum total value of all the other players, when i does not participate, which is

An incentive compatible auction mechanism

Initialize:
LS, L^i, T_k^i, M, MT

```
01 WHILE LS ≠ NULL DO
02     OMAX = NULL; MT = NULL; P^i = NULL;
03     PARFOR each S^i? LS DO
04         FOR each O_k? L^i DO
05             T_k^i = compute (B_k^i);   /*Compute the valuation corresponding to the desired object*/
06         ENDFOR
07         t^i = argmax_k(T_k^i);
08         SEND t^i to M; RECEIVE at M t^i in MT;
09     ENDPARFOR
10     OMAX = argmax_k(MT);   /*Choose the global dominate valuation*/
11     DELETE k from MT;
12     P^i = argmax_k(MT);         /*Calculate the Vickrey payment*/
13     BROADCAST OMAX;
14     SEND P^i to S^i; RECEIVE at S^i      /*Ask the winning agent to pay this amount*/
15     SEND P^i to M; RECEIVE at M         /*Send the required payment*/
16     Replicate O_OMAX;
17     b^i=b^i - o_k;              /*Update capacity*/
18     L^i = L^i - O_k;            /*Update the list*/
19     IF L^i = NULL THEN SEND info to M to update LS = LS - S^i;   /*Update mechanism players*/
20     PARFOR each S^i? LS DO
21         Update NN^i_OMAX       /*Update the nearest neighbor list*/
22     ENDPARFOR                  /*Get ready for the next round*/
23 ENDWHILE
```

FIGURE 45.1 Pseudo-code for an incentive compatible auction mechanism for the DRP.

$\max_{j \neq i} v^i$. The second number is the total value of all the other players when i wins, which is zero. Thus, when i wins, it pays $\max_{j \neq i} v^i$, which is equal to the second highest valuation. This is exactly the Vickrey payment [71].

Bids: Each player i reports a bid that is the direct representation of the true data that it holds. Therefore, a bid b^i is equivalent to $1/\{R_k^i + W_k^i\}$. That is, the lower the replication cost the higher is the bid and the higher are the chances for the bid b^i to win.

In essence, the I-CAM $(o(b), p(b))$, takes in the vector of bids b from all the players, and selects the highest bid. The highest bidder is allocated the object k that is added to its allocation set o^i. The auction mechanism then pays the bidder p^i. This payment is equivalent to the Vickrey payments and compensates the cost incurred (due to the entertainment of access requests for object k by users) by the player to host the object at its server. A pseudocode for an I-CAM is given in Figure 45.1.

Description of pseudo-code: We maintain a list L^i at each server. This list contains all the objects that can be replicated by player i onto server S^i. We can obtain this list by examining the two constraints of the DRP. List L^i would contain all the objects that have their size less than the total available space b^i. Moreover, if server S^i is the primary host of some object k', then k' should not be in L^i. We also maintain a list LS containing all servers that can replicate an object (i.e., $S^i \in LS$ if $L^i \neq NULL$). The algorithm works iteratively. In each step, the auction mechanism asks all the players to send their preferences (first **PARFOR** loop). Each player i recursively calculates the true data of every object in list L^i. Each player then reports the dominant true data (line 08). The auction mechanism receives all the corresponding entries, and then chooses the best dominant true data. This is broadcast to all the players, so that they can update their nearest neighbor table NN_k^i, which is shown in Line 21 (NN_{OMAX}^i). The object is replicated and payments made to the player. The auction progresses forward till there are no more players interested in acquiring any data for replication.

45.5.3 Further Discussion on the Casting Process

The mechanism described in Section 45.5.2 illustrates the usage of the auction theory as a possible solution toward the DRP with the property of utility maximization. This same process with minor modifications can be used to guarantee other auction properties applied to the DRP. We give a brief description of some of the properties in the subsequent text, but for details, the readers are encouraged to see some of the work performed by the authors that explicitly detail these properties.

Pareto optimality: Implementing an outcome that is not pareto dominated by any other outcomes, so no other outcomes make one player better off while making other players worst. Details on a pareto optimal auction applied to DRP can be found in [34].

Maximum utility to a particular player: Maximizing the expected utility to a single player, typically the central decision-making body, across all possible scenarios. This type of setting is very useful when considering revenue maximization scenarios. Details on a maximum utility to a particular player auction applied to DRP can be found in [33].

Deliberate discrimination of allocation: Maximize the system utilization by revoking allocations if deemed necessary. This type of property is very useful when considering dynamic scenarios, where it often warrants revoking a decision since the system parameters may change drastically during the computation of a decision. Details on a deliberate discrimination of allocation auction applied to DRP can be found in [35].

Budget balance: A budget balanced auction is when the total payments made or received by the players exactly equals zero. This property is important since the money is not injected or removed from the system. If the payments made or received by the players equal to zero, then the auction is termed as a strict budget balance auction. On the other hand, if the payments made or received by the players does not equal to zero but it is nonnegative, then the auction is termed as a weak budget balance auction. (In a weak budget balance auction, the auction does not run at a loss.) One can also consider an ex ante budget balance auction, in which the auction is balanced on average, and an ex post budget balance auction, in which the auction is balanced at all times. Details on a budget balance auction applied to DRP can be found in [36].

Budget balance is especially important in systems that must be self-sustaining and require no external benefactor to input money or central authority to collect payments [16]. For instance, a distributed system should always be a budget balanced system, since money has no literal meaning in the system—it is there just to drive the optimization process, and not the system as a whole.

45.6 Experimental Setup and Discussion of Results

We performed experiments on a 440 MHz Ultra 10 machine with 512 MB memory. The experimental evaluations were targeted to benchmark the placement policies.

Performance metric: The solution quality in all cases, was measured accordingly in terms of the OTC percentage that was saved under the replication scheme found by the technique, compared to the initial one (i.e., when only primary copies exist).

45.6.1 Network Topologies

To establish diversity in our experimental setups, the network connectivity was changed considerably. We used four types of network topologies, which we explain later. All in all, we employed 80 various topologies.

45.6.1.1 Flat Models

In flat random methods, a graph $G = (V, E)$ is built by adding edges to a given set of nodes V subject to a probability function $P(u, v)$, where u and v are arbitrary nodes of G.

Pure random model: A random graph $G(M, P(\text{edge} = p))$ with $0 \leq p \leq 1$ contains all graphs with nodes (servers) M in which the edges are chosen independently and with a probability p. Although this approach is extremely simple, yet it fails to capture significant properties of web-like topologies [20]. The five pure random topologies were obtained using GT-ITM [75] topology generator with $p = \{0.4, 0.5, 0.6, 0.7, 0.8\}$.

Waxman model: The shortcomings of pure random topologies can be overcome by using the Waxman model. In this method, edges are added between pairs of nodes (u, v) with probability $P(u, v)$ that depends on the distance $d(u, v)$ between u and v. The Waxman model is given by [71]

$$P(u, v) = \beta e^{\frac{-d(u,v)}{L\alpha}},$$

where L is the maximum distance between any two nodes and $\alpha, \beta \in (0, 1]$. β is used to control the density of the graph. The larger the value of β the denser is the graph. α is used to control the connectivity of the graph. The smaller the value of α, the larger is the number of short edges [20]. The 12 Waxman topologies were obtained using the GT-ITM [75] topology generator with values of $\alpha = \{0.1, 0.15, 0.2, 0.25\}$ and $\beta = \{0.2, 0.3, 0.4\}$.

45.6.1.2 Link Distance Models

In pure random and Waxman Models, there is no direct connection among the communication cost and the distance between two arbitrary nodes of the generated graph. To complement these two models, we propose a class of graphs in which the distance between two nodes is directly proportional to the communication cost. In such methods, the distance between two serves is reversed mapped to the communication cost of transmitting a 1 kB of data, assuming that we are given the bandwidth. That is, the communication cost is equivalent to the sum of the transmission and propagation delay. The propagation speed on a link is assumed to be 2.8×10^8 m/s (copper wire). Thus, if we say that the distance between two nodes is 10 km and has a bandwidth of 1 Mbps, then it means that the cost to communication 1 kB of data between the two nodes is equivalent to 10 km/(2.8×10^8 m/s) + 1 kB/(1 Mbps) = 8.03 ms, and the cost would simply be 0.00803.

Random graphs: This method involves generating graphs with random: node degree (d^*), bandwidth (b), and link distance (d) between the nodes of the graph. We detail the steps involved in generating random graphs as follows. First, M (user input) nodes are placed in a plain, each with a unique identifier. Second, from the interval d^*, each node's out degree is generated. (At this moment, the links do not have weights or communication costs.) Third, each link is assigned bandwidth (in Mbps) and distance (in kilometer) on random. Finally, for each link the transmission and propagation delay is calculated, on the basis of on assigned bandwidth and distance. The 12 random topologies were obtained using, $d^* = \{10, 15, 20\}$, b = $\{1, 10, 100\}$, and $d = \{5, 10, 15, 20\}$.

Fully connected random graphs: This method is similar to the random graphs except that now we do not require the node degree since the entire graph is fully connected. The five random topologies were obtained using, $b = \{1, 10, 100\}$ and $d = \{d_1 = [1, 10], d_2 = [1, 20], d_3 = [1, 50], d_4 = [10, 20], d_5 = [20, 50]\}$. Note that d has five elements $d_1, \ldots d_5$. Each element was used to generate a particular graph. For instance, for the first graph, we choose the bandwidth randomly from the values of $\{1, 10, 100\}$, and the link distance randomly from the interval of $d_1 = [1, 10]$.

Fully connected uniform graphs: This method is similar to the fully connected random graphs except that the bandwidth and link distance are chosen uniformly and not randomly. The five random topologies were obtained using, $b = [1, 100]$ and $d = \{d_1 = [1, 10], d_2 = [1, 20], d_3 = [1, 50], d_4 = [10, 20], d_5 = [20, 50]\}$.

Fully connected lognormal graphs: This method is similar to the fully connected random graphs except that link distance is chosen log-normally and not randomly. Note that the bandwidth is still assigned on random. (Curious readers are encouraged to see [20] for an insight on the lognormal distribution functions.) The nine lognormal topologies were obtained using, $b = \{1, 10, 100\}$ and

$d = \{\mu = \{8.455, 9.345, 9.564\}, \sigma = \{1.278, 1.305, 1.378\}\}$, where μ and σ are the mean and variance parameters of the lognormal distribution function, respectively.

45.6.1.3 Power-Law Model

The power-law model [52] takes its inspiration from the Zipf law [77], and incorporates rank, out-degree, and eigen exponents. We used Inet [10] topology generator to obtain the power-law-based Internet topologies. Briefly, Inet generates autonomous system (AS) level topologies. These networks have similar if not the exact characteristics of the Internet from November 1997 to June 2000. The system takes in as input two parameters to generate topologies, namely (i) the total number of nodes and (ii) the fraction (k) of degree-one nodes. Briefly, Inet starts from the total number of desired nodes and computes the number of months t it would take to grow the Internet from its size in November 1997 (which was 3037 nodes) to the desired number of nodes. Using t, it calculates the growth frequency and the out-degree of the nodes in the network. This information is used to iteratively connect nodes till the required out-degree of nodes is reached. The 20 power-law topologies were obtained using $k = \{0.01, 0.05, 0.1, 0.15, 0.2, 0.25, 0.3, 0.35, 0.4, 0.45, 0.5, 0.55, 0.6, 0.65, 0.7, 0.75, 0.8, 0.85, 0.9, 0.95\}$.

45.6.1.4 Hierarchical Transit-Stub Model

The Internet model at the AS level can also be captured by using a hierarchical model. Wolfson et al. [75] derived a graph generation method using a hierarchical model in order to provide a more adequate router model of the Internet than the Waxman model. In their paper, each AS domain in the Internet was classified as either a *transit* domain or a *stub* domain, hence the name transit-stub model. In a stub domain, traffic between any two nodes u and v goes through that domain if and only if either u or v is in that domain. Contrarily, this restriction is relaxed in a transit domain. The GT-ITM topology generator [75] models a three-level hierarchy corresponding to transit domains, stub domain, and LANs attached to stub domains [20]. Using the GT-ITM topology generator, we generated 12 random transit-stub graphs with a total of 3718 nodes each, and then placed the primary server inside a randomly selected stub domain. In order to make the topologies as realistic as possible, we introduced routing delays to mimic routers' decision delays inside the core network. We set this delay to be equal to 20 ms/hop. In order to have a realistic upper bound on the self-injected delays, the maximum hop count between any pair of servers was limited to 14 hops.

45.6.2 Access Patterns

To evaluate the replica placement methods under realistic access patterns, we used the access logs collected at the Soccer World Cup 1998 website [3]. The Soccer World Cup access log has over 1.35 billion requests, making it extremely useful to benchmark a given approach over a prolonged high access rate. The only drawback with these logs is that the users' IP addresses (that can potentially give us their approximate geographical locations) are replaced with an identifier. Although, we can obtain the information as to who were the top, say 500 users of the website, yet we cannot determine where the clients were from. To negate this drawback, one can use an access log that does not mask the IP addresses; however, we limit our discussion to the Soccer World Cup access logs.

An important point to note is that these logs are access (or read) logs; thus, they do not relay any information regarding the write requests. However, there is a tedious way around this. Each entry of the access logs has, among other parameters, the information about the size of the object that is being accessed. The logs are processed to observe the variance in the object size. For each entry that returns the change in the object size, a mock write request is generated for that user for the object that is currently being accessed. This variance in the object size generates enough miscellanies to benchmark object updates.

We used 88 days of the Soccer World Cup 1998 access logs, that is, the (24 h) logs from April 30, 1998 to July 26, 1998. To process the logs, we wrote a script that returned: only those objects that were present in all the logs (from this we choose 25,000 data objects on random—the maximum workload for our experimental evaluations), the total number of requests from a particular client for an object, the average and the variance of the object size. From this log, we chose the top 3718 clients (maximum experimental setup). (We will describe in the subsequent text how we came up with the number 3718. For the time being, assume that this is a correct measure.) A random mapping was then performed of the clients to the nodes of the topologies. Note that this mapping is not 1-1, rather 1-M. This gave us enough skewed workload to mimic real world scenarios. It is also worthwhile to mention that the total amount of requests entertained for each problem instance using the Soccer World Cup access logs was in the range of 3–4 million. The primary replicas' original server was mimicked by choosing random locations. The capacities of the servers $C\%$ were generated randomly with range from *Total Primary Object Sizes* 2 to $1.5 \times$ *Total Primary Object Sizes*. The variance in the object size collected from the access was used to mimic the object updates. The updates were randomly pushed onto different servers, and the total system update load was measured in terms of the percentage update requests $U\%$ compared that to the initial network with no updates.

45.6.3 Further Clarifications on the Experimental Setup

Since the access logs were of the year 1998 and before, we first used Inet to estimate the number of nodes in the network. This number came up to be approximately 3718, that is, there were 3718 AS-level nodes in the Internet at the time when the Soccer World Cup 1998 was being played. Therefore, we set the upper bound on the number of servers in the system to be $M = 3718$. Moreover, every topology model that was used in this study had the network topologies generated for $M = 3718$. Owing to space limitations, we do not show the detailed results obtained using every topology. However, we do provide the averaged performance of all the comparative algorithms over all the 80 topologies and 88 (24 h) access log.

45.6.4 The Determination of the Relocation Period

As noted previously (in Sections 45.5), the time when to invoke the replica placement techniques (say t) requires high-level human intervention. Here, we will show that this parameter, if not totally, can at least partially be automated. The decision when to initiate the replica placement techniques depend on the past trends of the user access patterns. The experiments performed to test the mechanism used real user access patterns collected at the 1998 Soccer World Cup website [3]. This access log file has become a de facto standard over the number of years to benchmark various replica placement techniques. Works reported in [24,25], [31–38], [47], and [64] all have used this access log for analysis.

Figures 45.2 show the user access patterns. The two figures represent different traffic patterns, that is, Figure 45.2a shows the traffic recorded on the days when there was no scheduled match, while Figure 45.2b shows the traffic on the days when there were scheduled matches. We can clearly see that the website incurred soaring and stumpy traffic at various intervals during a 24-h time period (it is to be noted that the access logs have a time stamp of GMT+1). For example, on days when there was no scheduled match, the traffic was mediocre before 0900 h. The traffic increased after 0900 h till 2200 h. The two vertical dashed lines indicate this phenomenon. These traffic patterns were recorded over a period of 86 days (April 30, 1998 to July 26, 1998). Therefore, on the days when there was no scheduled match, a replica placement algorithm (in our case the mechanism) could be initiated twice daily: (i) at 0900 h and (ii) at 2200 h. The time interval t for 0900 h would be $t = (2200-0900) = 11$ h and for 2200 h would be $t = (0900-2200) = 13$ h. On the other hand, the days when there were scheduled matches, the mechanism could be initiated at 1900 h and 0100 h. It is to be noted that the autonomous agents can easily obtain all the other required parameters (for the DRP) via the user access logs and the underlying network architecture.

45.6.5 Comparative Techniques

For comparison, we selected three various types of replica placement techniques. To provide a fair comparison, the assumptions and system parameters were kept the same in all the approaches. The techniques studied include efficient branch- and bound-based techniques (Aε-Star [31]). The algorithms proposed in [32], [47], [48], and [64] are the only ones that address the problem domain similar to ours. We select from

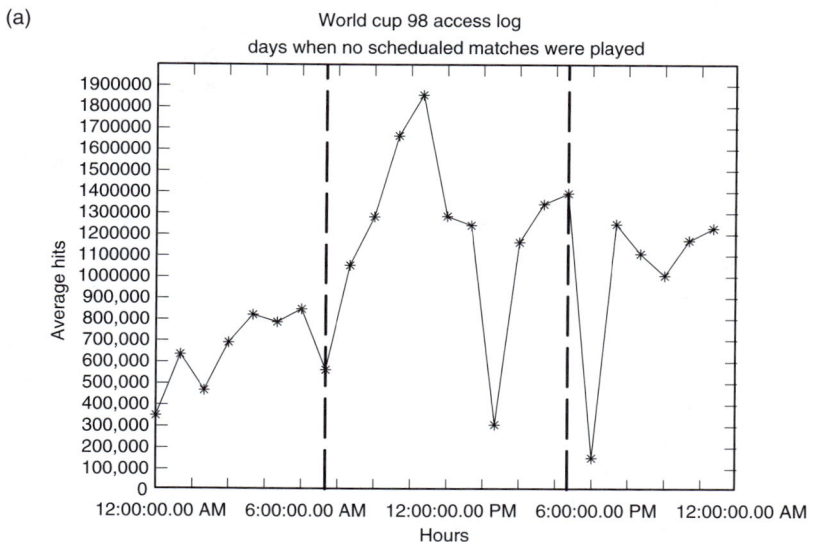

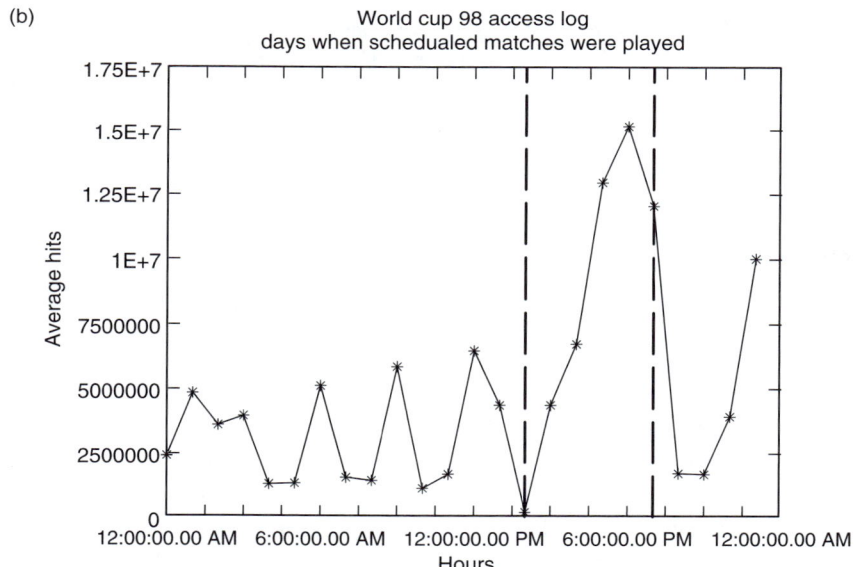

FIGURE 45.2 (a) Access on days with no matches. (b) Access on days with matches.

[64] the greedy approach (Greedy) for comparison because it is shown to be the best compared with four other approaches (including the proposed technique in [47]); thus, we indirectly compare with four additional approaches as well. From [48] we choose the genetic-based algorithm (GRA), which exhibits extreme robustness under various changing scenarios. Owing to space limitations we will only give a brief overview of the comparative techniques. Details for a specific technique can be obtained from the referenced papers.

1. Aε-Star: In [31], the authors proposed a $1 + \varepsilon$ ad missible A-Star-based technique called Aε-Star. This technique uses two lists: OPEN and FOCAL. The FOCAL list is the sublist of OPEN, and only contains those nodes that do not deviate from the lowest f node by a factor greater than $1 + \varepsilon$. The technique works similar to A-Star, with the exception that the node selection (lowest h) is done not from the OPEN but from the FOCAL list. It is easy to see that this approach will never run into the problem of memory overflow, moreover, the FOCAL list always ensures that only the candidate solutions within a bound of $1 + \varepsilon$ of the A-Star are expanded.

2. Greedy: We modify the greedy approach reported in [64], to fit our problem formulation. The greedy algorithm works in an iterative fashion. In the first iteration, all the M servers are investigated to find the replica location(s) of the first among a total of N objects. Consider that we choose an object i for replication. The algorithm recursively makes calculations on the basis of the assumption that all the users in the system request for object i. Thus, we have to pick a server that yields the lowest cost of replication for the object i. In the second iteration, the location for the second server is considered. On the basis of the choice of object i, the algorithm now would identify the second server for replication, which, in conjunction with the server already picked, yields the lowest replication cost. Observe here that this assignment may or may not be for the same object i. The algorithm progresses forward till either one of the DRP constraints is violated. The readers will immediately realize that derived I-CAM works similarly to the Greedy algorithm. This is true; however, the Greedy approach does not guarantee optimality even if the algorithm is run on the very same problem instance. Recall that Greedy relies on making combinations of object assignments and therefore, suffers from the initial choice of object selection (which is done randomly). This is never the case with the derived auction procedure, which identifies optimal allocations in every case.

3. GRA: In [48], the authors proposed a genetic algorithm-based heuristic called GRA. GRA provides good solution quality, but suffers from slow termination time. This algorithm was selected since it realistically addressed the fine-grained data replication using the same problem formulation as undertaken in this article.

From here onward, we will acronym the "incentive compatible auction mechanism" derived exclusively for the DRP as I-CAM.

45.6.6 Comparative Analysis

We record the performance of the techniques using the access logs and 80 topologies. Note that each point represents the average performance of an algorithm over 80 topologies and 88 days of the access log. Further, we detail our experimental findings.

45.6.6.1 Impact of Change in the Number of Servers and Objects

We study the behavior of the placement techniques when the number of servers increase (Figure 45.3), by setting the number of objects to 25,000, while in Figure 45.4, we study the behavior when the number of objects increase, by setting the number of servers to 3718. For the first experiment we fixed $C = 35\%$ and $R/W = 0.25$. We intentionally chose a high workload so as to see if the techniques studied successfully handled the extreme cases. By adding a server in the network, we introduce additional traffic due to its local requests, together with more storage capacity to be used for replication. I-CAM balances and explores these diverse effects, so as to achieve highest OTC savings. GRA showed the worst performance along all the techniques. It showed an initial gain, since with the increase in the number of servers, the population permutations increase exponentially, but with the further increase in the number of servers this phenomenon is not so observable as all the essential objects are already replicated. The top performing techniques (I-CAM, Greedy, and Aε-Star) showed an almost constant performance increase (after the initial surge in OTC savings). GRA also showed a similar trend but maintained lower OTC savings. This was in line with the claims presented in [31] and [48].

To observe the effect of increase in the number of objects in the system, we chose a softer workload with $C = 65\%$ and $R/W = 0.70$. The intention was to observe the trends for all the techniques under various workloads. The increase in the number of objects has diverse effects on the system as new read/write patterns (since the users are offered more choices) emerge, and also the strain on the overall storage

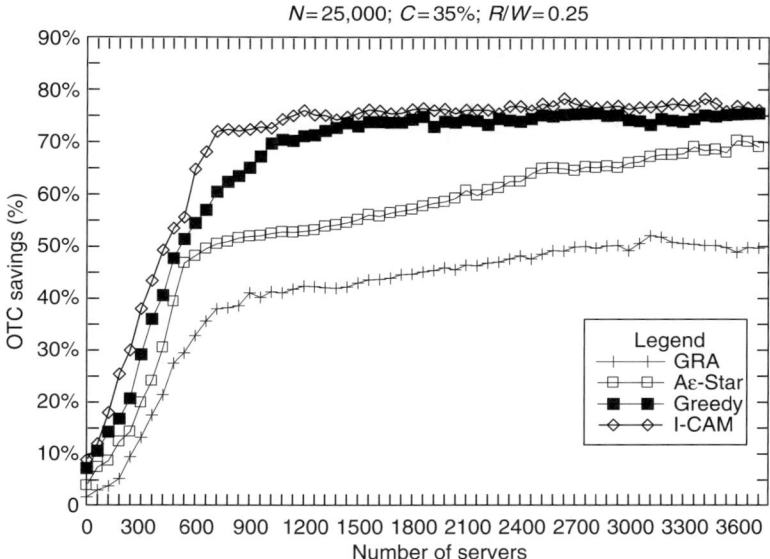

FIGURE 45.3 OTC savings versus number of servers.

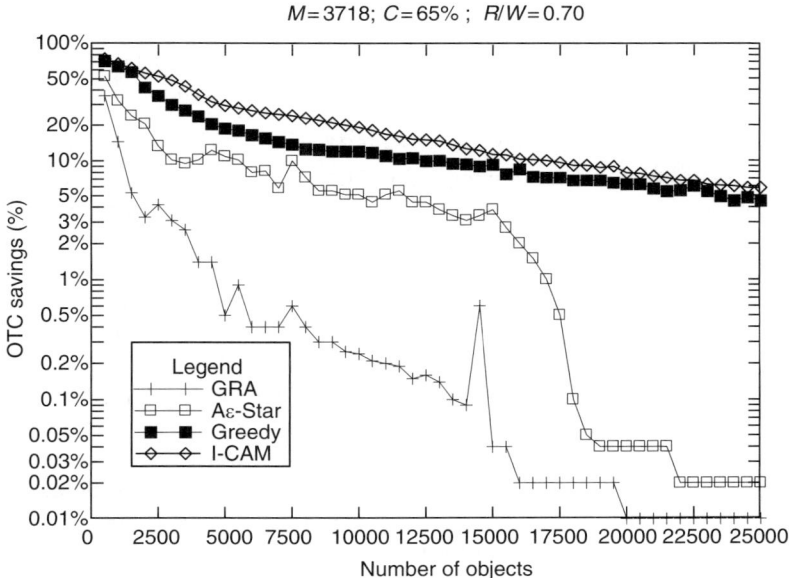

FIGURE 45.4 OTC savings versus number of objects.

capacity of the system increases (due to the increase in the number of replicas). An effective replica allocation method should incorporate both the opposing trends. From the plot, the most surprising result came from GRA. It dropped its savings from 47% to 0.01%. This was contradictory to what was reported in [48]. But there the authors had used a uniformly distributed link cost topology, and their traffic was based on the Zipf distribution [77]. While the traffic access logs of the Soccer World Cup 1998 are more or less double-Pareto in nature [3]. In either case, the exploits and limitations of the technique under discussion are obvious. The plot also shows a near identical performance by Aε-Star and Greedy. The relative difference among the two techniques was less than 7%. However, Greedy did maintain its

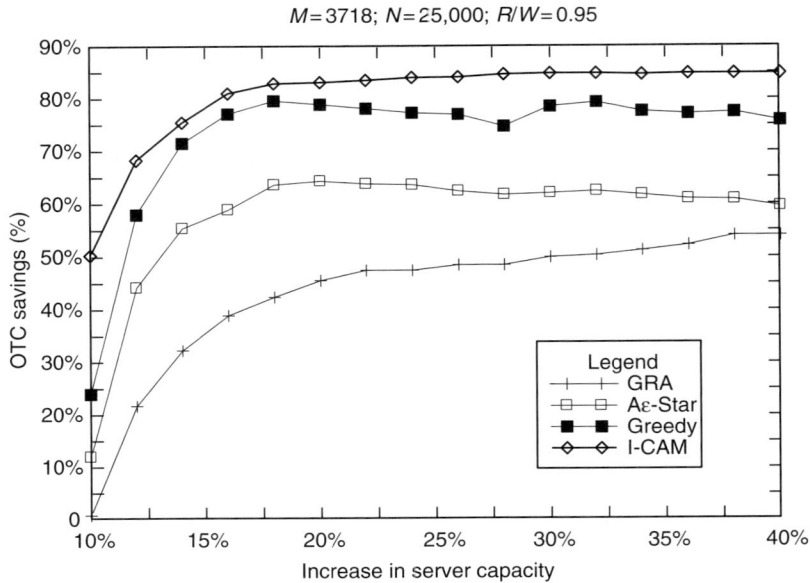

FIGURE 45.5 OTC savings versus capacity.

dominance. From the plots the supremacy of I-CAM is observable. [Figure 45.4 is deliberately shown with a log (OTC savings) scale to better appreciate the performances of the techniques.]

45.6.6.2 Impact of Change in the System Capacity

Next, we observe the effects of increase in storage capacity. An increase in the storage capacity means that a large number of objects can be replicated. Replicating an object that is already extensively replicated is unlikely to result in significant traffic savings as only a small portion of the servers will be affected overall. Moreover, since objects are not equally read intensive, increase in the storage capacity would have a great impact at the beginning (initial increase in capacity), but has little effect after a certain point, where the most beneficial ones are already replicated. This is observable in Figure 45.5, which shows the performance of the algorithms. GRA once again performed the worst. The gap between all other approaches was reduced to within 15% of each other. I-CAM and Greedy showed an immediate initial increase (the point after which further replicating objects is inefficient) in its OTC savings, but afterward showed a near constant performance. GRA although performed the worst but observably gained the most OTC savings (49%), followed by Greedy with 44%. Further experiments with various update ratios (5%, 10%, and 20%) showed similar plot trends. It is also noteworthy (plots not shown in this paper due to space restrictions) that the increase in capacity from 10% to 18%, resulted in 3.75 times (on average) more replicas for all the algorithms.

45.6.6.3 Impact of Change in the Read and Write Frequencies

Next, we observe the effects of increase in the read and write frequencies. Since these two parameters are complementary to each other, we describe them together. To observe the system utilization with varying read/write frequencies, we kept the number of servers and objects constant. Increase in the number of reads in the system would mean that there is a need to replicate as many object as possible (closer to the users). However, the increase in the number of updates in the system requires the replicas be placed as close to the primary server as possible (to reduce the update broadcast). This phenomenon is also interrelated with the system capacity, as the update ratio sets an upper bound on the possible traffic reduction through replication. Thus, if we consider a system with unlimited capacity, the "replicate everywhere anything" policy is strictly inadequate. The read and update parameters indeed help in drawing

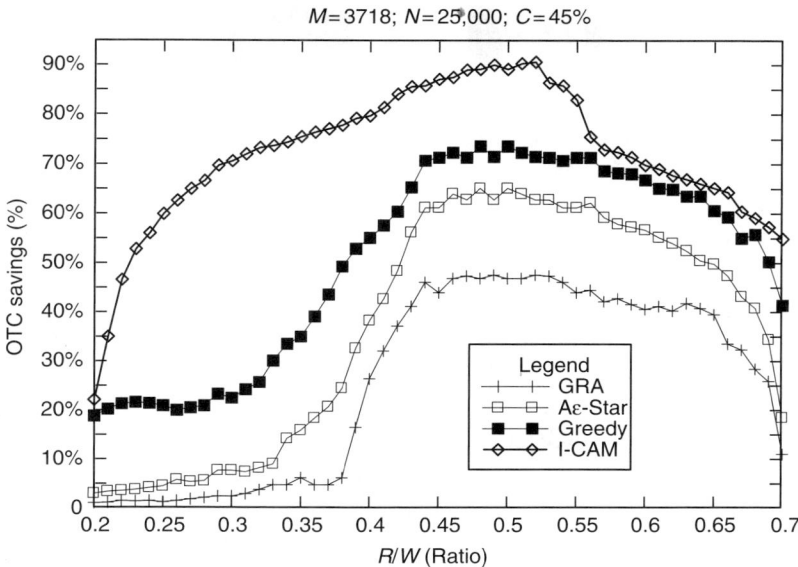

FIGURE 45.6 OTC savings versus read/write ratio.

TABLE 45.2 Running Time of the Replica Placement Methods in Seconds [$C = 45\%$, $R/W = 0.85$]

Problem size	Greedy	GRA	Aε-Star	I-CAM
$M = 2500, N = 15,000$	310.14	491.00	399.63	**185.22**
$M = 2500, N = 20,000$	330.75	563.25	442.66	**201.75**
$M = 2500, N = 25,000$	357.74	570.02	465.52	**240.13**
$M = 3000, N = 15,000$	452.22	671.68	494.60	**284.34**
$M = 3000, N = 20,000$	467.65	726.75	498.66	**282.35**
$M = 3000, N = 25,000$	469.86	791.26	537.56	**303.32**
$M = 3718, N = 15,000$	613.27	883.71	753.87	**332.48**
$M = 3718, N = 20,000$	630.39	904.20	774.31	**390.90**
$M = 3718, N = 25,000$	646.98	932.38	882.43	**402.23**

a line between good and marginal algorithms. The plot in Figure 45.6 shows the results of read/write ratio against the OTC savings. A clear classification can be made between the algorithms. I-CAM and Greedy incorporate the increase in the number of reads by replicating more objects, and thus, savings increased up to 88%, while GRA gained the least of the OTC savings of up to 42%. To understand why there is such a gap in the performance between the algorithms, we should recall that GRA specifically depends on the initial selection of gene population (for details see [48]). Moreover, GRA maintains a localized network perception. Increase in updates result in objects having decreased local significance (unless the vicinity is in close proximity to the primary location). On the other hand, I-CAM, Aε-Star, and Greedy never tend to deviate from their global (or social) view of the problem.

45.6.6.4 Running Time

Finally, we compare the termination time of the algorithms. Various problem instances were recorded with $C = 45\%$ and $R/W = 0.85$. The entries in Table 45.2 made bold represent the fastest time recorded over the problem instance. It is observable that I-CAM terminated faster than all the other techniques, followed by Greedy, Aε-Star, and GRA.

TABLE 45.3 Average OTC (%) Savings under Some Randomly Chosen Problem Instances

Problem size	Greedy	GRA	Aε-Star	I-CAM
$M = 100, N = 1000\ [C = 20\%, R/W = 0.75]$	71.46	85.77	86.28	**89.45**
$M = 200, N = 2000\ [C = 20\%, R/W = 0.80]$	84.29	78.30	79.02	**84.76**
$M = 500, N = 3000\ [C = 25\%, R/W = 0.95]$	68.50	70.97	67.53	**71.43**
$4M = 1000, N = 5000\ [C = 35\%, R/W = 0.95]$	88.09	67.56	78.24	**88.30**
$M = 1500, N = 10{,}000\ [C = 25\%, R/W = 0.75]$	89.34	52.93	76.11	**89.75**
$M = 2000, N = 15{,}000\ [C = 30\%, R/W = 0.65]$	67.93	51.02	52.42	**75.32**
$M = 2500, N = 15{,}000\ [C = 25\%, R/W = 0.85]$	77.35	71.75	73.59	**81.12**
$M = 3000, N = 20{,}000\ [C = 25\%, R/W = 0.65]$	76.22	65.89	73.04	**82.31**
$M = 3500, N = 25{,}000\ [C = 35\%, R/W = 0.50]$	66.04	59.04	67.01	**71.21**
$M = 3718, N = 25{,}000\ [C = 10\%, R/W = 0.40]$	76.34	63.19	76.02	**79.21**

45.6.6.5 Summary of Experimental Results

Table 45.3 shows the quality of the solution in terms of OTC percentage for 10 problem instances (randomly chosen), each being a combination of various numbers of server and objects, with varying storage capacity and update ratio. For each row, the best result is indicated in bold. The proposed I-CAM steals the show in the context of solution quality, but Greedy and Aε-Star do indeed give a good competition.

In summary, on the basis of the solution quality alone, the replica allocation methods can be classified into four categories: (i) high performance: I-CAM; (ii) medium–high performance: Greedy; (iii) medium performance: Aε-Star; and (iv) low performance: GRA. Considering the execution time, I-CAM and Greedy did extremely well, followed by Aε-Star and GRA.

45.7 Concluding Remarks and Future Directions

Replicating data across a distributed computing system can potentially reduce the user perceived access time, which in turn reduces latency, adds robustness, and increases data availability. Our focus here was to show how game theoretical auctions can be used to identify techniques for the DRP in distributed computing systems. A semidistributed technique based on a game theoretical auction was proposed for the DRP, which had the added property that it maximized the utility of all the players involved in the system—an I-CAM.

I-CAM is a protocol for automatic replication and migration of objects in response to demand changes. It aims to place objects in the proximity of a majority of requests while ensuring that no servers become overloaded. The infrastructure of I-CAM was designed such that each server was required to present a list of data objects that if replicated onto that server would bring the communication cost to its minimum. These lists were reviewed at the central decision body, which gave the final decision as to what objects are to be replicated onto what servers. This semidistributed infrastructure takes away all the heavy processing from the central decision-making body and gives it to the individual servers. For each object, the central body is only required to make a binary decision: (0) not to replicate or (1) to replicate.

To complement our theoretical results, we compared I-CAM with three conventional replica allocation methods, namely (i) branch and bound, (ii) greedy, and (iii) genetic. The experimental setups were designed in such a fashion that they resembled real-world scenarios. We employed GT-ITM and Inet to gather 80 various Internet topologies on the basis of flat, link distance, power-law, and hierarchical transit-stub models, and used the traffic logs collected at the Soccer World Cup 1998 website for mimicking user access requests. The experimental study revealed that the proposed I-CAM technique improved the performance relative to other conventional methods in four ways.

1. The number of replicas in a system was controlled to reflect the ratio of read versus write access. To maintain concurrency control, when an object is updated, all of its replicas need to be updated

simultaneously. If the write access rate is high, there should be few replicas to reduce the update overhead. If the read access rate is overwhelming, there should be a high number of replicas to satisfy local accesses.
2. Performance was improved by replicating objects to the servers based on locality of reference. This increases the probability that requests can be satisfied either locally or within a desirable amount of time from a neighboring server.
3. Replica allocations were made in a fast algorithmic turnaround time.
4. The complexity of the DRP was decreased multifold. I-CAM limits the complexity by partitioning the complex global problem of replica allocation, into a set of simple independent subproblems. This approach is well suited to the large-scale distributed computing systems that are composed of autonomous agents that do not necessarily cooperate to improve the system-wide goals.

All the aforementioned improvements were achieved by a simple, semidistributed, and autonomous I-CAM.

Our discussion only encircled the game theoretical auctions that had a central body to collect the information from the players, and based on that conclude a decision. However, there may be systems, such as grid computing and P2P system, that explicitly require a fully distributed mechanism. For instance, consider grid computing, which is predominately concerned with coordinated resource sharing in a dynamic, and sometimes in multiorganization structure. Consider also the P2P systems, which are similar to the grids but characteristically have more users with a wide spectrum of capabilities. Grids and P2P systems have distinct characteristics and stakeholders that require very efficient and effective resource allocation mechanisms, but there is no one central decision-making body. Thus, we need to consider applying distributed game theoretical auction mechanism, or the likes of it, which can consider to implement a social choice function under the constraint that no central decision-making body computes the outcome. This need can be owing to the following constraints:

1. The system has a structure that does not allow a central resource manager.
2. The system requires every entity to be a self-sufficient.

A distributed game theoretical auction mechanism would exhibit among others the following advantages over an ordinary game theoretical auction mechanism, and is a strong candidate for resource allocation and management techniques in grid and P2P computing:

1. A distributed game theoretical auction mechanism would transfer the computational workload from a central decision-making body in the mechanism to the players.
2. A distributed game theoretical auction mechanism would bring in robustness to the system, since in an ordinary game theoretical auction mechanism, the communications between the players and the central decision-making body are critical, and their malfunctioning may incapacitate the system. In distributed game theoretical auction mechanism, this communication structure simply does not exist, but at a cost—the system may only be able to attain suboptimal results.
3. Since in a distributed game theoretical auction mechanism no single entity would compute the outcome, a higher degree of trust would exist in the system.
4. Due to its distributed nature, the communication would never converge to a single point, thus, there would be no bottlenecks.

Briefly, a distributed game theoretical auction mechanism would distribute the mechanism's rules across the players so that they can perform computations (and eventually reach to an outcome) on the basis of the message sent and received from players in the system. Although, this setting is intriguing, yet it possesses several challenges, which we enumerate as follows:

1. The major challenge here would be to make these players play in a selfless manner, since they now have a firm control over the distributed structure of the underlying auction mechanism.
2. Another grand problem would be to reduce the complexity of messages passing in the communication network.

3. Computationally, we would seek to find social choice functions that can actually converge to solutions in a fully distributed fashion—something on the line of the distributed Vickrey auction implementation, which is a classical example of a canonically distributed convergence.
4. Theoretically, one needs to seek that the strategies applied by the players cater for cartel type behaviors. For instance, imagine a P2P system in which some servers only selectively (on personal preference) allow the sharing or resources. That kind of behavior has to be suppressed at all costs.

References

[1] I. Ahmad and A. Ghafoor, Semidistributed load balancing for massively parallel multicomputer systems, *IEEE Transaction on Software Engineering*, 17, 987–1004, 1991.
[2] P. Apers, Data allocation in distributed database systems, *ACM Transaction on Database Systems*, 13, 263–304, 1988.
[3] M. Arlitt and T. Jin, Workload characterization of the 1998 World Cup web site, Tech. report, Hewlett Packard Lab, Palo Alto, HPL-1999-35(R.1), 1999.
[4] R. Aumann and M. Maschler, Game theoretic analysis of a bankruptcy problem from the Talmud, *Journal of Economic Theory*, 36, 195–213, 1985.
[5] B. Awerbuch, Y. Bartal, and A. Fiat, Competitive distributed file allocation, *Proceedings of the 25th ACM STOC*, Victoria, B.C., Canada, 1993, pp. 164–173.
[6] B. Awerbuch, Y. Bartal, and A. Fiat, Distributed paging for general networks, *Journal of Algorithms*, 28, 67–104, 1998.
[7] I. Baev and R. Rajaraman, Approximation algorithms for data placement in arbitrary networks, *Proceedings of the 12th Annual ACM-SIAM Symposium on Discrete Algorithms*, 2001, pp. 661–670.
[8] M. Balinski, Integer programming: Methods, uses, computation, *Management Science*, 12, 253–313, 1965.
[9] R. Casey, Allocation of copies of a file in an information network, *Proceedings of the Spring Joint Computer Conference, IFIPS*, 1972, pp. 617–625.
[10] K. Chandy and J. Hewes, File allocation in distributed systems, *Proceedings of the International Symposium on Computer Performance Modeling, Measurement and Evaluation*, 1976, pp. 10–13.
[11] W. Chu, Optimal file allocation in a multiple computer system, *IEEE Transaction on Computers*, C-18, 885–889, 1969.
[12] I. Cidon, S. Kutten, and R. Soffer, Optimal allocation of electronic content, *Proceedings of IEEE INFOCOM*, April 2001, pp. 1773–1780.
[13] E. Clarke, Multipart pricing of public goods, *Public Choice*, 11, 17–33, 1971.
[14] V. Conitzer and T. Sandholm, Complexity of mechanism design, *Proceedings of International Conference on Uncertainty in Artificial Intelligence*, 2002, pp. 103–110.
[15] S. Cook, J. Pachl, and I. Pressman, The optimal location of replicas in a network using a READ-ONE-WRITE-ALL policy, *Distributed Computing*, 15, 57–66, 2002.
[16] R. Dash, N. Jennings, and D. Parkers, Computational mechanism design: A call to arms, *IEEE Intelligent Systems*, 18, 40–47, 2003.
[17] L. Dowdy and D. Foster, Comparative models of the file assignment problem, *ACM Computing Surveys*, 14, 287–313, 1982.
[18] K. Eswaran, Placement of records in a file and file allocation in a computer network, *Information Processing Letters*, 8, 304–307, 1974.
[19] A. Fiat, R. Karp, M. Luby, L. McGeoch, D. Sleator, and N. Young, Competitive paging algorithms, *Journal of Algorithms*, 12, 685–699, 1991.
[20] S. Floyd and V. Paxson, Difficulties in simulating the Internet, *IEEE/ACM Transaction on Networking*, 9, 253–285, 2001.
[21] T. Groves, Incentives in teams, *Econometrica*, 41, 617–631, 1973.

[22] S. Hakimi, Optimum location of switching centers and the absolute centers and medians of a graph, *Operations Research*, 12, 450–459, 1964.
[23] A. Heddaya and S. Mirdad, WebWave: Globally load balanced fully distributed caching of hot published documents, *Proceedings of the 17th International Conference on Distributed Computing Systems*, Baltimore, MD, 1997, pp. 160–168.
[24] S. Jamin, C. Jin, Y. Jin, D. Riaz, Y. Shavitt, and L. Zhang, On the placement of Internet instrumentation, *Proceedings of the IEEE INFOCOM*, 2000, pp. 295–304.
[25] S. Jamin, C. Jin, T. Kurc, D. Raz, and Y. Shavitt, Constrained mirror placement on the Internet, *Proceedings of the IEEE INFOCOM*, 2001, pp. 31–40.
[26] N. Jennings and S. Bussmann, Agent-based control systems, *IEEE Control Systems Magazine*, 23, 61–74, 2003.
[27] K. Kalpakis, K. Dasgupta, and O. Wolfson, Optimal placement of replicas in trees with read, write, and storage costs, *IEEE Transactions on Parallel and Distributed Systems*, 12, 628–637, 2001.
[28] J. Kangasharju, J. Roberts, and K. Ross, Object replication strategies in content distribution networks, *Proceedings of Web Caching and Content Distribution Workshop*, 2001, pp. 455–456.
[29] M. Karlsson and M. Mahalingam, Do we need replica placement algorithms in content delivery networks? *Proceedings of the International Workshop on Web Content Caching and Distribution*, August 2002.
[30] M. O'Kelly, The location of interacting hub facilities, *Transportation Science*, 20, 92–106, 1986.
[31] S. Khan and I. Ahmad, Heuristic-based replication schemas for fast information retrieval over the Internet, *Proceedings of 17th International Conference on Parallel and Distributed Computing Systems*, San Francisco, CA, 2004.
[32] S. Khan and I. Ahmad, Internet content replication: A solution from game theory. Technical report, Department of Computer Science and Engineering, University of Texas, Arlington, CSE-2004-5, 2004.
[33] S. Khan and I. Ahmad, A powerful direct mechanism for optimal WWW content replication, *Proceedings of 19th IEEE International Parallel and Distributed Processing Symposium*, 2005, p. 86.
[34] S. Khan and I. Ahmad, A game theoretical extended Vickrey auction mechanism for replicating data in large-scale distributed computing systems, *Proceedings of International Conference on Parallel and Distributed Processing Techniques and Applications*, 2005, pp. 904–910.
[35] S. Khan and I. Ahmad, Data replication in large distributed computing systems using discriminatory game theoretic mechanism design, *Proceedings of 8th International Conference on Parallel Computing Technologies*, 2005.
[36] S. Khan and I. Ahmad, RAMM: A game theoretical replica allocation and management mechanism, *Proceedings of the 8th International Symposium on Parallel Architectures, Algorithms, and Networks*, 2005, pp. 160–165.
[37] S. Khan and I. Ahmad, A pure Nash equilibrium guaranteeing game theoretical replica allocation method for reducing Web access time, *Proceedings of the 12th International Conference on Parallel and Distributed Systems*, 2006.
[38] S. Khan and I. Ahmad, Replicating data objects in large-scale distributed computing systems using extended Vickrey auction, *International Journal of Computational Intelligence*, 3, 14–22, 2006.
[39] M. Korupolu and C. Plaxton, Analysis of a local search heuristic for facility location problems, *Journal of Algorithms*, 37, 146–188, October 2000.
[40] C. Krick, H. Racke, and M. Westermann, Approximation algorithms for data management in networks, *Proceedings of the Symposium on Parallel Algorithms and Architecture*, 2001, pp. 237–246.
[41] V. Krishna. *Auction Theory*, Academic Press, San Diego, CA, 2002.
[42] P. Krishnan, D. Raz, and Y. Shavitt, The cache location problem, *IEEE/ACM Transactions on Networking*, 8, 568–582, October 2000.

[43] H. Kuhn, Excerpt from Montmort's Letter to Nicholas Bernoulli, in *Precursors in Mathematical Economics: An Anthology*, ser. Reprints of scarce works on political economy, W. Baumol and S. Goldfeld, eds., vol. 19, pp. 3–6, 1968.

[44] J. Kurose and R. Simha, A microeconomic approach to optimal resource allocation in distributed computer systems, *IEEE Transactions on Computers*, 38, 705–717, 1989.

[45] Y. Kwok, K. Karlapalem, I. Ahmad, and N. Pun, Design and evaluation of data allocation algorithms for distributed database systems, *IEEE Journal on Selected Areas in Communication*, 14, 1332–1348, 1996.

[46] A. Leff, J. Wolf, and P. Yu, Replication algorithms in a remote caching architecture, *IEEE Transactions on Parallel and Distributed Systems*, 4, 1185–1204, 1993.

[47] B. Li, M. Golin, G. Italiano, and X. Deng, On the optimal placement of Web proxies in the Internet, *Proceedings of the IEEE INFOCOM*, 2000, pp. 1282–1290.

[48] T. Loukopoulos and I. Ahmad, Static and adaptive distributed data replication using genetic algorithms, *Journal of Parallel and Distributed Computing*, 64, 1270–1285, 2004.

[49] T. Loukopoulos, I. Ahmad, and D. Papadias, An overview of data replication on the Internet, *Proceedings of ISPAN*, 2002, pp. 31–36.

[50] C. Lund, N. Reingold, J. Westbrook, and D. Yan, Competitive online algorithms for distributed data management, *SIAM Journal of Computing*, 28, 1086–1111, 1999.

[51] S. Mahmoud and J. Riordon, Optimal allocation of resources in distributed information networks, *ACM Transaction on Database Systems*, 1, 66–78, 1976.

[52] B. Maggs, F. Meyer auf der Heide, B. Vocking, and M. Westermann, Exploiting locality for data management in systems of limited bandwidth, *Proceedings of the Symposium on Foundations of Computer Science*, 1997, pp. 284–293.

[53] A. Mascolell, M. Whinston, and J. Green, *Microeconomic Theory*, Oxford University Press, 1995.

[54] R. McAfee and J. McMillan, Actions and bidding, *Journal of Economics Literature*, 25, 699–738, 1987.

[55] F. Meyer auf der Heide, B. Vocking, and M. Westermann, Caching in networks, *Proceedings of the 11th ACM-SIAM Symposium on Discrete Algorithms*, 2000, pp. 430–439.

[56] R. Myerson, *Game Theory: Analysis of Conflict*, Harvard University Press, 1997.

[57] B. Narebdran, S. Rangarajan, and S. Yajnik, Data distribution algorithms for load balancing fault-tolerant Web access, *Proceedings of the 16th Symposium on Reliable Distributed Systems*, 1997, pp. 97–106.

[58] J. Nash, The bargaining problem, *Econometrica*, 18, 155–162, 1950.

[59] J. Nash, Noncooperative games, *Annals of Mathematics*, 54, 286–295, 1951.

[60] J. Nash, Equilibrium points in N-person games, *Proceedings of the National Academy of Sciences*, 36, 48–49, 1950.

[61] J. Nash and L. Shapley, A simple three-person poker game, *Annals of Mathematical Studies*, 24, 105–106, 1950.

[62] N. Nisan and A. Ronen, Algorithmic mechanism design, *Proceedings of 31st ACM STOC*, 1999, pp. 129–140.

[63] M. Osborne and A. Rubinstein, *A Course in Game Theory*, MIT Press, 2002.

[64] L. Qiu, V. Padmanabhan, and G. Voelker, On the placement of Web server replicas, *Proceedings of the IEEE INFOCOM*, 2001, pp. 1587–1596.

[65] M. Rabinovich, Issues in Web content replication, *Data Engineering Bulletin*, 21, 21–29, 1998.

[66] P. Radoslavov, R. Govindan, and D. Estrin, Topology-informed Internet replica placement, *Computer Communications*, 25, 384–392, March 2002.

[67] J. Rosenchein and G. Zolotkin, *Rules of Encounter*, MIT Press, 1994.

[68] T. Sandholm, Distributed rational decision making, in *Multiagent Systems: A Modern Approach to Distributed Artificial Intelligence*, G. Weiss, ed., MIT Press, p. 201, 1999.

[69] R. Tewari and N. Adam, Distributed file allocation with consistency constraints, *Proceedings of the International Conference on Distributed Computing Systems*, 1992, pp. 408–415.

[70] A. Venkataraman, P. Weidmann, and M. Dahlin, Bandwidth constrained placement in a WAN, *Proceedings of ACM Symposium on Principles of Distributed Computing*, August 2001.

[71] W. Vickrey, Counterspeculation, auctions and competitive sealed tenders, *Journal of Finance*, 16, 8–37, 1961.

[72] J. von Neumann, Zur Theorie der Gesellschaftsspiele, *Mathematische Annalen*, 100, 295–320, 1928.

[73] M. Wellman, A market-oriented programming environment and its application to distributed multicommodity flow problem, *Journal of Artificial Intelligence Research*, 1, 1–23, 1993.

[74] O. Wolfson and S. Jajodia, Distributed algorithms for dynamic replication of data, *Proceedings of ACM Symposium on Principles of Database Systems*, June 1992, pp. 149–163.

[75] O. Wolfson, S. Jajodia, and Y. Hang, An adaptive data replication algorithm, *ACM Transaction on Database Systems*, 22, 255–314, 1997.

[76] W. Yuen, C. Sung, and W. Wong, Optimal price decrement strategy for Dutch auctions, *Communications in Information and Systems*, 2, 2002.

[77] E. Zegura, K. Calvert, and M. Donahoo, A quantitative comparison of graph-based models for Internet topology, *IEEE/ACM Transactions on Networking*, 5, 770–783, 1997.

46
Effectively Managing Data on a Grid

46.1	Introduction..	46-1
46.2	Related Work...	46-3
	Grid Computing • Data Grid Initiatives • The Advanced Computational Data Center Grid • The Advanced Computational Data Center Grid Monitoring Infrastructure	
46.3	The Scenario Builder	46-7
	Anatomy of a Scenario • Virtual Network Topology • Scenario Generation • Scenario Simulation • Scenario Visualization and Results • Further Remarks	
46.4	Intelligent Migrator..	46-16
	Anatomy of a Virtual Data Grid • Agent Infrastructure • Anatomy of an Agent • Generating a Virtual Data Grid • Simulating a Virtual Data Grid • Intelligent Migrator Visualization and Results • Further Remarks	
46.5	The Advanced Computational Data Center Data Grid	46-25
	Data Grid File Virtualization • Storage Network Architecture • Web Tools for the Advanced Computational Data Center Data Grid • Results • Further Remarks	
46.6	Conclusions	46-32
	Enabling Simulations with the Scenario Builder • Enabling Simulations with the Intelligent Migrator • Concluding Remarks	
	Acknowledgments..	46-34
	References	46-34

Catherine L. Ruby
Russ Miller
State University of New York

46.1 Introduction

Advances in parallel computing have resulted in the emergence of *grid computing*, a field that strives to integrate multiple distributed and heterogeneous resources, which are typically under distinct administrative domains, into a single, coherent system that can be efficiently and effectively used to solve complex and demanding problems in a fashion that is transparent to the end user [1–3]. Such resources include compute systems, storage systems, sensors, visualization devices, rendering farms, imaging systems, and a wide variety of additional Internet-ready instruments. Grids of various sorts are quite fashionable and are being deployed in many fields, including structural biology, computational chemistry, cancer research, biomedical informatics, astronomy, environmental science, and high-energy physics, to name a few.

Computing with limited resources is a pervasive theme in the field of computational science and engineering, which typically focuses on large-scale problems that rely on simulation and modeling. We live in a digital, data-driven society, where demands for the analysis, storage, maintenance, networking, and visualization of data is increasing at an extraordinary rate. Therefore, advances in scalable grid computing are critical to twenty-first-century discovery.

Although inherently scalable in its distributed design, grid computing is in its infancy. That is, it is common for grids to operate under well-controlled environments that predominantly include large computational clusters, data repositories, and high-end networking. Designing and deploying reliable data systems even within these controlled environments is an essential next step in terms of providing support for the ultimate ubiquitous grid.

A data grid consists of a geographically distributed network of storage devices intended to house and serve data. Data grids provide the control and management of data used and produced by grid applications, and are often, but not always, closely coupled to computational grids [1,4,5]. Distributing files across multiple storage elements can present load issues as usage patterns change over time. Data grids are dynamic systems, and it is essential that as storage network configurations and data usage patterns evolve, load issues can be predicted, analyzed, and addressed to maintain storage network performance. Overcoming the complexities associated with high-utilization distributed data systems, through examining the relationship between data grid performance and its file utilization patterns and storage network architecture, is a key to delivering a reliable and robust data grid.

The Advanced Computational Data Center Grid (ACDC Grid) is a grid deployed through Cyberinfrastructure Laboratory at the State University of the New York at Buffalo (SUNY-Buffalo) [6–9]. The ACDC Grid integrates several compute elements from SUNY-Buffalo and from other institutions across the country. This system supports users running a variety of grid applications, including the Shake-and-Bake (SnB) [10] application for molecular structure determination and the Princeton Ocean Model (POM) [11] for tracking harmful algal-blooms in the Great Lakes. The ACDC Grid supports many data-intensive grid applications, requiring the deployment of a reliable data grid solution to support the grid and enable continued growth [8].

In this chapter, we will discuss three initiatives that are designed to present a solution to the data service requirements of the ACDC Grid. We present two data grid file utilization simulation tools, the Scenario Builder and the Intelligent Migrator, which examine the complexities of delivering a distributed storage network. We then discuss the ACDC Data Grid, currently in use in conjunction with the ACDC Grid, and the ways in which the Scenario Builder and Intelligent Migrator enable us to simulate the performance of the ACDC Data Grid over time so that the system can be appropriately tuned for performance.

Deploying a distributed storage element solution to a system such as the ACDC Grid requires an analysis of storage load as data is utilized for computational jobs. The Scenario Builder, developed in conjunction with the ACDC Grid, is a dynamic, Web-based tool for creating, simulating, and analyzing virtual data grid environments in an effort to relate simulated file usage to potential production data grid scenarios and predict file utilization across a storage network over time. Through the Scenario Builder, highly customizable data grid scenarios may be generated and simulated over variable lengths of time in order to examine the demand on the virtual data grid storage network and predict the behavior of a physical data grid solution under similar circumstances.

Data files that are used to execute computational jobs, such as configuration files or data sets for analysis, can present key challenges when located across a distributed storage system. Trafficking files between computational resources as users submit grid jobs can become a difficult task, and users who submit jobs repeatedly can use a significant amount of time and bandwidth to move their files to the correct locations before job execution. Improving on the design of the Scenario Builder, we present the Intelligent Migrator, a system that consists of routines to generate, simulate, and visualize proposed data grid scenarios in an effort to learn about projected patterns of data usage over time. This allows for the effective management and replication of data to remote locations so that data files are present in advance when they are needed. By studying the performance of the Intelligent Migrator on generated scenarios, we can evaluate how the learning procedures would perform in a similar physical data system.

The ACDC Data Grid is a data grid serving the needs of the ACDC Grid. In this chapter, we present the ACDC Data Grid, its internal storage infrastructure, and its user interfaces, which provide a seamless environment for file access over its distributed design. The ACDC Data Grid, implemented across several storage repositories at SUNY-Buffalo and integrated with compute elements across Western New York, New York State, the Open Science Grid (OSG) [12] and TeraGrid [13], serves the data needs of the grid applications run on the ACDC Grid. The Scenario Builder and the Intelligent Migrator are based on the design and architecture of the ACDC Grid and enable the construction of tailored virtual data grid configurations for the examination of its potential performance in a wide variety of usage scenarios.

The development of data storage solutions is an essential component of ongoing grid research and advances in grid computing. Grid applications demonstrate a growing need for data services, and infrastructures for robust and reliable data grids are a necessity. The Scenario Builder, the Intelligent Migrator, and the ACDC Data Grid are three initiatives designed to meet the complex challenges of creating a distributed, heterogeneous storage framework and serve the ongoing needs of the ACDC Grid.

In this chapter, we will discuss the three data grid initiatives developed for the ACDC Grid. In Section 46.2, we review related research in the field of grid computing and present an overview of the ACDC Grid initiative. Section 46.3 discusses the Scenario Builder, both in its design and implementation, as well as in the tools developed for generating tailored data grid scenarios for simulation and visual analysis. We discuss the Intelligent Migrator in Section 46.4, including the design of a virtual data grid, its simulation and algorithms for modeling user file utilization over time, and the visualization tools developed for the analysis of the underlying algorithm's performance. In Section 46.5, we present the design and implementation of the ACDC Data Grid, which is currently providing the data services for the ACDC Grid. In Section 46.6, we include our final thoughts and discuss the ways in which the Scenario Builder and Intelligent Migrator impact the design of the ACDC Data Grid and enable performance simulations.

46.2 Related Work

The Scenario Builder, Intelligent Migrator, and the ACDC Data Grid are part of a research initiative to develop critical grid infrastructure. In this section, we will present a background of computational and data grid initiatives. We will then present the ACDC Grid and the monitoring infrastructure designed and deployed by the Cyberinfrastructure Laboratory.

46.2.1 Grid Computing

As the scientific community embraces simulation and modeling as the third component of research, complementing theory and experimentation, high-end computing has become a key to solving an ever-expanding array of scientific questions. Scientific and engineering applications that rely on simulation and modeling in areas that include physics, chemistry, astronomy, biology, manufacturing, and life sciences, to name a few, have produced a significantly increased need for computing, storage, networking, and visualization, creating the necessity for powerful compute resources and increased collaborations between research facilities. Founded on the principles of parallel computing and built on the maturing public Internet architecture over the last 10 years, grid computing has entered the mainstream as a viable solution to the growing needs of academic and corporate research ventures. Grid research strives to integrate geographically dispersed sites to provide a single infrastructure for accessing compute and storage resources available throughout the world [3,14].

Providing scalable and seamless access to distributed compute resources is a key component of grid infrastructure. Significant challenges include providing a viable solution for supporting potentially large and data-intensive applications between grid-enabled endpoints, given unpredictable bandwidth and connectivity limitations. Authentication and security are also critical issues when operating across administrative domains. Furthermore, common computational problems such as resource allocation and scheduling, which are present in a homogeneous computing environment, become more significant challenges in a heterogeneous and distributed setting [3].

Numerous grid initiatives have been spawned in the last several years in an attempt to overcome these challenges and provide a robust and scalable grid solution. Early grid research produced the concept of metacomputing, that is, integrating large, geographically dispersed supercomputing sites. Endeavors such as the information wide area year (I-WAY) experimented with linking high performance supercomputers by integrating existing high bandwidth networks. Further grid research yielded the second generation of the grid and the recognition of three main grid computing issues, namely (i) heterogeneity, or the ability to link machines with many different architectures or administrative policies, (ii) scalability, in order for grids to grow both in their size and in their support of the expanding scientific applications that require grid services, and (iii) the adaptability of the grid infrastructure to varying degrees of quality-of-service (QoS) between its heterogeneous and distributed computational components. This second generation resulted in the creation and deployment of grid middleware, such as Globus [15], for authentication and data-transfer capabilities over distributed systems, and Legion [16], which provided seamless resource integration through an object-oriented approach. The second generation of grid research also produced core technologies like the common object request broker architecture (CORBA) [17] and Grid Web portals [18].

The current focus is on developing grids that tie together the middleware discussed earlier with the overall mission of producing user-friendly grids. The concept of a Virtual Organization (VO) has been established in response to the goal of facilitating collaboration, in which rules regarding the access and sharing of computational resources are established as a response to the definition of a common academic or industrial goal [14]. For example, we established the Grid Resources for Advanced Science and Engineering (GRASE) VO for research in science, including, but not limited to, biology, chemistry, and engineering [19]. Grid initiatives currently take a more service-oriented approach, with an increased emphasis on enabling distributed collaboration and autonomic methods of failure detection and resolution [18].

46.2.2 Data Grid Initiatives

As the field of grid computing matures and grid applications evolve, the demand for efficient and reliable storage solutions is growing significantly. Many ongoing computational grid initiatives involve data-intensive applications. Scientific projects in many areas, including CERN's Large Hadron Collider (LHC), is projected to produce petabytes of data per year [3]. Grid applications are requiring and producing growing volumes of data, calling for better storage solutions to facilitate further advances in the field of grid computing [1].

A data grid is a grid solution designed to meet the data storage and service needs of a computational grid by networking a series of distributed, heterogeneous compute elements to provide a reliable system for serving available data [20]. Data grids house and serve data to grid users, providing the necessary data services for their computational grid-based jobs. The distributed and dynamic nature of a data grid, and the heterogeneous nature of the storage elements that may belong to one of several administrative domains, can present complex challenges in ensuring the reliability and availability of a user's data files [1,21].

A common approach to meeting these challenges is to provide virtualization services to better manage the data in the storage network. Common abstraction techniques include generating metadata for data files to facilitate straightforward retrieval from a knowledge repository. Others involve constructing a logical namespace for data in order to organize file attributes, which may be stored in a database. Emulation and migration techniques serve to reproduce portions of data stored in the storage network and maintain the processes necessary for managing a digital entity. Other endeavors utilize the concept of data replication for managing grid data [1,5]. These virtualization services attempt to alleviate the difficulties of retrieving data across arbitrarily large and heterogeneous data grids [21].

There are many examples of these virtualization services in grid initiatives around the world. The Storage Resource Broker [22], developed at the San Diego Supercomputing Center, is integrated with an Extensible Metadata CATalog (MCAT) System [23] to provide a logical namespace in a database for the files stored within its infrastructure, in order to easily apply file system operations as well as other extended capabilities. Similarly, the European DataGrid (EDG) project [24] maintains a mapping of global names to local names, and supports caching and streaming as well as data replication to multiple storage devices within the storage network to maintain high data availability [21]. The Grid Physics Network (GriPhyN) [25] utilizes the concept of Virtual Data and the creation of Petascale Virtual Data Grids (PVDGs) [26], which catalog data and virtualize information in order to serve needs of data-intensive grid applications [27]. These and other virtualization techniques are solutions to the challenges arising from creating robust and reliable data storage networks to serve data-intensive grid applications.

46.2.3 The Advanced Computational Data Center Grid

The ACDC Grid [6] is an implementation of a computational grid infrastructure that integrates high-performance compute elements at SUNY-Buffalo with computational resources across the country. It encompasses computing elements with a variety of queue managers, operating systems and Wide Area Network (WAN) connections, and focuses on seamlessly integrating highly heterogeneous compute elements to facilitate computational research [8,9,28].

The development of the ACDC Grid utilizes a series of application programming interfaces (APIs) to create a robust architecture for integrating compute hardware with grid middleware and user interfaces to the job submission system. The implementation of the ACDC Grid also includes the development of grid-enabling application templates (GATs) [29], which define the procedures for executing a scientific application in a grid environment. Authentication, data transfer, and remote execution are supported by the Globus Toolkit [8,9,15,28].

The ACDC Grid initiative encompasses the development of both the infrastructure for submitting and executing jobs, as well as the ACDC Grid Web Portal, the interface through which users access the ACDC Grid. Served by an Apache HTTP Server [30] and written in the PHP hypertext preprocessor scripting language [31] along with JavaScript [32], the Web portal provides real-time MySQL [33] database access that supports job submission and portal architecture [8,9]. Figure 46.1 shows the main ACDC Grid Web Portal page and the point of entry for ACDC Grid users.

The ACDC Grid is a general purpose grid and supports a wide variety of scientific applications. Grid applications include those in computational chemistry, structural biology, and environmental engineering, to name a few SnB [10], for example, is one grid application run on the ACDC Grid that utilizes the grid infrastructure to execute computationally intensive molecular structure determinations [8,9,28].

The ACDC Data Grid serves as the data storage network serving the grid users and grid applications run on the ACDC Grid infrastructure. Section 46.5 includes a detailed discussion of the design and implementation of this component and its service to the ACDC Grid.

46.2.4 The Advanced Computational Data Center Grid Monitoring Infrastructure

The Cyberinfrastructure Laboratory has created and deployed an extensive monitoring system for grid systems around the world, including the ACDC Grid, the OSG [12], and TeraGrid [13]. This lightweight grid monitoring system collects site functional and operational information for over 150 compute elements. Information is collected through a series of scripts designed to pull site environment information every few minutes from grid resources and use this as a bootstrap for collecting information, including the status of current running and queued jobs, the state of worker nodes, GridFTP [34] activity, bandwidth and network latency statistics, and a variety of other information. The data collected through this system provides the foundation for the historical job information and resource utilization statistics for the Intelligent Migrator, discussed in more detail in Section 46.4.

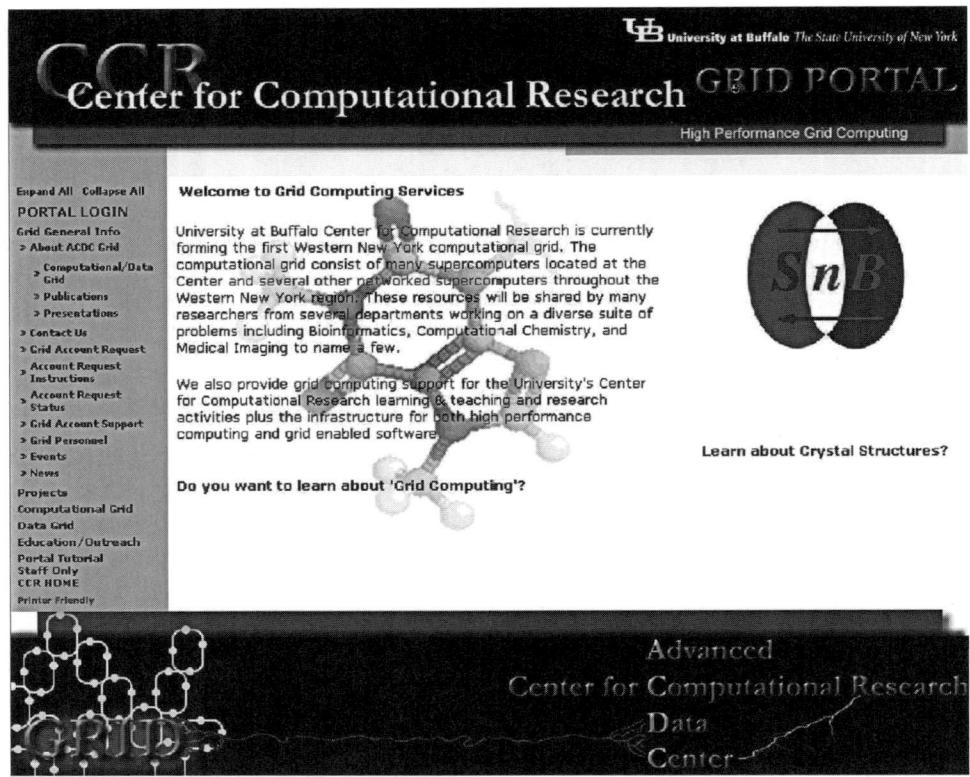

FIGURE 46.1 The ACDC Grid Web portal.

Data collected by the ACDC Grid Monitoring Infrastructure is aggregated into an online interface called the ACDC Grid Dashboard [35]. The main Grid Dashboard page provides an overview of the current status of a computational grid, job profiles, gatekeeper load information, GridFTP statistics, and resource storage load data. The presentation of this information is in the form of a variety of dynamic graphics. Clicking on any region of the Grid Dashboard brings the user to a new set of interactive plots with more detailed information on the grid environment, including more in-depth information on specific compute elements and historical information collected from the monitoring infrastructure. The main page of the ACDC Grid Dashboard is featured in Figure 46.2.

The ACDC Operations Dashboard [36], developed as an adjunct to the ACDC Grid Dashboard, provides an interactive and dynamic interface for viewing the operational status of compute elements available to grid users. The Operations Dashboard organizes operational status information in a Site Resource Service Matrix, showcasing the compute elements and their corresponding results from over 30 Site Functional Tests designed to test services offered by a compute element for the VOs it supports. These services range from simple connectivity tests and authentication verifications to the results of GridFTP tests and collecting and presenting VO support information from configuration files. Implemented in Perl [37], and based heavily on the site verification script written by Dr. Craig Prescott of the University of Florida [38], these tests are designed to provide detailed information regarding the functionality of a variety of services on a compute element for the VOs that are being monitored. Figure 46.3 shows a screenshot of the main ACDC Operations Dashboard page.

The Operations Dashboard provides an interactive subset of capabilities through Action Items. Organized in a multitiered authentication scheme using user certificates; these Action Items provide a pool of resources for administrators to detect, troubleshoot, and resolve operational issues on their compute elements. These and other popup displays are featured in Figure 46.4.

Effectively Managing Data on a Grid   46-7

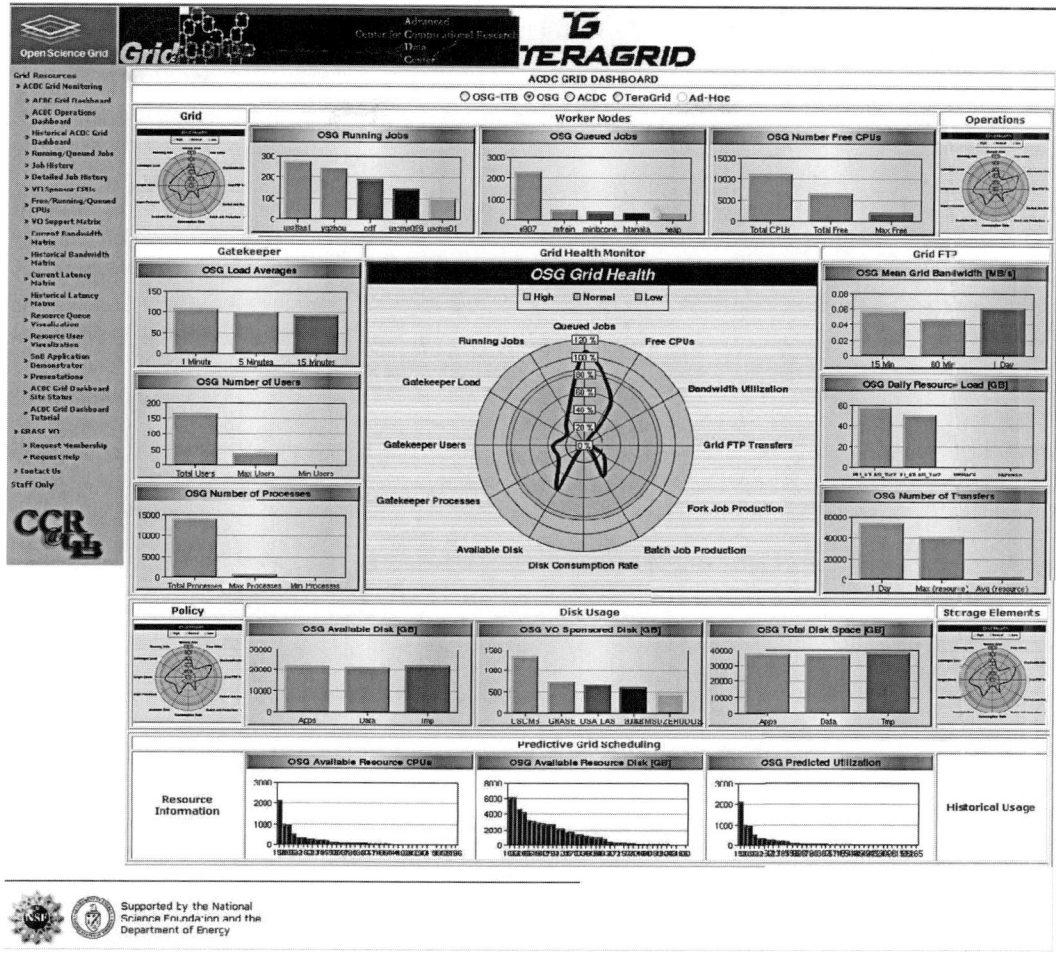

FIGURE 46.2 The ACDC Grid Dashboard.

Work on the ACDC Grid Monitoring Infrastructure serves to provide a wide range of accurate, up-to-date monitoring tools and interactive interfaces to facilitate the management of compute elements in a widely distributed grid setting. The underlying data collection routines and dynamic front-end interfaces provide a collaborative and interactive environment for publishing and responding to changes across monitored grid infrastructures. Plans have been made to apply the Operations Dashboard architecture to an ACDC Storage Elements Dashboard, which will track the functional status of machines dedicated as storage devices and running storage resource managers (SRMs) [39], a grid middleware component for file management and storage space allocation.

46.3 The Scenario Builder

In an early effort to model data grid infrastructure and examine storage load as it relates to file utilization over time, the Scenario Builder was created as a dynamic Web-based tool for constructing virtual data grid scenarios and simulating file accesses by grid users in order to explore how these would impact the performance of a distributed storage network. The Scenario Builder presents the user with a blank template for generating data grid usage scenarios. It contains simulation routines to model file accesses

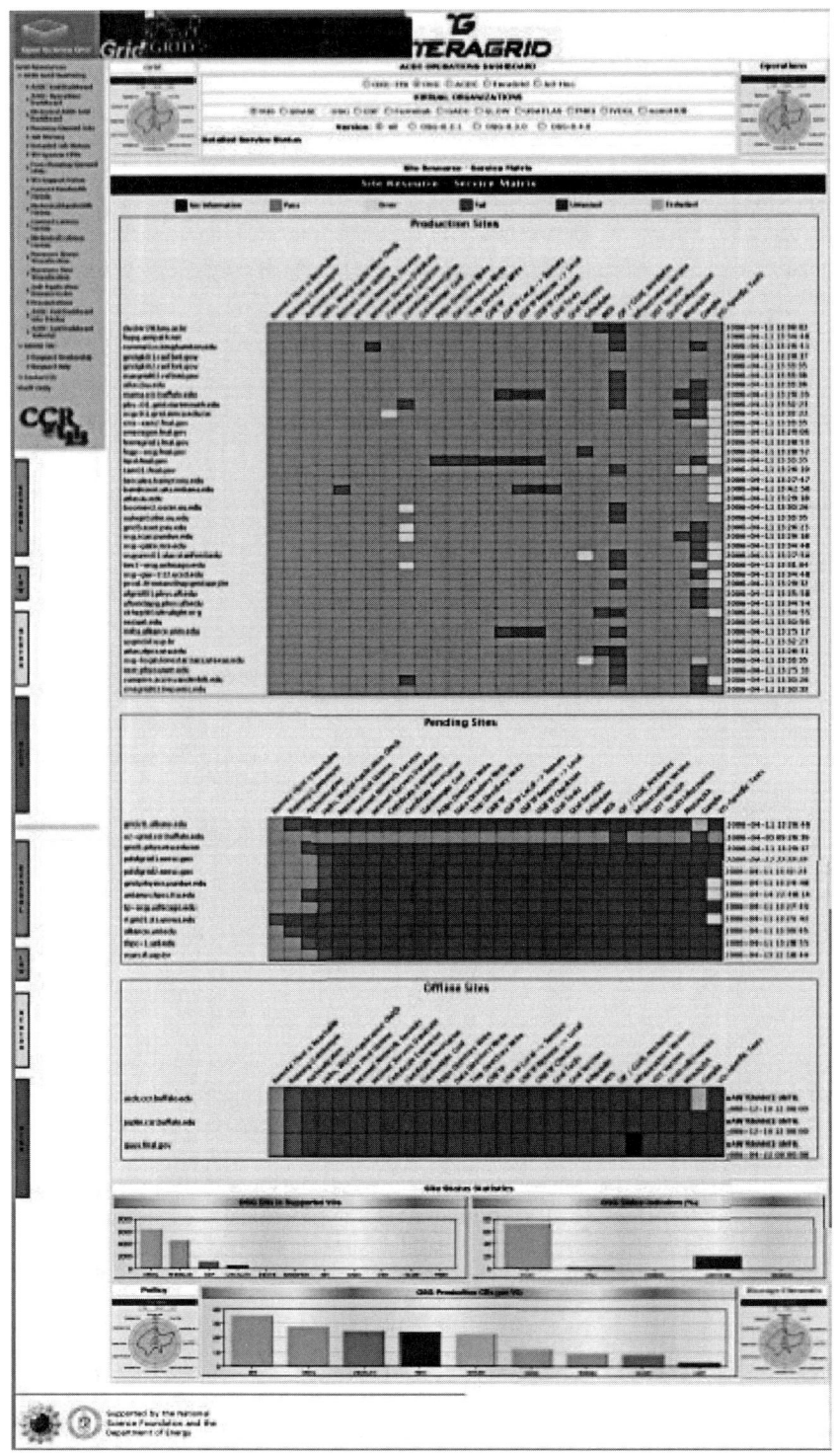

FIGURE 46.3 The ACDC Operations Dashboard.

Effectively Managing Data on a Grid

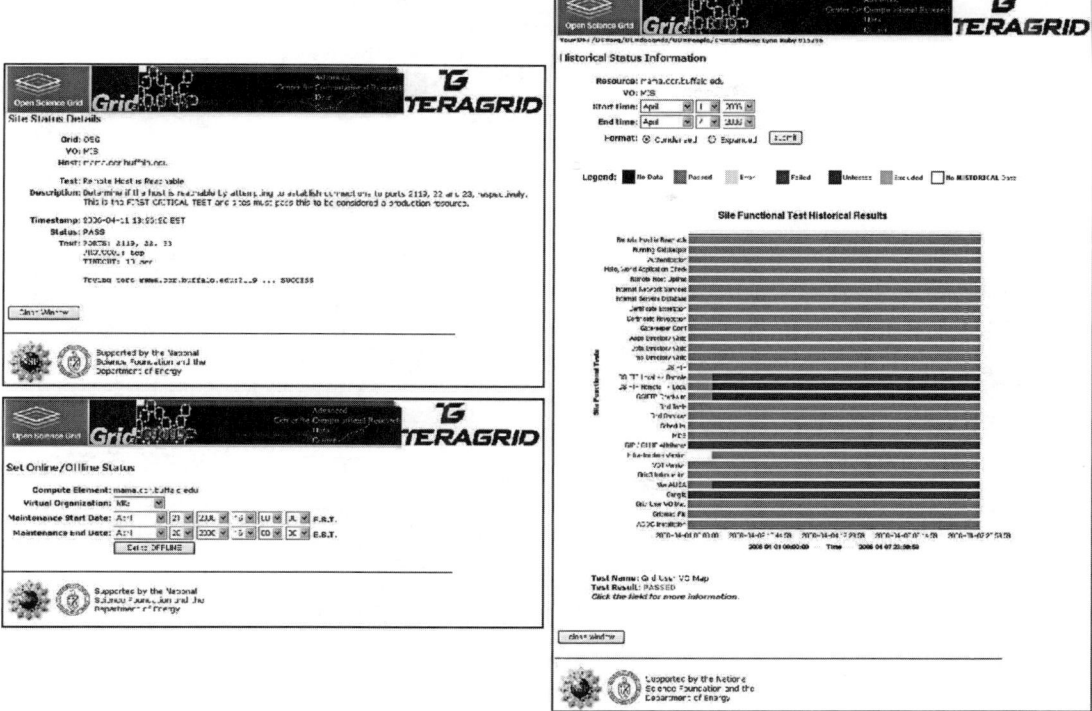

FIGURE 46.4 Select Action Items and popup displays from the ACDC Operations Dashboard.

and the system will present the user with records of bandwidth and patterns of file distribution based on a user-defined scenario, so that the user may evaluate scalability issues that could arise in such a setting.

In this section, we present the design and implementation of the Scenario Builder and its fundamental components, the concept of a scenario, and the architecture of the virtual storage network. We then discuss the Scenario Builder's three usage modes, namely, (i) generation, (ii) simulation, and (iii) visualization. These modes coupled with our Web-based interface provide a concise and simple manner for generating models, simulating usage, and evaluating the performance of the virtual data grid and its response to simulated file utilization. Finally, we will present results and further remarks.

46.3.1 Anatomy of a Scenario

The foundation of the Scenario Builder is built on the concept of a scenario, or a virtual user and file utilization model, which is a self-contained representation of a data grid. A scenario consists of users, groups, directories, and files in a virtual data grid. Virtual data grid users belong to groups and own directories that contain files. Files have permission levels (user, group, or public), sizes, creation times, and other common file attributes. Figure 46.5 illustrates these components and their relationships within a virtual data grid scenario.

Internally, a scenario is implemented as a series of MySQL database tables with entries that maintain the elements and their state within the virtual data grid. All scenario components are virtual entities within this database, and any location changes or migrations are simulated by reading or manipulating records in the database tables. This approach presents a lightweight and flexible design, where complex simulations can be run without maintaining physical files or constructing physical storage network architecture.

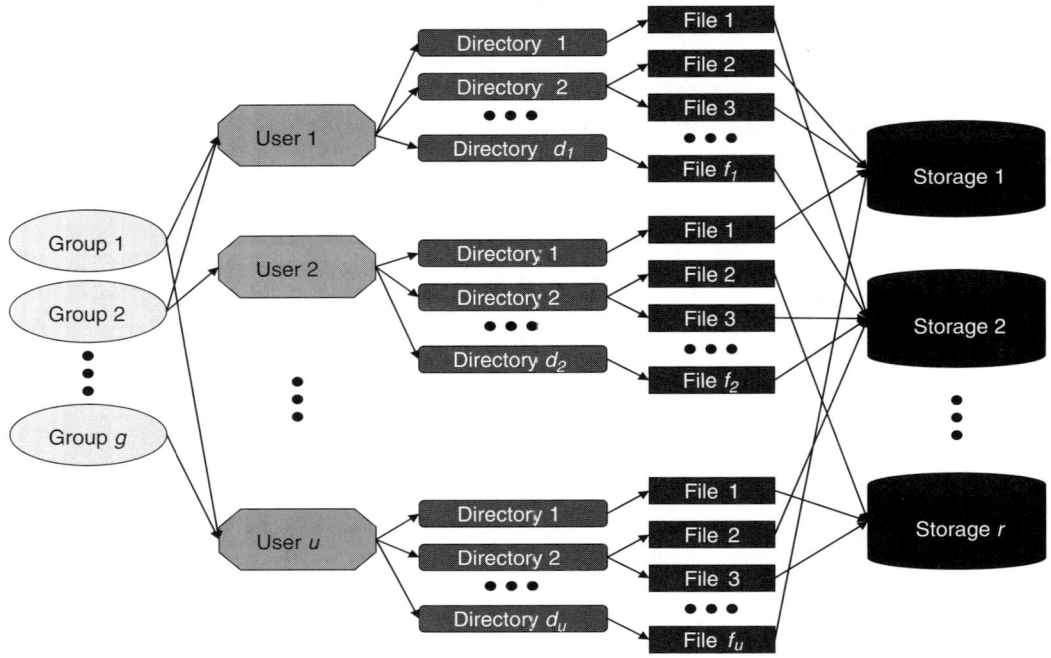

FIGURE 46.5 The anatomy of a scenario.

46.3.2 Virtual Network Topology

Another key element of the Scenario Builder is the virtual storage network. This represents the set of storage devices and their associated files, which are available to scenario users. The virtual storage network of the Scenario Builder is itself contained completely within the database, although it is designed to model a physical storage network with the capacity and connectivity constraints found in a physical system. The storage network architecture of the Scenario Builder can be altered to add or remove storage devices or modify system limits by updating its representation in the MySQL database, providing a more lightweight and flexible implementation than by creating a network of physical storage devices for simulation.

The virtual storage network of the Scenario Builder consists of a grid portal node and 11 storage devices. The grid portal node is the main storage repository in the model and represents the starting point of jobs run on the computational grid. When a file is accessed by a user during the simulation routines, its location is updated to the designated grid portal node. As the grid portal node has limited storage capacity, these files must be periodically migrated off of the grid portal node and onto the other storage repositories available. Figure 46.6 represents the topology of the virtual storage network for the Scenario Builder.

The virtual storage network grid portal node assumes a full duplex gigabit connection to the other 11 storage repositories in the simulation, with no network activity assumed outside of the Scenario Builder simulation. The bandwidth consumed during the periodic migrations (i) to free storage from the grid portal node and (ii) to distribute files across the storage network is considered to be significant, and must be shared between the eleven other storage devices. The bandwidth utilized is calculated on the basis of the random sizes bound to the virtual files and assumes that storage is distributed evenly over the other eleven repositories in proportion to their predesignated capacities. These imposed storage network limitations, and the topology of the storage network and its utilization, create the storage and

Effectively Managing Data on a Grid

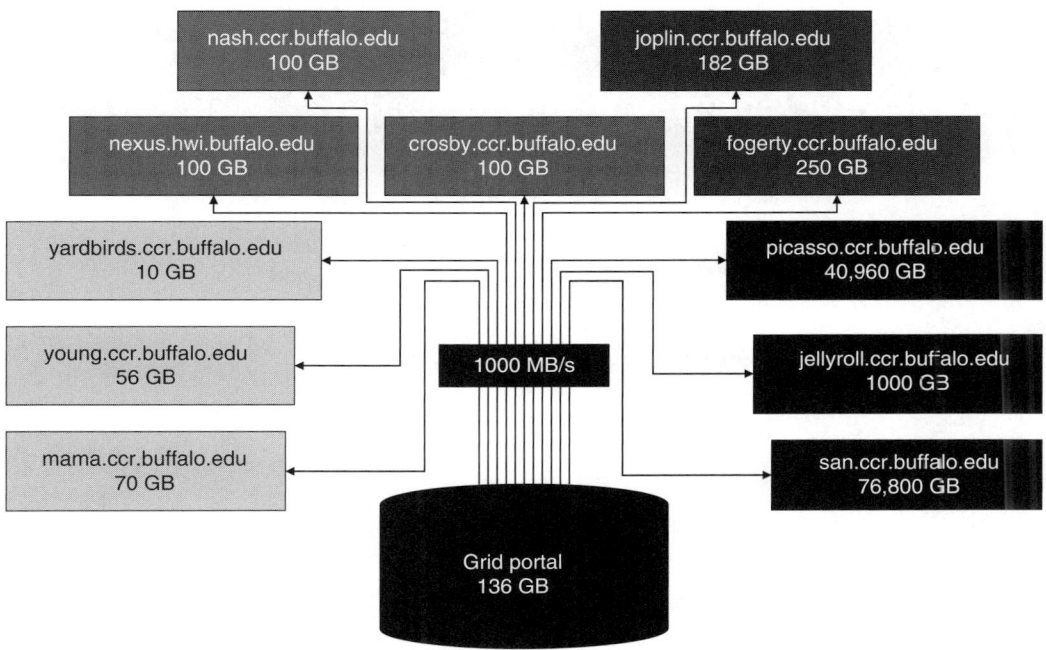

FIGURE 46.6 Scenario Builder network topology.

network load scenarios that we monitor. Further discussion is included with the presentation of results in Section 46.3.5.

46.3.3 Scenario Generation

Scenario generation takes place through a series of Web tools designed to facilitate the simple creation of tailored data grid scenarios through a dynamic and interactive interface. Accessible through the ACDC Grid Web Portal architecture discussed in Section 46.2, the Scenario Builder generation tool is built using the PHP and Apache architecture used for the portal. The generation suite maintains real-time access to the MySQL database where the scenarios are stored. A scenario is generated by creating the group, user, directory, and file components, which comprise the scenarios.

Figure 46.7 features a screenshot from the Scenario Builder generation tool, where 5000 files were created for a virtual user in the scenario on the grid portal node, with creation times ranging over the year previous to the time of generation. The attributes randomly selected for the file are shown below the form for selecting generation parameters.

46.3.4 Scenario Simulation

Simulating a virtual data grid scenario represents the core functionality of the Scenario Builder, which models file utilization by the virtual grid users over time. Built through the same interface used for scenario generation, the results of the simulation are used in the evaluation of the performance of a scenario. Users are prompted to enter the user, group, and public access probabilities for scenario data files, as well as the number of years back to initiate the simulation.

A simulation maintains the modified access times of data files, as well as their current locations on the virtual storage network. It also maintains records of bandwidth utilization during migration cycles. Upon creation, a virtual user is randomly assigned two parameters that model file utilization and dictate

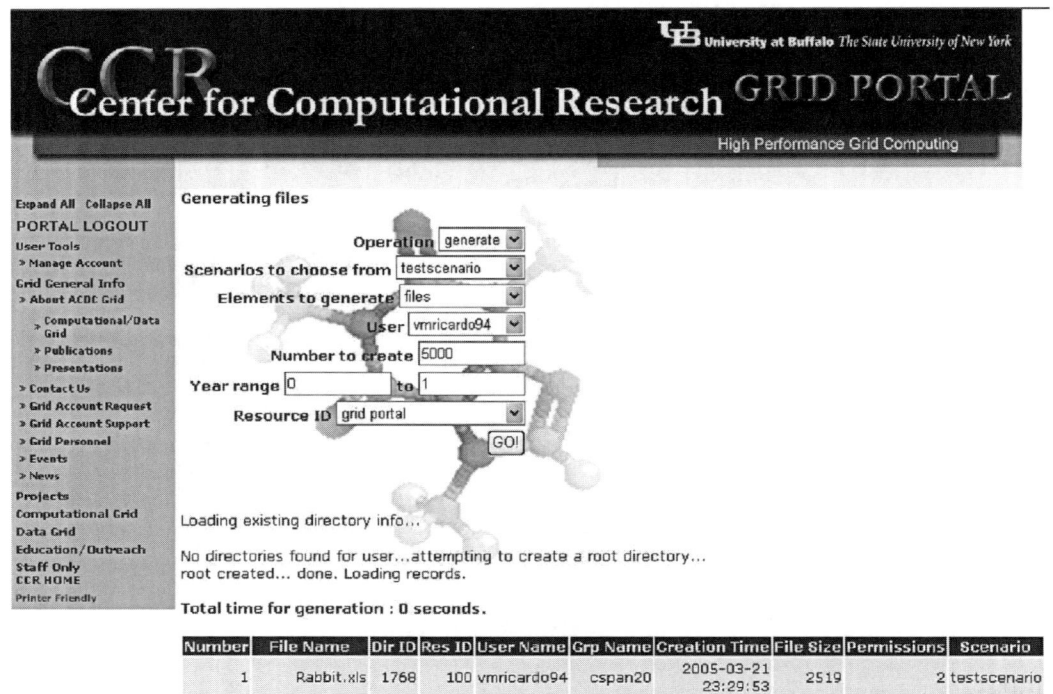

FIGURE 46.7 Example of scenario generation.

migration behaviors. The virtual grid user is given a user-aging parameter, which represents the probability that the user will access data files within the scenario. The virtual grid user is also given a global migration parameter, which indicates how long files should exist on the grid portal node before being flagged for migration to the other storage elements. These values shape the file utilization and migration patterns observed by the scenario simulation. A virtual user with a high user aging parameter and a low global migration parameter, for example, will access large numbers of files during the simulation while at the same time files will be flagged for migration more frequently, inflating the transfer loads and using more bandwidth than other virtual users possessing different parameters.

Simulations run once per virtual day from the starting point to the ending point of the designated time frame. Files are accessed at random according to the specified and randomly chosen access probabilities and placed on the grid portal node. Migrations are used to free space on the grid portal node on a virtual weekly basis or when the grid portal node's utilization exceeds 60%. During a migration cycle, files are flagged if they have gone more days without utilization than the virtual owner's global migration parameter, and are distributed evenly across the other storage devices by their assigned capacities. The bandwidth allocated to a storage device during a migration cycle is proportional to its transfer load during that cycle, and is reallocated as storage devices complete their transfers.

Figure 46.8 is a screenshot from the simulation of a 25,000 file scenario over one year with the default simulation parameters.

Effectively Managing Data on a Grid   46-13

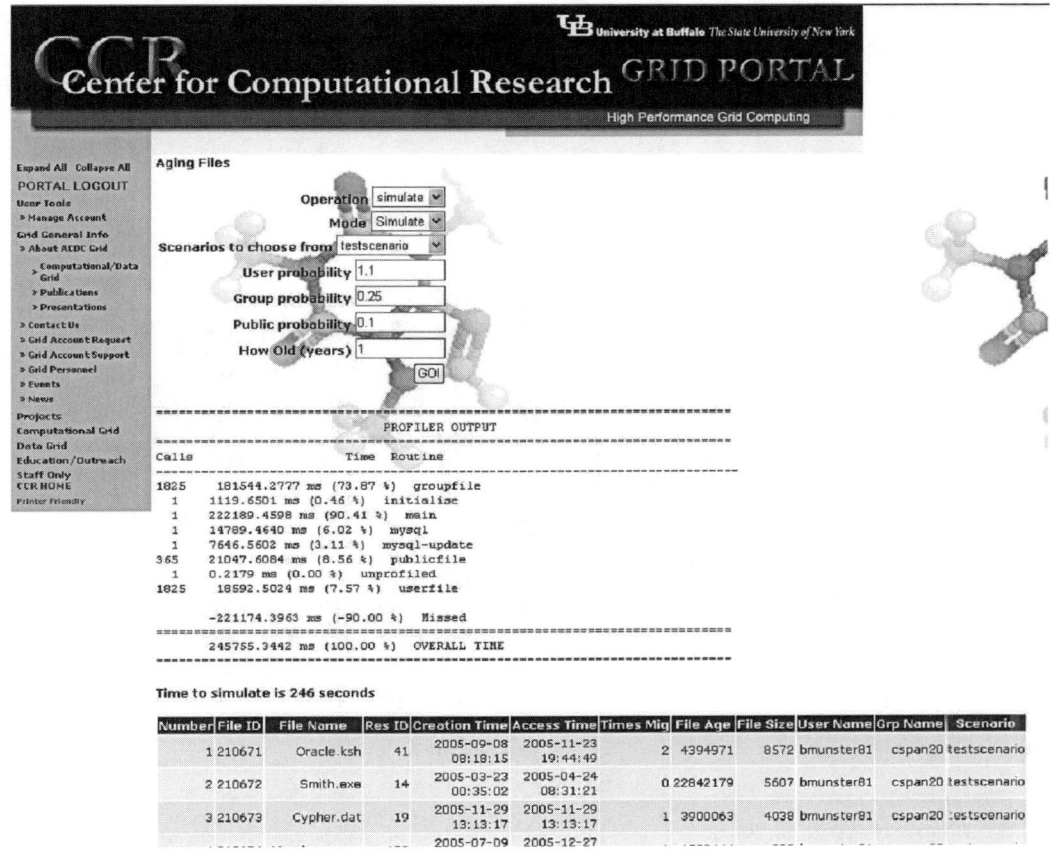

FIGURE 46.8 Example of scenario simulation.

46.3.5 Scenario Visualization and Results

The Scenario Builder is comprised of a series of visualization tools, designed to provide an in-depth view of the configuration of a scenario and the results of the simulation routines. This dynamic Web interface, available with the Scenario Builder generation and simulation tools, provides hundreds of interactive plots, pulling data from the MySQL database tables, and organizing it into clickable charts for the evaluation of scenario performance.

Using the visualization suite, we can explore the distribution of storage load across the Scenario Builder virtual storage network. Figure 46.9 shows the utilization of storage devices across the virtual storage network after a year-long simulation. The last bar represents the grid portal node, which in this example only 15% full, still within reasonable limits, whereas the other eleven bars, representing the other storage devices, are using percentages of their capacities all within 0.03% of one another, indicating an even storage load distribution.

We can also examine the network statistics from our scenario simulation to gain an understanding of the bandwidths experienced across the storage network. Figure 46.10 is a screenshot of a visualization chart detailing the bandwidths versus transfer loads over 52 migration cycles of the simulation. The bars represent the total megabytes migrated to each storage element over the time frame of the cycles selected, and the whisker layer indicates the minimum, maximum, and average bandwidths achieved by the storage devices over the same time frame. It is evident from this chart that the storage elements with the highest transfer load also achieved the highest average bandwidth, where the two storage elements with the largest

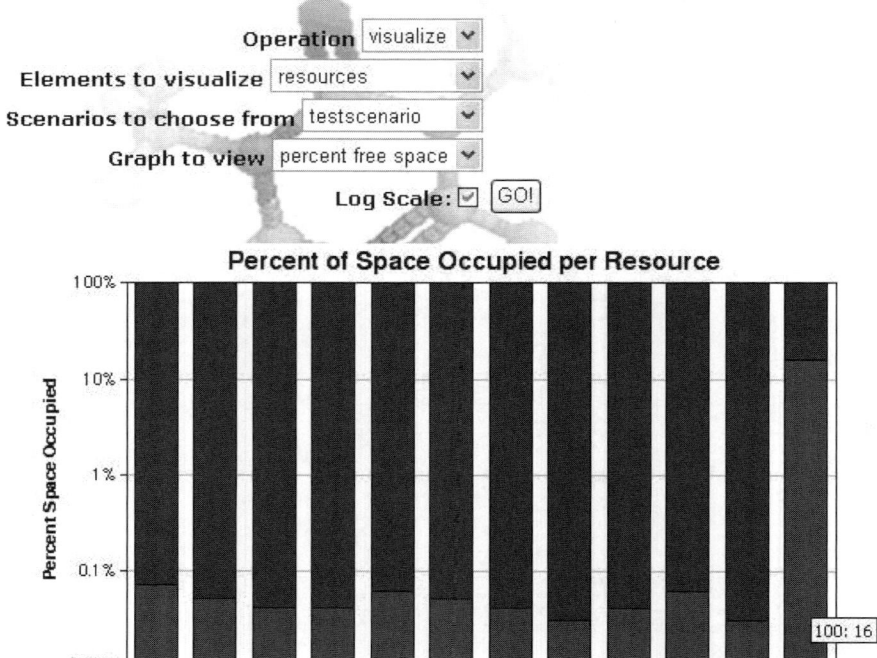

FIGURE 46.9 Resource capacity chart.

numbers of megabytes transferred observed bandwidths averaging approximately 125 MB per second, or the highest possible bandwidth given the gigabit connection. The simulation process is indeed allocating bandwidths proportional to migration transfer loads, and it is evident that for the scenario configuration simulated in this instance that the average bandwidths observed by all storage devices over the entire simulation period are within reasonable limits, or roughly half of the available bandwidth in spite of the being shared between up to ten other storage devices.

The distribution of files according to their ages, or the amount of time elapsed since their last access by a grid user within the simulation, is also an important statistic for scenario evaluation. Figure 46.11 shows scenario files by file ages on the grid portal node after a simulation over a virtual year. Each bar represents the number of files accessed on each virtual day, beginning from the leftmost bar, which is the last day of the simulation. It is clear from the figure that the largest scenario global migration parameter value of 42 days was enforced, as no files exist on the grid portal node that are more than 42 days old. It is also evident from this chart that the file access pattern is not linear. File access appears to occur in bursts for this usage scenario.

Consider a usage scenario where individual files tend to go roughly 50 days between utilization by the virtual users in the scenario. These files would be accessed, moved to the grid portal node, idle on the grid portal node for 42 days, waste valuable storage space, and consume bandwidth in their migration off the grid portal node, only to be called back to the grid portal node upon being accessed again shortly

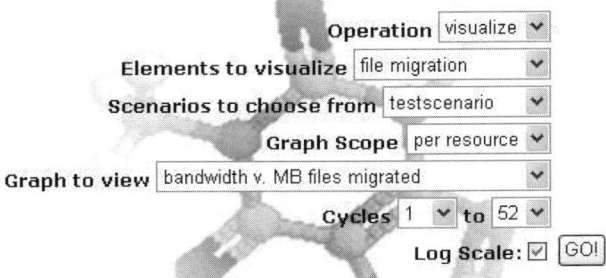
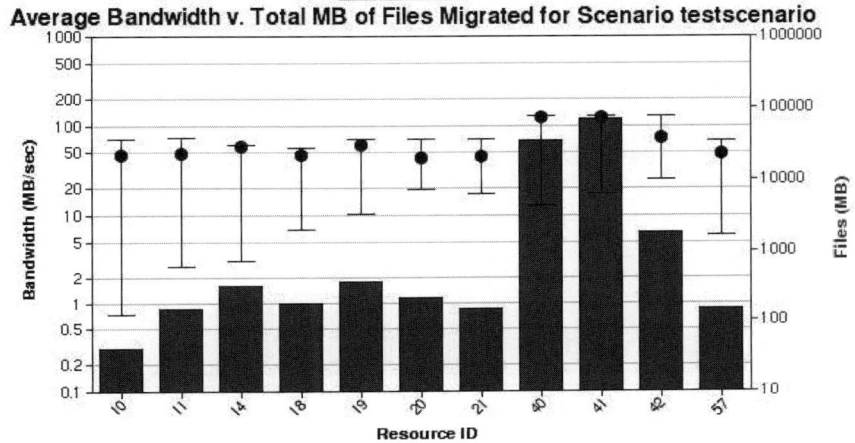

FIGURE 46.10 Bandwidth versus transfer load.

after. An examination of these usage trends and the application of curve-fitting techniques could serve to reduce the wasted storage on the grid portal node and the transfer load to other storage devices, improving the observed performance of the simulated data grid. This type of file usage scenario, and its relation to storage and bandwidth waste, is the basis for the Intelligent Migrator and is discussed in Section 46.4.

46.3.6 Further Remarks

The Scenario Builder represents early work for the ACDC Grid initiative in the area of utilizing simulations to examine file utilization and its relation to the performance of a general purpose data grid. The design of generation, simulation, and visualization routines for examining usage patterns, as well as the concept of a virtual data grid representing an independent data grid scenario completely within a MySQL database, serve as the foundation of the Intelligent Migrator discussed in Section 46.4. The Scenario Builder's relationship with the design of the ACDC Data Grid, and the ways in which it enables the simulation of its performance, are explored in Section 46.6.

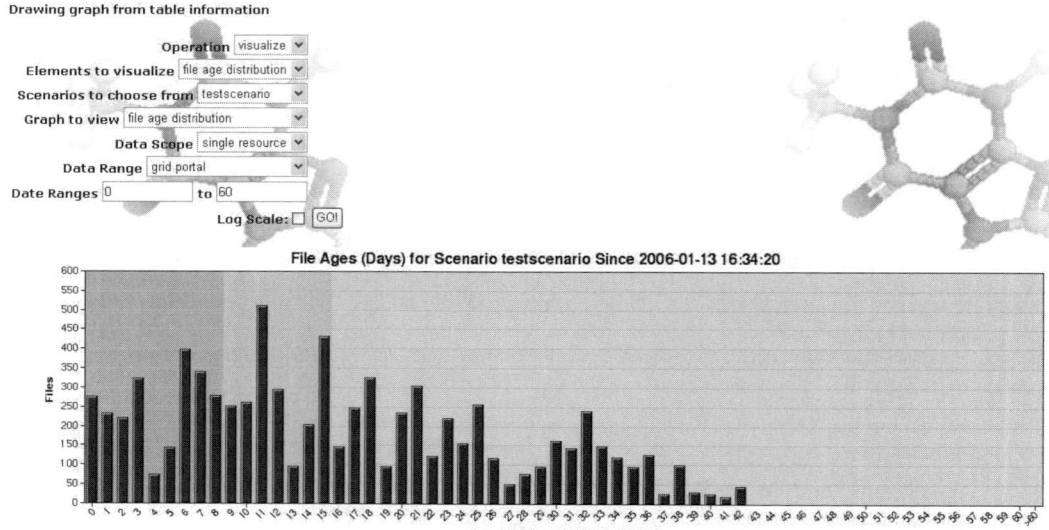

FIGURE 46.11 File age distributions for a simulated scenario.

46.4 Intelligent Migrator

In this section, we present the Intelligent Migrator, an agent-based system designed to minimize wasted storage and bandwidth on storage devices in a virtual data storage network. On the basis of the underlying design of the Scenario Builder, and using historical information collected by the monitoring infrastructure discussed in Section 46.2, the Intelligent Migrator is designed to monitor and observe user file usage and make intelligent decisions on where to place data files such that they will have maximum utilization on remote computational resources. The Intelligent Migrator consists of a series of tools for generating, simulating, and visualizing grid environments and file utilization over computational jobs across several remote resources. By studying the performance of the Intelligent Migrator, we can gain an understanding of the performance of a similar physical system.

46.4.1 Anatomy of a Virtual Data Grid

The Intelligent Migrator is based on the concept of a virtual file system. This component designates how file systems will be constructed to model real data grid scenarios, and provides the groundwork for simulating and evaluating its performance in a physical setting. Like the scenario of the Scenario Builder, the virtual file system of the Intelligent Migrator is implemented as a series of tables within a single MySQL database.

A storage element in the Intelligent Migrator is defined as remote storage space in a well-defined location that is accessible to a computational resource. Storage elements hold copies, or replicas, of files with a reference to the copy on a central storage repository or network of central storage devices local to the implementation of the data grid. By storing a data file on a storage element bound to a grid-enabled compute element, computational jobs scheduled to run on the compute element using the data file could utilize the file present on the storage element without staging the file to the compute element initially, saving time and bandwidth to the remote host. In the implementation of the Intelligent Migrator, a central data repository for a virtual data grid consists of a single MySQL database table, housing information on every file in the virtual file system. A storage element is represented by a table where file information can be written when a file is replicated to a remote storage element during the simulation. Upon the generation of a virtual file system for the Intelligent Migrator, one storage element is created for each resource represented by an agent.

A virtual file system generated for an Intelligent Migrator agent also maintains a set of virtual grid users. These users, like users in a physical grid environment, own files and submit computational jobs using their data files. User templates define the number and sizes of files that users own, the number of jobs they submit, the types of jobs submitted, and the number of files required for computational jobs, as well as the degree to which users deviate from the file utilization patterns established within the user template.

Grid users in a physical grid do not submit jobs at random, and often adhere to project schedules or demands particular to individual users over varying spans of time. The frequencies and durations of computational jobs submitted by grid users, as well as the compute elements that jobs that are submitted to, can have a significant impact on the usage of the associated data files and the patterns in which typical usage will follow and be captured by the Intelligent Migrator. The job history collected through the ACDC Grid Monitoring infrastructure serves as the pool of jobs from which virtual file system job histories are generated. Spanning several years of data collection for 68 compute resources, this repository holds valuable computational grid usage information on over 3.5 million jobs, including runtimes, user and group information, and other job attributes. Jobs for the compute elements supported by defined agents in the Intelligent Migrator are pulled at random from this monitored information and used as the computational jobs are submitted within the simulation routines.

To preserve the natural user job submission patterns present in the real monitored job histories, each user generated in the virtual file system is bound upon creation to a username from the job history pool, where a username in the job history repository represents all grid users who map to that username on the compute element to which the job was submitted. Files are bound to these jobs in the virtual data grid, translating the utilization of a computational grid to the data grid file utilization patterns we wish to examine.

There are several benefits to the architecture outlined in this section and to creating a completely simulated file system environment. A virtual data grid is lightweight, where its properties are housed in a small number of database tables of no more than a few megabytes each and are generated in a manner of minutes. The lightweight design of the file system is also highly flexible. Storage system components, distributions, and usage scenarios change at a rapid rate as grid initiatives are born, grow, and mature. The simulations of existing data grid storage scenarios and those that may develop in the future are essential in order to fully analyze the effectiveness of the optimization routines of the Intelligent Migrator. Providing an examination of the performance of the Intelligent Migrator in several usage scenarios is an important step in verifying the underlying intelligence of the system. The virtual file system architecture of the Intelligent Migrator allows for the generation of side-by-side comparisons and analysis of the performance of the optimization routines. Its design is also closely aligned with the design of the ACDC Data Grid such that the Intelligent Migrator can be easily deployed on the live system. The anatomy of a virtual data grid and its components is illustrated in Figure 46.12.

46.4.2 Agent Infrastructure

Methods of distributed artificial intelligence have long been investigated as solutions to such optimization problems. Such methods come from fields that include cooperative systems, game theory, and distributed computing. A multiagent system is an implementation of several small components, or agents, which interact to solve a problem or subproblems in a large system. Multiagent systems typically employ four main principles in their design and implementation. First, agents must have some level of autonomy and have some control over their actions and their internal state. Second, agents must be able to gather information on their environment and react appropriately. Third, they must be able to interact with other agents, typically with a predefined agent communication language or otherwise established protocol. Finally, they must also be goal-oriented, and their actions, as a part of a whole system, should move toward that goal as the system runs [40,41].

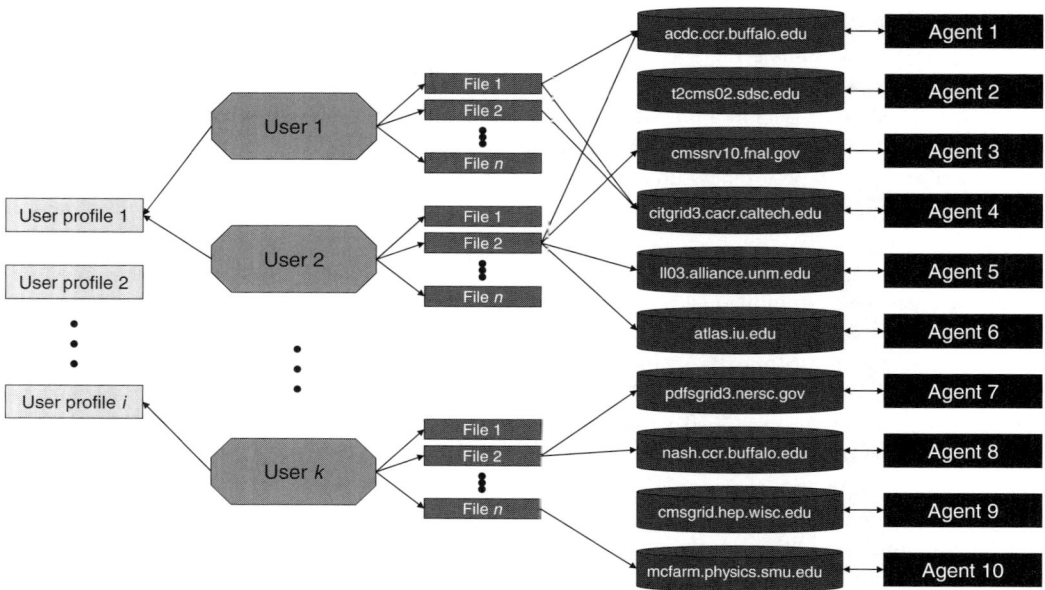

FIGURE 46.12 The anatomy of a virtual data grid.

These multiagent systems are often designed so that each agent has incomplete information regarding the overall problem to be solved and that there is no global control of the entire system. Similarly, in many multiagent systems, data is decentralized and the actions of agents are asynchronous with regard to the entire system. Inter-agent communication is often facilitated through defined techniques such as marketplace, auction, or blackboard dialogue styles [40–42].

Multiagent systems present attractive architectures for system design for several reasons. Partitioning responsibilities among individual agents results in a robust system, often eliminating single points of failure and providing a system more adaptable to failures and new scenarios. On a similar note, they may be easily distributed across parts of the system or even across many machines, lending themselves to growth in the development of distributed systems in many research sectors and even to grid development initiatives [40,41].

Although there are many advantages to deploying a multiagent system in a response to a difficult learning problem, there are many challenges that can arise in this type of distributed architecture. The methods of agent communication, for one, must be carefully planned as agent interdependencies may require the sharing of limited resources and a well-defined choreography of handoffs to prevent agent starving due to the lack of global control of the system. Keeping agent information consistent is another challenge, as agents may attempt to solve their subproblems with out-of-date information from surrounding agents due to limited communication opportunities or the dynamic nature of the system [40,41].

Multiagent systems vary widely in their practical applications. Problems such as managing control systems for satellites and communications systems are just two examples for which multiagent systems are well suited [43]. Multiagent systems have been used to solve scheduling problems and manage distributed resources [41]. They have also been applied to many problems relating to the field of grid research. One such project is the development of AGEGC [44], a grid computing model based on the Multi-AGent Environment (MAGE) platform, designed to facilitate matching resource requests with available grid-enabled compute elements. In the AGEGC system, resource agents represent high-performance compute elements and act as service providers, with registration and lookup mechanisms in place for locating resource agents and the services they provide.

The Intelligent Migrator uses independent agents to determine where to place files among a pool of remote compute resources with limited storage, given a set of user files. Each learning agent is privy to information regarding the past history of files on the resource it is assigned to, and is capable of making decisions regarding the distribution of data grid files over the limited storage available on its compute element. Agents submit requests to replicate certain files to their storage elements or to remove them, in order to ensure that the files with the highest likelihood of utilization are present for grid users and their computational jobs.

There are many benefits to the multiagent implementation of the Intelligent Migrator. Most importantly, using an agent-based method simplifies the problem of determining where to place a file to achieve the highest utilization, including the determination of whether a file should be replicated on a per compute element basis. As data files are copied to remote compute elements and not shared among them, optimizing data utilization across n compute elements is, in fact, equivalent to n completely independent subproblems, as optimal utilization on each resource constitutes optimal utilization across an entire computational grid. Agents are, thus, free to solve their subproblems in a self-interested manner, optimizing utilization on their own storage elements, and contributing to the solution of the entire problem. As this particular optimization problem is, in fact, a composite of independent optimization problems between agents, there is no need for agents in this system to communicate, and thus, many of the challenges associated with constructing a multiagent system as discussed previously are avoided.

In addition to simplifying the optimization problem, this type of system is extremely scalable as the underlying computational grid grows and evolves. Agents can be generated or removed as new compute elements are added to or retired from the computational grid. The asynchronous and decoupled nature of multiagent systems and especially of the architecture proposed in this section lends itself well to parallelization, and agent routines and computations could be integrated into computational grid jobs of their own and processed within the grid infrastructure they support.

46.4.3 Anatomy of an Agent

Artificial neural networks (ANNs) provide a robust method of solving complex machine learning problems. This supervised hill-climbing technique, commonly used for problems such as approximating real-valued or discrete-valued functions, has been proven to be highly effective in many practical applications. Neural networks are ideal for problems where supervised training sets are available, including training sets that contain noisy or erroneous entries. In addition, neural networks do not require a human understanding of the final learned target function. Changes to a neural network's design, such as its perceptron connection architecture, the encoding scheme of network inputs and outputs, and the energy function used for determining weight delta rules, are often tailored to meet the specific needs of a problem [45].

The Intelligent Migrator is an ideal candidate for the application of ANNs in terms of modeling file utilization in individual agents. Agents within the Intelligent Migrator each possess a layered ANN that represents the latest learned model of file utilization for the compute element assigned to the agent. Observed file utilization parameters are fed through the neural network file by file, and the magnitude of the output dictates the perceived strength of the file on the remote storage element assigned to the agent. Supervised training sets can be constructed on-the-fly from immediate past observed file utilizations, and data is inherently noisy from natural user deviations from past usage patterns. A fast evaluation of file parameters for the agent decision process is essential to the efficiency of the system, and an understanding of the learned user utilization model is not required for the execution of individual agents within the Intelligent Migrator.

ANNs have been applied to problems ranging from visual and speech recognition tools to scheduling systems and autonomous vehicle control. Pomerleau's autonomous land vehicle in a neural network (ALVINN) system at Carnegie Mellon uses visual sensory input to train a neural network to making steering decisions [46]. Neural networks have also been applied to the knapsack problem and similar optimization problems [47]. In Reference 48, an interesting discussion focuses on dynamically constructing generic neural networks of clusters of interlinked perceptrons to represent a system's state and possible solutions, with some configurations achieving convergence on a solution on the order of 10^8 times faster than sequential heuristic search techniques to solve similar problems.

The anatomy of an Intelligent Migrator agent is contained within the definition of the parameters of its agent type, including, but not limited to, the dimensions of the neural network model within the agent, the amount of simulated time over which to gather file parameters for calculating input values, and the training parameters used to tune the model on observed file utilizations. Agent types bring flexibility to the system, providing a means of customizing agents as new types of compute elements enter the already heterogeneous system. The agents deployed in this discussion of the Intelligent Migrator possess a single-layered neural network with four inputs, five hidden nodes, and a single output node. The input arguments fed to the neural network for a single file include the number of times the file was utilized on the remote resource, the number of times a job of the type most frequently associated with the file has been submitted by the user, the number of days since the file was last used by the user on the remote resource, and the current value of the file on the storage associated with the resource, or its value on the main storage repository if it is not present on storage local to the compute element. Figure 46.13 illustrates the topology of the agent's internal ANN.

46.4.4 Generating a Virtual Data Grid

Virtual data grid generation is the first step in utilizing the Intelligent Migrator to simulate file utilization and determine the accuracy and effectiveness of the underlying algorithms. Implemented as a PHP script

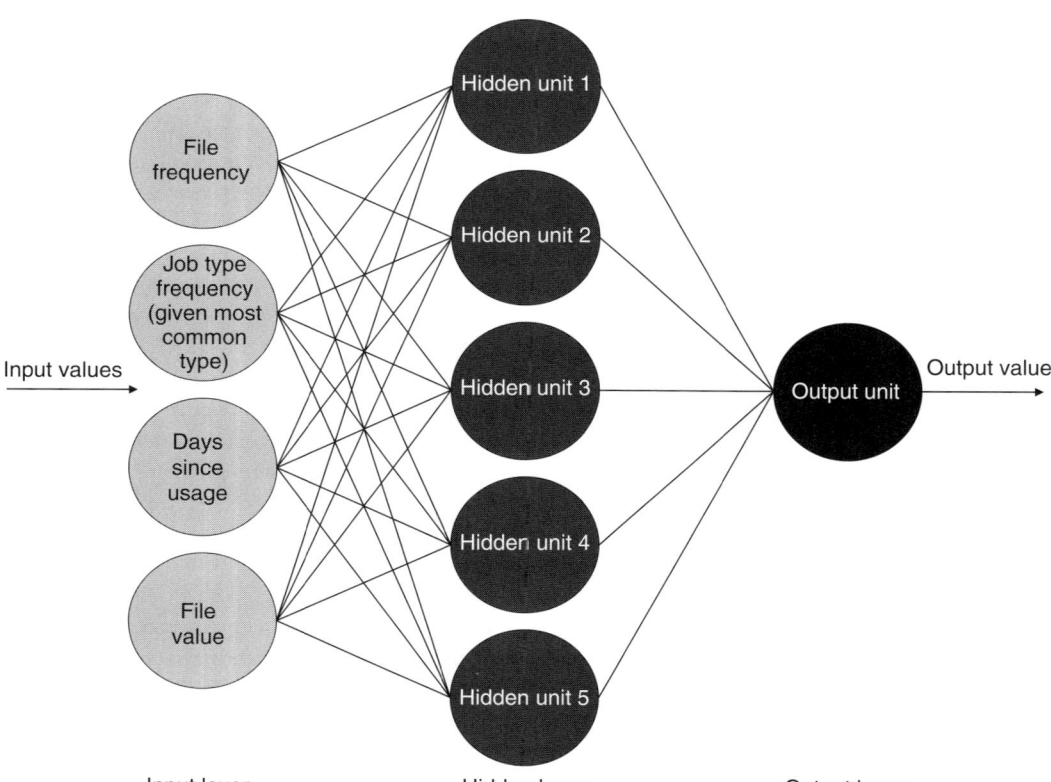

FIGURE 46.13 Agent neural network topology.

through a command line prompt, this procedure constructs the components that comprise a complete virtual file system. Upon the generation of a virtual data grid, a MySQL database is created on the basis of the user designated virtual file system name. Intelligent Migrator users then specify, in pairs, the user profile template number, which is discussed in Section 46.4.1, and the number of users to generate for each user profile template type. Information regarding the number and attributes of files to create, the number of jobs to select, and file binding information is pulled from the user template definitions in the MySQL database and used in generating the virtual file system.

The computational jobs run for virtual users in the generated data grid are pulled from job information collected by the ACDC Grid Monitoring infrastructure, as discussed in Section 46.2. The virtual users created are bound at random to usernames from the job pool, and jobs are drawn from the pool for each virtual grid user, although the order of the jobs is maintained to preserve the natural grid utilization patterns present in our job history records. As jobs are pulled from the job history pool available to the Intelligent Migrator, they are assigned to a job type, representing a grid application, according to user profile template information.

Grid users in a real grid setting do not access their files at random. There are often sets of files that a user tends to select for use with certain computational jobs, and these may or may not overlap with files submitted with other computational jobs. File utilization is not uniformly random, as there are files accessed for computational job submission with a much higher frequency than others. We model this in the Intelligent Migrator by ordering a virtual user's files and choosing a random starting point among them for each job type they will submit. For each file bound to a job, we walk randomly in either direction a small distance from this point and select files. The file furthest from our starting point becomes the starting point in the event of the next job submission of the same type. Choosing files in this manner approximates a multimodal function for file utilization, where small pools of files are utilized for computational jobs of the same type and represent the same scientific grid application. An example of generated file utilization for a virtual user submitting jobs of six types is shown in Figure 46.14, charted using the Intelligent Migrator visualization tools discussed in Section 46.4.6. In order to ensure that it is possible to correctly select and place all files required during a day onto the correct network storage element, the virtual data grid is examined and repaired such that on any given day for any given compute element, the total size of the files required does not exceed the capacity of the compute element. Any file that exceeds this limit is discarded.

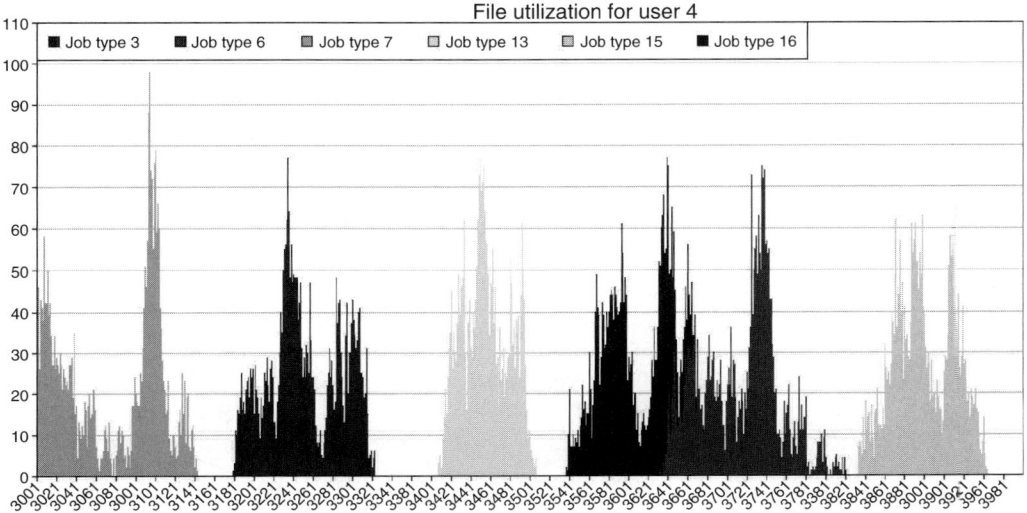

FIGURE 46.14 Intelligent Migrator file utilization example.

After the virtual data grid is constructed, the neural network is created for each agent present in the Intelligent Migrator system. The dimensions of the layered acyclic neural network, discussed in more detail in Section 46.4.3, are pulled from the definition of the agent type possessed by each agent and initialized to random values and stored in the MySQL database definition of the virtual data grid. These random weights represent the starting point of the neural network for each agent upon the initialization of the simulation procedure.

46.4.5 Simulating a Virtual Data Grid

The simulation process of the Intelligent Migrator uses the virtual data grids tailored to a specific storage scenario and simulates file utilization over time. The simulation procedure is implemented as a series of PHP scripts initiated through the command line, with full access to the MySQL database where agent information, user profile templates, and the virtual data grids are defined. The Intelligent Migrator simulation routine is designed to run in two modes. The first mode, or base case, is the most simplistic and replicates data to compute elements in a purely reactionary fashion. The second mode, built upon the functionality of the first mode, initiates the intelligent agents that perform additional replications based on observed past utilization and predicted future file usage on Intelligent Migrator storage elements. Simply by specifying the virtual data grid to run, simulations modify the virtual data grids to keep performance histories, submit and perform replication requests, and train the neural networks for each agent as file usage patterns are observed over time.

The generic simulation procedure begins at a random point in the job history of the virtual data grid and runs until the last submitted computational job is processed. The simulation proceeds once per virtual day and examines the computational jobs submitted over the next 24 h. A submitted job for a resource is associated with data files, and these files must be present on the storage element of their associated compute elements in order to run successfully with the job execution. All files that are not present on storage elements for the compute elements they are required on are copied to these storage elements during the current simulation round. If a storage element is at its capacity, the oldest data on the storage element is removed to free space for the required files. This step proceeds once daily until the simulation is complete.

This reactionary implementation provides the base case for comparison of the Intelligent Migrator's performance. Replicating files only as they are needed provides no analysis of past utilization or any attempt at replicating files before they are needed. Files are only present before they are required if there is high overlap in file utilization day-to-day, and any files that are removed from the storage element to make room for new required data must be re-replicated to that storage if needed in the future. The agent-based approach for learning file usage patterns is applied in the second mode of virtual data grid simulations.

There are two key differences between the intelligent mode of virtual data grid simulations and the generic simulation procedure. The intelligent mode utilizes assessed file values to assign weight and importance to files and to build a rough model of file utilization over time. File values appreciate as they are used by their owners and depreciate over time as they are passed over for other files. It is assumed that files with higher values are more valuable to their owners for submission with computational jobs. Appreciation and depreciation parameters are defined in the specification of an agent's type and are a function of recent file utilization on the file repository or on remote storage elements. In the intelligent mode, the file value is used in assessing which files to remove from remote storage when there is insufficient storage space on a compute element's storage device. In the generic mode of simulations, files are removed from a storage element on the basis of their age, where the oldest files are removed with the highest priority.

Second, and more importantly, agents are deployed in the simulation of a virtual file system using the intelligent mode. These agents, possessing neural networks accepting file parameters and returning an assessment of future file worth on a given compute resource, are utilized during the virtual days in the simulations to boost the performance of the Intelligent Migrator over the generic simulation mode. Before the daily replication of required files to storage elements, each agent in the storage network is consulted and given the opportunity to replicate files to the storage element beforehand. The agents gather past file utilization information and determine the pool of files to replicate by considering previous file utilization as well as file values on the repositories and remote storage devices. At the end of a simulated day, each agent's accuracy is measured over the previous day and stored in the database. The agent is only permitted to replicate files to its associated storage element if its latest test accuracy is over 70%.

As file utilization and job submission patterns change over the course of the simulated time period, agents are retrained to represent the latest model of perceived file usage to maintain the accuracy of their neural networks. After agents make their replications and the required files are replicated to the necessary storage elements, agents are retrained if their accuracy decays to below the 70% threshold.

The application of file values and agent learning models to the general simulation routines presents two models for the evaluation of the algorithms designed for the Intelligent Migrator. The generic case provides a basis for comparison with the intelligent approach, allowing for an assessment of the accuracy of the agents on a virtual data grid and the benefits of applying the intelligent model to a similar live data system.

46.4.6 Intelligent Migrator Visualization and Results

The Intelligent Migrator encompasses a suite of visualization charts designed for an analysis of the results of the internal learning algorithms. These Web tools, utilizing dynamic charts similar to those found in the Scenario Builder, provide information on-the-fly from the MySQL databases for the virtual data grids created. Graphs displaying the correct responses of the underlying implementation, and the test accuracy of the neural network, to name a few, can be generated by specifying the virtual data grid to evaluate.

In this discussion of results, we present four generated virtual data grids, each possessing different attributes for comparison. Each virtual data grid is associated with the same ten agents, representing ten different compute elements from across the grid initiatives monitored by the ACDC Grid Monitoring Infrastructure. Three user profile templates were established for these examples and applied to the virtual data grids created. Each user profile template creates users who possess 1000 files randomly sized roughly between 100 kB and 1 MB. The users submit up to 10,000 jobs chosen from six job types. The three user profile templates differ in that among user profiles 1, 2, and 3, users are limited to submitting a maximum of 10, 15, or 20 files per job, respectively. Each user is given a 5% chance of deviating from these figures and submitting anywhere between 0 and twice the number of files in their individual limit. We generated four virtual data grids such that one grid has ten users of profile type one, the second experiment has ten users of profile type two, the third has ten users of profile type three, and the fourth has fifteen users, five of each profile type. Each virtual data grid was run both in the general simulation mode as well as in the intelligent mode using multiple agents with ANNs.

Many factors indicate the performance of a simulation of the Intelligent Migrator. For example, in the intelligent case, it is important to examine the accuracy of the neural networks. However, most importantly, we wish to look at the hit rate for a simulated system. A hit is when a file is requested for submission with a computational job and it already exists on the storage system associated with the compute resource, such that the file does not need to be copied to the compute resource with the job, and time and bandwidth are spared in the transaction. The more hits during a simulation, the more efficient the storage network, and the more intelligent the system simulated by the Intelligent Migrator. A hit rate is the number of hits versus the number of requested files over a period of time.

The results of simulations of the four virtual data grid systems consistently show that the general mode simulation is slightly more effective than the intelligent mode in correctly replicating data to remote storage. Performance in the first three virtual data grids yielded hit rates within or close to 1% of one another between the two simulated scenarios, with a steadily increasing hit rate percentage as the number of files submitted per user increases, between 16% and 17% hit rate in the first virtual data grid and an 18–19% hit rate in the third virtual data grid. This may be attributed to an increased instance of repetition in file use, as more files are chosen for each job over the three user profile types while the number of files available to each user remains constant. The fourth virtual data grid, consisting of a mix of the three user profiles, and with the lowest instance of file repetition, faired worst of the four, yielding a 12–13% hit rate between the intelligent and general cases.

When broken down over compute resources or users, the intelligent mode in some cases exhibits an improved performance over the general mode. Virtual users may be the sole users of a compute element, or closely match the learned neural network models of the compute resources to which jobs are submitted. Similarly, compute elements represented by agents may experience low variability of file utilization over time or a generated file usage pattern otherwise conducive to an increased performance by the intelligent agent infrastructure, where neural network input arguments are more closely related to future file usage. Figure 46.15 shows the hit rates for the same compute resource between the intelligent and general modes. The intelligent mode, featured on the top of the graphic, sustains similar but slightly improved overall performance over the bottom general case.

In spite of the reduced performance of the intelligent replication mode over the general mode, the neural networks implemented and trained for each agent received high-test accuracies over the duration of the simulations. Average test accuracies for the neural networks, taken in the 24 virtual hours after

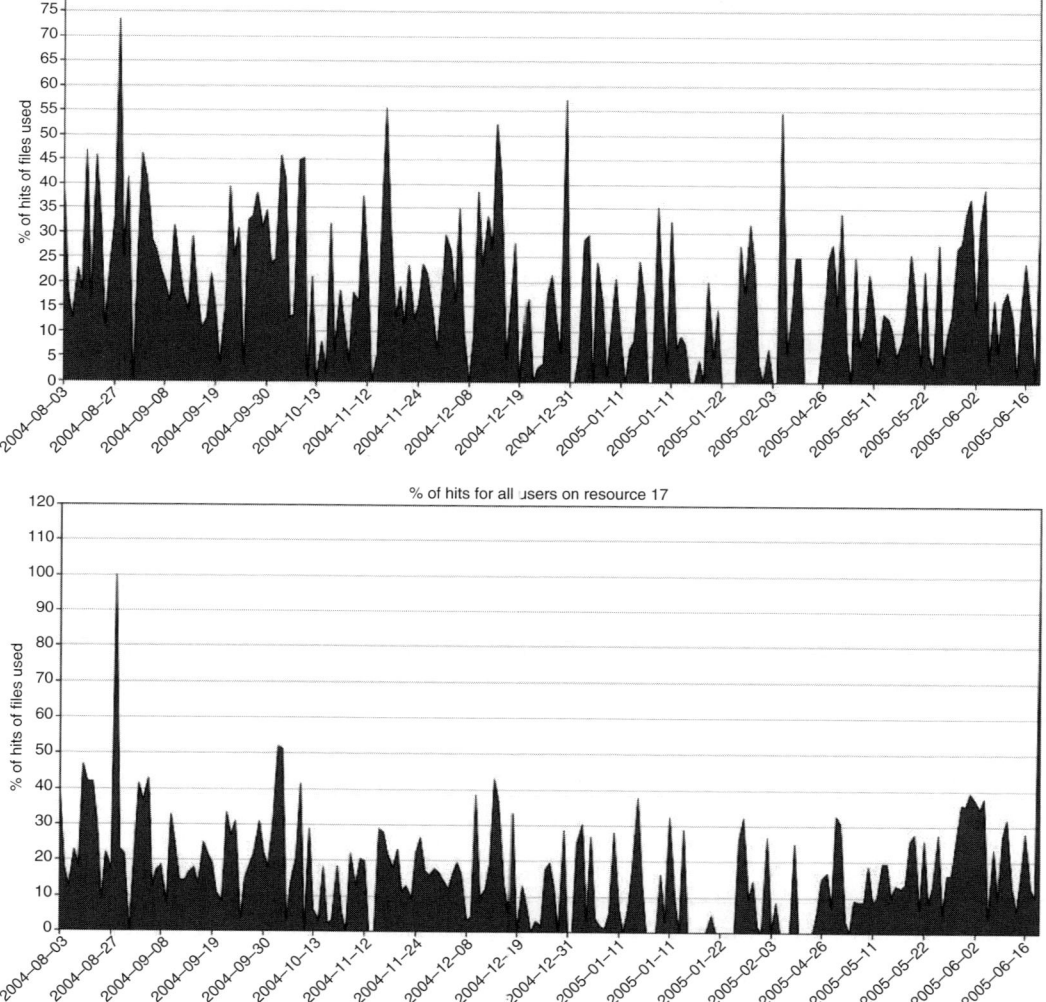

FIGURE 46.15 Intelligent Migrator hit rates example comparison. Intelligent mode (a), and general mode (b).

the neural network is consulted during the simulation, yield average test accuracies ranging from 30 to 98%, consistent over all virtual data grids utilized in this analysis in spite of their differing parameters. There are three measures of neural network accuracy, namely, (i) the percentage of positives marked, or the percentage of files submitted to a compute resource, which are marked as likely to be required on the remote storage device, (ii) the percentage of negatives marked, or the percentage of files not used, which are marked by the neural network as not likely to be required on the remote storage element and (iii) the overall classification accuracy of the neural network over measures (i) and (ii). Figure 46.16 shows example plots of these three neural network accuracy measures for an agent after a simulation.

Findings conclude, however, that high neural network accuracy does not always correspond with a high hit rate for an individual agent. The resulting low performances can be attributed to several factors. Flagging too many files for replication and having only the capacity on a remote storage device to replicate a few files that happen to be incorrect can lead to reduced performance. As there are many available files in the system and only a handful of them will be used on any given day, remote storage capacity is reached quickly and files may be left off of the storage element even if they are correctly marked for replication. In addition, as only a handful of files are utilized at a given time, a neural network state that marks no files for replication will have an artificially high overall accuracy as the majority of files, which were not used in actuality, were marked correctly. More closely tuned neural network training and an increased scrutiny over test accuracies could yield an improved performance and correlation between neural network performance and overall system efficiency.

Although the general mode of the Intelligent Migrator performs better than the neural network implementation of the Intelligent Migrator in most cases, the infrastructure presented here represents a framework for the development of learning algorithm enhancements or new approaches to modeling user file utilization. A further examination of usage patterns, job selection, and the learning model could yield substantial improvements in the Intelligent Migrator's performance. For example, changes to the input arguments for the neural network or to its internal structure and training could lead to better results. If the attributes being given for each file to the neural network are not good indicators of the file's future utilization, the neural network will frequently fail to flag the correct data. In addition, smaller training windows or an adjustment to the classification criteria when preparing the supervised training set may more closely tune the learned models. A closer look at the virtual data grid, the simulation routines, and in the ways in which files are chosen, could yield improved performance as well. The selection of files for submission, for example, may be biased toward supporting the general model for file replication.

The multiagent learning framework of the Intelligent Migrator represents an attempt to apply a distributed intelligent system to a distributed storage system. Although its initial performance does not yield improvements over the more general file placement case, the Intelligent Migrator presents a solid framework for enhancing the learning models and producing improved results. Further experimentation in the development of these improvements promises avenues for providing reliable and scalable data service to a computational grid and a means for efficiently utilizing storage and network resources.

46.4.7 Further Remarks

The Intelligent Migrator represents an effort in the ongoing initiative to present scalable and robust data service to the ACDC Grid. The examination of user utilization patterns and algorithms to model them is an important step in efficiently making use of limited storage and boosting the performance of a physical data grid and facilitating improved data service to a computational grid. In Section 46.6, we will discuss the relationship between the Intelligent Migrator and the design of the ACDC Data Grid.

46.5 The Advanced Computational Data Center Data Grid

The ACDC Data Grid, which serves the ACDC Grid and provides data storage services to grid users, is a network of interconnected and distributed storage resources that are logically linked to provide a single continuous storage environment. In this section, we present the ACDC Data Grid, including the

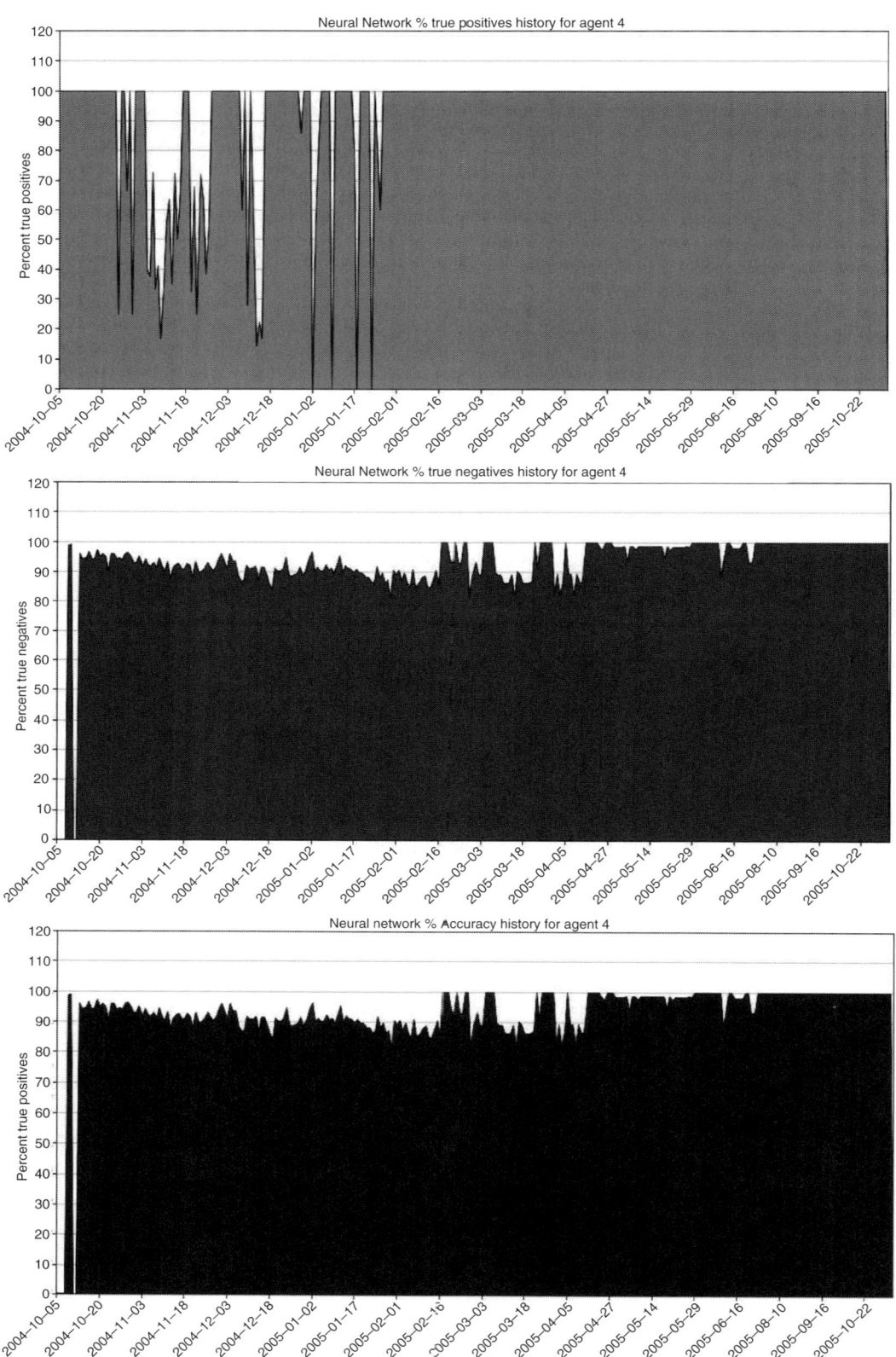

FIGURE 46.16 Intelligent Migrator neural network accuracy plots for a single agent.

Effectively Managing Data on a Grid

data file virtualization methods designed to provide seamless data access over multiple heterogeneous storage elements. We then discuss the implementation of a storage element and the architecture of the storage network for the ACDC Data Grid. We present the user and administrative Web tools that provide a user-friendly interface with the ACDC Data Grid. Finally, we include current results and present some additional remarks.

46.5.1 Data Grid File Virtualization

Data grids consist of a network of geographically distributed heterogeneous storage systems, where storage devices move in to and out of the network and the location of individual files is dynamic. Creating a robust and scalable data grid to provide the necessary data services to a computational grid, and, in particular, to grid users, requires that files be transparently accessible, regardless of their physical location in the storage network. This distributed storage environment should appear as a single homogeneous service to the grid user, and must implement seamless and scalable maintenance of user-imposed naming conventions and directory hierarchies similar to those found on traditional file systems. Maintaining such a flexible file system across an arbitrary storage network is a significant challenge.

In our design of the ACDC Data Grid, we have made a distinction between the information necessary for identifying and locating a data file within a distributed storage network, and the user-imposed naming conventions and directory hierarchy required to provide the impression of a single continuous storage space to grid users. By utilizing a MySQL database, we maintain the file organization imposed by grid users through the Web interfaces discussed in Section 46.5.3, while binding this information to file location records for referring to data internally. Files entering the ACDC Data Grid are stripped of their user-imposed names and directory locations and permanently assigned to a Globally Unique IDentifier (GUID), which is used as its physical file name. A record in the database is created for each file, mapping its GUID to its user-imposed name, directory, and its location on the storage element in the distributed storage network.

This process, in essence, creates two entities for each data file entering the data grid, namely, a physical file with a GUID, and a user-based identifier that only exists within the database. The virtual data grid, implemented through calls to the database, creates an image superimposed on the physical storage. The physical data grid can consist of an arbitrary distributed network of directories that are completely transparent to grid users, while the database provides a mapping to the higher-level virtual data grid that provides the complexity and flexibility users require in order to impose their own file naming conventions and organization.

There are several benefits to this approach. The decoupled architecture provides simplified access to files on distributed storage devices. Queries to the database reproduce the user-imposed directory hierarchies as a single directory tree accessible through the Web portal. APIs in place query the database and can retrieve a GUID for a user-defined name and directory location, and the mapping in the database for this GUID yields its physical location on the disk of a storage element. This process is illustrated in Figure 46.17. Files may be added or removed arbitrarily without needing to maintain complex user-imposed directory structures within the storage devices. Users are free to update the names and directory locations of their virtual files in a homogeneous representation within the Web portal with a series of simple, low cost updates to the database without visiting or modifying their physical data files. Creating this virtual image of physical storage provides a seamless interface to an arbitrarily distributed storage network, overcoming the complexities of managing a seamless interface to the data grid over multiple heterogeneous storage devices.

46.5.2 Storage Network Architecture

A scalable data grid solution must be equipped to grow with evolving demands by grid users and their applications. New storage devices must be easily integrated into the existing storage network, regardless of their architectures or platforms, so that the data grid can scale as demands for data services increase over

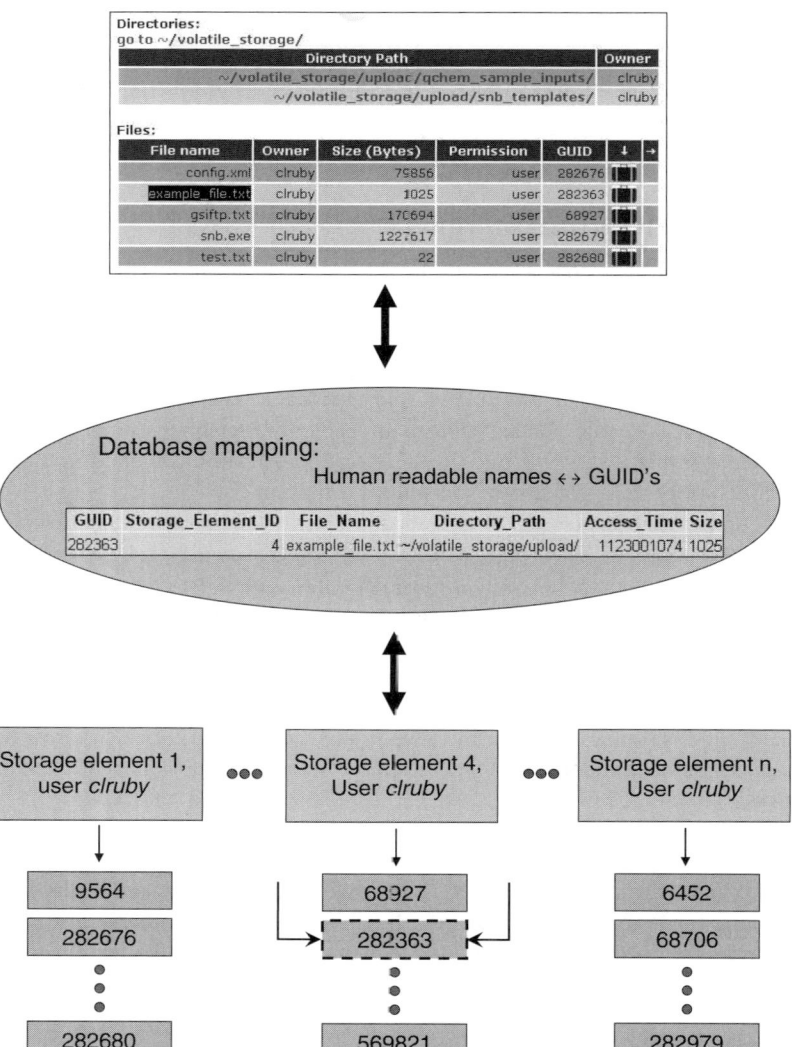

FIGURE 46.17 Virtual and physical database mapping.

time. Physical storage devices in the ACDC Data Grid storage network are implemented as designated directories on a grid-enabled compute element. These directories contain a directory for each VO users of the ACDC Data Grid, which at this time consists of the GRASE VO. Each grid user possesses a GUID directory within the VO directory of the VO the user is associated with, and all files belonging to the grid user on the storage element are stored directly within this GUID directory. As no two data grid files share a GUID, there is no chance of namespace overlap within this flat directory. An illustration of the structure of a storage element in the ACDC Data Grid storage network is shown in Figure 46.18.

Storage element information, including the names of the machines hosting them, the directories in which they are located, and the ways in which they are accessed, is stored within a MySQL database. These grid-enabled storage devices are accessed using the Globus Toolkit, and although storage elements within our administrative domain are mounted and accessed directly to enhance internal performance, these can be changed to use Globus through modifications to database entries. Given the GUID of a file to access, the full physical path to this GUID can be constructed given the owner of the file, the owner's

Effectively Managing Data on a Grid   46-29

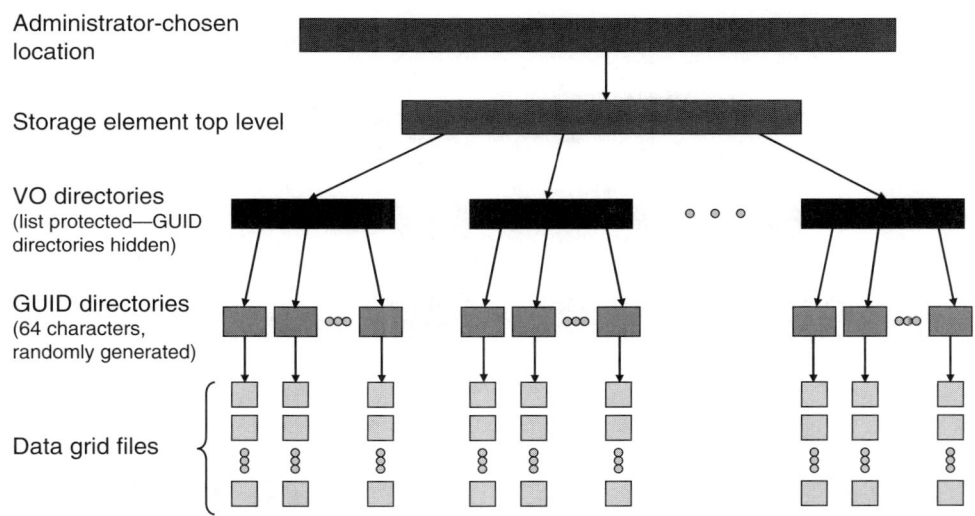

FIGURE 46.18 The anatomy of a storage element.

associated VO, and the storage element it is located on, allowing for seamless access in spite of arbitrarily distributed data files. The directory implementation of a storage element provides simple access in spite of the differing underlying architectures of the physical storage devices.

There are three types of storage elements implemented within the ACDC Data Grid storage network, namely, (i) persistent, (ii) volatile, and (iii) replica storage devices. Persistent and volatile storage elements are located within our administrative domain and the ACDC Grid. Persistent storage elements are directories that are recoverable in the event of a device failure, and are intended for user files that are critical and are not easily reproduced. Volatile storage, which is more abundant than physical storage, is implemented on devices where data recovery is not guaranteed. A data file exists exactly once on one of the persistent or volatile storage devices within our domain.

Replica storage space, on the other hand, is implemented on grid-enabled compute elements beyond the scope of ACDC Grid administrative control. Inherently volatile, the data files stored in these storage elements are copies, or replicas, of data files existing on persistent or volatile storage devices within our domain. This storage network architecture introduces the potential for saving storage space and bandwidth used in repeatedly submitting files with computational jobs.

Figure 46.19 illustrates the persistent, volatile, and replica storage space design of the ACDC Data Grid.

46.5.3 Web Tools for the Advanced Computational Data Center Data Grid

A series of Web tools has been developed to facilitate user data management as part of the ACDC Data Grid implementation. Users may upload, manage, and retrieve data from the data grid using the ACDC Grid Portal. This Web interface is designed to deliver seamless access to data files, concealing the distributed storage network and providing the impression of a single, coherent data repository.

Data files are added to a user's home space automatically as results from computational jobs, or by uploading them through the data grid upload interface. Users may upload single files or drag-and-drop entire directory structures from their local machines through the Rad Inks Rad Upload [49] Java applet [50], which is embedded in the Web tool. Users also have the ability to view and manage the files they own on the data grid. Files are displayed using the user-imposed virtual directory structure that is navigated by a series of links to all subdirectories of the current viewing directory. Files may be selected for a number of different tasks, including renaming, copying, removing, permissions modification, viewing, and editing.

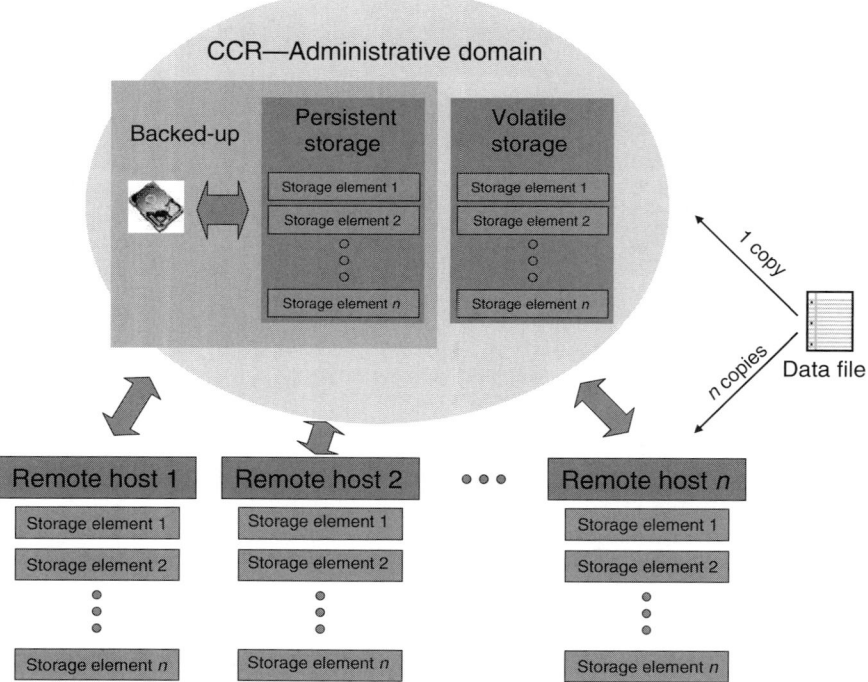

FIGURE 46.19 Persistent, volatile, and replica storage organization.

Users may also replicate their files to a number of remote storage elements available to them and manage their replicas through a similar file browse interface.

In addition to being able to view and manipulate files in the data grid, users may use the Web tools in place to select files to be used in scientific grid applications. Also available is a data grid file download interface, or a tool for extracting data grid files from the ACDC Grid Portal by selecting files or a directory to be zipped and made available to download to a user's local machine. Figure 46.20 includes several screenshots of user tools available for managing data files on the ACDC Data Grid.

Administrators have access to other Web tools designed for managing the ACDC Data Grid behind the scenes. These tools give data grid administrators access to many capabilities, including the ability to ensure the consistency between the virtual and physical data grid representations, physically pull files into the data grid, manually migrate files, and view file distributions across the available storage repositories, to name a few. In addition, replica storage elements can be installed, managed, or removed on storage devices remote to our domain. Figure 46.21 shows examples of the administrative tools available for managing the ACDC Data Grid.

46.5.4 Results

The ACDC Data Grid is used in conjunction with the ACDC Grid and serves the data requirements of grid users submitting computational jobs to more than ten grid applications supported by the ACDC Grid in a variety of scientific fields, including environmental engineering, computational chemistry, and structural biology, to name a few. The ACDC Data Grid consists of a production version and a development version, and currently supports 24 grid users and nearly 500 GB of data in over 1,000,000 data files. Since August of 2005, nearly 400,000 files totaling 70 GB of data have been staged for submission with computational jobs.

There are currently nine persistent and volatile storage elements installed throughout our domain. Four 2 TB volatile storage elements and four 500 GB persistent storage elements are installed on each of four

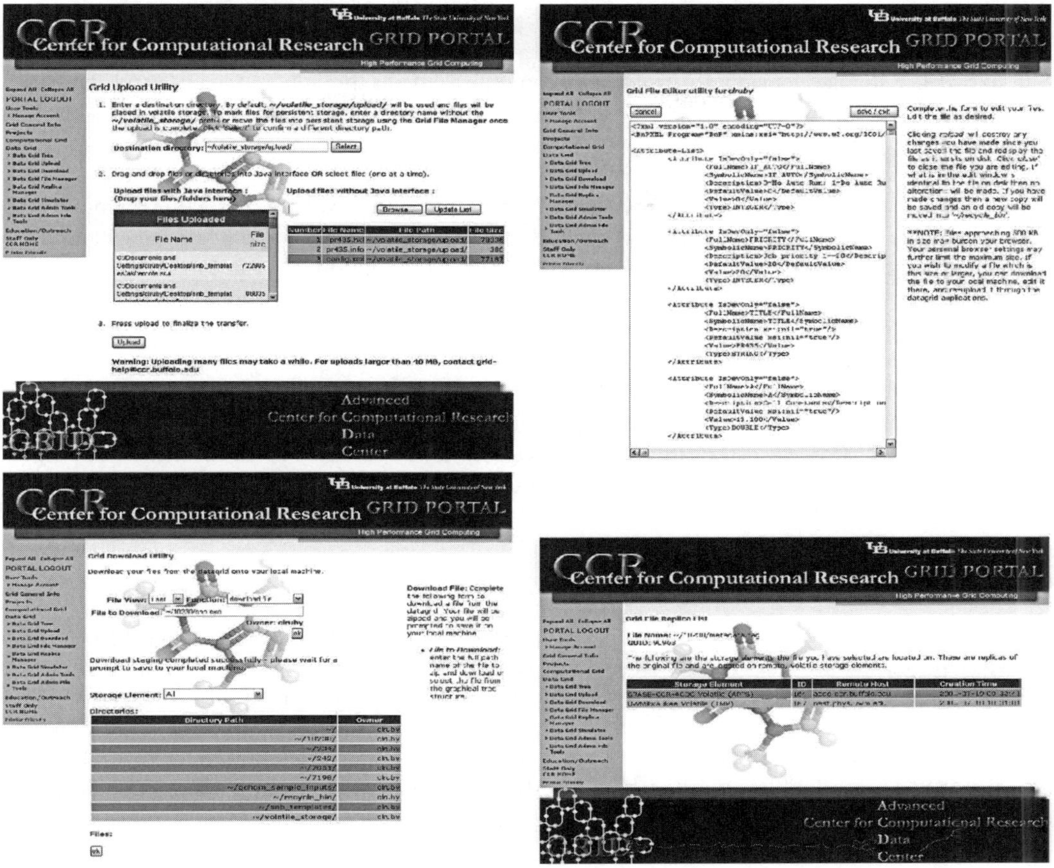

Effectively Managing Data on a Grid

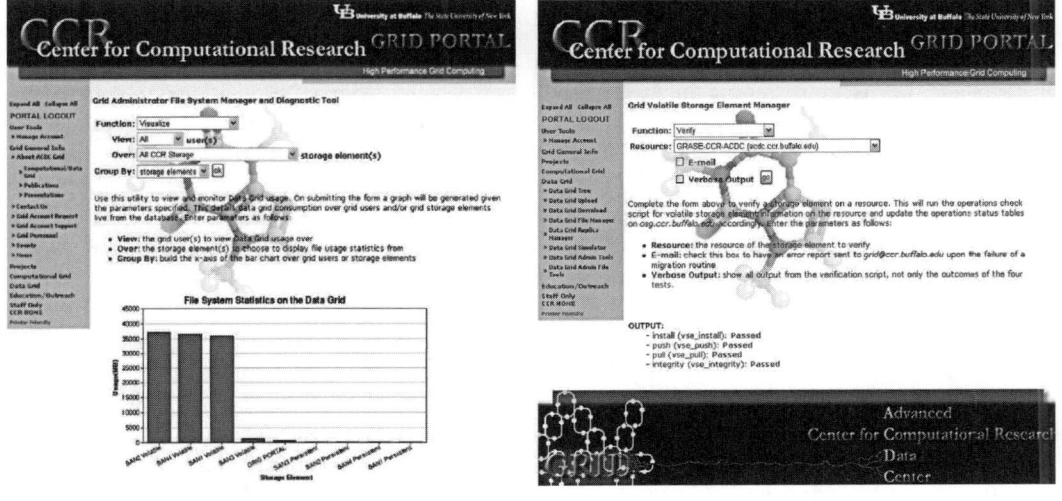

FIGURE 46.20 Select ACDC Data Grid user interface tools.

FIGURE 46.21 Select ACDC Data Grid administrative interface tools.

Hewlett-Packard storage area network (SAN) AlphaServer GS1280s. A ninth 380 GB persistent storage element is installed on the Dell four processor hyper-threaded machine running the ACDC Grid Portal. Multiple replica storage elements are installed on compute elements throughout the OSG and TeraGrid.

46.5.5 Further Remarks

The ACDC Data Grid is a part of the ongoing initiative within the Cyberinfrastructure Laboratory to provide scalable and reliable data services to the ACDC Grid. Improvements to its implementation are often made to strengthen its design and ability to serve the needs of grid users and their computational jobs. One key improvement planned for the ACDC Data Grid is to further integrate replicated data files with the job submission process. Automating a procedure for utilizing replicated data files present on a remote computational resource, and thus obviating the need to include them when staging a job and its associated files for submission to the resource, would enhance the performance of the ACDC Data Grid and fully integrate the remote replica storage repositories with the storage established within our domain. Moving the ACDC Data Grid forward in this direction is an important step in further developing its distributed design and in ensuring scalable data service to the ACDC Grid.

46.6 Conclusions

The Scenario Builder, the Intelligent Migrator, and the ACDC Data Grid represent three ongoing initiatives for developing a robust and reliable data system for the ACDC Grid. The Scenario Builder and the Intelligent Migrator are two applications that explore file utilizations and their effects on the performance of a data grid through generated models and simulations, whereas the ACDC Data Grid is the current implementation of the production data grid serving the ACDC Grid. In this section, we present the ways in which the Scenario Builder and the Intelligent Migrator relate to the design of the ACDC Data Grid and enable simulations of its performance. We then present our concluding remarks.

46.6.1 Enabling Simulations with the Scenario Builder

The Scenario Builder represents an early model of the ACDC Data Grid and contains elements from the data grid's early design. There are many similarities between these two data service initiatives, as well as a few key differences.

The virtual storage elements implemented in the Scenario Builder represent the persistent and volatile storage local to our domain in the ACDC Data Grid. The grid portal node of the Scenario Builder is analogous to the machine running the ACDC Grid Portal, which is where computational jobs are staged with the data grid files they require. As in the virtual storage network of the Scenario Builder, the ACDC Data Grid utilizes the grid portal node as a central location for files accessed for use in computational jobs, where data may be initially present on any storage device within our domain.

There are, however, important differences between the implementations of the Scenario Builder and the ACDC Data Grid. Files brought to the grid portal for submission with computational jobs are copied to the grid machine, not migrated as is done in the Scenario Builder. As such, these copies may be removed from the staging areas of the ACDC Grid portal machine without damaging grid user data integrity, as the original data files remain intact in their original storage locations. Thus, there is no need for periodic migrations of data from the grid portal node to the outlying data repositories managed by our domain.

The Scenario Builder represents early work in the initiatives of the Cyberinfrastructure Laboratory to provide robust data service to the ACDC Grid. In order to enable accurate simulations of the ACDC Data Grid, the Scenario Builder's underlying virtual storage network must be updated to accurately represent the current approach of the data grid. The storage network must be rearranged to include the ACDC Data Grid storage devices that are currently in use, and the storage and network limitations of the ACDC Data Grid must be updated within the Scenario Builder network definition in the MySQL database tables.

Although the current implementation of the ACDC Grid uses copied data and independent staging procedures for submitting files with computational jobs, and does not require migration cycles to free storage from the grid portal node, the study of these scenarios in the Scenario Builder still provides a useful examination of a storage network response to data access. As the utilization of the ACDC Data Grid grows with the ACDC Grid initiative, it may become important in the future to examine how to efficiently distribute files to evenly utilize storage devices as the data grid nears the capacity of its networked storage repositories. If utilization continues to evolve and grow, scalability will become a greater concern in ensuring reliable data service for the ACDC Grid. The results produced by the Scenario Builder and its power for producing storage network projections may prove to be even more relevant in the future as the need for evenly distributing larger volumes of files with respect to limited bandwidth becomes an issue in the growing ACDC Data Grid.

46.6.2 Enabling Simulations with the Intelligent Migrator

The Intelligent Migrator is intended to model the design and implementation of the ACDC Data Grid by providing a service for simulating the performance of generated virtual data grid usage scenarios and testing the learning and prediction accuracy of agents for the storage network topology. The Intelligent Migrator is based on the replica storage network implemented for the ACDC Data Grid. Recall that the ACDC Data Grid implements storage on compute elements external to our domain, where computational jobs may be submitted. These compute elements are simulated within the Intelligent Migrator, where an agent represents a compute element and an associated storage element. It is assumed within the Intelligent Migrator simulation design that a data file located on a storage element for a compute element is accessible to computational jobs submitted to the compute element, obviating the need to submit the data file to the computational resource when staging the job for submission.

The availability of this infrastructure presents the opportunity for efficiently utilizing remote storage elements with files repeatedly submitted with computational jobs. If the files required for jobs on a compute element are stored on the storage element ahead of time, time and bandwidth are saved in submitting the job as the files do not need to be staged. The Intelligent Migrator seeks to take advantage of this design by implementing learning algorithms that utilize past file usage patterns to predict which files will be required on a compute element in advance. The structure of the MySQL database supporting the virtual data grids of the Intelligent Migrator closely mirrors the database and implementation of the ACDC Data Grid. Agents utilizing the learning algorithms developed within the Intelligent Migrator could be deployed to learn ACDC Grid user file usage patterns and submit physical replication requests on their behalf with few changes to the Intelligent Migrator implementation.

Although the ACDC Data Grid implements remote storage and replicated data across storage elements bound to compute elements, the ACDC Data Grid is still under development in order to fully utilize file replicas located on remote storage elements. Remote storage can be installed on compute elements using the ACDC Data Grid administrative tools discussed in Section 46.5. However, jobs submitted to compute elements from the ACDC Grid are typically staged using the persistent or volatile data files stored locally within our domain. As the ACDC Data Grid is a part of an ongoing initiative to deliver a robust data system to the ACDC Grid, the capabilities utilized in the Intelligent Migrator are a part of future developments for improving the current system.

As the ACDC Data Grid matures, the utilization of data replicas on remote storage elements will serve to improve the efficiency of the data grid system and take advantage of the capabilities examined in the Intelligent Migrator. The learning agent infrastructure of the Intelligent Migrator could serve as a significant contribution to the performance of the ACDC Data Grid as it is enhanced to provide more robust and efficient data service to the ACDC Grid.

46.6.3 Concluding Remarks

Ongoing initiatives in the Cyberinfrastructure Laboratory to provide robust data service to the Advanced Computational Data Center and have led to the development of the Scenario Builder, the Intelligent

Migrator, and the ACDC Data Grid. The Scenario Builder and Intelligent Migrator represent two tools for generating a variety of tailored virtual data grid scenarios for simulation and evaluation. Both examine file utilization and its relationship to data grid performance by simulating file accesses and their effects on the storage and network load of the generated data environments. The Scenario Builder, representing early research in developing data services for the ACDC Grid, focuses on the direct impact of file accesses on a virtual storage network, whereas the Intelligent Migrator builds on the Scenario Builder by seeking to minimize waste by learning usage patterns and projecting future file utilization on remote storage elements. The Scenario Builder and the Intelligent Migrator each provide tools for the generation and simulation of tailored virtual data grid usage scenarios as well as for the interpretation of results with dynamic, interactive, charts through a convenient online interface.

The ACDC Data Grid serves the data requirements of the ACDC Grid by providing users with seamless access to their files distributed across multiple heterogeneous storage devices located both within our domain and across compute elements in the Western New York Grid, the emerging New York State Grid, the OSG, and TeraGrid. The ACDC Data Grid implements data virtualization and simple storage element installation procedures and provides a scalable and robust system serving the ACDC Grid. In addition, the ACDC Data Grid provides a set of online tools for grid users. Built on components of the design of the ACDC Data Grid, the Scenario Builder and the Intelligent Migrator provide insight into potential performance issues by enabling the generation and simulation of specific usage scenarios that could arise in a physical setting.

Institutions and organizations around the world are working on grid solutions to problems with significant computational and/or data components. The growth of grid applications has spurred the need for robust and reliable data management services that are typically used to facilitate computation on a large scale. Grid research in the Cyberinfrastructure Laboratory, as well as research in providing data service to the ACDC Grid, represents our role in an international initiative to develop robust and scalable grid infrastructures, facilitating further computational research and supporting science in the twenty-first century.

Acknowledgments

We would like to thank members of the Cyberinfrastructure Laboratory and the Center for Computational Research for their contributions to this work, including Dr. Mark L. Green, Steven M. Gallo, Jonathan J. Bednasz, and Anthony D. Kew. The development of the Scenario Builder, the Intelligent Migrator, and the ACDC Data Grid are supported by NSF grant ACI-0204918 and were developed during the second author's tenure as (founding) director of the Center for Computational Research, SUNY-Buffalo. The web site for the Cyberinfrastructure Laboratory is http://www.cse.buffalo.edu/faculty/miller/CI/

References

[1] Chervenak, A. et al., The data grid: Towards and architecture for the distributed management and analysis of large scientific data sets. *Journal of Network and Computer Applications*, 2000. **23**(3): pp. 187–200.
[2] Grid Computing. [Web page] 2006 April 5, 2006 [cited 2006 April 6]; Available from: http://en.wikipedia.org/wiki/Grid_computing
[3] Foster, I., The grid: A new infrastructure for 21st century science, in *Grid Computing—Making the Global Infrastructure a Reality*, F. Berman, G.C. Fox, and A.J.G. Hey, Editors. 2002, John Wiley & Sons Ltd.: West Sussex. pp. 50–63.
[4] Data Grid. [webpage] 2006 March 16, 2006 [cited 2006 April 6]; Available from: http://en.wikipedia.org/wiki/Data_grid
[5] Allcock, B. et al., Data management and transfer in high-performance computational grid environments. *Parallel Computing*, 2002. **28**(5): pp. 749–771.

[6] Grid Computing Services at SUNY-Buffalo. [Web page] 2006 April 1, 2006 [cited 2006 April 6, 2006]; Available from: https://grid.ccr.buffalo.edu

[7] The Cyberinfrastructure Laboratory at SUNY-Buffalo. [Web page] 2004 November 3, 2004 [cited 2006 April 6]; Available from: http://www.cse.buffalo.edu/faculty/miller/CI/

[8] Green, M.L. and R. Miller, Molecular structure determination on a computational and data grid. *Parallel Computing*, 2004. **30**(9–10): pp. 1001–1017.

[9] Green, M.L. and R. Miller, Evolutionary molecular structure determination using grid-enabled data mining. *Parallel Computing*, 2004. **30**(9–10): pp. 1057–1071.

[10] Miller, R. et al., SnB: crystal structure determination via Shake-and-Bake. *Journal of Applied Crystallography*, 1994. **27**(1994): pp. 631–621.

[11] The Princeton Ocean Model. [Web page] 2004 [cited 2006 April 25]; Available from: http://www.aos.princeton.edu/WWWPUBLIC/htdocs.pom/

[12] Open Science Grid. [Web page] 2006 April 13, 2006 [cited 2006 April 13]; Available from: http://www.opensciencegrid.org

[13] TeraGrid. [Web page] 2006 April 13, 2006 [cited 2006 April 13]; Available from: http://www.teragrid.org

[14] Foster, I., C. Kesselman, and S. Tuecke, The Anatomy of the Grid—Enabling Scalable Virtual Organizations. *International Journal of Supercomputer Applications*, 2001. **15**(3): pp. 200–222.

[15] Globus Toolkit. [Web page] 2006 April 3, 2006 [cited 2006 April 6]; Available from: http://www.globus.org/toolkit/

[16] Legion Worldwide Virtual Computer. [Web page] 2001 [cited 2006 April 30]; Available from: http://legion.virginia.edu

[17] CORBA (R) Basics. [Web page] 2006 January 5, 2006 [cited 2006 May 1]; Available from: http://www.omg.org/gettingstarted/corbafaq.htm

[18] Roure, D.D. et al., The evolution of the Grid, in *Grid Computing—Making the Global Infrastructure a Reality*, F. Berman, G.C. Fox, and A.J.G. Hey, Editors. 2003, John Wiley & Sons Ltd.: West Sussex. pp. 65–100.

[19] Science Grid This Week. Image of the Week [Web page] 2006 February 22, 2006 [cited 2006 April 14]; Available from: http://www.interactions.org/sgtw/2006/0222/

[20] Kunszt, P.Z. and L.P. Guy, The open grid services architecture, and data grids, in *Grid Computing—Making the Global Infrastructure a Reality*, F. Berman, G.C. Fox, and A.J.G. Hey, Editors. 2002, John Wiley & Sons Ltd: West Sussex. pp. 384–407.

[21] Moore, R.W. and C. Baru, Virtualization services for data grids, in *Grid Computing—Making the Global Infrastructure a Reality*, F. Berman, G.C. Fox, and A.J.G. Hey, Editors. 2002, John Wiley & Sons Ltd: West Sussex. pp. 409–435.

[22] The San Diego Supercomputing Center Storage Resource Broker. [Web page] 2006 April 7, 2006 [cited 2006 May 1]; Available from: http://www.sdsc.edu/srb/index.php/Main_Page

[23] MCAT. [Web page] 1998 May 11, 1998 [cited 2006 May 1]; Available from: http://www.sdsc.edu/srb/index.php/MCAT

[24] The DataGrid Project. [Web page] 2006 2004 [cited 2006 April 17]; Available from: http://eu-datagrid.web.cern.ch/eu-datagrid/

[25] Grid Physics Network. [Web page] 2006 [cited 2006 May 1]; Available from: http://www.griphyn.org/

[26] Grid Physics Network—Part 2: Petascale Virtual-Data Grids. [Web page] 2006 [cited 2006 May 1]; Available from: http://www.griphyn.org/projinfo/intro/petascale.php

[27] Bunn, J.J. and H.B. Newman, Data-intensive grids for high-energy physics, in *Grid Computing—Making the Global Infrastructure a Reality*, F. Berman, G.C. Fox, and A.J.G. Hey, Editors. 2002, John Wiley & Sons Ltd.: West Sussex. pp. 859–905.

[28] Green, M.L. and R. Miller, Grid computing in Buffalo, New York. *Annals of the European Academy of Sciences*, 2003: pp. 191–218.

[29] Green, M.L. and R. Miller, A client-server prototype for application grid-enabled template design. *Parallel Processing Letters*, 2004. **14**(2): pp. 1001–1017.
[30] Apache HTTP Server. [Web page] 2006 [cited 2006 April 6]; Available from: http://www.apache.org/
[31] PHP: Hypertext Preprocessor. [Web page] 2006 April 6, 2006 [cited 2006 April 6]; Available from: http://www.php.net
[32] JavaScript.com (TM)—The Definitive JavaScript Resource. [Web page] 2006 [cited 2006 April 6]; Available from: http://www.javascript.com/
[33] MySQL AB: The World's Most Popular Open Source Database. [Web page] 2006 April 6, 2006 [cited 2006 April 6]; Available from: http://www.mysql.com/
[34] GridFTP. [Web page] 2006 [cited 2006 April 13]; Available from: http://www.globus.org/grid_software/data/gridftp.php
[35] ACDC Grid Dashboard. [Web page] 2006 [cited 2006 April 19]; Available from: http://www.cse.buffalo.edu/faculty/miller/CI/
[36] ACDC Operations Dashboard. [Web page] 2006 [cited 2006 April 19]; Available from: http://www.cse.buffalo.edu/faculty/miller/CI/
[37] perl.com—The Source for Perl. [Web page] 2006 [cited 2006 April 28]; Available from: http://www.perl.com
[38] Prescott, C., *Site Verify*. 2005. p. Perform grid site verification checks on remote host(s) and report results.
[39] Shoshani, A., A. Sim, and J. Gu, Storage resource managers: Essential components for the grid, in *Grid Resource Management: State of the Art and Future Trends*, J. Nabrzyski, J.M. Schopf, and J. Weglarz, Editors. 2003, Kluwer Publishing: New York, pp. 329–347.
[40] Sycara, K.P., Multiagent systems. *Artificial Intelligence Magazine*, 1998. **10**(2): pp. 79–92.
[41] Lesser, V.R., Cooperative multiagent systems: A personal view of the state of the art. *IEEE Transactions on Knowledge and Data Engineering*, 1999. **11**(1): pp. 133–142.
[42] Hu, J. and M.P. Wellman, Online learning about other agents in a dynamic multiagent system. *Autonomous Agents*, 1998: pp. 239–246. http://portal.acm.org/citation.cfm?id=280765.280839
[43] Wolpert, D.H., K.R. Wheeler, and K. Tumer, General principles of learning-based multi-agent systems. *Autonomous Agents*, 1999: pp. 77–83. http://portal.acm.org/citation.cfm?id=301136.301167
[44] Shi, Z. et al., Agent-based grid computing. *Applied Mathematical Modelling*, 2006. **30**(7): pp. 629–640.
[45] Mitchell, T.M., *Machine Learning*. McGraw-Hill Series in Computer Science, C.L. Liu, Editors. 1997, McGraw-Hill: Boston, MA.
[46] Pomerleau, D.A., Efficient training of artificial neural networks for autonomous navigation. *Neural Computation*, 1991. **3**(1): pp. 88–97.
[47] Ohlsson, M., C. Peterson, and B. Soderburg, Neural networks for optimization problems with inequality constraints—the Knapsack problem. *Neural Computation*, 1993. **5**: pp. 331–339.
[48] Tsang, E.P.K. and C.J. Wang, A generic neural network approach for constraint satisfaction problems. *Neural Network Applications*, in Taylor, J.G. Editor. 1992, Springer-Verlag, pp. 12–22.
[49] Rad Inks—Rad Upload. [Web page] 2006 April 2006 [cited 2006 April 14]; Available from: http://www.radinks.com/upload/
[50] Sun Developer Network. Applets [Web page] 2006 [cited 2006 April 14]; Available from: http://java.sun.com/applets/

47
Fast and Scalable Parallel Matrix Multiplication and Its Applications on Distributed Memory Systems

47.1	Introduction...	**47**-1
	Motivation • New Development	
47.2	Preliminaries	**47**-3
	Distributed Memory Systems • Scalability of Parallel Algorithms	
47.3	Matrix Multiplication on DMSs	**47**-7
	Fast Matrix Multiplication • Scalable Matrix Multiplication	
47.4	Applications	**47**-17
	Matrix Chain Product • Matrix Powers	
47.5	Summary	**47**-23
	References	**47**-23

Keqin Li
State University of New York

47.1 Introduction

47.1.1 Motivation

Matrix multiplication is an important mathematical operation that has extensive applications in numerous research, development, and production areas. Furthermore, matrix multiplication is a building block in performing many other fundamental matrix operations such as computing the powers, characteristic polynomial, determinant, rank, inverse, LU-factorization and QR-factorization of a matrix, and in solving other mathematical problems such as graph theory problems. Owing to its fundamental importance in sciences and engineering, much effort has been devoted to finding and implementing fast matrix multiplication algorithms. Many sequential algorithms have been proposed for matrix multiplication. The standard sequential algorithm has $O(N^3)$ time complexity in multiplying two $N \times N$ matrices. Since Strassen's remarkable discovery of his $O(N^{2.8074})$ algorithm [43], successive progress has been made to develop fast sequential matrix multiplication algorithms with time complexity $O(N^\alpha)$, where $2 < \alpha < 3$. The current best value of α is less than 2.3755 [7].

The standard algorithm for matrix multiplication, Strassen's algorithm and Winograd's algorithm (see [36, p. 169]), both with $\alpha < 2.8074$, are used extensively in practice [15]. Although other known asymptotically fast algorithms are not practically useful because of large overhead constants hidden in the big-O notation (this applies to all the known algorithms with $\alpha < 2.78$), it is still plausible that faster practical algorithms for matrix multiplication will appear, which should motivate theoretical importance of parallelization of matrix multiplication algorithms, and in the case of Strassen's and Winograd's algorithms, such a parallelization should already have practical value.

As for parallel implementation of matrix multiplication algorithms, the standard algorithm can be easily parallelized, and Strassen's algorithm has been parallelized on shared memory multiprocessor systems [5]. Furthermore, it is well known that any known sequential $O(N^\alpha)$ matrix multiplication algorithm can be parallelized on a CREW PRAM in $O(\log N)$ time by using $O(N^\alpha/\log N)$ processors [3, 37, 39]. This significant result has been used extensively in many other parallel algorithms. Unfortunately, the PRAM model as well as large-scale shared memory multiprocessor systems are practically limited owing to their difficulty in realization.

On distributed memory multicomputer systems that are considered more practical, research has essentially focused on the parallelization of the standard algorithm. It was reported that matrix multiplication can be performed by a reconfigurable mesh with N^4 processors in constant time [41]. Such an implementation, though very fast, is far from cost optimal. It was also shown that matrix multiplication can be performed by a hypercube with p processors in $O(N^3/p + \log(p/N^2))$ time, where $N^2 \leq p \leq N^3$ [9]. This parallelization achieves linear speedup and cost optimality with respect to the standard algorithm when $p \leq N^3/\log N$.

However, before the recent breakthrough made in References 24, 29, and 30, it was not clear at all how nonstandard $O(N^\alpha)$ sequential matrix multiplication algorithms with $\alpha < 3$ can be parallelized and implemented efficiently on distributed memory systems (DMSs) with commonly used electronic interconnection networks such as hypercubes and meshes that have limited network connectivity and communication capability, or on DMSs with optical interconnection networks that have more network connectivity, flexibility, reconfigurability, and more powerful communication capability. In fact, none of the $O(N^\alpha)$ sequential algorithms with $\alpha < 3$ had been efficiently parallelized on any DMSs, especially for those systems with static or dynamic interconnection networks that do not seem to have sufficient capability to efficiently support the complicated communication patterns involved in nonstandard matrix multiplication algorithms.

Nevertheless, some initial attempt has been made. For instance, in Reference 9, it was shown that matrix multiplication can be performed by a hypercube with p processors in $O(N^\alpha/p^{(\alpha-1)/2})$ time, where $1 \leq p \leq N^2$, and α is the exponent of the best available $O(N^\alpha)$ sequential algorithm. This implementation is valid only in a small interval of p, and is not cost optimal with respect to the $O(N^\alpha)$ sequential algorithm. When $p = N^2$, we get the shortest execution time $O(N)$, which is very slow. The reason is that the $O(N^\alpha)$ algorithm is invoked sequentially to calculate submatrix products and not parallelized at all.

47.1.2 New Development

In this chapter, we present our recent contributions to fast and scalable parallel matrix multiplication and its applications on DMSs [24, 27, 28, 30]. Our main contributions are as follows:

- We develop a unified parallelization of any sequential $O(N^\alpha)$ matrix multiplication algorithm on any DMS. We show that for all $1 \leq p \leq N^\alpha/\log N$, multiplying two $N \times N$ matrices can be performed by a DMS with p processors in

$$T_{\mathrm{mm}}(N,p) = O\left(\frac{N^\alpha}{p} + T(p)\left(\frac{N^2}{p^{2/\alpha}}\right)(\log p)^{(\alpha-2)/\alpha}\right)$$

time, where $T(p)$ is the amount of time for the system to do a one-to-one communication.

- We show that for the class of distributed memory parallel computers (DMPC), where one-to-one communications can be realized in $T(p) = O(1)$ time, multiplying two $N \times N$ matrices can be performed in $T_{mm}(N, p) = O(N^\alpha/p)$ time, for all $1 \leq p \leq N^\alpha/\log N$. Such a parallel matrix multiplication algorithm achieves $O(\log N)$ time complexity by using only $N^\alpha/\log N$ processors and exhibits full scalability (i.e., the ability to attain linear speedup and cost optimality in the largest range of processor complexity). The above performance of matrix multiplication on DMPC matches the performance of matrix multiplication on CREW PRAM. Our result also unifies all known algorithms for matrix multiplication, standard or nonstandard, sequential or parallel.
- We instantiate our general result to specific interconnection networks. For instances, we show that matrix multiplication on DMSs with the class of hypercubic networks is highly scalable. We give the range of processor complexity such that matrix multiplication on DMSs with d-dimensional k-ary mesh or torus networks is scalable. We point out a condition for full scalability and a condition for high scalability. We also develop a compound algorithm and show how to achieve sublogarithmic parallel time complexity on a linear array with reconfigurable pipelined bus system (LARPBS) by using $o(N^3)$ processors.
- We demonstrate that by using our matrix multiplication algorithm as a building block, many matrix computation problems can be solved with increased speed and scalability. Owing to space limitation, we only demonstrate two applications, namely, solving the matrix chain product problem and the matrix powers problem.

To summarize, we have obtained currently the fastest and most processor efficient parallelization of the best known sequential matrix multiplication algorithm on a DMS, and have made significant progress in fast and scalable parallel matrix multiplication (as well as solving many other important problems) on DMSs, both theoretically and practically.

47.2 Preliminaries

47.2.1 Distributed Memory Systems

A DMS consists of p processors $P_0, P_1, P_2, \ldots, P_{p-1}$ (see Figure 47.1). Each processor has its own local memory and there is no global shared memory. Processors communicate with each other through message passing. The computations and communications in a DMS are globally synchronized into steps. A step is either a computation step or a communication step. In a computation step, each processor performs a local arithmetic/logic operation or is idle. A computation step takes constant amount of time.

In a communication step, processors send and receive messages via an interconnection network. A communication step can be specified as

$$((\pi(0), v_0), (\pi(1), v_1), (\pi(2), v_2), \ldots, (\pi(p-1), v_{p-1})),$$

where for all $0 \leq j \leq p-1$, processor P_j sends a value v_j to processor $P_{\pi(j)}$, and π is a mapping $\pi : \{0, 1, 2, \ldots, p-1\} \to \{-1, 0, 1, 2, \ldots, p-1\}$. If processor P_j does not send anything during a

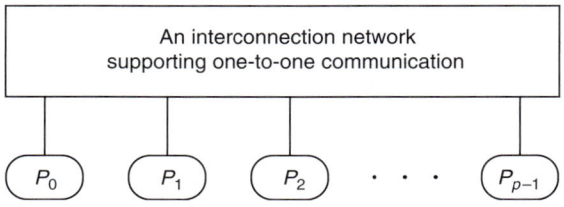

FIGURE 47.1 A distributed memory system (DMS).

communication step, then $\pi(j) = -1$ and v_j is undefined. It is required that for all $0 \leq i \leq p-1$, there is at most one j such that $\pi(j) = i$, that, each processor can receive at most one message in a communication step. Such a communication step is essentially *one-to-one communication*. It is assumed that the interconnection network connecting the p processors in a DMS can realize any one-to-one communication in $T(p)$ time. In a busiest communication step, every processor sends a message to another processor, and $(\pi(0), \pi(1), \pi(2), \ldots, \pi(p-1))$ is a permutation of $(0, 1, 2, \ldots, p-1)$.

On the basis of the above discussion, if a parallel computation on a DMS takes T_1 computation steps and T_2 communication steps, then the time complexity of the computation is $O(T_1 + T_2 T(p))$. From an algorithmic point of view, a DMS is characterized by the function $T(p)$ that measures the communication capability of the interconnection network.

In this chapter, we consider several typical types of DMS.

47.2.1.1 Distributed Memory Parallel Computers

In the class of DMPC, one-to-one communications can be realized in $T(p) = O(1)$ time [24]. We mention the following types of DMPC.

- Module Parallel Computer. The *module parallel computer* (MPC) model was used to study the capability of DMSs in simulating parallel random access machines (PRAM) [34]. The interconnection network in an MPC can be implemented by an electronic interconnection network such as a static completely connected network or a dynamic crossbar network, both can support a one-to-one communication step and arbitrary permutation in constant time. However, such implementations using electronic interconnection networks are very expensive and unrealistic when p is large.
- Optical Model of Computation. Owing to recent advances in optical interconnection technologies, the interconnection network in a DMPC can be implemented using optical interconnection networks. For instance, the *optical model of computation* (OMC) was proposed in Reference 11 (see References 1, 12, and 14 for more study on this model). In such a system, there is a processor layer and a deflection layer, both embedded in Euclidean planes. Interprocessor communications are performed using free space optical beams. An optical beam can carry and transmit information in constant time, independent of the distance covered. There are various techniques to implement the deflection layer (see Reference 11 for detailed discussions). The optical beams function properly as long as each processor receives at most one message in a communication step, thus supporting arbitrary one-to-one communication.
- Linear Array with Reconfigurable Pipelined Bus Systems. Recently, fiber optical buses have emerged as promising networks [2, 10, 17, 32, 45]. In a LARPBS, processors are connected by a reconfigurable pipelined optical bus that uses optical waveguides instead of electrical signals to transfer messages among electronic processors [30, 40]. In addition to the high propagation speed of light, there are two important properties of optical signal transmission on an optical bus, namely, unidirectional propagation and predictable propagation delay. These advantages of using waveguides enable synchronized concurrent accesses of an optical bus in a pipelined fashion [6, 23, 42]. Pipelined optical buses can support massive volume of data transfer simultaneously and can implement various communication patterns efficiently. In addition to one-to-one communication, a reconfigurable pipelined optical bus can also support broadcast, multicast, multiple multicast, and global aggregation in constant time. Hence, LARPBS is a more powerful computational model than DMPC.

47.2.1.2 Hypercubic Networks

It is well known that the Beneš, wrapped butterfly, and shuffle-exchange networks connecting p processors can support any one-to-one communication in $T(p) = O(\log p)$ time, provided that a mapping $\pi : \{0, 1, 2, \ldots, p-1\} \rightarrow \{-1, 0, 1, 2, \ldots, p-1\}$ is known in advance, which is really the case for matrix computations, where all communication steps are well defined (see Theorems 3.10, 3.12, 3.16 in Reference 22). The above networks belong to the class of *hypercubic networks*, including the hypercube network and many of its constant-degree derivatives such as cube-connected cycles, generalized shuffle-exchange, de Bruijn, k-ary de Bruijn, butterfly, k-ary butterfly, Omega, Flip, Baseline, Banyan, and Delta networks. All these networks are computationally equivalent in the sense that they can simulate each other with only a constant factor slowdown [22]. Thus, all these hypercubic networks can support one-to-one communications in $T(p) = O(\log p)$ time.

47.2.1.3 Meshes and Tori

A d-dimensional k-ary mesh network (i.e., d-dimensional k-sided array) or d-dimensional k-ary torus network (i.e., k-ary d-cube network) has d dimensions and each dimension has size k. The number of processors is $p = k^d$. When $d = \log p$ and $k = 2$, a d-dimensional k-ary mesh or torus network becomes a hypercube. It is well known that one-to-one communication can be realized in $T(p) = O(dp^{1/d})$ time on a d-dimensional k-ary mesh or torus network (see Section 1.9.4 and Problems 1.289 and 1.299 in Reference 22).

47.2.2 Scalability of Parallel Algorithms

47.2.2.1 Definitions

An important issue in parallel computing is the scalability of a parallel algorithm on a parallel system. Though there is no unified definition, scalability essentially measures the ability to maintain speedup that is linearly proportional to the number of processors.

To be more specific, when the number of processors is less than the maximum required, the parallel time complexity can generally be represented as

$$O\left(\frac{T(N)}{p} + T_{\text{comm}}(N, p)\right),$$

where N is the problem size, p is the number of processors available, $T(N)$ is the time complexity of the best sequential algorithm (or the time complexity of a sequential algorithm $\mathbb{A}$ to be parallelized), and $T_{\text{comm}}(N, p)$ is the overall communication overhead of a parallel computation.

Definition 47.1 *A parallel algorithm is* scalable *(or* scalable with respect to the sequential algorithm $\mathbb{A}$*) in the range $1 \leq p \leq p^*$ if the parallel time complexity is $O(T(N)/p)$, that is, linear speedup and cost optimality can be achieved, for all $1 \leq p \leq p^*$.*

Using Definition 47.1, we know that a parallel algorithm is scalable in the range $1 \leq p \leq p^*$, where p^* satisfies the following equation:

$$p^* = O\left(\frac{T(N)}{T_{\text{comm}}(N, p^*)}\right).$$

Clearly, p^* should be as large as possible, since this implies that the parallel algorithm has the ability to be scalable over a large range of processor complexity.

Definition 47.2 *A parallel algorithm is* highly scalable *(or* highly scalable with respect to the sequential algorithm $\mathbb{A}$*) if p^* is as large as*

$$O\left(\frac{T(N)}{T^*(N)(\log N)^c}\right)$$

for some constant $c \geq 0$, where $T^*(N)$ is the best possible parallel time complexity.

It is clear that not every sequential algorithm can be parallelized by using sufficient processors so that constant parallel execution time is achieved. If $T^*(N)$ is the best possible parallel time, the largest possible value for p^* is $O(T(N)/T^*(N))$. Definition 47.2 simply says that a parallel algorithm is highly scalable if p^* is very close to the largest possible except a polylog factor of N.

Definition 47.3 *A highly scalable parallel algorithm is* fully scalable *(or* fully scalable *with respect to the sequential algorithm* $\mathbb{A}$*) if $c = 0$.*

A fully scalable parallel algorithm implies that the best sequential algorithm (or the sequential algorithm $\mathbb{A}$) can be fully parallelized and communication overhead $T_{\text{comm}}(N, p)$ in parallelization is negligible.

47.2.2.2 Examples

As illustrative and motivating examples, we present the scalability of known parallel matrix multiplication algorithms on meshes and hypercubes, two most successful and popular types of interconnection networks ever developed and commercialized in parallel computing.

Example 47.1 *It is known that by using the block 2-D partitioning of matrices and parallelizing the standard matrix multiplication algorithm, multiplying two $N \times N$ matrices can be performed by a mesh network with p processors in*

$$O\left(\frac{N^3}{p} + t_s \log p + 2t_w \frac{N^2}{\sqrt{p}}\right)$$

time, where

- t_s *is the communication startup time required to handle a message at the sending and receiving processors;*
- t_w *is the per-word transfer time that is the reciprocal of link bandwidth (words per second) [16].*

According to Definition 47.1, the above parallel computation is scalable with respect to the standard algorithm with $p^* = O(N^2)$. However, since the best possible parallel time complexity of matrix multiplication is $T^*(N) = O(\log N)$, the above parallel computation is not highly scalable.

Example 47.2 *Again, on a mesh network with p processors, Cannon's algorithm [4] has time complexity*

$$O\left(\frac{N^3}{p} + 2\sqrt{p}t_s + 2t_w \frac{N^2}{\sqrt{p}}\right).$$

According to Definition 47.1, the above parallel computation is scalable with respect to the standard matrix multiplication algorithm with $p^* = O(N^2)$, but not highly scalable.

Example 47.3 *On a hypercube with p processors, the DNS algorithm [9] can multiply two $N \times N$ matrices in*

$$O\left(\frac{N^3}{p} + t_s \log p + t_w \frac{N^2}{p^{2/3}} \log p\right)$$

time. According to Definition 47.1, the above parallel computation is scalable with respect to the standard algorithm with $p^* = O(N^3/(\log N)^3)$. According to Definition 47.2, the above parallel computation is highly scalable with respect to the standard algorithm with $c = 2$. However, as mentioned earlier, the DNS algorithm is not scalable at all with respect to any nonstandard matrix multiplication algorithm.

Note that all the scalability results obtained in the following sections are with respect to the fastest sequential algorithms.

Step (B) (Basis) If $n = 0$, i.e., the matrices are of size 1×1, computer the product $C = A \times B$ directly, and return; otherwise, do the following steps.

Step (D) (Division) Calculate the $2R$ linear combinations of submatrices

$$L_u = \sum_{1 \leq i, j \leq m} f(i, j, u) A_{ij},$$

and

$$L_u^* = \sum_{1 \leq j, k \leq m} f^*(j, k, u) B_{jk},$$

for all $1 \leq u \leq R$, where L_u and L_u^* are $m^{n-1} \times m^{n-1}$ matrices.

Step (R) (Recursion) Calculate the R matrix products $M_u = L_u \times L_u^*$, for all $1 \leq u \leq R$.

Step (C) (Combination) Compute

$$C_{ik} = \sum_{j=1}^{m} A_{ij} \times B_{jk} = \sum_{u=1}^{R} f^{**}(k, i, u) M_u,$$

for all $1 \leq i, k \leq m$.

FIGURE 47.2 The recursive bilinear algorithm.

47.3 Matrix Multiplication on DMSs

All existing sequential matrix multiplication algorithms over an arbitrary ring fall into the category of *bilinear algorithms* (see Reference [3, pp. 315–316] and References [37–39]). Let $A = (A_{ij})$, $B = (B_{jk})$, and $C = (C_{ik})$ be $N \times N$ matrices, where $N = m^n$. Assume that m is a fixed constant, and $n \to \infty$. Each of these matrices are partitioned into m^2 submatrices A_{ij}, B_{jk}, C_{ik} of size $m^{n-1} \times m^{n-1}$. A bilinear algorithm for computing $C = A \times B$ can be recursively described in Figure 47.2. In the above computation, all the $f(i, j, u)$s, $f^*(j, k, u)$s, and $f^{**}(k, i, u)$s are constants. The value R is called the rank of the algorithm. For all the known algorithms for $N \times N$ matrix multiplication running in $O(N^\alpha)$ time for $\alpha \leq 3$ (including the standard algorithm for $\alpha = 3$, Strassen's algorithm for $\alpha = \log_2 7 < 2.8074$ [43], Pan's algorithm for $\alpha = \log_{70} 143640 < 2.7952$ [35], and the ones of Reference 7 for $\alpha < 2.3755$, which are currently asymptotically fastest), we may yield $R = m^\alpha$ in the associated bilinear construction. Note that for fixed m and α, R is finite.

The above recursive algorithm has $n = \lceil \log_m N \rceil$ levels. (A few extra dummy rows and columns are introduced if N is not a power of m.) The recursion reaches its base case (see Step (B)) when the size of the submatrices is 1×1. In general, a bilinear algorithm has three steps. If sufficiently many processors are available, we can calculate the L_us in parallel, and then all the L_u^*s in parallel in the division step (see Step (D)). After the recursion step (see Step (R)), all the C_{ik}s are also computed in parallel in the combination step (see Step (C)). Hence, each level takes constant time, and the overall time complexity is $O(\log N)$, which is the best possible. The execution time cannot be further reduced owing to the recursive nature of the class of bilinear algorithms, and not to the communication constraints on a parallel system.

For all so far available sequential algorithms for matrix multiplication, there is a parallelization under the CREW PRAM model, which requires $O(N^\alpha / \log N)$ processors (see p. 317 in Reference [3]).

47.3.1 Fast Matrix Multiplication

The recursive bilinear algorithm can be unrolled into an iterative algorithm, as shown in Figure 47.3. Let us label Steps (D), (R), and (C) in the lth level of the recursion as Steps (D_l), (R_l), and (C_l), respectively,

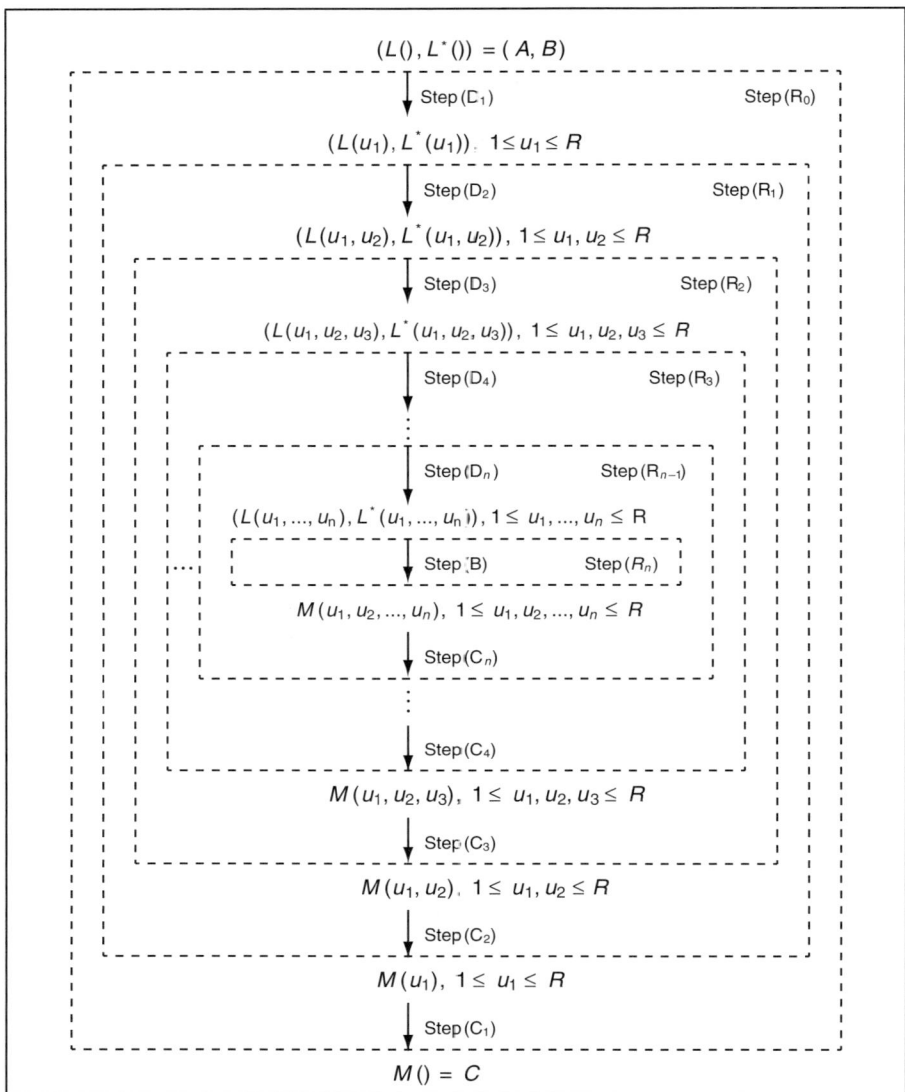

FIGURE 47.3 Unrolling of the recursive bilinear algorithm.

where $1 \leq l \leq n$. The computation proceeds in the following way:

$$\text{Step } (D_1) \to \text{Step } (D_2) \to \text{Step } (D_3) \to \cdots \to \text{Step } (D_n)$$
$$\to \text{Step } (B) \to \text{Step } (C_n) \to \cdots \to \text{Step } (C_3) \to \text{Step } (C_2) \to \text{Step } (C_1).$$

The recursive step (R_l), $1 \leq l \leq n$, is actually

$$\text{Step } (D_{l+1}) \to \text{Step } (D_{l+2}) \to \text{Step } (D_{l+3}) \to \cdots \to \text{Step } (D_n)$$
$$\to \text{Step } (B) \to \text{Step } (C_n) \to \cdots \to \text{Step } (C_{l+3}) \to \text{Step } (C_{l+2}) \to \text{Step } (C_{l+1}).$$

Given matrices $L(u_1, u_2, \ldots, u_l)$ and $L^*(u_1, u_2, \ldots, u_l)$ of size $N/m^l \times N/m^l$, Step (R_l) obtains their product

$$M(u_1, u_2, \ldots, u_l) = L(u_1, u_2, \ldots, u_l) \times L^*(u_1, u_2, \ldots, u_l),$$

for all $1 \leq u_1, u_2, \ldots, u_l \leq R$. Therefore, we can imagine that the entire bilinear algorithm is actually Step (R_0) which, given two matrices $L() = A$ and $L^*() = B$, computes their product $C = M() = L() \times L^*() = A \times B$.

To compute the product $M(u_1, u_2, \ldots, u_l)$ in Step (R_l), where $0 \leq l \leq n-1$, we first execute Step (D_{l+1}), that is, partitioning $L(u_1, u_2, \ldots, u_l)$ and $L^*(u_1, u_2, \ldots, u_l)$ into submatrices $L_{ij}(u_1, u_2, \ldots, u_l)$ and $L^*_{jk}(u_1, u_2, \ldots, u_l)$ of size $N/m^{l+1} \times N/m^{l+1}$, and calculating the following matrices:

$$L(u_1, u_2, \ldots, u_{l+1}) = \sum_{1 \leq i,j \leq m} f(i, j, u_{l+1}) L_{ij}(u_1, u_2, \ldots, u_l),$$

and

$$L^*(u_1, u_2, \ldots, u_{l+1}) = \sum_{1 \leq j,k \leq m} f^*(j, k, u_{l+1}) L^*_{jk}(u_1, u_2, \ldots, u_l),$$

for all $1 \leq u_{l+1} \leq R$. Then, all the matrix products

$$M^*(u_1, u_2, \ldots, u_{l+1}) = L(u_1, u_2, \ldots, u_{l+1}) \times L^*(u_1, u_2, \ldots, u_{l+1})$$

are computed by Step (R_{l+1}). Finally, in Step (C_{l+1}), we get $M(u_1, u_2, \ldots, u_l)$, which consists of submatrices $M_{ik}(u_1, u_2, \ldots, u_l)$, where

$$M_{ik}(u_1, u_2, \ldots, u_l) = \sum_{u_{l+1}=1}^{R} f^{**}(k, i, u_{l+1}) M(u_1, u_2, \ldots, u_{l+1}),$$

for all $1 \leq i, k \leq m$. When $l = n$, Step (R_n) is actually Step (B), and the $M(u_1, u_2, \ldots, u_n)$s are calculated directly. The complete iterative version of the bilinear algorithm is illustrated in Figure 47.3.

We now discuss how the above iterative bilinear algorithm can be implemented on a DMS with $p = N^\alpha$ processors.

The size of the matrix $L(u_1, u_2, \ldots, u_l)$ is $(N/m^l) \times (N/m^l)$, where $1 \leq l \leq n$. Since there are R^l $L(u_1, u_2, \ldots, u_l)$ matrices, the total number of elements in the $L(u_1, u_2, \ldots, u_l)$s is $R^l(N/m^l)^2 = N^2(R/m^2)^l = N^2 m^{(\alpha-2)l}$, which is the number of processors used in Steps (D_l) and (C_l). As the computation proceeds from Steps (D_1) to (D_n), the number of processors required increases, which reaches its maximum, that is, N^α, when $l = n$. Hence, the total number of processors required in Steps (D_1) to (D_n) is $p = N^\alpha$. The number of processors required in Step (C_l) is the same as Step (D_l), for all $1 \leq l \leq n$. It is clear that as the computation proceeds from Steps (C_n) to (C_1), the number of processors required decreases.

We use the notation $P[s : t]$ to represent the group of t processors with consecutive indices starting from s, that is, $P_s, P_{s+1}, P_{s+2}, \ldots, P_{s+t-1}$. All the matrices and their submatrices are of size m^l, where $0 \leq l \leq n$. A matrix X of size $m^l \times m^l$ is stored in a processor group $P[s : m^{2l}]$ in the shuffled-row-major order. That is, $X = (X_{ij})$ is partitioned into m^2 submatrices of size $m^{l-1} \times m^{l-1}$, and X_{ij} is stored in the processor group

$$P[s + ((i-1)m + (j-1))m^{2(l-1)} : m^{2(l-1)}]$$

in the shuffled-row-major order, where $1 \leq i, j \leq m$.

On the basis of the assumption that the interconnection network in a DMPC can support one-to-one communication in one step, it is not difficult to see the following lemmas [24].

Lemma 47.1 *A processor group $P[s : m^{2l}]$ that holds an $m^l \times m^l$ matrix X can send X to another processor group $P[s' : m^{2l}]$ in one communication step. Furthermore, let $P[s_u : m^{2l}]$, where $1 \leq u \leq R$ be R disjoint processor groups, and processor group $P[s_u : m^{2l}]$ holds an $m^l \times m^l$ matrix X_u, where $1 \leq u \leq R$. Let $P[s'_u : m^{2l}]$, where $1 \leq u \leq R$ be R disjoint processor groups. Then, in one communication step, $P[s_u : m^{2l}]$ can send X_u to $P[s'_u : m^{2l}]$, for all $1 \leq u \leq R$ in parallel.*

Lemma 47.2 *Let $P[s_u : m^{2l}]$, where $1 \leq u \leq R$ be R disjoint processor groups, and processor group $P[s : m^{2l}]$ holds an $m^l \times m^l$ matrix X. The matrix X can be broadcasted to all $P[s_u : m^{2l}]$, $1 \leq u \leq R$, in $(\lceil \log R \rceil + 1)$ communication steps.*

Lemma 47.3 *If a processor group $P[s : m^{2l}]$ holds d $m^l \times m^l$ matrices $X_1, X_2, \ldots, X_d$, the linear combination $f_1 X_1 + f_2 X_2 + \cdots + f_d X_d$, where $f_1, f_2, \ldots, f_d$ are constants, can be obtained in $(2d - 1)$ computation steps.*

Lemma 47.4 *Let $P[s_u : m^{2l}]$, where $1 \leq u \leq R$ be R disjoint processor groups. If processor group $P[s_u : m^{2l}]$ holds an $m^l \times m^l$ matrix X_u, where $1 \leq u \leq R$, the linear combination $f_1 X_1 + f_2 X_2 + \cdots + f_R X_R$, where $f_1, f_2, \ldots, f_R$ are constants, can be obtained in $\lceil \log R \rceil$ communication steps and $\lceil \log R \rceil + 1$ computation steps, such that the result is saved in $P[s_1 : m^{2l}]$.*

Let us first examine how the $L(u_1, u_2, \ldots, u_l)$s can be computed in Step (D_l), where $1 \leq l \leq n$. The $L(u_1, u_2, \ldots, u_l)$s can be arranged in the lexicographical order, where $0 \leq l \leq n$. Each $L(u_1, u_2, \ldots, u_l)$ is assigned an order number $\beta(u_1, u_2, \ldots, u_l)$ in the range $[0..R^l - 1]$:

$$\beta(u_1, u_2, \ldots, u_l) = (u_1 - 1)R^{l-1} + (u_2 - 1)R^{l-2} + \cdots + (u_{l-1} - 1)R + (u_l - 1).$$

Matrix $L(u_1, u_2, \ldots, u_l)$ is then stored in

$$P\left[\beta(u_1, u_2, \ldots, u_l)\left(\frac{N}{m^l}\right)^2 : \left(\frac{N}{m^l}\right)^2\right]. \tag{47.1}$$

In Step (D_l), where $1 \leq l \leq n$, we basically calculate the $L(u_1, u_2, \ldots, u_l)$s from the $L(u_1, u_2, \ldots, u_{l-1})$s. Let $L(u_1, u_2, \ldots, u_{l-1})$, which has size $(N/m^{l-1}) \times (N/m^{l-1})$, be partitioned into m^2 submatrices of size $(N/m^l) \times (N/m^l)$, that is, the $L_{ij}(u_1, u_2, \ldots, u_{l-1})$s, where $1 \leq i, j \leq m$. The submatrix $L_{ij}(u_1, u_2, \ldots, u_{l-1})$ is stored in

$$P\left[\beta(u_1, u_2, \ldots, u_{l-1})\left(\frac{N}{m^{l-1}}\right)^2 + ((i-1)m + (j-1))\left(\frac{N}{m^l}\right)^2 : \left(\frac{N}{m^l}\right)^2\right], \tag{47.2}$$

which is made available to the processor groups specified in Equation 47.1, for all $1 \leq u_l \leq R$. When the processor group specified in Equation 47.1 collects all the $L_{ij}(u_1, u_2, \ldots, u_{l-1})$s, $L(u_1, u_2, \ldots, u_l)$ can be computed using the following equation:

$$L(u_1, u_2, \ldots, u_l) = \sum_{1 \leq i,j \leq m} f(i, j, u_l) L_{ij}(u_1, u_2, \ldots, u_{l-1}). \tag{47.3}$$

Our algorithm for Step (D_l) is given in Figure 47.4.

It is clear that Lines 3–4 can be implemented in $(\lceil \log R \rceil + 1)$ communication steps (see Lemma 47.2). Thus, the sequential for-loop in Lines 2–5 takes $m^2(\lceil \log R \rceil + 1)$ communication steps. The computation in Line 7, which involves a linear combination of matrices defined in Equation 47.3, can be done in $(2m^2 - 1)$ computation steps (see Lemma 47.3). Thus, the parallel for-loop in Lines 6–8 takes $(2m^2 - 1)$ computation steps. The overall parallel for-loop in Lines 1–9 requires $O(m^2(\lceil \log R \rceil + 1)T(p) + (2m^2 - 1))$ time. Since

1. **for all** $(u_1, u_2, \ldots, u_{l-1}), 1 \leq u_1, u_2, \ldots, u_{l-1} \leq R$, **do in parallel**
2.   **for** $1 \leq i, j \leq m$ **do**
3.   The processor group specified by Equation 47.2 broadcasts $L_{ij}(u_1, u_2, \ldots, u_{l-1})$
4.   to the processor groups specified by Equation 47.1, for all $1 \leq u_l \leq R$.
5.   **end do**
6.   **for all** $1 \leq u_l \leq R$ **do in parallel**
7.   The processor group specified by Equation 47.1 calculates $L(u_1, u_2, \ldots, u_l)$ using Equation 47.3.
8.   **end do in parallel**
9. **end do in parallel**

FIGURE 47.4 Details of Step (D_l) to compute the $L(u_1, u_2, \ldots, u_l)$s.

m and R are constants, Step (D_l) has $O(T(p))$ running time to compute all the $L(u_1, u_2, \ldots, u_l)$s, for all $1 \leq l \leq n$. The total running time of Steps (D_1) to (D_n) is $O(T(p) \log N)$.

The computation of the $L^*(u_1, u_2, \ldots, u_l)$s in Step ($D_l$) can be done in a similar way, which requires the same amount of execution time and the same number of processors.

It is clear that Step (B) performs one local computation and takes constant time.

As shown in Figure 47.3, in Step (C_l), where $1 \leq l \leq n$, we calculate $M(u_1, u_2, \ldots, u_{l-1})$ on the basis of the $M(u_1, u_2, \ldots, u_l)$s, for all $1 \leq u_1, u_2, \ldots, u_{l-1} \leq R$, using the following equation,

$$M_{ik}(u_1, u_2, \ldots, u_{l-1}) = \sum_{u_l=1}^{R} f^{**}(k, i, u_l) M(u_1, u_2, \ldots, u_{l-1}, u_l). \tag{47.4}$$

Our algorithm for Step (C_l) is given in Figure 47.5. Again, for all $0 \leq l \leq n$, the $M(u_1, u_2, \ldots, u_l)$s are arranged in the lexicographical order, and $M(u_1, u_2, \ldots, u_l)$ is stored in the processor group specified in Equation 47.1. Lines 3–5 involves a linear combination in Equation 47.4, which can be obtained in $\lceil \log R \rceil$ communication steps and $\lceil \log R \rceil + 1$ computation steps (see Lemma 47.4). Hence, the sequential for-loop in Lines 2–6 requires $m^2 \lceil \log R \rceil$ communication steps and $m^2(\lceil \log R \rceil + 1)$ computation steps. Lines 7–8 only perform a one-to-one communication, that is, sending $M(u_1, u_2, \ldots, u_{l-1})$ in the processor group

$$P\left[\beta(u_1, u_2, \ldots, u_{l-1}, 1)\left(\frac{N}{m^l}\right)^2 : \left(\frac{N}{m^{l-1}}\right)^2\right] \tag{47.5}$$

to the processor group

$$P\left[\beta(u_1, u_2, \ldots, u_{l-1})\left(\frac{N}{m^{l-1}}\right)^2 : \left(\frac{N}{m^{l-1}}\right)^2\right], \tag{47.6}$$

which can be done in one communication step (see Lemma 47.1). The entire Step (C_l), that is, the parallel for-loop in Lines 1–9, requires $O((m^2 \lceil \log R \rceil + 1) T(p) + m^2(\lceil \log R \rceil + 1))$ time. Since m and R are constants, Step (C_l) has $O(T(p))$ running time to compute all the $M(u_1, u_2, \ldots, u_{l-1})$s, for all $1 \leq l \leq n$. The total running time of Steps (C_n) to (C_1) is $O(T(p) \log N)$.

Summarizing the discussion in this section, we obtain the following theorem that generalizes Theorem 1 in Reference 24.

Theorem 47.1 *Multiplying two $N \times N$ matrices can be performed by a DMS with $p = N^\alpha$ processors in $T_{\text{mm}}(N, p) = O(T(p) \log N)$ time.*

1. **for all** $(u_1, u_2, \ldots, u_{l-1}), 1 \leq u_1, u_2, \ldots, u_{l-1} \leq R$, **do in parallel**
2.   **for** $1 \leq i, k \leq m$ **do**
3.   The processor groups specified by Equation 47.1, where $1 \leq u_l \leq R$, compute
4. $M_{ik}(u_1, u_2, \ldots, u_{l-1})$ using Equation 47.4, and save the result in
5.

$$P\left[\beta(u_1, u_2, \ldots, u_{l-1}, (i-1)m + k)\left(\frac{N}{m^l}\right)^2 : \left(\frac{N}{m^l}\right)^2\right]. \quad (47.7)$$

6.   **end do**
7.   The processor group specified by Equation 47.5 sends $M(u_1, u_2, \ldots, u_{l-1})$ consisting of
8.   the $M_{ik}(u_1, u_2, \ldots, u_{l-1})$s in Equation 47.7 to the processor group specified by Equation 47.6
9. **end do in parallel**

FIGURE 47.5 Details of Step (C_l) to compute the $M(u_1, u_2, \ldots, u_l)$s.

47.3.1.1 Processor Complexity Reduction

In this section, we show that the number of processors in Theorem 47.1 can be reduced to $N^\alpha / \log N$, while the time complexity is still $O(T(p) \log N)$.

The size of the matrix $L(u_1, u_2, \ldots, u_l)$ is $(N/m^l) \times (N/m^l)$, where $1 \leq l \leq n$. Since there are R^l $L(u_1, u_2, \ldots, u_l)$ matrices, the total number of elements in the $L(u_1, u_2, \ldots, u_l)$s is $R^l(N/m^l)^2 = N^2(R/m^2)^l = N^2 m^{(\alpha-2)l}$. Let l^* be defined as the largest integer such that

$$N^2 m^{(\alpha-2)l^*} \leq \frac{N^\alpha}{\log N}.$$

Then, we have

$$l^* = \left\lfloor \frac{(\alpha - 2) \log N - \log \log N}{(\alpha - 2) \log m} \right\rfloor.$$

In Step (D_l), $l^* + 1 \leq l \leq n$, several passes are required to compute the $L(u_1, u_2, \ldots, u_l)$s. The number of $L(u_1, u_2, \ldots, u_l)$ matrices that can be accommodated by $p = N^\alpha / \log N$ processors in one pass is

$$K_l = \left\lfloor \frac{p}{(N/m^l)^2} \right\rfloor.$$

For all $0 \leq l \leq n$, matrices $L(u_1, u_2, \ldots, u_l)$ and $M(u_1, u_2, \ldots, u_l)$ are stored in

$$P\left[\left(\beta(u_1, u_2, \ldots, u_l) \bmod K_l\right)\left(\frac{N}{m^l}\right)^2 : \left(\frac{N}{m^l}\right)^2\right].$$

The number of passes required to compute all the $L(u_1, u_2, \ldots, u_l)$s in Step (D_l) is

$$\left\lceil \frac{R^l}{K_l} \right\rceil.$$

Hence, the total number of passes in Steps (D_1) to (D_n) is

$$\sum_{l=1}^{n} \left\lceil \frac{R^l}{K_l} \right\rceil \leq \sum_{l=1}^{n} \left(\frac{R^l}{K_l} + 1 \right)$$

$$\leq \sum_{l=1}^{n} \left(\frac{R^l}{p/(N/m^l)^2 - 1} + 1 \right)$$

$$= \sum_{l=1}^{n} \left(\frac{R^l (N/m^l)^2}{p - (N/m^l)^2} + 1 \right)$$

$$\leq \sum_{l=1}^{n} \left(\frac{R^l (N/m^l)^2}{p - N^2} + 1 \right)$$

$$= \sum_{l=1}^{n} \left(\frac{1}{p/N^2 - 1} \left(\frac{R}{m^2} \right)^l + 1 \right)$$

$$= \left(\frac{1}{p/N^2 - 1} \right) \sum_{l=1}^{n} \left(\frac{R}{m^2} \right)^l + \log_m N$$

$$= \left(\frac{1}{p/N^2 - 1} \right) \left(\frac{(R/m^2)^{n+1} - R/m^2}{R/m^2 - 1} \right) + \log_m N$$

$$= O\left(\left(\frac{\log N}{N^{\alpha-2}} \right) (m^{\alpha-2})^n + \log N \right)$$

$$= O(\log N).$$

Taking the $L^*(u_1, u_2, \ldots, u_l)$s into consideration, the running time of Steps (D_1) to (D_n) is at most doubled. The number of passes in Steps (C_n) to (C_1) is also $O(\log N)$, which can be obtained in a similar way. The number of passes in Step (B) is $\lceil R^n/K_n \rceil = \lceil \log N \rceil$. Since each pass performs constant numbers of communication and computation steps and takes $O(T(p))$ time, the overall time complexity of our parallelized bilinear algorithm is $O(T(p) \log N)$. The following theorem generalizes Theorem 2 in Reference 24.

Theorem 47.2 *Multiplying two $N \times N$ matrices can be performed by a DMS with $p = N^\alpha / \log N$ processors in $T_{mm}(N, p) = O(T(p) \log N)$ time.*

47.3.1.2 Time Complexity Reduction

Recall that the best parallel time complexity for matrix multiplication using the class of bilinear algorithms is $T^*_{mm}(N) = O(\log N)$. As indicated in Theorem 47.2, such a lower bound for parallel matrix multiplication can be achieved by a DMPC with $T(p) = O(1)$. To further reduce the execution time of parallel matrix multiplication to sublogarithmic time, we need the following algorithmic and architectural features:

- A different algorithmic approach
- Number of processors larger than N^α
- Additional communication/computation capabilities of the interconnection network connecting the processors in a DMS

It is well known that a reconfigurable pipelined optical bus in an LARPBS not only provides communication channels among the processors to support various communication patterns such as one-to-one communication, broadcasting, multicasting, and multiple multicasting, but also acts as an active component and agent for certain computations, for example, prefix sums, extraction and compression, and global aggregation operations. By parallelizing the standard matrix multiplication algorithm without recursion and fully exploiting the communication/computation capabilities of a reconfigurable pipelined optical bus, the following result was proven in Reference 30.

Lemma 47.5 *Multiplying two $N \times N$ matrices can be performed by an LARPBS with N^3 processors in $O(1)$ time.*

As suggested in Reference 30, a *compound algorithm* can be used for matrix multiplication. A compound algorithm combines two matrix multiplication algorithms, for example, the Strassen's algorithm and the standard algorithm [30]. In fact, any bilinear algorithm can be combined with the standard algorithm as follows. A compound algorithm proceeds in the same way as the bilinear algorithm for the first r levels of recursion, such that the submatrix size is reduced to N/m^r. Then, the recursive step (R_r) is replaced by the application of the algorithm in Lemma 47.5, that is, the product

$$M(u_1, u_2, \ldots, u_r) = L(u_1, u_2, \ldots, u_r) \times L^*(u_1, u_2, \ldots, u_r)$$

is computed directly by the algorithm in Lemma 47.5 without further recursion, for all $1 \leq u_1, u_2, \ldots, u_r \leq R$. Since there are R^r such products and each product requires $(N/m^r)^3$ processors, the number of processors required by the compound algorithm is

$$p = R^r \left(\frac{N}{m^r}\right)^3 = \frac{N^3}{(m^{3-\alpha})^r}.$$

Since each level of the recursion as well as Step (R^r) all take constant time, the overall execution time of the compound algorithm is $O(r)$. By letting $r = (\log_m N)^\delta$, where $0 \leq \delta \leq 1$, we get the following theorem, which generalizes Theorem 5 in Reference 30.

Theorem 47.3 *Multiplying two $N \times N$ matrices can be performed by an LARPBS with*

$$p = \frac{N^3}{(m^{3-\alpha})^{(\log_m N)^\delta}}$$

processors in $T_{\text{mm}}(N, p) = O((\log N)^\delta)$ time, where $0 \leq \delta \leq 1$.

47.3.2 Scalable Matrix Multiplication

It is clear that in the implementation mentioned in Theorem 47.2, processors perform computation and communication at the matrix element level. That is, processors

- Calculate one element level addition or multiplication during a computation step in Steps (D_1) to (D_n) and Steps (C_1) to (C_n)
- Calculate one element level multiplication during a computation step in Step (B)
- Send/receive one element during a communication step in Steps (D_1) to (D_n) and Steps (C_1) to (C_n)

When there are fewer processors, it is necessary for processors to perform computation and communication at the block level. The idea is to partition the input/output matrices A, B, and C into blocks. When a matrix $X = (X_{ij})$ of size $N \times N$ is divided into blocks, that is, the X_{ij}s, of size $s \times s$, X is treated as a supermatrix of size $q \times q$, such that $N = qs$, and the blocks are treated as superelements. Since the problem size is changed from N to q, the number of processors required in Theorem 47.2 is changed from $N^\alpha / \log N$ to $q^\alpha / \log q$. Let $r = \lceil \log_m q \rceil$. There are r levels of recursion. Then, processors perform computation and communication at the superelement (block) level, that is, processors

- Calculate one superelement level addition or a scalar-block multiplication during a computation step in Steps (D_1) to (D_r) and Steps (C_1) to (C_r)
- Calculate one superelement level multiplication (i.e., multiplying matrices of size $s \times s$) during a computation step in Step (B)
- Send/receive one superelement during a communication step in Steps (D_1) to (D_r) and Steps (C_1) to (C_r)

For supermatrices with superelements, the running times in Lemmas 47.1–47.4 are increased by a factor of $O(s^2)$. This implies that the total running time of Steps (D_1) to (D_r) and Steps (C_1) to (C_r) is $O(T(p)s^2 \log q)$. Step (B) requires $O(s^\alpha \log q)$ time (recall that the number of passes in Step (B) is $\lceil \log q \rceil$). Hence, multiplying two $q \times q$ supermatrices, where superelements are $s \times s$ submatrices, can be performed by a DMS with p processors in $O(s^\alpha \log q + T(p)s^2 \log q)$ time. Let q be the largest integer such that $q^\alpha / \log q \leq p$. This implies that $q^\alpha = \Theta(p \log q)$, $q = \Theta((p \log q)^{1/\alpha})$, and $\log q = \Theta(\log p)$. Therefore, the overall running time is

$$O(s^\alpha \log q + T(p)s^2 \log q)$$
$$= O\left(\left(\frac{N}{q}\right)^\alpha \log q + T(p)\left(\frac{N}{q}\right)^2 \log q\right)$$
$$= O\left(\left(\frac{N^\alpha}{p \log q}\right) \log q + T(p)\left(\frac{N^2}{(p \log q)^{2/\alpha}}\right) \log q\right)$$
$$= O\left(\frac{N^\alpha}{p} + T(p)\left(\frac{N^2}{p^{2/\alpha}}\right)(\log p)^{1-2/\alpha}\right).$$

The above discussion leads to the following theorem [24].

Theorem 47.4 *For all $1 \leq p \leq N^\alpha / \log N$, multiplying two $N \times N$ matrices can be performed by a DMS with p processors in*

$$T_{\mathrm{mm}}(N, p) = O\left(\frac{N^\alpha}{p} + T(p)\left(\frac{N^2}{p^{2/\alpha}}\right)(\log p)^{(\alpha-2)/\alpha}\right)$$

time. In particular, the above time complexity is $O(T(p) \log N)$ when $p = N^\alpha / \log N$.

47.3.2.1 Results of Scalability

Theorem 47.4 can be instantiated on different DMSs. Corollary 47.1 shows that matrix multiplication on DMPC is fully scalable.

Corollary 47.1 *For all $1 \leq p \leq N^\alpha / \log N$, multiplying two $N \times N$ matrices can be performed by a DMPC with p processors in $T_{\mathrm{mm}}(N, p) = O(N^\alpha / p)$ time. In particular, the time complexity is $O(\log N)$ when $p = N^\alpha / \log N$. According to Definition 47.3, matrix multiplication on DMPC is fully scalable.*

The significance of Corollary 47.1 is twofold. First, it shows that the performance of matrix multiplication on DMPC matches the performance on CREW PRAM. Second, it unifies all known algorithms for matrix multiplication on DMPC, standard or nonstandard, sequential or parallel.

Corollary 47.2 shows that matrix multiplication on DMS with hypercubic networks is highly scalable.

Corollary 47.2 *For all $1 \leq p \leq N^\alpha / \log N$, multiplying two $N \times N$ matrices can be performed by a DMS with p processors connected by a hypercubic network in*

$$T_{mm}(N,p) = O\left(\frac{N^\alpha}{p} + \left(\frac{N^2}{p^{2/\alpha}}\right)(\log p)^{2(\alpha-1)/\alpha}\right)$$

time. The above $T_{mm}(N,p)$ is a decreasing function of p and reaches its minimum $O((\log N)^2)$ when $p = N^\alpha / \log N$. Furthermore, if

$$p = O\left(\frac{N^\alpha}{(\log N)^{2(\alpha-1)/(\alpha-2)}}\right),$$

linear speedup and cost optimality can be achieved. According to Definition 47.2, with

$$c = 2\left(\frac{\alpha-1}{\alpha-2}\right) - 1 = \frac{\alpha}{\alpha-2},$$

matrix multiplication on DMS with hypercubic networks is highly scalable.

Corollary 47.3 shows the performance of matrix multiplication on DMS with mesh and torus networks.

Corollary 47.3 *For all $1 \leq p \leq N^\alpha / \log N$, multiplying two $N \times N$ matrices can be performed by a DMS with p processors connected by a d-dimensional mesh or torus network in*

$$T_{mm}(N,p) = O\left(\frac{N^\alpha}{p} + \left(\frac{dN^2}{p^{2/\alpha-1/d}}\right)(\log p)^{(\alpha-2)/\alpha}\right)$$

time. (The above time complexity reduces to that in Corollary 47.2 when $d = \log p$.) When $d = 1$, $T_{mm}(N,p)$ reaches its minimum

$$O\left(N^{\alpha^2/(2(\alpha-1))}(\log N)^{(\alpha-2)/(2(\alpha-1))}\right)$$

when

$$p = O\left(N^{\alpha(\alpha-2)/(2(\alpha-1))}/(\log N)^{(\alpha-2)/(2(\alpha-1))}\right).$$

When $d \geq 2$, $T_{mm}(N,p)$ is a decreasing function of p and reaches its minimum

$$O(dN^{\alpha/d}(\log N)^{(d-1)/d})$$

when $p = N^\alpha / \log N$. Furthermore, for $d \geq 1$, if $p \leq p^*$ where

$$p^* = O\left(\frac{N^{\alpha(\alpha-2)d/((\alpha-2)d+\alpha)}}{d^{\alpha d/((\alpha-2)d+\alpha)}(\log N)^{(\alpha-2)d/((\alpha-2)d+\alpha)}}\right),$$

linear speedup and cost optimality can be achieved.

The value p^* given by Corollary 47.3 is displayed in Table 47.1, assuming that $\alpha = 2.3755$. Clearly, for any constant $d \geq 1$, matrix multiplication on DMS with mesh and torus networks is not highly scalable.

47.3.2.2 Conditions for Scalability

Corollary 47.4 gives a condition for matrix multiplication on DMS to be fully scalable and a condition for matrix multiplication on DMS to be highly scalable.

Corollary 47.4 *Matrix multiplication on DMS is fully scalable if and only if $T(p) = O(1)$ and highly scalable if and only if $T(p) = O((\log p)^\gamma)$ for some constant $\gamma \geq 0$.*

TABLE 47.1 The Value p^* in Corollary 47.3

d	p^*
1	$O(N^{0.3242}/(\log N)^{0.1365})$
2	$O(N^{0.5706}/(\log N)^{0.2402})$
3	$O(N^{0.7641}/(\log N)^{0.3217})$
4	$O(N^{0.9202}/(\log N)^{0.3874})$
5	$O(N^{1.0487}/(\log N)^{0.4415})$
6	$O(N^{1.1563}/(\log N)^{0.4868})$
7	$O(N^{1.2478}/(\log N)^{0.5253})$
8	$O(N^{1.3265}/(\log N)^{0.5584})$
9	$O(N^{1.3950}/(\log N)^{0.5872})$
10	$O(N^{1.4550}/(\log N)^{0.6125})$

47.4 Applications

Given N matrices $A_1, A_2, \ldots, A_N$ of size $N \times N$, the matrix chain product problem is to compute $A_1 \times A_2 \times \cdots \times A_N$. Given an $N \times N$ matrix A, the matrix powers problem is to calculate the first N powers of A, that is, $A, A^2, A^3, \ldots, A^N$. Both problems are important in performing many fundamental matrix operations such as computing the characteristic polynomial, determinant, rank, inverse, eigenvalues, LU-factorization, and QR-factorization of a matrix, in solving linear systems of equations and in general scientific computations [8, 20–22, 26, 31, 33, 37, 44]. Both problems can be solved by using a matrix multiplication algorithm as a subroutine. For instance, sequentially, both the matrix chain product and the matrix powers problems can be solved in $O(N^{\alpha+1})$ time. We notice that the matrix chain product problem can be defined for matrices of different sizes. In sequential computing, the main concern is the optimal order of the $N-1$ matrix multiplications [13, 18, 19].

We are interested in solving the two problems on DMSs. It is clear that for high performance parallel computation of matrix chain product and matrix powers on DMS, a fast and scalable parallel matrix multiplication algorithm is required. In this section, we use the fast and scalable parallel matrix multiplication algorithm developed in the last section as a building block to solve the matrix chain product and the matrix powers problems. We also show that computing matrix chain product and matrix powers are fully scalable on DMPC, highly scalable on DMS with hypercubic networks, and not highly scalable on DMS with mesh and torus networks.

Before we move to our parallel algorithms for matrix chain product and matrix powers, we would like to point out that our algorithms use dynamic system partitioning and processor allocation during their execution, that is, processors are dynamically allocated to solve subproblems. We always divide a DMS with p processors into g subsystems of equal size p/g. Though the subsystems manipulate different data, these subsystems are allocated to do the same computation (i.e., matrix multiplication) using the same algorithm. In a computation step, subsystems are separate from each other, each performing its own computation. However, in a communication step, even though processors send/receive data to/from processors in the same subsystem, such communication is done in the entire DMS with the help from processors in other subsystems. Thus, a communication step still takes $T(p)$ time, not $T(p/g)$ time. Theorem 47.4 can be generalized to the following [28].

Theorem 47.5 *For all $1 \leq p \leq N^\alpha/\log N$, if $p^* \leq p$ processors are allocated, multiplying two $N \times N$ matrices can be performed by a DMS with p processors in*

$$T^*_{\text{mm}}(N, p, p^*) = O\left(\frac{N^\alpha}{p^*} + T(p)\left(\frac{N^2}{(p^*)^{2/\alpha}}\right)(\log p^*)^{1-2/\alpha}\right)$$

time. (Note that the above equation contains $T(p)$, not $T(p^*)$.)

The above matrix multiplication algorithm will be repeated and used as a building block in our algorithms for matrix chain product and matrix powers to compute partial products. It turns out that transfer of these intermediate partial results are straightforward and can be easily implemented on a DMS. We will thus focus on strategy description, processor allocation, and time complexity analysis.

47.4.1 Matrix Chain Product

47.4.1.1 Algorithm

Given N matrices $A_1, A_2, \ldots, A_N$ of size $N \times N$, the matrix chain product $A_1 \times A_2 \times \cdots \times A_N$ can be obtained by using the standard binary tree algorithm. The leaves are input matrices $A_1, A_2, \ldots, A_N$, and an internal node represents a task (i.e., matrix multiplication) that calculates a partial product. The root task of the tree computes the final result. Without loss of generality, we assume that $N = 2^n$ is a power of two. The computation is organized into n levels as follows:

- Level 1: The N matrices are grouped into $N/2$ pairs and $N/2$ matrix multiplications are performed.
- Level 2: The $N/2$ partial products from level 1 are grouped into $N/4$ pairs and $N/4$ matrix multiplications are performed.
  $\vdots$
- Level l: The $N/2^{l-1}$ partial products from level $(l-1)$ are grouped into $N/2^l$ pairs and $N/2^l$ matrix multiplications are performed.
  $\vdots$
- Level n: The 2 partial products from level $(n-1)$ are multiplied to get the matrix chain product $A_1 \times A_2 \times \cdots \times A_N$.

Altogether, the above algorithm performs $2^{n-1} + 2^{n-2} + \cdots + 2^0 = (N-1)$ matrix multiplications, and compared to the best sequential algorithm, there is no redundant computation. Figure 47.6 illustrates the binary tree algorithm with $N = 16$, where the notation $A_{i,j}$ means the partial product $A_i \times A_{i+1} \times \cdots \times A_j$.

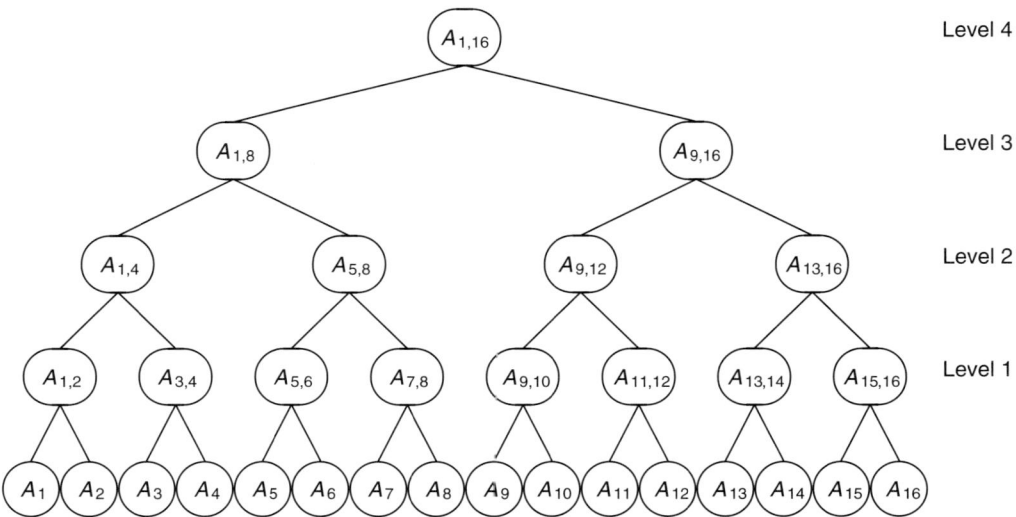

FIGURE 47.6 Illustration of the binary tree algorithm for matrix chain product ($N = 16$).

47.4.1.2 Analysis

We define the function $\log^+$ as follows: $\log^+ x = \log x$ if $x \geq 1$ and $\log^+ x = 1$ if $0 < x < 1$. The following theorem gives the time complexity of the binary tree algorithm for computing matrix chain product [28].

Theorem 47.6 *For all* $1 \leq p \leq N^{\alpha+1}/(\log N)^2$, *the product of N matrices of size $N \times N$ can be computed by a DMS with p processors in*

$$T_{\text{chain}}(N,p) = O\left(\frac{N^{\alpha+1}}{p} + T(p)\left(\left(\frac{N^{2(1+1/\alpha)}}{p^{2/\alpha}}\right)\left(\log^+ \frac{p}{N}\right)^{1-2/\alpha} + \log^+\left(\frac{p \log N}{N^\alpha}\right)\log N\right)\right)$$

time. In particular, the above time complexity is $O(T(p)(\log N)^2)$ *when* $p = N^{\alpha+1}/(\log N)^2$.

Corollary 47.5 instantiates Theorem 47.6 for DMPC and shows that matrix chain product on DMPC is fully scalable.

Corollary 47.5 *For all* $1 \leq p \leq N^{\alpha+1}/(\log N)^2$, *the product of N matrices of size $N \times N$ can be computed by a DMPC with p processors in* $T_{\text{chain}}(N,p) = O(N^{\alpha+1}/p)$ *time. In particular, the time complexity is* $O((\log N)^2)$ *when* $p = N^{\alpha+1}/(\log N)^2$. *According to Definition 47.3, matrix chain product on DMPC is fully scalable.*

Corollary 47.6 instantiates Theorem 47.6 for DMS with hypercubic networks.

Corollary 47.6 *For all* $1 \leq p \leq N^{\alpha+1}/(\log N)^2$, *the product of N matrices of size $N \times N$ can be computed by a DMS with p processors connected by a hypercubic network in*

$$T_{\text{chain}}(N,p) = O\left(\frac{N^{\alpha+1}}{p} + \left(\frac{N^{2(1+1/\alpha)}}{p^{2/\alpha}}\right)\log p\left(\log^+ \frac{p}{N}\right)^{1-2/\alpha} + \log^+\left(\frac{p \log N}{N^\alpha}\right)\log p \log N\right)$$

time. In particular, the time complexity is $O((\log N)^3)$ *when* $p \geq N^{\alpha+1}/(\log N)^{\alpha/2+1}$. *Furthermore, if*

$$p = O\left(\frac{N^{\alpha+1}}{(\log N)^{2(\alpha-1)/(\alpha-2)}}\right),$$

linear speedup and cost optimality can be achieved. According to Definition 47.2, with

$$c = 2\left(\frac{\alpha-1}{\alpha-2}\right) - 2 = \frac{2}{\alpha-2},$$

matrix chain product on DMS with hypercubic networks is highly scalable.

Corollary 47.7 instantiates Theorem 47.6 for DMS with mesh and torus networks.

Corollary 47.7 *For all* $1 \leq p \leq N^{\alpha+1}/(\log N)^2$, *the product of N matrices of size $N \times N$ can be computed by a DMS with p processors connected by a d-dimensional mesh or torus network in*

$$T_{\text{chain}}(N,p) = O\left(\frac{N^{\alpha+1}}{p} + dp^{1/d}\left(\left(\frac{N^{2(1+1/\alpha)}}{p^{2/\alpha}}\right)\left(\log^+ \frac{p}{N}\right)^{1-2/\alpha} + \log^+\left(\frac{p \log N}{N^\alpha}\right)\log N\right)\right)$$

time. (The above time complexity reduces to that in Corollary 47.6 when $d = \log p$.) *When* $d = 1$, $T_{\text{chain}}(N,p)$ *reaches its minimum* $O(N^{(\alpha+3)/2})$ *when* $p = O(N^{(\alpha-1)/2})$. *When* $d \geq 2$, $T_{\text{chain}}(N,p)$ *is a*

TABLE 47.2 The Value p^* in Corollary 47.7

d	p^*
1	$O(N^{0.6878})$
2	$O(N^{0.9170})$
3	$O(N^{1.0858}/(\log N)^{0.3217})$
4	$O(N^{1.3075}/(\log N)^{0.3874})$
5	$O(N^{1.4901}/(\log N)^{0.4415})$
6	$O(N^{1.6431}/(\log N)^{0.4868})$
7	$O(N^{1.7731}/(\log N)^{0.5253})$
8	$O(N^{1.8849}/(\log N)^{0.5584})$
9	$O(N^{1.9822}/(\log N)^{0.5872})$
10	$O(N^{2.0675}/(\log N)^{0.6125})$

decreasing function of p and reaches its minimum $O(dN^{(\alpha+1)/d}(\log N)^{2(d-1)/d})$ when $p = N^{\alpha+1}/(\log N)^2$. Furthermore, if $p \leq p^*$ where

$$p^* = O(N^{d(\alpha-1)/(d+1)}), \quad \text{for } d \leq 2,$$

and

$$p^* = O\left(\frac{N^{(\alpha-2)(\alpha+1)d/((\alpha-2)d+\alpha)}}{d^{\alpha d/((\alpha-2)d+\alpha)}(\log N)^{(\alpha-2)d/((\alpha-2)d+\alpha)}}\right), \quad \text{for } d \geq 3,$$

linear speedup and cost optimality can be achieved.

The value p^* given by Corollary 47.7 is displayed in Table 47.2, assuming that $\alpha = 2.3755$. Clearly, for any constant $d \geq 1$, matrix chain product on DMS with mesh and torus networks is not highly scalable.

Corollary 47.8 gives a condition for matrix chain product on DMS to be fully scalable and a condition for matrix chain product on DMS to be highly scalable.

Corollary 47.8 *Matrix chain product on DMS is fully scalable if and only if $T(p) = O(1)$ and highly scalable if and only if $T(p) = O((\log p)^\gamma)$ for some constant $\gamma \geq 0$.*

47.4.2 Matrix Powers

47.4.2.1 Algorithm

Given an $N \times N$ matrix A, our parallel algorithm to calculate the first N powers of A consists of the following two stages. Let $N = 2^n$ be a power of two.

- Stage 1: The powers $A^{2^1}, A^{2^2}, \ldots, A^{2^n}$ are calculated one by one, such that $A^{2^d} = A^{2^{d-1}} \times A^{2^{d-1}}$, where $1 \leq d \leq n$.
- Stage 2: For all $0 \leq q \leq N-1$, if $q = c_{n-1}2^{n-1} + c_{n-2}2^{n-2} + \cdots + c_0 2^0$, $A^q = M_{n-1} \times M_{n-2} \times \cdots \times M_0$ is calculated, where $M_d = A^{2^d}$ if $c_d \neq 0$, and $M_d = I_N$ if $c_d = 0$, and I_N is the $N \times N$ identity matrix.

Whereas the first stage is straightforward, the second stage needs more elaboration.

It is noticed that based on the results of the first stage, that is, $A^{2^0}, A^{2^1}, A^{2^2}, \ldots, A^{2^{n-1}}$, the second stage essentially computes $A^{2^{i_1}} \times A^{2^{i_2}} \times \cdots \times A^{2^{i_k}}$ for all subsets $S = \{i_1, i_2, \ldots, i_k\}$, where $S \subseteq \{0, 1, 2, \ldots, n-1\}$. In other words, the computation in the second stage can be considered as the following problem: given n matrices $M_0, M_1, M_2, \ldots, M_{n-1}$, to compute $M_{i_1} \times M_{i_2} \times \cdots \times M_{i_k}$, for all subsets $S = \{i_1, i_2, \ldots, i_k\}$. Let us assign to each subset $S = \{i_1, i_2, \ldots, i_k\}$ an index $I(S) = 2^{i_1} + 2^{i_2} + \cdots + 2^{i_k}$, such that there is a one-to-one correspondence between the power set of $\{0, 1, 2, \ldots, n-1\}$ and $\{0, 1, 2, \ldots, 2^n - 1\}$. If we

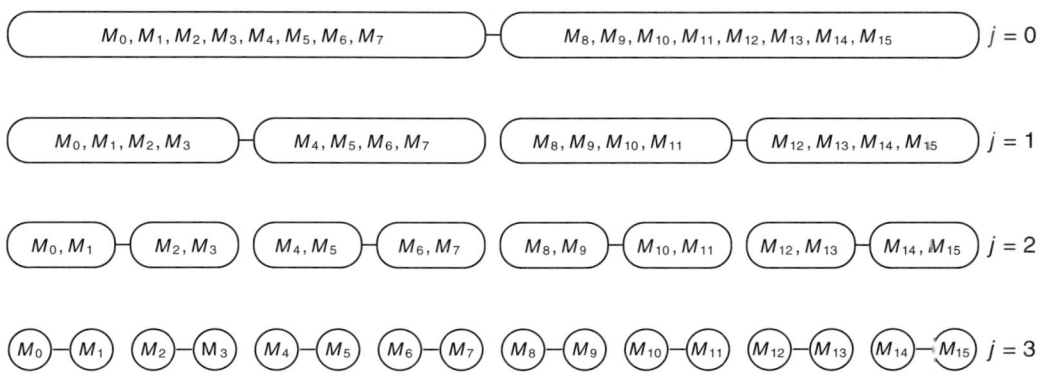

FIGURE 47.7 Unrolling of the recursive algorithm for the second stage ($n = 16$).

define $X_{I(S)} = M_{i_1} \times M_{i_2} \times \cdots \times M_{i_k}$, where $S = \{i_1, i_2, \ldots, i_k\}$, then the second stage computes X_0, X_1, X_2, ..., X_{2^n-1}, for all the 2^n subsets.

We assume that n is a power of two without loss of generality. Given M_0, M_1, M_2, ..., M_{n-1}, the following recursive algorithm can be employed to calculate X_0, X_1, X_2, ..., X_{2^n-1}. We divide the n input matrices (M_0, M_1, M_2, ..., M_{n-1}) into two parts, that is, (M_0, M_1, M_2, ..., $M_{n/2-1}$), and ($M_{n/2}$, $M_{n/2+1}$, $M_{n/2+2}$, ..., M_{n-1}). For convenience, let us rename ($M_{n/2}$, $M_{n/2+1}$, $M_{n/2+2}$, ..., M_{n-1}) as (M'_0, M'_1, M'_2, ..., $M'_{n/2-1}$), where $M'_j = M_{n/2+j}$, for all $0 \leq j \leq n/2 - 1$. These two partitions are regarded as two subproblems. In other words, for (M_0, M_1, M_2, ..., $M_{n/2-1}$), all the $2^{n/2}$ possible products, say, Y_0, Y_1, Y_2, ..., $Y_{2^{n/2}-1}$, are computed, where $Y_I = M_{i_1} \times M_{i_2} \times \cdots \times M_{i_k}$, if $I = I(S)$ and $S = \{i_1, i_2, \ldots, i_k\}$. Similarly, all the $2^{n/2}$ possible products, say, Z_0, Z_1, Z_2, ..., $Z_{2^{n/2}-1}$, for (M'_0, M'_1, M'_2, ..., $M'_{n/2-1}$), are also calculated, where $Z_I = M'_{i_1} \times M'_{i_2} \times \cdots \times M'_{i_k} = M_{n/2+i_1} \times M_{n/2+i_2} \times \cdots \times M_{n/2+i_k}$, if $I = I(S)$ and $S = \{i_1, i_2, \ldots, i_k\}$.

On the basis of the outputs of the two subproblems, that is, Y_0, Y_1, Y_2, ..., $Y_{2^{n/2}-1}$ and Z_0, Z_1, Z_2, ..., $Z_{2^{n/2}-1}$, the results of the original problem, that is, X_0, X_1, X_2, ..., X_{2^n-1}, can be obtained easily. Assume that $S \subseteq \{0, 1, 2, \ldots, n-1\}$, where $S = \{i_1, i_2, \ldots, i_{k_1}\} \cup \{n/2 + j_1, n/2 + j_2, \ldots, n/2 + j_{k_2}\}$, such that $S_1 = \{i_1, i_2, \ldots, i_{k_1}\} \subseteq \{0, 1, 2, \ldots, n/2 - 1\}$ and $S_2 = \{j_1, j_2, \ldots, j_{k_2}\} \subseteq \{0, 1, 2, \ldots, n/2 - 1\}$. Then, it is easy to see that $X_{I(S)} = Y_{I(S_1)} \times Z_{I(S_2)}$.

We notice that the recursive algorithm can be unrolled into $\log n$ levels: $\log n - 1$, $\log n - 2$, ..., 1, 0 (see Figure 47.7 for an illustration when $n = 16$). In level j, where $\log n - 1 \geq j \geq 0$, the n matrices M_0, M_1, M_2, ..., M_{n-1} are divided into 2^{j+1} groups (each group is shown by an oval in Figure 47.7), with $n/2^{j+1}$ matrices in each group. All the $2^{n/2^{j+1}}$ possible products of the matrices in each group have been computed in level $j + 1$. The 2^{j+1} groups are divided into 2^j pairs and we need to compute $2^{n/2^{j+1}} \times 2^{n/2^{j+1}} = 2^{n/2^j}$ products for each pair of groups, where each product is actually a matrix multiplication. The total number of products in level j is $w_j = 2^{n/2^j+j}$. It is easy to see that $2 \log N = w_{\log n-1} < w_{\log n-2} < \cdots < w_0 = N$.

47.4.2.2 Analysis

The following theorem gives the time complexity of the above algorithm for computing matrix powers [28].

Theorem 47.7 *For all $1 \leq p \leq N^{\alpha+1}/(\log N)^2$, the first N powers of an $N \times N$ matrix can be computed by a DMS with p processors in*

$$T_{\text{power}}(N, p) = O\left(\frac{N^{\alpha+1}}{p} + T(p)\left(\left(\frac{N^{2(1+1/\alpha)}}{p^{2/\alpha}}\right)\left(\log^+ \frac{p}{2\log N}\right)^{1-2/\alpha} + (\log N)^2\right)\right)$$

time. In particular, the above time complexity is $O(T(p)(\log N)^2)$ when $p = N^{\alpha+1}/(\log N)^2$.

Corollary 47.9 instantiates Theorem 47.7 for DMPC and shows that computing matrix powers on DMPC is fully scalable.

Corollary 47.9 *For all $1 \leq p \leq N^{\alpha+1}/(\log N)^2$, the first N powers of an $N \times N$ matrix can be computed by a DMPC with p processors in $T_{\text{power}}(N, p) = O(N^{\alpha+1}/p)$ time. In particular, the time complexity is $O((\log N)^2)$ when $p = N^{\alpha+1}/(\log N)^2$. According to Definition 47.3, computing matrix powers on DMPC is fully scalable.*

Corollary 47.10 instantiates Theorem 47.7 for DMS with hypercubic networks.

Corollary 47.10 *For all $1 \leq p \leq N^{\alpha+1}/(\log N)^2$, the first N powers of an $N \times N$ matrix can be computed by a DMS with p processors connected by a hypercubic network in*

$$T_{\text{power}}(N, p) = O\left(\frac{N^{\alpha+1}}{p} + \left(\frac{N^{2(1+1/\alpha)}}{p^{2/\alpha}}\right)\left(\log^+ \frac{p}{2\log N}\right)^{1-2/\alpha} \log N + (\log N)^3\right)$$

time. In particular, the time complexity is $O((\log N)^3)$ when $p \geq N^{\alpha+1}/(\log N)^{\alpha/2+1}$. Furthermore, if

$$p = O\left(\frac{N^{\alpha+1}}{(\log N)^{2(\alpha-1)/(\alpha-2)}}\right),$$

linear speedup and cost optimality can be achieved. According to Definition 47.2, with

$$c = 2\left(\frac{\alpha-1}{\alpha-2}\right) - 2 = \frac{2}{\alpha-2},$$

computing matrix powers on DMS with hypercubic networks is highly scalable.

Corollary 47.11 instantiates Theorem 47.7 for DMS with mesh and torus networks.

Corollary 47.11 *For all $1 \leq p \leq N^{\alpha+1}/(\log N)^2$, the first N powers of an $N \times N$ matrix can be computed by a DMS with p processors connected by a d-dimensional mesh or torus network in*

$$T_{\text{power}}(N, p) = O\left(\frac{N^{\alpha+1}}{p} + dp^{1/d}\left(\left(\frac{N^{2(1-1/\alpha)}}{p^{2/\alpha}}\right)\left(\log^+ \frac{p}{2\log N}\right)^{1-2/\alpha} + (\log N)^2\right)\right)$$

time. (The above time complexity reduces to that in Corollary 47.10 when $d = \log p$.) When $d = 1$, $T_{\text{power}}(N, p)$ reaches its minimum

$$O(N^{\alpha(\alpha+1)/(2(\alpha-1))}(\log N)^{(\alpha-2)/(2(\alpha-1))})$$

when

$$p = O(N^{(\alpha+1)(\alpha-2)/(2(\alpha-1))}/(\log N)^{(\alpha-2)/(2(\alpha-1))}).$$

When $d \geq 2$, $T_{\text{power}}(N, p)$ is a decreasing function of p and reaches its minimum

$$O(dN^{(\alpha+1)/d}(\log N)^{2(d-1)/d})$$

when $p = N^{\alpha+1}/(\log N)^2$. Furthermore, for $d \geq 1$, if $p \leq p^*$ where

$$p^* = O\left(\frac{N^{(\alpha-2)(\alpha+1)d/((\alpha-2)d+\alpha)}}{d^{\alpha d/((\alpha-2)d+\alpha)}(\log N)^{(\alpha-2)d/((\alpha-2)d+\alpha)}}\right),$$

linear speedup and cost optimality can be achieved.

TABLE 47.3 The Value p^* in Corollary 47.11

d	p^*
1	$O(N^{0.4607}/(\log N)^{0.1365})$
2	$O(N^{0.8108}/(\log N)^{0.2402})$
3	$O(N^{1.0858}/(\log N)^{0.3217})$
4	$O(N^{1.3075}/(\log N)^{0.3874})$
5	$O(N^{1.4901}/(\log N)^{0.4415})$
6	$O(N^{1.6431}/(\log N)^{0.4868})$
7	$O(N^{1.7731}/(\log N)^{0.5253})$
8	$O(N^{1.8849}/(\log N)^{0.5584})$
9	$O(N^{1.9822}/(\log N)^{0.5872})$
10	$O(N^{2.0675}/(\log N)^{0.6125})$

The value p^* given by Corollary 47.11 is displayed in Table 47.3, assuming that $\alpha = 2.3755$. Clearly, for any constant $d \geq 1$, computing matrix powers on DMS with mesh and torus networks is not highly scalable.

Corollary 47.12 gives a condition for computing matrix powers on DMS to be fully scalable and a condition for computing matrix powers on DMS to be highly scalable.

Corollary 47.12 *Computing matrix powers on DMS is fully scalable if and only if $T(p) = O(1)$ and highly scalable if and only if $T(p) = O((\log p)^\gamma)$ for some constant $\gamma \geq 0$.*

47.5 Summary

We have developed a unified parallelization of any known sequential algorithm for matrix multiplication over an arbitrary ring on any DMS. We have obtained currently the fastest and most processor efficient parallelization of the best known sequential matrix multiplication algorithm on a distributed memory system. We showed that for the class of DMPC, where one-to-one communications can be realized in constant time, our parallel matrix multiplication algorithm achieves $O(\log N)$ time complexity by using only $N^\alpha/\log N$ processors and exhibits full scalability. Such performance of matrix multiplication on DMPC matches the performance of matrix multiplication on CREW PRAM. Our result also unifies all known algorithms for matrix multiplication, standard or nonstandard, sequential or parallel. We instantiated our general analytical result to distributed memory systems with specific interconnection networks, such as the class of hypercubic networks and mesh and torus networks. We pointed out a condition for full scalability and a condition for high scalability. We also developed a compound algorithm and showed how to achieve sublogarithmic parallel time complexity on a linear array with reconfigurable pipelined bus system with subcubic processor complexity.

We have designed and analyzed parallel algorithms on DMS for the matrix chain product and the matrix powers problems. We demonstrated that by using our fast and scalable matrix multiplication algorithm as a building block, our algorithms can achieve linear speedup and cost optimality in a wide range of p, especially on DMPC and DMS with hypercubic networks. Further research should be directed toward revealing the impact of matrix multiplication on other important and interesting matrix computation problems [27].

References

[1] R. J. Anderson and G. L. Miller, Optical communication for pointer based algorithms, TR CRI 88-14, Computer Science Dept., University of Southern California, 1988.

[2] A. F. Benner, H. F. Jordan, and V. P. Heuring, Digital optical computing with optically switched directional couplers, *Optical Engineering*, vol. 30, pp. 1936–1941, 1991.

[3] D. Bini and V. Pan, *Polynomial and Matrix Computations, Vol. 1, Fundamental Algorithms*, Birkhäuser, Boston, 1994.

[4] L. E. Cannon, A cellular computer to implement the Kalman filter algorithm, PhD thesis, Montana State University, Bozman, MT, 1969.

[5] A. K. Chandra, Maximal parallelism in matrix multiplication, Report RC-6193, IBM T. J. Watson Research Center, Yorktown Heights, New York, October 1979.

[6] D. Chiarulli, R. Melhem, and S. Levitan, Using coincident optical pulses for parallel memory addressing, *IEEE Computer*, vol. 30, pp. 48–57, 1987.

[7] D. Coppersmith and S. Winograd, Matrix multiplication via arithmetic progressions, *Journal of Symbolic Computation*, vol. 9, pp. 251–280, 1990.

[8] L. Csanky, Fast parallel matrix inversion algorithms, *SIAM Journal on Computing*, vol. 5, pp. 618–623, 1976.

[9] E. Dekel, D. Nassimi, and S. Sahni, Parallel matrix and graph algorithms, *SIAM Journal on Computing*, vol. 10, pp. 657–673, 1981.

[10] P. W. Dowd, Wavelength division multiple access channel hypercube processor interconnection, *IEEE Transactions on Computers*, vol. 41, pp. 1223–1241, 1992.

[11] M. M. Eshaghian, Parallel algorithms for image processing on OMC, *IEEE Transactions on Computers*, vol. 40, pp. 827–833, 1993.

[12] M. Geréb-Graus and T. Tsantilas, Efficient optical communication in parallel computers, *Proceedings of 4th ACM Symposium on Parallel Algorithms and Architectures*, pp. 41–48, 1992.

[13] S. S. Godbole, On efficient computation of matrix chain products, *IEEE Transactions on Computers*, vol. 22, no. 9, pp. 864–866, 1973.

[14] L. A. Goldberg, M. Jerrum, T. Leighton, and S. Rao, Doubly logarithmic communication algorithms for optical-communication parallel computers, *SIAM Journal on Computing*, vol. 26, pp. 1100–1119, 1997.

[15] G. H. Golub and C. F. Van Loan, *Matrix Computations*, Johns Hopkins University Press, Baltimore, MD, 1996.

[16] A. Grama, A. Gupta, G. Karypis, and V. Kumar, *Introduction to Parallel Computing*, 2nd ed., Addison-Wesley, Harlow, England, 2003.

[17] Z. Guo, R. Melhem, R. Hall, D. Chiarulli, and S. Levitan, Pipelined communications in optically interconnected arrays, *Journal of Parallel and Distributed Computing*, vol. 12, pp. 269–282, 1991.

[18] T. C. Hu and M. T. Shing, Computation of matrix chain products. Part I, *SIAM Journal on Computing*, vol. 11, no. 2, pp. 362–373, 1982.

[19] T. C. Hu and M. T. Shing, Computation of matrix chain products. Part II, *SIAM Journal on Computing*, vol. 13, no. 2, pp. 228–251, 1984.

[20] O. H. Ibarra, S. Moran, and L. E. Rosier, A note on the parallel complexity of computing the rank of order n matrices, *Information Processing Letters*, vol. 11, no. 4,5, p. 162, 1980.

[21] H. Lee, J. Kim, S. J. Hong, and S. Lee, Processor allocation and task scheduling of matrix chain products on parallel systems, *IEEE Transactions on Parallel and Distributed Systems*, vol. 14, no. 4, pp. 394–407, 2003.

[22] F. T. Leighton, *Introduction to Parallel Algorithms and Architectures: Arrays, Trees, Hypercubes*, Morgan Kaufmann Publishers, San Mateo, CA, 1992.

[23] S. Levitan, D. Chiarulli, and R. Melhem, Coincident pulse techniques for multiprocessor interconnection structures, *Applied Optics*, vol. 29, pp. 2024–2039, 1990.

[24] K. Li, Scalable parallel matrix multiplication on distributed memory parallel computers, *Journal of Parallel and Distributed Computing*, vol. 61, no. 12, pp. 1709–1731, 2001.

[25] K. Li, Fast and scalable parallel algorithms for matrix chain product and matrix powers on reconfigurable pipelined optical buses, *Journal of Information Science and Engineering*, vol. 18, no. 5, pp. 713–727, 2002.

[26] K. Li, Fast and scalable parallel matrix computations with reconfigurable pipelined optical buses, *Parallel Algorithms and Applications*, vol. 19, no. 4, pp. 195–209, 2004.

[27] K. Li, Fast and scalable parallel matrix computations on distributed memory systems, *Proceedings of the 19th IEEE International Parallel and Distributed Processing Symposium*, Denver, CO, April 2005.
[28] K. Li, Analysis of parallel algorithms for matrix chain product and matrix powers on distributed memory systems, *IEEE Transactions on Parallel and Distributed Systems*, vol. 18, no. 7, pp. 865–878, 2007.
[29] K. Li and V. Y. Pan, Parallel matrix multiplication on a linear array with a reconfigurable pipelined bus system, *IEEE Transactions on Computers*, vol. 50, no. 5, pp. 519–525, 2001.
[30] K. Li, Y. Pan, and S. Q. Zheng, Fast and processor efficient parallel matrix multiplication algorithms on a linear array with a reconfigurable pipelined bus system, *IEEE Transactions on Parallel and Distributed Systems*, vol. 9, no. 8, pp. 705–720, 1998.
[31] K. Li, Y. Pan, and S. Q. Zheng, Parallel matrix computations using a reconfigurable pipelined optical bus, *Journal of Parallel and Distributed Computing*, vol. 59, no. 1, pp. 13–30, 1999.
[32] Y. Li, Y. Pan, and S. Q. Zheng, Pipelined TDM optical bus with conditional delays, *Optical Engineering*, vol. 36, no. 9, pp. 2417–2424, 1997.
[33] S.-S. Lin, A chained-matrices approach for parallel computation of continued fractions and its applications, *Journal of Scientific Computing*, vol. 9, no. 1, pp. 65–80, 1994.
[34] K. Mehlhorn and U. Vishkin, Randomized and deterministic simulations of PRAMs by parallel machines with restricted granularity of parallel memories, *Acta Informatica*, vol. 21, pp. 339–374, 1984.
[35] V. Pan, New fast algorithms for matrix operations, *SIAM Journal on Computing*, vol. 9, pp. 321–342, 1980.
[36] V. Pan, *How to Multiply Matrices Faster*, LNCS, 179, Springer Verlag, Berlin, 1984.
[37] V. Pan, Complexity of parallel matrix computations, *Theoretical Computer Science*, vol. 54, pp. 65–85, 1987.
[38] V. Pan, Parallel solution of sparse linear and path systems, in *Synthesis of Parallel Algorithms*, J. H. Reif, ed., pp. 621–678, Morgan Kaufmann, San Mateo, CA, 1993.
[39] V. Pan and J. Reif, Efficient parallel solution of linear systems, *Proceedings of 7th ACM Symposium on Theory of Computing*, pp. 143–152, May 1985.
[40] Y. Pan and K. Li, Linear array with a reconfigurable pipelined bus system—concepts and applications, *Journal of Information Sciences*, vol. 106, no. 3-4, pp. 237–258, 1998.
[41] H. Park, H. J. Kim, and V. K. Prasanna, An $O(1)$ time optimal algorithm for multiplying matrices on reconfigurable mesh, *Information Processing Letters*, vol. 47, pp. 109–113, 1993.
[42] C. Qiao and R. Melhem, Time-division optical communications in multiprocessor arrays, *IEEE Transactions on Computers*, vol. 42, pp. 577–590, 1993.
[43] V. Strassen, Gaussian elimination is not optimal, *Numerische Mathematik*, vol. 13, pp. 354–356, 1969.
[44] S.-T. Yau and Y. Y. Lu, Reducing the symmetric matrix eigenvalue problem to matrix multiplications, *SIAM Journal of Scientific Computing*, vol. 14, no. 1, pp. 121–136, 1993.
[45] S. Q. Zheng and Y. Li, Pipelined asynchronous time-division multiplexing optical bus, *Optical Engineering*, vol. 36, no. 12, pp. 3392–3400, 1997.

Index

Note: Page numbers in *italics* refer to illustrations

A

Absolute deadline, **24**-2
Accelerated single-program, multiple data (ASPMD), **11**-8
Accelerating machines
　for time-varying computational complexity, **1**-9
Active buffering, **35**-15
Activity-based model, **33**-8
Ad hoc wireless networks
　fault-tolerant ad hoc wireless networks, **39**-17
　issue in, **39**-2
　minimum energy communication in, **39**-1–**39**-18, *see also* Minimum energy communication
Adaptive bitonic sorting, **5**-32–**5**-34
　nonpipelined, **5**-33
　pipelined, **5**-33
Adaptive protocol, **33**-13, **33**-14
AdaptiveTkC, **24**-12
Additional-edge, **36**-15
Advanced computational data center (ACDC) grid, **46**-5–**46**-7
　ACDC data grid, **46**-25
　　data grid file virtualization, **46**-27
　　storage network architecture, **46**-27–**46**-29
　　web tools, **46**-29–**46**-30
　dashboard, **46**-7–**46**-9
　monitoring infrastructure, **46**-5–**46**-7
　web portal, *46*-6
Aggregators, **35**-11

Algorithmic optimizations, in multicare and multiprocessors, **26**-5–**26**-6
All-to-All broadcasting, **19**-3
Almost k-wise independence, **18**-11–**18**-14
　applications, **18**-13
Always-update strategy, **33**-6
AND/OR application model, **40**-13–**40**-14
Anonymous agents, **8**-2
Anonymous tree exploration, **8**-8
Antiport/symport system, **3**-17
Aperiodic task scheduling, **24**-18
AppLeS, **21**-9–**21**-10
Application Level Scheduler (AppLeS), **21**-9–**21**-10
Application view, **12**-4
Application-specific integrated circuits (ASICs), **11**-2
1.5-Approximation, for broadcast time minimization, **19**-23–**19**-26
Arbitrary networks, message dissemination over, **36**-18–**36**-19
Arbitrary-deadline system, **24**-2
Array processors with pipelined buses (APPB), **16**-1, *16*-*11*
Array processors with pipelined buses using switches (APPBS), **16**-1, *16*-*12*
Array structure with synchronous optical switches (ASOS), **16**-1, *16*-*12*
Array with reconfigurable optical buses (AROB), **16**-1
Art, **7**-13–**7**-15
Artificial neural networks (ANNs), **46**-19

Assert parallel directive, **31**-9
Association rule mining (ARM), parallel, **32**-2–**32**-12
　comparison, **32**-11–**32**-12
　mining by bitmaps, **32**-8–**32**-11
　pattern-growth method, **32**-6–**32**-8, *see also* Pattern-growth method
　priori-based algorithms, **32**-2–**32**-4
　vertical mining, **32**-4–**32**-5, *see also* Vertical mining
Association rules
　parallel data mining algorithms for, *see* Association rule mining, parallel
Asymmetric rendezvous, **8**-4–**8**-5
　agent asymmetry, **8**-5
　network asymmetry, **8**-4–**8**-5
Asymmetric wireless networks, **39**-7–**39**-13
　bidirected connectivity requirements, **39**-12–**39**-13
　multicasting, broadcasting, and group communication, **39**-8–**39**-12
Asynchronous task system, **24**-2
Atomic selfish routing in networks, **20**-1–**20**-28
Auctions/Auction theory, **45**-6
　game theoretical auction theory, properties, views, and extensions, **45**-7–**45**-9
　Nash equilibrium, **45**-6
　types, **45**-6
　　Dutch, **45**-6, **45**-7
　　English, **45**-6, **45**-7

I-1

Auctions/Auction theory (continued)
   first-price sealed-bid, **45**-6, **45**-7
   progressive auctions, **45**-6
   second-price sealed-bid, **45**-6, **45**-7
Automata fooling technique, *see* Fooling automata
Autonomous land vehicle in a neural network (ALVINN) system, **46**-19
Auxiliary buffers, **34**-5
Average case execution time (ACET), **40**-5
Aε-Star algorithm, **45**-15

B

Backward segments identification, in buffer vertices matching, **23**-11–**23**-13
Bandwidth allocation, for cellular networks, **33**-3–**33**-5
Basic patterns, **15**-1
Basis crashing, **18**-35
Benchmarks, scaling properties and, **43**-11
Benes network messages, **36**-4
Betweenness centrality, **31**-18
Biconnected components
   in efficient parallel graph algorithms, **26**-21–**26**-29
   improved algorithm for, **26**-27
   overheads, algorithm to reduce, **26**-25
   performance results and analysis, **26**-28–**26**-29
   Tarjan–Vishkin biconnected components algorithm, **26**-22–**26**-24
Bidirected connectivity requirements, **39**-12–**39**-13
Binary digits/Bits, for time-varying variables, **1**-5
Binary prefix sums, **16**-14–**16**-15
Binary representations, **18**-2–**18**-3
Binary search, **1**-10
Binary swap technique, **27**-14
Bin-packing heuristics, **24**-8
   Best Fit (BF), **24**-8
   First Fit (FF), **24**-8
   Next Fit (NF), **24**-8
Bit pairs PROFIT/COST problem, **25**-2–**25**-4
Bit vector, **32**-8
Bitstring broadcasting, **10**-13–**10**-14
   bitstring timeslicing, **10**-13–**10**-14
   iterated length timeslicing, **10**-14
   length timeslicing, **10**-14
Bitstring timeslicing, **10**-13–**10**-14
Black hole search (BHS), **8**-17

Black holes, **8**-16–**8**-17
Bor approach, **26**-14, **26**-32
   Bor-AL approach, **26**-14, **26**-16
   Bor-EL approach, **26**-14, **26**-16
   Bor-FAL approach, **26**-16
   flexible adjacency list, **26**-14
Borùvka's algorithms, **26**-1–**26**-15, **26**-17, **26**-32
   Bor approach, **26**-14
   Bor-AL approach, **26**-14
   compact-graph step, **26**-13
   connect-components step, **26**-13
   find-min step, **26**-13
   parallel Borùvka's algorithms, data structures for, **26**-1–**26**-15
Bounded degree model, **19**-6, **19**-13–**19**-14
Bounded parallelism, **3**-15–**3**-16
Bounded size matching model, **19**-6, **19**-13, **19**-14–**19**-15
Bounding global transfers, **19**-13–**19**-15
   bounded degree model, **19**-13–**19**-14
   bounded size matching, **19**-14–**19**-15
Branch decomposition, **8**-16
Branch patterns, **15**-2
Branch-On-Need Octree (BONO), **27**-12
Brandes' sequential algorithm, **31**-20
Brane calculi, **3**-16
Breadth-First Search (BFS) frontier, **5**-27–**5**-29, **8**-7, **31**-8
   algorithm execution, **5**-30
   comparison, **5**-3
   flattened BFS, **5**-30–**5**-31
   k-spawn BFS, **5**-28
   single-spawn BFS, **5**-27–**5**-28
   multithreaded approach to, **31**-9–**31**-10
   parallel BFS algorithms, programming, **5**-27–**5**-29
   flattened BFS, **5**-3
   nested spawn BFS, **5**-25–**5**-29
   single-spawn BFS, **5**-29
   performance results, **31**-15
Brent's scheduling principle, **26**-3
Bricks, as grid simulation tool, **21**-14
   scheduling unit, **21**-14
   simulated grid environment, **21**-14
Broadcasting, **2**-7, **36**-2
   Broadcast incremental algorithm (BIP), **39**-15
   in clusters of workstations, **19**-5–**19**-18, *see also* Clusters of workstations
   in heterogenous networks, **19**-19–**19**-29, *see also* Heterogenous networks

logP model, **19**-2–**19**-5, *see also* LogP model
   on networks of workstations (NOWs), **19**-1–**19**-29, *see also* Networks of workstations
   problem, **36**-18–**36**-19
Buffer vertices matching, in optimal parallel scheduling algorithms, **23**-9–**23**-15
   direct insertion, **23**-9
   forward and backward segments, identifying, **23**-11–**23**-13
   path search augmentation, **23**-9–**23**-11
   reachable set in parallel, expanding, **23**-13
Buffering, packet, **37**-2–**37**-9, *see also* Packet buffering
BUILD1 algorithm, **30**-14–**30**-16
BUILD2 algorithm, **30**-16–**30**-18
Bulk synchronous parallel (BSP), **2**-11
Bunch–Kaufman factorization, **29**-5
Buses, computing on, **28**-6–**28**-9
Byzantine faults, **10**-8–**10**-9

C

Cache and memory access patterns, **26**-5
Cache searching, **12**-15
Cache updating, **12**-15–**12**-16
Caching problems in networks, **37**-17–**37**-21
   connection caching, **37**-19–**37**-21
   document caching, **37**-17–**37**-19
Caching, **35**-15
Call blocking probability (CBP), **33**-10
Call dropping probability (CDP), **33**-10
Candidate distribution methods, **32**-2–**32**-4
Cannon's algorithm, **43**-11
Canonical schedule
   in power aware mapping, **40**-6–**40**-8
   and actual execution for independent tasks, **40**-9
   SPM and, **40**-6–**40**-8
   shifting schedule sections in, **40**-14–**40**-15
Carrier-to-Interference Ratio (CIR), **33**-4
Carry lookahead generator, **6**-7
Casting, **45**-10–**45**-12
   DRP casting into algorithmic output, **45**-9
   bids, **45**-10
   budget balance, **45**-11
   communications, **45**-9
   components, **45**-9

Index

deliberate discrimination of allocation, **45**-11
description of pseudocode, **45**-10
incentive compatible game theoretical auction, **45**-7–**45**-11
maximum utility to a particular player, **45**-11
monetary cost, **45**-9
Pareto optimality, **45**-11
PARFOR loop, **45**-10
payments, **45**-9
Catalytic system, **3**-17
C-competitive, **37**-1
Cells, **44**-14
  cell space, **44**-16
  in cellular networks, **33**-3–**33**-4
Cellular networks, **33**-2–**33**-10
  bandwidth allocation, **33**-3–**33**-5
  cells, **33**-3–**33**-4
  channel allocation, **33**-3
  channel borrowing with locking, **33**-5
  channel borrowing without locking (CBWL), **33**-5
  and corresponding graph model, **33**-3
  frequency reuse, **33**-4
  and interference graph, **38**-2
  load coping schemes, **33**-5
  location management, **33**-5–**33**-10
    arbitrarily, **33**-9
    cost, **33**-7–**33**-8
    geographical knowledge, **33**-9
    location inquiry, **33**-7
    location update, **33**-7
    queries, **33**-9–**33**-10
  mobile hosts, **33**-5
  mobility tracking, **33**-5
  one-dimensional network, **33**-4
  two-dimensional, mesh-configuration network, **33**-4
  user mobility patterns, **33**-8, *see also* User mobility patterns
Center for Computational Research (CCR), **46**-2
Centrality metrics, **31**-2, **31**-7–**31**-18
  betweenness centrality, **31**-18
  closeness centrality, **31**-18
  degree centrality, **31**-17
  preliminaries, **31**-17
  stress centrality, **31**-18
Centroid, **28**-17
Channel allocation, for cellular networks, **33**-3
Channel borrowing with locking, in cellular networks, **33**-5
Checkpointing, JavaSplit-FT **42**-13–**42**-14

Chernoff bound, **18**-4
Chernoff–Hoeffding bounds, **8**-10, **18**-22, **18**-48
Chinese remainder theorem, **15**-8
Cholesky factorization, **29**-5
Chord, **17**-3–**17**-7
  adding and removing nodes concurrently, **17**-5–**17**-6
  handling node failures, **17**-6–**17**-7
  structure, **17**-3–**17**-5
Classify and randomly select-based algorithms (CRS-A), **38**-11–**38**-15
  5-coloring used by, **38**-*14*
  4-coloring used by, **38**-*12*
  16-coloring used by, **38**-*15*
CLIQUE, **32**-17
Cloning, data migration with, **36**-9
Closeness centrality, **31**-18
Clustering, **5**-21–**5**-22, **28**-14
  cluster-aware virtual machines, **42**-17–**42**-19
  clustered bitmap, **32**-8
  compiler optimizations
    for k-spawn, **5**-22
    for single-spawn, **5**-22
    without nested spawns, **5**-22
  parallel data mining algorithms for, *see* Clustering algorithms, parallel
  prefix sums, **5**-26
    optimal k crossover, **5**-26–**5**-27
    optimal parallel-serial crossover, **5**-26–**5**-27
Clustering algorithms, parallel, **32**-12–**32**-19
  high-dimensional data, clustering, **32**-17–**32**-19
  k-means algorithm, **32**-12–**32**-13
  parallel hierarchical clustering, **32**-13–**32**-14
Clusters of workstations, broadcasting, **19**-5–**19**-18
  analysis, **19**-8–**19**-11
  bounded degree model, **19**-6
  bounded size matching model, **19**-6
  bounding global transfers, **19**-13–**19**-15, *see also* Bounding global transfers
  communication model, **19**-6
  interleaved LCF, **19**-6
  modified algorithm, **19**-8
  multicasting, **19**-11–**19**-13
  postal model, **19**-5–**19**-18, *see also* Postal model
  problem definition, **19**-7
Coarse-grain multicomputer, **9**-6–**9**-15
Coincident pulse technique, **16**-7, **16**-10
Columnsort, **14**-12–**14**-13
  on the R-Mesh, **14**-*13*

Combinatorial problems, multithreaded algorithms design for, **31**-1–**31**-24, *see also* Multithreaded algorithms design
Combined Free/Demand Assignment (CFDAMA) scheme, **33**-14
Combined random/reservation schemes, **33**-13
Combining broadcast problem, LogP model, **19**-4
Communication faults, **10**-1–**10**-2
  addition, **10**-2
  corruption, **10**-2
  omission, **10**-2
Communication model, **19**-6, **36**-3
  HiPer-D System, **41**-19–**41**-20
  communication setup time, **41**-19
  queuing delay, **41**-19
Communication primitives, message dissemination using, **36**-1–**36**-20, *see also* Message dissemination
Communication queue, **8**-10
Compiler-based DSM systems, **42**-19–**42**-20
Component labeling algorithms, **28**-13–**28**-14
Composite type faults, **10**-6–**10**-7
  additions and corruptions, **10**-7
  omissions and additions, **10**-7
  omissions and corruptions, **10**-6–**10**-7
Compression, **15**-4, **16**-15
Computation/Computational systems
  computational complexity, in robust resource allocation, **41**-13
  depth, **5**-11
  evolving, *see* Evolution
  meaning, **1**-3
  models, HiPer-D system, **41**-18–**41**-19
  parallel computation, **1**-3
  processes, **1**-3
Computer vision tasks, **28**-2, *see also* Mesh-based parallel algorithms
  high-level tasks, **28**-2
  low-level tasks, **28**-2, **28**-5
  mid-level tasks, **28**-2, **28**-5
Computing betweenness centrality, algorithms for, **31**-19
Computing on buses, **28**-5–**28**-9
Computing with deadlines, **1**-8–**1**-9
Concurrent-read concurrent-write (CRCW), **5**-5
Concurrent-read exclusive-write (CREW), **5**-5
Condition speculation table (CST), **7**-9

Conditional delay switch, **16**-6
Condor, **21**-10–**21**-11
Congestion model/games, for selfish routing in networks, **20**-2, **20**-3, *see also* Weighted Congestion games
　price of anarchy, **20**-4, **20**-7
　Pure Price of Anarchy (PPoA) in, **20**-27
　social cost, **20**-4
　social optimum, **20**-4
　symmetric congestion game, **20**-4
　unweighted congestion games, **20**-7–**20**-10
Congestion, **17**-11–**17**-12
Connected component labeling algorithms, **28**-13–**28**-14
Connected expansion number, **8**-16
Connected graph, **8**-6
Connected search number, **8**-16
Connection admission control (CAC) algorithm, **33**-18–**33**-19
　performance evaluation, **33**-19–**33**-21
　fixed reservation (FRES), **33**-19
　sequential probabilistic resource reservation (SPR), **33**-17
Connection caching, **37**-19–**37**-21
　Hyper Text Transfer Protocol (HTTP), **37**-19
Connectivity, **31**-2
Constrained triangulation scalable algorithms for, **9**-16–**9**-17
Constrained-deadline system, **24**-2
Content distribution networks (CDNs), **45**-3
Control flow graph (CFG), **7**-6
　nodes, **12**-11–**12**-13
　types, **12**-*12*
Conversion degree, **23**-3
Convex subdivision, mathematical transformations, **1**-18
Coordinate system, of wireless sensor networks, **33**-27–**33**-28
　clustering, **33**-28
Count distribution, **32**-2, **32**-11
Coverage mask, **27**-12
CPU reservations, **24**-22
Cray MTA-2, **31**-2–**31**-3
Cross-over pattern, **15**-2

D

DAChe [CCL+05], **35**-15
DARPA SSCA#2 benchmark (SSCA2) graphs, **31**-14
Data distribution, **32**-2
Data grid, effectively managing, **46**-1–**46**-32

ACDC grid, *see* Advanced computational data center (ACDC) grid
data grid initiatives, **46**-4–**46**-5
grid computing, **46**-3–**46**-4
intelligent migrator, **46**-16–**46**-26
　agent anatomy, **46**-19
　agent infrastructure, **46**-17–**46**-19
　file utilization, **46**-*21*
　neural network accuracy plots, **46**-*26*
　virtual data grid, *see* Virtual data grid
　visualization, **46**-23–**46**-25
scenario builder, **46**-7–**46**-16
　bandwidth vs. transfer load, **46**-15
　scenario anatomy, **46**-9
　scenario generation, **46**-11
　scenario simulation, **46**-11–**46**-13
　scenario visualization, **46**-13–**46**-15
　virtual network topology, **46**-10–**46**-11
Data libraries in parallel applications, **35**-1–**35**-6
Data locality, **2**-13
Data mining techniques, *see* Parallel association rule mining
Data remapping, **11**-10
Data replication problem (DRP), game theoretical solutions for, **45**-1–**45**-22
Data sieving technique, **35**-9–**35**-11
　factors to consider, **35**-*13*
Database mapping, **46**-*28*
Datatype I/O, **35**-18
DCLUB, **32**-8–**32**-11
　D-CLUB bitmap-based algorithm, **32**-8
Deadlines
　computing with, **1**-8–**1**-9
　deadline monotonic algorithm (DM), **24**-5
Deadlock, **26**-5
Decomposable bulk synchronous parallel (D-BSP) model, **2**-1–**2**-18
　basic algorithms, **2**-6
　broadcast and prefix, **2**-6–**2**-9
　vs. BSP, **2**-12–**2**-13
　definition, **2**-6
　effectiveness of, **2**-9–**2**-12
　　processor networks, with respect to, **2**-11–**2**-12
　　quantitative assessment methodology, **2**-10
　and memory hierarchy, **2**-13–**2**-18
　　PRAM simulation, **2**-9

sequential hierarchies models, **2**-14
　space locality, extension to, **2**-17–**2**-18
　submachine locality into temporal locality, **2**-15–**2**-17
Degree centrality, **31**-17
Degree of concurrency, **43**-10–**43**-11
Degree of parallelism, **2**-2
Demand Assignment Multiple Access (DAMA) schemes, **33**-13
Demand assignment protocols, **33**-13
Dense linear algebra (DLA) software library
　added functionality, **29**-7–**29**-10
　LAPACK into ScaLAPACK, **29**-7–**29**-8
　algorithmic improvements for linear systems solution, **29**-4–**29**-5
　better algorithms, **29**-3–**29**-7
　current functionality, extending, **29**-9–**29**-10
　eigenvalue solution, algorithmic improvements for, **29**-5–**29**-7
　EISPACK, **29**-1
　future architectures, challenges, **29**-2–**29**-3
　LAPACK, **29**-1, *see also* LAPACK library
　LINPACK, **29**-1
　motivation, **29**-2
　prospectus, **29**-1–**29**-16
　recursive data structures for, **29**-4
　ScaLAPACK, **29**-2, *see also* ScaLAPACK library
　software, **29**-10–**29**-16
　　ease of use, improving, **29**-10–**29**-11
　　improved software engineering, **29**-11–**29**-14
　　multithreading, **29**-14–**29**-16, *see also* Multithreading
　　performance, **29**-15–**29**-16
Depth-first search (DFS), **8**-7
Derandomization, **18**-6, **18**-49–**18**-53, **25**-1–**25**-2
　derandomization tree, **25**-*5*
　limitation, **25**-2
Destinations, **36**-6–**36**-8
　destination-forbidden colors, **36**-14
　messages with fixed number of, **36**-13–**36**-15
　multidestination messages, **36**-14–**36**-15
　two-destination messages, **36**-13–**36**-14

Index

Dhall effect, **24**-12, **24**-14, **24**-15
Diagonal patterns, **15**-1
Diffing technique, **42**-8
Dijkstra's algorithm, **31**-20
   for computing SPT, **39**-15
   Dijkstra's SSSP algorithm, **31**-19
Directed acyclic graphs, **34**-10
Directed Steiner Tree (DST), **39**-3
Discovery radius, **8**-12
Discrepancy theory, **18**-21
   2-color, **18**-21
   classical, **18**-21
   combinatorial, **18**-21
Discrete server model, in QoS scheduling, **22**-8–**22**-17
   advanced schedulers, **22**-17
   latency in, **22**-11
   performance analysis of, **22**-15–**22**-16
   self-clocked weighted fair queuing (S-WFQ), **22**-14
   service allocation, **22**-9–**22**-11
      cumulative service allocation, **22**-10
   weighted fair queuing (WFQ), **22**-12–**22**-14
      Finish Time Weighted Fair Queuing (F-WFQ), **22**-12, **22**-13
      Start Time Weighted Fair Queuing (S-WFQ), **22**-12–**22**-13
Distance matrix, **31**-20
Distributed computing, **10**-1–**10**-19
   boundaries of computability, **10**-3
      agreement, majority, and unanimity, **10**-3–**10**-4
      impossibility, **10**-4–**10**-5
      possibility, **10**-5–**10**-9
      tightness of bounds, **10**-9–**10**-10
   broadcast with mobile omissions, **10**-10–**10**-11
      general bounds, **10**-10–**10**-11
      specific bounds, **10**-11
   and communication faults, **10**-1–**10**-2
   environment, **10**-1
   fault-tolerant computations, **10**-2
   fractional, **10**-18–**10**-19
   function evaluation, **10**-12–**10**-18
      basic techniques applications, **10**-16–**10**-17
      basic tools, **10**-12–**10**-14
      input bits, **10**-14–**10**-15
      input strings, **10**-15–**10**-16
      omissions, **10**-12
      problem and basic strategies, **10**-12
      tightness of bounds, **10**-17–**10**-18

   game theoretical solutions for, **45**-1–**45**-22, *see also* Game theoretical solutions
   mobile faults, **10**-3
   probabilistic, **10**-18
Distributed database model, **12**-6
Distributed hash table (DHT), **3**-7
   chord, *see* Chord
Distributed KD tree, construction, **32**-15
Distributed memory model, in parallel isosurface extraction, **27**-8
Distributed memory parallel computers (DMPC), **47**-4
   applications, **47**-17–**47**-23
      matrix chain product matrix powers, **47**-20–**47**-23, *see also* Matrix powers
      matrix chain product, **47**-18–**47**-20
   hypercubic networks, **47**-4–**47**-5
   matrix multiplication, **47**-7–**47**-17
      fast matrix multiplication, **47**-7–**47**-14, *see also* Fast matrix multiplication
      scalable matrix multiplication, **47**-14–**47**-17, *see also* Scalable matrix multiplication
   meshes and tori, **47**-5
   parallel algorithms, scalability of, **47**-5–**47**-6
   types of, **47**-4
Distributed multimessage multicasting ($DMMF_C$), **36**-17–**36**-18
Distributed networks, **8**-1–**8**-17
   mobile agents, computing with, *see* Mobile agents
Distributed peer-to-peer data structures
   model and performance measures, **17**-2–**17**-3
   DHT, *see* Distributed hash table
   skip graphs, *see* Skip graphs
Distributed query processing, **12**-13
Distributed ray tracing, **27**-15
Distributed shared memory (DSM), **42**-2
   persistence, JavaSplit-FT, **42**-12–**42**-13
Distributed software systems
   hierarchical performance model
      control flow graph nodes, **12**-11–**12**-13
      module level, **12**-4, **12**-11–**12**-13
      operation level, **12**-4, **12**-14
      system level, **12**-3–**12**-8

      task level, **12**-3–**12**-4, **12**-10–**12**-11
   related work, **12**-2–**12**-3
Distributed synchronization, **42**-9–**42**-10
   distributed conditional waiting, **42**-10
   JavaSplit, **42**-3–**42**-10
Divide-and-conquer algorithm, **18**-39–**18**-44
Document caching, **37**-17–**37**-19
   bit model, **37**-17
   fault model, **37**-17
   general model, **37**-17
   landlord algorithm, **37**-17
   marking algorithm, **37**-18
Domain decomposition, *see* Spatial domain decomposition methods
Duels, **30**-4–**30**-5
Dynamic reconfiguration, **1**-2
Dynamic-priority scheduling, **24**-3, **24**-5–**24**-6

E

Earliest deadline first algorithm (EDF), **24**-6
   LR-heuristic-based solution with, **40**-3
   partitioned scheduling using, **24**-11
Early-release fair (ERfair) scheduling, **24**-18
Eclat algorithm, parallel, **32**-4–**32**-5, **32**-11
   Eclat–ParEclat, **32**-5
   ParClique, **32**-5
   ParMaxClique, **32**-5
   ParMaxEclat, **32**-5
Edge processor, **36**-10
Efficient parallel graph algorithms for multicore and multiprocessors, **26**-1–**26**-40
   algorithmic optimizations, **26**-5–**26**-6
   biconnected components, **26**-21–**26**-29, *see also* Biconnected components
   cache and memory access patterns, **26**-5
   designing, **26**-2–**26**-6
   experimental results **26**-33–**26**-38
   fast, parallel spanning tree algorithm, **26**-6–**26**-13, *see also* Spanning tree algorithm
   fine-grained mutual exclusion locks, **26**-30–**26**-31
   limited parallelism, **26**-3–**26**-4

Efficient parallel graph algorithms (*continued*)
  lock-free parallel algorithms, **26**-30–**26**-31
  lock-free protocols for resolving races among processors, **26**-31–**26**-32
  minimum spanning tree results, **26**-35, **26**-38
  minimum spanning tree, fast shared-memory algorithms for computing, **26**-13–**26**-21, *see also* Minimum spanning tree
  mutual exclusion and lock-free protocols, algorithms with, **26**-30–**26**-38
  network based algorithmic model, **26**-30
  parallel Borůvka minimum spanning tree algorithm, **26**-34
  Shiloach–Vishkin spanning tree implementation, **26**-35
  spanning tree results, **26**-35
  synchronization, **26**-4–**26**-5
  synchronous based algorithmic model, **26**-30
Eigenvalue solution, algorithmic improvements for, **29**-5–**29**-7
EISPACK library, **29**-1–**29**-3
Electrically Controlled Directional Coupler Switches (ECS), **28**-4
Embedded wireless multicast advantage, **39**-15
Erdös–Spencer method, **18**-1
Euler-tour step, **26**-24
Evolution, computational systems, **1**-1–**1**-21
  computational models, **1**-2–**1**-4, *see* Models, computational
  interacting variables, **1**-12–**1**-17, *see also* Interacting variables
  mathematical constraints, computations obeying, **1**-17–**1**-19, *see also* Mathematical constraints
  rank-varying computational complexity, **1**-9–**1**-12
  time-varying computational complexity, **1**-6–**1**-9
  time-varying variables, **1**-4–**1**-6, *see also* Time-varying variables
  universal computer, **1**-19–**1**-20
Exact potential game, **20**-2

Exclusive-read exclusive-write (EREW), **5**-32
Execution queue, **12**-5–**12**-8, **12**-13
Exploration algorithms, **8**-8
Extra-edge processor, **36**-10

F

Fair lexicographic scheduling, **22**-18–**22**-19
Fair queuing in the fluid flow model, **22**-5–**22**-7
Fast Fourier Transforms (FFTs), **43**-2
Fast matrix multiplication, **47**-7–**47**-14
  processor complexity reduction, **47**-12–**47**-13
  time complexity reduction, **47**-13
Fast multipole method (FMM), **44**-21
Fastest node first (FNF) technique, **19**-19, **19**-26
  bad example, **19**-25–**19**-26
Fault tolerance, **42**-10–**42**-15, **42**-21, *see also* JavaSplit-FT
  ad hoc wireless networks, **39**-17
  computations, **10**-2–**10**-3
  fault-tolerant distributed queue (FT-DQ), **42**-14
Feasible flow, **20**-21
  splittable feasible flow, **20**-21
  unsplittable feasible flow, **20**-21
FePIA procedure, **41**-3, **41**-7, **41**-10
FG to reduce latency effect in parallel programs running on clusters, **34**-1–**34**-15, *see also* Latency
  additional features, **34**-7–**34**-11
  directed acyclic graphs, **34**-10
  fork/join, **34**-9–**34**-10
  hard barriers, **34**-8–**34**-9
  intersecting pipelines, **34**-6–**34**-7
  macros, **34**-8
  multiple pipelines with FG, **34**-5–**34**-6
  pipeline-structured algorithms, **34**-2–**34**-4, *see also* Pipeline-structured algorithms
  soft barriers, **34**-9
  speeding up programs, **34**-11–**34**-15
  stage replication, **34**-10
  statistical hypothesis test, **34**-14
FI-cluster, **32**-8
Field-programmable gate arrays (FPGAs), **11**-1
  generic, **11**-3
  massively parallel FPGA array, **11**-5–**11**-6

File access in parallel applications, **35**-2–**35**-3
File allocation problem (FAP), **45**-2
File locking, **35**-14–**35**-15
Fine-grain parallelism, **11**-11, **26**-32–**26**-33, **26**-35
  code generation, **7**-3–**7**-8
  explicitly parallel code compilation, **7**-5–**7**-6
  internal code representation, **7**-6
  Inthreads-C language, **7**-4–**7**-5
  register allocation, **7**-6–**7**-8
  implementation, **7**-8–**7**-9
  misprediction handling, **7**-9–**7**-11
  performance evaluation, **7**-11–**7**-17
    benchmarks, **7**-13–**7**-15
    parallelization degree effect, **7**-15–**7**-16
    parallelization granularity, **7**-12–**7**-13
    register pressure effect, **7**-16–**7**-17
  programming model, **7**-2–**7**-3
  ISA extension, **7**-2–**7**-3
Fine-grain to coarse-grain transition, **9**-1–**9**-17
  model, **9**-5–**9**-7
  tools, **9**-7–**9**-11
    compaction, **9**-10
    library structure, **9**-10
    load balancing, **9**-7–**9**-8
    merging, **9**-8–**9**-9
    sorting, **9**-10
Finish Time Weighted Fair Queuing (F-WFQ), **22**-12, **22**-13
First Come First Served (FCFS), **22**-2
Fixed assignment protocols, **33**-13
Fixed Bandwidth Assignment (FBA), **33**-13
Fixed Memory Search, **8**-11
Fixed reservation (FRES), **33**-19
Flattened BFS, **5**-29, **5**-51–**5**-55
Floating-point cores, **11**-11
Flow intrinsic delay, **22**-11
Fluid flow server model, **22**-4–**22**-8
  fair queuing in the fluid flow model, **22**-5–**22**-7
  weighted fair queuing in, **22**-7–**22**-8
Fluid model, **33**-8
Fooling automata, **18**-34–**18**-44
  divide-and-conquer algorithm, **18**-39–**18**-44
  ε-fool constraints, **18**-35–**18**-37, **18**-39
  impact of, **18**-47–**18**-53, *see also* Srinivasan's parallel packing
  reducing support, **18**-35

transition probabilities, **18**-38–**18**-39
Footprint, **33**-10
Fork/join, **34**-9–**34**-10
Forward segments identification, in buffer vertices matching, **23**-11–**23**-13
Forwarding, messages, **36**-15–**36**-17
4D mesh simulation, **4**-8–**4**-9
FP-growth, **32**-11
Frame-based RT-applications, in power aware mapping, **40**-4–**40**-5
Frequency reuse, in cellular networks, **33**-4
Fully connected lognormal graphs, **45**-12
Fully connected random graphs, **45**-12
Fully connected uniform graphs, **45**-12
Fully dynamic-priority assignment, **24**-3
 algorithm, **24**-6, **24**-16
Fusing R-Mesh, **14**-3
FusionFS, **35**-16
Future parallelism, **31**-3

G

Game theoretical solutions
 comparative analysis, **45**-16–**45**-20
  number of servers and objects, change in, **45**-16–**45**-18
  read and write frequencies, change in, **45**-18–**45**-19
  running time, **45**-19
  system capacity, change in, **45**-18
 for data replication problem (DRP), **45**-1–**45**-22
  background material, **45**-6–**45**-11, *see also* Auctions
  problem formulation, **45**-4–**45**-6
  experimental setup and discussion, **45**-11–**45**-20
   access patterns, **45**-13–**45**-14
   comparative techniques, **45**-15–**45**-16
   flat models, **45**-11–**45**-12
   fully connected lognormal graphs, **45**-12
   fully connected random graphs, **45**-12
   fully connected uniform graphs, **45**-12
   hierarchical transit-stub model, **45**-13
   link distance models, **45**-12–**45**-13
   network topologies, **45**-11–**45**-13
   power-law model, **45**-13
   pure random model, **45**-12
   random graphs, **45**-12
   relocation period, determination, **45**-14
   Waxman model, **45**-12
  incentive compatible game theoretical auction, DRP casting into, **45**-7–**45**-11
   algorithmic output, **45**-9
   basics, **45**-7
   bids, **45**-8
   casting, **45**-8–**45**-10, *see also* Casting
   communications, **45**-7
   components, **45**-7
   desirable properties, **45**-8
   incentive compatible auction mechanism (I-CAM), **45**-8
   ingredients, **45**-7–**45**-8
   monetary cost, **45**-8
   payments, **45**-8
Gated-connection network (GCN), **28**-11, **28**-4
Gatekeeper array, **5**-25
Gauss–Markov model, **33**-8
General pairs PROFIT/COST (GPC) problem, **25**-4–**25**-7
General Parallel File System (GPFS), **35**-16
General purpose processors (GPPs), **11**-1
Generalized multiframe task model, **24**-22
Generalized robustness metric, **41**-2–**41**-6
 FePIA procedure steps, **41**-3
Genetic-based algorithm (GRA), **45**-15, **45**-16
Geometric flips, **1**-18
Geometric model
 in minimum energy communication, **39**-13–**39**-16, *see also* under Minimum energy communication
Global dynamic-priority scheduling, **24**-16
Global scheduling
 for identical multiprocessor systems, **24**-3
 on multiprocessors, **24**-12–**24**-15
  global dynamic-priority scheduling, **24**-14–**24**-15
  global static-priority scheduling, **24**-12–**24**-14
 in power aware mapping, **40**-2
Global static-priority scheduling, **24**-16
Glue, **34**-2
Gonzalez algorithm, **36**-10
Gossiping communication, **36**-5
Gossiping problem, **36**-18–**36**-19
 weighted gossiping problem, **36**-19
Gossiping, **36**-2
Graft-and-shortcut approach, **26**-6
Graph abstractions, **26**-1–**26**-40, *see also* Efficient parallel graph algorithms
Graph algorithm, **14**-14–**14**-20
Graph partitioning, **44**-2
Graph theoretic problems, **31**-2
Graph traversal, **31**-8–**31**-16
 Breadth-First Search (BFS), **31**-8
 st-connectivity and shortest paths, **31**-11–**31**-13
 performance results, **31**-13–**31**-16
  DARPA SSCA#2 benchmark (SSCA2) graphs, **31**-14
  Rand-ER graphs, **31**-14
  Rand-Hard graphs, **31**-14
  random graphs, **31**-13
  scale-free graphs (SF-RMAT), **31**-13
  synthetic sparse random graphs, **31**-14
Gravity model, **33**-8
Greedy algorithm, **45**-16
Greedy slack reclamation, **40**-9
Grid computing, **19**-5
Grid environments
 scheduling in, **21**-1–**21**-16, *see also* Scheduling
Grid scheduling, **21**-3–**21**-4
 abstract model of, **21**-3, **21**-4
 algorithms for, **21**-12–**21**-13
  list scheduling with round-robin order replication, **21**-12
  storage affinity (SA) algorithm, **21**-13
  Xsufferage, **21**-13
 centralized model, **21**-6
 decentralized model, **21**-6
 grid scheduler model, **21**-5–**21**-6
 grid scheduling systems, **21**-9–**21**-12
  application level scheduler (AppLeS), **21**-9–**21**-10
  Condor, **21**-10–**21**-11
  Moab grid scheduler (silver), **21**-11
  netsolve and gridsolve, **21**-1
  Nimrod/G, **21**-12
 hierarchical scheme, **21**-6
 scalability of, **21**-5
 scheduling procedure, **21**-6–**21**-9
  application properties, **21**-7
  execution preparation, **21**-8–**21**-9

Grid scheduling *(continued)*
  job analysis, **21**-7
  job description and submission, **21**-6–**21**-7
  monitoring, **21**-9
  postexecution, **21**-9
  preselection of resources, **21**-8
  resource allocation, **21**-8
  resource requirements, **21**-7
  time requirements, **21**-7
  traditional multiprocessor scheduling versus, **21**-3
    application diversity, **21**-3
    dynamicity of resource, **21**-3
    lack of control, **21**-3
Grid-enabling application templates (GATs), **46**-5
GridSim, as grid simulation tool, **21**-14
GridSolve project, **21**-11
Grouping, **32**-17
  group communication problems, **39**-2

H

Half lattice approximation, **18**-26–**18**-28
Hard barriers, **34**-8–**34**-9
Heisenberg's uncertainty principle, **1**-13
Helman–JáJá model/algorithm, **31**-4, **31**-5
Helper matrix, **5**-35
Heterogeneous distributed database system, **12**-18–**12**-19
Heterogeneous parallel and distributed systems, **41**-1–**41**-28
  robust resource allocation for, **41**-1–**41**-28, *see also* Robust resource allocation
Heterogenous networks broadcasting, **19**-19–**19**-29
  1.5-approximation for minimizing broadcast time, **19**-23–**19**-26
  completion times sum, minimizing, **19**-20–**19**-22
  fastest node first (FNF) technique, **19**-19
  multicast, **19**-29
  polynomial time approximation scheme, **19**-26–**19**-29
  problem definition, **19**-20
Heuristic descriptions, HiPer-D System, **41**-20–**41**-24
  duplex, **41**-23
  Fast greedy heuristic FGH-L, **41**-23
  max-min heuristic, **41**-23
  MCPF heuristic, **41**-20, **41**-23

MCTF heuristic, **41**-20–**41**-22
two-phase greedy X (TPG-X), **41**-23
upper bound, **41**-24
Hierarchical clustering algorithms, parallel, **32**-13–**32**-14
  for both shared and distributed memory multiprocessors, **32**-14
  parallel *HOP*, **32**-15–**32**-17, *see also* HOP clustering process
Hierarchical Data Format (HDF), **35**-4–**35**-6
Hierarchical memory model (HMM), **2**-14
Hierarchical performance modeling distributed software systems, *see* Distributed software systems
  reconfigurable computers, *see* Reconfigurable computers
Hierarchical transit-stub model, **45**-13
High-dimensional data, clustering, **32**-17–**32**-19
  CLIQUE, **32**-17
  pMAFIA algorithm, **32**-17
High-performance computing challenge (HPCC) benchmark, **34**-11
  application, **34**-11
    matrix multiplication, **34**-11
    matrix transpose, **34**-11
    three-dimensional FFT implementation, **34**-12
High-performance techniques for parallel I/O, **35**-1–**35**-22
  file access in parallel applications, **35**-2–**35**-3
  general MPI–IO usage and optimizations, **35**-6–**35**-15, *see also* MPI–IO usage and optimizations
  Hierarchical Data Format (HDF), **35**-4–**35**-6
  NetCDF, **35**-3–**35**-4
  Noncontiguous I/O methods, **35**-17–**35**-22, *see also* noncontiguous I/O methods
  parallel file systems, **35**-15–**35**-22
    FusionFS, **35**-16
    General Parallel File System (GPFS), **35**-16
    Lustre, **35**-15
    Panasas Active Scale File System (PanFS), **35**-16
    parallel virtual file system 1 (PVFS1), **35**-17
  parallel NetCDF, **35**-3–**35**-4
  portable file formats and data libraries, **35**-1–**35**-6

HiPer-D System, robustness metric of, **41**-8–**41**-10, **41**-16–**41**-18
  latency constraint, **41**-9
  robustness optimization, static heuristics design, **41**-18–**41**-24
    computation and communication models, **41**-18–**41**-20
    heuristic descriptions, **41**-20–**41**-24
  slack, **41**-16
  throughput constraint, **41**-9
Homeostatic principle, **1**-14
Homogenous distributed database system, **12**-16–**12**-17
HOP clustering process, parallelizing, **32**-15
  distributed KD tree, construction, **32**-15
  generating density, **32**-16
  grouping, **32**-17
  hopping, **32**-16
Hopping, **32**-16
Hough transform, computing, **28**-10–**28**-13
Hybrid MAC, **33**-13
Hybrid of RA and reservation protocols, **33**-13
Hypercubic networks, **47**-4–**47**-5
Hypergraphs, **18**-3

I

Identical multiprocessor systems, **24**-23
IEEE standard 754, **11**-11
Image segmentation, **28**-14–**28**-16
  clustering, **28**-14
  partition generated by, **28**-14
  quadtree, **28**-15
  region growing, **28**-15
  split and merge, **28**-15
Immediate subcell, **44**-15
Immediate supercell, **44**-15
Implicit-deadline task system, **24**-2
  synchronous periodic task systems, **24**-3
Incentive compatible auction mechanism (I-CAM), **45**-8
Input and output queuing switches combination, managing, **37**-8–**37**-9
  scheduling phase, **37**-8
Integer sorting, **13**-9–**13**-11
Integer summing, **15**-6–**15**-7
Intelligent paging method, **33**-9
Interacting variables, in computational systems, **1**-12–**1**-17
  disturbing the equilibrium, **1**-12–**1**-14

Index

quantum computing, distinguishability in, 1-16–1-17
solutions, 1-14–1-16, *see also* Solutions
uncertainty in measurement, 1-13–1-14
Intercluster activity, wireless sensor networks, 33-29
Interference graph, 7-8
Interleaved LCF, **19**-6, **19**-16, **19**-18
Intersecting pipelines, **34**-6–**34**-7
Inthreads computational model, **7**-2
Inthreads-C language, **7**-4–**7**-5
Intracluster activity, wireless sensor networks, 33-28
Intrasporadic (IS) model, **24**-18
Inverse quantum Fourier transform, 1-10–1-12
  parallel solution, 1-11–1-12
  sequential solution, 1-11
ISA extension, **7**-2–**7**-3
Isoefficiency metric of scalability, **43**-6–**43**-10
  isoefficiency function, **43**-7–**43**-10
  cost-optimality and, **43**-10
  degree of concurrency and, **43**-10–**43**-11
  lower bound on, **43**-10
  problem size, **43**-7
Isosurface/Isosurface extraction, **27**-1–**27**-5
  in 3D objects modeling, **27**-2
  interval tree, **27**-6
  parallel isosurface extraction strategies, **27**-7–**27**-8
  scalable and efficient parallel algorithm for, **27**-9–**27**-11
  simple optimal out-of-core isosurface extraction algorithm, **27**-5–**27**-7
  span space, **27**-6
  structured grid, **27**-3
  unstructured grid, **27**-3
  view-dependent isosurface extraction and rendering, **27**-11–**27**-15
    Branch-On-Need Octree (BONO), **27**-12
    full-resolution extraction, **27**-11
    multipass occlusion culling algorithm, **27**-13–**27**-14
    by ray tracing, **27**-15
    sequential algorithm for structured grids, **27**-12–**27**-13
Iterated length timeslicing, **10**-14

J

Java programs
  Addistant, 42-20
  Java-based distributed computing, **42**-17–**42**-21
    cluster-aware virtual machines, **42**-17–**42**-19
    compiler-based DSM systems, **42**-19–**42**-20
    standard JVMs, **42**-20–**42**-21
  JavaParty, 42-20
  JavaSplit-FT, system classes, 42-4
  jOrchestra, 42-20
  transparent distributed runtime for, **42**-1–**42**-21, *see also* Transparent distributed runtime for java
JavaSplit, **42**-3–**42**-10
  class file instrumentation, **42**-3–**42**-9
  class renaming transformation, 42-5
  distributed synchronization, **42**-9–**42**-10
  DSM, 42-3
  JavaSplit-pecific transformations, **42**-4–**42**-7
  shared memory management, **42**-7–**42**-9
  Lazy Release Consistency (LRC), 42-7
  twin class hierarchy (TCH), **42**-3–**42**-4
JavaSplit-FT, **42**-10–**42**-15
  checkpointing, **42**-13–**42**-14
  distributed shared memory persistence, **42**-12–**42**-13
  fault-tolerant distributed queueing, **42**-14–**42**-15
  conditional waiting, support for, **42**-15
  coordinator failures, recovery, **42**-14–**42**-15
  queue member failures, recovery, 42-14
  Java thread checkpointing, 42-11
  replication groups, **42**-11–**42**-12
Job-level dynamic-priority algorithm, **24**-3, **24**-6

K

k-Agreement problem, **10**-3–**10**-4
Karp–Miller–Rosenberg naming technique, **30**-21
K-ary summation, **5**-40–**5**-42
k-d Tree, **44**-7
k-*Means* clustering, **32**-12–**32**-13
k-wise independence method, **18**-2, **18**-7–**18**-11, **18**-14–**18**-34
  functions over Z_2^k, **18**-14–**18**-16
  lattice approximation, **18**-26–**18**-34, **18**-36,
  see also Lattice approximation
  low discrepancy colorings, **18**-21–**18**-26
  maximal independent sets in graphs, **18**-18–**18**-21
  multivalued functions, **18**-16–**18**-18
Kruskal's algorithm, **26**-20
k-Spawn BFS algorithm, **5**-57

L

Labeling algorithms, **28**-13–**28**-14
Labeling graph, **8**-7
Landlord algorithm, **37**-17
Language acceptor, **3**-17
LAPACK library, **29**-1–**29**-9
  data types supported in, 29-7
  LAPACK eigensolvers, algorithmic improvements to, **29**-5
  into ScaLAPACK, **29**-7–**29**-8
Large cell, **44**-15
Large-scale network analysis, **31**-16
Largest cluster first (LCF) algorithm, **19**-6, **19**-7, **19**-16–**19**-18
  modified LCF, **19**-8
Latency
  in discrete server model, **22**-11
  in parallel computing, **34**-1–**34**-2
Lattice approximation problem, **18**-2, **18**-26–**18**-34, **18**-36
  half lattice approximation problem, **18**-25–**18**-28
Layered networks, **20**-5
Lazy Release Consistency (LRC), 42-7
  home-based implementation, **42**-8–**42**-9
Le Châtelier's principle, 1-14
Leader election, **8**-6
Least laxity first (LLF), **24**-6
Leftmost finding, **15**-3
Length of sequence of round trips to memory (LSRTM), **5**-11
Length timeslicing, **10**-14
LexAS algorithm, **22**-18, **22**-20
Limited independence, **18**-6–**18**-14
  almost k-wise independence, **18**-11–**18**-14
  conditional probabilities method, **18**-6–**18**-7
Limited parallelism, **26**-3–**26**-4
Linear array with pipelined optical buses (LAPOB), **16**-1
Linear array with reconfigurable pipelined bus system (LARPBS), **16**-1, **16**-5–**16**-7, **47**-5
  structure, **16**-5

Linear mobility model, of LEO satellite networks, **33**-14, **33**-*15*
Linear pipelined bus (LPB), **16**-1
Linear R-Mesh, **14**-3
Linearization, **44**-9
Link distance models, **45**-12–**45**-13
LINPACK library, **29**-1–**29**-3
List I/O, **35**-17–**35**-18
List ranking, **14**-18–**14**-20, **31**-4–**31**-8
   classes
      ordered and random, **31**-5
   MTA list ranking code, **31**-7
   multithreaded implementation, **31**-5
   performance results, **31**-5–**31**-8
   scalable algorithms for, **9**-13
List scheduling, **40**-12
   with dual-processor system, **40**-*12*
(l,m)-Merge sort (LMM)
   merging algorithm, **6**-4–**6**-6
   on parallel disk model, **6**-6–**6**-8
Load balance index, **41**-14
Load coping schemes, in cellular networks, **33**-5
Load Sharing Facility (LSF), **21**-9
Localized faults, **10**-3
Locally essential tree, **44**-20
Location management, in cellular networks, **33**-5–**33**-10, *see also under* Cellular networks
Lock-free parallel algorithms, **26**-30–**26**-31
Lock-free protocols
   parallel algorithms with, **26**-30–**26**-38
   for resolving races among processors, **26**-31–**26**-32
Locks
   deadlock, **26**-5
   mutual exclusion locks, **26**-4
Logarithmic terms, in probabilistic method, **18**-3
Logarithmic-time prefix sums, **15**-6
LogP model broadcasting, **19**-2–**19**-5
   all-to-all, **19**-3
   combining broadcast problem, **19**-4
   single item, **19**-3
   summation, **19**-4–**19**-5
Look-up table (LUT), **11**-2
Loop future directive, **31**-9
Low- and mid-level vision tasks
   mesh-based parallel algorithms for, **28**-10–**28**-18
      connected component labeling algorithms, **28**-13–**28**-14
      Hough transform, computing, **28**-10–**28**-13
      image segmentation, **28**-14–**28**-16, *see also* Image segmentation
      moment computations, **28**-17–**28**-18
      region representation and description, **28**-16–**28**-17
Low discrepancy colorings, **18**-21–**18**-26
Low earth orbiting (LEO) satellite networks, **33**-10–**33**-17
   footprint, **33**-10
   medium access control (MAC) protocols, **33**-13–**33**-14, *see also* Medium access control (MAC) protocols
   mobility model and traffic parameters, **33**-14–**33**-17, *see also* Mobility model
   mobility models for, **33**-11
   Q-Win strategy, **33**-12
   real-time services, probabilistic resource reservation strategy for, **33**-12
   related work, **33**-10–**33**-12
   selective look-ahead (SILK) strategy, **33**-12
   spotbeams, **33**-10
   user location information, **33**-12
Low granularity parallelization, **7**-9
Lower bound, **38**-15–**38**-18
Lustre, **35**-15–**35**-16

M

Macros, **34**-8
Map coloring, **1**-18
Marching–Cubes (MC) algorithm, **27**-2, **27**-4
   active cells, **27**-4
   inactive cells, **27**-4
   interval tree, **27**-4
   metacell, **27**-5
   octree, **27**-4
   seed cells, **27**-4
   span space, **27**-4
   splitting value, **27**-4
Marking algorithm, **37**-18
Markovian model, **33**-8
Massively parallel field-programmable gate arrays, **11**-5–**11**-6
Massively parallel processor (MPP) systems, **21**-2
Mathematical constraints, computations obeying, **1**-17–**1**-19
   mathematical transformations, **1**-17–**1**-19
      convex subdivision, **1**-18
      geometric flips, **1**-18
      map coloring, **1**-18
      rewriting systems, **1**-18–**1**-19
   parallel solution, **1**-19
   sequential solution, **1**-19
Mathematical model solution, in computational systems, **1**-15–**1**-16
Matrices, **18**-2
Matrix chain product, **47**-18–**47**-20
   algorithm, **47**-18–**47**-19
   analysis, **47**-19–**47**-20
Matrix powers, **47**-20–**47**-23
   algorithm, **47**-20–**47**-21
   analysis, **47**-21–**47**-23
Matroid greedy algorithm, **23**-5
Maximal independent set (MIS), **18**-1, **25**-7–**25**-17
   algorithm overview, **25**-9–**25**-10
   in graphs, **18**-18–**18**-21
   initial stage, **25**-10–**25**-16
   lexicographically first MIS, **18**-18–**18**-19
   and maximal matching, parallel algorithms for, **25**-1–**25**-18
      bit pairs PROFIT/COST problem, **25**-2–**25**-4
      general pairs PROFIT/COST (GPC) problem, **25**-4–**25**-7
   speedup stage, **25**-16–**25**-17
Maximal matching, parallel algorithms for, **25**-17–**25**-18
Mcf benchmark, **7**-5, **7**-13–**7**-15
Medium access control (MAC) protocols, **33**-13–**33**-14
   groups of, **33**-13
      adaptive protocol, **33**-13–**33**-14
      Combined Free/Demand Assignment (CFDAMA) scheme, **33**-14
      combined random/reservation schemes, **33**-13
      Demand Assignment Multiple Access (DAMA) schemes, **33**-13
      demand assignment protocols, **33**-13
      fixed assignment protocols, **33**-13
      Fixed Bandwidth Assignment (FBA), **33**-13
      hybrid MAC, **33**-13
      hybrid of RA and reservation protocols, **33**-13
      random access (RA) protocols, **33**-13
Membrane systems, **3**-1–**3**-20
   cells to computers, **3**-3
   extensions and variants, **3**-7–**3**-8
   nonmaximal parallelism, *see* Nonmaximal parallelism
   P automata, **3**-8–**3**-9

Index

P systems, *see* P systems
P transducers, 3-8–3-9
polynomial time, solving in, 3-12–3-15
structure, 3-5
universality, 3-10–3-12
Memory access patterns, in efficient parallel graph algorithms, 26-5
Memoryless movement model, 33-8
Mesh-based parallel algorithms for ultra fast computer vision, 28-1–28-18
  basic building blocks, 28-5–28-6
  computer vision tasks, 28-2
  computing on buses, 28-6–28-9
  low- and mid-level vision tasks, 28-10–28-18, *see also* Low- and mid-level vision tasks
  parallel model of computation, 28-3–28-5, *see also* Parallel model of computation
Message dissemination using modern communication primitives, 36-1–36-20
  Benes network messages, 36-4
  broadcasting, 36-2
  communication model, 36-3
  definitions, 36-5–36-9
  distributed multimessage multicasting ($DMMF_C$), 36-17–36-18
  gossiping, 36-2, 36-5
  message communication patterns, 36-2
  message dissemination over arbitrary networks, 36-18–36-19
  messages with a fixed number of destinations, 36-13–36-15, *see also* Destinations
  multicasting, 36-2
  communication mode forwarding, 36-2
  multimessage multicasting, 36-9–36-20, *see also* Multimessage multicasting
  multiport communication mode, 36-3
  network, 36-2
  offline communication scheduling, 36-2
  offline multimessage multicasting problem, 36-4
  over arbitrary networks, 36-18–36-19
  pr-dynamic network, 36-4
  primitive operations, 36-2
  related work, 36-5–36-9
  single port communication mode, 36-3
  source and destination processors, 36-6
  splitting message destinations, 36-3
  unicasting, 36-2
  unsplittable message destinations, 36-7
Messages, *see also* Message dissemination
  exchange problem, 36-9
  with a fixed number of destinations, 36-13–36-15, *see also* Destinations
  message communication, 36-2, 36-7
Message passing interface (MPI) standard, 9-6
MicroGrid, 21-14–21-15
Milano theorem, 3-12
Minimal parallelism, 3-16
Minimax principle, 38-15
Minimax search trajectory, 8-12
Minimax value, 8-12
Minimum density directed tree (MDDT), 39-10
Minimum Energy Broadcast Tree (MEBT), 39-6, 39-7, 39-14
Minimum energy communication in ad hoc wireless networks, 39-1–39-18, *see also* Asymmetric wireless networks; Symmetric wireless networks
  extensions, 39-16–39-18
  geometric model, 39-13–39-16
  Broadcast Incremental Algorithm (BIP), 39-15
  linear case, 39-13–39-14
  multidimensional wireless networks, 39-14–39-16
  root-crossing, 39-13
Minimum Energy Inverse Broadcast Tree (MEIBT), 39-2, 39-14
Minimum Energy Inverse Multicast Tree (MEIMT), 39-2
Minimum Energy Multicast Tree (MEMT), 39-6–39-8, 39-10–39-11, 39-14
Minimum Energy Steiner Subgraph (MESS), 39-4–39-5
Minimum Energy Strongly Connected Subgraph (MESCS), 39-5, 39-17
Minimum Energy Subset Strongly Connected Subgraph (MESSCS), 39-5, 39-8
Minimum Spanning Arborescence (MSA), 39-2, 39-3
Minimum spanning tree (MST), 8-7, 14-17–14-18
fast shared-memory algorithms for computing, 26-13–26-21
  analysis, 26-15–26-17
  Borůvka's algorithms, 26-1–26-15, *see also* Borůvka's algorithms
  experimental results, 26-20–26-21
  new parallel minimum spanning tree algorithm, 26-17–26-20
Mining by bitmaps, 32-8–32-11
  bit vector, 32-8
  clustered bitmap, 32-8
  dCLUB, 32-8
  D-CLUB, 32-8
  FI-cluster, 32-8
  mining frequent itemsets by clustered bitmaps, 32-9
  PD-CLUB, 32-8
Min–max theorem, 45-5
Mixed radix representation technique, 15-9–15-10
Mixed strategies, 8-12
Moab grid scheduler (silver), 21-11
Mobile ad hoc networks (MANETs), 33-21–33-25
  applications, 33-22
  conventional network topologies and, 33-22
  operation, 33-22
  packet redirection, 33-24–33-25
  power-aware routing, 33-23–33-24
  cost-aware routing, 33-23–33-24
  routing, 33-22–33-25
Mobile agents
  black holes, 8-16–8-17
  definition, 8-2
  distributed networks, modeling in, 8-2–8-3
  game-theoretic approaches, 8-12–8-15
  search and rendezvous models, 8-12–8-15
  search games, 8-12
  graph exploration, 8-6–8-8
  anonymity, 8-7–8-8
  anonymous tree exploration, 8-8
  efficiency measures, 8-7
  labeling, 8-7
  problem of, 8-6–8-7
  underlying graph, 8-7
  network decontamination, 8-15–8-16
  rendezvous problem, 8-3–8-6
  asymmetric rendezvous, *see* Asymmetric rendezvous
  solvability, 8-4

Mobile agents *(continued)*
  symmetric rendezvous, *see* Symmetric rendezvous
  searching with uncertainty, 8-9–8-11
    complete network, 8-10–8-11
    network and search models, 8-9
    ring network, 8-9–8-10
Mobile computing algorithmics, 33-1–33-30
  call admission control algorithm, 33-17–33-21, *see also* Connection admission control (CAC) algorithm, 33-18–33-19
  cellular networks, 33-2–33-10, *see also* Cellular networks
  global integrated terrestrial and LEO satellite system, 33-2
  low earth orbiting satellite networks, 33-10–33-17, *see also* Low earth orbiting (LEO) satellite networks
  mobile ad hoc network (MANET), 33-21–33-25, *see also* Mobile ad hoc network (MANET)
  resource management, 33-17–33-21, *see also* Resource management strategy
  wireless sensor networks, 33-25–33-29, *see also* Wireless sensor networks
Mobile faults, 10-3
Mobile hosts (MHs), 33-1
  in cellular networks, 33-5
Mobile Switching Centre (MSC), 33-9
Mobility model, of LEO satellite networks, 33-14–33-17
  linear mobility model, 33-14
  mobility parameters, 33-16
  two-dimensional mobility model, 33-14–33-17
  user mobility parameters, 33-15
Mobility tracking, in cellular networks, 33-5
  always-update strategy, 33-6
  never-update strategy, 33-6
Models, computational, 1-2–1-4
  assumption, 1-4
  computation, meaning, 1-3
  parallel computation, 1-3–1-4, *see also* Parallel computation
  sequential model, 1-3
  time and speed, 1-2–1-3, *see also* Time unit
Modified LCF algorithm, 19-8
Module parallel computer, 47-5
Molecular dynamics simulation, 11-13–11-14

Moment computations, 28-17–28-18
Monotone search, 8-15
Most critical path first (MCPF) heuristic, 41-20, 41-23
Most critical task first (MCTF) heuristic, 41-20–41-22
MPI–IO usage and optimizations, 35-6–35-15
  current areas of research in, 35-14–35-15
    MPI–IO file caching, 35-15
    Persistent File Realm (PFR), 35-14
    portable file locking, 35-14–35-15
  MPI–IO interface, 35-6–35-9
    collective read and write, 35-9
    derived datatypes, 35-7
    file views, 35-7–35-8
    open, close, and hints, 35-6–35-7
    read and write, 35-8–35-9
  significant optimizations in ROMIO, 35-9–35-14, *see also* ROMIO implementation
MTA-2 compiler, multithreaded algorithms design for, 31-3–31-4, *see also under* Multithreaded algorithms
Multiagent rendezvous, 8-6
Multiagent systems, 46-18
Multibuffer problems with arbitrary packet values, 37-7–37-8
  algorithm transmit-largest (TL), 37-7
Multibuffer problems with uniform packet values, 37-4–37-7
Multicare and multiprocessors efficient parallel graph algorithms for, 26-1–26-40, *see also* Efficient parallel graph algorithms
Multicast acknowledgment, 37-15–37-17
  study models, 37-16
    asynchronous model, 37-16
    full information model, 37-16
    synchronous model, 37-16
Multicasting, 36-2, *see also* Multimessage multicasting
  clusters of workstations broadcasting, 19-11–19-13
    analysis, 19-11–19-13
  communication mode forwarding, 36-2
  communication primitive, 36-3
  patterns, 16-11
Multicore processors, 43-1
Multidestination messages, 36-14–36-15

Multidimensional bin packing, 18-44–18-47
  parallel bin packing algorithm, steps, 18-44–18-45
  derandomized rounding in NC, 18-45
  enlarged fractional bin packing, 18-45
  optimal fractional bin packing in parallel, 18-44
  randomized rounding, 18-45
Multidimensional wireless networks, 39-14–39-16
Multiframe task model, 24-22
Multilevel caching algorithm, 12-15–12-16
  cache searching, 12-15
  cache updating, 12-15–12-16
Multimessage multicasting, 36-2, 36-9–36-15
  edge processor, 36-10
  extra-edge processor, 36-10
  with forwarding, 36-15–36-17
    multisource MMF_C, 36-17
    NP-completeness, 36-15
  graph edge coloring instance, 36-10
  NP-completeness of, 36-9–36-10
  in optical networks, 36-19–36-20
  receiving buffers, 36-11–36-13
  upper and lower bounds for MM_C, 36-10–36-11
Multimessage unicasting, 36-8
Multipass occlusion culling algorithm, 27-13–27-14
  binary swap technique, 27-14
  modified coverage mask in, 27-13
    covered state, 27-13
    pending state, 27-13
    vacant state, 27-13
Multiple perturbation parameters, robustness against, 41-12–41-13
Multiple pipelines with FG, 34-5–34-6
Multiple Relatively Robust Representations (MRRR) algorithm, 29-5–29-6
Multiple uniform servers, 22-19–22-20
Multiple-writer protocol, 42-8
Multiport communication mode, 36-3
Multiprocessor (crew) parallelism, 31-3
Multiprocessor real-time scheduling, 40-2, *see also* Power aware mapping
Multiprocessor scheduling algorithms
  classes, 24-15
    global dynamic-priority scheduling, 24-16

Index

global static-priority scheduling, **24**-15–**24**-16
partitioned dynamic-priority scheduling, **24**-16
partitioned static-priority scheduling, **24**-16

Multiprocessor systems
real-time scheduling algorithms for, **24**-1–**24**-23, *see also* Real-time scheduling algorithms

Multistage interconnection network (MIN), **4**-5

Multithreaded algorithms design for combinatorial problems, **31**-1–**31**-24
for breadth-first search, **31**-9–**31**-10
assert parallel directive, **31**-9
loop future directive, **31**-9
Cray MTA-2, **31**-2–**31**-3
graph theoretic problems, **31**-2
graph traversal, **31**-8–**31**-16, *see also* Graph traversal
list ranking, **31**-4–**31**-8, *see also* List ranking
for MTA-2, **31**-3–**31**-4
complexity, analyzing, **31**-4
expressing parallelism, **31**-3–**31**-4
synchronization, **31**-4
in social network analysis, **31**-16–**31**-24, *see also* Social network analysis

Multi-threaded execution, **7**-*10*
Multithreading, **7**-2
in DLA software library, **29**-14–**29**-16
Multivalued functions in k-wise independence method, **18**-16–**18**-18

Mutual exclusion protocols
parallel algorithms with, **26**-30–**26**-38
compact-graph step, **26**-32
find-min step, **26**-33
fine-grained mutual exclusion locks, **26**-32–**26**-33

N

Naming technique, strings, **30**-9–**30**-12
Nash equilibrium, **20**-2, **20**-4
flows at, **20**-22
PNE construction, **20**-10–**20**-16
unweighted congestion games, **20**-7–**20**-10
weighted congestion games, **20**-13–**20**-16
PNE, existence and tractability of, **20**-5–**20**-6

Natural computing, **3**-2
N-body problem, **44**-19–**44**-22
Neighbor localization, **14**-6
Nested parallelism, **5**-8
Nested spawn BFS, **5**-27–**5**-29
NetCDF, **35**-3–**35**-4
NetSolve, **21**-11
Network based algorithmic model, parallel graph algorithms, **26**-30
Network congestion games, **20**-1
Network decontamination, **8**-15–**8**-16
Networks of workstations (NOWs), **19**-1–**19**-29, *see also* Broadcasting
Never-update strategy, **33**-6
Nimrod/G, **21**-12
No-busy-wait prefix sums, **5**-25
analysis, **5**-25–**5**-26
k-ary prefix sums, **5**-46–**5**-49
Node view, **12**-4
Node-Weighted Steiner Forest (NWSF), **39**-2, **39**-12
Node-Weighted Steiner Tree (NWST), **39**-3, **39**-6, **39**-7
Noncontiguous I/O methods for parallel file systems, **35**-17–**35**-22
datatype I/O, **35**-18
I/O suggestions for application developers, **35**-18–**35**-22
list I/O, **35**-17–**35**-18
possible I/O improvements, **35**-20–**35**-22
Nonmaximal parallelism, **3**-15–**3**-17
bounded parallelism, **3**-15–**3**-16
minimal parallelism, **3**-16
Nonoblivious search, **8**-11
Nonpartitioned or global scheduling, **24**-6
Nonpreemptive scheduling, **24**-2
NP-completeness, **36**-15
of MM_c, **36**-9–**36**-10
Null, **15**-1

O

Object transfer cost (OTC), **45**-5
Oblivious routing algorithms/scheme, **37**-10–**37**-14
simple, **37**-11–**37**-14
Octrees
compressed octrees, **44**-13–**44**-19
bit interleaving scheme for, **44**-*17*
constructing, **44**-16–**44**-19
data ownership, **44**-19
development, **44**-14
quality of partitioning, **44**-19
residual leaf node, **44**-21

space filling curves and, **44**-15–**44**-16
Offline communication scheduling, **36**-2
Offline multimessage multicasting problem, **36**-4
Online call admission control in wireless cellular networks, **38**-1–**38**-19
algorithms, **38**-3–**38**-5
p-random algorithms, **38**-5–**38**-11
classify and randomly select-based algorithms, **38**-11–**38**-15
extensions and open problems, **38**-18–**38**-19
interference graph G, **38**-2
lower bound, **38**-15–**38**-18
Online computation in large networks, **37**-1–**37**-21
caching problems, **37**-17–**37**-21
multicast acknowledgment, **37**-15–**37**-17
online routing, **37**-9–**37**-14, *see also* Online routing
packet buffering, **37**-2–**37**-9, *see also* Packet buffering
transmission control protocol acknowledgment, **37**-14–**37**-15
Online routing, **37**-9–**37**-14
model and known results, **37**-9–**37**-11
oblivious routing algorithms/scheme, **37**-10–**37**-14
Open problems, probabilistic method, **18**-53–**18**-54
parallel computation of low discrepancy colorings, **18**-53
parallelizing the random hyperplane method, **18**-53–**18**-54
practicability, **18**-54
sum of dependent random variables, **18**-53
Open System Interconnection (OSI) reference model, **33**-22
Optical buses, **16**-1–**16**-18
LARPBS, **16**-5–**16**-7
PR-Mesh, **16**-8–**16**-10
Optical networks, multimessage multicasting in, **36**-19–**36**-20
receiving fibers, **36**-20
sending fibers, **36**-20
Optical transpose interconnection system (OTIS)
algorithms, **4**-12–**4**-14
OTIS-hypercube, **4**-10–**4**-12

Optical transpose interconnection system (OTIS) (continued)
  diameter of, 4-10–4-11
  N^2-processor simulation, 4-11–4-12
  OTIS-Mesh, 4-7–4-10
    2D mesh simulation, 4-9–4-10
    4D mesh simulation, 4-8–4-9
    diameter of, 4-7–4-8
  parallel computers, 4-3–4-4
  permutation routing, 4-4–4-7
  routing, 4-14–4-15
Optimal algorithms, 30-2
Optimal parallel scheduling algorithms in WDM packet interconnects, 23-1–23-18
  formalization, 23-4–23-5
  implementation issues and complexity analysis, 23-15–23-17
  matching buffer vertices (phase two), 23-9–23-15, see also Buffer vertices
  matching output vertices (phase one), 23-6–23-9
  parallel segment expanding algorithm, 23-5–23-6, 23-16
  performance evaluation, 23-17–23-18
  wavelength conversion, 23-2–23-4
OptorSim, 21-15
ORB-Median of Medians (ORBMM), 44-8
Orthogonal recursive bisection (ORB), 44-6–44-9
  construction, 44-6–44-8
  k-d tree, 44-7
  data ownership, 44-8
  quality of partitioning, 44-8–44-9
Overheads, algorithm to reduce, 26-25

P

P systems
  basic model of, 3-5–3-7
  determinism vs. nondeterminism, 3-17–3-18
  essential ingredients, 3-2–3-3
  other classes, 3-18–3-19
  with proteins on membranes, 3-16–3-17
  with string objects, 3-9–3-10
Packet buffering, 37-2–37-9
  greedy algorithm, 37-3–37-6
  input and output queuing switches combination, managing, 37-8–37-9
  multibuffer problems with arbitrary packet values, 37-7–37-8
  with uniform packet values, 37-4–37-7
  single buffer problems, 37-3–37-4
Packet redirection, of MANETs, 33-24–33-25
Packet routing, 13-4–13-7
Packing integer programs (PIPs) parallel algorithms for, 18-44–18-53
  k-wise independence impact, 18-44–18-47, see also Multidimensional bin packing
Panasas ActiveScale File System (PanFS), 35-16
Parallel algorithms, 18-1–18-54
  for maximal independent set and maximal matching, 25-1–25-18, see also Maximal independent set
  scalability of, 47-5–47-23, see also Scalability
  in social network analysis, 31-19–31-22, see also under Social network analysis
  on strings, 30-1–30-21, see also Strings
  for volumetric surface construction, 27-1–27-15, see also Volumetric surface construction
  via probabilistic method, 18-1–18-54, see also Probabilistic method
Parallel association rule mining, see Association rule mining (ARM), parallel
Parallel betweenness centrality for unweighted graphs, 31-21
Parallel computation, 1-3–1-4
  interconnection network, 1-3
Parallel computing community challenge to, 43-4
Parallel data flow graph, 7-7
Parallel disk models, 6-1–6-16
  (l,m)-Merge Sort (LMM), 6-4–6-8
    merging algorithm, 6-4–6-6
    optimal integer sorting, 6-14–6-15
    optimal randomized sorting algorithm, 6-15–6-16
    practical realization of, 6-8–6-9
    sorting $M\sqrt{M}$ keys lower bound, 6-9
    three-pass algorithm, 6-9–6-10
  sorting results, 6-2–6-4
Parallel I/O
  high-performance techniques for, 35-1–35-22, see also High-performance techniques
Parallel isosurface extraction
  factors affecting, 27-8
  issues, 27-8
  single pass occlusion culling algorithm with random data partitioning, 27-14
  strategies for, 27-7–27-8
    distributed memory model, 27-8
    processor and main memory interconnection, 27-8
    shared-memory model, 27-8
Parallel matrix multiplication, 47-1–47-23
Parallel model of computation, 28-3–28-5
  reconfigurable bus system, 28-3
Parallel NetCDF (PnetCDF), 35-3–35-4
Parallel performance, metrics, 43-3–43-4
Parallel programs scalability, see Scalability
Parallel random access machine (PRAM) algorithms, 2-9, 16-17–16-18, 26-2, 26-4
Parallel segment expanding algorithm, 23-5–23-6
Parallel server models, 22-17–22-20
  fair lexicographic scheduling, 22-18–22-19
  multiple uniform servers, 22-19–22-20
Parallel solution
  to mathematical constraints, 1-19
  to time-varying variables of computational systems, 1-6
Parallel string matching, 30-2–30-9, see also under Strings
Parallel virtual file system 1 (PVFS1), 35-17
Parallelism
  levels of exploitation, 31-21
Parallelization schemes
  types, 31-3
    future parallelism, 31-3
    multiprocessor (crew) parallelism, 31-3
    single-processor (fray) parallelism, 31-3
Pareto optimality, 45-11
PARFOR loop, 45-10
Partition scheduling, in power aware mapping, 40-2
Partitioned dynamic-priority scheduling, 24-16
Partitioned optical passive star (POPS) network, 13-1–13-13
  integer sorting, 13-9–13-11
  packet routing, 13-4–13-7

preliminaries, 13-3–13-4
selection, 13-7–13-9
sparse enumeration sorting, 13-12–13-13
Partitioned scheduling, 24-6
for identical multiprocessor systems, 24-3
on multiprocessors, 24-7–24-11
advantage, 24-8
global scheduling versus, 24-8
scheduling algorithm, 24-7
task partitioning, 24-7
using earliest deadline first, 24-11
using rate monotonic, 24-8
Partitioned static-priority scheduling, 24-15–24-16
Paths
configuration paths and discrete dynamics graph in selfish routing, 20-5
best response dynamics graph, 20-5
best-reply improvement path, 20-5
closed path, 20-5
improvement path, 20-5
Nash dynamics graph, 20-5
simple path, 20-5
Pattern-growth method, 32-6–32-8
FP-growth algorithm, 32-6–32-7
local FP-trees construction, 32-7
PD-CLUB bitmap-based algorithm, 32-8–32-10
Pebbles, 8-8
Perfect skip list, 17-7, 17-8
Periodic task, 24-2
Permutation routing, 13-4, 14-5–14-6
on OTIS computers, 4-4–4-7
on an R-Mesh, 16-4
Perpetual exploration, 8-8
Persistent file realm (PFR), 35-14
Personal communication services (PCS), 33-10
Pfair scheduling, 24-16
Pipeline buffers, 34-5
Pipelined optical bus (POB), 16-1
structure, 16-11
Pipelined reconfigurable mesh (PR-Mesh), 16-1, 16-8–16-10
switch connections, 16-9
switch settings, 16-9, 16-10
Pipeline-structured algorithms, 34-2–34-4
FG overview, 34-4–34-5
for one pass of an out-of-core permutation on two processors, 34-3
operations of, overlapping, 34-3
pMAFIA algorithm, 32-17–32-18
Pointers, 2
Polymorphic torus, 14-2

Polynomial time, 3-12–3-15
approximation scheme, 19-26–19-29
Portable Batch System (PBS), 21-9
Portable file formats
and data libraries, 35-1–35-6
in parallel applications, 35-1–35-6
Portable file locking, 35-14–35-15
POSIX I/O interface, 35-9, 35-10
Postal model, 19-5–19-18
interleaved LCF, 19-18
largest cluster first (LCF) analysis, 19-16–19-18
Power aware mapping of real-time tasks to multiprocessors, 40-1–40-19
dynamic slack, 40-2, 40-8
fixed priority slack sharing for dependent tasks, 40-12–40-13, see also List scheduling
frame-based RT-applications and canonical schedule, 40-4–40-8
global scheduling, 40-2
partition scheduling, 40-2
practical considerations, 40-16
scheduling for parallel systems, 40-3
with slack sharing, 40-8–40-11
greedy slack reclamation, 40-9
slack sharing among processors, 40-9–40-11
slack reservation for speed adjustment overhead, 40-17–40-18
slack shifting for applications, 40-13–40-16
AND/OR application model, 40-13–40-14
shifting schedule sections in canonical schedule, 40-14–40-15
speculation for better performance, 40-15–40-16
static slack, 40-2
task scheduling for parallel systems, 40-5–40-6
variable speed processors and power management, 40-4
Power-aware routing, 33-23–33-24
Power-law model, 45-13
PRAM-on-chip platform
adaptive bitonic sorting, 5-32–5-34
BFS, see Breadth-first search (BFS)
compiler optimizations, 5-20–5-23
clustering, 5-21–5-22
nested parallel sections, 5-20–5-21

prefetching, 5-23
empirical validation, 5-14–5-20
performance comparison, 5-16–5-18
scalability, 5-19
model descriptions, 5-4–5-13
execution model, 5-8
high-level work-depth description, 5-6
PRAM model, 5-5
PRAM-on-chip programming model, 5-7–5-8
work-depth model, 5-6, 5-7
XMTC code, 5-8
prefix sums, 5-23–5-27
clustering, 5-26–5-27
no-busy-wait, 5-25–5-26
synchronous, 5-23–5-25
prefix-sums algorithms, comparing, 5-27
primitives
nested parallelism, 5-8
prefix-sum instruction, 5-7–5-8
spawn instruction, 5-7
thread-ID, 5-7
shared memory sample sort, 5-34–5-35
helper matrix, 5-35
sparse matrix, 5-35–5-36
p-Random algorithms, 38-5–38-11
Pr-Dynamic network, 36-4
Preemptive scheduling, 24-2, 24-3
Prefix remainders, 15-5–15-6
Prefix sums, 5-23–5-27
clustering, 5-26–5-27
instruction, 5-7–5-8
no-busy-wait, 5-25–5-26
synchronous, 5-23–5-25
Preliminaries, 18-1–18-6, 31-17
binary representations, 18-2–18-3
computational model, 18-6
graphs and hypergraphs, 18-3
logarithmic terms, 18-3
probabilistic tools, 18-3–18-6
vectors and matrices, 18-2
Price of anarchy of weighted congestion games, 20-20–20-27
bounding the price of anarchy, 20-25–20-27
flows and mixed strategies profiles, 20-21–20-22, see also Feasible flow
at Nash equilibrium, 20-22
maximum latency versus total latency, 20-23–20-24
upper bound on social cost, 20-24–20-25
Price of anarchy, 20-1
Price of stability, 20-1
Prim's algorithm, 26-17, 26-20
Prime numbers selection, 15-6

Primitive assessment table (PAT), **12**-14
Primitive operations, message dissemination using, **36**-2
Priori-based algorithms, association rule mining, **32**-2–**32**-4
  candidate distribution methods, **32**-2–**32**-4
  count distribution, **32**-2
  data distribution, **32**-2
Priority inversion, **26**-5
Probabilistic method, parallel algorithms via, **18**-1–**18**-54
  fooling automata, **18**-34–**18**-44, *see also* Fooling automata
  k-wise independence method, **18**-14–**18**-34, *see also* *k*-wise independence method
  limited independence, **18**-6–**18**-14, *see also* limited independence
  open problems, **18**-53–**18**-54, *see also* Open problems
  for packing integer programs, **18**-44–**18**-53, *see also* Packing integer programs
  preliminaries, **18**-1–**18**-6, *see also* Preliminaries
  tools, **18**-3–**18**-6
Processor array with a reconfigurable bus system (PARBS), **14**-2
Program count (PC), **7**-3
Proportionate-fair (Pfair) scheduling, **24**-3, **24**-6, **24**-15–**24**-18
Pseudodeadlines, **24**-6
Pure Price of Anarchy (PPoA), **20**-27
Pure random model, **45**-12

Q

QoS (Quality-of-Service) scheduling, in network and storage systems, **22**-1–**22**-21
  discrete server model, **22**-8–**22**-17, *see also* Discrete server model
  fluid flow server model, **22**-4–**22**-8
  parallel server models, **22**-17–**22**-20, *see also* Parallel server models
  queuing model, **22**-3–**22**-4, *see also* Queuing model
Quadtree, **28**-15, **44**-13
Quantum computing, distinguishability in, **1**-16–**1**-17
Quantum decoherence, **1**-5
Queuing delay, **5**-9

Queuing model, in QoS scheduling, **22**-3–**22**-4
  active flow, **22**-3
  backlogged flow, **22**-3
  fair queuing model, **22**-4
  flow queue, **22**-3
Q-Win strategy, **33**-12

R

Rainbow skip graph, **17**-13
Random access (RA) protocols, **33**-13
Random graphs, **8**-2, **31**-13, **45**-12
Range partitioning-based algorithms, in volumetric surface construction, **27**-9
Rank-varying computational complexity, **1**-9–**1**-12
  binary search, **1**-10
  inverse quantum Fourier transform, **1**-10–**1**-12
Rate monotonic (RM) algorithm, **24**-4
  partitioned scheduling using, **24**-8–**24**-10
Rate-based execution (RBE) model, **24**-22
Ray tracing, **27**-11
  distributed ray tracing, **27**-15
  view-dependent isosurface extraction by, **27**-15
Reachable set, parallel, expansion, **23**-13–**23**-14
  details, **23**-14
  proof for optimality of, **23**-14–**23**-15
Real-time scheduling algorithms for multiprocessor systems, **24**-1–**24**-23
  admission control, **24**-23
  CPU Reservations, **24**-22
  generalized multiframe task model, **24**-22
  global scheduling on multiprocessors, **24**-12–**24**-15, *see also* Global scheduling on multiprocessors
  identical multiprocessor systems, **24**-23
  implicit-deadline synchronous periodic task systems, **24**-3
  multiframe task model, **24**-22
  multiprocessor scheduling algorithms, **24**-6–**24**-7
  online multiprocessor scheduling, **24**-18–**24**-22
  partitioned scheduling, **24**-7–**24**-11, *see also* Partitioned scheduling

  proportionate-fair scheduling, **24**-15–**24**-18
  rate-based execution (RBE) model, **24**-22
  resource augmentation, **24**-22
  scope and map, **24**-3–**24**-4
  uniform multiprocessor model, **24**-23
  uniprocessor real-time scheduling, **24**-4–**24**-6, *see also* Uniprocessor real-time scheduling
  unrelated multiprocessor model, **24**-23
Real-time services, probabilistic resource reservation strategy for, **33**-12
Real-time tasks, **24**-2
  to multiprocessors, power aware mapping, **40**-1–**40**-19
Receiving buffers, in multimessage multicasting, **36**-11–**36**-13
Reconfigurable application specific computing technology (RASC), **11**-5
Reconfigurable array with spanning optical buses (RASOB), **16**-1
Reconfigurable bus system, **28**-3
Reconfigurable computers/computing
  element-level architectures, **11**-6–**11**-7
  fundamental algorithms
    binary prefix sums, **16**-14–**16**-15
    compression, **16**-15–**16**-16
    PRAM simulations, **16**-17–**16**-18
    sorting and selection algorithms, **16**-16–**16**-17
  massively parallel FPGA array, **11**-5–**11**-6
  with optical buses, **16**-1–**16**-18
    addressing techniques, **16**-7–**16**-8
    LARPBS, **16**-5–**16**-7
    other optical models, **16**-10–**16**-13
    PR-Mesh, **16**-8–**16**-10
    and pipelining, **16**-2–**16**-4
    relating optical models, **16**-13–**16**-14
  performance modeling, **11**-9–**11**-13
    GPP elements, **11**-10–**11**-11
    node-level, **11**-12–**11**-13
    RH element, **11**-11–**11**-12
  related work, **11**-18–**11**-19
  survey of, **11**-4–**11**-5
  system- and node-level architectures, **11**-7–**11**-9
    abstract model, **11**-8
    cluster architecture, **11**-8

node-level programming
model, **11**-9
shared-memory architecture,
11-7–**11**-8
system-level computing model,
11-8–**11**-9
Reconfigurable hardware,
11-2–**11**-3
abstract architecture, **11**-7
Reconfigurable Mesh (R-Mesh),
14-2–**14**-25
basic algorithmic, **14**-5–**14**-14
neighbor localization, **14**-6
partitioning R-Meshes,
14-6–**14**-7
permutation routing,
14-5–**14**-6
prefix sums, **14**-7–**14**-8
priority simulation,
14-8–**14**-11
sorting, **14**-11–**14**-14
computational complexity and
reconfiguration,
14-20–**14**-25
connection patterns, **15**-1–**15**-2
directed R-Mesh, **14**-5
multidimensional R-Meshes,
14-5
fundamental algorithms,
15-1–**15**-7
compression, **15**-4
integer summing, **15**-6–**15**-7
leftmost finding, **15**-3
logarithmic-time prefix sums,
15-6
prefix remainders, **15**-5–**15**-6
prime numbers selection, **15**-6
simple prefix sums, **15**-4
graph algorithms, **14**-14–**14**-20
connectivity, **14**-16–**14**-17
list ranking, **14**-18–**14**-20
minimum spanning tree,
14-17–**14**-18
mixed radix representation
technique, **15**-9–**15**-10
parallel prefix-remainders
technique, **15**-7–**15**-9
relations to
circuits, **14**-22
PRAMS, **14**-20–**14**-21
simulations with TM,
14-24–**14**-25
snake-like embedding technique,
15-11–**15**-13
summing integers algorithm,
15-13–**15**-15
variants based on
bus access, **14**-4
bus configuration, **14**-3–**14**-4
processor wordsize, **14**-4
Recursive bilinear algorithm, **47**-*8*
Reference frames, **16**-*7*
Region growing, **28**-15
Release consistency, **7**-11

Reliable neighbor transmission,
10-7-**10**-8
Rendezvous, **8**-4
asymmetric rendezvous, **8**-4, *see*
Asymmetric rendezvous
rendezvous search game,
8-14-**8**-15
symmetric rendezvous, *see*
Symmetric rendezvous
Replication algorithms, **45**-3
Reporting cells, **33**-6-**33**-7
Residual leaf node, **44**-21
Resource capacity chart, **46**-14
Resource management and call
admission control
algorithm, **33**-17-**33**-21
Resource management strategy,
mobile computing
algorithmics, **33**-17-**33**-18
nonreal-time (Class 2)
connections, **33**-17
real-time (Class 1), **33**-17
resource reservation strategy,
33-*18*
Restricted-domain searching
scalable algorithms for, **9**-11-**9**-13
Rewriting systems, to mathematical
constraints, **1**-18-**1**-19
Ring network, **8**-9
Robust resource allocation
for heterogeneous parallel and
distributed systems,
41-1-**41**-28
computational complexity,
41-13
generalized robustness metric,
41-3-**41**-6, *see also*
Generalized robustness
metric
related work, **41**-25-**41**-27
robustness metric for example
systems, **41**-6-**41**-12,
see also Robustness
metric
simulation experiments and
results, **41**-24-**41**-25
Robustness
against multiple perturbation
parameters, **41**-12-**41**-13
claim of, **41**-2
meaning, **41**-2
normalized, **41**-14, **41**-*15*
radius, **41**-5, **41**-13
Robustness metric, **41**-6-**41**-12
derivations, **41**-6-**41**-12
HiPer-D System, **41**-8-**41**-10
independent application
allocation system,
41-6-**41**-8
for machine failures requiring
reallocation
41-11-**41**-12
utility, demonstrating,
41-14-**41**-18

HiPer-D System, **41**-16-**41**-18
independent application
allocation system,
41-14-**41**-16
ROMIO implementation of
MPI–IO, **35**-9-**35**-14
common ROMIO hints,
35-12-**35**-14
data sieving, **35**-9-**35**-11
POSIX I/O interface, **35**-9
two-phase I/O, **35**-11-**35**-12
Root cell, **44**-14
Root-crossing, **39**-13
Round-robin order replication
(RR), **21**-12

S

Scalability of parallel programs,
43-1–**43**-15
analytical methods, **43**-2
analytical modeling, limitations,
43-14–**43**-15
characteristics, **43**-5–**43**-6
experimental approaches, **43**-2
heterogeneous composition of
applications, **43**-14
isoefficiency metric of scalability,
43-6–**43**-10, *see also*
Isoefficiency metric
metrics for, **43**-4–**43**-14
scaled speedup notions, **43**-12
metrics of parallel performance,
43-3–**43**-4
efficiency, **43**-3
parallel benchmarks and, **43**-11
scaled speedup, **43**-6
Scalable algorithms
constrained triangulation,
9-16–**9**-17
list ranking, **9**-13
restricted-domain searching,
9-11–**9**-13
Scalable matrix multiplication,
47-14–**47**-17
conditions for, **47**-16–**47**-17
results, **47**-15–**47**-16
Scalable visibility algorithms,
9-14–**9**-16
ScaLAPACK library, **29**-2–**29**-9
core of, **29**-12
data types supported in, **29**-7
drivers, **29**-12
F95 versions of, **29**-12
routing categories, **29**-3
Sca/LAPACK software
engineering (SWE)
approach, **29**-11–**29**-14
development plan, **29**-12
research plan, **29**-12, **29**-13
wrappers for, **29**-12
Scale-free graphs (SF-RMAT),
31-13

Scenario Builder, **46**-7–**46**-17
  bandwidth vs. transfer load, **46**-15
  scenario anatomy, **46**-9
  scenario generation, **46**-11
  scenario simulation, **46**-11–**46**-13
  scenario visualization, **46**-13–**46**-15
  virtual network topology, **46**-10-**46**-11
Scheduling in grid environments, **21**-1–**21**-16
  grid simulation tools, **21**-13–**21**-15
    bricks, **21**-14
    GridSim, **21**-14
    MicroGrid, **21**-14–**21**-15
    OptorSim, **21**-15
    SimGrid, **21**-15
  models, **21**-4–**21**-6
    application model, **21**-4–**21**-5
    grid model, **21**-5
    grid scheduler model, **21**-5–**21**-6
  scheduling problem, **21**-2–**21**-4
    grid scheduling, **21**-3–**21**-4, *see also* Grid scheduling
    traditional multiprocessor scheduling, **21**-2
Scientific computing
  spatial domain decomposition methods in **44**-1–**44**-22, *see also* Spatial domain decomposition methods
Search games, **8**-12
Searching phase, of parallel algorithms on strings, **30**-7–**30**-9
Segment switch, **16**-6
Select frames, **16**-7
Selection, **13**-7–**13**-9
Selective look-ahead (SILK) strategy, **33**-12
Self-clocked weighted fair queuing (S-WFQ), **22**-14
Selfish routing in networks, **20**-1–**20**-28, *see also* Atomic selfish routing in networks
  configuration paths and discrete dynamics graph, **20**-5, *see also* Paths
  layered networks, **20**-5
  model, **20**-2–**20**-5
    congestion model, **20**-3
    network congestion game, **20**-4
    symmetric congestion game, **20**-4, *see also* Symmetric congestion game
    weighted congestion model, **20**-3
  parallel links and player-specific payoffs, **20**-16–**20**-20

best-reply improvement paths, **20**-18
  players with distinct weights, **20**-20
  potential games, **20**-4–**20**-5
    b-potential, **20**-5
    exact potential, **20**-5
    ordinal potential, **20**-4
  pure Nash equilibrium (PNE) existence and tractability of, **20**-5–**20**-6, *see also* Nash equilibrium
  selfish behavior, dealing with, **20**-4–**20**-5
  strategic games, isomorphism of, **20**-5
Sequential algorithm for structured grids, **27**-12–**27**-13
  framework of, **27**-13
  mapping challenges, **27**-12
Sequential computer, **1**-3
Sequential hierarchies
  block transfer, **2**-14
  HMM, **2**-14
Sequential Probabilistic resource Reservation (SPR), **33**-17
Sequential solution, **1**-5–**1**-6
  to mathematical constraints, **1**-19
Serial prefix sums, **5**-50–**5**-51
Serial summation, **5**-50
Shared memory management, **42**-7–**42**-9
  JavaSplit, **42**-7–**42**-9
Shared object repository (SOR), **42**-12
Shared-memory architecture, **11**-7–**11**-8
Shared-memory model, in parallel isosurface extraction, **27**-8
Shifting schedule sections in canonical schedule, **40**-14–**40**-15
Shiloach–Vishkin parallel spanning tree algorithm, **26**-31, **26**-35
Shortest distance model, **33**-8
Silver, **21**-11
SimGrid, **21**-15
Simple prefix sums, **15**-4
SimpleScalar-PISA model, **7**-11
Single buffer problems, **37**-3–**37**-4
Single pass occlusion culling algorithm with random data partitioning, **27**-14
Single port communication mode, **36**-3
Single program multiple data (SPMD), **9**-6
Single shortest path (SSP), **8**-9
Single type faults, **10**-5–**10**-6
  additions, **10**-6
  corruptions, **10**-5–**10**-6
  omissions, **10**-6
Single-item broadcasting, **19**-3

Single-processor (fray) parallelism, **31**-3
Single-spawn BFS, **5**-29, **5**-55–**5**-57
Single-threaded execution, **7**-10
Six-standard-deviation theorem, **18**-22
Skip graphs, **17**-7–**17**-15
  achieving $O(|K|)$ space, **17**-12–**17**-13
  congestion, **17**-11–**17**-12
  fault tolerance, **17**-13–**17**-14
  hydra components, **17**-14–**17**-15
  inserting and deleting concurrently, **17**-11
  skip lists, **17**-7–**17**-9
  structure, **17**-9–**17**-11
Slack(s), **41**-16, **41**-26
  HiPer-D System, **41**-16–**41**-17
  reservation for speed adjustment overhead, **40**-17–**40**-18
  slack factor, **24**-12
  slack shifting applications, **40**-13–**40**-16
Slack sharing, power aware mapping with, **40**-8–**40**-11
  among processors, **40**-9–**40**-11
  fixed priority list scheduling with, **40**-13
  fixed priority slack sharing for dependent tasks, **40**-12–**40**-13
  greedy slack reclamation, **40**-9
  mapping algorithm, **40**-11
Sleep-waiting, **5**-21
Sliding window concept, **33**-12
Slowdown theorem, **1**-20
Small cell, **44**-15
Small sample spaces, **18**-7–**18**-11
Snake-like embedding technique, **15**-11–**15**-13
Social network analysis algorithms, **31**-16–**31**-24
  centrality metrics, **31**-7–**31**-18, *see also* Centrality metrics
  computing betweenness centrality, algorithms for, **31**-19
  straightforward way, **31**-19
  parallel algorithms, **31**-19–**31**-22
  closeness centrality, **31**-20
  degree centrality, **31**-20
  performance results, **31**-22–**31**-24
  test dataset characteristics, **31**-22
  stress and betweenness centrality, **31**-20–**31**-22
Soft barriers, **34**-9
Software, in DLA software library, **29**-10–**29**-16, *see also* Dense linear algebra: software
Solutions to interacting variables, in computational systems, **1**-14–**1**-16
  mathematical model, **1**-15–**1**-16

Index

parallel approach, 1-16
sequential approach, 1-16
simplifying assumptions, 1-15
Sorting
  coarse-grain multicomputer, 9-10
Sorting $\sqrt{M}$ keys, 6-9–6-10
  lower bound, 6-9
  three-pass algorithm, 6-9–6-10
Space filling curves (SFC), 44-9–44-13
  construction, 44-11
  data ownership, 44-11–44-12
  octrees and, 44-15–44-16
  quality of partitioning, 44-12–44-13
  range queries, 44-12
Spanning tree algorithm, fast, parallel, 26-6–26-13, 26-24, *see also* Minimum spanning tree
  experimental results, 26-10–26-11
  spanning tree results for multicare and multiprocessors, 26-35
  stub spanning tree, 26-6
  symmetric multiprocessor spanning tree algorithms, analysis, 26-9–26-10
  work-stealing graph traversal, 26-7
Sparse enumeration sorting, 13-12–13-13
Sparse matrix, 5-35–5-36
Spatial domain decomposition methods
  analyzing, 44-4–44-6
    computation of data ownership, 44-5
    orthogonal recursive bisection (ORB), 44-6–44-9
    quality of partitioning, 44-5–44-6
    runtime of, 44-4–44-5
  N-body problem, 44-19–44-22
  octrees and compressed octrees, 44-13–44-19, *see also* Octrees
  in parallel scientific computing, 44-1–44-22
    balanced graph partitioning, 44-2
  space filling curves, 44-9–44-13, *see also* Space filling curves
  spatial locality-based parallel domain decomposition, 44-3–44-6
    dynamic applications, 44-3
    motivating applications, 44-3–44-4
    particle-based methods applications, 44-4
    static applications, 44-3

Spawn instruction, 5-7
Speedup theorem, 1-20
Split and merge, 28-15
Splitting message destinations, 36-3
Sporadic task, 24-2
Spotbeams, 33-10
Srinivasan's parallel packing, 18-47–18-53
  derandomization, 18-49–18-53
  idea, 18-48–18-49
$S - t$ connectivity algorithm, 14-15
  and reachability, 14-15–14-16
Start Time Weighted Fair Queuing (S-WFQ), 22-12–22-13
Static power management (SPM), 40-2, 40-3
  canonical schedule and, 40-6–40-8
Static slack, 40-2
Static space multiplexing limitations, 22-2
Static-priority scheduling, 24-3–24-6
St-connectivity and shortest paths, 31-11–31-13, 31-16
Steiner forest (SF) problem, 39-3
Storage affinity (SA) algorithm, 21-13
Strategic games, isomorphism of, 20-5, 20-6
Stress and betweenness centrality, 31-20–31-22
Stress centrality, 31-18
Strings, parallel algorithms on, 30-1–30-21
  abaababaabaababaababab, 30-6
  naming technique, 30-9–30-12
    dictionary of basic subwords, 30-9
    Karp–Miller–Rosenberg naming technique, 30-21
    subwords, 30-9
  nonperiodic patterns, witnesses and duels, 30-4–30-5
  parallel construction of suffix arrays, 30-12–30-13, *see also* Suffix arrays
  parallel string matching, 30-2–30-9
    periodic case reduction to aperiodic, 30-2–30-3
    strongly aperiodic case, simple algorithm for, 30-3
    preprocessing the pattern, 30-5–30-7
    searching phase, 30-7–30-9
  suffix trees parallel construction by refining, 30-19–30-21
    local refining, 30-19
Stub spanning tree, 26-6

Subcell, 44-14
Subnets, *see* Clusters of workstations
Suffix arrays, strings
  parallel construction, 30-12–30-13
  skew algorithm, 30-13
  transformed into suffix trees, 30-13–30-18
    BUILD1 algorithm, 30-14–30-16
    BUILD2 algorithm, 30-16–30-18
Summation, logP model broadcasting, 19-4–19-5
Summing integers algorithm, 15-13–15-15
  two-stage summing method, 15-13
Supercell, 44-15
Supernode skip graph, 17-12
Switches, 37-2
  input-queued (IQ) switches, 37-2, 37-3
  output-queued (OQ) switches, 37-2
Symmetric multiprocessor (SMP) system, 11-6, 26-2, 26-5–26-7, 27-8
  architectures, 26-2
  features, 26-3
  spanning tree algorithms, analysis, 26-9–26-10
  TV adaptation on, 26-24
Symmetric rendezvous, 8-5
  leader election, 8-6
  randomized rendezvous, 8-5–8-6
  using tokens, 8-6
Symmetric wireless networks, 39-4–39-7
  multicasting and broadcasting, 39-6–39-7
  symmetric connectivity requirements, 39-4–39-6
Symport/antiport system, 3-17
Synchronization, 31-4
  among processors, 26-4–26-5
  synchronous based algorithmic model, parallel graph algorithms, 26-30
  synchronous k-ary prefix sums, 5-42–5-46
  synchronous prefix sums, 5-23
    analysis, 5-23–5-25
  synchronous task system, 24-2, 24-4
Synthetic sparse random graphs, 31-14
System-level computing model, 11-8–11-9

T

Tarjan–Vishkin (TV) biconnected components algorithm, **26**-22
  Euler-tour, **26**-23
    label-edge, **26**-23
    low-high, **26**-23
    spanning-tree, **26**-23
  implementation, **26**-24–**26**-25
    steps, **26**-23
      connected-components, **26**-23
  TV-filter, **26**-29
  TV-opt, **26**-29
  TV-SMP, **26**-24, **26**-29
Task scheduling for parallel systems, **40**-5–**40**-6
Task's period, **24**-2
Task-level parallelism, **11**-9
  constraining factors, **11**-11
Thread clustering, **5**-22
Thread control unit (TCU), **7**-8
Thread starting, **7**-11
Thread-level parallelism (TLP), **26**-3
Three-pass algorithm, **6**-9–**6**-10
Thumbnail, **34**-5
Time slice, **10**-7
Time unit, computational models, **1**-2–**1**-3
  definition, **1**-2
Time-cost expression deriver (TED), **12**-14
Time-Division Multiplexing (TDM), **22**-2
Time-varying computational complexity, **1**-6–**1**-9
  accelerating machines, **1**-9
  computing with deadlines, **1**-8–**1**-9
  increasing functions C(t), **1**-7–**1**-8
Time-varying variables, in computational systems
  evolution, **1**-4–**1**-6
  parallel solution, **1**-6
  quantum decoherence, **1**-5
  sequential solution, **1**-5–**1**-6
Tori, **47**-5
Traditional multiprocessor scheduling, **21**-2
Transaction Instance Report (TIR), **33**-28
Transmission Control Protocol (TCP), **37**-2
  and multicast acknowledgment, **37**-14–**37**-17
Transmit-largest (TL) algorithm, **37**-7
Transparent distributed runtime for java, **42**-1–**42**-21
  fault tolerance, **42**-10–**42**-15, see also JavaSplit-FT
  JavaSplit, **42**-3–**42**-10
  performance evaluation, **42**-15–**42**-17
  related work, **42**-17–**42**-21
  system model, **42**-3
Tree-based algorithms, in volumetric surface construction, **27**-9
Turing machine (TM), **3**-12, **14**-23–**14**-25
  simulations with, **14**-24–**14**-25
Twinning and diffing technique, **42**-8
2D mesh simulation, **4**-9–**4**-10
Two-dimensional mobility model, **33**-14–**33**-17
Two-phase greedy (TPG) heuristic, **41**-23
  TPG-X, **41**-23

U

Ultra fast computer vision
  mesh-based parallel algorithms for, **28**-1–**28**-18, see also Mesh-based parallel algorithms
Unbalanced recursive bisection (URB), **44**-8
Uncertainty in measurement, computational systems, **1**-13–**1**-14
  Heisenberg's uncertainty principle, **1**-13
  homeostatic principle, **1**-14
  Le Châtelier's principle, **1**-14
  reaction to stress, **1**-14
Unicasting, **36**-2
Uniform connection caching, **37**-20
Uniform multiprocessor model, **24**-23
Uniprocessor real-time scheduling, **24**-4–**24**-6
  dynamic-priority scheduling, **24**-5–**24**-6
  static-priority scheduling, **24**-4–**24**-5
Universal computer, **1**-19–**1**-20
Unrelated multiprocessor model, **24**-23
Unweighted congestion games, **20**-7–**20**-10
User mobility patterns, in cellular networks, **33**-8
  activity-based model, **33**-8
  fluid model, **33**-8
  Gauss–Markov model, **33**-8
  gravity model, **33**-8
  Markovian model, **33**-8
  memoryless movement model, **33**-8
  shortest distance model, **33**-8

V

Value, **8**-12
Variable speed processors and power management, **40**-4
Variable value transfer, **7**-11
Vectors, **18**-2
Vertical mining, **32**-4–**32**-5
  parallel *Eclat* algorithm, **32**-4–**32**-5, see also Eclat algorithm
Vickrey
  auction, **45**-6
  payment, **45**-9
Video allocation problem (VAP), **45**-3
View-dependent isosurface extraction and rendering, **27**-11–**27**-15, see also under Isosurface
Virtual data grid, **46**-17
  anatomy of, **46**-16–**46**-17, **46**-18
  generation of, **46**-20–**46**-22
  simulation, **46**-22–**46**-23
Virtual finish time, **22**-14
Virtual network topology, **46**-10–**46**-11
Virtual time concept, **22**-6
Voltage scaling (VS), **40**-2
Volumetric surface construction, parallel algorithms for, **27**-1–**27**-15, see also Isosurface
  Marching–Cubes (MC) algorithm, **27**-4
  simple optimal out-of-core isosurface extraction algorithm, **27**-5–**27**-7
  range partitioning-based algorithms, **27**-9
  tree-based algorithms, **27**-9

W

Wait buffers, **7**-8
Wait for mommy (WFM), **8**-5
Wavelength conversion, in WDM packet interconnects, **23**-2–**23**-4
Waxman model, **45**-12
WDM packet interconnects
  optimal parallel scheduling algorithms in, **23**-1–**23**-18, see also Optimal parallel scheduling algorithms
  output contention, **23**-2
Weighted congestion games
  for selfish routing in networks, **20**-3
  PNE existence and construction, **20**-13–**20**-16

Index

price of anarchy of, **20**-20–**20**-27
Weighted fair queuing (WFQ), **22**-12–**22**-14
   in fluid flow model, **22**-7–**22**-8
Wireless cellular networks, online call admission control in, **38**-1–**38**-19, *see also* Online call admission control
Wireless sensor networks, **33**-25–**33**-29
   network model, **33**-26–**33**-27
     advantage, **33**-26
     gateway, **33**-27
     sensor node, **33**-26
     sink, **33**-26–**33**-27
   network organization and clustering, **33**-27–**33**-28
     coordinate system, **33**-27–**33**-28
     output data assets, **33**-25
     sensor nodes, **33**-25
   work model, **33**-28–**33**-29
     intracluster activity, **33**-28
     workforce setup, **33**-29
     workforce size, **33**-29
Witnesses, **30**-4–**30**-5
Work-stealing graph traversal, **26**-7
Worst case execution time (WCET), **40**-2
Worst Fit (WF), **24**-8
Worst-Case Weighted Fair Queuing (WF^2Q), **22**-17

X

XMT PRAM-on-chip programming model, **5**-20
XMTC code, **5**-8
XSufferage, **21**-13